C000133419

ISBN 978-0-282-90626-9
PIBN 10872838

1 MONTH OF
FREE
READING

at

www.ForgottenBooks.com

By purchasing this book you are eligible for one month membership to ForgottenBooks.com, giving you unlimited access to our entire collection of over 1,000,000 titles via our web site and mobile apps.

To claim your free month visit:
www.forgottenbooks.com/free872838

English
Français
Deutsche
Italiano
Español
Português

www.forgottenbooks.com

Mythology Photography **Fiction** Fishing Christianity **Art** Cooking Essays Buddhism Freemasonry Medicine **Biology** Music **Ancient Egypt** Evolution Carpentry Physics Dance Geology **Mathematics** Fitness Shakespeare **Folklore** Yoga Marketing **Confidence** Immortality Biographies Poetry **Psychology** Witchcraft Electronics Chemistry History **Law** Accounting **Philosophy** Anthropology Alchemy Drama Quantum Mechanics Atheism Sexual Health **Ancient History** **Entrepreneurship** Languages Sport Paleontology Needlework Islam **Metaphysics** Investment Archaeology Parenting Statistics Criminology **Motivational**

CYCLOPÆDIA

OF

PRACTICAL RECEIPTS

AND COLLATERAL INFORMATION

COOLEY'S CYCLOPÆDIA

OF

PRACTICAL RECEIPTS

AND

COLLATERAL INFORMATION

IN THE

ARTS, MANUFACTURES, PROFESSIONS, AND TRADES

INCLUDING

Medicine, Pharmacy, Hygiene, and Domestic Economy

DESIGNED AS A COMPREHENSIVE

SUPPLEMENT TO THE PHARMACOPŒIA

AND

GENERAL BOOK OF REFERENCE

FOR THE MANUFACTURER, TRADESMAN, AMATEUR, AND
HEADS OF FAMILIES

SEVENTH EDITION

REVISED AND GREATLY ENLARGED BY

W. NORTH, M.A.Camb., F.C.S.

VOL. II

LONDON
J. & A. CHURCHILL
11, NEW BURLINGTON STREET
1892

A CYCLOPÆDIA

OF

PRACTICAL RECEIPTS, PROCESSES,

AND

COLLATERAL INFORMATION

VOLUME II

IRID'IUM. Ir. Atomic weight = 193. A rare metal, resembling osmium and platinum, and to a less degree rhodium, ruthenium, and palladium, in its properties. It was discovered by Descotils in 1803, and by Tennant in 1804, in the black powder left in dissolving crude platinum. This powder is an alloy of iridium with osmium. The metal is also found native and nearly pure amongst the Uralian platinum ores.

Prep. The residue, containing osmiridium, &c., that is left when native platinum is dissolved in aqua regia is fused in an earthen crucible with litharge, boracic acid, and lead, and the button which forms at the bottom of the crucible is dissolved in nitric acid, and the residue treated with aqua regia, and then fused with zinc in a crucible of gas-carbon, the temperature being finally raised so that the zinc volatilises and a spongy residue of osmiridium is left. To obtain pure iridium from this, the following method was adopted by MM. Deville and Debray, on behalf of the Paris Commission for the International Metric System. The spongy mass of osmiridium was ignited with barium nitrate, and the mass extracted with water; a residue of iridium oxide and barium osmate was thus formed. This was boiled with nitric acid in order to get rid of the osmium, which volatilised as the tetroxide. The solution was then treated with baryta, and the precipitated iridium salt redissolved in aqua regia and the iridium precipitated, by the addition of ammonium chloride, as the double chloride of iridium and ammonium. This was ignited, and a residue was thus obtained containing small quantities of platinum, ruthenium, and rhodium. It was ignited with potassium nitrate, and the mass treated with water, in which potassium ruthenate dissolved. The residue was then fused with lead, and the regulus treated with nitric acid and aqua regia. A residue of pure iridium was thus obtained.

Prop., &c. Brittle, white, very hard, only fusible by the strongest heat of Deville's gas furnace. In its pure state it is not acted upon by any of the acids, but it is oxidised by fusion with nitre, and when ignited in the finely divided state to redness in the air. MM. Deville and Debray prepared 25 kilos (55 lbs.) of pure iridium for the Paris Commission, to be used in the manufacture of standard meter measures. The raw material was obtained from Messrs Johnson and Matthey, which firm has since supplied the French Minister of War and the Minister of Agriculture and Commerce with some standard rules of osmiridium, which have been acknowledged as perfect in composition, manufacture, and physical properties. These standard measures are made of an alloy of 9 parts platinum with 1 part iridium. An alloy of iridium and osmium (artificial or native) has been employed for tipping the nibs of gold pens (everlasting pens).

Iridium, Chlorides of. SESQUICHLORIDE, Ir_2Cl_6. Formed by heating spongy iridium in chlorine; it is olive-green in colour, and is insoluble in water. With metallic chlorides it forms green double salts. — TETRACHLORIDE (BICHLORIDE), $IrCl_4$. Formed when finely divided iridium or either of its oxides is dissolved in aqua regia; it is yellowish-red in colour, and with alkaline chlorides gives double salts which are black when massive but red when powdered.

Iridium, Oxides of. SESQUIOXIDE, Ir_2O_3. Formed when finely divided iridium is ignited in air or oxygen. At a higher temperature (above 1000° C.) it loses its oxygen and the metal is re-formed.—DIOXIDE, IrO_2. Formed when the different chlorides of iridium are precipitated with potash in a hot solution in contact with air, and also when finely divided iridium is fused with potash and potassium nitrate.

IRITIS. Inflammation of the iris.

I'RON. Fe. Atomic weight = 56. *Syn.* FERRUM, L.; FER, Fr.; EISEN, Ger. A metallic element resembling to some extent cobalt, nickel, manganese, and chromium. The history of this most important metal extends to the remote past. The discovery of an iron rod in one of the Assyrian bronzes brought to England by Mr Layard established the interesting fact that this metal was known and commonly employed, where strength was required, nearly 3000 years ago. Rust of iron and scales of iron were used as medicines at a time equally remote.

Sources. Iron very rarely occurs in the metallic state (native iron)—perhaps never of terrestrial origin—but it occurs associated with nickel and other metals in one class of meteorites, which sometimes contain 90% of metallic iron. In combination with oxygen and other elements it occurs all over the world. The following are its chief ores:

1. *Magnetic Iron Ore, Magnetite, Loadstone,* Fe_3O_4. This ore, when pure, is the richest and most valuable one of iron, but is generally associated with more or less silica, &c., and contains from 66% to 45% of iron. It occurs crystalline, massive, and in small grains. It is difficult to reduce, but makes excellent iron and steel. It is found in Norway, Sweden, and Lapland, the Ural Mountains, Silesia, Elba, and the United States. The Swedish ore is very pure, and is reduced by means of charcoal; consequently the iron and steel manufactured from it is very pure, and free from sulphur, phosphorus, &c.

2. *Red Hæmatite or Specular Iron Ore,* Fe_2O_3. The latter is the crystalline variety, the former the massive, occurring generally in large mammillated masses. It contains from 66½% to 30½% of iron, according to its purity. The chief impurity is silica. It is found in England at Ulverston, in Lancashire, and on the Cumberland coast near Whitehaven; also in Westphalia, Elba, and at several localities in the United States.

3. *Brown or Yellow Hæmatite or Limonite,* $2Fe_2O_3 + 3H_2O$, a hydrated variety of the above. It usually occurs massive, and is found in the Forest of Dean, South Wales, and North Ireland; also in Sweden, Germany, France, Spain, and Canada. *Bog-ore* and *lake-ore* are varieties of it. Its impurities are water, silica, &c., and it contains from 63% to 35% of iron.

4. *Spathose Iron Ore or Siderite,* $FeCO_3$, is light in colour, and generally occurs massive. It is found in England in Somerset, Devon, and Yorkshire; also in Styria, Carinthia, and Prussia. It contains from 27% to 23% of iron. When mixed with clay or sand it is known as *clay-ironstone,* or *argillaceous iron ore.* This is the most important English ore of iron, and from it is obtained fully one half of the iron manufactured in this country. It occurs chiefly in nodules or beads in the coal measures, and is found in North and Central England, South Wales, and South Scotland. When it contains, in addition, 20% to 25% of carbonaceous matter it is known as *black-band ironstone,* and this is found in Lanarkshire, South Staffordshire, and South Wales; also in Westphalia, Silesia, South France, and several localities in the United States.

A *titaniferous iron ore* has been found at Taranaki, in New Zealand, in the form of black grains, of the size of rather coarse sand.

Iron also occurs in *iron* and *copper pyrites,* but these are useless as ores of iron, owing to the practical difficulty of getting rid of all the sulphur which they contain. Iron further occurs in minute quantities in the soil in some mineral (chalybeate) springs, and in all organisms, notably in the blood of higher animals; its presence is essential for the formation of chlorophyll in plants. It is also found in the atmosphere of the sun and some of the fixed stars.

Prep., &c. Iron is only prepared on the large scale, and an account of the manufacture would be out of place here. Those requiring detailed information must consult the elaborate works of Percy, Hunt, Fairbairn, Phillips, and other metallurgists.

Pure iron may be prepared by introducing fine iron wire, cut small, 4 parts, and black oxide of iron, 1 part, into a Hessian crucible, covering with a mixture of white sand, lime, and carbonate of potassium (in the proportions used for glass-making); and, after applying a closely fitting cover, exposing the crucible to a very high degree of heat. A button of pure metal is thus obtained, the traces of carbon and silicon present in the wire having been removed by the oxygen of the black oxide.

The stages involved in the manufacturing processes for the reduction of iron from its ores are very briefly the following:—The ore, unless a pure oxide, is first roasted, by which means carbonic acid, water, &c., are driven off. It is then heated with coal and limestone in a blast-furnace, the limestone being added to aid the formation of a slag. The ores are reduced as they descend through the furnace and combine with carbon; the iron then melts, and collects in the hearth at the bottom of the furnace; the melted slag floats on the top of the iron, and continually runs off through an opening made for that purpose. When sufficient iron has collected it is run off into sand-moulds, and forms 'pigs' of *cast iron,* which contain 3% or 4% of carbon with some silicon, and often phosphorus and sulphur; the presence of these last two substances is extremely prejudicial.

To convert the cast iron into *wrought iron* it is heated in a furnace, so as to oxidise the carbon and silicon, which form a slag with some of the iron that becomes oxidised; the sulphur and phosphorus are also oxidised away. The 'bloom' or mass of metal is worked about ('puddled') on the hearth of the furnace by a workman, or, in some cases, by the aid of a mechanical contrivance. The wrought iron only contains 0·15% of carbon. For the conversion of iron into steel see STEEL.

Prop., &c. The properties and uses of iron are too well known to require description. Its applications are almost universal. It is remarkably ductile, and possesses great tenacity, but is less malleable than many of the other metals. Its sp. gr. is 7·844. It is the hardest of all the malleable and ductile metals, and when combined with carbon (steel) admits of being tempered to almost any degree of hardness or elasticity. In dry air it does not oxidise at common temperatures; but at a red heat it soon becomes covered with a scaly coating of black oxide, and at an intense white heat burns brilliantly with the production of the same substance. Pure water, free from air or carbonic acid, does not tarnish the surface of polished iron, but the combined action of air and moisture, especially when a little acid vapour is present, causes its surface to be soon covered with rust, which is hydrated sesquioxide of iron. Nearly all acids attack iron; dilute sulphuric and hydrochloric acid do so with considerable energy, and the evolution of hydrogen gas. At a red heat iron decomposes water rapidly, hydrogen being evolved, and the black oxide of iron formed. Iron is magnetic up to a dull red heat, at which point it loses all traces of that property. It melts at about 1800° C. (3300° F.). With oxygen, chlorine, iodine, the acids, &c., it forms numerous important compounds. As a remedial agent, when properly exhibited, iron acts as a general stimulant, hæmatinic, and tonic, and generally proves beneficial in cases of chronic debility, unaccompanied with organic congestion or inflammation. The carbonate (ferrous carbonate), as it exists in mineral waters, held in solution by carbonic acid in excess, appears to be the form most congenial to the human body; and from its state of dilution is rapidly absorbed by the lacteals, and speedily imparts a ruddy hue to the wan countenance. Iron is undoubtedly one of the most valuable articles of the materia medica, and appears, from the antiquity of its introduction into medicine, and the number of its preparations, to have been deservedly appreciated.

Iron forms two classes of salts, namely, ferrous or proto-salts, in which iron exhibits a power of combining with two atoms of any monad element; and the ferric or per-salts, in which iron has a capacity of uniting with three atoms of any monad element.

Tests. The ferrous or proto-salts have generally a greenish colour, but yield nearly colourless solutions, except when concentrated. Their solutions are known by the following reactions:— They are not precipitated by hydrosulphuric acid when acid, and but incompletely when neutral. Sulphide of ammonium produces a black precipitate, becoming brown on exposure to the air, insoluble in alkalies, but easily soluble in the mineral acids. Ammonia and potassa give a greenish-white precipitate, gradually becoming green and then brown in the air. This precipitate occasionally is of a bluish black if excess of potassa is used. The presence of ammoniacal salts interferes with the action of these tests. Ferrocyanide of potassium gives a nearly white precipitate, becoming gradually blue in the air, and immediately so on the addition of a little weak nitric acid or chlorine water. Ferrocyanide of potassium produces a rich deep blue precipitate (*Turnbull's* blue), insoluble in hydrochloric acid. In highly dilute solutions the effect is only a deep bluish-green coloration. Phosphate of sodium produces a white precipitate, which after a time becomes green.

The ferric salts, which are also called the sesqui- or per-salts of iron, have for the most part a reddish-yellow colour, yielding deep-coloured solutions, which exhibit the following reactions: —They redden litmus-paper. Hydrosulphuric acid in acid solution reduces ferric to ferrous salts, giving a white or yellow precipitate of sulphur only. In alkaline solutions it yields a blackish precipitate, consisting of sulphur and ferrous sulphide. Sulphide of ammonium gives similar reaction. Ammonia and potassa produce bulky reddish-brown precipitates insoluble in excess. Ferrocyanide of potassium gives a rich blue precipitate (Prussian blue), insoluble in hydrochloric acid, and readily decomposed by potassa. Ferri-

cyanide of potassium deepens the colour, but does not give a blue precipitate, as it does with ferrous salts (proto-salts). Sulphocyanide of potassium gives an intense ruby-red colour to neutral or acid solutions. Tincture and infusion of galls strike a black colour. Phosphate of sodium gives a white precipitate, which becomes brown, and finally dissolves on the addition of ammonia.

Estim. 1. The solution must be boiled with hydrochloric acid and a crystal or two of potassium chlorate to ensure that any ferrous iron is oxidised to the ferric state. Ammonia is then added in excess, and the brown flocculent precipitate of ferric hydroxide is collected, dried, ignited, and weighed. Its weight multiplied by 0·7 gives the weight of iron in the sample taken.

2. The solution is made strongly acid with sulphuric or hydrochloric acid, and some pure zinc is added; this is to reduce any ferric salts present to the ferrous state. When the zinc is all dissolved the solution is boiled to expel hydrogen, and allowed to cool. It is then titrated with a standard solution of potassium permanganate, which is run in from a burette till a permanent colour is produced; or a standard solution of potassium bichromate may be used, and should always be employed when hydrochloric acid is present, as is generally the case. It is run into the iron solution till a drop of the latter ceases to give a blue coloration when touched against a drop of dilute potassium ferricyanide. If decinormal solutions are used, each 1 c.c. represents 0·0056 grm. iron in the sample taken.

Analysis of an Iron Ore. The ore is carefully sampled, and an average portion powdered.

Moisture. This is estimated by gently igniting for an hour 2 or 3 gr. of the powdered ore in a tube through which a current of air, dried by passing through sulphuric acid or through a calcium-chloride tube, passes, being then led through a second calcium-chloride tube, where it gives up the moisture it had taken from the ore. This last tube is weighed before and after the experiment; its gain in weight gives the amount of water in the sample. If necessary an estimation may also be made of the water lost at 100° C.

Carbonic acid is estimated by means of the apparatus described under CARBONIC ACID.

Silica, &c. Heat 10 gr. of the powdered ore in strong hydrochloric acid, mixed with a little nitric, in a porcelain basin until it is completely decomposed. Evaporate the whole to dryness, and repeatedly extract the residue with warm hydrochloric acid, each time filtering the supernatant liquid into a ½-litre flask. The residue is then thrown on to the filter, washed thoroughly with water, dried, ignited, and weighed. It consists of gangue and silica. The silica may be estimated, if required, by boiling the weighed residue with a solution of sodium carbonate in a platinum dish, filtering, and determining the weight of the gangue remaining; the difference between the two weights gives the amount of silica. These weights multiplied by 10 give the percentage quantities.

Sulphur. The filtrate from the silica, &c., contained in the ½-litre flask is diluted to 500 c.c., and well mixed by shaking. 100 c.c. are taken and evaporated nearly to dryness to expel excess of acid, diluted with water, boiled and treated with a few drops of barium chloride solution. After standing for 24 hours the barium sulphate is filtered off and weighed. Its weight in grms. multiplied by 6·87 gives the percentage of sulphur in the ore; to express as sulphuric acid multiply instead by 17·17.

Phosphoric Acid. To another 100 c.c. of the filtrate add a little clear solution of ammonium nitro-molybdate. After standing for 24 hours in a warm place, filter, wash the precipitate, treat it on the filter with ammonia, add magnesia mixture to the solution, and allow it to stand for 24 hours. Filter, and wash the precipitate, dry, ignite, and weigh it. Its weight multiplied by 21·85 gives the percentage of *phosphoric acid* in the ore.

Manganese, Alumina, Lime and Magnesia, Potash and Soda. Another 100 c.c. of the filtrate are boiled with a little nitric acid, ammonium carbonate is then added till the fluid is nearly neutral, and then to the *clear* red liquid ammonium acetate in excess. Boil and filter, dry, ignite, and weigh the precipitate, which consists of ferric oxide, alumina, phosphoric acid, and traces of silica. The silica is determined by fusing the precipitate with acid potassium sulphate, extracting the fused mass with water, and weighing the residual silica. The *alumina* is estimated by difference. To the filtrate from the basic acetates add a little bromine and warm, cork the flask, and let it stand for a few hours; then filter off, dry, ignite, and weigh the precipitate; its weight multiplied by 44·9 gives the percentage of *manganous oxide* in the ore. Evaporate the filtrate, ignite the residue, and treat it with a little water and about 1 grm. of oxalic acid. Again evaporate to dryness, ignite the residue, extract it with a little water, and filter off, dry, ignite, and weigh the *magnesia*. Its weight multiplied by 50 gives the percentage of magnesia in the ore. Treat the filtrate with hydrochloric acid, evaporate it and weigh the residue of mixed chlorides of potassium and sodium. The potassium in this may be estimated by precipitating it as the platino-chloride.

Iron. In 25 c.c. of the same filtrate the iron is estimated by means of a standard bichromate solution as described above. If the solution is decinormal, the number of c.c. multiplied by 1·12 gives the percentage of total iron (both ferric and ferrous) in the ore. To find the ferrous iron digest some of the ore with hydrochloric acid only, and estimate the iron in the solution with standard bichromate. If 1 grm. of the ore were taken, the number of c.c. used multiplied by 0·56 gives the percentage of ferrous iron in the ore; this number multiplied by 1·286 gives the percentage of ferrous oxide, FeO. The percentage of ferric iron is now found by subtracting ferrous from total iron; multiplied by 1·429 it gives the percentage of ferric oxide, Fe_2O_3, in the ore.

Iron, Preparations of:

Ferric Acetate. $Fe_2(C_2H_3O_2)_6$. *Syn.* PERACETATE OF IRON; LIQUOR FERRI ACETATIS FORTIOR (B. P.), L. Solution of persulphate of iron, 5 oz.; solution of ammonia, a sufficiency; glacial acetic acid (liquefied), 3 oz.; distilled water, a sufficiency. Mix 8 oz. of the ammonia with 1 pint of water; to this gradually add the persulphate of iron, previously diluted with a pint of

water; stir well, and keep the ammonia in excess. Set aside 2 hours, filter and wash precipitate well, press to remove remaining water; dissolve precipitate in the glacial acetic acid, and make up to 10 oz. with water.

It forms a dark ruby-red coloured fluid of acetous and astringent taste. Used as a hæmatin and astringent, acting like, but is milder than, perchloride of iron.—*Dose*, 2 to 8 minims.

Ferri Acetatis Liquor. SOLUTION OF ACETATE OF IRON (B. P.). Strong solution of acetate of iron, 5 oz.; water, 15 oz.—*Dose*, 5 to 30 minims.

Ferric Albuminate. *Syn.* LIQUOR FERRI ALBUMINATI, L. Dried egg-albumen, 30 parts; cinnamon water, 270 parts; solution of dialysed iron, 90 parts; caustic soda, 1·125 parts; rectified spirit, 150 parts; distilled, 1000 parts. Dissolve the albumen in the cinnamon water, then dilute the iron with 400 parts water, and add the spirit. Mix the solutions, add the soda, and set aside for several hours; then filter, and add water to 1000 parts.

Uses. Similar to dialysed iron; is easily borne by the stomach.—*Dose*, 1 to 4 dr.

Ferric Chlo'ride. Fe_2Cl_6. *Syn.* SESQUICHLORIDE OF IRON, PERCHLORIDE OF I., PERMURIATE OF I.; FERRI SESQUICHLORIDUM, L. *Prep.* 1. (Anhydrous.) By passing dry chlorine over heated iron filings. Brown scales.

2. (Hydrated.) Dissolve ferric hydrate in hydrochloric acid, evaporate to the consistence of a syrup, and crystallise. Yellow or red scaly crystals. The impure solution of this salt has been greatly used as a sewage deodoriser. See TINCTURE.

Ferric and Ammonium Chloride. Fe_2Cl_6. NH₄Cl. Aq. *Syn.* DOUBLE CHLORIDE OF IRON AND AMMONIUM, AMMONIO-CHLORIDE OF IRON; FERRI AMMONIUM CHLORIDUM, L.

Ferric oxide, 3 oz.; hydrochloric acid, ½ pint; digest in a sand-bath until dissolved, then add of ammonium chloride, 2¼ lbs., dissolved in water, 3 pints; filter the liquid, evaporate to dryness, and reduce the mass to coarse powder. Orange-coloured crystalline grains readily soluble in water.

Ammonio-chloride of iron is tonic, emmenagogue, and aperient.—*Dose*, 5 to 15 gr.; in glandular swellings, obstructions, &c.

Ferric Citrate. $Fe_2(C_6H_5O_7)_2$. *Syn.* PERCITRATE OF IRON, CITRATE OF SESQUIOXIDE OF I., CITRATE OF I.; FERRI CITRAS, L.

Prep. By saturating a solution of citric acid in an equal weight of water with freshly precipitated moist ferric hydrate, evaporating at 65° C. (150° F.) to the consistence of a syrup, and spreading on glass plates to dry.

By either of the methods adopted for the AMMONIO-CITRATE, merely omitting the addition of the ammonia. It much resembles the ammonio-citrate, but is only slightly soluble in water, and has a rather less agreeable taste.—*Dose*, 3 to 5 gr.

Ferric and Ammonium Citrate. *Syn.* AMMONIO-CITRATE OF IRON, AMMONIO-FERRIC CITRATE; FERRI ET AMMONII CITRAS, L.

There are several preparations to which the term 'citrate of iron' has been applied. That commonly known under this name is really a double citrate of iron and ammonia, and appears to be correctly called 'ammonio-citrate of iron.'

Prep. 1. (B. P.) Liquor Ferri Persulphatis (B. P.), 8 parts; liquor ammonia, 19½ parts; citric acid (in crystals), 4 parts; distilled water, a sufficiency; mix 14 parts of the solution of ammonia with 40 parts of water, and all gradually; the solution of ferric sulphate stir constantly and briskly; let the mixture stand 2 hours, and put into a calico filter and allow to drain. Wash well the precipitate until it no longer gives a precipitate with barium chloride. Dissolve the citric acid in 8 oz. of the water, and having applied the heat of a water-bath add the precipitate of ferric hydrate previously well drained, stir them together until the whole or nearly the whole of the hydrate has dissolved. Let the solution cool, then add 5½ parts of the ammonia, filter through flannel, evaporate to the consistency of syrup, and dry it in thin layers on flat porcelain or glass plates at a temperature not exceeding 100° F.

2. (Ph. L.) Ferrous sulphate, 12 oz.; carbonate of sodium, 12½ oz.; dissolve each separately in boiling distilled water, 6 pints; mix the solutions whilst still hot, and allow the precipitate to subside; after a time decant the supernatant liquor, wash the precipitate frequently with water (drain it), add of citric acid (in powder), 6 oz., and dissolve by the aid of a gentle heat; when the whole has cooled, add of liquor of ammonia (Ph. L.), 9 fl. oz., and gently evaporate to the consistence of a syrup; in this state spread it very thinly on flat earthenware dishes (or sheets of glass), dry by a gentle heat, and when dry keep it in well-stoppered bottles.

3. (Ph. D.) Citric acid, 4 oz.; distilled water, 16 fl. oz.; hydrated ferric oxide, obtained from the sulphate, 5 oz.; liquor of ammonia, 4 fl. oz., or q. s.

4. (Wholesale.) A mixture of iron filings and citric acid, in powder, with barely sufficient water to cover it, is kept in a warm situation for some days, occasionally stirring the mass, and replacing the water as it evaporates. A saturated solution is next made in distilled water, there being previously added more citric acid (about half the weight of the acid first used) as required; it is then neutralised with liquor of ammonia (about 1½ oz. of liquor of ammonia, sp. gr. ·882, to every gall. of the solution of sp. gr. 1·025), and the solution is concentrated by evaporation; the process is then completed as in No. 1. The first part of this process produces a salt of the protoxide of iron or ferrous citrate, which is afterwards converted, by exposure to the atmosphere, into a citrate of the magnetic acid, or ferri-ferro-citrate, and, lastly, into citrate of peroxide of iron, or ferric citrate.

Prop., &c. This beautiful salt is of a rich ruby colour, and forms glistening transparent scales, very soluble in aqueous menstrua, and the resulting solution is less easily decomposed by reagents than the solutions of most of the other salts of iron. It is 'compatible' with the alkaline bicarbonates and iodides, and several other salts, and is nearly tasteless, advantages which have been perhaps overrated by both prescriber and patient. It is doubtful whether this article has not obtained a larger sale from its pleasing appearance than from its medicinal virtues.

Ammonio-citrate of iron is soluble in water; the solution feebly reddens litmus; is not turned blue by ferrocyanide of potassium; but either potassium hydrate or lime-water being added, it throws down ferric hydrate, and ammonia is evolved. From 100 gr. by incineration about 30% of ferric oxide is·left.—*Dose*, 3 to 10 gr., in water, wine, or bitter infusions.

Ferric and Magnesium Citrate. *Syn.* CITRATE OF IRON AND MAGNESIA; FERRI MAGNESIO-CITRAS, FERRI ET MAGNESIÆ CITRAS, L. *Prep.* As the last, but using carbonate of magnesium instead of ammonia to neutralise the solution.— *Dose*, 2 to 10 gr. It has been recommended as a chalybeate in the dyspepsia of gouty and debilitated habits.

Ferric and Quinine Citrate. *Syn.* CITRATE OF QUININE AND IRON; FERRI ET QUINIÆ CITRAS, L.

(B. P.) Pure ferric hydrate is prepared from liquor ferri persulphatis, 4½ oz., and liquor ammoniæ, 8 oz., as in the ferric and ammonium citrate. Sulphate of quinine, 1 part, is mixed with water, 8 parts, and sulphuric acid, 1½ parts, and when dissolved, ammonia added until the quinine is precipitated. The precipitate is collected and washed with 30 parts of water. Citric acid, 3 parts, is dissolved in 8 parts of water by the aid of a water-bath, and the ferric hydrate, well drained, added; stir together until dissolved, and add the quinine, stirring well until all is dissolved, and allow to cool; add 1½ parts of solution of ammonia diluted with 2 parts of water, stirring the solution briskly until the quinine at first thrown down by the ammonia is redissolved; filter and evaporate to a syrup, drying in thin layers on flat porcelain or glass plates at a temperature of 100° F.

Ferric citrate, 4 parts; citrate of quinine, 1 part; distilled water, q. s.; dissolve, gently evaporate, and proceed as directed for ammonio-citrate of iron. Greenish golden-yellow scales when prepared by the B. P. process, soluble in 2 parts of water, and somewhat deliquescent; taste bitter as well as chalybeate.—*Dose*, 5 to 10 gr.; in cases where the use of both iron and quinine is indicated.

Ferric and Sodium Citrate. *Syn.* FERRI SODIO-CITRAS, FERRI ET SODÆ CITRAS, L. *Prep.* From citric acid, carbonate of sodium, and iron or the hydrate, as the ammonio-citrate or potassio-citrate.

Ferric and Strychnine Citrate. (U. S.) *Syn.* FERRI ET STRYCHNIÆ CITRAS, L. *Prep.* Citrate of iron and ammonia, 490 gr.; strychnia, 5 gr.; citric acid, 5 gr.; distilled water, 9 fl. dr. Dissolve the citrate of iron and ammonia in 1 oz. of the water, and the strychnia and nitric acid in 1 dr. of distilled water. Mix the two solutions, evaporate the mixture over a water-bath, at 140° F., to the thickness of a syrup, and spread on glass plates, so that the salt, when dry, may be obtained in scales.

Ferric Ferrocyanide. $Fe_4(FeCy_6)_3·18Aq$. *Syn.* SESQUIFERROCYANIDE OF IRON, PRUSSIAN BLUE; FERRI FERROCYANIDUM, F. SESQUIFERRO-CYANIDUM, L. *Prep.* Ferrous sulphate, 4 oz.; water, 1 pint; dissolve, add to the solution of nitric acid, 6 fl. dr., in small portions at a time, boiling for a few moments after each addition; next dissolve ferrocyanide of potassium, 4½ oz., in water, 1 pint, and add this last solution, by degrees, to the first liquid, stirring well each time; lastly, collect the precipitate, wash it with boiling water, drain, and dry it.—*Dose*, 3 to 5 gr., 3 or 4 times daily, as an alterative, febrifuge, and tonic, gradually increasing the quantity until some obvious effect is produced; in agues, epilepsy, and neuralgia. See PRUSSIAN BLUE.

Ferric Hydrate. $Fe_2(HO)_6$. See under FERRIC OXIDE.

Ferri Hypophosphis. *Syn.* HYPOPHOSPHITE OF IRON (U. S. P.). Dissolve in separate portions of water 480 parts of sulphate of iron and 326 parts of calcium hypophosphite, mix the solutions and filter; evaporate the filtrate to dryness. The solution first formed contains a ferrous salt; during evaporation it is changed to ferric salt. It forms a greyish-white powder, odourless and nearly tasteless, slightly soluble in water. Given in anæmia and cases of defective nerve nutrition.—*Dose*, 5 to 10 gr.

Ferric Iodide. Fe_2I_6. *Syn.* FERRI PERIODIDUM, L. *Prep.* Freely expose a solution of ferrous iodide to the air; or digest iodine, in excess, on iron, under water, gently evaporate, and sublime. A deliquescent, volatile, red compound, soluble in water and alcohol. It is rarely employed in *medicine*.

Ferric Nitrate. $Fe_2(NO_3)_6$. *Syn.* PROTONITRATE OF IRON, NITRATE OF SESQUIOXIDE OF IRON; FERRI PERNITRAS, L. By digesting nitric acid (diluted with about half its weight of water) on iron or ferric hydrate. A deep red liquid, apt to deposit a basic salt. It is used in dyeing, and has been recommended in dyspepsia, calculous affections, and chronic diarrhœa.—*Dose*, 5 to 10 or 12 drops.

Ferric Oxide. Fe_2O_3. *Syn.* PEROXIDE OF IRON, RED OXIDE OF I.; OXIDUM RUBRUM, L. This substance is found native under several forms, but employed in the *arts* is prepared by one or other of the following methods:—

From metallic iron :—From iron wire or clean iron filings cut into pieces, moistened with water, and exposed to the air until completely converted into rust; it is then ground with water, elutriated, and dried in a similar way to that adopted for chalk. For sale, it is usually made up into small conical loaves or lumps.

By calcination (BROWN - RED COLCOTHAR, CROCUS, INDIAN RED, ROUGE, JEWELLERS' R.; FERRI OXIDUM RUBRUM, L.):—Calcine ferrous sulphate until the water of crystallisation is expelled, then roast it with a strong fire until acid vapours cease to rise; cool, wash the residuum with water until the latter ceases to affect litmus, and dry it.

Ferrous sulphate, 100 parts; common salt, 42 parts; calcine, wash well with water, dry, and levigate the residuum. This process yields a cheap and beautiful product, which is frequently sold for the ferri sesquioxidum; but it is less soluble, and therefore unfitted for a substitute for that preparation.

Ferric Hydrate. *Syn.* FERRI PEROXIDUM HYDRATUM (Ph. D.), FERRUGO (Ph. E.), L.

Prep. (Ph. E.) Ferrous sulphate, 4 oz.; sulphuric acid, 3½ fl. dr.; water, 1 quart; mix, dissolve, boil, and gradually add off nitric acid, 9 fl. dr.; stirring well and boiling for a minute or two after each addition, until the liquor yields a yellowish-brown precipitate with ammonia; it must then be filtered and precipitated with liquor of ammonia (fort.), 3½ fl. oz., rapidly added and well mixed in; collect the precipitate, wash it well with water, drain it on a calico filter, and dry it at a heat not exceeding 180° F. When intended as an antidote for arsenic it should not be dried, but kept in the moist or gelatinous state.

FERRIC PEROXIDE, MOIST. *Syn.* FERRI PEROXIDUM HUMIDUM, L. *Prep.* Mix solution of persulphate of iron (B. P.), 4 fl. oz., with 1 pint of distilled water, and add it gradually to 33 fl. oz. of solution of soda (B. P.), stirring constantly and briskly. Let them stand for 2 hours, stirring occasionally; then put on a calico filter, and when the liquid has drained away, wash the precipitate with distilled water till what passes through ceases to give a precipitate with chloride of barium. Lastly, enclose the precipitate without drying it in a stoppered bottle, or other vessel, from which evaporation cannot take place.

DRY HYDRATE OF PEROXIDE OF IRON (B. P.). *Syn.* FERRI PEROXIDUM HYDRATUM, FERRI SESQUIOXIDUM, L.; PEROXIDE OF IRON. Fe_2O_3,H_2O. Dry the moist peroxide, 1 lb., at a temperature not exceeding 212° F., till it ceases to lose weight. Reduce to a fine powder.—*Dose*, 5 to 30 gr.

Prop. Ferric oxide, prepared by precipitation, is obtained in small hard grains, or as an impalpable powder, of a brownish-red colour, odourless, insoluble in water, freely soluble in acids, and possessing a slightly styptic taste, especially when recently prepared. When exposed to heat its colour is brightened, its sp. gr. increased, and it is rendered less easily soluble in acids. The oxide prepared by calcination is darker and brighter coloured, less soluble, and quite tasteless. It has either a scarlet or purplish cast, according to the heat to which it has been exposed. The finest Indian red, or crocus, usually undergoes a second calcination, in which it is exposed to a very intense heat. It is then known as 'purple brown.' The best jewellers' rouge is prepared by calcining the precipitated oxide until it becomes scarlet.

The hydrate is of a yellowish-brown colour, and though it can be dried without decomposition, it requires to be kept in a moist state. It is best preserved in a well-stoppered bottle, filled with recently distilled or boiled water.

Pur. Medicinal ferric oxide or sesquioxide of iron is soluble in dilute hydrochloric acid, and is again thrown down by potassa. The strained liquor is free from colour, and is not discoloured by the addition of either sulphuretted hydrogen or ferrocyanide of potassium.

The hydrate (FERRI PEROXIDUM HYDRATUM, Ph. D.; FERRUGO, Ph. E.) is entirely and very easily soluble in hydrochloric acid, without effervescence; if previously dried at 180° F., a stronger heat drives off about 18% of water.

Uses, &c. The precipitated oxide is employed in *medicine* as a tonic and emmenagogue, in doses of 10 to 30 gr.; and as an anthelmintic and in tic-douloureux, in doses of 1 to 4 dr., mixed up with honey. It is also employed to make some preparations of iron. The calcined oxide is employed as a pigment, as an ingredient in iron plaster, &c. The hydrate is used medicinally as a tonic in doses of 10 to 30 gr.; and in much larger doses as an antidote in cases of arsenical poisoning.

We are indebted to Bunsen and Berthold for the introduction of this substance as an antidote to arsenic. A table-spoonful of the moist oxide may be given every 5 or 10 minutes, or as often as the patient can swallow it (*Pereira*). When this preparation cannot be obtained, rust of iron or even the dry so-called carbonate (sesquioxide) may be given along with water instead. According to Dr Maclagan, 12 parts, and to Devergie, 32 parts, of the hydrate are required to neutralise 1 part of arsenious acid. Fehling says that the value of this substance as an antidote to arsenic is materially impaired by age even when kept in the moist state. The presence of potassium, sodium, ammonium, hydrates, sulphates, chlorides, or carbonates is not of consequence, and therefore, in cases of emergency, time need not be lost in washing the precipitate, which, in such cases, need only be drained and squeezed in a calico filter. The magma obtained by precipitating ferrous sulphate with magnesia in excess, and which contains free magnesia and magnesium sulphate, besides ferric hydrate, precipitates arsenious acid not only more quickly but in larger quantity than ferric hydrate does when alone. It will even render inert Fowler's solution, and precipitate both the copper and arsenic from solutions of Schweinfurt green in vinegar, which the pure gelatinous oxide alone will not do.

Soluble Saccharated Oxide of Iron. (G.) *Syn.* FERRUM OXIDATUM SACCHARATUM SOLUBILE, L. *Prep.* Solution of perchloride of iron (sp. gr. 1·480), 2 oz. (by weight); syrup, 2 oz. (by weight); mix, and add gradually solution of caustic soda (sp. gr. 1·330), 4 oz. (by weight), and set aside for 24 hours; then add to the clear liquid 30 fl. oz. of distilled hot water; agitate and set aside. Pour off the supernatant liquid from the precipitate which will have formed, and pour on fresh distilled water; then collect the precipitate on a filter and wash thoroughly with distilled water.

Put the drained precipitate into a porcelain vessel, and mix with it 9 oz. of sugar in powder, and evaporate to dryness with constant stirring over a water-bath, then mix in enough sugar in powder to make up 10 oz. by weight; reduce to powder and keep in a closed vessel. 100 parts contain 3 of metallic iron.

Ferric Phosphate. $Fe_2H_3(PO_4)_3$. *Syn.* FERRIC ORTHOPHOSPHATE (*Odling*); FERRI SESQUIPHOSPHAS, PHOSPHAS FERRICUS, L. A white powder obtained by precipitating ferric chloride by sodium phosphate.—*Uses* and *dose.* As the last.

Ferri Pyrophosphas. *Syn.* PYROPHOSPHATE OF IRON (U. S. P.). Citrate of iron, 9 parts; pyrophosphate of sodium, 10 parts; water, 18 parts. Dissolve the iron salt in hot water, add the sodium salt and stir constantly until it is dissolved. Evaporate at 60° C. to a thin syrup, spread on sheets of glass, and dry to scales.

Thin apple-green scales, freely soluble in water. One of the best mild iron tonic salts.—*Dose*, 5 to 10 gr.

Ferric Sulphate. $Fe_2(SO_4)_3$. *Syn.* PERSULPHATE OF IRON, SULPHATE OF SESQUIOXIDE OF IRON; FERRI PERSULPHAS, L. *Prep.* By adding to a solution of ferrous sulphate exactly half as much sulphuric acid as it already contains, raising the liquid to the boiling-point, and then dropping in nitric acid, until the liquid ceases to blacken by such addition. The solution evaporated to dryness furnishes a buff-coloured mass, slowly soluble in water.

Prop., &c. With the sulphates of ammonium and potassium it unites to form compounds to which the name 'iron alums' has been given. It forms the active ingredient in the 'liquor oxysulphatis ferri' of Mr Tyson, and is said by Dr Osborne to be a constituent of 'Widow Welch's pills.' This salt is also formed when ferrous sulphate is calcined with free exposure to the air. Dissolved in water, it is used as a test for hydrocyanic, gallic, and tannic acids. Used in making all the scale preparations of the B. P.

Ferric Sulphide. *Syn.* PERSULPHIDE OF IRON. This compound is prepared in the hydrated state (FERRI PERSULPHURETUM HYDRATUM) by adding, very gradually, a neutral solution of ferric sulphate to a dilute solution of potassium sulphide, and collecting, &c., the precipitate, as in the case of the hydrated ferrous sulphide. Proposed by Bouchardat and Sandras as a substitute for ferrous sulphide, to which they say it is preferable.

Ferric Tan'nate. *Syn.* FERRI TANNAS, FERRUM TANNICUM, L. *Prep.* From tannin, 1 part; boiling water, 150 parts; dissolve, add of freshly precipitated ferric hydrate (dried at 212° F.), 9 parts; evaporate by a gentle heat to one half, filter, add of sugar, 1 part, complete the evaporation, and at once put it into bottles.—*Dose*, 3 to 5 gr., thrice daily; in chlorosis, internal hæmorrhages, &c.

Double Ferric and Ammonium Tartrate. *Syn.* AMMONIO-TARTRATE OF IRON, DOUBLE TARTRATE OF IRON AND AMMONIUM, AMMONIO-FERRIC TARTRATE, FERRIC AMMONIO-TARTRATE; FERRI AMMONIO-TARTRAS, L.

Prep. 1. (*Aikin.*) Tartaric acid, 1 part; iron filings, 3 parts; digest in a sufficient quantity of hot water to barely cover the mixture for 2 or 3 days, taking care to stir it frequently, and to add just enough water to allow the evolved gas to escape freely; next add ammonia in slight excess, stir well, dilute with water, decant, wash the undissolved portion of iron, filter the mixed liquors, and evaporate to dryness; dissolve the residuum in water, add a little more ammonia, filter, and again gently evaporate to dryness, or to the consistence of a thick syrup, when it may be spread upon hot plates of glass or on earthenware dishes, and dried in a stove-room, as directed for the corresponding citrate.

2. Tartaric acid, 6½ oz.; water, 7 pints; dissolve, neutralise the solution with sesquicarbonate of ammonium, and add 6½ oz. more tartaric acid; to the solution heated in a water-bath further add moist hydrated oxide of iron (obtained from sesquioxide of iron, 53½ dr., dissolved in hydrochloric acid, and precipitated by ammonia);

when dissolved, filter and evaporate, &c., as before.

Prop., &c. Glossy, brittle lamellæ, or irregular pieces, of a deep garnet colour, almost black, very soluble in water, and possessing a sweetish and slightly ferruginous taste. By repeated re-solution and evaporation its sweetness is increased, probably from the conversion of a part of its acid into sugar. It contains more iron than a given weight of the sulphate of the same base. It is the most pleasant-tasted of all the preparations of iron except the ammonio-citrate, last noticed:—*Dose,* 3 to 10 gr.

FERRUM TARTARATUM, TARTARATED IRON (B. P.); FERRI POTASSIO-TARTRAS (Ph. L.), FERRUM TARTARIZATUM (Ph. E.), FERRI TARTARUM (Ph. D.), FERRI ET POTASSÆ TARTRAS (Ph. U. S.), L. *Prep.* (B. P.) Prepare ferric hydrate from 6 fl. oz. of liq. ferri persulphas (B. P.), as in making the double citrate, and add it to 2 oz. of the acid tartrate of potassium. Digest for 24 hours at 140° F., allow to cool, and decant off the clear solution, which is to be evaporated down and dried on glass plates (Ph. L.). Ferrous sulphate, 4 oz., is dissolved in water, 1 pint, previously mixed with sulphuric acid, ⅓ fl. oz.; heat is applied to the solution, and nitric acid, 1 fl. oz., gradually added; the solution is boiled to the consistence of a syrup and then diluted with water, 4 galls. (less the pint already used); liquor of ammonia, 10 fl. oz., is next added, and the precipitate washed and set aside for 24 hours; at the end of this time, the water being decanted, the still moist precipitate is added, gradually, to a mixture of bitartrate of potassium, 2 oz., and water, ½ pint, heated to 140° F.; after a time the undissolved oxide is separated by a linen cloth, and the clear solution either gently evaporated to dryness or treated in the same manner as the citrate. Lastly, preserve it in well-stoppered bottles. The formulæ of the Ph. E., D., & U. S. are essentially the same. The Ph. D. orders a heat not beyond 150° F. to be applied to the mixture of the oxide and bitartrate, with occasional stirring for 6 hours, and the desiccation to be conducted at the same temperature.

Obs. This preparation is a double salt of potassium and iron; it is therefore wrongly called 'tartrate of iron,' as is commonly done. It is totally soluble in water; the solution is unaffected by ferrocyanide of potassium, and not precipitated by acids with alkalies; on boiling it yields ferric hydrate. Heated with potassa, 100 gr. throws down about 30 gr. of sesquioxide of iron. Entirely soluble in cold water; taste freely chalybeate. That of commerce has generally a feebly inky taste, a slight alkaline reaction, is slightly deliquescent, dissolves in 4 parts of water, and is nearly insoluble in alcohol.

Potassio-tartrate of iron is an excellent ferruginous tonic.—*Dose,* 10 to 20 gr., dissolved in water or other convenient menstruum.

Ferric Valerianate. *Syn.* VALERIANATE OF SESQUIOXIDE OF IRON, VALERIATE OF IRON; FERRI VALERIANAS, L. (U. S. P.) By adding a solution of sodium valerianate to another of ferric sulphate, and collecting and washing the precipitate, which is to be dried by placing it for some days folded in bibulous paper on a porous brick; after which it is to be carefully kept from the air.

Prop., &c. A reddish-brown amorphous powder; nearly insoluble in water; soluble in rectified spirit, and in the dilute acids with decomposition.—*Dose,* 1 to 10 gr.; in anæmia and chlorosis complicated with hysteria.

Ferroso-ferric Hydrate. $Fe_3(HO)_6$. *Syn.* HYDRATED FERROSO-FERRIC OXIDE, HYDRATED MAGNETIC OXIDE. *Prep.* 1. Liquor ferri persulphas, 5½ parts; ferri sulphas, 2 parts; solution of soda, 80 parts; distilled water, a sufficiency. Dissolve the ferrous sulphate in 40 parts of water, add the solution of soda, stirring them well, boil the mixture, let it stand for 2 hours, put in a calico filter, wash with distilled water until the washing gives no precipitate with barium chloride, and dry at a temperature not exceeding 120° F.

2. Ferrous sulphate, 6 oz.; sulphuric acid, 160 minims; nitric acid, 4 fl. dr.; stronger solution of ammonia, 4½ fl. oz.; boiling water, 3 pints; dissolve half of the sulphate in half of the water, add the oil of vitriol, boil, add the nitric acid gradually, boiling after each addition for a few minutes; dissolve the remaining half of the sulphate in the rest of the boiling water; mix the two solutions, add the ammonia, stirring well (and boil for a short time); collect the precipitate on a calico filter, wash it with water until it ceases to precipitate a solution of nitrate of barium, and dry at a heat not exceeding 183° F. The formulæ of Gregory and Dr Jephson are similar.

3. Ferrous sulphate, 8 oz., dissolved in a mixture of water, 10 fl. oz., and sulphuric acid, 6 fl. dr., is converted by means of nitric acid, 4 fl. dr., diluted with water, 2 fl. oz., into ferric sulphate; this solution is then added to another, formed by dissolving ferrous sulphate, 4 oz., in water, ½ pint; the whole is then mixed with liquor of potassium hydrate, 2½ pints, and after being boiled for 5 minutes is collected on a calico filter, and washed, &c., as before, and is to be preserved in a well-stoppered bottle.

Prop., &c. The hydrate is a black sand-like substance, consisting of very minute crystals. When pure it is attracted by the magnet, and is entirely soluble in hydrochloric acid; and ammonia added to the solution throws down a black precipitate. The oxide is the chief product of the oxidation of iron at a high temperature in the air and in aqueous vapour. It is more permanent than ferrous oxide, but incapable of forming salts.—*Dose,* 5 to 20 gr. 2 or 3 times a day.

Ferroso-ferric Oxide. Fe_3O_4. *Syn.* MAGNETIC OXIDE OF IRON; FERRI OXIDUM NIGRUM, F. O. MAGNETICUM (Ph. D.), OXIDUM FERROSO-FERRICUM, L. This occurs native, but that used in *medicine* is prepared artificially.

From the black scales of iron that fall around the smith's anvil, by washing, drying, detaching them from impurities by means of a magnet, and then treating them by grinding and elutriation, as directed for prepared chalk. The product of this process is inferior as a medicine to the hydrate obtained as below, being less easily soluble in the juices of the stomach.

Ferrous Acetate. $F(C_2H_3O_2)_2$. *Syn.* FERRI ACETAS, L. *Prep.* 1. From freshly precipitated ferrous carbonate dissolved in dilute acetic acid.

2. By adding a solution of calcium acetate to another of ferrous sulphate, and evaporating the filtered liquid, out of contact with the air. Small, colourless, or pale greenish needles or prisms, very soluble and prone to oxidation.

Ferrous Arsenate. $Fe_3(AsO_4)_2$. *Syn.* FERRI ARSENIAS. *Prep.* 1. From a solution of sodium arseniate, added to a solution of ferrous sulphate, the precipitate being collected, washed in a little cold water, and dried.—*Dose*, $\frac{1}{16}$ to $\frac{1}{8}$ gr., made into a pill; in lupus, psoriasis, cancerous affections, &c. *Externally*, combined with 4 times its weight of ferrous phosphate and a little water, as a paint to destroy the vitality of cancerous formations. An ointment (20 to 30 gr. to the oz.) is also used for the same purpose. They are all dangerous remedies in non-professional hands.

2. (B. P.) Sulphate of iron, 20¾ oz.; arseniate of soda, dried at 300° F., 15¼ oz.; bicarbonate of soda, 4¼ oz. Dissolve the arseniate and the bicarbonate of soda in 5 pints, and the sulphate of iron in 6 pints, of boiling distilled water, mix the two solutions, collect the white precipitate which forms on a calico filter, and wash until the washings cease to be affected by a dilute solution of chloride of barium. Squeeze the washed precipitate between folds of strong linen in a screw-press, and dry it on porous bricks in a warm air-chamber whose temperature shall not exceed 100° F.—*Dose*, $\frac{1}{16}$ to $\frac{1}{2}$ gr.

Ferrous Arsenite. $Fe(AsO_2)_2$. *Syn.* FERRI ARSENIS, L. From the potassium arsenite, and ferrous sulphate, as the last. A yellowish-brown powder, occasionally used in medicine as a tonic, alterative, and febrifuge.—*Dose*, $\frac{1}{16}$ to $\frac{1}{8}$ gr.

Ferrous Bromide. $FeBr_2$. *Syn.* FERRI BROMIDUM, L. *Prep.* (*Moir.*) Bromine and iron filings, of each, 1 part; water, 3 parts; mix in a stoppered phial, set it aside, occasionally shaking it, for 2 or 3 days, and when the colour of the bromine has disappeared, and the liquid becomes greenish, filter and evaporate to dryness.—*Dose*, 1 to 6 gr., as a tonic, diuretic, and resolvent, in similar cases to those in which iodide of iron is given.

Ferrous Carbonate. $Fe(CO_3)$. *Syn.* PROTOCARBONATE OF IRON; FERRI CARBONAS, F. SUB-CARBONAS, L. This occurs in nature as SPATHOSE ORE, which is found alone, and also forms the chief constituent of CLAY IRONSTONE and BLACKBAND ORE, and in many CHALYBEATE WATERS.

Precipitate a solution of ferrous sulphate with a solution of sodium carbonate, well wash the green powder with water which has been boiled, and dry it out of contact with the air. On the slightest exposure to air it is converted into ferrous hydrate or oxide. This change is for the most part prevented by combining it with sugar, as in the following preparation.

FERRI CARBONAS SACCHARATA (B. P.); SACCHARINE CARBONATE OF IRON; FERRUM CARBONICUM SACCHARATUM, FERRI CARBONAS CUM SACCHARO (Ph. L.), L. *Prep.* (B. P.) Ferrous sulphate (sulphate of iron), 2 parts; ammonium carbonate, 1¼ parts; boiling distilled water, 320 parts; refined sugar, 1 part. Dissolve the sulphate and ammonium carbonate each in ¼ of the water, and mix; allow to stand for 24 hours and decant

off the clear solution, add the remainder of the water to the precipitate, stir well, allow to settle, and decant off. Collect the deposit in a calico filter, press, rub in the sugar in a porcelain mortar, and dry at a temperature not exceeding 212° F.

Prop., &c. A sweet-tasted greenish mass or powder, consisting chiefly of carbonate of iron. It is one of the best of the chalybeates.—*Dose*, 5 to 10 gr. When pure, it should be easily soluble in hydrochloric acid with brisk effervescence.

Ferrous Chloride. $FeCl_2$. *Syn.* PROTOCHLORIDE OF IRON; FERRI CHLORIDUM, L. *Prep.* 1. (Anhydrous.) By passing dry hydrochloric acid gas over ignited metallic iron. The chloride sublimes in yellowish crystals.

2. (Hydrated.) Dissolve iron filings or scale in hydrochloric acid, evaporate and crystallise. Soluble green crystals.

Ferrous Citrate. $Fe_3(C_6H_5O_7)_2$. *Syn.* PROTOCITRATE OF IRON, CITRATE OF PROTOXIDE OF IRON. This salt is easily formed by digesting iron filings or wire with citric acid, and evaporating the solution as quickly as possible out of contact with the air. It presents the appearance of a white powder, nearly insoluble in water, and rapidly passing to a higher state of oxidation by exposure to the air. Its taste is very metallic. It is exhibited under the form of pills, mixed with gum or syrup, to prevent it from being prematurely decomposed.

Ferrous Ferricy'anide. *Syn.* FERRICYANIDE OF IRON. *Prep.* By adding a solution of potassium ferricyanide ('red prussiate of potash') to a solution of ferrous sulphate (or any other soluble ferrous salt), and collecting and drying the precipitate. A bright blue powder. See TURNBULL'S BLUE.

Ferrous Hydrate. $Fe_2(HO)_2$. See under FERRIC OXIDE.

Ferrous Hydrate. $Fe_2(HO)_2$. May be precipitated from ferrous solutions as a white powder, by alkaline hydrates. It rapidly absorbs oxygen, and turns first green, and then red, by exposure to the air. Both the oxide and hydrate are very powerful bases, neutralising the acids, and forming stable salts, which, when soluble, have commonly a pale green colour, and a nauseous metallic taste.

Ferrous Hypophosphite. *Syn.* FERRI HYPOPHOSPHIS, L. From the double decomposition of hypophosphite of lime and sulphate of iron, as hypophosphite of potash.

Ferrous Iodide. FeI_2. *Syn.* PROTOIODIDE OF IRON, IODIDE OF I.; FERRI IODIDUM, L. *Prep.* Fine iron wire, 1 part; iodine, 2 parts; distilled water, 10 parts. Introduce the iron, iodine, and 8 parts of water into a flask, heat it about ten minutes, and boil until all the red colour is gone. Filter through paper into a polished iron dish, washing with the rest of the water, and boil until a drop of the solution taken out on iron wire solidifies on cooling. Pour on porcelain and cool. (Ph. L., 1836.) Iodine, 6 oz.; iron filings, 2 oz.; water, 4½ pints; mix, boil in a sand-bath until the liquid turns to a pale green, filter, wash the residuum with a little water, evaporate the mixed liquors in an iron vessel at 212° F. to dryness, and immediately put the iodide into well-stoppered bottles.

Iodine, 1 oz., and clean iron filings or turnings, ½ oz., are put into a Florence flask with distilled water, 4 fl. oz., and having applied a gentle heat for 10 minutes, the liquid is boiled until it loses its red colour; it is then at once filtered into a second flask, the filter washed with water, 1 fl. oz., and the mixed liquid is boiled down, until it solidifies on cooling.

SACCHARINE IODIDE OF IRON; SACCHARUM FERRI IODIDI, FERRI IODIDUM SACCHARATUM (U. S. P.), L. Iron (in powder), 6 parts; water, 20 parts; iodine, 17 parts; obtain a solution of iodide of iron, as above, and add to it of sugar of milk (in powder), 80 parts; evaporate at a temperature not exceeding 122° F., until the mass has a tenacious consistence, then further add of sugar of milk, 1 oz., reduce the mixture to powder, and preserve it in a well-stoppered bottle.—*Dose*, 2 to 10 gr.

From 'syrup of iodide of iron' exposed in a shallow vessel, in a warm place, until it crystallises; the crystals are collected, dried, and powdered. A simpler plan is to gently evaporate the whole to dryness, and to powder the residuum. The saccharine iodide may be kept for some time in a corked bottle without undergoing decomposition.

Obs. The preparation of the above compound needs care, exposure to air and excess of heat should be avoided. As soon as iodine and iron are mixed together under water much heat is evolved, and if too much water be not used, the combination is soon complete, and the liquor merely requires to be evaporated to dryness, out of contact with the air, at a heat not exceeding 212° Fahr. This is most cheaply and easily performed by employing a glass flask, with a thin broad bottom and a narrow mouth, by which means the evolved steam excludes air from the vessel. The whole of the uncombined water may be known to be evaporated when vapour ceases to condense on a piece of cold glass held over the mouth of the flask. A piece of moistened starch paper occasionally applied in the same way will indicate whether free iodine is evolved; should such be the case, the heat should be immediately lessened. When the evaporation is completed, the mouth of the flask should be stopped up by laying a piece of sheet india rubber on it, and over that a flat weight; the flask must be then removed, and when cold broken to pieces, the iodide weighed, and put into dry and warm stoppered wide-mouth glass phials, which must be immediately closed, tied over with bladder, and the stoppers dipped into melted wax.

Prop., &c. Ferrous iodide evolves violet vapours by heat, and ferric oxide remains. When freshly made it is totally soluble in water, and from this solution, when kept in a badly stoppered vessel, ferric hydrate is very soon precipitated; but with iron wire immersed in it, it may be kept clear in a well-stoppered bottle.—*Dose*, 1 to 3 gr., or more, as a tonic, stimulant, and resolvent; best given in the form of the syrup. It has been given with advantage in anæmia, chlorosis, debility, scrofula, and various glandular affections.

Ferrous Lactate. Fe(C₃H₅O₃)₂. *Syn.* PROTO-LACTATE OF IRON; FERRI LACTAS, FERRUM LACTICUM, L. *Prep.* Boil iron filings in lactic acid diluted with water until gas ceases to be evolved, and filter whilst hot into a suitable vessel, which must be at once closely stoppered; as the solution cools, crystals will be deposited, which after being washed, first with a little cold water and then with alcohol, are to be carefully dried. The mother liquor, on being digested, as before, with fresh iron, will yield more crystals.

Into sour whey, 2 lbs., sprinkle sugar of milk and iron filings, of each, in fine powder, 1 oz.; digest at about 100° F., until the sugar of milk is dissolved, then add a second portion, and as soon as a white crystalline powder begins to form boil the whole gently and filter into a clean vessel; lastly collect, wash, and dry the crystals as before.

Prop., &c. Ferrous lactate is a greenish-white salt; and when pure, forms small acicular or prismatic crystals, which have a sweetish ferruginous taste, and are soluble in about 48 parts of cold and in 12 parts of boiling water. It has been regarded by many high authorities as superior to every other preparation of iron for internal use, as being at once miscible with the lactic acid of the gastric juice, instead of having to be converted into a lactate at the expense of that fluid, as it is asserted is the case with the other preparations of iron.—*Dose*, 2 to 6 gr., frequently, in any form most convenient.

Ferrous Malate (Impure). *Syn.* FERRI MALAS IMPURUS, L. *Prep.* (P. Cod., 1839.) Porphyrised iron filings, 1 part; juice of sour apples, 8 parts; digest for 3 days in an iron vessel, evaporate to one half, strain through linen whilst hot, further evaporate to the consistence of an extract, and preserve it from the air.—*Dose*, 5 to 20 gr., where the use of iron is indicated.

Ferrous Nitrate. (FeNO₃)₂. *Syn.* PROTO-NITRATE OF IRON, NITRATE OF PROTOXIDE OF IRON; FERRI NITRAS, L. By dissolving ferrous sulphide in dilute sulphuric acid in the cold, and evaporating the solution *in vacuo*. Small green crystals, very soluble, and prone to oxidation.

Ferrous Oxalate (U. S.). *Syn.* FERRI OXALAS, L. *Prep.* Sulphate of iron, 2 oz.; oxalic acid, 396 gr.; distilled water, q. s. Dissolve the sulphate in 30 oz. (old measure), and the acid in 15 oz. (old measure) of distilled water. Filter the solutions, mix them, shake together, and set aside until the precipitate is formed. Decant the clear liquid, wash the precipitate thoroughly, and dry it with a gentle heat.

Ferrous Oxide. FeO. *Syn.* PROTOXIDE OF IRON; FERRI PROTOXIDUM, L. This substance is almost unknown in a pure state, from its extreme proneness to absorb oxygen and pass into the sesquioxide.

Ferrous Phosphate. *Syn.* PHOSPHATE OF IRON, NEUTRAL P. OF PROTOXIDE OF IRON, BIMETALLIC FERROUS ORTHOPHOSPHATE (*Odling*); FERRI PHOSPHAS (Ph. U. S.), L. A salt formed from ordinary or tribasic phosphoric acid.

Prep. (B. P.) Ferrous sulphate, 3 parts; sodium phosphate, 2½ parts; sodium bicarbonate ½ part; boiling distilled water, 60 parts; dissolve the sulphate and sodium salts, each in half the water, mix, and stir carefully; filter through calico, wash with hot distilled water until it ceases to give a precipitate with barium chloride, and dry at a heat not exceeding 120° F.

Prop., &c. A slate-coloured powder; insoluble in water; soluble in dilute nitric and hydrochloric acids.—*Dose*, 5 to 10 gr.; in amenorrhœa, diabetes, dyspepsia, scrofula, &c.; and *externally*, as an application to cancerous ulcers.

Ferrous Sulphate. $FeSO_4.7Aq.$ *Syn.* PROTOSULPHATE OF IRON, SULPHATE OF IRON, COPPERAS, GREEN VITRIOL, SHOEMAKER'S BLACK; FERRI SULPHAS (B. P., Ph. L. E. & D.), VITRIOLUM FERRI, L. The crude sulphate of iron or green vitriol of commerce (FERRI SULPHAS VENALIS, Ph. L.) is prepared by exposing heaps of moistened iron pyrites or native bisulphuret of iron to the air for several months, either in its unprepared state or after it has been roasted. When decomposition is sufficiently advanced, the newly formed salt is dissolved out with water, and the solution crystallised by evaporation. In this state it is very impure. The ferrous sulphate or sulphate of iron employed in *medicine* is prepared as follows:

Prep. (B. P.) Iron wire, 4 parts; sulphuric acid, 4 parts; distilled water, 30 parts. Pour the water on the iron, add the acid, and when the disengagement of gas has nearly ceased, boil for 10 minutes. Filter through paper. Allow to stand 24 hours, and collect the crystals. Sulphuric acid, 1 fl. oz.; water, 4 pints; mix, and add of commercial sulphate of iron, 4 lbs.; iron wire, 1 oz.; digest with heat and occasional agitation until the sulphate is dissolved, strain whilst hot, and set aside the liquor that crystals may form; evaporate the mother liquor for more crystals, and dry the whole.

Dissolve the transparent green crystals of the impure sulphate of iron in their own weight of water, acidulated with sulphuric acid, and recrystallise.

The formula of the Ph. U. S. is similar.

Dried: FERRI SULPHAS EXSICCATA (B. P.), FERRI SULPHAS EXSICCATUM (Ph. E.), F. S. SICCATUM (Ph. D.), L. From ferrous sulphate, heated in a shallow porcelain or earthen vessel, not glazed with lead, till it becomes a greenish-grey mass, and then reduced to powder. The heat should be that of an oven, or not exceeding 400° F. 5 parts of the crystallised sulphate lose very nearly 2 parts by drying.

Granulated: FERRI SULPHAS GRANULATA, L. (B. P.) A solution of iron wire, 4 oz., in sulphuric acid, 4 fl. oz., diluted with water, 1½ pints, after being boiled for a few minutes, is filtered into a vessel containing rectified spirit, 8 fl. oz., and the whole stirred until cold, when the granular crystals are collected on a filter, washed with rectified spirit, 2 fl. oz., and dried, first by pressure between bibulous paper, and next beneath a bell-glass over sulphuric acid, after which they are put into a stoppered bottle to preserve them from the air.

Prop., &c. Ferrous sulphate forms pale bluish-green rhombic prisms, having an acid, styptic taste, and acid reaction; it dissolves in 2 parts of cold and less than 1 part of boiling water; at a dull red heat it suffers decomposition; sp. gr. 1·82. It is perfectly soluble in water; a piece of iron put into the solution should not be covered with metallic copper. By exposure to the air it effloresces slightly, and is partly converted into a basic ferric sulphate.—*Dose*, ¼ to 4 gr., in pills or solution; *externally*, as an astringent or styptic. In the *arts*, as sulphate of iron (copperas), it is extensively used in dyeing, and for various other purposes. The dried sulphate (ferri sulphas exsiccatum) is chiefly used to make pills.

Crude sulphate of iron is frequently contaminated with the sulphates of copper, zinc, manganese, aluminium, magnesium, and calcium, which, with the exception of the first, are removed with difficulty. It also contains variable proportions of the neutral and basic ferric sulphates. The preparation obtained by direct solution of iron in dilute sulphuric acid should, therefore, be alone used in *medicine*.

In *commerce*, there are 4 varieties of crude sulphate of iron or copperas known: greenish-blue, obtained from acid liquors; pale green, from neutral liquors; emerald green, from liquors containing ferric sulphate; and ochrey-brown, which arises from age and exposure of the other varieties to the air. Even the first two of these contain traces of ferric sulphate, and hence give a bluish precipitate with ferrocyanide of potassium; whereas the pure sulphate gives one which is at first nearly white.

Ferrous Sulphide. FeS. *Syn.* SULPHIDE OF IRON, PROTOSULPHIDE OF I.; FERRI SULPHURETUM (Ph. E. & D.), L. *Prep.* 1. (Ph. E. & D.) From sublimed sulphur, 4 parts; iron filings, 7 parts; mixed together and heated in a common fire till the mixture begins to glow, then removing the crucible from the heat, and covering it up, until the reaction is at an end, and the whole has become cold.

2. Expose a bar of iron to a full white heat, and instantly apply a solid mass of sulphur to it, allowing the melted product to fall into water; afterwards separate the sulphide from the sulphur, dry, and preserve it in a closed vessel.

Hydrated: FERRI PROTOSULPHURETUM HYDRATUM, L. By adding a solution of ammonium sulphide or of potassium sulphide to a neutral solution of ferrous sulphate made with recently distilled or boiled water; the precipitate is collected on a filter, washed as quickly as possible with recently boiled water, squeezed in a linen cloth, and preserved in the pasty state, under water, as directed under FERRIC HYDRATE.

Prop., &c. The sulphide prepared in the dry way is a blackish brittle substance, attracted by the magnet. It is largely used in the *laboratory* as a source of sulphuretted hydrogen. The hydrated sulphide is a black, insoluble substance, rapidly decomposed by exposure to the air. Proposed by Mialhe as an antidote to the salts of arsenic, antimony, bismuth, lead, mercury, silver, and tin, and to arsenious acid; more especially to white arsenic and corrosive sublimate. A gargle containing a little hydrated sulphide of iron will instantly remove the metallic taste caused by putting a little corrosive sublimate into the mouth (*Mialhe*). On contact with the latter substance it is instantly converted into ferrous chloride and mercurous sulphide, two comparatively inert substances. It is administered in the same way as ferrous hydrate. When taken immediately after the ingestion of corrosive sublimate, it instantly renders it innocuous; but when the administration is delayed until 15 or 20 minutes after the poison has been swallowed, it is almost useless.

Ferrous Tar'trate. *Syn.* FERRI TARTRAS, FERRI PROTOTARTRAS, L. *Prep.* 1. From iron filings, 2 parts; tartaric acid, 1 part; hot water, q. s.; digest together until reaction ceases, agitate the liquid, pour off the turbid solution, and collect, wash, and dry the powder as quickly as possible, and keep it out of contact with the air.

2. Crystallised potassium tartrate, 132 parts; ferrous sulphate, 139 parts; dissolve each separately, mix the solutions, and collect the precipitate as before. A nearly insoluble powder; seldom used.

Obs. By dissolving the corresponding hydrates in a solution of tartaric acid, employing the former in slight excess, and evaporating, both the ferrous and ferric tartrate are easily obtained.

IRON AL'UM. See ALUMS.

IRON CEMENT. See CEMENTS.

IRON, DIALYSED. *Syn.* LIQUOR FERRI DIALYSATUS (B. P.), L. *Prep.* 1. Mix solution of perchloride of iron, 6 parts, with distilled water, 4 parts, and stir into the mixture sufficient diluted solution of ammonia to impart, after thorough agitation, a distinct ammoniacal odour. Collect the precipitated ferric hydrate on calico and wash it with distilled water, then squeeze to remove the superfluous water; add the precipitate to solution of perchloride of iron, 1 part, stir thoroughly, warm gently, and when complete or nearly complete solution is obtained filter if necessary, and place the liquid in a covered dialyser; then subject it to a stream of water in the usual manner until the solution on the dialyser is almost tasteless. The resulting solution should measure 28 parts. A clear dark reddish-brown liquid. Neutral to test-papers. Sp. gr. about 1·047.

Tests. The solution gives no precipitate with ferrocyanide of potassium or with nitrate of silver, but after being heated with hydrochloric acid it yields with ferrocyanide of potassium a blue precipitate. 100 gr. by weight affords a precipitate with a solution of ammonia, which, washed, dried, and ignited, weighs 5 gr.

2. ('American Journal of Pharmacy.') Take 10 gr. of liq. ferri perchlor. (B. P.), precipitate by liquor ammoniæ, and wash the precipitate thoroughly. Mix this with 12 parts of liq. ferri perchlor. (B. P.), and place in a dialyser. The dialyser is placed in a suitable vessel with distilled water, the water under it renewed every 24 hours. The operation is continued until no trace of chlorine exists, at which time the preparation is found to be neutral. It usually takes from 12 to 15 days to complete the process.

The resulting preparation, which should be of a deep dark red colour, contains about 5 per cent. of the oxide of iron. If the solution after completion of the operation should contain more than 5 per cent. of iron, it may be diluted with dialysed water till it reaches that point.

The above formula is said to furnish an article precisely similar to the original Bravais' dialysed iron.

3. (*E. B. Shuttleworth.*) Add ammonia to a solution of perchloride of iron as long as the precipitate formed is re-dissolved. A solution is

produced which contains ferric hydrate dissolved in ferric chloride, with free chloride of ammonium. Either the liquor ferri perchlor. fort. (B. P.), or the liquor ferri chloridi (U. S.), may be conveniently used, and the liquor ammoniæ, sp. gr. ·959 or ·960, of either Pharmacopœia will be found a convenient strength. If the ammonia be added to the strong solution of iron, considerable heat is evolved, and, on cooling, the preparation becomes gelatinised—often so much that the vessel containing it may be inverted. It is better to avoid this result, and to such end the solution of perchloride must be diluted until of a sp. gr. of about 1·300. This degree may be nearly enough approached by diluting 2 measures of the B. P. liquor with 1 of water, or adding 1 measure of water to 5 of the U. S. preparation. This solution will generally remain permanently bright and fluid. The amount of liquor ammoniæ required will of course vary with the acidity of the perchloride. The liquor ferri (B. P.) will sometimes bear as much as an equal volume. A gelatinised solution, even when made from the undiluted liquor, will often become fluid when put upon the dialyser, but, as has been said before, it is better to work with bright solutions.

4. (*Dr Pile*.) Dr Pile, noticing the fact that chloride of sodium is one of the most rapid crystalloids to dialyse, used a solution of carbonate of sodium to add to the solution of ferric chloride in place of the ammonia so generally recommended, and with great success. The solution of ferric chloride (U. S.) which has been neutralised by a cold solution of carbonate of sodium is poured into a floating dialyser. Starting with 1 pint of solution of ferric chloride, which on being treated with the sodium solution, and ready to dialyse, had a sp. gr. of 1·175, it had in 5 days increased to 5 pints. The water in which the dialyser floated was changed daily. At the end of five days it had passed through the membrane all the crystalloids, was free from taste of foreign substances, and owing to increase of bulk had now the sp. gr. of 1·0295, and on evaporation yielded 5% dry oxide of iron. Too long dialysation will cause the solution of iron to become gelatinous.

Mr Shuttleworth ('Canadian Pharmaceutical Journal,' Oct., 1877) says that an efficient dialyser may be made out of one of the flat hoops of an ordinary flour barrel, a bell-jar, or even an inverted glass funnel. He gives the preference to the former, and limits its diameter to 10 or 12 in.; if it exceeds this, the septum is liable to bulge in the centre, and to make the layer of liquid too deep at that point.

The parchment paper employed for the septum must be entirely free from holes; this is an essential condition, and if any should be discovered—by the simple process of sponging the upper surface of the paper with water, and then carefully examining the under surface—they must be stopped by means of a little white of egg, applied and coagulated by heat, or by a drop of collodion.

The parchment paper is not the kind ordinarily known under that name, but a less porous description, which has been made · by previous immersion in dilute sulphuric acid.

Well-washed bladder, deprived of its outer coat, also makes a good septum.

The septum should be tied around the hoop with twine, but not too tightly, and should be so arranged that its edges shall be left standing up around the hoop, so as to absorb any liquid escaping from the hoop at its junction with the septum. The dialyser being ready for use, the liquid intended for dialysis is poured into it to a depth of not more than half an inch, and the dialyser with its contents is then floated on the surface of some distilled water, contained in a suitable receptacle.

The hoop must only be allowed to sink just below the level of the water; if it gets below this point, it will be necessary to keep it up by some support.

It is necessary to change the water in the outer vessel daily. For the first 2 or 3 days distilled water should always be used. When this is not obtainable rain-water should be employed. When the water shows the absence of chlorides and the preparation ceases to have a ferruginous taste, the operation may be regarded as finished. The process generally occupies one or two weeks.

"A pig's bladder, completely filled with the iron solution, securely tied, and immersed in water frequently changed, answers well for making this preparation. The process requires a longer time than with a carefully regulated and properly conducted dialysis, but it entails considerably less trouble. I consider it an advantage to procure the bladder perfectly fresh, as it is then easily cleaned by pure water, and alkaline ley need not be used. Great care is necessary in tying the neck carefully. This can be best accomplished by a few turns of iron wire. Above this may be secured a piece of twine, to suspend the bladder, by means of a stick, or rod, placed on the edge of the vessel containing the water. The bladder should be perfectly full, and immersed altogether in water. The attraction of the solution for the water is so great that considerable pressure is manifested, and, should any parts or holes be in the bladder, the liquid will be forced out, water will take its place, and failure result."

Pretty general consent appears to have fixed the strength of the solution of dialysed iron at 5%. Where it exceeds this, the solution must be diluted with distilled water; and where it falls short of the amount, it will have to be reduced to the required volume by standing it in a warm and dry situation. The employment of much heat must be particularly avoided, as it very frequently leads to the destruction of the compound; hence every care should be taken to render the evaporation of the fluid unnecessary.

There seems little doubt that the so-called 'dialysed iron' is an oxychloride of the metal. Prof. Maisch believes it to be a very basic oxychloride of iron. On the supposition that the oxychloride and chloride of iron are both present in the liquid put into the dialyser, the origin of the oxychloride admits of easy explanation:—The chloride being a crystalloid, diffuses through the septum into the outer water, and thus becomes separated from the oxychloride, which being a colloid, and incapable of a passage through the membrane, remains in solution in the dialyser.

The comparative freedom from taste and easy assimilation of the oxychloride of iron render it a valuable therapeutic agent. The dose of the 5% solution is 15 to 20 drops daily, in divided doses. Syrup forms a pleasant vehicle for its administration.

Dialysed iron has been successfully employed in a case of arsenical poisoning. The 'American Journal of Pharmacy' for January, 1878, contains an interesting paper by Dr Mattison detailing a series of experiments, which conclusively prove its value as an antidote to arsenic. Dr Mattison recommends the administration of the iron to be immediately followed by a teaspoonful or more of common salt.

IRON FI"LINGS. *Syn.* FERRI RAMENTA (Ph. L. 1836), FERRI LIMATURA (Ph. E.), FERRI SCOBS (Ph. D.), L. The usual method of preparing iron filings for medical purposes has been already noticed; the only way, however, to obtain them pure is to act on a piece of soft iron with a clean file. The Fr. Cod. orders them to be forcibly beaten in an iron mortar, and to be separated from oxide and dust by means of a fine sieve, and from the grosser parts by means of a coarse hair-sieve. Iron filings can be purified by a magnet, as foreign metals and other impurities are not attracted to the magnet.—*Dose*, 10 to 30 gr., in sugar or honey, as a chalybeate; in larger doses it is an excellent vermifuge, especially for ascarides or the small thread-worm.

IRON LIQ'UOR. *Syn.* PYROLIGNITE OF IRON, DYER'S ACETATE OF I., BLACK LIQUOR, TAR IRON L.; FERRI ACETAS VENALIS, L. This article, so extensively used in dyeing, is a crude mixed acetate of the protoxide and sesquioxide of iron. It is usually prepared by one or other of the following methods:

1. Old scraps of iron (hoops, worn-out tin-plate, &c.) are left in a cask of pyroligneous acid, occasional agitation being had recourse to, until a sufficiently strong solution is obtained. By keeping the acid moderately warm in suitable vessels it will become saturated with the iron in a few days. With cold acid, on a large scale, forty days or more are required to complete the process.

2. A solution of pyrolignite or crude acetate of lime is added to another of green copperas as long as a precipitate is formed; after standing, the clear liquor is decanted.

IRONMOULD. To remove ironmould from linen soak the spots with a solution containing 1 gr. of ferrocyanide of potassium and 1 drop of sulphuric acid in each ounce, then wash with soft water and remove the stains, which will have become blue, with solution of potash.

IRON, REDUCED. *Syn.* QUEVENNE IRON; FERRUM REDACTUM (B. P.), FERRI PULVIS, L.; FER REDUIT, Fr. *Prep.* This preparation, which consists of metallic iron in a fine state of division mixed with a variable amount of magnetic oxide of iron, is made by passing perfectly dry hydrogen over peroxide of iron heated to redness in an iron tube.

Prop. A greyish-black powder, attracted by the magnet, and exhibiting metallic streaks when rubbed with firm pressure in a mortar. It dissolves in hydrochloric acid with evolution of hydrogen, and should not give an odour of sulphuretted hydrogen. 10 gr. added to an aqueous solution of 50 gr. of iodine and 50 gr. of iodide of potassium, and digested with them in a small flask at a gentle heat, should leave not more than 5 gr. undissolved, which should be entirely soluble in hydrochloric acid.

Uses. In *medicine*, it is chiefly given to restore the condition of the blood in all anæmic states of the system. There is no pulverulent state of iron so convenient as this for children, as it has no taste, and only a very small dose is required.— *Dose*, 1 to 5 gr. (children, ¼ to 1 gr.), in powder, pill, or between bread-and-butter.

Iron, to remove Rust from Polished. Rust of iron may be removed from a polished grate by means of emery paper, or by scraping some Bath brick to a fine powder, mixing it with a little oil and rubbing the spots well with a piece of flannel dipped in this mixture; after which some whiting should be applied by diligent friction. This operation requires daily repetition until the rust has disappeared. Steel fire-irons, fenders, &c., when put aside in the summer, should be previously smeared thinly over with soft paraffin, known to druggists by the name of 'vaseline' or 'cosmoline,' or with grease, mercurial ointment, &c.

Iron, to remove the Stains of, from Marble. Rub on very cautiously (confining it to the surface only occupied by the spot) some strong hydrochloric acid, removing it directly the spot disappears. Should this cause any diminution in the polish, this may be restored by means of emery paper and putty powder.

IRON WIRE. *Syn.* FERRUM IN FILA TRACTUM (Ph. L.), FERRI FILUM (Ph. E.), FERRI FILA (Ph. D.), L. This is the only form of metallic iron retained in the Ph. L. It is used to make preparations of iron.

ISATINE. $C_{16}H_{10}N_2O_4$. A yellow crystalline body obtained by the oxidation of indigo. When acted upon by potash it becomes converted into aniline. Isatine may be formed by heating indigo in a dilute solution of dichromate of potash and sulphuric acid, or by treating indigo under proper conditions with nitric acid.

ISCHU'RIA. In *pathology*, retention, stoppage, or suppression of the urine.

I"SINGLASS. *Syn.* ICHTHYOCOLLA, L. The finest kinds of isinglass are obtained from various species of the genus *Acipenser*, or sturgeon, that from the great sturgeon being perhaps the most esteemed. It is the air-bag, swimming bladder, or sound, dried without any other preparation than opening, folding, or twisting it. The picked or cut isinglass of the shops consist of the lumps of staple isinglass picked in shreds by women and children, or cut by machines.

Prop., &c. Good isinglass is the purest natural gelatin known. Its quality is determined by its whiteness, absence of the least fishy odour, and ready and almost entire solubility in boiling water; the solution forming a nearly white, scentless, semi-transparent, solid jelly when cold. It is soluble in weak acids, and this solution is precipitated by alkalies. The aqueous solution is not precipitated by spirit of the common strengths. 1 part of good isinglass dissolved in 25 parts of hot water forms a rich, tremulous jelly. It is very

commonly adulterated. Of the different varieties of isinglass, the Russian is the best and most soluble. See GELATIN.

ISOM'ERISM. In *chemistry*, identity of composition, with dissimilarity of properties. Isomeric compounds (isomerides) are such as contain the same elements in the same proportions, but which differ from each other in their chemical properties; thus, formate of ethyl ($H.COOC_2H_5$) and acetate of methyl ($CH_3.COOCH_3$) are isomeric, having precisely the same ultimate composition, though differing in the arrangement of their elements.

ISOMOR'PHISM. In *chemistry*, the quality possessed by different bodies, generally, however, having a similar molecular constitution, of assuming the same crystalline form. Isomorphous substances are found to be closely allied in their chemical nature; and the fact of two bodies crystallising in the same form has often led to the discovery of other points of similarity between them. The alums, for instance, no matter what their components, all crystallise in octahedra, and a crystal of potassium-alum, if transferred to a solution of chrome-alum, will continue to grow in it with perfect regularity.

Further, the molecular constitution of all the alums can be expressed by the general formula $RR'(SO_4)_2.12H_2O$, where R stands for an atom of such a metal as potassium or sodium (or ammonium), R' for such a metal as aluminium or chromium.

ISSUE. *Syn.* FONTICULUS, L. In *surgery*, a small artificial ulcer formed on any part of the body by means of caustic or the lancet, and kept open by daily introducing an ISSUE PEA covered with some digestive or stimulating ointment; the whole being duly secured by an appropriate bandage.

ISSUE PEAS. *Syn.* PISÆ PRO FONTICULIS, L. Those of the shops are the immature fruit of the orange tree (ORANGE BERRIES). They are usually smoothed in a lathe. Issue peas are also 'turned' from orris-root. The following compound issue peas are occasionally employed:

1. Orris-root (in powder) and Venice turpentine, of each, 1 part; turmeric, 2 parts; beeswax, 3 parts; melted together and made into peas whilst warm.

2. Beeswax, 3 parts; melt, add of Venice turpentine,1 part; mix, and further add, of turmeric, 2 parts; orris-root (in powder), 1 part; mix well, and form the mass into peas whilst warm. More irritating than the common pea.

3. (*Dr Gray.*) Beeswax, 12 parts; verdigris and white hellebore, of each, 4 parts; orris-root, 3 parts; cantharides, 2 parts; Venice turpentine, q. s. Used to open issues instead of caustic, but their employment requires care.

ISSUE PLASTERS. See PLASTERS.

ITCH. *Syn.* YOUK‡, SCOTCH FIDDLE‡; PSORA, SCABIES, L.; GALE, Fr. In *pathology*, a cutaneous disease, caused by a minute insect lodging under the skin, and readily communicated by contact. There are four varieties of itch, distinguished by nosologists by the names *Scabies papuliformis*, or rank itch; *Scabies lymphatica*, or watery itch; *Scabies purulenta*, or pocky itch; *Scabies cachectica*, a species exhibiting appearances resembling each of the previous varieties. Our space will not permit more than a general notice of the common symptoms, and the mode of cure which is equally applicable to each species, and will not prove injurious to other akin diseases simulating the itch.

The common itch consists of an eruption of minute vesicles, principally between the fingers, bend of the wrist, &c., accompanied by intense itching of the parts, which is only aggravated by scratching. The usual treatment is repeated applications of sulphur ointment (simple or compound), well rubbed in once or twice a day, until a cure is effected; accompanying its use by the internal exhibition of a spoonful or more of flowers of sulphur, mixed with treacle or milk, night and morning. Where the use of sulphur ointment is objectionable, a sulphur-bath, or a lotion or bath of sulphurated potash, or of chloride of lime may be employed instead.

In the 'Canadian Pharmaceutical Journal' for 1872 is a paper by Professor Rothmund recommending the employment of balsam of Peru in this objectionable disease. The writer states that one application generally effects a cure, and that its use does away with the necessity of baths. He recommends the balsam being rubbed all over the naked body. Carbolic acid is another and much cheaper remedy proposed by the same author. To obviate its caustic action he advises the acid to be mixed with glycerin or linseed oil, in the proportion of 1 scr. of the acid to 2 oz. of either excipient. He considers the objection to this remedy may be that it enters too rapidly into the circulation. Another agent employed by Professor Rothmund is a lotion composed of 1 part of carbolate of sodium dissolved in 12 parts of water. The affected parts of the skin are to be rubbed with this three times a day.

It is further recommended to continue this treatment 8 or 10 days after the cure, in order to kill any acari or their eggs that may have lurked among the clothes or bed-linen. See ACARUS, BATH, LOTION (Itch), OINTMENT, PSORIASIS, &c.

I'VORY. The osseous portion of the tusks and teeth of the male elephant, the hippopotamus, wild boar, &c. That of the narwhal or sea-horse is the most esteemed on account of its superior hardness, toughness, translucency, and whiteness. The dust or shavings (IVORY DUST, IVORY SHAVINGS) of the turner form a beautiful size or jelly when boiled in water. VEGETABLE IVORY is the hard albumen of the seed of the *Phytelephas macrocarpa*, one of the palm family.

Ivory may be dyed or stained by any of the ordinary methods employed for woollen, after being freed from dirt and grease; but more quickly as follows:

1. BLACK. The ivory, well washed in an alkaline lye, is steeped in a weak neutral solution of nitrate of silver, and then exposed to the light or dried and dipped into a weak solution of sulphide of ammonium.

2. BLUE. Steep it in a weak solution of sulphate of indigo which has been nearly neutralised with salt of tartar, or in a solution of soluble Prussian blue. A still better plan is to steep it in the dyer's green indigo-vat.

3. BROWN. As for black, but using a weaker solution of silver.

4. GREEN. Dissolve verdigris in vinegar, and steep the pieces therein for a short time, observing to use a glass or stoneware vessel; or in a solution of verdigris, 2 parts; and sal-ammoniac, 1 part, in soft water.

5. PURPLE. Steep it in a weak neutral solution of terchloride of gold, and then expose to the light.

6. RED. Make an infusion of cochineal in liquor of ammonia, then immerse the pieces therein, having previously soaked them for a few minutes in water very slightly acidulated with aquafortis.

7. YELLOW. *a.* Steep the pieces for some hours in a solution of sugar of lead, then take them out, and when dry immerse them in a solution of chromate of potassa.

b. Dissolve as much of the best orpiment in solution of ammonia as it will take up, then steep the pieces therein for some hours; lastly take them out and dry them in a warm place, when they will turn yellow.

Ivory is etched or engraved by covering it with an etching ground or wax, and employing oil of vitriol as the etching fluid.

Ivory is rendered flexible by immersion in a solution of pure phosphoric acid (sp. gr. 1·13), until it loses, or partially loses, its opacity, when it is washed in clean, cold, soft water, and dried. In this state it is as flexible as leather, but gradually hardens by exposure to dry air. Immersion in hot water, however, restores its softness and pliancy. According to Dr Ure, the necks of some descriptions of INFANTS' FEEDING BOTTLES are thus made.

Ivory is whitened or bleached by rubbing it with finely powdered pumice-stone and water, and exposing it to the sun, whilst still moist, under a glass shade, to prevent desiccation and the occurrence of fissures; observing to repeat the process until a proper effect is produced. Ivory may also be bleached by immersion for a short time in water holding a little sulphurous acid, chloride of lime, or chlorine, in solution; or by exposure in the moist state to the fumes of burning sulphur, largely diluted with air. Cloez recommends the ivory or bones to be immersed in turpentine, and exposed for three or four days to sunlight. The object to be bleached should be kept 1-8th or 1-4th of an inch above the bottom of the bath by means of zinc supports. For the preparation of ivory intended for miniature painting, Mr Ernest Spon, in his useful work, 'Workshop Receipts,' says: "The bleaching of ivory may be more expeditiously performed by placing the ivory before a good fire, which will dispel the wavy lines if they are not very strongly marked, that frequently destroy the uniformity of surface."

Ivory may be gilded by immersing it in a fresh solution of protosulphate of iron, and afterwards in solution of chloride of gold.

Ivory is wrought, turned, and fashioned in a similar manner and with similar tools to those used for bone and soft brass.

Obs. Bone for ornamental purposes is treated in a similar way to ivory, but less carefully, owing to its inferior value. The bones of living animals may be dyed by mixing madder with their food. The bones of young pigeons may thus be tinged of a rose colour in 24 hours, and of a deep scarlet in 3 or 4 days; but the bones of adult animals take fully a fortnight to acquire a rose colour. The bones nearest the heart become tinged the soonest. In the same way logwood and extract of logwood will tinge the bones of young pigeons purple (*Gibson*).

Ivory, Artificial. *Prep.* 1. Let a paste be made of isinglass, egg-shell in very fine powder, and brandy. Give it the desired colour, and pour it while warm into oiled moulds. Leave the paste in the moulds until it becomes hard.

2. (L'Union Pharmaceutique.) Two parts of caoutchouc are dissolved in 36 parts of chloroform, and the solution is saturated with pure gaseous ammonia. The chloroform is then distilled off at a temperature of 85° C. The residue is mixed with phosphate of lime or carbonate of zinc, pressed into moulds and dried. When phosphate of lime is used the product possesses, to a considerable degree, the nature and composition of ivory.

IVORY-BLACK. See BLACK PIGMENTS.

JABORANDI. *Syn.* IABORANDI, JAMBORANDI. The above names are given by the natives of Brazil, Paraguay, and other parts of South America, to any indigenous plants possessing strongly 'stimulant, diaphoretic, and sialogogue properties, which are principally employed in those countries as antidotes for the bites and stings of venomous snakes and insects.

As far as they have been examined, all the plants known under the generic name ' jaborandi' have been traced to the two Nat. Ord. RUTACEÆ and PIPERACEÆ. Those exercising the most marked physiological effects appear to belong to the former or the rutaceous division, and are very probably different species of *Pilocarpus*. The drug was first introduced into Europe by Dr Coutinho, of Pernambuco, who some four years since sent a sample of it to Dr Gubler, of Paris, by whom it was administered to some of the patients of the Beaujon Hospital there. The jaborandi with which these experiments were made was identified by Professor Baillon, of Paris, as belonging to the *Pilocarpus pinnatus* (*pinnatifolius*). 4 to 6 grms. of the bruised leaves and twigs were infused in a cup of water, and the patient, being put to bed, in ten minutes after taking the draught finds himself bathed in a perspiration lasting for four or five hours, this being so profuse as to render several changes of linen necessary during the time. Accompanying the diaphoresis are great salivary and bronchial secretions, which sometimes will not permit the patient to speak without his mouth becoming filled with water.

The quantity of saliva is stated to have sometimes equalled a litre in measure. These experiments have been repeated in this country by Dr Ringer with analogous effects—in one case reported with jaborandi obtained from the Beaujon Hospital, and in another from London—results the similarity of which strongly point to a corresponding composition in the two specimens of the plant used, if, as seems not improbable, they may have belonged to different species. A case

of impaired vision following the administration of jaborandi is also recorded; but this seems evidently to have been the effect of an overdose of the drug ('Pharm. Journal,' 3rd series, v, 364 and 561).

When jaborandi is administered in divided doses, instead of producing salivation or sweating, it acts as an active diuretic only, increasing the flow of urine to nearly double the usual amount. M. Albert Robins says : " The effect of jaborandi on animals is very marked ; guinea-pigs are seized with salivation, weeping, and diarrhœa, true ecchymoses being found in the intestines, and dogs become instantly salivated, their gastric secretion being also much increased " ('Medical Times and Gazette ').

Drs Coutinho and Gubler affirm they have employed jaborandi in dropsy, bronchitis, diabetes, and various other diseases, and that they have found it fully answer their expectations ; and in one case of albuminuria it is narrated that a permanent diminution of albumen from 14·40 to 12 grms. followed its use.

An alkaloid has been obtained from the piperaceous jaborandi by Parodi, and named by him *jaborandine*. Some short time afterwards Mr A. W. Gerrard succeeded in separating the alkaloid from the rutaceous jaborandi, to which, in accordance with Mr Holmes' suggestion, and because Parodi had anticipated him in the adoption of the previous title, he gave the name *pilocarpine*.

Mr Gerrard recommends the following process for the preparation of *pilocarpine*:—" Prepare a soft extract either with leaf or bark, with 50% alcohol. Digest this with water, filter and wash. Evaporate the filtrate to a soft extract, cautiously add ammonia in slight excess, shake well with chloroform, separate the chloroform solution, and allow it to evaporate ; the residue is the alkaloidal pilocarpine with probably a small amount of impurity." Mr Gerrard has also succeeded in preparing a crystalline nitrate and hydrochlorate of the alkaloid, both of which possess the medicinal powers of the jaborandi. A second base called *jaborine* has been obtained from the drug.

The abridged description of a sample of jaborandi from Pernambuco is from Mr. Holmes' paper in the 'Pharmaceutical Journal' (3rd series, v, 581). The *engr.* is from the last edition of Royle's 'Materia Medica.' "The specimens of the plant examined appear to belong to a shrub about 5 ft. high. The root is cylindrical, hardly tapering at all, nearly ¼ in. in diameter for the first 12 in., and very sparingly branched. Bark of root of a pale yellowish brown, about a line in thickness, and has a short fracture. The stem is ¼ in. in diameter near the root, narrowing to ¼ in. in the upper branches. The bark is thin, greyish brown, longitudinally striated, and in some specimens sprinkled over with a number of white dots. The wood of the stem is yellowish white and remarkably fibrous. The leaves (one of which is represented in the *engr.*) are imparipinnate, about 9 in. long, with from 3 to 5 pairs of opposite leaflets, which are articulated to the rachis, and have very short, slightly swollen petiolules. The rachis of the leaf is swollen at the base.

" The pairs of leaflets are usually about 1¼ in.

apart, the lowest pair being about 4 in. from the base of the rachis. The leaflets are very variable in size, even on the same leaf. Their general outline is oblong-lanceolate. They are entire,

Pilocarpus pinnatifolius. a, flower; b, flower with the petals removed; c, carpels.

with an emarginate or even retuse apéx and an unequal base, and texture coriaceous. The veins are prominent on both sides of the leaf, and branch from the midrib at an obtuse angle in a pinnate manner. When held up to the light the leaflets are seen to be densely pellucidly punctate. These pellucid dots, which are receptacles of secretion, are not arranged, as in another kind of jaborandi, in lines along the veinlets, but are irregularly scattered all over the leaf, and appear equally numerous in every part. The whole plant is glabrous."

Mr Holmes says there appear to be two varieties, if not species, of this *Pilocarpus*, the one being perfectly smooth in every part, as above described, and the other having the stems, petioles, and under surface of the leaves covered with a dense velvety pubescence composed of simple hairs.

The same author states that the jaborandi of commerce approximates more nearly to *P. seloanus* than to *P. pinnatifolius.*

Uses. Jaborandi is employed either in the form of extract, tincture, or infusion, as a diaphoretic and sialogogue, also galactogogue. It appears to exert a beneficial effect in chronic

deafness. It antagonises belladonna, and is used as an antidote to that drug.—*Doses.* Extract, 2 to 10 gr.; infusion, 1 to 2 oz.; tincture, ½ to 1 dr.

JACARANDA LANCIFOLIATA. Acts specially on the genito-urinary mucous membrane; appears to have no deleterious effect, and may be taken without nausea. Dr A. Wright considers that, when known, it will take the place of other drugs used internally for gonorrhœas. The fluid extract of the leaves has been introduced under the name of '*Salud.*'—*Dose,* 20 to 80 minims.

JAG'GERY. *Syn.* PALM SUGAR. A coarse brown sugar made in India by the evaporation of the juice of several species of palms. The following are the principal varieties of this product:

1. COCOA JAGGERY. From the juice of the Cocoa-nut palm (*Cocos nucifera*).
2. MALABAR JAGGERY. From the juice of the Gummut palm (*Saguerus saccharifer*).
3. MYSORE JAGGERY. From the juice of the wild Date palm (*Phœnix sylvestris*); 17 galls. yields 46 lbs.
4. PALMYRA JAGGERY. From the juice of the Palmyra palm (*Borassus flabelliformis*; 6 pints yield 1 lb.

JA'LAP. *Syn.* JALAPÆ RADIX, JALAPA, B. P. (Ph. L. & D.), CONVOLVULI JALAPÆ RADIX (Ph. E.), L. The dried tubercules of the *Ipomœa purga*, Royle. Jalap is a powerful stimulant and drastic purgative, producing copious liquid stools; but when judiciously administered, both safe and efficacious. It appears to be intermediate in its action between aloes and scammony.—*Dose,* 10 to 30 gr., in powder; in constipation, cerebral affections, dropsies, obstructed menstruation, worms, &c. Owing to its irritant properties, its use is contra-indicated in inflammatory affections of the alimentary canal, and after surgical operations connected with the abdomen and pelvis. It is usually administered in combination with sulphate of potass or bitartrate of potass and ginger; with mercurials, as the case may indicate. The powder is very generally adulterated.

Professor Flückiger has brought forward some new notes on jalap, from which it appears that for the last 20 years the yield of resin in the roots of this plant has become less than it was formerly. In 1842 Guibourt found 17·60%, though at that period the samples varied in their yield from 10% to 17%; but nowadays, says our author, the yield of resin is scarcely ever more than 12%. The plant itself has not deteriorated, for as much as 16·9% of resin was got from roots bought in the fields around Mexico, whereas those in Mexico itself gave only 7½%. It would seem, according to Flückiger, that the Mexican merchants let the roots steep for a time in spirit, and then dry them; in this manner their appearance is not altered, but a good deal of resin is extracted, and can be afterwards collected and sold as jalapin, or resin of jalap.

The jalap plants grown in the botanical gardens at Cassel and Munich, after desiccation, have given as much as 22·73% of resin, and 12% respectively, which indicates that, as far as the

yield of resinous matter goes, they can be cultivated successfully in Europe without much difficulty. In the species *Ipomœa orizabensis*, another resin called orizabine has been found. The
genera *Convolvulus* and *Ipomœa* are represented
in hot climates by about 50 different species, all
of which contain jalapin and orizabine, and
perhaps some other analogous substances. In
Asia there are several drastic species, such as
I. turpethum, which would be preferable to the
fraudulent Mexican root. The seeds of *I. hede-
racea* supply a very pure resin similar to that of
I. purgans. A somewhat less pure resin is
yielded by the seeds of *kaladana*, known as
jalapine or jalap resin, yet only used in India,
though the Arabs for the last 1000 years have
employed the seeds called *habb en nil*, which are
obtained from a plant very similar to the kaladana. Japan has two plants, *Ipomœa triloba* and
Phorbitis triloba, growing wild, which yield the
same resin of jalap.

Jalap Biscuits. *Prep.* 1. An oz. of jalap
mixed with 16 oz. of the materials for ginger-
bread or other kind of cake.

2. Pure resin of jalap, 56 grms., powdered
sugar and flour, 1000 grms.; tincture of vanilla,
10 grms., white of egg, No. 20, yelk of egg,
No. 40.

Let the resin be emulsified with the yelks of
the eggs, add successively the sugar, tincture, and
flour, and mix thoroughly into a paste, with
which thoroughly incorporate the whites of eggs,
previously beaten up. Let the mass be divided
into 144 biscuits and bake.

Jalap, Resin of. *Syn.* RESINA JALAPÆ.
Prep. 1. (B. P.) Jalap in No. 40 powder, 8 oz.,
rectified spirit and water a sufficiency. Digest
the jalap in warm spirit for 24 hours, then percolate with more spirit until nothing more is
dissolved. Add 4 oz. of water to the product
and distil off the spirit. The residue when cold
is washed 2 or 3 times with water, then dried.
Jalap should yield 10% resin.

2. (*Nativelle.*) Jalap root is digested in boiling water for 24 hours, and after being reduced
to thin slices more water is added, and the whole
boiled for 10 minutes, with occasional agitation;
the liquid is then expressed in a tincture press,
and the boiling and pressing repeated a second
and third time (these decoctions by evaporation
yield AQUEOUS EXTRACT OF JALAP); the pressed
root is next treated with rectified spirit, q. s.,
and boiled for 10 minutes, and then allowed to
cool; the tincture is then pressed out, and the
boiling with fresh alcohol and expression is repeated twice; a little animal charcoal is added to
the mixed tinctures, and, after thorough agitation, the latter are filtered; the liquid is now
distilled until nothing passes over, the supernatant fluid is poured off the fluid resin, and the
latter dried by spreading it over the surface of
the capsule, and continuing the heat. The product is a friable and nearly colourless resin,
which forms a white powder resembling starch.
—*Prod.* Fully 10% of pure resin.

3. (*Planche.*) Resinous extract of jalap is
dissolved in rectified spirit, the tincture agitated
with animal charcoal, and after filtration gently
evaporated to dryness.

Pur. The jalap resin of *commerce* is generally adulterated with scammony, gum, guaiacum or resin. When in a state of purity, it does
not form an emulsion with milk, like scammony
resin, but runs into a solid mass. It is insoluble
in fixed oils and turpentine, whilst the common
resins are freely soluble in those menstrua. Its
alcoholic solution, dropped on a piece of absorbent white paper, and exposed to the action of
nitrous gas, does not acquire a green or blue
colour; if it does, guaiacum resin is present.
2% of this adulteration may be thus detected
(*Gobley*). 10% only of the resin is soluble in
ether; but guaiacum resin, common resin, and
some others, are very soluble.

Powdered jalap resin placed in cold water does
not dissolve, but forms a semi-fluid transparent
mass, as if it had been melted. Dissolved in a
watch-glass with a little oil of vitriol, a rich
crimson-coloured solution is obtained, from which,
in a few hours, a brown viscid resin separates.
These last two characteristics distinguish it from
other resins.

Obs. Earthenware or well-tinned copper vessels
must alone be used in the above processes, as
contact with copper or iron turns the resin black,
and this tinge can only be removed by redissolving
the resin in alcohol, the addition of animal
charcoal, and re-evaporation.

Jalap resin is an energetic cathartic.—*Dose*,
1 to 5 gr. See JALAPIN.

Jalap, Factitious Resin of. *Syn.* RESINÆ JA-
LAPÆ FACTITIA, L. A substance frequently sold
for jalap resin is made by fusing a mixture of
pale yellow resin and scammony resin, and adding,
when it has cooled a little, but still semi-fluid, a
few drops of balsam of Peru or tolu; the mixture
is then poured into small paper capsules or tin
moulds. Its effects resemble those of jalap resin,
but it inflames less (*Landerer*).

Jalap, Soap of. *Syn.* SAPO JALAPÆ, SAPO
JALAPINUS, L. *Prep.* (Ph. Bor.) Resin of
jalap and Castile soap, of each, 1 part; rectified
spirit, 2 parts, or q. s. to dissolve the ingredients
softened by a gentle heat; subsequently evaporate
the mixture by the heat of a water-bath until
reduced to 4½ oz., or it has acquired the consistence of a pill-mass.

Prop., &c. A greyish-brown mass, soluble in
rectified spirit. Said to be milder in its action
than the resin alone.—*Dose*, 5 to 15 oz.

JALAPIC ACID. *Syn.* ODOROUS PRINCIPLE
OF JALAP. *Prep.* (*Pereira.*) Add an alcoholic
solution of acetate of lead to a similar solution of
jalap resin, collect the precipitate (jalapate of
lead), and throw down the lead by means of sulphuretted hydrogen. See ABSINTHIC ACID. A
brownish, soft, greasy substance, smelling strongly
of jalap, soluble in alcohol and alkali, and slightly
so in ether. Jalap resin contains about 13% of
this substance.

JALAPIN. $C_{34}H_{56}O_{16}$. *Syn.* JALAPINA. Jalap
resin is commonly sold under this name, but pure
jalapin is prepared by one or other of the following formulæ:

Prep. 1. The liquid filtered from the jalapate
of lead in preparing jalapic acid is a solution of
acetate of jalapin, which, after any trace of lead
is removed, by adding a few drops of dilute

sulphuric acid, and filtration, yields the whole of its jalapin, as a precipitate, on the addition of 5 or 6 times its volume of water; this is collected, washed with a little cold distilled water, and dried by exposure to a current of warm dry air.

2. (*Huss.*) Coarsely powdered jalap is digested in strong acetic acid for 14 days, the tincture filtered, ammonia added in excess, and the whole agitated strongly; the mixture is then filtered, the deposit washed in cold water, redissolved in acetic acid, re-precipitated by ammonia, and again washed and dried.

3. (*Kayser.*) Pure jalap resin, in powder, is digested for some time in boiling ether, by which means the jalapic acid is removed, and pure jalapin remains undissolved.

Prop., &c. A transparent, colourless, scentless, insipid resin, very soluble in alcohol, but insoluble in ether. It is the active purgative principle of crude jalap resin.

JAMAICA DOGWOOD. The bark of the root of the tree is the part employed in *medicine*; it is yielded by the *Picidia erythrina*, and used in the West Indies to intoxicate fish. It is a narcotic and sedative, relieves toothache, allays cough in bronchitis and phthisis. It dilates the pupil.—*Dose.* Of liquid extract, ½ to 2 dr.

JAMAI'CINE. *Syn.* JAMAICINA. A peculiar alkaloid obtained by Huttenschmidt from the bark of the cabbage-tree (*Andira inermis*).

Prep. The aqueous solution of cabbage-tree bark, treated with sulphuretted hydrogen and evaporated.

Prop. Yellow crystals soluble in water and, to a limited extent, in alcohol; fusible, and very bitter tasted. It forms salts with the acids, which, in small doses, produce restlessness and trembling; and in larger ones, purging. It is said to be vermifuge.

JAMBUL. The seeds or fruit-stones of *Eugenia jambolanum, Syzygium jambolanum.* Given in diabetes, diabetic ulceration, &c. Dr Kingsbury reports a case in the 'Lancet,' in which a patient had been suffering for six months, and was quite prostrate. 5 gr. of the powdered seeds were given six times in the 24 hours for a fortnight. The patient was then able to walk, had lost the abnormal thirst, &c., and was greatly relieved; sleeping well.—*Dose*, 5 to 10 gr.

JAMES'S POWDER. See POWDERS.

JAMS. *Syn.* PRESERVES. Conserves of fruit with sugar, prepared by boiling. In the latter respect they differ from the conserves of the apothecary.

Prep. The pulped or bruised fruit is boiled along with ½ to 2-3rds of its weight of loaf sugar until the mixture jellies, when a little is placed on a cold plate; the semi-fluid mass is then passed through a coarse hair sieve whilst hot to remove the stones and skins of the fruit, and as soon as it has cooled a little is poured into pots or glasses. It is usual to tie these over when cold with paper which has been dipped in brandy. The pots must then be placed aside in a dry and rather cold situation.

The following fruits are those from which jams are commonly prepared:—Apricots, cherries (various), cranberries, currants (black, red, and white), gooseberries (ripe and green), mulberries,

Orleans plums, raspberries, and strawberries. Red currants are commonly added to the last to remove insipidity.

JAPAN. See VARNISH, and *below.*

JAPAN'NING. The art of covering paper, wood, or metal, with a coating of hard, brilliant, and durable varnish. The varnishes or lacquers employed for this purpose in Japan, China, and the Indian Archipelago are resinous juices derived from various trees belonging to the Nat. Ord. ANACARDIACEÆ, especially *Stagmaria verniciflua, Holigarna longifolia, Semecarpus anacardium,* and species of *Rhus* (*Sumach*). For use, they are purified by a defecation and straining, and are afterwards mixed with a little oil, and with colouring matter as required. In this country varnishes of amber, asphaltum, or copal, or mixtures of them, pass under the names of 'JAPAN' and add 'JAPAN VARNISH.'

Prop. The surface is coloured or painted with devices, &c., as desired, next covered with a highly transparent varnish (amber or copal), then dried at a high temperature (135° to 165° F.), and, lastly, polished. Wood and paper are first sized, polished, and varnished. For plain surfaces, asphaltum, varnish, or japan, is used. See VARNISHING.

JAPON'IC ACID. $C_{12}H_{10}O_9$. When catechu is exposed to the air in contact with caustic alkalies black solutions (alkaline japonates) are formed; with carbonated alkalies, red solutions (alkaline rubates); the acid of the former may be separated. It is a black powder, insoluble in water, soluble in alkalies, and precipitated by acids. Rubic acid forms red insoluble compounds with the earths and some other metallic oxides.

JARAVE. The Spanish name for SARSAPARILLA BEER. See BEERS (in *pharmacy*).

JASPER. *Syn.* IASPIS, L. A mineral of the quartz family, occurring in rocky masses. It takes various shades of red, yellow, brown, and green, and is occasionally banded, spotted, variegated. It was formerly used as an amulet against hæmorrhages and fluxes. It is now extensively worked up into rings, seals, snuff-boxes, vases, &c., for which it is well suited from its extreme hardness and susceptibility of receiving a fine polish.

JATROPA CURCAS. The seed of the physic nut, *Curcas purgans*, is officinal in the Indian Pharmacopœia. It yields 30% of a purgative oil, acting like castor oil, and given in doses of 12 to 15 drops. *Externally* it is a stimulant, and used locally to increase the secretion of milk.

JATRO'PHIC ACID. *Syn.* CROTONIC ACID, IATROPHIC A. A peculiar fatty acid discovered by Pelletier and Caventou, and originally regarded by them as the cathartic principle of croton oil and croton seeds, but since shown by Redwood and Pereira to be nearly inert.

Prep. The oil is saponified by caustic potass, and the resulting soap is decomposed by tartaric acid; the fatty matter which floats on the surface of the liquid is then skimmed off the aqueous portion, and the latter submitted to distillation; the liquid in the receiver is a solution of jatrophic acid.

Prop., &c. Volatile; very acid; has a nauseous

odour; is solid at 23°, and vaporises at 35° F. It forms salts with the bases, none of which possess any practical importance.

JAU'MANGE. *Prep.* From isinglass, 1 oz.; boiling water, 12 oz.; dissolve; add of any sweet white wine, ½ pint; the yelks of 2 eggs beaten to a froth, and the grated yellow peel of 2 lemons; mix well, and heat the whole over the fire until sufficiently thickened, stirring all the time; lastly, serve it up or pour it into moulds.

JAUN'DICE. *Syn.* ICTERUS, MORBUS LUTE-OLUS, L.; ICTÈRE, FR.; GELBSUCHT, Ger. A disease characterised by a yellow colour of the eyes and skin, deep-coloured urine; and pale alvine evacuations.

Jaundice is of 2 kinds:—1. In which there is some impediment to the flow of bile into the small intestine, e. g. by reason of a gall-stone impacted in the bile-duct. 2. In which there is no such impediment.

The explanation of the disease in the first case is simple; the bile, unable to escape from the gall-bladder, is reabsorbed and thrown into the blood-vessels and lymphatics, and so distributed over the body, colouring the tissues.

The explanation in the second case is by no means clear, and authorities differ greatly as to the cause.

The number of diseases in which jaundice occurs, and the great variety of known causes of it, constitute it a *symptom* rather than a disease, and the treatment must be modified accordingly. Diuretics and purgatives are of great service, but must be used as the circumstances of each case appear to demand.

JEL'LY. *Syn.* GELATINA, L. A term now very loosely applied to various substances which are liquid or semi-liquid whilst warm, and become gelatinous on cooling.

Jellies are coloured by the addition of the usual stains used by confectioners, and are rendered transparent by clarification with white of egg.

Jelly, Almond. *Syn.* GELATINA AMYGDA-LARUM, L. *Prep.* From rich almond milk, ½ pint; thick hartshorn jelly, ½ pint; sugar, 2 oz.; with 2 or 3 bitter almonds and a little lemon peel to flavour, heated together, strained, and moulded.

Jelly, Ar'rowroot. *Syn.* GELATINA MA-RANTÆ, L. *Prep.* From arrowroot, 1½ oz., to water, 1 pint. *Tous les mois* jelly is made in the same way.

Jelly, Bis'cuit. *Prep.* From white biscuit (crushed beneath the rolling-pin), 4 oz.; cold water, 2 quarts; soak for some hours, boil to one half, strain, evaporate to 1 pint; and add of white sugar, ¾ lb.; red wine, 4 oz.; and cinnamon, 1 teaspoonful. In weakness of the stomach, and in dysentery and diarrhœa, and in convalescence combined with rich beef gravy or soup.

Jelly, Bladder-wrack. (*Dr Russell.*) *Syn.* GELATINA FUCI. *Prep.* Bladder-wrack (*Fucus vesiculosus*), 2 lbs.; sea-water, 2 lbs.; macerate for 15 days. Applied to glandular tumours.

Jelly Bread. *Syn.* PANADA; GELATINA PANIS, L. *Prep.* Cut a French roll into slices, toast them on each side, and boil in water, 1 quart, until the whole forms a jelly, adding more water if required; strain, and add sugar, milk, &c., to palate. It may be made with broth from which

the fat has been skimmed, instead of water. Used as the last.

Jelly Broth. *Syn.* SOUP JELLY. *Prep.* From broth or soup from which the fat has been skimmed, evaporated until it becomes gelatinous on cooling. A few shreds of isinglass are commonly added. See SOUP (Portable).

Jelly, Calves' Feet. *Prep.* For each foot take of water, 3 pints, and boil to one half; cool, skim off the fat, and again boil for 2 or 3 minutes with the peel of a lemon and a little spice; remove it from the fire, strain through a jelly bag (see FILTRATION), add the juice of a lemon and a glass of wine, and when it has cooled a little put it into glasses or 'forms.'

Obs. If this jelly is required to be very transparent it must be treated as follows:—After the fat is removed it should be gently warmed, just enough to melt it, next well beaten with the white of an egg and the seasoning, and then brought to a boil for a minute or two, when it will be ready for straining, &c. The calves' feet should not be bought ready boiled, but only scalded. Cows' feet ('COW HEELS') make nearly as good jelly as that from calves' feet, and are much more economical.

Jelly, Ceylon Moss. *Syn.* GELATINA GRACI-LARIÆ, L. *Prep.* (*Dr Sigmond.*) Boil Ceylon moss (*Granularia lichenoides*), ½ oz., in water, 1 quart, for 25 minutes, or till the liquid 'jellies' on cooling; strain and flavour. Very nutritious; recommended in irritation of the mucous membranes and phthisis.

Jelly, Codeine and Glycerin. *Prep.* Codeine, 72 gr.; citric acid, 720 gr.; refined gelatin, 6 oz.; glycerin, 36 oz.; oil of lemon, 1 dr.; balsam of tolu, and water, of each, a sufficiency. Boil the tolu in the water and filter, making up to 30 oz. Soak the gelatin in 25 oz. of the tolu water till dissolved, then add the glycerin. In the remaining 5 oz. of tolu water dissolve the codeine and citric acid, mix altogether, add the oil of lemon and stir well. Pour into wide-mouthed bottles to set. Useful in chronic laryngitis and phthisical cough.—*Dose*, 1 dr. (*S. Hardwick*).

Jelly, Copaiba. (*M. Caillot.*) *Syn.* GELA-TINA COPAIBÆ. *Prep.* Isinglass, 4 parts; water, 40 parts; dissolve in a water-bath, and add 20 parts of sugar; pour the clear liquid jelly into a warm mortar, and add copaiba, 60 parts; triturate, and pour in a vessel to jelly. Flavour with some aromatic essential oil or balsam of tolu.

Jelly, Cor'sican Moss. *Syn.* GELATINA HEL-MINTHOCORTI, L. *Prep.* (P. Cod.) Corsican moss (*Gracilaria helminthocorton*), 1 oz.; water, q. s.; boil 1 hour, and strain 8 fl. oz.; to this add of isinglass (previously soaked in a little water), 1 dr.; refined sugar, 2 oz.; white wine, a wine-glassful. Vermifuge. See DECOCTION.

Jelly, Fruit. Under this head we include those jellies made from the juices of fruits.

Prep. The strained juice mixed with ½ to ¾ its weight of refined sugar, until it 'jellies' on cooling, observing to carefully remove the scum as it rises. The process should be conducted by a gentle heat, and it is preferable not to add the sugar until the juice is somewhat concentrated, as by lengthened boiling the quality of the sugar is injured.

Obs. Jellies are sold in pots or glasses, like JAMS. Both jams and fruit jellies are refrigerant and laxative; they are, however, mostly employed as relishes, especially during fevers and convalescences. The principal fruit jellies are:— APPLE, BARBERRY, CHERRY (from either Cornelian or Kentish cherries), CURRANT (black, white, and red), ELDERBERRY, GOOSEBERRY, PLUM, QUINCE, RASPBERRY. See LEMON and ORANGE JELLY.

Jelly, Glycerin. *Syn.* GELATINA GLYCERINI. *Prep.* Mix glycerin to the required consistence with compound tragacanth powder. Or take powdered gum-arabic, ½ oz.; syrup, 4 oz. (3 oz. of sugar to 1 oz. of water); the yelks of 3 eggs; olive oil, 4 oz.; glycerin, 2 oz. Rub the gum and syrup well together, add the yelks, and when mixed add the oil and glycerin, previously triturated together. Applied to chapped hands, abrasions, &c. See GLYCERIN OF STARCH.

Jelly, Gra"vy. By evaporating meat gravies.

Jelly, Harts'horn. *Syn.* GELATINA CORNU CERVI, L. *Prep.* (P. Cod.) Hartshorn shavings, 8 oz.; wash it in water, then boil in clean water, 3 pints, till reduced to one half; strain, press, add of sugar, 4 oz., the juice of one lemon, and the white of an egg beaten up with a little cold water; mix well, clarify by heat, evaporate till it 'jellies' on cooling, then add the peel of the lemon, and set it in a cool place. It may be flavoured with wine, spices, &c. Very nutritious.

Jelly, Ice'land Moss. *Syn.* GELATINA LICHENIS, L. *Prep.* (P. Cod.) Iceland moss, 2 oz.; soak for 1 or 2 days in cold water, then boil for 1 hour in water, q. s. to yield a strong solution; strain, decant the clear after repose, apply heat, and dissolve therein of isinglass, 1 dr.; evaporate the whole to a proper consistence, put it into pots, and set them in a cool place. Nutritious. Recommended in phthisis. The jelly of Iceland moss and cinchona (GELATINA LICHENIS CINCHONA—P. Cod.) is made by adding to the above syrup of cinchona, 6 fl. oz.

Jelly, Ice'land Moss. Another formula. Iceland moss, 7½ oz.; water, 80 oz. Boil to 60 oz. and strain. Add lemon peel, 2½ oz.; isinglass, 2½ oz.; sugar, 40 oz. Boil to 40 oz. and strain.

Jelly, Ice'land Moss (Sweetened). (P. Cod.) *Syn.* GELATINA LICHENIS SACCHARATA, L. *Prep.* Saccharated Iceland moss (see ICELAND MOSS, SACCHARATED), 7½ oz.; sugar, 7½ oz.; water, 15 oz.; orange-flower water, 1 oz. Boil the first three substances and remove the scum which forms, and then let the jelly flow into a vessel which contains the orange-flower water.

Jelly, I"rish Moss. *Syn.* GELATINA CHONDRI, L. *Prep.* From Irish or Carrageen moss. See DECOCTION.

Jelly, I"singlass. *Syn.* CONFECTIONERS' JELLY; GELATINA ICHTHYOCOLLE, L. *Prep.* From isinglass dissolved in water by boiling, and evaporated till it 'jellies' on cooling, adding flavouring, as desired. 1½ oz. of good isinglass makes fully a pint of very strong jelly. See BLANCMANGE, ISINGLASS, CALVES-FEET JELLY, &c.

Jelly, Lem'on. *Prep.* From isinglass, 2 oz.; water, 1 quart; boil, add of sugar, 1 lb.; clarify,

and when nearly cold add the juice of 5 lemons, and the grated yellow rinds of 2 oranges and of 2 lemons; mix well, strain off the peel, and put it into glasses.

Jelly, No'yeau. As PUNCH JELLY, but strongly flavoured with bitter almonds.

Jelly, Orange. *Prep.* From orange juice, 1 pint; let it stand over the grated yellow rind of 3 or 4 of the oranges for a few hours, then strain, and add, of loaf sugar, ½ lb., or more, isinglass, ½ oz., dissolved in water, 1 pint; mix, and put it into glasses before it cools.

Jelly, Punch. *Prep.* From isinglass, 2 oz.; sugar, 1¼ lbs.; water, 1 pint; dissolve, add of lemon-juice, ½ pint; the peels of 2 lemons and of 2 oranges; rum and brandy, of each, ¼ pint; keep it in a covered vessel until cold, then liquefy it by a very gentle heat, strain, and pour it into moulds. A pleasant and deceptive way of swallowing alcohol.

Jelly, Quince. (Ph. E., 1744.) *Syn.* GELATINA CYDONIORUM, L. *Prep.* Juice of quinces, 3 lbs.; sugar, 1 lb.; boil to a jelly.

Jelly, Rice. *Syn.* CRÈME DE RIZ. From rice boiled in water, sweetened and flavoured.

Jelly, Sago. *Prep.* Soak sago in cold water for 1 hour, strain, and boil in fresh soft water until it becomes transparent; then add wine, sugar, clear broth, milk, or spices, to flavour. 1 oz. of sugar makes a pint of good jelly.

Jelly, Salep. (*Soubeiran.*) *Syn.* GELATINA SALEPE, L. *Prep.* Ground salep, 4 dr.; sugar, 4 oz.; water, q. s. Boil to 12 oz., and flavour to the taste.

Jelly, Tapio'ca. As SAGO JELLY, but using tapioca in lieu of sago.

JERVINE. $C_{30}H_{2}N_{2}O_{6}$. An alkaloid discovered in 1837, by Simon, in the root of the *Veratrum album*, and by Mitchell, in 1874, in the root of the *Veratrum viride*. Dr H. C. Wood, jun., describing the physiological effects of jervine, says they consist "in general weakness, lowering of arterial pressure, a slow pulse, profuse salivation, and finally convulsions." Jervine was analysed by Will, who ascribed to it the above composition.

JESUIT'S BARK. See CINCHONA.

JESUIT'S DROPS. There are various formulæ for this preparation, *e. g.* the old Dublin Pharmacopœia tr. benzoini co. was given for it. The following are typical formulæ:

1. Guaiacum, 7 oz.; bals. Peruv., 4 dr.; sarsaparilla, 5 oz.; rectified spirit, 2 pints. Digest for 14 days and filter.

2. Copaiba, 1 oz.; guaiacum, 2 dr.; oil of sassafras, 1 dr.; salt of tartar, ½ dr.; rectified spirit, 5 oz. Digest for a week and filter.

JESUIT'S POWDER. Powdered cinchona bark.

JET. A variety of mineral bituminous carbon, very hard, and susceptible of a fine polish.

JEWELRY. The gold in articles of jewelry, whether solid or plated, which are not intended to be exposed to very rough usage, is generally 'coloured,' as it is called in the trade. This is done as follows:

1. (RED GOLD COLOUR.) The article, after being coated with the amalgam, is gently heated, and, whilst hot, is covered with gilder's wax; it

is then 'flamed' over a wood fire, and strongly heated, during which time it is kept in a state of continual motion, to equalise the action of the fire on the surface. When all the composition has burned away, the piece is plunged into water, cleansed with the 'scratch-brush' and vinegar, and then washed and burnished. To bring up the beauty of the colour, the piece is sometimes washed with a strong solution of verdigris in vinegar, next gently heated, plunged whilst hot into water, and then washed, first in vinegar, or water soured with nitric acid, and then in pure water; it is, lastly, burnished, and again washed and dried.

2. (ORMOLU COLOUR.) This is given by covering the parts with a mixture of powdered hæmatite, alum, common salt, and vinegar, and applying heat until the coating blackens, when the piece is plunged into cold water, rubbed with a brush dipped in vinegar, or in water strongly soured with nitric acid, again washed in pure water, and dried. During this process, the parts not to be dried in 'or-molu colour' should be carefully protected.

The frauds practised in reference to the 'fineness' of the metal used in jewelry is noticed under GOLD (Jeweller's). See also ASSAYING, DIAMOND, GEMS, GILDING LIQUOR, GILDING WAX, &c.

JEWELS. See DIAMOND, EMERALD, GEMS, &c.

JEW'S PITCH. See ASPHALTUM.

JORDAN and VALENCIA ALMONDS. The kernel of the fruit of *Prunus amygdalus*, Baill. (*Amygdalus communis*, Linn.), a tree cultivated in the north of Africa, Italy, Spain, &c. Jordan (corruption of jardyne or garden, *i. e.* cultivated) and Valencia almonds are imported from Malaga without the shell, and differ from other sorts by their large size and oblong form.

JOURNAL BOXES, Alloy for. *Prep.* Copper, 24 parts; tin, 24 parts; antimony, 8 parts. First melt the copper, then add the tin, and lastly the antimony.

JUICE (Spanish). See EXTRACT and LIQUORICE.

JUJUBE. A fruit resembling a small plum, produced by various species of *Zizyphus*. Combined with sugar, it forms the JUJUBE PASTE of the shops, when genuine; but that now almost always sold under the name is a mixture of gum and sugar, slightly coloured and flavoured.

Jujubes, How to Make. 2 lbs. of picked gum-arabic, 1¼ lbs. of the finest sugar (sifted), 5 oz. of orange-flower water, and 1 pint of pure water. Powder the gum and then put it into a bright clean basin with 1 pint of water, and dissolve it over a slow fire, stirring constantly with a wooden spatula. When it is entirely dissolved, strain it through a towel or fine hair-sieve to free it from all sediment. Put the strained gum and the sugar into another clean bright basin, and stir it over a very moderate fire while it boils and reduces to the small pearl (or 30° by the saccharometer); then add the orange-flower water. Stir all together on the fire, take off the scum, and pour the mixture into very smooth clean tin pans that have been previously well rubbed with oil of almonds, or with olive oil; fill them with the mixture to the depth of a ¼ of an inch, and set them to dry in the drying-room (moderate heat). When sufficiently dried, so that on pressing the surface it proves to be somewhat elastic to the touch, remove them from the heat and allow them to become cold; the jujube may then be easily detached and removed from the pans, and is then to be cut up with scissors into strips, and then the strips into diamond-shaped pieces. The jujubes can be coloured with cochineal or ammoniated carmine solution, may be flavoured with vanilla, rose, &c., and may be medicated.

JULEP. *Syn.* JULAP; JULEPUM, JULEPUS, JULAPIUM, L. A term usually regarded as synonymous with 'MIXTURE'; but according to the best authorities, implying a medicine which is used as a vehicle for other forms of medicine. The word comes through the French, from a Persian expression, which signified 'sweet drink.' A julep, according to Continental writers, is a drink of little activity, generally composed of distilled waters, infusions, and syrups, to which mucilages and acids are sometimes added; "but never powders or oily substances, which could interfere with its transparency." In England the juleps of *old pharmacy* are now classed under 'MIXTURES.'

JULUS GUTTATUS, JULUS LONDINENSIS. The Thousand-legs. The hop-set, or young plants, are rugose and knotty, affording much shelter or cover for the eggs, grubs, and pupæ of insects. Planters usually plant two or three of these sets together to form one plant-centre. These, while keeping a separate or distinct existence, become much intertwisted, having many knots and cavities, hiding places, which are made use of by many species of the Julidæ or 'thousand-legs.' These are very frequently found in such cavities and in great abundance, especially where any decay has commenced. This they intensify, if they do not actually cause it, and if they contrive to penetrate into the softer, more sappy parts of the plant-centres, they rapidly occasion dangerous rotting. It is commonly held that these thousand-legs are merely attendants upon decay, and do not themselves create it; but the formation of their jaws, adapted for gnawing and biting, proves clearly that they are active sources of injury to plants. The thousand-legs (millipedes) must not be confounded with the species of another family of Myriapods, known as *Scolopendrida*, or, familiarly, centipedes, whose jaws are quite differently formed, and live on insects and animal matter. The two species commonly found injuring various crops in England are distinguished as *Julus Londinensis* and *Julus guttatus*. Similar species are known in France, Germany, and America, where they injure beans, peas, cabbages, many corn-crops, and hop-plants. The mischief, or rather the source of the mischief, which these creatures occasion to hop-plants is not at first apparent, and it would be desirable that planters should examine the roots of the plants closely when they flag or show symptoms of disease.

Life History. Strictly and scientifically speaking, the thousand-legs are not insects, though they are generally considered and may be treated here as such. They undergo no transformation like wireworms and other insects proper, and have only two stages of life, viz. the egg stage and the caterpillar or worm stage.

From the end of December to the beginning of May the female lays eggs in considerable numbers under stones, in decaying wood, and in vegetation, in the roots of the hop plants, and in other retreats where there is dampness. When the young emerge from the eggs they have at first only three pairs of legs, according to Taschenberg, but the number of legs increases. They are not full grown, Curtis says, until they are two years old, changing their skins or moulting five times during this period, and feeding actively throughout. It is believed that, like wireworms, they live four or five years from the time they come from the eggs.

Prevention. It is essential, for every reason, that hop land should be drained. For the prevention of thousand-legs this is most desirable, since they love moisture.

All vegetable rubbish and decaying matter should be removed from hop plantations which would serve as a harbour for them.

Frequent and thorough cultivation by digging round the plant-centres, and the application of nitrate of soda, soot, lime, or lime-ashes, to be dug deeply into the soil close round them, will be found very beneficial.

Remedies. Traps of pieces of carrot, turnip, mangel-wurzel, or vegetable marrow, put round the plant-centres, might be advantageously employed. This is done in Germany. Vegetable marrow is the best medium, being soft.

In the case of a serious attack the use of the paraffin-saturated materials recommended in many other cases would be beneficial. Curtis talks of lime-water being used with effect, as well as of nitrate of soda being washed in. This would hardly be practicable ('Reports on Insects Injurious to Crops,' by Chas. Whitehead, Esq., F.Z.S.).

JU"NIPER BERRIES. *Syn.* JUNIPERI BACCÆ, J. COMMUNIS BACCÆ (Ph. E.), JUNIPERUS (Ph. L.), L. The fruit of the *Juniperus communis*, or common juniper tree.

Donath obtained from 100 parts of the berries—

Water	29·44
Volatile oil . . .	·91
Formic acid . . .	1·86
Acetic acid . . .	·94
Malic acid (combined) . .	·21
Oxalic acid . . .	traces
Wax-like fatty matter . .	·64
Green resin (from ethereal solution) . . .	8·46
Hard brown resin (from alcoholic solution) . .	1·29
Bitter principle (called juniperin) . . .	·37
Pectine	·73
Albuminous substances .	4·45
Sugar	29·65
Cellulose . . .	15·88
Mineral substances .	2·83
Loss	2·89

In the old Ph. L. and D. both the tops and berries (JUNIPERI FRUCTUS ET CACUMINA, Ph. L.

1836; JUNIPERUS—BACCÆ, CACUMINA, Ph. D. 1862) were ordered. The berries are stomachic and diuretic, and have been long employed in dropsies, either alone or combined with foxglove and squills. The tops (SUMMITATES) have been highly praised in scurvy and certain cutaneous affections.—*Dose,* 1 to 2 dr., made into a conserve with sugar, or in the form of infusion or tea.

JUN'KET. *Syn.* DEVONSHIRE JUNKET, CURD JELLY. *Prep.* From warm milk put into a bowl, and then turned with a little rennet; some scalded cream and sugar are next added, with a sprinkling of cinnamon on the top, without breaking the curd. Much esteemed by holiday folk in the western counties during the hot weather of summer. Sometimes, very strangely, a little brandy finds its way into these trifles.

JUTE. This is the fibre yielded by the *Corchorus capsularis*, a lime tree growing in India and China. It is the material of which sacks, gunny bags, and coarse thread are made. It mixes even with linen or cotton, and hence may not improbably be employed as a sophisticant of these substances.

The *engravings* on page 920 exhibit the different microscopic appearances of the three substances.

KAIRIN. Hydrochlorate of oxychinoline-ethyl. Small white crystals, soluble in water. The solution gives a white precipitate with ammonia. Used in fevers and inflammation as an antipyretic and febrifuge; its taste is very unpleasant, so is best given hypodermically. *Dose,* 5 to 10 gr.

KALEIDOSCOPE. *Syn.* FLOWER-GLASS. A pleasing philosophical toy invented by Sir David Brewster, which presents to the eye a series of symmetrical changing views. It is formed as follows:—Two slips of silvered glass, from 6 to 10 in. long, and from 1 to 1¼ in. wide, and rather narrower at one end than the other, are joined together lengthwise, by one of their edges, by means of a piece of silk or cloth glued on their backs; they are then placed in a tube of tin or pasteboard, blackened inside, and a little longer than is necessary to contain them, and are fixed by means of small pieces of cork, with their faces at an angle to each other—that is, an even aliquot part of 4 right angles (as the ½, ⅓, ¼, &c.). The small end of the tube is then closed with an opaque screen or cover, through which a small eyehole is made in the centre; and the other end is fitted, first with a plate of common glass, and at the distance of about 1-8th of an inch, with a plain piece of slightly ground glass, parallel to the former; in the intermediate place or cell are placed the objects to form the images. These consist of coloured pieces of glass, glass beads, or any other coloured diaphanous bodies, sufficiently small to move freely in the cell, and to assume new positions when the tube is shaken or turned round. A tube so prepared presents an infinite number of changing and symmetrical pictures, no one of which can be exactly reproduced. This toy is so easily constructed, is so very inexpensive, and at the same time so capable of affording an almost inexhaustible fund of amusement to the young, that we advise our juvenile friends to try their hands at its construction. Any common

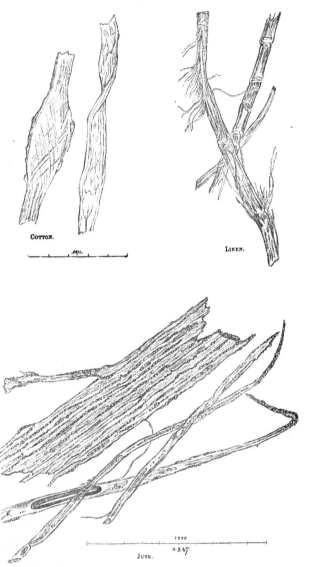

COTTON.

LINEN.

1000

× 247

JUTE.

tube of tin or pasteboard may be used, and strips of glass smoked on one side will answer for mirrors.

KA'LI. The name formerly applied to a species of *Salsola* employed for making BARILLA. It is sometimes used as a designation for the crude alkalies, and is the German synonym for 'potassa.'

Kali, Acid'ulated. *Syn.* LEMON AND KALI, LEMONIATED K. A common preparation of the shops for making a pleasant effervescing draught. It is sometimes incorrectly styled 'citrate of potash.' *Prep.* 1. Carbonate of soda and tartaric acid, of each, 5 oz.; lump sugar, 1 lb.; all in the state of fine powder, and separately dried by a very gentle heat, after which they are mixed together, flavoured with essence of lemon, 1 dr., rubbed through a gauze sieve in a warm dry situation, put into bottles and corked down immediately.

2. Finely powdered white sugar, 16 lbs.; tartaric acid, 4½ lbs.; carbonate of soda, 4 lbs.; essence of lemon, 1 oz.; as the last. Keeps well. A dessert-spoonful of either thrown into a glassful of water makes a pleasant effervescing draught.

KA'LIUM. [L.] Potassium.

KAL'YDOR. A cosmetic lotion; it resembles 'GOWLAND'S LOTION,' but is got up in a rather more pleasing style. See LOTION.

KAMALA. The red powder rubbed off the fruit capsules of *Mallotus philippinensis*, Muell. Arg. (*Rottlera tinctoria*, Roxb.), a plant belonging to the Nat. Ord. EUPHORBIACEÆ. Kamala is imported from India, where it is known under the name of kameela. The tree from which it is obtained is from 15 to 20 feet in height, indigenous to India and to many of the East Indian Islands.

Kamala has long been employed in India as a remedy for tapeworm, and within the last few years has been given for the same purpose in this country with fair success. It may be administered in doses of from 30 gr. to 3 dr., suspended in water, rubbed up with mucilage, or mixed with syrup. In large doses, such as 3 dr., it sometimes purges violently. After the third or fourth motion the worm is generally evacuated dead. A second dose may be taken in about 4 hours should the first fail to act; or, instead of a second dose, some castor oil may be given. Kamala is also used externally by the natives of India in various skin complaints, particularly in scabies. It is also said to have proved useful in herpetic ringworm.

Dr Anderson obtained from the resinous colouring matter, which is the principal constituent of kamala, a yellow crystalline substance, to which he gave the name *rottlerin*. The existence of rottlerin has since been confirmed by Mr Groves, who found that it becomes changed by exposure, a circumstance to which he attributes its non-detection in old specimens of the drug, and to which may very reasonably be attributed Leube's failure to find it.

The British Pharmacopœia ascribes the following 'characters' to kamala:—"A fine granular mobile powder, of a brick-red colour; it is with difficulty mixed with water, but when boiled with alcohol the greater part is dissolved, forming a red solution. Ether dissolves most of it, the

residue consisting principally of tufted hairs. It should be free from sand or earthy impurities."

Kamala forms a very considerable article of export from India, it being a valuable yellow dye.

KA'OLIN. *Syn.* CHINA CLAY, PORCELAIN C. A fine white clay, derived from the decomposition of the felspar of granitic rocks, and consisting almost entirely of hydrous silicate of aluminium, whereas ordinary clay contains quartz and iron oxide in addition. The potteries and porcelain works of this country are chiefly supplied with this substance from extensive tracts of it which occur near St Austell, Cornwall. See CLAY.

Eisner gives the following process for distinguishing kaolin from ordinary clay :—He agitates it in a test-tube with pure strong sulphuric acid till a uniform mixture is produced, decants the acid after subsidence, dilutes it carefully with 6 volumes of water, and supersaturates the cooled solution with ammonia.

Kaolin thus treated separates but slowly from the strong acid, and the diluted acid solution gives an immediate white precipitate with ammonia, whereas ordinary clay is but slightly attacked by the acid, separates quickly from it, and the acid after dilution gives but an insignificant precipitate with ammonia. When ground and washed it forms a powder often sold as fuller's earth. The powder with a little water is a good pill excipient for such substances as permanganate of potassium, chloride of gold, and silver nitrate. 1 oz. of powdered kaolin rubbed with 5 drops of creosote forms an excellent dusting powder for chafed or sore places, likewise for erysipelas. Kaolin ointment is a mixture of equal weights of vaseline, hard paraffin, and kaolin melted, mixed, and stirred till cold.

KAP'NOMOR. $C_{10}H_{11}O$. *Syn.* CAPNOMOR. A colourless oil obtained from crude creosote by distillation with potassa. It begins to boil at 180° C. (360° F.), but the greater part comes over between 200° and 208° C. (392° and 406° F.). It has a peculiar odour, and is insoluble in water, but readily soluble in an alkaline solution of creosote.

KATAL'YSIS. *Syn.* CATALYSIS, CONTACT ACTION. Terms applied to a class of chemical actions in which the decompositions and the recombinations of the elements of compound bodies are apparently excited by the mere presence of, or contact with, other bodies, which do not themselves suffer such a change. A good example of a katalytic agent is platinum-black (finely divided metallic platinum). When a mixture of hydrogen and oxygen are passed over this substance at the ordinary temperature combination takes place between them.

KAVA-KAVA (*Piper methysticum*). A plant growing in the Polynesian Islands, the root of which is employed by the natives to form an intoxicating drink. It contains a white crystalline principle known as *kavain*. In small doses it is tonic, stimulant, and diuretic. It has been highly recommended in gonorrhœa and in gout.—*Dose*, 30 to 60 minims of the fluid extract. See AVA.

KEFYR, or **KEPLIN.** A nutrient drink prepared from the Caucasian milk fungi and largely used in Germany and Switzerland under the name of kefyr-kumis. Following are methods for its preparation:—The dry fungi, after having been kept during 3 hours in tepid water and washed a few times with clean water, are put into new milk at 30°, which must be renewed daily. The fungi, which are originally of greater sp. gr. than milk, must be shaken frequently, but not too violently, for about 8 days, in order to acquire by increase of size change of colour to white, and gradual rising to the surface of the liquid, the qualities necessary for the preparation of kefyr and kefyr-kumis. New, cool skimmed milk is now poured upon the fungi, at the rate of 6 to 8 times their volume, the vessel closed with a cork, and left in a medium temperature for 24 hours, being frequently shaken meanwhile. The liquid is then strained, and the same procedure repeated once or twice again after washing the fungi with cold water. The beverage thus obtained is the kefyr of the mountaineers, which for appearance and taste may be compared to fresh sour cream.

The kefyr-kumis, called 'kapyr' by the natives, is obtained by pouring together in a champagne-bottle 1 part of kefyr and 2 of new, cool milk, corking the bottle tightly and leaving the compound to brew with frequent shaking for 24, 48, or 72 hours, according to the required strength of the kapyr. It is extensively prepared and consumed in Odessa and the ports round about. N. Seidemann, of Odessa, has analysed the product, and finds its percentage composition to be—

Casein	.	.	4·0
Albumen	.	.	0·8
Butter	.	.	8·0
Sugar of milk	.	.	2·0
Alcohol	.	.	0·6
Water and salts	.	.	88·0
Carbonic acid	.	.	1·0

KELP. *Syn.* VARECH, Fr. The alkaline ashes obtained by burning various species of sea-weed, formerly much used for the preparation of carbonate of soda. The weeds most valued for the purpose are the *Fucus vesiculosus, nodosus,* and *serratus,* and the *Laminaria bulbosa* and *digitata.*

Of late years the manufacture of kelp, like that of barilla, has been almost abandoned except as a source of iodine. Mr E. C. C. Stanford, by carefully collecting and compressing the weed, and afterwards submitting it to dry distillation, largely increases the yield of iodine and bromine, and obtains in addition various valuable hydrocarbons. See BARILLA, IODINE, SODA, &c.

KERATINE. A substance obtained from horn shavings by digesting them with pepsin and dilute hydrochloric acid which dissolves out the albuminous substances. The residue is now dissolved in ammonia, and evaporated to gum-like consistence. Keratine thus prepared is used for coating pills which are intended to pass the stomach and dissolve in the small intestine; the action of a pill can thus be localised.

KER'MES. *Syn.* KERMES-GRAINS, ALKERMES; GRANUM TINCTORIUM, L. The dried bodies of the female *Coccus ilicis* of Linnæus, a small insect of the Ord. HEMIPTERA, which flourishes on the Ilex oak. It has been used as a red and scarlet dye-stuff ever since the time of Moses; but is now superseded in this country by cochineal, which gives colours of much greater brilliancy.

KER'MES MIN'ERAL. *Syn.* KERMES MINERALE, K. MINERALIS, L. An amorphous ter-sulphide of antimony, containing a small admixture of teroxide of antimony and sulphuride of potassium. *Prep.* 1. IN THE HUMID WAY. *a.* (P. Cod.) Carbonate of soda (cryst.), 128 parts (say 21 parts), is dissolved in water, 1280 parts (say 210 parts), contained in a cast-iron pan; ter-sulphide of antimony (in fine powder), 6 parts (say 1 part), is next added, and the whole boiled for an hour, with constant agitation with a wooden spatula; the boiling liquid is then filtered into a heated earthen pan containing a small quantity of very hot water, and the solution is allowed to cool as slowly as possible; the red powder which is deposited is collected on a cloth, on which it is well washed with cold water, and the superfluous water being removed by pressure, the powder is dried by a gentle heat, and is, lastly, passed through a fine silk-gauze sieve, and preserved from light and air.

b. (Wholesale.) From black sulphide of antimony, 4 lbs.; carbonate of potassa, 1 lb.; boil in water, 2 galls., for ⅓ hour, filter, &c., as before. The undissolved portion of sesquisulphide of antimony may be boiled again several times with fresh potassa and water, until the whole is dissolved. Inferior to the last.

c. (CLUZELL'S KERMES.) From tersulphide of antimony, 4 parts; crystallised carbonate of soda, 90 parts; water, 1000 parts; boil, &c., as in 1, *a*, and dry the powder, folded up in paper, at a heat not exceeding 90° F.

2. IN THE DRY WAY. *a.* (P. Cod.) Carbonate of potassa, 100 parts; tersulphide of antimony, 50 parts; sulphur, 3 parts; mix, fuse in a Hessian crucible, pour the melted mass into an iron mortar, and when cold reduce it to powder; next boil it in water, 1000 parts, contained in an iron vessel, filter the solution, and otherwise proceed as before.—*Prod.*, large, but of inferior quality.

b. (*Fownes.*) From tersulphide of antimony, 5 parts; carbonate of soda (dry), 3 parts; water, 80 parts; fuse, &c., as before. Nearly equal to 1, *a*.

c. (*Berzelius.*) Carbonate of potassa (pure), 3 parts; tersulphide of antimony, 8 parts; water, q. s. Resembles the last.

Prop., &c. An odourless, tasteless powder, insoluble in both water and alcohol, and, when pure and carefully prepared, entirely soluble in hydrosulphate of ammonia. As prepared by the formulæ 1, *a*, and 1, *c*, it is a very dark crimson powder, of a velvety smoothness; but that from the other formulæ has a brownish-red colour, more or less deep. The secret of preparing this compound of a fine and velvety quality, like that imported from the Continent, consists simply in filtering the solution whilst boiling hot, and allowing it to cool very slowly, by placing the vessel in an appropriate situation for that purpose. Another important point, according to Rose, is to employ sufficient alkali to keep the whole of the teroxide of antimony in solution as the liquid cools, instead of allowing a part of it to be deposited with the kermes. This is the reason of the superior quality and mildness of that prepared according to the directions of the French Codex. The liquor decanted from the 'kermes mineral' yields the golden sulphide of antimony on the addition of an acid, for which purpose acetic acid is generally employed.

Dose, ¼ gr. to 3 or 4 gr., as a diaphoretic, cathartic, or emetic. It occupies in foreign practice the place of our James's Powder.

KETCH'UP. *Syn.* CATCHUP, CATSUP, KATCHUP. The juice of certain vegetables strongly salted and spiced, so as to be used as sauce; or a simple sauce made without the natural juice as a substitute for the true ketchup. The following are the principal varieties:

Ketchup, Camp. *Prep.* Take of good old beer, 2 quarts; white wine, 1 quart; anchovies, 4 oz.; mix, heat it to the boiling-point, remove it from the fire, and add of peeled shalots, 3 oz.; mace, nutmegs, ginger, and black pepper, of each, bruised, ½ oz.; macerate for 14 days, with frequent agitation, then allow it to settle, and decant and bottle the clear portion.

Ketchup, Cu'cumber. *Prep.* From ripe cucumbers, in the same way as mushroom ketchup. Very luscious. Mixed with cream or melted butter it forms an excellent white sauce for fowls, &c.

Ketchup, Marine. *Prep.* Take of strong old beer, 1 gall.; anchovies, 1½ lbs.; peeled shalots (crushed), 1 lb.; bruised mace, mustard-seed, and cloves, of each, ½ oz.; bruised pepper and ginger, of each, ¼ oz.; mushroom ketchup and vinegar, of each, 1 quart; heat the mixture to the boiling-point, put it into a bottle, and macerate for 14 days, frequently shaking; then strain through flannel, and bottle it for use. Excellent with anything; like the last, it makes good white sauce, and keeps well.

Ketchup, Mush'room. *Prep.* 1. Sprinkle mushroom flaps, gathered in September, with common salt, stir them occasionally for 2 or 3 days, then lightly squeeze out the juice, and add to each gallon cloves and mustard-seed, of each, bruised, ½ oz.; allspice, black pepper, and ginger, of each, bruised, 1 oz.; gently heat to the boiling-point in a covered vessel, macerate for 14 days, and decant or strain. Should it exhibit any indications of change in a few weeks, bring it again to the boiling-point, with a little more spice and a tablespoonful more salt.

2. Take of mushroom juice, 2 galls.; pimento, 2 oz.; cloves, black pepper, mustard-seed, and ginger, of each, bruised, 1 oz.; salt, 1 lb. (or to taste); shalots, 3 oz.; gently simmer for 1 hour in a covered vessel, cool, strain, and bottle.

3. Take of mushroom-juice, 100 galls.; black pepper, 9 lbs.; allspice, 7 lbs.; ginger, 5 lbs.; cloves, 1 lb.; all bruised; salt, q. s.; gently simmer in a covered tin boiler for 1 hour.

Ketchup, Oys'ter. *Prep.* Pulp the oysters, and to each pint add, of sherry wine, or very strong old ale, 1 pint; salt, 1 oz.; mace, ¼ oz.; black pepper, 1 dr.; simmer very gently for 10 minutes, strain, cool, bottle, and to each bottle add a spoonful or two of brandy, and keep them in a cool situation. COCKLE KETCHUP and MUSSEL KETCH-

UP are made in the same way. Used to flavour sauces when the fish are out of season; excellent with rump-steak, &c.

Ketchup, Pon'tac. *Prep.* Take of the juice of elderberries and strong vinegar, of each, 1 pint; anchovies, ½ lb.; shalots and spice, q. s. to flavour; boil for 5 minutes, cool, strain, and bottle. Used to make fish sauces.

Ketchup, Toma'to. *Prep.* Prepared from tomatoes or love-apples, like mushroom ketchup, except that a little very strong Chili vinegar is commonly added. An admirable relish for 'high' or rich-flavoured viands.

Ketchup, Wal'nut. *Prep.* 1. Take of the expressed juice of young walnuts, when tender, 1 gall.; boil 10 minutes, skim, add of anchovies, 2 lbs.; shalots, 1 lb.; cloves and mace, of each, 1 oz.; 1 clove of garlic, sliced; simmer in a covered vessel for 15 minutes, strain, cool, and bottle, adding a little fresh spice to each bottle, and salt, q. s. Will keep good in a cool place for 20 years.

2. Take of green walnut-shells, 16 galls.; salt, 5 lbs.; mix and beat together for a week, press out the liquor, and to every gallon add, of allspice, 4 oz.; ginger, 3 oz.; pepper and cloves, of each, 2 oz.; all bruised; simmer for half an hour, and set aside in a closed vessel and in a cool situation until sufficiently clear.

3. Take of walnut-juice, 1 gall.; vinegar, 1 quart; British anchovies (sprats), 3 or 4 lbs.; pimento, 3 oz.; ginger, ½ oz.; long pepper, ½ oz.; cloves, 1 oz.; shalots, 2 oz.; boil and bottle as before.

4. From the juice of walnut-shells, 30 galls.: salt, 1 bushel; allspice and shalots, of each, 6 lbs.; ginger, garlic, and horseradish, of each, 3 lbs.; essence of anchovies, 3 galls.; as before.

Ketchup, Wine. *Prep.* Take of mushroom or walnut ketchup, 1 quart; chopped anchovies, ½ lb.; 20 shalots; scraped horseradish, 2 oz.; spice, q. s.; simmer for 15 minutes, cool, and add of white and red wine, of each, 1 pint; macerate for 1 week, strain, and bottle.

General Remarks. In preparing the above articles vessels of glazed earthenware or stoneware, or well-tinned copper pans, should alone be used to contain them whilst being boiled or heated, as salt and vegetable juices rapidly corrode copper, and render the ketchup poisonous. Nothing in the shape of copper, lead, or pewter, should be allowed to touch them. Even a plated copper spoon left in a bottle of ketchup for some time will render its contents poisonous. Unpleasant and even dangerous fits of vomiting, colic, and diarrhœa have resulted from the neglect of this precaution. See SAUCE, &c.

KIBES. The vulgar name for ulcerated chilblains.

KID'NEYS. *Syn.* RENES, L. (In *anatomy*.) The kidneys are the organs which secrete the urine, and form the great channels by which the effete nitrogenous matter is removed from the blood. See URINE and URINARY AFFECTIONS.

Kidneys. (In *cookery*.) Soyer recommends kidneys to be dressed by gently broiling them, having previously split them, "so as nearly to divide them, leaving the fat in the middle," and "run a skewer through them, that they may remain open." After being rubbed with a little butter, and seasoned with salt and pepper, "they may be served on toast, or with any sauce." "You may also egg and bread-crumb them." "Five minutes suffice for a sheep or lamb's kidney of common size" (*Soyer*). 1 or 2 lamb's kidneys, plainly broiled and served up with the gravy in them, eaten along with a little dry-toasted bread, form a most excellent and appropriate luncheon or dinner for a dyspeptic or convalescent.

KIESERITE. A sulphate of magnesia found in the refuse salt (abraumsalz) of Stassfurt, near Magdeburg. It forms about 12% of the *abraumsals*. It is employed for washing wool and for the manufacture of 'permanent white' by treatment with chloride of barium; also for the preparation of Glauber salts, and of hypochlorite of magnesia for bleaching linen. See LINEN.

KING'S CUP. *Prep.* Yellow peel of 1 lemon; lump sugar, 1½ oz.; cold water, 1 pint; infuse 8 or 10 hours, and strain. The addition of a teaspoonful of orange-flower water is a great improvement. Used as a diluent in cases where acid liquors are inadmissible. See LEMONADE.

KING'S EVIL. See SCROFULA.

KING'S YELLOW. See YELLOW PIGMENTS.

KI'NIC ACID. $HC_7H_{11}O_6$. *Syn.* QUINIC ACID, CINCHONIC ACID. A monobasic acid occurring in the cinchona barks, in which it exists associated with the alkaloids.

Kinic acid is somewhat extensively diffused throughout the vegetable kingdom, being found in the bark of every species of the true cinchonas, as well as in the leaves of the oak, the elm, the ash, the ivy, the privet, and the coffee plant and berries. It occurs in the cinchona barks most probably combined with the alkaloids, which therefore exist in the plant as kinates.

It is readily obtained from kinate of lime by the action of dilute sulphuric acid; the filtered solution, evaporated to the consistence of a syrup, gradually deposits large crystals resembling those of tartaric acid.

Henry and Plisson give the following directions for the preparation of kinic acid:—Make a decoction of cinchona bark with water containing some sulphuric acid, and filter whilst hot, and to the filtrate add gradually freshly precipitated oxide of lead until the liquid becomes neutral and changes from a red to a pale yellow colour; care must be taken to add sufficient oxide. The filtrate is freed from lead by passing sulphuretted hydrogen through it, and filtered milk of lime is then added to precipitate the quinine and cinchonine; and the filtered liquid is evaporated to a syrup, which yields on cooling crystalline calcic kinate. To separate the acid from the calcic salt, Berzelius directs an aqueous solution of the salt to be made and to be precipitated by basic acetate of lead; the washed precipitate, suspended in water, is then decomposed by sulphuretted hydrogen, and the solution filtered and evaporated. Or the calcium kinate may be decomposed by an aqueous or alcoholic solution of sulphuric acid (*Watts*).

Kinic acid is, in the form of large tabular crystals, fusible at 161° C. These crystals dissolve in 2 parts of water; they are also soluble in spirits of wine, but scarcely, if at all, in ether.

It forms salts called kinates. Kinate of calcium is obtained from an acidulated infusion of cinchona bark, by adding an excess of lime, filtering, evaporating to a syrup, and setting the liquid aside to crystallise. These crystals are purified by re-dissolving them, treating the solution with a little animal charcoal, and crystallising the salt as before. The liquid from which the bark-alkaloids have been precipitated by hydrate of lime affords an almost inexhaustible supply of this salt. Mr H. Collier states that kinate of quinine is one of the most soluble and best salts of quinine for hypodermic use. See KINONE.

KI'NO. *Syn.* GUM-KINO; KINO (B. P., Ph. L. E. and D.) The juice flowing from the incised bark of the *Pterocarpus marsupium*, imported from the Malabar Coast, hardened in the sun.— *Dose*, 10 to 30 gr., in powder; as an astringent in chronic diarrhœa, &c.

Kino, Factitious, met with in the shops, is made as follows:—Logwood, 48 lbs.; tormentil root, 16 lbs.; madder root, 12 lbs.; exhaust by coction with water, q. s.; to the liquor add of catechu, 16 lbs.; dissolve, strain, and evaporate to dryness. *Prod.*, 24 lbs. Extract of mahogany is also commonly sold for kino.

KIRSCH'WASSER (-väs-ser). [Ger.] *Syn.* KIRSCHENWASSER. A spirituous liquor distilled in Germany and Switzerland from bruised cherries. From the rude manner in which it is obtained, and from the distillation of the cherry-stones (which contain prussic acid) along with the liquor, it has often a nauseous taste, and is frequently poisonous. When properly made and sweetened it resembles noyeau.

KISH. An artificial graphite occasionally produced in iron-smelting furnaces. It occurs in brilliant scales, and is said to possess peculiar efficacy in certain forms of anæmia and chlorosis.

KITCH'EN. The late Alexis Soyer set down as one of the crying faults of our countrymen the employment of an apartment for the kitchen which is either too small or inconveniently situated, and which, in general, is not sufficiently provided with 'kitchen requisites.' "As a workman cannot work properly without the requisite tools, or the painter produce the proper shade without the necessary colours, in like manner does every person wishing to economise his food and to cook it properly require the proper furniture wherewith to do it." The neglect of these matters, which is so general, is, undoubtedly, a mischievous and deceptive economy.

KNIVES, to Clean. After being used, all knives should be wiped on a coarse cloth, so as to ensure their freedom from grease previous to being cleaned. The practice of dipping the blades in hot water not only fails to remove any grease that may be on them, but is almost sure to loosen the handles. It is very essential to remove any grease from them, since if this remain it will spoil the knife-board.

For cleaning knives, a proper knife-cleaning machine, purchased of a good maker, is best. But where this is not used, the knife-board ought to be covered with very thick leather, upon which emery powder should be placed. The emery gives a good polish to the knives, and does not wear them out so quickly as Bath-brick. When the points of the knives become worn very thin, they should be rounded by the knife-grinder. Where the handles are good it will sometimes be worth while to fit them to new blades.

KNOX'S POWDER. *Prep.* From common salt, 8 parts; chloride of lime, 3 parts; mixed together. An ounce of it dissolved in a tumblerful of water furnishes a solution which is similar to Labarraque's disinfecting fluid.

KŒCHLIN'S LIQUID. *Prep.* From copper filings, 96 gr.; liquor of ammonia, 2 fl. oz.; digested together until it turns of a full blue colour, and then mixed with hydrochloric acid, 5 fl. dr.; distilled water, 5 lbs.—*Dose,* 1 to 2 teaspoonfuls daily; in scrofula. It is poisonous in large doses.

KOLA NUTS are the fruit of the *Sterculia acuminata,* a tree of Central Africa. There the kola is the remedy for all diseases, and is almost worshipped by the natives. It is sold at high prices, and no important bargain is ever concluded without a gift of kola. The natives have found this fruit to possess tonic, nutritive, stimulating, and aphrodisiac properties. They use an infusion of the roasted nut as well as the nut in its natural state. Analysis shows that kola contains a large proportion of caffeine with a little theobromine and tannin. It use is, therefore, indicated in the chronic diarrhœa of hot countries, where it has been successfully employed by naval surgeons in stomach complaints and in cachexia. Dr Dujardin Beaumetz has found it useful in chronic diarrhœa and in cardiac affections. He gives 15 grms. (about ½ oz.) in the course of the day in 2 cups of infusion of the roasted kola, or as an elixir, or as a chocolate.

M. Natton's formulæ are applicable either to the natural or the roasted kola. A tincture is made by macerating for 15 days 1 part of kola in 5 parts of alcohol 60°; a wine by macerating for 15 days 100 grms. in a litre; an extract by percolating 100 grms. with alcohol 60° and concentrating the percolate to the proper consistence; a syrup is made similarly, but instead of concentrating the percolate is made with sugar to weigh 1 kilogrm. Pills are made from the extract, 10 centigrms. in each with some powdered kola; an alcoholate by macerating 1 part of fresh scraped kola in 5 parts of alcohol 80° for 15 days; an elixir by mixing together equal parts of the alcoholate and of simple syrup; a saccharate, by rubbing together 1 part of fresh kola with 2 parts of sugar, sifting and drying; lozenges from the saccharate with 1 part of tragacanth and 6 parts of water to 100 parts of saccharate with any desired flavour; a chocolate with 60 grms. of the saccharate, 40 grms. of cocoa powder, and ½ grm. of cinnamon. Lastly, M. Natton gives the following form for a pleasant mixture of kola:—Alcoholate or tincture, 5 to 20 grms.; tincture of cinnamon, 1 grm.; brandy, q. v.; syrup of orange, 30 grms.; distilled water, q. s. to 150 grms. See CHOCOLATE.

KOOCH'LA NUT. See NUX VOMICA.

KOU'MISS. A liquor prepared by the Calmucs, by fermenting mare's milk previously kept until sour and then skimmed. By distillation it yields a spirit called rack, racky, or araka. 21 lbs. of fermented milk yield about ⅔ pint of low wines,

and this, by rectification, gives fully ¼ pint of strong alcohol. It has lately come into use as a remedy for phthisis and general debility.

The following formula from the 'Zeitschrift des Oesterr. Apoth. Ver.' (1876, 536), for the preparation of so-called KOUMISS EXTRACT, is said to be a good one:—Powdered sugar of milk, 100 parts; glucose (prepared from starch), 100 parts; cane-sugar, 300 parts; bicarbonate of potassium, 36 parts; common salt, 33 parts.

Dissolve these ingredients in 600 parts of boiling fresh whey of milk, allow the solution to cool, then add 100 parts of rectified spirit, and afterwards 100 parts of strained fresh beer-yeast. Stir the mixture well and put into bottles containing a ½ litre each. The bottles must be well corked and kept in a cool place.

For the preparation of koumiss add 5 to 6 table-spoonfuls of this extract to a litre of skimmed, lukewarm milk, contained in a bottle of thick glass; cork well, keep the bottle for ½ a day in a moderately warm room (at 16°—20° C.), and afterwards in a cool cellar, shaking occasionally. The bottle should be filled to within 3 to 4 cm. of the cork. After 2 days the koumiss is ready for use. See KEFYR.

KOUS'SO. *Syn.* CUSSO, KOSSO. This substance is the dried flowers of the *Hagenia abyssinica,* an Abyssinian tree which grows to the height of about 20 feet, and belongs to the Nat. Ord. ROSACEÆ. It is one of the most effective remedies known for both varieties of tapeworm. The dose for an adult is 3 to 5 dr., in powder, mixed with about ½ a pint of warm water, and allowed to macerate for 15 or 20 minutes. The method prescribed for its successful administration is as follows:—The patient is to be prepared by a purgative or a lavement, and the use of a very slight diet the day before. The next morning, fasting, a little lemon juice is to be swallowed, or a portion of a lemon sucked, followed by the dose of kousso (both liquid and powder), at 3 or 4 draughts, at short intervals of each other, each of which is to be washed down with cold water acidulated with lemon juice. The action of the medicine is subsequently promoted by drinking weak tea without either milk or sugar, or water flavoured with lemon juice or toasted bread; and if it does not operate in the course of 3 or 4 hours, a dose of castor oil or a saline purgative is taken.

The flavour of kousso is rather disagreeable and nauseating. Its operation is speedy and effectual; but at the same time it is apt to produce, in large doses, great prostration of strength, and other severe symptoms, which unfit it for administration to the delicate of both sexes, or during pregnancy or affections of the lower viscera. Care should be taken not to purchase it in powder, as, owing to its high price, it is uniformly adulterated. The powdered kousso of the shops is, in general, nothing more than the root-bark of pomegranate, coloured and scented. An infusion is contained in the B. P. 1 in 16.—*Dose,* 4 to 8 oz. taken without straining.

KRE'ASOTE. *Syn.* CREASOTE, CREOSOTE, KREOSOTE; CREASOTUM (B. P., Ph. L. & D.), CREASOTUM (Ph. E.), L. A peculiar substance, discovered by Reichenbach, and so named on account of its powerful antiseptic property. It is a product of the dry distillation of organic bodies, and is the preservative principle of wood-smoke and pyroligneous acid.

Prep. Kreasote is manufactured from wood-tar, in which it is sometimes contained to the amount of 20% to 35%, and from crude pyroligneous acid and pyroxilic oil.

Wood-tar is distilled till a black residue is obtained which solidifies on cooling. The distillate separates into two layers: an acid aqueous layer, and an oily one. This latter contains the kreasote, and is redistilled, only that part of the distillate being collected which is heavier than water. The product is now washed with a solution of carbonate of soda, and rectified in a glass retort, in order to separate any remaining oils that are lighter than water. A solution of potash is then added; this dissolves the kreasote with evolution of heat; while the hydrocarbons present remain for the most part undissolved. The alkaline solution is heated in contact with the air, when a foreign substance, which dissolved in the potash at the same time as the kreasote, separates out as a resin. The kreasote is then liberated by the addition of sulphuric acid, and further purified by repeated treatment with potash and sulphuric acid successively. Finally it is dried and rectified.

Prop. Kreasote is a colourless, transparent liquid, heavier than water, of a peculiar unpleasant penetrating odour resembling that of smoked meat, and a very pungent and caustic taste; its vapour irritates the eyes; it boils at 400° F., and is still fluid at—16·6° F.; it produces on white filter paper greasy spots, which disappear if exposed to a heat of 212°·F.; dissolves in 80 parts of water, and mixes in all proportions with spirit of wine, the essential and fatty oils, acetic acid, naphtha, disulphide of carbon, ammonia, and potassa; it dissolves iodine, phosphorus, sulphur, resins, the alkaloids, indigo-blue, several salts (especially the acetates and the chlorides of calcium and tin); reduces the nitrate and acetate of silver; is resinified by chlorine, and decomposed by the stronger acids. The aqueous solution is neutral, and precipitates solutions of gum and the white of eggs. It kindles with difficulty, and burns with a smoky flame. When quite pure, it is unaltered by exposure to the air. Sp. gr. 1·071, at 68° F. A slip of deal dipped into it and afterwards in hydrochloric acid, and then allowed to dry in the air, acquires a greenish-blue colour. It rotates a ray of polarised light to the right, whereas carbolic acid does not affect polarisation.

Pur. The fluid commonly sold in the shops for kreasote is a mixture of kreasote, picamar, and light oil of tar; in many cases it is little else than impure carbolic acid, with scarcely a trace of kreasote. Pure kreasote is perfectly soluble in both acetic acid and solution of potassa: shaken with an equal volume of water in a narrow test-tube, not more than the 1-80th part disappears; otherwise it contains water, of which kreasote is able to take up 1-10th without becoming turbid. If it can be dissolved completely in 80 parts by weight of water, at a medium temperature, it then forms a perfectly neutral liquid. An oily residue floating on the surface betrays the presence of other foreign products (EUPION, KAPNOMOR, PICAMAR), which are obtained at the same time with the kreasote during the dry distillation of organic substances.

Kreasote is "devoid of colour, has a peculiar odour, and is soluble in acetic acid. When it is dropped on bibulous paper, and a boiling heat is applied for a short time, it entirely escapes, leaving no transparent stain." (Ph. L.) "Entirely and easily soluble in its own weight of acetic acid." (Ph. E.) Sp. gr. 1·046 (Ph. L.), 1·066 (Ph. E. and D.). The density and boiling-point of absolutely pure kreasote is given above. When prescribed in pills with oxide of silver, the mass will take fire unless the oxide be first mixed with liquorice or other powder (*Squire*).

Uses. Kreasote has been recommended in several diseases of the organs of digestion and respiration, in rheumatism, gout, torpid nervous fever, spasms, diabetes, tapeworm, &c.; but its use has not, in general, been attended with satisfactory results. It is given in the form of pills, emulsion, or an ethereal or spirituous solution. Externally it has been employed in various chronic diseases of the skin, sores of different kinds, mortifications, scalds, burns, wounds (as a styptic), caries of the teeth, &c.; mostly in the form of an aqueous solution (1 to 80); or mixed with lard (5 drops to 1 dr.), as an ointment; dissolved in rectified spirit, it forms a useful and a popular remedy for toothache arising from decay or rottenness. In the *arts*, kreasote was extensively employed to preserve animal substances, either by washing it over them, or by immersing them in its aqueous solution. As an antiseptic, however, it is now nearly superseded by carbolic acid, but is still used to preserve wood. A few drops in a saucer, or on a piece of spongy paper, if placed in a larder, will effectually drive away insects, and make the meat keep several days longer than it otherwise would. A small quantity added to brine or vinegar is commonly employed to impart a smoky flavour to meat and fish, and its solution in acetic acid is used to give the flavour of whisky to malt spirit. See CARBOLIC ACID.

KRE'ATIN. $C_4H_9N_3O_2.Aq.$ *Syn.* CREATIN. A crystallisable substance obtained from the juice of the muscular fibre of animals. It was first obtained by Chevreul.

Prep. (*Liebig.*) Lean flesh is reduced to shreds, and then exhausted with successive portions of cold water, employing pressure; the mixed liquid is heated to coagulate the albumen and colouring matter of the blood, and is then strained through a cloth; pure baryta water is next added as long as a precipitate forms, the liquid is filtered, and the filtrate is gently evaporated to the consistence of a syrup; after repose for some days in a warm situation, crystals of kreatin are deposited; these are purified by redissolving them in water, agitating the solution with animal charcoal, and evaporating, &c., so that crystals may form.

Prop., &c. Brilliant, colourless, prismatic crystals; readily soluble in boiling water, sparingly so in cold water and in alcohol; the aqueous solution is neutral, bitter tasted, and soon putrefies.

KREAT'ININ. $C_4H_7N_2O.$ This substance exists in small quantities both in the juice of flesh and in conjunction with kreatin in urine. It is also produced by the action of the stronger acids on kreatin. It forms colourless prismatic crystals, which are soluble in water, and the solution has a strongly alkaline reaction. It is a powerful organic base, and produces crystallisable salts with the acids.

KRE'NIC ACID. See CRENIC ACID.

KRYSTAL'LINE. The name originally applied by Unverdorben to ANILINE.

KUSTITIEN'S METAL. *Prep.* Take of malleable iron, 3 parts; heat it to whiteness, and add of antimony, 1 part; Molucca tin, 72 parts; mix under charcoal, and cool. Used to coat iron and other metals with a surface of tin; it polishes without a blue tint, is hard, and has the advantage of being free from lead and arsenic.

KYANI'ZING. A method of preserving wood and cordage from decay, long known and practised; patented by Mr Kyan many years since. It consists in immersing the materials in a solution of corrosive sublimate, 1 part, and water, 50 or 60 parts, either under strong pressure or the contrary, as the urgency of the case or the dimensions of the articles operated on may require. See DRY ROT.

KY'ANOL. A substance obtained from coaltar oil, and at first thought to be an independent principle, but since shown to be identical with ANILINE.

LABARRAQUE'S FLUID. See SOLUTION OF CHLORIDE OF SODA.

LAB'DANUM. *Syn.* LADANUM. An odorous, resinous substance found on the leaves and twigs of the *Cystus creticus*, a plant growing in the island of Candia and in Syria. It was formerly much used for making stimulating plasters. The following compound is often vended for it:

Labdanum, Facti"tious. *Prep.* From gum-anime, resin, Venetian turpentine, and sand, of each, 6 parts; Spanish juice and gum-arabic, of each (dissolved in a little water), 3 parts; Canada balsam, 2 parts; ivory-black, 1 part; balsam of Peru, q. s. to give a faint odour.

LA'BELS capable of resisting the action of OILS, SPIRITS, WATER, SYRUPS, and DILUTE ACIDS, may be obtained as follows:—Lay a coat of strained white of egg over the label (an ordinary paper one), and immediately put the vessel into the upper portion of a common steam-pan, or otherwise expose it to a gentle heat till the albumen coagulates and turns opaque, then take it out and dry it before the fire, or in an oven, at a white heat of about 212° F.; the opaque white film will then become hard and transparent. The labels on bottles containing STRONG ACIDS or ALKALINE SOLUTIONS should be either etched upon the glass by means of hydrofluoric acid, or be written with incorrodible ink. A varnish made by dissolving gum-dammar in benzol is useful for preserving labels from the action of acids and other chemicals. See ETCHING and INK.

LAB'ORATORY. *Syn.* LABORATORIUM, L. A place fitted up for the performance of experimental or manufacturing operations in *chemistry, pharmacy,* and other sciences. For full in-formation respecting the best mode of fitting up a chemical laboratory the reader is referred to works especially devoted to chemical manipulation. Almost any well-lighted (a room with a good top or north light is to be preferred) spare room may be fitted up as a small laboratory at very little expense. The gas-furnaces and improved lamps introduced of later years have to a certain extent rendered chemists independent of brick furnaces. A strong working bench, fitted with drawers and cupboards, and having gas-pipes at intervals for attaching different kinds of jets, is an indispensable fixture. A close cupboard or closet, which is connected by a pipe with the chimney or the external air, is required to receive vessels emitting corrosive or evil-smelling vapours; the door of this closet should be of glass. A sink, with a copious supply of water, must be at hand for washing apparatus. A glass or stoneware barrel, with a tap of the same material, is required for holding distilled water. Shelves, supports for apparatus, and drawers, should be provided in abundance. The fine balances and other delicate instruments should be kept in a separate apartment. With regard to apparatus, we may state that the articles most frequently required in a laboratory are the gas or alcohol lamps; iron pans for sand-bath and water-bath; evaporating dishes; precipitating jars, funnels, and wash-bottles; retorts, flasks, and test-tubes; mortars and pestles; retort- and filtering-stands; rat-tail and triangular files, and glass rod and tubing.

LABURNINE. A poisonous alkaloid, found in the unripe seeds of the laburnum plant associated with another poisonous alkaloid called *Cytisine.*

LAC. *Syn.* LACCA, L. A resinous substance combined with much colouring matter, produced by the puncture of the female of a small insect, called the *Coccus lacca* or *ficus,* upon the young branches of several tropical trees, especially the *Ficus indica, Ficus religiosa,* and *Croton lacciferum.* The crude resinous exudation constitutes the STICK-LAC of *commerce.* SHELL-LAC or SHELLAC is prepared by spreading the resin into thin plates after being melted and strained. SEED-LAC is the residue obtained after dissolving out most of the colouring matter contained in the resin.

Shell-lac is the kind most commonly employed in the *arts.* The palest is the best, and is known as 'orange-lac.' The darker varieties—'liver-coloured,' 'ruby,' 'garnet,' &c.—respectively diminish in value in proportion to the depth of their colour.

Uses, &c. Lac was formerly much used in *medicine;* its action, if any, is probably that of a very mild diuretic. It is now chiefly used in DENTIFRICES, VARNISHES, LACQUERS, and SEALING-WAX.

Lac, Bleached. *Syn.* WHITE LAC; LACCA ALBA, L. By dissolving lac in a boiling lye of pearlash or caustic potassa, filtering and passing chlorine through the solution until all the lac is precipitated; this is collected, well washed and pulled in hot water, and, finally, twisted into sticks, and thrown into cold water to harden. Used to make pale varnishes and the more delicate coloured sealing-wax.

LAC-DYE. *Syn.* LAC, LAC-LAKE, INDIAN COCHINEAL. A colouring substance used to dye scarlet; imported from India.

Prep. By dissolving out the colour of ground stick-lac by means of a weak alkaline solution, and then precipitating it along with alumina by adding a solution of alum.

Obs. To prepare the lac for dyeing, it is ground and mixed with diluted 'lac spirit,' and the whole allowed to stand for about a week. The 'cloth' is first mordanted with a mixture of tartar and 'lac-spirit,' and afterwards kept near the boil for three quarters of an hour, in a bath formed by adding a proportion of the prepared lac-dye to the mixture used for mordanting. Lac-dye is only applicable to woollen and silk. The colours it yields are similar to those obtained from cochineal, but less brilliant.

LAC-SPIRIT. See TIN MORDANTS.

LACE. This decorative fabric is made by interweaving threads of linen, cotton, or silk, into various patterns and designs. Although in some instances lace is made by hand, the greater part is now manufactured by machinery worked by steam or water. The hand-made lace was called bone, pillow, or bobbin lace, these two latter names having been given it from its having been woven upon a pillow or cushion by means of a bobbin. The manufactured article is bobbin net. Lace and the machinery by which it is produced is of so complex a nature that Dr Ure says of one particular form of it: "It is as much beyond the most curious chronometer in the multiplicity of mechanical device as that is beyond a common roasting jack."

Owing to the improvements in machinery introduced of late years, it may be mentioned that a piece of lace which 20 years since could only be produced at a cost of 3s. 6d. for labour, may now be turned out for 1d., and a quantity of the fabric which sold for £17, now realises only 7s. A pair of curtains, each 4 yards long, may be made in 1 frame in 2 hours.

The following statistics relating to the British lace industry are of interest:—" In 1843 there were 3200 twist net and 800 warp frames, returning £2,740,000 that year; in 1851, 3200 bobbin net and 800 warp, giving a return of £3,846,000; and in 1866, 3552 bobbin and 400 warp, returning £5,180,000. There has since been no actual census, but about the same number is now at work, and the returns and profits are greatly increased by improved quality and patterns of goods produced. The returns of 1872 were certainly £6,000,000 at least; and from advancing wages and demand for Lever's laces, must still rapidly increase. Men are now earning by making them from £4 to £6 for 56 hours' weekly labour."

Lace, Gold and Silver, to Clean. Reduce to fine crumbs the interior of a 2-lb. stale loaf, and mix with them ¼ lb. of powder blue. Sprinkle some of this mixture plentifully on the lace, afterwards rubbing it on with a piece of flannel. After brushing off the crumbs rub the lace with a piece of crimson velvet.

Lace, to Scour. Take a perfectly clean wine bottle; wind the lace smoothly and carefully round it; then gently sponge it in tepid soap and water; and when clean, and before it becomes dry, pass it through a weak solution of gum and water. Next pick it out and place it in the sun to dry. If it be desired to bleach the lace, it should be rinsed in some *very weak* solution of chloride of lime, after removal from which it must be rinsed in cold water. Starch and expose it; then boil and starch, and again expose it if it has not become sufficiently white.

The following method is also said to whiten lace:—It is first ironed slightly, then folded and sewn into a clean linen bag, which is then placed for 24 hours in pure olive oil. Afterwards the bag, with the lace in it, is to be boiled in a solution of soap and water for 15 minutes, then well rinsed in lukewarm water, and finally dipped in water containing a small quantity of starch. The lace is then to be taken from the bag, and stretched on pins to dry.

To scour point lace proceed as follows:—" Fix the lace in a prepared tent, draw it tight and straight, make a warm lather of Castile soap, and with a fine brush dipped in rub over the lace gently, and when clean on one side do the same on the other; then throw some clean water on it, in which a little alum has been dissolved, to take off the suds; and, having some thin starch, go over with it on the wrong side, and iron it on the same side when dry; then open with a bodkin and set it in order. To clean the same, if not very dirty, without washing, fix it as before, and go over with fine bread, the crust being pared off, and when done, dust out the crumbs " (*Ernest Spon*).

Black lace may be cleaned by passing it through warm water containing some ox-gall, rinsing it in cold water, and then passing it through water in which a small quantity of glue has been previously dissolved by means of heat; it should then be taken out, clapped between the hands, and dried on a frame.

LAC'QUER. A solution of shell-lac in alcohol, tinged with saffron, annotta, aloes, or other colouring substances. It is applied to wood and metals to impart a golden colour. See VARNISH.

Lacquer, Burmese or Varnish. Obtained from *Melanorrhœa usitata*, Wall. It is used by the Burmese in lacquer work, both red and black, as size in gilding, and for covering buckets to make them watertight. It has been used as an anthelmintic. The wood is used for tool-handles, gunstocks, and railway sleepers.

LACQUER DYES. According to Metallarbeiten, these can be produced as follows:—½ dr. of boracic acid; white shell-lac, 15 dr.; mastic, 15 dr.; manilla copal, 15 dr., are dissolved in absolute alcohol, 1¾ pints. It is left at rest for at least 18 hours, and during the time is occasionally agitated. 3 dr. of Venetian turpentine is then added, and shaken until the latter has been thoroughly mixed and dissolved, after which the whole is thoroughly filtered. The colourless lacquer can be dyed with all kinds of non-acid colouring matter which are soluble in alcohol, and which should be previously dissolved.

LACTALBU'MEN. See CASEIN.

LACTATE. *Syn.* LACTAS, L. A salt of lactic acid. The lactates are characterised by yielding an enormous quantity of perfectly pure carbonic oxide gas when heated with 5 or 6 parts of oil of vitriol. Most of these salts may be directly formed by dissolving the hydrate or carbonate of the metal in the dilute acid.

LACTA'TION. See INFANCY, NURSING, &c.

LACTIC ACID. $C_3H_6O_3$. *Syn.* ACID OF MILK; ACIDUM LACTICUM, L. A sour, syrupy liquid discovered by Scheele in whey. It is also found in some other animal fluids, and in several vegetable juices, especially in that of beet-root.

Lactic acid is by no means an unimportant constituent of the human organism. It is contained in the gastric juice, and is frequently formed in the sweat. It has also been detected in the saliva of persons suffering from diabetes. A modification of the acid, termed sarkolactic acid, occurs in the fluids of the muscular tissue.

It is likewise a product of the fermentation of many vegetable juices, such as turnips, carrots, beet-root, and cabbage, which latter vegetable, after undergoing the lactic fermentation, becomes converted into the sauerkraut of the Germans.

In the form of calcic lactate it occurs in nux vomica.

Prep. 1. By decomposing any lactate with dilute sulphuric acid. As, however, the chief task is to get the lactate, a method of accomplishing this is described *below :*

2. Dissolve 6 kilos. (12 lbs.) of sugar and 30 grms. (1 oz.) of tartaric acid in 26 litres (5 galls.) of boiling water, and after some days add a mixture of 250 grms. (½ lb.) of putrid cheese with 8 kilos. (1¾ galls.) of sour skimmed milk and 3 kilos. (6 lbs.) of finely divided chalk. The mixture should then be placed in a spot where its temperature will be maintained at 30°—35° C. (86°—95° F.), and should be stirred up every day; after about a week the liquid becomes a thick magma of calcium lactate. This is then filtered through a cloth, and decomposed by sulphuric acid. The solution of lactic acid is filtered off from the precipitated calcium sulphate, and is neutralised with zinc carbonate, when zinc lactate is formed. A simpler plan is to form this salt at once by adding 2 kilos. (4 lbs.) of zinc-white instead of chalk. After 2 or 3 weeks a magma of zinc lactate is formed, which is purified by re-crystallisation. It is then dissolved in boiling water, and the zinc is precipitated with sulphuretted hydrogen. The filtrate is concentrated on a water-bath until mannite, deposited together with the zinc salt, separates out, forming a pasty mass. This is then dissolved in the smallest possible quantity of water and the lactic acid removed by shaking up with ether. By evaporating the ethereal solution lactic acid is obtained.

N.B.—In the above the quantities given in parentheses are not the *exact* equivalents of those expressed in the metric system.

3. If the acid is not required absolutely pure the calcium lactate, obtained by fermentation of sugar as described *above*, may be simply filtered off, recrystallised, and decomposed with sulphuric acid. The solution on evaporation yields aqueous lactic acid, but there will always be a little calcium present.

Obs. Lactic acid may be rendered quite pure by dilution with water, saturation with baryta, evaporation, crystallisation, re-solution in water,

and the careful addition of dilute sulphuric acid, as in No. 1 ; the liquid is, lastly, again filtered and evaporated. Another plan is to convert the acid into lactate of zinc by the addition of commercial zinc-white, and to redissolve the new salt in water, and then decompose the solution with a stream of sulphuretted hydrogen. In all cases the evaporation should be conducted at a very gentle heat, and, when possible, finished over sulphuric acid, or *in vacuo.* For particular purposes this last product must be dissolved in ether, filtered, and the ether removed by a very gentle heat. Care must also be taken to remove the solid lactate of calcium at the proper period from the fermenting liquid, as otherwise it will gradually redissolve and disappear, and on examination the liquid will be found to consist chiefly of a solution of butyrate of calcium.

Prop. An aqueous solution of lactic acid may be concentrated *in vacuo* over a surface of oil of vitriol until it appears as a syrupy liquid of sp. gr. 1·215 ; soluble in water, alcohol, and ether ; exhibiting the usual acid properties, and forming salts with the metals, called LACTATES. Heated in a retort to 130° C. (266° F.), a small portion distils over, and the residuum on cooling solidifies into a yellowish, compact, fusible mass of *lactic anhydride*, very bitter, and nearly insoluble in water. By long boiling in water this substance is reconverted into lactic acid. Heated to 250° C. (480° F.), it suffers decomposition, *lactide* (the anhydrous, concrete, or sublimed lactic acid of former writers) and other products being formed. This new substance may be purified by pressure between bibulous paper and solution in boiling alcohol from which it separates in dazzling white crystals on cooling. By solution in hot water and evaporation to a syrup, it furnishes common lactic acid.

Uses. Lactic acid has been given in dyspepsia, gout, phosphatic urinary deposits, &c. From its being one of the natural constituents of the gastric juice, and from its power of dissolving a considerable quantity of phosphate of calcium, it appears very probable that it may prove beneficial in the above complaints.—*Dose,* 1 to 5 gr. ; in the form of lozenges, or solution in sweetened water.

LACTIC FERMENTA'TION. The peculiar change by which saccharine matter is converted into lactic acid. Nitrogenous substances, which in an advanced state of putrefactive change act as alcohol-ferments, often possess, at certain periods of their decay, the property of inducing an acid fermentation in sugar, by which that substance is changed into lactic acid. Thus, the nitrogenised matter of malt, when suffered to putrefy in water for a few days only, acquires the power of acidifying the sugar which accompanies it ; whilst in a more advanced state of decomposition it converts, under similar circumstances, the sugar into alcohol. The gluten of grain behaves in the same manner. Wheat-flour, made into a paste with water, and left for 4 or 5 days in a warm situation, becomes a true lactic acid ferment ; but if left a day or two longer it changes its character, and then acts like common yeast, occasioning the ordinary panary or vinous fermentation. Moist animal membranes in a slightly decaying condition, often act energetically in developing lactic acid. The rennet employed in the manufacture of cheese furnishes a well-known example of this class of substances.

In preparing lactic acid from milk, the acid formed, after a time, coagulates and renders insoluble the casein, and the production of the acid ceases. By carefully neutralising the free acid by carbonate of sodium, the casein becomes soluble, and, resuming its activity, changes a fresh quantity of sugar into lactic acid, which may be also neutralised, and by a sufficient number of repetitions of this process all the sugar of milk present may, in time, be acidified. This is the *rationale* of the common process by which lactic acid is obtained. Cane-sugar (probably by previously becoming grape-sugar) and the sugar of milk both yield lactic acid; the latter, however, most readily, the grape-sugar having a strong tendency towards the alcoholic fermentation. If the lactic fermentation be allowed to proceed too far, the second stage of the process of transmutation commences, hydrogen gas and carbonic acid gas are evolved, and the butyric fermentation, by which oily acids are formed, is established.

Pasteur ascribes the lactic fermentation to the agency of a specific kind of ferment, which occurs in the form of a greyish layer deposited upon the surface of the sediment formed during the fermentation of the sugar in the manufacture of lactic acid.

If to a mixture of yeast or any nitrogenous substance and water sugar and then chalk be added, and finally a very small quantity of this greyish substance, taken from a portion of a liquid undergoing active lactic fermentation, lactic acid fermentation is almost immediately set up, the chalk disappears owing to the formation of calcic lactate, and the greyish substance is copiously deposited. When placed under the microscope this ferment is seen to be composed of "little globules, or very short articulations, either isolated or in threads, constituting irregular flocculent particles, much smaller than those of beer yeast, and exhibiting a rapid gyratory motion." If these little particles be washed thoroughly in pure water, and then placed in a solution of sugar, lactic acidification immediately commences in the saccharine liquid, and goes on steadily until stopped by the excess of free acid.

LAC′TIDE. See LACTIC ACID.

LAC′TIN. *Syn.* LACTOSE. See SUGAR OF MILK.

LAC′TOMETER. *Syn.* GALACTOMETER. An instrument for ascertaining the quality of milk. Milk may be *roughly* tested by placing it in a long graduated tube sold for the purpose, and allowing it to remain until all the cream has separated and measured, then decanting off the clear whey, and taking its sp. gr.; the result of the two operations, when compared with the known quantity of cream and the density of the whey of an average sample of milk, gives the value of the sample tested.

A little instrument called a 'milk-tester' is sold in London at a low price. It is essentially a hydrometer which sinks to a given mark on the stem in pure water, and floats at another mark at the opposite end of the scale in pure milk. The intermediate space indicates the quantity of water (if any) employed to adulterate the article. As the sp. gr. of pure milk varies, the indications of the 'tester' cannot be depended on. See MILK.

LACTOPEPTIN. A preparation of American origin, and stated to be composed of sugar of milk, 20 oz.; pepsin, 4 oz.; pancreatin, 3 oz.; diastase, 1 dr.; lactic acid, 2½ dr.; hydrochloric acid, 2½ dr.; mix well, dry and powder. Used as a remedy for dyspepsia.—*Dose*, 10 to 15 gr.

LAC′TOSE. See SUGAR OF MILK.

LACTUCA. (B. P.) *Syn.* LETTUCE. The leaves and flowering-tops of the wild indigenous plant *Lactuca virosa*. They are sedative, narcotic, and powerfully diuretic; also mildly laxative and diaphoretic. Given in dropsy and visceral obstructions. See LETTUCE, EXTRACT OF.

LACTUCA′RIUM. *Syn.* LETTUCE OPIUM. THRIDACE; LACTUCARIUM (Ph. E. & D.). The inspissated milky juice of the *Lactuca sativa* (common garden lettuce), or the *Lactuca virosa* (strong-scented wild lettuce), obtained, by incision, from the flowering stems, and dried in the air. The latter species yields by far the greatest quantity. M. Arnaud, of Nancy, adopts the following method of procuring this substance, which appears to be the most productive and simple of any yet published:—Before the development of the lateral branches, the stems of twelve plants are cut, one after another, a little below the commencement of these branches; returning to the first one, a milky exudation is found on the cut portion, and on that which remains fixed in the earth; this milky exudation is adroitly collected with the end of the finger (or with a bone knife), which is afterwards scraped on the edge of a small glass; the same operation is performed on twelve other heads, and so on; on the third day it is repeated on every portion of the plant remaining in the ground, a thin slice being first cut off the top; this is done every day until the root is reached. As soon as the lactucarium is collected it coagulates; the harvest of each day is divided into small pieces, which are placed on plates, very near each other, but without touching, and allowed to dry for two days, after which they are set aside in a bottle. In this way 15 or 20 times the ordinary product is obtained.

Prop., &c. Lactucarium is anodyne, hypnotic, antispasmodic, and sedative, allaying pain and diminishing the force of the circulation. It has been recommended in cases in which opium is inadmissible, and has been administered with advantage in chronic rheumatism, colic, diarrhœa, asthma, and troublesome cough of phthisis, the irritability and wakefulness in febrile disorders, &c.—*Dose*, 2 to 5 gr.; made into pills, lozenges, or tincture.

LACTU′CIN. *Syn.* LACTUCINUM, L. This is the active principle of lactucarium, and is found in the juice of several species of lettuce.

Prep. Exhaust lactucarium with hot rectified spirit, agitate the tincture with a little animal charcoal, filter, add a little milk of lime, and evaporate to dryness; digest the residuum in hot rectified spirit; filter, and evaporate by a gentle heat, so that crystals may form.

Prop., &c. A nearly colourless, odourless,

fusible, neutral, bitter substance; sparingly soluble in cold water and in ether, but freely soluble in alcohol. It possesses feeble basic properties. Good lactucarium contains fully 20% of this substance.

LAD'ANUM. See LABDANUM.

LAENNEC'S CONTRA-STIMULANT. See DRAUGHT.

LAKE. *Syn.* LACCA, L. Animal or vegetable colouring matter, precipitated in combination with oxide of tin or alumina, usually the latter. The term was formerly restricted to red preparations of this kind, but is now indiscriminately applied to all compounds of alumina and colouring matter. The term 'LAKE,' when unqualified by an adjective, is, however, understood to apply exclusively to that prepared from cochineal.

Prep. Lakes are made—1. By adding a solution of alum, either alone or partly saturated with carbonate of potass, to a filtered infusion or decoction of the colouring substance, and after agitation precipitating the mixture with a solution of carbonate of potash. 2. By precipitating a decoction or infusion of the colouring substance made with a weak alkaline lye, by adding a solution of alum. 3. By agitating recently precipitated alumina with a solution of the colouring matter, prepared as before, until the liquid is nearly decoloured or the alumina acquires a sufficiently dark tint. The first method is usually employed for acid solutions of colouring matter, or for those whose tint is injured by alkalies; the second for those that are brightened, or at least uninjured, by alkalies; the third for those colouring matters that have a great affinity for gelatinous alumina, and readily combine with it by mere agitation. By attention to these general rules, lakes may be prepared from almost all animal and vegetable colouring substances that yield their colour to water, many of which will be found to possess great beauty and permanence. The precise process adapted to each particular substance may be easily ascertained by taking a few drops of its infusion or decoction, and observing the effects of alkalies and acids on the colour. The quantity of alum or of alumina employed should be nearly sufficient to decolour the dye-liquor; and the quantity of carbonate of potassa should be so proportioned to the alum as to exactly precipitate the alumina without leaving free or carbonated alkali in the liquid. The first portion of the precipitate has the deepest colour, and the shade gradually becomes paler as the operation proceeds. A beautiful 'tone' of violet, red, and even purple may be communicated to the colouring matter of cochineal by the addition of perchloride of tin; the addition of arseniate of potassa (neutral arsenical salt) in like manner gives shades which may be sought for in vain with alum or alumina. After the lake is precipitated it must be carefully collected, washed with cold distilled water or the purest rain-water until it ceases to give out colour, and then carefully dried in the shade. In this state it forms a soft velvety powder. That of the shops is generally made up into conical or pyramidal drops (drop lake), which is done by dropping the moist lake through a small funnel on a clean board or slab, and drying it by a gentle heat as

before. A very little clear gum-water is commonly added to the paste to give the drops consistence when dry.

Lake, Blue. *Syn.* LACCA CŒRULEA, L. Prepared from some of the blue-coloured flowers; fugitive. The name is also applied to lump archil (lacca cœrulea), to moist alumina coloured with indigo, and to mixed solutions of pearlash and prussiate of potash, precipitated with another solution of sulphate of iron and alum. These are permanent and beautiful, but are seldom used, in consequence of indigo and Prussian blue supplying all that is wanted in this class of colours.

Lake, Brazil-wood. *Syn.* DROP LAKE; LACCA IN GLOBULIS, L. *Prep.* 1. Take of ground brazil-wood, 1 lb.; water, 4 galls,; digest for 24 hours, then boil for 30 or 40 minutes, and add of alum, 1½ lbs., dissolved in a little water; mix, decant, strain, and add of solution of tin, ¼ lb.; again mix well and filter; to the clear liquid add, cautiously, a solution of salt of tartar or carbonate of soda, as long as a deep-coloured precipitate forms, carefully avoiding excess; collect, wash, dry, &c., as directed above.

Obs. The product is deep red. By collecting the precipitate in separate portions lakes varying in richness and depth of colour may be obtained. The first portion of the precipitated lake has the brightest colour. An excess of alkali turns it on the violet, and the addition of cream of tartar on the brownish red. The tint turns more on the violet red when the solution of tin is omitted. Some persons use less, others more, alum.

2. Add washed and recently precipitated alumina to a strong and filtered decoction of Brazil-wood. Inferior to the last.

Lake, Carminated. *Syn.* COCHINEAL LAKE, FLORENCE L., FLORENTINE L., PARIS L., VIENNA L.; LACCA FLORENTINA, L. *Prep.* 1. The residuum of the cochineal left in making carmine is boiled with repeated portions of water until it is exhausted of colour; the resulting liquor is mixed with that decanted off the carmine, and at once filtered; some recently precipitated alumina is then added, and the whole gently heated, and well agitated for a short time; as soon as the alumina has absorbed sufficient colour the mixture is allowed to settle, after which the clear portion is decanted, the lake collected on a filter, washed, and dried, as before. The decanted liquor, if still coloured, is now treated with fresh alumina until exhausted, and thus a lake of a second quality is obtained. Very fine.

2. To the coloured liquor obtained from the carmine and cochineal as above, a solution of alum is added, the filtered liquor precipitated with a solution of carbonate of potassa, and the alum or alumina; this brightens the lake collected and treated as before. Scarcely so good as the last.

Obs. Some makers mix a little solution of tin with the coloured liquor before adding colour. The above lake is a good glazing colour with oil, but has little body. It may be made directly from a decoction of cochineal (see *below*).

Lake, Cochineal. *Prep.* 1. Cochineal (in coarse powder), 1 oz.; water and rectified spirit, of each, 2½ oz.; digest for a week, filter, and precipitate the tincture with a few drops of solution of tin, added every two hours, until the whole of the colouring matter is thrown down; lastly, wash the precipitate in distilled water and dry it. Very fine.

2. Digest powdered cochineal in ammonia water for a week, dilute the solution with a little water, and add the liquid to a solution of alum, as long as a precipitate falls, which is the lake. Equal to the last.

3. Coarsely powdered cochineal, 1 lb.; water, 2 galls.; boil 1 hour, decant, strain, add a solution of salt of tartar, 1 lb., and precipitate with a solution of alum. By adding the alum first, and precipitating the lake with the alkali, the colour will be slightly varied. All the above are sold as CARMINATED or FLORENCE LAKE, to which they are often superior.

Lake, Green. Made by mixing blue and yellow lake together. Seldom kept in the shops, being generally prepared extemporaneously by the artist on his palate.

Lake, Lac. *Prep.* Boil fresh stick-lac in a solution of carbonate of soda, filter the solution, precipitate with a solution of alum, and proceed as before.

Lake, Lichen. See ORCEIN.

Lake, Madder. *Syn.* LACCA RUBRÆ, L. COLUMBINA, L. *Prep.* 1. (*Sir H. C. Inglefield.*) Take of Dutch grappe or crop madder, 2 oz.; tie it in a cloth, beat it well in a pint of water in a stone mortar, and repeat the process with fresh water (about 5 pints) until it ceases to yield colour; next boil the mixed liquor in an earthen vessel, pour it into a large basin, and add of alum, 1 oz., previously dissolved in boiling water, 1 pint; stir well, and while stirring pour in gradually of a strong solution of carbonate of potassa ('oil of tartar'), 1¼ oz.; let the whole stand until cold, then pour off the supernatant yellow liquor, drain, agitate the residue with boiling water, 1 quart (in separate portions), decant, drain, and dry. *Prod.*, ½ oz. The Society of Arts voted their gold medal to the author of the above formula.

2. Add a little solution of acetate of lead to a decoction of madder, throw down the brown colouring matter, filter, add a solution of tin or alum, precipitate with a solution of carbonate of soda or of potassa, and otherwise proceed as before.

3. (*Ure.*) Ground madder, 2 lbs.; water, 1 gall.; macerate with agitation for 10 minutes, strain off the water, and press the remainder quite dry; repeat the process a second and a third time; then add to the mixed liquors, alum, ¼ lb., dissolved in water, 3 quarts; and heat in a water-bath for 3 or 4 hours, adding water as it evaporates; next filter, first through flannel, and when sufficiently cold, through paper; then add a solution of carbonate of potassa as long as a precipitate falls, which must be washed until the water comes off colourless, and, lastly, dried. If the alkali be added in 3 successive doses, 3 different lakes will be obtained, successively diminishing in beauty. See MADDER, MADDER (Red), &c.

Lake, Orange. *Prep.* Take of the best Spanish annotta, 4 oz.; pearlash, ¾ lb.; water, 1 gall.; boil for ½ hour, strain, precipitate with alum, 1 lb., dissolved in water, 1 gall., taking care not to add the latter solution when it ceases to procuce an effervescence or a precipitate; strain, and dry the sediment in small squares, lozenges, or drops. The addition of some solution of tin turns this lake on the LEMON YELLOW; acids redden it. See LAKE (Yellow).

Lake, Red. *Prep.* Take of pearlash, 1 lb.; clean shreds of scarlet cloth, 3¼ lbs.; water, 5 galls; boil till the cloth is decoloured, filter the decoction, and precipitate with a solution of alum, as before. See the LAKES noticed above (Brazil-wood, Carminated, and Madder).

Lake, Yellow. *Prep.* 1. Boil French berries, quercitron bark, or turmeric, 1 lb., and salt of tartar, 1 oz., in water, 1 gall., until reduced to 1 half, then strain the decoction, and precipitate with a solution of alum.

2. Boil 1 lb. of the dye-stuff with alum, ¼ lb.; water, 1 gall., as before, and precipitate the decoction with a solution of carbonate of potash. See LAKE (Orange), above.

LAMB in its general qualities closely resembles mutton, of which, indeed, it is merely a younger and more delicate kind. It is well adapted as an occasional article of food for the convalescent and dyspeptic; but it is unequal for frequent use, more especially for the healthy and robust, to the flesh of the adult animal.

LAMBS, DISEASES OF. Among other diseases, these animals are particularly prone to one affecting the lungs, in consequence of the existence of parasites (*Strongylus bronchialis*) in the air-passages. See PARASITES.

LAMP. A contrivance for producing artificial light or heat by the combustion of inflammable liquids. The term 'lamp' is also applied to a portable gas-burner (GAS-LAMP), and to a tubular candle-holder, which by the aid of a simple mechanical device, keeps the flame at one height (CANDLE-LAMP).

OIL LAMPS were employed for illumination among the nations of antiquity at the earliest period of which any record exists. The Assyrian, Greek, and Roman lamps preserved in our museums are generally noble specimens of art-workmanship. Though elegant in form, and rich in external embellishment, the ancient lamp was simply a vessel to contain the oil, with a short depression or spout on the one side, in which the wick is laid. Lamps of this rude construction are still in common use in many countries.

No important improvement in the principle and construction of lamps as a source of light occurred until a comparatively recent date; the smoke, dirt, and disagreeable odour of the common lamp having previously led to its disuse among the superior classes in favour of candles. At length, in 1789, M. Argand made a revolution in illumination by the invention and introduction of the well-known lamp which bears his name. In the ARGAND LAMP a hollow tubular wick of woven cotton replaces the solid bundles of fibres, and is so arranged that air passes through it into the interior of the flame. Over the burner is placed a cylindrical glass chimney, open at the bottom, and surrounding the flame at a short distance from it, by which another current of air is made to act on the exterior portion of the flame. In this way a due supply of oxygen is secured, and sufficient heat generated for the perfect combustion of the gaseous products of the oil, and the smoke and soot which escape from the ordinary lamp are converted into a brilliant and smokeless flame.

The earliest table-lamps constructed on Argand's principle had one serious defect – the oil vessels had to be placed almost on a level with the burners, in a position which caused them to cast objectionable shadows. This defect was almost entirely removed by making the oil vessel in the form of a flattish ring, connected by slender tubes with the burner. The more elegant contrivances, known as the MODERATOR LAMP and CARCEL LAMP, which are now so much used for burning colza and similar oils, cast no shadow. In these the oil, instead of being sucked up by the wick, or descending to it by the force of gravity, is driven up by mechanical means from the oil-reservoir contained in the foot or pedestal. A spiral spring, acting upon a piston, elevates the oil in the 'moderator,' while a little pump worked by clockwork does the same duty in the 'Carcel.' The burner and wick in each are formed on Argand's principle.

For burning the hydrocarbon oils distilled from coal and petroleum, lamps of very simple construction are used. These oils, in consequence of their diffusive character, rise to a considerable height up a wick, and therefore do not require mechanical lamps. The wicks of HYDROCARBON LAMPS are usually flat, but sometimes circular. To cause perfect combustion, a strong draught of air is created by placing over the flame a tall glass chimney, usually much contracted above the flame. A metallic cap, with an orifice the shape of the flame, is placed over the burner, its use being to deflect the currents of air upon the flame. The reservoirs of hydrocarbon lamps ought always to be constructed of some bad conductor of heat, as glass or porcelain.

For chemical operations, many forms of lamp are used. The ordinary glass SPIRIT-LAMP, fitted with a ground-glass cap, is quite indispensable for minor experiments. (See *engr.* 1.) Stone-

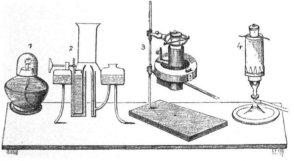

ware wick-holders are preferable to those of brass, which become greatly heated, and endanger the splitting of the glass. "An effective spirit-lamp may at any time be constructed out of a phial having a glass tube passing through the cork, a cover being formed from a test-tube inverted over the wick, and fitting with moderate tightness on the upper extremity of the cork" (*Greville Williams*). Alcohol or wood-spirit is the fuel used.

The ARGAND LAMP, when intended as a source of heat for chemical purposes, is so modified as to adapt it to burn either oil, spirit of wine, or wood-spirit, and the combustion is greatly aided by the chimney, which in this case is made of copper. (See *engr.* 2 and 3.) The lamp itself is also made of metal, and furnished with ground caps to the wick-holder and aperture by which the spirit is introduced, in order to prevent loss of spirit by evaporation when the lamp is not in use. When in use this aperture must always be left open, otherwise an accident is sure to happen, as the heat expands the air in the lamp, and the spirit is forcibly expelled.

In those situations in which coal-gas is cheap, it may be used with great economy and advantage as a source of heat in most chemical operations. Retorts, flasks, capsules, and other vessels can be thus exposed to an easily regulated and constant temperature for many successive hours. Small platinum crucibles may be ignited to redness by placing them over the flame on a little wire triangle. Of the various gas-lamps now used in the *laboratory*, the first and most simple consists of a common Argand gas-burner fixed on a heavy and low foot, and connected with a flexible gas-tube of caoutchouc or other material. (See *engr.* 4.) With this arrangement it is possible to obtain any degree of heat, from that of the smallest blue flame to that which is sufficient to raise a moderately large platinum crucible to dull redness. When gas mixed with a certain proportion of air is burnt, a pale blue flame, free from smoke, and possessing great heating power, is obtained. A lamp for burning the mixture may easily be made by fitting a close cover of fine wire gauze over the top of the chimney of the last-mentioned contrivance. The gas is turned on, and after a few minutes ignited above the wire gauze. The ingenious and useful burners of Bunsen and Griffin are so constructed that gas and air mixed in any proportions, or gas alone, may be burnt at pleasure. Bunsen's is a most efficient and convenient form of burner. (See *engr.*) It consists of a gas-jet surrounded by a metal tube about 6 to 9 in. high and about ½ in. in diameter; having at the bottom four large holes. On the admission of air, when the gas is turned on, the air rushes in by these orifices, and mingling with the gas, the mixture ascends to the top of the tube and is there ignited, giving rise to a flame of great heat, but without luminosity, owing to the simultaneous combustion of the carbon and the hydrogen. The burner, however, is so contrived that by shutting off the supply of air entirely or limiting it, the flame may be made more or less luminous at pleasure. To distribute the flame, a rosette burner is placed on the top of the tube.

An improved variety of this burner has been designed by Bunsen, and is figured *below.*

Fɪɢ. 1.

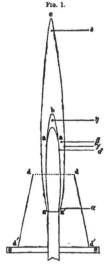

It is so contrived as to give a flame that is a very much better substitute for the flame of the

blowpipe than the ordinary Bunsen's burner, and may hence be employed for reducing, oxidising,

Fɪɢ. 2.

Improved Bunsen burner.

fusing, and volatilising, as well as for the observation of coloured flames. Fig. 1 is a sheath which, by turning round, regulates the admission of air. When it is used, the conical chimney, d d d' d', is placed in e e; it is of a size sufficient to allow of the flame burning tranquilly. In fig. 1 the flame is represented of half its natural size. This flame it will be seen consists of 3 divisions, viz.: 1, a a a' a', the dark zone, which is composed of cold gas mixed with about 62 per cent. of air; 2, a c a b, the mantle formed by the burning mixture of gas and air; 3, a b a, the luminous tip of the dark cone, which only appears when the orifices for the air are partially closed. Reductions may be performed in this part of the flame.

Bunsen, however, divides the flame into 6 parts, to which he attributes as many functions. These 6 divisions of the flame he names as follows:

1. *The base* at *a* has a relatively low temperature, because the burning gas is here cooled by the constant current of fresh air, and also because the lamp itself conducts the heat away. This part of the flame serves for discovering the colours produced by readily volatile bodies, when less volatile matters which colour the flame are also present. At the relatively low temperature of this part of the flame, the former vaporises alone instantaneously, and the resulting colour imparted to the flame is for a moment visible unmixed with other colours.

2. *The Fusing Zone.* This lies at β, at a distance from the bottom of somewhat more than one-third of the height of the flame, equidistant from the outside and the inside of the mantle, which is broadest at this part. This is the hottest part of the flame, viz., about 2300°, and it therefore serves for testing substances, as to their fusibility, volatility, emission of light, and for all processes of fusion at a high temperature.

3. *The Lower Oxidising Zone* lies in the outer border of the fusing zone at γ, and is especially suitable for the oxidation of oxides dissolved in vitreous fluxes.

4. *The Upper Oxidising Flame* at ε consists of the non-luminous tip of the flame. Its action is strongest when the air-holes of the lamp are fully open. It is used for the roasting away of volatile products of oxidation, and generally for all processes of oxidation, when the highest temperature is not required.

5. *The Lower Reducing Zone* lies at δ, in the inner border of the fusing zone next to the dark cone. The reducing gases are here mixed with oxygen, and, therefore, do not possess their full power, hence they are without action on many substances which are deoxidised in the upper reducing flame. This part of the flame is especially suited for reduction on charcoal or in vitreous fluxes.

6. *The Upper Reducing Flame* lies at η, in the luminous tip of the dark inner cone, which, as already explained, may be produced by diminishing the supply of air. This part of the flame must not be allowed to get large enough to blacken a test-tube filled with water and held in it. It contains no free oxygen, is rich in separated incandescent carbon, and therefore has a much stronger action than the lower reducing zone. It is used more particularly for the reduction of metals collected in the form of incrustations.

The subjoined is a drawing of the gauze burner, which is an open cylinder with wire gauze at the top.

Argand's lamp, with wire-gauze cap.

When this is placed over the gas burner, a supply of air is drawn in at the bottom by the ascending current of gas, and the mixture burns above the gauze, with a very hot flame, quite free from smoke, the metallic meshes preventing the flame from passing down to the gas below.

Lamp, Flame'less. *Syn.* GLOW LAMP. A coil of fine platinum wire is slipped over the wick of a spirit lamp, the greater part being raised above the cotton; the lamp is supplied with ether or alcohol, lighted for a moment, and then blown out. The coil continues to glow in the mixed atmosphere of air and combustible vapour, until the liquid in the lamp is exhausted.

Lamp, Monochromat'ic. A lamp fed with a mixture of a solution of common salt and spirit of wine. It gives a yellow light, and makes every object illuminated by it appear either yellow or black. The human features are changed in a remarkable degree; the countenance appearing truly ghastly and unearthly.

Lamp, Safety. *Syn.* MINER'S LAMP, DAVY, GEORDY. The safety lamps of Sir H. Davy and George Stephenson are similar in principle, and were independently invented about the same time. That of Sir H. Davy consists of a common oil lamp, surmounted with ██████ of wire gauze, the apertures of wh██████ greater than the 1-20th of an inch █████ the wire of which it is made to the 1████ ██ 1-60th of an inch in diameter. (Se███ █████ The fire-damp (carburetted hydrogen) along with air passes through the meshes into the interior of the gauze cylinder. Here it ignites, but the flame which is produced by its combustion cannot explode a mixture of fire-damp and air by which the lamp may be surrounded. The flame is prevented from passing to the exterior of the gauze by the cooling action of the metal of which it is constructed. When this lamp is taken into an explosive atmosphere, although the fire-damp may burn within the cage with such energy as sometimes to heat the metallic tissue to dull redness, the flame is not communicated to the mixture on the outside. These appearances are so remarkable, that the lamp becomes an admirable indicator of the state of the air in different parts of the mine, and if its admonitions are attended to, gives the miner time to withdraw before an explosion takes place.

Lamp, Telescope. This ingenious contrivance, invented by Messrs Murray and Heath, is intended for microscopic illumination. It consists of 3 brass tubes, sliding one within the other, the oil vessel being contained in the inner tube. The height of the lamp is regulated to the greatest nicety by simply turning one tube in the other, interior spiral guides preventing all chance of slipping. The great advantage of this arrangement is absence of the stand and bar usually employed for raising and lowering the lamp, which enables it to be used on all sides, and to be brought much closer to the microscope than other lamps. (See *engr.*)

LAMP-BLACK. See BLACK PIGMENTS.

LAMPREY. *Syn.* GREAT LAMPREY, SEAL. This fish is the *Petromizon marinus* of Linnæus. It generally quits the sea in the spring, for the purpose of spawning, and remains in our rivers for a few months. Its flesh is soft and glutinous, and though esteemed a delicacy, is extremely difficult of digestion, if not otherwise unwholesome. Potted lampreys are usually so highly seasoned as to become a dangerous article of food. Henry I is said to have lost his life from the effects of a surfeit of lampreys.

LAMPRONIA CAPITELLA, Linn. The Currant Borer. This is of very similar habits to the moth *Ægeria*. This small insect produces a caterpillar which injures currant bushes of all kinds, and raspberry canes as well, by burrowing into their shoots and killing them.

This *Lampronia* is mentioned by Kaltenberg as troublesome in Germany, but no mention is made of it by American writers.

Life History. The moth is a species of the genus *Lampronia* of the extensive family of *Tineina*. It is very small, being only about

8 lines across its expanded wings, and its body being only about three lines in length. Upon the brownish fore-wings there are markings or

tinges of purplish-yellow which glisten in the sun. The hinder wings are lighter in colour with delicate fringes. As it has a little dark yellow top knot of hairs, it is called *capitella*.

It is believed that the perfect insect places its eggs on the stems of the currant bushes in May, and that the larva makes its way into them and remains until the following spring.

Prevention. Carefully cutting away and burning infested branches and shoots should be adopted ('Reports on Insects Injurious to Crops,' by Chas. Whitehead, Esq., F.Z.S.).

LAMPRONIA RUBIELLA, Bjerkander. (*Lampronia*, from λαμπρος, splendid; *rubiella*, from *rubra*, a raspberry.) The Raspberry-shoot Borer. Raspberry and blackberry plants are liable to be injuriously affected by the larvæ of a very small and elegant moth throughout the month of May. They choose the tender shoots upon which the blossoms and fruit come, commencing their operations upon these when they are about an inch in length, and attacking them just under the whorls of leaves at their summits. Penetrating into the shoots, they work their way inside them down to their bases, revelling in the most abundant juices. The consequences to the shoots are fatal. They wither up after a while, and can produce no fruit.

Among several insects whose larvæ prey upon raspberry and blackberry plants it is very difficult for those unlearned in entomology, or those who have not studied carefully the habits of various species, to particularise, or, in plainer language, to 'spot' the offender. The terms bug, fly, and maggot serve almost as the sole, or at least as the most important, definition of distinction to busy cultivators, so that unless the actual culprit is presented, together with a description of its fashion of injury, identification is nearly impossible.

It may be said that the fashion of injury differs in each insect just as the form of each species of insect differs exceedingly. Thus the *Lampronia* has a mode of attack distinct altogether from that of the raspberry beetle, *Byturus*, or from the operation of the raspberry weevil, *Otiorynchus picipes*, and each of these latter has its own peculiar line of action. The *Lampronia rubiella* may be known at once by the brilliant red colouring of its larva, and its method of burrowing in the youngest shoots.

Fruit-growers who have been asked as to whether they had sustained harm and loss from this insect, have in many cases replied that they have certainly noticed that many of the young shoots of the raspberry plants have withered, but attributed this to natural decay or to unhealthiness.

A large fruit-grower near Orpington, in Kent, was much exercised as to this affection of his raspberry plants, and stated that he believed their roots had "got down to something they did not like." He subsequently and upon close investigation discovered "little red maggots in the shoots, which turned out to be the larvæ of a tiny moth, afterwards identified as the *Lampronia rubiella.*"

Injury to raspberry shoots of the same character was traced to the larvæ of this moth in fruit plantations near Lancaster in 1882, and from several other places complaints of harm to raspberry plants have been received in the last three or four years clearly caused by this insect.

Kaltenberg speaks of this moth as known in Germany, and says that the larvæ live in the shoots of the raspberry and blackberry plants. It does not appear that it is known in America.

Life History. This moth is readily distinguished by its lustrous brown wings, dotted over with yellow spots, two of which on the upper wings are large. The wing expanse is only about five lines, not ¼ an inch, and its body is not more than 1-3rd of an inch long.

It belongs to the LEPIDOPTERA and to the genus *Lampronia* of the numerous and interesting family *Tineidæ*. In moth form it escapes from the chrysalis in June and lays eggs upon the canes of the raspberry and blackberry under or among the folds of their fine cuticle or bark Caterpillars, or larvæ, are hatched in August and begin at once to feed upon the leaves

Westwood confirms this, and remarks that they feed for a time and then hybernate. Stainton also agrees with this in his description of this moth ('Lepidoptera-Tineina,' by H. Stainton). It is not quite clear whether the larvæ hybernate upon the stems under the bark or skin, or just under the ground sheltered by leaves or rubbish. The larva is bright red, about three lines, ⅓ in. in length, with a dark-coloured head. It begins feeding upon the succulent shoots of the raspberry and blackberry plants as soon as these are formed, boring into them as described above, and changes into a chrysalis towards the end of May.

Prevention. In order to prevent this insect from coming in a subsequent season it would be necessary to cut the canes down closer than usual at pruning time, and to remove every scrap of the cuttings. All dead leaves and grasses and weeds and rubbish should be cleared away from round the stocks. These should be well cleared out. After a very bad attack it would pay well to cut the stocks right back and let them make a fresh start. Where the canes are staked or trained, though this is not much practised in culture on a large scale, the stakes or supports should be removed. It may be remarked here that stakes with rind or bark left on frequently harbour insects in all stages, and the bands or ties which fasten the trees or canes to them serve as snug shelter for larvæ of all kinds during the winter. For example, in the winter 1888—89 it was seen that multitudes of larvæ of aphides had taken up their winter quarters in the straw bands put round damson trees to keep them from rubbing.

Good cultivation in the late autumn and in the early spring close to the stocks should be adopted, together with the application of lime, soot, or paraffin soaked substances, after a bad visitation.

Remedies. Those shoots which show signs of withering should be immediately cut off with shears and carried away to be burned ('Reports on Insects Injurious to Crops,' by Chas. Whitehead, Esq., F.Z.S.).

LANDANINE. $C_{20}H_{24}NO_4$. An alkaloid obtained by Hesse from the aqueous extract of opium. It is homologous with morphine and codeine. It dissolves in strong sulphuric acid with a rose-red colour, in strong nitric acid with an orange-red colour, and in ferric chloride with emerald-green colour.

LANOLIN. *Syn.* ADEPS LANÆ, L.; WOOL FAT. This substance is the natural fat of the skin and of epidermic tissues generally, such as hair, wool, hoofs, and feathers. It is mainly obtained from sheep's wool, and consists of cholesterin, stearic acid, and about 40% water. Prof. Liebreich, writing of lanolin, says:—"My researches have led me to believe that lanolin is always present wherever keratinous tissue is formed. As to the origin of this cholesterin fat, it may be asked whether it is secreted by glands, or formed simultaneously with the keratinous tissue. I can answer that the horny tissues are full of lanolin, even where there are no glands to produce it. The degree of brilliancy and elasticity of the horny tissue is in direct proportion to the amount of lanolin present. I must not forget to say that some researches made by French experimenters

were of great interest to me while I was working on the subject. Berthelot has shown, many years ago, that it is possible to combine synthetically cholesterine and fatty acids; he has also foreseen that this fat might be found in nature. I have found that the researches of Hartmann and Schultz have brilliantly confirmed those of Berthelot. I have, moreover, shown that lanolin is nearly always present in keratinous tissue. "From its use in over 400 cases in the hospital and private practice of Dr Lassar, the dermatologist, *no irritation of the skin was ever produced,* a result which my own experience, during the years in which I have been experimenting with it, confirms. For this reason alone it is to be highly recommended for massage. It was furthermore observed in Dr Lassar's clinic that the *most irritable skin could bear lanolin* when all other fats caused œdema and irritation."

Dr L. K. Pavlovsky writes that his experiments with lanolin enable him to arrive at the following conclusions:

1. Narcotic extracts, when combined with lanolin, are absorbed by the skin 'quite satisfactorily,' their analgesic action being obtained 'with almost an absolute certainty.' The dose used was only twice as large as that for internal use

2. Hydrochlorate of quinine is also absorbed very easily. This statement is based on four cases of intermittent fever in children, where lanolin and quinine inunctions rapidly gave the desired effects.

3. When an ordinary ointment, with iodide of potassium, is rubbed into the skin, iodine appears in the urine not sooner than two, four, or six hours after inunction, while Lassar *obtained iodine from the urine about three minutes after friction with a lanolin ointment.* Patschkowsky and Kaspar found traces of iodine in the urine 30 minutes after rubbing it into the skin with lanolin, whereas with the ordinary iodide of potash ointment results were *nil.*

4. In children, lanolin is better absorbed than in adults.

5. Washing the skin with ether considerably facilitates the absorption of lanolin ointments.

6. In general, lanolin is a substance which promises to supersede every other basis for ointments, and even, in certain cases, to supersede the internal administration of drugs.

Lanolinum Hydrargyri. *Syn.* MERCURIAL LANOLIN. *Prep.* Mercury, 100 parts; lanolin, 200 parts; mercurial ointment, 5 parts; mix in a warm mortar until the mercury is extinguished. Said to be superior to mercurial ointment.

LANTHANIUM. La = 92. A rare metal, discovered by Mosander, associated with oxide of cerium. Oxide of lanthanium is a pale salmon-coloured powder, unaffected by ignition in open vessels. According to Zschiesche, the atomic weight of lanthanium is 90·18. See CERIUM.

LANTHOPINE. $C_{23}H_{25}NO_4$. A base obtained by Hesse in small quantity, associated with other bases from the aqueous extract of opium. It is homologous with papaverine. Strong nitric acid dissolves it, giving rise to an orange-red colour. Strong sulphuric acid gives with it a faint violet colour.

LAPIS. [L.] A stone. The term was much

employed by the old chemists, and is still commonly applied to several preparations used in *medicine*.

Lapis Causticus. See POTASSIUM.

Lapis Divi'nus. *Syn.* DIVINE STONE; LAPIS OPHTHALMICUS, L.; PIERRE DIVINE, Fr. *Prep.* 1. (*Beer.*) Verdigris, nitre, and alum, equal parts, melted together.

2. (P. Cod.) Alum, nitre, and blue vitriol, of each, 3 oz.; camphor, 1 dr.; as last.

3. (*Woolfuse.*) Blue vitriol, nitre, alum, and camphor, equal parts, melted together, adding the camphor last. Astringent and detergent. 1 oz., dissolved in water, 1 pint, formed a once celebrated lotion. 1 dr. in water, 1 pint, is still used as a collyrium.

Lapis Inferna'lis. See NITRATE OF SILVER.

Lapis Lazu'li. See ULTRAMARINE.

Lapis Lydius. *Syn.* LYDIAN STONE. A siliceous slate, used as a touchstone by jewellers.

Lapis Medicamento'sus. *Syn.* MEDICINAL STONE; LAPIS MIRABILIS, L. *Prep.* (Ph. L. 1746.) Alum, litharge, and Armenian bole, of each, 6 oz.; colcothar of green vitriol, 3 oz.; vinegar, 4 fl. oz.; mix, and evaporate to dryness. Formerly used to make an astringent and detergent lotion:—1 oz. to water, 1 pint. Once a popular application to ulcers, and in other cases; now disused.

Lapis Vulnerar'ius. Very similar to LAPIS DIVINUS.

LARCH BARK. The inner bark of the *Pinus laris*, the common larch, has been lately introduced under the form of a tincture into the British Pharmacopœia.

Dr Stenhouse obtained from the bark a peculiar volatile constituent, possessed of acid properties for which the name of lariximic acid has been proposed. The other trees of the pine family are deficient in this acid. The young bark abounds most in it. Gum, starch, resin, and that variety of tannic acid which forms olive-green precipitates with the salts of iron, have also been found, in addition to other substances, in larch bark.

The inner bark, employed *internally*, has a special action on the mucous membranes, and acts as an astringent and mild stimulant. It is said to have been given with excellent results in hæmoptysis, as well as in bronchitis attended with copious expectoration, and in diseases of the urinary passages. *Externally* it has been found serviceable in psoriasis, chronic eczema, and some other skin diseases. It is best to combine its extract or tincture with glycerin when it is to be used outwardly. See TINCTURE OF LARCH BARK.

LARD. *Syn.* HOG'S LARD, AXUNGE; ADEPS (Ph. L.), AXUNGIA (Ph. E.), A. SUILLUS (Ph. D.), A. PORCI, A. PRÆPARATUS (B. P.), L. The fat of the pig (*Sus scrofa*, Linn.) melted by a gentle heat, and strained through flannel or a hair-sieve. The fat about the loins yields the whitest and hardest lard. "That which has been cured with chloride of sodium is not to be employed" (Ph. L.). "It is not to be used without being first carefully washed with water" (Ph. L. 1866). Used chiefly to make ointments, and in *cookery*. See ADEPS.

LARD'ING. By many this is regarded as belonging to the higher style of cookery only, and too troublesome and extravagant to be adapted to the kitchens of the middle classes and the poor. This, we are assured, is not the case. On the contrary, "it is an economical process, and will make lean meat go much farther than without it." The process of larding is as follows:—"Get what is called a larding needle—that is, a piece of steel from 6 to 9 inches long, pointed at one end, and having 4 slits at the other to hold a small strip of bacon when put between them. It will, perhaps, cost 10d. Cut the bacon into pieces 2 or 3 inches long, and ¼ to ⅛ an inch square; put each one after the other in the pin, insert it in the meat, and leave only about ⅛ an inch out; using 8 pieces to each pound" (*Soyer*).

LARK. The *Alauda arvensis* (SKYLARK) and the *Alauda cristata* (FIELD-LARK), with several other species of the same genus, form a light and nutritious article of food, by many esteemed a delicacy, though one which might well be dispensed with. The last, according to Galen and Dioscorides, eaten either roasted or boiled, 'helps the colic.' The heart, applied to the thigh, was also regarded to possess the same virtue. Larks are great insect-eaters, and the destruction of them should be prevented by every possible means.

LARYNGITIS. Inflammation of the larynx, or upper part of the windpipe. The symptoms that indicate this most dangerous malady are sore throat, accompanied with considerable pain in front of the throat, difficulty in breathing and swallowing, considerable hoarseness, change or loss of voice, a sense of suffocation, fever, restlessness, flushing of the face, and an eager desire for fresh air. Laryngitis may arise from numerous causes, and no one course of treatment will apply to all cases. Medical advice should be sought as soon as the malady is discovered.

LAUD'ANUM. This name is now understood to denote, exclusively, the common tincture of opium of the Pharmacopœia; but formerly the term was applied to several preparations of opium differing greatly from each other, both in their strength and mode of preparation (see *below*).

Laudanum, Dutchman's. From the flowers of bull's hoof or Dutchman's laudanum (*Passiflora merucuja*, Linn.) infused in rum. Narcotic. Used as a substitute for tincture of opium in the West Indies.

Laudanum, Ford's. This is merely the common tincture of opium aromatised with a little cloves and cinnamon.

Laudanum, Houlton's. *Prep.* From opium, 2½ oz.; distilled vinegar, 1½ pints; digested together for a week, the filtered tincture gently evaporated nearly to dryness, and then redissolved in weak spirit (1 of rectified spirit to 7 of water), 1 quart.—*Dose*, 10 to 60 drops.

Laudanum, Neumann's. A fermented infusion of opium evaporated to the consistence of honey.

Laudanum, Quince. *Syn.* EXTRACTUM OPII CYDONIATUM, LAUDANUM CYDONIATUM, L. *Prep.* 1. Extract of opium made with quince juice; a few drops of the oils of cinnamon, cloves, and mace being added before the mass cools. Now seldom used.

2. (LAUDANUM, LIQUID-QUINCE; LAUDANUM LIQUIDUM CYDONIATUM, L. L. C. PARATUM FER-

MENTATIONE, L.) A fermented infusion of opium prepared with quince juice, aromatised with cloves, cinnamon, aloes wood, and yellow sandal-wood, and evaporated so as to possess about twice the strength of the ordinary tincture. Now obsolete.

Laudanum, Rousseau's. Wine of opium prepared by fermentation. See WINE.

Laudanum, Smith's Concentrated. Resembles Battley's LIQUOR OPII SEDATIVUS, but possesses about 6 times its strength.

Laudanum, Swediaur's. Prep. From extract of opium, 2 parts, dissolved in a mixture of alcohol, 1 part, distilled water, 8 parts. Every 5 drops contain 1 gr. of opium.

Laudanum, Sydenham's Liquid. Syn. LAUDANUM LIQUIDUM SYDENHAMI, L. Similar to WINE OF OPIUM (Ph. L.), but rather stronger, and aromatised with a little cloves and cinnamon. Wine of opium is now always sold for it.

Laudanum, Tartarised. Syn. LAUDANUM LIQUIDUM TARTARIZATUM, L. A tincture of opium prepared with spirit alkalised with salt of tartar, and flavoured with aromatics. Obsolete.

LAUGH'ING GAS. See NITROUS OXIDE.

LAUR'EL. See CHERRY LAUREL, SWEET BAY, OIL, &c.

LA'VA. The matter thrown out by volcanoes. The beautiful ornamental vases, jugs, and other objects sold under the name are a superior sort of unglazed coloured porcelain.

LAVE'MENT. See ENEMA.

LAV'ENDER. The flowers or flowering tops of Lavandula vera, or common garden lavender, largely cultivated at Mitcham and Hitchin. An essential oil, spirit, and tincture, prepared from it, are officinal in the Pharmacopœias.

Lavender Dye (for COTTON). For 100 yards of material. Take 1 lb. of logwood, and 2 lbs. of sumach, and scald them separately. Then decant them into a proper-sized tub, let them cool to 150° F., and add 2 gills of vitriol. Winch the goods in this 20 minutes; lift, and run them slightly through acetate of iron; wash them in two waters; then give 1 lb. of logwood as before, raise with 1 pint of chloride of tin, wash in two waters; then in a tub of cold water put 4 oz. extract of indigo, enter and winch in this 15 minutes, lift; give one water and dry.

Lavender Dye (for WOOL). Boil 5½ lbs. of logwood with 2 lbs. of alum. Then add 10 oz. of extract of indigo. When cold put in the goods, and gradually raise to the boiling-point. For 50 lbs.

Lavender, Red. See TINCTURE.

Lavender, Smith's British. Prep. From English oil of lavender, 2 oz.; essence of ambergris, 1 oz.; eau de Cologne, 1 pint; rectified spirit, 1 quart. Very fragrant. See WATER, LAVENDER.

Lavender, to Dye Silk. (Mustpratt.) Into a vessel with warm water, as hot as the hand can bear, dissolve a little white soap, enough to raise a lather; then add 1 gill of archil liquor, and work the goods in this for 15 minutes; wring out and dry.

Boil 1 oz. of cudbear, and add the solution to the soap and water instead of archil, which will give a lavender having a redder tint than with the archil. If a still redder shade of lavender be required the soap may be dispensed with.

Lavender Water. See SPIRITS, PERFUMED.

LAX'ATIVES. Syn. LENITIVES; LAXATIVA, LAXANTIA, LENITIVA, L. Mild purgatives or cathartics. The principal of these are—almond oil, cassia pulp, castor oil, confection of senna, sulphur, cream of tartar, figs, grapes, honey, phosphate of soda, prunes, salad oil, tamarinds, cascara sagrada, &c.

LAY'ERS. Among gardeners, a mode of propagating plants by laying down the shoots of young twigs and covering a portion of them with the soil without detaching them from the parent plant. To facilitate the rooting of such layers, the part beneath the soil is fractured by twisting or bruising it, or it is partly cut through with a sharp knife immediately under a bud. When the layer has taken root it is divided from the parent stem, and transplanted or potted. In this way, with a little care, nearly all plants may be multiplied.

LEAD. Pb. Atomic weight=206·4. Syn. PLUMBUM, L.; PLOMB, Fr.; BLEI, Ger. A metallic element belonging to the same group as silicon, germanium, and tin. It does exhibit some analogies with those elements as regards its chemical properties, but is much more nearly related to the rare metal thallium. It has been known from the earliest times.

Sources. Lead rarely occurs native. The sulphide, galena (PbS) is by far the most important ore, and occurs in Cornwall, Derbyshire, S. Wales, N. of England, Scotland, and in various localities of Germany, Austria, Spain, United States, and Peru; some has also been obtained from Australia. Other less important naturally occurring compounds of lead are bournonite (CuPbSbS₃), the carbonate, cerussite (PbCO₃), and the sulphate, anglesite (PbSO₄).

Prep. Three distinct processes are made use of for smelting lead. These are:

1. The Air-reduction Process (Percy), which is employed when the ore consists mainly of galena, and is free from silica and the sulphides of other metals. The galena is heated in a reverberatory furnace, and some of it is converted into an oxide and some into the sulphate; at a certain stage the temperature is raised, and the rest of the sulphide is oxidised at the expense of the oxygen of the oxide and sulphate, according to the following equations, sulphur dioxide being evolved and metallic lead left behind.

$$2 PbO + PbS = SO_2 + 3Pb$$
$$PbSO_4 + PbS = 2SO_2 + 2Pb.$$

2. The Carbon-reduction Process is employed for less pure ores, and consists in the roasting of the ore and the subsequent reduction of the lead by carbonaceous matter in a 'slag-hearth,' or small blast-furnace.

3. The Precipitation Process, in which the reduction of the lead is effected by metallic iron, is chiefly practised in France and Germany, where the ore contains other metals, such as copper, antimony, and arsenic. Layers of the ore and the reducing material (at the present time a rich iron slag is generally used instead of cast-iron) are smelted together in a small blast ('schlieg') furnace. Much lead remains in the

slag that is formed, and this is again worked up, and finally is often heated to obtain the copper it contains.

On the small scale, chemically pure lead may be prepared thus (*Stas*):—Heat a solution of lead acetate in a leaden vessel in contact with thin sheet lead to about 40°—50° C. (105—120° F.) in order to precipitate silver and copper. Filter and pour the solution into very dilute pure sulphuric acid, collect the precipitate of lead sulphate, and carefully wash it with a solution of ammonium carbonate and ammonia; it is thus transformed into the carbonate. A portion of this is converted into lead oxide by carefully heating it in a platinum basin, whilst to the rest nitric acid is added in such quantity that a portion of the carbonate remains undissolved. Add the lead oxide to the boiling solution of the nitrate in order to precipitate any traces of iron, filter the solution, and pour it into a solution of pure ammonium carbonate. Reduce the lead carbonate thus formed by fusing it with potassium cyanide; fuse the metal thus obtained a second time with potassium cyanide, when it will assume a convex surface like mercury.

Purif. Lead obtained by the commercial processes described above contains small quantities of antimony, copper, zinc, silver, &c. If there is much antimony present the metal is hard, and is softened by melting and partially oxidising it in a reverberatory furnace with a cast-iron bottom. The antimony oxidises more readily than the lead, and so a scum of oxides is obtained, containing nearly all the antimony and some lead. This scum is reduced with coal, and yields a mixture of lead and antimony which can be used for type-metal. The silver in commercial lead, although present in very small quantities, pays for extraction, and this is carried out by either of the two following processes for desilverising lead.

Pattinson's Process. The metal is melted and carefully skimmed, and then allowed to cool. As it cools crystals of lead comparatively poor in silver form, and are fished out by means of a perforated ladle; a residue much richer in silver is left in the pot. A series of pots is often used, the first third of the fishings from each pot being placed in the pot above, the second third afterwards re-smelted in the same pot, and the residual third put into the pot below. By this means the silver is worked towards one end of the series, the lead towards the other. The richly argentiferous lead thus obtained is cupelled when the lead is oxidised and removed, and the silver remains.

Parkes' or Karsten's Process. Lead is melted with zinc, 11·2 lbs. of zinc being required for every 7 oz. of silver per ton of lead, and the mixture is allowed to cool, when an alloy of zinc and silver rises to the surface, and is fished out with a perforated ladle. To remove the small amount of zinc which remains dissolved in the lead, the mixture is heated to dull redness in a current of air, whereby the zinc is oxidised, whilst the main portion of the lead remains behind in the marketable state. Or superheated steam is passed into the melted alloy of lead and zinc, whereby the zinc is oxidised, together with a little of the lead, and the oxides are carried

over into chambers where they are condensed, while the lead remains behind in the pure state (*Corderié.*) The silver is obtained from the zinc-silver alloy by distilling off the zinc or dissolving it in hydrochloric acid.

The refined lead of *commerce* is very pure; an analysis of a sample from the Harz gave a total impurity of only 0·0164% —that is, only 164 parts per million.

According to Keith the refining of lead is now carried out by electrolysis in New York by the Electro Metal-Refining Company. In each of 30 wooden vessels about 1 yard high by 2 yards wide, dip 13 cylinders of thin brass plate, arranged concentrically at about 2 in. distance from each other, and serving as kathodes. The anodes are formed of plates of unrefined lead 2 ft. × 6 in. × ⅓ in., surrounded by muslin bags. A solution of lead sulphate and sodium acetate is constantly fed into the vessels at the bottom, and runs over at the top into a warming apparatus, where it is heated to 38° C. (100° F.), and is then conveyed back again into the electrolysing vessels. Arsenic, antimony, silver, &c., are deposited in the muslin bags surrounding the anodes, and are melted with saltpetre and soda to obtain the silver. Practically all the impurities are thus removed, and the refining of 1 ton of lead requires an expenditure of 148 lbs. coal.

In 1885 the lead produced in Europe amounted to 330,000 or 350,000 tons, in North America 120,000 tons. The average yearly consumption for recent years is, North America, 35,000; England, 115,000; France, 65,000; Germany, 45,000; in other parts of the world, 100,000; in all 460,000 tons. In France the consumption of lead plate and tubing has decreased enormously.

Prop. Lead is a white metal with a bluish-grey tinge; its sp. gr. is 11·25, or, after it has been poured into water, 11·36; it is but little increased by hammering. It is soft and tough, may be cut with a knife, and leaves a streak upon paper. It can be beaten into foil, but not drawn into wire. When melted repeatedly it becomes hard and brittle, probably owing to the formation of a small quantity of oxide; it is also rendered brittle by the presence of antimony, zinc, bismuth, arsenic, and silver. It crystallises in octahedra. If a piece of zinc be suspended in solution of sugar of lead (lead acetate), the lead separates out in an arborescent growth, known as the 'tree of Saturn,' Saturn being the old alchemistic name for lead. Lead melts at 334° C. (633° F.), but volatilises only at a bright red heat. It is scarcely at all attacked by sulphuric or hydrochloric acids, but dissolves in nitric acid. It oxidises superficially in the air, a bright, freshly-cut surface quickly tarnishing.

Salts. Lead forms a number of salts, which are mostly white in colour, and insoluble or only slightly soluble in water, though more so in acids. They are described in detail *below*.

Tests. The oxides and salts of lead, mixed with a little carbonate of soda, and exposed on a charcoal support to the reducing flame of the blowpipe, readily yield a soft and ductile globule of metallic lead, and the charcoal, at the same time, becomes covered with a yellowish incrustation of oxide of lead. Both metallic lead and its oxides are soluble in nitric acid, furnishing a solution which may be examined with ease.

Solution of lead salts may be recognised by the following reactions:—Sulphuretted hydrogen, sulphydrate of ammonium, and the alkaline sulphides, give black precipitates, insoluble in the cold dilute acids, alkalies, alkaline sulphides, and cyanide of potassium. Potassium and sodium hydrates give a white precipitate, soluble in excess. Ammonia (except with the acetate) gives a white precipitate, insoluble in excess. The carbonates of potassium, sodium, and ammonium, give a white precipitate, insoluble in excess. Dilute sulphuric acid (in excess), and solutions of the sulphates give a white precipitate, sparingly soluble in dilute acids, but soluble in a hot boiling solution of potassium carbonate. Chromate and bichromate of potassium give yellow precipitates, insoluble in dilute nitric acid, and soluble in solution of potassium hydrate. Iodide of potassium gives a yellow precipitate, soluble in great excess by heat, and separating in small, brilliant, golden-yellow scales as the liquid cools. A piece of polished zinc precipitates metallic lead in an arborescent form, hence called the lead tree. To prepare for these tests, a solid supposed to contain lead should be digested in nitric acid, when the solution, evaporated to dryness and redissolved in water, may be tested as *above*.

Estim. The ore of lead (galena) may be digested in nitric acid, when the solution may be treated with sulphuric acid, and the lead sulphate collected, washed with dilute sulphuric acid and then with alcohol, ignited, and weighed; its weight multiplied by 0·683 gives the equivalent weight of metallic lead. This is called an assay in the wet way. The method adopted by practical mineralogists is an assay in the dry way, and is conducted as follows:—A small but powerful air-furnace, charged with coke, is brought to as high a temperature as possible, and a conical wrought-iron crucible plunged into the midst of it; as soon as the crucible has attained a dull-red heat, 1000 gr. of the galena, reduced to powder, are thrown into it, and stirred gently with a long piece of stiff iron wire flattened at one end, in order to expose as large a surface of the powdered ore to the air as possible, observing now and then to withdraw the wire, to prevent it becoming red hot, in which case some of the ore would permanently adhere to it, and be reduced before the intended time; the roasting is completed in 3 or 4 minutes, and any portion of the ore adhering to the stirrer being detached by a knife, and returned into the crucible, the latter is covered up, and allowed to attain a full cherry-red heat, when about 2 or 3 spoonfuls of reducing flux are added, and the whole brought to a full white heat; in 12 or 15 minutes, the portion of metal and scoria adhering to the sides of the crucible are scraped down into the melted mass with a small stick of moist green wood, after which the crucible is again covered, and the heat urged for 2 or 3 minutes longer, so as to keep the mass in a perfectly liquid state during the whole time; the crucible is then removed from the fire with the crucible-tongs, and adroitly tilted so as to discharge its contents into a small, ingot-mould of brass, observing to rake the scoria

from the surface to the sides of the crucible, so as to allow the molten lead to be poured out without it; the scoria is then reheated in the crucible with about ⅓ spoonful of flux, and after being cleansed with a piece of green wood, as before, is at once poured into a second mould, which is instantly inverted; the little button of lead thus obtained is added to the lead in the other mould, and the whole is accurately weighed. The weight, divided by 10, gives the percentage of lead (including silver, if present) in the ore examined.

One half of the lead thus obtained is put into a dry cupel of bone-ash, and placed in the cupelling furnace, and treated as described in the article on assaying; the metallic button left on the cupel is then detached and weighed. The weight, divided by 5, gives the percentage of pure silver.

Obs. The flux commonly employed in the above assay is composed of red argol, 6 parts; nitre, 4 parts; borax, 2 parts; fluor-spar, 1 part; well pulverised and thoroughly mixed together. When the ore is very refractory, about a spoonful of carbonate of potassium should be added for each 1000 gr. of ore, in which case the roasting may be dispensed with. The quantity of silver in argentiferous galena varies from $\frac{1}{10000}$ to $\frac{1}{4}$ part of the whole. Whenever this ore contains above 2 parts of silver in the 1000, it is found to be profitable to extract the latter. Indeed, by Pattison's process it is found that as small a proportion as 1 in 8000 can be extracted with profit.

Uses. Lead is used to make pans for evaporating sulphuric acid, vitriol, and alum, for making leaden chambers of sulphuric acid works, for water and gas supply tubing, for retorts, bullets, plate, foil, shot (mixed with a little arsenic), extraction of metals, as silver and gold, alloys, lead acetate (sugar of lead), litharge, red-lead, white-lead, chrome yellow, &c. Some of its salts are employed in *medicine*.

Lead-poisoning. Administer an emetic of sulphate of zinc or sulphate of copper, and, if necessary, tickle the fauces with the finger or a feather, to induce vomiting. Should this not succeed, the stomach-pump may be had recourse to. Epsom or Glauber's salts, or alum dissolved in water, or water acidulated with sulphuric acid, followed by tea, water-gruel, or barley-water, are the proper antidotes, and should be taken as soon after the poison has been swallowed as possible. In poisoning by white-lead, Dr Alfred Taylor recommends the administration of a mixture of sulphate of magnesium and vinegar, as preferable to the sulphate alone. When the symptoms are those of painter's colic, the treatment recommended under that head should be adopted. In paralysis arising from lead, small doses of strychnine and its preparations may be cautiously administered. A symptom of poisoning by lead is the formation of a narrow leaden blue line, from 1-20th to 1-6th of an inch wide, bordering the edges of the gums, attached to the neck of two or more teeth of either jaw (*Dr Burton*). This discoloration may often be detected or rendered more conspicuous by rinsing the mouth out with water holding a little sulphuretted hydrogen or sulphydrate of ammonium in solution. Chevallier and Rayer recommend the use of sulphurous or hepatic mineral waters, or of artificial solutions of sulphuretted hydrogen or alkaline sulphides in water, both in cases of acute and chronic poisoning by lead; but the practical success of this plan does not appear to have been in proportion to theoretical anticipations. The moist and freshly precipitated sulphides of iron are said by their advocates to be infallible if taken sufficiently early.

Pure water put into a leaden vessel and exposed to the air soon corrodes it, and dissolves the newly formed oxide; but river and spring water have little action upon lead, provided there is no free carbonic acid present, the carbonates and sulphates in such water destroying their solvent powers. It has been found that a very small amount of phosphate of sodium or of iodide of potassium, dissolved in distilled water prevents its corrosive action on this metal. The lead in contact with such water gradually becomes covered with a superficial film of an insoluble salt of lead, which adheres tenaciously, and prevents further change. From this it appears that ordinary water ('hard water'), which abounds in mineral salts, may be more or less safely kept in leaden cisterns; but distilled water and rain water, and all other varieties that contain scarcely any saline matter, speedily corrode and dissolve a portion of lead when kept in vessels of that metal. When, however, leaden cisterns have iron or zinc fastenings or braces, a galvanic action is set up, the preservative power of saline matter ceases, and the water speedily becomes contaminated with lead, and unfit for consumption as a beverage. Water containing carbonic anhydride also acts on lead, and this is the reason why the water of some springs (although loaded with saline matter), when kept in leaden cisterns, or raised by leaden pumps, possesses unwholesome properties.

M. Fordos, in a communication to the 'Journal de Pharmacie et de Chimie,' xix, 20, states that in the course of some experiments on the applicability of lead for water-pipes and cisterns he could not detect a trace of lead in 10 litres of river water taken from the leaden cistern of one of the Paris hospitals. But upon shaking pure water with shot and air a coating of carbonate of lead was formed on the sides of the bottle, which almost rendered the glass opaque. On dissolving the film in nitric acid, and estimating the lead, it was found that 1 litre of water had produced 5 milligrms. of the carbonate. Wine and vinegar would also dissolve that film; and as shot is commonly used for cleaning wine-bottles, lead frequently finds its way into wines, a fact which may account for many of the cases of chronic poisoning by lead which occur in large towns. The detection of small quantities of lead in forensic investigation would afford, therefore, no proof of any intentional poisoning.

Orfila's erroneous statement that lead is a normal constituent of the human organism may also be accounted for in this way.

Free carbonic acid is evolved during the fermentation or decay of vegetable matter, and hence the absolute necessity of preventing the leaves of trees falling into water-cisterns formed of lead. The 'eau de rose' and the 'eau d'orange' of commerce, which are pure distilled water holding in solution small quantities of essential oil, and are imported in leaden canisters, always contain a small quantity of lead, and deposit a sediment, which is not the case when they are kept in glass or incorrodible vessels.

Lead and all its preparations are highly poisonous; and whether imbibed in almost infinitesimal quantities with our daily beverages and food, or swallowed in larger and appreciable doses, is productive of the most disastrous consequences, the real cause being unfortunately seldom suspected.

Lead in Aërated Water. Some time since Sir Robert Christison condemned the use of syphons for lemonade, owing to the action of free tartaric acid upon lead, and the rapidity with which waters containing any free acid become charged with lead in syphons. According to Professor Miller, 0·0175 gr. of lead per gallon is not an unusual amount for average cistern water. Mr John S. Thompson, however, reports to the Edinburgh University Chemical Society that, after such water has been aërated and put into a syphon, the amount of lead dissolved in it begins to rise in a rapid manner. Thus in potash water, drawn from a syphon, 0·0406 gr. of lead per gall. was found to be present, being nearly 25 times the quantity found in the same water before it entered the syphon. Pure aërated water again drawn in a similar manner from a syphon gave 0·0816 gr. of lead per gall., or exactly double the amount found in the potash water, showing at once the well-known protective action that salts of the alkalies and alkaline earths have on lead. "Although," says the 'Medical Journal,' "these results are sufficiently high and alarming, still, when the water is drawn off in small quantities at a time, as is frequently the case with invalids, the results are found to be still higher; thus, when potash water was so treated, 0·0455 gr. of lead per gall. was found, while aërated water, drawn off in small quantities, gave 0·0933 gr. of lead per gall., showing a very marked rise in both cases. The cause of this increase in quantity of the lead appears to be owing not so much to the lengthened period of contact between the liquid and the metal as to the fact that the nozzle of the syphon, being exposed to the atmosphere in a moist state, becomes rapidly oxidised or carbonated, and is left in the most suitable condition for entering into solution, so that, when merely small portions of the liquid are drawn off each time, a comparatively concentrated solution of lead is obtained. These results," continues the same authority, "compare accurately with those which were obtained by Messrs Savory and Moore in examining the contents of a series of syphons of aërated water for Dr George Owen Rees, F.R.S., whose attention was drawn to the subject by detecting symptoms of lead-poisoning in himself after he had been in the habit for some time of drinking such aërated water."

Lead, Acetate of. Pb($C_2H_3O_2$)$_2$,3H$_2$O. Syn. PLUMBIC ACETATE, SUGAR OF LEAD; PLUMBI ACETAS (B. P.), L. Prep. Litharge (in fine powder), 24 oz.; acetic acid, 2 pints; distilled water, 1 pint; mix the acetic acid and the water, add the litharge, and dissolve with the aid of a gentle heat, filter, evaporate until a pellicle forms, and crystallise. Drain and dry the crystal.

Acetic acid (sp. gr. 1·0643), 23 parts, is gently heated in a copper boiler rendered electro-negative by means of a large flat piece of lead soldered within it, and litharge (pure, and in fine powder), 13 parts, is sprinkled in; the heat is then continued, with constant stirring, until the acid is saturated, when the mother-waters of a former process, if any, are added, and the whole is heated to the boiling-point, and allowed to settle until cold; the clear portion is now decanted, and evaporated in a similar vessel until the liquor has the sp. gr. 1·266 or 1·267, when it is run into salt-glazed stone-ware vessels (the edges of which have been well smeared with candle-grease) and allowed to crystallise. The product is 38 to 38½ parts of crystallised sugar of lead. It is found to be advantageous to preserve a very slight excess of acid during the boiling and crystallisation, to prevent the formation of any basic acetate the presence of which impedes the formation of regular crystals.

From litharge, 112 lbs.; acetic acid (sp. gr. 1·057), 128 lbs. *Prod.*, 180 to 184 lbs.

Prop. Pure acetate of lead forms colourless, transparent, prismatic crystals, slightly efflorescent in dry air; it is soluble in 8 parts of alcohol and in 1¾ parts of cold water; the aqueous solution has a sweet astringent taste, and feebly reddens litmus, but turns turmeric and the juice of violets green; when gently heated, it melts in its water of crystallisation; by continuing the heat the whole of the water is expelled, and the dry acetate obtained; at a higher temperature the salt suffers decomposition, and acetic acid, acetone, &c., are given off. Commercial acetate of lead is in general a confused crystalline mass, somewhat resembling broken lump sugar. It is powerfully astringent and poisonous.

When pure it is completely soluble in distilled water acidulated with acetic acid, forming a transparent colourless solution. " 38 gr. dissolved in water require for complete precipitation 200 gr. measures of the volumetric solution of oxalic acid " (B. P.).

Uses, &c. Acetate of lead is extensively employed in dyeing and calico-printing. In *medicine*, it is used as an astringent, styptic, and hæmostatic, in pulmonary, uterine, and intestinal hæmorrhage, colliquative diarrhœa, phthisical sweats, &c. It is usually combined with morphia or opium, and with acetic acid to prevent it passing into the state of the poisonous carbonate in the stomach.—*Dose,* ½ to 2 gr. (*Collier*); 1 to 2 gr. to 8 or 10 gr., twice or thrice a day (*Pereira*); 3 to 10 gr., every 6 or 8 hours (*A. T. Thomson*). *Externally,* as a collyrium, 10 gr. to water, 8 fl. oz. (*A. T. Thomson*); as a lotion, 20 gr. (*A. T. Thomson*); 1 dr. (*Collier*) to water, 8 or 10 fl. oz.; as an injection, 40 gr. to rose-water, ½ pint. The lotion is cooling and sedative, and is commonly used in excoriations, local inflammations, &c.

Basic Acetates. There are several of these salts, but only one is of any importance:

Tribasic Lead Acetate, or Double Plumbic Acetate and Dioxide. $Pb(C_2H_3O_2)_2.2PbO.$ *Syn.* SUBACETATE OF LEAD, BASIC LEAD ACETATE, GOULARD'S ACETATE OF LEAD; PLUMBI SUBACETAS

(B. P.), L. *Prep.* Litharge, 7 parts; acetate of lead, 10 parts; and distilled water, 40 parts, are boiled ½ an hour, and evaporated down, and allowed to crystallise out of contact with air.

Used under the form of 'plumbi subacetas liquor' (B. P.).

Lead, Arse"niate of. $Pb_3(AsO_4)_2.$ *Syn.* ARSE-NIATE OF L.; PLUMBI ARSENIAS, L. *Prep.* By gradually adding a solution of acetate of lead to another of arseniate of sodium. A white, insoluble powder. Proposed as an external application in certain forms of cancer.

Lead, Bro"mide of. $PbBr_2.$ *Syn.* PLUMBI BROMIDUM, L. *Prep.* By precipitating a solution of neutral acetate or nitrate of lead with a solution of bromide of potassium. A white, crystalline powder, sparingly soluble in water. It fuses by heat into a red liquid, which turns yellow when cold. It has been used in the same cases as iodide of lead.

Lead, Car"bonate of. $PbCO_3.$ *Syn.* PLUMBI CARBONAS, L. *Prep.* By precipitating a cold solution of either acetate or nitrate of lead with a solution of an alkaline carbonate, taking care to well wash the precipitate and dry it in the shade. This preparation is seldom employed, the commercial (basic) carbonate (WHITE-LEAD) being substituted for it. See WHITE PIGMENTS.

Lead, Chloride of. $PbCl_2.$ *Syn.* LEAD CHLORIDE; PLUMBI CHLORIDUM (Ph. L. 1836), L. *Prep.* (Ph. L. 1836.) Dissolve acetate of lead, 19 oz., in boiling water, 3 pints; next dissolve chloride of sodium, 6 oz., in boiling water, 1 pint; mix the two solutions, and when cold wash and dry the precipitate. A white crystalline powder.

Dissolve finely powdered litharge in boiling dilute hydrochloric acid, and set aside the filtered solution to cool. Brilliant colourless needles.

Prop. Soluble in 135 parts of cold and in 22 parts of boiling water; it melts when heated, and solidifies on cooling, forming a horn-like substance (HORN LEAD; PLUMBI CORNEUM).

Uses, &c. In the Ph. L. 1836, chloride of lead was ordered to be employed in the preparation of 'hydrochlorate of morphia.' Mr Tuson highly recommends it in cancerous affections, to allay pain and restrain morbid action, either in the form of a lotion or ointment.

Various mixtures of lead chlorides and oxide are employed as a white pigment under the name of 'Pattison's white.' It is prepared by rapidly mixing a boiling solution of lead chloride with an equal volume of lime-water. Another similar compound is called 'patent yellow,' or 'Turner's yellow.'

Lead, Chromate of. $PbCrO_4.$ *Syn.* CHROME YELLOW, LEMON YELLOW, LEIPSIC YELLOW, PARIS YELLOW. *Prep.* By adding a filtered solution of acetate of nitrate of lead to a like solution of chromate of potassium as long as a precipitate forms, which is collected, washed with water, and dried. For information respecting the manufacture of this substance on the large scale, as a colouring substance (chrome yellow), see YELLOW PIGMENTS.

Lead, Dichromate of. *Syn.* CHROME ORANGE, CHROME RED. $PbCrO_4.PbO.$ *Prep.* By adding to a solution of nitrate or acetate of lead a solution of chromate of potassium, to which an equivalent of potassa has been added. This compound is of a splendid scarlet colour. See RED PIGMENTS.

Lead, Cy'anide of. PbCy$_2$. *Syn.* PLUMBI CYANIDUM, L. *Prep.* By adding hydrocyanic acid to a solution of acetate of lead as long as a precipitate forms, which, after being washed with distilled water, is dried by a very gentle heat, and preserved from the light and air. Sometimes used as a source of medicinal hydrocyanic acid.

Lead, Iodide of. PbI$_2$. *Syn.* LEAD IODIDE; PLUMBI IODIDUM (B. P., Ph. L. E. D.), L. *Prep.* (B. P.) Nitrate of lead, 4 parts; iodide of potassium, 4 parts; distilled water, a sufficiency. Dissolve with the aid of heat the nitrate of lead in 30 parts of water, and the iodide of potassium in 10 parts of water; mix, collect the precipitate, wash, and dry at a gentle heat.

Prop., &c. A rich yellow-coloured powder, soluble in acetic acid, alcohol, and boiling water; when heated, it fuses and volatilises in yellow vapour, but with a higher hegree of heat violet vapours of iodine are evolved, leaving a residuum (lead) which is wholly soluble in nitric acid.— *Dose,* ¼ to 4 gr. or more, made into a pill; as a deobstruent and resolvent in enlargements of the cervical, axillary, and mesenteric glands, and in scrofulous affections and scirrhous tumours.

Lead, Nitrate of. Pb(NO$_3$)$_2$. *Syn.* PLUMBI NITRAS, L. (B. P., Ph. E. D.) *Prep.* (Ph. D.) Litharge (in fine powder), 1 oz.; pure nitric acid, 2 fl. oz., diluted with water, ½ pint; mix, apply a sand-heat, and evaporate to dryness, occasionally stirring; boil the residuum in water, 2½ pints; filter, acidulate with a few drops of nitric acid, evaporate till a pellicle forms, and set the liquid aside to cool; lastly, dry the deposited crystals on bibulous paper, and preserve them in a well-closed bottle.

(*Commercial.*) By dissolving white-lead in dilute nitric acid, and crystallising.

Uses, &c. This salt is extensively used in calico-printing, and in the preparation of the iodide and other salts of lead. It was formerly much esteemed in asthmas, hæmorrhages, and epilepsy. It is now often used as an external application in cancer, ulcers, wounds, and various cutaneous affections. It is the basis of Liebert's celebrated 'cosmétique infallible,' and of Ledoyen's 'disinfecting fluid.' A very weak solution is an excellent application to chapped nipples, lips, hands, &c.—*Dose,* ½ to 1 gr.; in the form of pill or solution, washed down with a tablespoonful of water very slightly acidulated with nitric acid.

Lead, Nitro-sac'charate of. *Syn.* PLUMBI NITROSACCHARAS, L. *Prep.* (*Dr S. E. Hoskins.*) Nitric acid, 1 part; water, 19 parts; mix; in this dilute acid saccharate of lead (in fine powder) is to be dissolved, and set aside that crystals may form, which are to be dried by pressure between —the folds of bibulous paper. A weak solution of the salt, acidulated with saccharic acid, has been employed by Dr Hoskins as a solvent for phosphatic calculi with apparent success.

Lead, Oxides of. There appear to be four oxides of lead—the *monoxide* (PbO), *red-lead* (Pb$_2$O$_4$), the *sesquioxide* (Pb$_7$O$_3$), and the *dioxide* (PbO$_2$). Only the first of these forms stable salts.

Lead, Oxide of. PbO. *Syn.* MONOXIDE OF LEAD, PROTOXIDE OF LEAD, YELLOW OXIDE OF LEAD, MASSICOT, LITHARGE; PLUMBI OXYDUM (B. P.), L. *Prep.* This substance is obtained perfectly pure by expelling the acid from nitrate of lead by exposing it to heat in a platinum crucible; or, still better, by adding ammonia to a cold solution of nitrate of lead until the liquid becomes faintly alkaline, washing the precipitate with cold water, drying it, and heating it to moderate redness for 1 hour.

Prop., &c. Pure protoxide of lead has a lemon-yellow colour, and is the best of all the salts of lead. It is very heavy, slightly soluble in water, and freely so in acids, particularly when in the hydrated state; the aqueous solution has an alkaline reaction; at a red heat it melts, and assumes a semi-crystalline form on cooling; in the melted state it rapidly attacks and dissolves siliceous matter, with which it unites to form glass (flint glass); when heated along with organic substances of any kind it is easily reduced to the metallic state.

On the commercial scale this oxide is prepared by heating the grey film or dross that forms on the surface of melted lead when freely exposed to the air. When the process is arrested, as soon as the oxide acquires a uniform yellow colour, it is called massicot; when the heat is still further increased, until it fuses or partially vitrifies, it forms litharge, of which there are several varieties. See LITHARGE, MASSICOT.

Lead, Red Oxide of. Pb$_3$O$_4$. *Syn.* RED-LEAD, MINIUM. *Prep.* This is prepared by exposing unfused protoxide of lead to the air for a long time at a dull red heat. It is a very heavy powder, of a fine red colour, decomposed by a strong heat into protoxide of lead and oxygen gas, which is evolved. Somewhat uncertain in its composition, but this is generally Pb$_3$O$_4$ or PbO$_2$.2PbO. See RED PIGMENT.

Lead Dioxide. PbO$_2$. *Syn.* BINOXIDE OF LEAD, PEROXIDE OF LEAD, PUCE OXIDE OF LEAD. *Prep.* By digesting red oxide of lead in dilute nitric acid; or by fusing a mixture of protoxide of lead and chlorate of potassium at a heat a little below redness and washing the powdered mass in water; or by transmitting a current of chlorine gas through a solution of neutral acetate of lead. This oxide gives up half its oxygen at a red heat; acids also decompose it. Its chief use is in the chemical analysis of certain gaseous mixtures to separate sulphurous acid, which it converts into sulphuric acid, at the same time absorbing it, forming sulphate of lead. It has recently been employed as an oxidising agent in the manufacture of the ANILINE DYES.

Lead, Pyrolig'nite of. Sugar of lead made with crude pyroligneous acid. Used in dyeing, chiefly for the preparation of acetate of alumina.

Lead, Sac'charate of. *Syn.* PLUMBI SACCHARAS, L. *Prep.* (*Dr S. E. Hoskins.*) Nitric acid, 2 parts; water, 10 parts; mix in a porcelain capsule, add of sugar, 1 part, and apply heat until reaction ceases; then dilute the liquid with distilled water, neutralise it with powdered chalk, filter, and add to the filtrate a solution of acetate of lead as long as a precipitate (saccharate of lead) forms; lastly, collect the precipitate on a filter, wash and dry it. Used to make nitro-saccharate of lead, and as a source of saccharic acid.

Lead, Sul'phate of. $PbSO_4$. *Syn.* PLUMBI SULPHAS, L. This salt occurs native in transparent octohedra (lead vitriol), and is obtained in large quantities as a by-product in the preparation of acetate of aluminum for dyeing.

Prep. By adding dilute sulphuric acid to a solution of a soluble salt of lead. It is soluble in strong hydrochloric acid and bitartrate of ammonium, but almost insoluble in water and dilute sulphuric acid, though it dissolves to some extent in the strong acid. Commercial strong sulphuric acid contains small quantities of lead sulphate derived from the leaden pans in which it was concentrated; on diluting the acid the sulphate is precipitated as a white powder.

Lead, Sul'phide of. PbS. *Syn.* PLUMBI SULPHIDUM, L. This occurs abundantly in nature in the form of GALENA, see *above*.

Prep. By fusing metallic lead with sulphur, or by passing sulphuretted hydrogen through a solution of a salt of lead.

The naturally occurring compound is blue-grey in colour, with semi-metallic lustre and sp. gr. about 7·5. When precipitated by sulphuretted hydrogen it forms a black powder, insoluble in hydrochloric, soluble in nitric acid.

Lead, Tan'nate of. *Syn.* PLUMBI TANNAS, L. *Prep.* Precipitate a solution of acetate of lead with an infusion of galls, and wash and dry the precipitate. Astringent, sedative, and hæmostatic.—*Dose*, 1 gr. and upwards, made into a pill. It has been highly recommended in the form of ointment and cataplasms, in bedsores, chronic ulcers of the feet, white swellings, &c.

Lead, Tar'trate of. *Syn.* PLUMBI TARTRAS, L. *Prep.* By precipitating acetate of lead by tartrate of ammonium, washing and drying.

LEAD DUST. *Syn.* PULVIS PLUMBI, PLUMBUM DIVISUM, L. *Prep.* By melting new lead, adding bruised charcoal, mixing with violent agitation, which must be continued until the metal 'sets,' and then pounding and washing away the charcoal. Used by potters.

LEAD, GRANULATED. *Prep.* By melting new lead and pouring it in a small stream from an iron ladle with a hole drilled in its bottom into a pail of water. Used to make solutions and alloys.

LEAD PYROPH'ORUS. See PYROPHORUS.

LEAD, RED-. See RED PIGMENTS.

LEAD, WHITE-. See WHITE PIGMENT.

LEATH'ER. *Syn.* CORIUM, CORIUS, L. Leather is the skin of animals which has been prepared by one or other of several processes adopted for the purpose, having the common object of preventing its spontaneous destruction by putrefaction, besides other objects, which are more or less peculiar to each variety of this useful substance.

Leather is only prepared on the large scale, and primarily either by the process of 'TANNING' or 'TAWING,' in the manner briefly described under these heads.

CURRIED LEATHER is leather which has been tanned, and sold to the currier, who, after soaking it in water, and rubbing it to soften it, pares

it even with a broad, sharp knife, rubs it with a piece of polished stone or wood, and, whilst still wet, besmears it with oil or grease (DUBBING), which gradually penetrates the leather as the moisture evaporates. It next undergoes the operation of 'waxing,' which consists of first rubbing it on the flesh side with a mixture of oil and lamp-black; it is then 'black-sized' with a brush or sponge, and, when dry, is lastly 'tallowed' with a proper cloth, and 'slicked' upon the flesh side with a broad and polished lump of glass. Leather curried on the hair or grain side, termed 'black on the grain,' is blackened by wetting it with iron liquor, and rubbing it with an iron 'slicker' before applying the oil or grease. The grain is finally raised by the 'pommel' or 'graining board' passed over it in various directions.

Leather is dyed or stained by the application, with an ordinary brush, of any of the strong liquid dyes, in the cold or only gently heated, to the surface of the skin previously stretched on a board. The surface, when dry, is commonly finished off with white of egg and the pommel or smoothing stick. Bookbinders generally employ copperas water as a black stain or sprinkle, a solution of indigo as a blue one, and a solution of salt of tartar or common soda as a brown one.

Leather, before being japanned or varnished, as in the preparation of what is called 'ENAMELLED' and 'PATENT LEATHER,' is carefully freed from grease by the application of absorbent substances or hard pressure between rollers, and the surface is nicely shaved, smoothed and polished by appropriate tools; the varnish is then applied to the grain side for the former, and the flesh side of the skin for the latter, which is previously stretched out tight on a board to receive it. The whole is, lastly, submitted to a gentle stove-heat to harden the varnish; and the process is repeated if necessary.

Uses, &c. These are well known, and are all but universal. The leather manufacture of Great Britain is equal in importance and utility to any other department of our industry, and inferior in point of value and extent only to those of cotton, wool, and iron. " If we look abroad on the instruments of husbandry, on the implements used in most of the mechanic trades, on the structure of a multitude of engines and machines; or if we contemplate at home the necessary parts of our clothing—breeches, shoes, boots, gloves—or the furniture of our houses, the books on our shelves, the harness of our horses, or even the substance of our carriages; what do we see but instances of human industry exerted upon leather? What an aptitude has this single material in a variety of circumstances for the relief of our necessities, and supplying conveniences in every state and stage of life! Without it, or even without it in the plenty we have it, to what difficulties should we be exposed!" (*Dr Campbell*). Leather is a kind of natural felt, but of much closer and firmer texture than that of artificial origin. "The thinner and softer kinds of leather are sometimes used as body-clothing; but its special and proper purpose is the manufacture of coverings for the feet, to protect them from cold and water" (*Eras. Wilson*). See JAPANNING, VARNISH, &c.

60

Leather, Destruction of, by Gas. It is well known that the binding of books suffers considerable damage when the books are kept in apartments lighted by coal gas. That the cause of this deterioration is due, as was believed, to the combustion of the bisulphide of carbon contained in the gas, and its consequent oxidation into sulphuric acid, is exemplified by the following interesting communication from Professor Church, published in the 'Chemical News' for October 19th, 1877. He says, "Vellum seems unaffected; morocco suffers least; calf is much injured, and russia still more so. The disintegration is most rapid with books on the upper shelves of a library, whither the heated products of combustion ascend, and where they are absorbed and condensed." See BOOKBINDING.

By comparing specimens of old leather with specimens of new, it is quite clear that the destructive influence of gas is due mainly to its sulphur.

True, there are traces of sulphates in the dye and size of new leather bindings, but the quantity is insignificant, and there is practically no free sulphuric acid. That leather may be destroyed by the oil of vitriol produced by the burning of gas in a library is proved by the following observations and analyses:

The librarian of one of our public libraries forwarded to me the backs of several volumes, which had been 'shed' by the books on the upper shelves in an apartment lighted by gas. The leather of one of these backs (a volume of the 'Archæologia') was carefully scraped off so as to avoid any paper or size from underneath. This task of scraping was easy enough, for the leather was reduced to the consistency of Scotch snuff. On analysis of the watery extract of this leather, the following figures were obtained:

Free sulphuric acid in decayed
leather   .   .   .   .   6·21 per cent.
Combined   .   .   .   .   2·21   „
                         ———
                         8·42

LEAV'EN. Dough which has become sour or run into a state of incipient putrefaction. When a small quantity of it is added to recent dough it excites fermentation, but is apt to produce a disagreeable taste and odour in the bread. It is now superseded by yeast. Both these substances are used in the same way.

LEAVES (Medicated). Syn. FOLIA MEDICATA, L. On the Continent several preparations of this kind are in use. In many cases the leaves of tobacco deprived of nicotin, by soaking them in water, are dried, and then moistened or steeped in a tincture or infusion of the medicinal substance. In this way belladonna, camphor, and henbane are often administered. Cruveilhier recommends opiated belladonna leaves for smoking in troublesome coughs, phthisis, spasmodic asthmas, &c., to be prepared as follows:—Belladonna leaves, 1 oz., are steeped in an infusion of opium, 10 gr., in water, 1 fl. oz. (or less), and are then carefully dried in the shade. "MUSTARD LEAVES (Rigollot's) consist of mustard moistened with water, spread on paper, and dried" (Squire). See CIGARS (in pharmacy), and VEGETABLES.

Leaves, How to Dissect. "For the dissection of leaves," says Mrs Cussons, "I find the process of maceration too long and tedious, to say nothing of the uncertainty as to the results. I have therefore adopted the use of alkali in saturated solution, the specimens to be introduced while the liquid is heated to the boiling-point; the time of immersion to be regulated by the character of the various leaves and the nature of the epidermis to be removed. When the specimen is freed from epidermis and cellular tissue, it must be subjected to the action of chlorine to destroy the colouring matter. The introduction of peroxide of hydrogen not only serves to render the lace-like specimen purer in colour, but also preserves it. In destroying the colouring matter in ferns this also is invaluable; added to the chlorine it gives a solidity to the bleached fronds, and appears to equalise the action of the chlorine. For skeletonising capsules the slow process of maceration by steeping in rain-water is alone available; a moderate heat may be applied to hasten the process, but alkali is useless. The only known flower which can be dissected is the *Hydrangea japonica*. The fibrous nature of the petals renders it easy to skeletonise in the perfect truss in which it grows. Skeletonised leaves and capsules appear to gain in the process a toughness and durability not possessed by them in their natural state."

LEECH. Syn. HIRUDO (B. P.), L. The officinal leeches of the Pharmacopœias are the *Sanguisuga medicinalis* (*Hirudo medicinalis*, Cuv.) and *S. officinalis;* the first of these is familiarly known as the 'old English' or 'speckled leech.' It is also occasionally called the 'Hamburg grey' or 'Russian leech,' from being imported from those parts. Its characteristics are —Back, greenish or olive-green, sometimes almost black or intense brown, with 6 rusty-red or yellowish longitudinal stripes, which are mostly spotted with black. Belly, dirty yellow or light olive-green, spotted more or less with black. The spots are very variable in size and number; in some cases few, in others so numerous as to form the prevailing tint of the belly. This variety, which is the most valuable of the commercial leeches, is chiefly imported from Hamburg.

The *Sanguisuga officinalis*, familiarly known as the 'Hamburg' or 'French green leech,' is imported from Bordeaux, Lisbon, and Hamburg. Its characteristics are—Back, brownish olive-green, with 6 reddish or rusty yellow longitudinal bands. Belly, light dirty pea-green, or yellowish green, free from spots, but exhibiting two lateral stripes. This leech is vastly inferior to the preceding variety, and some of those imported from France and Portugal are absolutely useless, from their indisposition to bite, arising from the fraud practised by the collectors and dealers of gorging them with blood to improve their appearance before sending them to market. The above are the species of leech commonly employed in medicine in this country, but many others are noticed by writers on the subject.

Leeches are best preserved in water obtained from a pond, and occasionally changed; when kept in spring water they soon die. The introduction of a hand to which an ill-flavoured medicine or odour adheres into the water in which they are kept is often sufficient to poison them. The application of saline matter to the skin of leeches, even in very small quantities, immediately occasions the expulsion of the contents of the stomach; hence a few grains of common salt are frequently sprinkled over them to make them disgorge the blood which they have swallowed. The frequent changing of the water in which leeches are kept is injudicious. Once a month in winter and once a week in summer is deemed sufficiently often by the large dealers, unless the water becomes discoloured or bloody, when it should be changed every day, or every other day. When clean pond water cannot be obtained, clean rain water that has been well exposed to the air should alone be employed. Mr J. R. Kenworthy recommends placing in the water a few balls of irregular lumps of pure clay, about 2½ in. in diameter; a method which we can recommend as both simple and successful. The plan adopted by M. Fée is as follows:—Place 7 in. of a mixture of moss, turf, and charcoal in a marble or stone trough, over which sprinkle some small pebbles. At one end of the trough and about halfway up place a thin shelf of stone or marble pierced with small holes, on which put first some moss or portions of marsh horsetail (*Equisetum palustre*), and on this a layer of pebbles to keep it down; then pour in water sufficiently high just to moisten the moss and pebbles, put in the leeches, and tie over the mouth of the trough with a cloth. Another plan consists in keeping the leeches in a glass tank or aquarium, provided with a pebbly bottom and a few healthy aquatic plants.

Propag. According to Dr Wagner, an annual feast on living blood is necessary to render leeches able to grow and propagate. These conditions can only be fulfilled by restoring to the breeding cisterns those which have been already employed. All artificial methods of feeding them by bladders or sponges of blood have been found to fail. He recommends the employment of two tanks, with the bottom formed of loam, clay, or turf, surrounded by an inner border of a similar substance, and an outer one of sand—the one for leeches fit for medical use, and the other for breeding, or for such leeches as have been applied. No leeches are to be taken from the breeding-tank until a year has elapsed after their having been applied and fed with human blood; and their removal to the first tank should take place in September or October, as by this time the breeding season is over. By this plan all leeches that have been applied are to be carefully restored to the breeding-tank, without making them disgorge the blood they have swallowed.

LEECHING. This consists in the application of leeches to any vascular part of the body for the purpose of withdrawing blood from it, and thus allaying local inflammation, distension of vessels, &c. Leeches are most conveniently applied by means of a common pill-box or a wine-glass. The part should be previously washed perfectly clean, and if covered with hair should be closely shaved. Sometimes leeches are indisposed to bite; in such cases, allowing them to crawl over a piece of dry linen or calico, rolling them in porter, moistening the part with a little milk or sweetened milk, or drawing a little blood by a slight puncture or scratch will usually make them bite freely. To stop the bleeding from leech-bites various plans are adopted, among which the application of nitrate of silver or creasote, and gentle pressure for some hours with the finger, are the most successful. Of late years a piece of matico leaf or soldier's herb, applied in the same manner as a piece of lint, has been commonly adopted to stop the bleeding of leech-bites.

LEEK. *Syn.* PORRUM, L. The *Allium porrum*, Linn. Its general properties are intermediate between those of the onion and garlic. The juice is said to be powerfully diuretic, and capable of dissolving phosphate calculi.

LEGUMIN. Vegetable casein. It is found most abundantly in the seeds of leguminous (podded) plants, *e.g.* beans, peas, &c., as well as in the sweet and bitter almond.

In properties it closely resembles the casein of milk.

Legumin may be obtained from peas or from almonds as follows:—After digesting the crushed seeds for 2 or 3 hours in warm water, the undissolved portion is removed by straining through linen, and the strained liquid, after depositing the starch suspended in it, is next filtered and mixed with diluted acetic acid. The white flocculent precipitate which is thus produced is then collected on a filter and washed. It is afterwards dried, powdered, and digested, first in alcohol, and afterwards in ether.

Rochleder considered that, as thus obtained by Dumas and Cahours, it was not absolutely pure, since, as it was not entirely soluble in a cold concentrated solution of potash, he recommended the alkaline solution being decanted from the undissolved portion, and again precipitated by the addition of acetic acid.

Legumin as thus prepared was believed by Rochleder to be pure, and was found on analysis to give results analogous to those furnished by casein.

In the seed, legumin occurs associated with considerable quantities of the phosphates of calcium, magnesium, and potassium. Rennet coagulates it like it does the casein of milk, its similarity to which is exemplified by the manufacture of a kind of cheese from peas and beans by the Chinese.

Dried peas contain about a fourth of their weight of legumin.

LEMON. *Syn.* LIMO, L. The fruit of the *Citrus limonum*, or lemon tree. The juice, peel, and essential oil are officinal. See OIL, and *below.*

LEMON ACID. See CITRIC ACID.

LEMON FLAVOUR. See ESSENCE OF LEMON.

LEMON JUICE. *Syn.* LIMONIS SUCCUS (B. P.), SUCCUS LIMONUM (Ph. L. and D.), L. The juice of the lemon, obtained by squeezing and straining. When freshly expressed, it is turbid, owing to the presence of mucilage and extractive matter. These substances render the juice liable to decomposition, and various methods have from time to time been proposed for preserving it. Amongst these may be mentioned the addition to the fresh juice of 1% of bisulphite of calcium, or 10% of proof spirit.

"We have examined the juice expressed from two varieties of lemons, viz. Palermo and Messina, with the following results:

	Palermo.	Messina.
" Ounces of juice yielded by 100 lemons . .	108	96
Sp. gr. of juice . .	1044·85	1038·56
Percentage of citric acid	8·12	7·04
Percentage of ash . .	0·289	0·301

" 100 parts of the ash of the juice of Palermo lemons gave—

" Sulphuric acid	.	. 10·59
Carbonic acid	.	. 16·33
Chlorine	.	. 0·81
Phosphoric acid	.	. 6·74
Ferric phosphate	.	. 1·32
Lime	. .	. 8·89
Magnesia	.	. 3·02
Potash .	.	. 47·84
Soda	.	. 3·82
Silica	. .	. 0·72
Loss	. .	. 0·42
		100·00

"If lemons are kept a few months before squeezing, the yield of juice is slightly increased, but its specific gravity and percentage of citric acid remain unaltered. It is erroneous to suppose that the acid of the lemon is, by keeping, changed into sugar. We have kept lemons for 12 months, and found that the percentage of acid was not diminished. A certain proportion of sugar was formed, but at the expense of the soluble starch contained in the cell-walls of the lemon. Lemon juice on being kept is found to decrease in density, but the amount of acid remains the same " (*Harkness*).

Lemon juice may be preserved by heating it to 150° F., filtering, and setting it aside in bottles completely filled. If this process be performed in the winter, the juice, it is said, may be kept perfectly good for 12 months. Fresh lemon juice is prevented from decomposition and rendered fit for exportation by mixing it with 1-10th of alcohol (*Schweitzer*).

The Merchant Shipping Act of 1867 requires that after a ship has been at sea 10 days 1 oz. of lime or lemon juice, mixed with 1 oz. of sugar and ½ pint of water, shall be served out to each of the crew between the hours of 12 and 1 in the day.

Adult. Lemon juice is frequently adulterated, the adulterants being water, sugar, or gum, and sulphuric or acetic acid. The *modus operandi* is to dilute the genuine juice with water, and then bring up the density with the sugar or gum, and the percentage of acid with one or other of the above acids. The examination of lemon and lime juice supplied to the navy is now conducted in the Inland Revenue Laboratory, Somerset House, and it speaks well for that department when we say that cases of scurvy on board ships are now of very rare occurrence. No juice is passed unless it comes up to a certain standard in specific gravity and percentage of citric acid, and any sample containing any other acid is at once rejected.

Prop. Lemon juice is refrigerant and antiscorbutic, and has long been extensively employed in the preparation of cooling drinks and effervescing draughts, which are justly esteemed as wholesome summer beverages, as well as palliatives in fevers, nausea, &c. In scurvy there is no remedy equal to freshly expressed lemon juice; and in acute rheumatism and gout, according to the united testimony of Dr Owen Rees, Dr Babington, and numerous Continental practitioners, it has been exhibited with considerable success. In agues, dysentery, English cholera, nausea and vomiting, heartburn, putrid sore-throat, hospital gangrene, syphilis, and numerous skin diseases, it has proved most serviceable. See CITRIC ACID, GOUT, &c.

Lemon Juice, Facti"tious. *Syn.* SOLUTIO ACIDI CITRICI, SUCCUS LIMONUM FACTITIUS, L. *Prep.* 1. Citric acid, 1¼ oz.; carbonate of potassa, 45 gr.; white sugar, 2½ oz.; cold water, 1 pint; dissolve, add the yellow peel of a lemon, and in 24 hours strain through a hair-sieve or a piece of muslin.

2. As the last, but using 15 or 16 drops of oil of lemon, to flavour instead of the lemon peel.

Obs. The above is an excellent substitute for lemon juice, and keeps well in a cool place. Tartaric acid, and even vinegar, are sometimes used instead of citric acid; but it is evident that it then loses all claim to being considered as an imitation of lemon juice, and to employ it in lieu of which would be absurd.

LEM'ON-PEEL. *Syn.* CORTEX LIMONUM (B. P., Ph. L.), L. "The fresh outer part of the rind" (B. P.). "The fresh and the dried exterior rind of the fruit," the latter dried "in the month of April or May" (Ph. L.). Candied lemon-peel (CORTEX LIMONUM CONDITUS) is employed as a dessert, and as a flavouring ingredient by cooks and confectioners. It is reputed stomachic. See CANDYING.

LEMON PIC'KLE. See SAUCE.

LEMONADE'. *Syn.* LEMON SHERBET, KING'S CUP; LIMONADUM, L.; LIMONADE, Fr. *Prep.* 1. Lemons (sliced), 2 in no.; sugar, 2½ oz.; boiling water, 1½ pints; mix, cover up the vessel, and let it stand, with occasional stirring, until cold, then pour off the clear through a piece of muslin or a clean hair-sieve.

2. Juice of 3 lemons; yellow peel of 1 lemon; sugar, ¼ lb.; cold water, 1 quart; digest for 5 or 6 hours, or all night, and decant or strain as before.

3. Citric acid, 1 to 1½ dr.; essence of lemon, 10 drops; sugar, 2 oz.; cold water, 1 pint; agitate together until dissolved.

Obs. Lemonade is a pleasant, cooling summer beverage, and when made as above may be drunk in large quantities with perfect safety. It also forms an excellent refrigerant and antiseptic drink in fevers and putrid diseases generally. Tartaric acid is commonly substituted for citric acid, from being cheaper; it is, however, much inferior, being less wholesome and less agreeable. Lemonade for icing is prepared as above, only using a little more sugar. Orange sherbet, or orangeade for icing, is made in a similar way from oranges.

Lemonade, Aëra'ted. *Syn.* LIMONADUM AËRA-

TUM, L.; LIMONADE GAZEUSE, Fr. *Prep.* 1. (P. Cod.) Water, charged with 5 times its volume of carbonic acid gas, 1 pint; syrup of lemon, 2 oz.; mix.

2. (Without a bottling machine.) *a.* Into each bottle put lemon syrup, 1 to 1½ oz.; essence of lemon, 3 drops; sesquicarbonate of soda, ½ dr.; water, q. s. to nearly fill the bottle; have the cork fitted and ready at hand, then add of tartaric acid (cryst.), 1 dr.; instantly close the bottle, and wire down the cork; it should be kept inverted in a cool place, and, preferably, immersed in a vessel of ice-cold water.

b. As the last, but substituting lump sugar, ½ oz., for the lemon syrup.

c. From lump sugar, 1 oz.; essence of lemon, 3 drops; bicarbonate of potassa, 25 gr.; water, q. s., as No. 1; then add citric acid (cryst.), 45 gr., and cork, &c., as before. The last is most wholesome, especially for the scorbutic, dyspeptic, gouty, and rheumatic.

Obs. The best aërated lemonade of the London makers is prepared by putting 1½ fl. oz. of rich lemon syrup into each bottle, which is then filled up with aërated water at the bottling machine.

Lemonade, Antimo"niated. *Syn.* LIMONADUM ANTIMONIATUM, L. *Prep.* By adding tartar emetic, 1 gr., to each pint of ordinary lemonade. —*Dose.* A wine-glassful every ½ hour or hour, as a diaphoretic and expectorant. See ANTIMONY (Potassio-tartrate).

Lemonade, Ape"rient. *Syn.* LIMONADUM LAXATIVUM, L. *Prep.* 1. Sugar, 1 oz.; lemon juice, ¾ fl. oz.; sulphate of soda, 3 dr.; water, 8 fl. oz.; put them into a soda-water bottle without shaking, have the cork ready fitted, add of sesquicarbonate of soda (in cryst.), ½ dr., and instantly cork the bottle, wire it down, and keep it in a cool place, inverted. For a dose.

2. Heavy carbonate of magnesia, 1½ dr.; refined sugar, 1 oz.; essence of lemon, 5 or 6 drops; water, 8 fl. oz.; bottle as last, then add of citric acid (cryst.), 3 dr., and instantly cork, &c., as before. For a dose. It should be kept for at least 24 hours before being taken.

Lemonade, Lactic. *Syn.* LIMONADUM LACTICUM, L. *Prep.* (*Magendie.*) Lactic acid, 1 to 4 dr.; syrup, 2 oz.; water, 1 pint; mix. Recommended in dyspepsia, &c.

Lemonade, Milk. *Syn.* LIMONADUM LACTIS, L. *Prep.* Take of sugar, ½ lb.; water, 1 pint; dissolve, add the juice of 3 lemons; milk or whey, ½ pint; stir the whole together and strain through a hair-sieve. Some persons add a glassful of sherry.

Lemonade, Min'eral. *Syn.* LIMONADE MINÉRALE, Fr. On the Continent this name is applied to various drinks consisting of water acidulated with the mineral acids and sweetened with sugar. Thus we have limonade chlorhydrique, nitrique, phosphorique, sulphurique, &c., all of which are used as cooling drinks in fevers, inflammations, skin diseases, &c.

Lemonade, Port'able. See POWDERS.

LEMONADE POWDER. *Prep.* Bicarbonate of soda, 16 oz.; tartaric acid, 14 oz.; icing sugar, 32 oz.; essence of lemon, 80 drops; essence of pine-apple, 5 drops; mix.

LEMONADE POWDERS. See POWDERS.

LEMONADE WINE. *Prep.* Tartaric acid, 5 grms.; alcohol, 25 grms.; syrup of orange flowers, 50 grms.; sherry wine, 250 grms.; distilled water, 675 grms. Mix the liquids and dissolve the tartaric acid therein; filter into 3 12-oz. bottles, to each of which add 30 gr. of bicarbonate of soda, cork quickly and secure the cork with a string before shaking. The spirit can be substituted by cognac if a finer preparation is wanted.

LEMONATED KALI. See KALI, POTASSIUM (Citrate), &c.

LEN'ITIVES. In *medicine*, purgatives which act in a gentle manner, and have a soothing effect. See LAXATIVES.

LENS. In *optics*, a piece of glass or other transparent medium, having one or two curved surfaces, either convex or concave. A description of the different kinds of lenses belongs to a work on optics. It may, however, be useful to the chemical student to remark here that the CODDINGTON and STANHOPE LENSES, which may now be bought at any of the opticians, neatly mounted and of great power, for a few shillings, will be found of the greatest service in examining minute crystals, precipitates, &c.; and for all ordinary purposes offer a cheap and efficient substitute for more complicated microscopes. An extemporaneous instrument, possessing considerable power, may be made by simply piercing a small circular hole in a slip of metal, and introducing into it a drop of water, which then assumes a spherical form on each side of the metal, while the latter is held in a horizontal position. The ingenious little TOY MICROSCOPES sold about the streets of London, under the form of a perforated pill-box, at one penny each, consist of such a lens made with Canada balsam instead of water, which has the property of hardening without losing its transparency after exposure for a few hours to the air. A still simpler substitute for a lens is a piece of blackened card-paper with the smallest possible needle-hole pierced through it. Any very small object held in a strong light, and viewed through this hole at the distance of about an inch, will appear quite distinct, and from 10 to 12 times larger than its usual size.

Another method for the manufacture of an extemporaneous lens, by Mr Francis, is the following:—Procure a piece of thin platinum wire, and twine it once or twice round a pin's point, so as to form a minute ring with a handle to it. Break up a piece of flint glass into fragments a little larger than a mustard seed; place one of these pieces on the ring of wire, and hold it in the point of the flame of a candle or of a gas-light. The glass will melt and assume a complete lens-light or globular form. Let it cool gradually and keep it for mounting. It may be mounted by placing it between two pieces of brass which have corresponding circular holes cut in them of such a size as to hold the edge of the lens.

LEN'TIL. *Syn.* LENS, L. The seed of the *Ervum lens*, a plant of the Nat. Ord. LEGUMINOSÆ. The lentil is considerably smaller than an ordinary pea, and is of the shape of a double convex lens. Several varieties are cultivated on the continent of Europe and in many parts of Asia, where they are largely consumed as human food. Lentils are more nourishing than any other description of pulse, but are reputed difficult of digestion, apt to disorder the bowels, and injurious to the eyes. Several alimentary preparations, sold at high prices as cures for dyspepsia, constipation, &c., contain lentil flour as the principal ingredient.

Composition of Lentils.

Nitrogenous matter	.	.	25·2
Starch, &c.	.	.	56·0
Cellulose	.	.	2·4
Fatty matter	.	.	2·6
Mineral matter	.	.	2·3
Water	.	.	11·5
			100·00
			(Payen.)

Lentils, on account of their difficult digestibility, require to be very thoroughly cooked. See ERVALENTA and REVALENTA.

LEPROSY. *Syn.* LEPRA. A disease of the skin distinguished by circular scaly patches.

LEPTANDRIN. A peculiar crystalline principle obtained from the root of *Leptandra virginica*, a North American plant belonging to the Nat. Ord. SCROPHULARIACEÆ. Leptandrin is chiefly employed in American medical practice as a cathartic and cholagogue, in which latter function it has been recommended as a substitute for mercury; it excites the liver and promotes flow of bile without any irritation of the bowels. It is stated to be very serviceable in cases of duodenal indigestion and chronic constipation.—*Dose*, ¼ to 2 gr.

Mr Wayne obtained leptandrin by adding subacetate of lead to an infusion of the root, filtering, precipitating the excess of lead by carbonate of sodium, removing the carbonate of lead by filtration, passing the filtered liquid through animal charcoal to absorb all the active matter, washing the charcoal with water till the washings began to be bitter, then treating it with boiling alcohol, and allowing the alcoholic solution to evaporate spontaneously. By dissolving the powder thus obtained in water, treating this with ether and allowing the ether to evaporate, needle-shaped crystals were obtained, which had the bitter taste of the root. Leptandrin is soluble in water, alcohol, and ether.

Commercial leptandrin is a dark greenish powder consisting chiefly of resin, and is the substance intended to be used when ordered in prescriptions.

LETH'ARGY. *Syn.* LETHARGUS, L. A heavy, unnatural sleep, sometimes bordering upon apoplexy, with scarcely any intervals of waking, from which the patient is with difficulty aroused, and into which he again sinks as soon as the excitement is withdrawn. It frequently arises from plethora, in which case depletion is indicated; or from the suppression of some usual discharge or secretion, which it should then be our business to re-establish. It also often arises from over mental fatigue and nervous debility, when relaxation from business, the use of a liberal diet, and ammoniacal stimulants and antispasmodics are found useful. When depending on a determination of blood to the head, cupping may be had recourse to, and all sources of excitement avoided. In all cases the bowels should be moved as soon as possible by means of mild purgatives.

LETTUCE. *Syn.* LACTUCA, L. The early leaves or head of the *Lactuca sativa*, or garden lettuce, form a common and wholesome salad. They are reputed as slightly anodyne, laxative, hypnotic, and antaphrodisiac, and have been recommended to be eaten at supper by those troubled by watchfulness, and in whom there exists no tendency to apoplexy. The leaves and flowering tops of *L. virosa* are officinal in the B. P., the 'flowering herb' (LACTUCA) in the Ph. L., the 'inspissated juice' in the Ph. E., and the 'inspissated juice and leaves' in the Ph. D. The 'inspissated juice' of *Lactuca virosa*, or strong-scented wild lettuce, is also officinal in the Ph. E.; and both the 'leaves and inspissated juice' of the same variety are ordered in the Ph. D. The last species is more powerful than the cultivated lettuce. See EXTRACT and LACTUCARIUM.

LEUCORRHŒ'A. *Syn.* WHITES; CATARRHUS VAGINÆ, FLUOR ALBUS, L. The symptoms of this disease are well known to most adult females. The common causes are debility, a poor diet, excessive use of hot tea, profuse menstruation or purgation, late hours, immoderate indulgence of the passions, frequent miscarriages, protracted or difficult labours, or local relaxation. Occasionally it is symptomatic of other affections. The treatment must be directed to the restoration of the general health, and imparting tonicity to the parts affected. Tepid or sea bathing or shower-baths; bark, chalybeates, and other tonics; with local affusions of cold water and mild astringent injections, as those of black tea or oak bark, are generally found successful in ordinary cases.

LEVANT' NUT. See COCCULUS INDICUS.

LEVICO. An arseniated mineral water from the Austrian Tyrol. It has the following composition, according to Drs L. von Barth and Hugo Weidel:

Levico (Strong) springs from 4 rents at the bottom of the grotto named Vitriolo Cave, which is 450 feet deep; 19 pints are supplied per minute.

Per 10,000 Parts by Weight.

Arsenious acid	.	.	0·086879
Chloride of sodium	.	.	0·001781
Protosulphate of iron	.	25·675198
Persulphate of iron	.	13·019720
Sulphate of aluminium	.	6·239873
,, ,, manganese	.	0·002418
,, ,, calcium	.	3·724983
,, ,, magnesium	.	3·833451
,, ,, potassium	.	0·037031
,, ,, sodium	.	0·319031
,, ,, ammonium	.	0·032270
Silicic acid	.	.	0·310384
Carbon from organic matter	.	0·097825

Levico (Mild) springs from a grotto named the Ocker Cave with a supply of 38 pints per minute.

Per 10,000 *Parts by Weight.*

Arsenious acid	0·0095
Chloride of sodium . . .	0·0003
Protosulphate of iron . .	6·6278
Persulphate of iron . .	2·7272
Sulphate of aluminium . .	1·5919
„ „ copper . . .	0·0520
Protocarbonate of iron . .	0·1558
Sulphate of manganese . .	0·0003
„ „ magnesium . .	2·3648
„ „ calcium . . .	3·2477
„ „ sodium . . .	0·1579
„ „ potassium . .	0·0099
„ „ ammonium . .	0·0062
Silicic acid	0·2293

LEVIGATION. *Syn.* LEVIGATIO, L. The process of reducing substances to fine powder, by making them into a paste with water, and grinding the mass upon a hard smooth stone or slab, with a conical piece of stone having a flat, smooth under surface, called a 'muller.' Levigation is resorted to in the preparation of paints on the small scale, and in the elutriation of powders. The term is also sometimes incorrectly applied to the lengthened trituration of a substance in a marble or Wedgwood-ware mortar.

LEVORACEMIC ACID. See RACEMIC ACID.

LEYDEN JAR. *Syn.* LEYDEN PHIAL, ELECTRICAL JAR. An instrument for the accumulation of the electric fluid. Its simplest form is that of a wide-mouthed jar of rather thin glass, coated on both sides with tin-foil, except on the upper portion, which is left uncoated, and having a cover of baked wood, through which passes a brass wire terminating in a metallic knob, and communicating with the inner coating. To charge the jar the outer coating is connected with the earth, and the knob put in contact with the conductor of an electrical machine. The inner and outer surfaces of the glass thus become respectively positive and negative, and the particles of the glass become strongly polarised. On making connection between the two coatings with a conducting substance discharge takes place by a bright spark and a loud snap; and if any part of the body be interposed in the circuit a shock is felt.

LIBAVIUS'S LIQUOR. See TIN (TIN CHLORIDE).

LICHEN. In *pathology*, a dry papulous or pimply eruption of the skin, terminating in scurfy exfoliations. "Lichen exhibits great variety in its outward characters in different individuals; in one the pimples are brightly red; in another, of debilitated constitution, they are bluish and livid; in a third they are developed around the base of hairs; in a fourth they appear as circular groups and increase by their circumference, while they fade in the centre, forming so many rings of various size; in a fifth, a modification of the preceding, they have the appearance of flexuous bands; while in a sixth they are remarkable for producing intensity of suffering or unusual disorganisation of the skin. They are all occasioned by constitutional disturbance, sometimes referable to the digestive, and sometimes to the nervous system. In some instances, however, they depend upon a local cause. I have had a crop of lichenous pimples on the backs of my hands from rowing in hot weather; and in hot climates that annoying disorder called prickly heat is a lichen" (*Eras. Wilson*). The treatment of this affection is noticed under ERUPTIONS (Papular).

LICHENS. *Syn.* LICHENES (Juss.), LICHENALES (Lind.), L. In *botany*, these are cryptogamous plants, which appear under the form of thin, flat crusts, covering rocks and the barks of trees. Some of them, like Iceland moss (*Cetraria islandica*), are esculent and medicinal, and employed either as medicine or food; and others, when exposed in a moistened state to the action of ammonia, yield purple or blue colouring principles, which, like indigo, do not pre-exist in the plant. Thus the *Roccella tinctoria*, the *Variolaria orcina*, the *Lecanora tartarea*, &c., when ground to a paste with water, mixed with putrid urine or solution of carbonate of ammonia, and left for some time freely exposed to the air, furnish the archil, litmus, and cudbear of commerce, very similar substances, differing chiefly in the details of their preparation. From these the colouring matter is easily extracted by water or very dilute solution of ammonia. See ARCHIL, CUDBEAR, and LITMUS.

LIEBER'S HERBS OF HEALTH—Gesundheitskranter Liebersche—Blankenheimer Thee—Blankenheimer Tea—Herba Galeopsidis Grandiflorae Concisa (yellow hemp-nettle).

LIG'ATURE. In *surgery*, a small waxed piece of cord or string formed of silk or thread, employed for the purpose of tying arteries, veins, and other parts, to prevent hæmorrhage, or to cause their extirpation. To be safe and useful they should be round, smooth, and sufficiently strong to permit of being tied with security without incurring the danger of breaking or slipping. There are many cases recorded in which emigrants, soldiers, and travellers have lost their lives from the simple inability of those around them to apply a ligature.

LIGHT. *Syn.* LUMEN, LUX, L. Light acts as a vivifying or vital stimulus on organised beings, just as privation of light or darkness disposes to inactivity and sleep. "In maladies characterised by imperfect nutrition and sanguinification, as scrofula, rickets, and anæmia, and in weakly subjects with œdematous (dropsical) limbs, &c., free exposure to solar light is sometimes attended with very happy results. Open and elevated situations probably owe part of their healthy qualities to their position with regard to it." On the contrary, "in diseases of the eye, attended with local vascular or nervous excitement, in inflammatory conditions of the brain, in fever, and in mental irritation, whether attended or not with vascular excitement, the stimulus of light proves injurious, and in such cases darkness of the chamber should be enjoined. After parturition, severe wounds, and surgical operations, and in all inflammatory conditions, exclusion of strong light contributes to the well-doing of the patient" (*Pereira*).

LIGHT, ELECTRIC. Shortly after Faraday's discovery in 1830 of electrical induction, or the power of a bar of magnetised steel to set up in a certain direction a current of electricity in a coil of insulated wire when introduced into it, Pixü, reducing the result of Faraday's researches to practice, constructed an instrument which appears to have been the first dynamic magneto-electric machine. By Pixü's contrivance a current of electricity was generated by means of the poles of a permanent horseshoe magnet being made to revolve across those of an electro- or temporary magnet, the induced electricity set up in which in its turn established in the surrounding helix a current of electricity, which being made to escape by the terminals or ends of the wire coils could be applied to practical use.

The dynamic electro-magnetic machines of Saxton and Clarke, which succeeded Pixü's, may be regarded as modifications of this latter, since they differed only in the arrangement of their parts and mode of action. All three machines were chiefly in use in chemical and physical laboratories, whence they have gradually been supplanted by the far more useful Ruhmkorff's coil, a very powerful variety of the electro-magnetic instrument. In a small form Clarke's is now chiefly used for medical purposes. That electro-magnetic machines, as cheaper and more convenient sources of electric force, should have been applied to the purposes of telegraphy, will be an obvious inference.

Among the most important and effective of the various instruments for attaining this end, it will suffice to mention the magneto-electric machine of Messrs Siemens and Halske, first brought into use in 1854.

Except, however, in the case of short distances, or with telegraphs belonging to private persons or commercial firms, these instruments have not met with very general adoption. This is owing to the great tension of the induced current, and the consequent difficulty of insulating the wire, particularly for long distances, objections from which the old galvanic apparatus is in a much greater measure free. Mr Henley was the first to use the dynamic magneto-electric machine for working the electric telegraph soon after this instrument had been adopted in England; but, as we have seen, the method, except in the cases quoted, has been in great measure abandoned. A large magneto-electric machine has lately been invented by Wheatstone, the induced spark from which is used for firing mines.

The first electro-magnetic machine used for lighting purposes appears to have been one that was the joint invention of MM. Nollet and Van Malderen, of Brussels, a circumstance to which it probably owes its name of the 'Alliance Machine.'

Nollet, who brought out his invention (which is a modification of Clark's) in 1850, originally designed it for the electrolysis of water, the hydrogen resulting from which it was proposed to pass through camphine, or some other hydrocarbon illuminant, and to burn as gas. Additionally it was designed to use the hydrogen as a source of motive power by exploding it in a suitably constructed engine. Owing to the improvements, however, effected in the machine by Van Malderen, by which it became a powerful generator of magneto-electricity, this purpose was abandoned. The 'Alliance Machine' consists of a cast-iron frame, on the circumference of which 40 powerful horseshoe magnets, each capable of supporting a weight of 120 to 130 lbs., are fixed, in eight series of 5 magnets each. A number of circular metal discs, around the circumference of which are attached sixteen bobbins of insulated wire fixed to a horizontal shelf turned by a pulley, are in such a position with regard to the magnets, that with each revolution of the shaft each bobbin passes sixteen alternate poles of the magnets, and will have had sixteen alternate currents set up or induced in it. Until replaced by the later and smaller magneto-electric machine, the 'Alliance' has been the one mostly employed for the production of the electric light in France, and it is still in use in the lighthouses of Hève and Grisnez, as well as in those of many other places in that country. In 1856 Mr Holmes took out a patent for a machine, which differs from Nollet's in increasing the number of bobbins by arranging them in concentric circles between two brass discs. By this device the bobbins revolve more quickly in succession in front of the poles of the magnets, a plan which ensures the generation of a greater number of currents for every revolution.

Like the first application of Nollet's, Holmes' machine was used for lighthouse illumination. It was in work from December, 1858, until June, 1862, at the South Foreland lighthouse, since which time it has been removed to Dungeness, in the lighthouse of which station it has been in use ever since.

When applied to lighting purposes, both the the 'Alliance' and Holmes', and the other machines named, are worked in conjunction with the carbon points, which when arranged with proper machinery constitute the electric lamp.

Wild's and Ladd's are powerful dynamic magneto-electric instruments, capable of yielding large quantities of the electric fluid.

Artificial illumination by means of electricity has, however, been more or less occasionally practised for other than lighthouse purposes.

For instance, in 1854, during the building of the Napoleon Docks at Rouen, when 800 workmen were engaged nightly for 4 hours, the electric light was used for several nights with perfect success, the men being able to carry on their work at a distance of more than 100 yards from the source of the light.

In 1862 and 1863 it was frequently employed in Spain during the night in the construction of railways. During the late Franco-German war in 1870 it was applied to submarine illumination, and more lately it has been used in a series of street illuminations in St. Petersburg.

The electric light apparatus was placed on the tower of the Admiralty buildings of that city, and by means of it 3 of the larger streets were illuminated at night from 7 until 10 o'clock. In this latter case, as well as in that of the Rouen Docks, the lamps were supplied with the electric current generated in batteries.

It may be said, however, to have been only within the last two years that the question of electric lighting has developed into a burning one, and that the light itself has become so much more generally and extensively adopted.

This new era in the history of artificial illumination may be said to date from the introduction of two forms of dynamic magneto-electric apparatus, the one invented by Dr Siemens, the eminent telegraphic engineer, the other by M. Gramme, of Paris, who, from having been formerly a journeyman carpenter, has now become the head of a manufacture which forms a most important branch of scientific industry.

In the apparatus of Gramme and Siemens 3 marked features and improvements over the older machines have been achieved:

1. A great reduction in size, and, consequently, in cost and requisite space for the machine.

2. The method of generating large quantities of electricity by the mutual action between the different parts of the same machine, and the induction therein set up. This discovery was made independently and nearly simultaneously by Drs Siemens and Sir Charles Wheatstone.

3. The production of the electric current at a much less expenditure of motive power.

On this latter point Professor Tyndall, in his report to the Elder Brethren of the Trinity House, states that magneto-electric machines of old construction cost 10 times more, occupied 25 times the space, and weighed 14 times as much as the recent machines, while they produced only one fifth of the light with practically the same driving power; which in effect amounts to this— that, taking illuminating effect in each case into consideration, the new machines cost one fiftieth, and are, as regards space occupied, 125 times more advantageous than the earlier forms.

In all the older and larger machines the current of electricity, as it was given off from the wire and passed through the carbon points, was alternate, or first in one direction and then in the opposite—that is, it was a momentary current, first positive and then negative.

In Siemens' machine, and in one form of Gramme's, the current is direct—that is, it pursues one uniform course in its passage through the carbon points of the lamp, and in its circuit from the terminal of one wire to that of the other.

Scientific opinion is somewhat at variance as to the disadvantages of the indirect current; many electricians consider that it causes the partial destruction of the contacts, and sets up unnecessary heat in the machine. In magneto-electric machines employed in electro-metallurgic operations it is essential the current should be a direct one.

In the Gramme machine the electro-magnet consists of a ring composed of soft iron wire attached to a horizontal spindle or axis, which latter is turned by an endless strap revolving on a pulley. Around this iron ring are wound a number of coils, each having 300 turns, of insulated copper wire, each coil being bent inside the ring, and fixed to an insulated piece of brass. The wire being continuous, each coil is connected with the adjacent one, the whole of the coils thus forming a single conductor. The series of pieces of brass to which the wire is soldered are formed into a circle, which surrounds the axis of the machine, each piece of brass being insulated from its neighbour. The iron-wire ring with its attachments is so arranged, that when the shaft or axis to which it is fixed is turned, it revolves between the poles of a powerful horseshoe magnet in the same plane with it. As it turns the ring gives rise in the coils to two different and diverse currents of electricity, one in one half of the coils around the ring, and the other in the other half.

These currents are made to pass to the circle composed of the insulated pieces of brass, which are arranged radially to the axis of the machine.

Two brass brushes press against these insulated brass radii, one on each side.

These brushes are connected one to each terminal of the machine, and so contrived as always to be in contact with the coils, not becoming insulated from one coil until contact is established with the next one, an arrangement which gives rise to a continuous current of electricity always, and in the same direction.

The Gramme, although of very small dimensions, is an extremely powerful machine. It easily decomposes water, and will heat an iron wire 8 inches in length and a 25th of an inch in diameter to redness.

The following description of the Siemens magneto-electric machine is from a paper read some few months back at the Society of Arts by Dr Paget Higgs, and is extracted from the journal published by that body:

"In the latest form of construction of the Siemens magneto-electric machine the armature, as the revolving coil may be called, consists of several lengths of insulated copper wire, coiled in several convolutions upon a cylinder. The whole surface of the cylinder is covered with wire, laid on in sections, each convolution being parallel to its longitudinal axis. For about 2-3rds of its surface the wire cylinder is surrounded by curved iron bars, there being just sufficient space left between these curved iron bars and the wire cylinder to allow of its free rotation. The curved iron bars are prolongations of the cores of large, flat electro-magnets; the coils of these electro-magnets and the wire on the cylinder (from brush to brush) form a continuous electrical circuit. On revolving the cylinder (which is supported on a longitudinal axis in suitable bearings, the axis carrying a pulley) an initially weak current is generated into its wires by their passage through the magnetic field, formed by the residual magnetism of the iron coils of the electro-magnets, and the current being directed into the coils of the electro-magnets, increases the magnetism of the cores, which again induce a stronger current in the wire cylinder. This material action may continue until the iron has attained its limit of magnetisation. The maximum magnetic power acting upon each convolution is attained at every revolution of the armature, when the convolution passes through the centre of both magnetic fields, and gradually falls to zero as the convolution becomes perpendicular to that position. Each convolution has, therefore, a neutral position, and a convolution leaving that position on the one side of the axis and advancing towards the north pole of the electro-magnet would be subject to a direct induced current, and that portion of the convolution on the opposite side of the axis would be traversed by a current of opposite direction as regards a given point, but of the same direction as regards circuit. Each of the sections of wire coiled upon the cylinder consists of two separate coils, leaving four ends; two of these ends are connected to each of the segments of a circular commutator divided into parts. But all the coils are connected to the several segments of the commutator in such a manner that the whole of the double sections form a continuous circuit, but not one continuous helix. Two brushes placed tangentially to the segments of the commutator collect the electric currents; these brushes are connected one to each electro-magnet, and the two free ends of the electro-magnet coils are connected to the conducting wires leading to the lamp.

The dimensions, weight, number of revolutions made by the armature, light equivalent in normal candles, and horse-power required for driving, are for the 3 sizes of machines as follows:

Dimension in Inches.			Weight in lbs.	Revolutions of cylinder.	Candles' light.	Horse-power.
Length.	Width.	Height.				
25	21	8·8	298	1100	1,000	1½ to 2
29	26	9·5	419	650	6,000	3¼ to 4
44	28·3	12·6	1279	480	14,000	9 to 10

In the lamp which it is preferred to use with the Siemens machine, the points of the carbons after being separated are brought together again by the gravitation of the top carbon and its holder. The descent of the top carbon actuates by means of the straight rack it carries at its lower end, a large pinion, the spindle of which carries a small pinion, gearing into a second neck attached to the lower carbon holder, the superior weight of the top carbon and holder, in conjunction with the multiplying ratio of the two pinions, producing a continual tendency of the carbons to approach each other. The large and small pinions are connected to each other, and to the spindle that carries them, by an arrangement of friction discs, and the object of this construction is to allow of the two racks being moved equally and simultaneously up or down for the purpose of focussing the light when required. This movement is effected by means of bevelled gearing, and actuated by a milled head, which can be pressed into position when required. On the spindle carrying the large and small pinions and the friction discs is placed a toothed wheel, connected with the spindle by a pawl and ratchet. This wheel is the first of a train of wheels and pinions driving a regulating fly in the usual way. The pawl and ratchet are provided to allow of the rapid distancing of the carbon holders when it becomes necessary to introduce fresh carbons. The spindle of the fly also carries a small finely toothed ratchet wheel. This ratchet wheel is actuated by a spring pawl, carried at the end of a lever, which lever is the continuation of the armature of the electro-magnet, in such a manner that when the armature is attracted by the electro-magnet, the spring pawl engages in the teeth of the ratchet wheel, and causes the wheels in gearing therewith to act upon the racks of the carbon holders to draw them apart.

The action of the lamp is as follows:—The current passes from the conductor to the top carbon holder, thence through the carbons to the bottom carbon holder, then to the coils of the electro-magnet situated in the base of the lamp. From the coils of the electro-magnet the circuit is completed to the other conductor. Upon the current passing through the circuit, the armature of the electro-magnet is attracted, and the abutment from the armature lever caused to short-circuit the coils of the electro-magnet, releasing the armature. The armature being released, the short-circuit is removed from the coils of the electro-magnet, and the cycle of movement repeated; in this manner an oscillatory motion is given to the armature lever, which by the spring pawl actuates the ratchet wheel, the train of clockwork, and the racks of the carbon holders, forcing the carbons apart until the distance between their points sufficiently weakens the current, so that it no longer attracts the armature of the electro-magnet. Thus by the combined action of gravitation of the top carbon in drawing the carbons together, and of the current to separate the carbons when they approach too closely, a working distance is maintained between the points with perfect automatism.

Siemens' lamp is at the present time employed in the Lizard Lighthouse, in Messrs Siemens' Engineering Works in England and Wales, as well as in other localities or buildings requiring powerfully lighting up.

An interesting illustration of the value of the electric light to the sailor is furnished by the 'Telegraph Journal' of April 5th, 1878. This publication contains a letter from the captain of the s.s. 'Faraday,' narrating how that vessel was by its means prevented from running into another vessel during a dense fog.

Siemens' magneto-electric apparatus and lamp were used on the occasion above referred to.

In every form of contrivance for electrical illumination the lamp or lighting apparatus consists of carbon points separated by a very slight interval, through which the current of electricity passes by means of terminal wires attached to the dynamo-electrical machine.

The lighting effect is produced by the passage of the electric spark through the small gap which separates the carbon points, in which interval extremely minute but solid particles of carbon, given off by the points, are heated up to incandescence in the path of the spark, and thus give rise to the intensely luminous focus known as 'the electric light.' The brilliancy of the light of course depends upon the quantity of electricity employed.

A very large number and variety of designs and patents for electric lamps have made their appearance in England, America, France, and Russia within the period following the invention of the small, powerful, and economic dynamo-electric machines of Siemens and Gramme.

The lighting apparatus generally attached to and worked by that variety of Gramme's machine generating the continuous current is that known as the 'Serrin Lamp.' Two carbon electrodes placed vertically one above the other (the positive being the upper one) are fixed on brass holders, which are so connected by a suitably contrived clockwork movement, combined with the working of an electro-magnet in connection with the electric circuit, as to maintain the two carbon poles during their combustion at the necessary distance from each other. Serrin's lamp differs in detail from Siemens', but, like this latter, is automatic in principle. In Paris it was the one in general use until the introduction of the Jablochkoff candle, and, with the Duboscq lamp, may be looked upon as the precursor of the various lamps and regulators now employed in electric lighting. Serrin's lamp or regulator, with some slight modification in the machinery, is also used in the Lontin system of electric illumination, by which separate lights are supplied by separate circuits of electricity. The Jablochkoff candle is the invention of a Russian engineer, whose name it bears.

It consists of 2 sticks of gas carbon, about 9 inches long and ⅛ of an inch thick, which are placed vertically side by side, and insulated from one another by a very thin strip of kaolin or china clay (a silicate of alumina and potash), the whole forming a candle. Each carbon rod is connected with one of the terminal wires of a Gramme dynamo-electric machine, the electric current from which, however, not being continuous, sets up an alternate current between the tips or poles of the candles, which are gradually consumed like an ordinary taper, the only difference in action between Serrin's and Siemens' lamps, that whereas in these latter the spark passes from the top to the bottom carbon point, in the Jablochkoff candle it jumps from side to side. The inventor contends that the kaolin by becoming heated diminishes the resistance of the circuit, and thus permits of the passage of the electric spark more easily through the carbons; and also, we believe, asserts that the kaolin, being electrolytically decomposed as the carbons are consumed, becomes converted into silica, which melts and drops down, whilst the aluminium liberated contributes luminosity during combustion to the flame.

One of the chief advantages, however, claimed by M. Jablochkoff is, that he can divide the circuit into a number of different lights, as the resistance of the circuit is uniform.

A large number of Jablochkoff candles are employed in the celebrated 'Magasins du Louvre,' one of the most extensive commercial establishments in Paris for the sale of silks, ribbons, gloves, &c., and clothing of every description.

The pure white light diffused by electricity admirably adapts it for viewing colours of all kinds at night, whether seen in pictures or on fabrics and raiments, and more particularly blues and greens, the hues of which are frequently indistinguishable from each other by gaslight. The candle is also used to light the courtyard of the Hôtel du Louvre, a large building contiguous to, and with its apartments running over, the Magasins, as well as in several shops.

Jablochkoff's system is also in work in Paris in front of many public buildings, and by its means the Place and Avenue de l'Opéra, together occupying a space 900 yards long by 80 yards wide, are brilliantly illuminated every night.

That celebrated circus, so well known to every visitor to Paris, the Hippodrome, is also lighted by it.

Another form of electric lamp is that of M. Rapieff, now in use in the machine-room of the 'Times' newspaper office. In this lamp there are four carbon points instead of two. M. Rapieff, like M. Jablochkoff, states that by means of his system he is enabled to supply several lamps with the same electric current. In the Wallace-Farmer lamp slabs of carbon instead of points are had recourse to.

In the lamps of M. Regnier, in one variety two revolving carbon discs are used, whilst in another a rod of carbon descends upon a disc of the same material, an arrangement which the inventor states leads to the subdivision of the current and its separate utilisation by a number of such lights.

One of the latest and apparently most successful methods for dividing the electric current so that one and the same current shall be made simultaneously to supply and render incandescent a series of carbon points, and in so doing give rise to as many effective electric illuminators, is that of Mr Werdermann. Mr Werdermann, observing the disparity of consumption between the positive and negative poles of the electrodes, found by experiment that when the sectional area of the negative pole was 64 times greater than the positive one, the electric arc was so far reduced that the two carbons were in contact. Under these conditions the electric arc was infinitely small, the negative electrode was not consumed, whilst the positive one was incandescent. Two supplies of electric light, therefore, ensued, one by the electric arc, and the other by the incandescent carbon of the positive electrode. Under these circumstances, if it were possible to devise a plan by which the positive pole as it consumed should be kept in uniform contact with the negative pole, the difficulty which had hitherto proved the stumblingblock to using a series of lights from one current would be annihilated.

Mr Werdermann demonstrated the correctness of his premises by a practical illustration of his plan (November, 1878) at the British Telegraph Manufactory, 374, Euston Road. The current from a dynamo-electric Gramme machine of 2-horse power was conducted to two electric lamps, each having an illuminating value equal to 360 candles each. The light so produced is described by a spectator as "being soft and sun-like, and as being capable of being looked at without discomfort, though it was not shaded." These being extinguished, ten smaller lamps were ignited by means of the same current, each one having an illuminating power equal to 40 candles. "The lamps burned steadily with a beautiful soft and clear white light. First one of the ten lights was then extinguished, and afterwards a second, the only effect on the remainder being that they became slightly more brilliant" ('Daily News').

Unlike Mr Edison, Mr Werdermann does not believe in the indefinite divisibility of the electric light. It will be observed that the candle-power of the light becomes diminished by subdivision. Two lights gave a light equal to 700 candles, whereas the same current divided into ten lights gave an aggregate light of only 400 candles.

The following extract from the 'Times' of December 5th, 1878, illustrates the financial aspect of the electric light question:—

"At the usual weekly meeting of the Society of Arts, held last evening, Dr C. W. Siemens, F.R.S., in the chair, a paper on electric lighting was read by Mr J. N. Shoolbred, M.Inst.C.E. The object of the author was to present some results of the application of electric lighting to industrial purposes, especially as regards cost.

He noticed first the Holmes and the Alliance magneto-electric machines, giving alternating currents and single lights for lighthouses. Secondly, he referred to the dynamo-electric machines, producing single lights for general industrial purposes, as well as for lighthouses, and including the Siemens and the Gramme machines. In his third group the author included the machines used for producing divided lights, each group indicating a marked period representing a clearly defined stage of progress in electric lighting. With regard to cost, Mr Shoolbred stated that in every instance his figures and particulars were those afforded by the users of the various lights, and not by the inventors or their representatives. In the case of the Holmes machine the annual cost per lighthouse was about £1035, inclusive of interest, repairs, and wages. With the Siemens machine the annual cost was about £494 per lighthouse, including interest and the other expenses. With the Alliance machine as used at Havre the cost was about £474 per annum per lighthouse, interest, &c., included. The single-light Gramme machine has been in use in the Paris goods station of the Northern of France Railway for 2 years. Six machines have been kept going with one light each, and the cost is found to be 5d. per light per hour, or with interest on outlay at 10%, 8d. per hour. The same light at the ironworks of Messrs Powell at Rouen was stated to cost 4d. per light per hour, exclusive of interest and charge for motive power, the latter being derived from one of the engines on the the works. In 1877 a series of experiments were carried out with the Lontin light at the Paris terminus of the Paris, Lyons, and Mediterranean Railway. The passenger station was lighted, and the results were so satisfactory that the company have entered into a permanent contract with the proprietors of the Lontin light for lighting their Paris goods station with 12 lights, at a cost of

5d. per light per hour. The Western of France Railway Company have had 6 Lontin lights in the goods station at the Paris terminus, St Lazare, since May last, and 12 lights in the passenger station since June. Careful experiments have shown the cost to be 8d. per light per hour, inclusive of interest. Referring to the Jablochkoff light, Mr Shoolbred placed before the meeting some particulars with regard to its application in the Avenue de l'Opéra, Paris, which were afforded him by M. J. Allard, the chief engineer of the lighting department of the City of Paris. It appears that the authorities pay the Société Générale d'Électricité 37f. 2c. per hour for the 62 lamps in use there. These 62 lamps supersede 344 gas jets which were previously used, and which cost the authorities 7·244f. per hour. The electric illumination, however, is considered as equal to 682 gas jets, or about double the original illumination—that is, to a cost of 14·45f. per hour, as against 37·2f. for the electric light, the cost of which, therefore, is 2·6 times that of the gas. The contract for lighting by electricity was terminated by the City of Paris, and the authorities have declined to renew it except at the price paid for gas, namely, 7·224f. (or about 6s.) per hour, and that only until the 15th of January next. These terms have been accepted by the Société, so that the price paid to them will be at the rate of about 1½d. per light per hour. Mr Shoolbred stated that the Société place their expenses at 1·06f. (or just 11d.) per light per hour, which, however, they hope shortly to reduce by one half. A series of careful photometric experiments carried out by the municipal authorities with the Jablochkoff lights, above referred to, showed each naked light to possess a maximum of 800 candles of intensity. With the glass globe this was reduced to 180 candles, showing a loss of 40%, while during the darker periods through which the lights passed the light was as low as 90 candles. The foregoing were the only authenticated particulars which the author could obtain as regards the working of the various electric systems of electric lighting. In conclusion, Mr Shoolbred referred to the Rapieff light at the 'Times' office, which, he observed, worked fairly and with regularity, which could not be said of all others, and it might therefore be entitled to take rank as an established application of electric illumination. The paper was illustrated by the Siemens, Rapieff, Serrin, and other forms of electric light, which were shown in operation."

That the electric light is eventually destined to supplant coal-gas in illuminating the fronts of large buildings, open spaces, squares, assembly rooms, public halls, theatres, picture galleries, workshops and factories, &c., seems no very extravagant prediction. We have already seen that it has for some years been employed in one lighthouse; and we have the testimony of Mr Douglas, of the Trinity House, at a meeting of the Society of Arts, that at the Souter Point Lighthouse there had been only 2 stoppages in 8 years, once through a bad carbon breaking, and once through the lighthouse-keeper going to sleep.

In addition to places above specified, amongst other localities in which it is in work, we may mention the chocolate factories of M. Menier, at

Noiselle, his india-rubber works at Genelle, his sugar refinery at Nice, and Messrs Caille's works at Paris.

That electricity is more economical as a method of artificial lighing than coal-gas the figures previously given seem to demonstrate, and there can be no question as to the much greater luminosity and purity of the light over the gas flame, qualities which render it an admirable substitute for the sunlight, the absence of which it may be said to supply at night. One disadvantage urged against its employment in weaving rooms is that it casts such dark and distinct shadows that these are frequently mistaken for the threads themselves, an objection which is said to have been remedied by placing the light as near the ceiling as possible. The non-generation of carbonic acid and sulphurous products such as are given off by burning gas, although of slight importance when the light is employed in the open air, becomes a great advantage when it is used in crowded assembly rooms or theatres, since the atmospheric contamination caused by carbonic acid becomes of course considerably reduced. The absence of sulphur compounds especially qualifies the light for use in large libraries. If it be true that the light gives rise to an appreciable amount of ozone, this constitutes another point in its favour. Opinion is at variance as to the possibility of the practical application of the electric light for illuminating private houses and dwellings in such a manner as to supply the place of the gas we now burn in them. One serious impediment to the probable accomplishment of this result certainly seems to be the fact that electricity for lighting purposes can only practically be conveyed to short distances from its source, which would necessitate the establishment and supervision of a number of generating machines near the houses to be lighted. Another obstacle, which hitherto has not been overcome, is the circumstance that the current when subdivided yields proportionately a greatly diminished amount of light. For instance, one light which had a certain photometric candle value would yield when divided into two an aggregate amount of light considerably less than the one; and if divided into three still less, and so on. This has been pointed out when noticing Mr Werdermann's invention for the divisibility of the light. Mr Edison, the American inventor, asserts that he has conquered this difficulty, and additionally perfected a machine for measuring the current used in the electric light. He states that it consists of an apparatus placed in every house lighted by electricity, which registers the quantity of electricity consumed, and uses for the purpose 1-1000th part of the quantity employed in the building.

A matter of primary importance in connection with the successful working of the electric light is the quality of the carbon points. In their manufacture gas carbon obtained from the necks of the retorts used in gas-making, as being the hardest and purest, is employed.

Superior, however, as this form of carbon is to every other description of the substance, it is never chemically pure, and as any foreign substance imparts to the light the irregularity or flickering that sometimes accompanies it, it is necessary that the impurities should be removed. To effect their separation the carbon has to undergo several processes, such as soaking in caustic potash to remove the silica, treatment with strong acids, several washings, grinding, &c. It is then kneaded and put into moulds, in which it is subjected to a pressure as high as 12 tons to the square foot. Subsequently the points so made are baked.

Since the above article was written the electric light has made enormous strides, and is in constant use in numberless factories, exhibitions, and public places. Recent modifications in the law relating to the formation of Electric Lighting Companies has given a great impetus to its use, and at the present time the lighting of large areas in London and other cities is being undertaken. The experience of the past 10 years has carried electric lighting beyond the experimental stage, and the extended use of the light will probably serve to establish it firmly. This development has naturally led to the introduction of endless new apparatus, and the whole question is now far beyond the limits of even a long article.

LIGHTNING, PRECAUTIONS AGAINST. The object of a lightning conductor is to deliberately attract the lightning, and, by providing proper means, to conduct it safely and harmlessly to the earth. It is obvious, therefore, that a defective conductor is worse than none at all. In the case of houses placed among tall trees, it is probably the better plan to fix the conductor to one of the tallest, provided that in this way it is of a greater height than the highest point of the house or building. By this means, if the conductor should fail, the tree alone suffers. If it be remembered that electricity has a tendency to discharge itself through or to be attracted by points and projections, the danger of standing under an *isolated* tree during a thunderstorm will be readily understood. In the same way a person walking across open ground without trees constitutes himself an isolated point, and is, therefore, liable to be struck. In such cases it is probably safer to lie down than to continue walking. A wood is fairly safe provided the traveller do not halt under the tallest trees, which are, as before said, liable to be struck. The proximity of iron gates, railings, and the like should be avoided. In the house the centre of a carpeted room is practically safe unless there be a metal chandelier overhead.

LIGNIN. $C_8H_{10}O_4$. *Sys.* CELLULOSE. This is woody fibre deprived of all foreign matter. It forms about 95% of baked wood, and constitutes the woody portion of all vegetable substances. Fine linen and cotton are almost entirely composed of lignin, the associated vegetable principles having been removed by the treatment the fibres have been subjected to during the process of their manufacture.

Pure lignin is tasteless, inodorous, insoluble in water and alcohol, and absolutely innutritious; dilute acids and alkaline solutions scarcely affect it, even when hot; oil of vitriol converts it into dextrin or grape sugar, according to the mode of treatment. When concentrated sulphuric acid is added very gradually to about half its weight of lint, linen rag, or any similar substance shredded small, and contained in a glass vessel with constant trituration, the fibres gradually swell up and disappear, without the disengagement of any gas, and a tenacious mucilage is formed, which is entirely soluble in water. If after a few hours the mixture be diluted with water, the acid neutralised by the addition of chalk, and after filtration any excess of lime thrown down by the cautious addition of a solution of oxalic acid, the liquid yields, after a second filtration and the addition of alcohol in considerable excess, a gummy mass, which possesses all the characters of pure dextrin. If, instead of at once saturating the diluted acid solution with chalk, we boil it for 4 or 5 hours, the dextrin is entirely converted into grape sugar, which, by the addition of chalk and filtration as before, and evaporation by a gentle heat to the consistence of a syrup, will, after repose for a few days, furnish a concrete mass of crystallised sugar. By strong pressure between folds of porous paper or linen, redissolving it in water, agitation with animal charcoal, and recrystallisation, brilliant colourless crystals of grape sugar may be obtained. Hemp, linen, and cotton, thus treated, yield fully their own weight of gum, and 1% of their weight of grape sugar. During the above transformation the sulphuric acid is converted into sulpholignic acid, and may be procured in a separate state. A solution of oxide of copper in ammonia, or solution of basic carbonate of copper in strong ammonia, dissolves cotton, which may then be precipitated by acids in colourless flakes.

LIG'NITE. *Syn.* BROWN COAL. Wood and other matter more or less mineralised and converted into coal. The lignites are generally dark brown, and of obvious woody structure. They are distinguished from true coals by burning with little flame and much smoke. Those of Germany are largely used as a source of paraffin and burning oils.

LIG'NUM VITÆ. See GUAIACUM WOOD.

LIME. CaO. *Syn.* OXIDE OF CALCIUM; CHAUX, Fr.; KALK, Ger. Lime, when pure, and as a chemical and medical reagent, will be found treated of under CALCIUM (Oxide of). It is prepared on the large scale for commerce by calcining chalk, marble, or limestone, in kilns, and is called quicklime, caustic lime, burnt lime, stone lime, &c. The limekilns are usually of the form of an inverted cone, and are packed with alternate layers of limestone and fuel, and the burnt lime raked out from the bottom. The lime thus obtained is a pale yellow powder, combining eagerly with water, and crumbling to a light white powder—'slaked lime'—with the evolution of much heat. Lime which slakes well is termed 'fat lime,' while if it slakes badly it is termed 'poor lime.' The slaked lime—the CALCIS HYDRAS of the B. P.—is fresh lime sprinkled with water till it falls to powder.

Lime, Chloride of. *Syn.* BLEACHING POWDER, CHLORINATED LIME, HYPOCHLORITE OF CALCIUM. This article was formerly believed to be a compound of lime and chlorine (CaO.Cl) and consequently received the name of 'chloride of lime.' We now know, however, that it is not a definite substance, but a mixture of calcium hypochlorite, calcium chloride, and calcium hydrate. The value of this preparation is due to the readiness with which the calcium hypochlorite is decomposed by acids, even by the carbonic acid of the air, with the evolution of hypochlorous acid, which abstracts hydrogen from many vegetable colouring matters, badly smelling gases, &c.; the former are thereby bleached, and the latter deodorised.

Chloride of lime is most extensively used for bleaching linen, calico, and similar fabrics, thousands of tons being made near Newcastle alone every year. It is also largely employed as a deodoriser.

Prep. Freshly slaked lime is thinly spread out in a proper vessel, and exposed to an atmosphere of chlorine gas until it is saturated. Now included in the Materia Medica.

Slaked lime (fresh), 20 parts; common salt, 1 part, are mixed together, and the powder placed in long earthenware vessels, into which chlorine is passed until the mixture begins to grow damp, or until 1 part of it, dissolved in 130 parts of water, is capable of decolouring 4½ parts of sulphate of indigo (see CHLORIMETRY), when the whole is transferred to dry bottles.

(Wholesale.) The chlorine is generated from the usual materials mixed in leaden vessels, heated by steam, and the gas, after passing through water, is conveyed by a leaden tube into an apartment built of siliceous sandstone, and arranged with shelves or trays, containing dry fresh-slaked lime, placed one above another, about an inch asunder. The process, to produce a first-class article, is continued for 4 or 5 days. During this time the lime is occasionally agitated by means of iron rakes, the handles of which pass through boxes of lime placed in the walls of the chamber, which thus act as valves.

The successful manufacture of bleaching powder is dependent upon the careful observance of a number of conditions, such as the quality of the limestone, which should be free from iron; the presence of magnesia at the time is also very objectionable, since it gives rise to the formation and presence in the bleaching powder of deliquescent chloride of magnesium. The apportionment of the water in slaking the lime is also a matter of no inconsiderable importance, the lime forming into balls, which fail to properly absorb the gas if the water be insufficient, whilst if it be in excess it yields a powder deficient in chlorine. When slaked the lime is passed through a sieve to free it from small pebbles. After being slaked it is kept for 2 or 3 days before being used, as it is found that under these circumstances it absorbs chlorine more readily than when recently prepared. Previous to its entrance into the lime chamber the chlorine is passed through water, to free it from vapour of hydrochloric acid and solid particles of chloride of manganese.

The temperature of the chamber into which the chlorine is passed ought not to exceed 16° C. (62° F.). An excess of chlorine has been found to yield a powder deficient in hypochlorite.

Bleaching powder, unless protected from the air (carbonic acid), slowly parts with its chlorine. In summer it has been estimated that it loses as much as 86% of the gas, and in winter about 26%.

Prep., &c. Chloride of lime is a pale, yellowish-white powder, generally more or less damp, and evolving a chlorine-like odour of hypochlorous acid. Its soluble constituents dissolve in about 30 parts of water. It is decomposed by acids with the evolution of chlorine and oxygen (hypochlorous acid). Good chloride of lime should contain from 32% to 36% of chlorine, of which, however, but 25% to 30% can be easily liberated by an acid.

Estim. See CHLORIMETRY.

Uses. Chloride of lime is employed in *medicine* as a deodoriser and disinfectant. An ointment of chloride of lime has been used in scrofula, and a lotion or bath, moderately dilute, is one of the cleanest and readiest ways of removing the 'itch,' and several other skin diseases. It is also in great use as a disinfectant, and may be used either in substance or solution. A small quantity of the powder spread on a flat dish or plate, and placed on the chimney-piece, and a like quantity in an opposite part of the room will continue to evolve sufficient chlorine or hypochlorous acid to disinfect (?—Ed.) the air of an apartment for several days. The evolution of chlorine is promoted by occasionally renewing the exposed surface by stirring it with a piece of stick, and after it becomes scentless, by the addition of a little acid, as strong vinegar, or hydrochloric acid, or oil of vitriol, largely diluted with water. Of late, however, it has been partly superseded by sulphurous acid, carbolic acid, &c. The most extensive consumption of chloride of lime is, however, for bleaching textile fabrics. When employed for this purpose the goods are first immersed in a dilute solution of this substance, and then transferred to a vat containing dilute sulphuric or hydrochloric acid. The chlorine thus disengaged in contact with the cloth causes the destruction of the colouring matter. This process is generally repeated several times, it being unsafe to use strong solutions. White patterns may thus be imprinted upon coloured cloth; the figures being stamped with tartaric acid thickened with gum-water, the stuff is immersed in the chloride bath, when the parts to which the acid has been applied remain unaltered, while the printed portions are bleached white.

Concluding Remarks. Chloride of lime is now scarcely ever made on the small scale, as it can be purchased of the large manufacturer of better quality and cheaper than it could possibly be made by the druggist. The chief precaution to be observed in the manufacture of good bleaching powder is to maintain the ingredients at a rather low temperature.

Lime, Pyrolignite of. An impure acetate of calcium used for making mordants in dyeing and calico-printing, as a substitute for the more expensive acetate of lead.

Lime, Salts of. See under CALCIUM.

LIME. The fruit of *Citrus limetta.* It resembles the lemon, but is smaller and has a smoother skin. It is imported into Great Britain in a preserved state for use as a dessert. Its juice is also largely imported for the preparation of CITRIC ACID, and for the prevention of scurvy on board ship (see *below*).

LIME JUICE. *Syn.* LEMON JUICE. The juice of the fruits of various species of *Citrus,* principally LIMES, is known in *commerce* under these names. It is very variable as to quality, which depends upon the method of extraction, the quality of the fruit, and the honesty of the shipper.

We have examined the juice expressed from limes sent from the West Indies, from Jamaica, and from South Africa, with the following results: (*Tuson*)

	W. Indies.	Jamaica.	S. Africa.
Specific gravity of juice	1041·30	1044·18	1044·90
Per cent. of citric acid	7·96	8·66	8·50
Per cent. of ash	0·321	0·401	0·364

The yield from limes is very small, and the freshly expressed juice contains a large amount of pulp. This, however, on standing a few weeks, separates, and a clear sherry-coloured liquid is obtained.

A concentrated lime or lemon juice is used by calico printers. It is a "dark, treacly-looking fluid, marking from 48°—54° Twaddell," and contains about 30% of pure citric acid.

Adult. See LEMON JUICE.

Estim. Lime juice is only valuable on account of the citric acid it contains. If of good quality, 100 gr. will neutralise from 70 to 76 gr. of pure crystallised carbonate of soda. "For commercial purposes each grain of carbonate of soda neutralised may represent ½ gr. of crystallised citric acid (equal to 38 gr. of dry acid), and the value of the lime juice be calculated in proportion" (*O'Neill*). As commercial lime juice contains variable proportions of vegetable extractive matter, the indications of the hydrometer cannot be depended upon. See ACIDIMETRY, CITRIC ACID, &c.

LIMESTONE. A general term applied to a great variety of rocks in which carbonate of lime is the principal constituent; more or less silica is also invariably present.

Estim. The value of chalk, limestone, marble, &c., for hydraulic mortars and cements may be determined as follows:

A given weight (say 100 gr.) of the sample is reduced to powder and digested in hydrochloric acid diluted with about an equal weight of water, with frequent agitation for an hour or longer; the mixture is then diluted with thrice its volume of water, thrown upon a filter, and the undissolved portion washed, dried, ignited, and weighed. This weight indicates the percentage of clay and silica or sand, and the loss that of the lime or calcium oxide, magnesium oxide, and ferric oxide present in the substance examined. In most cases these results will be sufficient to show the quality of the limestone for the purpose of making mortar or cement.

The filtrate and the washings are mixed together, and ammonia is added in excess; the bulky, reddish-brown precipitate is collected, washed, dried, ignited, and weighed. This gives the percentage of ferric oxide.

The filtrate from last is then treated with oxalate of ammonium, and the quantity of lime determined in the manner described under the head of CALCIUM.

The liquid filtered from the precipitate in last

is boiled for some time with carbonate of potassium until ammoniacal fumes are no longer evolved; the precipitate is then collected on a filter, washed with hot water, dried, and strongly ignited for 3 or 4 hours, and, lastly, weighed. This gives the percentage of magnesium carbonate.

LIMETTIN. A crystalline substance which Professor Tilden and C. R. Beck extracted from the deposit which is found in oil of limes. It has the formula $C_{16}H_{14}O_5$. Essence of lemon yields a similar substance, $C_{14}H_{14}O_5$, but essence of bergamot yields a crystalline compound differing from both of these. The properties of the substance are described in a paper communicated to the Chemical Society.

LINCTUS. [L., Eng.] *Syn.* LOCH, LOHOCH, LINCTURE, LAMBATIVE; LOOCH, Fr. A medicine of the consistence of honey, intended to be licked off a spoon. This form of medicine is well adapted to females and children, but is not much used in England at the present time. Those employed in modern pharmacy and prescribing are included under the heads CONFECTION, CONSERVE, and ELECTUARY. The *dose*, when it is not otherwise stated, is a teaspoonful occasionally.

Linctus Acidus. *Prep.* 1. (Hospital for Consumption.) Dilute sulphuric acid, 5 minims, oxymel, 25 minims; simple linctus to 1 dr.

2. (University College Hospital.) Dilute sulphuric acid, 5 minims; oxymel, 20 minims; spirit of chloroform, 2 minims; treacle to 1 dr.

Linctus, Caca'o. *Syn.* LINCTUS CACAO, L.; CRÈME DE TRONCHIN, Fr. *Prep.* From cocoa butter, 2 oz.; white sugar (in powder), syrup of capillaire, and syrup of tolu, of each, 1 oz. Mix. Demulcent and pectoral; in coughs, sore throats, hoarseness, &c.

Linctus Communis. *Prep.* 1. Dilute sulphuric acid, 15 minims; syrup of squill, 30 minims; paregoric, 2 dr.; treacle, 1 dr.; anise water to 1 oz.; mix.—*Dose*, 1 dr. (King's Coll. Hospital).

2. Dilute sulphuric acid, 7½ minims; tincture of squill, 30 minims; syrup of extract of poppies, 1½ dr.; powdered tragacanth, 6 gr.; treacle, 2 dr.; water to 1 oz.—*Dose*, ½ to 2 dr. (Middlesex Hospital).

3. Solution of hydrochlorate of morphia, 3 minims; spirit of chloroform, 3 minims; glycerine and water, of each, ½ dr. (City Chest Hospital).

4. Dilute sulphuric acid, 5 minims; liquid extract of opium, 3 minims; syrup of squill, 15 minims; treacle, ½ dr.; water to 1 dr. (Charing Cross Hospital).

5. Acetic acid, 2 minims; vinegar of squill, 36 minims; syrup of poppies, 36 minims; confection of hips, 100 gr.; powdered tragacanth, 6 gr.; boiling water to 1 oz.—*Dose*, 1 to 2 teaspoonfuls (St. Bartholomew's Hospital).

6. Olive oil, 4 dr.; confection of hips, 6 dr.; vinegar of squill, 1½ dr.; tincture of opium, 7½ minims; treacle, 3 dr.—*Dose*, 1 to 2 dr. (St. George's Hospital).

7. Dilute sulphuric acid, 5 dr.; syrup of squill, 5 dr.; syrup of poppies, 5 dr.; comp. tincture of camphor, 5 dr.; ipecacuanha wine, 2 dr.; gum-arabic, 5 dr.; treacle, 2½ fl. oz.; water, 7½ oz.—*Dose*, 1 to 2 dr. (Westminster Ophthalmic Hospital).

8. Olive oil, 4 oz.; tartaric acid, 2 oz.; powdered gum, 4 oz.; powdered opium, 30 grs.; treacle, 5 lbs. (Women's Hospital).

9. Dilute sulphuric acid, 30 minims; vinegar of squill, 3 dr.; tincture of opium, 30 minims; powdered tragacanth, 40 gr.; treacle, 2 oz.; water, 2 oz. (London Ophthalmic Hospital).

Linctus, Cough. *Syn.* PECTORAL LINCTUS; LINCTUS PECTORALIS, L. *Prep.* (Dr. Latham.) Compound ipecacuanha powder (Dover's powder), ½ dr.; compound tragacanth powder, 2 dr.; syrup of tolu, confection of hips, and simple oxymel, of each, 1 oz.—*Dose*, 1 teaspoonful, 3 or 4 times a day. " This linctus has been extensively used, as a remedy for coughs, in the West End of London, having been found to be a safe and generally efficacious remedy " (*Redwood*). The preceding as well as the following are also useful preparations.

Linctus, Demulcent. *Syn.* LINCTUS DEMULCENS, L.; LOOCH DE TRONCHIN, Fr. *Prep.* From oil of almonds, syrup of capillaire, manna, and cassia pulp, of each, 2 oz.; powdered gum tragacanth, 20 gr.; orange-flower water, 2 fl. oz. As the last. The above is the quantity for 2 days, which is as long as it will keep.

Linctus of Egg. *Syn.* LINCTUS OVI, LOOCH OVI, L. *Prep.* Oil of almonds, ½ dr.; yolk of 1 egg; syrup of marsh-mallow, 1 oz. Mix.

Linctus, Emol'lient. *Syn.* OILY EMULSION; LOHOCH OLEOSUM, EMULSIO OLEOSA, L.; LOOCH HUILEUX, Fr. *Prep.* (P. Cod.) Oil of almonds, powdered gum, and orange-flower water, of each, 4 dr.; syrup of marsh-mallow, 1 oz.; water, 3 fl. oz. or q. s.; for an emulsion. In troublesome coughs.

Linctus, Expec'torant. *Syn.* LINCTUS EXPECTORANS, LOHOCH E., L. *Prep.* 1. Oxymel of squills, confection of hips, syrup of marsh-mallow, and mucilage of gum-arabic (thick), equal parts. Demulcent and expectorant.

2. (*Dr. Copland.*) Oil of almonds and syrup of lemons, of each, 1 fl. oz.; powdered ipecacuanha, 6 gr.; confection of hips, 1 oz.; compound powder of tragacanth, 3 dr.

Linctus, Green. *Syn.* LINCTUS VIRIDE, LOHOCH VIRIDE, L. *Prep.* Pistachio nuts (or sweet almonds), No. 14; syrup of violets, 1 oz.; oil of almonds, ½ oz.; gum tragacanth, 15 gr.; tincture of saffron, 1 scruple; orange-flower water, 2 dr.; water, 4 oz. Mix.

Linctus c. Ipecacuanhā. *Prep.* Ipecacuanha wine, 2 dr.; linctus to 1 oz.—*Dose*, 1 teaspoonful (St. Bartholomew's Hospital).

Linctus Limonis. *Prep.* Syrup of lemon, 6 dr.; solut. acet. morphia, 2 fl. dr.; water to 3 oz. —*Dose*, 1 dr. (Throat Hospital).

Linctus of Linseed. (E. 1744.) *Syn.* LINCTUS LINI, LOHOCH LINI, L. *Prep.* Fresh-drawn linseed oil, 1 oz.; syrup of tolu, 1 oz.; sulphur, 2 dr.; white sugar, 2 dr. Mix.

Linctus of Manna. (E. 1744.) *Syn.* LINCTUS MANNE, LOHOCH MANNE, L. *Prep.* Equal parts of manna, oil of almonds, and syrup of violets. Mix.

Linctus Morphinæ. *Prep.* 1. Solution of hydrochlorate of morphine, 3 minims; spirit of chloroform, 3 minims; glycerine, 1 dr. (Westminster Hospital).

2. **Hydrochlorate of morphine**, ¼ gr.; dilute hydrochloric acid, 2 minims; syrup of squill, 20 minims; dilute hydrocyanic acid, 2 minims; water to make 1 dr. (Royal Chest Hospital).

Linctus of Naphthalin (*Dupasquier*). *Syn.* LINCTUS NAPHTHALINI, LOHOCH NAPHTHALINI, L. *Prep.* To one common lohoch add from 8 gr. to 30 gr. of naphthalin. The latter must be well triturated with the gum.—*Dose.* One teaspoonful, as an expectorant.

Linctus Opiatus. *Prep.* Tincture of opium, 10 minims; dilute sulphuric acid, 12 minims; treacle, 5 dr.; water to 1 oz.—*Dose*, 1 to 2 dr. (Guy's Hospital).

Linctus, Pectoral. *Syn.* FOX LUNGS; LINCTUS PECTORALIS, LOHOCH à PULMONE VULPIUM, L. *Prep.* From spermaceti and Spanish juice, of each, 8 oz.; water, q. s. to soften the liquorice; make a thin electuary, and add of honey, 3 lbs.; oil of aniseed, 1 oz.; mix well. A popular and excellent demulcent in coughs. It formerly contained the herb 'fox lungs,' but spermaceti is now substituted for that article.

Linctus of Poppies (Th. Hosp.). *Syn.* LINCTUS PAPAVERIS. *Prep.* Compound tincture of camphor, syrup of poppies, and syrup of tolu, of each, equal parts; mix.—*Dose*, 1 fl. dr.

Linctus Potassæ Nitratis. *Prep.* Nitrate of potash, 1½ dr.; oxymel, 4 dr.; syrup of roses to 1½ oz.—*Dose*, 1 teaspoonful (Guy's Hospital).

Linctus Scillæ. *Prep.* 1. Syrup of squill, syrup of poppies, syrup of lemons, syrup of tolu, equal quantities.—*Dose*, 1 dr. (Throat Hospital).

2. Oxymel of squill, 20 minims; dilute sulphuric acid, 5 minims; tincture of opium, 2 minims; simple linctus to 1 dr. (Consumption Hospital).

3. Oxymel of squill, 1 dr.; paregoric, 15 minims; mucilage, 1 dr.—*Dose*, 2 dr. (St. Mary's).

Linctus Scillæ Co. (*vel c.* Opio). *Prep.* 1. Oxymel of squill, 5 minims; compound tincture of camphor, 2½ minims; spirit of nitrous ether, 2½ minims; water to 1 dr. (Westminster Hospital).

2. Oxymel of squill, 20 minims; compound tincture of camphor, 10 minims; ipecacuanha wine, 5 minims; mucilage to 1 dr. (Royal Chest Hospital).

3. Oxymel of squill, 2 dr.; paregoric, 1 dr.; ipecacuanha wine, ¼ dr.; mucilage to 5 dr. (University Hospital).

Linctus Simplex (*vel* Theriaca Preparata). *Prep.* Treacle, 20 minims; spirit of chloroform, 2 minims; water to 1 dr. (Consumption Hospital).

Linctus of Spermaceti. (E. 1744.) *Syn.* LINCTUS CETACHI, LOHOCH CETACHI, L. *Prep.* Spermaceti, 2 dr.; yolk of egg, q. s.; triturate, and add gradually oil of almonds, ½ oz.; syrup of tolu, 1 oz; mix.

Linctus of Syrup of White Poppies. (P. C.) *Syn.* LINCTUS SYRUPI PAPAVERIS ALBI, LOHOCH SYRUPUS PAPAVERIS ALBI. *Prep.* White lohoch, 5 parts; syrup of poppies (P. C.), 1 part. Mix.

Linctus, Turpentine. *Syn.* LINCTUS STIMULANS, L. TEREBINTHINÆ, LOHOCH ANTHELMINTICUM, L. *Prep.* (*Recamier.*) Oil of turpentine, 2 dr.; honey of roses, 3 oz.; mix.—*Dose.* A teaspoonful night and morning, followed by a

draught of any weak liquid; in worms, more especially tape-worm.

Linctus pro Tussi. *Prep.* Oxymel of squill, syrup of poppies, mucilage, of each, equal parts (St. Thomas's Hospital).

Linctus, White. *Syn.* LINCTUS ALBUS, MISTURA ALBA, LOHOCH ALBUM, L.; LOOCH BLANC, Fr. *Prep.* (P. Cod.) Jordan almonds, 4½ dr.; bitter almonds, ½ dr.; blanch them by steeping them in hot water and removing the skins; add of white sugar, ½ oz.; gum tragacanth, 20 gr.; beat to a smooth paste, and further add of oil of almonds and orange-flower water, of each, 4 dr.; pure water, 4 fl. oz. A pleasant demulcent in tickling coughs.

LINEN. *Syn.* LINTEUM, L. Linen is a textile fabric made of the liber-fibres of the *Linum usitatissimum*, or common flax, a plant which from time immemorial has been cultivated for this purpose. It is remarkable for the smoothness and softness of its texture, and is hence highly esteemed in temperate climates as an elegant and agreeable article of clothing to be worn next the skin. Its fibres are better conductors of heat, more porous, and more attractive of moisture than those of cotton, which render it less adapted for body-linen in cold weather, as well as in hot weather and hot climates, than calico. The latter, however, lacks the luxurious softness and freshness of linen, whilst the peculiar twisted and jagged character of its fibres renders it apt to excite irritation in extremely delicate skins. The common prejudice in favour of old linen and flax lint for dressing wounds is thus shown to have reason on its side, and, like many other vulgar prejudices, to be supported by the investigations of science.

Identif. Linen fabrics are commonly sophisticated with cotton, which is a much less costly and a more easily wrought material. Various plans have been proposed to detect this fraud, many of which are too complicated and difficult for practical purposes. The following commend themselves for their simplicity and ease of application:

1. A small strip (a square inch, for instance) of the suspected cloth is immersed for 2 or 3 minutes in a boiling mixture of about equal parts of hydrate of potassium and water, contained in a vessel of silver, porcelain, or hard glass; after which it is taken out and pressed between the folds of white blotting-paper or porous calico. By separating 8 or 10 threads in each direction their colour may be readily seen. The deep yellow threads are LINEN, the white or pale yellow ones are COTTON.

2. A small strip of the cloth, after having been repeatedly washed with rain-water, boiled in the water, and dried, is immersed for 1 or 2 minutes in sulphuric acid; it is then withdrawn, carefully pressed under water with the fingers, washed, immersed for a few seconds in ammonia, solution of carbonate of potassium, or solution of carbonate of sodium, again washed with water, and dried between filtering-paper. By this treatment the cotton fibres are dissolved, while the linen fibres are merely rendered thinner and more translucent according to the duration of the experiment; after a short immersion the

cotton fibres appear transparent, while the linen fibres remain white and opaque.

3. Böttger recommends the linen stuffs to be dipped into an alcoholic solution of rosolic acid, then into a concentrated solution of sodium carbonate, and finally washed with water. The linen fibre assumes a pink colour, whilst the cotton fibre remains unaltered.

4. (By the MICROSCOPE.) The indications afforded by both the previous tests, although quite visible to the naked eye, are rendered still more palpable by the use of a magnifying glass of small power, as the common pocket lens. Under a good microscope the presence of cotton in a linen tissue is very perceptible. The fibres of cotton present a distinctly flat and shrivelled appearance, not unlike that of a narrow, twisted ribbon, with only occasional joints; whilst those of flax are round, straight, and jointed. The

fibres of cotton, after being exposed to the action of strong alkaline lyes, untwist themselves, contract in length, and assume a rounded form, but still continue distinct in appearance from the fibres of linen. The cut represents a fibre of linen (1) and a fibre of cotton (2) as they appear when magnified 155 diameters. The difference between the two may be perceived, although less distinctly, through a good Stanhope or Coddington lens, provided the object be well illuminated.

Dyeing. Linen and cotton, from the similarity of their behaviour with dye-stuffs, are treated in nearly the same manner. The affinity of their fibres for colouring matter is very much weaker than that of the fibres of silk and woollen. On this account they are dyed with greater difficulty than those substances, and the colours so imparted are, in general, less brilliant and permanent under similar conditions. Linen shows less disposition to take dyes than cotton. The yarn or cloth, after being scoured and bleached in the usual manner, requires to have an additional tendency given to it by chemical means, to condense and retain the materials of the dye-bath in its pores. This is effected by steeping the goods in solutions (mordants) which have at once an affinity for both the fibres of the cloth and the colouring matter. A similar process is employed in dyeing most other substances; but with cotton and linen attention to this point is essential to the permanency of the dye. These matters are more fully explained under the heads DYEING and MORDANT.

The following process for bleaching linen, having been omitted from the article on "Bleaching," is inserted here :

Mr Hodges' process, which is known in Ireland as the 'chemico-mechanical process,' owing to the patentee turning to account the advantages derivable from the employment of mechanical contrivances driven by steam, combined with the introduction of a new method of obtaining the hitherto little used hypochlorite of magnesium, may

be said to date from the discovery of the substance known as *kieserite* (native sulphate of magnesia), which occurs as an essential constituent of the Abraumsalts of Stassfurth. For some time after the introduction of this substance into the market it was considered of little value except for the production of Epsom salts; but Mr Hodges, in the course of some investigations in bleaching jute, having had occasion to employ large quantities of hypochlorite of magnesia, it occurred to him that kieserite might be substituted for the more expensive crude sulphate of magnesia; and the importation into Ireland of the sample for this purpose was the first that was ever sent into that country for the manufacture of a bleaching liquor, or, indeed, for any other use. Mr Hodges, on experimenting with the kieserite, found that it not only supplied the place of the crude sulphate, but acted as a better precipitant for the lime of the bleaching powder, which is employed in the production of the hypochlorite of magnesia; and that it also produced a stronger and clearer solution. Without entering into a minute description of the process (which is at present successfully carried out in a factory erected for the purpose in the neighbourhood of Belfast), the following outline will be sufficient to show the nature of the methods adopted. The kieserite, which is imported from Germany in square blocks, on arriving at the works is conveyed to a house, on the ground-floor of which it is stacked until required, when it is ground to a fine powder placed in barrels, and drawn up by means of a crane to a room at the top of the building, at one end of which is a row of three tanks furnished with water-taps, agitators, and false bottoms. In one of the end tanks a definite quantity of the kieserite powder (varying according to its strength as ascertained by analysis) is placed and dissolved in a given quantity of water, the solution being assisted by agitators, and on settling the clear liquor is siphoned over into the middle tank. In the third tank bleaching powder (hypochlorite of lime), varying in quantity according to the strength of the kieserite solution, is placed. The bleaching powder, after being agitated with water, is allowed to settle, and the clear solution is siphoned over into the middle tank containing the clear kieserite solution, the agitator being kept in motion not only during the mixing of the liquids, but for some time after. The mixed liquids are then allowed to remain undisturbed all night, after which the clear hypochlorite of magnesia solution is siphoned into a large settling tank, which is situated in the room below. From this vessel it is conducted through wooden pipes (which are so contrived that they can be opened and cleansed at will) into a large cistern standing in the bleaching-house. This cistern is fitted with a ball-cock, by which arrangement the liquid can be drawn off by a system of wooden pipes as required. The bleaching-house in which the cistern is situated is fitted up in an original manner, and covers something more than an acre of ground; whilst the reeling-shed, which is the only part of the works our limits will permit us to describe, is 240 ft. long by 24 ft. broad, and contains 10 steeps and 12 reel-boxes. Each box is provided with water, a solution of the bleaching agent, and steam-pipes, and is capable of reeling at a time about 500 lbs. of yarn. Above the box is a line of rails or pillars. A travelling crane runs along the reels, and carries the reels from one box to another. Attached to this crane is a newly invented hydraulic pump, by means of which the reels with the yarn on them can be lifted in a few seconds from one box to another.

After the yarn has been boiled, washed, and passed through the squeezers in the usual manner, it is put on to a waggon, in which it is carried, by means of a line of rails, down to the first reel-box. Here it is placed on to the reels, which are made to revolve by means of steam, first in one direction and then in another, through a solution of carbonate of soda, previously heated by means of the steam-pipes before mentioned. The yarn having been sufficiently scalded, and so saturated with soda, the reels to which it is attached are raised by the hydraulic pump out of the box, and the yarn allowed to drain for a few minutes, after which the travelling crane carries it on to the next box. Into this box the yarn is again lowered by the pump, and made to revolve as before, but this time through a solution of the bleaching agent, which, immediately reacting on the carbonate of soda with which the yarn is charged, renders this bleaching agent free from the danger which attends the employment of chlorine, or the ordinary bleaching powder used in the older methods of bleaching. After the yarns have been brought to the desired shade in the solution of Hodges' bleaching agent they are either removed as before to a new box, and there washed before being scoured, or they are thrown into one of the steeps filled with water for the night. These operations are repeated with weaker solutions in the remaining reel-boxes, either once or twice, according to the shade required.

Mr Hodges claims as the chief features of his invention that it consists, first, in the employment of a bleaching agent which has not hitherto been practically employed, and a cheap method for its production; second, in the preparation of the yarn prior to being submitted to the action of the bleaching agent, this preparation setting free not only the imprisoned chlorine of the hypochlorite, but also another powerful bleaching agent, oxygen; third, in new and improved machinery, by which the work of bleaching the yarn is greatly shortened; fourth, in doing away with the tedious and expensive operation of exposing the yarn on the grass. If this last were the only feature in Mr Hodges' invention, the patentee would have greatly improved the process of bleaching; not only, however, does the new process supplant the old long and tedious one, but a great economy of time is additionally gained in other parts of the process. Added to these advantages it is stated that a superior finish is given to the yarns, and that in consequence a much greater demand for them has arisen.

Mr Hodges contends that the absence of caustic lime from his new bleaching compound gives it great advantages over the old bleaching powder, particularly in its application to finely woven fabrics, such as muslins, &c. He also says that fabrics bleached by it receive an increased capacity for imbibing and retaining colouring matter, a fact of considerable importance to the dyer and calico-printer, as they are thus enabled to communicate to the fabrics tints which have heretofore been considered impossible. See KIESERITE.

The domestic management of linen may here receive a few moments' attention. Fruit-stains, ironmoulds, and other spots on linen may, in general, be removed by applying to the part, previously washed clean, a weak solution of chlorine, chloride of lime, spirits of salts, oxalic acid, or salts of lemons in warm water, and frequently by merely using a little lemon juice. When the stain is removed the part should be thoroughly rinsed in clear warm water (without soap) and dried. Recent ironmoulds or ink spots on starched linen, as the front of a shirt, may be conveniently removed by allowing a drop or two of melted tallow from a common candle to fall upon them before sending the articles to the laundress. The oxide of iron combines with the grease, and the two are washed out together. If the spot is not entirely removed the first time, the process should be repeated. Linen that has acquired a yellow or bad colour by careless washing may be restored to its former whiteness by working it well in water to which some strained solution of chloride of lime has been added, observing to well rinse it in clean water both before and after the immersion in the bleaching liquor. The attempt to bleach unwashed linen should be avoided, as also using the liquor too strong, as in that case the linen will be rendered rotten.

LING. The *Gadus molva*, Linn., an inferior species of the cod-fish tribe, common in the northern seas, and used as a coarse article of food by the poor.

LINIMENT. *Syn.* LINIMENTUM, L. A fluid, semi-fluid, or soapy application to painful joints, swellings, burns, &c. The term is also occasionally extended to various spirituous and stimulating external applications. A preparation of a thinner consistence, but similarly employed, is called an 'EMBROCATION.' These terms are, however, frequently confounded together and misapplied. Liniments are generally administered by friction with the hand or fingers, or with some substance (as a piece of flannel) capable of producing a certain amount of irritation of the skin. Sometimes a piece of linen rag dipped in them is simply laid on the part. In most cases in which liniments are found beneficial, the advantage obtained from them is attributable rather to the friction or local irritation than to any medicinal power in the preparation itself. The greater number of cerates and ointments may be converted into liniments by simply reducing their consistence with almond or olive oil, or oil of turpentine.

Liniment, Ac'id. *Syn.* LINIMENTUM ACIDUM, L. ACIDI SULPHURICI, L. *Prep.* 1. (*Sir B. Brodie.*) Salad oil, 3 oz.; oil of vitriol, 1 dr.; mix, then add of oil of turpentine, 1 oz., and agitate the whole well together. As a counter-irritant, in rheumatism, stiff joints, &c. It closely resembles the 'GUISTONIAN EMBROCATION.'

2. (Hosp. F.) Olive oil, 3 oz.; oil of turpen-

tine, 2 oz.; sulphuric acid, 1 fl. dr. An excellent alterative, stimulant, discutient, and counter-irritant, in chronic rheumatism, stiff joints, indolent tumours, and various chronic diseases of the skin.

Liniment of Aconite. *Prep.* (B. P.) Aconite root in No. 40 powder, 20 oz.; camphor, 1 oz.; rectified spirit, enough to make 30 oz. Allow the spirit to percolate through the powder, then add the camphor and make up to 30 oz. with spirit. Painted on the face it relieves neuralgic pain.

Linimentum Aconiti. (B. P.) Aconite root, in powder, 20; camphor, 1; rectified spirit, to percolate, 30. Moisten the root for 3 days, then pack in a percolator, and pour sufficient rectified spirit upon it to produce with the camphor 30.

Strength, 1 in 1½. Applied with a camel-hair pencil, alone or mixed in equal proportions, with a soap liniment or compound camphor liniment, and rubbed on the part. Seven parts of this, and 1 part of chloroformum belladonnæ, sprinkled thinly on impermeable pilline, is the best application for neuralgia or lumbago.

Liniment of Am′ber Oil. *Syn.* LINIMENTUM SUCCINI, L. *Prep.* 1. From olive oil, 3 parts; oils of amber and cloves, of each, 1 part. Resembles 'ROCHE'S EMBROCATION.'

2. (Opiated: LINIMENTUM SUCCINI OPIATUM, L.) From rectified oil of amber and tincture of opium, of each, 2 fl. oz.; lard, 1 oz. Anodyne, antispasmodic, and stimulant. A once popular remedy in cramp, stiff joints, &c.

Liniment of Ammo′nia. *Syn.* AMMONIACAL LINIMENT, VOLATILE L., OIL AND HARTSHORN; LINIMENTUM AMMONIÆ (B. P., Ph. L. E. and D.), L. *Prep.* 1. (B. P.) Solution of ammonia, 1 part; olive oil, 3 parts; mix.

2. (Ph. L. and E.) Liquor of ammonia (sp. gr. ·960), 1 fl. oz.; olive oil, 2 fl. oz.; shake them together until they are mixed.

3. (Ph. D.) To the last add of olive oil, 1 fl. oz. Stimulant and rubefacient. Used in rheumatism, lumbago, neuralgia, sore throat, spasms, bruises, &c. When the skin is irritable more oil should be added, or it should be diluted with a little water.

4. (Camphorated: LINIMENTUM AMMONIÆ CAMPHORATUM, EMBROCATIO AMM. CAMPHORATA, L.) *a.* (HOSP. F.) Olive oil, 3 oz.; camphor, ? oz.; dissolve by a gentle heat, and when cold, add of liquor of ammonia, 1 fl. oz.

b. Soap liniment, 2 oz.; olive oil and liquor of ammonia, of each, 2 dr. As the last; more especially for sprains, bruises, chilblains, &c.

5. (Compound: DR GRANVILLE'S COUNTER-IRRITANT or ANTIDYNOUS LOTION; LINIMENTUM AMMONIÆ COMPOSITUM, L.) (Ph. E.) *a.* (STRONGER.) From liquor of ammonia (sp. gr. ·880), 5 fl. oz.; tincture of camphor, 2 fl. oz.; spirit of rosemary, 1 fl. oz.; mix. It should be kept in a well-stoppered bottle and in a cool situation.

b. (WEAKER.) Solution of ammonia (·880), 5 fl. oz.; tincture of camphor, 3 fl. oz.; spirit of rosemary, 2 fl. oz.

Obs. The above formulæ are nearly identical with the original ones of Dr Granville; the principal difference being in his ordering liquor of ammonia of the sp. gr. ·872, instead of ·880·

They are counter-irritant, rubefacient, vesicant, and cauterising, according to the mode and length of their application. The milder lotion is sufficiently powerful to produce considerable rubefaction and irritation in from 1 to 5 or 6 minutes, vesication in 8 or 10 minutes, and cauterisation in 4 or 5 minutes longer. For the latter purpose the stronger lotion is generally employed. According to Dr Granville, these lotions are prompt and powerful remedies in rheumatism, lumbago, cramp, neuralgia, sprains, swollen and painful joints, headache, sore throat, and numerous other affections in which the use of a powerful counter-irritant has been recommended. They are ordered to be applied by means of a piece of linen 6 or 7 times folded, or a piece of thick, coarse flannel wetted with the lotion, the whole being covered with a thick towel, and firmly pressed against the part with the hand. The stronger lotion is only intended to be employed in apoplexy, and to produce cauterisation. See COUNTER-IRRITANTS.

6. (From SESQUICARBONATE OF AMMONIA; LINIMENTUM AMMONIÆ SESQUICARBONATIS, Ph. L.) Solution of sesquicarbonate of ammonia, 1 fl. oz.; olive oil, 3 fl. oz.; shake them together until mixed. This preparation resembles ordinary liniment of ammonia in its general properties, but it is much less active, owing to the alkali being carbonated. It is the 'oil and hartshorn' and the 'volatile liniment' of the shops.

7. (WITH TURPENTINE.) (*Dr Copland.*) *Syn.* LINIMENTUM AMMONIÆ CUM TEREBINTHINÂ, L. *Prep.* Liniment of ammonia, 1½ fl. oz.; oil of turpentine, ½ fl. oz.; mix.

Liniment, An′odyne. See LINIMENTS OF BELLADONNA, MORPHIA, OPIUM, SOAP, &c.

Liniment, Antispasmod′ic. *Syn.* LINIMENTUM ANTISPASMODICUM, L. CAJEPUTI COMPOSITUM, L. *Prep.* (*Hufeland.*) Oils of cajeput and mint, of each, 1 part; tincture of opium, 3 parts; compound camphor liniment, 24 parts. Anodyne, stimulant, and rubefacient.

Liniment, Arceus's. Compound elemi ointment.

Liniment of Arnica. *Syn.* ARNICA OPODELDOC; LINIMENTUM ARNICÆ, L. *Prep.* Dissolve by heat Castile soap, 4 parts, and camphor, 1 part, in rectified spirit, 10 parts. Add tincture of arnica, 5 parts.

Liniment of Belladon′na. *Syn.* LINIMENTUM BELLADONNÆ (B. P.), L. *Prep.* 1. (B. P.) Prepared the same as LINIMENTUM ACONITI. Prescribed with equal parts of soap liniment or compound camphor liniment, and is an excellent topical application for neuralgic pain.

2. Extract of belladonna, 1 dr.; oil of almonds, 2 oz.; lime water, 4 fl. oz. In eczema, and some other cutaneous affections, to allay irritation, &c.

Liniment of Belladonna and Chloroform (*Mr. Squire*). *Syn.* LINIMENTUM BELLADONNÆ ET CHLOROFORMI, L. *Prep.* Belladonna liniment, 7 fl. dr.; belladonna chloroform (made by percolating the root with chloroform), 1 fl. dr.; sprinkled on pilline and applied to the loins, excellent in lumbago.

Liniment of Borax (*Swediaur*). *Syn.* LINIMENTUM BORACIS, L. *Prep.* Borax, 2 dr.; tincture of myrrh, 1 oz.; distilled water, 1 oz.; honey of roses, 2 oz.; mix.

Liniment, Bow's, is a Scotch remedy for chest complaints. *Prep.* Opium, 1 oz.; hard soap, 1½ oz.; compound camphor liniment, 8 oz. Digest for several days and filter. Instead of this a mixture of opium liniment, 2 parts; and solution of ammonia, 1 part, is sometimes sold.

Liniment of Caj'eput Oil. *Syn.* LINIMENTUM OLEI CAJEPUTI, L. *Prep.* 1. (*Dr Copland.*) Compound camphor liniment and soap liniment, of each, 1½ fl. oz.; oil of cajeput, 1 fl. oz.

2. (*Dr Williams.*) Oil of cajeput, ½ fl. dr.; castor oil, 1 fl. dr.; olive oil, 4½ fl. dr. A warm, antispasmodic, diffusible stimulant and rubefacient; in spasmodic asthma, colic, chronic rheumatism, spasms, chest affections, &c. See ANTISPASMODIC L. (*above.*)

Linimentum Calcis. (B. P.) Solution of lime, 1 part; olive oil, 1 part; mix. The best liniment for burns and scalds.

Liniment of Cam'phor. *Syn.* CAMPHORATED OIL, CAMPHOR EMBROCATION; LINIMENTUM CAMPHORÆ (B. P., Ph. L. E. & D.), OLEUM CAMPHORATUM, L. *Prep.* 1. (B. P.) Camphor, 1 part; olive oil, 4 parts; dissolve.

2. (Ph. L. & E.) Camphor, 1 oz.; olive oil, 4 fl. oz.; gently heat the oil, add the camphor (cut small), and agitate until dissolved. The Dublin College orders only ½ the above camphor. Stimulant, anodyne, and resolvent; in sprains, bruises, rheumatic pains, glandular enlargements, &c.

3. (Compound: LINIMENTUM CAMPHORÆ COMPOSITUM, B. P., Ph. L. & D.) *a* (B. P.) Camphor, 5 parts; English oil of lavender, ½ part; strong solution of ammonia, 10 parts; rectified spirit, 30 parts. Dissolve the oil and camphor in the spirit, and gradually add the ammonia.

b. (Ph. L.) Camphor, 2½ oz.; oil of lavender, 1 fl. dr.; rectified spirit, 17 fl. oz.; dissolve, then add of stronger liquor of ammonia, 3 fl. oz., and shake them together until they are mixed.

c. (Ph. L. 1836.) Liquor of ammonia, 7½ fl. oz.; spirit of lavender, 1 pint; distil off 1 pint, and dissolve in it camphor, 2½ oz. The formula of the Ph. D. 1826 was nearly similar.

d. (Wholesale.) Camphor (clean), 21 oz.; English oil of lavender, 3½ oz.; liquor of ammonia, 2½ lbs.; rectified spirit, 7 pints; mix, close the vessel, and agitate occasionally, until the camphor is dissolved. Powerfully stimulant and rubefacient. It closely resembles, and is now almost universally sold for, Ward's 'Essence for the Headache.'

e. (Ethereal.) *Syn.* LINIMENTUM CAMPHORÆ ETHEREUM. *Prep.* Camphor, 1 dr.; ether, 1 dr.; oil of vipers, 2 dr.; mix.

Liniment of Canthar'ides. *Syn.* LINIMENT OF SPANISH FLIES; LINIMENTUM LYTTÆ, LIN. CANTHARIDIS (Ph. D. & U. S.), L. *Prep.* 1. (*Dr Collier.*) Tincture of cantharides and soap liniment, equal parts.

2. (Ph. D.) Cantharides (in fine powder), 3 oz.; olive oil, 12 fl. oz.; digest for 3 hours over a water-bath, and strain through flannel with expression.

3. (Ph. U. S.) Spanish flies, 1 oz.; oil of turpentine, 8 fl. oz.; proceed as last. The above are irritant and rubefacient, but should be used cautiously, lest they produce strangury.

Liniment of Capsicum. *Syn.* LINIMENTUM CAPSICI. *Prep.* 1. (*Dr Copland.*) Compound camphor liniment, 1 fl. oz.; volatile liniment, 1 fl. oz.; tincture of capsicum, 3 fl. oz.; mix.

2. (*Dr Turnbull.*) Capsicums, 1 oz.; rectified spirit, 3 fl. oz. Macerate 7 days, and strain for use.

Liniment of Chlo"ride of Lime. *Syn.* LINIMENTUM CALCIS CHLORINATÆ, L. *Prep.* 1. Chloride of lime, 1 dr.; water (added gradually), 3 fl. oz.; triturate together in a glass mortar for 10 minutes, pour off the liquid portion, and add of oil of almonds, 2 fl. oz.

2. (*Kopp.*) Solution of chloride of lime (ordinary), 1 part; olive oil, 2 parts.

3. (*Waller.*) Chloride of lime (in fine powder), 1 part; soft soap, 2 parts; soft water, q. s. to make a liniment.

Obs. The above are cleanly and excellent applications in itch, scald-head, herpes, lepra, foul ulcers, &c.

Liniment of Chlo"roform. *Syn.* LINIMENTUM CHLOROFORMI, B. P. *Prep.* 1. (B. P.) Chloroform, 1 part; liniment of camphor, 1 part; mix. The oil in the camphor liniment prevents the evaporation of the chloroform. Stimulating on application to a tender skin.

2. Chloroform, 1 fl. dr.; almond oil, 7 fl. dr.: mix in a phial, and agitate it until the two unite.

3. (*Tuson.*) Chloroform, 1 fl. dr.; soap liniment, 2 fl. oz.; as the last. Used as an application in neuralgic pains, rheumatism, &c.

4. (*Peter Boa.*) Camphor, 1 oz.; chloroform, 5 fl. oz.; soft paraffin, q. s. Dissolve the camphor in the chloroform, and add enough of the soft paraffin to make 10 fl. oz. By increasing or diminishing the quantity of soft paraffin, a liniment of the desired consistency can be obtained, and the product will not stain the clothes, as is the case with the chloroform liniment which is made with olive oil, and which is wanting in consistency.

Liniment of Cod-liver Oil. *Syn.* LINIMENTUM OLEI MORRHUÆ, L. O. JECORIS ASELLI, L. *Prep.* (*Dr Brach.*) Cod-liver oil, 2 fl. oz.; liquor of ammonia, 1 fl. oz.; mix. Resolvent, dispersive; applied to glandular tumours, scrofulous enlargements, &c.

Liniment of Colchicum (Ear Infirmary). *Syn.* LINIMENTUM COLCHICI. *Prep.* Soap liniment, 1 fl. oz.; wine of colchicum seed, ½ fl. oz.; mix.

Liniment of Colocynth (*Heim*). *Syn.* LINIMENTUM COLOCYNTHIDIS. *Prep.* Tincture of colocynth, ½ fl. oz.; castor oil, 1½ oz.

Liniment of Cro'ton Oil. *Syn.* LINIMENTUM CROTONIS (B. P., Ph. D.), L. OLEI CROTONIS, L. O. TIGLII, L. *Prep.* 1. (B. P.) Croton oil, 1 part; oil of cajeput, 3½ parts; rectified spirit, 3½ parts; mix.

2. (Ph. D.) Croton oil, 1 fl. oz.; oil of turpentine, 7 fl. oz.; mix by agitation.

3. (*J. Allen.*) Croton oil and liquor of potassa, of each, 1 fl. dr.; agitate until mixed, then add of rose-water, 2 fl. oz.

4. (*Pereira.*) Croton oil, 1 part; olive oil, 5 parts.

Obs. The above are used as counter-irritants; in rheumatism, neuralgia, bronchial and pulmonary affections, &c. When rubbed on the skin, redness and a pustular eruption ensue, and in general the bowels are acted on.

Liniment, Diuretic (*Dr Christison*). Soap liniment, tincture of foxglove, and tincture of squills, equal parts. In dropsies; rubbed over the abdomen or loins twice or thrice a day.

Liniment, Emol'lient. *Syn.* LINIMENTUM ALBUM, L. EMOLLIENS, L. *Prep.* From camphor, 1 dr.; Peruvian balsam, ¼ dr.; oil of almonds, 1 fl. oz.; dissolve by heat, add of glycerin, ½ fl. oz.; agitate well, and, when cold, further add of oil of nutmeg, 15 drops. Excellent for chapped hands, lips, nipples, &c.

Liniment of Gar'lic. *Syn.* LINIMENTUM ALLII, L. *Prep.* From juice of garlic, 2 parts; olive oil, 3 parts; mix. In hooping-cough, infantile convulsions, &c.

Liniment of Glycerin. (*Mr. Startin.*) *Syn.* LINIMENTUM GLYCERINI. *Prep.* Soap liniment, 3 oz.; glycerin, 1 oz.; extract of belladonna, 1 oz.; mix. For gouty, rheumatic, and neuralgic pains. A little veratrine is sometimes added.

Liniment, Green. (*Dr Campbell.*) *Syn.* LINIMENTUM VIRIDE. Camphor, 1 oz.; olive oil, 6 oz.; extract of hemlock, 1 oz.; spirit of ammonia, 2 oz.; mix.

Liniment, Hunga''rian. *Syn.* LINIMENTUM HUNGARICUM, L. *Prep.* (*Soubeiran.*) Powdered cantharides and sliced garlic, of each, 1 dr.; camphor, bruised mustard seed, and black pepper, of each 4 dr.; strong vinegar, 6 fl. oz.; rectified spirit, 12 fl. oz.; macerate a week, and filter. An excellent rubefacient and counter-irritant.

Liniment of Hydrochlo''ric Acid. *Syn.* LINIMENTUM MURIATICUM, L. ACIDI MURIATICI, L. A. HYDROCHLORICI, L. *Prep.* 1. (Hosp. F.) Olive oil, 2 oz.; white wax, 2 dr.; dissolve by a gentle heat, add of balsam of Peru, 1 dr.; hydrochloric acid, 2 dr.; mix well. An excellent application to chilblains before they break.

2. (*W. Cooley.*) Olive oil, ¼ pint; white spermaceti (pure) and camphor, of each, ½ oz.; mix with heat, add of hydrochloric acid, ½ fl. oz., and proceed as before. Equal to the last, and cheaper. This was extensively employed among the seamen of the Royal Navy by Mr Cooley with uniform success.

Liniment of Iodide of Potas'sium. *Syn.* LINIMENTUM IODURETUM GELATINOSUM, L.; GELÉE POUR LE GOÎTRE, Fr. *Prep.* (*Foy.*) Iodide of potassium, 4 dr.; proof spirit, 2 oz.; dissolve, and add the liquid to a solution of curd soap, 6 dr., in proof spirit, 2 oz., both being at the time gently warmed; lastly, aromatise with rose or neroli, pour it into wide-mouthed bottles, and keep them closely corked. In goitre, &c.

Liniment of Iodide of Potassium with Soap. *Prep.* (B. P.) Curd soap (cut small), 16 parts; iodide of potassium, 12 parts; glycerin, 8 parts; oil of lemon, 1 part; water, 80. Dissolve the soap in glycerin and water by aid of heat, stir in the iodide; when cold stir in the oil of lemon.

Liniment of Iodide of Sulphur. (*Prof. E. Wilson.*) *Syn.* LINIMENTUM SULPHURIS IODIDI. *Prep.* Iodide of sulphur, 30 gr.; olive oil, 1 fl. dr.; triturate together.

Liniment of I'odine. *Syn.* IODURETTED LINIMENT; LINIMENTUM IODI (B. P.), L. IODINII, L. IODURETUM, L. *Prep.* 1. (B. P.) Iodine, 5

parts; iodide of potassium, 2; camphor,1; rectified spirit, 40; dissolve.

2. (*Dr Copland.*) Soap liniment, 1 oz.; iodine, 8 to 10 gr.

3. (*Guibourt.*) Iodide of potassium, 1 dr.; water, 1 fl. dr.; dissolve, and add to it white soap (in shavings) and oil of almonds, of each, 10 dr., previously melted together. Some perfume may be added. In scrofula, glandular enlargements, rheumatism, &c.

Liniment of Labdanum. (*Quincy.*) *Syn.* LINIMENTUM LABDANI, L. CRINISCANI, L. *Prep.* Labdanum, 6 dr.; bear's grease, 2 oz.; honey, ½ oz.; powdered southernwood, 3 dr.; oil of nutmeg, 1 dr.; balsam of Peru, 2 dr.; mix. To restore the hair.

Liniment of Lead. *Syn.* LINIMENTUM PLUMBI, L. *Prep.* (*Gooey.*) Acetate of lead, 40 gr.; soft water, 12 fl. oz.; olive oil, 6 oz.; mix, and agitate well. Astringent and refrigerant. Useful in excoriations, especially when accompanied with inflammation.

Liniment of Lime. *Syn.* LINIMENT FOR BURNS, CARRON OIL; LINIMENTUM CALCIS (Ph. L. E. & D.), L. AQUÆ CALCIS, OLEUM LINI CUM CALCE, L. *Prep.* 1. From olive oil (linseed oil—Ph. E.) and lime-water, equal parts, shaken together until they are mixed. Very useful in burns and scalds.

2. (Compound : LINIMENTUM CALCIS COMPOSITUM, L.) *a.* (Camphorated—*W. Cooley.*) Camphor liniment and lime water, equal parts.

b. (Opiated—*W. Cooley.*) Lime-water and camphor liniment, of each, 1 oz.; extract of opium, 5 gr.; mix. Both are used as anodynes to allay pain and irritation in severe burns, chilblains, &c., for which purpose they are excellent. All the above liniments with lime-water should be used as soon as possible after being prepared, as the ingredients separate by keeping.

Liniment of Mercury. *Syn.* MERCURIAL LINIMENT; LINIMENTUM HYDRARGYRI (B. P., Ph. L.), L. H. COMPOSITUM (Ph. L. 1836), L. *Prep.* 1. (B. P.) Ointment of mercury, 1 part; solution of ammonia, 1 part; liniment of camphor, 1 part. Melt the ointment in the liniment, add the ammonia, and shake them together.

2. (Ph. L.) Camphor, 1 oz.; spirit of wine, 1 fl. dr.; sprinkle the latter on the former, powder, add of lard and mercurial ointment (stronger), of each, 4 oz.; rub them well together, then gradually add of liquor of ammonia, 4 fl. oz.; and mix well. Stimulant and discutient. It resembles mercurial ointment in its effects; but though milder in its operation, it more quickly produces salivation.

Liniment of Mor'phia. *Syn.* LINIMENTUM MORPHIÆ, L. *Prep.* (*W. Cooley.*) Pure morphia, 3 gr.; put it into a warm mortar, add very gradually of oil of almonds (warm), 1 fl. oz., and triturate until the morphia is dissolved; then add of camphor liniment, 1 oz. An excellent topical anodyne and antispasmodic, which often allays pain when other means have failed.

Liniment of Mus'tard. *Syn.* LINIMENTUM SINAPIS, L. *Prep.* 1. Flour of mustard (best), 1 oz.; water, tepid, 2 fl. oz.; mix, and add of glycerin, liquor of ammonia, and olive oil, of each, 1 fl. oz.

2. (*Béral.*) Carbonate of ammonia (in fine powder), 1 part; camphor (in powder), 2 parts; oil of lavender, 4 parts; tincture of mustard, 6 parts; mix, dissolve by agitation, add of simple liniment (warm), 56 parts, and again agitate until the whole is perfectly incorporated.

3. Black mustard seed (ground in pepper-mill or otherwise well bruised), ¼ lb.; oil of turpentine, 1 pint; digest, express the liquid, filter, and dissolve it in camphor, ¼ lb. Stimulant and rubefacient. A popular and useful remedy in rheumatic pains, lumbago, colic, chilblains, &c. The last is a close imitation of Whitehead's ' Essence of Mustard.'

4. (LIN. OLEI VOLATILIS SINAPIS.) *a.* From volatile oil of black mustard seed, ½ dr.; oil of almonds, 1 fl. oz. As a rubefacient.

b. From volatile oil, 1 part; alcohol (sp. gr. ·815), 1 to 2 parts. As a vesicant.

Liniment of Mustard (Compound). *Syn.* LINIMENTUM SINAPIS COMPOSITUM (B. P.), L. *Prep.* Oil of mustard, 1 dr.; ethereal extract of mesereon, 40 gr.; camphor, 2 dr.; castor oil, 5 dr.; rectified spirit, 32 dr.; dissolve.

Liniment, Narcotic. (P. Cod.) *Syn.* LINIMENT CALMANT; LINIMENTUM NARCOTICUM, L. *Prep.* Anodyne balsam, 8 parts; compound wine of opium, cold cream, of each, 1 part; mix.

Liniment of Ni'trate of Mercury. *Syn.* CITRINE LINIMENT; LINIMENTUM HYDRARGYRI NITRATIS, L. *Prep.* (*Sir H. Halford.*) Ointment of nitrate of mercury and olive oil, equal parts, triturated together in a glass mortar, or mixed by a gentle heat. This liniment is stimulant, discutient, and alterative, and in its general properties resembles the ointment of the same name. For most purposes the quantity of oil should be at least doubled.

Liniment of Oleate of Mercury (5%, 10%). *Prep.* Made by dissolving 5 or 10 parts of yellow oxide of mercury in sufficient oleic acid to make 100 parts. The combination takes a few days, and it is best to avoid heat.

Liniment of Oleate of Mercury with Morphine. *Prep.* Pure morphine, 10 gr.; oleic acid, 5 dr.; dissolve, and add oleate of mercury (10%), 5 dr.

Liniment of O'pium. *Syn.* ANODYNE LINIMENT; LINIMENTUM OPII (B. P., Ph. L. and E.), L. OPII or L. ANODYNUM (Ph. D.), L. SAPONIS CUM OPIO, L. *Prep.* 1. (B. P.) Tincture of opium, 1 part; liniment of soap, 1 part; mix and filter.

2. Tincture of opium, 2 fl. oz.; soap liniment, 6 6 fl. oz.; mix.

3. (Ph. E.) Castile soap, 6 oz.; opium, 1½ oz. rectified spirit, 1 quart; digest for 3 days, then filter, add of camphor, 3 oz.; oil of rosemary, 6 fl. dr., and agitate briskly.

4. (Ph. D.) Soap liniment and tincture of opium, equal parts.

5. (Wholesale.) Soft soap, 1¼ lbs.; powdered opium and camphor, of each, ¼ lb.; rectified spirit, 1 gall.; digest a week.

Obs. This preparation is an excellent anodyne in local pains, rheumatism, neuralgia, sprains, &c.

Liniment of Phos'phorus. *Syn.* LINIMENTUM PHOSPHORATUM, L. *Prep.* (*Augustin.*) Phosphorus, 6 gr.; camphor, 12 gr.; oil of almonds, 1

oz.; dissolve by heat; when cold, decant the clear portion, and add of strongest liquor of ammonia 10 drops. A useful friction in gout, chronic rheumatism, certain obstinate cutaneous affections, &c.

Liniment de Rosen. (P. C.) *Prep.* Oil of mace, 4 parts, oil of cloves, 4 parts; oil of juniper, 9 parts; mix.

Liniment, Sim'ple. *Syn.* LINIMENTUM SIMPLEX (Ph. E.), L. *Prep.* (Ph. E.) White wax, 1 oz.; olive oil, 4 fl. oz.; melt together, and stir the mixture until it is cold. Emollient; resembles spermaceti ointment in all except its consistence.

Liniment of Soap. *Syn.* OPODELDOC, CAMPHORATED TINCTURE OF SOAP, BALSAM OF S.; LINIMENTUM SAPONIS (B. P., Ph. L. E. and D.), L. SAPONACEUM, TINCTURA SAPONIS CAMPHORATA, BALSAMUM-SAPONIS, L. *Prep.* 1. (B. P.) Hard soap (cut small), 2 oz.; camphor, 1 oz.; English oil of rosemary, 3 dr.; rectified spirit, 16 oz.; distilled water, 4 oz.; mix the water and spirit, add the other ingredients, digest at a temperature not exceeding 70° F., agitating occasionally for 7 days, and filter.

2. (Ph. L.) Castile soap (cut small), 2½ oz.; camphor (small), 10 dr.; spirit of rosemary, 18 fl. oz.; water, 2 fl. oz.; digest with frequent agitation until the solid substances are dissolved.

3. (Ph. E.) Castile soap, 5 oz.; camphor, 2½ oz.; oil of rosemary, 6 fl. dr.; rectified spirit, 1 quart.

4. (Ph. D.) Castile soap (in powder), 2 oz.; camphor, 1 oz.; proof spirit, 16 fl. oz.

5. (LINIMENT SAVONNEAU, P. Cod.) Tincture of soap (P. Cod.) and rectified spirit (·868, or 41 o. p.), of each, 8 parts; olive oil, 1 part.

Obs. This article, prepared according to the directions of the Pharmacopœia, from ' soap made of olive oil and soda ' (Castile soap), is apt to gelatinise in cold weather, and to deposit crystals of stearate of soda. This may be avoided, when expense is not an objection, by first well drying the soap, employing a spirit of at least 85%, and keeping the preparation in well-closed bottles. A cheaper and better plan is to substitute the ' soft soap ' of the Ph. L. (' soap made with olive oil and potassa ') for the Castile soap ordered by the College. The soft soap of commerce imparts to the liniment an unpleasant smell. The following formula, one of those commonly adopted by the wholesale druggists, produces a very good article, though much weaker than that of the Pharmacopœia.

6. (Wholesale.) Camphor (cut small), 1¼ lbs.; soft soap, 7 lbs.; oil of rosemary, 3 fl. oz.; rectified spirit of wine and water, of each, 3½ galls.; digest with occasional agitation for a week, and filter. This is the ' opodeldoc ' or ' soap liniment ' of the shops.

Uses. Soap liniment is stimulant, discutient, and lubricating, and is a popular remedy in rheumatism, local pains, swellings, bruises, sprains, &c.

7. (With opium.) See LINIMENT OF OPIUM.

8. (Sulphuretted: LINIMENTUM SAPONIS SULPHURETUM, L. SULPHURO-SAPONACEUM, *Jadelot,*

L.) Sulphuret of potassium, 3 oz.; soap, 12 oz.; water, q. s.; melt together, and add of olive oil, 12 oz.; oil of origanum, 1 fl. dr.; mix well. An excellent remedy for the itch, and some allied skin diseases.

Liniment of Sul'phide of Carbon. *Syn.* LINIMENTUM CARBONIS SULPHURETI, L. *Prep.* 1. From bisulphide of carbon, 1 dr.; camphorated oil, 1 oz.; mix.

2. (*Lampadius.*) Camphor, 2 dr.; bisulphuret of carbon, 4 fl. dr.; dissolve, and add of rectified spirit, 1 fl. oz. In rheumatism, gouty nodes, &c.

Liniment of Sulphu'ric Acid. See LINIMENT, ACID.

Liniment, Tripharm'ic. *Syn.* LINIMENTUM TRIPHARMICUM (Ph. L. 1746), L. *Prep.* Take of lead plaster and olive oil, of each, 4 oz.; melt, add of strong vinegar, 1 fl. oz., and stir until cold. Cooling and desiccative; in excoriations, burns, &c.

Liniment of Tur'pentine. *Syn.* KENTISH'S LINIMENT; LINIMENTUM TEREBINTHINÆ (B. P., Ph. L. & D.), L. TEREBINTHINATUM (Ph. E.), L. *Prep.* 1. (B. P.) Oil of turpentine, 16 parts; camphor, 1 part; soft soap, 2 parts; water, 2 parts; dissolve the camphor in the turpentine, then add the soap to the water, and rub till thoroughly mixed.

2. (Ph. L.) Soft soap, 2 oz.; camphor, 1 oz.; oil of turpentine, 10 fl. oz.; shake them together until mixed. Stimulant; in lumbago, cholera, colic, &c.

3. (Ph. L. 1824.) Resin cerate, 6 oz.; oil of turpentine, 4 fl. oz.; mix. An excellent application to burns.

4. (Ph. E.) Resin ointment, 4 oz.; camphor, 4 dr.; dissolve by a gentle heat, and stir in oil of turpentine, 5 fl. oz.

5. (Ph. D.) Oil of turpentine, 5 fl. oz.; resin ointment, 8 oz.; mix by a gentle heat. This forms Dr Kentish's celebrated application to burns and scalds. The parts are first bathed with warm oil of turpentine or brandy, and then covered with pledgets of lint smeared with the liniment.

6. Compound. *a.* (B. P.) LINIMENTUM TEREBINTHINÆ ACETICUM. Oil of turpentine, 4 parts; glacial acetic acid, 1 part; liniment of camphor, 4 parts; mix.

b. LINIMENTUM TEREBINTHINÆ COMPOSITUM, L. (Acetic: ST JOHN LONG'S LINIMENT; LINIMENTUM TEREBINTHINÆ ACETICUM, L.) Oil of turpentine, 8 oz.; rose-water, 2½ fl. oz.; acetic acid, 5 dr.; oil of lemons, 1 dr.; yolk of egg, 1; make an emulsion. As a counter-irritant in phthisis.

c. (Ammoniated, *Debreyne.*) Lard, 3 oz.; melt, and add of oil of turpentine and olive oil, of each, 1 oz.; when cold, further add of camphorated spirit, 4 fl. dr.; liquor of ammonia, 1 fl. dr. In sciatica, lumbago, &c.

d. (Opiated, *Recamier.*) Oil of turpentine, 1 fl. oz.; oil of chamomile, 2 fl. oz.; tincture of opium, 1 fl. dr. In neuralgia, &c.

e. (Sulphuric, Ph. Castr. Ruthena.) Oil of turpentine, 2 oz.; olive oil, 5 oz.; mix, and add of dilute sulphuric acid, 1½ dr. See ACID LINIMENT.

Liniment of Vera'trine. *Syn.* LINIMENTUM VERATRIÆ, L. *Prep.* (*Brande.*) Veratrine, 8 gr.; alcohol, ½ fl. oz.; dissolve, and add of soap liniment, ½ fl. oz. In neuralgic and rheumatic pains, gout, &c.

Liniment of Ver'digris. *Syn.* OXYMEL OF VERDIGRIS; LINIMENTUM ÆRUGINIS (Ph. L.), OXYMEL ÆRUGINIS (Ph. L. 1788), OXYMEL CUPRI SUBACETATIS (Ph. D. 1826), L. *Prep.* (Ph. L.) Verdigris (in powder), 1 oz.; vinegar, 7 fl. oz.; dissolve, filter through linen, add of honey, 14 oz., and evaporate to a proper consistence.

Obs. This preparation is wrongly named a 'liniment.' The College, after 'beating about the bush' for nearly a century, found a right name for it in 1788; but, as in many other cases, soon abandoned it for another less appropriate.

Oxymel of verdigris is stimulant, detergent, and escharotic. It is applied to indolent ulcers, especially of the throat, by means of a camel-hair pencil; and, diluted with water, it is used as a gargle. Care must be taken to avoid swallowing it, as it occasions vomiting and excessive purging.

Liniment, Vesicating. (*Dr Montgomery.*) *Syn.* LINIMENTUM VESICANS, L. For children. *Prep.* Compound camphor liniment, 4 fl. dr.; oil of turpentine, 2 fl. dr. To produce immediate vesication in adults. Mix 1 part of the strongest liquor ammoniæ with 2 of olive oil, and apply 6 drops on spongio-piline for 10 minutes.

Liniment, Ware's. *Prep.* From camphor liniment, 1 oz.; solution of carbonate of potassa, 1 dr. In amaurosis.

Liniment, White. *Syn.* LINIMENTUM ALBUM, L. *Prep.* Rectified oil of turpentine, 2 oz.; solution of ammonia, 2 oz.; soap liniment, 3 oz.; spirit of rosemary, 1 oz. Mix in the above order, and gradually add with continual agitation, distilled vinegar, 8 oz. For chapped hands.

Liniment, White's. The old name for spermaceti ointment.

Liniment, Wilkinson's. *Prep.* (Phœbus.) Prepared chalk, 20 gr.; sulphur, lard, and tar, of each ½ oz.; mix, and add of Boyle's fuming liquor, 10 or 15 drops. In certain chronic skin diseases, neuralgia, &c.

LINOLEIC ACID. $C_{16}H_{28}O_2$. This may be obtained by saponifying linseed oil. It is a liquid acid, and rapidly oxidises when exposed to the air, becoming converted into oxylinoleic acid, which is incapable of solidification even at low temperatures.

LIN'SEED. *Syn.* FLAX SEED; LINI SEMINA, L. The seed of *Linum usitatissimum*, Linn., or common flax. (Ph. L.) Oily, emollient, demulcent, and nutritive. Ground to powder (LINSEED MEAL; FARINA LINI), it is used for poultices. The cake left after expressing the oil (LINSEED CAKE) contains, when of average quality, in 100 parts, moisture, 12·70; oil, 11·32; albuminoids, 28·21; mucilage, &c., 29·42; indigestible fibre, 12·46; ash, 5·89. It is used for feeding cattle. Under the form of tea or infusion it is used as a diluent, and to allay irritation in bronchial, urinary, and other like affections. See INFUSION OF LINSEED.

LINSEED CAKE. See LINSEED.

LINT. *Syn.* LINTEUM, L. White linen cloth, scraped by hand or machinery, so as to render it soft and woolly. The hand-made lint is now little used; it was prepared from pieces of old linen cloth. The machine-made lint is prepared from a fabric woven on purpose. A lint made from cotton (cotton-lint) is now largely manufactured; it is much inferior to the true lint, being a bad conductor of heat. Lint is used for dressing ulcers and wounds, either alone or smeared with some suitable ointment or cerate.

Lints, Medicated. A large number of these are made and sold as antiseptic dressings. The chief forms are—

Lint, Benzoic. Lint soaked in an alcoholic solution of benzoic acid and dried. Strength, 4%.

Lint, Boric. Lint soaked in a hot saturated solution of boric acid, and dried without wringing.

Lint, Carbolic. Lint evenly sprayed with pure carbolic acid. Strength, 5%.

Lint, Corrosive Sublimate. Lint soaked in a watery solution of perchloride of mercury and dried. To contain 1%.

Lint, Iodoform. Dissolve iodoform in ether, saturate the lint with the solution, then dry. Strength, 10% iodoform.

Lint, Salicylic. Lint saturated with an alcoholic solution of salicylic acid and dried. Strength, 4% and 10% acid.

Lint, Thymol. Lint saturated with an alcoholic solution of thymol and dried. Strength, 5% thymol.

LIP SALVE. See SALVE.

LIQUATION. The process of sweating out by heat the more fusible metals of an alloy. Metallurgists avail themselves of this method in assaying and refining the precious metals, and procuring antimony and some other metals from their ores.

LIQUEFACIENTS. *Syn.* RESOLVENTS; LIQUEFACIENTIA, RESOLVENTIA, L. In *pharmacy*, substances or agents which promote secretion and exhalation, soften and loosen textures, and promote the absorption or removal of enlargements, indurations, &c. To this class belong the alkalies, antimony, bromine, chlorine, iodine, mercury, sulphur, &c., and their preparations.

LIQUEFACTION. The assumption of the liquid form. It is usually applied to the conversion of a solid into the liquid state, which may arise from increase of temperature (fusion), absorption of water from the atmosphere (deliquescence), or the action of a body already fluid (solution).

Liquefaction of Gases. Under the combined influence of pressure and cold, all the gases may be liquefied, and some even solidified. The first satisfactory experiments in this direction were made by Faraday, who succeeded in reducing to the liquid condition 8 bodies which had hitherto been regarded as permanent gases, namely, ammonia, carbonic anhydride, chlorine, cyanogen, hydrochloric acid, nitrous oxide, sulphuretted hydrogen, and sulphurous anhydride. His method of proceeding was very simple:—The materials were sealed up in a strong, narrow glass tube, bent so as to form an obtuse angle, together with a little 'pressure gauge,' consisting of a slender tube closed at one end, and having within it, near the open extremity, a globule of mercury. The gas, being disengaged by the application of heat or otherwise, accumulated in the tube, and by its own pressure brought about liquefaction. The force required for this purpose was judged of by the diminution of volume of the air in the pressure gauge. By employing powerful condensing syringes, and an extremely low temperature, Faraday subsequently succeeded in liquefying olefiant gas, hydriodic and hydrobromic acids, phosphuretted hydrogen, and the gaseous fluorides of silicon and boron. He failed, however, with oxygen, hydrogen, nitrogen, nitric oxide, carbonic oxide, and coal-gas, all of which refused to liquefy at the temperature of −166° F., while subjected to pressures varying in different cases from 27 to 58 atmospheres.

Toward the end of 1877 these hitherto refractory gases were reduced to the liquid, and, in the case of hydrogen, to the solid state. These results have been accomplished by subjecting the gases to a pressure considerably greater than that employed by Faraday, combined with the expedient of the sudden removal of this pressure, whereby the escaping gas (previously enormously reduced in temperature) in the act of expansion robs the remainder of so much of its heat as to leave it in the fluid condition.

The liquefaction of oxygen was accomplished independently by M. Cailletet, of Paris, and M. Pictet, of Geneva; the French chemist having effected it on December 2nd, 1877, and the Swiss one on the 22nd of the same month.

Simultaneously with Cailletet's announcement of the liquefaction of oxygen, that of carbonic oxide was made by the same chemist, who, about 3 weeks after, at a meeting in the Paris Academy of Sciences, stated that he had also reduced hydrogen, nitrogen, and atmospheric air to the fluid state.

In the previous November he had been equally successful in converting gaseous nitric oxide into a liquid.

M. Cailletet, in a communication to the Paris Academy of Sciences, read by M. Dumas at a meeting of that body on 24th December, 1877, thus describes the process by which he liquefied the gases oxygen and carbonic oxide:

"If oxygen or pure carbonic oxide be enclosed in a tube such as I have before described, and placed in an apparatus for compression like that which has already been worked before the Academy, and the gas be then lowered in temperature to − 29° C. by means of sulphurous acid, and at a pressure of about 300 atmospheres, the two gases preserve their gaseous state. (This apparatus, which consists of a hollow steel cylinder, to which is attached a strong glass tube, is described in the 'Comptes Rendus,' tome 85, p. 851. The gas is forced into it by means of an hydraulic pump with the intervention of a cushion of mercury.)

"But if they are allowed to suddenly expand, this expansion, according to the formula of Poisson, reducing them to a temperature at least 200° C. below their initial temperature, causes them immediately to assume the appearance of an intense fog, which is caused by the liquefaction and perhaps by the solidification of the oxygen or carbonic acid.

"The same phenomenon is also observed upon the expansion of carbonic acid, and of protoxide and binoxide of nitrogen, when under strong pressure.

" This fog is produced with oxygen, even when the gas is at the ordinary pressure, provided time is allowed for it to part with the heat it acquires in the mere act of compression.

" This I demonstrated by experiments performed on Sunday, the 16th December, at the Chemical Laboratory of the École Normale Supérieure, before a certain number of *savants* and professors, amongst whom were some members of the Academy of Sciences. I had hoped to find in Paris, together with the materials necessary for the production of a high degree of cold (protoxide of nitrogen or liquid carbonic acid), a pump capable of supplying the place of my compression apparatus at Châtillon-sur-Seine. Unfortunately a pump well fixed and suited to this sort of experiment could not be found in Paris, and I was obliged to send to Châtillon-sur-Seine for the refrigerating substances for collecting the condensed matters on the walls of the tube.

" To know whether oxygen and carbonic oxide are in a liquid or a solid state in the fog would necessitate an optical experiment more easy to imagine than to accomplish, because of the form and the thickness of the tubes containing them. Furthermore, chemical reactions will assure me that the oxygen is not transformed into ozone in the act of compression. I shall reserve the study of all these questions till the apparatus I am now having made is complete.

" Under the same conditions of temperature and pressure, even the most rapid expansion of pure hydrogen gives no trace of nebulous matter. There remains for me only nitrogen to study, the small solubility of which in water induces me to believe that it will prove very refractory to all change of condition " (' Comptes Rendus,' tome 5, p. 1213).

M. Pictet's process for liquefying oxygen, although differing in the method of working, is similar in principle to that of M. Cailletet. His paper, which was read at the same sitting of the Academy as M. Cailletet's, thus describes it :

" A and B, in the accompanying figure, are two double suction and force pumps, coupled together on the compound system, one causing a vacuum in the other in such a manner as to obtain the greatest possible difference between the pressures of suction and forcing."

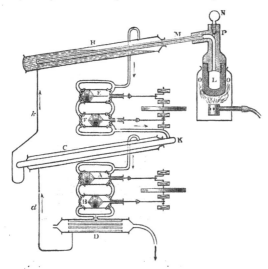

The pumps act on anhydrous sulphurous acid contained in the cylindrical receiver C. The pressure in this receiver is such that the sulphurous acid is evaporated from it at a temperature of 65° C. below zero.

The sulphurous acid is forced by the pumps into a condenser, D, cooled by a current of cold water; here it liquefies at the temperature of 25° above zero, and at a pressure of about 2½ atmospheres.

The sulphurous acid returns to the receiver C as it liquefies by the little tube d.

E and F are two pumps resembling the preceding, and coupled in the same manner. They act upon carbonic acid contained in a cylindrical receiver H.

The temperature in this latter receiver is such that the carbonic acid evaporates from it at a temperature of 140° C. below zero.

The carbonic acid forced on by the pumps is

driven into the condenser x, enclosed in the sulphurous acid receiver c, which has a temperature of 65° below zero; the carbonic acid here becomes liquefied at a pressure of 5 atmospheres.

The carbonic acid, in proportion as it liquefies, returns to the receptacle H by the small tube *k*.

L is a retort of wrought iron, sufficiently thick to resist a pressure of 500 atmospheres. It contains chlorate of potassium, and is heated in such a manner as to give off pure oxygen. It communicates by a tubulure with an inclined tube, M, made of very thick glass, one metre in length, which is enveloped by the receiver, H, containing carbonic acid at the temperature of 140° below zero.

A tap, N, situated upon the tubulure of the retort, permits of the opening of an orifice, P, which leads into the surrounding air.

After the four pumps have been worked for several hours by means of a steam-engine of 15-horse power, and when all the oxygen has been disengaged, the pressure in the glass tube is 320 atmospheres, and the temperature at 140° below zero.

Upon suddenly opening the orifice P the oxygen escapes with violence, producing, in doing so, so considerable an expansion and absorption of heat as to cause a liquefied portion to appear in the glass tube, and to spirt out from the orifice when the apparatus is sloped.

It ought to be stated that the quantity of liquefied oxygen contained in the tube 1 metre long, and 0·01 m. in internal diameter, occupied about a third of its length, and issued from the orifice, P, in the form of a liquid jet.

In a communication to M. Dumas, received two days after the above sitting, M. Pictet described his experiments more fully, prefacing the account by the following very interesting remarks:—

"The end to which I have been tending for the last 3 years has been to seek to demonstrate experimentally that molecular cohesion is a general property of bodies without exception.

"If the permanent gases cannot be liquefied, it must be concluded that their constituent particles do not attract each other, and are therefore independent of this law.

"To succeed experimentally in bringing the molecules of a gas into the closest possible proximity, and thus to obtain its liquefaction, certain indispensable conditions are necessary, which I thus sum up:

"1. To have a gas that must be perfectly pure and without a trace of foreign gas.

"2. To have at one's disposal very powerful means of compression.

"3. To obtain an intense degree of cold, and the abstraction of heat at these low temperatures.

"4. To have a large surface of condensation maintained at these low temperatures.

"5. To have the power of utilising the expansion of the gas under considerable pressure to the atmospheric pressure, which expansion added to the preceding means compels liquefaction.

"With these five conditions fulfilled we may formulate the following problem.

"When a gas is compressed at 500 or 600 atmospheres, and kept at a temperature of —100°

or 140°, and then let expand to the pressure of the atmosphere, one of two things must occur. Either the gas, obeying the action of cohesion, liquefies and yields its heat of condensation to the portion of the gas which expands and is lost in the gaseous form; or, under the hypothesis that cohesion is not a natural law, the gas passes beyond absolute zero—that is to say, it becomes inert, a dust without consistence. The work of expansion would be impossible, and the loss of heat absolute."

Spite of M. Cailletet's supposition that nitrogen would prove a very incoercible gas, his experiments showed the contrary, since he found that it easily condensed under a pressure of about 200 atmospheres and at a temperature of — 13° C., the conditions as to its sudden expansion being observed.

Hydrogen, the lightest of all the gases, which M. Cailletet could only procure in the form of mist, was unmistakably liquefied by M. Pictet within less than a fortnight afterwards, under a pressure of 650 atmospheres and 140° of cold.

The tap which confined the gas at this pressure being opened, a jet of a steel-blue colour escaped from the orifice, accompanied by a hissing sound, like that given off when a red-hot iron is dipped into cold water. The jet suddenly became intermittent, and a shower of solid particles of the hydrogen fell to the ground with a crackling noise. The hydrogen was obtained by the decomposition of formiate of potash by caustic potash, the gas thus yielded being absolutely pure.

Cailletet states that he succeeded perfectly in liquefying atmospheric air, previously deprived of moisture and carbonic acid, but he omits to mention the pressure and reduction of temperature to which the air was subjected. He liquefied nitric oxide at the pressure of 104 atmospheres and at a temperature of —11° C.

Carbonic anhydride is liquefied on the large scale by condensing it in strong vessels of gunmetal or boiler-plate. Thilorier was the first to procure it in a solid condition. It requires a pressure of between 27 and 28 atmospheres at 32° F. (*Adams*). The liquefied acid is colourless and limpid, lighter than water, and four times more expansible than air; it mixes in all proportions with ether, alcohol, naphtha, oil of turpentine, and sulphide of carbon, and is insoluble in water and fat oils. When a jet of liquid carbonic anhydride is allowed to issue into the air from a narrow aperture, such an intense degree of cold is produced by the evaporation of a part, that the remainder freezes to a solid (solid carbonic anhydride), and falls in a shower of snow. This substance, which may be collected, affords a means of producing extreme cold. Mixed with a little ether, and poured upon a mass of mercury, the latter is almost instantly frozen. The temperature of this mixture in the air was found to be —106° F.; when the same mixture was placed beneath the receiver of an air-pump, and exhaustion rapidly performed, the temperature sank to —166°. This degree of cold was employed in Faraday's last experiments on the liquefaction of gases.

LIQUEUR. [Fr.] *Syn.* CORDIAL. A stimulating beverage, formed of weak spirit, aromatised and sweetened. The manufacture of liqueurs constitutes the trade of the 'compounder,' 'rectifier,' or 'liquoriste.'

The materials employed in the preparation of liqueurs or cordials are rain or distilled water, white sugar, clean flavourless spirit, and flavouring ingredients. To these may be added the substances employed as 'finings' when artificial clarification is had recourse to.

The utensils and apparatus required in the business are those ordinarily found in the wine and spirit cellar; together with a copper still, furnished with a pewter head and a pewter worm or condenser when the method 'by distillation' is pursued. A barrel, hogshead, or rum puncheon, sawn in two, or simply 'unheaded,' as the case may demand, forms an excellent vessel for the solution of the sugar; and 2 or 3 fluted funnels, with some good white flannel, will occasionally be found useful for filtering the aromatic essences used for flavouring. Great care is taken to ensure the whole of the utensils, &c., being perfectly clean and 'sweet,' and well 'seasoned,' in order that they may neither stain nor flavour the substances placed in contact with them.

In the preparation or compounding of liqueurs, one of the first objects which engages the operator's attention is the production of an alcoholic solution of the aromatic principles which are to give them their peculiar aroma and flavour. This is done either by simple solution or maceration, as in the manufacture of tinctures and medicated spirits, or by maceration and subsequent distillation. The products, in this country, are called ESSENCES or SPIRITS, and by the French INFUSIONS, and are added to the solution of sugar (SYRUP or CAPILLAIRE) or to the dulcified spirit, in the proportions required. Grain or molasses spirit is the kind usually employed for this purpose in England. As before observed, it should be of the best quality; as, if this is not the case, the raw flavour of the spirit is perceptible in the liquor. Rectified spirit of wine is generally very free from flavour, and when reduced to a proper strength with clear soft water forms a spirit admirably adapted for the preparation of cordial liquors. Spirit weaker than about 45 o. p., which has been freed from its own essential oil by careful rectification, is known in trade under the title of 'pure,' 'flavourless,' 'plain,' or 'silent spirit.' Before macerating the ingredients, if they possess the solid form, they are coarsely pounded, bruised, sliced, or ground, as the peculiar character of the substance may indicate. This is not done until shortly before submitting them to the action of the menstruum; as, after they are bruised, they rapidly lose their aromatic properties by exposure to the air. When it is intended to keep them for any time in the divided state they should be preserved in well-corked bottles or jars. The practice of drying the ingredients before pounding them, frequently adopted by ignorant and lazy workmen for the sake of lessening the labour, is, of course, even more destructive to their most valuable qualities than mere exposure to the air. The length of time the ingredients should be digested in the spirit should never be less than 5 or 6 days, but a longer period is preferable when distillation is not employed. In either case the

time may be advantageously extended to 10 days or a fortnight, and frequent agitation should be had recourse to during the whole period. When essential oils are employed to convey the flavour, they are first dissolved in a little of the strongest rectified spirit of wine, in the manner explained under ESSENCE; and when added to the spirit, they are mixed up with the whole mass as rapidly and as perfectly as possible. In managing the still the fire is proportioned to the ponderosity of the oil or flavouring substance, and the receiver is changed before the faints come over, as these are unfitted to be mixed with the cordial. In many cases the addition of a few pounds of common salt to the liquor in the still facilitates the process and improves the product. Ingredients which are not volatile are, of course, always added after distillation. The stronger spirit is reduced to the desired strength by means of either clear soft water or the clarified syrup used for sweetening. The sugar employed should be of the finest quality, and is preferably made into capillaire or syrup before adding it to the aromatised spirit, and not until this last has been rendered perfectly 'fine' or transparent, by infiltration or clarification, as the case may demand. Some spirits or infusions, as those of aniseed, caraway, &c., more particularly require this treatment, which is best performed by running them through a clean wine bag, made of rather fine cloth, having previously mixed them with a spoonful or two of magnesia; but in all cases clarification by simple repose should be preferred. Under proper management, liqueurs or cordials prepared of good materials will be found perfectly 'clear' or 'bright' as soon as made, or will become so after being allowed a few days for defecation; but in the hands of the inexperienced operator, and when the spirit employed is insufficient in strength or quantity, it often happens that they turn out 'foul' or 'milky.' When this is the case, the liquid may be 'fined down' with the whites of 12 to 20 eggs per hogshead; or a little alum, either alone or followed by a little carbonate of sodium or potassium, both dissolved in water, may be added, in the manner described under FININGS.

An excellent and easy way of manufacturing cordial liquors, especially when it is inconvenient to keep a large stock on hand, is by simply 'aromatising' and 'colouring,' as circumstances or business may demand, spirit 60 or 64 u. p., kept ready sweetened for the purpose. To do this to the best advantage, two descriptions of sweetened spirit should be provided, containing respectively 1 lb. and 3 lbs. of sugar to the gallon. From these spirit of any intermediate sweetness may be made, which may be flavoured with any essential oil dissolved in alcohol, or any aromatised spirit or 'infusion' (see below), prepared either by digestion or distillation. As a general rule, the concentrated essences, made by dissolving 1 oz. of the essential oil in 1 pint of the strongest rectified spirit of wine, will be found admirably adapted for this purpose. These essences, which should be kept in well-corked bottles, are employed by dropping them cautiously into the sweetened spirit until the desired flavour is produced. During this operation the liquor should be frequently and violently shaken to produce complete admixture. If by any accident the essence is added in too large a quantity, the resulting 'milkiness' or excess of flavour may be removed by the addition of a little more spirit, or by clarification. In this way the majority of the liqueurs in common use may be produced extemporaneously, of nearly equal quality to those prepared by distillation. For those which are coloured, simple digestion of the ingredients is almost universally adopted. The 'process by distillation' should, however, be always employed to impart the flavour and aroma of volatile aromatics to the spirit, when expense, labour, and time are of less importance than the production of a superior article.

The French liqueuristes are famed for the preparation of cordials of superior quality, cream-like smoothness, and delicate flavour. Their success chiefly arises from the employment of very pure spirit and sugar (the former in a larger proportion than that adopted by the English compounder), and in the judicious application of the flavouring ingredients. They distinguish their cordials as 'eaux' and 'extraits' (waters, extracts), or liqueurs which, though sweetened, are entirely devoid of viscidity; and 'baumes,' 'crèmes,' and 'huiles' (balms, creams, oils), which contain sufficient sugar to impart to them a syrupy consistence. The greatest possible attention is given to the preparation of the aromatised or flavouring essences, in France called 'infusions.' These are generally made by macerating the aromatic ingredients in spirit at about 2 to 4 u. p. (sp. gr. ·922 to ·925), placed in well-corked glass carboys or stoneware jars or bottles. The maceration is continued, with occasional agitation, for 3, 4, or even 5 weeks, when the aromatised spirit is either distilled or filtered, generally the former. The outer peel of cedrats, lemons, oranges, limettes, bergamottes, &c., is alone used by our Continental neighbours, and is obtained either by carefully peeling the fruit with a knife, or by 'oleo-saccharûm,' by rubbing it off with a lump of hard white sugar. Aromatic seeds and woods are bruised by pounding before being submitted to infusion. The substances employed in France to colour liqueurs are—for blue, soluble Prussian blue, sulphate of indigo (nearly neutralised with chalk), and the juice of blue flowers and berries;—amber, fawn, and brandy colour, burnt sugar or spirit colouring;—green, spinach or parsley leaves (digested in spirit), and mixtures of blue and yellow;—red, powdered cochineal or brazil-wood, either alone or mixed with a little alum;—violet, blue violet petals, litmus, or extract of logwood;—purple, the same as violet, only deeper;—yellow, an aqueous infusion of safflower or French berries, and the tinctures of saffron and turmeric.

A frequent cause of failure in the manufacture of liqueurs and cordials is the addition of too much flavouring matter. Persons unaccustomed to the use of strong aromatic essences and essential oils seldom sufficiently estimate their power, and, consequently, are very apt to add too much of them, by which the liqueur is rendered not only disagreeably high-flavoured, but, from the excess of oil present, also 'milky' or 'foul,' either at once, or what is nearly as bad, on the addition of water. This source of annoyance, arising entirely from bad manipulation, frequently discourages the tyro, and cuts short his career as a manufacturer. From the viscidity of cordials they are less readily 'fined down' than unsweetened liquor, and often give much trouble to clumsy and inexperienced operators. The most certain way to prevent disappointment in this respect is to use too little rather than too much flavouring; for if the quantity proves insufficient, it is readily 'brought up' at any time, but the contrary is not effected without some trouble and delay.

A careful attention to the previous remarks will render this branch of the rectifier's art far more perfect and easy of performance than it is at present, and will, in most cases, produce at once a satisfactory article, 'fine, sweet, and pleasant.'

The cordials of respectable British 'compounders' contain fully 3 lbs. of white lump sugar per gallon, and are of the strength of 60 to 64 u. p. The baumes, crèmes, and huiles imported from the Continent are richer both in spirit and sugar than ours, and to this may be referred much of their superiority. Mere sweetened or cordialised spirits (eaux of the Fr.) contain only from 1 to 1½ lbs. of sugar per gallon.

The purity of liqueurs is determined in the manner noticed under BRANDY, WINE, &c.

The following list embraces nearly all the cordials and liqueurs, both native and imported, met with in trade in this country:

Absinthe. *Syn.* EXTRAIT D'ABSINTHE DE SUISSE; SWISS EXTRACT OF WORMWOOD. *Prep.* From the tops of *Absinthum majus*, 4 lbs.; tops of *Absinthum minus*, 2 lbs.; angelica root, *Calamus aromaticus*, Chinese aniseed, and leaves of dittany of Crete, of each, 15 gr.; brandy or spirit at 12 u. p., 4 galls.; macerate for 10 days, then add water, 1 gall.; distil 4 galls. by a gentle heat, and dissolve in the distilled spirit, of crushed white sugar, 3 lbs. Tonic and stomachic.

Alker'mes. This liqueur is highly esteemed in some parts of the south of Europe.

Prep. 1. Bay leaves and mace, of each, 1 lb.; nutmegs and cinnamon, of each, 2 oz.; cloves, 1 oz. (all bruised); cognac brandy, 3½ galls.; macerate for 3 weeks, frequently shaking, then distil over 3 galls., and add of clarified spirit of kermes, 18 lbs.; orange-flower water, 1 pint; mix well, and bottle. This is the original formula for the 'alkermes de Santa Maria Novella,' which is much valued.

2. Spice, as last; British brandy, 4 galls.; water, 1 gall.; macerate as before, and draw over 4 galls., to which add, of capillaire, 2 galls., and sweet spirit of nitre, ¼ pint. Cassia is often used for cinnamon. Inferior to the last.

An'iseed Cordial. *Prep.* 1. From aniseed, 2 oz. (or essential oil, 1½ dr.), and sugar, 3 lbs. per gall. It should not be weaker than about 45 u. p. as at lower strengths it is impossible to produce a full-flavoured article without its being milky, or liable to become so.

2. (ANISETTE DE BORDEAUX.) *a.* (Foreign.) Aniseed, 4 oz.; coriander and sweet fennel seeds, of each, 1 oz. (bruised); rectified spirit, ¼ gall.; water, 3 quarts; macerate for 5 or 6 days, then draw over 7 pints, and add of lump sugar, 2¼ lbs.

b. (English.) Oil of aniseed, 15 drops; oils of cassia and caraway, of each, 6 drops; rub them with a little sugar, and then dissolve it in spirit (45 u. p.), 3 quarts, by well shaking them together; filter, if necessary, and dissolve in the clear liquor, sugar, 1½ lbs. See PEPPERMINT (*below*).

Balm of Molucca. *Prep.* From mace, 1 dr.; cloves, ½ oz.; clean spirit (22 u. p.), 1 gall.; infuse for a week in a well-corked carboy or jar, frequently shaking, colour with burnt sugar, q. s., and to the clear tincture add of lump sugar, 4½ lbs.; dissolved in pure soft water, ½ gall. On the Continent this takes the place of the 'cloves' of the English retailer.

Bit'ters. These have generally from 1 to 1½ lbs. of sugar per gallon. See BITTERS.

Car'away Cordial. *Prep.* Generally from the essential oil, with only 2½ lbs. of sugar per gall. 1 fl. dr. of the oil is commonly reckoned equal to ½ lb. of the seed. The addition of a very little oil of cassia, and about half as much of essence of lemon or of orange, improves it. See BRANDY, CARAWAY.

Ce'drat Cordial. *Prep.* From essence (oil) of cedrat (½ oz.); pure spirit (at proof), 1 gall.; dissolve, add of water, 3 pints, agitate well; distil 3 quarts, and add an equal measure of clarified syrup. A delicious liqueur. See CRÈME and EAU (*below*).

Cin'namon Cordial. *Prep.* This is seldom made with cinnamon, owing to its high price, but with either the essential oil or bark of cassia, with about 2 lbs. of sugar to the gall. It is preferred coloured, and therefore may be very well prepared by simple digestion. The addition of 5 or 6 drops each of essence of lemon and orange peel, with about a spoonful of essence of cardamoms per gall., improves it. 1 oz. of oil of cassia is considered equal to 8 lbs. of the buds or bark. 1 fl. dr. of the oil is enough for 2½ galls. It is coloured with burnt sugar.

Cit'ron Cordial. *Prep.* From the oil or peel, with 3 lbs. of sugar per gall., as above (see *below*).

Citronelle. *Syn.* EAU DE BARBADES. *Prep.* 1. From fresh orange peel, 2 oz.; fresh lemon peel, 4 oz.; cloves, ½ dr.; corianders and cinnamon, of each, 1 dr.; proof spirit, 4 pints; digest for 10 days; then add of water, 1 quart, and distil ½ gall.; to the distilled essence add of white sugar, 2 lbs., dissolved in water, 1 quart.

2. Essence of orange, ½ dr.; essence of lemon, 1 dr.; oil of cloves and cassia, of each, 10 drops; oil of coriander, 20 drops; spirit (58 o. p.), 5 pints; agitate until dissolved, then add of distilled or clear soft water, 3 pints; well mix, and filter it through blotting-paper if necessary; lastly add of sugar (dissolved), q. s.

Clairet. *Syn.* ROSSALIS DES SIX GRAINES. *Prep.* From aniseed, fennel seed, coriander seed, caraway seed, dill seed, and seeds of the candy-carrot (*Athamantia cretensis*, Linn.), of each (bruised), 1 oz.; proof spirit, ½ gall.; digest for a week, strain, and add of loaf sugar, 1 lb., dissolved in water, q. s.

Cloves. *Syn.* CLOVE CORDIAL. *Prep.* From bruised cloves, 1 oz., or essential oil, 1 fl. dr., to every 3 galls. of proof spirit. If distilled, some common salt should be added, and it should be drawn over with a pretty quick fire. It requires fully 3 lbs. of sugar per gall., and is generally coloured with poppy flowers or burnt sugar. The addition of 1 dr. of bruised pimento, or 5 drops of the oil for every oz. of cloves improves this cordial. See BALM OF MOLUCCA (*above*).

Coriander Cordial. *Prep.* From corianders, as the last. A few sliced oranges improve it.

Crème d'Anis. As ANISEED CORDIAL, only richer.

Crème des Barbades. As CITRONELLE, adding some of the juice of the oranges, and an additional lb. of sugar per gall.

Crème de Cacao. *Prep.* Infuse roasted caracca-cacao nuts (cut small), 1 lb., and vanilla, ½ oz., in brandy, 1 gall., for 8 days; strain, and add of thick syrup, 3 quarts.

Crème de Cedrat. *Syn.* HUILE DE CEDRAT. *Prep.* From spirit of citron, 1 pint; spirit of cedrat, 1 quart; proof spirit, 3 quarts; white sugar, 16 lbs., dissolved in pure soft water, 2 galls.

Crème de Macarons. *Prep.* 1. From cloves, cinnamon, and mace, of each (bruised), 1 dr.; bitter almonds (blanched and beaten to a paste), 7 oz.; spirit (17 u. p.), 1 gall.; digest a week, filter, and add of white sugar, 6 lbs.; dissolve in pure water, 2 quarts.

2. Clean spirit (at 24 u. p., sp. gr. ·945), 2 galls.; bitter almonds, ¾ lb.; cloves, cinnamon, and mace, of each (in coarse powder), 1½ dr.; infuse for 10 days, filter, and add of white sugar, 8 lbs., dissolved in pure water, 1 gall.; lastly, give the liqueur a violet tint with infusion or tincture of litmus and cochineal. An agreeable, nutty-flavoured cordial, but, from containing so much bitter almonds, should be only drunk in small quantities at a time. The English use only ½ the above quantity of almonds.

Crème de Naphe. *Prep.* From sweetened spirit (60 u. p.) containing 3½ lbs. of sugar per gall., 7 quarts; orange-flower water (foreign), 1 quart. Delicious.

Crème de Noyeau. See NOYEAU.

Crème d'Orange. *Prep.* From oranges (sliced), 3 dozen; rectified spirit, 2 galls.; digest for 14 days; add, of lump sugar, 28 lbs. (previously dissolved in water, 4½ galls.); tincture of saffron, 1½ fl. oz.; and orange-flower water, 2 quarts.

Crème de Portugal. Flavoured with lemon, to which a little oil of bitter almonds is added.

Curaçao. *Prep.* From sweetened spirit (at 56 u. p.), containing 3½ lbs. of sugar per gall., flavoured with a tincture made by digesting the 'oleo-saccharum' prepared from Seville oranges, 9 in number; cinnamon, 1 dr.; and mace, ½ dr., in rectified spirit, 1 pint. It is coloured by digesting in it for a week or 10 days brazil-wood (in powder), 1 oz., and afterwards mellowing the colour with burnt sugar, q. s.

Delight of the Mandarins. From spirit (22 u. p.), 1 gall.; pure soft water, ½ gall.; white sugar (crushed small), 4½ lbs.; Chinese aniseed and ambrette or musk seed, of each (bruised), ½ oz.; safflower, ½ oz.; digested together in a carboy or stone bottle capable of holding double, and agitated well every day for a fortnight.

Eau de Cedrat. *Syn.* CEDRAT WATER. As CRÈME DE CEDRAT, but using less sugar.

Eau de Chasseurs. See PEPPERMINT (*below*).

Eau de Vie d'Andaye. *Syn.* EAU DE VIE D'ANIS; ANISEED LIQUEUR BRANDY. *Prep.* From brandy or proof spirit, 1 gall.; sugar, ¾ lb.; dissolved in aniseed water, 1 pint.

Gold Cordial. *Prep.* From angelica root (sliced), 1 lb.; raisins, ½ lb.; coriander seeds, 2 oz.; caraway seeds and cassia, of each, 1½ oz.; cloves, ½ oz.; figs and sliced liquorice root, of each, 4 oz.; proof spirit, 3 galls.; water, 1 gall.; digest 2 days, and distil 3 galls. by a gentle heat; to this add, of sugar, 9 lbs., dissolved in rose-water and clean soft water, of each, 1 quart; lastly, colour the liquid by steeping in it of hay saffron, 1¼ oz. This cordial was once held in much esteem. It derives its name from a small quantity of gold leaf being formerly added to it.

Huile d'Anis. See CRÈME D'ANIS (*above*).

Huile de Vanille. Flavoured with essence or tincture of vanilla. It is kept in a decanter, and used to flavour liqueurs, grog, &c.

Huile de Vénus. *Prep.* From the flowers of the wild carrot, 2½ oz., and sugar, 3 lbs. to the gall. It is generally coloured by infusing a little powdered cochineal in it.

Jargonelle. *Syn.* JARGONELLE CORDIAL. Flavoured with essence of jargonelle pear (acetate of amyl). Pine-apple cordial and liqueurs from some other fruits are also prepared from the new fruit essences. See ESSENCE.

Lem'on Cordial. *Prep.* Digest fresh and dried lemon-peel, of each, 2 oz., and fresh orange-peel, 1 oz., in proof spirit, 1 gall., for a week; strain with expression, add of clear soft water, q. s. to reduce it to the desired strength, and lump sugar, 3 lbs. to the gallon. The addition of a little orange-flower or rose water improves it.

Liquodilla. Flavoured with oranges and lemons, of each (sliced), 3 in number; with sugar, 2½ lbs. per gall.

Lov'age Cordial. *Prep.* From the fresh roots of lovage, 1 oz. to the gallon. A fourth of this quantity of the fresh roots of celery and sweet fennel are also commonly added. In some parts a little fresh valerian root and oil of savine are added before distillation. It is much valued by the lower classes in some of the provinces for its stomachic and emmenagogue qualities.

Oil of Ce'drat. See CRÈME DE CEDRAT (*above*).

Orange Cordial. Like LEMON CORDIAL or CRÈME D'ORANGE, from fresh orange-peel, ¾ lb. to the gallon.

Parfait Amour. *Syn.* PERFECT LOVE. *Prep.* Flavoured with the yellow rind of 4 lemons, and a teaspoonful of essence of vanilla to the gallon, with sugar, 3 lbs., and powdered cochineal, q. s. to colour.

Pep'permint. *Syn.* PEPPERMINT CORDIAL, SPORTSMAN'S C., X. MINT; EAU DE CHASSEURS, Fr. This well-known compound is in greater demand in every part of the kingdom than all the other cordials put together.

Prep. 1. From peppermint water and gin or plain spirit (22 u. p.), of each, 1 pint; lump sugar, ¾ lb.

2. (Wholesale.) English oil of peppermint, 5 oz., is added to rectified spirits of wine, 3 pints, and the mixture is agitated well together for some time in a corked bottle capable of holding 4 pints

or more; it is then emptied into a cask having a capacity of upwards of 100 galls., and perfectly white and flavourless proof spirit, 36 galls., is poured in, and the whole well agitated for ten minutes; a solution of the best double-refined lump sugar, 2½ cwt., in about 35 galls. of pure filtered rain-water, is then added, and the contents of the cask well 'rummaged up' in the usual manner for at least 15 minutes; sufficient clear rain-water to make up the whole quantity to exactly 100 galls., and holding in solution alum, 5 oz., is next added, and the whole is again well agitated for at least a quarter of an hour, after which the cask is bunged down, and allowed to repose for a fortnight before it is 'broached' for sale.

Obs. The last formula produces a beautiful article provided the ingredients are of good quality. Care on this point is particularly necessary in reference to the essential oil, which should only be purchased of some known respectable dealer. The sugar should be sufficiently pure to dissolve in a wine-glassful of clear soft water without injuring its transparency, and the cask should be a fresh-emptied gin pipe, or one properly prepared for gin, as, if it gives colour, it will spoil the cordial. When these particulars are attended to, the product is a bright transparent liquor as soon as made, and does not require fining. Should there be the slightest opacity, the addition of 2 oz. of salt of tartar, dissolved in a quart of hot water, will have the effect of 'clearing it down' in the course of a few days. The product is 100 galls. of cordial at 64 u. p.

Pimen'to. *Syn.* PIMENTO CORDIAL, PIMENTO DRAM. Rather strongly flavoured with allspice, or pimento. It has obtained a great repute in the West Indies in diarrhœa, cholera, and bowel complaints generally.

Rasp'berry Cordial. *Prep.* From raspberry brandy, capillaire, and water, equal parts. A similar article is prepared by flavouring sweetened spirit with the new 'raspberry essence.'

Ratafi'a. The numerous liqueurs bearing this name are noticed in another part of this volume. See RATAFIA.

Shrub. See the article SHRUB in another part of this work.

Sighs of Love. *Prep.* 1. From proof spirit flavoured with otto of roses and capillaire, equal parts.

2. From sugar, 6 lbs., pure soft water, q. s. to produce a gallon of syrup, to which add, of eau de rose, 1 pint; proof spirit, 7 pints. It is stained of a pale pink by powdered cochineal. A very pleasant cordial. A drop or two (not more) of essence of ambergris or vanilla improves it.

Tears of the Widow of Malabar. *Prep.* As BALM OF MOLUCCA, but employing cloves (bruised), ½ oz.; mace (shredded), 1 dr.; and a teaspoonful of essence of vanilla for flavouring. Some add of orange-flower water, ¼ pint. It is slightly coloured with burnt sugar.

Tent. From plain spirit (22 u. p.) and port wine, of each, 1 quart; sherry and soft water, of each, 1 pint; orange-flower water and lemon-juice, of each, ¼ pint; essence of ambergris, 2 drops (not more); sugar, 2 lbs. See WINE.

Us'quebaugh. See the article USQUEBAUGH in another part of this work.

LIQUEUR DE LA MOTTE. [Fr.] See DROPS, GOLDEN, and TINCTURE.

LIQUEUR DE PRESSAVIN. [Fr.] *Prep.* From oxide of mercury (freshly precipitated) and cream of tartar, of each, 1 oz.; hot water, 1 quart; dissolve and filter. For use 2 spoonfuls of this liquor are added to 1 quart of water.—*Dose.* A wine-glassful 3 or 4 times a day, avoiding the use of common salt. This is simply a solution of potassio-tartrate of mercury, and may be taken in the usual cases in which mercury is administered.

LIQUEUR DORÉE. [Fr.] *Prep.* Take of cinnamon, bitter orange-peel, and Peruvian bark, of each, ⅜ oz.; hay saffron, ¼ oz.; brandy and Malaga wine, of each, 3 quarts; digest for a week, strain, and add of lump sugar, 2 lbs. Tonic, stomachic, and stimulant; chiefly used as an agreeable alcoholic dram.

LIQUID-AMBAR. A fluid balsamic juice obtained from the *Liquidamber styraciflua*, an American tree. By keeping, it dries to a pale amber-coloured resin It closely resembles LIQUID STORAX in its properties, and may be applied to the same purposes. See STYRAX.

LIQUODIN'NA. See LIQUEUR.

LIQ'UOR. *Syn.* LIQUOR, L.; LIQUEUR, Fr. This term is given in the Pharmacopœia to those aqueous solutions commonly though improperly called ' WATERS;' ammoniæ, liquor potassæ, &c.

The term 'liquor' has also, of late years, been applied to certain concentrated preparations, most of which would be more correctly termed ' FLUID EXTRACTS,' as they merely differ from solid extracts in their consistence, and from ordinary extracts in containing less starchy matter, albumen, and gum. There is also usually a little spirit added to them to prevent decomposition. Liquors of this kind may be prepared of the finest quality by the same processes that are required for the preparation of good soluble extracts; observing to stop the evaporation as soon as the consistence of treacle is acquired, and when cold, to add 1-4th or 1-5th part of their volume (after evaporation) in rectified spirit. The liquors, which are merely concentrated infusions or decoctions, and which, in their consistence, do not even approximate to extracts, may be made in the manner directed under those heads. It is now the practice to prepare fluid extracts of such a strength that 1 fl. part of the finished extract is equal to 1 solid part of the dry drug.

Much confusion would be prevented if the terms 'concentrated decoction,' 'concentrated infusion,' &c., were adopted for those vegetable preparations possessing 8 times the usual strength; 'liquors' for those of a higher strength, but still sufficiently liquid to be treated as such in dispensing, &c.; and 'fluid extracts' for those possessing considerable consistence, and approaching the common extracts in their degree of concentration and mode of preparation. See DECOCTION, ESSENCE, EXTRACT, INFUSION, SOLUTION, &c.

⁎ The following formulæ present some illustrations of the preparation of this class of medicines:

Liquor of Ammonia. *Syn.* LIQUOR AMMONIÆ,

L. *Prep.* Strong solution of ammonia, 1 pint; distilled water, 2 pints; mix and preserve in a stoppered bottle. Sp. gr. ·959.

Liquor of Ammonia, Stronger. *Syn.* LIQUOR AMMONIÆ FORTIOR, L. *Prep.* Mix chloride of ammonium (in coarse powder), 3 lbs., and slaked lime, 4 lbs., and introduce the mixture into an iron bottle placed in a metal pot surrounded by sand. Connect the iron tube which screws airtight into the bottle in the usual manner, by corks, glass tubes, and caoutchouc collars, with a Woulf's bottle capable of holding a pint; connect this with a second Woulf's bottle of the same size, the second bottle with a matrass of the capacity of 3 pints, in which 22 oz. of distilled water are placed, and the matrass, by means of a tube bent twice at right angles, with an ordinary bottle containing distilled water, 10 oz. Bottles 1 and 2 are empty, and the latter and the matrass which contains the 22 oz. of distilled water are furnished each with a siphon safety-tube charged with a very short column of mercury.

The heat of a fire, which should be very gradually raised, is to be now applied to the metal pot, and continued until bubbles of condensable gas cease to escape from the extremity of the glass tube which dips into the water of the matrass.

The process being terminated, the matrass will contain about 43 fl. oz. of strong solution of ammonia. Bottles 1 and 2 will now include, the first about 16, the second about 10 fl. oz. of a coloured ammoniacal liquid.

Place this in a flask closed by a cork, which should be perforated by a siphon safety-tube containing a little mercury, and also by a second safety-tube bent twice at right angles, and made to pass to the bottom of the terminal bottle used in the preceding process. Apply heat to the flask until the coloured liquid it contains is reduced to 3-4ths of its original bulk. The product now contained in the terminal bottle will be nearly of the strength of solution of ammonia, and may be made exactly so by the addition of the proper quantity of distilled water, or of strong solution of ammonia. Density ·891, contains 32·5% of ammonia.

Ants. Vinegar and water followed by acidulated demulcent drinks.

Liquor, Anodyne. See SPIRIT OF ETHER.

Liquor, Antinephritic. *Syn.* LIQUOR ANTI-NEPHRITICUS, L. *Prep.* (*Adams.*) Poppy-heads, 6 oz.; water, 1½ pints; boil to 1-3rd, strain with pressure, and add of nitrate of potassa, 1 oz.—*Dose,* 1 to 2 teaspoonfuls night and morning; in gravel and painful affections of the kidneys and bladder.

Liquor, Antipodag'ric (Beguin's). *Syn.* HOFF-MANN'S GOUT LIQUID; LIQUOR ANTIPODAGRICUS HOFFMANNII, L. *Prep.* From Boyle's fuming liquor, 1 part; spirit of wine, 3 parts. Sudorific. —*Dose,* 20 to 30 drops; or externally, in gout and other painful affections, either alone or combined with camphor. See AMMONIUM, PERSULPHIDE OF.

Liquor, Bleaching. See SOLUTION OF CHLORIDE OF LIME.

Liquor, Blistering. (B. P.) *Syn.* LIQUOR EPISPASTICUS LINIMENTUM CANTHARIDES, L. *Prep.* Mix cantharides in powder, 5 oz., and acetic ether, a sufficiency. Pack in a percolator, and pass the ether slowly through until 20 fl. oz. are obtained. Keep in a stoppered bottle.

Liquor, Boyle's Fu"ming. The perhydrosulphate of ammonia.

Liquor of Calum'ba. *Syn.* LIQUOR CALUMBÆ, L. Same as CONCENTRATED INFUSION OF CALUMBA.

Liquor of Cam'phor. See ESSENCE.

Liquor of Chiret'ta. Same as CONCENTRATED INFUSION OF CHIRETTA.

Liquor Cinchonæ (B. P.). *Syn.* LIQUID EXTRACT OF CINCHONA. *Prep.* 1. Made by exhausting 20 oz. of red cinchona bark with a mixture of 5 pints water, 5 dr. hydrochloric acid, 2½ oz. glycerin ; then evaporating the fluid to 20 oz. 50 gr. of this liquid is shaken with caustic soda and benzolated amylic alcohol. Separate the alcohol, evaporate, dry, and weigh the residue, which multiplied by 2 will give the percentage of alkaloids present. From the result adjust the liquid, either by evaporation or addition of water, so that 85 gr. contains 5 gr. of alkaloids ; finally, add 12½ gr. rectified spirit, and enough water to make the product weigh 100 gr. The finished product will contain 5% alkaloids.

Uses. Tonic, febrifuge, astringent.—*Dose*, 3 to 10 mins.

2. Yellow cinchona bark (bruised), 56 lbs., and water holding in solution sulphuric acid, 1½ lbs., are macerated together, with occasional agitation, in a covered earthen vessel, for 48 hours, after which the liquor is expressed, and the residuum or marc is treated with fresh water ; the mixed strained liquid is then evaporated as rapidly as possible in earthenware to exactly 6 lbs. ; to this rectified spirit, 1½ lbs., is added, and the whole is set aside for a week or 10 days ; the clear portion is, lastly, decanted and preserved in well-closed bottles. The product is very rich in quinine. It is 96 times as strong as the DECOCTION OF CINCHONA (Ph. L.), and 12 times as strong as the above preparation of the Ph. L. This preparation resembles the 'LIQUOR CINCHONÆ' sold by certain houses in the trade at 24s. per lb. wholesale.

3. Exhaust the bark as above by maceration in 3 successive waters without acid, filter, evaporate the mixed liquors to 7 lbs., and proceed as before. Inferior to the last, and less rich in the cinchona alkaloids. Very thick, scarcely liquid.

4. From PALE BARK (LIQUOR CINCHONÆ PALLIDÆ ; INFUSUM CINCHONÆ SPISSATUM, Ph. L.). From pale bark, as the last. See INFUSION OF CINCHONA.

Liquor, Disinfect'ing. See SOLUTION (Chlorides of Lime, Soda, and Zinc), and DISINFECTING COMPOUNDS.

Liqueur du Docteur Laville. *Prep.* Alcohol, ½ oz. ; water, 3 oz. ; colchicin, 2 gr. ; quinine, 1½ gr. ; extract and staining material, 40 gr. ; mineral substances, 8 gr.

Liquor of Er'got. *Syn.* LIQUID EXTRACT OF ERGOT ; LIQUOR SECALE, L. *Prep.* Crushed ergot, 1 lb. ; distilled water, 6 pints ; rectified spirit, 6 fl. oz. Digest the ergot in 4 pints of the water for 12 hours, draw off the infusion, and repeat the digestion with the remaining water.

Press out, strain, and evaporate the liquors to 11 oz. ; when cold, add the spirit. Filter and make up to 16 fl. oz. with water.—*Dose*, ½ to 1 dr.

Liquor of Flints. See SOLUTION.

Liquor of Gutta Percha. *Syn.* LIQUOR GUTTA PERCHA. *Prep.* Gutta percha in thin slices, 1 oz. ; carbonate of lead in fine powder, 1 oz. ; chloroform, 8 fl. oz. Add the gutta percha to 6 fl. oz. of chloroform in a stoppered bottle, and shake them frequently till solution has been effected. Then add the carbonate of lead previously mixed with the remainder of the chloroform, and having several times shaken the whole together set the mixture aside, and let it remain at rest until the soluble matter has subsided. Lastly, decant the clear liquid and keep in a well-stoppered bottle.

Liquor, Libavius's. Bichloride of tin.

Liquor of Mat'ico. *Syn.* CONCENTRATED INFUSION OF MATICO ; LIQUOR MATICONIS, INFUSUM MATICONIS CONCENTRATUM, L. *Prep.* From matico leaves, 1 lb. ; rectified spirit, ½ pint ; distilled water, 32 fl. oz. ; digest 10 days, express, and filter. 1 fl. dr. added to 7 fl. dr. of water is equal to 1 fl. oz. of the common INFUSION.

Liquor of Myrrh. *Syn.* SOLUTION OF MYRRH ; LIQUOR MYRRHÆ, LOCO LIQUAMINIS MYRRHÆ, L. *Prep.* (Ph. Bor.) Extract of myrrh (Ph. Bor.), 1 oz. ; distilled water, 5 fl. oz. ; mix thoroughly, decant, and strain. It should be of a brownish-yellow colour, and turbid.—*Dose*, ½ to 1 fl. dr.

Liquor of O'pium. *Syn.* LIQUOR OPII, L. O. CONCENTRATUS, L. OPIATUS, L. See BLACK DROP (under DROPS).

Prep. 1. (*Messrs Smith.*) Opium, 4 oz., is made into an extract, and 'denarcotised' by ether ; it is then dissolved in alcohol, filtered, evaporated nearly to dryness, and redissolved in water, q. s. to furnish 12 oz. of solution ; to this is added, of rectified spirit, 2½ oz., with water, q. s. to make the whole up to 16 oz.—*Dose*, 8 to 12 drops.

2. (Acetic : LIQUOR OPII ACETICUS, L.) See LAUDANUM (*Houlton's*).

3. (Citric : LIQUOR OPII CITRICUS, L.) *a.* Powdered opium, 1½ oz. ; lemon-juice, 1½ pints ; evaporate to ½, cool, add of rectified spirit, 5 fl. oz., and the next day decant or filter ; same strength as 'LAUDANUM.'

b. (LIQUOR MORPHINÆ CITRATIS, L. ; *Dr Porter.*) Opium, 4 oz. ; citric acid, 2 oz. ; triturate, and add of boiling water, 15 fl. oz. ; digest with agitation for 24 hours, and filter. This last has above 3 times the strength of 'LAUDANUM.' It is sadly misnamed.

4. (Hydrochloric : SOLUTION OF MURIATE OF OPIUM ; LIQUOR OPII HYDROCHLORICUS, L. ; *Dr Nichol.*) Powdered opium, 1½ oz. ; distilled water, 1 pint ; hydrochloric acid, 1½ fl. oz. ; digest a fortnight, and strain with expression. Same strength as 'LAUDANUM.' According to Dr Nichol, this is preferable to every other preparation of opium.

5. (Sedative : BATTLEY'S SEDATIVE SOLUTION OF OPIUM ; LIQUOR OPII SEDATIVUS, L.) *a.* Hard aqueous extract of opium (bruised), 3 oz., is boiled in water, 1½ pints, until dissolved ; to the solution, when cold, rectified spirit, 6 oz., is added, together with water, q. s. to make the whole

measure exactly 1 quart; the liquor is, lastly, filtered.

b. From hard extract of opium, 22 oz.; boiling water, 13 pints; rectified spirit, 3 pints; as the last.

c. From extract of opium (Ph. L.), 4¼ oz.; water, 1 quart; boil till reduced to 34 fl. oz.; cool, filter, and add of rectified spirit, 5 fl. oz., and water, q. s. to make up exactly 1 quart.

Obs. The first two formulæ, which vary only in their quantities, are identical with that employed by Mr Battley. As hard extract of opium is not always at hand, we have introduced a formula in which the ordinary extract is ordered. It gives a precisely similar product to the others, provided the cold aqueous decoction is filtered before adding the spirit. Battley's LIQUOR OPII SEDATIVUS is an excellent preparation, less exciting than opium or laudanum.—*Dose,* 10 to 30 drops. Dr Christison states that 20 drops of Battley's solution are equal to 30 drops of the common tincture.

Liquor, Pancreatic. *Syn.* LIQUOR PANCREATICUS. *Prep.* Fresh pig's pancreas (well minced), 1 part; distilled water, 18 parts; rectified spirit, 2 parts. Macerate for 2 days with frequent stirring. Filter.—*Dose,* ½ to 2 dr.

Liquor of Pepsin (*Mr Squire*). *Syn.* LIQUOR PEPSINI. *Prep.* 1 dr. of Boudalt's pepsin in 1 oz. of distilled water. Salt must be added if it is to be preserved.—*Dose.* A teaspoonful.

Liquor of Rhu'barb. *Syn.* LIQUOR RHEI, INFUSUM RHEI CONCENTRATUM, L. *Prep.* 1. Rhubarb (well bruised), 6½ oz.; water, q. s.; rectified spirit, ½ pint; proceed as for INFUSION OF CALUMBA (conc.); to produce a quart. 8 times the usual strength.

2. See INFUSION OF RHUBARB (Concentrated).

3. See EXTRACT OF RHUBARB (Fluid).

Liquor of Sarsaparil'la. *Syn.* FLUID EXTRACT OF SARSAPARILLA; LIQUOR SARZÆ, ESSENTIA SARSAPARILLÆ, L. *Prep.* Sarsaparilla (in powder), 40 oz.; proof spirit, 2 pints; sugar, 5 oz.; water, 12 pints. Macerate the sarsaparilla and spirit for 10 days, press out 20 oz., and set aside. Mix the pressed residue with the water and macerate at 160° F. for 18 hours; then strain, and press and dissolve the sugar in the fluid; evaporate in a water-bath to 18 oz. Mix the two liquids and make up to 40 oz. with water.—*Dose,* 2 to 4 dr.

Liquor of Sen'na. *Syn.* LIQUOR SENNÆ, L. Both the FLUID EXTRACT and the CONCENTRATED INFUSION OF SENNA are called by this name, but more generally the former. The following are additional formulæ:

Prep. 1. (*Duncan.*) Senna, 15 lbs.; boiling water, 5 galls.; proceed by the method of displacement, evaporate the product to 10 lbs., add of molasses, 6 lbs. (previously concentrated over a water-bath until it begins to congeal on cooling), dissolve, and further add of rectified spirit, 1¼ pints, together with water, q. s. to make the whole measure exactly 12 pints. Every fl. oz. represents 1 oz. of senna.

2. (*Dr Tweedy.*) As the last, but using tincture of ginger (prepared with rectified spirit), 1½ pints, instead of the spirit there ordered.

Liquor of Soap. *Syn.* LIQUOR SAPONIS, L. See TINCTURE.

Liquor, Styp'tic. *Syn.* LIQUOR STYPTICUS, L. *Prep.* (Ph. Slevico-Holsat. 1831.) Alum and sulphate of copper, of each, 1½ oz.; sulphuric acid, 1 oz.; water, 1 lb.; dissolve and filter.

Liquor of Tarax'acum. *Syn.* FLUID EXTRACT OF DANDELION; EXTRACTUM TARAXACI FLUIDUM, LIQUOR TARAXACI, L. *Prep.* 1. Dry dandelion roots (in powder), 40 oz.; proof spirit, 4 pints; water, a sufficiency. Mix the dandelion with the spirit; macerate for 48 hours, then press out 20 oz. of fluid and set aside. Mix the pressed residue with the water, macerate 48 hours, press, and strain; evaporate the fluid to 18 oz. Mix the two liquids and make to 40 oz. with water.—*Dose,* ¼ to 2 dr.

2. The expressed juice of dandelion is heated to near the boiling-point, strained, and evaporated, as the last, to a proper consistence; 1-4th or 1-5th of rectified spirit is then added, and the liquid is otherwise treated as before. Very odorous and pale coloured.

3. Dried root (coarsely powdered), 1 lb.; water, 1½ pints; rectified spirit, ½ pint; digest a week, express the liquor, pass it through a hair-sieve into a bottle, and in 10 days decant the clear portion.

4. (Ph. Bor.) Extract of dandelion, 3 parts; water, 1 part (or q. s.); triturated together.

5. (*W. Procter.*) Fresh root, 2 lbs., is sliced and reduced to a pulp, and macerated with 1-6th of its bulk of rectified spirit for 24 hours; it is then subjected to strong pressure, the marc is treated with water containing a little spirit, 1 pint, and the liquid is again expressed; the mixed product is evaporated to 12 fl. oz., and when cold, rectified spirit, 4 fl. oz., is added, and the whole filtered.

Obs. Liquor of taraxacum has a very large sale. The dose is 1 to 2 fl. dr. See EXTRACT.

Liquor of Vale'rian. See EXTRACT. OF VALERIAN (Fluid).

Liquor of Vanil'la. *Syn.* FLUID EXTRACT OF VANILLA; LIQUOR VANILLÆ, EXTRACTUM V. FLUIDUM, L. *Prep.* 1. Vanilla (sliced), 1 lb.; rectified spirit, 3 pints; prepare a tincture either by displacement or maceration, and reduce it, by distillation at the lowest possible temperature, to 1½ lbs.; put this into a strong bottle whilst hot, add of white sugar-candy (in powder), ½ lb., cork down, and agitate the whole until it is nearly cold. Very fine. Used chiefly for its odour and flavour. It represents half its weight of vanilla.

2. (*W. Procter.*) Vanilla (cut into thin transverse slices), 1 oz.; sugar, 3 oz.; triturate until reduced to fine powder, put it into a strong pint bottle, along with syrup, ½ pint; water, 2 oz.; tie down the cork, and set the bottle for ½ an hour in boiling water; cool, strain, and treat the residue in a like manner with a mixture of water, 6 fl. oz., and rectified spirit, 1 fl. oz.; lastly, mix the two products. Greatly inferior to the last.

LIQUORICE. *Syn.* STICK LIQUORICE; LIQUORITIA, GLYCYRRHIZÆ RADIX (B. P.), GLYCYRRHIZÆ RADIX, GLYCYRRHIZA (Ph. L. and D.), G. GLABRA (Ph. E.), L. "The root or underground stem of the *Glycyrrhiza glabra*, fresh and dried, cultivated in Britain." "The recent and the dried root of *Glycyrrhiza glabra*," or common liquorice. "The fresh root is to be kept buried in dried sand for use" (Ph. L.). It has a sweetish taste, and is slightly aperient, expectorant, and diuretic. It is a popular demulcent and pectoral. Its extract and solution are much used as a domestic remedy for cough. As a masticatory it allays thirst and irritation.

Composition of the fresh root of liquorice:

Glycyrrhizin	. . .	8·60
Gum	. . . .	26·60
Matter soluble in alcohol,		
chiefly resin	. . .	0·75
Albumen	. . .	0·97
Starch .	. . .	22·91
Woody fibre	. .	13·36
Moisture	. . .	26·31
Ash, 3·07% .	. . .	—
		100·00
		(*Hassall.*)

Roussin asserts that the sweetness of liquorice root is not due to glycyrrhizin, as has been hitherto assumed, but to an ammoniacal compound of that substance. Glycyrrhizin, when purified 4 successive times by dissolving it in alcohol, and precipitating the foreign matter accompanying it by ether, is a yellowish substance, insoluble in cold water, and almost tasteless. Treated with dilute solution of potash or soda, it rapidly develops a sweet taste. In liquorice root, however, it is not contained in combination with either of these two alkalies, but appears to exist as an ammoniacal compound, for solutions of potash and soda liberate ammonia, both from the root and the extract. In its compounds with the alkalies glycyrrhizin plays the part of an acid, as it forms true salts capable of undergoing decomposition with most of the metallic salts, and also with the salts of the organic alkaloids. With ammonia it forms two compounds, a basic salt, which yields a deep yellow solution, and another containing less ammonia, the solution of which has an amber colour. The former is produced by dissolving glycyrrhizin in water with an excess of ammonia. Upon evaporating the resulting deep yellow solution to dryness it leaves a yellowish, scaly, shining, brittle, non-hygroscopic residue, which constitutes the second ammoniacal compound. This is readily soluble in cold water, to which it imparts a pale yellow colour and a very sweet taste. The solution turns deep yellow on the addition of a few drops of solution of ammonia, owing to the formation of the basic compound. The pale yellow solution possesses, in a marked degree, the taste of liquorice root, which, indeed, owes its sweetness to this glycyrrhizate of ammonia, or ammoniacal glycyrrhizin, as the author prefers to call it. 1 grm. of this compound imparts the sweet taste of the root to 2 litres of water.

The author gives the following process for the preparation of the ammoniacal glycyrrhizin in the pure state:—The carefully-selected roots, freed from all portions presenting a dark fracture, are scraped, and then well pounded, so as to reduce them to a kind of stringy tow. This substance is macerated in cold distilled water for some hours, pressed, and treated a second time in the same manner. The two liquors are mixed and allowed to stand for some time to deposit the starch. The supernatant liquor is then boiled and filtered, to separate the coagulated albumen. After cooling, sulphuric acid diluted with its weight of water is added gradually, with brisk stirring, until a precipitate is no longer formed. The precipitate, at first gelatinous and flocculent, after standing some time, forms a compact semisolid mass at the bottom of the vessel. The supernatant liquor is rejected, and after roughly washing the precipitate several times with pure water, it is finally kneaded repeatedly in distilled water until all trace of acidity has disappeared. The mass is then well drained and agitated in a flask with 3 times its weight of 90° alcohol until dissolved, when a similar quantity of 96° to 96° alcohol is added to the syrupy liquid so produced. A little pectic acid is thus precipitated, which is removed by filtration. Ether is then added to the alcoholic liquor as long as a precipitate is formed. After standing 24 or even 48 hours a blackish pitchy substance is deposited, which adheres to the glass, and allows of the clear liquor being decanted. To this clear liquor is added, in small quantities at a time, alcohol of 90° charged with gaseous ammonia, which determines the formation of a yellow, rather heavy, flocculent precipitate of glycyrrhizate of ammonia. The precipitate is washed rapidly on a fine cloth with a mixture of equal parts of alcohol and ether, pressed and dried in a current of warm air, or over sulphuric acid.

The author suggests the addition of ammoniacal glycyrrhizin to pill masses, powders, or mixtures, and states that its power of masking the taste of nauseous medicines is equal to 100 times its weight of sugar. Sulphate of quinine, sulphate of magnesia, iodide of potassium, and ipecacuanha, lose much of their taste by such an addition.

A dose of cod-liver oil or syrup of iodide of iron is rendered more palatable by being preceded by a small dose of the solid ammoniacal glycerin ('Journal de Pharmacie et de Chimie,' xii, 6–11). Its extract is the common LIQUORICE, SPANISH LIQUORICE, or SPANISH JUICE, of the shops. See EXTRACT, &c.

LISBON DIET DRINK. *Prep.* 1. (*Foy.*) Guaiacum wood (rasped), 1 oz.; sarsaparilla (bruised), 8 oz.; mezereon (sliced), ½ oz.; crude antimony (in a rag), 2 oz.; water, 12 pints. Boil down to 8 pints, and add, red sanders (rasped), white sandal (rasped), of each, 8 oz.; rosewood, rasped sassafras bark (sliced), of each, 1 oz.; liquorice root (sliced), ½ oz. Infuse for 4 hours, strain, and add syrup according to taste.—*Dose*, 1 to 2 pints a day.

2. (*Pearson.*) Sarsaparilla (bruised), 4 oz.; dried walnut peel, 4 oz.; guaiacum (rasped), 1¼ oz.; crude antimony (in a rag), ¼ oz.; water, 4 pints. Boil down to 8 pints.

LIST. The border or selvage torn off a piece of cloth. It is used by the French polishers and law stationers to form their rubbers, and for numerous other purposes.

LITHARGE. *Syn.* SEMI-VITRIFIED OXIDE OF LEAD; PLUMBI OXYDUM (Ph. L.), PLUMBI OXYDUM SEMI-VITREUM (Ph. E.), LITHARGYRUM (Ph. E.), L. The litharge of commerce is semi-vitrified monoxide of lead, PbO, obtained chiefly by scraping off the drops that form on the surface of melted lead exposed to a current of air (dross of lead ; plumbum ustum), and heating it to a full red, to melt out any undecomposed metal. The fused oxide, in cooling, forms a yellow or brownish semi-crystalline mass, which readily separates into scales ; these, when ground, constitute the 'powdered litharge' of the shops. The yellow variety is obtained when the metal is only moderately heated ; it is usually called massicot (q. v.). Litharge is also prepared by exposing red lead to a heat sufficiently high to fuse it, and ' English litharge' is obtained as a by-product, by liquefaction, from argentiferous lead ore, when it is often called ' silver stone.'

Prop. Litharge is a strong base; it forms a large class of salts often called plumbic salts. It is very readily soluble in dilute nitric and acetic acids, also in hot solutions of potash and soda. The acid solutions are blackened by sulphuretted hydrogen. It is easily reduced when heated with organic substances.

Pur. Digested with a cold solution of ammonium carbonate will remove any copper oxide. Heating in air will remove metallic lead. It is of great importance to the pharmaceutist to obtain pure litharge, as the slightest impurity will often colour and spoil his lead plaster (EMP. PLUMBI), and solution of diacetate of lead (LIQ. PLUMBI DIACETATIS).

Uses. Litharge is employed in *pharmacy*, to make plasters, &c. ; by painters as a 'drier' for linseed, poppy-seed and other oils ; in the manufacture of flint and crystal glass ; as a glaze for earthenware ; as a flux in glass and porcelain staining ; in very dilute solution as a hair dye ; for producing iridescent colours on brass, &c. ; and in the preparation of red lead, lead acetate, lead nitrate, white lead, putty, &c. ; and for various other purposes in the *arts*.

Obs. The litharge of commerce is distinguished by its colour as LITHARGE OF GOLD (LITHARGYRUM AURI, L. AURIUM, L. CHRYSITIS), which is dark coloured and impure, and LITHARGE OF SILVER (SILVER STONE ; LITHARGYRUM ARGENTI, L. ARGENTEUM, L. ARGYRITIS), which is purer, and paler coloured. Commercial litharge generally contains red oxide of lead and from 1% to 3% of the metal. Foreign litharge generally contains copper and iron oxides, and not infrequently a little silver and silica. These are readily detected by the usual tests. In grinding litharge, about 1 lb. of olive oil is usually added to each 1 cwt. to prevent dust. As it slowly absorbs carbon dioxide from the air, it generally effervesces slightly when treated with acids, and this effervescence is stronger in proportion to its age. The carbon dioxide and any absorbed water may be removed by ignition. When fused in a clay crucible at a red heat litharge forms an easily fusible silicate which perforates the sides.

LITHIUM. Li. At. wt. 7·01. Discovered by Arfvedson in 1817 in several Swedish minerals. It is widely distributed throughout the animal and vegetable kingdoms, and is found in the waters of many mineral springs, of which those at Karlsbad and Marienbad are typical. Lithium compounds have been detected by the spectroscope in sea and river water ; Lockyer has proved its presence in the solar atmosphere, and it has been found in meteorites. The chief sources of this metal are lepidolite, petalite, and spodumene.

Prep. Davy first obtained lithium by electrolysis, but the quantity was too small to allow of the properties being examined. Bunsen and Matthiesen in 1855 obtained lithium in considerable quantity and carefully investigated its properties. It is now obtained by fusing pure chloride of lithium in a small, thick porcelain crucible, and decomposing it while in a fused state by a current of electricity. For details of the process *vide* Bunsen's account in 'Chem. Soc. Journal,' viii, 143.

Prop. It is a white metal possessing a silver lustre, fusing at 180°, and having a sp. gr. of 0·59. It is the lightest solid known. It belongs to the 'alkali group,' of which potassium, sodium, cæsium, rubidium, and the hypothetical ammonium are the other members. It is not so oxidisable as potassium or sodium, but it soon tarnishes on exposure to the air. When thrown on water it oxidises without fusing. Heated in air above its melting-point it burns with a white light. Nitric acid has a very violent action on it ; dilute hydrochloric and sulphuric acids dissolve it readily.

Tests. Lithium forms salts analogous to those of sodium, but usually somewhat less soluble. They can be distinguished from those of potassium and sodium by the phosphate and carbonate being only sparingly soluble in water; from those of barium, strontium, and calcium by forming crystallisable and soluble salts with sulphuric acid and oxalic acid ; and from those of magnesium by the solution of its carbonate exhibiting an alkaline reaction. Heated on platinum, they tinge the flame of the blowpipe carmine-red. " To detect lithium in mineral waters evaporate part of the water to a small bulk, add baryta water, and, on cooling, ammonium carbonate, and filter; add sodium phosphate to the filtrate, evaporate to dryness, and treat the residue with a very small quantity of water. Lithium phosphate remains behind, and may be tested in the spectroscope " (*Thorpe* and *Muir*).

The salts of lithium may generally be formed by dissolving the hydrate or carbonate in dilute acids.

Lithium Benzoate. LiC7H5O2,H2O. (Paris Pharm. Society.) Benzoic acid, 122 grms. ; lithium carbonate, 37 grms. Suspend the benzoic acid in 10 parts of water, add the lithium carbonate, and heat. Solution takes place with effervescence, and upon evaporation, handsome, much flattened, more or less elongated prismatic crystals are obtained.

Lithium benzoate is very soluble in water. 1 gr. of the salt calcined, and then treated with slight excess of sulphuric acid and heated to redness, should give 0·876 grm. of lithium sulphate.

Lithium, Bromide of. LiBr. To 37 grms. of carbonate of lithium suspended in 200 grms. of distilled water, 80 grms. of bromide are added. A current of sulphuretted hydrogen is then passed through the mixture until the whole of the bromide has disappeared. Hydrobromic acid is thus formed, which decomposes the carbonate of lithium, bromide of lithium being produced and sulphur set free. The mixture is then gently heated to drive off the excess of sulphuretted hydrogen and to agglutinate the sulphur. After filtration the liquor is concentrated, and if it be desired to obtain the bromide in crystals, the desiccation is finished under a bell-jar by means of sulphuric acid.

Lithium, Carbonate of. Li_2CO_3. *Syn.* CARBONATE OF LITHIA; LITHIÆ CARBONAS (B. P.), L. *Prep.* To an aqueous solution of sulphate of lithium add a strong solution of carbonate of ammonium, collect the precipitate, drain and press, wash with a little rectified spirit, and dry.

Prop., &c. It resembles carbonate of magnesium in appearance; is but slightly soluble in cold water, and is insoluble in alcohol. The tests for its purity given in the B. P. are—in giving no precipitate with oxalate of calcium or lime-water, and leaving, when 10 gr. are neutralised with sulphuric acid and ignited, 14·86 gr. of dry sulphate. It has been proposed by M. Lipowitz, Dr Garrod, and others, as a solvent for uric acid calculi. According to Biswanger, 1 part of carbonate of lithia in 120 parts of water takes up, at blood-heat, nearly 4 parts of uric acid. Mr Alexander Ure recommends a dilute solution of this substance as an injection in lithic calculus, as it is a better solvent of uric acid than either borax or the alkaline carbonates. "Of all the various menstrua hitherto recommended, none appear to promise more favourably than the carbonate of lithia." "If by means of injections" (of this solution) "we can reduce a stone at the rate of a grain or more an hour, we shall not merely diminish the bulk of the calculus, but further loosen its cohesion, disintegrate it, so to speak, causing it to crumble down and be washed away in the stream of urine" (*Mr A. Ure*).—*Dose,* 2 to 5 gr., twice or thrice a day; as an injection, 1 gr. to water, 1 fl. oz.

Lithium Chloride. LiCl. One of the most deliquescent salts known. Prepared by dissolving the oxide or the carbonate in hydrochloric acid. It occurs naturally in the waters of the Mur spring at Baden-Baden. Miller found 372 milligrms. in 1 litre of water from a spring in the Wheal Clifford mine at Redruth, in Cornwall.

Lithium, Citrate of. $Li_3C_6H_5O_7$. *Syn.* LITHIÆ CITRAS (B. P.), L. A white deliquescent amorphous powder, made by acting on 50 grains of lithium carbonate with 100 of citric acid. It is readily soluble in 2½ parts of water.

Tests, &c. 20 gr. burnt at a low red heat until white leave 10·6 gr. of carbonate of lithium. Its medical properties are similar to those of the carbonate.—*Dose,* 5 to 16 gr., largely diluted.

Lithium, Citrate of, Effervescing. (Paris Pharm. Society.) Citric acid, 40 grms.; sodium bicarbonate, 50 grms.; lithium bicarbonate, 10 grms. Mix the powders and place them in a flat-bottomed vessel having a large surface; heat to about 100° C., stirring the powder continually until it takes the granular form, then by means of appropriate sieves obtain granules of suitable and uniform size, and preserve the preparation in well-closed bottles.

Lithium Nitrate. LiNO₃. Obtained from the carbonate and nitric acid.

Lithium, Oxide of. Li₂O. *Syn.* LITHIA. An alkaline earth found in petalite, &c., and in small quantities in most mineral waters.

Prep. Petalite (a silicate of aluminum and lithium) in powder mixed with twice its weight of fluor-spar is heated with strong sulphuric acid as long as acid vapours are given off. The residue is treated with ammonia, boiled and filtered, evaporated to dryness, and heated to redness. The residue consists of sulphate of lithium, from which the oxide is obtained by decomposing it with acetate of barium, filtering and heating after having evaporated the solution to dryness.

This yields the so-called oxide, which is in reality the hydrate, LiHO, a white, non-volatile, soluble, caustic solid. The true oxide is a white powder, sometimes coloured yellow by a small quantity of a higher oxide; it is decomposed by water, forming the hydrate, and is obtained by igniting the metal in oxygen. Dry oxygen does not act upon lithium at the ordinary temperature.

Lithium Phosphate. Normal, Li₃PO₄. A crystalline powder precipitated by adding phosphate of soda and caustic soda to any lithium salt.

Lithia, Effervescing Solution of. *Syn.* LIQUOR LITHIÆ EFFERVESCENS, L. *Comp.* Water charged with carbonic acid and holding in solution carbonate of lithium. 10 fl. oz. contain 5 gr. of the carbonate.—*Props.* Colourless liquid, possessing powerful diuretic properties.—*Use.* Antilithic, for dissolving calculi of uric acid.—*Dose,* 5 to 10 fl. oz.

LITHOFRACTEUR. See BLASTING POWDERS.

LITHOGRAPHY. The art of tracing letters, figures, and other designs on stone, and transferring them to paper by impression. Our notice of this beautiful and useful art must necessarily be brief.

There are two methods of lithography in general use. In the one, a drawing is made on the stone with a lithographic crayon, or with lithographic ink; in the other method the design is made on lithographic paper, which, on being moistened and passed through the press, leaves its design on the surface of the stone, reversed. In either method, water acidulated with nitrous acid, oil of vitriol, or hydrochloric acid, is poured over the stone, and this, by removing the alkali from the chalk or ink, leaves the design on it in a permanent form, at the same time that it 'etches' away a portion of the lights, and renders the exposed surface more absorbent of water, and therefore incapable of taking the ink.

The process of lithographic printing is as follows:—Water is thrown over the stone, the roller charged with printing ink is passed over the surface, the paper is applied, and a copy is obtained by the action of the lithographic press. The same process must be had recourse to for each copy. The nature of the stone is such that it retains with great tenacity the resinous and oily substances contained in the ink or crayon employed to form the design, and also absorbs water freely; this, combined with the peculiar affinity between resinous and oily substances, and their mutual power of repelling water, occasions the ink on the printing roller to adhere to the design, and to leave untouched the lights.

The stones are prepared for lithography by

polishing in the ordinary way, the style of work for which they are intended determining the degree of labour bestowed upon them. For crayon drawings the surface should have a fine grain, but the finish of the stone must depend upon the desired softness of the intended drawing; for writing or drawing on in ink the surface must receive a higher polish, and must be finished off with pumice-stone and water.

The best lithographic stones are obtained from Solenhofen, near Munich, and from Pappenheim, on the banks of the Danube. The white lias which lies immediately under the blue, near Bath, also yields good lithographic stones, and furnishes the principal portion of those employed in this country. If a gelatin mixture such as is used for the hektograph be written upon with a strong solution of alum or other salt which renders gelatin insoluble, the writing, after damping the surface, will be found to take lithographic ink in much the same way as the stone. A little patience is required, but the process is worthy of more extended use. See CRAYONS, INK, and PAPER.

LITHONTRYP'TICS. *Syn.* LITHOTRYPTICS; LITHONTRYPTICA, L. Under this head are intended numerous substances (LITHICS; LITHICA, L.) used to prevent the formation of urinary calculi, or to dissolve them when already formed. Those employed with the former intention are more correctly termed ANTILITHICS (ANTILITHICA, L.), and those with the latter, LITHONTRYPTICS, or LITHONLYTICS (LITHONTRYPTICA, LITHON-LYTICA, L.).

The following are the principal substances included under this head by pharmacological writers:—Alkalies and their carbonates, benzoic acid, borax, carbonate of lithia, effervescing solution of lithia, carbonic acid, cinnamic acid, diluents (generally), diuretics (generally), juniper, malic acid, Malvern waters, mineral acids, nitro-saccharate of lead, opium, phosphate of soda, phosphoric acid, poppies, turpentines, uva ursi, vegetable acids, vegetable astringents, vegetable bitters, Vichy waters, wall pellitory, water (pure).

LIT'MUS. *Syn.* TURNSOLE; LACMUS, LAOCA CÆRULEA, L. MUSIVA, L. MUSCI, L. A blue substance prepared by the united influence of water, air, ammonia, and either potassa or soda, from *Roccella tinctoria*, *Lecanora tartarea*, or any of the tinctorial lichens or seaweed, capable of yielding archil, by a process essentially similar to that adopted for the latter substance, except that gypsum or chalk is generally used to form the paste, which is moulded into cakes and dried.

Litmus is soluble in both water and alcohol. Its blue colour is reddened by acids, and is restored by the addition of alkalies. Hence it is much used as an indicator in alkalimetry.

The colouring matter of litmus is related to orcein, which is the chief constituent of the commercial orchil dye; when purified as much as possible it may be kept for an indefinite period unaltered in glycerin. Litmus is treated with hot water, and the solution, after concentration, is mixed with a sufficient quantity of alcohol (of 80 per cent.) to precipitate the colouring matter. After standing for 20 hours the alcohol is poured off, and carries with it a dirty blue foreign substance, which frequently occurs in litmus, and is not altered by acids. The sediment is treated with hot water, which dissolves it on account of the potassium carbonate which is present.

To remove this carbonate, sulphuric acid is added till the liquid assumes a faint wine tint; it is then heated to boiling for a few minutes, and again rendered blue by the addition of a few drops of lime-water. After the lapse of 24 hours the liquid is filtered and evaporated to a syrup, and left all night in a cool place, when the potassium sulphate crystallises out in the form of a crust. It is then filtered through moist cotton, mixed with glycerin, and carefully preserved from damp. Neutral litmus solution may be prepared for use in chemical analysis by the following method, which is due to Thorpe and Muir. "5 to 6 grms. of coarsely powdered litmus are digested with about 200 c.c. of distilled water for a few hours. The clear solution is decanted from the sediment, and very dilute nitric acid added drop by drop, until the colour changes to violet. The solution must neither be red nor blue, but between the two in colour. . . . . The solution should be kept in a wide-mouthed bottle, the cork of which is so cut that the air has ready access to the interior of the bottle, otherwise the liquid quickly loses its colour."

Slips of blotting-paper saturated with litmus solution are convenient test-papers for rough use.

Litmus has been extensively used by dyers, but the colour imparted to textile fabrics is rather fugitive. It has also been employed for imparting a bluish tinge to whitewash-lime, in the manufacture of confectionery, and for colouring champagne, &c., red. See ARCHIL, CUDBEAR, &c.

LIVE-LONG. Digestive candy. See CANDYING.

LIV'ER. *Syn.* HEPAR, L. A large abdominal organ situated on the right side of the body immediately beneath the diaphragm. The liver varies in weight from 4 to 5 lbs. Its functions are the secretion of bile and the formation of glycogen (animal starch). The liver probably takes an active part in the chemical changes by which the nitrogenous food-stuffs are broken down and reduced to less complicated forms. It is a common seat of disease, though by far the greater part of such disease is of the patient's own causing. Over-feeding, irregular feeding, rich and indigestible foods, and, above all, the abuse of alcohol, tend to produce a state of engorgement of the liver and a blocking of the portal circulation, as a result of which piles and other intestinal troubles arise, besides the effect produced on the general system by the failure of an important organ to do its work. Judicious abstinence from food and a saline purgative will generally cure the liver trouble which arises from any individual indiscretion in eating or drinking. Exercise is a sovereign remedy for those who suffer from what is commonly called "torpid liver," which is generally the result of over-feeding and sedentary habits. The livers of drunkards, especially spirit drinkers, are especially liable to inflammation of the interstitial connective or supporting tissue, which by enlarging crushes the proper liver cells out of place, and ultimately destroys their function.

Liv'er. *Syn.* HEPAR, L. In *chemistry* and *pharmacy*, a term formerly applied to numerous substances, on account of their colour; as liver of antimony (HEPAR ANTIMONII), liver of sulphur (HEPAR SULPHURIS), &c.

Liver, Edible. The livers of animals, such as the bullock, the calf, and the sheep, contain a large amount of nitrogenous matter (hence the instinct that leads man to cook it with a food rich in carbon, such as fat bacon), as may be seen from the following analysis by Payen:

Composition of Calf's Liver.

Nitrogenous matter	20·10
Fat	3·58
Carbo-hydrates (amyloid matter)	0·45
Saline matter	1·54
Water	72·33
	98·03

They are generally regarded as indigestible articles of diet, and as such should be avoided by dyspeptics.

It is of great importance to have the livers of animals thoroughly cooked, so as to ensure the destruction of a dangerous parasite—the *Distoma hepatica*, the liver fluke—that frequently infests them.

Liver used as food should be cut up into slices and carefully examined for signs of disease, local or general; cavities, hard lumps, white, chalky-looking particles, and discoloured patches should be regarded with suspicion, and the *whole* rejected unless of perfectly uniform texture, of a bright, healthy colour, and entirely free from blemish. This particularly applies to the livers of animals running wild or fed on wild pasture, but the precaution should not be neglected even with liver bought from the butchers in towns. Even healthy animals may have parasites in their livers capable of producing serious consequences if taken into the human body.

The *foie gras*, of which the celebrated Strasbourg pie is made, is the abnormally enlarged or, rather, diseased liver of the goose, brought to its enormous size and fatty condition by subjecting the bird to close confinement in a hot place and overfeeding it.

LIVER AND BACON. The liver must be washed, not soaked, then wiped dry and cut into slices. Flour each slice. Remove the rind from the bacon, and cut it into rashers. Let the bacon be fried first, then stand it in a hot dish before the fire during the time the liver is being fried in the melted fat from the bacon. When the liver is cooked place it on the bacon. Next mix a dessert-spoonful of flour into a smooth paste with a cupful of water, stir in it a pinch of pepper and salt, and pour it into the frying-pan; let it just boil, stirring it meanwhile, and, lastly, strain it over the liver and bacon.

LIXIVIATION. The process of dissolving out or extracting the saline matter of bodies, more especially of ashes, the residua of distillations, &c., by means of ablution or digestion in water. The solution so obtained is called a 'LYE,' 'LEY,' or 'LIXIVIUM,' and the salts resulting from the evaporation of such solutions 'LIXIVIAL SALTS.'

LLA'MA. *Syn.* GUANACO; LAMA, L. A genus of animals of the family *Bovidæ* and tribe *Camelina*. The llama is confined to South America, and may be regarded as the representative of the camel in the New World. The most important species are *Lama vicugna* (the VICUNA) and *L. guanacus* (the GUANACO). The wool of llamas is woven into stuffs for *ponchos*, and made into cords, sacks, &c. See ALPACA.

LOAD'STONE. *Syn.* LODESTONE, MAGNESIAN STONE, MAGNETIC IRONSTONE. Native magnetic oxide of iron (Fe_3O_4). It is often found massive, frequently crystallised, and occasionally in beds of considerable thickness. Its colour varies from reddish black to deep grey. Native magnets from Arabia, China, and Bengal are commonly of a reddish colour, and are powerfully attractive. Those found in Germany and England have the colour of unwrought iron; those from Macedonia are more black and dull.

LOAM. A native mixture of clay, sand, and oxide of iron, with more or less chalk. Loamy soils are of this description. They are called heavy or light, according to the proportion of clay; and sandy, calcareous, or gravelly, just as sand, gravel, or chalk forms a characteristic portion of them.

The term is also applied to the mixtures of earth, sand, and other materials used by metal founders.

LOBELIA. *Syn.* INDIAN TOBACCO; LOBELIA (B. P., Ph. L. E. & D.), L. "The flowering herb *Lobelia inflata*" (B. P.), or bladder-podded lobelia. The herb has an unpleasant odour, and an acrid, burning, nauseous taste, somewhat resembling that of tobacco. In small doses (1 to 3 gr.) it is expectorant and diaphoretic; in larger doses (5 to 15 gr.) nauseant and emetic; and in excessive doses poisonous. According to Dr. Pereira, its principal value is that of an antispasmodic. It has been highly recommended by Dr Elliotson in spasmodic asthma. He commences with small doses, and gradually increases them unless headache or nausea occurs. Others give a full dose at or before the commencement of the fit. It has been also tried in croup, hooping-cough, spasmodic asthma, and other diseases of the respiratory organs, with variable effect.

Lobelia is the panacea of Dr Coffin, the author of the pretended system of medicine irreverently called 'Coffinism.' Large doses of this drug are given by the Coffinites, sometimes with fatal results.

LOBE'LINE. *Syn.* LOBELINA, L. A light yellowish-brown oily substance, found in *Lobelia inflata*. It is volatile, soluble in alcohol, ether, and water; and in oil of turpentine, oil of almonds, and some other fixed oils; with the acids it forms crystallisable salts, which are soluble. It may be obtained from the seeds by the action of alcohol acidulated with acetic acid, evaporating, treating with magnesia and then with ether, and again evaporating. 1 oz. of the seeds furnishes 2 gr. When perfectly pure, 1 gr. will kill a large dog.

LOB'STERS. See SHELL-FISH.

LOCK'SOY. Rice boiled to a paste and drawn into threads. Used to thicken soups. It is imported from China.

LODGING-HOUSES. The following sections of the Public Health Act of 1875 embody the regulations in force with regard to *common* lodging-houses :

(S. 76.) Every local authority shall keep a register, in which shall be entered the names and residences of the keepers of all common lodging-houses within the district of such authority, and the situation of every such house, and the number of lodgers authorised according to this Act to be received therein.

A copy of any entry in such register, certified by the person having charge of the register to be a true copy, shall be received in all courts and on all occasions as evidence, and shall be sufficient proof of the matter registered without production of the register, or of any document or thing on which the entry is founded; and a certified copy of any such entry shall be supplied gratis by the person having charge of the register to any person applying at a reasonable time for the same.

(S. 77.) A person shall not keep a common lodging-house or receive a lodger therein until the house has been registered in accordance with the provisions of this Act, nor until his name as the keeper thereof has been entered in the register kept under this Act ; provided that when the person so registered dies his widow or any member of his family may keep the house as a common lodging-house for not more than 4 weeks after his death without being registered as the keeper thereof.

(S. 78.) A house shall not be registered as a common lodging-house until it has been inspected and approved for the purpose by some officer of the local authority ; and the local authority may refuse to register as the keeper of a common lodging-house a person who does not produce to the local authority a certificate of character in such form as the local authority direct, signed by 3 inhabitant householders of the parish respectively rated to the relief of the poor of the parish within which the lodging-house is situated, for property of the yearly rateable value of £6 or upwards.

(S. 79.) The keeper of every common lodging-house shall, if required in writing by the local authority so to do, affix and keep undefaced and legible a notice, with the words 'Registered common lodging-house,' in some conspicuous place on the outside of such house.

The keeper of any such house who, after requisition in writing from the local authority, refuses or neglects to affix or renew such notice, shall be liable to a penalty not exceeding £5, and to a further penalty of 10s. for every day that such refusal or neglect continues after conviction.

(S. 80.) Every local authority shall from time to time make bye-laws :

1. For fixing from time to time, varying the number of lodgers who may be received into a common lodging-house, and for the separation of the sexes therein ; and—

2. For promoting cleanliness and ventilation in such houses ; and—

3. For the giving of notices and taking precautions in the case of any infectious disease ; and—

4. Generally for the well-ordering of such houses.

(S. 81.) Where it appears to any local authority that a common lodging-house is without a proper supply of water for the use of the lodgers, and that such a supply can be furnished thereto at a reasonable rate, the local authority may by notice in writing require the owner or keeper of such house, within a time specified therein, to obtain such supply, and to do all works necessary for that purpose ; and if the notice be not complied with accordingly, the local authority may remove such house from the register until it is complied with.

(S. 82.) The keeper of a common lodging-house shall, to the satisfaction of the local authority, limewash the walls and ceilings thereof in the first week of each of the months of April and October in every year. Penalty for neglect, £2 or less.

(S. 83.) The keeper of a common lodging-house in which beggars or vagrants are received to lodge shall from time to time, if required in writing by the local authority or to such person as the local authority direct, every person who resorted to such house during the preceding day or night, and for that purpose schedules shall be furnished by the local authority to the person so ordered to report, which schedules he shall fill up with the information required, and transmit to the local authority.

(S. 84.) The keeper of a common lodging-house shall, when a person in such house is ill of fever or any infectious disease, give immediate notice thereof to the medical officer of health of the local authority, and also to the poor-law relieving officer of the union or parish in which the common lodging-house is situated.

(S. 85.) The keeper of a common lodging-house, and every other person having or acting in the care or management thereof, shall, at all times when required by any officer of the local authority, give him free access to such house or any part thereof. Penalty for refusing such access, £5 or less.

(S. 86.) Any keeper of a common lodging-house, or other person having or acting in the care or management thereof, who—

1. Receives any lodger in such house without the same being registered under this Act ; or—

2. Fails to make a report after he has been furnished by the local authority with schedules for the purpose, in pursuance of this Act, of the persons resorting to such house ; or—

3. Fails to give the notices required by this Act, where any person has been confined to his bed in such house by fever or other infectious disease—

shall be liable to a penalty not exceeding £5, and in the case of a continuing offence to a further penalty not exceeding £2 for every day during which the offence continues.

(S. 87.) In any proceedings under the provisions of this Act relating to common lodging-houses, if the inmates of any house or part of a house allege that they are members of the same family, the burden of proving such allegation shall lie on the persons making it.

(S. 88.) Where the keeper of a common lodging-house is convicted of a third offence against the provisions of this Act relating to common lodging-houses, the Court before whom the conviction for such third offence takes place may, if it thinks fit, adjudge that he shall not at any time within 5 years after the conviction, or within such shorter period after the conviction as the court thinks fit, keep, or have, or act in the care or management of a common lodging-house without the previous licence in writing of the local authority, who may withhold or grant on such terms or conditions as they think fit.

(S. 89.) For the purposes of this Act the expression 'common lodging-house' includes, in any case in which only part of a house is used as a common lodging-house, the part so used of such house.

Bye-laws as to Houses let as Lodgings.

(S. 90.) The Local Government Board may, if they think fit, by notice published in the 'London Gazette,' declare the following enactment to be in force within the district or any part of the district of any local authority, and from and after the publication of such notice such authority shall be empowered to make bye-laws for the following matter (that is to say):

1. For fixing the number, and from time to time varying the number, of persons who may occupy a house or part of a house which is let in lodgings, or occupied by members of more than one family, and for the separation of the sexes in a house so let or occupied.

2. For the registration of houses so let or occupied.

3. For the inspection of such houses.

4. For enforcing drainage and the provision of privy accommodation for such houses, and for promoting cleanliness and ventilation in such houses.

5. For the cleansing and limewashing at stated times of the premises, and for the paving of the courts and courtyards thereof.

6. For the giving of notices and taking of precautions in case of any infectious disease.

This section shall not apply to common lodging-houses within the provisions of this Act relating to such houses.

LOGWOOD. *Syn.* CAMPEACHY WOOD; HÆMATOXYLUM (Ph. L. E. & D.), HÆMATOXYLI LIGNUM (B. P.), LIGNUM CAMPECHENSE, L. CAMPECHIANUM, L. The heart-wood of *Hæmatoxylon campechianum*, a native of Central America, but now common in the West Indies and India. It is a valuable astringent, and its decoction, extract, and infusion are useful remedies in chronic diarrhœa and dysentery, and in hæmorrhages, &c. The extract is an efficient substitute for catechu and kino.

Logwood is extensively employed in dyeing and calico printing, for the production of reds, violets, purples, blacks, drabs, &c. The colouring matter which it contains is hæmatoxylin, $C_{16}H_{14}O_6\cdot 3H_2O$; it is deposited from a boiling aqueous solution in yellow needles, which are soluble in alcohol and ether. It resembles the phenols by dissolving in alkalies to a purple solution which absorbs oxygen, forming the red colouring matter hæmatin. The colouring matter requires a large quantity of water to dissolve it, but when dissolved can be concentrated to any degree by boiling down. Extract of logwood should be made in vacuum pans withdrawn from the oxidising action of the air. The infusion is of a fine red, turning on the purple or violet; acids turn it on the yellow, and alkalies deepen it. An intense black colour is yielded with potassium chromate, but it is fugitive. To stuffs mordanted with alum it gives various shades of violet and purple, according to the proportions of the materials. By using solution of tin as the mordant, various shades of red, lilac, and violet may be obtained. The addition of a little Brazil-wood is commonly made to brighten the red. With a mordant of sulphate or acetate of iron it dyes black; and with the addition of a little sulphate of copper greys of various shades. It is, however, chiefly employed, in conjunction with gall-nuts, for blacks, to which it imparts a lustre and velvety appearance. Silk is usually turned through the cold decoction, but for wool the decoction is used either hot or boiling. Logwood is one of the cheapest and most easily managed of the dye-stuffs. It is also used to make ink, and sometimes as an indicator in alkalimetry. See HÆMATOXYLIN, INK, MICROSCOPE, &c.

LO'HOCH. See LINCTUS.

LORICA. A species of lute applied as a coating to chemical vessels before exposing them to the fire. Its application is called LORICATION. See LUTE.

LOTION. *Syn.* LOTIO, L. An external application, or wash, consisting of water holding in solution or suspension medicinal substances. Lotions may be prepared of any soluble medicaments that are capable of exerting their action by contact with the skin. Writers on pharmacology have arranged them in classes, as sedative, anodyne, stimulant, &c., according to their effects. Sedative and refrigerant lotions are commonly employed to allay inflammation; anodyne and narcotic lotions, to relieve pain; stimulant lotions, to induce the maturation of tumours, &c.; detergent lotions, to clean foul ulcers; repellent and resolvent lotions, to discuss tumours, remove eruptions, &c.; counter-irritant lotions, to excite a secondary morbid action, with the intention of relieving one already existing. Lotions are usually applied by wetting a piece of lint or wool with them and keeping it on the part affected; or, in slight cases, by moistening the part with the fingers previously dipped into them. Lotions are more agreeable if made with rose-water, but are not thereby rendered more efficacious. In all cases distilled water, or filtered soft water, is alone admissible as the solvent.

As lotions are, in general, mere extemporaneous or magistral preparations, it will, of course, be only necessary here to give the formulæ for a few of those which are the most useful or the most frequently employed. These will serve as examples from which others may be prepared. As a general rule, the medium dose of any substance dissolved in a fluid ounce of distilled water forms a lotion of the proper strength under all ordinary circumstances; or, what is the same thing, the medium dose in grains, taken in scruples, is sufficient for a pint of such a lotion. Thus the dose of sulphate of zinc is 1 to 3 gr., therefore: $\frac{1+3}{2} = 2$ gr., which is the proportion of sulphate of zinc to be taken for 1 fl. oz. of water, or 40 gr. for 1 pint. Again, the dose of perchloride of mercury is $\frac{1}{12}$ to $\frac{1}{4}$ gr.; therefore: $\frac{\frac{1}{12}+\frac{1}{4}}{2} = \frac{1}{14}$ gr., or nearly 3 gr. per pint.

In this method extreme or unusual doses, as, for instance, those of sulphate of zinc, as an emetic, in poisoning, &c., are not taken into the calculation. In all cases in which lotions are intended for extremely susceptible parts it is proper to dilute them with an equal bulk of water. When intended for eye-waters (COLLYRIA) they should be diluted with at least 3 to 4 times their bulk of water. See EMBROCATION, LINIMENT, &c.

Lotion of Ac'etate of Ammo"nia. Syn. LOTIO AMMONIÆ ACETATIS, L. Prep. 1. Solution of acetate of ammonia, 1 part; water, 9 parts.

2. (Hosp. F.) Solution of acetate of ammonia, rectified spirit, and water, equal parts. Discutient and refrigerant. In ordinary inflammations.

Lotion of Ac'etate of Lead. Syn. LOTIO PLUMBI ACETATIS, L. Prep. 1. (Collier.) Acetate of lead, 1 dr.; distilled water, 8 fl. oz. Sometimes a little vinegar is added. In excoriations, burns, sprains, contusions, &c. See SOLUTION OF DIACETATE OF LEAD.

2. Acetate of lead, 2 gr.; distilled water, 1 oz. (Ophthalmic Hospital).

Lotion of Ac'etate of Mercury. Syn. LOTIO HYDRARGYRI ACETATIS, L. Prep. Acetate of mercury, 1 scruple; distilled water, 1 pint. Mix.

Lotion of Ac'etate of Zinc. Syn. LOTIO ZINCI ACETATIS, L. Prep. 1. (Béral.) Acetate of zinc, 1½ dr.; water, 1 pint. Astringent, similar to lotion of sulphate of zinc.

2. Acetate of zinc, 1 to 2 gr.; water, 1 oz. An astringent collyrium in ophthalmia, and as injection in gonorrhœa after the acute stage has passed. Neither tincture nor wine of opium gives a precipitate with this lotion.

Lotion, Acetic. Syn. LOTIO ACETI, L. Prep. 1. Vinegar, 1 part; water, 2 or 3 parts. For bruises, contusions, &c., and as a general refrigerant application to sound parts.

2. Vinegar, 1 fl. oz.; cold water, ½ pint; as a wash in chronic ophthalmia, &c.

Lotion, Acid. See LOTIONS OF ACETIC, NITRIC, and PHOSPHORIC ACID, &c.

Lotion of Acon'itine. Syn. LOTIO ACONITINÆ, L. Prep. (Turnbull.) Aconitine, 8 gr.; rectified spirit, 2 fl. oz. In neuralgia; applied by means of a small piece of sponge mounted at the end of a stick. It must never be employed when the skin is broken or abraded; and it would be wise, in most cases, to dilute it with double its volume of proof spirit.

Lotion, Al'kaline. Syn. LOTIO ALKALINA, L. POTASSÆ CARBONATIS, L. Prep. (P. Cod.) From salt of tartar, 1 oz.; water, 1 pint. Stimulant and detergent. Diluted with an equal bulk of water, it forms an excellent cosmetic wash to remove scurf from the hair. Sometimes it is made with almond milk instead of water.

Lotion, Almond, Alkaline. (Dr A. T. Thomson.) Syn. Solution of potash, 4 fl. oz.; emulsion of bitter almonds, 5½ fl. oz. To remove the scurf in porrigo furfurans, applied twice a day diluted with warm water.

Lotion of Al'um. Syn. LOTIO ALUMINIS, L. Prep. From alum, 1½ dr.; distilled or rose water, 1 pint. Astringent. For sore gums, nipples, excoriations, &c.

Lotion, Ammoni'acal. Syn. LOTIO AMMONIÆ, L. AMMONIACALIS, L. Prep. 1. Liquor of ammonia, 3 fl. dr.; cold water, 5 fl. oz. As a stimulant to indolent ulcers, and in certain skin diseases.

2. (Swediaur.) Liquor of ammonia, spirit of thyme, and spirit of camphor, equal parts. In headaches, applied to the forehead and temples, and in other cases, as a counter-irritant. In most cases it should be used diluted.

3. (Opiated—Dr Kirkland.) Sal volatile, 3½ fl. oz.; tincture of opium, ½ fl. oz.; water, 4 fl. oz. Anodyne, stimulant, and resolvent.

Lotion, Ammonio-camphorated. Syn. AQUA SEDATIVA, L.; EAU SÉDATIVE DE RASPAIL, EAU OU LOTION AMMONIACALE CAMPHRÉE, Fr. No. 1. Liquor ammoniæ ('923), 6 parts; camphorated spirit, 1 part; salt, 6 parts; water, 10 parts. No. 2 contains 8 parts, and No. 3 10 parts of ammonia.

Lotion, Antiphlogis'tic. Syn. LOTIO ANTIPHLOGISTICA, L. Prep. 1. (Copland.) Solution of diacetate of lead, 3 fl. dr.; solution of acetate of ammonia, 2 fl. oz.; distilled water, 1 pint. Refrigerant, sedative, and repellent. Used to allay inflammation, &c.

2. (A. T. Thomson.) Opium, 2 dr.; distilled vinegar, ½ pint. Anodyne and refrigerant; in swelled joints, &c.

Lotion of Ar'nica. Syn. LOTIO ARNICÆ, L. Prep. 1. Tincture of arnica, 1 fl. dr.; rosewater, 2½ fl. oz. In contusions, bruises, extravasations, &c.

2. (Niemann.) Arnica flowers, ½ oz.; hot vinegar, 3 fl. oz.; boiling water, 5 fl. oz.; infuse until cold, and strain. In acute hydrocephalus, or with water, q. s. to measure a pint, as a common lotion.

Lotion, Arsenical. Syn. LOTIO ARSENICALIS, L. ACIDI ARSENIOSI, L. Prep. 1. Arsenious acid, 5 gr.; water, 1 pint. In psoriasis, &c.

2. (Compound—M. le Febvre.) Arsenious acid, 8 gr.; boiling water, 16 fl. oz.; dissolve, and add of extract of hemlock, 1 oz.; solution of diacetate of lead, 3 fl. oz.; tincture of opium, 1 fl. dr. Every morning, in cancer.

Lotion, Astrin'gent. Syn. LOTIO ASTRINGENS, L. See LOTIONS OF ALUM, SULPHATE OF ZINC, &c.

Lotion, Barlow's. Prep. From sulphuret of potassium (in powder), 3 dr.; soap (sliced), 1½ dr.; lime-water, 7½ fl. oz.; proof spirit, 2 fl. oz. In itch, ringworm, &c.

Lotion, Bateman's. Prep. From perchloride of mercury, 2 gr.; compound spirit of lavender,

1 fl. oz.; dissolve, and add of distilled water, 4 fl. oz. In obstinate cutaneous eruptions, more especially those of a papular character.

Lotion of Belladon'na. *Syn.* LOTIO BELLADONNAE, L. *Prep.* (*Graefe.*) Extract of belladonna, ½ dr.; dilute solution of diacetate of lead, ½ pint. Applied to tumours and glandular enlargements.

Lotion of Benzoin. *Syn.* LOTIO BENZOINI, L. Tincture of benzoin, 1 part; rose water, 40 parts. A nice lotion to protect the face from the heat of the sun.

Lotion of Bismuth. *Syn.* LOTIO BISMUTHI, L. Nitrate of bismuth, 6 gr.; corrosive sublimate, ½ gr.; spirits of camphor, 1½ minims; water, 1 oz. A soothing lotion in chronic skin affections.

Lotion, Black. See LOTION, MERCURIAL.

Lotion of Borax. *Syn.* LOTIO BORACIS, L. BORACICA, L. *Prep.* 1. (*Dr. Abercrombie.*) Borax, 2½ dr.; distilled vinegar, ¼ pint. In ringworm.

2. (*Copland.*) Borax (in powder), 1 dr.; rose-water and orange-flower water, of each, 2 fl. oz.; dissolve. A fragrant and effective application to sore gums, sore nipples, excoriations, &c.

3. (*Dr Johnson.*) Borax, 2 dr.; precipitated chalk, 1 oz.; rose-water and rectified spirit, of each, 3 oz. For sore nipples.

4. (*Dr Meigs.*) Borax, ½ oz.; sulphate of morphia, 6 gr.; rose-water, 8 fl. oz. To allay itching and irritation, especially pruritus vulvae.

5. Borax, 1 part; rose-water, 24 parts. Cosmetic.

Lotion of Boric Acid. *Syn.* LOTIO ACIDI BORICI. A saturated solution of boric acid in water, about 1 in 23. Valuable as a mild antiseptic wash for ulcerated parts.

Lotion, Bro'mine. *Syn.* LOTIO BROMINII, L. *Prep.* (*Dr Glover.*) Bromine, 1 dr.; water, 1 pint. As an application to scrofulous ulcers.

Lotion for Burns. See LINIMENT.

Lotion of Calamine. *Syn.* LOTIO CALAMINAE, L. *Prep.* Calamine, 40 gr.; zinc oxide, 20 gr.; glycerin, 20 minims; water to 1 oz. Astringent and sedative; allays irritation in skin diseases, especially useful in eczema.

Lotion, Camphora'ted. See LOTION, EVAPORATING.

Lotion of Cap'sicum. *Syn.* LOTIO CAPSICI, L. *Prep.* (*Griffith.*) Tinctures of capsicum and camphor, of each, 4 fl. oz.; liquor of ammonia, 2 fl. oz. A powerful rubefacient and counter-irritant.

Lotion of Carbolic Acid. (*Sir J. Lister.*) *Syn.* LOTIO ACIDI CARBOLICI, L. *Prep.* 1 part of acid in 20 of water is used to promote the healing of wounds, abscesses, ulcers, and burns. A weaker solution of 1 in 40 is in common use in the London hospitals. 5 drops to 1 fl. oz. of glycerin forms a good application to eruptions of the skin.

Lotion of Car'bonate of So'da. *Syn.* LOTIO SODAE CARBONATIS, L. *Prep.* From carbonate of soda, ½ oz.; water, 1 pint. To allay itching and irritation. See LOTION, ALKALINE.

Lotion of Cher'ry Laurel. *Syn.* LOTIO LAUROCERASI, L. *Prep.* 1. Cherry-laurel water (distilled), 1½ fl. oz.; distilled water, ½ pint. Anodyne; useful to allay irritation, &c. Some

persons with delicate skin employ it as a wash after shaving.

2. Cherry-laurel water (distilled), 4 oz.; rectified spirit and ether, of each, 1 fl. oz.; extract of belladonna, 2 dr.; agitate well together in the cold. An excellent application in neuralgia, painful tumours, &c.

Lotion for Chilblains. See CHILBLAIN, LINIMENT, &c.

Lotion of Chlo'rate of Soda. *Syn.* LOTIO SODAE CHLORATIS, L. *Prep.* (*Darling.*) Chlorate of soda, 5 dr.; water, ½ pint. In pruritus, &c.

Lotion of Chlo''ride of Ammonium. LOTIO AMMONII CHLORIDI, L. Chloride of ammonium, 1 oz.; rectified spirit, 1 oz.; water, 10 oz. To this vinegar is sometimes added. Used as a dressing for bruises. See also LOTION OF HYDROCHLORATE OF AMMONIA.

Lotion of Chlo''ride of Lead. *Syn.* LOTIO PLUMBI CHLORIDI, L. *Prep.* (*Tuson.*) Chloride of lead, 1 dr.; hot distilled water, 1 pint; dissolve. In cancerous ulcerations, painful neuralgic tumours, &c.

Lotion of Chloride of Zinc. *Syn.* LOTIO ZINCI CHLORIDI, L. *Prep.* Chloride of zinc, 10 gr. (or solution, ½ fl. dr.); water, 1 pint. As a disinfectant and preventive lotion.

Lotion, Chlorina'ted. *Syn.* LOTIO CHLORINATA, L. *Prep.* 1. (LOTIO CALCIS CHLORINATAE.) a. From chloride of lime, 3 dr.; water, 1 pint; agitate together for some time, and strain through muslin.

b. (*Derheims.*) Chlorinated lime, 1 oz.; water, 1 quart; triturate and filter.

2. (LOTIO SODAE CHLORINATAE.) From chlorinated soda, as the last. They are both excellent washes for foul ulcers, the itch, &c.; and, when diluted for the teeth, to sweeten the breath, remove the smell of tobacco smoke, to prevent infection, and for various purposes. When intended for application to very tender or abraded surfaces, they must be largely diluted with water.

Lotion of Chlo''roform. *Syn.* LOTIO CHLOROFORMI, L. *Prep.* Chloroform (pure), 1½ fl. oz.; rectified spirit and cold distilled water, of each, ½ pint. Anodyne. A piece of oiled silk should be laid over the rag to prevent evaporation. The lotion made with water as commonly prescribed is inert.

Lotion for Corns. See CORN.

Lotion of Crea'sote. *Syn.* LOTIO CREASOTI, L. *Prep.* 1. Creasote, 2 fl. dr.; liquor of potassa, 3 fl. dr.; water, ½ pint.

2. Creasote, 3 fl. dr.; vinegar and water, of each, ½ pint. In burns, itch, phagedenic ulcerations, ringworm, chancre, &c.

Lotion of Cy'anide of Potas'sium. *Syn.* LOTIO POTASSII CYANIDI, L. *Prep.* 1. (*Casenave.*) Cyanide of potassium, 10 gr.; emulsion of bitter almonds, 6 fl. oz. In chronic eruptions and other cases attended with much itching or irritation.

2. (*Foy.*) Cyanide of potassium, 8 gr.; distilled water, 1 fl. oz. In neuralgia, acute rheumatism, &c.; applied by means of compresses of linen. Both the above are poisonous if swallowed, and should never be used except under medical supervision.

Lotion of Diac'etate of Lead. *Syn.* GOU-

LARD'S LOTION; LOTIO PLUMBI SUBACETATIS, L. The dilute liquor of diacetate of lead (LIQ. PLUMBI DIACETATIS DILUTUS, Ph. L.). See SOLUTION. Also solution of subacetate of lead (B. P.), 3 minims, with 7 minims to 1 oz. water.

Lotion, Evap'orating. *Syn.* LOTIO EVAPO-RANS, L. VAPORANS, L. SPIRITUS DILUTI, L. *Prep.* 1. (*Copland.*) Sulphuric ether, rectified spirit, and solution of acetate of ammonia, of each, 1½ fl. oz.; rose-water, 3½ fl. oz.

2. (Guy's Hosp.) Rectified spirit, 1 part; water, 5 parts.

3. (*Eras. Wilson.*) Rectified spirit, 1 part; water, 4 to 6 parts.

4. (CAMPHORATED, *Ware.*) Camphor, ½ dr.; elder flowers, ½ oz.; rectified spirit, 4 oz.; digest 24 hours and strain.

· *Obs.* The above are soothing and refrigerant if allowed to evaporate by free exposure; stimulant if the evaporation is prevented by covering the part with the hand or a piece of oiled silk. They are useful applications in nervous headaches, restlessness, itching and irritability of the skin, &c. " A little rose-water added to the simple water makes an agreeable addition, and sometimes camphor water (julep), or a little Goulard's extract, may be deemed advantageous when a greater degree of calming effect is required " (*Eras. Wilson*). Eau de Cologne, diluted with an equal quantity of water, is often used as an evaporating lotion.

Lotion of Gall-nuts. *Syn.* LOTIO GALLÆ, L. *Prep.* From gall-nuts (bruised), ½ oz.; boiling water, 1 pint; infuse until cold, and strain. Astringent. An excellent application to sore nipples, or to strengthen them before suckling; spirit of wine, 3 fl. oz., may be advantageously added to the cold infusion, and a like portion of water omitted. See DECOCTION.

Lotion of Glyc'erin. *Syn.* LOTIO GLYCERINI, L. GLYCERINIÆ, L. *Prep.* 1. Glycerin, 1 oz.; water, 1 pint. To allay itching, and remove dryness, &c., in various skin diseases; also in chaps of the nipples, lips, and hands. For the latter purpose the addition of 2 to 3 dr. of borax is recommended by some writers.

2. Glycerin, 1 oz.; thick mucilage, 2 oz.; lime-water, 7 oz. In burns, scalds, chaps, excoriations, &c.

3. (*Startin.*) Glycerin, 1 oz.; extract of belladonna, 1 dr.; soap liniment, 3 oz.; triturate together. In bruises, sprains, and swelled joints; gouty, neuralgic, and rheumatic pains, &c.

4. (*Startin.*) Trisnitrate of bismuth, ½ dr.; tincture of foxglove and dilute nitric acid, of each, 1 fl. dr.; glycerin, 4 dr.; rose-water, 8½ fl. oz. To allay the itching in prurigo, and some other skin diseases.

Obs. Various lotions may be prepared by dissolving active medicinal substances in glycerin.

Lotion, Goulard's. See LOTION OF DIACETATE OF LEAD.

Lotion, Gout. *Syn.* LOTIO ANTARTHRITICA, L. *Prep.* 1. Glycerin, 1 oz.; extract of belladonna, 3 dr.; veratrine, 10 gr., dissolved in rectified spirit, 2 fl. oz.; mix, and further add, of water, 17 fl. oz. It is poisonous if swallowed.

2. ('SCUDAMORE'S G. L.') From camphor mixture, 9 fl. oz.; rectified spirit, 3 fl. oz. The above

are applied on rags or compresses, or are poured on the surface of poultices.

Lotion, Gowland's. This celebrated nostrum is prepared as follows:—Take of Jordan almonds, 1 oz.; bitter almonds, ½ oz.; blanch them, and make an emulsion in soft water, 1 pint; to this add of bichloride of mercury, 15 gr.; previously dissolved in rectified spirit, 2 fl. dr., together with enough water to make the whole measure 1 pint, and put it into bottles.

Obs. This preparation is chiefly used as a cosmetic to improve the complexion; and also as a wash for obstinate eruptions and minor glandular swellings and indurations. As a beautifier of the complexion it is employed by simply wetting the skin with it, either by means of the corner of a napkin or the fingers dipped into it, after which it is gently wiped off with a dry cloth. Dr Paris represents this nostrum to contain ½ dr. of corrosive sublimate in every pint, which is not the case.

Lotion, Granville's Counter-irritant. See LINIMENT OF AMMONIA (Compound).

Lotion, Hem'lock. *Syn.* LOTIO CONII, L. *Prep.* (Mid. Hosp.) Extract of hemlock, 3 dr.; opium, 1 dr.; boiling water, 1 pint; digest until cold, and strain. Anodyne and resolvent; in glandular enlargements, painful ulcers, cancer, indurations, rheumatism, neuralgia, &c.

Lotion, Hooping-cough. (*Struve's.*) *Syn.* LOTIO ANTIPERTUSSICA, L. *Prep.* (Paris.) Potassio-tartrate of antimony, 1 dr.; tincture of cantharides, 1 oz.; water, 2 oz. This is a powerful counter-irritant, and should be used with caution, as it is apt to induce a troublesome eruption on the parts to which it is frequently applied.

Lotion of Hydrochlo"rate of Ammonia. *Syn.* LOTIO AMMONIÆ HYDROCHLORATIS, L. *Prep.* 1. (WEAKER.) From sal-ammoniac, 1 to 4 dr.; water, 1 pint. As a wash in itch, ulcers, tender feet, swelled joints, &c.

2. (STRONGER.) From sal-ammoniac, 1 to 2 oz.; water, 1 pint. In contusions, chronic tumours, extravasations, chilblains, &c., when the skin is not broken. Both are stimulant and resolvent or discutient. Vinegar is often substituted for the whole or part of the water, and sometimes a fifth or sixth part of rectified spirit is added. See also LOTION OF CHLORIDE OF AMMONIUM.

Lotion, Hydrochlo"ric. *Syn.* LOTIO ACIDI HYDROCHLORICI, L. *Prep.* 1. Hydrochloric acid, 1 fl. oz.; water, 1 pint. In lepra and several other skin diseases.

2. (*Foy.*) Hydrochloric acid, 1 part; water, 16 parts. In chilblains, when the skin is unbroken.

Lotion, Hydrocyan'ic. *Syn.* LOTIO HYDRO-CYANICI, L. ACIDI HYDROCYANICI, L. *Prep.* 1. (*Magendie.*) Medicinal hydrocyanic acid, 1 to 2 fl. dr.; lettuce water, 1 pint. In hepatic affections.

2. (*Sneider.*) Medicinal acid, 1½ fl. dr.; rectified spirit and water, of each, 6 fl. oz.

3. (*A. T. Thomson.*) Medicinal acid and rectified siprit, of each, 2 fl. dr.; acetate of lead, 16 gr.; distilled water, 7½ fl. oz. In impetigo, &c.

Obs. Lotions of prussic acid are employed to allay pain and irritation in various chronic skin diseases, especially the scaly and itchy eruptions; and in cancer, &c., with variable success. See HYDROCYANIC ACID.

Lotion of Hyposul'phite of Soda. *Syn.* LOTIO SODÆ HYPOSULPHITIS, L. *Prep.* (*Startin.*) Hyposulphite of soda and alum, of each, 1½ dr.; eau de Cologne, ½ fl. oz.; rose-water, 7½ fl. oz.; in the advanced stages of acne.

Lotion of I'odide of Ar'senic and Mer'cury. *Syn.* LOTIO ARSENICI ET HYDRARGYRI HYDRIODATIS, L. *Prep.* From Donovan's solution, 1 part; water, 9 parts. In lepra, psoriasis, and other scaly skin diseases. See SOLUTION.

Lotion of Iodide of Potas'sium. *Syn.* LOTIO POTASSII IODIDI, L. *Prep.* 1. From iodide of potassium, 1 to 2 dr.; water, 1 pint. In the usual cases in which ioduretted preparations are employed.

2. (*Dr O. Ward.*) Iodide of potassium, 1 dr.; water, ¾ pint. In itch. (See *below.*)

Lotion of Iodide of Zinc. *Syn.* LOTIO ZINCI IODIDI, L. *Prep.* (*Ross.*) Iodine, 1½ dr.; zinc filings, 1 dr.; water, 8 fl. oz.; digest with heat until the liquid becomes coloured, then filter. In enlarged tonsils.

Lotion of I'odine. *Syn.* LOTIO IODINII, L. *Prep.* From iodine, 2 gr.; rectified spirit, 1 fl. dr.; dissolve, well agitate the solution with distilled water, 1 pint, and filter. An excellent wash for scrofulous ulcers, and in chronic ophthalmia, cutaneous scrofula, and several chronic skin diseases, particularly in highly sensitive habits.

Lotion of Iodine, Compound. *Syn.* LOTIO IODI COMP., L. *Prep.* 1. Iodide of potassium, 80 gr.; iodine, 60 gr.; water, 1 oz.

2. (*Casenave.*) Iodide of potassium and iodide of sulphur, of each, 1 dr.; water, 1 oz. In itch, either alone, or diluted with an equal bulk of water.

3. (*Dauvergne.*) Iodine, 3 dr.; iodide of potassium, 6 dr.; water, 3 fl. oz.; dissolve, and label the bottle No. 1. Sulphuret of potassium, 4 oz.; water, 8 fl. oz.; dissolve. For use, a teaspoonful of No. 1 and a table-spoonful of No. 2 are to be added to about a pint of water. In itch and several other skin diseases.

4. (*Lugol.*) Iodine, 1 to 2 gr.; iodide of potassium, 3 to 6 gr.; water, 1 pint. In scrofulous ophthalmia, fistulas, &c.; and as a wash in numerous skin diseases.

5. (*Righini.*) Chloride of lime, 4 dr.; water, 2½ fl. oz.; triturate together, filter into a stoppered bottle, and add of tincture of iodine, 1 dr. With a pint of water it forms an effective application in itch.

6. (*Soubeiran.*) Iodide of potassium, 1 oz.; iodine, ½ oz.; water, 6 oz.; dissolve. Used as iodine paint; also as a caustic to touch the surfaces of scrofulous ulcers, and the eyelids in scrofulous ophthalmia.

7. Iodide of potassium, ½ dr.; iodine, 16 gr.; water, 1 pint. This is the common and best form of iodine lotion, but for certain purposes it is used much stronger (see *above*).

Lotion. *Syn.* LOTIO ANTIPSORICA, L. *Prep.* (*Casenave.*) Sulphuret of potassium, 1 dr.; soft soap, 2 dr.; water, 8 fl. oz.; dissolve.

An excellent remedy for the itch. It leaves little smell behind, and does not soil the linen. (See *above.*)

Lotion, Kirkland's. See LOTION OF MYRRH.

Lotion of Lemon Juice. *Syn.* LOTIO SUCCI LIMONIS, L. *Prep.* From the freshly expressed juice of lemon, diluted with 4 or 5 times its bulk of water. To render it more agreeable, rosewater may be employed, or a few drops of eau de Cologne added. It is cooling and detergent, and forms an excellent applicatition to foul ulcers, and to allay the itching in numerous cutaneous affections.

Lotion of Lime Wa'ter. *Syn.* LOTIO CALCIS SPIRITUOSA, L. *Prep.* (Ph. Chirur.) Rectified spirit, 4 oz.; lime water, 8 fl. oz. See EVAPORATING LOTION (*above*).

Lotion, Locock's, for the Hair. *Prep.* 1. (Ince's formula.) The external application called Locock's Lotion for the Hair was devised by Mr Alexander, the celebrated oculist, for the use of his wife. It was seen by Dr (afterwards Sir Charles) Locock, who recommended it to his friends, and thus it gained its name. The formula marked 2 was the original in Alexander's handwriting. The proportion of oil of mace, ʒss (½ oz.) to a 4-oz. lotion, was found too large, and was soon altered.

The hair lotion supplied to Sir Charles Locock and others was made according to the following working formula:—Ol. macis, 5 oz.; ol. olivæ, 20 oz.; aq. ammon. fort., 20 oz.; sp. rosemar., 50 oz.; aq. rosæ, ad 2 galls. imperial.

2. Ol. macis, ½ oz.; ol. olivæ, 2 dr.; aq. ammoniæ, ½ dr.; sp. rosmarini, 1 oz.; aq. rosæ, 2½ oz.

Lotion, Mercu"rial. *Prep.* 1. (BLACK WASH, BLACK LOTION, MILD PHAGEDÆNIC L.; LOTIO NIGRA (B. P.), L. HYDRARGYRI CINEREA, L. H. NIGRA, L. H. CHLORIDI CUM CALCE, L. MERCURIALIS N., AQUA PHAGEDÆNICA, MITIS, L.) *a.* (B. P.) From calomel, 3 gr.; lime water, 1 oz.; well shaken together.

b. (Mid. Hosp.) To the last add of thick mucilage, 1 fl. oz.

c. (Guy's Hospital.) From calomel, 1 dr., lime-water, 8 fl. oz.

Obs. Black wash is a favourite application to all kinds of syphilitic and scrofulous sores. The bottle should be well shaken before the lotion is applied.

2. (YELLOW WASH, Y. LOTION, PHAGEDÆNIC L.; LOTIO FLAVA, L. PHAGEDÆNICA, AQUA P., LOTIO HYDRARGYRI FLAVA, L. H. BICHLORIDI CUM CALCE, L.) *a.* (B. P.) Corrosive sublimate, 18 gr.; lime-water, 10 oz.; well shaken together.

b. (St B. Hosp.) Corrosive sublimate, 20 gr.; lime-water, 6 fl. oz. Used as the last, but it is stronger and more active, from containing a little undecomposed bichloride.

Lotion of Myrrh. *Syn.* KIRKLAND'S LOTION; LOTIO MYRRHÆ, L. *Prep.* 1. (*Dr Kirkland.*) Tincture of myrrh and lime-water, equal parts. In scorbutic ulcers and gums.

2. (Compound: LOTIO MYRRHÆ COMPOSITA, L., Ph. Chirur.) Honey of roses and tincture of myrrh, of each, 2 fl. dr.; lime-water, 2½ fl. oz. As No. 1; also used as a dentifrice.

Lotion of Ni'trate of Bis'muth. *Syn.* LOTIO BISMUTHI NITRATIS, L. *Prep.* (Cutan. Hosp.) Subnitrate of bismuth, ½ dr.; corrosive sublimate, 12 gr.; spirit of camphor, ½ fl. dr.; water, 1 pint. In itch, and some other eruptions.

Lotion of Nitrate of Sil'ver. *Syn.* LOTIO ARGENTI NITRATIS, L. *Prep.* 1. Nitrate of silver, 15 gr.; nitric acid, 10 drops; distilled water, ½ pint. As a wash for indolent ulcers, sore legs, &c.

2. (*Jackson.*) Nitrate of silver, 10 gr.; water, 1 fl. oz. For bedsores; applied, at first, twice or thrice a day.

3. (*Schreider.*) Nitrate of silver, ½ dr.; nitric acid, 10 drops; water, 1½ fl. oz. In chilblains, soft corns, &c.

Lotion of Nitrate of Sil'ver, Ethereal. *Syn.* LOTIO ARGENTI NITRATIS ÆTHEREA. Nitrate of silver, 20 gr.; distilled water, 1 dr.; spirit of nitrous ether, 1 oz.

Lotion of Nitrate of Silver (Strong) *Syn.* LOTIO ARGENTI NITRATIS FORTIS. Nitrate of silver, 60 gr.; distilled water, 1 oz.

Lotion of Ni'tre. *Syn.* LOTIO POTASSÆ NITRATIS, L. *Prep.* 1. Nitre, 3 dr.; vinegar, ½ pint; water, ½ pint.

2. Nitre, 2 dr.; sal-ammoniac, 1 dr.; vinegar and water, of each, ½ pint. In sprains, contusions, extravasations, tender feet, chilblains, &c. Diluted with an equal bulk of water, it is a popular application to 'black eyes.'

Lotion of Ni'tric Acid. *Syn.* LOTIO ACIDI, L. ACIDI NITRICI, L. *Prep.* 1. (*Collier.*) Nitric acid, ½ fl. oz.; water, 1 pint. In lepra, and other scaly skin diseases.

2. (*Phœbus.*) Nitric acid, 1 fl. dr.; laudanum, 1½ fl. dr.; rose-water, ½ pint. For venereal ulcers.

Lotion of Nitro-muriat'ic Acid. *Syn.* LOTION OF AQUA REGIA. *Prep.* (*Copland.*) Nitromuriatic acid, 1½ dr.; water, 1 pint. In gangrene and mortification.

Lotion of Nux Vom'ica. *Syn.* LOTIO NUCIS VOMICÆ, L. *Prep.* 1. Alcoholic extract of nux vomica, 10 gr.; rectified spirit and water, of each, 2½ fl. oz. In amaurosis.

2. (*Radius.*) Alcoholic extract of nux vomica, 8 gr.; liquor of ammonia (stronger), ½ fl. oz.; rectified spirit, 2 fl. oz. In paralysed limbs.

Lotion of O'pium. *Syn.* LOTIO OPII, L. OPIATA, L. *Prep.* 1. (*Christison.*) Opium, 40 gr.; water, ½ pint; infuse, add to the filtered liquid a solution of sugar of lead, 40, in water, ½ pint, and filter.

2. (St B. Hosp.) Opium, 1½ dr.; boiling water, 1 pint; triturate and strain. Anodyne; the first is also refrigerant and discutient.

Lotion of Ox'ide of Zinc. *Syn.* LOTIO ZINCI OXIDI, L. *Prep.* 1. (*Augustin.*) Oxide of zinc, 1 dr.; elder-flower water, 1½ fl. oz. In pustular erysipelas.

2. (Hosp. F.) Oxide of zinc, ½ dr.; mucilage, 2 fl. dr.; water, 6 fl. dr. As an astringent and desiccant in scrofulous eruptions, excoriations, moist chaps, &c.

Lotion of Perchloride of Mercury. *Syn.* LOTIO HYDRARGYRI PERCHLORIDI. A solution of perchloride of mercury in distilled water. Strengths, 1 in 1000, 2000, or 5000. A most powerful and valuable antiseptic, much used as a general dressing for wounds, by immersing lint or wool in the lotion.

Lotion, Phagedæn'ic. See MERCURIAL LOTION (*above*).

Lotion of Phosphor'ic Acid. *Syn.* LOTIO ACIDI PHOSPHORICI, L. *Prep.* (*Pereira.*) Dilute phosphoric acid (Ph. L.), 1 fl. oz.; water, ½ pint. In caries and fistula.

Lotion of Potas'sa. *Syn.* LOTIO POTASSÆ, L. *Prep.* From liquor of potassa, 1 fl. oz.; water, 1 pint. Detergent; in scorbutic eruptions and foul ulcers, and to prevent infection.

Lotion of Potas'sio-tar'trate of An'timony. *Syn.* LOTIO ANTIMONIALIS, L. ANTIMONII POTASSIO-TARTRATIS, L. RUBEFACIENS, L. *Prep.* 1. Tartar emetic, 1 dr.; tincture of camphor, 2 fl. dr.; water, 1 pint. As a local stimulant. Diluted with twice or thrice its weight of water, it is employed as a collyrium in chronic ophthalmia, and in specks on the cornea.

2. (*Sir Wm. Blizard.*) Tartar emetic, 20 gr.; boiling water, 1 fl. oz. Used to cleanse foul ulcers, to repress fungous growths and warts, and in ringworm, &c.

3. (*Pereira.*) Tartar emetic, 1 dr.; boiling water, 1½ fl. oz.; dissolve. Employed as a local irritant instead of the ointment. All the above are rubefacient and counter-irritant. See ANTIMONY.

Lotion of Quin'ine. *Syn.* LOTIO QUINÆ, EMBROCATIO Q., L. *Prep.* From disulphate of quinine, 1 dr.; rectified spirit, 5 fl. oz. Applied over the spine in intermittents.

Lotion, Sapona'ceous. *Syn.* LOTIO SAPONIS, L. SAPONACEA (Ph. L. 1746), L. *Prep.* From liquor of carbonate of potassa, ½ oz.; olive oil, 4 oz.; rose-water, 12 oz.; agitate together. Emollient; chiefly as a cosmetic.

Lotion, Saviard's. *Prep.* (*Foy.*) Caustic potassa, 1 dr.; camphor, 20 gr.; sugar, 1 oz.; water, 1 pint. As a wash for indolent ulcers.

Lotion, Struve's. See HOOPING-COUGH LOTION.

Lotion of Sul'phate of Cop'per. *Syn.* LOTIO CUPRI SULPHATIS, L. *Prep.* 1. Blue vitriol, 1 dr.; camphor julep, 1 pint. For phagedænic ulcers, and in itch, &c.

2. (Dr *Graves.*) Sulphate of copper, 10 gr.; water, 1 fl. oz. In chilblains, ringworm, &c.

3. (*Lloyd.*) Sulphate of copper, 1 oz.; water, 1 pint. In itch; either alone or diluted.

Lotion of Sul'phate of Iron. *Syn.* LOTIO FERRI SULPHATIS, L. *Prep.* Sulphate of iron, 2 gr.; water, 1 oz.

Lotion of Sul'phate of Zinc. *Syn.* LOTIO ZINCI SULPHATIS, L. *Prep.* 1. Sulphate of zinc, ½ dr.; water, 1 pint. Astringent; in some chronic skin diseases, as a wash for loose, flabby granulations, and for ulcers that discharge profusely, &c.

2. (*Collier.*) Sulphate of zinc, 2 dr.; water, 1 pint. As a counter-irritant in pains of the joints, periosteum, old sprains, &c.

Lotion of Sulphur. *Syn.* LOTIO SULPHURIS, L. *Prep.* Precipitated sulphur, 10 dr.; rectified spirit, 5 oz.; water, 10 oz. Stimulant and parasiticide.

Lotion of Sulphuret of Sodium. (Dr *Barlow.*)

Syn. LOTIO SODII SULPHURETI, L. *Prep.* Sulphide of sodium, 2 dr.; white soap, 2½ dr.; rectified spirit, 2 dr.; lime-water, 7 oz. For ringworm.

Lotion of Tannin. (*Mr Druitt.*) *Syn.* LOTIO TANNINI, L. *Prep.* Tannic acid, 5 gr.; distilled water, 1 oz.; mix. On lint, covered with oil-silk, to sore nipples.

Lotion of Tar. *Syn.* LOTIO PICIS LIQUIDÆ, L. *Prep.* (*Saunders.*) Quicklime, 6 oz.; water, 2½ pints; slake, add of tar, 4 oz., and boil to one half. This liquid may be advantageously employed in various chronic skin diseases, especially those affecting the heads of children. See INFUSION OF TAR.

Lotion of Ver'digris. *Syn.* LOTIO ÆRUGINIS, L. CUPRI CITRATIS, L. *Prep.* From verdigris, 3 dr.; vinegar, ½ pint; water, ½ pint. As a wash for indolent, scrofulous, and venereal ulcers.

Lotion of Vin'egar. See ACETIC LOTION (*above*).

Lotion, Yellow. See MERCURIAL LOTION (*above*).

LOUSE. *Syn.* PEDICULUS. There are several species of this offensive parasite infesting the bodies of man and domesticated animals. The three varieties of lice found on the human skin are—(1) The *Pediculus corporis*, (2) the *P. capitis*, (3) the *P. pubis*.

1. The *P. corporis*, the body louse, is of a dirty white colour, and varies from ½ to 2 lines in length. Its body is broad and elongated, with the margins divided into lobes, and covered with minute hairs; but it has a narrow thorax, furnished on each side with three legs, which terminate in claws. This creature produces great irritation of the skin, giving rise to a number of little pimples on it, which frequently discharge a watery fluid. It multiplies with extraordinary rapidity.

2. The *P. capitis*, the head louse, is much smaller than the above. It is devoid of hairs, with legs large in proportion to its body. It gives rise to a very troublesome eruption, attended with a watery discharge. It is propagated by means of the ova or nits, which are glued to the hairs of the head.

3. The *P. pubis*, the crab louse, is a small, round variety, which attaches itself with considerable tenacity to the hairs of the stomach and lower part of the body more particularly, and, like the preceding parasite, glues its eggs to the hairs.

Various applications have been recommended for the destruction of these loathsome parasites; amongst which we may mention sulphur, stavesacre, white precipitate, and cocculus indicus, in the form of ointments; carbolic acid and perchloride of mercury lotions, and tobacco. Benzoic acid has been found of service in allaying the irritation. Diligent washing with soap and water should be had recourse to previous to applying any of the above remedies, and should the head be infested, the hair should be cut short, and frequently combed with a small-tooth comb.

Pediculi are sometimes conveyed from filthy to cleanly persons by means of dirty water-closets, chairs, sheets, brushes and combs, and in various other ways.

School children frequently obtain them in consequence of their heads being brought into too close contact with the heads of other children infested by them.

LOZENGE. *Syn.* TROCHE; TROCHISCUS, TABELLA, L.; TABLETTE, Fr. A small cake, often medicated, consisting principally of powdered sugar, made into a mass with some glutinous liquid, without the aid of heat, and dried. The form given to lozenges (TROCHE TABELLÆ, TROCHISCI, TABLETTES) is generally that of a small round tablet or flattened cylinder; but originally they were exclusively made in the shape of a lozenge or rhomb, from which circumstance their familiar name is derived. LOZENGES are distinguished from DROPS or PASTILLES by the non-employment of heat in their preparation; and from PASTES, by the latter being formed of vegetable juice or pulp, and having a softer consistence. The lozenges of the Throat Hospital Pharmacopœia have a basis of either red or black currant paste.

They are intended to be used by placing them in the mouth and permitting them to remain until dissolved. They are valuable where prolonged local contact of a drug is required, as in disorders of the mouth and throat. Very powerful or disagreeable remedies should not be administered this way.

In the preparation of lozenges the dry ingredients, separately reduced to a very fine powder, are first perfectly mixed together, and then beaten into a stiff paste with the glutinous liquid employed to give them form; the mass is next rolled out to a desired thickness, and cut into pieces of the proper shape by means of a small cylinder or punch of steel or tin-plate, called a 'lozenge cutter.' The newly formed lozenges are lastly dried by placing them on an inverted sieve or frame covered with paper in a dry, warm, and airy situation, and are frequently turned until they become hard and brittle, due care being taken to preserve them from dust and dirt. To prevent the mass adhering to the fingers and utensils during the process of manufacture, a little finely powdered starch, or a very little olive oil, scented with the same aromatic as that contained in the lozenges, may be used. Mucilage of gum-arabic or of gum-tragacanth, thin isinglass size, and the strained white of egg, are the substances usually employed to make the pulverulent materials adhere together. A strained decoction of Irish moss is now frequently used for the same purpose, for inferior qualities. The larger the proportion of gum which enters into the composition of lozenges, the slower they dissolve in the mouth; hence powdered gum is frequently added to the other materials to increase their quality in this respect, as well as to give an additional solidity to those which, like chalk, for instance, are of a peculiarly dry or crumbly nature. Starch and potato flour are often added to lozenge-masses in lieu of a portion of the sugar, and even plaster of Paris is not unfrequently employed to give them weight—frauds which are readily detected in the manner noticed under GUM and SUGAR.

As a general rule, MEDICATED LOZENGES should weigh from 8 to 10 gr. each, and a medium dose of their active ingredient should be distributed through the bulk of 6 to 8 of them, in which case 3 to 5 of them may be safely taken as a dose, or sucked during the lapse of 3 or 4 hours. This will be useful in the preparation of those for which no established proportions are given. In ' sending out' compounds of this class containing active medicaments, as morphia or opium, the retailer as well as the manufacturer should be careful that the quantity contained in each lozenge is plainly marked on the label.

In lozenges intended for MOUTH COSMETICS or to perfume the breath, ambergris is generally regarded as the most appropriate perfume; but hard smokers frequently prefer cloves and cinnamon, and some ladies give the preference to roses, orange flowers, and orris or violets.

Lozenges are coloured with the same stains as are used for liqueurs and sweetmeats.

Lozenges, as well as all other similar articles of confectionery, should be preserved in well-closed glass bottles or jars, or in tin canisters, so as to be perfectly excluded from the air and damp.

Lozenges, Absor'bent. *Syn.* TROCHISCI ANT-ACIDI, L. *Prep.* 1. Take of precipitated chalk, ¼ lb.; gum-arabic, 2 oz.; double refined white sugar, 14 oz., all in impalpable powder; oil of nutmeg, ½ fl. dr.; pass the mixture through a fine sieve, beat it up with mucilage, q. s., roll the mass into a thin sheet, and cut it into lozenges; lastly, dry them by exposing them on a sheet of white paper to the air, out of contact with dust.

2. As the last, but substituting heavy carbonate of magnesium, 1½ oz., for an equal weight of chalk. In diarrhœa, heartburn, acidity, &c. See LOZENGES, CHALK, MAGNESIA, L., SODA, &c.

Lozenges, Aca'cia. See LOZENGES, GUM.

Lozenges, Acid'ulated. *Syn.* ACIDULATED LEMON LOZENGES, TARTARIC ACID L.; TROCHISCI ACIDI TARTARICI (Ph. E.), L. *Prep.* From tartaric acid, 2 dr.; oil of lemon, 10 drops; white sugar, 8 oz.; mucilage, q. s. to make a lozenge mass. The same ingredients mixed with heat form ACIDULATED or ACID DROPS. Both are useful in coughs, hoarseness, sore throats, &c. See LOZENGES, CAYENNE, CITRIC ACID, ROSE, &c.

Lozenges, Al'kaline. See LOZENGES, SODA, VICHY, &c.

Lozenges, Al'um. *Syn.* TROCHISCI ALUMINIS, L. Each lozenge contains 1½ gr. of alum. As an astringent. See LOZENGES, ASTRINGENT.

Lozenges of Ammonium Chloride. *Syn.* TROCHISCI AMMONII CHLORIDI, L. (Throat Hospital.) Ammonium chloride, 2 gr.; black currant paste, q. s.

Lozenges, An'iseed. *Syn.* TROCHISCI ANISI, L. *Prep.* From oil of aniseed, 1½ fl. dr.; finest white sugar, 1 lb.; mucilage, q. s. Carminative and stomachic. In colic, griping, &c.; and as a pectoral.

Lozenges, Anthelmin'tic. See LOZENGES, WORM.

Lozenges, Antimonial. *Syn.* TROCHISCI ANTIMONIALES, MORSULI STIBII KUNKELII, L.; TABLETTES DE KUNKEL, Fr. *Prep.* (P. Cod.) Levigated sulphuret of antimony and cardamom seeds, of each, 1 oz.; almonds (blanched), 2 oz.; cinnamon, ½ oz.; sugar, 18 oz.; mucilage of tragacanth, q. s.; to be divided into 15-gr. lozenges. As an alterative.

Lozenges, Ape'rient. *Syn.* TROCHISCI APERIENTES, L. Each lozenge contains 1 gr. each of calomel and scammony, and 2 gr. of jalap; or, instead of the last, ¼ gr. of jalapine. 2 to 3 for a dose.

Lozenges, Astrin'gent. *Syn.* TROCHISCI ASTRINGENTES, L. Each lozenge contains 1½ gr. of alum and 2 gr. of catechu. In spitting of blood, relaxed uvula, sore throat, &c. See LOZENGES, ALUM.

Lozenges, Bark. *Syn.* TROCHISCI CINCHONÆ, L. *Prep.* (P. Cod.) Cinchona, 2 oz.; cinnamon, 2 dr.; white sugar, 14 oz.; mucilage of gum tragacanth, q. s.; mix, and divide into 16-gr. lozenges. Tonic.

Lozenges, Bath. *Syn.* DAWSON'S LOZENGES. From extract of liquorice and gum-arabic, of each, 1½ oz.; sugar, 17 oz. It is both rolled into lozenges and formed into pipes. Demulcent; in tickling coughs, &c.

Lozenges, Benzoic Acid. (Th. Hosp.) *Syn.* TROCHISCI ACIDI BENZOICI. *Prep.* 1. Benzoic acid, in powder, 175 gr.; tragacanth, in powder, 70 gr.; refined sugar, in powder, 280 gr.; red currant paste, a sufficient quantity to make 1 lb. Divide into 350 lozenges, and dry at a moderate heat in a hot-air chamber. A valuable stimulant and voice lozenge in nervo-muscular weakness of the throat.

2. (B. P.) Benzoic acid, 360 gr.; sugar, 25 oz.; gum, in powder, 1 oz.; mucilage of acacia, 2 oz.; water, a sufficiency. Mix the first three ingredients; add the mucilage and water to form a mass; divide into 720 lozenges, and dry.

Lozenges, Bicarbonate of Soda. (B. P.) TROCHISCI SODÆ BICARBONATIS. Bicarbonate of soda, in powder, 3600 gr. (8½ oz.); refined sugar, 25 oz.; gum acacia, in powder, 1 oz.; mucilage, 3 oz.; distilled water, 1 oz.; mix, and form in 720 lozenges. Each lozenge contains 5 gr. of bicarbonate of soda.—*Dose*, 1 to 6 lozenges.

Lozenges, Bis'muth. *Syn.* TROCHISCI BISMUTHI. *Prep.* (B. P.) Subnitrate of bismuth, 1440 gr.; carbonate of magnesia, 4 oz.; precipitated chalk, 6 oz.; sugar, 29 oz.; gum acacia, 1 oz.; mucilage, 2 oz.; rose-water, a sufficiency; make 720 lozenges. Each lozenge contains 2 gr. of subnitrate of bismuth.—*Dose*, 1 to 6 lozenges.

Uses. Tonic and antispasmodic; in chronic dyspepsia, gastrodynia, nausea, cramp of the stomach, &c.

Lozenges, Black Cur'rant. TROCHISCI RIBIS NIGRI, L. *Prep.* From inspissated juice of black currants and sugar, of each, in powder, 1 lb.; tartaric acid, ¼ oz.; mucilage, q. s. In hoarseness, &c.

Lozenges, Bo'rax. *Syn.* TROCHISCI BORACIS, L. Each lozenge contains 3 gr. of borax. One occasionally in aphthous sore mouth, sore throat, &c.

Lozenges, Bromide of Ammonium. Each lozenge contains 2 gr. of bromide of ammonium.—*Dose*, 1 to 3 lozenges. In hooping-cough.

Lozenges, Burnt Sponge. *Syn.* TROCHISCI SPONGIÆ, T. S. USTÆ, L. *Prep.* (P. Cod.) Burnt sponge, 4 oz.; sugar, 12 oz.; mucilage of tragacanth, q. s.; divide into 12-gr. lozenges. In scrofula, glandular enlargements, &c.

Lozenges, Caca'o. *Syn.* TROCHISCI BUTYRI

CACAO, L. Each lozenge contains 1-3rd of its weight of pure cacao butter. In habitual constipation; and in phthisis, scrofula, &c., instead of cod-liver oil; taken *ad libitum*. They are usually scented with roses.

Lozenges, Caffe'ine. *Syn.* TROCHISCI CAFFEINÆ, L. Each lozenge contains ½ gr. of caffeine and ¼ gr. of citric acid. In hemicrania, hypochondriasis, &c.

Lozenges, Cal'omel. *Syn.* WORM LOZENGES; TROCHISCI CALOMELANOS, T. HYDRARGYRI CHLORIDI, L. *Prep.* (P. Cod.) Each lozenge contains 1 gr. of calomel. Alterative, &c. They afford a simple way of introducing mercury into the system. During their use salt food and acid liquors should be avoided. When given for worms they should be followed, in a few hours, by a purge.

Lozenges, Cam'phor. *Syn.* TROCHISCI CAMPHORÆ, L. Each lozenge contains ½ gr. of (finely powdered) camphor. They must be kept in a well-corked bottle.

Lozenges of Carbolic Acid. (Th. Hosp.) *Syn.* TROCHISCI ACIDI CARBOLICI, L. *Prep.* Carbolic acid, 350 gr.; gum-arabic, 220 gr.; refined sugar, 12½ oz.; mucilage, 1 oz.; distilled water, q. s. to make 1 lb. Divide into 350 lozenges, and finish as with benzoic acid lozenges.

Lozenges, Car'bonate of Lime. See LOZENGES, CHALK.

Lozenges, Cat'echu. *Syn.* CACHOU LOZENGES; TROCHISCI CATECHU (B. P.), T. DE TERRA JAPONICA, L.; TABLETTES DE CACHOU, Fr. *Prep.* 1. (Ph. E. 1744.) Catechu, 2 oz.; tragacanth, ½ oz.; white sugar, 12 oz.; rose-water, q. s.

2. (P. Cod.) Extract of catechu, 4 oz.; sugar, 16 oz.; mucilage of gum tragacanth, q. s.; for 10-gr. lozenges.

3. (TRO. CATECHU ET MAGNESIÆ, P. Cod.) Magnesia, 2 oz.; powdered catechu, 1 oz.; sugar, 13 oz.; mucilage of gum tragacanth (made with cinnamon-water), q. s. to mix.

4. (PERFUMED.) See CACHOU AROMATISÉ and PASTILS.

5. (B. P.) Pale catechu, in powder, 720 gr.; refined sugar, in powder, 25 oz.; gum-arabic, in powder, 1 oz.; mucilage, 2 oz.; distilled water, a sufficiency; divide into 720 lozenges. Each lozenge contains 1 gr. of catechu.—*Dose*, 1 to 3 lozenges.

Obs. All the above are taken in diarrhœa, in relaxation of the uvula, in irritation of the larynx, and as cosmetics to fasten the teeth and disguise a fetid breath. The one containing magnesia (No. 3) is also sucked in dyspepsia, acidity, and heartburn.

Lozenges, Cayenne'. *Syn.* TROCHISCI CAPSICI, L. Flavoured with essence or tincture of capsicum or cayenne, with a very concentrated Chili vinegar, or a little pure soluble cayenne pepper.

2. (ACIDULATED.) To each lb. add of tartaric acid, ½ oz. Both are used in dyspepsia, and to promote digestion and create an appetite. They have also been recommended in temporary deafness arising from exposure to cold. They are generally tinged of a light pink or red colour.

Lozenges, Chalk. *Syn.* HEARTBURN LOZENGES; TROCHISCI CRETÆ (Ph. E.), T. CARDIALGICI, TABELLÆ CARDIALGICÆ, L. *Prep.* (Ph. E.) Prepared chalk, 4 oz.; gum-arabic, 1 oz.; nutmeg, 1 dr.; white sugar, 6 oz.; rose or orange-flower water, q. s. Antacid and absorbent. 3 or 4 sucked *ad libitum*; in heartburn, dyspepsia, diarrhœa, acidity of the stomach and bowels, &c.

Lozenges, Char'coal. *Syn.* TROCHISCI CARBONIS, L. *Prep.* 1. (P. Cod.) Prepared charcoal, 4 oz.; white sugar, 12 oz., mucilage, q. s. to mix. In diarrhœa, cholera, dyspepsia, &c.

2. (TRO. CARBONAS CUM CHOCOLATÂ, *M. Chevallier*.) Charcoal and white sugar, of each, 1 oz.; chocolate, 3 oz.; mucilage of gum tragacanth, q. s. to mix. Nutritious; used as the last.

Lozenges, Ching's Worm. *Prep.* 1. (YELLOW.) From saffron, ½ oz.; boiling water, 1 pint; infuse, strain; add, of calomel, 1 lb.; powdered white sugar, 28 lbs.; mix well, make a mass with mucilage of tragacanth, and divide it into 7000 lozenges. Each lozenge contains 1 gr. of calomel.

2. (BROWN.) From calomel, 7 oz.; resinous extract of jalap, 3½ lbs.; white sugar, 10 lbs.; mucilage of tragacanth, q. s.; mix, and divide into 6125 lozenges. Each lozenge contains ½ gr. of calomel and 3½ gr. of resinous extract of jalap. 1 to 6 of the yellow lozenges, overnight, as a vermifuge, followed by an equal number of the brown ones the next morning fasting.

Lozenges, Chlo"rate of Potassium. *Syn.* TROCHISCI POTASSII CHLORATIS, L. *Prep.* 1. Each lozenge contains 1½ gr. of chlorate of potassa. In phthisis, sore throat, &c. 6 to 12 a day.

2. (B. P.) Chlorate of potash, in powder, 3600 gr. (8¼ oz.); refined sugar, in powder, 25 oz.; gum acacia, in powder, 1 oz.; mucilage, ½ oz.; distilled water, 1 oz., or a sufficiency; mix the powders, and add the mucilage and water to form a proper mass; divide in 720 lozenges. Each lozenge contains 5 gr. of chlorate of potash.—*Dose*, 1 to 6 lozenges.

Lozenges, Chloride of Ammonium. Each lozenge contains 2 to 3 gr. of chloride of ammonium. Used in bronchitis.—*Dose*, 2 to 4 lozenges.

Lozenges, Chlo"ride of Gold. *Prep.* 1. (TROCHISCI AURI CHLORIDI, L.) Each lozenge contains $\frac{1}{10}$ gr. of neutral chloride of gold. 2 to 4 daily; in scrofula, cancer, &c.

2. (With SODA: TROCHISCI AURI ET SODII CHLORIDI, T. SODII AURO-CHLORIDI, L.—*Chrestien*.) Each lozenge contains $\frac{1}{16}$ gr. of soda-chloride of gold. Two daily; as the last.

Lozenges, Chloride of Lime. *Syn.* TROCHISCI CALCIS CHLORIDI, T. C. CHLORINATÆ, L. Each lozenge contains ½ gr. of dry chloride of lime. They are frequently tinged with a little carmine. Used to sweeten the breath and whiten the teeth. They do not keep well.

Lozenges of Chlorinated Soda. *Syn.* TROCHISCI SODÆ CHLORINATÆ, L. *Prep.* Solution of chlorinated soda, 1 fl. dr.; sugar, 10 dr.; gum-arabic, 2 dr.; mucilage of tragacanth, q. s. (½ dr. of camphor may be added). To be held in the mouth during infection.

Lozenges, Choc'olate. *Syn.* TROCHISCI CHOCOLATÆ, L. From vanilla chocolate pressed into sheets, and cut into pieces whilst hot.

Lozenges, Cincho'na. *Syn.* TROCHISCI CIN-CHONÆ EXTRACTI, L. Each lozenge contains 1½ gr. of dry extract of bark. A little cinnamon or nutmeg is often added. See BARK LOZENGES.

Lozenges, Cin'namon. *Syn.* TROCHISCI CIN-NAMONI, L. From cinnamon (in fine powder), 1 oz., or the essential oil, 1 fl. dr. to each lb. of sugar. Carminative and stomachic. CASSIA LOZENGES are made in the same way, and are frequently substituted for them.

Lozenges, Cit'rate of Iron. *Syn.* TROCHISCI FERRI CITRATIS, L. Each lozenge contains 1½ gr. of ammonio-citrate of iron. As a mild chalybeate tonic. They are sometimes made with equal parts of sugar and vanilla chocolate.

Lozenges, Citrate of Magnesium. *Syn.* TRO-CHISCI MAGNESII CITRATIS, L. Each 15-gr. lozenge contains 5 gr. of pure citrate of magnesium. Laxative.

Lozenges, Cit'ric Acid. *Syn.* TROCHISCI ACIDI CITRICI, L. *Prep.* (P. Cod.) Citric acid, 3 dr.; sugar, 16 oz.; essence of lemon, 16 drops; mucilage of tragacanth, q. s.; mix, and divide into 12-gr. lozenges. In coughs, hoarseness, &c.

Lozenges, Clove. *Syn.* TROCHISCI CARYO-PHYLLI, L. From cloves (powdered along with sugar), 3 oz., or essential oil, 1 fl. dr., to each lb. of sugar. They are frequently coloured. Carminative and stomachic; also used as a restorative after fatigue, added to chocolate to improve its flavour, and sucked to sweeten the breath.

Lozenges, Cough. *Syn.* PECTORAL LOZENGES, PULMONIC L.; TROCHISCI ANTICATARRHALES, L. *Prep.* 1. Black currant lozenge mass, 1 lb.; ipecacuanha (in very fine powder), 2 dr. For 12-gr. lozenges.

2. To the last add of powdered opium and camphor, 1½ dr.

3. To either No. 1 or 2 add of oil of aniseed, 1½ fl. dr.

4. (TABLETTES DE TRONCHIN.) From powdered gum-arabic, 8 oz.; oil of aniseed, 16 drops; extract of opium, 12 gr.; kermes mineral, 1 dr.; pure extract of liquorice, 2 oz.; white sugar, 32 oz.; water, q. s.; mix, and divide into 10-gr. lozenges.

5. (TABLETTES DE VANDAMME.) From benzoic acid, 1 dr.; orris powder, 2 dr.; gum-arabic (powdered), 1 oz.; starch, 2 oz.; sugar, 16 oz.; water, q. s.; mix, and divide into 15-gr. lozenges.

6. Each lozenge contains ½ gr. of lactucarium, 1-8th gr. of powdered ipecacuanha, and 1-12th gr. of powdered squills, together with 1-3rd of their weight of pure extract of liquorice.

Obs. To render the above serviceable in coughs, hoarseness, &c., the bowels should be kept gently open with some mild aperient, and a light diet adopted, with abstinence from stimulating liquors. See LOZENGES, EMETINE, IPECACUANHA, &c.

Lozenges, Cro'ton Oil. *Syn.* TROCHISCI CRO-TONIS, L. *Prep.* (*Soubeiran.*) Croton oil, 5 drops; powdered starch, 40 gr.; white sugar, 1 dr.; chocolate, 2 dr.; divide into 30 lozenges; 5 or 6 generally prove cathartic.

Lozenges, Cu'bebine. *Syn.* TROCHISCI CUBE-BINI, L. *Prep.* (Ph. Hamb.) Copaiba and extract of cubebs, of each, 6 oz.; yolks of 3 eggs;

mix, add of powdered marsh-mallow root, 6 oz.; make it into pipes of 12 gr. each, and roll them in sugar. In gleet, &c., and in affections of the mucous membranes of the throat and fauces. Lablonye orders them to be made of sugar, and flavoured with oil of peppermint.

Lozenge of Cu'bebs. *Syn.* TROCHISCI CU-BEBÆ, L. *Prep.* (Throat Hospital.) Powdered cubebs, ½ gr.; black currant paste, q. s. Useful in bronchitis. Closely resembles Brown's bronchial troches.

Lozenges, Cu'bebs. *Syn.* TROCHISCI CUBEBÆ, L. *Prep.* 1. (*Spitta.*) Cubebs, 2 dr.; balsam of tolu, 6 gr.; mix, and add of extract of liquorice, 1 oz.; syrup of tolu, 1 dr.; powdered gum, q. s.; divide into 10-gr. lozenges. One of these, allowed to melt gradually in the mouth, is said to alleviate the obstruction in the nose in coryza.

2. (U. S.) Oleo-resin of cubebs, 50 gr.; oil of sassafras, 15 gr.; extract of liquorice in powder, 400 gr.; gum-arabic in powder, 200 gr.; syrup of tolu, q. s. Divide into 100 lozenges.

Lozenges of Cyanide of Gold. (*Chrestien.*) *Syn.* TROCHISCI AURI CYANIDI. *Prep.* Cyanide of gold, 2 gr.; chocolate paste, 1 oz. Made into 24 lozenges. From 1 to 4 in the day.

Lozenges, D'Arcet's. See LOZENGES, VICHY.

Lozenges, Diges'tive. See LOZENGES, RHUBARB, GINGER, CANDY, DIGESTIVE, &c.

Lozenges, Edinburgh. *Prep.* From extract of poppies, 2 oz; powdered tragacanth, 4 oz.; sugar, 10 oz.; rose-water, q. s. to form a lozenge mass.

Lozenges, Em'etine. *Syn.* TROCHISCI EME-TINÆ, L. *Prep.* (*Magendie.*) 1. From impure or coloured emetine, 32 gr. (or pure emetine, 8 gr.); white sugar, 2 oz.; mucilage, q. s. to mix; divide into 64 lozenges. Emetic.—*Dose,* 1 for a child, and 4 for an adult. They are generally tinged of a pink colour with carmine.

2. From impure or coloured emetine, 32 gr. (or pure emetine, 8 gr.); sugar, 4 oz.; mucilage, q. s.; divide into 256 lozenges. Pectoral. One every hour, or oftener, for an adult. The last are intended to take the place of ipecacuanha lozenges, but are rather stronger.

Lozenges, Escharot'ic. *Syn.* TROCHISCI ES-CHAROTICI, L. *Prep.* (P. Cod.) Corrosive sublimate, 2 dr.; starch, 4 dr.; mucilage of tragacanth, q. s.; mix, and divide into 8-gr. oat-shaped granules. For external use only. See CAUSTIC (Zinc).

Lozenges, Ferrocy'anide of Iron. *Syn.* TRO-CHISCI FERRI FERROCYANIDI, T. CÆRULEI, L. Each lozenge contains 1½ gr. of pure Prussian blue. Alterative, febrifuge, and tonic; in epilepsy, intermittents, diseases of the ganglionic system, &c.

Lozenges, Fruit. *Prep.* From juice of black currants (boiled to the consistence of an extract), 1 lb.; juice of red currants (similarly treated), ½ lb.; powdered gum tragacanth, ½ lb.; sugar, 3 lbs.; raspberry syrup, q. s.; pear essence, a few drops. Resemble black currant lozenges, but are more agreeable.

Lozenges, Garana'. See LOZENGES, PAUL-LINIA.

Lozenges, Gin'ger. *Syn.* TROCHISCI ZINGIBERIS, L. *Prep.* From the best unbleached Jamaica ginger and gum-arabic, of each, in very fine powder, 1½ oz.; double refined lump sugar, 1 lb.; rose-water (tinged with saffron), q. s. A still finer quality may be made by using an equivalent proportion of essence of ginger instead of the powder. Inferior qualities are prepared with coarser sugar, to which some starch is often added. Ginger lozenges are carminative and stomachic, and are useful in flatulency, loss of appetite, &c.

Lozenges, Gold. *Syn.* TROCHISCI AURI, L. Each lozenge contains 1-16th gr. of pulverulent gold.

Lozenge of Guaiacum. *Syn.* TROCHISCI GUAIACI, L. *Prep.* (Throat Hospital.) Guaiacum resin, 3 gr.; black currant paste, q. s. One every two hours in acute inflammation, and three times a day in chronic affections of the throat.

Lozenges, Gum. *Syn.* TROCHISCI ACACIÆ (Ph. E.), T. GUMMI ARABICI, T. GUMMOSI, L. *Prep.* 1. (Ph. E.) Gum-arabic, 4 oz.; starch, 1 oz.; white sugar, 12 oz. (all in very fine powder); rose-water, q. s.

2. (P. Cod.) Gum-arabic, 1 lb.; sugar, 3 lbs.; orange-flower water, 3 fl. oz.

3. (Transparent.) From the same materials, but employing a gentle heat. Demulcent; used to allay tickling coughs.

Lozenges, Gum Tra'gacanth. *Syn.* TROCHISCI TRAGACANTHÆ, T. GUMMI T., L. *Prep.* (Ph. E. 1744.) Compound powder of tragacanth, 3 oz.; sugar, 12 oz.; rose-water, 4 fl. oz. Resemble the last, but are more durable in the mouth.

Lozenges, Heart'burn. See LOZENGES, CHALK, &c.

Lozenges, Iceland Moss. *Syn.* TROCHISCI LICHENIS, L. (P. Cod.) Contain half their weight of dried and powdered lichen jelly. Resemble gum lozenges.

Lozenges, Indian Hemp. *Syn.* TROCHISCI CANNABIS, T. C. INDICI, L. (*Ebriard.*) Each lozenge contains $\frac{1}{16}$ gr. of extract of Indian hemp.

Lozenges, I'odide of Iron. *Syn.* TROCHISCI FERRI IODIDI, L. Each lozenge contains ½ gr. of dry iodide of iron. 12 to 20 daily; in amenorrhœa, chlorosis, scrofulous debility, &c. They are generally flavoured with a little nutmeg or cinnamon.

Lozenges, Iodide of Potassium. *Syn.* TROCHISCI POTASSII IODIDI, L. Each lozenge contains 1 gr. of iodide of potassium, flavoured with nutmeg or cinnamon. 10 to 15 daily; in scrofula, indurations, &c. One of the best ways of taking iodide of potassium.

Lozenges, Ipecacuan'ha. *Syn.* TROCHISCI IPECACUANHÆ, L. *Prep.* 1. (B. P.) Mix ipecacuanha in powder, 180 gr.; refined sugar in powder, 25 oz.; gum acacia in powder, 1 oz.; add mucilage of acacia, 2 fl. oz., and distilled water, 1 oz., or sufficient to form a proper mass. Divide into 720 lozenges, and dry in a hot-air chamber with a moderate heat. Each lozenge contains ¼ gr. of ipecacuanha.

2. (P. Cod., Hamb. do., and Ph. U. S.) Each lozenge contains ¼ gr. of ipecacuanha.

3. (TRO. IPECAC. CUM CAMPHORÂ.) Each lozenge contains ¼ gr. of camphor and ¼ gr. of ipecacuanha.

4. (TRO. IPECAC. CUM CHOCOLATÂ, P. Cod.) Each lozenge contains 1 gr. of ipecacuanha and 12 gr. of chocolate *à la vanilla*. The above are pectoral and expectorant, and are very useful in tickling and chronic coughs, hoarseness, &c.

Lozenges, Ipecacuanha and Morphia. *Syn.* TROCHISCI IPECACUANHÆ ET MORPHIÆ (B. P.). Each lozenge contains $\frac{1}{12}$ gr. ipecacuanha and $\frac{1}{36}$ gr. hydrochlorate of morphia.—*Dose*, 1 to 6 lozenges. See LOZENGES, MORPHIA and IPECACUANHA.

Lozenges, I'ron. *Syn.* TROCHISCI FERRI, T. CHALYBEATI, L. 1. Each lozenge contains 1 gr. of Quevenne's iron. See LOZENGES, REDUCED IRON.

2. (TRO. FERRI CARBONATIS.) Each lozenge contains 1½ gr. of saccharine carbonate of iron. They are both mild and excellent chalybeates. See LOZENGES, STEEL.

Lozenges, Ju'jube. See PASTE, JUJUBE.

Lozenges, Ker'mes Mineral. *Syn.* TROCHISCI KERMETIS, L. *Prep.* 1. (P. Cod.) Each lozenge contains ⅛ gr. of kermes mineral, and about ¾ gr. of gum, made up with sugar and orange-flower water. Diaphoretic and expectorant.

2. (Compound.) As the last, but with the addition of ⅛ gr. of opium, ¼ gr. of squills, and ½ gr. of ipecacuanha. Anodyne and expectorant; both are very useful in catarrhs.

Lozenges, Lactate of Iron. *Syn.* TROCHISCI FERRI LACTATIS, L. *Prep.* (Cap.) Each lozenge contains 1 gr. of lactate of iron. Tonic. Useful in debility accompanied by a diseased state of the organs of digestion.

Lozenges, Lac'tic Ac'id. *Syn.* TROCHISCI ACIDI LACTICI, L. Each lozenge contains 1 gr. of lactic acid to about 12 gr. of sugar. They are best flavoured with vanilla or nutmeg. In dyspepsia, &c., especially in gouty subjects. Those prepared by Magendie's formulæ contain a larger proportion of acid, but are much too sour for frequent use.

Lozenges, Lactuca''rium. *Syn.* TROCHISCI LACTUCARII, L. *Prep.* (Ph. E.) Prepared with lactucarium in the same manner as the opium lozenges, Ph. E. Each of these lozenges contains from ⅙ to ¼ gr. of lactucarium. Anodyne and demulcent. Used to allay tickling coughs, &c.

Lozenges, Lavender. *Syn.* TROCHISCI LAVANDULÆ, L. From ½ fl. dr. of Mitcham oil of lavender to each lb. of sugar, and tinged red with liquid lake or carmine, or violet with litmus or indigo. Used chiefly to scent the breath. Those of the shops are generally deficient in odour.

Lozenges, Lem'on. *Syn.* TROCHISCI LIMONIS, T. LIMONUM, L. *Prep.* 1. From 1½ fl. dr. of oil of lemon to each lb. of double refined white sugar.

2. (Acidulated.) See LOZENGES, CITRIC and TARTARIC.

Obs. Lemon lozenges and drops are agreeable sweetmeats, and those that are acidulated are often very useful to promote expectoration in coughs, &c. The last are also made into drops as well as lozenges, when they form the 'ACIDU-LATED LEMON DROPS' of the shops. Those that are made of citric acid are by far the most wholesome. Both lemon lozenges and drops are generally coloured with infusion of saffron or turmeric.

Lozenges, Lettuce. *Syn.* TROCHISCI LACTUCÆ, L. *Prep.* From extract of lettuce, extract of liquorice, gum, and sugar, equal parts. Anodyne and demulcent; in obstinate cough without expectoration. See LOZENGES, LACTUCARIUM.

Lozenges, Lichen. See LOZENGES, ICELAND MOSS.

Lozenges, Liquorice. *Syn.* BLACK LOZENGES; TROCHISCI GLYCYRRHIZÆ, T. G. GLABRÆ, T. BECHICI NIGRI, L. *Prep.* 1. (Ph. E.) Extract of liquorice and gum acacia, of each, 6 oz.; white sugar, 12 oz.; dissolve in water, q. s.; evaporate into a paste, and form into lozenges. Pectoral and demulcent. Useful to allay tickling coughs and remove hoarseness.

2. (With OPIUM.) See LOZENGES, OPIUM.

Lozenges, Magne'sia. *Syn.* HEARTBURN LOZENGES; TROCHISCI MAGNESIÆ (Ph. E.), L. *Prep.* 1. (Ph. E.) Carbonate of magnesium, 6 oz.; powdered white sugar, 3 oz.; oil of nutmeg, 20 drops; mucilage of tragacanth, q. s. to mix.

2. (Ph. U. S.) Calcined magnesia, 4 oz.; sugar, 12 oz.; nutmeg, 1 dr.; mucilage of tragacanth, q. s.; for 10-gr. lozenges.

3. (Wholesale.) Calcined magnesia, 3 oz.; powdered gum tragacanth, 1 oz.; double refined lump sugar, ¾ lb.; rose or orange-flower water, q. s. to make a lozenge mass.

Obs. Magnesia lozenges are very useful in heartburn, acidity, and indigestion. The confectioners generally omit the nutmeg, and make their mucilage with either rose or orange-flower water, or else add the dry gum to the mass, and then mix up the powders with one or other of these liquids. It is also an improvement to use calcined magnesia, which is about twice as strong as the carbonate, and consequently less need be employed.

Lozenges, Manna. *Syn.* TROCHISCI MANNÆ, L. *Prep.* (*Van Mons.*) Powdered tragacanth, 1 dr.; white sugar, 12 oz.; manna, 3 oz.; orange-flower water, q. s. to mix. Demulcent, and in large numbers slightly laxative.

Lozenges, Marsh-mallow. *Syn.* TROCHISCI ALTHÆÆ, L.; TABLETTES DE GUIMAUVE, Fr. *Prep.* (P. Cod.) Marsh-mallow root (decorticated and finely powdered), 2 oz.; sugar, 14 oz.; mucilage of tragacanth (made with orange-flower water), q. s. Demulcent and expectorant. Useful to allay the irritation in cough, &c. The preparations of marsh-mallow have always been highly esteemed as pectorals by the vulgar.

Lozenges, Morphia. *Syn.* TROCHISCI MORPHIÆ (Ph. E.), T. M. HYDROCHLORATIS, L. *Prep.* 1. (Ph. E.) Hydrochlorate of morphia, 20 gr.; tincture of tolu, ½ fl. oz.; powdered white sugar, 25 oz.; dissolve the hydrochlorate in a little warm water, mix it with the tincture and the sugar, make a mass with mucilage of gum tragacanth, q. s., and divide it into 15-gr. lozenges. Each lozenge contains about $\frac{1}{75}$ gr. of hydrochlorate of morphia. Used as opium lozenges, but are plea-

santer. The morphia lozenges of the shop generally contain $\frac{1}{12}$ gr. of hydrochlorate of morphia (*Pereira*).

2. (With IPECACUANHA: TROCHISCI MORPHIÆ ET IPECACUANHÆ, Ph. E.) As the last, adding of ipecacuanha, 1 dr. Each lozenge contains about $\frac{1}{12}$ gr. of hydrochlorate of morphia and $\frac{1}{12}$ gr. of ipecacuanha. Anodyne and expectorant; in tickling coughs, &c., and to allay pain.

3. (B. P.) Hydrochlorate of morphine, 20 gr.; tincture of tolu, ½ oz.; refined sugar, in powder, 24 oz.; gum-arabic, in powder, 1 oz.; mucilage, 2 oz., or a sufficiency; boiling distilled water, ½ oz. Divide the mass into 720 lozenges. Each lozenge contains $\frac{1}{36}$ gr. of hydrochlorate of morphine.—*Dose*, 1 or 2 occasionally, for cough.

Lozenges, Morphine and Ipecacuanha. *Syn.* TROCHISCI MORPHINÆ ET IPECACUANHÆ (B. P.). Hydrochlorate of morphine, 20 gr.; ipecacuanha, in fine powder, 60 gr.; tincture of tolu, ½ oz.; refined sugar, in powder, 24 oz.; gum-arabic, in powder, 1 oz.; mucilage, 2 oz., or a sufficiency; distilled water, ½ oz.; divide the mass into 720 lozenges. Each lozenge contains $\frac{1}{36}$ gr. of hydrochlorate of morphine and $\frac{1}{12}$ gr. of ipecacuanha.—*Dose*, 1 or 2 occasionally, for cough.

Lozenges of Naphthalin (*Dupasquier*). *Syn.* TROCHISCI NAPHTHALINI. *Prep.* Naphthalin, 5 scruples; sugar, 20 oz.; oil of aniseed to flavour; form a mass with mucilage of tragacanth, and divide into lozenges of 15 gr. each. Expectorant, and may be taken to the extent of 20 a day.

Lozenges, Ni'tre. *Syn.* TROCHISCI NITRICI, L. *Prep.* 1. (Ph. E. 1783.) Nitre, 3 oz.; white sugar, 9 oz.; mucilage of tragacanth, q. s. to mix. Diuretic; but chiefly sucked, without swallowing, to remove incipient sore throat.

2. (Camphorated: TROCHISCI NITRI CAMPHORATI, *Chaussier*, L.) Each lozenge contains ½ gr. of opium, ½ gr. of camphor, and 1 gr. of nitre. In hoarseness, sore throat, &c.

Lozenges, Nut'meg. *Syn.* TROCHISCI MYRISTICÆ, L. From oil of nutmeg, 1 fl. dr., to each lb. of sugar, and coloured with infusion of saffron. Carminative and stomachic; in colic, &c.

Lozenges, O'pium. *Syn.* TROCHISCI OPII (Ph. E.), T. GLYCYRRHIZÆ CUM OPIO, L. *Prep.* 1. (Ph. E.) Opium (strained), 2 dr.; tincture of tolu, ½ oz.; triturate together; add of powdered sugar, 6 oz.; extract of liquorice (soft) and powdered gum acacia, of each, 5 oz.; mix, and divide into 10-gr. lozenges. Each lozenge contains ½ to ½ gr. of opium. Used to allay tickling cough and irritation of the fauces, and as an anodyne and hypnotic.

2. (Ph. U. S.) Opium (in fine powder), 2 dr.; extract of liquorice, gum-arabic, and sugar, of each, 5 oz.; oil of aniseed, ½ fl. dr.; water, q. s.; divide into 6-gr. lozenges. Each lozenge contains $\frac{1}{10}$ gr. of opium. As the last.

3. Extract of opium, 72 gr.; tincture of tolu, ½ oz.; refined sugar (in powder), 16 oz.; gum (in powder), 2 oz.; extract of liquorice, 6 oz.; distilled water, a sufficiency. Divide the mass into 720 lozenges. Each lozenge contains 1-10th gr. of extract of opium.—*Dose*, 1 to 2 lozenges.

Lozenges, Or'ange. *Syn.* TROCHISCI AURANTII,

L. *Prep.* From oil of orange, 1½ fl. dr. to each lb. of sugar, and infusion of saffron for colouring. By adding nitric or tartaric acid, 3 dr., 'ACIDU-LATED ORANGE LOZENGES' will be formed.

Lozenges, Orange-flow'er. *Syn.* TROCHISCI AURANTII FLORUM, L. *Prep.* (P. Cod.) Powdered sugar, 1 lb.; neroli, 1 dr.; orange-flower water, q. s.; make it into drops (pastilli); or omit the water, and make it into lozenges with mucilage of tragacanth made with orange-flower water. Delightfully fragrant.

Lozenges, Or'ris-root. *Syn.* TROCHISCI IRIDIS, L. *Prep.* From orris-root (in very fine powder), 1 oz.; sugar, 1 lb.; mucilage of tragacanth, q. s. to mix. Used to perfume the breath.

Lozenges, Paregor'ic. *Syn.* TROCHISCI PARE-GORICI, L. *Prep.* Medicated with 2 fl. oz. of paregoric and 2 dr. of tartaric acid to each lb. of sugar, and tinged of a pink colour with lake or cochineal. As a pectoral in catarrhs, &c.

Lozenges, Pec'toral. *Syn.* TROCHISCI PEC-TORALES, T. BECHICI, L. *Prep.* 1. (*Dr Gruss.*) Powdered squills, 4 parts; extract of lettuce, 8 parts; ipecacuanha, 18 parts; manna, 125 parts; sugar, 250 parts; mucilage of tragacanth, q. s. to mix.

2. (*Magendie.*) See LOZENGES, EMETINE.

3. (BLACK: T. BECHICI NIGRI.) See LO-ZENGES, LIQUORICE.

4. (WHITE: T. BECHICI ALBI.) Orris-root, 4 dr.; liquorice powder, 6 dr.; starch, 1½ oz.; sugar, 18 oz.; mucilage of tragacanth, q. s. to make a lozenge mass.

5. (YELLOW: T. BECHICI FLAVI.) Powdered orris-root, 6 dr.; starch, 4 dr.; liquorice powder, 3 dr.; saffron, 2 dr.; sugar, 8 oz.; mucilage of tragacanth, q. s. to mix.

Obs. All the above are used as demulcents in coughs, colds, &c. Nos. 1 and 2 are anodyne as well as demulcent. For other formulæ see LO-ZENGES, COUGH, LIQUORICE, OPIUM, &c.

Lozenges, Pel'litory. *Syn.* TROCHISCI PY-RETHRI, L. *Prep.* From pellitory, mastic, and tragacanth, of each (in fine powder), equal parts; orange-flower water, q. s. to mix. In toothache.

Lozenges, Pep'permint. *Syn.* TROCHISCI MENTHÆ PIPERITÆ, L. *Prep.* 1. (P. Cod.) Oil of peppermint, 1 dr.; powdered sugar, 16 oz.; mucilage of tragacanth, q. s.

2. (Ph. U. S.) Oil of peppermint, 15 gr.; sugar, 1200 gr.; mucilage of tragacanth, q. s. to make 100.

3. (Wholesale.) 1 fl. dr. of the finest Mitcham oil of peppermint to each lb. of the finest double refined white sugar, with mucilage of either gum-arabic or tragacanth to mix.

Obs. The best peppermint lozenges are made of the very finest double refined sugar and of English oil of peppermint only, carefully mixed up with very clean mucilage. The commoner qualities are made by employing inferior lump sugar and foreign oil of peppermint, or, what is better, English oil of peppermint, but in a less proportion than for the better sorts. The addition of starch, in quantities varying from 1-6th to 2-9ths of the whole mass, is also commonly made to them; and in the cheapest varieties even plaster of Paris or chalk is occasionally introduced by

unprincipled makers. The addition of a very small quantity of blue smalts, reduced to an impalpable powder, is commonly made to the sugar, to increase its whiteness. 'TRANSPARENT' or 'SEMI-TRANSPARENT PEPPERMINT LOZENGES' are made from the same materials as the opaque ones; but the sugar is not reduced to quite so fine a powder, and the cake is rolled thinner before cutting it. A little oil of almonds or of olives is also occasionally mixed with the ingredients, to promote the transparency; but it tends to render the lozenges less white.

Peppermint lozenges and drops are useful in flatulency, nausea, and griping; and judging from the enormous and constantly increasing demand for them, they are more highly esteemed by the public than all other lozenges and confections.

Lozenges, Pontefract (POMFRET CAKES). These are made of the purest refined juice or extract of liquorice, and have long been esteemed as a demulcent.

Lozenges, Pop'py. *Syn.* TROCHISCI PAPAVERIS, L. *Prep.* From extract of poppies, 3 oz.; sugar, 15 oz.; powdered gum tragacanth, 2 oz.; rose-water, q. s. to mix. Used in coughs as an anodyne and demulcent, in lieu of opium lozenges.

Lozenges, Pulmon'ic. See LOZENGES, COUGH, PECTORAL, WAFERS, &c.

Lozenges, Quinine'. *Syn.* TROCHISCI QUI-NINÆ SULPHATIS, L. *Prep.* (*Soubeiran.*) Each lozenge contains about 1-10th gr. of sulphate of quinine. Tonic and stomachic, in dyspepsia, &c.; but to render them useful the quantity of the alkaloid should be doubled.

Lozenges of Red Gum. *Syn.* TROCHISCI GUMMI RUBEI. *Prep.* Red gum, 2½ gr.; tincture of capsicum, ½ min.; black currant paste, q. s. Useful for relaxed sore throat.—*Dose,* 2 to 6 daily.

Lozenges, Reduced Iron. *Syn.* TROCHISCI FERRI REDACTI (B. P.), L. *Prep.* Reduced iron, 720 gr.; refined sugar, in powder, 25 oz.; gum-arabic, in powder, 1 oz.; mucilage, 2 oz.; distilled water, 1 oz., or a sufficiency. Mix the iron, sugar, and gum, and add the mucilage and water to form a proper mass. Divide into 720 lozenges, and dry them in a hot-air chamber with a moderate heat. Each lozenge contains 1 gr. of reduced iron.—*Dose,* 1 to 6 lozenges.

Lozenges, Reduced Iron, with Chocolate (*Bouchardat*). *Syn.* TROCHISCI CHOCOLATÆ ET FERRI, L. *Prep.* Fine chocolate, 14 oz.; iron reduced by hydrogen, 1 oz. Soften the chocolate by heat, mix with the iron, and divide into lozenges of 15½ gr. each. Levigated iron filings are sometimes substituted for the reduced iron; others direct the peroxide.

Lozenges of Rhatany (Th. Hosp.). *Syn.* TRO-CHISCI KRAMERIÆ. *Prep.* Extract of rhatany, 3 gr.; black currant paste, q. s. Astringent.—*Dose,* 2 to 6 daily.

Lozenges, Rhu'barb. *Syn.* DIGESTIVE LO-ZENGES; TROCHISCI RHEI, L. *Prep.* (P. Cod.) Powdered rhubarb, 1 oz.; sugar, 11 oz.; mucilage of tragacanth, q. s.; divide into 12-gr. lozenges. Stomachic and laxative. Sucked before dinner they excite the appetite, and, after it, promote digestion. They are frequently aromatised with a little cinnamon or vanilla. See CANDY (Digestive).

Lozenges, **Rose**. *Syn.* TROCHISCI ROSÆ, L. *Prep.* 1. (ACIDULATED: T. R. ACIDÆ.) From otto, 5 to 10 drops; citric or tartaric acid, 3 dr.; sugar, 1 lb.; mucilage, q. s.

2. (Ph. E. 1746.) Red rose leaves (powdered), 1 oz.; sugar, 12 oz.; mucilage, q. s.

3. (PÂTE DE ROSE LOZENGES.) As No. 1, omitting one half of the acid.

4. (RED: T. R. RUBRI.) As No. 1, but coloured with liquid lake, or infusion of cochineal.

Obs. Some makers add of starch, 4 oz.; substitute oil of rhodium for otto of roses, and use mucilage made with rose-water; but the quality of course suffers. They are chiefly used to perfume the breath.

Lozenges, **Saffron**. *Syn.* TROCHISCI CROCI, L. *Prep.* From hay saffron (in fine powder), 1 oz.; white sugar, 1 lb.; mucilage of gum tragacanth, q. s. to mix. Anodyne, pectoral, and emmenagogue; but chiefly used to raise the spirits in hypochondriasis.

Lozenges, **Santonin**. *Syn.* TROCHISCI SANTONINI, L. *Prep.* Santonin, 720 gr.; refined sugar, 25 oz.; powdered gum, 1 oz.; mucilage of acacia, 2 oz.; distilled water, a sufficiency. Divide into 720 lozenges.—*Dose*, 1 to 6 daily, as a vermifuge.

Lozenges, **Scammony** (*Bouridres*). *Syn.* TROCHISCI SCAMMONII. *Prep.* Resin of scammony, 4 dr.; calomel, 4 dr.; sugar, 6 oz.; tragacanth, ½ dr.; tincture of vanilla, 40 minims. Make into 300 lozenges. 1 or 2 for a child; 2 to 4 for an adult.

Lozenges, **Soda**. *Syn.* TROCHISCI SODÆ BICARBONATIS (Ph. E.), L. *Prep.* 1. (Ph. E.) Bicarbonate of soda, 1 oz.; powdered gum-arabic, ½ oz.; sugar, 3 oz.; mucilage, q. s.

2. (Wholesale.) From bicarbonate of soda and powdered gum tragacanth, of each, 2 oz.; double refined lump sugar, ½ lb.; rose-water, q. s. to mix. In acidity, heartburn, &c. See LOZENGES, VICHY.

3. (With GINGER: TROCHISCI SODÆ ET ZINGIBERIS, L.) To the last, add of ginger (in very fine powder), 1½ oz.; powdered gum, ½ oz.

Lozenges of **Soluble Tartar** (*Guibourt*). *Syn.* TROCHISCI TARTARI SOLUBILIS, L. *Prep.* Borotartrate of potash, 1 oz.; sugar, 7 oz.; mucilage of tragacanth, q. s.; flavoured with lemon.

Lozenges, **Squills**. *Syn.* TROCHISCI SCILLÆ, L. *Prep.* 1. Each lozenge contains ½ gr. of powdered squills and 2 gr. of extract of liquorice.

2. (With IPECACUANHA: TROCHISCI SCILLÆ ET IPECACUANHÆ, L.) As the last, adding for each lozenge ½ gr. of powdered ipecacuanha. Both the above are useful cough lozenges.

Lozenges, **Starch**. *Syn.* TROCHISCI AMYLI, T. BECHICI ALBI, L. See PECTORAL LOZENGES.

Lozenges, **Steel**. *Syn.* TROCHISCI FERRI, T. CHALYBEATI, L. *Prep.* (P. Cod.) Levigated iron filings, 1 oz.; sugar, 10 oz.; cinnamon, 2 dr.; mucilage of tragacanth, q. s.; mix, and divide into 480 lozenges. Tonic. See LOZENGES, IRON.

Lozenges, **Sulphur**. *Syn.* TROCHISCI SULPHURIS, L. *Prep.* 1. (P. Cod.) From sulphur (pure precipitate), 2 oz.; sugar, 16 oz.; mucilage of tragacanth (made with rose-water), q. s. to mix. Useful in piles and some skin diseases.

2. (*Dr Garrod*.) Precipitated sulphur, 5 gr.; cream of tartar, 1 gr.; tincture of orange, q. s. Make one lozenge. Very useful in constipation.—*Dose*, 2 to 6 daily.

Lozenges, **Tannic Acid**. *Syn.* TROCHISCI ACIDI TANNICI (B. P.), L. *Prep.* Tannic acid, 360 gr.; tincture of tolu, ½ oz.; refined sugar, 25 oz.; gum acacia, 1 oz.; mucilage, 2 oz.; distilled water, 1 oz. Dissolve the tannic acid in the water; add first the tincture of tolu previously mixed with the mucilage, then the gum and the sugar, also previously well mixed. Form the whole into a proper mass, divide into 720 lozenges, and dry them in a hot-air chamber with a moderate heat. Each lozenge contains ½ gr. of tannic acid.—*Dose*, 1 to 6 lozenges.

Lozenges, **Tartaric Acid**. See LOZENGES, ACIDULATED.

Lozenges, **Tolu'**. *Syn.* BALSAMIC LOZENGES; TROCHISCI TOLUTANI, T. BALSAMICÆ, L. *Prep.* 1. (P. Cod.) Balsam of tolu and rectified spirit, of each, 1 oz.; dissolve, add of water, 2 fl. oz., heat the mixture in a water-bath, and filter; make a mucilage with the filtered liquid, and gum tragacanth (in powder), 80 gr.; add of sugar, 16 oz.; make a mass, and cut it into lozenges.

2. (Wholesale.) As the last, but using only one half the weight of balsam of tolu. Pectoral and balsamic.

Lozenges, **Tronchin's**. *Syn.* TABLETTES DE TRONCHIN, Fr. See LOZENGES, COUGH.

Lozenges, **Vanil'la**. *Syn.* TROCHISCI VANILLÆ, L. *Prep.* 1. Essence of vanilla, 3 fl. dr. to each lb. of sugar.

2. (*Guibourt*.) From vanilla triturated to a fine powder with 7 times its weight of sugar. Antispasmodic, nervine, and stomachic. Used to sweeten the breath, to flavour chocolate, &c.

Lozenges, **Vichy**. *Syn.* D'ARCET'S LOZENGES; TROCHISCI SODÆ, L.; PASTILLES DE VICHY, Fr. *Prep.* 1. (P. Cod.) Bicarbonate of soda, 1 oz.; powdered sugar, 19 oz.; mucilage of gum tragacanth, q. s.; mix, and divide into 20-gr. lozenges.

2. (*D'Arcet*.) As the last, adding a little oil of peppermint to give a slight flavour. Antacid or absorbent; in heartburn, &c.

Lozenges, **Vi'olet**. *Syn.* TROCHISCI VIOLÆ, T. VIOLARUM, L. *Prep.* Orris lozenges coloured with the juice of violets.

Lozenges, **Wistar's Cough**. *Prep.* Gum-arabic, extract of liquorice, and sugar, of each, 2½ oz.; powdered opium, 1 dr.; oil of aniseed, 40 drops; for 60 lozenges. One, three or four times a day.

Lozenges, **Worm**. *Syn.* TROCHISCI ANTHELMINTICI, MORSULI CONTRA VERMES, L. Most of the advertised nostrums under this name have a basis of calomel (about 1 gr. per lozenge), and require to be followed by a purge a few hours afterwards.

Prep. 1. (Ph. Austr. 1836.) Ethereal extract of wormseed, 1 dr.; jalap, starch, and sugar, of each, 2 dr.; mucilage of gum tragacanth, q. s.; divide into 60 lozenges.

2. (Ph. Dan. 1840.) Wormseed, 1 oz.; ethiops mineral and jalap, of each, 3 dr.; cinnamon, 2 dr.; sugar, 7 oz.; rose-water, q. s. See LOZENGES, CALOMEL, CHING'S, SANTONIN, &c. (*above*).

Lozenges, Zinc. *Syn.* TROCHISCI ZINCI, T. Z. SULPHATIS, L. *Prep.* (*Dr Copland.*) Each lozenge contains ½ gr. of sulphate of zinc. Antispasmodic, expectorant, and tonic, and in quantity emetic.

LUBRICATING COMPOUNDS. See ANTI-ATTRITION.

LU'CIFERS. See MATCHES.

LUMBA'GO. Rheumatism of the loins. It is distinguished from nephritis, or inflammation of the kidneys, by the pain being aggravated on stooping. The treatment consists of strong stimulant embrocations or liniments, or of blisters over the parts affected, with active aperients, warmth, and diaphoretics (as Dover's powder) at bedtime. The hot or vapour bath often gives almost immediate relief. The wearing of a broad flannel belt next the skin over the loins is of great service in protecting the individual against attacks of lumbago. The greatest care should be taken to avoid chill. See LINIMENT OF BELLADONNA and CHLOROFORM; RHEUMATISM.

LU'MINOUS PHIAL. See PHOSPHORUS.

LU'NA CORNEA. [L.] *Syn.* HORN SILVER. Fused chloride of silver.

LUNAR CAUSTIC. Fused nitrate of silver. See CAUSTIC and SILVER.

LUNCHEONS, HOT, by the River Side. We extract the following from 'Land and Water:'—
"In cold weather, by river side or on mountain or moor, when not too far from home, a hot lunch is often a *desideratum*, but one not easily accomplished without a more or less complicated apparatus and the trouble of lighting a fire—often an impossibility from the want of dry wood. A hot, substantial meal at the end of a hard day's work is often difficult to get when the time of return home may depend entirely on the humour of the fish; and for either purpose nothing will beat the homely Hot Pot, or 'Pâté de Lancashire,' as I have seen it pretentiously termed, though the latter name does not convey any of the comforting, cheering sensation to the inner man contained in the simpler denomination. I have never seen a good recipe for it, so append my own. Take a strong glazed earthenware jar of a cylindrical form, ten inches deep and twelve broad. At the bottom of this place a layer, about an inch thick, of potatoes cut into pieces, sprinkle with a little salt; on these place a layer of four or five mutton chops, season with salt and pepper, and a teaspoonful of Worcester sauce. Pour in enough broth, stock, or water to nearly cover the chops; then add another layer of potatoes (rather thicker than the first), on which place two or three chops and two kidneys, cut into smallish pieces for the sake of the gravy. If mushrooms are procurable, add a few with each layer of meat, or, in place of these, a few oysters. Season, and continue the meat and potatoes in alternate layers until within an inch of the top, when cover with small potatoes whole, or large ones cut into halves or quarters; bake slowly in the oven till the potatoes are quite soft inside and brown and well cooked at the top, when the dish is ready. If it is not wanted at once it may easily be kept hot, and the addition of a little stock will prevent its getting dry. To serve out of doors, wrap up in cloths, and carry in a small hamper lined with

straw, when it will keep steaming hot for an hour or more. One of the great excellences of this dish lies in the fact that all the aroma of the meat is retained, while the potatoes absorb any superfluous gravy. Sliced onions will improve the flavour for those who like them, especially when mushrooms cannot be got. I have tested the appreciation of this dish among a grouse-driving party on the Yorkshire moors on a raw December day, and there was no dissentient voice as to its merits when thankfully discussed over the subsequent pipe. I have found it not ungrateful, after a long day's fishing, nearly up to my waist in water, when the dinner ordered for six, with a view of taking an evening basket, would have been ruined before my arrival at eleven had it consisted of aught else; nay, I have assisted at more than one bachelor supper in chambers where it formed the dish of the evening, and mid-day, evening, or night I have always found it good."

LUNGS. In *anatomy*, the organ of respiration occupying the thorax or chest. See RESPIRATION.

LU'PULIN. *Syn.* LUPULINA, LUPULINE. Under this name two products are known, namely—1. (LUPULINIC GRAINS, L. GLANDS.) The yellow powder obtained from the dried strobiles or catkins of the hops, by gently rubbing and sifting them.—*Dose*, 5 to 10 gr.; as an anodyne, tonic, &c.

2. The aromatic bitter principle of hops.

Prep. The aqueous extract of the yellow powder or lupulinic grains of the strobiles, along with a little lime, are treated with rectified spirit; the filtered tincture is evaporated to dryness, redissolved in water, and the solution is again filtered, and evaporated to dryness; the residuum is, lastly, washed with ether, and allowed to dry.

Prop., &c. The latter product is a yellowish-white, bitter, uncrystallisable substance, soluble in 20 parts of water, very soluble in alcohol, and slightly so in ether. The yellow powder above alluded to (No. 1) is improperly called lupulin; a name which appears more appropriate to the pure bitter principle than to the lupulinic grains.

Adult. The lupulin sold to brewers is largely adulterated with quassia. In some samples, lately examined, the quassia amounted to 70 per cent.

LU'PUS. In *pathology*, a disease affecting the skin, remarkable for eating away the parts which it attacks with extreme rapidity. It is generally confined to the face, and commences with small, spreading ulcerations, which become more or less concealed beneath bran-like scabs, and end in ragged ulcers, which gradually destroy the skin and muscular tissue to a considerable depth.

LUSTRE. See PLUMBAGO.

LUTE. *Syn.* LUTING; LUTUM, CÆMENTUM, L. A composition employed to secure the joints of chemical vessels, or as a covering to protect them from the violence of the fire.

Prep. 1. Linseed meal, either alone or mixed with an equal weight of whiting, and made into a stiff paste with water. It soon becomes very hard and tough.

2. Ground almond cake, from which the oil has

been pressed, mixed up as the last. Both the above are much used for stills, retorts, and other vessels that are not exposed to a heat higher than about 320° F. They are capable of resisting the action of the fumes of volatile oils, spirits, weak acids, &c., for some time.

3. Fresh-slaked lime made into a paste with strained bullock's blood or size. As the last.

4. Plaster of Paris made into a paste with water, and at once applied. It bears a nearly red heat, but becomes rather porous and friable.

5. Powdered clay or whiting made into putty with water and boiled linseed oil. This is commonly known as 'fat lute.'

6. A mixture of powdered clay and ground bricks, made up with water or a solution of borax. For joining crucibles, &c., which are to be exposed to a strong heat.

7. Pipe-clay and horse-dung made into a paste with water. As a coating for glass vessels, to preserve them from injury from exposure to the fire. This composition is used by the pipe-makers, and will stand unharmed the extremest heat of their kiln for 24 hours. It is applied by spreading it on paper.

8. As the last, but employing shredded tow or plumbago for horse-dung.

Obs. For the joints of small vessels, as tubes, &c., especially those of glass or earthenware, pieces of vulcanised Indian tubing, slipped over and tied above and below the joint, are very convenient substitutes for lutes, and have the advantage of lasting for a long time, and bearing uninjured the heat at which oil of vitriol boils. Flat rings or ' washers ' of vulcanised rubber are also excellent for still heads, &c., whenever the parts can be pinched together by screws or clamps.

LYCOPO′DIUM. The fine powder known in commerce under this name consists of the minute spores of the common club-moss, or *Lycopodium clavatum.* It is exceedingly combustible ; thrown suddenly from a powder-puff or bellows across the flame of a candle, it produces the imitation flashes of lightning of our theatres. The powder is also employed as a ' dusting powder ' in excoriations, and to roll up boluses and pills.

According to M. Paul Cazeneuve, pine pollen is occasionally substituted for lycopodium.

MACABO′NI. This only differs from VERMI-CELLI in the size of the pipes, which are about as large as a goose-quill. When properly dressed it is very wholesome and nutritious. A pleasant dish may be made by boiling macaroni in water until soft, either with or without a little salt, draining off the water, and then stewing it with a little butter, cream, or milk, and grated cheese, adding spice to palate. It may be made into a ' form ' and browned before the fire.

MAC′AROONS (English). *Prep.* Take of sweet almonds, 1 lb. ; blanch and beat them to a paste, add of lump sugar, 1¼ lbs. ; whites of 6 eggs ; the grated yellow peel of 2 lemons ; mix well, make it into 'forms,' cover with wafer-paper, and bake in a moderate oven.

MACE. *Syn.* MACIS, L. The tough, membranous, lacerated covering (arillode) of the NUTMEG. It has a flavour and odour more agreeable than that of nutmeg, which in its general properties it resembles. It is used as a flavouring by cooks, confectioners, and liqueuristes ; and in medicine as a carminative. See OIL, &c.

MACERA′TION. *Syn.* MACERATIO, L. The steeping of a substance in cold water, for the purpose of extracting the portion soluble in that menstruum. The word is also frequently applied to the infusion of organic substances in alcohol or ether, or in water, either alkalised or acidulated.

MACHINERY, Electric Light, Belting for. Various kinds of belts have been used from time to time for the purpose of driving machinery, but all must yield the palm to leather, for there seems to be 'nothing like leather' as a material for driving-belts. It is most important to have the best belts for driving machinery, especially dynamos. The choice and care of driving-belts is a matter of consequence. The best belts are made of raw hide, with seamless joints, manufactured by an American firm. The joints are made by cementing the long chamfered edges of the leather together under pressure. These are sometimes made stronger, with a flat leather lace embedded in the leather. The joints are so neatly made as to present no additional thickness, and very little difference in suppleness from any other part of the belt. The leather is sent out oiled ready for use, and therefore the belts will retain their suppleness for many years whilst working in ordinary temperatures. Oiled belts take a better grip on the pulleys than dry belts, and therefore need not be run so tight as the latter. This lessens the strain on the grain of the leather, and conduces to the long life of the belt. Dry belts are apt to slip on the pulley, and the friction on the leather, caused by slipping, causes it to heat, and thus ' burns the life ' out of the belt. Belts should always present a clammy side to the pulley. In dry situations, such as in an engine-room or hot workshop, the clammy state of the belt should be kept up by giving it a dressing of dubbing and a coat or two of boiled linseed oil at least once a year. Always choose a belt wide enough to do the work without undue tightness. There is economy in using moderately wide belts running slightly slack, as against narrow ones put on as tight as they will bear. Run the flesh side of the leather next the pulley, and the grain side outside, because experience of both has shown that a belt run this way lasts longer than one run with the grain side next the pulley. It is also the natural bent of the leather. Small belts working light machinery run fairly well with butt-joints linked with double tee brass links (Green's patent belt fasteners) inserted in the leather, but these are apt to tear out if the belt has to do heavy work. These joints have the advantage of being easily and quickly made. Sewn lap-joints should be used for heavy driving-belts. Laced lap-joints with the laps well thinned down, and the lace-holes punched in diamond-shaped rows, do fairly well. All lumps accumulating on the pulleys or the inside faces of the belts should be promptly removed as soon as discovered, as they overstrain the belt and cause jerks in the machinery (' Work ').

MACKEREL. The *Scomber scombrus*, Linn., a well-known spiny-finned sea-fish, much esteemed at certain seasons for the table. Though nutritious it is very apt to disagree with delicate stomachs, and occasionally induces symptoms resembling those of poisoning.

MADDER. *Syn.* RUBIA, RUBIÆ RADIX, L. The root of *Rubia tinctorum*, Linn., or dyer's madder. The best madder has the size of a common goose-quill, a reddish-yellow appearance, and a strong odour. As soon as the roots are taken from the ground they are picked and dried, and before use they are ground in a mill. Levant, Turkey, and Smyrna madder is imported whole. It is obtained from the species *R. peregrina*. French, Dutch, and Zealand madder is imported ground. The finest quality of ground madder is called ' crop ' or ' grappe;' 'ombro' and 'gamene' are inferior sorts, and ' mull ' the worst.

Madder contains several distinct principles, *e.g.* madder red (alizarin), madder purple (purpurin), madder orange (rubiacin), madder yellow (xanthin), &c. The first of these (noticed below) is by far the most important. In addition to colouring matters madder contains a quantity of sugar; Stein found as much as 8%. From recent researches it appears that the *fresh* root only contains two colouring substances, viz. xanthin and purpurin. According to Dr Rochleder the alizarin is produced under the influence of a peculiar nitrogenous substance present in the root, and which converts part of the xanthin into alizarin and sugar.

' Flowers of madder ' is a commercial preparation made from the pulverised root by steeping it in water, inducing fermentation of the sugar, and washing the residue first with warm water, then with cold. Hydraulic pressure is used to remove the water, after which the substance is dried and again pulverised. This process eliminates the pectinous substances of the root, which otherwise become insoluble during the operation of dyeing.

Pur., &c. Madder is frequently adulterated with logwood, Brazil-wood, and other dye-stuffs of inferior value; and also, not unfrequently, with brickdust, red ochre, clay, sand, mahogany sawdust, bran, &c. These admixtures may be detected as follows:

1. When dried at 212° F., and then incinerated, not more than 10% to 12% of ash should be left.

2. It should not lose more than 50% to 56% by exhaustion with cold water.

3. When assayed for alizarin (see *below*), the quantity of this substance obtained should be equal to that from a sample of the same kind of madder which is known to be pure, and which has been treated in precisely the same manner. The operation may be conducted as follows :—500 gr. of the sample are weighed, and, after being dried by the heat of boiling water or steam, are gradually added to an equal weight of concentrated sulphuric acid, contained in a glass vessel, and stirred with a glass rod; after a few hours the charred mass is washed with cold water, collected on a filter, and dried by the heat of boiling water; the carbonised mass (' garacine ') is next powdered, and treated with successive portions of rectified spirit, to which a little ether has been added, at first in the cold, and afterwards with heat, until the liquid is no longer coloured by it, when the mixed tincture is filtered, and evaporated (distilled) to dryness; the weight of the residuum, divided by 5, gives the percentage of red colouring matter present. Or—The dried carbonised matter is exhausted by boiling it in a solution of 1 part of alum in 5 or 6 parts of water, and the decoction, after being filtered whilst in the boiling state, is treated with sulphuric acid as long as a precipitate falls, which is washed, dried, and weighed as before.

Uses, &c. Madder is principally employed as a dye-stuff. It has been given in jaundice and rickets, and as an emmenagogue.—*Dose,* ½ dr. to 2 dr., twice or thrice a day. See RED DYES, IVORY, PURPURIN, &c., also *below*.

MADDER RED. *Syn.* ALIZARIN. $C_{14}H_8O_4$. The most important constituent of the red dye of madder-root, first obtained in a separate form by Robiquet.

Prep. 1. The aqueous decoction of madder is treated with dilute sulphuric acid as long as a precipitate falls, which, after being washed, is boiled in a solution of chloride of aluminium as long as it gives out colour; the liquid is then filtered, precipitated with hydrochloric acid, and the precipitate washed and dried. It may be purified from any adhering purpurin by dissolving it in alcohol, again throwing it down with hydrate of aluminium, boiling the precipitate with a strong solution of soda, and separating the alizarin from its combination with alumina by means of hydrochloric acid; it is lastly crystallised from alcohol.

2. (*Meillet.*) Alum, 3 parts, is dissolved in water at 140° F., 30 parts, and madder, 13 parts, added to the solution; the whole is then gently boiled for 30 or 40 minutes, after which it is thrown upon a filter, and submitted to strong pressure; this treatment is repeated with fresh solutions a second and a third time; the mixed filtrates are then decanted, and when nearly cold, oil of vitriol, 1 part, diluted with twice its bulk of water, is added, care being taken to stir the liquid all the time; the supernatant fluid is next decanted, and the residuum well washed, and, lastly, dried in the air. If required quite pure, it is dissolved, whilst still moist, in a solution of 1¼ times its weight of potassium carbonate in 15 parts of water, and, after reprecipitation with sulphuric acid, is washed and dried as before.

3. (*Robiquet and Colin.*) Powdered madder is exhausted with water of a temperature not exceeding 68° F., and after being dried, 1 part of it is boiled for 15 or 20 minutes in a solution of alum, 8 parts, in water, 40 parts; the liquid is filtered whilst boiling, the marc well washed with a fresh solution of alum, the mixed liquids precipitated with sulphuric acid, and the precipitate washed and dried as before.

Obs. Alizarin has been produced artificially by Graebe and Liebermann from anthracene ($C_{14}H_{10}$), a hydrocarbon existing in coal-tar. For a description of the process see ALIZARIN, ARTIFICIAL.

4. Madder, exhausted by 2 or 3 macerations in 5 or 6 times its weight of cold water, is submitted to strong pressure, to remove adhering

water, and the marc, whilst still moist, is mixed with half its weight of oil of vitriol diluted with an equal quantity of water; the whole is kept at the temperature of 212° for an hour, and, after being mixed with cold water, is thrown on a linen strainer, well washed with cold water, and dried.

5. From powdered madder and oil of vitriol, equal parts, without heat, as described under MADDER.

6. (*F. Steiner.*) The 'used madder' of the dye-works is run into filters and precipitated with sulphuric acid; the matter thus obtained is put into bags and rendered as dry as possible by hydraulic pressure; the pressed cake is next crumbled to pieces, placed in a leaden vessel, and treated with 1-5th of its weight of oil of vitriol, afterwards assisting the action of the acid by introducing steam to the mixture; the resulting dark brown carbonised mass is, lastly, well washed, dried, powdered, and mixed with about 5% of carbonate of soda, when it is ready for sale.

Obs. The last three formulæ produce the 'GARANCE' or 'GARANCINE' of commerce, now so extensively used in dyeing.

Prop., &c. Pure anhydrous alizarin crystallises in magnificent orange-red crystals, which may be fused at 282° C., and sublimed; it is freely soluble in alkalies to a violet-red solution, and in oil of vitriol, giving a rich red colour; water throws it down from the last unchanged; it is also soluble in hot-alcohol, a hot solution of alum, and less freely in hot water. Hydrated alizarin occurs in small scales resembling mosaic gold. When impure it generally forms shining reddish-brown scales. Commercial 'garancine' is a deep brown or puce-coloured powder, and will probably ere long entirely supersede crude madder for dyeing. The properties of garancine as a dye-stuff are precisely similar to those of madder. A solution of alum added to a solution of alizarin, and precipitated by potassium carbonate or borax, furnishes a rose lake; the tin lake is a fine red colour, the iron lake violet-black, and the lime lake blue. The so-called madder lake is prepared by first making madder with water free from lime, and then proceeding as in the manufacture of the rose lake.

MAGILP'. *Syn.* MEGILLUP. A mixture of pale linseed oil and mastic varnish, employed by artists as a 'vehicle' for their colours. It is thinned with turpentine.

MAGISTERY. *Syn.* MAGISTERIUM, L. The old name of precipitates. The following are the principal substances to which this term has been applied:—MAGISTERY OF ALUM, hydrate of alumina; M. OF BISMUTH, subnitrate of bismuth; M. OF DIAPHORETIC ANTIMONY, washed diaphoretic antimony; M. OF OPIUM (*Ludolph's*), crude morphia; M. OF LAPIS CALAMINARIS or M. OF ZINC, hydrated oxide of zinc.

MAGNESIA. See MAGNESIUM, OXIDE OF.

Magnesia, Hydrate of. (P. Cod.) *Syn.* MAGNESIÆ HYDRAS. Obtained by boiling magnesia in 20 or 30 times its weight of water for 20 minutes, draining on a linen cloth, and drying. It contains 31% of water.

Magnesia, Lactate of. (Ph. Ger.) *Syn.* MAGNESIÆ LACTAS. *Prep.* Mix 1 oz. (by weight) of lactic acid in 10 oz. of distilled water, just made slightly warm, and add light carbonate of magnesia enough to neutralise it. Filter and evaporate till crystals form.

Magnesia Levis (B. P.). *Syn.* LIGHT MAGNESIA. *Prep.* (B. P.) Light carbonate of magnesium heated in a Cornish crucible until all the carbonic anhydride is driven off.

A bulky white powder, differing from the magnesia (B. P.) only in its density, the volume occupied by the same weight being 3½ to 1.

The properties of the two varieties of magnesium oxide are identical, and are used in medicine as antacids, laxatives, and antilithics, and much used in dyspepsia, heartburn, &c.—*Dose*, 10 to 20 gr. as an antacid, and 20 to 60 gr. as a purgative.

MAGNE'SIAN APE"RIENT (Effervescing). *Prep.* 1. Heavy carbonate of magnesia, 2 lbs.; tartaric acid and double refined lump sugar, of each, 1½ lbs.; bicarbonate of soda (dried without heat), 1 lb., each separately dried and in very fine powder; essential oils of orange and lemon, of each, ½ fl. dr.; mix well in a warm, dry situation, pass the whole through a sieve, put it into warm, dry bottles, and keep them well corked.

2. As the last, but substituting calcined magnesia, 1 lb., for the heavy carbonate, and adding sugar, ¾ lb. The preceding furnish a very pleasant effervescing saline draught.

3. (MOXON'S.) *a.* Take of sulphate of magnesia, 2 lbs.; dry it by a gradually increased heat, powder, add of tartaric acid (also dried and powdered), 1½ lbs.; calcined magnesia, ½ lb.; finely powdered white sugar, 3 lbs.; bicarbonate of soda (dried without heat), 1 lb.; essence of lemon, 1 dr.; mix, and proceed as before.

b. (*Durande.*) Carbonate of magnesia, 1 part; bicarbonate of soda, tartrate of soda and potassa (sel de Seignette), and tartaric acid, of each, 2 parts; mix as before.

c. ('Pharm. Journ.') Sulphate of magnesia and bicarbonate of soda, of each, 1 lb.; tartaric acid, ¾ lb.; mix as before. The last two are much less agreeable than the others.

4. Carbonate of magnesia, 2 parts; calcined magnesia, 4 parts; citric acid, 13 parts; lump sugar, 25 parts; essence of lemon, q. s. to flavour. Very agreeable. This is known as 'ROGERS PURGATIF.'

Obs. The above are very useful and popular medicines in indigestion, heartburn, nausea, habitual costiveness, dyspepsia, &c.—*Dose*, ½ to 2 dessert-spoonfuls, thrown into a tumbler ⅓ parts filled with cold water, rapidly stirred and drunk whilst effervescing, early in the morning fasting, or between breakfast and dinner.

MAGNESIAN LEMONADE'. See CITRATE OF MAGNESIA and LEMONADE (Aperient).

MAGNESIUM. Mg = 23·94. *Syn.* MAGNIUM, TALCIUM. The metallic radical of magnesia. The existence of this metal was demonstrated by Sir H. Davy in 1808, but it was first obtained in sufficient quantity to examine its properties by Bussy in 1830.

Prep. 5 or 6 pieces of sodium, about the size of peas, are introduced into a test-tube, and covered with small fragments of chloride of magnesium; the latter is then heated to near its point of fusion, when the flame of the lamp is applied to the sodium, so that its vapour may pass through the stratum of heated chloride; when the vivid incandescence that follows is over, and the whole has become cold, the mass is thrown into water, and the insoluble metallic portion collected and dried.

Commercial magnesium is prepared by evaporating solutions of the chlorides of sodium and magnesium, in the proportion of 1 to 3, to dryness, mixing with one quarter of its weight of fluor-spar and a like amount of sodium, and heating to bright redness in an iron crucible of proper construction.

On a larger scale it is prepared by heating to redness a mixture of chloride of magnesium, 9 parts; fused chloride of sodium, 1½ parts; fluoride of calcium, 1½ parts; and sodium in slices, 1½ parts.

The Magnesium Metal Company at Patricroft, near Manchester, and the American Magnesium Company at Boston, U.S.A., prepare the metal on the large scale.

Prop., &c. In colour and lustre it resembles silver, but in chemical properties is more like zinc; its sp. gr. is only 1·75; it is malleable, fusible at a red heat, and can be distilled like zinc; unaffected by dry air and by cold water; burns with brilliancy when heated to dull redness in air or oxygen gas, yielding oxide of magnesium; inflames spontaneously in chlorine, yielding chloride of magnesium; dissolves in the acids with the evolution of hydrogen gas, and the formation of pure salts of magnesium.

Tests. It is distinguished from the metals generally by the non-precipitation of its sulphide, and by the tendency of its salts, except the arsenate and phosphate, to form soluble compounds with the salts of ammonium. It is not precipitated by ammonium carbonate in presence of sal-ammoniac. Its presence is readily detected by the addition of sodium phosphate to any solution containing it. On standing a crystalline precipitate is deposited, the formation of which is greatly accelerated by scratching the sides of the test-tube with a glass rod.

Uses. It has been used somewhat extensively as an illuminating agent for photographing at night, for the light emitted by burning magnesium is capable of inducing chemical changes similar to those caused by sunlight, and also for the purpose of affording a brilliant light for microscopic, pyrotechnical, and magic-lantern effects. The metal is prepared for these purposes in the form of ribbon, wire, or powder; the latter is used for 'flash' lights, and should be handled with care. It was extensively used in the Abyssinian campaign for signalling at night. It has been suggested to alloy magnesium instead of zinc with copper; metallic magnesium is also used in toxicological investigations, in the estimation of nitrates and nitrites in drinking-water, and other chemical operations.

Magnesium Bromide. $MgBr_2$. *Syn.* MAGNESII BROMIDUM. To bromide of iron in solution add calcined magnesia in excess, heat the mixture, filter, and evaporate the clear solution to dryness.

It occurs free in sea-water and many brine springs.

Magnesium, Carbonate of (Heavy). *Syn.* HEAVY CARBONATE OF MAGNESIA; MAGNESII CARBONAS (B. P.), L. $3MgCO_3.MgO.5H_2O$. *Prep.* 1. (Apothecaries' Hall.) A saturated solution of sulphate of magnesium, 1 part, is diluted with water, 3 parts, and the mixture heated to the boiling-point; a cold saturated solution of carbonate of sodium, 1 part (all by measure), is then added, and the whole is boiled with constant agitation until effervescence ceases; boiling water is next freely poured in, and after assiduous stirring for a few minutes, and repose, the clear liquid is decanted, and the precipitate thrown on a linen cloth and thoroughly washed with hot water; it is, lastly, drained, and dried in an iron pot.

2. (Ph. D.) Dissolve sulphate of magnesium, 10 oz., in boiling distilled water, ½ pint; and carbonate of sodium (cryst.), 12 oz., in boiling distilled water, 1 pint; mix the two solutions, and evaporate the whole to dryness by the heat of a sand-bath; then add of boiling water, 1 quart, digest with agitation for ½ an hour, and wash the insoluble residuum as before; lastly, drain it, and dry it at the temperature of boiling water.

3. (B. P.) White granular powder precipitated from a boiling solution of sulphate of magnesium by a solution of carbonate of sodium, the whole evaporated to dryness, and the dry residue digested in water, collected on a filter, and washed.

Prop. The ordinary or light carbonate of magnesia is a white, inodorous, tasteless powder, possessing similar properties to calcined magnesia, except effervescing with acids, and having less saturating power. An ounce measure is filled by 45 to 48 gr. of the powder lightly placed in it. The heavy carbonate is sometimes fully thrice as dense (see *below*), but in other respects is similar.

Doses. As an antacid, ½ to a whole teaspoonful, 3 or 4 times daily; as a laxative, ½ to 2 dr. It is commonly taken in milk. It is apt to produce flatulence, but in other respects is preferable to calcined magnesia.

General Remarks. Although commonly called 'carbonate of magnesia,' the above substance, whether in the light or heavy form, appears to be a compound of carbonate with hydrate, in proportions which are not perfectly constant. In the preparation of these carbonates if the solutions are very dilute the precipitate will be exceedingly light and bulky; if otherwise, it will be denser. By employing nearly saturated solutions, and then heating them and mixing them together whilst very hot, a very heavy precipitate is obtained, but it is apt to be gritty or crystalline. The same occurs when cold solutions are mixed, and no heat is employed. The lightest precipitate is obtained from cold, highly dilute solutions, and subsequent ebullition of the mixture.

Mr Pattinson, a chemist of Gateshead, prepares a very beautiful and pure heavy carbonate from magnesian limestone. The latter is calcined at a dull red heat (not hotter) for some time, by which the carbonic anhydride is expelled from the carbonate of magnesium, but not from the carbonate of calcium, which hence continues insoluble. The calcined mass is next reduced to a milk with water in a suitable cistern, and carbonic anhydride resulting from its own calcination is forced into it under powerful pressure. The result is a saturated solution of carbonate of magnesia, the lime remaining unacted on so long as the magnesium is in excess. The solution by evaporation yields the heavy carbonate, whilst carbonic anhydride is expelled, and may be again used in the same manufacture. The heavy carbonate appears to be fully thrice as dense as the light carbonate. The bicarbonate of magnesium (MAGNESIÆ BICARBONAS, L.) exists only in solution. The so-called 'fluid magnesias' of Murray, Dinneford, Husband, &c., are solutions of this salt. The small prismatic crystals which are deposited when 'fluid magnesia' is exposed to the air for some time consist of hydrated neutral carbonate, and not bicarbonate, as is sometimes stated.

Magnesium, Carbonate of (Light). *Syn.* LIGHT CARBONATE OF MAGNESIA, CARBONATE OF MAGNESIA, MAGNESIA; MAGNESIÆ CARBONAS LEVIS (B. P.). $3MgCO_3.MgO.5H_2O$. *Prep.* 1. (Ph. L.) Sulphate of magnesium, 4 lbs., and carbonate of sodium, 4 lbs. 9 oz.; boiling distilled water, 4 galls.; dissolve the salts separately in one half the water, filter, mix the solutions, and boil for 2 hours, constantly stirring with a spatula, distilled water being frequently added to compensate for that lost by evaporation; lastly, the solution being poured off, wash the precipitated powder with boiling distilled water, and dry it.

2. (B. P.) Similar to the foregoing, except that precipitation takes place in the cold. The formula of this compound is $(Mg.CO_3)_3$. $Mg(HO)_2.4H_2O$.

3. (*Henry's.*) Ordinary carbonate of magnesia, the washing of which has been finished with a little rose-water.

4. Add a solution of carbonate of potassium or sodium to the bittern or residuary liquor of the sea-salt works, and well wash and dry the precipitate as before. This is known in commerce as 'Scotch magnesia.'

Obs. The carbonate of magnesia of commerce is usually made up into cakes or dice while drying; or it is permitted to drain and dry in masses, which are then cut into squares with a thin knife. It is powdered by simply rubbing it through a wire sieve. The presence of iron in the solution of the sulphate of magnesium, when the crude salt is employed, and which is destructive to the beauty of the preparation, may be got rid of by the addition of lime-water until the liquor acquires a slight alkaline reaction, and subsequent decantation after standing.

Magnesium, Chloride of. $MgCl_2$. *Syn.* MAGNESII CHLORIDUM, L. Occurs in sea-water, in many brine-springs and salt beds; it is at present prepared in large quantities at Stassfurt.

Prep. (*Liebig.*) By dissolving magnesia in hydrochloric acid, evaporating to dryness, adding an equal weight of chloride of ammonium, projecting the mixture into a red-hot platinum crucible, and continuing the heat till a state of tranquil fusion is attained. On cooling, it forms a transparent, colourless, and very deliquescent mass, which is anhydrous, and soluble in alcohol.

Obs. Without the addition of the chloride of ammonium it is impossible to expel the last portion of the water without at the same time driving off the chlorine, in which case nothing but magnesia is left. The fused mass should be poured out on a clean stone, and when solid broken into pieces, and at once transferred to a warm, dry bottle. The P. Cod. orders the solution to be evaporated to the sp. gr. 1·384, and to be put, whilst still hot, into a wide-mouthed flask to crystallise.—*Dose*, 1 to 4 dr.; as a laxative.

Magnesium, Cit'rate of. $Mg_3(C_6H_5O_7)_2$. *Syn.* MAGNESIÆ CITRAS, L. *Prep.* There is some difficulty in obtaining this salt in an eligible form for medicinal purposes. When precipitated from a solution it is insoluble. The following formulæ can be highly recommended:

1. (*Parrish.*) Dissolve crystallised citric acid, 100 gr., in water, 15 drops, and its own 'water of crystallisation' by the aid of heat; then stir in calcined magnesia, 35 gr.; a pasty mass will result, which soon hardens, and may be powdered for use.

Obs. The chief practical difficulty in this process results from the great comparative bulk of the magnesia, and the very small quantity of the fused mass with which it is to be incorporated. A part of the magnesia is almost unavoidably left uncombined, and the salt is consequently not neutral. The uncombined earth should be dusted off the mass before powdering the latter. A high temperature must be avoided.

2. (*Robiquet.*) Citric acid, 35¼ parts, is powdered and dissolved in boiling water, 10½ parts; when the solution is cold, and before it crystallises, it is poured into a wide earthen vessel, kept cold by surrounding it with water; then, by means of a sieve, carbonate of magnesium, 21¼ parts, is distributed evenly and rapidly over the surface without stirring; when the reaction ceases the mixture is beaten rapidly as long as it retains its pasty consistence. The salt should be dried at a temperature not exceeding 70° F.

3. (Effervescing: MAGNESIÆ CITRAS EFFERVESCENS, L.) *a.* Citric acid (dried and powdered), 7 parts; heavy carbonate of magnesium, 5 parts; mix, and preserve in well-corked bottles.

b. (*Ellis.*) Mix powdered citric acid, 2½ oz., with powdered sugar, 8 oz.; triturate to a fine powder, and drive off the water of crystallisation by the heat of a water-bath; add citrate of magnesium (prepared by fusion), 4 oz., and oil of lemons, 10 drops, and mix intimately; then add bicarbonate of sodium, 3 oz., and again triturate until the whole forms a fine powder, which must be preserved in stoppered bottles. From 1 to 3 table-spoonfuls, mixed in a tumbler of water, furnishes an effervescing draught in which the undissolved portion is so nicely suspended that it can be taken without inconvenience.

c. (Ph. Germ.) Light carbonate of magnesia, 25 oz.; citric acid, 75 oz.; distilled water, q. s; mix into a thick paste and dry at 86° F. With 14 oz. of the dried mass mix bicarbonate of soda, 13 oz.; citric acid, 6 oz.; sugar, 3 oz. Sprinkle over the mixture sufficient rectified spirit to make it moist enough to be granulated by rubbing it through a tinned iron sieve.

d. (Extemporaneous.) Citric acid (cryst.), 20

gr.; carbonate of magnesium, 14 gr.; mix in a tumbler of cold water, and drink the mixture whilst effervescing. A pleasant saline.

Obs. A dry white powder, sometimes sold as citrate of magnesia in the shops, is quite a different preparation from the above, and does not contain a particle of citric acid. The following formula is that of a wholesale London drug-house that does largely in this article:

Calcined magnesia (magnesium oxide), 1¼ lbs. (or carbonate, 2 lbs.); powdered tartaric acid, 1¼ lbs.; bicarbonate of sodium, 1 lb.; dry each article by a gentle heat, then mix them, pass the mixture through a fine sieve in a warm, dry room, and keep it in well-corked bottles. A few drops of essence of lemon and 3 lbs. of finely powdered sugar are commonly added to the above quantity. This addition renders it more agreeable.

Prop., &c. Citrate of magnesium is a mild and agreeable laxative; its secondary effects resemble those of the carbonate.—*Dose.* As a purgative, ½ to 1 oz. The dose of the effervescing citrate must depend on the quantity of magnesia present. A solution of this salt in water, sweetened and flavoured with lemon, forms magnesian lemonade.

Magnesium, Boro-cit'rate of. *Syn.* MAGNESIÆ BORO-CITRAS, L. *Prep.* (*Cadet.*) Boracic acid (in powder), 113 gr.; oxide of magnesium, 80 gr.; mix in a porcelain capsule, and add enough of a solution of citric acid, 260 gr., in water, 3½ pints, to form a thin paste; then add the remainder of the citric solution, and gently evaporate, with constant stirring, to dryness. A cooling saline, and in small doses, emmenagogue and lithontriptic.—*Dose.* As an aperient, 3 to 6 dr.

Magnesium, Oxide of. MgO. *Syn.* CALCINED MAGNESIA, MAGNESIA (B. P., Ph. L.).

Prep. Forms when the metal burns in the air. Magnesium carbonate, heated in a crucible until all the carbonic anhydride is driven off. It is also produced by the ignition of any magnesium salt containing a volatile acid.

Prop., &c. White heavy powder, scarcely soluble in water, but readily soluble in acids without effervescence. It is tasteless, but in the moist state turns litmus-paper blue. Its solution in hydrochloric acid, neutralised by a mixed solution of ammonia and ammonium chloride, gives a copious crystalline precipitate when sodium phosphate is added to it. See MAGNESIA LEVIS.

Magnesium, Phos'phate of. MgHPO₄.6Aq. *Syn.* MAGNESIÆ PHOSPHAS, L. *Prep.* From the mixed solutions of phosphate of sodium and sulphate of magnesium, allowed to stand for some time. Small, colourless, prismatic crystals, which, according to Graham, are soluble in about 1000 parts of cold water. Phosphate of magnesium exists in the grains of the cereals, and in considerable quantity in beer. It is also found in guano.

Magnesium and Ammo''nium, Phosphate of. MgNH₄.PO₄ + 6Aq. *Syn.* AMMONIO-PHOSPHATE OF MAGNESIA; MAGNESIÆ ET AMMONIÆ PHOSPHAS, L. This compound falls as a white crystalline precipitate whenever ammonia or carbonate of ammonium is added, in excess, to a solution of a salt of magnesium which has been previously

mixed with a soluble phosphate, as that of soda. It subsides immediately from concentrated solutions, but only after some time from very dilute ones.

Prop., &c. Ammonio-phosphate of magnesium is very slightly soluble in pure water; when heated, it is resolved into pyrophosphate of magnesium, and is vitrified at a strong red heat. It is found in wheaten bran, guano, potatoes, &c., and also frequently in urinary calculi.

Magnesium, Sil'icates of. There are several native silicates of magnesium, more or less pure, of which, however, none is directly employed in medicine. Meerschaum and steatite or soapstone are well-known varieties. Serpentine is a compound of silicate and hydrate of magnesium. Asbestos is a silicate of magnesium and calcium. The minerals augite and hornblende are double salts of silicic acid, magnesium, and calcium, with some ferrous oxide. The beautiful crystallised mineral called chrysolite is a silicate of magnesium, coloured with ferrous oxide. Jade is a double silicate of magnesium and aluminum, coloured with chromic oxide. Olivine and tourmaline are also silicates of magnesium.

Magnesium, Sulphate of. MgSO₄ + 7Aq. *Syn.* EPSOM SALT; MAGNESIÆ SULPHAS (B. P., Ph. L. E. and D.), SAL EPSOMENSIS, L. This compound was originally extracted from the saline springs of Epsom, Surrey, by Dr Grew, in 1695. It is now exclusively prepared on the large scale, and from either magnesian limestone, the residual liquor of the sea-salt works, or, as at Staesfurth, from kieserite, which is found in salt beds.

Prep. 1. From dolomite or magnesian limestone. *a.* The mineral, broken into fragments, is heated with a sufficient quantity of dilute sulphuric acid to convert its carbonates into sulphates; the sulphate of magnesium is washed out of the mass with hot water, and the solution, after defecation, is evaporated and crystallised.

b. The 'limestone,' either simply broken into fragments or else calcined, and its constituents quicklime and magnesia converted into hydrates by slaking it with water, is treated with a sufficient quantity of dilute hydrochloric acid to dissolve out all the calcium hydrate without touching the magnesium hydrate; the residuum of the latter, after being washed and drained, is dissolved in dilute sulphuric acid, and crystallised as before.

2. From bittern. *a.* The residual liquor or mother-water of sea-salt is boiled for some hours in the pans which are used during the summer for the concentration of brine; the saline solution is then skimmed and decanted from some common salt which has been deposited, after which it is concentrated by evaporation, and finally run into wooden coolers; in about 36 hours 1-8th part of Epsom salts usually crystallises out. This is called 'singles.' By re-dissolving this in water, and re-crystallisation, 'doubles,' or Epsom salts fit for the market, are obtained. A second crop of crystals may be procured by adding sulphuric acid to the mother-liquor and re-concentrating the solution, but this is seldom resorted to in England. Bittern yields fully 5 parts of sulphate of magnesia for every 100 parts of common salt that has been previously obtained from it.

b. A concentrated solution of sulphate of sodium is added to bittern, in equivalent proportion to that of the chloride of magnesium in it, and the mixed solution is evaporated at the temperature of 122° F. (*Ure*); cubical crystals of common salt are deposited as the evaporation proceeds, after which, by further concentration and repose, regular crystals of sulphate of magnesia are obtained.

c. A sufficient quantity of calcined and slaked magnesian limestone is boiled in bittern to decompose the magnesium salts, and the liquid is evaporated, &c., as before. This is a very economical process.

Prop. Small acicular crystals, or, by slow crystallisation from concentrated solutions, large four-sided rhombic prisms, which are colourless, odourless, transparent, slightly efflorescent, extremely bitter and nauseous. When heated, it fuses in its water of crystallisation, the larger portion of which readily passes off, but one equivalent of water is energetically retained; at a high temperature it runs into a species of white enamel. It dissolves in its own weight of cold water, and in 3·4ths of that quantity of boiling water; it is insoluble in both alcohol and ether. Sp. gr. 1·68. It is not deliquescent in air.

Pur., &c. Sulphate of magnesium is soluble in an equal weight of water at 60° F., by which it may be distinguished from sulphate of sodium, which is much more soluble.

An aqueous solution in the cold is not precipitated by oxalate of ammonium. The precipitate given by carbonate of sodium from a solution of 100 gr. should, after well washing and heating to redness, weigh 16·26 gr. (B. P.).

Digested in alcohol, the filtered liquid does not yield a precipitate with nitrate of silver nor burn with a yellow flame, and evaporates without residue. 10 gr., dissolved in 1 fl. oz. of water, and treated with a solution of carbonate of ammonium, are not entirely precipitated by 280 minims of solution of phosphate of sodium (Ph. E.).

Uses, &c. Sulphate of magnesium is an excellent cooling purgative, and sometimes proves diuretic and diaphoretic.—*Dose,* 1 dr. to 1 oz., as a purgative, or an antidote to poisoning by lead. Large doses should be avoided; instances are on record of their having proved fatal. Dr Christison mentions the case of a boy 10 years old who swallowed 2 oz. of salts, and died within 10 minutes. The best antidote is an emetic. A small quantity of Epsom salts, largely diluted with water (as a drachm to ½ pint or ½ pint), will usually purge as much as the common dose. This increase of power has been shown by Liebig to result rather from the quantity of water than the salt. Pure water is greedily taken up by the absorbents; but water holding in solution saline matter is rejected by those vessels, and consequently passes off by the intestines.

Epsom salt is also used as a dressing for cotton goods, and in dyeing with aniline colours.

Obs. Oxalic acid has occasionally been mistaken for Epsom salt, with fatal results. They may be readily distinguished from each other by the following characteristics:

EPSOM SALT.	OXALIC ACID.
Tastes extremely bitter and nauseous.	Tastes extremely sour.
Does not volatilise when heated on platinum foil.	Volatilises when heated on platinum foil.
Does not produce milkiness when dissolved in *hard* water.	Produces milkiness when dissolved in *hard* water.

Magnesium, Tar'trate of. *Syn.* MAGNESIÆ TARTRAS, MAGNESIA TARTARICA, L. *Prep.* By saturating a solution of tartaric acid with carbonate of magnesium, and gently evaporating to dryness. It is only very slightly soluble in water.—*Dose,* 20 to 60 gr , or more; in painful chronic maladies of the spleen (*Pereira,* ex *Radmacher*). The effervescing tartrate of magnesium, commonly sold under the name citrate, has already been noticed.

Magnesium and Potas'sium, Tartrate of. *Syn.* POTASSIO-TARTRATE OF MAGNESIA; MAGNESIÆ POTASSIO-TARTRAS, M. ET POTASSÆ TARTRAS, L. *Prep.* From acid tartrate of potassium (in powder), 7 parts; carbonate of magnesium, 2 parts; water, 165 parts; boiled until the solution is complete, and then evaporated and crystallised. A mild aperient.—*Dose,* 1 to 5 dr.; in scurvy, &c.

MAG'NET. *Syn.* MAGNES, L. Besides its application to the loadstone, this name was formerly given to several compounds used in medicine. ARSENICAL MAGNET (MAGNES ARSENICALIS), a substance once used as a caustic, consisted of common antimony, sulphur, and arsenious acid, fused together until they formed a sort of glass. MAGNES EPILEPSIÆ was native cinnabar.

MAGNOLIA BALM. The analysis of Hager and F. M. Clarke says that the balm consists of zinc oxide (coloured with carmine) in suspension in a little dilute glycerine, and perfumed with oil of bergamot, oil of lemon, and perhaps one other odour. The following formula makes a preparation substantially the same as the proprietary article:—Zinc oxide, 4 dr.; glycerine, 1½ fl. oz.; water, 2 fl. oz.; carmine, ½ gr.; oil bergamot, 1 minim; oil lemon, 1 minim.

MAGPIE MOTH, The (*Abraxas grossulariata,* Stephens). This is styled the 'Magpie' moth on account of its black and white markings, and it is known to fruit producers as mainly injurious to gooseberry and currant bushes. It is perhaps more destructive to black currants than to red currants. Sometimes it is found upon apricot trees and various forest trees, and it is especially fond of the blackthorn, *Prunus spinosa.*

Casual observers confound the attack of the *Abraxas* upon gooseberry and currant bushes with that of the gooseberry saw-fly, *Nematus grossulariæ.* In 1876 and in 1881 there was curious confusion between these insects, and it was necessary to request correspondents to send specimens of the foes that had come upon their gooseberry bushes in order to discover which was the culprit.

The attack of this *Abraxas* is not so serious as that of the *Nematus,* and is easily distinguished, as the two insects are utterly different. In the winged state there are no points of resemblance. In the larval state the caterpillar of the former differs essentially in size, colour, and conformation from the grub of the latter. The caterpillars of the *Abraxas*, as they pass the winter in this guise, are ready for action directly the weather invites them to quit their winter quarters, or as soon as there is a vestige of green upon the bushes. Thus they get a good start of the grubs of the *Nematus*, which are hatched from eggs laid by the female flies in the spring, and do not appear upon the scene until vegetation is far advanced.

Fortunately, however, the *Abraxas* is not nearly so abundant as the *Nematus*, at least in fruit plantations. When it gets a footing in these it is most important to take active steps to check its progress.

In the year 1876 there were many and grave complaints of injury caused by this insect to black and red currants in gardens as well as in plantations in Kent and in Cambridgeshire, where black currants are extensively produced. Also in 1876 several large gooseberry growers in Kent, in which county it is not by any means unusual for individual growers to have from thirty to eighty acres planted with gooseberry bushes, reported that large variegated caterpillars were at work in their plantations, having fixed upon bushes here and there, and that the area of their operations extended from day to day. They had come as soon as the leaves began to unfold, and cleared off these and the incipient blossoms so that the bushes were as bare as in winter. Without any difficulty these were declared to be the caterpillars of the Magpie moth.

Again in 1881 there were great outcries from fruit growers and gardeners from many counties in England and from several in Ireland, as to the destruction caused by it, though in some instances it was proved that the Saw-fly was the offender.

German cultivators suffer from this moth, which they term Harlequin (Harlekin) ('Praktische Insekten-Kunde,' von E. L. Taschenberg), though Köllar does not allude to it. In France it is well known, particularly in the central departments. Fitch speaks of it as very destructive to gooseberry and currant crops in America (third, fourth, and fifth 'Reports on the Noxious, Beneficial, and other Insects of the State of New York,' 1859, by Asa Fitch, M.D.).

Life History. The *Abraxas grossulariata* belongs to the Nat. Ord. LEPIDOPTERA, and to the family *Geometrida*. In its perfect, or moth form, it is about 20 lines across its expanded wings, with a length of body of from 12 to 14 lines. The typical Magpie moth has a yellow body and a black head, with a row of 6 black spots down its back. The fore-wings have a white ground with many black spots dotted irregularly upon them. Some specimens have these black spots in patches, while in others the black spots are very few and indistinct, so that the moth appears to be of a yellowish or creamy hue. As Stephens, Westwood, and Miss Ormerod have shown, there are great and unusual variations of colour and marks in different specimens of this pretty insect.

One of the distinctions between the sexes is that the antennæ of the male are slightly feathered, but those of the female are simple threads.

At the end of July, or in the beginning of August, the moth emerges from the chrysalis which has lain since May in the ground, and pairing takes place. Shortly after this the eggs, of a pale straw colour, are placed in little groups of 3 to 5, close to the midribs of the leaves of the gooseberry and currant bushes, or in the angles made by their nerves, and are hatched out within 10 days. At this time there is but little succulence in the leaves, and the caterpillars are obliged to put up with what they can get; but they quickly attain their full growth, and fall to the ground before the leaves come off, where they pass the winter under leaves, weeds, and rubbish, and actually under the surface. It is stated that some of the caterpillars remain snugly ensconced in the leaves during the winter, having fastened the under surfaces over their bodies with silken webs, and further, having bound the leaves tightly to the branches in order that they may not fall off or be blown away by wintry blasts. I have never found caterpillars in these aërial quarters. Mr. Newman, in his 'British Moths,' concludes that these are their natural and ordinary resorts for hibernation.

Although it is most exceptional for caterpillars to resemble their parents in distinctive markings and colourings, those of the *Abraxas grossulariata* are singularly like the moths in these respects. They have black heads and rows of black spots down their backs; while the bodies are yellowish, with a line of darker yellow on both sides. Like other caterpillars of the *Geometrida*, they have only pectoral and anal feet, and are therefore 'loopers,' making loops when they proceed.

At the beginning of April, or as soon as the leaves begin to show, the caterpillars come forth from their winter abodes, and crawl up the bushes under which they have been concealed. By this time their appetites are keen, and the young leaves and blossoms are more tempting and grateful to their taste than the withered-up foliage in September, and they rapidly clear up all vegetation before them, and change in due time to black chrysalids with three rings of golden colour at their extremities. Some of these are fixed to the leaf-stalks by means of threads. Other caterpillars fall or let themselves down to the ground, and their transformation takes place there under leaves or weeds or clods.

Prevention. Unusual premonitory indications are given of a coming attack from this moth. The strange arrival of its caterpillars in September upon gooseberry and currant bushes should serve to point out plainly what may be expected in the spring.

After this has been noticed, the ground for some distance round the bushes should be well covered with quicklime or gas lime, and dug in the early winter. Again, in the beginning of March, the soil should be well pulverised with pronged hoes, and another dressing of lime put on; or wood ashes or soot may be used instead of lime, and put thickly just round the bushes to prevent the caterpillars from crawling to the stems.

It is efficacious, but costly if carried out upon a large scale, to hoe away the surface soil round the bushes in November, and to put fresh tan close to them just before the approach of spring. Dressings of farmyard manure have also been put round the trees in this manner with good results. Ashes or finely powdered soil, or sawdust saturated with paraffin and water, have been put round the bushes in 3 or 4 cases.

Trial was made in a small plantation in Gloucestershire of Stockholm tar and cart-grease daubed round the stems near to the ground after a bad attack in 1881, and it was thought with considerable advantage, as the bushes were undisturbed.

Assuming that a part of the caterpillars do remain, as is alleged, upon the bushes, it would be desirable to have them hand-picked in the winter after the pruning has been accomplished. A deal of wood is cut away every year from these fruit bushes, especially from black currants, whose fruit is formed upon young wood. All these cuttings should be collected and burned if there is the slightest suspicion that any caterpillars are harboured upon them. I have asked many tree-cutters, as the pruners are called in Kent, whether they have ever noticed caterpillars upon the fruit bushes in the winter, and have been assured that they have never seen any. It must, however, be admitted that labourers are by no means observant.

Remedies. In gardens hand-picking or shaking the branches to dislodge the invaders may be practised in April or May when they are actively engaged. This could hardly be carried out upon a large scale in fruit plantations.

A mixture of lime and soot was thrown upon infested bushes in 1881, before the dew was off the leaves, by several growers, who expressed themselves well satisfied with the effect. Lime or soot by itself would, it is thought, be of equal advantage.

Powdered hellebore may be sprinkled over the trees, as it is for the grubs of the saw-fly. This is effectual when put on properly, but as it is a most deadly poison it is most dangerous to use it if the fruit is formed, though it be ever so small.

During the season of 1881 washing or syringing the bushes was tried successfully in Kent. By far the best mixture was soft soap and quassia with water, in the proportion of 7 lbs. of soap and 6 lbs. of quassia to 100 galls. of water. (The quassia chips must be well boiled in order that the bitter principle may be thoroughly extracted.) After syringing, which may be done with hand syringes and pails, or with hydronettes, the earth under the bushes must be hoed over and beaten down in order to kill the caterpillars which have been dislodged. It has been found, however, that they do not much relish the leaves and blossoms after these have been watered with infusions of quassia retained upon them by the soft soap. The fruit would not be injured, nor would it retain any flavour of quassia, unless it were ripe, or nearly ripe ('Reports on Insects Injurious to Crops,' by Chas. Whitehead, Esq., F.Z.S.).

MAHOG'ANY. This is the wood of *Swietenia mahagoni*, Linn., a native of the hotter parts of the New World. It is chiefly imported from Honduras and Cuba. The extract is astringent, and has been used in tanning, and as a substitute for cinchona bark. The wood is chiefly employed for furniture and ornamental purposes, and, occasionally, in shipbuilding.

Imitations of mahogany are made by staining the surface of the inferior woods by one or other of the following methods:

1. Warm the wood by the fire, then wash it over with aquafortis, let it stand 24 hours to dry, and polish it with linseed oil reddened by digesting alkanet root in it; or, instead of the latter, give the wood a coat of varnish, or French polish which has been tinged of a mahogany colour with a little aloes and annotta.

2. Socotrine aloes, 1 oz.; dragon's blood, ½ oz.; rectified spirit, 1 pint; dissolve and apply 2 or 3 coats to the surface of the wood, previously well smoothed and polished; lastly, finish it off with wax or oil tinged with alkanet root.

3. Logwood, 2 oz.; madder, 8 oz.; fustic, 1 oz.; water, 1 gall.; boil 2 hours, and apply it several times to the wood boiling hot; when dry, slightly brush it over with a solution of pearlash, 1 oz., in water, 1 quart; dry, and polish as before.

4. As the last, but using a decoction of logwood, 1 lb., in water, 5 pints. The tint may be brightened by adding a little vinegar or oxalic acid, and darkened by a few grains of copperas.

Stains and spots may be taken out of mahogany furniture with a little aquafortis or oxalic acid and water, by rubbing the part with the liquid by means of a cork till the colour is restored; observing afterwards to well wash the wood with water, and to dry it and polish it as before.

The best mahogany comes from the West Indies and America, and is yielded by the tree *Swietenia mahagoni*, one of the Nat. Ord. CEDRELACEÆ. Its growth is slow, and from the immense size of some of the trees, which often reach 80 to 100 feet in height, with a trunk 6 feet thick, it has been calculated that many of them must have been growing for approximately 200 years. As has been incidentally mentioned, those grown on low swampy ground produce inferior timber. Perhaps owing to the great difficulty of transportation from the place of growth to the port of shipment, there is an increasing tendency to deterioration in quality. The most accessible trees have been felled, and as distance from the coast increases there is less possibility of picking and choosing; the trees are now taken as they come.

East India mahogany is produced by the *Soymida febrifuga*, or Rohuna tree; as its name almost implies, it, or rather its bark, is occasionally used medicinally as a febrifuge. The bark of the mahogany tree has also been used for the same purpose.

African mahogany is the produce of the *Khaya senegalensis*. West Indian cedar, or, as it is often familiarly called, though erroneously, mahogany, is supplied by the *Cedrela odorata*.

All of them belong to the Nat. Ord. CEDRELACEÆ, which includes many other kinds of woods, that best known perhaps being satin-wood, which, apart from its distinctive colour, bears a very marked resemblance to mahogany in figure and general marking.

Everyone who is accustomed to handle it knows that mahogany is a reliable wood, pleasant to work, and susceptible of a high degree of finish. It is obtainable in large planks clean and sound, i. e. free from knots and shakes. Really fine mahogany is said to be difficult to procure nowadays, but there is no doubt whatever that by care, and the payment of sufficiently high prices, mahogany of the very finest figure is still to be purchased. Naturally, it will not have the fine dark colour of old mahogany, but that will come in time, unless, indeed, the stains which, at the request of ignorant purchasers, are so freely used to give an artificial appearance of age have a prejudicial effect. Stains may produce a pleasing appearance on new wood, but it may very reasonably be supposed that the benefit is only a temporary one, and that instead of improving as time goes on the colour will be anything but agreeable. If any mode of artificial darkening be resorted to, that by ammonia vapour is the least harmful, and gives a nearer approach to the colour of old mahogany than any other process. Of few woods can it be said that they improve with age, but mahogany is certainly one of them; the wood should be simply oiled, or at most French polished without any staining.

Mahogany, Polishing. The best way to finish mahogany is to French polish it if a bright glossy surface is required. The process embraces staining if necessary, to darken the colour, oiling, filling or stopping the grain of the wood, bodying it with polish, and finally 'spiriting off' to get a fine smooth surface without marks. For stain use a solution of either bichromate or permanganate of potash, the strength depending on the colour required. Rub down with fine glass-paper after staining to remove roughness caused by the moisture. Oil with raw linseed oil, rubbing it well in with a piece of rag, but not saturating the wood with it. Allow the work to stand by till the oil has become fairly dry, and as long as possible afterwards before beginning to fill in. The best filling is one composed of whiting and turpentine with a little rose pink to colour. Mix these into a stiffish paste, and then rub some of it well into the wood. When this has been sufficiently done to stop up the grain, wipe the surplus away before it gets hard with a clean cloth. The wood is then ready for 'bodying in' at any time, though it is always advisable not to hurry on too fast with any polishing process. To 'body in' use a pad formed of cotton wadding enclosed in a piece of soft rag. Moisten the wadding with French polish and cover it with one fold of the rag. Give this just the least touch of linseed oil, and go over the wood till there is a good body of polish on it. As the rubber dries add more polish, and be careful to cover the wood evenly, rubbing the polish till the spirit evaporates. If necessary, bodying in may be repeated several times at intervals of a day or two. At this stage the surface is smeared and dull-looking, and the final polish is got by 'spiriting off.' This is much the same as 'bodying in,' only spirit (methylated) alone is used instead of polish. Unless care be used, instead of getting

a highly finished surface, the previously laid body is apt to be removed. 'Spiriting off' is the most difficult part of the process, and requires considerable skill to manage it properly.

MAIZE. *Syn.* INDIAN CORN. The seeds of *Zea mays*, Linn. Like the other corn plants, it belongs to the Grass family (GRAMINACEÆ), and has albuminous grains sufficiently large and farinaceous to be ground into flour.

Maize is extremely nutritious, and although it is poorer in albuminoid matters than wheat, it is, of all the cereal grains, the richest in fatty oil, of which it contains about 9% (*Dumas* and *Payen*). It is remarkable for its fattening quality on animals, but is apt to excite slight diarrhœa in those unaccustomed to its use. Its meal is the 'POLENTA' of the shops. The peculiar starch prepared from it is known as 'CORN FLOUR.'

In America the young ears are roasted and boiled for food.

The centesimal composition of maize is as follows:—Albuminoid bodies, 9·9; starch, dextrin, and fat, 71·2; fibre, 4·0; ash, 1·4; water, 13·5.

Letheby says of maize: " The grain is said to cause disease when eaten for a long time, and without other meal—the symptoms being a scaly eruption upon the hands, great prostration of the vital powers, and death after a year or so, with extreme emaciation. See PELLAGRA.

" These effects have been frequently observed amongst the peasants of Italy, who use the meal as their chief food, but I am not aware of any such effects having been seen in Ireland, where it is often the only article of diet for months together."

Millions of bushels are grown every year in the United States of America, and large quantities are continually imported into England, where it is held in high esteem by cattle breeders. it being much cheaper than many of our home-grown productions. It is occasionally given to horses as a substitute for oats.

MALAG'MA. In *pharmacy*, a poultice or emollient application.

MA'LIC ACID. $C_4H_6O_5$. *Syn.* ACIDUM MALICUM, L. Hydroxysuccinic acid. This acid exists in the juice of many fruits and plants, either alone or associated with other acids, or with potash or lime. In the juice of cherries and the garden rhubarb it exists in great abundance, being associated with acid oxalate of potassium. Tobacco, gooseberries, currants, apples, pears, &c., contain malic acid.

Prep. 1. (*Everitt.*) The stalks of common garden rhubarb are peeled, and ground or grated to a pulp, which is subjected to pressure; the juice is heated to the boiling-point, and neutralised with carbonate of potassium, mixed with acetate of lime, and the insoluble oxalate of lime which forms is removed by filtration; to the clear and nearly colourless liquid, solution of acetate of lead is next added as long as a precipitate continues to form; this is collected on a filter, washed, diffused through water, and decomposed by sulphuric acid, avoiding excess, the last portion of lead being thrown down by a stream of sulphuretted hydrogen; the filtered liquid is, lastly, carefully evaporated to the consistence of

64

a syrup, and left in a dry atmosphere until it becomes converted into a solid and somewhat crystalline mass of malic acid. If perfectly pure malic acid is required, the malate of lead must be crystallised before decomposing it with sulphuretted hydrogen.

2. From the juice of the nearly ripe berries of the mountain ash (*Sorbus aucuparia*), as follows:—The juice is expressed, boiled and filtered, nearly neutralised with milk of lime, and again boiled. Calcium malate now forms in minute crystals; these are dissolved in hot aqueous nitric acid. On cooling hydrocalcium malate separates out in crystals, which are now dissolved in hot water and decomposed by lead acetate. Finally, the solution, which now contains lead malate, is treated with sulphuretted hydrogen, which precipitates lead sulphide, leaving malic acid still in solution, from which it may be obtained by crystallisation after evaporation.

3. Malic acid is also produced by the deoxidation of tartaric acid with hydriodic acid.

Prop., &c. Malic acid crystallises in groups of colourless prisms; it is slightly deliquescent, very soluble in water, soluble in alcohol, and has a pleasant acidulous taste. Two varieties exist, one being active, the other inactive with polarised light. The aqueous infusion soon gets mouldy by keeping. When kept fused for some time at a low heat, it is converted into fumaric acid; and when quickly distilled, it yields maleic acid, while fumaric acid is left in the retort. By the action of reducing agents, *e. g.* strong hydriodic acid, it is reduced to succinic acid. With the bases malic acid forms salts called malates. Of these the acid malate of ammonia is in large beautiful crystals; malate of lead is insoluble in cold water, but dissolves in warm dilute acid, from which it separates on cooling in brilliant silvery crystals; acid malate of lime also forms very beautiful crystals, freely soluble in water; neutral malate of lime is only sparingly soluble in water; the first is obtained by dissolving the latter in hot dilute nitric acid, and allowing the solution to cool very slowly.

MALLEABILITY. The peculiar property of metals which renders them capable of extension under the hammer.

MALT. *Syn.* BINA, BYNE, BRASIUM, MALTUM, L. The name given to different kinds of grain, such as barley, bere or bigg, oats, rye, maize, &c., which have become sweet from the conversion of a portion of their starch into sugar, in consequence of incipient germination artificially produced. Barley is the grain usually employed for this purpose.

Var. Independently of variations of quality, or of the grain from which it is formed, malt is distinguished into varieties depending on the heat of the kiln employed for its desiccation. When dried at a temperature ranging between 90° and 120° F. it constitutes 'PALE MALT;' when all the moisture has exhaled, and the heat is raised to from 125°—135°, 'YELLOW' or 'PALE AMBER MALT' is formed; when the heat ranges between 140° and 160°, the product receives the name of 'AMBER MALT;' at 160°—180°, 'AMBER-BROWN' or 'PALE BROWN MALT' is obtained. ROASTED, PATENT, or BLACK MALT, and CRYSTALLISED MALT,

are prepared by a process similar to that of roasting coffee. The malt is placed in sheet-iron cylinders over a strong fire, and the cylinders made to revolve at the rate of about 20 revolutions per minute if roasted malt is required, or 120 for crystallised malt. In the former case the finished malt has a dark brown colour; in the latter the interior of the grain becomes dark brown, whilst the husk assumes a pale amber hue. The temperature must never exceed 420°, or the malt will become entirely carbonised.

Qual. Good malt has an agreeable smell and a sweet taste. It is friable, and when broken discloses a flowery kernel. Its husk is thin, clean, and unshrivelled in appearance, and the acrospire is seen extending up the back of the grain, beneath the skin. The admixture of unmalted with malted grain may be discovered and roughly estimated by throwing a little into water; malt floats on water, but barley sinks in it. The only certain method, however, of determining the value of malt is to ascertain the amount of soluble matter which it contains by direct experiment. This varies from 62% to 70%, and for good malt is never less than 66% to 67%. If we assume the quarter of malt at 324 lbs., and the average quantity of soluble matter at 66%, then the total weight of soluble matter will be fully 213¾ lbs. per quarter; but as this, "in taking on the form of gum and sugar" during the process of mashing, "chemically combines with the elements of water, so the extract, if evaporated to dryness, would reach very nearly 231 lbs.; and this, reduced to the basis of a barrel of 36 galls., becomes in the language of the brewer 87 lbs. per barrel, which, however, merely means that the wort from a quarter of malt, if evaporated down to the bulk of a barrel of 36 galls., would weigh 87 lbs. more than a barrel of water" (*Ure*).

Assay. 1. A small quantity of the sample being ground in a coffee or pepper mill, 100 gr. are accurately weighed, and dried by exposure for about 1 hour at the temperature of boiling water. The loss in weight, in grains, indicates the quantity of moisture per cent. This, in good malt, should not exceed 6¼ gr.

2. A second 100 gr. is taken and stirred up with about ¼ pint of cold water; the mixture is then exposed to the heat of boiling water for about 40 minutes; after which it is thrown on a weighed filter, and the undissolved portion washed with a little hot water; the undissolved portion, with the filter, is then dried at 212° F., and weighed. The loss in weight, less the percentage of moisture last found, taken in grains, gives the percentage of soluble matter. This should not be less than 66 gr. The same result will be arrived at by evaporating the filtered liquid and 'washings' to dryness, and weighing the residuum.

3. A third 100 gr. is taken and mashed with about ¼ pint of water at 160° F., for 2 or 3 hours; the liquid is then drained off, the residue gently squeezed, and the strained liquid evaporated to dryness as before, and weighed. This gives the percentage of saccharine matter, and should not be less than about 71 gr., taking the above average of malt as the standard of calculation.

Uses, &c. Malt is chiefly employed in the arts

of brewing and distillation. Both roasted and crystallised malt are merely used to colour the worts produced from pale malt. 1 lb. of roasted malt mashed with 79 lbs. of pale malt imparts to the liquor the colour and flavour of 'porter.' The paler varieties of malt contain the largest quantity of saccharine matter. After the malt has been kiln-dried, the rootlets may be removed by means of a sieve. Before malt is mashed for beer it must be broken up, and the law requires that it be bruised or crushed by smooth metal rollers, and not ground by millstones. It has also been proposed to employ malt, instead of raw grain, for fattening domestic animals, and as food for their young and those in a sickly state. Infusion of malt (sweet wort, malt tea) is laxative, and has been recommended as an antiscorbutic and tonic. It has been given with great advantage in scurvy; but for this purpose good, well-hopped, mild beer is equally serviceable and more agreeable. See BREWING, DISTILLATION, FERMENTATION, &c.

For the processes of malting see under BREWING.

Malt, Extract of, and its Manufacture. The following extracts from a paper by J. L. Irwin read at the annual meeting of the Ohio State Pharmaceutical Association, Cincinnati, will be of interest:

Apparatus used. For the manufacture on a large scale two principal kinds of apparatus are at present in use, viz. the hot-air blast and the ordinary vacuum apparatus, the latter being the one with which the writer is more familiar. It includes an air-tight copper still, a cooler, a receiver or air-chamber, and a good air-pump, capable of producing a 27-inch vacuum. There are also required a thermometer, wood mash-tub, wood percolator, a good press, a platform for damping the malt, and the necessary buckets, dippers; &c.

Tests for Malt. Where the operator has not the facility to malt his own barley, some care in selecting a good article should be used. An ancient custom of determining a good sample was to take a glass nearly full of water, and put in some malt; if the grains swam the sample was considered good, but if any should sink to the bottom it was not considered true malt. I think the best test is the general appearance, &c. First notice whether the grains have a round body, break soft, are full of flour their whole length, smell well, and have a thin skin. Masticate some of the grains, and if sweet and mellow they should be considered good; but if hard and steely, and retaining something of the barley nature, the malt is not properly made, and weighs heavier than good malt.

To Grind the Malt. To obtain the extract, it is best to merely break the grain. For this purpose the malt is passed between revolving stones placed at such distances apart that each grain may be crushed without reducing it to a powder, for if ground too finely it thickens the solution and is difficult to percolate, while if not broken at all the extract is not all obtained. Pale malts are generally ground coarser than amber or brown. Malt should be used within 2 days after being ground to obtain the best results. Crushing

mills or iron rollers may be substituted for the revolving stones, and on a small scale the malt grains can be broken by wooden rollers, or even with an ordinary coffee mill.

Moistening the Malt. Take a convenient quantity of the ground selected malt, and having placed it upon a clean wooden platform, prepare a menstruum for it consisting of 1 vol. of 94% alcohol to 3 vols. of water. For a bushel of malt, 3 galls. of such menstruum is sufficient. After having thoroughly damped the malt with the menstruum cover it with a rubber blanket to prevent loss of alcohol, and allow it to stand about 12 hours, working it up with a shovel every 3 hours.

Extracting the Diastase. The writer has found that the best results are obtained when the diastase of the malt is first eliminated before preparing the starch for conversion into sugar by the former, so the damped grain is transferred to a conical percolator, cold water is gradually poured over it until the liquid begins to flow from the faucet of the percolator, then return the percolate and repeat until the liquor runs clear and free from starchy granules. Continue the percolation with fresh cold water until a quantity of liquid equal to 4 times the amount of the malt is obtained. Then by means of a rubber hose and air-pump transfer the percolate to the vacuum-still in order to have the spirit slowly recovered.

Recovering the Alcohol. Having exhausted the still-cooler and air-chamber by means of the air-pump until the vacuum-gauge registers 27 inches, open the steam-valve leading to vacuum-pan and evaporate off the alcohol, which will condense in a cooler and can be collected in the air-chamber. A temperature of 100° F. is generally enough to recover the alcohol with the vacuum. The time required for the recovery of the spirit varies according to the apparatus, amount of malt worked, &c., but for the working of 1 bushel of malt generally required is about ¾ an hour. In case the starch liquid is not ready for the still in time, it is best to allow some of the alcohol to remain with the diastase to preserve the latter until the starch liquid comes in contact with it. When the starch liquid is all prepared for evaporation, and having evaporated off all the remaining spirit, then stop the pump and remove the pressure by allowing the air to enter through the valve on top of the still. When the gauge registers zero, open the valve in the base of the air-chamber and remove the recovered spirit, which is usually quite weak.

Mashing the Malt. While the alcohol is being recovered from the diastatic percolate, the operator should give his attention to the separation of the starch liquid. To this end the malt is transferred from the percolator to the mash-tub, where the starchy matter must first be gelatinised before it can be acted on by the diastase now in the still. A volume of water about 4 times the weight of the malt employed is then added, and the whole mixed thoroughly. By means of the steam-valve at the base of the mash-tub the water is heated under constant stirring until it arrives at the boiling-point, at which temperature it is kept for 2 minutes under constant stirring, in order to insure the perfect coagulation of albuminoids not extracted in the first percolation, and to gelatinise and dissolve every particle of starch. The mixture is then allowed to cool, and an amount of water equal to ½ the weight of the malt is added, and the whole, after being well mixed, is allowed to stand for about 15 minutes. The outlet-valve of the tub is then opened, the liquid re-percolated until clear, and then collected in a receiver to be cooled down to at least 150° F. It is then transferred to the still by means of the hose and air-pump, and it is there converted by the diastase into malt sugar.

Evaporation. When the liquids are mixed in the still the whole is to be rapidly evaporated as soon as the vacuum-gauge registers 27 inches. The temperature necessary for rapid evaporation ranges from 110° to 140° F., the temperature rising above 110° as the extract begins to thicken. In no case should the temperature be allowed to rise above 150°, as it would impair the flavour, although the diastase may not be injured at a higher temperature, and the evaporation should never be stopped when once started until at least 10% of the liquid has been evaporated off.

The proper consistence for the extract depends upon the season of the year, that which would answer for summer not being suitable for winter; but the writer has found that a malt having a sp. gr. of 1·4, while a little difficult to handle in winter, will keep well in summer, and hence is of a desirable density.

Pressing the Mash. After the liquid has entirely drained from the mash-tub the malt is next transferred to a press, and the remaining liquid entirely pressed out and allowed to settle, the clear liquid poured off and mixed with the starch liquor in the still by means of the hose and air-pump, an equal volume of warm water is again added to the malt in the press, and the pressing repeated until the malt is entirely exhausted.

Precautions. While the operator should observe all the general precautions necessary to operate a vacuum apparatus, yet there are some in the manufacture of malt extract deserving special attention.

In the first place, unless the operator gives close attention to the apparatus while running on the malt, he will find that in nine cases out of ten the resulting product will be worthless as a digestive agent. From the time the malt is first moistened to the time it is in the form of a finished extract, quickness in detail and constant attention are essential for a good product.

The percolating process is quite often found troublesome. In the first percolation, after the process has proceeded far enough to drive out all the alcohol, it is possible that the malt has been so finely ground that when the water menstruum has been poured on, the latter will swell the small particles of the grain sufficiently to either cause the percolation to proceed so slowly as to impede progress, or to stop up altogether and thus incept fermentation; and again, even after this percolation has been successful, trouble may arise in the mash-tub, should the heat be allowed to continue longer than merely to coagulate the albuminoids and gelatinise the starch; especially when the malt has been finely ground

a paste is liable to form and again interfere with percolation; in such a case the best course to pursue is to transfer the whole to the press and squeeze out the liquid as quickly as possible. Care should also be taken that the starch liquid is cooled down below 150° F. before it be allowed to come in contact with the diastase.

Under no circumstances should the evaporation be stopped until the extract is of the consistence of simple syrup at least; even in that condition it should not be allowed to stand any length of time before the evaporation is completed and the extract is of the proper consistence.

Sufficient time should also be allowed in boiling, so that the albuminoids in the mass are entirely coagulated, and hence retained in the dregs after percolation, as with their presence in the finished extract there is liable to set up a fermentation.

In the process of evaporation the operator should note the rate of distillation by glancing every few minutes through the eye-piece of the air-chamber, and regulate the same by the time available to complete the process, and should keep the distillation at as constant a rate as possible. Especial care should here be used, as the writer has observed that a peculiar stony substance may at times be formed by the foam from the malt, and collect in the channel between the cooler and air-chamber in sufficient quantity either at least to stop up the latter entirely or diminish the rate of distillation. In such a case it will be readily seen that should the heat and pressure be allowed to continue, not only disaster to malt and apparatus might result, but even loss of life. The writer has noticed this trouble even with a high vacuum as shown by the gauge, and everything apparently going on all right except the rate of distillation.

The operator will find that there are many minor points which should be borne in mind while running extract of malt, and which experience alone can teach.

MALT LIQUORS. The qualities of ale, beer, and porter, as beverages, the detection of their adulteration, and the methods of preparing them, are described under their respective names, and in the article upon 'BREWING;' the present article will, therefore, be confined to a short notice of the cellar management, and the diseases of malt liquors generally.

AGE. The appearance and flavour to which this term is applied can, of course, be only given to the liquor by properly storing it for a sufficient time. Fraudulent brewers and publicans, however, frequently add a little oil of vitriol (diluted with water) to new beer, by which it assumes the character of an inferior liquor of the class 1 or 2 years old. Copperas, alum, sliced onions, Seville oranges, and cucumbers are also frequently employed by brewers for the same purpose.

BOTTLING. Clean, sweet, and dry bottles, and sound and good corks, should be had in readiness. The liquor to be bottled should be perfectly clear; and if it be not so it must be submitted to the operation of 'fining.' When quite fine and in good condition the bung of the cask should be left out all night, and the next day the liquor should be put into bottles, which, after remaining 12 or

24 hours, covered with sheets of paper to keep out the flies and dust, must be securely corked down. Porter is generally wired over. The wire for this purpose should be 'annealed,' and not resilient. If the liquor is intended for exportation to a hot climate the bottles should remain filled for 2 or 3 days, or more, before corking them. The stock of bottled liquor should be stored in a cool situation; and a small quantity, to meet present demands only, should be set on their sides in a warmer place to ripen. October beer should not be bottled before Midsummer, nor March beer till Christmas.

CLOUDINESS. Add a handful of hops boiled in a gallon of the beer, and in a fortnight fine it down.

FINING. See CLARIFICATION and BREWING.

FLATNESS. When the liquor is new or has still much undecomposed sugar left in it, a sufficient remedy is to remove it into a warmer situation for a few days. When this is not the case 2 or 3 pounds of moist sugar (foots) may be 'rummaged' into each hogshead. In this way a second fermentation is set up, and in a few days the liquor becomes brisk, and carries a head. This is the plan commonly adopted by publicans. On the small scale the addition of a few grains of carbonate of soda, or of prepared chalk, to each glass, is commonly made for the same purpose; but in this case the liquor must be drunk within a few minutes, else it becomes again flat and insipid. This may be adopted for home-brewed beer which has become sour and vapid.

FOXING or BUCKING. The spontaneous souring of worts or beer during their fermentation or ripening, to which this name is applied, may generally be remedied by adding to the liquor some fresh hops (scalded), along with some black mustard-seed (bruised). Some persons use a little made mustard, or a solution of alum or of catechu, and in a week or 10 days afterwards further add some treacle or moist sugar.

Frosted beer is recovered by change of situation, by the addition of some hops boiled in a little sweet wort, or by adding a little moist sugar or treacle to induce a fresh fermentation.

HEADING. This is added to thin and vapid beer to make it bear a frothy head. The most innocent, pleasant, and effective addition of this sort is a mixture of pure ammonio-citrate of iron and salt of tartar, about equal parts in the proportion of only a few grains to a quart.

IMPROVING. This is the trade synonym of 'ADULTERATION' and 'DOCTORING.' Nevertheless there are cases in which 'improvement' may be made without affecting the wholesome character of the liquor. Of this kind is the addition of hops, spices, &c., during the maturation of beer that exhibits a tendency to deteriorate. For this purpose some persons cut a half-quartern loaf into slices, and, after toasting them very high, place them in a coarse linen bag along with ¼ lb. of hops and 2 oz. each of bruised ginger, cloves, and mustard-seed, and suspend the bag by means of a string a few inches below the surface of the beer (a hogshead), which is then bunged close. The addition of a little ground capsicum in the same way is also a real improvement to beer when judiciously made.

MUSTINESS. To each hogshead, racked into clean casks, add 1 lb. of new hops boiled in a gallon of the liquor, along with 7 lbs. of newly burnt charcoal (coarsely bruised, and the fine dust sifted off), and a 4-lb. loaf of bread cut into thin slices and toasted rather black ; 'rouse up' well every day for a week, then stir in of moist sugar 3 or 4 lbs., and bung down for a fortnight.

RECOVERING. This is said of unsaleable beer when rendered saleable, by giving it 'head' or removing its 'tartness.'

RIPENING. This term is applied to the regular maturation of beer. It is also used to express the means by which liquors already mature are rendered brisk, sparkling, or fit and agreeable for immediate use. In the language of the cellars malt liquors are said to be 'up' when they are well charged with gaseous matter and bear a frothy head. These qualities depend on the undecomposed sugar undergoing fermentation, which, when active, can only be of comparatively short duration, and should, therefore, be repressed rather than excited in beers not required for immediate consumption. When we desire to give 'briskness' to these liquors, whether in cask or bottle, it is only necessary to expose them for a few days to a slight elevation of temperature, by removing them, for instance, to a warmer apartment. This is the plan successfully adopted by bottlers. The addition of a small lump of white sugar to each bottle of ale or beer, or a teaspoonful of moist sugar to each bottle of porter, just before corking it, will render it fit for drinking in a few days in ordinary weather, and in 2 or 3 days in the heat of summer. A raisin or a lump of sugar-candy is often added to each bottle with a like intention. The Parisians bottle their beer one day and sell it the next. For this purpose, in addition to the sugar as above, they add 2 or 3 drops of yeast. Such bottled liquor must, however, be drunk within a week, or else stored in a very cold place, as it will otherwise burst the bottles or blow out the corks.

ROPINESS. A little infusion of catechu or of oak-bark, and some fresh hops, may be added to the beer, which in a fortnight should be rummaged well, and the next day 'fined' down.

SOURNESS. Powdered chalk, carbonate of soda, salt of tartar, or pearlash, is commonly added by the publicans to the beer until the acidity is nearly removed, when 4 or 5 lbs. of moist sugar or foots per hogshead are 'rummaged' in, together with sufficient water to disburse double the amount of the outlay and trouble. Such beer must be soon put on draught, as it is very apt to get flat by keeping. Oyster-shells and egg-shells are also frequently used by brewers for the same purpose. To remove the acidity of beer on the small scale, a few grains of carbonate of soda per glass my be added just before drinking it.

STORING. The situation of the beer-cellar should be such as to maintain its contents at a permanently uniform temperature, ranging between 44° and 50° F., a condition which can only be ensured by choosing for its locality an underground apartment, or one in the centre of the basement portion of a large building.

VAMPING. Half fill casks with the old liquor,

fill them up with some newly brewed, and bung close for 3 weeks or a month.

MALTIN. A nitrogenous ferment obtained from malt, which it is believed by Dubrunfaut to be the active principle, and is more energetic than diastase. The above chemist states it may be precipitated from extract of malt by the addition of 2 molecules of alcohol at 90%. According to Dubrunfaut maltin exists in all cereal grains, and in the water of rivers and brooks; but not in the well water of Paris.

MALTING. The method of converting barley, wheat, oats, or any other description of grain into malt. There are four successive stages in the process of malting, viz. steeping, couching, flooring, and kiln-drying; for a description of which see BREWING.

MANCHINEEL TREE (*Hippomane mancinella*, Linn.). Native of tropical South America and the West Indies. Though of a poisonous character, its power, like that of the upas, has been much exaggerated.

MAN"GANESE. Mn = 55. *Syn.* MANGANESIUM, L. A hard, brittle metal, resembling iron in some of its physical and chemical characters, discovered by Gahn in the black oxide of manganese of commerce. In an oxidised state it is tolerably abundant in nature, entering into the composition of several interesting minerals. Traces of it have been found in the ashes of plants and in mineral waters. It chiefly occurs as pyrolusite, braunite, and manganese spar.

Prep. 1. Reduce manganous carbonate to fine powder, make it into a paste with oil, adding about 1-10th of its weight of calcined borax, place the mixture in a Hessian crucible lined with charcoal, lute on the cover, and expose it to the strongest heat of a smith's forge for 2 hours; when cold, break the crucible and preserve the metallic button in naphtha. The product is probably a carbide of manganese, just as steel is a carbide of iron.

2. Deville has lately prepared pure manganese by reducing the pure oxide by means of an insufficient quantity of sugar charcoal in a crucible made of caustic lime.

Prop. As prepared by Deville, metallic manganese is grey with a reddish lustre, like bismuth; it is very hard, brittle, and very difficult to fuse when powdered; it decomposes water, even at the lowest temperature. Dilute sulphuric acid dissolves it with great energy, evolving hydrogen. It scratches glass and hardened steel, and its sp. gr. is 7.13.

The salts of manganese may be easily prepared in a state of purity by dissolving the precipitated carbonate in the acids. Most of them are soluble, and several are crystallisable.

Tests. Manganous salts are distinguished as follows:—The hydrates of potassium and sodium give white precipitates insoluble in excess, and rapidly turning brown. The presence of ammonium salts interferes with these tests. Ammonia gives similar results.

Ferrocyanide of potassium gives a white precipitate. Sulphuretted hydrogen gives no precipitate in acid solutions, and precipitates neutral solutions only imperfectly; but in alkaline solutions it gives a bright, flesh-coloured, insoluble precipitate, which becomes dark brown on exposure to the air. Sulphide of ammonium, in neutral solutions, also yields a similar precipitate, which is very characteristic. A compound of manganese fused with borax in the outer flame of the blowpipe gives a bead, which appears of a violet-red colour whilst hot, and upon cooling acquires an amethyst tint; this colour is lost by fusion in the inner flame. Heated upon platinum foil with a little carbonate of sodium, in the outer flame, it yields a green mass whilst hot, which becomes bluish green when cold.

Uses. The metal itself has not been applied to any useful purpose. Spiegeleisen and ferro-manganese are alloys largely used in the production of Bessemer steel. Various ores of manganese are industrially employed in making oxygen, bromine, chlorine, and iodine, in the manufacture of glass and enamels, for producing mottled soaps, in puddling iron, and in dyeing and calico printing. Most of the manganese of commerce comes from Germany.

Manganate of Barium. $BaMnO_4$. Green insoluble powder, obtained by fusing barium hydrate, potassium chlorate, and manganic peroxide together, and washing the product. It forms the pigment known as Cassel green.

Manganate of Potassium. K_2MnO_4. Finely powdered manganic peroxide, potassium chlorate, and potassium hydrate, made into a thick paste with water, and heated to dull redness. The fused product is treated with a small quantity of water, and crystallised by evaporation *in vacuo*.

Prop. Dark green, almost black crystals, readily soluble in water, but decomposed by excess or by acids into manganic peroxide and potassium permanganate.

Manganate of Sodium. Na_2MnO_4. Prepared on the large scale by heating a mixture of manganese peroxide and sodium hydrate to redness in a current of air. Used in strong aqueous solution as a disinfectant under the name of 'Condy's green fluid.'

Manganic Acid. H_2MnO_4. This acid has not yet been obtained free, but some of its salts are extensively employed as disinfectants, as 'green Condy's fluid.' The chief compounds are the following:—

Manganic Hydrate. $Mn_2O_3(HO)_2$. *Syn.* HYDRATED SESQUIOXIDE OF MANGANESE. Found native as 'manganite,' in reddish-brown crystals. *Prep.* By passing a current of air through recently precipitated and moist manganous hydrate. It is a soft brown powder, and is converted into the oxide by heat. May be distinguished from MnO_2, with which it is often found associated, by its giving a brown instead of a black streak on unglazed porcelain.

Permanganic Acid. $H_2Mn_2O_8$. *Prep.* Obtained in a hydrated crystalline state by decomposing barium permanganate with sulphuric acid and evaporating *in vacuo*. Brown colour; dissolves in water, giving a red solution.

Permanganate of Barium. $BaMn_2O_8$. Black soluble prisms, formed by decomposing silver permanganate by means of barium chloride, and cautiously evaporating.

Permanganate of Potassium. $K_2Mn_2O_8$. *Prep.* With 4 parts of black oxide of manganese, 3½ parts of potassium chlorate, and 5 parts of caustic potash dissolved in a little water are mixed. The paste which results is dried by being heated to dull redness on an iron tray. This operation produces potassium manganate, which, when the cold mass is treated with water, forms a dark green solution. A stream of carbonic acid gas is now passed into this solution until no further change of colour is observed. The liquid now contains potassium permanganate, manganese dioxide, and potassium carbonate. The dioxide settles as a precipitate from which the liquid, which is red, is decanted. It is then concentrated and cooled. The permanganate crystallises out, whilst the more soluble carbonate remains in solution.

Prop. Dark purple, almost black, elongated rhombic prisms which are red by transmitted light, but reflect a dark green colour. Soluble in 20 parts cold water, forming a purple solution, which becomes green by contact with some substance capable of taking up oxygen. Thus an aqueous solution is easily decomposed and bleached by the action of sulphurous acid or a ferrous salt. Offensive emanations from putrescent organic matter are easily oxidised by potassium permanganate, which is extensively used for this purpose under the name of ' Condy's red disinfecting fluid.'

Uses. As a disinfectant (*vide supra*), in dyeing, and for staining wood. It is a most important reagent in volumetric analysis, and is especially useful in the estimation of iron. In preparing the solution about 5 grms. of pure crystallised permangate are dissolved in a small quantity of water, and then diluted to 1 litre; it is then standardised by means of pure iron wire, ferrous sulphate, or oxalic acid. The solution must be contained in a well-stoppered bottle which, when not in use, should be kept in a cool dark place.

Permanganate of Silver. $AgMnO_4$. *Prep.* Precipitate a strong solution of silver nitrate by means of a concentrated solution of potassium permanganate. Small black prisms, soluble in 100 parts of water, with a purple colour.

Permanganate of Sodium. $Na_2Mn_2O_8$. ' Obtained as a dark purple liquid by passing a current of carbonic anhydride through sodium manganate. It may also be made by heating the black oxide of manganese with caustic soda in a hollow vessel to dull redness for 48 hours. The mass is then boiled with water. It is often used as a disinfectant, being cheaper than the potassium salt. Condy's red fluid is chiefly sodium permanganate dissolved in water.

Manganous Ac'etate. $Mn(C_2H_3O_2)_2$. *Syn.* ACETATE OF PROTOXIDE OF MANGANESE; MANGANII ACETAS, L. *Prep.* By neutralising concentrated acetic acid with manganous carbonate, and evaporating the solution so that crystals may form.

Prop., &c. The crystals, when pure, are of a pale red colour, permanent in the air, soluble in alcohol, and 3½ parts of water, and possess an astringent and metallic taste.—*Dose,* 5 to 10 gr., as an alterative, hæmatinic, &c.

Manganous Car'bonate. $MnCO_3$. *Syn.* CARBONATE OF PROTOXIDE OF MANGANESE; MANGANESII CARBONAS, L. *Prep.* 1. Reduce the black oxide of manganese of commerce to fine powder, and after washing it in water acidulated with hydrochloric acid, dissolve it in strong hydrochloric acid, and evaporate the resulting solution to dryness; dissolve the residue in water, and add to the solution sufficient sodium carbonate to precipitate all the iron present; digest the mixed precipitate in the remainder of the liquid, filter, add ammonium sulphide until it begins to produce a flesh-coloured precipitate, then filter, and add sodium carbonate as long as a precipitate falls; lastly, well wash the newly formed carbonate in water, and dry it by a gentle heat.

2. By directly precipitating a solution of the chloride with sodium carbonate, and washing and drying the powder as before.

Prop., &c. A pale buff or cream-coloured powder, insoluble in water, freely soluble in acids; exposed to a strong heat, it loses its carbonic acid, absorbs oxygen, and is converted into the red oxide. It is chiefly employed in the preparation of the other salts of manganese.

Manganous Chlo"ride. $MnCl_2$. *Syn.* PROTOCHLORIDE OF MANGANESE, MURIATE OF M.; MANGANESII CHLORIDUM, L. *Prep.* 1. By saturating hydrochloric acid with manganous carbonate; the solution is greatly concentrated by evaporation, when crystals may be obtained, or it is at once evaporated to dryness; in either case the product must be placed in warm, dry, stoppered bottles, and preserved from the air.

2. From the dark brown residual liquid of the process of obtaining chlorine from binoxide of manganese and hydrochloric acid; this liquid is evaporated to dryness, and then slowly heated to dull redness in an earthen vessel, with constant stirring, and kept at that temperature for a short time. The greyish-looking powder thus obtained is treated with water, and the solution separated from the ferric oxide and other insoluble matter by filtration; if any iron still remains, a little manganous carbonate is added, and the whole boiled for a few minutes; the filtered solution is then treated as before. This is the least expensive and most convenient source of this salt.

Prop., &c. Rose-coloured tabular crystals; inodorous; very soluble both in water and alcohol; very deliquescent; when gradually heated to fusion the whole of the water is expelled, and at a red heat it slowly suffers decomposition. Astringent, tonic, hæmatinic, and alterative.—*Dose,* 3 to 10 gr.; in scorbutic, syphilitic, and certain chronic cutaneous affections; anæmia, chlorosis, &c.

Manganous Hydrate. $Mn(HO)_2$. *Syn.* HYDRATED PROTOXIDE OF MANGANESE. *Prep.* Formed by adding potassium hydrate to manganous sulphate, and filtering and drying the precipitate *in vacuo*. A white powder rapidly absorbing oxygen, and burning first green and then brown from formation of higher oxides.

Manganous I'odide. MnI_2. *Syn.* MANGANESII IODIDUM, L. *Prep.* By dissolving the carbonate in hydriodic acid and evaporating the filtered liquid *in vacuo* or out of contact with air.—*Dose,* 1 to 3 gr.; in anæmia, chlorosis, &c., occurring in scrofulous subjects.

Manganous Oxide. MnO. *Syn.* PROTOXIDE OF MANGANESE. *Prep.* By passing a current of hydrogen over manganous carbonate or manganese dioxide heated to whiteness in a porcelain tube. An olive-green powder, rapidly oxidising on exposure to air, and soluble in acids forming manganous salts. It has been found native in manganiferous dolomite.

Manganic Oxide. Mn_2O_3. *Syn.* SESQUIOXIDE OF MANGANESE. Found native as 'braunite,' and readily formed by exposing manganous hydrate to the action of air and drying, or by heating any of the oxides of manganese to redness in a current of oxygen.

Manganic Peroxide. MnO_2. *Syn.* PERMANGANIC OXIDE, BINOXIDE OF MANGANESE, PEROXIDE OF MANGANESE, BLACK OXIDE OF MANGANESE, OXIDE OF MANGANESE; MANGANESII OXIDUM NIGRUM (B. P.), MANGANESII BINOXYDUM (Ph. L.), MANGANESE OXYDUM (Ph. E.), L. Occurs massive and in prismatic crystals as pyrolusite; it is also found amorphous as peilomelane, and in the hydrated state as wad.

It is the only oxide of manganese that is directly employed in the *arts*. It is a very plentiful mineral production, and is found in great abundance in some parts of the west of England, in Germany and Spain. Manganese is prepared by washing, to remove the earthy matter, and grinding in mills. The blackest samples are esteemed the best. It is chiefly used to supply oxygen gas, which it evolves when treated to redness without fusing, in the manufacture of glass and bleaching powder, in dyeing, and in preparing the salts of manganese. It has been occasionally employed in medicine, chiefly externally in itch and porrigo, made into an ointment with lard. It has been highly recommended by Dr Erigeler in scrofula. Others have employed it as an alterative and tonic with variable success. When slowly introduced into the system during a lengthened period, it is said to produce paralysis of the motor nerves (*Dr Coupar*).—*Dose*, 3 to 12 gr., or more, thrice daily, made into pills.

Pur., &c. Native binoxide of manganese (pyrolusite) is usually contaminated with variable proportions of argillaceous matter, calcium carbonate, ferric oxide, silica, and barium sulphate, all of which lower its value as a source of oxygen, and for the preparation of chlorine. The richness of this ore can, therefore, be only determined by an assay for its principal ingredient.

Assay. There are several methods adopted for this purpose, among which the following recommend themselves as being the most accurate and convenient.

1. A portion of the mineral being reduced to very fine powder, 50 gr. of it are put into the little apparatus employed for the analysis of carbonates already described, together with about ½ fl. oz. of cold water, and 100 gr. of strong hydrochloric acid, the latter contained in the little tube (*b*); 50 gr. of crystallised oxalic acid are then added, the cork carrying the chloride of calcium tube fitted in, and the whole quickly and accurately weighed or counterpoised; the apparatus is next inclined so that the acid contained in the small tube may be mixed with the other contents of the flask, and the reaction of the ingredients is promoted by the application of a gentle heat; the disengaged chlorine resulting

from the mutual decomposition of the hydrochloric acid and the manganic peroxide converts the oxalic acid into carbonic acid gas, which is dried in its passage through the chloride of calcium tube before it escapes into the air. As soon as the reaction is complete, and the residual gas has been driven off by a momentary ebullition, the apparatus is allowed to cool, when it is again carefully and accurately weighed. The loss of weight in grains, if doubled, at once indicates the percentage richness of the mineral examined in manganic peroxide; or, more correctly, every gr. of carbonic anhydride evolved represents 1·982 gr. of the peroxide.

2. (*Fresenius* and *Will.*) The apparatus employed is the 'alkalimeter' figured at p. 70. The operation is similar to that adopted for the assay of alkalies, and is a modification of the oxalic acid and sulphuric acid test for manganese originally devised by M. Berthier. The standard weight of manganic peroxide recommended is 2·91 grms., along with 6·5 to 7 grms. of neutral potassium oxalate. The process, with quantities altered to adapt it for general employment, is as follows:—Manganic peroxide (in very fine powder), 50 gr.; neutral potassium oxalate (in powder), 120 gr.; these are put into the flask *A* along with sufficient water to about 1-4th fill it; the flasks *A* and *B* (the latter containing the sulphuric acid) are then corked air-tight, and thus connected in one apparatus, the whole is accurately weighed. The opening of the tube *a* being closed by a small lump of wax, a little sulphuric acid is sucked over from the flask *B* into the flask *A*; the disengagement of oxygen from the manganese immediately commences, and this reacting upon the oxalic acid present, converts it into carbonic anhydride, which passing through the concentrated sulphuric acid in the flask *B*, which robs it of moisture, finally escapes from the apparatus through the tube *d*. As soon as the disengagement of carbonic acid ceases, the operator sucks over a fresh portion of sulphuric acid, and this is repeated at short intervals until bubbles of gas are no longer disengaged. The little wax stopper is now removed, and suction is applied at *d* until all the carbonic acid in the apparatus is replaced by common air. When the whole has become cold it is again weighed. The loss of weight, doubled, indicates the amount of pure manganic peroxide in the sample, as before.

3. (*Otto.*) 50 gr. of the sample reduced to very fine powder are mixed in a glass flask, with hydrochloric acid, 1½ fl. oz., diluted with ¼ oz. of cold water, and portions of ferrous sulphate, from a weighed sample, immediately added, at first in excess, but afterwards in smaller doses, until the liquid ceases to give a blue precipitate with red prussiate of potash, or to evolve the odour of chlorine; heat being employed towards the end of the process. The quantity of ferrous sulphate consumed is now ascertained by again weighing the sample. If the peroxide examined is pure, the loss of weight will be 817 gr.; but if otherwise, the percentage of the pure peroxide may be obtained by the rule of three. Thus, suppose only 298 gr. of the sulphate were consumed, then

317 : 100 : : 298 : 94,

and the richness of the sample would be 94%. The percentage value of the oxide for evolving chlorine may be obtained by multiplying the weight of the ferrous sulphate consumed by 0·2688, which, in the above case, would give 76% of chlorine. For this purpose, as well as for chlorimetry, the ferrous sulphate is best prepared by precipitating it from its aqueous solution with alcohol, and drying it out of contact with air until it loses its alcoholic odour.

Obs. Before applying the above processes it is absolutely necessary that it be ascertained whether the peroxide examined contains any carbonates, as the presence of these would vitiate the results. This is readily determined by treating it with a little dilute nitric acid :—if effervescence ensues, one or more carbonates are present, and the sample, after being weighed, must be digested for some time in dilute nitric acid in excess, and then carefully collected on a filter, washed, and dried. It may then be assayed as before. The loss of weight indicates the quantity of carbonates present, with sufficient accuracy for technical purposes. The determination of this point is the more important, as these contaminations not merely lessen the richness of the mineral in pure manganic peroxide, but also cause a considerable waste of acid when it is employed in the manufacture of chlorine. For other methods of testing manganese, *vide* Crookes's 'Select Methods in Chemical Analysis.'

Manganous-manganic Oxide. Mn_3O_4, or MnO, Mn_2O_3. *Syn.* RED OXIDE OF MANGANESE, PROTOSESQUIOXIDE OF MANGANESE. Found native as 'hausmanite.' It is produced by igniting manganous carbonate, or manganic oxide, or manganic peroxide. Reddish-brown coloured crystals or powder, and communicates an amethyst colour to glass when fused with it. It seems probable that the true formula of this oxide is $2MnO.MnO_2$.

Manganous-manganic Peroxide. Mn_4O_7, or $MnO_2.Mn_2O_3$. *Syn.* INTERMEDIATE OXIDE OF MANGANESE. Found native as 'varvicite,' combined with water as a black hard crystalline mass. Decomposed, when heated, into a lower oxide and oxygen.

Manganous Phosphate. $MnH.PO_4 + 6Aq.$ *Syn.* PHOSPHATE OF PROTOXIDE OF MANGANESE; MANGANESII PHOSPHAS, L. *Prep.* By precipitating a solution of manganous sulphate with a solution of sodium phosphate. It must be preserved from the air.—*Dose*, 3 to 12 gr.; in anæmia, rickets, &c.

Manganous Sul'phate. $MnSO_4.7H_2O.$ *Syn.* SULPHATE OF PROTOXIDE OF MANGANESE; MANGANESII SULPHAS, L. *Prep.* 1. By dissolving manganous carbonate in dilute sulphuric acid, and evaporating the filtered solution so that crystals may form, or at once gently evaporating it to dryness. Pure.

2. (Commercial.) By igniting manganese peroxide (pyrolusite) mixed with about 1-10th of its weight of powdered coal in an iron crucible or gas retort, and digesting the residuum of the calcination in sulphuric acid, with the addition after a time of a little hydrochloric acid ; the solution of manganous sulphate thus obtained, after defecation, is evaporated to dryness, and heated to redness as before ; the mass, after ignition, is crushed small and treated with water ; the solution is nearly pure, the whole of the iron having been reduced to the state of insoluble peroxide.

Uses. Used by the calico printers and dyers in the production of black and brown colours. Cloth steeped in the solution, and afterwards passed through a solution of chloride of lime, is dyed of a permanent brown.

Prop., &c. Pale rose-coloured crystals of the formula $MnSO_4.7Aq.$; $MnSO_4.5Aq.$; or $MnSO_4.4Aq.$; according to the method of crystallising, readily yielding with water a solution of a rich amethyst colour. With sulphate of potassium it forms a double salt ('manganese alum').—*Dose.* As an alterative and tonic, 5 to 10 gr. ; as a cholagogue cathartic, 1 to 2 dr., dissolved in water, either alone or combined with infusion of senna. According to Ure, its action is prompt and soon over ; 1 dr. of it occasions, after the lapse of an hour or so, one or more liquid bilious stools. In large doses it occasions vomiting, and in excessive doses it destroys life by its caustic action on the stomach (*Dr G. C. Mitscherlich*). It has been administered with manifest advantage in torpor of the liver, gout, jaundice, syphilis, and certain skin diseases ; and, combined with iron, in anæmia, chlorosis, rickets, &c.

Manganous Sulphide. $MnS.$ Occurs as manganese blende in steel-grey masses. May be obtained as a greenish powder by heating any of the oxides of manganese in a current of $H_2S.$

Manganous Tar'trate. $MnC_4H_4O_6.$ *Syn.* MANGANESII TARTRAS, L. *Prep.* By saturating a solution of tartaric acid with most manganous carbonate. Alterative and tonic.—*Dose*, 4 to 12 gr.

MANGE. An eruptive, parasitic, contagious disease, common to several domestic animals, more especially the dog and horse. The causes are confinement, dirt, and bad living. The treatment should consist in the immediate removal of the cause, the frequent use of soft soap and water, followed by frictions with sulphur ointment, solution of chloride of lime or sporokton, the administration of purgatives, and a change to a restorative diet. Dun states that in India a very efficient remedy for mange is employed by the native farriers, which consists of castor-oil seeds well bruised, steeped for 12 hours in sour milk, and rubbed into the skin, previously thoroughly cleansed with soap and water. "The itchiness disappears almost immediately." A dressing consisting of 1 oz. of chloride of zinc (Burnett's disinfectant fluid) and 1 quart of water may also be applied with advantage. See Dog.

MAN'GEL-WUR'ZEL. *Syn.* MANGOLD-WURZEL, HYBRID BEET, ROOT OF SCARCITY. The *Beta vulgaris*, var. *campestris*, a variety of the common beet. The root abounds in sugar, and has been used in Germany as a substitute for bread in times of scarcity. In these countries it is chiefly cultivated as food for cattle. The young leaves are eaten as spinach. The percentage composition of mangold-wurzel is as follows :

Albuminoid bodies	. . .	1·54
Sugar, &c.	. . . .	8·60
Indigestible fibre	. . .	1·12
Ash	. . . . .	0·96
Water	. . . . .	87·78

MANGEL-WURZEL FLY. See ANTHOMYIA BETÆ.

MAN'HEIM GOLD. A gold-coloured brass. See GOLD (Dutch).

MAN'NA. *Syn.* MANNA (B. P., Ph. L., E., and D.), L. A concrete exudation from the stem of *Fraxinus ornus* and *F. rotundifolia*, obtained by incision (B. P.). "The juice flowing from the incised bark" of "*Fraxinus rotundifolia* and *F. ornus*, hardened by the air" (Ph. L.). The finest variety of this drug is known as flake manna, and occurs in pieces varying from 1 to 6 inches long, 1 or 2 inches wide, and ¼ to 1 inch thick. It has a yellowish-white or cream colour; an odour somewhat resembling honey, but less pleasant, a sweet, mawkish taste; and is light, porous, and friable. It is laxative in doses of 1 to 2 oz.

Manna, Factitious, made of a mixture of sugar, starch, and honey, with a very small quantity of scammony to give it odour and flavour, and to render it purgative, has been lately very extensively offered in trade, and met with a ready sale.

MAN'NACROUP. A granular preparation of wheat deprived of bran, used as an article of food for children and invalids (*Brande*).

MAN'NITE. $C_6H_8(OH)_6$. *Syn.* MANNITOL, MANNA SUGAR, MUSHROOM S.; MANNITA, L. A sweet, crystallisable substance, found in manna, in the sap of the common ash, larch, apple, cherry, &c.; in the leaves of the syringa and privet; in certain lichens, seaweeds, and fungi; in celery, asparagus, sugar-cane, olives, onions, rye-bread, and the root of the monkshood (*Aconitum napellus*). Mannite is also a product of the viscous or 'ropy' fermentation of sweet liquids, beetroot juice being especially liable to this change. It is now generally regarded as a hexhydric alcohol, and is an important substance in vegetable chemistry, and in several other vegetable productions. It has been formed artificially by the action of sodium amalgam upon an alkaline solution of cane-sugar.

Prep. 1. Digest manna in boiling rectified spirit, and filter or decant the solution whilst hot; the mannite crystallises as the liquid cools in tufts of slender, colourless needles.

2. (*Ruspini.*) Manna, 6 lbs.; cold water (in which the white of an egg has been beaten), 3 lbs.; mix, boil for a few minutes, and strain the syrup through linen whilst hot; the strained liquid will form a semi-crystalline mass on cooling; submit this to strong pressure in a cloth, mix the cake with its own weight of cold water, and again press it; dissolve the cake thus obtained in boiling water, add a little animal charcoal, and filter the mixture into a porcelain dish set over the fire; lastly evaporate the filtrate to a pellicle, and set the syrup aside to crystallise. Large quadrangular prisms, perfectly white and transparent.

3. Artificially from glucose or, still better, from fruit-sugar by treating an aqueous solution of it with sodium amalgam; the glucose takes up 2 atoms of hydrogen. The same transformation of glucose takes place under the action of certain ferments. This leads to the inference that grape-sugar is the aldehyde of mannite.

Prop., &c. Mannite has a moderately sweet and agreeable taste; dissolves in 5 parts of cold water and about half that quantity of boiling water; freely soluble in hot, and slightly so in cold alcohol; insoluble in ether; fuses at 166° without loss of weight. By oxidation in contact with platinum-black it is converted into mannitic acid, $C_6H_{12}O_7$, and mannitose, $C_6H_{12}O_6$, a sugar isomeric with glucose. By oxidation with nitric acid it yields saccharic acid, $C_6H_{10}O_8$, and ultimately oxalic acid. The nitrate is a crystalline body which explodes violently by percussion or when suddenly heated. Heated with organic acids mannite forms ethereal salts after the manner of alcohols; generally the resulting compound has a considerable resemblance to the fats. Heated to 200° C. it forms mannitane, a viscous substance very similar to glycerin. It is distinguished from the true sugars by its aqueous solution not being susceptible of the vinous fermentation, in not reducing an alkaline cupric solution, and not possessing the property of rotary polarisation. When pure, it is perfectly destitute of purgative properties. It is now extensively imported from Italy, and is chiefly used to cover the taste of nauseous medicines, and as a sweetmeat.

MANURES'. Substances added to soils to increase their fertility. The food of vegetables, as far as their organic structure is concerned, consists entirely of inorganic compounds; and no organised body can serve for the nutrition of vegetables until it has been, by the process of decay, resolved into certain inorganic substances. These are carbonic acid, water, and ammonia, which are well known to be the final products of putrefaction. But even when these are applied to vegetables, their growth will not proceed unless certain mineral substances are likewise furnished in small quantities, either by the soil or the water used to moisten it. Almost every plant, when burned, leaves ashes, which commonly contain silica, potassa, and phosphate of lime; often, also, magnesia, soda, sulphates, and oxide of iron. These mineral bodies appear to be essential to the existence of the vegetable tissues; so that plants will not grow in soils destitute of them, however abundantly supplied with carbonic acid, ammonia, and water. The carbon of plants is wholly derived from carbonic acid, which is either absorbed from the atmosphere, and from rain-water, by the leaves, or from the moisture and air in the soil, by the roots. Its carbon is retained and assimilated with the body of the plant, while its oxygen is given out in the gaseous form; this decomposition being always effected under the influence of light at ordinary temperatures. The hydrogen and oxygen of vegetables, which, when combined with carbon, constitute the ligneous, starchy, gummy, saccharine, oily, and resinous matters of plants, are derived from water chiefly absorbed by the roots from the soil. The nitrogen of vegetables is derived chiefly, if not exclusively, from ammonia, which is supplied to them in rain, and in manures, and which remains in the soil till absorbed by the roots.

According to the celebrated 'mineral theory' of agriculture advanced by Liebig, a soil is fertile or barren for any given plant according as it contains those mineral substances that enter into its composition. Thus "the ashes of wheat-straw contain much silica and potass, whilst the ashes of the seeds contain phosphate of magnesia. Hence, if a soil is deficient in any one of these, it will not yield wheat. On the other hand, a good crop of wheat will exhaust the soil of these substances, and it will not yield a second crop till they have been restored, either by manure, or by the gradual action of the weather in disintegrating the subsoil. Hence the benefit derived from fallows and from the rotation of crops.

"When, by an extraordinary supply of any one mineral ingredient, or of ammonia, a large crop has been obtained, it is not to be expected that a repetition of the same individual manure next year will produce the same effect. It must be remembered that the unusual crop has exhausted the soil probably of all the other mineral ingredients, and that they also must be restored before a second crop can be obtained.

"The salt most essential to the growth of the potato is the double phosphate of ammonia and magnesia; that chiefly required for hay is phosphate of lime; while for almost all plants potassa and ammonia are highly essential."

From these principles we "may deduce a few valuable conclusions in regard to the chemistry of agriculture. First, by examining the ashes of a thriving plant we discover the mineral ingredients which must exist in a soil to render it fertile for that plant. Secondly, by examining a soil, we can say at once whether it is fertile in regard to any plants the ashes of which have been examined. Thirdly, when we know the defects of a soil, the deficient matters may be easily obtained and added to it, unmixed with such as are not required. Fourthly, the straw, leaves, &c., of any plant are the best manure for that plant, since every vegetable extracts from the soil such matters alone as are essential to it. This important principle has been amply verified by the success attending the use of wheat-straw, or its ashes, as manure for wheat, and of the chippings of the vines as a manure for the vineyard. When these are used (in the proper quantity) no other manure is required. Fifthly, in the rotation of crops, those should be made to follow which require different materials; or a crop which extracts little or no mineral matter, such as peas, should come after one which exhausts the soil of its phosphates and potassa." (Liebig).

The experiments of Messrs Lawes and Gilbert have forced upon them opinions differing from those of Baron Liebig on some important points in relation to his 'mineral theory,' which endeavours to prove that "the crops on a field diminish or increase in exact proportion to the diminution or increase of the mineral substances conveyed to it in manure." The results obtained by the English investigators appear to prove that it is impossible to get good crops by using mineral manures alone, and that nitrogenous manures (farmyard manure, guano, ammoniacal salts, &c.) are fertilising agents of the highest order.

Of the chemical manures now so much used bone-dust is, perhaps, the most important, as it supplies the phosphates which have been extracted by successive crops of grass and corn, the whole of the bones of the cattle fed on these crops having been derived from the soil; its gelatin also yields ammonia by putrefaction. Guano acts as a source of ammonia, containing much oxalate and urate of ammonia, with some phosphates. Night-soil and urine, especially the latter, are most valuable for the ammonia they yield, as well as for the phosphates and potassa; but are very much neglected in this country, although their importance is fully appreciated in Belgium, France, and China. Nitrate of soda is valued as a source of nitrogen.

All organic substances may be employed as manures; preference being, however, given to those abounding in nitrogen, and which readily decay when mixed with the soil.

The analysis of manures, soils, and the ashes of plants, for the purpose of ascertaining their composition and comparative value, is not easily performed by the inexperienced; but a rough approximation to their contents, sufficiently accurate for all practical purposes, may be generally made by any intelligent person with proper care and attention. See BONEDUST, GUANO, &c.

Manures, Artificial. Various formulæ belonging to this head will be found dispersed, under their respective names, throughout this work. The following are additional ones:—

1. (Anderson.) Sulphate of ammonia, common salt, and oil of vitriol, of each, 10 parts; chloride of potassium, 15 parts; gypsum and sulphate of potassa, of each, 17 parts; saltpetre, 20 parts; crude Epsom salts, 25 parts; sulphate of soda, 33 parts. For clover.

2. (Huxtable.) Crude potash, 28 lbs.; common salt, 1 cwt.; bone-dust and gypsum, of each 2 cwt.; wood ashes, 15 bushels. For either corn, turnips, or grass.

3. (Johnstone.) Sulphate of soda (dry), 11 lbs.; wood ashes, 28 lbs.; common salt, ¼ cwt.; crude sulphate of ammonia, 1 cwt.; bone-dust, 7 bushels. As a substitute for guano.

4. (Lawes' 'Superphosphate.') See COPROLITE.

5. (Fertilising powder.) A mixture of very fine bone-dust, 18 parts; calcined gypsum and sulphate of ammonia, of each, 1 part. The seed is ordered to be steeped in the 'drainings' from a dunghill, and after being drained, but whilst still wet, to be sprinkled with the powder, and then dried. See FLOWERS, LIME (Superphosphate), &c.

The 'plant fertilisers' sold under various names require to be used with caution, or an artificial condition is produced which cannot be sustained unless the manure is continued. Gardeners 'get up' plants for market in this way, and the purchaser is frequently disappointed by their beginning to wither a few days after they are bought. Great care is required in order to save them, and more fertiliser must be used, gradually reduced till the plant reaches a normal state, when it will too often be found to have been hardly worth the trouble.

MANUSCRIPTS, Faded, to Restore. One of

the methods in use for the restoration of old or faded writing is to expose it to the vapours of hydrosulphate of ammonia (hydrosulphide of ammonium) until the ink becomes darkened by the formation of sulphide of iron. Another consists in carefully washing or sponging the faded manuscript over with a weak solution of the ammonic sulphide, and as soon as the characters become legible, soaking it in water so as to remove the remaining sulphide, and then drying it between folds of blotting-paper. A third plan, and one attended with less risk to the paper, is to brush over the manuscript with a moderately strong aqueous solution of gallo-tannic acid, to wash with water, and afterwards to dry it at a temperature of about 150° F.

The solution of gallo-tannic acid may be obtained by making a strong infusion of bruised nut-galls in boiling water, and when cold straining it. Some old and mediæval manuscripts are written in inks made of carbon. To such the above treatment is inapplicable, being suited only to those traced in ordinary writing ink. For parchments the latter method is preferable.

MAPS. These, as well as architects' and engineers' designs, plans, sections, drawings, &c., may be tinted with any of the simple liquid colours mentioned under 'VELVET COLOURS,' preference being given to the most transparent ones, which will not obscure the lines beneath them. To prevent the colours from sinking and spreading, which they usually do on common paper, the latter should be wetted 2 or 3 times with a sponge dipped in alum water (3 or 4 oz. to the pint), or with a solution of white size, observing to dry it carefully after each coat. This tends to give lustre and beauty to the colours. The colours for this purpose should also be thickened with a little gum water. Before varnishing maps after colouring them, 2 or 3 coats of clean size should be applied with a soft brush—the first one to the back.

MARASCHINO (-kēno). *Syn.* MARASQUIN, Fr. A delicate liqueur spirit distilled from a peculiar cherry growing in Dalmatia, and afterwards sweetened with sugar. The best is from Zara, and is obtained from the marasca cherry only. An inferior quality is distilled from a mixture of cherries and the juice of liquorice root.

MARBLE. *Syn.* LIMESTONE, HARD CARBONATE OF LIME; MARMOR, CALCIS CARBONAS DURUS, M. ALBUM (B. P., Ph. E. & D.), L. Marbles are merely purer and more compact varieties of limestone, which admit of being sawn into slabs, and are susceptible of a fine polish. White marble is employed for the preparation of carbonic acid and some of the salts of lime. It contains about 65% of lime. Sp. gr. 2·70 to 2·85. The tests of its purity are the same as those already noticed under CHALK.

Marble is best cleaned with a little soap and water, to which some ox-gall may be added. Acids should be avoided. Oil and grease may be generally removed by spreading a paste made of soft soap, caustic potash lye, and fuller's-earth over the part, and allowing it to remain there for a few days; after which it must be washed off with clean water. Or equal parts of American potash (crude carbonate of potash) and whiting are made into a moderately stiff paste with a sufficiency of boiling water, and applied to the marble with a brush. At the end of 2 or 3 days the paste is removed and the marble washed with soap and water. Any defect of polish may be brought up with tripoli, followed by putty powder, both being used along with water.

Marble is mended with one or other of the compounds noticed under CEMENTS.

Marble may be stained or dyed of various colours by applying coloured solutions or tinctures to the stone, made sufficiently hot to make the liquid just simmer on the surface. The following are the substances usually employed for this purpose:—

BLUE. Tincture or solution of litmus, or an alkaline solution of indigo.

BROWN. Tincture of logwood.

CRIMSON. A solution of alkanet root in oil of turpentine.

FLESH-COLOUR. Wax tinged with alkanet root, and applied to the marble hot enough to melt it freely.

GOLD-COLOUR. A mixture of equal parts of white vitriol, sal-ammoniac, and verdigris, each in fine powder, and carefully applied.

GREEN. An alkaline solution or tincture of sap green, or wax strongly coloured with verdigris; or the stone is first stained blue, and then the materials for yellow stain are applied.

RED. Tincture of dragon's blood, alkanet root, or cochineal.

YELLOW. Tincture of gamboge, turmeric, or saffron; or wax coloured with annotta. Success in the application of these colours requires considerable experience. By their skilful use, however, a very pleasing effect, both of colour and grain, may be produced.

Marble, Restoring. Take a rather firm linen pad, damp it, sprinkle it with rotten-stone or fine emery, and rub the marble until the gloss begins to appear. Finally, polish the whole with another linen pad, rouge and very finely ground emery being on it. After the marble is dry, give the finishing touch with a mixture of turpentine and wax or French polish, and polish with an old silk handkerchief until quite dry.

MARBLING (of Books, &c.). The edges and covers of books are 'marbled' by laying the colour on them with a brush, or by means of a wooden trough containing mucilage, as follows:—Provide a wooden trough, 2 inches deep, 6 inches wide, and the length of a super-royal sheet; boil in a brass or copper pan any quantity of linseed and water until a thick mucilage is formed; strain this into the trough, and let it cool; then grind on a marble slab any of the following colours in table-beer. For blue, Prussian blue or indigo;—red, rose-pink, vermilion, or drop lake;—yellow, king's yellow, yellow ochre, &c.;—white, flake white;—black, ivory-black, or burnt lamp-black;—brown umber, burnt u., terra di sienna, burnt s.; black mixed with yellow or red also makes brown;—green, blue and yellow mixed;—purple, red and blue mixed. For each colour provide two cups—one for the ground colours, the other to mix them with the ox-gall, which must be used to thin them at discretion. If too much gall is used the colours spread; when they keep their place on the surface of the trough, on being moved with a quill, they are fit for use. All things being in readiness, the prepared colours are successively sprinkled on the surface of the mucilage in the trough with a brush, and are waved or drawn about with a quill or a stick according to taste. When the design is thus formed, the book, tied tightly between cutting-boards of the same size, is lightly pressed with its edge on the surface of the liquid pattern, and then withdrawn and dried. The covers may be marbled in the same way, only the liquid colours must be allowed to run over them. The film of colour in the trough may be as thin as possible; and if any remains after the marbling, it may be taken off by applying paper to it before you prepare for marbling again. This process has been called FRENCH MARBLING.

To diversify the effect, a little sweet oil is often mixed with the colours before sprinkling them on, by which means a light halo or circle appears round each spot. In like manner spirit of turpentine, sprinkled on the surface of the trough, produces white spots. By staining the covers with any of the liquid dyes, and then dropping on them, or running over them, drops of the ordinary liquid mordants, a very pleasing effect may be produced. Vinegar black, or a solution of green copperas, thus applied to common leather, produces black spots or streaks, and gives a similar effect with most of the light dyes. A solution of alum or of tin in like manner produces bright spots or streaks, and soda or potash water dark ones. This style has been called EGYPTIAN MARBLE.—SOAP MARBLING is done by throwing on the colours, ground with a little white soap to a proper consistence, by means of a brush. It is much used for book-edges, stationery, sheets of paper, ladies' fancy work, &c.—THREAD MARBLE is given by first covering the edge uniformly of one colour, then laying pieces of thick thread irregularly on different parts of it, and giving it a fine dark sprinkle. When well managed the effect is very pleasing.—RICE MARBLE is given in a similar way to the last by using rice.—TREE MARBLE is done on leather book-covers, &c., by bending the board a little in the centre, and running the marbling liquid over it in the form of vegetation. The knots are given by rubbing the end of a candle on those parts of the cover.—WAX MARBLE is given in a similar way to thread marble, but using melted wax, which is removed after the book is sprinkled and dried; or a sponge charged with blue, green, or red may be passed over. This, also, is much used for stationery work, especially for folios and quartos. The 'vinegar black' of the bookbinders is merely a solution of acetate of iron, made by steeping a few rusty nails or some iron filings in vinegar. All the ordinary liquid colours that do not contain strong acids or alkalies may be used, either alone or thickened with a little gum, for marbling or sprinkling books.

SPRINKLING is performed by simply dipping a stiff-haired painter's brush into the colour, and suddenly striking it against a small stick held in the left hand over the work. By this means the colour is evenly scattered without producing 'blurs' or 'blots.'

PAPER, PASTEBOARD, &c., in sheets, are marbled and sprinkled in a similar manner to that above described, but in this case the gum trough must, of course, be longer.

MARGAR'IC ACID. This term was formerly applied to a mixture of palmitic and stearic acids, produced by decomposing the alkaline soaps of solid fats with an acid, but it is now given to a fatty acid which can only be obtained artificially.

MARG'ARIN. Syn. MARGARATE OF GLYCERYL. A constituent formerly supposed to exist in solid fats, but now regarded as a mixture of stearin and palmitin.

MARINE' ACID. See HYDROCHLORIC ACID.

MARL. A natural mixture of clay and chalk with sand. It is characterised by effervescing with acids. According to the predominance of one or other of its component parts, it is called argillaceous, calcareous, or sandy marl. It is very generally employed as a manure for sandy soils, more particularly in Norfolk. See SOILS.

MAR'MALADE. Originally a conserve made of quinces and sugar; now commonly applied to the conserves of other fruit, more especially to those of oranges and lemons.

Prep. Marmalades are made either by pounding the pulped fruit in a mortar with an equal or a rather larger quantity of powdered white sugar, or by mixing them together by heat, passing them through a hair-sieve whilst hot, and then putting them into pots or glasses. The fruit-pulps are obtained by rubbing the fruit through a fine hair-sieve, either at once or after it has been softened by simmering it for a short time along with a little water. When heat is employed in mixing the ingredients, the evaporation should be continued until the marmalade 'jellies' on cooling. See CONSERVES, CONFECTIONS, ELECTUARIES, JAMS, JELLIES, and below.

Marmalade, Apricot. From equal parts of pulp and sugar.

Marmalade, Mixed. From plums, pears, and apples, variously flavoured to palate.

Marmalade, Orange. Prep. 1. From oranges (either Seville or St Michael's, or a mixture of the two), by boiling the peels in syrup until soft, then pulping them through a sieve, adding as much white sugar, and boiling them with the former syrup and the juice of the fruit to a proper consistence.

2. By melting the confection of orange peel (Ph. L.), either with or without the addition of some orange or lemon juice, and then passing it through a sieve.

3. (CANDIED ORANGE MARMALADE.) From candied orange peel, boiled in an equal weight each of sugar and water, and then passed through a sieve.

4. (SCOTCH MARMALADE.) a. Seville orange juice, 1 quart; yellow peel of the fruit, grated; honey, 2 lbs.; boil to a proper consistence.

b. Seville oranges, 8 lbs.; peel them as thinly as possible, then squeeze out the juice, boil it on the yellow peels for ¾ of an hour, strain, add white sugar, 7 lbs., and boil to a proper consistence.

Marmalade, Quince. Syn. DIACYDONIUM. Prep. From quince flesh or pulp and sugar, equal parts; or from the juice (MIVA CYDONIORUM,

GELATINA C.), by boiling it to half, adding an equal quantity of white wine and 2-3rds of its weight of sugar, and gently evaporating the mixture.

Marmalade, Tomato. Like APRICOT MARMALADE, adding a few slices of onion and a little parsley.

MARMALADE PLUM (*Lucuma mammosa*, Griseb.).

MARMORA'TUM. Finely powdered marble and quicklime, well beaten together; used as a cement or mortar.

MARROW (Beef). This is extensively employed by the perfumers in the preparation of various pomades and other cosmetics, on account of its furnishing an exceedingly bland fat, which is not so much disposed to rancidity as the other fats. It is prepared for use by soaking and working it for some time in lukewarm water, and afterwards melting it in a water-bath, and straining it through a piece of muslin whilst hot. When scented it is esteemed equal to bear's grease for promoting the growth of the hair.

MARSH GAS. Light carburetted hydrogen.

MARSH-MALLOW. *Syn.* ALTHEA (Ph. L. and E.), L. The root (leaves and root, Ph. E.) of *Althæa officinalis*, Linn., or common marshmallow (Ph. L.). It is emollient and demulcent; the decoction is useful in irritation of the respiratory and urinary organs, and of the alimentary canal. The flowers as well as the root are reputed pectoral.

MARSH'S TEST. See ARSENIOUS ACID.

MARTIN'S POWDER. A mixture of white arsenic and the powdered stems of *Orobanche virginiana*, Linn., a plant common in Virginia. An American quack remedy for cancer.

MASS. *Syn.* MASSA, L. This term is commonly applied in pharmacy and veterinary medicine to certain preparations which are not made up into their ultimate form. Thus we have 'ball masses,' 'pill masses,' &c.; of which, for convenience, large quantities are prepared at a time, and are kept in pots or jars, ready to be divided into balls or pills, as the demands of business may require (see *below*).

MASSAGE. Dr Hale White gives the following instructions for conducting the operation of massage:—First grease the parts with vaseline or oil, and if the skin be very hairy it may be necessary to cut the hair; then stroke the muscles firmly several times with the edge of the hands in the direction of the venous flow. In places such as the back, where this is impossible, always stroke in the same direction; on the abdomen follow the course of the colon; next you may take up the skin between the thumb and forefinger, and rub it between them, one hand following the other in the same direction as the stroking, then thoroughly knead the muscles with one or both hands, according to the size of the part, in the same direction as already mentioned; after this move all the joints in every direction, then you may conclude by striking the muscles with several small blows, but this is not of much importance. The full time should be about an hour twice a day for the whole body. If there is a painful spot and the case be one of hysteria, particularly direct your energies to that part, avoiding the bones.

MASSARANDUBA. See MIMUSOPS.

MASSES (Veterinary). Reprinted from Tuson's 'Veterinary Pharmacopœia:'

Massa Aloes. MASS OF ALOES. *Syn.* CATHARTIC MASS. *Prep.* Take of Barbadoes aloes, in small pieces, 8 parts; glycerin, 2 parts; ginger, in powder, 1 part; melt together in a water-bath, and thoroughly incorporate by frequent stirring.—*Use.* Cathartic for the horse.—*Dose.* From 6 to 8 dr.

Massa Aloes Composita. COMPOUND MASS OF ALOES. *Syn.* ALTERATIVE MASS. *Prep.* Take of Barbadoes aloes, in powder, 1 oz.; soft soap, 1 oz.; common mass, 6 oz.; thoroughly incorporate by beating in a mortar, so as to form a mass.—*Use.* Alterative for the horse.—*Dose*, 1 oz.

Massa Antimonii Tartarata Composita. COMPOUND MASS OF TARTARATED ANTIMONY. *Syn.* FEVER BALL. *Prep.* Take of tartarated antimony, in powder, ½ dr.; camphor, in powder, ½ dr.; nitrate of potash, in powder, 2 dr.; common mass, a sufficiency; mix so as to form a bolus. Febrifuge for the horse.—*Dose.* The above mixture constitutes 1 dose.

Massa Belladonnæ Composita. COMPOUND MASS OF BELLADONNA. *Syn.* COUGH BALL. *Prep.* Take of extract of belladonna, ½ to 1 dr.; Barbadoes aloes, in powder, 1 dr.; nitrate of potash, in powder, 2 dr.; common mass, a sufficiency; mix so as to form a bolus.—*Use.* For the horse in chronic cough.—*Dose.* The above mixture constitutes 1 dose.

Massa Catechu Composita. COMPOUND MASS OF CATECHU. *Syn.* ASTRINGENT MASS. *Prep.* Take of extract of catechu, in fine powder, 1 oz.; cinnamon bark, in fine powder, 1 oz.; common mass, 6 oz.; mix.—*Use.* Astringent for the horse. —*Dose*, 1 oz., in the form of a bolus.

Massa Communis. COMMON MASS. *Prep.* Take of linseed, finely ground, and treacle, of each, equal parts; mix together so as to form a mass. —*Use.* An excipient for medicinal agents when they are to be administered in the form of bolus.

Massa Cupri Sulphatis. MASS OF SULPHATE OF COPPER. *Syn.* TONIC MASS. *Prep.* Take of sulphate of copper, finely powdered, 1 oz.; ginger, in powder, 1 oz.; common mass, 6 oz.; mix.—*Use.* Tonic for the horse.—*Dose*, 6 to 8 dr.

Massa Digitalis Composita. COMPOUND MASS OF DIGITALIS. *Syn.* COUGH BALL. *Prep.* Take of Barbadoes aloes, in powder, 2 oz.; digitalis, 1 oz.; common mass, 18 oz.; mix.—*Use.* For the horse in chronic cough.—*Dose*, 1 oz. once or twice a day.

Massa Ferri Sulphatis. MASS OF SULPHATE OF IRON. *Syn.* TONIC MASS. *Prep.* Take of sulphate of iron, in powder, 2 oz.; ginger, in powder, 1 oz.; common mass, 5 oz.; mix.—*Use.* Tonic for the horse.—*Dose*, 6 to 8 dr.

Massa Resinæ Composita. COMPOUND MASS OF RESIN. *Syn.* DIURETIC MASS. *Prep.* Take of resin, in powder, nitrate of potash, in powder, hard soap, of each, equal parts; mix.—*Use.* Diuretic for the horse.—*Dose*, 1 oz.

Massa Zingiberis Composita. COMPOUND MASS OF GINGER. *Syn.* CORDIAL MASS. *Prep.* Take of ginger, in powder, gentian root, in powder, treacle, of each equal parts, a sufficiency; mix so as to form a mass.—*Use.* Stomachic for the horse.—*Dose,* 1 oz.

MAS'SICOT. *Syn.* MASTICOT, YELLOW PROTOXIDE OF LEAD; PLUMBI OXYDUM FLAVUM, CERUSSA CITRINA, L. The dross that forms on melted lead exposed to a current of air, roasted until it acquires a uniform yellow colour. Artists often apply the same name to white-lead roasted until it turns yellow. Used as a pigment.

MASTIC. *Syn.* MASTICH, GUM MASTIC; MASTICHE, L. The "resin flowing from the incised bark of *Pistacia lentiscus,* var. *Chia*" (Ph. L.). It occurs in pale yellowish, transparent, rounded tears, which soften between the teeth when chewed, giving out a bitter, aromatic taste. Sp. gr. 1·07. It is soluble in both rectified spirit and oil of turpentine, forming varnishes. It is chiefly used as a 'masticatory,' to strengthen and preserve the teeth, and perfume the breath.

Mastic. Fine mortar or cement used for plastering walls, in which the ingredients, in a pulverulent state, are mixed up, either entirely or with a considerable portion of linseed oil. It sets very hard, and is ready to receive paint in a few days. See CEMENTS.

MASTICA'TION. The act of chewing food, by which it not only becomes comminuted, but mixed with the saliva, and reduced to a form fit for swallowing. It has been justly regarded as the highest authorities as the first process of digestion, and one without which the powers of the stomach are over-tasked, and often performed with difficulty. Hence the prevalence of dyspepsia and bowel complaints among persons with bad teeth, or who 'bolt' their food without chewing it.

MAS'TICATORIES. *Syn.* MASTICATORIA, L. Substances taken by chewing them. They are employed as intoxicants, cosmetics, and medicinals; generally with the first intention. The principal masticatory used in this country is tobacco. In Turkey, and several other Eastern nations, opium is taken in a similar manner. In India, a mixture of areca nut, betel leaf, and lime, performs the same duties; whilst in some other parts of the world preparations of caco are employed. As cosmetics, orris root, cassia, cinnamon, and sandal-wood are frequently chewed to scent the breath. Among medicinals, mastic and myrrh are frequently chewed to strenthen the teeth and gums; pellitory, to relieve the toothache; and rhubarb, ginger, and gentian, to relieve dyspepsia and promote the appetite.

Prep. 1. (*Augustin.*) Mastic, pellitory (both in powder), and white wax, of each, 1 dr.; mixed by heat and divided into 6 balls. In toothache, loose teeth, &c.

2. (*W. Cooley.*) Mastic, myrrh, and white wax, of each, 1 part; rhubarb, ginger, and extract of gentian, of each, 2 parts; beaten up with tincture of tolu, q. s., and divided into boluses or lozenges of 10 gr. each. One or two to be chewed an hour before dinner; in dyspepsia, defective appetite, &c.

3. (*Quincy.*) Mastic, 8 oz.; pellitory and stavesacre seed, of each, 2 dr.; cubebs and nutmegs, of each, 1 dr.; angelica root, ½ dr.; melted wax, q. s. to make it into small balls. As a stimulant to the gums, and in toothache.

4. Opium, ginger, rhubarb, mastic, pellitory of Spain, and orris root, of each, 1 dr.; melted spermaceti, q. s. to mix; for 6-gr. pills. As the last, and in toothache and painful gums.

MAS'TICOT. See MASSICOT.

MATCHES (Cooper's). *Syn.* SWEETENING MATCHES. These are made by dipping strips of coarse linen or canvas into melted brimstone. For use, the brimstone on one of them is set on fire, and the match is then at once suspended in the cask, and the bung loosely set in its place. After the lapse of 2 or 3 hours the match is removed and the cask filled with liquor. Some persons pour a gallon or two of the liquor into the cask before 'matching' it. The object is to allay excessive fermentation. The operation is commonly adopted in the western counties for cider intended for shipment, or other long exposure during transport. It is also occasionally employed for inferior and 'doctored' wines.

Matches (Instantaneous Light). Of these there are several varieties, of which the one best known, and most extensively used, is the common phosphorus match, known as the 'congreve' or 'lucifer.' The original 'LUCIFERS' or 'LIGHT-BEARING MATCHES,' invented in 1826, consisted of strips of pasteboard, or flat splints of wood, tipped first with sulphur, and then with a mixture of sulphide of antimony and chlorate of potassa, and were ignited by drawing them briskly through folded glass-paper. They required a considerable effort to ignite them, and the composition was apt to be torn off by the violence of the friction. The term 'lucifer,' having become familiar, was applied to the simpler and more effective match afterwards introduced under the names of 'CONGREVE' and 'CONGREVE LIGHT.' We need not describe the 'chemical matches,' 'phosphorus bottles,' and 'prometheans,' in use during the early part of the present century, as these are quite obsolete. We will simply sketch the general process of manufacture now in use for phosphorus matches.

Manuf. The wooden splints are cut by steam machinery from the very best quality of pine planks, perfectly dried at a temperature of 400° F. English splints are of two sizes—'large' and 'minnikins;' the former 2¼ inches longer, and the latter somewhat shorter. In the manufacture double lengths are used, so that each splint may be coated with the igniting composition at both ends, and then cut asunder in the middle to form two matches. In England the splints are usually cut square in form, but in Germany they are cylindrical, being prepared by forcing the wood through circular holes in a steel plate. The ends of the double splints having been slightly charred by contact with a red-hot plate, are coated with sulphur by dipping them to the requisite depth in the melted material. In some cases the ends are saturated with melted wax or paraffin instead of sulphur. The splints are then arranged in a frame between grooved boards in such a manner that the prepared ends project on each side of the frame. These projecting ends are then tipped with the phosphorus composition, which is spread to a uniform depth of about 1-8th inch on a smooth slab of stone, kept warm by means of steam beneath. When partially dry, the tipped splints are taken from the frames, cut through the middle, and placed in heaps of 100, ready for 'boxing.'

The different compositions for tipping the matches in use in different countries and factories all consist essentially of emulsions of phosphorus in a solution of glue or gum, with or without other matters for increasing the combustibility, for colouring, &c. In England the composition contains a considerable quantity of chlorate of potassa, which imparts a snapping and flaming quality to the matches tipped with it, and but little phosphorus, on account of the moisture of the climate. In Germany the proportion of phosphorus used is much larger, and nitre, or some metallic peroxide, replaces chlorate of potassa. The German matches light quietly with a mild lambent flame, and are injured quickly by damp. The following formulæ have been selected:

1. (ENGLISH.) Fine glue, 2 parts, broken into small pieces, and soaked in water till quite soft, is added to water, 4 parts, and heated by means of a water-bath until it is quite fluid, and at a temperature of 200° to 212° F. The vessel is then removed from the fire, and phosphorus, 1½ to 2 parts, is gradually added, the mixture being agitated briskly and continually with a 'stirrer' having wooden pegs or bristles projecting at its lower end. When a uniform emulsion is obtained, chlorate of potassa, 4 to 5 parts, powdered glass, 3 to 4 parts, and red-lead, smalt, or other colouring matter, a sufficient quantity (all in a state of very fine powder) are added, one at a time, to prevent accidents, and the stirring continued until the mixture is comparatively cool.

According to Mr G. Gore, the above proportions are those of the best quality of English composition. The matches tipped with it deflagrate with a snapping noise (see *above*).

2. (GERMAN.) *a.* (*Böttger.*) Dissolve gum-arabic, 16 parts, in the least possible quantity of water, add of phosphorus, in powder, 9 parts, and mix by trituration; then add of nitre, 14 parts; vermilion or binoxide of manganese, 16 parts, and form the whole into a paste, as directed above; into this the matches are to be dipped, and then exposed to dry. As soon as the matches are quite dry they are to be dipped into very dilute copal varnish or lac varnish, and again exposed to dry, by which means they are rendered waterproof, or at least less likely to suffer from exposure in damp weather.

b. (*Böttger.*) Glue, 6 parts, is soaked in a little cold water for 24 hours, after which it is liquefied by trituration in a heated mortar; phosphorus, 4 parts, is now added, and rubbed down at a heat not exceeding 150° F.; nitre, in fine powder, 10 parts, is next mixed in, and afterwards red ochre, 5 parts, and smalt, 2 parts, are further added, and the whole formed into a uniform paste, into which the matches are dipped, as before. Cheaper than the last.

c. (*Diesel.*) Phosphorus, 17 parts; glue, 21 parts; red-lead, 24 parts; nitre, 38 parts. Proceed as above.

Obs. Matches tipped with the above (*a*, *b*, and *c*) inflame without fulmination when rubbed against a rough surface, and are hence termed 'noiseless matches' by the makers.

3. (SAFETY MATCHES.) The latest improvement of note in the manufacture of matches is that of Landstrom, of Jonkoping, in Sweden, adopted by Messrs Bryant and May (patent). It consists in dividing the ingredients of the match-mixture into two separate compositions, one being placed on the ends of the splints, as usual, and the other, which contains the phosphorus, being spread in a thin layer upon the end or lid of the box. The following are the compositions used by the patentee:—*a*. (For the splints.) Chlorate of potassa, 6 parts; sulphuret of antimony, 2 to 3 parts; glue, 1 part.—*b*. (For the friction surface.) Amorphous phosphorus, 10 parts; sulphuret of antimony or peroxide of manganese, 8 parts; glue, 3 to 6 parts; spread thinly upon the surface, which has been previously made rough by a coating of glue and sand.

By thus dividing the composition the danger of fire arising from ignition of the matches by accidental friction is avoided, as neither the portion on the splint nor that on the box can be ignited by rubbing against an unprepared surface. Again, by using the innocuous red or amorphous phosphorus, the danger of poisoning is entirely prevented.

MATÉ. *Syn.* PARAGUAY TEA. This is the dried leaf of a small shrub, the *Ilex paraguayensis*, or Brazilian holly, growing in Paraguay and Brazil; by the inhabitants of which places, as well as South America generally, it is largely employed in the form of a beverage as tea. Its active ingredient, paraguaine, formerly supposed to be a distinct principle, has from further researches into its composition been discovered to be identical with theine and caffeine—the alkaloids of tea and coffee.

Mr Wanklyn ascribes the following composition to maté:

Moisture	. . .	.	6·72
Ash	. . .	.	5·86
Soluble organic matter	.		25·10
Insoluble organic matter	.		62·32
			100·00

MATE′RIA MED′ICA. A collective name of the various substances, natural and artificial, employed as medicines or in the cure of disease. In its more extended sense it includes the science which treats of their sources, properties, classification, and applications. The materia medica of the Pharmacopœia is a mere list, with occasional notes, "embracing the animal, vegetable, and chemical substances, whether existing naturally, prepared in officinal chemical preparations, or sold in wholesale trade, which we (the College) direct to be used either in curing diseases or in preparing medicines" (Ph. L.).

MAT′ICO. *Syn.* SOLDIER'S HERB; MATECO (B. P., Ph. D.); MATICA, HERBA MATICÆ, L. The dried leaves of *Piper angustifolium*, R. and P., a tropical American shrub, used as a mild aromatic. The leaves have been employed with considerable success as a mechanical external styptic; applied to leech-bites, slight cuts, and other wounds, &c., and pressed on with the fingers, they seldom fail to arrest the bleeding. Matico has also been much lauded as an internal astringent and styptic, in hæmorrhages from the lungs, stomach, bowels, uterus, &c.; but as it is nearly destitute of astringent properties, its virtues in these cases must have been inferred from its external action. As an aromatic, bitter stimulant, closely resembling the peppers, it has been proposed as a substitute for cubebs and black pepper, in the treatment of diseases of the mucous membranes, piles, &c.—*Dose*, ⅓ to 2 dr., in powder; or under the form of infusion, tincture, or boluses.

MAYONNAISE SAUCE. Powdered turmeric, 1 oz.; powdered tragacanth, 1 oz.; olive oil, 8 oz.; eggs, 8; water, 5¾ pints; ground mustard, 1½ oz.; salt, 8 oz.; acetic acid (glacial), 2 oz.; tincture of capsicum, ¼ oz., or according to taste; sugar, 1 lb. Mix the first three ingredients in a mortar capable of holding one gallon, then add the eggs, which have been whipped previously, and incorporate thoroughly until an emulsion is formed; next mix separately the mustard and water, allow to stand 10 or 15 minutes, or until the flavour is fully developed, then add the last 4 ingredients, mix and add the liquid gradually to the contents of the mortar. It should make a smooth, uniform emulsion; finally, strain through cheese-cloth.

MEAD. *Syn.* MELLINA, L. An old English liquor, made from the combs from which the honey has been drained, by boiling them in water, and fermenting the saccharine solution thus obtained. It is commonly confounded with metheglin. Some persons add 1 oz. of hops to each gallon; and, after fermentation, a little brandy. It is then called sack mead. See METHEGLIN.

MEAL. The substance of edible grain ground to powder, without being bolted or sifted. Barley meal and oatmeal are the common substances of this class in England. In North America the term is commonly applied to ground Indian corn, whether bolted or not (*Goodrich*). The four resolvent meals of old pharmacy (*quatuor farinæ resolventes*) are those of barley, beans, linseed, and rye.

MEALS. The "periods of taking food usually adopted, in conformity with convenience and the recurrences of hunger, are those which are best adapted to the purposes of health; namely, the morning meal, the midday meal, and the evening meal." "That these are the proper periods for meals is evident from the fact of their maintaining their place amid the changes which fashion is constantly introducing." "If we look at these periods in another point of view, we shall find an interval of four hours left between them for the act of digestion and sub-

sequent rest of the stomach. Digestion will claim between two and three hours of the interval; the remaining hour is all that the stomach gets of rest—enough, perhaps, but not too much, nor to be justly infringed " (*Eras. Wilson*).

MEA'SLES. *Syn.* RUBEOLA, MORBILLI, L. This very common disease is characterised by feverishness, chilliness, shivering, head-pains, swelling and inflammation of the eyes, shedding of sharp tears, with painful sensibility to light, oppressive cough, difficulty of breathing, and sometimes vomiting or diarrhœa. These are followed about the fourth day by a crimson rash upon the skin, in irregular crescents or circles, and by small red points or spots, which are perceptible to the touch, and which, after four or five days, go off with desquamation of the cuticle. The fever, cough, &c., often continue for sometime; and unless there have been some considerable evacuations, either by perspiration or vomiting, they frequently return with increased violence, and occasion great distress and danger.

Treat. When there are no urgent local symptoms, mild aperients, antimonial diaphoretics, and diluents should be had recourse to. The cough may be relieved by expectorants, demulcents, and small doses of opium; and the diarrhœa by the administration of the compound powder of chalk and opium; the looseness of the bowels, however, had better not be interfered with unless it be extreme.

Measles are most prevalent in the middle of winter, and though common to individuals of all ages, are most frequent amongst children. The contagion of measles will not travel far in the air, but is readily carried by clothing. The catarrhal stage is infectious, and often mistaken for a common cold, and neglected until too late to prevent spread of the disease; the period of incubation is ten or twelve days, and until this period have elapsed after contact with a patient suffering from the disease, a healthy person cannot be said to be safe. The value of isolation in preventing the spread of measles will be obvious.

Like the smallpox, the measles are contagious, and seldom attack the same person more than once during life.

MEASURE. *Syn.* MENSURA, L. The unit or standard by which we estimate extension, whether of length, superficies, or volume. The following tables represent the values and proportions of the principal measures employed in *commerce* and the *arts*:

TABLE I. *English Lineal Measures.*

Inches.	Feet.	Yards.	Poles.	Furlongs.	Miles.
1·	·063	·028	·00505	·00012626	·0000157828
12·	1·	·333	·06060	·00151515	·00018939
36·	3·	1·	·1818	·004545	·00056818
198·	16·5	5·5	1·	·025	·003125
7920·	660·	220·	40·	1·	·125
63360·	5280·	1760·	320·	8·	1·

⁎ The unit of the above table is the yard, of which no legal standard has existed since that established by the statute of 1824 was destroyed by the fire which consumed the two Houses of Parliament in 1834.

juicy appearance known as 'underdone.' Although a certain quantity of the gravy (which consists of the soluble and saline ingredients) escapes in the process, the greater part is retained. The brown agreeably sapid substance formed on the outside of the meat is known as *osmasome*, and is concentrated gravy. The melting fat which collects below forms the dripping. The loss in the meat is principally water.

The chemical effects of boiling are explained under the article devoted to that subject.

Meat generally loses from 30% to 40%, and sometimes as much as 60% in weight, by cooking. The better the quality the less the loss. Badly fed meat will lose twice as much weight as well-fed, and though often cheaper it is false economy to purchase it.

The amount of bone varies, being seldom less than 8%. It amounts in the neck and brisket to about 10%, and from ¼ to sometimes ⅓ the total weight in shins and legs of beef.

The most economical parts are the round and thick flank, next to these the brisket and sticking-piece, and lastly, the leg.

In choosing mutton and pork, selection should be made of the leg, after this of the shoulder (*Letheby*).

"Oxen," says M. Bizet, "yield of *best quality* beef 57% of meat and 43% waste. The waste includes the internal viscera, &c. *Second quality* of beef, 54% meat and 46% waste; *third quality* beef, 51% meat and 49% waste. In milking-cows, 46% meat and 54% waste. Calves yield 60% meat and 40% loss; and sheep yield 50% meat and 50% loss." Dr Parkes differs from Bizet as to the latter's value of the meat of the calf. He says the flesh of young animals loses from 40% to 50% in cooking.

It seems to be agreed, however, that animals when slaughtered should be neither too young nor too old. The flesh of young animals, although more tender, is less digestible than that of older ones; it is also poorer in salts, fat, and an albuminous substance called *syntonin*.

Consumption of Meat. Dr Letheby, writing in 1868, says that in London "the indoor operatives eat it to the extent of 14·8 oz. per adult weekly; 70% of English farm labourers consume it, and to the extent of 16 oz. per man weekly; 60% of the Scotch, 30 of the Welsh, and 20 of the Irish also eat it. The Scotch probably have a larger allowance than the English, considering that braxy mutton is the perquisite of the Scotch labourer; but the Welsh have only an average amount of 2½ oz. per adult weekly; and the Irish allowance is still less. It is difficult to obtain accurate returns of the quantity of meat consumed in London; but if the computation of Dr Wynter is correct, it is not less than 30½ oz. per head weekly, or about 4½ oz. per day for every man, woman, and child. In Paris, according to M. Armand Husson, who has carefully collected the *octroi* returns, "it is rather more than 49 oz. per head weekly, or just 7 oz. a day." Boudin states that throughout France the consumption is about 50 grms. daily, or under 1½ oz.

Dr Letheby, in his work 'On Food,' gives the following as the characteristics of good meat:

"1st. It is neither of a pale pink colour nor of a deep purple tint, for the former is a sign of disease, and the latter indicates that the animal has not been slaughtered, but has died with the blood in it, or has suffered from acute fever.

"2nd. It has a marked appearance from the ramifications of little veins of fat among the muscles.

"3rd. It should be firm and elastic to the touch, and should scarcely moisten the fingers—bad meat being wet, and sodden and flabby, with the fat looking like jelly or wet parchment.

"4th. It should have little or no odour, and the odour should not be disagreeable, for diseased meat has a sickly cadaverous smell, and sometimes a smell of physic. This is very discoverable when the meat is chopped up and drenched with warm water.

"5th. It should not shrink or waste much in cooking.

"6th. It should not run to water, or become very wet on standing for a day or so, but should, on the contrary, dry upon the surface.

"7th. When dried at a temperature of 212° or thereabouts, it should not lose more than from 70% to 74% of its weight, whereas bad meat will often lose as much as 80%.

"Other properties of a more refined character will also serve for the recognition of bad meat, as that the juice of the flesh is alkaline or neutral to test-paper, instead of being distinctly acid; and the muscular fibre, when examined under the microscope, is found to be sodden and ill-defined."

Unsound Meat—Diseased Meat. Dr Letheby, in his 'Lectures on Food,' published in 1868, states that the seizure and condemnation, in London, of meat unfit for human food, during a period extending over 7 years amounted to 700 tons per annum, or to about 1-750th of the whole quantity consumed. These 700 tons he dissects into lbs. as follows:—"805,653 lbs. were diseased, 568,375 lbs. were putrid, and 193,782 lbs. were from animals that had not been slaughtered, but had died from accident or disease. It consisted of 6640 sheep and lambs, 1025 calves, 2896 pigs, 9104 quarters of beef, and 21,976 joints of meat."

He admits, however, that this amount, owing to the difficulties and inefficiency of the mode of supervision, bears a very insignificant proportion to the actual quantity which escaped detection, and which was, therefore, partaken of as food. Professor Gamgee says that one fifth of the meat eaten in the metropolis is diseased. In 1863 the bodies of an enormous number of animals suffering from *rinderpest*, as well as from *pleuro-pneumonia*, were consumed in London; and we know that thousands of sheep die every year, in the country, of *rot*; the inference from which latter fact is that, since the carcases are neither eaten there nor buried on the spot, they are sent up to, and thrown upon, the London markets. The worst specimens find their way to the poorer neighbourhoods, where, as might be expected, their low price ensures a ready sale for them. These sales, it is said, mostly take place at night.

The above statements, which, if we exclude Professor Gamgee's figures, do not solve the problem as to the quantity of unsound meat consumed in London, not unreasonably justify the assumption that it is very considerable; and this being admitted, we should be prepared to learn that it was a fertile source of disease of a more or less dangerous character.

The flesh of tuberculous animals is now regarded as unfit for human food.

There is, however, such extensive divergence in the various data bearing upon this point, that no satisfactory solution of it can be said to be afforded. Thus, Livingstone states that, when in South Africa, he found that neither Englishmen nor natives could partake of the flesh of animals affected with *pleuro-pneumonia* without its giving rise to malignant carbuncle, and sometimes, in the case of the natives, to death; and Dr Letheby attributes the increased number of carbuncles and phlegmons amongst our population to the importation from Holland of cattle suffering from the same disease. On the contrary, Dr Parkes says he was informed, on excellent authority, that the Caffres invariably consume the flesh of their cattle that die of the same epidemic, without the production of any ill-effects. Again, there are numerous well-attested cases in which the flesh of sheep which have died·from *brasy* (a disease that makes great ravages amongst the flocks in Scotland) is constantly eaten without injurious results by the Scotch shepherd. The malady causes death in the sheep from the blood coagulating in the vital organs, and the sheep that so dies becomes the property of the shepherd, who, after removing the offal, is careful to cut out the dark congealed blood before cooking it (*Letheby*). Sometimes he salts down the carcass. In cases, however, where thorough cooking or an observance of the above precautions has been neglected, very dangerous and disastrous consequences have ensued. During the late siege of Paris large quantities of the flesh of horses with glanders appear to have been eaten with no evil consequences; and Mr Blyth, in his 'Dictionary of Hygiène,' quotes a similar case from Tardieu, who states that 300 army horses affected with glanders (*morve*) were lead to St Germain, near Paris, and killed. For several days they served to feed the poor of the town without causing any injury to health.

A similar exemption from any evil effect following the consumption of diseased flesh is recorded by Professor Brucke, of Vienna.

Not many years since, the cattle of a locality in Bohemia, being attacked by *rinderpest*, were ordered by the Government to be slaughtered, after which they were buried. The poor people dug up the diseased carcases, cooked the meat, and ate it, with no injurious result.

Parent Duchâtelet cites a case where the flesh of 7 cows attacked with rabies was eaten without injury; and Letheby states that pigs with scarlet fever and spotted typhus have been used for food with equally harmless results. The flesh of sheep with smallpox had been found to produce vomiting and diarrhœa, sometimes accompanied with fever.

One obvious suggestion of the immunity from disease recorded in part of the cases above given is that the injurious properties of the flesh had been destroyed by the heat to which it had been

subjected in the process of cooking, combined with the antiseptic and protective power of the gastric juice. The subject, however, has not been sufficiently examined to warrant the conclusion that every kind of unsound meat may be rendered innocuous by culinary means, for it must be remembered that a temperature far higher than that to which the inner parts of a joint are exposed in cooking, and much more prolonged, is required to kill many organisms, and that the spores remain unaffected by temperatures which would render the meat uneatable.

But even were this so, the idea of partaking of meat which had once been unsound, from whatever cause, and, as in the instances above quoted, with the pustules of smallpox, the spots generated by typhus, and the rash of scarlet fever upon it, becomes unspeakably repulsive and revolting. But we must not be misled because of the difficulty of reconciling the contradictory statements above given, nor by the evidence some of them appear to afford as to the innocuous character of diseased meat, since it is just possible that closer and more prolonged observation of the facts may have led to different conclusions. Thus, for example, pork infested with that formidable entozoon, the *Trichina spiralis*, had been partaken of for years, under the impression that it was a perfectly healthy food, until Dr Zencker, of Dresden, discovered that the parasite was the cause of a frightful disease, which he called *Trichinosis*, and which had hitherto baffled all attempts to find out its origin. Dr Letheby, writing on this subject, says : " I have often had occasion to investigate cases of mysterious disease which had undoubtedly been caused by unsound meat. One of these, of more than ordinary interest, occurred in the month of November, 1860. The history of it is this:— A fore-quarter of cow-beef was purchased in Newgate Market by a sausage-maker, who lived in Kingsland, and who immediately converted it into sausage-meat. Sixty-six persons were known to have eaten of that meat, and sixty-four of them were attacked with sickness, diarrhœa, and great prostration of vital powers. One of them died; and at the request of the coroner I made a searching inquiry into the matter, and I ascertained that the meat was diseased, and that it, and it alone, had been the cause of all the mischief " (Letheby, ' Lectures on Food,' Longmans and Co.).

Here are two instances in which but for subsequent investigation the evil effects narrated would not have been debited to diseased meat, but to some other cause.

One of the principal, and by far the most prolific sources of food poisoning is the sausage, the eating of which, in Germany more particularly, has caused the death of a number of persons.

The sausages in which these poisonous qualities occasionally develop themselves are the large kinds made in Wurtemburg, in which district alone they have caused the deaths of more than 150 out of 400 persons during the last fifty years. The poisonous character of the sausage is said to develop itself generally in the spring, when it becomes musty, and also soft in the interior. It is then found to be acid to test-paper, and to have a very disagreeable and tainted flavour.

Should it be eaten when in this state, after from

about 12 to 24 hours, the patient is attacked with severe intestinal irritation in the form of pains in the stomach and bowels, by vomitings, and diarrhœa.

To these symptoms succeed great depression, coldness in the limbs, weak and irregular pulse, and frequent fainting fits. Should the sufferer be attacked with convulsions and difficult respiration, the seizure generally ends in death. The nature of the poisonous substance that gives rise to these effects in the sausage has not yet been determined. Liebig believed them to be due to the presence in the meat of a particular animal ferment, which he conceived acted on the blood by catalysis, and thus rendered it diseased. Others have surmised that a poisonous organic alkaloid may have been produced in the decaying meat; and others again that the effects may have been caused by some deleterious substance of a fatty nature. M. Van den Corput was of opinion that the mischief was due to the presence in the meat of a poisonous fungus, which he calls a *Sarcina botulina*. This last theory receives support from the fact that a peculiar mouldiness is always to be observed in these dangerous sausages, and that this is coincident with the development of their poisonous qualities.

Several effects have been produced by other kinds of animal food—as veal, bacon, ham, salt beef, salt fish, cheese, &c., and the food has usually been in a decayed and mouldy condition. It would be tedious if I were to detail, or even to enumerate the cases recorded by medico-legal writers; but I may perhaps refer to a few of them. In 1839 there was a popular fête at Zurich, and about 600 persons partook of a repast of cold roast veal and ham. In a few hours most of them were suffering from pain in the stomach, with vomiting and diarrhœa; and before a week had elapsed nearly all of them were seriously ill in bed. They complained of shiverings, giddiness, headache, and burning fever. In a few cases there was delirium, and when they terminated fatally there was extreme prostration of the vital powers. Careful inquiry was instituted into the matter, and the only discoverable cause of the mischief was incipient putrefaction and slight mouldiness of the meat. A case is recorded by Dr Geisler of eight persons who became ill from eating bacon which was mouldy; and another by M. Ollivier of the death of four persons out of eight, all of whom had partaken of partially decomposed mutton.

If some of the foregoing statements fail to demonstrate that the act of partaking of diseased meat is a necessary source of danger to health, there can be no such doubt as to the pernicious and perilous consequences which ensue when meat is consumed containing in its tissues the ova and larvæ of certain parasitic creatures. If the fleshy part of a piece of measly pork be carefully examined it will be found to be more or less dotted about with a number of little bladder-like spots, in size about as large as a hemp-seed. See CYSTICERCI.

If now we carefully rupture one of these little bodies or cysts there will be found in it a minute worm, which under the microscope will be seen to have a head, from which proceed a number of little hooks that perform a very disagreeable office should the parasite be taken into the human stomach by any one making a meal off measly and undercooked pork; for then, being liberated from its sac, or nidus, by the action of the gastric juice of the stomach on this latter, the creature passes into the intestines. To these it attaches itself by means of the hooklets on its head, and instantly becomes a tapeworm, which grows by a succession of jointed segments it is able to develop, and each one of which is capable of becoming a separate and prolific tapeworm filled with countless eggs.

These eggs reach the land through the agency of manure (for they are found in the intestines of the horse), and from this source they get into the stomachs of pigs and oxen, where they hatch not into tapeworm or *tænia*, but, travelling through the animal's stomach, burrow into its muscular tissue. Here they establish and envelop themselves in the little cyst or small bladder-like substance, whose presence, as explained, constitutes the condition called 'measly' pork, and here they remain dormant until such time as, taken into the stomach, they may again become tapeworms, to be again expelled and to perpetuate by their ova the round of metamorphosis. From the circumstance of their being met with enclosed in little sacs or cysts, these parasites have been termed *Cysticerci*. The variety of them we have just been considering as occurring in pork is called the *Cysticercus cellulosæ*, whilst the tapeworm to which it gives rise is known as the *Tænia solium*.

Another variety of *Cysticercus* is met with in the flesh of the ox, the cow, and the calf. In the human body this also develops into a tapeworm called the *Tænia mediocanellata*. Tapeworm is a very common disease in Russia and Abyssinia, and its prevalence is no doubt due to the habit of giving the children in those countries raw meat to suck, under the impression that the child is strengthened in consequence. From experiments made by Dr Lewis it was found that a temperature of 150° F., maintained for five minutes, was sufficient to destroy these cysticerci.

Another and more formidable entozoon, communicable by unsound meat, is the *Echinococcus hominis* (see ECHINOCOCCUS HOMINIS), which represents one of the metamorphoses of the *Tænia echinococcus*, the tapeworm of the dog. In Iceland, where a sixth of the population are said to suffer from the ravages of the *Echinococcus hominis*, it is the custom to feed the dogs on the flesh of slaughtered animals affected with this parasite, which in the body of the dog develops into a tapeworm. The innumerable eggs which the worm produces are, however, incapable of being hatched in the dog's intestines. They have to find another and more suitable habitat, and this is secured for them as follows:—Segments of the tapeworm, with their countless ova, being voided with dog's excrement, fall into the running water, and on to the fields and pastures, whence they gain their entrance into the stomachs of human beings, oxen, and sheep. Here the eggs become hatched, not into tapeworms, but into *Echinococci hominis*, or respective tapeworms. Burrowing through the membranes of the stomach, the echinococcus establishes itself most commonly in the liver, but not unfrequently in the spleen, heart, lungs, and even the bones of man. In the animal economy they enclose themselves in little sacs or cysts, and give rise to the most alarming and painful diseases, which hitherto have proved incurable. They attack the brain in sheep, and are the cause of the disease known as 'staggers.' Sheep are also infested by another parasite, known as the *Distoma hepatica*, the ravages of which give rise in the sheep to that devastating disease, 'the rot.' The creature is also known by the name of the 'liver-fluke,' since it principally attacks this important organ in the animal. The liver-fluke is of constant occurrence in the livers of diseased sheep, and unless destroyed by thorough cooking will of course pass into the human economy. The embryo fluke gains admission to the sheep's body through the instrumentality of small snails, to the shells of which it attaches itself. In wet weather the snails crawl over the grass of the meadow which forms the pastures of the sheep, and are swallowed by it. Once in the sheep's stomach the embryo becomes a fluke, and commences its depredations on the animal's liver. After this the reason why the rot attacks sheep after a continuance of wet weather will be evident.

The most terrible of all the meat parasites is a minute worm about 1-30th of an inch long, found in the flesh of pork. This creature, which is named the *Trichina spiralis* (from the form it assumes when coiled up in the little cyst or capsule which encloses it), when it gets conveyed into the human stomach with improperly cooked or underdone pork, soon becomes liberated from its confinement owing to the destruction of its envelope by the gastric juice. Once in the stomach the parasite grows rapidly, giving birth to innumerable young *trichina*, which, by first boring through the membranes of the alimentary canal, pierce their way through the different parts of the body into the muscular tissue, where they become encysted, and where they remain until conditions favourable to their liberation again occur.

Until such time, however, as they have become enclosed in the cyst, their movements give rise to indescribable torture, and to a disease known as *trichinosis*, of which it has been estimated more than 50 per cent. of those attacked by it die. The symptoms of trichinosis commence with loss of appetite, vomiting, and diarrhœa, succeeded after a few days by great fever—resembling, according to Dr Aitken, that of typhoid or typhus. As might be expected, the pains in the limbs are extreme. Boils and dropsical swellings are not unusual concomitants of the malady.

Hitherto this frightful disease has been mostly confined to Germany, where there have been several outbreaks of it since its discovery in 1860 by Dr Zencker. Feidler says that only free *trichina* are killed by a temperature of 155° F.; and that when they are in their cysts a greater heat may be necessary. From what has been said the importance of efficient cooking must become manifest. There must always be risk in underdone pork, whether boiled or roasted. In the

pig, the trichina, if present, may always be found in the muscles of the eye. In Germany the makers of pork sausages are now said to have these muscles subjected to a microscopic examination previous to using the meat, which, of course, is rejected if the examination has been unfavourable.

The trichinæ, if present in the flesh of pork, may be seen as small round specks by the naked eye, the surrounding flesh itself being rather darker than usual owing to the inflammation set up in it. All doubt, however, on this point may be removed by having recourse to the microscope. Dr Parkes says a power of 50 to 100 diameters is sufficient, and that "the best plan is to take a thin slice of flesh, put it into liquor potassæ (1 part to 8 parts of water), and let it stand for a few minutes till the muscle becomes clear; it must not be left too long, otherwise the trichinæ will be destroyed. The white specks come out clearly and the worm will be seen coiled up. If the capsule is too dense to allow the worm to be seen, a drop or two of weak hydrochloric acid should be added. If the meat be very fat a little ether or benzine may be put on it in the first place."

Legislation relative to Meat Inspection and Seizure. The law recognising the importance of the supply of pure and wholesome meat gives considerable powers to the different sanitary officers who are appointed to inspect it. See FOOD, INSPECTION OF.

MEAT, AUSTRALIAN. See MEAT PRESERVING.

MEAT BISCUITS. *Prep.* 1. The flour is mixed up with a rich fluid extract of meat, and the dough is cut into pieces and baked in the usual manner.

2. Wheaten flour (or preferably the whole meal), 3 parts; fresh lean beef or other flesh (minced and pulped), 2 parts; thoroughly incorporate the two by hand-kneading or machinery, and bake the pieces in a moderately heated oven. Both the above are very nutritious, the last more especially so. 1 oz. makes a pint of good soup.

MEAT, COLD, to Stew. Let the cold meat be cut into slices about ¼ an inch thick. Take two large-sized onions, and fry them in a wine-glass of vinegar; when done, pour them on to the meat; then place the whole in a stewpan, and pour over sufficient water to cover it. After stewing about ¼ an hour add sufficient flour and butter to thicken the gravy, and also pepper, salt, and ketchup, to flavour; then let it simmer gently for another ¼ an hour. Serve up with a little boiled rice around it.

MEAT EXTRACTS. Some preparations of this nature have been already noticed under the heads ESSENCE and EXTRACT; the following are additional and highly valuable formulæ:

Prep. 1. (*Dr Breslau.*) Young ox-flesh (free from fat) is minced small, and well beaten in a marble mortar, at first alone, and afterwards with a little cold or lukewarm water; the whole is then submitted to the action of a press, and the solid residuum is treated in the same manner, with a little more cold water; the juice (reddish in colour) is now heated to coagulate the albumen, strained, and finally evaporated in a water-bath to the consistence of an extract. As ordinary flesh contains only 1% of kreatine, while that of the heart, according to Dr Gregory, contains from 1·37% to 1·41%, this is the part employed by Dr Breslau. The product possesses an agreeable odour and taste, and is easily soluble in water.

2. (*Falkland.*) Fresh lean beef (or other flesh), recently killed, is minced very fine, and digested, with agitation, in cold water, 1 pint, to which hydrochloric acid, 6 drops, and common salt, 1 dr., have been added; after about an hour the whole is thrown upon a fine hair sieve, and the liquid portion allowed to drain off without pressure, the first portions that pass through being returned until the fluid, at first turbid, becomes quite clear and transparent; when all the liquid has passed through, cold water, ½ pint, is gently poured on, in small portions at a time, and allowed to drain through into that previously collected. The product is about ¾ pint of cold extract of flesh, having a red colour, and a pleasant, soup-like taste. It is administered cold to the invalid—a teacupful at a time, and must on no account be warmed, as the application of even a very slight heat causes its decomposition and the separation of a solid mass of coagulated albumen. This cold extract of flesh is not only much more nutritious than ordinary beef-tea, but also contains a certain quantity of the red colouring matter of blood, in which there is a much larger proportion of the iron requisite for the formation of blood-particles. The hydrochloric acid also greatly facilitates the process of digestion. This formula is a modification of the one recently recommended by Liebig for the preparation of a highly nutritive and restorative food for invalids.

3. (EXTRACTUM SANGUINIS BOVIS—*Dr Mauthner.*) Pass fresh blood (caught from the slaughtered animal) through a sieve, evaporate it to dryness in a water-bath, and when cold rub it to powder.—*Dose*, 10 to 20 gr., or more, per diem, in a little water.

Obs. The above preparations are intended to supersede the inefficient compounds—beef-tea, meat soups, &c.—during sickness and convalescence. MM. Breslau and Mauthner describe their extracts of flesh and blood as being peculiarly advantageous in scrofulous exhaustion, anæmia, diarrhœa, &c. The extract of Falkland or Liebig is represented as having been employed both in the hospitals and in private practice at Munich with the most extraordinary success. It is said to be capable of assimilation with the least possible expenditure of the vital force.

Meat, Fluid. This preparation consists of lean meat, in which the albumen has been changed so as to be non-coagulable by heat, and the fibrin and gelatin from their normal insoluble condition to one admitting of their being dissolved in water.

In this soluble condition, the first stage effected in stomach digestion, the several bodies are known as peptones or albuminose, and the proportion of their simple constituents remains the same as in ordinary fibrin, albumen, and gelatin.

The alteration is effected by finely mincing meat and digesting it with peptone, hydrochloric acid, and water, at a temperature of about 100° F., until dissolved.

The solution is then filtered, the bitter principle, formed during the digestion, removed by the addition of a little pancreatic emulsion, and the liquor, which has been neutralised by the addition of carbonate of soda, evaporated to a thick syrup or extractive consistence.

Fluid meat is the only preparation which entirely represents, and yields the amount of nourishment afforded by, lean meat; it differs altogether from beef-tea and extracts of meat, as all these contain only a small portion of the different constituents of meat. A patent has been granted to its inventor, Mr Darby.

Meat, Liebig's Extract of. *Syn.* EXTRACT OF FLESH; EXTRACTUM CARNIS, L. This preparation is an aqueous infusion, evaporated to the consistence of a thick paste, of those principles of meat which are soluble in water, "altered as they be by the application of heat" (Deane and Brady, 'Pharmaceutical Journal,' Oct., 1866).

It is chiefly composed of alkaline phosphates and chlorides, a nitrogenous crystalline base, known as kreatine, various extractive matters, which it has been surmised may have originated in the decomposition of certain nitrogenous bodies, and possibly of a small quantity of lactic acid, as it contains neither albumen nor fibrin—two of the most important and nutritious ingredients of flesh; it must not, therefore, be regarded as a concentrated form of meat. Liebig says that it requires 34 lbs. of meat to yield 1 lb. of this extract—a statement which, as Dr Pavy justly remarks, shows how completely the substance of the meat which constitutes its real nutritive portion must be excluded. This absence of direct nutrient power, now admitted by physiologists, whilst disqualifying the extract as a substitute for meat, does not, however, preclude its use in certain cases of indisposition requiring the administration of a stimulant or restorative, in which circumstances it has been found a useful and valuable remedy, and has been suggested as a partial substitute for brandy where there is considerable exhaustion or weakness, accompanied with cerebral depression and lowness of spirits. In this latter respect its action seems analogous to strong tea.

In the vast pastures of Australia and the pampas of South America are countless herds of oxen and sheep, whose numbers far exceed the food requirements of the comparatively sparse population of those districts. The fat, horns, hoofs, bones, skins, and wool of these cattle, which form the chief part of the wealth of those countries, are exported principally to Europe. Until within a few years, however, no means had been adopted for the utilisation of the superfluous flesh of the animals, beyond employing it as a manure. By manufacturing it, however, into 'extract of meat' this waste has been remedied, and immense works for its preparation are now erected both in South America and Australia. The process followed by the different makers, although varying in some particulars, is essentially the same, and consists in extracting by water, either hot, cold, or in the form of the steam, those portions of the meat which are soluble in that fluid, and subsequently evaporating the solution so obtained until it becomes of a proper consistence to be put into jars. The extract so obtained keeps well (if all the fat and gelatin are removed), and is most conveniently adapted for exportation. It is said that the extract as being obtained from cattle that have had English progenitors possesses a flavour superior to that which comes from South America, where the animals are of a different and inferior breed.

The following interesting description of the manufacture of 'Liebig's Extract of Meat' is taken from the Buenos Ayres 'Standard' of September, 1867. The establishment, of which it is a description, is at Fray Bentos, on the Uruguay, South America. "The new factory is a building which covers about 20,000 square feet, and is roofed in iron and glass. We first enter a large flagged hall, kept dark, cool, and extremely clean, where the meat is weighed, and passed through apertures to the meat-cutting machines. We next come to the beef-cutting hall, where are four powerful meat-cutters, especially designed by the company's general manager, M. Geibert. Each machine can cut the meat of 200 bullocks per hour. The meat being cut is passed to 'digerators,' made of wrought iron; each one holds about 12,000 lbs. of beef; there are nine of these digerators, and three more have to be put up. Here the meat is digerated by high-pressure steam of 75 lbs. per square inch; from this the liquid which contains the extract and the fat of the meat proceeds in tubes to a range of 'fat separators' of peculiar construction. Here the fat is separated in the hot state from the extract, as no time can be lost for cool operation, otherwise decomposition would set in in a very short time.

"We proceed downstairs to an immense hall, 60 feet high, where the fat separators are working; below them is a range of 5 cast-iron clarifiers, 1000 galls. each, worked by high-pressure steam through Hallet's tube system.

"Each clarifier is provided with a very ingenious steam-tap. In the monstrous clarifiers the albumen, fibrin, and phosphates are separated. From hence the liquid extract is raised by means of air-pumps, driven by two 30-horse power engines, up to 2 vessels about 20 feet above the clarifiers;

thence the liquid runs to the other large evaporators. Now we ascend the staircase reaching the hall, where 2 immense sets of 4-vacuum apparatus are at work, evaporating the extract at a very low temperature; here the liquid passes several filtering processes before being evaporated in vacuo. We now ascend some steps and enter the ready-making hall, separated by a wire-gauze wall, and all the windows, doors, &c., guarded by the same to exclude flies and dust. The ventilation is maintained by patent fans, and the place is extremely clean. Here are placed 5 ready-making pans constructed of steel plates, with a system of steel discs revolving in the liquid extract.

"These 5 pans, by medium of discs, 100 in each pan, effect in 1 minute more than 2,000,000 square feet evaporating surface.

"Here concludes the manufacturing process. The extract is now withdrawn in large cans, and deposited for the following day.

"Ascending a few steps we enter the decrystallising and packing hall, where two large cast-iron tanks are placed, provided with hot water-baths under their bottoms; in these tanks the extract is thrown in quantities of 10,000 lbs. at once, and by decrystallising is made a homogeneous mass, and of uniform quality. Now samples are taken and analysed by the chemist of the establishment, Dr Seekamp, under whose charge the chemical and technical operations are performed.

"It may be mentioned that the company's butcher killed at the rate of 80 oxen per hour; separating by a small double-edged knife the vertebræ, the animal drops down instantaneously on a waggon, and is conducted to a place where 150 men are occupied dressing the meat for the factory, cutting each ox into 6 pieces; 400 are being worked per day."

Mr Tooth, at a meeting before the "Food Committee" held at the Society of Arts in January, 1868, said that he did not claim any difference in the composition of his article (which was made in Australia) as compared with that made by the South American company.

In the annexed table the composition of some of the extracts of meat of commerce is given:

	Liebig's Company.	Tooth, Sydney.	French Company, South America.	Whitehead.	Twentyman.
Water	18·56	16·00	17·06	16·50	24·49	20·81
Extractive, soluble in alcohol	45·43	53·00	51·28	28·00	22·08	13·37
Extractive, insoluble. .	13·93	13·00	10·57	46·00	44·47	59·10
Mineral matter . .	22·08	18·00	21·09	9·50	8·96	6·72
	100·00	100·00	100·00	100·00	100·00	100·00

The following are the characteristics of extract of meat of good quality. It should always have an acid reaction, its colour should be a pale yellowish brown, and it should have an agreeable meat-like odour and taste. It should be entirely soluble in cold water, and should be free from albumen, fat, and gelatin.

Meat Pie. Stew 2 lbs. of beef steak with 1 small onion, the gravy from which is to be thickened with flour, and flavoured with pepper and salt. Put it into a baking dish, and cover with a lard crust. It should be baked for 1 hour. The addition of 2 kidneys will greatly improve the pie.

Meat (Australian) Pie. Take 2 lbs. of Australian meat, or 1½ lbs. of meat and ½ lb. of kidney.

Season to taste, pour in a little water, cover with a lard crust, and bake *not more* than ½ an hour.

MEAT PRESERVERS. Dr Poleuske gives the following as the result of analyses of various compounds found in trade as preservers of meat and other articles of food:

Sozolith (concentrated meat preserver). This yields sulphate of soda, 37·27; oxide of sodium, 21; sulphurous acid, 39·68; water, 2·05 = 100·00 parts.

Berlinite (concentrated). Chloride of sodium, 7·46; boracic acid, 9·80; borax, 45·75; water of crystallisation, 36·80 = 99·81 per cent.

Chinese Preserving Powder. Chloride of sodium, 25·00; boracic acid, 17·70; sulphate of soda, 38·84; sulphite of soda, 9·20; water, 9·40 = 100·14 per cent.

Brockman Preserving Salt. Chloride of sodium, 34·32; hydrate of potash, 14·04; sulphate of potash, 15·00; crystallised borax, 24·86; boracic acid, 12·00 = 100·22 per cent.

Orthmann's Australian Salt. Chloride of sodium, 5·5; borax, 54·0; water of crystallisation, 40·8 = 103·3 per cent.

Rüger's Barmenite. Chloride of sodium, 49·95; anhydrous boracic acid, 27·00, with water of crystallisation, 22·50 = 99·45 per cent.

Magdeburg Preserving Salt. Lime, 0·46; chloride of sodium, 20·42; dry boracic acid, 33·45; borax, 15·00; water of crystallisation, 30·00 = 99·33 per cent.

Heydrick's Preserving Salt. Nitrate of potash, 15·50; chloride of sodium, 73·40; boracic acid, 9·45; water, 1·23 = 99·58 per cent.

Dreifache's Preserving Salt. Boracic acid, 55·5; water of crystallisation, 44·1 = 99·6 per cent.

Real Australian Meat Preserver. Lime, 9·50; sulphurous acid, 36·32; sulphuric acid, 8·00; oxide of iron and alumina, 0·60; silica, 0·40; magnesia and alkalies, 1·30 = a liquid of sp. gr. 1·034 at 19° C.

Ortmann's Real Australian Meat Preserve. This is also a liquid of the sp. gr. 1·046 at 19° C. It yields lime, 11·10, and sulphurous acid, 61·76.

Dalcendahl's Real Australian Meat Preserve (liquid). Sp. gr. 1·079 at 19° C. In 1 lb. (453 grms.) are found 20·7 grms. oxide of calcium and 100 grms. sulphurous acid.

MEAT PRESERVING. "The Belgian *Musée de l'Industrie* notes the following methods of preserving meats as the most deserving of attention amongst those communicated to the French Academy of Sciences, and published in the *Comptes Rendus.*

1. M. Bandet's method, by which the meat is kept in water acidulated with carbolic acid in the proportion of 1 to 5 parts of acid per 1000 parts of water. A series of experiments proved that all kinds of meat could thus be kept fresh for lengthened periods, without acquiring an ill taste or odour.

"The meat may be placed in barrels or air-tight tin cases, filled with acidulated water of the strength above specified, and headed up; or the pieces may be packed in barrels or cases in alternate layers with charcoal, pounded small, and saturated with water containing ¹⁄₁₀₀₀ of carbolic acid. The charcoal serves as a vehicle for the antiseptic fluid, and as an absorbent of any gaseous matters given off by the meat. The latter should be wrapped in thin linen covers to prevent the charcoal working its way into the tissues.

"This method, it is suggested, might be employed in curing pork in place of 'salting,' or of the more lengthy and costly process of 'smoking;' and also for the preservation of poultry, game, butter, eggs, &c.

"2. In the case of South American meat M. Baudet proposes the use of large sacks of caoutchouc. The meat should be packed in them, with alternate layers of charcoal as above described, and each sack, when filled, should be hermetically closed by drawing another empty caoutchouc sack, cap-wise, over it. The caoutchouc, it is supposed, would fetch enough in the market—its low price notwithstanding—to cover expenses of packing and freight, and so permit the meat to be sold in Europe at a very small advance on cost price. If intended for use a second time, the empty bags should be steeped in boiling water for a few minutes, to remove any organic impurities adhering to them.

"3. M. Gorge's method, which is in use in La Plata, consists in washing and drying the meat, and afterwards steeping in successive waters containing hydrochloric acid and sulphite of soda, and then packing it in air-tight cases holding 1, 5, or 10 kilog. each. Meat thus treated requires to be soaked in warm water for about ½ an hour before use.

"4. M. Léon Soubeiran has recommended braying and drying, in the fashion adopted by the Chinese and Mongols, as described by M. Simon, French consul in China, in a communication made by him to the Société d'Acclimatation. The *pemmican* of our Arctic voyagers and the *charqui* of South America are familar examples of meat preserved by analogous processes. The late M. Payen, a distinguished member of the Academy, insisted upon the great perfection to which this system might be carried by the aid of hot-air stoves and suitable apparatus."

Besides the foregoing, numerous patents have from time to time been taken out, and processes proposed for the preservation of meat; so as to enable it to be sent from those distant countries, such as South America, Australia, Canada, &c., where it is greatly in excess of the wants of the population, to other lands, in which the supply is as much below the demand, and the meat at such a price as to preclude its being regularly used as an article of food by the body of the people.

As the putrefactive changes set up in dead flesh are dependent upon the combined influences of moisture, air, and a certain temperature, it follows that most of the various methods of meat preservation resolve themselves into so many different efforts to remove the meat from the operation of one of the conditions above specified as necessary for its decomposition.

The *charqui* or jerked beef of South America affords an example of meat preserved by means of being deprived of moisture. It occurs in thongs or strips which have been prepared by placing freshly killed meat between layers of salt and drying them in the sun. *Charqui*, although it retains its soundness for a great length of time, and is rendered eatable by soaking in water and, prolonged cooking, is difficult of digestion and wanting in flavour, and if any fat be associated with it, this is liable to become rancid.

Pemmican is meat which, after being dried and powdered, is mixed with sugar and certain spices, both of which assist to preserve the meat as well as to improve its flavour, and to remove the tendency to rancidity caused by any fat that may be accidentally present.

Another process for the preservation of meat by means of desiccation is that of MM. Blumenthal and Chollet, who, in 1854, obtained a patent for preparing tablets composed of dried meat and vegetables, which, after being several times dipped into rich soup, were dried in warm air after each immersion.

At a meeting of the Food Committee, held at the Society of Arts, in May, 1868, specimens of dried beef and mutton in powder, from Brisbane, were shown by Mr Orr, who said they had been dried on tinned plates by means of steam. Dr A. S. Taylor, F.R.S., who examined the sample, found it perfectly fresh and good. It had been prepared at least 6 months previously.

At a subsequent meeting the Committee reported that the soup prepared from this desiccated meat, with the addition of a small quantity of vegetables, was considered very successful, and the Committee were of opinion that meat so preserved was likely to prove a valuable and cheap addition to the food resources of the people.

The specimen from which the soup was made had been in the Society's possession, and formed part of the contents of a tin opened upwards of 2 years ago. The preservation was perfect.

We have only space briefly to describe some of the more prominent of the processes which have been devised for the preservation of meat by excluding atmospheric air.

Mr Tallerman, a large importer of Australian meat, stated in evidence before the Food Committee of the Society of Arts, in May, 1870, that in the preservation of the meat he sent over to this country he had recourse to a very old practice, which was that of packing the meat in tins, the meat being previously salted or cured. Instead of the meat being packed in brine, the casks with the meat are filled up with melted fat.

In Mr Warrington's patent, which dates from 1846, it is proposed that animal substances shall be preserved by enveloping them in a layer of glue, gelatin, or concentrated meat gravy, or otherwise by dipping them in warm solutions of such substances, or by wrapping them in waterproof cloth, or by covering them with caoutchouc, gutta percha, or varnish, or thin cream of plaster of Paris, which when set was saturated with melted suet, wax, or stearin.

The patent of Prof. Redwood, which resembles Mr Warrington's in seeking to exclude atmospheric air by surrounding the meat with an impervious substance, claimed the use of paraffin for this purpose, the paraffin being afterwards coated with a mixture of gelatin and treacle, of gelatin and glycerin. The paraffin is easily removed from the meat by plunging this latter into boiling water, which dissolves the outer coating of gelatin mixture, and at the same time melts the paraffin and liberates the enclosed joint.

Messrs Jones and Trevethick's patent consisted in exhausting of air the vessel containing the meat, then forcing into it a mixture of nitrogen and sulphurous acids, and subsequently soldering the apertures. Dr Letheby says meat, fish, and poultry preserved in this manner have been found good after seven or eight years; and specimens of them were exhibited in the London Exhibition of 1862.

The removal, however, of atmospheric air from the vessels containing the meat it is designed to preserve is now principally accomplished by means of steam. The germ of this idea originated with M. Pierre Antoine Angilbert more than half a century ago, but the modification of Angilbert's process, which in principle is that generally adopted by the importers of Australian and South American *cooked* meat, as well as by the English preparers of the article, originated with Messrs Goldner and Wertheimer, nearly forty years since, and, briefly, is as follows:—The freshly killed meat is placed in tins, with a certain quantity of cold water. The tins and their contents are then securely soldered down, with the exception of a small opening not larger than a pin-hole, which is left in the lid. The tins are next placed in a bath of chloride of calcium, the effect of which is to heat the water in them up to the boiling-point, and after a certain time to more or less cook the meat contained in them. When the meat is thought to be sufficiently cooked, and whilst the steam arising from the boiling water is escaping from the aperture, this last is carefully soldered down, the steam not only having driven out all the atmospheric air from the vessel, but in the act of escaping having prevented the ingress of any from without. To still further guard against the entrance of air, the tins are covered over with a thick coating of paint.

Previously to their being allowed to leave the preserving works they are tested by being placed for some time in an apartment in which the temperature is sufficiently high to set up putrefactive action in the meat if any hair has been left in the tins, the evidence of which would be the bulging out of the tins, owing to the liberation of certain gaseous products of decomposition. When no distension from inside takes place, the result is considered satisfactory, and the vessels are regarded as properly and hermetically sealed. In some cases the vessels, instead of being heated in a bath of chloride of calcium, are exposed to the action of steam. If the operation be successfully performed, the meat so prepared will keep perfectly good and sound for years.

Mr Richard Jones effects the removal of the air from the vessels containing the meat as follows:—The meat is put into the tins and entirely soldered up, with the exception of a small tube about the size of a quill, which is soldered on the top of the tin. This tube is placed in connection with a vacuum chamber, and the air exhausted from the tin by means of it. In cooking the meat he also employs a chloride of calcium bath.

Dr Letheby, in one of his Cantor Lectures on

Food, delivered in 1865, speaking on this part of the subject, and on the above method of meat preservation, says:—"To-night, through the kindness of Messrs Crosse and Blackwell, I am able to show you a specimen of preserved mutton which has been in the case forty-four years, and you will perceive that it is in excellent condition. It formed part of the stores supplied by Messrs Donkin and Gamble, in 1824, to His Majesty's exploring ship *Fury*, which was wrecked in Prince Regent's Inlet in 1825, when the cases were landed with the other stores, and left upon the beach.

"Eight years afterwards, in August, 1833, they were found by Sir John Ross in the same condition as they were left; and he wrote to Mr Gamble at the end of that year, saying that 'the provisions were still in a perfect state of preservation, although annually exposed to a temperature of 92°(?) below and 80° above zero.' Some of the cases were left untouched by Sir John Ross; and after a further interval of sixteen years the place was visited by a party from Her Majesty's ship *Investigator*, when, according to a letter from the captain, Sir James Ross, 'the provisions were in excellent condition, after having lain upon the beach, exposed to the action of the sun, and all kinds of weather, for a period of nearly a quarter of a century.' Messrs Crosse and Blackwell have placed the original letters in my hands for perusal, and they show beyond all doubt that meat preserved in this manner will keep good for nearly half a century—in fact, the case of boiled mutton now before you has been preserved for forty-four years."

The generality of the samples of preserved meat from Australia are excellent in quality and flavour (the Food Committee of the Society of Arts, who have carefully and impartially examined numerous samples of Australian and South American preserved meat, say, "It is perfectly sweet and fresh, but somewhat insipid from overcooking, and it seems likely the flavour could be improved if the duration of exposure to heat could be shortened without endangering the preservation"), except that in most cases the meat has been overcooked, which has arisen from the too prolonged contact of the meat with the steam, which it is judged necessary shall be generated in such quantities as to ensure the certainty of the exclusion of the air. Another inconvenience attending the process, viz. the liability of the sides of the tin to collapse, owing to the vacuum formed in its interior, has been remedied by the introduction into the vessel of some inert gas, such as carbonic acid, or nitrogen.

Preserved meat at the present time forms a very considerable article of export both from Australia and South America. In the former country there are several establishments of a colossal character, where the work of tinning the meat is carried on, in many of which establishments hundreds of cattle are slaughtered daily. The largest establishments of the kind are at Sydney and Melbourne, whence extensive shipments are being constantly made. The following figures are taken from the Board of Trade returns:

Value of Meat preserved otherwise than by Salting.

	Imports from Australia.	Total Imports.
1871	£481,098	£610,228
1872	657,945	816,463
1873	557,552	733,331
1874	509,698	757,001
1875	249,611	592,196

Since 1876 tinned meat has been imported from North America.

Several methods have been proposed for the preservation of meat by subjecting it to such conditions that the surrounding temperature should be sufficiently low to arrest putrefaction. In Mr Harrison's process the reduction of temperature was effected by the application of melting ice and salt, made to run down the outside of the iron chambers containing the meat. It is affirmed that although the joints submitted to this treatment were solidly frozen, no loss of either flavour or immediate decomposition of the meat took place. Mr Harrison's experiment was perfectly successful in Australia, but broke down during the voyage of a large cargo of meat shipped from Australia in 1873, owing to a defect in the construction of the ice-chamber of the vessel and the failure of the supply of ice.

Of other forms of refrigeration applied for this purpose we may mention the process of M. Tellier, by which he proposes to place (on shipboard or elsewhere) joints of meat in a chamber through which a current of air charged with ether or other volatile substance may be passed, with a view to reduce the temperature to 30° F. Also that of M. Poggiale, from whose report to the Paris Academy of Medicine it appears that in chambers contrived on principles similar to M. Tellier's all kinds of butcher's meat and poultry have been hung for 10 weeks, at the end of which time they were found perfectly fresh and wholesome. The agent used in the latter case for the production of cold was methylic ether.

The process, however, of refrigeration which has proved not only the most, but in every respect successful, was first satisfactorily carried out in 1876, since which time large cargoes of dead meat have been constantly sent to our metropolitan markets, as well as to Glasgow, from New York. The following extract from the 'Dundee Advertiser' gives some interesting details of this process:

"As to dead meat, the first sale was held on the 5th June, when 100 carcases of beef and 72 of mutton were disposed of. Since then there has never been a smaller supply, and on the average about 150 carcases have been sold weekly. Last week 210 carcases were sold, and on Wednesday evening there were no fewer than 33 lorries, each laden with 3 tons of butcher's meat. The freight paid for carriage to Glasgow, Liverpool, and London, last week amounted to £1900. Altogether, since the importation began, a million and a quarter pounds of dead meat have been sold in Glasgow. The result of this importation has been a reduction in retail price of 1d. per lb., instead of an increase in price, which must have taken place without the increased supply.

"The oxen are collected chiefly in the States of Illinois and Kentucky. They are there reared in enormous numbers on the prairies. Before they reach New York they are driven over railway for fully a thousand miles. Those animals the carcases of which are to be sent to this country are killed the day before the departure of the steamer. As soon as the carcases are dressed they are put into a cooling room capable of containing 500, and subject to a constant current of cold air, supplied by means of a 25 h.-p. engine. This sets the beef and extracts the animal heat. Each carcass is next cut into quarters, and these are sewn up in canvas, and during the night transferred on board the vessel. Six of the Anchor Line mail steamers have been fitted up with refrigeration compartments, constructed on a patented principle specially for the conveyance of meat.

"After the quarters have been hung up in the room the door is hermetically closed. Adjoining the compartment is a chamber filled with ice. Air-tubes are connected with the beef room, and through them the animal heat ascends, and by means of a powerful engine it is blown across the ice, and returned to the beef room in a cold state. A temperature of about 38° is thus maintained in the beef room. If it were to get so low as 32°—freezing-point—the meat would be seriously injured [Mr Harrison's experiments make this statement doubtful]. The heat is, therefore, regulated by a thermometer, and when the temperature gets too low the speed of the engine is slackened, the normal degree of cold being thus maintained almost without variation during the voyage. Cattle killed on Thursday in New York are sold that day fortnight in Glasgow."

The first patent for the preservation of food by means of ice was granted to Mr John Ling in 1845.

Lastly, mention must not be omitted of another method for the preservation of meat, which consists in the application to it of certain antiseptic substances, the action of which is preventing putrefaction is due to their power of destroying minute parasitic organisms of low animal or vegetable life, that would otherwise attack and set up decomposition in the meat. Our ordinary salted meats owe their immunity from decay, as is well known, to the presence in their tissues of common salt. Meat preserved, however, by this means is tough, indigestible, and wanting in many of its most important soluble constituents, which, dissolving in the salt, run off from the meat and are lost.

Amongst other agents which have been found serviceable as antiseptics, and for which from time to time numerous patents have been taken out, are nitrate of potash, acetate and hydrochlorate of ammonia, the sulphates of soda and potash, and bisulphite of lime. The writer remembers partaking, some years since, of some Canadian turkey, which had been preserved by means of this latter substance, the turkey having been killed some 2 months before being eaten. It was perfectly sound and of excellent flavour. In this instance the bird had been sent from Canada, with several others, packed in waterproof casks, filled up with a weak solution of bisulphite of lime.

In some cases the saline solution is merely brushed over the outside of the meat, whilst in others it is injected into the substance of the flesh.

Thiebierge's process consists in dipping the joints for 5 minutes into dilute sulphuric acid, of the strength of about 10 of water to 1 of acid. The meat after being taken out is carefully wiped and dried, and is then hung up for keeping. Sulphurous acid also forms the subject of several patents for the preservation of meat. In the process of Laury, for which a patent was taken out in 1854, the gas was introduced into the vessels containing the food. In that of Belford, for which a provisional specification was granted the same year, the meat was soaked for 24 hours in a solution of sulphurous and hydrochloric acids (the latter being in the proportion of a hundredth of the volume of the former). The addition of the hydrochloric acid was made with the intention of decomposing any alkaline sulphites that might be formed by the combination of the alkaline salts of the meat with the sulphurous acid.

Dr Dewar's process, which is very similar to the foregoing, proposes, instead of exposing the meat to sulphurous acid fumigation, to immerse it in a solution of the acid of the same strength as that of the British Pharmacopœia. On being taken out of the liquid, the meat or other article is, as speedily as possible, dried at a temperature not exceeding 140° F., so that the albumen may be preserved simply in a desiccated, and not in a coagulated condition.

In the patent of Demait, which dates from 1855, the meat was directed to be hung up in a properly constructed chamber, and exposed for some time to the action of the gas. More recently, Professor Gamgee has taken out a patent which is a modification of Demait's, and which consists in hanging up the carcass of the animal, previously killed when under the influence of carbonic acid, in a chamber filled with this latter gas, to which a little sulphurous acid has been added, the chamber having been first exhausted of air. The carcass is allowed to remain in the chamber from 24 to 48 hours, after which it is hung in dry air. It is stated that meat subjected to the above treatment has been found perfectly sound and eatable after an interval of 5 months.

M. Lanjorrois proposes to preserve animal substances from decay by the addition to them of 1% of magenta. He states the process had been applied to slices of beef, which, after being kept for several months, yielded, after being washed and boiled, very good soup. Commenting on this suggestion for the preservation of meat, the 'Chemical News' very sensibly and properly remarks, "It is to be hoped the magenta employed will be free from arsenic."

The patent of M. de la Peyrouse (which dates from 1873) also consists in excluding the air by enveloping meat in fat. In this process, however, the fat is mixed, when in a melting condition, with a certain quantity of the carbonates of sodium, potassium, and ammonium, as well as with some chlorides of magnesium and aluminium, with the object of preventing the fat becoming rancid and decomposing, and thus imparting a disagreeable flavour to the meat.

In M. George's process the meat is partially dried, and then steeped in successive waters containing hydrochloric acid and sulphate of soda.

MECON'IC ACID. $H_2C_7HO_7.8H_2O.$ *Syn.* ACIDUM MECONICUM, L. A peculiar acid, first obtained by Sertuerner from opium, in 1804.

Prep. Meconate of lime is suspended in warm water, and treated with hydrochloric acid. Impure meconic acid crystallises on cooling, and may be purified by repeated treatment in the same way with hydrochloric acid. Its purity is ascertained by its leaving no residue when heated in a platinum or glass capsule.

Prop. Meconic acid forms beautiful pearly scales; possesses a sour astringent taste; is decomposed by boiling water; it is soluble in alcohol, and sparingly in cold water. With the acids it forms salts called 'meconates,' most of which are crystallisable. Meconate of lime is obtained by heating a solution of chloride of calcium with an infusion of opium made with cold water, and neutralised by powdered marble, and collecting the precipitate. Meconate of potassium is prepared by direct solution of the base in the impure acid obtained from meconate of lime till the liquor turns green, heat being applied, when the salt crystallises out as the liquid cools; it may be purified by pressure and recrystallisation.

Tests. Meconic acid is characterised by—1. Turning ferric salts red, and the red colour not being destroyed by the action of corrosive sublimate. 2. Precipitating a weak solution of ammonio-sulphate of copper green. 3. With acetate of lead, nitrate of silver, and chloride of barium it gives white precipitates, which are soluble in nitric acid. 4. It is not reddened by chloride of gold.

MEC'ONIN. $C_{10}H_{10}O_4.$ A white, crystalline, odourless, neutral substance, discovered by Couerbe in opium.

MECO'NIUM. See OPIUM.

MEDICINES. However skilful the medical practitioner may be, and however judicious his treatment, both are interfered with, and their value more or less neutralised, if the remedies he orders are not administered precisely according to his instructions. It is the duty of the attendant on the sick to follow implicitly the directions of the physician, as well in exactly complying with his orders as in doing nothing that she has not been ordered to do. At the same time there are exceptions to this rule, in which a suspension of the remedy, or a deviation from the order of the physician, is not only allowable, but is absolutely required. Thus, from idiosyncrasy or some other cause, the remedy in the doses ordered may have no effect, or may produce one widely different from that intended or expected. In such cases it is evident that a strict adherence to the direction of the physician would be productive of evil; but he should be immediately apprised of the circumstance. The common practice of neglecting to administer the doses of medicine at the prescribed time, or after prescribed intervals, and then, to compensate for the omission, giving the medicine more frequently or in larger doses, cannot be too severely censured, as destructive to the welfare of the patient and injurious to the credit of the physician.

For the purpose of disguising the taste of medicines, or lessening their nauseating properties, Dr Pollio has recommended a means founded on the physiological fact that a strong impression on the nerves (whether of vision, hearing, or taste) renders that which follows less perceptible than under the usual circumstances. Instead, therefore, of applying to the mouth agreeable substances after swallowing nauseous medicines, we should prepare it beforehand, in order that the taste of the medicine may not be perceived. Aromatic substances, as orange or lemon peel, &c., chewed just before taking medicine, effectually prevent castor oil, &c., being tasted. In preparing the mouth for bitters, liquorice is the only sweet that should be used, the others creating a peculiarly disagreeable compound taste. We have noticed already the effect of oil of orange peel in correcting the nauseating qualities of copaiba. See DOSE and PRESCRIBING.

MEDICINES FOR PASSENGER SHIPS. The annexed scale of medicines, medical stores, and instruments for ships clearing under the Passengers Acts, other than steamships engaged in the North Atlantic trade, has been issued and caused to be published by the Board of Trade, and is intended to supersede the scales hitherto in force.

The quantities mentioned in the scale are for every 100 passengers, when the length of the passage, computed according to the Passengers Act, is 100 days and upwards. Half the quantity of medicines indicated, but the same kind and quantity of medical stores, should be taken when the passage is less than 100 days.

N.B.—There is a separate scale for North Atlantic steam passenger ships.

The medicines are to be prepared according to the British Pharmacopœia, to be plainly labelled in English, and the average doses for an adult stated, according to the British Pharmacopœia.

All bottles are to be stoppered, and all medicines indicated thus (*) are to be marked with a red poison label. All fluid quantities are to be measured by *fluid* lbs., oz., or dr.

	lbs.	oz.	dr.
Acid, Acetic	0	6	0
* „ Carbolic	0	1	0
„ „ (a powder containing not less than 20% of pure carbolic or cresylic acid)	112	0	0
Acid, Citric	0	8	0
„ Gallic	0	1	0
„ Hydrocyanic Dil.	0	0	4
„ Nitric	0	1	0
„ Sulph. Dil.	0	6	0
Æther	0	1	0
Alumen	0	1	0
Ammon. Carb.	0	6	0
Amylum	1	0	0
Argent. Nit. (Stick)	0	0	2
Calx Chlorate	7	0	0
Camphor	0	6	0
Charta Epispastica, 4 sq. ft., in case			
*Chlor. of Zinc (Burnett's sol. of)	16	0	0
*Chloroform	0	8	0
Copaiba	0	8	0
Creosote	0	0	2
Cupri Sulph.	0	1	0

	lbs.	oz.	dr.
Emp. Cantharidis	0	1	0
Ferri et Quiniæ Cit.	0	1	0
„ Sulph.	0	0	4
Glycerin	0	5	0
„ Acid, Tannic	0	4	0
*Hydrat. Chloral	0	1	6
Hydrarg. cum Cretâ	0	0	4
„ Subchloridi	0	0	4
Lini Farina	6	0	0
Lin. Camph.	0	8	0
„ Opii	0	2	0
„ Saponis	1	0	0
*Liq. Atropiæ	0	0	1
„ Calcis	1	0	0
* „ Morphiæ Acetatis	0	1	0
* „ Plumbi Subacetatis	0	2	0
„ Potassæ	0	2	0
* „ „ Permanganatis (B.P. or Condy's Crimson Fluid)	3	0	0
Magnes. Sulph.	4	0	0
Mist. Sennæ Co. (omit Extract of Liquorice and substitute Aromatic Spirit of Ammonia, 1 oz. to 1 pint of the mixture)	3	0	0
Ol. Croton.	0	0	1
„ Lini	0	8	0
„ Menthæ Pip.	0	0	2
„ Morrhuæ	3	0	0
„ Olivæ	1	0	0
„ Ricini	2	0	0
„ Terebinthinæ	1	0	0
*Opium	1	0	0
*Plumbi Acetatis	0	1	0
Potassæ Bicarb. Pulv.	0	4	0
Potassii Iodid.	0	2	0
Pulv. Antimonialis	0	0	3
* „ Astringens (double the quantity indicated to be taken to all tropical ports. Pulv. Catechu Co., Pulv. Cretæ Arom: cum Opio—equal parts)	1	0	0
Pulv. Cretæ Arom. cum Opio	0	2	0
„ Ipecac.	0	2	0
„ „ Co.	0	2	0
„ Jalapæ Co.	0	3	0
„ Potassæ Nitratis	0	4	0
„ Rhei Co.	0	4	0
„ Scammon. Co.	0	0	6
Quiniæ Sulph. (double the quantity indicated to be taken to all tropical ports)	0	1	0
Soda Bicarb.	1	0	0
Sp. Æther. Nitrosi	0	8	0
„ Ammon. Arom.	0	8	0
„ Rectif.	0	8	0
„ Sulphur Sublimatum	3	0	0
Syr. Ferri Iodidi	0	4	0
*Sol. Morphiæ Acetat. (a neutral solution containing 4 grains in a drachm, and so marked. To be labelled—for hypodermic injection)	0	0	4
Tr. Arnicæ	0	6	0
„ Camphoræ Co.	0	8	0
„ Digitalis	0	0	6
„ Ergotæ	0	6	0
„ Ferri Perchloridi	0	4	0
* „ Opii	0	6	0
„ Scillæ	0	2	0
„ Valerian. Ammon.	0	8	0
Ung. Cetacei	1	0	0
„ Hydrargyri	0	2	0
„ „ Ox. Rub.	0	1	0
„ Sulph.	1	0	0
„ Zinci	0	2	0
Vin. Colchici	0	1	0
„ Ipecac.	0	1	0
Zinci Sulphatis	0	1	0
Desiccated Soup	4	0	0

All pills to be made and marked 5 grains.

Pil. Aloës cum Myrrhâ	2 dos.
„ Col. c. Hyoscy.	4 „
„ Hydrarg.	3 „
„ Ipecac. cum Scillâ	5 „
„ Quiniæ	6 „
„ Sapon. Co.	6 „

Medical Stores.

Lint	10 oz.
Tow	1 lb.
Adhesive Plaster	3 yds.
Male Syringe	1
„ Glass	1
Female „	1
Phials (assorted)	2 dos.
Phial corks	6 „
Sponges	3
Bed-pan	1
Paper of Pins	1
Hernia Truss, 36-in., reversible	1
Paper of Pill-boxes	1
Gallipots	6
Leg and Arm Bandages	6
Calico	3 yds.
Flannel Bandages, 7 yds. long, 6 in. wide	2
Flannel	2 yds.
Triangular Bandages, base 48 in., sides, 33 in. each	2
†Minim Measures	2
†1 oz. „	1
†2 oz. „	1
†Set of Splints	1
†Waterproof Sheeting	4 yds.
†Oiled Silk	1 yd.
†Enema Syringe and Stomach-pump	1
†Box of Small Scales and Weights	1
†Wedgwood Mortar and Pestle	1
„ Funnel	1
†Spatulas	2
†Authorised Book of Directions for Medicine Chests	1
†British Pharmacopœia	1

† One set only of these articles required, irrespective of number of passengers.

N.B.—Only one set of instruments required, without regard to the number of surgeons, passengers, or the length of the voyage.

Instruments.

In Pocket Case.

1 Tenaculum.	1 Skull Forceps.
1 Artery Forceps.	1 Trephine.
1 Operating ditto.	1 Elevator.
1 Finger Knife.	1 Hey's Saw.
1 Curve Bistoury, Probe Point.	1 Trephine Brush.
1 Ditto, Spear Point.	2 Scalpels.
2 Probes.	1 Hernia Knife.
1 Silver Director.	2 Trocars and Cannulas.
1 Caustic Case.	1 Aneurism Needle.
1 Scissors.	1 Hernia Director.
1 Spatula.	1 Tourniquet.
12 Needles.	2 Silver Catheters (Nos. 4 and 8).
1 Skein Ligature Silk.	4 Elastic Gum Catheters (Nos. 3, 5, and 7).
3 Lancets.	1 Clinical Thermometer.
1 Amputating Saw.	1 Hypodermic Syringe.
2 Ditto Knives.	1 dozen charged Tubes for Vaccination.
1 Bone Forceps.	1 Set of Midwifery Instruments.
3 Tooth ditto.	

MEDLAR (*Pyrus germanica*, Linn.). Common in many parts of Europe, and occurring in English hedgerows.

MELIGETHES ÆNEUS, Fabricius (from μέλι, honey; and γηθέω, to delight in). The TURNIP-FLOWER BEETLE. This is yet another beetle among many beetles injurious to plants of the *Brassica* tribe. It affects rape, turnip, mustard, and cabbage plants in its beetle form by feeding upon the pollen of their flowers, and thereby hindering fructification; while its larvæ, which are born in the buds and cradled in the flowers, live upon these and upon the seed-vessels developed later on. Rape is the plant which is especially liked by this insect. Much loss is often occasioned to growers of rape-seed in Lincolnshire, Northampton, Essex, Kent, and other seed-growing districts, by its operations. A large seed grower in Kent had 10 acres of rape plants literally beset with beetles of this species. The flowers were covered with them as fast as they came out, and were deprived of much of their pollen by their vigorous efforts. In course of time larvæ, tiny maggots, appeared upon the seed-pods that were produced, and lived upon them, biting out their substance with their hard jaws.

At first it was thought that they were the ordinary flea beetles, *Phyllotreta nemorum;* then some one suggested that they were mustard beetles, *Phædon betulæ;* but upon examination it was found that they were rather too large for the former, and that they did not jump like them; and were too brilliant in colour for the latter. They were unmistakable specimens of *Meligethes æneus.*

There was a bad attack of this insect near Wisbech in June, 1886. Also from a seed grower near Peterborough information was sent of mischief caused to mustard plants when in full blossom, about the 10th of June in the same year, by "little brassy bugs," which did not jump, but slipped off the flowers when disturbed. These bugs, it was stated, did not seem to eat the flowers, but rather to feast upon the pollen, as their rostra were thrust into the anthers.

Cabbage plants for seed also suffer considerably from the assaults, first of the beetles, then of their larvæ. Turnip plants, kohl rabi plants, and thousand-headed kale plants are also injured, as seed growers in Kent, Hants, Essex, and Bedfordshire have testified.

In Scotland many ravages have been committed by these beetles upon various cruciferous plants intended for seed during the last 4 or 5 years.

It seems clear that the work of destruction performed by this *Meligethes æneus* is sometimes very considerable, and that it is often attributed to other insects, and most commonly to the turnip flea beetle, which has quite enough sins of its own to account for.

This flower-loving beetle is well known in Germany, where it is termed *Raps-glanzkafer*, the brilliant rape-seed beetle. Nördlinger described it in 1855 as most troublesome to rape-seed cultivators. He adds that plants from seed drilled in rows were not so much affected by the beetle as those where the seed was sown broadcast—*breitwürfiger Saat* ('Die Kleinen Feinde der Landwirthschaft,' von Dr Nördlinger).

Taschenberg gives a long account of the beetle, and Kaltenbach says that the harm done by it in Germany is great, that the beetles assemble in quantities on the flowers of rape plants and eat the pollen, but their larvæ do infinitely more harm, and ruin the hopes of the seed harvest ('Die Pflanzen Feinde,' von J. H. Kaltenbach).

In France the rape-seed crop, which is a most important crop in that country, is often seriously diminished by the *Meligethes du Colza,* as it is styled there. Calwer, in his 'Käferbuch,' states that it is known in Germany, France, England, and Sweden.

Life history. The *Meligethes æneus* is a species of the genus *Meligethes*, of the family *Nitidulidæ*, and the Nat. Ord. COLEOPTERA. Its colour may be described as metallic green or brassy green, and it has red legs. In shape it is somewhat square, and only about 1¼ lines in length, having well-developed wings, club-shaped antennæ, and peculiarly shaped claws. Towards the middle of May the beetles emerge from their winter retreats in the ground, or under grass and weeds, and at the roots of cruciferous weeds, and are found upon these weeds directly they are in flower, and upon cultivated cruciferous plants as soon as their blossoms show. The female lays eggs in the flower-buds. These are hatched in 4 or 5 days, and produce larvæ which upon the attainment of full size are quite 2 lines, or the sixth of an inch, long, yellowish white in colour, with dark-coloured heads furnished with strong jaws well adapted for biting vegetable tissues, and pointed at their extremities. They have 6 legs on the upper part of the body, and 1 at the last or caudal segment. After a time, varying from 10 to 14 days, the larvæ fall to the ground, and burying themselves in it, make cells of earth and assume pupal form, in which they continue about 20 days.

Prevention. Cruciferous weeds must be kept down upon seed-growing farms, as upon all other farms and gardens. These serve to maintain the *Meligethes* beetles until cultivated plants are ready. Upon land subject to this beetle it would be desirable to cultivate early, and apply dressings of soot, lime, guano, or paraffin-saturated materials.

Remedies. A remedy is adopted in France and Germany which perhaps might be used here in the early stages of the attack. Boys and girls are sent to shake the beetles off the plants into bags. This would have to be done 3 or 4 times. It would be almost impossible to carry this out with mustard plants sown broadcast, but it might be managed with rape plants or turnip plants in drills ('Reports on Insects Injurious to Crops,' by Chas. Whitehead, Esq., F.Z.S.).

MELOLONTHA VULGARIS, Stephens. The COCKCHAFER or MAY BUG. This insect belongs to the Nat. Ord. COLEOPTERA, and to the family *Melolonthidæ*. It is endowed with an enormous appetite, and not of a discriminating or fastidious character. In its perfect state it eats the foliage of trees, shrubs. grasses, and corn plants. In its larval state it feeds upon the roots of corn plants, grasses, and other crops, and it is in this form that it is mainly injurious to agriculturists.

The perfect insect is known throughout this country, and called variously Cockchafer, May bug, Boom bug, Boomer bug. It feeds in this form for the most part upon the leaves of the oak, maple, thorn, beech. birch, apple, and pear trees. It flies and feeds in the twilight, and goes from tree to tree with heavy, awkward flight, and with a booming sound—"the shard-borne beetle's drowsy hum,"—and remains upon the under part of the leaves of trees and shrubs torpid and dormant during the day. This is the insect which is tortured by cruel boys to this day. Tormenting cockchafers is practised now as it was in the time of the ancient Greeks, as we read in Aristophanes' 'Comedy of the Clouds.' To trees in some districts and in certain seasons much destruction is occasioned by cockchafers. In France whole oak forests have been deprived of foliage by their attacks. Köllar says that in Germany they are often found in such numbers on oaks, willows, hazel, and fruit trees, that the branches bend with their weight. Occasionally in England they have been so numerous as to resemble a flight of locusts. Westwood, in his 'Introduction to the Classification of Insects,' remarks that 14,000 cockchafers were collected in a few days by children and men near Blois, in France. About 50 years ago the Council of the Society of Arts offered a premium for the best means of destroy-

ing this insect, but without any satisfactory results.

The larvæ are most destructive in grass land, devouring the roots of the grasses and destroying the herbage. In these cases the grasses lie withered on the ground, looking as if they had been violently pulled up. The rooks have been often accused of doing this by ignorant persons, as well as of divers other imaginary delinquencies, simply because they have congregated in meadows and have been actively engaged in digging for these large grubs, which are savoury morsels to them. It is not by any means infrequent to find acres of grass land destroyed by these grubs. The soil is honeycombed by them, and the grasses can be pulled up without any effort. Wheat, barley, and oats are frequently much injured by the cockchafer grubs, which weaken the plants by gnawing their roots, and in some cases kill them outright. Flax also suffers often from their attacks. They are very destructive in fir plantations, biting the roots so as to cause the death of young trees.

Life history. The cockchafer is very nearly an inch in length, of a brownish colour, with light-coloured scales. Its body is covered with a pubescence, or short down, like tiny scales. It is furnished with remarkable antennæ, having knobs at their extremities, which fold together like the divisions of a fan or the folds of a screen. In the male there are 7 of these folds, in the female only 6. It has very powerful jaws adapted for biting foliage. There are large hooks upon its claws to enable it to cling to leaves and branches, and its legs, 6 in number, are strong and well adapted for burrowing in the ground for the purpose of egg-laying. This takes place at the end of July. The female goes into the ground to the depth of 7 or 8 inches, and lays from 30 to 40 eggs of a dirty white colour, and a long oval shape. She then returns to the earth and resumes her ordinary life for a short period. The larvæ are hatched from the eggs in about 5 weeks. They are thick, fleshy, and more than an inch and a quarter long when full sized, of a whitish colour, with the head slightly yellow, having jaws fitted for gnawing roots, and 3 pairs of short dark feet. The last segment of the body is larger and more developed than the others, appearing to be filled with a substance of a violet hue. At first the larvæ grow slowly, as a rule congregating closely together just under the ground, feeding then upon the small and most tender roots. At the approach of frost and cold they go down to a depth of 9 to 12 inches for hibernation, coming up in the spring full-grown to attack roots of all kinds. In this state they remain in the land 3 years at least—Köllar says for 5 years and even longer. The pupa state is assumed in the autumn, and retained until the spring, when the pupa case is cast off, and the change is accomplished in about 14 days; then in perfect beetle guise the insects come from the earth, and commence their depredations. The larvæ cannot exist above ground, and soon die when exposed to the air.

Prevention. One special means of prevention is to make a raid upon the beetles when upon trees in the summer, and in other feeding-places. This is done in parts of France, and is called *le hannetonage général*, and might be performed in England in localities very much subject to incursions of cockchafers. It is such a dangerous and destructive insect that everyone's hand should be against it. Nets like those employed for sparrow catching at night, called bat-folding nets, only larger, lighter, and with very small meshes, might be used with advantage. The branches of infested trees being beaten with poles the insects would fly to the light of the lanterns held behind the nets, and become entangled in their meshes. Rooks, starlings, and jays are very fond of the beetles, and should on no account be driven from their haunts. These are farmers' friends, and should be encouraged. When they persistently visit fields under crops, or grass land, it may be assumed that there are larvæ or insects in some form or other at work pernicious to the crops, delicious to the birds. Moles, again, are wholesale devourers of the cockchafer grubs. Taschenberg speaks of the great services they rendered to farmers in this way in Germany, and French entomologists also speak highly of their invaluable benefits. In meadows moles are of great advantage in clearing off these grubs among many other insects. It is admitted that mole "heaves" are unsightly, and interfere much with mowing; but this is a slight disadvantage compared with the amount of good the moles do.

Meadows should be kept well rolled to prevent, if possible, the beetles from getting into the ground to lay eggs.

Remedies. As to remedial measures, it is somewhat difficult to apply these so as to be of direct and very apparent benefit. When meadows are badly affected, dressings of gas lime, or of earth, or wood or coal ashes carefully impregnated with kerosene or petroleum, should be tried. About 4 to 5 quarts of oil should be well mixed with a cart-load of earth, wood, or coal ashes or sawdust. Liquid manure copiously applied has been of much avail on light land. Kainite of potash put on at the rate of ½ a ton per acre has been found to answer. Rolling heavily and frequently tends to close the ground, and to keep the grubs from the roots in some degree. It may be said that meadows reserved for mowing and not regularly fed by sheep should be rolled more than they are, particularly upon light land, which is more subject to the attacks of cockchafers and other insects than land of clay or other adhesive composition. Folding sheep on grass land long and heavily, especially ewes and lambs, with plenty of artificial food and swedes or mangels carted on, is an admirable remedy against these grubs. The land is made firm so that they cannot work, and it is soaked with liquid which they cannot bear.

Corn crops are often attacked by these grubs after sainfoin leys, and clover leys that have been down longer than the usual period. In wheat land showing signs of loss of plant many grubs were discovered, though the finder had no idea what they were. He was advised to horse-hoe well, and to put on 5 cwt. of kainite of potash and 1 cwt. of nitrate of soda, and after this to roll heavily. This treatment was effectual. Soot has been also found very useful chopped in with hand hoes, and the land rolled down tightly afterwards. Nitrate of soda was usefully employed by itself in another case after ring rolling both ways, and a heavy plain roll put on finally (' Reports on Insects Injurious to Crops,' by Chas. Whitehead, Esq., F.Z.S.).

MEER'SCHAUM. *Syn.* ÉCUME DE MER, Fr. A native silicate of magnesia. It has a sp. gr. ranging between 2·6 and 3·4; is readily acted on by acids, and fuses before a powerful blowpipe into a white enamel. The finest qualities are found in Greece and Turkey. Its principal application is to the manufacture of tobacco pipes. The Germans prepare their pipes for sale by soaking them in tallow, then in white wax, and finally by polishing them with shave-grass. Genuine meerschaum pipes are distinguished from mock ones by the beautiful brown colour which they assume after being smoked for some time. Of late years some of the pipemakers have produced a composition clay pipe, which closely resembles meerschaum in appearance, and is 'warranted to colour well.' The composition, which is comparatively valueless, is made up into pipes of suitable patterns, which are frequently sold to the ignorant for 'meerschaums.' See CEMENTS, HYDRAULIC.

ME'GRIM. *Syn.* MEAGRIM, HEMICRANIA. A pain affecting one side of the head only, often periodic, like an ague, and generally of a nervous, hysterical, or bilious character. It is clavus when there is a strong pulsation, conveying the sensation of a nail piercing the part. See HEADACHE.

ME'GRIMS. *Syn.* MEAGRIMS, VERTIGO. In *veterinary medicine* this term is applied to horses which when at work reel, then stand for a minute dull and stupid, or fall to the ground, and lie partially insensible for a few minutes. "Horses subject to this affection should be driven with a breastplate or pipe collar, so as to prevent pressure on the veins carrying the blood from the head; the bowels should be kept in good order; an occasional laxative is advisable, and a course either of arsenic or quinine, or of arsenic and iron" (*Dun*).

MEL'ANCHOLY. See HYPOCHONDRIASIS and INSANITY.

MELIS'SIC ALCOHOL. A substance obtained by Brodie from beeswax. By oxidation it yields 'melissic acid.'

MELLA'GO. The old name for a medicine having the consistence of honey, with a somewhat sweetish taste. Mellago taraxaci is fluid extract of dandelion.

MELTING-POINT. The temperature at which solids assume the liquid form.

MENINGI'TIS, EPIZOOTIC. Cerebro-spinal fever, a disease peculiarly fatal to horses, and which has in recent years caused most serious losses in America. "Is a malignant, non-contagious epizootic fever of the zymotic class, occurring during the winter and early spring months, and affecting the coverings and surfaces of the brain and spinal cord " (*Williams*).

The *cause* is as yet unknown. "It attacks all classes of horses, but evidently prefers those that are rather of the superior order and well kept." Drainage and ventilation appear to have no effect upon it, and it is therefore probably due to a specific poison disseminated in the air.

Symptoms. The animal is dull and stupid; has a staggering gait and a gradually increasing paralysis, usually of the hinder extremities, which progresses for about 3 days, when coma comes on. If the symptoms appear and develop gradually the case will generally do well, but the paralysis is often obstinate. Relapse after 6 or 8 days is not uncommon. If the symptoms come on suddenly and violently there is little hope, and death will ensue in 12 to 72 hours.

Treatm. Sling the animal if possible; if not, lay it on a large thick bed of straw. "Have it well brushed, especially about the extremities; it must be well clothed and its legs dry-bandaged with flannel rollers" (*Williams*). The same author has found the subcutaneous injection of atropine, in conjunction with ergot in the food, "very efficacious if followed by stimulants and tonics during convalescence, but the stimulants must be used cautiously at first."

MEN'STRUUM. [L.] A solvent or dissolvent. The principal MENSTRUA employed in *chemistry* and *pharmacy*, to extract the active principles of bodies by digestion, decoction, infusion, or maceration, are water, alcohol, oils, and solutions of the acids and alkalies.

MEN'THOL. $C_{10}H_{20}O$. A stearopten obtained by cooling the oil obtained from the fresh herb of *Mentha arvensis* and of *M. piperita*; in colourless acicular crystals more or less moist, or in fused brittle masses. Melting-point 108°—110° F. Odour like peppermint, and taste pungent. Readily soluble in ether, alcohol, fixed and volatile oils; insoluble in water. It is a powerful antiseptic, having properties similar to those of its homologue thymol, and is extensively used as a local application in facial neuralgia, toothache, and sciatica, either painted in solution, or moulded into cones and gently rubbed over the painful part.—*Dose*, ½ to 2 gr.

MERCU'RIAL BAL'SAM. See OINTMENT OF NITRATE OF MERCURY.

MERCU'RIAL DISEASE'. *Syn.* MORBUS MERCURIALIS, HYDRARGYRIASIS, L. This results from the injudicious or excessive use of mercury, or exposure to the fumes of this metal. The common and leading symptoms are a disagreeable coppery taste; excessive salivation; sponginess, tumefaction, and ulceration of the gums; swollen tongue; loosening of the teeth; exfoliation of the jaws; remarkably offensive breath; debility; emaciation; ending (when not arrested) in death from exhaustion. Fever, cachexia, violent purging and griping, a species of eczema (ECZEMA MERCURIALE, LEPRA MERCURIALIS), and other forms of skin disease, are also phases of the same affection, the first of which occasionally proves fatal under the influence of sudden and violent physical exertion.

The treatment, in ordinary cases, may consist in free exposure to the open air, avoiding either heat or cold; the administration of saline aperients, as Epsom salts, phosphate of soda, &c.; the free use of lemon juice and water as a common drink; with weak gargles or washes of chloride of soda or chloride of lime to the gums, mouth, and throat.

MER'CURY. Hg = 199·8. *Syn.* QUICKSILVER; HYDRARGYRUM (B. P., Ph. L., E., and D.), L.;

MERCURE, VIF ARGENT, Fr.; QUECKSILBER, Ger. A remarkable metal, which has been known from a very early period, certainly since 300 B.C. The Romans employed it as a medicine externally, as did the Arabs; but the Hindoos were probably the first to prescribe it internally.

Sources. The most important are the mines of Idria, in Carniola; Almaden, in Spain; and New Almaden, in California, and at Wolfstein and Landsberg, in Bavaria; it is also imported from China and Japan, where it exists combined with sulphur, under the form of cinnabar.

Prep. From the ore the pure metal is obtained by distilling it with lime or iron filings in iron retorts, by which the sulphur it contains is seized and retained, while the mercury rises in the state of vapour, and is condensed in suitable receivers. Quicksilver is commonly imported in cylindrical iron bottles, containing ¾ to 1 cwt. each. It is also imported in small quantities from China, contained in bamboo bottles holding about 20 lbs. each.

Pur. Mercury, as imported, is usually sufficiently pure for ordinary purposes without any further preparation. Mere mechanical impurities, as floating dust, dirt, &c., may be got rid of by squeezing the metal through chamois leather or flannel, or by filtering it through a small hole in the apex of an inverted cone of paper. It can be further cleaned by shaking well with a little strong nitric acid, washing with distilled water, and drying by blotting-paper, or squeezing through warm chamois leather. The surest mode of freeing mercury from foreign metals is to redistil it, the surface being covered with iron filings.

Prop., &c. Mercury, at all common temperatures, is a heavy liquid, possessing a nearly silver-white colour, and a brilliant metallic lustre; solidifies at —39·5° C., and is then ductile, malleable, and tenacious; boils at 350° C., and escapes in colourless transparent vapour; its vapour density is 99·9; it also volatilises slowly at the ordinary temperature of the atmosphere. The presence of minute quantities of lead and zinc greatly retard its evaporation at its boiling heat. In chemical properties it much resembles silver. It unites with oxygen, chlorine, iodine, &c., forming numerous compounds. When reduced to a state of fine division as in some medicinal preparations it undergoes a partial oxidation. With the metals it unites to form AMALGAMS. The only acids which act directly on metallic mercury are the sulphuric and nitric, but for this purpose the former must be heated and concentrated. Nitric acid, however, even when dilute and in the cold, dissolves it freely. Pure mercury is unalterable in the air at ordinary temperatures. Sp. gr. 13·595 at 0° C.; 14·1932 when in the solid state.

Uses, &c. Mercury is applied to various purposes in the *arts;* as the amalgamation of gold and silver, in extracting these metals from their ores, 'wash gilding,' the silvering of looking-glasses, and the manufacture of barometers and thermometers. The zinc plates of certain galvanic batteries are amalgamated with mercury to prevent waste, and an amalgam of zinc and tin is used to promote the action of frictional electrical machines. Dentists employ gold amalgam and cadmium amalgam for stopping teeth. Copper amalgam is useful for sealing bottles, glass tubes, &c., when other plastic substances are undesirable; also for taking impressions of engraved metal work. Sodium amalgam in contact with water forms a convenient source of nascent hydrogen.

Mercury is used in the preparation of several very valuable medicines. In its metallic state it appears to be inert when swallowed unless it meets with much acidity in the alimentary canal, or is in a state of minute division; its compounds are, however, all of them more or less poisonous.

Mercury has been employed in one or other of its forms in almost *all* diseases; but each of its numerous preparations is supposed to have some peculiarity of action of its own, combined with that common to all the compounds of this metal. The mercurials form, indeed, one of the most important classes of the materia medica.

Tests. Metallic mercury is detected by its liquid condition and volatility; and, when in a finely divided or pulverulent state, by the microscope, or by staining a piece of copper white when the two are rubbed together.

Mercury, when present in combination, can be detected as under:

When intimately mixed with anhydrous sodium carbonate, and heated in a small test-tube, under a layer of the carbonate, decomposition ensues, and a crust of grey sublimate forms on the cooler portion of the tube. When examined by a lens this crust is seen to consist of minute metallic globules. By friction with a bright glass or iron rod these are united into globules, which are visible to the naked eye.

A perfectly clean and bright piece of copper, immersed in a slightly acid solution of mercury, becomes in a short time covered with a grey or whitish stain, which assumes a silvery lustre when gently rubbed with a piece of soft cork or leather, and is removed by the subsequent application of heat. A single drop of liquid may be tested on a bright copper coin in this way.

If copper foil be used in the previous test, after being washed with a weak solution of ammonia, and in distilled water, and dried by pressure between the folds of bibulous paper, it may be cut into small pieces, and heated in a test-tube, in order to obtain a sublimate of metallic globules. When the suspected solution contains organic matter, bright copper filings may be employed, and the process modified so as nearly to resemble Reinsch's test for arsenious acid. According to Orfila, 'scraped copper plate' is capable of detecting the presence of $\frac{1}{75000}$ part of corrosive sublimate in a solution. MM. Trousseau and Reveil state that a plate of brass is even more susceptible than one of red copper.

Smithson's Electrolytic Test. This consists in the use of a polished wire or plate of gold or copper, round which a strip or thread of zinc or tin is wound in a spiral direction. The suspected liquid is acidulated with a few drops of hydrochloric acid, and after immersion for a longer or a shorter period (half an hour to an hour or two) the gold will have become white if any mercury be present. The coil of zinc or tin is then removed from the gold, and the latter, after being washed and dried between folds of bibulous paper, is heated in a test-tube, to obtain metallic globules, as before.

An ingenious extemporaneous application of the electrolytic test may be made as follows:— Place a drop or two of the suspected liquid on a clean and bright gold or copper coin, and apply a bright key, so that it may at once touch the edge of the coin and the solution (see *engr.*). An

a. A gold or copper coin.
b. Drop of suspected solution.
c. A bright key.

electric current will then be established as before, and a white spot of reduced mercury will appear on the surface of the metal, which may be recognised in the manner already explained.

Mercury is best determined quantitatively by precipitating the solution with sulphuretted hydrogen. The sulphide is warmed with hydrochloric acid and nitric acid added drop by drop to separate the sulphur. Then the solution is diluted, almost neutralised with caustic soda, excess of potassium cyanide added, and the sulphide again thrown down with sulphuretted hydrogen; this is now quickly washed with water, dried at 100°, and weighed. Mercury can also be determined as mercurous chloride and as the metal (*Roscoe*).

The SALTS OF MERCURY are divided into two classes—mercur*ous*, where mercury is a monad element, and unites with one atom of chlorine; and mercur*ic* salts, where it plays the part of a diad element, and unites with two atoms of chlorine. The latter of these will be taken first.

Mercuric Salts. *Tests.* Sulphuretted hydrogen and ammonium sulphide, added in very small quantities, produce on agitation a perfectly white precipitate, which acquires successively a yellow, orange, and brownish-red colour as more of the test is added; and ultimately, when the test is added in considerable excess, an intensely black colour. This precipitate is insoluble in excess of the precipitant, potassium hydrate, potassium cyanide, hydrochloric acid, or nitric acid, even when boiling; but it dissolves readily and completely in potassium sulphide and in 'aqua regia' with decomposition. These reactions are characteristic.

Ammonia gives a white precipitate.

Potassium hydrate gives a reddish precipitate, turning yellow when the test is added in excess. The presence of ammonia causes the precipitate to be white, and when the solution contains much acid both reactions are imperfect.

Alkaline carbonates give a brick-red precipitate.

Potassium iodide gives a scarlet precipitate, which is soluble in excess and in alcohol, and solution of sodium chloride.

The alkaline bicarbonates either do not disturb the solution, or only cause a slight degree of turbidity.

Mercuric Acetate. $Hg(C_2H_3O_2)_2$. *Syn.* PROT-

ACETATE OF MERCURY. *Prep.* By dissolving mercuric oxide in warm acetic acid. It crystallises in brilliant micaceous laminæ, soluble in their own weight of cold water, and somewhat more soluble in boiling water. According to Robiquet, this is the basis of Keyser's antivenereal pills, which do not contain subacetate of mercury, as has been asserted.

Mercuric Bromide. HgBr₂. *Syn.* PROTOBROMIDE OF MERCURY; HYDRARGYRI BIBROMIDUM, L. *Prep.* Two equal parts of bromine and mercury and sublime. Soluble reddish mass; resembles the iodide in its action.—*Dose*, $\frac{1}{10}$ to $\frac{1}{4}$ gr.

Mercuric Chloride. HgCl₂. *Syn.* PROTOCHLORIDE OF MERCURY, PERCHLORIDE OF MERCURY, BICHLORIDE OF MERCURY, CORROSIVE SUBLIMATE ; HYDRARGYRI PERCHLORIDUM (B. P.), HYDRARGYRI BICHLORIDUM (Ph. L.), SUBLIMATUS CORROSIVUS (Ph. E.), SUBLIMATUM CORROSIVUM (Ph. D.), HYDRARGYRI CHLORIDUM CORROSIVUM (Ph. U. S.), HYDRARGYRI MURIAS CORROSIVUS, L. This is the 'corrosive sublimate' of the shops. It has been found native in one of the Molucca Islands.

Prep. 1. (Ph. L.) Mercury, 2 lbs.; sulphuric acid, 21½ fl. oz.; boil to dryness, and rub the residuum, when cold, with sodium chloride, 1½ lbs., in an earthenware mortar; lastly, sublime by a gradually increased heat.

2. (Ph. E.) Mercury, 4 oz.; sulphuric acid, 2 fl. oz. 3 fl. dr.; pure nitric acid, ½ fl. oz.; dissolve, add of sodium chloride, 3 oz., and sublime as before.

3. (Ph. D.) 'Persulphate of mercury' (mercuric sulphate), 2 parts; dried sodium chloride, 1 part; triturate, &c., as before.

4. (Ph. B.) Reduce sulphate of mercury, 20 oz., and chloride of sodium, 16 oz., each to fine powder, and, having mixed them, add black oxide of manganese, in fine powder, 1 oz., and thoroughly mix by trituration in a mortar; place the mixture in an apparatus adapted for sublimation, and apply sufficient heat to cause vapours of perchloride of mercury to rise into the less heated part of the apparatus arranged for their condensation.

Obs. In preparing corrosive sublimate, as well as calomel, by the common process, the solution of the mercury is usually made in an iron pot, set in a furnace under a chimney, to carry off the fumes ; and the sublimation is conducted in an earthen alembic placed in a sand-bath, or in an iron pot, covered with a semi-spherical earthen head. Corrosive sublimate may also be made by the direct solution of mercuric oxide in hydrochloric acid, or by bringing its constituents together in the state of vapour. The latter plan was patented by the late Dr A. T. Thomson.

Prop. The mercuric chloride of commerce occurs in white, semi-transparent masses, consisting of acicular or octahedral crystals of considerable density ; it possesses a sharp metallic taste, and is a violent poison ; it is soluble in about 16 parts of cold, and in 3 parts of boiling water ; the boiling solution deposits its excess of salt in long white prisms as it cools ; soluble in alcohol and ether, in the latter so much so that it has even the property of withdrawing it from its aqueous solutions ; the addition of hydrochloric acid, ammonious chloride, or camphor increases its solubility in all these menstrua. It forms a series of basic salts or oxychlorides. It is decomposed by contact with nearly all metallic bodies, and in solution by various organic substances, and by exposure to light. Sp. gr. 5·4. It melts at 265° C., and boils at 295° C., emitting an extremely acrid vapour which destroys the sense of smell for some time.

Tests. The presence of mercuric chloride may, under most circumstances, be readily detected by the general tests already given. To distinguish it from other salts, special tests for chlorine or hydrochloric acid must be applied. If on filtering the solution, acidulating it with dilute nitric acid, and testing it with silver nitrate, a cloudy white precipitate be formed, which is insoluble in excess of the precipitant, and in nitric acid, but soluble in ammonia water, and blackened by lengthened exposure to light, corrosive sublimate is shown to be present in the substance examined. Calomel, the only compound of mercury with chlorine besides corrosive sublimate, is an insoluble powder, which could not, therefore, be found in the filtered liquid. Calomel, or the white precipitate formed by the mercurous salt, with hydrochloric acid and the soluble chlorides, is soluble in excess of the precipitant, and is not only insoluble in liquor of ammonia, but is immediately blackened by it.

For the purpose of demonstrating the presence of corrosive sublimate in a highly coloured liquid, or one loaded with organic matter, it is necessary to agitate it for some minutes with an equal volume of ether. After standing for a short time the ethereal solution is decanted, and allowed to evaporate spontaneously. The residuum (if any) contains the corrosive sublimate, which, after being dissolved in distilled water, is readily recognised by the above characteristics.

When the substance under examination consists of food, or the contents of the digestive canal, or of animal tissue, it is in general necessary to destroy the organic matter in a nearly similar way to that described under ARSENIOUS ACID. The process adopted by Devergie for this purpose consists in dissolving the substance in concentrated hydrochloric acid, and passing a stream of chlorine through the liquid.—Flandin first carbonises the mass with ¼ or ½ its weight of concentrated sulphuric acid at 212° F., and then saturates the acid in the cold, with dry 'chloride of lime,' added in fragments, assisting the action by stirring, and further adding, by degrees, as the matter thickens and becomes white, a sufficient quantity of distilled water.—Lassaigne boils the suspected mixture for some time with a solution of sodium chloride—a method which, according to Orfila, is not sufficiently delicate to withdraw minute portions of mercury from flesh.—Millon agitates organic liquids (more especially blood, milk, &c.) in large flasks containing gaseous chlorine, which is frequently renewed.—Orfila either dissolves the matter in aqua regia, and passes a stream of chlorine through the liquid, or he carbonises it by means of concentrated sulphuric acid in close vessels. The apparatus consists of a matrass provided with a bent tube, the one end of which is plunged into a jar of cold distilled water. The corrosive sublimate is found both in the volatilised matter and in the carbonised residuum, and is extracted from the latter by boiling it for 15 or 20 minutes in aqua regia. —Personne proceeds by a similar method, but avoids raising the temperature of the substances operated on. In all cases it is advisable to operate in close vessels, on account of the volatility of the bichloride.

When the organic matter has been destroyed by any of the above processes, and a colourless and filtered solution in distilled water obtained, the usual tests may be at once applied. But in this way we can only detect the presence of mercury, and are unable to decide in what way it has entered the system, although we may infer it from other circumstances. It is, therefore, absolutely necessary, in all medico-legal investigations, to previously employ ether (see *above*), in order that we may be enabled to examine the deleterious matter in its original form, or that in which it was swallowed.

Uses, &c. Mercuric chloride is employed in dressing furs and skins; for the preservation of anatomical specimens; for preventing the decay of wood, and mixed with sal-ammoniac and water as an efficient bug poison, &c. " White precipitate," employed for destroying vermin, is deposited when a solution of corrosive sublimate is poured into an excess of solution of ammonia. In *medicine* mercuric chloride is used as an alterative, diaphoretic, and resolvent, in the chronic forms of secondary syphilis, rheumatism, scrofula, cancer, old dropsies, numerous skin diseases, &c.; and externally, as a caustic, in cancer, and made into an ointment, lotion, or injection, in a vast number of skin diseases, ulcers, gleet, &c., and as a preventive of contagion. It acts quicker than the other preparations of mercury, and it is less apt to induce salivation; but it has been said that its effects are less apparent.—*Dose*, $\frac{1}{16}$ to $\frac{1}{4}$ gr., either made into a pill or in solution. It is highly poisonous, and must be exhibited and handled with the greatest caution. Its use is contra-indicated in cases complicated with pulmonary affections or nervous derangement.

Pois. 1. *Symptoms.* Strong coppery or metallic taste; intense pain in the mouth, pharynx, oesophagus, stomach, and intestines; nausea, vomiting (often bloody), diarrhœa, and (sometimes) violent dysentery (these evacuations are generally more frequent than in poisoning by other metallic compounds). After a certain time there is generally an abatement of the severity of the symptoms; the circulation becomes slower, the pulse small and thready, the respiration gentle, and the skin cold; syncope then supervenes, and great general insensibility, always commencing at the pelvic extremities; and sometimes convulsions occur; the secretion of urine is generally diminished, sometimes even entirely suppressed; but the patients always urinate if the sublimate has been employed in a very diluted state, and if drinks have been administered. Death often appears to result from the shock to the nervous system, from intense exhaustion, or from mortification or intense inflammation of the primæ viæ. Poisoning by corrosive sublimate is distinguished from that by arsenic by the coun-

tenance being flushed, and even swollen, whereas in poisoning by arsenic it is wholly contracted and ghastly; and by the whitened condition of the epithelium of the mouth.

2. *Antidotes.* White of egg, hydrated ferric sulphide or ferrous sulphide, and gluten, are each of them powerful antidotes. White of egg has proved efficacious in numerous cases. It requires the white of 1 egg to decompose 4 gr. of corrosive sublimate (*Peschier*). The recently precipitated protosulphuret of iron is, however, according to Mialhe, the antidote *par excellence*, not only to corrosive sublimate, but to the salts of lead and copper. The gluten of wheat has also been recommended (*Taddei*); or, what is equally efficacious, wheat flour mixed up with water. When any of the above are not at hand, copious draughts of milk may be substituted. Iron filings have been occasionally used as an' antidote. All these substances should be taken in considerable quantities; the dose should be frequently repeated, and the general treatment similar to that in cases of poisoning by arsenic. Vomiting should be, in all cases, immediately induced, to remove, if possible, the poisonous matter from the stomach.

Mercuric-ammonium Chloride. HgNH$_2$Cl. *Syn.* AMMONIO-CHLORIDE OF MERCURY, AMMONIATED CHLORIDE OF MERCURY, WHITE PRECIPITATE, LEMERY'S W. P., COSMETIC MERCURY; HYDRARGYRI AMMONIATUM (B. P.), HYDRARGYRI AMMONIO-CHLORIDUM (Ph. L.), HYDRARGYRUM PRECIPITATUM ALBUM (Ph. E.). *Prep.* 1. (Ph. L.) Mercuric chloride, 6 oz.; distilled water, 3 quarts; dissolve with heat, and when the solution has cooled, add of liquor of ammonia, 8 fl. oz., frequently shaking it; lastly, wash the precipitate with water and dry it. The formulæ of the Ph. E. & D. are nearly similar.

2. Mercuric chloride and ammonium chloride, of each ½ lb.; water, 3 quarts; dissolve, and precipitate with solution of potassium hydrate, q. s.

Prop., &c. A white, inodorous, light mass or powder; insoluble in alcohol, partially soluble in boiling water, and wholly dissolved by sulphuric, nitric, and hydrochloric acids, without effervescence. It is totally dissipated by heat. When heated with solution of potash it exhales ammonia, and assumes a yellow colour. Used to make an ointment, which is employed in herpes, porrigo, itch, and other skin diseases, &c.; and by the lower orders as a dusting powder to destroy pediculi, an application which, from its liberal employment, is not always a safe one. It is highly poisonous, and must not be swallowed.

Mercuric and Ammonium Chloride. NH$_4$Cl, HgCl$_2$. *Syn.* CHLORIDE OF MERCURY AND AMMONIUM, SAL ALEMBROTH; HYDRARGYRI ET AMMONII CHLORIDUM, L. *Prep.* (P. Cod.) From mercuric chloride and ammonium chloride, equal parts, triturated together. " The object in adding the ammonium chloride here is to render the corrosive sublimate more soluble in water. The action of the latter is not otherwise altered " (*Redwood*). It is chiefly used for lotions and injections.

Mercuric and Quinine Chloride. *Syn.* CHLORIDE OF MERCURY AND QUININE; HYDRARGYRI ET QUINÆ CHLORIDUM, L. *Prep.* (*M'Dermott.*) From mercuric chloride, 1 part; quinine chloride,

3 parts; separately formed into saturated solutions with water and then mixed; the crystalline precipitate is collected and dried by a gentle heat.—*Dose*, ⅛ to ¼ gr., made into a pill with crum of bread; daily, as an alterative in debilitated habits; or combined with opium thrice daily, to produce salivation.

Mercuric Cyanide. HgCy₂ or Hg(CN)₂. *Syn.* CYANIDE OF MERCURY; HYDRARGYRI CYANIDUM, H. BICYANIDUM, H. CYANURETUM (Ph. U. S.), L. *Prep.* 1. (Ph. L., 1836.) Pure Prussian blue, 8 oz.; mercuric oxide, 10 oz.; distilled water, 4 pints; boil for ¼ an hour, filter, evaporate, and crystallise; wash what remains frequently with boiling distilled water, and again evaporate, that crystals may form. This is Proust's process. The formula of the Ph. U. S. is similar.

2. (Ph. D., 1826.) Prussian blue (pure), 6 parts; mercuric oxide, 5 parts; distilled water, 40 parts; as the last.

3. (*Desfosses.*) Potassium ferrocyanide, 1 part, is boiled for ¼ hour with mercuric sulphate, 2 parts, and distilled water, 8 parts; the deposit is separated by filtration, and the liquid evaporated to crystallising point.

4. (*Winckler.*) Saturate dilute hydrocyanic acid with mercuric oxide; evaporate and crystallise. Pure.

Prop., &c. Heavy, colourless, inodorous, square prisms; tasting strongly metallic; soluble in 8 parts of cold water; slightly soluble in alcohol. Those made by the first two formulæ are of a pale yellow colour. It is transparent and totally soluble in water. The solution, on the addition of hydrochloric acid, evolves hydrocyanic acid, known by its smell; and a glass moistened with a solution of nitrate of silver, and held over it, gives a deposit soluble in boiling nitric acid. When heated it evolves cyanogen, and runs into globules of metallic mercury. It has been administered in some hepatic and skin diseases, and has been proposed as a substitute for corrosive sublimate (*Parent*). It has been said to act directly on the skin and bones, and to have proved useful in allaying the pain of nodes and in dispersing them (*Mendoza*). It is, however, principally used as a source of cyanogen and hydrocyanic acid.—*Dose*, ¹⁄₁₆ to ¼ gr. (beginning with the smaller quantity), made into pills with crum of bread, or in alcoholic solution; as a gargle or lotion, 10 gr. to water, 1 pint; as an ointment, 10 or 12 gr. to lard, 1 oz. Dressings of mercuric cyanide are now used by surgeons as antiseptics.

Mercuric Fulminate. Hg[O(CN)₂]O. A detonating substance, for the manufacture of which consult 'Bloxam's Chemistry,' ed. 1890.

Mercuric Iodide. HgI₂. *Syn.* PROTIODIDE OF MERCURY, RED IODIDE OF MERCURY, IODIDE OF MERCURY, BINIODIDE OF MERCURY; HYDRARGYRI IODIDUM RUBRUM (B. P.), H. IODIDUM, H. BINIODIDUM (Ph. E.), H. IODIDUM RUBRUM (Ph. D.), L. *Prep.* 1. (B. P.) Mercuric chloride (corrosive sublimate), 4 parts; potassium iodide, 5 parts; boiling distilled water, 80 parts. Dissolve the mercuric chloride in 60 parts of water, and the potassium iodide in the remainder, and mix the two solutions. Allow to stand, decant the supernatant liquor, and collect the precipitate on a filter, wash twice with cold water, and dry at 212° F.

2. (Ph. L., 1836.) Mercury, 1 oz.; iodine, 10 dr.; rectified spirit, q. s. (2 or 3 fl. dr.); triturate until the globules of mercury disappear, and the mixture assumes a scarlet colour; dry it in the dark, and preserve it in a well-stoppered vessel.

3. (Ph. E.) Mercury, 2 oz.; iodine, 2 oz.; spirit, q. s.; triturate together as last, dissolve the product, by brisk ebullition, in concentrated solution of sodium chloride, 1 gall., filter whilst boiling hot, wash the crystals deposited on cooling, and dry them.

4. (Ph. D.) Mercuric chloride, 1 oz.; hot distilled water, 25 fl. oz.; dissolve; potassium iodide, 1½ oz.; water, 5 fl. oz.; dissolve; when the solutions are cold, mix them; filter off the precipitate, wash it with distilled water, and dry at 212° F.

Prop., &c. A bright scarlet powder consisting of microscopic octahedra insoluble in water, but soluble in alcohol and ether, and in the solutions of several of the iodides and chlorides. It is also soluble in cod-liver oil, and in several other fixed oils. Readily sublimed. When first heated it becomes yellow, then brown, then fuses, and is finally converted into a colourless vapour, which condenses in yellow crystals on a cold surface. These crystals when touched with a hard body instantly become red. The yellow crystals are rhombic.—*Dose*, ¹⁄₁₆ to ⅛ gr., dissolved in alcohol or made into a pill; in the same cases as the subiodides, from which it differs chiefly in its greater energy and poisonous qualities.

Mercuric Oleate. See OINTMENT OF OLEATE OF MERCURY.

Mercuric and Potassium Iodide. 2 (HgI₂.KI). 3Aq. *Syn.* IODIDE OF MERCURY AND POTASSIUM, IODO-HYDRARGYRATE OF POTASSIUM; HYDRARGYRI ET POTASSII IODIDUM, L. Yellow prisms, a solution of which mixed with potash forms Nessler's solution.—*Prep.* 1. (*M. Boullay.*) Mercuric iodide, potassium iodide, and water, equal parts; dissolve by heat, and crystallise by evaporation or refrigeration.

2. (*Puche.*) From mercuric iodide and potassium iodide, equal parts, triturated together.—*Dose*, q. s. to ½ gr., dissolved in water; in the same cases as the biniodide, and in chronic bronchitis, hooping-cough, inflammatory sore throat, &c.

Mercuric and Potassium Iodo-cyanide. *Syn.* HYDRARGYRI ET POTASSII IODO-CYANIDUM, L. *Prep.* To a concentrated solution of mercuric cyanide add a rather strong solution of potassium iodide, and dry the precipitate by a gentle heat.

Prop., &c. Small, white, pearly, crystalline plates or scales. It is chiefly used as a test of the purity of hydrocyanic acid. When put into this liquid it is instantly turned red if any mineral acid is present.

Mercuric Nitrate. Hg(NO₃)₂. *Syn.* PROTONITRATE OF MERCURY, PERNITRATE OF MERCURY. *Prep.* (NEUTRAL.) *a.* This is obtained by solution of mercuric oxide, in excess of nitric acid. The solution, evaporated in a bell-jar over sulphuric acid, yields large crystals. The same compound is obtained as a crystalline powder when the syrupy liquid is dropped into strong nitric acid.

b. By dissolving mercury in excess of nitric acid with heat, until the solution, when diluted with distilled water, ceases to give a precipitate with common salt.

(Basic.) 1st. $2Hg(NO_3)HO.H_2O$. *Prep.* By saturating hot dilute nitric acid with mercuric oxide. The salt, which is bibasic, crystallises on cooling.

The acid solution (before evaporation) is used as a caustic in cancerous, syphilitic, and other ulcerations; but it frequently produces intense pain, and occasionally the usual constitutional effects of mercury. It was formerly given in similar cases to those in which the bichloride is now employed.—*Dose* (of the dry salt), $\frac{1}{20}$ to $\frac{1}{6}$ gr. This is the preparation ordinarily referred to under the name 'pernitrate of. mercury.'

2nd. $2Hg(NO_3)HO.HgO$. *c.* By saturating strong nitric acid with mercury by heat, throwing the solution into cold water, and collecting and drying the precipitate. This salt, which is tribasic, is also formed when the preceding crystallised salts are put into hot water.

Obs. This preparation is a heavy, yellow powder, but the shade varies according to its basicity, which increases with the temperature of the water employed to effect the precipitation, until, at the boiling temperature, the colour is a dull red. It is extensively employed for the extemporaneous preparation of the ointment of nitrate of mercury, according to the formula on the following label which accompanies each bottle:—' Hydrarg. sub-nitras.' "Two scruples, mixed with one ounce of simple cerate, make the ung. hydrarg. nit. of the London Pharmacopœia." We need scarcely add that this statement, so unblushingly uttered, is a dangerous falsehood. An ointment so made possesses neither the quantity of mercury nor of nitric acid employed in the Pharmacopœia preparation, besides wanting many of its most sensible and valuable properties.

Mercuric Oxide. HgO. *Syn.* PROTOXIDE OF MERCURY, RED OXIDE OF MERCURY, RED PRECIPITATE, OXIDE OF M., BINOXIDE OF M., DEUTOXIDE OF M., PEROXIDE OF M.; HYDRARGYRI OXYDUM, H. O. RUBRUM (B. & L.). This substance is formed upon the surface of mercury when heated for a long time at its boiling-point in contact with air. The alchemists knew it as *mercurius præcipitatus per se.*

Prep. 1. Prepared on a large scale by heating an intimate mixture of mercury and mercuric nitrate until no more red fumes are evolved.

2. By precipitation (HYDRARGYRI BINOXYDUM —Ph. L. 1836.). Mercuric chloride (corrosive sublimate), 4 oz.; distilled water, 6 pints; dissolve and add of liquor of potass, 28 fl. oz.; drain the precipitate, wash it in distilled water, and dry it by a gentle heat.

Obs. A bright orange-red powder. It usually contains a little combined water; hence its readier solubility in acids than the oxide prepared by heat. When heated sufficiently it yields oxygen, and the mercury either runs into globules or is totally dissipated. It is entirely soluble in hydrochloric acid (Ph. L. 1836). The preparation

of the shops has frequently a brick-red colour, and contains a little oxychloride, arising from too little alkali being used.

3. By calcination of the nitrate (RED PRECIPITATE; HYDRARGYRI NITRICI OXYDUM—Ph. L.; HYDRARGYRI OXIDUM RUBRUM—B. P., Ph. L., Ph. D.). *Prep.* (B. P.) Mercury, by weight, 8 parts; nitric acid, 4½ parts; water, 2 parts. Dissolve half the mercury in the water and acid, evaporate to dryness, and triturate with the rest of the mercury until well blended. Heat in a porcelain capsule, repeatedly stirring, until acid vapours cease to be evolved.

Mercury, 3 lbs.; nitric acid, 18 fl. oz. (1½ lbs., Ph. L., 1836); water, 2 quarts; dissolve by a gentle heat, evaporate to dryness, powder, and calcine this in a shallow vessel, with a gradually increased heat, until red vapours cease to arise. The process of the Ph. E. and D. are similar, except that the Dublin College directs the evaporation and calcination to be performed in the same vessel, without powdering or stirring the mass.

Obs. Bright red crystalline scales, which usually contain a little undecomposed pernitrate of mercury; in other respects it resembles the last two preparations. It is more generally used as an escharotic and in ointments than the precipitated oxide. It is volatilised by heat without the evolution of nitrous vapours. According to Mr Brande it contains about 2½ per cent. of nitric acid. According to Mr Barker the process of the Ph. D. yields the finest coloured product; but Mr Brande states that the nitrate requires to be constantly stirred during the process. On the large scale the evaporation is generally conducted in a shallow earthen dish, and as soon as the mass becomes dry a second dish is inverted over it, and the calcination is continued, without disturbance, until the process is concluded. The heat of a sand-bath is employed.

Prop. Scarlet microscopic crystals, which are of a dark colour when hot. Decomposes at a red heat into mercury and oxygen. Explodes when heated with sulphur. Evolves light and heat when placed in contact with sodium. It is a powerful poison, possessing a metallic taste and an alkaline reaction. It is slightly soluble in water.

Uses. Mercuric oxide is valuable for various purposes in chemical analysis. It was formerly employed in medicine to induce salivation, but is now chiefly used as an escharotic, either in the form of powder or made into an ointment.— *Dose,* ⅛ to 1 gr., combined with opium. It is very poisonous.

Mercuric Sulphate. $HgSO_4$. *Syn.* PROTOSULPHATE OF MERCURY; HYDRARGYRI SULPHAS (Ph. B.), H. PERSULPHAS, H. BISULPHAS, L. *Prep.* 1. (Neutral.) *a.* By boiling together sulphuric acid and metallic mercury until the latter is wholly converted into a heavy, white, crystalline powder; the excess of acid is removed by evaporation. Equal weights of acid and metal may conveniently be employed.

b. (Ph. D. 1826.) Dissolve mercury, 6 parts, in a mixture of sulphuric acid, 6 parts, and nitric acid, 1 part, by boiling them in a glass vessel, and continue the heat until the mass becomes perfectly dry and white. Used to make calomel.

c. (Ph. B.) Place 20 oz. of quicksilver in a porcelain capsule with 12 fl. oz. of sulphuric acid, and apply heat until nothing remains but a white, dry, crystalline salt. Used to make perchloride and chloride of mercury.

2. (Basic.) $HgSO_4,2HgO$. *Syn.* TRIBASIC SULPHATE OF MERCURY, TURPETH MINERAL, TURBITH M., QUEEN'S YELLOW, SUBSULPHATE OF MERCURY†, TRIBASIC PERSULPHATE OF M.; HYDRARGYRI SUBSULPHAS, H. S. FLAVUS, TER-PETHUM MINERALE, L. *a.* Dissolve mercury in an equal weight of sulphuric acid by boiling them to dryness, fling the mass into hot water, and wash and dry the resulting yellow powder.

b. (Ph. D. 1826.) Mercuric sulphate, 1 part; warm water, 20 parts; triturate together in an earthen mortar, wash well with distilled water, drain, and dry it. Heavy lemon-yellow powder.

Prop., &c. The neutral mercuric sulphate is a white crystalline powder which becomes brown-yellow when heated and white again on cooling. Water decomposes it into a soluble acid sulphate, and into a yellow insoluble basic sulphate known as turbith.—*Dose.* As an alterative, $\frac{1}{8}$ to $\frac{1}{4}$ gr.; as an emetic, 3 to 5 gr.; as an errhine, 1 gr.; mixed up with a pinch of liquorice powder or fine snuff. It is a powerful poison, and one of the least useful of the mercurial preparations.

Obs. The temperature of the water employed to decompose the neutral sulphate influences the shade of colour of the resulting salt in a similar manner to that pointed out under the nitrate. It is now superseded as a pigment by chrome yellow and orpiment, which are not only more beautiful, but cheaper preparations.

Mercuric Sulphide. HgS. *Syn.* PROTOSUL-PHIDE OF MERCURY, RED SULPHURET OF MER-CURY, CINNABAR, VERMILION, SULPHURET OF MER-CURY, SULPHIDE OF M., BISULPHURET OF M.†; HYDRARGYRI BISULPHURETUM (Ph. B. and L.), CINNABARIS (Ph. E.), H. SULPHURETUM RUBRUM, L. Occurs in beds in slate rocks and shales, and more rarely in granite or porphyry. It is the chief ore of mercury.

Prep. 1. 100 parts of mercury and 38 parts sulphur, rubbed together for some hours, then mixed with 25 parts of potash dissolved in water at 45°; heated for 8 hours, then washed in water and dried (*Brunner*).

2. (Ph. L.) Quicksilver, 24 oz.; sulphur, 5 oz.; melt together, and continue the heat till the mixture swells up; then cover the vessel, remove it from the heat, and when cold, powder and sublime it. (Ph. B.) Quicksilver, 2 lbs.; sulphur, 5 oz. This is founded on the old Dutch process.

Prop., &c. Mercuric sulphide has a dark red semi-crystalline appearance in the mass, but acquires a brilliant scarlet colour by powdering. Sp. gr = 8·124. It is tasteless, odourless, and insoluble in most reagents, but it dissolves in aqua regia with liberation of sulphur and in cold concentrated or warm dilute hydriodic acid. It is commonly called vermilion, and is chiefly used as a pigment in the manufacture of paints, ink, and sealing-wax; but it is occasionally employed in *medicine* as a diaphoretic and vermifuge, and in some cutaneous diseases and gout.—*Dose,* 10 to 30 gr.; as a fumigation, about $\frac{1}{4}$ dr. is thrown on a plate of iron heated to dull redness. For the

last purpose it is inferior to mercurous oxide, owing to the more irritating nature of its vapour. Vermilion is sometimes adulterated with red-lead or red oxide of iron. The presence of these impurities can be readily ascertained, for the pure substance sublimes without leaving any residue.

Mercuric Thiocyanate. $Hg(SCN)_2$. A white crystalline precipitate, formed by the interaction of ammonium thiocyanate and corrosive sublimate solutions. Mixed with gum-water to a paste and made into balls this substance is used for producing the so-called 'Pharaoh's serpents.'

Mercurous Salts. *Tests.* Sulphuretted hydrogen and ammonium sulphide give black precipitates, insoluble in dilute acids, ammonium sulphide, potassium cyanide, and hot nitric acid, but slightly soluble in sodium sulphide, and decomposed by nitro-hydrochloric acid.

Potassium hydrate and ammonia give black-grey or black precipitates, which are insoluble in excess of the precipitant.

Hydrochloric acid and the soluble metallic chlorides occasion a precipitate, which assumes the form of a very fine powder of dazzling whiteness, insoluble in excess, but soluble in aqua regia. Potassium hydrate and ammonia turn it dark grey or black.

Potassium iodide gives a greenish-yellow precipitate, soluble in excess and in ether, and subliming in red crystals when heated.

Mercurous Acetate. $Hg(C_2H_3O_2)$. *Syn.* ACE-TATE OF MERCURY, SUBACETATE OF M. *Prep.* (P. Cod.) Dissolve mercurous nitrate, 1 part, in water (slightly acidulated with nitric acid), 4 parts, and precipitate the liquid with a solution of sodium acetate, gradually added, until in slight excess; carefully wash the precipitate with cold water, and dry it in the dark.

Prop., &c. Small, white, flexible scales; insoluble in alcohol; soluble in about 300 parts of water; blackened by light, and carbonised by a strong heat. It has been said to be one of the mildest of the mercurials; but this cannot be the case, as it occasionally acts with great violence on both the stomach and bowels, producing much pain and prostration.—*Dose,* $\frac{1}{2}$ to 1 gr., night and morning, gradually increased.

Mercurous Bromide. Hg_2Br_2. *Syn.* SUB-BROMIDE OF MERCURY; HYDRARGYRI BROMI-DUM, L. A whitish-yellow powder, insoluble in water. *Prep.* (*Magendie.*) By precipitating a solution of mercurous nitrate by another of potassium bromide. It closely resembles calomel in both its appearance and properties.—*Dose,* 1 to 5 gr.

Mercurous Chloride. Hg_2Cl_2. *Syn.* CALOMEL, SUBCHLORIDE OF MERCURY, MERCURY CHLORIDE; HYDRARGYRI SUBCHLORIDUM (B. P.), HYDRAR-GYRI CHLORIDUM (Ph. L.), H. C. MITE (Ph. U. S.), CALOMELAS (Ph. E. and D.), L. This substance is one of the best known, and probably the most valuable, of all the mercurials. It is found at Idria and Almaden, crystallised in rhombic prisms as 'horn silver.'

Prep. 1. (Ph. L.) Mercury, 2 lbs.; sulphuric acid, $21\frac{1}{2}$ fl. oz.; mix, boil to dryness (in a cast-iron vessel), and when the resulting mass has cooled, add of mercury, 2 lbs., and triturate the ingredients in an earthenware mortar until they are well mixed; then add of sodium chloride, 1½ lbs., and again triturate until the globules are no longer visible; next sublime the mixture, reduce the sublimate to the finest possible powder, diligently wash it with boiling distilled water, and dry it.

2. (Ph. F.) Mercury, 4 oz., is dissolved in a mixture of sulphuric acid, 2 fl. oz. 3 fl. dr., and nitric acid, ⅔ fl. oz., by the aid of heat; when cold, mercury, 4 oz., is added, and the remainder of the process is conducted as before.

3. (CALOMELAS SUBLIMATUM, Ph. D.) Sulphate of mercury, 10 parts; mercury, 7 parts; dry sodium chloride, 5 parts; triturate, &c., as before, and afterwards resublime it into a large chamber or receiver.

4. (Ph. B.) Same as Dublin.

5. (Apothecaries' Hall.) Quicksilver, 50 lbs., and sulphuric acid, 70 lbs., are boiled to dryness in a cast-iron vessel; of the dry salt, 62 lbs. are triturated with quicksilver, 40½ lbs., until the globules are extinguished, when sodium chloride, 34 lbs., is added, and after thorough admixture the whole is sublimed, &c., as before.

6. (Jewel's Patent.) The receiver, which is capacious, is filled with steam, so that the calomel vapour is condensed in it in a state of extremely minute division. The *engr.* represents the apparatus now usually employed when this plan is adopted. The product is extremely white, and of the finest quality. It is sometimes called ' hydro-sublimed calomel ' and ' hydrosublimate of mercury.' The 'flowers of calomel' of *old pharmacy* were prepared in a nearly similar manner.

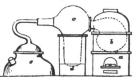

a. Furnace.
b. An earthenware retort, having a short and wide neck, containing the ingredients for making calomel.
c. An earthen receiver, having three tubulatures.
d. A vessel containing water.
e. A steam boiler.

7. (*Soubeiran*.) The crude calomel mixture is heated in an earthen tube in a furnace, and a current of air is directed uninterruptedly into the tube by means of a small ventilator. This sweeps away the vapours to the end of the recipient which is immersed in water, by which means the calomel is moistened and falls down. This plan, slightly modified, is now extensively adopted in this country.

8. A molecular weight of corrosive sublimate is intimately mixed with an atomic weight of metallic mercury, and a little water added to prevent dust; the mixture is then dried and sub-limed.

9. (PRECIPITATED CALOMEL; CALOMELAS PRÆCIPITATUM, L.) Digest pure quicksilver, 9 parts, in nitric acid (sp. gr. 1·02 to 1·25), 8 parts, until no more metal will dissolve, applying heat as the effervescence ceases; then mix the hot liquid quickly with a boiling solution of sodium chloride, 8 parts, dissolved in water (slightly acidulated with hydrochloric acid), 64 parts; lastly, well wash the precipitate in boiling distilled water, and dry it. The product, when the process is skilfully managed, is perfectly white and pure.

Prop. A heavy, white, tasteless powder, or semi-transparent fibrous mass with a slight yellow tinge. Light slowly decomposes it, turning it grey. It is insoluble in water, alcohol, and dilute acids; volatilises at a temperature below redness, and yields a white or yellowish-white sublimate; hot nitric acid oxidises and dissolves it; alkalies, the alkaline carbonates, and lime-water decompose it, with the production of the black oxide; ammonia converts it into a dark grey powder (BLACK PRECIPITATE, *Kane*). Sp. gr. 6·56.

Pur. Calomel is frequently contaminated with small quantities of corrosive sublimate, which may be detected by digesting a little in alcohol, decanting the clear portion, and testing it with a drop or two of potassium hydrate, when a reddish precipitate will be formed if any mercuric chloride be present. This impurity can be eliminated by washing repeatedly with water, in which the calomel is insoluble.

Uses. Calomel is one of the milder mercurials, and in this respect takes its position immediately after blue pill, mercury with chalk, and the grey oxide; but it probably ranks before all the other salts of mercury. Universal experience appears to show it to be a most valuable alterative when judiciously administered. With this intention it is given in doses of ⅓ to 1 gr., generally combined with antimonials, as in Plummer's pill, and repeated every night, or every other night, for some time, followed by a mild saline aperient in the morning. As a purgative, 2 to 5 gr., either combined with or followed by other purgatives, as jalap, rhubarb, senna, colocynth, Epsom salts, &c. As a vermifuge, 2 to 5 gr. overnight, followed by a sufficient dose of castor oil next morning. Combined with opium it is frequently used in various complaints to produce salivation, or bring the system under the influence of mercury. It is also employed as a sedative and errhine, and in a vast number of other indications. It is, indeed, more frequently used, and in a greater variety of complaints, than probably any other medicine.

Obs. Of the two methods of preparing calomel, that by precipitation is not only the best, but the most economical. That by sublimation is, however, the one generally adopted in England. Mr Brande states that "a small portion of sodium chloride is apt to remain combined with it, which might affect its medical uses." Such a contamination is not found in carefully prepared precipitated calomel, although we doubt whether the quantity of it which exists in any of the samples we have met with (being merely a trace) would at all interfere with its therapeutical action; more especially when it is recollected that alkaline chlorides are present in every part of the animal body. The late Mr Fownes once assured us that calomel was more easily and cheaply prepared of the best quality by precipitation than by sublimation, and that if, from careless manipulation, it occasionally contained a minute quantity of common salt, this was of much less importance than the contamination of corrosive sublimate, which was frequently present in samples of sublimed calomel.

To produce a superior quality of calomel in the dry way is a somewhat difficult task, and the process frequently fails in the hands of inexperienced operators. The solution of the mercury is best made in an iron vessel, and the sublimation should be conducted in an earthenware retort with a short but very wide neck, fitted in a spacious receiver, having a large flat bottom, also of earthenware, and containing a little cold water. The heat may be applied by means of a sand-bath. The apparatus, by precipitation, produces a large product, perfectly free from corrosive sublimate and mercurous nitrate.

"The form in which calomel sublimes depends much upon the dimensions and temperature of the subliming vessels. In small vessels it generally condenses in a crystalline cake, the interior surface of which is often covered with beautiful quadrangular prismatic crystals, transparent, and of a texture somewhat elastic or horny. In this state it acquires, by the necessary rubbing into powder, a decided yellow or buff colour, more or less deep, according to the degree of trituration it has undergone. If, on the contrary, the calomel be sublimed into a very capacious and cold receiver, it falls in an impalpable and perfectly white powder, which requires only one elutriation to fit it for use; it then remains perfectly colourless" (Brande).

The long-continued action of steam on calomel in a state of minute division is attended by the formation of a small quantity of corrosive sublimate (Bigbies). Boiling water, hot air, and light also produce a like effect.

Mercurous Iodide. Hg_2I_2. *Syn.* SUBIODIDE OF MERCURY, GREEN IODIDE OF MERCURY, MERCUROUS IODIDE; HYDRARGYRI IODIDUM VIRIDE (B. P.), HYDRARGYRI SUBIODIDUM, H. IODIDUM (Ph. L.), H. I. VIRIDE (Ph. D.), L. *Prep.* 1. (B. P.) Mercury, 1 oz.; iodine, 278 gr.; rectified spirit, a sufficiency. Rub the iodine and mercury in a porcelain mortar, moistening occasionally with a few drops of spirit, and continue the trituration until the mass assumes a uniform green colour, and no metallic globules are visible.

2. Precipitate a solution of mercurous nitrate by another of mercurous potassium iodide; wash the precipitate, first in a solution of sodium chloride, and then in pure water; dry it in the dark.

3. (Ph. L.) Mercury, 1 oz.; iodine, 5 dr.; triturate together, gradually adding of rectified spirit, q. s. (about 1 to 2 fl. dr.), until globules are no longer seen; dry the powder, by a gentle heat, in the shade as quickly as possible, and preserve it in a well-stoppered black glass vessel. The formula of the Ph. D. is similar. In this method, also in No. 1, mercuric iodide may be used in proper proportions instead of iodine.

Prep., &c. A heavy, unstable, greenish-yellow powder; slightly soluble in water, insoluble in alcohol and a solution of common salt; soluble in ether. Freshly prepared, it is yellowish. Heat being cautiously applied, it sublimes in red crystals, which turn yellow on cooling (Turon), and, on access of light, blacken. It is insoluble in a solution of chloride of sodium. The process of

the Ph. L. and F. P. does not answer when larger quantities than 4 or 5 oz. are prepared at once, owing to the great heat generated by the reaction of the ingredients, and the consequent volatilisation of a portion of the iodine, by which the colour of the product suffers.—*Dose,* ½ to 1 gr., made into pills; "in syphilis and scrofula, especially where they occur in the same individual." It is also used externally, in the form of ointment. It is very poisonous.

Mercurous Nitrate. $Hg_2(NO_3)_2$. *Syn.* SUBNITRATE OF MERCURY; HYDRARGYRI SUBNITRAS, H. NITRAS, H. PROTONITRAS, L. *Prep.* 1. (Neutral.) By digesting mercury in excess of cold dilute nitric acid, removing the short prismatic crystals soon after they are formed; these, when drained and redissolved in water slightly acidulated with nitric acid, furnish crystals of pure neutral mercurous nitrate by cautious evaporation.

2. (Basic.) A double salt deposited after some time, when excess of mercury has been employed as above. Long, thin, rhombic prisms of the formula $Hg_2(NO_3)_2 + Hg_2(OHNO_3)$.

Prop., &c. Both the above are decomposed by water, but the former may be dissolved in a very small quantity without decomposition; if there be excess of water the basic salt is formed. When the neutral salt is triturated with an excess of sodium chloride, and water subsequently added, the whole of the mercury is thrown down as calomel, and the filtered supernatant liquid does not contain corrosive sublimate. If this substance is detected, the salt examined contained mercurous nitrate, and if any basic mercurous nitrate was present, the newly formed calomel has a grey or black colour, due to presence of oxide.—*Dose.* Of the neutral salts, $\frac{1}{16}-\frac{1}{6}$ gr. It is seldom used internally. A solution is sometimes employed as a mild caustic to ulcers; and, more dilute, as a lotion in lepra, porrigo, psoriasis, &c.; or made into an ointment, in the same diseases.

Mercurous Oxide. Hg_2O. *Syn.* SUBOXIDE OF MERCURY, GREY O. OF M., BLACK O. OF M., DIOXIDE OF M., PROTOXIDE OF M.†; HYDRARGYRI SUBOXYDUM, H. OXYDUM, H. O. CINEREUM, H. O. NIGRUM, L. Easily obtained by the action of caustic alkalies on a mercurous salt. *Prep.* 1. (Ph. L., 1836.) Calomel, 1 oz.; lime-water, 1 gall.; mix, agitate well together, decant the clear liquid after subsidence, and well wash the sediment with distilled water; lastly, drain and dry it, wrapped in bibulous paper, in the air.

2. (Ph. D., 1826.) Sublimed calomel, 1 part; solution of potassium hydrate (warm), 4 parts; shake together, &c., as last.

3. Briskly triturate calomel in a mortar with pure potassium hydrate in excess; wash it with water, and dry it in the shade.

Prop., &c. A blackish-brown powder, suffering decomposition by exposure to light and air or gentle heat, becoming greyish from a portion being resolved into metallic mercury and red oxide. Digested for a short time in dilute hydrochloric acid, it remains undissolved, and the filtered liquid is not affected by potassium hydrate or by ammonium oxalate. It is totally soluble in acetic acid, and entirely dissipated by heat. As a medicine pure mercurous oxide is one of the mildest of the mercurials, and is used both internally and externally; but chiefly as a fumigant, or made into an ointment.—*Dose*, ⅓ gr. to 3 gr. twice a day.

Mercurous Phosphate. Hg_3PO_4. *Syn.* HYDRARGYRI PHOSPHAS, L. *Prep.* Add a solution of mercurous nitrate (slightly acidulated with nitric acid) to a solution of sodium phosphate; filter off, wash and dry the precipitate which forms. Soluble in excess of mercurous nitrate. In its physical characters it closely resembles calomel, than which it is said to be more appropriate in certain cases, especially in secondary syphilis. Alkalies turn it black.—*Dose*, ⅓ to 1 gr., made into a pill with sugar and aromatics.

Mercurous Sulphate. Hg_2SO_4. *Syn.* SUBSULPHATE OF MERCURY, SULPHATE OF THE SUBOXIDE OF M., PROTOSULPHATE OF M.†; HYDRARGYRI SUBSULPHAS, L. *Prep.* By adding sulphuric acid to a solution of mercurous nitrate. The salt falls as a white crystalline powder.

Mercurous Sulphide. Hg_2S. *Syn.* SUBSULPHATE OF MERCURY; HYDRARGYRI SUBSULPHURETUM CUM SULPHURE, H. S. NIGRUM, L. A very unstable substance, which Roscoe says does not exist. Most text-books, however, mention it.

Prep. 1. Falls as a black precipitate when a solution of mercurous acetate is treated with sulphuretted hydrogen.

2. (ETHIOPS MINERAL; HYDRARGYRI SULPHURETUM CUM SULPHURE, H. S. NIGRUM—Ph. L. 1824 and Ph. D. 1826; ÆTHIOPS MINERALIS—Ph. L. 1836 and Ph. D. 1826.) Quicksilver and sulphur, equal parts, triturated together in a stoneware mortar (Ph. D.) until globules are no longer visible.

Prop., &c. The last preparation of mercurous sulphide is alone employed medicinally. It is a heavy, insoluble, black powder. It is frequently met with imperfectly prepared, and sometimes adulterated. It is said to be a mixture of mercurous sulphide and sulphur, in variable proportions depending on the length of the trituration. It is said to be vermifuge and alterative, and has been used in some cutaneous and glandular diseases, but appears to be nearly inert.—*Dose*, 5 to 30 gr.

Mercurous Tartrate. (P. Cod.) *Syn.* PROTOTARTRATE OF MERCURY; HYDRARGYRI TARTRAS, L. Made by adding a solution of protonitrate of mercury in water, slightly acidulated with nitric acid, to a solution of tartrate of potash as long as a precipitate forms. Wash with distilled water, dry in the dark, and keep it in bottles covered with black paper.—*Dose*, 1 to 2 gr.

MERCURY, Other Preparations of.

Mercury, Hahnemann's. *Syn.* HAHNEMANN'S SOLUBLE MERCURY, H.'S BLACK OXIDE OF M., BLACK PRECIPITATE OF M.; HYDRARGYRI PRÆCIPITATUM NIGRUM, MERCURIUS SOLUBILIS HAHNEMANNI, L. *a.* By dropping weak ammonia into a solution of mercurous nitrate as long as the precipitate formed is of a black colour; the powder is washed, dried in the shade without artificial heat, and then preserved from the light and air.

b. (Ph. Bor. 1847.) Solution of mercurous nitrate (recent; sp. gr. 1·1), 9¼ oz.; distilled water, 2 lbs.; mix, filter, and add to the solution of ammonia (sp. gr. ·960), ½ oz., diluted with water, 4 fl. oz.; collect the powder immediately on a filter, wash it with water, 5 fl. oz., and dry it, &c., as before. A very black powder.—*Dose*, ⅓ to 1 gr.

Mercury, Precipitates of. 1. BLACK PRECIPITATE, Hahnemann's soluble mercury (basic mercurous and ammonium nitrate). 2. GREEN P. (MERCURIUS PRÆCIPITATUS VIRIDIS, LACERTA VIRIDIS), from equal parts of mercury and copper, separately dissolved in nitric acid, the solutions mixed, evaporated to dryness, and then calcined until red fumes cease to arise. Caustic. 3. RED P., mercuric oxide. 4. WHITE P., ammonio-chloride of mercury.

Mercury, Ward's. *Syn.* AMMONIO-NITRATE OF MERCURY; HYDRARGYRI AMMONIÆ NITRAS, L. To nitric acid, 4 parts, contained in a spacious bolt-head or matrass, add gradually ammonium sesquicarbonate, 2 parts; afterwards add of mercury, 1 part, and digest in a gentle heat until the solution is complete.

Mercury with Chalk. *Syn.* HYDRARGYRUM CUM CRETÂ; GREY POWDER (B. P.). *Prep.* Rub 1 oz. (by weight) of mercury, and prepared chalk, 2 oz., in a porcelain mortar, until metallic globules cease to be visible to the naked eye, and the mixture acquires a uniform grey colour.—From 3 to 8 gr.

A little water is said to aid in the extinction of the mercury. Mr Bottle suggests a slight departure from the Grey *modus operandi* followed by the British Pharmacopœia in the above preparation. He proposes to substitute for the tedious process of trituration in a porcelain mortar the agitation of the mercury with the chalk in a wide-mouthed glass bottle, by which means the metal may be minutely subdivided at a considerably less expenditure of time and labour.

Mercury with Magnesia. (Ph. D.) *Syn.* HYDRARGYRUM CUM MAGNESIÂ, L. Pure mercury, 1 oz.; carbonate of magnesia, 2 oz. Rub together in a porcelain mortar until the globules cease to be visible and the mixture acquires a uniform grey colour.—*Dose*, 3 to 8 gr.

MESLIN. A mixture of various kinds of grain (*Brande*).

METAL. *Syn.* METALLUM, L. See METALS.

METALLICA. [L.] Preparations of the metals. One of the divisions of the Ph. L.

METALLIC TREES. See VEGETATION (Metallic).

METALLOCHROMES. A name given by Nobili to extremely thin films of peroxide of lead deposited by electrolytic action upon plates of polished steel, so as to produce an iridescent play of colours. The effect is often very beautiful.

METALLOIDS. A name sometimes applied to the NON-METALLIC ELEMENTS.

METALLURGY. "The art of extracting metals from their ores, and adapting them to various processes of manufacture" (*Percy*).

"Notwithstanding the striking analogy which exists between common chemical and metallurgic operations, since both are employed to isolate certain bodies from each other, there are essential differences which should be carefully noted. In the first place, the quantity of materials being always very great in metallurgy, requires corresponding adaptations of apparatus, and often produces peculiar phenomena; in the second place, the agents to be employed for treating great masses must be selected with a view to economy, as well as chemical action. In analytical chemistry, the main object being exactness of result and purity of product, little attention is bestowed upon the value of the reagents, on account of the small quantity required for any particular process. But in smelting metals upon the large scale, profit being the sole object, cheap materials and easy operations are alone admissible" ('Ure's Dict. of Arts, Manufactures, and Mines,' 4th edit.).

The limits of this work do not permit of more than a general reference to the leading operations of metallurgy under this head. These are—*digging, picking* or *sorting, stamping* or *crushing,* and *washing,* included under the general term '*dressing ore;' roasting* or *calcination,* which expels water, CO_2, &c., volatilises or sublimes certain volatile substances, or oxidises some portion of the ore under treatment, and which is either performed with the fuel in contact with the ore, or in reverberatory furnaces; *reduction,* which brings oxides to the metallic state; *smelting,* or separation from the ore by fusion; *scorification,* which separates readily oxidisable metals from other metals with which they are associated; *cupellation,* which in a special way separates lead and other oxidisable metals from silver and gold; *sublimation,* by which substances are volatilised by heat and subsequently condensed in the solid state; *liquation,* by which substances having different melting-points are separated by subjecting them to a carefully regulated temperature; *lixiviation,* by which metallic salts are separated from metalliferous and other matters by the solvent action of water or saline solutions; *cementation,* in which process articles are embedded in certain powders or cementing materials and kept at below fusion temperature for several hours or days; and other less important operations. The application of these processes is noticed under the leading metals. Those who desire to study the subject minutely are referred to the treatises of Dr Percy, Robert Hunt, Karsten, and Le Play.

MET'ALS. *Syn.* METALLA, L. Metals are elementary bodies which are generally distinguished by their lustre and power of conducting heat and electricity. They form bases by combining with oxygen, have a powerful attraction for chlorine, and are but little disposed to combine with hydrogen. When their solutions are electrolysed the metals always appear at the electro-negative surface, and are hence termed electro-positive elements.

Formerly, when science was much less advanced than at present, the metals constituted a well-defined class. The properties which were regarded as specially characteristic were physical, and were not founded on chemical relations; thus lustre and high specific gravity were considered to be essential characters of all metals. But we are now acquainted with metals which have a lower specific gravity than water (lithium, sodium, &c.), and with so-called non-metallic elements which present a strong metallic lustre (carbon in the

state of graphite, crystallised silicon, &c.). It will therefore be seen that the term 'metal' is rather conventional than strictly scientific. By far the greater number of elementary bodies at present known are metals. Their physical characters and leading chemical properties are noticed under each of them in its alphabetical place. The following table exhibits some useful particulars :—

TABLE *of some of the properties of some of the metals.*

Names arranged in the order of their

Ductility.	Malleability.
Gold.	Gold.
Silver.	Silver.
Platinum.	Copper.
Iron.	Tin.
Nickel.	Platinum.
Copper.	Lead.
Zinc.	Zinc.
Tin.	Iron.
Lead.	Nickel.

Names arranged in the order of their

Power of conducting Heat.	Power of conducting Electricity.
Silver.	Silver.
Copper.	Copper.
Gold.	Gold.
Tin.	Zinc.
Iron.	Iron.
Lead.	Tin.
Bismuth.	Lead.
	Antimony.
	Bismuth.

The metals may be divided into ten groups, namely :

i. *Potassium group* = Lithium, sodium, potassium, rubidium, caesium.
ii. *Calcium group* = Calcium, strontium, barium.
iii. *Magnesium group* = Glucinum, magnesium, zinc, cadmium.
iv. *Aluminium group* = Aluminium, yttrium, gallium, zirconium, erbium, indium, lanthanium, didymium, thorium.
v. *Iron group* = Iron, cobalt, nickel, uranium, cerium.
vi. *Manganese group* = Vanadium, chromium, manganese, molybdenum.
vii. *Antimony group* = Antimony, bismuth.
viii. *Tin group* = Titanium, niobium, tin, tantalum, tungsten.
ix. *Silver group* = Copper, silver, mercury, thallium, lead.
x. *Platinum group* = Rhodium, ruthenium, palladium, gold, platinum, iridium, osmium.

METANTIMON'IC ACID. $H_4Sb_2O_7$. The name given by M. Fremy to that variety of antimonic acid obtained by decomposing pentachloride of antimony with excess of water. It should really be called pyro-antimonic acid. It differs from common antimonic acid in being tetra-basic, and forming different classes of salts with the acids. The acid metantimoniate of potassium is the only re-agent which yields a precipitate with the sodium salts, and is therefore of great value in chemical analysis.

Prep. By fusing antimonic acid with excess of potash in a silver crucible, dissolving the fused mass in a little cold water, and allowing it to crystallise *in vacuo.* The resulting crystals, by solution in pure water, are resolved into free potash and the acid salt. See ANTIMONY.

METAPEC'TIC ACID. See PECTIN.

METAPEC'TIN. See PECTIN.

METAPHOSPHORIC ACID. See PHOSPHORIC ACID.

METHEG'LIN. *Syn.* HYDROMELI, H. VINO-SUM, MELLIS VINUM, L. *Prep.* From honey, 1 cwt.; warm water, 24 galls.; stir well until dis-solved; the next day add of yeast, 1 pint, and hops, 1 lb., previously boiled in water, 1 gall.; along with water, q. s. to make the whole measure 1 barrel; mix well, and ferment the whole with the usual precautions adopted for other liquors. It contains on the average from 7% to 8% of alcohol. See MEAD.

METHYL. CH_3. The hypothetical radical of the methyl series. It forms a number of com-pounds analogous to those of ethyl, *e. g.* methyl alcohol, $CH_3.OH$.

METHYL ALCOHOL, Purification of Crude. Messrs Dittmar and Fawsitt have communicated to the 'Transactions of the Royal Society' of Edin-burgh a work upon the physical properties of methyl alcohol. It was of course necessary to procure this compound in a state of purity in order to carry out their observations. They effect this in a very simple manner, as follows:—100 c.c. of the crude methyl alcohol is digested with 150 grms. of pulverised hydrate of soda, after which it is distilled on a water-bath; 500 grms. of oxalic acid in crystals are now mixed with 200 c.c. of concentrated sulphuric acid, and then 400 c.c. of the methyl alcohol purified by the soda hydrate and distillation are added, and the mixture is cautiously heated on a water-bath. The methyl oxalate thus obtained is thoroughly dried by pres-sure, and the alcohol regenerated by digestion with water at 70° C. In order to dehydrate the alcohol thus obtained, digestion with baryta, lime, and dried sulphate of copper is proved to be necessary.

METHYLAMINE. *Syn.* AMIDOMETHANE. Occurs in herring-brine, the distillate from bones and wood, and in *Mercurialis perennis.* Produced in the decomposition of certain organic compounds, *e. g.* the alkaloids.

Prep. 1. By treating iodomethane with am-monia.

2. By the reaction of nascent hydrogen on prussic acid.

3. By heating methyl isocyanate with potash in a retort attached to a receiver cooled by a freezing mixture. The distillate is saturated with hydrochloric acid, evaporated to dryness, distilled with dry lime, and collected over mer-cury.

Prop. A colourless gas, having an ammoniacal and fishy odour; burns with a yellow flame. It is more strongly basic, and even more soluble in water, than ammonia. Closely resembles am-monia in its behaviour with acids, &c. Precipi-tates many metallic salts. It is condensed to a liquid at −18°. Most of its salts are very soluble in water.

METHYLATED SPIRIT. A mixture of 1 part of methylic alcohol (wood spirit) and 9 parts of ethylic alcohol (spirit of wine). See SPIRIT.

METHYLENE BLUE. $C_{12}H_4(CH_3)_4N_3SCl$. *Syn.* TETRAMETHYL - THIONINE CHLORIDE. Bronze-green crystals which dissolve in water to a fine blue liquid, employed in dyeing, being fixed on cotton with a mordant of antimony tannate. Prepared from dimethylaniline hydrochloride by treatment with sodium nitrate and then with hydrosulphuric acid. It has been found a useful staining agent in microscopic work. The forma-tion of this blue is one of the most delicate tests for sulphuretted hydrogen in solution; in apply-ing the test, the suspected liquid is mixed with hydrochloric acid, a little dimethyl-paradiamido-benzene sulphate added, followed by a drop of ferric chloride. If H_2S is present the blue colour will appear at once.

METHYLENE CHLORIDE. CH_2Cl_2. *Syn.* METHYLENE BICHLORIDE. There are various methods of obtaining this compound:—1. By heating chloroform with zinc filings and dilute sulphuric acid. 2. By acting on methylene iodide with chlorine. In this process prolonged treat-ment with chlorine, at ordinary temperatures, is required to remove the last traces of iodine (*Buttlerow*).

Prop. Chloride of methylene is a colourless mobile fluid, having a smell like chloroform, and a burning taste. It has been used as an anæs-thetic in place of chloroform. According to Dr Armstrong, the substance known as METHYLENE ETHER is a mechanical mixture of bichloride of methylene and ethylic ether. Dr Richardson says of this latter it is not so quick in its action as the methylene chloride, but that it is safer. See ANÆSTHETICS.

METHYLIC ALCOHOL. See WOOD SPIRIT.

MEZE'REON. *Syn.* GAROU, Fr.; MEZEREON BARK; MEZEREI CORTEX (B. P.), L.; MEZE-REON (Ph. L., E., & D.). The dried bark of the *Daphne mezereum,* mezereon; or *Daphne laureola,* spurge, or wood-laurel. The "bark of the root of *Daphne mezereum,*" or spurge-olive (Ph. L.). A stimulant and diuretic. It is em-ployed as a sudorific and alterative, in syphilis, rheumatism, scrofula, and chronic cutaneous dis-eases, usually in conjunction with sarsaparilla. It has also been used as a masticatory in tooth-ache, paralysis of the tongue, &c. On the Con-tinent it is used as a vesicant. For this purpose it is softened by soaking it in hot vinegar, and is then bound on the part, and renewed after intervals of some hours, until vesication is pro-duced.

MICE. See RATS.

MI'CROSCOPE. In the COMPOUND MICRO-SCOPE, which has quite superseded the 'simple microscope' as an instrument of research, the object is magnified in the first instance by the object-glass, and then remagnified by the eye-piece. It follows, therefore, that the magni-fying power of the instrument may be in-creased either by increasing the power of the object-glass or that of the eye-piece. It must be borne in mind, however, that in increasing the power of the eye-piece we do not magnify the object itself in a greater degree, but simply increase the image of the object formed by the object-glass. Any imperfections which may exist in the latter are thus greatly increased. At first the great drawback to the use of the compound microscope was its deficiency in achromatism; but the researches of Mr. Lester and Dr Goring led to the achromatising of the object-glass, which was the first of the rapid strides towards perfection made by this instrument during the last twenty years. The two most useful object-glasses are the 'quarter-inch,' which should magnify from 200 to 220 diameters, and the 'inch,' which should magnify from 30 to 40 diameters. The definition of these glasses should be good, and they should transmit plenty of light. Any lines in a structure examined by them should appear sharp and distinct, and there should be no coloured fringes around the object.

The following practical hints will be found useful to those using the microscope:

The instrument should always be chosen with regard to the work it is intended to do; and for the purposes of the student the simpler the instrument, consistent with efficiency, the better. Experience shows that large and costly microscopes are quite unfitted for the purposes of constant study, their bulk and the preparation required to make them ready for use, and the great care necessary to prevent their complicated construction suffering damage by dust, &c., prevent their being constantly at hand and always ready for use at a moment's notice; for this reason the Continental model has become the student's microscope, and English makers have not been slow to adopt it. One of the best in the market is Swift's student's microscope; whilst of the Continental makers those of Zeiss, of Jena, are in the first rank.

The following general directions for the use of the microscope will be especially useful to those who are but little acquainted with the instrument.

a. Always examine the object first with a *low power*, first adjusting the light with the mirror so that the field is evenly illuminated; place the object on the stage, and with the *coarse* adjustment bring the body (the tube which carries the eye-piece and objectives) to within a quarter of an inch from the object; then slowly raise the body, looking through the eye-piece the while, until the object is clearly visible; then focus accurately with the fine adjustment.

b. In using a high power always start with the objective some distance above the object, and bring it down very carefully to the proper position.

c. Never let the objective touch the object; dirt is sure to get on the front lens, and the cleaning of an objective should be avoided if possible. Glycerine may be removed by the use of a little *clean* water. Canada balsam and varnishes are best removed by the use of *clean* spirit and subsequent careful wiping with an old silk handkerchief. If this accident occur frequently it is exceedingly probable that the cement

of the front lens of the object-glass will be acted upon by the spirit, the glass loosened, and the object-glass seriously damaged.

d. Keep both eyes open, and get into the habit of using either eye; this saves much fatigue.

e. The higher the power the smaller the diaphragm required.

f. If the object appear dirty—(1) Turn the eye-piece round; if the dirt moves with it, take it out and clean it. (2) If after cleaning the object and the eye-piece the dirt still remains, it follows that it must be on the objective, which should be cleaned very carefully.

The object. The nature of the object will, to a large extent, determine the manner of its preparation, and the special treatment required by objects intended to be viewed by reflected light is so various as to preclude notice here.

The following methods are specially applicable to animal and vegetable tissues.

All objects should be mounted on *flat* slips of glass, measuring 3 inches by 1 inch, and should be covered by a piece of specially thin glass, called the cover-slip, which serves to protect the object from injury and prevent its coming in contact with the objective. The best cover-slips to use are ⅓ and ¼ inch *squares*, and they should be very thin; and whenever an object is mounted which will bear examination with a high power care should be taken to avoid a thick cover-slip, as it may, and often does, prevent the objective being brought sufficiently close to the object. Circular cover-slips are generally thick, as it is extremely difficult to cut circles out of very thin glass, and the price is high accordingly. New cover-slips are often very dirty and difficult to clean; they should be soaked for a time in strong nitric acid, then placed in a quantity of *clean* water, changed once or twice to get rid of all traces of acid, and, if in constant use, may be kept in a shallow glass dish with a good cover, under water; they may be cleaned between the finger and thumb by means of a very soft and well-washed silk handkerchief.

Methods of preparing Tissues. *Teasing.* Small portions of a tissue or organ are torn up with fine needles, in order to show the minute structure or the structural elements. Nerve and muscle are tissues to which this method is applicable.

Maceration greatly facilitates the process of teasing, and if properly carried out will preserve the individual cells, &c., whilst loosening their connections with one another. Suitable macerating fluids for animal tissues are—

(*a*) Ranvier's Alcohol. Strong spirit, 1 part; water, 2 parts. Fresh specimens may be soaked in this for 24 hours.

(*b*) Baryta Water. Useful for certain structures.

(*c*) Müller's Fluid. Bichromate of potash solution, ⅓% to 1% in water, with a little sodic sulphate.

(*d*) Schulze's Macerating Fluid for Vegetable Tissues. Potassium chlorate, 1 grm.; nitric acid, 50 c.c. The tissue is placed in a small quantity of this fluid, boiled in it for a short time, and subsequently well washed in water.

Hardening. This is a very necessary preliminary with all soft tissues, animal or vegetable, in order that sections of them may be readily cut.

The most-usual methods adopted for hardening *animal* tissues are the following:

Hardening of Organs. The following are to be hardened in spirit alone: lymphatic glands, heart, skin, thyroid, salivary glands, pancreas, suprarenals.

The following are to be placed in ½% chromic acid containing a very little osmic acid: tongue, liver, spleen, kidney, uterus, ovary, testis. The fluid must be changed the next day. After a week they may be transferred to spirit.

The following are to be slightly distended with the chromic acid and osmic solution, and immersed in a quantity of the same fluid: trachea and lungs, œsophagus, stomach and duodenum, ileum and large intestine, ureters and bladder. The next day pieces are to be cut from them and placed in fresh fluid. They are to be transferred in a week's time to spirit.

Many organs are well hardened by a 2% solution of bichromate of potash. They should remain in this a fortnight, and may then be well washed with water and transferred to spirit. The following can, if desired, be prepared in this way: tongue, liver, spleen, kidney, suprarenals.

The various parts of the central nervous system are best hardened in bichromate of ammonia (3%).

Almost any of the organs may be equally well hardened by immersion for two or three days in saturated solution of picric acid. They must be well washed by letting water flow over them from a tap for an hour or more, and the hardening is then completed by spirit. It is better to stain sections made from organs which have been hardened in picric acid in a solution of picro-carminate of ammonia (1%) instead of logwood.

Except in the case of those organs which are distended *in toto* with the hardening fluid, comparatively thin pieces should be taken for hardening, so that the fluid may rapidly penetrate to all parts of the tissue. A piece of filter-paper should be put at the bottom of the bottle, in order that the pieces of tissue may not rest against the glass. For bichromate of potash and bichromate of ammonia thicker pieces can be used than for chromic acid.

For vegetable tissues the following are recommended:

1. *Absolute Alcohol.* The specimen may be kept in this for an indefinite period. It becomes very brittle, but may be rendered less so by immersion, for 24 hours previous to cutting sections, in a mixture of equal parts of glycerine and absolute alcohol, exposed to the air so that the glycerine may evaporate. The sections cut from a preparation so treated must be mounted in glycerine.

2. *Picric Acid.* Saturated aqueous solution.

3. *Chromic Acid.* 0·1% to 0·5% solution in water. The time of immersion will vary with the nature of the material, from a few minutes to 24 hours. The preparation is removed from either fluid to a 50% solution of alcohol, next to a 70% solution, and finally to absolute alcohol; fresh alcohol is applied so long as it is discoloured by the preparation.

4. *Osmic Acid.* 0·1% to 1% solution in water. This reagent acts very rapidly, and in the case of simple structures, filamentous algæ, &c., 5 to 10 minutes is sufficient. The preparation should be washed with 50% alcohol, left in it for some time, and then removed to 70% alcohol. Osmic acid preparations must be mounted in glycerine.

Staining. The object of staining is twofold—to make transparent parts more visible, and more especially to enable the various elements of a structure to be distinguished from one another.

For this purpose a great number of colouring matters have been used by different investigators, of which the following are the most important:

1. *Hæmatoxylin (Kleinenberg's).* Prepared as follows:

(*a*) Saturated solution of crystallised calcic chloride in 70% alcohol; add alum to saturation.

(*b*) Saturated solution of alum in 70% alcohol: mix *a* and *b* in the proportion of 1 to 8.

(*c*) To this mixture add a few drops of a saturated solution of hæmatoxylin in absolute alcohol.

Specimens to be stained with this fluid must be free from acid.

2. *Borax Carmine.* Carmine, 2 parts; borax, 4 parts; water, 100 parts; add an equal volume of 70% alcohol; allow to stand for a day or two, and filter.

3. *Picro-carmine.* Carmine, 1 grm.; liquor ammoniæ fortis, 4 c.c.; distilled water, 200 c.c.; add 5 grms. of picric acid, shake well for some minutes, decant from excess of acid, allow to stand for some days, stirring occasionally, then evaporate to dryness, and to every 2 grms. of the dry residue add 100 c.c. of distilled water.

4. *Aniline Colouring Matters.* Alcoholic solutions of magenta, saffranin, fuchsin, methyl green, methyl violet, Hoffmann's blue, methylene blue, eosin, rosolic acid, and many others. These colours have the advantage that they stain very readily and are easy of application; but many of them are fugitive.

5. *Silver Nitrate.* This is a most useful reagent for many purposes, especially where it is desired to render distinct the outlines of cells. It stains the intercellular substance deeply, leaving the cells themselves almost unaffected. A 1% solution in water is the best strength. The tissue must be perfectly fresh, immersed for from 2 minutes to 1 hour, then well washed with *distilled* water and exposed to light.

Preparation of Sections. In order that the structure of tissues and organs may be satisfactorily made out it is generally necessary to prepare exceedingly thin sections. This may be done—

1. By holding a piece of the hardened material between the fingers and cutting with a keen razor moistened with spirit.

2. If the piece of tissue be small, by holding it between two pieces of carrot or potato grooved so as to hold it firmly.

3. By embedding it in paraffin, cacao butter, or mixtures of paraffin and lard, wax and castor oil, &c. This is best effected by making small paper boxes which are half filled with the melted embedding material, allowed to cool sufficiently to hold the tissue to be cut, and then filled up completely and allowed to solidify.

An excellent method, especially when only small pieces of tissue are available, is the following:

Small pieces of hardened tissues and organs may be embedded in cacao butter, preparatory to cutting sections from them. Supposing the hardening to have been completed by immersion in spirit, a thin piece of the tissue is removed and soaked first in oil of cloves to displace the spirit, and then in melted cacao butter, so as to fill all the interstices of the tissue with this fat. The tissue is then placed on a piece of wood or cork, and covered with an excess of cacao butter, which is allowed to set firmly, when thin sections can be made with a clean dry razor. The sections are placed in oil of cloves (in winter this must be slightly warmed) to dissolve out the cacao butter. They must be stained before mounting. The staining may be effected by immersion in a watch-glass of oil of cloves, coloured by a drop of a 1% solution of magenta in alcohol (this must be freshly prepared). When stained they are placed on a slide, the excess of coloured oil of cloves removed, and a drop of Canada balsam solution added. Or the sections may be stained with log-wood. With this end, the oil of cloves must be washed out by immersion in strong spirit, and from this the sections are transferred to the staining fluid (for oil of cloves will not mix with the water which the logwood solution contains). When sufficiently stained they are lifted out, dipped in water to remove the excess of the staining fluid, placed for a few minutes in strong spirit, then in oil of cloves, and finally may be mounted in balsam.

It is sometimes advantageous to stain the slices from which sections are to be prepared *in toto* before embedding. This may be done by immersing them for some hours in the staining fluid. The best fluid to choose for this purpose is a 1% solution of magenta in alcohol. The stained pieces are soaked first with oil of cloves, and then with cacao butter, in the same way as the unstained, and the sections which are made from them are placed in a watch-glass of oil of cloves to remove the cacao butter, and can then be mounted at once in Canada balsam solution.

The time which slices of the hardened tissues require to soak in oil of cloves and cacao butter depends chiefly upon the thickness of the slice. If this does not exceed that of a penny piece, from 15 to 20 minutes in oil of cloves and an hour in melted cacao butter is sufficient. Thick pieces would require to be left 2 or 3 hours in oil of cloves, and several hours in melted cacao butter. It is better, especially in summer, to mount the soaked slice of tissue upon the piece of wood or cork at least the day before the sections are to be cut from it, so that the cacao butter may be thoroughly set.

The parts of the central nervous system must not be *soaked* with cacao butter, but may be embedded simply, either in this fat, or in a mixture of wax and oil.

The processes of soaking and embedding may in great measure be dispensed with if a freezing microtome is available for use. These instruments enable larger sections to be prepared, and effect a considerable saving of time, but they are rather expensive.

Mounting. Whenever practicable Canada balsam should be used as a mounting medium. Solutions which require cells to contain them are troublesome, as the cells are constantly leaking, and the specimens are ruined by the admission of air.

Canada balsam (baked to expel turpentine and other volatile matters) may be dissolved in benzol, chloroform, or xylol, and the section, previously soaked in oil of cloves to remove spirit and render it transparent, is placed on the slide. Excess of oil of cloves is removed with filter- or blotting-paper. One drop of Canada balsam solution is allowed to fall on to it from a glass rod and a clean cover-slip placed over it and carefully pressed into position, care being taken to avoid air-bubbles. Glycerin diluted with its own bulk of water is a most useful mounting medium. Care must be taken that only just so much is put on the slide that none exudes from under the cover-slip. If this be the case it must be cleaned off. The cover is then cemented down with Canada balsam solution.

In addition to the above innumerable processes and methods have been devised to suit special cases. The student desiring further information should consult 'Marshall on the Frog,' 'Practical Histology' (Schäfer), 'Practical Botany' (Bower and Vines).

MICROSMIC SALT. NaNH₄HPO₄.4Aq. *Syn.* TRIBASIC PHOSPHATE OF SODIUM AND AMMONIUM. Occurs in putrid urine and in guano. *Prep.* Phosphate of sodium, 6 parts; water, 2 parts; liquefy by heat, and add of sal-ammoniac (in powder), 1 part; common salt separates, and after its removal the liquid is concentrated so that crystals may form. Purify by recrystallisation.

Prop., &c. Colourless prismatic crystals which are very soluble and fusible, leaving a glass of sodium metaphosphate which is valuable in blow-pipe assays for dissolving metallic oxides.

MILDEW. *Syn.* RUST, BLIGHT. The mouldy appearance on the leaves of plants produced by innumerable microscopic fungi. The hop, wheat, and the choicest garden fruit trees are those most commonly attacked. The causes are said to be excess of moisture, and absence of the free circulation of air and sunshine. On the small scale, finely powdered sulphur is occasionally dusted over the parts affected, as a remedy.

MILK. *Syn.* LAC, L. The value of milk as an article of food is clearly shown by the fact of it being sufficient to support, and to increase the growth of, the young of every species of the mammalia; at once supplying materials for the formation of the osseous, fleshy, and liquid portions of the body. "The substances present in milk are wonderfully adapted to its office of producing materials for the rapid growth and development of the animal frame. It contains an azotised matter, casein, nearly identical in composition with muscular flesh, fatty principles, and a peculiar sugar, and, lastly, various salts, among which may be mentioned phosphate of lime, held in complete solution in a slightly alkaline liquid.

"The white and almost opaque appearance of milk is an optical illusion. Examined by a microscope of even moderate power, it is seen to consist of a perfectly transparent fluid, in which float about numbers of minute transparent globules;

these consist of fat surrounded by an albuminous envelope, which can be broken mechanically, as in the churning, or dissolved by the chemical action of caustic potassa, after which, by agitating the milk with ether, the fat can be dissolved " (*Fownes*).

The fatty constituent of milk is not a simple chemical substance, but a mixture of various fats or glycerides, viz. olein, palmitin, stearin, and butyrin.

Comp. Cows' MILK of average quality contains from 10% to 12% of solid matter when evaporated to dryness by steam heat, and has the mean sp. gr. 1·03; while that of the skimmed milk is about 1·035; and of the cream, 1·0244 (*Ure*). The average CREAM of cows' milk contains 4·5% of butter, 3·5% of curd, and 92% of whey (*Berzelius*). The SKIMMED MILK consists of water, 92·9% ; curd, 2·8% ; sugar of milk, 3·5%; lactic acid, lactate of potash, and a trace of lactate of iron, 0·6%; chloride of potassium, phosphate of potash, and earthy phosphates (lime), 0·2% (*Berzelius*).

The following analysis of fresh milk is by M. Haidlen :

Water	873·00
Butter	30·00
Casein	48·20
Milk-sugar . . .	43·90
Phosphate of lime . .	2·31
,, magnesia . .	0·42
,, iron . .	0·07
Chloride of potassium .	1·44
,, sodium . .	0·24
Soda in combination with casein	0·42
	1000·00

The most important constituents of milk are milk-sugar and casein. It used to be thought that these substances existed in constant proportions, but recent researches have shown that they may vary widely, not only with the animal from which the milk is obtained, but with its food, general health, the time of the day and season of the year at which it is milked. The dietetic value of milk is not popularly recognised in this country. In Switzerland it forms the staple food of the entire peasant population, whilst in Kurdistan the average consumption of milk reaches from 4 to 6 pints per head daily. The poorer classes in England take very little, *e. g.* the masses of Spitalfields use about 7·6 oz. weekly, and those of Bethnal Green a fraction under 1½ oz. weekly per head (*Atcherley*).

Professor Wanklyn has devised and published in his excellent little manual, ' Milk Analysis ' (Trübner and Co.), a process by which a very thorough chemical examination of milk may be accomplished with great facility and expedition.

In his preliminary remarks he condemns, as utterly unreliable and misleading, the inferences to be drawn from those hydrometric instruments, the lactometer or lactodensimeter, and creamometer. " A very little consideration," he says, " will suffice to make intelligible the obliquity of the indications of the lactometer, and to show how untrustworthy it must be. The lactometer, as of

course will be understood, is simply the hydrometer applied to milk ; and readings of the instrument are neither more nor less than specific gravities. The more milk-sugar, and casein, and mineral matter there is in a given specimen of milk, the greater (other things being equal) will be its density or specific gravity, and the higher the lactometer reading. ·

" If, however, fat-globules (as happens in the instance of milk) be diffused through the fluid, then, because fat is lighter than water, the effect of the other milk solids on the gravity of the liquid will be more or less neutralised. The density of milk-fat is about 0·9, water being 1·0. Now, if a solution of casein and milk-sugar, of sp. gr. 1·03, be sufficiently charged with fat-globules, its specific gravity may be sent down even below the gravity of water. How much would be required to bring about such a result is a matter of simple calculation.

" This being understood, it will be obvious that if the specimens of milk differ in specific gravity, there must be two distinct and equally valid ways of accounting for the difference.

" The milk with the lower gravity may be milk let down with water, or let down with fat, *i. e.* milk let down by being enriched."

In support of this last assertion Professor Wanklyn quotes corroborative instances afforded by the examination of different specimens of milk known as ' strippings,' these being the last portions of milk yielded by the cow at the termination of the milking. All these ' strippings ' had a lower specific gravity than normal milk, through being richer in cream.

Further, Professor Wanklyn points out that the specific gravity of organic fluids is a fallacious index of the amount of solids they may contain, as is illustrated by the fact that whilst a 10% solution of chloride of potassium has a specific gravity of 1·066 at 15° C., a 10% solution of casein and milk-sugar has a specific gravity of only about 1·035.

The creamometer meets with equal condemnation in Professor Wanklyn's little book, since different specimens of milk vary considerably in their yield of cream, and a perfectly pure sample of milk may yield less cream than one which has been tampered with.

Dr Young stated in 1886 that " a very simple little instrument for the examination of milk by colour has recently been invented in Germany, viz. the Heeren patent milk tester. It is made of vulcanite, and on a raised portion of a raised disc a little milk is placed ; over this there is put a glass cover, which spreads out the milk. Round the glass are painted various shades of colour, representing cream, very fat milk, normal, less fat, poor, and very poor. . . . . It may be used as a rough and ready method of speedily testing milk."

A complete analysis of milk involves the determination of the water, the fat (the essential constituent of the cream), the casein, milk-sugar, and ash.

The following is an outline of Professor Wanklyn's neat and ingenious method of analysis :

By means of an accurately graduated pipette

he first places 5 c.c. of the milk in a small weighed platinum dish (about 14 grms. in weight), having just previously ensured that the sample from which the milk is taken is thoroughly mixed.

The dish is then placed over a water-bath (the water in which must be kept vigorously boiling the whole time) for 3 hours, at the end of which time all the water having been driven off, there will remain in the dish a completely dried-up residue.

The increase in weight between the empty dish and the residue will give the weight of the 'milk solids' from 5 c.c. of milk. Of course, if this weight be multiplied by 20, the yield from 100 c.c. of milk will be obtained.

To reduce this to a percentage statement it is necessary to remember that 100 c.c. of average milk weigh 102·9 grms.

The next proceeding consists in the determination of the fat. This is done by treating the dried milk solids resulting from the 5 c.c. of milk with ether. There are several important minutiæ necessary to be observed in connection with this part of the process, for the particulars of which the reader is referred to Professor Wanklyn's book. Suffice it to say that if properly performed the whole of the fat is dissolved by the ether, and, being separated from the non-fatty portion of the residue, is weighed and calculated as 'fat.' If then the amount found as fat be deducted from the whole of the milk solids previous to their treatment with ether, the 'milk solids, not fat,' will be arrived at. Professor Wanklyn estimates the casein (under the head 'Casein' Professor Wanklyn includes the entire nitrogenous materials of milk) as follows: He treats the milk solids, not fat, with hot alcohol, which dissolves out from them the milk-sugar and the soluble chlorides. The remaining residue, consisting of casein and phosphate of sodium (chemically combined with the casein), is dried on a water-bath until it ceases to lose weight. It is then weighed along with the vessel containing it, and ignited. The combined weight of the vessel and phosphate of sodium remaining after ignition being deducted from the weight previous to ignition, the difference is the casein.

Another and quicker method, recommended by Professor Wanklyn, for the determination of the casein, is to measure it by the amount of albuminoid ammonia it is capable of yielding when subjected to the 'albuminoid ammonia process,' invented by Messrs Wanklyn, Chapman, and Smith.

The alcoholic solution filtered off from the combined casein and phosphate of sodium contains the milk-sugar and soluble chlorides. It is evaporated to dryness on a water-bath, and the residue with the vessel containing it is weighed. It is then gently ignited, and the weight of the remaining residue, being deducted from the total weight before ignition, gives the yield of milk-sugar. Or the milk-sugar may be determined by titration with a standard copper solution.

For the determination of the ash it is only necessary to ignite the milk solids from 5 c.c. of milk in the small platinum dish, by which operation all the organic matter being burnt, that which remains behind constitutes the 'ash,' and is weighed as such.

It will be obvious that in order to determine with anything like rigid accuracy the quality of any sample of milk by analysis, not only must a normal standard for the purpose of comparison be adopted, but such normal standard must represent very closely and with but little variation the definite composition of all sound and genuine milk.

Professor Wanklyn says that "the following, which is the result of several concordant analyses of country-fed milk, may be taken as representing normal milk. In 100 grms. of milk—

Solids (dry at 100° C.) . . 12·5 grms.
Water . . . . . . 87·5 ,,
———
100·0

"The 12·5 grms. consist of 9·3 grms. of 'solids which are not fat,' and 3·2 grms. of fat." The above data, which are founded on the examination of a very large number of different samples of milk, are confirmed by the researches of Müller and Eisenstuck, who were employed by the Royal Agricultural Society of Sweden in a similar investigation. The labours of these chemists extended over a twelvemonth, and the result of them was to show that the milk yielded day by day, for a whole year, by a herd of cows was remarkably constant in composition.

Professor Wanklyn gives the following formulæ for the calculation and statement of the results of milk analysis. He says, "Treating the question quite rigidly, which I believe is the proper way of dealing with it, we arrive at the following:

"*Problem* 1. Given the percentage of 'solids, not fat' ($= a$), in a specimen of sophisticated milk (*i. e.* milk, either watered, or skimmed, or both)—required the number of grammes of genuine milk which was employed to form 100 grms. of it.

"*Answer.* Multiply the percentage of 'solids not fat' by 100, and divide by 9·3:

Or—
$$\frac{100}{9\cdot3}a.$$

"*Problem* 2. Given the percentage of 'solids, not fat' ($= a$), also the percentage of fat ($= b$), in a specimen of sophisticated milk—required the number of grammes of fat which have been removed by skimming from the genuine milk which was employed to form 100 grms. of it.

Answer :
$$\frac{3\cdot2}{9\cdot3}a - b.$$

"In translating fat into cream, the rule is that a removal of 0·2 grm. of fat equals a removal of 1·0 grm. of cream. This rule is directly founded on experiment. I do not, however, claim a high degree of accuracy for the measurement of the cream.

"Finally, a slight refinement may be noticed. If a specimen of sophisticated milk has been produced by both skimming and watering, it will be obvious, on consideration, that the extraneous waters employed in manufacturing 100 grms. of it is equal to the difference between 100 and the quantity of genuine milk employed to make 100 grms. of sophisticated milk, together with a quantity of water equal to that of fat removed by skimming."

$$\text{Extraneous water} = 100 \frac{100}{9 \cdot 3} a + \frac{3 \cdot 2}{9 \cdot 3} a - b$$

$$= 100 \frac{100 + 3 \cdot 2}{9 \cdot 3} a - b.$$

Save for the purpose of finding out the presence of matters other than an excess of water in the milk (a contingency regarded as very improbable), the estimation of the casein and milk-sugar is unnecessary. The determination of the ash is for the object of learning if foreign mineral matters, such as chalk or any other inorganic impurity, are present. Professor Wanklyn says he believes that such extraneous bodies are never employed. The chief, if not the sole forms of dishonesty are watering and skimming. It is doubtful whether mixing with watered condensed milk can pay, or indeed escape detection by the taste; such milk is always weaker than normal milk.

The amount of ash, however, is a good criterion as to the extent of dilution that has been practised, a deficient amount being, of course, confirmatory of a watered milk.

The determination of the amount of 'solids, not fat,' is, in almost every instance, all that is necessary to enable an opinion to be arrived at as to whether the sample of milk has had water added to it or not.

Dr Young considers that good milk should contain from 12% to 13% of solids. The Society of Public Analysts has fixed the percentage at 11·5, a very moderate standard indeed. In a prosecution, if the suspected milk barely comes up to this standard, the defendant can send a sample to Somerset House—the ultimate court of appeal in such cases. Now the authorities there have never yet declared what is this standard; all that is known of it is that it is a very low one. For the ends of justice and for the sake of the public it is time that some recognised standard should be declared, so that public analysts may have proper grounds for recommending a prosecution.

Out of fifty-six samples of milk supplied to the different London unions in 1873, Professor Wanklyn reports that he found only fifteen unwatered, or nearly unwatered. Of these fifteen samples nine had been skimmed, leaving only six that were at once unwatered and unskimmed. These figures, therefore, show that only about 10% of the milk supplied in the above year to the metropolitan unions was genuine. He adds, "It is curious to compare the language of the contract under which (as it appears from Mr Rowsell's report) the dealer supplied the various unions with milk, with the quality of the article as exhibited by the analysis. 'New unskimmed milk unadulterated,' 'genuine as from the cow,' 'best new unskimmed milk, to produce 10% of cream,' occur in these contracts."

This tale of sophistication is still practically true in the year 1890. Wigner says that "ten years' working of the Anti-adulteration Acts has brought us really to this point, that as regards milk our position is hopeless until the law is amended; no one can hope to get pure milk in London unless under other guarantees than these Acts afford."

Prop. Normal milk is opaquely white, and has no peculiar taste or smell; sp. gr. about 1·03; should not change its appearance by boiling. Perfectly fresh milk is slightly alkaline, but soon becomes acid on exposure to the air, and after a time white coagula of casein (CURDS) separate from it. This change is immediately effected by the addition of rennet or an acid. That from the first, when dried and pressed, constitutes cheese. This spontaneous acidification is due to the fermentation of the sugar of milk which results in the production of lactic acid. If it be now heated it coagulates, owing to the separation of the casein. When milk is kept for some time at 90° F. the result is different; the milk-sugar under the influence of the casein is converted into grape-sugar, and this then breaks down by fermentation into alcohol and carbonic acid.

Tests, &c. The common frauds practised by the milk dealers are the addition of water and the subtraction of part of the cream; the best methods for dealing with these adulterations have already been dealt with. Sometimes potato starch is added to the milk to give it a creamy or rich appearance, and this addition is still more frequently made to cream, to increase its consistence and quality.

The presence of potato starch may be determined by boiling some of the milk with a little vinegar, and after separating the coagulum by a strainer, and allowing the liquid to become cold, testing it with solution or tincture of iodine. If it turns blue, starch, flour, or some other amylaceous substance has been used to adulterate it. In most cases it will be sufficient to apply the test to the unprepared suspected milk.

Dr Young says that starch, dextrine, gum, and glycerin have been found in milk, having been added to make it thicker and richer looking; annatto and turmeric are sometimes added to give colour; nitre to take away the 'turnip taste;' magnesia, tragacanth, arrowroot, and yolk of egg to enrich the cream. But these are rarely resorted to, and when present require special methods of analysis, for which the reader must consult a text-book.

Mixtures of borax and carbonate of soda are sometimes added to preserve milk; these, however, can readily be recognised by ordinary tests.

It used to be frequently stated that chalk, plaster of Paris, gum, gelatin, sugar, flour, mucilage of hemp-seed, the brains of animals, and other similar substances are often added to London milk, but there is no reason to suppose there is any truth in these assertions, as some of these articles are too costly to be used, and the presence of others would so alter the flavour or appearance of the milk, or would so soon exhibit themselves by subsidence, as to lead to their detection.

The microscope is practically the only means by which diseased milk can be detected, but this instrument is only useful in highly skilled hands. The only reliable method of ascertaining the quality of milk is by means of a full chemical analysis (*Wigner*).

Pres. Milk may be preserved in stout bottles, well corked and wired down, by heating them, in this state, to the boiling-point in a water-bath, by which means the oxygen of the small quantity of enclosed air becomes absorbed. It must be afterwards stored in a cool situation. By this method, which is also extensively adopted for the preservation of green gooseberries, green peas, &c., milk will retain its properties unaltered for years. A few grains of carbonate of magnesia, or, still better, of bicarbonate of potassa or soda, may be advantageously dissolved in each bottle before corking it.

Under Bethel's patent the milk or cream is scalded, and, when cold, strongly charged with carbonic acid gas by means of a soda-water machine, and the corks are wired down in the usual manner. The bottles should be kept inverted in a cool place.

An excellent method of preventing milk from turning sour, or coagulating, is to add to every pint of it about 10 or 12 gr. of carbonate or bicarbonate of soda. Milk thus prepared may be kept for eight or ten days in mild weather. This addition is harmless, and, indeed, is advantageous to dyspeptic patients. According to D'Arcet, $\frac{1}{1000}$ part of the bicarbonate is sufficient for the purpose. An excess of alkali used in this manner may be detected by the milk turning turmeric paper brown, even after it has been kept some hours, and by the ash obtained by evaporating a little to dryness, and then heating it to dull redness, effervescing with an acid (see *below*).

*** Milk should not be kept in lead or zinc vessels, as it speedily dissolves a portion of these metals, and becomes poisonous.

Concluding Remarks. The principal difference between cows' milk and human milk consists in the former containing more casein and less sugar of milk than the latter. The remarkable indisposition to coagulate is another character which distinguishes human milk from cows' milk. Prof. Falkland, who has investigated the subject, has prepared a nutritive fluid for infants from cows' milk, closely resembling that of the healthy adult woman. His process is, however, unnecessarily complicated, and therefore unsuited to those who would have to employ it in the nursery. To remove this objection we have adopted the following formula:—Sugar of milk, 2 oz.; hot water, ¼ pint; dissolve, and when the liquor has become quite cold, add it to fresh cows' milk, ½ pint, and stir them together. This quantity, prepared morning and evening, will constitute the proper food for an infant of from 5 to 8 months old. More may be allowed if the child 'craves' for it; but there must be no 'cramming.' At first it will be advisable to remove a little of the cream from the milk before adding the saccharine solution; but after a few days this will be found to be unnecessary, and, indeed, may be injurious. One very important particular to be attended to is the employment of pure cows' milk, obtained from a healthy grass-fed animal only. With this precaution, and the use of a good FEEDING-BOTTLE, the infant will thrive nearly as well as on the breast of any human female, excepting its mother's (see *below*).

ASSES' MILK closely resembles human milk in colour, smell, and consistence, but it contains rather less cream. Though not an appropriate food for a healthy infant, it is easier of digestion than cows', containing as it does more water, less curd and butter, and an excess of sugar and salts. These latter sometimes cause diarrhœa. It is valuable as a change, but its use requires care and judgment.

EWES' MILK closely resembles cows' milk, than which, however, it is slightly richer in cream.

GOATS' MILK, for the most part, resembles cows' milk, but its consistence is much greater, and it contains much more solid matter.

MARES' MILK, in consistence, is between that of cows and human milk. Its cream is not converted into butter by agitation. See BUTTER, CHEESE, LACTIC ACID, &c.

The following table, compiled from Boussingault's analyses, shows the relative composition of the milk of woman, the cow, ass, and goat:

	Cow.	Ass.	Goat.	Woman.
Water	87·4	90·5	82·0	88·4
Butter	4·0	1·4	4·5	2·5
Milk-sugar	} 5·0	6·4	4·5	4·8
Soluble salts				
Casein	} 3·6	1·7	9·0	3·8
Insoluble salts				

Milk as a Cause or Carrier of Disease.—Dr Tacher, in his pamphlet on the 'Transmission of Disease by Milk,' sums up the ways in which a possible disease may be transmitted under the following heads:

1. It may be derived from an animal suffering from a specific epizootic disease.

2. It may be derived from a tuberculous animal.

3. It may be drawn from an inflamed udder.

4. It may have undergone chemical or fermentative change.

5. It may have been infected with the contagium of an animal disease.

6. It may have become infected with the contagium of a human disease.

Professor Brown says that the 'foot-and-mouth disease' (*aphtha epizootica*) may be communicated to man through the milk of an animal suffering from this disease. When a cow is very bad the milk contains numerous pus-like bodies, bacteria, vibriones, &c., and it rapidly undergoes putrefaction. The symptoms of this disease in man are fever, vomiting, swelling of the glands of the neck and throat, and ulcerations of the tongue and mouth.

Anthrax is another bovine disease which has been communicated to man through milk. The disease is propagated by a bacillus, according to Koch and Pasteur, which thrives well in milk.

With regard to the communication of tuberculosis there are various opinions. Professor Bang, of Copenhagen, has found, however, that animals fed upon the milk of tuberculous cows have themselves developed the disease, and a bovine form of tuberculosis has been found in man.

Milk of a mother labouring under strong mental emotion is, as is well known, capable of seriously endangering the health of the suckling babe. Payne narrates the case of a woman suffering under a nervous affection whose milk, two hours after an attack of the disease, became viscid, like the white of an egg. Similarly, a deterioration and consequent alteration in properties is induced in the milk of the cow if she be over-driven, exhausted, or harassed. The food of the animal likewise exercises an influence on the quality of its milk, often altering both taste and appearance; thus when cows are fed on turnips, wormwood, decayed leaves, and plants of the cabbage or onion family, the flavour of these substances is imparted to their milk; madder and saffron will colour it. The milk of animals that have fed on poisonous or deleterious plants is capable of setting up toxic symptoms in human beings partaking of it. In June, 1875, the inhabitants of a certain quarter of Rome were attacked with an epidemic, distinguished by great gastro-intestinal irritation. The cause of the outbreak was traced to the use of milk yielded by goats that had eaten of the meadow saffron, the *Colchicum autumnale*. It also appears that in the western States of America the milk of cows that have fed on the poison-oak, the *Rhus toxicodendron*, has on several occasions given rise to attacks of illness in children, marked by extreme weakness, vomiting, fall in bodily temperature, swollen and dry tongue, and constipation. Boiling seems to remove the dangerous properties of the milk.

Milk, as has been shown by Fuchs, is sometimes infested by a fungus, the *Oidium lactis* or *Penicillium*, which is capable of giving rise to gastric irritation, and sometimes to severe febrile gastritis (*Parkes*). This fungus turns milk blue, and yellow cream being mixed with it produces green milk. There is also a yellow milk resulting from a *vibrio*.

Although the evidence as to the power of the milk of animals affected with epizootic diseases to convey the particular affection to human beings is contradictory, there is little reason to doubt that soured milk may become a carrier of infection from the ailing or convalescent subject to the healthy one.

Typhoid, diphtheria, and scarlet fever have been known to have originated in this manner. The epidemics of these diseases due to milk occurring during the last 12 years have been tabulated by Mr Ernest Hart, who finds that of typhoid there were 50, of scarlatina 15, and of diphtheria 7.

The outbreak of the former malady in Marylebone in 1874 was traced to the contamination of milk by the remains of the water which had been used in rinsing the milk-pans. This water had been obtained from a well into which the excreta from a typhoid patient had percolated from a privy.

At Leeds a similar outbreak was caused owing to the absorption, by milk, of the typhoid effluvium. In the case of scarlet fever the malady has been conveyed by means of the throat discharges and particles of cuticle falling into the milk from the persons of servants and others employed in dairies.

The epidemic of diphtheria which broke out in the north of London in 1878 was definitely traced to the milk, and many others have been traced to the same cause.

Milk may acquire medicinal properties by cows feeding on certain plants, and this method of treating diseases has been suggested.

Milk, Al'mond. See EMULSION and MIXTURE.

Milk, Arrowroot. *Prep.* From arrowroot, 1 table-spoonful, first wetted and stirred with a little cold water, afterwards adding, gradually, of boiling water, ¼ pint; and, lastly, of boiling milk, ¼ pint; with sugar, spice, wine, &c., to taste. Very nutritious, and excellent in chronic diarrhœa. Some persons employ all milk.

Milk, Choc'olate. *Prep.* Dissolve chocolate (scraped), 1 oz., in boiling new milk, 1 pint. Nutritious, but apt to offend delicate stomachs.

Milk, Cof'fee. *Prep.* 1. Coffee, 1 oz.; boiling water, ¼ pint; infuse for 10 or 15 minutes in a warm situation, and add the strained liquid to boiling milk, ¾ pint.

2. Coffee, 1 oz.; tie it loosely in a piece of muslin, and simmer it for 15 minutes in milk, 1 pint. Both the above have been recommended for persons of spare habit, and for those disposed to affections of the lungs, more especially for the asthmatic.

Milk, Facti"tious. *Syn.* ARTIFICIAL MILK. Of the numerous compounds which have been proposed as substitutes for natural milks, the following are examples:

1. (FACTITIOUS ASSES' MILK; LAC ASININUM FACTITIUM, LAC A. ARTIFICIALE, L.) *a.* Cows' milk, 1 quart; ground rice, 1 oz.; oringo root (bruised), 1 dr.; boil, strain, and add sugar-candy (or white sugar), 1 oz.

b. Whites of 2 eggs; lump sugar, 1 oz.; cows' milk (new), ¾ pint; mix, then add syrup of tolu, ½ oz.

c. Water, 1 pint; hartshorn shavings, 1 oz.; boil to a jelly; then add lump sugar, 2 oz.; cool, add new milk, 1 pint; syrup of tolu, ½ oz. Used as substitutes for asses' milk, taken freely as a beverage. A cupful, with or without a spoonful of rum, 3 or 4 times daily, is a popular remedy in consumption and debility.

2. (F. GOATS' MILK—*A. T. Thomson.*) Fresh mutton suet (minced), 1 oz.; tie it in a muslin bag, and boil it in cows' milk, 1 quart; lastly, add of sugar-candy, 2 gr. In scrofulous emaciation, and in the latter stages of phthisis. The proportion of suet in the above may be advantageously increased a little. The LAC CUM SEBO of Guy's Hospital is a similar preparation.

3. (F. HUMAN MILK; LAC HUMANUM FACTITIUM, L.) *a.* See above.

b. (*Rosenstein.*) Almonds (blanched), 2 in number; white sugar, 1 dr.; water, 4 fl. oz.; make an emulsion, strain, and add of fresh cows' milk, 6 fl. oz. As a substitute for the breast in nursing.

Milk, Preserved'. *Syn.* MILK POWDER; LACTIS PULVIS, LAC PULVERATUM, L. *Prep.* 1. Fresh skimmed milk, 1 gall.; carbonate of soda (in very fine powder), 1½ dr.; mix, evaporate to 1-3rd by the heat of a water-bath, with constant agitation, then add of powdered white sugar, 3½ lbs., and complete the evaporation at a reduced temperature; reduce the dry mass to powder, add the cream (well drained) which was taken from the milk, and after thorough admixture put the whole into well-stoppered bottles or tins, which must be at once hermetically sealed.

2. (*Legrip.*) Carbonate of soda, ½ dr.; water, 1 fl. oz.; dissolve, add of fresh milk, 1 quart; sugar, 1 lb.; reduce it by heat to the consistence of a syrup, and finish the evaporation on plates by exposure in an oven.

Obs. About an ounce of the powder agitated with a pint of water forms an agreeable and nutritious drink, and a good substitute for milk. It may also be used for tea or coffee in a solid form. This process, which is very old, has been recently patented. See MILK (*above*).

The condensed or preserved milk, now in such general use, and which is met with in tins, is milk which has been more or less deprived of water by evaporation *in vacuo.* It occurs in the market in two forms—in one simply as condensed milk, and in the other as condensed milk mixed with a large quantity of sugar. Milk preserved as above without sugar will not keep long; whereas with sugar it may be preserved for an almost indefinite time. Either variety mixed with the proper quantity of water becomes normal milk again, the sweetened kind being, of course, milk with the addition of cane-sugar. Professor Wanklyn says he has examined the principal brands of preserved and condensed milk sent to the London market, and finds they contain their due proportion of fat. He gives the following analyses of the produce of the English Condensed Milk Company:

PRESERVED MILK.

In 100 parts by weight.

Water	20·5
Fat	10·4
Casein	11·0
Ash	2·0
Cane and milk sugar	56·1
	100·0

CONDENSED MILK.

Water	51·12
Fat	12·11
Casein	13·64
Milk-sugar	20·36
Ash	2·77
	100·00

Milk of Roses. *Syn.* LAC ROSAE, L. *Prep.* 1. (ENGLISH.) *a.* Almonds (blanched), 1 oz.; oil of almonds and white soft soap, of each, 1 dr.; rose-water, 1 pint; make an emulsion.

b. From liquor of potassa and oil of almonds, of each, 1 fl. oz.; hot water, 2 fl. oz.; agitate together until mixed, then add of rose-water and distilled or filtered soft water, of each, ¼ pint, and again agitate well.

c. As the last, but using ½ a teaspoonful of salt of tartar for the liquor of potassa.

d. (*Redwood.*) Blanched almonds, 8 oz.; rose-water, 3 pints; make an emulsion, add of white Windsor soap, white wax, and oil of almonds, of each, ½ oz.; previously melted together by a gentle heat; triturate until united, and strain; lastly, add a solution of oil of bergamot, ½ oz.; oil of lavender, 1 dr.; and attar of roses, ½ dr.; (dissolved in) rectified spirit, 12 oz.

2. (FRENCH.) *a.* From rose-water, 1 quart;

tinctures of benzoin and styrax, of each, 1 fl. oz.; spirit of roses, ½ fl. oz.; rectified spirit, ½ fl. oz.; mix.

b. (*Augustin.*) Tincture of benzoin, ½ fl. oz.; liquor of carbonate of potassa, 2½ fl. dr.; rose-water, ½ pint; agitate well together. As a lotion in acne.

c. (*Giannini.*) Tincture of benzoin, 1 dr.; tincture of balsam of Peru, 20 drops; rose-water, 1 pint; as the last.

d. (*Schubarth.*) Almond paste, 3 dr.; rose-water, ½ pint; tincture of benzoin, ½ fl. oz. As before. The addition to the last three of a little rectified spirit is an improvement.

3. (GERMAN.) From dilute solution of diacetate of lead (Goulard water) and spirit of lavender, of each, 1 fl. oz.; rose-water, 6 fl. oz.; soft water, 1 pint.

Obs. All the above are used as cosmetic washes, and to remove scurf, pimples, and eruptions in slight cases.

Milk, Sa'go. *Syn.* LAC SAGO, L. *Prep.* (*Dr A. T. Thomson.*) Sago, 1 oz.; cold water, 1 pint; macerate half an hour, pour off the water, add of milk, 1½ pints, and boil slowly until the sago is dissolved. Very nutritious; also in lieu of arrow-root milk.

Milk of Sulphur. See SULPHUR (Precipitated).

Milk, Thick. Mix one table-spoonful of flour with a pint of milk, and boil for ten minutes, stirring it well the whole time. It may be flavoured either with a little salt or sugar.

Milk, Vanil'la. *Syn.* LAC VANILLAE, L. *Prep.* 1. Essence of vanilla, 12 drops; lump sugar, 1 oz.; triturate, and add gradually, new milk, 1 pint.

2. (*Béral.*) Vanilla sugar, ½ oz.; milk, 16 oz.; dissolve.

MILK FEVER. *Syn.* FEBRIS LACTEA, L. A febrile condition of the system that sometimes occurs at the time the milk begins to be secreted after parturition. It often assumes a malignant character. See PUERPERAL FEVER.

MIL'LET. *Syn.* MILIUM, L. Several varieties of grain are known by this name. That commonly referred to under the name is the produce of *Panicum miliaceum* ('Indian millet').

In the subjoined table is given the composition of three different samples of millet meal, free from bran.

	Panicum miliaceum, common millet.	Penicillaria spicata, a kind of millet; much used in India under the name of Bajra.	Sorghum vulgare, Dharra of the Arabs, Goar of India.
Water	12·22	11·8	11·95
Nitrogenous substances	9·27	10·13	8·64
Dextrin	9·13	...	3·82
Sugar	1·80	...	1·46
Fat	7·43	4·62	3·90
Starch	59·04	71·75	70·23[1]
Silicin	0·11	...	...

[1] With husks.

The husked seeds (MILIUM MUNDATUM) are used to make gruel, and are ground for flour. 'Turkish millet,' or 'Guinea corn,' is produced by *Sorghum vulgare;* and the 'German' and 'Italian millet' by species of *Setaria.* In some parts of the world millet flour is used for bread, but it is chiefly cultivated as food for domestic animals.

Letheby says millets are a little more nutritious than rice.

MIMUSOPS ELATA, Allen. Massaranduba or Cow-tree of Para. The milk, resembling good cream in consistence, exudes slowly from the wounded bark. It is too viscid to be a safe article of diet.

MIMUSOPS ELENGI, Linn. The fruit is largely eaten in Guiana and elsewhere, the fragrant flowers are used for making garlands, and the bark yields a tonic and febrifuge.

MIMUSOPS GLOBOSA, Gaertn. The inspissated juice (*balata*) has been introduced from British Guiana as a substitute for gutta percha.

MINCEMEAT. *Prep.* From stoned raisins, currants, sugar, and suet, of each, 2 lbs.; sultana raisins and boiled beef (lean and tender), of each 1 lb.; apples, 4 lbs.; juice of two lemons; the rind of 1 lemon, chopped very fine; mixed spice, ¼ lb.; candied citron and lemon peel, of each, 2 oz.; brandy, a glassful or two; the whole chopped very fine. It may be varied by adding other spice or flavouring, and by the addition of eggs, or the substitution of chopped fowl or veal for beef, according to the taste of the cuisinier.

MINCE PIES. Take 3 apples, 3 lemons, 1 lb. of raisins, ½ lb. of currants, 1 lb. of suet, ½ lb. of raw beef, 2 lbs. of moist sugar, ¼ lb. of mixed candied peel, ¼ of a rind of a fresh orange, 1 teaspoonful of powdered mixed spice, composed of equal parts of cloves, cinnamon, and nutmeg, ½ pint of brandy, and 1 glass of port wine. Peel the apples and cut out the cores very carefully, and then bake the pieces until they are quite soft. The raisins must be carefully stoned, and the currants well washed, dried, and picked. Chop the suet very finely, as well as the raw meat and lemon-peel. Mix all the ingredients thoroughly together, add the brandy last of all, and press the whole down into a stone jar, and place a piece of paper soaked in brandy on the top. Remove the paper and stir up the mixture thoroughly every three days, replacing the paper. If this is done the mincemeat will keep a long time. To make the pies, roll out some thin puff paste, butter a small round tin, and line it with a piece of paste, then put in a generous quantity of the mincemeat, cover it over with a similar piece of puff paste, and bake it in a moderate oven. Mince pies are none the worse for being warmed up, but pray take care they are sent to table hot (*Cassell*).

MINDERE'RUS' SPIRIT. See AMMONIA (Acetate of), and SOLUTION.

MINERAL CHAME'LEON. *Prep.* From a mixture of binoxide of manganese and potassa and nitre, equal parts, heated to redness. It must be preserved in a well-corked bottle until required for use.

Prop., &c. When dissolved in water, its solution, at first green, passes spontaneously through all the coloured rays to the red, when, if potassa be added, the colour retrogrades until it reaches the original green. The addition of oil of vitriol, or of chlorine, renders the solution colourless. The addition of a weak acid, or even boiling or agitating the liquid, turns it from green to red. See MANGANIC ACID.

MINERALISERS. Substances which, by association with metallic bodies, deprive them of their usual properties, and impart to them the character of ores. Their removal belongs to metallurgy. The term 'MINERALISED' has been applied to caoutchouc, gutta percha, bitumen, &c., which have been combined with sulphur, silica, or metallic matter.

MINIM. *Syn.* MINIMUM, L. A measured drop, of which 60 are equal to a fluid drachm. The size of drops varies so greatly with different liquids, and is also much influenced by the size and shape of the vessels from which they are poured, that they afford no reliable measure of quantity for medicinal purposes. The poured drop has, in some cases, only ⅓ the volume of the measured drop, or minim; whilst in others it is nearly 3 times as large. According to Mr Durande. "liquids which contain a small proportion of water afford a small drop; while, on the contrary, liquids containing a large quantity of water furnish a large drop." "Among liquids containing a large proportion of water, those which are not charged with remedial substances give a larger and heavier drop than the same liquids when containing extraneous bodies in solution." In all cases in which the word 'drop' is mentioned in this work a minim is intended, and the quantity should be determined by means of a graduated minim measure.

MINIUM. See RED PIGMENTS.

MINT. *Syn.* SPEARMINT, GREEN M.; MENTHA VIRIDIS (Ph. L.), L. "The recent and the dried flowering herb" of *Mentha viridis.* It is aromatic and carminative, but its flavour is less agreeable than that of peppermint. It is employed in flatulence, colic, nausea, diarrhœa, &c.; also to make sauce.

MIRRORS. See AMALGAM (Silvering), SILVERING, SPECULUM METAL, &c.

MITES. See ACARI.

MITHRIDATE. *Syn.* DAMOCRATE'S CONFECTION; MITHRIDATIUM, CONFECTIO DAMOCRATIS, L. "This composition originally consisted of but few ingredients; viz. 20 leaves of rue, 2 walnuts, 2 figs, and a little salt. Of this we are informed that Mithridates took a dose every morning to guard himself against the effects of poison. It was afterwards altered, and the number of the ingredients increased to sixty-one. In this more complex form it contained opium, and was, in effect, an aromatic opiate, of which the confection of opium of the present day may be considered as a simplification. The 'mithridate' is still prepared in some shops, and is occasionally, though very rarely, prescribed" (Med. Lex.). "The formulæ for CONFECTION or ELECTUARY OF CATECHU may be considered as the representatives, in our modern Pharmacopœias, of the once celebrated recipes for CONFECTIO DAMOCRATIS and THERIACA ANDROMACHI" (*Redwood*). Mithridate was formerly conceived to be good for nearly every disease, and an antidote for every known poison.

MIXTURE. *Syn.* MISTURA, L. A compound medicine, either a clear fluid or containing matter in suspension, taken in divided doses. Mixtures are usually extemporaneous preparations, and in prescribing them care should be taken not to bring together substances that decompose each other, nor to order heavy powders that speedily separate from the body of the liquid by subsidence. EMULSIONS, JULEPS, and DRAUGHTS are special forms of mixtures.

Mixtures are usually dispensed in flat octagonal 6- or 8-oz. bottles with long necks, or in regular 'octagons' with short necks, having the doses marked on the glass, to which the strength of the medicine is made to correspond. Any mixture containing insoluble substances, such as bismuth, salts, oils, or balsams, must be labelled 'Shake the bottle.'

Our remarks respecting 'DRAUGHTS' equally apply here. By putting the active ingredients of six draughts into a 6-oz. mixture bottle, and filling it up with distilled water, a mixture will be made of corresponding properties, of which the dose will be 2 table-spoonfuls. When the formula for the draughts includes a decoction or infusion as the vehicle, instead of water, four of them only must be taken, which will then fill the 6-oz. bottle, and the proper dose will be 3 table-spoonfuls, or a small wine-glassful.

The following formulæ embrace the whole of the 'MIXTURE' of the British Pharmacopœia, as well as a few others in general use. These will serve as examples for the like preparations of medicines which are not included in the list. An extensive collection of mixtures will be found in 'Squire's Pharmacopœias of the London Hospitals.' See also DRAUGHT, EMULSION, JULEP, WATER, &c.

Mixture, Absorbent. See MIXTURE, ANTACID.

Mixture, Aca'cia. See MIXTURE, GUM.

Mixture, Ac'etate of Ammo"nia. *Syn.* MINDERERUS'S MIXTURE; MISTURA AMMONIÆ ACETATIS, L. *Prep.* From solution of acetate of ammonia, 1½ fl. oz.; nitre, 40 gr.; camphor mixture, 6 fl. oz.; rose syrup, ½ fl. oz.—*Dose,* 1 to 3 table-spoonfuls, every third or fourth hour, as a diaphoretic in inflammatory fevers, &c.

Mixture of Acetic Acid. *Syn.* MISTURA ACIDI ACETICI. *Prep.* Distilled vinegar, 2 fl. dr.; syrup, 4 fl. dr.; water, 2 fl. oz. A fourth part every 3 hours. For children with scarlatina.

Mixture of Aconite. (*Mr Fleming.*) *Syn.* MISTURA ACONITI. *Prep.* Tincture of aconite, 1 fl. dr.; carbonate of soda, 1½ dr.; sulphate of magnesia, 1½ oz.; water, 6 oz. A table-spoonful when the pain is urgent. In gastralgia this should only be administered under medical supervision or advice.

Mixture, Al'kaline. See MIXTURE, ANTACID.

Mixture, Al'mond. *Syn.* EMULSION OF ALMONDS, MILK OF A.; MISTURA AMYGDALÆ (B. P., Ph. L., E., & D.), LAC AMYGDALÆ, L. *Prep.* 1 (Ph. L.) Confection of almonds, 2½ oz.; distilled water, 1 pint; gradually add the water to the confection while triturating until they are mixed; then strain the liquid through linen.

2. (Ph. E.) From almond confection, 2 oz., and water, 1 quart; as the last. Or from sweet almonds (blanched), 10 dr.; white sugar, 5 dr.;

mucilage, ½ fl. oz. (or powdered gum, 3 dr.); made into an emulsion with water, 1 quart.

3. (Ph. D.) Sweet almonds (blanched), 5 dr.; refined sugar, 2 dr.; powdered gum, 1 dr.; distilled water, 8 fl. oz.; as the last.

4. (B. P.) Compound powder of almonds (sweet), 1 part; water, 8 parts; triturate and strain.

Obs. The last formula produces the article usually employed in dispensing in the shops. The addition of a little more sugar renders it more pleasant; and 2 or 3 bitter almonds, as in the formula of the Ph. D. 1826, or 1 or 2 fl. dr. of rose or orange-flower water, may occasionally be added to diversify the flavour.—*Dose,* 2 or 3 table-spoonfuls, *ad libitum;* as a demulcent and emollient in coughs and colds, or as a vehicle for more active medicines.

Mixture, Ammoni'acum. *Syn.* EMULSION OF AMMONIACUM, MILK OF A.; MISTURA AMMONIACI (B. P., Ph. L. & D.), LAC AMMONIACI, L. *Prep.* 1. (Ph. L.) Prepared ammoniacum, 5 dr.; distilled water, 1 pint; rub the ammoniacum with the water, gradually added, until they are perfectly mixed.

2. (Ph. D.) Ammoniacum, ¼ oz.; water, 8 fl. oz.; as the last, but straining through muslin.

3. (B. P.) Ammoniacum, ½ oz.; rubbed down with water, 8 oz., and strain.—*Dose,* ½ to 1 gr.

Obs. The last formula produces the best and most effective mixture, owing to the use of the raw instead of the strained drug.—*Dose,* 1 to 2 table-spoonfuls, either alone or combined with squills or ipecacuanha; as an expectorant and demulcent in chronic coughs, humoral asthma, &c.

Mixture, An'odyne. *Syn.* MISTURA ANODYNA, JULEPUM CALMANS, L. *Prep.* 1. Prepared chalk, 2 dr.; syrup of poppies, 1 oz.; fœtid spirit of ammonia, 1½ dr.; oils of dill and aniseed, of each, 3 drops; water, 4½ fl. oz.—*Dose.* A teaspoonful 3 or 4 times a day; in the diarrhœa and colic of infancy.

2. (P. Cod.) Syrup of opium, 2 dr.; syrup of orange flowers, 6 dr.; lettuce water, 4 fl. oz. To allay pain, induce sleep, &c. *Dose,* 1 table-spoonful.

3. (*Vicat.*) Ammoniated alcohol, ¾ fl. oz.; powdered opium, 1 dr.; powdered camphor, ½ dr.; proof spirit, 1½ fl. oz.; digest, with agitation, for 3 or 4 days, and filter. In toothache arising from caries, and as a lotion to the temples in headache.

Mixture, Antac'id. *Syn.* ABSORBENT MIXTURE, ALKALINE M.; MISTURA ALKALINA, M. ANTACIDA, L. *Prep.* 1. Liquor of potassa and spirit of nutmeg, of each, 2 fl. dr.; tincture of rhubarb, 3 fl. dr.; tincture of opium, 1 fl. dr.; water, 5 fl. oz. In dyspepsia, heartburn, &c., accompanied with flatulence.

2. Spirit of sal volatile and orange-flower water, of each, 1 fl. oz.; simple syrup, 1½ fl. oz.; water, 2½ fl. oz. In acidity, &c., accompanied with languor and low spirits.

3. Sesquicarbonate of ammonia, 2 dr.; syrup of orange peel and tincture of gentian, of each, 1 fl. oz.; water, 4 fl. oz. In dyspepsia, heartburn, &c., arising from excessive indulgence in spirituous or fermented liquors. It also pos-

sesses considerable stimulating properties, and will partially remove the fit of drunkenness.

4. (*Collier.*) Prepared chalk, 2 dr.; tincture of ginger, 2 fl. dr.; compound tincture of cardamoms, 1½ fl. oz.; pimento water, 6 fl. oz. In diarrhœa accompanied with acidity.

5. (*Collier.*) Chalk mixture, 5 fl. oz.; tinctures of catechu and cinnamon, of each, ½ fl. oz. In chronic diarrhœa. .

6. (*Ryan.*) Liquor of potassa, 2 fl. dr.; tincture of opium, 1 fl. dr.; calcined magnesia, 1 dr.; oil of peppermint, 5 drops; lime-water, 8 fl. oz. In dyspepsia accompanied with acidity, flatulence, and constipation.

Mixture, Anticroup'al. *Syn.* MISTURA SENE-GÆ, L. *Prep.* (*Jadelot.*) Infusion of senega, 4 oz.; syrup of ipecacuanha, 1 oz.; oxymel of squills, 3 dr.; tartarised antimony, 1½ gr.; mix. By spoonfuls, in croup.

Mixture, Antieme'tic. *Syn.* MISTURA ANTI-EMETICA, L. *Prep.* 1. Creasote, 12 drops; acetate of morphia, 1½ gr.; camphor, 10 gr.; rectified spirit, ½ fl. oz.; syrup of orange peel, 1½ fl. oz.; distilled vinegar, 4 fl. oz. In sea-sickness, &c. —*Dose*, 1 table-spoonful on the approach of vomiting, and repeated at intervals of half an hour until the vomiting ceases.

2. (*Dr Barker.*) Compound tincture of camphor, 1 fl. dr.; burnt brandy, 1 fl. oz.; sugar, ½ oz.; infusion of mint, 6 fl. oz.—*Dose*, ½ to 1 table-spoonful, every ¼ hour, until the vomiting ceases.

Mixture, Antiepilep'tic. *Syn.* MISTURA ANTI-EPILEPTICA, L. *Prep.* (*M. Lemoine.*) Liquor of ammonia, 12 drops; syrup of orange flowers, 1 oz.; distilled water of linden flowers, 2 oz.; do. of cherry laurel, ¼ oz.; mix. According to M. Lemoine, this is a specific in epilepsy.— *Dose*, 1 table-spoonful, or more.

Mixture, Antihyster'ic. *Syn.* MISTURA ANTI-HYSTERICA, L.; POTION ANTIHYSTÉRIQUE, Fr. *Prep.* 1. (*Dr Josat.*) Cyanide of potassium, 1½ gr.; distilled lettuce water, 4½ fl. oz.; syrup of orange flowers, 1½ fl. oz.—*Dose*, 1 or 2 teaspoonfuls every 10 minutes, when the fit is expected; during the fit it may be given in double doses. Dr Josat declares its efficacy to have been indisputably proved in upwards of 55 cases.

2. (*Magendie.*) Cyanide of potassium, 2 gr.; lettuce water (distilled), 4 oz.; syrup of marshmallow, 2 oz. Resembles the last.

3. (*Dr Paris.*) Assafœtida, 1 dr.; peppermint water, 5 fl. oz.; make an emulsion, and add of ammoniated tincture of valerian, 2 fl. dr.; tincture of castor, 3 fl. dr.; sulphuric ether, 1½ fl. dr. —*Dose*, 1 table-spoonful, 3 or 4 times a day, or oftener.

4. (P. Cod.) Syrup of wormwood, 1 oz.; tincture of castor, ½ dr.; valerian water and orange-flower water, of each, 2 oz.; ether, 1 dr. As the last.

Mixture, Antimo"nial. See MIXTURE, CONTRA-STIMULANT.

Mixture, Antipertus'sic. *Syn.* MISTURA ANTI-PERTUSSIENS, L. *Prep.* 1. Cochineal (powdered), 2 dr.; carbonate of potassa, 1 dr.; boiling water, 8 fl. oz.; infuse for 1 hour, strain, and add of lump sugar, 1½ oz.

2. (*Dr Bird.*) Extract of hemlock, 12 gr.; alum, 25 gr.; syrup of red poppies, 2 fl. dr.; dill water, 3 fl. oz.

3. (*Dr Reece.*) Tincture of assafœtida, 1 fl. dr.; tincture of opium, 10 or 12 drops; powdered ipecacuanha, 10 gr.; water, 2 fl. oz.—*Dose*. A teaspoonful every 3 hours, in hooping-cough, for a child 2 or 3 years old, and other ages in proportion.

Mixture, Antiscrof'ulous. *Syn.* MISTURA AN-TISCROFULOSA, L. *Prep.* From tincture of bichloride of gold, 30 drops; tincture of iodine, 40 drops; tincture of gentian, 1 fl. dr.; simple syrup, 7 fl. dr.; rose-water, 5 fl. oz.—*Dose*. A dessert-spoonful, 3 or 4 times daily, in a wine-glassful of water; observing to shake the bottle before pouring out the liquid. Mr Cooley states that he has seen repeated instances of the excellent effects of this medicine in scrofula, syphilis, and various glandular diseases, even under all the disadvantages of a salt-meat diet and confinement on shipboard.

Mixture, Antispasmod'ic. *Syn.* MISTURA AN-TISPASMODICA, L. *Prep.* 1. Tincture of castor, 6 fl. dr.; sulphuric ether and laudanum, of each, 1 fl. dr.; syrup of saffron, 1 fl. oz.; cinnamon water, 4 fl. oz.

2. (*Dr Collier.*) Assafœtida and camphor mixtures, of each, 2½ fl. oz.; tincture of valerian, 1 fl. oz.

3. (P. Cod.) Lime or linden-flower water and orange-flower water, of each, 2 oz.; syrup of orange flowers, 1 oz.; ether, ½ dr.—*Dose* (of each of the above), 1 to 2 tablespoonfuls.

Mixture, Ape"rient. *Syn.* MISTURA APERIENS, L. *Prep.* 1. (*Abernethy.*) Sulphate of magnesia, 1 oz.; manna, ½ oz.; infusion of senna, 1½ fl. oz.; tincture of senna, ½ fl. oz.; mint-water, 2 fl. oz.; distilled water, 4 fl. oz.; mix. This is the true 'ABERNETHY BLACK DRAUGHT.'

2. (*Dr Christison.*) Sulphate of magnesia, 1½ oz.; water, 4 fl. oz.; dissolve, and add of tincture of senna, 1 fl. oz.; infusion of roses, 4 fl. oz. —*Dose*. A wine-glassful hourly, until it begins to operate.

3. (*Dr Collier.*) Sulphate of iron, 20 gr.; Epsom salts, 1 oz.; pennyroyal water, 1 pint; dissolve.— *Dose*. A wine-glassful twice a day, in atonic amenorrhœa.

Mixture, Aromat'ic. *Syn.* MISTURA AROMA-TICA, L. *Prep.* (P. Cod.) Syrup of clove gilliflowers, 1 oz.; spirit of cinnamon, ½ oz.; confection of hyacinth, 2 dr.; peppermint water and orange-flower water, of each, 2 oz.

Mixture, Aromatic Iron. *Prep.* Red bark (in powder), 4 parts; calumba (in powder), 2 parts; cloves (bruised), 1 part; iron wire, 2 parts; compound tincture of cardamoms, 12 parts; tincture of orange peel, 2 parts; peppermint water, 50 parts; macerate the first four ingredients in the last one for 3 days, agitating occasionally; filter, add the tinctures, and make up to 50 parts. Used as a tonic.—*Dose*, 1 to 2 oz.

Mixture, Arsen'ical. *Syn.* MISTURA ARSENI-CALIS, L. *Prep.* From liquor of arsenite of potassa (Ph. L.), 2 fl. dr.; compound tincture of cardamoms, 4 fl. dr.; cinnamon water, 2 fl. oz.; pure water, 3 fl. oz.; mix.—*Dose*. A small table-spoonful, twice a day, after a full meal; in agues, periodic headaches, lepra, psoriasis, chronic rheumatism, &c. It should be exhibited with caution, and its effects watched; and after 5 or 6 days the dose should be reduced to half the quantity.

Mixture, Astrin'gent. *Syn.* MISTURA ASTRINGENS, L. *Prep.* 1. (*Pradel.*) Tannin, 12 gr.; tincture of rhatany, 1 dr.; simple syrup, 7 dr.; mucilage, 1 oz.; camphor mixture, 4 oz.

2. (*A. T. Thomson.*) Extract of catechu, 2 dr. (or tincture, 1 oz.); cinnamon water, 8 oz.; dissolve.—*Dose,* 1 to 3 table-spoonfuls, after every liquid dejection, in diarrhœa and dysentery.

Mixture, Atroph'ic. *Syn.* MISTURA ATROPHICA, L.; POTION ATROPHIQUE, Fr. *Prep.* (*Magendie.*) Iodide of potassium, 4 dr.; lettuce water, 8 oz.; peppermint water, 2 dr.; syrup of marsh-mallow, 1 oz.—*Dose,* 1 table-spoonful, twice a day; in hypertrophy (enlargement) of the heart. Sometimes 1 to 2 dr. of tincture of foxglove is added to the mixture.

Mixture, Balsamic. (P. Cod.) *Syn.* MISTURA BALSAMICA. *Prep.* Balsam copaiba, 2 oz.; rectified spirit, 2 oz.; syrup of tolu, 2 oz.; peppermint water, 4 oz.; nitric ether, 2 dr. Mix the alcohol and ether, add the balsam, then the syrup and water.

Mixture, Bar'ley. *Syn.* MISTURA HORDEI, L. See DECOCTION.

Mixture of Bisulphide of Carbon. (*Clarus.*) *Syn.* MISTURA BISULPHURETI CARBONIS. *Prep.* Bisulphide of carbon, 20 minims; sugar, 2 dr.; milk, 6 oz.—*Dose,* ½ oz., 4 times a day.

Mixture of Boracic Acid. (*Chaussier.*) *Syn.* MISTURA ACIDI BORACICI. *Prep.* Camphor mixture, 4 oz.; boracic acid, 60 gr.; syrup of orange peel, 1 oz.

Mixture, Brandy. *Syn.* MIXTURE OF SPIRIT OF FRENCH WINE, EGG-FLIP‡; MISTURA SPIRITUS VINI GALLICI (B. P., Ph. L.), L. *Prep.* 1. (Ph. L.) Brandy and cinnamon water, of each, 4 fl. oz.; yolks of 2 eggs; white sugar, ½ oz.; oil of cinnamon, 2 drops; mix. A valuable stimulant and restorative in low fevers, and in extreme exhaustion from hæmorrhages, &c.; but scarcely a fitting subject for the labours of the College of Physicians, since almost every cook and housewife could produce a better compound than the product of the College formula.

2. (B. P.) Brandy, 4 oz.; cinnamon water, 4 oz.; the yolks of 2 eggs; sugar, ½ oz.; mix.—*Dose,* ½ to 1½ oz., in prostration and last stages of fever.

Mixture of Burnt Hartshorn. See DECOCTION, WHITE.

Mixture of Caffein. (*Vanden-Corput.*) *Syn.* MISTURA CAFFEINÆ. *Prep.* Caffein, 7 gr.; distilled water, 3 oz.; hydrochloric acid, 2 drops; syrup of orange-flower water, ½ oz.; mix.—*Dose,* 1 table-spoonful.

Mixture, Cam'phor. *Syn.* CAMPHOR JULEP, CAMPHOR WATER; MISTURA CAMPHORÆ (Ph. L. & D.), EMULSIO CAMPHORÆ (Ph. E.), MISTURA CAMPHORATA, L. *Prep.* 1. (Ph. L.) Camphor, ½ dr.; rectified spirit, 10 drops; triturate together, gradually adding of water, 1 pint; and strain through linen.

2. (Ph. D.) Tincture of camphor, 1 fl. oz.; distilled water, 3 pints; agitate well together, and after 24 hours filter through paper.

3. (Ph. E.) See EMULSION.

Uses, &c. Camphor julep is chiefly used as a vehicle for other medicines.—*Dose,* ½ to 1 wine-glassful.

4. (With MAGNESIA: MISTURA CAMPHORÆ CUM MAGNESIÂ—Ph. E. & D., AQUA CAMPHORÆ—Ph. U. S.) *a.* (Ph. E.) Camphor, 10 gr., carbonate of magnesia, 25 gr.; triturate together, then gradually add of water, 6 fl. oz., still continuing the trituration.

b. (Ph. D.) Camphor, 12 gr.; carbonate of magnesia, ½ dr.; water, 6 fl. oz.; as last.

c. (Ph. U. S.) Camphor dissolved in alcohol, 16 parts, then pour on to 16 parts of cotton. When the cotton is nearly dry, pack in a percolator and pour on distilled water till 1000 parts are obtained. Antacid, antispasmodic, and anodyne.—*Dose,* 1 to 2 tablespoonfuls. Used without straining. It is stronger than the simple mixture.

d. CARBONATED CAMPHOR MIXTURE. *Syn.* MISTURA CAMPHOR CARBONICA. Water strongly charged with carbonic acid gas, agitated with powdered camphor, and strained.

Mixture, Carmin'ative. *Syn.* MISTURA CARMINATIVA, L. *Prep.* (*Dr Paris.*) Calcined magnesia, ½ dr.; peppermint water, 2½ fl. dr.; compound tincture of lavender, ½ fl. dr.; spirit of caraway, 4 fl. dr.; syrup of ginger, 2 fl. dr.; mix. Antacid and carminative. For 1 or 2 doses.

Mixture of Cassia. (Fr. Hosp.) *Syn.* MISTURA CASSIÆ, L.; EAU DE CASSE, Fr. Cassia pulp, 2 oz.; hot water, 1½ pints. May be taken by the wine-glass. Laxative.

Mixture of Cassia, Antimoniated. (*Foy.*) *Syn.* MISTURA CASSIÆ ANTIMONIATA, L.; EAU DE CASSE, Fr. Emitisée. *Prep.* Pulp of cassia, 1 oz.; boiling water, 1½ pints. Macerate, strain, and add sulphate of magnesia, 1 oz.; emetic tartar, 3 gr. By cupfuls. In painters' colic.

Mixture, Cathar'tic. See MIXTURE, APERIENT; M., SENNA, &c.

Mixture, Chalk. *Syn.* CRETACEOUS MIXTURE; MISTURA CRETÆ (Ph. L. E. & D.), M. CRETACEA, L. *Prep.* 1. (Ph. L.) Prepared chalk, ½ oz.; sugar, 3 dr.; mixture of acacia (mucilage), 1½ fl. oz., triturate together, then add of cinnamon water, 18 fl. oz.

2. (Ph. E.) Prepared chalk, 10 dr.; white sugar, 5 dr.; mucilage, 3 fl. oz.; spirit of cinnamon, 2 fl. oz.; water, 1 quart; as the last.

3. (Ph. D.) Prepared chalk, 2 dr.; syrup and mucilage, of each, ½ oz.; cinnamon water, 7 fl. oz.

4. (B. P.) Prepared chalk, 1 part; gum-arabic, in powder, 1 part; syrup, 2 parts; cinnamon water, 30 parts; mix by trituration.—*Dose,* 1 to 2 gr., with astringent tinctures and opium.

Obs. The above are antacid and absorbent.—*Dose,* 1 to 3 table-spoonfuls, either alone or combined with aromatic confection; in heartburn, and in diarrhœa, after every liquid motion. In the latter affection a little tincture of catechu or laudanum is often added; and when there is vomiting or nausea, either peppermint or spearmint water is generally substituted for the whole or a part of the simple water ordered in the above formulæ.

Mixture of Chlorine. *Syn.* MISTURA CHLORINII. Dr Watson prescribes 2 fl. dr. of the solution to 1 pint of water. The dose of the mixture is 4 fl. dr. every 3 hours, according to age, in scarlatina.

Mixture of Chloroform. (Ph. U. S.) *Syn.* MISTURA CHLOROFORMI. *Prep.* Chloroform, 8 oz.; camphor, 2 dr.; yolks of 10 eggs; water, 80 oz. Rub the yolks first by themselves, then with the camphor, previously dissolved in the chloroform, and lastly, with the water gradually added.

Mixture, Cincho'na. *Syn.* BARK MIXTURE; MISTURA CINCHONÆ, L. *Prep.* (*Copland.*) Confection of roses, ½ oz.; boiling decoction of bark, 1 fl. oz.; triturate, in 10 minutes strain, and add diluted sulphuric acid, 1½ fl. dr.; spirit of nutmeg, 4 fl. dr. Febrifuge, tonic, and stomachic.—*Dose*, 1 to 3 table-spoonfuls, 2 or 3 times a day.

Mixture of Citrate of Caffein. *Syn.* MISTURA CAFFEINÆ CITRATIS, POTION CONTRE MIGRAINE. Syrup of citrate of caffein, 1 fl. oz.; water (or any agreeable diluent), 5 oz. A table-spoonful frequently.

Mixture, Col'chicum. *Syn.* GOUT MIXTURE; MISTURA ANTARTHRITICA, M. COLCHICI, L. *Prep.* (*Sir S. Scudamore.*) Magnesia, 1½ dr.; vinegar of colchicum and syrup of orange-peel, of each, 4 fl. dr.; peppermint water, 8 fl. oz. A table-spoonful every 3 or 4 hours during the fit of gout.

Mixture, Contra-stim'ulant. *Syn.* MISTURA CONTRA-STIMULANS, JULEPUM C., M. ANTIMONII POTASSIO-TARTRATIS, L. *Prep.* (*Laennec.*) Tartar emetic, 3 gr.; infusion of orange leaves, 8 fl. oz.; syrup of do., 1 fl. oz.—*Dose.* A wine-glassful, or more, every 2 hours; in inflammation of the lungs, whooping-cough, &c.

Mixture, Cough. *Syn.* MISTURA BECHICA, L. *Prep.* 1. Almond mixture, 4 fl. oz.; oxymel of squills, 1 fl. oz.; ipecacuanha wine and syrup of tolu, of each, ½ fl. oz.

2. Tincture of tolu, ½ fl. oz.; paregoric elixir and tincture of squills, of each, 1 fl. oz.; syrup of poppies, 3 fl. oz.; water, 3½ fl. oz.

3. Mixture of ammoniacum, 4 fl. oz.; syrup of squills, 2 fl. oz. In the coughs of old persons.

4. Antimonial wine, 3 fl. dr.; syrup of poppies, 1½ fl. oz.; water, 4 fl. oz. In dry, husky coughs. —*Dose* (of each of the above), 1 table-spoonful, 2 or 3 times a day, or oftener.

5. (*Dr Monro.*) Paregoric, ½ fl. oz.; sulphuric ether and tincture of tolu, of each, ½ oz.—*Dose.* A teaspoonful in water, night and morning, or when the cough is troublesome.

6. (*Dr Radcliff.*) Syrup of poppies, syrup of squills, and paregoric, equal parts.—*Dose.* As the last. In all cases the bowels should be kept gently moved by some mild aperient.

7. (Dr Wood's Brown Mixture.) Extract of liquorice, 2 dr.; powdered gum-arabic, 2 dr.; boiling water, 4 oz.; dissolve, and add antimonial wine, 2 dr.; laudanum, 20 minims.—*Dose.* A table-spoonful occasionally. A popular American remedy.

Mixture, Cre'asote. *Syn.* MISTURA CREASOTI, M. CREAZOTI (Ph. E.), L. *Prep.* 1. (Ph. E.) Creasote and acetic acid, of each, 16 drops; mix, then add of compound spirit of juniper and syrup, of each, 1 fl. oz.; water, 14 fl. oz.; and agitate well together.—*Dose*, ½ to 1 wine-glassful, in nausea and vomiting, especially to prevent or relieve sea-sickness.

2. (B. P.) Creasote, 15 minims; glacial acetic acid, 15 minims; spirit of juniper, ½ dr.; syrup, 1 oz.; distilled water, 15 oz.; mix.—*Dose*, 1 to 2 oz.

Mixture of Cubebs. *Syn.* MISTURA CUBEBÆ. Powder of cubebs, 1 oz.; sugar, 2 dr.; mucilage, 2 oz.; cinnamon water, 6 oz.—*Dose*, ½ oz. to 1 oz.

Mixture, Demul'cent. *Syn.* MISTURA DEMULCENS, L. See ALMOND MIXTURE, GUM M., &c.

Mixture, Diaphoret'ic. *Syn.* MISTURA DIAPHORETICA, L. *Prep.* 1. Solution of acetate of ammonia, 3 fl. oz.; antimonial wine, 2 fl. dr.; tincture of henbane, 1½ fl. dr.; camphor mixture, 3 fl. oz.—*Dose*, 1 table-spoonful every 3 or 4 hours; in fevers, &c.

2. To the last add of sweet spirit of nitre, ½ fl. oz. As above.

Mixture for Diarrhœa. (Board of Health.) *Syn.* MISTURA PRO DIARRHŒÂ. Aromatic powder, 3 dr.; compound spirits of ammonia, 3 dr.; tincture of catechu, 10 dr.; compound tincture of cardamoms, 6 dr.; tincture of opium, 1 dr.; chalk mixture to make 20 oz.—*Dose.* For an adult, 1 oz.; for a child of 12 years of age, ½ oz.; for 7 years, ¼ oz.; after each liquid stool. See MIXTURE, CHALK, &c.

Mixture, Diuret'ic. *Syn.* MISTURA DIURETICA, L. *Prep.* 1. Nitrate of potassa, 2 dr.; sweet spirit of nitre, 3 fl. dr.; syrup of squills, 1½ fl. oz.; peppermint water, 4 fl. oz.

2. (*A. T. Thomson.*) Infusion of foxglove, 5½ fl. oz.; tincture of foxglove, ½ fl. dr.; acetate of potassa, 2 dr.; spirit of juniper, ½ fl. oz.; tincture of opium, ½ fl. dr. In dropsy.—*Dose*, 1 to 2 table-spoonfuls, every 2 or 3 hours. The last must be used with caution.

Mixture, Effervescing. (P. Cod.) *Syn.* MISTURA EFFERVESCENS; POTION GAZEUSE DE RIVIÈRE. *Prep.* Dissolve ½ dr. of bicarbonate of potash in 2 oz. of water, and add 4 dr. of syrup. Mix also ½ dr. of citric acid with ½ oz. of syrup of citric acid and 2 oz. of water. Mix an equal quantity of each, and give it while effervescing.

Mixture of Elaterium. (*Dr Ferriar.*) *Syn.* MISTURA ELATERII. *Prep.* Elaterium, 1 gr.; spirit of nitric ether, 2 fl. oz.; tincture of squills, ½ oz.; oxymel of colchicum, ½ oz.; syrup of buckthorn, 1 fl. oz.—*Dose*, 1 dr. 3 times a day in water.

Mixture, Emet'ic. *Syn.* MISTURA EMETICA, L. *Prep.* 1. (*Copland.*) Sulphate of zinc, 40 gr.; ipecacuanha wine and tincture of serpentary, of each, 4 fl. dr.; tincture of capsicum, 40 drops; oil of chamomile, 12 drops; peppermint water, 4½ fl. oz. As an excitant emetic; in cases of poisoning by narcotics, &c.

2. (*Magendie.*) Coloured emetine, 4 gr. (or white emetine, 1 gr.); acetic acid, 8 drops; mix, and add of infusion of orange leaves or lime flowers, 3½ fl. oz.; syrup of marsh-mallows, 1 fl. oz.

3. (*A. T. Thomson.*) Ipecacuanha, ½ dr.; tartar emetic, 1 gr.; tincture of squills, 1 fl. dr.; water, 6 fl. oz.—*Dose*, 1 to 2 table-spoonfuls, followed by half the quantity every 10 or 15 minutes, until vomiting is produced; at the same time assisting the action of the medicine by drinking copiously of warm water.

Mixture, Emmen'agogue. See MIXTURE, STEEL, &c.

Mixture, Expec'torant. *Syn.* MISTURA EXPECTORANS, L. *Prep.* 1. (*Collier.*) Oxymel of squills and mucilage, of each, 1 oz.; syrup of marsh-mallows, 2 oz.; camphor julep, 3 fl. oz.—*Dose*, 1 to 2 table-spoonfuls, 2 or 3 times a day; in coughs, hoarseness, asthma, &c.

2. (*A. T. Thomson.*) Almond mixture, 5 fl. oz.; ipecacuanha wine and tincture of squills, of each, 1 fl. dr.; syrup of tolu, 6 fl. dr.—*Dose*, 1 table-spoonful; in humoral asthma, catarrh, &c., when the cough is urgent.

Mixture, Feb'rifuge. *Syn.* MISTURA FEBRIFUGA, L. See ACETATE OF AMMONIA MIXTURE, DIAPHORETIC M., &c.

Mixture of Gentian. (Ph. B. 1867.) *Syn.* MISTURA GENTIANÆ, L. *Prep.* Macerate gentian root (sliced), ¼ oz.; bitter orange peel (cut small) and coriander fruit (bruised), of each, 30 gr., in proof spirit, 2 fl. oz., for 2 hours. Add distilled water, 8 fl. oz.; macerate again for 2 hours, and strain through calico.—*Dose*, 1 oz.

Mixture, Gentian (Compound). *Prep.* Gentian (bruised), 1½ parts; bitter orange peel (bruised), ¾ part; cardamom seeds (bruised), ½ part; proof spirit, 20 parts; macerate for 48 hours with 15 parts of the spirit, agitating occasionally, pack in a percolator, let it drain, and then pour on the remaining spirit; when it ceases to drop, wash the marc with spirit to make up 20 parts.—*Dose*, 1 to 2 dr.

Mixture, Gregory's. See POWDERS.

Mixture, Griffith's. See MIXTURE, STEEL.

Mixture, Guai'acum. *Syn.* EMULSION OF GUAIACUM, MILK OF G.; MISTURA GUAIACI (B. P., Ph. L. and E.), LAC G., L. *Prep.* (Ph. L.) Gum guaiacum, 3 dr.; white sugar, ½ oz.; gum acacia, 2 dr. (all in powder); triturate together, and to these, whilst rubbing, gradually add of cinnamon water, 1 pint.

2. (Ph. E.) Guaiacum, 3 dr.; sugar, ½ oz.; mucilage, ½ fl. oz.; cinnamon water, 19½ fl. oz.; as before.—*Dose*, 1 to 3 table-spoonfuls, 2 or 3 times a day; in chronic rheumatism, gout, &c.

3. (B. P.) Guaiac resin (in powder), 2 parts; sugar, 2 parts; gum-arabic (in powder), 1 part; cinnamon water, 80 parts; triturate, adding the cinnamon water gradually.—*Dose*, ½ to 2 oz.

Mixture, Gum. *Syn.* MUCILAGE; MISTURA ACACIÆ (Ph. L.), MUCILAGO (Ph. E.), M. ACACIÆ (Ph. D.), M. ARABICI GUMMI, L. *Prep.* 1. (Ph. L. Gum acacia (in powder), 13 oz.; boiling distilled water, 1 pint; rub the gum with the water, gradually poured in, until solution is complete.

2. (Ph. E.) Gum, 9 oz.; cold water, 1 pint; macerate, with occasional stirring, until dissolved, then strain through linen or calico.

3. (Ph. D.) Gum (coarsely powdered), 4 oz.; water (cold), 6 fl. oz.; dissolve, and strain through flannel.

Uses, &c. Mucilage of gum acacia is chiefly employed to render oily and resinous substances miscible with water. "Oils require about three-quarters their weight; balsams and spermaceti, equal parts; resins, 2 parts; and musk, 5 times its weight" for this purpose (*Montgomery*). The GUM MIXTURE, Ph. E., will be found under 'EMULSION.'

Mixture of Horseradish, Compound. (*Dr Paris.*) *Syn.* MISTURA ARMORACIÆ COMPOSITA, L. *Prep.* Horseradish root, ½ oz.; mustard seed, ½ oz.; boiling water, 1 pint. Macerate for an hour, and to 7 oz. of the strained infusion add aromatic spirit of ammonia, 1 fl. dr.; spirit of pimento, ½ oz. In paralysis.

Mixture, Hydrocyan'ic. *Syn.* MIXTURE OF PRUSSIC ACID; MISTURA ACIDI HYDROCYANICI, L. *Prep.* From medicinal prussic acid, 15 drops; simple syrup (pure), 1 fl. oz.; distilled water, 5 fl. oz.—*Dose*, 1 table spoonful, 2 or 3 times daily. Each dose contains 1¼ drops of medicinal prussic acid. The bottle should be shaken before pouring out the dose. Magendie's formulæ for this mixture are omitted, because the acid which he orders is not kept in the shops in England.

Mixture of Iodine with Sarsaparilla. (*Magendie.*) *Syn.* MISTURA IODINII CUM SARZÂ, L. *Prep.* Decoction of sarsaparilla, 1½ pints; iodide of potassium, 1 dr.; syrup of orange, 2 oz.

Mixture, I'ron. See MIXTURE, STEEL.

Mixture, I'ron (Compound). See MIXTURE, STEEL.

Mixture, Marsh-mallow. *Syn.* MISTURA ALTHÆÆ (Ph. E.), L. *Prep.* 1. (Ph. E.) Marsh-mallow root (dried), 4 oz.; stoned raisins, 2 oz.; water, 5 pints; boil to 3 pints, strain through linen, and after the sediment has subsided, decant the clear portion.

2. (Ph. D.) See DECOCTION. Demulcent.—*Dose.* A few spoonfuls *ad libitum*, so as to take 1 to 3 pints in the 24 hours; in strangury, calculus, coughs, fevers, &c.

Mixture of Monesia. (*Neligan.*) *Syn.* MISTURA MONESIÆ, L. *Prep.* Extract of monesia, 2 scr.; water, 7½ oz.; compound tincture of cardamoms, ½ oz.

Mixture of Musk. (Ph. L.) *Syn.* MISTURA MOSCHI, L. *Prep.* Musk, 3 dr.; triturate it with white sugar, 3 dr.; gum acacia, 3 dr.; and gradually add rose-water, 1 pint.—*Dose*, 1 to 2 oz.

Mixture of Musk-seed. (*Dr Reece.*) *Syn.* MISTURA ABELMOSCHI, L. *Prep.* Tincture of musk-seeds, 1 oz.; aromatic spirit of ammonia, 3 fl. dr.; compound spirit of lavender, 4 fl. dr.; camphor mixture, 6 oz.—*Dose*, ½ oz. to 1 oz.

Mixture, Myrrh. *Syn.* EMULSION OF MYRRH; MISTURA MYRRHÆ, L. *Prep.* (*Copland.*) Myrrh, 1½ dr.; add to it gradually, triturating all the time, decoction of liquorice, 6 fl. oz., and strain.—*Dose*, 1 to 2 table-spoonfuls, twice or thrice a day, combined with carbonate of soda, dilute hydrochloric acid, or paregoric; in debility, and diseases of the digestive organs.

Mixture, Narcot'ic. *Syn.* MISTURA NARCOTICA, M. FEBRIFUGA, L. *Prep.* 1. Tincture of henbane, 2 fl. dr.; solution of acetate of ammonia, 3 fl. oz.; water, 2½ fl. oz.; mix.—*Dose*, 1 to 2 table-spoonfuls, to relieve pain, procure sleep in fevers, &c.

2. (*W. Cooley.*) Laudanum, 1½ fl. dr.; syrup of poppies, sulphuric ether, and spirit of cinnamon, of each, 1 oz.; tincture of henbane, 2½ fl. dr.; tincture of capsicum, 4 fl. dr.; water, 2 fl. oz.—*Dose*, 1 to 2 table-spoonfuls, at the commencement of the hot fit of ague.

Mixture, Oleo-balsam'ic. *Syn.* MISTURA

OLEO-BALSAMICA, L. *Prep.* (Hamb. Cod.) Oils of cedrat, cinnamon, cloves, lavender, mace, and marjoram, of each 20 drops; oil of rue, 10 drops; balsam of Peru, ½ dr.; rectified spirit, 10 oz.; digest and filter.

Mixture of Oxalic Acid. (*Nardo.*) *Syn.* MISTURA ACIDI OXALICI. *Prep.* Oxalic acid, 8 gr.; mucilage, 3 oz.; syrup, 1 oz. In inflammation of the fauces and digestive tube.

Mixture of Phosphorus. (*Soubeiran*). *Syn.* MISTURA PHOSPHORICI. *Prep.* Phosphorated oil, 2 dr.; powdered gum acacia, 2 dr.; peppermint water, 3 oz.; syrup, 2 oz. Mix the gum with 10 dr. of water, and thin with the oil, and gradually add the others. Contains 1 gr. of phosphorus.—*Dose*, ½ fl. oz.

Mixture of Platinum Chloride. (*Hoeffer.*) *Syn.* MISTURA PLATINI CHLORIDI. *Prep.* Perchloride of platinum, 1½ gr.; gum juleps, 6 oz.

Mixture of Potassium Iodide. (*Cazenave.*) *Syn.* MISTURA POTASSII IODIDI, L. *Prep.* Iodide of potassium, 2 dr.; distilled water, 16 oz.; syrup, 2 fl. oz. 2 or 3 table-spoonfuls per diem.

Mixture, Pur'gative. *Syn.* MISTURA CATHARTICA, M. LAXATIVA, M. PURGANS, L. *Prep.* 1. From any of the purging salts (Epsom, Glauber, tasteless, &c.), 2 oz.; infusion of senna, 5 fl. oz.; syrup of orange peel, 1 fl. oz.; tincture of ginger, ½ fl. oz.; spirit of pimento, 2 fl. dr.; mix.—*Dose*, 1 to 3 table-spoonfuls, early in the morning; as an aperient in stomach complaints, &c.

2. (*Dr Copland.*) Manna, 1½ oz.; cream of tartar, ½ oz.; whey, 1 quart. By wine-glassfuls, as an aperient drink, in fevers, &c.

3. (*Corvisart.*) Boro-tartrate of potassa (soluble tartar), 1 oz.; tartar emetic, ½ gr.; sugar, 2 oz.; water, 1½ pints; dissolve. By wine-glassfuls, untils it begins to operate. This has been called 'NAPOLEON'S MEDICINE,' from its having been frequently taken by Napoleon I. See MIXTURES OF SCAMMONY, SENNA, &c.

Mixture of Quinine and Coffee. *Syn.* MISTURA QUINIÆ ET CAFFEÆ, L.; CAFÉ QUININÉ, Fr. Prepare 5 oz. of infusion from 4 dr. of ground coffee by percolation, and add 24 gr. of neutral sulphate of quinine and 4 dr. of sugar.—*Dose.* A tablespoonful. The coffee conceals the bitterness of the quinine.

Mixture of Quinine with Iron. *Syn.* MISTURA QUINIÆ CUM FERRO, L. Sulphate of quinine, 1 gr.; sulphate of iron, 2 gr.; dilute sulphuric acid, 5 minims; water, 1 oz. For 1 dose.

Mixture, Saline. *Syn.* MISTURA SALINA, L. See DRAUGHT and LEMONADE.

Mixture, Scam'mony. *Syn.* SCAMMONY MILK; MISTURA SCAMMONII (B. P., 1867), L. *Prep.* 1. (Ph. E.) Resin of scammony, 7 gr.; unskimmed milk, 3 fl. oz.; gradually mix, triturating all the time, so as to form an emulsion. Purgative.—*Dose.* One half.

2. (*Planche's* PURGATIVE POTION.) To the last add of white sugar, ¼ oz.; cherry-laurel (or bitter-almond) water, 4 or 5 drops. The above are the most tasteless and pleasant purgatives of an active character known.

3. (B. P.) Scammony in powder, 6 gr.; fresh milk, 2 oz.; triturate and form an emulsion.—*Dose.* The quantity of the formula for an adult, half for a child.

Mixture, Sen'na (Compound). *Syn.* BLACK DRAUGHT, ABERNETHY'S D., CATHARTIC MIXTURE; MISTURA SENNÆ COMPOSITA, L. *Prep.* 1. Infusion of senna, ¾ pint; tincture of senna, 1½ fl. oz.; Epsom salts, 4 oz.; carbonate of ammonia, ½ dr.; sugar, 3 oz.; agitate until the solids are dissolved.

2. Senna, 13 oz.; boiling water, 2 quarts; digest for 4 hours in a hot place, then press out the liquor in a tincture press, and add of compound tincture of senna, ¾ pint; Epsom salts, 1 lb.

3. East India senna, 2 lbs.; boiling water, 9 quarts; tincture of senna and Epsom salts, of each, 3½ lbs.; as the last.

4. Senna, 8 lbs.; boiling water, 9 galls.; Epsom salts, 16 lbs.; tincture of senna, 1½ galls.; treacle and colouring, of each, 1 quart.

5. (Guy's Hosp.) Senna and mint, of each, 1½ oz. (say 1½ oz.); boiling water, 1 quart; Epsom salts, 7½ oz. (say ½ lb.).

6. (*Redwood.*) Infusion of senna, 18 oz.; tincture of senna, 3 oz.; sulphate of magnesia, 6 oz.; extract of liquorice and spirit of sal volatile, of each, ¾ oz.; oil of cloves, 6 drops.

7. (B. P.) Infusion of senna, 15 oz.; sulphate of magnesia, 4 oz.; liquid extract of liquorice, 1 oz.; tincture of senna, 2½ oz.; compound tincture of cardamoms, 1½ oz.; dissolve and mix.—*Dose*, 1 to 1½ oz.

Obs. As the above mixture contains very little spirit, and from its great consumption, being made in large quantities at a time, it frequently spoils before the whole is sold, especially in hot weather. To avoid this, 1½ dr. of cloves and 3 dr. of mustard seed, both bruised, may be added to every gall. of the strained liquor at the same time with the salts, spirit, and colouring, after which it must be shaken up repeatedly for a few days, and then allowed to repose for a few days more, when it will become quite clear. It may be filtered through a flannel bag, but there is much loss and delay, owing to the consistence of the liquid. It is purgative in doses of 1 to 1½ fl. oz.

Mixture, Steel. *Syn.* MISTURA FERRI COMPOSITA (B. P.), MISTURA CHALYBEATA, L. Two compounds of this class are officinal:—

1. (GRIFFITH'S MIXTURE, COMPOUND IRON M.; MISTURA FERRI, M. F. PROTOXYDI, M. F. COMPOSITA—Ph. L. E. & D.) *Prep.* *a.* (Ph. L. & E.) Carbonate of potassa, 1 dr.; powdered myrrh, 2 dr.; spirit of nutmeg, 1 fl. oz.; triturate together, and whilst rubbing, add gradually, of sugar, 2 dr.; rose-water, 18 fl. oz.; mix well; then add of sulphate of iron (powdered), 50 gr.; and place it at once in a bottle, which must be kept closely corked.

b. (Ph. D.) Powdered myrrh and sugar, of each, 1 dr.; carbonate of potassa, ½ dr.; essence of nutmeg, 1 fl. dr.; rose-water, 7 fl. oz.; sulphate of iron, ½ dr.; (dissolved in) rose-water, 1 fl. oz. —*Dose*, 1 to 2 oz., 3 or 4 times a day, as a mild and genial chalybeate tonic and stimulant; in amenorrhœa, chlorosis, debility, &c., when there is no determination of blood to the head.

2. (HEBERDEN'S MIXTURE, H.'S INK; ATRA-

MENTUM HEREDII, MISTURA FERRI AROMA-
TICA — Ph. D.) Red cinchona bark, 1 oz.;
calumba root, ½ oz. (both in coarse powder);
cloves (bruised), 2 dr.; iron filings, ½ oz.; peppermint water, 16 fl. oz.; digest in a close vessel for
3 days, agitating frequently, then strain, and add
of tincture of cardamoms (comp.), 3 fl. oz.; tincture of orange peel, ½ oz. Bitter, stomachic, and
aromatic.—*Dose*, 1 or 2 table-spoonfuls, or more,
3 or 4 times a day. It is very slightly chalybeated. See also MIXTURE, AROMATIC IRON.

3. (B. P.) Sulphate of iron, 25 gr.; carbonate
of potash, 30 gr.; myrrh, 60 gr.; sugar, 60 gr.;
spirit of nutmegs, 4 dr.; rose-water, 9½ oz. Reduce the myrrh to powder, add the carbonate
of potash and sugar, and triturate them with
a small quantity of rose-water so as to form
a thin paste, then gradually add more rose-water,
and the spirit of nutmegs, continuing the trituration and further addition of rose-water until
about 8 fluid ounces of milky liquid is formed,
then add the sulphate of iron previously dissolved
in the remainder of the rose-water, and cork the
bottle immediately.—*Dose*, 1 to 2 oz., as a stimulating tonic.

Mixture of Sulphuric Acid. (Ph. G.) *Syn.*
MISTURA ACIDI SULPHURICI; HALLER'S ELIXIR.
Prep. To 3 oz. (by weight) of rectified spirit add
gradually 1 oz. (by weight) of pure sulphuric acid.
—*Dose*, 5 to 20 drops diluted.

Mixture, Ton'ic. *Syn.* STRENGTHENING MIX-
TURE; MISTURA TONICA, L. *Prep.* 1. Infusion
of cascarilla, 5 fl. oz.; tincture of orange peel,
7 fl. dr.; aromatic sulphuric acid, 2 fl. dr.

2. (*Collier.*) Decoction of bark, 5½ fl. oz.;
tincture of do., 3 fl. dr.; aromatic confection,
20 gr.; aromatic spirit of ammonia, 1 fl. dr.

3. (*Thomson.*) Infusion of calumba, 5½ fl. oz.;
compound tincture of cinnamon and syrup of
orange peel, of each, 2 fl. dr.—*Dose*, 1 to 3
table-spoonfuls, 2 or 3 times a day; in debility
of the digestive organs, loss of appetite, to check
nausea and vomiting, &c.

Mixture, Worm. *Syn.* MISTURA ANTHELMIN-
TICA, M. VERMIFUGA, L. *Prep.* 1. (*Collier.*)
Sulphate of iron, 20 gr.; infusion of quassia,
8 fl. oz.—*Dose*, 2 table-spoonfuls every morning
fasting.

2. (*Copland.*) Valerian, 2 dr.; wormseed,
4 dr.; boiling water, 8 fl. oz.; macerate 1 hour,
strain, and add of assafœtida, 1 dr., previously
triturated with the yolk of one egg. As the
last.

3. (*Richard.*) Root of male fern, 1 oz.; water,
9 fl. oz.; boil to 6 fl. oz., strain, and add of sulphuric ether, 1 dr.; syrup of tansy, 1 fl. oz. In
tapeworm; as above.

Mixture, Zinc. *Syn.* MISTURA ZINCI, M. Z.
SULPHATIS, L. *Prep.* (*Collier.*) Sulphate of
zinc, 5 gr.; sulphate of quinine, 10 gr.; compound infusion of roses, 2 fl. oz. Tonic.—*Dose*.
A teaspoonful 2 or 3 times a day, in a glass of
water. Said to be very efficacious in the cure of
coughs of a spasmodic character.

MIXTURES (Arithmetic of). The constantly
recurring necessity in business and chemical
manipulations of determining the value of mixtures, and of producing articles and preparations
of different strengths or prices from those already
in stock, has rendered a ready means of making
such calculations an indispensable qualification in
almost every department of trade and industrial
art. As we address ourselves to the intelligent
operative and busy tradesman, as well as to those
more blessed by education and leisure, we feel we
are bestowing a boon on many of our readers in
giving a short but sufficient outline of this useful
branch of commercial arithmetic, which is most
intimately connected with the objects of the
present work.

1. To determine the price of a mixture from
the value and quantity of each ingredient of
which it is composed.—RULE. Divide the 'gross
value' by the 'gross saleable' or 'useful quantity;' the quotient is the value or cost per gallon,
pound, &c., as the case may be.—*Example*. Required the value per gallon of a hogshead of wine
containing—

		s.	*d.*			£	*s.*	*d.*
30 gallons	@	10	6	. . . . .		15	15	0
20	„ „	12	6	. . . . .		12	10	0
13	„ „	14	6	. . . . .		9	8	6
63					)	37	13	6

Cost per gallon 0 11 11¼

2. To determine the proportions of substances
or articles of different values or strengths which
must be taken to prepare a mixture of any other
value or strength.—RULE. Arrange the 'prices'
or 'strengths of the ingredients' in a column, and
link them together in pairs; each of those above
the required price being always connected with
another below it. Then set the difference between
the required price and these numbers alternately
against those they are linked to, when they will
indicate the quantities to be taken, as in the following examples:—*a*. Required the proportions of
tea at 3*s.*, 4*s.*, 6*s.*, and 7*s.*, that must be taken to
produce a mixture 5*s.* the pound. Here—

$$5 \begin{cases} 3 & \cdots & 1, \text{ or } 1 \text{ lb. } @ \ 3s. \\ 4 & \cdots & 2, \text{ „ } 2 \text{ lbs. „ } 4s. \\ 6 & \cdots & 2, \text{ „ } 2 \text{ „ „ } 6s. \\ 7 & \cdots & 1, \text{ „ } 1 \text{ lb. „ } 7s. \end{cases}$$

b. (When the number of the ingredients or
prices is odd.) Required the proportions of teas
at 3*s.*, 5*s.*, and 6*s.* the pound, to sell at 4*s.* Here
the odd number must be taken a second time:—

$$4 \begin{cases} 3 & \cdots & 1 + 2, \text{ or } 3 \text{ lbs. } @ \ 3s. \\ 5 & \cdots & 1, \text{ „ } 1 \text{ lb. „ } 5s. \\ 6 & \cdots & 1, \text{ „ } 1 \text{ „ „ } 6s. \end{cases}$$

c. (When the number of the ingredients is not
merely odd, but the prices are unequally distributed either above or below the required price.)
A dealer having wines of the same name at 7*s.*,
9*s.*, 11*s.*, 12*s.*, and 14*s.* per gallon, wishes to produce a mixture of them worth 10*s.* per gallon:—

$$10 \begin{cases} 7 & \cdots & 1 + 4, \text{ or } 5 \text{ galls. } @ \ 7s. \\ 9 & \cdots & 2, \text{ „ } 2 \text{ „ „ } 9s. \\ 11 & \cdots & 3, \text{ „ } 3 \text{ „ „ } 11s. \\ 12 & \cdots & 1, \text{ „ } 1 \text{ gall. „ } 12s. \\ 14 & \cdots & 3, \text{ „ } 3 \text{ galls. „ } 14s. \end{cases}$$

It will be seen that by varying the manner
of linking the numbers, different answers may
often be obtained to the same question. It also
often happens that the dealer or operator desires

to use a given quantity of one particular article, or to produce a certain quantity only of the mixture instead of those indicated by the above calculations. In these instances he has simply to apply the common rule of 'practice' or the 'rule of three,' as the particular case may demand.

In the above manner the proportions of the constituents of a compound may be determined from their specific gravity, when no change of volume has arisen from their admixture; but when this is the case, as in alloys, alcoholic mixtures, &c., it is either quite inapplicable or the results obtained are mere approximations to the truth. It may, however, be conveniently employed for calculations connected with the 'mixing' and 're-duction' of spirits and other liquids, by substituting their percentage value in 'proof gallons' or other corresponding denomination, for the prices in the above examples; water, when introduced, be reckoned = 0. Thus: A spirit merchant having two puncheons of rum of the strengths of 17 and 21 o. p., wishes to know what proportions of each and of water he must take to form a spirit 10 u. p. The proof values of 100 gallons of these spirits are respectively equal to 121, 117, 90, and 0 (water). Therefore—

$$90 \begin{cases} 0 \quad \rule{1cm}{0.4pt} & 27+31, \text{ or } 58 \text{ g. water.} \\ 117 \rule{0.5cm}{0.4pt} . & . \quad 90, \,,, 90 \,,, \text{rum @ 117 o. p.} \\ 121 \rule{0.8cm}{0.4pt} . & . \quad 90, \,,, 90 \,,, \,,, 121 \,,, \end{cases}$$

Suppose the dealer required to use different proportions of the spirits referred to, instead of equal measures, he has only to take such aliquot parts of the quantities thus found referring to the smaller proportion, or such multiples of those referring to the larger one, as he wishes them to bear to each in the new mixture. Numerous other applications of this rule will occur to the ingenious reader.

Questions in 'alligation,' as the department of arithmetic above referred to is called, are very easily resolved by the 'method of indeterminate analysis,' even by persons but slightly conversant with rudimentary algebra, of which, indeed, they form a simple class of problems, often admitting of an almost indefinite number of solutions. See ALLIGATION.

MO'HAIR. The hair of a goat indigenous in Asia Minor. It is dyed and manufactured by similar materials and in a similar manner to wool.

MOIL. See CIDER.

MOIRÉE MÉTALLIQUE. [Fr.] A beautiful crystalline appearance produced on the surface of tin plate by acids. The tin plate is submitted for a few seconds, whilst gently heated, to the action of dilute aqua regia, by which it acquires a variegated primrose appearance. It is afterwards washed in hot water, dried, and lacquered. The degree of heat and dilution of the acid modifies the beauty and character of the surface. The effect is also varied by employing dilute sulphuric acid, either alone or mixed with a portion of nitric or hydrochloric acid; or by using a solution of citric acid or caustic potassa. According to Herberger, the best metal for the purpose is plate iron, which has been coated by dipping it into a tin bath composed of pure tin, 200 parts; copper, 3 parts; arsenic, 1 part. The varnish should consist of copal in highly rectified spirit. Moirée

métallique is in much less demand now than formerly.

MOLASSES. See TREACLE.

MOLES. The small, soft excrescences and discolorations of the skin which are popularly known under this name may, when slight, be removed by touching them every day with a little concentrated acetic acid by means of a hair pencil, observing due care to prevent the application from spreading to the surrounding parts. This does not discolour the skin. The application of lunar caustic is also very effective, but it turns the spot temporarily black. A solution of 2 parts caustic potash and 1 part of water will convert small moles into a gelatinous mass in a few minutes (Eras. Wilson). In the pure mole there is always a considerable production of hair.

MOLUC'CA BALM. See LIQUEUR.

MOLYBDATE OF AMMONIUM. $(NH_4)_2MoO_4$. Syn. MOLYBDENIC ACID, PEROXIDE OF MOLYBDENUM; ACIDUM MOLYBDICUM, L. Prep. Native sulphide of molybdenum, after being well roasted, is reduced to fine powder, digested with ammonia, and the mixture filtered, and the filtrate evaporated to dryness; the residue, molybdate of ammonium, is then dissolved in water, purified by crystallisation; and, lastly, decomposed by heat.

Prop., &c. Small white scales, soluble in 570 parts of water; the solution reddens litmus-paper, dissolves in the alkalies, forming alkaline molybdates, from which it is again precipitated by strong acids. It is used in the preparation of molybdenum blue, and in calico printing, but its scarcity precludes its extensive employment in the arts. Molybdate of ammonium is the salt principally used in dyeing. Silks and cottons passed through a solution of this salt, then through a bath soured with hydrochloric acid, and lastly (without washing), through another of protochloride of tin, are dyed of a rich and permanent blue colour. A solution of molybdate of ammonia in excess of nitric acid forms a valuable agent as a test for phosphates, with which it gives a beautiful yellow precipitate (phospho-molybdate of ammonia). See PHOSPHORIC ACID.

MOLYBDE'NUM. Mo. A very rare metal, having a white colour, discovered by Hielm in 1782.

Prep. By exposing molybdic acid, mixed with charcoal and placed in a covered crucible, to the strongest heat of a smith's forge.

Prop., &c. It is brittle and very infusible; when heated in contact with the air it is converted into molybdic anhydride, MoO_3.

MOMOR'DICINE. See ELATERIN.

MOMRAUGHAN FOR HARNESS, &c. A correspondent of the 'Field' newspaper gives the following formulæ for momraughan, a substance used in India for preserving saddles and every description of leather. It is made as follows:—1 lb. white wax, 3 lbs. mutton fat, 1 pint spirits of turpentine; melt, and mix well together while liquid. The saddle or leather should be rubbed well with a lime in the sun, then scrubbed with a brush with soap and water; when thoroughly dry, rub it well with the momraughan (letting it soak in) in the sun. One tablespoonful will be enough for a saddle. Another recipe is—1 pint neatsfoot oil, 2 oz. beeswax, 2 oz. spirits of turpentine; other directions as above. This latter mixture, with the addition of 1 oz. Burgundy pitch, makes a very good waterproof composition for boots.

MONE'SIA. *Syn.* MONESIA BARK, BURANHEIM B.; CORTEX MONESIÆ, L. The bark of *Chrysophyllum Buranheim*, a tree growing in the Brazils. The rough imported extract of this drug also commonly passes under the name of **MONESIA**. It is astringent, and possesses no advantage over rhatany or catechu.—*Dose* (of the latter), 18 to 20 gr.

MONE'SIN. A peculiar acrid principle, analogous to saponin, found in monesia bark to the extent of 4·7%.

MONOBROMATED CAMPHOR. See CAMPHOR, MONOBROMATED.

MONOMA'NIA. See INSANITY.

MOR'DANT. In *dyeing* and *calico printing*, any substance employed to fix the colouring matter of dye-stuffs in the fibres of organic bodies, and to give it brilliancy and permanency. This it effects either by serving as a bond of union between the two, owing to its attraction for each of them; or it acts by uniting with the colouring particles in the minute pores of the fibres, and rendering them insoluble in the alkaline, soapy, and other liquids, to the action of which they will subsequently be exposed. When an infusion of some dye-stuff, as cochineal or madder, for example, is mixed with alum or acetate of alumina, and a little alkali, a precipitate immediately forms, consisting of alumina in combination with colouring matter, constituting a LAKE. It is by a similar reaction occurring within the fibres that the permanent dyeing of the cloth is effected. Here the colouring matter of the dyeing materials not only passes from the soluble to the insoluble form, but it enters into chemical combination with other substances, and in the new compounds it assumes greater brilliancy and permanency than it previously possessed. Annotta and safflower afford instances of the second mode of action above referred to, by which substances operate as mordants. The colouring matter of these dye-stuffs is soluble in alkaline lyes, and into a solution of this kind the cloth is dipped. It has now received an extremely fugitive colour only; but by passing it through acidulated water the alkaline solvent is abstracted, and the tinctorial matter is precipitated in an insoluble and minutely divided state within its pores, and it becomes permanently dyed. A similar reaction takes place in dyeing with the 'indigo vat,' in which atmospheric oxygen performs the part of a mordant. It is believed that even in these cases the colouring principle, during its transition from the liquid to the solid form, enters into combination with the fibres of the organic substance, and that, in proportion to the affinity existing between the two, is the integrity and excellence of the dye. In wool and silk the affinity between their filaments and the tinctorial particles of the dye-bath is, in general, so considerable, that a permanent stain is very easily communicated to them; but with cotton and flax, the materials of which calico and linen goods are made, the reverse is the case, and the intervention of a third material, in the shape of a mordant,

is absolutely necessary to dye them of a permanent colour.

"Experience has proved that, of all the bases, those which succeed best as mordants are alumina, tin, and oxide of iron; the first two of which, being naturally white, are the only ones which can be employed for preserving to the colour its original tint, at least without much variation. But whenever the mordant itself is coloured, it will cause the dye to take a compound colour quite different from its own. If, as is usually said, the mordant enters into a real chemical union with the stuff to be dyed, the application of the mordant should obviously be made in such circumstances as are known to be most favourable to the combination taking place; and this is the principle of every day's practice in the dye-house.

"In order that a combination may result between two bodies, they must not only be in contact, but they must be reduced to their ultimate molecules. The mordants to be united with stuffs are, as we have seen, insoluble in themselves, for which reason their particles must be divided by solution in an appropriate vehicle. Now, this solvent or menstruum will exert in its own favour an affinity for the mordant, which will prove to that extent an obstacle to its attraction for the stuff. Hence we must select such solvents as have a weaker affinity for the mordants than the mordants have for the stuffs. Of all acids which can be employed to dissolve alumina, for example, vinegar (acetic acid) is the one which will retain it with the least energy, for which reason the acetate of alumina is now generally substituted for alum, because the acetic acid gives up the alumina with such readiness that mere elevation of temperature is sufficient to effect the separation of these two substances. Before the substitution of the acetate, alum alone was employed; but without knowing the true reason, all the French dyers preferred the alum of Rome, simply regarding it to be the purest; it is only within these few years that they have understood the real grounds of this preference.

"The two principal conditions, namely, extreme tenuity of particles and liberty of action, being found in a mordant, its operation is certain. But as the combination to be effected is merely the result of the play of affinity between the solvent and the stuff to be dyed, a sort of partition must take place, proportioned to the mass of the solvent, as well as to its attractive force. Hence the stuff will retain more of the mordant when its solution is more concentrated—that is, when the base diffused through it is not so much protected by a large mass of menstruum; a fact applied to very valuable uses by the practical man. On impregnating, in calico printing, for example, different spots of the same web with the same mordant in different degrees of concentration, there is obtained in the dye-bath a depth of colour upon these spots intense in proportion to the strength of their various mordants. Thus, with solution of acetate of alumina in different grades of density, and with madder, every shade can be produced from the fullest red to the lightest pink, and with acetate of iron and madder, every shade from black to pale violet" (*Ure*).

Besides the salts of aluminium, tin, and iron,

other substances are used as mordants, viz. soap, acids, albumen, tannin, &c.

In the employment of mordants in the ordinary processes of dyeing the goods are passed through the solution for a period varying, under different circumstances, according to the object in view. The cloth is subsequently aired, dried, and well rinsed, before immersing it in the colouring bath. In *calico printing* the mordant is applied partially or topically to the cloth by means of wooden blocks, or some similar contrivance; or certain parts of the cloth are stopped out by a suitable preparation, or 'resist,' by which means a pattern is produced, as the colouring matter of the dye-bath is removed from the other portions by the washing or scouring to which it is subsequently subjected. The substances used to thicken the mordant by the calico printers, to prevent them spreading, are gum, albumen, paste, starch, and dextrine. The first is preferred for neutral solutions; the others for acidulous ones. The removal of the undecomposed particles of the mordant, so as to preserve the other portion of the cloth from their action, is effected by the process of DUNGING (which *see*), or by the chalk-bath, bran-bath, &c. Chalk acts simply by precipitating the alumina or other oxide in the mordant. The action of cow-dung, which is especially used for madder goods, has been ascribed to a peculiar acid, also to phosphates, silicates, and other salts. That the latter is the more correct view is proved by the fact that the dung-bath is now almost wholly superseded by the solution of certain salts, viz. the double phosphate of soda and lime, arsenite and arsenate of soda, and silicate of soda—all of which act by precipitating the base of the mordant in the form of an insoluble salt, which will not unite with the colouring matter or with the fibre (*Watts*).

The process of GALLING or SOOTING, commonly employed as a preparation of cotton and linen for fast dyes, consists in working the stuff for some time, at a good hand heat, in a decoction of galls or an infusion of sumach. In this case the astringent matter plays the part of a mordant. About 2½ oz. of galls, or 5 oz. of sumach, and 3 or 4 pints of water, are commonly taken for every lb. of cotton. See CALICO PRINTING, DYEING, and the respective dye-stuffs and mordants.

Mordant. In *gilding*, any sticky matter by which gold-leaf is made to adhere. *Prep.* 1. Water or beer, rendered adhesive by the addition of a little gum, sugar, or honey, and tinged with a little gamboge or carmine to mark the parts to which it is applied. Used to attach gold-leaf to paper, taffety, vellum, &c.

2. (Mixtion.) From asphaltum, 1 part; mastic, 4 parts; amber, 12 parts; fused together, and then mixed with hot boiled oil, 1 pint. Used in gilding wood, &c. See GOLD SIZE.

MORISON'S PILLS. See PATENT MEDICINES.

MORPHINE. $C_{17}H_{19}NO_3.H_2O$. *Syn.* MORPHINA, MORPHINUM, L. The chief active principle of opium. Morphine was discovered by Ludwig in 1688, but it was first obtained pure and its precise nature pointed out by Sertuerner in 1804. It is peculiar to the PAPAVERACEÆ, or poppy order.

Prep. 1. (Ph. D.) Turkey opium (cut into thin slices), 1 lb., is macerated for 24 hours in water, 1 quart, and the liquid portion decanted; the residuum is macerated for 12 hours with a second quart of water, and the process is repeated with a third quart of water, after which the insoluble portion is subjected to strong pressure; the mixed liquids are evaporated by water or steam heat to a pint, and filtered through calico; to the filtrate is added a solution formed of chloride of calcium, 6 dr., dissolved in distilled water, 4 fl. oz., and the liquid is further evaporated until it is so far concentrated that nearly the whole of it becomes solid on cooling; this is enveloped in a couple of folds of strong calico, and subjected to powerful pressure, the dark liquid which exudes being preserved for subsequent use; the squeezed cake is next treated with about ½ pint of boiling water, and the undissolved portion is washed on a paper filter; the filtered solution is again evaporated, and the solid portion thus obtained submitted to pressure as before; if the product is not quite white, this process is repeated a third time; the squeezed cake is now dissolved in boiling water, 6 fl. oz., and the solution filtered through animal charcoal (if necessary); to the clear solution is added ammonia in slight excess; the crystalline precipitate which forms as the liquid cools is collected on a paper filter, washed with cold distilled water, and, lastly, the filter is transferred to a porous brick, in order that the morphine which it contains may become dry. (From the liquids reserved from the expressions more morphine may be obtained by dilution with water, precipitation with ammonia, re-solution in boiling water, and treatment with a little animal charcoal, &c., as before.)

2. (Ph. L. 1836.) Hydrochlorate of morphine, 1 oz., is dissolved in distilled water, 1 pint; and ammonia, 5 fl. dr. (or q. s.), previously diluted with water, 1 fl. oz., is added with agitation; the precipitate is well washed in distilled water, and dried by a gentle heat. By a similar process morphine may be obtained from its other salts.

3. (*Merck*.) A cold aqueous infusion of opium is precipitated with carbonate of sodium in excess; the precipitate washed, first with cold water, and then with cold alcohol of sp. gr. 85; the residuum is dissolved in weak acetic acid, the solution filtered through animal charcoal, and precipitated with ammonia; the precipitate is again washed with cold water, dissolved in alcohol, and crystallised. A good process where spirit is cheap.

4. (*Mohr*.) Opium, 4 parts, is made into a strong infusion with water, q. s.; lime, 1 part, reduced to a state of milk with water, is then added; the mixture is next heated to boiling, at once filtered through linen, and treated, whilst still hot, with chloride of ammonium, in fine powder, in slight excess (about 1 oz. to each lb. of opium); the morphine is deposited as the liquid cools, and may be purified by a second solution in lime and precipitation by chloride of ammonium. This process is remarkably simple, and in many points is preferable to any other, either on the small or large scale.

5. (PURE.) A filtered solution of opium in tepid water is mixed with acetate of lead in

excess; the precipitate (meconate of lead) is separated by a filter, and a stream of sulphuretted hydrogen is passed through the nearly colourless filtrate; the latter is warmed, to expel excess of the gas, once more filtered, and then mixed with a slight excess of ammonia, which throws down narcotine and morphine; these are separated by boiling ether, in which the former is soluble. For the B. P. process see OPIUM.

Prop. The morphine of commerce is a white crystalline powder; but when crystallised from alcohol it forms brilliant prismatic crystals of adamantine lustre, and the formula $C_{17}H_{19}NO_3$. H_2O. It exerts an alkaline reaction on test-paper; imparts a perceptible bitter taste to water; requires 1160 parts of cold water (Squire, 1 in 1000), and 94 parts of boiling water, for its solution; insoluble in ether; dissolves in 90 parts of cold and about 30 parts of boiling alcohol; it also dissolves in the fixed and volatile oils, and in solutions of the alkalies; heated in close vessels it forms a yellow liquid, like melted sulphur, which becomes white and crystalline on cooling; heated in the air it melts, inflames like a resin, and leaves a small quantity of charcoal behind. With the acids it forms salts, which are mostly soluble and crystallisable. These may all be made by the direct solution of the alkaloid in the dilute acid. The only ones of importance are the acetate, hydrochlorate, and sulphate.

Pur. Commercial morphine and its preparations are often contaminated with codeine, narcotine, and colouring matter. The proportion of the first two may be estimated by the loss of weight which the sample suffers when digested in ether; or by dissolving out the morphine by digestion in weak liquor of potassa. Pure morphine "is scarcely soluble in cold water, sparingly so in boiling water, and readily so in alcohol. This solution is alkaline to test-paper, and by evaporation leaves crystals, which are wholly dissipated by heat. It is soluble in pure potassa" (Ph. L. 1836).

Tests. 1. Potassium hydrate and ammonia precipitate morphine from solution of its salts, under the form of a white crystalline powder, which is very soluble in excess of hydrate of potassium, and, with somewhat more difficulty, in excess of ammonia. The solution formed by excess of the first is precipitated on the addition of bicarbonate of potassium. The precipitate in either case is soluble in a solution of chloride of ammonium, and in dilute acetic acid, and is insoluble in ether. A careful inspection of the precipitate through a lens of small power shows it to consist of minute acicular crystals; and seen through a glass which magnifies 100 times, these crystals present the form of right rhombic prisms.—2. The carbonates of potassium and sodium produce the same precipitate as hydrate of potassium, which is insoluble in excess of the precipitant.—3. The bicarbonates of potassium and sodium also give similar precipitates from neutral solutions, insoluble in excess. In each of the above cases stirring with a glass rod and friction on the sides of the vessel promote the separation of the precipitate.—4. If to a mixture of morphine and oil of vitriol a minute fragment of bichromate of potassium be added,

oxide of chromium is set free, and a fine green colour developed.—5. A drop or two of solution of terchloride of gold added to a weak solution of morphia gives a yellow precipitate, which is mostly redissolved on agitating the liquid, which then assumes various hues (green, blue, violet, purple) on the addition of a drop of liquor of potassa.—6. A minute fragment of terchloride of gold and of hydrate of potassium very gently dropped into the liquid occasion purple clouds or streaks in dilute solutions, followed by a precipitate, which is violet, purple, or blue-black, according to the strength of the liquid.

Another test, given by Siebold ('Year-book of Pharmacy,' 1873), is the following:—"Heat the substance which is believed to be, or to contain, morphine, gently with a few drops of sulphuric acid, add a very small quantity of pure perchlorate of potassium. The liquid immediately surrounding the perchlorate will at once assume a deep brown colour, which will soon spread and extend over the greater part of the acid. Warming increases the delicacy of the test. 0·0001 grm. of morphine can be distinctly recognised in this way, and no other alkaloid is acted upon in a similar way by the substances named. It is indispensable, however, for the success of the experiment that the perchlorate of potassium be absolutely free from chlorate." See ALKALOIDS.

The above are the most reliable tests for morphine; the first two may, indeed, be regarded as characteristic, and the remainder as almost so. The following are often referred to by medical writers, but are less exclusive and trustworthy:—Morphine and its salts are—7. Reddened by nitric acid, and form orange-red solutions, darkened by ammonia in excess, and ultimately turning yellow, with the production of oxalic acid.—8. They are turned blue by ferric chloride, either at once or on the addition of an alkali, and this colour is destroyed by water and by alkalies, or acids in excess.—9. Iodic acid added to their solutions turns them yellowish brown by setting iodine free, and the liquid forms a blue compound with starch.

Uses. Morphine and its salts are exhibited either in substance, made into pills, or in solution, generally the latter; or externally, in fine powder, applied to the dermis denuded of the cuticle. They are principally employed as anodynes and hypnotics in cases in which opium is inadmissible, and are justly regarded as the most valuable medicines of their class. "In cases wherein both opium and the morphine salts are equally admissible I prefer the former, its effects being better known and regulated; moreover, opium is to be preferred as a stimulant and sudorific, and for suppressing excessive mucous discharges" (*Pereira*).—*Dose.* Of pure morphine, $\frac{1}{16}$ to $\frac{1}{4}$ gr.; of its salts, $\frac{1}{8}$ to $\frac{1}{4}$ gr.; externally, $\frac{1}{8}$ to $1\frac{1}{4}$ gr. Morphine is chiefly used for the preparation of the acetate and some of its other salts.

Good opium yields from 10% to 13% of morphine. See OPIUM.

Morphine, Ac'etate of. $C_{17}H_{20}NO_3 \cdot C_2H_3O_2 \cdot 3H_2O$. *Syn.* MORPHIÆ ACETAS (Ph. L., E., & D.), L. *Prep.* 1. (Ph. L., 1836.) Morphine, 6 dr.; acetic acid (Ph. L.), 3 fl. dr.; distilled water, 4 fl. oz.; dissolve, gently evaporate, and crystallise.

2. (B. P.) Hydrochlorate of morphine, 1 part, is dissolved in water, 10 parts; and the solution is precipitated with ammonia in slight excess, the precipitate is washed in cold water, and dissolved by means of acetic acid, in excess, in warm water, 12 parts; from the solution crystals are obtained as before.

Pur. Soluble in about 3 parts water, sparingly soluble in rectified spirits, has a slight odour of acetic acid, which it evolves slowly; hence the salt often gives an alkaline reaction. 20 gr. of the salt form with 1 dr. of water a slightly turbid solution, which with ammonia in excess yields a white precipitate, which, after washing with a little cold water and drying, weighs 15 gr.

Obs. The acetate of morphine of commerce is usually in the form of a whitish powder, and is prepared by the mere evaporation of the solution to dryness by a gentle heat. During the process a portion of the acetic acid is dissipated, and hence this preparation is seldom perfectly soluble in water, unless it has been slightly acidulated with acetic acid.

Morphine, Hydrochlo″rate of. $C_{17}H_{19}NO_3, HCl, 3H_2O.$ *Syn.* MURIATE OF MORPHINE; MORPHIÆ HYDROCHLORAS (Ph. L. & Ph. B.), MORPHIÆ MURIAS (Ph. E., D., & U. S.), L. *Prep.* 1. (Ph. L., 1836.) Macerate sliced opium, 1 lb., in water, 4 pints, for 30 hours; then bruise it, digest it for 20 hours more, and press it; macerate what remains a second and a third time in water until exhausted, and as often bruise and press it; mix the liquors, and evaporate at 140° F. to the consistence of a syrup; add of water, 3 pints, and after defecation decant the clear portion; gradually add to this liquid crystallised chloride of lead, 2 oz. (or q. s.), dissolved in boiling water, 4 pints, until it ceases to produce a precipitate; decant the clear liquid, wash the residuum with water, and evaporate the mixed liquids, as before, that crystals may form; press these in a cloth, then dissolve them in distilled water, 1 pint, add freshly burnt animal charcoal, 1½ oz., digest at 120°, and filter; finally, the charcoal being washed, cautiously evaporate the mixed liquors, that pure crystals of hydrochlorate of morphia may form. To the decanted liquor from which the crystals were first separated, add of water, 1 pint, and drop in liquor of ammonia, frequently shaking, until all the morphine is precipitated; wash this precipitate with cold distilled water, saturate it with hydrochloric acid, digest with animal charcoal, 2 oz.; filter, wash the filtrate as before, and evaporate the mixed liquors, cautiously, as above, that pure crystals may be obtained.

2. (Ph. E.) Opium, 20 oz., is exhausted with water, 1 gall., in the quantity of a quart at a time, and the mixed liquors are evaporated to a pint; chloride of calcium, 1 oz., dissolved in water, 4 fl. oz. is added, and, after agitation, the liquid is placed aside to settle; the clear decanted liquid, and the washings of the sediment, are next evaporated, so that they may solidify on cooling; the cooled mass, after very strong pressure in a cloth, is redissolved in warm water, a little powdered white marble added, and the whole filtered; the filtrate is acidulated with hydrochloric acid, the solution again concentrated for crystallisation, and the crystals submitted to

powerful pressure, as before; the process of solution, clarification with powdered marble and hydrochloric acid, and crystallisation, is repeated until a snow-white mass is obtained. This is the process of Gregory and Robertson, and is one of the easiest and most productive on the large scale. To procure the salt quite white, 2 to 4 crystallisations are required, according to the power of the press employed. The Edinburgh College recommends, on the small scale, the solution, after two crystallisations, to be decoloured by means of animal charcoal; but, on the large scale, to purify the salt by repeated crystallisations alone.

3. (Ph. B.) Macerate opium, sliced, 1 lb., for 24 hours with distilled water, 2 pints, and decant. Macerate the residue for 12 hours with distilled water, 2 pints, decant, and repeat the process with the same quantity of water, subjecting the insoluble residue to strong pressure.

Unite the liquors, evaporate on a water-bath to the bulk of 1 pint, and strain through calico. Pour in now chloride of calcium, ¼ oz., previously dissolved in 4 fl. oz. of distilled water, and evaporate until the solution is so far concentrated that upon cooling it becomes solid. Envelop the mass in a double fold of strong calico, and subject it to powerful pressure, preserving the dark fluid which exudes. Triturate the squeezed cake with about ½ pint of boiling distilled water, and, the whole being thrown upon a paper filter, wash the residue well with boiling distilled water. The filtered fluids having been evaporated as before, cooled, and solidified, again subject the mass to pressure, and if it be still much coloured, repeat this process a third time, the expressed liquids being always preserved. Dissolve the pressed cake in 6 fl. oz. of boiling distilled water, add purified animal charcoal, ¼ oz., and digest for 20 minutes; filter; wash the filter and charcoal with boiling distilled water, and to the solution thus obtained add solution of ammonia in slight excess. Let the pure crystalline morphine which separates as the liquid cools be collected on a paper filter, and washed with cold distilled water until the washings cease to give a precipitate with solution of nitrate of silver acidulated with nitric acid.

From the dark liquids expressed in the above process an additional product may be obtained by diluting them with distilled water, precipitating with solution of potash added in considerable excess, filtering, and supersaturating the filtrate with hydrochloric acid. This acid liquid, digested with a little animal charcoal, and again filtered, gives upon the addition of ammonia a small quantity of pure morphine. Diffuse the pure morphine obtained as above through 2 oz. of boiling distilled water placed in a porcelain capsule, kept hot, and add, constantly stirring, dilute hydrochloric acid, 2 fl. oz. (or q. s.), proceeding with caution, so that the morphine may be entirely dissolved, and a neutral solution obtained. Set aside to cool and crystallise. Drain the crystals and dry them on filtering paper. By further evaporating the mother liquor, and again cooling, additional crystals are obtained.—*Doss.* From ½ to ½ gr.

4. (*Mohr.*) By dissolving the precipitate of morphine (see MORPHIA, *Prep.* 4) in dilute hydrochloric acid, and by crystallisation as before.

Pur., &c. It "is completely soluble in rectified spirit, and in water. What is precipitated from the aqueous solution by nitrate of silver is not entirely dissolved, either by ammonia, unless added in excess, or by hydrochloric or nitric acid" (Ph. L.). "Snowy white; entirely soluble; solution colourless; loss of weight at 212° F. not above 13%; 100 measures of a solution of 10 gr., in water, ⅓ fl. oz., heated to 212°, and decomposed with agitation by a faint excess of ammonia, yield a precipitate which, in 24 hours, occupies 12½ measures of the liquid" (Ph. E.). It takes 20 parts of cold and about its own weight of boiling water to dissolve it. The hydrochlorate of morphia of the shops is usually, like the acetate, under the form of a white crystalline powder.

Obs. Of all the salts of morphia, this one appears to be that most suitable for medical purposes, from its free solubility, and from its solution not being liable to spontaneous decomposition, at least under ordinary circumstances. "The opium which yields the largest quantity of precipitate by carbonate of sodium yields muriate of morphia, not only in the greatest proportion, but also with the fewest crystallisations" (Ph. E.). Smyrna opium contains the most morphine.

Morphia and Code'ia (Hydrochlorate of). *Syn.* GREGORY'S SALT; MORPHIÆ ET CODEIÆ HYDRO-CHLORAS, L.; SEL DE GREGORY, Fr. This is commercial HYDROCHLORATE OF MORPHINE prepared according to Dr Gregory's process.

Morphia, Mec'onates of. $(C_{19}H_{20}NO_3)_2, C_7H_2O_7$. *Prep.* 1. (NEUTRAL MECONATE OF MORPHINE; MORPHIÆ MECONAS, L.) By saturating an aqueous solution of meconic acid with morphia, and evaporating the solution by a gentle heat, so that crystals may be obtained.

2. (BIMECONATE OF MORPHINE; MORPHIÆ BIMECONAS, L.) $C_{19}H_{19}NO_3, HC_7H_3O_7$. Meconic acid, 11 parts; morphia, 14 parts; dissolve each separately in hot water, q. s.; mix the solutions, and either gently evaporate and crystallise, or at once evaporate to dryness.

Obs. Morphia exists in opium under the form of bimeconate, and hence this preparation of that drug has been preferred by some practitioners. A solution of this salt for medical purposes may be directly prepared from opium, by treating its solution in cold water with a little animal charcoal, filtering, gently evaporating to dryness, redissolving the residuum in cold water, filtering, and repeating the treatment with animal charcoal. The dose of the dry bimeconate is ½ gr. or more, and of the meconate rather less. "A powder is also sold, called 'bimeconate of morphia,' which is of the same strength as powdered opium, and is given in similar doses. It is obviously incorrect to apply this name to a powder which consists principally of foreign matter. It is to be hoped that physicians will not prescribe this powder under the above name, as such a practice might lead to fatal results if the prescription should be prepared with the substance which the name strictly indicates" (*Redwood*).

Morphine, Nitrate of. (*A. T. Thomson.*) *Syn.* MORPHINÆ NITRAS, L. Add morphia in slight excess to very dilute nitric acid, filter, concentrate by gentle evaporation, and set aside that crystals may form.

Morphine, Phosphate of. *Syn.* MORPHINÆ PHOSPHAS, L. As the nitrate, substituting dilute phosphoric for nitric acid.

Morphine, Sulphate of. *Syn.* MORPHINÆ SULPHAS, L. *Prep.* Saturate very dilute sulphuric acid with morphine, evaporate to a small volume, and set aside to crystallise. It is decomposed by-driving off the water of crystallisation. Sulphate of morphine is included in the Ph. U. S. According to Magendie, this salt sometimes agrees with patients who cannot bear the acetate.

Morphine, Tartrate of. (*A. T. Thomson.*) *Syn.* MORPHINÆ TARTRAS, L. *Prep.* Saturate a solution of tartaric acid with morphine, concentrate by evaporation, and set aside that crystals may form. By using an excess of acid an acid tartrate may be formed.

MORPHIOM'ETRY. A name given to the process of determining the richness of opium in morphine. See OPIUM.

MORSU'LI. An old name applied to lozenges and masticatories. It is still retained in some foreign Pharmacopœias.

MORTAR is the well-known cement, made of lime, sand, and water, employed to bind bricks and stones together in the construction of walls, buildings, &c.

In the composition of mortar stone lime is preferred to that obtained from chalk, and river sand to pit or road sand. Sea sand is unfitted for mortar until it has been well soaked and washed in fresh water. Sifted coal ashes are frequently substituted for the whole or a part of the sand.

HYDRAULIC MORTARS or CEMENTS are those which, like Roman cement, are employed for works which are either constantly submerged or are frequently exposed to the action of water. The poorer sorts of limestone are chosen for this purpose, or those which contain from 8% to 25% of alumina, magnesia, and silica. Such limestones, though calcined, do not slake when moistened; but if pulverised, they absorb water without swelling up or heating, like fat lime, and afford a paste which hardens in a few days under water, but in the air they never acquire much solidity.

"The essential constituents of every good hydraulic mortar are caustic lime and silica; and the hardening of this composition under water consists mainly in a chemical combination of these two ingredients through the agency of the water, producing a hydrated silicate of lime. But such mortars may contain other ingredients besides lime, as, for example, clay and magnesia, when double silicates of great solidity are formed; on which account dolomite is a good ingredient in these mortars. But the silica must be in a peculiar state for these purposes, namely, capable of affording a gelatinous paste with acids; and if not so already, it must be brought into this condition by calcining it along with an alkali or an alkaline earth at a bright red heat, when it will dissolve and gelatinise in acids. Quartzose sand, however fine its powder may be, will form no water mortar with lime; but if the powder be ignited with the lime, it then becomes fit for hydraulic cement. Ground felspar or clay forms with slaked lime no water cement; but when they are previously calcined along with the lime, the mixture becomes capable of hardening under water.

"All sorts of lime are made hydraulic, in the humid way, by mixing the slaked lime with solutions of common alum or sulphate of alumina; but the best method consists in employing a solution of the silicate of potash, called liquor of flints or soluble glass, to mix in with the slaked lime, or lime and clay. An hydraulic cement may also be made which will serve for the manufacture of architectural ornaments, by making a paste of pulverised chalk with a solution of the silicate of potash. The said liquor of flints likewise gives chalk and plaster a stony hardness by merely soaking them in it after they are cut or moulded to a proper shape. On exposure to the air they get progressively indurated. Superficial hardness may be readily procured by washing over the surface of chalk, &c., with liquor of flints, by means of a brush. This affords an easy and elegant method of giving a stony crust to the plastered walls and ceilings of apartments; as also to statues and busts cast in gypsum mixed with chalk."

Under Professor Kuhlman's patent, dated April, 1841, "instead of calcining the limestone with clay and sand alone, as has been hitherto commonly practised, this inventor introduces a small quantity of soda, or, preferably, potash, in the state of sulphate, carbonate, or muriate; salts susceptible of forming silicates when the earthy mixture is calcined. The alkaline salt, equal in weight to about 1-5th that of the lime, is introduced in solution among the earths" (*Ure*).

The hardening of the common mortars and cements is in a great measure due to the gradual absorption of carbonic acid; but even after a very great length of time this conversion into carbonate is not complete. Good mortar, under favourable circumstances, acquires extreme hardness by age.

Attempts have been made at various times to introduce the use of bituminous cements into this country, and thus to restore both to land and submarine architecture a valuable material which has now lain neglected for a period of fully 30 centuries; but, unfortunately, owing to the interest of our great building and engineering firms lying in another direction, these attempts have been hitherto unsuccessful. See ASPHALTUM, CEMENT, LIME, &c.

MORTIFICA'TION. *Syn.* GANGRENE; GANGRENA, MORTIFICATIO, L. Local death; the loss of vitality in one part of the animal body, whilst the rest continues living. "The terms gangrene and mortification are often used synonymously; but gangrene properly signifies the state which immediately precedes mortification, while the complete mortification or absolute death of a part is called sphacelus. A part which has passed into the state of sphacelus is called a slough."

MOSA'IC GOLD. See BRASS, GOLD, &c.

MOS'SES. *Syn.* MUSCI, L. Several vegetables of the Nat. Ords. ALGÆ, FUNGI, LICHENES, and MUSCI commonly pass under this name with the vulgar. Of these the following are the principal:

BOG-MOSS (*Sphagnum palustre*). Very retentive of moisture. Used to pack up plants for exportation.

CEYLON MOSS (*Gracilaria candida*). Very nutritive; made into a decoction or jelly, which is highly esteemed as an article of diet for invalids and children, more especially for those suffering under affections of the mucous membranes or phthisis.

CLUB-MOSS (*Lycopodium clavatum*). See LYCOPODIUM.

CORSICAN MOSS, C. WORM M. (*Gracilaria helminthocorton*).—*Dose*, ½ to 2 dr., in powder, mixed up with sugar; as a vermifuge.

CUP MOSS, C. LICHEN (*Cladonia pyxidata*). Astringent and febrifuge. A cupful of the decoction, taken warm, generally proves gently emetic. Used in hooping-cough, &c.

FIR CLUB-MOSS (*Lycopodium selago*). Violently emetic and purgative. It is also irritant and narcotic.

ICELAND MOSS (*Cetraria islandica*). Highly nutritious and easy of digestion. The decoction is a favourite alimentary substance in affections of the lungs and digestive organs. In Iceland, after the bitter has been removed by soaking it in hot water, it is made into jelly, or dried, ground to flour, and made into bread.

IRISH MOSS, PEARL M., CARRAGEEN M. (*Chondrus crispus*). Very nutritious. The decoction or jelly is a useful and popular demulcent and emollient in pulmonary affections, dysentery, scrofula, rickets, &c. It is often employed by cooks and confectioners instead of isinglass, and by painters to make their size.

REINDEER MOSS (*Cladonia rangiferina*). Esculent, very nutritious.

MOTHER-OF-PEARL. See PEARL.

MOTHER-WATER. See CRYSTALLISATION.

MOTH EXTERMINATORS. *Prep.* 1. Lupulin 1 dr.; snuff, 2 oz.; camphor, 1 oz.; cedar sawdust, 4 oz.; mix. This is to be used for sprinkling where the moths frequent.

2. Carbolic acid and gum camphor, of each, 1 oz.; benzine, 1 pint. Dissolve the gum and carbolic acid in the benzine. Apply by saturating a piece of blotting-paper, or use it in the form of spray by means of an atomiser.

The following is recommended for sprinkling among furs, clothes, &c., to prevent the ravages of moths:

3. Patchouly herb, 100 parts; valerian, 50 parts camphor, 40 parts; orris and sumbul, of each, 50 parts; oil of patchouly and otto of roses, of each, 1 part. The various ingredients are broken up as small as possible, passed through a wide sieve to separate the coarser pieces, and freed from dust by a fine sieve. The oils are mixed with the orris root, and all the ingredients are then combined.

4. Powdered cloves, 50 parts; powdered black pepper, 100 parts; powdered quassia, 100 parts; sprinkled with oil of cassia and oil of bergamot, of each, 2 parts; camphor, 5 parts, previously dissolved in ether, 20 parts; then mix with carbonate of ammonium, 20 parts; powdered orris, 20 parts ('National Druggist').

MOULDS. Numerous materials and compositions are employed for the purpose of taking moulds, among which are the following:

1. (COMPO'.) *a.* From spermaceti, stearine, or hard tallow, and white wax, equal parts, melted together. For fine work, as medals, small casts, &c.

b. From black resin, ¾ lb.; hard tallow, ½ lb.; beeswax, 6 oz.; as the last. For coarse work, as architectural ornaments, &c. The above are poured on the objects to be copied (previously oiled) whilst in the melted state. Articles in plaster of Paris are first soaked in water, observing that none of it remains on the surface so as to interfere with the design.

2. (ELASTIC.) *a.* Flexible or elastic moulds may be made of gutta percha softened in boiling water, and after being freed from moisture, pressed strongly against the object to be copied by means of a screw press. A ring or support should be employed to prevent undue lateral spreading.

b. By the use of gelatin or glue, elastic moulds are formed capable of reproducing with accuracy, and in a single piece, the most elaborately sculptured objects, of exquisite finish and delicacy. Casts from these are now common in the streets. The credit of the application of this substance to this purpose is due to M. H. Vincent. The process of casting consists in simply dissolving a certain quantity of gelatin in hot water until it is reduced to the state of liquid paste, when it is run over the object, previously oiled, intended to be reproduced. As it cools, the gelatin assumes a consistency offering a considerable degree of resistance, and is highly elastic, which latter quality enables it to be easily detached from the work on which it has been fitted. In the hollow formed by the gelatin the finest plaster, mixed to a thick cream with water, is next run; and when the plaster has acquired the requisite hardness, the gelatin mould is detached in the same manner as from the original. From this apparently fragile mould as many as 6 copies may be taken, all reproducing the original with unerring fidelity.

3. (METALLIC.) *a.* From fusible metal. See FUSIBLE ALLOYS.

b. (CLICHÉS MOULDS.) From a fusible alloy formed of bismuth, 8 parts; lead, 5 parts; tin, 4 parts; antimony, 1 part, repeatedly melted together. The above are poured out in the melted state on a plate or slab, and after being stirred until in a pasty state, the object to be copied is strongly pressed on the alloy at the moment it begins to solidify. They are chiefly used for medals and other like objects.

c. (Chameroy's Patent.) By melting together 1 part of some easily fusible metal in a crucible, and then mixing with it 4 parts of a metal far less readily fusible, steeped in ammonia and reduced to powder. Such a compound is stated to be of great solidity, hardness, facility of soldering, melts at a low temperature, and has great tractability in moulding to any form, and in casting takes the sharpest impressions, whilst in its nature it is peculiarly unchangeable. See ELECTROTYPE.

MOUTH COSMETICS. See BREATH, TEETH, LOZENGE, PASTE, POWDER, &c.

MOX'AS. Substances burnt upon the body, for the purpose of acting as counter-irritants, and allaying deep-seated pains and inflammation. They have been used in gout, rheumatism, &c. The small cone constituting the moxa is placed upon a part, lighted, and allowed to burn to its base. The CHINESE and JAPANESE moxas are made of the downy portion of the leaves of a species of wormwood (*Artemisia sinensis*); but various other substances, as the pith of the sunflower, cotton, or paper, soaked in a weak solution of nitrate, chlorate, or chromate of potassium, answer as well. Larrey's moxas consist of lycopodium, 4 oz.; nitre, 2 oz.; formed into small cones with alcohol, and dried for some days. Dr Osborne used quicklime enclosed in a hoop of card, and moistened with water. The actual cautery is said to be preferable to any of them.

MUCILAGE. *Syn.* MUCILAGO, L. An aqueous solution of gum, or other like substance, that gives a considerable consistency to water. See DECOCTION, MIXTURE, &c.

Mucilage, Acacia. (Ph. B.) *Syn.* MUCILAGO ACACIÆ. Put gum acacia, in small pieces, 4 oz., and distilled water, 6 oz., into a covered earthen jar, and stir frequently until the gum is dissolved. If necessary, strain through muslin.

Mucilage, Fenugreek. *Syn.* MUCILAGO FŒNUGRECI. Digest 1 oz. of fenugreek seed with ¼ pint of water for 12 hours, boil, and strain with pressure.

Mucilage, Linseed. (P. Cod.) *Syn.* MUCILAGO LINI. Linseed, 1 oz.; warm water, 6 oz. Digest for 6 hours, stirring now and then, and strain.

Mucilage, Liquorice. *Syn.* MUCILAGO GLYCYRRHIZÆ. From liquorice root, as MARSH-MALLOW MUCILAGE.

Mucilage, Marsh-mallow. (P. Cod.) *Syn.* MUCILAGO ALTHÆÆ. Marsh-mallow root, 1 oz.; boiling water, 6 oz.; digest for 6 hours, and strain.

Mucilage, Quicksilver. *Syn.* MUCILAGO MERCURIALIS PLENKII. Quicksilver, 1 dr.; gumarabic, 3 dr.; syrup of poppies, 4 oz.; mix.—*Dose*, ½ dr.

Mucilage, Sassafras. (Ph. U. S.) *Syn.* MUCILAGO SASSAFRAS. Infuse 2 dr. of pith of sassafras in 16 oz. (old measure) of boiling water for 3 hours, and strain.

Mucilage, Slippery Elm Bark. (Ph. U. S.) *Syn.* MUCILAGO ULMI. Slippery elm bark, sliced and bruised, 1 oz.; boiling water, 16 oz. Infuse for 2 hours.

Mucilage, Starch. (Ph. B.) Same as DECOCTION OF STARCH (Ph. L.), which *see.*

Mucilage, Tra'gacanth. *Syn.* MUCILAGO TRAGACANTHÆ (B. P., Ph. E., & Ph. D., 1826), L. *Prep.* 1. (Ph. E.) Tragacanth, 2 dr.; boiling water, 9 fl. oz. (8 fl. oz.—Ph. D.); macerate for 24 hours, triturate, and press through linen.

2. (B. P.) Tragacanth, in powder, 60 gr.; distilled water, 10 oz. To the water contained in a pint bottle add the tragacanth, agitate briskly for a few minutes, and again at short intervals, until the tragacanth is perfectly diffused, and has finally formed a mucilage.—*Dose*, 1 oz. (Should be made as required. 1 part of tragacanth gives more viscosity to water than 25 parts of gumarabic—*Squire*.) Used in *medicine* as a demulcent, and as an application to burns, &c., and in

pharmacy in making up pills, and to suspend heavy powders in liquids.

MU'DARIN. *Syn.* MADARINE. A peculiar substance, possessing powerful emetic properties, extracted from the root-bark of *Calotropis gigantea*, in which it exists to the extent of 11% (*Duncan*). It is soluble in water and in alcohol, and its aqueous solution, unlike that of most other substances, gelatinises by heat, and becomes fluid again on cooling.

MUFFINS. *Prep.* Take of fine flour, ¼ peck; warm milk-and-water, 1 quart; yeast, a wine-glassful; salt, 2 oz.; mix for 15 minutes, then further add of flour, ¼ peck, make a dough, let it rise 1 hour, roll it up, pull it into pieces, make them into balls, put them in a warm place, and when the whole dough is made into balls, shape them into muffins, and bake them on tins; turn them when half done, dip them into warm milk, and bake them to a pale brown.

MUFFLE. See ASSAYING.

MUL'BERRY. *Syn.* MORUM, L. Mulberries (MORA, MORI BACCÆ) are the fruit of *Morus nigra*, or the black mulberry tree. They are cooling and laxative, but when eaten too freely are apt to disorder the stomach and bowels. Mulberry juice (*mori succus*) is officinal in the Ph. L. A syrup (SYRUPUS MORI) is made of it. It is also occasionally made into wine.

MULTUM. A mixture of extract of quassia and liquorice, used by fraudulent brewers instead of malt and hops.

MUM. A beverage prepared from wheat malt, in a similar way to ordinary beer from barley malt. A little oat and bean meal is frequently added. It was formerly much drunk in England; but its use at the present day is chiefly confined to Germany, and to Brunswick more particularly.

MUMPS. *Syn.* PAROTITIS, L. Inflammation of the parotid gland, which is situated under the ear. There is little constitutional derangement, but the cheeks become swollen and painful, and there is some difficulty in opening the mouth, and in swallowing. The treatment consists in simply keeping the part warm with flannel, and the use of warm fomentations, at the same time that the bowels are kept freely open with some mild laxative.

The disease is infectious, and may become epidemic in schools and large institutions. It rarely attacks the same person twice. The incubation period varies from 8 days to 3 weeks.

MUREX'ID. $C_8N_6H_8O_6$. *Syn.* PURPURATE OF AMMONIUM.

Prep. (*Gregory.*) Alloxan, 7 parts; alloxantin, 4 parts; boiling water, 240 parts; dissolve, and add the solution to a cold and strong solution of carbonate of ammonia, 80 parts; crystals of murexid will separate as the liquid cools.

Obs. Murexid can be obtained directly from uric acid by the action of nitric acid and subsequent treatment with ammonia. This process is, however, very precarious, and often fails altogether.

Prop., &c. It is only very slightly soluble in cold water; freely soluble in solutions of ammonia and the fixed alkalies; the first, by exposure to the air, becomes purple, and deposits brilliant crystals of murexid. These compounds are the purpurates of Dr Prout. It forms iridescent crystals, having a metallic lustre, of a magnificent green colour by reflected light, and an equally beautiful reddish purple by transmitted light. It is soluble in boiling water, only very slightly soluble in cold water, and insoluble in alcohol and ether. A few years ago murexid was extensively used in dyeing; it is now almost superseded by rosaniline or magenta. An analogous substance, formed as above, by treating amalic acid with ammonia, is called ' caffein-murexid.'

MU''RIATE. An old name for hydrochlorate and chloride.

MURIATIC ACID. *Syn.* HYDROCHLORIC ACID, which *see*.

MURIDE. The name originally given to bromine by M. Balard.

MUR'RAIN. See ANTHRAX.

MUSCARIN. A very poisonous alkaloid, prepared from *Amanita muscaria*. Antidotes, atropin and digitalis. See MUSHROOMS.

MUSH'ROOMS. Edible fungi. The species commonly eaten in England are the *Agaricus campestris*, or common field or garden mushroom, used to make ketchup, and eaten either raw, stewed, or broiled; the *Morchella esculenta*, or morel, used to flavour soups and gravies; and the *Tuber cibarium*, or common truffle, also used as a seasoning.

Several fungi, which to the inexperienced closely resemble the common edible mushroom, possess poisonous narcotic properties, and their use has not unfrequently been productive of serious, and in some cases even fatal results. Unfortunately, no simple tests exist by which the edible and poisonous varieties can be distinguished from each other. So strongly was the late Professor L. C. Richards, the eminent botanist, impressed with this feeling, that though no one was better acquainted with the distinctions of fungi than he was, yet he would never eat any except such as had been raised in gardens, in mushroom beds.

" This difficulty of distinguishing edible from poisonous and noxious fungi must not be ignored. If only one out of a hundred, or for the matter of that a thousand, species were poisonous or noxious, it would not be sound advice to say that we should eat all that come to hand, and stand the chance of baneful results. Unfortunately it is the case that some of the most poisonous fungi are the most common, and there is scarcely a field, and perhaps not a single wood, which does not abound with varieties of *Coprinus*, the *Agaricus fascicularis*, and the beautifully coloured *Russula emetica*, and several other very undesirable species. Some writers, and among them, if we remember rightly, the learned and enthusiastic mycologist, Dr Badham, deny the existence of any poisonous fungi in our islands, and they account for the effects which are often produced by eating varieties different from our common mushroom by stating that some people, through idiosyncrasy of constitution, are injuriously affected by all fungi; and in support of this statement they instance the well-known fact that some people experience the most unpleasant effects after eating the common edible mushroom, which chemically contains no noxious ingredients. We all know that idiosyncrasy of constitution may account for much and for very strange phenomena; for instance, oysters are almost poison to some persons, while roast beef will cause hysterics in other cases; and to not a few certain odours, harmless in themselves, are causes of serious attacks of illness; but the fact remains that persons who can eat with impunity and greatly enjoy the common mushroom are unpleasantly affected by other species of fungi. Not a year passes but deaths are recorded of persons—sometimes of whole families—after eating noxious fungi, though they had no idiosyncrasy of constitution; and shortly prior·to the writing of this article a learned botanist and enthusiastic mycologist, and a friend, in experimenting on some specimens of fungi sent to him, narrowly escaped death, while another person who partook of the dish prepared actually succumbed. A thousand and one tests have been given in writing from time to time whereby our ordinary mushroom is to be distinguished from species which resemble it—and one species is to be distinguished from another; but we fear that practically they are not to be depended upon. Fungi differ in appearance according to the localities in which they grow, and according to their age. The common belief that the edible species never change colour when cut or bruised is untenable, for three varieties at least are perfectly edible, and yet assume different tints when injured in any way. The test of taste, too, which is applied under the idea that those with a pleasant savour and an inoffensive smell are always wholesome, is fallacious, for a raw mushroom is quite a different thing from the stewed or grilled one, and often what has an acrid taste when raw becomes perfectly savoury when cooked; and, *vice versâ*, a tasteless fungus may be poisonous, but only develop its latent flavour when submitted to the cook. Dr Christison declares that a sure test of poisonous fungus is an astringent, styptic taste, and a disagreeable pungent odour; but this, again, cannot always be depended on. Nor, again, is the popular idea that a mushroom which will skin easily is wholesome altogether based on fact. What, then, is to be done to enlarge the field of our mushroom gatherers, and to bring about the utilisation of food now suffered to run to waste? or, in other words, how is a knowledge of our fungi to be obtained? The only answer is that knowledge on this matter is to be got, generally speaking, as knowledge on other matters—partly from books, but more especially from oral instruction and demonstration. Such eminent authorities as Dr Badham, the Rev. M. Berkeley, Mr Cooke, and Mr Worthington Smith may be consulted with profit; and works such as that on 'Domestic Economy,' in which coloured plates bring accurately before the eye the different species of our fungi. And here we may mention that the plates prepared by Mr Worthington Smith, which were once at the South Kensington Museum, but now, we believe, at Bethnal Green, have done much to help the Londoner when in search for mushrooms in the country to distinguish between the good and bad species of fungi. It might be well that in our schools, where so many practically useless branches of knowledge

are crammed into children both in town and country, practical lessons on fungi should be given. Those, too, who wish to learn what is to be learned on this subject should avail themselves of opportunities now often given at exhibitions and botanical meetings. At Paris, in 1876, there was an exhibition of edible and poisonous fungi, in a fresh and dry state, together with books and drawings; and a similar exhibition took place in Aberdeen two years before; and, as most of our readers are probably aware, there exists a Fungus Club, or, rather, a botanical society which makes fungi a special study. This is the Woolhope Club, which has its headquarters at Hereford, and embraces in its scientific investigations all the district between Shropshire and the Bristol Channel. One day in each autumn is devoted to a fungus hunt, and the numbers that are gathered by this enthusiastic band are something enormous. The labours of the day are closed by a dinner, at which the main dishes are composed of the spoils of the chase, dressed in the most epicurean fashion, and of other good things flavoured with the most appetising (fungus) sauces. In the annual volume published of the transactions of the club there is a description of the fungi of the district, and the best modes of cooking them. It would be a great gain to the public if at least that part dealing with fungi were generally obtainable.

"Gastronomically the ordinary mushroom, and a large number of our British fungi, are most estimable, and ketchup produced from them—not the ordinary ketchup 'of commerce,' which is often innocent of any fungi whatever—is to the cultivated taste of the gourmet the best of sauces. Many an epicurean has been heard to aver that after that of an oyster that of a mushroom is the finest in the whole world of gastronomy. Bacon, in his 'Naturall Historie,' says of mushrooms, 'They yield a delicious meat;' and to these commendations it may be added that they can be cooked in almost as many ways as the French can cook eggs. Dr Letheby says that 'the edible varieties are highly nutritious;' and the late Dr Edward Smith, who was very chary of commending anything, also had a good word for them.

"Our word 'mushroom' is evidently an adaptation of the French *mousseron*, which, of course, is from *mousse*, 'moss' (Lat. *muscus*); but the suggestion of the learned Salmasius, that the French gave this name to the edible fungus 'because it grows only where the grass is the shortest and there is little else but moss,' strikes one as rather weak. The mushroom, like the moss, is a cryptogamous plant; but there is little connection in any way between the two. Perhaps, then, we must look to the Greek word *mucos*, though only used by the grammarians, for the origin of the French word and so of our own. This was one of the terms which signified a 'sponge,' and was probably applied to the 'fungi' because of their sponge-like growth. It is evident that some of our more exact botanists, or etymologists who compounded the word for them, consider the Greek word and not the Latin as the origin of the *mus* in the English word and the *mos* in the French, though, according to analogy, the *u* should have been changed into a *y*, for the study of 'fungi' is termed by them *mycology*. It is nardly necessary to add that the words 'fungology' and 'fungologist' are hybrid compounds of Greek and Latin, which are simply intolerable to ears correct, as are many other words similarly compounded, and recently introduced into our language. The Latin *fungus* is plainly a weakened form of the Greek *spongos*, and goes to show that the idea of a 'sponge' was from the first associated with the *fungi*, and that the Greek *mucos* must be taken as the origin of the French *mousseron* and the English 'mushroom.' It is curious that the Greek, Latin, and English 'fungous' terms have all been used in a sense reflecting on some of our species. The Greek *mucos* represented a silly, stupid fellow, and the Plautus couples the fungi—'soft-pated'—with the 'fools,' 'stolid,' and 'fatuous.' In like manner, in our own language, Bacon speaks of certain persons as 'mushrooms and upstart weeds' because of their sudden growth from a lowly origin. South, in one of his sermons, reflects on 'mushroom divines who start up of a sudden,' and whose success is 'not so good as to recommend their practice.' Carrying out the same analogy, the late Albert Smith, if we recollect rightly, spoke of 'stuck-up people' as springing like mushrooms suddenly into notice, and, like them, from very questionable soil" ('Daily Telegraph').

In cases of poisoning by fungi, vomiting should be immediately induced by an emetic and tickling the fauces with the finger or a feather; after which a purgative clyster or a strong cathartic should be administered, with ½ to 1 fl. dr. of ether in a glassful of water or weak brandy. As an antidote, a solution of tannin, ½ dr., in water, 1½ pints, or a decoction of ½ oz. of powdered galls, or of 1 oz. of powdered cinchona bark, in a like quantity of water, has been strongly recommended by M. Chansarel. Atropin or digitalis in small doses should be given as an antidote in poisoning by *Amanita muscaria*.

Alexis Soyer recommended the excellent method of cooking mushrooms by baking them under a glass or basin on toast, along with scalded or clotted cream, or a little melted butter, with 1 clove, and salt, pepper, &c., to taste. They take about ½ of an hour in a gentle oven or before the fire. When they are taken up, do not remove the glass for a few minutes, by which time the vapour will have become condensed and gone into the bread; but when it is, the aroma, which is the essence of the mushroom, is so powerful as to pervade the whole apartment.

MUSK. *Syn.* MOSCHUS (B. P., Ph. L., E., and D.), L. "A secretion deposited in a follicle of the prepuce of *Moschus moschiferus*, Linn." (Ph. L.), an animal inhabiting the mountains of Eastern Asia. It is imported from Bengal, China, and Russia; and, latterly, from the United States of America. That known as TONQUIN MUSK is the most esteemed for its odour; but that from Russia is the only kind which reaches us in perfect bags, or which has not been tampered with. POD MUSK (MOSCHUS IN VESICIS) is the bag in its natural state, containing the musk. The average weight of the pods is about 6 dr.; that of the grain musk which it contains, about 2½ dr.

Pur., &c. The musk of the shops is generally adulterated. Dried bullock's blood or chocolate is commonly employed for this purpose, along with a little bone-black. The extent of these additions varies from 25% to 75% of the gross weight of the mixture. The blood is dried by the heat of steam or a water-bath, then reduced to coarse powder, and triturated with the genuine musk in a mortar along with a few drops of liquid of ammonia. It is then either replaced in the empty pods, or it is put into bottles, and sold as grain musk. There are only 3 certain ways of detecting this fraud, viz.—by the inferiority of the odour, by an assay for the iron contained in the blood, or by the microscope. Genuine musk often becomes nearly inodorous by keeping, but recovers its smell on being exposed to the vapour of ammonia, or by being moistened with ammonia water. The perfumers sometimes expose it to the fetid ammoniacal effluvia of privies for the same purpose.

Pure musk, by trituration or digestion with boiling water, loses about 75% of its weight, and the boiling solution, after precipitation with nitric acid, is nearly colourless. A solution of acetate of lead, and a cold decoction of galls, also precipitate the solution; but one of corrosive sublimate does not disturb it. The ashes left after the incineration of pure musk are neither red nor yellow, but grey, and should not exceed 5% to 6%. The Chinese appear to be the most skilful and successful adulterators of musk. One of the best solvents for musk is ether.

Uses, &c. Musk is chiefly employed for its odour. As a *medicine* it is a powerful stimulant and antispasmodic, and is a valuable remedy in various diseases of a spasmodic or hysterical character, or attended with low fever.—*Dose,* 5 to 10 gr. made into an emulsion.

Musk, Factitious. *Syn.* RESIN OF AMBER; RESINA SUCCINI, MOSCHUS ARTIFICIALIS, M. FACTITIUS, L. *Prep.* 1. Oil of amber, 1 fl. dr.; nitric acid, 3½ fl. dr.; digest in a cold tumbler, and after 24 hours, wash in cold water the orange-yellow resinous matter which has formed and carefully dry it.

2. (*Elsner.*) From oil of amber, 1 part; fuming nitric acid, 3 parts; as the last, but employing artificial cold to prevent any portion of the oil being carbonised.

3. A remarkable oily liquid, having a brown colour, and smelling so like musk that, it is said, very few noses are able to detect the difference between the natural product and the artificial body, is obtained by a new process. 2 parts of isobutyl alcohol, 3 parts of metaxylol, and 9 parts of chlorate of zinc are heated together for 8 or 9 days at a temperature of about 440° or 450° F. in a strong vessel, the pressure inside which speedily rises to nearly 30 atmospheres, but gradually declines to about a quarter of that degree of tension, when the whole is allowed to cool gradually. The crude product so obtained is purified by distillation once or twice repeated, until an oily fluid is the result, which comes over between 220° and 260°; this when rendered slightly alkaline is the 'musk' in question, and it may be diluted with alcohol, for the use of the perfumer, to any desired degree of odoriferous strength.

Obs. Resin of amber smells strongly of musk, and is said to be antispasmodic and nervine. A tincture (TINCTURA RESINÆ SUCCINI) is made by dissolving 1 dr. of it in rectified spirit, 10 fl. dr., of which the dose is 1 fl. dr.; in hooping-cough, low fevers, &c.

Dr Collier mentions an artificial musk, prepared by digesting for 10 days nitric acid, ½ oz., on fetid animal oil, obtained by distillation, 1 oz.; then adding of rectified spirit, 1 pint, and digesting the whole for a month.

MUSK-SEED. *Syn.* GRAINS D'AMBRETTE, Fr. The seed of *Abelmoschus moschatus*, or musk mallow. They are chiefly used for their odour, in perfumery, hair powder, coffee, &c.

MUS'SEL. See SHELL-FISH.

MUS'SEL SCALE. See MYTILASPIS.

MUST. *Syn.* MUSTUM, L. The expressed juice of ripe grapes, before fermentation. When boiled to 2 to 3 dr. it is called CARENUM; when boiled to ½ it is called SAPA. On further concentration, it yields a species of granular sugar (grape-sugar).

Must, Facti"tious. *Syn.* MUSTUM FACTITIUM, L. *Prep.* Dissolve cream of tartar, ½ oz., in boiling water, 7 pints; when cold, add of lump sugar, 2¼ lbs.; raisins (chopped small), ½ lb.; digest for 3 or 4 hours, strain through flannel as quickly as possible, and add of lemon-juice, ⅓ pint.

MUSTARD. *Syn.* SINAPIS, L. "The seed of *Sinapis nigra* and *S. alba*" (Ph. L.). "Flour of the seeds of *Sinapis nigra*, generally mixed with those of *S. alba*, and deprived of fixed oil by expression" (Ph. E.). "The flour of the seeds" (Ph. D.). "The seeds of the *Sinapis nigra* and *S. alba* reduced to powder and mixed" (B. P.). That of the shops is very frequently adulterated with wheat flour. When this is the case it does not readily make a smooth paste with water, but exhibits considerable toughness, and somewhat of a stringy appearance, especially when little water and much heat is employed. The common proportions taken by some grocers are—dried common salt, wheat flour, and superfine mustard, equal parts; with turmeric, to colour, and cayenne, q. s. to give it piquancy and fire.

Uses, &c. Pure flour of mustard is used in medicine to make stimulating poultices, pedi-luvia, &c. As a condiment it is useful in torpor and coldness of the digestive organs. A few years since the use of mustard seed, by spoonfuls, *ad libitum*, was a common and fashionable remedy in torpor or atony of the digestive organs. The practice was a revival of that recommended by Dr Cullen; but it has now again sunk into disuse. Sir John Sinclair also approved of the use of mustard seed in this way, especially for the preservation of the health of the aged ('Lancet,' Jan., 1834). See POULTICES, &c.

Mustard for the Table. The common practice of preparing mustard for the table with vinegar, or still more, with boiling water, materially checks the development of those peculiar principles on which its pungency or strength almost entirely depends. To economise this substance we should use lukewarm water only; and when flavouring matter is to be added to it, this is better deferred until after the paste is made. The following forms for 'made mustard' are much esteemed for their flavour:

Prep. 1. Mustard (ground), 3¼ lbs.; water, q. s. to form a stiff paste; in ¼ hour add of common salt (rubbed very fine), 1 lb.; with vinegar, grape juice, lemon juice, or white wine, q. s. to reduce it to a proper consistence.

2. To the last add a little soluble cayenne pepper or essence of cayenne.

3. (*Lenormand.*) Best flour of mustard, 2 lbs.; fresh parsley, chervil, celery, and tarragon, of each, ½ oz.; garlic, 1 clove; 12 salt anchovies (all well chopped); grind well together, add of salt, 1 oz.; grape juice or sugar, q. s. to sweeten; with sufficient water to form the mass into a thinnish paste by trituration in a mortar. When put into pots, a red-hot poker is to be thrust into each, and a little vinegar afterwards poured upon the surface.

4. (MOUTARDE À L'ESTRAGON.) From black mustard seed (gently dried until friable, and then finely powdered), 1 lb.; salt, 2 oz.; tarragon vinegar, q. s. to mix. In a similar way the French prepare several other 'mustards,' by employing vinegars flavoured with the respective substances, or walnut or mushroom ketchup, or the liquors of the richer pickles.

5. (MOUTARDE SUPERBE.) Salt, 1½ lbs.; scraped horse-radish, 1 lb.; garlic, 2 cloves; boiling vinegar, 2 galls.; macerate in a covered vessel for 24 hours, strain, and add of flour of mustard, q. s.

6. (Patent.) Black ginger (bruised), 12 lbs.; common salt, 18 lbs.; water, 15 galls.; boil, strain, and add to each gallon flour of mustard, 5 lbs.

Mustard Leaves (*Rigollot's*) are made by spreading moistened mustard on paper, and drying.

MUSTINESS. See MALT LIQUORS and WINES.

MU'TAGE. The term applied to the 'matching' of grape must to arrest the progress of fermentation. See ANTIFERMENT, MATCHES, &c.

MUTTON. The flesh of sheep. That of the first quality is "between 4 and 5 years old; but at present it is rarely to be obtained above 3, and is often under 2. The flesh ought to be of a darkish, clear, red colour, the fat firm and white, the meat short and tender when pinched, and it ought not to be too fat." The flesh of the 'Southdown wether' is esteemed the finest flavoured. Mutton is one of the most wholesome of the 'red meats,' and in commercial importance is second only to beef. See MEAT.

MY'COSE. A peculiar variety of sugar, extracted by alcohol from ergot of rye. It crystallises in colourless prisms, and is distinguished from cane-sugar by not reducing the acetate of copper, when boiled with a solution of that salt.

MYDRIATIC ALKALOIDS. These are a small but important group of bases, which have the power, when placed in the eye, of dilating the pupil, destroying the power of accommodating vision to near objects. The best known are obtained from plants of the order SOLANACEÆ; they are atropine, daturin, hyoscyamine. Solutions of these bases in the pure state are distinguished from all other alkaloids by yielding a red precipitate when warmed with a weak alcoholic solution of mercuric chloride (*Gerrard*).

MYLABRIS. *Syn.* MYLABRIS CICHORII; CHINESE BLISTERING FLY. An insect found on the flowers of the succory plant in India and China. It is about $1\frac{1}{4}$ inches in length; sheath-wings black, each presenting anteriorly 2 almost quadrate, brownish-yellow spots; behind these, 2 brownish-yellow bands, each of which equals about 1-6th of the length of the sheath-wings. Its vesicant properties are due to the presence of cantharidin.

Its physiological actions are the same as those of cantharides, except that it is said not to affect the kidneys when topically applied.

MYRICIN. The portion of beeswax which is least soluble in alcohol, and saponified with difficulty.

MYRISTIC ACID. $HC_{14}.H_{27}O_2$. A monobasic fatty acid, obtained by the saponification of myristin. It melts at 120° F.

MYRISTIN. $C_{45}H_{86}O_6$. Syn. SERICINE. The white, solid portion of the expressed oil of nutmegs, which is insoluble in cold alcohol. See MYRISTIC ACID.

MYROBALANS. The fruits of *Terminalia chebula*, Retz, and *T. belerica*, Roxb., large deciduous trees common in India, Ceylon, and the Malay Islands. The hard woody fruits of both species are imported in large quantities for the use of tanners from various parts of the East Indies. Astringent galls are often formed on the twigs of *T. chebula*, used in India for making ink, as well as for dyeing and tanning. The hard woods of both species are used for a variety of purposes in India. The wood, gums, and bark of various species of *Terminalia* are met with in commerce. Amongst others, *T. tomentosa*, Bedd. —when polished the wood resembles walnut, and is considered one of the best woods for making stethoscopes at the Government Medical Store Depôt, Bombay; *T. paniculata*, Roth.; *T. myriocarpa*, Heurck. and Muell. Arg.; and *T. procera*, Roxb.

MYROLES. In French pharmacy, solutions of oleaginous or resinous substances in the volatile oils.

MYRONIC ACID. $HC_{10}.H_{18}NS_2O_{10}$. Bussy has given this name to an inodorous, bitter, non-crystallisable acid, obtained by him from black mustard, in which it exists as myronate of potassium. It is soluble in water and alcohol.

MYROSIN. Syn. EMULSION or BLACK MUSTARD. A name given by Bussy to a peculiar substance, soluble in water, and which possesses the power of converting myronic acid, in the presence of water, into the volatile oil of mustard seed.

MYROSPERMIN. The name given by Richter to the portion of the oil of balsam of Peru which is soluble in alcohol.

MYROXILIN. The name given by Richter to the portion of the oil of balsam of Peru which is insoluble in alcohol. By oxygenation it forms myroxilic acid.

MYRRH. Syn. MYRRHA (B. P., Ph. L., E., & D.), L. " Gum resin exuded from the bark of *Balsamodendron myrrha*" (B. P., Ph. L.).

Pur. 1. Triturate a small quantity of the powder of the suspected myrrh with an equal amount of chloride of ammonium, adding water, gradually: if the whole is readily dissolved, the myrrh is genuine; otherwise it is sophisticated with some inferior substance (*Righini*). 2. When incinerated it should not leave more than $3\frac{1}{2}\%$ to 4% of ashes.

Uses, &c. Myrrh is a stimulating aromatic bitter and tonic, and is given in several diseases accompanied by relaxation and debility; especially in excessive secretions from the mucous membranes, and in disorders of the digestive organs. *Externally*, as an ingredient in dentifrices and washes, in caries of the teeth, spongy and ulcerated gums, &c.—*Dose*, 10 to 30 gr.; either alone or combined with aloes or chalybeates.

MYTILASPIS (ASPIODOTUS) POMORUM, Bouché (μυτῖλος, a mussel). The MUSSEL SCALE. An affection is often discovered upon apple and pear trees of excrescences or scale-like coverings on the bark, chiefly upon the smaller branches, and where the bark is most smooth, and upon the main stems of particular sorts of these fruit trees, as well as of young trees. In orchards where the trees have been neglected there are often numerous groups of these scales upon both apple and pear trees, which anyone wholly ignorant of entomology would hardly notice, or distinguish from rugosities or lichenous growths upon the bark. Most usually the scales are upon the north side of trees, or upon those parts of their stems and branches shaded from the sun, and it seems that they like dampness and gloom. It will be understood, therefore, that in the primeval orchards of Devon, Herefordshire, Somersetshire, and Worcestershire, where the trees are close together, and their branches form an almost impenetrable shade, the scale insects find favourable conditions for their increase, and cause much harm by sucking out the juices upon which they feed.

The mussel scale is so named because it is shaped like the shell of the mussel. It might also be named after the limpet, as it sticks to the trees with as much tenacity as the limpet adheres to a rock. It is provided with powerful suckers, with which it extracts the sap from the trees, and when it is present in numbers the respiration of their surfaces is much impeded.

Apples and pear trees with smooth bark suffer more from the scale than those with thick rough coverings. The various kinds of Codlin, the Ribston Pippin, Margel, Pearmain, the Rennets (Reinette), Cox's Pomona, Hawthorndean, Wellington, Blenheim Orange, are especially liable to receive injuries from this insect. Young trees have been killed outright, having been literally covered with scale. Bush trees and half-standards are also subject to its attack.

Pear trees in orchards, notably in perry-making districts, are occasionally subject to this scale, and espalier trees, and trees grown against walls are troubled by it. Beurré Diel, Jargonel, and Marie Louise have been seen to be materially weakened by it. In Kentish orchards, Bergamot, Duchesse d'Angoulême, and Beurré de Capiaumont have been noticed to be suffering from its attacks.

There are several species of scale insects in America which do great harm to fruit trees of various kinds. Mr Matthew Cooke says, "No variety of fruit is exempt from their attacks, and in certain localities many trees have been seriously injured or even killed outright by them" ("A

Treatise on Insects Injurious to Fruit and Fruit Trees in the State of California,' by Matthew Cooke). In the southern States the species known as *Mytilaspis Gloveri*, introduced from China in 1840, almost entirely ruined the orange trees in Florida and Louisiana. Happily, however, parasitic flies which were not forthcoming for the first few years after the scale insects arrived, appeared in quantities a few years ago, and have since served to check their spread considerably.

This insect is known in France and Germany as hurtful to apple and pear trees and to currant bushes.

Life History. The mussel scale, or brown scale, is placed in the order HOMOPTERA, in its section *Monomera*, and in its family *Coccidæ*, "one of the most anomalous tribes of insects," says Westwood, "with which we are acquainted." The scale or shell of this insect is a horny covering, which is part and parcel of the insect, growing with it pretty much after the manner of the shells of snails, and serving after the completed growth of the larvæ as a protection, at least in the case of the females, and as a shelter for the eggs and young for a time. This scale is about ⅓ of an inch in length, and is shaped, as Taschenberg describes it, exactly like a comma. When the larvæ come forth from the eggs laid in the protecting shell and after a time leave its shelter, they roam about seeking for a desirable and comfortable spot in which to pitch their tents, or rather to set their shells. At this time they are not more than the nineteenth part of an inch in length, having antennæ and cornicles and six legs, together with a most serviceable apparatus wherewith to suck out the juices of the fruit trees. It appears that as soon as the larvæ have fairly established themselves they insert this apparatus into the bark, and after this they become fixed and cannot detach themselves or be detached without some difficulty. After this settlement has been gained the shells are soon visible, being formed from exudations, or secretions, from the bodies of the larvæ together with their exuviæ or moulted skins. Some of these larvæ are males; others are females. In the case of the males these undergo two moults or castings of their skins under the shells. They are similar to the female up to this time, but soon a change comes over them. They become pupæ, and in the course of a few days they shake off their shelly coats and appear in winged forms as long-winged flies. Pairing takes place then in the extraordinary manner described by Réaumur ('Mémoires pour servir à l'Histoire des Insectes,' par M. de Réaumur, tome iv, p. 34) and illustrated by many figures, the body of the male being peculiarly elongated in order that impregnation may be effected under the shelly covering of the female. Professor Comstock corroborates Réaumur's description in his elaborate treatises upon scale insects. Then the male quickly disappears from the scene, being, as is the case with the male insects of many of the *Amphididæ*, unprovided with mouths, and therefore unable to feed.

But the females remain still glued to the spot, having previous to their fecundation cast off their legs, antennæ, and cornicles with their skins in the course of several moults as useless appendages in their stationary condition. In due time eggs are laid which are arranged in the narrow parts of the shells in admirable order, the female keeping in the wider quarters until all the eggs are laid, and gradually getting smaller and smaller, and then finally dying. From 20 to 50 eggs are laid by each female. At first the eggs are whitish and opaque, and afterwards become darker, almost purple. They are hatched in about 10 days, and the larvæ, as described above, leave the parent shell and go forth on their own account.

The various transformations of this insect are completed in about 6 weeks. In hot countries there is more than one generation in a season, but in England it is believed that there is only one.

Prevention. It is very necessary to keep apple and pear trees free from lichenous and mossy growths, as these serve as harbours for scale insects and many others. Lime put on hot in damp weather in the autumn is a perfect cure for this. The bark should be kept scraped. The trees should not be planted too thickly in new orchards, and the branches of old trees in old orchards must be thinned out periodically to let in air and light. Young trees should be thoroughly overhauled before they are planted, in order to discover if they have scales upon them.

Remedies. Owing to the hard shells of the scale insects, syringings with even the most disagreeable compositions hardly make any impression upon them, except when they are taken, as Miss Ormerod has pointed out ('A Manual of Injurious Insects,' by E. A. Ormerod, 1881), just as the larvæ escape from the parental abode. If applied at this time syringing with strong soft soap and quassia concoctions, in the proportion of 12 lbs. of soft soap and 8 lbs. of quassia to 100 galls. of water, would be efficacious. Painting the trees with a wash compound of quicklime of about the consistency of whitewash, with soft soap added at the rate of about ⅓ lb. to the gallon, is most useful in the case of a bad attack, as well as for young trees. Or the stems of infested trees after having been scraped may be scrubbed over with a mixture of soft soap and water, in the proportion of ⅓ lb. of soft soap to a gallon of water, and ⅓ lb. of the finest flowers of sulphur, stirred well together. The mixture of soft soap and petroleum might also be advantageously used for brushing into the bark of the trees.

After trees have been scraped it is most essential that the scrapings of bark should be burned at once.

With regard to young trees and small trees, and all trees where it is practicable, it would be very advantageous to scrub the stems and branches with housemaid's scrubbing-brushes and a composition, as described above, of soft soap, sulphur and water, or of ⅓ lb. of soft soap and 8 wine-glasses full of paraffin oil to a gallon of water, or with the petroleum soap ('Reports on Insects Injurious to Crops,' by Charles Whitehead, Esq., F.Z.S.).

MYZUS CERASI, Passerini. The CHERRY APHIS. Fortunately this aphis does not often cause very much injury in large cherry orchards, as in those of East and Mid Kent, for instance, although it is frequently the source of a considerable amount of harm and annoyance in small orchards and gardens, especially upon half-standards, pyramids, and bushes. It is also particularly troublesome occasionally to Morello trees, whose large juicy subacid fruit is so valuable for making cherry brandy, and is largely produced throughout Kent. Sometimes, however, in blighting years, like that of 1885, when almost all the cultivated plants under the sun and many forest trees were infested with their own familiar aphides, the trees in the large orchards do not escape.

The aphides during a severe attack swarm upon the under surfaces of the leaves and pump out their life-juices with their siphon-like apparatus, and seal up their pores with filth, which also falls upon the upper surface of the leaves and prevents respiration. In these circumstances the fruit cannot fill out properly. If it become fully formed and ripen in due course it is of poor quality, and is naturally injured for sale by the black mixture of honey-dew and excreta that is sprinkled upon it.

This aphis is also found constantly upon black and red currant bushes. In 1885 these bushes, more particularly those of the black currant, were covered with these insects, which finally ruined the crop of fruit in some cases, and in others made it unfit to send to market. Mr Buckton, in his monograph of British Aphides, speaks of the *Myzus cerasi* as having been seen by him upon currant bushes, and the experience of recent years quite confirms Mr Buckton's statement.

In a neighbouring fruit plantation in June in 1885 there were to be found the *Aphis mali* upon almost every apple tree, the *A. pruni* actively engaged in ruining the plum and damson crop, and the *Myzus cerasi* hard at work upon the leaves of the black-heart cherry trees and upon the black and red currant bushes. Besides all these, side by side with the larvæ of the *M. cerasi* occasionally could be seen the larvæ of the *Aphis ribis*, engaged in throwing up red galls on the leaves of both kinds of currant bushes.

These currant bushes received more injury from the *Myzus cerasi* than the cherry trees in 1885, and the larvæ were innumerable and persistent.

No kind of migration from the cherry trees to the currant trees was noticed during the season. Their appearance upon each was nearly simultaneous.

Kaltenberg and Taschenberg both describe the *Myzus cerasi* as a plague upon cherry trees in Germany. Professors Asa Fitch and Lintner tell us that it is common in America, while Saunders speaks of an insect in Canada which seems to be exactly similar. It would appear from the statements of Professor Fitch that this *Myzus* is more formidable in America than in England, for he remarks that upon a cherry tree 10 feet high, reckoning that it had 17,000 leaves upon it, there were at least 12,000,000 of these creatures ('First and Second Reports upon the Insects of New York,' 1856, by Professor Asa Fitch).

Life History. This insect belongs to the family *Aphididæ*, to the tribe *Aphidinæ*, and the genus *Myzus*, so called from the Greek verb meaning to suck.

In colour the viviparous, apterous, or wingless female, bringing forth living larvæ, is dark, almost black, with dark yellow legs. Its body is very broad at its lower extremity. It comes from the egg in April, and soon after brings forth living larvæ, or lice, which at once begin to feed upon the juices of the leaves. After the larvæ have put on the pupa stage the winged viviparous females come forth and fly away to infest other trees and bushes. They have black bodies with yellow legs, and large wings measuring about 3 lines when expanded. Like the pupæ, they have red eyes.

Later on winged males are generated, whose bodies are yellow with brown or dark markings, and not so broad as those of the females. At or about the same time from the latter generations, produced by the winged viviparous females, wingless, egg-laying (oviparous) females come upon the scene, with which are found the winged males towards the middle of September. These wingless females are for the most part brown; their bodies shine and are squat in shape, and not so large as the foundress or Altmütter.

From 2 to 4 eggs are laid by each female upon the shoots of cherry trees and the currant bushes towards the end of September or in the beginning of October. Hatching takes place when the first warm days of spring arrive.

Prevention. After an attack of these aphides upon cherry trees, a close examination should be made in September to discover if there are egg-laying females upon their branches and side shoots. In case these are detected washing or syringing the trees with a mixture of soft soap, quassia, and water, or petroleum soap and water, may be adopted with good results. Care obviously must be taken to seize upon the right period for this operation, so as to remove the females before they have laid their eggs.

When black currant bushes have been badly infested they should be pruned in November, and the cuttings carried away and burned at once. Black currant bushes can hardly be pruned too closely, as the fruit comes upon the first year's wood. If the attack of the *Myzus* were continuous and persistent the bushes might be cut down close to the ground without any injurious consequences resulting to them.

Red currant bushes, on the other hand, are not pruned hard. It might, therefore, be well to brush the stems and shoots over with a thick solution of soft soap with paraffin oil in it, in the proportion of 20 lbs. of soft soap to 100 galls. of water and 2 quarts of paraffin oil, mixed well together, if it were suspected that there were any eggs upon them; or the petroleum soap, slightly diluted, may be employed.

Morello cherry trees are principally grown against walls and buildings, and it would be well after an attack of the *Myzus* to take down the branches from the walls and syringe them well with soft soap, quassia, and water; or to brush them over with the same composition as that prescribed for red currant bushes.

Remedies. Syringing with soft soap and quassia

mixed with water, by means of hop-washing engines, is the only remedy available in the case of cherry trees, and this must not be done before the cherries are well set and clear from the remains of the calyces, nor when they show the faintest tinge of colouring. This operation is arduous and costly, as it is in the case of all large fruit trees, and in all probability it might have to be repeated, so that it would not be undertaken unless there were special conveniences for carrying it out, and the prospect of remunerative prices. In the case of large cherry trees the ordinary hop engines would perhaps not have power enough to force the wash well up to the topmost boughs; but machines with stronger pumps could be made if the attack were serious and recurrent ('Reports on Insects Injurious to Crops,' by Chas. Whitehead, Esq., F.Z.S.).

NAILS (The) should be kept clean by the daily use of the nail-brush and soap-and-water. After wiping the hands, but whilst they are still soft from the action of the water, the skin, which is apt to grow over the nails, should be gently loosened and pressed back, which will not only preserve them neatly rounded, but will prevent the skin cracking around their roots (agnails, nail-springs) and becoming sore. The free ends or points of the nails should be pared about once a week; and biting them should be particularly avoided, as being at once destructive to their beauty and usefulness. "The (free) edge of the scarf-skin should never be pared, the surface of the nail never scraped, or the nails cleaned with any instrument whatever saving the nail-brush" (*Eras. Wilson*).

The consequences of wearing a shoe that is obviously too short for the foot are thus described by the above authority:—"In this case Nature gives us warning, by means of her agent, pain, that such a proceeding is contrary to her laws. We stop our ears, and get accustomed to the pain, which, perhaps, is not severe, and soon goes off; the shoes get a scolding for their malice, and we forget all about it for a time. But does Nature check her course to suit the convenience of thoughtless men? No, no. In a short time we find that the nail, intercepted in its forward course, has become unusually thick and hard, and has spread out so much upon the sides that it is now growing into the flesh, and so makes a case for the doctor. Or, perhaps, the continuance of pressure may have inflamed the sensitive skin at the root, and caused a sore and painful place there. And instances are by no means infrequent in which the power of production of the nail at the root becomes entirely abrogated, and then it grows in thickness only."

When the nails are stained or discoloured, a little lemon juice, or vinegar-and-water, is the best application. Occasionally a little pumicestone, in impalpable powder, or a little 'putty powder,' may be used along with water and a piece of soft leather or flannel for the same purpose. The frequent employment of these substances is, however, injurious to the healthy growth of the nail.

NANKEEN. The coloured cotton cloth which bears this name was originally brought from Nankin, the ancient capital of China, and was prepared from a native cotton, of a brownish-yellow hue. It is now successfully imitated in England, and at the present time the English manufacturers supply the Canton market. In this country the colour is generally given to the cloth by successive baths of sulphate of iron and crude carbonate of soda or lime water.

NANKEEN DYE. The liquid sold under this name in the shops is a solution of annotta. It is employed to dye white calicoes of a nankeen colour, but chiefly to restore the colour of faded nankeen clothing.

NAPH'THA. *Syn.* MINERAL NAPHTHA, ROCK OIL; NAPHTHA, L. A name given to the limpid and purer varieties of PETROLEUM (which *see*), which exudes from the surface of the earth in various parts of the world.

Prop. Naphtha possesses a penetrating odour and a yellow colour, but may be rendered colourless by distillation; it usually begins to boil at a temperature of about 180° F., but, being a mixture of several different hydrocarbons, it has no fixed boiling-point; it is very inflammable; it does not mix with water, but imparts to that fluid its peculiar taste and smell; mixes with alcohol and oils, and dissolves sulphur, phosphorus, camphor, iodine, most of the resins, wax, fats, and spermaceti; and forms with caoutchouc a gelatinous varnish, which dries with very great difficulty.

Pur. Mineral naphtha is very frequently adulterated with oil of turpentine, a fraud which may be detected by—1. The addition of some oil of vitriol, which will in that case thicken and darken it. 2. Hydrochloric acid gas passed through the liquid for an hour will occasion the formation of hydrochlorate of camphine, either at once or after a few hours' repose, even if only 5% of oil of turpentine is present (*Dr Bolley*). 3. If a few grains of iodide of potassium and a little water are rubbed with the suspected sample, the colour of the water should continue unchanged; the presence of 1-300th part of oil of turpentine will cause it to assume a red or orange colour (*Saladin*).

Uses, &c. Naphtha is chiefly employed for the purposes of illumination, as a solvent for india-rubber, and in the preparation of a very superior black pigment. It has been highly spoken of as a remedy for cholera, by Dr Andreosky, a Russian physician. The term naphtha has recently been extended so as to include most of the inflammable liquids produced by the dry distillation of organic substances. See PETROLEUM, and *below*.

Naphtha, Boghead. *Syn.* PHOTOGEN. Obtained by distilling Boghead coal, or any cannel coal or bituminous shale, at as low a temperature as possible.

Naphtha, Bone. *Syn.* BONE OIL, DIPPEL'S ANIMAL OIL. A mixture of hydrocarbons obtained in the distillation of bones.

Naphtha, Coal-tar. *Syn.* NAPHTHA, COAL N., LIGHT OIL. A mixture of volatile hydrocarbons, obtained by distilling coal-tar. It is one of the first products which comes over, and flows from the still as crude coal naphtha. To obtain rectified coal naphtha this crude liquid is distilled, and the product agitated with 10% of concentrated sulphuric acid; when cold the mixture is treated with 5% of peroxide of manganese, and the upper portion is submitted to further distillation. The specific gravity of this purified product is 0·850. It is extensively used as a solvent of caoutchouc and other allied substances, also of resins for the preparation of varnishes. By repeated purification and fractional distillation benzol, the chief and most important constituent of coal naphtha, is obtained. See BENZOL.

Naphtha, Wood. See PYROXYLIC SPIRIT.

NAPHTHALENE. $C_{10}H_8$. *Syn.* NAPHTHALINE. A colourless, crystallisable, volatile substance, possessing an odour of coal-gas. It is a common product of the action of heat upon substances rich in carbon, like coal, wood, alcohol, &c. Burmese petroleum and Rangoon tar contain it. It is found occasionally deposited in gas-pipes in cold weather.

Prep. The last portion of the volatile oily product in the distillation of tar is collected separately, and allowed to repose, when crude naphthalene separates in the solid state. By pushing the distillation until the residuum in the still begins to char, a further portion of dark-coloured naphthalene may be obtained. It is purified by resublimation a second or even a third time.

Prop., &c. Soluble in alcohol, benzene, and ether; very slightly soluble in boiling water; melts at 80° C.; boils at 217° C.; highly inflammable, burning with a red and smoky flame; heated with sulphuric acid it unites to form two naphthalene sulphonic acids. By the action of nitric acid upon naphthalene numerous substances may be formed, the most interesting being nitronaphthalene. Naphthalene has lately been extensively employed as a stimulating expectorant. With picric acid it behaves in a characteristic way; hot alcoholic solutions of these substances when mixed deposit stellate tufts of yellow needles on cooling. In its chemical relations naphthalene closely resembles benzene.—*Dose,* 5 to 20 gr.; or, preferably, ½ gr, frequently. *Externally,* made into an ointment, in dry tetters, psoriasis, &c., 30 gr. may be mixed with 1 oz. of lard.

NAPHTHOL (β *Naphthol*). $C_{10}H_7.HO.$ A derivative of coal-tar, recommended by Professor Kaposi, Vienna, in scabies, psoriasis, eczema, and other skin diseases. A simple naphthol ointment, 1 dr. to 1 oz. lard, was found very efficacious in psoriasis, and as it does not stain the skin and hair, it is especially suitable for psoriasis of the scalp, face, and hands. Produces internal antisepsis, given in 2 to 5 gr. doses for diarrhœa.

NAPLES YELLOW. See YELLOW PIGMENTS.

NARCEINE. $C_{23}H_{29}NO_9$. *Syn.* NARCEINA, NARCEIA. A peculiar substance discovered by Pelletier in opium. It is obtained from the aqueous solution of opium, after it has been freed from morphine and narcotine by ammonia, by adding to it hydrate of lime, or preferably baryta. On boiling the filtered solution to expel the ammonia, and evaporating the liquid, crystals of narceine are gradually deposited. It may be purified by solution in hot alcohol and recrystallisation.

Prop., &c. White, silky, acicular prisms; neutral; inodorous; bitter; pungent; soluble in 375 parts of water at 60°, and in 330 parts at 212° F.; insoluble in ether; does not neutralise the acids, and is destitute of basic properties. It is distinguished from morphia by its easier fusibility (190°), and by forming a blue liquid with the dilute mineral acids, which on gradual dilution changes to violet and rose-red, and ultimately becomes colourless. It does not strike a blue colour with ferric chloride, like morphia, but forms a blue compound with iodine, which is decomposed by boiling water. It appears to be inert, and has not been applied to any useful purpose.

NARCOTICS. Remedies which promote or artificially imitate the natural physiological process of sleep, but which in large quantity produce complete insensability. Narcotics may be divided into (1) indirect and (2) direct; the former have no primary effect on the cerebral circulation, but act by supplying warmth, quiet, and other tranquillising elements, or by removing some disturbing cause which renders sleep impossible, *e. g.* many soothing and hygienic conditions, anodynes, conium, &c. The latter have some direct effect upon the central nervous system or its blood supply, *e. g.* opium, chloral hydrate, croton chloral, potassium bromide, hyoscyamus, stramonium, belladonna, hop, Indian hemp, alcohol, digitalis, and the anæsthetic vapours.

NARCOTINE. $C_{23}H_{23}NO_7$. *Syn.* NARCOTINA, L.; SEL D'OPIUM, MATIÈRE DE DEROSNE, Fr. A peculiar crystalline substance, found by Derosne in opium, and on which its stimulant property was at first supposed to depend.

Prep. 1. From opium exhausted of soluble matter by cold water, by treating it with water acidulated with acetic or hydrochloric acid, filtering, neutralising with ammonia, and dissolving the washed precipitate in boiling alcohol; the narcotine is deposited as the liquid cools, and may be purified by solution in ether.

2. By acting on opium, previously exhausted by cold water, with ether.

Prop., &c. White, inodorous, fluted, or striated prisms; neutral to test-paper; insoluble in cold water, sparingly soluble in boiling water, freely soluble in boiling alcohol and in ether. It is only feebly basic.

Narcotine is distinguished from morphine by its insipidity, solubility in ether, insolubility in alkalies, giving an orange tint to nitric acid, and a greasy stain to paper when heated on it over a candle. Another test for narcotine, said by Orfila to be characteristic, is to add to a little of the suspected substance a drop or two of oil of vitriol, and then to add a very small fragment of nitrate of potassium; the liquid speedily acquires a deep blood-red colour if narcotine is present. Morphine treated in the same way strikes a brown or olive-green colour.

Obs. The physiological action of narcotine is differently stated by different authorities. 1 gr. of it, dissolved in olive oil, killed a dog in 24 hours; but 24 gr. dissolved in acetic acid were given with impunity (*Magendie*). In the solid state it is inert; 120 gr. at a dose scarcely produce any obvious effects (*Bally*). Scruple doses have been given without injury (*Dr Roots*). It has been recently proposed as a substitute for quinine in the cure of agues. For this purpose the sulphate or hydrochlorate is preferable. 200 cases of intermittent and remittent fevers have been thus successfully treated in India (Dr O'Shaughnessy).—Dose, 3 to 10 gr., as an antiperiodic, sedative, &c.

Turkey opium contains about 1%, and East Indian opium about 3% of narcotine.

NA'TRIUM. See SODIUM.

NA'TRON. Native carbonate of soda.

NAU'SEA. See SICKNESS.

NAU'SEANTS. Syn. NAUSEANTIA, L. Substances which induce an inclination to vomit without effecting it. See EMETICS.

NA'VEL, Starting of. To remedy this, take a slice of cork, about the circumference of a shilling, and a little thicker; and having covered the projecting navel with a small circular piece of clean, soft linen, place the cork on the linen, strapping it into position by means of cross strips of white sticking-plaster (simple lead plaster), over which the usual roller is to be adjusted. Be careful to have the plaster of sufficient length, and to see that it adheres tightly to the skin.

NEB-NEB. See BABLAH.

NEC'TAR. The fabled drink of the mythological deities. The name was formerly given to wine dulcified with honey; it is now occasionally applied to other sweet and pleasant beverages of a stimulating character. The following LIQUEURS are so called:

Prep. 1. Chopped raisins, 2 lbs.; loaf sugar, 4 lbs.; boiling water, 2 galls.; mix, and stir frequently until cold, then add 2 lemons, sliced; proof spirit (brandy or rum), 3 pints; macerate in a covered vessel for 6 or 7 days, occasionally shaking, next strain with pressure, and let the strained liquid stand in a cold place for a week to clear; lastly, decant the clear portion, and bottle it.

2. Red ratafia, 3 galls.; oils of cassia and carraway, of each, 25 drops (dissolved in); brandy, ½ pint; orange wine, 1 gall.; sliced oranges, 6 in number; lump sugar, 2 lbs.; macerate for a week, decant and bottle. See ARRACK (Factitious).

NE'GUS. A well-known beverage, so named after its originator and patron, Colonel Negus. It is made of either port or sherry wine, mixed with about twice its bulk of hot water, sweetened with lump sugar, and flavoured with a little lemon juice and grated nutmeg, and a small fragment only of the yellow peel of the lemon. The addition of about 1 drop of essence of ambergris, or 8 or 10 drops of essence of vanilla, distributed between about a dozen glasses, improves it.

NEMATUS GROSSULARIÆ, Westwood (from the Greek word νῆμα, the thread of a web); Nematus grossularis, Dahlbom. The GOOSEBERRY AND CURRANT SAW-FLY. Growers of gooseberries and red currants suffer exceedingly from this insect, whose larvæ clear off the leaves from these fruit bushes with surprising rapidity. They are very troublesome to gooseberry and currant bushes in gardens, but they can be removed generally by hand picking and other means, which it would be almost impracticable to adopt and carry out in large plantations of from 10 to 80 acres, such as may be seen in various parts of Kent and in other counties.

Gooseberry bushes are much more infested by the larvæ of this saw-fly than red currant bushes, while black currant bushes are not affected by them.

Although the methods of the Nematus grossulariæ in its campaign upon these fruit bushes resemble those of the gooseberry moth, Abraxas grossulariata, whose history is given in another place, it differs considerably in many essential points if close comparison is made between them. However, the Nematus is a far more common and dangerous enemy than the Abraxas.

In some seasons suitable for their propagation the larvæ or grubs of the Nematus, as they may be termed to distinguish them from caterpillars proper, or the larvæ of lepidopterous insects (Réaumur terms these grubs fausses chenilles, false caterpillars), cause the fruit bushes in May to look as in the middle of winter, without leaves or any sign of vegetation, except perhaps a few of the nerves or ribs of the leaves left upon the shoots. There is no leaf tissue; there are no fruits. These have been nipped in the bud.

During the spring in the years 1876, 1879, and 1881 grave complaints came up from fruit growers in many parts of the gooseberry and currant producing districts of Cambridge, Gloucester, Kent, and Worcestershire, and from many gardeners in all parts of the country. Many inquiries were made as to the habits and history of the grubs that were causing this destruction, and as to remedies to be used. In some instances it was reported that the bushes were actually killed by the onslaughts upon them continued for two years.

From all accounts it appears that the Nematus grossularia is known in all European countries where gooseberry and currant bushes or other species of the Ribes grow. It is certainly very injurious in France, especially near Paris, and in the fruit lands near Troyes, and in the more central departments. In Germany a good deal of mischief is caused by it. Reports of serious injury were made from various parts of Würtemberg, where fruit is extensively grown. Only within the last 30 years has the Nematus grossularia been noticed in America. Both in the United States and in Canada, particularly in Ontario, it is now an established pest upon gooseberry and currant bushes, having evidently been imported with cuttings or young bushes from Europe.

Life History. This insect belongs to the family Tenthrinidæ of the order HYMENOPTERA.

The perfect insect, the saw-fly, has four wings, translucent and beautiful when it is darting about in the sunshine. Between the tips of its extended wings it measures very nearly ½ an inch, or 5½ lines. Its body is 3 lines in length, and in colour yellow. The thorax is marked with black spots; the legs are yellow with dark-coloured extremities. The male is not quite so large as the female, and its body is narrower.

Pairing takes place in April. The flies may be seen in the first warm days of spring hovering over gooseberry and currant bushes in preparation for egg laying. This the female accomplishes by means of the wonderful saw-like apparatus, similar to that of the *Cephus pygmæus* (see 'Insects Injurious to Corn, Grass, Pea, Bean, and Clover Crops'), with which it makes slits in the leaves of the bushes (Réaumur gives a most interesting and elaborate description of this ; he says, "Cet instrument est une véritable scie qui ne diffère de celles que nous nous servons pour couper le bois, qu'en ce qu'elle est faite avec beaucoup plus d'art que les nôtres,' 'Mémoires,' tome v, p. 108), and places the long whitish eggs singly in each slit all down the ribs of the leaves. They are arranged most carefully and precisely, there being about half an egg's distance between each egg in the rows. After about 7 days the grub comes from the egg and begins at once to gnaw a tiny round hole in the thick part of the leaf. At first it is nearly transparent or slightly tinged with slate-colour; when it has commenced feeding it acquires a greenish hue. Differing from the caterpillar of the gooseberry moth this grub has 20 feet, viz. 6 pectoral and 12 abdominal feet, and 2 at the end of the body. In a week or 8 days it attains its full length of 9 lines, or ¾ of an inch. At this time these grubs are extremely voracious and destructive, and as it is not uncommon to find 500 or 600 and even more upon one fruit bush, it may be understood that they quickly clear off all the foliage. After the first casting of the skin, or moult, which takes place before the grub is fully grown, the colour is again very light, but becomes soon green again after a little feeding. In due course, or after 4 or 5 days, the second and last casting of skin occurs, and the grub crawls down the stem, or lets itself drop by means of threads of web to the ground, in which it buries itself some inches deep, and forming a kind of cell it makes a cocoon and assumes the chrysalis form, remaining in this until tempted to burst its bonds by spring weather. Dahlbom says that the cocoons are grouped and joined together in the earth by means of threads of web ('Clavis novi Hymenopterorum systematis,' p. 23, Gustavo Dahlbom). There are two broods of these insects, or at least of those which emerge earliest from their winter habitation.

Prevention. There can be no doubt that the best mode of prevention is to destroy the grubs or chrysalids while in the ground, and this may be done by deep cultivation round the fruit bushes with a spud, and by the application of copious dressings of fresh lime, or gas lime, or pure pungent soot, which should be worked well into the soil. The clods thus dug up should be well knocked to pieces with the large eyes of 'prong hoes' so as to dislodge the cocoons within them. This operation may be performed between October and the 1st of March, and after this the ground may be beaten down hard with spades, or trodden down hard, to prevent possibly the escape of some of the insects which have survived the liming and triturating process.

All this would only be done of course after a severe attack of grubs in the previous spring.

In garden or small plantations other means may be adopted, such as soaking the ground around the fruit bushes with liquid manure and removing the soil near them. These methods can hardly be carried out in large plantations.

Remedies. Quicklime powdered upon the fruit bushes early in the morning before the dew is off the leaves is a very useful remedy. Syringing the bushes with a strong wash of water and soft soap, consisting of from 10 to 12 lbs. of soft soap to 100 galls. of water, is an admirable remedial measure. The essence from ¼ lb. of tobacco may be mixed with this, or better still, the bitter extract from 4 or 5 lbs. of quassia chips.

Petroleum soft soap may also be used at the rate of ¼ gall. or ½ gall. to 100 galls. of water.

Paraffin oil in the proportion of a wine-glass to 3 galls. of water has been found to remove the grubs, but if applied when the young gooseberries are formed, this is said, or fancied, to have imparted some of its flavour to them. Washing or syringing a large plantation would be a tedious work. Fortunately the grubs generally appear here and there in patches, and not simultaneously upon a large area of fruit land. They should be taken in time. Directly a bush is seen to be infested active measures should be adopted, and when it has been limed or syringed the ground beneath must be hoed or well stamped down to kill the grubs which have fallen off. As there are two broods in some cases watchful care will be required that none of the grubs that fall escape.

Hellebore, *Veratrum*, sprinkled in the form of powder upon the fruit bushes, has a good effect in clearing off the grubs. This is a deadly poison, and if any of it remained upon the fruit most serious consequences might ensue. There are records of persons having been made seriously ill from having partaken of fruit after the bushes had been dusted with powdered hellebore. Gooseberries are picked very young and green for tarts and preserves, and it frequently happens that a portion of the crop of each bush in large plantations is picked green for these purposes if the price is good, so that it would be highly dangerous to apply hellebore even in these early stages. Hellebore is used extensively in America as a remedy against this and other insects.

Natural enemies have been created against this insect, as against many other insects that are destructive to crops. Among these may be cited the ladybirds, *Coccinella*, which eat the eggs, and have been seen attacking the grubs in their earliest stages. Also the larvæ of the *Chrysopa perla*— the Golden Eye, or Lacewing, a fly of the order NEUROPTERA and the family *Hemerobiidæ*—have been noticed devouring the grubs just after they have come from the eggs. There is also an ichneumon fly of some species which deposits its eggs in the eggs of the *Nematus*, as may be evidently seen by the dark colour under their transparent skins.

In America, Professor Riley discovered a similar parasite upon the *Nematus ventricosus*, a species allied to the *Nematus ribesii*. This he called *Trichogramma pretiosa*. Professor Lintner also confirms this, and relates that eggs of the currant saw-fly parasitized by the *Trichogramma* have been sent for distribution to various American States and to Canada ('Reports on Insects

Injurious to Crops,' by Charles Whitehead, Esq., F.Z.S.).

NEPEN'THE. A drink calculated to banish the remembrance of grief. In the 'Odyssey' Homer describes Helen as administering it to Telemachus. Nothing is known respecting the composition of the ancient nepenthe. The name is applied to a preparation of opium by many old writers, and is now employed by a Bristol firm to designate a preparation resembling in all essential points Battley's ' LIQUOR OPII SEDA-TIVUS.'

NESSLER'S TEST for ammonia, &c. This, the most delicate test for ammonia, was devised by Nessler. It is capable of detecting 1 part of ammonia in 20,000,000 parts of water. The test is based upon the fact that an alkaline solution of mercuric iodide produces a brown coloration with ammonia, due to the formation of the iodide of tetramercurammonium. It is prepared by saturating a solution of iodide of potassium with the biniodide of mercury, and then adding a weak solution of hydrate of sodium. The addition of a few drops of this solution to one containing ammonia produces a yellowish tint when only a trace of ammonia is present, but a dark brown precipitate when the ammonia is present in larger quantity. A modification of this test is applied to the detection of wood spirit in common alcohol. A dilute solution of the iodides in question in pure alcohol is formed, in the proportion of 2 or 3 gr. of the salts to 100 c.c. of alcohol. About 4 c.c. of the suspected alcohol are taken, to which are added 2 or 3 drops of the test solution, a few drops of alcoholic ammonia, and, lastly, a little alcoholic potash; if wood spirit be present, the solution will remain clear, but if the alcohol be pure, the characteristic reddish-brown precipitate will appear. This precipitate is soluble in acetone, which is always present in wood spirit.

Wanklyn gives the following formulæ for the preparation of the Nessler test:—Mercuric chloride in powder, 35 grms.; iodide of potassium, 90 grms.; water, 1¼ litres; heat gently till dissolved (say 20 minutes) in a large basin. Then add of stick caustic potash, 320 grms., and 50 c.c. of saturated solution of mercuric chloride. The above will be ready for use in 2 hours, and gives maximum colour in 3 minutes.

For quantitative chemical analysis Thorpe and Muir recommend the following method of preparation:—"Dissolve 35 grms. of potassium iodide in 120 c.c. of water, transfer 5 c.c. of the solution to a clean beaker, and add, little by little, a cold concentrated solution of mercuric chloride to the remainder until the mercuric iodide ceases to be redissolved on stirring. Add the 5 c.c. of the potassium iodide to redissolve the remaining mercuric iodide, and cautiously continue the addition of the corrosive sublimate solution until a very slight precipitate only remains. Now add an aqueous solution of potash, prepared by dissolving 100 grms. of 'stick' potash in 200 c.c. of water, and dilute the mixture to 500 c.c. The liquid should be allowed to stand for a short time, and a portion decanted into a small bottle for use." The rest is placed in a large bottle, from which the smaller

one is replenished by decantation as required. This solution is widely used in conjunction with standard ammonium chloride solution in the estimation of ammonia in potable waters; the method is colorimetric—that is, the tints produced by given quantities of Nessler's solution and water, and water treated with a known quantity of the ammonium chloride solution, are compared. *Vide* Wanklyn and Chapman's ' Water Analysis ' for further particulars.

NESTS, EDIBLE. These dietetic curiosities, which are esteemed as great gastronomic luxuries by the Chinese, are formed by several species of swallows frequenting the Indian seas. The so-called nests chiefly abound in Java, Borneo, and Celebes, being found in the caverns both inland and on the sea-shores of those islands.

They are not in reality birds' nests, but merely supports, by which the bird is enabled to sustain and also to attach its nest to the rock. The nests themselves consist of grass, leaves, and sea-weed; the last of which substances it was for a long time erroneously considered formed the esculent, whereas it is the support which exclusively constitutes this Eastern table luxury.

This in great part consists of a peculiar mucus, of a gelatinous nature, which it has been ascertained the bird secretes and discharges from its mouth in large quantities. The Chinese mostly use it in the form of soup, and believe it to be possessed of considerable nutrient power. As many as 8,400,000 of edible nests are said to be annually imported into Canton. "The finest and whitest kind sells for £5 or £6 the lb.; but it requires about 50 nests to make up 1 lb. The brackets or supports are moved three times, the best being obtained in July and August" (*Church*).

NETTLE RASH. See RASH.

NEURAL'GIA. Lit., pain in a nerve. This term is applied to a disease of the nervous sensory apparatus, marked by paroxysmal pain, which is for the most part unilateral and in the course of nerves. Neuralgia may, as is well known, manifest itself in almost any part of the body. The varieties of it are so numerous, and its causes and treatment so varied, as to preclude any detailed account. Those who suffer from it should seek medical advice and carefully follow out the directions given. Apart from all local treatment, a plain but generous diet, abundance of fresh air and exercise, regular habits, and a generally healthy mode of life will do much to assist the patient. Over-exertion, close and badly ventilated rooms and workshops, dyspepsia, late hours, and irregular habits are fertile causes of neuralgia, and should be carefully avoided by those who are liable to this distressing malady. The proper use of tonics, particularly quinine, arsenic, and iron, and the avoidance of anything approaching constipation, will do much to relieve the symptoms in most cases.

NEUTRALISA'TION. The admixture of an alkali or base with an acid in such proportions that neither shall predominate. A neutral compound neither turns red litmus-paper blue, nor blue litmus-paper red.

NEUTRALISING PROPORTIONS, Table of.

Table of the Neutralising Proportions of some of the Acids and Alkaline Carbonates, omitting minute fractions. The best commercial preparations must be used.

Tartaric Acid.	Citric Acid.	Lemon Juice.	Cr. Carb. of Soda.	Bicarb. of Soda and Carb. of Potash.	Bicarb. of Potash.	Carbonate of Magnesia.	Sesquicarbonate of Ammonia.	Bicarbonate of Ammonia.
Grs.	Grs.	Grs.	Grs.	Grs.	Grs.	Grs.	Grs.	Grs.
10	9¼	2¼	19	11	13½	6½	8½	10¼
10¼	10	2½	20¼	12	14½	7	8½	11¼
13	12	2¾	25	14½	17½	8½	10	13¼
15	14	3½	29	17	20¼	9½	12	16
15½	14½	3½	30	17½	21	10	12¼	16¼
18	17	4	34½	20	24½	11½	14	19
20	18½	4½	38½	22½	27	12½	15½	21
20½	19	4½	40	23	27½	13	16	21½
26	24	5½	50	29	35	16½	18½	27
27	25	5⅜	52	30	36	17	21	26½
32	30	7	61	36	43	20½	25	33½
36	33½	7⅝	69	40	48½	23	28	38
47	44	10¼	90	52½	68	30	37	49½
52	48¼	11½	100	58	70	33	41	55
62	58	13¼	120	69	84	40	49	65½
73	68	15⅝	140	82	98	46½	57	77
75	70	16¼	144	84	101	48½	59	79
90	84	19½	172	101	121	57½	71	94½
92	86	20	177	103	124	59	72	97
100	93	21½	192	112	134	64	78	105½
108	100	23½	206	120	145	69	84	113
180	168	39½	344	202	242	115	141	190

NEW BERLIN SANITARY LIQUEUR—Gesundheits-Liqueur, neuer Berliner (*Apotheker Emil Trotz*). An unpleasantly tasting bitter spicy schnapps, containing 18% of sugar. Leaves an after-taste of aloes (*Hager*).

NICK'EL. Ni = 58·6. *Syn.* NICKELIUM, L. A metal obtained from kupfernickel, NiAs, a native arsenide of nickel found in the Saxon mines in Styria, at Leadhills, and in Connecticut; from nickel-glance, Ni(AsS), nickel-blende, NiAs, and pentlandite, (NiFe)8; from magnetic pyrites in Pennsylvania; also from nickel speiss, an impure arsenio-sulphide of nickel left after the manufacture of cobalt blue from its ores. An important source has lately been opened up in New Caledonia, where large quantities of a silicate of nickel called garricerite occur.

Prep. The powdered arsenical ore is roasted first by itself, and next with charcoal powder, until all the arsenic is expelled, and a garlic odour ceases to be evolved; the residuum is mixed with sulphur, 3 parts, and potassium hydrate, 1 part; and the compound is melted in a crucible with a gentle heat; the fused mass when cold is reduced to powder, lixiviated with water, dissolved in sulphuric acid mixed with a little nitric acid, and precipitated with potassium carbonate; the precipitate (nickelous carbonate) is washed, dried, mixed with powdered charcoal, and, lastly, reduced by the heat of a powerful furnace.

When nickel predominates in the ore, after the arsenic, iron, and copper have been separated, ammonia is digested with the mixed nickelous and cobaltous oxides, and the resulting blue solution, after dilution with boiled pure water, is treated with potassium hydrate until the colour disappears, when the whole is put into an air-tight vessel, and set aside for some time. The powder (nickelous hydrate) which subsides, after washing, is mixed with charcoal, and reduced by fusion in a crucible containing some crown glass.

On the small scale, for chemical purposes, pure nickel is best obtained by moderately heating nickelous oxalate in a covered crucible lined with charcoal.

Pwr. Kruss and Schmidt have recently discovered that a new metal, which they have named gnomium, constantly occurs, associated with nickel and cobalt, as an impurity; this accounts for many of the irregular results which have been observed in dealing analytically with these metals.

Prop. White with steel-grey tinge; hard; malleable; magnetic; capable of receiving the lustre of silver; can be rolled into thin plates and drawn into wire; sp. gr. 8·9; fusibility between that of manganese and iron; it is oxidised with difficulty even on heating in the air; is little attacked by dilute hydrochloric or sulphuric acids, but easily soluble in dilute nitric acid. It decomposes steam slowly at a red heat. With the acids, &c., it forms numerous compounds, most of which may be prepared by the direct solution of the carbonate. When the metal contains carbon it is less malleable and more readily fusible than when pure.

Tests. The salts of nickel in the anhydrous state are for the most part yellow; when hydrated, green—and furnish solutions possessing a pale green colour. Solutions of its salts exhibit the following reactions:—Alkaline hydrates give a pale apple-green precipitate, insoluble in excess, but soluble in a solution of carbonate of ammonium, yielding a greenish-blue liquid. Ammonia gives a similar precipitate, soluble in excess, yielding a deep purplish-blue solution. The presence of ammonium salts or free acids interferes with this reaction. Cyanide of potassium produces a green precipitate, soluble in excess, forming an amber-coloured liquid, which is reprecipitated by hydrochloric acid. This last precipitate is scarcely soluble in excess of the acid in the cold, but readily so upon boiling the liquid. Ferrocyanide of potassium gives a greenish-white precipitate. Sulphuretted hydrogen occasions no change in solutions of nickel containing free mineral acid, but in alkaline solutions gives a black precipitate. Sulphide of ammonium in neutral solutions gives a black precipitate, soluble with difficulty in hydrochloric acid; but freely soluble in aqua regia.

Estim. Nickel may be thrown down from its ore in the form of either carbonate or hydrate, and after ignition may be weighed as oxide, each grain of which is equal to 7-8ths gr. of pure nickel; or, more accurately, 0·7871 gr.

Nickel may be separated from other metals in the same way as cobalt, but if both these metals be present the operation may be troublesome, and is then effected by the different reactions of their cyanides.

According to Rose, nickel may be separated from cobalt as follows:—The mixed metals are dissolved in considerable excess of hydrochloric acids, and the solution is diluted with a very large quantity of water; a current of chlorine is then passed through the liquor for several hours, and the upper part of the flask is left filled with the gas after the current has ceased; barium carbonate is next added in excess, the whole digested together with frequent agitation for 15 or 18 hours, and then thrown on a filter. The filtrate yields pure nickelous oxide by precipitation with hydrate of potassium; whilst the residuum on the filter, after being washed in water, dissolved in hot hydrochloric acid, and the barium precipitated with sulphuric acid, furnishes, with hydrate of potassium, a precipitate of cobaltous hydrate, free from nickel, which, when washed and dried, is reduced in a platinum or porcelain crucible by hydrogen gas.

Another simpler, and for all practical purposes sufficiently accurate, method of separating cobalt from nickel depends upon the precipitation of potassium cobalt nitrate by a solution of potassium nitrite. The preparation is dried at 100° and weighed. The nickel remains in the filtrate, and may be precipitated with caustic potash; the precipitate after being boiled and washed is converted into the monoxide by ignition (*Roscoe*).

Uses. Nickel is chiefly employed in the manufacture of German silver. Some of its salts have been recently introduced into medical practice, and appear likely to prove most valuable additions to the materia medica. It has also been much used recently for coating iron and steel by galvanic deposition; in this process it is used as the positive pole. If the coating be well deposited it scarcely undergoes any oxidation. This process of nickel plating is applied to firearms, surgical instruments, various parts of machines, harness, &c., to prevent them from rusting. The best bath for nickel plating is a solution of pure nickel ammonium sulphate saturated at 20°—25° C. Dishes and crucibles are made of nickel as substitutes for those of silver and platinum; such vessels are very useful in the chemical laboratory. Sheets of nickel can be welded upon iron and steel plates; and culinary vessels, &c., have been made of such plates, which are not liable to rust.

Alloys of nickel are used in the coinage of America, Belgium, Switzerland, &c., the proportion being about 25% of nickel to 75% of copper. The so-called 'German silver' and Chinese 'packfong' are alloys of nickel with copper and zinc in variable proportions. Several useful alloys, *e.g.* 'Webster's metal,' are made by combining nickel with aluminium bronze. The Steel Company of Scotland have recently produced some remarkable alloys of nickel and iron. Of these some are non-magnetisable, others magnetisable; and their properties have been investigated by Dr Hopkinson, who contributed several papers to the proceedings of the Royal Society in the spring of 1890.

Nickelic Oxide. Ni_2O_3. *Syn.* SESQUIOXIDE OF NICKEL, PEROXIDE OF NICKEL. *Prep.* By passing chlorine through water holding the hydrate in suspension; or by mixing a salt of nickel with bleaching powder; or by gently igniting the nitrate or carbonate in the air. An insoluble black powder, which is decomposed by heat.

Nickelous Acetate. $Ni(C_2H_3O_2)_2$. *Syn.* NICKELII ACETAS, L. *Prep.* By neutralising acetic acid with nickelous carbonate, and gently concentrating by evaporation, so that crystals may form. Small green crystals, soluble in 6 parts of water.

Nickelous Carbonate. $NiCO_3$. *Syn.* NICKELII CARBONAS, L. *Prep.* This salt may be obtained in the manner described above in connection with the preparation of metallic nickel, or by simply adding carbonate of sodium to a solution of nickelous chloride, but in this case the crystals contain 6 molecules of water. The following is another formula which produces a nearly pure carbonate, but one which may still contain a little cobalt, the entire separation of which is a matter of extreme difficulty, and can best be effected in the manner recommended by Rose, described above:

The mineral (crude speiss or kupfernickel) is broken into small fragments, mixed with from 1-4th to half its weight of iron filings, and the whole dissolved in aqua regia; the solution is gently evaporated to dryness, the residue treated with boiling water, and the insoluble ferrous arseniate removed by filtration; the liquid is next acidulated with hydrochloric acid, treated with sulphuretted hydrogen, in excess, to precipitate the copper, and, after filtration, is boiled with a little nitric acid, to bring back the iron into the ferric state; to the cold and largely diluted liquid a solution of bicarbonate of sodium is gradually added, and the ferric oxide separated by filtration; lastly, the filtered solution is boiled with carbonate of sodium in excess, and the pale green precipitate of carbonate collected, washed, and dried.

Uses, &c. It is freely soluble in the acids, and is chiefly employed to prepare the salts and other compounds of nickel.

Nickelous Chlo"ride. NiCl₂. *Syn.* NICKELII CHLORIDUM, L. *Prep.* From nickelous carbonate or oxide and hydrochloric acid. Small green crystals, of the formula NiCl₂,6Aq, which are rendered yellow and anhydrous by heat, unless they contain cobalt, when the salt retains a tint of green.

DOUBLE CHLORIDES. Nickelous chloride unites with the chlorides of ammonium, potassium, and sodium,' to form pale green crystallisable salts, which have been used for depositing nickel or iron, lead, copper, &c.

Nickelous Hy'drate. Ni(HO)₂. *Prep.* By precipitating a soluble solution of nickel with caustic potash. Green crystalline powder, freely soluble in acids, forming the ordinary salts of nickel.

Nickelous Ox'alate. NiC₂O₄. *Syn.* NICKELII OXALAS, L. *Prep.* By adding a strong solution of oxalic acid to a similar solution of nickelous sulphate, and collecting the pale bluish-green precipitate which forms after a time. Used to prepare metallic nickel and its oxide for laboratory purposes.

Nickelous Oxide. NiO. *Syn.* PROTOXIDE OF NICKEL. Occurs as bunsenite in Saxony. *Prep.* By heating the nitrate, carbonate, or hydroxide, to redness in open vessels. Green crystalline powder.

Nickelous Sulphate. NiSO₄. *Syn.* SULPHATE OF NICKEL. *Prep.* Dissolve nickelous carbonate or hydroxide in dilute sulphuric acid, evaporate down, and crystallise. Pale green prismatic crystals, and of the formula NiSO₄,7Aq, or small pale green octahedrons, when crystallised at a temperature, from a very acid solution, containing NiSO₄,6Aq.

Nickelous and Potassium Sulphate. NiSO₄, K₂SO₄,6Aq. *Syn.* DOUBLE SULPHATE OF NICKEL AND POTASSIUM. *Prep.* By crystallising a mixture of nickelous and potassium sulphates. Pale green crystals, readily soluble in water. Sodium and ammonium sulphates form similar compounds with nickelous sulphate.

Nickelous and Ammonium Sulphate. (NH₄)₂ SO₄ + NiSO₄,6Aq.

Prep. By dissolving pure nickel in dilute sulphuric acid, concentrating the solution, and then adding ammonium sulphate; re-crystallise.

Uses, &c. Employed for making the bath solution in nickel-plating.

According to Link, 100 parts of water dissolve of this salt, at 16°, 5·8 parts; at 20°, 5·9 parts; at 30°, 8·3 parts; at 40°, 11·5 parts; at 50°, 14·4 parts; at 85°, 28·6 parts.

Nickel Plating. A new process of nickel plating has recently come into use in Belgium, by which a thick plating may be deposited on any metal by a feeble electric current in a very short space of time. The bath is composed of 10 parts sulphate of nickel, 7¼ parts of neutral tartrate of ammonia, 0·5 parts of tannic acid, and 20 parts of water. The sulphate of nickel is dissolved in 3 to 4 part of water, carefully neutralised, the other ingredients added, and the solution boiled for ¼ of an hour; the rest of the water is added, and the liquid filtered or decanted. By adding

the materials in the same proportion the strength of the bath may be kept constant. It is said that the deposit is brilliantly white, soft, and homogeneous, and has, even when of great thickness, no tendency to scale.

Nickel Silver. See GERMAN SILVER.

Nickel Sulphides. The monosulphide, NiS, occurs as millerite; it is formed when the metal is heated with sulphur, and in the hydrated condition when ammonium sulphide is added to a solution of a nickel salt.

The disulphide, NiS₂, is obtained by heating nickel carbonate with sulphur and potassium carbonate, and then dissolving out with water.

A subsulphide, Ni₄S, is also known.

NIC'OTINE. C₁₀H₁₄N₂. *Syn.* NICOTINA, NICOTIA, L. A volatile base, discovered by Reiman and Posselt in tobacco.

Prep. 1. Infuse tobacco leaves 4 hours with warm water slightly acidified with hydrochloric acid, strain, and evaporate to a syrupy fluid. To the fluid add carbonate of sodium in excess, and shake out the alkaloid with ether. Separate the ether and shake it with a dilute solution of tartaric acid; remove the acid solution and evaporate to a small volume. Finally, add excess of lime to the solution and distil in a current of hydrogen. On cooling the distillate the nicotine separates in oily drops.

2. (*Ortigosa.*) Infuse tobacco leaves for 24 hours in water acidulated with sulphuric acid, strain, evaporate to a syrup, add ⅛ of its volume of a strong solution of potassa, and distil in an oil-bath at 288°, occasionally adding a little water to assist the process, and prevent the too great concentration of the solution of potassa in the retort; next saturate the distilled product with oxalic acid, evaporate to dryness, digest in boiling absolute alcohol, evaporate the resulting tincture to a syrup, and decompose the oxalate of nicotine thus obtained by adding potassa to it in a close vessel, and agitate the mass with ether, repeating the process with more ether until all the nicotine is dissolved out; lastly, distil the mixed ethereal solution in an oil-bath. At first ether comes over, then water, and, lastly, nicotine, which, towards the end of the process, assumes a yellowish tint.

3. (*Schloesing.*) This chiefly differs from the preceding by directing the concluding distillation to be conducted in a retort, by the heat of an oil-bath, at the temperature of 284° F., in a current of hydrogen, for 12 hours, after which, by raising the heat to 356° F., the nicotine distils over pure, drop by drop.

4. (*Kirchmann.*) A tin vessel provided with two tubulures is filled with tobacco, which is previously damped with sodium carbonate. One of the tubulures admits a glass tube reaching nearly to the bottom of the vessel; the other is provided with a glass tube merely penetrating the cork.

The vessel is made air-tight, placed in a boiling hot steam-bath, and a rapid stream of carbonic acid gas passed through it, entering the vessel by the longer and leaving it by the shorter tube; the latter dips into a mixture of alcohol and dilute sulphuric acid.

In this manner a large yield of perfectly

colourless nicotine is obtained. In order to obtain the pure alkaloid, caustic baryta is added to the solution, the latter evaporated to dryness, and the pure nicotine extracted with ether.

To estimate nicotine, weigh out 15 gr. of tobacco, digest for 24 hours with alcohol of 85%, acidified with 15 drops of sulphuric acid, so as to make 150 c.c. Evaporate 50 c.c. of the filtered liquid, and add iodohydrargyrate of potassium to the residue. The number of cubic centimetres employed, multiplied by 0·00405 (0·001 of the equivalent of nicotine), gives the quantity of alkaloid contained in 5 grms. of tobacco (*Linoff-sky*).

Prop., &c. Nicotine is a colourless, volatile liquid; highly acrid and pungent; smelling strongly of tobacco; boiling at 250° C.; soluble in water, ether, alcohol, and oils; and combining with the acids, forming salts, many of which are crystallisable. Quickly assumes a brown colour on exposure to light and air. It is a frightful poison; ¼ of a drop will kill a rabbit; a single drop will kill a large dog. Nicotine is the substance which was employed by the Count Bocarmé for the purpose of poisoning his brother-in-law, Gustave Fougnies, the particulars of which were developed in the celebrated trial, in Belgium, of that nobleman, in 1851. Good Virginia and Kentucky tobacco, dried at 212° F., contain from 6% to 7% of nicotina; Havannah tobacco (*cigars*), less than 2% (*Schlœsing*).

NIGER, or **Ramtil Seeds** (*Guizotia abyssinica*, Cass.). The plant is a native of tropical Africa, but is cultivated in many parts of India for the sake of the small black seeds, from which an oil is expressed, used as a lamp oil and as a condiment.

NIGHT'MARE. *Syn.* INCUBUS, EPHIALTES, L. The common causes of nightmare are indigestion and the use of narcotic and intoxicating substances. Its prevention consists in the selection of proper food, and in duly attending to the state of the stomach and bowels. Heavy and late suppers should be particularly avoided, as well as all articles of diet that are of difficult digestion, or apt to induce flatulency. When it arises from strong drink, tobacco, or opium, these should be abandoned, or employed in smaller quantities. A teaspoonful of aromatic spirits of ammonia, magnesia, or bicarbonate of soda, taken in a glass of cold water on going to bed, is a good and simple preventive. In cases accompanied by restlessness, a few drops of laudanum or tincture of henbane may be added. An occasional aperient is also excellent. See CHAMOMILE.

NIGHT'SHADE (Deadly). *Syn.* BELLADONNA (B. P., Ph. L., E., & D.). "The leaf, fresh and dried (leaves and root, Ph. D.), of *Atropa belladonna*, Linn." "The fresh leaves and branches to which they are attached; also the leaves separate from the branches, carefully dried, of *Atropa belladonna*, gathered, when the fruit has begun to form, from wild or cultivated plants in Britain" (B. P.). "Oval, acute, very perfect, glabrous, when bruised exhaling a disagreeable odour. The herb which grows spontaneously in hedges and uncultivated places is to be preferred to that which is cultivated in gardens" (Ph. L.). Belladonna is a powerful narcotic, and is used

as an anodyne, antispasmodic, and discutient, in a variety of diseases—neuralgia, arthritic pains, migratory rheumatic pains, spasmodic rigidity and strictures, angina pectoris, whooping-cough, fevers, phthisis, &c.; also as a prophylactic of scarlet fever, as a resolvent in enlarged and indurated glands, to produce dilatation of the pupil, &c.—*Dose.* Of the powder, commencing with 1 gr., gradually and cautiously increased until dryness of the throat or dilatation of the pupil occurs, or the head is affected. See ATROPIA.

NIGHTSHADE (Woody). *Syn.* BITTER-SWEET; DULCAMARA (B. P., Ph. L., E., & D.), L. The "new shoots (caules) of *Solanum dulcamara*, Lind." "The dried young branches of the *Solanum dulcamara* (bitter-sweet), from indigenous plants which have shed their leaves" (B. P.). "It is to be collected in autumn, after the leaves have fallen" (Ph. L.). Diaphoretic, diuretic, and (in large doses) narcotic. See INFUSION OF DULCAMARA.

NIM BARK. See AZADIRACHTA INDICA.

NIO'BIUM. See TANTALUM.

NIP'PLES (Sore). The most common form of this affection is that termed "chapped nipples" by nurses. As a preventive measure, the part may be moistened morning and evening, for some weeks before the period of lactation, with a little rum or brandy, which is more effective if slightly acidulated with a few drops of dilute sulphuric acid. Some persons employ tincture of tolu, or compound tincture of benzoin (Friar's balsam) for this purpose.

When chaps, cracks, or like sores, arising from lactation, are once developed, one of the safest and most effective remedies is tincture of catechu, applied 3 or 4 times a day, by means of a camel-hair pencil.

The celebrated nostrum of Liebert for cracked nipples, '*Cosmétique infaillible et prompt contre les gerçures ou crevasses aux seins et autres*,' is a lotion formed of 10 gr. of nitrate of lead dissolved in 4 fl. oz. of rose-water, and tinged with a little cochineal. The parts are moistened with the liquid, and are then covered with fine leaden nipple-shields, two of which are provided for the purpose. This is repeated soon after each time the child leaves the breast; and the nipple is carefully washed with a soft sponge and lukewarm water, and gently dabbed dry with a very soft towel, before the infant is again applied to it. This remedy is very successful, and has acquired great popularity and patronage in Brussels, Paris, Frankfort, and other parts. It must be recollected, however, that all applications of an active or poisonous nature should be employed with the greatest possible caution, as, unless unusual care is taken, a portion of the remedy may remain concealed within the delicate pores of the skin, and be sucked off by the infant, to the serious disturbance of its health.

The 'Medical Press' gives the following as a good application for fissures of the nipples :—1. Salol, 1 dr.; ether, 1 dr.; cocaine, 4 gr.; collodion, 5 dr. Pure cocaine should be used, not the hydrochlorate, and in compounding the application advantage should be taken of the solvent properties of the ether.

2. The nipples should be cleaned with a little

warm water, to which has been added a small amount of borax, before applying ;—Balsam Peru, ½ dr.; Tinct. Arnicæ, ½ dr.; Ol. Amygdalæ, ½ oz.; Aquæ Calcis, ½ oz. Shake well and apply to the nipples with a camel-hair brush.

All medicaments must be thoroughly removed before an infant is put to the breast.

NITRANILINE. This substance is obtained by acting on nitrobenzene with a mixture of fuming nitric acid and oil of vitriol; dinitrobenzene is formed, which is dissolved in alcohol, and the resulting solution subjected to the reducing action of ammonia and sulphuretted hydrogen, as described under ANILINE. Nitraniline forms yellow, acicular crystals, little soluble in cold water, but freely soluble in alcohol and ether. Three forms of this substance are known, viz. ortho-, meta-, and para-nitraniline. There are also 2 dinitranilines, and 1 trinitraniline.

NITRATE. *Syn.* NITRAS, L. A salt of nitric acid (*e. g.* Ag.NO₃, nitrate of silver). The nitrates are very easily prepared by the direct solution of the metal, or its oxide or carbonate, in nitric acid, which, in most cases, should be previously diluted with water. By evaporation, with the usual precautions, they may be obtained either in the pulverulent or crystalline form.

Tests. The nitrates are characterised by (1) deflagrating when thrown on red-hot charcoal; for the feeble attraction existing between oxygen and nitrogen, and the disposition of these elements to assume the gaseous state, cause nitrates to be readily decomposable by heat; (2) by an aqueous solution, after being mixed with half its bulk of strong sulphuric acid and thoroughly cooled, yielding a brown cloudy layer when a freshly made solution of ferrous sulphate is poured on to the surface; (3) mixed with a few drops of hydrochloric acid and a little indigo solution and boiled, the blue colour is discharged; (4) when mixed with a few drops of dilute sulphuric acid and potassium iodide solution to which a drop of starch paste has been added, the immersion of a strip of zinc-foil in the mixture will cause the nitric acid to be reduced to nitrous acid; this will liberate iodine, which will turn the starch blue. See NITRIC ACID, and the respective metals.

NITRE. Nitrate of potassa. See POTASSIUM.

NITRIC ACID. HNO₃. *Syn.* AZOTIC ACID; ACIDUM NITRICUM (B. P., Ph. L., E., & D.), AQUAFORTIS.

Prep. 1. (Ph. E. and Ph. L., 1836.) Purified nitre (dried) and sulphuric acid, equal parts; mix in a glass retort, and distil with a moderate heat, from a sand-bath (or naked gas flame, Ph. E.) into a cool receiver, as long as the fused materials continue to evolve vapours. "The pale yellow acid thus obtained may be rendered nearly colourless (if desired) by gently heating it in a retort" (Ph. E.). Sp. gr. 1·500. In the present Ph. L. this acid is included in the materia medica (see *below*).

2. (Ph. D.) The nitrate of potassa is dissolved in water, the solution treated with a little nitrate of silver, filtered, evaporated to dryness, weighed, and then treated as above.

3. Nitrate of soda (cubic nitre, Chili saltpetre) is introduced, in quantities varying between 4 and 10 lbs., into a cylindrical iron retort, which it

will only half fill, and after the lid is luted on and the connection made with the condensers, an equivalent of oil of vitriol is poured in through an aperture provided for the purpose, and the charge is worked off with a gradually increased heat. The condensing apparatus consists of a series of 5 or 6 salt-glassed stoneware receivers, about 1-6th part filled with cold water. The product of this process, the strongest brown and fuming 'NITROUS ACID' of commerce (AQUAFORTIS, FUMING NITRIC ACID; ACIDUM NITROSUM, ACIDUM NITRICUM FUMANS), has usually the sp. gr. of about 1·45. It contains about 46% of HNO₃. It is rendered colourless by gently heating it in a glass retort, when it forms COMMERCIAL NITRIC ACID (sp. gr. 1·37 to 1·4).

4. (PURE NITRIC ACID.) By mixing the strongest commercial acid with about an equal quantity of oil of vitriol; redistilling; collecting apart the first portion which comes over, and exposing it in a vessel slightly warmed and sheltered from the light, to a current of dry air made to bubble through it until the nitrous acid with which it is contaminated is completely removed.

Prop. Pure liquid nitric acid is colourless, highly corrosive, and possesses powerful acid and oxygenising properties. Phosphorus, sulphur, and even charcoal are oxidised by it. All the metals in common use are acted upon by nitric acid except gold and platinum, but tin and antimony are not dissolved. It forms nitro- substitution compounds with many organic substances, *e. g.* nitrobenzene, C₆H₅(NO₂). The sp. gr. of the strongest liquid acid has the sp. gr. 1·517 at 60° F., and contains about 67% of HNO₃. "On boiling nitric acid of different degrees of concentration at the ordinary atmospheric pressure, a residue is left boiling at 246° F., and 29 in. barometer, having a sp. gr. 1·414 at 60° F." (*Fownes*). Acid of less density than 1·414 parts with water gradually becomes stronger by boiling, but acid of less sp. gr. than 1·414 is weakened by exposure to heat. It begins to boil at 184° F., but cannot be distilled unchanged, for heat partially decomposes it into oxygen, water, and nitric peroxide. When exposed to intense cold, liquid nitric acid freezes. It is rapidly decomposed, with loss of oxygen, by contact with most organic and many metallic and non-metallic bodies. In many cases these reactions occur with considerable violence, and the production of light and heat. It stains the skin yellow.

Pur., &c. The nitric acid of commerce is generally contaminated by hydrochloric acid, nitrous acid, sulphuric acid, or chlorine, or by their soda or potash salts, and, occasionally, iodine, together with an excess of water. The last is readily detected by the sp. gr., and the others by the appropriate tests. "90 gr. by weight, mixed with ½ oz. of distilled water, require for neutralisation 1000 grain measures of the volumetric solution of soda. Evaporated, it leaves no residue. Diluted with six volumes of distilled water, it gives no precipitate with chloride of barium or nitrate of silver—indicating absence of sulphuric and hydrochloric acids" (B. P.). 5 measures of acid, sp. gr. 1·5, mixed with 2 of water, condenses into 6½ measures, and makes the sp. gr. 1·42. "Free from colour. Exposed to the air, it emits very acrid vapours. Totally volatilised by heat. Diluted with 3 times its volume of water, it gives no precipitate with either nitrate of silver or chloride of barium. 100 gr. of this acid (sp. gr. 1·42) are saturated by 161 gr. of crystallised carbonate of soda" (Ph. L.). The Ph. E. states that the density of commercial nitric acid is 1·380 to 1·390. "If diluted with distilled water it precipitates but slightly, or not at all, with solution of nitrate of baryta or nitrate of silver." The best 'double aquafortis' of the shops (aquafortis duplex) has usually the sp. gr. 1·36; and the single aquafortis (aquafortis simplex), the sp. gr. 1·22; but both are commonly sold at much lower strengths.

The nitric acid of commerce may be freed from impurities by one or other of the following methods:

1. By the addition of a little nitrate of silver, as long as it produces any cloudiness, and after repose, decanting the clear acid, and rectifying it at a heat under 212° F. To ensure a perfectly colourless product, a small portion of pure black oxide of manganese should be put into the retort (*Murray*).

2. By agitating the acid with a little red oxide of lead, and then rectifying it, as before.

3. By adding 1% of bichromate of potash to the acid before rectifying it. This answers well for acid not stronger than sp. gr. 1·48.

4. By rectification at a gentle heat, rejecting the first portion that comes over, receiving the middle portion as genuine acid, and leaving a residuum in the retort (*Ure*).

Tests. 1. It stains most organic colouring matters yellow, but it merely reddens litmus. 2. When mixed with a little hydrochloric acid or chloride of ammonium, it acquires the power of dissolving gold leaf. 3. Morphia, brucia, and strychnia give it a red colour, which is heightened by ammonia in excess. 4. When placed in a tube, and a fresh solution of ferrous sulpate is cautiously added, a dark colour is developed at the line of junction, which is distinctly visible when only $\frac{1}{12600}$ part of nitric acid is present. This test may be often conveniently modified by dropping into the liquid a crystal of ferrous sulphate; the fluid immediately surrounding this crystal then acquires a dark brown colour, which disappears upon simple agitation of the fluid, or by heating it. 5. When mixed with a weak solution of sulphate of indigo, and heated, the colour of the latter is destroyed and a yellow liquid is left. 6. When saturated with carbonate of potassium or sodium, and evaporated to dryness, the residuum deflagrates when thrown on burning coals. 7. When the mixture of a nitrate with cyanide of potassium, in powder, is heated on a piece of platinum, a vivid deflagration follows, attended with detonation (*Fresenius*). It is stated that sulphate of aniline is an extremely delicate test for nitric acid. The following is the method of its application: About a cubic centimetre of pure concentrated sulphuric acid (sp. gr. 1·84) is placed in a watch-glass; half a cubic centimetre of a solution of sulphate of aniline (formed by adding 10 drops of commercial aniline to 50 c.c. of diluted sulphuric acid in the proportion of 1 to 6) is poured on, drop by drop; a

glass rod is moistened with the liquid to be tested, and moved circularly in the watch-glass. By blowing on the mixture during the circular agitation, when a trace of nitric acid is present, circular *striæ* are developed of a very intense red colour, tinting the liquid rose. With more than a trace of nitric acid the colour becomes carmine, passing to a brownish red. This process serves to detect the presence of nitric acid in the sulphuric acid of commerce. It will also reveal the presence of nitrates in water (' Pharmaceutical Year-book '). 8. Put a very small piece of diphenylamine into a test-tube, and pour a little sulphuric acid over it, and then add a drop or two of water, so as to increase the temperature sufficiently to affect the solution of the diphenylamine. Now add very gently the solution to be tested, and if only a trace of nitric or nitrous acid be present, a beautiful and very permanent blue coloration is produced at the junction of the two liquids; but if there be any quantity of the nitrogen compound, the colour becomes almost black. This reaction is so delicate and certain that, in the case of a solution of nitric acid containing about 1 part B. P. acid in 10,000 of water, it is most distinct; 1 part of nitrite of potassium in 30,000 of water gives also almost unmistakable evidence of the presence of the nitrogen acid.

Estim. The strength of nitric acid may be roughly estimated by its sp. gr.; but more accurately by ascertaining the amount of carbonate of sodium, or other salt of known composition, which is required to neutralise it. To render this assay trustworthy, it must be, in all cases, also tested to detect the presence of impurities.

The following process for the quantitative estimation of nitric acid is by Fischer (' Dingl. Polyt. Journ.,' ccxiii, 423—427) :—Indigotin prepared by reduction of indigo by means of grape-sugar, alcohol, and caustic soda, oxidation in the air, and solution in sulphuric acid, may be kept unchanged for years. 5 c.c. of such a solution, diluted with water and mixed with 30 c.c. of pure sulphuric acid, is titrated by adding a standard nitric acid solution until the blue colour gives place to a light green; the indigo solution is then diluted, so that 1 c.c. shall be equal to 0·0025 milligramme-equivalent of nitric acid, or 0·2525 milligramme of potassium nitrate. If a water is being examined it is run into 4 c.c. of the titrated indigo solution, mixed with 20 c.c. of sulphuric acid, until the blue colour changes to light green. 10, divided by the number of c.c. of water used, expresses the milligramme-equivalents of nitric acid per litre; thus, if 4 c.c. of water are used, there are 2·5 milligramme-equivalents of nitric acid, equal to 252·5 milligrammes of potassium nitrate per litre. If a preliminary test with brucine has shown that the water contains very little nitric acid, 2 c.c. only of the indigo solution must be used, or sometimes as little as 1 c.c. If more than 8 c.c. of water is required to destroy the blue colour, 100 c.c. must be evaporated down to the volume of 8 c.c. and then titrated. The volume of sulphuric acid must be at least double the sum of the volumes of indigo and water; the temperature must not sink under 110°.

The nitrates may all be tested as above by first

adding a small quantity of pure sulphuric acid, which will liberate the nitric acid of the salt.

Ant., &c. See ACIDS.

Uses. Nitric acid is employed in assaying, in dyeing silk, &c., in etching on copper, in the preparation of gun-cotton, oxalic and sulphuric acids, &c. In chemical operations it is very valuable as an oxidising agent. In *medicine* it is used as a caustic to corns and warts; and in doses of 1 to 10 drops, in a tumbler of water, in liver complaints, fevers, dyspepsia, syphilis, to remove the effects of mercury, or as a substitute for that drug, &c. Externally it is employed in the form of baths, lotions, and ointment. Dr Collier states that a strong lotion of nitric acid is almost a specific in lepra and several other kindred skin diseases.

Concluding Remarks. The common laboratory source of nitric acid is nitrate of potassium, but it may also be obtained from other nitrates by a similar process. Nitrate of sodium is frequently used instead of nitrate of potassium, it is cheaper and is more convenient in some respects, for the residuum is more easily dissolved out of the retort or cylinder. The residuum of the common process with nitre ('sal enixum') is chiefly employed as a flux by the glasshouses, and as a source of potash in the manufacture of alum.

By proper management nitre yields more than ⅔ of its weight of pure nitric acid, sp. gr. 1·5; and nitrate of soda, its own weight of acid, sp. gr. 1·4.

By the patent process of M. Mallet, dried nitrate of soda is decomposed by dried or monohydrated boracic acid, on heating the two together. The products are nitric acid, which distils over, and biborate of soda (borax), which remains in the retort.

The crude coloured nitric acid of commerce (aquafortis) was originally prepared by distilling a mixture of nitre and copperas, and is still sometimes obtained in this way.

According to Apjohn and others, the strongest nitric acid, sp. gr. 1·520, is a monohydrate; that of the sp. gr. 1·500, a sesquihydrate; that of 1·486, a binhydrate; and that of 1·244, a quadrihydrate; or containing respectively 1, 1½, 2, and 4 atoms of water. (See *below*.)

Nitric Acid, Anhydrous. N_2O_5. *Syn.* NITRIC ANHYDRIDE. This interesting substance was first obtained in a separate form by M. Deville, in 1849.

Prep. (*Deville.*) Nitrate of silver is dried by exposure to a current of dry carbonic acid at a temperature of 356° F., and the tube containing it is then immersed in a water-bath heated to 203° F.; pure dry chlorine gas is next passed through the apparatus, and, as soon as the reaction commences, the temperature is reduced not lower than 154° F.; the production of crystals in the receiver, which must be cooled by a powerful freezing mixture, soon commences; lastly, the liquid portion of the product is removed by a current of dry carbonic acid gas.

Prop., &c. Colourless prismatic crystals, which melt at 85° F., boil at about 113°, and at a slightly higher temperature begin to suffer decomposition. Added to water, much heat is generated; it rapidly attacks organic bodies, even caoutchouc; sometimes it explodes spontaneously. When brought in contact with water nitric acid is produced with evolution of heat.

Nitric Acid, Dilute. *Syn.* ACIDUM NITRICUM DILUTUM (B. P., Ph. L., E., and D.), L. *Prep.* 1. (Ph. L.) Nitric acid (sp. gr. 1·42), 3 fl. oz.; distilled water, 17 fl. oz.; mix. Sp. gr. 1·082. "1 fl. oz. is saturated by 154 gr. of the crystals of carbonate of soda." It contains about 12% of pure anhydrous acid.

2. (Ph. E.) Nitric acid (1·500), 1 fl. oz.; distilled water, 9 fl. oz. Or, commercial nitric acid (1·390), 1 fl. oz. 5½ dr.; water, 9½ fl. oz. Sp. gr. 1·077. It contains 11·16% of pure dry nitric acid.

3. (Ph. D.) Nitric acid (1·500), 4 fl. oz.; water, 29 fl. oz. Contains about 9·7% of pure acid. The above are used for convenience in dispensing.—*Dose*, 15 drops to ½ fl. dr., or more. The above must not be confounded with the acidum nitricum dilutum, Ph. D. 1826, which had the sp. gr. 1·280; nor with the following:

4. (*Henry's.*) Sp. gr. 1·143; equal in saturating power to hydrochloric acid sp. gr. 1·074, and sulphuric acid 1·135. Used in assaying.

5. (B. P.) Nitric acid, 6 parts; distilled water, sufficient to make the mixture, when cooled to 60° F., measure 31 parts. Contains 15% of anhydrous nitric acid.—*Test*. Sp. gr. 1·101. 6 fl. dr. (361·3 gr.) by weight require for neutralisation 1000 gr. measures of the volumetric solution of soda, and therefore contain exactly one equivalent in grains of anhydrous acid, namely, 54 gr.—*Uses*. Tonic, astringent, lithonlytic.—*Dose*, 10 to 30 minims.

Nitric Acid, Fuming. *Syn.* NITROUS ACID‡; ACIDUM NITRICUM FUMANS, L. The red fuming nitrous or nitric acid of commerce is simply nitric acid loaded with nitric peroxide (which *see*). That of the Ph. Bor. is distilled from nitre, 2 parts; oil of vitriol, 1 part.

NITRIC ANHYDRIDE. See NITRIC ACID, ANHYDROUS.

NITRIC OXIDE. See NITROGEN, OXIDES OF.

NITRITE. A salt of nitrous acid; *e. g.* KNO_2, nitrite of potassium.

Tests. 1. White precipitate with silver nitrate; soluble in excess of water.

2. Grey precipitate (metallic mercury) with mercurous salts.

3. Brownish-black coloration with ferrous sulphate.

4. Blue coloration with a little potassium iodide, a drop of starch paste, and a few drops of dilute sulphuric acid. This test serves to detect the presence of small quantities of nitrite in potable waters. Fresenius recommends that the sample be acidified with acetic acid and then distilled; the first few drops which pass over are collected in a beaker containing the solution of potassium iodide, starch, and sulphuric acid.

NITRO-BENZENE. $C_6H_5NO_2$. *Prep.* By treating benzene with strong fuming nitric acid, with heat. The vessel must be kept cool. After the violence of the reaction is over, the liquid is diluted with water, and the heavy oily fluid which separates is collected, washed with water, then with caustic soda, and finally distilled with steam. A modification of this process is now used on the large scale. Mansfield patented a process in 1874 for its preparation from coal-tar.

Prop., &c. Light yellow, very sweet, but burning taste; smells of bitter almonds; scarcely soluble in water, but readily soluble in alcohol and ether; little affected by reagents; boils at 210° C., and at low temperature solidifies to needles, melting at 3° C.; sp. gr. 1·2. It is very poisonous, a quality which Letheby asserts it acquires owing to its conversion in the animal economy into aniline. Heated with an alcoholic solution of caustic potash, and the mixture submitted to distillation, it yields a red oily liquid, from which large red crystals of azobenzene separate. These are nearly insoluble in water, freely soluble in alcohol and ether, melt at 149° F., and boil at 559·4° F. DINITRO-BENZENE is made by dissolving benzene in a mixture of equal volumes of the strongest nitric and sulphuric acids, and boiling the liquid for a few minutes; the crystals which form as it cools are insoluble in water, but are freely soluble in alcohol. Several other nitro-benzenes are known.

Uses. Nitro-benzene is extensively used as a substitute for the essential oil of bitter almonds in *perfumery*. It is much more extensively used in the manufacture of pure aniline for the colouring matters known as aniline blue, aniline black, and magenta. Ferrand states that the presence of nitro-benzene in essence of bitter almonds may be detected as follows:—Heat to ebullition, in a test-tube, 3 or 4 c.c. of a 20% alcoholic solution of potash, together with 10 drops of the suspected essence. If nitro-benzene be present, the mixture takes a red colour; if the essence of bitter almonds be pure, it becomes a pale straw-colour.

NITROGEN. N=14. *Syn.* AZOTE; NITRO-GENIUM, AZOTUM, L. A gaseous elementary substance, discovered by Rutherford in 1772, and found to be a constituent of the atmosphere by Lavoisier, 1755. It is found combined both in the organic and inorganic kingdoms of nature; it forms about 4-5ths, or 79·19% by volume, or 76·99% by weight of the atmosphere, enters largely into the composition of most animal substances, and is a constituent of gluten, the alkaloids, and other vegetable principles.

Prep. 1. A small piece of phosphorus is placed in a capsule floating on the surface of the water of the pneumatic trough, and after setting it on fire a bell-jar is inverted over it; as soon as the combustion is over, and the fumes of phosphoric anhydride have subsided, the residual gas is washed by agitation with water, and with a solution of potash. It may be dried by either letting it stand over fused chloride of calcium, or by passing it through concentrated oil of vitriol.

2. A porcelain tube is filled with copper turnings, or, preferably, with spongy copper (obtained by reducing the oxide with hydrogen), and is then heated to redness, a stream of dry atmospheric air being at the same time directed through it. By repeating the process with the same air, and finally passing it over fragments of pumice moistened with strong solution of potash to absorb carbonic anhydride, the product is rendered quite pure.

3. Chlorine gas is passed into a solution of pure ammonia, care being taken to employ a *considerable excess* of the latter; the evolved gas, after being dried, is pure nitrogen. There is some danger of producing the explosive compound, chloride of nitrogen, with this process.

4. (*Corenwinder.*) From solution of nitrate of potassium, 1 vol.; concentrated solution of chloride of ammonium, 3 vols.; gently heated together in a flask, and the evolved gas passed through sulphuric acid. Pure.

5. By boiling a solution of nitrite of ammonium, or, which amounts to the same thing, a mixture of one measure of a solution of nitrate of potassium and 8 measures of a solution of chloride of ammonium. Both solutions must be concentrated. This is the easiest method of preparing nitrogen and of obtaining the gas in a pure state.

Note. The nitrite of potassium to be employed in this process is best prepared by passing nitrous anhydride, evolved from starch and nitric acid, into a solution of potash (sp. gr. 1·88) till it imparts an acid reaction to test-paper, and then neutralising by the addition of potash.

6. From lean flesh digested in nitric acid at a gentle heat.

Prop., &c. Pure nitrogen is a colourless, odourless, tasteless gas, neither combustible nor capable of supporting combustion or respiration. It is neutral to test-paper, does not affect lime-water, and is only slightly absorbed by pure water. Its sp. gr. is 0·9713. It is recognised by its purely negative qualities. It is, however, capable of directly combining with boron and silicon, and more readily with magnesium and titanium at high temperatures. With hydrogen it forms ammonia, NH_3. Forms very unstable compounds with the halogens, and enters into the composition of gun-cotton, the fulminates, nitro-glycerin, &c. Has been liquefied at a pressure of 300 atmospheres at a temperature of 13° C.

Nitrogen, Chloride of. NCl_3. *Syn.* NITROGEN TRICHLORIDE, TERCHLORIDE OF NITROGEN. This compound was discovered by Dulong in 1811, but its nature was first accurately determined by Sir H. Davy. Some chemists regard it as possessing the composition $NCl_2(NHCl_2)$.

Prep. (*Liebig.*) Dissolve chloride of ammonium, 1 oz., in hot water, 12 or 14 oz., and as soon as the temperature has fallen to 90° F. invert a wide-mouthed glass bottle full of chlorine over it. The gas is gradually absorbed, the solution acquires a yellowish colour, and in the course of 15 to 20 minutes yellow oil-like globules of chloride of nitrogen form upon the surface of the liquid, and ultimately sink to the bottom. The globules, as they descend, should be received in a small leaden saucer, placed under the mouth of the bottle for the purpose.

Prop., &c. Chloride of nitrogen should only be prepared in very small quantities at a time. Both its discoverer and Sir H. Davy met with severe injuries while experimenting on it. Its sp. gr. is 1·65; it volatilises at 160° F., and between 200° and 212° fulminates violently. Contact with combustible bodies at ordinary temperatures immediately causes detonation. *The explosive power of this compound seems to exceed that of every known substance, not even excepting fulmi-*

nating silver. A minute globule, no larger than a grain of mustard seed, placed on a platina spoon, and touched with a piece of phosphorus stuck on the point of a penknife, immediately explodes, and shivers the blade into fragments, at the same time that the vessel that contains it is broken to pieces. Olive oil, naphtha, and oil of turpentine have a similar effect.

Nitrogen, I'odide of. NHI₂. A dark brown or black insoluble powder, which is most safely and conveniently prepared by saturating alcohol (sp. gr. 0·852) with iodine, adding a large quantity of the strongest pure solution of ammonia, and agitating the mixture; water must now be added, when iodide of nitrogen will be precipitated, which must be carefully washed with cold distilled water and filtered off. The filter containing the precipitate should be spread out on a sheet of glass and torn into small pieces while the iodide is still moist. The precipitate should be simply exposed to dry in the air.

Prop., &c. It detonates violently as soon as it becomes dry, by the slightest pressure or friction, even that of a feather, and often spontaneously; but this explosion is scarcely so powerful as that of the chloride of nitrogen. It also explodes whilst moist, though less readily. It should only be prepared in very small quantities at a time. Recent researches have shown that it contains hydrogen.

Nitrogen, Ox'ides of. Nitrogen forms five distinct compounds with oxygen.

1. **Nitrous Ox'ide.** *Syn.* PROTOXIDE OF NITROGEN, LAUGHING GAS; NITROGENII PROTOXYDUM, L. *Prep.* (1) From fused nitrate of ammonium, introduced into a glass retort, or a flask furnished with a bent tube, and thence exposed, over a spirit-lamp or charcoal-chauffer, to a temperature of about 389° F.; the temperature must not be too high, or the gas will contain nitric oxide and nitrogen : the evolved gas may be collected in bladders, gas-bags, a gasometer, or in the pneumatic trough over *warm* water. The gas may be purified by passing it through three washbottles, one containing water, one a solution of sulphate of iron, and the other a solution of potash.

(2) Nitrous oxide may also be made in the same way, from crystallised nitrate of ammonia, or by exposing nitric oxide for some days over iron filings moistened with water, but, without great care, the product is not always fit for respiration.

Prop., &c. Colourless; possesses an agreeable odour and a sweetish taste, and when pure does not affect a solution of nitrate of silver; at 45° F., under a pressure of 40 atmospheres, it is liquid ; this, when exposed under the receiver of a powerful air-pump, changes into a snow-like solid ; at —150° F. it is a transparent, colourless, crystalline body; it supports combustion, and is absorbed by cold water. Sp. gr. 1·520. Its most remarkable property is its action on the system when inspired. A few deep inspirations are usually succeeded by a pleasing state of excitement, and a strong propensity to laughter and muscular exertion, which soon subside, without being followed by languor or depression. Its effects, however, vary with different constitutions. From 4 to 12 quarts may be breathed with safety. It

produces temporary insensibility to pain, like chloroform or ether; but its use is dangerous when affections of the heart, lungs, or brain are present. This gas is successfully and extensively employed as an anæsthetic in dental surgery. It can now be bought in a liquid state in wrought-iron vessels.

Obs. No particular caution is required in preparing the above compound, except the use of too much heat. The temperature should be so arranged as to keep the melted mass in a state of gentle ebullition, and should not be allowed, under any circumstances, to exceed about 500° F. Should white fumes appear within the retort after the evolution of the gas has commenced, the heat should be at once lowered, as, when heated to about 600°, nitrate of ammonia explodes with violence.

2. **Nitric Oxide.** NO. *Syn.* DEUTOXIDE OF NITROGEN, NITROUS GAS, BINOXIDE OF NITROGEN; NITROGENII BINOXYDUM, L. *Prep.* By pouring nitric acid, sp. gr. 1·2, on metallic copper, in the form of turnings, clippings, or wire. Effervescence ensues, and nitric oxide is evolved, and may be collected over water or mercury in the pneumatic trough. The residual liquid yields crystals of nitrate of copper on evaporation. A gentle heat assists the action.

Prop., &c. A colourless, tasteless, inodorous, irrespirable, and incombustible gas. In contact with free oxygen it produces dense orange or red vapours, chiefly consisting of nitric peroxide (NO₄), which are freely absorbed by water. Nitric oxide is absorbed by a solution of ferrous sulphate, which it turns of a deep brown or nearly black colour, which is removed by boiling. Sp. gr. 1·039. In the presence of water and excess of oxygen it is converted into nitric acid.

3. **Nitrous Anhydride.** N₂O₃. *Syn.* NITROGEN TRIOXIDE, ANHYDROUS NITROUS ACID. *Prep.* (1) Heat 1 part of powdered starch with 8 parts of nitric acid of sp. gr. 1·25, and pass the evolved gases first through a drying tube 2 feet long containing fused chloride of calcium, and then into a dry and empty U-tube cooled to 20° F. by surrounding it with a mixture of pounded ice and crystallised chloride of calcium.

(2) Heat nitric acid (sp. gr. 1·3) with an equivalent of white arsenic; pass the gas which comes off through a U-tube surrounded with cold water, then into another containing chloride of calcium, and finally collect in another cooled with ice and salt.

Prop., &c. Nitrous anhydride is a green liquid which boils at 14° C. and emits red fumes, and which on admixture with water at ordinary temperatures is decomposed, producing nitric acid and nitric oxide. If nitrous anhydride be mixed with water at temperatures below 0° F. the two combine, and a blue solution is formed which (probably) contains nitrous acid (HNO₂). See NITROUS ACID.

4. **Nitrogen Pentoxide.** N₂O₅. *Syn.* NITRIC PENTOXIDE, NITRIC ANHYDRIDE, ANHYDROUS NITRIC ACID. See NITRIC ACID (ANHYDROUS).

5. **Nitrogen Peroxide.** NO₂. *Syn.* NITRIC PEROXIDE, PEROXIDE OF NITROGEN, NITROGEN TETROXIDE, HYPONITRIC ANHYDRIDE. This compound forms the chief constituent of the red

fumes which develop on mixing nitric oxide with air or oxygen.

Prep. By heating thoroughly dried nitrate of lead in a retort, and conducting the evolved gases into a U-tube surrounded with a freezing mixture of ice and salt.

Prop., &c. If the U-tube be perfectly dry, and the cold intense, the nitric peroxide obtained assumes the form of transparent crystals which melt at 10° F., but the presence of the slightest trace of moisture prevents their formation, and produces instead a colourless liquid which, as the temperature rises, acquires a yellow and ultimately an orange-red colour. Nitric peroxide dissolves in nitric acid and turns it of a yellow or red hue. The so-called '*nitrous acid*' or '*fuming nitric acid*' of commerce owes its deep red colour

to the presence of this compound. At very low temperatures water converts nitric peroxide into nitric and nitrous acids; at ordinary temperatures it transforms it into nitric acid, nitrous acid, and nitric oxide. It boils with decomposition at 71° F. It is believed that the molecule at low temperatures is N_2O_4, which decomposes into $2NO_2$.

NITRO-GLYCERIN. $C_3H_5(NO_3)_3$. *Syn.* GLONOIN, NITRATE OF GLYCERYL, TRINITRITE, NITROLEUM, FULMINATING OIL, TRINITRO-GLYCERIN. This dangerously explosive compound, from the use of which in mining, quarrying, and such like operations so many fatal accidents have occurred, is glycerin in which 3 atoms of hydrogen have been replaced by 3 molecules of nitroxyl (NO_2), as illustrated by the following formulæ:

$$\underset{\text{Glycerin.}}{\left. C_3H_5 \atop H_3 \right\}O_3} + 3\left({H \atop NO_2}\right\}O\right) = \underset{\text{Nitro-glycerin.}}{\left. C_3H_5 \atop (NO_2)_3 \right\}O_3} + 3\left({H \atop H}\right\}O\right).$$

It was discovered in 1847 by Dr Sobrero, a pupil of Pelouze.

Prep. 1. Kopp prepares nitro-glycerin by mixing 3 parts of sulphuric acid, of sp. gr. 1·767, with 1 part of fuming nitric acid. 2800 grms. of the mixed acids are added to 350 grms. of glycerin, great care being necessary to avoid any elevation of temperature, which would lead to a violent reaction, resulting in the conversion of the glycerin into oxalic acid.

After standing 5 or 10 minutes, the mixture is poured into 4 or 6 times its bulk of very cold water to which a rotatory motion has been imparted. The nitro-glycerin falls to the bottom of the vessel as an oily-looking liquid, which is washed by decantation. 1% of magnesia is sometimes added to the nitro-glycerin in order to neutralise any acid arising from decomposition.

2. Böttger has devised a process for the preparation of nitro-glycerin, which being, as he affirms, entirely free from danger, adapts it for lecture experiments :—A few grains of pure glycerin, free from water, are poured into a test-tube which is surrounded by a freezing mixture, and containing a mixture of 1 vol. of the most concentrated nitric acid (1·52 sp. gr.), and 2 vols. of the strongest sulphuric acid (1·83 sp. gr.). Then, as quickly as possible, the whole is poured into a larger quantity of cold water. The nitro-glycerin, which has formed like oil drops, sinks rapidly to the bottom. It is then washed several times by decantation with fresh water, and, lastly, with a weak solution of soda. Remove the water with a few pieces of fused chloride of calcium.

If nitro-glycerin is not sufficiently purified it is liable, on being kept, to decompose and become dangerous. Nitro-glycerin is extensively used for blasting. Mixed with various inert substances it is the explosive principle in dynamite, kieselguhr, glyoxylin, lithofracteur, duolin, nitromagnite, blasting gelatin, gelatin dynamite, &c.

Prop., &c. Nitro-glycerin is a heavy yellow or brownish oily liquid; sp. gr. 1·6. It is very poisonous. It dissolves in alcohol, ether, and wood naphtha, from all of which it may be recovered by the addition of water, in which it is insoluble. Dissolved in either of these solutions it becomes converted into a crystalline mass when exposed

to a low temperature. If subjected to a blow it explodes with fearful violence, a single drop placed upon paper, and struck upon an anvil, giving rise to a report that is almost deafening. Neither a spark nor the application of a lighted body is said to cause its ignition, which takes place with difficulty even if it be applied to a thin layer of the substance. 100 parts of nitro-glycerin yield on combustion :

Water	.	.	20	parts.
Carbonic acid	.	.	58	„
Oxygen	.	.	3·5	„
Nitrogen	.	.	18·5	„
			100·0	

(*Wagner.*)

As the sp. gr. of nitro-glycerin is 1·6, 1 part by bulk will yield by combustion :

Aqueous vapour	.	.	554	vols.
Carbonic acid	.	.	469	„
Oxygen	.	.	39	„
Nitrogen	.	.	236	„
			1298	

(*Wagner.*)

Other experimenters affirm that, instead of free oxygen, nitrous oxide is one of the products of the combustion of nitro-glycerin. According to Nobel, the heat liberated when nitro-glycerin is exploded causes the expansion of the gases to be 8 times their original bulk; therefore 1 vol. of the substance will yield 10,384 vols. of gas, whilst 1 part by bulk of gunpowder only yields 800 vols. of gas. If these data be correct the explosive force of nitro-glycerin is 13 times greater than that of powder, bulk for bulk, and 8 times greater weight for weight. The manufacture of nitro-glycerin is attended with considerable danger, since very slight friction or pressure is sufficient to determine its explosion. Hence many methods have been suggested for guarding against accidents from it during storage. One of these consists in mixing it with finely powdered glass.

Wurtz advises the nitro-glycerin to be mixed with solutions of nitrate of lime, zinc, or magnesia, the solutions to have a sp. gr. equal to the nitro-glycerin. By this means a harmless emulsion would be formed, and the nitro-glycerin would be recoverable when required for use by simply adding water. Nobel's plan consists in dissolving it in wood spirit.

NITRO-HYDROCHLORIC ACID. *Syn.* NITROMURIATIC ACID; AQUA REGIA, ACIDUM NITROHYDROCHLORICUM (B. P.), A. NITRO-MURIATICUM, L.; EAU RÉGALE, Fr. *Prep.* 1. (B. P.) Nitric acid, 3 parts; hydrochloric acid, 4 parts; water, 25 parts. Mix the acids 24 hours before adding the water. (This precaution is necessary to allow of the development of the chlorine, and the chloronitrous and chloronitric gases which result from the mutual decomposition of the two acids, and upon which the therapeutic activity of the agent depends.) Colourless. Keep the mixture in a cool and dark place.

2. (Ph. D. 1826.) Nitric acid, part; hydrochloric acid, 2 parts (both by measure); mix in a refrigerated bottle, and keep the mixture in a cold and dark place. Used to dissolve gold and platinum; and in *medicine*, in liver complaints, syphilis, the exanthemata, &c., either internally, in doses of 5 to 15 drops in water, or externally, as a foot-or knee-bath. It is also occasionally employed as a caustic.

3. (AQUA REGIA WITH SAL-AMMONIAC.) Nitric acid (sp. gr. 1·2), 16 fl. oz.; sal-ammoniac, 4 oz.; dissolve. Occasionally used by dyers; does not keep well.

4. (DYERS' AQUAFORTIS.) Colourless nitric acid (sp. gr. 1·17), 10 lbs.; hydrochloric acid (sp. gr. 1·19), 1 lb.; mix. Used by dyers.

NITRO-PRUSSIDES. A series of salts discovered by Dr Playfair, and obtained by the action of nitric acid on the ferrocyanides and ferricyanides. The most important of these salts is the nitro-prusside of sodium, $Na_4Fe_2Cy_{10}(NO)_2.4Aq.$

Prep. Dissolve 2 parts of powdered ferrocyanide of sodium in 5 parts of common nitric acid, previously diluted with its own volume of water. When the evolution of gas has ceased, digest the solution on a water-bath until it no longer yields a blue but slate-coloured precipitate with ferrous sulphate. Cool the liquid, filter, neutralise the filtrate with carbonate of sodium, and again filter. This filtrate, on evaporation, yields crystals consisting of a mixture of nitro-prusside of sodium and nitrate of potassium; the former, which may be recognised by their rhombic shape and their fine ruby colour, should be picked out and preserved.

Use. As a test for soluble sulphides, with which nitro-prusside of sodium strikes a beautiful violet tint. According to Playfair this is the most delicate test for alkaline sulphides. The sulphur in an inch of human hair may be detected by it.

NITROUS ACID. See NITROUS ANHYDRIDE, under NITROGEN, OXIDES OF.

NITROUS OXIDE. See NITROGEN, OXIDES OF.

NOCTUA SEGETUM, Westwood; *Agrotis segetum,* Ochsenheimer. The COMMON DART MOTH. The large plump caterpillars of this moth are known as surface caterpillars, because they work mischief to plants just at or just under the surface of the ground. There are several species of caterpillars which are also termed surface caterpillars, as the caterpillar of the Heart

and Dart moth, *Agrotis exclamationis,* for example, among others. Most of these are very injurious to root crops of all descriptions, and the common Dart moth caterpillars especially attack turnips of all kinds, and mangel-wurzel, though they by no means despise wheat and other corn plants, and they are found frequently in celery and parsnip beds and in cabbage plots.

Not only does this caterpillar eat the young leaves of turnips and mangel-wurzel, gnawing them off close to the crown, and thus killing the plants outright, it also bores holes in the bulbs, causing them to decay. Much loss of weight is often occasioned in this way, and bulbs that are intended for clamping or storing will not keep when bored in many places by these caterpillars.

White turnips are even more subject to be burrowed into than swedes, as they are softer and their skins are not so thick. Much complaint came as to injury done by these caterpillars to white and red 'Tankard' turnips in Hants and Wilts, in which counties many of these early sorts are sown. The plants are decimated to begin with, and it was remarked by a careful observer that directly the plants were out of 'rough leaf' the caterpillars came up from the ground to feed upon the leaves, and that they dragged pieces of leaf down with them. Curtis, quoting Le Keux, speaks of this habit of carrying off leaves for food down with them into their holes ('Farm Insects,' by J. Curtis, p. 124). When the turnip leaves became old and hard the caterpillars fastened upon the bulbs, and with such vigorous onslaught that when the sheep were put upon them hardly a bulb had escaped from many perforations and consequent rottenness.

In 1884, in which year the ravages of the insect were very great, many mangel-wurzel growers remarked that their young plants got to a certain point when the tap-roots were about the size of a slight skewer, and then withered and died. Examination was made, and it was seen that the tap-roots were bitten completely through just beneath the surface of the soil. Wireworms again were blamed, but it was soon discovered that large caterpillars, proving to be those of *Agrotis segetum,* were swarming in the ground, and were the cause of the mischief.

Much damage is constantly occasioned in what are known as 'seed-beds,' or nurseries for cabbage, broccoli, cauliflower, and other plants, in the market gardens of Kent, Essex, Bedford, and other counties. A market-garden farmer, working nearly 1000 acres of land in Essex, estimated his losses, directly and indirectly, from injury to his 'seed-beds' at over £100 in 1884.

These caterpillars were very abundant in England and in Scotland in 1879, also in 1884 and 1885 heavy losses were sustained by root growers in various parts of the kingdom.

The *Agrotis segetum* is known upon the Continent and in America.

In Germany its attacks are as severe as in the United Kingdom. Taschenberg writes of it as very destructive to vegetation, and says it is known in Asia, South Africa, and North America ('Praktische Insekten Kunde,' von Dr E. Taschenberg). According to Kaltenbach, it has committed ravages in Silesia, Pomerania, and

Hungary ('Die Pflanzen Feinde,' von J. H. Kaltenbach). Nördlinger holds that it is spread all over Europe, and has done infinite mischief in Prussia, Poland, and Russia ('Die kleinen Feinde des Landwirthschaft,' von Dr H. Nördlinger).

Westwood cites a notice of it in the 'Annals' of the Entomological Society of France in 1834. It is called in France *la noctuelle des moissons*, and is particularly troublesome to sugar-beet, tobacco plants, and maize.

There are several species of *Agrotis* in America, which hurt crops of various kinds, whose caterpillars are styled 'cut-worms.' The *Agrotis segetum*, Harris says, is known in America as *Agrotis messoria*, or rather, this insect is its representative there.

Life History. The common Dart moth belongs to the Nat. Ord. LEPIDOPTERA, and to the genus *Agrotis* of the family *Noctuidæ*, which has, as Westwood points out, 400 species in Great Britain.

The perfect insect, or moth, is from 10 lines to an inch in the length of its body, and measures an inch and a half across the wings when expanded fully. In colour its body varies very much from shades of grey to shades of brown. Stephens says it is almost impossible to obtain two specimens precisely similar ('Illustrations of British Entomology, Haustellata,' by J. F. Stephens). The males differ from the females, their general colour being lighter than that of the females. The fore or anterior wings of the male are greyish brown. Those of the female are brown, or reddish grey, as some entomologists say. Curtis says they are nearly clay-coloured in some specimens.

Upon the fore-wings there are peculiar pointed marks from which it takes the name of 'Dart' moth.

Its hind or posterior wings are white, with dark divisions or nervures. Those of the female are pearly white or light grey.

As the name of its family implies, it flies by night, or in the twilight or dusk, remaining during the day upon trees, palings, hedges, plants, and weeds. It rests with its wings folded down its back in penthouse fashion, being inconspicuous on account of its colour.

Eggs are laid singly by the moth, and fastened to the under sides of leaves, the stems or stalks of plants of the *Brassica* tribe if conveniently near, or of other plants when these are not obtainable. Moths of this species have been seen flying about late in October in some seasons. The eggs are like poppy seeds. Caterpillars come from them in about 10 days and fall at once to the earth and go into it, beginning to feed at once upon leaves of plants near to them and congenial to their tastes. In about 8 weeks they are full grown—that is, an inch and a half in length, and thick in proportion. At this time their appetites are enormous, and their power of consumption of vegetation is most wonderful.

With their strong jaws shaped like a spoon or scoop, and furnished with 5 teeth, they both bite and gnaw leaves and stems, and scoop out burrows in the hard bulbs.

They are smooth and shiny, dusky grey in colour, not infrequently with a slight pink shade on their backs, having spots or freckles, as Curtis calls them, upon their skins, and a double line of dark colour down their bodies, and one line on either side of them. They have dark or brown heads, and 6 pectoral, 8 abdominal, and 2 anal feet.

The caterpillars remain in this form, feeding greedily, until the food fails, or the frosts drive them down deep into the earth, where they remain in oval-shaped cases of earth until the spring, and then become brown chrysalides and soon change into moths.

Prevention. When swedes and turnips have been attacked by the caterpillars of this moth, they should not be pulled, but fed off early by sheep folded on the land. The treading of the sheep and their manurial matters would kill them, or drive them down below and starve them out.

After an attack upon mangel-wurzel the ground should be ploughed up deeply. If wheat is put in after this crop it would be very advisable to apply a dressing of lime or lime ashes. But it would be far safer not to put wheat or any crop, and to fallow the land and keep it well and deeply stirred in the spring, and then to take spring tares.

A crop of turnips, swedes, mangel-wurzel, or cabbage should never be taken directly after either of these crops.

As these caterpillars feed upon corn plants if they cannot get plants of the *Brassica*, it is not desirable to put oats or barley in infested fields in the spring following an attack.

All weed growth must be carefully kept down, especially charlock, or cadlock, as it is called in some places, both in fields and in the sides of fields.

Remedies. Frequent stirring with horse-hoes between the drills is calculated to disturb the caterpillars and to kill a certain number of them. Side hoeing will also check them.

Soot scattered on both sides of the plants and chopped lightly in by hoes has been proved to be of much benefit. Guano also dry and well triturated, sprinkled close to the plants and hoed in, has been adopted in several cases with advantage. Agricultural salt sown close to the plants at the rate of 3 or 4 cwt. per acre is stated to be useful, care being taken not to put the salt on the plants. A mixture of quicklime and sulphur—black sulphur, or *sulphur vivum*—is an effective dressing put in close to the rows and covered in by means of hoes. About 7 lbs. of black sulphur to a bushel of quicklime is the proportion of this mixture.

Natural Enemies. Rooks, starlings, peewits, partridges, and moles are greedy devourers of these fleshy caterpillars, and should be encouraged in fields where these are at work ('Reports on Insects Injurious to Crops,' by Chas. Whitehead, Esq., F.Z.S.).

NOLI-ME-TANGERE. See LUPUS.

NOMENCLATURE (Chemical). The spoken language of chemistry; notation is the symbolic written language of the science. The following information will doubtless prove useful to many of our readers, as serving to explain terms which are necessarily of frequent occurrence in this work:

ACIDS. *a.* When a substance produces only one acid compound, the name of this acid is formed by adding the termination -IC to that of the radical, or to the leading or characteristic portion of it; as *sulphur*IC acid, an acid of sulphur. This is Latinised by changing -IC into -ICUM; as *acidum sulphur*ICUM. *b.* When a body forms two acid compounds containing oxygen, the name of the one containing the smaller proportion of that substance ends in -OUS; as *nitr*OUS acid, which contains 1 atom of nitrogen and 2 of oxygen; *nitr*IC acid, containing 1 atom of nitrogen and 3 of oxygen. In this case the Latin name ends in -OSUM; as *acidum nitr*OSUM. *c.* When a substance forms more than two acids with oxygen the Greek preposition HYPO- (below or under) is prefixed to the name of the acid in -OUS or -IC next above it; as HYPO*chlorous acid.* *d.* When a new acid compound of a substance is discovered, containing more oxygen than another acid of the same substances already known, the name of which ends in -IC, the prefix PER- or HYPER- is added; as PER*iodic acid.* This may be illustrated by the oxygen acids of chlorine:

Hypochlorous acid	(*acidum hypochlorosum*)		HClO
Chlorous	„ („	*chlorosum*) . .	HClO$_2$
Chloric	„ („	*chloricum*) . .	HClO$_3$
Perchloric *or* } Hyperchloric }	„ („	*perchloricum*) .	HClO$_4$

OXIDES. The names of these have, in general, reference to the number of atoms of oxygen which they contain. When a metal forms only one basic compound with oxygen, this compound is simply called the oxide of such a metal; but as most substances form more than one compound with oxygen, certain prefixes are introduced to express the proportions. In such cases it is generally found that one out of the number has a strongly marked basic character, and contains one atom of each of its constituents. This is called the oxide, protoxide, or monoxide, and forms the standard to which those both above and below it are referred. Thus, supposing M to be the metal, we may have:

Suboxide or dioxide (*suboxydum, dioxydum*) .		M$_2$O
Oxide, protoxide, or monoxide (*oxydum, pro-* *toxydum*)		MO
Sesquioxide (*sesquioxydum*)		M$_2$O$_3$
Binoxide, dioxide, or deutoxide (*binoxydum,* *deutoxydum*)		MO$_2$
Teroxide or trioxide (*teroxydum, trioxydum*)		MO$_3$
Peroxide (*peroxydum*) . .	{ That contain- ing the *largest* proportion of oxygen.	

The anhydrous oxides (such as N$_2$O$_5$ and SO$_3$), from which the acids are derived, may be best termed *acid-forming oxides;* whilst the lower oxides, because they have the power of acting as bases and of forming salts when brought into contact with acids, are termed *basic oxides* (*Roscoe* and *Schorlemmer*).

SALTS. *a.* Acids having names ending in -IC give rise to salts whose names end in -ATE; thus *nitr*IC acid yields *nitr*ATES, e. g. *nitrate of silver.* -ATE is Latinised by -AS, e. g. *nitrate of silver* becomes *argenti nitr*AS.

b. Acids possessing names ending in -OUS form salts having names ending in -ITE; thus *sulphur*OUS *acid* produces *sulph*ITES, e. g. *sulphite of sodium.* -ITE is Latinised by -IS; e. g. *sulphite of sodium* becomes *sulph*IS.

c. The preceding names are presumed to refer to neutral compounds. In *acid* salts the prefixes noticed above are added to express the preponderance of the acid radical over the metal. KHSO$_4$ is called *acid sulphate of potassium,* BI*sulphate of potassium,* or BI*sulphate of potash,* the neutral sulphate being K$_2$SO$_4$.

d. In *basic* salts, or those in which the metal is in excess of the acid radical, the prefixes SUB- and DI- are employed; e. g. the formula of *neutral* acetate of lead is PbA$_2$. This salt, when boiled with oxide of lead (a base), furnishes [PbA$_2$.PbO] and [PbA$_2$.2PbO]. They are both, therefore, *basic* acetates; and to distinguish one from the other the former is called DI*acetate* and the latter TRI*acetate* of lead; *di-* referring to the presence of two atoms of lead, and *tri-* to three.

Formerly the salts of the metals of the alkalies and alkaline earths received names which indicated the existence in them of the oxides of such metals. Thus the terms carbonate of soda, nitrate of potash, carbonate of lime, sulphate of magnesia, names by which these fluids are still designated by some chemists, are now substituted by the more systematic and less speculative names of carbonate of sodium, nitrate of potassium, carbonate of calcium, and sulphate of magnesium. Another, and still better system of nomenclature is that in which the metallic or basic radical is mentioned first; e. g. calcium sulphate instead of sulphate of calcium, ammonium chloride for chloride of ammonium. When the *same* radicals form more than one series of salts, each series is distinguished by appending the terminations -IC and -OUS to that part of the name which refers to the basic radical; e. g. *mercur*OUS *chloride* (HgCl), *mercur*IC *chloride* (HgCl$_2$); *ferr*OUS *sulphate* (FeSO$_4$), *ferr*IC *sulphate* (Fe$_2$[SO$_4$]$_3$).

NON-METALLIC BODIES, &c. The names of the compounds formed by the union of the non-metallic elements, and certain other bodies, with the metals and with each other, either terminate in -IDE, Latinised by -IDUM, or in -URET, Latinised by -URETUM; as *arsen*IDE or *arsen*IURET (*arsen*IDUM, *arsen*IURETUM), *brom*IDE, *carb*IDE or *carb*URET, *chlor*IDE, *cyan*IDE, *fluor*IDE, *hyd*rIDE, *iod*IDE, *sulph*IDE or *sulph*URET, &c. The first of these terminations now prevails among English scientific chemists. The prefixes already noticed are also employed here.

METALS. The names of the metals (those of them, at least, that have been given during the present century) end in -IUM or (less frequently) in -UM; as *potass*IUM, *sod*IUM, *platin*UM. The Latin names of several of the non-metallic elementary bodies also end in -IUM; as *iod*INIUM, *nitrog*ENIUM, &c.

ALKALOIDS. The names of the organic bases which resemble the alkalies in their properties end either in -IA, -NA, or -INE; as *morph*IA, *quin*A, *strych*NINE. These terminations are now limited, as much as possible, to substances exhibiting basic properties, but were formerly very loosely applied.

Many chemists reject the first two terminations, and apply -INE to every substance of this class; as *morph*AINE, *quin*INE, *anil*INE, &c.

OTHER ORGANIC SUBSTANCES. The names of organic radicals generally terminate in -YL; as *eth*YL, *meth*YL, *benz*OYL, &c.: they mostly contain carbon, hydrogen, and oxygen. Compounds corresponding to the electro-negative elements have the termination -OGEN; as *cyan*OGEN, *amid*OGEN. Neutral compounds of carbon and hydrogen, mostly liquid, have the termination -OL; as *glycer*OL, *pyrr*OL: such substances are usually alcoholic in character. Other neutral substances, generally solid, have the termination -IN; as *paraff*IN, *naphthal*IN. Compounds resembling ammonia, and generally considered as 'substitution compounds' of that body, terminate in -AMINE; as *ethyl*AMINE, *propyl*AMINE.

The Latin genitive or possessive of the above compounds in—

-as	is	-atis
-is	„	-itis
-icum	„	-ici
-osum	„	-osi
-idum	„	-idi
-etum	„	-eti
-ium	„	-ii
-um	„	-i
-ia		
-a	„	-æ
-na		

Ex. Acetas (acetate), acetatis, of acetate; arsenis, arsenitis; citricum, citrici; arseniosum, arseniosi; iodidium, iodidi; sulphuretum, sulphureti; sodium, sodii; platinum, platini; morphia, morphiæ; quina, quinæ; narcotina, narcotinæ. The genitives of common names vary with the termination. Most of those ending in -a make -æ, and most of those in -us and -um make -i; but there are many exceptions, among which *cornu* (a horn) and *spiritus* (spirit), which are unaltered in the genitive singular, may be mentioned as examples.

NORFOLK FLUID. *Prep.* Take of linseed oil, 3 pints; black resin, ¼ lb.; yellow wax, 12 oz.; melt, and add of neat's-foot oil, 1 quart; oil of turpentine, 1 pint. Used to preserve and soften leather.

NORTUM. An unexamined metal, the oxide of which, according to Svanberg, exists in certain varieties of ZIRCON.

"NORMAL" SOLUTIONS. This system was first adopted by *Mohr*, and is now almost universally followed, on account of its simplicity and convenience.

A so-called "normal" (or "N.") solution is one which, at a temperature of 16° C., contains per litre the hydrogen equivalent of the active reagent weighed in grammes (H=1). Thus a normal solution of the (monobasic) hydrochloric acid contains 36·4 grms. of the pure compound HCl in one litre (the molecular and also the equivalent weight of HCl being 35·4+1=36·4); a normal solution of the (dibasic) sulphuric acid contains 49, *i. e.* 98/2 grms. of the pure compound, H₂SO₄, in one litre (the molecular weight of H₂SO₄ being 98, but the equivalent weight 49); one of the (mono-acid) caustic potash contains 56 grms. KOH (the molecular and also the equivalent weight of KOH being 56); one of the

(di-acid)sodic carbonate contains 53, *i. e.* 106/2 grms. of Na₂CO₃ (the molecular weight of Na₂CO₃ being 106, but the equivalent weight 53), and so on.

It is obvious that an amount of caustic potash expressed by the formula KOH is equivalent to an amount of hydrochloric acid expressed by the formula HCl, thus:

$$KOH + HCl = KCl + HOH.$$

Again, an amount of sodic carbonate expressed by the formula Na₂CO₃ is equivalent to an amount of hydrochloric acid expressed by the formula 2HCl, thus:

$$Na_2CO_3 + 2HCl = 2NaCl + H_2O + CO_2.$$

It must, however, be borne in mind that the first thing to be considered with regard to any particular solution for use in volumetric analysis is not necessarily its equivalent hydrogen weight, but its reaction in the analysis in question. Thus tin is a tetravalent metal, but when a solution of stannous chloride is used as a reducing agent in the estimation of iron, a number (in grammes) corresponding to the half, and not to the fourth, of its molecular weight is required, as is shown by the equation—

$$Fe_2Cl_6 + SnCl_2 = 2FeCl_2 + SnCl_4.$$

(See Sutton's 'Volumetric Analysis,' 5th edition, p. 23.)

Semi-normal $\left(\frac{N}{2}\right)$, quintinormal $\left(\frac{N}{5}\right)$, decinormal $\left(\frac{N}{10}\right)$, and centinormal $\left(\frac{N}{100}\right)$ solutions are likewise frequently employed.

Where the 1000 gr. measure is used as the standard in place of the litre, the weight of the compound in grains is taken instead of that in grammes. Since 1000 gr. measures occupy but a small volume, it is found convenient in practice to prepare solutions of 10,000.

NOSTRUMS. See PATENT MEDICINES, &c.

NOTICES. The following sections of the Public Health Act refer to serving and delivery of notices under that statute:

(S. 266.) Notices, orders, and other such documents under the Public Health Act may be in writing or print, or partly in writing and partly in print; and if the same require authentication by the local authority, the signature thereof by the clerk to the local authority or their surveyor or inspector of nuisances shall be sufficient authentication.

(S. 267.) Notices, orders, or any other documents required or authorised to be served under the said Act may be served by delivering the same to or at the residence of the person to whom they are respectively addressed, or where addressed to the owner or occupier of premises, by delivering the same or a true copy thereof to some person on the premises, or if there is no person on the premises who can be so served, by fixing the same on some conspicuous part of the premises; they may also be served by post by a prepaid letter, and if served by post shall be deemed to have been served at the time when the letter containing the same would have been delivered in the ordinary course of post, and in proving such service it shall be sufficient to prove that the notice, order, or other document was properly addressed and put into the post.

Any notice required to be given to the owner or

occupier of any premises may be addressed by the description of the 'owner' or 'occupier' of the premises (naming them) in respect of which the notice is given, without further name or description.

Enforcing the Drainage of Houses.

(S. 23.) Notice is to be given to the owner or occupier, but in case of the failure of either to comply, and the authority having to do the work, the expenses fall on the owner.

Insufficient Privy Accommodation.

(SS. 36 and 37.) The same procedure as under the above section.

The Cleansing and Whitewashing of Houses.

(S. 46.) Notice to the owner or occupier.—The person on whom the notice is served is liable to a penalty if it is not complied with.

The Removal of Manure or Filth, &c., in an Urban District.

(S. 49.) Notice to be served on the person to whom the manure belongs, or to the occupier of the premises whereon it exists. If the urban authority have to remove it themselves, the expense of removal falls upon the owner of the manure, &c., or the occupier of the premises, or where there is no occupier, the owner of the premises.

In the case of Nuisances.

(S. 94.) Notice is to be served upon the person causing or permitting the nuisance to remain, or, if he cannot be found, on the owner or occupier of the premises on which the nuisance arises; but if the nuisance arises from the want or defective construction of any structural convenience, or where there is no occupier, notice is to be served on the owner.

In the case of Houses, &c., requiring Disinfection.

(S. 120.) Notice is to be given to the owner or occupier, and in case of non-compliance, the person on whom the notice is served is liable to penalties, and the expenses of the authority doing the necessary works falls upon that person (with certain exceptions in case of poverty).

NOVAR'GENT. *Prep.* From recently precipitated chloride of silver by dissolving it in a solution of either hyposulphite of sodium or of cyanide of potassium. Used chiefly to restore old plated goods. The liquid is rubbed over the metal to be coated with a little prepared chalk, and the part is afterwards polished off with a piece of soft leather. A powder recently sold under the same name is formed by mixing the preceding article with chalk, and drying the mass. It is made into a paste with a little water, spirit of wine, or gin, before applying it.

NOVAUR'UM. From a solution of neutral trichloride of gold, as the last.

NOXIOUS TRADES. See OFFENSIVE TRADES.

NOYAU. *Syn.* CRÈME DE NOYAU. This is a pleasant nutty-tasting liqueur; but from the large proportion of prussic acid which it contains, a small quantity only should be taken at a time.

Prep. 1. Bitter almonds (bruised), 3 oz.; spirit (22 u. p.), 1 quart; sugar, 1 lb.; (dissolved in) water, ¾ pint; macerate for 10 days, frequently shaking the vessel; then allow it to repose for a few days, and decant the clear portion.

2. As the last, but substituting apricot or peach kernels (with the shells, bruised) for the almonds.

3. To either of the above add of coriander seed and ginger, of each, bruised, 1 dr.; mace and cinnamon, of each, ¼ dr.

4. (Wholesale.) To plain cordial, at 54 to 60 u. p., containing 3 lbs. of sugar per gallon, add, gradually, essence of bitter almonds, q. s. to flavour.

5. (CRÈME DE NOYAU DE MARTINIQUE.) Loaf sugar, 24 lbs.; water, 2½ galls.; dissolve, add of proof spirit, 5 galls.; orange-flower water, 3 pints; bitter almonds (bruised), 1 lb.; essence of lemons, 2 dr.; as above. See LIQUEURS.

NUISANCE. The following are the chief clauses of the Public Health Act respecting nuisances:

Definition of Nuisances.

1. Any premises in such a state as to be a nuisance or injurious to health.

2. Any pool, ditch, gutter, watercourse, privy, urinal, cesspool, drain, or ashpit, so foul as to be a nuisance or injurious to health.

3. Any animal so kept as to be a nuisance or injurious to health.

4. Any accumulation or deposit which is a nuisance or injurious to health.

5. Any house, or part of a house, so overcrowded as to be dangerous or injurious to the health of the inmates, whether or not members of the same family.

6. Any factory, workshop, or workplace (not already under the operation of any general Act for the regulation of factories or bakehouses) not kept in a cleanly state, or not ventilated in such a manner as to render harmless as far as practicable any gases, vapours, dust, or other impurities generated in the course of the work carried on therein that are a nuisance or injurious to health, or so overcrowded while work is carried on as to be dangerous and injurious to the health of those employed therein.

7. Any fireplace or furnace which does not, as far as practicable, consume the smoke arising from the combustible used in such fireplace or furnace, and is used for working engines by steam, or in any mill, factory, dyehouse, brewery, bakehouse, or gaswork, or in any manufacturing or trade process whatsoever; and—

Any chimney (not being the chimney of a private dwelling-house) sending forth black smoke in such quantity as to be a nuisance;

Shall be deemed to be nuisances liable to be dealt with summarily under the Public Health Act: Provided—

First. That a penalty shall not be imposed on any person in respect of any accumulation or deposit necessary for the effectual carrying on any business or manufacture, if it be proved to the satisfaction of the court that the accumulation or deposit has not been kept longer than is necessary for the purposes of the business or manufacture, and that the best available means have been taken for preventing injury thereby to the public health.

Secondly. That where a person is summoned before any court in respect of a nuisance arising from a fireplace or furnace which does not consume the smoke arising from the combustible used in such fireplace or furnace, the court may hold that no nuisance is created within the meaning of this Act, and dismiss the complaint, if it is satisfied that such fireplace or furnace is constructed in such a manner as to consume as far as practicable, having regard to the nature of the manufacture or trade, all smoke arising therefrom, and that such fireplace or furnace has been carefully attended to by the person having the charge thereof. (P. H., s. 91.)

The Act also defines and specifies—1. The duty and powers of a local authority to inspect a district with the view to an abatement of any nuisance. 2. The process of information to be pursued in representing a nuisance to any local authority. 3. Procedure on failing to comply with notice. 4. The power of the Court to make an order dealing with such nuisance. 5. The penalty for neglecting to obey such order. 6. The power of complaint by private individuals. 7. The power of the police to proceed in certain cases. 8. The cost and expense of executing the provisions relating to nuisances. 9. The power of sale of manure, &c. 10. The supervision of nuisances caused by drains, privies, &c. 11. The proceedings to be taken in certain cases against nuisances in ships, &c.

NURSING. Milk is the natural food of the mammalia during the earlier period of their existence. It contains all that is necessary for the nourishment of their bodies, and on it they thrive and grow. Its secretion only actively commences at the time when it is required for the sustenance of the offspring, and it either materially lessens in quantity, or wholly disappears, as soon as the necessity for its existence has passed away, and the little being who depended on it has acquired sufficient age and strength to exist on cruder aliment. The nursing mother, when in a state of perfect health, and properly supplied with a sufficiency, without excess, of nutritious food, elaborates this secretion in the fittest condition to ensure the health and vigour of her offspring.

The milk of woman varies with the food, health, age, &c., of the nurse. That produced from a mixed animal and vegetable diet neither acesces nor coagulates spontaneously, like cow's milk; and when gently evaporated in an open vessel, "the last drop continues thin, sweet, and bland." Acids and rennet, however, coagulate it readily; and so does the gastric juice of the infant, as shown by the condition in which it is often ejected by the latter. The milk of a woman who lives wholly on vegetable food acesces and coagulates with equal readiness and in a precisely similar manner to cow's milk. The quality of the milk also varies with the progress of the digestion. Within the first hour or two after a meal it is thin and serous, and then gradually improves in richness and flavour, until at about the 4th or 5th hour it possesses these qualities in the highest degree. This, then, is the period at which the infant should be applied to the breast, which, according to the present habits of society, would be during the hour immediately preceding each meal except the breakfast. After about the 5th or 6th hour the milk gradually loses its peculiar colour and odour, until towards the 10th or 12th hour after eating food it becomes yellowish, bitter, and often nauseous, and in this condition is frequently refused by the infant. This points out the impropriety of a nurse fasting longer than 4 to 5 hours, except during the night, when the period may be extended to 7 or 8 hours, but never longer. The time after accouchement is another matter that influences the character of human milk in respect of its wholesomeness for the infant. The milk secreted soon after delivery is very thin and serous, but in the course of a few days it becomes thicker, richer, and more nutritious; and a gradual change in the same direction proceeds during the usual period of suckling. When the mother suckles her own infant, or the 'age of the milk,' as the nurses say, corresponds to that of the child, all goes on well; but when the former much exceeds the latter, the reverse is the case. Thus it is found that an infant is incapable of completely digesting the milk of a nurse whose own child is much older than itself; and that an infant of a few weeks old will often starve on the milk intended by nature for one several times its age. It is, therefore, necessary, in selecting a wet-nurse, to be certain that her condition, in this respect, closely corresponds to that of the mother of the infant, or that it does not differ on this point more than 3 or 4 weeks. In respect of the use of high-flavoured or improper food and beverages, medicine, &c., it appears that all these substances immediately affect the milk, and impart to it more or less of their peculiar flavour and properties; and, except with remedies administered under medical advice, in nearly all cases prove injurious to the infant. The diet of a nurse should be nutritious and succulent, and its healthy digestion should be promoted by exercise and pure air. Strong liquors, more especially spirits, act like slow poisons on the infant, and their habitual use by a nurse should, therefore, be considered as a positive disqualification for the duties of her office. The care of the mother or wet-nurse should be particularly directed to the maintenance of her own health and equanimity, by which both the health and good temper of the infant will be, as far as possible, ensured. A grieving, irritable, or angry mother forces her bad qualities on her offspring, in the shape of fits, convulsions, or hopeless marasmus. See INFANCY, MILK, INFANTS, FOOD FOR, &c.

NUTMEG. *Syn.* MYRISTICÆ NUCLEUS, NUCISTA, NUX MOSCHATA, N. MYRISTICA, N. AROMATICA, MYRISTICA (B. P., Ph. L.), L. "The shelled seed of *Myristica officinalis*, Linn. (*M. moschata*, Thunberg), or nutmeg tree." It is chiefly used as a spice and condiment, but it is also esteemed as an aromatic in flatulency and diarrhœa.—*Dose.* Half a teaspoonful, or more, grated. The distilled and expressed oils (OLEUM MYRISTICÆ) are also officinal.

Of the different varieties of nutmegs met with in commerce, those known as Penang are the most valuable. Next to these rank the Dutch or Batavian kind, and after these the Singapore nutmegs. In the Dutch or Batavian variety the exterior is composed of a number of white furrows, with brown projections, which aspect is caused by their having been dusted over with lime previous to their exportation. Besides the above, there is also a very inferior description, known as the long or wild nutmeg, which are met with either in the shell, out of the shell, or in the shell with the mace attached.

Nutmegs are subject to the ravages of a worm which would seem to devour or destroy their aromatic principle, since when attacked by this parasite they lose both their odour and taste.

In 100 parts sound nutmegs contain—

Volatile oil	.	.	.	.	.	6·0
Liquid fat	.	.	.	.	.	7·6
Solid fat	.	.	.	.	.	24·0
Acid	.	.	.	.	.	0·8
Starch	.	.	.	.	.	2·4
Gum	.	.	.	.	.	1·2
Ligneous fibre	.	.	.	.	.	54·0
Loss	.	.	.	.	.	4·0
						100·0

(Bonastre.)

NUTRI"TION. The phenomena of life are accompanied by the constant and unceasing waste of the materials of which the animal body is composed. Every act of volition, every exertion of muscular power, every functional action of the organism, whether perceptible or imperceptible and involuntary, every play of chemical affinity and decomposition, even thought itself, occasions the disorganisation and destruction, as living matter, of a portion of ourselves. But the process of respiration, and the various important changes with which it is connected, tend, more than all the other vital functions, to waste the substance of the body, the temperature of which it is its special office to support. This loss, this change, which commences with life and terminates only with death, is compensated for by the constant renewal of the whole frame by the deposition and assimilation, or organisation, of matter from the blood, which thus becomes gradually thinner and impoverished, unless, in its turn, it receives a corresponding supply of its vital elements. This it does from the food, which, by the function of digestion, is converted into a 'chyle,' and after being taken up by the 'lacteals,' passes into the blood, of which it then becomes a part, and attaches itself to those organs or tissues, the loss of which it is intended to supply. This constitutes nutrition.

NUTS, Cob (Jamaica). *Omphalea triandra,* Linn., a small tree exuding a white juice, which dries black, and bearing a yellow globose furrowed drupe, called Noisettier in the French W. Indies, and known in Jamaica as pig or hog nut. When ripe the seeds burst from the pericarp; they are eaten raw or roasted. By compression they yield a fine flavoured oil.

Nuts (Hickory). *Carya alba,* Nutt., and *C. tomentosa,* Nutt., the former species affording the principal supply. They are natives of North America, and the woods are both tough and elastic, especially that of *C. alba,* which is much used for spokes for carriage wheels, shafts, &c.

Nuts (Pecan). *Carya olivaformis,* Nutt., occasionally to be found in English fruit-shops; the kernels are sweeter than those of the former.

NUT TREE (of Eastern Australia). *Macadamia ternifolia,* F. Muell. The seeds of this tree are edible.

NUX VOMICA. *Syn.* KOOCHLA NUT, POISON N., VOMIT N.; NUCES VOMICÆ, NUX VOMICA (B. P., Ph. L., E., and D.). L. "The seed of *Strychnos nux-vomica,* Linn." (Ph. L.), imported from the East Indies (B. P.). This drug is chiefly known as a violent excitant of the cerebro-spinal system. In small doses, frequently repeated, it is tonic, diuretic, and, occasionally, laxative; in slightly larger ones it is emetic; and in large doses it is an energetic and fearful poison.—*Dose,* 1 to 3 gr.; in paralysis, nervous affections, impotence, chronic dysentery, chronic diarrhœa, &c. Its frequent use is said to render the system proof against the poison of serpents. See STRYCHNINE, its active principle.

OAK. The British oak is the *Quercus robur* of Linnæus, of which there are two varieties, Q. *pedunculata* and Q. *sessiliflora.* The wood of the oak is more durable than that of any other tree, and "for at once supporting a weight, resisting a strain, and not splintering by a cannon-shot, it is superior to every other kind." It nevertheless "warps and twists much in drying, and in seasoning shrinks about 1-32nd of its width." Foreign oak is less durable, but more brittle and workable. The bark (OAK-BARK; QUERCÛS CORTEX, QUERCUS, B. P., Ph. L., E., & D.) is used as an astringent and febrifuge, in doses of 30 to 120 gr. frequently; an astringent decoction is also made of it, but its chief employment is in tanning leather. The peculiar appearance of old oak or 'wainscoting' is given to the new wood by exposing it, whilst very slightly damp, to the fumes of ammonia.

Oak, Cork. *Quercus suber,* Linn. The cork tree grows in Spain, South France, Italy, and Algeria. Cork is the thick outer bark, which may be removed from the same tree at intervals of 6 to 10 years after it attains an age of about 30 years. The cork collected previously is of inferior quality. The bark is heated, loaded with weights to flatten it, and then slowly dried. The operation of removing the cork does not interfere with the healthy growth of the tree; it is said rather to favour it. A cork box called a 'tarro' is used in the province of Alentejo, Portugal, by agricultural labourers for carrying their food in, and to keep it cool.

OAT. *Syn.* AVENA, L. The common cultivated oat is the *Avena sativa,* Linn., a gramineous plant, of which there are several varieties, as the *Avena sativa alba,* or white oat; *A. s. nigra,* or black oat; the potato oat, &c. Other species are also cultivated, as *Avena nuda,* Linn., pilcorn, or naked oat; *A. strigosa,* or Spanish oat, &c. The seed (OATS; CARYOPSIDES, SEMINA AVENÆ CRUDA) form the common horse-corn of this country, but in the northern parts of the country it is extensively used as food for man. The husked grain constitutes GROATS, and its meal OATMEAL. The latter does not form a dough with water, as wheaten meal or flour does.

Oats consist of from 24% to 28% of husk, and 74% to 78% of grain. According to M. Payen, they contain of starch, 60·59%; azotised matter, 14·39%; saccharine and gummy matter, 9·25%; fatty matter, 5·50%; cellulose, 7·60%; silica and saline matter, 3·25%. The husks contain between 6% and 7% of saline matter (*Prof. Norton*). The ash amounts to 2·18%, and consists of potassa and soda, 26·18% ; lime, 5·95% ; magnesia, 9·95% ; oxide of iron, ·40% ; phosphoric acid, 43·84% ; sulphuric acid, 10·45% ; chlorine, ·26% ; silica, 2·67% ; alumina, ·06% (*Johnston*).

The yield of oats is from 20 bushels per acre in poor soils, up to 60, 70, and even 80 bushels per acre in rich soils. The weight per bushel varies from 35 to 45 lbs. and the product in meal is about one half the weight of the oats.

A large proportion of the oats given to horses

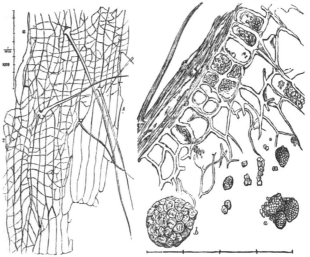

White oat.—Long. sect., 2nd and 3rd coats not separable. *a*. Compound grains × 100; *b*. One do. × 500.

passes off undigested. It has hence been proposed to prevent this loss by either coarsely bruising them in a mill, or by pouring boiling water over them, and allowing them to macerate till cold, when they are to be given to the horses without straining off the water. It is stated on good authority that oats thus treated will not only fatten quicker, but go twice as far as without preparation. Oat bruisers are now manufactured by most agricultural implement makers.

Under the microscope the oat is seen to consist of 2 or 3 envelopes ; the outer being composed of longitudinal cells ; the second obliquely transverse and not very clearly seen ; in this the cells are wanting in part, or pass into the cells of the third coat ; the third envelope consists of a layer, usually single, of cells, like wheat. Before the envelopes are searched for the husks must be removed. The starch-cells are small, many-sided, and cohere into round composite bodies, which are very characteristic, and which, by pressure, may be divided into separate grains. A high power is necessary for the examination of these latter. The starch of the oat does not polarise light.

OATMEAL. *Syn.* AVENÆ FARINA, F. EXSEMINIBUS AVENÆ (Ph. D.), L.

Oatmeal is the grain of the oat deprived of the skin, kiln-dried, and afterwards ground. It is regarded as one of the most nutritious of our

cereals, being rich in nitrogenous matter, fat, starch, and sugar. According to Letheby it contains in 100 parts—

Nitrogenous matter	12·6
Carbo-hydrates	63·8
Fatty matter	5·6
Saline matter	3·0
Water	15·0
	100·0

Kreusler has shown that the nitrogenous principle of oatmeal contains gluten-casein, a substance very similar to the legumin of peas and beans. Letheby points out that, although it contains more nutrient material than wheat, its higher price renders it less economical as an article of diet. Oatmeal forms the staple of the food of the farm labourer both in Scotland and in England, being consumed more largely by the Scotch than the English peasant. Scotch oatmeal is superior to English in nutritive value. Oatmeal, when mixed with water, does not possess sufficient tenacity to enable it to be made into bread. It can, however, be baked into excellent cakes, which, when made in Yorkshire, are leavened, and when in Scotland, unleavened.

The qualities of indigestibility and a tendency to produce irritability of the bowels and skin, have been ascribed to oatmeal; before it was so prepared as to effectually remove from it the husk and hairs by efficient screening, it was in Scotland a frequent source of intestinal concretions. These concretions, the nature of which was unravelled by Dr Wollaston, consisted principally of phosphate of lime mixed with the hairs and husks of the oats.

Of 30 samples of oatmeal examined by the 'Lancet' Sanitary Commissioner, no fewer than 16 samples, or more than one half, were adulterated. The substance generally used for this purpose is barley meal, which is only half the price of oatmeal. Husks of barley, wheat, and of the oat itself, are also frequently used. Rice and maize are also sometimes added. That supplied to the army, navy, and the workhouses, was very commonly adulterated with whiting, plaster of Paris, or ground bones. The mineral sophisticant may be detected by the excess of ash, which should not exceed 2·36%. These frauds are readily detected by the microscope.

Grits or *groats* are the decorticated grain of the oat, which when bruised or crushed constitute Embden groats. Flummery (known in Scotland as *sowans*) is made by steeping the husks of the grain in water, until they become slightly sour, the strained liquid being boiled down to the consistence of gruel. Oatmeal soon becomes sour and rancid. It should be purchased at such shops as have a quick sale for it. See ACARI, STIRABOUT.

OBE″SITY. *Syn.* OBESITAS, POLYSARCA, L. Unhealthy or troublesome fatness or corpulency. Sometimes the secretion of fat, and its accumulation in the adipose membrane, is almost as rapid as that of water in anasarca, on which account some of the old writers have called obesity a dropsy of fat. Persons in easy circumstances, of indolent habits, who live freely, and who are of a cheerful and contented disposition, are those most liable to obesity. The treatment consists in the very gradual reduction of the diet, until it falls rather below the average quantity required by a healthy adult; the very gradual disuse of fermented liquors, more especially beer; the gradual abridgment of the time devoted to repose, until it does not exceed 5 or 6 hours; the employment of several hours daily in exercise in the open air, at first moderate, but increased day by day in energy, until it becomes laborious; and, lastly, arousing the mind from a state of lethargy to one of active or even harassing employment.

In some cases the accumulation of fat has been enormous. Bright, of Maldon, weighed 728 lbs.; Daniel Lambert, of Leicester, 739 lbs.; a girl, 4 years old, noticed in the 'Phil. Trans.,' 1813, weighed 256 lbs.

Persons affected with obesity are generally short-lived.

The system known as 'Banting' is a very rational and physiological cure for obesity. It consists essentially in the reduction of the fats and carbo-hydrates of the food to a minimum, and the adoption of a chiefly animal diet with exercise.

OBSTRUCTION OF LOCAL AUTHORITY. Various penalties are mentioned in different sections of the Public Health Act for the offence of obstructing officers, &c., representing the local authority, in carrying out the Act. The follow-

ing section, which we select, deals with the subject generally:

SEC. 306. "Any person who wilfully obstructs any member of the local authority, or any person duly employed in the execution of this Act, or who destroys, pulls down, injures, or defaces any board on which any bye-law, notice, or other matter is inscribed, shall, if the same was put up by authority of the Local Government Board or of the local authority, be liable for every such offence to a penalty not exceeding £5.

"Where the occupier of any premises prevents the owner thereof from obeying or carrying into effect any of the provisions of this Act, any justice, to whom application is made in this behalf, shall by order in writing require such occupier to permit the execution of any works required to be executed, provided that the same appear to such justice to be necessary for the purpose of obeying or carrying into effect the provisions of this Act; and if within 24 hours after the making of the order such occupier fails to comply therewith, he shall be liable to a penalty not exceeding £5 for every day during the continuance of such non-compliance.

"If the occupier of any premises, when requested by or on behalf of the local authority to state the name of the owner of the premises occupied by him, refuses or wilfully omits to disclose, or wilfully mistakes the same, he shall (unless he shows cause to the satisfaction of the court for his refusal) be liable to a penalty not exceeding £5."

O'CHRES. These are native earthy compounds of clay, coloured with oxide of iron, with frequently a little chalk or magnesia. The differences in the colour arise partly from the quantity of iron present, and partly from the state of oxidation in which the iron is found. Several varieties are known in *commerce*—BROWN OCHRE, FRENCH O., OXFORD O., RED O., ROMAN O., YELLOW O. All these, with the exception of the first and fourth, have a yellow colour. ARMENIAN BOLE, INDIAN RED, VENETIAN R., and SPANISH BROWN are also ochres.

All the ochres are darkened by calcination. The yellow ochres acquire a red or reddish-brown colour by this treatment. The pigment called 'light red' is thus prepared from yellow ochre.

ODONTAL'GIA. See TOOTHACHE.

O'DORAMENTS. *Syn.* ODORAMENTA, L. Substances employed in *medicine* on account of their odour. They differ from disinfectants in only disguising, but not destroying, noxious vapours, &c. AMMONIA, STRONG VINEGAR, and PASTILLES furnish the most familiar examples of this class of substances. See DISINFECTANTS, PERFUMES, &c.

O'DOUR. The emanation of an odoriferous or scent-giving body. See PERFUMES.

ŒNAN'THIC ETHER. See ETHER (Œnanthic).

OFFENSIVE TRADES. These are declared by Section 112 of the Public Health Act to be that of a

Blood boiler,	Soap boiler,
Bone boiler,	Tallow melter,
Fellmonger,	Tripe boiler,

and

"Any other noxious or offensive trade, business, or manufacture shall be liable to a penalty not exceeding *fifty pounds* in respect of the *establishment* thereof, and any person carrying on a business so established shall be liable to a penalty not exceeding *forty shillings* for every day on which the offence is continued, whether there has or has not been any conviction in respect of the establishment thereof."

OFFICINAL. *Syn.* OFFICINALIS, L. A term applied to substances or medicines ordered in the Pharmacopœia.

OIL. *Syn.* OLEUM, L.; HUILE, Fr. This name is given to numerous liquid or semi-liquid substances, expressed or drawn from animal or vegetable bodies; to various products of the distillation of bituminous minerals; and to several unctuous mixtures in *perfumery* and *pharmacy*.

To facilitate reference, we have grouped the principal substances generally called 'oils' into classes, under the following heads:—OILS (Drying); OILS (Empyreumatic); OILS (Fixed); OILS (Medicated); OILS (Mineral); OILS (Mixed); OILS (Perfumed); OILS (Volatile). See these articles also *below*.

Oil, Consol'idated. *Syn.* CAMPTICON, FACTITIOUS CAOUTCHOUC. A substance having most of the properties of india rubber, prepared by oxidising boiled linseed oil, or any other oil that hardens on exposure to the atmosphere. To obtain the solid oil, plates of glass are dipped into linseed oil, the films are then allowed to dry, and the process is repeated again and again until the plates are coated with many layers of perfectly oxidised oil. Instead of plates of glass, extensive surfaces of prepared cloth are employed when the manufacture is carried out on a large scale.

The solid oil, having been scraped or peeled off the surfaces, is worked with a small proportion of shell-lac, by means of a mixing machine with hot rollers, until a material singularly like caoutchouc is produced. The consolidated oil can be rolled on to fabrics, so as to form a waterproof cloth, having the finish and flexibility of rubber-cloth. By the action of heat the consolidated oil may be converted into a hard substance resembling vulcanite and ebonite. Its useful applications appear to be very numerous, but its manufacture has not as yet made much progress.

OIL-GAS. A mixture of several gaseous hydrocarbons, obtained by passing common whale-fat resin, the heavy petroleum or shale oil, or the tarry residues left after the distillation of these two latter substances, or other cheap animal oil, through red-hot tubes, or by allowing it to fall in drops on red-hot stones or bricks arranged in an iron retort, or other suitable apparatus. The gas has great illuminating power, requires no purification, and is quite free from the ammoniacal and sulphur compounds which vitiate coal-gas.

The sp. gr. of oil-gas varies with the heat employed in its production. It averages from 0·76 to ·90, but it may rise as high as 1·1.

The composition of oil-gas, as given by Payen, is as follows:

	Oil-gas.	Gas from Petroleum residues.
Oleflant gas and homologues.	22·5	17·4
Marsh gas	50·3	58·3
Hydrogen	7·7	24·3
Carbonic oxide	15·5	—
Nitrogen	4·0	—

OILS, BLEACHING OF. According to a German chemist (*Puscher*) this is usually effected by means of 2% of concentrated sulphuric acid, and a subsequent washing with water. In many cases a perfectly pure oil is spoilt by an incomplete washing after the action of the sulphuric acid. Such is often the case in the colza oil, largely used as a source of illumination. Unless the purified oil is completely washed, the wick of the lamp is blackened in a very short time, and frequent cutting and trimming become requisite. By Puscher's new method this washing is greatly facilitated; instead of using pure sulphuric acid, he takes a mixture of equal parts of spirit at 96% and ordinary concentrated sulphuric acid. Half the quantity of acid usually employed is thus replaced by spirit. The effect is most satisfactory. No resinification of the oil is produced as when sulphuric acid alone is used, and from the first the mixture is perfectly homogeneous. Gradually the mass becomes cloudy, takes a green colour, and finally becomes black. In the course of one or two days, during which the vessels are left quite quiet, a comparatively small quantity of a black deposit has found its way down as a sediment; it occupies only a small space in proportion to the bulk of the oil so treated.

In testing the efficacy of this new process, experiments were made on 200 quarts of oil at a time. Walnut and colza oils become by this treatment as clear as water; linseed oil still retains a distinct yellow tinge when seen through a certain thickness.

Before being sent out oils treated in this manner must be submitted to a thorough washing with cold water, used in liberal quantities. It has been found in practice that the addition of the spirit economises one half of the sulphuric acid, and yields a much superior product.

Oils (Drying). All the fixed oils have an attraction more or less powerful for oxygen, and, by exposure to the air, they either become hard and resinous, or they only thicken slightly and become sour and rancid. Those which exhibit the first property in a marked degree, as the oils of linseed, poppy, rape, and walnut, are called 'drying oils,' and are used as vehicles for colours in painting. The others are frequently termed 'glutinous' for 'non-drying oils.' Chemically speaking, the drying oils are the glycerides of linoleic and allied acids. The principal vegetable drying oils are linseed, poppy-seed, grape-seed, and nut oils. Castor and cotton-seed oils seem to be intermediate between the drying and non-drying oils, and are sometimes classed with the latter (*Cameron*).

Light exerts a considerable influence upon the absorption of oxygen by the drying oils; while the process is very slow in the dark; it is most quickly accomplished in a blue or colourless light, and less quickly in a red, yellow, and green light (*Braunt*).

The resinifying or drying property of oils is greatly increased by boiling them, either alone or along with some litharge, sugar of lead, or white vitriol, when the product forms the 'boiled oil' or 'drying oil' (OLEUM DESICCATIVUM) of *commerce*. The oxalate and oxide of manganese have recently been used with some success. The efficacy of the process, according to Liebig, depends on the elimination of substances which impede the oxidation of the oil. The following formulæ are adopted for this purpose:

1. Linseed oil, 1 gall.; powdered litharge, ½ lb.; simmer, with frequent stirring, until a pellicle begins to form; remove the scum, and when it has become cold and has settled decant the clear portion. Dark coloured; used by house-painters.

2. Linseed oil and water, of each, 1 quart; white vitriol, in powder, 2 oz.; boil to dryness. Paler than the last.

3. Pale linseed or nut oil, 1 pint; litharge or dry sulphate of lead, in fine powder, 2 oz.; mix, agitate frequently for 10 days, then set the bottle in the sun or a warm place to settle, and decant the clear portion. Very pale.

4. Linseed oil, 100 galls.; calcined white vitriol ('sulphate of zinc'), in fine powder, 7 lbs.; mix in a clean copper boiler, heat the whole to 285° F., and keep it at that temperature, with constant stirring, for at least one hour; then allow it to cool; in 24 hours decant the clear portion, and in 3 or 4 weeks more rack it for use. Used for varnishes.

5. (*Liebig.*) Sugar of lead, 1 lb., is dissolved in rain-water, ¼ gall.; litharge, in fine powder, 1 lb., is then added, and the mixture is gently simmered until only a whitish sediment remains; levigated litharge, 1 lb., is next diffused through linseed oil, 2¼ galls., and the mixture is gradually added to the lead solution, previously diluted with an equal bulk of water; the whole is now stirred together for some hours, with heat, and is, lastly, left to clear itself by exposure in a warm place. The lead solution which subsides from the oil may be used again for the same purpose, by dissolving it in another lb. of litharge, as before.

6. (*Wilks.*) Into linseed oil, 236 galls., pour oil of vitriol, 6 or 7 lbs., and stir the two together for 3 hours; then add a mixture of fuller's-earth, 6 lbs., and hot lime, 14 lbs., and again stir for 3 hours; next put the whole into a copper with an equal quantity of water, and boil for about 3 hours; lastly, withdraw the fire, and when the whole is cold, draw off the water, run the oil into any suitable vessel, and let it stand for a few weeks before using it.

7. ('Allg. Polytech. Zeitung.') Binoxide of manganese (in coarse powder, but not dusty), 1 part; nut or linseed oil, 10 parts; mix, and keep the whole gently heated and frequently stirred for 24 to 36 hours, or until the oil begins to turn reddish. Recommended for zinc paint, but is equally adapted for other purposes for which boiled oil is employed.

Obs. There is often a difficulty in obtaining the oils 'bright' after boiling or heating them with the lead solutions; the best way, on the small scale, is either to filter them through coarse woollen filtering-paper, or to expose the bottle for some time to the sun or in a warm place. On the large scale, the finer oils of this kind are often filtered through Canton flannel bags. The litharge and sulphate of lead used in the above processes may be again rendered available for the same purpose by washing them in hot water, to remove adhering mucilage.

OILS (Empyreumat'ic). *Syn.* OLEA EMPYREUMATICA, L. The 'empyreumatic oils' of the old pharmaceutical writers were oily fluids obtained by the dry distillation of various substances, animal, vegetable, and mineral. But few of them are in use at the present day, though formulæ are given for them in some of the foreign pharmacopœias. Two or three have useful applications in the *arts*, and it is therefore necessary that we should briefly describe their preparation. When the ingredients are of a liquid or pasty nature, or become so when heated, they are usually mixed with about twice their weight of sand, powdered glass, or other like substance, to divide them, and thus expose them more effectually to the action of the fire. Care must also be taken to provide a well-cooled receiver, which must be furnished with a tube to carry off the non-condensable gases liberated at the same time as the oil. The products of the first distillation are generally purified by rectification, either alone or along with water. In general, they require to be preserved from the light and air.

The following are the principal substances belonging to this class:

Oil of Al'oes. *Syn.* ALOETIC OIL; OLEUM ALOETICUM, L. *Prep.* 1. From Socotrine or hepatic aloes distilled along with sand.

2. (Batavian—Cadet de Gassicourt.) Olive oil, 1 lb.; hepatic aloes and myrrh, of each, in powder, 2 oz.; olibanum, ¼ oz.; distil in a sand-bath, from a stoneware retort. Used as an external vermifuge for children; a portion is rubbed two or three times a day over the umbilical regions.

Oil of Am'ber. *Syn.* OLEUM SUCCINI, L. *Prep.* From coarse pieces of amber, distilled in an iron retort, either alone or reduced to powder and mixed with sand. The oil is separated from the fetid liquor and succinic acid which passes over, and rectified along with about 6 times its volume of water, by a gentle heat. It then forms 'RECTIFIED OIL OF AMBER' (OLEUM SUCCINI—Ph. L. 1836, O. S. RECTIFICATUM—Ph. D. 1826, O. S. PURISSIMUM—Ph. E. 1841).—*Prod.*, 20%.

Prop., &c. It has a pale yellow colour, a strong, ungrateful odour, and a hot, acrid taste; heat and air blacken and thicken it; it boils at 186° F. Sp. gr. ·758 at 75° F. It is antispasmodic, rubefacient, and stimulant.—*Dose*, 5 to 12 drops, made into an emulsion with mucilage; in hysteria, epilepsy, and convulsive affections. Externally, as a friction, either alone or combined with laudanum or sweet oil, in rheumatism, tic-douloureux, hooping-cough, &c.

Oil of Amber, Oxidated (*Artificial Musk*).

Prep. Put into a cup 1 dr. of oil of amber, and add to it, drop by drop, 3½ fl. dr. of strong nitric acid; let it stand for 36 hours, then separate and wash the resinous matter. Antispasmodic and nervine.—*Dose*, 5 to 10 gr. For children, ⅓ to 1 gr.

Oil, An'imal. *Prep.* 1. (Empyreumatic or Fetid: OIL OF HARTSHORN, DIPPEL'S O.; OLEUM ANIMALE EMPYREUMATICUM, O. CORNU CERVI, O. DIPPELII, L.) Chiefly obtained as a secondary product in the manufacture of bone-black. Fetid and dark-coloured. Used chiefly to make lamp-black. It contains—

2. (Ethereal: RECTIFIED OIL OF HARTSHORN; OLEUM ANIMALE ÆTHEREUM, O. CORNU CERVI RECTIFICATUM, LOCO OLEI ANIMALIS DIPPELII, L.) *a.* A finer kind of animal oil, made by slowly distilling oil of hartshorn, and collecting only the first portion that comes over. Pale and limpid. Exposure to light discolours it.

b. (Ph. Bor.) Fetid animal oil distilled in a sand-bath, and the product rectified with 4 times its volume of water. White, limpid, fragrant. Light discolours it.

Oil of Birch. *Syn.* OLEUM BETULÆ, L. *Prep.* From the inner bark of the birch, by heating it in an earthen pot with a hole in the bottom, to allow the oil to flow through into another jar sunk in the ground and luted to it. Thick, balsamic, fragrant. Used chiefly to dress russia leather.

Oil of Box-wood. *Syn.* OLEUM BUXI, O. E. EMPYREUMATICUM (Ph. L. 1746), L. *Prep.* From box-wood sawdust. Reputed resolvent; anodyne, antispasmodic, and diaphoretic.—*Dose*, 5 to 20 drops; in convulsions, epilepsy, gonorrhœa, &c. Externally, in toothache, &c.

Oil of Bricks. *Syn.* OLEUM LATERITIUM (Ph. L. 1746), L. *Prep.* From olive oil, mixed with brick-dust, and distilled; or from hot bricks steeped in olive oil, then broken to pieces, and distilled.

Oil of Bricks (Factitious). *Syn.* OLEUM LATERITIUM FACTITIUM, L. *Prep.* From linseed oil, 1 lb.; oil of turpentine, ½ lb.; oil of bones or of hartshorn and Barbadoes tar, of each, 1 oz.; simply stirred well together. This is generally substituted for the preceding in the shops.

Oil of Cade. *Syn.* OLEUM CADINUM, L.; HUILE DE CADE, Fr. *Prep.* From the *Juniperus oxycedrus*, or Languedoc juniper. Used as oil of tar, which is commonly sold for it.

Oil of Gua'iacum. *Syn.* OLEUM GUAIACI, O. G. EMPYREUMATICUM, L. *Prep.* From guaiacum shavings or raspings. Reputed balsamic, pectoral, and resolvent.

Oil of Harts'horn. Bone oil and rectified bone oil are commonly sold for it, but are inferior to it. See OIL, ANIMAL (*above*).

Oil, Paper. *Syn.* RAG OIL, PYROTHONIDÆ; OLEUM CHARTÆ, L. *Prep.* On the small scale, by burning paper on a cold tin plate, and collecting the oil; on the large scale, by the destructive distillation of paper or linen rags. In baldness, toothache, earache, &c.

Oil of Soot. *Syn.* OLEUM FULIGINIS (Ph. L. 1746), L. *Prep.* From wood soot. Fetid; reputed antispasmodic and nervine.

Oil of Tar. *Syn.* SPIRIT OF T.; OLEUM PINI,

O. P. RUBRUM, O. TEDÆ, O. PICIS LIQUIDÆ, L. *Prep.* By simple distillation from wood-tar. Reddish and strong-scented. By one or more rectifications it becomes colourless and limpid. It soon gets thick. Used in ringworm and several other skin diseases, made into an ointment with lard. It is poisonous if swallowed in large doses.

Oil of Tobac'co (Empyreumatic). *Syn.* OLEUM TABACI EMPYREUMATICUM (Ph. U. S.), L. *Prep.* From tobacco, in coarse powder, gradually heated in a green glass retort to dull redness, and kept at that temperature as long as any oil passes over; the oily portion is then separated from the water in the receiver, and kept for use. Highly narcotic and poisonous.

Oil of Wax. *Syn.* OLEUM CERÆ, L. *Prep.* From beeswax and sand distilled together; the product is rectified once or oftener. Reputed diuretic.—*Dose*, 3 to 6 drops.

OILS (Fixed). *Syn.* FAT OILS, UNCTUOUS O.; OLEA FIXA, O. EXPRESSA, L.; HUILES GRASSES, Fr. The fixed oils are compounds of carbon, hydrogen, and oxygen, obtained from the organic kingdoms, and characterised by their insipidity, unctuosity, insolubility in water, and being lighter than that fluid. Olive oil, which is obtained from the vegetable kingdom, and spermaceti oil, which is obtained from the animal kingdom, may be taken as types of the rest.

The fixed oils are chiefly found in the fruit and seeds of plants, and in thin membranous cells, forming what is called the adipose tissue, in the bodies of animals. According to their consistence, they may be classed into 'OILS,' 'BUTTERS,' and 'TALLOWS.'

Prop., &c. Among the best known properties of the fixed oils are—the permanent stain they give to paper, which they render translucid; their non-volatility at the ordinary temperature of the atmosphere, or at that of boiling water, or, indeed, at any temperature insufficient for their decomposition; their constantly floating on the surface of water when added to it; and, lastly, their inability to mix with that fluid. Some of them, as palm oil and cocoa-nut oil, are solid at ordinary temperatures; but the majority are fluid, unless they have been considerably cooled, when they separate into two portions—the one solid, consisting chiefly of stearin, or some analogous substance; and the other liquid, consisting chiefly of olein. Nearly all of them, when exposed to the air, absorb oxygen rapidly, and either gradually harden or become rancid and nauseous. From the first are selected the 'drying oils,' used by painters; the last are used as food, in cookery, and for machinery, lamps, &c. All of these oils, when heated to their boiling-points (500°—600° F.), suffer decomposition, yielding various hydrocarbons; and when suddenly exposed to a red heat they furnish a gaseous product (oil-gas), which has been employed for illumination. It is owing to this property of oil and liquid fats that candles and lamps give their light. With the caustic alkalies and water the fixed oils unite to form soap. When some of these oils are absorbed by porous bodies, and thus expose a vastly increased surface to the air, they absorb oxygen with such rapidity as to generate a considerable degree of heat.

Paper, tow, cotton, wool, straw, shavings, &c., moistened with oil, and left in a heap, freely exposed to the air or sun, often spontaneously inflame. In this way many extensive fires have arisen. The above is more particularly the case with linseed, rape, nut, and olive oil. The first, made into a paste with manganese, rapidly becomes hot, and ultimately inflames spontaneously.

The specific gravities of the fixed oils range between 0·865 and 0·970.

Prep. The fixed oils, except where otherwise directed, are obtained from the bruised or ground fruit or seed, by means of powerful pressure, in screw or hydraulic presses, and are then either allowed to clarify themselves by subsidence or are filtered. Both methods are frequently applied to the same oil. In some cases the impurities are removed by ebullition with water, and subsequent separation of the pure oil. Heat is frequently employed to increase the liquidity of the oil, and thus lessen the difficulty of its expulsion from the mass. With this object the bruised mass, placed in bags, is commonly exposed to the heat of steam, and then pressed between heated plates of metal. This is always necessary with the ' butyraceous oils.'

Another method is by boiling the bruised seed in water, and skimming off the oil as it rises to the surface. This is the plan adopted for castor oil in the West Indies.

In a few cases, for medicinal purposes, the bruised mass is mixed with half its weight, or an equal weight, of alcohol or ether, and after 24 hours' digestion the whole is submitted to pressure, and the alcohol or ether removed by distillation at a gentle heat. The first menstruum is commonly employed for croton oil on the Continent ; the second for that of ergot of rye.

Purif. Several methods are adopted for refining or purifying the fixed oils, among which are the following :

1. The oil is violently agitated along with 1½% to 2% of concentrated sulphuric acid, when it assumes a greenish colour, and, after about a fortnight's standing, deposits much colouring matter, becomes paler, and burns with greater brilliancy, particularly if well washed with steam or hot water, and clarified by standing or by filtration. This answers well for most recently expressed vegetable oils. It also greatly improves most of the fish oils.

2. A modification of the last method is to well mix the acid with the oil, then to blow steam through the mixture for some time, and afterwards to proceed as before.

3. FISH OIL (WHALE, SEAL, &c.) is purified by—

a. Violently agitating it with boiling water, or by placing it in a deep vessel with perforated bottom, through which high-pressure steam is forced for some time ; it is afterwards clarified by repose, and filtered through coarse charcoal.

b. The oil is violently agitated with a boiling hot and strong solution of oak-bark, to remove albumen and gelatin, and next with high-pressure steam and hot water ; it is, lastly, dried and filtered.

c. The oil, gently heated, is stirred for some time with about 1% of good chloride of lime, previously made into a milk by trituration with water ; about 1½% of oil of vitriol, diluted with 20 times its weight of water, is then added, and the agitation renewed and maintained for at least 2 hours ; it is, lastly, well washed with steam or hot water.

d. Mr. Davidson treats the oil first with a strong solution of tan, next with water and chloride of lime, then with dilute sulphuric acid, and lastly, with hot water.

e. Mr. Dunn's method, which is very effective, and admirable on account of its simplicity, is to heat the oil by steam to from 180° to 200° F., and then to force a current of air of corresponding temperature through it, until it is sufficiently bleached and deodorised ; it is, lastly, either at once filtered, or is previously washed with steam, or hot water.

f. Another method is to violently agitate the oil for some time with very strong brine, or with a mixed solution of blue vitriol and common salt, and then either to allow it to clarify by standing or filtering it through freshly burnt charcoal.

4. ALMOND, CASTOR, LINSEED, NUT, OLIVE, RAPE, and some other vegetable oils are readily bleached by either of the following processes :

a. Exposure in glass bottles to the sun's rays, in some suitable position, open to the south-east and south. This is the method employed by druggists and oilmen to whiten their castor and linseed oils. 14 to 21 days' exposure to the sun in clear weather during summer is usually sufficient for castor oil when contained in 2 to 4-quart pale green glass bottles (preferably the former), and capped with white gallipots inverted over them. The oil is filtered before exposing it to the light, as, if only in a slight degree opaque, it does not bleach well. Almond and olive oil are, when thus treated, apt to acquire a slight sulphurous smell ; but this may be removed by filtration through a little animal charcoal, or, still better, by washing the oil with hot water.

b. Heat the oils in a wooden, tinned, or well-glazed earthen vessel along with some dry 'filtering powder' (1 to 2 lbs. per gall.), with agitation for some time, and then filter them in the usual manner through an oil-bag. In this way the West-end perfumers prepare their ' WHITE ALMOND OIL' (OLEUM AMYGDALE ALBUM) and their ' WHITE OLIVE OIL' (OLEUM OLIVÆ ALBUM). Formerly, freshly burnt animal charcoal was used for this purpose, and is still so employed by some houses.

5. Mr Bancroft refines OILS FOR MACHINERY AND LUBRICATING PURPOSES generally, by agitating them with a lye of caustic soda of the sp. gr. 1·2. A sufficient quantity is known to have been added when, after standing, a portion begins to settle down clear at the bottom. About 4% to 8% is commonly required for lard oil and olive oil. After 24 hours' standing the clear supernatant oil is decanted from the soapy sediment, and filtered.

6. Not only the oils above referred to, but all other oils and fats, may be rendered perfectly colourless by the use of a little chromic acid ; or, what is the same, by a mixture of a solution of

bichromate of potash and sufficient sulphuric, hydrochloric, or nitric acid to combine with the alkali, and thus liberate the chromic acid.

7. PALM OIL and COCOA-NUT OIL are generally refined and bleached by either chromic acid or chlorine, or by heat.

a. The 'butyraceous oil' is liquefied by heat in a wooden vessel, and 7% to 9% of good chloride of lime, previously made into a smooth cream with water, is added, and the whole assiduously stirred until the ingredients appear united; the mixture is then allowed to cool and solidify. It is next cut up into small lumps, which are exposed to a free current of air for 2, 3, or even 4 weeks; these are melted in a wooden vessel heated by high-pressure steam circulating through leaden pipes, or in a cast-iron boiler lined with lead, and an equal weight of oil of vitriol (diluted with about 20 times its weight of water) is poured in, and the whole gently boiled until the oil is discoloured and runs clear; the fire is then moderated, and the whole allowed to settle; lastly, the fire is removed, and the oil is left to cool very slowly.

b. The process with chromic acid has been already noticed, but is more fully explained *below*.

c. The oil, heated to the temperature of about 250° F., is exposed to the action of high-pressure steam, which is continuously 'blown' through it for 10 or 12 hours, or even longer. The process is greatly facilitated by the introduction of some chromic acid.

8. Mr Watt's methods of purifying fats and oils are very effective, more especially for those intended for illumination. They are as follows:

a. (For FISH OILS.) Each ton is boiled for ⅓ an hour with caustic soda, ½ lb., previously made into a weak lye with water; or steam is blown through the mixture for a like period; oil of vitriol, ⅓ lb., diluted with 6 times its weight of water, is next added, the whole again boiled for 15 minutes, and allowed to settle for an hour or longer, when the clear oil is run off from the water and sediment into the bleaching tubs; here solution of bichromate of potash, 4 lbs., in oil of vitriol, 2 lbs., previously diluted with water, q. s., together with a little nitric acid and some oxalic acid, are added, and after thorough admixture of the whole, by blowing steam through it, strong nitric acid, 1 lb., diluted with water, 1 quart, is poured in, and the boiling continued for ½ an hour longer; a small quantity of naphtha or rectified spirit of turpentine is then mixed in, and the oil is, finally, well washed with hot water, and left to settle.

b. (For PALM OIL.) The oil is melted by the heat of steam, and after it has settled and cooled down to about 130° F., is carefully decanted from the water and sediment into the steaming tubs; here a mixture of a saturated solution of bichromate of potash, 25 lbs., and oil of vitriol, 8 or 9 lbs., is added, and after thorough admixture, hydrochloric acid, 50 lbs., is poured in; the whole is then constantly stirred until it acquires a uniform greenish colour, or is sufficiently decoloured, a little more of the bleaching materials being added if the latter is not the case, after which it is allowed to stand

for half an hour to settle; it is next run into a wooden vat, where it is washed, &c., as before.

c. (For VEGETABLE OILS.) These are treated with a solution of chromic acid, or with a solution of bichromate of potash, or some mineral acid, as noticed at No. 6. For COLZA, LINSEED, MUSTARD, NUT, and RAPE OIL, a little hydrochloric acid is added; but for ALMOND, CASTOR, OLIVE OIL, and POPPY OIL no such addition is required.

9. RANCID OILS and FATS are recovered by boiling them for about 15 minutes with a little water and calcined magnesia, or by filtering them through freshly burnt charcoal.

Obs. In reference to the above processes, it may be useful to remark that chlorine, the common bleacher and deodoriser of other substances, cannot be well employed directly in the purification of oils, as certain chemical reactions occur when these substances are brought together, which increase the colour instead of removing it, and are often otherwise injurious. The same remarks apply to the use of the 'chlorides,' which frequently fails in unskilful hands, and is, indeed, of questionable utility, except, perhaps, in the case of palm oil. Even charcoal exerts little of its usual energy on the oils, and whilst it removes or lessens their offensive odour, sometimes increases their colour. The addition of 1% or 2% of very pure and recently rectified naphtha or oil of turpentine to lamp oil is a real improvement, since it increases its combustibility and its illuminative power.

OILS FOR MEDICAL PURPOSES, as CASTOR OIL, COD-LIVER OIL, &c., must not be subjected to any process beyond mere clarification by subsidence, filtration though Canton flannel or porous paper, or, at the utmost, washing with warm water, as otherwise their active and valuable properties, if not wholly removed, will be considerably lessened. See FILTRATION.

Pur. The fixed oils vary greatly in their value, and hence the constant inducement to the unprincipled dealer to adulterate the more expensive ones with those of a similar character, but of an inferior kind or grade. Various methods are adopted to detect these frauds, among which the following are the most valuable of those capable of general application. Others referring to individual oils will be found under the respective heads.

1. (From the odour.) The method of applying this test is to heat a few drops of the oil under examination in a small porcelain, platinum, or silver spoon or capsule (a watch-glass answers well), and to carefully compare the odour evolved with that arising from a known pure sample of the same kind and quality of the oil similarly treated. The odour of the two, when each is pure, is precisely alike, and immediately suggests the plant or animal from which it has been obtained. The presence of LINSEED, NUT, RAPE, SEAL, TRAIN, or WHALE OIL is thus readily detected, and the imperfections of the sample, even if pure, rendered much more perceptible.

2. (From the density.) a. According to M. Penot, every oil supposed to come from the same plant, or the same animal, has its own particular density, which, at the same temperature, never

deviates more than a few thousands. To apply this test, the relative density or sp. gr. of the sample must be determined. This may be done by means of a 1000-gr. bottle or an ordinary 'AREOMETER;' or, more conveniently, by an 'ELAIOMETER' or 'OLEOMETER,' constructed and graduated for the purpose. ' Fischer's ELAIOMETER' or 'OIL-BALANCE' is much employed on the Continent for this purpose, and is a very useful instrument. On the large scale, the weight of an accurately measured imperial gallon of the oil may be taken.

b. M. Lauret, an eminent Parisian chemist, a short time since observed that the variations of the density of an oil from adulteration are rendeted much more apparent when it is examined in a heated state. To render this discovery practically available, he plunges an 'elaïometer,' graduated for the given temperature, into a small tin cylinder nearly filled with the oil, and then places this in a vessel containing boiling water; as soon as the whole has acquired a uniform temperature, he observes the point on the scale of the instrument at which it floats. This point for—

Colza oil is	0°
Fish oil	83°
Poppy oil	124°
Hemp-seed oil	136°
Linseed oil	210°

c. By employing a sp. gr. bottle or small glass globe, fitted with a stopper in which is hermetically fixed a capillary tube of about 8 or 9 inches in length, we may apply the above principle of M. Lauret with the greatest accuracy. This little apparatus is filled with the oil, and then immersed in boiling water for a sufficient length of time for it to acquire that temperature; it is then removed and weighed. The smallest adulteration is, it is said, in this way immediately detected.

When the density of the given sample has been taken and the name of the oil used to adulterate it is known, the quantity of the latter present may be approximately determined from the specific gravities by the common method of alligation. See MIXTURES, ARITHMETIC OF.

3. (Sulphuric acid test.) *a.* Heidenreich found that when oil of vitriol is mixed with the fatty oils, very intense chemical action commences, the temperature of the mixture rises, and the mass becomes coloured. These changes are sufficiently varied in the case of the different oils to furnish us with the means of identifying many of them, and of determining their purity. The method of Heidenreich is to lay a plate of white glass over a sheet of white paper; on the glass he places 10 or 15 drops of oil, and then adds to it a small drop of concentrated sulphuric acid. The appearances which follow differ with the character of the fatty oil examined, whether the acid is allowed to act on the oil undisturbed, or the two are stirred together with a glass rod. In many cases, as with tallow oil, a peculiar odour as well as a change of colour is developed, and a further means of detection supplied. Heidenreich has minutely described these reactions, which, for the most part, closely resemble those given in the table below. It is necessary, however, in order to ensure great accuracy, to compare the effect of the reagent on the sample with those which it produces on pure oil of the same kind and character under precisely similar circumstances.

b. Penot, who has followed up the researches of Heidenreich with considerable success, recommends the employment of 20 drops of oil, instead of only 10 or 15; and the use of a small capsule of white porcelain, instead of a plate of glass. He also employs a saturated solution of bichromate of potash in sulphuric acid, which he uses in the same proportion as before; but in this case the oil and the reagent are always stirred together.

The observations of M. Penot have been repeated in many cases by Mr Cooley, and the results, with additions, and rearranged, are given in the table.

" By perusing this table," writes M. Penot, " it will be observed that the same oil does not, under all circumstances, yield precisely similar results with the same reagent. This depends on the place of growth, the age and the manner of pressing. If, however, any oil be examined comparatively with a perfectly pure one, the proof of adulteration may be rendered if not certain, at least probable, by noting the difference. Thus I obtained, by adding 1 part of either whale, train, or linseed oil, or oleic acid, or 10 parts of rapeseed oil, the following results :

NAME OF OIL.	REAGENTS.		
	Sulphuric Acid.		Solution of Bichromate of Potash.
	Not Stirred.	*Stirred.*	*Stirred.*
Rape oil with whale-train oil	More red ground than with rape oil	Brownish-olive coloured	Small reddish lumps on a grey ground.
Rape oil with linseed oil	No perceptible difference from the rape oil	Olive coloured	Small and more numerous red lumps on a very dark green ground.
Rape oil with olein or oleic acid	No perceptible difference from the rape oil	Greenish brown	Small brownish lumps on an olive-coloured ground.

" The adulteration being ascertained as far as is possible, the oil is then tested by endeavouring to discover the adulterating oil, either by reagents or by its odour when gently heated, as before described. This having been found out, small quantities of the suspected oil are added to a perfectly pure oil of the kind under examination. Every mixture is then tested by the reagents until precisely similar results are obtained as those yielded by the oil under examination. Thus the proportions of the two mixed oils will be discovered by approximation."

4. (From increase of temperature.) M. Maumené proposed the increase of temperature arising from the admixture of mono-hydrated sulphuric acid (oil of vitriol, sp. gr. 1·845) with the fatty oils as a test of their purity, but a sufficient number of observations have not yet been made to furnish data for a general application of this method. According to MM. Faisst and Knauss, who have re-examined the subject, the following are the results when 15 grms. of oil are mixed with 5 grms. of the acid:

		Rise of Temperature.
Almond oil	. . .	72·5°
Olive „	. . .	68·0°
Poppy „	. . .	127·0°
Rape or colza oil	. . .	100·0°
Linseed oil (with Nordhausen or fuming acid only)	. . .	133·0°

The above method is less liable to error when a larger quantity of the substances are used.

5. The presence of FISH OIL in the vegetable oils may be readily detected by passing a stream of chlorine through them; the pure vegetable oils are not materially altered, but a mixture of the two turns dark brown or black.

6. Mr Coleman states that the presence of mineral oils in animal or vegetable oils can be easily detected by two characteristic tests: (1) The fluorescent properties they impart to all animal or vegetable oils. (2) The strongly marked aromatic burning flavour they communicate to mixtures containing them. The first-mentioned property is brought out by smearing a metallic surface, such as tin plate or steel, with the oil, and then viewing it at different angles in the open air or sunlight. Mr Coleman suggests that, in examining a dark-coloured oil, it may first be necessary to refine the sample by successive treatments with concentrated sulphuric acid and weak soda solution or lime-water; so small a quantity as 2½% may then be detected by the bluish colour noticed on viewing the oil at certain angles and by tasting it.

The absence of resin oil must also be proved. Nitric acid is said to be a good test, as the colour developed is much greater than in pure oils. Sometimes it may be detected by the smell. The presence of 10% of resin or mineral oil in non-drying oils delays their solidification with the nitrate of mercury test ('Journal of Applied Chemistry,' Dec., 1874).

7. Miss Kate Crane ('American Journal of Pharmacy,' iv, 406) states that the cohesion figures of oils may be usefully employed as tests of the identity and purity of the oils. She says, "A number of experiments on this subject have led me to the conclusion that a little patient practice will teach the eye of the observer in a short time to detect the characteristic differences of the figures. To make these perfect it is necessary to observe the time in forming, for at different periods some varieties form figures very like; but with this precaution each is entirely characteristic.

"It is essential that the dish used, &c., be perfectly clean, so that when filled with water no dust or lint floats upon the surface, as this materially interferes with the perfect formation of the figure.

"A single drop is let fall from a burette or glass rod held steadily above the water upon the centre of the surface. The experiments made with fixed oils are as follows:—*Poppy-seed oil* spreads instantly to a large figure, retaining an entire outline, and for a few seconds the surface is unbroken, except the bare intimation of a beaded edge.

"In a few moments little holes appear round the edge, and soon the whole surface is broken in like manner; these increase in size very slowly. In fifteen minutes the edge begins to open, forming indentations, which gradually work their way across the figure. As they increase in length these begin to curve, and in three quarters of an hour have doubled themselves two or three times.

"*Cod-liver oil* spreads in a large film; a little way from the edge a row of small holes appears, and in a minute or two the surface is covered with them; these gradually enlarge, assuming irregular shapes, soon separated by branching lines.

"*Cod-liver oil* with *lard oil* spreads very like the former, but in a few moments the edge opens, and the film separates partly across; in a moment one of the projecting points begins to curve itself towards the centre, bending more and more until it forms a coil; meanwhile a few holes have appeared, which spread irregularly, throwing out projecting points.

"*Castor oil* spreads instantly, the edge remaining entire; openings appear quickly in thirty seconds, and increase gradually, but unevenly, those nearer the edge being larger, and lengthening out irregularly as they spread. The figure lasts some time.

"*Castor* with a little *lard oil* makes a smaller figure, and not nearly so much broken; in five minutes the holes open into each other, and the figure breaks up from the edge.

"A mixture of *castor* and *poppy-seed oils* spreads to form a lacework border, but smooths out to an entire edge soon, and within a few seconds openings appear. The figure, in size and general appearance, is more like castor oil alone, but the holes spread less uniformly in a given time, a few being larger, but the greater portion much smaller. In fifteen minutes there is a general tendency to break up.

"*Castor* with a little *croton oil* throws out a spray, which in a few moments unites into a thin film. The spray, as it spreads, draws out the inner portion into radiate points, which open into a beautiful network, the centre cohering closely.

"*Croton oil* throws out, in spreading, a fine spray in advance of the more closely cohering portion, which follows quickly. The outer edge breaks up unevenly into little indentations, the border of the inside portion being quite broken, but gradually becomes nearly entire. The surface, too, has openings, which increase quite rapidly in size, the outer ones being much the larger. In the final breaking up, before the holes open one into another, the outlines are beautifully fringed."

8. "Spontaneous combustion ensues when a handful of cotton waste is embued with oil and placed in an air-bath at 130° to 200° F. Boiled linseed oil required 1½ hours; raw linseed oil, 4 hours; lard oil, 4 hours; refined rape, about 9 hours" (J. J. Coleman, 'Journal of Applied Chemistry,' Dec., 1874).

Mr. Gellatly found that an admixture of 20% of mineral oil retarded combustion, and 50% prevented it completely.

9. M. Burstyn (Ure's 'Dictionary of Arts,' &c.), believing that the value of a fatty oil as a lubricant depends on the amount of acid it contains, has invented a method for volumetrically determining the acidity. The process is as follows:—A tall cylindrical vessel provided with a ground-glass stopper, and having two marks on it to indicate respectively 100 c.c. and 200 c.c., is filled to the first mark with the oil to be tested, and to the second mark with 88 to 90 per cent. alcohol. The cylinder is then closed and well shaken. Equal quantities other than 100 c.c. can be employed without any other change in the process. After standing two or three hours the oil settles, and the clear alcohol, which contains in solution the free acids and a little of the oil, rises to the top perfectly clear; 25 c.c. of the clear solution is taken from the top by means of a pipette. A few drops of alcoholic turmeric is added, and the acid determined by means of a standard solution of potash, as in acidimetry. The change from yellow to brownish red takes place with great sharpness when neutralisation is reached.

The number of cubic centimètres of potash employed, multiplied by 4, gives the quantity of normal solution requisite to neutralise the free acid in 100 c.c. of oil. As it is not an individual acid, but a variable mixture of acids, it is not possible to calculate the percentage of acids present. These numbers, however, may be taken as degrees of acidity. For instance, an oil of 3° of acidity is one which contains enough free acid to neutralise 3 c.c. of normal alkali.

If we assume that oleic acid predominates, which in most cases is the fact, 1° of acidity corresponds to 0·28% by weight of oleic acid. The olive oil of commerce has an acidity ranging from 0·4° to 12°. The first passes as very fine, and is called free from acid or salad oil, while the latter is known by smell and taste as very rancid. Oil that has 4°—6° of acidity has been found to answer very well as a lubricator.

The relation which exists between the degree of acidity and any injurious effect upon metals is shown by the following experiments:—4 shallow vessels of sheet brass, having a surface of 40 sq. cm. each at the bottom, were filled to the depth of 2 mm. with oils of different acidity, and exposed to the air at the ordinary temperature. The vessels were soon more or less covered with green fatty salts, and the oil too acquired a green colour. Oil and vessel No. 1 were the only ones in which no change could be perceived. At the end of three days the vessels were cleaned with ether and weighed. The following table shows the amounts of action:

Vessel No. 1, filled with oil of 0·8° lost 0·03 gr.
　　" No. 2,　　　" 　　　4·6° " 0·22 "
　　" No. 3,　　　" 　　　7·8° " 0·36 "
　　" No. 4,　　　" 　　　8·8° " 0·04 "

9. *Hübl's Iodine Absorption Method* (Dingler's 'Polytechnisches Journal,' t. 253, p. 281). Make a solution of 25 grms. iodine and 30 grms. mercuric chloride, each in ½ litre of 95% alcohol; mix the two solutions, stand 12 hours, and titrate against standard sodium hyposulphite. Weigh out from 0·2 to 0·8 gr. of the oil; dissolve in 10 c.c. chloroform; add excess of the iodine solution; allow to stand two hours, and then determine the excess of iodine by the sodium hyposulphite solution. The number of grammes of iodine taken up by 100 grms. of the oil is then found. This number is compared with a standard number. The following is a list of numbers for pure substances:

	Iodine. Hübl.	Degrees. Möller.
Shark-liver oil	—	268·2
Manhaden oil	—	170·8
Porpoise oil	—	131·2
Seal oil	—	103·4
Linseed oil	158·0	175·7
Walnut oil	143·0	—
Poppy-seed oil	136·0	—
Cotton-seed oil	106·0	107·9
Rape-seed oil	100·0	99·4
Almond oil	98·4	—
Castor oil	84·4	—
Olive oil	82·8	81·3
Lard oil	59·0	47·2
Palm oil	51·5	48·6
Tallow oil	40·0	—
Cocoa-nut oil	8·9	6·8
Oleic acid	—	86·2
Olein	—	82·3

The quantity of metal destroyed, in equal times and under equal conditions, increases with the acidity of the oil.

The table on p. 1119, by Mr Bottome, describes the most striking physical properties of some of the principal fixed oils.

*** The following are the principal fixed oils met with in commerce, or which are objects of interest or utility:

Oil of Al'monds. *Syn.* OLEUM AMYGDALÆ (B. P., Ph. L.), O. AMYGDALARUM (Ph. D.), O. AMYGDALIA COMMUNIS (Ph. E.), L. *Prep.* "Bruise the fresh almonds in a stone mortar, then put them into a hempen sack, and express the oil, without heat" (Ph. E.). The oil of almonds B. P. and of commerce is obtained from either the bitter or sweet almond, but chiefly from the first, on account of their less value, and the marc being employed in the manufacture of essential oil.

Prop., &c. Oil of almonds is bland, demulcent, emollient, and nutritious; possesses a purely oleaginous taste, and is one of the most agreeable of the fixed oils; when taken in quantity it is mildly laxative; it is little affected by cold, and congeals with difficulty; is soluble in 35 parts of cold and 6 parts of boiling alcohol; ether dissolves it freely. Sp. gr. 0·915 to 0·918.—*Av. prod.* Sweet almonds, 46%; bitter a., 41%.

Pur. It is extensively adulterated with poppy,

Name of Oil	Specific Gravity at 15°C., Water=1000.	Combustibility. Grms. consumed per hour in a Lamp with Wick.	Freezing-point in Degrees Centigrade.	Colour.	Taste.	Smell.	Limpidity. Time (in seconds) required to trickle a given distance.	Drying Power.
Plum kernel	0·9187	68	− 9	Brownish yellow	Amygdalaceous	Very slight	98	Non-drying.
Rape seed	0·9198	80	− 4	Yellow	Nauseous	Nauseous	159	Non-drying.
Colza	0·9136	40	− 9·25	Yellow	Nauseous	Nauseous	162	Non-drying.
Cabbage seed	0·9139	48·5	− 8	Yellow	Nauseous	Nauseous	148	Non-drying.
White mustard	0·9142	39·8	− 16·25	Light yellow	Pleasant	Very slight	157	Non-drying.
Ground nut	0·9163	?	− 8	Pale greenish yellow	Like peas	Very slight	Not tested	Non-drying.
Black mustard	0·9170	25	− 17·5	Yellow	Pleasant	Peculiar	141	Non-drying.
Olive	0·9176	63	− 6[1]	Yellow	Sweet	None	195	Non-drying.
Sweet almond	0·9180	52·8	− 21·5	Amber	Agreeable	None	150	Non-drying.
Horseradish seed	0·9187	43	− 16·25	Yellowish brown	Pleasant	None	143	Non-drying.
Grape seed	0·9203	37	− 17·5	Gold yellow	Sweet	None	99	Dries slowly.
Beech nut	0·9236	50	− 15	Amber	Very sweet	None	158	Non-drying.
Pumpkin	0·9231	43	− 15	Pale brown yellow	Sweet	Disagreeable	185	Dries slowly.
Land cress	0·9240	42	− 10	Brownish yellow	Acrid	None	108	Dries slowly.
Hazel nut	0·9243	58·4	− 18[2]	Amber	Sweet	None	166	Non-drying.
Poppy	0·9243	31	− 18	Pale yellow	Peculiar	Peculiar	123	Drying.
Camelina	0·9252	34	− 18	Yellowish	Flat	None	119	Drying.
Walnut	0·9260	45	− 16	Light yellow	Sweet	None	88	Dries slowly.
Sunflower	0·9263	51·8	− 27·5	Colourless	Disagreeable	Disagreeable	114	Drying.
Hemp seed	0·9276	46	− 27·5	Dark greenish yellow	Strong	None	87	Drying.
Cotton seed	0·9316	?	− 2·5	Reddish brown	Pleasant, slightly piquant	Disagreeable	Not tested	Drying.
Sesame	0·9320	?	− 5	Bright yellow		None	Not tested	Non-drying.
Linseed	0·9347	88	− 27	Dark greenish yellow	Strong	Disagreeable	88	Drying.
Wood	0·9358	44	Not noted	Green	Unpleasant	None	73	Drying.
Spindle	0·9360	61	− 20	Reddish brown	Acrid	Slight	148	Non-drying.
Castor	0·9611	47	− 18	Colourless	Sickly	Very slight	1880	Dries slowly.

[1] Though these oils do not become quite solid till the point indicated is reached, yet they begin to become grainy at + 4° C.

[2] Once solidified, this oil does not liquefy until the temperature reaches 2° C.

TABLE *giving the reactions of various* OILS *with* SULPHURIC ACID *and with a saturated solution of* BICHROMATE OF POTASH *in sulphuric acid.* Rearranged from M. PENOT's table, with additions, by Mr COOLEY.

⁎ *The result indicated is obtained in each case by the action of* one drop *of the* REAGENT *on twenty drops of* OIL.

NAME OF OIL.	REAGENTS.		
	Sulphuric Acid.		Saturated Solution of Bichromate of Potash in Sulphuric Acid.
	Not stirred.	*Stirred.*	*Stirred.*
Almond oil . . .	Greenfinch yellow, with orange spots	Dirty green	Yellowish small lumps.
Castor oil . . .	Yellow, with slight spots	Little reaction	Slightly green.
Cod-liver oil (*fine sample of pale oil*)	Deep purple in the centre, rapidly turning brown, whilst violet or purple clouds or streaks spread out towards the circumference, the colour of which remains unaltered for some minutes after the central portion has turned nearly black	Deep purple, passing into purple brown, reddish brown, and gradually deepening to an intense brown, approaching black	Reddish - brown clots, changing to a clear bright green.
Hemp-seed oil . .	Small brown lumps or clots on a yellow ground	Greenish brown	Small yellow lumps or clots on a green ground.
Linseed oil (*from the Upper Rhine*)	Dark reddish brown	Brown small lumps on a grey ground	Small brown lumps on an almost colourless ground.
,, (*from Paris*)	Reddish brown, less dark coloured	Brown clots on a green ground	Small brown lumps on a green ground.
,, (*English*)	Chestnut brown	Brown clots on a greenish-grey ground	Brown lumps on a greenish-grey ground.
Liver-train oil . .	Dark red	Dark red	Dark red.
Madia-sativa oil .	Slightly reddish brown underneath a thin greyish film	Olive green	Light brown small lumps on an olive-coloured ground.
Black-mustard oil .	Bluish green	Olive green	Olive brown.
Neat's-foot oil . .	Yellow slight spots	Dirty brown	Brown spots on a brownish ground.
Nut oil (*recent*) .	Yellowish brown	Clotted, dark brown	Small brown lumps or clots.
,, (*one year old*)	Yellow	Dirty brown, less dark coloured	Small brown lumps.
,, (*still older*) .	Orange yellow	Dirty brown	Small brownish lumps.
Olein, oleic acid, lard, or tallow oil	Reddish spots, with reddish circles	Reddish brown	Bright chestnut colour.
Olive oil . . .	Yellow	Dirty brown	Olive brown.
,, (*another sample*)	Orange yellow	Brownish grey	Brown.
,, (*from fermented olives*)	Orange yellow	Brownish grey	Brown.
Poppy oil (*recent cold drawn*)	Yellow spots	Olive brown	Small yellow lumps on a white ground.
,, (*recent, as-pressed with slight heat*)	Greenish-yellow spots	Olive brown, turning more on the green	Small yellow lumps on a greenish-grey ground.
,, (*one year old, expressed with heat*)	Greenish spots	Olive green	Small yellow lumps on a green ground.
Rape or colza oil (*trade*) .	Yellowish-brown streaks, surrounded by a bluish-green ring	Brownish, turning on the olive green	Small yellow lumps on a green ground.
,, (*recent*) .	Green	Bluish green	Small yellow lumps on a green ground.
,, (*one year old*)	Green	Bluish green	Yellow lumps on a brighter green ground.
,, (*one year old, rough hot-pressed*)	Green	Olive green	Small yellow lumps, more numerous, on an olive-green ground.
Whale-train oil .	Small reddish lumps on a brownish ground	Resembles wine lees	Small, bright, chestnut-coloured lumps on a brown ground.

nut, and teel oil, and not unfrequently with refined rape or colza oil, and lately with nitro-benzene, q. v. If adulterated, the sp. gr., boiling-point, solubility, taste, and odour will be altered (see *above*).

Detection of Nitro-benzene in Bitter Almond Oil. Warm a specimen with manganese dioxide and sulphuric acid. Nitro-benzene does not lose its odour, but after a time smells like oil of cinnamon, while bitter almond oil develops a disagreeable odour, which soon disappears, leaving the oil odourless (Morpurgo, in ' Chem. Centralblatt,' 1890, i, 879).

Oil of Bay. *Prep.* 1. (EXPRESSED O. OF B.; OLEUM LAURI, O. LAURINUM, L.) By expression from either fresh or dried bayberries, as CASTOR OIL. Limpid ; insipid.

2. (By decoction : BUTTER OF B.; OLEUM LAURI NOBILIS, O. L. VERUM, L.) From the berries, by boiling them in water, and skimming

off the oil. Green, buttery; chiefly imported from Italy. Used in bruises, sprains, rheumatism, deafness, &c.—*Prod.*, 20%.

Oil of Beech. *Syn.* OLEUM FAGI, L. *Prep.* From the nuts of *Fagus sylvatica*, Linn., or beech mast. Clear; keeps well; when washed with hot water, it is used for salads, and burnt in lamps. Sp. gr. 0·9235.—*Prod.*, 16%.

Oil of Belladon'na. *Syn.* OLEUM BELLADONNÆ SEMINUM, O. E. BACCÆ, L. *Prep.* From the seeds or berries of *Atropa belladonna*, or deadly nightshade. Yellow; insipid. Used for lamps in Swabia and Wurtemberg, and as an application to bruises. The marc is poisonous. It freezes at 34° F. Sp. gr. 0·9250.

Oil of Ben. *Syn.* OIL OF BEHEN; OLEUM BALATINUM, L. *Prep.* From the seeds of *Moringa pterygosperma* (ben nuts). Scentless, colourless; keeps long without growing rank; by standing, it separates into two parts, one of which freezes with difficulty, and is hence much used in perfumery. Sp. gr. 0·912 to 0·915 at 15·5° C.

Oil of Benne Seed. See OIL OF GINGELLY.

Oil of Brazil Nuts. *Syn.* OLEUM BERTHOL-LETIÆ, L. *Prep.* From the kernels of the fruit of *Bertholletia excelsa*, or Brazil nuts. An oil of a bright amber colour, congealing at 24° F. Sp. gr. 0·917. It has been used as a substitute for olive oil in plasters and ointments.

Oil of Caca'o. *Syn.* BUTTER OF C.; OLEUM CACAO CONCRETUM, BUTYRUM CACAO, L. *Prep.* From the seeds of *Theobroma cacao*, or chocolate nuts, gently heated over the fire, and then decorticated, and pressed between hot iron plates. Sp. gr. 0·945 to 0·952. It has lately been used in the adulteration of American lard (*Allen*), and in Germany it is refined and used instead of butter.

Oil, Castor. *Syn.* RICINI OLEUM (B. P.), OLEUM CASTOREI, O. RICINI (Ph. L., E., and D.), L. "The oil prepared by heat, or by pressure, from the seeds of *Ricinis communis*, Linn." (Ph. L.), the *Palma Christi*, or Mexican oil-bush.

The best castor oil (COLD-DRAWN CASTOR OIL; OLEUM RICINI SINE IGNE) is prepared by pressing the shelled and crushed fruit (seed) in hemp bags in an hydraulic press, and heating the oil thus obtained along with water in well-tinned vessels, until the water boils and the albumen and gum separate as a scum; this is carefully removed, and the oil, as soon as it has become cold, is filtered through Canton flannel. The commoner kinds are of a darker colour, and are prepared by gently heating the crushed seeds, and pressing them whilst hot. Another method, sometimes adopted, is to put the crushed seed into loose bags, to boil these in water, and to skim off the floating oil.

Prop. It is the most viscid of all the fixed oils; when pure it mixes in all proportions with alcohol and ether, and also dissolves, to a certain extent, in rectified spirit, but a portion of the oil separates on standing. Camphor and benzoic acid increase its solubility in spirit. By long exposure to the air it becomes rancid, thick, and is ultimately transformed into a transparent yellow mass; light hastens these changes. Exposed to cold, a solid, white crystalline fat separates from the liquid portion, and when cooled to 0° it con-

geals into a yellow transparent mass, which does not again liquefy until the temperature rises to about 18° F. Sp. gr. 0·9611 to 0·9612, at 60°; 0·9690, at 55° (Saussure); 0·9575, at 77° (Saussure).—*Prod.*, 38% to 46%.

Pur., &c. Pure castor oil rotates a ray of polarised light (x) = + 12·15°. This behaviour may be used as a test of its purity. Castor oil is sometimes adulterated with rape oil or with lard oil, a fraud which may be detected by its diminished density; and, when the added oil exceeds 33%, by its insolubility in its own weight of alcohol of 0·820. In many cases croton oil is added to increase the purgative quality of the mixture. A compound of this kind is vended in gelatine capsules under the name of 'CONCENTRATED CASTOR OIL,' the use of which is fraught with danger. "I have heard of several cases in which very violent and dangerous effects were produced by these capsules" (*Pereira*). The best is imported from the East Indies in tin canisters. The oil obtained from the seeds of *Ricinus viridis*, Willd., or lamp-oil seeds, is often mixed with or sold for castor oil.

Uses, &c. Castor oil is an exceedingly useful mild purgative, particularly when abdominal irritation should be avoided, as in inflammations of the stomach and bowels, pregnancy, surgical operations, &c.—*Dose*, 2 fl. dr. to 1 fl. oz.

Oil, Cocoa-nut. *Syn.* COCOA-NUT BUTTER; OLEUM COCOIS NUCIFERÆ, L. By expression from the kernels of the cocoa-nut, or fruit of the *Cocos nucifera*.

Oil, Cod-liver. *Syn.* MORRHUÆ OLEUM (B. P.), COD-FISH OIL; OLEUM JECORIS ASELLI, O. GADI, O. E. MORRHUÆ, OLEUM MORRHUÆ (Ph. L.), L. "The oil extracted from the fresh liver of the *Gadus morrhua* by a steam heat or waterbath not exceeding 180° F. Yellow." The oil prepared from the liver of *Gadus morrhua*, Linn." (Ph. L.).

The common cod-liver oil of commerce drains from the livers of the cod-fish when freely exposed to the sun, and just beginning to putrefy. It is dark coloured, strong, and nauseous, and is now chiefly employed in this country by the curriers, for dressing leather. It is the 'OLEUM JECORIS ASELLI FUSCUM' of Continental writers. Formerly, the less fetid varieties of this crude oil, after the impurities were removed, either by subsidence or filtration, constituted the only cod-liver oil used in medicine. As its employment as a remedy increased, its revolting flavour, and its great tendency to permanently disorder the stomach and bowels, were found to be serious obstacles to its general use. It was observed that the oil as it exists in the liver of the cod is bland and nearly colourless, and has only a slight fishy, but not a disagreeable flavour. The attention of persons interested was therefore immediately directed to the subject, and improved methods of obtaining the oil were adopted on the large scale.

The methods of preparing cod-liver oil are noticed in another part of this work, but we think it advisable to add to these a description of the plan adopted by Messrs Charles Fox and Co., of Newfoundland, Scarborough, and London, the well-known manufacturers and importers of cod-liver oil:

"The Newfoundland fisheries are entirely carried on in small boats, principally by the hand-line system, and quite close to the shore. The boats go out early in the morning, and return about four o'clock in the afternoon. The fish, on landing, are handed over to a 'fish-room keeper,' whose duty it is to split and open the fish, and to deposit the livers in small tubs holding 17 or 18 galls. each. The tubs are soon afterwards collected from the different 'fish-rooms,' and conveyed to the manufactory. The livers are here thrown into tubs filled with clean cold water, and, after being well washed and jerked over, are placed on galvanised iron-wire sieves to drain. They are next put into covered steam-jacket-pans, and submitted to a gentle heat for about three quarters of an hour, after which the steam is turned off, cold air again admitted, and the whole allowed to repose for a short time, during which the livers subside, and the oil separates and floats on the top. The oil is then skimmed off into tin vessels, and passed through flannel strainers into tubs, where it is left to subside for about 24 hours. From these the purer upper portion of oil is run into a very deep, galvanised-iron cistern, and again left to clarify itself by defecation for a few days. It is now further refined by carefully passing it through clean and very stout mole-skin filters, under pressure. The transparent filtered oil is received in a clean, galvanised-iron cistern containing a pump, from which the casks are filled for exportation. The latter, before being filled, are carefully seasoned and cleaned, to prevent their imparting either flavour or colour to the pure oil."

The superiority of the oil prepared as above consists essentially in every part of the process of extraction being performed whilst the livers are fresh, and in no chemical means being adopted to give the oil a factitious appearance. Its natural pale colour is thus preserved from contamination, and its medicinal virtues maintained intact.

Much of the light brown oil of commerce is obtained from *Gadus callarius* (the dorse), *G. carbonarius* (the coal-fish), and *G. pollachius* (the pollack).

Pur., &c. "The finest oil," remarks Dr Pereira, "is that which is most devoid of colour, odour, and flavour. The oil, as contained in the cells of the fresh liver, is nearly colourless, and the brownish colour possessed by ordinary cod-liver oil is due to colouring matters derived from the decomposition (putrefying) of hepatic tissues and fluids, or from the action of the air on the oil (age). Chemical analysis lends no support to the opinion, at one time entertained, that the brown oil was superior, as a therapeutic agent, to the pale oil. On the other hand, the disgusting odour and flavour and nauseating qualities of the brown oil preclude its repeated use. Moreover, there is reason to suspect that, if patients could conquer their aversion to it, its free use, like that of other rancid and empyreumatic fats, would disturb the digestive functions, and be attended with injurious effects" ('Elem. Mat. Med.,' &c., 3rd edit., iii, 2239).

Tests. Among the tests of purity, that generally relied on is known as the 'sulphuric acid test.' See OILS (Fixed): *Purity.* DORSE OIL and other FISH OIL, sold as 'LIGHT-BROWN COD-LIVER OIL,' exhibit with this test much lighter reactions, which closely resemble those of LIVER-TRAIN and WHALE-TRAIN OIL.

Boudard adds fuming nitric acid to a portion of oil; if pure it becomes rose-coloured, but this effect is interfered with by the presence of other fish oils.

To detect the presence of combined iodine, upon which, by some, the therapeutic value of cod-liver oil is thought to depend, the sample is saponified by trituration with a little caustic potash and hot water, the resulting soap cautiously incinerated, the ashes digested with water, and the whole thrown on a filter. The usual tests for iodine may be then applied to the filtered liquid.

The presence of iodine artificially added is best detected by agitating the oil with a little rectified spirit, and then testing this last for iodine. Or, a little solution of starch and a few drops of sulphuric or nitric acid may be at once added to the oil, when a blue colour will be developed if iodine, or an iodide, has been mixed with the sample.

The sp. gr. of the pale oil is 0·9231 to 0·9238; of the light brown oil, 0·924 to 0·9245; of the dark brown oil, 0·929 to 0·9315. The density is, however, apt to vary a little with the quantity of moisture present.

Uses, &c. Cod-liver oil is a most valuable medicine in a great variety of diseases, more especially in glandular indurations and enlargements, scrofula, phthisis, rheumatism, gout, certain cutaneous diseases, amenorrhœa, chlorosis, caries, rickets, &c. To be of service, however, its use must be continued for several weeks, and the oil must be recent.—*Dose,* 1 to 2 table-spoonfuls, 3 or 4 times daily, or oftener.

Oil, Col'za. From the seeds of *Brassica campestris,* var. *oleifera,* or *colza de printemps,* a variety of *Brassica campestris,* Linn. It may be regarded as a superior sort of rape oil. Burns well in lamps, especially after being refined. Used also for lubricating purposes and in the manufacture of india-rubber. Sp. gr. 0·9136, at 60° F.—*Prod.,* 39%. The term 'colza oil' is commonly applied to ordinary refined rape.

Oil, Cotton-seed. *Syn.* OLEUM GOSSYPII SEMINUM, L. From the seeds of *Gossypium barbadense.*

The yield is about 10%. Sp. gr. crude oil =0·928 to 0·93, refined oil = 0·92 to 0·923 (*Gilmour*). It possesses slight drying properties.

Uses. It is used for paints, lamps, lubrication, soap-making, and especially in the adulteration of olive, linseed, sperm, and lard oils. In the 'blown' condition it has recently been much used for adulterating American lard.

Test. One of the simplest is that proposed by Leone ('Gazetta,' 19, 355). It is applicable to the detection of cotton-seed oil in fats (lard and olive oil. Add a few c.c. of a 1% solution of silver nitrate in alcohol, acidified with 0·5% of nitric acid, to a few c.c. of the fat, and heat on the water-bath for 5 or 6 minutes. If the adulterant be present, a brownish-yellow ring is formed at the surface of separation of the two liquids. The reaction is sufficiently delicate to detect the presence of 5% of cotton-seed oil in lard. In the case of olive oil heat must be applied for 10—12 minutes. With other oils a white ring is observed, which changes on prolonged heating to green.

Oil, Croton. *Syn.* CROTONIS OLEUM (B. P.), OLEUM CROTONIS (Ph. E.), O. TIGLII (Ph. L.), L. From the shelled seeds of *Croton tiglium* or Molucca grains. Imported chiefly from the East Indies. It is one of the most powerful cathartics known, and acts when either swallowed or merely placed in the mouth. Externally, it is a rubefacient and counter-irritant, often causing a crop of painful pustules, like tartar emetic.—*Dose,* 1 to 2 drops, on sugar; in apoplexy, &c. It is poisonous in larger doses. Sp. gr. 0·942 to 0·953. —*Prod.* Unshelled seeds, 22% to 25%; shelled do., 32% to 35%.

Pure croton oil is soluble in an equal volume of alcohol of 0·796, but in 2 or 3 days about 96% of the oil separates. In France the marc is exhausted with alcohol, and the oil thus obtained is added to that previously obtained from the same seeds by expression. The East Indian oil (OLEUM CROTONIS EXOTICUM) is usually of a pale yellow; that pressed in England (O. CROTONIS ANGLICANUM) is much darker.

Oil of Cu'cumber. *Syn.* OLEUM CUCURBITÆ, L. From the seeds of *Cucurbita pepo* or squash, and the *C. melopepo* or pumpkin. Pale, used sometimes as a soothing application to piles. Sp. gr. 0·92.

Oil of Eggs. *Syn.* OLEUM OVI, O. O. VITELLI, O. OVORUM, L. From the yolks of eggs, gently heated until they coagulate and the moisture has evaporated, and then pressed or broken up, digested in boiling rectified spirit, the tincture filtered whilst hot, and the spirit distilled off. Bland; emollient. The common plan is to fry the yolks hard; but the oil is then darker coloured and stronger. The P. Cod. orders them to be exhausted with ether by displacement. Formerly used to 'kill' quicksilver, and still held in great esteem in some parts of England for sore nipples and excoriations.—*Prod.* 10 to 12 eggs yield 1 oz. See MIXED OILS.

Oil of Garden Cress. *Syn.* OLEUM LEPIDII SATIVI, L. From the seed. Drying. Sp. gr. 0·924.—*Prod.*, 54%.

Oil of Gar'den Spurge. *Syn.* OLEUM LATHYRIS, O. EUPHORBIÆ L., L. From the seeds of *Euphorbia lathyris*, or garden spurge. Cathartic.—*Dose,* 3 to 8 drops. Sp. gr. 0·9281. —*Prod.*, 30% to 41%. Croton oil mixed with 6 times its weight of nut or rape oil is usually sold for it.

Oil of Gingel'ly. *Syn.* OIL OF SESAMUM or SESAME, BENNE OIL, TEEL O., TIL O.; OLEUM SESAMI, L. From the seeds of *Sesamum orientale*, Willd., or gingelly. Pale; bland. Used in salads, paints, &c.; also to adulterate oil of almonds.—*Prod.*; 46%.

Oil, Gourd. See OIL OF CUCUMBER.

Oil of Ground Nuts. From the nuts of *Arachis hypogæa.* Colourless. Sp. gr. 0·916. Used in making soap, as a lubricant, and in the adulteration of expensive oils.

Oil of Gurgun. See BALSAM, GURGUN,

Oil of Hemp. *Syn.* OLEUM CANNABIS, L. From the seed of *Cannabis sativa,* Linn., or common hemp. Mawkish. Sometimes used for frying, but chiefly for paints, soaps, &c. Freely soluble in boiling alcohol; does not thicken until cooled to 5° F. Sp. gr. 0·9276.—*Prod.*, 18% to 24%.

Oil of Jatro'pha. *Syn.* OIL OF WILD CASTOR SEEDS; OLEUM JATROPHÆ, L. From the seeds of *Jatropha purgans.* Somewhat resembles CROTON OIL. Used for lamps in the East Indies.

Oil, Kundah. *Syn.* TALLICOONAH O.; OLEUM TOULOUCOUNÆ, L. From the fruit of *Carapa Touloucouna.* Rancid, nauseous, vermifuge, rubefacient, emetic, and purgative. Chiefly used in lamps.

Oil, Lard. *Syn.* TALLOW O., CRUDE OLEIN, C. OLEIC ACID; OLEUM ADIPIS, L. By separating the olein of lard from the stearin by means of boiling alcohol. Only applicable where spirit is cheap. The product is, however, excellent. The crude oleic acid, or lard oil of commerce, is chiefly obtained as a secondary product in the manufacture of stearin. It is purified by agitation with sulphuric acid, and subsequently by steaming it, or washing it with hot water. Burns well in lamps if the wick-tube is kept cool. Superior to olive oil for greasing wool. Sp. gr. 0·9008.

Oil, Linseed. *Syn.* OLEUM LINI (B. P., Ph. L., E. & D.), L. *Prep.* 1. (COLD-DRAWN LINSEED OIL; OLEUM LINI SINE IGNE.) From the seed of *Linum usitatissimum,* Linn., or common flax, bruised or crushed, and then ground and expressed without heat. Pale, insipid, viscous; does not keep so well as the next.—*Prod.*, 17% to 22%.

2. (ORDINARY LINSEED OIL.) As the last, but employing a steam heat of about 200° F. Amber-coloured; less viscous than the last; congeals at 2°—4° F.; soluble in 5 parts of boiling and 40 parts of cold alcohol. Both are drying and cathartic.—*Dose,* 1 to 2 oz.; in piles, &c. Chiefly used in making paints, printing inks, varnishes, floor-cloths, &c. Sp. gr. varies from 0·93—0·935. —*Prod.*, 22% to 27%.

3. (BOILED LINSEED OIL.) See OILS (Drying).

Oil of Mace (Expressed). See OIL OF NUT-MEG (Expressed).

Oil of Male Fern. See EXTRACT OF MALE FERN.

Oil of Mustard. *Syn.* OLEUM SINAPIS, L. *Prep.* 1. (OIL OF WHITE MUSTARD.) From *Sinapis alba,* or white mustard, but chiefly from *Sinapis arvensis, S. chinensis, S. dichotoma, S. glauca, S. ramosa,* and *S. tori.* Sweet. Used for the table. Sp. gr. 0·9142.—*Prod.*, 36%.

2. (OIL OF BLACK MUSTARD; OLEUM SINAPIS NIGRI, L.) From the 'hulls' of black mustard seed. Viscid, stimulant. Used in rheumatism. Sp. gr. 09168 to 09170. See OILS (Volatile).

3. (OIL OF WILD MUSTARD; OLEUM RAPHANI, L.) From the seed of *Raphanus raphanistrum,* Linn., or jointed charlock, or wild mustard. —*Prod.*, 30%.

Oil, Neat's-foot. *Syn.* NERVE OIL, TROTTER O.; OLEUM BUBULUM, O. NERVINUM, AXUNGIA PEDUM TAURI, L. Yellow or colourless. From neat's-feet by toiling them in water, and skimming off the oil. Does not thicken by age. Used to soften leather, to clean fire-arms, as a 'low-temperature' lubricant, &c.

Oil, Nut. *Syn.* HAZEL-NUT O.; OLEUM NUCIS, O. CORYLI, L. From the kernels of *Corylus avellana*, Linn., or hazel-nut tree. Pale, mild-tasted, drying; superior to linseed oil for paints and varnishes. It is employed to adulterate oil of almonds, &c. Walnut oil is also frequently sold for nut oil. Sp. gr. 0·9260.—*Prod.*, 63% (*Ure*).

Oil of Nut'meg (Expressed). *Syn.* EXPRESSED OIL OF MACE, BUTTER OF M.; OLEUM MYRISTICÆ (CONCRETUM—Ph. L.), MYRISTICÆ ADEPS (Ph. E.), M. BUTYRUM, O. MYRISTICÆ EXPRESSUM (H. P.), O. MOSCHATÆ, O. NUCISTÆ, L. "The concrete oil expressed from the seed of *Myristica officinalis*," Linn. (Ph. L.), or common nutmeg. The nutmegs are beaten to a paste, enclosed in a bag, exposed to a vapour of hot water, and then pressed between heated iron plates. Orange-coloured, fragrant, spicy; butyraceous, or solid. It is a mixture of the fixed and volatile oils of the nutmeg. When discoloured and hardened by age, it is called 'BANDA SOAP' (OL. MACIS IN MASSIS). When pure it is soluble in 4 parts of boiling alcohol and in 2 parts of ether. It has been used in rheumatism and palsy, but is now chiefly employed for its odour and aromatic qualities. From the East Indies.—*Prod.*, 17% to 28%.

Oil, Olive. *Syn.* SALAD OIL, SWEET O.; OLIVÆ OLEUM (B. P.), OLEUM OLIVARUM, O. OLIVÆ (Ph. L., E., & D.), L. The "oil expressed from the fruit" of "*Olea europæa*, Linn." (Ph. L.), or common olive. Five different methods are employed to obtain the oil from the fruit:

1. (VIRGIN OIL; O. O. VIRGINEUM, L.; HUILE VIERGE, Fr.) From olives, carefully garbled, either spontaneously or only by slight pressure, in the cold. That yielded by the pericarp of the fruit is the finest.

2. (Ordinary 'FINE OIL.') This is obtained by either pressing the olives, previously crushed and mixed with boiling water, or by pressing, at a gentle heat, the olives from which the virgin oil has been obtained. The above processes furnish the finer salad oils of commerce. The cake which is left is called 'GRIGNON.'

3. (SECOND QUALITY.) By allowing the bruised fruit to ferment before pressing it. Yellow, darker than the preceding, but mild and sweet-tasted. Much used for the table.

4. ('GORGON.') By fermenting and boiling the pressed cake or marc in water, and skimming off the oil. Inferior.

5. OIL OF THE INFERNAL REGIONS (OLEUM OMPHACINUM) is a very inferior quality of oil, which is skimmed off the surface of the water in the reservoirs in which the waste water which has been used in the above operations is received, and allowed to settle. The last two are chiefly used for lamps, and in soap-making, &c.

Of the principal varieties of olive oil known in commerce, and distinguished by the place of their production, 'PROVENCE OIL' is the most esteemed; 'FLORENCE OIL' and 'LUCCA OIL' are also of very fine quality; 'GENOA OIL' comes next, and then

'GALLIPOLI OIL,' which forms the mass of what is used in England; 'SICILY OIL,' which has a slightly resinous flavour, is very inferior; and 'SPANISH OIL' is the worst imported.

Prop., &c. Olive oil is a nearly inodorous, pale greenish-yellow, unctuous fluid, with a purely oleaginous taste, peculiarly grateful to the palate of those who relish oil. It does not suffer active decomposition at a heat not exceeding 600° F., and when cooled to 32° it congeals into a granular solid mass. It is very slightly soluble in alcohol, but its solubility is increased by admixture with castor oil. It is soluble in 1½ parts of ether. When pure it has little tendency to become rancid. Sp. gr. 0·914 to 0·918, at 60° F.—*Prod.*, 32%, of which 21% is furnished by the pericarp, and the remainder, which is inferior, by the seed and woody matter of the fruit.

Pur. Olive oil, with the exception of that of almonds, being the most costly of the ordinary fixed oils of commerce, is, consequently, the one most subject to adulteration. Nut, poppy, rape, and lard oil are those most commonly used for this purpose. The addition of any other oil to olive oil renders it far less agreeable to the palate, and, by increasing its tendency to rancidity, much more likely to offend and derange the stomach and bowels of those who consume it. When pure, and also fresh, olive oil is most wholesome as an article of food or as a condiment.

The detection of the sophistication of salad oil is a matter of no great difficulty. The palate of the connoisseur will readily perceive the slightest variation in the quality of his favourite condiment. Other methods, however, of a more accurate and certain description, and of more general application, are adopted. Amongst these, in addition to those mentioned above, are the following:

a. When pure olive oil is shaken in a phial, only half filled, the 'bead' or bubbles rapidly disappear; but if the sample has been mixed with poppy or other oil the bubbles continue longer before they burst.

b. Olive oil begins to solidify at 32°—50° F., and is completely solidified when a small bottle containing it is surrounded by ice; but when mixed with poppy oil it remains partly liquid, even when the latter forms only 1-4th of the mass; if more than 1-3rd of poppy oil is present it does not solidify at all, unless cooled much below the freezing-point of water.

c. (Ph. E.) When pure olive oil is "carefully mixed with 1-12th part of its volume of a solution of 4 oz. of mercury in 8 fl. oz. 6 dr. of nitric acid (sp. gr. 1·5), it becomes in 3 or 4 hours like a firm fat, without any separation of liquid oil."

d. M. Pontet recommends the mercurial solution to be made by dissolving 6 parts of mercury in 7½ parts of nitric acid (sp. gr. 1·35), without heat; of this solution he adds 1 part to every 48 parts of the oil, and well shakes the mixture every 30 minutes, until it begins to solidify. This it does after about 7 hours in summer and 4 or 5 hours in winter, and when the oil is pure it will have formed in 24 hours a hard mass. The other edible oils do not furnish a hard mass with nitrate of mercury. The solidity of the mass is exactly in proportion to the quantity of foreign oil present. When the sophistication is equal to 1-8th of the whole a distinct liquid layer separates; when the mixture contains half its volume of an inferior oil, one half only of the mixture becomes solid, and the other half continues liquid. A temperature of about 90° F. is the best to cause the oil and coagulum to separate perfectly from each other. When the oil has been adulterated with animal oil the mixture solidifies in about 5 hours; but in this case the coagulum consists of the animal oil, whilst the olive oil floats on the surface, and may be decanted for further examination. This coagulum, on being heated, exhales the well-known odour of rancid fat or melted tallow.

e. The following is Dr Langlies' process for proving that olive oil does not contain any seed oil:

He mixes 3 grms. of the oil to be tested with 1 grm. of nitric acid (3 parts acid to 1 part water) in a test-tube, or a small stoppered flask, and heats the liquid in a water-bath. If the oil is pure the mixture becomes clearer, and takes a yellow colour like purified oil; if it is adulterated with seed oil it acquires the same transparency as the pure oil, but becomes red. With 5% of seed oil the reddish colouring is characteristic; with 10% it is decided. The reaction does not require more than from 15 to 20 minutes. The colouring of the oils lasts for 3 days. A large number of other tests have been proposed during the last few years, but all the above have been tested and proved.

Uses, &c. The dietetical uses of olive oil are well known. In Spain and Italy it is commonly employed as a substitute for butter. It is highly nutritious, but is digested with difficulty by some persons, and hence should be avoided by the dyspeptic. Like almond oil, it is occasionally employed as a laxative and vermifuge, and is, perhaps, one of the mildest known. In *pharmacy* it is extensively employed in the preparation of cerates, liniments, ointments, and plasters.—*Dose.* For an adult, ½ to 1 wine-glassful as a mild aperient; for an infant, ½ to 1 teaspoonful, mixed up with an equal quantity of honey, syrup of roses, or syrup of violets. The white fibrous sediment which forms in the recently expressed oil is the 'AMURCA' of Pliny, and was formerly highly esteemed in medicine.

Oil, Olive, Droppings. *Syn.* SWEET-OIL D. The 'foots' or 'deposits,' and the 'drippings' of the casks, cisterns, and utensils. Used for machinery, making soap, &c.

Oil, Olive (Oxygenated). *Syn.* OLEUM OLIVÆ OXYGENATUM (Ph. Batav.), L. Olive oil, 16 oz., is placed in a receiver surrounded with ice or very cold water, and chlorine is slowly transmitted through it for several days, or until it becomes thick and viscid, after which it is well washed with warm water.

Oil, Palm. *Syn.* PALM BUTTER; OLEUM PALMÆ, L. From the fruit of *Elais guineensis* and *E. melanococca*, the Guinea oil palms. Orange or red coloured; butyraceous or solid; smells of violets; unchanged by alkalies; bleached by sunlight, age, exposure, chlorine, chromic acid, and oil of vitriol; melting-point varies between 76° and 95° F. Sp. gr. 0·968. Demulcent. Used to colour and scent ointments, pomades, &c.; but chiefly to make soap and candles.

Oil, Palm Nut. *Syn.* PALM-NUT KERNEL O. Extracted from the kernels of the palm fruit. Primrose-yellow. Used in soap-making.

Oil, Pi'ney. *Syn.* PINEY TALLOW, P. DAMMAR, P. RESIN. From the seeds of *Vateria indica*, Linn., or peenoe tree. Resinous flavoured, fragrant; made into candles. Sp. gr. 0·926.

Oil, Pop'py. *Syn.* OLEUM PAPAVERIS, L.; OLIETTE, HUILE BLANCHE, Fr. From the seeds of *Papaver somniferum*, Linn. or white poppy. Sweet; pale; dries and keeps well. Used for salads, paints, and soaps, and to adulterate almond oil. It does not freeze until cooled to 0° F. Sp. gr. 0·913 to 0·9240.—*Prod.*, 48% to 54%.

Oil of Pumpkin. *Syn.* OLEUM CUCURBITÆ, L. Expressed from the seeds of the pumpkin; a soothing application to piles.

Oil, Rape. *Syn.* COLZA OIL, BROWN O.; OLEUM RAPÆ, L. From the seed of *Brassica napus*, Linn. (cole or rape), and from *B. campestris*, Linn. (wild navew or rape). Glutinous; buttery at 25° F. Dries slowly; makes soft soaps and good ointments, but bad plasters. Smokes much in burning, unless well refined. Sp. gr. 0·9130 to 0·915.—*Prod.*, 32%.

OIL, REFINED or **PALE RAPE (OLEUM RAPÆ REFINUM, OL. R. ALBUM),** is prepared from crude rape oil by agitating it with about 2% of oil of vitriol, previously diluted with about twice its weight of water, and, after 10 or 12 days' repose, decanting the clear oil, and filtering it through Canton flannel or felt. The quality is improved by washing it with hot water or steam before filtration. Used for lamps, blacking, and machinery; also extensively employed to adulterate both almond and olive oil. It forms the common 'SWEET OIL' of the oilmen and druggists.

Oil, Seal. *Syn.* OLEUM PHOCÆ, L. From the hood seal and harp seal, and other species of PHOCIDÆ. PALE SEAL OIL is that which drains from the blubber before putrefaction commences, and forms about 60% of the whole quantity of oil obtained. It is very clear, free from smell, and, when recently prepared, not unpleasant in its taste. REFINED SEAL OIL is the last, washed and filtered. Ranks close after sperm oil. BROWN or DARK SEAL OIL is that which subsequently drains from the putrid mass. It is very strongly scented and nauseous, and smokes in burning. Used for lamps and dressing leather. A full-grown seal yields 8 to 12 galls. of oil; a small one, 4 to 5 galls.

Oil of Ses'amum. *Syn.* OIL OF GINGELLY (*above*).

Oil, Shark-liver. Prepared from the livers of various species of shark. Used in tanneries to adulterate cod-liver oil. The lightest of the fixed oils. Sp. gr. 0·865 to 0·876.

Oil, Skate. *Syn.* OLEUM RAIÆ, L. From the livers of *Raia batis*, Linn., or common skate, as cod-liver oil; also from *Raia rhinobatus*, or white skate, and *Raia clavata*, or thornback. Often mixed with cod-liver oil.

Oil, Spermace'ti. *Syn.* SPERM OIL; OLEUM CETACEI, L. From the 'head matter' of *Physeter* *macrocephalus*, or spermaceti whale; a species once common in all the principal seas, but now chiefly confined to the Southern Ocean. It is very limpid, smells little, and burns well; and has long been reputed the best oil for lamps and machinery, as it does not thicken by age or friction. The solid portion is refined for candle-making. It is frequently adulterated with refined seal oil. Sp. gr. 0·875.

Oil, Sun'flower. *Syn.* OLEUM HELIANTHI, L. From the seeds of *Helianthus annuus* and *H. perennis.* Clear, pale yellow, tasteless; thickens at 60° F. Used for salads and lamps. Sp. gr. 0·926. —*Prod.,* 15%.

Oil, Teel. See OIL, GINGELLY.

Oil, Tobac'co-seed. *Syn.* OLEUM TABACI (EX-PRESSUM), L. From the seeds of *Nicotiana tabacum,* Linn., or true tobacco plant. Pale; dries well; equal to nut oil. Its production has recently been carried on with considerable success in some parts of Russia. Sp. gr. 0·923.

Oil of Touloucou'na. See OIL, KUNDAH.

Oil, Train. See OIL, WHALE.

Oil, Turkey-red. The soluble product obtained by the intersection of various oils with sulphuric acid.—*Prep.* Mix castor oil with sulphuric acid diluted with ¼ its bulk of water; stand. Wash with salt and water, and saponify with caustic alkali. — *Use.* As a mordant in Turkey-red dyeing.

Oil, Walnut. *Syn.* OLEUM JUGLANDIS, O. NUCIS J., L. From the kernels of the nuts of *Juglans regia,* Linn., or common walnut tree. Soon gets rank; dries well. Used in paints, and occasionally in plasters. When ' cold drawn' and washed it is sometimes eaten with salad. Sp. gr. about 0·926.—*Prod.,* 48% to 52%.

Oil of Wax. *Syn.* BUTTER OF WAX; OLEUM CERÆ, L. From beeswax, by quick distillation in a close vessel. Butyraceous. By rectification along with quicklime it yields a liquid oil.

Oil, Whale. *Syn.* TRAIN OIL, WHALE TRAIN O.; OLEUM BALENÆ, O. CETI, L. From the blubber of the *Balæna mysticetus,* Linn., or the common or Greenland whale, by heat. Coarse; stinking. SOUTHERN WHALE OIL is the best. Used for lamps, machinery, &c. Sp. gr. 0·923. —*Prod.* per fish, about 1¼ tons for each foot of bone.

Oil of Wheat. *Syn.* OLEUM TRICITI, L. From bruised Colne wheat, with heat. In chilblains, ringworm, and several other skin diseases.

Oil of Wine-seed. *Syn.* GRAPE-STONE OIL; OLEUM VITIS VINIFERÆ LAPIDUM, L. From the seeds of grapes, separated from the marc. Pale yellow, bland, emollient. Used for salads and lamps. Sp. gr. 0·918 to 0·92.—*Prod.,* 14% to 18%.

*** The numbers given above, under ' products,' unless when otherwise stated, refer to the respective fruits, kernels, nuts, seeds, &c., deprived of their husks, pods, shells, and every other portion destitute of oil.

OILS (Medicated). *Syn.* OLEA COCTA, O. INFUSA, O. MEDICATA, L. These are prepared by infusion or decoction. The bruised ingredients are either simply digested in 2 to 4 times their weight of olive oil for some days, or they are gently boiled in it until they become dry or crisp, great care being taken that the heat towards the end of the process is not greater than that of boiling water. As soon as the process is complete, the oil is allowed to drain from the ingredients, which are then (if necessary) submitted to the action of the press. The product is commonly run through flannel or a hair sieve whilst still warm, after which it is allowed to repose for a week or ten days, when the clear portion is decanted from the dregs. The green or recent plants are usually employed for this purpose, but, in many cases, the dried plants, reduced to powder, and digested for 6 or 8 hours in the oil, at the heat of hot water, with frequent agitation, yield a much more valuable product. They are nearly all employed as external applications only.

*** The following are the most important preparations of this class :

Oil of Adder's-tongue. *Syn.* OLEUM OPHIO-GLOSSI, L. From the herb, as OIL OF BELLA-DONNA. A popular vulnerary.

Oil of Ants. *Syn.* OLEUM FORMICARUM. Digest 4 oz. of ants in 16 oz. (by weight) of olive oil with a gentle heat, and strain.

Oil of Bal'sam Apple. *Syn.* OLEUM BALSA-MINE. *Prep.* Balsam apple (deprived of seeds), 1 oz.; oil of almonds, 4 oz.; digest and strain.

Oil of Belladon'na. *Syn.* OLEUM BELLA-DONNÆ (P. Cod.), L. *Prep.* From the fresh leaves, bruised, 1 part; olive oil, 4 parts; digested together at a gentle heat until the moisture is evaporated; the oil is then strained off with pressure, and filtered.

Oil of Cantha'rides. *Syn.* OLEUM CANTHA-RIDIS, O. CANTHARIDIBUS, L. *Prep.* (P. Cod. 1839.) From Spanish flies (powdered), 1 part; olive oil, 8 parts; as OIL OF BELLADONNA. Stimulant and rubefacient. Used as a dressing to indolent sores, blisters, &c.; and in dropsy, rheumatism, gout, &c. OIL OF THE OIL BEETLE (*Meloe proscarabæus,* Linn.) is prepared in a similar manner.

Oil of Cap'sicum. *Syn.* OLEUM CAPSICI, L. *Prep.* (Dr Turnbull.) From powdered capsicum or Cayenne pepper, 4 oz.; olive oil, 1 pint; digested together for 6 hours, with heat, and strained. Stimulant; rubefacient in colic, cholera, &c.

Oil of Cham'omile. *Syn.* OLEUM ANTHEMIDIS, OL. CHAMEMELI, L. From the dried flowers (rubbed to pieces), 1 part; olive oil, 8 parts; digested together, with heat, for 6 hours. Stimulant, emollient, and vermifuge.

Oil of Col'ocynth. *Syn.* OLEUM COLOCYNTHI-DIS, L. From the pulp, as OIL OF CHAMOMILE. Diuretic. In dropsy, neuralgia, rheumatism, worms, &c.

Oil of Earth'worms. *Syn.* OLEUM LUMBRI-CORUM. (E. Ph. 1744.) Washed earthworms, ½ lb.; olive oil, 1½ pints; white wine, ½ pint. Boil gently till the wine is consumed, and press and strain.

Oil of Elder Flowers. *Syn.* WHITE OIL OF ELDER; OLEUM SAMBUCI ALBUM, O. SAMBU-CINUM (P. Cod.), L. *Prep.* From the flowers, as OIL OF CHAMOMILE. Emollient and discussive.

Oil of Elder Leaves. *Syn.* GREEN OIL, GREEN OIL OF ELDER, OIL OF SWALLOWS; OLEUM VI-RIDE, O. SAMBUCI VIRIDE, L. *Prep.* 1. Green elder leaves, 1 lb.; olive oil, 1 quart; boil gently until the leaves are crisp, press out the oil, and again heat it till it turns green.

2. As before, but by maceration, at a heat under 212° F. More odorous than the last.

3. Elder leaves, 1 cwt.; linseed oil, 3 cwt.; as No. 1.

Obs. The last form is the one usually employed on the large scale. It is generally coloured with verdigris, ¼ lb. to the cwt., just before putting it into the casks, and whilst still warm; as, without great skill and a very large quantity of leaves, the deep green colour so much admired by the ignorant cannot be given to it. The oil is got from the leaves by allowing them to drain in the pan or boiler (with a cock at the bottom), kept well heated. Emollient; in great repute among the vulgar as a liniment, in a variety of affections.

Oil of Fen'ugreek. *Syn.* OLEUM FŒNUGRÆCI, L. *Prep.* (P. Cod.) From the seeds, as OIL OF CANTHARIDES or of CHAMOMILE. Emollient and resolvent.

Oil of Fox'glove. *Syn.* OLEUM DIGITALIS, L. *Prep.* (P. Cod.) From the fresh leaves, as OIL OF BELLADONNA. Used as an application to chronic ulcers and indurations, painful swellings, &c. As usually met with it is nearly inert.

Oil of Garden Night'shade. *Syn.* OLEUM SOLANI, L. *Prep.* (P. Cod.) From the leaves, as OIL OF BELLADONNA. Anodyne and discussive.

Oil of Gar'lic. *Syn.* OLEUM ALLII INFUSUM, L. From garlic, as OIL OF BELLADONNA. Used as a liniment in deafness, diarrhœa, infantile convulsions, palsy, rheumatism, &c.

Oil, Green. *Syn.* OLEUM VIRIDI, L. From bay leaves, origanum, rue, sea-wormwood, and elder leaves, of each, 2¼ oz.; olive oil, 1 quart; as OIL OF ELDER. Detergent, stimulant, and resolvent. Green oil of elder is now usually sold for it.

Oil of Hem'lock. *Syn.* OLEUM CONII, L. *Prep.* (P. Cod.) As OIL OF BELLADONNA. Anodyne and emollient; in painful ulcers, glandular tumours, &c.

Oil of Hen'bane. *Syn.* OLEUM HYOSCYAMI, L. *Prep.* (P. Cod.) As OIL OF BELLADONNA. Used as the last, in various painful local affections.

Oil, Iodised, Marshall's. *Syn.* OLEUM IODATUM. *Prep.* Oil of almonds, 15 parts; iodine, 1 part. Triturate and digest till dissolved.

Oil of Ju"niper (by Infusion). *Syn.* OLEUM JUNIPERI INFUSUM, L. From the crushed berries, as OIL OF BELLADONNA. Diuretic and vulnerary; in frictions, &c.

Oil of Lil'ies. *Syn.* OLEUM LILIORUM, L. From white lilies, 1 lb.; olive oil, 3 lbs.; as OIL OF BELLADONNA. Emollient; used to soften and ripen tumours, indurations, &c.

Oil of Mel'ilot. *Syn.* OLEUM MELILOTI, L. As the last, avoiding much heat. Emollient and resolvent.

Oil of Mu'cilage. *Syn.* OLEUM MUCILAGINUM, O. CUM MUCILAGINIBUS, L. *Prep.* 1. (Ph. L. 1746.) Marsh-mallow root, ½ lb.; linseed and fenugreek seed, of each, bruised, 3 oz.; water, 1 quart; boil 1 hour, add of olive oil, 2 quarts, and boil until the water is consumed.

2. Fenugreek seeds, 8 oz.; linseed oil, 1 quart;

infuse a week, and strain. Once a highly popular emollient application in various local affections.

Oil of Mu'dar. *Syn.* OLEUM MUDARIS, L. From mudar bark (in coarse powder), 1 dr.; warm olive oil, ⅓ pint; digest 24 hours and strain. Used as an application to cutaneous ulcers, the bites of venomous animals, &c., and as a friction in worms.

Oil of O"pium. *Syn.* ANODYNE OIL, OPIATED O.; OLEUM OPIATUM, L. *Prep.* From opium (in powder), 1 dr.; olive oil, 2½ fl. oz.; digest at a gentle heat, with frequent agitation, for 5 or 6 hours. The powder should be rubbed in a mortar with a few drops of the oil before adding the remainder. As a local anodyne. The above is the only reliable formula for this preparation. Others are extant; but whilst the products of several are much stronger, those from others have only 1-5th or 1-6th the strength.

Oils, Ozonised. (*Dr Thompson.*) *Syn.* OLEA OZONATA. *Prep.* Pass oxygen gas into the oil (cocoa-nut, sunflower, cod-liver oil, &c.) until it will dissolve no more. Then expose for a considerable time in the direct rays of the sun. Used in phthisis.

Oil of Pel'litory. *Syn.* OLEUM PYRETHRI, L. From bruised pellitory root, as OIL OF BELLADONNA. Used as the last.

Oil of Black Pep'per (by Infusion). *Syn.* OLEUM PIPERIS INFUSUM, L. From black pepper, in coarse powder, as OIL OF CAPSICUM. Stimulant and rubefacient; in frictions.

Oil of Poison Oak. *Syn.* OLEUM RHOIS TOXICODENDRI, L. *Prep.* From the leaves, as OIL OF BELLADONNA. Externally; in paralysis, &c.

Oil of Rhu'barb. *Syn.* OLEUM RHEI, L. *Prep.* From rhubarb (in powder), 1 part; oil of almonds, 8 parts; digested together in a gentle heat for 4 hours, and strained, with expression. As an application to indolent ulcers, and as a friction over the abdomen in diarrhœa, English cholera, &c., or as a laxative when the stomach will not bear medicine.

Oil of Ro"ses. *Syn.* OLEUM ROSÆ, O. ROSACHUM, O. R. INFUSUM, O. ROSATUM, L. *Prep.* From the fresh petals, pulled to pieces, crushed, and digested for 2 or 3 days in the sun, or a warm situation, in 4 times their weight of olive oil, and then pressed; the process being repeated with fresh roses. Ph. E. 1744 and P. Cod. are nearly similar. ALMOND, BEN, or OLIVE OIL, coloured with ALKANET, and scented with attar of roses, is now almost universally sold for it. Used for the hair.

Oil of Rue. *Syn.* OLEUM RUTÆ (INFUSUM) L. *Prep.* (P. Cod.) From fresh rue, bruised, as OIL OF CHAMOMILE. Reputed antispasmodic, emmenagogue, stimulant, and vermifuge. In frictions.

Oil of St John's-wort. *Syn.* OLEUM HYPERICI (Ph. L. 1746), O. H. SIMPLEX, BALSAMUM H., L. *Prep.* From the flowers, 1 part; olive oil, 6 parts; digested together until the oil is well coloured. Antispasmodic, stimulant, and resolvent. A mixture of equal parts of RAPE OIL and GREEN ELDER OIL is usually sold for it.

Oil of Scam'mony. *Syn.* OLEUM SCAMMONII, O. PURGANS, L. *Prep.* (*Van Mons.*) From scammony (in powder), 1 dr.; hot oil of almonds,

3 fl. oz.; triturate together until cold, and the next day decant the clear portion.—*Dose*, ½ to 1 table-spoonful.

Oil of Stramo′′nium. *Syn.* OLEUM STRAMONII, L. *Prep.* (P. Cod.) From the leaves of thornapple or stramonium, as OIL OF BELLADONNA. Anodyne and discussive; as an application to painful tumours, joints, &c.

Oil of Tobac′co (by Infusion). *Syn.* OLEUM TABACI, O. T. INFUSUM, L. *Prep.* From fresh tobacco leaves (bruised), like OIL OF CHAMOMILE. As an application in ringworm, irritable ulcers, pediculi, &c.; and as a friction in itch, neuralgia, painful indurations, &c. It must be used with extreme caution, as it is poisonous.

Oil of Tooth′wort. *Syn.* OLEUM SQUAMARIÆ, L. *Prep.* From the herb of *Lathræa squamaria*, Linn., as OIL OF ST JOHN'S-WORT. Astringent and vulnerary. This must not be confounded with another preparation sometimes called 'OIL OF TOOTHWORT' (OLEUM PLUMBAGINIS EUROPEÆ), and which has been occasionally used in itch, as the latter is acrid and apt to cause much irritation.

Oil of Turpentine, Sulphurated. *Syn.* OLEUM TEREBINTHINÆ SULPHURATUM. *Prep.* Sulphurated linseed oil, 1 part; oil of turpentine, 3 parts.

Oil of Turpentine (for acoustic use). *Syn.* OLEUM TEREBINTHINÆ ACOUSTICUM. (*Mr Maule*.) Oil of almonds, 4 dr.; oil of turpentine, 40 minims.

Oil of Worm′wood. *Syn.* OLEUM ABSINTHII, L. *Prep.* From the fresh herb, as OIL OF LILIES. The P. Cod. and Ph. Wurtem. order only 1 part of the herb to 8 parts of oil. Applied to the abdomen in dyspepsia, diarrhœa, heartburn, worms, &c. It is seldom used in this country.

OILS (Mineral). *Syn.* HYDROCARBON OILS. An important class of liquids, consisting solely of carbon and hydrogen—the elements of ordinary coal-gas, and obtained by the distillation of coal, lignite, petroleum, and other bituminous substances. For the purposes of illumination, many of these oils are in most respects superior to the fixed or fatty oils containing oxygen. They give a whiter and more brilliant light, and are produced at a much lower cost. The lamps in which they are burnt, when properly constructed, are less liable to get out of order than those adapted for the combustion of fatty oils, and require less attention when in use. The experiments of Dr Frankland on the relative value of the ordinary illuminating agents (see ILLUMINATION) prove that the mineral oils are cheaper than all other portable illuminating agents in common use, and that they give the largest amount of light with the least development of heat, and the smallest production of carbonic acid. Some oils adapted for burning in lamps are very volatile and highly inflammable, and their safety depends on their proper extraction. These volatile liquids are used in the arts as substitutes for spirits of turpentine, as solvents for various substances, and to increase the illuminating power of coal-gas. Others are of a greasy nature, and are too heavy to be conveniently used in lamps. These, however, are well adapted for lubricating fine ma-

chinery, and are extensively employed instead of sperm oil by the cotton manufacturers of Lancashire. When the more volatile ingredients are separated from the burning oils, the latter are perfectly safe. Most of the mineral burning oils now in use are, we believe, free from danger in this respect. See *Tests* (*below*).

Hist. For many years the manufacture of burning oils by the distillation of bituminous shales has been extensively carried out on the Continent, but the discovery which formed the foundation of the modern manufacture was made nearly 30 years ago by Mr James Young. This gentleman took the lease of a spring of petroleum in 1847, and after numerous experiments succeeded in obtaining two useful oils from the crude liquid; the one being adapted for lubricating machinery, and the other for burning in lamps. The almost total cessation of the flow of petroleum terminated the business after 2 years' working, and led Mr Young to institute a series of experiments with a view to obtaining artificially by the destructive distillation of coal. These experiments resulted in the discovery of an oil which Mr Young named 'paraffin oil,' as it had many of the chemical properties of the solid body of paraffin, discovered 20 years before by Reichenbach in beech-wood tar. Young's patent (dated Oct. 7, 1850) involved the slower distillation of coals, at a lower temperature than had hitherto been employed for the purpose, and this novelty in practice resulted in a copious production of liquid hydrocarbons. The gas or cannel coals were found to yield these liquids in largest quantities, that variety known as Boghead coal or Torbane Hill (this species of coal is now exhausted—Ed.) mineral being specially adapted for the patented process (see PARAFFIN OIL, *below*). Soon after Young's discovery native petroleum was brought from Rangoon, and purified by distillation, so as to produce oils very similar to the coal products. During the last few years, however, rich sources of petroleum have been discovered in North America, whence are imported the greater part of the vast quantities of petroleum oil, both for burning and lubricating purposes, together with the paraffin spirit, or naphtha, which are consumed in this country.

Tests, Precautions. The Sanitary Commission of the 'Lancet' took as the limit of safety an oil that gave off inflammable vapour when heated to 130° F., and this has been generally accepted by dealers. If an oil gives off inflammable vapours before being heated up to 130°, it is considered unsafe for domestic use.

1. A rough-and-ready method of testing the inflammability of a sample is to pour a little out on a dry flat board, and try whether it can be ignited readily by a lighted paper. If it catches fire like turpentine or brandy, the oil is dangerous.

2. The following plan, proposed by Mr Tegetmeier, requires no scientific knowledge and no apparatus but what is to be found in every house, while it is sufficiently accurate for all practical purposes:

"Take an earthenware dish, holding about half a pint (a breakfast-cup will do), fill the cup full from a kettle of boiling water, pour this into an earthenware quart jug, then fill the same cup again with boiling water from the kettle, and pour it also into the quart jug, then fill the cup with cold water, put it into the jug, shake the jug to mix the hot and cold water, then pour the tepid water from the jug into the cup till the cup is half full, then pour about a table-spoonful of the oil to be tested on the tepid water in the cup, take the oil-can with the oil out of the room, then touch the surface of the oil in the cup with a lighted splinter of wood, or a match without sulphur. If the match causes a flash of flame to appear on the surface of the oil, the oil is below the standard of safety, and should not be used; if no flame appears, the oil is up to the standard. We may mention that in this trial no time should be lost after pouring the boiling water from the kettle, as the water may get too cold, but the whole may be gone through in from 2 to 3 minutes. It is well to have a saucer at hand, and if the oil should be a bad oil, and ignite with the match, place the saucer on the mouth of the cup, and the flame is extinguished. This trial should be done by daylight, and at a distance from a fire, and the directions must be followed exactly in the order as given above."

3. Provided that the oils to be examined have been produced by careful fractional distillation, their relative volatility, as indicated by their specific gravity, shows to a great extent the facility with which they ignite. The lightest oils are more volatile and more easily inflamed than those which are heavier. Oils much under 0·80 inflame directly a lighted match is thrown into them, whereas oils at about 0·815 to 0·823 (if unmixed products) cannot be set on fire in this manner. The specific gravity test cannot, however, be depended on to determine the inflaming point of any commercial oil. A heavy oil, badly rectified, may contain a proportion of very volatile vapour, and have a low inflaming point; whereas a much lighter oil may be perfectly safe, from its having the more volatile portions carefully removed.

4. (*Van der Weyde.*) The oil to be tested is placed in a graduated tube closed at one end; the open end is then closed with the finger, and is then placed mouth downwards in a vessel of water that is heated from 43°—44° C. The vapour from the portion volatilised at this temperature then collects in the upper part of the tube, and expels a corresponding quantity of oil. See PETROLEUM.

In Great Britain petroleum is defined by Act of Parliament as being any oil which gives off an inflammable vapour at a temperature less than 100° F.

To prevent accidents with paraffin or petroleum lamps, the following precautions ought to be observed:

The lamps should be filled and trimmed by daylight.

They should never be over-filled; the oil should not be allowed to come into contact with the metal work of the burner.

Any portion of oil spilled on the outside of the lamp should be carefully wiped away.

When not in use the wick should be turned down into the wick-holder.

⁎ The principal products noticed below rank high among the numerous varieties of mineral oil now in the market, but there are many others

equally good and safe. Their properties are described in accordance with the results obtained by Mr W. B. Tegetmeier, who has carefully examined the mineral oils:

Oil, Al′bertite. From 'Albertite,' a lustrous black mineral found in New Brunswick. A sample was shown in the Colonial Department of the International Exhibition of 1862, but the oil has not yet appeared in the English market.

Prop. Odour very slight; illuminating power high; boiling-point 838° F., or 126° above that of water.

Oil, American. See PETROLEUM OIL (*below*).

Oil, Apyroæ′tic. *Syn.* NON-EXPLOSIVE OIL. A burning oil, and prepared, we believe, from American petroleum.

Prop. Slightly coloured; perfectly limpid; odour slight, but not perceptible during combustion. The most remarkable property of this oil is that, in spite of its limpidity, the point at which it gives off inflammable vapour is 180° F., or 80° above the requirements of the Petroleum Act.

Oil, Bel′montine. From Rangoon tar, or Burmese petroleum, by distillation; superheated steam being employed as the heating agent.

Prop. Colourless; odour not unpleasant; sp. gr. 0·847; but although so heavy, the oil is altogether free from viscosity, and will rise rapidly in a comparatively long wick; inflaming point 134° F.; burns with an exceedingly white light, and possesses a very high illuminating power.

Obs. Besides the above lamp oil, several beautiful and useful products are obtained. At first there comes over a very volatile liquid, termed SHERWOOD OIL, used for removing grease from fabrics, cleaning gloves, &c.; then comes the BELMONTINE OIL, already noticed; then two lubricating oils, the one light and the other heavy; and, last of all, when the temperature is considerably elevated, the beautiful white, translucent solid known as BELMONTINE distils over. This last is a kind of paraffin, and is used for making ornamental candles.

Oil, Caz′eline. An excellent burning oil, prepared from American petroleum.

Prop. Bright, limpid, with scarcely a trace of colour; odour very slight, and quite free from any objectionable character; sp. gr. 0·805; lowest point of ignition 144° F.; burns with a pure white light, free from smoke and smell.

Oil, Col′zarine. A heavy hydrocarbon oil, adapted for burning in lamps constructed from the old 'Moderators' and 'Carcels,' formerly so much used for the fatty oils.

Prop. Limpid; quite inodorous; of a pale amber colour; sp. gr. about 0·838; temperature at which the vapour can be permanently ignited, 250° F. Tested in the altered moderator it gives an intense white light, without smoke or smell. Compared with vegetable colza oil, its illuminating power is in the proportion of 3 to 2.

Obs. This oil is suitable for burning in lamps where 'colza' and other vegetable and animal oils have been usually consumed. Similar oils are prepared by other firms.

Oil, Machin′ery. *Syn.* LUBRICATING OIL, SHAFTING O., SPINDLE O. The heavier hydrocarbon oils obtained in distilling coal, shale, and petroleum have almost superseded the fatty oils for lubricating purposes. They have very little chemical action on the ordinary metals, and are not affected by cold. The lightest of these comparatively heavy oils are used for spindles, or other kinds of rapid machinery; the heaviest for the bearing parts of heavy machinery; and those of an intermediate character for such machines as printing-presses, agricultural steam-engines, &c. In America and on the Continent this oil is also used for making gas. See OIL, BELMONTINE (above), and OIL, PARAFFIN (below).

Oil, Par'affin. *Syn.* PARAFFINE OIL. This name was given by Mr Young to the oil produced by the distillation of cannel coal, Boghead coal, &c., at a temperature considerably lower than that employed in the manufacture of illuminating gas. The following is a brief outline of Mr Young's process:

Manuf. (Young's patent.) Coal, bituminous schists, or shales of the lower carbouiferous formation broken into small fragments, are introduced into perpendicular tubes or retorts, about 11 feet in height, by conical hoppers at their upper extremities. Four of these tubes constitute a set, being built into one furnace, and charged by a single workman. They pass completely through the furnace, and are closed below by dipping into shallow pools of water, while the openings into the hoppers above are shut by valves. The coal in each tube is gradually heated as it descends to that part which passes through the furnace, and when it reaches the bottom of the tube it has parted with its volatile constituents, and is raked away as refuse, the coal from above descending as it is removed. Thus the action of these perpendicular retorts is continuous, and the distillation goes on uninterruptedly both day and night. The vapours produced are conducted by iron tubes to the main condensers, which consist of a series of syphon pipes freely exposed to the air. The quantity of uncondensable gas formed is inconsiderable; and it is this result, so different from that obtained in the ordinary gasworks, that marks the great value of Young's process. The crude oil, a dark-coloured, thick liquid, is then distilled to dryness in large iron cylindrical stills, and is thus freed from the excess of carbon which is left behind as coke. The oil, after distillation, is further purified by the action of strong sulphuric acid, which chars the principal impurities, and causes them to subside in the form of a dense, black, heavy acid tar. To separate the remaining impurities, and that portion of the sulphuric acid which remains in the oil, it is next subjected to the action of caustic soda. As thus purified, the paraffin oil, which is also called shale oil and Scotch oil, contains four distinct commercial products. To effect their separation, the process of fractional distillation is first employed. The first elevation of temperature drives over the lighter and more volatile portions, which, when purified by a subsequent distillation, yield the fluid known as 'paraffin naphtha,' 'petroleum spirit,' or 'benzoline.' This product is used as a substitute for 'turps,' as a solvent for india rubber, for cleaning gloves, and for burning in those naphtha lamps so much employed by cos-

termongers, and workmen in railway tunnels, &c. On the perfect separation of this naphtha the safety of the burning oil depends. This burning oil, the 'paraffin oil' of commerce, comes over at a much higher temperature than the naphtha. It is a perfectly safe lamp oil, and has a greater illuminating value than any other oil in the market. Its properties are noticed *below*. The third product in point of volatility is a comparatively heavy liquid (machinery oil), largely used for lubricating purposes. From this oil, and others which come over at a very high temperature, the fourth commercial product is separated by the action of artificial cold. This last product is the beautiful translucent solid paraffin, now much used in candle making, for which purpose it is specially adapted, being a most elegant substance, and surpassing all other candle materials, even spermaceti, in illuminating power. The softer kinds, when dissolved in naphtha and mixed with a little vegetable oil, are, according to Stenhouse, excellent for waterproofing wood, *e. g.* matches, barrels, sleepers, &c.; also for waterproofing hose, cloth, linen, leather, &c. To these fabrics they also impart greater tensile strength. The naphtha solution also makes a good lubricating 'cream.' For a detailed account of the processes carried on at the Bathgate works, see Mr Tegetmeier's paper in 'England's Workshops' (Groombridge and Sons). See OIL, PARAFFIN, PETROLEUM.

In Germany and other countries the extraction of the crude oil is effected in ovens of special construction, but neither the yield nor the quality of the output is so good.

Young's method has been improved upon in detail by various patents; that of Henderson is, perhaps, the most important.

"Lignite or brown coal is extensively used on the Continent for preparing paraffin and paraffin oil. The following are the final products of the distillation:"

a. Volatile oil, called photoform and solar oil.
b. Paraffin.
c. Volatile spirit, called benzol.
d. Phenol, or carbolic acid.

In the preparation of paraffin oil from native petroleum, the oil is obtained by direct distillation from the petroleum, and subsequently separated from the more or less volatile hydrocarbons (the paraffin naphtha, the lubricating oils, and the solid paraffin) that are associated with it, by fractional distillation as in Young's process; whereas, when procured from bituminous minerals, it is derived from the *tar* or *crude oil*, which has to be previously extracted from the bituminous matters by destructive distillation. There are various methods for obtaining this tar or crude oil, which, although differing in detail, are in general principles very similar to that described in Young's patent. Thus, whilst in many works *closed* horizontal retorts are employed, in other establishments vertical ones, to the bottoms of which are attached receptacles for the receipt of the exhausted coal or other material as it falls from the retort, the same as in Young's apparatus, are extensively adopted. When horizontal retorts are employed they are made of cast iron, and vary in length from 8 to 10 feet, being from

28 to 34 inches wide and from 9 to 14 inches deep. The charge is introduced by an opening in the end of the retort, by which aperture the exhausted residue is removed when necessary. This aperture is closed by a tightly fitting cast-iron cover while the distillation is going on. At the other end of the retort is a pipe for carrying off the products of distillation. This communicates with a larger pipe, and this latter with the condensing apparatus. A number of these retorts are set together in a row, with a furnace at one end, and flues extending beneath the retorts, while the upper parts of the retorts are covered with brickwork, to prevent the oil vapours from being decomposed by the heat of the waste furnace gas passing to the chimney through the flues above the retorts.

The gaseous products of the distillation of the tar, leaving the retort by the exit tube already described, are cooled by being made to pass through a number of iron pipes exposed to the air, or surrounded by water, and thus becoming condensed, pass into a reservoir in the form of the oil, which forms the material from which the various hydrocarbons are separated by fractional distillation. Accompanying the oil vapours are certain uncondensable gases; these escape through a properly contrived outlet which is made in the condensing pipes; in some works these escaping gases are utilised as fuel, and in others for purposes of illumination.

In other works superheated steam is driven into the retorts during the process of distillation; but although this has the effect of sweeping the oil vapour more quickly out of the retort into the condenser, it is questionable whether this advantage covers the extra cost of the production of the steam (Payne's 'Industrial Chemistry,' edited by Dr Paul).

In many parts of Germany the extraction of the crude oil or tar from bituminous substances is effected in ovens. In these ovens the bituminous body is thrown upon a layer of burning fuel, which covers the bottom of the oven, the result being that the bituminous matter is resolved into gaseous bodies which are lost, and tar which flows downwards toward the burning fuel, which, being covered with a layer of clay, is prevented from entering into violent combustion. This method, however, is only resorted to on a small scale, since it is found that in most cases the tar obtained by means of it is not of a kind suited for yielding paraffin and paraffin oils.

The preparation of the tar or crude oil from coal, shale, &c., of the character already specified, constitutes one of the most delicate and difficult branches in the manufacture of paraffin oils and paraffin, &c. The chief sources of failure to be avoided are the overheating of the oil vapour, its consequent decomposition into useless gaseous products, and its inefficient condensation.

It has been shown by Vohl that even when the construction of the retorts is not of the best, an average yield of tar may be obtained by the proper condensation of the vapours. "The complete condensation of the vapours of the tar is one of the most difficult problems the mineral oil and paraffin manufacturer has to deal with; while the means usually adopted for condensation, such as large

condensing surfaces, injection of cold water, and the like, have proved ineffectual. It has often been attempted to condense the vapours of tar in the same manner as those of alcohol, but there exist essential differences between the distillation of fluids and dry distillation. In the former case the vapours soon expel all the air completely from the still and from the condenser, and provided, therefore, that, in reference to the size of the still and bulk of the boiling liquid, the latter be large and cool enough, every part of the vapour must come into contact with the condensing surfaces. In dry distillation the process is entirely different, because with the vapours—say of tar—permanent gases are always generated. On coming into contact with the condensing surfaces a portion of the vapours is liquefied, leaving a layer of gas as a coating, as it were, on the condensing surface. The gas being a bad conductor of heat prevents to such an extent the further action of the condensing apparatus, that a large proportion of the vapours are carried on, and may be altogether lost. A sufficient condensation of the vapours of tar can be obtained only by bringing all the particles of matter which are carried off from the retorts into contact with the condensing surface, which need neither be very large nor exceedingly cold, because the latent heat of the vapours of tar is small, and consequently a moderately low temperature will be sufficient to condense those vapours to the liquid state. The mixture of gases and vapours may be compared to an emulsion such as milk, and as the particles of butter may be separated from milk by churning, so the separation of the vapours of tar from the gases can be greatly assisted by the use of exhausters, acting in the manner of blowing fans. It is of the utmost importance in condensing the vapours of tar that the molecules of the vapours be kept in continuous motion, and thus made to touch the sides of the condenser. The condenser should not be constructed so that the vapours and gases can flow uninterruptedly in one and the same direction" (R. Wagner).

An important condition for the safe and quiet distillation of the tar or crude oil is that it should be free from water. Unless the removal of the water is effectually accomplished the tar may boil over, and, coming into contact with the fire under the still, may give rise to an alarming conflagration. The dehydration of the tar is effected in an apparatus constructed for the purpose, consisting of an iron tank placed within a larger tank; a space of about two inches intervening between the two tanks is filled with water, which is heated to, and kept at a temperature of, between 60° and 80° C. for 10 hours, by the end of which time the ammoniacal water, having separated from the lighter tar, is drawn off by a stopcock placed at the bottom of the tank, whilst the tar is decanted through a valve at the top.

In America the distillation of the natural petroleum oils is carried out in cylindrical stills capable of holding as much as 1600 galls. each. The retorts employed in the distillation of the tar, or crude oils obtained from shale and other bituminous compounds, are often constructed of large cast-iron flanged pans, each capable of containing from 1½ to 3 tons of the oil, "and forming the body of the retort. The pan is set in brickwork with flues running round the upper portion, and beneath it is a perforated dome of brickwork, through which the flame and hot gas from the furnace pass up round the bottom of the pan before entering the flues by which the upper portion of the pan is heated. To the flange of the pan is fitted a flanged cover, having on one side a discharge pipe through which the vapour is passed to the worm of the condenser. In the centre of the cover is a manhole. The oil condensed in the worm is discharged through a pipe into a receiver, and the uncondensable gas escapes through an ascending pipe" (*Palen*).

The processes to which the crude oil or tar and the natural petroleum are next submitted differ only in the degree of treatment with certain agents to which these products are subjected when, after similar methods of fractional distillation, they have been isolated from each other. The benzoline and paraffin oils (both for burning and lubricating purposes) separately yielded by the natural oils seldom require purification, or if so, in a minor degree only, whilst the same bodies as obtained from the crude shale oil or tar must be submitted to various processes of depuration before they are fit for the market. Thus the crude petroleum or burning oil derived from tar is characterised by a more or less dark colour and disagreeable smell—properties which are partly due to the presence of carbolic acid and its homologues. By agitating the paraffin oil with a solution of caustic soda these objectionable substances are removed.

The oil, having been next separated from the alkali by subsidence, and any remains of the soda having been removed from it by washing with water, is next mixed with an aqueous solution of sulphuric acid in the proportion of 5% of acid (sp. gr. 1·7). The acid removes from the oil certain basic substances derived from the tar, which, like the carbolic acid, give to it a bad odour and a dark colour. In this operation thorough admixture of the acid with the oil is important, and this is generally effected by mixing the two in vessels furnished with paddles. After a time, and when the mixture has separated into two layers, the upper one, i. e. the paraffin oil, is drawn off from the lower or acid layer, and well washed with water; in some instances lime-water is used for the washing, in others the water is impregnated with caustic alkali. With some samples of crude paraffin oil the above operations have to be repeated 2 or 3 times, and even redistilled before the oil becomes sufficiently pure and colourless for sale. When redistilled, the last portions which come over are often found to yield some solid paraffin in addition to that furnished by the first fractional distillation. The 'paraffin,' 'naphtha,' 'petroleum,' 'spirit,' or 'benzoline' (by all of which names it is known), which forms the more volatile portion of the tar, and which is the first to pass over from the retort, is subjected to the same treatment as that used for burning oil; as for the denser lubricating oil, which passes over after the burning oil has collected, this being freed from any of the latter, is set aside in a cool place, in order that any solid paraffin it contains may crystallise out, and be separated from it.

The waste carbolate of soda resulting from the treatment of the oil with the caustic alkali, having been decomposed by sulphuric acid, the liberated carbolic acid is utilised either as a disinfectant, or for saturating railway sleepers; and sometimes as a source of certain tar colours; or it may be used in the manufacture of gas, the soda which remains in the coke being extracted by lixiviation. The waste sulphuric acid combined with the ammoniacal liquors that always accompany the first stages of the distillation of the tar is made into sulphate of ammonia.

The following conspectus of operations and quantities (variable with the oil and the state of the markets) will render the whole modern process of refining more intelligible (Dr Mills' 'Destructive Distillation'). See next page.

"Within recent times considerable attention has been bestowed on the production of a highly illuminating gas from the less valuable liquid products of the paraffin industry. Thus, 'Green' oil of sp. gr. 0·894, from acid tar, has been found to yield 87 c. ft. per gall. of such gas. An oil of sp. gr. 0·844 has, however, furnished 88 c. ft. per gall.; a gravity of 0·822 corresponds to 90 c. ft. with less tar, and that of a thinner quality. The produce of tar from the lighter oils is in general about ½ to 1 gall. of tar of sp. gr. 1·081 for every 5 galls. of oil; from the heavier oils, about 1½ galls. It is, of course, neither acid nor alkaline. After passing through condensers and a washer, the gas traverses two purifiers containing layers of chopped straw, sawdust, and lime. It is admirably adapted for compression, the original compression being 30, the working pressure 6 to 10 atmospheres. Before such treatment it has the sp. gr. 0·7; during the process it deposits 1 gall. of light 'gasoline' per 1000 c. ft., the eventual lighting power being 25·9 candles, and the consumption (in a railway carriage lamp) 0·78 c. ft. per hour" (Mills' 'Destructive Distillation').

Prop. The paraffin oil of commerce is of a very pale amber colour, or it may be quite colourless, but possessing a strong bluish fluorescence; is bright, perfectly transparent, and remarkably limpid. Its sp. gr. is 0·823. Its point of temporary ignition is 150° F., that of permanent ignition being a few degrees higher. Its odour is very slight. Its rate of combustion is slow, as may be inferred from the absence of the lighter oils, as indicated by its high sp. gr. and inflaming point. At the same time its limpidity proves the absence of the heavier oils, and accounts for its rising through a long wick with freedom, and burning without charring the cotton.

Oil, Petro'leum. *Syn.* KEROSENE OIL, REFINED PETROLEUM, PARAFFIN OIL. Petroleum consists chiefly of a mixture of fatty hydrides, and occurs abundantly in the Upper Devonian and Carboniferous Limestone formations. It is found in all parts of the world, and is probably not confined to any one geological formation. The American petroleums vary greatly in properties, and numerous methods of refining are employed by the manufacturers. The Canadian petroleum is richer in aromatic compounds and poorer in gaseous paraffin. It contains sulphuretted hydrogen,

Operations and Quantities.

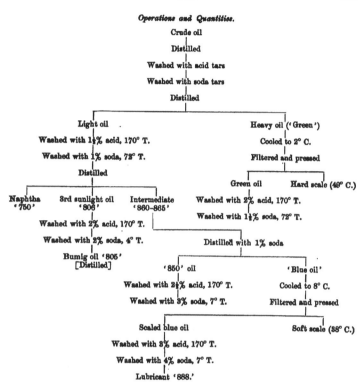

which imparts to it a very disagreeable smell, which is difficult of removal. In purifying this oil, some make use of both acids and alkalies, others employ alkalies alone, and steam is applied at various degrees of heat. Some of the oils produced are of excellent quality, but others are inferior, and do not ascend the wick in sufficient quantity to afford a constant light.

The petroleum of the United States is chiefly obtained in Pennsylvania by boring operations. The oil wells discharge large volumes of gas containing hydrogen, marsh gas, and ethane, which are used for heating and lighting purposes in the neighbouring district.

The liquid which spouts, flows, or is pumped out of the wells consists chiefly of members of the paraffin series, of which the following is a list:

Methane	CH_4	gas.
Ethane	C_2H_6	,,
Propane	C_3H_8	,,
Butane	C_4H_{10}	,,

		Boils at
Pentane	C_5H_{12}	88° C.
Hexane	C_6H_{14}	70° ,,
Heptane	C_7H_{16}	98° ,,
Octane	C_8H_{18}	125° ,,
Nonane	C_9H_{20}	148° ,,
Decane	$C_{10}H_{22}$	168° ,,
Dodecane	$C_{12}H_{26}$	202° ,,
Hexadecane	$C_{16}H_{34}$	278° ,,

When the petroleum is distilled, the hydrocarbons as far as butane are evolved in a gaseous state; these are collected and subjected to the action of a condensing pump, which liquefies a portion of them, yielding the liquid sold as *cymogene*, which is used in freezing machines on account of the cold produced by its rapid evaporation. It consists chiefly of butane. The liquid constituents of the petroleum are separated by the process of fractional distillation, which depends upon the difference in their boiling-points. The portion which distils over below 76° C. consists chiefly of pentane and hexane, and is sold as *petroleum spirit* or *petroleum ether*, and used for dissolving india-rubber and for making varnishes. The next fraction of the distillate is chiefly heptane, and is sold for burning in paraffin lamps under the names of benzoline, paraffin oil, and mineral sperm oil. The oils boiling below 76° C. are not safe for burning in ordinary lamps, because they so easily evolve vapour, which forms an explosive mixture with air. That portion which distils over between 150° C. and 200° C. consists chiefly of nonane and dodecane, and is used for lubricating machinery. At still higher temperatures the liquid which distils over consists of hexadecane and other hydrocarbons richer in carbon. These form soft solids like vaselin; those containing most carbon form the wax-like crystalline solid originally termed paraffin (Bloxam's 'Chemistry').

Mr Boverton Redwood says, in his 'Cantor Lectures on Petroleum,' that "of the producing wells in the United States the great majority furnish only a few barrels a day, but some are stated to have yielded for a short time as much as 260,000 galls. per 24 hours." The production of oil in Pennsylvania is now on the decline; the wells have to be bored deeper every year, in some instances to a depth of 5000 feet, and the yield is not so good. Thus the famous Bradford field is steadily drying up, and the Richburg field is regarded as not of a durable character.

For more than 2000 years Baku, on the Caspian Sea, has been famous for its marvellous springs of petroleum, and there is historical evidence that for nearly 1000 years its oil resources have been drawn upon. The Zoroastrian fire-worshippers resorted to Baku 1000 years B. C. to pay their devotion to the perpetual flames of fire which burnt around the natural oil springs. Oil is found not only at Baku but throughout the whole of the Caucasus, covering an area of about 1200 miles across. It exudes in places 9000 feet above the level of the sea and 600 feet below it. The area worked at Baku is 1836 acres; the wells average 350 feet by 10 inches, and yield an average of 1,000,000 galls. a day, frequently under enormous pressure. At Tagieff's wells a fountain commenced playing at the rate of 500 tons of petroleum per hour on Oct. 5, 1886. Its height was 224 feet. In a few days it reached a maximum of 2,750,000 galls. per diem. Since the Russian Government threw open the oil industry to foreign capital the production of petroleum has been increasing by millions of gallons yearly.

Hitherto the want of transport facilities has impeded the development of the trade, but now the Caspian Sea and the Black Sea are united by rail and pipe lines, and tank steamers are coming into use, the marvellous abundance and cheapness of the oil must cause it to prove a formidable competitor with the produce of America, especially as from Baku petroleum can be extracted a better and safer kerosene and an incomparably better lubricating oil of high sp. gr., whilst the refuse furnishes an inexpensive fuel (*Marvin*).

The Baku petroleum is, according to Mendlejeff, strongly characterised by the presence of olefines. Its sp. gr. for a given boiling-point is greater than that of American or Scotch oil. It contains no solid paraffin. The residues from the rectification process are used as fuel at the oil works and on board the oil steamers.

There is nothing whatever in the geological history of Baku and the Caucasus to warrant the belief that the supply is not of a practically inexhaustible character (*Marvin*).

Petroleum is also worked in Japan, California, the Argentine Republic, Italy, Bavaria, Hanover, Roumania, Alsace, and in the Limogne Valley, Turkestan, the Punjab, Beluchistan, Egypt, &c.

A great future is predicted for the oil-fields recently acquired by England in her annexation of Burmah. The so-called Rangoon tar yields a splendid heavy illuminating oil (30·38%, sp. gr. 0·832) and an excellent lubricating oil (51·24%, sp. gr. 0·901).

1889—*Imports.*

Petroleum.	Gallons.	£.
From Russia . .	31,582,385 .	628,833
„ Germany . .	207,637 .	10,592
„ Holland . .	77,484 .	6,207
„ Belgium . .	111,817 .	4,220
„ France . .	119,311 .	4,206
„ United States of America	70,739,663 .	1,932,850
„ other foreign countries .	39,777 .	1,975
Total from foreign countries .	102,878,574 .	2,588,883
„ from British possessions	2,682 .	64
Grand total . .	102,881,256 .	2,588,947

None of the native petroleums contain carbolic acid and other impurities which exist in the oils distilled from coals and shales; hence their purification is simple and comparatively cheap. "The oil prepared from petroleum is almost colourless; it has a sp. gr. of about ·810, and when of good quality only a slight and rather aromatic odour" (*Payen*). See PETROLEUM and *above.*

Oil, Shale. As we have stated, products analogous to those derived from cannel coal are obtained by the destructive distillation of bituminous shales and schists, and lignites or brown coals. On the Continent the production of shale oils has of late years declined considerably, owing to their unsuccessful competition, in point of price, with the American petroleum oils. The oil obtained from bituminous shale or from coal is generally of higher specific gravity than that procured from petroleum; it is deeper in colour, and not so pleasant in smell.

OILS (Mixed). *Syn.* COMPOUND OILS; OLEA COMPOSITA, OLEA MIXTA, L. Under these names are commonly included various mixtures of oils and other substances that possess an unctuous appearance. When not otherwise stated, they are prepared by simply agitating the ingredients together, and, after a sufficient time, decanting the clear portion, which, in some cases, is then filtered. A few of them only possess any importance. Some of them are highly esteemed as remedies among the vulgar, and the use of others is confined to veterinary medicine.

The following include the principal mixed oils of the shops, to which the names of a few other compounds, which are frequently called 'oils' by the ignorant, are added, for the purpose of facilitating a reference to them.

Oil, Acou'stic. *Syn.* EAR OIL; OLEUM ACOUSTICUM, O. TEREBINTHINÆ ACOUSTICUM, L. *Prep.* From oil of turpentine, 1 part; oil of almonds, 6 parts; mix. In atonic deafness, accompanied with induration of the wax. 1 or 2 drops are poured into the ear, or on a piece of cotton wool, which is then gently placed in it.

Oil for Bicycle Lamps. Camphor, ¼ oz.; sperm oil, 2 oz.; paraffin oil, 6 oz.

Oil, Black. *Syn.* OLEUM NIGRUM, L. *Prep.* 1. Oil of turpentine, 1 pint; rape oil, 3 pints; oil of vitriol, ¼ lb.; agitate well together with care; then add of Barbadoes tar, 3 oz.; again agitate well, and in 10 days decant the clear portion. Linseed oil is preferred for the above by many persons.

2. (*Percivall.*) Sweet oil, 1 pint; oil of turpentine, 2 oz.; mix, add gradually of oil of vitriol, 1¼ oz.; again mix, and leave the bottle open until the next day. Detersive, stimulant. Used by farriers for mange, &c.

Oil, British. *Syn.* COMMON OIL OF PETER; OLEUM BRITANNICUM, O. PETRÆ VULGARE, L. *Prep.* From oil of turpentine, 1 quart; Barbadoes tar, 1 lb.; oils of rosemary and origanum, of each, 1 oz. Stimulant. Formerly reputed to possess the most astonishing virtues.

Oil, Cam'phorated. Liniment of camphor.

Oil, Carbolised. *Syn.* OLEUM CARBOLATUM. Pure carbolic acid in crystals, 1 part; olive oil, 10, 20, or 40 parts; mix, and dissolve by the aid of heat. Used as a local antiseptic application, also for oiling catheters.

Oil, Carron. Liniment of lime.

Oil, Chabert's. *Syn.* CHABERT'S EMPYREUMATIC OIL; OLEUM CHABERTI, O. CONTRA TÆNIAM CHABERTI, L. Oil of turpentine, 3 parts; Dippel's animal oil, 1 part; mix, and distil 3 parts. It must be preserved from the air and light. Used in tapeworm.—*Dose*, 1 to 2 teaspoonfuls, in water, night and morning, until 5 or 6 fl. oz., or more, have been taken; a cathartic being given every third day.

Oil, Cologne. This is a convenient mixture for the ready preparation of eau de Cologne and for perfuming dental and other preparations. Oil of bergamot, 8 oz.; oil of lemon, 4 oz.; oil of orange peel (sweet), 2 oz.; oil of bitter almonds, 2 oz.; oil of lavender, 4 oz.; oil of rosemary, ½ oz.; oil of neroli, 1 oz.; oil of cloves, ½ oz.; extract of musk, 8 oz.; alcohol, to make 64 oz. Use 8 oz. of the 'oil' to 1 gall. of alcohol (Druggists' Circular).

Oil, Exeter. *Syn.* OLEUM EXCESTRENSE (*Gray*). Green oil, 16 lbs.; euphorbium, mustard seed, castor, pellitory, of each, 1 oz.; digest and strain. The original form is more complex. The following is also used. Rape oil, 1½ pints; green oil, ½ pint; oils of wormwood, rosemary, and origanum, of each, half a dr.

Oil, Fur'niture. *Syn.* MAHOGANY OIL, OIL STAIN. *Prep.* 1. From refined linseed oil, 1 pint; alkanet root, ¼ oz.; digested together in a warm place until the former is sufficiently coloured, when it is poured off and strained.

2. Pale boiled oil, 1 pint; beeswax, ¼ lb.;

melted together, and coloured as before. Gives a superior polish, which becomes very tough by age.

3. Linseed or boiled oil, 1 pint; Venice turpentine (pure), 6 oz.; as before. The above are used for mahogany and other dark-coloured woods.

4. Linseed oil, 8 oz.; vinegar, 4 oz.; oil of turpentine, mucilage, rectified spirit, of each, ½ oz.; butter of antimony, ½ oz.; hydrochloric acid, 1 oz. Mix.

5. Linseed oil, 16 oz.; black resin, 4 oz.; vinegar, 4 oz.; rectified spirit, 3 oz.; butter of antimony, 1 oz.; spirit of salts, 2 oz.; melt the resin, add the oil, take it off the fire, and stir in the vinegar; let it boil for a few minutes, stirring it; when cool put it into a bottle, and add the other ingredients, shaking all together. The last two are specially used for reviving French polish.

6. (Pale.) *a.* As the preceding, omitting the alkanet.

b. From nut oil, ½ pint; beeswax (finest), 3 oz.; melted together.

c. To the last add of copal varnish, 3 or 4 oz. The last three are employed for pale woods. They are all applied by means of a rag, and are 'polished off' with a 'woollen rubber' or 'furniture brush.' A little strong vinegar, or a few drops of hydrochloric acid, are sometimes added. See POLISH.

Oil, Hair. See OIL (Perfumed).

Oil and Hartshorn. Liniment of ammonia.

Oil, I'ron. *Syn.* OLEUM FERRI, O. MARTIS, L. The old name for the liquid formed when perchloride of iron is allowed to deliquesce by free exposure to the air. It is excessively caustic and corrosive.

Oil, Lime. See CALCIUM (Chloride).

Oil, Macas'sar. See OILS (Perfumed).

Oil, Mar'row. *Prep.* From clarified beef marrow, 1 part; oil of almonds, 3 parts; melted together and strained through muslin. It is usually scented with ambergris, cassia, or mace, and slightly tinged with palm oil or annotta. Used for the hair.

Oils, Marshall's. *Prep.* From linseed oil and rape oil, of each, 1 lb.; green oil and oil of turpentine, of each, ½ lb.; oil of origanum, ½ fl. oz.; oil of vitriol, ½ oz.; well shaken together.

Oils, Mixed. *Syn.* OLEUM MIXTUM COMMUNE, L. *Prep.* From linseed oil and green oil, of each, 1 lb.; oil of turpentine, ½ lb.; Barbadoes tar and balsam of sulphur, of each, 2 oz.; oils of spike and origanum, of each, 1 oz. Stimulant and rubefacient. Used by farriers for sprains, &c. See OILS, STAMFORD'S (*below*).

Oils, Newmarket. *Prep.* From oils of linseed, turpentine, and St John's-wort, of each, 3 lbs.; oil of vitriol, 1½ oz.; well shaken together, and the clear portion decanted in a few days. A favourite remedy for sprains in horses.

Oils, Nine. *Syn.* OLD MIXED OILS; OLEUM EX OMNIBUS, L. *Prep.* From train oil, 1 gall.; oil of turpentine, 1 quart; oil of amber and oil of bricks, of each, 5 oz.; oil of spike and oil of origanum, of each, 2 oz.; Barbadoes tar, 2½ lbs.; oil of vitriol, 2 oz.; camphorated spirit, ½ pint; mixed together as the last. A favourite remedy with provincial farriers.

Oil of Petre. See OIL, BRITISH (*above*).

Oil, Phos'phorated. *Syn.* OLEUM PHOSPHORA-
TUM, L. *Prep.* 1. (Ph. Bor.) Phosphorus (dried
and sliced small), 6 gr.; oil of almonds, 1 oz.;
mix, place the phial in hot water, agitate for some
time, and, when cold, decant the clear oil from the
undissolved phosphorus.

2. (*Magendie.*) Phosphorus (sliced), ½ dr.; almond oil, 1 oz.; macerate in the dark, with frequent agitation, for 14 days, then, after repose,
decant the clear portion, and aromatise it with a
little essence of bergamot.

3. (B. Ph.) *Prep.* Take of phosphorus and oil
of almonds, of each, q. s. Heat the oil in a porcelain dish to 300° F., and keep it at this temperature for about 15 minutes, then let it cool and
filter it through paper. Put 4 fl. oz. of this oil
into a stoppered bottle capable of holding 4½ fl.
oz.; then add to it 16 gr. of phosphorus. Immerse the bottle in hot water until the oil has acquired the temperature of 180° F., removing the
stopper two or three times to allow the escape of
expanded air; then shake the oil and phosphorus
together until the latter is entirely dissolved.—
Dose, 5 to 10 minims.

Obs. A fl. oz. of oil dissolves rather less than
5 gr. of pure phosphorus. The large excess ordered in the second formula must be merely for
the purpose of increasing the extent of surface
acted on. It is, however, with the other precautions
given, quite unnecessary. The products of both
formulæ have the same strength.—*Dose,* 5 to 10
or 12 drops, in milk, barley-water, or gruel, or
made into an emulsion; in chronic rheumatism,
gout, &c., and as a powerful diffusible stimulant
in various diseases, with debility and general prostration of the vital powers, &c. Externally, as a
friction. It is chiefly to the presence of phosphorus that cod liver owes its wonderful remedial
power in these affections.

Oil, Quit'ter. *Prep.* 1. Red precipitate, 2 dr.;
aquafortis, 1 oz.; dissolve, add of olive oil, oil of
turpentine, and rectified spirit, of each, 2 oz.;
and agitate well and frequently for 3 or 4
hours.

2. Ointment of nitrate of mercury (Ph. L.),
1 part; nut oil, 3 parts; melt together, and stir
until the mixture is cold. Used by farriers for
quitters, &c.

Oils, Radley's. From Barbadoes tar, ½ lb.; linseed oil and oil of turpentine, of each, ½ pint;
gently warmed, and shaken together.

Oil, Shav'ing. See ESSENCE OF SOAP.

Oil, Sheldrake's. *Prep.* From pale boiled nut
oil and copal varnish, equal parts, melted together
by the heat of hot water, and, when perfectly
mixed, placed aside in a bottle for a week to
settle, after which the clear portion is decanted.
Used by artists to grind their colours in to
brighten them.

Oil of Spike. 1. (FARRIER'S.) From oil of
turpentine, 1 quart; Barbadoes tar, 1½ oz.; alkanet root, ½ oz.; digested together for a week.
Used as a stimulating liniment by farriers.

2. (PAINTER'S.) *a.* From rectified oil of turpentine, 3 pints; oil of lavender, 1 pint; mix.

b. Oil of turpentine (warm), 5 parts; lavender
oil bottoms (genuine), 3 parts; agitate well together, and in a fortnight decant the clear away.
Used by artists and enamellers.

Oils, Stamford's. *Syn.* LORD STAMFORD'S
MIXED OILS. *Prep.* Dissolve camphor, 1 oz., in
rectified spirit of wine, ½ pint; add oil of origanum, 2 oz.; oil of turpentine, ½ pint; green
elder oil, 2 lbs.; and agitate until mixed. The
rectified spirit is now generally omitted, the camphor being dissolved in the green oil by aid of heat
before adding the other ingredients. Stimulant.
Used by farriers.

Oil of Stone. This very old-fashioned preparation is, says the 'Druggists' Circular,' of variable
composition, its chief ingredient being crude
petroleum. Some recipes direct equal parts of
American petroleum and Barbadoes tar, and again
this mixture is diluted with turpentine as follows:—American petroleum, 1 part; Barbadoes
tar, 1 part; spirit of turpentine, 3 parts. Another
more complex petroleum mixture, usually called
British oil, was also sometimes sold as ' oil of
stone.'

Oil, Sul'phurated. *Syn.* BALSAM OF SULPHUR;
OLEUM SULPHURATUM, BALSAMUM SULPHURIS,
L. *Prep.* 1. (Ph. L. 1746.) Flowers of sulphur,
1 part; olive oil, 4 parts; boil together in a vessel
lightly covered, until they assume the consistence
of a thick balsam.

2. (Ph. L. 1824.) Olive oil, 16 fl. oz.; heat it
in a sand-bath, and gradually add of washed sulphur, 2 oz.; stirring until they combine.

Prop., &c. Balsam of sulphur is a dark reddish-brown, viscid fluid, having an extremely
disagreeable and penetrating odour, and a strong,
nauseous taste. The local action of balsam of
sulphur is that of an acrid and irritant; its
remote effects those of a stimulant, expectorant,
and diaphoretic. Externally, it is occasionally
used as an application to foul ulcers; and was
formerly commonly employed internally in
chronic pulmonary affections, in doses of 20 to
50 drops. It is now seldom given internally
except in veterinary practice.

Oils, Three. *Syn.* OLEUM DE TRIBUS (*Van
Mons*), L. Oils of brick, lavender, and turpentine, equal parts. As a stimulant liniment.

Oil of Vit'riol. Sulphuric acid.

Oils, Ward's. *Syn.* WARD'S WHITE OILS.
From powdered camphor, rape oil, oil of turpentine, rectified spirit, and liquor of potassa, equal
parts, agitated together for some time, and again
before use. Beef brine was formerly used instead
of liquor of potassa.

Oil, Watchmaker's. Prepared by placing a
clean strip or coil of lead in a small white glass
bottle filled with pure almond or olive oil, and
exposing it to the sun's rays at a window for
some time till a curdy matter ceases to be deposited, and the oil has become quite limpid and
colourless. Used for fine work; does not become
thick by age.

Oil, Wedell's. *Syn.* BEZOAR OIL; OLEUM
BEZOARDICUM, L. From nut oil, ½ pint; camphor, ½ oz.; dissolve by a gentle heat, and, when
cold, add of essence of bergamot, 1 dr., and let it
stand over a little alkanet root until sufficiently
coloured.

Oils, White. *Syn.* WHITE EGG-OILS. *Prep.*
1. Yolks of eggs, 4 in number; oil of turpentine,
½ pint; mix, add of liquor of ammonia, 3 fl. oz.;
oil of origanum, ½ oz.; soaper's lye, ½ pint; water,

¼ pint; agitate well, and strain through a coarse hair sieve.

2. Rape oil, ⅜ pint; liquor of ammonia and oil of turpentine, of each, 3 oz.; agitate until they form a milk.

3. (*Redwood.*) Whites and yolks of 2 eggs; oil of turpentine, 1½ oz.; triturate together, add of Goulard's extract, ¼ oz.; mix, next add of distilled vinegar, 1½ pints, and, lastly, of rectified spirit, 1½ fl. oz. Stimulant and detergent. Used by farriers.

Oil, Worm (Canine). *Syn.* OLEUM VERMIFUGUM CANINUM. *Prep.* From oil of turpentine and castor oil, equal parts; tinged yellow with a little palm oil or annotta.—*Dose.* For a middle-sized dog, ¼ oz., repeated in 2 or 3 hours if it does not operate.

OILS (in Perfumery). *Syn.* SCENTED OILS; OLEA FIXA ODORATA, L. The oils which usually form the basis of these articles are those of almonds, ground nut, cotton seed, ben, or olives; but others are occasionally used. The methods adopted for their preparation vary with the nature of the substances whose fragrance it is intended to convey to the oil. The Continental perfumers employ three different processes for this purpose, which they technically distinguish by terms indicative of their nature. These are as under:

1. A sufficient quantity of the essential oil of the plant, or of the concentrated essence of the substance, if it does not furnish an oil, is added to the fixed oil which it is desired to perfume, until the latter becomes agreeably fragrant; the whole is then allowed to repose for a few days, and, if any sediment falls (which should not be the case when the ingredients are pure), the clear portion is decanted into another bottle. When alcoholic essences are thus employed, the fixed oil should be gently warmed, and the admixture made in a strong bottle, so as to permit of it being corked and well agitated with safety; and in this case the agitation should be prolonged until the whole has become quite cold. In this way all the ordinary aromatised and perfumed oils of the English druggists and perfumers, as those of bergamot, cassia, cloves, lavender, lemon, millefleurs, neroli, nutmeg, oranges, roses, &c., are made; but those of a few of the more delicate flowers, and of certain other substances, can only be prepared of the first quality by one or other of the processes described below.

In general, 1 to 1½ dr. of the pure essential oil, or 3 to 4 fl. dr. of the alcoholic essences, are found sufficient to render 1 pint of oil agreeably fragrant. ½ dr. of pure attar of roses is, however, enough for this purpose, owing to the very powerful character of its perfume; but even a less quantity than this is commonly employed, on account of its costliness, the deficiency being made up by a mixture of the oils of rhodium, rosemary, and bergamot. Most of the oils of this class are intended for hair cosmetics.

2. (By INFUSION.) Dry substances, after being reduced to powder, or sliced very small—flowers or petals, after being carefully selected, and picked from the stems and other scentless portions—and soft or unctuous matters, as ambergris, civet, or musk, after being rubbed to a

paste with a little oil, either with or without the addition of about twice their weight of clean sand or powdered glass, to facilitate the reduction, are digested in the fixed oil for about 1 hour, at a gentle heat obtained by means of a water-bath, continual stirring being employed all the time; the mixture is then removed from the heat, covered up, and left to settle until the next day, when the clear portion is decanted into clean bottles. When flowers are employed, the free oil is drained off, and the remainder obtained by the action of a press. The process is then repeated with fresh flowers, five or six times, or even oftener, until the oil is sufficiently perfumed. For ambergris, musk, or civet, the digestion is generally continued for 15 to 20 days, during which time the vessel is either freely exposed to the sunshine, or kept in an equally warm situation.

The first quality of the oils of ambergris, balsam of Peru, benzoin, cassia, cinnamon, civet, orange flowers, orris, roses, styrax, and vanilla is made by infusion.

3. (By THE FLOWERS.) *a.* Upon an iron frame a piece of white, spongy cotton cloth is stretched, and then moistened with almond or olive oil, usually the latter; on the cloth is placed a thin layer of the freshly plucked flowers; another frame is similarly treated, and in this way a pile of them is made. In 24 or 30 hours the flowers are replaced by fresh ones, and this is repeated every day or every other day, until 7 or 8 different lots of flowers have been consumed, or the oil is sufficiently loaded with their odour. The oil is then obtained from the cotton cloth by powerful pressure, and is placed aside in bottles to settle, ready to be decanted into others for sale. Sometimes thin layers of cotton wool, slightly moistened with oil, are employed instead of cotton cloth.

The oils of honeysuckle, jasmine or jessamine, jonquil, may blossom, myrtle blossom, narcissus, tuberose, violet, and, in general, of all the more delicate flowers, are prepared in the above manner.

b. The native perfumers of India prepare their scented oils of bela, chumbul, jasmine, &c., in the following manner:—A layer of the scented flowers, about 4 inches thick and 2 feet square, is formed on the ground; over this is placed a layer of moistened tel or sesamum seeds, 2 inches thick, and on this another 4-inch layer of flowers. Over the whole a sheet is thrown, which is kept pressed down by weights attached round the edges. The flowers are replaced with fresh ones after the lapse of 24 hours, and the process is repeated a third and even a fourth time, when a very highly scented oil is desired. The swollen sesamum seeds, rendered fragrant by contact with the flowers, are then submitted to the action of the press, by which their bland oil is obtained strongly impregnated with the aroma of the flowers. The expressed oil is then set aside in dubbers (bottles made of untanned hides) to settle. We have employed poppy seed in this country, in a similar manner, with great success.

c. The flowers are crushed in a mortar or mill, with one half their weight of blanched sweet almonds, and the next day the mass is gently

72

heated and submitted to the action of a powerful press; the liquid thus obtained is allowed to repose for a week, when the upper portion of oil is decanted and filtered. This plan is occasionally adopted in this country for the oils of roses and of a few other flowers. (See *below*.)

The solution of a few grains of benzoic acid, or of gum benzoin (preferably the first), in any of the above oils, will materially retard the accession of rancidity, if it does not prevent it altogether.

The oils of the last two classes (2 and 3) are chiefly used to impart their respective odours to the simple oils, pomades, &c.; and in the manufacture of scented spirits or esprits. The following formulæ are given as examples of both classes of preparations:

Oil of Am'bergris. From ambergris, 2 dr.; oil, 1 pint; by infusion.

Oil of Ben'zoin. From gum benzoin, 7 dr.; oil, 1 pint; by infusion.

Oils for the Hair. *Syn.* HUILES ANTIQUES, Fr. These are numerous. All those scented with the simple perfumes are prepared in the way explained under class 1 (*above*). The selection depends entirely upon the judgment of the operator or the fancy of the purchaser. In general, a mixture of two or three perfumes is preferred in these countries to the pure fragrance of any single flower, and a grossness of taste is exhibited in these matters which surprises our Continental neighbours, and the inhabitants of Italy more particularly. Some of these oils are coloured. A red tinge is given to them by allowing the oil to stand for a few hours over a little alkanet root (2 dr. to the pint) before scenting it. The application of a gentle heat facilitates the process. Yellow and orange are given by a little annotta or palm oil; and green, by steeping a little green parsley or lavender in them for a few days; or by dissolving 2 or 3 dr. of gum guaiacum in each pint by the aid of heat, and, when cold, decanting the clear portion. Huile antique au jasmin, Huile antique à la fleurs d'oranges, Huile antique à la rose, Huile antique à la tubéreuse, Huile antique à la violette, &c., are simple oils flavoured with the respective perfumes or their preparations.—Huile antique à la rose is the ordinary oil of roses coloured with alkanet root.—Huile antique verte is simple oil coloured green, as above, and scented.—Huile antique aux mille-fleurs is so scented with several perfumes that none predominate. A mixture of bergamot, lemons, lavender, neroli, pimento, and ambergris or musk, is commonly employed for the purpose.

Oil, Macas'sar. *Syn.* HUILE DE MACASSAR. *Prep.* 1. (*Rowland's*.) Oil of ben or almonds (reddened by alkanet root), 1 pint; oils of rosemary and origanum (white), of each, 1 dr.; oil of nutmeg and attar of roses, of each, 15 drops; neroli, 6 drops; essence of musk, 3 or 4 drops.

2. (*De Naquet.*) Oil of ben, 1 quart; nut oil, 1 pint; rectified spirit, ⅓ pint; essence of bergamot, 3½ dr.; tincture of musk and esprit de Portugal, of each, 2 dr.; attar of roses, ⅓ dr.; alkanet root, q. s. to colour.

Oil, Mar'row (Perfumed). 1. Simple marrow oil, scented at will.

2. (FLUIDE DE JAVA.) Marrow oil, coloured with a little palm oil and scented.

3. (HUILE COMOGÈNE.) Marrow oil, 4 oz.; spirit of rosemary, 1½ oz.; oil of nutmeg, 12 drops.

4. (HUILE PHILOCOME D'AUBRIL.) Cold-drawn nut oil and marrow oil, equal parts; scent at will, q. s.

5. (HUILE DE PHÉNIX.) Clarified beef marrow, lard, pale nut oil, and expressed oil of mace, of each, 4 oz.; melt together by the heat of hot water, strain through linen into a warm stone mortar, add of oils of cloves, lavender, mint, rosemary, sage, and thyme, of each, ⅓ dr.; rectified spirit, 1 oz., in which has been dissolved by a gentle heat balsam of tolu, 4 dr.; camphor, 1 dr.; triturate until the whole is cold, and then put it into bottles. All the above are used to make the hair grow, and to prevent it falling off.

Oil of Musk. *Prep.* From grain musk, 1 dr.; ambergris, ½ dr.; oil of lavender, 20 drops; oil, 1 pint, by infusion. A second quality is made by working the same ingredients, after the oil is poured from them, with ¾ pint of fresh oil. This also applies to OIL OF AMBERGRIS and HUILE ROYALE.

Oil of Musk and Am'bergris. *Syn.* HUILE ROYALE. *Prep.* From ambergris, 2 dr.; grain musk, ⅓ dr.; oils of cassia, lavender, neroli, and nutmeg, of each, 10 drops; oil, 1 pint; by infusion. See *above*.

Oil of Sty'rax. *Prep.* From liquid styrax (pure), 5 dr.; oil of nutmeg, 10 drops; ambergris, 6 gr.; oil, 1 pint; by infusion.

Oil of Vanil'la. *Syn.* HUILE À LA VANILLE. *Prep.* From purest olive or almond oil, 1½ pints; vanilla (finest, in powder), 2 oz.; oil of bergamot, 1 dr.; attar of roses (finest), 15 drops; by infusion.

OILS (Volatile). (Although essential oils are volatile oils, volatile oils are not always essential ones as the term is understood. This is the case with the petroleum and paraffin oils obtained by the distillation of native petroleum and bituminous bodies. To describe the two as synonymous is therefore incorrect.—ED.) *Syn.* OLEA DESTILLATA, OLEA DISTILLATA, OLEA ESSENTIALA, OLEA VOLATILIA, L.; HUILES VOLATILES, Fr. The volatile oils are an extensive and important class of bodies, derived from the vegetable kingdom, and found in almost every part of the majority of the plants which produce them, except the cotyledons of the seeds, in which, in general, the fixed oils are exclusively stored up. Their presence confers upon flowers, leaves, fruit, seeds, roots, bark, and woods their peculiar and characteristic odours; but among these they are not equally distributed in the same individual, and are often altogether absent from some of them. To them we are indebted for our most delightful perfumes, and our choicest spices and aromatics. Some of them are found to possess valuable medicinal properties, and others are invested with the highest possible interest on account of their peculiar chemical constitution, and the reactions which occur when they are brought into contact with other bodies.

The volatile oils are often called 'essences,' and the same loose and unmeaning term is also commonly applied to their alcoholic solutions.

Prop. The volatile or essential oils are usually more limpid and less unctuous than the fixed oils; but some of them are butyraceous or crystalline. Nearly all of them consist of two or more oils, differing in their sp. gr. and boiling-points, one of which is generally liquid, the other, in some cases, crystalline. All of them, when perfectly pure, are colourless, though before rectification nearly the whole of them have as pale yellow tint, and some of them are brown, blue, or green. Their odour is that of the plants which yield them, and is usually powerful; their taste is pungent and burning. They mix in all proportions with the fixed oils, dissolve freely in both alcohol and ether, and are sparingly soluble in water, forming 'perfumed' or 'medicated waters.' Their boiling-point usually ranges between 310° and 325° F., and is always considerably higher than that of water. They resist saponification, and (excepting oil of cloves) do not combine with the salifiable bases. Their density fluctuates a little on either side of water. The lightest oil is that of citrons (sp. gr. ·847), and the heaviest that of sassafras (sp. gr. 1·096). When cooled sufficiently, they all solidify. The common temperature of the atmosphere is sufficient for this with some of them, as the oils of roses and aniseed; whilst others require to be cooled below the freezing-point of water before they assume the solid form. In this state they appear to consist of a crystalline or semi-crystalline substance (stearopten, stearessence), and a fluid portion (eleopten, oleiessence). The two may be separated by pressing the concrete oil between the folds of bibulous paper, in the cold. By exposure to the air the volatile oils rapidly absorb oxygen, and become partially converted into resin. This is the cause of the deposit that usually forms in them (especially in the expressed oil of orange) when kept in an ill-corked vessel. The solid crystalline matter which separates from them when kept in closed vessels is stearopten, as mentho and thymol.

Class. Chemically considered, the essential oils may be divided into three great classes:

1. Oils composed of carbon and hydrogen only (binary volatile oils, carbo-hydrogens, hydrocarbons, terebenes, camphenes), of which oil of turpentine may be regarded as the type. These are characterised by being, as a class, less soluble in rectified spirit and in water than the other essential oils. The oils of bergamot, capivi, cubebs, elemi, hops, juniper, lemons, orange peel, pepper, the grass oil of India, the laurel oil of Guiana, and some others, belong to this class.

2. Oils containing carbon, hydrogen, and oxygen (oxygenated oils), including most of those used in medicine and perfumery. These, as a class, are more soluble in rectified spirit and in water than those containing carbon and hydrogen only. To this class belong the oils of almonds, aniseed, cassia, cedar-wood, cinnamon, cumin, jasmine, lavender, meadow-sweet (*Spiræa ulmaria*), orange flowers, penuyroyal, peppermint, spearmint, rosemary, rose-petals, valerian, winter-green (*Gaultheria procumbens*), and others too numerous to mention. A few of these oxygenated oils contain nitrogen.

3. Oils containing sulphur (sulphuretted oils). These are characterised by their extreme pungency, suffocating odour, vesicating power, property of blacking silver, and being decomposed by contact with most other metallic bodies. The oils of assafœtida, black mustard-seed, garlic, horse-radish, and onions are of this kind. Some sulphuretted oils contain nitrogen.

Prep. The volatile oils are generally procured by distilling the odoriferous substances along with water; but in a few instances they are obtained by expression, and still more rarely by the action of alcohol.

According to the common method of proceeding, substances which part freely with their oil are put into the still resting on a perforated diaphragm, along with about an equal weight of water, and are at once submitted to distillation. Those substances which give out their oil with difficulty are first soaked for 24 hours, or longer, in about twice their weight of water, to each gallon of which 1 lb. of common salt has been added in order to raise its boiling-point. The distillation is conducted as quickly as possible, and, when one half the water has come over, it is returned into the still, and this cohobation is repeated when necessary, until the distilled water ceases to be mixed with oil. The heat of steam or a salt-water bath should be preferably employed. When a naked fire is used the still should be deep and narrow, by which means the bottom will be better protected by the gradually decreasing quantity of water towards the end of the process, and empyreuma prevented. When the distilled water is to be repeatedly cohobated on the ingredients a convenient and economical plan is to so arrange the apparatus that, after the water has separated from the oil in the receiver, it shall flow back again into the still. An ordinary worm-tub, or other like condensing apparatus, may be employed; but in the case of those oils which readily solidify the temperature of the water in the condenser must not fall below about 55° F.

The mixed vapours which pass over condense and fall as a milky-looking liquid into the receiver. This separates after a time into two portions, one of which is a solution of a part of the newly eliminated oil in water, and the other is the oil itself. The latter either occupies the upper or the lower portion of the receiver, according as its specific gravity is less or greater than that of distilled water. The separation of the oil and water is effected by allowing the mixed liquids to drop into a 'Florentine receiver' (see *engr.*) when the oil is the lighter of the two, by which means the latter accumulates at *a*, and the water flows over by the spout (*b*).

The same receiver may be employed for oils heavier than water by reversing the arrangement;

but a glass 'separator' (see *engr*.) is, in general, found more convenient. In this case the oil accumulates at the bottom of the vessel, and may be drawn off by the stopcock provided for the purpose.

The essential oils of lemons and oranges of commerce, and of some other fruits, are chiefly obtained by submitting the yellow rind to powerful pressure; but in this way they are not so white, nor do they keep so well, as when distilled, although in the case of the fruits referred to the oils are more fragrant than when prepared by any other method.

Chevallier gives the following rules for the distillation of essential oils:

1. Operate upon as large quantities as possible, in order to obtain a greater product, and one of finer quality.

2. Conduct the distillation rapidly.

3. Divide the substances minutely in order to facilitate the extrication of the oil.

4. Employ only sufficient water to prevent the matter operated on from burning, and the product from being contaminated with empyreuma.

5. For substances whose oil is heavier than water saturate, or nearly saturate, the water in the still with common salt to raise the boiling-point, and thus to enable the vapour to carry over more oil.

6. Employ, when possible, water which has been already distilled from off the same substances, and has thus become saturated with oil.

7. For oils naturally fluid keep the water in the refrigerator cool; but for those oils which easily become solid preserve it at 80° to 90° F. (?).

To the above may be added:

8. Collect the oil as soon as possible after it separates from the water with which it passes over, and in its subsequent treatment keep it, as much as possible, from free contact with the air.

Dr Ure remarks, "The narrower and taller the alembic is, within certain limits, the greater will be the proportion of oil, relative to that of the aromatic water, from like proportions of aqueous and vegetable matter employed." "Some place the plants in baskets, and suspend these immediately over the bottom of the still under the water, or above its surface in the steam; but the best mode, in my opinion, is to stuff an upright cylinder full of the plants, and drive down through them steam of any desired force, its tension and its temperature being further regulated by the size of the outlet orifice leading to the condenser. The cylinder should be made of strong copper, tinned inside, and encased in the worst conducting species of wood, such as soft deal or sycamore."

The newly distilled oils may be separated from adhering water, which frequently renders them partially opaque or 'cloudy,' by repose in a temperature between 60° and 70° F., and subsequent decantation; but to render them quite dry (anhydrous) it is necessary to let them stand over some fragments of fused chloride of calcium. This is not, however, required with the commercial oils.

The rectification of the volatile oils is commonly performed without water by the careful application of a heat just sufficient to make them flow over pretty rapidly, so that they may be kept heated for as short a time as possible. One half, or at most two thirds only, is drawn off, that left in the retort being usually mixed with the raw oil intended to be sold in that state. This method often leads to much loss and disappointment, and we have known more than one rather dangerous explosion result from its use. A better plan is to rectify the oil from strong brine, and then to separate any adhering water, either by repose or chloride of calcium.

Pres. Volatile oils should be preserved in well-closed and nearly full bottles, in the shade, and should be opened as seldom as possible. By age they darken, lose much of their odour, increase in density, and become thick and clammy. It is then necessary to distil them, by which the undecomposed portion is separated from the resin. Agitation along with animal charcoal will restore their clearness and original colour, but nothing more.

Pur., Tests. The essential or volatile oils of commerce are very frequently adulterated with the fatty oils, resins, spermaceti, or alcohol, or with other essential oils of a cheaper kind or lower grade. The presence of the first three of these may be readily detected by placing a drop of the suspected oil on a piece of white paper, and exposing it for a short time to heat. If the oil is pure it will entirely evaporate, but if adulterated with one of these substances a greasy or translucent stain will be left on the paper. These substances also remain undissolved when the oil is agitated with thrice its volume of rectified spirit.

The presence of alcohol may be detected by agitating the oil with a few small pieces of dried chloride of calcium. These remain unaltered in a pure essential oil, but dissolve in one containing alcohol, and the resulting solution separates, forming a distinct stratum at the bottom of the vessel. When only a very little alcohol is present the pieces merely change their form, and exhibit the action of the solvent on their angles or edges, which become more or less obtuse or rounded.

Another test for alcohol in the essential oils is the milkiness occasioned by agitating them with a little water, as well as the loss of volume of the oil when it separates after repose for a short time.

A more delicate test of alcohol in the essential oils than either of the preceding is potassium, as employed by M. Beral:—12 drops of the oil are placed on a perfectly dry watch-glass, and a piece of potassium, about the size of an ordinary pin's head, set in the middle of it. If the small fragment of metal retains its integrity for 12 or 15 minutes, no alcohol is present; but if it disappears after the lapse of 5 minutes, the oil contains at least 4% of alcohol; and if it disappears in less than 1 minute, it contains not less than 25% of alcohol.

Boettger states that anhydrous glycerin possesses the property of dissolving in alcohol, without mixing with the volatile oils. The mode of applying the glycerin is as follows:—The oil to be examined is well shaken in a graduated tube, with its own volume of glycerin (sp. gr. 1·25). Upon being allowed to settle, the mixture separates into two layers. The denser glycerin separates rapidly, and if the essence has been mixed with alcohol, this is dissolved in the glycerin, the augmentation in the volume of glycerin showing the proportion of alcohol present.

This species of adulteration is very common, as it is a general practice of the druggists to add a little of the strongest rectified spirit to their oils, to render them transparent, especially in cold weather. Oil of cassia is nearly always treated in this way.

The admixture of an inferior essential oil with one more costly may be best detected by pouring a drop or two on a piece of porous paper or cloth, and shaking it in the air, when, if occasionally smelled, the difference of the odour at the beginning and the end of the evaporation will show the adulteration, especially if the added substance is turpentine. The presence of the latter may also be detected by agitating the oil with rectified spirit, when it will remain undissolved.

The following method, which may also be used as a test for the presence of turpentine, is based upon its power of dissolving fats:—Take about 50 gr. of oil of poppy in a graduated glass tube, and add an equal quantity of the sample of essential oil. Shake the mixture up thoroughly and then allow it to stand; if the essential oil be pure, the mixture becomes milky, and does not clear until after several days have passed, whereas it will remain transparent if even so little as 5% of essence of turpentine be present.

Turpentine may be detected as an adulterant in lemon oil by heating a portion of the sample in a dry test-tube, with a piece of copper butyrate about the size of a pin's head; the temperature is slowly raised to 170° but must not exceed 180°. If the oil of lemon is pure the copper salt dissolves and colours the oil green. If turpentine is present the oil becomes turbid, is coloured yellow, and reddish-yellow copper protoxide is separated.

The purity of essential oils may likewise, in many cases, be determined by taking their sp. gr.; or, with still greater accuracy and convenience, by measuring their index of refraction, as suggested by Dr Wollaston. A single drop of oil is sufficient for the application of the last method.

The adulteration of a heavy oil with a light one, or the reverse, may be detected by agitating the suspected oil with water, when, in most cases, the two will separate and form distinct strata.

Miss Crane believes that the cohesion figures afforded by the volatile oils, like those of the fixed ones, will be found useful indications of their purity. The application of her method is precisely similar to that followed in her examination of the fixed oils as already described. She finds that—

Oil of Turpentine, by itself, spreads instantly to the whole size of the plate (a common soup plate), and almost immediately the edge begins to break into irregular shapes, when a rapid motion takes place over the surface of the film, and there seems to be a contest between the cohesion of the oil particles and the adhesion between them and the water. The oil makes repeated efforts to gather itself closer together, when the water instantly reacts, giving a wavy appearance to the whole figure.

The play of colours at this point is beautiful, and serves to bring out the lines more perfectly. In a few seconds innumerable little holes appear over the surface, which soon are separated only by threaded lines, and the figure is like the most exquisitely fine lace.

Oil of Cinnamon forms a figure not more than half the size of the last-named. In a few seconds small portions are detached, and shortly separate into distinct drops, four or five larger, and a number of smaller ones, scattered about. With mixtures in different proportions of oil of turpentine, the figures formed differently, taking more of the characteristics of the adulterant as it predominated.

Oil of Nutmeg forms a large figure instantly, the edge showing a beaded line. It gathers itself together and spreads again, very like oil of turpentine, but the surface presents more the appearance of watered silk. Within sixty seconds some holes appear, and in eighty more the surface is covered with them; these scarcely spread to more than a sixteenth of an inch in diameter, but from the first each is bordered with a dotted edge. The figure lasts some time without changing materially, except the openings lengthen out into an oblong shape, remaining entirely distinct. The play of colours is very fine. With the addition of one third of the oil of turpentine, the first spreading is little different, but openings appear in half the time, and the dotted border does not come as soon; in about four minutes the figure is most characteristically marked, and soon breaks up entirely, this being the distinctive difference between the pure oil and the mixture.

Oil of Peppermint spreads instantly to a large figure, and in 10 or 15 seconds openings appear, which increase rapidly in size. At first they look somewhat like the last-named, but are not nearly so numerous, and the border soon is more like tiny drops. In 1½ or 2 minutes they begin to run together, and the figure breaks up.

With the addition of turpentine oil the figure forms more slowly, and the breaking up is less rapid, but in five minutes the outlines only remain.

Since the demand for menthol has extended, it is no uncommon circumstance to meet with specimens of peppermint oil, supposed to be genuine, from which the menthol has been abstracted. To detect this fraud, the following test may be found useful:—A test-tube, partially filled with the oil and corked, is placed in a freezing mixture of snow and salt, for 10 or 15 minutes. At the end of that time, if the oil has not been tampered with, it will have become cloudy, thick, or of a jelly-like consistence. If 4 or 5 crystals of menthol be then added, and the tube be replaced in the freezing mixture, the oil will, after a short time, form a solid frozen mass of crystals. If, on the other hand, the oil remains limpid, it may be concluded that the menthol has been removed.

Oil of Bergamot spreads instantly; in 30 seconds tiny openings appear, not very abundant, and increase in size slowly; in 5 minutes they are not larger than *oil of nutmeg* at 1½ minutes. At first they have a dotted border, but as they increase in size this changes to a scalloped film, which spreads, until, in 8 or 10 minutes, they are joined together over the whole surface. This, with the *turpentine oil*, gives a watered surface in spreading, much more marked, and with a fine play of colours.

R. H. Davies, 'Pharm. Journ.,' 3rd, xix, estimates the purity of essential oils by their power of absorbing iodine; for the details of the method the original communication should be referred to.

Uses, &c. The volatile oils are chiefly used by perfumers and rectifiers, and in medicine. Some of the cheaper kinds are largely employed as vehicles for colours, and in the manufacture of varnishes. The dose of the aromatic and carminative oils is from 1 to 10 drops, on sugar, or dissolved in a little weak spirit. This does not apply to oil of bitter almonds, the dose of which is ¼ to ½ a drop.

*** The following list includes short notices of nearly all the volatile oils which have been examined, as well as of some other substances of a similar character which commonly pass under the name:

Oil of **Allia'ria**. From the flowers of *Alliaria officinalis*, or sauce-alone. Identical with the oil of black mustard.

Oil of **All'spice.** See OIL OF PIMENTO.

Oil of **Al'monds.** See OIL OF BITTER ALMONDS.

Oil of **American Arbor Vitæ.** *Syn.* HUILE CÈDRE BLANC, Fr. From the fresh tops of *Thuja occidentalis*, or American arbor vitæ tree. Yellow; fragrant; stimulant. Used in frictions for rheumatism.—*Prod.*, 1½% to 2% (nearly).

Oil of **Angel'ica.** From the dried root of *Angelica archangelica.*—*Prod.*, 25% (fully).

Oil of **An'iseed.** *Syn.* OLEUM ANISI (Ph. L., E., and D.), O. ESSENTIALE ANISI, L. From the fruit (seeds) of *Pimpinella anisum*, or of *Illicium anisatum*. Nearly colourless. It is very frequently adulterated with one or other of the cheaper oils, in which case spermaceti or camphor is added to it, to make it 'candy.'

Prop., &c. When pure it congeals into a solid crystalline mass on being cooled to 50° F., and does not melt again until heated to about 63°. Treated with iodine, it quickly congeals into a solid hard mass, with a perceptible increase of temperature, and the development of orange-coloured and grey fumes. Sulphuric acid, with heat, turns it of a rich purple-red colour, and the compound soon afterwards becomes inspissated

and hard (resinified). In alcohol of ·806 it is soluble in all proportions, but rectified spirit (·838) dissolves only 42% of this oil. Sp. gr. (recent) ·0768; (one year old) ·9853 to ·9855; (old) ·9856 to ·9900. The foreign oil is generally the heaviest.

Oil of aniseed is carminative and pectoral; and both itself and preparations have long been in favour with the masses in coughs, colds, &c. In preparing it care must be taken that the temperature of the water in the receiver and refrigerator does not fall lower than about 68° F.—*Prod.* (From the dried fruit of commerce) av. 2% (nearly). See OIL OF STAR-ANISE.

Oil, **Ap'ple.** See AMYL (Valerianate of), and ESSENCE OF APPLE.

Oil of **Ar'nica.** *Syn.* OLEUM ARNICÆ, O. A. RADICUM, L. From the roots of *Arnica montana*. Yellowish brown. Sp. gr. ·940.—*Prod.* 16 lbs. yielded 1 oz. of oil. The oil from the flowers of arnica is blue.

Oil of **Asarabac'ca.** *Syn.* OLEUM ASARI, O. A. LIQUIDUM, L. From the roots of *Asarum europæum*. Yellow; glutinous. Two butyraceous oils pass over at the same time.

Oil of **Assafœt'ida.** *Syn.* OLEUM ASAFŒTIDA, L. From the gum-resin. Contains sulphur. Very fetid and volatile.

Oil of **Balm.** *Syn.* OLEUM MELISSÆ, L. From the herb (*Melissa officinalis*). Pale yellow; fragrant. Sp. gr. ·970 to ·975.—*Prod.* 100 lbs. of the fresh flowering herb yielded ¼ oz. of oil (*M. Raybaud*). A mixture of oil of lemons and rosemary is commonly sold for it.

Oil of **Balsam of Peru.** See CINNAMEINE.

Oil of **Ber'gamot.** *Syn.* BERGAMOT, ESSENCE OF B.; OLEUM BERGAMII, O. BERGAMOTÆ, L. By expression from the yellow portion of the rind of the fruit of *Citrus bergamia*, or bergamot orange. Pale greenish yellow; highly fragrant. It is obtained purer by distillation, but its perfume is then slightly less delicate. Sp. gr. ·875 to ·885. —*Prod.* The rind of 100 bergamot oranges yielded by distillation nearly 3 oz. of oil (*M. Raybaud*).

Oil of bergamot is frequently adulterated with rectified spirit, or with the oils of lemons, oranges, or turpentine. The presence of these substances may be detected in the manner explained under OILS (Volatile), *Purity and Tests* (*antè*), as well as by the altered density of the oil. Pure bergamot oil is much more soluble in rectified spirit than either of the others, and is further distinguished from them by its free solubility in solution of potass, forming a clear solution.

Oil of **Bit'ter Almonds.** *Syn.* ESSENCE OF B. A.; OLEUM AMYGDALÆ AMARÆ, O. A. ESSENTIALE, L. From the ground cake of bitter almonds from which the fixed oil has been expressed. The common plan is to soak the cake (crumbled to fragments) for about 24 hours in twice its weight of water, to which ⅓ or ½ of its weight of common salt has been added, and then to submit the whole to distillation, allowing the first half of the water that passes over to deposit its oil, and to run back again into the still. Pale golden yellow; colourless when rectified; tastes and smells strongly nutty, like peach kernels. It consists of 85% to 90% of hydride of benzoyl and 8% to 12% of hydrocyanic acid with a variable quantity of benzoic acid and benzoin. The density varies a little with the age of the oil, and the temperature and rapidity with which it has been distilled. Sp. gr. (recent) 1·0525; (trade crude oil) 1·079 (*G. Wippel*); (old) 1·081 (1·0836—*Pereira*). "Essential oil of almonds, free from adulteration, should have a sp. gr. at most of 1·052" (*Ure*). According to Professor Redwood, the density may vary from 1·0524 to 1·0822. The light oil contains the most hydride of benzoyl, and the heavy oil the most benzoin.—*Prod.* From less than ·2% to ·5% .

Pur. This oil is generally adulterated with cheaper oils, and in nearly every case with alcohol. When it is pure—mixed with oil of vitriol, it strikes a clear crimson-red colour, without visible decomposition—mixed with an alcoholic solution of potassa, crystals are eliminated. Iodine dissolves only partially and slowly in it, without further visible results. Chromate of potassa does not affect it. Nitric acid (sp. gr. 1·42) causes no immediate reaction, and in the course of 3 or 4 days crystals of benzoic acid begin to appear; but if only 8% or 10% of alcohol or rectified spirit is present, a violent effervescence speedily commences, and nitrous fumes are evolved. By using nitric acid, sp. gr. 1·5, the smallest quantity of alcohol may be detected.

Obs. This oil does not pre-exist in the almond, but is formed by the action of water and emulsion on a peculiar crystallisable substance, called amygdalin. It is essentially the hydride of benzoyl, but it always contains a portion of hydrocyanic or prussic acid, to which it owes its very poisonous properties. It is occasionally employed as a substitute for hydrocyanic acid in medicine; but its principal consumption is as a flavouring ingredient and a perfume by cooks, confectioners, liquorists, and perfumers. For this purpose it is dissolved in rectified spirits. See ESSENCE.—*Dose*, ½ to 1 drop.

An oil closely resembling that from bitter almonds is obtained by distillation from the leaves of the peach and cherry-laurel, the bark of the plum-tree, the bruised kernels of cherries, plums, and peaches, the pips of apples, and from several other vegetable substances that possess a nutty odour and flavour.

A NON-POISONOUS OIL OF ALMONDS has been introduced. This is simply the ordinary oil of commerce freed from hydrocyanic acid, and is intended to be substituted for the crude, poisonous oil for domestic purposes. Unfortunately, the purified essence does not keep well, and is often converted after a few months into little else than a solution of benzoic acid, almost devoid of the usual odour and flavour of the bitter almond. "No wonder, then, under such circumstances, that the public preferred the preparations they had been accustomed to, which were not so liable to change" (*Redwood*). The following methods have been adopted for this purpose.

1. (*Liebig.*) Agitate the crude distilled oil with red oxide of mercury in slight excess, and after a few days' contact, rectify the oil from a little fresh oxide of mercury. The product is quite pure when the process is properly managed. The cyanide of mercury thus formed may be either

employed as such or reconverted into mercury and hydrocyanic acid.

2. (*Mackay.*) Commercial oil of almonds, 1 lb.; fresh-slaked lime, q. s. to form a milk-like liquid; afterwards add of solution of potassa, 1½ lbs.; water, 3 pints; agitate occasionally for 48 hours, then distil over the oil, and rectify it from a fresh mixture of lime and potassa.

3. (*Redwood.*) The oil is mixed with an equal quantity of water, and the mixture is digested in a water-bath with red oxide of mercury and small quantities of fresh-slaked lime and protochloride of iron, with as little access of air as possible; as soon as decomposition of the acid has taken place, the whole is introduced into a copper retort, and submitted to distillation. The product is perfectly free from hydrocyanic acid. The first process is, however, the simplest, cheapest, and best.

The sp. gr. of this non-poisonous oil is 1·051 (*G. Whippel*). That of pure colourless hydride of benzoyl is 1·043; it boils at 356° F., is soluble in 35 parts of water, and in all proportions in alcohol and ether. Exposed to the air, it greedily absorbs oxygen, and becomes converted into a mass of crystallised benzoic acid. The purified oil of almonds does the same, only less rapidly.

Oil of Almonds (Facti"tious). *Syn.* ESSENCE OF MIRBANE, NITRO-BENZOL. The preparation of this article on the small scale is explained under NITRO-BENZOL. It is now extensively prepared as a substitute for the oil of almonds obtained by distillation. The following is Mansfield's process. The apparatus consists of a large glass worm, the upper end of which is divided into two branches, gradually dilating so as to form two funnel-shaped tubes. Into one of these concentrated nitric acid is poured, and into the other benzol, which need not, for this purpose, be chemically pure. These bodies meet at the point of junction of the two tubes, and the rate of their flow is regulated by any appropriate means. Chemical reaction instantly takes place, and the new compound is cooled by its passage through the worm, which is refrigerated for the purpose. It has then only to be washed with water or a very weak solution of carbonate of soda for the process to be complete. The product has the sp. gr. 1·209, boils at 415° F., has an intensely sweet taste, and an odour closely resembling, but not actually identical with, that of oil of bitter almonds. Unlike genuine oil of almonds or hydride of benzol, it is insoluble in water, and does not distil without suffering partial decomposition. It is chiefly used to scent soaps, and to adulterate the genuine oil. The benzol for this purpose is obtained from coal-tar. See BENZOL and NITRO-BENZOL.

Oil of Boxwood. (Ph. L. 1746.) *Syn.* OLEUM BUXI EMPYREUMATICUM. Distilled from fragments of boxwood in a retort, with a sand-bath gradually increased in heat. Anodyne, antispasmodic, and diaphoretic.—*Dose*, 10 to 20 drops (Jourdan says 4 to 5 drops in gonorrhœa). It relieves toothache.

Oil, Brandy. See OIL OF GRAPE.

Oil of Bu'chu. OLEUM BAROSMÆ, O. DIOSMÆ, L. From the leaves of *Barosma crenata.* Yellow; lighter than water; smells of the leaves.

Oil of Caj'eput. *Syn.* CAJEPUTI OIL, KYA-POOTIE O.; CAJEPUTI OLEUM (B. P.), OLEUM CAJAPUTI (Ph. L., E., & D.), L. From the dried leaves of the *Melaleuca leucadendron*, Linn., var. *minor.* Colourless when pure (that of commerce is usually green); odorous; aromatic; taste hot and penetrating. Its odour has been compared to a mixture of those of camphor and cardamoms. It boils at 343° F. Sp. gr. ·925 to ·927. When rectified about 3-4ths of the quantity passes over colourless, and has the density ·897; the remaining portion is green, and has the density ·920 to ·925. Its green colour is derived from chloride of copper, the presence of which may be recognised by the red precipitate occasioned by agitating the oil with a solution of ferrocyanide of potassium (*Guibourt*). From the East Indies.

Pure oil of cajeput is slightly soluble in water, entirely and freely soluble in alcohol, dissolves iodine, and when dropped on water rapidly diffuses itself over the surface, and soon completely evaporates. A spurious kind (FACTITIOUS OIL OF CAJEPUT), made of oil of rosemary, flavoured with camphor and the oils of peppermint and cardamoms, and coloured with verdigris, is occasionally met with in the shops.

Oil of cajeput is a powerful antispasmodic and diffusible stimulant.—*Dose*, 3 to 6 drops, on sugar; in cholera, colic, epilepsy, hysteria, rheumatism, spasms, toothache, &c.

Oil of Cam'phor. *Syn.* LIQUID CAMPHOR; OLEUM CAMPHORÆ, O. C. VOLATILE, L. Obtained from incisions in the wood of the camphor tree of Borneo and Sumatra (*Dryobalanops aromatica*), in which it exists in cavities in the trunk; also by distillation from the branches of the *Camphora officinarum*, or laurel camphor tree. Colourless when rectified. Sp. gr. ·910.—*Prod.* 60 lbs. of the crude brown oil yield 40 lbs. of pure white oil and 20 lbs. of camphor. It rapidly oxidises in the air. Used to scent soap. See CAMPHOR (Liquid).

Oil of Car'away. *Syn.* OLEUM CARUI (B. P., Ph. L., E., & D.), O. C. ESSENTIALE, L. From the fruit of *Carum carui* (caraway seeds). Nearly colourless; aromatic; carminative. Sp. gr. ·940; (old) ·946 to ·950.—*Prod.* Av. 5% (nearly). It is frequently adulterated with oil of cumin. Added to purgative medicines to prevent griping.

Oil of Car'damoms. *Syn.* OLEUM CARDAMOMI, O. C. ESSENTIALE, L. From the seed of *Elettaria cardamomum*, or true cardamom. Colourless; fragrant; carminative. Sp. gr. ·948.—*Prod.*, 5% (nearly). The capsules ('lesser cardamoms') yield only about 1% of oil (*M. Raybaud*).

Oil of Cascaril'la. *Syn.* OLEUM CASCARILLÆ, L. From the bark of *Croton eleuteria*, Swartz, or cascarilla tree. Very fragrant.—*Prod.*, ·4% to ·75%.

Oil of Cas'sia. *Syn.* OIL OF CHINA CINNAMON; OLEUM CASSIÆ (Ph. E.), L. From cassia buds, or from cassia bark. Golden yellow; aromatic; fragrant. It is generally adulterated with rectified spirit. Nitric acid converts the pure oil into a crystalline mass. Sp. gr. 1·071 to 1·073; (old) 1·078 to 1·090.—*Prod.* From the buds 1% (barely); from the bark of commerce, ·75%. It is frequently sold for oil of cinnamon.

Oil of Ce'dar-wood. From the wood of a species of *Cedrus.* It consists of two hydrocarbons; one a volatile liquid (cedrene), and the other a solid crystalline compound containing oxygen. Used in preparing objects for the microscope.—*Prod.*, ·2% to 25%.

Oil of Ce'drat. *Syn.* ESSENCE OF CEDRA; OLEUM CEDRI, O. CITRI FINUM, L. From the exterior yellow rind of the fruit of *Citrus medica*, Risso, or citrons, either by expression or distillation, as oil of bergamot. The first portion of oil that comes over is colourless, the latter portion greenish. Very fragrant. Sp. gr ·847.—*Prod.* 100 citrons yield nearly 1 fl. oz. of pale and ½ fl. oz. of green oil. See OIL OF CITRON (*below*).

Oil of Cel'ery-seed. *Syn.* OLEUM APII, L. From the fruit (seed) of *Apium graveolens*. Diuretic; stimulant.—*Prod.*, ½% to 1% (nearly).

Oil of Cham'omile. *Syn.* OIL OF ROMAN CHAMOMILE; OLEUM ANTHEMIDIS (Ph. L., E., & D.), O. CHAMÆMELI, O. C. FLORUM, O. ESSENTIALE EX FLORIBUS C., L. From the flowers of *Anthemis nobilis.* In the Ph. L., English oil of chamomile (ANTHEMIDIS OLEUM ANGLICUM) is ordered. Blue; turns yellow and brown by exposure and age; odour characteristic. Sp. gr., English (from the flowers), ·9083; foreign, ·9289.—*Prod.* Fresh flowers, ·1% (barely); recently dried (finest commercial), 5%; av. of 6 dried samples, ·25% (nearly). If much water is employed, even the above small quantities of oil will not be obtained.

Oil of chamomile is reputed antispasmodic, tonic, and stomachic. 1 to 3 drops on a lump of sugar, taken just before retiring to rest, is an excellent preventive of nightmare, and will frequently induce quiet sleep where more active substances have failed. Unfortunately, the oil of the shops is generally either adulterated or old, and commonly both, in which case the oil acts as an irritant. A common plan is to mix it with old oil of lemons, a fraud which may be detected by the lessened density of the oil, and by its diminished solubility in rectified spirit.

Oil of Cherry-laurel. *Syn.* OLEUM LAURO-CERASI, L. From the leaves of *Cerasus lauro-cerasus*, or common laurel. Closely resembles oil of almonds, but is said to be weaker. Like that substance, it is powerfully poisonous.—*Prod.*, 100 lbs. fresh leaves (undeveloped, June), 10·13 oz.; do. (half-grown, June), 7·2 oz.; do. (full-grown, 8 weeks on tree, July), 4·96 oz.; do. (do., 3 months on tree, Sept.), 7·04 oz.; do. (15 months on tree), 2·24 oz. (*Christison*).

Oil of Cher'vil. *Syn.* OLEUM CHÆROPHYLLI, L. From the bruised fresh herb, macerated for 2 or 3 days in salt and water, and then distilled.

Oil of Cin'namon. *Syn.* HYDRIDE OF CINNA-MYL; OLEUM CINNAMOMI (B. P., Ph. L., E., and D.), O. C. VERI, L. From the bark of *Cinnamomum zeylanicum*, macerated for several days in salt and water, and then distilled. Yellowish or red; very aromatic; both odour and taste resemble that of the bark. Sp. gr. 1·035.—*Prod.* 11 lbs. yielded 1 oz.; 100 lbs. yielded 1·56 oz. (*M. Raybaud*).

Pur. Oil of cinnamon, owing to its high price, and the consequent premium for its adulteration, can scarcely be obtained pure from the shops of this country. Oil of cassia and highly rectified spirit are the substances usually employed for this purpose. The increased sp. gr. resulting from the first, and the diminished sp. gr. from the second, afford ready means of detecting these frauds. The presence of oil of cassia may also be detected by an experienced person by the odour, which differs a little from that of pure oil of cinnamon. Oil of cassia is less limpid than oil of cinnamon, and it stands a greater degree of cold without becoming turbid or congealing. "Wine-yellow, when recent; cherry-red, when old; odour purely cinnamonic; nitric acid converts it nearly into a uniform crystalline mass" (Ph. E.). During this reaction the odour of bitter almonds is perceptible. Both oil of cassia and oil of cinnamon are thus converted into a brown balsam; with oil of cassia, however, a brisk decomposition occurs sooner, and at a slighter heat. It also forms a crystalline compound with ammonia. These reactions, unfortunately, are not characteristic. "The most distinguishing characteristic of the cinnamon oils is, perhaps, their relation to the alcoholic solution of caustic potash. Both dissolve in it readily and clear, with a reddish, yellowish-brown colour; after some time, however, the solution becomes very turbid, and a rather heavy undissolved oil precipitates, when the solution gradually becomes clear again" (*Ure*). The palest oil is considered the best.

Obs. Oil of cinnamon is chiefly imported from Ceylon, where it is distilled from bark that is unfit for exportation. The dark coloured oil is usually rectified, when two pale oils are obtained, one lighter, and the other heavier, than water; but 10% of oil is lost by the process. The oil distilled from the root of the tree (O. CINNAMOMI RADICIS) is much weaker than that from the bark. The oil from the leaves (O. C. FOLIORUM), also imported from Ceylon, smells of cloves, but has a less density than oil of cloves.

Oil of cinnamon consists essentially of hydride of cinnamyl, but unless when very recently prepared, it also contains a variable proportion of cinnamic acid formed by the oxidation of the hydride.

Oil of Cit'ron. *Syn.* ESSENCE OF CITRON; OLEUM CITRI, L. From the lees of citron juice; or from the peels, as oil of lemons or bergamot. The last generally goes by the name of oil of cedrat. Both are fragrant (see *above*).

Oil of Citron Flowers. *Syn.* OLEUM CITRI FLORUM, L. Amber-coloured; highly fragrant. —*Prod.* 60 lbs. yield 1 oz.

Oil of Cloves. *Syn.* ESSENCE OF CLOVES; OLEUM CARYOPHYLLORUM, O. CARYOPHYLLI (B. P., Ph. L., E., and D.), O. EUGENIÆ C. (Ph. D., 1826), L. From the unexpanded flowers (cloves) of the *Eugenia caryophyllata*, or Molucca clovetree, soaked for some time in salt and water, and then submitted to distillation; the distilled water, after having deposited its oil, being returned three or four times into the still, and again 'worked off' from the same materials. Nearly colourless when recent, gradually becoming pale yellow and ultimately light brown, by age; highly aromatic, with the characteristic odour and flavour of cloves. It is the least volatile of all the essential oils. Sp. gr. 1·055 to 1·061.—*Prod.*, 16% to 22%.

Pur. Oil of cloves is frequently adulterated with inferior essential oils, especially with those of pimento, pinks, and clove-gillyflowers, and, occasionally, with castor oil. 1. Pure oil of cloves forms a butyraceous coagulum when shaken with pure liquor of ammonia, which crystallises after fusion by a gentle heat. 2. Treated with an alcoholic solution of potassa, it entirely congeals into a crystalline mass, with total loss of its characteristic odour. 3. Shaken with an equal volume of strong caustic soda lye, it forms, on repose, a mass of delicate lamellar crystals. 4. Solution of chromate of potassa converts it into brown flakes, whilst the salt loses its yellow colour. 5. Chlorine turns it first green, and then brown and resinous. 6. Nitric acid turns it red, and a reddish-brown solid mass is formed; with heat, it converts it into oxalic acid. 7. It dissolves freely in sulphuric acid (oil of vitriol), yielding a transparent, deep reddish-brown solution, without any visible decomposition. 8. Mixed, gradually, with about 1-3rd of its weight of oil of vitriol, an acid liquor is formed, together with a resin of a rich purple colour, which, after being washed, is hard and brittle, and forms a red tincture with rectified spirit, which is precipitated of a blood-red colour by water. 9. It dissolves iodine freely, without any marked reaction. 10. It dissolves santaline freely. 11. Mix 1 drop of the oil with a small trace of solution of aniline by means of a glass rod, and then shake with 5 or 6 c.c. of distilled water. By the addition of a few drops of sodium hypochlorite to the mixture the characteristic blue coloration due to phenol will be developed in a few minutes, if the adulterant be present; whereas with the pure oil nothing but the purplish-violet colour of aniline will be perceived. Stirring or shaking must be avoided after the addition of the hypochlorite. The presence of 1% of phenol can thus be demonstrated in 1 drop of the oil.

Obs. Clove oil contains a heavy oil, sp. gr. 1·079 (caryophylic acid), and a light oil, sp. gr. ·918 (clove hydrocarbon); by rectification, much of the light oil is lost, and the product becomes denser (1·361—*Bonastre*).

Oil, Cog'nac. See OIL, GRAPE (*below*).

Oil of Copai'ba. *Syn.* OIL OF CAPIVI; OLEUM COPAIBÆ (B. P., Ph. L. and E.), L. *Prep.* 1. (Ph. E.) Balsam of capivi, 1 oz.; water, 1¼ pints; distil, returning the water into the still, until oil ceases to pass over.

2. (Wholesale.) From the crude oil which separates during the manufacture of 'specific solution of copaiba' and 'soluble capivi,' by distillation along with a little salt and water.

Pur., &c. Colourless when pure; that of commerce has frequently a greenish tinge, derived from the copper utensils; odour, not disagreeable when recent. Sp. gr. ·876 to ·878.—*Prod.*, 50% to 55%. When adulterated with oil of turpentine, its solubility in rectified spirit is greatly diminished, and the solution is turbid.—*Dose*, 10 to 15 drops, in sugar; in the usual cases in which copaiba is ordered. 20 to 60 minims, three times a day (B. P.).

Oil of Corian'der. *Syn.* OLEUM CORIANDRI, L. From the fruit (seeds) of *Coriandrum sativum*. Yellowish; aromatic; carminative.—*Prod.* (dried fruit), 5½% to 6%.

Oil, Corn. The name given by Mulder to a peculiar fatty compound found in the fusel oil of the distilleries of Holland. It has a very powerful odour, resembling that of some of the umbelliferous plants, and is unaffected by caustic potassa. See OIL, FUSEL.

Oil of Cu'bebs. *Syn.* OLEUM CUBEBARUM, O. CUBEBÆ (B. P., Ph. E. and D.), L. From the fruit of *Cubeba officinalis*, or cubebs, coarsely ground. Aromatic, hot, and bitter tasted; odour, that of the fruit; faintly green, colourless when pure. Sp. gr. ·129.—*Prod.*, 9% to 11%.

Pur., &c. When pure, iodine has little action on this oil, and immediately gives it a violet colour, without any very marked reaction; nitric acid turns it opaque, and the mixture changes to a pale red when heated; sulphuric acid turns it of a crimson red. When adulterated with oil of turpentine, its viscidity, solubility in rectified spirit, and its density are lessened; when mixed with castor oil it leaves a greasy stain on paper.—*Dose*, 10 to 15 drops, in the usual cases in which cubebs in substance is given. 5 to 20 minims (B. P.), suspended in water by mucilage and sugar.

Oil of Cumin. *Syn.* OLEUM CUMINI, O. CYMINI, L. From the fresh fruit (seed) of *Cuminum cyminum*, or cumin. Pale yellow; smells and tastes strongly of the seeds. Sp. gr. ·975.—*Prod.*, 2¼% to 3%.

Obs. Oil of cumin is a mixture of two oils differing in volatility, and which may be separated by careful distillation. The more volatile one has been named cymol; the other cuminol.

Oil of Dill. *Syn.* OLEUM ANETHI (Ph. L. and E.), L. From the bruised fruit (seed) of *Anethum graveolens*. Pale yellow; odour, that of the fruit; taste, hot and pungent; carminative. Sp. gr. ·188 to ·882.—*Prod.*, 4% (nearly).

Oil of El'der. *Syn.* ATTAR OF ELDER FLOWERS; OLEUM SAMBUCI, L. From elder flowers (*Sambucus nigra*). Butyraceous; odour not very marked.

Oil of El'emi. *Syn.* OLEUM ELEMI, L. From the resin. Isomeric with oil of turpentine.

Oil of Er'got. *Syn.* ETHEREAL O. OF E.; OLEUM ERGOTÆ, O. E. ÆTHEREUM, O. SECALIS CORNUTI, L. Prepared by evaporating the ethereal tincture at a very gentle heat, and, preferably, allowing the last portion of the ether to escape by spontaneous evaporation. Brownish-yellow; lighter than water; soluble in ether and solution of potassa; only partly soluble in alcohol. It appears to be a mixture of volatile and fixed oil, with some resinous matter.—*Dose*, 10 to 20 drops, in hæmorrhages; 10 or 12 drops every 3 or 4 hours, in diarrhœa; 20 to 50 drops, as a parturifacient, &c. Externally, in rheumatism, toothache, &c.

Obs. The above is the oil of ergot now employed in medicine. It must not be confounded with other preparations occasionally called by the same name, but which differ from it in character. Among the latter are the following:

a. A fixed oil obtained by distilling off the spirit from the alcoholic tincture. It has the odour of rancid fish oil, and the distilled spirit has also a putrid odour.

b. A fixed oil, obtained from coarsely powdered ergot by strong pressure between iron plates, at a heat of about 212° F. It is fluid, coloured, smells strongly of the drug, but is nearly destitute of its leading qualities. Both the preceding contain some volatile oil and resinous matter.

c. An empyreumatic oil obtained by distilling ergot *per se*. It is light brown, viscid, acrid, and nauseous.

d. A volatile oil obtained by digesting powdered ergot in solution of potassa at 125° F., diluting the saponaceous mass thus formed with one half to an equal weight of water, neutralising the alkali with dilute sulphuric acid, and then submitting the whole to distillation in a chloride of sodium or oil bath. It is white, adhesive, butyraceous, and tasteless. It appears a product, rather than a simple educt.

e. This is the ethereal oil, first described, in its purest form. It is colourless, translucent, oily, and acrid-tasted, with the odour of ergot; it has a high boiling-point, at which it suffers partial decomposition, but may be volatilised at a lower temperature, like the other oils. By long exposure to heat, it thickens and partly solidifies; light and air darken it; it is lighter than water, very slightly soluble in water, but sufficiently so to impart to it its peculiar odour; it is soluble in pure alcohol, in ether, the volatile and fixed oils, alkaline lyes, liquor of ammonia, creasote, and naphtha. The dilute mineral acids clear it but do not produce any marked reaction.

Oil, Ethe''real. See OIL OF WINE (*below*).

Oil of Eucalyptus. *Syn.* OLEUM EUCALYPTI GLOBULI. See EUCALYPTUS.

Oil of Fen'nel. *Syn.* OLEUM FŒNICULI (Ph. D.), O. F. OFFICINALIS (Ph. E. & D.), O. F. DULCIS, L. From the fruit or seed of *Fœniculum dulce*, or sweet fennel (Ph. L.). Colourless; odour that of the plant; tastes hot and sweetish; congeals at 50° F.; carminative and stomachic. It consists of two oils; the one solid and identical with that of oil of aniseed. When treated with nitric acid, it affords benzoin. Sp. gr. ·997.—*Prod.* Dried fruit (of commerce), 3% to 3·5%. The flowering herb yields ·35% of a similar oil.

Obs. The oil of fennel of the shops is the product of the fruit of *Fœniculum vulgare*, or common, wild, or bitter fennel. It closely resembles that of sweet fennel, but is scarcely so agreeable either in taste or smell. It is chiefly used to scent soaps.

Oil of Fir-wood. *Syn.* OLEUM PINI SYLVESTRIS. An essential oil, distilled from the leaves of *Pinus sylvestris*. Much used, as in inhalation in sore throat and laryngeal catarrh.

Oil of Fleabane. (Ph. U. S.) *Syn.* OLEUM ERIGERONTIS CANADENSIS. An essential oil, distilled from *Erigeron canadensis*.—*Dose*, 5 minims, in hæmorrhage.

Oil, Fu'sel. See page 773.

Oil of Galbanum. *Syn.* OLEUM GALBANI (Ph. Bor.), L. From galbanum, 2 lbs.; water, 16 fl. oz.; distilled together. Yellow; resembles oil of assafœtida, but milder.

Oil of Garlic. *Syn.* SULPHIDE OF ALLYL. From the bruised bulbs or 'cloves' of *Allium sativum*, or garlic. It possesses the peculiar odour, taste, and other properties of the bulbs, in a highly exalted degree.

Obs. When a mixture of oil of black mustard and sulphide of potassium is exposed in a sealed glass tube to a temperature above that of 212° F., sulphocyanide of potassium and garlic oil are formed. On the other hand, when the compound of garlic oil and chloride of mercury (formed by adding to an alcoholic solution of the oil a like solution of the chloride) is gently heated with sulphocyanide of potassium, mustard oil, with all its characteristic properties, is called into existence.

Oil of Gaulthe"ria. See OIL OF PARTRIDGE-BERRY (*below*).

Oil of Gera"nium. *Syn.* OIL OF GINGER-GRASS, O. OF SPIKENARD. The oil of commerce which passes under this name, and which was formerly imported from the East Indies, was not obtained from any species of *Geranium* or *Pelargonium*, but probably from a species of *Andropogon*. Of recent years, however, genuine geranium oil, obtained from the rose geranium (*Pelargonium roseum*), has been and continues to be met with in our markets. This essential oil is manufactured in immense quantities at La Trappe de Staonelli, not far from the Bay of Sidi Ferruch, in Algiers, where about 40 acres of the plant are in cultivation. "Three harvests are gathered every year, and each yields from 170 to 200 kilograms of oil, or equal to 500 to 600 kilograms per annum. The value of this product never falls below 40 francs the kilogram, the average gross value being therefore from 20,000 to 25,000 francs, or at least £20 per acre. Seven distillatory apparatus are employed in this manufactory " (" The Paris Exhibition," 'Pharmaceutical Journal,' 3rd series, No. 433). A finer oil is yielded by the rose geranium when grown in France, but it is much dearer. It is often employed to adulterate otto of roses. See OIL, GRASS (*below*).

Oil of Gin'ger. *Syn.* OLEUM ZINGIBERIS, L. From the dried root (rhizome) of *Zingiber officinale*, or ginger of commerce. Bluish green; possesses a less agreeable odour than that of good ginger, without any pungency.—*Prod.*, $\frac{1}{4}$ to $\frac{1}{2}$ of 1% (*M. Raybaud*).

Oil of Goosefoot. (Ph. U. S.) *Syn.* OLEUM CHENOPODII. Distilled from the seeds of *Chenopodium anthelminticum.*—*Dose.* From 4 to 8 drops, with treacle or milk, for 3 nights in succession, for children. For adult, $\frac{1}{2}$ dr. Vermifuge.

Oil of Grain-spirit. *Syn.* GRAIN OIL. Two distinct substances are found in spirit distilled from fermented grain; one of which is butyraceous and highly offensive (corn oil of Mulder—?), the other liquid (crude fusel oil). The relative proportions of these substances to each other, and to the spirits which they contaminate, vary with the materials and the management of the process. The 'GRAIN OIL' of the London rectifiers consists chiefly of fusel or potato oil, mixed with alcohol and water, and with small and variable proportions of solid ethyl and amyl-compounds of certain fatty acids (œnanthic and margaric). The latter are said to be similar to the butyraceous matter before referred to, as well as the solid fat of the whisky distilleries conducted on the old plan. According to Mr Rowney, the fusel oil of the Scotch distilleries contains capric acid. See OIL OF CORN (*above*), and FUSEL OIL.

Oil, Grape. *Syn.* BRANDY OIL, COGNAC O. This is essentially the sulphate of amyl. It is prepared by dissolving the fusel oil of marc brandy in strong rectified spirit, and then adding concentrated sulphuric acid; alcohol and excess of acid are removed by washing the newly formed compound with water. Dissolved in rectified spirit, it forms 'BRANDY ESSENCE,' which is used to impart the cognac flavour to plain spirit. See SULPHATE OF AMYL and AMYL-ETHER, also OIL OF MARC BRANDY (*below*).

Oil, Grass. Several of the grasses (*Graminaceæ*) yield fragrant volatile oils. See OIL OF GERANIUM, GRASS OIL (of Namur), OIL OF LEMON-GRASS, OIL OF SPRING-GRASS, &c.

Oil, Grass (of Namur). *Syn.* INDIA GRASS OIL. From *Andropogon calamus aromaticus*, Royle, supposed to have been the 'sweet cane' and 'rich aromatic reed from a far country' of Scripture; formerly supposed to be obtained from *Andropogon Iwarancusa*. Stimulant and highly fragrant. See OIL OF SPIKENARD.

Oil of Hops. *Syn.* OLEUM LUPULI, L. From commercial hops, by distillation along with water. It may also be collected during the brewing of beer. Odorous; acrid; narcotic; soluble in water; becomes resinous by exposure and age. Sp. gr. ·910. Chiefly used to increase the aroma and flavour of old or damaged hops.

Oil of Horse-mint. *Syn.* OLEUM MONARDÆ (Ph. U. S.), L. From the fresh herb of *Monarda punctata*, a plant indigenous in the U. S. of America. Dark amber-coloured; fragrant; pungent; carminative; rubefacient; and vesicant. It is a source of thymol.

Oil of Horse-rad'ish. *Syn.* OLEUM ARMO-RACIÆ, L. From the fresh roots of *Cochlearia armoracia*, Linn., or common horse-radish. Pale yellow; heavier than water; acrid; vesicant; identical with that from black mustard.—*Prod.*, ·5% (nearly).

Oil of Hyssop. *Syn.* OLEUM HYSSOPI, L. From the flowering herb of *Hyssopus officinalis*. Aromatic; stimulant.—*Prod.*, ·25% to ·33%.

Oil of Jargonelle Pear. See AMYL (Acetate).

Oil of Jas'mine. *Syn.* OIL OF JESSAMINE; OLEUM JASMINI, O. J. VOLATILE, L. From the flowers of *Jasminum grandiflorum* and *J. fragrans*, carefully picked, by placing them in alternate layers with cotton wadding imbued with olive oil, in any suitable vessel, and renewing the flowers till the fixed oil becomes strongly odorous, and then distilling the wadding along with a little water. The volatile oils of hyacinths, jonquil, tuberose, violets, and most of the more delicate flowers are obtained in the same way. Used in perfumery. From the East Indies.

Oil of Ju'niper. *Syn.* JUNIPERI OLEUM (B. P.), OLEUM JUNIPERI (Ph. L., E., and D.), O. à BACCIS J., O. ESSENTIALE à B. J., L. From either the wood, tops, or berries, preferably the last. The berries should be chosen fully grown, but still slightly green, and should be bruised before being placed in the still. In the Ph. L., English oil of juniper (O. JUNIPERI ANGLICUM) is ordered. Colourless, or very pale greenish.

yellow; odour and taste, sweet and terebinthinate; rather viscid; soluble in rectified spirit; rendered opaque and resinous by exposure and age. It is reputed carminative and diaphoretic, and possesses powerful diuretic properties. Sp. gr. ·911 (English, ·8688; foreign, ·8834—*Brande*).—*Prod.* Green berries, ·25% ; ripe do. (one year old), ½ to 1% (fully).

Pur. It is frequently adulterated with oil of turpentine, a fraud readily discovered by the lessened density, viscidity, and solubility, in rectified spirit, of the oil.

Obs. Oil of juniper consists of two oils—one, white and most volatile, sp. gr. ·8393 ; the other, dark-coloured and less volatile, sp. gr. ·8784; together with some resin left in the retort.

Oil, Krumholz. *Syn.* OLEUM TEMPLINUM, L. From Hungarian balsam, a terebinthinate exudation from the *Pinus pumilio*, or mountain pine of Southern Europe. Fragrant; golden yellow; tastes oily, acidulous, and resinous.

Oil of Laurel. *Syn.* OIL OF SWEET BAY; OLEUM LAURI VOLATILE, O. L. ESSENTIALE, L. From either the berries or leaves of *Laurus nobilis*, Linn., or sweet bay-tree. Pale yellow, clear, odorous, aromatic, stimulant, and narcotic. Sp. gr. ·871.—*Prod.* From the leaves, ½% to 1% (fully).

Oil of Lavender. *Syn.* ESSENCE OF L.; LAVANDULÆ OLEUM (B. P.), OLEUM LAVANDULÆ (Ph. L.), O. L. VERE (Ph. E. and D.), O. L. SPICÆ, O. L. ESSENTIALE, O. L. FLORUM, L. The oil (OLEUM LAVANDULÆ ANGLICUM) distilled from the flowers of *Lavandula vera* (Ph. L.). Very pale lemon-yellow; highly fragrant; taste, warm and not disagreeable; carminative, antispasmodic, and stimulant. Sp. gr. ·877 to ·905. According to Brande, the sp. gr. of the oil obtained from the flowers only is ·8960; that from the whole plant, ·9206. The lightest is esteemed the best.—*Prod.* Flowers, 1¼% to 2% (nearly). The whole of the flowering herb is commonly distilled. According to Raybaud, the herb, after flowering (Sept.), yields the most oil.

Pur. Alcohol is the substance commonly used to adulterate this oil; but, occasionally, oil of bergamot is used for the same purpose. If the density is below ·87, there is reason to suspect adulteration. When pure—1. Sulphuric acid turns it reddish brown, and the reaction is accompanied by strong inspissation. 2. It fulminates quickly and violently with iodine, and the thick syrupy residue possesses a pungent, acid, balsamic odour. The oils of the other labiate plants fulminate much less powerfully with iodine. The presence of alcohol weakens, but does not destroy, the action of this test, unless it is added in an equal volume, when only a lively effervescence and a disengagement of orange-coloured vapours are produced by the iodine, without fulmination. 3. Santaline is nearly insoluble in pure oil of lavender, and exerts no marked action on it, but is freely soluble in oil of lavender adulterated with alcohol or rectified spirit.

Obs. English oil of lavender possesses the purest fragrance; and of this, the variety known as 'MITCHAM OIL OF LAVENDER,' from the place of its preparation, is esteemed the best. The foreign oil of lavender is inferior.

This last is improved by rectification. See OIL OF SPIKE.

Oil of Lem'on-grass. *Syn.* ESSENCE OF L.-G., INDIAN GRASS OIL, OIL OF VERBENA. Probably from *Andropogon citratum*, the Indian lemon-grass. Pale yellow; powerfully fragrant. CITRONELLE OIL is also the product of this or of an allied species of *Andropogon*.

Oil of Lem'ons. *Syn.* ESSENCE OF L.; OLEUM LIMONIS (B. P.), OLEUM LIMONIS, O. LIMONUM (Ph. L., E., and D.). From the yellow portion of the rind, grated, placed in hair bags, and exposed to powerful pressure; also by distillation, but the product is then less agreeably fragrant and sweet, but keeps better. Nearly colourless; odour, that of the fruit. Sp. gr. ·8752 to ·8785. Expressed oil, ·8517; distilled do., ·845, at 72° F. (*Ure*).—*Prod.* 100 lemons yield, by expression, 1½ to 2 oz. (nearly) ; by distillation, 1¼ to 1½.

Pur. Commonly adulterated with oil of turpentine, and occasionally with nut or poppy oil. These may be detected in the manner already explained. When pure it is soluble in all proportions in absolute alcohol, but rectified spirit only dissolves 16% of it. It also boils at 148° F., whereas oil of turpentine boils at 312°, and mixtures of the two at intermediate temperatures depending on the proportions.

Oil of Lemon Thyme. *Syn.* OLEUM SERPYLLI, L.; HUILE DE TAIN, Fr. From the fresh flowering herb of *Thymus serpyllum*, the lemon or wild thyme of our hills and pastures. Very fragrant. Used to scent soaps, &c. Sp. gr. ·867.—*Prod.* 100 lbs. yield 2½ to 5½ oz. of oil. When pure, it is scarcely affected by iodine, but solution of chromate of potassa acts on it with energy.

Oil of Let'tuce. *Syn.* OLEUM LACTUCÆ VIROSÆ, L. From *Lactuca virosa*, Linn., or strong-scented wild lettuce. Closely resembles the odorous matter of opium.

Oil of Limes. *Syn.* OLEUM LIMETTÆ, L. From the rind of the fruit of *Citrus limetta*, or lime, as OIL OF LEMONS, which it somewhat resembles.—*Prod.* 100 limes yield 2½ to 2½ oz. of oil.

Oil of Lev'age. *Syn.* OLEUM LEVISTICI, L. From the leaves and fruit of *Levisticum officinale*, lovage. Pale yellow, aromatic, carminative.—*Prod.* Fresh herb, ·1% to ·15%.

Oil of Mace. *Syn.* OLEUM MACIDIS, O. M. ESSENTIALE, O. M. STILLATITIUM, L. From the arillus of *Myristica officinale* (commercial mace). Nearly colourless; fragrant; lighter than water; closely resembles oil of nutmeg. Sp. gr. ·945.—*Prod.*, 4½% to 9%.

Oil of Marc Brandy. *Syn.* FUSEL OIL OF M. B., O. OF GRAPE-SPIRIT. Obtained after the spirit (marc brandy) has passed over during the distillation of the fermented residuum of expressed grapes. Limpid; odorous; acrid; offensive; soon turns yellow in the air; soluble in 1000 parts of water, and in all proportions in rectified spirit; 6 or 7 drops will spoil a hogshead of brandy. According to M. Balard, this oil is a mixture of potato oil and œnanthic ether.

Oil of Mar'joram. *Syn.* OIL OF SWEET M.; OLEUM MARJORANÆ, O. ORIGANI M. (Ph. E.), L. From the fresh flowering herb of *Origanum marjorana*, or sweet or knotted marjoram. Pale yellow ; odorous; tonic ; stimulant. Sp. gr. ·925

(·940—*Baumé*).—*Prod.*, ·33% to ·35%. See OIL OF ORIGANUM.

Oil of Mea'dow-sweet. *Syn.* OLEUM SPIRÆÆ ULMARIÆ, L. From the flowers or flowering tops of *Spiræa ulmaria*, Linn., or common meadow-sweet. This oil is a native hydride of salicyl. It is yellow, sweet-scented, and slightly soluble in water, which then strikes a deep violet colour with the persalts of iron. It boils at 385° F. Sp. gr. 1·172 (see *below*).

Oil of Mea'dow-sweet (Facti"tious). This is prepared as follows:—Salicin, 1 part, is dissolved in distilled water, 10 parts, and being placed in a glass retort, bichromate of potassa (in powder), 1 part, is added, followed by oil of vitriol, 2½ parts, previously diluted with 4 times its weight of water; a gentle heat is next applied to the retort, and after the first effervescence resulting from the mutual reaction of the ingredients is over, the heat is increased, and the mixture is distilled for the oil in the usual manner. The product is absolutely identical with the natural oil of meadow-sweet (see *above*).

Oil of Mil'foil. *Syn.* OLEUM MILLEFOLII, L. From the flowers of *Achillæa millefolium*, Linn., or yarrow. Blue. Sp. gr. ·852.—*Prod.* 14 lbs. of the dried flowers yield 3 dr. of oil.

Oils, Mixed (Essential). *Syn.* OLEA MIXTA ESSENTIALIA. From the oils of bergamot and lemons, of each, 1 oz.; oils of lavender and pimento, of each, ½ oz. Used to scent 'sal volatile drops,' smelling-bottles, &c.

Oil of Mus'tard (Volatile). *Syn.* SULPHO-CYANIDE OF ALLYL; OLEUM SINAPIS NIGRÆ, O. S. ESSENTIALE, L. From the seeds of *Sinapis nigra*, Linn., or black mustard, as oil of bitter almonds. Nearly colourless; intensely acrid, pungent, rubefacient, and vesicant; slightly soluble in water; boils at 289° F. It contains sulphur. Sp. gr. 1·035 to 1·088; 1·015 at 68° F.—*Prod.* Av. ·6% (fully).

Obs. This oil, like that of bitter almonds, does not pre-exist in the seed, but is the result of the action of a peculiar substance, myrosin, in the presence of water upon myronate of potassium contained in the seeds of black mustard. Oil of black mustard has been used as a stimulant or counter-irritant in palsy, &c.; and the distilled water, or a solution of the oil in water, is said to be an excellent and cleanly remedy for the itch.

Oil of Myrrh. *Syn.* OLEUM MYRRHÆ, O. M. ESSENTIALE, L. Colourless; thin; heavier than water; stimulant; smells strongly of the drug.

Oil of Myr'tle (Volatile). *Syn.* ESSENCE OF M.; OLEUM MYRTÆ ESSENTIALE, L. From the flowers and leaves of *Myrtus communis*. 100 lbs. of the fresh leaves yield 2½ to 5 oz.

Oil of Namur Grass. See OIL OF GRASS (NAMUR).

Oil of Narcis'sus. *Syn.* ESSENCE OF JONQUIL; OLEUM NARCISSI, L. As OIL OF JASMINE. Delightfully odorous.

Oil of Nero'li. See OIL OF ORANGE FLOWERS.

Oil of Nut'meg (Volatile). *Syn.* OLEUM MY-RISTICÆ (B. P., Ph. E.), O. M. MOSCHATÆ (Ph. D.), L. From the officinal nutmeg or kernel of the fruit of *Myristica fragrans*. Nearly colourless; odour and flavour that of the fruit, but more

powerful. By agitation with water, it is separated into two oils—one lighter, the other heavier, than water; the last is butyraceous. Sp. gr. ·948.—*Prod.*, 4½% to 7%. It is reputed to make the hair grow, and prevent baldness.

Oil of On'ions. From the bulbs of *Allium cepa*, or common onion. Contains sulphur, and smells strongly of the herb.

Oil of Orange. *Syn.* ESSENCE OF O.; OLEUM AURANTII, O. AURANTIORUM, O. A. CORTICIS, L. From the yellow portion of the rind of either the Seville or sweet orange, preferably of the latter; as oil of bergamot or lemons. Closely resembles oil of lemons, but is more agreeably fragrant. The expressed oil is very apt to become opaque, and deposit a stearopten, especially in cold weather, unless well kept from the air. Sp. gr. ·875.—*Prod.* 100 fruits yield 4 to 5 oz. (See *below*.)

Oil of Orange Berries. *Syn.* OLEUM AU-RANTII BACCÆ, L. From the small unripe fruit of the orange-tree. Does not keep well. (See *below*.)

Oil of Orange Flowers. *Syn.* NEROLI, OIL OF N., ESSENCE OF N.; OLEUM NAPHÆ, O. AU-RANTII FLORUM, AURANTII OLEUM (Ph. E. and D.), L. From the flowers of either the bitter (Seville) or sweet orange (*Citrus vulgaris* or *C. aurantium*), by distillation with water. That from the fruit is said to be preferred, but there does not appear any actual difference between the two. Very fluid; lighter than water, in which it is slightly soluble; it is delightfully aromatic and fragrant, but the odour differs slightly from that of the flowers.—*Prod.* 100 lbs. of flowers gathered in May or December yield 3 to 6 oz. of oil; 6 cwt. of the fresh flowers yield 1 lb. of oil.

Pur. Neroli is commonly adulterated with alcohol or essence de petit grain, and generally with both of them. The presence of the first is easily determined (see *above*); that of the second can only be discovered by comparing the odour evolved during the evaporation of a drop of the suspected oil, placed on a piece of white paper, with a like drop of pure neroli similarly treated. (See *above* and *below*.)

Oil of Orange Leaf. *Syn.* OLEUM AURANTII FOLII, L.; ESSENCE DE PETIT GRAIN, Fr. From the leaves of either the bitter or sweet orange, that from the first being preferred. Delightfully fragrant. Extensively used to adulterate oil of neroli, and is itself commonly sophisticated with both alcohol and oil of orange berries. (See *above*.)

Oil of Orig'anum. *Syn.* OLEUM ORIGANI, O. O. ESSENTIALE, L. From the flowering herb of *Origanum vulgare*, or common or winter marjoram. Pale yellow colour; fragrant; acrid, pungent, and rubefacient. Sp. gr. ·927 (·940—Baumé).—*Prod.*, ·5% to ·75%. The dark-coloured oil of origanum of the shops is obtained from *Thymus vulgaris*. The oil of origanum (Ph. E.) is oil of *Origanum marjorana*. See OILS OF MARJORAM, THYME, and LEMON THYME.

Oil of Or'ris. *Syn.* ESSENCE OF VIOLET; OLEUM IRIDIS, L. From the dried rhizomes of *Iris Florentina*, or Florentine orris-root. Fragrant. Sold for oil and essence of violets.

Oil of Parsley. *Syn.* OLEUM PETROSELINI,

L. From the fresh herb or dried fruit (seed) of *Apium petroselinum*, or garden parsley. Yellowish; smells strongly of the plant. It consists of two oils, separable by agitation with water, one of which is concrete, and melts at 80° F.; the other, liquid.—*Prod.* Herb, ·50% to 1% (nearly).

Oil of Par'tridge-berry. *Syn.* OIL OF WINTERGREEN, METHYLO-SALICYLIC ETHER, SALICYLATE OF OXIDE OF METHYL; OLEUM GAULTHERIÆ (Ph. U. S.), L. From the leaves or the whole plant of *Gaultheria procumbens*, a herb common in North America, and otherwise known by the names box-berry, chequer-berry, partridge-berry, mountain tea, winter-green, &c. Pale yellow, growing brown by exposure and age; aromatic; sweet; highly pungent; when diluted, agreeably fragrant; mixed with a dilute solution of potassa, it solidifies to a crystalline mass (salicylate of potassium), from which the oil may be again separated by the addition of an acid. It is the heaviest of all the essential oils. Sp. gr. 1·173. Boils at 412°, and, when purified, at 435° F.

Oil of partridge-berry, dissolved in rectified spirit, is in common use in the United States of America as an antispasmodic, carminative, diuretic, emmenagogue, and stimulant; chiefly as an adjunct to mixtures, &c.; and also with the view of increasing the flow of milk during lactation. It is likewise extensively used in perfumery, and is an object of great interest to the organic chemist, on account of its peculiar constitution and reaction. It is the chief source of natural salicylic acid.

Oil of Partridge-berry (Facti"tious). See SALICYLIC ACID.

Oil, Pearl. See AMYL (Acetate of), and ESSENCE OF JARGONELLE PEAR.

Oil of Pennyroy'al. *Syn.* OLEUM PULEGII (Ph. L.), O. MENTHÆ P. (B. P., Ph. E. & D.), O. P. ESSENTIALE, L. From the flowering herb of *Mentha pulegium*, or the common pennyroyal of our gardens. Pale yellow, growing reddish yellow by age and exposure; antispasmodic, carminative, and emmenagogue. Boils at 395° F. Sp. gr. ·925 to ·931.—*Prod.*, ½% to 1%. (See below.)

Oil of Pennyroyal (American). *Syn.* OLEUM HEDEOMÆ (Ph. U. S.), L. From *Hedeoma pulegioides*, as the last. Light yellow; closely resembles oil of pennyroyal, for which it passes in the U. S. Sp. gr. ·945 to ·948.

Oil of Pepper. *Syn.* OIL OF BLACK P.; OLEUM PIPERIS, O. P. NIGRI, L. From bruised black pepper (*Piper nigrum*). Colourless, turning yellow; odorous; pungent; not so hot as the spice. Sp. gr. ·9932.—*Prod.*, 1·25% to 1·5%. White pepper (of commerce), 1% (barely).

Oil of Pep'permint. *Syn.* OLEUM MENTHÆ PIPERITÆ (B. P., Ph. L., E., and D.), O. ESSENTIALE M. PIPERITIDIS, L. From the fresh flowering herb of *Mentha piperita*, or garden peppermint. Nearly colourless, or at most a very pale greenish yellow; powerfully odorous; tastes pungent, at the same time imparting a sensation of coldness to the tongue and palate. Boils at 365° F. Sp. gr. ·902 to ·905.—*Prod.* Fresh flowering herb, ·25% to ·4%; dried do., 1% to 1·25% (fully). In a warm dry season, 5 lbs. of the fresh flowering herb yield 1 oz. of oil; in a wet and unfavourable one, 11 lbs. yield barely the same quality.

Pur. The oil of commerce usually contains fully a third part of rectified spirit, and is also frequently adulterated with the oils of rosemary, spearmint, and turpentine. When pure—1. It is soluble in its own weight of rectified spirit. 2. Mixed with 1-4th its volume of nitric acid, a rich purple-red colour is developed. 3. Chromate of potash, in solution, turns it of a deep reddish-brown colour, and converts it into a soft coagulum, which assumes a flaky form when divided with a glass rod, whilst the solution of the salt loses its yellow colour or becomes greenish yellow. 4. With iodine it forms a homogeneous mass, without fulmination. If it explodes with iodine, it contains turpentine. The yellowish, resinous oil, sold under the name of 'American' or 'crude oil of peppermint,' consists chiefly of oil of turpentine, and on evaporation leaves a residuum of pine resin.

Obs. English oil of peppermint is the best, a fact clearly shown by its price in the market being so greatly above that of the imported oil. The oil distilled at Mitcham, in Surrey (Mitcham oil of peppermint), is the most esteemed. It has usually a very pale greenish colour, which is often imitated by steeping a leaf or two of green mint or parsley in the oil. Old dark-coloured oils are commonly bleached by exposure to the light, to the destruction of a portion of their other properties.

According to a recent and valuable report upon those articles in the Paris Exhibition of 1878, more particularly interesting to the pharmacist, the chemical manufacturer, the perfumer, &c., which lately appeared in the 'Pharmaceutical Journal,' the above statement is open to question. Of late years it seems that a considerable industry has sprung up at Arzim in the Département du Nord, in France, where large quantities of labiate plants are cultivated, and subsequently submitted to distillation.

An acre of land generally yields every year from 3 to 4 tons of the peppermint plant; and from 500 parts of this, one part of essential oil is usually obtained, which it is alleged by M. Hanart, the distiller of the oil in question, after being carefully bottled and kept for some years, successfully rivals the English oil both in quality and price.

Of late years an essential oil of peppermint manufactured by Messrs Hotchkiss, of New York, has lately come into considerable demand. This, which is said to be a very pure article, differs from the other peppermint oils in becoming thick when first mixed with spirit of wine. After a short time, however, the mixture clears and becomes perfectly bright.

Oil of peppermint is stimulant, antispasmodic, and carminative, and has always been a favourite remedy in flatulence, nausea, vomiting, loss of appetite, cramp of the stomach, colic, griping pains, diarrhœa, the early stage of cholera, &c.—*Dose*, 1 to 3 drops, on sugar.

Oil of Petro'leum. See NAPHTHA, OILS (Mineral), PETROLEUM, &c.

Oil of Pimen'to. *Syn.* OIL OF ALLSPICE; OLEUM

PIMENTÆ (B. P., Ph. L., E., and D.), L. From the bruised fruit of *Eugenia pimenta*, allspice, or Jamaica pepper. Pale yellow, growing reddish-brown by age; odour, a combination of cloves and cassia; taste, pungent. Sp. gr. 1·021.—*Prod.*, 5% to 8%.

Obs. Oil of pimento contains two oils similar to those found in clove oil. When pure, nitric acid turns it red, with active effervescence and the assumption of a rusty brown colour. It combines with the salifiable bases in a nearly similar manner to oil of cloves. It is much used in perfumery, especially in hair cosmetics.

Oil of Pim'pernel. *Syn.* OLEUM PIMPINELLÆ, L. From the root of *Sanguisorba officinalis*, or pimpernel. Blue; carminative.

Oil, Pine-apple. This artificial essential oil dates its commercial importance from the Great Exhibition of 1851. It is essentially butyric ether, and may be regarded as simply the crude form of that substance. On the large scale it is prepared by saponifying butter or crude butyric acid with a strong lye of caustic potassa, and dissolving the resulting soap in the smallest possible quantity of hot alcohol; to the solution is added a mixture of alcohol and oil of vitriol in excess, and the whole is then submitted to distillation as long as the product has an aromatic fruity odour; the product is rectified from dried chloride of calcium and a little litharge. Dissolved in rectified spirit it is much used as a flavouring substance by confectioners and liquorists. See ETHER (Butyric) and ESSENCE OF PINE-APPLE, &c.

Oil of Pota'to Spirit. See FUSEL OIL.

Oil of Pumilio Pine. *Syn.* OLEUM PINI PUMILIO. From the acicular leaves of the *Pinus pumilio*. It is a nearly colourless, very fragrant pine oil, which, in the refined condition, is sold under the proprietary name of Pumiline. Employed as an inhalation in laryngeal catarrh and sore throat, also as a liniment in bronchitis and rheumatism. See PUMILINE.

Oil of Ravens'ra. *Syn.* OLEUM RAVENSARÆ, L. From the roots of *Ravensara aromatica*. Chiefly used to adulterate oil of cloves, which it somewhat resembles.

Oil of Rho'dium. *Syn.* OLEUM RHODII, L. Said to be derived from the wood of a species of *Rhodoriza*. Very fluid and limpid; pale yellow; soon darkens by age and exposure; tastes bitter and aromatic; has a modified odour of roses. Chiefly used as a substitute for otto of roses in cheap perfumery, and to adulterate it. Oil of sandal-wood is frequently sold for it.—*Prod.*, 1% to 16%. See OIL OF ROSES (*below*).

Oil of Rose'mary. *Syn.* ROSMARINI OLEUM (B. P.), OLEUM ANTHOS, O. ROSISMARINI, O. ROSMARINI (Ph. L., E., and D.), O. ROSISMARINI ESSENTIALE, L. From the flowering tops of *Rosmarinus officinalis*. Colourless; strongly fragrant, but scarcely agreeable unless compounded; carminative and stimulant. Boils at 365° F. Sp. gr. ·910; recent, ·897; rectified, ·8887.—*Prod.*, ¾% to 1% (nearly).

Pur., &c. It is frequently adulterated with oil of turpentine. When pure it dissolves in all proportions in spirit of ·830. By age it deposits a crystalline stearopten, and acquires a terebinthinate odour. It is chiefly used as a stimulant in liniments, hair oil, pomatums, &c.

Oil of Ro''ses. *Syn.* OLEUM ROSÆ, L. *Prep.* 1. From the petals of *Rosa sempervirens*, Linn., or the musk rose, as OIL OF CLOVES, observing to keep the water in the worm-tub at 85° F., and afterwards subjecting the water in the receiver to refrigeration. Resembles otto of roses, of which it is merely a variety.—*Prod.*, 1/10 to 1/8 of 1%.

2. (ATTAR OF ROSES, OTTO OF R.; OLEUM ROSÆ.) From the petals of *Rosa damascena*, and probably other varieties of rose. Fluciger and Hanbury state, "The rose is cultivated by Bulgarian and Turkish peasants in gardens and open fields, in which it is planted in rows as hedges, 3 to 4 feet high. The flowers attain perfection in April and May, and are gathered before sunrise. Those not wanted for immediate use are spread out in cellars, but are always used for distilling the same day. The apparatus is a copper still of the simplest description, connected with a straight tin tube, cooled by being passed through a tube fed by a stream of water. The charge for a still is 25 to 50 lbs. of roses, from which the calyces are not removed. The first runnings are returned to the still; the second portion is received in glass flasks and kept for a day or two, by which time most of the oil, bright and fluid, will have risen to the surface. From this it is skimmed off. The produce is about ·04%."

Roses are also cultivated for the making of attar in the South of France about Grasse, Cannes, and Nice; likewise in India, at Ghazipur, Lahore, and Amritsar.

Prop., &c. A light yellow fluid or semi-solid; sp. gr. ·87 to ·89; odour intense and diffusive, most pleasant when diluted. At low temperature it concretes, separating platy crystals of a stearopten, the proportion of which differs with locality and period of production. Turkish attar fuses at from 16°—18° C.; Indian, 20° C. 1000 parts of alcohol of ·806 dissolve only 7 parts of otto at 57° F., and only 33 parts at 72°. Sp. gr. ·832 at 90°, to water 1·000 at 60° F.—*Prod.* 100 lbs. of roses yield 2 to 3 dr.

Pur. Otto of roses is frequently adulterated with the oils of rhodium, sandal-wood, and geranium, and with camphor; and occasionally with spermaceti, to give the spurious compound the usual crystalline appearance. The oil of geranium, also known as oil of Indian grass, is imported into Turkey and sprinkled on the roses before distillation. The following are reliable tests:—1. Pure otto has a bland, sweet taste; if it is bitter, it contains oil of rhodium or sandal-wood; if it is pungent, or 'bites' the palate, it contains either oil of geranium or camphor, and probably both; if it imparts an unctuous sensation, it contains spermaceti. 2. Exposed for some hours to the fumes of a small quantity of iodide under a bell-glass in the cold, pure otto remains white, and continues so when exposed to the air; an adulterated sample, on the contrary, becomes yellow or brown, and afterwards, on exposure to the air, continues to darken in colour, until it becomes of a deep brown, or even perfectly black, according to the quantity of foreign oil present. A single drop may be thus tested. 3. (*Guibourt.*)

One or two drops of the suspected oil are put into a watch-glass; the same number of drops of concentrated sulphuric acid are added, and the two fluids are mixed with a glass rod. All the oils are rendered more or less brown by this proceeding, but otto of roses retains the purity of its odour; oil of geranium acquires a strong and disagreeable odour, which is perfectly characteristic; the odour of the oil of rhodium is increased, and becomes somewhat unctuous, and, in general, it acquires an odour distinctly like that of cubebs.

Oil of Rose'wort. *Syn.* OIL OF ROSE-ROOT; OLEUM RHODIOLÆ, L. From the roots of *Rhodiola rosea*. Yellowish; odour resembles that of oil of rhodium, for which it is often sold, as well as the distilled water for rose-water. 1¼ lbs. yield about 1 dr.

Oil of Rue. *Syn.* RUTÆ OLEUM (B. P.), OLEUM RUTÆ. The "oil distilled from the fresh herb of *Ruta graveolens*" (B. P.), or common rue. Pale yellow, turning brown by age, and depositing a brownish, resinous sediment; congeals at about 40° F.; acrid, bitter; odour that of the plant; stimulant, antispasmodic, and emmenagogue. Sp. gr. ·909 to ·911.—*Prod.*, ¾% to 1% (nearly). According to Raybaud, the recent dried seeds yield fully four times as much oil as the flowering herb.

Pur. Nearly always adulterated. When pure—1. It forms a clear solution with rectified spirit. 2. It does not form a camphor with gaseous hydrochloric acid. 3. Iodine dissolves in it slowly, without any apparent reaction, beyond a darkening and a slight increase of viscidity. 4. It is unaffected by a solution of chromate of potassa. 5. Nitric acid very slowly changes it into a greenish-yellow liquid balsam. 6. If it forms a reddish-brown solution with liquor of potassa and a still darker one with oil of vitriol, or if it fulminates with iodine, it is adulterated with the oil of some labiate plant. It is more soluble in both rectified spirit and water than any of the oils used to adulterate it.

Oil of Saf'fron. *Syn.* OLEUM CROCI, L. From the pistils of *Crocus sativus* (saffron). Yellow; heavier than water; acrid, pungent, and narcotic; decomposed by exposure to light and age, with the formation of a white solid matter, which is lighter than water.

Oil of Sage. *Syn.* OLEUM SALVIÆ, L. From the herbaceous portion of *Salvia officinalis*, or common sage.

Oil of San'dal-wood. *Syn.* OLEUM SANTALI, O. S. FLAVI, L. From the wood of *Santalum album*, or sandal-tree, and preferably from that of Malabar. It has an odour somewhat resembling that of oil of rhodium, for which it is commonly used; also used to adulterate otto of roses.—*Prod.* 9 lbs. yield 1 oz.; 100 lbs. yield 5 oz. Given in doses of 10 to 30 min. for gonorrhœa.

Oil of Sarsaparil'la. *Syn.* OLEUM SARZÆ, L. From the root bark, distilled along with salt-and-water. Acrid; odour and flavour same as the root.

Oil of Sas'safras. *Syn.* VOLATILE OIL OF S.; OLEUM SASSAFRAS (Ph. E.), O. LAURI S., O. OFFICINALIS, L. From bruised sassafras chips, the sliced root of *Sassafras officinale*, as oil of cloves. Pale yellow; highly odorous; hot; pungent, rubefacient, and stimulant; reputed alterative, sudorific, and diuretic, and, as such, occasionally given in rheumatism, cutaneous affections, &c. Sp. gr. 1·094 to 1·096.—*Prod.*, 1½% to 2% (fully).

Pur., &c. 1. If the density is lower than 1·094, it is adulterated. 2. Nitric acid acts on this oil, at first slowly, merely turning it of an orange-red, but afterwards with violence, and a reddish-brown resin is formed. 3. Mixed with about one half its weight of sulphuric acid, a green colour is at first developed, which, by heat, is changed to a blood-red. A large quantity of sulphuric acid acts at once violently, white fumes are given off, and mere charcoal is left. 4. With iodine it forms a permanently clear solution, or at least one that remains so for some time. 5. By agitation with water, it separates into two oils—one lighter, the other heavier, than that fluid.

Oil of Sav'ine. *Syn.* OLEUM SABINÆ (B. P.), OLEUM JUNIPERI SABINÆ, L. From the fresh top or leaves of *Juniperus sabina*, or common savine. Pale yellow; limpid; acrid, pungent, and stimulant. It possesses the general properties of the plant in a highly exalted degree. Sp. gr. ·915.—*Prod.* Fresh herb, 1·25% to 1·5%; dried ditto (recent), 2½% to 3%.—*Dose*, 2 to 6 drops; as an anthelmintic, diaphoretic, and emmenagogue. Its use must be carefully avoided during pregnancy or disease of the abdominal viscera.

Pur., &c. It is less frequently adulterated than the other volatile oils. Its high sp. gr. and free solubility in rectified spirit offer the means of detecting the presence of either oil of turpentine or alcohol, the substances occasionally added to it. A mixture of equal parts of oil of savine and oil of vitriol, by distillation from milk of lime, furnishes an oil apparently identical with oil of thyme (*Winckler*).

Oil of Spear'mint. *Syn.* ENGLISH OIL OF SPEARMINT (B. P.), OIL OF MINT, OIL OF GREEN M.; MENTHÆ VIRIDIS OLEUM (B. P.), OLEUM MENTHÆ VIRIDIS, O. M. SATIVÆ, O. ESSENTIALE MENTHÆ S., L. From the fresh flowering herb of *Mentha viridis*, Linn., or garden or spearmint. Pale yellow, reddened by age; odour and general properties resemble those of oil of peppermint, but it is less grateful. It boils at 320° F. Sp. gr. ·915 (·9394, *Brande*).—*Prod.*, ·2% to ·25%. Its common adulterants are alcohol and oil of turpentine.

Oil of Spike (True). *Syn.* FOREIGN OIL OF LAVENDER; OLEUM SPICÆ, O. S. VERUM, O. STŒCHADIS, O. LAVANDULÆ S., L.; HUILE D'ASPIC, Fr. Chiefly from *Lavandula spica* and *L. stœchas*, or French and Alpine lavenders. It differs from English oil of lavender by its darker green colour and inferior odour. From France. Used by artists to mix their colours in, and to make varnishes. Oil of turpentine scented with lavender is commonly sold for it.—*Prod.* From *L. spica* (fresh), ¾% to 1½%; *L. stœchas* (dried), ¾% to 1% (fully).

Oil of Spikenard. *Syn.* OLEUM NARDI, L. The precious oil mentioned under this name in Scripture is supposed to have been derived from *Andropogon Iwarancusa*. The commercial oil of geranium (see *above*) is also called by this name.

Oil of Spring Grass. *Syn.* OLEUM ANTHOXANTHI ODORATI, L. From *Anthoxanthum odoratum*, or sweet-scented vernal grass. It is this oil that gives the very agreeable odour to new hay.

Oil of Star-an'ise. *Syn.* BADIAN OIL; OLEUM BADIANI, O. ANISI STELLATI, L. From the capsules of *Illicium anisatum*, or star-anise. It continues liquid at 35½° F. At 35° F. it congeals. This, and its weaker reaction with iodine, distinguish it from the preceding compound.—*Prod.*, 2% (fully).

Oil of Sweet Fennel. See OIL OF FENNEL.

Oil of Sweet Flag. *Syn.* OLEUM ACORI, O. A. AROMATICA, L. From the rhizomes or roots of *Acorus calamus* (Linn.), or sweet flag. Yellow; agreeably fragrant. Used to scent snuff, aromatic vinegar, &c.—*Prod.* Fresh rhizomes, ½% to 1%; dried (recent), 1 to 1·25%.

Oil of Tan'sy. *Syn.* OLEUM TANACETI, L. From the flowering herb of *Tanacetum vulgare* (Linn.), or tansy. Pale greenish yellow; very odorous; bitter; aromatic. Sp. gr. ·946 to ·950.—*Prod.* Fresh, ·25% to 5%; dried (recent), ⅛% to 1% (fully).

Oil of Thyme. *Syn.* OLEUM THYMI; OIL OF ORIGANUM; OLEUM ORIGANI (of the shops). From the flowering herb of *Thymus vulgaris* (Linn.), or garden thyme. Nearly colourless; the imparted oil has a reddish colour, which it loses by rectification; very fragrant; acrid; hot tasted, stimulant, and rubefacient; boils at 354° F. Sp. gr. ·867 to ·875.—*Prod.*, 5% to ·75%.

Obs. This is the dark-coloured 'OIL OF ORIGANUM' of the shops. It is frequently adulterated with oil of turpentine. It is occasionally used in toothache and in stimulating liniments; but its chief consumption is in perfumery, more particularly for hair-oils, pomatums, and hair-washes, as it is reputed to make the hair grow and to prevent baldness.

Oil of Tobacco (Volatile). From the leaves of *Nicotiana tabacum* (Linn.), or the tobacco plant. Concrete.

Oil of Turpentine. *Syn.* SPIRIT OF T., ESSENCE OF T., TURPS, CAMPHENE, CAMPHINE; TEREBINTHINÆ OLEUM (B. P.), SPIRITUS TEREBINTHINÆ, ESSENTIA T., OLEUM TEREBINTHINÆ, O. T. PURIFICATUM, L. The oil of turpentine of commerce is obtained by distilling strained American turpentine along with water. The residuum in the still is 'resin' or 'rosin.' The product in oil varies from 14% to 16%. The colleges order it to be rectified before being employed for medicinal purposes. This is effected by redistilling it along with 3 or 4 times its volume of water, observing not to draw over quite the whole. The portion remaining in the retort (balsam of turpentine) is viscid and resinous. A better plan is to well agitate it with an equal measure of solution of potassa or milk of lime before rectifying it. This is the plan adopted for the camphine used for lamps. By agitating crude oil of turpentine with about 5% of sulphuric acid, diluted with twice its weight of water, and after repose and decantation rectifying it from 5 or 6 times its volume of the strongest lime water, a very pure and nearly scentless oil may be obtained. Dr Nimmo recommends oil of turpentine to be purified by agitation with ⅛th part of rectified spirit, after repose to decant the spirit, and to repeat the process 3 or 4 times. The product retains, however, fully ⅛th part of spirit in solution, and hence this method is objectionable, except for medicinal purposes, for which, according to Dr Garrod, it is better than the oil purified by rectification. The sweet spirits of turpentine (SPIRITUS TEREBINTHINÆ DULCIS), vended of late years in the shops, is simply the common oil which has been agitated with, and rectified from, somewhat dilute sulphuric acid.

Prop. Pure oil of turpentine is colourless; limpid; very mobile; neutral to test-paper; has an odour neither powerful nor disagreeable when recently prepared, but becoming so by exposure to the air; dissolves ⅛th part of alcohol of ·830; is soluble in 8½ parts of ether and in 6½ parts of rectified spirit; hot strong alcohol dissolves it freely, but the greater part separates in globules as the liquid cools. Oil of vitriol chars it, and strong nitric acid attacks it violently, even with flame. It congeals at 14°, and boils at 312° F. Sp. gr. ·867; that of the oil of the shops varies from ·872 to ·878. It possesses a very high refractive power. At 72° it absorbs 163 times its volume of hydrochloric-acid gas (if kept cool), and in 24 hours from 26% to 47% of crystals of terpene mono-hydrochloride (KIND'S CAMPHOR) separate. These have a camphoraceous odour, and, after being washed with water, and sublimed along with some dry chalk, lime, or charcoal, assume the form of a white, translucent, flexible, crystalline mass, which is volatile, soluble in alcohol, and possesses a considerable resemblance to camphor. A nearly similar substance is produced by the action of oxygen gas on oil of turpentine.

By continued agitation of turpentine oil with water and air, peroxide of hydrogen and camphine acid are produced. It is in this way 'sanitas' preparations are made.

Uses, &c. Oil of turpentine is extensively used in the manufacture of varnishes and paints. Under the name of 'camphine' it is occasionally employed for burning in lamps. For the last purpose it must be newly rectified and preserved from the air. By exposure it rapidly absorbs oxygen, resin is formed, its density increases, and it gives a dull fuliginous flame. In medicine, it is employed as a diaphoretic, stimulant, vermifuge, &c.—*Dose*, 6 to 30 or 40 drops; in rheumatism, hemicrania, &c., 1 fl. dr. every 4 hours, in combination with bark or capsicum; in tapeworm, 3 fl. dr. to 1 fl. oz., either alone or combined with a little syrup of orange peel, every 8 hours, until the worm is expelled. The common symptoms of large doses of this oil are dizziness and a species of temporary intoxication, and occasionally nausea and sickness, which subside after two or three alvine evacuations, leaving no other effect, when the oil is pure, than a certain degree of languor for a few hours. In tapeworm, a little castor oil may be advantageously combined with the second and subsequent doses. Oil of turpentine imparts a violent odour to the urine. To prevent loss by evaporation and resinification, this oil should be kept in tin cans or glass bottles.

For store vessels, closely covered tin cisterns are the best. To *prevent accidents*, it is proper to caution the operator of the extremely penetrating and inflammable nature of the vapour of this oil, even in the cold. During the process of its distillation, without the greatest precautions are taken, an explosion is almost inevitable.

Oil of Vale″rian. *Syn.* OLEUM VALERIANÆ (Ph. Bor.), L. From the root of *Valeriana officinalis* (Linn.), or wild valerian. Yellowish; viscid; lighter than water; smells strongly of the plant. By exposure to the air it is partly converted into valerianic acid, and more readily so under the influence of an alkali. In its usual form it consists of valerol, a neutral oily body; borneene, a volatile liquid hydrocarbon; and valerianic acid. It is powerfully antispasmodic, emmenagogue, tonic, and stimulant, and, in large doses, narcotic.—*Dose*, 2 to 6 drops; in epilepsy, hysteria, hemicrania, hypochondriasis, low fevers, &c.—*Prod.*, 1½% to 2% (nearly).

Oil of Ver′bena. *Syn.* OLEUM VERBENÆ, L. From the fresh flowering herb of *Verbena odorata*.—*Prod.*, 2% to 5%. The 'OIL OF VERBENA' of the shops is imported from India, and is obtained from *Andropogon citratum*. See OIL OF LEMON-GRASS.

Oil of Wine. *Syn.* HEAVY OIL OF WINE, ETHEREAL OIL, OILY ETHEREAL LIQUOR, SULPHATE OF ETHER AND ETHEROLE; OLEUM ÆTHEREUM (Ph. L.), OLEUM VINI, LIQUOR ÆTHEREUS OLEOSUS, L. This is an artificial production which, for convenience, may be included under this head.

Prep. 1. (Ph. L.) Rectified spirit, 2 pints, and sulphuric acid, 36 fl. oz., are cautiously mixed together in a glass retort, and submitted to distillation until a black froth appears, when the retort is immediately removed from the fire (sand heat); the lighter, supernatant liquor is next separated from the fluid in the receiver, and exposed to the air for 24 hours; it is then agitated with a mixture of solution of potassa and water, of each, 1 fl. oz., or q. s., and, when sufficiently washed, is, lastly, separated from the aqueous liquid from which it has subsided. The formula of the Ph. L. 1836 is nearly similar.

2. (Ph. D.) Rectified spirit and oil of vitriol (commercial), of each, 1¼ pints; as the last, employing a Liebig's condenser, and a capsule for the exposure to the air; the oil is then transferred to a moistened paper filter, and washed with a little cold water to remove any adhering acid.

3. (Ph. D. 1826.) From the residuum in the retort after the process of preparing ether, distilled to one half, by a moderate heat, and the oil treat as before.

4. From rectified spirit (sp. gr. ·833), 2 parts; oil of vitriol, 5 parts; mix and distil, as before; wash the product with distilled water, and free it from adhering water and undecomposed alcohol by exposure in the vacuum of an air-pump, between two open capsules, the one containing fragments of solid potassa, and the other concentrated sulphuric acid. Pure.

5. By distilling a mixture of ether, and oil of vitriol, and treating the product as before.

6. By the destructive distillation of dry sulphovinate of calcium; the product is freed from alcohol, &c., by washing it. This process yields the largest product.

Prop., &c. An oily liquid, nearly colourless, neutral, with an aromatic taste, and an odour resembling that of oil of peppermint. It is insoluble in water, but freely soluble in both alcohol and ether; boiling water converts it into sulphovinic acid, and a volatile liquid called light or sweet oil of wine; with an alkaline solution, this effect is produced with even greater facility. Sp. gr. 1·05 (*Hennel* and Ph. L.); 1·13 (*Serullas*). Boils at 540° F. "Dropped into water, it sinks, the form of the globule being preserved" (Ph. L.).—*Prod.*, 1·25% to 1·5%; 33 lbs. of rectified spirit, and 64 lbs. of oil of vitriol, yield 17 oz. of this oil (*Hennel*).

Uses. Oil of wine is reputed anodyne, but is only used in the preparation of other compounds. See SPIRIT OF ETHER (Compound), &c.

Oil of Wine (Light). *Syn.* SWEET OIL OF WINE. See ETHERIN, ETHEROLE, and *above*.

Oil, Wood (of India). From the *Chloroxylon Swietenia* (De Cand.), the tree which yields the satin-wood of the cabinet-makers. Another wood oil (GURJUN BALSAM) is obtained by incision from various species of *Dipterocarpus*. This balsam yields about 38% of a volatile oil by distillation, which in its general properties closely resembles OIL OF COPAIBA (*O'Shaughnessy*).

Oil of Worm'seed. *Syn.* OLEUM CHENOPODII (Ph. U. S.), L. From the seeds of *Chenopodium anthelminticum*, or Jerusalem oak (American wormseed). Light yellow, or greenish; powerfully anthelmintic. Sp. gr. ·908—*Dose*. For an adult, 25 to 30 drops, in sugar, honey, or milk, night and morning, for 3 or 4 days, followed by a good dose of castor oil, or some other suitable purgative.

Oil of Worm'wood. *Syn.* OLEUM ABSINTHII, L. From the herbaceous portion of *Artemisia absinthium*, or common wormwood; green or brownish-green; odorous; acrid; bitter; stomachic. Sp. gr. ·9703 (*Pereira*); ·9725 (*Brande*).—*Prod.* Fresh herb (picked), ¾% to ¾%, dry herb (a year old), ¾% (fully); do. (recent), ¾% to 1% (fully).

Pur. That of the shops is nearly always either adulterated or partly spoiled by age; hence the discrepancies in the densities given for this oil by different authorities. A specimen of this oil distilled by Mr Cooley from the green plant had the sp. gr. ·9712; but after being kept for 12 months it had increased to ·9718. Nitric acid of 1·25 colours the pure oil first green, then blue, and, lastly, brown. The positive character of these reactions is in direct proportion to the purity and freshness of the sample.

OILY EMUL′SION. See LINCTUS (Emollient).

OILY ETHE′REAL LIQ′UOR. See OIL OF WINE (*above*).

OINT′MENT. *Syn.* UNGUENTUM, L. Any soft, fatty substance applied to the skin by inunction. The term is now commonly restricted to those which are employed in medicine.

Ointments (unguenta) differ from 'cerates' chiefly in their consistence, and in wax not being a constant or essential constituent; and they are made and used in a nearly similar manner to that class of preparations. Their proper degree of solidity is that of good butter, at the ordinary temperature of the atmosphere. When the active ingredients are pulverulent substances, nothing can be more suitable to form the body of the ointment than good fresh lard, free from salt; but when they are fluid or semi-fluid, prepared suet, or a mixture of suet and lard, will be necessary to give a due consistence to the compound. In some instances wax is ordered for this purpose. Another excellent 'vehicle' for the more active ingredients is a simple ointment, formed by melting together 1 part of pure white wax with about 4 parts of olive oil. The use of the last excludes the possibility of the irritation sometimes occasioned by the accession of rancidity, when inferior lard is employed. In a few cases butter is employed to form the body of the ointment.

Soft and hard paraffins sold under the fancy names of vaseline, cosmoline, petroline, &c., have of late years largely superseded fats as ointment bases. In certain particulars they have advantages as they keep well, never going rancid or bad. On the other hand, they are said to act more like a varnish than an ointment to the skin, protecting the skin in such a way that remedies cannot be absorbed. Lanolin is a valuable ointment base, made from sheep's wool, and contains about 30% of water, it appears to be the most rapidly absorbed of all the ointment bases.

Some ointments are made from recent vegetable substances by infusion or coction, in the manner adopted for medicated oils. See OILS, MEDICATED.

Ointments are best preserved by keeping them in salt-glazed earthen or stoneware jars, covered with tin-foil, in a cool situation.

The accession of rancidity in ointments and other unctuous preparations may be greatly retarded, if not wholly prevented, by previously dissolving in the fat about 2% of gum-benzoin, in fine powder, or rather less quantity of benzoic acid by the aid of heat. This addition renders the ointment peculiarly soothing to irritable or highly sensitive skins. Poplar buds act in a similar manner.

₊ The formulæ for all the more useful and generally employed ointments are given below. Those not included in the list may be prepared of the proper strength for all ordinary purposes, by combining about 12 to 15 times the medium dose of the particular medicinal with 1 oz. of lard or simple ointment. For substances which possess little activity, ⅛ to 1 dr. per oz., or even more, may be taken. See CERATE, FAT, &c.

Ointment of Ac'etate of Lead. Syn. UNGUENTUM PLUMBI ACETATIS (P. B., Ph. E. and D.), L. Prep. 1. (Ph. E.) Acetate of lead, in fine powder, 1 oz.; simple ointment, 20 oz.; mix them thoroughly (by trituration).

2. (Ph. D.) Ointment of white wax, 1 lb.; melt by a gentle heat, then add, gradually, of acetate of lead, in very fine powder, 1 oz., and stir the mixture until it concretes.

3. (B. P.) Acetate of lead; in fine powder, 12 gr.; benzoated lard, 1 oz.; mix.

Obs. A useful, cooling, astringent, and desiccative ointment. For the formula of Ph. L., see CERATE.

Ointment, Ace'tic. See OINTMENT, VINEGAR.

Ointment of Ac'onite. Syn. UNGUENTUM ACONITI, L. Prep. 1. (Dr Turnbull.) Alcoholic extract of aconite, 1 part; lard, 2 parts; carefully triturated together. In neuralgia, &c.

2. (Ammoniated; UNGUENTUM ACONITI AMMONIATUM—Turnbull.) Ammoniated extract of aconite, 1 part; lard, 3 parts. In neuralgia, paralysis, old rheumatic affections, &c. The use of the above preparations of aconite requires the greatest caution. They are intended as substitutes for OINTMENT OF ACONITINE, a still more dangerous preparation.

Ointment of Acon'itine. Syn. UNGUENTUM ACONITINÆ, L. Prep. 1. (Dr Garrod.) Pure aconitine, 1 gr.; lard, 1 dr.; mix by careful trituration.

2. (Dr Turnbull.) Aconitine, 2 gr.; rectified spirits, 6 or 7 drops; triturate together, then add of lard, 1 dr., and mix well.

3. (B. P.) Aconitia (aconitine), 8 gr.; rectified spirit, ⅓ dr.; dissolve and add lard, 1 oz.; mix.

Use, &c. As a topical benumber in neuralgic affections, rheumatic pains, &c. Its application generally occasions considerable tingling, and sometimes redness of the part to which it is applied, followed by temporary loss of sensation in the skin and the cessation of the pain. For slight cases Dr Paris formerly employed only 1 gr. to the oz. Owing to the intensely poisonous nature of aconitine this ointment must be both prepared and used with great caution, and must never be applied to an abraded surface. It is seldom employed, owing to its extreme costliness. See ACONITIA and above.

Ointment, Ague. See OINTMENT, ANTIPERIODIC.

Ointment, Albinolo's. See PATENT MEDICINES.

Ointment, Alkaline. Syn. UNGUENTUM ALKALINUM, L. Prep. 1. (Biett.) Carbonate of soda, 2 dr.; fresh-slaked lime, 1 dr.; powdered opium, 2 gr.; lard, 2 oz.; mix by trituration. In prurigo, ringworm, and some other cutaneous affections.

2. (Casenave.) Carbonate of potassa, 1 dr.; lard, 1 oz. In psoriasis, lepra, and scorbutic eruptions.

3. (Devergie.) a. From carbonate (not sesquicarbonate) of soda, 10 to 15 gr.; lard, 1 oz. In lichen.

b. From carbonate of soda, 20 to 30 gr.; lard, 1 oz. In ichthyosis, lepra, psoriasis, and some other scaly skin diseases.

c. From carbonate of soda, ½ to 1 dr.; lard, 1 oz. In porrigo favosa, especially when occurring in adults.

4. (Soubeiran.) Carbonate of soda, 1 to 2 dr.; wine of opium, 1 fl. dr.; lard, 1 oz. In any of the above affections when there is much pain or irritation.

Obs. Carbonate of potassa is thought to be preferable to carbonate of soda when the above affections occur in scorbutic habits. A little camphor is also occasionally added.

Ointment of Aloes. See OINTMENT FOR WORMS.

Ointment of Aloes (Compound). See OINTMENT FOR WORMS.

Ointment of Al'um. *Syn.* UNGUENTUM ALUMINIS, L. *Prep.* 1. Alum, in very fine powder, 1 dr.; lard, 1½ oz. In piles.

2. To the last add of powdered opium, 7 gr. In piles, when there is much pain. See OINTMENT, BANYER'S.

Ointment, Ammoni'acal. *Syn.* UNGUENTUM AMMONIACALE, U. AMMONIÆ, L.; LIPAROLE D'AMMONIAQUE, POMMADE DE GONDRET, Fr. *Prep.* 1. (P. Cod.) Suet and lard, of each, 1 oz.; melt in a strong wide-mouthed bottle, add of liquor of ammonia (sp. gr. ·923), 2 oz., at once close the bottle, and agitate it until its contents concrete. As little heat as possible should be employed, to prevent unnecessary loss of ammonia.

2. (*Gondret.*) Lard, 3 parts; suet, 2 parts; almond oil, 1 part; strong solution of ammonia, 6 parts; mix as before. Rubefacient, vesicant, and counter-irritant. Smeared over the skin and covered so as to prevent evaporation, it raises a blister in 5 or 6 minutes. Its general effects and uses are similar to those of compound liniment of ammonia.

Ointment of Car'bonate of Ammo"nia. *Syn.* UNGUENTUM AMMONIÆ CARBONATIS, U. A. SESQUICARBONATIS, L. *Prep.* From carbonate of ammonia, 1 dr.; lard, 9 dr. An excellent application to painful joints, indolent tumours, scrofulous sores, &c.

Ointment of Ammo'niated Mercury. *Syn.* UNGUENTUM HYDRARGYRI AMMONIATI (B. P.). Ammoniated mercury, 1 part; simple ointment, 9 parts; mix. See next preparation.

Ointment of Ammo"nio-chloride of Mercury. *Syn.* WHITE PRECIPITATE OINTMENT; UNGUENTUM HYDRARGYRI AMMONIO-CHLORIDI (Ph. L.), U. H. PRECIPITATI ALBI, U. PRECIP. A. (Ph. E.), U. H. SUBMURIATIS AMMONIATI (Ph. D. 1826), L. *Prep.* 1. (Ph. L.) Ammonio-chloride of mercury, 2 dr.; lard, 3 oz.; triturate together.

2. (Ph. E.) As the last, but employing heat.

Uses, &c. Alterative; detergent; stimulant. In itch, scald-head, and various other skin diseases; in inflammation of the eyes; as an application to scrofulous and cancerous tumours; to destroy vermin on the body, &c. It "may be safely used" (in small quantities) "on infants" (*A. T. Thomson*).

Ointment, An"glo-Saxon. *Prep.* Heat olive oil, 1 pint, and beeswax, ¼ lb., until the mixture acquires a reddish-brown colour; then add red lead (levigated), ¼ lb., and continue the heat, with constant stirring; when the union appears complete add of amber and burnt alum, of each, in fine powder, ½ oz.; lastly, when considerably cooled add of powdered camphor, 3 dr. As a dressing to foul ulcers.

Ointment, An'odyne. See OINTMENT OF OPIUM, HEMLOCK, &c.

Ointment of An'thracoka'li. *Syn.* POMADE DE ANTHRACOKALI, Fr. *Prep.* (*Dr Polya.*) Anthracokali, in very fine powder, 1 part; lard, 30 parts. See ANTHRACOKALI.

Ointment, Antiherpet'ic. *Syn.* UNGUENTUM ANTIHERPETICUM, L. *Prep.* 1. (*Alibert.*) Red sulphide of mercury, 3 dr.; powdered camphor, 1 dr.; lard, 3 oz.

2. (*Chevallier.*) 'Subsulphate of mercury; (Turpeth mineral), 2 dr.; chloride of lime, 3 dr.' almond oil, 6 dr.; lard, 2 oz. In herpes or tetters.

Ointment, Antimo'nial. See OINTMENT OF POTASSIO-TARTRATE OF ANTIMONY.

Ointment of Araroba. See ARAROBA.

Ointment, Aromat'ic. *Syn.* BALSAMUM STOMACHALE WACKERI, UNGUENTUM AROMATICUM, L. *Prep.* (Ph. Austr. 1836.) Simple ointment, 2½ lbs.; yellow wax and oil of laurel, of each, 3 oz.; melt together, and, when considerably cooled, add of oils of juniper, mint, lavender, and rosemary, of each, 2 dr. Anodyne, balsamic, and stimulant.

Ointment of Arse"niate of I'ron. *Syn.* UNGUENTUM FERRI ARSENIATIS, L. *Prep.* 1. (*Carmichael.*) Arseniate of iron, ½ dr.; phosphate of iron, 3 dr.; spermaceti ointment, 6 dr.

2. (*Dr Pereira.*) Arseniate of iron, ½ dr.; lard, 1½ oz. In cancer.

Ointment of Arseniate of Soda. *Syn.* UNGUENTUM SODÆ ARSENIATIS, L. *Prep.* Arseniate of soda, 1 dr.; lard, 2 oz. Mix.

Ointment, Arse'nical. *Syn.* OINTMENT OF WHITE ARSENIC; UNGUENTUM ARSENICALE, U. ARSENICI, U. ACIDI ARSENIOSI, L. *Prep.* 1. Arsenious acid (levigated), 3 gr.; lard or simple ointment, 1 oz. In lepra, psoriasis, malignant whitlows, &c.

2. (Hosp. F.) Levigated white arsenic, 15 to 20 gr.; lard, 1 oz. As a dressing for cancerous sores.

3. (*Soubeiran.*) White arsenic, 1 dr.; lard and spermaceti ointment, of each, 6 dr. In malignant cancer. The above must be carefully prepared, and used with great caution. See CERATE.

Ointment, Astrin'gent. *Syn.* UNGUENTUM ASTRINGENS, L. *Prep.* Triturate powdered catechu, 1½ dr., with boiling water, 2 fl. dr.; add, gradually, of spermaceti ointment (melted), 1½ oz., and continue the trituration until the mass concretes. An excellent dressing for ill-disposed sores and ulcers, especially during hot weather. See the several LEAD OINTMENTS, OINTMENT OF GALLS, &c.

Ointment of Atro"pia. *Syn.* UNGUENTUM ATROPIÆ (B. P.), L. *Prep.* 1. Atropia, 1½ gr.; simple ointment, 1 dr.; mix by careful trituration.

2. (*Dr Brookes.*) Atropia, 5 gr.; lard, 3 dr.; otto of roses, 1 drop. In neuralgia, rheumatic pains, &c., when the affection is not deeply seated.

3. (B. P.) Atropia, 8 gr.; rectified spirit, ½ dr.; lard, 1 oz.; dissolve the atropia in the spirit and mix with the lard.

Ointment, Bailey's. See OINTMENT, ITCH.

Ointment of Bal'sam of Peru. *Syn.* UNGUENTUM BALSAMI PERUVIANI, L. *Prep.* 1. Lard or spermaceti ointment, 1 oz.; balsam of Peru, 1 dr.; melt together by the heat of boiling water, stir for 5 or 6 minutes, allow it to settle, and pour off the clear portion. In chaps and abrasions.

2. (Compound: UNG. B. P. COMPOSITUM—*Copland.*) Lard, 1 oz.; white wax, ½ oz.; balsam of Peru, 1 dr.; melt as before, and when nearly

cold, add of oil of lavender, 10 or 12 drops. As the last, and to restore the hair.

Ointment, Banyer's. *Syn.* COMPOUND ALUM OINTMENT; UNGUENTUM ALUMINIS COMPOSITUM, U. CALOMELANOS, U. BANYERI, L. *Prep.* From burnt alum and calomel, of each, 1½ oz.; carbonate of lead, or litharge (levigated), 2 oz.; Venice turpentine, ½ lb.; lard, 2 lbs.; carefully triturated together. In milk-scald, porrigo, &c.

Ointment of Bark. See OINTMENT OF CINCHONA.

Ointment, Basil'icon. *Syn.* BASILICON, YELLOW B.; UNGUENTUM BASILICUM, U. B. FLAVUM, L. *Prep.* (Ph. L. 1746.) Olive oil, 16 fl. oz.; yellow wax, yellow resin, and Burgundy pitch, of each, 1 lb.; melt, remove the vessel from the fire, and stir in of common turpentine, 3 oz. This form is still occasionally employed in some shops, but is generally superseded by the resin cerate and resin ointment of the Pharmacopœias. A nearly similar preparation, under the name of ' basilicon ointment,' is contained in the Ph. Bor. 1847 (see *below*).

Ointment, Basilicon (Black). See OINTMENT OF PITCH.

Ointment, Basilicon (Green). *Syn.* UNGUENTUM BASILICUM VIRIDE, L. *Prep.* (Ph. L. 1746.) Prepared verdigris, 1 oz.; yellow basilicon, 8 oz.; olive oil, 3 fl. oz. Detergent. Used to keep down fungous growths, to dress syphilitic ulcers, &c. See CERATE and OINTMENT OF VERDIGRIS.

Ointment, Bateman's. See OINTMENT, ITCH.

Ointment of Bay-leaves. See OINTMENT, LAUREL.

Ointment of Belladon'na. *Syn.* UNGUENTUM BELLADONNÆ, L. *Prep.* 1. (Ph. L.) Extract of belladonna (deadly nightshade), 1 dr.; lard, 1 oz.; mix by trituration.

2. (*Soubeiran*.) Fresh belladonna leaves (bruised), 1 part; lard, 2 parts; simmer together until the leaves become crisp, and, after digestion for a short time longer, drain with pressure.

3. (B. P.) Alcoholic extract of belladonna, 1 part; rubbed with benzoated lard, 9 parts.

Uses, &c. As a local anodyne, in painful and indolent tumours, nervous irritations, &c. Also as an application to the neck of the uterus in cases of rigidity (*Chaussier*).

4. (Compound: UNGUENTUM BELLADONNÆ COMPOSITUM, L.) *a.* (*W. Cooley.*) Compound iodine ointment, 7 dr.; extract of belladonna, 1 dr. Powerfully discutient. A most excellent application to all glandular tumours and indurations, buboes, &c., which it is desirable to disperse instead of mature, more especially where there is much pain. It is particularly suitable to cases occurring on shipboard; and when its application (at least twice a day) is accompanied with the internal use of the mixture of iodine and gold (see ANTISCROFULOUS MIXTURE), this treatment has seldom failed, even when the patients were dieted chiefly on salt food.

b. (*Debreyne*.) Extract of belladonna and lard, of each, 3 dr.; powdered opium, ½ dr. As an external anodyne and benumber, more especially in neuralgia, painful cancerous tumours, &c. A small piece is to be applied to the part, and the friction continued for 6 or 8 minutes. The above prepara-

tions are useless unless the extract employed is recent and of good quality.

Ointment of Benzoin. (Ph. U. S.) *Prep.* Tincture of benzoin, 2 oz.; lard, 16 oz.; melt the lard over a water-bath and add the tincture, stirring constantly, and when the spirit has evaporated, remove from the water-bath, and stir whilst cooling.

Ointment of Bismuth. *Syn.* UNGUENTUM BISMUTHI, L. *Prep.* 1. Nitrate of bismuth (' white bismuth '), 1 dr.; simple ointment, 1 oz.

2. (*Fuller*.) Nitrate of bismuth, 1 dr.; spermaceti ointment, 19 dr. In itch and some chronic cutaneous diseases.

Ointment, Blist'ering. See OINTMENT OF CANTHARIDES and VESICANTS.

Ointment, Blue. This is the vulgar name in England of mercurial ointment. On the Continent an ointment made of smalts and Goulard water is commonly so called.

Ointment of Bo'rax. *Syn.* UNGUENTUM BORACIS, L. *Prep.* From borax (in very fine powder), 1 dr.; simple ointment or lard, 7 dr. In excoriations, chaps, &c.

Ointment of Boric Acid. (B. P.) *Syn.* UNGUENTUM ACIDI BORICI, L. *Prep.* Boric acid in powder, 1 part; soft paraffin, 4 parts; hard paraffin, 2 parts. Melt, mix, and stir till cold. This ointment was devised by Sir J. Lister. It is a mild antiseptic. Used for dressing ulcers and burns.

Ointment of Bromide of Potas'sium. *Syn.* UNGUENTUM POTASSII BROMIDI, U. POTASSÆ HYDROBROMATIS, L. *Prep.* (*Magendie*.) Bromide of potassium, ½ dr.; lard, 1 oz. Resolvent; in bronchocele, scrofula, &c.

Ointment of Bro'mine. *Syn.* UNGUENTUM BROMINII, U. B. COMPOSITUM, L. *Prep.* (*Magendie*.) Bromide of potassium, 20 gr.; bromine, 6 to 12 drops; lard, 1 oz. As the last, but more active.

Ointment, Brown. *Syn.* FRENCH POOR MAN'S FRIEND; UNGUENTUM FUSCUM, U. HYDRARGYRI F., L. *Prep.* (P. Cod.) Nitric oxide of mercury (levigated), ½ dr.; resin ointment, 1 oz. In ophthalmia (cautiously), after the inflammatory stage is over; as an application to sore legs, &c.

Ointment of Cad'mium. *Syn.* UNGUENTUM CADMII, U. C. SULPHATIS, L. *Prep.* (*Radius*.) Sulphate of cadmium, 1 to 2 gr.; pure lard, 1 dr.; carefully triturated together. In specks on the cornea.

Ointment of Cadmium, Iodide of. (B. Ph.) *Syn.* UNGUENTUM CADMII IODIDI. *Prep.* Mix thoroughly iodide of cadmium in fine powder, 62 gr., with simple ointment, 1 oz.

Ointment of Caffeine. *Syn.* UNGUENTUM CAFFEINÆ, L. *Prep.* Citrate of caffeine, 8 gr.; lard, 10 oz. Mix.

Ointment of Cal'amine. (B. P.) *Syn.* UNGUENTUM CALAMINÆ, L. *Prep.* Prepared calamine, 1 part; benzoated lard, 5 parts. This is known as Turner's cerate.

Ointment of Cal'omel. *Syn.* UNGUENTUM HYDRARGYRI SUBCHLORIDI (B. P.), UNGUENTUM CALOMELANOS, U. HYDRARGYRI CHLORIDI, L. *Prep.* 1. From calomel, 80 gr.; benzoated lard, 1 oz.

Obs. " Were I required to name a local agent

pre-eminently useful in skin diseases generally, I should fix on this. It is well deserving a place in the Pharmacopœia " (*Pereira*). Dr Underwood uses elder-flower ointment as the vehicle.

2. (Compound: UNGUENTUM CALOMELANOS COMPOSITUM—*Dr A. T. Thomson*.) Calomel, 1 dr.; tar ointment, 4 dr.; spermaceti ointment, 1 oz.

Ointment of Cam'phor. *Syn.* UNGUENTUM CAMPHORÆ, L. *Prep.* 1. Camphor, 1 to 2 dr.; lard, 1 oz.; dissolve by a gentle heat and stir until the mass is nearly cold. Stimulant and anodyne; in prurigo, psoriasis, &c.

2. (Compound.) From powdered opium, ½ dr.; powdered camphor, 1½ dr.; lard, 1½ oz.; mix by trituration. As an anodyne friction in rheumatic pains, swelled joints, colic, &c.

Ointment of Canthar'ides. *Syn.* UNGUENTUM CANTHARIDIS (B. P., Ph. L., D., and U. S.), U. LYTTÆ, L. *Prep.* 1. (Ph. L.) Cantharides (in very fine powder), 3 oz.; distilled water, 12 fl. oz.; mix, boil to one half; to the strained liquid add of resin cerate, 1 lb., and evaporate to a proper consistence.

2. (Ph. D.) Liniment of Spanish flies, 8 fl. oz.; white wax, 3 oz.; spermaceti, 1 oz.; melt together with a gentle heat, and stir until it concretes.

3. (Ph. E.) *a.* (UNGUENTUM INFUSI CANTHARIDIS—Ph. E.) Powdered cantharides, 1 oz. boiling water, ½ pint; infuse one night (12 hours), strain with expression, add of lard, 2 oz., and boil until the water is expelled; then add beeswax and resin, of each, 1 oz., and when these are liquefied, remove the vessel from the fire, and further add of Venice turpentine, 2 oz.

b. (UNGUENTUM PULVERIS CANTHARIDIS—Ph. E.) Resin ointment, 7 oz.; melt, add of cantharides (in fine powder), 1 oz., and stir until the whole is nearly cold.

4. (B. P.) Cantharides, in fine powder, 1 part; olive oil, 6 parts; yellow wax, 1 part; digest the cantharides in the oil for 12 hours, and for ½ hour at 212°; strain, add the melted wax, and stir till cold.

Obs. The above preparations are frequently called 'blister ointment' or 'epispastic ointment.' They are used to keep blisters open after they have been produced by stronger compounds. The first three compounds are regarded as milder than the last (3, *b*), which contains the flies in substance. The P. Cod. contains an ointment (UNG. EPISPASTICUM FLAVUM) which is weaker than the above, prepared by digesting the bruised flies in lard, for 3 hours, over a warm bath; about 1-6th part of wax is next added to the strained fat, which is then coloured with turmeric, and scented with oil of lemon. See CERATE, POMMADE, VESICANTS, and *below*.

Ointment of Cantharides, Extract of. (M. Cap.) *Syn.* UNGUENTUM CUM EXTRACTO CANTHARIDIS. *Prep.* Alcoholic extract of cantharides, 8 gr.; oil of roses, 1 dr.; beef marrow, 2 oz.; oil of lemon, 40 minims. To promote the growth of the hair.

Ointment of Cantharides with Mercury. *Syn.* UNGUENTUM CANTHARIDIS CUM HYDRARGYRO. *Prep.* Lard, 65 parts; Spanish flies, 29 parts; strong mercurial ointment, 6 parts. Mix. Used in Normandy to indolent tumours.

Ointment of Canthar'idine. *Syn.* UNGUENTUM CANTHARIDINÆ, L. *Prep.* (*Soubeiran.*) Cantharidine, 1 gr.; white wax, 1 dr.; lard, 7 dr.; mix thoroughly. See *above.*

Ointment of Cap'sicum. *Syn.* UNGUENTUM CAPSICI, L. *Prep.* (*Dr Turnbull.*) Tincture of capsicum (pure), q. s.; gently evaporate it until it begins to gelatinise, then mix the extract with twice its weight of lard. As a powerful stimulant and rubefacient. When very freely used, it vesicates.

Ointment of Carbolic Acid. (B. P.) *Syn.* UNGUENTUM ACIDI CARBOLICI, L. *Prep.* Carbolic acid, 1 part; soft paraffin, 18 parts; hard paraffin, 9 parts. Melt and mix.

Ointment of Car'bonate of Ammo"nia. See OINTMENT, AMMONIACAL.

Ointment of Car'bonate of Lead. *Syn.* WHITE-LEAD OINTMENT; UNGUENTUM PLUMBI CARBONATIS (P. B., Ph. E. and D.), U. CERUSSÆ, L. *Prep.* 1. (Ph. E.) Carbonate of lead, 1 oz.; simple ointment, 5 oz.; mix thoroughly.

2. (Ph. D.) Carbonate of lead, 3 oz.; ointment of white wax, 1 lb.; mix with heat.

3. (B. P.) Carbonate of lead, in fine powder, 1 part; simple ointment, 7 parts. Mix.

4. UNGUENTUM PLUMBI CAMPHORATUM (E., 1744). Add to the last 2 scruples of camphor ground with a little oil.

Uses, &c. Cooling, desiccative. Useful to promote the healing of excoriated parts and slight ulcerations. The camphorated white ointment of old *pharmacy* (UNG. ALBUM CAMPHORATUM—Ph. L., 1744) was made by adding 40 gr. of camphor to the first of the above.

Ointment of Cat'echu. *Syn.* UNGUENTUM CATECHU, L. *Prep.* From alum, 1 oz.; catechu, 3 oz. (both in very fine powder); added to olive oil, ½ pint, and yellow resin, 4 oz., previously melted together. Used to dress ulcers in hot climates, where the ordinary fat ointments are objectionable; also in this country during hot weather. See OINTMENT, ASTRINGENT.

Ointment of Chalk. *Syn.* UNGUENTUM CRETÆ, L. *Prep.* Prepared chalk, 1 oz.; lard, 4 oz. Mix.

Ointment of Chamomile. (*M. Basin.*) *Syn.* UNGUENTUM ANTHEMIDIS, L. *Prep.* Freshly powdered chamomile flowers, olive oil, and lard, in equal quantities. For the cure of itch.

Ointment of Char'coal. *Syn.* UNGUENTUM CARBONIS, L. *Prep.* 1. Resin ointment, 10 dr.; recently burnt charcoal (levigated), 3 dr. As a dressing to foul ulcers, especially those of the legs.

2. (*Caspar.*) Lime-tree charcoal and dried carbonate of soda, of each, 2 dr.; rose ointment, 1 oz., or q. s. In scald-head.

3. (*Radius.*) Animal charcoal (recent), 1 part; mallow ointment, 2 parts. As a friction in glandular enlargements and indurations, as a dressing to fetid ulcers, &c.

Ointment of Chaulmoogra. *Syn.* UNGUENTUM GYNOCARDIÆ. Chaulmoogra oil, 1 part; petroleum cerate, 3 parts. Used in leprosy, lupus, and eczema.

Ointment of Cherry Laurel. *Syn.* UNGUENTUM LAURO-CERASI, L. *Prep.* (*Soubeiran.*) Essential oil of cherry laurel, 1 dr.; lard, 1 oz.

To alleviate the pain in cancer, neuralgia, and other local affections.

Ointment, Chil'blain. *Syn.* UNGUENTUM AD PERNIONES, L. *Prep.* 1. From made mustard (very thick), 2 parts; almond oil and glycerin, of each, 1 part; triturated together. To be applied night and morning.

2. (*Cottereau.*) Acetate of lead, camphor, and cherry-laurel water, of each, 1 dr.; tar, 1½ dr.; lard, 1 oz.

3. (*Devergie.*) Creasote and Goulard's extract, of each, 12 drops; extract of opium, 1½ gr.; lard, 1 oz. Twice or thrice daily.

4. (*Giacomini.*) Sugar of lead, 2 dr.; cherry-laurel water (distilled), 2 fl. dr.; lard, 1 oz.

5. (*Linnæus.*) Balsam of Peru, 1 dr.; hydrochloric acid, 2 dr.; spermaceti ointment, 2½ oz.

Obs. For Swediaur's, Vance's, and Wahler's ointments, see article CHILBLAIN.

Ointment of Chloral Hydrate. (*Dowault.*) *Syn.* UNGUENTUM CHLORALIS HYDRAS. *Prep.* Chloral hydrate, 2 parts; lard, 20 parts. Stimulant; stronger if required as a rubefacient.

Ointment of Chlo"ride of Cal'cium. *Syn.* UNGUENTUM CALCII CHLORIDI, U. CALCIS MURIATIST, L. *Prep.* (*Sundelin.*) Chloride of calcium (dry), 1 dr.; strong vinegar, 40 gr.; foxglove (recent, in fine powder), 2 dr.; lard, 1 oz. In bronchocele, scrofulous tumours, &c.

Ointment of Chloride of Lead. *Syn.* UNGUENTUM PLUMBI CHLORIDI, L. *Prep.* (*Tuson.*) Chloride of lead, 1 part; simple cerate, 8 parts; carefully triturated together. In painful cancerous ulcerations and neuralgic tumours. See LEAD (Chloride).

Ointment of Chloride of Lime. See OINTMENT OF HYPOCHLORITE OF LIME.

Ointment of Chloride of Mercury. *Syn.* See OINTMENTS of CALOMEL and CORROSIVE SUBLIMATE.

Ointment of Chlo"rine. *Syn.* UNGUENTUM CHLORINII, L. *Prep.* (*Augustin.*) Chlorine water, 1 part; lard, 8 parts; well triturated together. In itch, lepra, ringworm, fœtid ulcers, &c.

Ointment of Chlori'odide of Mercury. *Syn.* UNGUENTUM HYDRARGYRI CHLORIODIDI, L. *Prep.* (*M. Recamier.*) Chloriodide (iodo-chloride) of mercury, 3 gr.; lard, 5 dr. Recommended as a powerful discutient or resolvent. See OINTMENT OF IODO-CHLORIDE OF MERCURY.

Ointment of Chlo"roform. *Syn.* UNGUENTUM CHLOROFORMI, L. *Prep.* (*M. Louis.*) Chloroform, 1 dr.; simple ointment, 1 oz. In neuralgia and rheumatic pains, &c. It must be kept in a stoppered, wide-mouthed phial.

Ointment of Chrysarobin. (B. P.) *Syn.* UNGUENTUM CHRYSAROBINI, L. Chrysarobin, 1 part; benzoated lard, 24 parts. Melt, and stir whilst hot, so as to promote solution.

Ointment of Cincho'na. *Syn.* OINTMENT OF BARK; UNGUENTUM CINCHONÆ, L. *Prep.* (*Biett.*) Red cinchona bark (in very fine powder) and almond oil, of each, 1 part; beef marrow (prepared), 3 parts. In the variety of scald-head termed porrigo decalvans. A little oil of mace or tar is a useful addition.

Ointment, Cit'rine. See OINTMENT OF NITRATE OF MERCURY.

Ointment of Cobalt, Oxide of. (Amst. Ph.) *Syn.* UNGUENTUM OXIDI COBALTI. *Prep.* Simple cerate, 16 oz.; liquid subacetate of lead, 4 oz.; powdered smalt, 4 oz.

Ointment of Cocaine. *Syn.* UNGUENTUM COCAINÆ. Cocaine hydrochlorate, 1 part; lanoline, 30 parts. Used in neuralgia, shingles, urticaria, eczema, and pruritus.

Ointment of Coc'culus In'dicus. *Syn.* UNGUENTUM COCCULI (Ph. E.), L. *Prep.* (Ph. E.) Kernels of *Cocculus indicus*, 1 part; beat them to a smooth paste in a mortar, first alone, and next with a little lard; then further add of lard, q. s., so that it may be equal to 5 times the weight of the kernels. Used to destroy pediculi, and in scald-head, &c.

Ointment of Cod-liver Oil. *Syn.* UNGUENTUM OLEI MORRHUÆ, U. O. JECORIS ASELLI, L. *Prep.* Cod-liver oil (pale and recent), 7 parts; white wax and spermaceti, of each, 1 part; melted together. In ophthalmia and opacity of the cornea, either alone or combined with a little citrine ointment; as a friction or dressing for scrofulous indurations and sores; in rheumatism, stiff joints, and in several skin diseases. It often succeeds in porrigo or scald-head when all other remedies have failed. Scented with oil of nutmeg and balsam of Peru, it forms an excellent pomade for strengthening and restoring the hair.

Ointment of Col'ocynth. *Syn.* UNGUENTUM COLOCYNTHIDIS, L. *Prep.* (*Chrestien.*) Colocynth pulp (in very fine powder), 1 part; lard, 8 parts. Used in frictions on the abdomen as a hydragogue purgative, in mania, dropsy, &c.

Ointment of Corrosive Sub'limate. *Syn.* OINTMENT OF PERCHLORIDE OF MERCURY; UNGUENTUM HYDRARGYRI PERCHLORIDI, L. *Prep.* 1. From corrosive sublimate, 2 to 5 gr.; rub it to powder in a glass or wedgwood-ware mortar; add of rectified spirit, 6 or 7 drops, or q. s.; again triturate; lastly add, gradually, of spermaceti ointment (reduced to a cream-like state by heat), 1 oz., and continue the trituration until the whole concretes. Used as a stimulant, detergent, and discutient application in various local affections; in lepra, porrigo, acne, &c., and as a dressing to syphilitic and some other ulcers.

2. (Ph. Chirur.) Corrosive sublimate, 10 gr.; yelk of 1 egg; lard, 1 oz. As a dressing.

3. (POMMADE DE CIRILLO, P. Cod.) Corrosive sublimate, 1 dr.; lard, 1 oz. Caustic; must not be confounded with the preceding.

Ointment, Cosmet'ic. *Syn.* UNGUENTUM COSMETICUM, L.; POMMADE DE LA JEUNESSE, Fr. *Prep.* (*Quincey.*) Spermaceti, 3 dr. (better, 4½ dr.); oil of almonds, 2 oz.; melt together, and, when cooled a little, stir in of nitrate of bismuth ('white bismuth'), 1 dr.; and, lastly, of oil of rhodium, 6 drops. In itch and some other cutaneous eruptions, but chiefly as a pomade for the hair. Its frequent use is said to turn the latter black.

Ointment of Cre'asote. *Syn.* UNGUENTUM CREASOTI (B. P., Ph. L., E., D. & U. S.), L. *Prep.* 1. (Ph. L.) Creasote, ½ fl. dr.; lard, 1 oz.; triturate together.

2. (Ph. E.) Lard, 3 oz.; melt it by a gentle heat; add of creasote, 1 dr., and stir the mixture until it is nearly cold.

3. (Ph. D.) Creasote, 1 fl. dr.; ointment of white wax, 7 dr.; as the last.

4. (B. P.) Creasote, 1 part; simple ointment, 8 parts. Mix.

Uses, &c. In several skin diseases, especially ringworm; as a friction in tic-douloureux; a dressing for scalds and burns; an application to chilblains, &c.

Ointment of Cro'ton Oil. *Syn.* UNGUENTUM CROTONIS, L. *Prep.* 1. Croton oil, 15 to 30 drops; lard (softened by heat), 1 oz.; mix well. This is the usual and most useful strength to prepare the ointment. Rubefacient and counter-irritant; in rheumatism and various other diseases. When rubbed repeatedly on the skin it produces redness and a pustular eruption. It also often affects the bowels by absorption. The only advantage it possesses over other preparations of the class is the rapidity of its action.

2. (RUBEFACIENT POMADE—*Caventou.*) White wax, 1 part; lard, 5 parts; melt together, and, when quite cold, mince it small, add of croton oil, 2 parts, and mix by trituration. Stronger than the last.

Ointment of Cucumber. *Syn.* UNGUENTUM CUCUMERIS, L. Cucumber juice, 1200 parts; lard, 1000 parts; veal suet, 600 parts; balsam of tolu, dissolved in spirit of wine, 2 parts; rose-water, 10 parts. Used as a cooling ointment like cold cream.

Ointment of Cy'anide of Mer'cury. *Syn.* UN-GUENTUM HYDRARGYRI CYANIDI, L. *Prep.* 1. (*Casenave.*) Cyanide of mercury, 8 gr.; lard, 1 oz.; carefully triturated together.

2. (*Pereira.*) Cyanide of mercury, 10 to 12 gr.; lard, 1 oz. As a dressing for scrofulous and syphilitic ulcers, &c.; as an application in psoriasis, moist tetters, and some other skin diseases, &c. Biett orders the addition of a few drops of essence of lemon.

Ointment of Cyanide of Potas'sium. *Syn.* UN-GUENTUM POTASSII CYANIDI, L. *Prep.* (*Case-nave.*) Cyanide of potassium, 12 gr.; oil of almonds, 2 dr.; triturate, add of cold cream (dry), 2 oz., and mix by careful trituration. As an anodyne in neuralgia, rheumatism, swelled joints, &c.; also as a friction over the spine in hysteria, and over the epigastrium in gastrodynia, &c. The greatest possible care must be used in the employment of this compound.

Ointment of Del'phinine. *Syn.* UNGUENTUM DELPHINIÆ, L. *Prep.* (*Dr Turnbull.*) Delphinine or delphinia, 10 to 30 gr.; olive oil, 1 dr.; lard, 1 oz.; mix as the last. Used as a friction in rheumatism, and the other cases in which veratrine is employed.

Ointment, Depil'atory. *Syn.* UNGUENTUM DE-PILATORIUM, L. See DEPILATORY (*Casenave's*).

Ointment, Desic'cative. *Syn.* DRYING OINT-MENT; UNGUENTUM DESICCATIVUM, U. EXSIC-CANS, L. See the OINTMENTS of CALAMINE, LEAD, ZINC, &c.

Ointment, Deter'gent. *Syn.* UNGUENTUM DE-TERGENS, L. The OINTMENTS of NITRATE OF MERCURY, NITRIC OXIDE OF MERCURY, TAR, VER-DIGRIS, &c., when not too strong, come under this head.

Ointment, Diges'tive. *Syn.* UNGUENTUM DI-GESTIVUM, L. *Prep.* 1. (P. Cod.) Venice tur-

pentine, 2 oz.; yolks of 2 eggs; mix, and add of oil of St John's-wort, ½ oz.

2. (DIGESTIF ANIMÉ, P. Cod.) As the last, with an equal weight of liquid styrax.

3. (DIGESTIF MERCURIEL, P. Cod.) As No. 1, with an equal weight of mercurial ointment.

4. (UNG. D. VIRIDE, *Dr Kirkland.*) Beeswax, gum elemi, and yellow resin, of each, 1 oz.; green oil, 6 oz.; melt them together, and, when considerably cooled, add of oil of turpentine, 2 dr.

Ointment, Edinburgh. Two compounds are known under this name. 1. (BROWN.) From black basilicon, 6 parts; milk of sulphur, 2 parts; sal-ammoniac, 1 part.

2. (WHITE.) From white hellebore, 3 oz.; sal-ammoniac, 2 oz.; lard, 1 lb. Both are used in itch.

Ointment of Eggs. *Syn.* UNGUENTUM OVO-RUM, L. *Prep.* 1. Yolk of 1 egg; honey and fresh linseed oil, of each, 1 oz.; balsam of Peru, ½ dr.; mix well.

2. (*Soubeiras.*) Beeswax, 4 dr.; oil of almonds, 1½ oz.; yolk of 1 egg. As an emollient and soothing dressing to excoriations, irritable ulcers, &c.

Ointment, Egyp'tian. *Prep.* (*Giordano.*) Burnt alum, 1 part; verdigris, 10 parts; strong vinegar, 14 parts; purified honey (thick), 32 parts; mix by heat and agitation. As a detergent application to foul ulcers. It is a modification of the 'UNGUENTUM ÆGYPTIACUM' of old pharmacy.

Ointment of Elder Flowers. *Syn.* WHITE ELDER OINTMENT; UNGUENTUM SAMBUCI FLO-RUM, U. SAMBUCI (Ph. L.), L. *Prep.* 1. (Ph. L.) Elder flowers and lard, of each, 1 lb.; boil them together until the flowers become crisp, then strain, with pressure, through a linen cloth. The same precautions must be observed as are necessary in the preparation of the medicated oils by infusion. Emollient; less white and odorous than the following.

2. (Wholesale.) Take of lard (hard, white, and sweet), 25 lbs.; prepared mutton suet, 5 lbs.; melt them in a well-tinned copper or earthen vessel, add of elder-flower water, 3 galls.; agitate briskly for about ½ an hour, and set it aside; the next day gently pour off the water, remelt the ointment, and add of benzoic acid, 5 dr.; otto of roses, 20 drops; oil of bergamot and oil of rosemary, of each, 1 dr.; again agitate well, let it settle for 10 minutes, and then pour off the clear portion into pots for sale. Very agreeable, and keeps well.

Obs. The last formula is the one now generally adopted by the large wholesale houses.

Ointment of Elder Leaf. *Syn.* ELDER OINT-MENT, GREEN E. O.; UNGUENTUM VIRIDE, U. SAMBUCI VIRIDE, U. SAMBUCI (Ph. D. 1826), L. *Prep.* 1. (Ph. D. 1826.) Fresh elder leaves (bruised), 3 lbs.; suet, 4 lbs.; lard, 2 lbs.; boil together as above.

2. (Wholesale.) Good fresh lard, 1 cwt.; fresh elder leaves, 56 lbs.; boil till crisp, strain off the oil, put it over a slow fire, add hard prepared mutton suet, 14 lbs., and gently stir it until it acquires a bright green colour.

Obs. The above ointment is reputed to be emollient and cooling, and has always been a great favourite with the common people. Both elder-flower and elder-leaf ointment are, however, unnecessary preparations. "They are vestiges of the redundant practice of former times" (*A. T. Thomson*). The above formulæ are those now almost exclusively employed in trade. The ointment should be allowed to cool very slowly; and after its temperature has fallen a little, and it begins to thicken, it should not be stirred, in order that it may 'grain' well, as a granular appearance is much admired. It is a common practice to add powdered verdigris to deepen the colour, but then the ointment does not keep well. This dangerous fraud may be detected in the manner noticed under CERATE, SAVINE.

Ointment of Elecampane. *Syn.* UNGUENTUM INULÆ. *Prep.* Fresh elecampane root (boiled till soft and pulped), 1½ oz.; lard, 1 oz. Mix.

Ointment of El'emi. *Syn.* BALSAM OF ARCŒUS†; UNGUENTUM ELEMI (B. P., Ph. L. & D.), L. *Prep.* 1. (Ph. L.) Elemi, 8 oz.; suet, 6 oz.; melt them together, remove the vessel from the fire, and stir in of common turpentine, 2½ oz.; olive oil, ½ fl. oz.; lastly, strain the whole through a linen cloth.

2. (Ph. D.) Resin of elemi, 4 oz.; ointment of white wax, 1 lb.; melt them together, strain through flannel, and stir the mixture constantly until it concretes.

8. (B. P.) Elemi, 1; simple ointment, 4; melt and strain.

Uses, &c. Stimulant and digestive. It is frequently employed to keep open issues and setons, and as a dressing for old and ill-conditioned sores. The 'UNG. ELEMI CUM ÆRUGINE' of St George's Hospital is made by adding 1 dr. of finely powdered verdigris to every 6 oz. of the ointment.

Ointment, Escharot'ic. *Syn.* UNGUENTUM ESCHAROTICUM, L. *Prep.* (Sir B. Brodie.) Corrosive sublimate, 1 dr.; nitric oxide of mercury, sulphate of copper, and verdigris, of each 2 dr. (all in very fine powder); lard, q. s. See OINTMENT, and CERATE, ARSENICAL.

Ointment of Eucalyptus. (B. P.) *Syn.* UNGUENTUM EUCALYPTI. Oil of eucalyptus, 1 part; hard and soft paraffin, of each, 1 part. Melt the paraffins, add the oil, and stir till cold.

Ointment of Euphorbium. (*Dr Neligan.*) *Syn.* UNGUENTUM EUPHORBII. *Prep.* Powdered euphorbium, 25 to 30 gr.; lard, 1 oz.; mix. To keep up a discharge from issues.

Ointment, Eye. 1. (*Dessault.*) Nitric oxide of mercury, carbonate of zinc, acetate of lead, and dried alum, of each, 1 dr.; corrosive sublimate, 10 gr.; rose ointment, 1 oz. In chronic ophthalmia, profuse discharges, &c.; generally diluted.

2. (*Dupuytren.*) Red oxide of mercury, 10 gr.; sulphate of zinc, 20 gr.; lard, 2 oz. For chronic inflammation of the eyelids, chronic ulcers, &c.

8. (*Regent.*) Acetate of lead and red precipitate, of each, 1 dr.; camphor, 6 gr.; washed fresh butter, 2½ oz. As the last, and in chronic ulcerations.

4. (Singleton's GOLDEN OINTMENT.) According to Dr Paris, this compound consists of lard medicated with orpiment (native yellow sulphuret of arsenic). There appears, however, to be some mistake in this, as that sold us under the name had nearly the same composition as the OINTMENT

OF NITRIC OXIDE OF MERCURY of the Pharmacopœia. It did not contain even a trace of either arsenic or sulphur. The action of this nostrum, and the reputation which it has acquired, fully justify this conclusion.

5. (*Smellome.*) From verdigris (levigated), ½ dr.; olive oil, 1 fl. dr.; triturate together; add of yellow basilicon, 1 oz., and again triturate until it begins to concrete. A popular nostrum, sometimes useful in chronic inflammation and ulcerations of the eyelids, &c., especially in those of a scrofulous character.

6. (*Spielmann.*) Acetate of lead, 20 gr.; spermaceti cerate, 5 dr.; compound tincture of benzoin, 40 gr. Cooling, desiccative. In inflamed eyelids, excoriations, &c.

7. (*St Yves.*) Fresh butter (washed), 1 oz.; white wax, 1 dr.; camphor, 15 gr.; melt by a gentle heat, and, when cooled a little, add of red precipitate (levigated), ½ dr.; oxide of zinc, 20 gr. In chronic inflammation of the coats of the eye or of the eyelids, specks on the cornea, &c.

8. (*Thomson.*) Levigated oxide of zinc, 1 dr.; lard, 9 dr.; wine of opium, 20 drops. In chronic ophthalmia depending on want of tone in the vessels and integuments of the eye.

9. (*Ware.*) Wine of opium, 1 fl. dr.; simple ointment, 8 dr. In ophthalmia, after the inflammatory symptoms have subsided, and the vessels remain red and turgid.

Obs. The ingredients entering into the composition of all the above ointments must be reduced to the state of impalpable powder before mixing them; and the incorporation should be made by long trituration in a wedgwood-ware mortar, or, preferably to those that contain substances that are very gritty, by levigation on a porphyry slab with a muller. The most serious consequences, even blindness, have resulted from the neglect of these precautions. They should all be employed in exceedingly small quantities at a time, and they should be very carefully applied by means of a camel-hair pencil or a feather; and, in general, not until acute inflammation has subsided. The stronger ones, in most cases, require dilution with an equal weight to twice their weight of lard or simple ointment, and should only be used of their full strength under proper medical advice. Various other formulæ for OPHTHALMIC OINTMENTS will be found under the names of their leading ingredients.

Ointment of Fig'wort. See OINTMENT OF SCROPHULARIA.

Ointment of Fox'glove. *Syn.* UNGUENTUM DIGITALIS, L. *Prep.* 1. From fresh foxglove as ointment of hemlock (Ph. L.). As an application to chronic ulcers, glandular swellings, &c.

2. (*Rademacher.*) Extract of foxglove, 2 dr.; lard, 1 oz. In croup; spread on lint, and applied as a plaster to the throat.

Ointment of Fuligoka'li. See FULIGOKALI.

Ointment of Galls. *Syn.* UNGUENTUM GALLÆ (B. P., Ph. D.), L. *Prep.* 1. (Ph. D.) Gallnuts (in very fine powder), 1 dr.; ointment of white wax, 7 dr.; rub them together until a uniform mixture is obtained.

2. (B. P.) Galls, in very fine powder, 80 gr.; benzoated lard, 1 oz. Mix. An excellent application to piles, either alone or mixed with an equal

quantity of zinc ointment; also highly useful in ringworm of the scalp.

Ointment of Galls with Camphor. *Syn.* UN-GUENTUM GALLÆ CUM CAMPHORÂ. *Prep.* Galls, 2 dr.; camphor, ¼ dr.; lard, 1 oz. Mix them.

Ointment of Galls with Morphia. *Syn.* UN-GUENTUM GALLÆ ET MORPHIÆ (*Dr Paris*). Morphia, 2 gr.; olive oil (hot), 2 fl. dr.; triturate; add of zinc ointment (Ph. L.), 1 oz.; powdered galls, 1 dr.; mix thoroughly. In piles. The quantity of galls should be doubled.

Ointment of Galls and Opium. UNGUENTUM GALLÆ CUM OPIO (B. P.), UNGUENTUM GALLÆ OPIATUM, U. GALLÆ COMPOSITUM (Ph. L.), U. GALLÆ ET OPII (Ph. E.). *Prep.* 1. (Ph. L.) Gall-nuts (very finely powdered), 6 dr.; powdered opium, 1½ dr.; lard, 6 oz.; rub them together.

2. (Ph. E.) Galls, 2 dr.; opium, 1 dr.; lard, 1 oz.; as the last.

3. (B. P.) Ointment of galls, 1 oz.; opium (in powder), 32 gr. Mix.

Uses, &c. A most valuable astringent and anodyne in blind piles, slight cases of prolapsus ani, &c. Some practitioners add 1 dr. of camphor. The ointment of the Ph. E. is much the strongest.

Ointment of Garlic. *Syn.* UNGUENTUM ALLII, L. *Prep.* 1. Fresh garlic (bruised), 2 parts; lard, 3 parts; simmer together for ¾ an hour, and then strain with expression. Rubbed on the abdomen in chronic diarrhœa and colic, and over the chest and spine in hooping-cough.

2. (*Beasley.*) Fresh garlic and lard, equal parts; beaten together. Applied to the feet in hooping-cough.

Ointment, Giacomini's. See OINTMENT, CHILBLAIN.

Ointment of Glycerin. *Syn.* UNGUENTUM GLYCERINI. *Prep.* Glycerin, 4 fl. oz.; oil of almonds, 8 fl. oz.; wax and spermaceti, of each, ¼ oz.

Ointment of Glycerin of Subacetate of Lead. (B. P.) *Syn.* UNGUENTUM GLYCERINI PLUMBI SUBACETATIS, L. Glycerin of subacetate of lead, 1 part; soft paraffin, 4 parts; hard paraffin, 1½ parts. Melt, mix, and stir till cold.

Ointment of Gold. *Syn.* UNGUENTUM AURI, L.; POMMADE D'OR, Fr. *Prep.* 1. (*Legrand.*) Gold (in powder), 12 gr.; lard, 1 oz. As a dressing for syphilitic ulcers, and as a friction in glandular indurations, &c.; also endermically.

2. (*Magendie.*) Amalgam of gold, 1 dr.; lard, 1 dr. For endermic use chiefly. When the surface becomes dry, the ointment of terchloride of gold is to be substituted as a dressing. In rheumatic pains, neuralgia, &c.

Ointment, Gold'en. See OINTMENT, EYE; CITRINE O., &c.

Ointment, Gondret's. See OINTMENT, AMMONIACAL.

Ointment, Goulard's. *Syn.* UNGUENTUM GOULARDI, U. LITHARGYRI ACETATIS, L. *Prep.* (Ph. Chirur.) Goulard's extract, 1 dr.; simple ointment, 2 oz. See CERATE (Lead).

Ointment, Green. See OINTMENT, ELDER.

Ointment of Hamamelis. *Syn.* UNGUENTUM HAMAMELIDIS, L. (B. P.) Liquid extract of hamamelis, 1 part; simple ointment, 9 parts. Mix well.

Uses. Astringent and sedative in piles.

Ointment of Hel'lebore. *Syn.* OINTMENT OF WHITE HELLEBORE; UNGUENTUM VERATRI, L. *Prep.* 1. (Ph. L. 1836.) White hellebore (in very fine powder), 2 oz.; lard, 8 oz.; oil of lemons, 20 drops. In itch, lepra, ringworm, &c.; and to destroy insects in the hair of children. It should be used with caution, and, preferably, diluted with an equal weight of lard.

2. (Compound: UNGUENTUM VERATRI COMPOSITUM.) *a.* (*Rayer.*) White hellebore, 1 oz.; sal-ammoniac, ½ oz.; lard, 8 oz. Used as the last.

b. See SULPHUR OINTMENT (Compound), Ph. L.

Ointment of Hemlock. *Syn.* UNGUENTUM CONII (B. P.), L. *Prep.* 1. Juice of hemlock, 2 oz.; hydrous wool fat, ¾ oz.; boric acid (in fine powder), 10 gr.; evaporate the juice to 2 dr. at a temperature not exceeding 60° C.; then mix all well together.

2. (Ph. L.) Fresh hemlock leaves and lard, of each, 1 lb.; boil them together (very gently) until the leaves become crisp, then strain through linen, with pressure. See OILS (Medicated).

3. Extract of hemlock, 1 dr.; lard, 9 dr.; triturate together.

Uses, &c. As a local anodyne in neuralgic and rheumatic pains, glandular enlargements, painful piles, &c.; and as a dressing to painful and irritable ulcers, cancerous sores, &c.

Ointment of Hen'bane. *Syn.* UNGUENTUM HYOSCYAMI, L. *Prep.* 1. Fresh henbane leaves, 1 lb.; lard, 2 lbs.; boil until nearly crisp.

2. (*Taddei.*) Extract of henbane, 1 dr.; lard, 1 oz. Anodyne; in painful piles, sores, &c., as the last.

Ointment, Holloway's. See PATENT MEDICINES.

Ointment of Hops. *Syn.* UNGUENTUM LUPULI, L. *Prep.* (*Swediaur.*) Hops (commercial), 2 oz.; lard, 10 oz.; as extract of hemlock, Ph. L. In painful piles and cancerous sores.

Ointment of Hydri'odate of Ammo''nia. *Syn.* UNGUENTUM AMMONIÆ HYDRIODATIS, L. *Prep.* From hydriodate of ammonia (iodide of ammonium), ½ dr.; simple ointment, 1 oz. Used chiefly as an application to scrofulous tumours and ulcers in irritable subjects.

Ointment of Hydrochlo''ric Acid. *Syn.* UNGUENTUM ACIDI HYDROCHLORICI, L. *Prep.* (*Dr Corrigan.*) Hydrochloric acid, 1 dr.; simple ointment, 1 oz. As a dressing for scald-head, after the scabs have been removed by emollient liniments or poultices.

Ointment of Hypochlo''rite of Lime. *Syn.* OINTMENT OF CHLORIDE OF LIME; UNGUENTUM CALCIS HYPOCHLORITIS, U. C. CHLORINATÆ, L. *Prep.* 1. From chlorinated lime (chloride of lime), 1 dr.; lard, 1 oz.; carefully triturated together. In scrofulous swellings, goitre, chilblains, indolent glandular tumours, &c.

2. Chlorinated lime, 1 dr.; powdered foxglove, 2 dr.; simple ointment, 2 oz. As an application to fetid and malignant ulcers, &c.

Ointment of Hypochlo''rite of Sul'phur. *Syn.* UNGUENTUM SULPHURIS HYPOCHLORITIS, L. *Prep.* (*Dr Copland.*) Hypochlorite of sulphur, 1 dr.; simple ointment, 1 oz. It is generally

-scented with oil of almonds. Used in psoriasis inveterata, and some other skin diseases.

Ointment of I'odide of Ar'senic. *Syn.* UN-GUENTUM ARSENICI IODIDI, L. *Prep.* (*Biett.*) Iodide of arsenic, 2 to 3 gr.; lard, 1 oz.; carefully triturated together. In lepra, psoriasis, &c.; and in corroding tubercular diseases. It should be used with caution, and not more than ½ dr. applied at once.

Ointment of Iodide of Ba"rium. *Syn.* UN-GUENTUM BARII IODIDI, L. *Prep.* (*Magendie.*) Iodide of barium, 3 to 4 gr.; lard, 1 oz. As a friction to scrofulous swellings and indurations. The usual proportions are now 5 gr. to the oz.

Ointment of Iodide of Iron. (*Pierquin.*) *Syn.* UNGUENTUM FERRI IODIDI, L. *Prep.* Iodide of iron, 1 dr.; lard, 1 oz. Mix them.

Ointment of Iodide of Lead. *Syn.* UNGUEN-TUM PLUMBI IODIDI (B. P., Ph. L. and D.), L. *Prep.* 1. (Ph. L.) Iodide of lead, 1 oz.; lard, 8 oz.; rub them together.

2. (Ph. D.) Iodide of lead (in fine powder), 1 dr.; ointment of white wax, 7 dr.

3. (B. P.) Iodide of lead (in fine powder), 62 gr.; simple ointment, 1 oz. Mix. An excellent application to scrofulous tumours and swelled glands, especially when accompanied with pain.

Ointment of Green Iodide of Mer'cury. *Prep.* 1. (OINTMENT OF SUBIODIDE OF MERCURY, O. OF PROTIODIDE OF M.*; UNGUENTUM HYDRARGYRI IODIDI—Ph. L.) White wax, 2 oz.; lard, 6 oz.; melt them together, add of iodide (green iodide) of mercury, 1 oz., and rub them well together.

2. (*Magendie.*) Green iodide of mercury, 28 gr.; lard, 1½ oz.

Uses, &c. In tubercular skin diseases, as a friction in scrofulous swellings and indolent granular tumours, and as a dressing for ill-conditioned ulcers, especially those of a scrofulous character.

Ointment of Red Iodide of Mercury. *Syn.* O. OF BINIODIDE OF M.*; UNGUENTUM HYDRARGYRI IODIDI RUBRI (B. P.), UNGUENTUM HYDRARGYRI BINIODIDI*, U. H. IODIDI RUBRI (Ph. D.), L. *Prep.* 1. (Ph. D.) Red iodide of mercury 1, dr.; ointment of white wax, 7 dr.; mix by careful trituration.

2. (*Soubeiran.*) Red iodide of mercury, 20 gr.; lard, 1½ oz.

· 3. (B. P.) Red iodide of mercury (in very fine powder), 16 gr.; simple ointment, 1 oz. Mix.

Uses, &c. Similar to those of the preceding, but it is much more stimulant, and is regarded as better adapted for obstinate syphilitic sores. Largely diluted with lard or almond oil, it is applied to the eyes in like cases.

Ointment of Iodide of Potas'sium. *Syn.* UN-GUENTUM POTASSII IODIDI (B. P., Ph. L. and D.), L. *Prep.* 1. (Ph. L.) Iodide of potassium, 2 dr., dissolved in boiling distilled water, 2 fl. dr.; lard (softened by heat), 2 oz.; triturate together until united.

2. (Ph. D.) Iodide of potassium, 1 dr.; distilled water, ½ fl. dr.; ointment of white wax, 7 dr.; as before.

3. (*Magendie.*) Iodide of potassium, 1 dr.; lard, 12 dr.

4. (*Le Gros.*) Iodide, 1½ dr.; lard, 1 oz.

5. (B. P.) Iodide of potassium, 64 gr.; carbonate of potash, 4 gr.; distilled water, 1 dr.; benzoated lard, 1 oz.; dissolve the carbonate and the iodide in the water, and mix thoroughly with the lard.

Uses, &c. As a friction in scrofula, bronchocele, glandular enlargements, indurations, &c.; as a dressing to scrofulous ulcers, as an application in scrofulous ophthalmia, and in most of the other applications in which the employment of iodine is indicated. The last formula has been successfully employed by M. Le Gros in itch.

Obs. The strength of this ointment as prescribed by different physicians varies greatly, the proportions of the iodide ranging from $\frac{1}{24}$ to $\frac{1}{6}$ of the whole, to adapt it to particular cases. When other ingredients are added, the iodide must be used in a perfectly dry state, and in fine powder, instead of being dissolved in water. This is particularly necessary when it is to be mixed with mercurial ointment.

Ointment of Iodide of Sulphur. *Syn.* UN-GUENTUM SULPHURIS IODIDI (B. P.), L. *Prep.* Iodide of sulphur, 5 parts; hard paraffin, 18 parts; soft paraffin, 55 parts. Powder the iodide, and mix with the melted paraffins.

Uses, &c. As a local stimulant and alterative in the chronic forms of lepra, lupus, porrigo, psoriasis, itch, &c.; also a remedy for acne punctata. A few drops of oil of cloves or nutmeg are commonly added.

Ointment of Iodide of Zinc. *Syn.* UNGUEN-TUM ZINCI IODIDI, L. *Prep.* 1. From iodide of zinc, 12 gr.; simple ointment, 1 oz. In scrofulous excoriations, and in the chronic ophthalmia of scrofulous subjects, arising from a relaxed state of the tissues and vessels.

2. (*Dr Ure.*) Iodide of zinc, 1 dr.; lard, 1 oz. As a friction to glandular tumours and indurations, and as a dressing to flabby and obstinate scrofulous ulcers.

Ointment of I'odine. *Syn.* UNGUENTUM IODI (B. P.), UNGUENTUM IODINII (Ph. U. S.), L. *Prep.* 1. (B. P.) Iodine, 32 gr.; iodide of potassium, 32 gr.; glycerin, 1 dr.; rub together and add prepared lard, 2 oz. See OINTMENT OF IODINE (Compound).

2. (Ph. U. S.) Iodine, 20 gr.; rectified spirit, 20 drops; rub them together, then add of lard, 1 oz.

Ointment of Iodine (Compound). *Syn.* OINT-MENT OF IODURETTED IODIDE OF POTASSIUM; UNGUENTUM POTASSII IODIDI IODURETUM, U. IODINII COMPOSITUM (Ph. L. and D.), U. IO-DINII (Ph. E.), L. *a.* (Ph. L.) *Prep.* Iodide of potassium (in very fine powder), 1 dr.; lard, 2 oz.; mix, then add of iodine, ½ dr.; dissolved in rectified spirit, 1 fl. dr., and mix all together. See OINTMENT OF IODINE (B. P.).

b. (Ph. E.) Iodine, 1 dr.; iodide of potassium, 2 dr.; rub them together, then gradually add of lard, 4 oz.

c. (Ph. D.) Pure iodine, ½ dr.; iodide of potassium, 1 dr.; rub them well together in a glass or porcelain mortar, then gradually add of ointment of white wax, 14½ dr., and continue the trituration until a uniform ointment is obtained.

Uses, &c. The compound ointment is an excellent friction in goitre, and in enlarged or indurated glands or tumours, more especially those of a scrofulous character; in the quantity of ½ to 1 dr., night and morning. It may be advantageously combined with extract of belladonna in the incipient bubo of scrofulous subjects, and in the early stages of cancer; and, with an equal weight of mercurial ointment, as a friction in cases of enlarged liver and spleen, and ovarian dropsy. The simple ointment of the Ph. U. S. is generally regarded as weaker and less efficacious than the compound.

Ointment of Iodo-chlo"ride of Mercury. *Syn.* UNGUENTUM HYDRARGYRI IODO-BICHLORIDI*, L. *Prep.* From iodo-chloride of mercury, 16 gr.; simple ointment, 1 oz. Discutient; probably one of the most powerful known in syphilitic cases complicated with scrofula. See OINTMENT OF CHLORIODIDE OF MERCURY.

Ointment of Iodoform. *Syn.* UNGUENTUM IODOFORMI (B. P.), L. *Prep.* Iodoform, 1 part; benzoated lard, 9 parts. Mix.

Ointment, Iodo-hydrar'gyrate of Potassa. *Syn.* UNGUENTUM POTASSÆ IODO-HYDRARGYRATIS, L. *Prep.* I. (*Lamothe.*) Iodo-hydrargyrate of potassa, 20 gr.; lard, 1 oz.

2. (*Puche.*) Red iodide of mercury and iodide of potassium, of each, 8 gr.; lard, 1 oz. As a powerful stimulant discutient; in tumours, inflammatory sore throat, &c.

Ointment, Iodo-narcot'ic. *Syn.* UNGUENTUM IODO-NARCOTICUM, L. *Prep.* (*Purvis.*) Iodine, 20 gr.; iodide of potassium, 2 dr.; oil of tobacco (by infusion), 1½ dr.; lard, 8 dr. To relax rigid muscles.

Ointment, Is'sue. *Syn.* UNGUENTUM ADFONTICULOS, L. *Prep.* (*Golding-Bird.*) Ointment of cantharides (Ph. L.), 1½ oz.; tartar emetic (in impalpable powd'r), 8 gr.; spermaceti ointment, 2 oz. As a stimulating application to issues, to promote the discharge. See ELEMI OINTMENT, CERATE, PLASTER, &c.

Ointment, Itch. *Syn.* UNGUENTUM ANTIPSORICUM, L. Several excellent formulæ for itch ointments will be found under the names of their leading ingredients. The following are additional ones, including some nostrums:

1. (*Bailey.*) From alum, nitre, and sulphate of zinc, of each, in very fine powder, 1½ oz.; vermilion, ½ oz.; mix, add gradually of sweet oil, ½ pint; triturate together until perfectly mixed, then further add of lard (softened by heat), 1 lb., with oils of aniseed, lavender, and origanum, q. s. to perfume.

2. (*Bateman.*) Carbonate of potassa, ½ oz.; rose-water, 1 fl. oz.; red sulphuret of mercury, 1 dr.; oil of bergamot, ½ fl. dr.; sublimed sulphur and hog's lard, of each, 11 oz.; mix them (Bateman, 'Cutaneous Diseases'). The nostrum vended under the name is made as follows:—Carbonate of potash, 1 oz.; vermilion, 3 dr.; sulphur, 1 lb.; lard, 1½ lbs.; rose-water, 3 fl. oz.; oil of bergamot, 1½ dr.

3. (French Hosp.) Chloride of lime, 1 dr.; rectified spirit, 2 fl. dr.; sweet oil, ½ fl. oz.; common salt and sulphur, of each, 1 oz.; soft soap, 2 oz.; oil of lemon, 20 drops. Cheap, effectual, and inoffensive.

4. (*De la Harpe.*) Sulphur, 2 oz.; powdered white hellebore, ½ oz.; sulphate of zinc, ½ oz.; soft soap, 4 oz.; lard, 8 oz.

5. (*Jackson.*) From palm oil, flowers of sulphur, and white hellebore, of each, 1 part; lard, 2 parts.

6. (*Nugent.*) From white-lead, 2 oz.; orris root, 1 oz.; corrosive sublimate, in very fine powder, ½ oz.; palm oil, 4 oz.; lard, 1½ lbs.

7. (Ph. E. 1744.) Elecampane root and sharppointed dock (*Rumex acutus*, Linn.), of each, bruised, 3 oz.; water, 1 quart; vinegar, ½ pint; boil to one half, add of water-cress, 10 oz.; lard, 4 lbs.; boil to dryness, and strain with expression; to the strained liquid add of beeswax and oil of bays, of each, 4 oz., and stir the mixture until nearly cold.

8. (UNG. A. COMP.—Ph. E. 1744.) To each lb. of the last add of strong mercurial ointment, 2 oz.

9. (*Robertson.*) Soft soap, 1 oz.; rum, 1 tablespoonful; chloride of lime (dry and good), ¼ oz.; mix, and add of lard, 2 oz.

10. (*Swediaur.*) Stavesacre (in powder), 1 oz.; lard, 3 oz.; digest with heat for 8 hours, and then strain. The formula of the Ph. Bruns. is nearly similar. Very useful in itch; also to destroy pediculi.

11. (*Thomson.*) Chloride of lime and common salt, of each, in fine powder, 1 dr.; soft soap, 1 oz.; rectified spirit, 2 fl. dr.; mix, add of lard, 1 oz.; and lastly, of strong vinegar, 2 fl. dr. Very cleanly and effective; but should not be made in quantity, as it does not keep well.

12. (*Vogt.*) Chloride of lime (dry), 2 dr.; burnt alum, 3 dr.; lard, 9 dr. To be mixed with an equal quantity of soft soap at the time of fusing it.

Obs. The products of the preceding formulæ are used by well rubbing them into the part affected, night and morning, as long as necessary, the number of applications required depending greatly on the manner in which this is done.

Ointment of I'vy. *Syn.* UNGUENTUM HEDERÆ, L. *Prep.* From the leaves of common ivy, by infusion, as ointment of henbane. Used as an application to soft corns, in itch, and as a dressing to indolent ulcers and issues.

Ointment of Jatropha (PHYSIC-NUT). The milky juice of the English physic-nut (*Jatropha curcas*) mixed with half its weight of lard. In piles.

Ointment of Juniper. *Syn.* UNGUENTUM JUNIPERI, L. *Prep.* Juniper leaves, 1 part; resin ointment, 6 parts; boil gently and strain.

Ointment of Ju"niper-tar. *Syn.* UNGUENTUM OLEI PYROLIGNI JUNIPERI, U. CADINUM, L. *Prep.* (*Eras. Wilson.*) Lard and suet, of each, 6 parts; beeswax, 4 parts; liquefy by heat, and add of pyroligneous oil of juniper ('huile de cade'), 16 parts; with a few drops of any fragrant essential oil, to conceal the smell. In ringworm, and as a stimulant ointment in some other skin diseases.

Ointment of Kaolin. *Syn.* UNGUENTUM KAOLINI. *Prep.* Vaseline, 1 part; paraffin, 1 part; melt, and add kaolin (in powder), 1 part. Used as a protective to allay irritation of the skin; also as a pill excipient for permanganate of potassium.

Ointment, Kirkland's. See LEAD OINTMENT (Compound).

Ointment of Labdanum. (*Quincy.*) *Syn.* UNGUENTUM CRINISCUM. *Prep.* Labdanum, 6 dr.; bear's grease, 2 oz.; powdered southernwood, 3 dr.; oil of mace, 1 dr.; balsam of Peru, 2 dr.

Ointment of Lard. *Syn.* UNGUENTUM ADIPIS, L. *Prep.* (Ph. L. 1788.) Prepared lard, 2 lbs.; melt, add of rose-water, 3 fl. oz.; beat the two well together, then set the vessel aside, and when the whole is cold, separate the congealed fat. A simple emollient. See OINTMENT, ELDER.

Ointment of Lau'rel. *Syn.* LAURINE OINTMENT; UNGUENTUM LAURINUM, U. LAURI NOBILIS, L. *Prep.* 1. (Ph. Lusit.) Suet (softened by heat), 8 oz.; laurel oil (expressed oil of bay), 1 lb.; oil of turpentine, 1½ oz. This is the 'nervine balsam' and 'nervine ointment' of the shops in the Peninsula, and in some other parts of Southern Europe. The Ph. Bat. 1805 added ½ oz. of rectified oil of amber.

2. (P. Cod.) Fresh bay leaves and berries (bruised), of each, 1 lb.; lard, 2 lbs.; as hemlock ointment (Ph. L.). Highly esteemed on the Continent as a stimulating friction, in bruises, strains, stiff joints, &c., and in deafness.

3. (Trade.) From fresh bay leaves, 2 lbs.; bay berries, 1 lb.; neat's-foot oil, 5 pints; boil as last; to the strained oil add of lard suet, 3 lbs.; true oil of bay, ¼ lb., and allow it to cool very slowly, in order that it may 'grain' well. Sold for laurel ointment and common oil of bay.

Ointment of Lavender. (*Baume.*) *Syn.* OLEUM LAVANDULÆ, L. *Prep.* Lard, 2½ lbs.; lavender flowers, 10 lbs.; white wax, 3 oz. Melt the lard, digest with 2 lbs. of the flowers for 2 hours, and strain; repeat this with fresh flowers till all are used; melt the ointment and leave it at rest to cool; separate the moisture and dregs, and melt the ointment with the wax.

Ointment of Lead. *Syn.* UNGUENTUM PLUMBI, U. LITHARGYRI (P. Cod.), L. *Prep.* 1. Litharge, 3 oz.; distilled vinegar, 4 oz.; olive oil, 9 oz.; mix with heat, and stir until they combine. Camphor, morphia, and opium are common additions to lead ointment when an anodyne effect is desirable.

2. (Compound: NEUTRAL OINTMENT, HIGGINS' O., KIRKLAND'S O.; UNGUENTUM NEUTRALE, U. PLUMBI COMPOSITUM—Ph. L.) Lead plaster, 2 lbs.; olive oil, 18 fl. oz.; mix by a gentle heat, and add of prepared chalk, 6 oz.; lastly, add of dilute acetic acid, 6 fl. oz., and stir well until the mass has cooled. As a dressing in indolent ulcers, "but its utility is doubtful" (*Dr Garrod*).

Obs. It will be observed that the College has already modified the old formula of this ointment. The vinegar is now the last ingredient added to the mass. "Gradually add the chalk, separately mixed with the vinegar, the effervescence being finished, and stir," &c. (Ph. L. 1836). See ACETATE OF LEAD, CARBONATE OF L., CHLORIDE OF L., IODIDE OF L.; EYE, GOULARD'S, LE MORT'S, and other OINTMENTS containing lead.

Ointment, Le Mort's. *Prep.* Carbonate of lead, corrosive sublimate, litharge, and Venice turpentine, of each, 1 oz.; alum, ¼ oz.; lard, ½ lb.; vermilion, q. s. to colour.

Ointment of Mace. *Syn.* UNGUENTUM MACIDIS, L. *Prep.* From mace (beaten to a paste) and palm oil, of each, 1 lb.; purified beef marrow, 3 lbs.; gently melted together and strained. Emollient and stimulant; chiefly used as a pomade for the hair. Sold for 'common oil of mace.'

Ointment of Marsh-mal'low. *Syn.* UNGUENTUM ALTHÆÆ, DIALTHÆÆ, L. *Prep.* 1. (Ph. L. 1746.) Oil of mucilages, 2 lbs.; beeswax, ½ lb.; yellow resin, 3 oz.; melt them together, then add of Venice turpentine, ½ oz., and stir the mixture until it concretes.

2. (Wholesale.) From palm oil, ½ lb.; yellow resin, 1½ lbs.; beeswax, 2½ lbs.; pale linseed oil, 9 lbs. (say 1 gall.); melt together and stir until it is nearly cold.

Uses, &c. Emollient and stimulant; seldom used in regular practice, but in great repute amongst the common people. Linseed oil is now almost universally substituted for the oil of mucilages.

Ointment of Masterwort. *Syn.* POMMADE ANTI-CANCÉREUSE DE MILIUS; UNGUENTUM IMPERATORIÆ, L. *Prep.* (*Beasley.*) Powdered masterwort (*Imperatoria ostruthium*), 1½ oz.; tincture of masterwort, 1 oz.; lard, 2 oz.

Ointment of Matico (*Mr Young*). *Syn.* UNGUENTUM MATICO, L. *Prep.* Powdered matico, 3 dr.; opium, 3 gr.; lard, 1 oz.

Ointment, Mayer's. *Prep.* To olive oil, 2½ lbs., add white turpentine, ½ lb.; beeswax and unsalted butter, of each, 4 oz.; melt them together and heat to nearly the boiling-point. Then add, gradually, red-lead, 1 lb., and stir constantly until the mixture becomes black or brown; then remove from the fire, and when it has become somewhat cool, add to it a mixture of honey, 12 oz., and powdered camphor, ½ lb. Lard may be used instead of butter.

Ointment, Mercu''rial. *Syn.* STRONG MERCURIAL OINTMENT, BLUE O., NEAPOLITAN O.; UNGUENTUM HYDRARGYRI (B. P., Ph. L., E., and D.), U. H. FORTIUS, U. CÆRULEUM, L. *Prep.* 1. (B. P.) Mercury, 16 parts; prepared lard, 16 parts; prepared suet, 1 part; rub together until metallic globules cease to be visible. See also OINTMENT, MERCURIAL (Compound).

2. (Ph. L. and E.) Mercury, 1 lb.; lard, 11½ oz.; suet, ½ oz.; rub the mercury with the suet and a little of the lard until globules are no longer visible; then add the remaining lard, and triturate altogether.

3. (Ph. D.) Pure mercury and lard, of each, 1 lb.; as before.

Pur., &c. The 'stronger mercurial ointment' of the shops is usually made with a less quantity of mercury than that ordered by the Colleges, and the colour is brought up with finely ground blue-black or wood charcoal. This fraud may be detected by its inferior sp. gr., and by a portion being left undissolved when a little of the ointment is treated first with ether or oil of turpentine, to remove the fat, and then with dilute nitric acid, to remove the mercury. When made according to the instructions of the Ph., its sp. gr. is not less than 1·781 at 60° F. It "is not well prepared so long as metallic globules may be seen in it with a magnifier of 4 powers" (Ph. E.). When rubbed on a piece of bright copper or gold, it should immediately give it a coating of metallic mercury and a silvery appearance.

The *Ung. hyd. fort.* of the wholesale houses is generally made of mercury, 12 lbs., suet, 1½ lbs., and lard, 16½ lbs. It thus contains only ⅓ instead of ½ its weight of mercury. That of the same houses labelled '*Ung. hyd. partes aquales*' is prepared with mercury, 12 lbs.; suet, 1½ lbs.; lard, 13½ lbs.

Uses. This ointment is chiefly used to introduce mercury into the system when the stomach is too irritable to bear it; in syphilis, hepatic affections, hydrocephalus, &c. For this purpose ½ to 1 dr. is commonly rubbed into the inside of one of the thighs until every particle of the ointment disappears. This operation is repeated night and morning until the desired effect is produced, and should be, if possible, performed by the patient himself. During its administration the patient should avoid exposure to cold, and the use of fermented or acidulous liquors, and his diet should consist chiefly of toast, broth, gruel, milk-and-water, and other inoffensive matters. This ointment has been employed to prevent the 'pitting' in smallpox; and, diluted with 3 or 4 times its weight of lard, in several skin diseases, as a dress-for ulcers, to destroy pediculi, &c. Camphor is often added to this ointment to increase its activity. With the addition of a little extract of belladonna, or hydrochlorate of ammonia, it forms an excellent anodyne and resolvent friction in painful syphilitic tumours and glandular enlargements.

Obs. The preparation of mercurial ointment according to the common plan is a process of much labour and difficulty, and usually occupies several days. The instructions in the Pharmacopœias are very meagre and unsatisfactory, and, so far as details go, are seldom precisely carried out. Employers grumble, and operatives become impatient, when they find the most assiduous trituration apparently fails to hasten the extinction of the globules. To facilitate matters various tricks are resorted to, and various contraband additions are often clandestinely made. Among the articles referred to, sulphur and turpentine are those which have been longest known, and, perhaps, most frequently employed for the purpose; but the first spoils the colour, and the other the consistence, of the ointment; whilst both impart to it more or less of their peculiar and respective odours. On the Continent oil of eggs was formerly very generally used for the purpose, and is even now occasionally so employed. Nearly half a century ago Mr W. Cooley clearly showed that the difficulty might be satisfactorily overcome by simply triturating the quicksilver with ⅛ to ¼ of its weight of old mercurial ointment before adding the lard; and that the effective power of this substance was in direct proportion to its age, or the length of time it had been exposed to the air. His plan was to employ the 'bottom' and 'scraping' of the store pots for the purpose. At a later period (1814-15) Mr Higginbottom, of Northampton, repeated this recommendation, and at length the plan has been imported into the Pharmacopœia Borussica. About 20 years since, "we reopened an investigation of the subject, which extended over several months, during which we satisfied ourselves of the accuracy of the assertion of M. Roux, that

the mercury in mercurial ointment exists entirely, or nearly so, in the metallic state, and not in the form of oxide, as was generally assumed. We succeeded in preparing an excellent sample of mercurial ointment by agitating washed suet and quicksilver together *in vacuo*. The quantity of oxide present at any time in this ointment is variable and accidental, and is largest in that which has been long prepared; but in no case is it sufficient to materially discolour the fat after the metallic mercury is separated from it. We were led to conclude that the property alluded to, possessed by old ointment, depends solely on the peculiar degree of consistence or viscidity of the fat present in it, and on the loss of much of the thoroughly greasy 'anti-attritive' character possessed by the latter in a recent state. In practically working out this idea we obtained pure fats (MAGNETIC ADEPS; SEVUM PRÆPARATUM), which, without any addition, were capable of reducing in a few mintues 8, 16, 32, and even 48 times their weight of mercury. We also found that the formula of the Pharmacopœia might be adopted, and that a perfect ointment might be readily obtained by skilful management in from half an hour to an hour, even without these resources. All that was necessary was to employ a very gentle degree of heat by either performing the operation in a warm apartment or by allowing the mortar to remain filled with warm water for a short time before using it. Suet or lard, reduced either by gentle warmth or by the additiou of a little almond oil to the consistence of a thick cream, so that it will hang to the pestle without running from it, will readily extinguish 7 or 8 times its weight of running mercury by simple trituration. The exact temperature must, however, be hit upon, or the operation fails. This fact was afterwards noticed in the 'Ann. de Chim.' and some other journals" (*A. J. Cooley*).

Professor Remington recommends the following as a rapid and convenient mode of preparing the ointment :—Mercury, 50 parts; lard, 25; suet, 25; mercurial ointment, 10; comp. tincture benzoin, 4. Mix the mercury with the tincture in a mortar, add the mercurial ointment, and stir till the globules of mercury cease to be visible, then add the suet and lard, previously mixed and melted; stir till uniform.

M. Pomonti has proposed a method of preparing strong mercurial ointment, which, modified to suit the English operator, is as follows :—Fresh lard, 8 parts; solution of nitre (see *below*), 1 part; mix by trituration; add of mercury, 32 parts, and again triturate. The globules disappear after a few turns of the pestle, but reappear in a few minutes, and then again disappear to return no more. When this happens, the trituration is to be continued for a few minutes longer, when lard, 24 parts, is to be rubbed in, and the ointment at once put into pots. It is said that the globules are so completely extinguished as to escape detection, even when the ointment is examined by a microscope of low power. The SOLUTION :—Nitre, 100 gr. ; water, 1 fl. oz.; dissolve. This quantity is sufficient for a kilogramme of mercury.

M. Lahens strongly recommends for the rapid

preparation of mercurial ointment the application of oil of almonds in the following proportions:—Mercury, 1000 parts; oil of almonds, 20 parts; lard, 980 parts. The mercury is first triturated with the oil for about fifteen minutes, after which its globules are said to be no longer discernible by the naked eye; 200 parts of the melted lard are now added, and the trituration continued to the complete extinction of the metal, which is generally accomplished within an hour. The ointment is then mixed with the remainder of the lard. See OINTMENT OF OXIDE OF MERCURY.

Ointment, Mercurial (Milder). MILDER BLUE OINTMENT, TROOPER'S O., UNCTION; UNGUENTUM HYDRARGYRI MITIUS, U. CÆRULEUM MITIUS, L. Prep. 1. Stronger mercurial ointment, 1 lb.; lard, 2 lbs.

Dose, &c. In the itch and several other cutaneous diseases, as a dressing to syphilitic ulcers, to destroy pediculi on the body, &c. Each drachm contains 10 gr. of mercury. That of the shops generally contains considerably less.

2. (With soap: UNGUENTUM HYDRARGYRI SAPONACEUM; SAVON MERCURIEL). a. (Draper.) Mercurial ointment (softened by a gentle heat), 1 oz.; hydrate of potassa, 1 dr.; dissolved in water, ½ fl. oz.; triturate them together until the mass solidifies.

· b. (Swediaur.) Milder mercurial ointment, 8 parts; soft soap, 2 parts; camphor, 1 part. In periostitis, engorgements of the testicles, soft corns, &c. See OINTMENT OF NITRATE OF MERCURY, &c.

Ointment, Mercurial (Compound) (B. P.). Mercurial ointment, 6 parts; yellow wax, 3; olive oil, 3; camphor, 1½. Melt the wax and oil, and when the mixture is nearly cold add the camphor in powder and the mercurial ointment, and mix.

Ointment, Mercurial, with Hydrochlorate of Ammonia (Dupuytren). Syn. UNGUENTUM HYDRARGYRI CUM AMMONIÆ MURIATE, L. Prep. Stronger mercurial ointment, 2 oz.; hydrochlorate of ammonia, 1 dr.

· Ointment, Mercurial, with Soda. (F. H.) Syn. UNGUENTUM HYDRARGYRI CUM SODA, L.; SAVON MERCURIEL, Fr. Prep. Mercurial ointment, 3½ oz.; solution of soda, 3 oz.; triturate until they combine.

Ointment of Mercury, Oleate of. (U. C. Hosp.) Syn. LINIMENTUM HYDRARGYRI OLEATIS, UNGUENTUM HYDRARGYRI OLEATIS, L. (10%.) Prep. Yellow peroxide of mercury, 1 dr.; pure oleic acid, 10 dr. To the oleic acid kept agitated in a mortar sprinkle in the peroxide gradually, and triturate frequently during 24 hours, until the peroxide is dissolved, and a gelatinous solution is formed. 20% as above, using double the quantity of yellow oxide. To be applied with a brush, or spread lightly over the part with the finger. In persistent inflammation of the joints, Professor Marshall adds to 1 dr. of the above preparation 1 gr. of morphia—the pure alkaloid —not one of its salts, which are insoluble in oleic acid.

In the preparation of ointment of oleate of mercury it is of the utmost importance that the mercuric oxide should be thoroughly dry, and further that it should be sifted in small portions at a time upon the surface of the oleic acid, each fresh portion being well incorporated before another is added. Solution should be promoted by frequent stirring at ordinary temperatures, since experience has shown that all heating is positively injurious (C. Rice).

Ointment, Mezereon. Syn. UNGUENTUM MEZEREI, L. Prep. 1. (Hamb. Cod.) Alcoholic extract of mezereon, 2 dr.; dissolve in rectified spirit, q. s.; add it to white wax, 1 oz.; lard, 8 oz., and mix by a gentle heat.

2. (P. Cod.) Mezereon (dried root-bark), 4 oz.; moisten it with rectified spirit, bruise it well, and digest it for 12 hours, at the heat of boiling water, in lard, 14½ oz.; then strain with pressure, and allow it to cool slowly; lastly, separate it from the dregs, remelt it, and add of white wax, 1½ oz. Used as a stimulating application to blistered surfaces and indolent ulcers.

3. (P. Cod.) Ethereal extract of mezereon, 176 gr.; lard, 9 oz.; white wax, 1 oz.; rectified spirit, 1 oz.; dissolve the extract in the alcohol, add the lard and wax, heat moderately, stir until the spirit is driven off, strain and stir till cold.

Ointment of Monesia. Syn. UNGUENTUM MONESIÆ, L. Prep. Oil of almonds, 4 parts; white wax, 2 parts; extract of monesia, 1 part; water, 1 part.

Ointment of Mustard. Syn. UNGUENTUM SINAPIS, L. Prep. 1. Flour of mustard, ½ oz.; water, 1 fl. oz.; mix, and add of resin cerate, 2 oz.; oil of turpentine, ¼ oz. Rubefacient and stimulant. As a friction in rheumatism, &c.

2. (Frank.) Flour of mustard, 8 oz.; oil of almonds, ½ fl. oz.; lemon juice, q. s. In sunburn, freckles, &c.

Ointment of Naph'thalin. Syn. UNGUENTUM NAPHTHALINÆ, L. Prep. (Emery.) Naphthalin, ¼ dr.; lard, 7½ dr. In dry tetters, lepra, psoriasis, &c.

Ointment, Narcotic and Balsamic (G. Ph.). Syn. UNGUENTUM NARCOTICO-BALSAMICUM HELIMUNDI. Prep. Acetate of lead, 10 dr.; extract of hemlock, 30 dr.; wax ointment, 33 oz.; balsam of Peru, 30 dr.; wine of opium, 5 dr.

Ointment, Neapolitan. See OINTMENT, MERCURIAL.

Ointment, Nervine. Syn. BALSAMUM NERVINUM, UNGUENTUM N., L.; BAUME NERVAL, Fr. Prep. (P. Cod.) Expressed oil of mace and ox marrow, of each, 4 oz.; melt by a gentle heat, and add, of oil of rosemary, 2 dr.; oil of cloves, 1 dr.; camphor, 1 dr.; balsam of tolu, 2 dr.; (the last two dissolved in) rectified spirit, 4 dr. (In rheumatism, &c. A somewhat similar preparation was included in the Ph. E. 1744.

Ointment, Neu'tral. See OINTMENT OF LEAD (Compound).

Ointment of Ni'trate of Mercury. Syn. CITRINE OINTMENT, YELLOW O., MERCURIAL BALSAM; UNGUENTUM HYDRARGYRI NITRATIS (B. P., Ph. L. & D.), U. H. N., or U. CITRINUM (Ph. E.), L. Prep. 1. (Ph. L.) Mercury, 2 oz.; nitric acid (sp. gr. 1·42), 4 fl. oz.; dissolve, and mix the solution, whilst still hot, with lard, 1 lb., and olive oil, 8 fl. oz., melted together. (For the milder ointment see below.)

2. (Ph. E.) Mercury, 4 oz.; nitric acid (sp. gr. 1·500), 8 fl. oz. 6 fl. dr.; dissolve by a gentle

heat, add the liquid to lard, 15 oz.; olive oil, 32 fl. oz.; melted together and whilst the whole are still hot, and mix them thoroughly. "If the mixture does not froth up, increase the heat a little until this takes place. Keep the ointment in earthenware vessels, or glass vessels, secluded from the air." This admirable formula is a modification of that originally introduced into pharmacy by the late Dr Duncan, of Edinburgh. (For the milder ointment see *below*.)

3, (Ph. D.) Mercury, 1 oz.; nitric acid (1·500), 1 fl. oz.; (diluted with) water, ¼ fl. oz.; dissolve by a gentle head, and add the liquid to lard, 4 oz.; olive oil, 8 fl. oz.; melted together, and still hot; next " let the temperature of the mixture be raised so as to cause effervescence, and then, withdrawing the heat, stir the mixture with a porcelain spoon until it concretes on cooling."

4. (P. Cod.) Mercury, 3 parts; nitric acid (1·321), 6 parts; lard and oil, of each, 24 parts; as above.

5. (Ph. U. S.) Mercury, 1 oz.; nitric acid (1·42), 14 fl. oz.; lard, 3 oz.; fresh neat's-foot oil, 9 fl. oz.; mix the mercurial solution with the melted fat and oil at 200° F.

6. (B.P.) Mercury, 4 parts; nitric acid, 12; prepared lard, 15; olive oil, 32; dissolve the mercury in the nitric acid with the aid of a gentle heat; melt the lard in the oil by a steam or water bath in a porcelain vessel capable of holding six times the quantity, and while the mixture is about 212° F. add the solution of mercury, also hot, and mix them together thoroughly. If the mixture does not froth up, increase the heat till this occurs. (The heat required for this is from 170° to 180° F.)

Uses, &c. Detergent and stimulant. In ringworm, herpes, itch, porrigo, psoriasis, and some other chronic skin diseases; in various chronic affections of the eyes, especially chronic inflammation and ulceration of the eyelids, 'blear eye,' &c. It " may be almost regarded as specific in psorophthalmia, in the purulent ophthalmia of infants producing ectropium (eversion of the eyelids), and in ulcerations of the tarsi (edges of the eyelids)" (*A. T. Thomson*). As a dressing to old ulcers, more especially those of a syphilitic character, it is superior to all the other ointments containing mercury; in sore legs, assisted by the internal use of the pill of soap with opium (PIL SAPONIS CUM OPIO), it often acts like a charm when all other modes of treatment have failed. For most of these purposes it should be diluted with from twice to seven times its weight of some simple fatty matter. One of the principal reasons why this ointment is in less general use than its merits deserve is the very inferior quality of that vended in the shops under the name, arising from almost every druggist preparing some mess of his own, instead of adhering to the College formulæ.

Obs. Ointment of nitrate of mercury, faithfully prepared according to the instructions in the Pharmacopœia, possesses a rich golden-yellow colour, and a buttery consistence, and keeps well. Unfortunately, clumsy and careless operators, who regard the Pharmacopœia as a foolish book, which it is quite unnecessary to look into, often fail in their attempts to produce an article of good quality. The difficulty is immediately surmounted by employing pure ingredients, in the proportions ordered, and mixing them at the proper temperature. The acid should be of the full strength, or, if somewhat weaker than that directed, an equivalent quantity should be employed. A slight excess of acid is not injurious, rather the contrary; but a deficiency of acid, in all cases, more or less damages the quality of the product. If, on stirring the mercurial solution with the melted lard and oil, the mixture does not froth up, the heat should be increased a little, as, unless a violent frothing and reaction take place, the ointment will not turn out of good quality, and will rapidly harden and lose its colour. The most favourable temperature for the union of the ingredients is from 185° to 200° E., and in no case should it exceed 212°; whilst below 180° F. the reactions are feeble and imperfect.

Stoneware or glass vessels must alone be employed in the preparation of this ointment, and the stirrers or spatulas should be either of glass or white deal. The best plan is to keep the whole exclusively for the purpose, and when out of use to preserve them from dust and dirt. (See *below*.)

Ointment of Ni'trate of Mercury (Milder). *Syn.* MILDER CITRINE OINTMENT; UNGUENTUM HYDRARGYRI NITRATIS MITIUS (Ph. L.), U. H. N. M., or U. CITRINUM M. (Ph. E.), L. *Prep.* 1. (Ph. L.) Ointment of nitrate of mercury, 1 oz.; lard, 7 oz.; rub them together. " This ointment is to be used recently prepared."

2. (Ph. E.) As the stronger ointment, Ph. E., but using a triple proportion of oil and lard.

Uses, &c. See the STRONGER OINTMENT (*above*).

Ointment of Ni'trate of Mercury (Diluted). (B. P.) *Syn.* UNGUENTUM HYDRARGYRI NITRATIS DILUTUM, L. *Prep.* Nitrate of mercury ointment, 1 part; soft paraffin, 2 parts; mix.

Ointment of Ni'trate of Sil'ver. *Syn.* UNGUENTUM ARGENTI NITRATIS, L. *Prep.* 1. (*M. Jobert.*) Nitrate of silver, 2, 4, or 6 parts; lard, 20 parts. These ointments are respectively numbered 1, 2, and 3, and are used in white swelling.

2. (*Macdonald.*) Nitrate of silver, 1 part; lard, 7 to 8 parts. To smear bougies, in gonorrhœa, &c.

3. (*Mackensie.*) Nitrate of silver, 5 gr.; lard, 1 oz. In purulent and chronic ophthalmia, ulcers on the cornea, &c.

4. (*Velpeau.*) Nitrate of silver, 1 gr.; lard, 1 dr. In acute ophthalmia, &c. The above compounds require to be used with caution.

Ointment of Ni'tric Acid. *Syn.* OXYGENISED FAT; UNGUENTUM OXYGENATUM, U. ACIDI NITRICI, L.; POMMADE D'ALYON, Fr. *Prep.* (Ph. D. 1826.) Olive oil, 1 lb.; lard, 4 oz.; melt them together, add, gradually, of nitric acid (sp. gr. 1·500), 5½ fl. dr., and stir the mixture constantly with a glass rod until it concretes.

Uses, &c. In itch, porrigo, and some other chronic skin diseases; and as a dressing for syphilitic and herpetic ulcers, old sores, &c. It is frequently employed as a substitute for the ointment of nitrate of mercury, wh ich it some what resembles in appearance; but it is less active and useful.

Ointment of Ni'tric Ox'ide of Mer'cury. *Syn.* OINTMENT OF RED OXIDE OF MERCURY (B. P.), RED PRECIPITATE OINTMENT; UNGUENTUM HY DRARGYRI NITRICO-OXYDI (Ph. L.), U. H. OXYDI (Ph. E.), U. H. O. RUBRI (B. P., Ph. D.), L. *Prep.* 1. (Ph. L.) White wax, 2 oz.; lard, 6 oz.; mix by heat, add of nitric oxide of mer cury, in very fine powder, 1 oz., and rub them together.

2. (Ph. E.) Nitric oxide of mercury, 1 oz.; lard, 8 oz.; mix by trituration.

3. (Ph. D.) Red oxide of mercury (nitric oxide), 1 dr.; ointment of white wax, 7 dr.; as the last.

4. (B. P.) Red oxide of mercury, in very fine powder, 62 gr.; hard paraffin, ⅓ oz.; soft paraffin, ⅔ oz.

Uses, &c. An excellent stimulant application to indolent and foul sores, ulcers, &c.; and, when diluted, as an eye ointment in chronic inflamma tion and ulceration of the eyes and eyelids, and especially in psorophthalmia; also in specks on the cornea, and the other affections noticed under OINTMENT OF NITRATE OF MERCURY. It forms the basis of numerous quack medicines. See also OINTMENT OF OXIDE OF MERCURY (*below*).

Ointment, Obstet'ric. *Syn.* UNGUENTUM OB STETRICUM, L.; POMMADE OBSTETRICALE, Fr. *Prep.* 1. (*Chaussier.*) Extract of belladonna, 1 dr.; water, 2 dr.; lard, 1 oz. To promote the dilatation of the os uteri.

2. (POMMADE POUR LE TOUCHER.) From yel low wax and spermaceti, of each, 1 oz.; olive oil, 16 oz.; melt them together, strain, add of solu tion of caustic soda, 1 fl. oz., and stir until the whole is nearly cold.

Ointment of Oleate of Copper. *Syn.* UNGUEN TUM CUPRI OLEATIS, L. Oleate of copper, 1 part; lanolin or soft paraffin, 4 parts. Valuable in ring worm; destroys warts and removes freckles.

Ointment of Oleate of Zinc (B. P.). *Syn.* UN GUENTUM ZINCI OLEATI, L. Oleate of zinc, 1 part; soft paraffin, 1 part. Melt and mix.

Ointment of Oleo-resin of Capsicum. *Syn.* UN GUENTUM OLEO-RESINÆ CAPSICI. (Unofficial for mulary.) Take of oleo-resin of capsicum, 1 oz.; yellow wax, ½ oz.; benzoated lard, 4 oz. Melt the wax and lard at a low temperature, add the oleo-resin, mix thoroughly, and, if necessary, strain through muslin. Stir until cold.

Ointment of O'pium. *Syn.* UNGUENTUM OPIA TUM, U. OPII (Ph. L.), L. *Prep.* 1. (Ph. L.) Powdered opium, 20 gr.; lard, 1 oz.; mix by trituration. As a simple anodyne friction or dressing.

2. (*Augustin.*) Opium, 2 dr.; ox-gall, 2 ox.; digest 2 days, strain, and add of melted lard, 2 oz.; oil of bergamot, 10 drops.

3. (*Brera.*) Opium, 1 dr.; gastric juice of a calf, ½ oz.; digest 24 hours, and add of melted lard, 1 oz.

Ointment of Oxide of Lead. See OINTMENT, LEAD.

Ointment of Oxide of Man"ganese. *Syn.* UN GUENTUM MANGANESII OXYDI, U. M. BINOXYDI, L. *Prep.* 1. Black oxide of manganese (levi-

gated), 1 dr.; lard, 1 oz.; mix by patient tritura tion. As a friction in scrofulous swellings and indurations; and in itch, scald-head, chil blains, &c.

2. (*W. Cooley.*) Binoxide of manganese, 1 dr.; sulphur, 2 dr.; lard, 9 dr.; cajeput oil, 15 drops. As the last; also as a friction in rheumatism, swelled joints, &c., and in porrigo and some other skin diseases.

Ointment of Oxide of Mer'cury. Under this name the two ointments noticed below are often confounded, owing to the different opinions held respecting the atomic weight of mercury.

i. **Ointment of Grey Oxide of Mer'cury.** *Syn.* OINTMENT OF SUBOXIDE OF MERCURY, O. OF PROTOXIDE OF M.†; UNGUENTUM HYDRARGYRI OXYDI, U. H. SUBOXYDI, U. H. O. OINERRI, L. *Prep.* 1. (Ph. E. 1817.) Grey oxide of mer cury, 1 oz.; lard, 8 oz.; triturate together. For merly proposed as a substitute for mercurial ointment, but in practice it has been found use less as a friction, owing to the unctuous matter only being absorbed, whilst the oxide is left on the surface. This objection does not apply to the following preparations.

2. (*Donovan.*) Grey oxide of mercury, 20 gr.; lard, 1 oz.; mix, and expose them to the tempera ture of 320° F. for 2 hours, constantly stirring. Grey coloured. It may also be made from the nitric or red oxide in the same way, by keeping the ointment heated to about 300° for some hours. Cleaner and stronger than *Ung. hyd. fort.* (Ph. L.).

3. (*Tyson.*) Black oxide of mercury (prepared by decomposing precipitated calomel with liquors of potassa and ammonia), 2 oz.; lard, 1 lb.; tri turate together. Inferior in activity to the last. It closely resembles in appearance a fine sample of mercurial ointment.

ii. **Ointment of Red Oxide of Mercury.** *Syn.* UNGUENTUM HYDRARGYRI BINOXYDI†, U. H. OXYDI RUBRI, L. *Prep.* (*Casenave.*) Red oxide of mercury, 30 gr.; camphor, 5 gr.; lard, 1 oz. Closely resembles ointment of nitric oxide of mer cury, over which it, perhaps, possesses some ad vantage from the oxide being in a more minutely divided state.

Ointment of Oxide of Sil'ver. *Syn.* UNGUEN TUM ARGENTI OXYDI, L. *Prep.* (*Serre.*) Oxide of silver, 16 to 20 gr.; lard, 1 oz. As a dressing for scrofulous and syphilitic sores, &c.

Ointment of Oxide of Zinc. *Syn.* ZINC OINT MENT, NIHIL ALBUM OINTMENT†; UNGUENTUM ZINCI (B. P., Ph. L., E., & D.), U. OXYDI ZINCI, L. *Prep.* 1. (Ph. L.) Oxide of zinc, 1 oz.; lard, 6 oz.; mix them together.

2. (Ph. E.) Oxide of zinc, 1 oz.; simple lini ment (Ph. E.), 6 oz.

3. (Ph. D.) Ointment of white wax, 12 oz.; melt it by a gentle heat, add of oxide of zinc, 2 oz., and stir constantly until the mixture con cretes.

4. (B. P.) Oxide of zinc, in very fine powder, 1 part; benzoated lard, 5½ parts; mix.

Uses, &c. Astringent, desiccative, and stimu lant; in excoriations, burns, various skin diseases attended by profuse discharges, in chronic inflam mation of the eyes depending on relaxation of the vessels, in sore nipples, indolent sores, ringworm of the scalp, &c. It is an excellent and very useful preparation. See OINTMENT OF TUTTY (*below*).

Ointment, Pagenstecher's. *Prep.* Yellow oxide of mercury, 30 gr.; vaseline, 1 oz. Used for inflamed eyelids. Often ordered one fourth the above strength.

Ointment of Pep'per. *Syn.* UNGUENTUM PIPERIS NIGRI, L. *Prep.* 1. Black pepper (bruised), 1 oz.; lard, 2 oz.; suet, 1 oz.; digest together in a covered vessel, by the heat of a water-bath, for 6 hours, then strain with pressure, add of expressed oil of mace, 2 dr., and stir until the mixture concretes. In piles, itch, as a friction in rheumatism, &c.

2. (Ph. D. 1826.) Black pepper (in fine powder), 4 oz.; lard, 1 lb.; mix. In scald-head, &c.

Ointment of Petro'leum. *Syn.* UNGUENTUM PETROLEI, L. *Prep.* Yellow wax, 1 part; vaseline, 13 parts.

Ointment of Phosphor'ic Acid. *Syn.* UNGUENTUM ACIDI PHOSPHORICI, L. *Prep.* (*Soubeiran.*) Phosphoric acid, 1 dr.; lard (softened by heat), 1 oz.; triturate carefully together. As a friction in caries, osseous tumours, &c.

Ointment of Phos'phorus. *Syn.* UNGUENTUM PHOSPHORI, U. PHOSPHORATUM, L. *Prep.* (P. Cod.) Phosphorus, 1 dr.; lard, 6 oz. 3 dr.; melt together (in a wide-mouthed bottle) by the heat of a water-bath, remove the vessel from the heat, and shake it briskly until the ointment concretes. As a friction in gout, chronic rheumatism, and several skin diseases.

Ointment of Picrotox'in. *Syn.* UNGUENTUM PICROTOXINE, L. *Prep.* (*Jäger.*) Picrotoxin, 10 gr.; lard, 1 oz. In ringworm of the scalp, and to destroy pediculi. It should be used with care.

Ointment for Piles. *Syn.* UNGUENTUM HEMORRHOIDALE, U. ANTI-HEMORRHOIDALE, L. *Prep.* 1. Burnt alum and oxide of zinc, of each, ⅓ dr.; lard, 7 dr.

2. (*Bories.*) Acetate of lead, 15 gr.; freshly burnt cork, ⅓ oz.; washed fresh butter, 2 oz.; triturate well together.

3. (*W. Cooley.*) Morphia, 8 gr.; melted spermaceti ointment, 1 oz.; triturate together until solution is complete, then add of galls (in impalpable powder), 1½ dr.; essential oil of almonds (genuine crude), 12 to 15 drops, and stir until the mass concretes. In painful piles, prolapsus, &c. It is not only very effective, but does not soil the linen so much as most other ointments.

4. (*Dr Gedding.*) Carbonate of lead, 4 dr.; sulphate of morphia, 15 gr.; stramonium ointment, 1 oz.; olive oil, q. s. When there is much pain and inflammation.

5. (*Sir H. Halford.*) Ointment of nitrate of mercury and oil of almonds, equal parts, triturated together.

6. (*Mazzini.*) Nitrate of morphia, 15 gr.; citrine ointment, 1 dr.; fresh butter, 1 oz. As the last.

7. (*Valles.*) Extract of elder leaves, ⅓ dr.; burnt alum, 16 gr.; poplar ointment, 1 oz. For other formulæ, see the respective names of their leading ingredients.

8. (*Ware.*) Camphor, 1 dr.; simple ointment, 1 oz.; dissolve by heat, add of powdered galls, 2 dr.; mix well, further add of tincture of opium,

2 fl. dr., and stir until the whole is cold. In flabby mucous and painful piles.

9. (*Zanin.*) Spermaceti ointment, 1 oz.; powdered galls, 1 dr.; powdered opium, 18 gr.; solution of diacetate of lead, 1 fl. dr. When there is both pain and inflammation.

10. (From ' New Remedies.') Yellow wax, 8 parts; resin, 4 parts; lard, 12 parts; oil of sassafras, 2 parts. Melt the wax, resin, and lard, remove from the fire, add the oil of sassafras, and stir until the mass is solid. This is said to be a most excellent application for painful or itching piles.

Ointment of Pitch. *Syn.* BLACK BASILICON, OINTMENT OF BLACK PITCH; UNGUENTUM PICIS (B. P., Ph. L.), U. PICIS NIGRE, L. *Prep.* 1. (Ph. L.) Black pitch, resin, and beeswax, of each, 11 oz.; olive oil, 1 pint; melt together, strain through a linen cloth, and stir until the mass concretes.

2. (B. P.) Tar, 5 parts; yellow wax, 2 parts; melt together, and stir till cold.

Uses, &c. Stimulant and detergent; very useful in indolent ulcerations, scald-head, and various foul eruptions. In itch and psoriasis, and other scaly skin diseases, a little sulphur is commonly added to it.

Ointment of Plat'inum. *Syn.* UNGUENTUM PLATINI, L. *Prep.* (*Hager.*) Bichloride of platinum, 15 gr.; extract of belladonna, ⅓ dr.; lard, 1 oz. As a dressing for painful indolent ulcers.

Ointment of Plumbago. *Syn.* OINTMENT OF GRAPHITE; UNGUENTUM GRAPHITIS, U. PLUMBAGINIS, L. *Prep.* From pure plumbago (' black-lead '), 1½ dr.; lard, 1 oz. As a dressing to ulcers, and in certain skin diseases.

Ointment, Plunket's. *Prep.* (Original formula.) Crowsfoot, 1 handful; dog's-fennel, 3 sprigs; pound well, add of flowers of sulphur and white arsenic, of each, 3 thimblefuls; beat them well together, form the mass into boluses, and dry them in the sun. For use, powder them; and mix the powder with yolk of egg, spread a little on a small piece of pig's bladder (size of half a crown), and apply it to the sore, where it must remain until it falls off by itself. Poisonous; in cancer, with great caution.

Ointment, Poma'tum. See OINTMENT, LARD.

Ointment of Pop'lar Buds. *Syn.* UNGUENTUM POPULEUM, L. *Prep.* 1. Fresh poplar buds (bruised), 1 part; lard, 4 parts; boil until crisp, and strain. It never gets rancid. Emollient and stimulant.

2. (Compound—P. Cod.) Poplar buds, 12 oz.; fresh leaves of belladonna, common nightshade (*Solanum nigrum*), henbane, and poppies, of each, 8 oz.; lard, 4½ lbs.; as the last. Emollient, stimulant, and anodyne.

Ointment of Potas'sio-tar'trate of An'timony. *Syn.* ANTIMONIAL OINTMENT, TARTAR EMETIC O.; UNGUENTUM ANTIMONII TARTARATI (B. P.), UNGUENTUM ANTIMONII POTASSIO-TARTATRIS (Ph. L.), U. A. TARTARIZATI (Ph. D.), U. ANTIMONIALE (Ph. E.), U. TARTARI EMETICI, L. *Prep.* 1. (Ph. L. and E.) Potassic tartrate of antimony (rubbed to a very fine powder), 1 oz.; lard, 4 oz.; mix by trituration.

2. (Ph. D.) Tartar emetic (in very fine powder), 1 dr.; ointment of white wax, 7 dr.

3. (B. P.) Tartrated antimony (in fine powder), 1 part; simple ointment, 4 parts; mix.

Uses, &c. Counter - irritant; in phthisis, chronic rheumatism, certain liver affections, and other deep-seated pains and diseases. A portion about the size of a nut is rubbed on the skin night and morning, until a crop of pustules is produced. The part should be well rubbed with a coarse towel, so as to be reddened, before applying the ointment. The product of the Dublin formula is of only half the strength of those of the other Colleges.

Obs. Before adding the tartar emetic to the lard it should be reduced to the state of an impalpable powder. The precipitated salt is the best for this purpose. As the pustules formed by this ointment permanently mark the skin, it should only be applied to those parts of the person which are covered by the dress.

Ointment, Purgative. See OINTMENT OF COLO-CYNTH, WORM O., &c.

Ointment of Quinine'. *Syn.* UNGUENTUM QUININE, U. QUINIE SULPHATIS, L. *Prep.* 1. Sulphate of quinine, 1 dr.; lard, 3 dr. In the agues of children.

2. (*Beasley ex Antonini.*) Sulphate of quinine, 1 dr.; alcohol (rectified spirit), 2 dr.; sulphuric acid, 10 drops; dissolve, and mix it with lard, ½ oz. In malignant intermittents; 2 to 4 dr. at a time, rubbed into the groin or axilla.

Ointment of Red Oxide of Mer'cury. (B. P.) *Syn.* UNGUENTUM HYDRARGYRI OXIDI RUBRI, L. Red oxide of mercury in powder, 1 part; hard paraffin, 1¼ parts; soft paraffin, 5¼ parts. Melt, mix, and stir till cold.

Ointment of Red Sul'phuret of Mer'cury. *Syn.* UNGUENTUM HYDRARGYRI BISULPHURETI, U. H. SULPHURETI RUBRI, L. *Prep.* 1. (*Alibert.*) Red sulphuret of mercury, 1 dr.; camphor, 20 gr.; simple ointment, 1 oz. In herpes, applied twice a day.

2. (*Collier.*) Bisulphuret of mercury, 1½ dr.; sal-ammoniac, ½ dr.; lard, 1 oz.; rose-water, 1 fl. dr. In several skin diseases, to diminish the itching, destroy pediculi, &c.

3. (*Radius.*) As the last, with 1 oz. more lard.

Ointment of Res'in. *Syn.* YELLOW BASILICON; UNGUENTUM RESINÆ (Ph. D.), U. RESINOSUM (Ph. E.), L. *Prep.* 1. (Ph. D.) Yellow wax, ½ lb.; yellow resin, in coarse powder, ½ lb.; prepared lard, 1 lb.; melt them together by a gentle heat, strain the mixture, whilst hot, through flannel, and stir it constantly until it concretes.

2. (Ph. E.) Beeswax, 2 oz.; resin, 5 oz.; lard, 8 oz.

Obs. A useful stimulant dressing to foul and indolent ulcers. For the corresponding preparation of the Ph. L., see CERATE, RESIN.

Ointment, Resol'vent. See OINTMENT, DIS-CUTIENT.

Ointment of Rhatany. (*Trousseau.*) *Syn.* UNGUENTUM RHATTINIÆ, L. *Prep.* Extract of rhatany, 1½ dr.; cacao butter, 5 dr. Mix.

Ointment, Ring'worm. UNGUENTUM CONTRA-TINEAM, L. *Prep.* 1. Carbonate of soda, 1 part; fresh-slaked lime, 4 parts; lard, 120 parts.

2. Ointment of nitrate of mercury, 1 dr.; tar ointment and lard, of each, ½ oz.

3. (*Henke.*) Hydrochloric acid, 1 fl. dr.; juniper-tar ointment, ½ oz.; marsh-mallow do., 1 oz.

4. (*Pereira.*) Tar, 3 dr.; lard, 1½ oz.; melt them together, and stir in of acetic acid (Ph. L.), 2 fl. dr.

5. (*Thompson.*) Carbonate of soda and sulphuret of potassium, of each, 1 dr.; creasote, ½ dr.; lard, 1½ oz.

Obs. The hair must be cut off close, and the part washed clean before each application. For other forms see *above.*

Ointment of Rose. *Syn.* ROSE POMMADE, ROSE LIP-SALVE; UNGUENTUM ROSÆ, U. ROSA-TUM, L. *Prep.* 1. (P. Cod.) Washed lard (melted) and roses (centif.), of each, 2 lbs.; mix, and in 2 days remelt the mass, and press out the fat; to this last add of fresh roses, 2 lbs., and repeat the process; lastly, colour it with alkanet root if required red.

2. (UNG. AQUÆ ROSÆ—Ph. U. S.) This is spermaceti ointment melted and beaten up with about two thirds of its weight of rose-water until they congeal. Both the above are simple emollients. The last is an officinal 'cold cream.'

Ointment of Rosemary (Compound). (Ph. G.) *Syn.* UNGUENTUM ROSMARINI COMPOSI-TUM, L. *Prep.* Lard, 16 oz.; suet, 8 oz.; yellow wax, 2 oz.; oil of mace, 3 oz.; liquefy in a vapour-bath, and when nearly cold, add oil of rosemary and oil of juniper, of each, 1 oz. by weight.

Ointment of Rue. (Span. Hosp.) *Syn.* UN-GUENTUM RUTÆ, L. *Prep.* Fresh rue, 2 oz.; wormwood, 2 oz.; nitre, 2 oz.; lard, 16 oz.; boil till the moisture is expelled.

Ointment, Rust's. *Prep.* Calcined alum, 1½ dr.; camphor, ½ dr.; powdered opium, 20 gr.; balsam of Peru, 1 dr.; lead ointment, 5 dr.; triturate together. In chilblains, frostbites, frosted limbs, &c.

Ointment of Sabadil'line. *Syn.* UNGUENTUM SABADILLINÆ, L. *Prep.* (*Dr Turnbull.*) Sabadilline, 15 to 20 gr.; lard, 1 oz. Intended as a substitute for ointment of veratrine.

Ointment of Salicylic Acid. (B. P.) *Syn.* UNGUENTUM ACIDI SALICYLICI, L. Salicylic acid, 1 part; soft paraffin, 18 parts; hard paraffin, 9 parts. Melt and mix.

Ointment of Sav'ine. *Syn.* UNGUENTUM SABINÆ (Ph. P., Ph. L. and D.), CERATUM SABINÆ, L. *Prep.* 1. (Ph L.) White wax, 3 oz.; benzoated lard, 1 lb.; melt them together, mix in of fresh savine (bruised), ½ lb., and press through a linen cloth.

2. (Ph. D.) Savine tops, dried and in fine powder, 1 dr.; ointment of white wax, 7 dr.; mix by trituration. For the formula of the Ph. E., the uses, &c., see CERATE.

Ointment of Scrophula"ria. *Syn.* UNGUEN-TUM SCROPHULARIÆ, L. *Prep.* (Ph. D. 1826.) Green leaves of knotted-rooted figwort and lard, of each, 2 lbs.; prepared suet, 1 lb.; boil till crisp and strain with pressure. In ringworm, 'burnt holes' (*pemphigus gangrænosus* of children), impetigo, and some other cutaneous diseases; also as an application to piles, painful swellings, &c. In the second it is said to be almost specific.

Ointment, Simple. *Syn.* OINTMENT OF WHITE WAX, SIMPLE DRESSING; UNGUENTUM SIMPLEX (B. P., Ph. E.), U. CERÆ ALBÆ (Ph. D.), L. *Prep.* 1. (Ph. E.) Olive oil, 5½ fl. oz.; white wax, 2 oz.; melted together, and stirred whilst cooling.

2. (Ph. D.) Prepared lard, 4 lbs.; white wax, 1 lb.; as the last.

3. (B. P.) White wax, 2 parts; benzoated lard, 3 parts; almond oil, 3 parts; melt together, and stir till it becomes solid.

Obs. The above are mild emollients, useful in healthy ulcers, excoriations, &c.; but chiefly as forming the basis for other ointments. The corresponding preparation of the Ph. L. is spermaceti ointment. See *below;* also OINTMENT, LARD, &c.

Ointment, Singleton's. See OINTMENTS, EYE.

Ointment, Smallpox. *Syn.* UNGUENTUM BOTROTICUM, L. *Prep.* 1. Mercurial ointment, 1½ oz.; beeswax and black pitch, of each, ½ oz.; expressed oil of mace, 2 dr.; mixed together by a very gentle heat.

2. (*Briquet.*) Mercurial ointment, 4 parts; powdered starch, 1 part.

3. (*Tourrière.*) Iodide of potassium (dry and in fine powder), 1 part; expressed oil of mace, 2 parts; black resin, 4 parts; mercurial ointment, 8 parts. Used to prevent the 'pitting of the pustules.' See SMALLPOX.

Ointment, Smellome's. See OINTMENTS, EYE.

Ointment of Soap. 1. See CERATE.

2. (Camphorated: UNGUENTUM SAPONIS CAMPHORATUM—Hamb. Cod.) White soap (scraped), 1 lb.; water, ½ lb.; dissolve by heat; add of olive oil, 5 oz.; and when the mixture has partly cooled, further add of camphor, 1 oz., previously dissolved by heat in olive oil, 1 oz.; lastly, stir until the mass concretes. As an anodyne and stimulating friction in various local affections, as chaps, chilblains, rheumatism, &c.

Ointment of So'dio-chlo"ride of Gold. *Syn.* UNGUENTUM AURI SODIO-CHLORIDI, L.; POMMADE DE MURIATE D'OR ET DE SOUDE, Fr. *Prep.* (*Magendie.*) Sodio-chloride of gold, 10 gr.; lard, 4 dr. In scrofulous and syphilitic swellings, indurations, ulcers, &c.

Ointment of Spermace'ti. *Syn.* EMOLLIENT DRESSING, SIMPLE OINTMENT, WHITE O.; UNGUENTUM CETACEI (B. P., Ph. L. and D.), U. SPERMATIS CETI, L. *Prep.* 1. (Ph. L.) Spermaceti, 5 oz.; white wax, 14 dr.; olive oil, 1 pint, or q. s.; melt them together by a gentle heat, and stir the mixture until cold.

2. (Ph. D.) White wax, ½ lb.; spermaceti, 1 lb.; prepared lard, 3 lbs.; as the last.

3. (B. P.) Spermaceti, 5 parts; white wax, 2 parts; almond oil, 20 parts, or a sufficiency; stir constantly until it cools.

Uses, &c. As an emollient and healing application or dressing to abrasions, excoriations, blistered surfaces, healthy ulcers, chilblains, chaps, &c. In trade the Dublin formula, with double the amount of lard, is commonly employed. See OINTMENT, LARD, SIMPLE O., &c.

Ointment of Squills. *Syn.* UNGUENTUM SCILLÆ, L. *Prep.* 1. (*Brera.*) Squills (in very fine powder), 1 dr.; mercurial ointment, 2 dr.

2. (*Hufeland.*) Squills, 1 oz.; liquor of potassa, 2 fl. oz.; reduce to a mucilage by boiling, then add of lard, 2 oz., or q. s. As a resolvent friction to indolent tumours and indurations.

Ointment of Staves'acre. *Syn.* UNGUENTUM STAPHISAGRIÆ, L. *Prep.* 1. (*Swediaur.*) Powdered stavesacre, 1 oz.; lard, 8 oz.; melt together, digest 3 or 4 hours, and strain. A very cleanly remedy for itch, and to destroy pediculi on the person. A similar ointment is much used by farriers.

2. (B. P.) Stavesacre seed, 1 part; benzoated lard, 2 parts. Crush the seeds and macerate them in the melted lard, over a water-bath for two hours. Strain and cool. Less powerful parasiticide, useful in itch and to kill lice.

Ointment of Storax. *Syn.* UNGUENTUM STYRACIS, L. *Prep.* Strained storax, 1 part; lanolin or lard, 3 parts. Melt together and stir till cold. A good remedy for scabies.

Ointment of Stramo"nium. *Syn.* UNGUENTUM STRAMONII, L. *Prep.* 1. Fresh thornapple leaves, 1 part; lard, 4 parts; as ointment of hemlock.

2. (*Pereira.*) Powdered leaves, 1 oz.; lard, 4 oz.; mix by trituration.

3. (Ph. U. S.) Extract of stramonium, 1 dr.; lard, 1 oz.; as the last.

Uses, &c. To dress irritable ulcers, and as an application to painful piles.

Ointment of Strych'nine. *Syn.* UNGUENTUM STRYCHNIÆ, L. *Prep.* 1. (*Bouchardat.*) Strychnine, 16 gr.; lard, 1 oz.; carefully triturated together.

2. (*Wendt.*) Nitrate of strychnine, 6 dr.; lard, 1 oz.; as last. Both are used as a friction in paralysed parts, &c. From the extremely poisonous character of strychnine it should be used with caution.

Ointment of Subac'state of Cop'per. See OINTMENT OF VERDIGRIS.

Ointment of Subchloride of Mercury. See OINTMENT OF CALOMEL.

Ointment of Subsul'phate of Mercury†. *Syn.* UNGUENTUM HYDRARGYRI SUBSULPHATIS, L. *Prep.* 1. (*Alibert.*) Turpeth mineral, ½ dr.; lard, 1 oz.

2. (*Biett.*) Turpeth mineral, 1 dr.; sulphur, 2 dr.; lard, 2 oz.; oil of lemons, 15 drops. In herpes, porrigo, and the scaly diseases.

Ointment of Sulphate of I'ron. *Syn.* UNGUENTUM FERRI SULPHATIS, L. *Prep.* (*Velpeau.*) Sulphate of iron, 1½ dr.; simple ointment, 1 oz. In erysipelas.

Ointment of Sulphate of Zinc. *Syn.* UNGUENTUM ZINCI SULPHATIS, L. *Prep.* (*Scarpa.*) Sulphate of zinc (in very fine powder), 1 dr.; lard, 1 oz. In some chronic skin diseases attended with a lax state of the tissues, and as a dressing to scrofulous tumours after they have suppurated and the abscess has been discharged.

Ointment of Sulphur. *Syn.* UNGUENTUM SULPHURIS (B. P., Ph. L., E., & D.), L. *Prep.* 1. (Ph. L.) Sulphur, ½ lb.; lard, 1 lb. In the Ph. L. 1836 oil of bergamot, 40 drops, were added. See 5, Compound.

2. (Ph. E.) Sulphur, 1 oz.; lard, 4 oz.

3. (Ph. D.) Sulphur, 1 lb.; lard, 4 lbs.

4. (B. P.) Sublimed sulphur, 1 part; benzoated lard, 4; mix.

Uses, &c. In itch, scald-head, &c., in the first of which it is specific. It should be well rubbed in every night until the disease is cured; "but not more than one fourth part of the body should be covered with it at a time" (*A. T. Thomson*).

5. (Compound: ITCH OINTMENT; UNGUENTUM SULPHURIS COMPOSITUM—Ph. L.) *a.* (Ph. L.) Nitrate of potassa (powdered), 40 gr.; white hellebore (powdered), 10 dr.; sulphur and soft soap, of each, 4 oz.; lard, 1 lb.; rub them together.

b. (P. Cod.) Alum and sal-ammoniac, of each, ½ oz.; sulphur, 8 oz.; lard, 16 oz.

Uses, &c. In itch, as the simple ointment (1, 2, and 3). They are more efficacious, but, owing to the presence of white hellebore, the Ph. L. preparation is apt to cause irritation in persons with delicate skins. See OINTMENT, ITCH.

Ointment of Sulphurated Potash. *Syn.* UNGUENTUM POTASSÆ SULPHURATÆ (B.P.), L. *Prep.* 1. Sulphurated potash, 30 gr.; triturate, and add hard paraffin, ½ oz.; soft paraffin, ½ oz.; mix.

2. Sulphurated potash, 5 parts; hard paraffin, 18 parts; soft paraffin, 55 parts. This ointment should be freshly made.

Ointment of Sulphuret of Mercury. See OINTMENT OF RED SULPHURET OF MERCURY.

Ointment of Sulphuret of Potas'sium. Sulphuret of potassium, 2½ dr.; lard and soft soap, of each, 1 oz.; olive oil, ½ oz. In several chronic skin diseases, as itch, psoriasis, ringworm, lepra, eczema, &c.

Ointment of Sulphuret of So'dium. *Syn.* UNGUENTUM SODII SULPHURETI, L. *Prep.* (*Swediaur.*) Sulphuret of sodium, 3 dr.; lard, 1½ oz. In itch, for which it is very cleanly and effective. The last two ointments are most powerful when recently prepared.

Ointment of Sulphuric Ac'id. *Syn.* UNGUENTUM ACIDI SULPHURICI, L. *Prep.* 1. (*Dr Duncan.*) Sulphuric acid, 1 dr.; lard, 2 oz.

2. (Ph. D. 1826.) Sulphuric acid, 1 dr.; lard, 1 oz.; mix.

Uses, &c. Black, fœtid; in itch. It is now seldom used. With oil of turpentine it has been used as a stimulating liniment in rheumatism. An ointment made of 1½ dr. of dilute sulphuric acid to 1 oz. of lard is a good application in prurigo.

Ointment, Sulta'na. Spermaceti and white wax, of each, ½ oz.; oil of almonds and butter of cacao, of each, ½ lb.; melt together, add of balsam of Peru, 1 dr., stir constantly for a few minutes, and after it has settled pour off the clear portion; to this add of orange-flower water, 2 fl. dr., and stir the mixture constantly until it concretes. A very agreeable species of cold cream.

Ointment of Tan'nate of Lead. *Syn.* UNGUENTUM PLUMBI TANNATIS, L. *Prep.* 1. Tannate of lead, 1½ dr.; powdered camphor, 20 gr.; spermaceti ointment, 7 dr. In inflamed piles, &c.

2. (*Swadelin.*) Decoction of oak bark, 6 fl. oz.; solution of diacetate of lead, 1½ oz.; mix, collect, and drain the precipitate, and mix it, whilst still moist, with lard, 1 oz.; camphor, 10 gr. In bedsores.

Ointment of Tan'nin. *Syn.* UNGUENTUM TANNINI, U. ACIDI TANNICI, L. *Prep.* (*Richard.*) Tannin, 2 dr.; water, 2 fl. dr.; triturate them together, then add of lard, 1½ oz. Astringent

and hæmostatic. In piles, prolapsus, &c. It is a very cleanly and effective application.

Ointment of Tar. *Syn.* UNGUENTUM PICIS LIQUIDÆ (Ph. L., E., & D.), L. *Prep.* 1. (Ph. L.) Tar and suet, of each, 1 lb.; melt them together, and press the mixture through a linen cloth.

2. (Ph. B.) Tar, 5 oz.; beeswax, 2 oz.; melt together, and stir the mixture briskly until it concretes.

3. (Ph. D.) Tar, ½ pint; yellow wax, 4 oz. as the last.

Uses, &c. As a detergent application in ringworm, scald-head, scabby eruptions, foul ulcers, &c. It should be, in general, at first diluted with half of its weight of lard or oil. See also OINTMENT OF PITCH.

Ointment of Tartar Emet'ic. See OINTMENT OF POTASSIO-TARTRATE OF ANTIMONY.

Ointment of Tobac'co. *Syn.* UNGUENTUM TABACI, L. *Prep.* 1. (*Chippendale.*) Extract of tobacco, 1 dr.; lard, 1 oz. As a friction in neuralgia.

2. (Ph. U. S.) Fresh tobacco leaves, 1 oz.; lard, 12 oz.; as ointment of hemlock. As an anodyne application in irritable ulcers, ringworm, prurigo, and some other skin diseases.

Ointment, Tripharm'ic. *Syn.* OINTMENT OF THREE THINGS; UNGUENTUM TRIPHARMACUM, L. *Prep.* From lead plaster, 4 oz.; olive oil, 2 fl. oz.; distilled vinegar, 1 fl. oz.; melt together, and stir until they combine, and a proper consistence is obtained. Cooling and desiccative; formerly greatly esteemed as a dressing.

Ointment, Trooper's. See OINTMENT, MERCURIAL.

Ointment of Turpentine. *Syn.* UNGUENTUM TEREBINTHINÆ (B. P.), L. *Prep.* 1. (Guy's Hosp.) Camphor, 1 dr.; oil of turpentine, 1 to 2 fl. dr.; dissolve, and add of resin of cerate, 1 oz. As a stimulant and anodyne friction in nephritic and rheumatic pains, engorgements, &c.

2. (Ph. Austr.) Turpentine, 2 lbs.; simple ointment, 1 lb.; mix by a gentle heat. As a stimulant dressing.

3. (B. P.) Turpentine, 1 oz.; resin, 54 gr.; yellow wax, ½ oz.; prepared lard, ½ oz. Melt, and stir till cold.

Ointment of Tut'ty. *Syn.* UNGUENTUM ZINCI OXYDI IMPURI, U. TUTLÆ, L. *Prep.* From prepared tutty, 1 part; simple ointment, 5 parts; mix by trituration. Formerly in great repute in ophthalmic practice, more particularly in inflammation, &c., of the eyelids. See OINTMENT OF OXIDE OF ZINC.

Ointment of Vera'trine. *Syn.* UNGUENTUM VERATRINÆ (B. P.), L.; POMMADE DE VERATRINE, Fr. *Prep.* 1. (*Magendie.*) Veratrine, 4 gr.; lard, 1 oz.; mixed by careful trituration.

2. (*Pereira.*) Veratrine, 30 gr.; lard, 1 oz.

3. (*Turnbull.*) Veratrine, 10 to 20 gr.; olive oil, 1 dr.; triturate, and add of spermaceti ointment, 1 oz.

4. (B. P.) Veratrine, 8 gr.; hard paraffin, ½ oz.; soft paraffin, ½ oz.; olive oil, 1 dr. Rub the veratrine smooth with the oil, melt the remaining ingredients, and mix.

4a. Veratrine, 1 part; hard paraffin, 14 parts;

soft paraffin, 41 parts; olive oil, 7 parts; rub the veratrine and the oil to a smooth condition, then mix with the melted paraffins.

Uses, &c. As a friction in neuralgia, neuralgic rheumatism, gout, dropsy, &c. A piece about the size of a hazel nut is to be rubbed for 10 or 15 minutes over the seat of pain twice a day. It must not be applied where the skin is unsound, nor to a large surface at a time, and the greatest caution must be used, on account of the extremely poisonous character of veratrine.

Ointment of Ver'digris. *Syn.* OINTMENT OF SUBACETATE OF COPPER; UNGUENTUM ÆRUGINIS (Ph. E.), U. CUPRI SUBACETATIS (Ph. D.), L. *Prep.* 1. (Ph. E.) Resinous ointment, 15 oz.; melt by a gentle heat, sprinkle into it of verdigris (in very fine powder), 1 oz., and stir the mixture briskly until it concretes.

2. (Ph. D.) Prepared subacetate of copper, ½ dr.; ointment of white wax, 7½ dr.; mix by trituration.

Uses, &c. Detergent and escharotic; as an occasional dressing to foul and flabby ulcers, to keep down fungous flesh, and, diluted with oil or lard, in scrofulous ulceration and inflammation of the eyelids.

Ointment of Vin'egar. *Syn.* ACETIC OINTMENT; UNGUENTUM ACETI, U. ACIDI ACETICI, L. *Prep.* 1. (*Dr Cheston.*) Olive oil, 1 lb.; white wax, 4 oz.; melt them together by a gentle heat, add of strong vinegar, 2 fl. oz., and stir until the mixture concretes. As a cooling astringent dressing, and as an application in chronic ophthalmia.

2. (*W. Cooley.*) Acetate of morphine, 6 gr.; acetic acid (Ph. L.) and water, of each, 1½ fl. dr.; dissolve, add the solution to simple ointment (melted), 1½ oz., and stir the mixture briskly until nearly cold. In chronic ophthalmia, painful inflamed piles, &c.; also to remove freckles, and to allay itching and irritation in several skin diseases.

Ointment of Walnut Leaves (*Negrier*). *Syn.* UNGUENTUM JUGLANDIS. *Prep.* Extract of walnut leaves, 3 dr.; lard, 4 dr.; oil of bergamot, 1 drop. Mix.

Ointment, White. Both SPERMACETI OINTMENT and OINTMENT OF CARBONATE OF LEAD were formerly so called, but the name is now obsolete. The CAMPHORATED WHITE OINTMENT of the Ph. L. of 1746 (UNG. ALBUM CAMPHORATUM) was spermaceti ointment to which a little camphor had been added.

Ointment of White Precipitate. *Syn.* OINTMENT OF AMMONIATED MERCURY; UNGUENTUM HYDRARGYRI AMMONIATI (B. P.), L. *Prep.* Ammoniated mercury, 1 part; simple ointment, 9 parts. Mix. See OINTMENT OF AMMONIO-CHLORIDE OF MERCURY.

Ointment of White Wax. See OINTMENT, SIMPLE.

Ointment of Witch Hazel. *Syn.* UNGUENTUM HAMAMELIDIS. *Prep.* Liquid extract of hamamelis, 1 part; benzoated lard, 9 parts. Astringent for piles.

Ointment of Wolfsbane. See OINTMENT OF ACONITE.

Ointment of Wood Soot. *Syn.* UNGUENTUM FULIGINIS, L. *Prep.* Wood soot and lard, of each, equal parts. Mix.

Ointment of Wood Soot (Compound). *Syn.* UNGUENTUM FULIGINIS COMPOSITUM, L. *Prep.* Acetic extract of wood soot, 4 dr.; dried salt, 10 dr.; lard, 14 oz. For ringworm.

Ointment for Worms. *Syn.* UNGUENTUM ANTHELMINTICUM, U. VERMIFUGUM, L. *Prep.* 1. (*Boerhaave.*) Aloes and ox-gall, of each, 1 part; marsh-mallow ointment, 8 parts.

2. (Fr. Hosp.) Aloes and oil of tansy, of each, 1 part; dried ox-gall, 2 parts (both in fine powder); lard, 8 parts.

3. (Ph. Bat.) Aloes, 1 dr.; dried ox-gall and petroleum, of each, 1½ dr.; lard, 1½ oz.

4. (*Soubeiran.*) Powdered aloes, 2 dr.; lard, 1 oz. *Uses, &c.* The above are purgative and vermifuge, applied as frictions to the abdomen. They are chiefly employed for children and delicate females. See OINTMENT, COLOCYNTH.

Ointment of Yel'low Wax. *Syn.* UNGUENTUM CERÆ FLAVÆ, L. *Prep.* (Ph. D. 1826.) Beeswax, 1 lb.; lard, 4 lbs.; melt them together. A mild emollient dressing. Some parties regard it as more 'healing' than the OINTMENT OF WHITE WAX.

Ointment of Zinc. See OINTMENT OF OXIDE OF ZINC.

Ointment of Zinc Cyanide (*Cussier*). *Syn.* UNGUENTUM ZINCI CYANIDI, L. *Prep.* Cyanide of zinc, 12 gr.; lard, 5 dr.; butter of cacao, 5 dr.; mix.

OINTMENTS (Flower of). *Syn.* FLOS UNGUENTORUM, L. *Prep.* From resin, thus, wax, and suet, of each, ½ lb.; olibanum and Venice turpentine, of each, 2½ oz.; myrrh, 1 oz.; wine, ½ pint; boil them together, and, lastly, add of camphor, 2 dr. Suppurative; warming.

OLEATES. *Syn.* OLEATA, L. These are chemical compounds of oleic acid with a base. Of recent years they have come much into use as local applications, as rubbed into the skin they are readily absorbed. Two methods of making them are followed:

1. A metallic oxide or a pure alkaloid is dissolved in proper quantity in pure oleic acid, using heat in most cases.

2. A soluble salt of a metal, such as sulphate of zinc, is mixed with a solution of olive-oil soap, washing the precipitate well with water, and drying. This is the better of the two processes.

Oleate of Aconitine. *Prep.* Pure aconitine, 10 gr.; oleic acid, 1 oz.; dissolve.

Uses. Local anodyne to relieve the pain of neuralgia and rheumatism.

Oleate of Atropine. *Prep.* Pure atropine, 8 gr.; oleic acid, 1 oz.; dissolve.

Uses. Local anodyne, also to dilate the pupil.

Oleate of Bismuth. *Prep.* Nitrate of bismuth (in crystals), ½ oz.; dissolve in glycerin, 4 oz.; add slowly solution of oleate of sodium with constant stirring until it ceases to precipitate. Decant, wash with water, and dry.

Uses. Sedative, astringent; applied to pustular eruptions and hyperæmia of the skin.

Oleate of Copper. *Prep.* Sulphate of copper, 1 part; water, 40 parts; dissolve. Add slowly a solution of oleate of sodium until a precipitate ceases to form, stirring constantly. Decant the water, and wash the semi-fluid oleate several times with hot water.

Uses. Antiseptic and antiparasitic agent. Useful in ringworm.

Oleate of Lead. Same process as for OLEATE OF COPPER, using acetate of lead in place of sulphate of copper.

Oleate of Mercury. *Syn.* OLEATUM HYDRARGYRI (B. P.), L. *Prep.* 1. Yellow oxide of mercury, 1 part; oleic acid, 9 parts. Mix well, and stir till dissolved.

2. Yellow oxide of mercury, 150 gr.; nitric acid, 180 gr.; water, 2 oz.; dissolve, and dilute to 40 oz. with water. To this add a weak solution of pure soft soap, until a precipitate ceases to form. Wash the precipitate with warm water and dry.

Uses. Oleate of mercury was first introduced by Professor Marshall, and is used of 3 strengths, 5%, 10%, 20% oxide of mercury. Strongly recommended as an application for chronic inflammation in the joints, or rubbed into the axilla for syphilis. Applied to the head it destroys pediculi.

Oleate of Mercury and Morphine. *Prep.* Pure morphine, 1 gr.; oleate of mercury (5%), 1 dr.; mix and dissolve.

Oleate of Zinc. *Syn.* OLEATUM ZINCI (B. P.), L. *Prep.* 1. Oxide of zinc, 1 part; oleic acid, 9 parts; mix well, and dissolve by the aid of heat.

2. Same process as for OLEATE OF COPPER, using sulphate of zinc for sulphate of copper. The product should be a hard mass, and is usually sold in the form of powder.

Uses. In form of ointment or dusting powder, as an astringent in eczema ulcerations and burns. See OLEIC ACID.

OLEFIANT GAS. C₂H₄. *Syn.* ETHYLENE, HEAVY CARBONETTED HYDROGEN, HEAVY CARBURETTED H., ELAYL, ETHENE; GAZ HUILEUX, Fr. A substance discovered by some associated Dutch chemists in 1795. This gas occurs as an important constituent amongst the products of the action of heat upon coal and other substances rich in carbon.

The name *olefiant gas* is derived from its property of uniting with halogens to form oily liquids, a fact which is applied to the estimation of the proportion of this gas present in coal gas, upon which part of the illuminating value of the latter depends.

Prep. 1. One measure of alcohol (rectified spirit) is gradually added to two measures of oil of vitriol, and the mixture is heated in a retort until it blackens, and sulphurous acid begins to be evolved; the product is then passed first through a wash-bottle containing a solution of caustic potash, or milk of lime, and next through a Woulff's bottle containing concentrated sulphuric acid, the last being furnished with a tube dipping into the water of a pneumatic trough: here the gas is collected in tall glass cylinders.

2. The vapour of boiling alcohol is passed into a mixture of oil of vitriol diluted with rather less than one half its weight of water, and so heated as to be in a state of tranquil ebullition (320°—330° F.); the gaseous product is chiefly olefiant gas and the vapour of water, from which it may be separated as above. No sulphurous acid is formed, nor does the acid blacken as in the last process.

Prop., &c. Colourless; neutral; possessing a peculiar ethereal odour; nearly insoluble in water; alcohol, ether, and the volatile and fixed oils absorb a portion of it; burns with a brilliant white flame; at a full red heat or under the action of a strong electric spark it suffers decomposition, with deposit of carbon and liberation of light carburetted hydrogen gas; mixed with three times its volume of oxygen it explodes with extreme violence when ignited; it detonates powerfully when brought in contact with strongly ozonised oxygen; mixed with twice its volume of chlorine and inflamed, hydrochloric acid is formed, and the carbon of the gas is precipitated in the form of dense black soot; if the mixture (best in equal volumes), instead of being kindled, be left standing over water, it soon condenses into a heavy oily liquid (chloride of olefiant gas, Dutch liquid). Liquefies to a colourless liquid at 103° C. Sp. gr. 0·978; 100 cubic inches weigh 30·57 gr.

Olefiant Gas, Bromide of. C₂H₄Br₂. *Syn.* BROMIDE OF ETHYLENE. From bromine and olefiant gas as Dutch liquid (q. v.). A colourless liquid, with an ethereal odour, boiling at 265°, and solidifying at 0° F. Sp. gr. 2·16.

Olefiant Gas, Chloride of. C₂H₄Cl₂. *Syn.* DUTCH LIQUID, ETHYLENE DICHLORIDE. This substance, referred to above, may be easily prepared in any quantity by the following process:—Chlorine and olefiant gas (the latter a little in excess) are conveyed by separate tubes (passing through the same cork) into a glass globe, having a narrow funnel-shaped neck at its lower part, dipping into a small bottle destined to receive the product of their mutual reaction: the newly formed liquid trickles down the sides of the globe into the receiver, and when a sufficient quantity is collected, it is purified by agitating it first with water, and then with sulphuric acid, and, lastly, submitting it to distillation.

Prop., &c. Colourless; sweet-tasted; agreeably fragrant, the odour approaching that of oil of chloroform; slightly soluble in water, freely so in alcohol and ether; it sinks in water; boils at 180° F.; burns with a smoky greenish flame; is unaffected by oil of vitriol, but decomposed by solution of caustic potash. It combines with chlorine, forming new compounds. See CHLORIDES OF CARBON.

OLEIC ACID. C₁₇H₃₃.CO₂H. *Syn.* OLEINE, ELAIC ACID. One of the fatty acids discovered by Chevreul, and produced by saponifying oils, and then separating the base from the resulting soap by means of a dilute acid. It occurs as trioleine in most liquid and solid fats. It now forms an important secondary product in the manufacture of stearic acid and stearin candles, in which its presence would be injurious by lowering the melting-point. Perfectly pure oleic acid may be obtained as follows:

1. By saponifying triolein, as just noticed.

2. Pure almond, lard, or olive-oil soap is decomposed by a dilute acid, and the resulting oily acid is digested in a water-bath with half its weight of litharge (in very fine powder) for some hours, constantly stirring; the mixture is then agitated

with twice its volume of ether in a close vessel, and in 24 hours the clear ethereal solution is decanted, and decomposed with dilute hydrochloric acid; the oleic acid separates, and the ether mixed with it is expelled by evaporation. To render it colourless, the acid is again saponified with caustic soda, and the soap thus obtained is repeatedly dissolved in a solution of soda, and as often separated by adding common salt; this soap is, lastly, decomposed by dilute hydrochloric acid as before.

3. (H. N. Fraser's method.) Oil of cotton seeds, deprived of most of its stearin by chilling and pressure, is first saponified with potash, using a slight excess of the base. The soap is then treated with tartaric acid, or any other acid which will make a soluble salt with potash, until the base is completely neutralised: the residue is washed until a mass is left about the consistence and colour of cerate, free from any of the salt; this is heated for several hours with nearly its weight of litharge, and three or four times its bulk of water; the resulting compound is shaken up while yet warm with ether, and allowed to stand until all the soluble matter separates. This removes the stearate, and leaves a nearly pure oleate of lead.

The clear liquor is decanted and briskly shaken with dilute muriatic acid for a few minutes to precipitate all the chloride of lead, the lighter liquid washed to remove traces of muriatic acid, and filtered; the filtrate is finally heated slowly in a water-bath, and the ether distilled until the residue ceases to have an ethereal odour. The product is about 50% of the bulk of the oil.

Crude oleic acid may be purified as follows:—
1. Expose it repeatedly to a temperature of about 45° F., and as often express the liquid portion. With this mix an equal bulk of solution of sulphurous acid, place the mixture in the light, and shake it frequently until no more colour is discharged. After separation the oleic acid is to be washed repeatedly with cold distilled water, and put into bottles, which should be kept filled up and in a cool place.

2. Heat with litharge over boiling water for several hours; extract the oleate with ether; shake the solution with muriatic acid, which precipitates the lead as chloride, the oleic acid remaining dissolved in the ether, which forms the upper layer; distil off the ether; dissolve the remaining oleic acid in ammonia, and then precipitate with chloride of barium; recrystallise the barium oleate from an alcoholic solution of it, and finally decompose and separate by adding tartaric acid.

Prop., &c. A colourless oily acid, which on cooling solidifies to brilliant, colourless, tasteless needles, melting at 14° C., insoluble in water, soluble in alcohol, ether, and oil: with the bases it forms salts called oleates. The best kind of oleic acid is known as 'pale cloth oil.'

It is used in greasing the wool in the process of spinning; olive oil used to be employed, but oleic acid is much more readily removed by alkalies, and, therefore, more suitable.

The following are the most important compounds of oleic acid

Ammonium Oleate. Employed as a mordant for aniline dyes on cotton.

Barium Oleate. A crystalline powder insoluble in water, and slightly soluble in boiling water.

Lead Oleate. A light, white powder, melting at 80° C. to a yellow oil, and cooling to a brittle translucent mass; it forms the chief part of lead plaster.

Potassium Oleate. A transparent jelly, which can be decomposed by water into caustic potash, and the insoluble acid salt. The soft soap made by saponifying whale and seal oils with potash chiefly consists of this substance.

Sodium Oleate. A constituent of hard soap; it can be crystallised from absolute alcohol. J. Lightfoot introduced the use of this salt in calico printing as a 'prepare' for cloth for steam colours, the effect being to heighten and brighten the tints.

By fusing oleic acid with caustic potash it is resolved into acetic and palmitic acids. This first is taken advantage of in the utilisation of the large quantities obtained in the manufacture of candles.

Impurities, Tests, &c. Pure oleic acid has no acid reaction; if it reddens litmus, products of oxidation are present.

Mineral and rosin oils are sometimes used to adulterate oleic acid, and their presence greatly interferes with the adaptability of the latter for greasing wool. Such admixture reduces the power of being readily saponified, for which oleic acid is chiefly valued.

To detect such hydrocarbons they may be dissolved out from the dry soap (the sample having been saponified), mechanically divided by admixture with sand, by the use of suitable solvents, such as ether, chloroform, carbon disulphide, benzene, or petroleum spirit (*A. H. Allen*). This method requires very careful manipulation, for the details of which the reader should consult Allen's 'Commercial Organic Analysis,' vol. ii, pp. 165, 166.

Determination of Oleic Acid in Insoluble Fatty Acid. 1. Heat with finely powdered litharge; dissolve out the oleate of lead formed by digesting with warm ether repeatedly and filtering; decompose the filtrate with muriatic acid; decant the solution containing the liberated oleic acid; evaporate, and then weigh the residue in a capsule.

2. (Muter's method—'Analyst,' ii, 78.) Saponify about 1·5 grms. of the fatty matter with alcoholic potash, and dilute well with boiling water. Treat the solution with acetic acid till slightly acid, and then carefully neutralise it with weak potash. Precipitate with slight excess of lead acetate and stir until the soap settles. Decant the supernatant liquid; wash the soap once with a large quantity of water, and decant again. The process so far has yielded lead oleate, lead palmitate, and lead stearate; the first of these salts is soluble in ether, while the other two are insoluble.

The soap is now transferred to a flask (capacity = 100 c.c.), the basin which contained it being well rinsed with pure ether and the washings placed in the flask along with the soap. The flask is then filled up with pure ether, corked, and shaken at intervals for several hours, after which it is allowed to subside. The contents of the flask are now filtered, and the precipitate washed with ether until the washings cease to blacken with ammonium sulphide. The filtrate and washings now contain the lead oleate only. This solution is now transferred to a tube (capacity = 250 c.c.) graduated from the bottom upwards, and furnished with a well-ground stopper and a stopcock which is placed at 50 c.c. from the bottom. About 20 c.c. of dilute hydrochloric acid (1 part acid, 2 parts water) are then added, the stopper replaced, the tube well shaken, and then set to subside. Lead chloride will form and settle, and a clear solution of oleic acid will rise to the top. A definite volume of the solution is then drawn off through the stopcock into a tared platinum dish, the ether evaporated, and after being dried at 100° C. the oleic acid is weighed and calculated on the whole bulk.

This method yields the most accurate results of any that have been hitherto proposed.

O'LEIN. $C_3H_5(C_{18}H_{33}O_2)_3$. Syn. TRIOLEIN, ELAIN; HUILE ABSOLUE, Fr. It is the principal component of olive and almond oils; it occurs also in most of the fixed oils and fats. By saponification it yields oleic acid, but it is less easily decomposed by alkalies than palmitin or stearin. It is one of the three glycerides of oleic acid obtained by Berthelot ('Ann. Chim. Phys.' [3], xli, 243); the other two, monolein and diolein, are oily liquids, which, on cooling, solidify to a crystalline mass; these are not important commercially.

Prep. 1. Olive oil or almond oil is digested for 24 hours with a quantity of caustic soda lye, only sufficient to saponify one half of the oil, and the undecomposed oily portion (olein) is then separated from the alkaline solution and newly formed stearin soap.

2. The saponified mixture of oil and alkali (see No. 1) is digested with proof spirit until all the soap is dissolved out, and the olein separates and floats on the surface; the latter, after repose, is decanted.

3. Almond or olive oil is agitated in a stout bottle with 7 or 8 times its weight of strong alcohol (sp. gr. 0·798) at nearly the boiling-point, until the whole is dissolved; the solution is next allowed to cool, after which the clear upper stratum is decanted from the stearin which has been deposited, and, after filtration, the spirit is removed by distillation at a gentle heat; by exposure at a very low temperature it deposits any remaining stearin, and then becomes pure.

4. Olein can be made artificially by heating pure glycerin with oleic acid in a closed vessel.

5. (Kerwyck's method.) From cold-pressed olive oil, by allowing it to stand 24 hours over a solution of caustic soda with frequent agitation. The soap produced is removed with dilute alcohol, and the olein is then decolourised by animal charcoal.

Prop., &c. The products of methods 2 and 3 have only a very slight yellow colour, but may be rendered quite limpid and colourless by digestion for 24 hours with a little pure, freshly burnt animal charcoal, and subsequent filtration. In this state the olein is devoid of taste and smell, is perfectly neutral to test-paper, does not in the slightest degree affect metallic bodies immersed in it, and does not thicken by exposure to the greatest cold. Olein is used by watchmakers for their fine work. Some years ago the product of the last formula was sold by a certain metropolitan house as 'watchmakers' oil,' at 1s. 6d. a drachm. Commercial olein is generally lard oil. The refined oleic acid of the stearin works also commonly passes under the same name. Olein burns well in lamps; but oleic acid does not do so unless when well refined, and unless the wick tube is so formed as to remain cool. See LARD OIL and OLEIC ACID.

OLEITE. Syn. RICINOL-SULPHONATE OF SODA. Chemically this substance is essentially ricinol-sulphonate of soda. The following descriptive outline of the method of its production is furnished by Mr Kilmer:—"It is prepared from castor oil by treating with sulphuric acid at a low temperature, when a compound of sulphuric and ricinoleic acid is formed. The free sulphuric acid being removed by washing, and any unchanged oil by ether, the resulting sulphoricinoleic acid is then neutralised by sodium hydrate, the finished product being a jelly-like liquid, with a little odour, acrid taste, soluble in water, alcohol, chloroform, and essential oils." Mr W. A. H. Naylor says, "This description is characterised by brevity and vagueness, while the latter part of it is unfortunately so worded as to invite, if not literally to compel, the deduction of an erroneous inference. In the absence of particular knowledge of the action of sulphuric acid upon certain oils, one would conclude that the product of the reaction between the castor oil and the acid—sulphoricin-oleic acid—was not sensibly soluble in water or in ether, while as a matter of fact the reverse is the case.

"My present object is simply to supply a working formula for the soda compound, one that I have used and can recommend. Take 1 lb. of castor oil, and add to it gradually, with continuous stirring, 2 oz. by weight of sulphuric acid (B. P.). This part of the process will occupy several hours, and should be timed so as to be finished towards the end of the working day. In the morning introduce in the same manner 1 oz. by weight of the acid, or a sufficiency. The point of finality is reached when the product remains clear, or, as is generally the case, is only faintly opalescent when diluted with about 40 times its volume of distilled water.

"The temperature of the mixed oil and acid may be allowed to reach 110° F., and may, without detriment, even rise to 120° F. When chemical combination is complete, the product is at once intimately mixed with 1½ times its weight of distilled water, and allowed to stand until separation into two distinct portions has ensued. The supernatant and oily layer is then removed and neutralised with a 10% aqueous solution of caustic soda. This soda compound is shaken up with 5 times its volume of proof spirit and set aside, when any free oil will rise to the surface. The lower and spirituous portion is evaporated on a water-bath to a thick jelly, the liquid being kept faintly alkaline by the addition of soda solution if necessary.

"The resulting product usually contains a small

proportion of sulphate of soda, but the quantity is insufficient to rank as a serious objection in view of the uses to which oleite is likely to be applied. If, however, in any case it is deemed necessary to eliminate traces of alkaline sulphate, the ricinol-sulphonate of soda must be treated with alcohol, in which the latter is soluble and the former practically insoluble.

"The free acid (ricinol-sulphonic acid) may be readily obtained by decomposing the soda compound with hydrochloric acid."

OLEOM'ETER. *Sys.* ELAÏOMETER, ELÆOMETER, OIL-BALANCE. A delicate areometer or hydrometer, so weighted and graduated as to adapt itself to the densities of the leading fixed oils. As the differences of the specific gravities of these substances are inconsiderable, to render it more susceptible the bulb of the instrument is proportionately large, and the tube or stem very narrow. The scale of the oleometer in general use (Gobby's) is divided into 50 degrees, and it floats at 0° or zero in pure poppy oil, at 38° or 38·5° in pure almond oil, and at 50° in pure olive oil. The standard temperature of the instruments made in this country is now 60°; those made on the Continent, 54·5° F. The oil must therefore be brought to this normal temperature, before testing it, by plunging the glass cylinder containing it into either hot or cold water, as the case may be; or a correction of the observed density must be made. The last is done by deducting 2 from the indication of the instrument for each degree of the thermometer above the normal temperature of the instrument, and adding 2 for every degree below it. Thus: suppose the temperature of the oil at the time of the experiment is 60°F., and the oleometer indicates 61°; then—

> 60·0° Actual temperature.
> 54·5 Normal temperature.
> ———
> 5·5 Difference.

> Indication of the oleometer . . 61·0
> The difference 5·5 × 2 = . . . 11·0
> ———
> Real density . . . . . 50·0

Suppose the temperature observed at the time of the experiment is 52°, and the oleometer indicates 45°; then—

> 54·5° Normal temperature.
> 52·0 Actual temperature.
> ———
> 2·5 Difference.

> Indication of the oleometer . . 45·0
> The difference 2·5 × 2 = . . . 5·0
> ———
> Real density . . . . . 50·0

The oil is, therefore, presumed to be pure. Excellent results at high temperatures have been obtained by using a hydrostatic balance made by

G. Westphal, of Celli, Hanover. The bulb, or plummet, suspended from the balance, is immersed in the test-tube (1¼ in. × 5 in.) containing the melted fat or oil. The desired temperature is obtained by placing the test-tube in a paraffin-bath. The latter is heated by an outer water-bath, and, when it arrives at a constant

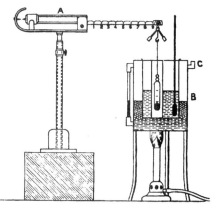

temperature (206°—208° F.), the weights on the arm of the balance are exactly adjusted, and the sp. gr. of the oil under examination may then be read off.

The diagram is taken from the 'Chemical News,' vol. xxxviii, p. 267.

A represents the Westphal balance with bulb immersed in the test-tube, containing the fat whose gravity it is desired to obtain. B is a vertical section of the bath, the outer casing of which is filled with water; the outlet and inlet for water is marked C.

The only precaution needful is to see that the plunger of the balance does not rest either on the bottom or the side of the test-tube.

The apparatus is described in detail in the 'Chemical News,' vol. xxxiv, p. 254. See HYDROMETER, OILS (Fixed), and SPECIFIC GRAVITY.

O'LEO-PHOSPHOR'IC ACID. An acid compound found by Frémy in the brain and nervous matter. The sodium salt occurs in almost all parts of the animal body, its quantity increasing with the age of the animal and differing in amount in different species (Valenciennes and Frémy, 'Ann. Ch. Phys.' [3], i, 172).

OLEO-RES'INS. The natural compounds of resin and essential oil forming the vegetable balsams and turpentines. Copaiba, Canada balsam, and Venice turpentine are examples. Certain extracts prepared with ether, as the fluid extracts of cubebs and pepper in the Ph. U. S., may be regarded as oleo-resins. See EXTRACT.

OLEO-SAC'CHARUM. *Sys.* ELÆOSACCHARUM. Sugar aromatised or medicated by being rubbed up with an essential oil. The oleo-sacchara of aniseed, caraway, cinnamon, peppermint, pennyroyal, and the other like essential oils are made by rubbing 15 to 20 drops of the respective oils with white sugar, 1 oz. The Ph. Græca, 1837, prescribes 1 part of oil to 20 parts of sugar. The Ph. Austr. 1836 and Ph. Bor. order the same proportions, or 3 drops of oil to the dr., and 24 drops to the oz., of powdered sugar. When intended for making extemporaneous distilled waters, 1 dr. of magnesia is a common addition. The oleosacchars of citrons, lemons, oranges, &c., are made from the peels, as follows:—After cleaning off any specks in the outer rind of the fruit, rub a large piece of loaf sugar on it until the yellow rind is completely removed. Those parts of the sugar which are impregnated with the essence are, from time to time, to be cut away with a knife, and put into an earthen pot. The whole being thus taken off, the sugared essence (oleosaccharum) is to be closely pressed down in the pot, tied over with bladder, and preserved in a cool place for use.

OLIB'ANUM. *Syn.* OLIBAN, INDIAN OLIBANUM, FRANKINCENSE. This gum-resin is of uncertain origin, but ascribed to several varieties of *Boswellia*, notably *B. sacra* and *B. Carteri*, natives of Asia and Africa. Olibanum occurs in the form of fragments, which are sometimes pale yellow, and at others of a reddish colour; these fragments have a splintery fracture, a mealy surface, a faintly balsamic odour, and a bitter taste. "It consists of about 4% or 5% of a volatile oil, 56% of a resinous acid, 30% to 36% of gum, and 6% of bassorin" (*Payen*). Incense, of which olibanum forms one of the ingredients, owes much of its pleasant balsamic odour when burning to its presence. It is also used in *pharmacy*.

OLIVE. *Syn.* OLEA, OLIVA, L. The *Olea europæa*, Linn., a native of the south of Europe. The unripe fruit is preserved in brine (SPANISH OLIVES, FRENCH OLIVES); the ripe fruit furnishes olive oil; the bark is bitter, astringent, and febrifuge, and has been used as a substitute for cinchona bark; it yields a gum-like substance (OLIVE GUM), which was formerly reported vulnerary, and contains olivine. The olive tree has in all ages been held in peculiar estimation. It is remarkable for yielding a fixed oil from the pericarp instead of from the seed.

OLIVINE. *Syn.* CHRYSOLITE. A crystallised double silicate of magnesia and iron, found in basaltic and volcanic rocks, and very frequently in masses of meteoric iron. It is sometimes yellow in colour, but mostly green.

Olivine (*Dr Landerer*). *Syn.* OLIVINA. *Prep.* Treat olive leaves with acidulated water, concentrate, precipitate with ammonia, re-dissolve the washed precipitate in a diluted acid, purify with animal charcoal, filter, and re-precipitate with ammonia.

OLLIVIER'S BISCUITS. See PATENT MEDICINES.

OMBENE. A native name for Kola nuts. They are also known as 'gura.'

OM'ELET. *Syn.* OMELETTE, Fr. A variety of pancake or fritter made of eggs and other ingredients. Omelets may contain bacon, ham, herbs, fish, shell-fish, cold meat, cold game, fruit, or anything else at hand at the pleasure of the cook. 'Spirit omelets' are made by pouring a

little brandy, rum, or whisky over them on serving them up, and setting it on fire for a moment just before placing the dish on the table. "Where is the man or woman cook but says they know how to make an omelette, and that to perfection? But this is rarely the case, It is related of Sarah, the Duchess of Marlborough, that no one could cook a 'fraise,' as it was then called, for the great duke but herself. The great point is, if in an iron pan, it should be very clean and free from damp, which sometimes comes out of the iron when placed on the fire. The best plan is to put it on the fire with a little fat, and let it get quite hot, or until the fat burns; remove it, and wipe it clean with a dry cloth, and then you will be able to make the omelette to perfection." (*Soyer*).

The following formula for a plain omelet is by the above culinary authority:—"Break four eggs into a basin, add ½ teaspoonful of salt, and ¼ do. of pepper, and beat them up well with a fork; put into the frying-pan 1½ oz. of butter, lard, or oil, place it on the fire, and, when hot, pour in the eggs, and keep on mixing them quickly with a spoon until they are delicately set; then let them slip to the edge of the pan, laying hold by the handle, and raising it slantways, which will give an elongated form to the omelette; turn in the edges, let it rest a moment to set, turn it over on to a dish, and serve." "It ought to be of a rich yellow colour, done to a nicety, and as light and delicate as possible." "2 table-spoonfuls of milk and 1 oz. of the crum of bread, cut into thin slices, may be added."

MIXED and FANCY OMELETS are made by simply dropping the ingredients, cut into dice or fragments, into the above. ANCHOVY, OYSTER, and SHRIMP OMELETS are generally prepared by placing a few spoonfuls of the respective sauces in the centre of each when nearly dressed.

ONABINE. An alkaloid obtained from the roots of the onabaio plant, closely related to the *Carissa Schimperi*. The poisoned arrows of the Somalis, East Africa, owe their deadly character to being steeped in the juice of the onabaio plant.

ONGUENT (de la Mère). A stimulant and digestive ointment, very popular in French pharmacy.

Prep. (P. Cod.) Black pitch, 1 part; butter, lard, litharge, suet, and yellow wax, 4 parts; olive oil, 8 parts.

ON'ION. *Syn.* CEPA, L. The bulb of *Allium cepa*. The onion is diuretic, expectorant, rubefacient, and stimulant. The juice, made into a syrup with sugar (SYRUPUS CEPÆ), has been given in chronic catarrh, diarrhœa, croup, dropsy, and calculus. Roasted and split open, onions have been applied as poultices to suppurating tumours, and applied to the pubes to relieve suppression of urine in children. As an article of diet for those undergoing severe bodily labour onions are by no means to be despised. They also possess antiscorbutic properties.

Onions, to Chop. "Few persons know how to chop onions properly. In the first place all the dry skin must be removed, then a thin slice off the top and bottom, or they will be bitter, then cut them into thin slices, dividing the onion, and cut crossways to form dice. If a very slight flavour is required and the onion is strong, like in the north of England—for it must be remembered that the further north you go the stronger the flavour of the root, and if French receipt books are exactly copied, it is no wonder complaints are made of the preponderance of the flavour of the onion; in which case, when chopped, put them in the corner of a napkin or cloth, wash them in water, squeeze them dry, then put them back on the board, and chop finer, or sometimes only rubbing the pan or the meat with the onion is quite sufficient" (*Soyer*).

O'NYX. A sub-species of quartz often wrought into small ornamental articles. Among jewellers any stone exhibiting layers of two or more colours, strongly contrasted, is called an ' onyx.' A regularly and richly banded agate of this class is much prized for cameos. The *sardonyx* of the ancients is a variety of onyx. It is so called from *sard*, or *sarda*, a rich dark red variety of carnelian, stripes of which in the stone alternate with white stripes. See GEMS.

O'OLITE. A variety of limestone composed of a number of collections of small round particles, bound together by a calcareous cement. The resemblance the mass is supposed to bear to the roe of a fish has caused oolite to be called *roe-stone*. When the grains are of larger dimensions it is called *peastone*. The little spherical bodies of which the stone is composed are mostly formed of concentric layers of carbonate of lime arranged round a grain of sand, a fragment of shell, or some other nucleus.

The building stones of Caen, Portland, and Bath are all oolitic limestones. When first quarried they are mostly soft, a circumstance which admits of their being easily sawn and carved. They harden by exposure to the air.

In *geology* the term ' oolite' has a wider signification, and is applied to an important division of the middle secondary rocks of England, underlying the chalk formation, and rich in interesting fossil remains.

O'PAL. A mineral allied to agate and chalcedony, but distinguished by its peculiar resinous lustre. The variety most admired as a gem is the precious or noble opal, which is remarkable for its beautiful play of colours. The finest opals come from Hungary, and although usually very small, will, if of good quality, realise as much as £5 each. Their value increases in a much greater ratio than their increase in size. They are also found in Saxony and in North America. The largest known specimen of an opal is that in the Imperial Cabinet at Vienna, the dimensions of which are five inches by two and a half. *Girasol*, *Cacholong*, *Hyalite*, and *Menilite* are varieties of opal. See GEMS and PASTES.

OPHTHAL'MIA. *Syn.* OPHTHALMITIS, L. Inflammation of the eye. The term is generally applied in the present day to the various forms of inflammation of the conjunctiva or cornea; inflammation of other parts of the eye being distinguished by special names, *e. g.* retinitis, sclerotitis, iritis, &c.

Conjunctivitis is perhaps the most common form of ophthalmia. There are three or four varieties of the disease, which, in their earlier stages, closely resemble one another. There is

heat, redness, swelling, and pain in the part, and a discharge which is often profuse, and which may at first be mucous, becoming afterwards muco-purulent or truly purulent in character. Sometimes the exudation is of such a character as to raise the conjunctiva from the sclerotic, and form a swollen ridge around the cornea. The redness is variable, and the pain is not severe except in those cases in which the subconjunctival swelling is very dense. As a rule the transparency of the cornea is not affected, and vision is not interfered with.

In mild cases conjunctivitis is an unimportant disorder, but certain secondary results may ensue which are of serious consequence. The inflammation in its acute stage may cause partial or even complete destruction of the cornea, resulting in impairment of vision or even blindness. When the disease is chronic there is often great enlargement of the papillæ of the mucous membrane lining the lids, and these act as hard points, irritating the cornea, and causing very great distress.

The chief varieties of conjunctivitis are—(1) *Infantile*, (2) *Simple* or *Catarrhal*, (3) *Contagious*, and (4) *Diphtheritic*.

1. Infantile Conjunctivitis. *Syn.* OPHTHALMIA NEONATORUM, L. This disease is probably due in most cases to inoculation in the act of birth with some of the secretions of the vagina. It shows itself about the third day after birth, and rapidly takes on the purulent form. There is great swelling of the lids, and a thick discharge which often glues them together. If neglected, blindness may result from damage to the cornea, but if properly treated it is not difficult of cure.

Treatment. The discharge must be carefully and frequently washed away, and an astringent lotion applied to the surface (Dr Brudenell Carter recommends a solution of 2 gr. of nitrate of silver in 1 oz. of distilled water) every four hours, or less frequently when improvement begins. Some simple ointment should be applied to the edges of the lids to prevent their sticking together. The food of the child should be carefully regulated, and "cod-liver oil combined in an emulsion with two-minim doses of liquor cinchonæ" may be given if the child be very feeble.

2. Simple or Catarrhal Conjunctivitis. This is usually due to cold or to chemical or mechanical irritation. The discharge is usually mucous, and does not tend to become purulent.

Treatment. Foreign bodies should first be carefully removed, and any chemical or other irritating material washed away by the free use of tepid water. In many cases rest is all that is required to complete the cure. If there be no foreign body the 2-grain solution of nitrate of silver may be used with the best results. The eye should be protected from cold, dust, and bright light.

3. Contagious Conjunctivitis. *Syn.* PURULENT OPHTHALMIA. In this form of the disease the discharge rapidly becomes purulent. It is especially common where there is over-crowding and general insanitary conditions of life. The conjunctiva is beset with granular semi-transparent bodies known as ' sago grains,' which are collections of

lymph-corpuscles. In the more severe forms there is great swelling of the lids and lifting of the ocular conjunctiva, producing the characteristic elevation round the cornea, which has a great tendency to slough.

Treatment. Dr Carter recommends in the worst cases that the eyelids should be everted and "carefully touched with a stick composed of 1 part of nitrate of silver fused with 4 parts of nitrate of potash. The caustic should be neutralised by a drop or two of solution of common salt, applied by means of a camel's-hair pencil, before the lid is suffered to return into contact with the cornea; and the cauterisation must be done carefully and with a light hand, so that the resulting eschar may include only the epithelium; for if the basement membrane be destroyed, there will be danger of subsequent adhesions between the eyelids and the eyeball. The cauterisation should be repeated about every 8 hours, or as soon as the eschar falls; and in the intervals, if the patient be awake, the conjunctiva should be gently syringed every hour with a weak alum lotion at a comfortable temperature."

The patient's health and strength must be sustained by a good diet and the use of tonic medicines. It may be necessary to give anodynes to allay the pain. Contagious ophthalmia is a very serious disorder, and if it breaks out in schools, barracks, or other places in which a number of persons are collected together, every possible means should be taken, by isolating the patients, extreme cleanliness, good food, &c., to prevent the spread of the disease. It is exceedingly common in the poorer quarters of the large towns in Southern and Eastern countries, and seems to be inseparably connected with dirt and over-crowding.

4. **Diphtheritic Conjunctivitis.** This is a rare disease in England, but has from time to time prevailed extensively in Berlin. The subjects of it are chiefly feeble and ill-fed children, and the cases are regarded by the German physicians as hopeless unless treated in their earliest stages.

Ophthalmia, Strumous or Scrofulous. This form of ophthalmia is generally met with in children of scrofulous habit of from four to ten or eleven years of age. Its most distinctive characteristic is the inability of the sufferer to bear the light, the effect of which is that the eyes are kept spasmodically partially closed. If the eyes are examined, a slight fulness of the vessels, usually stopping at the edge of the cornea, is observable; and about the line dividing the cornea and sclerotic coat small opaque pimples or pustules are visible. This variety of ophthalmia, being the outcome of a constitutional taint, is frequently very obstinate, and yields with difficulty to medical treatment, besides being very likely to reappear. It is not unfrequently accompanied with a troublesome cutaneous affection known as *Crusta lactea,* which occurs on the cheeks, and arises from the irritation caused by the flow down the cheeks of the acrid lachrymal secretion. The usual treatment consists in improving the general health and strength of the patient by means of tonics, such as quinine, quinine and iron, cod-liver oil, or syrup of iodide of iron. The diet should be nutritious and easy of digestion, and there should be no stint of fresh air.

O''PIATES. *Syn.* OPIATA, L. Preparations containing opium or active principles of opium. The word is often applied in a general sense to anodynes and soporifics. In French pharmacy the name is commonly used synonymously with confections, as in the following preparations:

ANTIDYSENTERIC OPIATE—*Quarin.* Purified opium, 4 gr.; ipecacuanha, ⅓ dr.; tormentilla, 1 dr.; syrup of whortleberries and conserve of red roses, of each, 6 dr.—*Dose.* A teaspoonful every hour.

ANTIHYSTERICAL OPIATE — *Trousseau* and *Reveil.* Powdered indigo, 1 oz.; white honey, 3 oz.—*Dose,* 1 table-spoonful daily, gradually increased until the whole is taken in a day. In hysteria, epilepsy, and nervous affections of an epileptic character.

BALSAMIC OPIATE — *Trousseau* and *Reveil.* Oleo-resin (balsam) of copaiba, 1 oz.; cubebs (in powder), 3 oz.; potassio-tartrate of iron, 2½ dr.; syrup of quince, q. s. In gleet.—*Dose,* 3 boluses the size of a nut, thrice daily.

CHARCOAL OPIATE—*Ratier.* Willow charcoal (recent), 1 oz.; prepared chalk, 1 dr.; powdered white sugar, 2 oz.; rose-water, q. s. to form an electuary. In diarrhœa and incipient cholera, in dysentery with fetid stools, and in gastralgia, flatulence, &c. By substituting calcined magnesia for chalk it becomes an excellent remedy for habitual constipation.

CUBEB OPIATE—*Deyeaux.* Powdered cubebs, 4 dr.; powdered camphor, 1 dr.; mix, and divide it into 18 powders.—*Dose.* One, 3 or 4 times daily, in gleet, painful and scalding micturition, &c.

O''PIUM. *Syn.* OPIUM (B. P., Ph. L., E., and D.), L. The juice inspissated by spontaneous evaporation, obtained by incision from the unripe capsules of the *Papaver somniferum,* grown in Asia Minor.

Hist. The milky juice of the poppy has been known from remote times. Theophrastus, who lived in the third century B. C., was acquainted with the substance. About the year 77 A. D. Dioscorides speaks of the juice of the poppy capsules as being more active than an extract of the entire plant; he also alludes to the adulteration of the drug with the juices of *Glaucium* and *Lactuca.* Celsus, in the first century, speaks of the drug as *Lacrima papaveris.* The Arabs, who call the drug *Afyun,* transmitted its use to the natives of the East. It is believed that the nepenthe of Homer was opium, for he speaks of it as "a pain-assuaging drink, a grief-allaying remedy, causing obliviousness of all evil."

The first mention of opium in connection with India is during the fifteenth century, when Pyras, in a letter to Manuel, King of Portugal, says, "It fetches a good price, that the kings and lords eat of it, and even the common people." During the early and middle ages opium was prepared in many forms, especially in the form of confections, the chief of these being Mithridates' elixir and theriaca.

Var. 1. EGYPTIAN; in roundish flattened lumps; inferior to Turkish opium. 2. ENGLISH; often equal to the best Smyrna. 3. FRENCH; resembles the last. 4. GERMAN; similar to English opium. 5. INDIAN:—*a.* BENARES; in large balls;—*b*, MALWA; in roundish flattened cakes, of 9 or 10 oz. in weight each;—*c*. PATNA; in balls or square cakes; inferior to Turkey opium. 6. LEVANT; same as Smyrna opium. 7. PERSIAN; in rolls or sticks, 6 × ½ inch; inferior; resembles hepatic aloes in appearance. 8. SMYRNA; in irregular, rounded, flattened pieces, varying in weight from 2 or 3 lbs. to only as many oz. It forms the best variety of Turkey opium, and is particularly rich in morphia. It is the only one adapted for the manufacture of the salts of morphia, as it contains on the average from 7% to 9% of that alkaloid, and usually yields about 12% to 12·5% of hydrochlorate of morphia, which is more than can be obtained from any other variety of opium. Of five kinds of Smyrna opium examined by Merk, the worst were found to yield 3% to 4% of morphia, and the best from 13% to 13·5%. 9. TURKEY; of which two varieties are known in commerce, viz. Constantinople opium and Levant or Smyrna opium, noticed above. Constantinople opium is generally in small, flattened, roundish cakes, 2 to 2½ inches in diameter, and covered with poppy leaves. It is more mucilaginous and less esteemed than Smyrna opium, from which it may be distinguished by the last being always covered with the reddish capsules of a species of *Rumex*.

The following account of the method of opium collection adopted in Asia Minor is extracted from a paper in the 'Pharmaceutical Journal,' contributed by Messrs Maltass and Wilkin (first series, vol. xiv). About the end of May the plants arrive at maturity, and the flowers expand. A few days after the petals have fallen the capsule is ready for incision.

This operation is performed in the afternoon of the day, and in the following manner:—A transverse incision is made with a knife in the lower part of the capsule, the incision being carried round until it arrives nearly at the part where it commenced; sometimes it is continued spirally to halfway down its starting-point. The greatest nicety is required to avoid cutting too deep, and penetrating the interior coating of the capsule, as this would cause the exuding milky juice to flow into the inside.

The following morning those engaged in collecting the opium lay a large poppy leaf on the palm of the left hand, and having a knife in the right hand, they scrape the opium which has exuded from the incision in each capsule, and then transfer it from the knife to the leaf, until a mass of sufficient size has been formed, when a second poppy leaf is placed over the top of the mass. If the dew has been heavy during the night the yield is greater, but the opium is dark in colour; if, on the contrary, there has been no dew, the yield is less, but the opium is of a lighter colour. A high wind is prejudicial, as the dust raised from the pulverised soil adheres to the exudation, and cannot be separated. The poppy capsules are cut but once, but as each plant will from one stem produce several branches, and each branch produce a flower, it is usual to pass over the field a second or a third time, to cut such capsules as were not ready at the first cutting. After the opium is collected it is dried in the shade.

The proceeds arising from the sale of the opium crop in British India form a considerable item in the revenues of our Eastern Empire; hence the poppy as the source of this valuable export, almost the whole of which goes to China, is very extensively cultivated in India.

The cultivation of opium in India appears to have existed as a monopoly so far back as the sixteenth century, for it is mentioned in the Ain-i-Akbari that the produce of this monopoly at that time amounted to 1000 chests. Under the British Government the cultivation of the poppy was at first in the hands of contractors, who held the monopoly on payment of a fixed sum; but in 1797 the Benares Opium Agency was established under a covenanted officer, and, with some changes in administrative details, the system of control and executive management is the same now as eighty years ago.

The Ghazipore Opium Agency is, however, under the Board of Revenue of Lower Bengal after a certain fashion, but there is reason to believe that the entire department will be brought directly under the Government of India.

The present opium agent (Mr H. Rivett Carnac) at Benares has done much to raise the pay and position of the department under him, and has systematised the details of cultivation, collection, and manufacture in an admirable manner. The poppy is cultivated under a system of advances, which are made to selected representatives of the cultivators known as *lumberdars*, who make over the whole of their produce to the State. The rate of advance is from 4 to 8 rupees per bigha (⅓ of an acre), according to the known productive capabilities of the village. These advances are made in September, and help the cultivator to pay his autumn instalment of rent and prepare his land for rearing the somewhat delicate opium poppy. During the cold weather strict supervision is exercised over the cultivation, and a large amount of information is collected by the officers of the department, who are also authorised to make advances for the construction of wells on very reasonable terms. In February the poppy is in flower, and then commences the collection of the petals, which are utilised as flower leaves (known as *chupatti*), in which the balls of opium exported to China are encased.

In March the drug is extracted by a rude method of incision, and collected in earthen vessels; and in April it is taken into central stations to be graded, weighed, paid for, and packed for despatch to the agency. The poppyseed is also collected, and forms an article of commerce, whilst in the neighbourhood of Ghazipore even the stalks are taken to the factory, and used as 'trash' for packing the chests of opium for China.

When the opium has been tested at Ghazipore, the cultivator's accounts are finally settled up at the rate of 5r. per seer (2 lbs.) for opium of 70° consistence. The elaborate processes by which the opium is finally prepared for the Chinese market are not of sufficient interest to call for detailed notice. The extent of the present operations of the factory may be conceived when we compare the 1,900 maunds produced in 1795 with the 51,000 maunds which is now the average amount treated annually, whilst in 1877 no less than 14,000 maunds were manufactured at the Ghazipore Agency. Besides the export opium, excise, or *abkari*, opium is prepared for local consumption. This consists of pure opium dried in the sun to the consistency of 90°, and divided into cakes of nominally a seer each.

Opium produced in Northern India is known as the Bengal opium, while the produce of Central India is known in commerce as the Malwa opium; the latter is of an inferior quality. Good opium contains about 4 per cent. of morphia and 8 per cent. of narcotine. Opium is generally used as an intoxicant in the form of pills, or smoked in a prepared form, known as the *madak* and *chandu*. Europeans generally have an aversion to opium eaters, as it is said to stupefy the persons using it. It is, however, asserted by many authorities that excess in opium is less injurious than excess in spirituous drink, and that the moderate use of opium after the age of forty prevents waste of tissue, prolongs life, and makes the system less liable to the influences of malarial and other poisons which vitiate the atmosphere of tropical countries. Opium is chiefly exported to China, and yields an annual revenue of over 10½ millions sterling, or more than a seventh part of the revenue of India.

The following figures give the gross revenue derived from opium in India in the two presidencies, with the changes in the last ten years, in millions of pounds, omitting the last three figures:

	Bengal.		Bombay.
	Revenue.	Charges.	Revenue.*
1871	£5644	£2012	£2401
1872	6898	1592	2355
1873	6069	1809	2615
1874	5583	1998	2742
1875	5608	2338	2954
1876	5922	2216	2550
1877	6174	2840	2948
1878	6433	2657	2750
1879	7006	1695	2398
1880	7042	2065	3141

* The charges in this Presidency range from £2000 to £3000 per annum.

The following shows the quantity and value of the exports of opium from India:

		Quantity.	Value.
1866–77	cwt.	130,775	£12,404,748
1877–78	„	126,789	12,374,355
1878–79	„	125,765	12,993,978
1879–80	„	144,638	14,323,314
1880–81	„	127,484	16,660,147
1881–82	„	123,918	12,432,142
1882–83	„	126,789	11,481,376

We may add to the foregoing the statistics of our imports and re-exports of opium for a series of years, which appear to be largely on the increase, for whilst the average annual receipts of opium in the last eight years have been 543,419 lbs., the annual imports in the preceding nine years were only 360,559 lbs.

	Imports.		Exports.
1866	lbs. 196,223	lbs.	124,154
1867	„ 273,522	„	148,519
1868	„ 322,309	„	123,965
1869	„ 219,495	„	107,279

	Imports.		Exports.
1870	lbs. 371,665	lbs.	150,414
1871	„ 591,466	„	307,399
1872	„ 356,211	„	308,273
1873	„ 400,469	„	250,577
1874	„ 514,372	„	306,374
1875	„ 536,291	„	298,556
1876	„ 400,303	„	237,700
1877	„ 607,362	„	251,267
1878	„ 558,840	„	314,380
1879	„ 572,411	„	232,388
1880	„ 400,374	„	195,510
1881	„ 793,146	„	401,883
1882	„ 478,624	„	191,316

The chief imports are from Turkey, the quantity of Turkish opium received in 1882 being 359,560 lbs., and from Persia 54,442 lbs. In some years a little comes in from Bombay and China. The exports of opium are principally made to the United States, Holland, Peru, British Guiana, and the West India Islands, and a little is sent to Hong-Kong, there being a considerable demand in Queensland and some other of the Australian colonies for the Chinese ('Chemist and Druggist').

The yield of morphia from East Indian opium is usually very small, a circumstance which Messrs Flückiger and Hanbury conceived to be partly due to the climate and partly to the defective method of cultivation.

He believed that the period, three or four weeks, during which the juice was allowed to remain in the wet state was much too long, and exercised a destructive influence on its constituents.

Since 1879 Bulgaria has given attention to the cultivation of the opium poppy, which is chiefly carried on in the districts of Kuestendil, Lowtscha, and Hatits. The opium from Kuestendil is in hemispherical cakes weighing from 120 to 300 grms. The opium is formed into balls, laid upon grape-vine leaves, and covered with the same leaves, so as to leave the sides free. The cakes have an exceptionally strong opium odour, externally brown, internally lighter, very dry, and show upon the surface a number of small tears. Calculated for dry opium, 100 parts yielded—ash, 3·63; morphine, 20·73; principles soluble in water, 47·64; insoluble in water, 31·73. Its morphine value is thus shown to be very high.

Opium from England, France, and Germany is occasionally met with, but never in considerable quantity. The cultivation of the opium poppy, however, in these three countries is chiefly carried on for the sake of the capsules, which are largely employed in medicine; and the oil extracted from the seed, which is highly valued and extensively employed by artists.

Pur. The opium of commerce is not unfrequently adulterated with extract of poppies, extract of lettuce, lactucarium, mucilage of gum tragacanth, dried leaves, starch, water, clay, sand, gravel, and other substances, in order to increase its weight. This fraud is readily detected by inspection, by chemical analysis, and the microscope; and indirectly, with the greatest certainty, by a simple assay of the sample of its morphia (morphiometry). This may be effected by one or other of the following methods:

1. (*Couerbe.*) Opium, 4 parts, and quicklime, 1 part, made into a milk with water, q. s., are boiled together, and the solution filtered whilst hot ; the filtrate is then saturated with dilute hydrochloric acid and the morphia precipitated by the addition of ammonia, any excess of the latter being expelled by heat ; the precipitate is then collected, dried, and weighed. If 100 gr. have been operated on, the given weight will represent (nearly) the percentage richness of the sample in morphia.

2. (*Guilliermond.*) 100 gr. of opium are triturated for some time in a mortar along with 4 times its weight of rectified spirit, and the tincture strained through linen, with expression, into a wide-mouthed bottle ; the marc is triturated a second time with about 3 times its weight of alcohol, and the tincture strained into the bottle as before ; to the mixed tincture is added a fl. dr. of liquor of ammonia, and the whole is agitated for a short time. In about 12 hours the morphia spontaneously separates, accompanied with some narcotina and meconate of ammonium ; the morphia covering the interior of the vessel with large, coloured, and gritty crystals, feeling like sand, and the narcotina crystallising in very light, small, white and pearly needles. These crystals are washed with water, either through a paper filter or linen, to free them from the meconate of ammonia which they contain ; after which the narcotina is separated from the morphia by decantation in water, which removes the narcotina, which is the lighter of the two. According to M. Mialhe, however, the morphia is more effectually removed by washing the crystals with 1 to 1½ fl. dr. of ether, by triturating the two together, when the morphia is left in an insoluble state, and may then be dried and weighed.

3. (B. Pharm., 1885.) Take of powdered opium, dried at 212° F. (100° C.), 240 gr. ; lime (freshly slaked), 60 gr. ; chloride of ammonium, 40 gr. ; rectified spirit, ether, and distilled water, of each, a sufficiency.

Triturate together the opium, lime, and 400 grain-measures of distilled water in a mortar until a uniform mixture results ; then add 1000 grain-measures of distilled water and stir occasionally during half an hour (meconates and sulphates precipitated as lime salts, morphine held in solution of ' lime-water '). Filter the mixture through a plaited filter about 3 inches in diameter into a wide-mouthed bottle or stoppered flask (having the capacity of about 6 fluid ounces, and marked at exactly 1040 grain-measures) until the filtrate reaches this mark. To the filtered liquid (representing 100 gr. of opium) add 110 grain-measures of rectified spirit and 500 grain-measures of ether, and shake the mixture (resins and fat dissolved out) ; then add the chloride of ammonium, shake well and frequently during half an hour, and set it aside for 12 hours (2AmCl + Ca2HO = CaCl₂ + 2AmHO ; morphine crystallises out). Counterbalance two small filters ; place one within the other in a small funnel, and decant the ethereal layer as completely as practicable upon the inner filter. Add 200 grain-measures of ether to the contents of the bottle and rotate it ; again decant the ethereal layer upon the filter, and afterwards wash the latter with 100 grain-measures of ether added slowly and in portions. Now let the filter dry in the air, and pour upon it the liquid in the bottle in portions, in such a way as to transfer the greater portion of the crystals to the filter. When the fluid has passed through the filter, wash the bottle and transfer the remaining crystals to the filter, with several small portions of distilled water, using not much more than 200 grain-measures in all, and distributing the portions evenly upon the filter. Allow the filter to drain, and dry it, first by pressing between sheets of bibulous paper, and afterwards at a temperature between 131° and 140° F. (55° and 60° C.), and, finally, at 194° to 212° F. (96° to 100° C.). Weigh the crystals in the inner filter, counterbalancing by the outer filter. The crystals should weigh 10 gr., or not less than 9½ and not more than 10½ gr., corresponding to about 10% of morphine in the dry powdered opium.

4. (Société de Pharmacie of Paris.) Mix 15 grms. of the sample to be tested with 9 grms. calcium hydrate and 150 c.c. water, rub them well together and shake for about half an hour. The mass is then thrown upon a filter, and 100 c.c. of the filtrate accurately measured and placed in a stoppered bottle. To this 20 c.c. of ether are added with constant shaking, then 6 grms. of ammonium chloride are dissolved in the solution, the whole agitated and allowed to remain at rest for two hours. After that period the ether may be drawn off and replaced by a second quantity of fresh ether, after which removal the precipitate of morphine which forms in the fluid may be collected on a tared filter, washed very carefully with distilled water, dried, and weighed. The weight of dry precipitate multiplied by 10 shows the percentage of morphine in the sample.

5. G. Loof, in the ' Apotheker Zeitung,' recommends the following method :—Five grms. of the finely ground opium are carefuly rubbed with water, and diluted to 78 c.c. At the end of one or two hours, during which the mixture is shaken frequently, 60·8 c.c., corresponding to four grms. of opium, are filtered off; 0·2 grm. of oxalic acid is added, and at the end of half an hour 5·2 c.c. of potash (1 : 2) are added, the mixture well shaken, and 16·5 c.c. of it filtered through a dry filter into an Erlenmeyer flask of 30 c.c. capacity, this quantity corresponding with one grm. of opium. Five grms. of ether, free from alcohol, are then added, and the mixture shaken briskly for ten minutes in the closed flask. The excess of the ether is volatilised by blowing a current of air into the flask, after which the separated morphine is collected upon a filter, and washed with water saturated with ether. The morphine on the dried filter may be transferred back to the portion remaining in the flask, and the weight of the whole obtained by drying until the weight of the flask and its contents remain constant. In the case of tincture of opium 50 c.c. are used for each experiment ; and in case it is opium extract that is required to be analysed, 2·5 grms. are a suitable quantity to take. The whole of the operations in the two latter cases are carried out exactly as described above.

6. (*Prollius.*) This is a very simple process,

and is said to give very exact results. It is as follows:—The opium is exhausted with 9 or 10 times its weight of spirit of 34 per cent. strength. Of the resulting tincture, 100 parts are well shaken with 5 parts of ether and 2 parts of solution of ammonia in a stoppered bottle, and then allowed to stand from 12 to 24 hours. The liquids separate slowly, and retain, partly in the ether, partly in the alcoholic liquid, the colouring matter, narcotine, and other crystallisable constituents of opium; while the morphia separates in crystals between the two layers, and finally sinks to the bottom. The fluid portion is decanted, the crystals are washed with diluted alcohol, dried, and weighed.

7. (*Teschemacher,* 'Chemical News,' xxxv, 47.) In employing the following method the use of alcohol to extract the morphia is avoided, and meconic acid is separated at an early stage, which prevents the formation of a basic meconate on precipitation of the morphia. Two special reagents are required for this process: the one prepared by mixing 1 part of ammonia, sp. gr. 0·880, with 20 parts of methylated alcohol, and digesting in this mixture a large excess of morphine; this, when filtered, is termed "*morphiated spirit;*" the other, *morphiated water,* is water saturated with excess of morphine, and contains 0·04 per cent. of this alkaloid. 1000 gr. of opium are macerated for 12 to 24 hours in about 4000 gr. of cold distilled water, together with 300 gr. of lead acetate, stirring the mixture from time to time. This separates the meconic acid as lead meconate, whilst the morphia is dissolved in the acetic acid set free.

After this maceration the opium may be readily ground in a mortar to a paste, and so much more cold distilled water added, raising the pestle and mortar with successive portions of it, as to fill with the mixture a measure=20,250 gr. of distilled water; experience has shown that the space occupied by the insoluble matters measures from 200 to 300 gr., so that the limit of possible error, by averaging and allowing 250 gr. for the insoluble portion, amounts to 0·05% in opium containing 10% of morphia. The mixture is to be filtered, and 15,000 measured gr.=750 gr. of opium, of the clear solution are to be evaporated to an extract on a water-bath, and this extract to be drenched with 3090 gr. of boiling alcohol or methylated spirit, and the whole digested with frequent stirring for about 10 minutes.

This separates the gum, &c., of the opium which is insoluble in alcohol, and so far frees the solution of morphia from impurity.

At this stage of the process it is well to get rid of the excess of lead salts, and to accomplish this sulphuric acid is preferable to sulphuretted hydrogen. So much diluted sulphuric acid as may be equal to 30 gr. of oil of vitriol will almost always be sufficient for this purpose, any excess of acid being converted into sulphate of ammonia by the subsequent addition of so much solution of ammonia as shall be equivalent to the 30 gr. of oil of vitriol, thus forming a salt but slightly soluble in the alcoholic solution. This mixture may now be transferred to a beaker and allowed to settle for 12 hours, after which it is to be filtered, and

the filter and insoluble residue thoroughly washed with alcohol or methylated spirits. This alcoholic filtrate is then distilled, or evaporated on a water-bath, to about 1000 gr.; and mixed, while still hot, with 400 gr. of solution of ammonia, sp. gr. 0·880, stirring rapidly and continuously for at least 20 minutes, whilst the beaker or evaporating dish should be cooled as rapidly as possible by immersion in an external vessel filled with cold water. The rapid and continuous stirring is most important, as the precipitation of the whole of the morphia *in fine powder* is thereby effected, instead of the granular or mammillated condition so frequently met with, and it thus permits of the easy and thorough separation of all the narcotine which may be mixed with the morphine. When the cooling of the mixture and precipitation of the morphia is thus attained, transfer it quickly and completely to a filter of sufficient capacity to hold the whole, and when the liquid portion has passed through, wash the remainder of the precipitated morphia adhering to the dish or beaker on to the filter, using for this purpose the morphiated spirit already described, and continuing the washing of the precipitate until it is completely freed from the mother-liquor. To do this effectually requires some little care; thus the morphia on the filter must be kept in a spongy condition and never allowed to cohere, which is easily effected by pouring the morphiated spirit round the edges of the filter, so as not to disturb the precipitate, which must not be permitted to drain or solidify until this washing is completed.

The precipitate is now to be washed from off the filter-paper with the morphiated water previously described, and digested therein for a few minutes, which removes some more colouring matter, together with any salts soluble in water, but insoluble in alcohol, which may have adhered to the precipitated morphia; then once more collect the precipitate on a filter, washing it with morphiated spirit, after this once with ether, and finally thrice or more with benzine; this completely frees it from narcotina, which is very soluble in benzine; morphia, on the contrary, being insoluble in this liquid. It now remains to drain and dry at a low temperature, say 100° F., the resulting pure and white morphia, the weight of which will indicate the amount of this alkaloid present in 750 gr. of the opium under examination.

Tests. These depend chiefly on the chemical and physical characters of morphia and meconic acid, the tests for which have been already noticed. In operating upon the contents of the stomach, or upon solid organs, in cases of suspected poisoning, the best method of proceeding is that already described under ALKALOID.

Another method is to boil the substances in water slightly acidulated with acetic acid, next to evaporate the solution to the consistence of a thick syrup, and then to treat it twice with boiling rectified spirit; the tincture thus obtained is to be filtered when cold, and again evaporated to the consistence of a syrup; it is now re-dissolved in distilled water, the filtrate treated with solution of subacetate of lead, and the precipitate of meconate of lead separated by filtration and carefully preserved. A current of sulphuretted hydrogen is then passed through the solution to precipitate excess of lead, and after again filtering it the liquid is evaporated, at first in a waterbath, and afterwards under the receiver of an airpump. The shapeless mass of crystals thus obtained present all the characters of morphine, if the substance examined contained opium. In the meantime the precipitate of meconate of lead is to be boiled with water acidulated with sulphuric acid, and the insoluble sulphate of lead separated by filtration; the filtered liquid, by evaporation, furnishes meconic acid, either under the form of crystals or an amorphous powder, the solution of which precipitates ferric salts of a deep bloodred colour.

The following are additional tests to those already noticed:

1. From the peculiar odour of opium, often perceptible when the drug has been taken only in very small quantities.

2. A solution containing crude opium is turned of a deep red colour, or if coloured, it is turned of a reddish brown, and is darkened by tincture of ferric chloride.

3. (*Hare.*) A portion of the suspected liquid is poured into a beaker glass, and a few drops of solution of acetate of lead are added to it; the whole is stirred frequently for 10 or 12 hours, and then allowed to settle, after which the supernatant liquid is decanted; 20 or 30 drops each of dilute sulphuric acid and solution of ferric sulphate are next poured on the precipitate (meconate of lead), when a deep and beautiful red colour will be developed if the original liquid contained opium.

4. (*Dr Rieget.*) The suspected substance is mixed with some potassa, and is then agitated with ether; a strip of white unsized paper is next several times moistened with the solution, and when dry it is re-moistened with hydrochloric acid, and exposed to the steam of hot water. The paper assumes a red colour, more or less deep, if opium is present.

Uses, &c. Opium is one of the most valuable substances employed in medicine. Its general uses are to lessen pain; produce sleep; to lessen irritation in various organs. In small doses it acts as a powerful and diffusible stimulant; in somewhat larger ones it is narcotic, and in excessive doses it proves an active narcotic poison. It is also anodyne, antispasmodic, diaphoretic, soporific, and sedative, its peculiar action being greatly modified by the dose and the condition of the patient. Its action as a stimulant is followed by sedative effects, which are, in general, much more marked than could be expected from the degree of previous excitement it induces. It is employed to fulfil a variety of indications—to procure sleep, to lull pain, allay irritation, check morbid discharges, alleviate cough and spasm, &c. It also, when judiciously administered, renders the body less susceptible of external impressions, as those of cold, contagion, &c.; but it is injurious when the pulse is high, the heat of the body above the natural standard, and the skin dry, or when there is a disposition to local inflammation or congestion. In peritonitis it is valuable both taken internally and applied externally. When applied externally, in the form of frictions, liniments, ointments, &c., it is absorbed, and produces similar effects to those produced by swallowing it, but in this way it requires to be used in larger quantities.—*Dose.* As a stimulant, ¼ gr., every 2 or 3 hours; as an anodyne and antispasmodic, ½ to 1 gr.; as a soporific, ½ to 2 gr.; in violent spasms, neuralgia, acute rheumatism, &c., 2 to 4 gr., increased in delirium tremens, hydrophobia, mania, tetanus, &c., to several times that quantity, according to circumstances.

The use of opium as a stimulant and intoxicant is common among the nations of the East. The Turks chew it, and the Chinese smoke a watery extract of it, under the name of 'chundoo,' the preparation of which from the crude article constitutes a special business. Messrs Flückiger and Hanbury, in their 'Pharmacographia,' published in 1874, say this particular business is not confined to the Celestials, since, in 1870, a British firm at Amoy opened an establishment for preparing chundoo for the consumption of the Chinese in California and Australia.

The qualities most valued by the Chinese in opium are its fulness and peculiarity of aroma, and its degree of solubility. The amount of morphine it contains is a secondary consideration.

The practice of opium-smoking yearly increases in China. It appears to be openly followed, and no odium attaches to it, provided it is not carried so far as to intoxicate or incapacitate the smoker.

In the larger cities and towns adjacent to Amoy the proportion of opium-smokers, according to Mr Hughes, Commissioner of Customs at Amoy, is estimated at from 15% to 20% of the adult population.

In the country districts 5% to 10% of the population are believed to be opium-smokers.

In many of the western States of America the practice has become so notoriously common that in 1872 the Legislature of Kentucky passed a bill by which any person who, through the excessive use of opium, is incapacitated from managing himself or his affairs, may, upon the affidavit of two citizens, be confined in an asylum, and subjected to the same restraint as lunatics or habitual drunkards (*Blythe*).

Mr J. Calvert, of San Francisco, gives the following account of how the Chinese prepare smoking opium:—The essential apparatus employed consists of two charcoal fire-clay furnaces, about 15 in. high and of about the same width, open on three sides; several brass pans, a brass ladle, and several tin ones; a large spoon for skimming; a gridiron, two pairs of pincers for lifting the pans, some fibre brushes, buckets, basket strainers, muslin for straining, heavy sticks to be used as pestles, several spatulas about a foot long and 3 in. wide made of oak or ash, and a steel-bladed scraper. Using Turkey opium, the balls are first steeped in water to soften the surface, from which the leaves and grit are then removed by the hands. The opium is then gently heated in water, being constantly kneaded with the wooden pestle until it is homogeneous, when it is uniformly spread over the inner surface of the brass pan, and the heat continued until the opium is so solid that the pan can be turned up. The direct heat of a small fire is now applied to the opium until it is hard. The drying process is not yet complete, however, for the opium is now scraped off the pan, and in thin layers dried upon the gridiron until it is crisp and crusty. The dried opium is now steeped in warm water overnight, the infusion strained off in the morning, and the residue again treated with warm water. These infusions are used for the extract, subsequent washings being employed for the extraction of the next batch. There seems to be no precise rule as to the quantity of water for making the infusion, the crusts are merely covered. The infusion is then mixed with some egg albumen, and a part of it is placed in the largest of the brass pans over the naked charcoal fire, and is heated, skimmed, and boiled constantly; fresh portions of warm infusion containing albumen are added from time to time as the bulk diminishes. When all the infusion has been added, and the evaporation has proceeded as far as is considered to be necessary, the pan is removed from the fire, and the extract constantly stirred by means of a wooden spatula in a current of air produced by fanning until cool and uniformly mixed. The yield of extract varies, 18 lbs. of the first quality Turkey opium generally giving about 10 lbs. of this extract. Mr Calvert states—contrary to the common belief—that there is no appreciable difference in the yield of morphine when opium has gone through this barbarous process. Whatever changes may take place among the other proximate constituents is not known, but the natural morphine salts, protected by extractive, are not decomposed, or only to a very small extent, by such a heat as is necessary for the desired alteration of the valueless or inert matters contained in opium.

Of late years opium-eating and laudanum-taking have, unfortunately, been greatly on the increase in this country, and the employment of this drug as a soporific for infants and young children has become so general amongst the poor and dissipated as to call for the interference of the Legislature.

According to Dr Chevers the practices of opium-eating and opium-smoking are very common among the natives of India. The same authority also states that in that country a large number of female infants are purposely poisoned by it, by introducing the drug into the child's mouth, and in various other ways.

The first effect of opium as a stimulant is to excite the mental powers and to elevate those faculties proper to man; but its habitual use impairs the digestive organs, induces constipation, and gradually lessens the energy of both the mind and body. In excessive quantities it destroys the memory, induces fatuity and a state of wretchedness and misery, which after a few years is mostly cut short by a premature death. In this respect the effects of the excessive use of opium closely resemble those of fermented liquors.

Opium is somewhat uncertain in its action; some persons being able, sometimes from idiosyncrasy, but more frequently from previous indulgence in it, to take a much larger dose than others. The smallest quantity which is said to have proved fatal with an adult is 4 gr. of the crude opium. In contrast with this may be quoted the statement of Dr Garrod, of a young man who not only swallowed 60 gr. of Smyrna opium night and morning, but very frequently, in addition to this, 1 to 1½ oz. of laudanum during the day.

Dr Chapman also cites the case of a patient to whom a wine-glass of laudanum had to be administered several times in 24 hours.

Pois.—Symptoms. Headache; drowsiness; stupor; frightful reveries; vertigo; contracted pupil (generally); scanty urine; pruritus or dry itching of the skin, often accompanied by a papular eruption; thirst; dryness of mouth and throat; weak and low pulse; vomiting; respiration generally natural. Sometimes the drowsiness or sleep is calm and peaceful.—*Ant., &c.* Vomiting must be induced as soon as possible, by means of a strong emetic and tickling the fauces. If this does not succeed the stomach-pump should be applied. The emetic may consist of a ½ dr. of sulphate of zinc dissolved in ½ pint of warm water, of which one third should be taken at once, and the remainder at the rate of a wineglassful every 5 or 10 minutes, until vomiting commences. When there is much drowsiness or stupor, 1 or 2 fl. dr. of tincture of capsicum will be found a useful addition; or one of the formulæ for emetic draughts may be taken instead. Infusion of galls, cinchona, or oak-bark should be freely administered before the emetic, and water soured with vinegar and lemon juice after the stomach has been well cleared out. To rouse the system, spirit-and-water or strong coffee or pure caffeine may be given. To keep the sufferer awake, rough friction should be applied to the skin, an upright posture preserved, and walking exercise enforced if necessary. When this is ineffectual, cold water may be dashed over the chest, head, and spine, or mild shocks of electricity may be had recourse to. To allow the sufferer to sleep is to abandon him to destruction. Bleeding may be subsequently necessary in plethoric habits, or in threatened congestion. The costiveness that accompanies convalescence may be best met by aromatic aperients, and the general tone of the habit restored by stimulating tonics and the shower-bath. The smallest fatal dose of opium in the case of an adult within our recollection was 4½ gr. Children are much more susceptible of the action of opium than of other medicines, and hence the dose of it for them must be diminished considerably below that indicated by the common method of calculation depending on the age. See DOSES, &c.

According to Mulder, 100 parts of ordinary Smyrna opium contain—

Morphine	.	.	.	10·842
Codein	.	.	.	·678
Narcotine	.	.	.	6·808
Narceine	.	.	.	0·662
Meconine	.	.	.	·804
Meconic acid	.	.	.	5·154
Resin	.	.	.	3·582
Gummy matter	.	.	.	26·342
Mucus	.	.	.	19·086
Fatty matter	.	.	.	2·166
Caoutchouc	.	.	.	6·012
Water	.	.	.	9·846
Matter undetermined and loss	.			2·118
				100·

The sp. gr. of Smyrna opium is 1·336.

Concluding Remarks. Opium is a very complicated substance, and contains a number of alkaloids and other proximate vegetable principles, besides a certain portion of saline matter. The substances already detected in it are caoutchouc, codein, fatty matter, lignin, meconic, acetic, sulphuric acids, meconia, morphia, narceia, narcotia, odorous matter, opiania, papaveria, pseudo-morphia (?), porphyroxin, resin, saline matter, &c. It is doubtful, however, whether some of these substances are not generated from other principles existing in opium during the process adopted to obtain them.

The following chart, showing the natural alkaloids of opium and a few of their artificial derivatives, is taken from the ' Pharmacographia ' of Messrs Flückiger and Hanbury :

Discovered by		C.	H.	N.	O.
Wöhler, 1844 . .	COTARNINE . . . Formed by oxidising narcotine, soluble in water.	12	13	1	3
Hesse, 1871 . . .	1. Hydrocotarnine . . Crystallisable, alkaline, volatile at 100°.	12	15	1	3
Matthiesen & Wright, 1869	APOMORPHINE . . . From morphine by hydrochloric acid, colourless, amorphous, turning green by exposure to air, emetic.	17	17	1	2
Wright, 1871 . .	DESOXYMORPHINE . . .	17	19	1	2
Sertürner, 1816 . .	2. Morphine . . . Crystallisable, alkaline, levogyre.	17	19	1	3
Pelletier & Thibouméry, 1835	3. Pseudo-morphine . . . Crystallises with H_2O, does not unite even with acetic acid.	17	19	1	4
Matthiesen & Burnside, 1871	APOCODEINE . . . From codeine by chloride of zinc; amorphous, emetic.	18	19	1	2
Wright, 1871 . .	DESOXYCODEINE . . .	18	21	1	2
Robiquet, 1832 . .	4. Codeine . . . Crystallisable, alkaline, soluble in water.	18	21	1	3
Matthiesen & Foster, 1868	NORNARCOTINE . . .	19	17	1	7
Thibouméry, 1835 .	5. Thebaine . . . Crystallisable, alkaline, isomeric with buxine.	19	21	1	3
Hesse, 1870 . .	THEBENINE . . .	19	21	1	3
Hesse, 1870 . .	THEBAICINE . . . From thebaine or thebenine by hydrochloric acid.	19	21	1	3
Hesse, 1871 . . .	6. Protopine . . . Crystallisable, alkaline.	20	19	1	5
Matthiesen & Foster, 1868	METHYLNORNARCOTINE . . .	20	19	1	7
Hesse, 1871 . . .	DEUTEROPINE . . . Not yet isolated.	20	21	1	5
Hesse, 1870 . .	7. Laudanine . . . An alkaloid which, as well as its salts, forms large crystals; turns orange by hydrochloric acid.	20	25	1	4
Hesse, 1870 . .	8. Codamine . . . Crystallisable, alkaline, can be sublimed; becomes green by nitric acid.	20	25	1	4
Merck, 1848 . .	9. Papaverine . . . Crystallisable, also its hydrochlorate; sulphate in sulphuric acid precipitated by water.	21	21	1	4
Hesse, 1865 . .	10. Rhœadine . . . Crystallisable, not distinctly alkaline, can be sublimed; occurs also in *Papaver rhœas*.	21	21	1	6
Hesse, 1865 . .	RHŒAGENINE . . . From rhœadine; crystallisable, alkaline.	21	21	1	6

Discovered by		C.	H.	N.	O.
Armstrong, 1871. . . . . DIMETHYLNORNARCOTINE . . .		21	21	1	7
Hesse, 1870 . . . . 11. Meconidine . . . . Amorphous, alkaline, melts at 58°, not stable, the salts also easily altered.		21	23	1	4
T. & H. Smith, 1864 . . . . . 12. Cryptopine . . . . Crystallisable, alkaline, salts tend to gelatinise, hydrochlorate crystallises in tufts.		21	23	1	5
Hesse, 1871 . . . . 13. Laudanosine . . . . Crystallisable, alkaline.		21	27	1	4
Derosne, 1803 . . . . . . 14. Narcotine . . . . Crystallisable, not alkaline, salts not stable.		22	23	1	7
Hesse, 1870 . . . . . . 15. Lanthopine . . . . Microscopic crystals, not alkaline, sparingly soluble in hot or cold spirit of wine, ether, or benzol.		23	25	1	4
Pelletier, 1832 . . . . . . 16. Narceine . . . . Crystallisable (as a hydrate), readily soluble in boiling water, or in alkalies, levogyre.		23	29	1	9

₄ The following preparations, once famous, are now nearly obsolete in this country. Those that are made with cold water or by fermentation are supposed to be milder than crude opium, and in this respect to be similar to 'BLACK DROP.'

Opium, Homberg's. Opium exhausted by repeated coction in 10 or 12 times its weight of water, and the mixed liquors evaporated to one third, and kept boiling for two or three days, adding water from time to time, and then straining and evaporating to a pilular consistence. BAUMÉ'S PURIFIED OPIUM is similar.

Opium, Launcelotte's. Prep. Opium, 1 lb.; quince juice, 1 gall.; pure potassa, 1 oz.; sugar, 4 oz.; ferment for some time, evaporate to a syrup, digest in rectified spirit, filter and evaporate the tincture.

Opium, Let'tuce. Lactucarium.

Opium, Newman's. Infusion of opium, strained, mixed with a little sugar, and fermented for some months in a warm place; and, lastly, strained and evaporated to an extract, or preserved in the liquid form.

Opium, Powell's. Opium exhausted by coction with water, the residuum treated with spirit of wine, and the mixed tincture and decoction evaporated to an extract.

Opium, Pu"rified. Syn. OPIUM PURIFICATUM, L. The purified opium of the old pharmacy is now represented by the aqueous extract of the Pharmacopœias. (See EXTRACT.) Formerly, picked opium, beaten to a pilular consistence, with the addition of a little water or proof spirit, was called 'SOFT PURIFIED OPIUM' (OPIUM PURIFICATUM MOLLE); and picked opium, dried in a water-bath until brittle enough for powdering, was called 'HARD PURIFIED OPIUM' (O. P. DURUM). CORNETTE'S and JOSSE'S PURIFIED OPIUM are similar to the extract of Ph. L.

Opium, Quercetan's. Vinegar of opium evaporated to an extract.

Opium, Strained. Syn. EXTRACTUM THEBAICUM, OPIUM COLATUM, OPIUM PURIFICATUM,

L. Opium dissolved or softened in an equal weight of water, passed through canvas, and evaporated to the consistence of an extract. It is now superseded by the aqueous extract.

Opium, Tor'refied. Syn. ROASTED OPIUM; OPIUM TORREFACTUM, L. Opium dried, cut into thin slices, and roasted on an iron plate, at a low heat, as long as it emits vapours, care being taken not to burn it.

OPODEL'DOC. 1. See LINIMENT OF SOAP.

2. (STEER'S OPODELDOC.) This, which differs from common opodeldoc chiefly in containing more soap, is prepared as follows:

a. White Castile soap (cut very small), 2 lbs.; camphor, 5 oz.; oil of rosemary, 1 oz.; oil of origanum, 2 oz.; rectified spirit, 1 gall.; mix, and digest in strong bottle (closely corked), by the heat of a water-bath, until solution is complete; when the liquid has considerably cooled, add of liquor of ammonia, 11 oz., and immediately put it into wide-mouthed bottles (Steer's), cork them close, and tie them over with bladder. Very fine, solid, and transparent when cold.

b. Soap, 4 oz.; camphor, 1 oz.; oil of rosemary and origanum, of each, 1 dr.; rectified spirit, 1 pint; liquor of ammonia, 1½ fl. oz. Mix.

c. (Phil. Coll. of Pharm.) White soap, 28 oz.; camphor, 8 oz.; rectified spirit, 6½ pints; dissolve, suffer the impurities to subside, add of liquor of ammonia, 4 fl. oz.; oils of rosemary and horsemint, of each, 1 fl. oz.; and pour it into phials, as before.

OPOPONAX. A resinous substance obtained from the roots of the Opoponax chironium. It occurs in lumps of a reddish-yellow or brown colour, and has a waxy fracture. It has a powerful odour, which somewhat resembles garlic, and a bitter taste.

Opoponax is only partially soluble in alcohol. According to Payen it consists of a little volatile oil, a resin that melts at 100° C., gum, inorganic and organic salts, and mechanical admixtures. It is used in French pharmacy, and was held in great esteem by Hippocrates, Theophrastus, and Dioscorides, all of whom employed it therapeutically.

ORANGE. *Syn.* AURANTIUM, L. The common SWEET ORANGE is the fruit of *Citrus aurantium*. The SEVILLE or BITTER ORANGE is produced by *Citrus vulgaris* or *bigaradia*.

Oranges are probably about the most wholesome and useful of all the subacid fruits. Their juice differs from that of lemons chiefly in containing less citric acid and more sugar. In their general properties the two are nearly similar.

Oranges are imported into England from many countries; the best come from Denia, on the east coast of Spain. Malta oranges are almost the latest in the market, arriving in spring. Orange-growing in Florida has become a profitable business on a large scale. The fruit from the Canary Islands commands, perhaps, the best price. The small Tangerine orange has suffered in quality in recent years from the efforts of the cultivators to increase the yield of the tree.

FACTITIOUS ORANGE JUICE is made by dissolving citric acid, 1 oz., and carbonate of potassa, 1 dr., in water, 1 quart, and digesting the solution on the peel of half an orange until sufficiently flavoured; Narbonne honey or white sugar is then added to impart the necessary sweetness. Instead of orange peel, 5 or 6 drops of oil of orange peel, with ½ fl. oz. of tincture of orange peel may be used.

ORANGE PEEL (CORTEX AURANTII) is an agreeable stomachic, bitter tonic, especially useful as an adjunct to more active medicines. That ordered to be used in medicine is the exterior (yellow) rind of the *Citrus bigaradia*, or bitter orange, dried in the months of February, March, and April. See CANDYING, INFUSION, ISSUE PEAS, OILS (Volatile), &c.

ORANGEADE′. *Syn.* ORANGE SHERBET. *Prep.* 1. Juice of 4 oranges, thin peel of 1 orange, lump sugar, 4 oz.; boiling water, 3 pints.

2. Juice and peel of 1 large orange, citric acid, 15 gr.; sugar, 3 oz.; boiling water, 1 quart.

Orangeade, Effervescing or Aërated. *Prep.* 1. Mix 1 lb. of syrup of orange peel, a gallon of water, and 1 oz. of citric acid, and charge it strongly with carbonic acid gas with a machine.

2. Syrup of orange juice, ¾ fl. oz.; aërated water, ½ pint.

3. Simple syrup, ½ fl. oz.; tincture of orange peel, ½ dr.; citric acid, 1 scruple; fill the bottle with aërated water.

4. Put into a soda-water bottle ½ oz. to 1 oz. of syrup of orange peel, 30 gr. of bicarbonate of potash, 8 oz. of water, and, lastly, 40 gr. of citric acid, in crystals, and cork immediately.

5. Put into each bottle 2 or 3 dr. of sugar, 2 drops of oil of orange peel, 30 gr. of bicarbonate of potash, or 25 gr. of bicarbonate of soda water to fill the bottle, and 40 gr. of citric acid, as before.

ORANGE CHROME. *Prep.* 1. From a solution of chromate of potash and diacetate of lead, as chrome yellow.

2. From chrome yellow or chromate of lead, by acting on it with a weak alkaline lye until sufficiently darkened. Used as a pigment.

ORANGE DYES. These are produced from mixtures of red and yellow dyes in various proportions; or by passing the cloth, previously dyed yellow, through a weak red bath. 1. A very good fugitive orange may be given with annotta, by passing the goods through a solution made with equal parts of annotta and pearlash; or, still better, through a bath made of 1 part of annotta, dissolved in a lye of 1 part each of lime and pearlash and 2 parts of soda. The shade may be reddened by passing the dyed goods through water acidulated with vinegar, lemon juice, or citric acid, or through a solution of alum. The goods are sometimes passed through a weak alum mordant before immersion in the dye-bath.

2. (For COTTON.) For 40 lbs. 2½ lbs. annotta, 24 lbs. of bark, 3 quarts of chloride of tin. Boil the annotta, put off the boil, enter and rinse until it has a good body. Then wring out, wash well, wring again, and shake out. Next, in a clean boiler, boil the bark in a bag for a quarter of an hour, add the chloride of tin, and enter, rinse till the required shade is got.

3. (For SILK.) For 10 yards. Annotta, 1¾ oz.; bark, 1¼ oz.; chloride of tin, 1½ oz. Give a good body of annotta at 212° F.; wash in one water, then top with the bark and chloride of tin.

4. (For WOOL.) For 50 lbs. Boil 10 lbs. of bark and 1½ lbs. of cochineal; add 2 lbs. of tartar, 2½ quarts of yellow spirits. Enter at 200° F.; boil 30 minutes. See ANNOTTA, DYEING, &c.

The following list includes the more recent orange dyes.

Orange-alizarin (for wool). Mordant with from 5% to 8% of stannous chloride to the same quantity of cream of tartar. Dye with 10% of alizarin (20%).

Orange-α-naphthol. *Syn.* TROPÆOLIN 000 No. 1, ORANGE No. 1 (*Poirrier*). Dyes a reddish-orange shade.

Orange-β-naphthol. *Syn.* TROPÆOLIN 000 No. 2, ORANGE No. 2 (*Basf, Poirrier*), ORANGE EXTRA (*L. Casella & Co.*), &c. Dyes a bright reddish-orange shade.

Orange-dimethylaniline. *Syn.* HELIANTHIN, GOLD ORANGE, ORANGE III, &c. Will dye cotton, wool, and silk; it is not fast to washing on cotton.

Orange-diphenylamine orange. *Syn.* TROPÆOLIN 00, ORANGE IV, ORANGE N., &c. The presence of free acid induces a deeper tint.

Orange-palatine (*Basf*). The ammonium salt of tetra-nitro-γ-diphenol. It dyes wool and silk in a bath acidulated with sulphuric or acetic acid (*Hummel*).

ORANGE RED. *Syn.* SANDIX. From white-lead, by calcination, in a nearly similar manner to that by which red-lead is prepared from the protoxide. Brighter than red-lead. Used wholly as a pigment.

OR′ANGERY. The gallery, building, or enclosure in a garden in which orange trees are preserved or cultivated, to shield them from the effects of the external winter, or to assist their growth by artificial heat.

OR′CEIN. $C_7H_7NO_3$. *Syn.* LICHEN LAKE. A brownish-red powder, obtained by dissolving orcin in ammonia, exposing the solution to the air, and then precipitating with dilute acetic acid. It is sparingly soluble in water, but dissolves freely in alcohol, in solutions of ammonia and the fixed alkalies with the production of a rich purple or violet colour; such solutions are reddened by acids. It constitutes the leading tinctorial ingredient in ARCHIL, CUDBEAR, and LITMUS. (See *below*.)

OR'CHARD. See CIDER.

OR'CHIL. See ARCHIL.

Orchella Weeds. The species of *Roccella*, used in the manufacture of orchil, archil, or cudbear. They are, according to Pereira—

Angola orchella, *Roccella fuciformis*.

Madagascar „	„	„
Mauritius „	„	„
Canary „	„	*tinctoria.*
Cape de Verde „	„	„
Madeira „	„	„ and *fuciformis.*
Linia „ (large and round)	„	„
Linia „ (small and flat)	„	*fuciformis.*
Cape of Good Hope „	„	*hypomecha.*
Barbary „	„	*tinctoria.*
Corsican and Sardinian „	„	„

OR'CIN. $C_7H_8O_2$. *Syn.* ORCINOL, DIHYDRO-TOLUENE. The general product of the decomposition of the acids obtained from the tinctorial lichens under the influence of heat or the alkaline earths.

Prep. 1. The powdered lichen is treated with boiling alcohol, the tincture whilst hot, and again after it has become cold; the alcohol is then removed by distillation, and the remainder evaporated to the consistence of syrup; this is redissolved in water, the solution again filtered and evaporated to a syrup; it is then set aside some days in a cool place, and the crystals of orcin which form are collected, and dried by pressure in bibulous paper. Impure.

2. Leanoric or orsellinic acid (impure will do) is boiled in baryta water, and the excess of baryta is precipitated by carbonic acid; the filtered liquid is then evaporated to a small bulk, and set aside to crystallise, as before. Lime and water may be used instead of baryta.

3. By the action of fused potash or aloes.

4. Artificially from toluene by converting it into orthochlorotoluene-sulphuric acid, fusing this with excess of potash.

Prop., &c. Large, square, prismatic crystals; colourless; sweet; very soluble in ether, water, and alcohol; melt to a syrupy liquid, and then distil unchanged. Alkalies decompose it; when exposed to the air it gradually reddens. (See *above*.)

OREIDE. A variety of brass, in appearance very much like gold. The following, according to MM. Meurier and Valient, its inventors, is the composition of this alloy:—Copper, 100 parts; zinc, 17 parts; magnesia, 6 parts; sal-ammoniac, 3·6 parts; quicklime, 1·8 parts; tartar of commerce, 9 parts. The copper being first melted, the other ingredients are added by small portions at a time, the whole being kept in fusion for about half an hour, and well skimmed. The oreide has

a fine grain, is malleable, is capable of being brilliantly polished, and has its lustre restored by the use of acidulated water.

It is used for the cases of cheap watches and for ornamental castings. It resembles gold in colour, and forms a good basis for electro-plating with gold. Sometimes tin is substituted for the zinc.

OR'ELLIN. A yellow colouring matter contained together with bixin in annotta. It is soluble in water and in alcohol, slightly soluble in ether, and dyes alumed goods yellow. Also the name sometimes given to purified annotta. The commercial annotta is dissolved in an alkaline solution, either caustic or carbonated, and then precipitated by an acid. See ANNOTTA.

ORES. The mineral bodies from which metals may be obtained. The processes adopted for this purpose constitute OPERATIVE METALLURGY; those by which their value is determined, MINERAL ASSAYING.

A very small proportion only of the metals are met with in nature in the free or elementary condition, by far the greater number found being united with some non-metallic element or elements, in definite atomic proportions, and as such forming true chemical compounds, in which, in almost every instance, the physical and chemical properties of the metal are obliterated. In these bodies, which, when they are used as sources of the metals commonly employed by man, are called ORES, the metal is mostly combined with oxygen or sulphur, sometimes with carbonic acid, and less frequently with chlorine and other negative elements. Thus we have the native combinations of iron and oxygen constituting the minerals known as red hæmatite iron ore (Fe_2O_3), brown hæmatite ($2Fe_2O_3.3H_2O$), and magnetic iron ore ($Fe_2O_3.3FeO$), of tin and oxygen in tinstone (SnO_2), and of copper and oxygen known as red copper ore (Cu_2O). Of the principal ores into which sulphur enters as a chemical ingredient, we may mention native sulphide of antimony (Sb_2S_3); the two native sulphides of arsenic, realgar (As_2S_2) and orpiment (As_2S_3); galena, or native sulphide of lead (PbS), blende, or native sulphide of zinc (ZnS), and cinnabar, or native sulphide of mercury (HgS). Besides the above there are also certain double native sulphides, such as the double sulphide of iron and copper, known as Peacock ore, and having the composition $Fe_2S_3.3Cu_2S$; iron and copper pyrites ($Fe_2S_3.Cu_2S$); and red silver ore ($Sb_2S_3.3AgS$).

In the state of carbonate, ores occur—as malachite, native carbonate of copper, $CuCO_2.Cu(HO)_2$; as calamine, or native carbonate of zinc ($ZnCO_2$); and as spathose iron ore, or native carbonate of iron ($FeCO_2$). Horn silver or horn lead, the former having the composition AgCl, and the latter $PbCO_3 + PbCl_2$, are illustrations of ores containing chlorine.

The process of obtaining the metal from the ore of course varies with the nature and character of the latter. Before, however, this operation can be undertaken, the ore itself is subjected to certain mechanical operations, in order to remove the gangue or the adhering earthy, rocky, stony, and other matters with which it is always more or less mixed up. The amount of attention which is given to this preparatory treatment of the ores greatly depends upon their value; those, for instance, of copper and lead, as commanding a higher market price than those of zinc and iron, being submitted to commensurate treatment. This process of freeing the ores from the gangue, which is termed dressing, is generally conducted as follows, usually near the pit entrance of the mine whence the ores have been extracted.

If the material brought up to the pit's mouth is a lead or copper ore, it mostly contains a number of lumps, which are considered sufficiently pure for the smelting even, and these are set aside without being dressed. Generally, however, the ore is first broken by hammers into pieces about as large as a walnut, and the best pieces are then selected for smelting.

The remaining or inferior portions are then crushed under the large and horizontal cylinders of a grinding mill, to which they are supplied by hoppers. After being ground the ore is separated by being made to pass through coarse sieves, the coarser portions being set aside for the stampers, whilst the finer ones are subjected to the operation of jigging. This consists in a workman separating the contents of the sieve under water by imparting to them such a movement that the bits of ore (particularly if they are of a friable nature like galena) become broken, and thus pass through the meshes of the sieve to the bottom of the water; whilst the less friable and specifically lighter matter, mostly consisting of gangue, remains behind on the sieve. This residue being mixed with the coarser portions resulting from the first siftings, and which have not been subjected to the jigging process, is transferred to the stamping mill, whilst those portions of ore found at the bottom of the well are reserved for smelting. If the ore be one containing tin, it does not undergo the above processes, but passes at once to the stamping apparatus.

This stamping apparatus may consist of five or six large wooden beams, each weighing 1-8th of a ton. Each beam is covered at the bottom with iron, and is made to rise and fall in succession by means of projections from a horizontal axle, caused to revolve either by water or steam power. Behind the stampers is an inclined board, upon which are placed the residue and coarser portions of the ore already described, and when the stampers are in motion the ore slides down the inclined plane under them, and thus gets crushed. When it is thought the ore has been sufficiently crushed, it is, by means of a current of water running through the mill, carried away through a grating in front of the mill into a channel in which there are two pits, with the result that the more valuable and heavier portion of the ore becomes deposited in the first pit, whilst the inferior portion is carried on, and falls into the second one.

The crushed ore has, however, to undergo other operations before it is considered sufficiently pure for the furnace. That part (the purer portion, called the crop by the Cornish miner) which has been deposited in the first pit, after removal therefrom, is subjected to a series of further washings, the different apparatus by which these are effected being known in Cornish language as a buddle and a kieve.

"The crop is first subjected to washing in the buddle; this is a wooden trough about 8 feet long, 3 wide, and 2 deep, fixed in the ground with one end somewhat elevated. At the upper end a small stream of water enters, and is reduced to a uniform thin sheet by means of a distributing board, on which a number of small pieces of wood are fastened to break the stream. The ore to be washed is placed in small quantities at a time on a board just below the distributing board, and somewhat more inclined than the body of the buddle, and as the ore is spread out into a thin layer the water carries it forward.

"The richer portions subside near the head of the trough, and the light ores are carried further down. 'The heads' are then tossed into the kieve, a covered wooden tub, which is filled with water, and ore added by a workman, who keeps the contents of the kieve in continual agitation by turning an agitator, the handle of which projects through the lid of the tub. When the vessel is nearly full the agitation is stopped; the kieve is struck sharply upon the side several times, and its contents are allowed to subside; the upper half of the sediment is again passed through the buddle. Various modifications of the washing process are resorted to, but they are all the same in principle" (*Miller*).

The water which has been used in washing the ore on the buddle, as well as that in the kieve, contains in addition to the *débris* of the gangue more or less of small pieces of the ore itself. Hence this water is not allowed to escape, but conveyed into a narrow channel cut at the end of the buddle, where it deposits the solid minerals. These being then removed undergo a second washing on an inclined stage, a process which is followed in Cornwall, and by which any remaining mineral is recovered.

The above is the method of dressing the ores of lead and tin, and, with some modifications, those of copper.

Some metals, as, for example, certain iron and zinc ores, previous to being dressed, require a preliminary exposure for some time to the atmosphere. This operation, which is called 'weathering,' has the effect of aiding the subsequent removal by water of certain materials of a clayey, slaty, or marly nature, which sometimes adhere very tenaciously to the ores in question.

Again, in some cases weathering is had recourse to for obtaining a metallic compound in a soluble form. It is by this means that iron pyrites, if exposed to the air, after a time becomes converted into a sulphate of the metal.

Large quantities of commercial sulphate of iron or green vitriol are manufactured from this natural sulphate after it has been dissolved by the rain, and then crystallised. Sometimes the ores after dressing, and previous to roasting or smelting, are subjected to a process of calcination without excess of air, with the object of depriving them of water, carbonic acid, and bituminous matters, and also of rendering the ore softer and in a favourable condition to be acted upon by the subsequent metallurgic operations.

The ores, having been by these various processes sufficiently freed from extraneous matters, are next, according to their composition, either submitted to the operation of roasting or smelting, and in many cases to both.

Roasting. This operation is mostly carried out in a reverberatory furnace. The result of the process upon the ores containing sulphur, which are those chiefly subjected to it, varies with the nature of the ore. Thus when the sulphides of antimony, arsenic, or zinc are roasted, the sulphur escapes as sulphurous anhydride with the formation of the volatile oxides of arsenic, antimony, or zinc, which sublime, and are afterwards collected and purified.

With cinnabar or native sulphide of mercury sulphurous anhydride is evolved along with the vapours of metallic mercury, these being at the same time condensed by cooling.

When copper pyrites (the double sulphide of sulphide of copper and iron) is placed in the reverberatory furnace, the copper and iron become converted into oxides.

When galena or lead sulphide is exposed to the roasting process, lead oxide and sulphate, with the copious escape of sulphurous acid, are at first formed. The oxide and sulphate become eventually decomposed, leaving behind metallic lead, with a small portion of a subsulphide of the metal. In most cases, however, the effect of roasting on an ore is to convert it into an oxide.

Clay ironstone, which is that from which the greater part of the iron is manufactured in Great Britain, and that known as the black band of the Scotch coal-fields, are impure carbonates of iron, and these when roasted yield ferric oxide. The roasting in the case of these minerals is sometimes effected in kilns, but more frequently in the open air; in the latter case by the firing of stacks composed of alternate layers of the ore and of small coal. Calamine or native carbonate of zinc is converted into oxide sometimes by being roasted in kilns, but more frequently in a reverberatory furnace.

Smelting. Except in those cases in which the ore is directly reduced from the state of a sulphide to that of a metal, it is, as has been shown, converted into an oxide. If, therefore, it be required to procure the metal *per se*, some method must be adopted for the removal of the oxygen from its oxide.

This process, which is called smelting, and is applied to most metallic oxides, whether of natural or artificial origin, consists in heating the oxide with a substance which has a stronger attraction for oxygen than the metal has. Such bodies are coal, coke, and charcoal, which, when raised to very high temperatures in contact with certain metallic oxides, rob them of their oxygen, and thus reduce them to the state of metals, carbonic oxide or carbonic anhydride being at the same time formed and carried off. A mechanical impediment, however, to the reducing action of the fuel upon the ore exists in the rocky, earthy, and other impurities mostly present in large quantities, even after the dressing; these envelop the mineral, and afford it a protective covering. To remove them it is not only necessary that some substance should be added which has the power of combining with them, but one which is capable of forming a compound which shall become fusible by the heat of the furnace, so that

the molten metal as it sinks through it by reason of its greater specific gravity, and falls to the bottom of the furnace, shall be protected in doing so from contact with the air. Many substances, varying with the nature of the gangue accompanying them, are thus employed as fluxes, such as limestone, fluor spar, gypsum, heavy spar, &c., and they act by combining with the silicious compounds contained in the gangue attached to the ore, and forming a fusible silicate known as slag, which is from time to time run off by an aperture at the side of the furnace. Considerable knowledge and experience are required in the selection of suitable fluxes.

The smelting-furnaces in which the deoxidation of iron is accomplished are of considerable size. The following description of one, together with the engraving, are from Professor Bloxam's able work, 'Chemistry, Inorganic and Organic.'

"Great care is necessary in first lighting the blast-furnace, lest the new masonry should be cracked by too sudden a rise of temperature; and when once lighted, the furnace is kept in constant work for years, until in want of repair.

"When the fire has been lighted the furnace is filled up with coke, and as soon as this has burnt down to some distance below the chimney, a layer of the mixture of calcined ore with the requisite quantity of limestone is thrown upon it; over this there is placed another layer of coke, then a second layer of the mixture of ore and flux, and so on in alternate layers, until the furnace has been filled up; when the layers sink down fresh quantities of fuel, ore, and flux are added, so that the furnace is kept constantly full.

"As the air passes from the tuyère-pipes into the bottom of the furnace, it parts with its oxygen to the carbon of the fuel, which it converts into carbonic acid; the latter passing the red-hot fuel as it ascends in the furnace is converted into carbonic oxide by combining with an additional quantity of carbon. It is this carbonic oxide which reduces the calcined ore to the metallic state when it comes in contact with it at a red heat in the upper part of the furnace, for carbonic oxide removes the oxygen at a high temperature from the oxides of iron, and becomes carbonic acid, the iron being left in the metallic state.

"But the iron so reduced remains disseminated through the mass of ore until it has passed down to a part of the furnace which is more strongly heated, where the iron enters into combination with a small proportion of carbon to form cast iron, which fuses or runs down into the crucible or cavity for its reception at the bottom of the furnace.

"At the same time the clay contained in the ore is acted upon by the lime of the flux, producing a double silicate of alumina and lime, which also falls in the liquid state into the crucible, where it forms a layer of slag above the heavier metal. This slag, which has five or six times the bulk of the iron, is allowed to accumulate in the crucible and to run over its edge down the incline upon which the blast-furnace is built; but when a sufficient quantity of cast iron is collected at the bottom of the crucible, it is run out through a hole provided for the purpose, either into channels made in a bed of sand, or into iron moulds where it is cast into rough semi-cylindrical masses, called pigs, cast iron being spoken of as pig-iron.

"The temperature of the furnace is, of course, highest in the immediate neighbourhood of the tuyères. The reduction of the iron to the metallic state appears to commence at about two thirds of

the way down the furnace, the volatile matters of the ore, fuel, and flux being driven off before this point is reached.

"Some idea may be formed of the immense scale upon which the smelting of iron ores is carried out, when it is stated that each furnace consumes in the course of 24 hours about 50 tons of coal, 30 tons of ore, 6 tons of limestone, and 100 tons of air.

"The cast iron is run off from the crucible once or twice in 12 hours, in quantities of 5 or 6 tons at a time. The average yield of calcined clay-iron stone is 35% of iron.

"The gases escaping from the chimney of the blast-furnace are highly inflammable, for they contain, beside the nitrogen of the air blown into the furnace, a considerable quantity of carbonic oxide and some hydrogen, together with the carbonic acid formed by the action of the carbonic oxide upon the ore. Since the carbonic oxide and hydrogen confer considerable heating power upon these gases, they are employed in some iron-works for heating steam boilers, or for calcining the ore, or for raising the temperature of the blast.

"The composition of the gas issuing from a hot blast-furnace (fed with uncoked coal) may be judged of from the following table:

"*Gas from Blast furnaces.*

Nitrogen	.	.	.	55·35 vols.
Carbonic oxide	.	.	.	25·97 ,,
Hydrogen	.	.	.	6·73 ,,
Carbonic acid	.	.	.	7·77 ,,
Marsh gas	.	.	.	3·75 ,,
Olefiant gas	.	.	.	0·43 ,,
				100·00 ,,

"The carbonic oxide, of course, renders these gases highly poisonous, and fatal accidents occasionally happen from this cause. Although the bulk of the nitrogen present in the air escapes unchanged from the furnace, it is not improbable that a portion of it contributes to the formation of the cyanide of potassium which is produced in the lower part of the furnace, the potassium being furnished by the ashes of the fuel." See METALLURGY.

Assay. Three general methods are adopted for this purpose:

1. (MECHANICAL.) This consists in pulverising the ore by any convenient method, and expertly washing a given weight of it (say 1000 gr.) in a wooden bowl or capsule with water, so as to remove the earthy gangues from the denser and valuable metallic matter in such a way that none of the latter may be lost. This is the common plan adopted with auriferous sands, the ores of

tin after they have passed the stamping-mill, galena, grey antimony, &c., and may either be employed as an independent process or merely as preparatory to more exact investigations. When galena is thus tested, the product is a nearly pure sulphide of lead, of which every grain is equivalent to 0·8666 of metallic lead, the rest being sulphur. The results with grey antimony ore are still more direct.

2. (HUMID.) Assays in the 'humid way' are true chemical analyses, and are described under the head 'Estim.' attached to most of the more important minerals noticed in this work. This plan offers greater facilities and gives more accurate results than either of the other methods.

3. (DRY.) Of the methods of assay in the 'dry way' the following are the most accurate, generally useful, and easily applied:

a. (Dr Abich.) The mineral is reduced to powder, and mixed with five or six times its weight of carbonate of barium, also in powder; this mixture is fused at a white heat in a platinum crucible, and the resulting slag, after being powdered, is exhausted with hydrochloric acid. This process answers well with both stony and metallic minerals, the most refractory of which give way under this treatment.

b. (Liebig.) Into a crucible containing commercial cyanide of potassium, a weighed quantity of the ore, in the state of fine powder, is sprinkled, when the metallic oxides and sulphides which it contains are almost immediately reduced to the metallic state, and may be separated from the scoria by lixiviation with water. When the oxides and sulphides of antimony and tin this reduction occurs at a dull red heat; with the compounds of copper it occurs with the disengagement of light and heat; but an ore of iron requires to be mixed with a little carbonate of potassium or of sodium before throwing it into the fused cyanide, and to be then submitted to a full red heat for a short time, before it is reduced to the regulus state. In this case any manganese present in the ore of iron is left under the form of protoxide. A mixture of about equal parts of dry carbonate of sodium and cyanide of potassium answers better for the crucible than the cyanide alone.

c. By cupellation, a method applicable to the assay of gold, silver, lead, &c. In assaying gold, for example, a small piece of metal is wrapped in tissue-paper together with 3 times its weight of pure silver: it is then added to 12 times its weight of pure lead fused in a crucible or cupel made of bone-ash (see engr.), and placed

in a muffle. When the lead, copper (if any), &c., are oxidised, the fused oxide of lead dissolves that of copper, and both are absorbed by the cupel. After a time the button of metal ceases to diminish in size; it is then allowed to cool, hammered into a flat disc, annealed by heating it to redness, rolled into a thin plate, and then between the finger and thumb twisted into a cornette. This is now boiled with nitric acid (sp. gr. 1·18), to extract

the silver; the gold which remains is washed with distilled water and boiled with nitric acid (sp. gr. 1·28) to extract the last traces of silver, after which it is again washed, heated to redness in a small crucible, and finally weighed. See ALLOYS, METALLURGY, &c.; also 'Percy's Metallurgy' and 'Mitchell's Manual of Practical Assaying.'

ORGANIC BASES. These interesting bodies may be divided into two classes; the first comprising those which occur ready formed in nature (ALKALOIDS), and the second those produced by artificial processes in the laboratory (ARTIFICIAL ALKALOIDS, ARTIFICIAL ORGANIC BASES). They all contain the element NITROGEN. The natural bases have already been described under ALKALOID. Until recently none of them have been produced by artificial means. The bases of entirely artificial origin are mostly volatile, and their constitution is, as a rule, simpler than that of the native bases. Of the vast number which have been formed the following are, perhaps, the most interesting:—ETHYLAMINE, METHYLAMINE, AMYLAMINE, ANILINE, NAPHTHYLAMINE, CHINOLINE, and PICOLINE. These and other bodies of the class are noticed under their respective heads.

By Berzelius the natural organic bases (owing to the invariable presence in them of hydrogen and nitrogen) were regarded as compound ammonias, or combinations of ammonia with a variety of neutral principles.

He conceived that the greater part of these neutral bodies were incapable of isolation, and further that the closest union existed between them and the ammonia. Thus it was his opinion that quinine, $C_{20}H_{12}NO_2$,3HO (halving the modern formula), was a compound of the group $C_{20}H_6O_2$ with oxide of ammonium and water of crystallisation, thus: $(C_{20}H_6O_2H_4NO)_2HO$. He believed that the organic base owed its basicity to the ammonia. Berzelius's opinion carried weight at the time, from the circumstance that certain neutral substances when directly combined with ammonia were capable of forming a number of artificial bases very similar in qualities and also in composition to the natural ones, or those obtained from living plants. Thus a base may be artificially obtained from the union of oil of bitter almonds with ammonia.

Liebig, who was one of the first chemists to dispute the correctness of Berzelius's hypothesis, by showing that the natural organic bases never gave any indication of the presence in them of ready-formed ammonia, replaced it by the suggestion that they might be bodies into the composition of which amidogen (H_2N) entered, and that these, instead of being compounds of ammonia and an organic group, might be derivatives from ammonia, or ammonia in which an atom of hydrogen had been displaced by an equivalent organic radical.

The labours of subsequent chemists, notably those of Messrs Wurtz and Hofmann, have developed Liebig's theory, and have proved the analogy in structural arrangement between ammonia and the greater number of organic bases; whilst they have further shown that not only one, but all three of the hydrogen atoms in ammonia may be substituted by certain compound radicals.

ORGANIC SUBSTANCES. We have reserved a notice of the method of estimating the quantity of carbon, hydrogen, oxygen, and nitrogen, in organic compounds, until now, in order to present them to the reader in a more useful and connected form. The operation essentially consists, in respect of the first three, in causing the complete combustion of a known quantity of the substance under examination, in such a manner that the carbonic acid and water thus produced shall be collected, and their quantity determined. From these the proportions of their elements are easily calculated. The estimation of the quantity of nitrogen (as is also the case with chlorine, phosphorus, sulphur, &c) requires a separate operation. The two great classes of organic bodies (azotised and non-azotised) are readily distinguished from each other by heating a small portion with some solid hydrate of potassium in a test-tube. If nitrogen is present, it is converted into ammonia, which may be recognised by its characteristic odour and its alkaline reaction.

1. *Estimation of the* CARBON, HYDROGEN, *and* OXYGEN. *a.* The method of Professor Liebig, now almost exclusively adopted for this purpose, is as follows:—The substance under examination, reduced to powder, is rendered as dry as possible, either by the heat of a water-bath or by exposure over concentrated sulphuric acid *in vacuo*; 5 or 6 gr. of it are then weighed in a narrow open test-tube, 2 or 3 inches long, and to ensure accuracy this tube and any little adhering matter is again weighed after its contents have been removed—the difference between the two weights being regarded as the true quantity of the substance employed in the experiment. A 'combustion-tube,' of hard white Bohemian glass (0·4 to 0·5 inch diam., 14 to 18 inches long), is next taken, and about 2-3rds filled with black oxide of copper, prepared by the ignition of the nitrate, and which has been recently re-heated to expel moisture. Nearly the whole of this oxide, whilst still warm, is then gradually poured from the tube and triturated with the organic sample in a dry and warm mortar, after which the mixture is transferred to the combustion-tube, and the mortar being rinsed out with a little fresh oxide, which is then added to the rest, the tube is, lastly, nearly filled with some warm oxide fresh from the crucible. The contents of the tube are next arranged in a proper position by a gentle tapping, so as to leave a small passage for the evolved gases from the one end of the tube to the other. (See *engr.*)

The 'combustion-tube' with its 'charge' is next placed in a 'furnace' or 'chauffer' of thin sheet iron lined with fireclay (see figure *below*).

Its open end is then connected with a 'drying-

tube' filled with fragments of fused chloride of calcium, and carefully weighed.

This tube is, in its turn, connected with a series of small glass bulbs ('Liebig's potash bulbs') containing solution of pure potash of sp. gr. 1·27, also carefully weighed. The junction with the first is made by means of a perforated cork; that with the second by means of a small tube of india-rubber tied with silk, the whole being made quite air-tight. The apparatus is then tested by sucking a few bubbles through the liquid with dry lips, when, if the level of the solution of potash in the two legs continues unequal for some minutes, the joints are regarded as perfect. The whole arrangement being complete (see *engr.*), burning charcoal is now placed in the furnace around the front part of the combustion-tube, and when this has become red-hot

the screen is slowly moved back, and more burning charcoal is added, until the furthest extremity of the tube has been exposed to its action. (Gas burned in furnaces specially contrived for the purpose is now usually employed instead of charcoal.) The heat is so regulated that the gas enters the potash apparatus in bubbles easily counted, without any violence, and it is kept up as long as gas is given off. As soon as the apparatus is complete, and the slightest retrograde action of the potash is observed, the charcoal is removed from the combustion-tube or the gas is turned off, and the extreme point of this last is broken off. A little air is then sucked through the apparatus in order to seize on any remaining carbonic acid gas and moisture. The potash apparatus and the chloride of calcium tube are lastly detached, and again accurately weighed. The increase in the weight of the first gives the weight of the carbonic acid formed during the combustion; that of the second the weight of the water.

The numbers equivalent to any given number of grains, found as above, are converted into the proportions per cent. by simply dividing them by the weight of the organic substance which has been employed in the experiment, and moving the decimal point of the result two figures to the right.

b. In applying the preceding method to volatile liquids, it is necessary to enclose them in a small bulb with a narrow neck, instead of mixing them directly with the protoxide of copper. The bulb with its contents is introduced into the combustion-tube, and after some 6 or 8 inches of the protoxide is heated to redness, heat is applied near where the bulb is situated, so that the liquid which it contains may be slowly volatilised and passed through the heated mass in the state of vapour, being thus completely burned. For further information consult Fresenius's ' Chemical Analysis.'

2. An improved and more complex apparatus is thus described in the last edition of Bloxam's ' Chemistry: '

" The substance to be analysed having been carefully dried and weighed (about 0·5 grm.) is placed in a small boat-shaped tray of porcelain or platinum, which is introduced into one end of a glass tube about 30 inches long, of which about 24 inches are filled with small fragments of carefully dried cupric oxide. The end of the tube

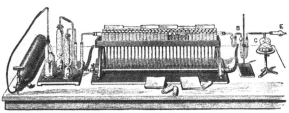

where the boat is placed is connected with an apparatus for transmitting air or oxygen, which has been purified from CO_2 by passing through potash, and from H_2O by calcium chloride. To the other end of the tube is attached, by a perforated cork, a weighed tube (B) filled with small fragments of calcium chloride to absorb H_2O, and to this is joined by a caoutchouc tube a bulb apparatus (C), containing strong potash to absorb CO_2, and a small guard-tube (E) with calcium chloride to prevent loss of water from the potash. The potash bulbs and guard-tube are accurately weighed. The combustion-tube is supported in a charcoal or gas furnace, and that portion which contains the cupric oxide is heated to redness. The end containing the boat is then gradually heated, so that the organic substance is slowly vaporised or decomposed. The vapour or the products of decomposition in passing over the red-hot cupric oxide will acquire the oxygen necessary to convert the C into CO_2, and the H into H_2O, which are absorbed in the potash bulbs and calcium chloride tube. At the end of the process, which commonly occupies about an hour, a slow stream of pure air or oxygen is passed through, whilst the entire tube is red-hot, in order to burn any charcoal which may remain in the boat, and to carry forward all the CO_2 and H_2O into the absorption apparatus. The weight of the CO_2 is given by the increase in weight of the potash bulbs, and that of H_2O by that of the calcium chloride."

The following calculation will serve to illustrate the manner in which the result is obtained in an analysis of sugar. The figures are those of a real experiment.

Quantity of sugar taken . . . 0·2375 grm.
Weight of potash bulbs after experiment . . . . 39·0565 „
Ditto ditto before experiment . 38·6910 „
                                            ————
Carbon dioxide = 0·3655 „
Weight of calcium chloride tube
after experiment . . . . 11·3025 „
Ditto ditto before experiment . 11·1650 „
                                            ————
Water = 0·1375 „

0·3655 grm. CO_2 = 0·0997 grm. C.
0·1375 „ H_2O = 0·0153 „ H.
Hence in 100 grms. sugar—
     Carbon . . . . . . 41·98
     Hydrogen . . . . . 6·43
     Oxygen by difference . 51·59
                              ————
                              100·00

3. *Estimation of the* NITROGEN. *a.* Several methods are employed for this purpose, but the only one of general application, and adapted to the non-scientific operator, is that of Varrentrap and Will, described under GUANO. To ensure correct results the caustic soda must be pure, and the lime of good quality and well burnt. The last, having been properly slaked with a little water, holding the former in solution, the mixture is thoroughly dried in an iron vessel, and then heated to full redness in an earthen crucible. The ignited mass is rubbed into powder in a warm dry mortar, and either used at once or carefully preserved from the air. The best quantity of the organic substance to operate on is, in this case, about 10 gr., which must be dried and accurately weighed with the usual precautions. Bodies very rich in either nitrogen or hydrogen are best mixed with about an equal weight of pure sugar before triturating them with the soda-lime. The nitrogen is weighed under the form of double chloride of platinum and ammonium, dried at 212° F. This salt contains 6·272% of nitrogen.

b. Péligot has modified the preceding plan by conducting the gaseous matter extricated during the operation into a three-bulb tube charged with a standard solution of sulphuric acid. This he subsequently pours into a beaker glass, and after tingeing it with a single drop of tincture of litmus, he tests it with either a standard aqueous solution of soda or one of lime in sweetened water, after the common method of alkalimetry. The difference between the saturating power of the acid in its normal condition and after its exposure in the condenser indicates the amount of ammonia formed (see GUANO). Each grain of ammonia contains 0·82353 gr. of nitrogen.

c. (*Kjeldahl's method.*) This consists in oxidising the substance with potassium permangenate and sulphuric acid, by which ammonium sulphate is produced. Boiling with an alkali liberates the ammonia, which is then absorbed by hydrochloric acid, and determined with standard alkali or with platinic chloride.

Estimation of the Sulphur and Phosphorus. These are converted into sulphuric and phosphoric acids respectively by the action of strong oxidising agents, such as chloric and nitric acids, bromine, &c. The acids are then determined by the usual methods.

Concluding Remarks. The successful application of the above processes requires considerable care and some aptitude in manipulating, as well as the employment of a very delicate balance for determining the weights. A greater error in the weighings than the $\frac{1}{110}$ gr. cannot be tolerated when exact results are desired. The method of Varrentrap and Will for the determination of nitrogen answers admirably for all organic compounds containing it, except those in which it exists under the form of hyponitrous, nitrous, and nitric acids; for which, however, it is not required. When extreme accuracy is aimed at, the atmospheric air in the apparatus, and that absorbed during the preliminary operations by the substances employed, must be expelled before the application of heat to the combustion-tube. See WATER, ANALYSIS OF.

ORGYIA ANTIQUA, Hübner. (From the Greek word to extend, as the moths extend their feet when sitting—'British Moths,' *Westwood*.) THE COMMON VAPOURER MOTH. This moth is termed 'Vapourer' for the reason that the male is continually darting about, always on the wing, flying hither and thither evidently in search of the female, which, like the female of the winter moth, *Cheimatobia brumata*, is without wings, and is insignificant in appearance.

Although somewhat indiscriminate in its selection of food, and attacking many kinds of trees, it appears to have a predilection for the ROSACEÆ and for the fruit trees of this order, especially apple, pear, and plum trees, and it occasionally does much harm to these by eating their leaves. Fortunately it is not very often that serious injury is caused by it. It is well, however, to give a short sketch of its habits and history.

It is known in America. Harris speaks of it, and Lintner also alludes to it. In Germany it is injurious to fruit trees and other trees. Réaumur describes it at length, and gives admirable illustrations of it in all stages. He remarks that its caterpillars live on the leaves of plum trees ('Histoire des Insectes,' par M. de Réaumur, tome i, p. 321).

Life History. According to Westwood's classification the *Orgyia antiqua* belongs to the seventh family of the LEPIDOPTERA, the *Arctiidæ*.

As has been mentioned above, the female moth is wingless. It is of a dark ash colour, and is not easily detected upon the trunks and branches of trees which it inhabits. It comes from the pupa state between July and September, and after having paired it lays many eggs upon the web within which the pupal stage was passed.

The male is a rather pretty insect, whose body is seven lines in length, with a wing expanse of fourteen or fifteen lines, and with antennæ much fringed or pectinated. In colour it is tawny or chestnut, and has a white spot, by which it may readily be identified, towards the end of each fore-wing.

From this plain mother and vapouring sire a most brilliant caterpillar is procreated, which is hatched from the egg in the spring, and immediately proceeds to feed upon the foliage near to it.

It is twelve lines in length, the ground colour is dark with red spots upon it, and there are four long yellowish tufts upon its body at intervals, and two dark-coloured tufts at its head and tail. Réaumur calls these tufts *grandes aigrettes de plumes.*

Prevention. After an attack upon fruit trees search should be made for the eggs upon the webs placed upon the stem and branches, and these should be brushed off with stiff brushes. The females also may be found upon the stems between July and late in September, and may be easily killed.

Remedies. Nothing but washing, or syringing, can be suggested as remedial in this case, if practicable.

Birds are very fond of the eggs of this moth. They also eat numbers of the females, which are an easy prey ('Reports on Insects Injurious to Crops,' by Charles Whitehead, Esq., F.Z.S.).

OR-MOLU'. [Fr.] This name is given to gold-coloured brass or bronze, so finished off as to have the appearance of gold, or of being gilt; but it is often applied in a more general sense. The French more particularly excel in working in or-molu, and the products of this branch of their industry hold an important position in the art manufactures of France.

To give or-molu its richest appearance, "it is not unfrequently brightened up after 'dipping' (that is, cleaning in acid) by means of a scratch-brush (a brush made of very fine brass wire), the action of which helps to produce a very brilliant gold-like surface. It is protected from tarnish by the application of lacquer " (*Ure*).

Ure says or-molu contains more copper and less zinc than ordinary brass, and that although, in many of its applications, the colour is heightened by means of a gold lacquer, in some cases the true colour of the alloy is best preserved after it has been properly developed by means of dilute sulphuric acid.

OR'PIMENT. Native yellow sulphide of arsenic. The finest samples used by artists (golden orpiment) come from Persia. See ARSENIC (Tersulphuret).

OR'RIS. *Syn.* ORRIS ROOT, FLORENTINE R.; RADIX IRIDIS, L. The dried rhizome of *Iris florentina*, *pallida*, and *germanica*. Sialogogue, irritant, subacrid, and errhine. Chiefly employed to impart a violet-like odour to oils, tooth powder, snuffs, spirits, &c.; and when cut into peas to keep open issues.

ORSE'DEW. Dutch leaf-gold.

ORSEL'LIC ACID. Two compounds pass under this name—α-ORSELLIC ACID and β-ORSELLIC ACID. They closely resemble each other, and are obtained in a similar manner; the first from the South American variety of *Roccella tinctoria*, the last from that grown at the Cape.

ORSELLINIC ACID. A compound formed along with picro-erythrin by boiling erythric acid for some time with water. It is also formed by boiling a-orsellic acid with water. In both cases, if the ebullition is too long continued, the new acid is wholly or in part converted into orcin.

Prop., &c. Crystallisable; bitter; soluble in water; its aqueous solution, by exposure to the air, assumes a beautiful purple colour.

ORTHOCLASE. *Syn.* POTASSIUM FELSPAR. This material, which is a double silicate of potassium and aluminium, enters into the composition of many rocks, and is a common ingredient in granite. It has the following composition:—Silica, 64·8 parts; alumina, 18·4 parts; and potash, 16·8 parts. Part of the potassium is frequently replaced by small quantities of calcium, magnesium, and sodium.

Orthoclase is used for glazing the finest varieties of porcelain, a very intense heat being necessary to effect its fusion in the porcelain furnace. By the Chinese potters it is called *pe-tun-tse.* "The name 'orthoclase' is generally restricted to the subtranslucent varieties, there being many subvarieties (founded on variations of lustre, colour, and other differences), of which the following are some of the principal, viz. *adularia,* a transparent or translucent felspar, met with in granitic rocks (frequently in large crystals); *moonstone; sunstone; erythrite;* glassy felspar or *lanadine,* a transparent variety found in volcanic rocks, containing 4% of soda or upwards" (*Ure*).

ORTHOPÆ'DIA. In *surgery,* the straightening, correcting, or curing deformities of children. See SURGERY.

OSCINIS VASTATOR, Curtis. THE FRIT FLY. This is a small fly, of the family *Oscinidæ,* as defined by Westwood, which works in a somewhat similar manner to the *Cephus pygmæus.* Curtis gives it the designation *vastator* because of its serious injuries, and he considered it a far worse enemy to cereal crops than the Cephus or the *Chlorops tæniopus.* He says that the ten or twelve stalks of corn he opened were filled only with powder at the base, every portion of the young ear being consumed; the destruction was complete.

The larvæ of this fly burrow within the stems of cereal plants, and live upon their parenchyma or internal tissues, and utterly prevent the development of the ears.

Serious injuries are often occasioned by this insect in America, France, Germany, and Sweden.

Specimens of wheat plants were sent me from Worcestershire in 1883 in the first week in June, in which it was seen that the inner leaves or blades were yellow or light brown, and were manifestly dying. The farmer who sent these reported that many of the plants in the field from which these were taken were similarly affected. Upon searching a tiny yellowish maggot was discovered at the lower part of the stem. The blades were yellow or brownish at the tips, and could be easily pulled away from the stem. It was clear that in a short time the whole of the stem would have been rendered unfruitful and useless. Other affected plants were sent later on, in which the larvæ of the Oscinis had completely

destroyed the nascent ear, and had left nothing within the stems but a little dust.

Life History. The perfect fly is greenish black, with a somewhat shiny appearance. It comes first at the beginning of May, and deposits eggs upon the under side of the leaves. When the larvæ are hatched they make their way speedily into the hearts of the stems. They are whitish maggots. As the pupæ have been found in wheat stems in the middle of June, it is supposed that there are two broods during the year, and that the second brood again attack the wheat plants or grasses.

Prevention. As it is believed that the pupæ hibernate in the stems of wheat plants and grasses, it is important that all stubble, weeds, and rubbish should be burnt or ploughed in deeply under the soil ('Reports on Insects Injurious to Crops,' by Charles Whitehead, Esq., F.Z.S.).

OSIER. The osier, which is a species of willow (*Salix*), and is largely used in the construction of baskets and other wicker work, is extensively cultivated at Nottingham and on the level lands of Cambridgeshire and Huntingdonshire, as well as on the banks of the Thames, Severn, and other rivers. The small islands in these rivers, when planted with osiers, are known as osier HOLTS. But large as is the supply of shoots afforded by the English osier beds, it is insufficient for home consumption; hence great quantities of osier rods are imported into this country from Holland, Belgium, and France. There are a great variety of osiers, and it is found that those which have been the most highly cultivated yield the toughest and finest wood, and are best adapted for the superior kinds of basket work. The branches of the wilder and less domesticated kind are more liable to break, and are used for making hoops and coarse baskets. This last variety, which is known as the COMMON OSIER (*Salix viminalis*), grows on the alluvial grounds of Britain, and in other European countries; it is often planted on the banks of rivers to prevent their being washed away.

The following are the principal varieties of osier indigenous to this country, and which yield the most valuable wood:—1. THE FINE BASKET OSIER (*Salix Forbyana*). 2. THE GREEN-LEAVED OSIER, or ORNARD (*Salix rubra*). 3. THE SPANISH ROD (*Salix triandra*). 4. THE GOLDEN OSIER, or GOLDEN WILLOW (*Salix vitellina*).

The osier requires plenty of water, and hence it thrives best in those localities and low grounds which are washed by a river. The soil best adapted for it is a rich but not clayey one. In planting an osier bed an important condition is that the trees should be placed sufficiently closely together, since it is found that with too much space the shoots do not develop into those long and slender branches which are so much sought after. The shoots are cut once a year, at any time between the fall of the leaf and the rising of the sap in spring. After being cut they are divided into those destined for brown, and those for white baskets. In the latter case the rods have to be peeled, but as this operation cannot be performed at once, and the removal of the bark would be difficult were they allowed to dry, the shoots are placed upright and sustained in that position in wide shallow trenches in about four inches of water, where they are kept until they begin to bud and blossom in the spring, which they do as if they were attached to the parent plant. The peeling is easily done by passing them through an instrument known as a *break*. If the spring has been a cold one, they have, previous to peeling, to be laid for some time under a layer of litter.

When they have been peeled they are stacked, preparatory to being sold. With the rods intended for brown baskets, no peeling is of course necessary. They are therefore carefully stacked in some place protected from the rain, and diligently watched to see that no heat is set up in them, as is sometimes the case with freshly stacked hay, and which, if not stacked, would cause the rods to rot and render them useless.

In England, besides the native produce, 5000 tons of osiers are annually imported, valued at about £40,000. Of late years the Australian colonists have turned their attention to the cultivation of the osier, in the hopes of supplying the demand for it in Great Britain.

OSMAZOME. The substance on which the peculiar odour and flavour of boiled meat and broth are supposed to depend. Nothing is really known of its true nature.

Prep. From lean meat, minced and digested in cold water, with occasional pressure; the filtered infusion is gently evaporated nearly to dryness, and then treated with alcohol; the alcoholic tincture is, lastly, evaporated. The product has a brownish-yellow colour, is soluble in water, and its aqueous solution may be precipitated by an infusion of galls and the mineral astringent salts.

OSMIUM. Os. A rare metal found associated with the ores of platinum by M. Tennant, in 1803.

These ores contain an alloy of rhodium, osmium, ruthenium, and iridium, together with platinum and palladium. When they are treated with aqua regia, the insoluble residue which remains chiefly consists of the alloy. This alloy is also found associated with native gold, and being very heavy it accumulates at the bottom of the crucible during the melting operations. To separate the osmium from the other metals, Fremy takes advantage of its easy oxidability, and of the volatility of its tetroxide.

In the first part of this process (which is a great improvement upon the methods previously followed) the above residue or alloy is heated to redness in a platinum or porcelain tube. In that part of the tube which projects from the furnace some fragments of porcelain are placed, and the tube is connected with a series of glass flasks, in which the tetroxide of osmium is condensed as it distils over, any tetroxide that may have escaped condensation being retained by a solution of caustic potash, placed in the last flask of the series. This last flask is connected with an aspirator, by means of which a current of air is drawn through the apparatus.

Before being allowed to enter the heated tube the air is dried by passing it through tubes filled with pumice-stone moistened with sulphuric acid. During the operation the osmium and ruthenium become oxidised, the tetroxide of osmium con-

denses in needles in the flasks, and mechanically carries forward the oxide of ruthenium, which is deposited upon the pieces of porcelain. The vapours of tetroxide of osmium are very dangerous to the eyes.

Prep. 1. By treating the volatile tetroxide of osmium obtained by Fremy's method, as above described, with hydrochloric acid and metallic mercury in a closed vessel at 140° C. The mercurous oxide, which is first formed at the expense of the oxygen contained in the tetroxide of osmium, is decomposed by the hydrochloric acid, and calomel is produced, together with metallic osmium. The water and excess of acid are removed by evaporation to dryness, and on heating the residue in a small porcelain retort the excess of mercury and calomel are drawn off, pure osmium being left behind in the form of a fine powder.

2. Deville and Debray procure it in the metallic form by passing the tetroxide of osmium in a current of nitrogen, over carbon which has been obtained from the vapour of benzine by passing it through a porcelain tube at a high temperature.

Prop. Crystalline in cubes or obtuse rhombohedra, of a bluish-white colour with violet lustre, and harder than glass. The specific gravity of osmium in the pulverulent form is about 10; but after having been heated to the fusing-point of rhodium in the oxyhydrogen jet, it acquires a density of 21·4, and in the crystalline state it has a sp. gr. of 22·477. Osmium has not yet been fused.

There are five known oxides of osmium:

1. **Osmium Protoxide.** OsO. The anhydrous protoxide is of a greyish-black colour. It is insoluble in acids; is obtained from the corresponding sulphite ignited with sodium carbonate in a current of CO₂. Its bluish-black hydrate, which dissolves in hydrochloric acid, forms a solution of osmium dichloride of a deep indigo-blue colour. The solution absorbs oxygen readily, and becomes converted into the tetrachloride (OsCl₄).

2. **Osmium Sesquioxide.** Os₂O₃. Black powder, insoluble in acids, obtained by heating its salts with carbonate of soda in a current of CO₂. Of its salts, the osmic chloride of potassium and ammonium have been most accurately examined.

3. **Osmium Dioxide.** OsO₂. This is dark-coloured, but has a coppery lustre; it is obtained from its salts like the foregoing oxides.

4. **Osmium Tetroxide.** *Syn.* OSMIC ACID, OSMIC ANHYDRIDE. OsO₄. This oxide may be obtained by operating, according to Fremy's process, on the ores of platinum, as already described. It is also formed when metallic osmium is heated with potassic nitrate, or roasted in air. It crystallises in colourless, transparent, flexible needles, which fuse easily, and dissolve readily in water. Its aqueous solution, however, does not redden litmus. Tetroxide of osmium is converted into vapour at about 100° C. The fumes are excessively irritating and dangerous, and have an odour somewhat like that of chlorine. As an antidote to the effects of osmic acid Claus recommends the cautious inhalation of sulphuretted hydrogen. This oxide unites with alkalies, but not with acids. It is given off as tetroxide when the alkaline solution which contains it is boiled. If applied to the skin this oxide becomes partially reduced,

causing a painful eruption, and imparting a permanent black colour to the skin, due to the deposition of metallic osmium. With tincture of galls its solutions give a distinctive blue precipitate.

There are four chlorides of osmium, the best known of which are the dichloride and the tetrachloride, which are formed by direct combination.

1. Osmium Dichloride. OsCl₂. *Syn.* OSMIOUS DICHLORIDE, OSMIUM PROTOCHLORIDE. This is green, and sublimes in green needles, which give a blue solution with water. It may be obtained by heating metallic osmium in a current of dry chlorine gas. It forms double salts, which are of a green colour.

2. Osmium Tetrachloride. OsCl₄. *Syn.* OSMIC TETRACHLORIDE, OSMIUM DICHLORIDE. This may be procured in the same manner as the dichloride, using, however, an excess of chlorine. It occurs as a red, crystalline, fusible, deliquescent powder, which yields a yellow solution with water. It is more volatile than the dichloride.

OSTEOCOL'LA. A rough sort of glue or gelatin obtained from bones by digestion in dilute hydrochloric acid, to remove their earthy matter, and afterwards acting on the residuum with water at a high temperature, until it is wholly dissolved.

OTAL'GIA. Pain in the ear. See EARACHE.

O'THYL. In *chemistry*, a radical having the formula C_2H_5O, assumed by Professor Williamson to exist in acetic acid.

OTIORYNCHUS PICIPES, Curtis. THE RASPBERRY WEEVIL. This is one of a genus of weevils which prey upon cultivated crops, upon fruit trees, fruit bushes, corn, turnips, and other plants. Curtis calls this the pitchy-legged weevil and the night-feeding weevil, and speaks of its injury to many plants, and especially to raspberry canes. This has been, unfortunately, the experience of many fruit-growers in parts of Kent, Worcester, Bedfordshire, and Gloucestershire, and other places, who have made loud complaints of much damage to this fruit. In some instances even the bark, or thin tissue-like outer rind or cuticle of the cane has been eaten away in patches, so that an escape of sap was occasioned, while the leaves were pierced or bitten through in many holes. But the main and preferred subjects of the attack of this weevil are the fruit blossoms and the embryonic fruit-buds.

Raspberry culture is important and increasing in many districts, as it is profitable and not materially affected by foreign competition, and cultivators of this fruit were therefore rather alarmed at the onslaughts of this unknown foe. A foe unknown because the weevil feeds only at night, and remains concealed in the earth during the day. By watching the canes closely in the late eventide it was discovered that troops of 'little brown bugs' came forth from under the clods and stones around the canes, and swarmed up to these to feed upon the juices of their tender leaves and buds, just at a period when they are very full of sap and succulence, and when punctures and suctions by numerous snouts are calculated to do infinite mischief.

In a large raspberry plantation in Kent, upon a light 'stone-shattery' soil, serious harm accrued to young raspberry canes in their second

VOL. II.

year. Again, near Evesham, upon rather heavy land, though friable and in good cultivation, considerable damage was done to canes in their fourth season. As no invaders were ever seen, the evil was put down to flea beetles, to some species of the genus *Haltica*, by the labourers who knew that hop plants and turnip plants are constantly and seriously ravaged by these insects in a somewhat similar manner. Closer observation showed that the injury was different, and after a while the *Otiorynchus picipes* was seen by the light of a lantern in the very act of feeding upon the buds.

On the 24th May, 1886, at the time this was written, a lamentation concerning the action of this weevil upon raspberry canes has come in from a large fruit producer in Kent, whose land is a clay loam of medium texture upon the London clay beds overlying the chalk. This correspondent stated that they were more common in some fields than in others, and that if they were present in one season they nearly always came again in greater numbers the next.

Grave complaints of harm occasioned to raspberry canes have been made from time to time by growers in Cornwall, whose soil, or that upon which fruit is produced, appears to be favourable to the spread of the weevil.

This weevil also attacks blackberry canes in a similar manner. Blackberries are grown to some extent for market, and their culture is increasing, as they make admirable jam, alone or mixed with apples. A large prolific species has been introduced from the United States, where they are largely cultivated.

Near Ightham, in Kent, where cobnuts of the finest quality are grown, considerable injury was caused to the nut trees by insects biting the small twigs or spurs upon which the bunches of nuts are formed. Some of these were captured and found to be the raspberry weevil, *Otiorynchus picipes.*

At Hunton, also in Kent, where fruit trees of all kinds flourish exceedingly, it was reported that 'little bugs' were biting the red currant fruit-bearing spurs and those of the filbert trees, thus doing serious damage. These little bugs were *Otiorynchi.*

It is also not unfrequently found upon apple blossoms and leaves, whose juice it exhausts in the same manner as those of the raspberry and blackberry canes.

Life History. The raspberry weevil belongs to the extensive family of *Curculionidæ,* and to the genus *Otiorynchus,* comprising a number of species.

It is a very small insect, hardly three lines—the fourth of an inch—in length, having a short rostrum or snout, somewhat dilated, or spatulate —spoon-like—at the extremity, with lobes of an ear-like shape at both ends of the snout, and the under lip a little projecting. In colour it is light brown, and is on this account difficult to distinguish in a clayey soil. Upon its back there are spots and lines of a dark colour, and the ends of its six feet are black, from which its name of *picipes* is derived. Like all the species of the genus it is without wings, though it has elytra or wing-cases.

76

A great part of its existence is passed in the ground. It spends the day there during the period of its weevil form, and only comes forth at night to search for food. When it is discovered in its subterranean retreat it remains perfectly motionless, with legs folded up, counterfeiting death, after the crafty manner of many of the *Curculionidæ*.

Towards the end of the summer, when food grows scarce, the weevil lays eggs in the ground. From these in a short time whitish grubs are produced, legless, rather elongated, having brown heads. They feed for some time upon the roots of the plants whose leaves and buds their weevil progenitors have destroyed, and put on the pupal form upon the advent of spring, appearing as perfect weevils about the beginning of May.

Prevention. After an attack upon raspberry and blackberry canes, hot or quick lime, or lime ashes, or *pure* soot, should be put thickly round the canes in the autumn and dug in. Another good dressing of caustic substance may be given again in March, and well hoed in with prong-hoes directly the soil is dry enough. The clods should be well knocked about and pulverised. All stones, rubbish, and weeds should be removed.

Remedies. It has been found to be of some service to send labourers at night having tarred boards which they hold on either side of the rows of canes, while the canes are shaken violently, in order to dislodge the weevils and precipitate them into the tar. They stick fast in this, and many are killed thus, just as in the hop plantations the jumpers, *Euacanthus interruptus*, are trapped and slain. But these beetles are very wide awake, and fall to the ground on the slightest suspicion of danger and the first glimmer of a light. It is better, therefore, not to take lanterns in these expeditions.

Knowing that these enemies are but a little way under the ground during the day, it is easy to make raids upon them while they are napping. This may be done by chopping round the plants with prong-hoes put in deeply and smartly, and by applying at the same time mixtures of a caustic character or of a pungent odour. Fine earth, or dry ashes, or sawdust, or sand, saturated with a solution of carbolic acid in the proportion of about a pint to a bushel of either of the above media, might be used with great advantage.

Or a pint and a half of paraffin oil to a bushel of either of these would be equally efficacious in routing the weevils by making their headquarters unbearable.

Water containing three quarters of a pint of carbolic acid or a pint and a half of paraffin oil to ten gallons of water would have the same result. This might be put round the plants with garden engines, care being taken to direct the hose steadily and not too near the plants.

It need hardly be said that this operation would require great accuracy in making the mixture as well as in applying it.

Curtis speaks of natural enemies of this weevil in the shape of insects known commonly as sand wasps, of the order HYMENOPTERA, the family *Crabronidæ*, and genus *Cerceris*. These are like the common wasp, *Vespa vulgaris*, in colour, but have longer though narrower bodies, and a larger wing expanse. They make nests in sand-banks, gravel-pits, and other places, and carry home enormous quantities of weevils, especially those of the *Otiorynchus picipes*, *Otiorynchus sulcatus*, and the *Balaninus nucum*, for their young to feed upon ('Reports on Insects Injurious to Crops,' by Chas. Whitehead, Esq., F.Z.S.).

OTIORYNCHUS SULCATUS, Fabricius (from two Greek words signifying ear-snouted). THE STRAWBERRY WEEVIL. Strawberry plants often suffer considerably from this insect in its perfect state, as well as from its grubs or larvæ; though it by no means limits its attention to these plants, but also injures vines, raspberry and blackberry canes, and various plants and flowers.

The operations of this weevil, like those of many other insects, are very frequently unsuspected, and its effects are attributed to other causes. All who have strawberry plants obviously failing from root affection or attack should closely examine their roots; while if the runners are bitten through or the young blossoms nipped in the bud careful watch should be set to discover the origin of the evil. In the former case investigation will in all probability show that grubs are working hard among the roots, gnawing them with their horny jaws, and living on their succulent parts; while in the latter case a patient and discreet look-out will prove that weevils batten upon the plants in the stilly night.

There is no doubt that this weevil preys upon raspberry and blackberry plants in the same manner as its congener of the pitchy legs, which is described under O. *picipes*, but not to the same extent, as it evidently prefers strawberry plants if it can get them.

This weevil, as well as the *Otiorynchus picipes*, is known and dreaded in France and Germany as destructive to vines, strawberry, raspberry, and blackberry plants, to root crops, and to cultivated flowers. It is not known in America—at least Harris, Fitch, Lintner, Saunders, and other entomologists do not mention it.

Life History. The strawberry weevil is of the family *Curculionidæ* and the genus *Otiorynchus*. It is wingless, and rather longer than the raspberry weevil, or a little more than four lines, the third of an inch long. In colour it is dark, nearly black, and its six legs are clear-coloured and long. Its rostrum is short and stout, with a deep wide furrow, from which it derives its affix *sulcatus*, furrowed or grooved. Eggs are laid in the earth in the summer, from which grubs are speedily hatched. These are white, or of a slightly creamy-white hue, hairy, legless, and a little larger than the grubs of *Otiorynchus picipes*. They feed upon the roots of various plants, their transformation taking place in the first spring days, so that the weevils are fully grown and ready to seize upon the early leaves and buds directly they appear.

Prevention. A practice prevails of putting short straw or farmyard or stable manure between and under strawberry plants. This should not be done, as weevils of all kinds are undoubtedly encouraged, and capital shelter is thus afforded them. Good cultivation is most essential, both by digging and hoeing in spring and autumn, to disturb them, and to prevent them from egg-laying near fruit plants. Caustic substances should also be put on, and other applications, as suggested under *O. picipes*, to make their homes obnoxious to them.

When these weevils attack raspberry and blackberry canes the same measures of prevention should be adopted as recommended in respect of the raspberry weevil.

Remedies. Having discovered that the affection of the strawberry plants is due to weevils, the soil all round the plants should be forked deeply, and at the same time as delicately as possible, so as not to interfere with the blossom and the forming fruit. The earlier the weevils are detected the easier it will be to rout them by cultivation, by forking close round the plants, and by digging or horse-hoeing between the rows, and it may be by putting lime on if the attack be very bad. Strong-smelling remedies, such as paraffin-saturated earth or sawdust, would obviously be out of the question in the case of strawberry plants ('Reports on Insects Injurious to Crops,' by Chas. Whitehead, Esq., F.Z.S.).

OTIORYNCHUS TENEBRICOSUS. THE RED-LEGGED GARDEN WEEVIL. This is another species of the same genus of weevils, whose habits are exactly similar to those of the weevil which have already been described. It feeds upon many of the same plants, and the same modes of precaution and the same remedies should be adopted to check it. Strawberry plants are frequently much infested by it, whose leaves it pierces in innumerable holes. It also bites the runners and blossom-bearing joints.

It is about the same size as the *Otiorynchus sulcatus*, shiny black in colour, with reddish legs. Stephens says that it is slightly variable in colour, being sometimes of a reddish black, 'rufo-piceous.' This is probably the result of immaturity, Stephens adds ('Illustrations of British Entomology—*Mandibulata*,' by J. F. Stephens).

Moles are extremely fond of the grubs of this insect. It was observed that these animals were making 'heaves' in a strawberry-field, as it appeared, from mere wantonness. Some of the strawberry plants were dying, and the blossoms, just changing into fruit, were withering upon many. Upon searching it was discovered that at the roots and in the roots of the plants there were many grubs of this weevil, as well as many of the perfect insects, in the first week of June ('Reports on Insects Injurious to Crops,' by Chas. Whitehead, Esq., F.Z.S.).

OTTO OF ROSES. See OILS (Volatile).

OVALBU'MEN. White of egg; to distinguish it from seralbumen, or the albumen of the serum of the blood.

OVENS. A very ingenious and useful improvement in the apparatus for baking was introduced some years ago by Mr Sclater, of Carlisle. It consists in causing the articles to be baked to traverse a heated earthenware tube. This tube forms the oven. It is of considerable length, and the biscuits or other articles are slowly traversed through it, from end to end, at such a rate as will allow of the baking being completed

during the passage. The biscuits are carried on trays, set on travelling chains; or the trays are made into an endless web or chain. The oven is thus entirely self-acting, and the articles demand no attention whatever from the attendants, whilst the system combines superior economy with the best results. A 'pyrometer,' or heat indicator, is attached externally, so that the attendant can regulate the heat with great facility. The object of these improvements is to reduce the cost of baking, and to improve the appearance of the baked articles. The apparatus is applicable as well to the baking of articles of clay or earthenware as to bread or biscuits.

Of the ovens now in common use by the bakers, that known as the 'hot-water oven' is perhaps the best; not merely in reference to economy, but also with reference to its superior cleanliness, and the ease with which the articles operated on may be turned out of that delicate yellowish-brown tint for which the bread of the Viennese and Parisian bakers is so celebrated. See BAKING, BREAD, &c.

OWNER. For the purposes of the Public Health Act this term is thus defined:—"'Owner' means the person for the time being receiving the rack-rent of the lands or premises in connection with which the word is used, whether on his own account, or as agent or trustee for any other person, or who would so receive the same if such lands or premises were let at a rack-rent."

OX. The *Bos taurus*, Linn., one of the RUMINANTIA. In its more limited sense the word is restricted to the emasculated animal. The flesh, milk, skin, horns, bones, and blood of this animal are all serviceable to man. Goldbeater's skin is prepared from the peritoneal membrane of its cæcum. Its blood, fat, horns, and excrement were among the simples of the Ph. L., 1618. See BEEF, GALL, MILK, and *below*.

Ox-gall. *Syn.* OX-BILE; FEL BOVINUM, F. BOVIS, F. TAURI, L. Crude ox-gall is noticed under GALL. Refined ox-gall (*Fel bovinum purificatum*) is prepared as under:

1. Fresh ox-gall is allowed to repose for 12 or 15 hours, after which the clear portion is decanted, and evaporated to the consistence of a thick syrup by the heat of a water-bath; it is then spread thinly on a dish, and exposed in a warm situation near the fire, or to a current of dry air, until nearly dry; it is, lastly, put into wide-mouthed bottles or pots, and carefully tied over with bladder. In this state it will keep for years in a cool situation. For use a little is dissolved in water.

2. Fresh gall, 1 pint; boil, skim, add powdered alum, 1 oz.; boil again till the alum is dissolved, and when sufficiently cool pour it into a bottle, and loosely cork it down. In a similar manner boil and skim another pint of gall, add to it 1 oz. of common salt, and again boil, cool, and bottle it, as above. In three months decant the clear from both bottles, and mix them in equal quantities; the clear portion must then be separated from the coagulum by subsidence or filtration.

Uses, &c. Both the above are employed by artists to fix chalk and pencil drawings before tinting them, and to remove the greasiness from ivory, tracing-paper, &c. The first is also used in medicine.

OX'ALATE. *Syn.* OXALAS, L. A salt of oxalic acid. The soluble oxalates are easily formed by directly neutralising a solution of oxalic acid with a metallic hydrate, carbonate, or oxide; and the insoluble oxalates by double decomposition. See OXALIC ACID and the respective bases.

OXAL'IC ACID. $H_2C_2O_4$. *Syn.* ACIDUM OXA-LICUM, L. 'Essential salt of lemons.' This substance was discovered by Bergman in 1776. It occurs both in the mineral and organic kingdoms, and is produced artificially by the action of nitric acid on sugar, starch, woody fibre, &c. It abounds in wood-sorrel and other plants, in which it exists in combination with potassium or calcium. With few exceptions all starchy and saccharine substances yield oxalic acid when treated with nitric acid at a somewhat elevated temperature or by fusion with caustic alkalies.

Prep. 1. From sugar:

a. Nitric acid (sp. gr. 1·42), 5 parts, diluted with water, 10 parts, is poured on sugar, 1 part, and the mixture is digested at a gentle heat as long as gaseous products are evolved; the liquid is then concentrated by evaporation until it deposits crystals on cooling; the crystals, after being drained and freed from superfluous moisture, are redissolved in the smallest possible quantity of boiling water, and the solution is set aside to crystallise. The residuary 'mother-water' is treated with a little fresh nitric acid (say 1¼ parts) at a gentle heat, after which it is evaporated, as before, for a second crop of crystals. This process is repeated until the solution is exhausted. The brownish-coloured crystals thus obtained are allowed to effloresce by exposure to dry air, and are then redissolved and recrystallised. By repeating this treatment they yield pure colourless oxalic acid at the third crystallisation.

b. (*Schlesinger.*) Sugar (dried at 257° F.), 4 parts, and nitric acid (sp. gr. 1·38), 33 parts, are digested together, as before; and as soon as the evolution of gas ceases the liquid is boiled down to one sixth of its original volume, and set aside to crystallise. The whole process may be completed in about 2 hours, and in one vessel, and yields of beautifully crystallised oxalic acid, at the first crystallisation, a quantity equal to 56% to 60% of the weight of the sugar employed.

c. (*Ure.*) Nitric acid (sp. gr. 1·4), 4 parts, and sugar, 1 part, are digested together over a water-bath, and as soon as gas ceases to be evolved the vessel is removed from the bath, and set aside to cool and crystallise. The use of a little sulphuric acid along with the nitric acid contributes to increase the product.

2. From POTATO- or DEXTRIN-SUGAR. (*Nyren.*) From the washed pulp of potatoes, boiled for some hours with water in a leaden vessel, with about 2% of oil of vitriol, until the fecula of the pulp is converted into saccharine matter, shown by the liquid being no longer turned blue by iodine; the whole is then filtered through horse-hair bags or strainers, and the filtrate is evaporated until its density is such that a gallon of it weighs 14 to 14½ lbs.; in this state it is converted into oxalic acid by treatment with nitric acid in the way already described. A similar process was patented some years ago by Messrs Davy, Macmurdo, and Co.

3. From SAWDUST: (*Roberts, Dale, & Co.'s* patent.) This process is the one now usually employed for the manufacture of oxalic acid on the large scale. It is based on Gay-Lussac's discovery that wood and similar substances are converted into oxalic acid by fusion with caustic alkali. The practical details of the process are thus given by Dr Murray Thomson, of Edinburgh:—(1) Hydrate of sodium and hydrate of potassium, mixed in the proportion of 2 equivalents of the former to 1 equivalent of the latter, are dissolved, and the solution evaporated until of specific gravity 1·85; sawdust is now stirred in until a thick paste results. (2) This paste is then heated on iron plates, during which it is constantly stirred; water is first given off; the mass then swells; inflammable gases, hydrogen, and carburetted hydrogen are evolved, along with a peculiar aromatic odour. When the temperature has been maintained at 400° for one or two hours this stage of the process is complete. The mass has now a dark colour, and contains only 1% to 4% of oxalic acid, and about 0·5% of formic acid. The bulk, therefore, of the mass at this stage consists of a substance whose nature is not yet known, but which is intermediate between the cellulose of the sawdust and oxalic acid. (3) The next stage consists in a simple extension of the last, in which the mass is heated till quite dry, care being taken that no charring takes place. It now contains the maximum quantity of oxalic acid, 28% to 30%. (4) This oxalic acid now exists as oxalate of potassium and sodium in the grey powder resulting from stage 3. This powder is now washed on a filter with solution of carbonate of sodium, which seems to have the singular and unexpected power of decomposing the oxalate of potassium, and converting it into oxalate of sodium. At all events, it is quite true that all traces of potash are washed out with the solution of carbonate of sodium. The only explanation that occurs to account for this unusual decomposition is that oxalate of sodium is a more insoluble salt than oxalate of potassium, and therefore may be formed by preference. (5) This oxalate of sodium is now decomposed by boiling milk of lime. Oxalate of calcium falls as a precipitate, and soda remains in solution. The soda is boiled down, and again made use of with fresh sawdust. This recovery of alkali is also practised with the potassium salt which filters through in the last stage. (6) The oxalate of calcium is now decomposed in leaden vessels with sulphuric acid. Sulphate of calcium is precipitated, and oxalic acid forms in solution, which is now evaporated; the acid separates in crystals, which now need only to be re-crystallised to make them quite pure, and fit the acid for all the purposes for which it is employed. By this process 2 lbs. of sawdust are made to yield 1 lb. of oxalic acid.

Prop., &c. Colourless, transparent, prismatic crystals, possessing a powerful, sour taste and acid reaction; these effloresce in warm dry air, with loss of 28% (2 eq.) of water, and then form a white powder, which may be sublimed in part without decomposition; the crystals are soluble in 9 parts of cold water, in their own weight or less of boiling water, and in about 4 parts of alcohol; with the acids it forms salts called oxalates.

Tests. 1. Solution of chloride of barium occasions a white precipitate in neutral solutions of oxalic acid (oxalates), which is soluble in both nitric and hydrochloric acid. 2. Solution of nitrate of silver, under like circumstances, gives a white precipitate, which is soluble in nitric acid and in ammonia, and which, when heated to redness, yields pure silver. 3. Lime-water and solutions of all the soluble salts of calcium produce white precipitates, even in highly dilute solutions of oxalic acid or of the oxalates, which are freely soluble in both nitric and hydrochloric acid, but are nearly insoluble in either acetic or oxalic acid, and are converted into carbonate of calcium upon ignition. 4. Oxalic acid (or an oxalate), when heated, in the dry state, with oil of vitriol in excess, is converted into carbonic anhydride and carbonic oxide; the former produces a white precipitate with lime-water, and the latter, when kindled, burns with a faint blue flame. Of the above tests solution of sulphate of calcium (*vide* No. 3) is the most delicate and characteristic. 5. It is distinguished from Epsom salt by its acid reaction, its solubility in rectified spirit, its complete dissipation by heat, and by emitting a slight crackling noise during its solution in water. See MAGNESIA (Sulphate).

Uses, Pois., &c. Oxalic acid is chiefly used in the arts of dyeing, calico printing, and bleaching; to remove ink-spots and ironmoulds from linen, and to clean boot-tops and brass. It is extremely poisonous. The treatment, in cases of its having been swallowed, is to promote vomiting, and to administer chalk, whiting, or magnesia, mixed up with water, in considerable quantities. The use of the alkalies or their carbonates must be avoided, as the compounds which these form with oxalic acid are nearly as poisonous as the acid itself. The remaining treatment is noticed under ACIDS. In poisoning by oxalic acid the nervous system is almost always affected, and the patients experience numbness, formication of the extremities, and sometimes convulsions, so that the symptoms somewhat approach those produced by strychnia, from which it is distinguished by its corrosive action on the tissues, and its effect upon the heart and circulatory system.

Concluding Remarks. The manufacture of oxalic acid is an important one, and the process of Roberts, Dale, and Co. has so cheapened the price that more than half the amount of oxalic acid used all over the world is now made from sawdust. In manufacturing the acid from sugar, on the large scale, the first part of the process is either conducted in salt-glazed stoneware pipkins of the capacity of 3 to 5 quarts each (which are about two thirds filled and set in a water-bath), or in wooden troughs lined with lead, and heated by means of a coil of steam-pipe. On the small scale, a glass retort or capsule is commonly employed: The most appropriate temperature appears to be about 125° F., and the best evidence of the satisfactory progress of the decomposition is the free but not violent evolution of gas, without the appearance of dense red fumes, or, at all events, any marked quantity of them. When these are disengaged with violence and rapidity, a greater quantity of the newly formed acid suffers decomposition, and flies off in a gaseous form. The sp. gr. of the nitric acid commonly used on the large scale ranges from 1·22 to 1·27, equivalent quantities being taken. The evaporation is preferably conducted by the heat of steam. The evolved nitrous vapours are usually allowed to escape, but this loss may be in part avoided by conveying them into a chamber filled with cold damp air, and containing a little water, when they will absorb oxygen, and be recondensed into fuming nitric acid. Various modifications of this plan have been patented. That of Messrs McDougall and Rawson, which is one of the simplest and best, consists in passing the mixed nitrous fumes through a series of vessels containing water, and connected together by tubes, so that the fumes which collect at the top of one vessel are conveyed to nearly the bottom of the next one, and then, bubbling up through the water, mix with air, a supply of which is provided for the purpose. The nitrous fumes are thus brought alternately into contact with air and water, and by the time they reach the last vessel are reconverted into nitric acid. Another plan is to pass the mixed nitrous vapours through a vessel stuffed with some porous substance such as pumice-stone or pounded glass, conjointly with a supply of steam from a boiler and a supply of oxygen by a blowing machine.

The products obtained by skilful manipulation are—from good dry sugar, 128% ; from good treacle, 107%. "One cwt. of good treacle will yield about 116 lbs. of marketable oxalic acid, and the same weight of good brown sugar may be calculated to produce about 140 lbs. of acid." "As a general rule, 5 cwt. of saltpetre, or an equivalent of nitrate of soda, with 2½ cwt. of sulphuric acid, will generate sufficient nitric acid to decompose 1 cwt. of good sugar, and yield, as above, 140 lbs. of fair marketable oxalic acid, free from superfluous moisture " (*Ure*). On the small scale, 5 parts of sugar yield nearly 6 parts of crystallised acid.

Chemically pure oxalic acid is best prepared by precipitating a solution of binoxalate of potassium with a solution of acetate of lead, washing the precipitate with water, decomposing it, whilst still moist, with dilute sulphuric acid or sulphuretted hydrogen, and gently evaporating the filtrate so that crystals may form as it cools.

OXALURIA. Also known as the oxalic acid diathesis. An abnormal condition of the system, marked by the presence in the urine of crystals of oxalate of lime. The crystals occur as minute transparent octahedra, and sometimes in the form of dumb-bells. They can be easily recognised under a microscope with a power of from 200 to 250 diameters, when they present a very beautiful appearance. They differ from phosphatic deposits in being insoluble in acetic acid. Their presence is mostly indicated by the appearance in the urine of a cloud of mucus, which forms after the urine has stood some little time.

Oxaluria most generally affects persons of dyspeptic and sedentary habits and of nervous temperament; those suffering from skin affections and neuralgia are also occasionally attacked by it. In ordinary cases the treatment consists in the administration of the nitro-hydrochloric acid in infusion of gentian two or three times a day, or of a course of quinine and iron, aided by plenty of exercise in the open air, care being taken to avoid fatigue. If it can be borne, the shower-bath should also be had recourse to. Rhubarb tarts and tomatoes, which contain oxalic acid, must be excluded from the diet; so also should aërated water and too much sugar.

If after a short time the oxalates should not disappear from the urine under this treatment, the patient should seek proper medical advice; since the persistent presence of this deposit is of very serious significance, as indicating the existence in the bladder of that dangerous form of urinary concretion known as 'mulberry calculus.'

OXIDA'TION. The combination of bodies with oxygen, forming oxides; the operation or process adopted to induce or facilitate such conversion. Some familiar examples of oxidation are the tarnishing of metals in air, the drying of oils in paints, the formation of vinegar from alcohol, the respiration of animals, and combustion.

OX'IDE. *Syn.* OXYD; OXYDUM, L. A compound formed by the union of oxygen with another body.

OXYCHLO'RIDE. *Syn.* OXICHLORIDE; OXY-CHLORIDUM, L. A term often loosely applied to compounds of an oxide and chloride, whether in definite or variable proportions. See ANTIMONY (Oxychloride), &c.

OX'YCRATE. *Syn.* OXYCRATUM, L. The old name of a mixture of vinegar and water, dulcified with honey.

OXYCRO'CEUM. See PLASTERS.

OX'YGEN. O. *Syn.* OXYGEN GAS, DE-PHLOGISTICATED AIR†, EMPYREAL A., VITAL A.†; OXYGENIUM, L. An elementary body discovered independently by Scheele and Priestley in 1774. It is remarkable that, although this substance forms a large proportion of our atmosphere (nearly one fourth), and confers upon it the power of supporting respiration and combustion, and also constitutes the principal portion of the water of our rivers and seas (eight ninths), and enters largely into the composition of the majority of the various mineral bodies that form the bulk of our globe, its existence should have remained unsuspected, or at least undetermined, until a comparatively recent date. Oxygen is an essential constituent of all living organisms. It is absorbed by animals during respiration, and evolved in a free state by growing vegetables when exposed to sunlight. The oxygen gas of the atmosphere is mechanically mixed, not chemically combined, with the nitrogen.

Prep. 1. From red oxide of mercury, heated over a spirit lamp or a few pieces of ignited charcoal. The operation is usually performed in a small green glass retort, or in a short tube of hard Bohemian glass, closed with a perforated cork furnished with a piece of bent glass tube of small bore, to convey the liberated gas to the vessel arranged over a pneumatic trough to receive it. Pure. 1 oz. yields about 100 cubic inches.

2. From chlorate of potassium, as the last. Pure. 100 gr. yield nearly 100 cubic inches. This is the plan adopted in the P. Cod. The decomposition occurs at a heat below that of redness.

3. From a mixture of chlorate of potassium

(in coarse powder), 3 parts; powdered binoxide of manganese, 1 part; both by volume. Pure. 100 gr. of the mixture yield about 110 cubic inches of oxygen. This method, which has received the approval of Faraday, is exceedingly convenient. The gas is evolved with a rapidity which is entirely at the command of the operator by simply increasing or lessening the heat. The residuum in the retort may be kept for another operation, if not exhausted; or it may be at once washed out with a little warm water, and the manganese, which is uninjured by the process, reserved for future use. Red-lead, black oxide of copper, red oxide of iron, and several other substances, will do nearly as well as binoxide of manganese.

4. From a mixture of bichromate of potassium, 3 parts; oil of vitriol, 4 parts; gently heated, as before. Yields pure oxygen very freely (*Balmain*).

5. From binoxide of manganese and oil of vitriol, equal parts; as the last. 44 gr. of pure binoxide of manganese yield 8 gr., or 24 cubic inches, of oxygen; 1 oz. yields 88 gr., or 256 cubic inches (*Liebig*).

6. (On the large scale.) *a.* From nitre exposed to a dull red heat in an iron retort or gunbarrel. 1 lb. yields about 1200 cubic inches of gas, contaminated, more or less, with nitrogen (*Ure*).

b. From binoxide of manganese, as the last. 1 oz. of the pure binoxide yields 44 gr., or 128 cubic inches, of oxygen (*Liebig*); 1 lb. of good commercial binoxide yields from 1500 to 1600 cubic inches, or from 5 to 6 galls.

c. M. Boussingault has reinvestigated a process, long known, although not usefully applied, by which pure oxygen gas may be obtained from the atmosphere at a trifling cost, so as to enable it to be collected in unlimited quantities and preserved in gasometers, like coal-gas, for application in the arts, manufactures, and sanitation. This process depends upon a peculiar property possessed by the earth baryta, of absorbing atmospheric oxygen at one temperature and evolving it at another. Thus if baryta, BaO, be heated gently in the air to dark redness it takes up another atom of oxygen, and becomes barium dioxide, BaO₂; but when the temperature is raised to a bright red heat this additional atom of oxygen is given off, and baryta is re-formed. Brin's patent process depends upon this property of baryta; by it large quantities of oxygen are very cheaply prepared. It consists in passing air through earthenware retorts filled with baryta which are kept at a dull red heat. The baryta absorbs the oxygen and the nitrogen passes on. When no more oxygen is absorbed by the baryta the current of air is turned off. The retorts are connected with the gas-holders and the temperature is raised, the pressure within them being at the same time slightly lowered. The absorbed oxygen under these conditions is given off. By thus alternately varying the temperature and the other conditions of the process a regular production of gas can be obtained from a small quantity of baryta. The original baryta can thus be used over and over again.

The oxygen obtained by this method is usually

stored in iron bottles under great pressure, in which condition it can be conveniently transported to factories, laboratories, &c.

d. From ferrate of potassium, prepared on the large scale. When exposed to moisture or thrown into water, pure oxygen is evolved. This method has been successfully adopted to maintain the air of diving-bells, and of other confined spaces, in a state fit for respiration.

e. The decomposition of sulphuric acid has been recommended by Deville and Debray as a means whereby large quantities of oxygen gas may be obtained at a low price. Into a tubulated retort are put fragments of fire-brick, and upon these, when raised to a full red heat, sulphuric acid is made to fall drop by drop, through an iron tube, which is luted to the tubulure and reaches to the bottom of the retort, the acid being poured into it through a bent funnel. The sulphuric acid becomes decomposed into sulphurous anhydride, oxygen, and water. The volatilised products are sent through a spiral condenser, by which the water and any undecomposed acid become liquefied; whilst the sulphurous acid is removed by subsequent washing with water, the oxygen being collected in the usual manner. Sulphuric acid yields 15·68% of its weight of oxygen.

f. A process for obtaining oxygen on a large scale has been devised by Tessié du Motay. It consists in heating in a current of steam the manganates, permanganates, chromates, and ferrates of the alkalies and alkaline earths, and regenerating the residue by passing air over it at a red heat. This plan gives good results if the steam be kept dry.

7. OXYGEN GAS AT THE ORDINARY TEMPERATURE. Boettger states that when a mixture is made of equal weights of the peroxides of lead and barium, and dilute HNO_3 (9° Beaumé) is poured thereon, a current of pure O, free from ozone, is given off abundantly. This mixture of the two peroxides may be kept dry in a stoppered bottle for any length of time. Boettger also prepares pure oxygen, free from ozone, by submitting permanganate of potassium to a gentle heat.

8. Fleitman (Watts) has found that when chloride of lime in solution is heated with a small quantity of freshly prepared peroxide of cobalt, it is completely resolved into chloride of calcium and oxygen. A concentrated solution consisting of 35% of chloride of lime, which must be previously filtered to prevent frothing, yields, when heated with $\frac{1}{10}$ to $\frac{1}{8}$ per cent. of peroxide of cobalt, a volume of oxygen from 25 to 30 times as great as that of the liquid, and always rather more than the calculated quantity, probably in consequence of the absorption of oxygen from the air. The remaining peroxide may always be employed again. A like result follows if, instead of the peroxide, an ordinary salt of cobalt be used. Fleitman explains the reaction on the supposition that there are several peroxides of cobalt, and that the effects produced depend upon the alternate formation and partial reduction of a higher oxide; or on the formation of a cobaltic and a percobaltic hypochlorite, which is subsequently decomposed into cobaltous chloride and oxygen.

Prop. Oxygen gas is colourless, tasteless, inodorous, and incombustible; the sp. gr. is 1·057 (Dumas; 1·1026—Berzelius and Dulong; 1·111—Thomson); according to Regnault 100 cubic inches at 60° F., and 30 inches of the barometer, weigh 34·19 gr. (Dumas; 34·109 gr. —Berz; 34·6 gr.—Brande). Its density to that of atmospheric air is, therefore, as about 11 to 10. It is a powerful supporter of combustion, and its presence is essential to the existence of both animal and vegetable life. It forms 21% (20·81%) by volume, and 23% (23·01%) by weight, of the atmosphere (Dumas). Water dissolves about 5% by volume of oxygen, and by pressure a much larger quantity, forming oxygenated water (AQUA OXYGENII). Oxygen has been liquefied at a pressure of 320 atmospheres and −140° C.

Tests. 1. It is distinguished from other gases by yielding nothing but pure water when mixed with twice its volume of hydrogen and exploded, or when a jet of hydrogen is burnt in it. 2. A recently extinguished taper, with the wick still red-hot, instantly inflames when plunged into this gas. 3. A small spiral piece of iron wire ignited at the point, and suddenly plunged into a jar of oxygen, burns with great brilliancy and rapidity. Charcoal, sulphur, and phosphorus do the same.

Estim. The estimation of the quantity of oxygen in an organic compound is generally made by difference, and has already been described. For determining the quantity present in atmospheric air, and other like gaseous mixtures, Döbereiner has proposed the use of pyrogallic acid. The air under examination (freed from moisture) is measured into an accurately graduated tube over mercury, capable of holding about 30 c.c., and which it should 2-3rds fill. A solution formed of 1 part of dry hydrate of potassium and 2 parts of water, and in volume about 1-35th that of the air, is next introduced by means of a pipette with a curved point, and is gently agitated therewith in the gas for a short time; the decrease of volume gives the proportion of carbonic anhydride present. A solution of pyrogallic acid (1 grm. in 5 or 6 cm. of water), equal in volume to one half that of the solution of potash already used, is then introduced by means of another pipette, and the mixed liquids are cautiously shaken together over the inner surface of the tube. When absorption ceases (which it does in a few minutes), the quantity of residual gas (nitrogen) is read off from the graduations; the difference in volume before and after the introduction of the pyrogallic acid indicates the proportion of oxygen. This is a modification of Prof. Liebig's method. 1 grm. of pyrogallic acid in conjunction with hydrate of potassium is capable of absorbing about 189 c.c. of oxygen. Other methods employed for the analysis of air, depending on the increase or loss of weight when the air is passed over finely divided copper heated to redness, the loss of volume when the air is exploded in a eudiometer with half its bulk of hydrogen, or when a stick of phosphorus is left in it for some hours, are described at length in every elementary work on chemical analysis. The last method, although less accurate than the others, has the advantage of extreme simplicity.

Uses. Oxygen has been employed to increase the illuminative and heating power of lamps, and to render vitiated air respirable, &c.; and when largely diluted with atmospheric air, or condensed in water, as a remedial agent in asphyxia arising from the inhalation of carbonic anhydride and carbonic oxide.

Dr Ringer says that if oxygen be administered as a gaseous bath for an hour or two at a time, and the bath repeated six or eight times a day, it is of great service in senile gangrene.

Concluding Remarks. Oxygen gas may be collected in the usual way, either over water, mercury, or in bags; or, on the large scale, in gasometers. The purity of the products of the several processes given above depends on the substances from which the gas is obtained being themselves pure. For particular experiments the first portion of gas should be allowed to escape, or be received apart, as with this, as with the other gases, it is contaminated with the atmospheric air of the apparatus. The gas procured from manganese or nitre may be purified by passing it through milk of lime or a solution of caustic potash; it will still, however, retain some traces of nitrogen. Limousin ('Pharm. Centralhalle,' xiv, 318) has devised an apparatus for the preparation of oxygen by the attendants of hospitals, which obviates the risk of bursting of the retorts, attending its preparation by the old method. The apparatus consists of two cast-iron hemispheres, whose edges, which are well polished and about two centimetres thick, can be fitted hermetically upon each other, and fastened by three screws. The mixture of chlorate of potash and peroxide of manganese is placed in the lower hemisphere, which rests upon a tripod; the upper hemisphere, from which projects an iron tube, is now screwed on, and the iron tube connected by india-rubber and glass tubing with a Woulfe's wash-bottle, from which the gas after being washed passes through a second glass tube, and is thus ready for use. It may be conducted into an air-tight bag, in which it will keep for several weeks. Such a bag when supplied with a tube and stop-cock will afford a ready means for inhalations. See ORGANIC SUBSTANCES, OZONE, GASES, LIQUEFACTION OF, &c.

OXYGENATION. The act or process of combining with oxygen. Formerly it was of more general application than the word 'oxidation,' with which it has been regarded as synonymous. 'Oxygenation' is, however, at the present day practically obsolete.

OXYGENISED LARD. *Syn.* OXYGENATED AXUNGE; AXUNGIA OXYGENATA, L. *Prep.* (Ph. Bat. 1805.) From prepared lard, 16 parts, melted over a slow fire, and then mixed with nitric acid, 1 part, the combination being promoted by constant stirring with a glass rod until it ceases to affect litmus paper. It should be extremely white, and should be kept in the dark. See OINTMENT OF NITRIC ACID.

OXYHYDROGEN BLOWPIPE. See BLOWPIPE. Deville and Debray ('Ann. Ch. Phys.' [3], lvi, 385) employ the oxyhydrogen blowpipe in the following manner for effecting the fusion of pla-

tinum and the refractory metals which accompany it. The apparatus consists of the blowpipe *C* (*see below*), a furnace *ABD*, and a crucible *GHI*. The blowpipe is composed of a copper tube about half an inch in diameter, terminating below in a slightly conical platinum jet about 1½ inches long. Within this tube, which is supplied with hydrogen or coal-gas through the stopcock *H*, is a second copper tube *C'* for supplying oxygen, which is supplied through the stopcock *O*, terminated also by a platinum nozzle with an aperture of about a twelfth of an inch in diameter.

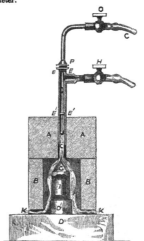

The furnace *ABD* consists of three pieces of well-burnt lime of slightly hydraulic quality, which may be turned at a lathe with ease. The cylinder *A* is about 2½ inches thick, and is perforated by a slightly conical hole, into which the blowpipe fits accurately, passing about halfway through the thickness of the mass. A second somewhat deeper cylinder of lime, *B*, is hollowed into a chamber wide enough to admit the crucible, and leave an interval of not more than a sixth of an inch clear around it. *KK* are four apertures, two of which are shown in the *engr.*, for the escape of the products of combustion.

The outer crucible *HH* is also made of lime, but it contains a smaller crucible *I* of gas coke, provided with a cover of the same material; and in this the substance to be used is placed, the crucible resting on the tube support *D'*. The conical cover *G* is made of lime, and its apex should be placed exactly under the blowpipe jet, at a distance from it of ¾ to 1¼ inches.

The different pieces of the furnace must be bound round with iron wire to support them should they crack. The oxygen is admitted under a pressure equal to that of a column of 16 inches of water. The temperature is gradually raised to the maximum, and in about eight minutes from this time the operation is complete.

By employing a jet of mixed coal-gas and oxygen (*EQ*—see *engr.*) in a furnace of lime

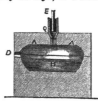

Deville and Debray succeeded, at an expense of about 43 cubic feet of oxygen, in melting and refining, in 42 minutes, 25·4 lbs. avoirdupois of platinum, and casting it into an ingot in a mould of gas coke; much larger masses have since been melted by this method. Lime is so bad a conductor of heat that if a cup of lime not more than 0·8 inch thick be filled with melted platinum the exterior scarcely rises beyond 300° F. (Miller's 'Elements of Chemistry,' 3rd ed., pt. ii, p. 825).

OXYHYDROGEN LIGHT. The following hints as to the use of the oxyhydrogen light for the optical lantern will be found useful to the inexperienced :

1. If the apparatus is frequently moved about from one place to another, it should be contained in a box which will hold the lantern and all accessories in compartments, so that it may be seen at a glance whether all that is required for an exhibition is in its place.

2. The 'blow-through' jet is the simplest and safest, but the 'mixed' jet gives a more powerful light, and is said to be more economical. Any form of jet should have an arrangement by which the lime cylinder can be turned round and raised or lowered at will while the light is burning and the lamp is in position.

3. The simplest, safest, and most convenient way of using the gases is from steel cylinders; bags are cumbrous, short-lived, and very liable to accident. Coal-gas may often be obtained from the mains in the room in which the exhibition is to take place. This is convenient, as the supply is practically unlimited and requires no attention.

4. Each cylinder of gas should be fitted with a regulator. Beard's is an excellent form for the oxygen. It is best to have some other make for the coal-gas or hydrogen cylinder; the terrible consequences of mixing the gases and exploding them under pressure will thus be effectually avoided. *The regulators must always be used for the same gas, never interchanged.* These regulators are generally sent out with differential screws for tightening them on to the gas cylinder. This is objectionable, as the threads are very liable to be torn and are difficult to adjust. A simple union fly nut, rather heavy and with strong arms to fit the wrench, is altogether preferable.

5. It is well to paint the words OXYGEN, HYDROGEN, COAL-GAS, in large letters on the respective cylinders, and to use one coloured

rubber tubing for oxygen and another for the coal-gas or hydrogen in making connections.

6. Pressure gauges are sold for testing the amount of gas in the cylinders. One only is really necessary, viz. for the oxygen. In using a gauge *never on any account* oil or grease any part of it, or a terrible explosion may result. Further, turn the tap of the cylinder *very slowly ;* if turned rapidly the great pressure suddenly put upon the gauge may blow it to pieces. A strong gauze cover is advisable. As gauges are expensive instruments, it is well to know that the weight of the cylinder is an excellent guide to the quantity of gas left, a cubic foot of oxygen weighing almost one ounce.

7. The best lime cylinders are the cheapest in the end, and with care one will last a long time. They should at all times, when not in use, be kept in quicklime in a tin case, or better in a well-stoppered bottle. When used the cylinder should be carefully heated in the coal-gas flame, and frequently turned, before turning on the oxygen. This saves the lime from splitting. During an exhibition, at convenient times, the lime should be turned round a little in order to prevent pitting.

8. Before commencing an exhibition the lantern, condensers, front lens, and other optical parts of the lantern should be carefully wiped with a hot handkerchief to remove moisture, and in cold weather it is very desirable to warm all the apparatus before the fire, as this prevents the deposition of moisture and the appearance of drops, &c., on the screen.

9. When in use the oxyhydrogen light should not hiss or make more than a hardly perceptible noise, otherwise waste of gas is going on. A skilled operator will not use more than 3 feet of oxygen and the same quantity of coal-gas, or a little more, per hour.

OXYMEL. *Syn.* OXYMEL, L. An acidulous syrup made of honey and vinegar. There are only two oxymels in the last Ph. B. The ingredients in an oxymel should be of such a character and in such proportions as to produce a mixture of the proper consistence without evaporation.

Oxymel of Col'chicum. *Syn.* OXYMEL COLCHICI CORMI, OXYMEL COLCHICI, L. *Prep.* (Ph. D. 1826.) Fresh corms (roots) of meadow saffron, 1 oz.; distilled vinegar, 1 pint (wine-measure); macerate for two days, press out the liquor, filter, add of clarified honey, 2 lbs., and boil down the mixture to the consistence of a syrup, frequently stirring.—*Dose,* 1 to 3 dr., twice or thrice a day; in gout, rheumatism, dropsy, &c.

Oxymel of Gar'lic. *Syn.* OXYMEL ALLII, L. *Prep.* (Ph. L. 1746.) Sliced garlic, 1½ oz. ; caraway seed and sweet fennel seed, of each, 2 dr. ; boiling vinegar, 8 fl. oz. ; infuse, strain, and add of clarified honey, 10 oz. In hooping-cough, chronic diarrhœa, rheumatism, &c.

Oxymel of Narcissus. (*Van Mons.*) *Syn.* OXYMEL NARCISSI. *Prep.* Vinegar of narcissus (made with 1 part of fresh flowers of daffodil to 8 of vinegar), 1 part ; honey, 4 parts. Dissolve. —*Dose,* ½ teaspoonful. In hooping-cough and spasmodic asthma.

Oxymel, Pec'toral. *Syn.* OXYMEL PECTORALE, O. INULÆ COMPOSITUM, L. *Prep.* (Ph. Br.)

Elecampane, 1 oz.; orris root, ½ oz.; water, 1½ pints; boil to half a pint, strain, add of honey, 16 oz.; ammoniacum, 1 oz.; (dissolved in) vinegar, 3 fl. oz.; lastly, boil to an oxymel.—*Dose*, 1 spoonful occasionally; in coughs, humid asthma, &c.

Oxymel, Sim'ple. *Syn.* VINEGAR SYRUP, ACETATED HONEY; OXYMEL (Ph. L. & D.), OXYMEL SIMPLEX, MEL ACETATUM, L. *Prep.* 1. (Ph. L.) Acetic acid (sp. gr. 1·048), 7 fl. oz.; distilled water, 8 fl. oz.; mix and add them to honey, 5 lbs., previously made hot. This contains only one half the acid ordered in the Ph. L. 1836.

2. (Ph. D.) Clarified honey, 1 lb.; acetic acid (sp. gr. 1·044), 3 oz.; as before. Stronger than the last.

3. (B. P.) Clarified honey, 8 parts; acetic acid, 1 part; distilled water, 1 part. Liquefy the honey by heat, mix in the acid and water.

4. (Wholesale.) From honey (thick and good), 12 lbs.; melt it by a gentle heat, add of distilled vinegar (of fully 5%), 2 quarts, and strain the mixture through flannel. No evaporation is required.

Uses, &c. Demulcent and refrigerant.—*Dose*, 1 to 4 fl. dr., either gradually sucked from the spoon or dissolved through some simple liquid. Dissolved in water, it forms a useful and pleasant cooling drink or gargle in fevers, sore throats, hoarseness, &c.; but in some individuals it occasions griping. It is commonly used as an adjunct, in mixtures, &c.

Oxymel of Squills. *Syn.* HONEY OF SQUILLS; OXYMEL SCILLÆ (B. P.), MEL SCILLÆ (Ph. L.), O. SCILLITICUM, L. *Prep.* 1. (B. P.) Mix and evaporate on a water-bath vinegar of squills, 1 pint, and clarified honey, 2 lbs., till the product when cold has a specific gravity of 1·32.

2. Vinegar of squills, 2½ pints; gently evaporate it to 12 fl. oz., and add of honey (previously made hot), 5 lbs.

3. (Ph. L. 1836.) Strained honey, 3 lbs.; vinegar of squills, 1½ pints; boil to a proper consistence. The formula of the Ph. D. 1826 was similar.

Uses, &c. Expectorant, and in large doses nauseant.—*Dose*, ½ to 2 fl. dr.; in chronic coughs, hoarseness, humoral asthma, &c.

Oxymel of Ver'digris. See LINIMENT OF VERDIGRIS.

OXYR'RHODYNE. *Syn.* OXYRRHODINON. An old compound formed of 1 part of vinegar of roses and two parts of oil of roses.

OXYSAC'CHARUM. A syrup acidulated with vinegar. See SYRUP.

OXYSUL'PHIDE. A name given to certain compounds or mixtures of metallic oxides and sulphides. See ANTIMONY OXYSULPHIDE, &c.

OYS'TER. *Syn.* OSTREA, L. This well-known shell-fish is the *Ostrea edulis*, Linn.

"The oyster is a genus of lamellibranchiate molluscs of the section with a single adductor muscle. The shell consists of two unequal and somewhat irregularly shaped valves of laminated and closely foliated structure, and the hinge is without tooth or ridge, the valves being held together by a ligament lodged in a little cavity in each. The animal is in its organisation among the lowest and simplest of lamellibranchiate molluscs. It has no foot, and, except when very young, no power of locomotion, or organ of any kind adapted to that purpose. Its food consists of animalcules, and also of minute vegetable particles, brought to it by the water, a continual current of which is directed towards the mouth by the action of the gills. The gills are seen in four rows when the valves of the shell are separated, a little within the fringed edge of the mantle. In the most central part is the adductor muscle; and between the adductor muscle and the liver is the heart, which may be recognised by the brown colour of its auricle. The mouth—for, as in the other *Lamellibranchiata*, there is no head—is situated beneath a kind of hood formed by the union of the two edges of the mantle near the hinge. It is jawless and toothless. The ovaries are very large during the season of reproduction, which extends over certain months when oysters are out of season for the table. Oysters are hermaphrodite" (Chambers' 'Encyclopædia').

The fecundity of the oyster is amazing. Leeuwenhoek estimated that an oyster, when full of spawn, contained from 3000 to 4000 of its offspring, and it has also been computed that one oyster alone produces nearly a million and a quarter of eggs. The eggs are hatched and the young produced within the shell and mantle of the parent, where they continue floating or swimming about in the vicinity of the gills in a creamy-looking kind of mucus or fluid until expelled. Their expulsion is preceded by a change of appearance in the fluid to a brownish or muddy colour; a circumstance that may possibly indicate an alteration of composition in the liquid unfavourable to the infant oyster, and thus lead to its departure. Their departure or expulsion from all the parent molluscs of the oyster bank or bed takes place at the same time.

When they leave the parent shell the young oysters, which in this condition are called *spat*, are not more than 1⁄140 of an inch in length; and two millions of them when closely packed do not occupy a space of more than a cubic inch. Thus cast adrift they are carried away by currents, their multitudinous numbers being considerably diminished by their falling a prey to numerous fish, as well as from their frequent inability to find a suitable resting-place. This obtained, the young oyster or spat attaches itself to it, and makes it the permanent home on which it eats, grows, and breeds, and, debarred of locomotion, passes its existence, unless, of course, removed by external causes. Pending its obtaining a suitable locality the young oyster is provided with a powerful swimming apparatus which, it has been surmised, becomes absorbed or otherwise disappears when its function is rendered unnecessary by the stationary life of the oyster after it has secured a habitat.

The objects to which it attaches itself are numerous. The *Ostrea parasitica*, a species of oyster found in warm climates, fixes itself to the roots and branches of trees growing within reach of and washed by the tide. Again, in some of the southern States of North America, large oyster-beds, which are sometimes of such magnitude as to form buttresses against the force of the tides and winds, originate from the habit of young oysters attaching themselves to the shells of old ones. Similarly the banks of some of the rivers of Georgia, which run up some few miles inland from the sea, are composed of masses of living oysters attached to each other. These banks, which are so massive as to make a channel for the river, are known as *racoon banks*, because this animal is one amongst others which frequents them for the sake of devouring the oysters. In some of the French *parcs*, or artificial oyster-beds, the young oysters attach themselves to large unglazed tiles, or to fagots or other solid bodies which are placed in suitable situations for the purpose; in the English, artificial beds of hurdles are frequently employed, upon which the spat become deposited. It appears the young oysters select dark objects, such as slate or black stone, in preference to bodies of a lighter colour to fix themselves to, and that they choose, where practicable, the inner side of the object, or that portion of it away from the light. After a time the young oysters are removed from the breeding beds, placed in the fattening beds, from whence they are removed when they have attained a sufficient size, and sent upon the market. In England oysters are not regarded as fit to be eaten until they are at least three years old; whereas in France they are served up to table about a year earlier. The chief enemy of the young oyster is a species of whelk, known in France as the *bigourneau*, *dog-whelk*, or *piercer*. These creatures, which are found in immense quantities in the celebrated oyster-beds at Arcachon, near Bordeaux, cause great destruction amongst the bivalves. Part of their anatomy consists of a boring apparatus, with which they pierce the shell of the oyster; whatever of the dead oyster is left by the whelk is devoured by the crabs, which creep into the aperture in the shell made by the former.

We have already alluded to the abundance of oysters in parts of Georgia, where, we may add, they are not only confined to the alluvial shores of the rivers, but are also found in large numbers amongst the long grass of the adjoining low lands.

In these districts it is by no means an uncommon practice for the inhabitants to improvise a meal by picking up a bunch of oysters and roasting them over a fire kindled on the spot. In many of these localities the oysters occur in quantities so immense that a vessel of 100 tons might be loaded within three times her own length (Chambers' 'Encyclopædia').

There are also many other parts of America in which the yield of the oyster-beds is enormous. In the State of Maryland 6000 persons are said to be employed dredging, and nearly 11,000,000 bushels of oysters were taken in 1870-71.

In Baltimore as many as 10,000 persons are employed in tinning this bivalve. Comparing the plenteousness of the oyster in America with its great scarcity of late years in our country, and the consequent much lower price of the foreign bivalve, we should be prepared to learn that considerable supplies of oysters, both alive and preserved in tins, come to us from America.

The bulk of those consumed in Britain are a small variety, and come from Maryland and Virginia.

In 1872, owing to the diminished yield of the English oyster-beds, an attempt was made to introduce the American oyster into British waters; and we believe the depôt for this purpose still exists at Cleethorpes, at the mouth of the Humber, where operations in this branch of oyster culture are being carried on by the Conway Company. If, however, the opinion of an eminent pisciculturist be correct, viz. that the American oyster will not breed in our waters, we should conceive the experiment will be abandoned, since nothing will be gained by relaying them that cannot be attained by simply importing them and sending them to the market, since it is asserted they are kept alive out of water for a month.

A few years back a Select Committee appointed by Parliament to inquire into the causes of the scarcity of oysters, issued in 1876 a report in which, endorsing the opinion of previous authorities on oyster culture, they attributed the diminished yield of our oyster-beds to continual over-dredging for them in open waters, without allowing sufficient 'close time.' The Committee found that, in France, where the stringent observance of the 'close season' was enforced, the supply of oysters had increased concurrently. The Committee, therefore, recommended the establishment of a 'general close time,' extending from May 1st to September 1st, subject to certain exceptions under the supervision of the Board of Trade; the levying of penalties for buying or selling oysters for consumption during the 'close season' being also recommended. The Committee further recommended that no oysters should be sold from the deep-sea fisheries under 2½ or 3 inches in diameter. Commenting upon the above report, 'Nature' very sensibly remarks :—" What is really wanted for the protection of the oyster is the assurance that these animals shall not be sold before they have a chance of reproducing their kind. Since the introduction of the railway system, the demand for oysters in distant places has become so great and the price has risen so high, that oyster culturists are tempted to send immature animals to market, and it is this fact, more than any failure of spat, that is leading to the scarcity. There are not, in consequence of the unceasing demand and consequent high price, so many full-grown oysters left to spat as there ought to be; hence the scarcity. Any Act of Parliament that decrees two oysters to grow where only one grew before will be greedily welcomed both by oyster culturists and by the public, and we hope that the issue of the present report will lead to some effective measures being taken for the preservation of this delicious creature ere it be too late." Previous to 1846 the wholesale price of best English natives was £2 2s. a bushel; since then the price has risen rapidly to £4 4s. in 1865, in 1866 to £5, and in 1869 it had advanced to £8; that is, they had risen nearly 300% in 8 years, which is equivalent to an advance of from ⅝d. to 2d. each. At the present time they are, we believe, sold at from 3s. to 3s. 6d. a dozen by the retail dealer.

Oysters are nutritious and easy of digestion when fresh, but are apt to prove laxative to those unaccustomed to their use. It is generally believed that they are in season each month of the year the name of which contains the letter R. Whitstable in Kent, and Colchester and other places in Essex, are the great nurseries or feeding-grounds for supplying the metropolis, and, indeed, the whole of England, with the most esteemed variety (NATIVES) of this shell-fish. The shells (TESTÆ PREPARATÆ, T. OSTRE-ARIÆ) were formerly used in medicine as an absorbent.

Of the various species of oysters, that which holds the foremost place in the estimation of the *gourmet* is the ‘English native;’ now, alas! owing to the unwise rapacity of the collector, nearly dredged out of existence.

The native has an historic reputation, too, since it appears it was eagerly sought after by the old Romans, and was a frequent dish at their tables. The enthusiasm of the celebrated Dr Kitchener for this particular oyster was very intense. He is very particular in directing its shell to be opened with the greatest care, so that it may be eaten alive and ‘*tickled to death by the teeth.*’

The green oyster of Ostend is also prized by epicures; it acquires its colour from its food, which consists chiefly of green monads and confervæ. Some of the American oysters are excellent in flavour, and are said to be without the copper taste occasionally to be met with in English oysters. They smack a little of the mussel.

Payen gives the following as the composition of the oyster:

	Mean of two Analyses.
Nitrogenous matter	14·010
Fatty matter	1·515
Saline matter	2·695
Non-nitrogenous matter and loss	1·395
Water	80·385
	100·000

See SHELL-FISH, SAUCES, &c.

Oyster, Scalloped. Put them with crumbs of bread, pepper, salt, nutmeg, and a bit of butter, into scallop-shells or saucers, and bake them before the fire in a Dutch oven.

Oysters, Fried (to garnish boiled fish). Make a batter of flour, milk, and eggs, add a little seasoning to it, dip the oysters into it, and fry them a fine yellow-brown. A little nutmeg should be put into the seasoning, and a few crumbs of bread into the flour.

Oysters, Stewed. Open them, and separate the liquid from them, then free them from grit by washing, strain the liquor, and add to the oysters a small piece of mace and lemon-peel, and a few white peppercorns. Simmer very gently, and add some cream and a little flour and butter. Let them be served with sippets.

Oysters, to Feed. Put them into water, and wash them with a birch broom till quite clean. Then place them bottom downwards in an earthen-ware pan; sprinkle them with flour, oatmeal, and salt, and then cover with water. Repeat this treatment every day, taking care to make the water pretty salt.

OZOKERIT. *Syn.* FOSSIL WAX, MINERAL WAX, CERITE, CERESIN. This substance, which has within the last few years been utilised as a source of paraffin and the mineral hydrocarbon oils, is found in various localities in the tertiary strata, mostly occurring in close proximity to petroleum springs, and often associated with bituminous sandstones, clay schist, gypsum, and sodium chloride. But although extensive deposits of it are to be met with in Galicia, on the slopes of the Carpathian mountains, it is by no means an abundant body. In the Austrian empire there are many large manufactories for its conversion into paraffin and the mineral oils. The products of this conversion are benzine, naphtha, heavy oils, solid paraffin, and coke. Ozokerit is usually met with as a brown and compact substance, occasionally yellow, but sometimes black. It melts at a temperature varying from 60° to 80° C., but some inferior kinds are fusible at 100° C.

NEFT-GIL is a variety of ozokerit, and is found on the island of Swätoi-Ostrow, in the Caspian Sea. According to Rossmässler, neft-gil is treated in the following manner:—15 cwt. of the crude material is put into iron stills provided with a leaden worm, and submitted to fractional distillation, yielding 8% of distillate, consisting of 8% of oil and 60% of crude paraffin. The oil thus obtained is yellow, opalescent, possesses an ethereal odour, and a sp. gr. of 0·75 to 0·81. Each distillation yields a quantity of a light oil, boiling below 100° C., which is used for the purpose of purifying the paraffin. The crude paraffin obtained by the first distillation is tolerably pure, has a yellow colour, and can at once be treated by the hydraulic press and centrifugal machine; the oil from these operations is again submitted to fractional distillation in order to obtain more paraffin. The pressed paraffin is melted and treated at 170° to 180° C. with sulphuric acid, which is next neutralised by means of lime, and the paraffin again rapidly distilled, then again submitted to strong pressure, and the material obtained yielding 68% of the light oil; it is then again melted, again pressed, and finally treated with steam for the purpose of eliminating the last trace of oil. The material obtained by this treatment is a perfectly pure, colourless material, free from smell, transparent, and so hard as to exhibit in large blocks almost a metallic sound. The fusing-point is 63° C.

Uses, Tests, &c. Ozokerit is imported from Galicia, Hungary, and Russia, for the manufacture of candles. It is a source of illuminating and lubricating oil, and is often used for adulterating beeswax. From the latter substance it may be distinguished by the fact that warm concentrated oil of vitriol scarcely attacks it, whereas beeswax is completely decomposed.

In all its chemical properties ozokerit resembles solid paraffin.

OZONE (Greek ὄζω, I smell) is a variety of oxygen characterised by its greater weight, its peculiar chlorous smell, its intensely active oxidising powers, and, finally, by the ease with which it passes into common oxygen. The history of ozone may be summed up as follows:—In 1785 Van Marum observed the production of a peculiar smell when electric sparks were passed through oxygen. This smell, which every one who has worked with an electric machine must have noticed, Van Marum regarded as the 'smell of electricity,' thinking that electricity was a substance. In 1840 Schönbein, of Basle, proved the existence of a definite substance, to which he assigned the name of ozone; he also discovered several methods of producing it, a delicate test for it, and several of its most striking properties. He subsequently added many new facts, but to the time of his death he never held a correct theory with regard to its nature. Later researches by Marignac and De la Rive, Becquerel, and Frémy, Andrews and Tait, Soret, Brodie, and others, have established the true nature of this remarkable body. It is now generally admitted that it only differs from common oxygen in containing three atoms of oxygen in each molecule instead of two. In fact, as the formula for oxygen is O_2, that of ozone is O_3. It follows that ozone is half as heavy again as oxygen, and it has been demonstrated that its specific gravity is 24 (H = 1), that of oxygen being 16. All the known reactions of ozone are easily explained in accordance with this view.

Ozone may be generated in several ways.

1. By the action of electricity on oxygen or air, the silent or 'slow' discharge being the most efficacious. The best apparatus is the induction-tube of Siemens. This consists of two tubes, one inside the other. The inner side of the inner and the outer side of the outer tube are coated with tinfoil, and these coatings are connected with the terminals of a powerful induction coil. Dry air or oxygen streams between the tubes and passes out, strongly charged with ozone.

2. Boillot has proposed a modification of Siemens' apparatus, which consists of two glass tubes, one fitting within the other, and each coated externally with powdered coke, which is made to adhere by means of gelatin. The coatings of the two tubes are connected with the terminals of an induction coil, and a stream of oxygen is made to pass between the tubes, and becomes thus exposed to the influence of the silent discharge, as in Siemens' contrivance.

3. Houzeau has invented an apparatus which he calls an 'ozoniser,' by means of which ozone is produced in considerable quantities. In an ordinary straight gas delivery-tube is placed a wire of copper, lead, or, better, platinum, 4 to 6 decimètres long, with one of its extremities passing through the side of the upper portion of the tube. On the exterior of the tube is coiled a similar wire over the path of the preceding. When the two are placed in communication with a Ruhmkorff's coil, giving a 2 or 3 centimètre spark, a slow stream of oxygen passing through the tube will be strongly charged with ozone. By this apparatus Houzeau has prepared oxygen containing 60 to 120 (once 188) milligrams of ozone per litre. Electrolysis of water furnished only 3 to 5 milligrams, barium peroxide and sulphuric acid 10 milligrams per litre ('Comptes Rendus,' 'Watt's Dictionary,' second supplement, lxx, 1286).

4. During certain processes of oxidation. A piece of phosphorus, half covered with water in a bottle of air, absorbs a portion of the oxygen,

while another portion becomes partially ozonised.

5. By plunging a clean glass rod heated to about 260° C. into a jar containing a few drops of ether.

6. By mixing very gradually 3 parts of strong sulphuric acid and 2 of permanganate of potash.

7. It has been shown that ozone is formed in small quantity during the burning of hydrogen at a jet, and in several analogous reactions.

8. During the liberation of oxygen at low temperatures. When barium dioxide is moistened with sulphuric acid, the odour of ozone is at once apparent, and the evolution proceeds for a considerable time.

9. In the electrolysis of water the oxygen evolves a small quantity of ozone, especially if the poles are small.

10. Linder has suggested an easy method for the production of ozone for hygienic purposes, which is as follows:—Make a mixture of manganese peroxide, potassium permanganate, and oxalic acid. Two spoonfuls of this powder, if placed on a dish and gradually mixed with water, will generate ozone sufficient for a room of medium size; more water is added in small portions from time to time; the powder may be kept in a bottle ready for use.

But Schönbein has shown that many essential oils possess the property of absorbing it without decomposing it. By the use of Siemens' apparatus, oxygen containing, as a maximum, twenty volumes per cent. of ozone may be obtained. This represents a contraction of about 1-11th during formation. But it is at present impossible to separate the one from the other. Ozone is entirely converted into oxygen by a temperature of 237° C. The conversion is effected more slowly at lower temperatures. Silver, iron, copper, when moistened, are oxidised on the surface immediately at ordinary temperatures by ozone; organic substances are destroyed.

Silver even becomes converted into a peroxide, although it will not combine with ordinary oxygen, either when moist or dry. Little or no absorption of ozone takes place when the metals are perfectly dry, except with dry mercury and dry iodine, both of which remove it immediately. It was conclusively shown by Andrews and Tait that little or no contraction followed the absorption of ozone by these or any other agents. Hence, as suggested by these observers, it seems probable that the ozone is resolved into a quantity of ordinary oxygen equal in bulk to itself, which is liberated at the moment when another portion of its oxygen enters into combination with the metal or the iodine.

Ozone has been condensed into an indigo-coloured liquid, which boils at 106°.

Ozonised air becomes deozonised when passed over cold manganese dioxide, silver dioxide, or lead dioxide. When ozone is mixed with peroxide of hydrogen, water and oxygen are formed. In these cases the ozone is converted into ordinary oxygen, and the peroxides into monoxides.

Antozone, which Schönbein surmised to be oxygen in an oppositely electrified condition to ozone, has been shown by Van Babo to be peroxide of hydrogen.

From the ease with which it gives up its third atom of oxygen, ozone has been proposed, when mixed with air, as a means of decolourising wax, stearin, and other organic substances which cannot be subjected to the fumes of sulphurous acid or chlorine, or at any rate only partially so. On account of its oxidising properties, ozone is used to bleach engravings discoloured by age; these are rolled into the neck of a large glass balloon, in which a stick of phosphorus is suspended, and which contains a little water in the bottom. It has also been employed in oxidising alcohol to aldehyde in the manufacture of the well-known aniline green dye.

Being one of the most energetic oxidising agents known, it is not surprising that the claims of ozone as a disinfectant should have found many supporters. One of its strongest advocates for this purpose is Dr Cornelius Fox, who says, "Ozone should be diffused through fever wards, sick-rooms, the crowded localities of the poor, or wherever the active power of the air is reduced and poisons are generated. Its employment is especially demanded in our hospitals, situated as they mostly are in densely populated districts, where the atmosphere is almost always polluted by rebreathed air, decomposing substances and their products, and where no mere ventilation can be fully effective. If practicable, it would be highly advantageous to direct streams of sea air, or air artificially ozonised, into the fever and cholera nests of our towns. Ozone may be easily disseminated through public buildings, theatres, and other confined atmospheres, where numbers of people are accustomed to assemble, in order to maintain the purity of the air."

Another ardent believer in the hygienic value of ozone is Lender, who is also a strong advocate for its medical application, and recommends it, both in the form of ozonised air and water, in tuberculosis, rheumatism, asthma, and many other diseases.

The contention of those who assert that it is impossible to convey such an unstable body as ozone into the blood without the ozone becoming decomposed into ordinary oxygen is denied upon the authority of Lebone and Houzeau, who state that it is less liable to change than is generally supposed, for they found, after working with it, that its peculiar odour remained on their hands and garments for some time. These views, largely shared by many others, as to the beneficial effects of ozone have, however, not been allowed to pass unchallenged. P. Thénard considered it important that both the public and medical men should be apprised of the erroneous character of the opinions generally entertained respecting the action of ozone on the organism. Ozone, he says, so far from exerting a beneficial effect, is one of the most energetic of poisons; and the serious accidents which have occurred in his own laboratory do not leave the slightest room for doubt in the matter.

Writing to the 'Comptes Rendus,' lxxxii, p. 1857, Thénard narrates the case of a guinea-pig, in which the beats of the pulse, normally 148 per minute, fell to $\frac{1}{10}$ after the exposure of the animal for a quarter of an hour to an atmosphere charged with ozone. He states that under the influence of ozone, even when very largely diluted, the blood-corpuscles rapidly cohere and change their form. Other instances are recorded in which the blood, contrary to anticipation, has been found in the venous condition.

Drs Dewar and M'Kendrick found that ozone acted as a very powerful irritant upon the mucous membranes. Further, an experiment was made by placing some small birds in a mixture of oxygen and ozone, containing 10% of the latter. In two minutes the birds were dead.

Ozone is frequently present in the atmosphere, formed by electricity and perhaps by other means. Gorup-Besanez has shown (vide 'Ann. Chem. Pharm,' clxi, 232) that ozone is formed when water evaporates, and he ascribes its occurrence in the atmosphere to this cause rather than to the influence of electrical discharges. Payen states that it does not amount to more than $\frac{1}{170000}$ by weight, and $\frac{1}{70000}$ by volume of atmospheric air. Other observers state that it varies in amount according to height, locality, temperature, electricity, &c. Dr Buchanan says it is more abundant "on the sea-coast than inland, in the west than in the east of Great Britain, in elevated than in low situations, with south-west than with north-east winds, in the country than in towns, and on the windward than on the leeward side of towns." According to the Scottish Meteorological Society ozone is most prevalent in the atmosphere from February to June, when the average amount is 6·0, and least from July to January, when the average is 5·7. The maximum 6·2 is reached in May, and the minimum 5·3 in November.

These results are said to be in accordance with the conclusions arrived at by Berigny and Houzeau.

Although there appears no ground for doubting that artificially prepared ozone, by reason of its actively disinfectant properties, may prove a valuable auxiliary in checking the spread of certain diseases, it seems far from satisfactorily established that the same quality is possessed by the ozone in the atmosphere, or, on the contrary, as has been asserted, that certain ailments are caused by it. During an outbreak of influenza at Berlin, Schönbein states that the air contained a large quantity of ozone; a circumstance confirmed by Dr Pietra-Santa during the prevalence in another locality of the same epidemic, which it was imagined might be caused by the irritating effect of the ozone on the organs of respiration. Billard, Wolf, Bœckel, and Strambis all state that, during the prevalence of cholera at Strasbourg, Berlin, and Milan, ozone was absent from the atmosphere, and that the decline of the malady was marked by its reappearance. Uhle ascribes the accumulation of malaria at night to the non-formation of ozone by solar heat.

The above facts have, however, been disputed by some observers, whilst others have refused to regard them as anything more than coincidences, and have indeed cited evidence of a totally opposite character; thus Grellois has stated that he found more ozone in a marsh than elsewhere.

Kingzett has shown the incorrectness of Schönbein's statement that, when oil of turpentine and other essential oils are oxidised by exposure to the air, ozone is formed. Schönbein was misled because from the oxidised oil and the air in its vicinity he obtained the ozone reaction with potassium iodide.

Kingzett has demonstrated that the compound can be neither ozone nor hydrogen dioxide, because it is destroyed at the boiling-point of oil of turpentine, viz. 160°, at which temperature ozone and hydrogen dioxide are permanent; besides which it resists to a certain extent the action of sodium thiosulphate, and its solution in water retains its properties after long-continued boiling. Kingzett believes that the active properties of the oxidised turpentine oil are due to the formation of monohydrated terpene oxide.

One of the most delicate tests for ozone is potassium iodide, either alone or mixed with starch. A brown colour in the former case, a blue in the latter, indicates the liberation of iodine. In the ozometer, strips of paper saturated with starch and potassium iodide are exposed to the action of a definite volume of air in a dark chamber. The comparative quantities of ozone in different samples of air are judged of by the intensity of the colour compared with a fixed scale on which 1 is the lightest and 10 generally the darkest shade. See OZONOMETER.

Ozone acts as a reducing agent in certain curious cases. Thus hydrogen peroxide and ozone reduce one another, water and oxygen being the sole products; and some substances, such as platinum black and manganese peroxide, convert it into oxygen without suffering change themselves, being probably oxidised and reduced alternately.

OZONIC ETHER. Ether containing in solution peroxide of hydrogen and a little alcohol. Employed as a test for blood, especially in urine, it changes the colour of the blood to blue when mixed with fresh-made tincture of guaiacum.

OZONOM'ETER. This name has been given to paper prepared with a mixed solution of starch and iodide of potassium. It is white, but is turned blue by ozonised air when exposed to it in a slightly moistened state.

The following are the proportions given by Schönbein for the preparation of the paper:—1 part of *pure* iodide of potassium, 10 parts of starch, and 200 of water. Lowe gives 1 part of iodide to 5 parts of starch; Moffatt, 1 part to 2½ parts. The starch must be treated with warm water and filtered, so that a clear solution is obtained.

The iodide is dissolved in another portion of water, and gradually added. The paper, cut in slips and previously soaked in distilled water, is placed in the mixed iodide and starch for several hours; and, lastly, slowly dried in a cool dark place, the slips being hung horizontally. Schönbein's papers require moistening with water after exposure before the trial is taken.

Payen's ozonometer, which is an improvement on the above, is made of red litmus paper with half its surface impregnated with a 1% solution of potassium iodide. The portion of the paper becomes blue by contact with air containing ozone, in consequence of oxidation and the formation of potash. The unimpregnated portion of the paper undergoes no change unless the air

contains ammoniacal vapours, and then the paper becomes blue over its entire surface.

Böttger has suggested the use of papers impregnated with thallious oxide, as this substance is not changed by the action of nitrogen oxides.

Davy states that he has obtained very satisfactory results in the estimation of ozone in the atmosphere by employing a mixture of iodide of potassium and arsenite of potassium.

The value of the ozonometer as an indicator of atmospheric ozone must be looked upon as uncertain, when it is borne in mind that there are other bodies besides ozone frequently present in the air, such as nitrous acid, chlorine, &c., which give similar reactions with the above reagents.

PACK'FONG. *Syn.* PAKFONG, PACKTONG, CHINESE WHITE COPPER. An alloy of copper, zinc, and nickel, containing also traces of iron. It has been manufactured for several hundred years in China and the East Indies. An analysis by Fyfe is as follows:—Copper = 40·4, zinc = 25·4, nickel = 31·6, iron = 2·6. Total = 100·0.

Prep., &c. White, slightly ductile, and permanent at ordinary temperatures; at a temperature below that of redness it suffers decomposition, with the extrication of fumes of arsenious acid. Formerly much used for the scales of thermometers and other instruments, dial-plates, candlesticks, &c. It is now almost superseded by the alloy of nickel and copper called German silver, to which the name is also applied by some recent writers.

PACK'ING. As there is considerable art in packing brittle hollow-ware, as glass, china, &c., in such a way that it will stand exposure to the jolting, blows, and agitation of land carriage, it is better, when it is of much value, or in quantity, to employ a person qualified for the job. A man accustomed to packing such articles may be readily procured at any glass-works or china warehouse for a trifling consideration. When this cannot be done, it must be recollected that the great secret of safe packing consists in the articles being carefully preserved from undue pressure or contact with each other, yet so firmly arranged, and so surrounded with some material as hay, straw, sawdust, &c., that they cannot be shaken into such a condition by the ordinary contingencies of transport. Loose packing must always be avoided.

PAD'DING. Among calico printers this term is applied to the operation of impregnating the pores of their cloth with a mordant. It is now almost exclusively performed by means of a simple piece of machinery (padding machine), which essentially consists of a 'large reel,' around which the unprepared cloth is wound; a 'guide-roller,' over which it passes to smooth and adjust it before entering the liquor; a copper cylinder, or 'dip-roller,' nearly at the bottom of the 'mordant-trough,' under which it is carried from the guide-roller; a half-round polished 'stretched bar,' to give it equal tension; a pair of 'padded cylinders,' to remove superfluous moisture; and, lastly, a 'reel' to receive the mordanted ('padded') cloth. The degree of tension is regulated by a weight suspended on a lever, and motion is given to the whole by an endless band from the driving shaft. This machine is also applicable to many of the operations of dyeing, bleaching, and starching textile fabrics.

PAINTER'S CREAM. *Prep.* Take of pale nut oil, 6 oz. ; mastic, 1 oz. ; dissolve, add of sugar of lead, ¼ oz., previously ground in the least possible quantity of oil; then further add of water q. s. gradually, until it acquires the consistence of cream, working it well all the time. Used by painters to cover their work when they are obliged to leave it for some time. It may be washed off with a sponge and water.

PAINTING. The art or employment of laying on colour. In the *fine arts,* the production of a picture or a resemblance in colours on a flat surface. The artistic and mechanical consideration of this subject does not come within the province of our volume; but notices of the leading materials employed by both artists and house-painters are given under the respective names. See the various pigments, COLOURS, OILS, VARNISHES, &c., and *below.*

Painting, Distem'per. A method of painting generally adopted by the ancients. Water was the principal medium, but various gelatinous and albuminous 'binders' were added to fix the pigments. Of these the most important were glue, size, and white of egg. In modern distemper, as executed by the painters of theatrical scenery, panoramas, &c., spirit of turpentine is largely employed as a medium.

Painting, Elydor'ic. A method of painting invented by M. Vincent, of Montpelier, having for its object to combine the fresh appearance and finish of water-colours with the mellowness of oil painting. The liquid employed as a vehicle for the pigments is an emulsion formed of oil and water by the intervention of certain portions of gum or mucilage.

Painting, Enam'el. In this variety of painting vitrifiable colours are laid on thin plates of metals and fused into them. The outline is first burnt in, after which the parts are filled up gradually, with repeated fusions at an enameller's lamp, to the most minute finishing touches. " The enamel painter has to work, not with actual colours, but with mixtures which he only knows from experience will produce certain colours after the operation of the fire " (*Aikin*).

Painting, Encaus'tic. This method is very ancient, but is now seldom practised. According to Pliny, the colours were made up into crayons with wax, and the subject being traced on the ground with a metal point, they were melted on the picture as they were used. A coating of melted wax was then evenly spread over all, and when it had become quite cold was finally polished.

The art of encaustic painting, after lying dormant for about fifteen centuries, was revived by Count Caylus in 1753. In its new form the wood or canvas to be painted on is first well rubbed over with wax, and then held before the fire, so that the wax may penetrate and fill up all the interstices, and form a perfectly even surface. The coloured pigments are next mixed with the powder noticed below, which is then rubbed smooth with some thick gum-water, and applied

with brushes in the same manner as ordinary water-colours. When the painting is finished, and quite dry, it is brushed over with pure white wax in a melted state, the surface being equalised by the skilful application of heat; it is, lastly, polished off, as before.

The Powder. To white wax, melted in an earthen pipkin, add, in small portions at a time, an equal weight of powdered mastic, stirring continuously until the whole is incorporated ; then pour it into cold water, and afterwards reduce it to powder in a wedgwood-ware mortar. A small quantity only of this powder is used with light colours; but more is required with the darker ones, until, on approaching black, the two may be mixed in almost equal proportions.

Painting, Fres'co. This method of painting was known to the ancient Egyptians, and was commonly practised by the Greeks and Romans. It is confined to the decoration of the walls of buildings, and is executed by incorporating the colours with the still moist plaster, or gesso. The pigments employed are entirely mineral or vitreous. As it is extremely difficult to alter the work after the colours are once absorbed, or after the ground has hardened, the whole must be carefully designed before commencing the picture, and no more commenced at once than can be executed during the day.

Of all the varieties of painting, fresco is " undoubtedly the most virile, most sure, most resolute, and most durable " (*Vasari*), and the one most adapted for the purposes of historical painting in its grandest and most exalted forms. In comparison with it, it has been said that even oil painting is " employment fit only for women and children " (*Michael Angelo*).

Painting, Glass. See STAINED GLASS.

Painting, Oil. This well-known and much-practised method of painting takes its name from the vehicle employed for the colours. The last may be any of those of a permanent character, and whose natural tint is not altered by admixture with oil. Linseed, nut, and poppy oil are those which are principally employed. The first requires the addition of 'driers,' and hence is generally used under the form of 'boiled oil.' Spirit of turpentine is commonly used to thin down the prepared colours, and the finished picture is frequently covered with a coat of varnish.

Painting, Por'celain. See POTTERY, STAINED GLASS, &c.

Painting, Vel'vet. Any of the ordinary non-corrosive pigments or liquid colours, thickened with a little gum, may be employed in this art; preference being, however, given to those that possess the greatest brilliancy, and which dry without spreading. See STAINS, &c.

Painting, Water-colour. In its strictest and modern sense, 'water-colour painting' means the painting on paper with colours diluted with water. The English school of water-colour painting has produced works which bear comparison with the great masterpieces in oil, and even surpass them in the delicacy of atmospheric effects. The old practice of making the entire drawing in light and shade by washes of Indian ink or neutral tints, and then adding the various local colours in transparent washes, has given place to the more healthy system of painting every object in its appropriate local colour at the outset.

PAINTINGS. Many valuable paintings suffer premature decay from the attacks of a microscopic insect, a species of acarus or mite. The best method of preventing this variety of decay is to add a little creasote (dissolved in brandy or vinegar), or a few grains each of corrosive sublimate and sal-ammoniac (dissolved in a little water) to the paste and glue used to line the picture, as well as to add a few drops of pure creasote or of an alcoholic or ethereal solution of corrosive sublimate to the varnish, when any is to be applied. If the destruction alluded to has already commenced the painting should be at once carefully cleaned and relined, observing to employ one or other of the remedies just mentioned.

The most appropriate and only safe situation in which to keep paintings is where there is a pure and moderately dry atmosphere. To protect pictures from the effects of damp it has been suggested to dip the canvas into a solution of silicate of potash, and afterwards dry it, previous to its being used. Impure air abounds in carbonic acid and sulphuretted hydrogen. It is the presence of the last in the air that blackens the 'lights,' and causes most of the 'middle tints' and 'shades' to fade; and it is exposure to damp that produces mouldiness and decay of the canvas. For this reason valuable paintings should not be kept in churches, nor suspended against heavy walls of masonry, especially in badly ventilated buildings. Excess of light, particularly the direct rays of the sun, also acts injuriously on paintings, since it bleaches some colours and darkens others.

The blackened lights of old pictures may be instantly restored to their original hue by touching them with peroxide of hydrogen, diluted with 6 or 8 times its weight of pure water. The part must be afterwards washed with a clean sponge and water. The most astonishing results have been produced in this way. See PEROXIDE OF HYDROGEN.

Pettenkofer, observing the colours of many of the oil paintings in the Munich galleries apparently fading, discovered that the dim and grey appearance they then presented was not really due to any decay of colour, but to a discontinuity of the molecules of the vehicle, and the resinous substances mixed with the pigments, the effect of which was to break up and lessen the mass of transparent colour, and to diminish its intensity. This separation from each other of the alternate particles he conceived was owing to the shrinking and contraction they underwent after long years of exposure to a moist atmosphere. To remedy it Pettenkofer subjected the affected picture to two simple processes, which he is said to have found absolutely successful. The first, which he terms the 'regeneration' process, consists in enclosing the picture in a flat box, where it is exposed to the vapours of alcohol, part of which being absorbed by the resinous molecules, restores them to their original volume. Hence it follows that the gaps between the molecules being thus filled up, there is presented to the eye a continuous mass of transparent colour, as when the picture was freshly painted.

In the previous operation the resinous constituents only of the picture have been acted upon and restored to their normal condition. The hardened molecules of the oil which have been employed as a vehicle have likewise diminished in bulk from the same causes, and in so doing have contributed to the lessening of the brightness of the picture. In cases where it is found the increased volume of resinous particles has failed to fill up the intervals between the shrunken oil molecules, Pettenkofer subjects the picture to a further process. In this, which he terms 'nourishing it,' the picture is simply rubbed over with balsam of copaiba.

Oil, which was formerly employed for this purpose, is very strongly condemned by Pettenkofer.

Oil paintings, as probably most of our readers are aware, are mostly executed either on wood ('panel') or canvas, now principally on the latter. Both these substances have to undergo a preliminary operation, known as 'priming,' the priming being, in short, the ground on which the paint is placed. This priming may consist either of a number of layers composed of a mixture of chalk or plaster with paste or glue, or else of a series of coats of oil colour. When a canvas or panel is prepared with the former it is called 'distemper priming;' when with the latter, 'oil priming.' The distemper is the more quickly prepared, but is open to the objection of being easily broken, and of a liability to absorb moisture, which renders it apt to separate from the canvas. If the priming be of oil colour it is desirable that the chief pigment used in making it should be white-lead, and that if any other colours are added they should be in comparatively small quantities. Dr H. Liebreich cites an example in which a departure from this precaution, persevered in from the middle of the 16th to that of the 17th century, by a celebrated school of Italian painters (the Bologna), has resulted in the destruction in their works of all the glazing of the picture, "so that those colours only can be recognised which either contain white, or are glazed on white." Furthermore, that the dark priming used by these artists has caused the dark parts of their pictures to become still darker.

This priming, which was of a reddish-brown colour, was composed of a mixture of bole Armenian and umber; and it is conjectured it was employed with the object of modifying or softening too violent contrasts of light and dark colours, and thus of easily securing effective chiaroscuro, and of aiding rapid execution.

The Dutch and Flemish painters mostly employed a light-coloured priming; sometimes it was of a light oak colour. Vandyke is said to have used grey grounds for his pictures, and in some few instances dull red ones; and since his pictures are free from the objectionable qualities met with in the works of the Bologna artists, it has been surmised that in this method of working he had recourse to impasto colouring.

In the selection of wood, which is subsequently to be used for the picture, considerable judgment and experience are required, that from the toughest and soundest oaks, nut trees, or cedar being sought after. The cutting it into boards, and seasoning it, are also points exacting a great amount of time and care.

The backs of pictures, if made of wood, in addition to their liability to attacks from insects, not unfrequently warp, or fissures form in them, or they may become hopelessly rotten.

When the picture warps, it should be moistened with water at the back, on which it should be laid for 24 hours, at the end of which time, or sometimes less, it becomes perfectly straight. Fissures may be filled up by pieces of wood cut to the required size. Small pieces of rotten wood, if not too near the painting, may be cut out, and the gaps filled up with wedge-shaped pieces of wood. Where the loss is insignificant it may be stopped up with cement. When the panel is very rotten and decayed, it may be necessary to remove the picture from it altogether, and to place it either on a new panel, or upon what Dr Liebreich regards as better still, a piece of canvas.

This is by no means so formidable and astonishing an operation as it may at first sight appear; in short, as will be directly shown, the picture may, if necessary, be freed from its priming even, without any difficulty.

Hacquin, of Paris, was one of the first to remove an oil painting from its base, and to place it upon a new one. He did this with one of Raphael's Madonnas, in the gallery of the Louvre; and the same treatment has since been extended to the 'Resurrection of Lazarus,' by Sebastian del Piombo, one of the pictures in our National Gallery. This process is generally accomplished as follows:

"First of all the surface of the picture is pasted over with gauze and paper; after that the wood is made straight by moistening, or, if necessary, by making incisions with the saw, into which cuneiform pieces of wood are driven. By means of a tenon-saw the panel is to be sawn into little squares, which must be removed by a chisel, and in this way the thickness of the wood is reduced to half an inch; it is then planed until it becomes no thicker than paper, and the rest is removed by means of a knife and with the fingers.

"The painting being thus severed from its basis, it can be fixed on canvas if the priming is sufficiently preserved. In the opposite case a mixture made of chalk and glue, or something of the kind, must be put on first, and very evenly smoothed after being dry. This done, the new canvas has to be fixed upon it by means of a mixture of glue, varnish, and turpentine, and the substance of the picture pressed tightly and evenly against it by means of warm irons" (*Liebreich*).

Defects in the priming of an oil painting, when they are confined to a slight separation of the priming of a canvas, may be remedied by pouring into the gap caused by the severance a little solution of size, and then pressing the separated surfaces gently together. Slight cracks must be filled up with fresh priming.

For paintings in which the whole of the priming seems insecure, or has extensively separated from the canvas, it is recommended to remove them entirely from the old basis and to transfer them to new panels or canvas.

The property of unchangeableness or indisposition to fade, as exemplified in the retention of its freshness of colour by a picture, is one which, it is asserted, is very much more generally met with in the pictures of the Italian (from the Italian school must be excepted that of Bologna) and Dutch painters of the 15th, 16th, and 17th centuries, than in those of the French and English schools of the last hundred years. Opinions have been advanced in explanation of this circumstance. One is, that the older masters used pigments and vehicles of much greater purity and freedom from adulteration than the latter generations of painters; another, that they worked by a method and prepared their colours by a process unknown since their time,—in fact, that they were possessed of a technical secret, which, as they never divulged it, has died with them; a third, that they had choice of many colours unknown in the present day.

One of the later and most valuable contributions to our knowledge of these points has been made by Dr R. Liebreich, in his lecture 'On the Deterioration of Oil Paintings,' delivered at the Royal Institution, March 1st, 1878, which also embraces the practical deductions to be drawn from the results of his investigations. The plan adopted by Liebreich for unravelling the so-called secret by which the old masters so generally contrived to secure permanency for their colours was ingenious and logical; it consisted in dissecting the structure and chemically analysing the pigments, vehicles, &c., of the pictures of the pupils of the great masters; for "fortunately they painted with the same material and by the same methods as the masters, and thousands of pictures by the pupils, well preserved and in different stages of decay, may be easily secured."

The third explanation, previously given as a reason for the superior durability of the colouring of the old over the later oil paintings, is thus disposed of by him. He says:

"We meet very often with the idea that the old masters had been in possession of colours, that is, pigments, the knowledge of which has been lost, and that this accounts principally for the difference between the oil painting of the 15th and 16th centuries, on the one hand, and that of the 18th and 19th on the other. But this is a great mistake. We know perfectly well the pigments used by the old masters; we possess the same and a considerable number of new ones, good as well as bad, in addition."

He adds, "In using the expression of good and bad, I am thinking principally of their durability. From this point of view the pigments can be placed under three headings:

"1. Those that are durable in themselves, and also agree well with the other pigments with which they have to be mixed.

"2. Such as when sufficiently isolated remain unaltered, but when in contact with certain other pigments change colour, or alter the others, or produce a reciprocal modification.

"3. Those which are so little durable that, even when isolated from other pigments, the mere contact of the vehicle, the air, or the light makes them in time fade, darken, or disappear altogether.

"The old masters used without reserve only those belonging to the first of these three categories. For those belonging to the second they imposed on themselves certain limits and precautions. Those belonging to the third they did not use at all.

"That some of the modern masters have not followed these principles is not owing to a lost secret, but to the fact that they disregarded those well-known principles, and even consciously acted against them. In Sir Joshua Reynolds' diary, for instance, we read that in order to produce certain tints of flesh he mixed orpiment, carmine lake, and blue-black together.

"Now, orpiment is one of the colours of the second category, carmine lake one of the third. That is to say, orpiment, as long as it remains isolated, keeps its brilliant yellow or reddish-orange colour; but when mixed with white-lead it decomposes, because it consists of sulphur and arsenic; and it moreover blackens the white-lead, because the sulphur combines with it. Carmine lake, even if left isolated, does not stand as an oil colour, and therefore has been superseded by madder lake.

"Unfortunately some of the most brilliant colours are perishable to such a degree that they ought never to be used; yet it seems to me that just in one branch of art, in which of late remarkable progress has been made—I mean landscape painting—the artists, in order to obtain certain effects of colour not easy to be realised, do not always resist the temptation to make use of a number of pigments, the non-durability of which is proved beyond doubt."

Another point which Dr Liebreich regards as of much more importance even than the selection and treatment of their pigments, and in which he says the old masters exercised great discretion, was the more sparing use of the vehicles and liquids they mixed with their colours.

He points out that there are certain pigments which, when mixed with the oil, impede its drying, whilst others there are which hasten it. "Supposing now," he says, "we should add to each of the different pigments the same quantity of oil, the drying of it would progress at different rates. But in reality this difference is very greatly increased by the fact that the different pigments require very different quantities of oil, in order to be ground to the consistency requisite for painting."

Pettenkofer quotes the following figures given to him by one of the colour manufacturers:

100 parts (weight)	White-lead	require 12 parts of oil.
	Zinc white	„ 14 „
„ „	Green chrome	„ 15 „
„ „	Chrome yellow	„ 19 „
„ „	Vermilion	„ 25 „
„ „	Light red	„ 31 „
„ „	Madder lake	„ 62 „
„ „	Yellow ochre	„ 66 „
„ „	Light ochre	„ 72 „
„ „	Camel's brown	„ 75 „
„ „	Brown manganese	„ 87 „
„ „	Terre-verte	„ 100 „
„ „	Parisian blue	„ 106 „
„ „	Burnt terre-verte	„ 112 „
„ „	Berlin blue	„ 112 „
„ „	Ivory black	„ 112 „
„ „	Cobalt	„ 125 „
„ „	Florentine brown	„ 150 „
„ „	Burnt terra sienna	„ 181 „
„ „	Raw terra sienna	„ 140 „

According to this table a hundred parts of the quick-drying white-lead are ground with 12 parts of oil; and on the other hand, slow-drying ivory black requires 112 parts of oil.

It is very important that artists should have an exact knowledge of these matters. But it seems to me that they are insufficiently known to most of them. All, of course, know perfectly how different the drying quality of different colours is. But that these different colours introduce into the picture so different a quantity of the oil, and how large the quantity is in the colours they buy, and, further, that the oil as well as the mediums or siccatives they add to dry the colours are gradually transformed into a caout-chouc-like opaque substance, which envelops and darkens the pigments, and, moreover, that the oil undergoes, not in the beginning, but much later on, when it is already completely dry, changes of volume, and so impairs the continuity of the picture—all this is not sufficiently known; otherwise the custom of painting with the ordinary oil colours, to be bought at any colourman's, would not have been going on for nearly a hundred years, in spite of all the clearly shown evil results—results due chiefly to the principal enemy of oil painting, that is to say, the oil.

A close optical examination and accurate study of the pictures of the French and English masters of the last hundred years have revealed to Dr Liebreich their principal defects, which he says are—

1. Darkening of the opaque bright colours.
2. Fading of the transparent brilliant colours.
3. Darkening, and, above all, cracking of the transparent dark colours. He states that these cracks are so characteristic and distinctive of the pictures of this period that they might be used as a test as to whether or not a picture really belonged to this school, or was only a copy.

This peculiar cracking in the paint is, according to Dr Liebreich, particularly observable in Guericault's 'Wreck of the Medusa' in the Louvre, and also in Ingres' 'Portrait of Cherubini;' and as the same effect is not to be seen in the works of the Dutch and Italian artists, the very rational inference to be drawn is that the methods followed by these schools were sounder than those adopted by their English and French successors. Dr Liebreich believes the cracks were owing to the practice of painting over one colour with another before the first was perfectly dry.

"The study of the alterations," says Dr Liebreich, "already fully developed within the last hundred years only, and their comparison with the works of the old masters, would suggest the following rules for the process of painting:

"1. That the oil should in all colours be reduced to a minimum, and under no form should more of it than absolutely necessary be introduced into a picture.

"2. All transparent colours which dry very slowly should be ground, not with oil at all, but with a resinous vehicle.

"3. No colour should be put on any part of a picture which is not yet perfectly dry, and, above all, never a quick-drying colour upon a slowly drying one which is not yet perfectly dry.

"4. White and other quick-drying opaque

colours may be put on thickly. On the contrary, transparent and slowly drying colours should always be put on in thin layers. If the effect of a thick layer of these latter is required it must be produced by laying one thin layer over another, taking care to have one completely dry before the next is laid on. If transparent colours are mixed with sufficient quantity of white-lead they may be treated like opaque ones."

Dr Liebreich concludes his interesting lecture with some judicious advice on the subject of picture cleaning, and points out that, since different pictures require to be differently operated upon, all universal agents and methods suggested for the purpose are open to suspicion, and should be discarded.

For pictures the varnish of which has become cracked or dim he recommends Pettenkofer's treatment with alcoholised vapour, already described. For those in which the varnish may have become dark yellow, brown, or dirty, he advises its removal altogether, being very careful to specify the conditions under which this should be accomplished, and the risk the picture may run of being spoiled if entrusted to an unintelligent and ignorant manufacturer. "If a picture," he says, "is throughout painted in oil, if its substance has remained sound and even, and it has been varnished with an easily soluble mastich or dammar varnish, there will be neither difficulty nor danger in removing the varnish. This can, in such a case, be done either by a dry process—that is, by rubbing the surface with the tips of the fingers, and thus reducing the varnish by degrees to a fine dust—or by dissolving the varnish by application of liquids which, when brought only for a short time into contact with the oil painting, will not endanger it. We have, however, seen that the works of the old masters are not painted with oil colours like those used by modern painters, but, on the contrary, that certain pigments, and especially the transparent colours used for glazing, were ground only with resinous substances. These latter have in the course of time been so thoroughly united with the layer of varnish spread over the surface of the picture that there no longer exists any decided limit between the picture and the varnish. It is in such pictures that a great amount of experience and knowledge of the process used for the picture, as well as precaution, are required, in order to take away from the varnish as much only as is indispensable, and without interfering with the picture itself.

"Numberless works of art have been irreparably injured by restorers, who, in their eagerness to remove dirt and varnish, attacked the painting itself. They then destroyed just that last finishing touch of the painting without which it is no longer a masterpiece."

"The cleaner is, then, reminded that if the removal from the pictures of their varnish, when this is known to consist of a spirituous solution of the gum mastich or dammar, requires the amount of discretion and judgment before specified, still greater care and prudence are necessary when dealing with pictures whose surfaces have been covered with oil, oil varnish, or oleo-resinous varnish. All these substances, which in time

more or less obscure the picture, form on its face a dark and opaque film, and this frequently requires for its removal the application of some agent, which, in dissolving the layer of varnish, is very liable at the same time to dissolve the substance of the picture also."

As a recent instance of the injurious effects of injudicious cleaning, Dr Liebreich mentions the case of a valuable picture in the Pitti Palace, at Florence, the 'St. John of Andrea del Sarto.' The softness of the outline of the face of the figure, which he remembers previous to its attempted restoration, had been entirely destroyed, which disastrous result Dr Liebreich conceived had been caused by the entire removal of the glazing.

A new method for cleaning pictures is described by E. Von Bibra in the 'Journal für Praktische Chemie.' A very indistinct oil painting was freed from dust with a feather, washed with a sponge and water, and then covered for eight minutes with a layer of shaving soap. The soap was then washed off with a brush and then left to dry. It was next thoroughly cleaned with linen cloth soaked in nitro-benzol. The picture was now distinct, but the colours dull. Finally, it was treated with olive oil, and a coating of quick-drying varnish laid on ('Academy,' May 6th, 1878). In giving insertion to the above, we do not venture to give an opinion as to its value or the reverse. We would recommend it to be read side by side with Dr Liebreich's advice on picture cleaning, given above.—ED. See WATER-COLOURS, EFFECT OF LIGHT ON.

PAINTS. In trade, this term is commonly applied to pigments ground with oil to a thick paste, ready to be 'thinned down' with oil or turpentine to a consistence adapted for application with a brush.

Paints are prepared on the small scale by grinding the dry pigments with the oil by means of a stone and muller; on the large scale they are ground in a colour mill. There are several pigments, as King's yellow, Scheele's green, verdigris, white-lead, &c., which from their poisonous character cannot be safely ground by hand, except in very small quantities at a time, and then only by the exercise of extreme caution.

In mixing or thinning down paints for use it may be useful to mention that, for outdoor work, boiled oil is principally or wholly employed, unless it be for the decorative parts of houses, when a portion of turpentine and pale linseed oil is often added. For indoor work, linseed oil, turpentine, and a little 'driers' are generally used in the same way. The smaller the proportion of oil employed for the purpose, the less will be the gloss, and the greater the ultimate hardness of the coating. For 'flatted white,' &c., the colour being ground in oil, requires scarcely any further addition of that article, as the object is to have it 'dead' or dull. The best driers are ground litharge and ground sugar of lead; the first for dark and middle tints, and the last for light ones.

To preserve mixed paints in pots from 'skinning over' or drying up, they should be kept constantly covered with water; or, what is better, with a thin film of linseed oil.

Brushes, when out of use, may be preserved in a similar manner to mixed paints. When dirty, or required for a paint of another colour, they may be cleaned with a little oil of turpentine, which may be either preserved for the same purpose another time, or may be allowed to deposit its colour, and then used to thin down paints as usual. In no case, however, should it be thrown back into the cistern or pan with the pure 'turps.'

Paints, Flexible. *Prep.* Take of good yellow soap (cut into slices), 2½ lbs.; boiling water, 1½ galls.; dissolve, and grind the solution whilst hot with good oil paint, 1½ cwt. Used to paint canvas.

Paints, Vitrifi'able. See ENAMEL, GLAZE, STAINED GLASS, &c.

PALAMOND. Chocolate, 1 oz.; rice flour, 4 oz.; potato arrowroot, 4 oz.; red sanders, in fine powder, 1 dr. Mix. (In the above, by chocolate is meant the cacao beans roasted and pulverised without addition. Indian arrowroot, or toasles-moi, may be substituted for the potato arrowroot.)

PALLA'DIUM. Pd=106·2. A rare metal discovered by Dr Wollaston in the ore of platinum, in 1803. It occurs in a fairly pure condition along with Brazilian platinum ore, and to some extent in most ores of platinum. It is also found associated with gold in the Hars, and in several parts of South America.

Prep. 1. A solution of the ore of platinum in aqua regia, from which most of the metal has been precipitated by chloride of ammonium, is neutralised by carbonate of sodium, and then treated with a solution of cyanide of mercury; the white insoluble precipitate (cyanide of palladium) is next washed, dried, and heated to redness; the residuum of the ignition (spongy palladium) is then submitted to a gradually increased pressure, and welding at a white heat, so as to form a button, in a similar manner to that adopted with platinum.—*Prod.* Columbian ore of platinum, 1% ; Uralian do., 0·25% to 0·75% .

2. The native alloy of gold and palladium (from the Brazils) is submitted to the operations of quartation and parting, the nitric acid employed being of the density of 1·3; the silver is next precipitated from the solution by means of a solution of common salt or dilute hydrochloric acid, and the decanted supernatant liquid, after evaporation to one half, is neutralised with ammonia, and concentrated so that crystals may form; these (chloride of palladium and ammonium) are cautiously washed in a little very cold water, dried, mixed with borax, and exposed in a crucible to the strongest heat of a powerful blast-furnace, when a solid button of pure palladium is formed.

Prop., &c. Palladium closely resembles platinum in appearance, fusibility, malleability, and ductility ; but it is less dense, and has a rather more silvery colour than that metal; it is freely soluble in aqua regia, and is slowly attacked by nitric acid, but the other acids exert little or no action on it; heated to redness in the air, a very superficial blue or purple film of oxide forms on the surface, which is again reduced at a white heat. It melts at 156° (*Wedgwood*). Sp. gr. 11·4 at 22·5° (*Deville* and *Debray*). It readily

unites with copper, silver, and some other metals, by fusion. *Tests.* 1. The hydrochloric acid solution is precipitated by potassium cyanide as yellowish-white cyanide of palladium, soluble in both hydrochloric acid and ammonia. 2. The neutral solutions of palladium are precipitated in the metallic state by ferrous sulphate, dark brown by sulphuretted hydrogen, and yellowish white by cyanide of mercury. 3. A drop of tincture of iodine placed on the surface of metallic palladium, and then evaporated by the heat of a spirit lamp, leaves a black spot. By the last two tests palladium is readily distinguished from platinum.

Uses. It has been employed to form the scales of mathematical and astronomical instruments, and is used in dentistry. Its alloy with silver is a very valuable white metal. It is also used for making the smaller divisions of grain and gramme weights. Palladium is not tarnished by sulphuretted hydrogen. An alloy of 1 part of palladium and 100 parts of steel is well adapted for cutting instruments which require to be perfectly smooth on the edge. The palladious and palladic salts are not of commercial interest.

PALMIT'IC ACID. $C_{16}H_{31}.CO_2H$. It is the first of the fatty acids which occur as glycerides in vegetable and animal fats, and form true soaps with the alkalies; palm oil largely contains it as palmitin, which is also found in notable quantities in spermaceti and beeswax.

Prep. 1. By the action of potash on oleic acid (new commercial process).

2. (Small scale.) Palm oil is boiled with potash; dilute sulphuric acid is added to the solution, which precipitates palmitic and oleic acids. The precipitate is washed and dried, dissolved in hot alcohol, from which solution the palmitic acid crystallises out.

3. (Large scale.) Palm oil is decomposed by superheated steam in a still; the volatile products of the reaction condense in the receiver and separate into two layers, of which the lower consists of glycerine and water, and the upper oily layer of palmitic acid. These layers are separated, and on cooling the upper one forms a white crystalline solid, which is used in candle-making.

Prop., &c. Pure palmitic acid crystallises in fine white needles which fuse at 62° C., and congeal to a scaly crystalline mass, having a foliated fracture. It is odourless, tasteless, friable, and lighter than water, in which it is insoluble; it dissolves freely in boiling alcohol or ether. It burns well, and is used in the manufacture of candles.

Estim. It is usually required to determine palmitic acid in the presence of stearic acid. For all practical purposes the approximate proportion of each may be found by Muter's method. The mixed acids are melted, and a little is drawn up into a pair of glass tubes, drawn out at one end to a long thin point, until the drawn-out parts are completely filled. After having been allowed to cool these tubes are suspended in water with a thermometer between them. Heat is applied and the temperature noted at the moment when the contents of the tube become transparent. They are again allowed to cool gradually and the temperature read off at the instant of re-solidification.

A sufficiently approximate result is obtained by reference to the following table by Heintz:

Proportion by weight of		Mixture			
		Melts at		Solidifies at	
Stearic acid.	Palmitic acid.	Cent.	Fahr.	Cent.	Fahr.
90	10	67·2	153·00	62·5	144·50
80	20	65·3	149·50	60·3	140·50
70	30	62·9	145·25	59·3	138·75
60	40	60·3	140·50	56·5	133·75
50	50	56·6	133·75	55·0	131·00
40	60	56·3	133·25	54·5	130·00
35	65	55·6	130·25	54·3	129·75
30	70	55·1	131·00	54·0	129·25
20	80	57·5	135·50	53·8	128·80
10	90	60·1	140·90	54·5	130·10

PAL'MITIN. *Syn.* TRIPALMITIN. C_3H_5 $(C_{16}H_{31}O_2)_3$. The solid portion of palm oil purified by first pressing out and then repeatedly treating it with hot alcohol and crystallising from ether. It is also formed when monopalmitin is treated with excess of palmitic acid and the mixture kept at 250°—270° for about eight hours. It is produced artificially by the action of palmitic acid on glycerine at a high temperature. Berthelot has obtained and examined a mono-and a di-palmitin.

Prop., &c. Pearly white glistening crystals, which melt at 50·5° C., and which, according to Mackelyne, after further heating again crystallise and then melt at 66·5°. These crystals are very soluble in ether and moderately soluble in hot alcohol. By saponification it is converted into palmitic acid. (See *above.*)

PALPITA'TION. *Syn.* PALPUS, PALPITATIO CORDIS, L. A violent and irregular beating or action of the heart, either temporary or occasional. When it does not arise from sudden or violent agitation or distress of mind, it may be regarded as a symptom of a disturbance of the nervous functions by disease, in which case attention should be directed to the removal of the primary affection, which in a very large number of cases consists in anæmia and dyspepsia. Good plain food, regular habits, and proper exercise will cure most cases of palpitation. Some suitable tonic is often a useful adjunct to the treatment.

PAL'SY. See PARALYSIS.

PANACE'A. A term formerly applied to those remedies which were supposed to be capable of curing all diseases, and still applied to some quack medicines.

PANA'DA. See BREAD JELLY (under JELLY).

PAN'ARY FERMENTATION. The vinous fermentation as developed in the dough of bread.

PAN'CAKES. These are essentially fried batter, variously enriched and flavoured, according to the taste of the cook. When they contain fruit, fish, meat, or poultry, or are highly seasoned or ornamented, they are commonly called FRITTERS.

Prep. (*M. Soyer.*) Break 2 to 4 eggs into a basin, add 4 small table-spoonfuls of flour, 2 teaspoonfuls of sugar, and a little salt; beat the whole well together, adding, by degrees, ½ pint of milk, or a little more or less, depending on the size of the eggs and the quality of the flour, so as to form a rather thick batter; next add a little

ginger, cinnamon, or any other flavour at will; lastly, put them into the pan, and when set, and one side brownish, lay hold of the frying-pan at the extremity of the handle, give it a sudden but slight jerk upwards, and the cake will turn over on the other side; when this is brown, dish up with sifted sugar, and serve with lemon. See FRITTERS.

PANCREAS. This gland, known popularly as the sweetbread, or stomach sweetbread, resembles the salivary glands very closely in structure, but differing greatly from them in the nature of its secretion. It lies just below the stomach on the left side of the body, with its broad end or head in the horseshoe bend of the duodenum, in the centre of which its duct opens, generally by an opening common to the pancreatic and bile ducts.

The pancreatic secretion produces four distinct and separate effects upon the food, due probably to the action of as many distinct ferments, as follows:

1. **Diastatic Action,** or power of converting starch into sugar, is much more energetic than that of the salivary ferment acting upon *raw* as well as boiled starch.

2. **Tryptic Action,** or power of converting proteids into peptones, depends upon the presence of a ferment called *Pancreatin* (*Corvisart*) or *Trypsin* (*W. Kühne*). This ferment is exceedingly energetic, and if a fresh and still warm pancreas be rubbed up with an equal volume of a 1% solution of acetic acid and then extracted with glycerine, the extract will be found to act powerfully on proteids. If the action be prolonged the peptones are further altered, and *leucin* ($C_6H_{13}NO_2$) and *tyrosin* ($C_9H_{11}NO_3$) with a number of other bodies are produced.

[Putrefactive Phenomena.] If the action of the pancreatic fluid be still further prolonged, and if the fluid be alkaline, a number of foul-smelling bodies are produced; *indol* (C_8H_7N), *skatol* (C_9H_9N), also *phenol* (C_6H_6O). These changes are due to putrefaction, and may be prevented by the addition of calomel, salicylic acid, or thymol to the fluid.

3. **Action on Neutral Fats.** This is twofold:
   (*a*) The formation of a *fine permanent emulsion*.
   (*b*) The splitting of the fats into glycerine and the corresponding fatty acid, *e.g.*
   $$(C_{57}H_{110}O_6) + 3(H_2O) = (C_3H_8O_3) + 3(C_{16}H_{32}O_2)$$
   Tristearin.   Water.   Glycerine.   Stearic acid.

This latter action is due to a special ferment, the alkaline juice combining with the fatty acids to form soaps, and thus by saponification and emulsification the fats are absorbed.

4. According to Kühne and W. Roberts the pancreas also contains a MILK-CURDLING FERMENT.

Preparation of Peptonised Food. The following directions for the use of 'liquor pancreaticus' (*Benger*) in the preparation of peptonised or partially digested foods are mainly reprinted from Sir Wm. Roberts's 'Lumleian Lectures,' delivered before the Royal College of Physicians, London:

1. *Peptonised or Partially Digested Foods.*— Peptonised Milk. Mix three quarters of a pint of fresh milk with a quarter of a pint of water, and warm in a saucepan to a temperature of about

140° F., that is, as hot as it can be tasted without burning the mouth; then pour into a jug or basin, add 2 teaspoonfuls of liquor pancreaticus and half a level teaspoonful of bicarbonate of soda, stir, and place near the fire to keep warm. In a few minutes a considerable change will have taken place in the milk, but in most cases it is best to allow the digestive process to go on from 10 to 20 minutes, according to the degree of peptonisation or predigestion desired. Partially peptonised milk is scarcely distinguishable in taste from ordinary new milk, though it is very much more digestible. As the process of peptonisation or digestion goes on a slight bitterness is developed, which is unobjectionable to many palates; a few trials will, however, indicate the limit most acceptable to the individual patient; and as soon as this is reached the milk must, if not required by the patient at once, be boiled up to prevent the further action of the liquor pancreaticus. It will then keep like ordinary milk. The addition of a little coffee to peptonised milk effectually covers the slightly bitter taste. If peptonised milk is consumed at the period indicated, that is to say, at the end of 10 to 20 minutes, it need not undergo any final boiling; it is better, indeed, to use it without boiling, because the half-finished process of digestion will go on for a time in the stomach.

Peptonised Gruel, Arrowroot, &c. Gruel may be prepared from any of the numerous farinaceous articles which are in common use—wheaten flour, oatmeal, arrowroot, sago, pearl barley, pea or lentil flour. The gruel should be very well boiled, and made thick and strong. It is then poured into a covered jug, and allowed to cool to a lukewarm temperature. Liquor pancreaticus is then added in the proportion of 2 teaspoonfuls to the pint of gruel, and the jug is kept warm as before. After standing half an hour to an hour the product is boiled and strained. The action of liquor pancreaticus on gruel is twofold; the starch of the meal is converted into sugar, and the albuminoid matters are peptonised. The conversion of the starch causes the gruel, however thick it may have been at starting, to become quite thin. This is often an advantage to the weak invalid, who can then drink the product with ease, when thick foods could not be swallowed. Peptonised gruel is useful as a basis for peptonised soups, jellies, blanc-manges, &c.

Peptonised Milk-gruel. This may be regarded as an artificially digested bread and milk, and as forming by itself a complete and highly nutritious food for weak digestions. It is very readily made. First a good thick gruel is prepared from any of the farinaceous articles above mentioned. The gruel, while still boiling hot, is added to an equal quantity of *cold* milk. The mixture will then be of the required temperature. To each pint of this mixture 2 teaspoonfuls of liquor pancreaticus and a pinch of bicarbonate of soda are added. It is kept warm in a covered jug for half an hour, and then boiled for a few minutes and strained. The slight bitterness of the digested milk is almost completely covered in the peptonised milk-gruel, and invalids take it without the least objection. Those who fail to peptonise milk-gruel so as to make it acceptable to the

palate and stomach of the patient invariably allow the peptonising process to go on too far, and in such cases the mixture should be boiled after standing a shorter time.

Peptonised Custard Pudding. A delicious and highly nutritive pudding for invalids may be made as follows:—Take half a pint of peptonised milk-gruel prepared as above, and allowed to cool somewhat after the final boiling; add to this two eggs well beaten, with sugar and flavouring to taste, and bake in a very slow oven.

Peptonised Soups, Jellies, and Blanc-manges. To give variety to peptonised dishes, soups, jellies, and blanc-manges containing peptonised aliments may be prepared, and these, whilst containing a large amount of digested starch and digested proteids, possess excellent flavour, which the most delicate palate could not accuse of having been tampered with. Soups are prepared in two ways. The first way is to add what cooks call 'stock' to an equal quantity of peptonised gruel or peptonised milk-gruel. A second and better way is to use peptonised gruel, which is quite thin and watery, instead of simple water, for the purpose of extracting shins of beef and other materials employed for the preparation of soup. Jellies are prepared simply by adding the due quantity of gelatin or isinglass, previously soaked in water, to hot peptonised gruel, and flavouring the mixture according to taste. Blanc-manges are made by treating peptonised milk in the same way, and then adding cream. In preparing all these dishes it is absolutely necessary to complete the operation of peptonising the gruel or the milk, even to the final boiling, before adding the stiffening ingredient; for if liquor pancreaticus be allowed to act on the gelatin, the gelatin itself undergoes a process of digestion, and its power of setting on cooling is utterly abolished. N.B. Flavouring agents may be added to any of the above preparations if thought desirable.

Peptonised Beef-tea. Half a pound of finely minced lean beef is mixed with a pint of water. This is simmered for an hour and a half. When it has cooled down to a lukewarm temperature (about 140° F.) a tablespoonful of the liquor pancreaticus is added, and it is then kept warm for two hours, and occasionally stirred. At the end of this time it is boiled for five minutes, and the liquid portion, measuring about half a pint, is strained off. Beef-tea prepared in this way is rich in peptone, highly nutritious, and of very agreeable flavour.

2. *Liquor Pancreaticus as an Addition to Food shortly before it is eaten.* Certain dishes commonly used by invalids—farinaceous gruels, milk, bread and milk, milk flavoured with tea or coffee or cocoa, and soups strengthened with farinaceous matters or with milk—are suitable for this mode of treatment. A teaspoonful or two of the liquor pancreaticus should be stirred up with the warm food as soon as it comes to table. And such is the activity of the preparation that, even as the invalid is engaged in eating—if he eat leisurely, as an invalid should—a change comes over the contents of the cup or basin: the gruel becomes thinner; the milk alters a shade in colour, or perhaps curdles softly; and the pieces of bread soften. The transformation thus begun goes on for a time

in the stomach, and, before the gastric acid puts a stop to the process, the work of digestion is already far advanced.

This mode of using liquor pancreaticus is simple and convenient. No addition of alkali is required, and of course no final boiling. The only precaution to be observed is that the temperature of the food, when the liquor is added, does not exceed 150° F. (55° C.). This point is very easily ascertained, for no liquid can be tolerated in the mouth, even when taken in sips, which has a temperature above 140° F. (60° C.). If, therefore, the food is sufficiently cool to be borne in the mouth, the liquor pancreaticus may be added to it without any risk of injuring the activity of the ferments.

3. *Liquor Pancreaticus with or after Meals.* One or two teaspoonfuls of liquor pancreaticus may be mixed with a wine-glassful of water, and sipped *during* meals consisting of starchy or farinaceous foods. It may also be taken 2 or 3 hours *after* meals; in the latter case it is well to add about 15 gr. (half a level teaspoonful) of bicarbonate of soda, to protect the pancreatic ferments for a time against the action of the acid juices of the stomach.

4. *Liquor Pancreaticus as an Addition to Nutritive Enemata.* Liquor pancreaticus is peculiarly adapted for administration with nutritive enemata. The enema may be prepared in the usual way with milk-gruel and beef-tea, and a dessert-spoonful of liquor pancreaticus should be added to it just before administration.

NOTE. In peptonising or partially digesting food by means of 'liquor pancreaticus' (*Benger*) it is important to remember that the liquor must not be added to food of any kind at a higher temperature than 140° F. This temperature can be estimated with sufficient accuracy, should no suitable thermometer be at hand, by tasting. If too hot to sip without burning the mouth it would entirely destroy the activity of the liquor pancreaticus, and must be allowed to cool somewhat before such addition is made. Artificial digestion, like cooking, must be regulated as to its degree, and it is easy to regulate it by the length of time during which the process is allowed to go on. The practical rule for guidance in peptonising articles of food containing milk is to allow the process to go on until a perceptible bitterness is developed, but not unpleasantly pronounced, and not longer. As soon as this point is reached the milk or milk-gruel should be consumed, or, if not required at once, should be boiled for a minute, so as to put a stop to further changes which would render the product less palatable. The extent of the peptonising action can be regulated either by increasing or diminishing the quantity of the liquor pancreaticus, or by increasing or diminishing the time during which it is allowed to act on the food. The bitter taste referred to is only produced in articles of food containing milk. In peptonising these, therefore, it is important not to carry the process so far as to render them unpalatable.

PANCREATIN. (*Griffith.*) It is obtained from the pancreas of recently killed animals by treating the colourless viscous juice with alcohol, and drying the precipitate *in vacuo.*

Dr Dobell's 'Crude Pancreatic Emulsion' is prepared as follows:—After freeing from fat and all foreign matters the pancreas of a freshly killed pig, 2½ lbs. of purified pancreas are bruised in a marble mortar, and to it are added 2½ lbs. of lard; these are well beaten together, and then to the mixture 3 lbs. of water are added, very gradually, so as to ensure the perfect absorption of the latter.

The pancreatised fat is prepared by shaking up 1 part of the 'crude emulsion' with 3 parts of ether, allowing the mixture to stand, drawing off the ethereal solution, and carefully distilling off the ether. The pancreatised fat remains. Dr Dobell says that pancreatised fat, unlike the crude fat, has no tendency to putrefy. His 'purified pancreatic emulsion' is made by mixing very carefully together 5 parts of pancreatised fat, 7½ parts of distilled water, and 2½ parts of rectified spirit, and flavouring with oil of cloves.

SACCHARATED PANCREATIN. Mr Mattison ('American Journal of Pharmacy') adopts the following process for the preparation of this substance:—The pancreas is dissected and macerated in water acidulated with hydrochloric acid for about forty-eight hours, then separated, and the acidulated solution of pancreas passed through a pulp filter until it is perfectly clear. To this clear solution is then added a saturated solution of chloride of sodium, which is allowed to stand until the pancreatin is separated. This is carefully skimmed off and placed upon a muslin filter, and allowed to drain, after which it should be washed with a less concentrated solution of sodium chloride and then put under the press. When all the salt solution has been removed, and the mass is nearly dry, it is rubbed with a quantity of sugar of milk and dried thoroughly without heat, after which it is diluted until ten grains emulsify two drachms of cod-liver oil.

PANIFICATION. The changes which occur in flour-dough under the influence of the fermentative process and heat, by which it is converted into bread.

PAPAINE (Papayotine). A ferment prepared from the juice of the papaw fruit (*Carica papaya*) which has the property of digesting albumen and fibrin. It is a whitish amorphous powder, soluble in water, said to be capable of peptonising 200 times its weight of blood-fibrin. The fruit of the papaw tree has long been used, both in the East and West Indies, for rendering tough meat and poultry tender. Prof. Finkler and Dr Schoffer recommend a 5 per cent. solution of papaine as the best solvent for diphtheritic and croupous membrane. The surface is painted with the solution every five or ten minutes; the membranes are said to be thus removed in a few hours, and the fever to disappear. Mr E. Hurry Fenwick has used papaine, in combination with cocaine, with marked benefit in syphilitic ulcers of the tongue and throat. The ulcers and white patches rapidly clean and begin to skin over.—*Dose,* 2 to 8 grains.

PAPA'VERINE. *Syn.* PAPAVERINA. An alkaloid discovered by Merck in opium. It crystallises in needles; is insoluble in water; is slightly soluble in cold alcohol and in ether; and forms crystallisable salts with the acids which possess little solubility. The hydrochlorate, one of the most characteristic of these compounds, crystallises in beautiful colourless prisms, which possess a high refractive power, and are only very slightly soluble in hydrochloric acid. Flückiger states that papaverine is much less active than thebaine, that it is not soporific either with men or animals, that it does not arrest diarrhœa, and is but slightly analgesic.

PAPER. *Syn.* CHARTA, PAPYRUS, L.; PAPIER, Fr. The limits of this work preclude the introduction of a description of the manufacture of this well-known and most useful article, which is now almost exclusively made by machinery of an elaborate and most ingenious description. We must, therefore, content ourselves with a short notice of a few of the preparations of the manufactured article. (See *below*.)

Good white paper should be perfectly devoid of odour, and when burnt it should leave a mere nominal amount of ash; digested in hot water, the liquid should be neutral to test-paper, and not affected by sulphuretted hydrogen or the alkaline sulphurets, or by tincture of iodine. Coloured papers should not contain any deleterious matter.

Paper, Antiasthmatic. (P. Codex.) *Syn.* CHARTA FUMIFERA, L.; CARTON ANTIASTHMATIQUE, Fr. Unsized grey filtering paper, 12 oz.; nitre, 6 oz.; belladonna, stramonium, digitalis, lobelia inflata, phellandrium, all in powder, ½ oz. of each; myrrh and olibanum, in powder, 1 oz. each. Tear the paper in pieces and soak it in water till quite soft; drain off the greater part of the water, and beat it into a paste; incorporate with it the powders previously mixed. Then put into tinned iron moulds, and dry by a stove.

Paper, Antirheumatic. *Syn.* CHARTA ANTIRHEUMATICA, L. (M. Berg.) Euphorbium, 30 parts; cantharides, 15 parts; alcohol, 150 parts. Digest eight days, filter, and add resin, 60 parts, and turpentine, 50 parts. Thin paper is to be brushed over two or three times with this varnish.

Paper, Atropine. *Syn.* CHARTA ATROPIÆ, L. Paper is impregnated by steeping in solution of sulphate of atropia in such a manner that a piece ¼ of an inch square shall contain ₁/₁₅ of a grain of the salt; a square of ₁/₁₅ of an inch the ₁/₁₀₀₀ of a grain. This square inserted between the eyelids will dilate the pupil.

Paper, Atropine, Gelatinised. Tablets of gelatin are impregnated with sulphate of atropia, as above.

Paper, Blistering. See VESICANTS.

Paper, Cloth. This is prepared by covering gauze, calico, canvas, &c., with a surface of paper pulp in a 'Foudrinier machine,' and then finishing the compound sheet in a nearly similar manner to that adopted for ordinary paper.

Paper, Col'oured. For those papers which are merely coloured on one side the pigments, ground up with gum water or size, or the stains thickened with a little of the same, are applied with a brush, after which the sheets are suspended on a line to dry.

For paper coloured throughout its substance the tinctorial matter is usually mixed with the pulp in the process of manufacture; or the manufactured paper is dipped into a bath of the colouring substance, and then hung up to dry.

Paper, Cop'ying. *Prep.* Make a stiff oint-

ment with butter or lard and black-lead or lampblack, and smear it thinly and evenly over soft writing-paper by means of a piece of flannel; the next day wipe off the superfluous portion with a piece of soft rag. *Use, &c.* Placed on white paper and written on with a style or solid pen, a copy of the writing is left on the former. By repeating the arrangement, two, three, or more copies of a letter may be obtained at once. This paper, set upon a case, forms the ordinary 'manifold writer' of the stationers. The copying or transfer paper used for obtaining fac-similes of letters written with 'copying ink' is merely a superior quality of bank-post paper.

Paper, Emery. See EMERY.

Paper, Glass. *Prep.* From powdered glass, as emery paper. Used to polish wood, &c. See GLASS (Powdered).

Paper, Gout. *Syn.* CHARTA ANTIARTHRITICA, L.; PAPIER FAYARD, Fr. *Prep.* 1. Euphorbium, 1 part; cantharides, 2 parts (both in powder); rectified spirit, 8 parts; ether, 3 parts; digest in a stoppered bottle, with frequent agitation, for a week; to the strained tincture add of Venice turpentine, 1 part; lastly, dip thin white paper into it, and dry the sheets in the air.

2. (*Mohr.*) Euphorbium, 1 dr.; cantharides, 4 dr.; rectified spirit (strongest), 5 oz.; make a tincture, to which add of Venice turpentine, 1½ oz., previously liquefied with resin, 2 oz.; and spread the mixture, whilst warm, very thinly on paper. Used as a counter-irritant in gout, rheumatism, &c.

Paper, Hydrograph'ic. An absurd name given to paper which may be written on with simple water or with some colourless liquid having the appearance of water.

Prep. 1. A mixture of nut-galls, 4 parts, and calcined sulphate of iron, 1 part (both perfectly dry and reduced to very fine powder), is rubbed over the surface of the paper, and is then forced into its pores by powerful pressure, after which the loose portion is brushed off. Writes black with a pen dipped in water.

2. From persulphate of iron and ferrocyanide of potassium, as the last. Writes blue with water.

3. As the last, but using sulphate of copper instead of sulphate of iron. Writes reddish brown with water.

4. The paper is wetted with a colourless solution of ferrocyanide of potassium, and after being dried is written on with a colourless solution of persulphate of iron. Writes blue.

Obs. The above applications, we need scarcely say, are more amusing than useful. See SYMPATHETIC INK.

Paper, Incombus'tible. See INCOMBUSTIBLE FABRICS.

Paper, Irides'cent. *Prep.* (*Beasley.*) Sal-ammoniac and sulphate of indigo, of each, 1 part; sulphate of iron, 5 parts; nut-galls, 8 parts; gum-arabic, ½ part; boil them in water, and expose the paper washed with the liquid to (the fumes of) ammonia.

Paper, Issue. *Syn.* CHARTA AD FONTICULOS, L. *Prep.* (*Soubeiran.*) Elemi, spermaceti, and Venice turpentine, of each, 1 part; white wax, 2 parts; melt them together by a gentle heat, and spread the mixture on paper. Used to keep issues open.

Paper, Lithograph'ic. *Prep.* 1. Starch, 6 oz.; gum-arabic, 2 oz.; alum, 1 oz.; make a strong solution of each separately, in hot water, mix, strain through gauze, and apply it whilst still warm to one side of leaves of paper, with a clean painting-brush or sponge; a second and a third coat must be given as the preceding one becomes dry; the paper must be, lastly, pressed, to make it smooth.

2. Give the paper 3 coats of thin size, 1 coat of good white starch, and 1 coat of a solution of gamboge in water; the whole to be applied cold, with a sponge, and each coat to be allowed to dry before the other is applied. The solutions should be freshly made.

Use, &c. Lithographic paper is written on with lithographic ink. The writing is transferred by simply moistening the back of the paper, placing it evenly on the stone, and then applying pressure a reversed copy is obtained, which, when printed from, yields corrected copies resembling the original writing or drawing. In this way the necessity of executing the writing or drawing in a reversed direction is obviated. See LITHOGRAPHY, INK, &c.

Paper, Oiled. *Prep.* Brush sheets of paper over with 'boiled oil,' and suspend them on a line till dry. Waterproof. Extensively employed as a cheap substitute for bladder and gut skin to tie over pots and jars, and to wrap up paste blacking, ground white-lead, &c.

Paper, Parch'ment. *Syn.* PAPYRIN, VEGE-TABLE PARCHMENT. *Prep.* 1. (*Poumarède* and *Figuier.*) Dip white unsized paper for half a minute in strong sulphuric acid, sp. gr. 1·842, and afterwards in water containing a little ammonia.

2. (*W. E. Gaine,* Patent 1857.) Plunge unsized paper for a few seconds into sulphuric acid diluted with half to a quarter its bulk of water (this solution being of the same temperature as the air), and afterwards wash with weak ammonia. This process, now extensively worked by Messrs De la Rue and Co., produces a much better material than does that of Poumarède and Figuier.

Prop. A tough substance, resembling animal parchment, and applicable to the same purposes. It is largely used for covering pots of pickles and preserves, and by the chemist for the intervening membrane in experiments in diffusion. See DIALYSER, DIALYSIS, &c.

Parchment paper in the form of tubes of various diameters and of any length may be obtained of Karl Brandegger, Ellwangen, Wurtemburg, Bavaria. Portions of this tube are far superior for dialysing to the ordinary hoop arrangement, as there is no risk of leakage.

Paper, Paste. Boil white paper in water for five hours; then pour off the water, and pound the pulp in a mortar; pass it through a sieve and mix with some gum-water, isinglass, or glue. It is used in modelling by artists and architects.

Paper, Protective. Various attempts have from time to time been made to prepare paper which might make the fraudulent alterations of cheques and other documents difficult or impossible. These attempts have taken two different directions, which may be briefly described.

The first and best known method consists in printing, in some delicate and easily destroyed colour, a complicated pattern on the face of the paper. Any reagent which will remove the writing will, of course, destroy the pattern below, and so render the alteration evident. The cheques used by Messrs Coutts and Co. are fine examples of this kind of protection, the whole of the paper being printed over with the name of the firm in characters so delicate, that they can scarcely be read without the assistance of a lens.

The obvious objection to this method is, that it is possible for a skilful forger to replace the printed design before the completion of the alteration.

The other method consists in the introduction into the paper during its manufacture of some substance or mixture of substances which shall strike a characteristic colour when chemical agents are applied to the ink.

One of the earliest attempts of this kind was that of Stephenson, who introduced ferrocyanide of potassium into the pulp. When any acid was applied to the writing, Prussian blue was formed with the aid of the iron of the ink. In another process iodide of potassium and starch were introduced into the paper, the application of chlorine then producing a blue stain (iodide of starch), while in a third (*Robson's*) the pulp was stained with the ingredients of common writing ink. None of these methods gave, however, any very efficient protection against fraud, for in each case it was tolerably easy to restore the paper to its original condition. But another process which followed upon the others has proved more successful, and, when properly applied, gives a paper which is practically secure. This process was patented by Barclay, and consists in the introduction into the pulp of ferrocyanide of manganese. When any acid is applied to the writing on this paper the blue stain of Prussian blue appears. This can, it is true, be removed by alkalies, but in that case the manganese is precipitated as the brown peroxide, an effect also produced by bleaching powder. This brown stain can be removed by sulphurous acid, but in that case Prussian blue appears simultaneously, so that the forger has merely a choice between a brown and a blue stain.

When such paper is printed with a delicate design in some fugitive ink (common writing ink would be the best), the greatest attainable safety is obtained.

Ferrocyanide of manganese is easily formed by adding to the pulp pure crystallised chloride of manganese, and rather more than an equal weight of ferrocyanide of potassium, both in solution (*Heaton*).

Paper, Ra"zor. Smooth unsized paper, one of the surfaces of which, whilst in a slightly damp state, has been rubbed over with a mixture of calcined peroxide of iron and emery, both in impalpable powder. It is cut up into pieces (about 4 × 3 inches), and sold in packets. Used to wipe the razor on, which thus does not require stropping.

Paper, Razor-strop. From emery and quarts (both in impalpable powder), and paper pulp (estimated in the dry state), equal parts, made into sheets of the thickness of drawing-paper, by the ordinary process. For use, a piece is pasted on the strop and moistened with a little oil.

Paper, Re'sin. Syn. POOR MAN'S PLASTER; CHARTA RESINOSA, L. Prep. 1. Beeswax, 1 oz.; tar and resin, of each, 3 oz.; melted together and spread on paper.

2. (Ph. Bor.) Paper thinly spread over with black pitch. Calefacient, stimulant, and counter-irritant; in rheumatism, chest affections, &c.

Paper, Rheu'matism. See PAPERS, GOUT and RESIN.

Paper, Safe'ty. Syn. PAPIER DE SURETÉ, Fr. White paper pulp mixed with an equal quantity of pulp tinged with any stain easily affected by chlorine, acids, alkalies, &c., and made into sheets as usual.

Paper, Test-. Syn. CHARTA EXPLORATORIA, L. Under this head may be conveniently included all the varieties of prepared paper employed in testing. For this purpose sheets of unsized paper or of good ordinary writing-paper (preferably the first) are uniformly wetted with a solution of the salt, or with a cold infusion or decoction of the tinctorial substance in distilled water, and are then hung up to dry in a current of pure air; they are, lastly, cut into pieces of a convenient size, and preserved in closed bottles or jars. For use, a small strip of the prepared paper is either dipped into or moistened with the liquid under examination, or it is moistened with distilled water and then exposed to the fumes. A single drop, or even less, of any liquid may be thus tested.

The following are the principal test-papers and their applications:

PAPER, BRAZIL-WOOD. From the decoction. Alkalies turn it purple or violet; strong acids, red.

PAPER, BUCKTHORN. From the juice of the berries. Reddened by acids.

PAPER, CHERRY-JUICE. As the last.

PAPER, DAHLIA; GEORGINA P. From an infusion of the petals of the violet dahlia (Georgina purpurea). Alkalies turn it green, acids red; strong caustic alkalies turn it yellow. Very delicate.

PAPER, ELDERBERRY. From the juice of the berries. As the last.

PAPIER FAYARD. See PAPER, GOUT.

PAPER, INDIGO. From a solution of indigo. Decoloured by chlorine.

PAPER, IODIDE OF POTASSIUM. a. From the solution in distilled water. Turned blue by an acidulated solution of starch.

b. From a mixture of a solution of iodide of potassium and starch paste. Turned blue by chlorine, ozone, and the mineral acids, and by air containing them.

PAPER, LEAD. From a solution of either acetate or subacetate of lead. Sulphuretted hydrogen and hydrosulphuret of ammonia turn it black.

PAPER, LITMUS. In general this is prepared from infusion of litmus, without any precaution, but the following plan may be adopted when a superior test-paper is desired:

a. (Blue.) Triturate commercial litmus, 1 oz., in a wedgwood-ware mortar, with boiling water, 3 or 4 fl. oz.; put the mixture into a flask, and add more boiling water until the liquid measures fully ½ pint; agitate the mixture frequently until it is cold, then filter it, and divide the filtrate into two equal portions; stir one of these with a glass rod previously dipped into very dilute sulphuric acid, and repeat the operation until the litmus infusion begins to look very slightly red; then add the other half of the filtrate, and the two being mixed together, dip strips of unsized paper into the liquid in the usual manner, and dry them. Acids turn it red; alkalies blue. The neutral salts of most of the heavy metals also redden this as well as the other blue test-papers that are affected by acids.

b. (Red.) The treatment of the whole quantity of the infusion (see above) with the rod dipped in dilute sulphuric acid is repeated until the fluid begins to look distinctly red, when the paper is dipped into it as before. The alkalies and alkaline earths, and their sulphides, restore its blue colour; the alkaline carbonates and the soluble borates also possess the same property. Very sensitive. An extemporaneous red litmus paper may be prepared by holding a strip of the blue variety over a pot or jar into which 2 or 3 drops of hydrochloric acid have been thrown.

PAPER, MALLOW. From an infusion of the purple flowers of the common mallow. Affected like 'dahlia paper.'

PAPER, MANGANESE. From a solution of sulphate of manganese. Ozonised air blackens it.

PAPER, RHUBARB. From a strong infusion of the powdered root. Alkalies turn it brown, but boracic acid and its salts do not affect it. Very sensitive.

PAPER, ROSE. From the petals of the red rose. As the last. Alkalies turn it bright green. Dr A. S. Taylor recommends the infusion to be very slightly acidulated with an acid before dipping the paper into it. More sensitive than turmeric paper.

PAPER, STARCH. From a cold decoction of starch. Free iodine turns it blue.

PAPER, SULPHATE OF IRON. From a solution of ferrous sulphate. As a test for hydrocyanic acid and the soluble cyanides.

PAPER, TURMERIC. From decoction of turmeric (2 oz. to the pint). It is turned brown by alkalies, and by boracic acid and the soluble borates. It is not quite so susceptible as some other tests, but the change of colour is very marked and characteristic.

PAPER, ZINC OLEATE. Tissue-paper saturated with a hot solution of pure oleate of zinc. Used as a healing astringent application.

Paper, Tra"cing. Prep. 1. Open a quire of smooth unsized white paper and place it flat upon a table, then apply, with a clean 'sash tool,' to the upper surface of the first sheet, a coat of varnish made of equal parts of Canada balsam and oil of turpentine, and hang the prepared sheet across the line to dry; repeat the operation on fresh sheets until the proper quantity is finished. If not sufficiently transparent, a second coat of varnish may be applied as soon as the first has become quite dry.

2. Rub the paper with a mixture of equal parts of nut oil and oil of turpentine, and dry it immediately by rubbing it with wheaten flour; then hang it on a line for 24 hours to dry.

Obs. Both the above are used to copy drawings, writing, &c. If washed over with ox-gall and dried, they may be written on with ink or water-colours. The first is the whitest and clearest, but the second is the toughest and most flexible. The paper prepared from the refuse of the flax-mills, and of which bank-notes are made, is also called 'tracing paper,' and sometimes 'vegetable paper.' This requires no preparation; but, though very flexible, it has little strength.

Paper, Varnished. Before proceeding to varnish paper, card-work, pasteboard, &c., it is necessary to give it two or three coats of size, to prevent the absorption of the varnish, and any injury to the colour or design. The size may be made by dissolving a little isinglass in boiling water, or by boiling some clean parchment cuttings until they form a clear solution. This, after being strained through a piece of clean muslin, or, for very nice purposes, clarified with a little white of egg, is applied by means of a small clean brush, called by painters a *sash* tool. A light, delicate touch must be adopted, especially for the first coat, lest the ink or colours be started or smothered. When the prepared surface is perfectly dry, it may be varnished in the usual manner. See MAPS, VARNISH, &c.

Paper, Wa″fer. See WAFERS.

Paper, Waxed. *Prep.* Place cartridge paper, or strong writing-paper, on a hot iron plate, and rub it well with a lump of beeswax. Used to form extemporaneous steam or gas pipes, to cover the joints of vessels, and to tie over pots, &c.

For the various photographic papers see PHOTOGRAPHY.

PAPER-HANGINGS. The ornamental paper used to cover the walls of rooms, &c. Under the old system, the paper, after being sized and prepared with a ground colour, had the pattern produced on it by the common process of 'stencilling,' a separate plate being employed for each colour that formed the pattern. To this succeeded the use of wooden blocks, the surface of which bearing the design in relief, and being covered with colour, was applied by simple hand pressure on the paper, in a precisely similar manner to that adopted in the block-printing of calicoes. The cylinder calico-printing machine has now been successfully applied to the manufacture of paper-hangings.

The colours employed for paper-hangings are—

BLACKS. Frankfort, ivory, and blue black.

BLUES. Prussian blue, verditer, and factitious ultramarine.

BROWNS. Umber (raw and burnt), and mixtures.

GRAYS. Prussian blue and blue-black, with Spanish white.

GREENS. Brunswick green, Scheele's green, Schweinfurt green, and green verditer; also mixtures of blues and yellows.

REDS. Decoctions of Brazil-wood (chiefly), brightened with alum or solution of tin; the red ochres; and, sometimes, red lake.

VIOLETS. Decoction of logwood and alum; also blues tempered with bright red.

WHITES. White-lead, sulphate of baryta, plaster of Paris, and whiting, and mixtures of them.

YELLOWS. Chrome yellow, decoction of French berries or of weld, terra di sienna, and the ochres.

The vehicle employed to give adhesiveness and body to the colours is a solution of gelatin or glue, sufficiently strong to gelatinise on cooling. The satiny lustre observable in some paper-hangings (SATIN PAPERS) is produced by dusting finely powdered French chalk over the surface, and rubbing it strongly with a brush or burnisher. The ground for this purpose is prepared with plaster.

FLOCK and VELVET PAPERS are produced by covering the surface of the pattern with a mordant formed with boiled oil thickened with white-lead or ochre, and then sprinkling powdered woollen flocks on it. These are previously dyed, and ground to the required fineness in a mill.

PAPIER-MÂCHÉ. Pulped paper moulded into forms. It possesses great strength and lightness. It may be rendered partially waterproof by the addition of sulphate of iron, quicklime, and glue or white of egg to the pulp; and incombustible by the addition of borax and phosphate of soda. The papier-mâché tea-trays, waiters, snuff-boxes, &c., are prepared by pasting or glueing sheets of paper together, and then submitting them to powerful pressure, by which the composition acquires the hardness of board when dry. Such articles are afterwards japanned, and are then perfectly waterproof.

The refuse of the cotton and flax mills, and numerous other substances of a like character, are now worked up as papier-mâché, and the manufactured articles formed of them are indistinguishable from those prepared directly from paper.

A practical paper on moulding, inlaying, decorating, and varnishing papier-mâché will be found in 'Work,' September, 1889 (Cassell and Co.).

PAPIN'S DIGESTER is a strong, closed, iron vessel, in which water can be heated above 212° F., thereby acquiring a temperature that adds considerably to its solvent powers. This apparatus is put to many useful applications in the arts, of which one is the speedy extraction of gelatin from the earthy matter of bones. The bones may be boiled for hours at 212° without any such effect being produced. The high temperature acquired by the water is effected by the confinement of the steam, the internal pressure of which can be regulated by means of a safety-valve attached to the vessel. By this arrangement the water may be kept at any uniform temperature above 212° at pleasure. Professor Junichen ('Chemical News') recommends the use of the digester for the purpose of boiling meat and other food. It appears from the author's experiments that the time for cooking various articles of daily consumption is much shorter when effected under strong pressure, while a great saving of fuel is also effected.

PAPRIKA. *Syn.* HUNGARIAN RED PEPPER.

Prepared from the fruit of *Capsicum annuum*. The *Szegediner paprika* is the sort most esteemed. It is much employed as a condiment, and placed on the table in a salt-cellar.

PAPY'RIN. See PAPER, PARCHMENT.

PAR'ACHUTE. In aërostation, an instrument or apparatus having for its object to retard the descent of heavy bodies through the air. The only form of parachute which has been hitherto adopted with success is that of the common umbrella when extended. The materials of which the apparatus is made are canvas and cord, both light but strong, and carefully put together. The car to contain the adventurer resembles that of the balloon, only smaller.

It is estimated that a circular parachute, to descend in safety with an adult, weighing, with the apparatus, 225 lbs., must have a diameter of at least 30 ft. Its terminal velocity would then be at the rate of 12 to 18 ft. per second, or about 6½ miles per hour; and the shock experienced on contact with the earth would be equal to that which the aëronaut would receive if he dropped freely from a height about 2½ ft.

Several descents from balloons, after they have acquired a great elevation, have been effected without accident by means of parachutes. Unfortunately, however, any want of integrity in the machine, or any accident which may happen to it after its detachment from the balloon, is irreparable and fatal.

PARACYAN'OGEN. The brown solid matter left in the retort when cyanide of mercury is decomposed by heat. It is polymeric with cyanogen.

PARAFFIN. *Syn.* TAR-OIL STEARIN. This remarkable solid hydrocarbon is one of the several substances discovered by Reichenbach in WOOD-TAR. It may be obtained from coal, bituminous shale, &c., by distillation. It exists also in the state of solution in many kinds of petroleum. From a chemical point of view the minerals known as fossil wax, ozocint, hatchettin, &c., may be regarded as solid paraffin.

Prep. 1. (From WOOD-TAR—*Reichenbach*.) Distil beech-tar to dryness, rectify the oily portion of the product, which is heavier than water, until a thick matter begins to rise, then change the receiver, and moderately urge the heat as long as anything passes over; next digest the product in the second receiver, in an equal measure of alcohol of 0·833, gradually add 6 or 7 parts more of alcohol, and expose the whole to a low temperature; crystals of paraffin will gradually fall down, which, after being washed in cold alcohol, must be dissolved in boiling alcohol, when crystals of pure paraffin will be deposited as the solution cools.

2. (From COAL—*James Young*.) The details of this process for obtaining paraffin and its congeners by the slow distillation of coal (preferably 'Boghead') are given in our article on PARAFFIN OIL. The solid paraffin is separated from the last products, or 'heavy oils,' by artificial cold; it is then melted and run into moulds.

3. (From RANGOON PETROLEUM—Patent.) In this process, which is worked by Price's Candle Company, superheated steam is employed as the heating agent. The paraffin, or 'BELMONTINE,'

as it is called, is the last product which distils over.

4. (From PEAT.) The various processes which have been suggested for obtaining paraffin from peat, &c., are similar in principle to Young's. The great point is to conduct the distillation at as low a temperature as possible.

Prop. A white, hard, tasteless, inodorous, translucent body, melting at 110° F. and upwards, according to its source, and burning with a bright white flame. It has great stability—sulphuric acid, chlorine, and nitric acid below 212° exerting no action upon it. Dr Anderson states that its composition and properties vary with the source from which it is derived. With respect to the melting-point, this variation is very remarkable. Thus, Young's paraffin, from Boghead coal, melts at 114°, while that from Rangoon petroleum ('belmontine') melts at 140°, and that from peat at 116°.

Uses, &c. Paraffin is now largely used for making candles, for which purpose it is specially adapted, being a most elegant substance, and surpassing all other candle materials, even spermaceti, in illuminating power. Its property of not being acted upon by acids or alkalies renders it suitable for preparing the stoppers for vessels holding chemical liquids; also for electrotype moulds. It is not acted upon by ozone, so that it has been employed with great advantages in experiments on this body for rendering air-tight all joints in the apparatus. As it contains no oxygen, it might be employed to protect oxidisable metals like sodium and potassium from contact with the air. One use of paraffin candle-ends will commend them to the ladies of the household—a small piece of paraffin added to starch will be found to give a gloss and brilliancy of surface to starched linen that can be obtained by no other addition.

Imports of Paraffin in 1889.

	Paraffin.	cwt.	£
From Germany	. . . .	1,804	8,181
" United States of America		306,338	361,363
" other foreign countries	.	36	58
Total from foreign countries	.	808,178	364,602
From British East Indies	. .	7,422	10,277
" Australasia	. . . .	6	8
Total from British Possessions		7,428	10,285
Grand total	. . . .	315,606	374,887

PARAFFIN OILS. See OILS.

PARALDEHYDE. $C_6H_{12}O_3$. A colourless liquid of peculiar odour and pungent taste. Sp. gr. ·996. A new hypnotic. In physiological action it strongly resembles chloral, but differs from it in its action on the circulatory system, strengthening the heart's action while diminishing its frequency. It has also a well-marked action on the kidneys, greatly increasing the flow of the urine. The skin is not at all affected. The drug is said not to give rise to digestive disturbances, no headache, or to any other unpleasant symptom. It has been found a valuable remedy in mania, melancholia, and other nervous affections, as well as in the sleeplessness that accompanies acute bronchial catarrh, lobar pneumonia, and heart diseases. When prescribed in mixtures, syrup of orange or orange-flower water may be added, to disguise the disagreeable taste of the drug.—*Dose*, 30 to 60 minims.

PARAL'YSIS. *Syn.* **PALSY.** A loss or considerable diminution of power of voluntary motion, or functional action, of any part of the body. In its most usual form one side only of the body is affected. It not uncommonly seizes the lower extremities, or all parts below the pelvis; sometimes the arms only; and occasionally a part, as one side of the face, one eyelid, the tongue, or the muscles of deglutition. In these cases the speech frequently becomes indistinct and incoherent, and the memory and judgment impaired, whilst the features become drawn and distorted.

The causes of paralysis are various. It may be occasioned by pressure on particular parts of the brain, the spinal cord, or the nerves; by poisons, the long-continued use of sedatives, local injuries, the sudden suppression of profuse and habitual evacuations. It may also be a consequence of an attack of apoplexy, or it may be symptomatic of other diseases, as scrofula, syphilis, and worms. When it is of a distinctly local character it may arise from excessive use or undue employment of the part or organ. That of old age is, probably, a mere consequence of the failing nervous energy of the system being unequally distributed.

Palsy usually comes on with a sudden and immediate loss of the motion and sensibility of the parts; but in a few instances it is preceded by a numbness, coldness, and paleness; and sometimes by slight convulsive twitches. If the disease affects the extremities, and has been of long duration, it not only produces a loss of motion and sensibility, but likewise a considerable flaccidity and wasting away of the muscles of the parts affected.

The treatment of paralysis depends upon a careful consideration of its cause, and requires in all cases skilled advice. Much harm may be done by the reckless use of violent remedies.

PARANAPH'THALIN. *Syn.* **ANTHRACEN.** See **ANTHRACEN.**

PARAPEC'TIN. See **PECTIN.**

PAR'ASITES. The parasitical animals that infest the human body are referred to under the heads **ACARI** and **PEDICULI.**

Parasites, Animal. The following list is given in Williams' 'Veterinary Medicine:'

1. Nat. Ord. NEMATODA.

Genus *Ascaris*.
A. megalocephala, *horse and ass*, small intestine.
A. lumbricoides, *pig, cattle*, small intestine.
A. mystax, *cat, dog*, small intestine.

Genus *Eustrongylus*.
E. gigas, *dog, horse, cattle*, kidneys and bladder.

Genus *Filaria*.
F. lachrymalis, *horse and ox*, lachrymal ducts.
F. papillosa, *horse, ox, and ass*, eye, brain.
F. immitis, *dog*, heart and blood.
F. trispinulosa, *dog*, capsule of lens.

Genus *Spinoptera*.
S. megastoma, *horse*, tumours in stomach.
S. sanguinolenta, *dog and wolf*, tumours in stomach.
S. strongylina, *pig*, stomach.
S. scutata, *ox*, œsophagus.
S. hamulosa, *common fowl*.
S. cincinnata, *horse*, foot ligaments.

Genus *Oxyuris*.
O. curvula, *horse and ass*, large intestine.

Genus *Dochmius* (*Strongylus* of some authors).
D. hypostomus, *sheep, goats, &c.*, intestine.
D. tubæformis, *cat*, duodenum.
D. trigonocephalus, *dog*, stomach and intestine.
D. cernuus, *sheep*, intestine.
D. duodenalis, *man*, duodenum.

Genus *Strongylus*.
S. armatus, *horse*, intestine.
S. tetracanthus, *horse*, intestine.
S. dentatus, *pig*, large intestine.
S. syngamus, *fowls*, trachea and bronchi.
S. radiatus, *ox*, intestine.
S. venulosus, *goat*, intestine.
S. micrurus, *cattle, horse, ass*, trachea and bronchi.
S. filaria, *sheep, goat, camel, &c.*, trachea and bronchi.
S. paradoxus, *pig*, trachea and bronchi.
S. filicollis, *sheep*, small intestine.
S. ventricosus, *cattle*, small intestine.
S. inflatus, *cattle*, colon.
S. contortus, *sheep and goat*, abomasum.
Stephanurus dentatus, *pig*, kidneys.

Genus *Trichina*.
T. spiralis, *man, pig, ox, rabbit, rat, &c.*, muscle.

Genus *Tricocephalus*.
T. dispar, *man*.
T. affinis, *sheep and goat*, cæcum.
T. depressiusculus, *dog*, cæcum.
T. crenatus, *pig, wild boar*, large intestine.

2. Nat. Ord. TREMATODA.

Genus *Distoma*.
D. hepaticum, *sheep, cattle, goat, and pig*, rarely in *horse, ass, cat*, and very rare in *man*, gall-bladder.
D. lanceolatum, *sheep, cattle, goat, pig*, gall-bladder.
D. campanulatum, *dog*, liver.
D. conjunctum, *Indian dogs*, bile-ducts.

Genus *Amphistoma*.
A. conicum, *cattle*, paunch.
A. truncatum, *cat*.

Genus *Hemistoma* (*Holostoma*).
H. alatum, *dog, wolf, fox*, small intestine.
H. cordatum, *cat*.

Genus *Gastrodiscus*.
G. polymastos, *Egyptian horses*.

3. Nat. Ord. CESTODA.

See **TÆNIA** and **TAPEWORM.**

Parasites, Human. The following is a list of the principal parasites infesting man. It is extracted from the 'Dictionary of Hygiene,' of Wynter Blyth, who states that he has arranged it, with some slight alterations, from a table in Dr Aitken's 'Science and Practice of Medicine.' The first two divisions include animal parasites, the third vegetable ones. No. 1, or *Entozoa*, are animal parasites found inside the human body; No. 2, those found outside; No. 3, consisting of vegetable parasites, comprises *Entophyta* and *Epiphyta*, the former existing in the interior, and the latter on the exterior of the human body. Some of the principal parasites have already been described and figured in these pages.

I. *Entozoa.*

Acephalocystis endogena, *liver.*
A. multifida, *brain.*
Anchylostomum seu Sclerostoma duodenale, *intestines.*
Anthomia canicularis, *intestines.*
Ascaris alata, *intestines.*
A. lumbricoides, *intestines.*
A. mystax, *intestines.*
Bilharzia seu Distoma hæmatobia, *portal and venous system.*
Bothriocephalus cordatus, *intestines.*
B. latus, *intestines.*
Cysticercus cellulosæ, seu telæ cellulosæ (C. of Tænia solium), *muscles.*
Cysticercus of Tænia marginata (C. tenuicollis), *intestines.*
Dactylius aculeatus, *urinary bladder.*
Diplosoma crenatus.
Distoma seu Distomum crassum, *duodenum.*
D. hepaticum, seu Fasciola hepatica, *gall-bladder.*
D. heterophryes, *intestines.*
D. lanceolatum, *hepatic duct.*
D. oculi humani, seu ophthalmobium, *capsule of crystalline lens.*
Ditrachycerus rudus, *intestines.*
Echinococcus hominis (hydatid of Tænia echinococcus), *liver, spleen, and omentum.*
Filaria bronchialis, seu trachealis, *bronchial glands.*
F. seu Dracunculus medinensis, *skin and areolar tissue.*
F. oculi, seu lentis, *eye.*
F. sanguinis hominis, *blood.*
Hexathrydium pinguicola, *ovary.*
H. venarum, *venous system.*
Monostoma lentis, *crystalline.*
Œstrus hominis, *intestines.*
Oxyuris vermicularis, *intestines.*
Pentastroma constrictum, *intestines and liver.*
P. denticulatum, *intestines.*
Polystroma pinguicola, *ovary.*
P. sanguicola, seu verarum, *venous system.*
Spiroptera hominis, *urinary bladder.*
Strongylus seu Eustrongylus bronchialus, *bronchial tubes.*
S. seu Eustrongylus gigas (Acarus renalis), *kidney and intestines.*
Tænia acanthotrias, *intestines.*
T. elliptica, *intestines.*
T. flavopuncta, *intestines.*
T. lophosoma, *intestines.*
T. mediocanellata, *intestines.*
T. nana, *intestines and liver.*
T. solium, *intestines.*
Tetrastoma renale, *kidney.*
Trichina spiralis, *muscles.*
Tricocephalus dispar, *intestines.*

II. *Ectozoa.*

Demodex seu Acarus folliculorum, *sebaceous substance of cutaneous follicles.*

Pediculus capitis (head louse).
P. corporis, seu vestimenti (body louse).
P. palpebrarum (brow louse).
P. pubis (Phthirius inguinalis) (crab louse).
P. tabescetium, *phthiriasis (lousy disease).*
Pulex penetrans (chigoë), *skin, cellular tissue.*
Sarcoptes seu Acarus scabiei (itch insect), *scabies.*

III. *Entophyta and Epiphyta.*

Achorion Lebertii (Tricophyton tonsurans), *Tinea tonsurans.*
A. Schönleinii, *Tinea favosa.*
Chionyphe Carteri (fungus of Mycetoma), *deep tissues, bones of hands and feet.*
Leptothrix buccalis (alga of the mouth).
Microsporon Audouini, *Tinea decalvans.*
M. furfur, *Tinea versicolor.*
M. mentagrophytes, *follicles of hair in sycosis or mentagra.*
Oïdium albicans (thrush fungus), *mouth, mucous and cutaneous surfaces.*
Puccinia favi, *Tinea favosa.*
Sarcina ventriculi, *stomach.*
Torula cerevisiæ (Cryptococcus cerevisiæ, yeast plant), *stomach, bladder, &c.*
Tricophyton sporuloïdes, *Tinea polornia.*

PARATARTARIC ACID. See RACEMIC ACID.
PARCHMENT. See VELLUM, and PAPER (Parchment).
PARCHMENT PAPER. See PAPER.
PAREGORIC. See TINCTURE OF CAMPHOR (Compound).
Paregoric, Scotch. See TINCTURE OF OPIUM (Ammoniated).
PAREIRA BRAVA. See VELVET LEAF.
PARKESINE. An old name for xylonite and celluloid, from its original discoverer, Alex. Parkes, of Birmingham, patented by him in the year 1855.
PARR. A name applied to the salmon until near the end of its second year, when it loses its dark lateral bars by the superaddition of a silvery pigment. It was formerly regarded as a distinct species.
PARSLEY. *Syn.* PETROSELINUM, L. This well-known herb is the *Apium petroselinum.* The root is diuretic; the fruit (seed) carminative; the leaves are a pleasant stimulating salad and condiment, and are much used to flavour broth and soup. "The fruit is a deadly poison to parrots" (*Lind. ex Burnett*).
PARSNIP. The root of *Pastinaca sativa.* The parsnip is native to England and Ireland, but does not grow in Scotland. It is likewise met with in many parts of Europe and in Northern Asia. In the wild state the root is somewhat acrid, and injurious effects have been known to follow its use as food. By cultivation, however, it loses both its acridity and dangerous properties, and forms a table vegetable not in universal favour.

In the Channel Islands parsnips constitute the winter food of cows; and these animals when fed upon them are said to yield butter of a better quality than can be obtained from them when partaking of any other fodder.

The flesh of cattle fed on the parsnip is also highly commended. In the north of Ireland the juice of the root, mixed with hops and yeast, is made into a fermented liquor. Parsnip wine is an agreeable alcoholic beverage.

Composition of the Parsnip.

Nitrogenous matter	.	.	.	.	1·1
Starch	.	.	.	.	9·6
Sugar	.	.	.	.	5·8
Fat	.	.	.	.	0·5
Salts	.	.	.	.	1·0
Water	.	.	.	.	82·0
					100·0

PASTE. *Syn.* PASTA, L.; PÂTE, Fr. This word is very loosely applied to substances and preparations differing so widely from each other, that it would be scarcely possible to class them together. We shall, therefore, refer the reader to the individual articles. The pastes (pâtes) of French pharmacy are compound medicines of the consistence of hard dough, and which do not stick to the fingers. They are formed of sugar and gum, dissolved in water or in some medicated liquid. They are evaporated so as to unite these principles by degrees, and give them the pliancy and the firmness of paste. They are employed internally in doses more or less variable in a similar manner to lozenges. "Pâtes, properly so called, are divided into transparent, or such as are made without agitation, like jujubes of brown liquorice; and opaque, or such as are made with agitation, like the pâtes of marsh-mallow, lichen, &c." (*Trousseau* and *Reveil*). See PASTES (Artificial Gems), PASTRY, and *below*.

Paste, Adhesive. 1. Let 4 parts, by weight, of glue soften in 15 parts of cold water for 15 hours, after which the mixture must be moderately heated until it becomes quite clear. To this mixture 65 parts of boiling water are to be added without stirring. In another vessel 30 parts of starch paste are stirred up with 20 parts of cold water, so that a thin milky fluid is obtained without lumps. Into this the boiling glue solution is poured, with constant stirring, and the whole is kept at the boiling temperature. When cooled 10 drops of carbolic acid are to be added to the paste. This paste possesses great adhesive power, and may be used for leather, paper, or cardboard with great success. It must be preserved in closed bottles to prevent evaporation of the water, and will in this way keep good for years ('Dingler's Journal').

2. The paste used by the United States Government for gumming postage stamps is made by the formula given below. It has the properties of being very adhesive, does not become brittle or scale off, and is well adapted for sticking paper labels to tin and other metals. Take of starch, 2 dr.; white sugar, 1 oz.; gum-arabic, 2 dr.; water, q. s. Dissolve the gum, add the sugar, and boil until the starch is cooked.

Paste, Al'mond. *Syn.* PASTA AMYGDALINA, P. AMYGDALARUM, P. REGIA, L.; PÂTE ROYALE, Fr. *Prep.* 1. (MOIST.) *a.* Take of blanched Valentia almonds, 4 oz.; reduce them to a very smooth paste by patient pounding in a clean mortar, adding, towards the last, a little rose-water with some eau de Cologne, or 3 or 4 drops of otto of roses or neroli, or an equivalent quantity of any other perfume, according to the fancy of the artiste.

b. From bitter and sweet almonds (blanched), equal parts; rose-water, q. s. It requires no other perfume.

c. To either of the preceding add spermaceti, ½ oz. The white of an egg, or ¼ oz. of white soap, is added by some makers. With about ¼ dr. of powdered camphor to each oz. of the above it forms the 'camphorated almond paste' of the shops.

d. Take fine Narbonne honey and white bitter paste (see *below*), of each, 1 lb.; beat them to a smooth paste, then add, in alternate portions, of oil of almonds, 2 lbs.; yolks of 5 eggs; and reduce the whole to a perfectly homogeneous pasty mass. Much esteemed. It is commonly sold under the name of 'honey paste,' 'pâte royale,' &c. In a similar manner are made nosegay, orange, rose, vanilla, and other like pastes having almonds for a basis, by merely adding the respective perfumes.

2. (PULVERULENT.) *a.* (Grey.) Prepared from the cake of bitter almonds from which the oil has been thoroughly expressed by drying, grinding, and sifting it.

b. (Bitter white.) As the last, but the almonds are blanched before being pressed.

c. (Sweet white.) As the last, but using sweet almonds.

Obs. All the above are used as cosmetics, to soften and whiten the skin, prevent chaps, abrasions, chilblains, &c. The honey paste, and the sweet and bitter white pastes, are those most esteemed. (See *below*.)

Paste, Almond (*in confectionery*). *Prep.* 1. Take of Valentia almonds, 3 lbs.; bitter do., ¼ lb.; blanch them, and reduce them to a very smooth paste by pounding, then put them into a clean copper pan along with white sugar and good gum-arabic, of each, 1 lb. (the last previously dissolved in about a pint of water); apply a gentle heat, and stir until the whole is mixed and has acquired a proper consistency, then pour it out on a smooth, oiled, marble slab, and when cold cut it into squares.

2. As the last, but when the mixture has acquired the consistence of thick honey, setting it aside to cool; when nearly cold the whites of six eggs are to be added, and heat being again gradually applied, the whole is to be stirred until it acquires the proper consistence, as before.

3. Blanched sweet almonds and white sugar, of each, 1 lb.; blanched bitter almonds and powdered gum, of each, 3 oz.; beat them, in the cold, to a perfectly smooth paste, with orange-flower water or rose-water, q. s., so that it may be sufficiently stiff not to stick to the fingers, and then cut the mass into squares, as before. The above are eaten as confections.

Paste, Ancho'vy. *Prep.* Remove the larger bones from the fish, and then pound them to a smooth paste in a marble mortar, adding a little bay-salt and cayenne pepper at will; next rub the pulp through a fine hair sieve, and about 3-4ths fill the pot with it; lastly, cover the surface of each to the depth of about ¼ inch with good butter in a melted state. It should be kept in a cool situation. Other fish pastes, as those of bloaters, lobsters, shrimps, caviare, &c., are made in a similar manner.

Paste, Arsen'ical. See CAUSTICS, PATENT MEDICINES, and POWDERS.

Paste, Baudry's. See PASTE, PECTORAL (below).

Paste, Bird. See GERMAN PASTE.

Paste, Black Currant. As black currant lozenges, but simply cutting the mass into dice or squares.

Paste, Car'rageen. *Prep.* From Irish moss, as the lichen paste of the P. Cod. (see below).

Paste, Chilli. *Prep.* Powdered capsicum, 8 oz.; olive oil, 32 oz.; spermaceti, 6 oz. Macerate the capsicum in the oil for three days, strain, press, filter, and warm. Melt the spermaceti and add it to the oil, stirring the mixture until cold.

Paste, Chinese'. *Prep.* From bullock's blood, 10 lbs, reduced to dryness by a gentle heat, then powdered and mixed with quicklime, also in fine powder, 1 lb. It is used as a cement, made into a paste with water, and at once applied.

Paste of Chlo''ride of Zinc. See CAUSTICS.

Paste of Dates. *Syn.* PASTA DATYLIFERÆ, P. DACTYLORUM, L.; PÂTE DE DATTES, Fr. From dates (stoned), as jujube paste. Pectoral, and slightly astringent. Paste of gum Senegal is usually sold for it.

Paste, De Handel's. *Prep.* From opium, ½ dr.; camphor, 1 dr. (both in powder); extracts of belladonna and henbane, of each, 1 dr.; oil of cajeput and tincture of cantharides, of each, 10 or 12 drops; distilled water of opium (or of lettuce), q. s. In toothache.

Paste, Depil'atory. *Syn.* PASTA EPILATORIA, L. Several preparations of this character are noticed at pages 552-3. 1. A mixture of slacked lime, 2 parts, and water, 3 parts, saturated with sulphuretted hydrogen, is said to be so powerful, that "a layer a line in thickness denudes the scalp in three minutes" (*Beasley*).

2. (*Payès.*) Powdered sulphate of copper made into a soft paste with yolk of egg.

Paste of Figs. *Syn.* PASTA CARICARUM, P. FICARIA, L. *Prep.* 1. From figs, as jujube paste.

2. (*Soubeiran.*) Pulp of figs, 1 part; press it through a sieve, mix it with powdered sugar, 4 parts, concentrated by a gentle heat (if necessary), roll the mass out, and cut it into squares or lozenges.

Paste, Flour. *Syn.* COLLE DE PÂTE, Fr. From wheaten flour. Paper-hangers, shoemakers, &c., usually add to the flour ½ to ½ of its weight of finely powdered resin. It is then sometimes called 'hard paste.' The addition of a few drops of creasote or oil of cloves, or a little powdered camphor, colocynth, or corrosive sublimate (especially the first two and the last), will prevent insects from attacking it, and preserve it in covered vessels for years. Should it get too hard, it may be softened with water. See CEMENTS.

Paste, Fruit. *Prep.* 1. To each pint of the strained juice add of gum-arabic, 1 oz.; gently evaporate to the consistence of a syrup, and add an equal weight of bruised white sugar; as soon as the whole is united, pour it out on an oiled slab, and, when cold enough, cut it into pieces.

2. Citric acid, ¾ oz.; gum-arabic, 6 oz.; white sugar, ¾ lb.; water, q. s.; dissolve, and flavour with any of the fruit essences. It may be coloured with any of the stains used for confectionery or liqueurs.

3. As fruit lozenges.

Paste, Fur'niture. See POLISH.

Paste, Glove. See GANTRINE.

Paste of Gum-arabic. *Syn.* PASTA GUMMI, L.; PÂTE DE GOMME, P. DE G. ARABIQUE, Fr. *Prep.* 1. As marsh-mallow paste, omitting the mallow roots.

2. Gum-arabic (picked), 1 lb.; water, 1 pint; dissolve, add of white sugar, 1 lb.; evaporate by a gentle heat to a very thick syrup, then add the whites of 3 eggs, previously beaten up with orange-flower water, 1 fl. oz., and strain through muslin, and continue the heat with constant stirring, until of a proper consistence on being cooled. The last two are commonly sold for marsh-mallow paste (pâte de guimauve).

3. (Transparent.) From gum-arabic (picked), 1 lb.; cold water, 1 pint; white sugar, 1¼ lbs.; proceed as the last, adding orange-flower water, 1 fl. oz., towards the end. Often sold under the name of 'white jujubes.'

Paste of Gum Senegal. *Syn.* PÂTE DE GOMME SÉNÉGAL, Fr. As JUJUBE PASTE, without the fruit.

Paste, Hon'ey. See PASTE, ALMOND.

Paste, Ju'jube. *Syn.* JUJUBES, JUJUBE LOZENGES; PASTA JUJUBÆ, L.; PÂTE DE JUJUBES, Fr. *Prep.* (P. Cod.) Jujubes (the fruit), 1 lb.; water, 4 lbs.; boil ½ hour, strain with expression, settle, decant the clear portion, and clarify it with white of egg; add a strained solution of gum-arabic, 6 lbs., in water, 8 lbs., and to the mixture add of white sugar, 5 lbs.; gently evaporate, at first constantly stirring, and afterwards without stirring, to the consistence of a soft extract, then add of orange-flower water, 6 fl. oz., and place the pan in a vessel of boiling water. In 12 hours carefully remove the scum, pour the matter into slightly oiled tin moulds, and finish the evaporation (hardening) in a stove heated to 104° F. It is commonly coloured with beetroot, cochineal, or saffron. Expectorant; in coughs, &c. Paste of gum-arabic is usually sold for it.

Paste, Lassar's. *Prep.* Salicylic acid, 35 gr.; oxide of zinc, 1 oz.; starch, 1½ oz.; vaseline, 2½ oz. Melt the vaseline, and mix with the powders in the proper manner.

Paste, Li'chen. *Syn.* PASTA LICHENIS, L.; PÂTE DE LICHEN, Fr. *Prep.* (P. Cod.) Iceland moss, 1 lb.; water, q. s.; heat them to nearly the boiling-point, strain with pressure, reject the liquor, and boil the moss in fresh water, q. s., for 1 hour; strain, press, add of gum-arabic, 6 lbs.; white sugar, 4 lbs., and evaporate to a proper consistence, as above. Pectoral. With the addition of ½ gr. of extract of opium to each oz., it forms the opiated lichen paste (P. Cod.).

Paste, Liquorice. *Syn.* LIQUORICE JUJUBES; PASTA GLYCYRRHIZÆ, L.; PÂTE DE RÉGLISSE, P. DE E. NOIRE, Fr. *Prep.* 1. (P. Cod.) Refined juice and white sugar, of each, 1 lb.; gum-arabic, 2 lbs.; water, 3 quarts; dissolve, strain, evaporate considerably, add of finely powdered orris-root, ½ oz.; oil of aniseed or essence of cedrat, a few drops, and pour the paste upon an oiled slab, or into moulds, as before.

2. (Brown: PASTA G. FUSCA, L.; PÂTE DE R. BRUNE, Fr.) Refined juice, 4 oz.; white sugar, 2 lbs.; gum-arabic, 8 lbs.; water, 4 pints; proceed as last.

3. (Opiated: PÂTE DE R. OPIACÉ, Fr.—P. Cod.) To the last add of extract of opium, 15 gr.

4. (White: PÂTE DE R. BLANCHE, Fr.) As No. 2, substituting the powder of the decorticated root for the extract. All the above are pectoral; the second is also slightly anodyne. They are useful in tickling coughs, hoarseness, &c.

Paste, London. *Syn.* PASTA LONDINENSIS, L. *Prep.* Equal parts of caustic soda and unslaked lime. Reduce to a fine powder in a warm mortar, and mix intimately. Keep it in well-closed bottles, and when required for use take as much as is sufficient, and make it into a paste with water.

Paste, Marsh-mallow. *Syn.* PASTA ALTHEAE, L.; PÂTE DE GUIMAUVE, Fr. *Prep.* (P. Cod. 1816.) Decorticated marsh-mallow root (French), 4 oz.; water, ½ gall.; macerate for 12 hours, strain, add white sugar and gum-arabic, of each, 2½ lbs.; dissolve, strain, evaporate without boiling to the thickness of honey, constantly stirring, and add, gradually, the whites of 12 eggs, well beaten with orange-flower water, 4 fl. oz., and strain; continue the evaporation and constant stirring until the mass is so firm as not to adhere to the fingers, then proceed as before.

Obs. It should be very light, white, and spongy. In the P. Cod. of 1889 the marsh-mallow root is omitted, and the name is changed to that of 'pâte de gomme,' a compound long sold for it in the shops. Both are agreeable pectorals. See PASTE OF GUM-ARABIC.

Paste, Odontal'gic. *Syn.* PASTA ODONTAL-GICA, L. *Prep.* 1. Pellitory (in powder), 1 dr.; hydrochlorate of morphia, 3 gr.; triturate; add of honey, 2 dr.; and oil of cloves, 6 drops.

2. Powdered mastic, pellitory, and white sugar, of each, 1 dr.; chloroform, q. s. to form a paste. It must be kept in a stoppered bottle. See TOOTH-ACHE, and *below.*

Paste, Or'ange. *Prep.* From orange flowers, 2 lbs.; bitter and sweet almonds, of each, blanched, 2½ lbs., beaten to a perfectly smooth paste. An agreeable cosmetic. See PASTE, ALMOND.

Paste, Or'geat. *Prep.* From blanched Jordan almonds, 1 lb.; blanched bitter a. and white sugar and honey, of each, ¾ lb.; beaten to a paste, with orange-flower water, q. s. (or neroli, a few drops), and put into pots. As a cosmetic, or to make orgeat milk. For use, rub 1 oz. with ⅛ pint of water, and strain through muslin.

Paste, Pec'toral. *Syn.* PASTA PECTORALIS, L. *Prep.* 1. (PÂTE PECTORALE DE BAUDRY.) Take of gum-arabic and white sugar, of each, 7 lbs.; water, q. s.; dissolve, add of extract of liquorice, 3 oz.; evaporate, add extract of lettuce, 2 dr.; balsam of tolu, 1½ oz.; orange-flower water, 4½ fl. oz.; whites of 4 eggs; oil of citrons, 5 or 6 drops.

2. (PÂTE PECTORALE BALSAMIQUE DE REG-NAULT.) From the flowers of coltsfoot, cudweed, marrow, and red poppy, of each, 1 oz.; water, 1

quart; boil, strain; add of gum-arabic, 30 oz.; white sugar, 20 oz.; dissolve, concentrate, add of tincture of tolu, 3 fl. dr., and pour the mixture on an oiled slab.

3. (ANISATED COLTSFOOT PASTE; PÂTE DE TUSSILAGE À L'ANIS.) From a strong decoction of coltsfoot flowers, 1 quart; Spanish juice, ½ lb.; dissolve, strain, evaporate, as before, and towards the end add of oil of aniseed, 1 dr. All the above are useful in hoarseness, coughs, &c.

Paste, Phos'phor. See RATS.

Paste, Phosphorous. A phosphorous paste that will keep a long time may be made by the following process:—Shake 9 parts of phosphorus in 90 parts of warm syrup, and pour the still warm mixture into a pan in which 90 parts of wheat-flour have been previously put; agitate rapidly, and add 60 parts of ivory-black, 60 parts of water, and 120 parts of lard.

Paste, Pol'ishing. *Prep.* 1. (For copper and brass.) See BRASS PASTE.

2. (For iron and steel.) From emery (in fine powder) and lard, equal parts.

3. (For pewter.) From powdered Bath brick, 2 parts; soft soap, 1 part; water, q. s. to make a paste. Used with a little water, and afterwards well rinsed off.

4. (For furniture.) See POLISH.

Paste, Ra'zor. *Prep.* 1. From jewellers' rouge, plumbago, and suet, equal parts, melted together, and stirred until cold.

2. From prepared putty powder (levigated oxide of tin), 3 parts; lard, 2 parts; crocus martis, 1 part; triturated together.

3. Prepared putty powder, 1 oz.; powdered oxalic acid, ¼ oz.; powdered gum, 20 gr.; make a stiff paste with water, q. s., and evenly and thinly spread it over the strop, the other side of which should be covered with any of the common greasy mixtures. With very little friction this paste gives a fine edge to the razor, and its action is still further increased by slightly moistening it, or even breathing on it. Immediately after its use the razor should receive a few turns on the other side of the strop.

4. Diamond dust, jewellers' rouge, and plumbago, of each, 1 part; suet, 2 parts. Powdered quartz is generally substituted for diamond dust, but is much less effective.

5. (*Mechi's.*) Emery (reduced to an impalpable powder), 4 parts; deer suet, 1 part; well mixed together.

6. (*Pradier's.*) From powdered Turkey stone, 4 oz.; jewellers' rouge and prepared putty powder, of each, 1 oz.; hard suet, 2 oz.

Obs. The above (generally made up into square cakes) are rubbed over the razor strop, and, the surface being smoothed off with the flat part of a knife or a phial bottle, the strop is set aside for a few hours to harden before being used.

Paste, Regnault's. See PASTE, PECTORAL.

Paste, Ricord's. Wood charcoal and strong sulphuric acid, equal parts by weight; mix well. Used as an application to cancer and phagedenic growths.

Paste, Rubefa"cient. *Syn.* PASTA RUBEFA-CIENS, L. *Prep.* (*Clarus.*) From acetate of lead, 1 oz.; bisulphate of potassa, 3 oz.; water, q. s. It acts powerfully and quickly on the skin.

Paste, Rust's. *Prep.* From powdered opium and extract of henbane, of each, 10 gr.; powdered pellitory and extract of belladonna, of each, 20 gr.; oil of cloves, 10 drops. In toothache.

Paste, Sha'ving. *Prep.* 1. Naples soap (genuine), 4 oz.; powdered Castile soap, 2 oz.; honey, 1 oz.; essence of ambergris and oils of cassia and nutmegs, of each, 5 or 6 drops.

2. White wax, spermaceti, and almond oil, of each, ½ oz.; melt, and, whilst warm, beat in 2 squares of Windsor soap previously reduced to a paste with a little rose-water.

3. White soft soap, 4 oz.; spermaceti and salad oil, of each, ½ oz.; melt them together and stir until nearly cold. It may be scented at will. When properly prepared these pastes produce a good lather with either hot or cold water, which does not dry on the face. The proper method of using them is to smear a minute quantity over the beard, and then to apply the wetted shaving-brush, and not to pour water on them, as is the common practice.

Paste, Styptic, of Gutta Percha. *Syn.* PASTA GUTTÆ PERCHÆ STYPTICA, L. (*Mr Beardsley.*) Gutta percha, 1 oz.; Stockholm tar, 1½ to 2 oz.; creosote, 1 dr.; shellac, 1 oz., or q. s. to render it sufficiently hard. To be boiled together with constant stirring till it forms a homogeneous mass. For alveolar hæmorrhage, and as a stopping for teeth in toothache. To be softened by moulding with the fingers.

Paste, Swediaur. See CHILBLAIN.

Paste, Tooth. *Syn.* PASTA DENTIFRICIA, ELECTUARIUM DENTIFRICUM, L. Various preparations are known under this name. They consist, for the most part, of the ordinary substances used as dentifrices, reduced to the state of a very fine powder, and mixed with sufficient honey, sugar, or capillaire, to give them the required consistence. Honey of roses is often used for this purpose, with some agreeable perfume at will. A little eau de Cologne or rectified spirit is a useful addition. The following are a few examples:

1. (CARBON PASTE; OPIAT CARBONIQUE.) The chippings of Turkey stone, cylinder charcoal, and prepared chalk, of each, 2 oz.; cochineal and cloves, of each, 1 dr.; honey, 5 oz.; eau de Cologne, q. s. It should not be put into the pots until the next day, and should be afterwards well preserved from the air. Much prized by smokers and by persons troubled with a fetid breath from rotten teeth.

(CHERRY TOOTH PASTE.) Precipitated chalk, 2 lbs.; rose pink, 1 lb.; powdered orris, 4 oz.; glycerine, 4 oz.; honey, 8 oz.; English oil of lavender, 2 dr.; oil of cinnamon, ½ dr.; oil of bergamot, 2 dr. Powder and sift the solids, beat well with the other ingredients, and allow to stand a month. If at that time the paste is too stiff, reduce with water only.

2. (CORAL PASTE; OPIAT DENTIFRICE ROUGE.) From prepared coral, 8 oz.; cuttle-fish bone, 4 oz.; mastic, 2 oz.; cochineal, ½ oz.; honey, ¾ lb.; essence of ambergris, 1 fl. dr.; oil of cloves, ½ fl. dr., dissolved in rectified spirit, 1 fl. oz. As the last. Cleanse the teeth rapidly.

3. (Dyon's CHARCOAL PASTE.) From chlorate of potassa, 1 dr.; mint water, 1 fl. oz.; triturate until dissolved, then add of powdered charcoal, 2 oz.; honey, 1 oz.

4. (MAGIC PASTE.) From white marble dust, 4 oz.; pumice-stone (in impalpable powder), 8 oz.; rose pink, 1 oz.; honey, ½ lb.; otto of roses, 15 drops. Rapidly whitens the teeth, but it should not be used too freely, nor too frequently.

5. (P. Cod.) Prepared coral, 4 oz.; bitartrate of potassa, 2 oz.; cuttle-fish bone and cochineal, of each, 1 oz.; alum, ½ dr.; Narbonne honey, 10 oz.; with essential oil, q. s. to aromatise the mixture.

6. (Pelletier's ODONTINE.) This is stated to be a mixture of pulverised sepia-bone, butter of cacao, and honey, with essential oil.

7. (ROSE PASTE.) Coral paste scented with roses, or the following:—Cuttle-fish bone, 1 oz.; prepared chalk, 2 oz.; cochineal, ½ dr.; honey of roses, 8 oz.; otto of roses, 6 drops.

8. (SOLUBLE PASTE, SALINE DENTIFRICE.) From bitartrate of potassa or sulphate of potassa (in fine powder), 3 oz.; honey of roses, 2 oz.

9. (SPANISH DENTIFRICE, CASTILIAN TOOTH CREAM.) From Castile soap (in fine powder) and cuttle-fish bone, of each, 2 oz.; honey of roses, 5 oz. An excellent preparation. It is superior to all the other pastes for removing tartar and animalcula from the teeth.

10. (VANILLA PASTE.) From red cinchona bark, 2 dr.; vanilla, 1 dr.; cloves, ½ dr.; (the last two reduced to powder by trituration with) white sugar, 1 oz.; cuttle-fish bone and marble dust, of each, ½ oz.; syrup of saffron, q. s.

11. (VIOLET PASTE.) From prepared chalk and cuttle-fish bone, of each, 3 oz.; powdered white sugar, 2 oz.; orris root, 1 oz.; smalts, ½ oz.; syrup of violets, q. s. to mix.

12. (Winckler's ROSEATE DENTIFRICE.) From cuttle-fish bone, 1 part; conserve of roses (Ph. L.), 3 parts; white otto of roses, 2 drops to the oz.

13. Chalk, 8 oz.; myrrh and rhatany root, of each, 2 oz.; orris root, 1 oz.; honey of roses, q. s. to mix. In foul and spongy gums.

Paste, Tooth'ache. See PASTE, ODONTALGIC.

Paste, Tor'mentil. *Syn.* PASTA TORMENTILLÆ, L. *Prep.* (*Morin.*) Powdered tormentil root made into a paste with white of egg. In whitlow; applied on linen. Mixed with an equal weight of simple syrup, it has also been recommended in dysentery and diarrhœa.

Paste, Unna's. *Prep.* White gelatin, 3 parts; zinc oxide, 3 parts; glycerin, 5 parts; water, 9 parts. Dissolve the gelatin in the water and glycerin, then add the zinc oxide and stir well. Used to paint on the skin in eczema, also as a dressing for ulcerated legs. Before use it must be melted by standing near a fire, or in a hot-water bath. It is painted on with a brush.

Paste, Vienna. See CAUSTIC POTASSA WITH LIME.

Paste, Vohler's. *Prep.* From dragon's blood, 1 dr.; powdered opium, 2 dr.; powdered gums of mastic and sandarach, of each, 4 dr.; oil of rosemary, 20 drops; tincture of opium, q. s. to form a paste. In toothache.

Paste, Ward's. See CONFECTION OF PEPPER.

PASTES. *Syn.* ARTIFICIAL GEMS, FACTI-

TIOUS G.; PIERRES PÉRCIEUSES ARTIFICIELLES, Fr. Vitreous compounds made in imitation of the gems and precious stones. The substances which enter into their composition, and the principles on which their successful production depends, have been already briefly noticed. The present article will, therefore, be confined to giving the reader a few original formulæ, together with several others carefully selected from the most reliable English and Continental authorities. Like enamels, the artificial gems have for their basis a very fusible, highly transparent and brilliant, dense glass, which is known under the name of 'frit,' 'paste,' 'strass,' 'flux,' 'fondant,' or 'Mayence base,' and which in its state of greatest excellence constitutes the 'artificial diamond.' For convenience, this will be noticed here under its last synonym. (See *below*, also Ure's 'Dictionary of Arts,' &c.)

Amethyst. 1. Paste or strass, 500 gr.; oxide of manganese, 3 gr.; oxide of cobalt, 24 gr.

2. (*Douault-Wieland*.) Strass, 4608 gr.; oxide of manganese, 36 gr.; oxide of cobalt, 2 gr.

3. (*Lançon*.) Strass, 9216 gr.; oxide of manganese, 15 to 24 gr.; oxide of cobalt, 1 gr.

Aqua Marina. From strass, 4800 gr.; glass of antimony, 30 gr.; oxide of cobalt, 1½ gr. See BERYL, of which this is merely a variety.

Aventurine. 1. From strass, 500 gr.; scales of iron, 100 gr.; black oxide of copper, 50 gr.; fuse until the black oxide of copper is reduced to the reguline form, then allow the mass to cool very slowly, so that the minute crystals of metal may be equally diffused through it. Has a rich golden iridescence.

2. As the last, but substituting oxide of chromium for the protoxide of copper. Appears brown, filled with countless gold spangles; or, when mixed with more paste, of a greenish grey, filled with green spangles.

Beryl. (*Douault-Wieland*.) Strass, 3456 gr.; glass of antimony, 24 gr.; oxide of cobalt, 1½ gr. See AQUA MARINA.

Carbuncle. See GARNET.

Chrysolite. From strass, 7000 gr.; pure calcined sesquioxide of iron ('trocus martis'), 65 gr.

Cornelian. 1. (RED.) From strass, 7000 gr.; glass of antimony, 3500 gr.; calcined peroxide of iron, 875 gr.; binoxide of manganese, 75 gr.

2. (WHITE.) From strass, 7200 gr.; calcined bones, 250 gr.; washed yellow ochre, 65 gr.

Diamond. 1. From rock crystal (purest), 1600 gr.; borax, 560 gr.; carbonate of lead (pure), 3200 gr.; oxide of manganese, ½ to 1 gr.; powder each separately, mix them together, fuse the mixture in a clean crucible, pour the melted mass into water, separate any reduced lead, and again powder and remelt the mass.

2. Pure silica, 150 gr.; pure litharge, 250 gr.; borax and nitre, of each, 50 gr.; arsenious acid, 21 gr.

3. (*Douault-Wieland*.) a. From rock crystal, 4056 gr.; minium, 6300 gr.; potash, 2154 gr.; borax, 276 gr.; arsenic, 12 gr.

b. From rock crystal, 3600 gr.; ceruse of Clichy (pure carbonate of lead), 3608 gr.; potash, 1260 gr.; borax, 360 gr.

4. (*Fontanier*.) Pure silica, 8 oz.; salt of tartar, 24 oz.; mix, bake, cool, treat the fused mixture with dilute nitric acid until effervescence ceases, and afterwards with water as long as the washings affect litmus paper; next dry the powder, add to it of pure carbonate of lead, 12 oz., and to every 12 oz. of the mixture add of borax, 1 oz.; triturate in a porcelain mortar, melt in a clean crucible, and pour the fused mass into cold water; dry, powder, and repeat the process a second and a third time in a clean crucible, observing to separate any revived lead. To the third frit add of nitre, 5 dr., and again melt. The product is perfectly limpid and extremely brilliant.

5. (*Lançon*.) Litharge, 100 gr.; pure silica, 75 gr.; white tartar or potash, 10 gr.

6. (*Loysel*.) Pure silica, 100 parts; red oxide of lead (minium), 150 parts; calcined potash, 30 to 35 parts; calcined borax, 10 parts; arsenious acid, 1 part. This produces a paste which has great brilliancy and refractive and dispersive powers, and also a similar specific gravity to the Oriental diamond. It fuses at a moderate heat, and acquires the greatest brilliancy when remelted, and kept for two or three days in a fused state, in order to expel the superabundant alkali, and perfect the refining ('Polytech. Journ.'). The products of the above formulæ are not only employed to imitate the diamond, but they also form the basis of the other factitious gems. (See *above*.)

7. (YELLOW DIAMOND.) Strass, 500 gr.; glass of antimony, 10 gr.

Eagle Marina. From strass, 3840 gr.; copper stain, 72 gr.; pure zaffre, 1 gr.

Emerald. 1. From strass, 7000 gr.; carbonate of copper, 65 gr.; glass of antimony, 7 gr.

2. Paste, 960 gr.; glass of antimony, 42 gr.; oxide of cobalt, 3½ gr.

3. (*Douault-Wieland*.) Paste, 4608 gr.; green oxide of copper, 42gr.; oxide of chrome, 2 gr.

4. (*Lançon*.) Paste, 9612 gr.; acetate of copper, 72 gr.; peroxide of iron, 1½ gr.

Garnet. 1. Paste or strass, 1200 gr.; glass of antimony, 580 gr.; purple of Cassius and binoxide of manganese, of each, 3 gr.

2. (*Douault-Wieland*.) Paste, 513 gr.; glass of antimony, 256 gr.; purple of Cassius and oxide of manganese, of each, 2 gr.

3. (VINEGAR GARNET.) From paste, 7000 gr.; glass of antimony, 3460 gr.; calcined peroxide of iron, 56 gr.

Lapis Lazuli. From paste, 7000 gr.; calcined horn or bones, 570 gr.; oxides of cobalt and manganese, of each, 24 gr. The golden veins are produced by painting them on the pieces with a mixture of gold powder, borax, and gum-water, and then gently heating them until the borax fluxes.

Opal. 1. From strass, 960 gr.; calcined bones, 48 gr.

2. (*Fontanier*.) Paste, 1 oz.; horn silver, 10 gr.; calcined magnetic ore, 2 gr.; absorbent earth (calcined bones), 26 gr.

Ruby. 1. Paste, 45 parts; binoxide of manganese, 1 part.

2. Paste, 1 lb.; purple of Cassius, 3 dr.

3. (*Douault-Wieland*.) a. From paste, 2880 parts; oxide of manganese, 72 parts.

b. Topas paste that has turned out opaque, 1 part; strass, 8 parts; fuse them together for 30 hours, cool, and again fuse it in small pieces before the blowpipe. Very fine.

4. (*Fontanier.*) Strass, 16 oz.; precipitate of Cassius, peroxide of iron, golden sulphide of antimony, and manganese calcined with nitre, of each, 168 gr.; rock crystal, 2 oz. or more.

5. Paste and glass of antimony, of each, 8 oz.; rock crystal, 1 oz.; purple of Cassius, 1½ dr. Turns on the orange.

Sapphire. 1. From strass, 3600 gr.; oxide of cobalt, 50 gr.; oxide of manganese, 11 gr.

2. (*Douault-Wieland.*) Paste, 4608 gr.; oxide of cobalt, 68 gr.; fuse in a little Hessian crucible for 30 hours.

3. (*Fontanier.*) Paste, 8 oz.; oxide of cobalt, 49 gr.

Topas. 1. From strass, 1050 gr.; glass of antimony, 44 gr.; purple of Cassius, 1 gr.

2. (*Douault-Wieland.*) Paste, 3456 gr.; calcined peroxide of iron, 36 gr.

Turquoise. From blue paste, 20 to 24 parts; calcined bones, 1 part.

Concluding Remarks. It is absolutely necessary for the successful application of the preceding formulæ that the substances employed should be perfectly free from impurities, more particularly those of a mineral kind. The litharge, oxide of lead, and carbonate of lead, above all things, must be entirely free from oxide of tin, as the smallest particle of that substance may impart a 'milkiness' to the paste. All the ingredients must be separately reduced to powder, and, after being mixed, sifted through lawn. The fusion must be carefully conducted and continuous, and the melted mass should be allowed to cool very slowly, after having been left in the fire from 24 to 30 hours at the least. Hessian crucibles are preferred for this purpose, and the heat of an ordinary pottery or porcelain kiln is sufficient in most cases; but a small wind-furnace devoted exclusively to the purpose is, in general, more convenient. It is found that the more tranquil, continuous, and uniform the fusion, the denser and clearer is the paste, and the greater its refractive power and beauty.

All the coloured vitreous compounds noticed under GLASS may be worked up as ornamental stones, in the same way as those just referred to.

The following method of obtaining artificial rubies and emeralds, first pointed out by Boëttger, is exceedingly simple and inexpensive, and deserves the serious attention of those interested in this ingenious art:—Recently precipitated and well-washed hydrate of aluminum is moistened with a few drops of neutral chromate of potassium, and kneaded so that the mass assumes a tinge scarcely perceptible; it is then rolled up into small sticks, about the thickness of a finger, and slowly dried, taking the precaution to fill the fissures (if any) that form during desiccation with fresh hydrate of aluminum. When perfectly dry, and after having been submitted to a gentle heat, one end of these sticks is brought into the termination of the flame of an oxyhydrogen blowpipe, until a portion of the mass is fused into a small globule. After the lapse of a few minutes, several minute balls form, having a diameter of some millimetres, and of such intense hardness that quartz, glass, topaz, and granite may be easily and perceptibly scratched with them. These, when cut and polished, appear, however, slightly opaque. By employing nitrate of nickel in lieu of chromate of potassium, green-coloured globules, closely resembling the emerald, are obtained.

By the substitution of oxide of chromium for chromate of potassium, Mr Booley produced factitious gems of considerable hardness and beauty, though slightly opaque in some portion of the mass. The addition of a very little silica prevented, in a great measure, this tendency to opacity.

It may be observed that the beauty of pastes or factitious gems, and especially the brilliancy of mock diamonds, is greatly dependent upon the cutting, setting up, and the skilful arrangement of the foil or tinsel behind them. See ENAMEL, FOILS, GEMS, GLASS, &c.

PAS'TELS. [Fr.] Coloured crayons.

PASTIL. *Syn.* PASTILLE; PASTILLUS, PASTILLUM, L. A lozenge or confection. The pastilles (PASTILLI) of French pharmacy are merely 'confectionery drops' aromatised or medicated. The name is also given to mixtures or odorous substances made up into small cones and burnt as incense. (See *below.*)

The following very useful notes are abstracted from a paper by Mr Wyatt:

Pastils are soft jelly-like jujubes, variously medicated, made from a gelatin and glycerin base, called in the Throat Hospital Pharmacopœia 'Glyco-gelatine,' which is made according to the following form:—Gelatin, 1 oz.; glycerin, 2½ oz. (by weight); orange-flower water, 2½ oz. (by weight); ammoniacal solution of carmine, a sufficiency.

Cut the gelatin into shreds and soak in the orange-flower water for two hours; then transfer to a water-bath and heat with the glycerin until the gelatin is dissolved. Colour with the carmine solution, and pour into an oiled tray to cool.

The Pharmacopœia gives no formula for the solution of carmine, but 30 minims of one made as follows is enough for 6 oz. of glyco-gelatin:—Carmine, 30 gr.; solution of ammonia, a sufficiency. Dissolve the carmine in 6 dr. of the ammonia, filter, and wash the filter with more ammonia until 1 fl. oz. has been collected.

The medication of the pastils is accomplished by melting an ounce of glyco-gelatin on a water-bath, adding the medicine, previously rubbed to a thick syrup with glycerin if a powder, stirring until nearly cool, and pouring into an oiled mould, cutting the mass into 24 pastils when cold. A suitable mould for small quantities is one with sides soldered on, square in shape, 3 in. by 3 in., divided into 36 squares by means of deeply impressed lines on the under side, these causing the finished pastils to have a slightly rounded surface, the lines leaving a series of deep grooves which serve as a guide to cutting.

Orange-flower water being distasteful to many persons, Mr Wyatt recommends as a pleasant variety other flavoured waters, fruit-juice, tolu, and glycyrrhizin.

Of the flavoured waters rose or cinnamon water

may be used instead of orange-flower water, and in the same proportion, whilst 2 fl. dr. of cherry laurel water with 2½ oz. of distilled water impart a pleasant almond flavour.

He has also used raspberry and lime-fruit juices, the raspberry in the same proportion as the orange-flower water, the lime-juice in the proportion of half juice and half distilled water.

A tolu mass was made by using in the place of glycerin toluinated glycerin, made by heating 1¼ parts of tolu with 30 of glycerin and 6 of water over a water-bath for an hour, filtering on cooling, and making up 36 fl. parts by the addition of glycerin.

Glycyrrhizin, 24 gr., dissolved in the water used to soak the gelatin, imparted an excellent liquorice flavour, very useful to hide the taste of ammonium chloride.

In addition to the formula of the Throat Hospital Pharmacopœia, the following have proved useful in many cases :

Pastilli Apomorphinæ. Apomorphine hydrochlorate trituration (1 in 8), 9 gr.; glycerin, 10 minims; glyco-gelatin (lime-juice), 1½ oz. Rub the trituration smooth with the glycerin, add to the previously melted glyco-gelatin, stir until nearly cool, and pour in au oiled mould, cutting into 36 pastils on cooling.

Pastilli Cocainæ. Cocaine hydrochlorate, 1 gr.; glyco-gelatin (raspberry), 1½ oz. Melt the glyco-gelatin, stir in the cocaine, allow to stand until the air-bubbles have risen, pour into an oiled mould, and cut into 36 pastils when cold. Each contains ₁⁄₃₆ gr. cocainæ hydrochlor.

Pastilli Codeinæ. Codeine, 4½ gr.; glycerin, 2 fl. dr.; glyco-gelatin (liquorice), 1½ oz. Dissolve the codeine in the glycerin with heat, add to the melted glyco-gelatin, stir, and pour into an oiled mould, cutting, when cold, into 36 pastils.

Each contains ⅛ gr. codeine, and is equal to 1 fl. dr. of the syrupus codeinæ (*Martindale*).

Pastilli Salol. Salol, 90 gr.; glycerin, 90 minims; glyco-gelatin (toluinated), 1½ oz. Rub the salol with the glycerin, add this to the previously melted glyco-gelatin, and heat until the salol is dissolved; stir until nearly set, pour into an oiled mould, and cut into 36 pastils when cold. Each contains 2½ gr. salol.

Various other formulæ can be devised, according to the dispenser's ingenuity, taking the precaution, of course, to use no ingredients with which gelatin is incompatible.

The plain glyco-gelatin of the raspberry, lime-fruit, and the liquorice flavours, when cast in a large mould (such as the lid of an ordinary 3 lb. jujube tin), and cut into square or diamond-shaped lozenges with a pair of scissors, is an excellent substitute for the glycerin (?) jujubes usually sold, containing, as it does, more than a homœopathic dose of glycerin.

Pastils, Explo'sive. Fumigating pastilles, containing a little gunpowder. Used to produce diversions, but they often prove far from harmless.

Pastils, Fu'migating. *Syn.* AROMATIC PASTILLES, INCENSE P.; PASTILLI FUMANTES, P. ODORATI, L. *Prep.* 1. Benzoin, 4 oz.; cascarilla, ¼ oz.; nitre and gum-arabic, of each, 3 dr.; myrrh, 1 dr.; oils of nutmeg and cloves,

of each, 25 drops; charcoal, 7 oz.; all in fine powder; beat them to a smooth ductile mass with cold water, q. s.; form it into small cones with a tripod base, and dry them in the air.

2. (*Henry* and *Guibourt.*) Powdered gum-benzoin, 16 parts; balsam of tolu and powdered sandal-wood, of each, 4 parts; a light charcoal (Linden), 48 parts; powdered tragacanth and true labdanum, of each, 1 part; powdered nitre and gum-arabic, of each, 2 parts; cinnamon water, 12 parts; as above.

3. (P. Cod.) Benzoin, 2 oz.; balsam of tolu and yellow sandal-wood, of each, 4 dr.; nitre, 2 dr.; labdanum, 1 dr.; charcoal, 6 oz.; mix a solution of gum tragacanth, and divide the mass into pastilles, as before.

4. (PASTILLES À LA FLEUR D'ORANGE.) For powdered roses in the next formula substitute pure orange powder, and for the essence of roses use pure neroli.

5. (PASTILLES À LA ROSE.) Gum-benzoin, olibanum (in tears), and styrax (in tears), of each, 12 oz.; nitre, 9 oz.; charcoal, 4 lbs.; powder of pale roses, 1 lb.; essence of roses, 1 oz.; mix with 2 oz. of gum tragacanth, dissolved in rose-water, 1 quart.

6. (PASTILLES À LA VANILLE.) Gum-benzoin, styrax, and olibanum (as last), of each, 12 oz.; nitre, 10 oz.; cloves, 8 oz.; powdered vanilla, 1 lb.; charcoal, 4½ lbs.; oil of cloves, ½ oz.; essence of vanilla, 7 or 8 fl. oz.; as before.

Obs. The products of the above formulæ are all of excellent quality. They may be varied to please the fancy of the artiste by the substitution of other perfumes or aromatics. Cheaper pastilles may be made by simply increasing the quantity of the charcoal and saltpetre. The whole of the ingredients should be reduced to fine powder before mixing them. The use of musk and civet, so often ordered in pastilles, should be avoided, as they yield a disagreeable odour when burned. The addition of a little camphor renders them more suitable for a sick chamber. The simplest and most convenient way of forming the mass into cones is by pressing it into a mould of lead or porcelain.

Pastilles are burned either to diffuse a pleasant odour, or to cover a disagreeable one. For this purpose they are kindled at the apex and set on an inverted saucer or a penny piece to burn. Persons who use them frequently employ a small china or porcelain toy (' pastille house ') sold for the purpose.

Pastils, Mouth. *Syn.* BREATH PILL, CACHOU LOZENGES; PASTILLI COSMETICI, L.; CACHOU AROMATISÉ, C. AROMATIQUE, C. DE BOLOGNA, GRAINS DE CACHOU, Fr. *Prep.* 1. Soft extract of liquorice, 3 oz.; gum catechu and white sugar, of each, 1 oz.; gum tragacanth (powdered), ½ oz.; oil of cloves, 1 dr.; oil of cassia, ¼ dr.; essence of ambergris and oil of nutmeg, of each, 12 drops; make a firm mass with rose or orange flower water, q. s., and divide it into 1-gr. pills; when these are dry, cover them with gold or silver leaf.

2. Solazzi juice (dried by a gentle heat and powdered), of each, 1 oz.; lump sugar, 3 oz.; powdered catechu, 2 oz.; powdered tragacanth, 1 oz.; oil of cloves, 2 fl. dr.; oil of cassia, 1 fl. dr.; white of

egg or rose-water, q. s. to form a pill-mass; as before.

3. Powdered catechu, 1 oz.; Solazzi juice, 4 oz.; lump sugar, 12 oz.; oils of cloves, cassia, and peppermint, of each, 1 fl. dr.; mucilage of tragacanth, q. s. to mix; as before.

4. Extract of liquorice (soft), 2 oz.; white sugar, 3 oz.; powdered tragacanth and cascarilla (or orris root), of each, ½ oz.; oil of cloves, ½ fl. dr.; oil of cassia, 12 drops; water, q. s.; as before.

5. (*Chevallier.*) Powdered coffee, chocolate, and sugar, of each, 1½ oz.; powdered vanilla and freshly burnt charcoal, of each, 1 oz.; mucilage of tragacanth, q. s.

6. Chloride of lime (dry and good), 1 dr.; white sugar, 3 oz.; powdered tragacanth, 1 oz.; oil of cloves, 30 drops; rose-water, q. s. To disinfect the breath.

Obs. Almost every maker employs his own forms for these articles. The objects to be aimed at are the possession of rather powerful and persistent odour, and a toughness to prevent their too rapid solution in the mouth. The original Italian formula included liquorice, mastic, cascarilla, charcoal, orris root, oil of peppermint, and the tinctures of ambergris and musk, but is now seldom employed in this country. The flavour of peppermint does not, indeed, appear to be approved of by English smokers. Sometimes, instead of being made perfectly spherical, they are flattened a little.

CACHOU À L'AMBRE GRIS, CACHOU À LA CANELLE, CACHOU À LA FLEUR D'ORANGE, CACHOU MUSQUÉ, CACHOU À LA ROSE, CACHOU À LA VANILLE, CACHOU À LA VIOLETTE, &c., are merely flavoured and scented respectively with the essences or oils of ambergris, cinnamon, neroli, musk, rose, vanilla, violets, &c. See BREATH, CACHOU AROMATISÉ, LOZENGES, PILLS, &c.

PA"STRY. Articles of food made of 'paste' or dough, or of which 'paste' forms a principal and characteristic ingredient. The word is popularly restricted to those which contain puff paste, or such as form the staple production of the modern pastrycook; but it is, in reality, of much more general signification.

Several varieties of paste are prepared for different purposes, of which the following are the principal:

PUFF PASTE. The production of a first-class puff paste is commonly regarded as a matter of considerable difficulty, but by the exercise of the proper precautions it is, on the contrary, an extremely simple affair. This paste, before being placed in the oven, consists of alternate laminæ of butter or fat and ordinary flour dough, the latter being, of course, the thicker of the two. During the process of baking, the elastic vapour disengaged, being in part restrained from flying off by the buttered surfaces of the dough, diffuses itself between these laminæ, and causes the mass to swell up, and to form an assemblage of thin membranes or flakes, each of which is more or less separated from the other. Individually, these flakes resemble those of an ordinary rich unleavened dough when baked; but collectively, they form a very light crust, possessing an extremely inviting appearance and an agreeable flavour.

The precautions above referred to are—the use of perfectly dry flour, and its conversion into dough with a light hand, avoiding unnecessarily working it; the use of butter free from water or buttermilk, and which has been reduced to precisely the same degree of plasticity as the dough between which it is to be rolled; conducting the operation in a cool apartment, and, after the second or third folding of the dough, exposing it to a rather low temperature before proceeding further with the process; and, lastly, baking the paste in a moderately smart but not too hot an oven. The following are examples:

1. (Rich.) Take of flour, 1 lb.; butter, ½ lb.; cold spring water, q. s.; make a moderately soft flexible dough, then roll in (as described above) of dry fresh butter, ½ lb.

2. (Ordinary.) Take of flour, 1 lb.; cold water, q. s.; make a dough, and roll in, as before, of butter, 6 oz.

3. (*Rundell.*) Take ½ peck of flour, rub into it 1 lb. of butter, and make a 'light paste' with cold water, just stiff enough to work well; next lay it out about as thick as a crown piece; put a layer of butter all over it, sprinkle on a little flour, double it up, and roll it out again; by repeating this with fresh layers of butter three or four times or oftener, a very light paste will be formed. Bake it in a moderately quick oven.

4. (*Soyer.*) Put 1 lb. of flour upon your pastry-slab, make a hole in the centre, into which put a teaspoonful of salt, mix it with cold water into a softish flexible paste with the right hand, dry it off a little with flour until you have well cleared the paste from the slab, but do not work it more than you can possibly help; let it remain for 2 or 3 minutes upon the slab, then take 1 lb. of fresh butter from which you have squeezed all the buttermilk in a cloth, and brought to the same consistency as the paste, upon which place it; press it out flat with the hand, then fold over the edges of the paste, so as to hide the butter, and reduce it with the rolling-pin to the thickness of about ½ an inch, when it will be about two feet in length; fold over one third, over which again pass the rolling-pin; then fold over the other third, thus forming a square; place it with the ends top and bottom before you, shaking a little flour both under and over, and repeat the rolls and turns twice again as before; flour a 'baking-sheet,' upon which lay it on ice, if handy, or otherwise in some cool place, for about half an hour; then roll it twice more, turning it as before, and again place it upon ice or in the cold for half an hour; next give it two more rolls, making seven in all, and it is ready for use. "You must continually add enough flour while rolling to prevent your paste sticking to the slab."

HALF-PUFF PASTE. As the preceding, using only one half the quantity of butter, and giving the paste only three or four folds.

SHORT PASTE, SHORT CRUST. 1. Flour (dry and warm), 1 lb.; sugar, 3 oz.; butter, ½ lb.; 2 eggs; water, ½ pint; make a light dough. If one half of 'Jones's patent flour' be used no eggs will be required.

2. (*Soyer.*) Put on the 'paste-slab' or 'pieboard' 1 lb. of flour, 3 oz. of pounded sugar, 6 oz. of butter, 1 egg, ½ teaspoonful of salt, and ½ pint

of water; mix the sugar and water well together, add them with the water by degrees to the flour, and form a paste, but firmer than puff paste.

PIE PASTE. That commonly used is 'short paste,' varied at will, but at good tables the upper crust of the pie is generally made of 'puff paste,' and the remainder of 'short paste.'

PUDDING PASTE. This for baked puddings may resemble the last. For boiled puddings (or, indeed, for any) the paste may be either ordinary 'short paste,' or one made with 2 to 6 oz. of butter or lard, or 3 to 8 oz. of chopped beef suet, to each lb. of flour, with or without an egg, and a little sugar, according to the means of the parties. The first is most appropriate for those containing fresh fruit, and that with suet for meat puddings, and those containing dried fruit, as grocers' currants, plums, &c. Milk or milk and water is often used instead of simple water to make the dough. Ginger, spices, savoury herbs, &c., are common additions to the crusts of puddings. Where economy is an object, especially among the lower classes, kitchen fat is frequently substituted for suet, and lard for butter. When 'Jones's patent flour' is employed an excellent plain pudding paste may be made by simply mixing it up with very cold water, and immediately putting it into the water, which should be boiling, and kept in that state until the pudding is dressed.

PATENT MEDICINES. Syn. MEDICAMENTA ARCANA, L. The majority of the preparations noticed under this head are the nostrums popularly termed 'quack medicines,' and which are sold with a Government stamp attached to them. The term patent, as applied to these preparations, is confusing and misleading, the majority of the public believing them to be inventions which have received the seal of the Patent Office, whereas very few are real patents. The application of a Government stamp to a medicine has, by usage, given the right to use the word patent. A few other secret or proprietary remedies are also, for convenience, included in the list. An alphabetical arrangement, based on the names of the reputed inventors or proprietors of the articles, has been adopted, as being the one best suited for easy reference. The composition of a number of them is given from careful personal inspection and analysis (by Mr Cooley), and that of the remainder on the authority of Gray, Griffith, Paris, Redwood, the members of the Philadelphia College of Pharmacy, and other respectable writers. A variety of articles, not included in the following lists, is noticed along with other preparations for the class to which they belong, or under the names of their proprietors. See BALSAM, CERATE, DROPS, ESSENCE, TINCTURE, OINTMENT, PILLS, &c.

Abernethy's Pills. See ABERNETHY MEDICINES.

Albinolo's Ointment. See HOLLOWAY'S OINTMENT (below).

Ali Ahmed's Treasures of the Desert. There are three preparations included under this name:

a. (ANTISEPTIC MALAGMA.) From lead plaster, 3 parts; gum, thus and salad oil, of each, 2 parts; beeswax, 1 part; melted together by a gentle heat, and spread upon calico.

b. (PECTORAL, ANTIPHTHISIS, or COUGH PILLS.) From myrrh, 3½ lbs.; squills and ipecacuanha, of each, 1 lb. (all in powder); white soft soap, 10 oz.; oil of aniseed, 1½ oz.; treacle, q. s. to form a pill mass.

c. (SPHAIROPEPTIC or ANTIBILIOUS PILLS.) From aloes, 28 lbs.; colocynth pulp, 12 lbs.; rhubarb, 7 lbs.; myrrh and scammony, of each, 3½ lbs.; ipecacuanha, 3 lbs.; cardamom seeds, 2 lbs. (all in powder); soft soap, 9 lbs.; oil of juniper, 7 fl. oz.; treacle, q. s. This, as well as the last, is divided into 3½ gr. pills, which are then covered with tin-foil or silver leaf. An excellent aperient pill, no doubt, and one likely to prove useful in all those cases in which the administration of a mild diaphoretic and stomachic purge is indicated. Unlike many of the advertised nostrums of the day, there is nothing in their composition that can by any possibility prove injurious; but beyond this they are destitute of virtue.

Anderson's Scott's Pills. See PILLS.

Atkinson's Infant Preservative. From carbonate of magnesia, 6 dr.; white sugar, 2 oz.; oil of aniseed, 20 drops; spirit of sal volatile, 2½ dr.; laudanum, 1 dr.; syrup of saffron, 1 oz.; caraway water to make up 1 pint.

Balm of Rackasiri. See BALSAM.

Balsam of Life. Syn. BAUME DE VIE, Fr. Several compound medicines of this name are noticed on page 261. The following are well-known nostrums:

1. (Hoffmann's.) a. Of the oils of cinnamon, cloves, lemon, lavender, and nutmegs, and balsam of Peru, of each, 2 dr.; essence of ambergris, oil of amber, and oil of rue, of each, 1 dr.; cochineal, 12 gr.; strongest rectified spirit, 3½ pints; mix.

b. (Ph. Dan. 1840.) Oils of cinnamon, cloves, lavender, and nutmegs, of each, 20 gr.; purified oil of amber, 10 drops; balsam of Peru, 30 gr.; rectified spirit (tinged with alkanet root), 10 oz.

2. (Gabius's.) Nearly similar to Hoffmann's.

3. (Turlington's.) Benzoin and liquid styrax, of each, 13 oz.; balsam of tolu and extract of liquorice, of each, 4 oz.; balsam of Peru, 2 oz.; aloes, myrrh, and angelica root, of each, 1 oz.; highly rectified spirit of wine, 7 pints; digest, with frequent agitation, for 10 days, and filter. Externally, the above are rubefacient and corroborant; internally, stimulant, cordial, and pectoral.

Betton's British Oil. From oil of turpentine, 1 pint; Barbadoes tar, ½ lb.; oil of rosemary, 1 fl. oz.

Blake's Green Mountain Ointment. We are told that the active ingredient in this compound is Arnica montana, with a basis of soap cerate. It is very useful as an external application in several affections. The chief objection to its use is that it is a secret preparation.

Blake's Toothache Essence. From alum, in fine powder, 1 dr.; sweet spirit of nitre, 5 dr.

Boerhaave's Odontalgic Essence. From opium, ½ dr.; oil of cloves, ½ dr.; powdered camphor, 5 dr.; rectified spirit, 1½ fl. oz.

Bouchardat's Tasteless Aperient. From phosphate of soda, ½ oz., placed in a soda-water bottle, which is then filled up with carbonated water at the bottling machine. For a dose.

Brand's Tooth Tincture. From pellitory of Spain (bruised), 1 oz.; camphor, ¾ oz.; opium, ¼ oz.; oil of cloves, 1 dr.; digest for 10 days in rectified spirit, ½ pint.

Brodum's Nervous Cordial. *Prep.* 1. "Originally it consisted simply of an infusion of gentian root in English gin, coloured and flavoured with a little red lavender (compound spirit of lavender). After a time the doctor added a little bark to the nostrum, and subsequently made other additions" ('Anat. of Quackery').

2. (*Paris.*) Tinctures of gentian, calumba, cardamoms, and cinchona, compound spirit of lavender, and steel wine, of each equal parts. "It is tonic, stomachic, and stimulant; but, beyond these, possesses no curative properties" ('Anat. of Quackery').

Chlorodyne. This nostrum, which was first introduced as 'a combination of perchloric acid with a new alkaloid,' has become a popular anodyne and sedative. Several preparations are sold under this name, and the claims of the rival makers have occasioned some expensive lawsuits. The name was undoubtedly invented by Dr J. Collis Browne, but Mr Freeman, pharmaceutical chemist, claims to be the inventor of the preparation. Whether Browne's and Freeman's 'chlorodynes' are essentially the same, we are not able to determine, but we know that there is not the slightest foundation for the statements made by each manufacturer respecting the new vegetable principle contained in his medicine. Chlorodyne, in every one of its forms, is simply a mixture of certain well-known materials, some of which are rather dangerous ingredients for a popular nostrum. According to the analysis of Dr Odgen, Browne's chlorodyne is composed as follows:

Chloroform, 6 dr.; chloric ether, 1 dr.; tincture of capsicum, ½ dr.; oil of peppermint, 2 drops; hydrochlorate of morphine, 8 gr.; Scheele's hydrocyanic acid, 12 drops; perchloric acid, 20 drops; tincture of Indian hemp, 1 dr.; treacle, 1 dr. 'Towle's chlorodyne' is prepared according to this formula, the ingredients being named on the label.

Clarke's Conglutinum. See CONGLUTINUM.

Cochrane's Cough Remedy. Acidulated syrup of poppies.

Corn Nostrums. See CORN.

Cottereau's Odontalgic Essence. A nearly saturated ethereal solution of camphor, mixed with about 1-12th of its volume of strong liquor of ammonia.

Curtis's Antivenereal Lotion. A mixture of Beaufoy's solution of chloride of lime, 2 fl. oz., with cold soft water, 8 fl. oz. For use, 1 to 2 table-spoonfuls are put into a wine-glassful of water.

Dalby's Carminative. *Prep.* 1. (*Dr Paris.*) Carbonate of magnesia, 40 gr.; tincture of castor and compound tincture of cardamoms, of each, 30 drops; tincture of assafœtida and spirit of pennyroyal, of each, 15 drops; laudanum, 5 drops; oil of aniseed, 3 drops; oil of nutmeg, 2 drops; oil of peppermint, 1 drop; peppermint water, 2 fl. oz.—*Dose*, ½ to 1 teaspoonful. The bottle should be well shaken before pouring it out.

2. (Wholesale.) Carbonate of magnesia, 1 oz.; tincture of castor, 5 fl. dr.; tincture of assafœtida, 3 fl. dr.; oils of aniseed and pennyroyal, of each, ¼ fl. dr.; oil of nutmeg, 15 drops; syrup of poppies, 7 oz.; rectified spirit, 3½ fl. oz.; peppermint water, 1 pint; as before.

Davidson's Cancer Remedy. A mixture of arsenious acid and hemlock, both in powder (*Dr Paris*).

Davis's Calorific. The 'LIQUID' is commercial acetic acid (sp. gr. 1·048), diluted with about an equal volume of water, and coloured with burnt sugar or spirit colouring. The 'SHIELD' consists of a piece of red flannel backed with oil-skin, to prevent evaporation. A few drops of calorific are sprinkled on the flannel, which is then bound over the affected part. The heat of the body gradually volatilises the acetic acid, and the escape of the vapour being prevented by the oil-skin, a strongly counter-irritant action is set up.

Derbyshire's Embrocation. *Prep.* From opium and mottled soap, of each, 2 oz.; extract of henbane, 2 dr.; and mace, ½ dr.; boiled for 30 minutes, in water, 3 pints; to the cold liquor, rectified spirit, 1 quart, and liquor of ammonia, 1 fl. oz., are added, and, after repose, the clear portion is decanted. A preventive of sea-sickness.

Deshler's Cerate. Yellow basilicon.

Duncan's Gout Medicine. See GOUT.

Dutch Ague Remedy. *Prep.* A mixture formed of Peruvian bark and cream of tartar, of each, 1 oz.; cloves, ½ dr.; reduced to fine powder. —*Dose*, 1½ dr., every 3 hours (*Dr Paris*).

Godfrey's Cordial. *Prep.* 1. (Original formula.) Opium (sliced), ¾ oz.; sassafras chips, 1 oz.; English brandy, 1 quart; macerate for 4 or 5 days, then add of water, 1 quart, treacle, 3½ lbs., and simmer the whole gently for a few minutes; the next day decant the clear portion.

2. (*Dr Paris.*) Aniseed, caraways, and coriander, of each (bruised), 1 oz.; sassafras chips, 9 oz.; water, 6 pints; simmer gently until reduced to 4 pints, then add of treacle, 6 lbs.; and when nearly cold, further add of tincture of opium, 3 fl. oz.

3. (Phil. Coll. of Phar.) Carbonate of potassa, 2½ oz.; water, 26 pints (old wine measure); dissolve, add of sugar-house molasses (treacle), 16 pints (o. w. m.); simmer the mixture, remove the scum, and when it has considerably cooled, add of tincture of opium, 24 fl. oz.; oil of sassafras, ½ fl. oz.; (dissolved in) rectified spirit, 1 quart (o. w. m.). It contains about 16 drops of laudanum (= 1¼ gr. of opium) in each fl. oz.

The following forms are also current in the wholesale trade:

4. From molasses, 16 lbs.; distilled water, 2½ galls.; oil of sassafras, 1 fl. oz.; (dissolved in) rectified spirit, ½ gall.; bruised ginger, ¾ oz.; cloves, ½ oz.; laudanum, 8 fl. oz.; macerate for 14 days, and strain through flannel.

5. Sassafras chips, 1 lb.; ginger (bruised), 4 oz.; water, 3 galls.; simmer until reduced to 2 galls.; then add of treacle, 16 lbs.; rectified spirit, 7 lbs.; laudanum, 1 pint.

6. Opium, ½ oz.; treacle, 7 lbs.; boiling water, 1 gall.; dissolve, and add of rectified spirit, 1 quart; oil of sassafras, ½ dr.; cloves and mustard

seed, of each, ¼ oz.; coriander and caraway seeds, of each, 1 dr.; digest for a week.

7. Caraways, corianders, and aniseed, of each, 1 lb.; water, 6 galls.; distil 5 galls., and add of treacle, 23 lbs.; laudanum, 1 quart; and oil of sassafras, 1 fl. oz., previously dissolved in rectified spirit, 1 gall.

Obs. This preparation is anodyne and narcotic, and, amongst the lower classes, is commonly given to children troubled with wind and colic. Its frequent and excessive use has sent many infants prematurely to the grave. Gray says, "It is chiefly used to prevent the crying of children in pain or starving." The dose is ¼ teaspoonful and upwards, according to the age and susceptibility of the child.

Grave's Gout Preventive. *Prep.* A tincture prepared by steeping, for a week, dried orange peel and hiera picra, of each, 1 oz., and rhubarb, ¼ oz., in brandy, 1 pint.

Grinrod's Remedy for Spasms. From acetate of morphia, 1 gr.; spirit of sal volatile and sulphuric ether, of each, 1 fl. oz.; camphor julep, 4 fl. oz.; for a mixture. It should be kept closely corked, in a cool place, and should be well shaken before use.—*Dose.* A teaspoonful in a glass of cold water or wine, as required. It is a really valuable preparation.

Holloway's Ointment. The original formula of ALBINOLO'S OINTMENT, of which this pretends to be a reproduction, contained the 'graisses de serpent et de vipère,' and other pharmaceutical curiosities. The principal ingredients, however, in the HOLLOWAY'S OINTMENT of the present day are very homely substances. In the case of Sillen *v.* Holloway, tried at the Court of Common Pleas in January, 1863, the plaintiff's counsel asserted that, on the ointment being received by the agent in Paris, it was submitted to the authorised Government chemists to be analysed, in accordance with the laws of France prohibiting the sale of secret remedies, and was found by them to contain butter, lard, Venice turpentine, white wax, yellow wax, and nothing else. In a letter to the 'Times' Mr Holloway stated that the French analysis was incorrect, for three of the ingredients named were not in the ointment, while there were other components which the analysts had not discovered. The formula adopted by those who prepare an imitation ointment on the large scale, and which closely resembles, if it be not actually identical with, that employed by Mr Holloway, is as follows:—Fresh butter (free from water), ¼ lb.; beeswax (good), 4 oz.; yellow resin, 3 oz.; melt them together, add of vinegar of cantharides, 1 fl. oz., and simmer the whole, with constant agitation, for 10 or 12 minutes, or until the moisture is nearly evaporated; then add of Canada balsam, 1 oz.; expressed oil of mace, ½ dr.; balsam of Peru or liquid styrax, 10 or 12 drops; again stir well, allow the mixture to settle, and when it is about half cold (not before) pour it into the pots, previously slightly warmed, and allow it to cool very slowly. The label will do the rest. No two samples of Holloway's ointment are precisely of the same colour or consistence.

Holloway's Pills. From aloes, 4 parts; jalap, ginger, and myrrh, of each, 2 parts; made into a

mass with mucilage, and divided into 2-gr. pills, of which about 4 dozen are put into each 1*s*. 1½*d*. box.

Jackson's Bathing Spirit. A species of soap liniment, made of soft soap, 1 lb.; camphor, 6 oz.; oils of rosemary and thyme, of each, ½ fl. oz.; rectified spirit, 1 gall.

Kaye's Infant's Preservative. A preparation partaking of the joint properties of Atkinson's nostrum and Godfrey's cordial, but more powerful than either, as indicated by the doses in which it is directed to be given during early infancy, viz. "two, three, or more drops."

Keating's Cough Lozenges. These are said to be composed of—lactucarium, 2 dr.; ipecacuanha, 1 dr.; squills, ½ dr.; extract of liquorice, 2 oz.; sugar, 6 oz.; made into a mass with mucilage of tragacanth, and divided into 20-gr. lozenges.

King's Sarsaparilla Pills. From the compound extract. "Instead of two pills being equivalent to ½ fl. oz. of the concentrated decoction or essence of sarsaparilla, as asserted, it takes about 32 of them to represent the given quantity, and about 4 of them to be equal in strength to the common decoction of the Pharmacopœia." "Instead of one 2*s*. 9*d*. box of these pills being equal to a pint of the costly concentrated fluid preparation, it would take nearly 1½ lbs. of them for that purpose" ('Med. Circ.,' ii, 493).

Kitchener's Peristaltic Persuaders. See PILLS.

Lambert's Asthmatic Balsam. The active ingredients in this compound are said to be squills and aqueous extract of opium.

Lemasurier's Odontalgic Essence. From acetate of morphia, 1 gr.; dissolved in cherry-laurel water, 1 oz. For use, a teaspoonful is added to ½ a wine-glassful of warm water, and the mouth well rinsed out with the mixture.

Leroy's Purgative. *a.* (No. 1.) Vegetable turpeth, 6 dr.; scammony, 1½ oz.; jalap, 6 oz.; brandy, 10 pints; digest for 24 hours, and add a syrup made of senna, 6 oz.; water, 1½ pints; sugar, 32 oz.

b. (No. 2.) As the last, only ½ stronger.

c. (No. 3.) Twice as strong as No. 1.

Lewis's Balsamic Ointment. This preparation, which is declared by its proprietor to be 'utterly unsurpassable,' for the most part resembles Holloway's ointment ('Med. Circ.,' ii, 493).

Lewis's Electuarium. A liquid nostrum, said to be alterative and to contain a small quantity of both antimony and mercury.

Lewis's Silver Cream. This nostrum is said to depend for its efficacy on white precipitate and a salt of lead.

Locock's Pulmonic Lozenges. See WAFERS.

Mahomed's Paste. See ELECTUARY.

Mardant's Norton's Drops. A mixture of the tinctures of gentian and ginger, holding in solution a little bichloride of mercury, and coloured with cochineal.

Marriott's Dry Vomit. A mixture of equal parts of tartar emetic and sulphate of copper.

Marsden's Drops. A coloured solution of corrosive sublimate (*Dr Paris*).

Matthieu's Vermifuge. *a.* (To destroy the worms.) Tin filings, 1 oz.; male fern root, 6 dr.; worm-seed, 4 dr.; resinous extract of jalap and sulphate of potass, of each, 1 dr.; honey, q. s. to form an electuary.—*Dose.* A teaspoonful, repeated every third or fourth hour for 2 or 3 days, when the following are to be substituted, and continued until the bowels are well acted on.

b. (To expel the worms.) Jalap and sulphate of potassa, of each, 40 gr.; scammony, 20 gr.; gamboge, 10 gr.; honey, q. s. as before.

McKinsey's Golden Cerate. This appears to resemble Poor Man's Friend.

McKinsey's Katapotia. This notorious nostrum is compounded of aloes, 5 oz.; soap, 1½ oz. (both in powder); beaten up with syrup of saffron and a little essential oil, and divided into pills varying in weight from 2 to 2½ gr. each ('Med. Circ.,' iv, 86).

McKinsey's Medicinal Powder. *Syn.* REV. T. SMITH'S M. P. From dried lavender flowers and rosemary tops, of each, 2½ oz.; assarbacca, 1 oz.; reduced to powder, and further disguised with a little perfume. A very small quantity of subsulphate of mercury is also most probably added. 2 or 3 pinches of this powder, taken 3 or 4 times a day as snuff, is said by the proprietor to be sufficient to cure almost any known disease. See ASARABACCA.

Morison's Adhesive Paste. See PLASTER.

Morison's Aperient Powder. A mixture of cream of tartar and lump sugar, in nearly equal proportions, with sufficient powdered cassia to give it an aromatic flavour. See PILLS.

Ollivier's Biscuits. Take of the white of 2 eggs; water, ¼ pint; beat them together, strain the mixture, and add to it a solution of bichloride of mercury, 76 gr.; collect the precipitate, wash, dry, powder, and carefully weigh it; next add to it such a quantity of flour, &c., that each 2-dr. biscuit may contain exactly ¼ gr.

Papier Fayard. See PAPER (Gout).

Pâte Arsenicale. A powder composed of arsenious acid, 8 gr.; dragon's blood, 22 gr.; cinnabar, 70 gr. It is to be made into a paste with the saliva at the time of applying it. A favourite remedy in cancer on the Continent (*Dr Paris*).

Perry's Balm of Syriacum. From English gin, 1 pint; moist sugar, ½ lb.; (dissolved in) water, 4 oz.; mix, and add of paregoric (Tinct. Camph. Co.—Ph. L. 1836), 1 oz.; tincture of tolu, ½ oz.; tincture of cantharides, q. s.; together with a few drops each of the oils of aniseed and spearmint; agitate well together, and the next day filter, or decant the clear portion.

Perry's Preventive Lotion. This is said to be a solution of sal alembroth, 2 dr., in water, 1 pint. For use, it is diluted with 4 or 5 times its bulk of water.

Pisate's Toothache Essence. From liquor of ammonia, 2 parts; laudanum, 1 part. It is applied on lint.

Pilules Angéliques. *Syn.* GRAINS DE SANTÉ. Take of aloes and juice of roses, of each, 4 oz.; juices of borage and chicory, of each, 2 oz.; beat them together, and when they are reduced to the consistence of a soft pill-mass, add of powdered rhubarb, 2 dr.; powdered agaric, 1 dr.; and divide the mixture into 1½-gr. pills. A good purgative. —*Dose,* 4 to 12.

Poor Man's Friend (*French*). See OINTMENT (Brown).

Poor Man's Friend (*Dr Roberts's*). This consists chiefly of ointment of nitric oxide of mercury.

Pringle's Remedy for Typhus (*Dr Paris*). Pale cinchona (bruised), ½ oz.; water, 12 fl. oz.; boil them together for 10 minutes, adding, towards the end, Virginian snake-root (bruised), 2 dr.; macerate for an hour in a covered vessel, and to the strained liquid add of dilute sulphuric acid, 2 fl. dr., and when the mixture is cold, further add of spirit of cinnamon, 1 fl. oz. The dose is 2 table-spoonfuls every six hours.

Reynolds' Gout Specific. Wine of colchicum disguised by some unimportant additions.

Righini's Odontalgic Drops. A solution of creasote in an equal weight of the strongest rectified spirit, coloured with cochineal, and disguised by the addition of a few drops of oil of peppermint.

Ruspini's Styptic. A strong solution of gallic acid in spirit of roses. Dr A. T. Thomson says that it also contains sulphate of zinc.

Rust's Toothache Paste. See PASTE.

Scott's Drops. *Syn.* TINCTURE OF SOOT. From wood-soot, 2 oz.; assafœtida, 1 oz.; brandy or proof spirit, 1 pint.—*Dose,* 1 to 2 table-spoonfuls; in hysteria, &c.

Smith's Powder. See MCKINSEY'S POWDER.

Solomon's Anti-impetigines. A solution of bichloride of mercury disguised by the addition of a little flavouring and tinctorial matter ('Med. Circ.,' ii, 69, 70).

Standert's Red Mixture. Take of carbonate of magnesia, 1 oz.; powdered Turkey rhubarb, ½ oz.; tincture of rhubarb, 8 fl. oz.; tincture of opium, 2 fl. dr.; oils of aniseed and peppermint, of each, ¼ dr.; (dissolved in) gin or proof spirit, 5 fl. oz.; agitate the whole together, then further add of soft water, 1½ pints. In colic and diarrhœa.— *Dose.* A wine-glassful. The spirit is frequently omitted, but then the mixture soon spoils.

Standert's Stomachic Candy. Take of lump sugar, 1 lb.; water, 3 fl. oz.; dissolve by heat; add cardamom seeds, ginger, and rhubarb, of each, 1 oz.; and when the mixture is complete, pour it out on an oiled slab or into moulds.

Storey's Worm Cakes. Take of calomel and cinnabar, of each, 24 gr.; powdered jalap, 72 gr.; ginger, 1 dr.; white sugar, 1½ oz.; syrup, q. s.; mix and divide into a dozen cakes. Resemble 'Ching's lozenges' in their action. (See page 1007.)

Struve's Lotion. See LOTION, HOOPING-COUGH.

Succession Powder. A mixture of powdered quartz and diamond dust, chiefly the first. Used as an escharotic.

Tasteless Ague Drops. A solution of arsenite of potassa. It is the common ague medicine in the fen counties of England.

Turlington's Balsam. See BALSAM OF LIFE (*above*).

Valangin's Solution of Solvent Mineral. From arsenious acid, ½ dr., dissolved in hydrochloric acid, 1½ dr., and the solution diluted with distilled water, 1½ pints. In ague, &c. It has rather less than half the strength of the solution of arsenite of potassa, Ph. L.

Vance's Cream. See CHILBLAIN.

Wahler's Ointment. See CHILBLAIN.

Ward's Purging Powder. A mixture of jalap and cream of tartar, equal parts, coloured with a little red bole.—*Doss.* A teaspoonful, or more, in broth or beer, twice or thrice daily; in dropsy.

Webster's Diet Drink. A sweetened decoction of betony, dulcamara, guaiacum wood, liquorice root, sarsaparilla, sassafras, thyme, and turmeric.

Wilson's Gout Tincture. This is said to be wine of colchicum.

Wright's Pearl Ointment. Take of white precipitate, 8 oz.; Goulard's extract, 1 pint; rub them to a cream, and add the mixture to white wax, 7 lbs., and olive oil, 10 lbs., previously melted together by a gentle heat; lastly, stir the whole until it is nearly cold ('Pharm. Journ.').

Young's Aperient Drink. From carbonate of soda, 2½ dr.; bitartrate of potass, 3 dr. (both in crystals); throw them into a soda-water bottle containing cold water, 8 fl. oz., and immediately cork it down securely, and keep it inverted, in a cool place, until required for use.

Zanhetti's Bohemian Restorative Tincture. From crushed raisins, ½ lb.; hay saffron, 2 oz.; aqueous extract of opium, 3 dr.; powdered cochineal, 2 dr.; capillaire and orange-flower water, of each, ½ pint; Proof spirit, 3 pints; digested together for a week, and then strained, with expression.

PAULLIN'IA. See GUARANA.

PAYN'IZING. The name given to Mr Payne's process for preserving and mineralising wood. See DRY-ROT.

PEACH. *Syn.* PERSICUM, L. The fruit of *Prunus Persica.* Two varieties are known in our gardens—CLINGSTONE PEACH and FREESTONE PEACH, terms which explain themselves. The fruit is wholesome; but the flowers and kernels contain prussic acid, and are poisonous.

Dr Fresenius has analysed this fruit, and found its composition to be—

Soluble matter:

	Large Dutch.
Sugar	1·580
Free acid (reduced to equivalent in malic acid)	0·612
Albuminous substances	0·463
Pectous substances	6·313
Ash	0·422

Insoluble matter:

Seeds	4·639
Skins	} 0·991
Pectose	
[Ash from soluble matter included in weights given	0·042]
Water	84·990

　　　　　　　　　　　100·000

It will be seen from the above that the peach contains a very small amount of sugar.

The peach, the original habitats of which were Persia and the north of India, is now very generally grown in many parts of Europe, in many parts of the East, and very largely in the more temperate portions of North and South America; more particularly in Pennsylvania, New Jersey, and Maryland, where there are extensive orchards of peach trees. This fruit is also extensively cultivated by the Mormon community at Utah. The fruit of the NECTARINE, which is a variety of the peach, differs from that of the latter in having a smooth skin. When stewed, the fruit of the peach is said to be useful in habitual constipation.

PEACH'WOOD. The produce of a species of *Cæsalpinia*, now extensively used in calico printing.

PEAR. *Syn.* PYRUS, L. The fruit of *Pyrus communis*, Linn., one of the ROSACEÆ. Its general qualities resemble those of the apple.

COMPOSITION OF THE PEAR.

Soluble matter:

Sugar	7·000
Free acid (reduced to equivalent in malic acid)	0·074
Albuminous substances	0·260
Pectous substances, &c.	3·281
Ash	0·285

Insoluble matter:

Seeds	0·390
Skins	} 3·420
Pectose	1·340
[Ash from insoluble matter included in weights given	0·050]
Water	83·950

　　　　　　　　　　　100·000
　　　　　　　　　　　(*Fresenius.*)

PEARL. *Syn.* MARGARITA, MARGARITUM, PERLA, UNIO, L. The most beautiful and costly pearls are obtained exclusively from the pearl oyster (*Meleagrina margaritifera*) of the Indian seas. The principal fisheries are on the coast of Ceylon, and at Olmutz, in the Persian Gulf. An inferior description of pearl is procured from a fresh-water shell-fish (*Unio margaritifera*) in the neighbourhood of Omagh, county of Tyrone. A similar quality is also procured from the river Ythan, Aberdeenshire. It is probable that pearls may have given rise to the statement by Tacitus, in his 'Life of Agricola,' of pearls "not very orient, but pale and wan," being among the indigenous products of Great Britain.

Pearls are composed of membrane and carbonate of calcium; or, in other words, of substances similar to bladder and chalk, in alternate layers. The cause of the production of pearls is highly curious and interesting. When any foreign body gains a permanent lodgment within the shells of any of the mollusca which are lined with pearly matter, or nacre, the pearly secretion of the animal, instead of being spread in layers on the inside of its habitation, is accumulated around the offending particles in concentric films of extreme tenuity, and more or less spherical, forming a pearl.

Pearls were formerly used in medicine as absorbents or antacids; and among the ancients they were occasionally taken, dissolved in acid, both as a remedy and for the purpose of displaying the careless opulence and luxury of their possessors. A perfect pearl, large, truly spherical, highly iridescent, and reflecting and decomposing the rays of light with vivacity, claims to rank with the most costly of the gems, and in some parts of the East is, with justice, more highly prized than even the diamond. In Europe, however, the present estimation of their value is somewhat different. "A handsome necklace of Ceylon pearls, smaller than a large pea, costs from £170 to £300; but one of pearls about the size of pepper-corns may be had for £15. The pearls in the former sell at a guinea each, and those in the latter at about 1s. 6d." (*Milburn*). Seed pearls are of little value, however beautiful.

Pearl, Artificial. These are hollow spheres or beads of glass, perforated with two holes at opposite sides to permit of their being strung into necklaces. A small portion of essence d'orient is introduced into each, by suction, and is then spread over the inner surface of the glass. When this has become dry and hard, the globe is filled up with white wax, spermaceti, or gum-arabic. The glass of which the beads are formed is slightly bluish and opalescent, and very thin. The latest improvement consists in removing the glassy appearance of the surface of the prepared bead by exposure to the fumes of hydrofluoric acid, highly diluted.

Pearl, Mother of. *Syn.* UNIONUM CONCHÆ, L.; NACRE DE PERLE, Fr. This is the internal or nacreous layer of those shells which produce the pearls for ornamenting the persons; hence the term 'mother of pearl' is by no means inappropriate. It is also derived from several other species, known in trade as ear-shells, green snail-shells, Bombay shells, &c.

The brilliant hues of mother of pearl do not depend so much upon the nature of the substance as on its structure. Its surface is covered by minute corrugations or furrows, which give a chromatic appearance to the reflected light. Sir David Brewster was the first to show that this substance is capable of imparting its iridescent appearance to fusible metal or fine black wax.

Mother of pearl is cut and wrought with nearly similar tools to those used for ivory, but its treatment, owing to its more fragile nature and delicate structure, requires considerably greater care. It is polished with colcothar or putty powder.

The numerous applications of mother of pearl, for buttons and knife-handles, boxes, inlaying work, &c., are well known.

PEARL/ASH. This is prepared by calcining crude potashes on a reverberatory hearth, dissolving the calcined mass in water, and, after repose, decanting the clear solution, and evaporating it to dryness in flat iron pans, the product being constantly stirred towards the end to reduce it to a semi-granular state. Although purer, its richness in absolute alkali is less than that of the potashes from which it is prepared, being only from 47% to 51%. This exists almost entirely under the form of carbonate. The commercial value of this substance is determined by the ordinary processes of ALKALIMETRY.

PEARL BARLEY. See BARLEY.

PEARL FLAVOUR. See ESSENCE.

PEARL WHITE. This is a subchloride of bismuth; but the name is now commonly applied to trisnitrate of bismuth, which is sold for it.

PEARLS (Rose). *Syn.* ROSE BEADS. The petals of red roses beaten in an iron mortar for some hours, until they form a smooth black paste,

then rolled into beads and dried. Hard; very fragrant; take a fine polish.

Pearls, to Polish. Take very finely pulverised rotten-stone, and make it into a thick paste by adding olive oil; then add sulphuric acid, a sufficient quantity to make into a thin paste.

This is to be applied on a velvet cork, rub quickly, and as soon as the pearl takes the polish wash it. This mixture when properly applied will give to pearl a brilliant polish.

PEARS, Wooden (*Xylomelum pyriforme*, Knight), so called from the extreme hardness and form of the fruit.

PEAS. *Syn.* GARDEN PEAS, MOTOR P.; PISA, L. The seed of *Pisum sativum*, Linn., Poggiale found in 100 parts of common green peas, dried and shelled, 57 of starch, 21·7 of a nitrogenous substance (legumin), 1·9 of fatty matter, 8·2 of cellulose, 2·8 of ash, and 13·7 of water. In the fresh state (GREEN PEAS) they are nutritive, and, with the pods which contain them, are highly serviceable in scurvy. The last have been used for making beer. The dried seeds are still more nutritious, but are heavy and flatulent unless well cooked. For kitchen use 'SPLIT PEAS' should be chosen, and after having washed them in a little clean soft water, and allowed them to drain, they should be left to soak in cold soft water for at least 12 hours before applying heat to them, and should then be dressed in the same water in which they have been soaked, and be only gently simmered until they are reduced to a pulp. Additions of meat, vegetables, &c., should not be made until they have nearly arrived at this condition. 'WHOLE PEAS' require soaking for at least 18 or 20 hours.

A substitute for green peas in winter may be obtained by placing the dried seeds on a flat dish, sprinkling them with water, and keeping them in a warm situation. In a few days germination commences, and, after it has proceeded sufficiently far, the whole is dressed in the usual manner. An easier and simpler plan is to preserve the green

peas, when they are in season, by the common method adopted for gooseberries and other like fruit.

Pea flour. is sometimes used to adulterate ordinary flour. It is never added to this latter to a greater extent than 4%, as, if this quantity be exceeded, it makes the bread heavy and dark. It is also used as a sophisticant for other substances, sometimes for butter.

Peas, Issue. *Syn.* PISA PRO FONTICULIS, L. Orange berries, or the small unripe fruit of the orange trees, dried, and smoothed in a lathe. See ISSUE.

PEB'BLE. The trade name for the transparent colourless variety of rock crystal or quartz used for the lenses of spectacles instead of glass, over which, from its extreme hardness, it has the advantage of being little apt to be scratched.

PEC'TIC ACID. The name given by Braconnot to an acid which is found very generally diffused throughout the vegetable kingdom.

Prep. From carrot roots, from which the juice has been pressed out, by boiling them with $\frac{1}{11}$ part of their weight of carbonate of potassium, and about 6 times their weight of water, until the liquid becomes gelatinous when neutralised with an acid. A pectate of potassium is formed, from which the acid may be obtained by neutralising the alkali with a stronger acid, or by carefully adding a solution of chloride of calcium as long as a gelatinous precipitate (pectate of calcium) falls, and after washing this with water, decomposing it with dilute hydrochloric acid.

Prop., &c. A colourless jelly, having an acid reaction; scarcely soluble in cold water, more so in hot water; and precipitated by acids, alkalies, alcohol, salts, and even sugar. Its compounds with the bases are called pectates. By long boiling with solution of caustic alkali it is converted into metapectic acid, which does not gelatinise (see below).

PEC'TIN. *Syn.* VEGETABLE JELLY. Obtained by adding alcohol to the juice of ripe currants or other fruit, until a gelatinous precipitate forms, which must be drained, washed with a little weak alcohol, and dried.

Prop., &c. In the moist state it forms a neutral, tasteless, soluble, transparent jelly; when dried, a translucent mass, closely resembling isinglass; boiled with water, or with dilute acids, it is converted into parapectin and metapectin; in the presence of alkalies, these, as well as pectin, are changed into pectic acid, and by continuing the ebullition for some time longer, into metapectic acid, which is not gelatinous. See PECTIC ACID.

PECTORAL BALSAM. The same as BALSAM OF HONEY, which *see.* The reference to 'Pectoral Balsam,' which occurs at the end of the article 'Balsam of Honey,' conveys the impression that it is a different medicine. This is an error.

PEC'TORALS. Under this head are popularly included all the various remedies employed in breath or chest diseases.

PEDIC'ULI. See LOUSE and ACARUS.

PELLAGRA. A skin disease accompanied by nausea, vertigo, diarrhœa, cramps, and neuralgic pains, which, when once acquired, returns every summer with increasing force, the patient becoming demented and even insane and maniacal, with suicidal tendencies. The disease is indigenous in hot countries, and is common in Italy, Spain, and the south of France. The Italian Government,

believing it to result from the use of badly harvested maize, has caused kilns to be erected in several of the affected districts for drying the grain, the usual open-air methods being prohibited. Some benefit is said to have resulted, but the cause of the disease is doubtful, and as good food and tonics and avoidance of exposure to the sun are among the chief remedies, the improvement in these districts may be due rather to the extra care taken to provide the staple food in good condition than to the removal of any specific cause.

PELLETIERINE. *Syn.* PELLETIERINA. An alkaloid ($C_8H_{15}NO$) discovered by Tauret in the bark of *Punica granatum.* Pelletierine is a colourless liquid; dissolves in 20 parts water; freely soluble in alcohol and ether. When exposed to oxygen it is rapidly changed to a resinous mass. Salts of the alkaloid when heated become acid in reaction.

A sulphate and tannate of pelletierine are used medicinally for the removal of tapeworm.—*Dose,* 3 to 8 gr., followed in 2 hours with 1 oz. castor oil.

PELLETS, TOOTHACHE (*Dieterich*). *Prep.* Cocaine hydrochlorate, 16 gr.; powdered opium, 64 gr.; menthol, 16 gr.; althœa (powdered), 48 gr.; mucilage of acacia, 9 gr. Make into $\frac{1}{2}$-gr. pills and keep in well-stoppered vials. For use, one of these pellets is to be inserted in the hollow tooth.

PEL'LICLE. See CRYSTALLISATION.

PEL'LITORY. *Syn.* PELLITORY OF SPAIN, PELLITORY ROOT; PYRETHRI RADIX (B. P.), PYRETHRUM (Ph. L. and E.), L. The root of *Anacyclus pyrethrum.* It is a powerful topical excitant. It is chiefly employed as a masticatory in headache, toothache, palsy of the tongue, and facial neuralgia and rheumatism; and made into a tincture with rectified spirit, it is a common remedy among dentists for the toothache. Internally, it has been given as a gastric stimulant, and in intermittents, &c. Half to 1 dr. may be chewed at a time.

PEL'TRY. The name applied to fur skins in the state in which they are received from the hunters. To prepare them as furs, the inside of them is generally first 'tawed' by the application of a solution of alum. They are next well dusted over and rubbed with hot plaster of Paris or whiting, and are, lastly, thoroughly dried and brushed clean. When it is desired to change or modify their colour, the grease being removed by lime-water or a weak soda lye, they are stretched out on a table or board, and the ordinary liquid mordants and dyes are applied to them hot by means of a painter's brush.

The furs of the rabbit and hare are rendered fit for the purposes of the felt and hat manufacturers by a process called by the French '*secretage.*' This consists in thoroughly moistening the hair with a solution of quicksilver, 1 part, in aquafortis, 16 parts, diluted with half to an equal bulk of water. This is applied with a brush, and the moistened skins being laid together, face to face, are dried as rapidly as possible in a stove room. See FURS.

PEMPHIGUS. A somewhat rare disease of the skin, in which large vesicles or blisters filled with a serous fluid develop themselves. In the mild form of the disorder the blisters vary in size from a pea to a chestnut. They chiefly attack the extremities, and break after three or four days, when they then give rise to a thin scab, which soon heals and disappears without causing any bodily derangement.

In the acute form, however, there is a considerable constitutional disturbance, which shows itself in the shape of more or less fever and inflammation; the blisters too are larger, and the scabs very irritable and obstinate. Children during teething, or owing to injudicious diet, are frequently subject to this kind of pemphigus. There is also a chronic variety of the disease, which varies but slightly from the acute form, except that it continues longer. Old people are those who principally suffer from this chronic pemphigus.

A mild attack seldom calls for any treatment; the best course to pursue in the case of an acute one is to administer some saline aperient, to adopt a moderately low diet, and to protect the exposed parts caused by the breaking of the blister by applying to them some simple dressing, such as spermaceti ointment.

When the case becomes chronic it will be advisable to consult the medical practitioner.

PENALTIES. The following sections of the Public Health Act of 1875 refer to various offences for which penalties may be inflicted under the statute:

BUILDING or re-erecting a house in an urban district without proper drains, &c., £50 (s. 25). For building or re-erecting a house in any district without proper sanitary conveniences (privies, &c.), £20, or less (s. 35).

Unauthorised building over sewers or under streets in an urban district, £5 penalty, and 40s. per day during continuance of offence (s. 26).

BURIAL. For obstructing a justice's order with regard to the burial of a person who has died from an infectious disease, &c., £5, or less (s. 142).

BYE-LAWS. Penalties may be imposed by local authorities for the contravention of bye-laws; such penalties are not to exceed £5, and for continuing offences further penalties of sums not exceeding 40s. a day (s. 183). Penalty for injury or defacement of any board, &c., on which a notice or bye-law of any authority is inscribed by the authority of the Government Local Board, or of the local authority, £5, or less (s. 306).

CELLARS, unauthorised occupation of, 20s. per day (s. 73).

CLEANSING AND WHITEWASHING, &c. Failure to comply with notice to cleanse and whitewash a house, 10s. per day (s. 46).

CONTRACTS. All contracts are to specify some pecuniary penalty (s. 174). Officers or servants being concerned or interested in contracts, accepting fees, are liable to a penalty of £50, recoverable with full costs of suit.

DISINFECTION. Failure to comply with notice to disinfect and cleanse articles and premises, not less than 1s. and not more than 10s. per day. Expenses of local authority doing the work may also be recovered (s. 120). Failure to disinfect public conveyances after conveying infected per-

sons, £5, or less (s. 127). For letting infected houses without proper disinfection, £20, or less (s. 128).

DRAINS, &c. Unauthorised connection of a drain with a sewer, £20, or less (s. 21). For neglecting to comply with notice for the construction of privies, &c., for factories, £20, or less, and 40s. per day. For non-compliance with notice for the construction of drains, privies, &c., 10s. per day (s. 41).

EPIDEMIC DISEASES. For violation or obstruction of the regulations of the Local Government Board with regard to epidemic diseases, £5, or less (s. 140).

EXPOSURE of infected persons or things, £5, or less (s. 126).

HOUSES OR ROOMS. Making false statements with regard to infectious diseases for the purpose of letting, £20, or less, or imprisonment for one month with or without hard labour (s. 129).

LODGING-HOUSES. Receiving lodgers in unregistered houses, failure to make a report, failure to give notice of infectious diseases, £5, or less, and 40s. per day during continuance of offence. Refusal or neglect to affix or renew notice of regulation in common lodging-houses, £5, or less, and 10s. a day during continuance of offence after conviction (s. 79). For neglecting the lime-washing and cleansing of lodging-houses according to the Act, 40s., or less (s. 82).

MANURE. Failure to comply with a notice of urban authority to periodically remove manure, &c., 20s. a day (s. 50).

MEAT. For exposing for sale or having in possession unsound meat and other articles of food specified in the Act, £20, or less, for each carcass or piece of meat, or fish, &c., or three months' imprisonment with or without the option of a fine (s. 117). For obstruction of officer inspecting the food, £5, or less (s. 118).

MORTGAGE OR RATE. Refusal of custodian of register to permit inspection, £50, or less. Neglect or refusal of clerk to register transfer of mortgage, £20, or less.

NUISANCE. The court may impose a penalty of £5, or less, with regard to nuisances generally (s. 98). For want of diligence in carrying out the order to abate nuisance, 10s. per day; for contravention of order, if wilful, 20s. per day during such contrary action, besides the expenses of the local authority in abating the nuisance (s. 98).

For nuisance of pigs, pigsties, and the contents of cesspools, &c., overflowing, 40s., or less, and 5s. per day during continuance of offences (s. 98).

OBSTRUCTION. For wilful obstruction of member of, or person authorised by, local authority, £5, or less (s. 306). Obstruction of owner by occupier in carrying out any of the provisions of the Act, £5 per day, commencing twenty-four hours after non-compliance with the justice's order (s. 306).

OFFICES. Certain offices are not to be held by the same person. Penalty for offence, £100, recoverable with full costs of suit (s. 192).

ORDER OF JUSTICES. Refusal to obey order for admission of local authority, £5, or less (s. 103).

RATES. Refusal of officers in custody of rate-books, valuation lists for the relief of the poor, &c., to permit inspection, £5, or less (s. 212).

Refusal of person to permit inspection of rate, £5, or less (s. 219).

SCAVENGING. Obstruction of the contractor or local authority in scavenging the streets or in removal of refuse, £5, or less (s. 42). Neglect of local authority to scavenge after undertaking to do so, 5s. per day (s. 43).

STREETS. Wilful unauthorised displacement or injury of pavement stones, injury to fences, &c., of streets vested in urban authority, £5, or less, and a further penalty of 5s. or less for every square foot of pavement injured, &c. Compensation may also be awarded by the court for injury to trees (s. 149).

For building or bringing forward buildings beyond the general line of the houses in the street in an urban district, 40s. per day after written notice (s. 156).

TRADE, OFFENSIVE. Unauthorised establishment of, in an urban district, £50, and 40s. per day during continuance of offence (s. 112). Nuisance arising from offensive trade is punishable by penalty—for first offence, not less than 40s., and not exceeding £5 ; for second or any subsequent offence, double the amount of the last penalty which has been imposed, but in no case to exceed £200 (s. 114).

WATER. Pollution of, by gas, £200 ; and when offence is continued at the end of 24 hours' notice, £20 per day (s. 68).

For injuring water-meters, 40s., or less, and the damage sustained may also be recovered (s. 60).

WORKS. For wilful damage of works or property belonging to a local authority, in cases where no other penalty is provided, £5, or less (s. 307).

⁎ All penalties, forfeitures, costs, and expenses directed to be recovered in a summary manner, or not otherwise provided for, may be prosecuted and recovered under the 'Summary Jurisdiction Acts' before a court of summary jurisdiction (P. H. S. 251) ; but proceedings for the recovery of penalties are only to be taken by the person aggrieved, or by the local authority of the district, except the consent in writing of the Attorney-General be obtained. But this restriction does not apply to the proceedings of a local authority with regard to nuisances, offensive trades, houses, &c., without their district, in cases in which the local authority are authorised to take proceedings with respect to any act or default (s. 258).

Unless otherwise provided for, the penalty is thus applied :—One half goes to the informer, and the remainder to the local authority of the district in which the offence was committed ; but if the local authority be the informer they are entitled to the whole of the penalty recovered.

All penalties and sums recovered by a local authority are paid to the treasurer, and carried to the account of the fund applicable to the general purposes of the Public Health Act.

(The justices or court have power to reduce penalties imposed by 6 Geo. IV, c. 78. P. H. Part III.)

PENCILS. This name is applied to the small brushes made of camel's hair used by artists, as well as to the plumbago crayons familiarly known as black-lead pencils. The last are prepared by one or other of the following methods :

1. The blocks of plumbago are exposed to a bright red heat in a closely covered crucible, and are afterwards sawn into minute sticks, and mounted in cases of cedar or satin-wood.

2. The plumbago, in powder, is calcined as before, and then mixed with an equal, or any other desired proportion of pure washed clay, also in powder, after which the mixture is reduced to a plastic state with water, and pressed into grooves cut on the face of a smooth board, or into well-greased wooden moulds, in which state it is left to dry. When dry, the pieces are tempered to any degree of hardness by exposing them, surrounded by sand or powdered charcoal, in a closely covered crucible to various degrees of heat. The crucible is not opened until the whole has become cold, when the prepared ' slips' are removed and mounted as before. This method was invented by M. Conté in 1795.

3. The dough or paste, prepared as last, is reduced to the required form by forcing it through a perforated plate (in a similar manner to that adopted for coloured crayons), or into minute metallic cylinders, from which it may be readily shaken after it has become partially dry.

Obs. The leads for some varieties of drawing-pencils are immersed for a minute in very hot melted wax or suet before mounting them. To the composition for others a little lamp-black is added, to increase and vary the degree of blackness. The pencils for asses' skin books and prepared paper are tipped with ' fusible metal.' Numerous improvements in pencil cases and pencil mounts have been patented of late years by Stevens and others.

Pen'cils, Medical. (Codex.) Under the term Crayons medicamenteux the Codex includes Crayons d'azotate d'argent mitigé, which are composed of nitrate of silver with 10, 50, 66, or 75% of nitrate of potash. The crystals are to be melted in a silver or porcelain crucible and poured into moulds.

Pencils, Sulphate of Copper, are to be prepared in the same manner, but in melting them the crystals should be first broken small, and the heat employed must be gentle.

Pen'cils, Tannin, are thus prescribed :— Powdered tannin, 10 grms. ; powdered gum, 50 centigrms. ; distilled water and glycerine, q. s., as little as possible. Having mixed the tannin and the gum, this powder is to be made into a mass of pilular consistence, and rolled and cut into cylindrical strips of the size required.

Pen'cils, Iodoform, are to be prepared in the same manner.

PENNYROY'AL. Syn. PULEGIUM (Ph. L. and E.), MENTHA P. (Ph. D.), L. " The recent and dried flowering herb of Mentha pulegium, Linn." (Ph. L.). PENNYROYAL TEA is a popular emmenagogue, expectorant, and diaphoretic, and is in common use in asthma, bronchitis, hooping-cough, hysteria, suppressions, &c. Water, essence, oil, and spirits of pennyroyal are officinal. They are now chiefly used as mere adjuncts or vehicles.

PENTASTOMATA. There are two varieties of this entozoon—the *Pentastoma denticulatum*, which Leuckart has shown to be the larvæ of the *Pentastoma tænoides*, and the *Pentastoma constrictum*. The *P. denticulatum* infests the human liver and small intestines. The *P. constrictum* does not appear to be known in this country. The latter appears to have caused death by setting up peritonitis. According to Dr. Aitken these parasites are provided with two pairs of hooks or claws, placed on each side of a pit or mouth, on a flattened head. He says : " These claws appear to be implanted in socket-like hollows or depressions, surrounded by much loose integument. These socket-like hollows appear to be elevated on the summit of the mass of tissues which lies underneath the folds of integuments surrounding the base of the hooks. These parts are regarded as the feet of the parasite, and the hooks are the fore claws. The pit or mouth is of an oval shape, the long axis of the oval lying in the direction of the length of the worm.

" The less or outer margin of the pit is marked by a well-defined thin line. There are no spines nor hooks on the integument of the elongated body."

PEPPER (Black). *Syn.* PEPPER; PIPER, B.P.; NIGRI BACCÆ, PIPER NIGRUM (Ph. L., E., and D.). L. " The immature fruit (berry) of *Piper nigrum*, Linn., or the black pepper vine." (Ph. L.)

Pur. The ground black pepper of the shops is universally adulterated; in fact, the public taste and judgment are so vitiated that the pure spice is unsaleable. A most respectable London firm, on commencing business, supplied their customers with unadulterated ground pepper, but in three cases out of every four it was returned on their hands and objected to, on account of its dark colour and rich pungency, which had induced the belief that it was sophisticated. The house alluded to was therefore compelled by the customers to supply them with an inferior, but milder and paler, article. The substances employed to lower black pepper are known in the trade as—' P. D.,' ' H. P. D.,' and ' W. P. D.'— abbreviations of pepper dust, hot p. d., white p. d. The first is composed of the faded leaves of autumn, dried and powdered; the second, the ground husks (hulls) of black mustard, obtained from the mustard mills; and the third is common rice, finely powdered. Equal parts of black peppercorns, H. P. D., and W. P. D., form the very best ground pepper sold. The ordinary pepper of the shops does not contain more than ⅔th to ¾th of genuine pepper, or 2 to 2½ oz. in the lb. Very recently, ground oil-cake or linseed meal has been chiefly employed as the adulterant, instead of the old ' P. D.'

Dr. Parkes ('Practical Hygiene') says : " The microscopic characters of pepper are rather complicated. There is a husk composed of four or five layers of cells and a central part. The cortex has externally elongated cells, placed verti-

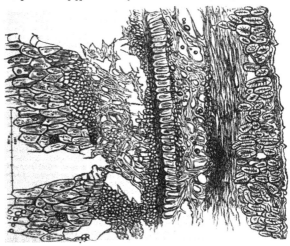

cally, and provided with a central cavity, from which lines radiate towards the circumference; then comes some strata of angular cells, which, towards the interior, are larger and filled with oil. The third layer is composed of woody fibre and spiral cells. The fourth layer is made up of large cells, which, towards the interior, become smaller and of a deep red colour; they contain most of the essential oil of the pepper. The central part of the berry is composed of large angular cells, about twice as long as broad. Steeped in water, some of these cells become yellow; others remain colourless. It has been supposed that these yellow cells contain piperine, as they give the same reaction as piperine does, namely, the tint is deepened by alcohol and nitric acid, and sulphuric acid applied to a dry section causes a reddish hue" (*Hassal*).

Uses, &c. Black pepper is a powerful stimulant, carminative, and rubefacient. Its use in moderation, as a condiment, is peculiarly serviceable to persons who are of cold habit, or who suffer from weak digestion; but in inflammatory habits, and in affections of the mucous membranes, it is generally highly injurious. As a medicine it is often serviceable in nausea, vomiting, chronic diarrhœa, and agues. In North America a common remedy for the last is ½ oz. of ground pepper stirred up with a glassful of warm beer; or a like quantity made into a tincture by steeping it in five or six times its weight of gin, rum, or whisky, for a few days.

Prepared black pepper is made by steeping the berries for three days in three times their weight of vinegar, and then drying and grinding them. It is milder than common pepper. See CONFECTIONS, PIPERINE, &c.

Pepper, Cayenne. *Syn.* BIRD PEPPER, CHILI P., GUINEA P., INDIAN P., RED P.; PIPER CAPSICI, P. CAYENNE, L. This is prepared from chillies, or the pods of *Capsicum frutescens*, or from *Capsicum baccatum*, or bird pepper, but generally from the first, on account of its greater pungency and acrimony; and, occasionally, from *Capsicum annuum*, or medicinal capsicum.

Prep. 1. From the dried pods (powdered), 1 lb.; and wheaten bread or captain's biscuits (heated until they are perfectly dry and brittle, and begin to acquire a yellow colour throughout, and then powdered), 7 lbs.; mixed and ground together. Colouring matter and common salt are frequently added, but are unnecessary.

2. As the last, but making the mixture into a dough with water, then forming it into small cakes, drying these as rapidly as possible at a gentle heat, and then grinding them.

3. (Loudon.) The ripe pods, dried in the sun, are stratified with wheaten flour in a dish or tray, and exposed in a stove-room or a half-cold oven until they are quite dry; they are then removed from the flour, and ground to fine powder; to every oz. of this powder 1 lb. (say 15 oz.) of wheaten flour (including that already used) are added, and the mixture is made into a dough with a little tepid water and a teaspoonful of yeast; after fermentation is well set up, the dough is cut into small pieces, and baked in a slow oven until it is perfectly hard and brittle; it is then beaten or ground to powder, and forms 'cayenne pepper.'

Pure cayenne pepper, when burnt, leaves a scarcely perceptible quantity of white ash; a red-coloured ash indicates the presence of red ochre, brick-dust, Armenian bole, or other earthy colouring matter. If red lead is present, it will be left behind under the form of a dark-coloured powder, or a small metallic globule.

Pur. The 'cayenne pepper' of the shops is often a spurious article, made by grinding a mixture of any of the reddish woods or saw-dust with enough red pods or chillies to render the mixture sufficiently acrid and pungent. Common salt, colcothar, red bole, brick-dust, vermilion, and even red lead, are also common additions.

Uses, &c. The capsicums resemble the peppers, except in their greater energy and their pungency being unmodified by the presence of essential oil. As a condiment, under the form of cayenne pepper, and in all diseases in which the employment of a powerful stimulant or rubefacient is indicated, their uses are well known. In medicine the fruit of *Capsicum annuum* (Linn.—Ph. E. and D.; *C. fastigiatum* Blume—B. P., Ph. L.), or annual capsicum is ordered (CAPSICUM—Ph. L., E., and D.). The London College directs the fruit to be that of 'Guinea,' less than one inch long, oblong, cylindrical, and straight. See ESSENCE OF CAYENNE.

Pepper, Prepared Cayenne, is the residuum of cayenne—vinegar, essence, or tincture, dried and ground (see *below*).

Pepper (Soluble) Cayenne. *Syn.* CRYSTALLISED SOLUBLE CAYENNE. *Prep.* 1. Capsicum pods (recent, ground in a pepper mill), 1 lb.; rectified spirit, 2½ pints; proceed by percolation so as to obtain 2½ pints; from this distil one half of the spirit by the heat of a water bath; to the residuum add of fine dry salt, 5 lbs.; mix them well together, and dry the mixture at a very gentle heat, frequently stirring; lastly rub it through a sieve, and put it into warm dry bottles. It is usually coloured with a little vermilion or rouge (sesquioxide of iron), but it possesses an agreeable colour without it.

2. Essence of cayenne (No. 1, page 652), 6 pints; distil off 3 pints, add to the residual liquor of dry salt, 12 lbs.; mix well, dry by a gentle heat, and otherwise proceed as before.

3. Capsicums (ground), 3 lbs.; red sanders or Brazil wood (sliced or rasped), 10 oz.; rectified spirit, 1 gall.; macerate for 14 days, then express the tincture, filter, distil off one half, add of dry salt, 15 lbs., and proceed as before.

4. As the first formula, with the addition of a strong decoction of saffron, q. s. It gives a beautiful colour to soups, &c.

Obs. The above formulæ are those actually employed by the houses most celebrated for their 'soluble cayenne.' The products are of the very finest quality, and are perfectly wholesome. We speak from an extensive experience in the manufacture. The spirit distilled from the essence forms a most suitable menstruum for making fresh essence or tincture of cayenne.

Pepper, Cu'beb. See CUBEBS.

Pepper, Jamai'ca. See PIMENTO.

Pepper, Kit'chen. See SPICE.

Pepper, Long. *Syn.* PIPERIS LONGI FRUCTUS, PIPER LONGUM (Ph. L. and E.), L. "The immature fruit (dried female spikes) of *Piper Longum*, Linn." (Ph. L.), or long-pepper vine. The spikes are about 1½ inches in length, with an indented surface, and are of a dark-grey colour. In its general properties it resembles black pepper, but it is less aromatic, though equally pungent. Elephant pepper is merely a larger variety of this species (*Gray*). The root and stems, sliced and dried, form the 'pippula moola' of the East Indies (*Roxburgh*).

Pepper, Red. See CAYENNE.

Pepper, White. *Syn.* PIPER ALBUM, L. This is made by either soaking ordinary black pepper in a solution of common salt until the outside skins are soft, and then rubbing them off in the hands, or by merely rubbing off the skins of the over-ripe berries that fall from the vines. An inferior quality is made by bleaching black pepper with chlorine.

Obs. The use of white pepper instead of black is an instance of the sacrifices made to please the eye. Pure white pepper has only about 1·4th of the strength of pure black pepper, whilst it is nearly destitute of the fine aroma of the latter. It also contains a mere trace of piperina or piperine, one of the most valuable constituents of black pepper.

PEPPER PODS. Capsicums. See CAYENNE PEPPER.

PEPPERMINT. *Syn.* MENTHA PIPERITA (Ph. L., E., and D.), L. "The recent and dried flowering herb of *Mentha piperita*" (Ph. L.), or garden peppermint. The flavour and odour of this herb are well known. It is the most pleasant and powerful of all the mints. Peppermint water and the essential oil have long been employed in nausea, griping, flatulent colic, hysteric, diarrhœa, &c.; but in regular practice chiefly to cover the taste of nauseous medicines, or as an adjunct or vehicle for more active remedies. See OILS (Volatile), WATERS, &c.

PEPSIN. *Syn.* GASTERACE, CHYMOSIN. A peculiar principle found in the gastric juice, and which, in conjunction with hydrochloric acid, also present in the stomach, confers upon it the power of digesting the albuminous portions of the food.

Prep. 1. (*Beale*, 'Med. Times and Gaz.,' February 10th, 1872, p. 152.) "The mucous membrane of a perfectly fresh pig's stomach is carefully dissected from the muscular coat, and placed on a flat board. It is then lightly cleansed with a sponge and a little water, and much of the mucus, remains of food, &c., carefully removed. With the back of a knife, or with an ivory paper-knife, the surface is scraped very hard, in order that the glands may be squeezed and their contents pressed out. The viscid mucus thus obtained contains the pure gastric juice with much epithelium from the glands and surface of the mucous membrane. It is to be spread out upon a piece of glass, so as to form a very thin layer, which is to be dried at a temperature of 100° over hot water, or *in vacuo* over sulphuric acid. Care must be taken that the temperature does not rise much above 100° F., because the action of the solvent would be completely destroyed. When dry the mucus is scraped from the glass, powdered in a mortar, and transferred to a well-stoppered bottle. With this powder a good digestive fluid may be made as follows: of the powder, 5 grs.; strong hydrochloric acid, 18 drops; water, 6 oz. Macerate it at a temperature of 100° for an hour. The mixture may be filtered easily, and forms a perfectly clear solution very convenient for experiment.

"If the powder is to be taken as a medicine, from two to five grains may be given for a dose, a little diluted hydrochloric acid in water being taken at the same time. The pepsin powder may be mixed with the salt at a meal. It is devoid of smell, and has only a slightly salt taste. It undergoes no change if kept perfectly dry, and contains the active principle of the gastric juice almost unaltered.

"The method of preparing this pepsin was communicated to Mr Bullock, of the firm of Messrs Bullock and Company, 3, Hanover-street, Hanover Square, who at once adopted it for the preparation of medicinal pepsin, and soon improved upon it in some particulars. The dose is from 2 to 4 or 5 grs.—*Test.* ¼ths of a grain of this pepsin, with 10 drops dilute hydrochloric acid and an ounce of distilled water, dissolve 100 grs. of hard-boiled white of egg in from 12 to 24 hours. In the body probably twice this quantity of white of egg or even more would be dissolved in a comparatively short space of time. The digestive powder prepared from the pig's stomach retains its activity for any length of time if kept dry. The solution made with this pepsin and hydrochloric acid was nearly tasteless and inodorous. One pig's stomach, which costs sixpence, will yield about 45 grs. of the powder prepared as above described.

"Gradually the usefulness of this preparation of pepsin of the pig was found out, and it had to be prepared in increasing quantities. I should be afraid to say how many pigs' stomachs have been used of late years during the winter season.

"In 1857 Dr Pavy carefully examined the pepsin prepared and sold by many different firms, and found that this dried mucus of the pig's stomach was the most active of them all ('Medical Times and Gazette,' 1857, vol. i, p. 336). In 1870 Professor Tuson instituted a still more careful comparative examination, and with a similar result ('Lancet,' August 13th, 1870); for he found that this preparation was *twenty-five times stronger than some others that he obtained for examination.*"

2. (*Scheffer*, 'Pharm. Journ.,' March 23rd, 1872, p. 761.) "Of the well-cleaned fresh hog stomach the mucous membrane is dissected off, chopped finely and macerated in water acidulated with muriatic acid for several days, during which time the mass is frequently well stirred. The resulting liquid, after being strained, is, if not clear, set aside for at least 24 hours in order to allow the mucus to settle. To the clarified liquid the same bulk of a saturated solution of sodium chloride is added, and the whole thoroughly mixed. After several hours the pepsin, which, by the addition of chloride of sodium, has separated from its solution, is found floating on the surface, from whence it is removed with a spoon and put upon cotton cloth to drain; finally it is submitted to strong pressure, to free it as much as possible from the salt solution.

"The pepsin, when taken from the press and allowed to become air-dry, is a very tough substance, and presents, according to thickness, a different appearance, resembling in thin sheets parchment paper, and in thick layers sole leather; its colour varies from a dim straw yellow to a brownish yellow. Besides a little mucus, it contains small quantities of phosphate of lime and chloride of sodium, which, however, do not inter-.

fere with its digestive properties, as they are found also in normal gastric juice.

In order to get a purer article I redissolve the pepsin, as obtained after expression, in acidulated water, filter the solution through paper and precipitate again with a solution of sodium chloride; the precipitate, after draining and pressing, is now free of phosphate of lime and mucus, but still contains salt. In the freshly precipitated state the pepsin is very readily soluble in water, and cannot therefore be freed from adhering salt by washing.

" By allowing the pressed sheet of pepsin to get perfectly air-dry—whereby it becomes coated with a white film and small crystals of chloride of sodium—and by immersing it then in pure water for a short time, the greater part of sodium chloride can be extracted, but it has to be done very rapidly, as the pepsin swells up considerably and loses its tenacity. By operating in this matter I have obtained a pepsin which dissolves in acidulated water to quite a clear colourless liquid, but as it still contains traces of salt, I prefer to call it purified pepsin."

3. (B. P.) A preparation of the mucous lining of a fresh and healthy stomach of the pig, sheep, or calf. The stomach of one of these animals, recently killed, having been cut open and laid on a board with the inner surface upwards, any adhering portions of food, dirt, and other impurity, are to be removed and the exposed surface slightly washed with cold water; the cleansed mucous membrane is then to be scraped with a blunt knife or other suitable instrument, and the viscid parts thus obtained are to be immediately spread over the surface of glass or glazed earthenware, and quickly dried at a temperature not exceding 100° F.; the dried residue is to be reduced to powder, and preserved in a stoppered bottle.—*Dose*, 2 to 5 gr.

Pepsin, Saccharated. To work it into saccharated pepsin ('American Journal of Pharmacy,' January, 1871) the damp pepsin, as it is taken from the press, is triturated with a weighed quantity of sugar of milk to a fine powder, which, when it has become air-dry, is weighed again, the quantity of milk sugar subtracted, and so the amount of pepsin found. The strength of this dry pepsin is now ascertained by finding how much coagulated albumen it will dissolve at a temperature of 100° F. in five or six hours, and after this sufficient milk sugar is added to result in a preparation of which 10 gr. will dissolve 120 gr. of coagulated albumen, and this preparation I have called saccharated pepsin.

Pepsin with Starch. Pepsin mixed with starch ·is the *medicinal pepsine* of M. Boudault; the *Poudre nutrimentive* of M. Corvisart.

Pepsin, Glycerite of. *Syn.* GLYCERITUM PEPSINI, L. *Prep.* Pepsin (N. F.), 640 gr.; hydrochloric acid, 80 minims; purified talcum, 120 gr.; glycerin, 8 fl. oz.; water, enough to make 16 fl. oz. Mix the pepsin with 7 fl. oz. of water and the hydrochloric acid and agitate until solution has been effected. Then incorporate the purified talcum with the liquid, filter, returning the first portions of the filtrate until it runs through clear, and pass enough water through the filter to make the filtrate measure 8 fl. oz. To

this add the glycerin and mix. Each fl. dr. represents 5 gr. of pepsin (N. F.).

Note. For filtering the aqueous solution of pepsin first obtained by the above formula, as well as for filtering other liquids of a viscid character, a filter-paper of loose texture (preferably that known as ' Textile Filtering Paper '), or a layer of absorbent cotton placed in a funnel, or percolator, should be employed.

Pepsin, Acid Glycerin of. *Prep.* Pure pepsin, 1 oz.; hydrochloric acid, 2 dr.; glycerin, 8 oz.; water, 12 oz. Mix the acid with 2 oz. of water, and rub up with the pepsin; add the rest of the water, digest for 24 hours, then add the glycerin, and after 2 days decant or filter.

PERCENT'AGE. Literally, 'by the hundred.' In commerce the term is applied to an allowance, duty, or commission on a hundred (*Webster*).

PERCHLO'RATE. *Syn.* PERCHLORAS, L. A salt of perchloric acid.

The perchlorates are distinguished from the chlorates by their great stability, and by not turning yellow when treated with hydrochloric acid. Like the chlorates, they give off oxygen when heated to redness. They may be prepared by directly neutralising a solution of the acid with a solution of the base. See POTASSIUM (Perchlorate of) and CHLORINE.

PERCHLO'RIC ACID. See CHLORINE.

PERCOLA'TION. *Syn.* METHOD OF DISPLACEMENT. A method of extracting the soluble portion of any substance in a divided state, by causing the menstruum to filter or strain through it. The 'sparging' of the Scotch brewers is an example of the application of this principle on the large scale. In *pharmacy*, the 'method of displacement' is frequently adopted for the preparation of tinctures, infusions, &c., and is, in some respects, superior to digestion or maceration. " The solid materials, usually in coarse or moderately fine powder, are moistened with a sufficiency of the solvent to form a thick pulp. In 12 hours, or frequently without delay, the mass is put into a cylinder of glass, porcelain, or tinned iron, open at both ends, but obstructed at the lower end by a piece of calico or linen, tied tightly over it as a filter; and the pulp being backed by pressure, ranging as to degree with different articles, the remainder of the solvent is poured into the upper portion of the cylinder, and allowed gradually to percolate. In order to obtain the portion of the fluid which is absorbed by the residuum, an additional quantity of the solvent is poured into the cylinder, until the tincture which has passed through equals in amount the spirit originally prescribed. The spirit employed for this purpose is then recovered, for the most part, by pouring over the residuum as much water as there is spirit retained in it, which may be easily known by an obvious calculation in each case. The method of percolation is now preferred by all who have made sufficient trial of it to apply it correctly " (Ph. E.).

The first portion of liquid obtained by the method of displacement is always in a state of high concentration. In general it is a simple solution of the soluble ingredients of the crude drug in the fluid employed. But sometimes the solvent, if compound, is resolved into its component parts, and the fluid which passes through at any given time is only one of these, holding the soluble parts of the drug in solution. Thus if diluted alcohol be poured over powder of myrrh, in the cylinder of the percolator, the fluid which first drops into the receiver is a solution of an oily consistence, chiefly composed of resin and volatile oil, dissolved in alcohol. In like manner, when the powder of gall-nuts is treated in the same way by hydrated sulphuric ether, two layers of fluid are obtained, one of which is a highly concentrated solution of tannin in the water of the ether, and the other a weak solution of the same principle in pure ether. In all cases, therefore, in which it is not otherwise directed it is absolutely necessary to agitate the several portions of the liquid obtained by percolation together, in order to ensure a product of uniform strength or activity.

Several forms of displacement apparatus are employed by different operators. A simple and useful one is that figured in the margin. It has, also, the advantage of being inexpensive, and may be made by any worker in tin plate.

In operating on some substances it is found advantageous to hasten the process by pressure. This may be effected by any of the methods adopted for that purpose, and already described under FILTRATION. An ingenious little apparatus, which is well adapted for small quantities, is shown in the *engr.* By pouring mercury or water

a. Percolator.

b. Stand.

c. Receiver.

d. Menstruum.

e. Substance operated on.

f. Calico strainer.

through (*e*), into the bottle (*c*), the air in the latter suffers compression, and acts in a corresponding manner on the percolating liquor in (*a*). The whole of the joints must be made air-tight.

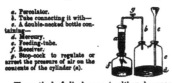

a. Percolator.
b. Tube connecting it with—
c. A double-necked bottle containing—
d. Mercury.
e. Feeding-tube.
f. Receiver.
g. Stop-cock to regulate or arrest the pressure of air on the contents of the cylinder (*a*).

The method of displacement, although apparently simple, requires for its successful application no inconsiderable amount of experience and skill in manipulation. The principal points to be attended to are—the reduction of the substance to the proper state of comminution (neither too coarse nor too fine),—the due regulation of the period of maceration according to the hardness,

density, and texture of the substance; and, more important still,—the proper packing of the ingredients in the cylinder. On the correct performance of the last the success of the process mainly depends. Some substances require considerable pressure to be used, whilst others, when even lightly packed, scarcely permit the fluid to pass through them. When the material is too loosely packed, the menstruum passes through quickly, but without exerting its proper solvent action; when too great pressure is employed, percolation either progresses very slowly or not at all. On the whole, the firmness of the packing should be inversely as the solvent and softening power of the menstruum upon the solids exposed to its action; but to this rule there are many exceptions, and each substance may be said to require special treatment. An excellent plan, applicable to all substances, and especially to those of a glutinous or mucilaginous nature, is to mix the powder with an equal bulk of well-washed siliceous sand before rubbing it up with the menstruum. In reference to the coarseness of the powder it must be observed that substances which readily become soft and pappy when wetted by the menstruum, should not be used so fine as those that are more woody and fibrous, and not of a glutinous or resinous nature.

The 'method of displacement' has the advantage of expedition, economy, and yielding products possessing considerable uniformity of strength; but the difficulties attending its application by the inexperienced are serious obstacles to its general adoption in the laboratory. It answers admirably for the preparation of all tinctures that are not of a resinous nature, and for most infusions of woody and fibrous substances, as roots, woods, barks, leaves, seeds, insects, &c., and particularly when cold or tepid water is used as the solvent. It is also especially adapted for the preparation of concentrated infusions and essences, as they may thus be obtained of any required strength without loss, or requiring concentration by heat, which is so destructive to their virtues.

" When (ordinary) tinctures are made in large quantities, displacement is never likely to supersede maceration, on account of any practical advantages it may possess. If the prescribed directions be duly attended to, the process of maceration is unexceptionable. The process is more simple than the other; the mode of operation is more uniform, it is, in fact, always the same; it requires less of skill and dexterity in conducting it; it requires less constant attention during its progress, which, in operating on large quantities, is a consideration; and, finally, the apparatus required is less complicated. When, however, only small quantities of tincture are made at a time, and kept in stock, the adoption of the process of displacement will often be found convenient and advantageous. It offers the means of making a tincture in two or three hours, which, by the other process, would require as many weeks " (*Mohr and Redwood*).

Another useful application of the method of displacement is to the manufacture of extracts on the large scale. Here it is superior to any other plan. By the simple and inexpensive forms of apparatus in block-tin, stoneware, or glass, which have recently been designed for the purpose, not merely a first-class product is ensured, but a great saving in fuel and labour is at the same time effected. The reader is referred to the last edition of the 'United States Pharmacopœia,' and to papers by Messrs Saunders and Schweitzer in the 'Pharmaceutical Year Book for 1878,' and by Mr Campbell in the same publication for 1874, for additional information in the subject of "Percolation." See BREWING, EXTRACT, TINCTURE, &c.

PERCUS'SION. *Syn.* PERCUSSIO, L. In *medicine*, the act of striking any part of the body with the fingers, or any instrument, to ascertain its condition.

PERCUS'SION CAPS. The composition employed to prime these articles is noticed under FULMINATING MERCURY.

PERFECT LOVE. See LIQUEUR (Parfait amour).

PERFUME. A substance that emits or casts off volatile particles which, when diffused through the atmosphere, agreeably affect the organs of smell. The term is also applied to the volatile effluvia so perceived. The principal source of perfumes is the Vegetable Kingdom. Its flowers, seeds, woods, and barks furnish a rich variety, from which the most fastidious connoisseur may select his favourite bouquet. A few perfumes, as musk, ambergris, and civet, are derived from the Animal Kingdom; but none of these evolve an aroma comparable in freshness to that of the rose, or in delicacy to that of the orange-blossom, or even the unpretending jasmine. The Inorganic Kingdom yields not a single perfume, so called; nor has the science of chemistry yet been able to produce a single odoriferous compound from matter absolutely inorganic.

PERFU'MERY. Perfumes in general; also the art of perfuming them. In its commercial application, this word embraces not merely perfumes, but also cosmetics, and other articles of a closely allied character employed at the toilet, the manufacture and sale of which constitute the trade of the modern perfumer. In addition to those given here, many formulæ met with in trade, both simple and compound, will be found under the heads COSMETICS, DEPILATORY, ESSENCE, HAIR DYES, OILS, PASTES, PASTILS, POMADE, SPIRIT, WATERS, &c., to which we refer the reader.

PERFUMES. For most of the following formulæ we are indebted to the "Chemist and Druggist" and the "British and Colonial Druggist":

Bouquet d'Amour. Rose triple, 10 oz.; extract, of rose, 20 oz.; of violet, 20 oz.; of cassie, 20 oz.; of jasmine, 20 oz.; of civet, 3 oz.; of musk, 3 oz.; of ambergris, 1 oz.

Bouquet des Fleurs. Extract, of jasmine, 20 oz.; of rose, 20 oz.; of violet, 20 oz.; of tuberose, 20 oz.; of orris, 20 oz.; of orange, 10 oz.; of musk, 4 oz.; Ol. French geranium, ½ oz.

Bouquet du Roi. Extract, of jasmine, 20 oz.; of violet, 20 oz.; of rose, 20 oz.; of vanilla, 8 oz.; of vitivert, 5 oz.; of musk, 2 oz.; of ambergris,

½ oz.; essence, of bergamot, 60 minims; Ol. cloves, 60 minims; otto of rose, 50 minims.

Clove Pink, Essence of. Rose triple, 7 oz.; extract, of rose, 10 oz.; of orange, 5 oz.; of cassie, 5 oz.; of vanilla, 3 oz.; of musk, ½ oz.; Ol. cloves, 18 minims.

Court Bouquet. Extract, of rose, 20 oz.; of violet, 20 oz.; of jasmine, 20 oz.; of tuberose, 10 oz.; of cassie, 8 oz.; of musk, 2 oz.; of civet, 1 oz.; of ambergris, 1 oz.; essence, of bergamot, ½ oz.; of lemon, ¼ oz.; Ol. citron, ¼ oz.; Ol. neroli, 69 minims.

Essence of White Rose. Take of oil of patchouly, 10 minims; essence of musk, 10 minims; otto of rose, 20 minims; alcohol, 90%, 20 fl. oz. Mix.

Extrait d'Ambre, for the Handkerchief. Rose triple, 10 oz.; extract, of ambergris, 20 oz.; of musk, 5 oz.; of vanilla, 2 oz.; aq. rosæ, 7 oz.

Extract of Flowers. Extract, of rose, 20 oz.; of tuberose, 20 oz.; of violet, 20 oz.; of storax, 1½ oz.; of musk, ½ oz.; essence, of bergamot, ½ oz.; of lemon, ½ oz.; Ol. citron, 33 minims.

Fleur d'Italie. Extract, of rose, 40 oz.; of jasmine, 20 oz.; of violet, 20 oz.; of cassie, 10 oz.; of tuberose, 10 oz.; of triple, 16 oz.; of musk, 3 oz.; of ambergris, 1 oz.

Gardenia. Extract, of jasmine, 20 oz.; of tuberose, 55 oz.; of cassie, 5 oz.; of musk, 2½ oz.; of storax, 2½ oz.; of civet, 1 oz.; Ol. Ylang-ylang, 30 minims; Ol. neroli, 40 minims; Ol. French geranium, 60 minims; otto of rose, 30 minims.

Helitrope Extract. Take of heliotropine, 60 grs.; oil of orange flowers, 2 drops; essence of musk, 15 minims; alcohol, 14 fl. oz. Mix.

Note. The essence of musk is made by rubbing 1 dr. of musk with 1 dr. of milk sugar to a uniform powder in a warm mortar, adding thereto 2½ fl. oz. of water, 5 fl. oz. of 90% alcohol, and ½ fl. dr. of liquor ammoniæ. The whole is digested eight days and filtered.

Holy Basil. Extract, of tonquin, 40 oz.; of vanilla, 50 oz.; of geranium, 40 oz.; S.V.R., 20 oz.; extract, of storax, 16 oz.; of musk, 4 oz.; of orange, 20 oz.; of cassie, 20 oz.; of jasmine, 20 oz.; of tuberose, 20 oz.; of rose, 25 oz.; Ol. citron, 2 dr.; Ol. bergamot, 1 dr.; otto of rose, 45 minims.

Hovenia, Essence of. S.V.R., 40 oz.; essence of lemon, ½ oz.; Ol. French geranium, 30 minims; Ol. cloves, 30 minims; Ol. neroli, 10 minims; otto of rose, 40 minims; aq. rosæ, 10 oz.; extract of musk, 1 oz.

Honeysuckle. Extract, of rose, 20 oz.; of violet, 20 oz.; of tuberose, 20 oz.; of cassie, 20 oz.; of vanilla, 7 oz.; of storax, 4 oz.; of musk, 1 oz.; Ol. neroli, 12 minims; Ol. almonds, 7 minims; otto of rose, 10 minims.

Jockey Club. Essence of jasmine, 4 oz.; otto of rose, 1 dr.; essence, of musk, 1 oz.; of tonquin, 2 oz.; S.V.R., 9 oz.

Kew Gardens Bouquet. Extract, of neroli, 20 oz.; of cassie, 10 oz.; of tuberose, 10 oz.; of jasmine, 10 oz.; of violet, 10 oz.; of rose, 10 oz.; of geranium, 10 oz.; of musk, 3 oz.; of civet, 1 oz.

Leap-year Bouquet. Extract, of tuberose, 20 oz.; of jasmine, 20 oz.; of rose, 16 oz.; of santal, 10 oz.; of patchouly, 10 oz.; of verbena, 2 oz.; of civet, 2 oz.; rose triple, 6 oz.

Lebanon Cedar-wood, for the Handkerchief. Oil of cedar, 1 oz.; S.V.R., 25 oz.; rose triple, 6 oz.

Lime-tree Blossom Bouquet. S.V.R., 40 oz.; Ol. lign aloe, ½ oz.; extract, of rose, 10 oz.; of jasmine, 10 oz.; of orris, 10 oz.; of musk, 3 oz.; rose triple, 10 oz.; essence of lemon, 30 minims.

Moss Rose. Extract, of rose, 40 oz.; of orange, 20 oz.; of violet, 20 oz.; of jasmine, 20 oz.; rose triple, 20 oz.; extract, of musk, 4 oz.; of ambergris, 3 oz.

Mousseline Bouquet. Extract, of maréchale, 20 oz.; of cassie, 10 oz.; of jasmine, 10 oz.; of tuberose, 10 oz.; of rose, 30 oz.; Ol. santal, 45 minims.

Narcissus. Otto of rose, 10 minims; extract, of tuberose, 60 oz.; of jonquille, 40 oz.; of violets, 10 oz.; of storax, 5 oz.; of musk, 1 oz.

Naval Bouquet. Rose triple, 8 oz.; extract, of rose, 10 oz.; of violet, 10 oz.; of jasmine, 10 oz.; of santal, 6 oz.; of vitivert, 6 oz.; of patchouly, 4 oz.; of verbena, 2 oz.

Oxford and Cambridge Bouquet. Rose triple, 10 oz.; extract, of jasmine, 10 oz.; of tuberose, 10 oz.; of cassie, 10 oz.; of vanilla, 5 oz.; of santal, 5 oz.; of violet, 20 oz.; of geranium, 4 oz.; of rose, 10 oz.; of patchouly, 3 oz.; of storax, 1 oz.; essence, of bergamot, 60 minims; of lemon, 40 minims; extract of musk, 3 oz.

Reseda. Extract, of orris, 120 oz.; of orange, 50 oz.; of cassie, 90 oz.; of jasmine, 40 oz.; of rose, 40 oz.; of violet, 20 oz.; of storax, 6 oz.; of musk, 4 oz.; aq. flor. aur., 10 oz.; aq. rosa, 10 oz.

Royal Hunt Bouquet. Rose triple, 10 oz.; extract, of rose, 10 oz.; of cassie, 7 oz.; of neroli, 5 oz.; of orange, 7 oz.; of orris, 6 oz.; of civet, 1½ oz.; of musk, ½ oz.; of tonquin, 7 oz.; Ol. citron, 60 minims; essence of lemon, 60 minims.

Sweet Pea, Essence of. Extract, of orange, 10 oz.; of tuberose, 18 oz.; of rose, 10 oz.; of violet, 7 oz.; of vanilla, 2 oz.; of storax, 1 oz.; otto of rose, 7 minims.

Tea Rose, Essence of. Rose triple, 15 oz.; extract, of rose, 25 oz.; of geranium, 20 oz.; of santal, 10 oz.; of neroli, 5 oz.; of orris, 5 oz.; of violet, 10 oz.; of storax, 3 oz.

Volunteer's Garland. S.V.R., 20 oz.; Ol. neroli, ¼ oz.; Ol. lavandula ang., ¼ oz.; Ol. bergamot, ¼ oz.; Ol. French geranium, 60 minims; Ol. cloves, 10 minims; otto of rose, 50 minims; extract, of orris, 20 oz.; of jasmine, 7 oz.; of cassie, 7 oz.; of violet, 10 oz.; of musk, 2 oz.; of civet, ½ oz.; of ambergris, ½ oz.

White Lilac, Essence of. Extract, of tuberose, 20 oz.; of rose, 10 oz.; of orange, 7 oz.; Ol. almonds, 5 minims; extract civet, ½ oz.

Yellow Roses. Rose triple, 16 oz.; extract, of rose, 30 oz.; of tonquin, 5 oz.; of tuberose, 30 oz.; of violet, 10 oz.; verbena, 4 oz.; of musk, 3 oz.

Ylang-Ylang Essence. 1. Take of alcohol, 90%, 8 fl. oz.; oil of ylang-ylang, 8 minims; otto of rose, 4 minims; oil of orange flowers, 2 minims; vanillin, 1 gr.; tincture of tolu, 2 oz.; rose water, 1 oz. Digest together several days and filter through carbonate of magnesia.

Note. The tincture of tolu is made by digesting 1½ oz. of tolu balsam with 15 oz. of 85% alcohol for 5 days, with frequent agitation, allowing to stand, and filtering.

2. Take of alcohol, 90%, 8 fl. oz.; oil of ylang-ylang, 8 minims; oil of orange flowers, 2 minims; rose water, 6 fl. dr. Mix.

Eaux de Cologne. The author of the first on the list was awarded the prize of a free trip to Paris by Messrs Gosnell.

Prep. 1. Essence, of bergamot, 2 dr.; of lemon, 1 dr.; oil, of neroli, 20 drops; of origanum, 6 drops; of rosemary, 20 drops; S.V.R. treble-distilled, 1 pint; orange-flower water, 1 oz.

2. Oil, of bergamot, 150 minims; of lemon, 60 minims; of Portugal, 50 minims; of neroli, 20 minims; of petit-grain, 10 minims; of lavender (Eng.), 20 minims; of rosemary, 10 minims; of melissa, 5 minims; finest spirit, 30 oz.; rose-water, 14 dr.; orange-flower water, 14 dr.

3. Oil, of bergamot, 100 minims; of lemon, 50 minims; of Portugal, 30 minims; of petit-grain, 10 minims; of lavender, 20 minims; of rosemary, 15 minims; finest spirit, 30 oz.; rose-water, 9 dr.; orange-flower water, 9 dr.; distilled water, 9 dr.

The above formulæ are for preparing the perfume by the cold method. The proper plan is to add the oils to the spirit in the order in which they are set down, shake well, and set aside for a few days, shaking occasionally before adding the waters. After these are added, again set aside for some time, and, if not perfectly clear, filter.

4. Oil, of Portugal, 180 minims; of bergamot, 180 minims; of cedrat, 120 minims; of lemon, 120 minims; of neroli, 190 minims; of petit-grain, 120 minims; of rosemary, 240 minims; of lemon, 240 minims; finest spirit, 10 oz.

This formula is for the preparation of a concentrated eau de Cologne, which will bear dilution with 10 times its volume of fine spirit. Dissolve the oils in 10 oz. of the spirit, and set aside for 14 days, shaking 4 times a day. Then distil the mixture twice, when the result will be 10 oz. of an exceedingly strong perfume, which improves in odour the longer it is kept, and is specially suited for exportation. It is of good odour when freshly diluted with spirit, but in this case also the odour improves on keeping.

5. Oil, of bergamot, 375 minims; of cedrat, 60 minims; of lemon, 60 minims; of lavender, 30 minims; of Portugal, 60 minims; of thyme, 4 minims; of neroli, 75 minims; of rosemary, 75 minims; finest spirit, 62 oz. Mix and distil, then add to the distillate 2½ oz. of melissa water and 5 oz. orange-flower water, and distil again. The product is a very fine eau de Cologne, the formula dating as far back as 1821, but the following goes even farther, viz. to 1813.

6. Oil, of neroli, 10 minims; of lemon, 40 minims; of bergamot, 50 minims; of cedrat, 15 minims; of lavender, 18 minims; of rosemary, 10 minims; melissa water, 4½ oz.; finest spirit, 30 oz. Dissolve the oils in the spirit contained in a retort, giving the mixture a thorough shaking, then close the retort and keep the contents just warm for 48 hours, whereby perfect blending of the oils with the spirit is ensured. Then place it for 24 hours in a cool place, after which filter it through paper until it is obtained per-

fectly clear. With the filtrate mix the melissa water.

7. Oil, of bergamot, ½ oz.; of lemon, ½ oz.; of rosemary, 1 dr.; of citronella, 20 drops; of neroli, 1 dr.; rectified spirit, 32 oz.

8. Oil, of bergamot, 12 minims; of lemon, 12 minims; of neroli, 12 minims; of orange peel, 12 minims; of rosemary, 12 minims; cardamom seeds (in powder), 60 gr.; rectified spirit, 1 pint.

9. Oil, of lemon, 2 dr.; of neroli, 2 dr.; of orange peel, 2 dr.; English oil of lavender, 20 minims; English oil of rosemary, 10 minims; oil of bergamot, 7 dr. 30 minims; rectified spirit, 66½ oz.; orange-flower water, 17½ oz.

10. Essential oils of bergamot, lemon, cedrat, of each, 100 grms.; essential oils of lavender, neroli, rosemary, of each, 50 grms.; essential oil of cinnamon, 25 grms.; alcohol (90°), 12,000 grms.; alcoholate of balm (eau des carmes), 1500 grms.; spirit of rosemary, 1,000 grms. Dissolve the essential oils in the alcohol, add the two alcoholates, and let stand for 8 days. Distil in a sand-bath ⅞ of the mixture (Codex).

11. Siam benzoin, 15 gr.; oil, of lavender, 30 gr.; of rosemary, 15 gr.; of neroli, 80 gr.; of petit-grain, 80 gr.; of cedrat, 80 gr.; of Portugal, 160 gr.; of lemon, 150 gr.; of bergamot, 150 gr.; of rose geranium, 15 gr.; alcohol (95%), 68 fl. oz. The essential oils (all by weight) are dissolved, in the order given above, in the spirit, and then the finely powdered benzoin is added. Allow to stand with frequent agitation for at least 4 weeks; place in a still, add an equal quantity of water and distil over about 64 fl. oz., having previously collected and set aside the first ounce. Allow the distillate (64 fl. oz.) to stand another 4 weeks in a glass vessel which is exposed to sunlight or diffused daylight; the longer the water is kept the better it is.

12. Oil, of bergamot, 14 minims; of lemon, 34 minims; of petit-grain, 20 minims; of neroli bigaradia, 7 minims; of rosemary, 14 minims; spirit of wine (genuine), 12½ fl. oz.; mix.

Note. The addition of a little amber or musk essence makes the perfume more persistent. This eau de Cologne obtained the gold medal at the Sydney Exhibition.

Lavender Water. *Prep.* 1. Oil of lavender, 4½ oz.; tonquin beans, 7 oz.; oil of bergamot, 2 oz.; otto of rose, 160 minims; musk, 32 gr.; rectified spirit, 3 galls. (old measure of 128 oz. to the gall.).

2. Ambergris, 12 gr.; oil of bergamot, 6 oz.; English oil of lavender, 1½ oz.; oil of cloves, 6 dr.; English oil of santal, 4 dr.; otto of rose, 4 dr.; musk, 4 dr.; rectified spirit, 3 galls. (old measure, 128 oz. to the gall.)

3. English oil of lavender, 16 oz.; oil of bergamot, 4 oz.; rectified spirit of wine, 4½ galls. (old measure); distilled water, 5 pints (old measure) musk, 30 gr.; sugar (powdered lump), 1 oz.; orange-flower water, 5 pints (old measure).

4. English oil of lavender, 3 dr.; essence of ambergris (1 dr. in 16 oz.), 1 dr.; oil of bergamot, ½ dr.; orange-flower water, 1 oz.; rose-water, 1 oz.; rectified spirit, 1 oz.

5. English oil of lavender, 2½ dr.; oil of bergamot, ½ dr.; musk, 10 gr.; rectified spirit, 16 oz.

6. English oil of lavender, 2 oz.; foreign oil of lavender (good), 1 oz.; oil of bergamot, 1 oz.; essence of musk (1 dr. in 16 oz.), 6 dr.; essence of tonquin beans (1 in 10) 1 oz.; rectified spirit, 96 oz.

7. English oil of lavender, 3 oz.; oil of bergamot, 1½ oz.; essence of tonquin beans (1 in 10), 1 oz.; triple rose-water, 12 oz.; rectified spirit, 80 oz.

8. Musk, 40 gr.; oil of bergamot, 1 oz.; English oil of lavender, 5 oz.; French essence of millefleur, 8 oz.; pulv. rad. iridis, 2 oz.; otto of roses, 20 minims; essence of ambergris (1 dr. in 16 oz.), 2 oz.; distilled water, 40 oz.; rectified spirit, 6 pints.

9. English oil of lavender, ½ oz.; oil of bergamot, 30 minims; essence of tonquin bean (1 in 10), 30 minims; essence of musk, (1 dr. in 16 oz.), 60 minims; rectified spirit, 16 oz.

10. English oil of lavender, 10 dr.; oil of bergamot, 1½ dr.; essence of musk (1 dr. in 16 oz.), ½ oz.; oil of neroli, 4 drops; oil of geranium, 6 drops; English oil of sandal-wood, 7 drops; rectified spirit, 30 oz.; water, 30 oz.

11. Oil, of lavender, 4 dr.; of bergamot, ½ dr.; of lemon, ½ dr.; musk, 2 gr.; rose-water, 2 oz.; rectified spirit, 18 oz.

12. English oil of lavender, ½ oz.; oil of neroli, 10 drops; essence of ambergris (1 dr. in 16 oz.), 1 oz.; essence of musk (1 dr. in 16 oz.), 1 oz.; rectified spirit, to make 30 oz.

13. Alcohol (pure, 90%), 25 fl. oz.; oil of lavender (Mitcham), 6 dr.; essence of musk, 3 dr.; essence of amber, 1 dr.

Perfumes, Ace'tic. See VINEGAR.

PERIGEE. In astronomy that point in the orbit of the moon where she is nearest to the earth, or the point in the earth's orbit where our globe is nearest to the sun. It is also used as a general term to denote the least distance of a body from the earth.

PERIHELION. That point in the orbit of a planet or comet which is nearest to the sun.

PERIODIC ACID. *Syn.* HYDRIC PERIODATE. (HIO_4). 1. By passing a current of chlorine gas through a solution of sodic iodate, containing caustic soda, in the proportion of 3 atoms of the latter to 1 atom of sodic iodate. The hydrated basic sodic periodate, which crystallises out, is dissolved in diluted nitric acid, and precipitated by the addition of argentic nitrate; a normal argentic periodate crystallises as the liquid cools, and this salt being treated with water, is decomposed into a basic argentic periodate, which is insoluble, and periodic acid, which is dissolved. By evaporating the solution, the periodic acid may be obtained in deliquescent, oblique, rhombic prisms, which are somewhat soluble in alcohol and in ether.

2. From perchloric acid by the action of iodine. See IODINE.

PERISTAL'TIC PERSUA'DERS. See PILLS (Kitchener's).

PER'MANENT WHITE. See BARIUM (Sulphate) and WHITE PIGMENTS.

PERNAMBU'CO WOOD. *Syn.* PEACH WOOD. The wood of *Cæsalpinia echinata*. It constitutes the paler variety of Brazil wood used by the dyers.

PERONEA COMPARANA or **COMARIANA. THE STRAWBERRY MOTH.** Injury is occasioned to strawberry plants by the caterpillars of this little moth fastening up their leaves and blossoms, and the small immature strawberries later on, with their webs, and feeding upon them.

A report was received from Perthshire that strawberry plants were suffering in this manner, and in some extensive beds near Southampton much loss was sustained from the action of these caterpillars, and in other parts of the country. Complaints have been received from a few places in Kent with regard to this attack. On the other hand two large growers, near London, in this county say that they have not noticed that any trouble has been caused by this insect. Most gardeners will remember to have seen its work in their strawberry beds, though perhaps ignorant of the origin of the evil.

Life History. It belongs to the family *Tortricidæ* of the *Lepidoptera*, and the genus *Peronea*, so called, as Westwood says, from the Greek word signifying a button, because the typical species of this genus have a tuft of raised scales upon each of their fore wings, resembling a button.

There is some little doubt as to the exact name of this strawberry moth. Miss Ormerod styles it *comariana*, or *comparana*. Dr Ellis, whom Miss Ormerod quotes, terms it *comariana*, probably after the weed which bears a small inconspicuous pseudo-fruit rather like a small strawberry, called *comarum palustre*. Neither Stephens nor Westwood speaks of this species, though Hübner styles a somewhat similar moth *comparana*, and Zeller *comariana*. It is most probable that it is a distinct species, and might be termed *fragaria*.

In colour the moth is brownish with yellowish wings, having a large patch of black upon each fore wing, whose expanse is about nine lines. The length of its body is five lines. It is presumed, for unfortunately the life history of this moth is most incomplete, that the perfect insect emerges from the chrysalis state, which is passed in the ground, in May, and places eggs upon the bases of the strawberry plants, from which large green caterpillars come, and drawing the leaves together with their webs feed upon them in snug seclusion.

Prevention. Digging quicklime in between the rows of strawberry plants in the autumn when the runners and leafage have been cleared away, would kill many of the caterpillars, if done after an attack in the preceding summer.

Good cultivation between the rows is also desirable when the fruit is picked. This may be done by hand, or by horses, as the plants are now set wide enough apart by modern cultivators for horse hoeing.

Remedies. Women and children, the future strawberry pickers, might much check the caterpillars by cutting off the infested leaves with scissors or shears, where they could do this without injury to the large fruit (Report on ' Insects Injurious to Crops,' by Charles Whitehead, Esq., F.Z.S.).

PERRY. *Syn.* PYRACEUM, L. A fermented liquor prepared from pears in the same way as cider is from apples. The red rough-tasted sorts are principally used for this purpose. The best

perry contains 9% of absolute alcohol; ordinary perry from 5 to 7%.

Perry is a very pleasant-tasted and wholesome liquor. When bottled 'champagne fashion,' we have seen it frequently passed off for champagne without the fraud being suspected.

PERSIAN BERRIES. See FRENCH BERRIES.

PERSPIRATION. The liquid or vapour secreted from the surface of the body by the sweatglands, small tubular glands situated deeply in the true skin, and communicating with the surface by means of a spiral tube, the opening of which on the skin (pore) may readily be seen by the use of an ordinary hand-magnifier. The number of these glands is very great, and has been estimated at two-and-a-half millions, with a total secreting surface of about twelve hundred square yards. Under ordinary circumstances the water excreted by these glands passes off in the form of vapour, and we are unconscious of its presence, the perspiration is then said to be "insensible;" but under the influence of great heat, or severe bodily labour, or even of the emotions, terror, pain, &c., the glands secrete so rapidly that the fluid accumulates in drops on the surface of the skin, such perspiration is said to be "sensible." The total amount lost by the skin under ordinary circumstances is about three pounds weight per diem; but, as the result of severe exertion, this amount may be doubled—the former loss is replaced by the water of the food and drink almost as rapidly as it takes place; the latter, only partially, and any great loss of weight as the result of bodily labour generally requires a day or two for its complete reparation. The amount of such loss is an excellent guide to the condition of the individual, and one of the great objects of "training" is to bring the organism to such a state that the excessive loss of water through the skin shall not occur.

The secretion of the sweat is one of the most important functions of the skin, as by it the temperature of the body is to a large extent regulated and maintained at a constant level (about 98·4° F.). This is accomplished by the incessant evaporation of the sweat from the surface of the body. The sweat glands are abundantly supplied with minute capillary blood-vessels, from which they derive the material of their secretion. These blood-vessels are under the control of the nervous system, and dilate or contract under suitable stimuli, thus increasing or diminishing the supply of blood to the gland. Heat and cold are such stimuli, the former causing the vessels to dilate, and the supply of blood being greatly augmented, the glands secrete more freely, the temperature is reduced by the increased evaporation, and the effect of excessive heat is thus neutralised. Cold produces the converse effect, causing the vessels to contract, diminishing the blood-supply to the skin and therefore the activity of the glands, the diminished evaporation preventing too great a loss of heat from the body. The importance of the sweat cannot be exaggerated; and it is a matter of great consequence that the skin should be kept clean, and every care taken to prevent the blocking of the pores of the sweat glands. The extra quantity of blood which finds its way to the vessels of the skin as the result of a warm or hot bath is of great therapeutic value, enabling us as it does, by very simple means, to relieve the tension on the internal organs, heart, brain, &c., and in many cases thereby save life.

PERU'VIAN BALSAM. See BALSAM OF PERU.

PERU'VIAN BARK. See CINCHONA.

PES'SARY. *Syn.* PESSUM, PESSARIUM, L. An instrument made of caoutchouc, gutta percha, box-wood, or ivory, inserted into the vagina to support the mouth and neck of the uterus. They are variously formed, to meet the prejudices of the individual or the necessities of the case. The cup, conical, globe, and ring pessaries (pessi) are those best known.

Pessaries prepared by the pharmacist are of a conical shape, weighing from ¼ to 2 drachms, cast in gun-metal moulds. They have, as a basis, either pure cocoa-butter, or what is called ' Gelatine Moss,' made as follows: Gelatine, 1 oz., immerse in water a few minutes, pour off the water, and in half-an-hour dissolve in 4 oz. of glycerine. Those most commonly used are : Tannic acid, 10 gr.; glycerine, 30 gr.; atropine, 1/40 gr.; belladonna extract, ¼ to 1 gr.; cocaine, ½ gr.; iodoform, 5 to 10 gr.; potassium bromide or iodide, 10 gr.; zinc oxide, 10 gr.

The different formulæ are given below :

Pessary, Alum. *Syn.* PESSUS ALUMINIS. Alum, catechu, wax, of each 1 dr.; lard, 5½ dr.

Pessary, Belladonna. *Syn.* PESSUS BELLADONNE. Extract of belladonna, 10 gr.; wax, 22½ gr.; lard, 1½ dr.; in each pessary.

Pessary, Mercurial. *Syn.* PESSUS HYDRARGYRI. Strong mercurial ointment, ½ dr.; wax, ½ dr.; lard, 1 dr. Mix.

Pessary, Lead. *Syn.* PESSUS PLUMBI. Acetate of lead, 7½ gr.; white wax, 22½ gr.; lard, 1½ dr.

Pessary, Iodide of Lead. *Syn.* PESSUS PLUMBI IODIDI. Iodide of lead, 5 gr.; wax, 25 gr.; lard, 1½ dr.

Pessary, Tannin. *Syn.* PESSUS TANNINI. Tannin, 10 gr.; wax, 25 gr.; lard, 1½ dr.

Pessary, Zinc. *Syn.* PESSUS ZINCI. Oxide of zinc, 15 gr.; white wax, 22½ gr.; lard, 1½ dr.

PESTILENCE. See PLAGUE.

PESTILENTIAL DISEASES. All those diseases which are epidemic and malignant and assume the character of a plague. See CHOLERA, &c.

PETONG'. Same as *packfong.*

PET'ROLENE. The pure liquid portion of mineral tar. It has a pale yellow colour, a penetrating odour, and a high boiling point; is lighter than water, and is isomeric with the oils of turpentine and lemons. In its general proportions it resembles rectified mineral naphtha.

PETRO'LEUM. *Syn.* ROCK OIL, LIQUID BITUMEN, OIL OF PETRE; OLEUM PETRÆ, BITUMEN LIQUIDUM, L. PETROLEUM is an oil found oozing from the ground or obtained on sinking wells in the soil. To a limited extent it is met with in most countries of Europe and in the West India islands, but occurs in abundance in the district of the Caucasus, in Pennsylvania and other parts of the United States, in Canada and Burmah. It varies in colour from slight yellow to brownish black, in consistence from a

thin mobile liquid to a fluid as thick as treacle, in specific gravity from 800 to 1100 (water being 1000), and is either clear and transparent or turbid and opaque. Petroleum is essentially a volatile oil, and when submitted to distillation yields gases homologous with light carburetted hydrogen of marsh-gas, liquids of similar constitution, and solid paraffin-like bodies. Commercially petroleum is distilled so as to yield petroleum-spirit or mineral naphtha, which is used as a substitute for turpentine and for burning in sponge-lamps and costermongers' barrow-lamps; petroleum oil, which is used all over the world as mineral lamp-oil for illuminating purposes; and a heavy oil employed for lubricating machinery. The value of a sample of rock-oil is roughly determined by distilling a weighed quantity in a small glass retort and weighing the products. The petroleum or middle product must be of such a character as to have a specific gravity not higher than 810 or 820, and to contain so little petroleum spirit that it only evolves inflammable vapour when heated to 100° Fahr. in the manner prescribed in the Petroleum Act, 1871 (see *below*). Any petroleum product or mineral oil which will not stand this test, and which is kept in larger bottles than one pint, and in larger total quantity than three gallons, cannot be stored or sold except by licence of the local authorities.

Directions for Testing Petroleum to ascertain the temperature at which it gives off inflammable vapour.

The vessel which is to hold the oil shall be of thin sheet iron; it shall be two inches deep and two inches wide at the opening, tapering slightly towards the bottom; it shall have a flat rim, with a raised edge one quarter of an inch round the top; it shall be supported by this rim in a tin vessel four inches and a half deep and four and a half inches in diameter; it shall also have a thin wire stretched across the opening, which wire shall be so fixed to the edge of the vessel that it shall be a quarter of an inch above the surface of the flat rim. The thermometer to be used shall have a round bulb about half an inch in diameter, and is to be graduated upon the scale of Fahrenheit, every ten degrees occupying not less than half an inch upon the scale.

The inner vessel shall be filled with the petroleum to be tested, but care must be taken that the liquid does not cover the flat rim. The outer vessel shall be filled with cold, or nearly cold water; a small flame shall be applied to the bottom of the outer vessel, and the thermometer shall be inserted into the oil so that the bulb shall be immersed about one and a half inches beneath the surface. A screen of pasteboard or wood shall be placed round the apparatus, and shall be of such dimensions as to surround it about two thirds and to reach several inches above the level of the vessels.

When heat has been applied to the water until the thermometer has risen to about 90° Fahr., a very small flame shall be quickly passed across the surface of the oil on a level with the wire. If no pale blue flicker or flash is produced, the application of the flame is to be repeated for every rise of two or three degrees in the thermometer.

When the flashing-point has been noted, the test shall be repeated with a fresh sample of the oil, using cold, or nearly cold water as before; withdrawing the source of heat from the outer vessel when the temperature approaches that noted in the first experiment, and applying the flame test at every rise of two degrees in the thermometer. See NAPHTHA, OILS (Mineral), &c.

PEWTER. This is an alloy of tin and lead, or of tin with antimony and copper. The first only is properly called pewter. At least three varieties are known in the trade :—

Prop. 1. (PLATE PEWTER.) From tin, 79% ; antimony, 7% ; bismuth and copper, of each, 2% ; used to make plates, teapots, &c. Takes a fine polish.

2. (TRIFLE PEWTER.) From tin, 79% ; antimony, 15% ; lead, 6% ; as the last. Used for minor articles, syringes, toys, &c.

3. (LEY PEWTER.) From tin 80% ; lead, 20%. Used for measures, inkstands, &c.

Obs. According to the report of the French commission, pewter containing more than 18 parts of lead to 82 parts of tin is unsafe for measures for wine and similar liquors, and, indeed, for any other utensils exposed to contact with our food or beverages. The legal sp. gr. of pewter in France is 7·764; if it be greater it contains an excess of lead, and is liable to prove poisonous. The proportions of these metals may be approximately determined from the sp. gr. ; but correctly only by an assay for the purpose. Britannia metal, Queen's metal, &c., are varieties of pewter much used for making teapots, cream-jugs, &c. Articles of pewter used to be cast in moulds, but the process of 'spinning' is now more commonly resorted to ; this consists in bringing the sheet of pewter against a rapidly revolving tool by which it is gradually fashioned. See BRASS, GERMAN SILVER, LEAD, and TIN.

PHÆDON BETULÆ (Linn.). THE MUSTARD BEETLE. "Black Jack" is the name given to this destructive little beetle in the Fen district, where it does much harm to white and brown mustard crops. It is known pretty generally where mustard plants are cultivated for seed, as well as where turnip seeds of all kinds are produced. The mustard plant, however, seems to be its chief attraction. Kohl rabi and thousand-headed kale are also often much injured by it, but it draws a line at mangel wurzel plants, and leaves them uninjured. It has been known as infesting cruciferous plants for a long while. Curtis states that he often found it upon turnip leaves ('Farm Insects,' by J. Curtis, 1859), and Professor Westwood gave an account of it in the 'Gardeners' Chronicle,' in 1884. Without doubt this beetle has been mistaken for the common turnip flea, *Phyllotreta nemorum*, as it is nearly the same size and of a somewhat similar colour, though the mustard beetle does not spring or jump like it, and its injury to plants is almost identical, at least in the early stages of the attack. Those who know this particular beetle, the *Phædon betulæ*, have taken notes of its habits, relate that it advances in battalions and invades fields of mustard, rape, turnips, and kohl rabi, clearing off the leaves, and leaving nothing but the stems and stalks. Lately the onslaughts

of this beetle have increased so much that the Seeds and Plants Diseases Committee of the Royal Agricultural Society of England advised the Council to issue a circular to agriculturists, inviting information to be sent to Miss Ormerod, the Consulting Entomologist of the society, to prepare for publication. This will contain interesting and valuable facts as to the habits of this insect and means of prevention and remedies.

In the neighbourhood of Peterborough, Whittlesea, Ely, Wisbech, and other Fen districts, where mustard seed is extensively grown, infinite harm has been caused. One grower of rape seed estimated his losses from this insect at £1,000.

Growers of turnip, rape, and mustard seeds in Romney Marsh, in Kent, and near Sandwich, in Kent, and in other parts of Kent where these seeds are grown, have noticed from time to time the ravages of a beetle upon these crops, but they had considered that it was a species of flea beetle, and were surprised to find that it belonged to an altogether different family.

It is not known in America. It is common in France, and destructive there to turnip and mustard plants, and especially to rape plants which are extensively cultivated for oil. In Germany it is well known. Calwer says it is found in Northern Europe, and in that part of Europe where the climate is moderate (C. G. Calwer's 'Käferbach'). Kaltenbach says the spring larvæ are found in May and June, and there is a second generation later on ('Die Pflanzenfeinde,' von J. H. Kaltenbach).

It appears as if there were more than one generation in this country also, as beetles are to be seen throughout the summer.

Life History. The *Phædon betulæ* belongs to the family *Chrysomelidæ*, of the division *Phytophaga*, of the order *Coleoptera*, and to the genus *Phædon.*

The beetle is about one and a half lines long, of an oval shape, and in colour is of various tints of dark blue, dark violet, and dark green, with black legs and antennæ, having large wings. The beetles pass the winter in the perfect state in the ground, in the stems of plants, in pieces of straw, dead charlock stems, pieces of dead mustard stems, and under grasses and weeds. They attack mustard, turnip, rape, kohl rabi, and thousand-headed kale plants, directly their leaves appear, and the females deposit eggs upon these, from which larvæ, or grubs, come in a few days and live on the leaves for a short time and, descending to the earth, become chrysalides. They change again to beetles in about a fortnight, according to Kaltenbach, which go forth to destroy in countless swarms.

The larva, or grub, a little more than two lines long, has six feet, and is of a smoky yellow colour, as Curtis has it, with black spots.

Prevention. After these beetles have been abundant, all pieces of stems of mustard, rape, and turnip plants should be got off the land, or ploughed in very deeply. Weeds, especially charlock, *Sinapis arvensis*, and grasses should be kept from the land and the outsides of fields, as the beetles shelter under them during the winter. Ditches, water-courses, and drains should be kept cleared out, because these beetles like such places, and shelter there during the winter.

When mustard, rape, and other cruciferous plants are grown for seed, it is advisable to burn the stalks, as the beetles are harboured in these. It is stated that the beetles have been found alive in the seed after it has been stored in sacks for two years; therefore it is desirable that seed from fields that have been infested should be examined before it is sown, and run through the seed-winnowing machine with fine screens.

Remedies. When young mustard plants intended for seed are infested with this beetle in the spring, top dressings of lime, or of soil, or of finely triturated guano, should be put on early in the morning, before the dew is off the plants. Unless the beetles are checked and discomfited at the beginning of their attack, they will increase and multiply with astonishing rapidity, and concentrate their energies upon the plants when these require all their power to develop seed. The same remarks apply to rape and turnip plants for seed, which should be watched carefully, and early endeavours made to stop their advance.

It does not appear that there are any remedies that can be applied with much advantage, when mustard, rape, or turnip plants for seed are attacked by this beetle in the late stages of their growth. ('Reports on Insects Injurious to Crops,' by Chas. Whitehead, Esq., F.Z.S.)

PHARAOH'S SERPENTS. 1. The chemical toy sold under this name consists of the powder of sulphocyanide of mercury made up in a capsule of tin foil in a conical mass of about an inch in height.

Ignited at the apex an ash is protruded, long and serpentine in shape. The fumes evolved are very poisonous.

2. (NON-POISONOUS.) Bichromate of potassium, 2 parts; nitrate of potassium, 1 part; and white sugar, 3 parts. Pulverise each of the ingredients separately, and then mix them thoroughly. Make small paper cones of the desired size, and press the mixture into them. They will then be ready for use, but must be kept from light and moisture.

PHARMACY ACT. The following are the principal clauses of the Pharmacy Act of 1860 (31 and 32 Victoria cap. cxxi). We have separated and placed last, those provisions of the Act which relate to the sale of poisons:

Whereas it is expedient for the safety of the public that persons keeping open shop for the retailing, dispensing, or compounding of poisons, and persons known as chemists and druggists should possess a competent practical knowledge of their business, and to that end, that from and after the day herein named all persons not already engaged in such business should, before commencing such business, be duly examined as to their practical knowledge, and that a register should be kept as herein provided, and also that the Act passed in the 15th and 16th years of the reign of her present Majesty, intituled 'An Act for Regulating the Qualification of Pharmaceutical Chemists,' hereinafter described as the Pharmacy Act, should be amended: Be it enacted, by the Queen's most excellent Majesty, by and with the advice and consent of the Lords Spiritual and Temporal and Commons in this present Parliament assembled, and by authority of the same, as follows:

From and after the 31st day of December, 1868, it shall be unlawful for any person to sell or keep open shop for retailing, dispensing, or compounding poisons, or to assume or use the title 'Chemist and Druggist,' or chemist or druggist, or pharmacist, or dispensing chemist, or druggist, in any part of Great Britain, unless such person shall be a pharmaceutical chemist or a chemist and druggist, within the meaning of this Act, and be registered under this Act, and conform to such regulations as to the keeping, dispensing, and selling of such poisons as may from time to time be prescribed by the Pharmaceutical Society with the consent of the Privy Council (Clause 1).

Chemists and druggists within the meaning of this Act shall consist of all persons who at any time before the passing of this Act have carried on in Great Britain the business of a chemist and druggist in the keeping of open shop for the compounding of the prescriptions of duly qualified medical practitioners, also of all assistants and associates, who before the passing of the Act shall have been duly registered under or according to the provisions of the Pharmacy Act, and also of all such persons as may be duly registered under this Act (Clause 3).

All such persons as shall from time to time have been appointed to conduct examinations under the Pharmacy Act shall be, and are hereby declared to be, examiners for the purposes of this Act, and are hereby empowered and required to examine all such persons as shall tender themselves for examination under the provisions of this Act (see above), and every person who shall have been examined by such examiners, and shall have obtained from them a certificate of competent skill, and knowledge, and qualification, shall be entitled to be registered as a chemist and druggist under this Act, and the examination aforesaid shall be such as is provided under the Pharmacy Act for the purposes of a qualification to be registered as assistant under that Act, or as the same may be varied from time to time by any bye-law to be made in accordance with the Pharmacy Act as amended by this Act, provided that no person shall conduct any examination for the purposes of this Act until his appointment has been approved by the Privy Council (Clause 6).

No name shall be entered in the register, except of persons authorised by this Act to be registered, nor unless the registrar be satisfied by proper evidence that the person claiming is entitled to be registered; and any appeal from the decision of the registrar may be decided by the council of the Pharmaceutical Society; and any entry which shall be proved to the satisfaction of such council to have been fraudulently or incorrectly made may be erased from or amended in the register, by order in writing of such council (Clause 12).

The registrar shall, in the month of January in every year, cause to be printed, published, and sold, a correct register of the names of all pharmaceutical chemists, and a correct register of all persons registered as chemists and druggists, and in such registers, respectively the names shall be in alphabetical order, according to the surnames, with the respective residences, in the form set forth in schedule (B) to this Act, or to the like effect, of all persons appearing on the register of pharmaceutical chemists, and on the register of chemists and druggists, on the 31st day of December last preceding, and such printed registers shall be called 'The Registers of Pharmaceutical Chemists and Chemists and Druggists,' and a printed copy of such registers for the time being, purporting to be so printed and published as aforesaid, or any certificate under the hand of the said registrar, and countersigned by the president or two members of the council of the Pharmaceutical Society, shall be evidence in all courts and before all justices of the peace and others, that the persons therein specified are registered according to the provisions of the Pharmacy Act or of this Act, as the case may be, and the absence of the name of any person from such printed register shall be evidence, until the contrary shall be made to appear, that such person is not registered according to the provisions of the Pharmacy Act or of this Act (Clause 13).

From and after the 31st day of December, 1868, any person who shall sell or keep an open shop for the retailing, dispensing, or compounding poisons, or who shall take, use, or exhibit the name or title of chemist and druggist, or chemist or druggist, not being a duly registered pharmaceutical chemist, or chemist and druggist, or who shall take, use, or exhibit the name or title pharmaceutical chemist, pharmaceutist, or pharmacist not being a pharmaceutical chemist, or shall fail to conform with any regulation as to the keeping or selling of poisons, made in pursuance of this Act, or who shall compound any medicines of the British Pharmacopœia, except according to the formularies of the said Pharmacopœia, shall for every such offence be liable to pay a penalty or sum of £5, and the same may be sued for, recovered, and dealt with in the manner provided by the Pharmacy Act for the recovery of penalties under that Act; but nothing in this Act contained shall prevent any person from being liable to any other penalty, damages, or punishment to which he would have been subject if this Act had not been passed (Clause 15).

Clauses of the Pharmacy Act relating to the sale of Poisons.

It shall be unlawful to sell any poison either by wholesale or retail, unless the box, bottle, vessel, wrapper, or cover in which such poison is contained be distinctly labelled with the name of the article and the word poison, and with the name and address of the seller of the poison; and it shall be unlawful to sell any poison of those which are in the first part of schedule (A) to this Act, or may hereafter be added thereto under section II of this Act, to any person unknown to the seller, unless introduced by some person known to the seller; and on every sale of any such article the seller shall, before delivery, make or cause to be made an entry in a book to be kept for that purpose, stating, in the form set forth in schedule (F) to this Act, the date of the sale, the name and address of the purchaser, the name and quantity of the article sold, and the purpose for which it is stated by the purchaser to be required, to which

entry the signature of the purchaser and of the person, if any, who introduced him, shall be affixed; and any person selling poison otherwise than is herein provided, shall, upon a summary conviction before two justices of the peace in England or the sheriff in Scotland, be liable to a penalty not exceeding £5 for the first offence, and to a penalty not exceeding £10 for the second or any subsequent offence; and for the purposes of this section the person on whose behalf any sale is made by any apprentice or servant shall be deemed to be the seller, but the provisions of this section, which are solely applicable to poisons in the first part of the schedule (A) to this Act, or which require that the label shall contain the name and address of the seller, shall not apply to articles to be exported from Great Britain by wholesale dealers, nor to sales by wholesale to retail dealers in the ordinary course of wholesale dealing, nor shall any of the provisions of this section apply to any medicine supplied by a legally qualified apothecary to his patient, nor apply to any article when forming part of the ingredients of any medicine dispensed by a person registered under this Act, provided such medicine be labelled in the manner aforesaid with the name and address of the seller, and the ingredients thereof be entered, with the name of the person to whom it is sold or delivered, in a book to be kept by the seller for that purpose, and nothing in this Act contained shall repeal or affect any of the provisions of an Act of the Session holden in the fourteenth and fifteenth years in the reign of her present Majesty, intituled 'An Act to regulate the Sale of Arsenic' (Clause 17).

SCHEDULE (A).
Part 1.

Arsenic and its preparations.
Prussic acid.
Cyanide of potassium and all metallic cyanides.
Strychnine and all poisonous vegetable alkaloids and their salts.
Aconite and its preparations.
Emetic tartar.
Corrosive sublimate.
Cantharides.
Savin and its oil.
Ergot of rye and its preparations.

Part 2.

Oxalic acid.
Chloroform.
Belladonna and its preparations.
Essential oil of almonds, unless deprived of its prussic acid.
Opium and all preparations of opium or of poppies.

By virtue and in exercise of the powers vested in the council of the Pharmaceutical Society of Great Britain, the said council do hereby resolve and declare that each of the following articles, viz.—

Preparations of prussic acid,
Preparations of cyanide of potassium and of all metallic cyanides,
Preparations of strychnine,
Preparations of atropine,
Preparations of corrosive sublimate,

Preparations of morphine,
Red oxide of mercury (commonly known as red precipitate of mercury),
Ammoniated mercury (commonly known as white precipitate of mercury),
Chloral hydrate and its preparations,
Nux vomica and its preparations,
Every compound containing any poison within the meaning of 'The Pharmacy Act, 1868,' when prepared or sold for the destruction of vermin,
The tincture and all vesicating liquid preparations of cantharides,
—ought to be deemed a poison within the meaning of the 'Pharmacy Act, 1868;' and also that of the same each of the following articles, vis.—
Preparations of prussic acid,
Preparations of cyanide of potassium and of all metallic cyanides,
Preparations of strychnine,
Preparations of atropine,
—ought to be deemed a poison in the first part of the schedule (A) to the said 'Pharmacy Act, 1868.'
And notice is hereby also given, that the said Society have submitted the said resolution for the approval of the Lords of Her Majesty's Council, and that such approval has been given.
By order,
ELIAS BREMRIDGE,
Secretary and Registrar of the Pharmaceutical Society of Great Britain. .

And whereas the council of the Pharmaceutical Society of Great Britain did, on the 17th day of November, 1877, resolve and declare in the words following :
" That by virtue and in exercise of the powers vested in the council of the Pharmaceutical Society of Great Britain, the said council does hereby resolve and declare that *Chloral Hydrate and its preparations* ought to be deemed poisons within the meaning of the 'Pharmacy Act, 1868,' and ought to be deemed poisons in the second part of the schedule (A) of the said 'Pharmacy Act, 1868.' "
And whereas the said Society have submitted the said resolution for the approval of the Privy Council, and the Lords of the Privy Council are of opinion that the said resolution should be approved.
Now, therefore, their Lordships are hereby pleased to signify their approval of the said resolution.   C. L. PEEL.

Further information on the laws affecting pharmacy will be found in the Calendar of the Pharmaceutical Society.
Tardieu states that of late years the criminal administration of phosphorus has increased considerably in France. For example, from 1851 to 1872, in 793 cases of poisoning, 287 or 36·2% were due to arsenic, and 267 or 31·1% to phosphorus; whilst in the years 1872 and 1874, in 141 criminal poisonings by arsenic and phosphorus, only 74 were due to arsenic. The explanation of these facts may reasonably be ascribed to the much greater facility with which phosphorus, in the form of matches or vermin pastes, can be procured than arsenic.

PHENACETINE. *Syn.* PHENACETINUM PARA-ACETPHENITIDIN. $C_{10}H_{13}NO_2$. Is an acetyl compound of phenitidin and analogous to antifebrin (acetanilide). It was first prepared by Dr O. Hinsberg, of Elberfeld, who together with Professor Kast submitted it to physiological investigation as to its antipyretic properties.
According to Dr Koller of Vienna who has made extensive experiments with this new body, phenacetin is undoubtedly an antipyretic, but action takes place less promptly but lasts longer, than in the case of other antipyretics. He usually administered it in doses of from 4 to 7 gr. and found that single large doses were more serviceable than successive small ones. He found that it was not followed by any disagreeable after-effects.
Mr Grenfell, in the 'Practitioner' for May, 1888, gives some interesting detailed accounts of the use of phenacetin in cases of pyrexia. " The action of the drug," he says, " begins within half an hour after administration, the patient generally prespires freely and feels drowsy. Sleep often follows, pain is relieved, while the patient always says that he feels more comfortable after it." He finds the most suitable dose for an adult is about 8 gr. He has used it extensively both as an antipyretic and as an analgesic in neuralgia with good results. Phenacetin is in white shining laminar crystals, tasteless and without smell, only slightly soluble in water and glycerin, readily soluble in hot alcohol. It is most easily administered in the form of capsules.—*Dose*, 5 to 15 gr.
PHENOL. C_6H_6O. See CARBOLIC ACID.
PHENOL SODIQUE. Mr Beringer found that the following formula yielded a preparation very similar :—Coal tar, 2 troy oz.; soda, 190 gr.; water, sufficient to make 16 fl. oz. Dissolve the soda in 4 fl. oz. of water and warm, add the coal tar, and thoroughly agitate the mixture for a few minutes. Then add the remainder of the water, and set aside in a covered vessel in a warm place, frequently agitating for 7 days. Decant the aqueous solution, and filter through a moistened filter, washing the residue with sufficient water to make the finished product measure 16 fl. oz.
No. 2 phenol, 8 parts; caustic soda, 4 parts; water, 100 parts.
PHENYL. C_6H_5. The hypothetical compound radical of the phenyl-series. Carbolic acid is said to be its hydrate.
PHENYLAMINE. $C_6H_5H_2N$. Aniline is sometimes so named on account of its relation to the phenyl series.
PHIALS. The ordinary green moulded phials used by the pharmaceutist are made of a glass obtained from common river sand and soapboilers' waste. In the manufacture of the glass for the white phials purer materials (and these as free from iron and alumina as possible) are used. Decolourising agents are also employed. The following is given as the composition of a white glass for apothecaries' phials in 'Chemistry: Theoretical, Practical, and Analytical' (*Mackensie and Co.*) :
100 lbs. white sand.
30— 26 „ potash, impure.
       17 „ lime.
110—120 „ ashes.
      ·25—·5 lbs. binoxide of manganese-cullet.

Phials, Bologna. Small flasks or phials of unannealed glass, which fly to pieces when their surface is scratched by a hard body. Thus, if a small piece of flint be dropped into them they are shivered; whereas if a bullet be used they remain uninjured.

PHILO'NIUM. The ancient name of an aromatic opiate, reputed to possess many virtues, invented by Philo. See CONFECTION OF OPIUM.

PHILOS'OPHER'S STONE. *Syn.* LAPIS PHILOSOPHORUM, L. A wonderful substance, the discovery of which formed the day dreams of the alchemists. It was supposed to be capable of converting all the baser metals into gold, and of curing all diseases. Some of the alchemists appear to have laboured under the delusion that they had actually discovered it. The last of these enthusiasts was the talented and unfortunate Dr Price, of Guildford. Speaking of the age of alchemy, Liebig says:—" The idea of the transmutability of metals stood in the most perfect harmony with all the observations and all the knowledge of that age, and in contradiction to none of these. In the first stage of the development of science, the alchemists could not possibly have any other notions of the nature of metals than those which they actually held. . . . We hear it said that the idea of the philosopher's stone was an error; but all our views have been developed from errors, and that which to-day we regard as truth in chemistry may, perhaps, before to-morrow, be regarded as a fallacy."

PHILOSOPHIC CANDLE. An inflamed jet of hydrogen gas.

PHILOSOPHIC WOOL. Flowers of zinc.

PHIL'TRE. *Syn.* PHILTRUM, L. A charm or potion to excite love. The ancients had great faith in such remedies. Nothing certain is now known respecting their composition; but there is sufficient evidence that recourse was frequently had to them by the ancients, and that "their operation was so violent that many persons lost their lives and their reason by their means." The Thessalian philtres were those most celebrated (Juv., vi, 610, &c.). At the present day the administration of preparations of the kind is interdicted by law.

PHLORE'TIN. $C_{14}H_{14}O_5$. A crystallisable, sweet substance, formed along with grape sugar, when phloridzin is acted on by dilute acids.

PHLORID'ZIN. $C_{21}H_{24}O_{10}$. *Syn.* PHLORIZINE; PHLORIDZINUM, L. *Prep.* By acting on the fresh root-bark of the apple, pear, or plum tree, with boiling rectified spirit the spirit is distilled off, and the phloridzin crystallises out of the residual liquor as it cools.

Prop., &c. Fine, colourless, silky needles, freely soluble in rectified spirit and in hot water, but requiring 1000 parts of cold water for its solution; its taste is bitter and astringent. When its solution is boiled with a little dilute sulphuric acid or hydrochloric acid, it is changed into grape sugar and phloretin.

Phloridzin bears a great likeness to salicin. It is said to be a powerful febrifuge.—*Dose,* 3 to 15 gr.

PHOCE'NIC ACID. See DELPHINIC ACID.

PHŒNICINE. See INDIGO PURPLE.

PHONO'GRAPH. A review of the history of the art of recording and reproducing sound shows that Dr. Hooke, in 1681, exhibited some experiments before the Royal Society, demonstrating how musical notes and other sounds could be produced by means of toothed wheels rapidly rotated.

In 1854, Charles Bourseuil proposed to use two diaphragms connected by an electric wire, and, by speaking into one of them, reproduce the spoken sounds at any distance in the other. This idea was actually carried out by Philip Reis five years later.

The phonautograph was patented by Leon Scott in 1857; and Faber constructed a complicated speaking machine which pronounced a few words and sentences most unsatisfactorily. The complex mechanism by which this was effected was contrived upon the principles of the human organs of speech, for the machine possessed an india-rubber tongue and lips, and an artificial larynx, made out of a thin vibrating tube of ivory. Faber's automaton, although of much greater scientific interest than the automatic flute and flageolet players of Vancanson, the trumpeter of Drox, and similar exhibitions of curious workmanship, was, like these, only a mechanical curiosity, without any promise of a useful application.

In 1876 appeared the Bell Telephone, the first really good instrument for the transmission of speech.

In April, 1877, Charles Cros deposited a paper at the Academy of Science in Paris on 'A process of recording and reproducing audible phenomena,' in which he proposed to obtain tracings of sound-waves by means of a vibrating membrane. Then, by going over these tracings with a stylus attached to another membrane the sounds would be reproduced. Consequently, to M. Cros belongs the credit of having suggested a means of mechanically recording and reproducing spoken sounds.

Later in the year, Thomas Alva Edison realised this idea in his phonograph. It was described in a report to the "Times" on February 17th, 1878, and shortly afterwards it was exhibited for the first time by W. H. Preece, at the Royal Institution.

The following description and diagram will serve to illustrate the principle of the phonograph. B is a brass cylinder, through whose centre passes a metal shaft, the arms of which rest on upright supports, one of which is shown in the engraving. The arm of the shaft, obscured from view, corresponding in length with the part of it which is visible, is screw-turned, and it works in a nut bored out of the support. Attached to the screw-end of the shaft or axle is a crank C, by turning which a double movement, viz. a rotatory and a horizontal one, may be simultaneously imparted to the cylinder. Round the surface of the cylinder is cut a spiral groove corresponding in dimensions with the threads of the screw part of the shaft. Covering the whole of the cylinder is a sheet of tin-foil, which is secured to its edges by means of shell-lac varnish. In front of the cylinder, resting on a proper support, is a mouthpiece, A, at the bottom of which (the end nearest the cylinder) is a very thin plate or diaphragm of metal, and to this diaphragm is attached a steel point, or stylus, which when not in use does not touch the foil. Previously to using the apparatus this steel point has to be accurately adjusted opposite to that part of the foil lying over the spiral groove. If now the lips be applied to the

mouth-piece, and any sentence be spoken, the crank being at the same time turned, the vibrations imparted to the metal plate by the voice will cause the steel point to come into contact with that part of the foil overlying the groove in the cylinder, and to make on the foil a number of indentations, as it revolves, and is carried forward laterally before the mouth-piece. Furthermore, these indentations will be found to vary in depth and sectional outline according to the nature of the vibrations which have produced them; and, as experiment proves, are the specific and infallible caligraphy of those vibrations.

It might be said that at this point the machine has already become a complete phonograph or sound writer, but it yet remains to translate the remarks made. Now, by much practice, and the aid of a magnifier, it might be possible to read phonetically the marks made on the foil (Mr. Edison does not appear to have yet solved the problem of reading the phonograph record by sight. He states that, although a specific form exists for each articulated sound, the chief difficulties arise from the varying indentations or marks caused by the same sound. Amongst the circumstances giving rise to these results are: the same sound uttered by different people, the manner in which it is spoken, the distance of the mouth from the instrument, the force with which it is spoken, or the speed with which the barrel is rotated); but he saves us that trouble by literally making it read itself. The distinction is the same, as if, instead of perusing a book ourselves, we drop it into a machine, set the latter in motion, and behold! the voice of the author is heard repeating his own composi-

tion. The reading mechanism is nothing but another diaphragm, held in the tube D, on the opposite side of the machine, and a point of metal, which is arranged against the tin-foil on the cylinder by a delicate spring.

"It makes no difference as to the vibrations produced, whether a nail moves over a file or a file moves over a nail, and, in the present instance, it is the foil or indented foil-strip which moves, and the metal point is caused to vibrate as it is affected by the passage of the indentations. The vibrations, however, of this point must be precisely the same as those of the other points which made the indentations, and these vibrations transmitted to a second membrane, must cause the latter to vibrate similar to the first membrane, and the result is a synthesis of the sounds, which in the beginning we saw, as it were, analysed" ('Scientific American,' December, 1877).

In later instruments, that section of the apparatus shown at D is dispensed with, and the reproduction of the spoken words or sentences is effected by bringing the cylinder back to its original starting point, opposite to the little steel projection attached to the metal disc at the end of the mouthpiece A. The steel point is then brought by means of a screw into contact with the foil, and as the cylinder moves onward in its former track, the metal point retraces the indentations on the foil from beginning to end, in doing which it communicates the vibrations it thus receives to the metal diaphragm in precisely the same manner, and with the same results as were shown with D. For the diaphragm, more particularly when employed as a resonator or reproducer of the words which have been spoken into the mouth-piece, other substances than metal, such as glass and paper, have been tried, with, it is said, more satisfactory results.

The crank, C (shown in the figure), by which the cylinder is turned is very frequently supplanted by an apparatus consisting of weights and wheels, or else by clockwork, whereby the cylinder is put in motion. The advantage of the working of these arrangements over that of the crank are, that a regularity of movement of the cylinder is ensured, and it is thus made to advance at the same rate whilst the words are being reproduced as when they are being spoken. This uniformity tends to preserve the pitch of the voice of the speaker.

Edison's instrument created a great sensation, and glowing anticipations were entertained of its future application, but it was found that its articulation was far too imperfect, and its general performance too crude, to admit of its being used for practical purposes; and the inventor, himself, gave it up, applying himself to other work, even allowing his two English patents to lapse.

In 1881, Professor Graham Bell, the inventor of the telephone, with Dr Chichester Bell and Charles Sumner Tainter, formed the Volta Laboratory Association, in Washington, for the purpose of investigating the art of transmitting, recording, and reproducing sound. They conducted many elaborate experiments, and, among other things, sought for and discovered the cause of the failure of the Edison Phonograph. They found that tin-foil, as used in the instrument, was far too pliable for the purpose, as it always had a tendency to pucker, and destroy the symmetry of the sound-waves. They perceived that no good result could be obtained by merely indenting a pliable material; it was necessary to engrave a record in a solid resisting substance; and this discovery enabled them to produce a really practical instrument, which they termed the Graphophone. Instead of tin-foil, Tainter employed wax, ploughing out, by means of the vibrating stylus, a narrow undulating groove, which constitutes the sound-record. When this

groove was retraced by another stylus and diaphragm, the original sounds were reproduced with a fidelity undreamed of by those only acquainted with the tin-foil method.

In the new phonograph, or "graphophone," as it is called, the general principle already described and illustrated is retained: the differences are difference of detail. The chief feature is the "recording cylinder," six inches long by an inch-and-a-quarter broad, formed of cardboard coated with wax. This is placed in a small lathe, and rotated by a treadle in contact with the "recorder," which consists of a metal frame supporting a thin mica diaphragm, in the centre of which is a steel point that cuts a narrow groove on the surface of the cylinder, according to the quality and intensity of the sound spoken against it.

The "recorder" is then removed, and replaced by the "reproducer," a light feather of steel that travels along the grooves made on the cylinder, and transmits their undulations to a small mica diaphragm, which in its turn communicates its vibrations, as sound-waves, to the ears of the auditor by means of two india-rubber tubes, for it was found best to reduce the size of the record, and concentrate the sound in this way, on account of the greater distinctness that was thus secured.

The manipulation of the graphophone is simplicity itself. It requires no adjustment, no electric motor, no galvanic battery. The foot supplies the motive power, and the machine regulates its own speed by means of an ingenious, but simple, governor. The secondary sound, however, is much less powerful than the original

one, the difference between the two being, as a writer in 'Nature' said, "very similar in effect to the feeling produced when looking upon a worn print and an early wood engraving."

Amongst the predictions as to the ultimate capabilities of the phonograph we may notice the following, that—the phonograph will be able to record and reproduce at a future time any air sung to it, so that the vocal triumphs of some of our most accomplished singers may be preserved and resung after their death; that by its means may also be conserved and respoken, likewise after death, a speech delivered by a great statesman or orator; that a dying testator by breathing into his last wishes may have these securely registered, to be expressed after his demise, if need be, in a court of justice; and that the contents of a book or novel may be read to us in the very accents of its author, long after he has passed away. All these predictions have been practically realised.

The grooves are cut very closely together, so as to give a great total length to each inch of surface—a close calculation gives as the capacity of each cylinder upon which the record is made as about 1,000 words.

The practical application of this form of phonograph for communications is very simple. A cylinder is placed in the phonograph, which is then set in motion, and the matter dictated into the mouthpiece, without other effort than when dictating to a stenographer. It is then removed, placed in a suitable form of envelope, and sent through the ordinary channels to the correspondent for whom designed. He, placing it upon his phonograph, sets it moving, *listens* to what his correspondent has to say. Inasmuch as it gives the tone of voice of his correspondent it is *identified*. As it may be filed away as other letters and at any subsequent time reproduced, it is a perfect *record*.

The phonograph letters may be dictated at home or in the office, the *presence* of a stenographer *not being required*. The dictation may be as rapid as the thoughts can be formed, or the lips utter them. The recipient may listen to his letters being read at the rate of 150 to 200 words per minute, and at the same time busy himself about other matters. Interjections, explanations, emphasis, exclamations, &c., may be thrown into such letters *ad libitum*.

Journalists and reporters may dictate their articles and reports, leaving others to transcribe them. The principal of a firm can speak his day's correspondence into the machine, which will repeat it, sentence by sentence, to be written down in proper form by pen or type writer. All these applications are now in active operation in America, where the instrument has achieved great success.

The advantages of such an innovation upon the present slow, tedious, and costly methods are obvious, while there are no disadvantages which will not disappear coincident with the general introduction of the new method.

A vast number of patents have been taken out since the labours of the Volta Electric Association ended in success, showing that invention was stimulated in many quarters. Many of these

80

refer to modifications in the nature of the diaphragm, the recording cylinder, and to details in the working of the instrument. Mr. Edison, evidently encouraged by the results obtained with the graphophone, took again to experimenting with his old phonograph, and after trying wax covered with tin-foil for indentation, he abandoned that mode of recording, and also settled upon a cylinder of wax and the graving-out process, thus confirming the correctness of Messrs. Bell and Tainters' conclusions, and as Emile Berliner states in his interesting paper read before the Franklin Institute at Philadelphia on the 'Gramophone' (a modified phonograph), May 16th, 1888, the new Edison phonograph and the graphophone appear to be practically the *same apparatus, differing only in form and motive power.*

found in all soils upon which plants will grow. It is an important constituent of most plants, especially cereals, while bones largely consist of phosphates. These also enter into the composition of almost every solid and liquid in the animal body. Phosphorus is likewise found in small quantities in meteoric stone, a fact which indicates its wide cosmical distribution.

Prep. This is now only conducted on the large scale : Bone-ash (in powder), 12 parts, and water, 24 parts, are stirred together in a large tub until the mixture is reduced to a perfectly smooth paste ; oil of vitriol, 8 parts, is then added in a slender stream, active stirring being employed during the whole time, and afterwards until the combination appears complete. The next day the mass is thinned with cold water, and, if con-

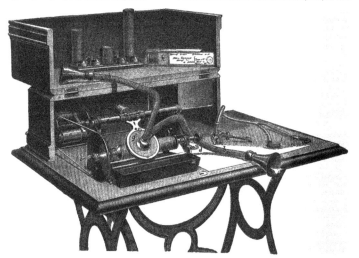

The lettering refers to a detailed description which may be consulted on p. 247, 'Engineering,' September 14th, 1888.

PHOS'GENE GAS. See CHLOROCARBONIC ACID,

PHOS'PHATE. *Syn.* PHOSPHAS, L. A salt of phosphoric acid. See PHOSPHORIC ACID and the respective metals.

PHOS'PHIDE. See PHOSPHURET.

PHOS'PHITE. *Syn.* PHOSPHIS, L. A salt of phosphorous acid. See PHOSPHOROUS ACID.

PHOS'PHORUS. P. =30·96.

This substance appears to have been first prepared by the Alchemist Brand from urine, early in the 17th century. Being a very easily oxidisable substance it never exists free in water, but it is found abundantly in the form of phosphate of lime ; it occurs also in coprolites, and in the minerals apatite, sombrerite, and phosphorite. In small proportions phosphorous compounds are

venient, heated in a leaden pan or boiler until it has entirely lost its granular character ; it is now transferred to one or a series of tall casks (according to the extent of the batch), and further diluted with a large quantity of water. After repose, the clear liquid is decanted, the sediment washed with water, and the 'washings' and 'decanted liquor' evaporated in a leaden or copper boiler until the white calcareous deposit (gypsum) becomes considerable ; the whole is then allowed to cool, the clear portion decanted, and the sediment thoroughly drained on a filter. The liquid thus obtained is evaporated in an iron pot to the consistence of a thick syrup (say 4 parts), when dry charcoal (in powder), 1 part, is added, and the desiccation continued until the bottom of the pot becomes nearly red hot, after which it is covered over and allowed to cool ; the dry mixture, when cold, is put into one or more earthen retorts well covered with 'luting,' which must be properly dried. Heat is then applied (sideways rather than at the bottom) by means of a good air-furnace; after a short time the beak of the retort is connected with a copper tube, the other end of which is made to dip about one fourth of an inch beneath the surface of some luke-warm water placed in a trough or wide-mouthed bottle.

Pur. The distilled product is purified from the small quantities of carbon, which are mechanically carried over, by squeezing it through chamois leather under warm water; or more frequently by mixing the crude melted material with sulphuric acid and bichromate of potash, 8½ parts of each being used for every 100 parts of phosphorus, when the impurities rise as a scum to the surface. It is then moulded for sale by melting it under water heated to about 145° F., and sucking it up to any desired height in slightly tapering, but perfectly straight, glass tubes, previously warmed and wetted. The bottom of the tube being now closed with the finger, it is withdrawn, and transferred to a pan of cold water to congeal the phosphorus, which will then commonly fall out, or may be easily expelled by pressure with a piece of wire.

Prop., &c. Phosphorus in its normal condition is a pale yellow, semi-transparent, and highly combustible solid; soft and flexible at common temperatures; melts at 44·3° C., and boils in an atmosphere free from oxygen at 290° C.; it takes fire in the air when heated slightly above its melting point, and oxidises at all temperatures above the freezing point. Exposed to the air, its surface is slowly converted into phosphorous acid. It is apparently insoluble in water, but it conveys its peculiar flavour and odour to that fluid when agitated with it; it is slightly soluble in ether, naphtha, and the fixed and volatile oils, and more freely so in bisulphide of carbon. It unites with oxygen, forming oxides, and with oxygen and hydrogen, forming acids, and with the metals, forming phosphides. It crystallises in octahedra, and its sp. gr. at 10° C. is 1·83.

Phosphorus is remarkable for assuming several allotropic forms. In one of these forms (amorphous phosphorus) its properties are so altered that they might be those of a distinct element.

Uses. The principal consumption of phosphorus is in the manufacture of lucifer matches. When swallowed, it acts as a powerful corrosive poison; but small doses of its ethereal and oily solutions are occasionally administered in cases of chronic debility, extreme prostration of the nervous powers, impotency, &c. Its action is that of a powerful diffusible stimulant and diuretic; it is also aphrodisiac. Its use requires great caution, and the effects must be narrowly watched. The treatment of poisoning by phosphorus consists of the administration of a powerful emetic and the copious use of mucilaginous drinks. The French practitioners recommend oil of turpentine as the most effective antidote. They administer about a teaspoonful of the turpentine every four hours.

Concluding Remarks. From the great inflammability of phosphorus it can only be safely preserved under water. In commerce, it is always packed in tin cylinders filled with water and soldered up air-tight. The leading points to be observed in order to ensure success in this manufacture are chiefly connected with the firing. The heat of the furnace should be *very* slowly raised at first, but afterwards equably maintained in a state of bright ignition. After 3 or 4 hours of steady firing, carbonic and sulphurous anhydrides are evolved in considerable abundance, provided the materials have not been well dried in the iron pot; then sulphuretted hydrogen makes its appearance, and next phosphuretted hydrogen, which last should continue during the whole of the distillation. The firing should be regulated by the escape of this remarkable gas, which ought to be at the rate of about two bubbles per second. If the discharge becomes interrupted, it is to be ascribed either to the temperature being too low, or to the retort getting cracked; and if, upon raising the heat sufficiently, no bubbles appear, it is a proof that the apparatus has become defective, and that it is needless to continue the operation. We may infer that the process approaches its conclusion by the increasing slowness with which the gas is disengaged under a powerful heat; and when it ceases to come over we may cease firing, taking care to prevent reflux of water into the retort (and consequent explosion) from condensation of its gaseous contents, by admitting air into it through a recurved glass tube, or through the tube of the copper adapter. The usual period of the operation, upon the large scale, is from 24 to 30 hours.

Phosphorus, Amor'phous. *Syn.* RED PHOSPHORUS, ALLOTROPIC PHOSPHORUS; PHOSPHORUS FUSCUS, P. RUBER, L. This is noncrystalline phosphorus in that peculiar condition to which Berzelius has applied the term 'allotropic.' The honour of its discovery is due to Dr Shrötter, of Vienna.

Prep. The ordinary phosphorus of commerce, rendered as dry as possible, is placed in a shallow vessel of hard and well-annealed Bohemian glass, fitted with a safety tube just dipping beneath the surface of a little hot water contained in an adjacent vessel; heat is then applied by means of a metallic bath (a mixture of lead and tin), the temperature of which is gradually raised until it ranges between 464° and 482° F., and bubbles of gas escape from the end of the safety tube and catch fire as they come in contact with the air; this temperature is maintained until the amorphous condition is produced, the length of the exposure being regulated by a miniature operation with tubes conducted in the same bath. As soon as this point is reached, the apparatus is allowed to cool, and the amorphous phosphorus, which still contains some unconverted phosphorus, detached from the glass; it is then reduced to powder by careful trituration under water, drained on a calico filter, and, whilst still moist, spread thinly on shallow trays of iron or lead; in this state it is exposed, with frequent stirring, to heat in a chloride of calcium bath, at first gentle, but then gradually increased to its highest limit, the heat being continued until no more luminous vapour escapes; the residuum on the trays is then cooled, washed with water until the washings cease to affect test-paper, and is, lastly, drained and dried. To render it absolutely free from unaltered phosphorus, it may be washed with bisulphide of carbon.

On the small scale, common phosphorus may be converted into amorphous phosphorus, by simply exposing it for 50 or 60 hours to a temperature of about 473° F., in any suitable vessel from which the air is kept excluded by a stream of carbonic acid, or any other gas which is unable to act chemically on the phosphorus.

By keeping common phosphorus fused at a high temperature, under the above conditions, for fully 8 days, compact masses of amorphous phosphorus may be obtained.

Prop., &c. A reddish brown, infusible, inodorous, solid substance, which is reconverted into ordinary phosphorus by simply exposing it to a heat a little above 500° F. It is unaltered by atmospheric air; is insoluble in bisulphide of carbon, alcohol, ether, or naphtha; is non-luminous in the dark below about 390° F.; and does not take fire at a lower temperature than that necessary for its reconversion into the common or crystalline form. The sp. gr. ranges between 2·089 to 2·017. Its properties render it an admirable substitute for the common phosphorus in the composition for tipping matches, both as regards security from spontaneous ignition, and the health of the manufacturers who, when exposed to the fumes of ordinary phosphorus, were very liable to be attacked with caries of the lower jaw.

Phosphorus, Metallic or Rhombohedral. *Prep.* By heating ordinary yellow phosphorus in sealed tubes in contact with metallic lead for about 12 hours at a red heat a third modification is formed. When the tube is cooled and broken the lead will be found to be permeated with small crystals which can be separated by dissolving the matrix in dilute nitric acid and purified afterwards by boiling them in strong hydrochloric acid. This variety is also formed when red phosphorus is heated to 580° under pressure.

Prop. Bright, lustrous, dark rhombohedral crystals which, when in thin plates, possess a red colour. Sp. gr. 2·34 at 15° C. This form reverts to yellow phosphorus and takes up the octahedral form when heated to 358° C.

Phosphorus, Black. *Prep.* From melted phosphorus which contains foreign matters, especially mercury and other metals, when cooled down (*Thénard*).

Phosphorus, Trichloride of. PCl₃. Syn. **PHOSPHORUS TERCHLORIDE, PHOSPHORUS CHLORIDE.** By gently heating phosphorus, in excess, in dry chlorine gas; or by passing the vapour of phosphorus through a stratum of powdered mercuric chloride, strongly heated in a glass tube. It is limpid, colourless, and highly pungent liquid which fumes in the air, and is slowly resolved by water into phosphorous acid and hydrochloric acid. Sp. gr. 1·62.

Phosphorus, Pentachloride of. PCl₅. Syn. **PHOSPHORIC CHLORIDE, PERCHLORIDE OF PHOSPHORUS.** *Prep.* By the spontaneous combustion of phosphorus in an excess of dry chlorine; or by passing a stream of dry chlorine into the liquid trichloride. By the first method it is obtained as a white crystalline sublimate; by the second, as a solid crystalline mass. It is volatile; water resolves it

into phosphoric acid and hydrochloric acid; heat into PCl₃ and Cl₂. It is of great use in effecting certain transformations in organic substances.

Phosphorus, Oxychloride of. PCl₃O. Syn. **PHOSPHORIC OXYCHLORIDE, PHOSPHORIC MONOXYCHLORIDE.** *Prep.* By heating phosphoric chloride with phosphoric anhydride. It is a colourless, fuming liquid, having the sp. gr. 1·7.

Phosphorus, Hydride of. PH₃. Syn. **PHOSPHORETTED HYDROGEN, PHOSPHURETTED HYDROGEN.** Three of these are known, viz. PH₃ gas, P₂H₄ liquid, P₄H₂ solid; the former need only be noticed here.

Prep. 1. Phosphorous acid is gently heated in a retort, and the first portion of the gas collected.

2. From phosphorus (in small lumps) boiled in a solution of hydrate of potassium or milk of lime, contained in a small retort, as before. Take a very small thin retort, capable of holding not more than 1 oz. or 1½ oz. of water; place in this 3 or 4 fragments of the sticks of fused hydrate of potassium, each being about ¼ inch in length; add as much water as will barely cover them, and then drop in a small fragment of phosphorus, about the size of a horse-bean; apply a very gentle heat with the small flame of a spirit lamp, agitating the retort continually. A pale lambent flame will first appear in the interior, and when this reaches the orifice, and burns in the open air, the retort should be placed on the stand with its beak about an inch under water. Care must be taken not to withdraw the flame of the lamp. When the bubbles of the gas rise to the surface they spontaneously inflame, forming the well-known 'smoke- or vortex-rings.'

3. From phosphide of calcium and dilute hydrochloric acid, as above; or simply from the phosphide thrown into the water.

Obs. The gas obtained by methods 2 and 3 is contaminated with the vapour of a liquid phosphide of hydrogen, PH₂, which gives to it the property of spontaneous inflammability, producing clouds of phosphorous pentoxide.

Prop., &c. Colourless; very fetid; slightly soluble in water; burns with a white flame; decomposed by light, heat, and strong acids; as commonly prepared, inflames on contact with air, at ordinary temperatures, but when pure, only at the heat of boiling water. It is slightly heavier than air. Sp. gr. 1·19. It is rendered quite dry by standing over fused chloride of calcium.

Phosphorus, Suboxide of. P₄O. (*Odling*.) A reddish-brown powder, formed when a stream of oxygen is forced upon phosphorus, melted beneath the surface of hot water. To purify it from phosphoric acid and free phosphorus, it is washed on a filter with water, then dried by bibulous paper, and finally digested with bisulphide of carbon.

Hypophosphorous Acid. H₃PO₂. By cautiously decomposing a solution of hypophosphite of barium with sulphuric acid, filtering from the precipitate (sulphate of barium), and evaporating. Dissolve hypophosphite of calcium, 480 gr. in distilled water, 6 fl. oz.; dissolve crystallised oxalic acid, 350 gr., in another portion of distilled water, 3 fl. oz.; mix the solutions and filter the mixture through white filtering paper. Add distilled water carefully to the filtrate till it measures 10 fl. oz., and evaporate this to 8¼ fl. oz. The solution thus prepared contains about 10% of terhydrated hypophosphorous acid.

Prop. A viscid, uncrystallisable liquid having a strong acid reaction. It is a powerful drying agent, and forms salts called hypophosphites.

Ammonium, Hypophosphite of. $(NH_4)_2PO_2$. *Prep.* Dissolve hypophosphite of calcium, 6 oz., in water, 4 pints; and dissolve sesquicarbonate of ammonium, 7·23 oz. (barely 7½), in water, 2 pints; mix the solutions; filter, washing out the solution retained by the carbonate of lime with water, q. s.; evaporate the filtrate to dryness with great care; dissolve it in alcohol, q. s.; filter, evaporate, and crystallise. Very soluble in both alcohol and water.

Barium, Hypophosphite of. $BaH_4(PO_2)_2$. *Prep.* Boil phosphorus in a solution of hydrate of barium till all the phosphorus disappears and the vapours have no longer a garlic odour. Filter, evaporate, and set aside to crystallise.

Calcium, Hypophosphite of. $Ca_2(PO_2)_2$. *Prep.* Slake recently burnt lime, 4 lbs., with water, 1 gall., and mix it with water, 4 galls., just brought to the boiling temperature in a deep open boiler, stirring until a uniform milk of lime is formed; then add phosphorus, 1 lb., and keep up the boiling constantly, adding hot water from time to time, so as to preserve the measure as nearly as may be until all the phosphorus is oxidised and combined, and the strong odour of the gas has disappeared; then filter the solution through muslin, wash out that portion retained by the calcareous residue with water, and evaporate the filtrate to 6 pints; re-filter, to remove the carbonate of calcium resulting from the action of the air upon the solution; evaporate again until a pellicle forms, and set aside to crystallise—or continue the heat with constant stirring until the salt granulates.

Obs. As spontaneously inflammable phosphuretted hydrogen is given off during the boiling, the process must be conducted under a hood, with a strong draught or in the open air. Smaller proportions than those given may be used.

Prop. Hypophosphite of calcium is a white salt, with pearly lustre, crystallising in flattened prisms; soluble in 6 parts of cold water, and slightly soluble in dilute alcohol. It is the most important of these compounds, and when introduced into the stomach it is supposed to be converted into phosphate of calcium. It has been termed 'chemical food.' By decomposition it readily furnishes the other hypophosphites.

Ferric, Hypophosphite. $FePO_2$. *Prep.* By precipitating a solution of hypophosphite of sodium or ammonium, with solution of ferric sulphate, washing the gelatinous precipitate with care (it being somewhat soluble); and, finally, drying it into an amorphous white powder. This is freely soluble in hydrochloric and hypophosphorous acids.

Potassium, Hypophosphite of. K_2PO_2. *Prep.* From hypophosphite of calcium, 6 oz., dissolved in water, 4 pints; and granulated carbonate of potassium, 5¾ oz., dissolved in water, ⅓ pint. Mix, filter, and wash the precipitate till the fil-

trate measures 5 pints. Evaporate till a pellicle forms, then stir constantly, continuing the heat till the salt granulates. A white, opaque, deliquescent body, very soluble in water and alcohol.

Quinine, Hypophosphite of. Dissolve sulphate of quinine, 1 oz., in very dilute sulphuric acid; precipitate the alkaloid with ammonia; wash the precipitated quinine and digest it in hypophosphorous acid with heat, the quinine being in excess; after filtering the solution, allow it to evaporate spontaneously till the required salt crystallises. It forms elegant tufts of soft, feathery crystals, which are soluble in 60 parts of water.

Sodium, Hypophosphite of. Na_2PO_2. *Prep.* From hypophosphite of calcium, 6 oz., dissolved in water, 4 pints; and crystallised carbonate of sodium, 10 oz., dissolved in water, 1½ pint. Proceed as in making hypophosphite of potassium, but allowing 6 pints as the measure of the filtrate. If required in crystals, the granulated salt may be dissolved in alcohol sp. gr. 0·835, evaporated till syrupy, and set by in a warm place. Crystallises in rectangular tables, with a pearly lustre; is very soluble in water and ordinary alcohol, and deliquesces when exposed to the air.

Phosphorus, Trioxide of. P_2O_3. *Syn.* PHOSPHOROUS ANHYDRIDE; ANHYDROUS PHOSPHORIC ACID.

Prep. By burning phosphorus in a limited supply of air. White flaky powder, with an odour of garlic, and rapidly absorbing water to form phosphorous acid.

Phosphorous Acid. H_3PO_3. *Syn.* HYDRATED PHOSPHOROUS ACID. Pure phosphorus is volatilised through a layer of powdered mercuric chloride, contained in a glass tube; trichloride of phosphorus comes over, which, on being mixed with water, is resolved into hydrochloric acid and phosphorous acid; by evaporating the mixed liquid to the consistence of a syrup, the first is expelled, and the residuum forms a crystalline mass of hydrated phosphorous acid on cooling.

Prop., &c. It is a powerful deoxidising agent. Heated in a closed vessel, it is resolved into phosphoric anhydride and free phosphorus. With the bases it forms salts, called phosphites, which possess little practical importance.

Phosphorous Tetroxide. P_2O_4. Very little is known of this substance itself, but from it hypophosphoric acid is derived, which has lately been prepared by Salzer, and has a composition represented by the formula $P_2O_2(OH)_4$.

Phosphorous Pentoxide. P_2O_5. *Syn.* ANHYDROUS PHOSPHORIC ACID; PHOSPHORIC ANHYDRIDE; PHOSPHORIC OXIDE. Obtained by the vivid combustion of phosphorus in a stream of dry atmospheric air, or under a bell-jar, copiously supplied with dry air. The product is pure anhydrous phosphoric acid in the form of snowlike flakes. It must be immediately collected and put into a warm, dry, well-stoppered bottle. In this state it exhibits an intense attraction for water, and when thrown into it combines with explosive violence; exposed to moist air for only a few seconds, it deliquesces to a syrupy-looking liquid. It is often used in the laboratory as a dessicating agent.

Phosphoric Acid. There are three distinct acids usually grouped under this head, namely, META-PHOSPHORIC ACID, HPO_3; PYROPHOSPHORIC ACID, $H_4P_2O_7$; and ORTHOPHOSPHORIC ACID, H_3PO_4.

Metaphosphoric Acid. HPO_3. Syn. MONO-BASIC PHOSPHORIC ACID; GLACIAL PHOSPHORIC ACID.

Prep. 1. Bones (calcined to whiteness and powdered), 3 parts, are digested for several days in oil of vitriol, 2 parts, previously diluted with water, 6 parts, the mixture being frequently stirred during the time; a large quantity of water is next added, the whole thrown upon a strainer, and the residual matter washed with some hot water; the mixed liquors are then precipitated with a solution of carbonate of ammonium, in slight excess, filtered from the insoluble and finally ignited in a platinum crucible.

2. By acting upon the anhydride with cold water.

When phosphoric acid is added to a strong solution of phosphate of sodium, and the mixture, after concentration, is exposed to a low temperature, prismatic crystals are deposited. These, after being strongly heated to expel their basic water, are pure metaphosphate of sodium. From the solution of this salt in cold water, a solution of pure metaphosphoric acid may be obtained, as above, by means of nitrate or acetate of lead and sulphuretted hydrogen.

Obs. This acid precipitates the salts of silver white, and is distinguished from the other modifications of phosphoric acid by the property which its solution possesses of coagulating albumen.

Pyrophosphoric Acid. $H_4P_2O_7$. Syn. DIBASIC PHOSPHORIC ACID. Prep. By strongly heating common phosphate of sodium. The water of crystallisation only is at first expelled, and the salt becomes anhydrous; but as the temperature reaches that of redness the salt loses water and is decomposed. By solution of the altered salt in water, crystals of pyrophosphate of sodium may be obtained. A solution of this last compound, treated with nitrate of lead, and the resulting precipitate, suspended in cold water, and decomposed by sulphuretted hydrogen, yields a solution of pure pyrophosphoric acid.

Obs. Heat resolves this into a solution of the ordinary acid. Pyrophosphoric acid precipitates the salts of silver of a white colour. The salts of this acid are called pyrophosphates.

Orthophosphoric Acid. H_3PO_4. Syn. TRI-HYDRIC PHOSPHATE. TRIBASIC PHOSPHORIC ACID. Prep. 1. Ordinary nitric acid is heated in a tubulated retort connected with a receiver, and small fragments of phosphorus are dropped into it, singly and at intervals. As soon as the oxygenation of the phosphorus is complete, the heat is increased, the undecomposed acid distilled off, and the residuum evaporated to the consistency of a syrup. In this state it forms the phosphoric acid of the shops.

2. Commercial phosphate of sodium is dissolved in water and the solution precipitated with another of acetate of lead; an abundant white precipitate (phosphate of lead) falls; this is collected on a filter, well washed, and, whilst still moist, is suspended in distilled water, and sulphuretted hydrogen gas passed into it, in excess; a black insoluble precipitate forms, while pure tribasic phosphoric acid remains in solution, and is easily deprived of the residual sulphuretted hydrogen by a gentle heat. By concentration in vacuo over sulphuric acid, it may be obtained in thin crystalline plates.

The solution of this acid may be boiled without change, but when concentrated and heated to about 400° F. it is converted into pyrophosphoric acid, and at a red heat into metaphosphoric acid. Its salts are the ordinary phosphates, or ortho-phosphates, and they give a yellow precipitate with nitrate of silver.

Tests. The following reactions characterise the ordinary or ortho-phosphates:—1. Chloride of barium produces in aqueous solutions of the neutral and basic phosphates a white precipitate, which is insoluble in either hydrochloric or nitric acid, and with difficulty soluble in a solution of chloride of ammonium.—2. Solution of sulphate of calcium produces in neutral and alkaline solutions of the phosphates a white precipitate, freely soluble in acids, even the acetic.—3. Sulphate of magnesium produces in solutions of the phosphates, to which some chloride of ammonium and free ammonia has been added, a white, crystalline, and quickly subsiding precipitate of the phosphate of ammonium and magnesium, which is insoluble in a solution of either ammonia or chloride of ammonium, but readily soluble in acids, even in acetic.—4. Nitrate of silver, with neutral and basic alkaline phosphates, gives a light yellow precipitate. If the fluid in which the precipitate is suspended contained a basic phosphate, it does not affect test-paper; if it contained a neutral phosphate, the reaction will be acid. If the phosphate examined has been heated to redness before solution, it then, as a metaphosphate, gives a white precipitate with nitrate of silver.—5. Hydrochloric acid is added to the solution until an acid reaction is produced, and afterwards 1 or 2 drops of a concentrated solution of ferric chloride; a solution of acetate of potassium is next added in excess, when a flocculent, gelatinous, white precipitate will be formed if phosphoric acid or any phosphate was present in the original liquid. This test is highly characteristic, and of general applicability.

Obs. The insoluble phosphates must be first treated with diluted hydrochloric or sulphuric acid, and the resulting solution filtered and neutralised with an alkali, before applying the reagents. When the substance under examination consists of a very small quantity of phosphoric acid or phosphate, with a large quantity of sesquioxide of iron, it should be fused with some carbonate of sodium, the residuum of the ignition exhausted with water, and the tests applied to the filtered solution. Arsenious acid, if present, should be removed by sulphuretted hydrogen before applying the tests. When phosphate of aluminum is present, the solution in hydrochloric acid is neutralised with carbonate of sodium; carbonate of barium is next added in excess, followed by the addition of hydrate of potassium, also in excess, after which the whole is boiled. An insoluble phosphate of barium is formed, which may be decomposed by sulphuric acid, as before. See MOLYBDATE OF AMMONIUM.

Estim. Pure solutions of phosphoric acid may be tested by the common methods of acidimetry. When in a state of combination, it may be separated and weighed in either of the forms noticed under GUANO.

Uses, &c. This acid is the common form, and is the compound alluded to when 'phosphoric acid' is spoken of. The commercial variety is usually contaminated with arsenic acid. It is extensively employed by the bleacher, dyer, calico-printer, and enameller. Unlike sulphuric acid and the other strong acids, it does not coagulate albumen nor injure vegetable fibre, and is not decomposed by contact with organic matter. In combination with alumina and a large quantity of boracic acid, it is said to be capable of producing a glaze for earthenware of extreme beauty and durability, and perfectly innocuous. It is also used in medicine.

PHOSPHORIC ACID, DILUTED. (B. Ph.) Put 6 fl. oz. of nitric acid (sp. gr. 1·42), diluted with 8 oz. of distilled water, into a tubulated retort connected with a Liebig's condenser, and having added 413 gr. of phosphorus, apply a very gentle heat until 5 fl. oz. of liquid have distilled over. Return this to the retort, and renew and continue the distillation until the phosphorus has entirely dissolved.

Transfer the contents of the retort to a porcelain capsule and evaporate the liquid until it is reduced to 4 fl. oz. Transfer to a platinum vessel and evaporate to about 2 fl. oz., and until orange vapours cease to form. Mix when cool in such an amount of distilled water that the volume shall become 1 pint. (It contains 10% by weight of anhydrous acid. Sp. gr. 1·08.)—*Dose*, 10 to 30 minims properly diluted.

PHOSPHORUS, BALDWIN'S. Recently fused nitrate of calcium. For this purpose it must be broken into fragments whilst still warm, and at once placed in dry and well-stopped phials. After exposure for some time to the direct rays of the sun it emits sufficient light in the dark to render visible the figures on the dial-plate of a watch.

PHOSPHORUS, BOLOGNIAN. *Syn.* KIRCHER'S PHOSPHORUS, BOLOGNIAN STONE. This substance was accidentally discovered by a shoe-maker of Bologna, and excited much interest about the middle of the 17th century. The following is said to have been the formula employed by the Logani family, who were particularly successful in its preparation, and acquired wealth by its sale to the curious throughout Europe:

Prep. Reduce recently calcined native sulphate of barium to powder, make it into a paste with mucilage of gum tragacanth, and roll the mass into pieces about ¼ inch thick and 1 to 2 inches long; dry these slowly by a moderate heat, and then expose them to ignition in a wind furnace, by placing them loosely among the charcoal; lastly, allow them to cool slowly, and at once place the pieces in well-stopped phials. Like the preceding substance, it phosporesces in the dark after exposure to the sun's rays.

PHOSPHORUS, CANTON'S. *Prep.* From cal-cined oyster shells, 3 parts; flowers of sulphur, 1 part; placed in alternate layers in a covered crucible, and exposed to a strong heat for about an hour. It is preserved and used like the above.

PHOSPHORUS, HOMBERG'S. Recently ignited chloride of calcium.

PHOSPHORUS BOTTLES. *Prep.* 1. Phosphorus, 12 gr.; olive oil, ½ oz.; mix in an oz. phial, and place the latter, loosely corked, in a basin of hot water; as soon as the phosphorus is melted, remove the phial, cork it securely, and agitate it until nearly cold. On being uncorked it emits sufficient light in the dark to see the time by a watch, and will retain this property for some years if not too frequently employed. These are frequently called 'luminous phials.'

2. (BRIQUETS PHOSPHORIQUES.) *a.* From phosphorus, 8 parts; white wax, 1 part; cautiously melted together by the heat of hot water; as the mixture begins to cool, the bottles are turned round so that it may adhere to the sides.

b. (*Bendix.*) Cork (rasped small, and dry) and yellow wax, of each, 1 part; phosphorus, 4 parts; petroleum, 8 parts; mixed, by fusion, and placed in bottles in the same way as in *a.* Used as instantaneous-light bottles. A sulphur match rubbed against the composition immediately inflames on exposure to the air. They should be only unstoppered at the instant of introducing the match, and should be handled with caution.

PHOSPHORUS MATCHES. See MATCHES, and above.

PHOSPHORUS PASTE. *Syn.* ANTI-ARSENICAL RAT-POISON, PHOSPHOR-PASTE. *Prep.* 1. Phosphorus, 1 oz.; warm water, 1 pint; place them in a bottle, cork it, and agitate them well together until the phosphorus is reduced to a minute state of division, adding towards the end moist sugar, ¼ lb.; next add of lard (melted by a gentle heat), 1 lb., and repeat the agitation until the whole is nearly cold; when cold, form it into a stiff dough with oatmeal or barley meal, and make this into small balls or cakes; lastly, dry these in the air, without artificial heat.

2. (*Simon.*) Phosphorus, 8 parts; water (lukewarm), 180 parts; mix in a mortar, and add of rye meal, 180 parts; when cold, further add of butter or lard, 180 parts; sugar, 125 parts; and mix the whole thoroughly together. This is the formula authorised by the Prussian Government.

Obs. Rats, mice, &c., eat the above composition with avidity, after which they soon die. It is said that the best method of using it is to place small pieces of it in and about the holes, with some water in a shallow vessel for them to drink. It has the advantage of retaining its efficacy for many years, and is less dangerous to human beings than compositions containing arsenic, whilst it is even more effective for the purpose for which it is employed. Some persons recommend the addition of a little oil of rhodium or oil of aniseed. See RATS, &c.

PHOSPHURET. *Syn.* PHOSPHIDE; PHOSPHURETUM, PHOSPHIDUM, L. A compound of phosphorus with a metal or other basic radical. See the respective METALS, &c.

PHOSPHURETTED HYDROGEN. *Syn.* PHOSPHORETTED HYDROGEN. See HYDROGEN.

PHOTOGRAPHY

PHOTOGRAPHY. The art of producing pictures by the action of light.

HISTORICAL.

The action of light upon silver chloride was known to the alchemists of the 16th century but not understood by them. Scheele, in 1777, investigated the properties of the body formed, and Ritter, of Jena, in 1801 carried Scheele's studies still further by discovering that the rays of the spectrum beyond the extreme violet darkened the chloride rapidly. Josiah Wedgwood and Sir Humphry Davy, in 1802, obtained pictures upon leather covered with silver chloride, and in 1803 Dr Wollaston discovered the action of light upon gum guaiacum. In 1814 Joseph Nicéphore de Niépce, experimenting with resins, found that light rendered them insoluble, and that pictures upon polished metal plates could be produced in this way. In 1829 Daguerre, a French painter, went into partnership with Niépce, and discovered the action of light upon a silver plate which had been exposed to the vapour of iodine (1839). In the same year Mr Talbot read his paper on photogenic drawings before the Royal Society, with which he exhibited prints made upon paper which had been dipped into a solution of common salt, then dried, and brushed over with a solution of silver nitrate. In this way he obtained a negative from which positives in any quantity could be produced. In 1841 Talbot patented the calotype process by which an invisible image formed by exposing paper, covered with a film of iodide of silver, to the light, was developed by a solution of gallic acid. In 1839 Mungo Ponton had discovered the action of light upon chromium salts, and in 1843 Sir J. Herschell succeeded in obtaining a picture of his 40 feet telescope on a plate of glass covered with silver chloride. Niépce de St Victor used albumen as a vehicle for the silver salts, and Le Gray suggested the use of collodion, applying the collodion process in a practical form was due to Scott-Archer and Dr Hugh Diamond in 1851. From these beginnings the whole of our modern photographic processes take their origin.

The object of this article being practical rather than theoretical, the reader who desires information on the chemistry of the action of light on various chemical compounds is referred to Abney, 'A Treatise on Photography' (Longmans), and Meldola, 'The Chemistry of Photography' (Macmillan's Nature Series).

THE DAGUERREOTYPE PROCESS.

A plate of copper silvered on one side is cleaned on the plated surface with tripoli and alcohol, using a Canton flannel rubber until it is quite free from scratches and fairly smooth; it is then polished with a leather buff and jewellers' rouge. The clean plate is at once placed in a box, at the bottom of which iodine in crystals is strewn, so arranged that it can rest face downwards at some distance above the iodine. After a time the silvered surface is attacked by the iodine, and a thin film of silver iodide produced, the action is allowed to go on until a reddish colour is reached. This accomplished the plate is transferred to a similar box, at the bottom of which is placed a mixture of bromine and quicklime, the bromine now attacks the silver, and with the iodide already formed, produces a coating of bromo-iodide of silver, this is allowed to go on till the colour is steel-grey or violet, and the plate is once more transferred to the iodine box and allowed to remain for one third of the time of the first operation. The plate is now ready for exposure in the camera, and is very sensitive, the exposure completed, development is effected by allowing the vapour of mercury to act on the plate by placing it face downwards over a tray of mercury heated to about 150° F. The image is fixed by immersion in a 10% solution of sodic hyposulphite.

THE WET COLLODION PROCESS.

The discovery that certain salts of silver were sensitive to light, and that pictures could be produced by exposing a thin film of these salts in the camera, naturally suggested to workers the desirability of having some method by which the salts could be spread over any required surface as wanted, and kept in position by some inert material, which should serve as a carrier, and above all, some means by which a plate of glass could be rendered sensitive to light, so that a negative being obtained positive pictures could be made from it in any required number. The required conditions are satisfied in a remarkable degree by the wet collodion process, in which a plate of glass is first carefully cleaned and then coated with a solution of nitro-cellulose in a mixture of alcohol and ether, to which small quantities of certain bromides and iodides have been added. The film of iodised collodion so formed is then immersed for a moment in a solution of silver nitrate, which is decomposed *in the film,* and an exceedingly fine and even deposit of bromide and iodide of silver upon glass is thus obtained, which is exceedingly sensitive to light. After exposure the plate is developed by a solution of pyrogallic acid and ferrous sulphate.

Formulæ, &c., in Connection with the Wet Process.

Preparation of Pyroxyline. (*Hardwich*.) Take of—

Sulphuric acid, sp. gr. 1·842 at 15° C.	. 500 c.c.	
Nitric acid ,, 1·456 ,,	. 166·6 c.c.	
Water	. 145·7 c.c.	

Mix the nitric acid and water thoroughly in a porcelain dish, and then add the sulphuric acid, stirring well all the time; allow the mixture to cool to 65° C., and have ready a dozen balls of cotton wool, weighing about 1·5 grammes each. (The wool should first be well steeped in soda and water and then *thoroughly* washed and *completely* dried.) Immerse the balls quickly, and assist the soaking of the wool by pressing with a glass rod or spatula. Allow them to remain in the mixture ten minutes to a quarter of an hour, then lift them out with the spatula, squeezing out as much acid as possible against the sides of the dish, and throw them one by one into a *large quantity of water,* and continue the process of washing in abundance of clean water until a piece of blue litmus paper is no longer affected by the wet cotton. The strength of the acids and the temperature is of the utmost importance, and should be accurately determined, otherwise failure will result.

Another formula:
Sulphuric acid, sp. gr. 1·842 . 170 c.c.
Dried potassium nitrate (pure) 110 grms.
Water . . . . . 28·3 c.c.
Best dried cotton wool . . 4 grms.
Proceed as before.
Collodions (Abney):
No. 1 for cold weather.
Pyroxyline (Hardwich's formula) 12 to 14 grms.
Alcohol, sp. gr. ·820 . . . 450 c.c.
Ether „ ·725 . . . 550 „
No. 2 for warm weather.
Pyroxyline (Hardwich's formula) 12 to 14 grms.
Alcohol, sp. gr. ·820 . . . 500 c.c.
Ether „ ·725 . . . 500 „
Iodo-Bromide Collodion (Abney). 1. Ammonium iodide, 7 grms.; cadmium bromide, 4 grms.; plain collodion, 1000 c.c.
2. Ammonium iodide, 8 grms.; cadmium bromide, 2·5; plain collodion, 1000 c.c.
3. Cadmium iodide, 9 grms.; cadmium bromide, 4 grms.; plain collodion, 1000 c.c.
Nos. 1 and 2 are soon ripe enough for use, and with a little alcoholic tincture of iodine added, they may be used immediately. No. 3 requires keeping.
Iodised Collodion (for Negatives):
Ether, sp. gr. ·725 . . . 10 fl. oz.
Alcohol „ ·805 . . . 8 „
Pyroxyline . . . . 120 grs.
Ammonium iodide . . . 12 „
Cadmium „ . . . 20 „
Bromo-iodised Collodion (for Negatives):
Ether, sp. gr. ·725 . . . 10 fl. oz.
Alcohol „ ·805 . . . 10 „
Pyroxyline . . . . 120 grs.
Ammonium iodide . . . 40 „
Cadmium „ . . . 40 „
„ bromide . . . 20 „
Bromo-iodised Collodion (for Positives or Ferrotypes). Ether, 10 fl. oz.; alcohol, 10 fl. oz.; pyroxyline, 100 grs.; cadmium iodide, 50 grs.; ammonium bromide, 20 grs.

THE NITRATE BATH (for Negatives).
1. Nitrate of silver (pure recryst.), 6 oz.; distilled water, 80 fl. oz.; nitric acid (pure), 12 minims. Saturate with iodide of silver and filter.
2. Nitrate of silver recryst., 80 grms.; potassium iodide, ·25 grm.; water, 1000 c.c. Dissolve the silver nitrate in 250 c.c. of the water, and the iodide in as small a quantity as possible. Mix, shake well, add the remaining water, filter, and acidify with a few drops of a 5% solution of nitric acid (Abney).

(For Positives or Ferrotypes.)
Nitrate of silver (recryst.), 5 oz.; distilled water, 80 fl. oz.; nitric acid (pure), 12 minims. Saturate with silver iodide and filter.

DEVELOPER.
For Negatives. 1. Protosulphate of iron, ½ oz.; glacial acetic acid, ½ oz.; alcohol, ½ oz.; water, 8 oz. ('British Journal Photographic Almanac,' 1891).
2. Ammonio-sulphate of iron, 75 gr.; glacial acetic acid, 75 gr.; sulphate of copper, 7 gr.; water, 3 oz. ('British Journal Photographic Almanac,' 1891).

For Collodion Positives or Ferrotypes. Protosulphate of iron, 1½ oz.; nitrate of baryta, 1 oz.; water, 1 pint; alcohol, 1 oz.; nitric acid, 40 drops ('British Journal Photographic Almanac,' 1891).
For Collodion Transfers. Pyrogallic acid, 5 gr.; citric acid, 3 gr.; acetic acid, 45 minims; water, 1 oz.; alcohol, q. s. ('British Journal Photographic Almanac,' 1891).
Developers. (Abney.) 1. Pyrogallic acid, 1 grm.; glacial acetic acid, 20 c.c.; alcohol, q. s.; water, 500 c.c.
2. (Weak.) Ferrous sulphate, 10 grms.; glacial acetic acid, 30 to 40 c.c.; alcohol, q. s.; water, 1000 c.c.
3. (Strong.) Ferrous sulphate, 100 grms.; glacial acetic acid, 40 c.c.; alcohol, q. s.; water, 1000 c.c.

INTENSIFICATION.
Any method by which the apparent density or blackness of the image when viewed by reflected light may be increased, or by which its power of transmitting light may be diminished, is termed intensification. There are many methods, of intensification; the simplest, perhaps, is to flood the plate with a solution of mercuric chloride, then thoroughly wash for some time and treat with a weak solution of ammonia, or we may use a solution of bromide of copper; wash, and again treat with a solution of silver nitrate, by which the silver forming the image is very greatly increased; or, after development with iron and thorough washing, take of—
(a) Pyrogallic acid . . 4 grms. . . 2 grs.
Citric acid . . 4 to 8 „ . . 2 „
Water . . . 1000 c.c. . . 1 oz.
and whilst the plate is still wet, flood it with either of the above solutions, to which a few drops of a solution of silver nitrate, 4%, has been added immediately before use. Or (b) ferrous sulphate, 10 grms.; citric acid, 20 grms.; water, 1000 c.c.; using the silver solution as before. The following can be used after fixing:
(1) Iodine . . . . . ·1 grm. or 10 parts by weight.
Potassic iodide . ·2 „ 20 „ „
Water . . . 50 c.c. or 500 „ „
(2) Mercuric chloride ·2 grm. or 2 „ „
Water . . . 750 c.c. or 7500 „ „
and
Potassic iodide . ·1 grm. or 1 „ „
Water . . . 50 c.c. or 50 „ „
Add the latter solution to the former until the precipitate begins to become permanent, then flood the negative with the filtered solution.

FIXING THE IMAGE.
After thorough removal of developer by washing, place the negative in either—
(a) Sodium hyposulphite 100 grms. or 1 part.
Water . . . . 500 c.c. or 5 „
or—
(b) Potassium cyanide . 30 grms. or 6 „
Water . . . . 500 c.c. or 100 „
and wash very thoroughly. The cyanide is objectionable because of its great activity and the possibility of destruction of the image by the use of too strong a solution, and further because it is excessively poisonous.

VARNISHING THE FILM.

Captain Abney recommends—

Unbleached lac . . . . 65 grms.
Sandarach . . . . . . 65 „
Canada balsam . . . . . 4 „
Oil of thyme or lac acetic . 32 c.c.
Alcohol ·830· . . . . . 500 „

or—

Seed lac . . . . 120 grms.
Methylated spirit . . . . 1000 c.c.

POSITIVES BY THE WET PROCESS.

The process is essentially the same as for negatives, but the silver bath should be of the strength of 65 grms. to the litre (65 parts per 1000 by weight), silver iodide added as for the negative bath, and slightly acidified with nitric acid. Captain Abney recommends as developer— Ferrous nitrate, 7 grms.; ferrous sulphate, 3 grms.; nitric acid, 1·45, 1·25 c.c.; alcohol, q. s.; water, 1000 c.c.

See also above, under DEVELOPER, COLLODION, &c.

DRY PLATE PROCESSES WITH THE BATH.

A collodion film on glass if allowed to dry will not bear subsequent wetting without leaving the support and peeling off. The apparatus necessary for the wet process renders it cumbrous and unsuitable under many circumstances, and a method by which a sensitive collodion film, dry and ready for use at any time, could be prepared was an obvious step in advance of the wet process requiring the apparatus to be taken into the field.

The process is in principle briefly as follows:— A clean glass plate is coated with some material which will cause the collodion to adhere firmly to it. Gelatin, india-rubber, and albumen solutions are those most generally used. The plate is then covered with collodion as for the wet process, all excess of the bath solution is washed off by the liberal use of distilled water, and the plate is then coated with some material which shall serve as a protection to the collodion film against atmospheric or other influences without at the same time interfering with the sensitiveness of the plate of the action of the developers. Innumerable 'preservatives' or organifiers have been used; the following are among the most approved:

1. Tannic acid, 5 to 10 gr.; sugar, 1 gr.; water, 1 oz.
2. A strong infusion of tea or coffee.
3. A decoction of malt, 4 oz., to 1 pint of water.

(From the 'British Journal Photographic Almanac,' 1891:)

For Landscape Work.

4. Tannin, ¼ oz.; gallic acid, 60 gr.; water, 20 fl. oz.
5. Tannin, 300 gr.; water, 20 fl. oz.

For Landscapes or Transparencies (warm brown tone).

6. Freshly ground coffee, 1 oz.; boiling water, 1 pint.

For Transparencies (brownish-black tone).

7. Tannin, 80 gr.; pyrogallic acid, 60 gr.; water, 30 fl. oz.

Developers for Collodion Dry Plates (Abney).

1. (a) Ferrous sulphate, 2 grms.; water, 20 c.c.

(b) Gelatin, 4 grms.; glacial acetic acid, 60 c.c.; water, 400 c.c.

Dissolve the gelatin in the water and add the glacial acetic acid, then mix the solution in the proportion of 3 parts by measure of a to 1 part by measure of b; filter, and the developer is ready for use. To every 4 c.c. of the mixed developer add 1 drop of a 60% solution of silver nitrate, and use at once.

2. (a) Pyrogallic acid, 10 grms.; alcohol, 60 c.c.

(b) Silver nitrate, 4 grms.; citric acid, 4 grms.; water, 85 c.c.

1 part by measure of a is mixed with ½ part of b, and 30 parts of water added.

3. (a) Pyrogallic acid, 1 grm.; water, 40 c.c.

(b) Ammonium hydrate (·880), 1 part; water, 4 parts.

(c) Citric acid, 4 grms.; acetic acid (glacial), 2 c.c.; water, 30 c.c.

(d) Silver nitrate, 1 grm.; water, 20 c.c.

This is specially useful for albumen beer plates. To each 50 c.c. of a add 10 drops of b, mix well and flood the plate with the mixture. When the image begins to appear pour off the developer and add 7 more drops of b, and flood the plate again. Then pour off and add 20 drops of c, and repeat the process. Rinse the plate once more with solution a, and intensify with a few drops of d. Lastly, wash and fix.

COLLODION EMULSION PROCESSES.

In the processes just described the plate was sensitized *after* coating with collodion. It is possible, however, to prepare excellent dry plates for many purposes by impregnating the collodion in bulk with the sensitive silver salts, and then coating the plates; the use of the bath is thus dispensed with. Collodion emulsions are of two kinds, viz. washed and unwashed according as they are or are not submitted to a process of washing with pure distilled water previous to coating the plates. The washing may be accomplished *after* coating, but the process is more tedious and the plates more liable to defects than when the washing is effected *before* coating.

Collodion dry plates are not by any means so rapid as gelatin plates, though for many purposes this is a distinct advantage. Some recent experiments by Mr Wellington lead to the belief that it may be possible to prepare such plates of a sensitiveness equal to that of certain gelatin emulsion plates.

The object of the washing is to remove excess of soluble salts, which would otherwise crystallise out and crack the silver, rendering it useless.

Formulæ for pyroxyline for dry collodion processes. ('British Journal Photographic Almanac,' 1891.)

For Collodio-Bromide or Unwashed Emulsion.

Nitric acid, sp. gr. 1·45, 2 fl. oz.; suphuric acid, sp. gr, 1·845, 4 fl. oz.; water, 1 fl. oz.; cotton (cleaned and carded), 100 grms.; temperature, 150° F.; time of immersion, 10 minutes.

For Washed Emulsion.

1. Nitric acid, sp. gr. 1·45, 2 fl. oz.; sulphuric acid, sp. gr., 1·845, 6 fl. oz.; water, 1 fl. oz.; cotton (cleaned and carded), 100 gr.; temperature, 140° F.; time of immersion, 10 minutes.

2. Nitric acid, sp. gr. 1·45, 2 fl. oz.; sulphuric acid, sp. gr., 1.845, 3 fl. oz.; white blotting paper, 145 grs.; temperature, 100° F.; time of immersion, 30 minutes.

Collodio-Bromide Emulsion.

Ether, sp. gr. ·720, 5 fl. oz.; alcohol, sp. gr., .820, 3 fl. oz.; pyroxyline, 50 grs.; bromide of cadmium and ammonium, 80 grs.; or bromide of zinc, 76 grs.

Sensitise by adding to each oz. 15 *gr.* of silver nitrate dissolved in a few drops of water and 1 dr. of boiling alcohol. This is suitable for slow landscape work or for transparencies.

Washed Emulsion (for Landscapes).

1. Ether, sp. gr. ·720, 4 fl, oz.; alcohol, sp. gr., ·820, 2½ fl. oz.; pyroxyline, 40 grs.; Castile soap (dissolved in alcohol), 30 grs.; bromide of ammonium and cadmium, 84 grs.

Sensitise with 100 gr. silver nitrate dissolved in 1 oz. of boiling alcohol; and after standing ten days, add a further 20 gr. of silver dissolved as before in 2 dr. of alcohol.

2. Rapid. Ether, sp. gr., ·720, 4 fl. oz.; alcohol, sp. gr., ·820, 2½ fl. oz.; pyroxyline, 40 grs.: Castile soap, 30 grs.; bromide of cadmium and ammonium, 56 grs.

Sensitise with 125 gr. silver nitrate, dissolved as before in 1 oz. of alcohol with the aid of heat. In 12 hours' time add 30. gr. more of the double bromide of ammonium and cadmium dissolved in ½ oz. of alcohol.

For Washed Emulsion (for Transparencies).

Ether, 5 fl. oz.; alcohol, 3 fl. oz.; pyroxyline or papyroxyline, 60 gr.; bromide of cadmium and ammonium, 100 gr.; or, bromide of zinc, 96 gr.; hydrochloric acid, sp. gr. 1·2, 8 minims.

Sensitise with 20 gr. of silver nitrate to each oz., dissolved in a minimum of water, with 2 dr. of boiling alcohol. Allow to stand two or three days.

N.B.—In the last three formulæ, the emulsion, after being allowed to ripen for the time stated, should be poured into a dish and allowed to become thoroughly dry. The mass of dry emulsion is then washed to remove all soluble salts, and is then again dried and redissolved in equal parts of ether and alcohol, at the rate of 20 to 24 gr. to each oz. of the solvents.

Developing Solutions for Collodion Emulsion.

A. Pyrogallic acid, 96 gr.; alcohol, 1 fl. oz.;
B. Potassium bromide, 10 gr, ; water, 1 fl. oz. ;
C. Liquor ammoniae, sp. gr. ·880, 1 fl. dr.; water, 15 fl. dr.; or D. Ammonium carbonate, 2 gr.; water, 1 fl. oz.

For each drachm of developer take, for a normal exposure, 5 minims of A., 1 or 2 minims of B , and 1 or 2 minims of C., or if D. be used, add the above quantities of A, B, and C to 1 dr. of D. When the details of the image are out add double the quantities of B. and C.

Intensifying Solution for Collodion Emulsion.

Silver nitrate, 60 gr.; citric acid, 30 gr.; nitric acid, 30 minims; water, 2 fl. oz. To each dr. of a 3-dr. solution of pyrogallic acid add 2 or 5 minims of the above, and apply until sufficient density is attained.

GELATIN DRY PLATES.

In these a film of gelatin is used to carry the sensitive salts instead of collodion. The commercial dry plates of the present day are almost all prepared with gelatin, and being made in enormous quantities are of wonderfully uniform quality, easy to develope, and with proper treatment can be made to serve almost any purpose, though as will be seen later there are still some branches of photography in which the old wet collodion process holds its own. The use of gelatin in the preparation of plates has of late years revolutionised photography, reducing the cost and the trouble and skill required for picture taking to a minimum; while the relative cheapness of the necessary apparatus has put the art within the reach of all. The processes are cleanly in the extreme, and "photographer's fingers" are unknown except to the professionals, who are still obliged to use the wet process. Photography can now no longer be said to be a "black art."

GELATIN DRY PLATE PROCESSES.

Gelatino-Bromo-Iodide Emulsion (Abney).
1. Potassium iodide, 5 gr.
2. Potassium bromide, 135 gr.
3. Nelson's No. 1 photographic gelatin, 30 gr.
4. Silver nitrate, 175 gr.
5. Autotype gelatin, 240 gr.

Or the same quantity of a mixture of Nelson's No. 1, 3 parts, with a hard gelatin, such as Heinrich's, 1 part.

Cover Nos. 3 and 5 with water, stir well, and pour off to get rid of dust; dissolve 1 and 2 in 1 dr., and 1½ oz. of water respectively. To the solution of No. 2 add 1 minim of strong hydrochloric acid, and enough of a solution of iodine in alcohol to make it a deep sherry colour. No. 3 is swelled for 10 minutes in 1 oz. of water, and then dissolved by heat. No. 4 is dissolved in ¾ oz. of water and heated to about 120° F.

In the Dark Room.

No. 3 and No. 4 are mixed and shaken in a bottle till a perfect mixture is secured; three quarters of the solution of No. 2 is then dropped in little by little and shaken well, and then No. 1 is added to the remaining quarter of the solution of No. 2, and the mixture added as before. The emulsion should appear of a ruby colour, when a thin film of it is examined by a gas-light.

Boiling the Emulsion.

The bottle containing the emulsion is set in a saucepan of water, which is brought to the boiling-point, and kept there for 45 minutes, and shaken occasionally during the process.

Cooling and Washing.

The gelatin No. 5 having been washed to get rid of dust, swelled in 2 oz. of cold water, and then melted at a temperature of 100° F.; *the emulsion in the bottle and the gelatin No.* 5 are cooled to 70° or 80° F., and then well shaken and mixed; the mixture, after the froth has subsided, is poured into a flat porcelain dish, and allowed to set. A piece of very coarse canvas or mosquito netting is now prepared by washing and boiling, so as to be perfectly clean, and the emulsion is scraped out of the dish by means of a piece of *clean* glass, placed in the canvas, and squeezed through it by twisting *under* water; the shreds are then again put on the canvas, stretched over a sieve or jar, and washed by pouring a couple of gallons of water over the mass; it is then again squeezed through the canvas, and again washed as before. The operation is repeated once more, and the whole is then left for a short time in a jar of water, which is changed a few times. Captain Abney regards the repetition of the squeezing as equivalent to 12 hours' washing.

Draining the Emulsion.

Allow the shreds to drain for 2 or 3 hours on the canvas through which it has been squeezed.

Dissolving the Emulsion.

Melt in a clean jar or pot at about 120°, and add ¼ gr. of chrome alum dissolved in 1 dr. of water, and stir well during the addition. Now add 6 dr. of absolute alcohol, and filter through wet chamois leather or two thicknesses of swansdown calico previously well boiled and washed. The filtered material should be received into a glass flask which will bear heat, and is then ready for coating the plates. The above résumé of Captain Abney's account of the method of making a washed gelatin emulsion will serve as a general outline of the processes necessary. Those who wish to make their own plates are strongly recommended to consult his work, 'Photography with Emulsions.' The plates will require to be carefully dried in a light and dust-tight apparatus, such as that described in 'Burton's ABC of Photography.'

The object of the washing is to remove the soluble salts and to increase the sensitiveness of the product.

Captain Abney found that regarding the unwashed emulsion as 1, the first squeezing and washing increased the sensitiveness to 1¼, the second to 2½, and that the third gave the same result, *i. e.* two squeezings and washings are sufficient.

Formula for Gelatin Emulsions (Bennet's).

Ammonium bromide, 70 gr.; pure silver nitrate, 110 gr.; gelatin, 200 gr.; distilled water, 6 oz.

Paget's Prize Emulsion.

Pure hydrochloric acid, 1 dr.; distilled water, 12½ oz. Put into a 20 oz. bottle; of the above, dilute acid, 20 minims; distilled water, 3 fl. oz.; ammonium bromide, 210 gr.; Nelson's No. 1 gelatin, 80 gr. Dissolve 330 gr. pure silver nitrate in 3 oz. of distilled water. Pour about 2 dr. of this solution into another vessel and dilute with an equal bulk of water. Heat the 20 oz. bottle and contents *gradually* until the gelatin is dissolved; pour in the 4 dr. dilute silver solution, and shake well for half a minute, adding the other ¼ oz. at a time, shaking well after each addition. Now bring the whole to a high temperature by immersion in boiling water for fifty-five minutes, and then cool as quickly as possible to 90° F. Now take 1 oz. of Nelson's No. 1 gelatin and soak in 10 oz. of clean water till 4 oz. are absorbed; this should be done previously. Pour off the unabsorbed 6 oz. of water and add the rest after melting to the contents of the 20 oz. bottle; shake well and mix thoroughly, then pour into a clean beaker, and wash and squeeze as before, but using 3 oz. of a saturated solution of potassium bichromate to 3 pints of cold

water; in this the shreds are left for one hour, it is next squeezed again twice into clean cold water. Then strain and put cloth and all into a clean beaker, immersed in hot water till all is melted. With a clean hand take out the cloth and squeeze again; now add 2 oz. of alcohol and make up to 20 oz. with clean water. Filter and coat the plates.

W. K. Burton's Gelatin Emulsion ('British Journal Photographic Almanac,' 1891.)

a. Potassium bromide, 260 gr.; potassium iodide, 20 gr.; gelatin (Nelson's No. 1), 80 gr.; distilled water, 10 oz.

b. Fused silver nitrate, 200 gr.

c. Silver nitrate, 200 gr.; distilled water, 1 oz. Converted to ammonio-nitrate.

d. Gelatin, hard (dry), 600 gr.

Burbank's Gelatin Emulsion.

Water, 1 oz.; ammonium bromide, 15 to 20 gr., or potassium bromide, 18 to 25 gr.; silver nitrate (proportioned to the amount of bromide), 25 to 30 gr.; gelatin, 30 to 40 gr.

Developing Formula for Gelatin Dry Plates.

Almost every maker of dry plates in the present day sends out with each box a formula or formulæ for their development, by which it may be supposed the best possible results may be obtained, and it is not altogether fair to complain of a certain brand of plate until the maker's formula for development has been thoroughly tried. The number and variety of these special developers is so great that it is quite impossible to give them in detail. Those who wish information on these developers should consult a table prepared by Messrs Lyonel Clark and E. Ferrero, in the 'British Journal Photographic Almanac,' 1891, in which 65 different developers are analysed and presented in a tabular form, so that the composition of any given developer may be seen at a glance. Those who practise photography on a large scale and as a business, confine themselves as a rule to one make of plate and one developer. Experience acquired in this way generally gives better results than constant change of plates and developing solutions.

General Principles of Development.

A sensitised plate which has been exposed in the camera under proper conditions exhibits to the eye no change whatever, but that some change has taken place, profoundly modifying the silver salts in the film in those parts of it on which the light has acted, becomes evident at once on the application of certain chemical reagents which, as they result in the production of an image or picture where there was none before, are called developers. We may therefore define a developer as a substance which acts upon those portions of a compound, sensitive to light, which have been exposed to the action of light, in a manner different from its action upon the same body which has not been so exposed.

The simplest case is the action of light on ferric salts. A piece of paper which has been coated with a solution of ferric chloride and subsequently exposed to light, one portion of it being covered by some opaque object of definite form such as a coin, will be found to be slightly altered, sufficiently at all events for the outline of the coin to be just visible, the salts on the parts of the paper exposed to light being converted into the ferrous state, that protected by the coin remaining in the ferric condition. If now the paper be washed with a solution of potassium ferricyanide, the area exposed to the light becomes blue, the area protected by the coin remaining white and an image in white on a blue ground is thus produced or *developed* by the ferricyanide thus:

$$3Fe_2Cl_4 + 2K_3Fe_2Cy_{12} = 3Fe_2(Fe_2Cy_{12}) + 12KCl.$$

similarly paper treated with a *uranic* salt will yield a brown print.

The chemistry of the reactions is more or less simple and comprehensible. What exactly occurs in the case of the silver bromide is by no means so clear, but in the result the silver compounds in the plate are so acted on by light that upon the application of the developer reduction takes place and metallic silver is deposited, and this in proportion to the amount of action which has taken place. The highest lights in the original are represented in the developed plate by the greatest amount of reduction, *i. e.* the darkest shadows and *vice versâ*, *i. e.* the developed plate is a *negative* from which the original or *positive* picture has to be obtained by a repetition of the process, viz. the exposure of a sensitive film *under* the negative to the action of light.

Two general methods of development are in universal use, (1) acid and (2) alkaline, a reducing agent in acid solution in the first case and in alkaline solution in the second.

Acid Development. This is usually effected by means of a solution of ferrous oxalate prepared as follows:

1. Saturated solution of ferrous sulphate.
2. Saturated solution of potassio oxalate (neutral), add a trace of oxalic acid to the oxalate solution to prevent possible alkalinity and pour the *ferrous sulphate into the* oxalate in the proportion of 3 parts of the latter to 1 of the former. Less of the sulphate may and often will suffice, and it will be found that if much more be added a precipitate will form which will cause great trouble.

This developer requires care in use, as it is very energetic and not very easily controlled in its action. The addition of a drop or two of a 2% solution of potassium bromide slows the action and enables the operator to watch the process and, if necessary, remove the plate and wash it before the action has gone too far and resulted in " fog," *i. e.* the general reduction of the silver all over the plate. It is a good plan to begin by flooding the plate with a strong developer diluted with its own bulk or rather less of water, conducting development with this till all detail appears and then finishing the operation with some fresh undiluted ferrous oxalate to which a few drops of the bromide solution have been added to give density. The development should be continued until the image is visible on the back of the plate, which is then to be well washed in water and transferred to a saturated solution of alum which hardens the film and decomposes any calcium oxalate which might have been formed on the plate by the action of the developer on the wash water. After two or three minutes' immersion it is again well washed

and placed in the fixing bath (hyposulphite of soda). As soon as all traces of the white film have disappeared the plate may be removed and thoroughly rinsed under the tap to get rid of excess of hyposulphite; it should then be soaked for several hours in water which is frequently changed to remove the last traces of hyposulphite, and lastly set on its edge to dry in a current of air, but protected from dust. If required in a hurry, the negative, after removal from the fixing-bath, may be well rinsed, allowed to drain, and then put into a dish full of clean methylated spirit. Five or ten minutes' immersion will suffice to remove the water from the film, and, on draining the plate, will be found to dry very rapidly and will be ready for printing from in a few minutes, especially if it be gently warmed. Negatives so treated should as soon as possible be soaked in water for several hours, so as to ensure the complete removal of the hyposulphite.

Any well-made gelatin plate should bear ferrous oxalate development if due care be used. The resulting negatives are exceedingly clean and bright, if the exposure has been correct, and are specially suited for platinum printing on account of their vigour and density. In unskilled hands they are apt to be a little hard and to show too great contrasts. The process is still very largely used on the Continent, though other developers have taken the place of iron in England and America almost entirely.

Captain Abney recommends a ferrous-citro-oxalate developer made as follows:

1. Potassium citrate, 700 gr.; potassium oxalate, 200 gr.; water, 3½ oz. 2. Ferrous sulphate, 300 gr.; water, 3½ oz. Mix in equal proportions.

Alkaline Developers for Gelatin Plates.

The variety of formulæ published for these developers is so great, every maker and almost every user having some modification of his own, that a few general examples must suffice.

PYROGALLOL DEVELOPERS.

The editor has found the following to work exceedingly well:

1. Pyrogallol, 1 dr.; water, 10 oz. 2. Potassium bromide, 1 dr.; liquor ammoniæ, 1 dr.; water, 10 oz. Mix in equal proportions.

The addition of a little sodic sulphite is advantageous, preventing stains, and tending to preserve the pyrogallic acid solution.

The Britannia Works Co. recommend for their Ilford plates:

1. Pyrogallic acid, 1 oz.; ammonium bromide, 600 gr.; water, 6 oz.; pure nitric acid, 20 drops.
2. Liquor ammoniæ ('880), 3 dr.; water, 1 pint.
3. No. 1 solution, 1 oz.; water, 19 oz.

For developing, mix No. 2 and 3 in equal proportions just before using.

The ' British Journal Photographic Almanac,' 1891, gives the following:

1. (*a*) Sulphite of soda, 6 oz.; hot water, 32 oz.; pyrogallic acid, 1 oz.; citric acid to acid reaction.

(*b*) Carbonate of soda, 3 oz.; carbonate of potash, 1 oz.; water, 32 oz.

Mix just before using in equal proportions, and then add water to twice the bulk of the mixture.

2. (a) Sodium sulphite, 4 oz.; warm distilled water, 4 oz.; when cooled to 70° F. add sulphurous acid (strongest obtainable), 3½ oz.; pyrogallol, 1 oz.

(b) Potassium carbonate, 3 oz.; water, 4 oz.; sodium sulphite, 2 oz.; water, 4 oz.

Dissolve separately and then mix in one solution.

For 2 oz. of developer, take 1 dr. of a (= 6 gr. pyro.), and 20 minims of b, add water to 2 oz. If development has to be pushed, add 20 minims more of b, but the total of b used should not exceed 2½ dr. in 2 oz.

After using pyrogallic acid developers it is well to immerse the plate in a strong solution of alum containing a small quantity of citric acid. This hardens the film and also greatly tends to prevent staining of the plate. After a thorough rinsing under the tap, the plate should be soaked for a while (10 minutes) in clean water, and then placed in the fixing bath. The object of the soaking is to remove the last trace of acid in the plate, which would otherwise decompose the hyposulphite, cause the formation of sulphuretted hydrogen, and consequent staining of the negative. Sufficient ammonia added to the fixing bath to make it smell perceptibly is an advantage, as decomposition is thereby effectually prevented.

Hydroquinone Developers.

The editor has found the following very reliable and uniform in its action:

Hydroquinone, 2½ dr.; potassium carbonate, 7½ dr.; sodium sulphite, 10 dr.; water to 30 oz.

A dilute solution of potassium bromide (2%) should be kept at hand for use when a restrainer is necessary.

The following are from the 'British Journal Photographic Almanac,' 1891:

1. Hydroquinone, 1 part; sodium sulphite, 2 parts; sodium carbonate, 10 parts; water, 67 parts.

2. (a) Hydroquinone, 4 gr.; meta-bisulphite of potash, 4 gr.; potassium bromide, 1 gr.; distilled water, 1 oz.

(b) Potassium hydrate, 10 gr.; distilled water, 1 oz.

Of each equal parts.

3. (a) Hydroquinone, 80 gr.; citric acid, 10 gr.; sodium sulphite (recryst), 80 gr.; distilled water, 20 oz.

(b) Caustic potash (fused), 160 gr.; sodium sulphite, 160 gr.; distilled water, 20 oz.

(c) Potassium bromide, 24 gr.; distilled water, 1 oz.

(d) Caustic potash, 160 gr.; distilled water, 20 oz.

For normal exposures use equal parts of a and b, adding 5 minims of c to every ounce of solution. For over exposed plates use d instead of b, with an extra quantity of c. For under exposed plates omit c, and in extreme cases add 6 or 8 gr. more of sulphite of soda to each ounce of the developer. The object of increasing or decreasing the quantity of sulphite is to give greater or lesser density.

4. (a) Hydroquinone, 160 gr.; sodium sulphite, 2 oz.; citric acid, 60 gr.; ammonium bromide, 20 gr.; water to 20 oz.

(b) Potassium carbonate, 2 oz.; sodium carbonate (cryst), 2 oz.; water to 20 oz.

Of each equal parts.

5. (a) Hydroquinone, 15 gr.; sodium sulphite, 75 gr.; water, 5 oz.

(b) Potassium carbonate, 90 gr.; water, 5 oz.

(c) A 10% solution of potassium bromide.

Use equal parts of a and b, with 2 or 3 drops of c.

6. Potassium bitartarate, 90 gr.; potassium sulphite, 45 gr.; potassium carbonate, 4 oz.; water, 16 oz. Filter, and add hydroquinone, ½ oz. For use, one part diluted with sixteen of water.

For Chloride Plates.

7. Hydroquinone, 2 gr.; sodic sulphite, 10 gr.; potassium or ammonia carbonate, 10 gr.; potassium bromide, ¹⁄₁₀ gr.; water, 1 oz.

Maker's Formula for Ilford Plates.

a. Hydroquinone, 160 gr.; potassium bromide, 30 gr.; sodic sulphite, 2 oz. avoir.; water, 20 oz.

b. Sodio hydrate, 100 gr.; water, 20 oz.; for use take equal parts of each and mix.

Hydroquinone is a trifle slower in its action than pyrogallic acid, but does not require such constant rocking during development. The negatives have a characteristic black colour, are very clean, and print well. Those developers containing caustic alkali are apt to give rather grey negatives, and require more attention in use than those made with alkaline carbonates.

Eikonogen Developers.

Eikonogen is an amido-β-naphthol-β-monosulphonate of sodium, introduced about three years ago by Dr Andriessen as a developer for dry plates. He claims for it that it gives a greater range of half-tone than other developers, and that the negatives are softer and more delicate.

Maker's Formula.

1. a. Sodic sulphite (crystallised), 40 grms.; eikonogen, 3 grms.; distilled water, 500 c.c.

b. Potassium carbonate or calcined soda, 60 to 75 grms.; distilled water, 500 c.c. For use of each, equal parts; mix.

2. a. Sodic sulphite, 4 parts; eikonogen, 1 part; water, 10 parts. Boil and stir till solution is effected, then pour into a flask containing 50 parts cold water.

b. Sodic carbonate, 3 parts; water, 20 parts. Mix 3 parts of a with one of b immediately before use.

Fixing Bath for Negatives Developed with Eikonogen.

Sodium hyposulphite, 4 parts; sodium bisulphite, 1 part; water, 20 parts.

Fixing Bath for Gelatin Dry Plates.

Sodium hyposulphite, 1 oz.; water, 10 oz.

Washing Negatives.

Where one or two only have to be washed it will be sufficient to rinse thoroughly after removal from the hyposulphite, and then place in a large basin full of water for 8 or 10 hours, changing the water two or three times during this period. When a large number of negatives are to be washed it is best to employ a tank with a moveable frame inside to hold the plates and an arrangement by which a constant stream of water shall pass in at the bottom and overflow at the top. A great number of contrivances are sold by the dealers in photographic apparatus for washing negatives, some of which are very ingenious and effective. Before setting a plate in the rack to dry after removal from the washing tank it is well to examine its surface for deposits of lime salts from the water, which if allowed to dry on the plate would damage it seriously by causing minute holes in the film. A good flooding with water under the tap (or better a rose) should be given, and if necessary the film should be lightly rubbed with a tuft of cotton wool, and again flooded before setting aside to dry. Dust should be carefully excluded from places in which plates are drying, or irremediable damage may result. Artificial heat, beyond that of a warm room, should not be employed unless the plates have been previously treated with alcohol.

Intensifying Negatives.

It frequently happens that a plate is sufficiently over exposed to give a print, thin negative wanting in contrast, or it may not be up to the standard of density required for printing in quantity, *i. e.* it requires too much looking after during the process. In these cases intensification is resorted to in order to strengthen the image and give the required density. There are many methods in use, but the simplest, and one of the best is to immerse the plate, *after thorough washing*, in a solution of bichloride of mercury, 20 gr.; ammonium chloride, 20 gr.; water; 1 oz. (*J. England*). The surface gradually becomes greyish white, and when the deposit is judged to be sufficient the plate is again *thoroughly washed*, placed in a clean dish and treated with a solution of ammonia, 10 drops to the ounce of water; the film first turns brown and then black, it is again *thoroughly washed* and set aside to dry. Mr B. J. Edwards uses the following:

a. Mercuric chloride, 60 gr.; water, 6 oz.
b. Potassium iodide, 90 gr.; water, 2 oz.
c. Sodium hyposulphite, 120 gr.; water, 2 oz.

Add *b* to *a*, and then add *c* to the mixture, immerse the plate and wash. Captain Abney says that negatives thus intensified turn yellow after a time.

Dr Eder's Intensifier.

Uranium nitrate, 15 gr.; potassium ferricyanide, 15 gr.; water, 4 oz.; immerse the plate. Allow the action to proceed as far as necessary, then remove from the solution and wash. Every trace of hyposulphite must be removed before applying the intensifier or the shadows will be veiled. Simple and permanent.

Over intensification with mercury can be reduced or entirely removed by immersing the plate in a solution of sodium hyposulphite.

To Reduce over Dense Negatives. Immerse in a solution of potassium ferricyanide, 3 gr.; 5% solution of sodium hyposulphite, 1 oz.; use as soon as prepared and with caution. A *very weak* solution of ferric chloride may also be used for the same purpose, but the greatest care must be used to prevent the disappearance of the picture. In some cases it is best to allow the reduction to go on to a considerable extent; wash thoroughly and intensify to the desired pitch. Local reduction may be effected by the cautious use of a weak solution of ferric chloride, but it must be borne in mind that this is a very active

reagent, and valuable negatives should not be tampered with except by persons skilled in its use.

VARNISH FOR NEGATIVES.

1. A saturated solution of seed-lac in methylated spirit, thinned down to a proper consistency, makes an excellent negative varnish.

2. Sandarac, 4 oz.; alcohol, 28 oz.; oil of lavender, 8 oz.; chloroform, 5 dr.

3. (Cheap.) White hard varnish, 15 oz.; methylated spirit, 25 oz.

4. Shellac, 1¼ oz.; mastic, ¼ oz.; oil of turpentine, ¼ oz.; sandarac, 1¼ oz.; Venice turpentine, ¼ oz.; camphor, 10 gr.; alcohol, 20 fl. oz.

5. Sandarac, 90 oz.; turpentine, 36 oz.; oil of lavender, 10 oz.; alcohol, 500 oz.

6. Sandarac, 2 oz.; seed-lac, 1 to 1½ oz.; castor oil, 3 dr.; oil of lavender, 1½ dr.; alcohol, 18 fl. oz.

NEGATIVE NETORICHING VARNISH.

7. Sandarac, 1 oz.; castor oil, 80 gr.; alcohol, 6 oz.

Dissolve the gum in the alcohol, then add the oil.

GROUND-GLASS VARNISH.

8. Sandarac, 90 gr.; mastic, 20 gr.; ether, 2 oz.; benzol, ½ to 1½ oz.

The benzol determines the character of the surface. Nos. 2 to 8 are from the 'British Journal Photographic Almanac.'

EXPOSURE OF PLATES.

Elaborate tables have been constructed by which the exact duration of the exposure with a given lens, stop, and plate under all conditions of time, place, weather, and season of the year may be calculated. These tables are interesting, and may be of use to the beginner, but the judgment which comes of experience is a far more reliable guide. It is well to note the stop used and the exposure given under certain conditions for reference in case the same object has to be photographed again.

STORING NEGATIVES.

All manner of contrivances have been devised for this purpose—cupboards fitted with racks, grooved boxes, envelopes, files, &c. The simplest plan is to place the negatives carefully *without paper or other packing*, one on the top of the other in the empty plate boxes. If these are labelled, and the plates numbered and a catalogue kept, there is no difficulty in finding a negative required, nor is there any risk of damage if ordinary care be used in handling them. Specially valuable negatives should be kept in grooved boxes. It is sometimes forgotten by photographers that glass is a heavy article, and that the weight of some hundreds of large negatives packed closely together is not merely considerable but enormous. Studios are often on the upper floors of houses, not perhaps too well built, and never intended to bear a ton weight concentrated on a very small surface.' Old negatives, especially if of small size below 'whole plate,' are almost valueless; the larger sizes can sometimes be got rid of for glazing greenhouses if there be a nursery near the studio. If not, the cost of carriage is greater than that of new glass specially made for the purpose.

THE CALOTYPE PROCESS.

This is a process by which negatives may be obtained upon paper, and in spite of the introduction of such materials as Eastman's negative paper, celluloid films, and other excellent means of securing lightness and portability, the calotype process is so simple and the results so satisfactory that it is worthy of being better known and more used than is actually the case. It is one of the oldest negative processes, and has been but little improved since its introduction by Fox Talbot.

The Paper.—Any good, hand made, white paper, free from grit and chalk will answer. What is known as medium Saxe is recommended by Captain Abney. If this cannot be obtained it is well to treat an untried paper with dilute hydrochloric acid, brushing it over the surface, and then thoroughly washing and drying, taking care that it dries *perfectly flat.* When dry and ready for coating, silver iodide is prepared as follows: Take of (a) Silver nitrate, 3 grms.; distilled water, 20 c.c. (b) Potassium iodide, 3 grms.; distilled water, 20 c.c.; and dissolve each separately. Pour b into a; allow the precipitate to settle and pour off the supernatant liquid. Wash the precipitate thoroughly with several lots of distilled water by stirring, allowing to settle and decanting, and dissolve it in the following: Potassium iodide, 30 grms.; water, 60 c.c.; this will not completely dissolve the iodide, and crystals of potassium iodide must be added till after much stirring a semi-transparent and milky solution is obtained.

The paper is cut to a proper size and pinned on a flat board. Next prepare a brush by taking a tuft of clean cotton wool, putting a loop of thread round it and drawing it through a piece of glass tube, leaving enough of the wool outside to form a brush. The paper on the board is now evenly coated with the iodide and allowed to partially dry, it is then immersed in three changes of distilled water in a dish, and after two or three hours' washing, to remove potassium iodide, it is hung up to dry in a dark room. When dry it may be stored between the leaves of a clean copying-book for future use.

When required for use, pin on the board as before, and with another cotton-wool brush cover the surface with:

1. Silver nitrate, 5 grms.; glacial acetic acid, 8 c.c.; water, 50 c.c.

2. Saturated solution of gallic acid in distilled water.

To every cub. cent. of No. 1 add 60 c.c. of distilled water, next 1 c.c. of No. 2, and finally 30 c.c. of distilled water. Apply the mixture plentifully but lightly to the iodised paper with a cotton-wool brush, and take up all excess with the best filtering paper. Two sheets are placed back to back, with blotting-paper between, between two sheets of glass in the dark slide of the camera. The paper is most sensitive when moist.

To develop, pin out the paper as before, and apply equal parts of Nos. 1 and 2 as before. When the mixture begins to fail in its action use No. 2 alone until the deep shadows begin to get dim by transmitted light, the sheet should then be at once immersed in a 6% solution of sodium

hyposulphite and washed for several hours in running water. When thoroughly washed and dried, the whole of the paper except the sky, if there be any, should be impregnated and rendered translucent by wax worked into it by a moderately hot iron.

Captain Abney advises beginners to use the process for making prints from negatives at first, and to try it in the camera when some experience of manipulation has been obtained. He strongly recommends the process to travellers, as the apparatus required is very small. "A few dozen sheets of iodised paper, and a chest containing silver nitrate, gallic acid, and a bottle of acetic acid and sodium hyposulphate being all the chemicals required; scales and weights, the camera and its legs, a couple of pieces of clean glass the size of the slides, a few drawing pins, a folding dish, a cotton-wool brush-holder, and a candle-shade, complete the apparatus."

PHOTOGRAPHIC APPARATUS.

We have thus far discussed only the various processes by which sensitive media may be prepared on which a photograph may be taken, and the means by which the action of the light may be made apparent, and the negative fixed and preserved. Without entering into great detail it will be well to point out some of the essential features of the apparatus required for various kinds of photography, but excluding the photomechanical processes which will be treated separately.

THE DARK ROOM.

The construction and arrangement of the dark room will very greatly depend upon the nature and amount of the work done in it. The accommodation required for the mere development of dry plates is small, and the appliances required exceedingly simple; but if emulsions are to be mixed, plates coated, and several different processes worked side by side; in fact, if the dark room is to be what it should be, a photographic laboratory, the conditions to be satisfied are much more complicated, and the need for plenty of space with properly constructed working benches, and an ample supply of water, with corresponding sinks and appliances for heating, drying cupboards, abundant shelf space for chemicals and apparatus; in fact, a chemical laboratory arranged for photographic work, and lighted in such a way that the sensitive materials used may be manipulated therein without risk. The plan and arrangement of such a laboratory hardly comes within the scope of this work, but some general instructions as to the conversion of existing premises may be of service.

Lighting.—The best light for any studio, laboratory, or workshop is undoubtedly a north light, and in a photographic laboratory this is perhaps of more importance than in others, as the direct rays of the sun are often difficult to control, and unless the control is practically perfect it is useless. The whole of the light admitted to the dark room must be made to pass through some coloured medium which is capable of absorbing or cutting off as much as possible of the actinic rays; for this purpose, deep orange stained glass or flashed ruby glass or a combination of the two should be used, and in order to avoid risk there should be no more glazing in the room than is required to give a good light for conducting the operations, and even then no glass, or other material, paper, cloth, stained varnishes, &c., *should be trusted until it has been thoroughly tried* by leaving a sensitive plate exposed for a reasonable time, e.g. ten minutes, to its action. This plate, a portion of which should be covered with some opaque material, is then to be developed as if it were a negative, and if after fixing it should be possible to detect the boundary of the covered part, it is proof that the light is unsafe, and the glass should be rejected or some additional precautions taken. There is hardly any medium which can be trusted implicitly for emulsion making where the exposure to light is of necessity prolonged, and accordingly it is best to employ artificial illumination for this purpose. Instead of glazing a window sash with ruby glass, the existing panes may be covered with ruby or canary fabric, orange or ruby tissue-paper, or even varnished with some coloured medium; the last plan is not to be recommended, as no varnish will long stand exposure to light without fading, and the light then slowly becomes unsafe, the cause perhaps remaining unsuspected until some valuable negative is spoiled. It is a great mistake to work in the dark, as with care it is possible to have an abundance of a perfectly safe and comfortable light. Where *artificial light* is used care should be taken to secure proper ventilation, and the free escape of the products of combustion, especially where gas is used, as the sulphuric acid produced by the traces of sulphur in the gas will find its way into everything kept in the room; and if there be an open water tank, the addition of a little barium chloride to a sample of the water from it, after the gas has been burning for a short time, will yield a heavy precipitate of barium sulphate, showing the presence in the water of either free sulphuric acid or sulphates, in this case probably sulphate of ammonia. The water tank should therefore be *outside* the dark room if the supply cannot be obtained direct from the mains. Every imaginable form of lamp which ingenuity can devise is in the market. A biscuit tin with a gas elbow soldered into the lid and a hole at one end with a small stove elbow set on it to carry off the hot air and prevent the escape of light, the bottom being cut out and replaced by two folds of canary fabric, makes a most excellent lamp. If gas is not to be had, a small paraffin lamp or a carriage candle set inside will answer every purpose. A cylinder made of two thicknesses of canary fabric set over a lamp or candle makes a good portable light for developing. The light which streams out of the top may be neglected if the plates be quickly immersed in the solution. The dark slides should be filled by the operator turning his back on the light in the furthest corner of the room. Very sensitive plates may be changed in this way *with care.* Experience and observation alone will show what precautions are necessary.

It is desirable that the walls of the dark room be covered with a varnished paper or be painted, or otherwise rendered smooth, so that as little

dust as possible may be deposited—to descend in a cloud, when some unusual vibration occurs, upon a plate. The dust of a laboratory is always chemical, *i. e.* contains particles of salts, &c., which, if they fall on a plate, are capable of producing minute spots, pin-holes, and other troubles. Reference has already been made to the arrangement of the water-supply where tanks are used. The tap of the developing sink should have a rose attached to it by a piece of flexible tubing, a fine spray of water washing far more effectually and quickly than a heavy stream.

Bottles and jars containing chemicals and solutions should be well closed by good stoppers and corks, and should be very *distinctly labelled,* and they should, if possible, always be kept in the same place, so that in the obscure light no mistakes may be made. The same applies to the dishes and other vessels used for the development, fixing, and washing of plates ; every care should be taken to prevent their being interchanged; for this reason it is well to use dishes of different materials for each purpose. There is, perhaps, nothing of so much consequence in photography as cleanliness, and it is well to acquire the habit of washing the hands after dipping them in any of the solutions used, especially the hyposulphite bath. Common yellow soap, plentifully used, will remove the "hypo" as effectually as anything. It is a mistake to use any other dishes for development than those made for the purpose to the size of the plates employed. The initial cost of a dish may be saved, it is true, but the waste of developer will more than compensate for this in a very short time.

Portable Dark Tents. These, which are an absolute necessity when the wet process is worked in the field, may be obtained in many different patterns from the dealers in photographic apparatus. It is not difficult to construct a folding frame large enough to admit the head and shoulders, and covered with some opaque material, e. g. two thicknesses of Turkey red calico and one of black, with a window of ruby or canary fabric to admit light, is easily constructed. The base should be about 21 inches by 13, and there should be enough material in the cover that when the frame rests on a table the operator may sit down with the cover tied round his waist, thus excluding all light. By developing at night all necessity for a tent is done away with, but the changing of plates in the field requires some such device. Changing boxes and bags, of every imaginable design, may be obtained of the makers, and require no notice here beyond the remark that a little ingenuity will save much expense.

The choice of a camera and lenses is a matter which calls for remark. The number of types and patterns in the market and the great range of price are such as to bewilder the inexperienced. The best advice that can be given to the would-be purchaser is first of all to make up his mind clearly the purpose for which he requires the apparatus, field, or studio, and the size of the picture for field work, and whether the apparatus has to be carried constantly. A ¼-plate, 5 × 4, or ½-plate apparatus will be the most convenient, as the weight increases rapidly beyond these sizes. The modern amateur photographer will not apparently entertain the purchase of a camera where weight (even whole plate) may not be reckoned by *ounces*; and the skill and ingenuity of the makers has proved equal to the task of providing cameras of extraordinary lightness—some of them marvels of workmanship. Much depends on the uses to which the instrument is to be put. For hard rough work and constant exposure to the weather in all climates the old, albeit somewhat heavy, pattern, made by Hare and Meagher, with rigid front, solid base board, and folding tail-piece, brass-bound, is still unsurpassed; and the amount of rough usage which one of these cameras will stand is simply astonishing. The modern type with taper bellows and folding front, in which every particle of material, which is not absolutely necessary, has been removed, are, as before said, wonderful examples of cabinet and brass work, but absolute rigidity cannot be secured without a certain amount of weight, and the folding front, which carries the lens, is, to practical workers, a serious objection, as any unsteadiness of the lens is fatal to the production of a picture. Price is another very important consideration. Really good work, either light or heavy, is always expensive; and though machinery has done much to lessen the cost of production, and cameras and lenses are to be had at very low prices, and not by any means to be despised on that account, it should be remembered that the camera is an optical instrument as much as the microscope, and that there is a certain necessary relation between price and quality.

The Lens.—The choice of a lens must be entirely governed by the use to which it is to be put, as the variety in photographic work is very great, and there is no such thing as a really universal lens, though there is one form which makes a considerable approach to this desideratum.

If we arrange the first simple lens which comes to hand in such a way that we can obtain an image on the ground glass of the camera its defects are at once apparent. The image will probably be very ill-defined excepting just at the centre. It will only occupy a small portion of the screen, will exhibit great distortion at the margins of the picture which will be very unequally illuminated, and there will be a distinct tendency to fringes of colour at the edges. For photographic as for any other purposes we require a lens, corrected as far as possible for spherical and chromatic aberration, and which shall at the same time cover the whole of the plate sharply to the corners. In certain cases, *e.g.* in portraiture, an additional quality is required in an extreme degree, viz. depth of focus as it is called, *i.e.* the capacity for bringing into one plane the various parts of a solid object.

The simplest form of photographic lens is *the landscape lens* (Fig. A) which is practically a simple meniscus, made of two or more pieces of glass cemented together for the purpose of correcting aberration; such a lens will define sharply distant objects in very different planes, and according to the angle of the lens will yield a sharp image of objects lying at a considerable angle to its axis; near objects, however, will appear blurred and distorted to a greater or less extent, and though it is possible to construct a lens of this kind of such an angle as to define near objects, the effect

produced is apt to be strained and unnatural, and the use of these wide angle lenses requires considerable discretion, though in cramped situations they are often the only means of securing a picture of any kind.

FIG. A.

Landscape Lens.

The Rectilinear Lens (Fig. B) is a combination or doublet, *i.e.* a pair of compound lenses separated from one another by a considerable interval; each combination consists of a compound meniscus, and the two are mounted with their con-

FIG. B.

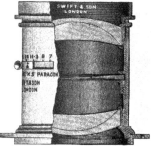

Rectilinear Lens.

cave surfaces facing one another with a slit in the mount between them for the introduction of the diaphragm. These lenses, if properly constructed, may be used for almost any purpose. They are, as their name implies, rectilinear, *i.e.* the pictures produced show no distortion, and they may be used for landscape, architecture, or copying.

The Portrait Lens (Fig. C) resembles the rectilinear in general construction, but the curves are worked specially for portraiture, and the combinations separated by a considerable interval in order to secure the depth of focus required. This reduces the field very greatly, so much so, that a good portrait lens will not produce a picture much larger than the area of its own surface, hence the great size and cost of portrait combinations intended for taking large pictures, one such, 6½ inches in diameter, by a maker of repute, costing between £70 and £100.

FIG. c.

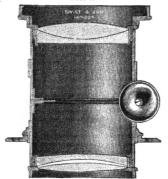

Portrait Lens.

The Triplet-Lens. This is a form of lens containing three sets of combinations for the purpose of preventing distortion. These lenses are now obsolete, but those made by Ross and Dallmeyer still command a good price second hand, as for certain purposes, especially copying drawings, maps, letterpress, &c., they can hardly be surpassed. The increase in the number of surfaces, and consequent loss of light, render them slow in action, but in many cases this is a distinct advantage.

. *The Diaphragm.* The chief use of the diaphragm is to cut off the oblique rays of light passing through the lens and thus further correct the spherical aberration and increase the sharpness of the picture. This sharpness is, however, obtained at the obvious expense of the illumination, and in order that the exposure necessary for a given make of plate with diaphragms of different sizes may be known with some degree of accuracy, the size and position of the diaphragm must be adjusted according to certain rules. In a doublet lens, if the combinations have the same focal length, the diaphragm should be exactly midway between them in order to diminish distortion in the highest degree. In some of the older lenses a circular patch of light always made its appearance in the centre of the picture called a "flare spot;" this was due to the surface of the lens reflecting the image of the aperture in the diaphragm. By altering the position of the stop very slightly, the defect may be overcome, but the remedy may result in the reappearance of distortion. Something further may be done by altering the distance between the combinations, but there is reason to believe that the defect is only distributed over the whole plate, and may in some cases result in general "fog" over the developed negative.

Relative Sizes of Diaphragms. In order that the aperture of the diaphragms to be used with a given lens many have some numerical value, Mr Dallmeyer has proposed that they should bear a definite relation to the focal length; thus if the focal length of the lens be twelve inches, a diaphragm with an aperture two inches in diameter would be marked ⅙th, *i. e.* one sixth of the focal length, and so on with other apertures generally expressed *f*6.

The Photographic Society of Great Britain has established a uniform system of diaphragms for which a lens with an aperture one quarter its focal length has been taken as the unit of measurement and is marked "1," a stop of half the area which, when used, will require double the exposure to be given is marked 2, the next requiring double this exposure again is marked 4, and so on, 1, 2, 4, 8, 16, 32, 64, 128, 256, the last being the smallest stop in general use. Lenses with apertures greater than one fourth of their focal length are not common, when such exist, the apertures are marked by decimal fractions thus ·5 ·25,&c. The latter is the largest aperture obtainable in lens construction. These numbers in terms of the focal length of the lens may be written—

$$\frac{\cdot25}{2}, \quad \frac{\cdot5}{2\cdot828}, \quad \frac{1}{4}, \quad \frac{2}{5\cdot657}, \quad \frac{4}{8}, \quad \frac{8}{11\cdot314}, \quad \frac{16}{16}, \quad \frac{32}{22\cdot627}, \quad \frac{64}{32}, \quad \frac{128}{44\cdot25}, \quad \frac{256}{64},$$

It may happen that the stops supplied with a particular lens are not the society standard stops. In this case it will not be difficult to determine approximately the relation of the given set of stops to the standard. The focal length of the lens being known, a table can readily be calculated, showing what the exact diameters of a series of standard stops should be, and by comparing these figures with those obtained by measuring carefully the unknown stops, the relative values of these may be known with sufficient accuracy for most purposes.

Beginners in photography would do well to master this question of stops thoroughly, as success greatly depends upon it. A very clear explanation of the system, with tables to facilitate judgment of exposure, will be found in Burton's 'Modern Photography' ('Piper and Carter's Photographic Handy Books,' No. viii).

Focal Length of Lenses. For making small pictures from which enlargements are intended to be made there is a distinct optical advantage in the use of a lens of short focus, the picture being theoretically sharper than if taken the same size direct. In practice this will be found to be of but limited application, and it is now recognised that in order to obtain true artistic effect the focal length of the lens must be suited to the subject, and lenses of long focus are now being largely used for landscape work as giving a more truthful picture. This is especially the case in hilly or mountainous country. A lens of focal length exactly equal to the longest edge of the plate used will be found generally useful, but where distant views of mountain tops are required a focus equivalent to two, or even three times the length of the long edge of the plate will be required.

To find the Focus of a Lens. The following simple method is given by Mr T. R. Dallemeyer, F.R.A.S., in the 'British Journal Photographic Almanac' for 1890:

Having marked on the base-board of the camera the position of the screen when the given lens is focussed on some very distant object, "mount the camera on a board covered with white paper; mark two fine vertical lines on the ground glass at a definite known and accurately measured distance apart; place a naked candle flame a considerable distance off in the room, and as nearly as possible in a line with the plane of the axis of the lens. First focus the flame in the centre of the ground glass, and mark a pencil line by the side of the camera on the paper on the board below; then, without using the board below, turn the camera on its axis until the candle flame is on the other vertical line, and make a second line by the side of the camera ; remove the camera now, and find the angle included between these lines; also mark the position of the camera back on the base. We have now all the data for finding the focus for parallel rays.

"Let the distance between the lines on the ground glass B C = a inches, the angle BAC = ϕ; the difference in the positions of the camera back

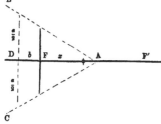

for the focus for the candle flame, and that registered from parallel rays (*i. e.* distant object) DF = b inches, and the focus for parallel rays AF (to be found) = x inches. Then—

$$\tan \frac{\phi}{2} = \frac{BD \text{ or } CD}{AD} \text{ or } \frac{\frac{a}{2}}{x+b}$$

from which x is easily found."

The focussing screen should be made of the finest ground glass. A piece of patent plate is best, set in its frame with the ground surface towards the lens. For very delicate work, *e. g.* microphotography, it is better to use plain glass, and focus with a suitable lens. Accurate focussing can only be accomplished in this way. Substitutes for focussing screens, in case of accident, may be made from an undeveloped plate or other piece of glass, using a lens for focussing as before. Care should be taken that the glass is close up against the rebate of the frame, and, if plain glass be used, a few very fine scratches should be made on the surface next the lens, there to be brought into focus with the picture, which will

then be sharp. For micro-photographic work fine parallel lines crossing one another to form small squares is the best marking.

Speed of Lenses.—There is considerable confusion in the text-books as to what constitutes a rapid lens, and why. Technically one lens is said to be more rapid than another, if with the same stop and the same plate, turned upon the same object at the same time ; the exposure necessary to obtain a normal result is shorter in the one case than the other. This may arise from several causes. A badly-corrected lens will require a great deal of stopping down in order to obtain a picture free from distortion ; consequently there will be great loss of light and longer exposure will be necessary. The quality of the glass of which lenses are made varies somewhat in its power of absorbing light, hence another cause of difference ; but the chief effect is produced in doublet lenses by the arrangement of the combinations. This is best illustrated by the portrait lens, in which both the combinations are widely separated. These are carefully corrected for distortion, so as to give a good picture without any diaphragm if necessary, and, as the image produced is only a little larger than the surface of the lens, it follows that more light reaches a given point on the plate than would be the case if the combinations were brought closer together, and the image spread over a larger area.

Instantaneous Photography. — The extreme sensitiveness of modern gelatin emulsions, and the skill of the makers in the construction of lenses, have led to an enormous development of the photography of moving objects. Hundreds of different makes of so-called "detective cameras" are in the market, with the result that in the majority of hands vast quantities of inartistic and technically poor negatives are produced. In portraiture, especially children's portraits, good lenses and rapid plates are invaluable, and when put to its proper use instantaneous photography is of great service. We have learned from it the details of the movements of men and animals, and the work of Marey and Muybridge in this direction has not been without its effect upon art and painters, teaching them that the conventional method of representing animals in motion, especially horses, was entirely wrong, and affording a curious example of how the constant uncritical use of something which is in itself wrong and purely conventional, may lead to its being regarded as correct.

There is no doubt that many of our greatest artists who paint subjects in motion study instantaneous photographs of these subjects with great care, and are thus enabled to avoid absurdities, in the representation of such subjects, as breaking waves, moving animals, &c., &c. The great French artist, lately dead, M. Meissonnier, paid great attention to the results of photographing animals in motion, and embodied the results of his studies in his painting.

Under the head of instantaneous photography, the question of *shutters* requires some mention. By this term is implied an apparatus by which the lens may be uncovered and covered again in a shorter period of time than is possible by uncapping and capping the lens with the hand. The number and variety of shutters in the market at the present time is literally legion, and it is quite impossible to enter here into a discussion as to which is the best form or type. A shutter constructed to work between the combinations of a lens, and which during the greater part of the exposure allows the lens to be fully open is, in the opinion of Professor Burton, the best form to use. But excellent work may be done with very simple apparatus, and a "drop shutter," constructed on the plan given in his manual already referred to, will be found, at first at all events, to serve most purposes, and the experience gained with it will assist the choice of a better instrument.

Substitutes for Glass as the Support. Space will not permit a detailed notice of all the various appliances which have been introduced for diminishing the weight of the sensitive plates, Negative paper is now an article of commerce, and various contrivances for using it either in single sheets or as a continuous roll are sold which will be found figured in the catalogues of most dealers in photographic apparatus. Films of celluloid, coated with emulsion, are now manufactured and are rapidly coming into use as substitutes for the heavy glass plates, and possess the additional advantages that they occupy far less space and are not liable to breakage.

Stereoscopic Photography. In order to obtain a stereoscopic negative, twin lenses of exactly equal focal length are used in the camera which is generally divided by a moveable partition, the axis of the lenses being set at about 2½ inches apart, in this way a double negative is obtained from which a print is made by any suitable process. This must be cut in two and mounted on a card, the right-hand print on the left of the card, and the left-hand on the right. Thus mounted and placed in the stereoscope the well-known effect of solidity is obtained. Stereoscopic photography is deservedly becoming popular again after some years of neglect.

PRINTING PROCESSES.

The production of a negative is, though perhaps the most important, only the first stage in any photographic process, which cannot be said to be complete until a positive print in some form or other has been obtained.

It will be obvious that a direct photograph of a *negative* taken in the camera will yield a *positive*, and this process is frequently resorted to especially where enlargement or reduction is required. These processes will be described later.

Fox Talbot's calotype process has already been mentioned as applicable to the production of positive prints on paper, and some of the earliest paper positives in existence actually produced by Fox Talbot were made upon a paper treated with silver chloride. Prints so made were dull and wanting in detail, and it was found better to coat the paper with albumen before sensitising in order to keep the image as near the surface as possible, and thus the modern silver print and albumenised paper originated.

Silver Printing. For this purpose a good paper is required which will bear soaking in water

and return afterwards to its original size without puckering. The papers known as Saxe and Rives are generally used for preparing sensitised paper, as they possess these qualities in a high degree.

Plain Salted Paper. Make a solution of Nelson's No. 1 gelatin in water (1 gr. to the ounce) by soaking and the subsequent use of heat, then add 3 gr. of ammonium chloride to every oz. of the solution, good Saxe paper is soaked in this solution, taking care to avoid air bubbles, and is then carefully dried. When dry it is floated on a solution of silver nitrate containing 50 to 60 gr. to the ounce of water, or it may be coated *by means of a brush* with a solution of ammonio-nitrate of silver made as follows : " Dissolve 60 gr. of silver nitrate in ½ oz. of water, and drop in ammonia until the precipitated oxide is exactly redissolved. Then divide this solution of ammonio-nitrate of silver into two equal parts, to one of which add nitric acid cautiously until a piece of immersed litmus paper is reddened by the excess of the acid ; then mix the two together, fill up to 1 oz. with water, and filter from the milky deposit of chloride or carbonate of silver, if any be found " (Hardwick). The solution must be kept in the dark. Paper coated with this ammonio-nitrate of silver solution keeps badly and should be used soon after it is made. It prints much more quickly than paper sensitised with silver nitrate alone. Plain paper may be fixed, washed, and toned in the usual way, but a more dilute toning bath is required, about 1 gr. of gold to 16 oz. of water is sufficient.

Albumenising Paper. According to the size and quantity of the paper to be albumenised eggs are taken and the yolks very carefully separated from the whites (each egg will coat about two sheets of paper, 22 × 17 in.). The whites are then mixed and measured and 8 gr. of ammonium chloride dissolved in the smallest possible quantity of water, added to each ounce. After " salting " the albumen is beaten into a very fine froth and set aside for 24 or 48 hours, it is then filtered through two thicknesses of fine muslin. The filtered albumen is poured into a flat dish and the paper is floated on it, every care being taken to avoid air bubbles. As soon as the paper has ceased to curl and lies flat on the surface of the albumen it is drawn off by one end and hung up to dry over a roller of wood 2 or 2½ in. in diameter. This operation may be performed twice, the paper after the first drying being passed through a bath of methylated spirits 4 parts, water 1 part, to coagulate the first layer of albumen ; it is then dried again and floated once more ; when dry it is ready for use.

The Sensitising Bath. A 50 gr. to the ounce solution of silver nitrate. As each sheet sensitised takes away silver from the bath it gets weaker and must be replenished, this is best done by adding ½ oz. of a solution of silver nitrate 100 gr. to the oz., after each sheet if the bath be a small one ; 2 oz. after every eight sheets will suffice if the bath contain a gallon or more of solution.

Floating the Paper. The albumenised and salted paper is first damped and made limp by keeping for a few hours in a damp cellar, or by careful exposure in a box or cupboard to the action of steam from a bowl of boiling water. The paper is floated on the bath in just the same way as on the albumen, but for a longer time, viz. about half an hour; the time varies with the strength of the bath, temperature, and other conditions. Burton recommends as a test to brush a little solution of potassium chromate *on the back* of the sheet in one corner or on a separate piece of the same paper, and to continue the floating until this spot has become of a deep orange colour. The paper is then removed carefully from the bath, drained a little and hung over a roller to dry in the dark in a warm, well ventilated room. Care should be taken to provide for catching the solution which drips from the corners; a small piece of good white blotting-paper is perhaps the best, as it attaches itself readily and can be afterwards burned and the silver recovered.

Fuming the Paper.—If the sensitised paper be exposed to the action of the fumes of ammonia for from three to twenty minutes before printing, the resulting prints will be more brilliant, the paper will print more quickly and will tone to a purple colour much more easily. Experiment alone will determine the amount of fuming desirable.

Printing on Albumenised Paper. The 'printing frame' is an apparatus so familiar nowadays as hardly to require description. It resembles an ordinary picture-frame with a hinged back, kept in its place by springs. The negative takes the place of the glass in the picture-frame (film inwards), on this lies the sheet of sensitised paper, and is kept in its place by the hinged back, which is usually padded with cloth or felt. By releasing one of the springs one half of the back can be lifted up, and the progress of the printing examined without shifting the paper. For sizes larger than whole plate it is better to use a deep and heavy frame with a plate-glass front, against which the negative lies, and is thus saved from a strain which might otherwise cause its fracture. As a general rule the printing is continued twice as long as is necessary to produce a pleasing picture in the frame; it is then ready for the next process. The printing should rarely, if ever, be done in direct sunlight, and if the negatives are very thin it is well to cover them with a sheet of tissue paper. Cracked negatives may be printed from by attaching the frame to an ordinary "bottle jack," and allowing it to spin during the exposure, or by printing at the bottom of a deep box not much larger than the frame.

Washing Prints from the Frames. The prints after removal from the frames are placed in water which is constantly changed until it is no longer rendered milky. It is then advisable to let them lie in a bath consisting of ½ oz. of common salt to the pint of water for five minutes. They are then again washed in several changes of water, and are ready for toning.

Toning the Prints. This consists in the substitution of gold for silver on the surface of the print, thereby giving it a much richer tone, and adding considerably to its permanency.

Various formulæ for toning baths are in use; some of the following will be found reliable:

1. Chloride of gold, 1 gr.; borax, 60 gr.; water, 10 oz.

2. Chloride of gold, 1 gr.; acetate of soda, 20 to 30 gr.; water, 12 oz.
Dissolve the acetate in the water, and add the gold. Keep a week before use.

3. Gold chloride, 1 gr.; phosphate of soda, 20 gr.; water, 12 oz.
To be made as wanted; will not keep.

4. Gold chloride, 1 gr.; sodæ bicarb., 5 gr.; water, 12 oz.
To be made as wanted; will not keep.

5. Take a solution of gold chlorate, 1 gr. to the ounce, shake with a little prepared chalk, and allow to settle; pour off the clear liquid. Take of this clear liquid, 10 dr.; calcium acetate, 20 gr.; chloride of lime, 1 gr.; tepid water, 20 oz. For use, take 2 oz. of this mixture and make up to 10 oz. with tepid water; this will suffice for one full-sized sheet of paper.

6. Toning and fixing in one bath, 'British Journal Photographic Almanac.' Gold chloride, 1 gr.; sodic phosphate, 15 gr.; ammonium sulphocyanide, 25 gr.; sodium hyposulphite, 240 gr.; water, 2 oz.
Dissolve the gold separately and add it last. Not in general use.

Experience alone can determine the extent to which the toning should be carried. A good purple colour is desirable if the paper will bear it. A rich deep brown is as much as most papers can be made to give without general degradation of the prints.

Washing the Toned Prints. The prints are taken out of the toning bath one by one and placed in clean water, and then washed in several changes of water to get rid of all traces of the toning bath. They are then ready for

Fixing. A bath of sodium hyposulphite, 2 oz. to the pint of water, is made up, and the prints carefully immersed in this. The fixing bath should be made to smell slightly of ammonia, decomposition is thereby avoided, and a double hyposulphite of silver and sodium, which is not very soluble in the hyposulphite alone, is thereby removed from the prints, greatly increasing their permanency. Immersion for twenty minutes face downwards is sufficient to remove all the soluble silver salts. A fresh bath should be made for each batch of prints, and the old bath saved for recovery of the silver it contains (¼ oz. or more of metallic silver for each quire of paper fixed).

Washing the Prints. The prints are taken out of the fixing bath one by one and put into a porcelain dish full of water, in which they lie for a moment to get rid of the bulk of the hyposulphite. This preliminary washing may be repeated once or twice. The whole batch is then transferred to a 'washer,' in which they are kept constantly moving by a stream of water for several hours. If a washer be not at hand, let the prints lie in a clean tub of water changed frequently for three or four hours, then take each separately, let it face downwards on a clean piece of glass, and allow a stream of water (from a rose, if possible) to play on it for a few minutes; then turn it face upwards and repeat the process.

Drying the Prints. Take them from the wash water, allow to drain as much as possible, then lay them face downwards on a thick pad of *clean* blotting paper (or blotting boards), cover with more paper, then a layer of prints, then more blotting, and so on, till all have been treated; put a board on the top and load with a weight for an hour or more, and are ready for mounting.

Mounting Media. 1. Thin starch paste carefully made as required from pure starch (without blue) and laid on with the finger.

2. One oz. of hard gelatin dissolved in 10 oz. of water.

3. ('British Journal Photographic Almanac.') Nelson's gelatin (No. 1), 4 oz.; water, 16 oz.; glycerine, 1 oz.; methylated spirit, 5 oz. Dissolve the gelatin in water, add the glycerine, and lastly, the spirit.

RECOVERY OF RESIDUES FROM SILVER PRINTING PROCESSES.

Paper Cuttings, Spoiled Prints, &c. These should be carefully kept and burned in a crucible or pipkin, the ashes being preserved.

Washings. The first washings of silver prints should be kept and poured into a large vessel, a little potassium chromate solution is added, and then common salt, till the red colour changes to white; allow to settle, pour off the clear liquor and repeat the process in the same vessel till enough silver chloride is obtained to be worth removing.

Fixing Baths for plates or paper should be set aside and treated with a solution of potassium sulphide 'liver of sulphur.' This converts all the silver into sulphide. All residues may thus be converted into ashes, chlorides or sulphides.

Treat the ashes with nitric acid and precipitate the solution with common salt, adding this to the chloride residues. Digest the residue, after washing, with a little hyposulphite of soda, and add to the fixing bath residues. We thus reduce all the residues to two forms, chloride and sulphide.

Now take a Stourbridge clay crucible with a cover and put into it the dried chloride thoroughly mixed with twice its weight of sodium carbonate and heat slowly to bright redness; in a short time the chloride of silver is reduced to the metallic state. Now add the dry sulphide and continue the heating, pushing it to a full whiteness; in 20 minutes after this is obtained the whole process will be complete, and the molten silver may be poured out into a mould or allowed to run into water to form granulated silver.

CARBON PROCESSES.

These are all based upon the fact that organic matter in the presence of a bichromate when exposed to light is rendered more or less incapable of absorbing water. The simplest of these processes is that known as

The Powder or Dusting-on Process, in which a plate of glass or other material is coated with a suitable mixture, exposed to light under a negative, and then dusted over with a very fine black powder which adheres to those parts protected from the light by the negative.

The following formulæ are given by Burton:

1. Dextrine, 4 dr.; grape sugar, 4 dr.; ammonium bichromate, 4 dr.; water, 10 oz.

2. Gum-arabic, 7 dr.; grape sugar, 8 dr.; potassium bichromate, 5 dr.; water, 10 oz.

3. Honey, 90 gr.; albumen (filtered), 90 min.; ammonia bichromate (saturated solution), 150 min.; water, 10 oz.

4. Honey, 2 dr.; glucose, 4 dr.; albumen, (filtered), 8 dr.; dextrine, 90 gr.; potassium bichromate, 4 dr.; water, 10 oz.

A little consideration will show that by this process a negative yields a negative, and a positive a positive, so that for the production of positives a positive on glass will be required. Opal plates are generally used, the roughened side is cleaned with whitening and water, and then well washed under a tap; whilst still wet the mixture is poured over it from one edge, driving the water before it. When coated the plate may be dried in a hot oven or before a fire. As the plate is very sensitive when dry, the oven is best; in brilliant sunshine an exposure of a quarter of a minute suffices, a little longer in diffused light, but as the plate cannot be examined during the printing, the method of trial and error, or comparison with a scrap of sensitised paper exposed at the same time, must be adopted; before development warm plate, transparency and frame together before a fire. Open the frame when warm by gas or candle light, and put a small quantity of fine black-lead or lamp-black in the middle of the plate, and with a large round camel's-hair brush about 1½ in. diameter, spreading it as quickly and as lightly as possible all over the surface; if the image is not fully out at the end of a minute, gently blow on the plate and continue to rub in the powder. Breathing on it is apt to make it take the colour all over and foul the high lights. When finished flood the plate with methylated spirit 8 parts, water 4 parts, sulphuric acid 1 part; this should remove all the yellowness due to the bichromate in five minutes, if not add a little more water and flood again; the extra water should be added cautiously. When the colour is quite gone, soak the plate for some time in spirit and set on edge to dry; mount under a glass cover or coat with plain collodion and varnish. The process is useful for the production of reversed negatives.

THE AUTOTYPE PROCESS.

Gelatin containing bichromate of potash or ammonia (or any of several other salts) remains soluble so long as it is kept in the dark, but becomes insoluble when exposed to light. So that if a mixture of gelatin, pigment, and the bichromate be spread on glass or paper and exposed to light under a negative, and the film be then soaked in warm water, a print would be produced by the water, dissolving away those parts on which the light had not acted. But a little consideration will show that in the case assumed the picture could not be developed, because the surface acted upon by the light would be the outer one, and as *some* action probably takes place all over it, no part of a negative being absolutely opaque, the warm water would be met by a layer of insoluble gelatin. This was the first difficulty encountered, and it has been met in a very ingenious manner.

The paper on which the sensitive material is spread is of tough quality, and retains its toughness when wet. The sheet which has been printed is taken out of the frame and soaked in cold water until it is quite flat and limp, and is then squeezed on to a plate of glass or metal or a sheet of prepared paper, which is put under the print (upside down) in the water. The two are drawn out together, and with a few strokes of the squeegee are brought into hermetic contact. The two are then allowed to rest awhile between sheets of blotting paper to absorb superfluous moisture. When somewhat dry the transfer paper as it is called and the print are put into water at 100° F. or 110° F., and kept under the surface until the pigmented gelatin begins to ooze out between the two sheets of paper. In a short time it will be found that the original paper support can be stripped off with ease, and the surface of the pigmented gelatin exposed, *i. e. the back of it to which the light has not penetrated, and which consequently remains soluble.* The warm water is now diligently splashed on to the film until the lighter parts of the picture begin to appear, and the process is continued carefully until the whole of the detail is out. The picture is then transferred to a dish of cold water and well rinsed, and from this to a saturated solution of common alum where it remains until every trace of the yellow colour produced by the bichromate has disappeared. It is then well washed in water and hung up to dry. A print so produced will obviously be reversed unless a reversed negative be used (*e.g.* one made by the powder process described above). This reversal is especially objectionable in portraiture, and the method of 'double transfer' as it is called is used to overcome the difficulty. The print is developed upon *waxed paper*, and to this it will adhere when wet; whilst wet it is squeegeed again on to a paper to which it will adhere when dry. The two are allowed to dry together when the waxed sheet (*or temporary support*) peels off, leaving the finished picture no longer reversed, but correct as regards right and left upon the permanent support, which may be then treated as a finished print and mounted or otherwise as required.

The temporary support may be a sheet of zinc finely ground and coated with a mixture of beeswax, 3 dr.; yellow resin, 6 dr.; oil of turpentine, 1 pint.

The Autotype Company prepare a flexible support of prepared paper coated first with insoluble gelatin, and then with a mixture of various lacs so as to present a smooth surface quite impervious to water. The autotype final support consists of paper coated with gelatin which is rendered insoluble just before use by soaking in alum. In order to produce good carbon prints only vigorous, brilliant negatives should be used.

Gelatin mixture for coating paper (*Burton*). Nelson's flake gelatin, 2½ lbs.; Coignet's gold medal gelatin, ½ lb.; liquor ammoniæ ·880, ½ oz.; 5% solution of phenol, 2 oz.; sugar (white loaf), 1½ lbs.; water, 6 pints. Soak the gelatin in the water, then melt with heat, and add the

other ingredients, stirring briskly all the time with an egg beater.

For colour use Indian ink broken into small pieces, soaked for 24 hours in water and then rubbed smooth in a porcelain or glass mortar. Enough is to be added to the jelly to give a certain opacity. A drop of the jelly and ink allowed to set on a piece of glass should be almost opaque for prints, quite opaque for transparencies. To sensitise the above quantity of jelly, 6 oz. of potassium bichromate should be used, and added just before coating the paper; either dissolved in the smallest possible quantity of boiling water, or better incorporated with the mass in fine powder, and dissolved by constant agitation. The tissue when sensitised will not keep for long especially in warm weather, and it is best to buy unsensitised tissue, and when required immerse it in a bath consisting of potassium bichromate, 3 oz.; strongest liquor ammonia, ½ oz.; water, 80 oz. After sensitising it must be dried rapidly in a warm well ventilated room in which no gas is burned.

Carbon prints may be intensified by flooding with a strong solution of potassium permanganate.

THE PLATINOTYPE PROCESS.

This beautiful process depends upon the following chemical facts :

Ferric oxalate is reduced to ferrous oxalate by the action of light. Ferrous oxalate in solution reduces chloro-platinate of potassium to metallic platinum. Ferrous oxalate is soluble in oxalate of potassium. If then a sheet of paper be coated with a mixture of ferric oxalate and potassium chloro-platinate allowed to dry in the dark, and then exposed to light under a negative, the first of the above reactions takes place, and on floating the paper upon a hot solution of neutral potassic oxalate, the second, which results in the production of a picture in metallic platinum. All that is required is to wash this in several changes of dilute hydrochloric acid, then in plenty of water, and dry, and the print is ready for use, and is so far as is known absolutely permanent.

The sensitised paper is difficult to prepare, and must be kept *perfectly dry*. Special tin cases are constructed for holding it with a false bottom in which fused calcium chloride is placed to absorb moisture.

1. THE BLUE PROCESS.

Prepare the following solutions *separately* :—
a. Potassium ferricyanide, 5 oz.; water, 20 oz.
b. Ammonio-citrate of iron, 5 oz.; water, 20 oz.

When required mix equal parts of the above and coat good white paper with the mixture, using a clean sponge. Expose under a negative (or architectural, or other drawing, on tracing paper or cloth) to a good light till the picture appears bronzed, then wash in plenty of cold water. Blue picture on a white ground if a negative be used, white on a blue ground if a plan be used as a negative.

2. PIZZIGHELLI'S PROCESS.

Prepare—*a.* Gum-arabic, 3 oz.; water, 15 oz.
b. Citrate of iron and ammonia, 1½ oz.; water, 3 oz.

c. Ferric chloride, 1 oz.; water, 2 oz.

d. Potassium ferricyanide, 2 oz.; water, 20 oz.

Mix *a, b,* and *c,* and coat the paper as soon as possible and expose as before, but for only one half or one third the time required for sensitised albumenised paper. When exposure is complete apply *d* with a brush till the image appears blue on a white or bluish ground, then dip in

e. Hydrochloric acid, 2 oz.; water, 20 oz.

This clears the ground and darkens the lines; then wash and dry.

3. PELLET'S PROCESS.

a. Oxalic acid, 5 grms.; ferric chloride, 10 grms.; water, 100 c.c.

b. Potassium ferricyanide, 3½ oz.; water, 20 oz.

c. Hydrochloric acid, 2 oz.; water, 1 pint.

Sensitise with *a,* develop with *b,* clear in *c,* then wash and dry.

GELATINO-BROMIDE PRINTING PROCESSES.

These consist essentially in the use of paper coated with a thin film of a slow gelatino-bromide emulsion, exposed behind a negative in the usual way, but to a gas or lamp flame. The prints are developed with a weak solution of ferrous oxalate. There are several firms who make these papers, and as the fullest possible instructions are issued with them they do not require repetition here. Professor Burton gives the following developer as one which will yield good results with almost any brand of paper :—Saturated solution of ferrous sulphate, 1 oz.; saturated solution of potassic oxalate, 4 oz.; water, 5 oz.; citric acid, 40 gr. Negatives for this process should be rather thin and full of detail, with a short exposure and slow development. Such negatives yield beautiful prints.

Enlargements from negatives can be made on this paper by the aid of a magic lantern. Details of the process will be found in most of the works referred to in this article.

Gelatino-Chloride Paper. This is paper coated with a gelatin emulsion of bromide *and* chloride of silver. It is by no means so sensitive as the gelatino-bromide paper and possesses this curious property, that if an exposure several hundred time longer than that necessary to give all the details of the picture be given, and if the paper be developed with a very weak developer containing excess of bromide, red prints are obtained which may be toned with gold to produce all the effects of albumenised paper. Full directions are sent out by the makers.

Gelatino-citro-chloride Paper for Printing out. In this paper the sensitive surface is composed of an emulsion containing citrate and chloride of silver. The paper known as 'Aristotype' is a good example. The advantages of this class of sensitised paper are considerable, and with thin weak negatives, vigorous prints, full of detail can be obtained. The necessary exposure is much shorter than with the ordinary albumenised paper. The toning can be carried to a much greater extent, and thus the permanency of the print is probably greater.

The following toning-bath is recommended :— Chloride of gold, 1 gr.; hyposulphite of soda, 1 gr.; ammonium sulphocyanide, 20 gr.; water,

2 oz. Dissolve the sulphocyanide and hyposulphite in 2 oz. of the water, the gold chloride in 1 oz., and pour the latter into the former, stirring well the while.

PHOTO-MECHANICAL PRINTING PROCESSES.

There is probably no department of photography which has made such rapid advances as the reproduction of pictures originally taken in the camera by purely mechanical means, and the enormous number of cheap and well-illustrated papers which are now published almost owe their existence to one or other of the numerous photo-mechanical processes. To attempt to give anything approaching a complete account of even the more important would far exceed the limits allowed by this work, but it may be well to indicate briefly the general principles upon which some of the best known of these processes are based, and to make clear to the reader who is not technically informed the meaning of certain terms which are constantly met with, and which relate to these processes.

THE WOODBURY PROCESS.

From the description given of the method of development of a carbon print it will be clear that the film of bichromated gelatin yields a picture in which the shadows are represented by elevations, and the lights by depressions, the amount of colour being dependent upon the thickness of the film. It is conceivable that a cast might be made from such a developed film from which a metal mould could be made and the picture thus reproduced indefinitely. This in principle is the Woodbury process, carried out in practice somewhat as follows :

A thick film of slightly pigmented or unpigmented gelatin is prepared from a suitable mixture of gelatin and bichromate of potassium and dried very rapidly in a special stove. This film is exposed to light under a good negative, full of contrast; the exposed film is then attached by a solution of rubber to a glass plate, and is then immersed for several hours in water at about 106° F. to develop, *i. e.* dissolve any of the gelatin which has not been acted on by light. The film is then rinsed in cold water and soaked for ten minutes in a 4% solution of chrome alum; again washed in cold water, allowed to drain, then set in a dish of methylated spirit for an hour, drained and set on edge to dry like any other negative. The film is then removed from the glass by inserting a penknife blade under it and stripping. After this it is kept for some hours in a dry place, so that it may contract as far as it will and become thoroughly hard. The next process is the production direct of a mould in lead by forcing this dry gelatin film into the metal, under a *pressure of nearly four tons on the square inch,* in an hydraulic press. It seems wonderful that any result whatever should be obtained beyond a crushed film of gelatin and an indented plate, but as a matter of fact the finest lines are accurately reproduced in the metal, and a number of moulds may be made from one film. The moulds are then set in special screw presses, a proper quantity of pigmented gelatin is spread over them, and a sheet of prepared paper laid on the top. The lever or screw of the press is now worked, and the ink is crushed between the prepared paper and the mould, filling all the interstices; it is allowed to set, the press is opened, and the cast removed in the form of a thin pellicle of varying thickness, giving all the lights and shades of the original picture. The films are soaked in alum, trimmed and mounted. There is only one objection to the process, viz. that prints more than 8 in. square cannot well be produced; the difficulty being the construction of a press which shall exert the enormous pressure required evenly over a large surface.

STANNOTYPE.

This is a modification of the Woodbury process, in which the use of the hydraulic press is dispensed with. A print on thick gelatin is prepared *from a positive*, the film is attached to glass, developed as before, and dried. It is then covered with a special steel-faced tinfoil by painting the film with rubber varnish, laying the foil on it, and passing both through a pair of rubber rollers; the foil-covered film is then printed from in a press as before described.

PHOTO-LITHOGRAPHY.

A subject in line or one without any half-tone may be reproduced by photo-lithography as follows:

1. A very good negative is prepared, preferably by the wet process.

2. A print on photographic transfer paper (paper coated with bichromated gelatin) is made and inked up with lithographic ink; a thin, even coating is required.

3. The inked print is floated on water heated to 100° F. till the lines show a depression.

4. It is removed to a stone or other level surface, and developed by the aid of warm water and a sponge. The ink leaves all parts excepting the lines of the drawing, and the gelatin which has not been acted on by light is dissolved and removed.

5. The print is well washed in cold water and dried. It may then be laid upon the stone and passed through the press; the ink lines are thus transferred to the stone, and the process of printing proceeds in the ordinary way. There are a number of details which require attention in order to obtain good results, but the above is an outline of the principles of the process.

PHOTO-ZINCOGRAPHY.

A similar transfer is laid down on a sheet of zinc, which is then etched with the following:—Decoction of nut-galls, ¾ pint; solution of gum (consistence of cream), ¼ pint; solution of phosphoric acid, 3 dr. After the etching, printing proceeds in the ordinary way.

COLLOTYPE.

When a film of bichromated gelatine is developed after exposure under a negative, those parts on which the light has acted refuse to take water, and if such a plate be rolled up with a greasy ink, it will be found that the ink follows the lines of the action of light, and that the parts which are still soluble and take up water refuse the ink. It will be obvious that an impression in ink may be obtained by setting such a plate on the bed of an ordinary press and working it as type. Collotype printing, though so exceedingly

simple in theory, is much more complicated in practice. The preparation of the plates requires skill and judgment on the part of the operator. The 'machining' must be conducted with care in order to obtain good impressions and prevent the destruction of the plate. The best work can only be done on hand-presses, though for ordinary purposes, advertisements, trade lists, &c., power presses are used, and the results are fairly satisfactory. Collotype is a process which tends to produce excess of contrast in the resulting prints, the negatives for reproduction should therefore be thin and full of detail, *i.e.* somewhat over-exposed.

HELIOTYPE.

This process resembles collotype in principle, but the sensitised gelatin is treated as a film stripped from the glass plate on which it was poured and allowed to set, and after exposure is cemented to a metal plate and treated as a collotype plate.

PHOTO. ZINC BLOCKS FOR PRINTING WITH TYPE.

Blocks for this purpose must be so prepared that the picture is represented by raised lines, and the method of production is theoretically exceedingly simple. If a design be drawn upon zinc with some medium or varnish which is not acted on by acids, and the whole plate be then dipped in some etching fluid, the part not protected by the varnish will be dissolved away, leaving the design in relief. The varnish is cleaned off, the plate mounted on wood 'type high,' and the block is then ready for setting up with type and printing from in the ordinary way. In practice the process is by no means so simple, as there is great difficulty in preventing the undermining of the lines by the lateral action of the etching fluid.

Three principal methods are employed for obtaining an image on the zinc plate in a material not acted on by acids.

1. By laying a lithographic transfer on the plate and passing the two through the press.

2. By coating the plate with bitumen and exposing to light under a negative, and then developing with turpentine and benzol. This process has already been described under BITUMEN.

3. By coating the plate with bichromated gelatine or albumen, exposing under a negative, and developing with water, *i.e.* removing the soluble gelatine or albumen and then etching as before, or by inking up the whole plate before development, and then proceeding as in photo-lithography, the ink forming a stronger 'resist,' as it is called, than the albumen or gelatin.

To prevent the undercutting of the lines by the etching fluid, the plate is slightly etched, then inked up and dusted over with powdered asphalte; this adheres to the ink, and on warming the plate melts with it and runs down the side of the line. By repeating this process with care and skill the line is etched in steps, thus ▲ instead of ▼ which would result in a line so undermined and rotten that it would be destroyed in the press.

One other process is that of casting from gelatin reliefs, obtained as already described; the plaster cast is dried and dipped in water at 120°

F., and a cast made from it in stearine 1 inch thick. This is allowed to harden, rubbed over with bronze powder, and an electrotype made from it which is backed up with type metal, and mounted on a block of wood as before. Sharp and delicate lines cannot be obtained by this process.

PRODUCTION OF INTAGLIO PLATES.

In these plates the lines to be reproduced are engraved instead of being in relief, and the plate resembles an etching. It has the advantage over a 'process block' that much finer lines can be represented accurately, but it cannot be printed with type, but must be treated as an etching or engraving. The process is in principle much the same as those above described, a positive being used instead of a negative.

PROCESSES FOR MECHANICAL REPRODUCTION OF HALF-TONES.

Careful consideration of the conditions under which the various mechanical processes already described are conducted will show that in photo-lithography, in photo-zincography, and in all the processes for the preparation of metal blocks for printing with type, the production of half-tones, i. e. the degrees of light and shade which exist in an ordinary negative, is, by the methods described, an impossibility; and that only in the case of line drawings, or drawings done in stipple, can reproductions be made in which the half-tone will be faithfully represented. This involves a very serious restriction of the application of these processes, and one which the ingenuity of experimenters has been sorely taxed to remove. Fortunately their efforts have not been unattended with success, but there is still great room for improvement. The case of photo-lithography will serve as an illustration. Wherever the stone takes the ink it does so all over alike, and not in varying shades or degrees, and the effect of half-tone can only be produced by drawing on the stone in lines or dots of varying size or numbers. Each point or dot prints absolutely black (if black ink be used), but the total effect is half-tone. In copying plans, engravings, or even some kinds of pencil drawings, the difficulty does not exist, as the originals are produced in such a way as to give the necessary lines or dots; but in a photograph of, say, a piece of machinery, or a sepia drawing, or a landscape from nature, it is possible that there may be no absolutely white parts, and the result would be that the stone would take ink everywhere. The problem, therefore, resolves itself into the introduction of some sort of grain into the transfer, or the production of a negative which shall be itself grained or broken up into dots.

The methods actually in use are very numerous, and are, in their detail at all events, for the most part trade secrets. A brief account of two methods must therefore suffice.

Asser's Process. A sheet of unsized paper is coated with starch, and floated (when dry) coated side upwards, on a solution of potassium bichromate; it is then dried in the dark. This paper is now sensitive to light, and a print from the required negative is made upon it, thoroughly washed in cold water to remove the bichromate, and then dried between blotting-paper and finished in the open air. When dry the print is heated evenly by placing it on a slab of hot stone or ironing with a hot iron; this causes the starch which has been acted on by light to take up lithographic ink greedily, in addition to which the surface of the paper is raised into a number of granular points. By careful inking of this print a transfer may be prepared and laid upon the stone.

Collotype prints have a grain which is due to reticulation of the gelatin, and it is possible so to prepare and ink up a gelatin film that a grained impression may be taken from it and transferred to stone. Sprague and Co. use this process successfully. The same grain may be transferred to zinc and etched in the usual way.

Use of Grained Screens. In 1866 Messrs E. and J. Bullock patented a process for obtaining half-tone by exposing a sensitive plate under a positive with a grained screen superposed, the resulting negative when developed showing the grain of the screen as well as the picture. These grained screens are obtained by photographing fine netting or muslin, and preparing a positive with great care.

The process is worked at present somewhat as follows for the reproduction of a picture, e. g. for such an illustrated book as ' Academy Notes.' A negative of the desired size is made, and from this a transparency by contact. The transparency and grained screen are placed film to film, and a negative reproduced from the two in the camera. The grained negative is then used for printing with bitumen on a zinc plate by the usual process. The modification of the method known as the Meisenbach process consists in giving the grained screen a slight movement once or twice during the process of making the negative.

Ives' Process. A relief in bichromated gelatin is made from a transparency, and from this a plaster cast. On this a rubber surface, or plate of rubber which is covered with fine V-shaped grooves and ridges, is (after inking) pressed. The rubber does not touch the hollows, but is quite flattened where it touches the highest part of the relief. A stipple of varying degree is thus produced all over the relief. The image so obtained is transferred to zinc and etched.

The above sketch of the photo-mechanical processes is, of necessity, very brief and incomplete. The reader who wishes further information should consult—

' Practical Guide to Photographic and Photo-mechanical Printing Processes,' by W. K. Burton (Marion and Co.); papers in the ' Photographic News,' 1883–4; ' British Journal of Photography,' 1884 ; ' Instruction in Photography,' by Captain Abney (Piper and Carter); ' The Grammar of Lithography,' by W. D. Richmond (Wyman and Sons) ; ' Der Licht Druck und die Photolithographie,' von Dr Julius Schnauss (Düsseldorf : Ed. Liesegang's Verlag) ; Bolas, ' Cantor Lectures ' ; ' The Application of Photography to the Production of Printing Surfaces and Pictures in Pigment,' 1878 ; ' Modern Methods of Illustrating Books,' by H. Trueman

Wood; 'Photo-engraving and Photo-lithography,' by W. T. Wilkinson.

PREPARATION OF TRANSPARENCIES FOR THE LANTERN.

The size of an English lantern slide is $3\frac{1}{4}$ in. square, and as a negative is rarely or ever taken upon a plate so small, although cameras are made for this special purpose, it follows that some process of reduction must be resorted to in order to obtain the picture on an ordinary lantern plate. Quarter-plate negatives, being the same size in one direction as the lantern plate, may be and usually are printed from by direct contact in a frame, the important parts of the picture generally falling within the required limits. With larger negatives the process of copying in the camera must be resorted to. The simplest way of effecting this is to have a table of sufficient length set with one end against a window having a north aspect; this window should either be glazed with ground glass or have a sheet of tissue-paper stretched over it in order to diffuse the light. The negative to be copied is set up, either in a printing frame or some contrivance made for the purpose, with its film side towards the camera which is set on the table; great care should be taken that the negative is perfectly parallel with the front of the camera, and where much of this kind of work is to be done it will be found advisable to secure this by some permanent mechanical arrangement. A framework, temporary or otherwise, is arranged in such a way that it may be covered with cloth or paper so as to secure that no light shall reach the lens except that passing through the negative; but, as this is not desirable in all cases, it is well to make the cover moveable. A carrier is fitted to the dark slide of the camera of the proper size to take a lantern plate, and the picture accurately focussed on the centre of the screen. It will be found convenient to mark the space to be occupied by the image on the glass by means of strips of gummed paper, and if the slide and carrier are in perfect register, the picture will be found when the plate is developed to be properly placed upon it. Attention to these details is very necessary, as otherwise great annoyance is caused by the picture being askew. There is a great variety of lantern plates in the market, all equally good if the makers' instructions for development are carefully followed; they therefore require no further consideration here. Lantern slides made by the wet collodion process are perhaps still to be regarded as the best, but almost equally good results may be obtained on dry plates. Slides and transparencies made by the Woodbury and carbon processes are exceedingly beautiful, and in some cases, especially when the negatives are not very vigorous, better results can be obtained. The lens used for making a transparency in the camera should be of the rectilinear or portrait type, and capable of covering a very much larger plate than that used for the transparency; with such a lens and the use of a comparatively small stop, the most minute detail is accurately reproduced.

For lecture purposes the optical lantern is rapidly taking the place of diagrams; the latter are cumbrous and costly compared with the lantern slide, and necessarily indifferent representations of the subjects in most cases. It is a matter of every-day occurrence with lecturers who use the lantern and make their own slides, to require a copy of a drawing or engraving—in some both; and, though at first sight this might seem to be a simple matter, there are difficulties in its execution which have to be overcome, and which to the inexperienced often prove very serious.

To Copy a Drawing or Engraving. Use a lens of the rectilinear type, and one capable of covering a much larger plate than the one to be used for the negative. Take every possible care that the sheet to be copied and the sensitive plate shall be absolutely parallel, and use a rather small stop. Before making the experiment take care that every part of the sheet to be copied is *equally illuminated*, and this by a light coming from the side and not from the front, otherwise reflected light will pass through the lens and fog the picture. In copying silver prints or others having an equally glazed and bright surface the lighting is of great consequence, also when there is much grain in the paper. In this latter case a compromise must be effected, and a top light used with discretion. The exposure for a line subject should be long enough to give a rather thin negative; indeed, it is often very difficult to obtain anything else, and should err on the side of under-exposure rather than over. Experience alone will teach the exact point. The negative so obtained from a map or other subject full of fine lines will be found to be so thin that it is almost impossible to obtain from it a lantern slide showing dense black lines on a clear ground, and for such subjects the wet collodion process has many advantages. If, however, the negative be *very carefully intensified* with bichloride of mercury, so that none of the deposit falls on the lines, a negative of sufficient density to yield a good lantern slide by contact may be obtained. For line subjects hydroquinone is an excellent developer both for negative and transparency, especially if the alkaline carbonates be used with a liberal allowance of potassium bromide.

Care should be taken in handling the transparencies during development, fixing, washing, and drying, to avoid dust and scratches, as almost invisible markings on the plate become very prominent on the screen.

SPECIAL METHODS FOR THE PHOTOGRAPHY OF COLOURED OBJECTS.

The plates prepared by any of the processes given above are almost insensitive to the colours red and yellow, so much so that the presence of these colours to any considerable extent in the original constitutes a serious obstacle to its correct rendering in the photograph. If an attempt be made to photograph a bunch of flowers of varied colours ranging over the whole spectrum, and the print be compared with the original, this defect of the plates will come out very markedly, the yellows and reds of the flowers, perhaps the brightest and most prominent of them all, being represented by almost perfect black. The defect is even more marked in photographs of certain paintings, e.g. a rosy sunset will appear more like a gathering storm. The correct representation of an illuminated manuscript is rendered very difficult for the same reason. Various devices have been adopted in order to overcome this difficulty; and though it cannot yet be said that the relative tones of a coloured subject are correctly rendered, some considerable approach to nature has been made.

The simplest plan, and one which for certain purposes yields a result not altogether to be despised, is to prolong the exposure to such a degree that the reds and yellows shall affect the plate. A certain equalisation of tone is brought about in this way, but the method is unscientific, and is only of real use in a certain very limited number of cases. The second plan is to use coloured screens, e.g. a film of tinted collodion placed in front of the lens; here, again, the conditions are so altered that the result is but little nearer the truth than if nothing of the kind were used. The third and best plan is to what is called "isochromatise" the plates. An ortho- or iso-chromatic plate is one whose sensitiveness to yellow, and possibly red, has been greatly increased. This increase, great though it is, does not make the plate by any means equally sensitive to all colours, so that there is still much to be done before a perfect plate can be prepared; nevertheless the improvement in the direction named is great and undeniable. The method of their preparation is the subject of several patents, and depends upon the fact that certain dyes and colouring matters possess the power—why is not altogether known—of increasing the sensitiveness of the silver salts to yellow and red light. Ammoniacal solutions of eosin, erythrosin, rose Bengal, cyanin, and others are generally used. Prof. W. K. Burton gives the following formula:

Erythrosin, 1 part; water, 1000 parts. Take of this solution, 1 part; ammonia (10%), 1 part; and water, 8 parts; bathe the plate for two minutes in liquor ammoniæ, 1 part; water, 100 parts; and then, without washing, immerse in the alkaline solution of the dye, and dry in absolute darkness.

These plates, however prepared, are excessively sensitive to light and to every sort of noxious vapour, and must, therefore, be manipulated in the smallest amount of deep ruby light. The best results are obtained when the exposure is made through a yellow screen.

PHOTOGRAPHY WITH THE MICROSCOPE.

The great complexity of microscopic structures and the difficulty of making accurate and reliable drawings of them, even by the aid of the camera lucida and similar appliances, render the application of photography peculiarly valuable, especially as many very competent microscopists are not equally good draughtsmen, and knowledge of the particular structure under investigation is absolutely essential to the production of a really reliable drawing. Unfortunately it is not quite so easy as might at first sight appear to obtain a good negative of an object as seen under the microscope.

The apparatus required is comparatively simple, and consists of a camera body without the lens, a dark slide, the microscope and its objectives, and a source of light.

The first essential is that the apparatus should be so mounted that the source of light may be in the axis of the objective, and that the sensitive plate should be in a plane perfectly perpendicular to the axis, as very slight deviations will lead to failure in the result. This is best accomplished by mounting the whole upon one stout and rigid board, determining the proper position by actual experiment, and so arranging the different parts of the apparatus that if removed they may be replaced in exactly the same position at some future time.

Unless a specially constructed "projection eye-piece" be employed it is best to dispense with this part of the microscope altogether, and to place the instrument in a horizontal position with its body projecting slightly into the lens opening, or into a brass tube screwed on to the front of the camera in the place of the lens; this tube should be carefully blacked inside, and the junction between it and the microscope body covered with a bag of velvet to prevent the entrance of light into the camera between the two. The best source of light is a very broad-wicked paraffin lamp (1½ in.). Having placed these in position and a low-power objective on the microscope, the light is arranged so as to give an evenly illuminated disc on the focussing screen. Some simple object, such as the tongue of a blowfly, is now placed on the stage and very carefully focussed, first by the eye and then by the use of a lens mounted in a tube. A simple shutter may be arranged in the tube which takes the place of the lens of the camera. If a dry plate be now placed in the slide, an exposure given, to be determined by experiment, and the plate developed, a negative of the object considerably magnified will be obtained. Simple though these manipulations may appear, the results will at first be, in all probability, highly unsatisfactory, and the following will be among the causes of failure:—Unequal illumination, due to wrong position of the light; too near to or too far from the object, or not in the axis of the objective. Failure to use the proper size of diaphragm beneath the object; this leads to loss of definition. Incorrect focussing; this is a very common cause of trouble, and occurs in this way. The finest ground glass is much too coarse for focussing such delicate lines as those of microscopic objects, and being so delicate, unless they are accurately focussed on the sensitive film, the blurring which results is very serious; it is therefore of great importance that the lens used for focussing should be of such a focal length that when the lens mount is placed close against the back of the focussing screen the image on the ground surface should be accurately in focus. Even this precaution will give unreliable results, and in practice it is found best to focus roughly on the ground glass, and then to insert in its place a piece of plain glass with a few very fine lines ruled on one side of it. The image should now be focussed as before, until it and these fine lines appear sharp at the same time. A dark slide of the American pattern answers well for carrying this second screen, which must of course be in perfect register. It will be necessary to arrange a brass rod parallel with the camera and projecting beyond it to the level of the fine adjustment screw of the microscope. By means of a milled pulley fixed to the board and a thick rubber ring slipped over the end of the rods so as to bite on this pulley and turn it when the rod is turned on its axis, the fine adjustment of the microscope may be moved from the rear end of the camera by tying a silk thread round the pulley and fine adjustment screw as a driving band. Having taken all these precautions, the result may still appear blurred and out of focus. This is due to an optical defect in the objective, the chemical and visual foci not being coincident. This can be corrected by trial and error, throwing the object very slightly out of focus by means of the rod acting on the fine adjustment; or it may be done away with by the use of specially constructed " apochromatic" objectives, such as those of Professor Abbé, of Jena. If these are not used, and objectives for micro-photography are to be bought, the purchaser should go to some respectable maker, explain the uses to which the objectives are to be put, and ask him to select those whose chemical and visual foci are practically coincident. This the makers are generally willing to do for a small extra charge. For low powers the paraffin lamp mentioned will answer very well, but if an objective of higher power than ⅓ in. be used it will be found necessary to use a sub-stage condenser, and further to collect the light by means of another condenser, and focus it upon the diaphragm. Again, the greatest care must be taken to secure the accurate centring of all these parts.

A further difficulty arises when still higher powers are used; a more powerful light, e.g. the oxyhydrogen light, is required, and great patience is often necessary to secure even illumination of the object without projecting an image of the surface of the lime cylinder upon the lens. An alum trough must also be interposed between the light and the object, in order to prevent dislocation of the mounting material by the great heat evolved. These are difficulties which can only be properly understood and appreciated by those who have actually attempted to take a photograph with the microscope.

Yet another difficulty, and that a serious one, will present itself in the coloration of the objects. Staining with logwood, though most useful for simple examination under the microscope, is not suited to photographic reproduction; blurred and foggy images seem almost always to result. Blue stains are still more unsuitable; the reds, blacks, and browns, which can be obtained with carmine and some of the aniline dyes, appear to be the best, and on the whole it is desirable to use orthochromatised plates.

The above necessarily brief outline of the arrangements desirable for successful micro-photography will perhaps be made more clear to the reader by a careful study of the annexed drawing of one of the largest and most complete pieces of apparatus for the purpose which has ever been constructed. It was recently made by Messrs Swift & Son, of London, from the designs of Andrew Pringle, Esq., for the laboratory of the Royal

Veterinary College. The whole instrument is of exceedingly solid construction, and supported on heavy metal castings, so as to ensure great rigidity and freedom from vibration, a most essential quality. The oxyhydrogen light is collected by an achromatic bull's eye upon a sub-stage condenser, which is provided with focussing screws and a fine adjustment. The stage of the microscope has rectangular movements actuated by micrometer screws which are fitted with verniers, and the fine adjustment of the microscope itself is so constructed that one whole turn of the screw head only moves the tube 1/150 of an inch. The microscope is arranged on a massive revolving table for convenience in arranging the object before photographing it, and the whole constitutes an optical instrument of the very first quality. It is possible to produce most excellent work without these complicated and costly appliances, but where money is no object they undoubtedly present great advantages, and effect a very material saving of time and trouble.

PHOTOGRAPHY IN NATURAL COLOURS.

It is impossible to close this article without some reference to the reproduction of the colours of nature by photographic processes. Orthochromatic plates have enabled us to obtain accurate representations of the variations in *tone* present in nature, but thus far all attempts to reproduce the varieties of *colour* have entirely failed.

In 1802 Sir Humphry Davy found that if the image of the spectrum were allowed to fall on paper coated with silver chloride, a coloured image corresponding in some degree to the spectrum was obtained, but it could not be fixed. In 1810 Dr. Seebeck, of Jena, obtained variously coloured images in the same way. In 1839 Sir J. Herschel described similar experiments, and Fox Talbot recorded the observation that the red portions of a coloured print copied of a red colour on paper prepared with silver chloride.

Between 1840 and 1843 Robert Hunt obtained a coloured image of certain parts of the spectrum on paper coated with silver fluoride, and in 1843, on a paper prepared with silver bromide and gallic acid, he obtained by prolonged exposure a picture in which the sky was crimson, the houses slaty blue, and the green fields of a brick-red tint. In 1848 M. Becquerel (died May, 1891), by using silver plates on which a layer of chloride was deposited electrolytically, obtained images of brightly dressed dolls in colours bearing some relation to the originals.

The failure of all these attempts was caused by the fact that no means has ever been found of fixing the images.

Niépce de St. Victor sent to the Exhibition of 1862 a number of photographs in colour obtained by a modification of Becquerel's method.

In 1868 Poitevin repeated Herschel's experiments. Chloride of silver paper was exposed to light, then dipped into a solution of potassium bichromate and copper sulphate, and dried; such paper exposed under coloured glass yielded coloured prints. In 1874 St. Florent described a process by which similar results might be obtained.

Microphotographic Apparatus constructed for the Royal Veterinary College from the designs of Andrew Pringle, Esq.

Recent Reported Discoveries of Photography in Colours. From time to time reports are spread abroad that the means of obtaining photographs in colours have been discovered. Most of these are impudent frauds intended to mislead the ignorant, and may be at once dismissed. The rest are unfortunately only exaggerated and distorted accounts of experiments by competent persons, written by newspaper reporters. Some recent investigations by M. Lippmann which yielded curious results were announced in this way. The theoretical difficulties in the way of the production of a plate which shall reproduce the colours of nature or their complements are very great, and beyond the observed facts mentioned above we appear to have made no nearer approach to its accomplishment than the point reached nearly eighty years ago.

PHOTOMETRY. The art of determining the relative intensities of different lights. Various methods have been adopted, at different times, for this purpose, among which, however, a few only are sufficiently simple for general application. The principle adopted by Bouguer and Lambert depends on the fact that, though the eye cannot judge correctly of the proportional intensity of different lights, it can generally distinguish with great precision when two similar surfaces or objects presented together are equally illuminated or when the shadows of an opaque object produced by different lights are equally dark. Now, as light travels in straight lines, and is equally diffused, it is evident that its intensity will progressively lessen as the distance of its source increases. This diminution is found to be in the duplicate ratio of the distance. To apply this principle to candles, lamps, gaslights, &c., we have only to arrange two of them so that the light or shadow resulting from both shall be of equal intensity, after which we must carefully measure the distance of each of them from the surface on which the light or shadow falls. The squares of

these distances give their relative intensity. In general some known light, such as that from a wax candle (4 to the lb.), is taken as the standard of comparison.

In London the standard is a sperm candle of 6 to the lb., and burning 120 grs. in an hour ; Harcourt adopts as a unit the light obtained on burning a mixture of 7 volumes of pentane gas and 20 volumes of air at the rate of ½ c. ft. per hour in a specially constructed burner, which yields a flame of a certain height. The absolute unit of light adopted by the International Congress of Electricians is that given out by a square centimetre of melted platinum at the moment of its solidification. This light is equivalent to that emitted by 15 standard candles.

Dr Ritchie's 'photometer' consists of a rectangular box, about 2 inches square, open at both ends, and blackened inside to absorb extraneous light. In this, inclined at angles of 45° to its axis, are placed two precisely similar rectangular plates of plain silvered glass, which are fastened so as to meet at the top, in the middle of a narrow slit about an inch long and the eight of an inch broad, and which is covered with a strip of tissue or oiled paper. In employing this instrument, the "lights must be placed at such a distance from each other, and from the instrument between them, that the light from each shall fall on the reflector next it, and be reflected to the corresponding portion of the oiled paper. The photometer is then to be moved nearer to the one or the other, until the two portions of the oiled paper corresponding to the two mirrors are equally illuminated, of which the eye can judge with considerable accuracy."

In Bunsen's photometer, a circular spot is made on a paper screen with a solution of spermaceti in naphtha ; on one side of the screen is placed the standard light. The light, the intensity of which is to be examined, is then so arranged that it can be moved in a straight line to such a

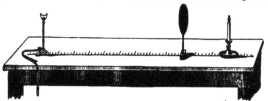

distance on the other side of the screen that the eye is unable to detect any difference in brightness between the greased spot and the rest of the paper. The distance of the lights from this screen is then measured, and then their relative illuminating powers are respectively as the squares of their distance from the screen. This method is the most generally resorted to, and it answers exceedingly well for all ordinary practical purposes.

In Prof Wheatstone's photometer the relative intensity of the two lights is determined by the relative brightness of the opposite sides of a revolving silvered ball illuminated by them.

In the method of photometery usually, but erroneously, ascribed to Count Rumford, the shadows of an opaque object formed by different lights, and allowed to fall on a white wall or screen, are contrasted. A wire about $\frac{1}{16}$ of an inch thick, and about a foot in length, with the one end bent so as to form a handle, is commonly used to form the shadows. The method of proceeding is similar to that first noticed above.

It is supposed by some that the equality of two shadows can be appreciated with greater certainty than that of two lights, hence several methods involving this principle has been proposed.

PHOTO'XYLIN. A variety of nitro-cellulose made from wood pulp, and used in making collodion. *Prep.* Nitrous acid (48° Baume), 8½ lbs.; sulphuric acid, 4½ lbs.; potassium nitrate (granular), 8 oz.; wood pulp, 4 oz. Mix the acids in an earthenware jar, and when the temperature of the mixture has fallen to 90° F. add the potassium nitrate, stirring well all the time; then immerse the wood pulp in the mixture and allow it to soak for 12 hours. At the end of the period remove the pulp and wash it well with water, to which a few drops of ammonia solution have been added. Dry carefully in the same way as gun-cotton is dried. The resulting photoxylin is soluble in equal parts of ether (s.g. ·725) and alcohol (s.g. ·820). Three parts of the photoxylin to 100 parts of this mixture make a collodion sufficiently thick for all practical purposes, and 5 drops of castor oil to 1 oz. make it flexible.

PHRATORA VITELLINÆ. THE WILLOW BEETLE. A box full of specimens of this beetle was received in May, with a report that the leaves and young shoots were being fast eaten off in extensive willow, or osier, beds. It was stated that for two or three years past this beetle had been fearfully destructive in willow beds in the whole of the districts in which willows are extensively cultivated and form a most profitable and labour-employing industry. A little later on other accounts to the same effect were forwarded from Lancashire. No beetles were forwarded in these cases, but from the descriptions of the mischief done there could be no doubt that it was the work of the willow beetle. The owners of the willow beds remarked that they were in despair, that they had tried many remedies without any mitigation of the evil.

These beetles, together with their larvæ, not only clear off the leaves from the willows, or so riddle them that they can serve the plants in no way, but they also eat the shoots and the rind of the willows, and completely ruin a valuable crop.

Life History. The beetle is rather more than a sixth of an inch in length. It is somewhat variable in colour, from blue to green, with metallic lustre, having faint spots upon the wing-cases. The body beneath is of a reddish hue; the antennæ are black. It is most tenacious of life, and difficult to kill with water, and pungent and poisonous solutions and fumes. It comes forth in May from its winter retreats in the earth, in rubbish, under the bark of trees, in the chinks and crannies of buildings, posts, and rails. Fences, especially "made" fences of "brush" woven between stakes, form admirable shelters for it. In short, any refuge near the willow beds seems to be suitable to keep the beetles from birds and from weather, for they are not affected by cold. Having strong wings, they can fly considerable distances.

The eggs are placed under the leaves in groups, and without any regular arrangement. They are white, long, and cylindrical. Many eggs were laid in the boxes in which the beetles were put. Larvæ, however, were not hatched out in these artificial conditions. In ordinary circumstances they are found on the willow plants towards the end of June. These larvæ are about half an inch

long, dirty-white in colour, having black heads and rows of black spots along their bodies; they have six feet. Westwood says: "I have also traced the transformation of the *Chrysomela vitellina*, the larvæ of which feed on the willow, arranged in a single row six or seven abreast, eating only the surface of the leaf, and leaving their exuviæ attached to its surface. They were found at the beginning of September, shortly after which they descended into the earth and assumed the pupa state, and appeared in the perfect state in the beginning of October" ('An Introduction to the Modern Classification of Insects,' by J. O. Westwood, F.L.S., vol. i, p. 389). Kaltenberg says that the larvæ attack both sides of the leaves.

In this country, as in Germany, and according to the reports received during the year, there are two attacks, one in the spring, and the other in September and October.

Methods of Prevention. These consist in flooding the willow beds where this can be done artificially. Though they require a deal of drowning this tends to decrease them, or at least those below the water-level. Many are ensconced under the bark of trees, in posts, and hedges above the water-mark. Upon some sewage farms willow beds have been made. Flooding with sewage has been found to be far more effectual than flooding with water.

As far as possible rubbish, and any other possible refuges for the beetles, should be removed from the willow beds and their neighbourhood.

Remedies. Many things have been tried to dislodge these insects, such as soot, sulphur, and other unpleasant materials. Those who have seen willow plants growing luxuriantly in beds will appreciate the difficulties of applying insecticides or insectifuges, either in dry or liquid form. Paris green and London purple have been experimented with, and found of some benefit. This requires to be done early, upon the first appearance of the beetles, and before the plants have made too much headway. The wash can be put on with the Vermorel machines strapped upon the backs of workmen. Care must be taken not to make the arsenical washes too strong, as the willow leaves are tender. Not more than one ounce to twenty gallons of water should be used at first.

In despair some willow planters have taken to picking the insects off by hand, and shaking the beetles into vessels held beneath the plants ('Reports on Insects Injurious to Crops,' by Chas. Whitehead, Esq., F.Z.S.).

PHYLLOTRETA NEMORUM, Chevrolat; HALTICA NEMORUM, Linn. (from φυλλόν, a leaf; and τετραίνω, to bore). THE TURNIP BEETLE (the turnip flea, or 'fly'). This insect, known generally as the 'fly,' is by far the most destructive to turnip and swede crops of any in the long list of their enemies. Every farmer has had painful experiences of its evil influences which have entailed losses that can hardly be estimated. Large breadths of land have been sown and resown in the same season and all the successive plants have been cleared off as fast as they appeared. As all practical persons know full well this entails not only the expense of seed and of cultivation, but the loss of a valuable crop essential in rotation, and upon which the maintenance of the sheep and cattle upon the farm depends. The failure of turnips and swedes throws the whole system out of gear. The full extent of the inconvenience and loss can only be recognised by those who live in a turnip-growing district, or by those who have large flocks of breeding ewes dependent upon the turnip crop.

The turnip flea has been known as destructive to turnips almost from the time when they were first cultivated. In the 'Annals of Agriculture,' edited by Arthur Young, published 100 years ago, there are many references to the baneful effects of this insect, and remedies prescribed to counteract them. The loss in Devonshire alone, in 1786, was estimated by Arthur Young at £100,000. Marshall speaks of great ravages committed by this insect in various counties, whose agriculture he described towards the end of the last century in his 'Practice of Agriculture and Rural Economy in the Six Agricultural Departments of England.'

Curtis states that it was not known in Scotland until 1826.

During the last 20 years the depredations of this insect from time to time have been seriously intensified, particularly in hot and dry seasons, and appear to have culminated in 1881, in which year in many of the most important turnip-growing districts this crop was completely ruined. On account of this unprecedented calamity an inquiry was instituted by the Council of the Royal Agricultural Society and conducted by Miss E. Ormerod, the Consulting Entomologist, as to the extent of the injury, the circumstances connected with it, and as to means of prevention and remedies that have been found in any way effectual.

In very many cases in England and Scotland it was ascertained by this inquiry that turnips and swedes were sown three times over, without any crop after all. The estimated loss for seed, expenses of sowing and re-sowing, in 22 English and 11 Scotch counties, in this season of 1881, amounted to over half a million of money, quite independent of the enormous losses and inconveniences sustained from the failure of the crop entirely in many parts of this area.

A typical instance, given in a graphic manner by a practical Kentish farmer, of the far-spreading results of the pertinacious onslaughts of the turnip beetle may be cited from this inquiry. "Mainly owing to the fly, the turnip crops were a complete failure. On the observer's own ground the plants no sooner showed than the fly attacked them and cleared them off. A second sowing, part swedes and part turnips, was swept away; a third sowing was made as soon as practicable, but it was too late for turnips, and the only substitute, rape and mustard, did not have time to produce half a crop. The fact that our turnip crops were ruined had the effect of lowering the prices of store sheep and lambs from 5s. to 10s. each at the sheep fairs and sales. Many of the large buyers from Essex, Surrey, &c., did not put in an appearance at all, and those who did come only made limited purchases, and this owing to the failure of the root crops. The bearings of the question are

so extensive that one hardly knows what interest is affected and what not."

A farmer having a flock of 1000 breeding ewes in Wiltshire found himself in November, 1881, with not nearly a quarter of his usual supply of turnips and swedes from the pertinacious attack of this insect. His ewes were due to lamb down at the end of January, and his 250 ewe tegs had to be kept well during the winter and spring. There was a breadth of short mustard and rape sown late in August, but no good pieces of turnips, swedes, and rape to 'hold' the ewes and lambs, and to keep the tegs upon in February, March, and April. It was calculated that the extra cost of keeping these sheep until the water meadows were fit, and the trifolium was ready, of which he was provident enough to have a good supply, was £480.

This beetle is not only destructive to turnips, swedes, rape, and mustard, but to cabbages and Kohl rabi also.

Its main and most dangerous attack is undoubtedly when the plants have just started, and until they are fairly established. At this time it constantly happens that the plants are eaten completely up in a day or two, or are so crippled and despoiled of leaf surface that they cannot grow, while there is a continuous supply of beetles ready to take any vegetation that struggles out. But it is by no means unusual to find, if the plants manage to get away from the first onslaughts of the enemy, that they are so steadily beset by them and their leaves so riddled that they never make good roots. Even when the roots are formed and are of some size the late generation of beetles pertinaciously stick to them, so that even in September sportsmen notice the sound made by the insects disturbed by their feet as they alight upon other leaves.

Life History. The *Phyllotreta nemorum*, literally the leaf-fretter of the groves, is a *Coleopterous* insect of the great family *Galeracidæ* and the genus *Phyllotreta*, according to Chevrolat. It is very small, only about 1¼ lines long, but it has large wings expanding more than the fourth of an inch, which enable it to take extensive flights and convey it to congenial food. According to some it scents it afar off, and flies forthwith to it. The shape of the beetle is somewhat oval. In colour it is black, having a broad sulphurous band upon each elytra. Its thighs, or, rather, its hinder thighs, are very stout and made for leaping. Curtis states that it can jump 18 inches, or about 216 times its own length ('Farm Insects,' by J. Curtis). It passes the winter in the perfect shape under clods and tufts of grass, and under weeds, on the outskirts of woods, by the sides of fields, hedgerows, and ditches, and under clods in the ground. It is thus sheltered, and it is sustained during the early days of spring until the turnips have sprouted, upon the wild cruciferous plants, such as charlock, the wild radish, hedge mustard, and others. When the turnip plants are in 'rough leaf,' this beetle lays eggs upon the under sides of these leaves, distinguished from the seed leaves by having hairs or bristles upon them.

A female lays only a few eggs, and only one daily, as Curtis says. In about 10 days yellow larvæ come forth, and piercing the leaves, make burrows in them, living upon their tissues. They are 2¼ lines long, having 6 feet and a caudal proleg, with dark marks upon the anterior and the posterior joints of their bodies. In the course of a few days, from 5 to 7 days, they leave the leaves and fall to the ground, in which they ensconce themselves close to the turnip plants, and change to chrysalides. From these the perfect beetles come in 11 or 12 days, and make furious raids upon the seed leaves. Not much injury is done to the leaves by the larvæ, at least compared with that done by the beetles. These arrive in a rapid succession of generations throughout the summer, if it is hot and dry, and if other circumstances are favourable, when it is believed that there are as many as 6 generations.

Prevention. A 'stale furrow' is calculated to prevent the attacks of this beetle. One reason for this is that a stale furrow implies in most cases what is known as a 'good season,' or a fine tilth; whereas land freshly ploughed up does not, as a rule, work down well, but is 'knubby' or cloddy. Besides this the moisture evaporates much more quickly from fresh ploughed land that harrows down cloddy, than from land stale ploughed. Beetles object to moisture, and nature naturally helps the young plants to grow away quickly from their foes.

If it is not convenient to provide autumn-ploughed land for the turnip crop, or if it is necessary to move autumn-ploughed land that may have been beaten down by heavy rains and snow, it is far better to work the land with cultivators rather than to bring up wet, unkind, and sticky furrows, which it would be difficult to pulverise, and from which the moisture would quickly depart.

Rolling down the land immediately after the drill should be adopted, as it tends to keep in the moisture and to level the earth in the drills, so that the seed may come away as rapidly as possible.

Finely comminuted manure, mixed with fine ashes or mould, should be drilled in with the seed, in order that it may be close to the plants to help them along out of the way of the earliest onslaughts of the beetle. Superphosphate is a good manure for this, at about 5 cwts. per acre; or guano at 2 cwts. per acre. Care must be taken that the ashes and mould should not be too dry so as to hinder vegetation. The ashes or mould may be advantageously moistened with paraffin oil, at the rate of 2 pints to a cwt. of material.

A water drill is of certain advantage upon some soils, and should be used where it is not too costly, and where the beetles are really troublesome. One objection to the water drill is that the small amount of moisture from it is very soon evaporated in a dry season, and though it starts the germination of the seed rapidly, this is liable to be checked, and the vitality of the seed destroyed unless rain comes soon.

Plenty of seed of the preceding year's harvest should be used, carefully examined as to its germinating powers, and as to its freedom from other and worthless seeds. From 3 to 4 lbs. per acre may be put in. The great importance of having seed of full germinating power cannot be too strongly insisted upon.

The growth of cruciferous weeds, such as charlock, encourages the beetles and furnishes them with food until the turnip plants are ready for them. After the fearfully wet season of 1879, all kinds of weeds were rampant, especially charlock, which encouraged the beetles. In Miss Ormerod's report, alluded to before, the clear connection between a prolific charlock crop and a fatal 'fly attack' succeeding it next year, is abundantly shown. One correspondent remarks that it has often been noticed, when charlock was abundant on any part of a field, that it was on this spot that fly-attack began, and also that hedges and other surroundings where weeds of this and similar kinds are allowed to grow neglected and unchecked, are spots from which the fly comes forth and spreads over the crop.

Therefore, all endeavours must be made to keep charlock down on farms, and to have the outsides of fields brushed and free from weeds as far as possible.

If it can be managed land that is intended for turnips or swedes should be allowed to lie a few days after it has been stirred and before it is sown, that the charlock seeds may germinate and be dragged up by the harrows when the sowing takes place.

Remedies. Dressings of soot are frequently of great service when the beetles are numerous and thick upon the young plants. Soot should be put on before the dew has gone. Wood ashes, and ashes from burnt earth, turf, and rubbish, and peat moss, all well powdered, have been moistened with paraffin oil at the rate of 2½ pints to a cwt. and have been found valuable dressings. A mixture of wood ashes with a little finely powdered sulphur has been tried with considerable benefit. Lime is useful as a dressing, but it must be put on very hot and while dew is on the leaves. Rolling the land with a light roller very frequently proves serviceable, especially if it is rough. This operation disturbs the beetles and presses the soil round the plants, keeping in the moisture.

Drawing a light wide framework of tarred boards upon wheels just over the plants is a means of catching many beetles, as they jump instinctively as the machine goes over them and alight in the tar. Many acres can be got over in a day by a man pushing this machine. The tar requires renewing occasionally, and the beetles which accumulate in masses must be scraped off. Driving flocks of sheep over beetle-infested plants has been tried. This does not appear to be any more efficacious than rolling, unless it is done very early in the morning, that the dust may remain on the plants, and it is not beneficial by any means to the sheep. Still some practical farmers adopt it. A Kentish hop planter tried washing the plants with soft soap and quassia, employing his hop washing machines for this purpose. He considered that this saved the plant, though it was a somewhat costly process.

Horse-hoeing should be done early and often, and side hoeing also, directly the plants are at all out of the way. It is important to keep on disturbing the beetles, and the roller may be applied at once, or soon after, to close the ground again ('Reports on Insects Injurious to Crops,' by Chas. Whitehead, Esq., F.Z.S.).

PHYLLOXERA VASTATRIX. In 1866 M. Delorme, of Arles, in the South of France, was the first to suggest that a peculiar disease which had manifested itself the previous year amongst the vines growing in the plateau of Pujaut on the west bank of the river Rhone, in the Department of the Gard, was of a new and specific character.

Shortly afterwards a commission appointed by the Herault Agriculture Society visited one of the infested localities, and one of its members, M. Planchon, confirmed M. Delorme's conjectures by discovering the cause of the vine malady. This he conclusively showed was due to the presence of a peculiar and hitherto unknown description of *Aphis*, belonging to the genus *Phylloxera*, which, as illustrative of its devastating qualities, he named *P. vastatrix*.

A full-grown *Phylloxera vastatrix* does not exceed more than the 33rd or 40th of an inch in length. Examined under a microscope, in addition to short pointed legs, it is seen to be furnished with a proboscis nearly half as long as its body. Upon examination this proboscis seems to be composed of three tongues, of which the centre one is the longest, and these are united at their base into a kind of flat, sharp-pointed blade, which is the boring or puncturing apparatus, by the aid of which the insect pierces into the roots, from which it sucks the juices that constitutes its food. About half the proboscis or sucker is inserted into the bark of the root, and the creature can not only attach itself to the root by means of it, but can also turn on it, as on a pivot, when engaged in the depredations.

These are continued from April to October, by which month the insect has lost the yellow colour that distinguishes it in the summer months, and assumed a copper-brown shade.

From October to April the Phylloxera hybernate, or rather, such of them do as have laid no eggs during the period of their active existence, for the egg-laying females die, and young phylloxerae only are preserved during the winter months.

With the return of April they awake from their winter sleep, and recommence their devastating career. They then increase rapidly in size and begin to lay unimpregnated eggs, for there are at that time no males. "These bring forth females, which in their turn develop and lay unimpregnated eggs, and the virginal reproduction continues for five or six generations, the development increasing in rapidity with the heat, but the prolificacy or the number of the eggs decreases.

"In July some of the individuals show little wing-pads at the sides, and begin to issue from the ground and acquire wings. These winged individuals become very numerous in August, and continue to appear in diminishing numbers thereafter till the leaves have all fallen. They are all females and carry in their abdomen from three to eight eggs of two sizes, the larger ones about ${}_{10}^{3}$ths of an inch long and half as wide; the smaller ⅔ths as long. These eggs are also unimpregnated and are laid by preference on the under side of the more tender leaves, attached by one end, amid the natural down. They increase somewhat in size, and give birth in about ten

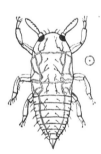

Male Phylloxera; dot in circle showing natural size.

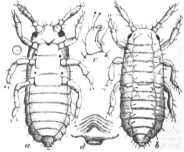

True female Phylloxera; *a*, ventral view, showing obsolete mouth and solitary egg, occupying nearly the entire body; *b*, dorsal view; *c*, tarsus; *d*, contracted anal joints after the egg is laid; dot in circle showing natural size.

days to the true sexual individuals, the larger producing females, the smaller, males.

"Anomalous as it may seem these creatures are born perfect, though without mouth, and with no other than the reproductive function.

"A most remarkable fact, discovered by Babiani, is that some of the females never acquire wings, but always remain on the roots, also produce the few different sized eggs from which these true, mouthless males and females hatch. The sexes pair soon after hatching, and the

female is delivered on the 3rd or 4th day of a solitary egg, and then perishes. This egg is never laid on the leaf, but always on the wood, either under the bark, or in sheltered situations above ground, or on the roots underground. The young hatching from it is the normal agamous mother, which, with increased vigour and fertility, lays a large number of eggs, and recommences the virginal reproduction and the cycle of the species' curious life. The impregnated eggs laid early in the season doubtless hatch the same year, though some of the later deposited ones may pass the winter before hatching." (*Riley*.)

The parts of the vine attacked by the Phylloxera are the rootlets, which, in a diseased plant, may be seen more or less covered with what appears to the naked eye a yellowish powder, but upon a microscopic examination reveals itself as a mixture of phylloxeræ of different sizes, and of their eggs.

During August and September, the Phylloxera invest the rootlets in countless numbers, and are so abundant as to entirely obscure the colour of the roots, and to cause them to appear yellow from the enormous number of their minute organisms.

The effect of the attacks of the parasite upon the rootlets is to give rise in it to the formation of a number of little tumefactions or enlargements. These in course of time decay, and their destruction results in the death of the plant.

Exposure to air and sunlight acts fatally to the Phylloxera, shrivelling and drying it up. Hence its instinct of self-preservation, no less than its search after its food, leads it to bury itself beneath the surface of the soil. But, as the insect does not possess an organisation that fits it for burrowing, the character of the soil has a great deal to do in affording facilities or the reverse favorable to its existence.

If the soil be of such a nature that it splits easily into fissures or cracks, which better lead to or serve to expose the vine roots, it will, of course, afford a much more easy means of access to the parasite than if it be compact and close.

Hence it is that clayey and chalk soils, from their liability to split up on the surface, afford much more congenial habitats for the Phylloxera than sandy ones, which, being dry and closely-knit, afford a much more impenetrable barrier to the entrance of the insect, or to its subterranean movements.

These statements are borne out by the fact that where the disease has shown itself, it has been found to vary in extent and intensity in proportion as the soil of the vineyard is more or less clayey; and many instances are known in which patches of a vineyard have continued unaffected amidst the surrounding devastation, owing to the absence in those particular parts of the soil of the argillaceous element.

A forcible illustration of this malign influence of clay in the soil is afforded by the following analysis of two specimens of earth taken from the same vineyard. The specimen marked 'healthy' was from a small plot of ground in which the vines were perfectly sound; that distinguished as 'unhealthy' formed by far the greater portion of the soil of the vineyard, the plants growing in which were all suffering from the ravages of the parasite:

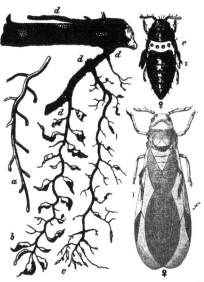

a, healthy root; *b*, root on which the lice are working, showing the knots and swellings caused by their punctures; *c*, root deserted by them, on which the rootlets have begun to decay; *d, d, d*, lice on the larger roots, natural size; *e*, female pupa, dorsal view; *f*, winged female, dorsal view, greatly enlarged.

	Healthy.	Unhealthy
Water	2·25	3·20
Nitrogen	0·11	0·12
Sulphate of calcium	0·62	0·42
Chloride of sodium	1·15	0·18
Carbonate of calcium	49·00	42·00
Siliceous sand	23·50	10·20
Clay	17·75	37·50
Organic substances and error of analysis	5·61	6·38
	100·00	100·00

From the locality already indicated, where it had first developed itself in 1865, the vine disease gradually extended until, in the year 1873, it was ravaging the vineyards of the Gard, Vaucluse, Isère, Herault, Drôme, Bouches du Rhone, Ardèche, Basses-Alpes, Var, the Gironde, and the Charentes, since which time it has gradually continued to spread into the adjacent districts.

"We may gain a more precise idea than can be afforded by a mere observation of the geographical extension of the disease, of the disastrous nature of the ravages of the *Phylloxera*, by the examination of some of the statistics of the grape-crop in successive years, in some of the departments attacked. Thus, in the Department of Vaucluse, where the disease showed itself in 1866, there were in 1866, according to the results obtained by the departmental commission instituted at Avignon to observe on the new vine-disease, 6000 hectares absolutely dead or dying, and a much larger number already attacked, which have since succumbed to the parasite. Out of the 30,000 hectares of vineyards comprised in this department, 25,000, or five sixths of the total area, have been destroyed. In the Gard, where the vine flourishes better than in the above-mentioned department, the ravages of the disease are yet most terrible; for in 1871, in the Arrondissement of Uzes, but one half of the average crop was produced, and in the Arrondissement of Nismes a tenth part of the crop was destroyed. These proportions, moreover, have increased since that year.

"If we examine the mischief done in the less extended areas of the communes, we shall obtain a still clearer idea of the rapid spread of the disease:—

"COMMUNE OF GRAVESON.

1865-66-67 mean crop 10,000 hectolitres.

1868	„	5,500	„
1869	„	2,200	„
1870	„	400	„
1871	„	205	„
1872	„	100	„
1873	„	50	„

"In the Commune of Maillanne the crop in 1868 was only 40 per cent. of the average of the three preceding years, while in 1869 it was only 10 per cent. In the Commune of Eyragues the crop in 1868 was about 33 per cent. of the average of the three preceding years, and in 1869 there was a further falling off of about 10 per cent. In 1870 the crop in the three above-named communes was almost entirely destroyed. From instances such as these, fairly selected from many others equally tragic in their stern figures, we may form some idea of the magnitude of the disaster. Indeed, it is difficult to see, so rapid is the extension of the disease, how, unless some potent and effective remedy can be soon applied, any vine-bearing district in France, can escape the visitation of the Phyllosera." ('Nature,' vol. x.)

The French Government, fully alive to the peril threatening the staple product of their country, shortly after the appearance of the Phyllosera in the vineyards of France, offered through their Minister of Commerce and Agriculture a reward of 300,000 francs for the discovery of a means of arresting and stopping its ravages; and in 1871 the Academy of Sciences at Paris appointed a commission, presided over by the celebrated chemist M. Dumas, to investigate the biology, habits, &c., of the parasite, together with the nature of the injuries it inflicted upon the vine, the area of its depredations, &c. From amongst the members of this commission three gentlemen were chosen to visit the infected districts, so as to be afforded an opportunity of studying the Phyllosera at its destructive work, and its environments of soil, situation, temperature, &c.

The delegates selected by the Commission were

MM. Balbiani, Cornu, and Duclaux, respectively amongst the most distinguished living representatives of zoology, botany, and chemistry in France, and the results of their labours was the issue, some few years back, of a most exhaustive and valuable report to the Academy of Sciences on the subject of the Phyllosera.

The vines of other countries besides those of France have also suffered from the attacks of the Phyllosera. Thus it has made its appearance in the vineyards of Algiers, Italy, Germany, Spain, Portugal, Switzerland, Australia, and North America, on all of which it has been productive of more or less serious injury to the vintage.

Amongst the numberless remedies that have been suggested and tried, with varying but by no means uniform or satisfactory results, for the destruction of the Phyllosera, may be mentioned sulphur, the sulphites, tobacco, caustic soda, and potash, bisulphide of carbon, coal-tar, soft soap, lime, the immersion of the vine in sulpho-carbonate of potassium, and the application around the roots of sand.

A certain amount of success, it has been said, has attended the employment of the sulpho-carbonate of potassium and sand.

An American botanist, Mr. Riley, recommends the importation into French vineyards of the American vines, which he suggests should be employed as stocks on which to graft the French ones.

The American plant being of a hardy nature he believes its incorporation with the more susceptible French ones, would give rise to a vine sufficiently vigorous to resist, or at any rate not to be injured by the ravages of the parasite.

PHYSIC BALLS. See VETERINARY MEDICINES.

PHYSOSTIGMATIS FABA. See CALABAR BEAN.

PHYSOSTIGMINE. Syn. PHYSOSTIGMINA, ESERINA. $C_{15}H_{21}N_3O_2$. An alkaloid obtained from Calabar bean. To prepare it an alcoholic extract of Calabar bean is dissolved in water, bicarbonate of sodium added in excess, the freed alkaloid shaken out with ether, and the ethereal liquid removed and evaporated.

Characters. It is colourless when fresh and carefully made, by exposure to the air it becomes pink. The crystals dissolved in water soon darkens to red, which change is facilitated by adding solution of potash. It causes contraction of the pupil of the eye.

Both sulphate and salicylate of physostigmine are used in medicine, and prepared by neutralising pure physostigmine with sulphuric or salicylic acid, evaporating, and setting aside to form crystals.

Uses. In eye diseases, as ulcers of the cornea; it removes dilatation of the pupil after the use of atropine. Internally it has been given in chorea, tetanus, and hysteria. Dose, $\frac{1}{12}$ gr.

Physostigmine, Neutral Hydrobromate of. This body is prepared with colourless hydrobromic acid in the same manner as the sulphate. The solution evaporated to a syrupy consistence, crystallises in the course of a few days in fibrous masses, rarely colourless and non-deliquescent.

The neutral hydrobromate of physostigmine is

employed like the sulphate and in the same doses, although it contains a little less physostigmine. (From 'Formulæ for New Medicaments,' adopted by the Paris Pharmaceutical Society.) See CALABAR BEAN.

Physostigmine, Neutral Sulphate of. The salt is obtained by saturating directly and exactly a known quantity of physostigmine with dilute sulphuric acid (1 in 10); or better still, by shaking a solution of the physostigmine with a titrated solution of sulphuric acid so as not to exceed the point of saturation. The filtered solution of sulphate of physostigmine is evaporated rapidly to dryness by the aid of a gentle heat.

Sulphate of physostigmine can be crystallised in long prismatic needles, combined in radiating groups, but it is very difficult. It is preferable to preserve it in the amorphous state, and in well-stoppered bottles, as it is very deliquescent.

Sulphate of physostigmine is employed like physostigmine internally under the form of granules containing up to 1 milligram. It is employed also for the eyes as a solution, containing 2 to 5 centigrams of the salt to 10 grams of distilled water.

Solutions containing physostigmine, pure or combined, alter rapidly in contact with the air, becoming red; they should only be prepared in small quantities as required. (From 'Formulæ for New Medicaments,' adopted by the Paris Pharmaceutical Society.) See CALABAR BEAN.

Phytolaccin. A powdered extract obtained from poke-root, *Phytolacca decandra*. It acts as an alterative, cathartic, and emetic; given in syphilis and rheumatism. *Dose*, 1 to 5 gr. in pill.

PHYTOMYZA NIGRICORNIS, Macquart (from φύτον a plant, and μύξω, to suck), THE BLACK-HORNED TURNIP-LEAF MINER. In some seasons the leaves of swedes and turnip plants are seen to be much punctured on their under sides. Upon examination it will be found that there are maggots under the cuticle, which have mined in the parenchyma, and have made long burrows therein. Though they do not cause a great amount of injury to the plants they effect them in a degree, and tend to make them unhealthy. Upon one swede plant as many as 80 maggots, or larvæ, of this insect have been counted. This attack is very often overlooked, and its consequences are attributed to other causes, because the maggots are always on the under side of the leaf, and cannot be seen at all under the cuticle of the upper part of the leaf. A large farmer in Oxfordshire remarked that some swede plants did not get on as they should; but nothing could be discovered until by chance he pulled off a leaf of a plant that was specially flagging, and saw the mines made by the maggots of the *Phytomyza* between the ribs or veins. It was discovered by further examination that many of the plants in a field of considerable extent were more or less affected by this leaf miner.

An account was sent of a somewhat mysterious and important change in the appearance of the turnip plants in a piece of white Tankards, towards the latter part of July. The weather had been hot and dry, and the turnips had been put in very early for feeding at the beginning of September. The leaves that were sent were full of

mines or burrows, and several pupæ of the *Phytomyza* were found within these. It was reported that a large quantity of the leaves were in this condition.

German entomologists do not speak of this insect as injuring turnip plants, but Kaltenbach says that it mines the leaves of Monkshood ('Die Pflanzenfeinde,' von J. H. Kaltenbach, page 15). American writers do not allude to it. Meigen states that it is known in the northern part of France ('Systematische Beschreibung der bekannter Europaischen Zweiflugeligen Insekten,' von J. W. Meigen, vol. vii).

Life History. As a two-winged fly, the black-horned turnip-leaf miner is naturally of the order *Diptera*, and belongs to its family *Muscidæ*, and is a species of the section *Phytomyzides*. It is dusky grey, or dark slate coloured, being rather more than a line, the twelfth of an inch in length, with a wing expanse of from two to two and a half lines. The head is yellowish, while its poisers and parts of its legs are white, and the wings translucent. The fly lays eggs on the under side of the leaf of the turnip plant, as well as of other plants, in May. From these eggs maggots are hatched, and make burrows invariably in the cuticle of the lower part of the leaf, in this respect differing entirely from the *Drosophila flava*, feed upon the soft substances of the leaf, turning into pupæ in about three weeks. The pupæ are dark brown. It is supposed that the pupæ pass the winter in the ground, as well as in the decayed leaves, if these by any chance remain.

Prevention. . Turnip plants attacked by this fly should be fed off by sheep, or care should be taken to clear away the tops that are cut off when swedes or turnips are stored. These should not be put in heaps to rot on the outside of fields, but spread on the land and ploughed deeply in, or taken away and burnt (' Reports on Insects Injurious to Crops,' by Chas. Whitehead, Esq., F.Z.S.).

PHYTOPTHORA INFESTANS. THE POTATO DISEASE, or BLIGHT. The disorder affecting potatoes, commonly known as the potato disease, or potato blight, and as potato ' rot ' in the United States, is caused by a fungus named by De Bary, the distinguished German mycologist, *Phytopthora infestans*. This disorder first appeared in Great Britain in 1844, and simultaneously in other parts of Europe. It spread rapidly in that year from the south of England, through the Midland Counties, to Scotland and Ireland. In 1844, and in succeeding years, very much loss was sustained by potato-growers, particularly in Ireland where the distress consequent upon the potato disease was most calamitous.

Outbreaks of this disease have occurred with varying frequency since it first came to these shores. In 1870 it was specially considered by the Royal Agricultural Society of England, and experiments were made to determine whether there were varieties of potatoes proof against disease. The elaborate report of these experiments written by Mr Carruthers, the Society's Consulting Botanist, showed that no varieties that had been tried were disease-proof. Incidentally it was shown that no special manures nor mode of cultivation seemed to have any influence upon the course of the disorder.

In 1880 a select committee of the House of Commons was appointed to inquire into the best means of diminishing the frequency and the extent of failures in the potato crop, and made a report founded on the interesting evidence of scientific and practical witnesses including Mr Carruthers, Mr Thistleton Dyer, Mr Worthington Smith, Dr Voelcker, and Professor Baldwin. This report indicated that the failures in the potato crop were due to the action of a fungus, and that all soils and climates were pretty nearly alike liable to be attacked by it. The results of the Royal Agricultural Society's experiments as to manures and cultivation were generally confirmed, though it was adduced that there were certain varieties of potatoes which were better able to resist the fungus than others, yet none able to resist it altogether.

During a few years previous to 1800, the attack of this fungus had not been of a very serious nature, but it then showed itself in very many parts of England and Scotland, and it was feared that it would cause much loss.

The Board of Agriculture made special inquiry into the position and prospects of the potato crop in the autumn of 1890, when a summary of the reports received from inspectors under the Drainage and Improvement Acts showed that no serious loss of the potato crop in Great Britain was to be apprehended from disease, as the late, or main crop, was regarded as comparatively sound.

It is curious to note that the early sorts of potatoes were more affected by the disease in 1890 than those of later habit which constitute the most important part of the crop, and are mainly relied on for storing purposes. The reverse has usually been the case. Early sorts have escaped, the tubers having been fit to dig before the fungus had spread in an important degree, and generally consumed before it could show itself within them. Thus the main crop has generally suffered in previous years, the latest sorts being most affected.

It is considered that the reason for this is that the weather in June, 1890, was eminently suited for the development and rapid increase of the fungus. It was wet and warm, with much electrical disturbance and occasional fitful periods of scorching sun-heat,—in short, typical potato-blight meteorological conditions. Later on, in July and part of August, the downpour of wet and the low temperature were not favourable for the dissemination of the spores of the fungus. At the end of August brilliant hot dry weather set in and continued for six weeks. This was fatal to this special fungoid growth, and at the same time gave healthful vigour to the plants.

Both in England and Scotland the potato plants were much more affected by the disease in gardens and small holdings than those in fields, and those cultivated upon a large scale. This has been noticed in former years of potato blight, and is due to potatoes being grown more frequently upon the same land in the former case, so that infection is carried on.

The continual cropping of the same land with potatoes in Ireland would seem to account for the greater prevalence of disease in that part of the United Kingdom, together with the less care that is there exercised in changing and selecting seed potatoes.

Life History. This fungus belongs to a genus of the family PERONOSPOREÆ named by De Bary *Phytopthora* (plant devourer), to distinguish it from other PERONOSPOREÆ, as it differs from these in respect of peculiarities in the form of its conidia, or spores.

Propagation of this fungus is carried on in two ways, first by means of oospores, termed 'resting spores,' because they rest through the winter; and, secondly, by the mycelium, or centre, in the tubers, which is supposed to be passive until influences of moisture, raised temperature, or exposure to air render it active. In an active condition it extends filaments—hyphæ—which proceed to destroy the cells of the tubers and to set up decay within them so that they rot in the clamps, or stores. If the mycelium remains dormant within the tubers until they are planted, it may be taken up in the plants, and by the plants as they grow, and produce conidia in due time to infect neighbouring plants.

De Bary has shown that the mycelium of this fungus when in the tubers and in the ground, is able to send forth conidiophores bearing conidia or spores directly from the tubers, which might be conveyed with the growing plants, or by insects; though he thinks there should be but little weight attached to this source of infection. The same writer observes that tubers infected and containing the mycelium of the *Phytopthora* may infect tubers in the same clamp, or store, by means of the conidiophores bursting their way through the skin, or eyes, and the conidia finding their way to sound tubers and attaching themselves to them. "If these quite healthy tubers should then be planted in the ground, the conidia will germinate, the germs penetrate some of the tubers, and the mycelium develop itself in them. All this is obvious from simple experiments which have been well known for a long time."

The oospores, or 'resting spores,' which also carry on the life of the fungus through the winter, as discovered by Mr Worthington Smith, remain in the haulm, foliage, and decaying tubers left on the earth, or lie upon the earth, or upon weeds and rubbish. Mr Worthington Smith states, "We have secured potato oospores direct from the ground by observing water filtered through earth on which diseased potato material has been allowed to decay" ('Diseases of Field and Garden Crops,' by J. Worthington Smith, F.L.S.). Mr Smith adds that these 'resting spores' can remain dormant for three years. Swarms of conidia—spores—come from the oospores in the early summer, and are conveyed to the potato plants by the wind, insects, and other agencies. These spores, ovoid in shape, and not more than the eight-hundredth part of an inch in length, fall on the leaves and send out tube-shaped shoots, if there is moisture present and other conditions are suitable, which find their way into the tissues. From these are formed mycelia, serving as the centres of the fungus, as it were, and putting forth many branches or filaments, called hyphæ, which run between the cells of the leaves and affect the formation of starch in the leaves, and its ultimate supply to the tubers.

It is not quite clear how this is done, whether by breaking down the walls of the cells and so preventing the manufacture of starch, or by the direct absorption of starch by the fungus, but the starch, or the starch manufactory, as put by Mr Marshall Ward ('Diseases of Plants,' by H. Marshall Ward, F.R.S.), is the object of the invader. Upon filaments sent up from the mycelia through the stomata or pores of the leaves, conidia are generated, by means of which the infection is conveyed to neighbouring plants.

In conditions favourable to the progress of the fungus within the leaf tissues it proceeds from the leaves to the stems, and finally descends to the tubers.

Prevention. It is almost obvious from the life history of the cause of the potato disease that the greatest possible care should be taken to destroy every particle of haulm and leafage from infected fields and gardens, as well as every infected tuber that may be left decaying, or decayed, in or upon the ground.

It would also follow that potatoes should not be grown upon the same land for some time after an infected crop. If Mr Worthington Smith's theory, that the resting spores can retain vitality for three years, is correct, though it must be said that it has not been proved, potatoes should not be taken again for three years. In any case it would be right to plough infected land deeply, and to treat it with a good dressing of quicklime, or gas lime. Garden land and allotment land which must be cropped frequently with potatoes should be dug deeply, and also should be dressed with lime or gas lime. With regard to allotments it is, of course, highly important that there should be unity of action. Where there has been potato disease in any part of an allotment, it should be the duty of all the holders to insist upon the carrying out of the simple precautionary measures of burning, or otherwise destroying, infected haulm and tubers.

There are other precautions that should be taken, such as keeping outsides, and ditches, and corners of fields and gardens free from weeds and rubbish which might harbour the oospores.

Above all things it is essential that the seed potatoes should be free from infection. As I have shown, the mycelium of the fungus may be in a dormant state within the tubers planted for seed, and assume virulent activity when moisture is given, and the tuber begins to shoot. Seed potatoes should therefore not be planted if disease has been prevalent in the crop from which they were selected.

Potato-growers who save their own seed can well arrange this. Those who buy seed potatoes must examine them, cutting whole seed through occasionally to see if there are traces of the disease within them, and examining the portions where cut seed potatoes are employed. By the time seed potatoes are planted the mycelium will show pretty plain indications of its presence, in the form of brown patches scattered irregularly throughout the tubers.

Special Investigation of Diseased Potatoes. In November last potatoes were sent to the Board

of Agriculture for examination as to the occasion of brown patches pervading their internal tissues. The tubers externally were particularly clear skinned and healthy-looking, and would have been accepted by anyone as first-rate sound seed. Upon cutting them open the brown patches were clearly seen; they were small and by no means continuous, and in all cases unconnected with the outside of the tubers. For some time it was difficult to make out their nature, but by exposing pieces of tuber to moisture under glass in a warm temperature it was seen that the brown spots were the mycelia of the potato fungus, as its filaments were sent forth. Portions of uninfected tubers were infected by placing mycelium upon them. Had these tubers been planted as seed they would assuredly have been most dangerous sources of infection; yet ninety-nine growers out of one hundred would have planted them without the slightest suspicion.

Remedies. It has not yet been demonstrated that there is any remedy for this disorder. Sulphur has been tried, but not systematically. A grower in Kent, knowing that powdered sulphur is largely used for hop mildew, put some on a number of rows of potato plants showing signs of infection, leaving other rows near without any sulphur. At digging time the sulphured rows were certainly more free from disease than those unsulphured, and at this stage the record ceased, as the potatoes were all sold.

In the United States experiments have been made with the sulphate of copper washes, such as the *Bouillie bordelaise* and *Eau celeste*, used extensively and most advantageously in France against the vine mildew.

Mr Galloway, the Chief of the section of Vegetable Pathology of the United States Department of Agriculture, reports upon trials made with the *Bouillie bordelaise* in the Report of the Secretary of Agriculture for 1889. He says, "It is well known that a very small quantity of sulphate of copper will prevent the spores from germinating and consequently from infecting healthy plants, and the treatment was made with this fact in mind. Bordeaux mixture containing 6 lbs. of copper sulphate and 4 lbs. of lime to 22 gallons of water, was used for the experiment. The first application was made when the plants were a foot high, there being no signs of blight at the time, and the sprayings were repeated every two weeks until the 10th of September. The variety treated was the Peach Blow, and for convenience the field was divided into three plats of 75 hills each. On November 5th the potatoes were dug, the yield of each plat being as follows:

Plat.	Treatment.	Yield.
		lbs.
1	Bordeaux mixture	346
2	No application	164
3	Bordeaux mixture . . .	283

"Diameter of the largest tuber on treated plats, 5 inches. Diameter of largest tuber from untreated plat, 3 inches. The treated vines kept green until killed by frost, November 5th, while the untreated were killed by the blight a month previously. Plat three grew alongside a row of trees, which probably accounts for the falling off of its yield.

"The results in this case are certainly very satisfactory, and it is hoped that another year more extended experiments can be undertaken, from which further and more important deductions can be made. To those wishing to test the remedy we will say that it is of the utmost importance the mixture be applied early. The fact that the treatment is entirely preventive must constantly be kept in mind, as on this hinges the whole secret of success."

In the course of last summer attention was drawn to the interesting series of experiments carried on by M. Aimé Girard, in France, on the use of sulphate of copper as a remedy for potato disease in 1888 and 1889. These showed that much benefit had been derived from the application of this substance. The experiments were brought in August last to the notice of the Irish Government, and the feasibility of making similar experiments in Ireland was suggested.

It is to be hoped that the sulphate of copper washes will be tried in Great Britain and Ireland, not in a half-hearted manner and when the disease is established, but as a preventive measure adopted as the Continental wine producers adopt it, and as the English hop-growers adopt sulphur, as almost a necessary part of cultivation. The washes are inexpensive. They can be put on in the fields with the Strawsonizer. In gardens and allotments the cheap 'Knapsack' form of engine as used in the United States and in the smaller vineyards on the Continent would serve admirably to distribute them.

With regard to disease-proof varieties of potatoes, the experiments instigated by Earl Cathcart, and conducted by the Royal Agricultural Society of England, through the Seeds and Plants Diseases Committee, showed that there were no disease-proof varieties. It is hardly reasonable that there should be, considering the cause and nature of the disease. Some varieties, however, are better able to resist it than others, as the Champion, for example, whose stems are stout and high, and the Magnum Bonum, though neither the Champion nor any other potato is disease-proof ('Reports on Fungi,' by Chas. Whitehead, Esq., F.Z.S.).

PHYTOPTUS RIBIS. THE CURRANT MITE. This mite gets more troublesome year by year in black currant plantations. Many complaints were made last year from all parts of the country, both in England and Scotland. A large black currant grower in Kent especially commented upon the heavy losses he had sustained for some seasons, stating that the infested area steadily increased, and young bushes were as badly attacked as those of mature age.

Many fruit-growers are quite unconscious of the origin of this injury, and set down the disorder of the bushes to unsuitable soil or to other causes. The mites are so tiny, being only about the two-hundredth part of an inch long, that they cannot be distinguished without a good pocket lens, and even with this their characteristics are

not detected. They affect the buds by feeding upon them and sucking up their sap, beginning directly there is the least sign of swelling in the buds, and probably even before this. On account of their action many of the leaves and blossoms within the whorls of leaves do not come out at all or do not come out properly.

There are mites of this family peculiar to pear trees, birch trees, nut trees, peach trees, lime trees, and many others, each differing in some characteristics and being distinct species, as this *Phytoptus ribis* is a distinct species.

Upon examining the infested twigs of currant bushes in April it was noticed that some of the buds were partly out on one side, and the part on the other side was evidently unable to get out at all, and much of the fruit blossom within the leaves was nipped in the bud. Upon stripping off the outer bracts of the buds little white specks were discernible with a strong pocket glass.

Many have certainly remarked that after the tiny currants have been formed and have just grown out from the leaf whorls which surround them they fail suddenly, shrivel up, and fall off. This is the result of the injury occasioned to the buds by the continuous action of the mites, which begin to work directly there is any indication of swelling in the buds. It is indeed quite probable, though it has not been and can hardly be ascertained, that they are feeding upon the buds in the winter, even before they swell at all. They are present in them throughout the winter, as Mr. Andrew Murray relates ('Economic Entomology,' by Andrew Murray).

Life History. This mite is of the order *Arachnoidea*, and belongs to the family *Phytoptidæ*. It is not so large as the point of a pin, and of a whitish colour, not milky white, but translucent. In form it is long, and has four legs close to the thorax. At its extremity near the end segment there are two bristles, which it is supposed aid the mite in moving about. The egg is nearly oval and is transparent. It is laid towards the end of the summer, and the mite is soon afterwards hatched out and remains in the bud during the winter, but it is not clear whether it feeds during this season or remains in a state of torpor.

Prevention. Black currants in well-managed plantations and gardens are cut very closely back in the autumn, as the fruit comes on young shoots. Where infestation exists they should be cut harder than usual and all the cuttings removed at once and burnt. In a very bad case I advised the grower to cut the bushes down close to the ground, to burn every fragment and to give a good dressing of lime, thrown over the stocks and dug in close round them. Directly after infested bushes have been cut in the autumn, and it is best to cut such early, the stems left should be brushed over with solutions of soft soap and paraffin oil, or quassia, well worked in. A painter's brush would answer for this. It will be seen that there must be a considerable amount of soap in this mixture to make it hang well to the brush and the stems. A little Paris green paste, or London purple fluid might safely be employed mixed with the soft soap.

Dead leaves and all the rubbish round the

stocks should be raked up and burnt when the bushes are cut. A mixture of soft soap and sulphur has been used with effect.

With regard to remedies it would be difficult to apply these in the summer when the attack was bad and the bushes full of leafage. In the early spring where infestation was noted, early syringing with soft soap and quassa, paraffin oil, or Paris green might be useful. This could be put on with the Vermorel machine, or other 'Knapsack' machines for distributing liquids ('Reports on Insects Injurious to Crops,' by Charles Whitehead, F.Z.S.).

PI'CA. Depraved appetite.

PIC'AMAR. One of the peculiar principles discovered by Reichenbach in beech-tar, and described by him as a viscid, colourless, oily liquid, only feebly odorous, but intensely bitter; insoluble in water; freely soluble in alcohol, ether, and oils; boiling point 520° F.; sp. gr. 1·095. See KREASOTE.

PICALIL'LI. See PICKLES.

PICK'LE. The liquor in which substances used as food are preserved. That for flesh is commonly brine; that for vegetables, vinegar; both of which are commonly flavoured with spices, &c.

Prep. 1. (FOR MEAT.)—*a.* From bay salt, 3 lbs.; saltpetre, 2½ oz.; moist sugar, 1 lb.; allspice and black pepper, of each (bruised), 1 oz.; water, 9 pints; simmer them together in a clean covered iron or enamelled vessel for 7 or 8 minutes: when the whole has cooled, remove the scum, and pour it over the articles to be preserved. Used for hams, tongues, beef, &c., to which it imparts a fine red colour and a superior flavour.

b. From bay salt and common salt, of each 2 lbs.; moist sugar, 1 lb.; saltpetre ½ lb.; allspice (bruised), ½ oz.; water, 1 gall.; as before. Used chiefly for pork and hams. Common salt may be substituted for bay salt, but it is less powerfully antiseptic, and the flavour is less grateful.

2. (FOR VEGETABLES.)—*a.* Strong distilled vinegar, to each quart of which 1½ oz. of good salt has been added.

b. Good distilled vinegar, 4 pints; common salt, 2½ oz.; black pepper, ½ oz.; unbleached Jamaica ginger, 2½ oz. (the last two bruised but not dusty); mace (shredded), ¼ oz.; simmer in an enamelled iron or stoneware vessel, as above, and strain through flannel. Sometimes a little capsicum is added. Used either hot or cold, according to the vegetable it is intended to preserve.

Pickle, Lemon. See SAUCES.

PICKLES. These well-known articles are easily prepared of the finest quality. The vegetables and fruit, selected of the proper quality and at the proper season, after being well cleansed with cold spring water, are steeped for some time in strong brine; they are then drained and dried and transferred to the bottles or jars; the spice (if any) is then added, the bottles filled up with hot, strong, pickling vinegar, and at once securely corked down and tied over with bladder. As soon as the bottles are cold the corks are dipped into melted wax, the more surely to preserve them air-tight. Good wood or distilled vinegar is commonly used for this purpose; but the best malt or white wine vinegar of the strength known as No.

22 or 24 is exclusively employed for the finer pickles which are not spiced. In those for early use the 'steep' may be made in hot or boiling brine, by which the product will be ready for the table in a much shorter period; but with substances of a succulent and flabby nature, as cabbage, cauliflower, some fruit, &c., or in which crispness is esteemed a mark of excellence, this is inadmissible. To such articles the vinegar should also be added cold, or, at the furthest, should only be slightly warmed. As a general rule, the softer or more delicate articles do not require so long soaking in brine as the harder and coarser kinds; and they may be often advantageously pickled by simply pouring very strong pickling vinegar over them without applying heat. It must also be observed that beetroot, and other like substances which are sliced, as well as certain delicate fruits, must not be steeped at all. The spice is commonly added whole to the bottles, but a more economical plan is to steep it (bruised) for some time, or to simmer it in the vinegar before using the latter, as in the forms given under PICKLE (*above*).

The spices and flavouring ingredients employed for pickles are—allspice, black and white pepper, capsicums or red pods, cloves, garlic, ginger, horseradish, lemon peel, mace, mustard, shallots, and turmeric. These are chosen with reference to the particular variety of the pickle, or the taste of the consumer.

A good SPICED VINEGAR for pickles generally is the following :—Bruise in a mortar 2 oz. of black pepper, 1 oz. of ginger, ½ oz. of allspice, and 1 oz. of salt. If a hotter pickle is desired, add ½ dr. of cayenne, or a few capsicums. For walnuts, add also 1 oz. of shallots. Put these into a stone jar, with a quart of vinegar, and and cover them with a bladder wetted with the pickle, and over this place a piece of leather. Set the jar on a trivet near the fire for three days, shaking it 3 times a day, then pour it on the walnuts or other vegetables. For walnuts it is used hot, but for cabbage, &c., cold. To save time it is usual to simmer the vinegar gently with the spices; which is best done in an enamelled saucepan.

In the preparation of pickles it is highly necessary to avoid the use of metallic vessels, as both vinegar and brine rapidly corrode brass, copper, lead, &c., and thus become poisonous. These liquids may be best heated or boiled in a stoneware jar by the heat of a water-bath or a stove. Common glazed earthenware should be avoided, either for making or keeping the pickles in, as the glazing usually contains lead. Pickles should also be kept from the air as much as possible, and should only be touched with wooden or bone spoons. They are also better prepared in small jars, or bottles, than in large ones, as the more frequent opening of the latter exposes them too much. Copper or verdigris is frequently added to pickles to impart a green colour, or the vinegar is boiled in a copper vessel until sufficiently 'greened' before pouring it on the vegetables. This poisonous addition may be readily detected by any of the tests mentioned under COPPER. If a green colour be desired, it may be imparted to the vinegar, and ultimately to the pickles, by steeping vine leaves, or the leaves of parsley or spinach, in it. A teaspoonful of olive oil may be advantageously added to each bottle to keep the pickles white, and to promote their preservation.

*** The following list includes the leading pickles of the shops, and some others:

Barberries. From the ripe fruit without heat.

Beans. From the young green pods of the scarlet bean, and the French or kidney bean, with heat.

Beetroot. From the sliced root, without steeping in brine, and with cold spiced vinegar. When wanted for immediate use the vinegar may be used boiling hot.

Broccoli. As CAULIFLOWERS.

Cabbage. This, either red or white, is cut into thin slices, and steeped in strong brine or sprinkled with common salt, and allowed to lie for one or two days; after which it is drained for 10 or 12 hours in a warm room, and then put into jars or bottles, with or without a little mace and white peppercorns, and at once covered with cold strong white vinegar. Another plan is to steep the sliced cabbage in alum water for 10 or 12 hours, and, after draining and drying it, to pour the vinegar upon it as before. The product of the last formula eats very fresh and crisp, but takes longer to mature than that of the other. Some persons add a little salt with the vinegar; and others mix slices of red beet with the cabbage.

Capsicums. As GHERKINS.

Cauliflowers. As CABBAGE (nearly). Or, they may be steeped in hot brine for 1 or 2 hours before pouring the vinegar over them.

Cherries. From the scarcely ripe fruit, bottled, and covered with strong and colourless pickling vinegar.

Codlins. As BEANS.

Cucumbers. As GHERKINS.

Elderflowers. From the clusters, just before they open, as RED CABBAGE. A beautiful pickle.

English Bamboo. From the young shoots of elder, denuded of the outer skin, pickled in brine for 12 or 14 hours; then bottled with a little white pepper, ginger, mace, and allspice, and pickled with boiling vinegar. Excellent with boiled mutton.

Eschalots. With boiling spiced vinegar, or spices added to each bottle.

French Beans. See above.

Garlic. As ESCHALOTS.

Gherkins. From small cucumbers (not too young), steeped for a week in very strong brine; this last is then poured off, heated to the boiling point, and again poured on the fruit; the next day the gherkins are drained on a sieve, wiped dry, put into bottled or jars with some spice (ginger, pepper, or cayenne), and at once covered with strong pickling vinegar, boiling hot. Several other pickles may be prepared in the same way.

Gooseberries. From the green fruit, as either CABBAGE or CAULIFLOWERS.

Indian Mango. From green peaches (see below).

Indian Pickle. Syn. PICCALILLI. This is a mixed pickle which is characterised by being highly flavoured with curry-powder, or turmeric, mustard, and garlic. The following form is commonly used:—Take 1 hard white cabbage (sliced), 2 cauliflowers (pulled to pieces), some French beans, 1 stick of horseradish (sliced), about 2 dozen small white onions, and 1 dozen gherkins; cover them with boiling brine; the next day drain the whole on a sieve, put into a jar, and add, of curry-powder or turmeric, 2 oz.; garlic, ginger, and mustard seed, of each, 1 oz.; capsicums, ½ oz.; fill up the vessel with hot pickling vinegar, bung it up close, and let it stand for a month, with occasional agitation. See MIXED PICKLES (below).

Lemons. From the fruit, slit half way down into quarters, and cored, put into a dish, and sprinkled with a little salt; in about a week the whole is placed in jars or bottles with a little turmeric and capsicums, and covered with hot vinegar.

Limes. As the last.

Mangoes. As LEMONS, adding mustard seed and a little garlic, with spices at will. ·ENGLISH MANGOES are made from cucumbers or small melons, split and deprived of their seeds.

Melons. As LEMONS (nearly).

Mixed Pickles. From white cabbage, cauliflowers, French beans, cucumbers, onions, or any other of the ordinary pickling vegetables, at will (except red cabbage or walnuts), treated as GHERKINS; with raw ginger, capsicum, mustard seed, and long pepper, for spice, added to each bottle. A little coarsely bruised turmeric improves both the colour and flavour.

Mushrooms. From the small button mushrooms, cleansed with cold spring water, and gently wiped dry with a towel, then placed in bottles, with a blade or two of mace, and covered with the strongest white pickling vinegar, boiling hot.

Myrobalans. The yellow myrobalan preserved in strong brine. Gently aperient.

Nasturtiums. From the unripe or scarcely ripe fruit, simply covered with cold strong vinegar; or, as CABBAGE or GHERKINS.

Onions. From the small button or filbert onion, deprived of the outer coloured skin, and either at once put into bottles and covered with strong white pickling vinegar, or previously steeped for a day or two in strong brine or alum water. When required for early use, the vinegar should be poured on boiling hot.

Peaches. From the scarcely ripe fruit, as GHERKINS.

Peas. As BEANS or CAULIFLOWERS.

Piccalilli. See INDIAN PICKLE.

Radish Pods. As BEANS or GHERKINS.

Samphire. From the perennial samphire (*Erythmum maritimum*), covered with strong vinegar, to each pint of which ½ oz. of salt has been added, and poured on boiling hot. Said to excite the appetite.

Tomatoes. From the common tomato or love apple, as GHERKINS.

Walnuts. From the young fruit of *Juglans regia*, or common walnut:—1. Steep them in strong brine for a week, then bottle them, add spice, and pour on the vinegar boiling hot.

2. On each pint of the nuts, spread on a dish,

sprinkle 1 oz. of common salt; expose them to the sun or a full light for 10 or 12 days, frequently basting them with their own liquor; lastly, bottle them, and pour on the vinegar, boiling hot.

3. (*Dr Kitchener.*) Gently simmer the fruit in brine, then expose it on a cloth for a day or two, or until it turns black; next put it into bottles or jars, pour hot spiced vinegar over it, and cork down immediately. In this way the pickle becomes sufficiently mature for the table in half the time required for that prepared by the common method. Dr Kitchener also recommends this parboiling process for several other pickles. Some persons pierce the fruit with an awl or stocking-needle in several places, in order to induce early maturation. The spices usually employed are mustard seed, allspice, and ginger, with a little mace and garlic.

PIC'OLINE. C_6H_7N. An oily substance discovered by Dr Anderson, associated with aniline, chinoline, and some other volatile bases, in certain varieties of coal-tar naphtha. It is isomeric with aniline, and similar in properties to pyridine.

PIC'RIC ACID. $C_6H_2(NO_2)_3OH$. *Syn.* CARBAZOTIC ACID, NITROPHENISIC ACID, TRINITROPHENOL. A peculiar compound formed by the action of strong nitric acid on indigo, aloes, wool, phenol, and several other substances.

Prep. 1. Add, cautiously and gradually, 1 part of powdered indigo to 10 or 12 parts of hot nitric acid of the sp. gr. 1·48; when the reaction has moderated and the scum has fallen, add an additional quantity of nitric acid, and boil the whole until red fumes are no longer evolved; re-dissolve the crystals of impure picric acid that are deposited in boiling distilled water, and remove any oily matter found floating on the surface of the solution by means of bibulous paper; a second time redissolve in boiling water the crystals which form as the liquid cools, saturate the new solution with carbonate of potash, and set it aside to crystallise; the crystals of picrate of potassium thus obtained must be purified by several re-solutions and re-crystallisations, and next decomposed by nitric acid; the crystals deposited as the liquid cools yield pure picric acid, when they are again dissolved in boiling water, and re-crystallised.

2. Dissolve the yellow resin of *Xanthorrhœa hastilis* (Botany Bay Gum) in a sufficient quantity of strong nitric acid. Red vapours are evolved, accompanied by violent frothing, and a deep red solution is produced, which turns yellow after boiling. Evaporate this solution over a water-bath. A yellow crystalline mass is deposited, which consists of picric acid with small quantities of oxalic and nitrobenzoic acids. The picric acid is purified by neutralising the yellow mass with potash, and crystallising twice out of water. The pure picrate of potassium thus obtained is decomposed by hydrochloric acid, and the liberated picric acid is purified by two crystallisations. This process, devised by Stenhouse, is one of the best, and yields a quantity of the acid amounting to 50% of the resin employed.

The Manufacture and Storage of Picric Acid. In consequence, doubtless, of the serious explosion at the works of Messrs Roberts and Dale, Cornbrook, Manchester, picric acid and the picrates are now, by an Order in Council, declared to be explosives coming within the provisions of the Explosives Act. Hitherto they have been regarded as explosives only when manufactured for military or engineering purposes, but not when for use in dyeing or printing. This distinction was not very logical, since they were equally dangerous whatever the purpose for which it or they were ultimately designed. The only exceptions are picric acid wholly in solution and picric acid manufactured or stored within a place exclusively devoted to such storage and in such a manner as to prevent it from coming in contact with any basic metallic oxide or oxidising agent, or with any detonator or other article capable of exploding picric acid, or with any fire or light capable of igniting it. We may add that the use of picric acid in dyeing has much declined.

Prop. Brilliant yellow scales, sparingly soluble in cold water, but very soluble in boiling water, alcohol, and ether; fusible at 122° C.; volatile; taste insupportably bitter, and very permanent. It forms salts with the bases (picrates, carbazotates), mostly possessing a yellow colour, and exploding when heated.

Prep. The picrate of lead has been proposed as a fulminating powder for percussion caps. A solution of picric acid in alcohol is an excellent test for potash, if there be not too much water present, as it throws down a yellow crystalline precipitate with that alkali, but forms a very soluble salt with soda. Most of the picrates may be made by the direct solution of the carbonate, hydrate, or oxide of the metal, in a solution of the acid in hot water. The picrate of silver forms beautiful starry groups of acicular crystals, having the colour and lustre of gold. The principal use of crude picric acid is for dyeing silk and wool yellow. It is said to be largely employed as a hop-substitute in beer. It is, however, highly poisonous. According to Prof. Rapp, it acts deleteriously both when swallowed and applied to the unsound skin. Five grains seriously affected a large dog, and killed it within 24 hours. It induces vomiting, feebleness, and general loss of nervous tone. The tissues of animals poisoned by it (even the white of the eye) were tinged of a yellow colour. The picrate of potassium has been given with advantage in intermittent fevers. See PORTER, &c.

PICROTOX'INE. Professor E. Schmidt has definitely settled the chemical composition of picrotoxine, the active and poisonous principle of *Cocculus indicus*. Pure picrotoxine may be represented by the formula $C_{30}H_{34}O_{13}$. It crystallises in stellate groups of needles, which have an intensely bitter taste and are very poisonous. They melt at 190° C.—200° C. By means of sundry reagents, picrotoxine may be split up into picrotoxinine ($C_{15}H_{18}O_6$) and picrotine ($C_{15}H_{18}O_7$).

PICROTOX'IN. $C_{12}H_{14}O_5$. *Syn.* PICROTOXINE, PICROTOXIA, PICROTOXINA. A poisonous principle discovered by Boullay in the fruit of *Anamirta paniculata*, or *Cocculus indicus*. It is a vegetable principle which strongly resembles the glucosides.

Prep. 1. Precipitate a decoction of *Cocculus indicus* with a solution of acetate of lead, gently evaporate to dryness, redissolve the residuum in alcohol of 0·817, and crystallise by evaporation; repeat the solution and crystallisation a second and a third time. Any adhering colour may be removed by agitating it with a very little water; or by animal charcoal, in the usual manner.

2. (*Kane.*) Alcoholic extract of *Cocculus indicus* is exhausted with the smallest possible quantity of water, and the mixed liquors filtered; to the filtrate hydrochloric acid is added, and the whole set aside to crystallise. The product may be purified as before.

Prop., &c. It forms small, colourless, stellated needles; soluble in alcohol, ether, and acetic acid, and feebly so in water; boiling water dissolves it freely; taste of solutions inexpressibly bitter; reaction neutral. It does not combine with acids, as formerly asserted, but it forms feeble combinations with some of the bases. It is a powerful intoxicant and narcotico-acrid poison. It acts powerfully on the spinal cord and nervous system generally, occasioning an increase of temperature, and peculiar movements, similar to those described by Flourens as resulting from sections of the cerebellum. It is frequently present in malt liquors, owing to their common adulteration with *Cocculus indicus.*

PICTURES, OIL. To clean. See PAINTINGS, OIL.

PIERIS BRASSICÆ, Linnæus; PONTIA BRASSICÆ, Latreille. Although the mischief of the caterpillars of this large butterfly is more noticed in gardens than in the fields, it is important that the economy of the insect should be known to agriculturists, as all crops of the species of *Brassica* grown by them are liable to be attacked by it. Occasionally it happens that field cabbages, which are now extensively cultivated and thousand-headed kale, a most invaluable farm plant, and turnips of all descriptions, mustard, and rape, are much damaged, especially in small fields, and fields surrounded with hedge-rows, shaws, or shaves and spinneys.

The most important harm, however, is occasioned by it when plants of mustard, rape, turnips, kale, and cabbage are in seed. In some districts, as in Romney Marsh in Kent, in Essex, Lincolnshire, and other places where seed growing is largely adopted by farmers and market-garden farmers, the caterpillars attack the seed pods just after they are formed, and soon clear them off the plants.

A large seed grower in Essex wrote of this kind of injury in 1880 as follows: "The caterpillars of the Dart moth, *Agrotis segetum,* much damaged the plants in my seed beds, and now the large green caterpillars of the white butterfly are devouring the seed pods of my white mustard, rape, and turnips."

Serious complaints were received in 1882, 1883, and 1884 of swedes having been stripped of their leaves by large green caterpillars in small fields in East Sussex, also in fields near the chalk cliffs that abound the coast between Hastings and Brighton. These proved to be the caterpillars of the large white butterfly, whose chrysalides had most probably passed the winter in the rubbish near the hedge-rows, and in the chalk cliffs which are very harbours and refuge for them, as it appears that generally these insects are abundant in localities where there are chalk cliffs, chalk pits, sand banks, and railway cuttings, all of which afford admirable refuges for the chrysalides during the winter.

There are stories told of large flights of these butterflies having apparently come from France, but it is most likely that these had come from the chalk cliffs fringing the south-eastern shores of England.

Curtis relates that "the caterpillars of the white cabbage butterfly greatly injured some swedish turnips in 1841, and no doubt frequently assist in reducing the foliage very considerably" ('Farm Insects,' by J. Curtis). Kirby also speaks of it as destructive to turnips ('An Introduction to Entomology,' by Kirby and Spence).

Westwood states that this butterfly is very common throughout Europe, and is found in Egypt, Barbary, Siberia, and Nepaul ('British Butterflies and their Transformations,' by J. O. Westwood, Esq., F.L.S), Köllar, Taschenberg, Kaltenbach, and Nordinger describe it as very injurious to the *Brassicæ* in Germany.

There is a butterfly somewhat similar in America, called *Pieris oleracea,* by Harris. Its habits of destruction are the same as those of *Pieris brassica* in England, but it is not exactly the same, as it does not have black spots upon its wings.

Life History. This butterfly belongs to the family of *Papilionida,* and the sub-family *Pierides.*

It is rather more than 2½ in. across the wings in the male, and nearly 3 in. in the female, and its body is about an inch in length, and black. The ground colour of the wings is white. At the ends of the fore wings there are broad black bands, wider in the case of the female. In both sexes there is a small dark patch in each corner of the anterior, or hind, wings. The females have two black spots on each fore wing.

The perfect insect appears first between the 7th and the 20th of May, according to the weather. In 1880 it was first seen on the 8th of May, while in 1886 it was not noticed until the 15th. It lays numerous eggs after a few days upon the under sides of the leaves of various plants of the *Brassica* tribe. The egg is yellow skittle-shaped, as Buckler has it, and is fastened in clusters of from 60 to 100 with a glutinous substance. Under the microscope it is seen that the egg has many longitudinal and transverse lines upon it, and forms an interesting microscopical object. The caterpillars are hatched in six or seven days. Their first step is to devour the egg shells. Curtis and Westwood both note this as a peculiarity of these creatures. They feed in companies for three or four weeks, the time being regulated by the food supply and the weather, changing finally to chrysalides.

The caterpillar is greenish yellow, having three yellow lines upon its body and black spots with pale hairs. It has 16 feet. When full grown it is as thick as a small goose quill, being 1½ in. in length.

In the first generation, for there are at least

two generations during the summer, the caterpillars, as has been said, soon become chrysalides. Before this change they crawl away to sheltered places upon palings, trunks of trees, stems of plants, and weeds, to assume the chrysalis stage from which they emerge in imago form in the course of ten days, and commence again the cycle of their life changes. Sometimes the transformation is accomplished upon the plants on which the caterpillars are feeding, at least in the first generation. The chrysalides of the last generation, it should be observed, do not change to butterflies until the following May, remaining in their retreats during the winter in pale green cocoons.

Curtis has it that there is a succession of broods during the summer. Probably there are more than two if the weather is suitable. Nördlinger remarks that in Germany young caterpillars are found upon *Brassica* plants even so late as November (' Die Kleinen Feinde der Landwirthschaft').

Mr. Buckler notes that he found eggs of this butterfly on the 15th of September, which produced larvæ on the 25th of September, but many of these were killed by the frost and rain ('The Larvæ of the British Butterflies and Moths,' by W. Buckler, edited by H. T. Stainton, F.R.S.)

Prevention. Attention to plants that are infested with the caterpillars of the first generation, in June, may prevent a much more serious and general attack upon them, as well as upon others near. When it is possible in these circumstances the leaves with caterpillars on them should be pulled off, or cut off by women and children. It would certainly pay to do this where plants of turnips, mustard, rape, and cabbage were intended for seed. Though a costly process it would, it is believed, pay well in the case of these valuable seed crops.

Outsides of fields should be kept clear of weeds, and hedge-sides and hedge-row sides brushed from time to time. Cruciferous weeds particularly should be extirpated. In market gardens and market-garden farms all open lodges, sheds, and out-buildings should be periodically cleansed, as the walls of these are favourite resorts of the caterpillars for pupation.

Frost unfortunately does not kill the chrysalides. Miss Ormerod states that "during the severe winter of 1878-79, chrysalides of the large white butterfly, which I had the opportunity of examining, appeared perfectly uninjured by cold, which ranged at various temperatures between 10° F. and 30° F." (' Manual of Injurious Insects,' by E. A. Ormerod, Consulting Entomologist, R.A.S.E.)

One hears frequently, "Splendid frost! it has got deep into the ground, it will kill the insects." It is a common error to imagine that frost invariable kills insects either in the perfect, the egg, the larval, or the pupa state. Very many insects are so constituted as to be frost-proof. This is a provision of nature. Many are overtaken and killed by abnormally early or abnormally late frosts before they can assume the form in which they can naturally resist them; but as a rule insects are unharmed by the ordinary winter frosts. As a fact, chrysalides which have been exposed to sharp frosts, and frozen so hard that they would snap asunder like pieces of stick, nevertheless have preserved vitality.

Remedies. From the nature of this attack it is most difficult to apply remedial measures.

I have seen some good done by broad-casting lime, and soot, and guano over swedes very early in the morning, while the dew was on the leaves, or after a shower.

An application of either of these dressings has been found beneficial when plants intended for seed have been attacked. A few pounds of powdered black sulphur—sulphur vivum—mixed with the dressing will be useful if put on when the plants are wet.

In gardens, small plots, and market gardens, syringing the plants with quassia infusion and soft soap and water, in the proportion of 7 lbs. of quassia and 6 lbs. of soft soap to 100 gallons of water is very efficacious.

Natural Enemies. Several parasites feed upon this insect, keeping its numbers down as a rule. Sometimes circumstances transpire in which the butterflies get the upper hand.

The principal among these parasites is one shown in the illustration (Nos. 5 and 6)—a fly belonging to the *Chalcidiæ Pteromalus brassica*, which places from 200 to 300 eggs in the chrysalide. From these eggs larvæ come and quickly eat up the interior of the cocoons. Another deadly parasite is the *Microgaster glomeratus*, a very tiny fly, which deposits quantities of eggs within the bodies of the caterpillars. In course of time the larvæ of the *Microgaster*, when they have literally "cleaned out their victims," turn to chrysalides enwrapped in yellow silken cocoons. Curtis gave a graphic account of these, and Miss Ormerod says that they should not be destroyed.

It need hardly be reiterated that it is of the utmost importance that agriculturists should know their insect friends and be able to distinguish them from their foes, in order that they may be preserved and encouraged ('Reports on Insects Injurious to Crops,' by Charles Whitehead, Esq., F.Z.S.)

PIERIS NAPI. Latreille. **THE GREEN-VEINED WHITE BUTTERFLY.** Yet another species of *Pieris* has to be described, which is common enough everywhere, and in a degree mischievous in turnip, rape, and cabbage fields. It is known throughout England and on the Continent. Köllar speaks of it as living on the leaves of turnips, cabbages, and mignonette in Germany. Taschenberg says it is the least common of the three white cabbage butterflies, but even then it is quite common enough.

Life History. The green-veined butterfly is of the family *Papilionidæ*, and the sub-family *Pierides.* The wings of the butterfly are white, with black or dusky tips, and it has a black back. The male has one black spot on each of its fore wings. The female has two spots. It is distinguished from *Pieris rapæ*, as the ribs or nervures on the under side of its hind wings are of a green colour. The wing expanse is from 1½ to 2 inches.

The egg is flask-shaped, of a pale green hue, deeply furrowed with longitudinal and transverse lines, and is laid singly towards the first week in May. From this in about six days the caterpillar comes. It is an inch in length, dark green, with yellow lines on either side, and escapes observation as it lies under the leaves from being so like them in colour. There are at least two generations of this insect during the summer. In the earlier generations the chrysalis frequently remains and is transformed on the plant; but in the later generations the caterpillar seeks shelter under leaves, on walls, palings, the trunks and stems of trees, the sides of cliffs, and cuttings.

Prevention. The method of preventing the attacks of this butterfly and remedies for them are the same as those recommended in respect of *Pieris brassicæ* and *Pieris rapæ*.

Natural Enemies. This butterfly has a very formidable enemy in the shape of an Ichneumon fly described as *Hemiteles melanarius*, a black four-winged fly, about the fifth of an inch long, which lays its eggs in the chrysalis. Curtis relates that he found the pupæ of the *Pieris napi* with largish holes in them, from which this parasite had issued, and that he bred an incredible number of male and female parasites from one pupa ('Farm Insects,' by J. Curtis). ('Reports on Insects Injurious to Crops,' by Charles Whitehead, Esq., F.Z.S.)

PIERIS RAPÆ. Latreille. **THE SMALL WHITE BUTTERFLY.** Westwood says that this butterfly is sometimes mistaken by persons ignorant of entomology for the young of the large white butterfly, *Pieris brassicæ.* It is as distinct from this as a wren is from a sparrow. There is some resemblance between the two species in point of colour and markings, and in their modes of attacking crops. In other respects they differ extremely.

It is known in many of the southern and south-eastern and western counties as the 'turnip butterfly,' on account of the harm which it does to turnips and swedes. Some also call it the 'cabbage' butterfly, and the 'heart' butterfly, because its caterpillars get into the hearts of cabbages and between the rings of their leafage, and on this account it is considered a great pest by farmers who grow cabbages for stock.

This butterfly is well known in Germany and France; in the latter country its caterpillar is called *ver de cœur.*

Life History. This small white butterfly belongs to the family *Papilionida,* and the sub-family *Pierides.* In measurement of its wing expanse it is close upon two inches, and of its body three parts of an inch. The colour of the body is dark. That of the wings is white, or faint creamy white. Upon the fore wings of the male there is a small black spot, while there are two black spots on each wing of the female.

This butterfly comes out earlier than *Pieris brassicæ,* generally towards the 27th of April, and lays its eggs singly, fastening them on the under surface of the leaves of *brassicæ,* or other cruciferous plants. The egg is light yellow, getting deeper yellow in time. It is skittle-shaped, as Mr. Buckler says ('Larvæ of British Butterflies and Moths,' by W. Buckler, edited by H. T. Stainton, F.R.S.) In six to eight days the caterpillars emerge from the eggs and eat the shells. The full grown caterpillar is about one inch and a quarter in length, dull green in colour,

and therefore very difficult to distinguish when upon leaves. It has two faint yellow lines down its body, with yellow spots upon the lines. There are two or more generations or broods, and the winter is passed in the chrysalis state in pretty much the same conditions as those of the *Pieris brassicæ.*

Prevention. The same precaution must be adopted in respect of this butterfly as in the case of its congener, the large white butterfly, *Pieris brassicæ.*

Remedies. These also are similar to those recommended for the large white butterfly. I may add that when field cabbages are attacked, agricultural salt may be advantageously broadcasted over them before they have become too 'hearty.' This dressing must be applied very judiciously.

PIERRE DIVINE. *Syn.* **CUPRUM ALUMINATUM.** See **LAPIS DIVINUS.**

PIES. Alexis Soyer gives the following instructions for making pies:—

To make a pie to perfection—when your paste (half-puff or short) is carefully made, and your dish or form properly full, throw a little flour on your paste-board, take about a ⅛ lb. of your paste, which roll with your hand until (say) an inch in circumference ; then moisten the rim of your pie-dish, and fix the paste equally on it with your thumb. When you have rolled your paste for the covering or upper crust, of an equal thickness throughout, and in proportion to the contents of your pie (½ inch is about the average), fold the cover in two, lay it over one half of your pie, and turn the other half over the remaining part; next press it slightly with your thumb round the rim, cut neatly the rim of the paste, form rather a thick edge, and mark this with a knife about every quarter of an inch apart; observing to hold your knife in a slanting direction, which gives it a neat appearance; lastly, make two small holes on the top, and egg-over the whole with a paste-brush, or else use a little milk or water. Any small portion of paste remaining may be shaped to fanciful designs, and placed as ornaments on the top.

For meat pies, observe that, if your paste is either too thick or too thin, the covering too narrow or too short, and requires pulling one way or the other, to make it fit, your pie is sure to be imperfect, the covering no longer protecting the contents. It is the same with fruit; and if the paste happens to be rather rich, it pulls the rim of the pie to the dish, soddens the paste, makes it heavy, and, therefore, indigestible, as well as unpalatable.

Meat pies require the addition of either cayenne, or black pepper, or allspice; and fruit pies, of enough sugar to sweeten, with mace, ginger, cloves, or lemon peel, according to taste and the substance operated on. See **PASTRY,** &c.

PIG. The pig or hog (*Sus scrofa*—Linn.), one of the common pachydermata, is now domesticated in all the temperate climates of the world. Its flesh constitutes pork, bacon, ham, &c.; its fat (lard) is officinal in the Pharmacopœias. The skin, bristles, and even the blood and intestines of this animal, are either eaten as food or turned to some useful purpose in the arts. See **PORK, LEATHER,** &c.

PIG'MENTS. These are noticed under the *respective colours.*

PIG-STYE. In order that a pig-stye may not become a nuisance and a danger to health it is essential that the liquid excrement of the pig should be carried off by means of an effective and well-covered drain, and that the solid matters should be frequently removed.

Should it come to the knowledge of the sanitary inspector of the district that a pig-stye is deficient in this particular, the inspector has power to compel the owner of the stye to construct proper drainage.

Urban authorities have full powers in the matter of pig-styes, since under section 26 of the Public Health Act it is enacted "that the owner of any swine or pig-stye kept in a dwelling-house, or so as to be a nuisance to any person, is liable to a penalty of 40s. or less, and to a further penalty (if the offence is continued) of 5s. a day. The authority can also, if they choose, abate the nuisance themselves, and recover the expenses of such action from the occupier of the premises in a summary manner.

A rural authority has power to deal with the matter under provision 3, section 91, of the Public Health Act, which defines as a nuisance "any animal so kept as to be a nuisance or injurious to health."

PIKE. The *Esox lucius* (Linn.), a fresh-water fish. It is remarkable for its voracity, but is highly esteemed by epicures. Various parts of it were formerly used in medicine. The fat (OLEUM LUCII PISCIS) was one of the simples of the Ph. L. of 1618, and was esteemed as a friction in catarrhs. It is even now used in some parts of Europe to disperse opacities of the cornea.

PIL'CHARD. The *Clupea pilchardus*, a fish closely resembling the common herring, than which, however, it is smaller, but thicker and rounder and more oily. It abounds on the coasts of Devon and Cornwall, where it is not only consumed as food, but pressed for its oil.

PILES. *Syn.* HEMORRHOIDS; HÆMORRHOIDES, L. A painful disease occasioned by the morbid dilatation of the veins at the lower part of the rectum and surrounding the anus.

Piles are principally occasioned by costiveness and cold; and, occasionally, by the use of acrid food. They have been distinguished into—BLIND PILES, or a varicose state of the veins without bleeding,—MUCOUS PILES, when the tumours are excoriated, and mucus or pus is discharged,—BLEEDING PILES, when accompanied with loss of blood, and—EXCRESCENTIAL PILES, when there are loose fleshy excrescences about the verge of the anus and within the rectum.

The treatment of piles consists in the administration of mild aperients, as castor oil, or an electuary of sulphur and cream of tartar. When there is much inflammation or bleeding, cold and astringent lotions, as those of sulphate of zinc or alum, should be applied; and when the pain is considerable, fomentations of decoction of poppy heads may be used with advantage. To arrest the bleeding, ice is also frequently applied, but continued pressure is more certain. When the tumours are 'large and flaccid, the compound

ointment of galls is an excellent application; and if there is a tendency to inflammation, a little liquor of diacetate of lead may be added. In confirmed piles the internal use of copaiba, or, still better, of the confection of black pepper, should be persevered in for some time, together with local applications. In the early stages external piles may be very effectually treated by the local application of tincture of hamamelis. In severe cases the protruded tumours are removed by surgeons, by the knife or ligature. See OINTMENTS, ELECTUARIES, &c.

PILL COCHIA. See COMPOUND COLOCYNTH PILLS (*below*).

PILL RUFI. See PILLS OF ALOES WITH MYRRH (*below*).

PILLS. *Syn.* PILULÆ (Ph. E. & D.), PILULA Ph. L.; PILULES, SACCHAROLÉS SOLIDES, Fr. Pills are little balls, of a semi-solid consistence, composed of various medicinal substances, and intended to be taken whole. The facility with which they are made and administered, their comparatively little taste, their power of preserving their properties for a considerable length of time, and, lastly, their portability and inexpensiveness, have long rendered them the most frequently employed and the most popular form of medicine.

The rapid and skilful preparation of pills, from all the numerous substances of which they are composed, is justly considered to demand the highest qualifications in the practical dispenser. The medicinals employed must be made into a consistent and moderately firm mass, sufficiently plastic to be rolled or moulded into any shape, without adhering to the fingers, knife, or slab, and yet sufficiently solid to retain the globular form when divided into pills. A few substances, as certain extracts, &c., are already in this condition; but the others require the use of an excipient to give them the requisite bulk or consistence. As a general rule, all the constituents of a pill that can be pulverised should be reduced to fine powder before mixing them with the soft ingredients which enter into its composition; and these last, or the excipient, should next be gradually added, and the mixture triturated and beaten until the whole forms a perfectly homogeneous mass. It is then ready to be divided into pills. This is effected by rolling it on a slab, with a pill or bolus knife, into small pipes or cylinders, then dividing these into pieces, of the requisite weight; and, lastly, rolling them between the thumb and finger to give them a globular form. A little powdered liquorice-root or starch is commonly employed to prevent the pills adhering to the fingers, or to each other, after they are made. Magnesia, so frequently used for this purpose, is unsuited for pills containing metallic salts or the alkaloids, or other remedies, which are exhibited in very small doses.

Instead of forming the mass into pills by hand, in the manner just referred to, a convenient and simple instrument called a 'pill-machine,' is now generally used by the druggists for the purpose. This consists of two pieces. The first (see fig. 1) is divided into three compartments:—*c* is a vacant space to receive the divided mass, which is to be rolled into pills; *b* is a grooved brass plate, which assists in dividing the mass into pills; and a is a box for containing the powder for covering the pills, and to receive them as they are formed. The second (see fig. 2) consists of a brass plate (a), grooved to match the plate b in fig. 1, and bounded

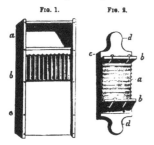

FIG. 1.   FIG. 2.

at both ends by movable projecting plates (b b), containing each two wheels under the ledge of the plate (b); and a wooden back (o), with two handles (d d), to which this plate is affixed. In using this machine, the pill-mass is rolled into a cylindrical form on the front part of it, by means of fig. 2 inverted; the small roll is then laid on the cutting part of the instrument (1, b), and divided by passing fig. 2 over it, the little wheels enabling the latter to run easily on the brass plate which forms the margin of the bed of the machine. The pills, thus formed, are then drawn forward on to the smooth bed on which the mass was first rolled, and receiving a finishing turn or two with the smooth side of the 'cutter,' by which they are rendered more nearly spherical. They are, lastly, thrown over into 1, o, ready to be transferred to the pill-box.

The nature of the excipient should be suited to that of the active ingredients in pills, as well as in all other forms of medicine. Furthermore, it should be of such a nature "that," to quote Dr Redwood, "it will modify as little as possible the action of the pills, either by causing them to become hard, or in any other way, and will not unnecessarily or inconveniently increase their size." Soft extracts, and other substances of a like character, may be rendered more consistent by the addition of any simple powder, as that of liquorice or sugar. Vegetable powders are generally beaten up with syrup or treacle, and heavy powders with conserve of roses or extract of liquorice. Castile soap (made of olive oil and soda) and medicinal soft soap (made of olive oil and potash) are commonly employed for fatty and resinous matters, as well as for many others which are not decomposed by alkalies. When the chief ingredient of the mass is resin rectified spirit is frequently used to soften it, either with or without the addition of soap to increase its solubility in the stomach. For many substances no excipient is required. Thus, most of the gum-resins and stiff extracts may be at once made into pills, or, at all events, after being slightly softened by heat. Mucilage, formerly so much used in the preparation of pills, is now only employed for those which

are to be taken within a day or two after being made; as pills containing it become so hard and insoluble when kept for some time as to resist the action of the stomach, and frequently to pass through the bowels without even losing their form. Glycerin as well as tragacanth are also employed as pill-excipients.

The use of glycerin as a pill-excipient has been advocated on account of its generally neutral properties, and also because its substitution for syrup, the conserves of roses, treacle, and such like substances, results in the production of a pill of smaller bulk. It seems to be suited for pills containing chemical substances, such as quinine, tannic acid, &c. It is said, however, to possess the disadvantage of attracting moisture, making the pills damp on their surfaces.

Quinine, 3 parts, with 1 of the glycerin mass, P. B. strength, works well. Compound powder of ipecacuanha, 5 gr., with ½ gr. of the mass, makes a good pill. Oxide of zinc, too, 4 gr. with 1, makes a good mass.

But for most mineral and insoluble powders it is too moist, and will not form with them a firm mass; some additional absorbent is necessary, and for this purpose I found nothing better than flour; equal parts of the glycerin mass and flour form a tolerably firm, solid, adhesive paste, somewhat resembling dough, but it is not so elastic; this I call bread mass. It possesses great capacity for the absorption of insoluble powders, such, for example, as calomel (3 gr. with 1½ gr. of this mass makes a good pill), nitrate and carbonate of bismuth, arsenic, &c. Of reduced iron, 3 parts with 2 of it, form a good mass, in which the iron is not liable to oxidation. Carbolic acid, too, of which it is a good solvent, is readily made into a pill with the bread mass, a little additional flour being necessary for this substance. Then again, substances which are given in minute doses, as the salts of morphia, resin of podophyllum, and other active principles to partially dilute their action, or where an excipient is needed to partially increase the bulk of the pill, it is well adapted for use. And among the official pill masses an equal quantity of it can with great advantage be used to supplant confection of roses in all these, with the exception of pilula aloes cum ferro, for which the glycerin masses is needed, and pilula ferri carbonatis. This, too, requires the glycerin mass, with which it mixes well, but after a time the pills have a tendency to become moist. Mercurial pill I have not tried with it. The same quantity of this bread mass will replace the treacle in pilula scillæ composita. Equal parts of it and powdered soap, in place of powdered soap alone (if this might be permitted), form a much better mass than the official one of pilula saponis composita.

This pill mass, made strictly according to the Pharmacopœia, soon becomes set into a condition resembling a piece of soap, in which state much beating is necessary to make it again plastic.

Of the glycerin mass to be added to the Pharmacopœia quantities of—

Pil. cambogiæ comp. (vice syrup), 1 oz., makes a good mass.

Pil. colocynth comp. (vice water), 3 dr., makes a good mass, and does not get so hard.

Pil. hydrarg. subchlor. comp. (vice castor oil),

1¼ oz., makes a good mass, but becomes slightly moist.

Pil. ipecac. cum scilla (vice treacle), 1 oz., makes a good mass, which does not crumble.

Pil. rhei comp. (vice 4 oz. of treacle), 2 oz., makes a good mass, and keeps tolerably plastic.

Among the other official pill masses which I have not tried with these excipients are pilula colocynthidis et hyoscyami and pilula conii composita. These I find do not generally require any excipient, and pilula ferri iodidi, the starch contained in the flour, with that would not form an elegant preparation.

Nitrate of silver is generally recommended in works on materia medica to be made into a pill with bread crumb, but this contains common salt, with which it is incompatible. I recommend the following formula, which is a modification of the bread mass :

℞   Nitrate of silver .   .   6 gr.
    Distilled water .   .   6 minims.
Dissolve, and add—
    Glycerin mass .   .   12 gr.
    Flour .   .   .   .   24 gr.

Mix to form a mass which may be divided into 2-gr. pills, each containing ¼ gr. of nitrate of silver. The mass rolls out well. Keep them from exposure to the air and light.

For *Perchloride of Mercury* Pills :

℞   Perchloride of mercury .   6 gr.
    Distilled water .   .   48 minims.
Heat in a test-tube till dissolved, and add to it—

    Glycerin mass .   .   48 gr.
    Flour .   .   .   .   96 gr.

Mix well, and divide into 96 2-gr. pills, each of which will contain a sixteenth of a grain of perchloride of mercury.

Substances like nitrate of silver and perchloride of mercury may form different combinations with the albuminoid principles contained in the flour, but in such state they will probably be quite as readily assimilated, and have a similar medicinal action, as physiologists affirm that most metallic substances enter into the blood as albuminates. I have had some fear lest the gluten contained in the flour might favour some decomposition similar to fermentation, but such, from nearly two years' use of them, I have never yet seen take place ; the glycerin seems to check anything of the kind.

The crude gluten obtained in the moist condition from flour I find is nearly entirely soluble in glycerin, the solution does not appear to undergo any change when kept.

A mixture of glycerin and tragacanth is often used, and produces very similar results to those I have obtained from the glycerin mass. I have not had much experience with such a mixture, but I find that it makes a more elastic paste, which is often a disadvantage, as it causes the pills to have a certain amount of springiness, and renders them difficult to form perfectly globular.

For dry sulphate of iron, of which a large quantity is sometimes ordered in a pill, I find syrup the best excipient. By this means 5 gr. of this can be thus made into a pill ('Pharmaceutical Year Book').

It may be further remarked that no deliquescent salt should enter into the composition of pills not intended for immediate use ; and that when efflorescent salts are so employed they should be first freed from their water of crystallisation.

When the mixed ingredients are made into a mass (pill-mass), which it is not intended at once to divide into pills, it should be preserved in a piece of bladder or gut-skin placed in a covered stoneware or earthenware pot. In this state it may be occasionally moistened with a little weak spirit to prevent its getting hard.

The weight (size) of pills varies from ½ gr. to 6 gr. If heavier than this, they are called 'boluses.' Formerly, as a general rule, they were made of 5 gr. each ; but pills of this weight are, in general, so large that some persons find a difficulty in swallowing them. Another disadvantage of large pills is the trouble of nicely apportioning the dose,—one pill being, perhaps, too small a quantity, and two pills the reverse. Hence, 2 to 3-gr. pills are now the favourite size with both pill-takers and dispensers, notwithstanding that 5-gr. pills are still ordered in one of the authorised Pharmacopœias.

Pills are occasionally coated with gold, silver, gelatin, and other substances, to render them more agreeable to the eye, or to prevent the taste of nauseous ingredients affecting the palate during deglutition. They are gilded and silvered by placing them, in the moist state, on a leaf or two of the metal in a small gallipot, and covering them in a similar manner with another leaf of metal ; over the mouth of the gallipot is laid a piece of smooth writing paper, and on this the palm of the hand ; a sudden and rapid circular motion is then given to the whole. If the pills are not sufficiently moist or sticky, they should be rendered so by rubbing them between the fingers very slightly moistened with mucilage, before proceeding to silver them. Another method is to shake them in a similar manner with a little gold or silver dust.

Mr Haselden recommends a varnish composed as follows :—Resin (this is the substance which is left in the preparation of syrup of tolu) of tolu, 8 parts ; rectified spirit, 6 parts ; methylated ether, 2 parts. Well shake until all that will is dissolved. Use the clear solution. Mr Haselden says iodide of iron pills are well preserved by this coating, and also granules of secale cornutum.

When pills are to be covered with gelatin, each pill, being stuck on the point of a very thin wire 4 or 5 inches in length, is dipped into a solution of gelatin, so as to coat it completely, and the wire is then inserted into a pin-cushion, or a vessel containing fine sand, and left until the gelatin is firm, which occurs in about a quarter of an hour ; the pins may then be easily removed by simply warming them, by placing the centre of each wire for a second or two in the flame of a spirit lamp or candle. 'Sugar-coated pills' are prepared in nearly the same way, but substituting hot and highly concentrated syrup, to which a little gelatin has been added, for a simple solution of gelatin.

The following details for coating pills with sugar

are taken from the 'Chemist and Druggist,' Dec. 15th, 1871 :—"The pills are first varnished with the following liquids—

Ether	.	.	.	.	100 parts.
Bals. tolu	.	.	.	.	10 „
Colophonium	.	.	.	1 „
Absolute alcohol	.	.	.	10 „

By first rolling them in a mortar with this ethereal solution, and then transferring to a sheet of writing paper with the sides bent upwards, shaking being continued till they are perfectly dry. Then to a small quantity of the saccharated albumen (see next receipe) add a few drops of water, at the same time beating for a short while, so that a thick paste will be formed. Into this mass the pills are stirred, and when moistened on all sides, quickly poured into a wooden pill-box, which has previously been filled about one third with the finest powdered sugar obtainable, and immediately shaken, or rather rolled in a lively way with great force, separating from time to time those cohering. When no more sugar will adhere they are dried over a gentle fire, taking care not to bring them too near the stove lest they should crack. Shaking, of course, must be continued till dryness is effected."

Albumen cum Saccharo. Take the white of an egg, and in an evaporating dish beat with it as much powdered sugar, passed through a sieve, as will make rather a thick fluid. Then place it in a water-bath and evaporate to dryness, stirring constantly that no sugar may be deposited. Pulverise and set aside for further use.

The following notes on pill-coating are by Dr Hughes Davies, 'Pharm. Journ.,' January, 1891 :

"There are various ways of coating pills, but the first I was ever able to perform was the tolut. and creta gall., which consists in dissolving ʒiij of residue from making syrup tolut. in ʒiss of ether, varnish the pills with this solution, and when dry rub over with a little powdered French chalk; there are, however, improvements upon that process which I will try and define.

1. *The Gelatine Process.* I will not make any comments upon the many methods of gelatin coating I have tried, but will simply define the one with which I have obtained by far the best results, viz. make the solution from gelatin, 1 oz.; water, 8 oz. Dissolve at a gentle heat, then add the white of an egg, and heat until the albumen coagulates, strain through flannel into a water bath kept at a low temperature, add 2 dr. glycerin, 2 dr. S.V.R., and acid. boric., gr. vi. A beautifully clear solution is thus obtained; the clearer the solution the better the polish. When gelatin coating is carried out on a small scale, it is the usual custom to coat the pills singly, but I have adopted another plan and find it answer equally well and occupy considerably less time.

I have a rounded piece of thin wood with a thick layer of cork stuck round the edge, and in the centre a small hole, through which I have a little ferrule, which enables me to place the concern on a small iron peg fastened in a wooden stand.

The cost of making the whole apparatus would amount to about 9d. It is convenient to have three or four boards at hand; the stand, of course,

would be adaptable to any of them. I have the boards with good needles firmly fastened in the cork to the number of 6, 12, 24, and 48. Now attach the pills to be coated to the points of the needles and dip in the solution, taking care not to keep them in too long, as a thick coating is undesirable. Place the board with the pills on back on the peg, revolve in a gentle manner to render the coating even, and give it an occasional turn round. By doing the coating in the evening the pills are ready to be taken off the needles and stored away in bottles the next morning.

2. *Pearl Coating.* To do this successfully several conditions are of great importance, without attention to which the French chalk will fail to shine. Care must be taken in the selection of a proper excipient for working the mass, for although pills when pearl coated are not within view of the naked eye, they nevertheless must be properly made to be properly coated.

Glycerin being hygroscopic is not an excipient that should be used.

The pills ought to be as nearly round as possible and moderately hard and dry; it is best to keep them exposed on trays for at least a day before the coating is proceeded with. Should the mass be crumbly the condition may be considered to be one of the most bitter enemies of successful coating; the operation will necessarily be a failure, as the pills will most likely crack, and when that takes place the attempt may be given up. Another difficulty that has to be overcome is with pills containing essential oils. Unless these are varnished previous to the coating the oil will work through and spoil the appearance. It is best to dilute the pill varnish in common use to half strength and allow the pills a day's rest before clothing in white.

I use two covered gallipots and a round tin box in the process. The pots should be perfectly smooth, and have well-fitting lids, and should be large enough to hold double the quantity of pills for coating. The tin corresponds in size to the pots.

Place some French chalk in the tin and the pills in one of the pots, damp with a solution. The one I use is equal parts of mucilago, acaciæ, syr. simp. and aqua; too much solution should not be used. A 2 dr. measure is convenient for the purpose. The pills should all be damped, but if too much solution be used too much chalk is taken up. Now turn them out of the pot into the tin containing the chalk, shake sharply and empty out on to a proper receptacle (I use the lid of a cardboard box), keep moving and separate the loose chalk. They are now ready for the polishing pot, being the second gallipot, which should be kept as a polisher only.

Repeat the operation, but this time removing as much of the loose chalk as possible before using the polisher, never forgetting that these small things are the tedious puzzles of pill coating. Repeat the operation once again and it is complete. It is necessary to give particular attention to the washing of the pots between each course, and to keep the polisher perfectly dry. I generally after washing and wiping the pot hold it over a spirit lamp and polish out with a soft cloth."

As pill-masses are likely to get hard and brittle by keeping, an excellent plan is to keep the dry ingredients powdered and mixed together in well-corked bottles or jars, when a portion may at any time be beaten up with syrup, conserve, soap, &c., according to the formula, and as wanted for use. The mixed ingredients in this state are technically known as 'species' or 'powder' for the respective pills.

Pills, Abernethy's. See ABERNETHY MEDICINES (page 6).

Pills of Ac'etate of Lead. *Syn.* PILULÆ PLUMBI ACETATIS, L. *Prep.* 1. Acetate of lead, 20 gr.; powdered camphor, 15 gr.; conserve of roses, q. s.; mix and divide into 12 pills.

2. (*Radius.*) Acetate of lead and powdered mallow or liquorice root, of each, ½ dr.; simple syrup. q. s.; divide into 18 pills.—*Dose*, 1 to 5 daily, washed down with water soured with vinegar; as a powerful astringent in hæmorrhages, diarrhœa, the night-sweats in phthisis, &c. See OPIATED LEAD PILLS.

Pills of Acetate of Mercury. *Syn.* PILULÆ HYDRARGYRI ACETATIS, L. *Prep.* 1. Sub-acetate of mercury, 18 gr.; sugar of milk (or manna), 1 dr.; mucilage, q. s.; divide into 24 pills.—*Dose.* As an alterative, 1 daily; as a sialogogue, one every four or five hours, or oftener in syphilis, &c. See KEYSER'S PILLS.

3. (Opiated—Carmichael.) Acetate of mercury, camphor, and opium, of each 30 gr.; syrup of poppies to mix. For 30 pills. Less apt to affect the stomach and bowels than the last.

Pills of Acetate of Mor'phine. *Syn.* PILULÆ MORPHIÆ ACETATIS, L. *Prep.* 1. Acetate of morphine, 2 gr.; sugar of milk, 15 gr.; conserve of roses, 20 gr.; for 12 pills. Anodyne, sedative, and soporific.—*Dose*, 1, as required.

2. (*Dr A. T. Thomson.*) Acetate of morphine, 1 gr.; powdered foxgloves, 6 gr.; powdered camphor, 10 gr.; powdered gum arabic, 8 gr.; syrup of tolu, q. s.; to be divided into 6 pills. Sedative and antispasmodic.—*Dose.* One every 3 or 4 hours; in phthisis, palpitation, spasms, &c. The hydrochlorate of morphine may be used instead of the acetate, with advantage.

Pills of Ac'onite. *Syn.* PILULÆ ACONITI, P. EXTRACTI A., L. *Prep.* (*Dr Turnbull.*) Alcoholic extract of aconite, 1 gr.; liquorice powder, 12 gr.; simple syrup, q. s.; mix, and divide the mass into 6 pills.—*Dose*, 1 pill every 3 or 4 hours; as a powerful anodyne and sedative in excessive action of the heart, acute rheumatism, gout, neuralgia, &c. The utmost care should be taken both in their preparation and administration.

Pills, Alibert's. See PILLS, APERIENT.

Pills of Aloes. *Syn.* PILULA ALOES SOCOTRINE (B. P.), PILULÆ ALOETICÆ, PILULÆ ALOÈS (Ph. E.), L. *Prep.* 1. (Ph. E.) Socotrine aloes (in powder) and Castile soap, equal parts; conserve of red roses, q. s. to form a pill-mass.

2. (B. P.) Socotrine aloes, 16 parts; hard soap, 8 parts; oil of nutmeg, 1 part; confection of rose, 8 parts. Form into a mass.

3. (B. P.) PILULA ALOÈS BARBADENSIS. *Prep.* Barbadoes aloes (in powder), 2 oz.; hard soap, in powder, 1 oz.; oil of caraway, 1 fluid dr.;

confection of roses, 1 oz. Beat all together until thoroughly mixed.

Obs. "This pill may be also correctly made with the finer qualities of East Indian aloes, as the (true) Socotrine variety is very scarce; and many, not without reason, prefer (pure) Barbadoes aloes." (Ph. E.) The dose, as a laxative, is 5 to 10 gr.; as a purgative, 12 to 20 gr., or more. See PILLS OF ALOES AND SOAP.

Pills of Aloes and Assafœtida. *Syn.* PILULA ALOÈS ET ASSAFŒTIDA (B. P.). *Prep.* Socotrine aloes, in powder, 1; assafœtida, 1; powdered hard soap, 1; confection of roses, 1 (½ confection sufficient—Squire). Mix. Cathartic and anti-spasmodic.—*Dose*, 5 to 10 gr.

Pills of Aloes (Compound.) *Syn.* PILULÆ ALOETICÆ COMPOSITÆ, PILULA ALOÈS COMPOSITA (Ph. L.), PILULÆ A. COMPOSITÆ (Ph. D.), L. *Prep.* 1. (Ph. L.). Socotrine aloes (in powder), 1 oz.; extract of gentian, ½ oz.; oil of caraway, 40 drops; treacle, q. s.; the whole to be beaten together until they form a mass proper for making pills.

2. (Ph. D.). Hepatic aloes (in powder), 2 oz.; extract of gentian and treacle, of each 1 oz.; oil of caraway, 1 fl. dr.; as the last.

Obs. The above is a very valuable purgative in habitual costiveness and indigestion, in all cases in which the use of aloes is not contra-indicated. The dose is from 5 to 15 gr., or more.

Pills of Aloes (Diluted.) *Syn.* PILULÆ ALOÈS DILUTÆ, L. *Prep.* 1. (Dr Marshall Hall.) Barbadoes aloes, Castile soap, extract of liquorice and treacle, equal parts; water, q. s.; dissolve, with heat, strain, and evaporate to the consistence of a pill-mass. Resembles the PILULA ALOÈS CUM SAPONE—Ph. L.

Pills of Aloes and Assafœti'da. *Syn.* PILULÆ ALOÈS ET ASSAFŒTIDA (Ph. E.), L. *Prep.* (Ph. E.). Aloes (Socotrine or East Indian, powdered), assafœtida, and Castile soap, equal parts; beat them with conserve of red roses to a proper pill-mass.—*Dose*, 5 to 10 gr., once or twice daily, as a stomachic tonic and laxative, in dyspepsia, flatulence, &c.; and 12 to 20 gr., as a purgative in similar cases. It is extremely useful in costiveness, with flatulency, occurring in hysterical and hypochondriacal subjects. The B. P. preparation is the same as this, except that hard soap is used instead of Castile soap.

Pills of Aloes and Gin'ger. *Syn.* PILULÆ ALOÈS ET ZINGIBERIS, L. *Prep.* (Ph. D. 1826.) Aloes, 1 oz.; Castile soap, ½ oz.; ginger, 1 dr.; oil of peppermint, ½ dr.; beaten to a mass. A useful laxative in cold habits.—*Dose.* As the last.

Pills of Aloes and Ipecacuanha. *Syn.* DE BAILIE'S DINNER PILLS; PILULÆ ALOÈS ET IPECACUANHÆ, L. *Prep.* (Dr Bailie.) Powdered aloes, 30 gr.; powdered ginger (finest), 45 gr.; ipecacuanha, 12 gr.; syrup of orange peel q. s. to mix. For 24 pills.—*Dose.* One, about an hour before dinner.

Pills of Aloes and I'ron. *Syn.* PILULÆ ALOÈS ET FERRI (Ph. E.), L. *Prep.* 1. (B. P.) Barbadoes aloes, 2; sulphate of iron, 1½; compound powder of cinnamon, 3; confection of roses, 4; mix (6 of confection required—Squire). —*Dose*, 5 to 10 gr.

2. (Ph. E.) Sulphate of iron, 3 parts; Barbadoes aloes, 2 parts; aromatic powder, 6 parts; conserve of red roses, 8 parts; powder the aloes and sulphate of iron separately, beat the whole to a mass, and divide this into 5-gr. pills. An excellent medicine in chlorosis, hysteria, and atonic amenorrhœa.—*Dose*, 1 to 3 pills daily.

Pills of Aloes and Mas'tic. See PILLS, DINNER.

Pills of Aloes and Mercury. *Syn.* PILULÆ ALOËS CUM HYDRARGYRO, L. See PILLS, APERIENT (8).

Pills of Aloes and Myrrh. *Syn.* RUFUS'S PILLS; PILULA ALOËS CUM MYRRHÂ (Ph. L. & D.), PILULÆ RUFI or COMMUNES (Ph. L. 1720), P. ALOËS ET MYRRHÆ (B. P., Ph. E.), L. *Prep.* 1. (Ph. L.) Socotrine or hepatic aloes (in powder, ½ oz.; saffron, myrrh powdered), and soft soap (Ph. L.), of each 2 dr.; treacle q. s. to form a pill-mass.

2. (Ph. D.) Hepatic aloes, 2 oz.; myrrh, 1 oz.; dried saffron, ½ oz.; all in powder; treacle, 2¼ oz.

3. (Ph. E.) Aloes (Socotrine or East Indian), 4 parts; myrrh, 2 parts; saffron, 1 part; beat them to a pill-mass with conserve of red roses, q. s.

4. (Ph. L. 1836 and Ph. D. 1826.) Aloes (in powder), 2 oz.; saffron and powdered myrrh, of each 1 oz.; syrup, q. s. to form a pill-mass.

5. (B. P.) Socotrine aloes, 2; myrrh, 1; dried saffron, ½; treacle, 1; glycerine, a sufficiency. Mix. Stimulant and cathartic.—*Dose*, 5 to 10 gr.

Obs. This compound is a most excellent stomachic purgative and emmenagogue, when there are no febrile symptoms present. It is said to have been employed ever since the time of Rhazes, and is still in extensive use.—*Dose*, 10 to 20 gr.

Pills of Aloes and Rhubarb. *Syn.* PILULÆ ALOËS ET RHEI, P. R. CUM RHEO, L. *Prep.* Powdered Socotrine or hepatic aloes, powdered rhubard, and soft soap (Ph. L.), of each, ½ oz.; oil of chamomile, 10 drops; for 30 pills.—*Dose*, 1 to 5, either as a stomach tonic or laxative; especially in dyspepsia, with loss of appetite.

Pills of Aloes and Rose-juice. *Syn.* PILULÆ ALOËS ROSATÆ, L.; PILULES ANGÉLIQUES, GRAINS DE SANTÉ, FR. *Prep.* Take aloes and rose-juice, of each 4 oz.; juice of borage and chicory, of each 2 oz.; dissolve with heat, evaporate to an extract; add, of rhubarb, 2 dr.; agaric, 1 dr.; and divide the mass into 1½-gr. pills.—*Dose*, 4 to 12, as a purge.

Pills of Aloes with Soap. *Syn.* PILULÆ ALOËS CUM SAPONE (Ph. L.), L. *Prep.* (Ph. L.) Powdered extract of Barbadoes aloes, soft soap, and extract of liquorice, equal parts; treacle, q. s. to form a pill-mass.—*Dose*, 10 to 20 gr.; in the usual cases in which aloes is administered. It is more readily soluble in the juices of the primæ viæ, and is milder than most of the aloetic pills without soap. See PILLS OF ALOES (Diluted).

Pills, Aloes and Turpentine. (*Bois.*) *Syn.* PILULÆ ALOËS ET TEREBINTHINÆ. *Prep.* Boiled turpentine, 2 dr.; aloes, ½ dr. Divide into 40 pills.

Pills, Al'terative. *Syn.* PILULÆ ALTERANTES,

L. See PILLS, CALOMEL, MERCURIAL, and PLUMMER'S, &c.

Pills of Al'um. *Syn.* PILULÆ ALUMINIS, P. A. COMPOSITÆ, L. *Prep.* 1. (*Augustin.*) Alum, 20 gr.; benzoic acid, 6 gr.; powdered gum and white sugar, of each 10 gr.; water, q. s. to form a mass. For 36 pills. In phthisis and atonic mucous discharges. The whole to be taken in the course of 2 or 3 days.

2. (*Capuron.*) Catechu, 1 dr.; alum, ½ dr.; opium, 10 gr.; syrup of red roses, q. s.; divide into 5-gr. pills.—*Dose*, 1 to 3; in chronic diarrhœa and leucorrhœa.

3. (*Radius.*) Alum and catechu, equal parts; extract of gentian, q. s. to mix; divide into 2 or 3-gr. pills.—*Dose*, 2 to 4, every four hours; in passive hæmorrhages, mucous discharges, and chronic diarrhœa.

Pills of Ammoni'acum. *Syn.* PILULÆ AMMONIACI, L. *Prep.* 1. Gum ammoniacum, 1 dr.; powdered sugar, ½ dr.; conserve of hips, q. s. In old coughs and hysterical affections.

2. (Compound.)—*a.* (*Ainslie.*) Ammoniacum, 1 dr.; mercurial pill, 15 gr.; powdered squills, 6 or 8 gr.; simple syrup, q. s. For 16 pills. In asthmatic coughs, with deranged action of the liver.—*Dose*, 1, two or three times a day.

b. (*W. Cooley.*) Ammoniacum and sagapenum, of each 1 dr.; dried sulphate of iron, ½ dr.; conserve of hips, q. s. In obstructed menstruation, and in the chronic diarrhœa of hysterical subjects.

Pills of Ammo''niated Cop'per. *Syn.* PILULÆ CUPRI AMMONIATI (Ph. E.), P. C. AMMONIURETI, L. *Prep.* (Ph. E.) Ammoniated copper (in fine powder), 1 part; bread-crumb, 6 parts; solution of carbonate of ammonia, q. s. to make a mass, which is to be divided so that each pill may contain ½ gr. of ammoniated copper. In epilepsy, and in some other spasmodic diseases.—*Dose*, 1 pill, night and morning, gradually increased to 5 or 6.

Pills of Ammoniated I'ron. *Syn.* PILULÆ FERRI AMMONIATI, P. F. AMMONIO-CHLORIDI, L. *Prep.* 1. (*Dr Copland.*) Ammoniated iron, 1 dr.; aloes and extract of gentian, of each ½ dr.; for 30 pills. In scrofula, chlorosis, amenorrhœa, &c.

2. (*Radius.*) Ammoniated iron and galbanum, of each 1 dr.; assafœtida, 2 dr.; castor, 20 gr.; tincture of valerian, q. s. For 3-gr. pills.—*Dose*, 2 pills, night and morning; in atonic nervous disorders, epilepsy, &c.

Pills of Ammo''nio-cit'rate of Iron. *Syn.* PILULÆ FERRI AMMONIO-CITRATIS, L. *Prep.* (*Beral.*) Ammonia-citrate of iron, 1 dr.; white sugar, 3 dr.; mucilage, q. s. to mix. For 3-gr. pills.—*Dose*, 1 to 3, or more; as a mild chalybeate tonic.

Pills, Analep'tic. See PILLS, JAMES', ANALEPTIC, &c.

Pills, Anderson's Scot's. Various formulæ for these pills are extant, the products of which differ widely from the genuine article. Dr Paris, some years since, declared that they consisted of Barbadoes aloes, jalap, and oil of aniseed. "A careful examination of the proprietary article, with other facts that have come to our knowledge, leads us to believe that the first of the following formulæ is the one now employed in the preparation of the 'Grana Angelica,' or 'Anderson's True Scot's Pills,' of the present day." (*Cooley.*) *Prep.* 1. From Barbadoes aloes, 7 lbs.; jalap (in fine powder), 2½ lbs.; treacle, ½ lb.; soap, 6 oz.; melted together by the heat of a warm bath, and, when partly cold, aromatised by stirring in oil of aniseed, 1 oz. The mass is divided into about 3½-gr. pills, of which 26 or 27 are placed in each 1s. 1½d. box. A mild and useful aperient.—*Dose*, 5 to 15 gr., or more.

2. (Original formula.) Socotrine aloes, 1 oz.; best myrrh, ½ oz.; saffron, 1 dr.; separately pounded very fine; mix them in an earthen pipkin, with a spoonful each of water and sweet oil, by the heat of a slow fire, and form the mass into "common-sized pills." From a copy of the original document in the Chapel of the Rolls.

3. (*P. Cod.*) Aloes and gamboge, of each 6 dr.; oil of aniseed, 1 dr.; syrup, q. s.; mix, and divide into 4-gr. pills. Much more powerful than the preceding, and closely resembling Morison's 'No. 2 pills.'

4. (Phil. Coll. of Pharm.) Barbadoes aloes (in powder), 3 lbs.; Castile soap, ½ lb.; colocynth and gamboge (both in fine powder), 2 oz.; oil of aniseed, 1 oz.; beat to a mass with water, q. s., and divide it into 3-gr. pills. Less active than the last, but more so than the 'True Scot's Pills.'

Pills, An'odyne. *Syn.* PILULÆ ANODYNÆ, L. *Prep.* 1. (Hosp. F.) Opium (in powder), 6 gr.; camphor, 15 gr.; conserve of roses, q. s.; divide into 12 pills.—*Dose*, 1 to 3, as required.

2. (*A. T. Thomson.*) Calomel, potassio-tartrate of antimony, and opium, equal parts; syrup of saffron, q s.; divided in 3½-gr. pills. In acute rheumatism and neuralgia.—*Dose*, 1 pill, at bedtime.

Pills, Antibil'ious. All the ordinary aperient and stomachic pills may be classed under this head. See the names of their proprietors or reputed inventors, or those of their leading ingredients.

Pills, Antichlorot'ic. *Syn.* PILULÆ ANTICHLOROTICÆ, L. *Prep.* (*Radius.*) Aloes and carbonate of iron, of each, ½ dr.; gum ammoniacum, 1 dr.; extract of taraxacum, q. s. For 3-gr. pills.—*Dose*, 2 to 6, night and morning; in chlorosis, amenorrhœa, &c.

Pills, Antimonial (Compound). *Syn.* PILULÆ ANTIMONIALIS COMPOSITÆ, P. ANTIMONII CO., L. *Prep.* Antimonial powder, ½ dr.; calomel, camphor, and powdered opium, of each, 6 gr.; conserve of roses, q. s.; divide into 4-gr. pills.—*Dose*, 2, at night; in acute rheumatism, neuralgia, chronic coughs, &c.

Pills, Antineuralgic. *Prep.* Sulphate of quinine, 2 gr.; sulphate of morphine, ₁/₁₆ gr.; strychnine (alkaloid), ₁/₂₀ gr.; arsenious acid, ₁/₂₀ gr.; extract of aconite leaves (Ph. U. S. 1870), ½ gr.

Note. When 'Antineuralgic Pills,' or 'Neuralgia Pills,' without other specification, are prescribed, it is recommended that the above preparation be dispensed. Sometimes the sulphate of morphine is directed to be omitted (N. F.).

Pills, Antispasmod'ic. *Syn.* PILULÆ ANTISPASMODICÆ, L. *Prep.* 1. (*Dr A. T. Thomson.*)

Opium, 1 gr.; Russian castor, 13 gr.; powdered digitalis, 2 gr.; syrup, to mix; divide into 4 pills.—*Dose*, 1 or 2, two or three times a day; in spasmodic asthma, difficulty of breathing, &c. Several other formulæ for antispasmodic pills will be found both above and below.

2. (*Trousseau* and *Reveil.*) Musk, 15 gr.; extract of valerian, ½ dr.; liquorice powder, q. s. For 20 pills.—*Dose*, 1 every 2 hours, until there is a marked improvement in the symptoms; in pneumonia, accompanied by delirium, especially in drunkards; in spasms of the uterus, and in various other spasmodic affections.

Pills, Ape"rient. *Syn.* PILULÆ APERIENTES, L. *Prep.* 1. Hepatic aloes, 2 dr.; rhubarb and Castile soap, of each, 1 dr.; scammony, ½ dr. (all in powder); essential oil (at will), 10 or 12 drops; beaten to a smooth mass, and divided into pills.

2. Compound extract of colocynth (Ph. L. 1836), 1½ dr.; extract of gentian, ½ dr.; powdered ipecacuanha, 20 gr.; oil of cloves, caraway, or cassia, a few drops. In dyspepsia, loss of appetite, &c.

3. (*Abernethy's.*) See page 6.

4. (*Alibert's.*) From calomel, resin of jalap, and Castile soap, of each, 1 dr.; oil of orange peel or citron, 6 or 8 drops. For 60 pills. As an occasional mild purgative, especially in bilious habits and worms.

5. (*Sir B. Brodie.*) Compound extract of colocynth and mercurial pill, of each, ½ dr.; scammony and Castile soap, of each, 15 gr.; oil of caraway, 6 or 7 drops. For 24 pills. As the last.

6. (*W. Cooley.*) Aloes, 1½ dr.; jalap and Castile soap, of each, 1 dr.; rhubarb and cardamoms, of each, ½ dr. (all in powder); oil of juniper, 12 drops. For 3-gr. pills. A useful mild aperient, for either frequent or occasional use.

7. (*Dr Copland.*) Compound extract of colocynth (Ph. L. 1836), 40 gr.; extract of henbane, 30 gr.; Castile soap, 12 gr.; ipecacuanha, 6 or 7 gr. For two dozen pills.—*Dose*, 2, on retiring to rest. As an aperient in nervous affections and irritable habits.

8. (*Harvey.*) Mercurial pill and powdered aloes, of each, ½ dr.; ginger, 20 gr. For 24 pills. In constipation, attended with a deficiency of bile.

9. (*Dr Neligan.*) Compound colocynth pill and soap of jalap, equal parts; either with or without a few drops of some aromatic essential oil. For 4 or 5-gr. pills. As an aperient for general use.

10. (*Sir C. Scudamore.*) Compound extract of colocynth, 40 gr.; extract of rhubarb, ½ dr.; scammony and soap, of each, 12 gr.; oil of caraway, 5 or 6 drops. For 20 or 24 pills.

11. (*Stahl's :* PILULÆ APERIENTES STAHLII—Ph. Hannov.) Powdered aloes, 1 oz.; compound extract of colocynth, ½ oz.; iron filings, 2 dr.; mucilage, q. s. In amenorrhœa, low habits, and worms.

12. (*Vance.*) Compound extract of colocynth, 80 gr.; extract of rhubarb, 12 gr.; Castile soap, 6 or 8 gr.; oil of cinnamon, 4 or 5 drops.

Obs. The products of the above formulæ may be divided into pills of any size deemed most agreeable to the patient, and they may be aromatised by the addition of any essential oil at will. The dose varies, according to circumstances, from 5 to 10 or 12 gr., or more. Those containing aloes or mercurials are best taken at bedtime. For other formulæ see the various officinal and other pills containing aloes, colocynth, gamboge, rhubarb, scammony, &c.

Pills of Arse″niate of Iron. *Syn.* PILULÆ FERRI ARSENIATIS, L. *Prep.* (*Biett.*) Arseniate of iron, 3 gr.; extract of hops, 2 dr.; powdered mallow-root, ½ dr.; syrup, q. s. For 48 pills.—*Dose*, 1 to 2, daily; in cancerous, scrofulous, and herpetic affections. See PILLS, ARSENICAL.

Pills of Arseniate of So′da. *Syn.* PILULÆ SODÆ ARSENIATIS, L. *Prep.* (*Erasmus Wilson.*) Arseniate of soda, 2 gr.; distilled water, the smallest possible quantity to dissolve it; powdered gum guaiacum, ¼ dr.; oxysulphuret of antimony, 20 gr.; mucilage, q. s. For 24 pills.—*Dose*, 1 pill, as the last; in herpes, &c. See PILLS, ARSENICAL.

Pills, Arsen′ical. *Syn.* ASIATIC PILLS, CARMATIC P., EAST INDIAN P., TANJORE P.; PILULÆ ARSENICI, P. ARSENICALIS, P. ASIATICÆ, P. ACIDI ARSENIOSI, L. *Prep.* (P. Cod.) Arsenious acid, 1 gr.; black pepper (in fine powder), 12 gr.; rub them together for some (considerable) time in an iron mortar, then add, of powdered gum, 2 gr.; water, q. s. to make a mass; which is to be accurately divided into 12 pills. Each pill contains ¹⁄₁₂ gr. of white arsenic.

Obs. This compound is commonly employed in the East Indies in syphilis, elephantiasis, intermittents, the bites of venomous snakes, &c.; and as a preventive of hydrophobia. The common practice in England is to employ 16 gr. of pepper to 1 gr. of arsenious acid, and to divide the mass into 16 instead of 12 pills. The dose is 1 or 2 pills daily, taken *after* a meal. The use of all compounds containing arsenic demands great caution.

Pills, Arsenical (Opiated). *Syn.* PILULÆ ARSENICI CUM OPIO, L. *Prep.* (*A. T. Thomson.*) Arsenious acid, 2 gr.; powdered opium, 8 gr.; Castile soap, 20 gr.; simple syrup, q. s. For 34 pills. *Dose*, as the last; in intermittents, herpes, lepra, psoriasis, periodical headaches, neuralgia, &c. (See *above*.)

Pills, Asiat′ic. See PILLS, ARSENICAL.

Pills of Assafœt′ida. *Syn.* PILULÆ ASSAFŒTIDÆ (Ph. E. and U. S.), L. *Prep.* 1. (Ph. E.) Assafœtida, galbanum, and myrrh, of each 3 parts; conserve of red roses, 4 parts, or q. s.; mix, and beat them to a proper pill-mass.

2. (Ph. U. S.) Assafœtida, 1½ oz.; Castile soap, ½ oz.; water, q. s.; divide into 240 pills. *Obs.* The above (particularly the last) are stimulant and antispasmodic.—*Dose*, 5 to 10 gr.; twice or thrice daily; in hysterical affections, &c. (See *below*.)

Pills of Assafœtida (Compound). *Syn.* PILULÆ ASSAFŒTIDÆ COMPOSITÆ (B. P., Ph. D.) *Prep.* 1. (Ph. D.) Assafœtida, 2 oz.; galbanum, myrrh, and treacle, of each 1 oz.; mix in a capsule, by the heat of steam or a water bath, and stir until it becomes a uniform mass.—*Dose*, &c. As the last. The B. P. directs the quantity of galbanum to be double the above.

2. (Hosp. F.) Assafœtida, 1 dr.; soft soap

(Ph. L.), 20 gr.; ipecacuanha and squills, of each (in powder), 12 gr.; syrup, q. s.—*Dose*, 5 to 10 gr.; in chronic asthmas, coughs, &c.

Pills of Assafœtida with Iron. *Syn.* PILULÆ ASSAFŒTIDÆ CUM FERRO, L. *Prep.* (*W. Cooley.*) Assafœtida, 1 dr.; extract of chamomile, ½ dr.; mix with a slight heat; add, of dried protosulphate of iron, 15 gr.; oil of cajeput, 10 drops; and divide into 36 pills. In hypochrondriasis, hysteria, amenorrhœa, chlorosis, &c., after an aperient.

Pills, Asthma. *Syn.* PILULÆ ANTIASTHMATICÆ, L. *Prep.* 1. (Expectorant.) From compound squill pill, 20 gr.; calomel, 5 gr.; powdered opium, 3 gr.; made into 6 pills.—*Dose*, 1 or 2, at bedtime. Expectorant, and sometimes laxative.

Pills, Astringent. *Syn.* PILULÆ ASTRINGENTES, L. See PILLS OF ACETATE OF LEAD, ALUM, GALLIC ACID, NITRATE OF SILVER, SULPHATE OF IRON, SULPHATE OF COPPER, TANNIN, &c.

Pills of Atropine. (*P. Cod.*) *Syn.* PILULÆ ATROPIÆ. *Prep.* Atropia, 1½ gr.; sugar of milk, 1 dr.; gum Arabic, 12 gr.; syrup of honey, q. s. Triturate the atropia for a long time with the sugar of milk, and make into 100 granules and silver them. Granules of arsenious acid, digitalin, and strychnia, are prepared in the same way.

Pills, Dr. Baillie's. *Prep.* (*Cooley*). Aqueous extract of aloes and compound extract of colocynth, of each, 3 dr.; Castile soap, 1 dr.; oil of cloves, 15 drops. For 4-gr. pills. A good occasional aperient.—*Dose*, 1 to 3, at bedtime, or early in the morning. See PILLS, DINNER.

Pills, Barbarossa's. These are supposed to have been the first mercurial preparation employed in medicine. They consisted of quicksilver, rhubarb, musk and amber.

Pills, Rev. D. Barclay's. *Prep.* (*Cooley.*) Resinous extract of jalap, 1 dr.; almond or Castile soap, 1½ dr.; extract of colocynth, 2 dr. (or powdered colocynth, 3 dr.); gum guaiacum, 3 dr.; potassio-tartrate of antimony, 10 gr.; oil of juniper, 8 or 10 drops; oils of caraway and rosemary, of each 4 drops; make a mass with syrup of buckthorn (the smallest possible quantity), and divide into 4-grain pills. A diaphoretic aperient. —*Dose*, 1 to 3; at bedtime.

Pills, Dr. Baron's. *Prep.* From compound rhubarb pills, 30 gr.; compound extract of colocynth, 20 gr.; powdered ipecacuanha, 6 gr. For 3-gr. pills. An excellent stomachic aperient.— *Dose*, 1 to 3 pills, at bedtime; in dyspepsia, loss of appetite, &c.

Pills, Barther's. *Prep.* From myrrh, 1 dr.; aloes, ½ dr.; musk, 15 gr.; camphor, 12 gr.; balsam of Peru, q. s. to form a mass. For 3½-gr. pills.—*Dose*, 2, thrice daily; in hysteria, amenorrhœa, chlorosis, &c.

Pills, Bath Digestive. *Prep.* (*Cooley.*) Rhubarb, 2 oz.; ipecacuanha and Castile soap, of each ½ oz.; capsicum, ginger, and gamboge, of each ¼ oz. (all in powder); syrup of buckthorn, q. s. For 4-gr. pills.—*Dose*, 1, as a dinner pill; 2 or 3 as an aperient.

Pills of Be′beerine. *Syn.* PILULÆ BEBEERINÆ, L. *Prep.* From sulphate of bebeerine, ½ dr.;

aromatic confection, q. s.; oil of cajeput, 5 or 6 drops. For 18 pills.—*Dose*, 1 to 3, every four hours; as an antiperiodic, instead of bark or quinine.

Pills, Be'chic. Pilulæ Bechicæ, L. *Prep.* (*Trousseau* and *Reveil*.) Extract of digitalis, 15 gr.; white oxide of antimony, 30 gr.; extract of liquorice, 40 gr.; mix carefully, and divide into 40 pills. Expectorant and sedative.—*Dose*, 2 to 12, or more; in cases of irritating coughs, catarrh of the pulmonary capillaries or bronchia, &c. See Pills, Cough.

Pills, Beddoe's. *Prep.* From dried (effloresced) carbonate of soda, 1 dr.; soap, 1½ dr.; oil of juniper, 12 drops; sugar of ginger, q. s.; divide into 30 pills. In gravel, stone, &c.—*Dose*, 2 to 5.

Pill's of Belladon'na (Compound). *Syn.* Pilulæ Belladonnæ compositæ, L. *Prep.* 1. (*Ainslie*.) Extract of belladonna, mercurial pill, and powdered ipecacuanha, equal parts. For 3-gr. pills.—*Dose*, 1 night and morning, in cancerous and glandular affections.

2. (*Debreyne*.) Camphor and assafœtida, of each, 1 dr.; extract of belladonna, 20 gr.; extract of opium, 5 gr.; syrup, q. s. For 48 pills.—*Dose*, 1 pill, gradually increased to 6, daily. In hysteria, amenorrhœa, &c.

Pills, Bellosto's. See Pills Mercurial.

Pills, Bennet's. See Pills Fuller's.

Pills, Benzoic. (*Dr Paris*.) *Syn.* Pilulæ benzoæs. *Prep.* Benzoic acid, 12 gr.; extract of poppies, 18 gr. Mix, for 6 pills.—*Dose*, 1 pill. Expectorant.

Pills of Bichlo'ride of Mercury. Pills of corrosive sublimate.

Pills of Bichlo'ride of Plat'inum. *Syn.* Pilulæ platini bichloridi, L. *Prep.* (*Dr Hoefer*.) Bichloride of platinum, 7½ gr.; extract of guaiacum, 1 dr.; liquorice powder, q. s. For 24 pills. —*Dose*, 1 pill, twice or thrice daily; as an alterative in syphilis, &c.

Pills, Bicker's. *Prep.* From rust (carbonate) of iron, 2 dr.; aloes, myrrh, and sulphur, of each, 1 dr.; ox-gall, q. s. to mix. For 4-gr. pills.— *Dose*, 1 to 6, morning and evening; in debility, chlorosis, &c.

Pills of Bit'tersweet. *Syn.* Pilulæ dulcamaræ, L. *Prep.* (*Radius*.) Extract of bittersweet (dulcamara), 1 dr.; crude antimony and bittersweet (in powder), of each ½ dr. For 3-gr. pills.—*Dose*, 6 to 12, twice or thrice a day; in obstinate skin diseases.

Pills, Bland's (Pil. Ferri B. P.) *Syn.* Pilulæ antichloroticæ, L. *Prep.* (*Trousseau Reveil*.) Sulphate of protoxide of iron, 2 parts; reduce it to powder, and dry it in a stove at 104° F.; add to this dry carbonate of potassa, 2 parts; honey, 1 part; and form the mass into 50 pills. Tonic and emmenagogue.—*Dose*, 1 to 10 daily; in debility chlorosis, &c.

2. Pure ferrous sulphate (dried and powdered), 30 grms.; pure potassic carbonate (dried), 30 grms.; powdered gum arabic, 5 grms.; distilled water, 30 grms.; simple syrup, 15 grms.

3. (B. P. additions, 1890.) Sulphate of iron, 120 parts; carbonate of potassium, 72 parts; sugar, 24 parts; tragacanth powder, 8 parts; glycerine, 4½ parts; water, a sufficiency. Mix

the iron in powder with the sugar and tragacanth in a mortar. Mix the potash with the glycerine in another mortar, transfer this to the first mortar, beat into a mass suitable to form pills.—*Dose*, 1 to 4 pills of 5 gr. each.

Pills, Blue. See Pills, Mercurial.

Pills, Bontius's. *Syn.* Pilulæ hydrogogæ, P. H. Bontii, L. *Prep.* (B. Cod.) Socotrine aloes, gamboge, and gum ammoniacum, of each, 1 dr.; white-wine vinegar, 6 dr.; dissolve by heat at twice, press out the liquor, evaporate to a pilular consistence, and divide into 4-gr. pills.—*Dose*, 1 to 3; as a strong cathartic, in dropsy.

Pills, Brigg's Gout and Rheumatic. This nostrum closely resembles in appearance, odour, and properties, the Plummer's Pill of the Pharmacopœia; the two are probably identical (*Cooley*).

Pill of Bro'mide of I'ron. *Syn.* Pilulæ ferri bromidi, L. *Prep.* (*Magendie*). Bromide of iron and powdered gum-arabic, of each, 12 gr.; conserve of roses, 20 gr.; mix, and divide into 20 pills. They should be kept in a dry, corked phial. Tonic and alterative.—*Dose*, 1 to 2, night and morning; in debility, especially that of scrofulous habits, in chlorosis, &c.

Pills of Bru'cine. *Syn.* Pilulæ brucilæ, L. *Prep.* (*Magendie*.) Brucine, 12 gr.; confection of roses, ½ dr.; carefully mixed and divided into 24 pills, which are recommended to be silvered. The quantity of the confection may be advantageously doubled.—*Dose*, 1 pill night and morning; in the same affections as those for which strychnine is administered. The acetate hydrochlorate, or sulphate of brucine may be substituted for the alkaloid in the above formula, in a slightly larger quantity.

Pills of Calomel. *Syn.* Pilulæ calomelanos, P. E calomelane, P. hydrargyri subchloridi, P. H. cloridi†, P. H. c. mitis (Ph. U. S.), L. *Prep.* 1. Calomel, 24 dr.; powdered gum-arabic, 1 dr.; simple syrup, q. s.; mix and divide into 240 pills. Each pill contains 1 gr. of calomel. A convenient form of exhibiting this drug when uncombined with other remedies.—*Dose*, 1 to 5 pills, according to the indication.

Pills of Calomel (Compound). *Syn.* Plummer's pills, Red P.; Pilula hydrargyri subchloridi composita, Pilulæ calomelanos compositæ (Ph. E. and D.), Pilulæ plummeri, Pilula hydrargyri chloridi composita, L. (Ph. L.). *Prep.* 1. (Ph. L.) Chloride of mercury (calomel) and oxysulphide of antimony, of each, 2 dr.; rub them together, add of guaiacum (in powder) and treacle, of each, 4 dr., and form the whole into a pill-mass.

2. (Ph. E.) Calomel and golden sulphide of antimony, of each, 1 part; guaiacum (in powder) and treacle, of each, 2 parts; beat the whole to a pill-mass, and divide it into 6-gr. pills.

3. (Ph. D.) Calomel and precipitated sulphide of antimony, of each, 1 dr.; triturate them together, then add, of guaiacum resin (in powder), 2 dr.; castor oil, 1 fl. dr.; and beat the whole to a uniform mass.

4. (B. P.) Calomel, 1; sulphurated antimony, 1; guaiac resin (in powder), 2; castor oil, 1; mix.—*Dose*, 5 to 10 gr.

Obs. An excellent alterative pill; very useful

in lepra, in secondary syphilis affecting the skin, and in various other chronic cutaneous diseases; also in dyspepsia and liver complaints.—*Dose*, 3 to 10 gr., night and morning.

Pills of Calomel and Opium. *Syn.* PILULÆ CALOMELANOS ET OPII (Ph. E.), L. *Prep.* Ph. E.) Calomel, 3 parts; opium, 1 part; conserve of red roses, q. s.; divide the mass so that each pill may contain 2 gr. of calomel.—*Dose*, 1 or 2 pills, in rheumatism, facial neuralgia, and various inflammatory affections. They offer a convenient form for gradually introducing mercury into the system, and, if continued, induce salivation.

Pills of Cam'phor. *Syn.* PILULÆ CAMPHORÆ, P. CAMPHORATÆ, L. *Prep.* Camphor and sugar, of each (in powder), 2 parts; conserve of hips, 1 part. For 3-gr. pills. Anaphrodisiac, sedative, diaphoretic, and nervine.—*Dose*, 1 to 5, twice or thrice a day.

Pills of Camphor (Compound). *Syn.* PILULÆ CAMPHORÆ COMPOSITÆ, P. CAMPHORATÆ C., L. *Prep.* 1. (*Dupuytren.*) Camphor, 24 gr.; pure musk, 8 gr.; opium, 2 gr.; syrup, q. s.; divide into 12 pills.—*Dose*, 1 to 4, three or four times daily; in putrescent sores, hospital gangrene, &c.

2. (Fr. Hosp.) Gum ammoniacum, 40 gr.; camphor, 30 gr.; musk, 10 gr.; opium, 5 gr.; tincture of valerian, q. s.; divide into 4-gr. pills.—*Dose*, 2 to 6 pills, daily; in nervous and hysterical affections, &c.

3. (*Ricord.*) Camphor and lactucarium (or extract of lettuce), equal parts; divide into 4-gr. pills.—*Dose*, 3 to 6 pills daily; as an anaphrodisiac.

Pills of Canthar'ides. *Syn.* PILULÆ CANTHARIDIS, P. O. COMPOSITÆ, L. *Prep.* 1. Cantharides (in very fine powder), 8 gr.; extract of gentian, ½ dr.; liquorice powder, 10 gr. For 12 pills.—*Dose*, 1 to 4 daily; as a diuretic, emmenagogue, &c.

2. (*Ellis.*) Cantharides (in very fine powder), 18 gr.; opium and camphor, 36 gr.; mix, and divide into 36 pills.—*Dose*, 1 pill, at bedtime; as an aphrodisiac, in parties labouring under general debility. They should be used with extreme caution, and but seldom.

Pills of Capsicum. *Syn.* CAYENNE PEPPER PILLS; PILULÆ CAPSICI, L. *Prep.* 1. (Guy's Hosp.) Capsicum, 1 part; rhubarb, 2 parts, (both in powder); treacle, q. s.; mix, and divide into 3½-gr. pills.—*Dose*, 1 to 3, an hour before dinner, to create an appetite and promote digestion.

2. (*Radius.*) Powdered capsicum, 20 gr.; extract of gentian, 1 dr.; powdered gentian, q. s. to form a mass. For 60 pills.—*Dose*, 2 to 4 pills, thrice daily; in chronic dyspepsia, especially in the loss of tone of the stomach arising from intemperance.

Pills of Carbolic Acid. *Syn.* PILULÆ ACIDI CARBOLICI. *Prep.* Carbolic acid, 3 drops; soap powder, ·60 grm.; lycopodium, ·06 grm.; powdered tragacanth, q. s. For 6 pills. The two first ingredients form a semi-fluid mass, which the lycopodium does not absorb, but which is solidified by means of the tragacanth.

Pills of Car'bonate of I'ron. *Syn.* VALLET'S PILLS; PILULÆ FERRI CARBONATIS (Ph. E), L.

Prep. (B. P., Ph. E.) Saccharated carbonate of iron, 4 parts; conserve of red roses, 1 part; mix, and divide the mass into 5-gr. pills.—*Dose*, 1 to 3, or more; as a mild chalybeate and antichlorotic. 5 to 20 gr.., B, P. For another formula, see PILLS, BLAUD'S (*above*).

Pills, Carbolic Acid. (*Mr Morson.*) *Syn.* PILULÆ ACIDI CARBONICI. *Prep.* Mix ¼ dr. of bicarbonate of soda and 25 gr. of tartaric acid, coarsely powdered, with the smallest possible quantity of syrup and mucilage to form a mass. Divide into 12 pills.

Pills, Catarrh'. *Syn.* PILULÆ ANTICATARRHALES, L. *Prep.* 1. (*Trousseau* and *Reveil.*) Turpentine, 4 dr.; ammoniacum, 1 dr.; balsam of tolu, ½ dr.; aqueous extract of opium, 5 gr.; liquorice powder, q. s.; mix, and divide into 80 pills.—*Dose*, 5 or 6 daily; in chronic catarrh of the bronchi and bladder.

2. (*Trousseau* and *Reveil.*) Alcoholic extract of aconite, 30 gr.; sulphuret of calcium, 16 gr.; powdered sugar, q. s. For 24 pills.—*Dose*, 1 pill, three or four times daily; in chronic pulmonary catarrh.

Pills, Cathartic. *Syn.* PILULÆ CATHARTICÆ, L. *Prep.* 1. (*Dr Collier.*) Calomel, 10 gr.; powdered jalap and prepared chalk, of each ½ dr.; oil of caraway, 10 drops; syrup of buckthorn, to mix; divide into 5-gr. pills.—*Dose*, 1 to 4.

2. (*Dr A. T. Thomson.*) Scammony, 4 gr.; extract of taraxacum, 16 gr.; divide into 6 pills. *Dose*, 3 pills, twice daily; in hypochondriasis and chronic inflammation of the liver.

3. (*A. T. Thomson.*) Calomel, 15 gr.; powdered jalap, 45 gr.; mucilage, q. s. to mix. For 18 pills.—*Dose*, 1 to 3, at night, to empty the bowels, in bilious affections. Other formulæ for cathartic pills will be found both *above* and *below*.

Pills, Cathartic (Compound). *Syn.* PILULÆ CATHARTICÆ COMPOSITÆ, L. *Prep.* (Ph. U. S.) Compound extract of colocynth, 4 dr.; powdered extract of jalap and calomel, of each, 3 dr.; powdered gamboge, 40 gr.; water, q. s.; mix, and divide into 180 pills. An excellent purgative, especially in bilious affections, dyspepsia, &c.—*Dose*, 1 to 3 pills.

Pills of Cetrarine. (*Dr Neligan.*) *Syn.* PILULÆ CETRARINÆ. Cetrarine, 24 gr.; extract of calumba, ½ dr.; make into 12 pills; one every four hours as a febrifuge.

Pills, Chamberlain's Restor'ative. A nostrum composed of cinnabar and milk of sulphur, equal parts; beaten up with conserve of hips.

Pills of Cham'omile. *Syn.* PILULÆ ANTHEMIDIS, P. FLORUM CHAMÆMELI, L. *Prep.* Extract of gentian, 1 dr.; powdered aloes, ½ dr.; powdered rhubarb, 20 gr.; oil of chamomile, 10 drops. A tonic and stomachic aperient.—*Dose*, 5 to 15 gr. This forms the 'chamomile pills' of the shops. They should be kept in a corked phial. (See *below*.)

Pills of Chamomile (Compound). *Syn.* PILULÆ ANTHEMIDIS COMPOSITÆ, L. *Prep.* 1. (*Ainslie.*) Extract of chamomile, 1 dr.; assafœtida, ½ dr.; powdered rhubarb, 20 gr.; divided into 30 or, better, 36 pills.—*Dose*, 1, as a dinner pill; or 2 to 3, twice a day, in flatulent dyspepsia.

2. (*Beasley.*) Aqueous extract of aloes, 12 gr.;

extract of chamomile, 36 gr.; oil of chamomile, 3 drops. For 12 pills.—*Dose*, 2 at night, or twice a day; in dyspepsia, loss of appetite, &c. See PILLS, NORTON'S CHAMOMILE.

Pills, Chapman's. *Prep.* Mastic, 12 gr.; aloes, 16 gr.; rhubarb, 24 gr. For 12 pills. An excellent stomachic aperient.—*Dose*, 2 to 4.

Pills of Chirat'ta. *Syn.* DR REECE'S PILLS; PILULÆ CHIRATTÆ, L. *Prep.* From chiratta, 2 dr.; dried carbonate of soda, 20 gr.; powdered ginger (best), 15 gr.; divided into 36 pills.—*Dose*, 2, twice a day. In acidity, flatulence, and dyspepsia, especially when complicated with gout or debility.

Pills of Chlo''ride of Ba''rium. *Syn.* PILULÆ BARII CHLORIDI, L. *Prep.* 1. (*Pierquin.*) Chloride of barium, 1 dr.; resin of guaiacum, 4 dr.; conserves of fumitory, q. s.; divided into 188 pills.—*Dose*, 1 pill, morning and evening, afterwards increased to 2; in tapeworm, and in the rheumatism of scrofulous subjects.

2. (*Walsh.*) Chloride of barium, 15 gr.; powdered marshmallow or liquorice root and mucilage of tragacanth, of each, q. s. to make 200 pills.—*Dose*, 3, gradually increased to 10 or 12, daily; in cancer, scrofula, goitre, syphilis, &c.

Obs. The above are very poisonous, and their exhibition demands great caution.

Pills of Chloride of Cal'cium. *Syn.* PILULÆ CALCII CHLORIDI, L. *Prep.* 1. As the last.

2. (*Gräfe.*) Chloride of calcium, 1 dr.; extract of opium, 10 gr.; mucilage, q. s. For 54 pills.—*Dose*, 1, every two or three hours, gradually increased until 10, or even 12, are taken every hour; in gonorrhœa, more especially when occurring in scrofulous subjects.

Pills of Chloride of Gold. *Syn.* PILULÆ AURI CHLORIDI, L. *Prep.* From terchloride of gold, 3 gr.; powdered liquorice, 1 dr.; syrup, q. s. For 48 pills.—*Dose*, 1 pill, twice or thrice daily.

Pills of Chloride of Gold and So'dium. *Syn.* PILULÆ AURI ET SODI CHLORIDI, P. A. SODIO-CHLORIDI, L. *Prep.* (*Magendie.*) Soda-chloride of gold, 1 gr.; extract of mezereon, 2 dr.; divide into 60 pills.

Pills of Chloride of Lime. *Syn.* PILLS OF CHLORINATED LIME; PILULÆ CALCIS HYPO-CHLORITIS, L. *Prep.* 1. Chloride of lime, 12 gr.; starch powder, 24 gr.; conserve of hips, q. s.; divide into 36 pills.

2. (*Dr Copland.*) Chloride of lime, 15 gr.; compound powder of tragacanth, 90 gr.; syrup, q. s. For 24 pills.—*Dose*, 1 to 3, twice or thrice daily; in various putrid affections, fevers, &c.

Pills of Chloride of Mercury. Pills of calomel.

Pills, Chol'era. *Syn.* PILULÆ ANTICHOLE-RICÆ, E. *Prep.* 1. Powdered camphor, 15 gr.; powdered capsicum (pure), ½ dr.; bicarbonate of soda, 1 dr.; conserve of roses, q. s. For 36 pills. —*Dose*, 2 to 4, every fifteen minutes, washed down with a wine-glassful of cold water containing half a tea-spoonful of ether; repeated every fifteen or twenty minutes until reaction ensues. They should be freshly made.

2. (PILULA ANTICHOLERICA ARABICA.) *Prep.* Assafœtida, asclepias gigantea, and opium, of

each, 1½ gr. in each pill. One every half or three quarters of an hour, broken down in a spoonful of brandy and water, till the symptoms yield. After vomiting and purging have ceased, if prostration and spasms are urgent, give ¼ or ½ doses. Black pepper is substituted for asclepias in this country.

3. (PILULÆ CAMBOGIÆ COMPOSITÆ, B. P.) *Prep.* Gamboge, aloes beds, and compound cinnamon powder, of each, 1 part; soap, 2 parts; syrup, q. s.—*Dose*, 5 gr. to 10 gr.

Pills of Ci'trate of I'ron and Quinine'. *Syn.* PILULÆ FERRI CITRATIS CUM QUINÂ, L. *Prep.* From citrate of iron and quinine, 1 dr.; powdered citric acid, 20 gr.; conserve of hips, q. s. For 36 pills. An excellent tonic in debility, chlorosis, &c.—*Dose*, 1 to 3, twice or thrice daily.

Pills, Sir C. Clark's. See DINNER PILLS.

Pills, Coindet's. See PILLS OF IODIDE OF MERCURY.

Pills of Col'chicum. See PILLS, GOUT.

Pills of Col'ocynth. *Syn.* PILULÆ È DUOBUS, P. EX COLOCYNTHIDE SIMPLICIORES, L. *Prep.* (Ph. L. 1746.) Colocynth and scammony, of each, 2 oz.; oil of cloves, 2 dr.; syrup of buckthorn, q. s. An active hydragogue cathartic.—*Dose*, 2 to 12 gr.

Pills of Colocynth (Compound). *Syn.* PILL OF COCHIA; PILULÆ COCCIÆ, P. COCHIÆ, PILULA COLOCYNTHIDIS COMPOSITA (B. P.), P. COLOCYNTHIDIS COMPOSITÆ (Ph. L. and D.), P. COLOCYNTHIDIS (Ph. E.), L. *Prep.* 1. (Ph. L.) Extract of colocynth (simple), 1 dr.; powdered extract of aloes, 6 dr.; powdered scammony, 2 dr.; powdered cardamoms, ¼ dr.; soft soap (Ph. L.), 1½ dr.; mix, and beat them altogether, so that a mass may be formed. This is intended as a substitute for the compound extract of colocynth of the Ph. L. 1836.

2. (Ph. E.) Socotrine or East Indian aloes and scammony, of each, 8 parts; sulphate of potassa, 1 part; beat them together; add of colocynth (in fine powder), 4 parts; next add of oil of cloves, 1 part; and with the aid of a little rectified spirit, beat the whole to a mass, and divide this into 5-gr. pills.

3. (Ph. D.) Colocynth pulp, scammony, and Castile soap, of each 1 oz. (in powder), 1 oz.; hepatic aloes, 2 oz.; treacle, 10 dr.; oil of cloves, 1 fl. dr.; mix, and beat them into a mass of uniform consistence.

4. (Ph. L. 1746.) Socotrine aloes and scammony, of each, 2 oz.; pulp of colocynth, 1 oz.; oil of cloves, 2 dr.; syrup of buckthorn, q. s. to form a pill-mass. This is the original formula published by Galen for 'pilulæ cochiæ minores,' and under various slight modifications, it has continued in use ever since.

5. Aloes, 1½ lbs.; colocynth, ½ lb.; jalap, 6 oz. (all in powder); oil of cloves, 1½ oz.; syrup or treacle, q. s. to mix.—*Prod.* About 4¼ lbs. This forms the common 'pil. cochiæ' of the druggists. A few, more conscientious than the rest, add to the above, scammony, 6 oz. It is greatly inferior to the Ph. pill.

6. (B. P.) Colocynth (in powder), 1 part; Barbadoes aloes (in powder), 2 parts; scammony (in powder), 2 parts; sulphate of potash (in powder),

¼ part; oil of cloves, ¼ part; distilled water, a sufficiency (about ¼ part); mix. Dr Gregory's favourite pill.—*Dose*, 5 to 10 gr.

Obs. Compound colocynth pill is a cheap and excellent cathartic, more powerful than the other officinal aloetic pills, and well adapted to cases of habitual costiveness. It has long been extensively used by the poorer classes, and in domestic medicine generally.—*Dose*, 5 to 15 gr.

Pills of Colocynth and Hen'bane. *Syn.* PILULÆ COLOCYNTHIDIS ET HYOSCYAMI (B. P., Ph. E.), L. *Prep.* (Ph. E.) Colocynth pill-mass, 2 parts; extract of henbane, 1 part; beat them up with a few drops of rectified spirit (if necessary), and divide them into 5-gr. pills.—*Dose*, 1 to 3 pills; as an anodyne purgative, in irritable bowels.

Pills of Copai'ba. *Syn.* PILULÆ COPAIBÆ, L. *Prep.* (Ph. U. S.) Pure balsam of copaiba, 2 oz.; recently prepared calcined magnesia, 1 dr.; mix thoroughly, then set the mixture aside until it acquires a pillular consistence, and lastly, divide it into 200 pills.

Obs. Unless the magnesia has been very recently calcined, the copaiba hardens very slowly or not at all. It is said that "lime produces the affect more completely and uniformly than magnesia," and that "specimens of copaiba which are old and contain the most resin harden quickest" (*Redwood*). For present use, the quantity of magnesia may be at least doubled. Dr Pereira orders copaiba, 1 oz.; magnesia, 5 or 6 dr.—*Dose*, 10 to 30 gr., frequently; in diseases of the mucous membranes of the urinary organs. Cubebs are often added.

Pills, Dr Copland's. See PILLS, APERIENT and PECTORAL.

Pills of Corro'sive Sub'limate. *Syn.* PILLS OF CHLORIDE OF MERCURY, P. OF BICHLORIDE OF M.†, HOFFMANN'S P.; PILULÆ SUBLIMATIS CORROSIVI, P. HYDRARGYRI BICHLORIDI†, P. MAJORES HOFFMANNI, L. *Prep.* 1. Corrosive sublimate, 3 gr.; white sugar, ½ dr.; triturate together in a glass mortar for some time, then add of powdered gum-arabic, 20 gr., and beat the whole to a mass with dilute hydrochloric acid, q. s. For 36 pills, each containing 1/12 gr. of corrosive sublimate.

2. (*Brera.*) Corrosive sublimate, 3 gr.; rectified spirit, the smallest possible quantity to dissolve it; bread-crumb, q. s. to form a mass. For 24 pills, each containing ⅛ gr. of the corrosive sublimate.

3. (*Dr Paris.*) Corrosive sublimate and sal-ammoniac, of each, 5 gr.; water, ½ fl. dr.; triturate together until solution is complete, then add of honey, ½ dr.; liquorice powder, 1 dr. (or q. s.), and divide into 40 pills. Each pill contains ⅛ gr. of corrosive sublimate.

4. (Ph. Hannov.) Corrosive sublimate, 15 gr.; distilled water, ¼ fl. dr.; crumb of bread, q. s. to form a mass. For 120 pills, each containing ⅛ gr.

5. (PILULÆ HYDRARGYRI BICHLORIDI CUM GUAIACO—*Dupuytren.*) *Prep.* Perchloride of mercury in subtle powder, 3 gr.; extract of opium, 6 gr.; extract of guaiacum, 12 gr. Make into 20 pills.

Obs. The above formulæ are among those most usually employed. Other authorities order pills containing 1/12 of a gr. Dzondi orders 1/30 gr., and Hüfeland only 1/80 gr., in each pill. The commencing dose should not exceed 1 pill containing the 1/12 of a gr., twice or thrice a day. It may afterwards be safely kept at ⅛ of a gr. They are chiefly employed in syphilis, but are also occasionally exhibited with great advantage in glandular indurations and enlargements, and in cancer; due caution being observed.

Pills, Cough. See PILLS, PECTORAL, EXPECTORANT, &c.

Pills of Cre'asote. *Syn.* PILULÆ CREASOTI, L. *Prep.* 1. (*Pitschaft.*) Creasote, 6 gr.; powdered henbane, 24 gr.; conserve of hips, q. s. For 24 pills.—*Dose*, 1 three times daily; in sea-sickness, the vomiting during pregnancy, &c.

2. (*Riecke.*) Creasote, 1 dr.; extract of liquorice and gum galbanum, of each, ½ dr.; powdered mallow-root, 2 dr.; to be divided into 2-gr. pills.—*Dose*, 3 to 6, four times a day; in acute rheumatism, bronchitis, neuralgia, phthisis, &c.

Pills, Crespigny's. See PILLS, DINNER.

Pills of Cro'ton Oil. *Syn.* PILULÆ CROTONIS, P. TIGLII, L. *Prep.* 1. Croton oil, 3 drops; oil of cloves, 4 drops; bread-crumb, q. s. For 3 pills, one of which is a dose.

2. (*Dr Copland.*) Croton oil, 6 drops; pill of aloes and myrrh, 1½ dr.; soap, 20 gr.; liquorice powder, q. s. For 30 pills.—*Dose*, 2 to 4.

3. (*Dr Reece.*) Croton oil, 6 drops; Castile soap, ½ dr.; oil of caraway, 8 drops; liquorice powder, q. s. For 12 pills.—*Dose*, 1 to 3. In dropsy, visceral obstructions, &c. See CROTON OIL.

4. (With MERCURY—*Dr Neligan.*) Croton oil soap, 3 gr.; extract of henbane and mercurial pill, of each, 24 gr.; oil of pimento, 12 drops; divide into 12 pills.—*Dose*, 2 at bedtime (see *above*).

Pills of Cy'anide of Mer'cury. *Syn.* PILULÆ HYDRARGYRI CYANIDI, P. H. CYANURETI, L. *Prep.* (*Guibourt.*) Cyanide of mercury, 6 gr.; opium, 12 gr.; bread-crum, 60 gr.; honey or syrup, q. s. For 96 pills.—*Dose*, 1 night and morning; in syphilis, chronic inflammation of the viscera, &c.

Pills of Cyanide of Potas'sium. *Syn.* PILULÆ POTASSII CYANIDI, L. *Prep.* (*Golding Bird.*) Cyanide of potassium, 2 gr.; arrowroot, 20 gr.; simple syrup, q. s. For 18 pills.—*Dose*, 1 twice or thrice a day; as a sedative in hysteria, gastrodynia, extreme nervous excitability, &c. See DRAUGHT and MIXTURE, HYDROCYANIC.

Pills of Dande'lion. See PILLS, TARAXACUM.

Pills, De Haen's. *Prep.* (*St Marie.*) Gum ammoniacum and pill aloes with myrrh, of each, 1 dr.; extracts of hemlock and Castile soap, of each, 1½ dr. For 2-gr. pills.—*Dose*, 3 to 6 daily; in painful or obstructed menstruation, chlorosis, &c.

Pills of Del'phine. *Syn.* PILULÆ DELPHINIÆ, L. *Prep.* (*Dr Turnbull.*) Delphine, 1 gr.; extracts of henbane and liquorice, of each, 12 gr. For 12 pills.—*Dose*, 1 to 3 twice a day; in dropsy, gout, rheumatism, &c., instead of veratrine.

Pills, Deobstruent. (L. Ph., 1746.) *Syn.*

PILULÆ ECPHRACTICA. *Prep.* Aromatic pill, 3 oz.; rhubarb, 1 oz.; extract of gentian, 1 oz.; sulphate of iron, 1 oz.; carbonate of potash, ¼ oz.; syrup of roses, q. s.

Pills of Deuto-iodide (Biniodide) of Mercury. *Syn.* PILULÆ HYDRARGYRI DEUTO - IODIDI. (*Magendie.*) *Prep.* Deuto-iodide (biniodide) of mercury, 7½ gr.; extract of juniper, 75 gr.; powdered liquorice, q. s. for 100 pills.

Pills, Diaphore'tic. *Syn.* PILULÆ DIAPHO-RETICÆ, L. *Prep.* 1. Antimonial powder, ½ dr.; opium, 10 gr.; calomel, 5 gr.; confection of opium, q. s. to mix; divide into 10 pills.—*Dose,* 1 at bedtime; in coughs and bronchial irritability after an aperient.

2. Guaiacum, 19 gr.; emetic tartar and opium, of each, 1 gr.; simple syrup, q. s. to mix; divide into 3 pills.—*Dose,* 1 to 2, in acute rheumatism, &c.

3. Camphor and antimonial powder, of each, ½ dr.; opium, 10 gr.; aromatic confection, q. s. to mix. For 12 pills. In fevers, and in some spasmodic diseases.—*Dose,* 1 pill.

4. Powdered guaiacum, 10 gr.; compound powder of ipecacuanha, 5 gr.; confection of roses, q. s. to mix; for a dose. As a diaphoretic, in inflammatory affections and rheumatism.

Pills, Diarrhœ'a. *Syn.* PILULÆ ANTIDI-ARRHOALES, L. *Prep.* (*Trousseau* and *Reveil.*) Soft extract of opium, 1½ gr.; calomel and powdered ipecacuanha, of each, 3 gr.; conserve of hips, q. s.; divide into 10 pills.—*Dose,* 1, two or three times daily; in chronic and choleraic diarrhœa.

Pills, Diges'tive. Under this head are generally classed all the stomachic and milder aperient pills. See PILLS, BATH; PILLS, DINNER, &c.

Pills of Digita'line. *Syn.* PILULÆ DIGI-TALINÆ, L. *Prep.* Digitaline, 1 gr.; powdered sugar, ½ dr.; thick mucilage, q. s. For 24 pills. —*Dose,* 1 to 4 daily, watching the effects; as a sedative to reduce the force of the circulation, in phthisis, enlargement of heart, &c. See PILLS, FOXGLOVE.

Pills, Din'ner. *Syn.* PILULÆ DICTÆ ANTE-CIBUM, L.; GRAINS DE SANTÉ, Fr. *Prep.* 1. Aloes, 1 dr.; rhubarb and extract of gentian, of each, ½ dr.; ipecacuanha and capsicum, of each, 12 gr.; syrup of ginger, q. s. to mix. For 3½-gr. pills.

2. (*Dr Baillie's.*) See above.

3. (BATH DIGESTIVE PILLS.) See *above.*

4. (PILLS OF ALOES AND MASTIC; LADY CRESPIGNY'S PILLS, LADY HESKETH'S P., LADY WEBSTER'S P., DIGESTIVE P., STOMACH P., PILULÆ ALOËS ET MASTICHES, P. A. CUM MASTICHE, P. STOMACHICÆ MESUES; GRAINS DE VIE, GRAINS DE MESUES.) From aloes (powdered), 6 dr.; powdered mastic and petals of red roses, of each, 2 dr.; syrup of wormwood, q. s. to form a pill-mass. For 3-gr. pills. In small doses they excite the appetite; in larger ones they produce a bulky and copious evacuation. This is the formula of the old Paris Codex. Rhubarb is now frequently substituted for the rose petals.

5. (*Sir C. Bell's.*) From sulphate of quinine, 4 gr.; mastic, 6 gr.; rhubarb, 50 gr.; syrup of orange peel, q. s. to mix. For 12 or, preferably, 18 pills.

6. (*Sir Chas. Clarke's.*) From extract of chamomile, ½ dr.; myrrh and rhubarb (in powder), of each, 20 gr.; powdered Socotrine aloes, 10 gr.; oil of chamomile, 8 drops; mucilage, q. s. to form 20 pills. "These pills, which were originally prescribed by Sir Chas. Clarke, are much used in London" (*Redwood*).

Extract of Socotrine aloes, 1 gr.; mastiche, ½ gr.; spirits of wine, q. s. One at dinner.

Socotrine aloes, 1 gr.; powdered rhubarb, 1 gr.; mastiche, 1 gr.; spirits of wine, q. s. One at dinner.

Pills, Diuret'ic. *Syn.* PILULÆ DIURETICÆ, L. *Prep.* 1. From powdered foxglove, 12 gr.; calomel, powdered squills, and opium, of each, 4 gr.; conserve of hips, q. s. For 12 pills.

2. (*Dr A. T. Thompson.*) Mercurial pill, 1 dr.; powdered squills, 20 gr.; confection of roses, q. s.; divided into 20 pills. The dose of either of the above is 1 pill, twice or thrice daily; in dropsy, &c.

3. (St. Mary's Hospital.) Blue pill, 1 gr.; powdered digitalis, 1 gr.; powdered squill, 2 gr. One or two for a dose in dropsy.

Pills, Dixon's. According to Dr Paris these pills consist of aloes, scammony, rhubarb and a little tartar emetic, beaten up with syrup. "The following formula produces a pill precisely similar to this nostrum:—Take of compound extract of colocynth (Ph. L. 1836), 4 dr.; powdered rhubarb, 2 dr.; potassio-tartrate of antimony, 8 gr.; syrup of buckthorn, q. s.; mix, and divide into 120 pills. Aperient and diaphoretic.—*Dose,* 2 or 3 at bedtime." (*Cooley.*) Although a nostrum it is really an excellent medicine, adapted for numerous cases.

Pills, Duchesne's. *Prep.* From aloes and gum ammoniacum, of each, 30 gr.; mastic and myrrh, 10 gr.; carbonate of potass and saffron, of each, 3 gr.; syrup, q. s. In the dyspepsia of hysterical patients, in engorgements of the abdominal viscera, following intermittent fevers, &c.

Pills, Dys'entery. *Syn.* PILULÆ DYSENTE-RICÆ, L. *Prep.* Pure alumina and tannic acid, of each, 20 gr.; antimonial powder, 15 gr.; castor oil, ½ dr.—*Dose,* 5 to 10 gr.; frequently.

Pills of Elate'rium. *Syn.* PILULÆ ELATERII, L. *Prep.* (*Radius.*) Elaterium, 6 gr.; extract of gentian and Castile soap, of each, 9 gr.; mix, and divide into 12 pills.—*Dose,* 1 to 4; in obstinate constipation, and as a purge in dropsy, &c.

Pills of Ergotine. *Syn.* PILULÆ ERGOTINÆ, L. *Prep.* (*Bonjean.*) Ergotine (*Bonjean's*), 24 gr.; liquorice powder, 40 gr.; syrup, q. s. For 24 pills.—*Dose,* 3 to 6 daily; as an internal hæmostatic, &c.

Pills, Everlast'ing. *Syn.* PERPETUAL PILLS; PILULÆ ÆTERNÆ, P. PERPETUÆ, L. Small spheres of metallic antimony. They possess the property of purging as often as swallowed, but have now long fallen into disuse.

Pills, Expec'torant *Syn.* PILULÆ EXPEC-TORANTES, L. *Prep.* 1. Myrrh, 1½ dr.; powdered squills, ½ dr.; extract of henbane, 2 dr.; syrup, q. s.; divide into 30 pills.—*Dose,* 2, night and morning.

2. (*A. T. Thomson.*) Powdered squills and extract of hemlock, of each, ½ dr.; ammoniacum, 1½ dr.; divide into 30 pills.—*Dose,* 2, twice or

thrice a day. In chronic coughs, asthma, &c., after an aperient. See PILLS, PECTORAL, &c.

Pills, Family Antibil'ious. *Syn.* ALOE PILLS; ALOËS ROSATA, PILULÆ ALOËS ROSATÆ, L. *Prep.* Socotrine or hepatic aloes, 3 oz.; juice of roses, 1 pint; dissolve by heat, strain through a piece of coarse flannel, evaporate to a proper consistence, and form it into pills. Purgative, in doses of 5 to 15 gr.

Pills, Fe'ver. *Syn.* PILULÆ FEBRIFUGÆ, L. Of these the principal are those containing antimonials, bark, quinine, and salicine (which *see*).

Pills, Fordyce's. An active purgative, closely resembling in composition the compound gamboge pill of the Ph. L.

Pills, Dr Fothergill's. *Prep.* (*Cooley*). Aloes, 4 dr.; extract of colocynth and scammony, of each, 1 dr.; diaphoretic antimony, 30 gr.; syrup, q. s. For 3½-gr. pills. A diaphoretic aperient.—*Dose*, 1 to 4 pills at bedtime.

Pills of Fox'glove and Hen'bane. *Syn.* PILULÆ DIGITALIS ET HYOSCYAMI, L. *Prep.* (*Dr A. T. Thomson.*) Powdered foxglove, 4 gr.; powdered camphor, 12 gr.; extract of henbane, 18 gr. For 6 pills.—*Dose*, 1 or 2 at bedtime; as a sedative in maniacal and spasmodic affections, &c.

Pills of Fox'glove and Squills. *Syn.* PILULÆ DIGITALIS ET SCILLÆ (Ph. E.), L. *Prep.* (Ph. E.) Powdered foxglove and squills, of each, 1 part; aromatic electuary (Ph. E.), 2 parts; conserve of red roses, q. s.; divide into 4-gr. pills. A valuable diuretic in dropsies.—*Dose*, 1 to 2 pills.

Pills, Frankfort. These are the Pilules Angeliques noticed among PATENT MEDICINES, formed into 2-gr. pills, and silvered.

Pills, Franks'. See PILLS, DINNER.

Pills of Fuligoka'li. *Syn.* PILULÆ FULIGOKALI, L. *Prep.* (*Deschamps*.) Fuligokali, 5 dr.; starch, 2½ dr.; powdered tragacanth, 10 gr.; syrup, q. s. For 100 pills, which must be covered with two or three coats of gum, and preserved from the air. The pills of sulphuretted fuligokal (Pilulæ Fuligokali Sulphurati) are prepared in a similar manner.

Pills, Fuller's. *Syn.* BENNET PILLS; PILULÆ BENEDICTÆ, L. *Prep.* (*Cooley*.) Aloes and sulphate of iron, of each, ¼ dr.; myrrh and senna, of each, 20 gr.; assafœtida and galbanum, of each, 10 gr.; mace and saffron, of each, 6 gr.; syrup, q. s.; mix, and divide into 4-gr. pills. Antispasmodic, emmenagogue, and tonic, and slightly aperient.—*Dose*, 1 to 4, according to the object in view.

Pills, Gairthorn's Mild Provi''sional. *Prep.* (*Cooley*.) Compound gamboge pill, 60 gr.; aqueous extract of aloes, 40 gr.; sulphate of potassa and extract of senna, 30 gr.; compound scammony powder, 15 gr.; balsam of Peru, 6 or 8 gr.; emetic tartar, 3 gr.; mix, and divide into 36 pills. Purgative.—*Dose*, 1, 2, or more, when required.

Pills of Gal'banum (Compound). *Syn.* PILULA GALBANI COMPOSITA (Ph. L.), PILULÆ G. COMPOSITÆ, L. *Prep.* 1. (Ph. L.) Myrrh and prepared sagapenum, of each, 3 dr.; prepared galbanum and soft soap, of each, 2 dr.; pre-

pared assafœtida, 1 dr.; treacle, q. s. to form a pill-mass.

2. (Ph. L. 1836.) As the last, omitting the soap.

3. (Ph. D. 1826.) As the Ph. L., except that treacle is substituted for syrup.

Obs. These pills are stimulant, expectorant, antispasmodic, and emmenagogue.—*Dose*, 10 to 20 gr.; in hysteria, chronic coughs, chlorosis, amenorrhœa, &c.

Pills of Galbanum with Iron. *Syn.* PILULÆ GALBANI CUM FERRO, L. *Prep.* (Guy's Hosp.) Compound galbanum pill, 2 parts; precipitated sesquioxide of iron, 1 part; water, q. s. to form a mass. For 4½-gr. pills. An excellent tonic, emmenagogue.—*Dose*, 10 to 20 gr.; in chlorosis, amenorrhœa, &c., when chalybeates are not contra-indicated.

Pills of Gam'boge (Compound). *Syn.* GAMBOGE PILLS, FORDYCE'S P.; PILULÆ CAMBOGIÆ COMPOSITA (Ph. L.), P. CAMBOGIÆ (Ph. E.), L. *Prep.* 1. (Ph. L.) Powdered Socotrine or hepatic aloes, 3 dr.; powdered gamboge, 2 dr.; powdered ginger, 1 dr.; soft soap (Ph. L.), 4 dr.; mix, and beat them to a pill-mass. The formulæ of the Ph. L. 1836 and Ph. D. 1826 are precisely similar.

2. (Ph. E.) Gamboge, East Indian or Barbadoes aloes, and aromatic powder, of each (in powder), 1 part; Castile soap, 2 parts; syrup, q. s.

Obs. All the above are active cathartics.— Dose, 5 to 15 gr.; at bedtime; in obstinate constipation, &c.

Pills of Gen'tian (Compound). *Syn.* PILULÆ GENTIANÆ COMPOSITÆ, L. *Prep.* (*W. Cooley*.) Extract of gentian, 1 dr.; powdered rhubarb and cardamom, of each, ½ dr.; ipecacuanha, 12 gr. For 3-gr. pills. Stomachic.—*Dose*, 2 or 3, twice or thrice daily, to improve the appetite and digestion.

Pills, Gout. *Syn.* PILULÆ ANTARTHRITICÆ, L. *Prep.* 1. (*Bouchardat*.) Extract of colchicum and compound extract of colocynth, of each, 1 dr.; aqueous extract of opium, 32 gr.; mix, and divide into 3-gr. pills.—*Dose*, 1 or 2, according to their purgative action, as required.

2. (*Sir H. Halford's*.) From acetic extract of colchicum, ½ dr.; Dover's powder and compound extract of colocynth, of each, 18 gr. For 12 pills. —*Dose*, 1 pill.

3. (*Lartigue's*.) From compound extract of colocynth, 20 gr.; alcoholic extract of colchicum seeds and alcoholic extract of digitalis, of each, 1 gr. For 2-gr. pills.—*Dose*, &c. As the last.

4. (St George's Hosp.) Acetic extract of colchicum, 12 gr.; Dover's powder, 30 gr. For 12 pills.—*Dose*, 2 pills.

5. (*Sir C. Scudamore's*.) From acetic extract of colchicum, 1 dr.; powdered marshmallow root, q. s. to form a mass. For 40 pills.—*Dose*, 1 to 3, or more, with caution, as required.

6. (*Trousseau and Reveil*.) Powdered colchicum seeds, ½ dr.; powdered digitalis and sulphate of quinine, of each, 15 gr.; calomel and extract of colocynth, of each, 8 gr.; syrup, q. s. For 20 pills.—*Dose*, 1 to 4, during the day, at the commencement of an attack of gout. Other formulæ for gout pills will be found under their respective names.

Pills, Dr Griffith's. Powdered rhubarb, 1½ dr.; sulphate of iron, ½ dr.; Castile soap, 40 gr.; water, q. s. to form a mass. For 48 pills. An excellent remedy in costiveness, with loss of tone of the bowels.—*Dose*, 2 to 4 at bedtime.

Pills of Gu'aiacum (Compound). *Syn.* PILU-LÆ GUAIACI COMPOSITÆ, L. *Prep.* 1. Powdered resin of guaiacum, 1 dr.; oxysulphide of antimony, 40 gr.; oil of cajeput, 12 drops; extract of gentian, q. s. to form a mass. For 4-gr. pills.—*Dose*, 3 to 6, thrice daily; in gout, rheumatism, secondary syphilis, various obstinate cutaneous affections, &c.

2. (St B. Hosp.) Guaiacum, 30 gr.; ipecacuanha and opium, of each, 3 gr.; syrup, q. s. For 12 pills.—*Dose*, 1 to 3; as the last.

Pills, Halford's. See PILLS, GOUT.

Pills, Hall's Dinner. *Prep.* Aloes, 1 gr.; extract of glycyrrhiza, 1 gr.; soap (in powder), 1 gr.; molasses, 1 gr. (N. F.).

Pills, Dr Hamilton's. The same as the colocynth and henbane pill of the Ph. E. The compound pills of gamboge, now vended under the title of 'MORISON'S No. 2 PILLS,' were long known in Scotland as Dr Hamilton's Pills.

Pills, Head'ache. *Syn.* CEPHALIC PILLS; PIL-ULÆ CEPHALICÆ, P. ANTICEPHALAIGICÆ, L. *Prep.* 1. Caffeine, 15 gr.; aloes, 20 gr.; conserve of hips, q. s. For 12 pills.—*Dose*, occasionally; when only one side of the head is affected.

2. (*Broussais*.) Extract of opium, 6 gr.; extracts of belladonna and henbane, of each, 15 gr.; extract of lettuce, 30 gr.; butter of cacao, 4 dr. For 120 pills.—*Dose*, 1, twice or thrice daily; in headache, accompanying spasmodic affections, &c.

3. (*Dr Wilson Philip*.) Powdered nutmeg and rhubarb, of each, 20 gr.; extract of chamomile, 20 gr.; oil of peppermint, 10 or 12 drops. For 30 pills.—*Dose*, 1 to 3, thrice daily; in nervous headaches.

Pills, Heim's. *Prep.* Powdered digitalis, ½ gr.; powdered opium, ½ gr.; quinine, 1 gr.; powdered ipecacuanha, ½ gr.—*Dose*, 1, three times a day; as a sedative and tonic in heart disease.

Pills, Helvetius's. *Syn.* PILULÆ ALUMINIS HELVETII, L. *Prep.* Alum, 2 dr.; dragon's blood, 1 dr.; honey of roses, to mix. For 48 pills. Astringent.

Pills of Hem'lock (Compound). *Syn.* PILULÆ CONII COMPOSITA (B. P., Ph. L.), L. *Prep.* (Ph. L.) Extract of hemlock, 5 dr.; powdered ipecacuanha, 1 dr.; treacle, q. s. Antispasmodic, expectorant, and narcotic.—*Dose*, 4 to 8 gr. (B. P. 5 to 10 gr.), twice or thrice a daily; in hooping-cough, bronchitis, incipient phthisis, &c.

Pills of Henbane (Compound). *Syn.* PILULÆ HYOSCYAMI ET ZINCI, L.; PILULES DE MEGLIN, Fr. *Prep.* (P. Cod.) Extracts of henbane and Valerian, and oxide of zinc, equal parts. For 3-gr. pills.—*Dose*, 1 to 10; as an anodyne or sedative in neuralgia, nervous attacks, &c.

Pills, Lady Hesketh's. See PILLS, DINNER.

Pills, Hoffmann's. See PILLS OF CORROSIVE SUBLIMATE.

Pills, Holloway's. See PATENT MEDICINES.

Pills, Hooper's Female. *Prep.* 1. (*Gray*.)

Sulphate of iron and water, of each, 8 oz.; dissolve, add, Barbadoes aloes, 2½ lbs.; white camella, 6 oz.; myrrh, 2 oz.; opopanax, 1 oz.

2. (Phil. Coll. of Pharm.) Barbadoes aloes, 8 oz.; dried sulphate of iron, 2½ oz.; myrrh, extract of black hellebore, and Castile soap, of each, 2 oz.; canella and ginger, of each, 1 oz.; water, q. s.; divide the mass into 2½- or 3-gr. pills, and put 40 in each box. Cathartic and emmenagogue.—*Dose*, 2, or more. "If we omit the soap, lessen the quantity of extract of hellebore, and increase that of the aloes, we think the form will be nearer that of the original" (*Cooley*).

Pills of Hound's-tongue. (P. Cod.) *Syn.* PIL-ULÆ CUM CYNOGLOSSÓ, L. *Prep.* Root-bark of hound's-tongue, 4 dr.; henbane seeds, 4 dr.; extract of opium, 4 dr.; myrrh, 6 dr.; olibanum, 4 dr. and 48 gr.; saffron, 96 gr.; castor, 96 gr.; syrup of honey, 14 dr.; mix. Contains 1 gr. of extract of opium in 10 gr. The original form of NICOLAUS contained styrax, and seems to have been the origin of the compound styrax pill, as well as of this compound.

Pills, Humphrey's. See PILLS, PECTORAL.

Pills, Hunter's. See PILLS, RENAL.

Pills, Hydragogue. See BONTIUS'S PILLS, &c.

Pills, Hydrophobia. *Syn.* PILULÆ AD RABIEM, L. *Prep.* (*Werlhoff*.) Cantharides (in very fine powder), 2 gr.; belladonna and calomel, of each, 4 gr.; camphor, 8 gr.; mucilage, q. s. For 12 pills.—*Dose*, 2 to 3, twice daily.

Pills of In'dian Hemp. *Syn.* PILULÆ CAN-NABIS INDICÆ, L. *Prep.* From alcoholic extract of Indian hemp, ½ dr.; sugar of milk, 1 dr.; mucilage, q. s. For 48 pills. An excellent pill for soothing pain and quieting the system, acting without causing headache or constipation of the bowels.—*Dose*, 1 pill, increased to 2 or more, as necessary.

Pills of I'odide of Arsenic. *Syn.* PILULÆ ARSENICI IODIDI, L. *Prep.* 1. (*Dr Neligan*.) Iodide of arsenic, 2 gr.; manna, 40 gr.; mucilage, q. s.; mix, and divide into 12 pills.

2. (*Gardner*.) Iodide of arsenic, 1 gr.; extract of hemlock, 20 gr. For 20 pills.—*Dose*, 1 pill, twice or thrice daily; in lepra, psoriasis, and some other scaly skin diseases.

Pills of Iodide of Iron. *Syn.* PILULÆ FERRI IODIDI, L. *Prep.* 1. Unoxidised iron filings (recently levigated), 20 gr.; iodine, 40 gr.; distilled water, ½ dr.; mix in a cold wedgwood-ware mortar, and triturate them together until the red colour of the mixture has entirely disappeared; then add, of powdered gum, 20 gr.; powdered sugar, 1 dr.; liquorice powder, q. s. to form a mass, and divide it into 48 pills. Each pill contains 1 gr. of dry iodide of iron.—*Dose*, 1 to 6 pills, twice or thrice a day.

2. (B. P.) Fine iron wire, 40 gr.; iodine, 80 gr.; refined sugar (in power), 70 gr.; liquorice root (in powder), 140 gr.; distilled water, 20 minims. Agitate the iron with the iodine and the water in a strong stoppered ounce phial until the froth becomes white. Pour the fluid upon the sugar in a mortar, triturate briskly, and gradually add the liquorice.—*Dose*, 3 to 8 gr.

Obs. The above pills are reputed alterative, tonic, and emmenagogue, and are found peculiarly useful in indurations, scrofula, chlorosis, leucorrhœa, &c., when the administration of chalybeates is not contra-indicated.

Pills of Iodide of Lead. *Syn.* PILULÆ PLUMBI IODIDI, L. *Prep.* From iodide of lead, 15 gr.; powdered sugar, 1½ dr.; mucilage, q. s. For 60 pills.—*Dose*, 1 pill, gradually increased to 3, or more, twice a day; in scrofula, scirrhus, &c.

Pills of Iodide of Mar'cury. *Syn.* PILULÆ HYDRARGYRI IODIDI, L. *Prep.* 1. (Ph. L. 1836.) Green iodide of mercury and powdered ginger, of each, 1 dr.; conserve of hips, 3 dr.—*Dose*, 2 to 5 gr., twice or thrice daily, as an alterative in scrofula and scrofulous syphilis, &c.

2. (COINDET'S PILLS.) From green iodide of mercury, 1 gr.; extract of liquorice, 20 gr.; mix, and divide into 8 pills.—*Dose*, 2 to 4, as the last. Pills of red iodide of mercury are made in the same way, but, owing to its greater activity, only one fourth of the above quantity of iodide must enter into their composition.

Pills of Iodide of Potas'sium. *Syn.* PILULÆ POTASSII IODIDI, L. *Prep.* 1. Iodide of potassium and powdered starch, of each, ½ dr.; conserve of hips, q. s. For 36 pills.—*Dose*, 1 to 6, thrice daily; in glandular indurations and enlargements, goit, scrofula, &c.

2. (*Vogt.*) Iodide of potassium, 15 gr.; burnt sponge and extract of dulcamara, of each, 5 dr.; water, q. s. For 180 pills.—*Dose*, 4 to 6, twice a day, as the last.

Pills of Iodide of Silver. *Syn.* PILULÆ ARGENTI IODIDI, L. (*Dr Patterson.*) *Prep.* Iodide of silver, nitrate of potash, of each, 10 gr., rub together into a very fine powder, and add, liquorice powder, ¼ dr.; white sugar, 20 gr.; mucilage, q. s. to form a mass, to be divided into 40 pills; 1 three times a day.

Pills of I'odine. *Syn.* PILULÆ IODINII, L. *Prep.* (*Radius.*) Iodine, 6 gr.; extract of gentian, 1 dr.; powdered gum, q. s. For 24 pills.—*Dose*, 1 to 3; in scrofula, &c.; also in mercurial and scorbutic salivation.

Pills of Iod'oform. *Syn.* PILULÆ IODOFORMI, L. *Prep.* (*Bouchardat.*) Iodoform, ¼ dr.; extract of wormwood (or gentian), 1 dr.; mix, and divide into 36 pills.—*Dose*, 1, twice or thrice daily; in scrofula, &c.

Pills of Ipecac'uanha (Compound). *Syn.* PILLS OF IPECACUANHA WITH SQUILLS, P. OF I. AND OPIUM; PILULÆ IPECACUANHÆ CUM SCILLÂ (Ph. L.), P. IPECACUANHÆ ET OPII (Ph. E.), L. *Prep.* 1. (Ph. L.) Compound powder of ipecacuanha (Dover's powder), 3 dr.; powdered ammoniacum and squills (freshly powdered), of each, 1 dr.; treacle, q. s. to form a pill-mass. Anodyne, sudorific, and expectorant.—*Dose*, 5 to 10 gr.; in chronic coughs and asthma, &c.

2. (Ph. E.) Dover's powder, 3 parts; conserve of red roses, 1 part; mix, and divide into 4-gr. pills. Resembles Dover's powder in its effects. It is hence regarded by many as a useless preparation.

3. (B. P.) PILULA IPECACUANHÆ CUM SCILLÂ. Compound ipecacuanha powder, 3 oz.; fresh-dried squill, 1 oz.; ammoniacum (in powder), 1 oz.; treacle, q. s. Beat all together.—*Dose*, 5 to 10 gr.

Pills of I'ron (Compound). *Syn.* PILULÆ FERRI COMPOSITÆ (Ph. L.), P. F. CUM MYRRHA,

L. *Prep.* (Ph. L.) Myrrh (in powder), 2 dr.; carbonate of soda, 1 dr.; rub them together in a warm mortar, then add of sulphate of iron, 1 dr., and again triturate; lastly, add of treacle, 1 dr., and beat all together to form a pill-mass. An excellent mild chalybeate tonic and emmenagogue, similar in its properties to 'Griffith's Mixture.'—*Dose*, 5 to 15 gr., two or three times a day.

Pills, Italian Black. *Syn.* PILULÆ ITALICÆ NIGRÆ, P. ALOETICÆ FERRATÆ, L. *Prep.* (Ph. Bor.) Powdered aloes and dried sulphate of iron, equal parts; beaten up with rectified spirit, q. s., and divided into 2- or 2½-gr. pills. See PILLS OF ALOES AND IRON.

Pills of Jal'ap. *Syn.* PILULÆ JALAPÆ, L. *Prep.* 1. (Ph. E. 1788.) Extract of jalap, 2 dr.; aromatic powder, 1 dr.; syrup, q. s.

2. (Ph. Bor.) Soap of jalap, 3 parts; powdered jalap, 1 part; beat them to a pill-mass.—*Dose* (of either), 10 to 15 gr.

Pills, James's Analep'tic. *Prep.* 1. Antimonial powder, guaiacum, and pill of aloes with myrrh, equal parts; syrup, q. s.

2. (*Cooley.*) Antimonial powder (James's), pill aloes with myrrh, and compound aloes powder, of each, 2 parts; powdered ammoniacum, 1 part; beaten up with tincture of castor, q. s., and divided into 3½-gr. pills. A diaphoretic purge.—*Dose*, 2 to 4 pills.

Pills, Dr J. Johnson's. *Prep.* From compound extract of colocynth, 2 dr.; calomel, ½ dr.; potassio-tartrate of antimony, 2 gr.; oil of cassia, 12 drops. For 4 dozen pills. An excellent alterative and diaphoretic aperient.—*Dose*, 1 to 3 pills.

Pills, Kaye's. See PILLS, WORSDELL'S.

Pills, King's. See PATENT MEDICINES.

Pills, Kitchener's. *Syn.* Dr KITCHENER'S PERISTALTIC PERSUADERS; PILULÆ RHEI ET CARUI, L. *Prep.* From powdered Turkey rhubarb, 2 dr.; simple syrup, 1 dr.; oil of caraway, 10 or 12 drops. For 40 pills. An admirable stomachic, dinner, or laxative pill, according to the quantity taken.—*Dose*, 2 to 6. "From 2 to 4 will generally produce one additional motion within twelve hours. The best time to take them is early in the morning."

Pills, Klein's. *Prep.* From ammoniacum and extract of centaury, of each, ½ dr.; Castile soap, 1 dr.; oil of amber, 3 drops. For 2-gr. pills. Stomachic, emmenagogue, and pectoral.—*Dose*, 2 to 6 pills.

Pills of Lac'tate of Iron. *Syn.* PILULÆ FERRI LACTATIS, L. *Prep.* (*Cap.*) Lactate of protoxide of iron and powdered marshmallow root, equal parts; clarified honey, q. s. For 3-gr. pills. One of the most valuable of the chalybeates.—*Dose*, 1 to 2, three or four times a day.

Pills of Lactuca'rium. *Syn.* PILULÆ LACTUCARII, L. *Prep.* 1. (*Brera.*) Lactucarium, 18 gr.; conserve of elder-berries and extract of liquorice, of each, q. s. For 12 pills. *Dose*, 1 to 2 pills, every three or four hours; in dry asthma, obstinate coughs without expectoration, &c.

2. (*Dr Duncan.*) Lactucarium, 12 gr.; liquorice powder, 20 gr.; simple syrup, q. s. For 12 pills.—*Dose*, 1 to 2 pills, every hour, as an anodyne, or to induce sleep.

Pills, Lartigue's. See PILLS, GOUT.

Pills, Laxative. POST PARTUM. Laxative pills after confinement (*Barker's Post Partum Pills*). Each pill contains, compound extract of colocynth, 1½ gr.; aloes, ⅔ gr.; extract of nux vomica, $\frac{1}{18}$ gr.; resin of podophyllum, $\frac{1}{16}$ gr.; ipecac., in fine powder, $\frac{1}{16}$ gr.; extract of hyoscyamus, 1½ gr. *Note.*—This is the formula generally employed by Dr Fordyce Barker, except where special circumstances render modifications necessary. The formula usually quoted in manufacturers' lists and some formularies is not correct (N. F.).

Pills of Lead. *Prep.* 1. See PILLS OF ACETATE OF LEAD.

2. (Opiated; PILULÆ PLUMBI OPIATÆ—Ph. E.; PILULA PLUMBI CUM OPIO—B. P.) Acetate of lead, 6 parts; opium, 1 part; conserve of red roses, about 1 part; beat them to a proper mass, and divide this into 4-gr. pills. "This pill may also be made with twice the quantity of opium." In hæmorrhages, obstinate diarrhœa, dysentery, spitting of blood, and other cases demanding the use of a powerful astringent. It has also been highly extolled in cholera.—*Dose*, 1 to 3 pills, twice or thrice daily, washed down with water soured with pure vinegar.

Pills, Lee's Antibil'ious. *Prep.* ('Amer. Journ. of Pharm.') Aloes, 12 oz.; scammony, 6 oz.; calomel, 5 oz.; gamboge, 4 oz.; jalap, 3 oz.; Castile soap and syrup of buckthorn, of each, 1 oz.; mucilage, 7 oz.; beat them together, and divide the mass in 5-gr. pills. A powerful cathartic, and, from containing mercury, not adapted for frequent use. See WYNDHAM'S PILLS.

Pills, Lewis's Al'terative and Liver. These "for the most part resemble SCOTT'S BILIOUS AND LIVER PILLS. They are, however, of a more drastic and powerful character, and frequently operate with considerable violence."

Pills, Lockstadt's. *Prep.* (*Phœbus.*) Sulphate of quinine, 3 gr.; aromatic powder, 10 gr.; essential oil of almonds, 1 drop; extract of gentian, q. s. For 10 pills.—*Dose*, 1 to 2, thrice daily, as a stomachic tonic; or the whole at once, before an expected attack of an ague or intermittent.

Pills, Dr. Lynn's. *Prep.* From pill of aloes with myrrh and compound extract of colocynth, of each, 1 dr.; calomel, ⅔ dr. For 4 dozen pills. Aperient and antibilious.—*Dose*, 1 to 3; in costiveness, biliousness, &c.

Pills of Manganese, Carbonate. (*Hannon.*) *Syn.* PILULÆ MANGANESII CARBONATIS. *Prep.* Dissolve separately, 17 oz. of crystallised sulphate of manganese, and 19 oz. carbonate of soda, in water, q. s. Mix the solutions and add to every 17 oz. of the liquid, 1 oz. of syrup, and allow the precipitate to subside in a well-closed bottle. Pour off the supernatant liquid, wash the precipitate with sugared water, express, mix it with 10 oz. of honey, and evaporate rapidly to a pill consistence.—*Dose*. From 4 to 10 4-gr. pills daily, in anæmia, chlorosis, &c.

Pills of Manganese, Iodide. (*Hannon.*) *Syn.* PILULÆ MANGANESII IODIDI. *Prep.* Iodide of potassium, 1 oz.; dried sulphate of manganese, 1 oz.; mix with honey, q. s. to form a pill mass; divide into 4-gr. pills.—*Dose.* From 1 pill daily, gradually increased.

Pills of Manganese, Malate. (*Hannon.*) *Syn.* PILULÆ MANGANESII MALATIS. Malate of manganese, 15 gr.; powdered cinchona bark, 15 gr.; honey, q. s. for 30 pills. 3 to 5 or 6 daily.

Pills of Manganese, Muriate. (*Niemann.*) *Syn.* PILULÆ MANGANESII MURIATIS. *Prep.* Chloride of manganese, 2 scruples; gum-arabic, 2 scruples; liquorice, 1 scruple. Mix.

Pills of Manganese, Phosphate. (*Hannon.*) *Syn.* PILULÆ MANGANESII PHOSPHATIS. *Prep.* Phosphate of manganese, 1½ dr.; cinchona bark, ⅔ dr.; syrup of catechu, q. s. Make into 4-gr. pills.

Pills of Manganese, Tartrate. *Syn.* PILULÆ MANGANESII TARTRATIS. As Pills of Malate Manganese.

Pills of Manganese and Sulphate of Iron. (*Hannon.*) *Syn.* PILULÆ MANGANESII ET FERRI SULPHATIS. *Prep.* Sulphate of iron, 18 oz.; sulphate of manganese, 8½ oz.; carbonate of soda, 17½ oz.; honey, 10 oz.; syrup, q. s. to make a mass to be divided into 4-gr. pills.

Pills, Mar'tial. *Syn.* PILLS OF IRON AND WORMWOOD; PILULÆ FERRI CUM ABSINTHIO, P. MARTIALES, L. *Prep.* (*Sydenham.*) Levigated iron filings, 1 dr.; extract of wormwood, q. s. Tonic and hæmatinic.—*Dose*, 5 to 10 gr., twice a day.

Pills, Matthew's. *Syn.* PILULÆ MATTHÆI, P. PACIFICÆ, L. *Prep.* 1. (*Dr Paris.*) Black hellebore, Castile soap, liquorice, opium, saffron, and turmeric, equal parts; made into pills with oil of turpentine.

2. (Ph. E. 1744.) Opium and saffron, of each, 1 dr.; castor, 2 dr.; soap of turpentine, 3 dr.; balsam of copaiba (or oil of turpentine), q. s. to form a mass. Alterative and anodyne.—*Dose*, 3 to 10 gr.

Pills, McKinsey's. See McKINSEY'S KATAPOTIA, among PATENT MEDICINES.

Pills, Meglin's. *Syn.* PILULÆ DE MEGLIN, Fr. See PILLS OF HENBANE (Compound).

Pill, Mercu"rial. *Syn.* BLUE PILL; PILULA HYDRARGYRI (B. P.), PILULÆ HYDRARGYRI (Ph. L., E. & D.), P. MERCURIALES, L.; PILULÆ MERCURIELLES, Fr. *Prep.* 1. (Ph. L.) Mercury, 4 dr.; confection of roses, 6 dr.; rub them together until globules can no longer be seen; then add of liquorice powder, 2 dr., and beat the whole together, so that a proper mass may be formed.

2. (Ph. E.) As the last; afterwards dividing the mass into 5-gr. pills.

3. (Ph. D.) As the Ph. L. formula, but taking four times the quantity of the respective ingredients.

4. (B. P.) Mercury, 2; confection of roses, 3; decorticated liquorice root, in fine powder, 1; rub the mercury with the confection of roses, until metallic globules are no longer visible, then add the liquorice, and mix the whole well together.—*Dose*, 3 to 6 gr. as an alterative, 10 gr. as a purgative.

Obs. The remarks under 'MERCURIAL OINTMENT' (p. 1166), for the most part also apply here. This pill, when properly prepared, presents no globules of mercury when moderately rubbed on a piece of white paper, and immediately communicates a white stain to a piece of bright gold or copper. It possesses considerable density, and has a dark blue or slate colour. It contains one third of its weight of mercury, which may be ascertained from its sp. gr.; or, more exactly, by an assay for the metal. It is the mildest and most extensively used of all the mercurial preparations—*Dose.* As an alterative, 1 to 3 gr.; as a purgative, 10 to 15 gr.; and as a sialogogue, 5 or 6 gr., or more, twice or thrice daily. To prevent it affecting the bowels, it is commonly combined with either rhubarb or opium. A blue-pill taken overnight, and a black draught in the morning, is a popular remedy in bilious complaints. See ABERNETHY MEDICINES.

5. (*Collier.*) Mercury, 2 dr.; sequioxide of iron, 1 dr.; confection of red roses, 3 dr.; triturated, as before, until the globules disappear. An excellent extemporaneous substitute for the common mercurial pill. The addition of only a few gr. of the sesquioxide of iron to 1 or. of conserve renders the latter capable of rapidly killing a large quantity of quicksilver.

6. (*Tyson.*) Grey oxide of mercury (prepared by decomposing calomel with liquor of potassa to which a little liquor of ammonia has been added), 2 dr.; confection of roses, 6 dr.; powdered chamomiles, 1 dr.; mix. As a substitute for the College pill.

7. ('Pharm. Journ.') Stearin, 1 dr.; rub it in a warm mortar till it assumes the consistence of thick cream, then add of mercury, 4 dr., and again triturate until the globules disappear; next further add, of confection of roses and wheaten flour, of each 3 dr., powdered gum, 1 dr., and form the whole into a pill-mass. As a substitute for the College pill.

8. (PILULÆ HYDRARGYROSÆ—P. Cod.) Mercury and honey, of each, 6 dr.; triturate till the globules are extinguished, then add of aloes, 6 dr.; rhubarb, 3 dr.; scammony, 2 dr.; black pepper, 1 dr.; and make a pill-mass as before. Contains 1-4th part of quicksilver. Alterative and aperient.—*Dose,* 5 to 10 gr. BELLOSTE'S BARBAROSSA'S, SÉDILLOT'S, and MORELOT'S PILLS are nearly similar compounds. See PILLS OF CALOMEL and CORROSIVE SUBLIMATE, &c.

9. (PILULÆ UNGUENTI HYDRARGYRI—*Biett.*) *Prep.* Mercurial ointment, powdered sarsparilla, 1 dr. Mix, and divide into 48 pills. From 1 to 4 daily.

10. (PILULÆ HYDRARGYRI CUM SAPONE—P. Cod.) *Prep.* Mercurial ointment, 2 dr.; soap, 4 scruples; liquorice powder, 2 scruples. Make into 3-gr. pills.

Pills, Mercurial (Arabic). *Syn.* PILULÆ MERCURIALES ARABICÆ, L. *Prep.* Take of quicksilver and corrosive sublimate, of each, ½ dr.; triturate them patiently together until the globules disappear; then add, of agaric, pellitory, and senna, of each 1 dr.; honey, q. s. to make a pill-mass. For 3½-gr. pills.—*Dose,* 2 a day. Employed in the 'traitement arabique' for the cure of obstinate cutaneous diseases.

Pills, Mercurial, Hahnemann's. (F. H.) *Syn.* PILULÆ HYDRARGYRI HAHNEMANNI. *Prep.* Hahnemann's soluble mercury, 20 gr.; gum-arabic, 30 gr.; sugar, 30 gr. Mix, and divide into 30 pills.

Pills, Mitchell's. *Prep.* Aloes, ½ dr.; rhubarb, 1 dr.; calomel, 6 gr.; emetic tartar, 2 gr. For 36 pills. An alterative aperient.—*Dose,* 2 to 4 pills.

Pills, Moat's. Similar to MORISON'S PILLS.

Pills, Morison's. *Prep.* a. (No. 1 pills.) From aloes and cream of tartar, equal parts, made into a mass with either syrup or mucilage. A mild aperient.

b. (No. 2 pills.) From colocynth, 1 part; gamboge, 2 parts; aloes, 3 parts; and cream of tartar, 4 parts; made into a mass with syrup, as the last. An active purgative, often acting with great violence. Both No. 1 and No. 2 are divided into 3-gr. pills, of which 4 dozen are put into each 1s. 1½d. box. The proper dose of either is 1 to 3 or 4 pills; but they are often given by the Morisons in doses of 12, 20, 30, or even more daily. For the history of these pills and their proprietors, see 'Anat. of Quackery,' or 'Med. Circ.,' ii, 9—27.

Pills of Mor'phia. *Syn.* PILULÆ MORPHIÆ, L. *Prep.* (*Magendie.*) Morphine, 1 gr.; conserve of roses (stiff), q. s. For 6 (or, better, 8) pills.—*Dose,* 1 pill; as an anodyne or soporific.

Pills of Morphia (Compound). (*Rougier.*) *Syn.* PILULÆ MORPHIÆ COMPOSITÆ. *Prep.* Sulphate of morphia, 2 gr.; cyanide of potassium, 4 gr.; mucilage, q. s. Make into 24 pills; one every six hours, in neuralgia.

Pills, Moseley's. *Prep.* Finest Turkey rhubarb, 60 gr.; Jamaica ginger, 30 gr.; sugar, 20 gr. (all in powder); tincture of rhubarb, q. s. to form a mass. For 4-gr. pills. A mild and excellent medicine, closely resembling KITCHENER'S PERISTALTIC PERSUADERS.

Pills of Musk. *Syn.* PILULÆ MOSCHI, L. *Prep.* (*Dupuytren.*) Opium, 2 gr.; musk (genuine), 8 gr.; camphor (in powder), 24 gr.; syrup, q. s. For 8 pills. Antispasmodic and stimulant.—*Dose,* 1 to 3 thrice daily, in low nervous affections; or the whole during the day, in hospital gangrene, &c.

Pills of Myrrh. See PILLS OF ALOES AND MYRRH.

Pills, Napier's Neureton'ic. Of these, like MORISON'S PILLS, there are No. 1 and No. 2. The first is a simple stomachic aperient; the other, a stimulant tonic. They both owe their sale and reputed virtues to extensive advertising ('Anat. of Quackery').

Pills, Napoleon's. See PECTORAL PILLS.

Pills, Neural'gic. *Syn.* PILULÆ ANTINEURALGICÆ, L. *Prep.* 1. (*Marchal De Calvi.*) Aqueous extract of opium, 4 gr.; sulphate of quinine, 16 gr.; powdered cinnamon, powdered orange leaves, and extract of valerian, of each, 20 gr.; syrup of belladonna, q. s. For 3 dozen pills.—*Dose,* 1 hourly.

2. (*Trousseau* and *Reveil.*) Extracts of opium and stramonium, of each, 8 gr.; oxide of zinc, 2 dr.; syrup, q. s. For 40 pills.—*Dose,* 1 every two or three hours, gradually increased in frequency until there is some considerable disorder of vision, &c. Both of the above should be used with care. See PILLS, MEGLIN'S, &c.

3. Butyl chloral hydrate, 3 gr.; hydrochlorate of gelsemine, ₁₀₀ gr.—*Dose,* 1 every four hours.

Very useful in toothache and neuralgia affecting the fifth nerve (*Dr Ringer*).

Pills of Nitrate of Bismuth. *Syn.* PILULÆ BISMUTHI TRISNITRATIS, L. *Prep.* From trisnitrate of bismuth and powdered rhubarb, equal parts; syrup of orange peel, q. s. to form a mass. For 8-gr. pills.—*Dose*, 1 to 2 every two hours; as a tonic, stomachic, and antispasmodic, in dyspepsia, debility, spasms, &c.

Pills of Nitrate of Mercury. *Syn.* PILULÆ HYDRARGYRI PROTO-NITRATIS, L.; PILULES DE SAINTE MARIE, Fr. *Prep.* Powder of proto-nitrate of mercury, 7½ gr.; extract of liquorice, ½ dr. Mix accurately, and divide into 60 pills.—*Dose*, 1, four times a day.

Pills of Nitrate of Sil'ver. *Syn.* PILULÆ ARGENTI NITRATIS, L. *Prep.* 1. (St B. Hosp.) Nitrate of silver (crystallised), 12 gr.; liquorice powder, 24 gr.; treacle, q. s. For 12 pills.—*Dose*, 1 pill twice or thrice a day; in chronic epilepsy and some other spasmodic disorders.

2. (*Dr A. T. Thomson.*) Nitrate of silver, 6 gr.; crumb of bread, q. s. (say ½ dr.); mix, and divide into 12 pills.—*Dose*, one every six hours.

Obs. To prevent the blue or slate-coloured tinge of the skin, so often produced by the continued use of the salts of silver, 8 drops of diluted nitric acid in 1 fl. oz. of water should be taken after each pill.

Pills, Norton's Chamomile. *Prep.* From aqueous extract of aloes, 1 dr.; extract of gentian, 3 dr.; mix, and drive off the excess of moisture by the heat of a water bath; then add of essential oil of chamomile, 20 drops, and divide the mass into 60 pills. To preserve their aromatic properties, they should be kept in a dry glass bottle or a well-covered earthenware pot. —*Dose*, 1, as a dinner pill; or 2 night and morning, as a stomachic tonic (' Anat. of Quackery ').

Pills of Nux Vom'ica. *Syn.* PILULÆ NUCIS VOMICÆ, L. *Prep.* 1. Nux vomica and aloes (both in powder), equal parts; syrup, q. s. For 3-gr. pills.—*Dose*, 1 to 3, twice or thrice daily, carefully watching the effects in the nervous derangement, general debility, impotence, paralysis, &c.

2. Alcoholic extract of nux vomica, 1 part; powdered sugar, 2 parts; beaten up with rectified spirits, q. s. For 2½-gr. pills.—*Dose*, 1 to 2; as the last.

3. (*Moudière.*) Alcoholic extract, 6 gr.; levigated black oxide of iron, 1 dr.; syrup, q. s. In atonic incontinence of urine, amenorrhœa, &c.

Pills, Odontal'gic. *Syn.* PILULÆ ODONTALGICÆ, L. *Prep.* (Ph. Bor.) Powdered opium and extracts of belladonna and henbane, of each, 10 gr.; oil of olives, 20 drops; powdered pellitory of Spain, ½ dr.; beat them to a mass, and divide it into 1-gr. pills; keep them in a corked phial.

Pills, Opiated Lead. See PILLS OF LEAD.

Pills of O"pium. *Syn.* ANODYNE PILLS, NIGRT F., THEBAIC P.; PILULÆ OPII (Ph. U. S.), P. O. or THEBAIOæ (Ph. E.), L. *Prep.* 1." (Ph. E.) Opium and conserve of red roses, of each, 1 part; sulphate of potash, 3 parts; rub them together to a proper mass, and divide into 5-gr. pills.—*Dose*, 1 to 2 pills, as an anodyne or soporific. Each pill contains 1 gr. of opium, or double the quantity in the same pill of the previous edition of the Ph. E.

2. (Ph. U. S.) Powdered opium, 1 dr.; Castile soap, 12 gr.; water, q. s. For 60 pills. As the last.

Pills of Ox-gall. *Syn.* BILE PILL; PILULÆ FELLIS BOVINI, P. BILIS, L. *Prep.* 1. From inspissated ox-gall formed into pills by the addition of any simple powder; or the harder extract beaten up with a little proof spirit. Powdered rhubarb is frequently used for the purpose. For 3-gr. pills.—*Dose*, 1 to 6; for constipation, flatulence, &c., arising from a deficiency of bile.

2. (Compound.) From inspissated ox-gall, 1 dr.; powdered rhubarb, ½ dr.; powdered ipecacuanha and capsicum, of each, 15 gr.; oil of caraway, 12 drops. For 48 pills.—*Dose*, 1 to 4; in loss of appetite, and dyspepsia, with torpor of the bowels, &c. See CONSTIPATION, GALL, &c.

Pills of Ox'ide of Gold. *Syn.* PILULÆ AURI OXYDI, L. *Prep.* (*Magendie.*) Teroxide of gold, 5 gr.; extract of mezereon, 2 dr.; mix and divide into 60 pills. Each pill contains ¹⁄₁₂ gr. of teroxide.—*Dose*, 1 to 3; in scrofula, syphilis, malignant fevers, &c.

Pills of Oxide of Mercury. See PILLS, MERCURIAL.

Pills of Oxide of Sil'ver. *Syn.* PILULÆ ARGENTI OXYDI, L. *Prep.* From oxide of silver, 6 gr.; powdered rhubarb and extract of gentian, of each, 12 gr. For 1 dozen pills.—*Dose*, 1 pill, twice or thrice daily; in gastralgia, hæmorrhages, nervous affections, &c. Milder than the pills of nitrate of silver.

Pills of Oxide of Zinc. *Syn.* PILULÆ ZINCI OXYDI, L. *Prep.* From oxide of zinc, powdered cascarilla, and conserve of hips, equal parts. For 3½-gr. pills. Tonic and antispasmodic.—*Dose*, 1 to 3, thrice daily; in dyspepsia, gastric or spasmodic coughs, epilepsy, chorea, &c.

Pills, Parr's Life. *Prep.* Aloes, 7 lbs.; rhubarb and jalap, of each, 5 lbs. (all in powder); extract of gentian, 3½ lbs.; soft soap, ½ lb.; liquorice powder, treacle, and moist sugar, of each, 4½ lbs.; oil of cloves, 10 oz.; oil of caraway, 3½ oz.; mix, and beat the whole to a proper mass with syrup bottoms, q. s., and divide it into 3½-gr. pills. "There are about 4 dozen in each 1s. 1½d. box, weighing (dry) barely 3 gr. each." A good stomachic and aperient pill, but possessing none of the extraordinary virtues ascribed to it by its proprietors (' Med. Circ.,' ii, 146, 167, &c.).

Pills of Paullin'ia. *Syn.* GUARANA PILLS; PILULÆ GUARANÆ, P. PAULLINIÆ, L. *Prep.* 1. Paullinia mixed up with syrup of orange-peel, and the mass divided into 2½-gr. pills.—*Dose*, 2 to 3.

2. (*Dr Gavrelle.*) Extract of guarana, 1 dr. liquorice powder, q. s. For 40 pills.—*Dose*, 3 to 6 daily.

Obs. These pills are highly esteemed on the Continent as a tonic and astringent, in diseases of the bowels and bladder, in chlorosis, debility, &c.

Pec'toral Pills. *Syn.* BREATH PILLS; PILULÆ PROTORALES, L. *Prep.* 1. Compound squill pill, 1 dr.; gum benzoin, ½ dr.; powdered ipecacuanha and extract of henbane, of each, 15 gr.; syrup, q. s. For 3-gr. pills.—*Dose*, 2 to 4, three or four times a day; in asthmas, chronic bronchial affections, coughs, &c.

2. (*Dr Copland*.) Camphor (in powder), 10 gr.; ipecacuanha, 15 gr.; extract of hemlock, 1 dr.—*Dose*, 3 to 6 gr.; in irritating and spasmodic coughs, &c.

3. (*Haggart*.) Powdered ipecacuanha and squills, of each, ½ dr.; acetate of morphia, 6 gr.; Castile soap, 3 dr.; mix, and divide into 72 pills. A most excellent medicine, at once soothing and expectorant.—*Dose*, 1 to 2, thrice a day, or oftener.

4. (HUMPHRIES' COUGH PILLS.) From powdered ipecacuanha, 15 gr.; compound squill-pill, 1 dr.; compound extract of colocynth, ½ dr.; For 3½-gr. pills.—*Dose*, 2 pills, night and morning.

5. (*Dr Latham*.) Compound powder of ipecacuanha, 1 dr.; fresh squill and gum ammoniacum, of each, 20 gr.; calomel, 4 gr. For 20 pills. A most valuable pectoral and expectorant.—*Dose*. 1 pill, thrice daily; in bronchitis, coughs, &c., after the more active inflammatory symptoms have subsided.

6. (*Napoleon's*.) From ipecacuanha, 30 gr.; squills and ammoniacum, of each, 40 gr. (all in powder); mucilage, q. s. to mix. For 24 pills. It is said that this was a favourite remedy with the Emperor Napoleon I for difficulty of breathing, bronchitis, and various affections of the organs of respiration.—*Dose*, 2 pills, night and morning.

7. (*Dr Paris*.) Powdered squills, 1½ dr.; powdered myrrh, 1½ dr.; extract of henbane, 40 gr.; water (or simple syrup), q. s. to mix. For 4-gr. pills.—*Dose*, 2 pills, night and morning. As No. 2.

8. (Ph. L. 1746.) Gum ammoniacum, 4 dr.; gum benzoin, 3 dr.; gum myrrh, 2 dr.; saffron, 1 dr.; anisated balsam of sulphur, ½ dr.; syrup of tolu, q. s. to mix.—*Dose*, 5 to 15 gr.

9. (*Richter*.) Assafœtida and valerian, of each, ½ dr.; castor, 15 gr.; powdered squills and sesquicarbonate of ammonia, of each, 8 gr.; extract of aconite (alcoholic), 3 gr. For 4-gr. pills.—*Dose*, 1 to 3 pills, night and morning; in spasmodic affections of the respiratory organs.

Pills, Perpet'ual. See PILLS, EVERLASTING.

Pills, Peter's. *Prep.* (*Cooley*.) Aloes, 3 dr.; gamboge, jalap, and scammony, of each, 2 dr. (all in powder); calomel, 1 dr.; beaten up with rectified spirit, q. s. A powerful cathartic.—*Dose*, 1 to 3 pills.

Pills of Phosphorus. (B. P.) *Syn.* PILULÆ PHOSPHORI. *Prep.* Phosphorus, 3 gr.; balsam of tolu, 120 gr.; yellow wax, 57 gr.; curd soap, 90 gr. Put the phosphorus and balsam into a Wedgwood mortar half full of hot water, and when the phosphorus has melted and the balsam become sufficiently soft, rub them together beneath the surface of the water until no particles of phosphorus are visible, the temperature of the water being maintained at or near 140° F. Add now the wax, and as it softens mix it thoroughly with the other ingredients. Allow the mass to cool without being exposed to the air, and keep it in a bottle immersed in cold water. It may

be softened with a few drops of rectified spirit when made into pills.—*Dose*, 2 to 4 gr.

Pills of Pip'erine. *Syn.* PILULÆ PIPERINÆ, L. *Prep.* From piperine, ½ dr.; extract of cinchona, q. s. For 30 pills.—*Dose*, 1 pill, every two hours, during the intermission of an ague; also as an aphrodisiac and a remedy in piles.

Pills of Pitch. *Syn.* PILULÆ PICIS NIGRÆ, L. *Prep.* From black pitch and powdered black pepper, equal parts; beaten together in a warm mortar, and divided into 4-gr. pills.—*Dose*, 2 pills, night and morning; in piles, &c.

Pills, Pitschaft's Eccoprot'ic. *Prep.* From strained aloes and disulphate of quinine, equal parts; made into 2-gr. pills. A tonic and stomachic aperient.—*Dose*, 2 to 4, at bedtime; in torpor of the large intestines, the dyspepsia of the debilitated, &c.

Pills of Podophyllin. *Syn.* PILULÆ PODO-PHYLLINI. *Prep.* Resin of podophyllin, ½ gr.; extract of henbane, 1 gr. To make one pill. One or two for a dose.

Pills, Plummer's. See PILLS OF CALOMEL (Compound).

Pills, Pur'gative. *Syn.* PILULÆ PURGANTES, L. *Prep.* 1. (*Dr Robinson*.) Aqueous extract of aloes, 1 dr.; powdered scammony, ½ dr.; balsam of Peru, 10 or 12 gr.; oil of caraway, 9 or 10 drops; mix, and divide into 30 pills. A warm, stimulating aperient, highly recommended to excite the peristaltic action of the bowels of the aged, sedentary, and debilitated.—*Dose*, 1 to 4 pills, as required.

2. (*Trousseau* and *Reveil*.) Resin of jalap, 1 dr.; scammony, ½ dr.; extract of colocynth, 6 dr.; excipient, as required. For 20 (or, better, 24) pills.—*Dose*, 1, "every two hours, in the morning, fasting, until they operate." For other formulæ see PILLS, APERIENT and CATHARTIC, and PILLS OF ALOES, JALAP, COLOCYNTH, &c.

Pills of Quinine'. See PILLS OF SULPHATE OF QUININE.

Pills, Reece's. See PILLS, CHIRATTA.

Pills, Re'nal. *Syn.* PILULÆ RENALES, L. *Prep.* 1. Squills, myrrh, and digitalis, of each (in powder), 10 gr.; extract of rhubarb and mercurial pill, of each, 15 gr.; powdered nitre, 20 gr.; oil of juniper, 10 or 12 drops. For 24 pills. Alterative, diuretic, and tonic.—*Dose*, 3 to 6, thrice a day. Hunter's Renal Purifying Pills are similar, but omitting the mercurial pill. De Roos' Renal Pills contain a preparation of copaiba.

Pills, Rheu'matism. *Syn.* PILULÆ ANTIRHEU-MATICÆ, L. *Prep.* 1. Gum guaiacum, 1 dr.; nitrate of potassa, 1½ dr. (both in powder); soft soap (Ph. L.), ½ dr.; oil of cajeput, 16 drops. For 4 dozen pills.—*Dose*, 2 to 6, night and morning; in chronic rheumatism,'and rheumatic gout. Their action is accelerated by the copious use of lemon juice during the day.

2. (*Beasley*.) Extract of artichoke, ½ dr.; powdered sarsaparilla, 20 gr.; oil of sassafras, 1 drop. For 12 pills.—*Dose*, 1 pill, thrice daily.

Pills of Rhu'barb. *Syn.* PILULÆ RHEI (Ph. E.), L. *Prep.* 1. (Ph. E.) Powdered rhubarb, 9 parts; acetate of potassa, 1 part; conserve of

red roses, 5 parts; mix, and divide into 5-gr.
pills. A stomachic and gentle aperient, particularly useful in atonic dyspepsia.—*Dose*, 2 to 4
pills.

2. (Ph. U. S.) Powdered rhubarb, 6 dr.; Castile soap, 2 dr.; beaten up with water, q. s., and
divide into 120 pills. As the last.

Pills of Rhubarb (Compound). *Syn.* ARO-
MATIC PILLS, BALSAMIC LAXATIVE P., EDIN-
BURGH P., STOMACHIC P.; PILULA RHEI COM-
POSITA (B. P., Ph. L.), PILULE R. COMPOSITE
(Ph. E. and D.), P. STOMACHICE, P. AROMATICE,
L. *Prep.* 1. (Ph. L.) Powdered rhubarb, 4
dr.; powdered Socotrine aloes, 3 dr.; pow-
dered myrrh, 2 dr.; soft soap (Ph. L.), ½ dr.;
oil of caraway, 15 drops; treacle, q. s. to form a
mass.

2. (Ph. L. 1836.) Powdered rhubarb, 1 oz.;
aloes, 6 dr.; myrrh, 4 dr.; Castile soap, 1 dr.;
oil of caraway, ½ fl. dr.; syrup, q. s.

3. (Ph. E.) Powdered rhubarb, 12 parts;
aloes, 9 parts; myrrh and Castile soap, of each,
6 parts; conserve of red roses, 5 parts; oil of
peppermint, 1 part; mix, and divide into 5-gr.
pills. The oil of peppermint may be omitted
when so preferred.

4. (Ph. D.) Rhubarb, 1½ oz.; hepatic aloes,
9 dr.; myrrh and Castile soap, of each in fine
powder, 6 dr.; oil of peppermint, 1 fl. dr.;
treacle, 2 oz.; mix, and beat the whole to a
uniform mass.

5. (Ph. U. S. and Ph. E. 1817.) Rhubarb, 8
dr.; aloes, 6 dr.; myrrh, 4 dr.; oil of peppermint, ½ fl. dr.; syrup of orange peel, q. s.; mix,
and divide into 240 pills.

6. (B. P.). Rhubarb, in fine powder, 3 oz.;
Socotrine aloes, in fine powder (some physicians
prefer the aqueous extract—*Squire*), 2½ oz.;
myrrh in fine powder, 1½ oz.; hard soap, 1½ oz.;
English oil of peppermint, 1½ dr.; glycerine, 1 oz.,
treacle about 3 oz.; reduce the soap to fine powder
and triturate it with the rhubarb, aloes, and
myrrh; add the treacle, glycerine, and oil, and
beat into a mass.—*Dose*, 5 to 10 gr.

Obs. The above are tonic, stomachic, and
gently laxative; extremely useful for obviating
costiveness and giving tone to the stomach and
bowels.—*Dose*, 6 or 8 to 20 gr. The London
pill is not only the most agreeable, but it keeps
the best.

Pills of Rhubarb and Car'away. See KIT-
CHENER'S PERISTALTIC PERSUADERS (Patent
medicines).

Pills of Rhubarb and Chamomile. *Syn.*
SPEEDIMAN'S PILLS; PILULE RHEI ET ANTHE-
MIDIS, L. *Prep.* From aloes, myrrh, rhubarb,
(each in powder), and extract of chamomile, of
each, 1 dr.; essential oil of chamomile, 10 or 12
drops. For 4-gr. pills. An excellent tonic and
stomachic aperient, particularly useful in the
dyspepsia and loss of appetite of hard drinkers.
—*Dose*, 1 to 3 pills, either before dinner or at
bedtime.

Pills of Rhubarb and Copaiba. *Syn.* PILULE
RHEI ET COPAIBE, P. R. BALSAMICE, L. *Prep.*
(*Swediaur*.) Powdered rhubarb and gum, equal
parts; balsam of copaiba, q. s.

Pills of Rhubarb and Gin'ger. *Syn.* STOMACH
PILLS; PILULE RHEI ET ZINGIBERIS, L. *Prep.*

From powdered rhubarb, 1 dr.; powdered ginger,
½ dr.; Castile soap, 20 gr.; tincture of essence of
ginger, q. s. to form a mass. For 80 pills.—*Dose*,
1 to 6.

Pills of Rhubarb and Ipecacuan'ha. *Syn.*
PILULE RHEI ET IPECACUANHE, L. *Prep.* From
rhubarb, ½ dr.; ipecacuanha, 15 gr.; opium, 5 gr.
(each in powder); oil of cinnamon, 6 drops;
syrup, q. s. For 18 pills.—*Dose*. In loss of
appetite and spasmodic dyspepsia, 1 to 3 pills,
twice a day; in dysentery, diarrhœa, &c., to relieve
tormina and tenesmus, 1 every two hours.

Pills of Rhubarb and I'ron. *Syn.* PILULE
RHEI ET FERRI (Ph. E.), L. *Prep.* (Ph. E.)
Dried sulphate of iron, 4 parts; extract of
rhubarb, 10 parts; conserve of red roses, 5 parts;
beat them to a proper mass, and divide this into
5-gr. pills.—*Dose*, 2 to 4 pills; in the atonic
dyspepsia of debilitated subjects, in chlorosis, &c.

Pills of Rhubarb and Ox-gall. *Syn.* PILULE
RHEI ET FELLIS BOVINI, L. *Prep.* From powdered rhubarb, gum ammoniacum, and inspissated
ox-gall, in equal parts beaten up with a little
tincture of ginger or proof spirit, and the mass
divided into 2½-gr. pills. In dyspepsia and constipation dependent on a torpid action of the liver.
—*Dose*, 2 to 6 pills.

Pills of Rhubarb and Soda. *Syn.* PILULE
RHEI ET SODE, P. R. COMP. CUM SODA, L. *Prep.*
(Guy's Hosp.) Dried carbonate of soda, powdered
rhubarb, and extract of gentian, equal parts.
For 4½-gr. pills.—*Dose*, 2 to 4 pills; acidity,
heartburn, diarrhœa, loss of appetite, &c.

Pills, Richter's. See PECTORAL PILLS.

Pills, Dr Robinson's. See PILLS, PURGATIVE.

Pills, Rudius's. *Syn.* RUDIUS'S EXTRACT;
PILULE RUDII, EXTRACTUM RUDII, L. *Prep.*
1. Colocynth pulp, 6 dr.; agaric, black hellebore,
and turpethum root. of each, 4 dr.; cinnamon,
mace, and cloves, of each 40 gr.; rectified spirit,
½ pint; digest for 4 days, express the tincture,
and evaporate it to a proper consistence for
making pills. Formerly esteemed one of the
most safe and certain cathartics in troublesome
constipation.—*Dose*, 5 to 20 gr.

2. (Ph. E. 1783.) Black hellebore and colocynth, of each, 2 oz.; water, 4 pints (o. w. m.);
boil to a quart, strain, evaporate to the consistence of honey, and add of aloes, 2 oz.; scammony
(powdered), 1 oz.; next remove the vessel from
the fire, and further add of sulphate of potassa,
2 dr.; oil of cloves, 1 dr.; and form the whole
into a pill-mass. Resembles the last (nearly).

Pills, Rufus's. See PILLS OF ALOES WITH
MYRRH.

Pills of Sabadilla. *Syn.* PILULE CEVADILLE.
Prep. Equal parts of sabadilla and honey; make
into 5-gr. pills.—*Dose*. For an adult, 4 to 6 pills;
for a child, 1 to 2. (Vermifuge.)

Pills of Saf'fron. *Syn.* PILULE CROCI, L.
Prep. 1. From hay saffron, 1 dr.; myrrh, ½ dr.;
oil of cajeput, 6 drops; syrup of saffron, q. s.
For 36 pills.—*Dose*, 1 to 3 or 4 occasionally; as
a stimulant in low spirits, hypochondriasis, &c.

2. (*Phœbus.*) Saffron, myrrh, and sulphur,
equal parts; inspissated bile, q. s. For 2-gr.
pills.—*Dose*, 2 to 12, daily; as an emmenagogue.

Pills of Sagapenum (Compound). *Syn.* PILULE

SAGAPENI COMPOSITÆ, L. *Prep.* (Ph. L. 1836.) Sagapenum, 1 oz.; aloes, ½ dr.; syrup of ginger, q. s.—*Dose*, 5 to 20 gr.; as a stimulant antispasmodic laxative, in dyspepsia with flatulence, flatulent colic, &c.

Pills of Sal'icin. *Syn.* PILULÆ SALICINÆ, L. *Prep.* From salicin, ½ dr.; powdered rhubarb, 20 gr.; extract of gentian, q. s. to mix. For 4-gr. pills.—*Dose*, 2 to 4, every three hours, during the apyrexia of intermittents.

Pills of Sandal-wood Oil. (*Ebert.*) *Syn.* PILULÆ OLEI SANTALI. *Prep.* Oil of yellow sandal-wood, ½ oz.; yellow wax, ½ oz. Melt the wax into a capsule, and weigh into it the oil of sandal-wood. Mix, and stir until cold, then roll out the mass and divide it into 80 pills, by means of the pill machine or pill-tite, in the same manner as in the ordinary mass, and sprinkle with marsh-mallow root powder. Each pill contains about three gr. or about 5 drops of the oil. The excipient is unobjectionable, as it is readily soluble in the juices of the stomach.

Pills of Scam'mony (Compound). *Syn.* PILULÆ SCAMMONII COMPOSITÆ, L. *Prep.* 1. (St. B. Hosp.) Scammony, 24 gr.; ginger, 20 gr.; aloes and gamboge, of each, 12 gr.; treacle, q. s.; mix, and divide into 12 pills. A powerful cathartic and vermifuge.—*Dose*, 1 to 3 pills.

2. (B. P.) Resin of scammony, resin of jalap, of each, 1 oz.; curd soap, in powder, 1 oz.; strong tincture of ginger, 1 fl. oz.; rectified spirit, 2 fl. oz. Add the tincture and spirit to the soap and resins, and dissolve by the aid of a gentle heat, then evaporate the spirit over a water-bath until the mass has a pilular consistence.—*Dose*, 5 gr, to 15 gr.

Pills, Scot's. *Prep.* From aloes, 9 lbs.; jalap, 3 lbs.; gamboge and ginger, of each, ½ lb.; beaten with treacle, q. s. See PILLS, ANDERSON'S SCOT'S.

Pills, Dr Scott's Bil'ious and Liver. *Prep.* (*Cooley.*) Compound extract of colocynth (Ph. L. 1836), 8 oz.; powdered rhubarb, 4 oz.; powdered myrrh, 2 oz.; soft soap, ½ oz.; oil of caraway, 2½ dr.; strong syrup of saffron, q. s. to form a pill-mass. "There are twenty-five 3½-gr. pills in each 1s. 1½d. box." "It has been stated that these pills contain a minute portion of antimony" ('Anat. of Quackery').

Pills, Sed'ative. *Syn.* PILULÆ SEDATIVÆ, L. *Prep.* Hydrochlorate of morphia, 6 gr.; powdered sumbul, 20 gr.; alcoholic extract of Indian hemp, ½ dr. For 2-gr. pills.—*Dose*, 1 to 3, twice or thrice daily; in excessive nervous irritability, painful menstruation, &c.

Pills, Sedillot's Febrifuge. *Prep.* From powdered opium, 3 gr.; sulphate of quinine, 12 gr.; confection of opium, 10 gr., or q. s. For 12 pills. —*Dose*, 1 to 2, every second hour, during the intermission of an ague.

Pills of Sen'na. *Syn.* PILULÆ SENNÆ, P. S. COMPOSITÆ, L. *Prep.* 1. Powdered senna, 1 dr.; extract of rhubarb, ½ dr.; powdered capsicum, 4 gr.; oil of juniper, 6 or 8 drops. For 3-gr. pills. An aperient well suited for females. —*Dose*, 5 to 8 pills.

2. (*Hufeland.*) Powdered senna, 1 dr.; extract of dandelion, q. s. to mix. For 80 pills. As the last.

Pills, Smith's. *Prep.* From powdered aloes, 4 dr.; jalap, 2 dr.; ginger and soft soap, of each, 1 dr.; oil of juniper, ½ dr.; emetic tartar, 6 gr. For 120 pills. Laxative and diuretic.—*Dose*, 1 to 4, at bedtime, or early in the morning.

Pills, Dr Hugh Smith's. See STOMACH PILLS.

Pills of Soap. *Syn.* PILULÆ SAPONIS, P. CUM SAPONE, L. *Prep.* (P. Cod.) White Castile soap, 32 parts; powdered marsh-mallow root, 4 parts; powdered nitrate of potash, 1 part; beat them to a mass, and divide this into 4-gr. pills. In habitual costiveness, calculary affections, &c.—*Dose*, 1 to 6 pills, twice or thrice a day.

Pills of Soap (Compound). *Syn.* PILLS OF SOAP AND OPIUM, LAUDANUM PILLS; PILULA SAPONIS COMPOSITA (Ph. L.), PILULÆ SAPONIS CUM OPIO, L. *Prep.* 1. (Ph. L.) Opium and liquorice, of each (in powder), 2 dr.; soft soap (Ph. L.), 6 dr.; beat them to a uniform mass.

2. (B. P.) Opium (in fine powder), ½ oz.; Castile soap, 2 oz.; glycerin, q. s.; reduce the soap to powder, mix it with the other ingredients, and beat the whole together, as before.—*Dose*, 3 gr. to 5 gr. See PILLS OF OPIUM.

Obs. The above pills contain 1-5th part of their weight in dry opium. The dose is 3 to 10 gr., in the usual cases in which the administration of opium is indicated. Mr Skey, the eminent surgeon of St Bartholomew's Hospital, has shown the great value of this pill in promoting the healing of obstinate ulcers, more especially those of the legs.

Pills of Soda. *Syn.* PILULÆ SODÆ CARBONATIS, L. *Prep.* (Ph. E. 1817.) Exsiccated carbonate of soda, 4 parts; Castile soap, 3 parts; syrup, q. s. to form a mass. Antacid and slightly laxative.—*Dose*, 10 to 20 gr. This pill was a great favourite of the once celebrated Dr Beddoes.

Pills, Speedman's. *Prep.* (*Cooley.*) Aloes, 3 dr.; rhubarb, myrrh (all in powder), and extract of chamomile, of each, 1 dr.; oil of chamomile, 20 drops. For 4-gr. pills. An excellent aperient, tonic, and stomachic.—*Dose*, 2 to 4 pills, as a purgative; 1, as a stomachic or dinner pill.

Pills, Splenet'ic. *Syn.* PILULÆ ANTISPLENETICÆ, L. *Prep.* (*Saunders.*) Strained aloes and gum ammoniacum, of each, 3 dr.; myrrh and bryony, of each, ½ dr. For 4-gr. pills.— *Dose*, 3 to 5. "Extolled in amenorrhœa and hypochondriasis" (*Dr R. E. Griffith*).

Pills of Squill (Compound). *Syn.* COUGH PILLS, PILLS OF SQUILLS AND GINGER; PILULA SCILLÆ COMPOSITÆ (B. P., Ph. L.), PILULÆ SCILLÆ COMPOSITÆ (Ph. D.), P. SCILLÆ (Ph. E.), L. *Prep.* 1. (Ph. L.) Freshly powdered squills, 1 dr.; powdered ginger and powdered ammoniacum, of each, 2 dr.; mix, add of soft soap (Ph. L.), 3 dr.; treacle, 1 dr.; and beat the whole together, so that a mass may be formed.

2. (Ph. E.) Squills, 5 parts; ammoniacum, ginger (all in fine powder), and Spanish soap, of each, 4 parts; conserve of red roses, 2 parts; mix, as before, and divide the mass into 5-gr. pills.

3. (Ph. D.) Squills (in fine powder), 2½ dr.

ammoniacum, ginger, and Castile soap, of each (in fine powder), 2 dr.; treacle, ½ oz.

4. (B. P.) Squill (in fine powder), 1¼ parts; ginger (in fine powder), 1 part; ammoniacum (in powder), 1 part; hard soap (in powder), 1 part; treacle (by weight), 2 parts, or a sufficiency; mix the powders, add the treacle, and beat into a mass.—*Dose*, 5 to 10 grains.

Obs. Compound squill pill is a most useful expectorant in chronic coughs, asthmas, bronchial affections, difficulty of breathing, &c.; and, combined with calomel and foxglove, and, occasionally, with croton oil, as a diuretic, &c., in dropsies. Unfortunately, however, it soon spoils; and, therefore, to be effective as a remedy it must be recently prepared. As an expectorant, it should not be administered until the inflammatory symptoms have been subdued by purgatives or bleeding. A little powdered opium, or extract of henbane, is occasionally added, to allay irritation.—*Dose*, 5 to 20 gr., twice or thrice a day, accompanied by an occasional aperient.

Pills, Stahl's. See PILLS, APERIENT.

Pills, Starkey's. *Prep.* (Original formula.) Extract of opium, 4 oz.; saffron and Virginian snake-root, of each, 2 oz.; Starkey's soap, ½ lb.; oil of sassafras, ½ oz.; tincture of antimony (Old Ph.), 2 fl. oz. Anodyne, diaphoretic, &c.—*Dose*, 3 to 10 gr. The formula already given under MATTHEW'S PILLS is erroneously assigned to this pill by some writers.

Pills, Mrs Stephen's. This once celebrated remedy for stone was prepared from the calcined shells of eggs and snails, made into 3-gr. pills with soft soap. Its active ingredients were, consequently, lime and potash.

Pills, Stim'ulant. *Syn.* PILULÆ STIMULANTES, L. *Prep.* 1. Capsicum, ½ dr.; nitrate of silver, 2 gr.; conserve of hips, q. s. For 12 pills.—*Dose*, 2 to 4, washed down with a spoonful of warm spirit and water, and repeated hourly until reaction ensues; in cholera, &c.

2. (*A. T. Thomson.*) Strychnine, 1 gr.; acetic acid, 1 drop; crum of bread, 20 gr.; mix very carefully, and divide the mass into 10 pills.—*Dose*, 1 every six hours; in paralysis arising from lead.

Pills, Stoerck's. *Syn.* PILULÆ CONII, P. CICUTÆ, L. *Prep.* From extract of hemlock, 1 dr.; powdered hemlock, q. s. to make a mass. For 2-gr. pills.—*Dose*, 1 to 4, twice a day; in various glandular and visceral enlargements, pulmonary affections, cancer, scrofula, neuralgia, &c.

Pills, Stomach. *Syn.* PILULÆ STOMACHICÆ, L. *Prep.* 1. Ipecacuanha, 10 gr.; sumbul and extract of rhubarb, of each, 30 gr.; powdered quassia, 20 gr.; oil of sassafras, 6 drops; beaten up with essence of ginger (strongest), q. s. For 3-gr. pills.—*Dose*, 1 to 3, thrice daily; in loss of appetite, flatulence, dyspepsia, &c.

2. (*Dr Hugh Smith's.*) From aloes, rhubarb, ginger (all powdered), and sagapenum, of each, 1 dr.; oils of peppermint and cloves, of each, 10 drops; balsam of Peru, q. s. to mix. For 5-gr. pills.—*Dose*, 2 or 3 nightly; or 1 to 2 before dinner. For other formulæ, see DINNER, APERIENT, COMPOUND RHUBARB, ALOES AND MASTIC PILLS, &c.

Pills of Sto'rax (Compound). *Syn.* STORAX PILLS; PILULA STYRACIS COMPOSITA (Ph. L.), PILULÆ STYRACIS (Ph. E.), L. *Prep.* 1. (Ph. L.) Prepared storax, 6 dr.; saffron and powdered opium, of each, 2 dr., beat them together to a uniform mass. Contains 1-5th of its weight of opium.

2. (Ph. E.) Opium and saffron, of each, 1 part; extract of styrax, 2 parts; beat them to a uniform mass, and divide this into 4-gr. pills. Contains 1-4th part of opium.

Obs. The storax is here chiefly employed to disguise the odour and taste of opium. The name of the preparation has been chosen so that the word 'opium' may not appear in the prescription, a point highly necessary with certain patients.—*Dose*, 3 to 10 gr.; as compound soap pill, and as an anodyne and expectorant in chronic coughs, &c.

Pills of Stramo"nium. *Syn.* PILULÆ STRAMONII, L. *Prep.* 1. Stramonium seeds (in powder), 12 gr. (or leaves, 25 gr.); powdered camphor and extract of seneka root, of each, 1 dr.; powdered savine, 1½ dr.; oil of cajeput, 15 drops. For 2½-gr. pills.—*Dose*, 2 to 4, thrice daily; in rheumatism, &c.

2. (*Sir H. Halford.*) Extract of stramonium and liquorice powder, of each, 1 dr.; powdered Castile soap, 2 dr.; mucilage, q. s. to mix. For 60 pills.—*Dose*, 1 night and morning; in asthmas, &c.

Pills of Strych'nine. *Syn.* PILULÆ STRYCHNIÆ, L. *Prep.* (*Magendie.*) Strychnine, 2 gr.; conserve of hips, 36 gr. (liquorice powder, q. s.); mix very carefully, divide the mass into 24 pills, and silver them.—*Dose*, 1 pill night and morning; in amaurosis, impotence, paralysis, &c.

Pills of Sulphate of Copper. (*Brande.*) *Syn.* PILULÆ CUPRI SULPHATIS, L. *Prep.* Sulphate of copper, 3 gr.; bread crum, 1 dr. Mix, for 24 pills; 1, three or four times a day.

Pills of Sul'phate of Iron. *Syn.* PILULÆ FERRI SULPHATIS (Ph. E.), L. *Prep.* 1. (Ph. E.) Dried sulphate of iron and conserve of red roses, of each, 2 parts; extract of dandelion, 5 parts. For 5-gr. pills. A useful chalybeate tonic.—*Dose*, 1 to 2, twice or thrice daily; in dyspepsia, chlorosis, amenorrhœa, &c.

2. (Ph. E. 1817.) Sulphate of iron (dried), 1 oz.; extract of chamomile, 1½ oz.; oil of peppermint, 1 dr.; syrup, q. s. As the last.

Pills of Sulphate of Quinine'. *Syn.* PILULÆ QUINIÆ SULPHATIS, P. Q. DISULPHATIS, L. *Prep.* 1. Sulphate of quinine, 20 gr.; extract of gentian, 40 gr. For 20 pills.

Pills of Sulphate of Zinc. *Syn.* PILULÆ ZINCI SULPHATIS, P. Z. S. COMPOSITÆ, L. *Prep.* 1. Sulphate of zinc, 12 gr.; extract of gentian, ½ dr.; liquorice powder, q. s. For 20 pills. In dyspepsia, epilepsy, and various convulsive diseases.

2. (*Dr Paris.*) Sulphate of zinc, 10 gr.; powdered myrrh, 1½ dr.; conserve of roses, q. s. For 30 pills.—*Dose*, 1 to 2, twice or thrice daily; in hooping-cough, &c.

Pills of Sulphuret of Iron. (*Biett.*) *Syn.* PILULÆ FERRI SULPHURETI. *Prep.* Sulphuret of iron, ½ dr.; marsh-mallow powder, 10 gr.;

syrup, q. s. Make into 20 pills; 1 to 4 pills daily, in scrofulous eruptions.

Pills, Syph'ilis. *Syn.* PILULÆ ANTISYPHILI-TICÆ, L. See the various pills of mercury, gold, &c. The pills of corrosive sublimate commonly pass under this name.

Pills, Tangore. See PILLS, ARSENICAL.

Pills of Tan'nic Acid. *Syn.* PILULÆ TANNINI, P. ACIDI TANNICI, L. *Prep.* From tannic acid or tannin and powdered sugar, of each, ½ dr.; conserve of roses, q. s. For 24 pills.—*Dose,* 1 or 2 pills thrice daily, in diarrhœa, or 2 every three hours, in internal hæmorrhages, spitting of blood, &c.

Pills of Tar. *Syn.* PILULÆ PICIS LIQUIDÆ, L. *Prep.* From tar, 1 dr.; powdered gentian, ½ dr., or q. s. For 24 pills. Stimulant, diuretic, and sudorific.—*Dose,* 1 to 4, thrice a day; in dropsies, worms, ichthyosis, and several other skin diseases, &c.

Pills of Tarax'acum. *Syn.* PILULÆ TARAXACI, L. *Prep.* 1. Extract of dandelion, 1 dr.; powdered rhubarb, q. s.; divide into 8½-gr. pills. In dyspepsia, &c., complicated with congestion of the liver.

2. (*St Marie.*) Extract of dandelion and Castile soap, equal parts; liquid acetate of potassa, q. s. to mix. For 4-gr. pills. As a diuretic in dropsy.

3. Extract of dandelion, 1 dr.; mercurial pill, 20 gr.; powdered digitalis, 15 gr.; liquorice powder, q. s. For 24 pills.—*Dose,* 1, afterwards increased to 2 or 3; in dropsy connected with liver disease.

Pills, Thomson's Stomach and Liver. *Prep.* From extract of dandelion, 1 dr.; scammony and rhubarb, of each, 15 gr. For 14 pills.—*Dose,* 2 pills, night and morning; in hysteria, hypochondriasis, and chronic inflammation of the liver or kidneys.

Pills of Tobacco. (*Augustin.*) *Syn.* PILULÆ TABACI. *Prep.* Powder of tobacco, 24 gr.; confection of roses, q. s. Mix, and form 72 pills.—*Dose,* 2 to 4 daily, till nausea is produced. In dropsy.

Pills, Tonic. *Syn.* PILULÆ TONICÆ, L. *Prep.* 1. Sulphate of iron, ginger, and myrrh (all in powder), equal parts; conserve of roses, q. s.; mix, and divide into 4-gr. pills.—*Dose,* 1, twice a day; in debility, chlorosis, &c.

2. Powdered myrrh and sulphate of iron, of each, 1 dr.; disulphate of quinine, ½ dr.; powdered capsicum, 15 gr.; conserve of roses, q. s. to mix. For 60 pills.—*Dose,* 1 or 2, twice or thrice a day; in debility, dyspepsia, ague, &c.

3. (*Dr Collier.*) Tartrate of iron and extract of gentian, of each, 1 dr.; oil of cinnamon, 2 drops. For 30 pills.—*Dose,* 3 to 6, three or four times a day. A good stomachic tonic.

4. (*Dr Collier.*) Oxide of zinc, ½ dr. (or sulphate of zinc, 20 gr.); myrrh, 2 dr.; camphor, 20 gr.; confection of hips, to mix. For 40 pills. —*Dose,* 1 or 2 pills, three times a day; in epilepsy, chorea, and other nervous disorders, debility, &c.

Pills of Turpentine. (P. Cod.) *Syn.* PILULÆ TEREBINTHINÆ. *Prep.* Venice turpentine, 1½ oz.; carbonate of magnesia, 1 oz. Make into 200 pills.

Pills of Vale'rian (Compound). *Syn.* PILULÆ

VALERIANÆ COMPOSITÆ, L. *Prep.* (*Dupuytren.*) Powdered valerian, ½ dr.; castor and white oxide of zinc, of each, 20 gr.; syrup, q. s. to mix. For 18 pills.—*Dose,* 2 or 3, thrice daily; in hysteria, hypochondriasis, chlorosis, hemicrania, &c.

Pills of Vale'rianate of Zinc. *Syn.* PILULÆ VALERIANAS, L. *Prep.* From valerianate of zinc and powdered gum, of each, 15 gr.; conserve of hips, q. s. to form a mass. For 18 pills.—*Dose,* 1 pill, twice daily; in nervous headache, neuralgia, hysteria, &c.

Pills, Vallet's. See PILLS OF CARBONATE OF IRON.

Pills, Vance's. See PILLS, APERIENT.

Pills of Veratrine. *Syn.* PILULÆ VERATRINÆ, L. *Prep.* 1. (*Magendie.*) Veratrine, ½ gr.; powdered gum-arabic and syrup of gum, of each, q. s. to form 6 pills (see *below*).

2. (*Turnbull.*) Veratrine, 1 gr.; extract of henbane and liquorice powder, of each, 12 gr.; mix, and divide into 12 pills.—*Dose,* 1 pill, every 3 hours; in dropsy, epilepsy, hysteria, paralysis, nervous palpitations, &c. This should be prepared and used with great caution.

Pills, Ward's Red. *Syn.* WARD'S ANTIMONIAL PILLS. *Prep.* From glass of antimony (finely levigated), 4 oz.; dragon's blood, 1 oz.; mountain wine, q. s. to form a mass. For 1½-gr. pills. Emetic. "They are recommended in obstinate rheumatic affections, in foulness of the stomach and bowels, &c. Their action is often of a very unpleasant character" ('Anat.' of Quackery').

Pills, Lady Webster's. See PILLS, DINNER.

Pills, Whitehead's Essence of Mustard. Balsam of tolu with resin (*Dr Paris*).

Pills, Whytt's. *Prep.* (*Radius.*) Aloes, chloride of iron, and extract of horehound, of each, ½ dr.; assafœtida, 1½ dr. For 2-gr. pills. —*Dose,* 2 to 5, thrice daily; in leucorrhœa, chlorosis, hysteria, &c., with constipation.

Pills, Worsdell's (Kaye's). *Prep.* (*Cooley.*) Powdered aloes, gamboge, and ginger, equal parts; together with a very small quantity of diaphoretic antimony, beaten into a mass with either syrup or treacle, and divided into 2½-gr. pills. "There are about 4½ dozen pills in each 1s. 1½d. box." "The dose, as given in the directions, is from 2 to 8 pills (or even 10 to 12) daily" ('Anat. of Quackery'). They frequently operate with great violence.

Pills, Worm. *Syn.* PILULÆ ANTHELMINTICÆ, P. VERMIFUGÆ, L. *Prep.* 1. Calomel, 1 oz.; sugar, 1½ oz.; mucilage, q. s.; mix, and divide into 240 pills.—*Dose,* 1 to 2, overnight, followed by a strong dose of castor oil early the next morning.

2. Gamboge, 6 gr.; calomel, 5 gr.; mucilage, q. s.; divide into 3 pills. For a morning's dose, fasting.

3. Extract of wormwood, calomel, and powdered scammony, equal parts. For 4-gr. pills.—*Dose,* 1 to 2, as the last. For ascarides, and other small worms.

4. (*Brecmer.*) Powdered aloes and tansy seed, of each, ½ dr.; oil of rue, 9 or 10 drops. For 12 pills.—*Dose,* 3 to 6, in the morning, fasting, and repeated in two or three hours.

5. (*Phœbus.*) Iron filings, ½ dr.; assafœtida,

1½ dr.; essential oil of tansy, 10 or 12 drops; extract of wormwood, q. s.; mix, and divide into 80 pills.—*Dose*, 6 pills, thrice daily.

6. (*Peschier.*) Ethereal extract of male fern, 30 drops; extract of dandelion, 1 dr.; powdered rhizomes of male fern, q. s. to mix. For 30 pills. In tapeworm.—*Dose*, 6 to 15 at bedtime; the dose being repeated in the morning, and then followed in an hour by a strong dose of castor oil.

Pills, Wyndham's (Lee's). *Prep.* (*Cooley.*) Aloes and gamboge, of each (in powder), 3 oz.; Castile soap and extract of cow-parsnip, of each, 1 oz.; nitre, ½ oz. For 5-gr. pills. A powerful drastic cathartic.—*Dose*, 1 to 3 pills.

Pills of Zinc. See PILLS OF OXIDE, SULPHATE, and VALERIANATE OF ZINC, &c.

PILOCARPINE. $C_{11}H_{16}N_2O_2$. An alkaloid discovered by Gerrard, and extracted from the leaves of jaborandi (*Pilocarpus pennatifolius*). *Prep.* Exhaust the leaves or bark of jaborandi with 80% alcohol, to which hydrochloric acid has been added in the proportion of 8 gr. per litre; distil and evaporate to the consistence of an extract. Redissolve the extract with a small quantity of distilled water and filter; treat with ammonia in slight excess, and a large quantity of chloroform. Distil off the chloroform, dissolve the residue in distilled water acidulated with hydrochloric acid, and filter. Treat afresh with chloroform and ammonia. The chloroformic solution is then shaken with water, to which hydrochloric acid is added, drop by drop, up to the quantity sufficient to saturate the pilocarpine. The foreign matters remain in the chloroform, and upon evaporation of the aqueous liquid the hydrochlorate is obtained, well crystallised, in long needles radiating from a common centre. The hydrochlorate dissolved in distilled water, and treated with ammonia and chloroform, yields the pilocarpine upon evaporation of the chloroform solution.

Pilocarpine appears under the form of a soft viscous substance; it is slightly soluble in water and very soluble in alcohol, ether, and chloroform. It presents all the chemical characters of an alkaloid, and rotates the plane of polarized light strongly to the right.

Two salts of pilocarpine are in common use, the nitrate and hydrochlorate, both of which form crystals, freely soluble in water.

Pilocarpine salts are powerful diuretics and sialogogues. A dose taken by the writer kept him in a continual perspiration for five hours, with saliva running from the month for two or three hours. It contracts the pupil of the eye.

Uses.—Antidote in cases of poisoning with belladonna or atropine, useful in asthma, diabetes, intermittent fever, and puerperal convulsions.—*Dose*. Nitrate or hydrochlorate, $\frac{1}{15}$ to ¼ gr.

PIMA'RIC ACID. A resin acid first obtained by Laurent from the turpentine of *Pinus maritima* (Bordeaux turpentine), by the action of hot alcohol.

PIMEN'TO. *Syn.* ALLSPICE, CLOVE PEPPER, JAMAICA P., PIMENTO BERRIES; PIMENTA (B. P., Ph. L., E., and D.), PIPER CARYOPHYLLATUM, P. JAMAICENSE, P. ODORATUM, PIMENTÆ BACCÆ, L. "The dried unripe berries of the allspice tree, *Eugenia pimenta*, from the West Indies"—B. P. "The immature fruit of *Eugenia pimenta* (*Myrtus pimenta*, Linn.)"—Ph. L. *Pimenta officinalis*, Lindl. A tree common in Jamaica, from whence large quantities are imported into this country. Pimento is very largely used as a spice, also in medicine for its aromatic and stimulant properties. Oil of pimento, obtained by distillation from the fruits, is often used for similar purposes as the oil of cloves, as well as in perfumery. Sticks of the pimento are imported in very large quantities for walking-sticks and umbrella handles.

From the leaves of an allied species (*P. acris*, Wight.) the oil of bay or bayberry is obtained, used in the manufacture of bay rum, employed in the United States as a refreshing perfume in faintness, or to sprinkle about sick rooms, as well as for hair washes.

It possesses a mixed odour of cinnamon, cloves, and nutmegs, which, with its other properties, it for the most part yields to alcohol, ether, and water. It is a stimulant and tonic, and is much esteemed as an adjuvant in medicines prescribed in dyspepsia, flatulence, gout, hysteria, &c.; and also to cover the taste of disagreeable medicines.—*Dose*, 5 to 30 gr., bruised or in powder. See ESSENCE, OILS (Volatile), SPIRITS, and WATERS.

PIM'PLES. See ERUPTIONS (Papular).

PINCH'BECK. A gold-like alloy of copper and zinc. See DUTCH GOLD.

PINE-APPLE. *Syn.* ANANAS. The fruit of *Ananassa sativa*, a plant of the Nat. Ord. BROMELIACEÆ. It is astringent, esculent, and possesses a rich flavour and odour. In Europe it is chiefly used as a delicacy for the table, but in tropical climates it is said to be valuable in renal diseases. See ESSENCE, &c.

PI'NEY TAL'LOW. *Syn.* PINEY RESIN, P. DAMMAR. An oleo-resinous substance obtained from the fruit of *Vateria indica*, a tree common in Malabar, by boiling it with water. It is intermediate between fat and wax, and makes good soap and excellent candles. It melts at 95° F. Sp. gr. ·9250 to ·9265.

PI'NIC ACID. The portion of common resin or colophony which is soluble in cold alcohol of sp. gr. ·833.

PINK. A well-known shade of light red. The name is also applied to several pigments, consisting of whiting stained with liquid dyes. See RED and YELLOW PIGMENTS, &c.

PINK DYE. *Prep.* From washed safflower, 2 oz.; salt of tartar, ½ oz.; cold water, 1 quart; digest for three hours, express the liquor, and strain it. Used as a cosmetic, and to dye silk stockings, &c., of a rose colour. The colour is brought out by afterwards applying to, or passing the articles through, water soured with lemon juice. See SAUCERS (Pink).

PIP'ERIN. $C_{17}H_{19}NO_3$. *Syn.* PIPERINA, PIPERINUM, L. *Prep.* Alcoholic extract of white pepper is treated with a weak solution of caustic potash (1 to 100), and the residuum, after being washed with cold water, is dissolved in alcohol; the solution is next agitated with a little animal charcoal, and the filtrate allowed to evaporate spontaneously; the product may be purified by re-solution in alcohol and re-crystallisation.

Prop., &c. Colourless, or only slightly yellow; tasteless; inodorous; fusible; crystallises in plates; insoluble in water; freely soluble in alcohol, ether, and in the acids; very feebly basic; a few definite compounds have, however, been obtained with difficulty; reddened by oil of vitriol. It has been much employed in Italy and on the Continent as a febrifuge.

Obs. An assay for its piperin is the only certain method of testing the quality of pepper. For this purpose a weighed quantity of the sample is reduced to powder, and is exhausted with alcohol of the *sp. gr.* 0·883; the mixed tinctures are then evaporated to an extract, which is treated as above. See PEPPER.

PIPES (in *confectionery*). These are formed from any of the common lozenge-masses, by rolling them into cylinders of about the thickness of a goose-quill. They are frequently medicated.

PIPETTE. A graduated glass instrument, in frequent use in the chemical laboratory, for conveying a measured quantity of fluid from one vessel to another. The pipette generally consists of a bulb, from each end of which proceeds a straight, slender, hollow stem, communicating with the bulb, and varying in length with the capacity of the instrument. Thus constructed, the lower end of the pipette can be dipped into a vessel containing a fluid, the required volume of which can be removed from it. The pipette varies in capacity from 10 to 200 cubic centimetres.

Dr Fresenius gives the following directions for its use:—"To fill a pipette with the fluid which it is intended to transfer from one vessel to another, the lower part of the instrument is dipped into the fluid, and suction applied to the upper aperture, either direct with the lips or through a caoutchouc tube until the fluid in the pipette stands a little above the required mark; the upper, somewhat narrowed, ground orifice is then closed with the point of the index of the right hand, which to that end had always better be moistened a little, and holding the pipette in a perfectly vertical direction, the excess over the quantity required is made to drop out by lifting the finger a little. When the fluid in the pipette has fallen to the required level, the drops which may happen to adhere to the outside of the pipette are carefully wiped off, and the contents of the tube are then fully transferred to the other vessel. In this process it is found that the fluid does not run out completely, but that a small portion of it remains adhering to the glass in the point of the pipette; after a time, as this becomes increased by other minute drops of fluid trickling down from the upper part of the tube, a drop gathers at the lower orifice, which may be allowed to fall from its own weight, or may be made to drop off by a slight shake; if, after this, the point of the pipette be laid against a moist portion of the inner side of the vessel, another minute portion of fluid will trickle out; and lastly, another trifling droplet or so may be got out by blowing into the pipette through the upper orifice. Now, supposing the operator follows no fixed rule in this respect, letting the fluid, for instance, in one operation simply run out, whilst in another operation he lets it drain afterwards, and in a third blows off the last particles of it from the pipette,

it is evident that the respective quantities of fluid delivered in the several operations cannot be quite equal. I prefer in all cases the second method, viz. to lay the point of the pipette whilst draining finally against a moist portion of the inner side of the vessel, which I have always found to give the most accurate corresponding measurements."

PISTA'CHIO NUTS. *Syn.* PISTACIA NUTS; NUCES PISTACIÆ, L. The kernels of the fruit of *Pistachia vera*, Linn., one of the turpentine trees. They closely resemble almonds, but are sweeter, and form a green emulsion with water. Used in confectionery and perfumery, and also as a dessert fruit.

PITCH. *Syn.* BLACK PITCH, BOILED P., STONE P., WOOD P.; PIX (Ph. L.), PIX NIGRA, L. The residuum from boiling tar in an open iron pot, or in a still, until the volatile and liquid portion is driven off. The volatile products principally consist of crude pyroligneous acid and oil of tar. Pitch is chiefly employed in shipbuilding. As a medicine it is a stimulant and tonic; it has been used internally in some skin diseases, and in piles. An ointment made of it is also extensively used in cutaneous affections of the scalp.—*Dose*, 10 gr. to ½ dr.

Pitch, Burgundy. *Syn.* WHITE PITCH, BURGUNDY PINE RESIN; PIX BURGUNDICA (B. P., Ph. L., E., & D.), L. This is an impure resin prepared from the turpentine of *Abies excelsa*, or Norway spruce fir, and from its concrete resinous exudations. It is chiefly used in plasters.

Obs. The importation of this substance has for some years past been gradually lessening in amount, in consequence of the substitution for it of a fictitious pitch, made by melting common resin with linseed oil, and colouring the mass with annotta or palm oil. The physiological action of the two articles is, however, considerably different, since Burgundy pitch acts upon the skin as a powerful local irritant, exciting a slight degree of inflammation, and not unfrequently producing a pimply eruption and an exudation of purulent matter. It is celebrated for its effects when employed as a plaster in all cases where warmth, support, and long adhesion to the skin are desirable; and in the latter quality no substance equals it. The fictitious Burgundy pitch has similar properties, but in an immensely less degree.

PREPARED BURGUNDY PITCH (PIX BURGUNDICA PREPARATA—Ph. L.) may be obtained in the same way as that adopted for strained ammoniacum. This plan is, however, seldom, if ever, adopted in trade.

Pitch, Burgundy (Facti"tious). *Syn.* PIX BURGUNDICA FACTITIA, L. *Prep.* By melting good yellow resin, 1 cwt., with linseed oil, 1 gall., and palm oil (bright), q. s. to colour. The mixture is allowed to cool considerably, and is then pulled with the hands in the same way as lead plaster is treated; after which it is placed in 'bladders' or 'stands' for sale.

Obs. The product of the above formula is the 'Burgundy pitch' of the shops. The 'pulling' or 'working' destroys the translucency of the resin, and imparts to it the peculiar semi-opacity of foreign Burgundy pitch. Cold water is commonly employed to cool it down. Annotta is often substituted for palm oil as a colouring substance. The addition of some of the 'droppings' or 'bottoms' of Canada balsam, Chio turpentine, oil of juniper, &c., renders this article nearly equal to foreign pitch; but in commerce this is never attempted, the aim being only the production of a good colour with moderate toughness. A common melting-pan and fire (if carefully managed) may be used, but, both for safety and convenience, steam pans are preferable, and on the large scale almost indispensable. A good workman can pull and put into stands or casks about 5 cwt. daily; or from 1½ cwt. to 3 cwt. in bladders, the latter quantity depending on the size of the bladders (see *above*).

Pitch, Can'ada. *Syn.* HEMLOCK GUM, H. PITCH. Similar to Burgundy pitch, but from the *Abies canadensis*, or hemlock spruce fir.

Pitch, Jews'. Asphaltum.

Pitch, Min'eral. Indurated mineral bitumen. See ASPHALTUM, BITUMEN, &c.

PIT'COAL. *Syn.* COAL; HOUILLE, Fr.; STEINKOHLE, Ger. This article has been truly described as the most valuable of all those mineral substances from which Great Britain derives its prosperity, and the one which may be regarded as the main support of the whole system of British production. It fuses the metals, it produces the steam which sets our machinery in motion, and, in short, it may be said to render all the resources of this country available for use.

Coal appears to have been formed by some process of decomposition or fermentation of buried vegetable matter, the result being that much of the hydrogen in this matter has separated in the form of marsh gas (the chief constituent of firedamp) and other analogous organic compounds; whilst the oxygen has for the most part passed off in the form of carbonic acid gas, leaving the carbon and other elements in the residual mass. By properly selecting samples from different neighbourhoods and analysing them, a series may be obtained showing the different stages of decomposition through which coal passes. Roscoe gives the following table:

	Carbon.	Hydro-gen.	Oxygen and Nitrogen.
Wood cellulose . . .	50·00	6·00	44·00
Irish peat	60·02	5·88	34·10
Lignite from Cologne .	66·96	5·25	27·76
Earthy coal from Dax .	74·20	5·89	19·90
Cannel coal from Wigan	85·81	5·85	8·34
Newcastle 'Hartley' .	88·42	5·61	5·97
Welsh anthracite . .	94·05	3·38	2·57

The fossilised vegetable remains found in coal embrace over 500 distinct species, of which the calamites (the representatives of the living Equisetums), Lepidodendra, Sigillaria, ferns, conifers, and cycads are the most important.

Coal occurs chiefly in the 'carboniferous system' of rock formations, but it is also worked to advantage in more recent strata.

The more important kinds of coal may be classified as follows:—1. Lignite or brown coal occurs in the tertiary deposits; it has a low specific gravity, a ligneous structure, and consists of the remains of recent plants. 2. Bituminous or caking coals. The most widely diffused and valuable of English coals. They are subdivided into—*a*. Caking coal. Splinters on heating, but the fragments then fuse together in a semi-pasty mass. The chief sources of this valuable variety of coal are the Newcastle and Wigan districts. *b*. Cherry coal or soft coal. Lustre very bright; does not fuse, ignites well and burns rapidly; it occurs in the Glasgow district, Staffordshire, Derbyshire, Nottingham, Lancashire, &c. *c*. Splint, rough, or hard coal. Black and glistening; does not ignite readily, but burns up to a clear hot fire. It constitutes the bulk of the produce of the great coal-fields of North and South Staffordshire, and occurs in the Glasgow district, in Shropshire, Leicestershire, Warwickshire, &c. *d*. Cannel or parrot coal. Dense and compact, having a shelly fracture, and taking a polish like jet. Splinters in the fire, and burns clearly and brightly; it occurs in Wigan and other parts of Lancashire, West Glasgow district, &c. The curious deposit at Bathgate, near Edinburgh, commonly known as 'Boghead cannel coal,' or 'Torbane Hill mineral,' differs considerably from the ordinary 'cannels;' by distillation it yields paraffin oils (q. v.), which are largely used for illuminating and lubricating purposes. 3. Anthracite or stone-coal. The densest, hardest, and most lustrous of all kinds of pitcoal. Burns with little flame or smoke, but gives great heat; it occurs in South Wales, Kilkenny, Devonshire, &c. 4. Steam coal. This approaches nearly to anthracite. Admirably adapted for steam-vessels; it occurs in South Wales, Tyne district, &c.

The quality of coal may be ascertained by either directly testing its heating power or by chemical analysis. In the investigations undertaken at the Museum of Economic Geology, under the directions of Sir H. De la Beche, which furnished the materials for the celebrated 'Admiralty Reports,' three different methods were adopted for this purpose. These consisted in the determination of the quantity of water which a given weight of the coal was capable of converting into steam, the quantity of litharge which it was capable of reducing to the metallic state, and, lastly, its ultimate analysis by combustion with oxide of copper. See ORGANIC SUBSTANCES.

The quantity of sulphur in coal is another matter of importance that may be determined by chemical analysis (see SULPHUR). The presence of more than 1% of sulphur renders coal unfit for the economical production of good illuminating gas, and more than 2% of sulphur renders it objectionable for use as domestic fuel. In like manner, coals containing mineral ingredients in excess are to be avoided, not merely on account of the quantity of ashes left by them, but for their tendency to vitrify upon the bars of the furnace, and to produce what is technically called 'clinkers.' The presence of much silica or alumina, and more particularly of any of the salts of lime, in 'steam coal,' is, on this account, highly objectionable.

For some further information connected with this subject see ANTHRACITE, CHIMNEYS, COKE, FUEL, GAS, LIGNITE, OILS (Mineral), ORGANIC SUBSTANCES, &c.

PITYRIASIS. [Πίτυρον = bran.] The technical name for dandriff, a superficial chronic inflammation of the skin without exudation or swelling, and accompanied by a disturbance of the nutrition of the epidermis causing it to desquamate.

PLAICE. The *Platessa vulgaris*, a well-known flat-fish, common to both the English and Dutch coasts. Its flesh is good and easy of digestion, but more watery than that of the flounder.

PLANTAIN. The plantain, which belongs to the Nat. Ord. MUSACEÆ, and is a native of the East Indies, is cultivated in all tropical and subtropical regions of the world, in many of which it constitutes the principal food of the inhabitants. There are a great many varieties of the plantain, in some of which the stem is 15 or 20 ft. high, whilst in others it does not exceed 6 ft. It is one of the largest of the herbaceous plants.

The fruit is sometimes eaten raw, but is more generally boiled or roasted. It contains both starch and sugar. Boiled and beaten in a mortar, it forms the common food of the negroes in the West Indies. It also constitutes the chief food of the Indians of North and South America.

Humboldt has calculated that the food produce of the plantain is 44 times greater than that of the potato, and 133 times that of wheat.

The banana is a species of plantain. See BANANA.

PLANTAIN, WATER-, or *Alisma plantago*. The use of the root of this plant as a remedy for hydrophobia is by no means recent, and was sanctioned by the College of Physicians of Moscow in the year 1820. Its value is, however, very doubtful. The root contains a very active principle. Cattle are frequently poisoned by it, and it is held in repute in some parts of America as a remedy for the bite of the rattlesnake. It has powerful sedative properties, and is best administered by scraping about an ounce of the solid root and letting it be eaten between two slices of bread (*Christy*).

PLASMA. The liquor sanguinis, in which the corpuscles float.

PLASTER. (In *boiling*, &c.) See MORTAR.

Plaster of Paris. Calcined sulphate of lime. See ALABASTER, GYPSUM, LIME, &c.

PLASTER. (In *pharmacy*.) *Syn.* EMPLASTRUM, L. Plasters (emplastra) are external applications that possess sufficient consistence not to adhere to the fingers when cold, but which become soft and adhesive at the temperature of the human body.

Plasters are chiefly composed of oils, fats, or fatty acids, united to metallic oxides, or mixed with powders, wax, or resin. They are usually formed, whilst warm, into ½-lb. rolls, about 8 or 9 inches long, and wrapped in paper. When required for use a little is melted off the roll by means of a heated iron spatula, and spread upon leather, linen, or silk. The less adhesive plasters, when spread, are usually surrounded with a margin of resin plaster, to cause them to adhere.

In the preparation of plasters the heat of a water-bath or of steam should alone be employed. On the large scale well-cleaned and polished copper or tinned copper pans, surrounded with iron jackets, supplied with high-pressure steam, are used for this purpose. The resins and gum-resins that enter into their composition are previously purified by straining. After the ingredients are mixed, and the mass has acquired sufficient consistence by cooling, portions of it are taken into the hands, and well pulled or worked under water until it becomes solid enough to admit of being formed into rolls; but this process must not, on any account, be practised on compound plasters containing odorous substances, or substances soluble in water. These should be suffered to cool on an oiled marbled slab until sufficiently 'stiff' to be formed into rolls. Many plasters, as those of lead and resin, derive much of their whiteness from the treatment just referred to. White plasters are not, however, always the best; but they are those which are most admired, and the most sought after in trade.

Plasters are preserved by enveloping the rolls with paper to exclude the air as much as possible, and by keeping them in a cool situation. A few, as those of belladonna and ammoniacum with mercury, are commonly placed in pots. When kept for any length of time they are all more or less apt to become hard and brittle, and to lose their colour. When this is the case they should be remelted by a gentle heat, and sufficient oil added to the mass to restore it to a proper consistence.

The operation of spreading plasters for use requires skill and experience on the part of the operator. Various fabrics are employed for the purpose, of which linen or cotton cloth, or leather, are those most generally employed. Silk and satin are used for 'court plaster.' The shape and size must be regulated by the part to which they are to be applied.

On the large scale plasters are spread by means of a 'spreading machine,' illustrations of which are found in works on pharmacy.

Compound plasters are now much less frequently employed in medicine than formerly. Those principally in use are such as afford protection to sores and abraded surfaces, and give support to the parts. A few, however, which contain acrid, stimulating, and narcotic substances, and operate as rubefacients, blisters, or anodynes, are still retained in the Pharmacopœias.

Alcoholic extracts are far more suitable for mixing with plasters than watery extracts.

Plaster of Aconite. *Syn.* EMPLASTRUM ACONITI, L. *Prep.* (*Curtis*.) Gently evaporate tincture of aconite to the consistence of a soft extract, then spread a very small portion over the surface of a common adhesive plaster, on either calico or leather. Mr Curtis has strongly recommended this plaster in neuralgia. A little of the alcoholic extract may be employed instead of that obtained fresh from the tincture.

Plaster, Adhe'rent. See PLASTER, SOAP (Compound).

Plaster, Adhe'sive. See PLASTER, RESIN, COURT P., &c.

Plaster, Adhesive Lime. *Syn.* EMPLASTRUM ADHESIVUM CALCAREUM. *Prep.* Soap of lime, 200 parts; boiled turpentine, 100 parts; suet, 25 parts.

Plaster, Ammoni'acal. *Syn.* DE KIRKLAND'S VOLATILE PLASTER; EMPLASTRUM AMMONIÆ, E. A. HYDROCHLORATIS, L. *Prep.* Take of lead plaster, 1 oz.; white soap (shaved fine), ½ oz.; melt them together, and, when nearly cold, add of sal-ammoniac (in fine powder), 1 dr. Stimulant and rubefacient. Dr Paris, who highly recommends it in pulmonary affections, employs double the above proportion of sal-ammoniac. Its efficacy depends on the gradual extrication of free ammonia by the decomposition of the sal-ammoniac, on which account it is proper to renew the application of it every twenty-four hours.

Plaster of Ammoni'acum. *Syn.* EMPLASTRUM AMMONIACI (Ph. L., E., & D.), L. *Prep.* 1. (Ph. L. & E.) Ammoniacum (strained), 5 oz.; dilute acetic acid (distilled vinegar), 8 fl. oz. (9 fl. oz.—Ph. E.); dissolve, and, frequently stirring, evaporate by a gentle heat to a proper consistence.

2. (Ph. D.) Gum ammoniacum (in coarse powder), 4 oz.; proof spirit, 4 fl. oz.; dissolve by the aid of a gentle heat, and evaporate, as before.

Obs. This plaster is adhesive, stimulant, and resolvent, and is employed in scrofulous and indolent tumours, white swellings, &c. In the Ph. D. 1826 vinegar of squills was ordered instead of distilled vinegar.

Plaster of Ammoniacum with Hem'lock. *Syn.* EMPLASTRUM AMMONIACI CUM CICUTÂ, L. *Prep.* (Ph. E. 1744.) Gum ammoniacum, 8 oz.; vinegar of squills, q. s. to dissolve; hemlock juice, 4 oz.; gently evaporate, as before. In cancerous and other painful tumours. A better plan is to add 1 dr. of extract of hemlock to 1½ oz. of strained ammoniacum (previously reduced to a proper consistence with a little distilled vinegar), melted by a very gentle heat.

Plaster of Ammoniacum with Mer'cury. *Syn.* EMPLASTRUM AMMONIACI CUM HYDRARGYRO (B. P., Ph. L., E., & D.), L. *Prep.* 1. (Ph. L. & E.) Olive oil, 56 gr.; heat it in a mortar; add of sulphur, 8 gr.; triturate; further add of mercury, 3 oz.; again triturate, and when the globules are extinguished add it to ammoniacum (strained), 1 lb. (12 oz.—B. P.), previously melted by a gentle heat, and mix them well together.

2. (Ph. D.) From ammoniacum plaster, 4 oz.; mercurial plaster, 8 oz.; melted together by a gentle heat, and then stirred constantly until nearly cold.

3. (Wholesale.) Take of mercury, 38 oz.; prepared suevm, 5 oz.; triturate as last, and add the mixture to strained ammoniacum, 10 lbs., previously sufficiently softened by a gentle heat. Possesses a fine blue colour, and is quickly made.

Obs. This plaster cannot be rolled till considerably cooled, and neither this nor the simple plaster must be put into water. It is powerfully discutient, and is applied to indurated glands, indolent tumours, &c.

Plaster, An'odyne. See PLASTER, OPIUM; PLASTER, BELLADONNA, &c.

Plaster, Arnica. (Ph. U.S.) *Syn.* EMPLASTRUM ARNICÆ. *Prep.* Alcoholic extract of arnica, 1½ oz.; resin plaster, 3 oz. Add the extract to the plaster previously melted over a water-bath, and mix it thoroughly.

Plaster, Aromatic. *Syn.* STOMACH PLASTER; EMPLASTRUM AROMATICUM, L. *Prep.* (Ph. D. 1826.) Strained frankincense (thus), 8 oz.; beeswax, ½ oz.; melt them together, and, when the mass has considerably heated, add of powdered cinnamon, 6 dr.; oils of allspice and lemons, of each, 2 dr. Stimulant; applied over the stomach in dyspepsia, spasms, nausea, flatulence, &c. Camphor, 1 dr., is commonly added.

Plaster of Assafœtida. *Syn.* ANTIHYSTERIC PLASTER, ANTISPASMODIC P.; EMPLASTRUM ASSAFŒTIDÆ (Ph. E.), E. ANTIHYSTERICUM, &c., L. *Prep.* (Ph. E.) From lead plaster and strained assafœtida, of each, 2 oz.; strained galbanum and beeswax, of each, 1 oz.; melted together. Antispasmodic; applied to the stomach or abdomen in spasms, hysteria, &c.; and to the chest in hooping-cough.

Plaster, Baynton's Adhesive. *Prep.* From yellow resin, 1 oz.; lead plaster, 1 lb.; melted together. Recommended for bad legs and other like sores.

Plaster of Belladon'na. *Syn.* EMPLASTRUM BELLADONNÆ (B. P., Ph. L., E., & D.), L. *Prep.* 1. (Ph. L.) Soap plaster, 8 oz.; melt it by the heat of a water-bath; add of extract of belladonna (deadly nightshade), 3 oz.; and keep constantly stirring the mixture until it acquires a proper consistence.

2. (Ph. E.) Resin plaster, 3 oz.; extract of belladonna, 1½ oz.; as the last.

3. (Ph. D.) Resin plaster, 2 oz.; extract of belladonna, 1 oz.

4. (B. P.) Alcoholic extract of belladonna, 1 part; resin plaster and soap plaster, of each, 2 parts; melt the plasters and the extract and mix.

Uses, &c. As a powerful anodyne and antispasmodic; in neuralgia and rheumatic pains, and as an application to painful tumours. With many persons this plaster produces a rash and dryness of the throat; its use should then be discontinued. The plaster of the shops is usually deficient in extract. The following formula is in common use in the wholesale trade:—Lead plaster and resin plaster, of each, 2½ lbs.; extract of belladonna, 1½ lbs. This plaster must not be 'pulled' in water.

Plaster, Berg's Antirheumatic. *Syn.* GOUT PAPER; EMPLASTRUM ANTIRHEUMATICUM, CHARTA ANTIRHEUMATICA, L. *Prep.* By digesting euphorbium, 2 parts, and cantharides, 1 part (both in powder), in rectified spirit, 10 parts, for 8 days; adding to the strained liquid, black resin and Venetian turpentine, of each, 4 parts; assisting the mixture by a gentle heat. Two or three coats of the product are successively spread over the surface of thin paper. Used in gout and rheumatism ('Anst. of Quackery').

Plaster, Black. *Syn.* EMPLASTRUM NIGRUM, L. *Prep.* Mr Sharp's black plaster was formed by boiling together olive oil, 18 oz.; wax, 3½ oz.; carbonate of lead, 10 oz.

Plaster, Black Diach'ylon. See COURT PLASTER.

Plaster of Black Pitch. *Syn.* EMPLASTRUM
PICIS NIGRÆ, L. *Prep.* (Ph. Wirtem.) Black
pitch, black resin, and beeswax, of each, 8 parts;
suet, 1 part; melted together. Rubefacient and
stimulant.

Plaster, Blistering. See PLASTER OF CAN-
THARIDES.

Plaster, Bree's Antiasthmatic. *Prep.* From
lead plaster, 1 oz.; olive oil, 1 dr.; melted together, and, when somewhat cooled, mixed with
powdered camphor, 2 dr.; powdered opium, 1 dr.,
and at once spread on leather.

Plaster, Brown. *Syn.* EMPLASTRUM FUSCUM,
L.; ONGUENT DE LA MÈRE, Fr. The butter, lard,
oil, suet, and wax should be first melted together,
and the heat gradually increased until they begin
to smoke; the litharge is then to be sifted in,
and the stirring and heat continued until the
mixture assumes a brown colour; the pitch is
next added, and the whole stirred for some time
longer.

Plaster, Brown Diach'ylon. See PLASTER OF
GALBANUM.

Plaster, Bryony. (*Boerhaave.*) *Syn.* EM-
PLASTRUM BRYONIÆ, L. *Prep.* Strained galbanum, 4 oz.; wax plaster, 9 oz.; olive oil, 1 oz.
Melt together, and add powdered bryony root, 2
oz.; flowers of sulphur, 1 oz.; Ethiops mineral, 2
dr.; stir till cold.

Plaster of Bur'gundy Pitch. *Syn.* CEPHALIC
PLASTER, BREAST P.; EMPLASTRUM PICIS (B. P.,
Ph. L. & E.), E. P. COMPOSITUM, E. P. BURGUN-
DICÆ, L. *Prep.* 1. (Ph. L.) Prepared (strained)
Burgundy pitch, 2 lbs.: prepared frankincense
(thus), 1 lb.; yellow resin and beeswax, of each,
4 oz.; melt them together, then add olive oil and
water, of each, 2 fl. oz.; expressed oil of nutmeg
(mace), 1 oz.; and, constantly stirring, evaporate
to a proper consistence.

2. (Ph. E.) Burgundy pitch, 1 lb.; resin
and beeswax, of each, 2 oz.; olive oil and
water, of each, 1 fl. oz.; oil of mace, ½ oz.; as
the last.

3. (B. P.) Burgundy pitch, 26 parts; common
frankincense, 13 parts; resin, 4½ parts; yellow
wax, 4½ parts; expressed oil of nutmegs, 1 part;
olive oil, 2 parts; water, 2 parts; add the oil and
the water to the other ingredients, previously
melted together; stir, and evaporate to a proper
consistency.

Uses, &c. Burgundy pitch plaster is stimulant, rubefacient, and counter-irritant. It is a
common application to the chest in pulmonary
affections, to the joints in rheumatism, and to
the loins in lumbago. Spread on leather, it
forms a good warm plaster to wear on the
chest during the winter. "When it produces
a serous exudation it should be frequently renewed."

The BURGUNDY PITCH PLASTER of the shops
is commonly made as follows:—Factitious Burgundy pitch (bright coloured), 42 lbs.; palm oil
(bright), ¾ lb.; beeswax (bright), 5 lbs.; melt,
and, when nearly cold, add of oil of mace, 6 oz.;
oil of nutmeg, 1 oz.

Plaster of Burgundy Pitch (Irritating). (Ph.
G.) *Syn.* EMPLASTRUM PICIS IRRITANS. *Prep.*
Burgundy pitch, 32 oz.; yellow wax, 12 oz.;
turpentine, 12 oz.; euphorbium, 3 oz.

Plaster, Calefa'cient. *Syn.* WARM PLASTER;
EMPLASTRUM CALEFACIENS (Ph. D.), L. *Prep.*
(Ph. D.) 1. Plaster of cantharides, ½ lb. (1 part);
Burgundy pitch, 5½ lbs. (11 parts); melt them
together by a gentle heat, and stir the mixture
as it cools until it stiffens. Stimulant, rubefacient, and counter-irritant; in a variety of
affections. In some persons, when long applied,
it blisters or produces a running sore.

2. (Ph. B.) Cantharides in coarse powder,
4 oz.; boiling water, 1 pint; expressed oil of
nutmeg, 4 oz.; yellow wax, 4 oz.; resin, 4 oz.;
soap plaster, 2 lbs.; resin plaster, 8½ lbs. Infuse
the cantharides in the boiling water for six hours;
squeeze strongly through calico, and evaporate
the expressed liquid by a water-bath till reduced
to one third. Then add the other ingredients and
melt in a water-bath, stirring well till the whole
is thoroughly mixed.

Plaster, Camphor. *Syn.* EMPLASTRUM CAM-
PHORÆ. Camphor is best applied by sprinkling
the powder on the warm surface of a spread
adhesive or other plaster. Blisters are treated in
this way to prevent strangury.

Plaster, Can'cer. *Syn.* EMPLASTRUM ANTI-
CANCROSUM, L. *Prep.* 1. Wax plaster, 1 oz.;
extract of hemlock, 1 dr.; levigated arsenious
acid, ½ dr.

2. (*Richter.*) Extract of hemlock, 1 oz.; extract of henbane, ½ oz.; powdered belladonna,
1 dr.; acetate of ammonia, q. s. to form a plaster.
Both the above must be used with great caution.
See CANCER, &c.

Plaster of Canthar'ides. *Syn.* BLISTERING
PLASTER, VESICANT P., PLASTER OF SPANISH
FLIES; EMPLASTRUM CANTHARIDIS (B. P., Ph.
L., E., & D.), E. O. VESICATORIÆ, E. LYTTÆ, L.
Prep. 1. (Ph. L.) Yellow wax and suet, of
each, 7½ oz.; lard, 6 oz.; resin, 3 oz.; melt them
together, remove the vessel from the fire, and, a
little before they concrete, sprinkle in of cantharides (in very fine powder), 1 lb. (12 oz.—B. P.),
and mix.

2. (Ph. E.) Cantharides, beeswax, resin, and
suet, equal parts; as the last.

3. (Ph. D.) Spanish flies, 6 oz.; prepared lard,
resin, and yellow wax, of each, 4 oz.; proceed as
before, and "stir the mixture constantly until the
plaster is cool."

4. (Wholesale.) From beeswax and good lard,
of each, 4 lbs.; flies and yellow resin, of each,
6 lbs.; suet, 10 lbs. A commonly used formula,
the product of which is, however, greatly inferior
to that of the Pharmacopœia.

Obs. All the above are used to raise blisters.
The plaster is spread on white leather or adhesive
plaster with a knife, and is surrounded with
a margin of resin plaster to make it adhere. A
piece of thin muslin or tissue-paper is sometimes placed between the plaster and the skin
to prevent absorption. A little powdered camphor is sometimes sprinkled on the surface of
the spread plaster, to prevent strangury. A better
mode of obviating the action on the urinary
organs is by the copious use of diluents. This
plaster should be rolled in starch powder, and not
with oil.

Plaster of Cantharides (Compound). *Syn.* EM-
PLASTRUM CANTHARIDIS COMPOSITUM, L. *Prep.*

(Ph. E.) Venice turpentine, 4½ oz.; cantharides and Burgundy pitch, of each, 3 oz.; beeswax, 1 oz.; verdigris (in fine powder), ½ oz.; powdered mustard and black pepper, of each, 2 dr.; mix at a heat under 212° F. Stronger than the last, and quicker in its action; but it causes more pain, and is much more apt to occasion troublesome ulcerations. Used in gout, spasms of the stomach, &c.

Plasters, Caoutchouc. Dr. Schneegans and Corneille have made known some caoutchouc plaster formulæ. The groundwork of these plasters is a mixture of lanoline, benzoated lard, and dammar resin with caoutchouc; they keep well, and are usually self-adhesive. A slight addition of glycerine prevents the product drying too rapidly by exposure to the air. The caoutchouc is previously dissolved in benzol (1 to 6 parts) by soaking for three or four days. The composition of the various plasters is as follows:

1. *Zinc Caoutchouc Plaster*, 20%. Resin dammar, 20 parts; benzoated lard, 25 parts; lanolin, 15 parts; caoutchouc, 8 parts; glycerine, 12 parts; and zinc oxide, 20 parts; total, 100 parts.

2. *Iodoform Caoutchouc Plaster*, 20%. Resin dammar, 15 parts; benzoated lard, 30 parts; lanolin, 20 parts; caoutchouc, 5 parts; glycerine, 10 parts; iodoform, 20=100 parts.

3. *Mercurial Caoutchouc Plaster*, 20%. Resin dammar, 25 parts; benzoated lard, 12 parts; yellow wax, 15 parts; caoutchouc, 8 parts; lanolin, 20 parts; quicksilver, 20=100 parts.

4. *Boracic Caoutchouc Plaster*, 20%. Resin dammar, 20 parts; benzoated lard, 25 parts; white wax, 15 parts; caoutchouc, 8 parts; lanolin, 12 parts; and boracic acid, 20 parts; total, 100 parts.

5. *Salicylic Acid Caoutchouc Plaster*, 20%, is prepared like that with boracic acid.

6. *Ichthyol Caoutchouc Plaster*, 20%. Resin dammar, 20 parts; benzoated lard, 30 parts; yellow wax, 20 parts; caoutchouc, 8 parts; lanolin, 12 parts; and ichthyol (ichthyolnatrium), 20 parts; total, 100 parts. The ichthyol is first heated with the lanolin in a water-bath, and the other materials added afterwards.

7. *Zinc and Quicksilver Caoutchouc Plaster*, 20% and 10%. Resin dammar, 20 parts; benzoated lard, 12 parts; yellow wax, 10 parts; caoutchouc, 8 parts; lanolin, 20 parts; quicksilver, 20 parts; and zinc oxide, 10 parts; total, 100 parts. The zinc oxide is mixed previously with the mercury, the two being well worked up together, and to the lukewarm mixture the other substances are added afterwards. Of course similar plasters can be made in like manner with other active ingredients, as may be prescribed. The authors say that these preparations can be produced at a cost of about threepence to sixpence a square yard.

Plaster, Capuchin'. See PLASTER OF EUPHORBIUM.

Plaster of Car'bonate of Lead. *Syn.* EMPLASTRUM PLUMBI CARBONATIS, E. CERUSSÆ, L. *Prep.* (P. Cod.) Carbonate of lead, 1 lb.; olive oil and water, of each, 2 lbs.; boil them together until they combine and form a plaster; lastly, remelt this with white wax, 8½ oz. Its properties resemble those of ordinary lead plaster. An excel-

lent emollient and defensive plaster. See PLASTER, MAHY'S.

Plaster, Cephal'ic. *Syn.* LABDANUM PLASTER; EMPLASTRUM CEPHALICUM, E. LABDANI, L. *Prep.* (Ph. L. 1788.) Labdanum, 3 oz.; frankincense (thus), 1 oz.; melt, and add to the mixture, when nearly cold, powdered cinnamon and expressed oil of mace, of each, ½ oz.; oil of mint, 1 dr. Applied to the forehead or temples, in headache; to the stomach, in colds, &c. See PLASTER OF BURGUNDY PITCH, &c.

Plaster, Chesselden's Stick'ing. *Syn.* EMPLASTRUM PLUMBI CUM PICE, L. *Prep.* From lead plaster, 2 lbs.; Burgundy pitch (genuine), 1 oz.; melted together.

Plaster, Corn. *Syn.* EMPLASTRUM AD CLAVOS, L. *Prep.* 1. Resin plaster, 5 parts; melt, stir in of sal-ammoniac (in fine powder), 1 part, and at once spread it on linen or soft leather.

2. (*Baudot's.*) Resin cerate, 40 parts; galbanum plaster, 40 parts; verdigris, 15 parts; turpentine, 5 parts; creasote, 3 parts.

3. (*Kennedy's.*) From beeswax, 1 lb.; Venice turpentine, 5 oz.; verdigris (in fine powder), 1¼ oz.; mixed by a gentle heat, and spread on cloth. It is cut into pieces and polished, and of these 1 dozen are put into each box.

4. (*Le Foret.*) Galbanum plaster, 2 oz.; melt by a very gentle heat; add sal-ammoniac and saffron, of each, ½ oz.; powdered camphor, 2 oz.; and, when nearly cold, stir in of liquor of ammonia, 2 oz. Applied, spread on leather, to the corn only, as it will blister the thinner skin surrounding its base.

5. (Ph. Sax.) Galbanum plaster, 1 oz.; pitch, ½ oz.; lead plaster, 2 dr.; melt them together, and add verdigris and sal-ammoniac (in fine powder), of each, 1 dr. For other formulæ see PLASTER OF VERDIGRIS (*below*), and CORNS.

Plaster, Court. *Syn.* STICKING PLASTER, ISINGLASS P.; EMPLASTRUM ICHTHYOCOLLÆ, E. ADHESIVUM ANGLICUM, L. *Prep.* 1. Isinglass, 1 part; water, 10 parts; dissolve, strain the solution, and gradually add to it of tincture of benzoin, 2 parts; apply this mixture, gently warmed, by means of a camel-hair brush, to the surface of silk or sarcenet, stretched on a frame, and allow each coating to dry before applying the next one, the application being repeated as often as necessary; lastly, give the prepared surface a coating of tincture of benzoin or tincture of balsam of Peru. Some manufacturers apply this to the unprepared side of the plaster, and others add to the tincture a few drops of essence of ambergris or essence of musk.

2. (*Deschamps.*) A piece of fine muslin, linen, or silk is fastened to a flat board, and a thin coating of smooth, strained flour paste is given to it; over this, when dry, two coats of colourless gelatin, made into size with water, q. s., are applied warm. Said to be superior to the ordinary court plaster.

3. (*Liston's.*) Soak isinglass, 1 oz., in water, 2½ fl. oz., until it becomes swollen and quite soft; then add of proof spirit, 3½ fl. oz., and expose the mixture to the heat of hot water, frequently stirring, until the union is complete; lastly, apply four coats of the solution to the surface of oiled silk nailed to a board, by means of a soft brush.

4. (*Dr Paris.*) Black silk or sarcenet is strained and brushed over ten or twelve times with the following composition: gum benzoin, ½ oz.; rectified spirit, 6 oz.; dissolve. In a separate vessel dissolve of isinglass, 1 oz., in as little water as possible; strain each solution, mix them, decant the clear portion, and apply it warm. When the last coating is quite dry, a finishing coat is given with a solution of Chio turpentine, 4 oz., in tincture of benzoin, 6 oz.

Obs. The common 'COURT PLASTER' of the shops is generally prepared without using spirit, and with merely sufficient tincture of benzoin, or other aromatic, to give it an agreeable odour. Formerly black silk or sarcenet was exclusively employed as the basis of the plaster, but at the present time chequered silk is also much in favour. 'FLESH-COLOURED COURT PLASTER' is likewise fashionable. 'TRANSPARENT COURT PLASTER' is prepared on oiled silk. 'WATER-PROOF COURT PLASTER' is simply the common plaster which has received a thin coating of pale drying oil on its exposed surface. The FINEST COURT PLASTER of the West-end houses is now prepared on goldbeaters' skin (or the prepared membrane of the cæcum of the ox), one side of which is coated with the isinglass solution as above, and the other with pale drying oil or a solution of either gutta percha or caoutchouc in chloroform, or in bisulphuret of carbon.

Plaster of Cro'ton Oil. *Syn.* EMPLASTRUM CROTONIS, E. OLEI TIGLII, L. *Prep.* (*Bouchardat.*) To lead plaster, 4 parts, melted by a very gentle heat, add of croton oil, 1 part. A powerful counter-irritant; it also generally acts powerfully on the bowels.

Plaster of Cum'in. *Syn.* EMPLASTRUM CUMINI (Ph. L.), E. CYMINI, L. *Prep.* 1. (Ph. L.) Burgundy pitch, 3 lbs.; beeswax, 3 oz.; melt, add of cumin seed, caraways, and bayberries, of each (in powder), 3 oz.; next add of olive oil and water, of each, 1½ fl. oz., and evaporate to a proper consistence.

2. (Wholesale.) From yellow resin, 7 lbs.; beeswax and linseed oil, of each, ½ lb.; powdered cumin and caraway seeds, of each, 7 oz.; mix.

Obs. This is a mere revival of the formula of the Ph. L. 1724. In that of the Ph. L. 1778 no water was ordered, and the powders simply stirred into the melted mass shortly before it cools—the common practice in all laboratories.

Cumin plaster is carminative, stimulant, and discutient. It is applied over the regions of the stomach and bowels in colic, dyspepsia, and flatulence, and is also applied to indolent tumours. It has long been a favourite remedy with the lower classes.

Plaster, Delacroix's Agglu'tinative. *Syn.* EMPLASTRUM GLUTINANS SANCTI ANDREÆ À CRUCE, E. FICIS CUM ELEMI, L.; EMPLÂTRE D'ANDRÉ DE LA CROIX, Fr. *Prep.* (P. Cod.) From Burgundy pitch, 25 parts; gum elemi, 6 parts; Venice turpentine and oil of bays, of each, 8 parts; melted together, and strained.

Plaster, Diach'ylon. See PLASTER OF LEAD.

Plaster, Diapal'ma. See PLASTER, PALM.

Plaster of El'emi. *Syn.* EMPLASTRUM ELEMI, L. *Prep.* From wax plaster, 3 parts; gum elemi, 1 part; melted together by a gentle heat. Stimulant and discutient. Used for issues, &c.

Plaster of Euphor'bium. *Syn.* EMPLASTRUM EUPHORBII, L. *Prep.* 1. (Guy's Hosp.) Burgundy pitch plaster, 8 oz.; melt, and add of euphorbium (in powder), 1 dr.

2. (CAPUCHIN PLASTER—Ph. Wirt.) Burgundy pitch and beeswax, of each, 3 oz.; Venice turpentine, 1 oz.; melt them together, add gum ammoniacum, olibanum, mastic, and lapis calaminaris, of each, 1 oz.; euphorbium, pyrethrum, and common salt, of each (in powder), 2 oz.; and stir until the mass concretes. Both of the above are stimulant, rubefacient, and counter-irritant.

Plaster, Fayard's. See PAPER (Gout).

Plaster of Flower of Ointments. *Syn.* EMPLASTRUM FLOS UNGUENTORUM DICTUM, L. *Prep.* From frankincense (thus), yellow resin, suet, and beeswax, of each, 1 lb.; olibanum, ¼ lb.; Venice turpentine, 5 oz.; gum myrrh, 2 oz.; white wine, 16 fl. oz.; boil to a plaster, adding, before the mass cools, of camphor, ¼ oz. Calorifacient and stimulant.

Plaster of Frank'incense. *Syn.* STRENGTHENING PLASTER; EMPLASTRUM THURIS, E. ROBORANS, L. *Prep.* (Ph. L. 1788.) To lead plaster, 2 lbs., melted by a gentle heat, add of frankincense (thus), ½ lb.; dragon's blood (in powder), 3 oz., and stir well. In muscular relaxations, weak joints, &c. Mr Redwood says that a "better-looking plaster is produced by melting the frankincense and dragon's blood together, and straining them through a cloth, then mixing these with the lead plaster previously melted." See PLASTER OF OXIDE OF IRON.

Plaster of Gal'banum. *Syn.* COMPOUND GALBANUM PLASTER, YELLOW DIACHYLON, GUM PLASTER, DIACHYLON WITH THE GUMS; EMPLASTRUM GALBANI (B. P., Ph. L.), E. GUMMOSUM (Ph. E.), L. *Prep.* 1. (Ph. L.) Take of strained galbanum, 8 oz.; common turpentine, 1 oz.; melt them together, then add of prepared frankincense (thus), 3 oz.; and next, of lead plaster, 3 lbs., previously melted over a slow fire.

2. (Ph. E.) Gum ammoniacum and galbanum, of each, ½ oz.; melt them together, strain, and add of litharge plaster, 4 oz.; beeswax, ½ oz. (both previously melted); and mix the whole thoroughly. These proportions are the same as those of the B. P.

3. (Wholesale.) From lead plaster, 48 lbs.; yellow resin, 12 lbs.; strained galbanum, 3 lbs.; strained assafœtida, 1 oz.

4. Galbanum, ammoniacum, yellow wax, of each, 1 part; lead plaster, 8 parts. Melt the galbanum and ammoniacum together and strain, then mix with the other melted ingredients.

Obs. Galbanum plaster is stimulant and resolvent, and is much used in indolent, scrofulous, and other tumours, painful gouty and rheumatic joints, in rickets, &c.

Plaster, Gaulthier's. (*Guibourt.*) Palm plaster, 12 parts; olive oil and white wax, of each, 1 part; melt, and add of Venice turpentine, 2 parts. More adhesive than the simple PALM PLASTER.

Plaster of Gin'ger. *Syn.* EMPLASTRUM ZINGIBERIS, L. See GINGER.

Plaster, Gout. *Syn.* EMPLASTRUM ANTAR-

TRRITICUM, L. See PLASTER OF GALBANUM, PITCH, &c.; PAPER, GOUT.

Plaster of Gum. See PLASTER OF GALBANUM.

Plaster of Hem'lock. *Syn.* EMPLASTRUM CONII, E. CICUTÆ, L. *Prep.* 1. Wax, 1 part; Burgundy pitch, 9 parts; melt them together, and add of extract of hemlock, 3 parts.

2. (Ph. Bat.) Lead plaster and beeswax, of each, 1 lb.; olive oil, 6 fl. oz.; melt, and add of powdered hemlock (recent), 1 lb.

Obs. Hemlock plaster is occasionally used as an application to painful and malignant ulcers and tumours, painful joints, &c. A spread plaster of it, with 6 or 8 gr. of tartar emetic (in very fine powder) sprinkled over its surface, has been highly extolled as a counter-irritant in hooping-cough, phthisis, &c.

Plaster of Hen'bane. *Syn.* EMPLASTRUM HYOSOYAMI, L. *Prep.* As the last, but using henbane instead of hemlock. As an anodyne, in various external affections.

Plaster of I'odide of Lead. (Ph. B.) *Syn.* EMPLASTRUM PLUMBI IODIDI. *Prep.* Add iodide of lead in fine powder, 1 oz., to lead plaster, 8 parts, and resin, 1 part, previously melted together. Mix thoroughly.

Plaster of Iodide of Potas'sium. *Syn.* EMPLASTRUM POTASSII IODIDI (Ph. L.), L. *Prep.* (Ph. L.) Iodide of potassium, 1 oz.; olive oil, 2 fl. dr.; triturate them together, then add of strained frankincense (thus), 6 oz.; wax, 6 dr.; and stir constantly until the mass cools. "This plaster is to be spread on linen rather than on leather." Used as a discutient or resolvent; more particularly as an application to scrofulous tumours and indurations.

Plaster of I'odine. *Syn.* EMPLASTRUM IODINII, L. *Prep.* Triturate iodine, 1 dr., in a warm mortar, with olive oil, 1 oz.; then add of beeswax, 1 oz.; yellow resin, ½ oz., previously melted together, and stir the whole until it concretes. It should be, preferably, spread at once on leather, and applied shortly after being prepared. Used as the last.

Plaster of Iodine (Compound). *Syn.* EMPLASTRUM IODINII COMPOSITUM, L. *Prep.* 1. Iodine, 1 dr.; iodide of potassium, 2 dr.; rub them to a fine powder, add this to lead powder, 2 oz.; Burgundy pitch, 1 oz., previously melted together, and just about to concrete. More active than either of the preceding.

2. (EMP. IOD. CUM BELLADONNÆ.) To belladonna plaster, 2 oz., melted by a very gentle heat, add iodine and iodide of potassium (in fine powder), of each, 1 dr., and stir the mixture until nearly cold. Powerfully resolvent and anodyne. Used in the same cases as the preceding, when there is much pain.

Plaster of I'ron. See PLASTER OF OXIDE OF IRON.

Plaster of I"singlass. See PLASTER, COURT.

Plaster, Is'sue. *Syn.* EMPLASTRUM AD FONTICULOS, SPARADRAPUM PRO FONTICULIS, L. *Prep.* 1. From beeswax, ½ lb.; Burgundy pitch and Chio turpentine, of each, 4 oz.; vermilion and orris powder, of each, 1 oz.; musk, 4 gr.; melted together and spread upon linen. This is afterwards polished with a smooth piece of glass moistened with water, and cut into pieces.

2. (Ph. Aust.) Yellow wax, 6 oz.; mutton suet, 2 oz.; lard, 1½ oz.; melt, add of turpentine, 1½ oz., and afterwards of red-lead, 4 oz.; dip pieces of linen into the melted mixture, pass these between rollers, and when cold polish them, as before, and cut them into squares. The issue plaster (issue paper; charta ad fonticulos) of the Ph. Suecica is a nearly similar compound, with the addition of about 1-48th part of verdigris in very fine powder, and being spread upon paper.

Plaster, Kennedy's. See PLASTER, CORN.

Plaster, Kirkland's. See PLASTER, AMMONIACAL.

Plaster of Lab'danum. See PLASTER, CEPHALIC.

Plaster of Lead. *Syn.* LEAD PLASTER, LITHARGE P., COMMON P., DIACHYLON, SIMPLE DIACHYLON, WHITE D.; EMPLASTRUM PLUMBI (B. P., Ph. L.), E. LITHARGYRI (Ph. E. & D.), E. COMMUNE, DIACHYLON SIMPLEX, L. *Prep.* 1. (Ph. L.) Oxide of lead (litharge), in very fine powder, 6 lbs.; olive oil, 1 gall.; water, 1 quart; boil them over a slow fire, constantly stirring to the consistence of a plaster, adding a little boiling water if nearly the whole of that used in the beginning should be consumed before the end of the process.

2. (Ph. E.) Litharge, 5 oz.; olive oil, 12 fl. oz.; water, 8 fl. oz.; as the last.

3. (Ph. D.) Litharge, 5 lbs.; olive oil, 1 gall.; water, 1 quart.

4. (*Otto Kohnke.*) For each lb. of litharge employed, add ¼ pint of colourless vinegar (each fl. oz. of which is capable of saturating ⅓ dr. of carbonate of potassa), add the oil, boil until all moisture is evaporated, and until only a few strise of litharge rise to the surface; then remove the vessel from the heat, add gradually 1-3rd to half as much vinegar as before, and boil the mixture to a proper consistence.

5. (Wholesale.) From Genoa oil, 7 galls. (or 65 lbs.); litharge (perfectly free from copper), 28 lbs.; water, 2½ galls.; boil to a plaster as before.

6. (B. P.) Oxide of lead, in very fine powder, 5 parts; olive oil, 10; water, 5; boil all the ingredients together gently by the heat of a steam-bath, and keep them simmering for 4 or 5 hours, stirring constantly until the product acquires the proper consistence for plaster, adding more water during the process if necessary.

Obs. The London College orders too little oil. The second, fourth, and fifth formulæ produce beautiful plasters, that keep well; those of the others, although very white, get hard and brittle much more rapidly. The proper proportion of oil is fully 2½ times the weight of the litharge,—2½ times appears the best quantity; and without this is used, the plaster speedily gets hard and non-adhesive. The process consists in putting the water and the litharge into a perfectly clean and well-polished tinned copper or copper pan, mixing them well together with a spatula, adding the oil, and boiling, with constant stirring, until the plaster is sufficiently hard when thoroughly cold. This process usually occupies from 4 to 5 hours, but by adopting the fourth formula an excellent plaster may be made in from 20 to 30 minutes. This plaster is generally cooled by immersion in cold water; and to render it very white, a quality highly prized in the trade, it is usual to submit it to laborious 'pulling,' in the manner already noticed. *Use.* As a simple defensive plaster or strapping; but principally as a basis for other plasters.

Plaster, Liston's. See PLASTER, COURT.

Plaster, Mahy's. *Syn.* EMPLASTRUM PLUMBI CARBONATIS, E. P. C. COMPOSITUM, L. *Prep.* (Ph. U. S.) Carbonate of lead (pure white-lead), 1 lb.; olive oil, 32 fl. oz.; water, q. s.; boil them together, constantly stirring until perfectly incorporated; then add of yellow wax, 4 oz.; lead plaster, 1½ lbs.; and when these are melted and the mass somewhat cooled, stir in of powdered orris root, 9 oz. A favourite application in the United States of America to inflamed and excoriated surfaces, bedsores, burns, &c.

Plaster of Mel'ilot. *Syn.* EMPLASTRUM MELILOTI, E. à MELILOTO, L. *Prep.* 1. (Ph. E. 1744.) Fresh melilot, chopped small, 6 lbs.; suet, 3 lbs.; boil until crisp, strain with pressure, and add of yellow resin, 8 lbs.; beeswax, 4 lbs., and boil to a plaster. Stimulant. Used to dress blisters, &c. The greater portion of this plaster in the shops is made without the herb, and is coloured with verdigris. (See the next formula.)

2. (Wholesale.) Take of yellow resin, 18 lbs.; green ointment, 4½ lbs.; yellow wax, 3 lbs.; finely powdered verdigris, q. s. to give a deep green colour.

Plaster of Menthol. *Syn.* EMPLASTRUM MENTHOL (B. P. additions, 1890), L. *Prep.* 1. Menthol, 2 parts; yellow wax, 1 part; resin, 7 parts. Melt the wax and resin, and as the mixture cools stir in the menthol.

2. Lead plaster, 75 parts; yellow wax, 10 parts; yellow resin, 5 parts; melt, strain, and add menthol, 10 parts. Mix well, and spread on cloth or leather.

Plaster, Mercu"rial. *Syn.* EMPLASTRUM MERCURIALE, E. HYDRARGYRI (B. P., Ph. L., E., & D.), L. *Prep.* 1. (Ph. L.) Add, gradually, of sulphur, 8 gr., to heated olive oil, 1 fl. dr., and stir the mixture constantly with a spatula until they unite; next add of mercury, 3 oz., and triturate until globules are no longer visible; lastly, gradually add of lead plaster (melted over a slow fire), 1 lb., and mix them all well together. (About 1 fl. dr. of balsam of sulphur may be substituted for the oil and sulphur ordered above.) These proportions are the same as those of the B. P.

2. (Ph. E.) Resin, 1 oz.; olive oil, 9 fl. dr.; mix by heat, cool, add of mercury, 3 oz., and triturate until its globules disappear; then add of litharge plaster, 6 oz. (previously liquefied), and mix the whole thoroughly.

3. (Ph. D.) Oil of turpentine, 1 fl. oz.; resin, 2 oz.; dissolve with the aid of heat; add of mercury, 6 oz.; triturate until the globules disappear, and the mixture assumes a dark grey colour; then add of litharge plaster (previously melted), 12 oz., and stir the whole until it stiffens on cooling.

4. (Wholesale.) Take of mercury, 7 lbs.; prepared serum, ¼ lb.; triturate until the globules disappear, and add the mixture to lead plaster

(melted by a gentle heat), 36 lbs.; stir them well together, and until they concrete. Very fine bluish-slate or lead colour. *Obs.* Mercurial plaster is used as a discutient in glandular enlargements and other swellings, and is also applied over the hepatic regions in liver complaints.

Plaster, Mercurial, with Belladon'na. *Syn.* EMPLASTRUM HYDRARGYRI CUM BELLADONNA, L. *Prep.* From mercurial plaster, 6 dr.; extract of belladonna, 2 dr.; olive oil, 1 dr.; mixed by a gentle heat. One of our most useful anodyne and discutient applications, in painful scirrhous, scrofulous, and syphilitic tumours. The Medico-Chirurgical Pharm. orders ½ fl. dr. of hydrocyanic acid to be added to every 2 oz. of the above.

Plaster of Mezereon and Cantharides. (Ph. G.) *Syn.* EMPLASTRUM MEZEREI CANTHARIDATUM, L. *Prep.* Cantharides in coarse powder, 3 oz.; mezereon, cut and dried, 1 oz.; acetic ether, 10 oz. by weight. Macerate for 8 days, filter, and dissolve in the filtered liquid 175 gr. of sandarac, 87 gr. of elemi, 87 gr. of resin, which spread on silk previously covered with the following solution:—Isinglass, 2 oz.; distilled water, 20 oz.; rectified spirit, 5 oz. by weight.

Plaster of Min'ium. *Syn.* EMPLASTRUM MINII, E. à MINIO, E. PLUMBI OXYDI RUBRI, L. *Prep.* (Ph. L. 1746.) Olive oil, 4 lbs.; minium (red-lead), in fine powder, 2½ lbs.; water, q. s.; proceed as for lead plaster (which it closely resembles). *Obs.* To ensure a good colour and the quality of keeping well, the quantity of oil should be increased about 1-3rd. When discoloured by heat it forms the 'brown minium plaster' (emp. à minio fuscum) of old pharmacy. Lead plaster, either alone or with the addition of a little red-lead, is usually sold for it.

Plaster of Minium (Compound). *Syn.* NUREMBERG PLASTER; EMPLASTRUM MINII COMPOSITUM, L.; EMPLÂTRE DE NUREMBERG, Fr. *Prep.* (Soubeiran.) Red-lead, 12 parts; olive oil, 8 parts; grind them together on a porphyry slab, and add the mixture to lead plaster, 50 parts; beeswax, 24 parts, melted together; lastly, when nearly cold, stir in of camphor, 1 part.

Plaster, Morrison's Adhesive. *Syn.* MORRISON'S ADHESIVE PASTE. From wheaten flour, 2 oz.; mild ale, ¼ pint; stir them together, and heat the mixture to the boiling-point; when cold, add of powdered resin, 3 oz.; and constantly stirring, again heat them to boiling. Used as a depilatory in ringworm, &c.

Plaster of Mus'tard. *Syn.* EMPLASTRUM SINAPIS, L. This is always an extemporaneous preparation. Flour of mustard is made into a stiff paste with lukewarm water, or with vinegar, and is then spread on a piece of calico or linen (folded two or three times); over the surface of the mustard is placed a piece of gauze or thin muslin, and the plaster is then applied to the part of the body it is intended to medicate. Its action is that of a powerful rubefacient and counter-irritant; but its application should not be continued long, unless in extreme cases. Its effects are often apparently wonderful. We have seen very severe cases of facial neuralgia, sore throat, painful joints, rheumatic pains, &c., relieved in a few minutes by means of a mustard plaster or 'poultice.'

Plaster, Nuremberg. See PLASTER, MINIUM (COMPOUND).

Plaster of Oak Mistletoe. (*Hardy.*) *Syn.* EM-PLASTRUM VISCI QUERCINI, L. *Prep.* To 2 parts of melted beeswax add gradually 1 part of juice of true oak mistletoe, and form a plaster. In neuralgic pains.

Plaster of O″pium. *Syn.* EMPLASTRUM ANO-DYNUM, E. OPII (B. P., Ph. L., E., & D.), L. *Prep.* 1. (Ph. L.) Lead plaster, 8 oz.; melt, and add of frankincense (thus), 2 oz.; next add of extract of opium, 1 oz., previously dissolved in boiling water, 1 fl. oz.; and, constantly stirring, evaporate the mixture over a slow fire to a proper consistence. This plaster is much stronger than that of the Ph. L. 1836 and of the other British Colleges.

2. (Ph. L. 1836.) Lead plaster, 1 lb.; melt, add of powdered thus, 3 oz.; mix, and further add of powdered opium, ½ oz.; water, 8 fl. oz., and boil to a proper consistence.

3. (Ph. E.) Litharge plaster, 12 oz.; Burgundy pitch, 3 oz.; liquefy by heat, then add, by degrees, of powdered opium, 2 oz., and mix them thoroughly. This and the preceding contains only 1-3rd part of the opium ordered in the present Ph. L. & D.

4. (Ph. D.) Resin plaster, 9 oz.; opium, in fine powder, 1 oz.; as the last. Same as B. P.

5. (Ph. B.) Powdered opium, 1 oz.; resin plaster, 9 oz.; melt the plaster and add the opium.

Obs. The above plaster is reputed anodyne, and useful in various local pains; but its virtues in this way have been greatly exaggerated. The formula of the Ph. L. 1836, from being less costly, is still often employed in place of that of the Ph. L. 1851. The following is commonly used:—Lead plaster, 14 lbs.; yellow resin, 2 lbs.; powdered opium, ¼ lb.

Plaster of Opium and Camphor. (*Dr Paris.*) *Syn.* EMPLASTRUM OPII ET CAMPHORÆ. *Prep.* Opium and camphor, of each, ½ dr. Lead plaster, q. s. Mix.

Plaster of Ox'ide of I'ron. *Syn.* IRON PLASTER, FRANKINCENSE P., STRENGTHENING P.; EMPLAS-TRUM ROBORANS, E. FERRI (B. P., Ph. L., E., & D.), E. THURIS, E. FERRI OXYDI RUBRI, L. *Prep.* 1. (Ph. L.) Lead plaster, 8 oz.; frankincense (thus), 2 oz.; melt them together over a slow fire, sprinkle into the mixture sesquioxide of iron, 1 oz., and mix the whole well together.

2. (Ph. E.) Litharge plaster, 3 oz.; yellow resin, 6 dr.; beeswax, 3 dr.; melt them together, then add of red oxide of iron, 1 oz., previously triturated with olive oil, 3½ fl. dr.

3. (Ph. D.) Litharge plaster, 8 oz.; Burgundy pitch, 3 oz.; peroxide of iron, in fine powder, 1 oz.; as No. 1. Same as B. P.

4. (Wholesale.) From lead plaster (quite dry), 84 lbs.; powdered yellow resin, 14 lbs.; 'crocus martis' (lively coloured), 14 lbs.; olive oil, 3 pints; as No. 2.

5. (B. P.) Add hydrated peroxide of iron in fine powder, 1 oz., to Burgundy pitch, 2 oz., and litharge plaster, 8 oz., previously melted together, and stir the mixture constantly till it stiffens on cooling.

Obs. Iron plaster is reputed strengthening and stimulant. It is employed as a mechanical support in muscular relaxation, weakness of the joints, &c., especially by public dancers. Its tonic action is probably wholly imaginary. No. 4 is the 'EMPLASTRUM ROBORANS' of the shops at the present time.

Plaster, Oxycro'ceum. *Syn.* EMPLASTRUM OXY-CROCEUM, L. *Prep.* 1. (Ph. E. 1744.) Beeswax, 1 lb.; black pitch and strained galbanum, of each, ¼ lb.; melt, and add of Venice turpentine, powdered myrrh, and olibanum, of each, 2 oz.; powdered saffron, 2 oz.

2. (Wholesale.) From black pitch, 9 lbs.; black resin, 11 lbs.; beeswax and lard, of each, 2¼ lbs., melted together. Warm, discutient. Still popular with the lower orders. The saffron of the original formula never finds its way into the oxycroceum plaster of the druggists.

Plaster, Palm. *Syn.* EMPLASTRUM DIAPAL-MUM, L.; DIAPALME, EMPLÂTRE DIAPALME, Fr. *Prep.* (P. Cod.) Lead plaster, 32 parts; yellow wax, 2 parts; melt them together, add of sulphate of zinc, 1 part; dissolve in a little water, and continue the heat, with constant agitation, until all the water is evaporated.

Obs. This plaster originally contained palm oil, and this ingredient is still ordered in the formulæ of Plenck and Reuss. Soubeiran directs white wax to be employed.

Plaster, Paracelsus's. *Syn.* EMPLASTRUM PARACELSI, E. STYPTICUM, L. *Prep.* From lead plaster, 28 lbs.; galbanum plaster, 2 lbs.; powdered white canella and gum thus, of each, 1½ lbs., melted together. The original formula, as well as that of the Ph. L. 1721, was similar, although much more complicated.

Plaster of Pitch. *Syn.* POOR MAN'S PLASTER, GOUT P., ANTIRHEUMATIC P.; EMPLASTRUM PAUPERIS, E. ANTIRHEUMATICUM, E. ANTARTHRI-TICUM, E. PICIS COMMUNE, L. This has been already noticed under the head of RESIN PAPER. It is also, but less frequently, spread on cloth and leather.

Plaster, Prestat's Adhe'sive. *Prep.* From lead plaster, 2½ lbs.; yellow resin, 5 oz.; Venice turpentine, 4 oz.; gum ammoniacum and mastic, of each, 1½ oz.; made into a plaster, and spread on linen or calico.

Plaster of Red-lead. See PLASTER OF MINIUM.

Plaster of Res'in. *Syn.* ADHESIVE PLASTER, RESINOUS P.; EMPLASTRUM ADHESIVUM, E. RE-SINÆ (B. P., Ph. L. & D.), E. RESINOSUM (Ph. E.), E. LITHARGYRI CUM RESINÂ, L. *Prep.* 1. (Ph. L.) To lead plaster, 3 lbs., melted by a gentle heat, add of resin, ½ lb., also liquefied by heat, and mix. The formula of the Ph. U.S. is similar.

2. (Ph. E.) Litharge plaster, 5 oz.; resin, 1 oz.; mix with a moderate heat.

3. (Ph. D.) To litharge plaster, 2 lbs., melted by a gentle heat, add of powdered resin, 4 oz.; Castile soap, in powder, 2 oz.; and mix them intimately.

4. (Wholesale.) Pale lead plaster (from a previous batch, and quite dry), 72 lbs.; olive oil (Genoa), 3 lbs.; melt them together in a bright and perfectly clean copper pan, and sift m of pale yellow resin (in powder), 12 lbs., stirring all the

while. The mixture is to be cooled, and 'pulled' or 'worked,' after the manner of lead plaster.

5. (B. P.) Resin (in powder), 2 parts; litharge plaster, 16 parts; curd soap, 1 part; melt the plaster with a gentle heat, add the resin and soap, first liquefied, and mix.

Obs. Resin plaster, spread upon calico, forms the well-known '**STRAPPING**' or '**ADHESIVE PLASTER**' so extensively used to protect raw surfaces, support parts, and for dressing ulcers, retaining the lips of recent cuts and wounds in contact, &c. It is gently stimulant, and is thought to assist the healing process. It is also employed as a basis for other plasters. The '**HOSPITAL PLASTER**' of certain houses is of this kind. See **PLASTER OF SOAP (COMPOUND).**

Plaster, Resol'vent. *Syn.* EMPLASTRUM RESOLVENS, E. EX MIXTIS QUATUOR, L. *Prep.* (P. Cod.) Galbanum, hemlock, mercurial, and soap plasters, equal parts, melted together.

Plaster, Roper's Royal Bath. *Prep.* (*Cooley.*) Strained black pitch, 16 oz.; Burgundy pitch, 10 oz.; tar and beeswax, of each, 1 oz.; melt, and, when considerably cooled, add of expressed oil of mace, 2 dr.; croton oil, 1 dr.; and spread the mixture upon heart-shaped pieces of white sheepskin, without remelting it. Stimulant and counter-irritant; recommended by its proprietor as a cure for all human ailments. The '**BATH-PLASTER PILLS**,' also prepared by Mr Roper, resemble several of the aperient pills already noticed. (See 'Anat. of Quackery.')

Plaster, Scott's. *Prep.* From lead plaster, 14 oz.; olive oil and white resin, of each, 1 oz.; melted together, and spread on calico.

Plaster, Sharp's Black. *Prep.* From olive oil, 5 parts; carbonate of lead, 4 parts; beeswax, 1 part; boiled to a plaster.

Plaster, Simple. See **PLASTER, WAX.**

Plaster of Soap. *Syn.* EMPLASTRUM è SAPONE, E. SAPONIS (Ph. L. E. and D.), L. *Prep.* 1. (Ph. L.) To lead plaster, 3 lbs., melted by a slow heat, add of Castile soap, sliced, ½ lb.; resin, 1 oz., both (also) liquefied by heat, and, constantly stirring, evaporate to a proper consistence.

2. (Ph. E.) To litharge plaster, 4 oz., gum plaster, 2 oz., melted together, add of Castile soap, in shavings, 1 oz., and boil a little.

3. (Ph. D.) To litharge plaster, 2½ lbs., melted over a gentle fire, add of Castile soap (in powder), 4 oz., and heat them together (constantly stirring) until they combine.

4. (B. P.) Curd soap, lead plaster, 36 parts; resin (in powder), 1 part; to the lead plaster, previously melted, add the soap and the resin, first liquefied, then, constantly stirring, evaporate to a proper consistence.

Obs. Care must be taken to evaporate all the moisture from the above compounds, as, if any is left in the plaster, it turns out crumbly, and does not keep well. Much heat discolours it. (See *below.*)

Soap plaster is emollient and resolvent, and is used in abrasions and excoriations, and as a dressing to soft corns, lymphatic tumours, &c.

Plaster of Soap (Camphorated). (P. Cod.) *Syn.* EMPLASTRUM SAPONIS CAMPHORATUM. Soap plaster, 10 oz.; camphor, 48 gr.

Plaster of Soap (Compound). *Syn.* EMPLAS-

TRUM SAPONIS COMPOSITUM, E. ADHAERENS, L. *Prep.* (Ph. D. 1826.) Resin plaster, 3 oz.; soap plaster, 2 oz.; melted together.

Obs. Less emollient, but more stimulant, than the simple plaster. The '**EMPLASTRUM è MINIO CUM SAPONE**' (Ph. E. 1744) was made by melting 1 part of soap with 5 parts of minium plaster. Neither of the above must be put into water. See **PLASTER OF RESIN, Ph. D.**

Plaster of Soap-ce'rate. *Syn.* EMPLASTRUM CERATI SAPONIS (B. P.), L. *Prep.* 1. From soap-cerate, heated by means of a water-bath until all the moisture is evaporated. Sometimes 2 or 3 dr. each of powdered mastic and gum ammoniacum are added for each pound of cerate. The product is generally spread whilst still warm. Said to be suppurative, resolvent, cooling, and desiccative. See **CERATE** (Soap).

2. (B. P.) Curd soap, 10 parts; beeswax, 12½ parts; oxide of lead (in powder), 15 parts; olive oil, 20 parts; vinegar, 160 parts; boil the vinegar with the oxide over a slow fire, or by a steam-bath, constantly stirring them until they unite; then add the soap and boil again in a similar manner until all the moisture is evaporated; lastly, mix with the wax previously dissolved in the oil, and continue the process till the product takes the consistence of a plaster.

Plaster of Squill (Compound). *Syn.* EMPLAS-TRUM SCILLÆ COMPOSITUM. *Prep.* Galbanum, ½ oz.; soap, ½ oz.; litharge plaster, 2 oz.; melt together, and add opium, 1 dr.; ammoniacum, ½ oz.; vinegar of squills, 3 oz., mixed together; keep them over the fire constantly stirred till they are incorporated.

Plaster, St Andrew's. *Prep.* From yellow resin, 8 oz.; gum elemi, 2 oz.; Bordeaux turpentine and oil of the bay laurel, of each, 1 oz.; melted together by a gentle heat. A stimulant, resolvent, and adhesive plaster, once supposed to possess extraordinary virtues.

Plaster, Stick'ing. See **PLASTER, COURT, PLASTER OF RESIN, &c.**

Plaster, Stom'ach. See **PLASTER, AROMATIC, &c.**

Plaster, Strength'ening. See **PLASTERS OF FRANKINCENSE** and **OXIDE OF IRON.**

Plaster, Styp'tic. See **PLASTER OF OXIDE OF IRON, PARACELSUS'S P., &c.**

Plaster of Thus. See **PLASTER OF FRANKIN-CENSE.**

Plaster of Ver'digris. *Syn.* EMPLASTRUM ÆRUGINIS, E. CUPRI SUBACETATIS, L. *Prep.* (P. Cod.) Beeswax, 4 parts; Burgundy pitch, 2 parts; melt, add of Venice turpentine and prepared verdigris (in powder), of each, 1 part, and stir until the mass is nearly cold. For other formulæ see **PLASTER, CORN, &c.**

Plaster, Vigo's. *Syn.* EMPLASTRUM VIGONIS, L. *Prep.* (P. Cod.) Lead plaster, 40 oz.; mercury, 12 oz.; liquid styrax, 6 oz.; beeswax, turpentine, and resin, of each, 2 oz.; ammoniacum, bdellium, myrrh, and olibanum, of each, 5 dr.; saffron, 3 dr.; oil of lavender, 2 dr.; made into a plaster s. a.

Plaster, Warm. See **CALEFACIENT PLASTER, BURGUNDY PITCH P., &c.**

Plaster of Wax. *Syn.* SIMPLE PLASTER; EM-PLASTRUM ATTRAHENS, E. SIMPLEX (Ph. E.), E.

OKRE, L. *Prep.* 1. (Ph. E.) Beeswax, 3 oz.; suet and yellow resin, of each, 2 oz.; melt them together, and stir the mixture briskly until it concretes by cooling.

2. (Ph. L. 1836.) Yellow wax and suet, of each, 8 lbs.; yellow resin, 1 lb.; as the last. Intended to be employed as a simple dressing, especially to blistered surfaces. It is now seldom used.

Plaster, White Diach'ylon. See PLASTER OF LEAD.

Plaster, Yellow Diach'ylon. See PLASTER OF GALBANUM.

Plaster, Zinco-lead. *Syn.* EMPLASTRUM ZINCO-PLUMBICUM, E. DIAPOMPHOLYGOS, L. *Prep.* (Ph. Succ.) Beeswax, 1 lb.; olive oil and graphite (black-lead), of each, 6 oz.; carbonate of lead, 4 oz.; oxide of zinc (impure), 3 oz.; olibanum, 1½ oz.; boil to a plaster. Astringent and desiccant. Other forms substitute an equal weight of litharge for the graphite.

PLASTERMULLS. (*H. Unna.*) This name has been given to a dressing or plaster consisting of a layer or sheet of gutta-percha fixed to muslin. On the gutta-percha side is spread a layer of soft material, of unknown composition, containing one or more active compounds. They are spread in strips 1 metre long and 20 cm. wide. Those in common use contain salicylic acid or a mixture of salicylic acid and creasote. They are prepared in Germany.

Uses. Mainly for the removal of hard skin, or as an application to lupus.

PLATE. The name is commonly given to gold and silver wrought into instruments or utensils for domestic use.

The cleaning of plate is an important operation in a large establishment, as its durability, and much of its beauty, depend on this being properly done. The common practice of using mercurial plate powder is destructive to both of these, as mercury not only rapidly erodes the surface of silver, but renders it soft, and, in extreme cases, even brittle. The only powder that may be safely used for silver is prepared chalk, of the best quality. For gold, the form of red oxide of iron known as *Jeweller's Rouge* is the most useful and appropriate.

In his 'Workshop Receipts' Mr Spon recommends the following:—"Take an ounce each of cream of tartar, common salt, and alum, and boil in a gallon or more of water. After the plate is taken out and rubbed dry it puts on a beautiful silvery whiteness. Powdered magnesia may be used dry for articles slightly tarnished, but if very dirty it must be used first wet and then dry."

Chamois leather, a plate brush, or very soft woollen rags should alone be used to apply them; and their application should be gentle and long continued rather than the reverse. Dirty plate, after being cleaned with boiling water, may be restored by boiling it in water, each quart of which contains a few grains of carbonate of soda, and about an ounce of prepared chalk, calcined hartshorn, or cuttle-fish bone, in very fine powder. The ebullition sets up a gentle friction, which effects its purpose admirably. The boiled plate, after being dried, is best 'finished off'

with a piece of soft leather or woollen cloth which has been dipped into the cold mixture of chalk and water, and then dried. The same method answers admirably with German silver, brass, pewter, and all the softer metals. See POWDER (Plate), &c.

PLATINA. See PLATINUM.

PLATING. The art of covering copper and other metals with either silver or gold.

Plating is performed in various ways. Sometimes the silver is fluxed on to the surface of copper by means of a solution of borax, and subsequent exposure in the 'plating furnace,' and the compound ingot is then rolled to the requisite thinness between cylinders of polished steel. The common thickness of the silver plate before rolling is equal to about the 1-40th of that of the compound ingot. Sometimes the nobler metal is precipitated from its solutions upon the copper by the action of chemical affinity, or, more frequently, by the agency of electro-chemical decomposition (electro-plating).

The metal employed for plating is a mixture of copper and brass, annealed or hardened, as the case may require. For electro-plated goods, 'nickel silver' is now almost invariably employed. See ELECTROTYPE, GILDING, PLATINISING, SILVERING, &c.

PLATINISING. Metals may be coated with platinum by nearly similar processes to those already referred to under PLATING. In . the 'moist way' vessels of brass, copper, and silver are conveniently platinised in the following manner:—Solid bichloride of platinum, 1 part, is dissolved in water, 100 parts, and to this solution is added of common salt, 8 parts; or, still better, 1 part of ammonio-chloride of platinum and 8 parts of chloride of ammonium are placed in a suitable porcelain vessel, with about 40 parts of water, and the whole heated to ebullition; the vessels or utensils, previously made perfectly bright, are then immersed in the boiling liquid. In a few seconds they generally acquire a brilliant and firmly adhering layer of platinum.

Silver plates for voltaic batteries are commonly platinised in order to make them last and to facilitate depolarisation by immersing them for a few seconds in a mixture of saturated solution of bichloride of platinum, 1 part; dilute sulphuric acid, 3 parts; water, 4 to 6 parts. Platinum battery plates are covered with a pulverulent deposit of platinum by means of the electrotype.

The electro-deposition of platinum has within the last few years become an important art. It is very difficult to produce a bright adherent deposit of platinum. Most of the processes which have been proposed give good results at the commencement, but the solutions deteriorate from a variety of causes, the chief among which is the insolubility of platinum as an anode, which necessitates occasional additions to the bath of fresh quantities of the platinum salt, whereby the electrolytes are continually altering in conductivity, and gradually becoming contaminated with secondary products. The character of the deposit of platinum is naturally impoverished by such alterations in the conditions.

All the solutions recommended for electro-plating, except Boettger's, which is the double chloride of ammonium and platinum in sodium citrate, are made by treating the platinic chloride with alkaline salts, the most favoured of which are the phosphates and oxalates. The result is a solution of the double phosphate or oxalate, as the case may be, together with the chloride of the alkali from the decomposition of the platinic chloride.

As the solution becomes impoverished it is strengthened by fresh additions of platinic chloride, with the result that the alkaline chloride accumulates until the bath is practically spoilt. Boettger maintains his bath by fresh additions of the original solution, but here again the accumulation of foreign substances must follow.

W. H. Wahl, who has been investigating the question of the best method for the electro-deposition of platinum, states in a recent issue of the 'Journal of the Franklin Institute' that with an alkaline platinate solution, an oxalate solution or a phosphate solution may be used.

Solution of platinic hydrate in caustic potash will give a good deposit of metal, and the bath may be kept up to a standard by additions of platinic hydrate without any deterioration due to the accumulation of foreign salts. He recommends a solution of platinic hydrate in the preparation and maintenance of the strength of the ordinary solution in use, and gives the following directions :

The Alkaline Platinate Solution. Platinic hydrate, 2 oz. ; caustic potash, 8 oz. ; distilled water, 1 gall.

One half of the caustic potash is dissolved in a quart of water and the platinic hydrate gradually added ; when solution is effected, the remainder of the caustic potash dissolved in another quart of water is stirred in, and the solution made up to a gallon.

A current of about two volts is the best, and there should be only a slight, if any, evolution of hydrogen at the cathode, but a liberal one of oxygen at the anode. The solution may be worked at half the above strength. A little acetic acid improves the working of the bath when a heavy deposit is required. Articles of steel, nickel, tin, zinc, or German silver are preferably thinly coated with copper in a hot cyanide bath.

An oxalate solution may be prepared by dissolving 1 oz. of platinic hydrate in 4 oz. of oxalic acid, and diluting to 1 gall. The best plan is to work with a saturated solution of the oxalate, keeping an undissolved excess always present. The addition of a small quantity of oxalic acid now and then is advantageous. The double oxalate may be prepared by saturating the alkaline oxalate with platinic hydrate, the strength of the bath being maintained by the presence of the single oxalate, as above. The deposits from these solutions are sensibly harder than those obtained with the alkaline bath, and will buff tolerably well.

The Phosphate Solution. Phosphoric acid, syrupy (sp. gr. 1·7), 8 oz. ; platinic hydrate, 1—1½ oz. ; distilled water, 1 gall.

The acid should be moderately dilute, and the solution of the hydrate effected at the boiling temperature, after which it is diluted to a gallon. The current in this case may be stronger than in the previous one. The strength is maintained by additions of platinic hydrate. The double alkaline phosphates may be used, and are prepared by neutralising the above with the alkali, and then adding an excess of phosphoric acid. The deposit is described as brilliant and adherent, with the same steely appearance as with the oxalate, but to a less pronounced degree.

Platinised asbestos is prepared by dipping asbestos into a solution of bichloride of platinum, or one of the double chlorides of that metal, and then gradually heating it to redness. It is used as a substitute for spongy platinum. See ELECTROTYPE, VOLTAIC ELECTRICITY.

PLATINOTYPE. See PHOTOGRAPHY.

PLATINUM. Pt = 194·4. *Syn.* PLATINA, WHITE GOLD ; PLATINUM, L. A heavy, greyish-white metal, occurring chiefly in certain of the alluvial districts of Mexico and Brasil, in the Ural Mountains of Russia, in Ceylon, Brasil, Australia, Peru, Borneo, and California. It occurs in nature only in the metallic state under the form of grains and small rolled masses, associated with palladium, rhodium, osmium, ruthenium, iridium, and a little iron ; gold usually accompanies it in the form of grains. It has only been known in Europe since 1748.

Prep. 1. The native alloy of this metal (crude platinum) is acted upon, as far as possible, by nitro-hydrochloric acid containing an excess of hydrochloric acid, and slightly diluted with water in order to dissolve as small a quantity of iridium as possible ; to the deep yellowish-red and highly acid solution thus produced ammonium chloride is added, by which nearly the whole of the platinum is thrown down in the state of the ammonio-chloride. This substance, after being washed with a little cold water, is dried and heated to redness ; the product is spongy metallic platinum. This is made into a thin uniform paste with water, introduced into a slightly conical mould of brass, and subjected to a graduated pressure, by which the water is squeezed out, and the mass rendered at length sufficiently solid to bear handling. It is next dried, very carefully heated to whiteness, and hammered, or subjected to powerful pressure by suitable means, whilst in the heated state. It will now bear forging into a bar, and may afterwards be rolled into plates, or drawn into wire, at pleasure.

2. The crude platinum is fused with six parts of lead, and the alloy treated with dilute nitric acid (1 : 8) ; the reagent dissolves most of the lead, and along with it any copper, iron, palladium, and rhodium that may be present, leaving a residue, which consists of lead, platinum, and iridium. This residue is now treated with dilute aqua regia, which dissolves the lead and platinum, but leaves the iridium ; the solution is separated off, the lead which it contains precipitated with sulphuric acid, and then the solution of platinic chloride is treated as in Prep. 1.

Prop., &c. Platinum is one of the heaviest substances known, its sp. gr. being 21·5. It is whiter than iron, harder than silver ; it is infusible in ordinary furnaces, and melts only when exposed to the highest temperature obtained by Deville's oxyhydrogen gas furnace, viz. about 2000° C. It is unaffected by air, water, and all the ordinary acids, and even its polish is uninjured by the strongest heat of a smith's forge; aqua regia, however, dissolves it, though with much more difficulty than gold; it is also superficially oxidised by fused hydrate of potassium; it is malleable and ductile. Spongy platinum, powdered platinum, and even perfectly clean platinum-foil possess the remarkable property of causing the union of oxygen and hydrogen gases, and of promoting the oxidation of other bodies, with more or less elevation of temperature. Platinum is precipitated from its solutions by deoxidising substances in the form of a black powder (platinum-black), which has the power of absorbing oxygen, and again imparting it to combustible substances, and thus causing their oxidation. In this way alcohol and pyroxylic spirit may be converted into acetic and formic acids, &c. The slight expansion which platinum undergoes when heated allows of its being sealed into glass without cracking by unequal contraction on cooling.

Platinum-black is simply platinum in a fine state of division; Berthelot, however, considers that it is really an oxide: it is readily obtained as follows:—1. A solution of platinic chloride, to which an excess of carbonate of sodium and a quantity of sugar have been added, is boiled until the precipitate which forms after a little time becomes perfectly black, and the supernatant liquid colourless; the black powder is then collected on a filter, washed, and dried at a gentle heat.

2. Platinic ammonium chloride, reduced to very fine powder, is moistened with strong sulphuric acid, and into the mixture a small piece of zinc is thrust; after a while it is reduced to a black powder; it is then washed, first with hydrochloric acid, then with pure water, and is, lastly, dried.

3. (*Zdrowkowitch.*) Platinum-black, in a highly active condition, can be obtained by adding 3 to 5 c.c. of solution of perchloride of platinum, drop by drop, to a boiling mixture of 15 c.c. of glycerin and 10 c.c. of solution of caustic potash of 1·08 sp. gr.

Platinum, in the state of platinum-black, possesses the property of condensing gases, more especially oxygen, of which it will absorb 800 times its volume, into its pores, and afterwards giving it out to various oxidisable substances. When placed in contact with a solution of formic acid it converts it, with effervescence, into carbonic acid; alcohol, dropped upon it, becomes changed by oxidation into acetic acid, the rise of temperature being often sufficient to cause inflammation; exposed to a red heat it shrinks in volume, assumes the appearance of spongy platinum, and, for the most part, loses these peculiarities. That prepared with zinc explodes, when heated, like gunpowder. The spongy platinum is obtained by igniting the ammonio-platinic chloride at a red heat.

Tests. If a platinum compound be heated on a carbonised match in a flame a grey spongy mass is obtained soluble only in aqua regia. The salts of platinum are recognised as follows:—Sulphuretted hydrogen throws down from neutral and acid solutions of the plantinic salts a blackish-brown precipitate, which is only formed after a time in the cold, but immediately on heating the liquid. Ammonium sulphide also gives a blackish-brown precipitate, which completely re-

dissolves in a large excess of the precipitant, provided the latter contains an excess of sulphur. Chloride of ammonium and chloride of potassium give yellow crystalline precipitates,' insoluble in acids, but soluble in excess of the precipitate upon the application of heat; these precipitates are decomposable by heat, with production of spongy platinum. Ammonia and potassium hydrate also give similar precipitates in solutions previously acidulated with hydrochloric acid.

Estim. This may be effected by throwing down the metal in the form of chloride of ammonium and platinum, which, after being washed on a filter with a little weak alcohol, to which a little of the precipitate has been added, and afterwards with alcohol alone, may be carefully dried at 212° F., and weighed. Or the precipitate may be ignited in a platinum crucible, and weighed in the state of spongy platinum. 198·25 gr. of the platinic and ammonium chlorides are equivalent to 98·75 gr. of metallic platinum.

Uses. Platinum is valuable for making crucibles, capsules, and other utensils or instruments intended to be exposed to a strong heat, or to the action of acids. In the form of basins, foil, wire, and crucibles, it is indispensable to the analytical chemist. Platinic chloride and the platinic and sodium chloride are much used in chemical analysis. Both of these are also used in medicines with the same intentions, and in the same doses, as the corresponding salts of gold. These compounds are poisonous. The antidotes and treatment are similar to those described under GOLD.

Concluding Remarks. Deville and Debray introduced a method of refining platinum which has already done much to extend the useful applications of the metal. The process consists in submitting the crude metal to the action of an intensely high temperature, obtained by the combustion of hydrogen (or coal-gas) with oxygen in a crucible of lime. By this means large quantities of platinum (50 lbs. or more) can be kept fused until the sulphur, phosphorus, arsenic, and osmium, generally occurring in crude platinum are oxidised and volatilised, and the iron and copper are oxidised and absorbed by the lime forming the crucible.

Platinic Chloride. PtCl₄. Syn. BICHLORIDE OF PLATINUM, CHLORIDE OF PLATINUM, PERCHLORIDE OF P.; PLATINI BICHLORIDUM (Ph. L.), PLATINI TETRACHLORIDUM, L. *Prep.* By dissolving scraps of platinum-foil in nitro-hydrochloric acid (4HCl,HNO₃), and evaporating the solution to dryness at a gentle heat.—*Prop., &c.* Reddish brown, deliquescent, and very soluble in both water and alcohol, yielding orange-coloured solutions. It combines with a variety of metallic chlorides to form 'double salts.' Used as a test in chemical analysis for browning gun-barrels, and as an alterative in secondary syphilis, &c.—*Dose*, ₁₆ to ¼ gr., dissolved in distilled water, or made into a pill with syrup and liquorice powder. Some persons prescribe much larger doses, but unsafely. Hoefer recommends an ointment made with it as an application to indolent ulcers. In doses of 5 gr. and upwards it acts as a violent caustic poison. This last salt is the 'chloride of platinum' of the shops, and the one used in the arts and medicine. It forms one of the tests included in the Appendix to the Ph. L.

Platinic Ammonium Chloride. PtCl₄,2NH₄Cl. *Syn.* AMMONIO-CHLORIDE OF PLATINUM, PLATINO-CHLORIDE OF AMMONIUM. *Prep.* A solution of chloride of ammonium is added to a strong solution of platinic chloride, and the precipitate washed with dilute alcohol.

Prop. Minute transparent, yellow, octahedral crystals, very feebly soluble in water, less so in dilute alcohol, and insoluble in acids; heat converts it into spongy platinum.

Platinic Potassium Chloride. PtCl₄,2KCl. *Syn.* PLATINO-CHLORIDE OF POTASSIUM, POTASSIO-CHLORIDE OF PLATINUM. *Prep.* A bright yellow crystalline precipitate, formed whenever solutions of the chlorides of platinum and of potassium are mixed; or a salt of potassium, acidulated with a little hydrochloric acid, is added to platinic chloride. In appearance, solubility, &c., it closely resembles ammonio-chloride of platinum.

Platinic Sodium Chloride. PtCl₄,2NaCl. *Syn.* CHLORIDE OF PLATINUM AND SODIUM, SODIO-CHLORIDE OF PLATINUM, PLATINO-BICHLORIDE OF SODIUM; PLATINI ET SODII CHLORIDUM, PLATINI SODIO-CHLORIDUM, &c., L. *Prep.* Platinic chloride, 17 parts; chloride of sodium, 6 parts; dissolve the two salts separately in water, q. s.; mix the solutions, and evaporate, that crystals may form. The crystals are large, transparent, and of a yellow-red colour.—*Dose*, ¹⁄₁₆ to ¼ gr.; in the same cases as the bichloride.

Platinic Oxide. PtO₂. *Syn.* BINOXIDE OF PLATINUM. *Prep.* 1. By exactly decomposing the platinic sulphate with nitrate of barium, and adding pure hydrate of sodium to the filtered solution, so as to precipitate only half the oxide (*Berzelius*). 2. By boiling platinic chloride with hydrate of sodium in considerable excess, and then adding acetic acid.

Prop., &c. As the hydrate, Pt(HO)₄, it is a bulky brownish powder; this, when gently heated, becomes the black anhydrous dioxide. It forms salts with the acids, and combines with some of the bases. The salts have a red or yellow colour, and a remarkable tendency to form double salts with the alkaline salts.

Obs. Both the oxides of platinum are reduced to the metallic state on ignition.

Platinous Oxide. PtO. *Syn.* OXIDE OF PLATINUM. *Prep.* 1. By heating to below redness the platinic chloride and digesting the residue with hydrate of potassium.

2. By carefully igniting the corresponding hydroxide.

Prop., &c. A dark grey powder, soluble in excess of alkali, and freely so in the acids, forming brown solutions of the platinous salts. These are distinguished from solutions of the platinic salts by not being precipitated by chloride of ammonium. Platinous oxalate, in fine copper-coloured needles, may be obtained by heating platinic oxide in a solution of oxalic acid.

Platinous Chloride. PtCl₂. *Syn.* PLATINUM DICHLORIDE. *Prep.* 1. By dissolving the metal in aqua regia and crystallising out the chloroplatinic acid, H₂PtCl₆, which forms: this heated to 300° C.

2. By heating spongy platinum in a current of dry chlorine to between 240° and 250° C. (*Schützenberger*).

Prop. A greenish-grey powder which is insoluble in water.

Platinum Gas. *Syn.* GAS-PLATINE; GILLARD'S GAS. In Paris this gas is employed by gold- and silversmiths and electro-platers because it gives rise to no sulphur product, and burns without giving off soot or smoke. It is free from smell. Steam is decomposed by being made to pass through a retort filled with red-hot charcoal. The hydrogen being free from the carbonic acid which is associated with it, by means of crystallised carbonate of soda, is burnt from an Argand burner provided with numerous small holes. The flame, which is not luminous in itself, is surrounded by a network of moderately fine platinum wire, which on becoming white-hot is luminous. It burns quite steadily, and its illuminating power is said to exceed slightly that of coal-gas.

Platinum, Spongy. *Prep.* 1. By heating ammonio-chloride of platinum to redness.

2. Crude bichloride of platinum and chloride of ammonium are separately dissolved in proof spirit, and the one solution is added to the other as long as a precipitate forms; this is collected, and, whilst still moist, formed into little balls or pieces, which are then dried, and gradually heated to redness.

Prop., &c. These have been noticed above. Small balls of spongy platinum are used for the hydrogen 'instantaneous-light' lamp (Döbereiner's lamp); but they are apt to absorb moisture from the atmosphere, and then lose their power of inflaming hydrogen until they are re-dried and heated.

PLEURISY. Inflammation of the pleura, or membrane covering the lungs. The symptoms of pleurisy are a sharp pain in the side, which is rendered more acute when a deep breath is taken; quick, short, difficult inspiration; cough; a quick pulse; and fever. Much pain is also experienced if the attempt be made to lie on the affected side.

Pleurisy sometimes accompanies pneumonia or inflammation of the substance of the lungs. If allowed to run on, the disease produces effusion of serum or of lymph into the cavity of the chest, in either case giving rise to adhesions, which cause embarrassment of breathing. On the contrary, it may terminate by resolution or complete recovery.

Pleurisy generally arises from exposure to the cold. A blow or a wound will also cause it, and a not uncommon origin is the splintered end of a broken rib. In every case the advice of the medical practitioner should be sought upon the first indications of the disease. A perfectly normal case of acute pleurisy will attain its maximum in about a week.

Treatment. Perfect rest, warmth, protection from chill, leeches applied to the affected side, followed by bandaging and, when the leech-bites will allow, strapping with plaster. James's powder, ipecacuanha, and liquor ammoniæ acetates are useful medicines. Alkaline effervescing drinks and *liquid* foods should be given. Solids and stimulants forbidden. Gentle blistering or painting with iodine over the affected part of the chest is often useful. The bowels should be kept gently acting by salines.

Epizootic Pleurisy in Horses. *Def.* An inflammation of the pleura and substance of the lungs, preceded and accompanied by a low typhoid or adynamic form of fever, which lasts from seven to fourteen days. It generally occurs but once in a season, but one attack does not render an animal exempt from a second or third (*Williams*).

This disease in an epizootic form raged in the north of England in 1861-2, and caused great mortality, especially among young horses, and those removed from pasture to stables.

Alternations of heat and cold in spring and early summer, and exposure of the animals, uncovered, to cold winds while waiting in carts and carriages, are common causes of pleurisy.

Symptoms. The animal is dull and stupid, off its food, is easily fatigued, and perspires freely on small provocation. The pulse is 60 to 80 per minute, and the temperature 103° to 104°; sometimes there is a cough, but often this does not appear in the first three or four days. The extremities are alternately hot and cold, the mucous membranes injected, the tongue foul, and the animal does not lie down, and has obvious pain and difficulty in breathing, indicating pleural inflammation with exudation; often there is pericarditis in addition.

Treatment. 1. Perfect rest on the first signs of illness. 2. Warm housing in a dry, light, well-ventilated loose box. 3. Protection from draught and cold. 4. Plenty of clothing to the body, and bandages on the legs.

Williams strongly condemns bleeding, purging, and counter-irritation, and advises plenty of cold water to drink, and warm or cold bran mashes, a boiled linseed mash every night, roots such as carrots, turnips, or potatoes, and a handful or two of the best hay. Two or three doses daily of spirits of nitrous ether in warm water may be given, and if the kidneys do not act half-ounce doses of nitrate of potash; 10-minim doses of Fleming's tincture of aconite in a ball three times a day are useful when the fever is high. Opium should be given as a tincture when the pain is great, with linseed oil to relieve the constipation caused by it. Warm fomentations applied to the sides for an hour three or four times a day give great relief. When the appetite is very bad give plenty of milk, or even eggs beaten in milk.

A similar treatment will prove useful in ordinary pleurisy. If the quantity of fluid in the chest be large relief must be given by tapping, followed by careful feeding with warm nourishing food.

PLEURISY ROOT. The root *Asclepias tuberosa*. A tincture of the root (1 in 10) is employed as a remedy for pleurisy and heart disease, acting as an expectorant and diuretic.—*Dose,* 5 to 40 minims.

PLEURO-PNEUMONIA CONTAGIOSA. *Def.* A contagious febrile disease peculiar to horned cattle, due to a contagion which gains access to the system by the lungs, and which, after an incubative period of from two to three weeks to as many months, induces complications in the form of extensive exudations within the substance of the lungs and on the surfaces of the pleura, finally resulting in consolidation of some portions of the lungs, occlusion of the tubes, embolism of the vessels, and generally adhesion of the pleural surfaces. In some cases there is extensive and rapid destruction of lung tissue, with death by suffocation; but most commonly the disease is of a lingering character, symptoms of great prostration manifesting themselves, with blood-poisoning from absorption of the degraded pulmonary exudates, and death from marasmus and apnœa (*Williams*).

Synonyms. Lung disease, pleura, new disease, new delight (Yorkshire), pulmonary murrain, epizootic pleuro-pneumonia, &c. Lungenseuche, Ger.; peripneumonie contagieuse, Fr.

Under the Contagious Diseases (Animals) Act, 1878, all animals suffering from pleuro-pneumonia are directed to be slaughtered. And by the Pleuro-pneumonia Slaughter Order of 1888, all cattle being or having been in the same field, shed, or other place, or in the same herd, or otherwise in contact, with cattle affected by pleuro-pneumonia, are to be slaughtered within ten days after the fact of their having been so in contact has been ascertained, or within such further period as the Privy Council may in any case direct.

PLUGGING. The introduction of a mass of lint, sponge, or other suitable material, into a wound or cavity, with the intention of arresting hæmorrhage. It is now seldom adopted, except in cases of bleeding from the nose, and that only after more approved methods have failed.

PLUM. A name applied to several varieties of the *Prunus domesticus*, Linn., or wild plum. Among the cultivated varieties, the damson, greengage, French plum, magnum bonum or Mogul p., mirabelle p., Orleans p., and prune, are those best known. Grocers' 'plums' are raisins, or dried grapes.

In the table on the next page will be found the composition of the principal varieties of plum.

PLUMBAGO. *Syn.* GRAPHITE, BLACK-LEAD. One of the native forms of carbon. The black powder known by the name of ' black-lead ' has no relation to lead, but probably received this name because pencils made of it caused a mark on paper resembling that made by lead, only blacker. This similarity, together with its metallic appearance, also gave it the name of plumbago, from the Latin *plumbum*, meaning lead. The name graphite is derived from a Greek source, and bears a reference to its use as a writing material. It is really a crystalline form of carbon found in the oldest sedimentary rocks. It is sometimes found associated with iron in its ores, and in some districts is found in the form of veins in the rocks. Its specific gravity varies from 2·15 to 2·35. It contains from 95 to 100% of pure carbon, has a metallic lustre, and conducts electricity nearly as well as the metals. It was formerly regarded as a carbide of iron, but the iron generally found is now known to be merely in a state of mixture. There are two distinct varieties of graphite—crystallised or foliated graphite, obtained chiefly from Ceylon; and amorphous graphite (the ordinary plumbago or black-lead), which is largely imported to this country from Germany. The Borrowdale mine

	Mirabelle, common yellow.	Greengage.		Black-blue, middle-sized Plums.	Dark black-red Plums.	Mussel Plums.	
		Yellow-green, middle size.	Large green, very sweet.			Common.	Italian, very sweet.
Soluble matter :							
Sugar	3·584	2·960	3·405	1·996	2·252	5·793	6·730
Free acid, reduced to equivalent in malic acid . . .	0·582	0·960	0·870	1·270	1·331	0·952	0·841
Albuminous substances . .	0·197	0·477	0·401	0·400	0·426	0·785	0·832
Pectous substances, &c. . .	5·772	10·475	11·074	2·313	5·851	3·646	4·105
Ash	0·570	0·318	0·398	0·496	0·553	0·734	0·590
Insoluble matter :							
Seeds	5·780	3·250	2·852	4·190	3·329	3·540	3·124
Skins, &c.	0·179	0·680	1·035	}0·509	1·020	{1·990	0·972
Pectose	1·080	0·010	0·245			0·630	1·534
[*Ash from insoluble matter included in weights given*] .	[0·082]	[0·089]	[0·037]	[0·041]	[0·063]	[0·094]	[0·066]
Water	82·256	80·841	79·720	88·751	85·238	81·930	81·272
	100·00	99·971	100·00	99·925	100·00	100·00	100·00

in Cumberland, from which the finest black-lead was formerly derived, is now nearly exhausted. The foliated graphite of Ceylon and other parts is the principal material employed for making plumbago crucibles and other fire-resisting goods. The amorphous graphite is used for making black-lead pencils, polishing powder for stoves and grates ('lustre,' 'servants' friend,' &c.), and to diminish friction in heavy machinery (anti-friction powder). Its powder is also used to give conducting surfaces to articles on which it is desired to deposit copper by the electrotype. In medicine plumbago has been used with apparent advantage in herpes and several chronic skin diseases as an ointment made with four times its weight of lard; and internally, in the form of pills.

Purification. For medical and chemical use graphite may be treated as follows :

1. Heat it to redness with caustic potash in a covered crucible, then wash it well with water, boil it in nitric acid and in aqua regia, again wash it with water, dry it, and expose it at a white heat to a stream of dry chlorine gas; lastly, wash it with water, and again heat it to dull redness.

2. Pure native plumbago, 1 lb., is boiled in water for one hour, then drained, and digested for twenty-four hours in a mixture of water, 8 oz.; nitric acid and hydrochloric acid, of each, 2 oz.; it is, lastly, well washed with water, and dried.

3. (*Brodie's Process.*) This is only applicable to the hard varieties of graphite, as that of Ceylon. It consists in introducing coarsely powdered graphite, previously mixed with 1–14th of its weight of chlorate of potash, into 2 parts of concentrated sulphuric acid, which is heated in a water-bath until the evolution of acid fumes ceases. The acid is then removed by water, and the graphite dried. Thus prepared, this substance, when heated to a temperature approaching a red

heat, swells up to a voluminous mass of finely divided graphite. This powder, which is quite free from grit, may be afterwards consolidated by pressure, and used for making pencils or other purposes.

This material is of great use to the electrotyper, since it enables him to coat a non-conducting surface of a mould with a conducting substance capable of reproducing the finest lines impressed thereon. For this purpose the very best graphite should be employed.

That which rubs into a very fine powder of a dead-black appearance when undisturbed, but having a metallic lustre when rubbed or brushed on a surface, is the best. Coarse graphite is useless, however much it may be lauded by the vendor as being "pure as it comes from the mines." Much of this native graphite is too impure to be used for black-leading moulds.

"Coarse impure graphite may be purified by heating the powder with sulphuric acid and potassium chlorate; a compound is thus obtained which, on being strongly heated, decomposes, leaving pure graphite in a bulky, finely divided powder" (*Roscoe*). Electrotyper's graphite may have its conducting power improved by mixing with it some tin or copper-bronze powder. Mr Watt gives the following recipe for improving the conductivity of plumbago :—"Dissolve 1 part of chloride of gold in 100 parts of sulphuric ether; this is then to be mixed with 50 parts of plumbago, and the mixture is exposed to sunlight, being frequently stirred until quite dry."

Black-leading or Plumbagoing. The process of applying plumbago or graphite to render their surfaces conductors of electricity. Small moulds of coins and medallions are black-leaded by brushing in the fine plumbago dust with a sable or camel-hair brush or pencil. Larger moulds require larger brushes, which should always be soft; whilst those of printing electrotypes are black-leaded by machinery, the mould being fixed to a travelling carriage and caused to move to and fro under a vibrating brush. Every part of the mould must be coated with the conducting material, and the coat must be nicely polished to produce good results. Some electrotypists, it should be said, dispense with the dry black-leading process and adopt Knight's wet process. By this method the mould is coated with a thin wash of plumbago in water squirted on to it from a rose nozzle.

PLUM'RIC ACID. Binoxide of lead occasionally receives this name on account of its combining with some of the bases to form compounds which have been called plumbates.

PLUMBUM COR'NEUM. See LEAD, CHLORIDE OF.

PLUMOSE AL'UM. The old name of the silky amianthine crystals of the double sulphate of aluminium and iron occasionally found on alum slate. Asbestos has also been so called.

PLUNKET'S CANCER REMEDY. See CAUSTIC, PLUNKET'S.

PLUSIA GAMMA, Linn. [From πλούσιος, and the Greek letter γ (*gamma*), which the wing-markings resemble in shape.] THE SILVER Y MOTH. This moth was very abundant in several districts of Kent in 1881, and in other parts of the country it was noticed to be unusually plentiful in this same year, as well as in 1882. In Scotland, also, it was remarked that it was much more numerous than usual, and did much mischief to swede and turnip plants. Irish turnip-fields do not escape, as several attacks have been reported in Ireland during the last eight years. The injury caused is not by any means so extensive or so serious as that of the Diamond Back turnip moth, but in some instances farmers have estimated their losses from it at from £4 to £6 per acre. An attack was reported from the neighbourhood of Derby in 1879, where swede and white turnip plants were infested somewhat badly, and others in the same season from Gosport, Andover, and Reigate.

The caterpillars are universal feeders, being found upon corn plants, clover plants, trees, weeds, swede, turnip, beet, and mangel-wurzel plants.

Stephens says that this is by far the most common species of this genus, frequenting every hedge and field where flowers abound ('Illustrations of British Entomology,' Haustellata, by J. F. Stephens). Sometimes in fields of clover in blossom one may see hundreds of these moths rising at every step that is taken, attracted evidently by the honey in the flowers, for the extraction of which the insect has a long tongue. Mr Whitehead says he has seen swarms upon sainfoin plants when in flower, as well as upon vetches and lucerne. It appeared, from observations made at this time, that the moths did not lay eggs upon these plants, for they were not noticed afterwards to have any caterpillars upon them during the summer, but plants of swedes and rape not far off were injured to some extent by the caterpillars of this moth.

A large quantity of moths were especially noticed swarming upon the flowers of the 'second cut' of clover plants towards the beginning of August in 1879, which was a very wet season.

These were evidently of the second, or even third generation, and had probably come from the turnip and rape fields hard by, or elsewhere, to feed upon the honey in the flowers.

The *Plusia gamma* is very common throughout the Continent. Both French and German writers speak of it as abundant and destructive to crops of most kinds. Nördlinger says it is common in Europe, and swarms from spring to autumn in abundance. Taschenberg and Köllar also treat of it as occasioning much loss to cultivators, especially in respect of sugar beet.

Life History. The silver Y moth belongs to the group *Noctuidæ* of the Nat. Ord. LEPIDOPTERA, and the genus *Plusia.* The moth is nearly ¾ of an inch in length of body, with a wing expanse of from 1¼ inches to 1¾ inches. The head and thorax and upper part of the body are dark grey. The fore-wings are silvery grey, with a tinge of purple in certain lights, and darkish markings and a brilliant gloss. In their centres there is a metallic spot shaped like the Greek leter γ, or the English Y. The stigmata, or spots, on the fore-wings are lustrous. The hind or posterior wings are pale ash-coloured, with brown marks. Although it is classified among the *Noctuidæ* it flies at all times of the day.

In the spring, between the 27th of April and the 7th of May according to the season, the female lays eggs singly (Curtis, in 'Farm Insects,' says that one female Y moth might become the progenitor of 16,000,000 caterpillars in the space of twelve months, viz. from the spring of one year to the following spring); but abundantly, fixing them upon the under side of the leaves of various plants. The egg, seen microscopically, is very beautiful, having, as Curtis says, a curiously sculptured shape. From the eggs the caterpillars proceed in ten days. The caterpillar is green, with pale yellow or white lines down its back, and a more dark yellow streak along the sides, having somewhat sparse hairs upon the body. It has only twelve legs, and moves in a similar manner to the 'Loopers,' *Geometridæ*, only that its 'loop' is not so pronounced. Before changing it spins a cocoon, in which the dark chrysalis is ensconced under the leaves of plants. This insect passes the winter in this caterpillar form under leaves, roots of plants, grass, and rubbish.

Prevention. Sainfoin leys and clover leys kept down for more than one season are without doubt harbours of refuge for the caterpillars, as they are for many other insects. Close feeding with sheep folded on the land is very desirable where these moths have been noticed in the summer and autumn, or where swedes, turnips, or rape have been attacked near. Outsides of fields must be carefully cleaned, and hedges also cleared out at the bottoms. Strips of grass land must not be allowed to skirt fields. Paring and burning the outsides of fields and ditches which divide fields is a good practice for the purpose of destroying insects as well as weeds.

Remedies. Soot broadcasted in the early morning upon infested plants may be of some benefit; also guano or lime; but the caterpillars keep under the leaves mainly, so that these applications do not in some cases prove of full advantage.

Dialodging the caterpillars with bunches of birch, furze, or green broom fastened to the horse-hoes is an excellent remedy, with another horse-hoe following to bury or kill them.

Natural Enemies. Birds, as rooks, starlings, peewits, thrushes, eagerly eat these caterpillars, and should be encouraged ('Reports on Insects Injurious to Crops,' by Chas. Whitehead, Esq., F.Z.S.).

PNEUMONIA. Inflammation of the substance of the lungs. When the inflammation extends to the pleura, or covering of the lungs, the disease is distinguished as PLEURO-PNEUMONIA. By most pathologists pneumonia is described under the three general heads of—(1) Croupous pneumonia, (2) catarrhal pneumonia, (3) chronic pneumonia, each of which has, by some medical writers, been subdivided into other forms and varieties.

1. ACUTE CROUPOUS PNEUMONIA. This first description of pneumonia is most common amongst persons of from twenty to thirty years of age, although no age escapes it, and it is generally very severe in character when it attacks the very young or old. It prevails more amongst men than women, since the former, from their more frequent exposure to the weather and to changes of temperature, run greater risk of being overtaken by a very fertile cause of croupal pneumonia, viz. a sudden chill when the body is unusually heated.

It frequently seizes those suffering from chronic or acute disorders, as well as those who are intemperate and drunken. It often assails patients suffering from contagious and acute maladies, such as measles, smallpox, pyæmia, puerperal fever, typhus, and, as appears from the accounts of the recent outbreak of Astrakan plague, in that disease also. It likewise frequently prevails amongst the poor and badly fed living in the overcrowded quarters of large towns and cities.

The following are the principal symptoms of acute croupous pneumonia, given by Dr Roberts ('Handbook of the Theory and Practice of Medicine,' by F. J. Roberts, M.D., &c.; Lewis, 1878):

"In some cases there are premonitory signs of general indisposition for a short time. In primary, or unmixed pneumonia, the attack sets in usually very suddenly, the invasion being attended with a *single, severe, more or less prolonged rigor.* There may be great prostration with fever; vomiting or nervous symptoms, viz. headache, delirium, restless stupor, or, in children, convulsions. The special symptoms are *local* and *general.*

"*Local Symptoms.* Pain in the side is usually present, commonly stabbing or piercing, increased by a deep breath. Difficulty of breathing. Cough also commences speedily; it does not come on in violent paroxysms, but is short and hacking and difficult to repress. Soon expectoration occurs, the expectorated matter presenting peculiar characters. It is scarcely at all frothy, but extremely viscid and adhesive, and the vessel which contains it may often be overturned without its escaping. The expectorated matter has a rusty colour or presents various

tints of red, from admixture of blood, and as the case progresses, changes of colour are observed through shades of yellow, until finally they become merely like the expectoration of bronchitis. In some cases of croupal pneumonia pain and other symptoms are sometimes very slight or absent, and the expectoration may be merely like that in bronchitis, absent, or in low cases present the appearance of a dark, offensive, thin fluid, resembling liquorice or prune juice.

"*General Symptoms.* These may be summed up generally as severe fever with great depression and prostration. The skin is hot, dry, and burning. The temperature rises with great rapidity to 102°, 103°, 105°, or sometimes higher. It has been known to reach 107° in cases which recovered, and in fatal cases it has attained to 109·4°. In a large number of instances it does not exceed 104°. There is usually considerable flushing of the cheeks. The pulse ranges generally from 90 to 120, or may be much above this."

In the majority of cases this variety of pneumonia has a favourable termination, but, however slight the form in which it shows itself, or the mildness of its attack, the properly qualified practitioner should be called in to combat it. We have described the nature and cause of the disease, and given the course to be followed in treating it, for the benefit only of the emigrant and others similarly situated. The above comments are meant to apply to the other descriptions of pneumonia, which will be adverted to in the course of the present article.

Treatment to be followed in croupous pneumonia. Bleeding was formerly had recourse to, but this treatment has either been abandoned of late years, or very rarely practised, the only case in which its moderate employment is recommended being that in which the patient is threatened with death from partial privation or suspension of breath.

Leeches may be applied to the spot in pain, and a large blister near it, but it is preferable to first try the effect of hot fomentations and poultices containing laudanum; or turpentine sprinkled on a warm damp flannel may be tried. A third of a grain of tartarised antimony, with a few drops of laudanum, or a third of a grain of hydrochlorate of morphia may be given every four hours.

"In all *low* forms of the disease the only chance is in *free stimulation.* At the same time full doses of carbonate of ammonia, with bark, spirits of chloroform, ether, camphor, and such remedies must be administered. In some cases quinine with iron is useful" (*Dr Roberts*). The best diet consists of milk and beef-tea. The patient, it is needless to say, should be kept in bed, and the temperature of his chamber should be maintained at about 60° F. It is also most essential that the room should be thoroughly ventilated, and all the expectorated matter, stools, &c., thoroughly disinfected before removal.

2. CATARRHAL PNEUMONIA. The acute variety of this form of pneumonia is that which principally attacks infants and children, and frequently complicates diphtheria, hooping-cough,

measles, and influenza; although it may occasionally occur when not associated with these diseases.

In the other variety—chronic catarrhal pneumonia—the greater number of cases arise from bronchitis. Many authorities look upon the last variety of pneumonia as the cause of a great proportion of the cases of pulmonary phthisis.

Symptoms. These differ, in the great majority of cases, from croupal pneumonia, in not being preceded by rigors. There is always fever and a rise of bodily temperature from 103° to 105°. There is often copious perspiration and increased pulse. As the disease progresses the breathing becomes more difficult and rapid, the cough changes its character, and "becomes short, harsh, hacking, and painful, the child endeavouring to repress it, and having an expression of pain or crying and diminished expectoration" (*Dr Roberts*).

The treatment of this form of pneumonia consists in keeping up the strength of the patient by means of good nourishing food, and stimulants judiciously administered. Ammonia and senega should be given if the sufferer is very weak. In ordinary cases ipecacuanha wine will be found useful. Poultices of linseed or mustard to the chest are also prescribed. During convalescence the patient requires careful watching; his diet should be generous, and should include wine, cod-liver oil, quinine, and iron, or other tonics.

3. CHRONIC PNEUMONIA. This disease, in which the substance of the lung is in a more or less abnormal or altered condition, is mostly the result of some previous pulmonary affection. It frequently follows successive attacks of the catarrhal variety of pneumonia and the bronchial irritation arising from the inhalation of small particles of dust given off by substances employed in certain occupations or manufactures, such as coal, steel, granite, &c.

The symptoms are pains in the side, cough, sometimes occurring in severe paroxysms, shortness of breathing, the patient meantime gradually becoming thinner and weaker. Sometimes night sweats occur, but generally there is little or no fever.

The best treatment is nourishing diet, combined with tonics and cod-liver oil.

Of late years the doctrine of the contagious nature of some forms of acute pneumonia (whether complicated with pleurisy or not) seems to have been gaining ground amongst medical practitioners. The well-known fact that the pleuro-pneumonia of cattle is propagated by contagion, if it does not prove this contention, is at any rate "worthy," as Dr Parkes remarks, "of all attention."

POACHING. Amongst cooks, a peculiar method of cooking small articles by a slight boiling or stewing process.

POACHED EGGS are prepared by breaking them into a small saucepan or stewpan containing about ½ pint of boiling water, to which a teaspoonful of common salt, and, occasionally, a little vinegar, are added, and gently simmering them for three or four minutes, or until sufficiently firm to bear removal with a spoon or 'slice.' Another me-

thod is to employ melted butter instead of water, and to dress them either with or without stirring.

Poached eggs are commonly served on toast, or with fried ham or bacon, with spice or vegetable seasoning at will. They form an excellent breakfast, or 'make-shift dinner.'

PODOPH'YLLIN. *Syn.* RESIN OF PODOPHYLLUM; RESINA PODOPHYLLI (B. P.), L. Obtained from the root of the *Podophyllum peltatum*, Linn., or may-apple.

Prep. 1. The alcoholic extract of may-apple is digested in cold ether to remove fatty matter, and is then dissolved in rectified spirit; the solution is decoloured with a little animal charcoal, and filtered; it is, lastly, allowed to evaporate spontaneously.

2. (B. P.) Podophyllum, in coarse powder, 1 part; rectified spirit, 3 parts, or a sufficiency; distilled water, a sufficiency; exhaust the podophyllum by percolation with the spirit; distil over the spirit; slowly pour the liquid remaining after distillation of the tincture into three times its volume of water, constantly stirring; let it stand 24 hours; collect the resin which falls, wash on a filter with distilled water, and dry in a stove. Cholagogue purgative; used as a substitute for calomel.—*Dose*, ⅙ to ¼ gr., or even 2 gr. It is best to begin with ½ gr. (*Squire*).

Prop., &c. An amorphous powder, varying in colour from a pale yellow to a deep orange, soluble in alcohol, and slightly soluble in water. It is a safe and certain cathartic, superior in activity to resin of jalap.—*Dose*, ¼ to 1 gr. See EXTRACT OF MAY-APPLE.

PODOPHYLLUM ROOT. *Syn.* PODOPHYLLI RADIX (B. P.), L. The dried rhizome of the *Podophyllum peltatum*; imported from North America. Active and certain cathartic.—*Dose*, 10 to 20 gr.

PODOSPHÆRA CASTAGNEI. THE HOP MILDEW, or MOULD. This disorder, termed the 'white blight' by hop planters, is due to a fungus. Considerable losses were occasioned by it in the last season, 1890, in certain hop-gardens. Patches of mildew appeared upon the leaves first; directly the burr—the incipient cone—showed itself it was affected and prevented from further development. Sulphur was used in enormous quantities, but on account of the wet, cold weather of July and August its operation was ineffectual to a great extent. Sulphur acts as a preventive of mildew, and as a remedy against it, by the disengagement of sulphurous acid gas, which is prejudicial to fungoid growths. Heat and some moisture are essential to the formation of this gas. In a dull, cold time there are but few fumes given off. In hot weather there is a strongly perceptible evolution, and even in ordinary summer temperature sulphurous acid gas can be readily smelt on passing through or by a recently sulphured hop-garden. It is well known that sulphur fumes are effectual in checking parasitic fungi in greenhouses and hothouses. The fumes then are generated by heat and moisture, and are confined; while in the open air they are slowly evolved and cannot be concentrated upon the fungus, so that it is not strange in these circumstances that sulphur often fails in its operation. The best possible conditions for its proper working are burning sun heat with slight rainfall. These conditions were entirely wanting in the last season. It is undoubtedly from the great uncertainty of the action of sulphur, even in a climate of greater heat than that of England, that the French wine producers have discontinued the use of sulphur for the vine mildew, and have had recourse to sulphate of copper washes, whose efficacy in keeping the fungus in check is very great. I have urged hop planters to use these remedies. A few experiments were made in July, 1890, with the *Bouillie bordelaise*, but they were not carried out thoroughly, so that no reliable records of the results have been obtained. It has been proved beyond doubt that sulphate of copper solutions, properly mixed, arrest or destroy parasitic fungi ; it only requires a little trouble and skill on the part of the hop planters to put it on the hop plants evenly and at right periods. The hop-washing engines are perfectly fitted to distribute sulphate of copper washes over hop plants, with the substitution of copper or lead for iron tanks.

The life history of this fungus has been described in previous reports, and it is not deemed necessary to repeat it here ('Reports on Fungi,' by Chas. Whitehead, Esq., F.Z.S.).

POISONS. *Syn.* POISONS, Fr.; GIFTE, Ger. Any substance may be said to be a poison which possesses an inherent deleterious property of such a nature as to render it capable of destroying life if introduced into the animal economy in any way soever. Those substances which act in a purely mechanical manner are excluded from this definition.

Poisons are now usually classed under three heads : 1, *corrosive poisons ;* 2, *irritant poisons ;* and 3, *neurotic poisons.*

Corrosive Poisons. The mineral acids—oxalic acid—the caustic alkalies and corrosive salts, such as potassium bisulphate and carbonate, the chlorides of zinc, tin, antimony, and mercury, and nitrate of silver.

Irritant Poisons. Poisons which cause inflammation of the parts to which they are applied, generally the alimentary tract. The irritant action is almost always combined with some more or less well-marked effect on the nervous system. The most important division of irritant poisons is into *metallic* and *vegetable* irritants, *animal* irritants being grouped with the latter. Arsenic is the most important metallic irritant, and the salts of antimony, zinc, and other metals. Elaterium, essential oils, and gamboge are examples of vegetable irritants, and cantharides of an animal irritant.

Neurotic Poisons. Poisons whose most important effect is produced upon the nervous system—morphia, chloral hydrate, hyoscyamus, digitalis, strychnia, prussic acid, nitro-benzol, phenol, alcohol, aconite, belladonna, and many others.

Treatment of Poisoning. Under the heading ANTIDOTE and under each separate drug some account will be found of the special symptoms produced in each case, and the mode of treatment. When corrosive poisons have been taken and there is great damage to tissues, the use of the stomach-pump is undesirable by reason of the risk of perforating the gullet or stomach.

In all other cases it can at least do no harm. Acids may be neutralised by lime-water or saccharated lime-water, or by frequently repeated doses of chalk or whiting and water, or the alkaline carbonates. Alkalies are similarly to be neutralised by dilute acids. The effects of corrosive and irritant poisons must be counteracted afterwards by the administration of oil and demulcents ; opiates may also be required to allay the intense pain, and this is particularly the case with carbolic acid. In poisoning by prussic acid artificial respiration is our only help, and this should be persisted in till all hope is gone. In alkaloidal poisoning emetics should be followed by tannin, tincture of galls, strong tea, and coffee, in the hope of reducing the alkaloids to an insoluble form.

POLARISATION (of Light). A change produced upon light by the action of certain media and surfaces, by which it ceases to present the ordinary phenomena of reflection and transmission, and, on the undulatory theory, instead of traversing all planes the beam of light is more or less perfectly restricted to one. Light thus affected is said to be 'plane polarised.' Instruments or apparatus employed to effect this change are called 'polariscopes.' Although the polarisation of light is frequently employed as a means of chemical investigation, and is of the utmost interest to the philosophical inquirer, its consideration scarcely comes within the province of this work. See 'Watt's Dict. of Chemistry,' 'Ganot's Physics,' &c.

PO-LIO-YO. A variety of oil of peppermint, prepared in China. It is sold as Japanese drops, in small bottles, with a label in Chinese. The oil is rich in menthol, and used by the Chinese and Japanese to paint on painful parts, especially to relieve toothache and neuralgia.

POLISH. Various substances, differing widely from each other, are popularly known under this name. See POWDERS, VARNISH, &c., and *below.*

Polish, French. See FRENCH POLISH (*below*).

Polish, French Reviver. *Prep.* 1. Linseed oil, ⅓ pint ; pale lac varnish and wood naphtha, of each, ⅓ pint ; well shaken together, and again every time before use.

2. Methylated rectified spirit, 3 pints ; linseed oil and French polish, of each, 1 pint ; as the last.

3. Linseed oil (pale), 1 quart ; strong distilled vinegar, ⅓ pint ; spirit of turpentine, ⅓ pint ; muriatic acid, 1 oz.

Furniture Cream. *Prep.* 1. Pearlash, 2 oz. ; soft soap, 4 oz. ; beeswax, 1 lb. ; water, 1 gall. ; boil until the whole is united and forms a creamy liquid when cold.

2. Beeswax, ⅓ lb. ; good yellow soap, ¼ lb. ; water, 5 pints ; boil to a proper consistence with constant agitation, then add of boiled oil and spirit of turpentine, of each, ⅓ pint. For use, the above are diluted with water, spread upon the surface with a painter's brush, and then polished off with a hard brush, cloth, or leather.

3. Boiled oil (pale), ⅓ pint ; beeswax, 1½ oz. ; mixed by heat. Applied by a 'rubber,' and at once polished off.

4. (For wooden furniture.) White wax, 8

parts; resin, 2 parts; true Venice turpentine, ½ pint; melt at a gentle heat. The warm mass, completely melted, is poured into a stone jar, agitated, and 6 parts of rectified oil of turpentine added thereto. After twenty-four hours the mass, having the consistency of soft butter, is ready for use. Before using the paste the furniture should be washed with soap and water, and then well dried ('Dingler's Journal').

Furniture Oil. See OILS, MIXED.

Furniture Paste. *Prep.* 1. Oil of turpentine, 1 pint alkanet root, ¼ oz.; digest until sufficiently coloured, then add of beeswax (scraped small), 4 oz.; put the vessel into hot water, and stir until the mixture is complete, then put it into pots. If wanted pale, the alkanet root should be omitted.

2. (White.) White wax, 1 lb.; solution of potassa, ½ gall.; boil to a proper consistence.

Furniture Polish. Linseed oil, 10 oz.; turpentine, 3 oz.; vinegar, 2 oz.; methylated spirit, 2 oz.; hydrochloric acid, ½ oz. Mix the oils with the hydrochloric acid, then add the vinegar and spirit. This should be applied sparingly, and the furniture well rubbed afterwards. It gives an excellent polish, and leaves no finger-marks if well rubbed.

German Furniture Polish. This semi-translucent white paste, known on the Continent as 'Moebel-politur-Pomade,' has been submitted to analysis, and the result shows that it is composed of 33 per cent. of hard paraffin, and 66 per cent. of turpentine. We are informed that it can be made by taking 8 oz. of hard paraffin to 1 pint of turpentine. The paraffin must be carefully melted, and the turpentine gradually added, great precaution being taken lest the mixture inflame. It is stirred until it assumes a creamy consistency, and is then poured into appropriate boxes.

Polish, Harness. See BLACKING, HARNESS.

Polish, Leather. See BLACKING.

Polish for Marble. Mr. W. C. Durkee (Boston, U.S.A.) gives the following formula for a marble dressing or polish :—Pure beeswax, 10 parts; Japan gold size, 2 parts; spirits of turpentine, 88 parts. The mixture is of creamy consistence, and should be applied in small quantities, with the aid of a piece of white flannel. If it is desired for use upon white marble, white wax may be substituted. The same preparation can be used with advantage on woodwork. The Japan size prevents the stickiness which exists when wax alone is used.

Polish for Shoes, Liquid. Lampblack, 1 dr.; oil of turpentine, 4 dr.; alcohol (methylated spirit), 12 oz.; shellac, 1½ oz.; white turpentine, 5 dr.; sandarac, 2 dr. Make a solution by digesting the mixture in a close vessel at a gentle heat and strain.

French Polish. Shellac, 1½ oz.; gum benzoin, ½ oz.; gum sandarac, ¼ oz.; methylated spirits, ½ pint. Dissolve.

POL'LARD. See FLOUR.

POL'YCHREST. *Syn.* POLYCHRESTUS, L. A term formerly applied to several medicines on account of the numerous virtues they were supposed to possess. Sal polychrestus is the old name for sulphate of potassa.

POLYCHRO'ITE. The name formerly given to the colouring matter of saffron, from the variety of colours which it exhibits with different reagents. Its alcoholic and aqueous solutions are of a golden yellow; nitric acid turns it green; sulphuric acid, first blue, and then lilac.

POLYDESMUS COMPLANATUS, Linn. THE THOUSAND LEGS (*Millipedes*). Though these are not insects in the strict scientific meaning of the term, as having no wings, nor undergoing any transformation, and not having bodies divided or cut, they must be described here as having the habits of insects and habits injurious to cultivation. Linnæus classified these among the order *Aptera*, and Mr Murray in his 'Handbook of Economic Entomology' follows this classification, and treats them as insects. These thousand legs are utterly distinct from, first, *Wireworms*, with which they are sometimes confounded; second, *Centipedes* (*Scolopendridæ*), of which there are many species not coming within the scope of this work, as living mainly upon animal substances.

The thousand legs eat wheat, oats, and barley plants, but they are not nearly so destructive as wireworms and several other root-eating insects. They do much harm also to bean and pea crops, and are most injurious to French beans and broad beans in market gardens, and market garden farms in Essex, Bedfordshire, Surrey, and Kent. The species named above, *Polydesmus complanatus*, is perhaps the most troublesome to farm crops generally, but all the species are more or less injurious to vegetation.

Life History. The female lays eggs in the spring in damp places under stones and decaying wood and leaf rubbish. From these tiny worms come, which do not attain their full growth and power of reproduction until two years. They have only three pairs of legs at first. In course of time these are multiplied even to as many as one hundred pairs. They live for five years, and always under ground. This species is about nine lines, or three quarters of an inch in length. Other species, as *Julus guttatus*, are an inch long.

Prevention. As the Polydesmi, as well as the Julidæ, like dampness and moisture, wet land, and boggy, marshy places should be drained. A good dressing of hot lime should be ploughed into land infested with them early in the spring to destroy their eggs. Rubbish and decaying matter must not be allowed to lie about in fields.

Remedies. When corn is attacked dressings of soot, lime, nitrate of soda, and guano may be used with some advantage, especially if soaking showers follow the applications. In cases where peas and beans are suffering from their onslaughts horse and hand hoeing should follow dressings of these manures. In market gardens pieces of swedes, mangels, or vegetable marrows, if procurable, should be put between the drills to attract the millipedes from the growing crops, as they burrow into these like wireworms, and can be taken from them and destroyed ('Reports on Insects Injurious to Crops,' by Charles Whitehead, Esq., F.Z.S.).

POM'ACE. See CIDER.

POMA'TUM. *Syn.* POMMADE, Fr. This term was originally applied to a fragrant ointment prepared with lard and apples; but is now wholly restricted, in this country, to solid greasy substances used in dressing the hair. The pomatums of French pharmacy (POMMADES, GRAISSES MÉDICAMENTEUSES—P. Cod.; LIPAROLÉS—Guibourt; LIPAROLÉS and LIPAROIDÉS—Béral; STEAROLÉS—Chéreau) are soft ointments, having a basis of lard or fat, without resinous matter. See OINTMENT and POMMADE.

POMEGRAN'ATE. The fruit of *Punica granatum*, Linn., cultivated from early antiquity for its fruit; naturalised in the Mediterranean region, but a native of Western Asia, south of the Caspian, and not of Carthage, as its name would denote (*Malum punicum*). It was known to the Hebrews under the name *Rimmon*, and is mentioned in Deuteronomy as a product of Palestine. The root is an excellent vermifuge; the bark gives the colour to yellow morocco leather, which is tanned with it. The dried rind of the fruit is valued as a remedy in India for diarrhœa and dysentery. Walking-sticks are made from the stems of young plants imported from Algeria. Fruit (POMEGRANATE; GRANATA, MALA PUNICA) is cooling and astringent; fruit-rind (POMEGRANATE PEEL; MALACORIUM, CORTEX GRANATI; GRANATUM—Ph. L.) and root-bark (GRANATI RADIX—B. P., Ph. L., E., and D.) are powerfully astringent, detersive, and anthelmintic; the last more particularly so. The double flowers of the wild tree (BALAUSTINES; BALAUSTIÆ), as well as those of the cultivated one (CYTINI), are tonic and astringent.—*Dose*, 15 to 20 gr. of the root-bark, repeated every 30 to 40 minutes, until four doses have been taken, followed by castor oil; in tape-worm. As an astringent, all the parts described are commonly given under the form of decoction.

POMMADE. [Fr.] The term applied by Continental perfumers to any soft fragrant ointment (POMATUM).

In the preparation of pommades one of the first objects of consideration is to obtain their fatty basis in as fresh and pure a state as possible. Lard, beef, and mutton suet, beef marrow, veal fat, and bear's fat are the substances commonly employed for this purpose, either singly or in mixtures of two or more of them. The fat, carefully selected from a young and healthy animal, after being separated from extraneous skin and fibre, is pounded in a marble mortar, in the cold, until all the membranes are completely torn asunder. It is next placed in a covered porcelain or polished metal pan, and submitted to the heat of a water-bath, which is continued until its fatty portion has liquefied, and the albuminous and aqueous matter, and other foreign substances, have completely separated and subsided. The liquid fat is then carefully skimmed, and at once passed through a clean flannel filter. In this state it may be aromatised or perfumed at will; after which, when it is intended that the pommade should be opaque and white, it is assiduously stirred or beaten with a glass or wooden knife, or spatula, until it concretes; but when it is desired that it should appear transparent or crystalline, it is allowed to cool very slowly, and without being disturbed. To prevent the accession of rancidity, a little benzoic acid, gum benzoin, or nitric ether may be added to the fat, whilst in the liquid state, as noticed under FAT and OINTMENT. Sometimes a small portion of white wax

or beeswax (according to the intended colour of the product) is melted with the fat to increase its solidity. Some makers employ a few grains of powdered citric acid per ounce, in a like manner, with the intention of increasing the whiteness of the compound; but the practice is not to be commended, as pommades so prepared prove injurious to the hair.

The French perfumers, who are celebrated for the variety and excellence of their pommades, divide them into four classes:

1. POMMADES BY INFUSION. These are made by gently melting in a clean pan, over a water-bath, 2 parts of hog's lard, and 1 part of beef suet (both of the finest quality, and carefully 'rendered'), and adding thereto one part of the given flowers, previously carefully picked and separated from foreign matter; or, if the odorous substance is a solid, then coarsely bruised, but not reduced to fine powder. The mixture is next digested at a very gentle heat for from 12 to 24 hours, with occasional stirring, the vessel being kept covered as much as possible during the whole time. The next day the mixture is reheated, and again well stirred for a short time, after which it is poured into canvas bags, and these, being securely tied, are submitted to powerful pressure, gradually increased, in a screw or barrel press. This operation is repeated with the same fat and fresh flowers, several times, until the pommade is sufficiently perfumed. A good pommade requires thrice to six times its weight in flowers to be thus consumed; or of the aromatic barks and seeds a corresponding proportion. The pommades of cassia, orange flowers, and several others kept by the French perfumers, are prepared in this manner.

2. POMMADES BY CONTACT (ENFLEURAGE). These are made by spreading with a palette knife simple pommade (made with lard and suet as above) on panes of glass or pewter plates, to the thickness of a finger, and sticking the surface all over with the sweet-scented flowers. These last are renewed daily for one, two, or three months, or until the pommade has become sufficiently perfumed. On the large scale, the panes are placed in small shallow frames, made of four pieces of wood nicely fitted together, and are then closely piled one upon another. On the small scale, pewter plates are generally used, and they are inverted one over the other. In some of the perfumeries of France many thousands of frames are employed at once. The pommades of jasmine, jonquil, orange flowers, narcissus, tuberose, violet, and some other delicate flowers are prepared in this manner.

3. POMMADES BY ADDITION. These are prepared by simply adding the fragrant essences or essential oils, in the required quantity, to the simple pommade of lard and suet to produce the proper odour. In this way the pommades of bergamotte, cédrat, cinnamon, lemons, lemon thyme, lavender, limettes, marjoram, Portugal roses, rosemary, thyme, verbena, and about forty others kept by the Parisian perfumers, are made.

4. MIXED POMMADES. Of these a great variety exists, prepared by the addition of judicious combinations of the more esteemed perfumes to simple pommade; or, by the admixture of the different

perfumed pommades whilst in the semi-liquid state. (See *below*.)

THE COLOURED POMMADES derive their respective tints from tinctorial matter added to the melted fat before perfuming it. GREEN is given by gum guaiacum (in powder), or by the green leaves or tops of spinach, parsley, lavender, or walnut;—RED, by alkanet root and carmine;—YELLOW and ORANGE, by annatto or palm oil. WHITE POMMADES are made with mutton suet instead of beef suet. The BROWN and BLACK hard pomatums, vended under the name of 'COSMETIQUE,' are noticed at page 553. A few compound pommades are used as skin cosmetics.

Pommade. *Sym.* POMATUM. *Prep.* 1. (PLAIN POMATUM, SIMPLE P.) *a.* From lard, 2 lbs.; beef suet, 1 lb.; carefully rendered as above. The ordinary consistence for temperate climates.

b. Lard and suet, equal parts. For warm climates. Both may be scented at will.

2. (SCENTED POMATUM.) *a.* Plain pomatum, 1 lb.; melt it by the least possible degree of heat, add of essence of lemon or essence of bergamot, 3 dr., and stir the mixture until it concretes. This forms the ordinary 'pomatum' of the shops.

b. Plain pomatum, 1½ lbs.; essence of bergamot, 1½ dr.; essence of lemon, 1 dr.; oils of rosemary and cassia, of each, ½ dr.; oil of cloves, 20 drops. More fragrant than the last.

Pommade, Castor Oil. *Prep.* 1. From castor oil, 1 lb.; white wax, 4 oz.; melt them together; then add, when nearly cold, of essence of bergamot, 3 dr.; oil of lavender (English), ½ dr.; essence of ambergris, 10 drops. Supposed to render the hair glossy.

2. (Crystallised.) From castor oil, 1 lb.; spermaceti, 3 oz.; melt them together by a gentle heat, add of essence of bergamot, 3 dr.; oil of verbena, lavender, and rosemary, of each, ½ dr.; pour it into wide-mouthed glass bottles, and allow it to cool very slowly and undisturbed.

3. Castor oil, 680 parts; vaseline, 170 parts; yellow wax, 100 parts. Perfume to fancy.

Pommade, Castor Oil and Glycerin. (American receipt.) White wax, 1½ oz.; glycerin, 2 oz.; castor oil, 12 oz.; essence of lemon, 5 dr.; essence of bergamot, 2 dr.; oil of lavender, 1 dr.; oil of cloves, 10 drops; annatto, 10 gr.; rectified spirit and distilled water, of each a sufficient quantity. By a moderate heat dissolve the wax in a small portion of the castor oil (one fourth), and triturate it with the remainder of the oil and glycerin till quite cool; then add volatile oils. Lastly, rub the annatto with a drachm of water till smoothly suspended; add a drachm of alcohol, and stir the colouring into the pommade until it is thoroughly mixed. Avoid much heat.

Pommade, Casenave's. *Prep.* From prepared beef marrow, 4 oz.; tincture of cantharides (P. Cod.), 3 to 4 dr.; powdered cinnamon, ½ oz.; melt them together, stir until the spirit has, for the most part, evaporated, then decant the clear portion, and again stir it until it concretes. Recommended as a remedy for baldness and weak hair. It is to be used night and morning, the head being washed with soap and water, and afterwards with salt and water, before applying it. Dr Cattell scents it with the oils of origanum and bergamot instead of cinnamon.

Pommade, Collante. *Prep.* 1. Oil of almonds, 3 oz.; white wax, ¾ oz.; melt them together, and add of tincture of mastic (strong), 1 oz.; essence of bergamot, ½ dr. Used to stiffen the hair and keep it in form.

2. Burgundy pitch (true), 3 oz.; white wax, 2 oz.; lard, 1 oz.; melt, and, when considerably cooled, stir in of tincture of benzoin, 1 oz.; essence of bergamot, ½ dr. Used to fasten false curls.

Pommade, Cowslip. *Prep.* From plain pommade, 2 lbs.; essence of bergamot, 3 dr.; essence of lemon and essence of orange peel, of each, 1 dr.; huile au jasmin and essence de petit grain, of each, ½ dr.; essence of ambergris, 6 drops.

Pommade, Crystallised. *Prep.* From olive oil and spermaceti, as crystallised castor oil pommade, with scent at will.

Pommade of Cucumbers. *Sym.* POMMADE DE CONCOMBRES, Fr.; UNGUENTUM CUCUMIS, L. *Prep.* Lard, 10 oz.; veal suet, 6 oz.; balsam of tolu, 9 gr.; rose water, 44 minims; cucumber juice, 12 oz. by weight. Melt the lard and the suet over a water-bath, and add the tolu, previously dissolved in a little alcohol, and then the rose water. When clear, decant it into a tinned basin, then add to a third of the cucumber juice, and stir continually for four hours; pour off the juice and add another third, stir as before, then pour off, and add the remainder of the juice; separate as much as possible the fat from the liquid, melt by a water-bath, and after some hours skim, and put into pots. (Beat when in a semi-liquid state with a wooden spatula, when it will become much lighter and nearly double in bulk.)

Pommade, Dandruff. Salicylic acid, 30 gr.; borax, 15 gr.; Peruvian balsam, 25 minims; oil of anise, 6 drops; oil of bergamot, 20 drops; vaseline, 6 drachms. Mix.

Pommade d'Alyon. See OINTMENT OF NITRIC ACID, and CUPS.

Pommade de Beauté. *Prep.* From oil of almonds, 2 oz.; spermaceti, 2 dr.; white wax, 1½ dr.; glycerin, 1 dr.; balsam of Peru, 1 dr.; mixed by a gentle heat. Used as a skin cosmetic as well as for the hair.

Pommade de Casse. *Prep.* From plain pommade, 1 lb.; palm oil, ½ oz.; melt, pour off the clear, and add oil of cassia and huile au jasmin, of each, 1 dr.; neroli, 20 drops; oil of verbena or lemon-grass, 15 drops; otto of roses, 5 drops; and stir until nearly cold. Very fragrant.

Pommade d'Hebe. *Prep.* To white wax, 1 oz., melted by a gentle heat, add of the juice of lily bulbs and Narbonne honey, each, 2 oz.; rose water, 2 dr.; otto of roses, 2 drops. Applied night and morning to remove wrinkles.

Pommade de Ninon de l'Enclos. *Prep.* Take of oil of almonds, 4 oz.; prepared lard, 3 oz.; juice of houseleek, 3 fl. oz. Used chiefly as a skin cosmetic. Said to be very softening and refreshing.

Pommade, Divine. *Prep.* 1. Washed and purified beef marrow, 2 lbs.; liquid styrax, cypress wood, and powdered orris root, of each, 2 oz.; powdered cinnamon, 1 oz.; cloves and nutmeg, of each (bruised), ½ oz.; digest the whole together by the heat of a water-bath for six hours, and then strain through flannel.

2. Plain pommade, 2 lbs.; essence of lemon and bergamot, of each, 2 dr.; oils of lavender and origanum, of each, 1 dr.; oils of verbena, cassia, cloves, and neroli, of each, 12 drops; huile au jasmin, 3 dr.; essence of violets, ½ oz.

Pommade, Dupuytren's. *Prep.* 1. Take of prepared beef marrow, 12 oz.; melt, add of baume nerval (see OINTMENT, NERVINE), 4 oz.; Peruvian balsam and oil of almonds, of each, 3 oz.; and lastly, of alcoholic extract of cantharides, 36 gr.; (dissolve in) rectified spirit, 3 fl. dr. This is the original formula for this celebrated pommade. The following modifications of it are now commonly employed:—

2. (Cap.) Beef marrow, 2 oz.; alcoholic extract of cantharides, 8 gr.; rose oil, 1 dr.; essence of lemons, 30 drops.

3. (*Guibourt.*) Beef marrow and 'baume nerval' (see page 1167), of each, 1 oz.; rose oil, 1 dr.; alcoholic (or acetic) extract of cantharides, 6 gr.; (dissolved in) rectified spirit, q. s. These compounds are used to promote the growth of the hair and to prevent baldness, for which purpose they are usually coloured and scented according to the taste of the manufacturer. To be useful, they should be well rubbed on the scalp, at least once daily, for several weeks, and the head should be occasionally washed with soap and water.

Pommade, East India. *Prep.* Take of suet, 3 lbs.; lard, 2 lbs.; beeswax (bright), ½ lb.; palm oil, 2 oz.; powdered gum benzoin, 3 oz.; musk (previously triturated with a little lump sugar), 20 gr.; digest the whole together in a covered vessel, by the heat of a water-bath, for two hours, then decant the clear portion, and add of essence of lemon, ½ oz.; oil of lavender, ½ oz.; oils of cloves, cassia, and verbena, of each, ½ dr. A favourite pommade in the East Indies.

Pommade for Freckles. ('New York Druggists' Circular.') *Prep.* Citrine ointment and oil of almonds, of each, 1 dr.; spermaceti ointment, 6 dr.; oil of roses, 3 drops. Mix well in a wedgwood mortar, using a wooden or bone knife.

Pommade, Hard. *Syn.* HARD POMATUM, ROLL P. *Prep.* 1. Take of beef suet, 2 lbs.; yellow wax, ½ lb.; spermaceti, 1 oz.; powdered benzoin, ¼ oz.; melt them together, then add of oil of lavender, 2 dr.; essence of ambergris, ½ dr. Before it concretes pour it into moulds of paper or tin-foil.

2. Mutton suet and lard, of each, 1 lb.; white wax, 6 oz.; melt, and add of essence of lemon, 2 dr.; oil of cassia, ½ dr. Other perfumes may be employed at will.

Hard pomatums are used to gloss and set the hair. They act both as 'pommade' and 'fixateur.' See COSMETIQUE.

Pommade, Macassar. *Prep.* From castor oil, 5 oz.; white wax, 1 oz.; alkanet root, ½ dr.; heat them together until sufficiently coloured, then strain, and add oil of origanum and oil of rosemary, of each, 1 dr.; oil of nutmeg, ½ dr.; otto of roses, 10 drops. Said to be equal in efficacy to MACASSAR OIL.

Pommade, Marechal. Plain pommade scented by digesting it with *poudre maréchale.*

Pommade, Marrow. *Syn.* MARROW POMA-

TUM. *Prep.* From prepared beef marrow, ½ lb.; beef suet, ½ lb.; palm oil, ¼ oz.; melted together and scented at will.

Pommade, Millefleur. *Prep.* From plain pommade scented with a mixture of essence of lemon and essence of ambergris, each, 4 parts; oil of lavender, 2 parts; oil of cloves and essence de petit grain, of each, 1 part; or with other like perfumes so proportioned to each other that no one shall predominate. Much esteemed.

Pommade, Roll. See POMMADE, HARD.

Pommade, Roman. See *below.*

Pommade, Rose. *Syn.* ROSE POMATUM. This is plain pommade or hard lard which has been well beaten with eau de rose, or, better still, scented with otto of roses. It is sometimes tinged with alkanet root.

Pommade, Soft. Plain pomatum scented at will.

Pommade, Soubeiran's. *Prep.* From beef marrow, 1½ oz.; oil of almonds, ½ oz.; disulphate of quinine, 1 dr. Recommended for strengthening and restoring the hair.

Pommade, Transparent. *Prep.* Spermaceti, 2 oz.; castor oil, 5 oz.; alcohol, ·5 oz.; oil of bergamot, ½ dr.; oil of Portugal, ½ dr.

Pommade, Transparent Brilliantine. Melt together on a water-bath 200 grms. of suet and 120 grms. of clear amber resin; while liquid and at a temperature of about 80° C. add to the resinous fat a solution of 150 grms. of caustic soda (40°) in 300 grms. of rectified spirit. Use a vessel for holding these ingredients which will enable them to be boiled. Heat the contents of the vessel until saponification is complete, and a transparent soap has formed. Meanwhile melt in a separate vessel 4 kilos. of vaseline in 5 kilos. of castor oil by the heat of a water-bath. Add by portions 590 grms. of the transparent soap mass, and 3 kilos. of rectified spirits. Heat the whole until bubbles rise to the surface, then pour out, colour with gamboge, and perfume with 100 grms. of oil of sweet orange or any other perfume.

Pommade, Vanilla. *Syn.* ROMAN POMMADE; POMMADE À LA VANILLE, POMMADE ROMAIN, Fr. From plain pommade and pommade à la rose, of each, 12 lbs.; powdered vanilla, 1 lb.; heat them together in a water-bath, stir constantly for one hour, let it settle for another hour, decant the clear, and add oil à la rose, 2½ lbs.; bergamot, 4 oz.

Pommade, Vaseline. In the following formulæ the fatty basis consists of 3 parts of white vaseline and 1 part of creasin (purified mineral wax). These substances should be melted together and placed in a warm porcelain vessel, the colouring matter added, and the whole diligently stirred until the mixture is of the consistence of thick cream; then add the perfumes, and pour into pots or bottles:

Pommade à la Rose. Fatty basis, 1000 parts; oil of rose geranium, 15 parts; oil of bergamot, 6 parts; oil of neroli, 2 parts. To be coloured a faint red with alkanet.

Pommade à l'Héliotrope. Fatty basis, 1000 parts; oil of cassia flowers, 7 parts; oil of bitter almonds, 3 parts; oil of cinnamon, 2 parts; Peruvian balsam, 9 parts.

Pommade au Riséda. Fatty basis, 1000 parts; oil of bergamot, 8 parts; oil of bitter almonds, 6 parts; oil of neroli, 4 parts; oil of ylang-ylang, 1 part. To be coloured green with spinach.

Pommade au Citron. Fatty basis, 1000 parts; oil of lemon, 10 parts; oil of bergamot, 2 parts; oil of citronella, 2 parts. To be coloured yellow with gamboge.

Pommade aux Oranges. Fatty basis, 1000 parts; oil of orange-peel, 10 parts; oil of bergamot, 2 parts; oil of rose geranium, 2 parts. To be coloured orange with annatto.

POND'S EXTRACT. An aromatic water distilled from the leaves of *Hamamelis virginica*, or winter bloom. It is also known as hazelina.

Uses. Valuable hæmostatic, very useful in piles, or to check mucous discharges, or as an application to bruises and wounds.

POPPY. *Syn.* WHITE POPPY; PAPAVER SOMNIFERUM, L. The capsules or fruit ("mature," Ph. L.; "not quite ripe," Ph. E.) form the poppies or poppy-heads of the shops (PAPAVERIS CAPSULÆ; PAPAVER, Ph. L., E., & D.). They are anodyne and narcotic, similar to opium, but in only a very slight degree. The seeds (MAW SEED), which are sweet, oleaginous, and nutritious, are used as a substitute for almonds in confectionery and mixtures, and are pressed for their oil. See EXTRACT, OPIUM, and SYRUP.

Poppy, Red. *Syn.* CORN POPPY, CORN ROSE; PAPAVER RHŒAS, L. The fresh petals or flowers (RHŒADOS PETALA; RHŒAS, Ph. L., E., & D.) are reputed pectoral, but are chiefly employed on account of their rich colour. See SYRUP.

POP'ULIN. *Syn.* POPULINUM, L. A peculiar neutral, crystallisable substance, formerly supposed to be an alkaloid, found, associated with SALICIN, in the root-bark of the *Populus tremula*, Linn., or aspen.

Prep. Concentrate the decoction by a gentle heat, and set it aside in a cool situation to crystallise; dissolve the crystals which are deposited in rectified spirit, decolour them by digestion with animal charcoal, filter, and again crystallise. To render them still purer they may be redissolved and crystallised a second and a third time, if necessary.

Prop., &c. It resembles salicin in appearance and solubility, but, unlike that substance, has a penetrating sweet taste. Dilute acids convert it into benzoic acid, grape sugar, and saliretin; and with a mixture of sulphuric acid and bichromate of potassa it yields a large quantity of salicylous acid. It appears to be tonic, stomachic, and febrifuge.

POR'CELAIN. See POTTERY.

PORK. The value of pork as an article of diet is well known. That from the young and properly fed animal is savoury, easy of digestion, and, when only occasionally employed, highly wholesome; but it is apt to disagree with some stomachs, and should, in such cases, be avoided. To render it proper for food it should be thoroughly but not overcooked. When salted it is less digestible. The frequent use of pork is said to favour obesity, and to occasion disorders of the skin, especially in the sedentary. See MEAT.

POR'PHYRIZED, PORPHYRIZA'TION. Words coined by recent pharmaceutical writers, and possessing similar meanings to LEVIGATED and LEVIGATION.

PORPHYROX'IN. A neutral crystallisable substance discovered by Merck in opium. It is soluble in both alcohol and ether, insoluble in water, and is characterised by assuming a purplish-red colour when heated in dilute hydrochloric acid.

PORRI'GO. See RINGWORM.

PORTER. This well-known beverage, now the common drink of the inhabitants of London, by whom it is generally termed 'beer,' originated with a brewer named Harwood in 1722. Previously to this date, 'ale,' 'beer,' and 'twopenny' constituted the stock in trade of the London publican, and were drunk, either singly or together, under the name of 'half-and-half' or 'three threads,' for which the vendor was compelled to have recourse to two or three different casks, as the case might demand. The inconvenience and trouble thus incurred led Mr Harwood to endeavour to produce a beer which should possess the flavour of the mixed liquors. In this he succeeded so well that his new beverage rapidly superseded the mixtures then in use, and obtained a general preference among the lower classes of the people. At first this liquor was called 'entire' or 'entire butt,' on account of it being drawn from one cask only, but it afterwards acquired, at first in derision, the now familiar name of 'porter,' in consequence of its general consumption among porters and labourers. The word 'entire' is still, however, frequently met with on the signboards of taverns about the metropolis.

The characteristics of pure and wholesome porter are its transparency, lively dark brown colour, and its peculiar bitter and slightly burnt taste. Originally these qualities were derived from the 'high-dried malt' with which alone it was brewed. It is now generally, if not entirely, made from 'pale' or 'amber malt,' mixed with a sufficient quantity of 'patent' or 'roasted malt,' to impart the necessary flavour and colour. Formerly this liquor was 'vatted' and 'stored' for some time before being sent out to the retailer, but the change in the taste of the public during the last quarter of a century in favour of the mild or new porter has rendered this unnecessary. The best 'draught porter,' at the time of its consumption, is now only a few weeks old. In this state only would it be tolerated by the modern beer-drinker. The old and acid beverage that was formerly sold under the name of porter would be rejected at the present day as 'hard' and unpleasant, even by the most thirsty votaries of malt liquor.

The 'beer' or 'porter' of the metropolitan brewers is essentially a weak mild ale, coloured and flavoured with roasted malt. Its richness in sugar and alcohol, on which its stimulating and nutritive properties depend, is hence less than that of an uncoloured mild ale brewed from a like original quantity of malt. For pale malt is assumed to yield 80 to 84 lbs. of saccharine per quarter; whereas the torrefied malt employed by the porter brewers only yields 18 to 24 lbs. per quarter, and much of even this small quantity is altered in its properties, and is incapable of undergoing the vinous fermentation. In the manufacture of porter there is a waste of malt which does not occur in brewing ale; and the consumer must, therefore, either pay a higher price for it or be content with a weaker liquor.

The hygienic properties of porter, for the most part, resemble those of other malt liquors. Some members of the faculty conceive that it is better suited to persons with delicate stomachs and weak digestion than either ale or beer. That there may be some reason for this preference, in such cases, we are not prepared to deny, but undoubtedly, when the intention is to stimulate and nourish the system, ale is preferable. Certain it is, however, that the dark colour and strong taste of porter render its adulteration easier than that of ale, whilst such adulteration is more difficult of detection than in the paler varieties of malt liquors. " For medical purposes, ' bottled porter ' (CEREVISIA LAGENARIA) is usually preferred to ' draught porter.' It is useful as a restorative in the latter stages of fever, and to support the powers of the system after surgical operations, severe accidents, &c." (Pereira, ii, 982). When ' out of condition' or adulterated, porter, more than perhaps any other malt liquor, is totally unfit for use as a beverage, even for the healthy; and when taken by the invalid, the consequences must necessarily be serious. Dr Ure says that pure ' porter,' " when drunk in moderation, is a far wholesomer beverage for the people than the thin acidulous wines of France and Germany."

The manufacture of porter has been described in our article on BREWING, and is also referred to above. It presents no difficulty or peculiarity, beyond the choice of the proper materials. A mixture of ' brown ' and ' black malt' is thought to yield a finer flavour and colour to the pale malt that gives the body to the liquor than when ' black ' or ' roasted malt' is employed alone. The proportion of the former to the latter commonly varies from 1-6th to 1-4th. When ' black malt' is alone used, the proportion varies from the 1-10th to 1-15th. 1 lb. of ' roasted malt,' mashed with about 79 lbs. of pale malt, is said to be capable of imparting to the liquor the flavour and colour of porter. The following formulæ were formerly commonly employed in London:

1. (DRAUGHT PORTER.) From pale malt, 8½ qrs.; amber malt, 3 qrs.; brown malt, 1½ qrs.; mash at twice with 28 and 24 barrels of water, boil with brown Kent hops, 56 lbs., and set with yeast, 40 lbs.—*Prod.*, 28 barrels, or 3½ times the malt, besides 20 barrels of table-beer from a third mashing.

2. (BOTTLING PORTER, BROWN STOUT.) From pale malt, 2 qrs.; amber and brown malt, of each, 1½ qrs.; mash at 3 times with 12, 7, and 6 barrels of water, boil with hops, 50 lbs., and set with yeast, 26 lbs.—*Prod.*, 17 barrels, or 1½ times the malt.

The purity and quality of porter as well as of other malt liquors may be inferred in the manner noticed under BEER, but can only be positively determined by a chemical examination. For this purpose several distinct operations are required:

1. *Richness in* ALCOHOL. This may be cor-

rectly found by the method of M. Gay-Lussac, or from the boiling-point. (See ALCOHOLOMETRY and EBULLIOSCOPE.) The method with anhydrous carbonate of potassa will also give results sufficiently near to the truth for ordinary purposes, when strong or old beer is operated on. The quantity of the liquor tested should be 3600 water-grains measure; and it should be well agitated, with free exposure to the air, after weighing it, but before testing it for its alcohol. The weight of alcohol found, multiplied by 1·8587, gives its equivalent in sugar. This may be converted into ' brewer's pounds' or density per barrel, as below.

2. *Richness in* SACCHARINE OF EXTRACTIVE MATTER. A like quantity of the liquor under examination, after being boiled for some time to dissipate its alcohol, is made up with distilled water, so as to be again exactly equal to 3600 water-grains measure. The sp. gr. of the resulting liquid is then taken, and this is reduced to ' brewer's pounds' per barrel by multiplying its excess of density above that of water (or 1000) by 360, and pointing off the three right-hand figures as decimals.

3. ACETIC ACID OF VINEGAR. This is determined by any of the common methods of ACIDIMETRY (which *see*; see also ACETIMETRY). Each grain of anhydrous acetic acid so found represents 1·6765 gr. of sugar.

4. *Gravity of* ORIGINAL WORT. This is obtained by the addition of the respective quantities of saccharine matter found in Nos. 1, 2, and 3 (above). These results are always slightly under the true original density of the wort, as cane sugar appears to have been taken by the Excise as the basis of their calculations. More correctly, 12% of proof spirit is equivalent to 19 lbs. of saccharine per barrel. 10½ lbs. of saccharine are equivalent to 1 gall. of proof spirit.

5. *Detection of* NARCOTICS. This may be effected either by the method described under ALKALOID, or by one or other of the following processes:

a. Half a gallon of the beer under examination is evaporated to dryness in a water-bath; the resulting extract is boiled for 30 or 40 minutes in a covered vessel with 10 or 12 fl. oz. of alcohol or strong rectified spirit, the mixture being occasionally stirred with a glass rod, to promote the action of the menstruum; the alcoholic solution is next filtered, treated with a sufficient quantity of solution of diacetate of lead to precipitate colouring matter, and again filtered; the filtrate is treated with a few drops of dilute sulphuric acid, again filtered, and then evaporated to dryness; it may then be tested with any of the usual reagents, either in the solid state, or after being dissolved in distilled water. Or the extract, obtained as above, may be boiled as directed with rectified spirit, the solution filtered, the spirit distilled off, and a small quantity of pure liquor of potassa added to the aqueous residue, which is then to be shaken up with about 1 fl. oz. of ether; lastly, the ethereal solution, which separates and floats on the surface, is decanted, evaporated, and the residuum tested, as before. The alkaline liquid, from which the ether has been decanted, is then separated from any precipitate which may have formed, and both of these separately tested for alkaloids.

b. From 2 to 8 oz. of purified animal charcoal is diffused through ½ gall. of the beer, and is digested in it, with frequent agitation, for from 8 to 12 hours; the liquor is next filtered, and the charcoal collected on the filter is boiled with about ½ pint of rectified spirit; the resulting alcoholic solution is then further treated as above, and tested. This answers well for the detection of strychnia or nux vomica.

6. PICRIC ACID. This substance, which was formerly employed to impart bitterness to London porter in lieu of hops, may be detected as follows:

a. A portion of the liquor agitated with a little solution of diacetate of lead loses its bitter flavour if it depends on hops, but retains it if it depends on picric acid.

b. Pure beer is decoloured and deodorised by animal charcoal; but beer containing picric acid, when thus treated, retains a lemon-yellow colour and the odour.

c. Unbleached sheep's wool, boiled for six or ten minutes, and then washed, takes a canary-yellow colour if picric acid be present. The test is so delicate that 1 gr. of the adulterant in 150,000 gr. of beer is readily detected.

d. (*Vitate*, 'Chemical News,' vol. xxxv, p. 75.) The author agitates 10 c.c. of the suspected beer in a test-tube with half its volume of pure amylic alcohol. If the mixture is left to settle, the amylic stratum separates entirely, and is drawn off with a pipette, evaporated to dryness at a convenient temperature in a porcelain capsule, and the residue is finally taken up in a little distilled water with the aid of heat. The aqueous solution is divided into portions, and submitted to the following reagents. One portion is treated with a solution of ammonio-sulphate of copper, which, in dilute solutions of picric acid, instantly produces a turpidity, due to the formation of very minute crystals of the ammonio-picrate of copper, of a greenish colour. Another portion may be treated with a concentrated solution of cyanide of potassium, which produces a blood-red colour, more or less intense, according to the quantity of picric acid present, in consequence of the formation of iso-purpuric acid. A third portion may be submitted to the action of sulphide of ammonium, rendered still more alkaline by the addition of a few drops of ammonia. Here also a blood-red colour is produced, which becomes more intense on the application of heat, and is due to the formation of picramic acid.

7. MINERAL MATTER. *a.* A weighed quantity of pure beer evaporated to dryness, and then incinerated, does not furnish more than from ·20% to ·85% of ash, the quantity varying within these limits with the strength of the liquor and the character of the water used in brewing it.

b. A solution of this ash, made by decoction with distilled water, should be only rendered slightly turbid by solutions of acetate of lead, bichloride of platinum, nitrate of baryta, nitrate of silver, oxalate of ammonia, and sulphuretted hydrogen.

c. If the beer contained common salt, the above solution will give a cloudy white precipitate with a solution of nitrate of silver. Each grain of this precipitate is equivalent to ½ gr. of common salt (nearly).

d. If GREEN COPPERAS (sulphate of iron) is present, ferridcyanide of potassium gives a blue precipitate, and ferrocyanide of potassium a bluish-white one, turning dark blue in the air; solution of chloride of barium gives a white precipitate, each grain of which, after being washed, dried, and ignited, represents 1·188 gr. of crystallised protosulphate of iron.

e. The ash digested in water slightly acidulated with nitric acid, and then boiled, yields a solution which, when cold, gives a black precipitate with sulphuretted hydrogen, and a white one with dilute sulphuric acid when lead is present.

8. *Wittstein's method for the detection of* ADULTERANTS *in beer.* ('Archiv der Pharmacie,' January, 1876; 'Pharm. Journal,' 3rd series, v.) One litre of the suspected beer is evaporated by a moderate heat to the consistence of a thick syrup. This is poured into a tarred glass cylinder capable of containing ten times its volume and weighed; five times its weight of 93° to 96° alcohol is added, and the whole frequently stirred, by means of a thick glass rod, during twenty-four hours.

By this means all the gum, dextrin, sulphates, phosphates, and chlorides are separated, and a comparatively small portion is obtained in solution. After clearing this solution is decanted, the residue is again treated with fresh alcohol, the two products mixed, filtered, and the alcohol driven off by a gentle heat.

a. Of the syrupy residue left after this evaporation, a small portion is diluted with three times its bulk of water, and tested for picric acid, according to the directions already given.

b. The remaining largest portion of the syrup is agitated for some time with six times its weight of pure colourless benzol (boiling-point 80° C.); this is decanted off, and the operation is repeated with fresh benzol, and the two liquors, the first of which has become yellow, the second having scarcely changed colour, are evaporated at a gentle heat. The pale yellow, resinous residue thus obtained may possibly contain brucine, strychnine, colchicine, or colocynthin. To ascertain this, three portions of the resin are placed on a porcelain capsule; one is treated with nitric acid (sp. gr. 1·33 to 1·40), another with concentrated sulphuric acid, and the third, after a few morsels of red chromate of potash have been added, also with sulphuric acid. A red colour, produced by the nitric acid, indicates brucine with certainty, and a violet colour colchicine; a red colour produced by sulphuric acid indicates colocynthin, and a purple-violet, produced by sulphuric acid and bichromate of potash, reveals strychnine. Resin in which one or other of these colorations is produced possesses an extremely bitter taste; that in which the coloration does not take place is also bitter, but the bitterness recalls the well-known hop flavour.

c. The syrup which has been treated with benzol is freed by gentle heating from the small quantity of benzol remaining, and agitated twice with pure colourless amylic alcohol (boiling-point

132° C.).   The first portion of the alcohol acquires a more or less wine or golden-yellow colour. It would take up any picrotoxin or aloes if present, and thereby acquire a strongly bitter taste.

If neither of these two substances be present, the amylic alcohol does not become bitter, because neither the hop bitter nor the remaining four bitter principles—absinthin, gentipicrin, menyanthin, and quassiin—are soluble in it.

In order to distinguish picrotoxin from aloes a portion of the first obtained amylic alcoholic solution is poured upon glass, and allowed to evaporate spontaneously. If a fine white crystallisation be formed picrotoxin is present, if not aloes is present, and can only be recognised by its peculiar saffron-like odour.

d. The syrup which has been treated with benzol and amylic alcohol is freed by means of blotting-paper from the small quantity of amylic alcohol adhering to it, evaporation by heat being impracticable in consequence of the high boiling-point of the alcohol, and shaken with anhydrous ether. This takes up the hop bitter yet present and absinthin. After evaporation the latter is easily recognised through its wormwood-like aroma; it also gives a reddish-yellow solution with concentrated sulphuric acid, which changes quickly to an indigo-blue colour.

e. After treating with ether the syrup has yet to be tested for gentipicrin, menyanthin, and quassiin. As it is now free from the hop bitter, a decidedly bitter taste points to one of these three substances. Any remaining ether is removed, and the syrup is dissolved in water and filtered; to one portion is added strong ammoniacal solution of silver, and it is then heated.

If it remains clear quassiin is present; if a silver mirror be formed it originates either with gentipicrin or menyanthin. Another portion is evaporated to dryness on porcelain, and concentrated sulphuric acid added. If, while cold, no change of colour takes place, but on heating it becomes carmine-red, gentipicrin is present; menyanthin would give a yellowish-brown colour, gradually changing to violet.

For further information connected with this subject, see ALCOHOLOMETRY, ALE, BEER, BREWING, MALT LIQUORS, &c.

PORT-FIRE. A paper tube, from 9 to 12 inches in length, filled with a slow-burning composition of metal powder, nitre, and sulphur, rammed moderately hard by a similar process to that adopted for small rockets. It is used in lieu of a touch-match to fire guns, mortars, pyrotechnical devices, &c.

PORTLAND CEMENT. A species of mortar formed by calcining a mixture of limestone and argillaceous earth, and grinding the calcined mass to powder, in which state it must be preserved from the air. It is characterised by absorbing a large quantity of water and then rapidly becoming solid, and after a time acquiring considerable hardness. See MORTAR and CEMENT.

POSOLOGY. See DOSE.

POSSET. *Syn.* POSSETUM, L. Milk curdled with wine or any other slightly acidulous liquor. It is usually sweetened with either sugar or treacle, and is taken hot.

Prep. From new milk, ½ pint; sherry or

marsala, 1 wineglassful; treacle, 1 or 2 tablespoonfuls, or q. s.; heat them together in a clean saucepan until the milk coagulates. This is called 'treacle posset' or 'molasses posset,' and taken on retiring to rest is highly esteemed in some parts of the country as a domestic remedy for colds. Lemon juice, strong old ale, or even vinegar, is occasionally substituted for wine, and powdered ginger or nutmeg added at will.

POT METAL. See COOK METAL.

POTASH. The 'potash' or 'potashes' of commerce is an impure carbonate of potassium, so first made. The 'potash' or 'potass' of the chemist is the hydrate of the metal potassium, which is more particularly referred to below. See CARBONATE OF POTASSIUM, &c.

Potash, which is in much demand for the manufacture of soap and glass, is now principally obtained from the following sources:

1. From carnallite, a hydrated double chloride of potassium and magnesium, which occurs associated with other salts of potassium and magnesium, as well as of sodium, in a bed of clay, at Stassfurt, near Magdeburg, in Prussia.

2. Feldspar and similar minerals.

3. Sea water, and the mother-liquor of salt works.

4. Native saltpetre.

5. The ashes of plants.

6. The calcined residue of the molasses of beet-root sugar remaining after distillation.

7. The seaweeds, as a by-product of the manufacture of iodine.

8. From the fleece of the sheep. Maumené and Rogelet state that a fleece weighing 9 lbs. contains about 6 oz. of pure potash.

The following is a process for obtaining alkali from seaweed, described in the 'Chemical News' for Nov. 10th, 1876:

At the chemical works at Aalbourg, in Jutland, Denmark, where about 30 tons of alkali are made per week by the ammonia process, Mr Theobald Schmidt, the director of the manufactory, works, in conjunction with this process, a method of treating seaweed so as to obtain iodine, potash, salts, and other marketable products therefrom.

In Denmark a very heavy duty is levied on the importation of common salt, whilst enormous quantities of seaweed rich in iodine and potash can be obtained at small cost in the neighbourhood of the works. Mr Schmidt's process is as follows:—After the seaweed is dried and burnt, a concentrated solution of the ash is made and added to the liquor, containing chlorides of sodium and calcium, left after the ammonia has been recovered in the ammonia-soda process by boiling with lime. The sulphates of potash, soda, and magnesia contained in the ash of the seaweed are thereby decomposed, and hydrated sulphate of lime and hydrated magnesia are precipitated in a form which is available for paper-making, as 'pearl-hardening.' The last traces of sulphates are got rid of by adding a small quantity of solution of chloride of barium. To the clear solution nitrate of lead is now added, until all the iodine is precipitated as iodide of lead, which is then separated by filtration and treated for the production of iodine or iodides. After filtration the liquid is boiled; nitrate of soda is added to convert the chloride of potassium present into nitrate of potash. The latter is separated by crystallisation. There remains a solution of common salt, containing traces of ammonia from the previous soda operation, and a trace of chloride of potassium. This solution is again treated by the ordinary ammonia-soda process for the production of bicarbonate of soda and white alkali. See CARBONATE OF POTASSIUM, &c.

POTASSIUM. K = 39·04. The metallic base of potash. It was discovered, in 1807, by Sir H. Davy, who obtained it by submitting moistened potassium hydrate, under a film of naphtha, to the action of a powerful voltaic current. It has since been procured by easier methods, of which the following, invented by Brunner, is the best.

Prep. An intimate mixture of carbonate of potassium and charcoal is prepared by calcining, in a covered iron pot, the crude tartar of commerce; when cold, it is rubbed to powder, mixed with 1-10th part of charcoal in small lumps, and quickly transferred into a retort of stout hammered iron; the latter may be one of the iron bottles in which quicksilver is imported, a short and somewhat wide iron tube having been fitted to the aperture; the retort, thus charged, is placed upon its side, in a furnace so constructed that the flame of a very strong fire, preferably fed with dry wood, may wrap round it, and maintain every part of it at a very high and uniform degree of heat. A copper receiver, divided in the centre by a diaphragm, is next connected to the iron pipe, and kept cool by the application of ice, whilst the receiver itself is partly filled with mineral naphtha, to preserve from oxidation the newly formed potassium as it distils over. The arrangement of the apparatus being completed, the fire is gradually raised until the requisite temperature, which is that of full whiteness, is reached, when decomposition of the alkali by the charcoal commences, carbonic acid gas is abundantly disengaged, and potassium distils over, falling in large drops into the liquid. To render the product absolutely pure, it is redistilled in an iron or green glass retort, into which some naphtha has been put, so that its vapour may expel the air, and prevent the oxidation of the metal. The pieces of charcoal are introduced for the purpose of absorbing the melted carbonate of potassium and preventing its separation from the finely divided carbonaceous matter.—*Prod.*, 8% to 4% of the weight of tartar acted upon.

Prop., &c. Pure potassium is a brilliant white metal, with a high lustre; at the common temperature of the air it is soft, and may be easily cut with a knife, but at 32° F. it is brittle and crystalline; it melts completely at 136° F., and in close vessels distils unaltered at a low red heat. Sp. gr. 0·865. It has an affinity for oxygen, which is so great that it takes it from many substances containing it. Exposed to the air, its surface is instantly tarnished, and quickly becomes covered with a crust of oxide or hydrate. It inflames spontaneously when thrown upon water, and burns with a beautiful purple or purple-red flame, yielding a pure alkaline solution. It can

only be preserved in naphtha, rock oil, or some other fluid hydrocarbon.

Tests. The salts of potassium are all soluble in water, the tartrate, periodate, and fluosilicate being the least so; they are usually colourless, unless the acid be coloured, crystallising readily, and forming numerous double compounds. They can be recognised as follows:

Sulphuretted hydrogen, sulphide of ammonium, and carbonate of ammonium do not affect them. A solution of tartaric acid, added in excess to moderately strong neutral or alkaline solutions of potassium salts, gives a quickly subsiding, crystalline, white precipitate, which is redissolved on heating the liquid, and again separates as it cools; and is also soluble in aqueous solutions containing free alkali, or free mineral acids. Platinic chloride produces, in neutral and acid solutions, a yellow crystalline precipitate. Alkaline solutions require to be first slightly acidulated with hydrochloric acid. The separation of the precipitate here, as well as that produced by tartaric acid, is promoted by violent agitation and friction against the sides of the vessel, and the delicacy of both is increased by the addition of some alcohol. When converted into carbonate by igniting with excess of carbonate of ammonium and alcohol, and treated with sulphuretted hydrogen solution and nitro-prusside of sodium, it gives a splendid violet colour, turning through red to green on standing.

Potassium salts give with sodium periodate and hydro-fluosilicic acid white precipitates soluble in much water.

Heated in the inner flame of the blowpipe on platinum wire, they impart a violet coloration, which must be observed through a piece of blue glass, for it is masked by a mere trace of sodium salts.

Estim. 1. The double chloride of platinum is formed in the separation of potassium from sodium. An excess of platinum tetrachloride is added to the mixed chlorides of potassium and sodium, the liquid evaporated on a water-bath, and the cooled residue mixed with strong alcohol in which the excess of platinic chloride and the sodium double salt easily dissolve; the potassium double salt is then collected and weighed.

2. The mixed chlorides of potassium and sodium are converted into sulphates by treatment with strong sulphuric acid, and the weight of these is then found. Then the amount of potassium (a) and of sodium (b) present can be calculated when the weights of the several salts are known, thus:

$$\frac{\text{Weight of}}{\text{mixed chlorides}} = \frac{\text{Md.wt.of KCl}}{(39\cdot04)} + \frac{\text{Md.wt.of NaCl}}{(22\cdot99)} a$$

$$\frac{\text{Weight of}}{\text{mixed sulphates}} = \frac{\text{Md.wt.of K}_2SO_4}{2(39\cdot04)} + \frac{\text{Md.wt.of Na}_2SO_4}{2(22\cdot99)} b$$

Potassium, Acetate of. $KC_2H_3O_2$. *Syn.* ACETATE OF POTASH, POTASSIC ACETATE; POTASSÆ ACETAS, L. *Prep.* Acetic acid, 26 fl. oz.; distilled water, 12 fl. oz.; mix, and add, gradually, carbonate of potassium, 1 lb., or q. s. to saturate the acid; next, filter the solution, and evaporate it by the heat of a sand-bath, gradually applied, until the salt is dried.

Prop., &c. Acetate of potassium, prepared as above, occurs in shining white masses, having a soft foliated texture, a slight but peculiar odour, and a warm, sharp taste; it deliquesces in the air; dissolves in rather less than its own weight of water, and in about twice its weight of alcohol; and by exposure to a red heat is converted into pure carbonate of potassium. It should be preserved in well-corked and sealed bottles. It is soluble in water and in alcohol. These solutions neither affect litmus nor turmeric, nor are they disturbed by either chloride of barium or nitrate of silver; but if from a stronger solution anything is thrown down by nitrate of silver, the same is again dissolved on the addition of water or dilute nitric acid. Sulphuric acid being added, the vapour of acetic acid is evolved.

Uses, &c. Acetate of potassium has been found useful in dropsies, febrile affections, jaundice, scurvy, calculus, and several chronic skin diseases. During its exhibition the urine becomes at first neutral, and then alkaline, owing to the salt being converted into carbonate of potassium in the system.—*Dose.* As a diaphoretic and antiscorbutic, 15 to 20 gr.; as a diuretic, 20 to 60 gr.; as an aperient, 2 to 3 dr.; in each case dissolved in some bland liquid, or in the infusion of some mild vegetable bitter.

Potassium Antimoniates. The normal potassium antimoniate ($KSbO_3$) may be obtained by heating, in an earthen crucible, 1 part of metallic antimony with 4 parts of nitrate of potash. The mass so obtained is reduced to powder, and afterwards washed with warm water to remove the excess of potash and potassium nitrate. The residue must be boiled in water for an hour or two; the insoluble anhydrous antimoniate is thus converted into a soluble gelatinous hydrated modification ($K_5Sb_5O_{16}.nH_2O$). The insoluble residue now consists chiefly of acid antimoniate of potassium. The normal salts possess the property of readily dissolving the acid antimoniate, which is precipitated when such a solution is mixed with any neutral salt of one of the alkalies. The normal antimoniate does not crystallise, and has an alkaline reaction.

Acid antimoniate of potassium, $K_2H_2(SbO_4)$, $6.9H_2O$, may be procured by passing a stream of carbonic anhydride through a solution of the normal antimoniate.

Normal potassium metantimoniate ($K_4Sb_2O_7$) is best obtained by fusing the soluble hydrated antimoniate with three times its weight of potash, dissolving the mass in water, and crystallising out by evaporation. It forms deliquescent crystals, which are decomposed by water into free alkali and an acid metantimoniate ($H_2K_2Sb_2O_7 + 6H_2O$), which is a slightly soluble crystalline powder. Its aqueous solution readily passes into the gelatinous antimoniate.

Potassium, Arsenite of. $KAsO_2$. *Syn.* PO-TASSIUM METARSENITE. A salt of arsenious acid. Very stable, but soluble easily in water.

Uses. An ingredient in 'sheep-dipping' liquids. In the manufacture of arsenical soap. Naturalists often use a soap composed of potassium arsenite, common yellow soap, and camphor, in order to preserve the skins of animals.

Potassium, Borate of. KBO_2. *Syn.* POTASSE

BORAS, L. *Prep.* From dry carbonate of potassium and dry boracic acids, equal parts, reduced to powder, and heated to redness in a covered crucible; the sublimed mass, when cold, being dissolved in boiling water, and the filtered solution concentrated by evaporation, and then set aside to crystallise; or at once completely evaporated to dryness.

Potassium, Boro-tartrate of. *Syn.* POTASSE BORO-TARTRAS, CREMOR TARTARI SOLUBILIS, L.; CRÈME DE TARTRE SOLUBLE, Fr. *Prep.* Crystallised boracic acid, 1 part; bitartrate of potassium, 4 parts; water, 24 parts; dissolve, by the aid of heat, in a silver basin, constantly stirring; evaporate the resulting solution either to dryness, and then powder it, or it may merely be evaporated to a syrupy consistence, spread upon plates, and dried by the heat of a stove. It must afterwards be preserved from the air.

Prop., &c. A white, deliquescent powder, freely soluble in water. It has been used as a solvent for lithic calculi, and in gout, &c.—*Dose*, 15 to 30 gr. In doses of 2 to 3 dr.; it is laxative, and is very popular as such on the Continent.

Potassium, Bromide of. KBr. *Syn.* POTASSII BROMIDUM (B. P.). *Prep.* Exactly as the iodide, which it resembles in its character, only being somewhat less soluble in water, and more so in alcohol. Employed in similar cases and given in similar doses to the iodide.

Potassium, Carbonate of. K_2CO_3. *Syn.* CARBONATE OF POTASSA, SUBCARBONATE OF POTASSA, SALT OF TARTAR; POTASSÆ CARBONAS (B. P., Ph. L., E., & D.). Impure or crude carbonate of potassium is chiefly imported from America and Russia, where it is obtained by lixiviating wood ashes, and evaporating the solution to dryness. The mass is then transferred into iron pots, and kept in a state of fusion for several hours, until it becomes quiescent, when the heat is withdrawn, and the whole is left to cool. It is next broken up and packed in air-tight barrels, and in this state, mixed as it is with much potassium chloride and some sulphate, constitutes the 'potashes' or 'potash' of commerce. Another method is to transfer the black salts, or product of the first evaporation, from the kettles to a large oven or furnace, so constructed that the flame is made to play over the alkaline mass, which is kept constantly stirred by means of an iron rod. The ignition is continued until the impurities are burned out, and the mass changes from a blackish tint to a dirty or bluish white. The whole is next allowed to cool, and is then broken into fragments, and packed in casks as before. It now constitutes 'pearlash.'

When pearlash is dissolved in cold distilled water, the solution depurated, filtered, and crystallised, or simply evaporated to dryness, it forms 'refined ashes,' or carbonate of potash sufficiently pure for most pharmaceutical and technical purposes. The granulated carbonate of potash, salt of tartar, or prepared kali, of the shops, is simply refined ashes which, during evaporation, and more especially towards the conclusion of the desiccation, have been assiduously stirred, so that they may form small white granules, instead of adhering together as an amorphous solid mass. In this state it constitutes the ordinary or carbonate of potass of the Pharmacopœias.

Pur. Ordinary potash or pearlash may be refined as follows:—Raw potash, 10 parts, is dissolved in cold water, 6 parts, and the solution allowed to remain for 24 hours in a cool place; it is then filtered, and somewhat concentrated by evaporation, crystallisation being prevented by continually stirring the mass until the whole is nearly cold; it is next decanted into a strainer, and the mother-liquor allowed to drip off; the residuum is evaporated to dryness at a gentle heat, and redissolved in an equal quantity of cold distilled water; the new solution, after filtration, is again evaporated to dryness. The product is quite free from potassium sulphate, and is nearly free from potassium chloride and any silicates that may be present.

Potassium, Pure Carbonate of. *Sys.* CARBONATE OF POTASSA (POTASSÆ CARBONAS PURUM, Ph. E. & D., and Ph. L., 1836).

Prep. 1. From bicarbonate of potassium, in crystals, heated to redness in a crucible.

2. As the last; or, more cheaply, by dissolving bitartrate of potassium in 30 parts of boiling water, separating and washing the crystals which form on cooling, heating them in a loosely-covered crucible to redness as long as fumes are given off; breaking down the mass, and roasting it in an oven for two hours with occasional stirring; lixiviating the product with cold distilled water, filtering the solution thus obtained, evaporating it to dryness, granulating the salt towards the close by brisk agitation; and, lastly, heating the granular salt nearly to redness.

3. Bitartrate of potassium, 2 lbs., is exposed to a red heat in an iron crucible as before; the powdered calcined mass is boiled for 20 minutes in water, 1 quart, the solution filtered, and the filtrate washed with water, 1 pint, to which ammonium sesquicarbonate, ½ oz., has been added; the mixed and filtered liquors are evaporated to dryness, and, a low red heat having been applied, the residuum is rapidly reduced to powder in a warm mortar, and at once enclosed in dry and well-stoppered bottles.

Prop. It exhibits most of the properties of hydrate of potassium, but in a vastly less degree; hence it is often termed 'mild' alkali. It is very deliquescent, effervesces with acids, exhibits an alkaline reaction with test-paper, is insoluble in alcohol, but dissolves in less than its own weight of water, its affinity for the last being so great that it is from alcoholic mixtures.

Pur., &c. Carbonate of potassium frequently contains an undue quantity of water, as well as silicic acid, sulphates, and chlorides. The water may be detected by the loss of weight the salt suffers when heated; the silica, by adding to it hydrochloric acid in excess, evaporating to dryness, and igniting the residuum, by which this contamination is rendered insoluble. The sulphates and chlorides may be detected by adding nitric acid in excess, and testing the liquid with nitrate of silver and chloride of barium; if the former produces a white precipitate a chloride is present, and if the latter does the same the contamination is a sulphate. Carbonate of potassium deliquesces in the air, and is almost entirely dis-

solved by water. It may be crystallised in prisms of the formula $2K_2CO_2.3H_2O$, which becomes $K_2CO_2.H_2O$ at 100° C. It changes the colour of turmeric brown. Supersaturated with nitric acid, neither carbonate of sodium nor chloride of barium throws down anything, and nitrate of silver very little. 100 gr. lose 16 gr. of water by a strong red heat; and the same weight loses 26·3 gr. of carbonate anhydride when placed in contact with dilute sulphuric acid.

Potassium, Bicarbonate of. KHCO₂. *Syn.* POTASSIUM HYDROGEN CARBONATE, BICARBONATE OF POTASSA; POTASSÆ BICARBONAS (B. P., Ph. L., E., & D.), L. *Prep.* 1. Carbonate of potassium, 6 lbs.; distilled water, 1 gall.; dissolve, and pass carbonic anhydride (from chalk and sulphuric acid diluted with water) through the solution to saturation; apply a gentle heat, so that whatever crystals have been formed may be dissolved, and set aside the solution that crystals may again form; lastly, the liquid being poured off, dry them.

2. Carbonic anhydride, obtained by the action of dilute hydrochloric acid on chalk (the latter contained in a perforated bottle immersed in a vessel containing the acid), is passed by means of glass tubes connected by vulcanised india-rubber to the bottom of a bottle containing a solution of carbonate of potassium, 1 part, in water, 2½ parts; as soon as the air is expelled from the apparatus the corks through which the tubes pass are rendered air-tight, and the process left to itself for a week; the crystals thus obtained are then shaken with twice their bulk of cold water, drained, and dried on bibulous paper, by simple exposure to the air. From the mother-liquor, filtered, and concentrated to one half, at a heat not exceeding 110° F., more crystals may be obtained. The tube immersed in the solution of carbonate of potassium will have to be occasionally cleared of the crystals with which it is liable to become choked, else the process will be suspended.

3. Potassium carbonate, 100 lbs.; distilled water, 17 galls.; dissolve, and saturate the solution with carbonic anhydride, as in No. 1, when 35 to 40 lbs. of crystals of bicarbonate of potassium may be obtained; next dissolve carbonate of potassium, 50 lbs., in the mother-liquor, and add enough water to make the whole a second time equal to 17 galls.; the remaining part of the operation is then to be performed as before. This plan may be repeated again and again for some time, provided the carbonate used is sufficiently pure.

4. Take of carbonate of potassium, 6 oz.; sesquicarbonate of ammonium, 3½ oz.; triturate them together, and, when reduced to a very fine powder and perfectly mixed, make them into a stiff paste with a very little water; dry this very carefully at a heat not higher than 140° F., until a fine powder, perfectly devoid of ammoniacal odour, be obtained, occasionally triturating the mass towards the end of the process.

5. (Commercial.) From carbonate of potassium, in powder, made into a paste with water, and exposed for some time on shallow trays in a chamber filled with an atmosphere of carbonic anhydride, generated by the combustion of either coke or charcoal, and purified by being forced

through a cistern of cold water; the resulting salt is next dissolved in the least possible quantity of water at the temperature of 120° F., and the solution filtered and crystallised.

Prop. It can be crystallised in large transparent monoclinic prisms. It is soluble in four times its weight of water at ordinary temperature; it is stable in the air, but loses carbonic acid below the temperature of a carbonate at a red heat. It possesses the general alkaline properties of carbonate of potassium, but in an inferior degree, having a saline or only a slightly alkaline taste, and, when absolutely pure, not affecting the colour of turmeric. When an aqueous solution of it is boiled it gives off CO_2.

Pur. and Tests. In a solution of pure bicarbonate of potassium a solution of mercuric chloride merely causes an opalescence, or very slight white precipitate; if it contains normal carbonate, a brick-coloured precipitate is thrown down. From 100 gr. of the pure crystals of bicarbonate, 30·7 gr. of water and carbonic acid are expelled at a red heat. In other respects it may be tested like the carbonate.

Uses, &c. Bicarbonate of potassium is the most agreeable of all the salts of potassium, and is much used as an antacid or absorbent, and for making effervescing saline draughts. It has also been successfully employed in rheumatism, scurvy, gout, dyspepsia, and various other diseases in which the use of potassium is indicated. The dose is from 10 gr. to ½ dr.

20 gr. bicarbonate, in crystals,

are equivalent to

14 gr. of crystallised nitric acid,
15 gr.   „   tartaric acid, and
½ oz. of lemon juice.

Potassium, Chlorate of. $KClO_3$. *Syn.* CHLORATE OF POTASH; POTASSAE CHLORAS (B. P., Ph. L. and D.), L. *Prep.* 1. Chlorine gas is conducted by a wide tube into a moderately strong and warm solution of hydrate or carbonate of potassium, until the absorption of the gas ceases and the alkali is completely neutralised; the liquid is then kept at the boiling temperature for a few minutes, after which it is gently evaporated if necessary until a pellicle forms on the surface; it is then set aside, so as to cool very slowly; the crystals which form are drained and carefully washed on a filter with ice-cold water, and are purified by re-solution and re-crystallisation. The mother-liquor, which contains much chloride of potassium mixed with some chlorate, is either evaporated for more crystals (which are, however, less pure than the first crop) or is preserved for a future operation.

Obs. The product of the above process is small, varying from 10% to 45% of the weight of the potassium consumed in it, according to the skill with which it is conducted; this apparent loss of potassium arises from a large portion of it being converted into chloride, a salt of comparatively little value. The following processes have been devised principally with the view of preventing this waste, or of employing a cheaper salt of potassium than the carbonate:

2. Mix slaked lime, 53 oz., with carbonate of potash, 20 oz., and triturate them with a few ounces of distilled water, so as to make the mixture slightly moist. Place oxide of manganese, 80 oz., in a large retort or flask, and having poured upon it hydrochloric acid, 24 pints, diluted with 6 pints of water, apply a gentle sand heat, and conduct the chlorine as it comes over, first through a bottle containing 6 oz. of water, and then into a large carboy containing the mixture of carbonate of potash and slaked lime. When the whole of the chlorine has come over remove the contents of the carboy and boil them for 30 minutes with 7 pints of distilled water; filter and evaporate till a film forms on the surface, then set aside to cool and crystallise.

The crystals thus obtained should be purified by dissolving them in three times their weight of boiling distilled water, and again allowing the solution to crystallise.

3. A solution of chloride of lime is precipitated with a solution of carbonate of potassium, and the liquid, after filtration, saturated with chlorine gas; it is then evaporated and crystallised as before. Dr Ure proposed the substitution of sulphate of potassium for the carbonate, by which the process would be rendered very inexpensive.

4. Carbonate of potassium, 69 parts of the dry or 82 parts of the granulated, hydrate of calcium (dry fresh slaked lime), 37 parts, both in powder, are mixed together, and exposed to the action of chlorine gas to saturation (the gas is absorbed with great rapidity, the temperature rises above 212° F., and water is freely evolved); the heat, with free exposure, is then maintained at 212° for a few minutes (to remove some trace of 'hypochlorite'); the residuum, consisting of chlorate of potassium and chloride of calcium, is treated with hot water, and the chlorate of potassium crystallised out of the resulting solution, as before. This process is an excellent one.

5. A solution of chloride of lime (18° to 20° Baumé) is heated in a leaden or cast-iron vessel, and sufficient of some salt of potassium added to raise the density of the liquid 3 or 4 hydrometer degrees; the solution is then quickly, but carefully, concentrated until the gravity rises to 30° or 31° Baumé, when it is set aside to crystallise. A good and economical process.

6. Chloride of potassium, 76 parts, and fresh calcium hydrate, 222 parts, are reduced to a thin paste with water, q. s., and a stream of chlorine gas passed through the mixture to saturation; chloride of calcium and chlorate of potassium are formed; the last is then removed by solution in boiling water, and is crystallised as before. This process, which has received the approval and recommendation of Liebig, has long been practised in Germany, and was originally introduced to this country by Dr Wagenmann. The product is very large, and of excellent quality.

7. The following process given in Dingler's 'Polytechnisches Journal,' clxxxix, p. 488, by Lunge, is stated to be the most efficient, and the one that will be employed on the large scale in the future. Into a solution of milk of lime (sp. gr. 1·04), chlorine gas is passed until the liquid is nearly saturated. The clear solution is then

evaporated until its sp. gr. = 1·18. Potassium chloride is now added, and the mixture reduced by evaporation to a sp. gr. 1·28. It is then allowed to cool, and the crystals of chlorate separate out.

Prop. White, inodorous, glassy, monoclinic tables, which, when of certain dimensions, exhibit iridescence and emit light on being rubbed in the dark; soluble in about 20 parts of cold and 2½ parts of boiling water; in taste it resembles nitre, at about 450° F. it fuses, and on increasing the heat almost to redness effervescence ensues, fully 39% of pure oxygen gas being given off, whilst the salt becomes changed into chloride of potassium. When mixed with inflammable substances such as sulphur, and triturated, heated, or subjected to a smart blow or strong pressure, or moistened with a strong acid, it explodes with great violence.

Pur., Tests. The usual impurity of this salt is chloride of potassium, arising from careless or imperfect manipulation. When this is present, a solution of nitrate of silver gives a curdy, white precipitate, soluble in ammonia; whereas a solution of the pure chlorate remains clear.

Uses. Chlorate of potassium is principally used in the manufacture of lucifer matches, fireworks, oxygen gas, &c., and as an oxidising agent in calico printing. It was formerly used to fill percussion caps, but was abandoned for fulminating mercury, on account of its disposition to rust the nipples of the guns. As a medicine it is stimulant and diuretic. It used to be given in dropsy, syphilis, scurvy, cholera, typhus, and other depressing affections, as it was believed to act as an oxidising agent on the blood. It has, however, lately been shown that the whole of the salt passes out undecomposed in the urine. It is still largely employed in the form of lozenges for allaying inflammation of the tonsils, and as a gargle.

Concluding Remarks. Formerly, chlorate of potassium was a salt which was made only on the small scale, and chiefly used in experimental chemistry; now it is in considerable demand, and forms an important article of chemical manufacture. The chlorate requires to be handled with great care. It should never be kept in admixture with any inflammable substance, more especially with sulphur, phosphorus, or the sulphides, as these compounds are exploded by the most trivial causes, and, not unfrequently, explode spontaneously.

Potassium, Chloride of. KCl. *Syn.* CHLORIDE OF POTASSA. This substance is an important natural source of potassium, being extracted from the ashes of seaweed, from sea water, and the refuse of beetroot sugar manufactories. In combination with magnesium chloride it forms the mineral camallite ($KCl,Mo_u,Cl_2,6H_2O$), which is found at Stassfrest in Saxony deposited in strata which overlie beds of rock-salt in the salt mines. Large deposits of a mineral consisting of the chloride and sulphate of potassium have also been found in East Galicia.

Prep. The chloride of potassium of commerce is usually a secondary product in the manufacture of chlorate of potassium and other substances. The mother-liquor of the former is

evaporated to dryness and heated to dull redness, the calcined mass is then dissolved in water, the solution purified by defecation and evaporated down for crystals.

It can also be well prepared by neutralising boiling solution of carbonate of potassium by dilute hydrochloric acid, evaporating down, and crystallising.

Prop., &c. It crystallises in four-sided tables, and closely resembles culinary salt in appearance; is anhydrous; dissolves in about 4 parts of cold and 2 of boiling water; has a slightly bitter, saline taste; fuses at a red heat; and is volatilised at a very high temperature. As a medicine it is diuretic and aperient. It was formerly in high repute as a resolvent and antiscorbutic, and, particularly, as a remedy for intermittents. It is now seldom used.

Potassium, Chromate of. K_2CrO_4. *Syn.* CHROMATE OF POTASSA, NEUTRAL CHROMATE OF P., MONOCHROMATE OF P., YELLOW C. OF P., SALT OF CHROME; POTASSE CHROMAS, P. C. FLAVA, L. The source of this salt is 'CHROME ORE,' a natural octahedral chromate of iron, found in various parts of Europe and America. For medicinal purposes the commercial chromate is purified by solution in hot water, filtration, and recrystallisation.

Prep. 1. The ore, previously assayed to determine its richness, and freed as much as possible from its gangue, is ground to powder in a mill, and mixed with a quantity of coarsely powdered nitre rather less than that of the oxide of chromium which it contains; this mixture is exposed for several hours to a powerful heat on the hearth of a reverberatory furnace, during which time it is frequently stirred up with iron rods ; the calcined mass is next raked out and lixiviated with hot water, and the resulting yellow-coloured solution evaporated briskly over a fire, or by the heat of high-pressure steam; chromate of potassium falls down in the form of a granular yellow salt, which is removed from time to time with a ladle, and thrown into a wooden vessel, furnished with a bottom full of holes, where it is left to drain and dry. In this state it forms the chromate of potassium of commerce. By a second solution and recrystallisation it may be obtained in large and regular crystals.

2. A mixture of pulverised chrome ore and chloride of potassium is exposed to a full red heat, on the hearth of a reverberatory furnace, with occasional stirring for some time, when steam at a very elevated temperature is made to act on it until the conversion is complete ; this is known by assaying a portion of the mass; the chromate is then dissolved out of the residuum, as before. Common salt or hydrate of calcium may be substituted for chloride of potassium, and then the chromates of sodium or calcium are respectively produced.

3. On the small scale this salt may be prepared from the bichromate by neutralising it with hydrate of potassium, or with potassium carbonate until the red colour changes to yellow; it is then evaporated and crystallised.

Prop. Yellow prismatic efflorescent crystals; tastes cool, bitter, and disagreeable; soluble in 2 parts of water at 60° F.

Pur. The salt of commerce is frequently contaminated with large quantities of sulphate or chlorate of potassium. To detect these Zueber adds tartaric acid, dissolved in 50 parts of water, to an aqueous solution of the sample. As soon as the decomposition is complete, and the colour verges towards green, the supernatant liquor should afford no precipitate with solutions of the nitrates of silver and barium, whence the absence of chlorides and sulphates may be respectively inferred. The proportions are 8 parts of tartaric acid to 1 part of the chromate. If saltpetre is the adulterating ingredient, the sample deflagrates when thrown upon burning coals.

Assay. 1. A solution of 50 gr. of the salt is treated with a solution of nitrate of barium, the precipitate digested in nitric acid, and the insoluble portion (sulphate of barium) washed, dried, and weighed. 117 gr. of this substance are equivalent to 89 gr. of sulphate of potassium.

2. The nitric acid solution, with the washing (see *above*), is treated with a solution of nitrate of silver, and the precipitate of chloride carefully collected, washed, dried, ignited, and weighed. 144 gr. of chloride of silver represent 76 gr. of chloride of potassium.

3. The nitric solution, with the washing (see *above*), after having any remaining barium precipitated by the addition of dilute sulphuric acid in slight excess, is treated with ammonia, and the resulting precipitate of chromic oxide collected on a filter, washed, dried, carefully ignited in a silver, platinum, or porcelain crucible, and weighed. 40 gr. of this oxide represent 100 gr. of pure chromate of potassium. Any deficiency consists of impurities or adulterants.

Uses. Chromate of potassium is used in dyeing, bleaching, the manufacture of chromic acid, bichromate of potassium, &c. It is the common source of nearly all the other compounds of chromium. A solution in 8 parts of water is occasionally used to destroy fungus; 1 in 30 to 40 parts of water is also used as an antiseptic and desiccant.

Concluding Remarks. The first process is undoubtedly the best when expense is not an object. To reduce this a mixture of ' potash ' or 'pearlash,' with about half of its weight of nitre, or 1-5th of its weight of peroxide of manganese, may be substituted without much inconvenience. The assay of the chrome ore, alluded to above, may be made by reducing 100 gr. of it to powder, mixing it with twice its weight of powdered nitre and a little hydrate of calcium, and subjecting the mixture to a strong red heat for 3 or 4 hours; the calcined mass may then be exhausted with boiling water, and the resulting solution, after precipitation with dilute sulphuric acid in slight excess, and filtration, may be treated with alcohol, when its chromium may be thrown down by the addition of ammonia (see *above*). In the conversion of chrome ore into chromate of potassium care should, in all cases, be taken that the proportion of nitre or alkali should be slightly less than what is absolutely required to saturate the ore, as the production of a neutral salt is thereby ensured; for should not the whole of the chromate be decomposed by the first calcination it may

easily be roasted a second time with fresh alkali. The nature of the furnace to be employed in the conversion is not of any great importance so long as carbonaceous matters from the fire are entirely excluded, and the required temperature is attainable.

Potassium, Bichromate of. $K_2Cr_2O_7$, or $K_2CrO_4.CrO_3$. *Syn.* BICHROMATE OF POTASSA, RED CHROMATE OF POTASH, ACID C. OF P.; POTASSE BICHROMAS, L. *Prep.* 1. To a concentrated solution of yellow chromate of potassium, acetic acid is added in quantity equal to one half that required for the entire decomposition of the salt; the liquid is then concentrated by evaporation and slowly cooled, so that crystals may form.

2. (*Jacquelain.*) Chrome ore, finely ground and sifted, is mixed with chalk; the mixture is spread in a thin layer on the hearth of a reverberatory furnace, and heated to bright redness, with repeated stirring, for about 10 hours. The yellowish-green product consists essentially of neutral chromate of calcium mixed with ferric oxide. Having been ground and stirred up with hot water, sulphuric acid is added till a slight acid reaction becomes apparent, a sign that the neutral chromate has been converted into bichromate. Chalk is now stirred in to precipitate the ferric sulphate, and after a while the clear solution is run off into another vessel, where it is treated with carbonate of potassium, which precipitates the lime, and leaves bichromate of potassium in solution. The solution is then evaporated to the crystallising point. This process, when carried out on a large scale, is very economical.

3. (*Stromeyer's* new method.) 4½ parts of finely-ground roasted chrome iron ore is mixed with 2½ parts of potassium carbonate and 7 parts of lime. The mixture is dried at 150°, and then heated to redness with an oxidising flame, being kept constantly stirred. The charge is then withdrawn from the furnace, cooled, and lixiviated with a small quantity of hot water. Should calcium chromate have been formed a hot solution of potassium sulphate is added; this precipitates the lime, and potassium chromate remains in solution. After being treated with the proper quantity of sulphuric acid and diluted with twice its volume of water, the liquid is allowed to cool. A precipitate forms, which is then collected and recrystallised. The mother-liquors are used for the lixiviation of more roasted mixture.

Prop., &c. It forms very beautiful garnet-red square tables, or flat four-sided prismatic crystals; permanent in the air; soluble in 10 parts of water at 60°, and in less than 3 parts at 212° F.; it has a metallic, bitter taste, and is poisonous. It is chiefly used in dyeing and bleaching, in the manufacture of chrome yellow, and as a source of chromic acid. The tests, &c., are the same as for the yellow chromate.

Potassium, Citrate of. $K_3C_6H_5O_7$. *Syn.* POTASSE CITRAS, L. *Prep.* From a solution of citric acid neutralised with carbonate of potassium, evaporated, and granulated, or crystallised; very deliquescent. Or in the form of solution, by adding carbonate, or bicarbonate, of potassium to lemon juice, as in the common effervescing draught.

Potassium Cyanate. KO.CN. *Prep.* By passing gaseous cyanogen chloride into an aqueous solution of potash, kept well cooled.

Prop. Crystallise in needles, which, when heated till they fuse, become changed into the isocyanate.

Potassium, Isocyanate of. KCyO, or KCNO. *Prep.* 1. By roasting, at a red heat, dry ferrocyanide of potassium, in fine powder, upon an iron plate, constantly stirring it until it becomes fused into one mass, which must be reduced to fine powder and digested in boiling alcohol, from which crystals of the cyanate will be deposited as the solution cools.

2. (*Liebig.*) A mixture of ferrocyanide of potassium with half its weight of peroxide of manganese is kindled by a red-hot body, and allowed to smoulder away, after which it is treated with alcohol, as before.

3. A mixture of ferrocyanide of potassium and litharge is heated as before, then dissolved out by alcohol, and crystallised.

Prop. Colourless or white salt, crystallisable in plates, readily soluble in alcohol and water, but decomposed when moist into bicarbonate of potassium and ammonia, or in solution into the carbonate of potassium and ammonium.

This salt is poisonous. The cyanates of silver, lead, and many other metals may be made by adding a solution of cyanate of potassium to another of a neutral salt of the base.

Potassium, Cyanide of. KCN, or KCy. *Syn.* CYANIDE OF POTASH, CYANURET OF POTASSIUM; POTASSII CYANIDUM, P. CYANURETUM, L. *Prep.* 1. *a.* A solution of pure hydrate of potassium, 2 parts, in alcohol, 7 parts, is placed in a receiver furnished with a safety tube, and surrounded with bruised ice; the beak of a tubulated retort, containing ferrocyanide of potassium, in powder, 4 parts, is then adapted to it in such a manner that any gas or vapour evolved in the retort must traverse the solution in the receiver; the arrangement being complete, sulphuric acid, 3 parts, diluted with an equal weight of water, and allowed to cool, is cautiously poured into the retort, and the distillation conducted very slowly, only a very gentle heat being applied; as soon as the force of ebullition in the retort has subsided the distillation is complete, and the connection between the retort and receiver is broken; the contents of the receiver, now transformed into a mixture of a crystalline precipitate of cyanide of potassium, and an alcoholic solution of undecomposed potassium hydrate, is carefully thrown on a filter, and the precipitate, after the mother-liquor has drained off, very cautiously washed with ice-cold and highly rectified spirit; then drained, pressed, and dried. The product is chemically pure, and equal to fully 10% of the ferrocyanide employed. This is a modification of what is commonly known as 'Wigger's process.'

b. Expose well-dried and powdered ferrocyanide of potassium to a dull red heat, in a closed vessel; when cold, powder the fused mass, place it in a funnel, moisten it with a little alcohol, and wash it with cold water; evaporate the solution thus formed to dryness, expose it to a dull red heat in a porcelain dish, then cool, powder, and digest it in boiling alcohol; as it cools, crystals of cyanide of potassium, nearly pure, will be deposited. The alcohol employed in both this and the preceding process may be recovered by distillation from calcined sulphate of iron.

2. (CRUDE or COMMERCIAL CYANIDE—*Liebig.*) Commercial ferrocyanide of potassium, 8 parts, rendered anhydrous by gently heating it on an iron plate, is intimately mixed with dry carbonate of potassium, 3 parts; this mixture is thrown into a red-hot earthen crucible, and kept in a state of fusion, with occasional stirring, until gas ceases to be evolved, and the fluid portion of the mass becomes colourless; the crucible is then left at rest for a few minutes, to allow its contents to settle, after which the clear portion is poured from the heavy black sediment upon a clean marble slab, and the mass, whilst yet warm, broken up, and placed in well-stoppered bottles.

Obs. A cheap and excellent process. The product, though not sufficiently pure for employment in medicine as potassium cyanide, is admirably adapted for various technical applications of this substance, such as in electro-plating, electro-gilding, photography, &c. It may also be advantageously substituted for the ferrocyanide in the preparation of hydrocyanic acid by the distillation of that substance with dilute sulphuric acid.

Prop., &c. When pure, this salt is colourless and odourless; it forms cubic or octahedral crystals, which are anhydrous; it is freely soluble in water and in boiling alcohol, but most of it separates from the latter as the solution cools; it is fusible, and undergoes no change even at a full red heat in close vessels; it exhibits an alkaline reaction; when exposed to the atmosphere it absorbs moisture, and acquires the smell of hydrocyanic acid. If it effervesces with acids, it contains carbonate of potassium, and if it be yellow it contains iron. It is employed in chemical analysis, and for the preparation of hydrocyanic acid; cyanide of sodium may be made in the same way. The dose is $\frac{1}{12}$ to $\frac{1}{4}$ gr., in solution; in the usual cases in which the administration of hydrocyanic acid is indicated.—*Antidotes.* The same as for hydrocyanic acid.

Potassium Ethylate. C_4H_5KO. Löweg and Weidmann obtained this compound by heating together acetate of ethyl and potassium. Dr B. W. Richardson, some few years ago, recommended the employment of the alkaline ethylates as caustics, and they are now frequently used as such in surgery. When first applied to the body the ethylates produce no action, but as they absorb water from the tissues they are decomposed, the potassium or sodium is oxidised, yielding caustic potash or soda in a nascent condition, while alcohol is re-formed from the recombination of hydrogen derived from the water. Dr Richardson believes the ethylates of potassium and sodium will be found the most effective and manageable of all caustics, and that in cases of cancer, when it is important to destroy structure without resorting to the knife, and in the removal of simple growths, they will be of essential service. The ethylates dissolve in alcohol of different strengths; the solution may either be applied with a glass brush ,or injected by the needle, and a slow or quick effect can be insured according to the wish of the operator.

Potassium, Ferricyanide of. $K_6(C_2N_3)_4Fe_2$. *Syn.* FERRIDCYANIDE OF POTASSIUM, FERRICYANURET OF P., RED PRUSSIATE OF POTASH; POTASSII FERRICYANIDUM, P. PRUSSIAS RUBRUM, L. This important and beautiful salt was discovered by L. Gmelin. At first it was merely regarded as a chemical curiosity, but it is now extensively employed in dyeing, calico printing, assaying, &c.

Prep. 1. Chlorine gas is slowly passed into a cold solution of ferrocyanide of potassium, 1 part, in water, 10 parts, with constant agitation, until the liquid appears of a deep reddish-green colour, or of a fine red colour by transmitted light, and ceases to give a blue precipitate, or even a blue tinge to a solution of ferric chloride, an excess of chlorine being carefully avoided; the liquor is next evaporated by the heat of steam or boiling water, until a pellicle forms upon the surface, when it is filtered, and set aside to cool; the crystals are afterwards purified by re-solution and re-crystallisation; or simply evaporate the original solution to dryness, by a steam heat, with agitation, then re-dissolve the residuum in the least possible quantity of boiling water, and, after filtration, allow the new solution to cool very slowly so that crystals may form.

2. Add nitric acid or some other oxidising agent very gradually to a cold solution of ferrocyanide of potassium, with constant agitation, until a drop of the mixture ceases to impart a blue colour to a solution of ferric chloride, carefully avoiding excess of acid. It may be at once used in solution, or evaporated, &c., as before.

Prop. &c. Magnificent monoclinic prismatic crystals, of a rich ruby-red tint, which are frequently mackled; permanent in the air; combustible; decomposed by a high temperature; soluble in 4 parts of cold water; yielding a yellowish-brown or, when diluted, lemon-coloured solution, which ultimately becomes changed into a ferrocyanide or a blue precipitate; insoluble in alcohol. It is a powerful oxidising agent. Colours ferric salts a pale brown, gives with ferrous salts a deep blue, and precipitates bismuth salts pale yellow; cadmium and mercuric salts, yellow; zinc salts, deep yellow; mercurous, cupric, molybdenic, silver, and uranic salts, reddish brown; cobalt salts, dark brown; manganous salts, brown; cupric salts, greenish; and nickelous salts, olive-brown.

Potassio-ferric Ferrocyanide. $K_2Fe_2(FeCN)_2$. *Syn.* SOLUBLE PRUSSIAN BLUE.

Prep., &c. By the addition of ferric chloride or ferric sulphate solution to potassium ferrocyanide solution, the latter being in excess. This blue is insoluble in the liquid containing saline matter, but dissolves as soon as the latter has been removed by washing; the addition of an acid or a salt reprecipitates it.

Potassium, Ferrocyanide of. $K_6(C_2N_3)_4Fe_2 + 6H_2O$. *Syn.* FERROCYANURET OF POTASSIUM, PRUSSIATE OF POTASH, YELLOW P. OF P.; POTASSÆ PRUSSIA FLAVA (B. P.), POTASSII FERROCYANIDUM (Ph. L., E., and D.), L. This valuable salt, the well-known 'prussiate of potash' of commerce, was discovered by Scheele about the middle of the 18th century.

Prep. (Large scale.) Good 'potash' or 'pearlash,' 2 parts, and dried blood, horns, hoofs, woollen rags, or other refuse animal matter, 5 or 6 parts, are reduced to coarse powder, and mixed with some coarse iron borings; the mixture is then placed in cased hemispherical cast-iron pots set in brickwork heated by fire; the mass is constantly stirred by means of vertical spindles which pass through the lids, so as to prevent it running together, and the calcination is continued until fetid vapours cease to be evolved, care being taken to exclude the air from the vessels as much as possible; during the latter part of the process the mass is stirred less frequently; it is then removed with an iron ladle, and excluded from the air until cold; it is next exhausted by lixiviation with boiling water, and the resulting solution, after filtration, is concentrated by evaporation, so that crystals may form as the liquid cools; these are redissolved in hot water, and the solution allowed to cool very slowly, when large yellow crystals of ferrocyanide of potassium are deposited.—*Product.* 1 ton of dried blood or woollen refuse, with 3 cwt. of pearlash, yields from 2 to 2½ cwt. of commercial ferrocyanide. The mother-liquor contains sulphate of potassium.

In this operation the sulphur contained in the animal matter, as well as that contained as sulphate in the 'potashes,' very quickly attacks the iron pots, and a further loss occurs owing to the formation of potassium thiocyanate.

A better yield is obtained when potash free from sulphate is employed, and when on lixiviating the mass with water some freshly precipitated ferrous carbonate (obtained by dissolving chalk in ferrous chloride solution) is added. The solution is then evaporated to a specific gravity 1·27, and allowed to crystallise; the crude salt is dissolved in warm water to form a solution of the same specific gravity as before, and again crystallised. The mother-liquors are reserved for use in dissolving fresh quantities of crude salt.

Prop. It forms large and very beautiful lemon-yellow quadratic pyramidal crystals, which are permanent in the air, and very tough and difficult to powder, sp. gr. 1·88; it is soluble in 4 parts of cold and 2 parts of boiling water; insoluble in alcohol; has a mild saline taste, and is not poisonous; at a gentle heat loses water; at a higher temperature, in closed vessels, it is for the most part converted into cyanide of potassium, and, when exposed to the air, into cyanate of potassium. Precipitates solutions of antimonious, bismuth, mercurous, and zinc salts, white; cadmium salts, pale yellow; cuprous salts, white, turning red; ferrous salts, white, turning blue; lead salts, white; manganous salts, white, turning red; mercuric salts, white, turning bluish; nickelous salts, white, turning green; silver salts, white; stannous salts, white; cobalt salts, green; cupric salts, chocolate-red; ferric salts, dark blue; palladous salts, green; stannic salts, yellow; uranic salts, reddish brown; and zinc salts, white.

The commercial salt frequently contains potassium sulphate, which can only be removed by repeated recrystallisation.

Uses, &c. Ferrocyanide of potassium is chiefly used in dyeing and calico printing, in the manufacture of Prussian blue, in electro-plating, and, in chemistry, as a test, and a source of hydrocyanic acid. As a medicine it is said to be sedative and astringent, and in large doses purgative, but appears to possess little action.—*Dose*, 10 gr. to ½ dr., dissolved in water; in hooping-cough, chronic bronchitis, night-sweats, leucorrhœa, &c. Potassio - ferrous Ferrocyanide. (K₂Fe₂CN.) *Prep.* Obtained as a white precipitate when a solution of a ferrous salt quite free from ferric salt is added to potassium ferrocyanide quite free from ferricyanide. When exposed to the air it absorbs oxygen and becomes blue; oxidising agents, like chlorine water and nitric acid, change it into Prussian blue.

Potassium, Hydrate of. KHO. *Syn.* CAUSTIC POTASH, POTASSA HYDRATE, HYDRATE OF POTASSA, POTASSA, HYDRATED OXIDE OF POTASSIUM; POTASSA CAUSTICA (B. P.), POTASSA (Ph. E.), P. CAUSTICA (Ph. D.), P. HYDRAS (Ph. L.), P. FUSA, L. This substance was considered to be an oxide of potash till Darcet, in 1808, showed that when ignited it contains some other ingredient, which he considered to be water. For a long time it was then supposed to be a compound of potassium oxide, and it was not till comparatively recent years that it was recognised to be a hydroxide.

Prep. 1. (Pure.) From the metal or the oxide by solution in water and subsequent evaporation.

2. From a dilute solution of carbonate of potassium by decomposition with slaked lime. Potassium carbonate, 1 part by weight is dissolved in water, 12 parts, and the solution heated in a covered iron or silver vessel to boiling; milk of lime is then gradually added till on heating a portion of the filtered liquid with hydrochloric acid, it evolves no carbonic anhydride. After settling the liquid is decanted into a well-stoppered vessel, then evaporated in a silver vessel till the residue begins to volatilise. It is then cast in sticks.

3. *Wöhler's Process.* (Pure.) Pure potassium nitrate, 1 part, and thin copper-foil, 3 parts, are arranged in thin layers alternately in a copper crucible, and exposed to a red heat for several hours. The mass is then lixiviated with water after it has cooled down, the liquid allowed to stand in a tall stoppered vessel, decanted, and finally evaporated down in a silver vessel and cast in sticks ('Ann. Chem. Pharm.,' lxxxvii, p. 273).

4. *Schubert's Process.* (Pure.) To hot concentrated baryta water powdered potassium sulphate is added till there is a slight excess of the latter. This is removed by carefully adding more baryta water. The solution is then treated as in No. 3 ('Journal Pract. Chem.,' xxvi, p. 117).

5. *Polacci's Process.* Very pure hydrate of potash may be obtained in a few minutes by the following process. In an iron vessel a mixture consisting of 1 part of nitrate of potash well triturated with 2 or 3 parts of iron filings is heated. The mass becomes red in a few minutes, and, after cooling, it is treated with water, left to settle, and then decanted. A more or less concentrated solution may thus be obtained, or it may be evaporated to produce the solid potash.

Pur. The hydrate, obtained as above, is dissolved in alcohol or rectified spirit, and, after repose for a few days in a closely stoppered green glass bottle, the solution is decanted, and cautiously evaporated in a deep silver basin, out

of contact with the air. By this method it may be obtained free from potassium sulphate and alumina, but it always contains traces of potassium chloride, potassium carbonate, and potassium acetate, the latter being formed by the action of the potash on the alcohol.

6. A quantity of potassium hydrate may be obtained from the liquor potassæ of the shops thus:—Evaporate 1 gall. in a clean iron or silver vessel over an open fire until the ebullition being finished the residuum liquefies. Pour this into proper moulds.

Prop. When perfectly pure it is a hard, white solid, very soluble in water and in alcohol; intensely acrid and corrosive, and exhibiting the usual signs of alkalinity in the highest degree. That of the shops has usually a greyish or bluish colour. Caustic potash often exhibits a fibrous structure; it melts below red heat to a clear oily liquid, and at higher temperature it volatilises at a white heat; the vapours decompose into potassium, oxygen, and hydrogen. It is very deliquescent, rapidly absorbing moisture from the air and also carbonic anhydride. 1 part water dissolves 2·18 parts caustic potash with evolution of heat.

Uses, &c. Since caustic potash destroys both animal and vegetable substance it acts as a powerful cautery, and is much used in surgery; for this reason aqueous or alcoholic solutions can only be filtered through glass or sand; they are best clarified by subsidence. It is largely used in chemical analysis for absorbing carbonic anhydride, and for drying certain gases and liquids; it is an indispensable adjunct to the laboratory. Its most important commercial use is in the manufacture of soft soap. The Liquor Potassæ of the Pharmacopœia has a sp. gr. = 1·058, and contains about 5% of KOH.

Gerlach (vide 'Zeitschrift für analytische Chemie,' vol. viii, p. 279) gives the following table of specific gravities of aqueous solutions at 15° C.:

Per cent. of KOH.	Sp. gr.	Per cent. of KOH.	Sp. gr.
1	1·009	40	1·411
5	1·041	45	1·475
10	1·083	50	1·539
15	1·128	55	1·604
20	1·177	60	1·667
25	1·230	65	1·729
30	1·288	70	1·790
35	1·349		

Solid caustic potash must always be kept in well-stoppered bottles, which when not in use may be waxed down.

Potassium, Iodate of. KIO₃. *Syn.* PO-TASSÆ IODAS, L. *Prep.* 50 gr. of iodine are digested with 50 gr. of potassium chlorate, both freely powdered, with ½ fl. oz. of nitric acid in a flask till all colour disappears; the liquid is then boiled for about 1 minute, poured out into a dish, evaporated to dryness, and then moderately heated. The product consists of potassium iodate and potassium chloride; the latter can be dissolved out with water.

Prop. Small colourless cubical crystals. It is a useful test for SO₃ which at once liberates iodine from it; if a drop or two of starch paste has been previously mixed with a solution of KIO_3, the addition of SO_2 generates the characteristic blue coloration.

Potassium, Iodide of. KI. *Syn.* POTASSII IODIDUM (B. P., Ph. L., E., and D.), L.

1. Take of iron filings, 2 oz.; distilled water, 2 quarts; iodine, 6 oz.; mix them, and heat the solution until it turns green, and then add of carbonate of potassium, 4 oz., dissolved in water, 1 quart; filter, wash the residuum on the filter with water, evaporate the mixed filtered liquors and crystallise.—*Product.* 1 oz. of iodine yields 1 oz. 45 gr. of iodide.

2. Add iodine to a hot solution of pure hydrate of potassium until the alkali is perfectly neutralised, carefully avoiding excess; evaporate the liquid to dryness, and expose the dry mass to a gentle red heat in a platinum or iron crucible; afterwards dissolve out the salt, gently evaporate the solution and crystallise. An excellent process, yielding a large product, but, if the ignition be not carefully managed, it is apt to contain a little undecomposed iodate. The addition of a little powdered charcoal to the mass before ignition will obviate this (*Scanlan*).

3. Iodine is treated with a small proportion of phosphorus in water, and is thus converted into hydriodic acid; hydrate of calcium is then added, and the iodide of calcium formed is separated, fused, and then decomposed by sulphate of potassium into sulphate of calcium, which is precipitated, and iodide of potassium, which remains in solution, and may be crystallised, as in the other processes. This is a modification of a method devised by Liebig.

4. Put solution of potash, 1 gall., into a glass or porcelain vessel, and add iodine, 29 oz., or q. s., in small quantities at a time, with constant agitation, until the solution acquires a permanent brown tint. Evaporate the whole to dryness in a porcelain dish, pulverise the residue, and mix this intimately with wood charcoal in fine powder, 3 oz. Throw the mixture, in small quantities at a time, into a red-hot crucible, and when the whole has been brought to a state of fusion, remove the crucible from the fire and pour out its contents. When the fused mass has cooled, dissolve it in 2 pints of boiling distilled water, filter through paper, washing the filter with a little distilled boiling water; add the washings and then evaporate till a film forms on the surface. Set aside to cool and crystallise. Drain the crystals and dry quickly with a gentle heat; more crystals may be obtained by evaporating the mother-liquor, and cooling. The salt should be kept in a stoppered bottle.

Prop. It crystallises in cubes, which in the pure salt are transparent if slowly deposited from a somewhat dilute solution, whilst if they are deposited from a hot solution they have an opaque porcelain-like appearance; these are anhydrous; fuse at 639° C., without decomposition; dissolve in less than an equal weight of water, at 60° F., and sparingly soluble in alcohol; do not deliquesce in moderately dry air unless they contain undecomposed hydrate of potassium. Its solution dissolves iodine freely, and also, less readily, several of the insoluble metallic iodides and oxides. Its aqueous solution alters the

colour of turmeric either not at all or but very slightly, nor does it affect litmus paper, or effervesce with acids. Nitric acid and starch being added together, it becomes blue. It is not coloured by the addition of tartaric acid with starch. No precipitate occurs on adding either a solution of hydrate of calcium or of chloride of barium. 100 gr., dissolved in water, by the addition of nitrate of silver, yield a precipitate of 141 gr. of iodide of silver.

Assay. The iodide of commerce frequently contains fully one half its weight of either chloride or carbonate of potassium, or both of them, with variable quantities of iodate of potassium, a much less valuable salt. The presence of these substances is readily detected. As the first of these is only very slightly soluble in cold alcohol, and the others insoluble in that liquid, a ready method of determining the richness of a sample in pure iodide, sufficiently accurate for ordinary purposes, is as follows:—Reduce 50 gr. of the sample to fine powder, introduce this into a test-tube with 6 fl. dr. of alcohol, agitate the mixture violently for 1 minute, and throw the whole on a weighed filter set in a covered funnel, observing to wash what remains on the filter with another fl. dr. of alcohol. The filtrate, evaporated to dryness, gives the quantity of pure iodide, and the increase of weight of the filter dried at 100° C., that of the impurities present in the sample examined, provided it contained no hydrate of potassium. The quantity of alkali, whether hydrate or carbonate, may be found by the common method of 'alkalimetry.'

Uses, &c. Chiefly in photography, medicine, and pharmacy.—*Dose,* 1 to 10 gr., twice or thrice daily, made into pills, or, better, in solution, either alone or combined with iodine; in bronchocele, scrofula, chronic rheumatism, dropsy, syphilis, glandular indurations, and various other glandular diseases. Also externally, made into a lotion or ointment.

Potassium, Nitrate of. KNO_3. *Syn.* NITRATE OF POTASH, NITRE, SALTPETRE; POTASSÆ NITRAS (B. P., Ph. L., E., and D.), NITRUM†, SAL NITRI†, SAL PETRÆ†, KALI NITRATUM†, L. This salt is found as an efflorescence on the surface of the soil and in certain porous felspathic rocks in various parts of the world, especially in the East Indies.

Prep. 1. On the Continent it has long been produced artificially, by exposing a mixture of calcareous soil and animal matter to the atmosphere, when calcium nitrate is slowly formed, and is then extracted by lixiviation. The solution is decomposed by the addition of wood ashes, or carbonate of potassium, by which carbonate of calcium is precipitated, and nitrate of potassium remains in solution. The places where these operations are performed are called 'nitrières,' or 'nitrières artificielles.' The salt of the first crystallisation, by either process, is called 'crude nitre' or 'rough saltpetre.' This is purified by solution in boiling water, skimming, and, after a short time allowed for defecation, straining (while still hot) into wooden crystallising vessels. The crystals thus obtained are called 'single refined nitre;' and when the process is repeated 'double refined nitre.'

2. (Artificial.) Equal molecular quantities of sodium nitrate and potassium chloride are dissolved in hot water until the sp. gr. = 1·5. Chloride of sodium forms and is thrown down as a precipitate. The clear solution is decanted off, and on being agitated and allowed to cool potassium nitrate crystallises out as 'saltpetre flour.'

Prop. White, pellucid, regular six-sided prisms; permanent in the air; soluble in 7 parts of water at 60° with absorption of much heat, and in 1 part at 212° F.; insoluble in alcohol; its taste is cool, saline, slightly bitter; at about 560° it fuses to an oily-looking mass, which concretes on cooling, forming 'sal prunella ;' at a red heat it gives out oxygen, and, afterwards, nitrous fumes; sp. gr. 2·1.

Pur. Commercial nitre generally contains chlorides, sulphates, or calcareous salts. The first may be detected by its solution giving a cloudy white precipitate with nitrate of silver; the second by chlorides of barium or calcium giving a white precipitate; and the third by oxalate of ammonium giving a white precipitate. It may be purified thus :—Commercial nitre, 4 lbs.; boiling distilled water, 1 quart; dissolve, withdraw the heat, and stir the solution constantly as it cools; the minute crystals, thus obtained, are to be drained, and washed, in a glass or earthenware percolator, with cold distilled water, until that which trickles through ceases to give a precipitate with a solution of nitrate of silver; the contents of the percolator are then to be withdrawn, and dried in an oven.

Assay. Of the numerous methods prescribed for this, few are sufficiently simple for our purpose. The proportion of chlorides, sulphates, and calcareous salts may be determined separately, and the general richness of the sample by the method of Gay-Lussac, modified as follows: —100 gr. of the sample (fairly chosen) are triturated with 50 gr. of lampblack and 400 gr. of common salt; the mixture is then placed in an iron ladle, and ignited or fused therein, due care being taken to prevent loss; the residuum is exhausted with hot water, and the solution thus obtained tested by the usual methods of alkalimetry for carbonate of potassium. The quantity of carbonate found, multiplied by 2·125 or 2⅛, gives the percentage richness of the sample in nitrate.

Uses, &c. Nitre is chiefly employed in the manufacture of gunpowder, fireworks, and nitric acid. It is used in curing meat, to which it imparts a red colour. It is used in many chemical operations, and in medicine as a sedative, refrigerant, and diaphoretic, and as a cooling diuretic. It has been recommended in active hæmorrhages (especially spitting of blood), in various febrile affections, in scurvy, and in herpetic eruptions; and it has been highly extolled by Dr Basham as a remedy in acute rheumatism.—*Dose,* 5 to 15 gr., every two hours. A small piece, dissolved slowly in the mouth, frequently stops a sore throat at the commencement. In large doses it is poisonous. The best treatment is a powerful emetic, followed by opiates.

Potassium, Nitrite of. KNO₂. *Syn.* NITRITE OF POTASSA; POTASSÆ NITRIS, L. This is formed when saltpetre is heated until one atom of oxygen is

evolved. *Prep.* 1. By heating nitre to redness, dissolving the fused mass in a little water, and adding twice the volume of the solution in alcohol; after a few hours the upper stratum of liquid is decanted, and the lower one, separated from the crystals, evaporated to dryness.

2. (*Corenwinder.*) Nitric acid, 10 parts, are passed through a solution of hydrate of potassium of the sp. gr. 1·38 to saturation; the liquid is then neutralised with a little potassium hydrate (if necessary), and at once evaporated.

Prop., &c. Small indistinctly formed crystals; deliquescent; insoluble in alcohol; decomposes, evolving red fumes, on being heated with dilute sulphuric acid. Used for the separation of cobalt and nickel, and in organic chemistry for effecting simultaneously the removal of 3 atoms of hydrogen from a compound and the insertion of 1 atom of nitrogen.

Potassium, Oxalate of. K₂C₂O₄.4H₂O. *Syn.* NEUTRAL OXALATE OF POTASSA, NORMAL O. OF P.; POTASSÆ OXALAS, L. *Prep.* Neutralise a solution of oxalic acid, or the acid oxalates, with carbonate of potassium, evaporate, and crystallise. Transparent colourless rhombic prisms, soluble in 3 parts of water.

Potassium, Hydrogen Oxalate of. KHC₂O₄.2H₂O. *Syn.* POTASSIUM BINOXALATE, SALT OF SORREL, ESSENTIAL SALT OF LEMONS; POTASSÆ BINOXALAS, L. *Prep.* By saturating a solution of oxalic acid, 1 part, with carbonate of potassium, adding to the mixture a similar solution of 1 part of oxalic acid, and evaporating for crystals. It may also be obtained from the expressed juice of rhubarb, sheep's sorrel, or other species of *Rumex,* by clarifying it with eggs or milk, and evaporating, &c., as before.

Prop., &c. Colourless rhombic crystals, soluble in 40 parts of cold and 6 parts of boiling water, yielding a very sour solution. A solution of this salt is often used for removing ink stains from paper.

Potassium, Trihydrogen Oxalate. of. KHC₂O₄.H₂C₂O₄.2H₂O. *Syn.* POTASSIUM QUADROXALATE, POTASSIUM ACID OXALATE. *Prep.* By neutralising 1 part of oxalic acid with carbonate of potassium, adding to the solution 3 parts more of oxalic acid, evaporating and crystallising.

Prop., &c. Resembles the last, but is less soluble, and more intensely sour, and forms triclinic crystals. The salt is occasionally sold under the names of 'sal acetosella,' 'salt of sorrel,' and 'essential salt of lemons.' It is used to remove ink and iron stains from linen, to bleach the straw used for making bonnets, and, occasionally, in medicine, as a refrigerant.

Potassium, Oxide of. *Syn.* P. MONOXIDE. K₂O. *Prep. &c.* Burn pure potassium in a current of oxygen. White powder or grey brittle mass, rapidly absorbing water and forming the hydrate.

Potassium Peroxide. K₂O₄. *Prep.* Heat clean potassium in a current of dry air, and then in dry oxygen, according to Harcourt's plan (*vide* 'Journal Chemical Society,' xiv, p. 257). Dark chrome-yellow coloured powder.

Potassium, Perchlorate of. KClO₄. *Syn.* POTASSÆ PERCHLORAS, L. Prepared by mixing

87

well-dried and finely powdered chlorate of potassium in small portions at a time with warm nitric acid. The salt is separated by crystallisation.

Potassium, Prus'siate of. See POTASSIUM FERRICYANIDE and FERROCYANIDE.

Potassium, Salicylite of. Formed by mixing salicylous acid with a strong solution of hydrate of potassium; it separates, on agitation, as a yellow crystalline mass, which, after pressure between bibulous paper, is recrystallised from alcohol. Golden-yellow crystals, soluble in both water and alcohol; damp air gradually converts them into acetate of potassium and melanic acid.

Potassium, Silicate of. K_2SiO_3. Syn. POTASSIUM SILICAS, P. METASILICATE. Prep., &c. Mix 1 part of powdered quartz or flint, or of fine siliceous sand, with 2 parts of carbonate of potash, and fuse them in a Hessian crucible. Dissolve the mass in water, filter the solution and evaporate; a glassy deliquescent mass.—Dose, 10 to 15 gr., in 6 or 8 oz. of water, twice a day. To dissolve gout concretions.

Potassium Tetrasilicate. $K_2Si_4O_9$. Syn. WATER GLASS. The 'soluble glass' of Freek. Prep. By fusing 45 parts quartz, 30 of potashes, and 3 of powdered charcoal for five or six hours; the greyish-black glass is then boiled with five times its weight of water, the volume somewhat reduced, and then 1-4th its bulk of strong alcohol added. After standing the mother-liquor is poured off and the residue dissolved in water. It is used in the preservation of stone; mixed with sand and lime as a cement, and in the soap manufacture, although the sodium silicate is superior to it for these purposes.

Potassium, Sulphate of. K_2SO_4. Syn. POTASSÆ SULPHAS (B. P., Ph. L., E., & D.). This salt has been known since the 14th century. It is found native in the lava of Vesuvius and in kainite. It is also obtained as a bye-product in several chemical manufactures, e. g. the preparation of potash, bichromate of potash, &c.

Prep. 1. From kainite. The mineral, after having been weathered by exposure to the atmosphere, becomes deliquescent; the soluble magnesium chloride which it also contains is then partially decomposed by treatment with boiling water, and on cooling the sulphate, which is only sparingly soluble, crystallises out.

2. The residuum of the distillation of nitric acid from nitre is dissolved in water, the solution neutralised with carbonate of potassium, and after defecation and evaporation until a pellicle forms, it is strained, or decanted, and set aside to crystallise. Or, the residuum is simply ignited, to expel excess of acid, and then dissolved and crystallised as before.

Prop., &c. Anhydrous, heavy, hard, rhombic, pyramidal crystals; permanent in the air; soluble in 12 parts of water at 60° and in 5 parts at 212° F.; insoluble in alcohol; extremely nauseous and bitter tasted. It crepitates on the application of heat; fuses at a red heat, but can be volatilised only at high temperature. It is used as a purgative in medicines, and in the manufacture of potash alum and potassium carbonate.

Potassium, Bisulphate of. $KHSO_4$. Syn. POTASSIUM-HYDROGEN SULPHATE, ACID POTAS-

SIUM SULPHATE; POTASSÆ BISULPHAS, L. Found native in the Grotto del Solfo, near Naples, in the form of long silky needles.—Prep. 1. (Anhydrous.) Neutral sulphate of potassium and sulphuric acid, equal parts; hot water, q. s. to dissolve; anhydrous bisulphate crystallises out, in long delicate needles, as the solution cools. If these are left for several days in the mother-liquor they are redissolved, and crystals of the ordinary hydrated bisulphate are deposited.

2. (Hydrated.) a. Salt left in distilling nitric acid, 2 lbs.; boiling water, 3 quarts; dissolve; add of sulphuric acid, 1 lb.; concentrate by evaporation, and set the liquid aside, so that crystals may form.

b. Powdered sulphate of potassium, 3 oz.; sulphuric acid, 1 fl. oz.; mix in a porcelain capsule, and expose to a heat capable of liquefying its contents, until acid vapours cease to be evolved; powder the residuum, and preserve it in a well-stoppered bottle.

Prop., &c. Sour and slightly bitter-tasted rhombic prisms; soluble in about 2 parts of cold and 1 part of boiling water, the solution exhibiting a strongly acid reaction. It is much employed in lieu of tartaric acid, for the production of carbonic acid, in 'gazogenes,' &c.; also to adulterate cream of tartar and tartaric acid. According to Dr Paris, it forms a "grateful adjunct to rhubarb."—Dose, 12 gr. to 1½ dr., in solution, combined with rhubarb or bitters, as the neutral sulphate.

Potassium, Sulphide of. Syn. SULPHURET OF POTASSIUM, LIVER OF SULPHUR†; POTASSA SULPHURATA, POTASSII SULPHURETUM (Ph. L., E., & U. S.), HEPAR SULPHURIS (Ph. D.), L. Liver of sulphur and Hepar sulphuris are the old names given to a mixture of potassium polysulphides with potassium sulphate or potassium thiosulphate. The true sulphides of potassium are the monosulphide K_2S, the trisulphide K_2S_3, the tetrasulphide K_2S_4, and the pentasulphide K_2S_5, none of which are of much importance.—Prep. 1. Sulphur, 1 oz.; carbonate of potassium, 4 oz.; mix, heat them in a covered crucible till they form a uniform fused liver-coloured mass; when cold, break into fragments, and preserve it in well-closed vessels.

2. Sublimed sulphur, 4 oz.; carbonate of potassium (from pearlash, first dried, and then reduced to powder), 7 oz.; mix in a warm mortar, heat in a Hessian crucible, pour the fused mass into an iron cup, over which immediately invert a second vessel to exclude the air, and, when cold, break the mass into fragments, and preserve it in a green glass-stoppered bottle.

Prop., &c. A hard, brittle, liver or greenish-brown coloured solid; inodorous whilst dry; soluble in water, forming a highly fetid solution; and, in acids, evolving strong fumes of sulphuretted hydrogen; reaction, alkaline; exposed to the air, it is gradually converted into sulphate of potash. As a medicine it is reputed diaphoretic, expectorant, and stimulant.—Dose, 2 to 6 gr., in solution, or made into pills with soap; in gout, rheumatism, liver affections, and various chronic skin diseases. Externally, made into a lotion and ointment. It is highly acrid and corrosive, and in large doses poisonous.

Potassium, Sulphocyanide of. KCNS. *Syn.* P. THIOCYANATE, P. SULPHOCYANATE, SULPHO-CYANURET OF POTASSIUM; POTASSII SULPHOCYA-NIDUM, P. SULPHOCYANURETUM, L. *Prep.* 1. Dried ferrocyanide of potassium, 46 parts; sulphur, 32 parts; pure carbonate of potassium, 17 parts; reduce them to powder, and very gradually heat the mixture to low redness in a covered iron crucible, which it will less than one half fill; remove the half-cooled but still soft mass, crush and exhaust it with water, then evaporate the aqueous solution to dryness; powder the residuum, and exhaust it with hot alcohol or rectified spirit; the alcoholic solution will yield beautiful white crystals as it cools, and the residuum or mother-liquor may be evaporated for the remainder of the salt.

2. Cyanide of potassium, 8 parts; sulphur, 1 part; water, 6 parts; digest them together for some time, add 8 parts more of water, filter, evaporate, and crystallise.

Prop., &c. Long, slender, colourless prisms or plates, which are anhydrous, bitter tasted, deliquescent, fusible, blue when hot, very soluble in both water and alcohol, and are non-poisonous. It is chiefly used as a test for ferric oxide, for which purpose it is preferable to all other substances.

Potassium, Tartrate of. K₂C₄H₄O₆. *Syn.* NEUTRAL TARTRATE OF POTASSIUM, TARTRATE OF POTASSA, NEUTRAL TARTAR, SOLUBLE T.; POTASSÆ TARTRAS (B. P., Ph. L., E., & D.), KALI TARTARIZATUM†, L. *Prep.* Carbonate of potassium, 8 oz.; distilled water, 2 quarts; dissolve, and to the solution, whilst boiling hot, gradually add of bitartrate of potassium, in fine powder, 1 lb., or q. s. until the liquid, after ebullition for a couple of minutes, ceases to change the colour of either blue or reddened litmus paper; next filter the liquid through calico, evaporate it until a pellicle forms on the surface, and set it aside to crystallise; after 12 hours collect the crystals, dry them on bibulous paper, and preserve them from the air.

Prop., &c. The crystals of this salt, which are obtained with difficulty, are right rhombic prisms, and are deliquescent. The salt of commerce is usually in the form of a white granular powder, which is obtained by simply evaporating the solution to dryness, with constant stirring. In this state it requires 4 parts of cold water for its solution.

The solution changes the colour neither of litmus nor turmeric. Almost any acid throws down crystals of bitartrate of potassium, which generally adhere to the vessel. The precipitate occasioned by either chloride of barium or acetate of lead is dissolved by dilute nitric acid.

As a medicine it acts as a gentle diuretic and aperient, and is valued for correcting the griping properties of senna and resinous purgatives. It is also antiscorbutic.—*Dose,* ½ dr. to ½ oz., in powder, or dissolved in water.

Potassium, Bitartrate of. KHC₄H₄O₆. *Syn.* ACID TARTRATE OF POTASSA, SUPERTARTRATE OF P., CREAM OF TARTAR; CREMOR TARTARI, POTASSÆ BITARTRAS (B. P., Ph. L., E., & D.), POTASSÆ SUPERTARTRAS, TARTARI CRYSTALLI, L. *Prep., &c.* This well-known salt is deposited during the fermentation of grape juice as a crust on the sides of the casks or vats. In its unprepared or crude state it is called white or red tartar or argol, according to the colour of the grape juice from which it has been obtained. It is purified by boiling it in water, crystallisation, resolution in water, and treatment with freshly burnt charcoal and clay, to remove the colour; the clear liquid is then decanted whilst still hot, and allowed to cool slowly; the resulting crystals form the 'cream of tartar' of commerce.

Prop., &c. Small, translucent, gritty, prismatic crystals, irregularly grouped together; permanent in the air; requiring fully 100 parts of cold water, and about 15 parts of boiling water, for their perfect solution; the solution has a harsh, sour taste, and, like that of the tartrate, suffers spontaneous decomposition by keeping. It dissolves easily in acids and alkalies, which convert it into the neutral tartrate. When heated it gives off the odour of burnt sugar and leaves a black mass of charcoal and potassium carbonate (salts of tartar). Its solution reddens litmus. It is much used to make a pleasant cooling drink ('Imperial'), and in tooth-powders.—*Dose.* As an aperient, 1 to 8 dr.; as a diuretic, ½ dr. to 1 dr.; as an antiscorbutic, 10 to 20 gr. frequently.

POTATO. This well-known and valuable article of food is the tuber of the *Solanum tuberosum* or *esculentum*, a plant which was introduced into this country by either Sir Francis Drake or Sir Walter Raleigh, towards the latter part of the 16th century. It is now extensively cultivated in all the temperate climates of the world. It yields a vast quantity of food on a small space of ground, but only about 1-7th part of the weight of the tuber is nutritious, and this is chiefly farinaceous. This farina or starch is, however, accompanied by no inconsiderable portion of saline matter, more especially of potassa, which renders it highly antiscorbutic, and a powerful corrective of the grossness of animal food. When forming part of a mixed diet, perhaps no substance is more wholesome than the potato, and, certainly, no other esculent hitherto discovered appears equally adapted for universal use.

Boussingault gives the following as the average composition of the tubers of the potato:

	Moist.	Dry.
Water	. 75·9	. —
Albumen	. 2·8	. 9·6
Oily matter	. 0·2	. 0·8
Fibre	. 0·4	. 1·7
Starch	. 20·2	. 88·8
Salts	. 1·0	. 4·1
	100·0	100·0

Manuring experiments on potatoes in 1867-9 show that on light soils a mixture of mineral superphosphate, crude potash salts, and ammonium sulphate produces very beneficial results; on stiff soils nitrogenous manures have little effect. Further proof is given that manuring with common salt tends to decrease the yield.

Analyses, conducted by A. Stockardt, of potatoes grown in eight different years, show that those manured with salt invariably contain less starch than those unmanured, the decrease being from 10% to 20% of starch; the same effect is produced when the salt is mixed with other manures.

Under the microscope the cells are seen to be very sparingly fitted with starch-grains.

Unmanured potatoes contained ·43% of sodium chloride in the dry substance, and those which had received a small dressing of common salt 1·84% (Voelcker, 'Roy. Agric. Society's Journal,' quoted in 'Journal of Chemical Society,' vol. xxv).

No certain rule can be laid down for 'dressing' potatoes. "If boiled, it may be that they require to be put into boiling water, or, may be, into cold, and either boiled quickly or slowly; but this you must find out. Choose them all about the same size, with a smooth skin, and when they are boiled and begin to crack, throw off the water immediately, as it only damages the root. When dressed, let them stand near the fire, with a cloth over them, and serve them in their skins. Salt may be put into the water at the beginning. A watery potato will require quick boiling, and sometimes to be put into boiling water" (Soyer).

To retain the highest amount of nourishment in potatoes they should be 'dressed' with their skins on them. The bruised or damaged parts, worm-holes, &c., being removed with a knife, the dirt should be carefully cleaned out of the 'eyes' and from the rough parts of the skins, by means of a brush and water, after which they should be well rinsed in clean water, and drained in a colander. If they are at all dry or shrivelled they may be advantageously left to soak for 3 or 4 hours in clean cold water before cooking them. Potatoes 'dressed' in the skins have been found to be nearly twice as rich in potass salts as those which have been first peeled. The skins are easily removed before sending them to table.

NEW POTATOES should have their loose outer skin rubbed off with a cloth or stiff brush before being dressed or cooked.

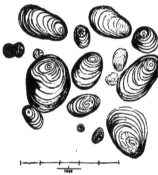

Potato starch granules.

MASHED POTATOES are prepared by crushing with the back of a spoon, or with a rolling-pin, the hot 'dressed' tubers, placed in a bowl or dish, or on a pie-board. A little milk, butter, and salt

may be added to them at will, and they may be either at once 'served up,' or pressed into 'forms,' and first 'browned off' in the oven. Potatoes, if

Potato starch granules swollen by liquor potassa.

not soft and mealy, and well masticated, frequently escape the solvent action of the stomach, and pass off undigested, often to the serious derangement of the health. By mashing them this inconvenience is removed. The delicate, the dyspeptic, and the aged should take them in no other form.

Potatoes may be preserved so as to stand the longest voyages unchanged, by thoroughly desiccating them in an oven, or by steam heat. For this purpose the roots, either raw or three parts dressed, are generally first cut into dice of above ½ inch square, to facilitate the operation. Under a patent granted to Mr Downes Edwards, Aug., 1840, the boiled potatoes are mashed and granulated by forcing them through a perforated plate before drying them. The granulated product, beaten up with a little hot milk or hot water, forms an excellent extemporaneous dish of mashed potatoes.

The microscopic detection of potato starch is easy. Instead of being round or oval, and with a central hilum, the starch grains are pyriform, with an eccentric hilum placed at the smaller end, and with well-marked concentric rings. A weak solution of liquor potassæ (one drop of the Pharmacopœia preparation to ten of water) swells them out greatly after a time, while wheat starch is little affected by potash of this strength; if the strength is 1 to 3 (as in the figs.) the swelling takes place very rapidly.

Potatoes, like many other plants, degenerate when propagation by tubers is constantly resorted to. Varieties become extinct, and when once this has occurred it can never be restored. The terrible ravages of the potato disease have led to careful inquiry as to the best means of raising healthy vigorous plants, and much valuable information will be found on the subject in 'An Essay on Raising New Varieties of Potatoes,' by Charles Lynch, 1886, threepence, published by Eyre and Spottiswoode; or in 'Potato Culture for the Million,' by H. W. Ward, by the same publishers, 1891.

POT-POURRI. [Fr.] A mixture of odorous flowers, roots, gums, &c., varied according to the taste of the operator, either mixed together dry, or in the fresh state preserved with salt. "The usual way of making it is to collect roses, lavender, and other sweet-scented flowers, as they blow; to put them into a large jar mixed (stratified) with salt, until a sufficient quantity has been collected; then to add to these such other odorous substances as may be required to form an agreeable perfume." Among the substances thus added are—ambergris, benzoin, calamus root, cascarilla, cassia, cassia buds, cinnamon, civet, cloves, musk, musk seed, orange berries and flowers, orris root, pimento, storax, vanilla, yellow sandal-wood, &c.

"Instead of the fresh flowers, dried roses are sometimes used, and, with the addition of some essential oils, these answer quite as well" (*Redwood*).

The following is a French formula:—Take the petals of the pale and red roses, pinks, violets, moss rose, orange flower, lily of the valley, acacia flowers, clove gilliflowers, mignonette, heliotrope, jonquils, with a small proportion of the flowers of myrtle, balm, rosemary, and thyme; spread them out for some days, and as they become dry put them into a jar, with alternate layers of dry salt mixed with orris powder, till the vessel is filled. Close it for a month and stir the whole up, and moisten it with rose water.

POTTED MEATS. See PUTREFACTION and POTTING (*below*).

POTTERY. The mechanical operations connected with the manufacture of pottery (CERAMIC ART) do not come within the province of this work. The materials employed in this country to produce the best kind of earthenware (STAFFORDSHIRE WARE) are the fine white clays of Devonshire and Cornwall, and powdered chert or flint. This is brought to a perfectly homogeneous plastic mass with water, and in this state is fashioned upon the potter's wheel and lathe, or by moulding, into all the varied objects of utility and ornament which are made in this material. After the newly formed vessels and other articles have been dried by exposure in heated rooms they are enclosed in clay cases (SEGGARS) and exposed to heat in a kiln, by which they arrive at a state (BISCUIT) which renders them fit for glazing; the patterns (if any) and, afterwards, appropriate vitreous materials having been applied to their surfaces, they are again placed in the seggars, and are again exposed in a kiln to a heat sufficiently high to fuse the newly applied compound, by which they acquire a uniform enamelled surface, and become fit for the market. PORCELAIN, or CHINA, as it is commonly called, is manufactured in a nearly similar manner, but the materials are selected and the various processes conducted with corresponding skill and care.

The ornamental patterns are produced upon both Staffordshire ware and porcelain by a number of ingenious processes, among which printing, painting, and moulding are the principal. The colours employed are those which have been already referred to under ENAMELS.

The metallic lustres now so common on stoneware, &c., are given as follows:

1. GOLD LUSTRE. Dissolve grain-gold, 1 dr., in aqua regia, ¼ oz.; to the solution add of metallic tin, 6 gr.; and when this is dissolved pour it very gradually, with constant stirring, into a mixture of balsam of sulphur, ¼ dr.; oil of turpentine, 20 gr.; when the mass begins to stiffen, ¼ dr. more of oil of turpentine must be added, and well mixed in. More gold deepens and brightens the lustre; more tin turns it on the violet or purple.

2. IRON LUSTRE. From a mixture of 'muriate of iron' (ferric chloride) and spirit of tar.

3. PLATINUM LUSTRE. To a solution of platinum in aqua regia (platinic chloride) is added, drop by drop, a mixture of spirit of tar and balsam of sulphur in equal proportions, until by a trial the composition is found to give the required result. This gives the appearance of polished steel.

4. SILVER LUSTRE. The ammonio-chloride of platinum is reduced to an impalpable powder, ground up to the requisite consistence with a little spirit of tar, and at once applied with a brush.

The above lustres are applied over an easily fusible glaze to the outer surfaces only of the vessels, after which adhesion is produced by exposing the pieces to a slight degree of heat in the muffle. To give them their full beauty they must be rubbed with cotton, wool, or chamois leather, after the firing. See ALUMINA, CLAY, ENAMELS, GILDING, GLAZES, KAOLIN, &c.

POTTING. A term commonly applied to the operation or practice of preserving animal substances in a state fit for immediate use, in small pots or jars. The method of proceeding is, first, to mince the substance (previously well dressed, and carefully deprived of bones, sinews, skin, &c.), and next to pound it in a clean polished marble or iron mortar, along with a little butter and some cayenne pepper, or other suitable spice or sauce, until it forms a perfectly smooth paste; this is pressed into the pots, so as to about 2-3rds fill them, and clarified melted butter is then poured in to the depth of about 1-8th of an inch; the pots are, lastly, closely covered over, and kept in a cool situation. In this state their contents may be preserved for a year or longer. Potted beef, ham, veal, poultry, game, anchovies, bloaters, salmon, &c., are commonly sold in the shops. They are all intended for relishes, and are spread upon bread in the same manner as butter.

POUDRE KOUSIQUE. [Fr.] A French nostrum, consisting of nitre and sulphur, of each, 50 parts; charcoal and antimony, of each, 1 part. It is divided into ¼-dr. doses, of which three are put into each packet. It is given to dogs in a ball of butter to prevent the disorders to which they are liable.

POUDRE MÉTALLIQUE. [Fr.] See TOOTH CEMENTS.

POUDRE SUBTILE. [Fr.] See DEPILATORY.

POUDRETTE. [Fr.] Dried night-soil. The manure sold under this name is a compound of night-soil with clay, charcoal, or gypsum, made into balls or cakes. Its extensive use in Belgium, France, the United States of America, and more particularly in China, where it was first employed, has shown it to be one of the most generally applicable of all the numerous substances used as manure; but as its nitrogen is very low compared with guano and other materials, the cost of carriage is very greatly against its use, 1 ton of guano being equal to 9 or 10 tons of poudrette.

Tardieu, speaking of the men engaged in making poudrette, says, "The action of the exhalations from the manure manufacturers is certainly not injurious. The workmen show actually no trace of sickness or disease which can be referred to the influence of these exhalations."

That 'poudrette' is, however, occasionally liable to set up fermentation seems undeniable; and when this is the case, and large quantities of poudrette are stored within a small space, serious consequences may ensue.

Parent Duchâtelet cites the case of a vessel carrying poudrette to Guadaloupe, in which half the crew died, and the remainder were in a very low state of health on the arrival of the vessel at Guadaloupe, owing, as he affirms, to the exhalations given off by the poudrette.

POULTICE. *Syn.* CATAPLASM; CATAPLASMA, L. An external application, generally extemporaneous, used to promote suppuration, allay pain and inflammation, and resolve tumours, by means of moisture, warmth, or certain remedies they may contain.

Poultices (cataplasmata) are generally prepared with substances capable of absorbing much water and assuming a pulpy consistence, so as to admit of their application to any surface, however irregular. Their curative action principally depends upon the liquids with which they are moistened, and the heat retained by the mass. With this object they should never be heavy or very bulky, and should be frequently repeated, and lightly but securely bandaged on to prevent displacement.

The addition of a little lard, olive oil, or, still better, glycerin, to a poultice, tends greatly to promote its emollient action and to retard its hardening.

As the continued medication of the part with warmth and moisture, or with substances applied in the moist way, is the principal object to be attained in the application of poultices, a fold or two of lint or soft linen dipped in hot water, either simple or medicated, and covered with a piece of thin sheet gutta percha or india-rubber cloth to prevent evaporation may be often conveniently applied in their stead. A very elegant and useful substitute of this kind is sold under the name of 'spongio-piline.' Its cleanliness, lightness, and ease of application has led to its extensive adoption by the medical profession.

The following formulæ embrace all the cataplasms of the Pharmacopœias:

Poultice of Al'um. *Syn.* CATAPLASMA ALUMINIS, L. *Prep.* (B. P., Ph. D. 1826.) Alum (in powder), 1 dr.; whites of 2 eggs; shake them together until they form a coagulum. Applied between the folds of fine linen to chilblains, sore nipples, inflamed eyes, &c.

Poultice, Anodyne. (P. Cod.) *Syn.* CATAPLASMA ANODYNUM, L. *Prep.* Poppy heads, 1 oz.; dried leaves of henbane, 2 oz.; water, 24 oz. Boil, strain, and add to the liquor 4 oz. of emollient meals, to form a poultice.

Poultice, Astringent. (*Foy.*) *Syn.* CATA-

PLASMA ASTRINGENS, L. *Prep.* Catechu, 1 oz.; powdered oak bark and barley meal, of each, 1 oz.; cold water, q. s.

Poultice of Belladonna. (*Dr Recce.*) *Syn.* CATAPLASMA BELLADONNÆ, L. *Prep.* Extract of belladonna, made *in vacuo*, 1 dr.; oatmeal, ½ lb.; boiling water, q. s.

Poultice, Bran. *Syn.* CATAPLASMA FURFURIS, L. *Prep.* Fine bran, with 1-10th of linseed meal, made into a poultice with boiling water. Mr Payne recommends, as a cheap hospital poultice, 3½ pecks of pollard, 14 lbs. of bruised meal, and ½ lb. of lard.

Poultice of Bread. *Syn.* CATAPLASMA PANIS, L. *Prep.* From crumb of bread, soaked in hot water, slightly pressed, and then beaten up with a little lard, butter, or oil. Emollient. See POULTICE, LINSEED MEAL (*below*).

Poultice of Car'rot. *Syn.* CATAPLASMA DAUCI, L. *Prep.* 1. From the common esculent carrot, scraped fine, so as to form a pulp.

2. (Ph. D. 1826.) From the cultivated carrot boiled with a little water until it becomes soft enough to form a poultice. Anodyne and antiseptic. Used in foul and painful ulcers, burns, contusions, &c. That from the first formula is the more stimulant.

Poultice of Char'coal. *Syn.* CATAPLASMA CARBONIS (B. P., Ph. L.), C. C. LIGNI, L. *Prep.* 1. (Ph. L.) Soak bread, 2 oz., in boiling water, ½ pint; to this add, by degrees, of linseed meal, 10 dr.; and, afterwards, of powdered (recently burnt) charcoal, 2 dr.; lastly, sprinkle on the surface of the poultice powdered charcoal, 1 dr. As an application to fœtid and gangrenous sores; frequently renewed.

2. (B. P.) Wood charcoal, ½ oz.; bread, 2 oz.; linseed meal, 1½ oz.; boiling water, 10 oz.; soak the bread in the water near the fire, add the linseed meal and half the charcoal, stirring to a soft poultice, sprinkling the remainder of the charcoal on the surface.

Poultice of Chlorine. *Syn.* CATAPLASMA SODÆ CHLORINATÆ (B. P., Ph. L.), L. *Prep.* 1. (Ph. L.) Boiling water, 6 fl. oz.; linseed meal, 4½ oz.; mix gradually, then add of solution of chlorinated soda, 2 fl. oz. Applied to foul ulcers, gangrenous parts, &c.

2. (B. P.) Solution of chlorinated soda, 1 part; linseed meal, 2 parts; boiling water, 4 parts; add the linseed meal gradually to the water, stirring constantly, then mix the solution of chlorinated soda.

Poultice, Compound Farina. *Syn.* CATAPLASMA FARINÆ COMPOSITUM, L. *Prep.* Rye flour, 1 lb.; old yeast, 4 oz.; salt, 2 oz.; hot water, q. s.

Poultice, Cummin. (L. Ph. 1788.) *Syn.* CATAPLASMA CUMINI. *Prep.* Cummin seeds, 1 lb.; bay berries, scordium leaves, serpentaria root, of each, 3 oz.; cloves, 1 oz.; to be powdered together and mixed with thrice their weight of honey.

Poultice, Discutient. *Syn.* CATAPLASMA DISCUTIENS, L. *Prep.* Barley meal, 6 oz.; fresh hemlock, 2 oz.; vinegar, q. s. Boil and add sal-ammoniac, ½ oz. (Fr. Hosp.) The same, with 2 dr. of acetate of lead added.

Poultice, Effervescing. *Syn.* CATAPLASMA EFFERVESCENS, L. *Prep.* Fresh wort thickened with oatmeal, and a spoonful of yeast added.

Poultice, Emetic. *Syn.* CATAPLASMA EMETI-
CUM, L. *Prep.* Bruised groundsel (*Senecio
vulgaris*) applied over the stomach produces
vomiting.

Cataplasma of Fig. *Syn.* CATAPLASMA FICI,
L. *Prep.* A dried fig, roasted or boiled (sometimes in milk), is frequently applied to gum-
boils, &c.

Poultice, Foxglove (*Mr Alland*). *Syn.* CATA-
PLASMA DIGITALIS, L. *Prep.* A strong decoction of foxglove, with bread crumb, or linseed
meal, q. s.

Poultice, Galbanum. *Syn.* CATAPLASMA GAL-
BANI, L. *Prep.* Lily roots, 4 oz.; figs, 1 oz.;
boil till soft, and bruise them with 1½ oz. of
onions and ¼ oz. of galbanum, triturated with
yolk of egg and a sufficient quantity of linseed
meal.

Poultice, Galvanic. (*Recamier.*) *Syn.* CATA-
PLASMA GALVANICUM, L. It consists of cotton
wadding containing a layer of very thin zinc
plates, and another layer of copper ones. This
pad, conveniently quilted, is enclosed in a bag,
one face of which is of quilted calico, the other
of impermeable tissue. The natural perspiration,
confined by the impermeable tissue, excites galvanic action between the metals.

Poultice of Hemlock. *Syn.* CATAPLASMA
CONII (B. P., Ph. L.), L. *Prep.* 1. (Ph. L.)
Boiling water, ½ pint; linseed meal, 4½ oz., or
q. s.; make a poultice, and on this spread of
extract of hemlock (Ph. L.), 1 oz., first softened
with a little hot water. Anodyne. In irritable
and painful cancerous, scrofulous, and syphilitic
sores, tumours, &c.

2. (B. P.) Juice of hemlock, 1 oz.; linseed
meal, 4 oz.; boiling water, 10 oz. Evaporate
the juice to half; mix well.

Poultice, Henbane. *Syn.* CATAPLASMA HYO-
SCYAMI. The same as POULTICE OF POPPY.

Poultice, Hop. (*Dr Trotter.*) *Syn.* CATA-
PLASMA HUMULI, L. Hops softened with hot
water. To foul ulcers.

Poultice, Iodine. *Syn.* CATAPLASMA IODURE-
TUM, L. To a common poultice add solution
or tincture of iodine.

Poultice, Lead. *Syn.* CATAPLASMA PLUMBI,
L. *Prep.* Goulard water, 1 lb.; bread crumb,
q. s.

Poultice, Lily. *Syn.* CATAPLASMA LILII, L.
The pulp of the white lily boiled and bruised.

Poultice, Lime. *Syn.* CATAPLASMA CALCIS,
L. *Prep.* Slaked lime, 2 oz.; oatmeal, 2 oz.;
lard, 4 oz. Formerly used at Bath Hospital.

Poultice of Linseed Meal. *Syn.* CATAPLASMA
LINI (B. P., Ph. L.), L. *Prep.* 1. (Ph. L.) To
boiling water, ½ pint, add gradually, constantly
stirring, of linseed meal, 4½ oz., or q. s. Emollient. Used to promote the suppuration or
'ripening' of tumours. A little oil or lard
should be added, and some smeared over the surface as well, to prevent its getting hard. For
small 'gatherings,' as of the fingers, a little
chewed bread and butter is an efficient and convenient substitute.

2. (B. P.) Linseed meal, 4 parts; boiling water,
10 parts; mix the linseed meal with the water,
constantly stirring.

Obs. Linseed meal prepared from the cake,

from which the oil has been expressed, is less
adapted for poultices than that prepared from the
unpressed, whole seed.

Poultice, Malt. (Guy's Hosp.) *Syn.* CATA-
PLASMA BYNES. *Prep.* Ground malt, with yeast,
q. s. to form a poultice; to be applied warm.

Poultice of Mustard. *Syn.* CATAPLASMA
SINAPIS (Ph. L.), L. *Prep.* 1. (B. P., Ph. L.)
Linseed meal and powdered mustard, of each, 2½
oz., or q. s.; boiling water and lukewarm water,
q. s. Mix the mustard with 2 or 3 oz. of lukewarm water; mix the linseed meal with 6 to 8
oz. boiling water; mix both together.

2. (Ph. L. 1836.) As the last, but substituting
boiling vinegar for water. Used as a powerful
counter-irritant, stimulant, and rubefacient; in
low fevers, apoplexy, coma, &c., where there is a
determination of blood to the head; in deep-seated
inflammatory pains, neuralgic pains, &c. It should
not be left on long enough to raise a blister. See
PLASTERS.

Poultice, Onion. *Syn.* CATAPLASMA CEPÆ, L.
Prep. Onions roasted and mashed.

Poultice of Pop'py. *Syn.* CATAPLASMA PA-
PAVERIS, L. *Prep.* (P. Cod. 1839.) A strong
decoction of poppies, thickened with crumb of
bread. Anodyne.

Poultice, Potash, Acetate of. *Syn.* CATAPLASMA
POTASSÆ ACETATIS, C. NEUTRALE, L. *Prep.*
Acetate of potash, 1 oz.; water, 1 pint; crumb of
bread, q. s. To ill-conditioned sores.

Poultice of Pota'to. *Syn.* CATAPLASMA
SOLANI TUBEROSI, L. *Prep.* From the raw
potato, scraped or grated fine. A popular application to fresh bruises, extravasations, burns,
scalds, &c.

Poultice of Potato Starch. (P. Cod.) *Syn.*
CATAPLASMA FECULÆ, L. *Prep.* Potato starch,
2 oz.; mix with a little cold water, add to it 20
oz. of boiling water, and boil for an instant.
Cataplasms of rice and wheat starch are made in
the same manner.

Poultice, Pradier's. *Syn.* PRADIER'S CATA-
PLASMA; CATAPLASME DE PRADIER, Fr. *Prep.*
Take of balsam of Mecca, 6 dr.; rectified spirit,
16 oz.; dissolve; next, separately, take red cinchona bark, sarsaparilla, and sage, of each, 1 oz.;
saffron, ½ oz.; rectified spirit, 32 oz.; digest for
48 hours, and filter; mix the two liquors; add to
them twice their weight of lime water. In gout;
2 fl. oz. are sprinkled on the surface of a hot
linseed-meal poultice sufficiently large to surround
the affected part. It is said that the Emperor
Napoleon gave £2500 for this receipt.

Poultice of Pyroligneous Acid. (*Dr Reece.*)
Syn. CATAPLASMA ACIDI PYROLIGNOSI, L. *Prep.*
Bran, 1 lb.; linseed meal, 1 oz.; impure pyroligneous acid, q. s. For scrofulous ulcers; occasionally 30 minims of tincture of perchloride of
iron, and 3 dr. of extract or powder of hemlock,
are added.

Poultice of Sea-wrack. (*Dr Russell.*) *Syn.*
CATAPLASMA FUCI, L. *Prep.* Fresh bladder
fucus (sea-wrack) bruised. Applied to glandular
tumours, &c.

Poultice, Simple. *Syn.* CATAPLASMA SIMPLEX,
L. *Prep.* (Ph. D. 1826.) Powder for a cataplasma and boiling water, of each, q. s. to form a
poultice, the surface of which is to be smeared

over with olive oil. Emollient. Bread poultice and linseed-meal poultice are now generally called by this name. See POWDER (Poultice).

Poultice of Slippery Elm. *Syn.* CATAPLASMA ULMI, L. The powdered bark of the slippery elm (*Ulmus fulva*) mixed with a sufficient quantity of hot water.

Poultice of Soap. *Syn.* CATAPLASMA SAPONIS, L. *Prep.* From white soap (scraped or sliced), 1 oz.; boiling water, ¼ pint; dissolve and add crumb of bread, q. s. As an application to scalds and burns.

Poultice, Sorrel. *Syn.* CATAPLASMA OXALIS, L. *Prep.* Bruised sorrel leaves, mixed with oatmeal and beer.

Poultice, Stimulating. (*Dr Hugh Smith.*) *Syn.* CATAPLASMA STIMULANS, L. *Prep.* Rye flour, 1 lb.; old yeast, 4 oz.; common salt, 2 oz.

Poultice, Sulphate of Lime. (*Blisard.*) *Syn.* CATAPLASMA CALCIS SULPHATIS, L. *Prep.* Paris plaster mixed with water to a soft paste, and applied before it hardens. Formerly applied to ulcers to form an artificial scab; now occasionally used to afford mechanical support in some surgical cases.

Poultice of Sulphate of Soda. (*Kirkland.*) *Syn.* CATAPLASMA SODÆ SULPHATIS, L. *Prep.* Sulphate of soda, 1 oz.; boiling water, ½ lb.; crumb of bread, a sufficient quantity.

Poultice to cause Suppuration. (E. Ph. 1774.) *Syn.* CATAPLASMA SUPPURANS. To an emollient cataplasm add bruised onions, 1½ oz.; basilicon ointment, 1 oz.

Poultice, Turnip. *Syn.* CATAPLASMA RAPI. *Prep.* Peel turnips, boil them till soft, beat them to a pulp, and apply warm.

Poultice, Turpentine. (*Dr Reeve.*) *Syn.* CATAPLASMA TEREBINTHINÆ. *Prep.* Oil of turpentine, 2 dr.; olive oil, 1 oz.; linseed meal, 1 oz.; oatmeal, 4 oz.; boiling water, q. s. To indolent ulcers, and, with more turpentine, to deep burns, scalds, and chilblains.

Poultice of Vin'egar. *Syn.* CATAPLASMA ACETI, L. *Prep.* From crumb of bread soaked in vinegar. Applied cold in bruises, extravasations, &c., especially black eyes. Verjuice is often employed in the same way.

Poultice of Walnut Leaves. (*Perfect.*) *Syn.* CATAPLASMA JUGLANDIS. *Prep.* The fresh leaves of walnut bruised and mixed with honey. Applied over the abdomen as a vermifuge.

Poultice of Yeast. *Syn.* CATAPLASMA FERMENTI (B. P., Ph. L.), C. F. CEREVISIÆ, L. *Prep.* 1. (Ph. L.) Beer yeast and water at 100° F., of each, 5 fl. oz.; mix, stir in flour, 1 lb., and place it near the fire until it rises. In gangrenous or foul ulcers; to correct the fetor of the discharge, and to hasten the sloughing.

2. (B. P.) Beer yeast, 6 parts; flour, 14 parts; water (at 100° F.), 6 parts; mix. Place the mass near the fire till it rises.

POULTRY. Domestic birds, which are propagated and fed for the table, and for their eggs, feathers, &c.

The poultry of this country are the common domestic fowl, the turkey, the duck, and the goose; to which some writers add the guinea-fowl and the peacock. Of these, the first is the most generally useful. Though a native of India,

it accompanies man through almost every gradation of civilisation and climate, and flourishes almost everywhere, when properly secured from the inclemency of the weather, and allowed an ample supply of fresh air, with proper food. For the production of abundance of eggs it must, however, be well fed and warmly lodged. The hen-roosts and poultry-houses should be well protected from the weather, and their temperature should be duly maintained by proximity to the stables, cowhouses, or dwelling-house, and, in cold weather, by the employment, when necessary, of artificial heat. The food should also contain an ample supply of nitrogenous matter, for without this how can it be expected that hens can produce abundance of eggs, which are peculiarly rich in nitrogen? The 'greaves' of the tallow-chandlers, and such like substances, are hence excellent additions to the ordinary food of poultry. But it is not sufficient merely to supply poultry with abundance of food and warmth; it is equally necessary that they should have ample space for exercise and recreation. This space "should always contain living plants of various kinds, and some gravelly or sandy soil; because worms, snails, as well as, occasionally, grass and herbage, form a part of the food of poultry; and sand or gravel is swallowed by them for the purpose of promoting digestion. Hence, no healthy poultry can ever be reared in towns, however much the natural food may be imitated by the supply of animal matters, herbage, and sand" (*London*).

Poultry-rearing in England is, except in a few hands, very far from being the science it ought to be, and it is little short of a disgrace that we should import annually more than £3,000,000 sterling worth of eggs from abroad, and nearly £500,000 worth of poultry of various kinds. In France poultry-rearing is a great industry, and a trade journal states that there are in the country (1890) 45,000,000 of hens, which, at an average price of 2 francs 50 centimes per head, represent a value of 112,050,000 francs. One fifth of the stock is annually consumed as food, and is sold for about 22,500,000 francs. Two millions of cocks, worth 5,000,000 francs, are also sold as food. The number of laying hens is put down at 35,000,000, and the annual value of their eggs is 188,000,600 francs.

The following quotations from a letter which appeared in the 'Standard' newspaper Aug. 24, 1890, puts the main facts of the case very clearly, and the writer's strictures on the careless, haphazard way in which poultry are reared in England are, unfortunately, only too well justified :

"Every one who knows anything of poultry must admit that a hen, if properly bred and managed well, will leave a handsome profit on her keep at the end of the year; and it is impossible for a chick, if kept growing from birth and sent to market as soon as ready, to die in debt. The fact is, the poultry department of the farm is looked upon by the owner as too trifling a matter to need his attention, and it is left for the other members of the household to attend to when they find it convenient to do so. In the course of a year I visit some 50 farms, ranging from 50 to 450 acres, and in no single instance can it be said that the poultry business is managed as it should be, although, as might be expected, some are better than others. I have good reason to believe that the farmers I have alluded to are fairly representative men of their class in their respective districts.

"As one who has had a considerable experience in poultry-keeping in various parts of the country, I contend that no class of stock pays the farmer better, and I further maintain that there is scarcely a holding in the country upon which a good stock of fowls may not be kept. Mr Weir, in his second letter, counsels his correspondents to 'avoid clay land.' It is, I admit, an unsatisfactory soil to deal with; still, by selecting the right breeds of fowls, and taking care to hatch them at the proper season, much might be done even in such a situation as this. But whatever the situation may be, the manager must have a proper system to work upon. It is not sufficient to let the hens manage the business for themselves. In a state of nature the hen may lay a batch of eggs every spring, and hatch them; no more will be forthcoming until the next season. In the farmyard the hen should be laying or sitting the best part of the year, and a good proportion of the eggs must come at midwinter, when prices are the best, or the balance-sheet will be an indifferent one. So with chickens. Any one can send chickens into the market when prices are at their lowest in consequence of the abundant supply, but only the careful breeder and skilful raiser can share in the profits obtainable during the first five months of the year. The average poultry-keeper does not think sufficiently about the future. To secure early chickens you must have winter eggs, and these can only be forthcoming in any quantity from March and April hatched pullets. The sisters of the brood mature more rapidly than the cockerels; hence it follows that the pullets are too often sent into the market for killing purposes, and the farmer depends on later-hatched birds, which by no system of feeding can be brought to lay until the end of January or February, when eggs are becoming plentiful. Again, no skill is shown in the selection of stock birds. Anything in the shape of a cock or hen is considered to be good enough to perpetuate their species, and in-breeding is very frequently indulged in. Under such circumstances as these it is not surprising that indifferent layers are the rule, or that deterioration of the stock in other respects is often seen. Two instances of this came under my notice not many months ago. In one case, the season's pullets proved poor layers and indifferent sitters, although they were bred from hens good in both points, the explanation being that the sire used was a bird of which the purchaser knew nothing. In the other case, the pullets, although fine birds for table use, laid wretchedly small eggs, and this must have been the outcome of injudicious mating. The eggs, too, were so few in number that the hens could not possibly be kept at a profit.

"Before the farm poultry of our country can be said to be in a healthy state there must be a weeding out of the worthless specimens, and their place must be filled by well-bred birds of their respective breeds. As Mr Weir remarks, a well-bred bird eats no more than a mongrel, and the returns from it are much better. It must not be supposed, however, that any breeds will answer the purpose, or that exhibition specimens are the best to procure. What the farmer wants is a class of fowl which is a good layer of large eggs, producing chickens suitable for the table, and at the same time hardy. It may be doubted whether there is any one breed which really answers these requirements. Mr Weir praises the Dorking, the prince of table fowls; but the hens of many strains are very poor layers, and some strains are unable to stand the cold and wet in exposed situations. For winter layers under such conditions one of the Asiatic breeds will be most suitable, and such hens crossed with a cock belonging to one of the table breeds will produce very useful chickens. Still no hard and fast line can be laid down, for whilst one farmer may find it most profitable to produce eggs only, another may be able to do best with chickens. The owner, then, should be careful to produce the article which is most in demand. Whilst advocating a good proportion of sitting hens in each yard, I am bound to admit it is desirable to keep a good number of non-sitters, for there is less danger of eggs going wrong during the summer months, when sitting hens are so fond of stealing their nests.

"Few farmers, I believe, ever think of working upon any recognised plan, and, in consequence, the supply of eggs or poultry fluctuates very considerably. Yet nothing is easier to manage when the concern is once got into working order. I have already alluded to the way in which winter eggs are obtained. The supply of chickens can be maintained by having two hatching seasons in each year; the first from January to May, the second from July to September. I am here speaking of Midland districts; in the southern counties hatching may be carried on much later. The spring-hatched chickens may be made to serve from April until the next batch is fit; and the latter will keep up the supply until the spring lot is again ready. In addition to table poultry, the sale of pullets almost ready for laying might be made a profitable business, and in sporting districts there would be a demand in the spring for sitting hens, which might be met by selling the pullets which have laid all the winter. Other kinds of poultry ought on no account to be neglected. Turkeys should be kept where the soil is suitable; geese whenever there is a good acreage of grass land; and ducks if the water supply is not too abundant.

"Poultry-farming on a large scale has often been recommended as likely to prove a lucrative business. Few people, however, who urge such schemes are aware of the enormous outlay required at the commencement in setting up houses and runs. Nor do they take into account the risk of disease when fowls are kept so thickly on the ground. The occupier of 50, 100, or more acres need not make this outlay, for if his fences be good he may make each field a poultry run, and disease from foul ground might be avoided by changing the fowls, with their house, into another field when one pasture becomes tainted. Purely corn farms are not so well adapted for the business. Still, by using the temporary pastures during the summer months, and removing the houses to the stubbles as soon as the corn is carted, much might be done."

POUNCE. Powdered gum sandarac generally passes under this name. It is used to prepare parchment for writing on, and to prevent ink from spreading upon paper after erasures. Powdered cuttle-fish bone is occasionally employed in the same way. Both are applied to the surface by means of a cylindrical roll of list called a 'rubber.' Packers rub the surface of porous and greasy woods, as the heads of boxes, cases, casks, &c., with whiting or powdered resin to make them bear the ink. The coloured powders used by pattern drawers, for sprinkling over pricked papers, are also called 'pounce.' For liquid pounce, see MARKING INK.

POWDER. Syn. PULVIS, L. Powders are divided by pharmaceutical writers into two classes—simple and compound. The first are prepared by simple pulverisation; the second by the admixture of two or more simple powders. For use the appropriate doses are separately weighed, and placed in separate papers. They are usually exhibited in a little honey, sugar, milk, or enclosed in a cachet, either taken from a spoon or made into an electuary or bolus, and swallowed in the semi-solid form. Metallic and other heavy powders are best taken in the latter state. Very active substances should be, in all cases, mixed with some inert powder, as that of starch, gum, sugar, liquorice, or marsh-mallow, at the time of 'dispensing' them.

"This form of preparing medicines is the simplest, and perhaps the least objectionable; but it is not applicable to all the articles of the Materia Medica. Those remedies which are very unpleasant to the taste; those which deliquesce rapidly when exposed to the air, or are very volatile; and those which require to be given in very large doses, or which are not diffused readily in water, cannot, with propriety, be administered in the form of powder. Some substances cannot be reduced to powder unless they be very much dried, and the heat necessary for that purpose alters their properties." Nor can we "be surprised that a great alteration should be effected in a short time by the action of the air on so great an extension of surface as takes place in the operation usually adopted for reducing drugs to a fine powder" (A. T. Thomson).

In this country compound powders appear to be a favourite form of medicine in the diseases of infancy and childhood.

"It is necessary that whatever we order to be reduced to powder should be rubbed through a fine sieve, so that the impurities and coarser parts may be separated; and it is needful that most powders should be recently prepared, and not too long kept" (Ph. L.).

As nearly all powders suffer by exposure to the air and light, they should be preserved in closely corked opaque or green glass bottles, or in tin canisters from which the external air is carefully excluded. See PULVERISATION, &c.

Powder, Algaroth's. See ANTIMONY OXY-CHLORIDE.

Powder of Al'oes (Compound). Syn. PULVIS

ALOËS COMPOSITUS (Ph. L.), P. ALOËS CUM GUAIACO, L. Prep. (Ph. L.) Socotrine or hepatic aloes (in powder), 1½ oz.; guaiacum (in powder), 1 oz.; compound cinnamon powder, ½ oz.; rub them together. A warm, sudorific purgative.—Dose, 10 to 20 gr.

Powder of Aloes with Canel'la. Syn. ALOETIC POWDER, HOLY BITTER†; HIERA PICRA†, PULVIS ALOËS CUM CANELLA, L. Prep. From powdered Socotrine or hepatic aloes, 4 parts; powdered white canella, 1 part. Uses and dose, as the last.

Obs. Once a highly popular remedy. It was originally made into an electuary with honey, and in this form was frequently called 'HIERA LOGADII.' It is still a favourite in domestic medicine and veterinary practice. The principal objection to both this and the preceding preparation is the nauseous flavour of the aloes, which is ill concealed by the aromatics. The 'HIERA PICRA' for farriers is usually made with the cheapest Cape aloes.

Powder of Aloes with Iron. (L. Ph. 1788.) Syn. PULVIS ALOETICUS CUM FERRO, L. Prep. Aloes, 1½ oz.; myrrh, 2 oz.; sulphate of iron, 1 oz.; dried extract of gentian, 1 oz.

Powder of Al'um (Compound). Syn. STYPTIC POWDER; PULVIS STYPTICUS, P. ALUMINIS COMPOSITUS (Ph. E.), L. Prep. (Ph. E.) Alum, 4 oz.; kino, 1 oz.; mix them, and reduce them to fine powder. Astringent and styptic.—Dose, 5 to 15 gr.; in diarrhœa, profuse menstruation, &c. Externally, in hæmorrhages, &c.

Powder, Alum, Opiated. (Bouchardat.) Syn. PULVIS ALUMINIS OPIATUS. Prep. Alum, 1 dr.; sugar, 1 dr.; opium, 4 gr.; mix for 12 powders. 2 or 3 daily in obstinate diarrhœas and passive hæmorrhages.

Powder of Alum with Capsicum. (Dr Turnbull.) Syn. PULVIS ALUMINIS CUM CAPSICO, L. Prep. Alum, 8 parts; concentrated tincture of capsicum, 1 part; mix, dry, and triturate again. Applied to the tonsils.

Powder of Alum with Gum. (Frankel.) Prep. Alum, gum tragacanth, of each, equal parts. Applied to sore breasts.

Powder of Alum with Starch. (St. Th. Hosp.) Syn. PULVIS ALUMINIS CUM AMYLO. Prep. Alum and starch, equal parts. In insufflation of rhinorrhœa.

Powder of Ambergris with Musk. (Bat. Ph.) Syn. PULVIS AMBERGRISEA MOSCHATUS, L. Prep. Ambergris, 6 dr.; musk, 1 dr.; oil of cinnamon, 2 scruples; refined sugar, 11½ oz.; mix.

Powder, Ammoniated Aromatic. Syn. PULVIS AMMONIATUS AROMATICUS, L.; LLAYSON's AMMONIACAL COLLYRIUM. Prep. Muriate of ammonia, 1 dr.; slaked lime, 1 oz.; charcoal, 15 gr.; cinnamon, 15 gr.; cloves, 15 gr.; bole, ½ dr. Put them into a bottle and moisten with a little water.

Powder of Ancho'vy. Syn. PULVIS CLUPEA ENCRASICOLI, L. Prep. Pound anchovies to a paste, then rub them through a sieve, and add enough flour to make a dough, which pour be rolled out into thin slices and dried by a gentle heat in a stove; it is, lastly, powdered and bottled. Colouring is frequently added. Chiefly used to make sauces. British anchovies are frequently substituted for the genuine fish.

Powder, Anthrakokali, Compound. *Syn.* PULVIS ANTHRAKOKALI, COMPOUND. *Prep.* Anthrakokali, 2 gr.; washed sulphur, 6 gr.; mix. For 1 dose.

Powder, Anthrakokali, Simple. (*Poyla.*) *Syn.* PULVIS ANTHRAKOKALI SIMPLEX, L. *Prep.* Anthrakokali, 2 gr.; liquorice powder, 6 gr.; mix. For 1 dose.

Powder, Antiepileptic. *Syn.* PULVIS ANTI-EPILEPTICUS (E. Ph., 1744), L. *Prep.* White dittany, pæony, valerian, mistletoe of the oak, equal parts.—*Dose,* 6 to 10 gr. (*Behrends*). *Prep.* Valerian, 4 dr.; magnesia, muriate of ammonia, oil of cajeput, of each, 1 scruple.—*Dose.* A teaspoonful three times a day. Dr Paris says the following was used successfully by a Dutch empiric:—Sulphur, 1 scruple; sulphate of potash, 10 gr.; rhubarb, 5 gr.; nutmeg, 2 gr.; mix (Germ. Hosp.). *Prep.* Oxide of zinc, 16 gr.; carbonate of magnesia, 48 gr.; oleo-saccharum of cajeput, 3 dr. Mix for 8 doses. *Poudre de Ragolo.* Oxide of zinc, 10 gr.; valerian, mistletoe, sugar, orange leaves, of each, 4 dr.; magnesia, 2 scruples; oil of cajeput, 2 scruples; a teaspoonful three times a day. Pasquier prescribes—Wall crop, 10 gr.; gum-arabic, 10 gr.; 1 to 4 powders daily for eight times. SOMMER'S SPECIFIC consists of—Wall crop, 6 to 10 gr.; oleo-saccharum of mint, 8 gr.; one morning and evening for six times. See PULVIS ARTEMISLE SACCHARATUS. The *Poudre de Gutiète* consists of mistletoe, 2 parts; white dittany, 2 parts; pæony root and seeds, 2 parts; prepared coral, 1 part; elk's hoof, 2 parts; seeds of orache, 2 parts. Given in doses of a few grains in convulsions of infants, or in larger doses for epilepsy.

Powder, Antihydrophobic. (*Dr Mead.*) *Syn.* PULVIS ANTILYSSUS. *Prep.* Ash-coloured ground liverwort (*Peltidea canina*), ½ oz.; black pepper, 2 dr.; mix, and give a fourth part every morning for four times.

Powder, Antimo″nial. *Syn.* FEVER POWDER, LISLE'S P., JAMES'S P.; PULVIS JACOBI, PULVIS ANTIMONIALIS (B. P., Ph. E. and D.), PULVIS ANTIMONII COMPOSITUS (Ph. L.), L. *Prep.* 1. (Ph. L.) A mixture of tersulphide of antimony, 1 lb., and hartshorn shavings, 2 lbs., is reduced to powder, thrown into a crucible heated to whiteness, and stirred constantly until vapour no longer rises; the calcined mixture is then rubbed to powder, again put into the crucible, and the heat gradually increased to whiteness, and maintained so for two hours; the residuum is, lastly, reduced to a very fine powder.

2. (Ph. E.) From sulphide of antimony and hartshorn shavings, equal weights; as the last.

3. (Ph. D.) Tartarised antimony, 4 oz., is dissolved in water, ½ gall., and added to solution of phosphate of soda, 4 oz., in water, 1 quart; a solution of chloride of calcium, 2 oz., in water, 1 quart, and to which solution of ammonia (Ph. D.), 4 fl. oz., has been added, is next poured in, and the whole boiled for 20 minutes; the precipitate is then collected on a calico filter, and washed with hot distilled water, until the liquid which passes ceases to give a precipitate with a weak solution of nitrate of silver; it is, lastly,

dried by a steam or water heat, and reduced to a fine powder.

4. (B. P.) Oxide of antimony, 1 part; precipitated phosphate of lime, 2 parts; mix.—*Dose,* 2 to 6 gr.

Uses, &c. Febrifuge and diaphoretic. Intended as a substitute for the proprietary and more expensive JAMES'S POWDER.—*Dose,* 3 to 10 or 12 gr., or more, repeated every fourth or fifth hour until diaphoresis is set up; in fevers, rheumatic affections, chronic skin diseases, &c. It is a very uncertain and variable compound, unless it has been carefully prepared. Dr Elliotson exhibited it in doses of 100 gr. without producing any sensible effect. A spurious article, made by triturating 1 oz. of tartar emetic with 18 or 19 oz. of burnt hartshorn, is frequently sold for it in the shops. See ANTIMONIOUS ACID and JAMES'S POWDER.

Powder, Antispasmodic. (*P. Cod.*) *Syn.* PULVIS ANTISPASMODICUS. *Prep.* Cyanide of zinc, 3 gr.; calcined magnesia, 24 gr.; cinnamon, 12 gr.; mix. For 6 doses.

Powder, Antispasmodic. (*Jourdan.*) *Syn.* PULVIS ANTISPASMODICUS. *Prep.* Valerian, 1 oz.; oxide of zinc, 1 scruple; musk, 8 gr.; mix.

Powder, Aromat′ic. See POWDER, COMPOUND CINNAMON.

Powder, Arsen′ical. See POWDER, ESCHA-ROTIC.

Powder of Asarabac′ca (Compound). See SNUFF (Cephalic).

Powder, Astrin′gent. *Syn.* PULVIS ASTRINGENS, P. STYPTICUS, L. *Prep.* From Aleppo galls and burnt alum, in fine powder, equal parts. Used in piles, soft polypi of the nose, chilblains, &c.

Powder, Ba″king. *Prep.* 1. Tartaric acid, ½ lb.; bicarbonate of soda and potato farina or British arrowroot, of each, ½ lb. (each in powder); separately dry them perfectly by a very gentle heat, then mix them in a dry room, pass the mixture through a sieve, and at once put it into packets, preparatory to press it hard, and to cover it with tinfoil or close-made paper, to preserve it as much as possible from the air and moisture.

2. (*Delforte's.*) Powdered tartaric acid, ½ lb.; powdered alum, ½ lb.; bicarbonate of soda, ⅜ lb.; farina, 1 lb.; dry separately, as before, mix, and further add of sesquicarbonate of ammonia (in powder), 3 oz.; lastly, closely pack it in tinfoil.

3. (*Green's.*) Tartaric acid, 35 lbs.; sesquicarbonate of soda, 56 lbs.; potato flour, 1 cwt.; mix as before.

Uses, &c. Baking powder is chiefly employed as a substitute for yeast. 1 or 2 teaspoonfuls are mixed with the dry flour and other ingredients, which are then made into a dough, as quickly as possible, with cold water, and at once baked or made as the case may be. By the addition of about ½ dr. of turmeric powder to each pound of the mixture it is converted into egg powder. When intended to be kept for any length of time it should be preserved in bottles or tins, so as to prevent the absorption of moisture. We have discovered traces of arsenic in some of the baking powders of the shops, which we refer to common washerwoman's soda being used in their composition, instead of the pure carbonate or sesquicarbonate.

Powder, Baking, Alum. Mr C. V. Petraeus, in an article on baking powders in the 'Pharmaceutical Record,' states that burnt alum is the most perfect acid element that can be used in baking powders, and for several reasons, viz. :—(1) when exposed to the air it does not become moist ; (2) when mixed with bicarbonate of soda and starch or flour, burnt alum evolves no gas at ordinary temperatures; therefore an alum baking powder does not deteriorate in the package like a cream of tartar powder—its keeping quality is far above the latter ; (3) though burnt alum does not dissolve in water, during the baking process it sets free the gas from bicarbonate of soda slowly, and with greater regularity than cream of tartar, and, therefore, does much better and more effective work. He shows further that 80 gr. of burnt alum decompose as much bicarbonate (84 gr.) as 188 gr. of cream of tartar, and while the dry residue in the latter case weighs 210 gr., in the case of the alum it is 110 gr. (71 gr. sulphate of soda, 22 gr. sulphate of ammonia, and 17 gr. alumina). The use of alum in baking powder must not be confounded with its use for 'improving' bad flour. In the one case the alum remains in the bread as alum, just as it was put into the flour; but when mixed with bicarbonate of soda, as in baking powders, it is entirely decomposed, and there remains in the bread only a few grains of insoluble alumina, which is quite as harmless as would be a few grains of white clay or any other inert material. For these and other reasons Mr Petraeus considers that alum baking powders are the best, not only because a given quantity will raise more bread than the same quantity of cream of tartar baking powder, but because of the small quantity and innocent character of the residue they leave in the bread. A suitable formula for alum baking powder based on the figures given above would be as follows :—Burnt alum (in fine powder), 8 oz.; bicarbonate of soda, 8 oz. 3 dr.; rice flour, 1 lb.

Powder, Basil'ic. *Syn.* ROYAL POWDER, CORNACHINI'S P.; PULVIS BASILICUS, P. CORNACHINI, L. *Prep.* From scammony, calomel, cream of tartar, and diaphoretic antimony, equal parts. This is the formula generally adopted for this compound, which has now long been omitted from the Pharmacopœias. It is still a favourite with many practitioners, as an alterative purgative and vermifuge for children.—*Dose.* For a child, 2 to 8 gr.; for an adult, 5 to 20 gr. Compound powder of scammony is now generally sold for it.

Powder, Belladonna, Saccharated. (*Wertsler.*) *Syn.* PULVIS BELLADONNÆ SACCHARATUS. *Prep.* Belladonna root, 15 gr.; pure sugar, 1 dr.; mix. For 72 powders. One twice a day, or oftener, according to the age. In hooping-cough.

Powder, Elaine's Distem'per. The basis of this preparation is the 'aurum musivum,' or bisulphuret of tin (*Dr Paris*).

Powder, Blancmange'. *Prep.* From sago meal, 1 lb.; essence of lemon, 15 drops ; mace, 12 gr.; mix.

Powder, Bleach'ing. Chloride or hypochlorite of lime.

Powder, Blue. See SMALTS.

Powder, Bronze. See STANNIC SULPHIDE, BRONZING, &c.

Powder of Burnt Hartshorn. *Syn.* PULVIS CORNU CERVINI USTI, L. *Prep.* From pieces of hartshorn calcined to whiteness, and powdered. It consists principally of phosphate of lime.—*Dose,* 10 to 30 gr. ; in rickets, &c.

Powder of Burnt Hartshorn with O'pium. *Syn.* PULVIS OPIATUS, P. CORNU USTI CUM OPIO, L. *Prep.* From powdered calcined hartshorn, 1 oz.; powdered opium and cochineal, of each, 1 dr.—*Dose,* 5 to 20 gr.

Powder of Camphor. Camphor may be readily pulverised by triturating it with the addition of a few drops of rectified spirit or ether.

Powder, Camphorated Nitre. (*Swediaur.*) *Syn.* PULVIS NITRO CAMPHORATUS, L. *Prep.* Nitre, 10 gr.; camphor, 4 gr.; gum-arabic, 24 gr.; mix. For two or three doses.

Powder, Capuchin'. *Prep.* From powdered cevadilla, parsley seed, stavesacre, and tobacco, equal parts. Used to destroy pediculi.

Powder, Castillon's. *Prep.* From sago meal, salep, and gum tragacanth, of each, 3 dr.; prepared oyster shells, 1 dr.; cochineal, q. s. to colour. Absorbent.—*Dose,* ½ to 1 dr., boiled in milk ; in diarrhœa, &c.

Powder of Cat'echu (Compound). *Syn.* PULVIS CATECHU COMPOSITUS (B. P., Ph. D.), L. *Prep.* 1. (Ph. D.) Take catechu and kino, of each, 2 oz. ; cinnamon and nutmeg, of each, ½ oz.; reduce each to a fine powder, mix, and keep the prepared powder in a well-stoppered bottle. Aromatic and astringent.—*Dose,* ½ dr. to 2 dr. ; in various affections.

2. (B. P.) Pale catechu, 4 parts ; kino, 2 parts ; rhatany, 2 parts ; cinnamon, 1 part ; nutmeg, 1 part ; mix.—*Dose,* 15 to 30 gr.

Powder of Chalk (Compound). *Syn.* PULVIS CRETÆ AROMATICUS (B. P.), P. C. COMPOSITUS (Ph. L., E., and D.), P. CARBONATIS CALCIS COMPOSITUS, L. *Prep.* 1. (Ph. L.) Prepared chalk, ½ lb.; cinnamon, 4 oz. ; tormentil and gum acacia, of each, 3 oz. ; long pepper, ½ oz. ; rub them separately to a very fine powder, and mix them.

2. (Ph. E.) Prepared chalk, 4 oz. ; cinnamon, in fine powder, 1½ dr. ; nutmeg, in fine powder, 1 dr.

3. (Ph. D.) Prepared chalk, 5 oz. ; cinnamon, 2½ oz. ; gum, 2 oz. ; nutmeg, ½ oz.

4. (AROMATIC POWDER OF CHALK — B. P.) Chalk, 11 parts ; cinnamon, 4 parts ; nutmeg, 3 parts ; saffron, 3 parts ; cloves, 1½ parts ; cardamom seed, 1 part ; refined sugar, 25 parts ; all in powder ; mix.—*Dose,* 30 to 60 gr.

Uses, &c. Aromatic, astringent, and antacid.—*Dose,* 10 to 30 gr. ; in acidity, flatulence, heart-burn, diarrhœa, &c. The following form is used by many wholesale houses :—Prepared chalk, 4 lbs.; powdered cassia, 2 lbs. ; powdered calamus aromaticus, ½ lb.; powdered gum, 1½ lbs. ; long pepper, ½ lb.

Powder of Chalk with Opium (Compound). *Syn.* OPIATED CHALK POWDER; PULVIS CRETÆ AROMATICUS CUM OPIO (B. P.), P. C. COMPOSITUS CUM OPIO (Ph. L.), P. C. OPIATUS (Ph. E. and D.), L. *Prep.* 1. (Ph. L.) Compound chalk powder, 6½ oz.; powdered opium, 80 gr.

2. (Ph. E.) Compound chalk powder, 6 oz. ; powdered opium, 80 gr.

5. (Ph. D.) Compound chalk powder, 4 oz. 7 dr.; opium, in fine powder, 1 dr.

4. (Wholesale.) Compound chalk powder, 21 oz.; powdered opium, ½ oz. Anodyne, antacid, and carminative.—*Dose*, 10 to 30 gr.; in the same cases as the preceding, than which it is more active. It has long been a favourite remedy in all cases of simple and even choleraic diarrhœa.

5. Aromatic powder of chalk (see POWDER OF CHALK (Compound) 4), 39 parts; opium, in powder, 1 part; mix thoroughly, and pass through a sieve.—*Dose*, 10 to 40 gr.

Powder, Chalk Mixture. *Syn.* PULVIS PRO MISTURA CRETÆ, L. *Prep.* (*Beasley.*) Powdered gum acacia, 5 oz.; prepared chalk, 4 oz.; white sugar, 3 oz.; oil of cinnamon, 1½ fl. dr.; mix. 40 gr. of this powder, triturated with 1 fl. oz. of water.

Powder, Chel'era (Saline). *Syn.* PULVIS SALINUS ANTICHOLERICUS, L. *Prep.* (*Dr O'Shaughnessy.*) Carbonate of soda, 5 gr.; chloride of sodium, phosphate of soda, and sulphate of soda, of each, 10 gr. For a dose.

Powder of Cin'namon (Compound). *Syn.* AROMATIC POWDER; PULVIS CINNAMOMI COMPOSITUS (B. P., Ph. L.), P. AROMATICUS (Ph. E.), L. *Prep.* 1. (Ph. L.) Cinnamon, 2 oz.; cardamoms, 1½ oz.; ginger, 1 oz.; long pepper, ½ oz.; rub them together so that a fine powder may be made.

2. (B. P., Ph. E.) Cinnamon, cardamom seeds, and ginger, equal parts; to be kept in a well-closed glass vessel.

3. (Ph. D.) Cinnamon and ginger, of each, 2 oz.; cardamom seeds (husked) and nutmegs, of each, 1 oz. Aromatic and carminative.—*Dose*, 10 to 30 gr. In the powder of the shops cassia is generally substituted for cinnamon.

Powder, Cla'rifying. For clarifying alcoholic drinks, Dieterich recommends a powder composed of egg albumen (dried), 40 parts; sugar of milk, 40 parts; starch, 20 parts. All the ingredients must be in impalpable powder. Use 5 grms. of the powder to a litre of the liquor (about 3 gr. to 1 oz.); let the mixture stand in a warm room several days, shaking it at intervals. Finally filter through paper.

Powder, Cock'le. From the well-known shell-fish *Cardium edule*, Linn., as oyster powder.

Powder, Colbatche's Specific. *Prep.* From solution of sesquichloride of iron and acetate of lead, of each, 4 oz.; mix, evaporate to dryness, powder the residuum, and preserve it from the air. Astringent and hæmostatic.—*Dose*, 3 to 8 gr.

Powder of Colocynth. *Syn.* PULVIS COLO-CYNTHIDIS, L. That of the shops is generally prepared from the whole of the peeled fruit, with the seeds, instead of merely from the pulp, by which its activity is greatly lessened. A factitious article is also met with in trade, made by grinding bryony root with about twice its weight of colocynth seeds and a very small quantity of gamboge.

Powder, Compound Almond. (B. Ph.) *Syn.* PULVIS AMYGDALÆ COMPOSITUS. *Prep.* Steep 8 oz. of Jordan almonds in warm water till their skins can be easily removed; and, when blanched, dry them thoroughly with a soft cloth, and rub them lightly in a mortar to a smooth consistence; mix gum-arabic in powder, 1 oz.; and refined sugar, in powder, 4 oz.; and adding them to the pulp gradually, rub the whole to a coarse powder. Keep it in a lightly covered jar.

Powder, Compound Bark. (Geneva Ph.) *Syn.* PULVIS CINCHONÆ COMPOSITUS. *Prep.* Peruvian bark, 1 oz.; rhubarb, 1½ dr.; muriate of ammonia, 1½ dr.; mix.

Powder, Compound Belladonna. *Prep.* 1. (*Hecker.*) Belladonna, 1 to 3 gr.; musk, 5 gr.; camphor, 5 gr.; white sugar, 30 gr.; mix. For 3 powders.

2. (*Kopp.*) Belladonna root, 2 gr.; ipecacuanha, 2 gr.; sulphur, 32 gr.; sugar of milk, 32 gr. Mix, and divide into 3 powders, three daily. In hooping-cough.

Powder, Compound Bismuth. (Ferrier's Snuff.) *Syn.* PULVIS BISMUTHI COMPOSITUS. Hydrochlorate of morphine, 2 gr.; powdered acacia, 2 dr.; subnitrate of bismuth, 6 dr. Used as a snuff for cold in the head.

Powder, Compound Ceruse. (Ph. L., 1788.) *Syn.* PULVIS CERUSÆ COMPOSITUS. *Prep.* Carbonate of lead, 5 oz.; sarcocol, 1½ oz.; tragacanth, ½ oz.; mix. For outward use.

Powder, Compound of Cyanide of Zinc. (*Gilbourt.*) *Syn.* PULVIS ZINCI CYANIDI COMPOSITUS. *Prep.* Cyanide of zinc, 2 gr.; calcined magnesia, 27 gr.; cinnamon, 10 gr.; mix. For 6 doses, in cramp of the stomach.

Powder, Compound Fennel. (Brunsw. Ph.) *Syn.* PULVIS FŒNICULI COMPOSITUS. *Prep.* Carbonate of magnesia, 1 oz.; fennel seeds, ½ oz.; orange peel, 2 dr.; white sugar, 2 dr. Reduce each to a fine powder and mix.

Powder, Compound Musk. (Russ. Ph.) *Syn.* PULVIS MOSCHI COMPOSITUS. *Prep.* Musk, 8 parts; valerian, 10 parts; camphor, 3 parts.

Powder, Compound Opium. (B. P.) *Syn.* PULVIS OPII COMPOSITUS. *Prep.* Mix thoroughly 1½ oz. of opium, in powder, with 2 oz. of powdered black pepper, 5 oz. of ginger, 6 oz. of caraways, and ½ oz. of tragacanth. Pass the powders through a fine sieve, rub them lightly in a mortar, and keep the powder in a stoppered bottle. This powder nearly represents the dry ingredients of confection of opium.—*Dose*, 2 to 5 gr.

Powder, Compound Silver. (*Serre.*) *Syn.* PULVIS ARGENTI COMPOSITUS. *Prep.* Chloride of silver, 1 gr.; washed orris powder, 2 gr. Used in frictions, the same as COMPOUND GOLD POWDER.

Powder, Compound of Sulphate of Soda. *Syn.* PULVIS SODÆ SULPHATIS COMPOSITUS; SEL DE GUINDRE. *Prep.* Dried sulphate of soda, 18 dr.; nitrate of potash, ½ dr.; potassio-tartrate of antimony, 1 gr. A third part to be taken in water or herb broth.

Powder, Compound Sulphur. *Syn.* PULVIS SULPHURIS COMPOSITUS. RATIER:—Sulphur, 1 oz.; cream of tartar, 1 oz.; white sugar, q. s. VAN MONS (antidysenteric powder):—Sulphur, 1 oz.; fennel seed, 1 dr.; white sugar, 2 oz.; gum-arabic, 2 oz.; mix. SWEDIAUR (pectoral powder): —Sulphur, ½ oz.; liquorice, 1 oz.; orris, 2 dr.; benzoic acid, 1 scruple; white sugar, 2 oz.; oil of anise and fennel, of each, 10 drops.

Powder of Contrayer'va (Compound). *Syn.*

PULVIS CONTRAYERVÆ COMPOSITUS, L. *Prep.* (Ph. L., 1824.) Powdered contrayerva root, 5 oz.; prepared oyster shells, 1½ lbs.; mix. A tonic absorbent or antacid.—*Dose*, 10 gr. to ½ dr., as required.

Powder, Cooling. (G. Ph.) *Syn.* PULVIS TEMPERANS. *Prep.* Nitrate of potash, 1 oz.; cream of tartar, 3 oz.; sugar, 6 oz.

Powder, Corn. See CORN SOLVENT and POWDER, WART (*below*).

Powder, Cosmetic. *Syn.* PULVIS COSMETICUS, L. *Prep.* (Ph. Hann. 1831.) Blanched sweet almonds and beans, of each, 18 oz.; orris root, 8 oz.; white Spanish soap, 6 oz.; spermaceti, 1½ oz.; dried carbonate of soda, 1 oz.; oils of lavender, bergamot, and lemon, of each, 6 dr.; mix, and beat them to a powder. See POWDER, HAIR, and HAND (*below*), PASTE (Almond), POWDERS (Scented), &c.

Powder, Creasote and Starch. *Syn.* PULVIS CREASOTI ET AMYLI. Creasote, 10 minims; starch, 1 oz. A useful dusting powder in cases of erysipelas.

Powder of Crystal. From quartz, like POWDERED GLASS (p. 796). Used to make fine glass; also for a drier for paints, and sold under the name of ' diamond dust' for razor strops.

Powder of Cubebs with Alum. (*Matthieu.*) *Syn.* PULVIS CUBEBÆ CUM ALUMINE. *Prep.* Cubebs, 2 oz.; alum, 4 dr.; mix. For 9 doses. Three daily, in gonorrhœa.

Powder, Curry. *Syn.* INDIAN CURRY POWDER. The samples of this compound prepared by different houses vary so greatly from each other in the proportions of the ingredients, that it is difficult to regard any one as a standard. The following are, therefore, merely given as examples :

Prep. 1. Corianders, 1 lb.; turmeric, ¾ lb.; black pepper, ¼ lb.; scorched mustard, ¼ lb.; ginger, 2 oz.; cumin seed, 1 oz.; capsicums, ¾ oz.; mace, ¼ oz. (all in powder); mix well.

2. Coriander seeds and black pepper, of each, 8 lbs.; turmeric and cumin seeds, of each, 4 lbs.; allspice, ¾ lb.; mace, 1 oz. (all in powder); mix. This receipt is employed by an eminent wholesale house that does very largely in curry powder.

3. See p. 565.

Used as a condiment and flavouring ingredient. The addition of a few heads of garlic gives it an increased zest for Indian veterans.

Powder, Cust'ard. *Prep.* From sago meal, 2 lbs.; powdered turmeric, ½ oz.; bitter-almond powder, cassia, and mace, of each, ½ dr.

Powder, Cyprus. From *Cladonia rangiferina*, or reindeer moss. It has a very agreeable smell, and, being extremely retentive of odours, is much used as a basis for scent-powders, sachets, &c. The lichen known as the ragged hoary evernia also possesses nearly similar properties, and is often substituted for it. See POWDERS (Scented).

Powder, Diapente. (Ed. Ph. 1744.) *Syn.* PULVIS DIAPENTE, L. *Prep.* Aristolochia root, gentian, bay berries, myrrh, ivory dust, of each, 2 oz.; mix.

Powder, Disinfect'ing. *Syn.* PULVIS DISINFECTANS, L. *Prep.* (*Keist.*) Bisulphate of potassa, 41 parts; sugar of lead, 7 parts; binoxide of manganese, 3 parts; reduce them separately to a fine

powder, and, when wanted for use, mix a proper quantity in any suitable vessel. For other formulæ, see DISINFECTING COMPOUNDS. The name is generally applied to hypochlorite of lime.

Powder, Diuretic. (P. Cod.) *Syn.* PULVIS DIURETICUS, L. *Prep.* Gum-arabic, 6 oz.; sugar of milk, 6 oz.; nitrate of potash, 1 oz.; marshmallow root, 1 oz.; liquorice root, 2 oz.; mix.

Powder, Dover's. *Syn.* PULVIS DOVERI, L. *Prep.* (Original formula.) Nitre and sulphate of potassa, of each, 4 oz.; melt them together in a red-hot crucible, reduce the cold fused matter to powder, and add powdered ipecacuanha, liquorice, and opium, of each, 1 oz. This is the formula adopted in the Paris Codex. COMPOUND IPECACUANHA POWDER is now sold under this name (see *below*).

Powder, Duke of Portland's. *Syn.* PULVIS ANTIARTHRITICUS, L. *Prep.* Round birthwort, gentian, tops of lesser centaury, tops of ground pine, and germander, of each, equal parts.—*Dose*, 1 dr.

Powder, Egg. See POWDER, BAKING.

Powder of Elaterin, Compound. (B. Ph.) *Syn.* PULVIS ELATERINI COMPOSITUS, L. *Prep.* Elaterin, 5 gr.; sugar of milk, 195 gr. Rub them together to fine powder.—*Dose*, ½ gr. to 5 gr.

Powder, Emmen'agogue. *Syn.* PULVIS EMMENAGOGUS, P. HÆMATINUS, P. CONTRA AMENORRHŒAM, L. *Prep.* 1. Saccharine carbonate of iron, 3 parts; powdered myrrh, ginger, and nutmeg, of each, 1 part; divide into ½-dr. papers. One for a dose, twice or thrice daily.

2. (*Augustin.*) Myrrh, 12 gr.; saffron, 3 gr.; oil of cloves, 1 drop. For a dose, as the last.

3. (*Klein.*) Calomel, 24 gr.; extract of yew, 10 gr.; powdered savine, 1 dr.; Quevenne's iron, ¼ dr.; loaf sugar, 2 dr. For 6 powders; as before.

Powder, Emulsive, of Gluten. (*Taddei.*) *Syn.* PULVIS GLUTENIS EMULSIVUS, L. *Prep.* Fresh vegetable gluten, 10 oz.; soap, 2 oz.; water, 1 pint. Dissolve, evaporate the solution, and reduce to powder. As an antidote to corrosive sublimate.

Powder, Escharot'ic (Arsenical). *Syn.* PULVIS ESCHAROTICUS ARSENICALIS, L.; POUDRE DU FRÈRES COSME, Fr. *Prep.* 1. (Original formula.) From white arsenic, 12 gr.; burnt hartshorn, ½ dr.; cinnabar, 1 dr.

2. (P. Cod.) Red sulphuret of mercury and powdered dragon's blood, of each, 2 parts; levigated arsenious acid, 1 part; carefully mixed together. See CAUSTIC, ARSENICAL.

Powder of Extract of Col'ocynth (Compound). *Syn.* PULVIS EXTRACTI COLOCYNTHIDIS COMPOSITI, L. *Prep.* From compound extract of colocynth (Ph. L. 1836), dried by a gentle heat, and powdered.

Obs. This, like many other articles employed by lazy dispensers, does not represent the preparation for which it is used as a substitute; whilst, from its peculiar character, it is very open to sophistication—a practice, we regret to say, very general with some druggists. Indeed, some of these persons make this article by simply throwing the ingredients of the extract into a pan along with a little water, and, when they have become soft, stirring them together with a spatula, after which they are desiccated and powdered. This is then labelled by certain houses, ' Pulv. Ext. Coloc. co.—P. L.,' and sold to their unfortunate customers as such, although no such an extract has been in the Ph. L. since that of 1836.

Powder, Faynard's. The charcoal of beech wood, finely powdered (*Paris*). Used in piles, and as a styptic.

Powder, Fe'ver. See POWDER, ANTIMONIAL (*above*).

Powder of Flint. *Syn.* SILEX CONTRITUS (Ph. L.), L. *Prep.* As powdered glass (see p. 796). It is ordered in the Ph. L. to be employed, instead of magnesia, for the purpose of mechanically dividing the essential oils used in the preparation of distilled water. It is also used as an escharotic.

Powder, Fly. *Prep.* From white arsenic, 5 oz.; white sugar, 6 lbs.; rose pink, 2 oz.; mix, and put 6 dr. in each paper. Used to kill flies. It is poisonous, and should be employed with great caution, particularly where there are children.

Powder, Fu"migating. *Syn.* PULVIS FUMALIS, L. *Prep.* (Ph. Russ.) Amber, mastic, and olibanum, of each, 3 parts; storax, 2 parts; benzoin and labdanum, of each, 1 part; reduce them to coarse powder, and mix them well. See FUMIGATION.

Powder of Galls (Compound). *Syn.* PULVIS GALLÆ COMPOSITUS, L. See POWDER, ASTRINGENT.

Powder, Gascoign's. *Syn.* PULVIS ECHELIS COMPOSITUS. *Prep.* Prepared crab shells, 1 lb.; prepared chalk, 3 oz.; prepared coral, 3 oz.; mix.

Powder, Goelis's Antihec'tic. *Prep.* From burnt hartshorn, powdered nutmeg, black pepper, and roasted laurel berries, of each, 1 part; liquorice powder, 3 parts.—*Dose,* ¼ to 1 dr.; in the hectic fever of scrofulous subjects.

Powder, Gold. *Syn.* PULVIS AURI. Triturate gold leaf with 10 or 12 times its weight of sulphate of potash till bright particles are no longer visible; pass it through a sieve; mix with boiling water, wash what remains on the filter, and dry in a stove.

Powder, Gold (Compound). *Syn.* PULVIS AURI COMPOSITUS. *Prep.* Auro-chloride of sodium, 1 gr.; lycopodium, starch, or washed orris powder, 1 scruple; mix. A ₁/₁₀th part, gradually increased to ¼th part, of this powder to be rubbed on the gums.

Powder of Gold and Iron. (*Buokler*.) *Syn.* PULVIS AURI ET FERRI. *Prep.* Pulverised gold, 2 scruples; clean levigated iron filings, 2 scruples; gum-arabic in powder, 30 gr.; mix. For one dose, to be given in water acidulated with a few drops of sulphuric acid, as an antidote for corrosive sublimate.

Powder, Goulard. Effloresced sugar of lead. Poisonous.

Powder, Gregory's. See POWDER OF RHUBARB (Compound).

Powder, Grey. Mercurial powder.

Powder of Guarana, Compound. (*Dr Gavrelle*.) *Syn.* PULVIS PAULLINIÆ COMPOSITUS. *Prep.* Guarana, 1 dr.; compound cinnamon powder, 4 dr. Mix.

Powder, Gum. (G. Ph.) *Syn.* PULVIS GUM-MOSUS. *Prep.* Gum-arabic, 3 oz.; liquorice, 2 oz.; refined sugar, 1 oz. Mix.

Powder, Hæmostat'ic. *Syn.* PULVIS HÆMOSTATICUS, L. *Prep.* 1. (*Guibourt.*) Charcoal and gum-arabic, of each, in powder, 1 part; powdered resin, 4 parts.

2. (*Mialhe.*) From powdered alum, gum tragacanth, and tannin, equal parts. Used to check local bleeding.

3. (P. Cod.) Resin, 4 dr.; gum-arabic, 1 dr.; powdered catechu, 1 dr. Mix.

Powder, Hair. *Syn.* PULVIS PRO CRINE, L. Starch reduced to a very fine powder, and then scented according to the fancy of the artist; it is lastly passed through a gauze sieve. In its simple form, without any addition, it constitutes 'plain hair powder.' In other cases it is distinguished by the name of the substance added to perfume it. Thus we have ' rose hair powder,' ' violet h. p.,' &c. Potato farina, well triturated, is now commonly used for hair powder. Amongst the lower classes, the contents of the ' flour dredger ' of the kitchen are frequently misappropriated to this purpose. See POWDERS, SCENTED (*below*).

Powder, Hand. *Prep.* From almond powder, 1 lb.; powdered cuttle-fish bone and white soap, of each, 4 oz.; orris powder, 1 oz.; mix. Used to clean the hands and to render them soft and white. See POWDER, COSMETIC.

Powder, Helvetius's. *Syn.* PULVIS HELVETII, L. A mixture of powdered alum and dragon's blood (*Dr Paris*).

Powder, Herrenschwand's Specific. See PATENT MEDICINES.

Powder, Hiera Picra. Powder of aloes with canella (see *above*).

Powder, Hufeland's. *Syn.* HUFELAND'S QUINQUINA TACTICE, PULVIS CINCHONÆ FACTITIUS, P. SALICIS COMPOSITUS, L. *Prep.* From bennet (the herb), calamus aromaticus, chestnut bark, gentian root, and willow bark, equal parts; reduced to powder.

Powder, Hunt's. See POWDER, BREAKFAST.

Powder, Hunter's. See POWDER, WART (*below*).

Powder of Ipecacuan'ha (Compound). *Syn.* DOVER'S POWDER, COMPOUND POWDER OF IPECACUANHA WITH OPIUM; PULVIS DOVERI, PULVIS IPECACUANHÆ COMPOSITUS (B. P., Ph. L., E., & D.), L.; POUDRE D'IPECACUANHA ET D'OPIUM, Fr. *Prep.* 1. (B. P., Ph. L.) Ipecacuanha and opium, of each, in fine powder, 1 dr.; sulphate of potassa, in fine powder, 1 oz.; mix them (thoroughly). The Edin. and Dublin formulæ are similar.

2. (P. Cod.) Nitrate and sulphate of potassa, of each, 4 oz.; ipecacuanha, liquorice root, and hard extract of opium, of each, 1 oz. This closely resembles the original formula.

3. (*Wholesale.*) From powdered ipecacuanha and opium, of each, 1 lb.; powdered sulphate of potassa, 8 lbs.—*Uses, &c.* ' Dover's powder is a powerful and valuable sudorific.'—*Dose,* 5 to 15 or 20 gr., followed by warm diluents; in inflammatory affections, rheumatisms, colds, &c.

Powder, Itch. *Syn.* PULVIS ANTIPSORICUS, L. *Prep.* 1. Sulphur and potato farina, of each, ½ lb.; essence of bergamot, ¼ oz.; mix.

2. (Poudre de Pihorel.) A mixture of finely

pulverised sulphuret of calcium and farina, in nearly equal quantities. Used either as a dusting powder or mixed with a little oil or fat, and rubbed into the affected part.

3. (Fr. Hosp.) Flowers of sulphur, 1 oz.; acetate of lead, 1 oz. Mix.

4. Equal parts of sulphur and charcoal.

Powder of Jal'ap (Compound). *Syn.* PULVIS JALAPÆ COMPOSITUS (B. P., Ph. L., E., & D.), L. *Prep.* 1. (Ph. L.) Jalap, 3 oz.; bitartrate of potassa, 6 oz.; ginger, 2 dr.; rub them separately into fine powder, then mix them.

2. (Ph. E.) As the last, omitting the ginger.

3. (Ph. D.) Jalap, 2 oz.; bitartrate of potassa, 3½ oz.; ginger, ½ oz. (all in fine powder); mix by careful trituration.—*Dose*, 20 to 60 gr.; as a purgative in habitual costiveness, dropsies, &c.; also in worms, the tumid bellies of children, &c.

4. (B. P.) Jalap, in powder, 5 parts; acid tartrate of potash, 9 parts; ginger, in powder, 1 part; mix.—*Dose*, 20 to 60 gr.

5. (PULVIS LENITIVUS; SUCRE ORANGE PURGATIF.) From refined sugar, ¾ lb.; jalap and cream of tartar, of each, 2 oz.; oil of orange peel, ¼ oz. A popular purgative on the Continent.—*Dose*, 1 to 3 dr.

Powder, James's. *Syn.* PULVIS JACOBI, P. FEBRIFUGUS JACOBI, L. The antimonial powder, or compound powder of antimony, of the Pharmacopœias (see *above*) is the preparation which usually passes under this name; but the true James's powder is a nostrum, the pretended secret of the preparation of which is claimed to be possessed by only two persons in the kingdom. The patent specification of the once celebrated Dr James runs as follows:

"Take of antimony, calcine it with a continued protracted heat in a flat, unglazed earthen vessel, adding to it, from time to time, a sufficient quantity of any animal oil and salt, well dephlegmated; then boil it in melted nitre for a considerable time, and separate the powder from the nitre by dissolving it in water." On this it has been remarked that it yields a product totally different from that which Dr James and his successors have sold under the name, and he has hence been charged with concealing the real formula for his powder, and publishing a false one in its stead.

According to Dr Robinson the original formula for this nostrum, and that still adopted by the vendors of the proprietary article at the present day, is—tartarised antimony, 1 part; prepared burnt hartshorn and calx of antimony, of each, 5 parts; carefully mixed together, and divided into 21-gr. powders ('Phil. Journ. Pharm.,' vi, 282).

From analyses recently made of three specimens of James's powder ('Newbery's,' 'Butler's,' and a sample of 60 years old obtained by Mr Squire), it appears that antimonious acid was present in different proportions, from about 45% to 33%, the amount being greatest in the old specimen; teroxide of antimony was also present to the extent of from 9% to less than 1%, the greatest quantity being again in the old preparation; the remainder in each specimen consisted chiefly of phosphate of lime; no trace of tartaric acid was discoverable in any of the samples.

Perhaps no nostrum ever received such extensive patronage from the faculty as James's powder. Dr James himself was remarkably successful in its use; but whether his success depended upon his powder or the mercurials and bark, which he commonly employed at the same time, is still undetermined.

Powder of Kermes with Camphor. (Germ. Hosp.) *Syn.* PULVIS KERMETIS CUM CAMPHORA. *Prep.* Kermes mineral, 3 gr.; camphor, 6 gr.; white sugar, 2 dr.; mix. For 12 doses.

Powder of Kermes with Ipecacuanha. (Fr. Hosp.) *Syn.* PULVIS KERMETIS CUM IPECACUANHA. *Prep.* Kermes, 2 gr.; ipecacuanha, 2 gr.; crabs' eyes, 2 scruples; gum-arabic, 2 scruples; mix. For 12 doses. In hooping-cough.

Powder of Ki'no (Compound). *Syn.* PULVIS KINO COMPOSITUS (Ph. L.), L. *Prep.* 1. (Ph. L.) Kino, 15 dr.; cinnamon, 4 dr.; dried opium, 1 dr.; reduce them separately to fine powder, and then mix them.—*Dose*, 5 to 20 gr.; in diarrhœa, pyrosis, &c.

2. (B. P.) Kino, 3½ oz.; cinnamon, 1 oz.; opium, ½ oz.; mix.—*Dose*, 5 to 20 gr.

Powder, Lansanne. *Prep.* From nitre, 1½ dr.; carbonate of magnesia, bitartrate of potassa, precipitated sulphur, and oleo-saccharum of peppermint, of each, 4 dr.; sugar of milk, 1 oz. Lenitive and antidysenteric.

Powder, Lax'ative. See SPECIES, LAXATIVE.

Powder of Liquorice (Compound). *Syn.* PULVIS GLYCYRRHIZÆ COMPOSITUS, L. *Prep.* 1. (Ph. Bor.) Liquorice root and senna leaves, of each, 6 oz.; fennel seed and milk of sulphur (pure), of each, 3 oz.; white sugar, 18 oz. (all in fine powder); mix. Pectoral and laxative.

2. (B. Ph.) Senna and liquorice root, both in fine powder, of each, 2 oz.; sugar, in fine powder, 6 oz.; sublimed sulphur, 1 oz.; mix thoroughly, and pass through a fine sieve.—*Dose*, ½ dr. to 1 dr.

Powder of Magne'sia and Rhu'barb. See POWDER OF COMPOUND RHUBARB (*below*).

Powder, Martin's Cancer. An American nostrum, composed of the powdered stems of the *Orobanche Virginiana*, Linn., combined with a very small quantity of arsenious acid. It is used as a sprinkle for open cancers and cancerous sores.

Powder, Mercu"rial. *Syn.* GREY POWDER (HYDRARGYRUM CUM CRETA, B. P.), MERCURY WITH CHALK. *Prep.* 1. (B. P.) Mercury, 1 part; prepared chalk, 2 parts; triturate till the globules disappear.—*Dose*, 3 to 8 gr.

2. Mercury, 3 oz.; powdered resin, ½ oz.; prepared chalk, 5 oz.; rectified spirit, q. s.; make a paste with the resin and a small quantity of the spirit; add the mercury, which may be extinguished in a short time; then the chalk and alcohol gradually, so as to keep up the pasty consistence; lastly, add sufficient spirit to dissolve out the resin, wash the powder on a filter, and dry it. Rectified oil of turpentine may be substituted for the spirit.

Powder, Morison's Ape"rient. See PATENT MEDICINES.

Powder of Mugwort, Saccharated. (*Breslaw.*) Powdered mugwort root, 3 oz.; sugar, 6 oz.;

a teaspoonful four times a day in chorea and epilepsy.

Powder of Mush'room. *Syn.* PULVIS AGARICI, P. A. ESCULENTI, L. From edible mushrooms, dried by a gentle heat, and then powdered along with a little white pepper, cloves, and mace. Some cayenne is frequently added.

Powder of Mus'sel. From the *Mytilus edulis*, Linn., or common mussel, in the same way as POWDER OF OYSTER.

Powder of Myrrh (Compound). *Syn.* PULVIS È MYRRHÀ COMPOSITUS, L. *Prep.* (Ph. L., 1788.) Myrrh, dried savine, dried rue, and Russian castor, equal parts, rubbed to powder, and then well mixed. Emmenagogue and antispasmodic.—*Dose,* 12 to 30 gr.

Powder, Nursery. See POWDER, VIOLET (*below*).

Powder of Nux Vomica (Compound). *Syn.* PULVIS NUCIS VOMICÆ COMPOSITUS; VOGT'S STOMACHIC POWDER. *Prep.* Nux vomica, 18 gr.; ipecacuanha, 24 gr.; rhubarb, 1 dr.; prepared oyster shell, 48 gr.; oleo-saccharum of mint, 1 dr. Mix, and divide into 12 powders.

Powder, O″piated. Powder of chalk with opium.

Powder, Opiated Guaiacum. (*Pareira.*) *Syn.* PULVIS GUAIACI OPIATUS. *Prep.* Guaiacum, 1 dr.; orange leaves, ½ dr.; acetate of morphia, ⅛ gr. Mix, and divide into 6 powders. One every two hours in articular rheumatism.

Powder of Oxide of Zinc with Starch. (*Casenave.*) *Syn.* PULVIS ZINCI OXYDI CUM AMYLO. *Prep.* Starch, 1 oz.; oxide of zinc, 1 dr.; camphor, in powder, 1 dr. For excoriations and bedsores.

Powder of Oys'ter. *Syn.* PULVIS OSTREÆ, L. *Prep.* From the common oyster (*Ostrea edulis*, Linn.), pulped through a sieve, made into a paste with wheaten flour and a little salt, and then rolled out into thin pieces, and dried; these are reduced to powder, sifted, and packed in well-corked bottles. Used to make sauce, about 1 oz., to water, 1 pint. Other shell-fish are treated in the same way.

Powder, Parturifacient. (E. Ph., 1744.) *Syn.* PULVIS AD PARTUM. *Prep.* Borax, 4 dr.; castor, 1½ dr.; saffron, 1½ dr.; oil of cinnamon, 8 drops; oil of amber, 6 drops. Mix.—*Dose,* 20 to 30 gr. (This name, and also that of PULVIS PARTURIFACIENS, has been given to powdered ergot.)

Powder, Pea. *Syn.* PEA FLOUR; FARINA PISORUM, L. *Prep.* From peas, in the usual manner. Used to make extemporaneous pea soup.

Powder, Pearl. *Prep.* From pure pearl white and French chalk (scraped fine by Dutch rushes), equal parts, triturated together. Some makers add more French chalk. Used as a skin cosmetic. This mixture is preferable to pearl white alone, from being more adhesive.

Powder, Pease. *Prep.* From dried mint and sage, of each, 4 oz.; celery seed and white pepper, of each, ½ oz.; turmeric powder, ¼ oz.; reduced to fine powder. Used as a condiment and kitchen spice.

Powder, Pec'toral. See POWDER OF LIQUORICE, &c.

Powder of Phosphate of Lime, Saccharated.

VOL. II.

Syn. PULVIS CALCIS PHOSPHATIS SACCHARATUS. *Prep.* Precipitated phosphate of lime, 15 gr.; white sugar, 85 gr.; triturate and divide into 20 packets. Two or more powders daily, according to age of child. In rickets.

Powder, Piles. *Syn.* PULVIS ANTIHÆMORRHOIDALIS, P. HÆMORRHOIDALIS, L. *Prep.* 1. (Fr. Hosp.) Precipitated sulphur, 3 oz.; cream of tartar and black pepper, of each, 1 oz.; oil of cubebs, ½ dr.—*Dose,* A teaspoonful, in milk or honey, thrice a day.

2. (External.) *a.* From Aleppo galls, in very fine powder, 2 oz.; opium, in fine powder, 1 dr. A pinch to be applied occasionally.

b. From sesquioxide of iron, 1 oz.; powdered acetate of lead, ½ dr. As the last.

Powder of Pitch, Compound. *Syn.* PULVIS PICIS COMPOSITUS; DISINFECTING POWDER OF CORNE AND DEMAUX. 100 parts of plaster of Paris are triturated thoroughly with 1 to 5 parts of coal tar. Used as an absorbent and disinfectant to fetid ulcers and wounds.

Powder, Plate. *Syn.* PULVIS PRO ARGENTO, L. *Prep.* 1. Jeweller's rouge, ½ lb.; prepared chalk or levigated burnt hartshorn, ¾ lb.; mix.

2. Levigated putty powder, ½ lb.; burnt hartshorn, ½ lb.; prepared chalk, 1 lb.; rose pink, 1 oz.

3. (MERCURIAL.) From quicksilver with chalk, 1 oz.; prepared chalk, 11 oz.; mix. Used to clean and polish plate. See PLATE.

Powder, Plate-boiling. *Prep.* From cream of tartar, common salt, and alum, equal parts. A little of this powder, added to the water in which plate is boiled, gives to it a silvery whiteness.

Powder, Plummer's Al'terative. See ANTIMONY, ETHIOPS OF.

Powder, Poul'tice. *Syn.* PULVIS PRO CATAPLASMATE (Ph. D. 1826), L. *Prep.* From linseed meal, 1 part; oatmeal, 2 parts; mixed together.

Powder, Purgative and Anthelmintic. (*Boerhaave.*) *Syn.* PULVIS PURGANS ANTHELMINTICUS, L. *Prep.* Jalap, 12 gr. (or agaric, 8 gr.); Ethiops mineral, 12 gr.; for one dose.

Powder of Quinine, Aërated. (*Dr Meireu.*) *Syn.* PULVIS QUINIÆ AËRATUS, L. *Prep.* Tartaric acid, 15 gr.; disulphate of quinia, 1½ gr. Mix, and add bicarbonate of soda, 18 gr.; refined sugar, 30 gr. Mix for one dose, between the fits of intermittent fever.

Powder of Quinine and Tobacco. (*Hug.*) *Syn.* PULVIS QUINIÆ SULPHATIS ET TABACI. *Prep.* Disulphate of quinine, 12 gr.; snuff, 1 oz. To be used as a snuff for nervous headaches.

Powder, Rats. See RATS.

Powder of Rhu'barb (Compound). *Syn.* GREGORY'S MIXTURE, GREGORY'S POWDER; PULVIS RHEI COMPOSITUS (B. P., Ph. E. and D.), L. *Prep.* 1. (Ph. E.) Calcined magnesia, 1 lb.; rhubarb, 4 oz.; ginger, 2 oz. (all in fine powder); mix, and preserve it from the air.

2. (B. P., Ph. D.) Calcined magnesia, 6 oz.; rhubarb, 2 oz.; ginger, 1 oz.

3. Calcined magnesia, 8 oz.; rhubarb, 3 oz.; chamomile, 2 oz.; ginger, 1 oz.

Obs. An excellent stomachic, antacid, and laxative.—*Dose,* 20 gr. to ½ dr. Some druggists

88

substitute the heavy carbonate for the calcined magnesia ordered above, but this alters the nature of the preparation, and requires the dose to be increased. Heavy calcined magnesia may, however, be employed with advantage.

Powder, Bach'et. See SCENTED POWDERS.

Powder, Saline' (Compound). *Syn.* PULVIS SALINUS COMPOSITUS (Ph. E.), L. *Prep.* (Ph. E.) Pure chloride of sodium and sulphate of magnesia, of each, 4 oz. ; sulphate of potash, 3 oz. ; each separately dried by a gentle heat, and pulverised, then triturated together, and preserved in well-closed vessels. An excellent saline purgative.—*Dose,* 2 to 6 dr., in ½ pint of water or table-beer, in the morning, fasting.

Powder of Scammony (Compound). *Syn.* PULVIS SCAMMONII COMPOSITUS (B. P., Ph. L., E., & D.), L. *Prep.* 1. (Ph. L.) Scammony and hard extract of jalap, of each, 2 oz. ; ginger, ½ oz. ; rub them separately to a very fine powder, and then mix them.—*Dose,* 5 to 15 gr.

2. (Ph. E.) Scammony and bitartrate of potassa, of each, in very fine powder, equal parts.—*Dose,* 7 to 20 gr.

3. (Ph. D.) Scammony, in fine powder, 1 oz. ; compound powder of jalap, 3 oz. ; mix.—*Dose,* 13 to 30 gr.

4. (B. P.) Scammony, 4 parts ; jalap, 3 parts ; ginger, 1 part ; mix, and reduce to fine powder.—*Dose,* 10 to 12 gr.

Obs. The above are favourite cathartics in worms, especially for children. They are commonly sold for basilic powder (see *above*).

Powder of Scammony with Cal'omel. *Syn.* PULVIS SCAMMONII CUM CALOMELANE, L. *Prep.* From scammony, ½ oz. ; calomel and white sugar, of each, 2 dr. An excellent vermifuge for children.—*Dose.* For an adult, 5 to 20 gr. ; for a child, 2 to 8 gr. Sold for basilic powder, to which it approaches nearer in composition than the preceding.

Powder of Scammony with Soot. *Syn.* PULVIS SCAMMONII CUM FULIGINE ; POUDRE D'AILHAUT. *Prep.* Scammony, 1 dr. ; wood soot, 1½ dr. ; resin, 2 dr. ; mix.—*Dose,* ½ dr. A once fashionable purgative.

Powder, Schmidt's Parturifa"cient. *Syn.* SCHMIDT'S POUDRE OCYTIQUE, Fr. *Prep.* From powdered ergot of rye, borax, and oleo-saccharum of camomile, of each, 8 gr. ; powdered sugar, q. s. For a dose ; to be repeated every quarter of an hour until some effect is produced.

Powder of Scordium (Compound). *Syn.* PULVIS E SCORDIO COMPOSITUS (L. Ph. 1746). *Prep.* Bole, 4 oz. ; scordium, 2 oz. ; cinnamon, 1½ oz. ; styrax, tormentil, bistort, gentian, dittany, galbanum, gum acacia, red rose petals, of each, 1 oz. ; long pepper, ½ oz. ; ginger, ½ oz. Make a powder.

Powder of Scordium with Opium. *Syn.* PULVIS E SCORDIO CUM OPIO. *Prep.* Add to the preceding, 3 dr. of dry strained opium, and powder it with the other ingredients.

Powder of Senna (Battley's Green). *Syn.* PULVIS SENNÆ VIRIDIS, L. *Prep.* From senna leaves, dried and heated until they turn yellow, then powdered along with a little (blue) charcoal, to give a green colour.

Powder of Sen'na (Compound). *Syn.* PUL-VIS SENNÆ COMPOSITUS, L. *Prep.* (Ph. L. 1824.) Senna and bitartrate of potassa, of each, 2 oz. ; scammony, ½ oz. ; ginger, 2 dr. ; all in fine powder ; mix.—*Dose,* 20 to 30 gr. or more ; as a purgative or anthelmintic.

Powder, Sil'vering. *Prep.* 1. Silver dust (fine), 20 gr. ; alum, 30 gr. ; common salt, 1 dr. ; cream of tartar, 3 dr. ; rub them together to a fine powder.

2. As the last, but substituting 35 gr. of nitrate of silver for the silver dust.

3. Chloride of silver is dissolved in a solution of hyposulphite of soda, and the solution made into a paste with levigated burnt hartshorn or bone dust ; this is next dried and powdered.

4. Silver dust, 1 oz. ; common salt and sal-ammoniac, of each, 4 oz. ; corrosive sublimate, ½ oz.

Obs. The above powders, made into a paste with a little water, are used to silver dial-plates, statuettes, and other articles in copper, previously well cleaned, by friction. The best silver powder for the purpose is that precipitated from its nitric acid solution by means of a copper plate. When the product of the last formula is used, the articles should be afterwards made red-hot, and polished.

Powder of Soap. *Syn.* SAPO CONTRITUS, PULVIS SAPONIS, L. Castile soap, sliced or cut small, dried by exposure to a warm atmosphere, or by a very gentle heat, and then powdered. Used in dispensing ; also as a hand, shaving, and tooth powder. As a cosmetic it may be scented at will.

Powder, Spermaceti. *Syn.* PULVIS CETACEI. Spermaceti is pulverised as camphor, by the aid of a few drops of spirit.

Powder of Spermaceti with Sugar. *Syn.* PULVIS CETACEI CUM SACCHARO. One part of powdered spermaceti with two of sugar. Pectoral.

Powder of Sponge. *Syn.* PULVIS SPONGIÆ, P. SPONGIÆ USTÆ, L. *Prep.* Let sponge, cut into small pieces, be beaten so as to free it from sand or stones ; then burn it in a covered iron vessel until it becomes black and friable ; finally, reduce it to powder. Dobstruent.—*Dose,* ½ to 3 dr. ; in glandular indurations and enlargements, &c. It should be of a brownish-black colour ; if over-burnt its efficacy is destroyed.

Powder of Squills. *Syn.* PULVIS SCILLÆ, L. *Prep.* Remove the membranous integuments from the bulb of the squill, cut it into thin slices, and dry it at a heat between 90° and 100° F. ; next reduce it to powder, and keep it in well-stoppered bottles.

Powder, Stahl's Resolvent. *Syn.* PULVIS RESOLVENS STAHLII. *Prep.* Antimonial powder, nitre, prepared crabs' eyes, in equal parts.

Powder of Starch with Soda. *Syn.* PULVIS AMYLI ET SODÆ ; DEVERGIE'S ALKALINE POWDER. *Prep.* Mix 1 part of carbonate of soda in fine powder with 10 of white starch. For external use in some skin diseases.

Powder, Sternu'tatory. See SNUFFS (Medicated).

Powder, Styp'tic. See POWDER, ASTRINGENT, FAYNARD'S P., &c.

Powder, Tonquin. *Syn.* PULVIS ANTILYSSICUS TONQUINENSIS ; SIR G. COBB'S TONQUIN POWDER.

Prep. Musk, 16 gr.; cinnabar, 48 gr.; to be mixed or washed down with arrack or other spirit. Three doses to be given on three alternate days, and three more on the three next changes of the moon.

Powder of Trag'acanth (Compound). *Syn.* PULVIS TRAGACANTHÆ COMPOSITUS (B. P., Ph. L. & E.), L. *Prep.* 1. (Ph. L.) Gum tragacanth, gum acacia, and starch, of each, in fine powder, 1½ oz.; powdered white sugar, 3 oz. The Edinburgh formula is similar. Demulcent.—*Dose,* ½ dr. to 2 dr., in water or any simple liquid; in hoarseness and catarrhs, combined with squills and henbane, to allay irritation; in dysentery, combined with ipecacuanha; in gonorrhœa, strangury, &c., combined with acetate of potassa or nitre.

2. (B. P.) Tragacanth, in powder, 1 part; gum-arabic, in powder, 1 part; starch in powder, 1 part; refined sugar, in powder, 3 parts; rub well together.—*Dose,* 10 to 60 gr.

Powder of Vanilla, with Sugar. (P. Cod.) *Syn.* PULVIS VANILLÆ CUM SACCHARO; POUDRE DE VANILLE SUCRÉE. Vanilla is reduced to powder by cutting it in pieces, and triturating it with 9 times its weight of refined sugar.

Powder of Verdigris with Calomel. *Syn.* PULVIS ÆRUGINIS CUM CALOMELANE. *Prep.* Prepared verdigris, 1 dr.; calomel, 1 dr.; mix. For external use.

Powder, Violet. *Syn.* NURSERY POWDER, SKIN P. This is simply starch, reduced to a very fine powder, and scented with orris powder or essence of violets. The best kinds are also perfumed with a little musk or ambergris, and are now generally made with potato farina. The commoner sort is only scented with a little essence of bergamot, or essence of lemon. 'Plain violet powder' is, of course, unscented.

Prep. 1. Powdered starch, 28 lbs.; powdered orris root, 1 lb.; essence of ambergris and essence of bergamot, of each, ½ oz.; oil of rhodium, ¾ dr.; mix, and pass the powder through a sieve.

2. Powdered starch, 14 lbs.; essence of bergamot, ½ oz.; oil of cloves, ¾ oz.; as last. Used as a dusting powder in excoriations, &c. See POWDERS, COSMETIC (*below*).

Powder, Ward's Sweating. Resembles DOVER'S POWDERS.

Powder, Wart. *Syn.* CORN POWDER, COSMETIC CAUSTIC, &c. *Prep.* 1. Ivy leaves ground to powder. A pinch is applied with a rag, the part being first moistened with strong vinegar. Useful for soft corns and warts.

2. (*Hunter's.*) From savine and verdigris, equal parts. See CORN SOLVENT.

Powder, Warwick's (Earl of). *Syn.* PULVIS COMITIS WARWICENSIS, L. *Prep.* From scammony, prepared with the fumes of sulphur, 2 oz.; diaphoretic antimony, 1 oz.; cream of tartar, ¾ oz. —*Dose,* 15 to 30 gr.

Obs. This is a modification of CORNACHINI'S POWDER. It is represented in the present Pharmacopœias by COMPOUND SCAMMONY POWDER. "Cornachini wrote a whole book about his powder, the proportions of the ingredients of which he varied according to circumstances" (' Med. Lex.').

Powder, Wash'ing. The numerous compounds vended under this name have for their basis the soda-ash of commerce, blended with common Scotch soda in variable proportions. The best of them consist either wholly or chiefly of the first of these substances. The alkaline matter is reduced to coarse powder, and stirred up with liquid size, or with a decoction of linseed, Irish moss, or British gum, and is then dried, and again crushed or powdered, and at once put into the packages, in which it is rammed tight, and covered up immediately. The object aimed at by the manufacturer is to keep his commodity from the air as much as possible, because exposure renders it less caustic, and consequently less detergent.

Powder of Yellow Bladder-wrack. (Ph. D.) *Syn.* PULVIS QUERCÛS MARINÆ. *Prep.* Yellow bladder-wrack, in flower, is dried, cleansed, and heated in a crucible with a perforated lid till vapours cease to be given off, and the carbonaceous residue reduced to powder.—*Dose,* 10 gr. to 2 dr.

POWDERS. The following preparations have been placed under this head instead of under ' POWDER,' because some are invariably spoken of in the plural number, and the others may be conveniently noticed in classes or groups :

Powders, Aërated Sherbet (IN ONE BOTTLE). Double refined sugar, 14½ oz.; powdered orange peel, 12 gr.; bicarbonate of soda, 3½ oz.; essence of cedrat, 12 drops; oil of orange peel, 60 drops; tartaric acid, 4 oz. The powders must be carefully dried, mixed quickly, and afterwards kept dry, in a bottle securely corked. A measure holding nearly 3 dr. of the powder should accompany each bottle.

Powders, Efferves'cing. *Prep.* 1. (PULVERES EFFERVESCENTES—Ph. E.) Take of tartaric acid, 1 oz.; bicarbonate of soda, 1 oz. 54 gr. (534 gr.), or bicarbonate of potassa, 1 oz. 2 dr. 40 gr. (640 gr.); reduce the acid and either bicarbonate separately to fine powder, divide each of these into 16 powders, and preserve the acid and alkaline powders in separate papers of different colours.

2. (PULVERES EFFERVESCENTES CITRATI—Ph. D.) Take of citric acid (crystallised), 9 dr.; bicarbonate of soda, 11 dr., or bicarbonate of potassa, 13 dr.; proceed as last, dividing each into 18 parts.

3. (PULVERES EFFERVESCENTES TARTARIZATI —Ph. D.) Take of tartaric acid (in crystals), 10 dr.; bicarbonate of soda, 11 dr., or bicarbonate of potassa, 13 dr.; reduce them to powder, and divide them into 18 parts, as before (see *below*).

Powders, Effervescing, with Iron. (P. Cod.) *Syn.* PULVERES EFFERVESCENTES CUM FERRO. *Prep.* Tartaric acid, 2½ oz.; bicarbonate of soda, 2 oz.; powdered sugar, 9 oz.; dried sulphate of iron, 46 gr. Mix the acid and the sulphate of iron (previously reduced to coarse powder), add the sugar, and lastly the soda, not in very fine powder. All the ingredients must be very dry. Half an ounce of this powder is to be quickly added to 2 pints of pure water (without air) contained in a bottle, which is to be immediately corked.

Powders for Gasogene. For 2 pints :—Powdered tartaric acid, 14 scruples; bicarbonate of soda, 17 scruples.

For 3 pints:—Powdered tartaric acid, 17 scruples; bicarbonate of soda, 21 scruples.

For 5 pints:—One each change of 2 and 3 pints.

Powders, Gin'ger Beer. *Syn.* PULVERES EFFERVESCENTES CUM ZINGIBERE, L. *Prep.* 1. Powdered white sugar, 1 to 2 dr.; bicarbonate of soda, 26 gr.; finest powdered Jamaica ginger, 6 gr.; essence of lemon, 1 drop; mix, and wrap it in blue paper. In the white paper put of powdered tartaric acid, 35 gr., or of powdered citric acid, 30 gr.

2. Finest Jamaica ginger, 1 dr.; bicarbonate of soda, 5 dr.; white sugar, 16 dr.; essence of lemon, 6 or 8 drops; mix and divide it between 12 papers (blue). For the white papers, divide tartaric acid, 6 dr., in the same way. By taking the drachms as ounces, the quantity will be sufficient for 8 dozen. For use dissolve one of each colour separately in somewhat less than half a glass of water, mix the two, and drink the mixture whilst effervescing.

3. (In one bottle.) *a.* The sugar and the saline ingredients are separately dried by a very gentle heat, then mixed in a dry room with the ginger and essence of lemon, and at once put into bottles.

b. By adding to the 'acidulated kali,' noticed at page 921, about 1-16th of its weight of the finest powdered Jamaica ginger (*i. e.* ½ dr. to each oz.; 1 oz. to each lb.) at the time of mixing the ingredients together. A dessert-spoonful, thrown into a tumbler 2-3rds filled with cold water, produces an excellent glass of ginger beer.

Powders, Ink. The article usually sold under this name is noticed under INK. Another formula, which we have adopted with considerable success, is as follows:—Good black ink, 3 pints; lump sugar, 1½ oz.; and gum-arabic, ½ oz., are put into a clean iron pan, and evaporated by the heat of boiling water, with occasional stirring, to dryness; the dried mass is reduced to powder, and divided into 12 parts, which are enveloped in either tin-foil or glazed paper, and kept dry. One of these papers dissolved in ½ pint of hot water forms that quantity of excellent black ink, without sediment, and which answers well with the copying press.

Powders, Lemonade'. *Syn.* LEMON SHERBET; LIMONADUM SICCUM, PULVIS PRO LIMONADO, L. *Prep.* 1. Powdered citric or tartaric acid, 12 gr.; powdered white sugar, ½ oz.; essence of lemon, 1 drop (or a little of the yellow peel of a lemon rubbed off on a piece of sugar); mix. For one glass.

2. White sugar, 4 lbs.; citric or tartaric acid, 1½ oz.; essence of lemon, ¼ oz.; mix well, and preserve it in a bottle for use. 1 to 2 dessert-spoonfuls make a glass of lemonade. It is also put up in papers containing about 2½ dr. each.

3. (EFFERVESCING.) *a.* For the blue papers, take of powdered white sugar, 1 lb.; bicarbonate of soda, ¼ lb.; essence of lemon, 1½ dr.; mix, and divide it between 6 dozen papers. Next divide tartaric or citric acid, 5 oz., between 6 dozen white papers. Or the two may be kept in bulk, in separate bottles.

b. (In one bottle.) As 'ACIDULATED KALI.' Some makers slightly increase the quantities of acid and essence of lemon there ordered.

Powders, Orangeade. *Syn.* AËRATED SHERBET. *Prep.* Powdered sugar, 14½ oz.; powdered orange peel, 12 gr.; oil of orange peel, 60 drops; essence of cedrat, 12 drops; bicarbonate of soda, 3½ oz.; mix, and put 145 gr. in each blue paper. In the white paper put 32 gr. of tartaric acid (or 30 gr. of citric acid). Or the alkaline and acid powders may be put into separate bottles, with a measure holding the proper proportions of each. The orange peel may be omitted if necessary.

Powders, Pol'ishing. *Prep.* 1. (For brass and copper.) *a.* From rotten-stone, 3 oz.; powdered soap, 1 oz.

b. From rotten-stone, 7 oz.; powdered oxalic acid, 1 oz. Both are used with a little water. See BRASS PASTE.

2. (For gold.) Jeweller's rouge. See SESQUIOXIDE OF IRON.

3. (For ivory.) Pumice-stone and putty powder.

4. (For plate.) See PLATE and POWDER, PLATE.

5. (For silver.) As the last.

Powders, Preservative. The German Imperial Health Department has ordered the examination of various powders offered to the public for the preservation of meat. The following formulæ are based upon the analytical results:

1. Chloride of sodium, 46 parts; nitrate of potassium, 34 parts; boracic acid, 20 parts.

2. Chloride of sodium, 25 parts; boracic acid, 20 parts; dried sulphate of sodium, 40 parts; sulphate of sodium, 15 parts.

3. Chloride of sodium, 6 parts; borax, 94 parts.

Powders, Scented. *Prep.* 1. COSMETIC POWDERS. *a.* (POUDRE DE CHIPRE.) Macerate oak moss in running water for 2 or 3 days, then dry and powder it. Used as a basis for other powders, on account of its being highly retentive of odours. Reindeer moss and ragged hoary evernia are also used for the same purpose. See CYPRUS POWDER (*above*).

b. (POUDRE DE CHIPRE DE MONTPELIER.) From poudre de chipre, 2 lbs.; musk, 30 gr.; civet, 20 gr. (the last two powdered by means of a little sugar); cloves, ½ oz.

c. (POUDRE DE FLEURS D'ORANGES.) From starch or cyprus powder, 25 lbs.; orange flowers, 1 lb.; mixed in a covered chest, and stirred twice or thrice daily; the process being repeated, with fresh flowers, a second and a third time. Or the plain powder is scented by the addition of a little neroli or essence of petit grain.

d. (POUDRE DE FRANGIPANI.) From poudre de fleurs d'oranges and poudre de chipre, of each, 6 lbs.; essence of ambergris, 1 oz.; civet (powdered with sugar), ½ dr. Ash-grey colour.

e. (POUDRE DE JASMINE.) As POUDRE DE FLEURS D'ORANGES, but using jasmine flowers.

f. (POUDRE À LA MARÉCHALE.) From poudre de chipre, 2 lbs.; starch powder, 1 lb.; calamus aromaticus, cloves, and cyperus perennis or rotundis, of each, 2 oz. Or starch powder, 28 lbs.; powdered cloves, ¾ lb.; powdered orris root, ½ lb.; essence of ambergris, 2 dr.

g. (POUDRE À LA MOUSSELINE.) From orris root, 1 lb.; coriander seed, 6 oz.; mace and violet ebony, of each, 2 oz.; musk seed, cassia, cloves, and sandal-wood, of each, 1 oz.

h. (POUDRE DE JONQUILLE.) From jonquils, as POUDRE DE JASMINE.

i. (POUDRE À L'OEILLET.) From plain powder, 2 lbs.; orris root and dried red rose leaves, of each, 1 lb.; cloves and musk seed, of each, 4 oz.; essence of bergamot and essence of petit grain, of each, ½ dr.

k. (POUDRE DE ROSES COMMUNES.) From pale roses, as POUDRE DE FLEURS D'ORANGES.

l. (POUDRE DE ROSES MUSQUÉES.) From musk roses, as the last.

m. (POUDRE À LA VANILLA.) From poudre de chipre or cyprus, 3 lbs.; vanilla, powdered by means of sugar, 2 dr.; oil of cloves and essence of ambergris, of each, 20 drops.

n. (POUDRE À LA VIOLETTE.) See POWDER, VIOLET (*above*).

The above are used as cosmetic powders for the skin and hair; also, but less frequently, for sachets, drawers, &c.

2. SACHET POWDERS. These are used, along with cotton-wool, to fill scent-bags, cassolettes, &c.; and as scent powder for boxes, drawers, and the like. The scent is added to the dry ingredients, separately reduced to powder, and the whole is then passed through a fine sieve to ensure perfect admixture.

(1) CASSIE SACHET. Cassie flowers, ground, 1 lb.; powdered orris, 1 lb.

(2) CHYPRE SACHET. Ground cedar wood, 1 lb.; ground santal, 1 lb.; ground vanilla beans, ½ lb.; ground Tonquin beans, 2 oz.; powdered orris, 1½ lbs.; ol. French geranium, 30 minims; ol. bergamot, 15 minims; otto rose, 25 minims; extract musk, 1 oz.; mix well.

(3) FRANGIPANI SACHET. Powdered orris, 3 lbs.; ground vitivert, ½ lb.; ground santal, ½ lb.; ground vanilla beans, ½ lb.; ground Tonquin beans, 2 oz.; ol. neroli, 60 minims; ol. santal, 40 minims; ol. bergamot, 60 minims; ol. French geranium, 60 minims; otto rose, 30 minims; extract musk, 1 oz.; extract civet, ½ oz.; mix well.

(4) HELIOTROPE SACHET. Powdered orris, 2½ lbs.; ground rose leaves, 1 lb.; ground vanilla beans, 6 oz.; ground Tonquin beans, 4 oz.; extract musk, 1½ oz.; extract civet, ½ oz.; ol. almonds, 7 minims; mix.

(5) LAVENDER SACHET. Ground lavender flowers, 2 lbs.; powdered gum benzoin, 2 oz.; ol. French lavender, 1 oz.; extract musk, 1 oz.; mix.

(6) ROSE SACHET. Powdered orris, ½ lb.; ground rose leaves, 1½ lbs.; ground santal wood, 4 oz.; ground patchouly, 2 oz.; extract civet, ½ oz.; ol. French geranium, 30 minims; otto rose, 20 minims; mix.

(7) MARÉCHALE SACHET. Ground santal wood, ½ lb.; ground rose leaves, ½ lb.; powdered orris, 1 lb.; ground vitivert, 2 oz.; ground cloves, ½ lb.; ol. bergamot, 60 minims; ol. French geranium, 60 minims; extract musk, 1 oz.

(8) MOUSSELAINE SACHET. Ground cloves, 2 oz.; ground vitivert, 1 lb.; ground santal wood, ½ lb.; ground rose leaves, ½ lb.; powdered orris, 1 lb.; ground cassie leaves, ½ lb.; powdered gum benzoin, 2 oz.; ol. neroli, 5 minims; ol. French geranium, 35 minims; extract musk, 2 oz.

(9) JOCKEY CLUB SACHET. Powdered orris, 3 lbs.; ground santal wood, ½ lb.; ol. bergamot, 1 oz.; otto rose, 30 minims; extract musk, 2 oz.; extract civet, 1 oz.

(10) ESS. BOUQUET SACHET. Powdered orris, 4 lbs.; ground cassie leaves (flowers), 1 lb.; ground rose leaves (flowers), 1 lb.; ground vanilla beans, 3 oz.; essence bergamot, 1 oz.; essence lemon, 1 oz.; ol. French geranium, 60 minims; extract musk, 2 oz.; extract ambergris, ½ oz.

(11) PATCHOULY SACHET. Ground patchouly leaves, 2 lbs.; powdered orris, ½ lb.; ol. patchouly, 30 minims; ol. French geranium, 30 minims.

(12) MILLEFLEUR SACHET. Ground lavender flowers, 1 lb.; ground cassie flowers, 1 lb.; ground rose flowers, 1 lb.; powdered orris, 2 lbs.; powdered benzoin, ½ lb.; ground Tonquin beans, ½ lb.; ground vanilla beans, 3 oz.; ground santal wood, ½ lb.; ol. bergamot, ½ oz.; extract civet, ½ oz.; extract musk, ½ oz.; ground cloves, 2 oz.; ground cinnamon, 2 oz.; ol. French geranium, 30 minims; ol. patchouly, 10 minims; mix.

(13) OPOPONAX SACHET. Powdered orris, 3 lbs.; ground rose leaves (flowers), 1 lb.; ground cassie leaves (flowers), 1 lb.; ground Tonquin, ½ lb.; ground vanilla, 3 oz.; ground musk pods, 1 oz.; ol. citronella, 16 minims; ol. citron, 30 minims; ol. bergamot, 120 minims; ol. patchouly, 30 minims; ol. French geranium, 60 minims; extract civet, ½ oz.; otto rose, 5 minims; mix.

(14) LIGN ALOE SACHET. Powdered orris, 3½ lbs.; ground santal wood, ½ lb.; ground vanilla, ½ lb.; ground rose leaves, 1 lb.; ol. lign aloe, 1 oz.; ol. French geranium, 40 minims; otto rose, 20 minims; extract civet, 1 oz.; extract musk, ½ oz.

(15) VERBENA SACHET. Powdered orris, 3 lbs.; ol. bergamot, 120 minims; ol. verbena, 180 minims; ol. French geranium, 30 minims; essence musk, ½ oz.

(16) POT-POURRI. Ground lavender flowers, 1 lb.; powdered orris, 1 lb.; ground rose leaves, 1 lb.; ground cloves, ½ lb.; ground cinnamon, ½ lb.; ground gum benzoin, ½ lb.; ground pimento, ½ lb.; ground table salt, ½ lb.; ol. lavender ang., 60 minims; ol. santal, 60 minims; ol. French geranium, 60 minims; ol. bergamot, 120 minims; essence lemon, 120 minims; otto rose, 10 minims; ground vanilla beans, 3 oz.; ground musk pods, 1 oz.; extract ambergris, ½ oz.; mix.

(17) YLANG-YLANG SACHET. Ground rose leaves, 1 lb.; ground cassia leaves, 1 lb.; ground pimento, ½ lb.; ground Tonquin beans, 2 oz.; ground vanilla beans, 2 oz.; powdered orris, 3 lbs.; ol. pimento, 60 minims; ol. bergamot, 120 minims; ol. French geranium, 60 minims; ol. ylang-ylang, 120 minims; otto rose, 20 minims; extract musk, 1 oz.; extract civet, ½ oz.; gum benzoin (ground), 1 oz.; mix.

(18) VIOLET SACHET. Powdered orris, 3 lbs.; essence bergamot, 30 minims; ol. French geranium, 20 minims; otto rose, 20 minims; extract musk, 1 oz.; mix.

(19) NEW-MOWN HAY SACHET. Powdered orris, 4 lbs.; ground Tonquin beans, ½ lb.; ground vanilla beans, ½ lb.; ol. almonds, 10 minims; ol.

French geranium, 120 minims; otto rose, 30 minims; ol. bergamot, 60 minims; extract musk, 1½ oz.; mix.

(20) SWEET-BRIAR SACHET. Powdered orris, 4 lbs.; ground santal wood, 1 lb.; ol. French geranium, 30 minims; ol. neroli, 55 minims; ol. verbena, 55 minims; ol. bergamot, 40 minims; essence lemon, 60 minims; otto rose, 30 minims; extract ambergris, 1 oz.; extract musk, ½ oz.; mix.

(21) RONDELETIA SACHET. Powdered orris, 3 lbs.; ground lavender flowers, 1½ lbs.; ol. French geranium, 30 minims; ol. bergamot, 120 minims; ol. cloves, 120 minims; ol. lavender ang., 2 dr.; otto rose, 20 minims; ground musk pods, 1 oz.; extract ambergris, 1 oz.; ground cloves, ½ oz.; mix ('Chemist and Druggist').

3. PARFUM POUR LES AUTRES POUDRES. From poudre d'ambrette, 12 lbs.; civette, 1½ oz.; musk, 1 dr.; reduce the last two to powder by grinding them with some dry lump sugar; then mix the whole together, and pass it through a sieve. Used to perfume hair powder, sachets, &c.

Powders, Seidlitz. *Syn.* PULVIS SODÆ TARTARATÆ EFFERVESCENS. *Prep.* 1. Potassio-tartrate of soda (Rochelle salt), 2 dr.; bicarbonate of soda, 40 gr.; mix, and put it in a blue paper; tartaric acid, 38 gr.; to be put in a white paper. For about ½ pint of water. Laxative.

2. (In one bottle.) From potassio-tartrate of soda, 12 oz.; bicarbonate of ditto, 4 oz.; tartaric acid, 3½ oz.; white sugar, 1 lb. (all in fine powder); dry each separately by a gentle heat, add of essence of lemon, ½ dr.; mix well, pass the mixture through a sieve, and put it at once into clean, dry bottles.—*Dose.* A dessert-spoonful, or more, to a tumblerful of water.

Obs. The above mixtures, though now universally sold as Seidlitz powder, do not, when dissolved, exactly resemble the natural water, which contains carbonates, sulphates, and chlorides of calcium and magnesium. However, the factitious article is equally effective, and much more agreeable.

Powders, Sher'bet. These are made of the same materials as lemonade powders, the flavouring ingredient being varied to suit the particular case.

Powders, So'da-water. *Syn.* EFFERVESCING POWDERS, E. SALINE P., SODAIC P., AËRATED SODA P.; PULVERES EFFERVESCENTES, L. *Prep.* 1. From bicarbonate of soda, 30 gr. in each blue paper; tartaric acid, 25 gr. (or citric acid, 24 gr.) in each white paper. One of each is dissolved separately in about half a glassful of water, and the two solutions mixed and drunk immediately. A cooling, wholesome summer beverage, but it should not be indulged in to excess.

2. (Chalybeated.) By adding 1 gr. of dried protosulphate of iron to each paper of acid. Tonic.

3. (*Midgsley's*.) Made by adding ½ gr. of tartarised antimony to each paper of acid. Refrigerant and diaphoretic. For the Ph. formulæ see POWDERS, EFFERVESCING (*above*).

Powders, Soup. See POWDER, CURRY; POWDER, PEA; SPICE, &c.

Powders, Spruce Beer. *Syn.* PULVERES EFFERVESCENTES CUM ABIETE, L. *Prep.* As

ginger-beer powders, but substituting essence of spruce, 3 to 6 drops, for the powdered ginger.

Powders, Toilet. *Syn.* FACE POWDERS. The following formulæ are the result of analyses, but must not be taken as the absolute formulæ from which the powders are made, as perfumes have to be added to suit the public taste (*W. H. Snow*).¶

Swan Down (manufactured by Henry Tetlow). Zinc oxide, 38·9%; orris root, 18·35%; French chalk, 42·75%.

Wright's. A harmless face powder manufactured by Alfred Wright, of Rochester, N.Y.; claimed by its manufacturer to be "entirely free from lead or other poisonous minerals, and no more hurtful in use than common starch." Upon examination it proved to be—French chalk, 25·48%; corn starch, 33·73%; bismuth oxide, 0·8%; calcium sulphate, 40·19%.

Saunders' Bloom of Ninon. Saunders' pure white face powder, or Bloom of Ninon, manufactured by J. T. Saunders, Oxford Street, London; claimed by its manufacturer to be a "delicate preparation for beautifying the complexion, free from anything which can possibly injure the skin." Each box holds 1 oz. 25 gr. We offer the following formula:—Precipitated chalk, 23·00 parts; French chalk, 23·76 parts; bismuth subcarbonate, 6·64 parts; zinc oxide, 16·60 parts; corn starch, 30·00 parts.

Pozzoni's (White). J. A. Pozzoni's complexion powder, manufactured in St. Louis, Mo., states on the label that it "imparts a brilliant transparency to the skin, removes all pimples, freckles, and discolorations, makes the skin delicately soft, perfectly harmless, containing no arsenic or other deadly material." Found upon examination to be—French chalk, 55·95%; calcium carbonate, 31·25%; bismuth oxychloride, 12·8%.

Palmer's Lily White Tablet for the complexion, prepared only by Solon Palmer, New York. Examination proved it to be—Precipitated chalk, 42·5%; French chalk, 57·5%.

Palmer's Invisible was found upon examination to be a silicate of alumina, magnesia, potash, and soda, coloured with carmine. The natural silicate is probably French chalk.

Powders, Tooth. *Syn.* PULVIS DENTIFRICII, L. The general principles which should be kept in view in the selection of the materials, and in the preparation of dentifrices, have been already fully noticed under DENTIFRICES, and need not, therefore, be repeated here. Care must be taken that all the dry ingredients be finely pulverised, and that the harder and gritty ones be reduced to the state of an impalpable powder, either by levigation or elutriation. The mixture of the ingredients must also be complete. This is the most readily effected by stirring them well together until they form an apparently homogeneous powder, and then passing this powder through a very fine sieve. Those which contain volatile substances should be preserved in closely corked wide-mouthed bottles, and those which contain acidulous or gritty matter should not be frequently employed. The selection of the tooth-brush likewise deserves attention. It should be sufficiently stiff to effect its purpose completely; but, at the same time, it should be so formed as not to cause irritation or injury to the gums during its use.

Prep. 1. Cuttle-fish bone and prepared chalk, of each, 2 oz.; oil of cloves, 20 drops. This may be perfumed at will, and medicated by any of the substances referred to under DENTIFRICES.

2. To the last add of powdered Castile soap, 2 oz.

3. Prepared chalk, 12 oz.; cuttle-fish bone, 8 oz.; orris root, 4 oz.; dragon's blood, 1½ oz.; oils of cloves and cassia, of each, ½ dr.

4. Prepared chalk, 1 lb.; pumice-stone in impalpable powder, ½ lb.; orris root, 2 oz.; pure rouge, ½ oz.; neroli, ½ dr.

5. Yellow cinchona bark and myrrh, of each, ½ oz.; recently burnt charcoal, 3 oz.; cloves, 1 dr.

6. Pumice-stone, red coral, and powdered rhatany root, of each, 2 oz.; orris root, ½ oz.; essence of vanilla, ½ dr.

7. (AROMATIC TOOTH POWDER.) From cuttle-fish bone, 4 oz.; calamus aromaticus, 2 oz.; powdered Castile soap, 1 oz.; oil of cloves, ½ dr.

8. (ASIATIC DENTIFRICE.) From prepared red coral, 8½ lbs.; Venetian red, ¾ lb.; prepared chalk and pumice-stone, of each, 1½ lbs.; China musk, 30 gr.

9. (*Cadet's.*) From lump sugar and charcoal, of each, 1 oz.; Peruvian bark, ½ oz.; cream of tartar, ½ oz.; cinnamon, ½ dr.

10. (Camphorated.) See CAMPHORATED CHALK.

11. (CHARCOAL DENTIFRICE.) From charcoal, preferably that from the willow or the areca nut, either alone or combined with twice its weight of prepared chalk. Scent or medicinals injure it (see 9, 19, and 26).

12. (CORAL DENTIFRICE.) See 16, 23, and 25 (*below*).

13. (Deschamps' ALKALINE DENTIFRICE.) From powdered talc, 4 oz.; bicarbonate of soda, 1 oz.; carmine, 6 gr.; oil of mint, 12 or 15 drops.

14. (FLORENTINE DENTIFRICE.) From prepared shells, 4 oz.; orris root, 1½ oz.; bitartrate of potassa, ¾ oz.; Florentine lake, q. s. to colour.

15. (GALVANIC DENTIFRICE.) From gold, 3 leaves; silver, 4 leaves; triturate them with alum and sulphate of potassa, of each, 1½ dr.; then add of dry common salt, pellitory of Spain, and Peruvian bark, of each, 1 dr.; prepared hartshorn, 1 oz.; mix, and either colour it blue with smalts or red with lake. A useless compound.

16. (*Grosvenor's.*) From red coral, 3 lbs.; prepared oyster-shells, 2½ lbs.; orris powder, ½ lb.; oil of rhodium, 25 drops. Rose pink is now commonly substituted for the coral.

17. (*Hemet's.*) From cuttle-fish bone, 6 oz.; cream of tartar, 1 oz.; orris root, ½ oz.

18. ('Lancet.') Red bark and Armenian bole, of each, 1 oz.; powdered cinnamon and bicarbonate of soda, of each, ½ oz.; oil of cinnamon, 2 or 3 drops.

19. (*Lardner's.*) From charcoal, in very fine powder, 1 oz.; prepared chalk, 3 oz.; mix.

20. (Mialhe's RATIONAL DENTIFRICE.) From sugar of milk, 3 oz.; pure tannin, 3 dr.; red lake, 1 dr.; oils of mint and aniseed, of each, 7 or 8 drops; neroli, 4 or 5 drops.

21. (MYRRH DENTIFRICE.) From cuttle-fish bone, 6 oz.; myrrh and orris root, of each, 2 oz.

22. (PEARL DENTIFRICE.) From heavy carbonate of magnesia, or precipitated chalk, 1 lb.; finest smalts, 3 dr.; essence de petit grain, ½ dr.

23. (Pelletier's QUININE DENTIFRICE.) From prepared red coral, 3 oz.; myrrh, 1 dr.; disulphate of quinine, 12 to 15 gr.

24. (Ph. Russ.) Cinchona bark, 4 oz.; orris root, 2 oz.; catechu and myrrh, of each, 1½ oz.; sal-ammoniac, 1 oz.; oil of cloves, 20 drops.

25. (POUDRE DENTIFRICE—P. Cod.) Red coral, red bole, and cuttle-fish bone, of each, 3 oz.; dragon's blood, 1½ oz.; cinnamon, ¾ oz.; cochineal, 3 dr.; cloves, 1 dr.; bitartrate of potassa, 4½ oz.; reduce them separately to very fine powder before mixing them. This is the 'coral dentifrice' of the French.

26. (*Rignini's.*) From charcoal, 1 oz.; yellow bark, ½ oz.

27. (ROSE DENTIFRICE.) From precipitated chalk, 6 oz.; cuttle-fish bone, 3 oz.; bicarbonate of soda, 2 oz.; red lake, ½ oz.; otto of roses, 20 drops.

28. (*Ruspini's.*) From cuttle-fish bone, 8 oz.; Roman alum and orris root, of each, 1 oz.; cream of tartar, 2 oz.; oil of rhodium, 6 or 8 drops.

29. (VIOLET TOOTH POWDER.) From orris root, 3 oz.; cuttle-fish bone and rose pink, of each, 5 oz.; precipitated chalk, 12 oz.; pure indigo, q. s. to give it a pale violet tinge.

30. (*Zister's.*) From finely powdered calcined hartshorn and cuttle-fish bone, of each, 6 oz.; calamus aromaticus, cassia, and pellitory of Spain, of each, 1 oz.; essence of vanilla, 1 dr.; essence of ambergris, 10 or 12 drops.

31. Chalk, carbonate of magnesia, and pale bark, of each, 1 oz.; oil of peppermint, 5 drops.

32. Cream of tartar, sugar of milk, of each, 2 oz.; carmine, 88 gr. (all in very subtle powder); oil of peppermint, 4 drops.

Powders, Worm. *Syn.* PULVERES ANTHELMINTICI, P. VERMIFUGI, L. *Prep.* 1. (*Bouchardat.*) Powdered Corsican moss and worm-seed, of each, 5 dr.; calomel, 40 gr.; rub them together.

2. (*Collier.*) From powdered jalap and scammony, of each, 1 dr.; cream of tartar, 2 dr.; Ethiops mineral, 3 dr.

3. (*Guibourt.*) Sulphate of iron, 1 dr.; tansy, 2 dr.; worm-seed, 3 dr.

4. (P. Cod.) Corsican moss and worm-seed, of each, 2 oz.; rhubarb, 1 oz.; rubbed to a fine powder, and carefully mixed.

POX. A corruption of a Saxon word, originally applied to pustules or eruptions of any kind, but now restricted to varicella, variola, vaccinia, and, in its unqualified form, to syphilis (see *below*).

Pox, Chick'en-. *Syn.* WATER-POX; VARICELLA, L. An eruptive disease, consisting of smooth, semi-transparent vesicles, of various sizes, which afterwards become white and straw-coloured, and about the fourth day break and scale off without leaving any permanent mark behind them. In hot weather the discharge sometimes becomes purulent, and at others the eruption is attended with considerable fever. Sometimes the vesicles assume a pointed form, and the fluid remains clear throughout the disease; it is then frequently called the 'swine-pox.' When the vesicles are large and globular, and their contents, at first whey-coloured, afterwards turn yellow, it is popularly known as 'hives.'

The treatment of chicken-pox consists in the adoption of a light vegetable diet, and in the administration of mild saline aperients and cooling drinks.

The chicken-pox, except in children of a very bad habit of body, is an extremely mild disease. Like the smallpox, it rarely attacks the same person more than once during life.

Pox, Cow-. *Syn.* VACCINIA, VARIOLA VACCINA, L. This disease was proposed as a substitute and a preventive of smallpox by Dr Jenner in 1798, and its artificial production (vaccination) has rendered smallpox a comparatively rare disease in Britain. There appears no reason to doubt that the pretensions of the advocates of vaccination have been fully justified by the experience of more than half a century; or that this disease, when actively developed, evinced by the completeness and maturation of the pustules, acts as a prophylactic of smallpox.

The process of vaccination is similar to that of inoculation for smallpox. The point of a lance is wetted with the matter taken from one of the pustules, and is then gently inserted under the cuticle, and the scratch afterwards rubbed over with the same. Hæmorrhage should be avoided, as the blood is apt to wash away the virus, or to form a cake, which shields the living tissue from its action.

Pox, Small-. *Syn.* VARIOLA, L. This disease comes on with the usual symptoms of inflammatory fever. About the third day red spots, resembling flea-bites, make their appearance on the face and head, and gradually extend over the whole body. About the fifth day small circular vesicles, depressed in the centre, surrounded by an areola, and containing a colourless fluid, begin to form, when the feverish symptoms abate; about the sixth day the throat becomes sore; about the eighth day the face is swollen; and about the eleventh day the pustules acquire the size of a pea, and cease to enlarge; the matter which they contain becomes opaque and yellow, a dark central spot forms on each, the swelling of the face subsides, and secondary symptoms of fever come on; the pustules become rough, break, and scab over, and a dark spot remains for some days, often followed by permanent indentations, popularly known as 'pock-marks.' At the end of the sixteenth or eighteenth day the symptoms usually disappear. In the confluent smallpox, a severer form of the disease, the pustules coalesce, the eruption is irregular in its progress, and the inflammatory symptoms are more severe.

The treatment of ordinary cases of smallpox resembles, for the most part, that mentioned above for chicken-pox. As soon as the febrile symptoms become marked the patient should not be suffered to lie in a hot bed, but on a mattress, in a cool and well-ventilated apartment, and antiseptic cooling drinks should be freely administered. When convulsions occur, or great irritability exists, small doses of morphine, opium, or camphor may be administered, and obstinate vomiting arrested by effervescing saline draughts. When the skin is pale and cold, the pulse weak, and the eruption languidly developed, the warm or tepid

bath is often serviceable. The assistance of a competent medical practitioner should always be sought, and his instructions carefully carried out.

Smallpox is an exceedingly infectious disease, and every precaution should be taken to prevent its spread through clothing or contact of the healthy with the sick. Fortunately vaccination has reduced the terrors of this disease. Vaccinated persons are rarely affected, and if attacked the disease generally takes a mild form, and leaves little or no trace behind. In the unvaccinated, years gone by, blindness and terrible disfigurement were common results of the disease, now happily rarely seen.

PRAYER BEADS. See ABRUS.

PRECIP'ITATE. Any substance which has separated from its solution in a solid and, usually, a pulverulent or flocculent form, not a mere turbidity, is strictly called a precipitate. The substance by which such a change is produced is called the 'precipitant;' and the act or operation by which it is effected is called 'precipitation.' The old chemists gave this name to several compounds. Red precipitate, or precipitate *per se*, is the red oxide of mercury prepared by heat. White precipitate is the AMMONIATED MERCURY of the B. P.

PRECIPITA'TION. The formation or subsidence of a precipitate (see *above*). When the precipitate is the chief object of the process, it is necessary to wash it, after it is separated, by filtration. This operation requires little attention when the substance thrown down is insoluble in water; but when it is in some degree soluble in that liquid, great care is required to prevent the loss which might result from the use of too much water.

Precipitates soluble in water, but insoluble in alcohol, are frequently, on the small scale, washed with spirit more or less concentrated.

The best precipitating vessel is a very tall glass jar, furnished with a lip and spout, and narrower at the bottom than at the mouth, so that the precipitate may readily collect by subsidence, and the supernatant liquor be decanted off with more ease.

Heavy precipitates may be separated from

FIG. 1.

FIG. 2.

liquids by decantation, and are also washed by the same process; the precipitate is shaken with distilled water, allowed to settle, and when the water has become quite clear it is poured off by allowing it to run gently down a wet glass rod which is pressed against the edge of the vessel (*vide* fig. 2), the precipitate being left in the vessel.

Precipitates in general are washed free from adhering liquid after they have been placed on a filter contained in a funnel according to the following directions :—"Support the funnel which contains the filter and the precipitate with its neck in a beaker or flask, and blow in a fine stream of distilled water from the wash-bottle (*vide* fig. 3), so directed, by moving the jet with the fingers, as to stir up the precipitate well; in this way fill the filter to within a short distance from its edge; let this water run through perfectly, then nearly fill the filter again in the manner just described; repeat this process two or three times, letting the liquid run through perfectly each time before putting in a fresh quantity; the water running through from the third or fourth washing will usually be quite tasteless, and the precipitate and filter will be freed from everything soluble in water.

FIG. 3.

"A precipitate is often required in a dry condition after it has been filtered off and washed. It is dried by placing the funnel in a hollow tin cone or cylinder called the filter-drier (*vide* fig. 4), and supporting this on a piece of wire gauze upon a tripod stand over the flame of a rose-burner turned very low; or the filter-drier may be placed on a gently heated sand-bath. The funnel is then heated by a current of hot air, and rapidly dries the filter and precipitate. Great care must be taken to regulate the heat so as not to char the filter. A more rapid method of drying a precipitate after it has drained for some time consists in spreading the filter upon a piece of wire gauze supported on a tripod stand; a small flame from a rose-burner is then placed beneath the gauze, and the filter carefully watched to avoid charring it. A precipitate is partially dried by opening out the filter upon several dry filter-papers; this process may precede those already mentioned.

FIG. 4.

"When a small quantity of a moist precipitate has to be taken from a filter to test its behaviour or closely examine its appearance, it is most readily removed by dipping the end of a glass rod into the precipitate; by touching a watch-glass or the

interior of a test-tube with the end of the rod a small quantity of the precipitate is deposited for examination.

"If the precipitate is to be removed from the filter as completely as possible several methods are available; one or other must be chosen according as circumstances render it suitable.

"a. The bottom of the filter may be pushed out through the neck of the funnel with a glass rod which is small enough to pass easily through the neck, and the precipitate may then be washed down into a vessel beneath with a fine stream of water or other liquid from the wash-bottle.

FIG. 5.

"b. Without breaking the filter the funnel may be held with its neck horizontal, and the rim just inside the edge of a porcelain dish (*vide* fig. 5); the precipitate is then washed out by directing a fine stream of water against the side of the filter.

"c. The filter and precipitate are allowed to stand for some time, so as to drain off as much water as possible; the filter is then carefully taken out of the funnel, partially dried if necessary by laying it upon several folds of filter-paper, and after removing the portions of paper which contain no precipitate together with the empty fold, it is spread out inside a porcelain dish; the liquid with which it is to be treated is poured upon it, and by shaking the dish so as to cause the liquid to move round and round, and occasionally carefully stirring the precipitate with a glass rod, the precipitate is washed off the paper without tearing the latter to pieces. The paper is then carefully removed by a glass rod.

"d. If it is undesirable to add a liquid to the precipitate upon the filter, the filter and precipitate, after draining for a short time, are removed from the funnel; the filter is spread upon a flat piece of glass and the precipitate carefully scraped off with a glass rod or a small spatula. If the precipitate is required dry, the filter, after removal from the funnel, may be carefully opened and spread upon several thicknesses of filter-paper to drain. When there is a large quantity of the precipitate a sufficient quantity may be removed on the end of a glass rod or spatula without taking the filter out of the funnel. This method is usually the most imperfect, but is frequently the best for other reasons.

"e. A precipitate has sometimes to be dissolved off the filter. The hot liquid used as a solvent may then be poured upon the precipitate; it will run through the filter into a vessel below, taking with it the precipitate in solution; the liquid after it has run through should be heated again, and once more poured upon the precipitate, if the latter is not entirely dissolved; this reheating and returning of the liquid to the filter should be continued as long as anything is dissolved; any remaining portion of the precipitate must then be removed by a little fresh solvent.

"f. A precipitate, if small in quantity, may also be rinsed off the filter with the liquid with which it is to be treated or dissolved. The funnel is placed with its neck in a test-tube, and the precipitate is quickly stirred up with the liquid with a glass rod thin enough to pass down through the neck of the funnel; the bottom of the filter is then pushed out through the neck by the glass rod, and the liquid carrying most of the precipitate will run through; if some of the precipitate remains on the filter, the same liquid is poured through the filter again into the other tube, and by thus pouring backwards and forwards from one tube to another all the precipitate may be removed" (Clowes' 'Practical Chemistry').

PREGNANCY. For the preservation of the health, and the prevention of the numerous discomforts and dangers which so frequently attend this condition, nothing is so effective as exercise. It is this that is so favourable to the humble peasant, and it is its absence that inflicts such calamities on the wealthier classes. Exercise, moderate and unfatiguing, when assisted by regular habits, and a diet nutritious, but not too liberal, is, indeed, capable of not only affording pleasure and increasing the comforts of existence, but is also generally sufficient to greatly lessen the severity of the sufferings, and to ward off the not unfrequently fatal results which terminate this interesting condition.

The sickness of pregnancy may be greatly ameliorated, if not removed, by the occasional use of a saline aperient, and by effervescing draughts formed with the bicarbonate of potass and citric acid. The oxalate of cerium is strongly recommended by Professor Simpson, of Edinburgh, as a remedy for obstinate vomiting in pregnancy.—*Dose*, 1 gr. to 2 gr. three times a day in pills.

PRESCRI''BING (Art of). Besides a knowledge of diseases and their treatment, much of the success of the physician depends on circumstances connected with the form in which the remedies are exhibited. In writing a prescription it is necessary to consider the age, sex, temperament, habits, and idiosyncrasy of the patient, as well as the conditions of climate and season, before the selection of the leading medicament and the apportioning of the dose. The most convenient form of exhibiting it, whether it should be given alone or in some simple form, or combined with other ingredients, the compatibility of the latter, and how far these are likely to assist, impede, or modify its operation, must also receive the consideration of the practitioner. Without a careful attention to all these circumstances the most valuable remedies may be rendered worthless, and the highest medical skill and the best intentions frustrated.

A prescription generally contains several medicinal substances, which are distinguished by medical writers by names indicative of the office which each of them performs. These are—1. The BASIS, which is the principal or most active ingredient;—2. The ADJUVANT, or that which is intended to promote the action of the base;—3. The CORRECTIVE, included to correct, modify, or control its action, or to cover its odour or taste, as when we add carminatives or diaphoretics to cathartics, or aromatics or liquorice to nauseous substances;—4. The EXCIPIENT, or that which gives the whole a commodious or agreeable form, and which, consequently, gives the prescription its peculiar character, as that of draught, mixture, pills, &c. To these, certain Continental writers add a 5th, the INTERMEDIUM, which is the substance employed to unite remedies which are not, by themselves, miscible with each other, or with the excipient. Of this character are the yolk of egg and mucilage, employed in the preparation of emulsions.

The medicinal substances, with the quantities to be taken, generally arranged as above, are said to form the 'inscription,'—the directions as to their combination or dispensing, which usually comes next, the 'subscription,'—and the orders for the exhibition of the compound medicine, which follow these, the 'instructions.' These distinctions are, however, in many cases more technical and useful.

In choosing the form of a prescription it should be recollected that solutions and emulsions generally act with more certainty and rapidity than powders diffused through water; and these, again, than the semi-solid and solid forms of medicine, represented by electuaries, boluses, and pills. On these matters, however, the taste and wishes of the patient should not be disregarded. For this purpose the taste of nauseous medicines should be disguised as much as possible by the judicious selection of an appropriate corrective or excipient. Thus the disagreeable flavour of Epsom salt may be in a great measure covered by dissolving it in peppermint water; that of aloes by liquorice; that of castor oil and copaiba by orange peel; and that of quinine by mixing it with milk immediately before taking it; whilst the bitterness of all bitter substances is concealed by strong coffee.

In order that a prescription may be well made it is not necessary to unite all the elements above referred to. The basis and the excipient are the only two which are absolutely necessary, since there are many medicines which have no need of an adjuvant. The agreeable flavour and odour of some, and the mild and harmless nature of others, often render the intervention of a corrigent unnecessary when they are employed. A single substance may also "be capable of answering two or more purposes. Thus the adjuvant may also act as a corrigent, as when the addition of soap to aloes, or to extract of jalap, lessens their griping properties, and at the same time promotes their action. In the same way neutral salts correct the colic which follows the use of resinous purgatives, and accelerate their action." According to Gaubius, the number of ingredients in a prescription should scarcely ever exceed three or four. See DOSE, MEDICINES, INCOMPATIBLES, PILLS, &c.

PRESCRIP'TIONS. Recipes or formulæ for the preparation and exhibition of medicines intended, generally, for immediate use. See PRESCRIBING (*above*).

PRESERVES. A general term, under which are included the various fruits and vegetables which are seasoned and kept in sugar or syrup, more especially those which are so preserved whole or in slices. See CANDYING, JAM, MARMALADE, &c.

PRESS (Correcting for the). See PROOFS.

PRESSURE, BAROMETRIC, Influence of, on the

Phenomena of Life. M. P. Bert contributed to the 'Comptes Rendus' ('Journal Chemical Society,' vol. xxv) (lxxiii, 213, 503; lxxiv, 617; lxxv, 29, 88) an account of the following experimental researches on the influence of changes in the barometric pressure on the phenomena of life:

He found that at pressures under 18 centimètres of mercury animals die from want of oxygen; at a pressure of one to two atmospheres, from want of oxygen and presence of carbonic acid; at 2—6 atmospheres, from the presence of carbonic acid alone; at 6—15 atmospheres, from the presence of carbonic acid and of excess of oxygen; and at 15—25 atmospheres, from the poisonous action of oxygen alone.

Animals die from want of oxygen when the amount contained in their arterial blood is not sufficient to balance a pressure of 3·5% of oxygen in the atmosphere. They die from poisoning by carbonic anhydride when the amount contained in their venous blood is sufficient to balance a pressure of 26% to 28% of carbonic anhydride in the atmosphere in the case of sparrows, of 28% to 30% for mammals, and of 15% or 16% for reptiles.

As the pressure of oxygen in the surrounding air depends on two factors, the percentage proportion and the barometric pressure, the barometric pressure may be reduced to 6 centimètres for sparrows, if the proportion of oxygen in the air is increased; and it may be raised to 23 atmospheres without causing death, if the proportion of oxygen is reduced by mixing the air with nitrogen. Aëronauts might, therefore, ascend higher than it has hitherto been possible to do by taking with them a bag of oxygen to inhale; and the danger that threatens divers of being poisoned by the oxygen in the compressed air might be averted by using a mixture of air and nitrogen.

From an examination of the gases in the blood of animals confined in rarefied air the author finds that both the oxygen and the carbonic anhydride in the blood diminish. The dyspnœa which is felt in ascending mountains is therefore due to want of oxygen in the blood. The diminution in oxygen becomes diminished at 20 centimètres pressure, yet this is the pressure under which the inhabitants of the elevated Mexican plateau of Anahuac live. The oxygen diminishes more quickly and more regularly than the carbonic anhydride. Although there are but very small quantities of gases simply dissolved in the blood, the chemical combinations in which they take part are dissociated very easily and in a progressive manner under the influence of diminished pressure, and this dissociation takes place more easily in the organisms than in experiments in vacuo.

PRICKLY ASH (*Xanthoxylum fraxineum*). The bark of this shrub is a stimulant, tonic, alterative, and sialogogue. It owes its virtues to a soft resin, a crystalline resin, a bitter principle, and an acrid green oil. The drug is not used in this country, but is officinal in the United States Pharmacopœia.

PRINCE'S METAL. One of the names for Dutch gold. See GOLD, DUTCH.

PRINTING (Anastatic). A method of xinco-graphy, patented in 1845, having for its object the reproduction of drawings, engravings, and letterpress, from copies however old. To describe briefly the preparation of a plate or cylinder, let us suppose a newspaper about to be reprinted by this means. The sheet is first moistened with dilute acid, and placed between sheets of blotting-paper, in order that the superfluous moisture may be absorbed. The ink resists the acid, which attacks the blanks only. In all cases where the letterpress is of recent date, or not perhaps older than half a year, a few minutes suffice for this purpose. The paper is then carefully placed upon the plate with which the letterpress to be transferred is in immediate contact, and the whole passed under a press, on removal from which, and on carefully disengaging the paper, the letters are found in reverse on the plate. A preparation of gum is then applied to the plate by means of a roller, after which the letters receive an addition of ink, which is immediately incorporated with that by which they are already formed. These operations are effected in a few minutes. The surface of the plate round the letters is next bitten in a very slight degree by dilute acid, and on the fresh application of the ink it is rejected by the zinc, and received only by the letters, which are charged with the ink by the common roller used in hand-printing. Each letter comes from the press as clear as if it had been imprinted by type-metal; and the copies are fac-similes, which cannot easily be distinguished from the original sheet.

When pen-and-ink drawings are to be reproduced, they are made on any paper free from hairs or filaments, and well sized. The ink used is a preparation made for the purpose, closely resembling lithographic ink, and may be mixed to any degree of thickness in pure distilled water. It should be used fresh, and slightly warm when a fine effect is to be given. In making or copying a design a pencil may be used; but the marks must be left on the paper, and by no means rubbed with india-rubber or bread. It is necessary to add that the paper should be kept quite clean and free from friction, and should not be touched by the fingers, inasmuch as it will retain marks of very slight touches.

Before closing this notice of anastatic printing it may be proper to remark that the great pretensions originally set up by the patentees have not been fulfilled by its extensive adoption in trade. The grave objection to the process is the practical destruction of the original by the acids used. Photographic methods have entirely superseded the anastatic process.

PRINTING (Letterpress). [The Editor is much indebted to Mr J. E. Adlard for this interesting article.] *Syn.* TYPOGRAPHY. The art of collecting together and arranging moveable types for the purpose of printing, in one or more colours, by pressure applied from a flat surface or by means of a cylinder biting the paper to be printed, and which is inserted between itself and the type.

In illustration of this section some specimen types are appended, the greater portion being from the well-known foundry of Messrs V. & J Figgins, and should now be carefully read down to render the further remarks intelligible.

The ordinary printing [1]
types are technically known as [2]
Book Founts—those more espe- [3]
cially adapted for newspapers ' are [4]
styled News Founts. Each fount is [5]
divided into two distinct portions—the [6]
roman or upright letters forming one part, [7]
and the *italic*, or *sloping*, the other. There is, [8]
moreover, an addition of SMALL CAPITALS to the roman [9]
section of the type-founders' bill for a complete fount of [10]
a particular weight. The height of a type is rather more [11]
than 7-8 of an inch, thus giving depth at the sides for locking-up the [12]
types. The smallest fount cut is "Brilliant."

🙼𝖍𝖊 🙼𝖊𝖝𝖙 🙼𝖔𝖓𝖘𝖎𝖉𝖊𝖗𝖆𝖙𝖎𝖔𝖓 [13]
is the Nomenclature [14]
FORMATION. [15]
Printing Types are [16]
technically named according to, [17]
first, the BODY, that is, [18]
how many lines, when [19]
PLACED IN CONSECUTIVE [20]
order, will make, by Measurement, the length of a [21]
FOOT. HAVING NOW DETERMINED [22]
THE NAME OF THE BODY, [23]
The SPECIAL CUT, or [24]
FACE, IS ADDED THERETO, [25]
thus completing the NAME by which [26]
Type-Founders and Typographers recognise [27]
EACH DISTINCTIVE SIZE AND STYLE [28]
OF THE MANY-VARIED TYPES USED IN THE PRODUCTION [29]
of that luxury which has now become an apparent [30]
NECESSITY OF THE AGE—PRINTING. [31]

References to the above types—by reading across.

BODY.	BOOKWORK FACE.	DISPLAY FACE.
1 Great Primer—Roman.	14 Gt. Primer Manuscript.	
	15 ,, Ext. Ornamented.	
	16 ,, Black.	
2 English—	Roman.	17 ,, Condensed Black.
3 Pica—	Roman.	18 Pica Antique.
	19 ,, Clarendon.	
	20 ,, Rustic.	
4 Small Pica—	Roman.	21 ,, Narrow Gauge.
5 Long Primer—	Roman.	22 Long Primer Condensed Sans-serif.
6 Bourgeois—	Roman.	
7 Brevier—	Roman.	23 Brevier Grotesque.
	24 ,, Extended.	
8 Minion—Rom. & Italic.	25 ,, Open Sanserif.	
9 Nonpareil—	Roman.	26 Nonpareil Egyptian.
	27 ,, Hair-line.	
	28 ,, Ornamented.	
	29 ,, Condensed Gro-	
10 Ruby—	Roman.	tesque.
11 Pearl—	Roman,	30 Pearl Clarendon.
12 Diamond—	Roman.	31 Diamond Grotesque.
13 Small Pica 2-line (No. 4 doubled) German Text.		

The key is contained in itself by reading the column as one continuous paragraph with the help of the foot-notes. Some idea may thus be formed of the vast number of distinct kinds of type necessary to carry out the requirements of the present system of printing.

Mention there has been made that the name of the body is determined by its number of lines to a foot; but this must be qualified. The imperial length in reality as well as in name. When one foundry was sufficient to supply all the types that were required for use in the early ages of printing, then a name and its dimensions could be taken as absolute. But with the increase of printing, type-founders also increased; and this has produced the variations of bodies which are so annoying to the typographer, for one single letter or space taken from a body larger than its own, yet of the same name, will be enough to throw the column of type out of a straight line all the way through. Still, when we look to the fact that, according to the ancient masters, the large-sized type called *Pica* (No. 3 and Nos. 18, 19, 20, and 21) requires 72½ lines to the foot, and that *Nonpareil*, half its size (No. 9, and Nos. 26, 27, 28, and 29), requires 145 lines to the foot, and recollecting that the slightest variation multiplied 145 times must produce a very sensible deviation, the wonder is that each of the founders should approach each other so closely as they do. An attempt was made some years ago to introduce a certain fixity of standard for each body throughout the trade, based on the French system; the difficulties of altering the standards and matrices of each foundry were seen to be so great that the effort was unavailing.

During the latter half of the present century there has been a growing disposition to return to the cut of the letters as used by the early printers. To meet this desire, nearly all the type-founders have introduced Old-style faces, but yet modernised as to their peculiarities. Considering that this article would not be complete without some notice thereof, as well as to show the contrast, the following is here introduced to the reader.

Thefe Old-faced Types
CUT BY THE CELEBRATED
WILLIAM CASLON, in or about the
years 1716-30, are even now viewed
with great satisfaction, and held in high
efteem, by judges of the typographic art
as mafter-pieces of fhape and finifh.

To the list of types presented, and which give a sufficient general view, may be added *Emerald* —between *Minion* and *Nonpareil*—for book-work, and also for borders and flowers to be used in neat and artistic work; *Gem* and *Semi-Nonpareil* for music; and *Minikin*, for music and Oriental work.

The larger sizes of type are, with very few exceptions, simple multiples of the Pica; for instance, 6-*line Roman* means a roman letter of the depth of six lines of Pica; 20-*line Antique*, an antique of the depth of twenty lines; and so on.

Very little more need be said on the names applied to the different faces. Letters used in title-pages are especially cut for, and styled *Titling*—*Square*, *Condensed*, and if very much condensed in width, *Compressed* or *Narrow-Gauge*. On the other hand, when the letters seem pulled out right and left, they are styled *Extended*.

If the reader will notice the type in which this volume is composed, he will observe that the bottoms of the *tail letters* are very close down upon the tops of the tall letters, and all but touch: this is termed *solid*. When a page of book is required to look light and less wearisome to the vision, the lines of type are removed from each other, and a space-line inserted between them: the page is now termed *leaded*. These space-lines used to be cut, by the compositor, from milled lead, first in strips of the necessary width, then of the required length; hence the term *leads*, by which name they are commonly known. However, they were but poor appliances at the best. Moulds are now used for casting the metal to the specified thickness in strips of about 9 inches long, then cut by a machine to a set gauge; by these means the thickness of the space-lines, or leads, is not only more uniformly secured, but far greater regularity obtained in the lengths cut. Here, as in the large type, as above mentioned, Pica is the standard which regulates the lead; in other words, leads are cast as 3 to a pica—that is, 3 leads form the solid measurement of the pica body; 4-to-pica requires 4 leads, and the body of the lead continues to decrease according to the prefixed figure, which simply denotes into how many parts the pica is to be divided. Leads are cast so delicately fine that 16 form the pica, but they are seldom used. In many of the News offices brass space-lines have superseded those cast from type-metal.

The method of manufacturing type is—

The face having been determined upon—light or heavy, round or narrow, as well as the thickness of the downstroke—a piece of prepared soft iron is taken, and, upon the tip-end thereof the proposed letter is cut in relief; when this cutting is finished it is case-hardened, and afterwards styled the punch. The strike is the next operation. The punch (the letter cut upon which, by-the-bye, is backward) is now punched, or struck, into an oblong piece of copper, about 3 inches long and ½ of an inch thick, the breadth such as the size of the letter may require: this is the matrix. A most particular part has now to be performed, called justifying; which means that the matrices shall, when placed in the mould, deliver the letters perfectly upright, and all to be true on a line as fine as a razor's edge. When the process of justifying is accomplished, the matrix is fixed at the bottom of a mould, of the shape of a parallelogram, of the size of the body one way, of the width of the letter the other, and the depth the standard height of the type; the molten metal is forced down this tube, either by hand or by a pump worked by hand or steam, the metal filling the matrix (the sunk letter upon which is now forward) receives the shape of the letter, which is once more reversed, or in a backward position, like as the original punch was cut.

The castings are released from the mould by a very ingenious method of opening from the two diagonal corners. The types as cast are forwarded on to the dressers to remove burrs and other superfluities; then are placed in long lines in a frame for finishing; next turned face downwards, and a grooving plane driven across the feet to insure correctness in height; finally looked over for blemishes, when all faulty letters are thrown out; the process is completed by ranging into lines of handy length, and tied up—ready for delivery to the typographer.

PRINTING INK. *Prep. a.* The VARNISH. Linseed or nut oil, 10 or 20 galls., is set over the fire in an iron pot capable of containing fully as much more; when it boils, it is kept stirred with an iron ladle, and, if it does not take fire of itself soon after the smoke begins to rise, it is kindled by means of a piece of burning paper, stuck in the cleft end of a long stick; the pot is shortly afterwards removed from the fire, and the oil is suffered to burn for about half an hour, or until a sample of the varnish cooled upon a palette knife may be drawn into strings of about ½ inch long, between-the fingers; the flame is now extinguished by the application of a closely-fitting tin cover, and, as soon as the froth of the ebullition has subsided, black resin is added, in the proportion of ¼ lb. to 1 lb. for every quart of oil thus treated; the mixture is next stirred until the resin is dissolved, when dry brown soap, cut into slices, 1½ lbs., is further added (cautiously), and the ingredients are again stirred with the spatula until the whole is united, the pot being once more placed over the fire to promote the combination; when this is effected, the varnish is removed from the heat, and, after a good stirring, is covered over and set aside.

b. The INK. Indigo and Prussian blue, of each, in fine powders, 2½ oz.; mineral lampblack (finest), 4 lbs.; vegetable lampblack, 3½ lbs.; stir them gradually into the warm varnish (a), and submit the mixture to careful grinding, either in a mill or by means of a slab and muller. On the large scale, steam power is now generally employed for this purpose.

An extemporaneous superfine black ink may be made by the following formula:—Take of balsam of copaiba (pure), 9 oz.; lampblack, 3 oz.; indigo and Prussian blue, of each, ¾ oz.; Indian red, ¾ oz.; yellow soap (dry), 3 oz.; grind the mixture to an impalpable smoothness by means of a stone and muller. Canada balsam may be substituted for balsam of copaiba where the smell of the latter is objectionable, but the ink then dries very quickly.

COLOURED PRINTING INKS are made in a similar way from the following pigments:—Carmine, lakes, vermilion, chrome yellow, red-lead, orange red, Indian red, Venetian red, for red; orange chrome, chrome yellow, burnt terra di sienna, gall-stone, Roman ochre, yellow ochre, for orange and yellow; verdigris, Scheele's green, Schweinfurt green, blues, and yellows mixed, for greens; indigo, Prussian blue, Antwerp blue, cobalt blue, charcoal blue, for blue; lustre, bronze powders, &c., for metallic colours; and umber, sepia, &c., for brown.

Obs. It is necessary to prepare two kinds of

[Proof.]

1 *Ital.*   As the <u>vine</u>, which has long
2   twined its grac∮ful foliage
3   about the oak⁄ and been
lifted by it into sunshine, will,
4   when the hardy plant is rift͜
5   ed by the thunder⏝bolt,
6   cling round∮ it with its
7 *Rom*   *caressing* tendrils, and bind
8   ∧its shattered boughs up⏝,
9   so is it ⌒ordered ∖beautifully⌣
10   by ⫽rovidence that woman,
12   who is the mere depend∦t
13   and ornament of man in t̶h̶e
14   happier hours, should ∧ his
15   stay and solace⌒
16   when smitten by
sudden calamity ⁄ winding
17 | herself ▬ into the rugged
18   recèsses of his ⫽ature, tenderly supporting the droop-
19   ing h̶e̶a̶d̶, and binding up
20   the broken heart. [It also
21   is/interesting to/notice how
22 *i.e.*   *some* MINDS seem almost to
23   create themselves, springing
24 ⸺ up un͟de͟r, and working their
∧
25   ∧solitary, | but irresistible way,
∧
26   through | a thousand obsta-
27   cles ⁄ | Nature seems, &c.
28   IR w̬ING.

29 ∖*every disadvantage*

[The same corrected.]

As the *vine*, which has long twined its graceful foliage about the oak, and been lifted by it into sunshine, will, when the hardy plant is rifted by the thunderbolt, cling round it with its caressing tendrils, and bind up its shattered boughs, so is it beautifully ordered by Providence, that WOMAN, who is the mere dependant and ornament of man in his happier hours, should be his stay and solace when smitten by sudden calamity; winding herself into the rugged recesses of his nature, tenderly supporting the drooping head, and binding up the broken heart.

It also is interesting to notice how *some* minds seem almost to create THEMSELVES, springing up under every disadvantage, and working their "solitary, but irresistible way," through a thousand obstacles. Nature seems, &c.

IRVING.

Explanation of the marks:

1. When a letter or word is to be in *italics*.
2. When a letter is turned upside down.
3. The substitution of a comma for another point or letter.
4. The insertion of a hyphen; also marked (-).
5. When letters should be close together.
6. When a letter or word is to be omitted.
7. When a word is to be changed to roman.
8, 9. Two methods of marking a transposition: when there are *several* words to be transposed, and they are much intermixed, it is a common plan to number them, and to put the usual mark in the margin.
10. Substitution of a capital for a small letter.
11. When a letter is to be changed from small letters to capitals.
12. The transposition of letters in a word.
13. The substitution of one word for another.
14. When a word or letter is to be inserted.
15. When a paragraph occurs improperly.
16. The insertion of a semicolon.
17. When a space or quadrat stands up, and is seen along with the type.
18. When letters of a wrong fount are used.
19. When words crossed off are to remain.
20. The mark for a paragraph, when its commencement has been neglected. Sometimes the sign [, or ¶, or the word '*break*' is used instead of the syllables '*New Par.*'
21. For the insertion of a space when omitted or insufficient.
22. To change capitals to small letters.
23. To change small letters to small capitals.
24. When lines or words are not straight.
25, 26. The insertion of inverted commas. The apostrophe is similarly marked.
27. The insertion of a period when omitted, or in place of another point or letter.
28. Substitution of one letter for another.
29. The method of marking an omission or insertion when too long for the side margin.

varnish, varying in consistence, from more or less boiling, to be occasionally mixed together as circumstances may require; that which answers well in hot weather being too thick in cold, and *vice versâ*. Large characters also require a thinner ink than small ones. Old linseed oil is preferable to new. Yellow resin soap is preferred for black and dark-coloured inks, and white curd soap for light ones.

A good varnish may be drawn into threads like glue, and is very thick and tenacious. The oil loses from 10% to 14% by the boiling. Mr Savage obtained the large medal of the Society of Arts for his black ink made as above.

A PRINTER'S INK EASILY REMOVED FROM WASTE PAPER. The following process for the preparation of a printer's ink that can be far more readily removed from waste paper than ordinary printer's ink has been patented by Kirscher and Ebner. Iron is dissolved in some acid —sulphuric, hydrochloric, acetic, &c., will answer, and half of the solution is oxidised with nitric acid and added to the other half, and the oxide precipitated from the mixture by means of soda or potash. The precipitate is thoroughly washed, and treated with equal parts of solutions of tannic and gallic acids, and the bluish-black or pure black pigment formed is thoroughly washed and dried, and mixed with linseed-oil varnish, and can then be immediately used for printing from type, copper, wood, steel, or stone. Waste paper printed with it can be bleached by digesting it for 24 hours in a lukewarm bath of pure water, and 10 per cent. of caustic potash or soda, and then grinding it well in the rag engine, and throwing the pulp upon cloth and allowing it to drain. It is then to be washed with pure water, containing 10 per cent. of hydrochloric, acetic, or oxalic acids, or of binoxalate of potass, and allowed to digest for 24 hours, and may then be worked up into paper, or it can be dried and used as a substitute in the manufacture of finer paper.

PRINTS (Ackerman's Liquor for). *Prep.* Take of the finest pale glue and white curd soap, of each, 4 oz.; boiling water, 3 pints; dissolve, then add of powdered alum, 2 oz. Used to size prints and pictures before colouring them.

PRINTS, To Bleach. Simple immersion of the prints in a solution of hypochlorous acid (the article remaining in the solution for a longer or shorter space, according to the strength of the solution) is generally all that is required to whiten them. See ENGRAVING.

PRIVIES. See WATERCLOSETS.

PROOF. See ACETIMETRY, ALCOHOLOMETRY, &c.

PROOFS (correcting). The specimen of corrected proof given on page 1406 has been so prepared as to include all the usual errors which are met with. It must not, moreover, be supposed that any printer would send out a proof so full of errors, but in any large work some or all of them may be met with, and the reader by referring to this example will be enabled to correct 'proof' in such a way that any printer will understand the exact nature of the corrections required. To those who contemplate the printing of a book, or even a small pamphlet, a brief outline of the process by which the manuscript is converted into

print may be useful, and it will be convenient to discuss it in its various stages.

THE MANUSCRIPT should in every case be written on one side of the paper only, and preferably on sheets of medium quarto size, in a clear and legible hand, with plenty of space between the lines; and if the original MS. contains in itself many alterations, additions, and corrections, a fair copy should always be made before putting it in the printer's hands. Otherwise, even in the most careful hands, mistakes will occur, which may require considerable labour and time to set right. Printers have the reputation of being able to read any handwriting whatever, and to unravel the most disorderly manuscript; but as their time and work have to be paid for by the author who employs them, it is well to make the task as easy as possible. Paper is cheap, and there is no excuse for an author who sends his work to the compositors written on small scraps of paper of every imaginable size and quality, in a cramped hand, and full of corrections and alterations, unless he is prepared to encounter considerable trouble in the correction of proof and heavy expense in the production of his work.

Supposing the manuscript to be complete, and a fair copy sent to the printers, the kind of type, the size and quality of paper determined on, an ESTIMATE OF THE COST OF PRINTING will be given, based upon a calculation of the number of words in the whole MS. This calculation is called 'casting off,' and unless the MS. be uniform, and clearly written, it is very difficult to form more than a very approximate estimate of the space the work will occupy without actually counting the words, which may be a serious and costly operation. It should also be remembered that it costs more as a rule to set small type than large; and that tabular work, *e. g.* tables of figures, &c., is more costly than mere letterpress, and that much of it may add considerably to the total expense of printing a book. The next thing to be done is to decide upon the way in which the PROOFS shall be sent out. The most usual plan is to set up the type in long slips the width of the intended page cut of variable length (called 'galley slips,' after the special press on which they are generally printed), each slip containing matter enough for two or three pages. These proof slips are read before they are sent to the author, and all gross errors corrected, doubtful words marked, and the author's attention called by the printer's 'reader' to any redundancies of expression or any sentences which are not apparently intelligible. The author now reads this proof very carefully, and makes all the necessary corrections, and, if he thinks fit, alters the wording of passages, or makes additions, following the directions given in the example on page 1406.

It is easy to make such alterations in galley slips, but all such as do not appear in the original MS. must of course be paid for as extras. In the most carefully prepared work some such alterations always occur, and it is not until an author has had some experience that he can realise how his manuscript will read in print. The corrected proof is returned to the printer, and the typographical errors are set right and any new matter inserted. This done, a fresh impression is taken and sent to the author, marked 'REVISE,' which he reads and corrects as before, and if perfect he will mark it legibly 'MAKE UP' and return it to the printers. When sufficient has been so returned it will be made up into SHEETS; i. e. if the book is a quarto each sheet will consist of eight pages, if an octavo of sixteen pages. Each sheet (2ND REVISE) is again examined by the author to see that no letters or words have been dropped in the process of making up, especially at the bottom of one page and the top of the next; also that the type has not shifted, and that the headings and numbering of the pages are correct. If there are no serious errors the author will correct such as there are, and return it to the printers marked 'PRESS;' it will then be finally corrected, and the required number of copies printed from it on the quality of paper decided upon originally, and the type will then be broken up and 'distributed.' This process goes on until the last sheet of the book is complete, when the whole is sent to the binders and put into such a cover as the author may wish. A book worked through the press in the manner above described is less likely to contain typographical errors than one in which the first proofs are sent out as made-up sheets, paged and titled, but it is somewhat more costly. The latter plan can, however, only be adopted when the author does not make any material alterations in the text, for such alterations may cause a great deal of trouble, especially when additions are made. Suppose, for example, that after sheet 86 of this work had been passed for press and worked off, and sheets 87 and 88 had been sent out by the printers in obedience to the order to 'make up' written at the bottom of the galley slips, the editor were to discover that a column and a half of matter had been omitted from the end of the article POTASSIUM which must be inserted. The only way in which this could be done would be by pulling to pieces all the columns from page 1372 onwards, putting in the new matter, and then reimposing sheets 87 and 88. In the simplest case this would cause page 1408 to end about this point, PROPIONIC ACID and PROPYLAMINE being thus thrust into sheet 89. If it were a question of only a word or two, the disturbance might not extend beyond a column or even less of this work; but in an ordinary book, if the type be closely set and but little broken up into paragraphs, an alteration of this kind, even of a few words added, might possibly make itself felt over many pages. The expense of such alterations is obviously serious, and an author who does not know his own mind, or who does not correct his proofs carefully, may find the printing of his volume a very costly luxury indeed. Type-written copy has this very great advantage, that it is perfectly legible, may be sent to the printers corrected as a proof, and the author is enabled to see what his work looks like in print before incurring the printer's bill.

The publication of a book even of very modest pretensions is a more or less serious undertaking to those who are not acquainted with the technicalities of printing, and as a consequence the printers are often blamed for what is after all the result of the author's ignorance of press work.

Much useless labour and annoyance will be saved by a careful study of the directions given for the correction of proof, and the author who will take the trouble to master the chief technicalities of printing, and who will take the advice and help which the printer will be only too glad to give him, will find his way smoothed and his labour lightened to an extent which he will perhaps at first hardly credit.

PROPIONIC ACID. $C_2H_5.CO_2H$. Formed in small quantity by the distillation of wood and by the fermentation of various organic bodies. *Prep.* 1. From ethyl cyanide and caustic potash.

2. By reducing lactic acid with hydriodic acid.

Prop., &c. Colourless liquid with a penetrating odour, somewhat resembling that of acetic acid; boils at 140° C.; yields simple substitution products with the halogens, &c.

PROPYL'AMINE (Normal). *Syn.* TRITYLAMINE.

C_3H_9N, or $\left. \begin{matrix} C_3H_7 \\ H \\ H \end{matrix} \right\} N$. This compound or substituted ammonia, in which one of the three atoms of hydrogen is displaced by the radical propyl or trityl (C_3H_7), is isomorphous with trimethylamine, which has been often mistaken for it.

It has been proposed as a remedy for acute and chronic rheumatism. Hence the commercial substance known under the name of 'propylamine,' which has been proposed and employed as a remedy for rheumatism, has been shown to be not propylamine, but its isomer, trimethylamine, or a mixture of this latter in varying proportions with ammonia.

Prep. Mendius's process: 36 grms. of cyanide of ethyl, 500 grms. of common alcohol, 200 grms. of water, and 50 grms. of 20% hydrochloric acid are allowed to act on excess of granulated zinc, and then distilled. The distillate is put back once, and 400 grms. of hydrochloric acid are added. The product is distilled to get rid of the alcohol, then excess of alkali is added to the residue, and the distillation continued, whereupon propylamine and water come over. 36 grms. of the cyanide of ethyl yield 9 grms. of pure propylamine. It is dried by distillation from solid potash.

It may also be prepared by the action of boiling potash on the mixture of propyl isocyanate and isocyanurate, obtained by distilling normal propyl iodide with silver cyanate.

Prop., &c. Propylamine is a bright, colourless, highly refracting, very mobile liquid, strongly alkaline, possessing a peculiar, strongly ammoniacal odour. It mixes with water, heat being generated by the mixture. It boils at 50° C., and has a sp. gr. of 0·7283 at 0° C.

Propylamine combines with acids, and forms crystallised salts. The chloride is a very deliquescent salt. The sulphate occurs in crystals, and is also deliquescent. Isopropylamine is a sweet ammoniacal liquid, boiling at 32° C.; it is liberated from the formate (produced by the action of hydrochloric acid on isopropyl cyanide) by successive treatment with hydrochloric acid and potash. See TRIMETHYLAMINE.

PROPYL'IC ALCOHOL. C_3H_8O. *Syn.* PROPYL ALCOHOL, HYDRATED OXIDE OF PROPYL, TRITYL

ALCOHOL. There are two isomeric modifications of the three-carbon alcohol, viz. normal propyl alcohol and isopropyl alcohol, pseudopropyls, or secondary propyls. Normal propyl alcohol is an oily liquid boiling at 96° C., sp. gr. 0·8305 at 0° C., obtained by repeatedly rectifying the first products of the distillation of the fused oil of marc brandy. It stands to ethylic alcohol (ordinary alcohol) in the same relation in which the latter stands to methylic alcohol (pyroxylic spirit). By oxidation with a mixture of sulphuric acid and potassium dichromate it is converted into propionic acid.

Isopropyl alcohol is a colourless liquid having a peculiar odour, a sp. gr. of 0·791 at 15° C., and boiling at 83°—84° C., under a pressure of 789 mm. It does not act on polarised light. It is prepared from acetone by the direct addition of hydrogen evolved by the action of water on sodium amalgam. It yields acetone by oxidation with dilute chromic acid.

PROTEIDS. See ALBUMEN.

PRO'TEIN. The name given by Mülder to a substance which he regarded as the original matter from which animal albumen, casein, and fibrin were derived ; but which is now considered as a product of the decomposition of those important principles by moderately strong caustic alkali.

PRO'TIDE. A soluble, straw-yellow substance, formed, along with other products, by the action of strong solution of potass on albumen, fibrin, or casein. See ERYTHROPROTIDE.

PROTO-. See NOMENCLATURE.

PROVI'SIONS (Preservation of). See PUTRE-FACTION.

PRUNE, VIRGINIAN. *Syn.* WILD CHERRY. The bark of the *Prunus Virginiana* is much employed as a remedy in the United States ; it contains amygdaline, and yields on distillation with water an essential oil rich in hydrocyanic acid. Both a syrup and tincture are used in this country as sedatives, to allay cough in phthisis and bronchitis.

PRUNES. [Fr.] The fruit of cultivated varieties of *Prunus domestica*, Linn. The dried fruit (FRENCH PRUNES or PLUMS; PRUNUM— B. P., Ph. L. ; PRUNA—Ph. E. & D.) is cooling and gently laxative, and, as such, is useful in habitual costiveness and fevers.

Prunes, Pulp of. *Syn.* PREPARED PRUNES; PULPA PRUNORUM, PRUNUM PRÆPARATUM (Ph. L.), L. *Prep.* The imported dried fruit is boiled gently for four hours with water, q. s. to cover them, and then pressed, first through a flue cane sieve, and afterwards through a fine hair sieve; the pulp is, lastly, evaporated by the heat of a water-bath to the consistence of a confection. A better plan is to use as little water as possible, by which the necessity of subsequent evaporation is avoided. Used in the preparation of confection of senna.

PRU'NING varies according to the kind of plant or tree operated on and the particular object in view, and its skilful performance must, therefore, greatly depend on the experience and knowledge of the gardener. " In the operation of pruning, the shoots are cut off close to the buds, or at a distance not greater than the diameter of

the branch to be cut off ; because without the near proximity of a bud the wounds would not heal over. In shoots which produce their buds alternately the cut is made at the back of the bud sloping from it, so that it may be readily covered by the bark in the same or in the following year ; but in the case of branches where the buds are produced opposite each other, either one bud must be sacrificed or the branch must be cut off at right angles to its line of direction, which is most conveniently done with the pruning shears " (*London*).

PRUSSIAN AL'KALI. Ferrocyanide of potassium.

PRUSSIAN BLUE. $(Fe_7)_4(Fe_2)_3(C_2N_3)_{12}$. *Syn.* BERLIN BLUE, INSOLUBLE P. B., WILLIAMSON'S B., PARIS B., FERROCYANIDE OF IRON, PRUSSIATE OF I., CYANURET OF I. This is the well-known blue pigment of the shops. It was discovered early in the 18th century by a colour maker named Diesbach.

Prep. 1. A clear solution of ferrocyanide of potassium is precipitated by a mixed solution of alum, 2 parts, and green sulphate of iron, 1 part ; the dingy greenish precipitate that falls gradually becomes blue by absorption of atmospheric oxygen, which is promoted by exposure and agitation of the liquor ; as soon as it has acquired its full colour, the sediment is repeatedly washed with water, and is then drained and dried, at first in a stove, but afterwards on chalk stones. Product large, but inferior in quality.

To obtain pure Prussian blue repeatedly digest and wash the precipitate obtained in process 1 in very dilute hydrochloric acid and then in pure water ; drain and dry it.

2. (Paris blue.) *a.* Neutralise a solution of ferrocyanide of potassium with dilute sulphuric acid, precipitate the liquid with a solution of any per-salt or sesqui-salt of iron (as the persulphate, nitrate, sesquichloride, or peracetate) ; well wash the precipitate with water, and dry it as before. A very rich and intense colour.

b. (*Hochstätter.*) Crystallised ferrocyanide of potassium and green sulphate of iron, of each, 6 parts, are each separately dissolved in water, 15 parts ; after the admixture of the solutions, add frequent agitation, oil of vitriol, 1 part, and fuming hydrochloric acid, 24 parts, are stirred in ; after some hours have elapsed a strained solution of chloride of lime, 1 part, dissolved in water, 80 parts, is gradually added, the addition being stopped as soon as an effervescence from the escape of chlorine is perceived ; the whole is now left for 5 or 6 hours, when the precipitate is thoroughly washed in pure soft water, drained and dried. The precipitate may be exposed to the air, treated with chlorine water, dilute nitric acid, or bleaching powder solution, and afterwards with hydrochloric acid to remove the ferric oxide which is formed. These modifications of the method are now most commonly employed on the large scale. The product is of the finest quality.

3. (*Williamson.*) By oxidising Turnbull's blue (*i. e.* ferrous ferricyanide, q. v.) by treating it with nitric acid or chlorine water.

4. (*Shrewp.*) By precipitating soluble Prussian blue with ferric chloride.

Prep. A deep blue powder, which on trituration

assumes a bright copper-like lustre. Insoluble in water and in dilute acids, except the oxalic, in solutions of which, and of ammonium tartrate, it dissolves freely when pure; oil of vitriol dissolves it to a white pasty mass, which is again precipitated of the usual blue colour by water; alkalies instantly decompose it, and so do red oxide of mercury and some other oxides when boiled with it; it burns in the air like tinder, leaving an ash of oxide of iron. It is not poisonous. *Pur., &c.* The quality of Prussian blue may be estimated by the richness of its colour, and by the quantity of potash or soda required to destroy this. It always contains a certain amount of water which cannot be driven off by heat with decomposition. If it effervesces with acids, it contains chalk; and if it forms a paste with boiling water, it is adulterated with starch. It is pure if, "after being boiled with dilute hydrochloric acid, ammonia throws down nothing from the filtered liquid" (Ph. L. 1836). It is distinguished from indigo by exhibiting a coppery tint when broken, which is removed by rubbing with the nails.

Concluding Remarks. The commercial Prussian blue is not pure ferrocyanide of iron, but a mixture of this salt with varying proportions of the ferrocyanide of iron and potassium, which also has a fine deep blue colour. The object in employing alum is to prevent or lessen the precipitation of oxide of iron by the free alkali in the ferrocyanide of potassium solution, but a portion of alumina is in consequence thrown down with the blue, and tends to make it paler and to increase the product. The quantity of alum employed may be varied according to the shades of the intended blue. Samples containing this contamination must not be employed medicinally. A solution of Prussian blue in oxalic acid was formerly much used as a blue ink; it has, however, now been replaced by the aniline colours.

Prussian Blue, Sol'uble. $K_2Fe_5(C_2N_2)_4Fe_3$. *Syn.* FERRIC POTASSIUM FERROCYANIDE. *Prep.* By precipitating a solution of a sesqui-salt or per-salt of iron (as the persulphate, pernitrate, peracetate, or sesquichloride) with a stronger solution of ferrocyanide of potassium, so that the latter may be in considerable excess. A blue precipitate is formed, which is treated as before. This variety is precipitated by alcohol. Both are freely soluble in pure water, but not in water which has the slightest saline contamination. Hence it is that lengthened exposure to the atmosphere and the use of the common steel pen causes the gradual precipitation of this substance from its solution when used as ink. See WRITING FLUIDS, POTASSIO-FERRIC FERROCYANIDE.

Prussian Blue, Soluble. The following is said to be a rapid and easy process for preparing this substance:—Pure Prussian blue, 5 dr.; ferrocyanide of potassium, 2½ dr.; distilled water, q. s. Rub the two salts to a fine powder in a mortar, add 2 to 4 pints of water, according to strength desired. Digest for ½ an hour with occasional agitation, then filter ('American Textile Record').

PRUSSIAN GREEN. $(Fe_2)_3(Fe_2)_3(C_2N_2)_{18}$. *Prep., &c.* (*Pelouze.*) A green hydrated precipitate obtained from a solution of potassium

ferricyanide or ferrocyanide by the action of chlorine gas in excess; then heating the liquid to boiling, separating the precipitate which forms and boiling it with prussic acid. Decomposed by caustic potash into a mixture of potassium ferri- and ferrocyanides and ferric hydroxide. Heated to 180° C.; when dry it yields a violet substance and gives off cyanogen.

PRUSSIC ACID. See HYDROCYANIC ACID.

PSEU'DO-MOR'PHIA. A substance of little importance, occasionally found in opium. It differs from morphine chiefly in not decomposing iodic acid. It is said to contain nitrogen.

PSILA ROSÆ, Fabricius. (From ψιλός, bald, as the head or forehead of this genus has only a very few hairs upon it.) THE CARROT FLY. Carrots are largely grown by farmers for horses, and are a useful and most valuable crop. They are also produced to a very large extent by market-garden farmers and by market gardeners proper, in Essex, Bedford, Surrey, Kent, Middlesex, and, in short, wherever vegetables are grown for market. To market-garden farmers and to market gardeners they are sometimes most remunerative. As many as from 30 to 40 acres are frequently cropped with carrots in a season upon some of the largest market-garden farms. As much as £70 per acre is often returned for a crop of carrots, from which, of course, there are heavy expenses to be deducted (Report upon the 'Market Garden and Market-Garden Farm Competition in connection with the Royal Agricultural Society's Show at Kilburn in 1879,' by Charles Whitehead, Esq., F.L.S., F.G.S.); it will be seen, therefore, that it is most important that these should be well grown, and without spots and blemishes. And carrots for feeding and for storing should also be free from injuries, or their sale is spoilt, and they will not keep.

The larvæ or maggots of the carrot fly, *Psila rosæ*, seriously damage carrot plants by working their way into the roots and feeding upon their substance. It is a very common occurrence to find deep marks upon the roots of carrots all round them. Some of them will be found to go quite into the centre if the root is cut down lengthways. When they are thus affected the roots lose their bright red clear colour, and become rusty—iron-mouldy, as the Germans have it. In these circumstances the roots get shrivelled after a time, and are no longer sweet and juicy. The tops also change their colour, appearing as if the plants were dead. There can be no mistaking the cause of this disorder, as upon pulling up a root the maggots will be seen within the holes, protruding from these in many cases. Complaints of injuries to carrots by this fly have been made from many parts of England, Wales, Scotland, and Ireland during the past six or seven years. Curtis spoke of this insect attacking carrots in 1845 in Ireland, but it does not appear that its effects were very serious until within the past ten years. Miss Ormerod describes it as having done much mischief in 1880, 1881, 1882, and 1883, in many parts of the United Kingdom ('Reports of Observations on Injurious Insects, for 1880, 1881, 1882, and 1888,' by Miss E. Ormerod).

When carrots are suffering from the attacks of this insect, it will be found that their roots are infested by other insects, such as millipedes, which delight to live in decayed vegetable matter, and are frequently accused of being the causes of the mischief. Other insects, as slugs, are also attracted by the unhealthy state of the plants. This insect appears to be known generally in Europe. Taschenberg and Nördlinger speak of it as troublesome to carrot plants in Germany. Köllar also describes it. Kaltenberg says it was very troublesome in Switzerland in 1851 (' Die Pflanzen Feinde,' von J. H. Kaltenbach). It is not known in America.

Life History. The *Psila rosæ* belongs to the extensive family *Muscidæ* of the order DIPTERA. It is nearly three lines, or a quarter of an inch, in length, and its wings are close upon five lines across; it is greenish black in colour, with a brassy tinge like that of the ' blue-bottle,' or meat fly. Its head is very round, and ochreous in colour, with the front part bare of hairs. The fly comes forth towards the middle of May, and goes down into the ground and places its eggs upon the roots of the carrots. From the eggs, maggots or larvæ are quickly hatched, which bore into the roots and thus injure the plant, as well as by living upon its juices. The maggot is yellowish white, and three lines in length. It appears to have no head, and its body is pointed or tapering at its fore-end. On close examination it will be seen that there are here two tiny instruments for boring. The other end of the body is rounded off somewhat unevenly. The maggots are found in the carrots up to the time they are dug, and some of them remain in this form during the winter in the ground and the roots when stored. But most of them change to pupæ, or, rather, the larvæ acquire *puparia* or cases in which, after a time, the transformation to the fly stage is finally accomplished. There are two or more broods during the summer. In the case of the earlier broods the insect remains in the pupal or semi-pupal state about three weeks. In that of the late broods the winter is passed in the *puparium.*

Prevention. It is very important to keep the ground firm round the carrot plants, so that the flies may not be able to get down to lay eggs upon the roots. After the process of hoeing out, or ' singling' the plants, it would be desirable to send men or boys to tread heavily on both sides of the drills, or rows. This might be done almost at ordinary walking pace. This is the most dangerous time, after ' singling,' as the ground is loosened, and the first broods of flies are actively hunting about for congenial sites for their eggs.

In localities where these flies cause much injury, ashes, sawdust, or wood ashes, or sand saturated with paraffin oil, at the rate of a quart of oil to a hundredweight of ashes and sand, and two quarts or more to a hundredweight of sawdust, should be put into the drills with the seed. Peat moss, as used for litter, well triturated, with the coarser fibre screened out, might be adopted for this purpose as a medium for the absorption and retention of paraffin or petroleum oils. A top dressing of an oil-saturated substance just after the plants have been hoed is efficacious. Soot put on at the rate of 15 bushels an acre has been tried

with considerable advantage. This should be sprinkled upon the plants in the drills or rows. Where the seed is broadcasted, as is sometimes the case upon market-garden farms and in market gardens, the paraffin-saturated dressings may be broadcasted before or just after the seed is sown, and harrowed in with it.

After an attack it is very essential that the carrots should be cleared away in the early autumn, and the ground well limed and deeply ploughed. Also that stored carrots from infested fields should be consumed before the spring comes, and not consumed upon the land.

Remedies. When the attack is established it may be modified by top dressings of soot, or guano, or nitrate of soda. These will stimulate the plants and keep them vigorous.

If flies are seen while the plants are young, the dressings should be at once put on, as these will be more efficacious at this stage than when the plants are older; and when the plants are large the maggots, dislodged by this application, merely move a little lower down.

In gardens the attack has been checked by the use of water in which quassia has been infused at the rate of 9 to 10 lbs. to 100 gallons. This infusion was poured from a water-pot close round the plants. Garden engines might be employed where there is an extensive breadth of plants (' Reports on Insects Injurious to Crops,' by Chas. Whitehead, Esq., F.Z.S.).

PSYLLIODES ATTENUATUS (or *Agromyza frontalis ?*). THE HOP CONE-STRIG MINERS. During the last few years the hop cones in many parts of the hop-yielding districts have become rapidly red or rust-coloured some days before they were ready to be picked, and after a short time they have dried up, and their bracts have fallen to pieces. This was at first attributed to red mould or to red rust, but upon careful examination it has been found that the strigs or stalks of the cones had been bored or mined by an insect throughout. Moreover, in many of these mines little white maggots, the larvæ of an insect, were found.

It is a moot point as to what kind of insect these larvæ belong to. Some are of opinion that they are the larvæ of a species of flea-beetle of the tribe Psylliodes, either *Psylliodes attenuatus* or *Psylliodes chrysocephalus,* which, to a casual observer, resembles the common hop flea-beetle, *Haltica concinna.* According to Taschenberg the larvæ of the latter commonly bore into bulbs or stalks of plants. Others hold that they are the larvæ of a species of fly, *Agromyza frontalis,* which are also known to be leaf and stalk miners (' Report on Insects Injurious to Crops,' by Charles Whitehead, Esq., F.Z.S.).

PTIS'AN. *Syn.* PTISANA, L. A decoction made of pearl barley, licquorice, raisins, and other like vegetable matters, either alone or so slightly medicated as to be taken as a common drink in fevers, catarrhs, &c. Those retained in English pharmacy have been already noticed. The French physicians often employ this form of medicine. The ' tisanes' of the P. Cod. are numerous. See DECOCTION, INFUSION, JULEP, TISANE, &c.

PTOMAÏNES. Bodies resembling alkaloids

and having many alkaloidal chemical reactions, prepared from decomposing animal matters. The poisonous characters of most of the ptomaïnes are intense, their chemistry is but little understood, but there is good reason for hope that re-cent researches will throw much light on these obscure but dangerous products of putrefaction. Ptomaïnes are divided into two classes, those which contain oxygen and those which do not. The following have been described :

Name.		Source.	Discoverer.
Parvoline . .	$C_9H_{13}N$ . .	Mackerel and horseflesh	Gautier & Etard.
Hydrocollidine	$C_9H_{13}N$ . .	„ „	„ „
Base	$C_{17}H_{38}N_4$ . .	—	—
Base	$C_{10}H_{15}N$ . .	Bullock fibrine and cuttle-fish . . .	Guaverchi & Merro.
Collidine . .	$C_8H_{11}N$ . .	Gelatine and ox pancreas	Neucki.
Neuridine . .	$C_5H_{14}N_2$ . .	Albuminoids	Brieger.
Cadaverine . .	$C_5H_{14}N_2$ . .	Bodies subjected to prolonged putrefaction .	Brieger.
Putrescine . .	$C_4H_{12}N_2$ . .	Flesh of mammifers and herring brine .	—
		Ptomaïnes containing oxygen.	
Neurine . . .	$C_5H_{13}N(OH)$	—	
	Hydrate of trimethylvinyl-ammonium.		
Choline . . .	$C_5H_{15}NO_2$	—	
	Hydrate of trimethylhydroxethylenine-ammonium.		
Muscarin . .	$C_5H_{15}NO_3$.	—	Brieger.
Gardinine . .	$C_7H_{17}NO_2$	—	Brieger.
Bases	$\begin{cases} C_7H_{15}N_2O_6 \ . \\ C_8H_{22}N_2O_4 \ . \end{cases}$	—	Brieger.

PTY'ALIN. A peculiar animal ferment, analogous to diastase, obtained from the saliva. It is soluble in water, but insoluble in alcohol.

Mialhe named ptyalin 'animal diastase,' and regarded it as the principal agent in effecting the digestion of starchy foods, by converting them into soluble glucose. One part of ptyalin, according to Mialhe, was capable of transforming 800 parts of insoluble starch into sugar. It has been computed that the average daily secretion of ptyalin by an adult amounts to 116 grains. It very quickly decomposes, and in properties somewhat resembles sodic albuminate.

PUCHÁ PÁT. *Syn.* PATCHOULI. Puchá pát is the dried foliaceous tops of *Pogostemon Patohouli,* an Indian species of *Labiatæ.* It is much used in perfumery, particularly for making sachets; but its odour, although very durable, is not so agreeable as that of many other substances, unless it is combined with lavender, bergamot, ambergris, musk, or some other like perfume.

PUD'DINGS. The instructions given under CAKES, PIES, &c., will be found, with some slight modifications, also to apply to puddings, and, therefore, need not be repeated here. Soyer tells us that every sort of pudding, if sweet or savory, is preferably dressed in a basin instead of in a cloth. If boiled in a basin the paste receives all the nutriment of the materials, which, if boiled in a cloth, are dissolved out by the water, when by neglect it ceases boiling. To cause them to turn well out, the inside of the basin should be thoroughly 'larded' or rubbed with butter.

In the preparation of meat puddings the "first and most important point is never to use any meat that is tainted; for in pudding, above all other dishes, it is least possible to disguise it by the confined progress which the ingredients undergo. The gradual heating of the meat, which alone would accelerate decomposition, will cause the smallest piece of tainted meat to contaminate all the rest. Be particular, also, that the suet and fat are not rancid, ever remembering the grand principle that everything which gratifies the palate nourishes."

"A pudding cloth, however coarse, ought never to be washed with soap; it should be simply dried as quickly as possible, and kept dry and free from dust, and in a drawer, or cupboard, free from smell" (*Soyer*).

PUD'DLING. See IRON.

PULMONITIS. Inflammation of the lungs.

PULP. *Syn.* . PULPA, L. The softer parts of plants, more particularly of fruits, separated from the fibrous and harder portions.

"Pulpy fruits, if they be unripe, or ripe and dried, are to be placed in a damp situation until they become soft; then the pulp is to be pressed out through a hair sieve; afterwards it is to be boiled with a gentle heat, frequently stirring; and finally, the (excess of) water is to be evaporated in a water-bath, until the pulp acquires proper consistence.

"Press the pulpy fruits which are ripe and fresh through a hair sieve, without boiling them" (Ph. L. 1836).

PULQUE. The national drink of the Mexicans. It is produced by the fermentation of the maguey, or *Agave Americana.* For ages pulque has been considered to have medicinal virtues in a high degree. Physicians use it as a tonic, stimulant, and antispasmodic; they recommend it to weak, infirm, anæmic, nursing mothers.

PULSATILLA (*Anemone pulsatilla,* Pasque-flower herb, Meadow Anemone, or Wind-flower). It contains *Anemorium* or pulsatilla camphor. Mr Gerard Smith ('Lancet,' January 15th, 1887) says this drug has a striking curative action in inflammatory states of the testicle, epididymis, and spermatic cord: the relief is so rapid that it is even unnecessary to employ morphine to subdue the pain, while the swelling and heat subside more rapidly than under any other drug. Pulsatilla has been used with success in nasal, bronchial, vaginal, vesical, and conjunctival catarrh, and is recommended in amenorrhœa and dysmenorrhœa. A tincture (in 10 of proof spirit) is given in 2 to 8 minim doses.

PULVERISATION. The reduction of any substance to dust or powder.

On the small scale, pulverisation is usually performed by means of a pestle and mortar; on the large scale, by stamping, grinding, or cutting the substance in a mill. A few soft substances, as carbonate of magnesium, carbonate of lead, &c., may be pulverised by simply rubbing them through a fine sieve, placed over a sheet of paper, whilst many hard, gritty substances can only be reduced to fine powder by porphyrisation or levigation. Elutriation, or 'washing over,' is adopted for several substances, as chalk, antimony, &c., which are required to be reduced to fine powder on the large scale. For some articles which are very tough, fibrous, or resisting, a rasp or file is employed. Whichever of these methods is adopted the body to be powdered must be very dry, and where spontaneous drying is insufficient, artificial desiccation in a stove or oven, gently heated, is employed. To facilitate this, the substance should be first cut into pieces or crushed small. On the other hand, a few substances, as rice, sago, nux vomica, and St Ignatius's bean, are often soaked in water, or steamed, before being further operated on. Whenever a substance cannot be dried completely, without an alteration of its properties, an intermedium is had recourse to, by which the moisture may be absorbed, or its state of aggregation modified. Thus, sugar is employed in pulverising civet, musk, nutmeg, and vanilla. When camphor is to be pulverised, the addition of a very small quantity of alcohol renders the operation easy. In other cases the intermedium is of so hard a nature as to assist in breaking down the substance to be powdered; thus, gold-leaf is reduced to powder by rubbing it with sulphate of potassa, and afterwards removing this last by means of water. Fusible metals, as zinc and tin, are powdered by pouring them into a mortar, and stirring them rapidly whilst cooling; or by briskly agitating them, in the melted state, in a wooden box covered with chalk or whiting. Phosphorus is powdered by melting it in urine or lime water, and then shaking the bottle until its contents have become quite cold. Glass, quartz, and silicated stone require to be heated red-hot, and in this state to be thrown into cold water, by which they become sufficiently friable to admit of pulverisation. Many salts which are reduced to fine powder with very great difficulty, and do not dissolve in spirit of wine, are easily obtained in a pulverulent form by agitating their concentrated aqueous solution with a considerable quantity of rectified spirit; the disengaged fine crystallised powder may then be dried, and further divided by trituration. Potassio-tartrate of antimony may be advantageously thus treated. A large number of salts, including nitre, sal-ammoniac, and carbonate of potash, may also be reduced to powder by keeping their solutions in a state of constant and violent agitation during their rapid evaporation.

The following rules should be observed in the preparation of powders:

1. If possible, perfectly dry articles should alone be operated on, and only in dry weather.

2. The nature of the mortar, and the mode of operating, should be adapted to the nature of the substance. Thus, woods and barks should be pulverised in an iron mortar; sugar, alum, and nitre in one of marble or wedgwood-ware; and corrosive sublimate, only in one of glass.

3. The mortar should be provided with a cover, to prevent loss and annoyance to the operator. If much powder escapes, or if it is dangerous or disagreeable when breathed, or if the substance is rare or costly, the mortar should be covered with a skin of leather, to which the pestle is attached, so that the latter may be freely moved without causing the slightest opening for the escape of the dust occasioned by the process. When aloes or gamboge is powdered, a few drops of olive oil are commonly added with the same intention.

4. The pulverised portions should be separated from time to time by aid of a sieve, the coarser particles being returned to the mortar to be again beaten and triturated; and this alternate pulverisation and sifting is to be repeated until the process is complete.

The size of a powder is regulated by passing through sieves; a sieve having 20, 40, or 60 meshes to the linear inch gives a powder called No. 20, 40, or 60. Powders for internal use, or for local dusting purposes, should be extremely fine, or what is termed impalpable.

PUMICE - STONE. Syn. PUMEX, LAPIS PUMICEUS, L. PUMICIS, L. Found in the neighbourhood of volcanoes. Used, in the solid form, to polish wood, paint, &c.; also, when pulverised, as a polishing powder for glass, bone, ivory, marble, metals, &c.

PUMILINE. STIRUS PUMILINE. (Oleum Pini Pumilionis, volatile oil distilled from the needles of the Mugho, or Mountain Pine.) Dr. Prosser James calls attention ('Lancet,' March 10th, 1888) to the value of this preparation in diseases of the respiratory mucous tracts. It is a very pure essential oil, possessing in a high degree the odour of the most fragrant variety of the pine, and is less irritating than other fur oils. Sprinkled or sprayed about a sick-room by means of a Siegel's inhaler, or handball atomiser, it imparts a lasting and grateful fragrance to the air, which is not oppressive, and it seems to be disinfectant; either of these methods may be utilised for maintaining an atmosphere laden with pine odour, as a substitute for that of Arcachon, Reichenhall, or Bournemouth. The oil is admirably adapted for inhalation by means of a respirator or steam inhaler, and it may be given internally in doses of 1 to 5 minims on sugar, or in lozenges. It is a very mild stimulant to the mucous membrane, and an agreeable remedy for inhalation in relaxation, congestion, and chronic catarrhal affections of the respiratory tract. It is its action on the bronchial membrane during excretion that renders it valuable in disease of this surface, being a stimulant, expectorant, and disinfectant; hence indicated in chronic bronchitis, dilatation of bronchi, bronchorrhœa, some states of phthisis, and other affections. Externally, sprinkled on flannel or spongio-piline, the oil is a cleanly, prompt, and useful stimulant and counter-irritant, and sometimes appears to possess slight anæsthetic properties.

PUMP (for use in Chemical, Paper, and other Works). The Perreaux Pump Valve is made of vulcanised india rubber, and is of the form of the valves in the human body. It is of the greatest, and, perhaps, the only really valuable improvement in valves applicable equally to the common hand or jack pump, and the most elaborate mechanical combinations for raising water.

The valve may be taken as the key of the pump; a perfect valve renders an indifferent pump a valuable and effective machine, whereas an imperfect valve, in an otherwise excellently constructed pump, renders it practically useless.

The pump which Simon the tanner, of Joppa, used for pumping his pits, nearly two thousand years ago, may be taken as the type of the common hand pump in use to this day. Various mechanical improvements have been made in its form and construction, but, practically, and effectively, the only real and valuable improvement is the Perreaux valve, now under consideration.

. Constructed of a flexible material, and made in form, as nearly as may be, to the valves of the human body, they may be said to be automatic in their action, or self-acting; upon the pump being actuated, the least motion of the pump ensuring a corresponding action of the valve, and the most rapid action of the pump being equally responded to by the pulsation of the valves.

Although the most perfect valves for pumping clear water, because, what is mechanically termed the duty of the pump is complete—in other words, the quantity displaced is discharged absolutely without loss,—yet their most valuable feature is that they pump semi-fluids equally well as clear water.

· For the pulps and stuffs in paper mills, for bleaches, dyes, and corrosive liquors, for liquid manures and other such semi-fluids, they stand alone, they are absolutely unchokable.

Used in conjunction with cylinders or barrels made of toughened glass, they form the most perfect pump where the fluid to be raised is of a caustic or corrosive nature, and where the fluid would be destructive to or destroyed by its action upon metals—such, for example, as the caustic bleach used in the manufacture of paper, &c. See VALVES.

PUNCH. An acidulous, intoxicating beverage, composed of water sweetened with sugar, with a mixture of lemon juice and spirit, to which some aromatic, as nutmeg, mace, or cinnamon, is occasionally added. Wine is sometimes substituted for spirit. It is much less drunk than formerly. Rum punch is the most popular amongst sailors, who are now the principal consumers of this beverage.

Prep. 1. Juice of 3 or 4 lemons; yellow peel of 1 lemon; lump sugar, ¾ lb.; boiling water, 3½ pints; infuse ½ an hour, strain, and add of bitter ale, ½ pint; rum and brandy, of each, ¾ to 1 pint (or rum alone, 1½ to 2 pints). More hot water and sugar may be added if the punch is desired either weaker or sweeter.

2. (COLD PUNCH.) From arrack, port wine, and water, of each, 1 pint; juice of 4 lemons; white sugar, 1 lb.

3. (GIN PUNCH.) From the yellow peel of ½ a lemon; juice of 1 lemon; strongest gin, ¾ pint; water, 1½ pints; sherry, 1 glassful.

4. (ICED PUNCH). From champagne or Rhenish wine, 1 quart; arrack, 1 pint; juice of 6 lemons; yellow peel of 3 lemons; white sugar, 1 lb.; soda water, 1 or 2 bottles; to be iced as cream.

5. (MILK PUNCH; VEEDER.) Steep the yellow rinds of 18 lemons and 6 oranges, for 2 days, in rum or brandy, 2 quarts; then add 3 quarts more of either spirit; hot water, 3 quarts; lemon juice, 1 quart; loaf sugar, 4 lbs.; 2 nutmegs, grated; and boiling milk, 3 quarts; mix well, and in two hours strain the liquor through a jelly-bag.

6. (NORFOLK PUNCH.) Take of French brandy, 20 quarts; yellow peels of 18 oranges and 30 lemons; infuse for 12 hours; add of cold water, 30 quarts; lump sugar, 20 lbs.; and the juice of the oranges and lemons; mix well, strain through a hair sieve, add of new milk, 2 quarts, and in 6 weeks bottle in. Keeps well.

7. (ORANGE PUNCH.) As No. 1, using oranges, and adding some orange wine, if at hand. A little curaçoa, noyau, or maraschino improves it.

8. (RASPBERRY PUNCH.) As the last, but using raspberry juice, or raspberry vinegar, for the oranges or lemons.

9. (REGENT'S PUNCH.) From strong hot green tea, lemon juice, and capillaire, of each, 1½ pints; rum, brandy, arrack, and curaçoa, of each, 1 pint; champagne, 1 bottle; mix, and slice a pineapple into it.

10. (TEA PUNCH.) From strong hot tea, 1 quart; arrack, ½ bottle; white sugar, 6 oz.; juice of 8 lemons, and the yellow rinds of 4 lemons; mixed together.

11. (WINE PUNCH.) From white sugar, 1 lb.; yellow peel of 8 lemons; juice of 9 lemons; arrack, 1 pint; port or sherry (hot), 1 gall.; cinnamon, ¼ oz.; nutmeg, 1 dr.; mix.

12. (YANKEE PUNCH.) Macerate sliced pine-apple, 3 oz.; vanilla, 6 gr.; and ambergris (rubbed with a little sugar), 1 gr., in the strongest pale brandy, 1 pint, for a few hours, with frequent agitation; then strain with expression; add of lemon juice, 1 pint; lemon syrup, and either claret or port wine, of each, 1 bottle; with sugar, ½ lb., dissolved in boiling water, 1½ pints. See SHRUB.

PURG'ATIVES. *Syn.* DEJECTORIA, PURGAN-TIA, PURGATIVA, L. These have been divided into five orders or classes, according to their particular actions. The following are the principal of each class:

1. (LAXATIVES, LENITIVES, or MILD CATHARTICS.) Manna, cassia pulp, tamarinds, prunes, honey, phosphate of soda; castor, almond, and olive oils; ripe fruit.

2. (SALINE or COOLING LAXATIVES.) Epsom salt, Glauber's salt, phosphate of soda (tasteless salt), seidlitz powders, &c.

3. (ACTIVE CATHARTICS, occasionally acrid, frequently tonic and stomachic.) Rhubarb, senna, aloes, &c.

4. (DRASTIC or VIOLENT CATHARTICS.) Jalap, scammony, gamboge, croton oil, colocynth, elaterium, &c.

5. (MERCURIAL PURGATIVES.) Calomel, blue pill, quicksilver with chalk, &c.

In prescribing purgatives regard should be had to the particular portion of the alimentary canal on which we desire more immediately to act, as well as to the manner in which the medicine effects its purpose. Thus, Epsom salt, sulphate of potass, and rhubarb act chiefly on the duodenum; aloes on the rectum; blue pill, calomel, and jalap on the larger intestines generally, and tartrate and bitartrate of potassa, and sulphur on the whole length of the intestinal canal. Again, others are stimulant, as aloes, croton oil, jalap, scammony, &c.; others are refrigerant, as most of the saline aperients; magnesia and its carbonate are both aperient and antacid; whilst another class, including rhubarb, damask roses, &c., are astringent. Further, some produce only serous or watery dejections, without greatly increasing the peristaltic action of the bowels; whilst a few occasion a copious discharge of the fæces in an apparently natural form. See DRAUGHT, MIXTURE, PILLS, PRESCRIBING, &c.

PURL. *Prep.* To ale or beer, ½ pint, gently warmed, add of bitters, 1 wine-glassful, or q. s. Some add a little spirit. A favourite beverage with hard drinkers early in the morning.

PURPLE. A rich compound colour, produced by the admixture of pure blue and pure red. This colour has always been the distinguishing badge of royalty and distinction. The celebrated Tyrian purple was produced from a shell-fish called murex.

Purple An'iline. *Syn.* PERKIN'S PURPLE, MAUVE. The sulphate of a base called mauvine, $C_{27}H_{24}N_4$. This valuable dye-stuff is prepared under W. H. Perkin's patent by mixing solutions of sulphate of aniline and bichromate of potash in equivalent proportions, and, after some hours, washing the black precipitate with water, drying it, digesting it repeatedly in coal-tar naphtha, and, finally, dissolving it in boiling alcohol. It may be further purified by evaporating the alcoholic solution to dryness, dissolving the residue in a large quantity of boiling water, reprecipitating by caustic soda, washing with water, dissolving in alcohol, filtering, and evaporating to dryness. Thus purified, mauve forms a brittle substance, having a bronze-coloured surface. It imparts a deep purple colour to cold water, though dissolving sparingly in that liquid; it is more soluble in hot water, and very soluble in alcohol. See PURPLE DYES (*below*), and TAR COLOURS.

Purple of Cassius. *Syn.* PURPLE PRECIPITATE OF CASSIUS, GOLD PURPLE, GOLD PREPARED WITH TIN; AURUM STANNO PARATUM, PURPURA MINERALIS CASSII, L. A compound of gold, tin, and oxygen, which is believed to be combined according to the formula $Au_2SnO_2.Sn.8nO_2.4H_4O$.— *Prep.* 1. Seven parts of gold are dissolved in aqua regia, and mixed with 2 parts of tin, also dissolved in aqua regia; the mixed solutions are largely diluted with water, and then a weak solution of 1 part of tin in hydrochloric acid is added till a fine purple colour is produced. The addition of salt assists the subsidence of the purple precipitate.

2. (*Frick.*) Dissolve pure grain tin in cold dilute aqua regia until the fluid becomes faintly opalescent, then take the metal out and weigh it; next, dilute the solution largely with water, and add, simultaneously, a dilute solution of gold and dilute sulphuric acid in such proportion that the tin in the one shall be to the gold in the other in the ratio of 10 to 36.

3. (P. Cod.) Terchloride of gold, 1 part, is dissolved in distilled water, 200 parts; another

solution is made by dissolving in the cold, pure tin, 1 part, in a mixture of nitric acid, 1 part, and hydrochloric acid, 2 parts; this last solution is diluted with distilled water, 100 parts, and is then added to the solution of terchloride of gold until precipitation ceases to take place; the powder is, lastly, washed by decantation, and dried by a very gentle heat.

4. Silver, 150 parts; gold, 20 parts; pure grain tin, 35 parts; fuse them together under charcoal and borax, cool, laminate, and dissolve out the silver with nitric acid.

Prop., &c. Purple of Cassius is soluble in ammonia, but the solution is decomposed by exposure to light, becoming blue and finally colourless; metallic gold being precipitated, and binoxide of tin left in solution. Heat resolves it into a mixture of metallic gold and binoxide of tin. It is used as a purple in porcelain painting, and to communicate a ruby-red colour to glass, when melted in open vessels.

PURPLE DYES. The purples now in vogue are the numerous shades of 'mauve' and 'magenta' obtained by the 'aniline colours.' (See *above*, also RED.) For silk and woollen goods no mordant is required. The proper proportion of the clear alcoholic solution is mixed with water slightly warm, any scum that may form is cleared off, and the goods are entered and worked until the required shade is obtained; a small quantity of acetic or tartaric acid is added in some cases. For dyeing on cotton with the aniline colours, the cloth or yarn is steeped in sumach or tannic acid, dyed in the colour, and then fixed by tin; or it may be steeped in sumach and mordanted with tin, and then dyed. Purples were formerly, and are still occasionally, produced by first dyeing a blue in the 'indigo vat,' and then dyeing a cochineal or lac scarlet upon the top. The purple dyes which are now most commonly used are known as Alizarin P., Ethyl P., and Regina P.

Purple, Alizarin. Fast shades of purple are obtained on cotton from alizarin, either with or without the use of oil. If prepared with oil, it is mordanted in a solution of ferrous sulphate (8°—4° Tw. Sp. gr. 1·015 to 1·02), and washed after remaining overnight; it is then dyed with 5% to 15% of alizarin (10%); it is afterwards washed and soaped at 60° C. When oil is not used, the cotton is worked in a cold solution of tannin (1 to 2 grms. tannic acid per litre), mordanted with a solution of pepolignite of iron (1° —3° Tw. Sp. gr. 1·005 to 1·015), washed, and then treated as before with the alizarin.

Purple, Ethyl. 6 B. (*Basf.*) The hydrochloride hexa-ethyl-para-rosaniline. It is the bluest shade of violet at present known. In dyeing cotton the fibre is prepared with tannic acid and tartar emetic, or with sulphated oil and aluminium acetate; it is then washed and dyed at 45° to 50° C. in a neutral bath. Wool is dyed at 60° to 80° C. in a neutral bath, to which 2% to 4% of soap has been added. Silk is dyed at 60° to 60° C. in a bath containing soap, then washed and brightened in a bath slightly acidulated with acetic or tartaric acid.

Purple, Regina, is closely related to the rosaniline violets. Cotton is dyed in a bath slightly acidulated with alum or sulphuric acid, after having been prepared with tannic acid and tartar emetic. Wool is dyed in a colour solution at 60° to 80° C., acidulated with 4% sulphuric acid (168° Tw. Sp. gr. 1·84). Silk is dyed at 60° to 80° C. in a bath containing 'boiled-off liquor,' slightly acidulated with sulphuric acid (*Hummel*).

PUR'PURATE OF AMMO'NIA. See MU-REXIDE.

PURPU'RIC ACID. See MUREXAN.

PURPURIN. $C_{14}H_4O_2(OH)_3$. *Syn.* MADDER PURPLE. The name given by Robiquet and Colin to a beautiful colouring principle originally obtained from madder. It is really trihydroxy-anthraquinone.

Prep. 1. Coarsely powdered madder is allowed to ferment with water, after which it is boiled in a strong solution of alum, which dissolves only the purpurin; the decoction is next mixed with sulphuric acid, and the resulting red precipitate is purified by one or more crystallisations from alcohol.

2. It is obtained artificially by oxidising alizarin with manganese dioxide and sulphuric acid.

Prop., &c. Crystalline red needles, insoluble in cold water, but soluble in hot water, and in alcohol, ether, and solutions of alum and alkalies. It differs from alizarin or madder red in containing 2 atoms less of carbon. In dyeing, it is used like alizarin. Commercial samples contain also anthrapurpurin and flavopurpurin.

PUR'REE. *Syn.* INDIAN YELLOW. A yellow substance imported from China and India, and now extensively used in both oil and water-colour painting. There has always been some doubt about its origin. It was believed to be the urinary sediment of the camel or buffalo, after the animal had fed on decayed and yellow mango leaves, but nobody was quite sure. Dr Hugo Müller made some inquiries at Kew in 1883, and the Kew authorities set the India Office to work, with the result that an official of the Revenue and Agricultural Department of the Government of India proceeded to Monghyr, a town in Bengal, to see how it was obtained. There is a purree—or, more correctly, 'piuri'—of mineral origin imported from London; this, of course, differs materially from the genuine article. The official found that the latter is really obtained from the urine of cows kept by a sect of gualas, or milkmen, residing in a suburb of Monghyr, who are the only people who manufacture the purree. They feed the cows solely with mango leaves and water, which increases the bile pigment and imparts to the urine a bright yellow colour. The cows are made to pass urine three or four times a day by having the urinary organ rubbed with the hand, and they become so habituated to this that they cannot urinate unless this is done. The urine is collected, and at night heated in earthen vessels, whereby the yellow principle is precipitated. It is collected, made into balls, and dried, first over a charcoal fire, and then in the sun. A cow yields about 2 oz. per day, this quantity being the product of 3 quarts of urine.

PURRE'IC ACID. *Syn.* EUXANTHIC ACID. This substance is obtained from purree. It crystallises in nearly colourless needles, which are only sparingly soluble in cold water, and forms rich yellow-coloured compounds with the alkalies and earths. Heat converts it into a neutral, crystallisable substance, called purrenone.

PUS. The cream-like, white or yellowish liquid secreted by wounded surfaces, abscesses, sores, &c.

PUTREFAC'TION. *Syn.* PUTREFACTIO, L. The spontaneous decomposition of animal and nitrogenised vegetable substances, under the joint influence of warmth, air, and moisture. The solid and fluid matters are resolved into gaseous compounds and vapours, which escape, and into earthy matters, which remain. The most striking characteristic of this species of decomposition is the ammoniacal or fetid exhalations that constantly accompany it.

The nature of putrefaction, and the conditions essential to its occurrence, have been briefly alluded to under fermentation, to which we must refer the reader. It may here, however, be useful to reiterate that this change can only be prevented by the abstraction or exclusion of the conditions essential to its occurrence. This may be effected by reduction of temperature, exclusion of atmospheric air, or the abstraction of moisture. The antiseptic processes in common use are effective in precisely the same degree as these preventive means are carried out. Frozen meat may be preserved for an unlimited period, while the same substance will scarcely keep for more than a few days at the ordinary heat of summer. Animal substances will also remain uninjured for a long period if kept in vessels from which the air is entirely excluded, as in the process now so extensively adopted for the preservation of fresh meat for the use of our army and marine. The third condition is fulfilled when nitrogenised matter is preserved in alcohol, brine, or any similar fluid, and when it is dried. In either case water is abstracted from the surface, which then loses its propensity to putrefy, and forms an impervious layer, which excludes atmospheric oxygen from the interior and softer portion of the substance. Creasote, and most of the antiseptic salts, also act in this way.

Among special antiseptic processes are the following:

APPLICATION OF COLD. The accession of putrefaction is prevented, and its progress arrested, by a temperature below that at which water freezes. In the colder climates of the world, butchers' meat, poultry, and even vegetables, are preserved from one season to the other in the frozen state. In North America millions are thus supplied with animal food, which, we can state from personal experience, is often superior in flavour, tenderness, and apparent freshness, to that from the recently killed animal. In temperate climates, and in cold ones during their short summer, ice-houses and ice-safes afford a temperature sufficiently low for keeping meat fresh and sweet for an indefinite period. Substances preserved in this manner should be allowed to gradually assume their natural condition before cooking them; and on no account should they be plunged into hot water, or put before the fire, whilst in the frozen state.

BUCANING. A rude kind of drying and smoking

meat, cut into thin slices, practised by hunters in the prairies and forests.

DESICCATION or DRYING. In this way every article of food, both animal and vegetable, may be preserved without the application of salt or other foreign matter. The proper method is to expose the substances, cut into slices or small fragments, in the sun, or in a current of warm dry air, the temperature of which should be under 140° F. Articles so treated, when immersed for a short time in cold water, to allow the albumen and organic fibres to swell, and then boiled in the same water, are nearly as nutritious as fresh meat cooked in the same manner. If a higher degree of heat than 140° be employed for animal substances, they become hard and insipid. Owing to the practical difficulties in the way of applying the above process to fresh meats, it is usually employed in conjunction with either salting or smoking, and, frequently, with both of them.

EXCLUSION OF ATMOSPHERIC AIR. This is effected by the method of preserving in sugar, potting in oil, and, more particularly, by some of the patented methods noticed below. Fresh meat may be preserved for some months in that state, by keeping it in water perfectly deprived of air. In practice some iron filings and sulphur may be placed at the bottom of the vessel, over which must be set the meat; over the whole is gently poured recently boiled water, and the vessel is at once closed, so as to exclude the external air.

IMMERSION IN ANTISEPTIC LIQUIDS. One of the commonest and most effective liquids employed for this purpose is alcohol of 60% to 70%, to which a little camphor, ammonia, sal-ammoniac, or common salt is occasionally added. A cheaper and equally efficient plan is to employ a weak spirit holding a little creasote in solution. A weak solution of sulphurous acid may be substituted for alcohol. Weak solutions of alum, or carbolic acid, with or without the addition of a few grains of corrosive sublimate, or of arsenious acid, are also highly antiseptic. These are chiefly employed for anatomical specimens, &c. A solution containing only $\frac{1}{500}$ part of nitrate of silver is likewise very effective; but from this salt being poisonous, it cannot be employed for preserving articles of food. Butchers' meat is occasionally pickled in vinegar. By immersing it for one hour in water holding $\frac{1}{500}$ part of creasote in solution, it may be preserved unchanged for some time even during summer.

INJECTION OF ANTISEPTIC LIQUIDS into the veins or arteries of the recently killed animal. It is found that the sooner this is done after the slaughter of the animal the more effective this becomes, as the absorbent power of the vessels rapidly decreases by age. See GANNAL'S PROCESS (below).

JERKING is a method of preserving flesh sometimes adopted in hot climates. It consists in cutting the lean parts of the meat into thin slices, and exposing these to the sunshine until quite dry and brittle, when they are bruised in a mortar, and pressed into pots.

PICKLING IN VINEGAR. In this method the substances, rendered as dry as possible by expo-sure to the air, are placed in glass or stoneware jars (not salt-glazed), or wooden vessels, when strong vinegar, either cold or boiling hot, is poured over them, and the vessel at once closely corked or otherwise covered up, and preserved in a cool situation. Meat is occasionally thus treated; vegetables frequently so. See PICKLE.

POTTING IN OIL. In this case salad or olive oil is substituted for vinegar (see above), and is always used cold.

SALTING acts chiefly by abstracting water from the albuminous portions of the meat, by which its disposition to change is lessened.

SMOKING. This process, which, as well as the last, is referred to further on, acts both by the abstraction of moisture and the antiseptic properties of certain substances (creasote, &c.) contained in wood smoke. Fresh meat and fish are occasionally smoked; but, in general, substances intended to be thus treated are first salted.

In Donkin and Gamble's patent process the substances, previously parboiled, are placed in small tin cylinders, which are then filled up with rich soup; the lids are next soldered on quite air-tight, and a small hole is afterwards made in the centre; the cylinders are then placed in a bath of strong brine, or a strong solution of chloride of calcium, which is at once heated to the boiling-point, to nearly complete the cooking process; after which the small hole in the lid is hermetically sealed by covering it with solder while the vessel still remains boiling hot; the tins are, lastly, again submitted to heat in the heated bath, the duration of which is proportioned to the quantity and character of their contents, the 'dressing' of which is to be perfected by this operation. The ends of the tins, on cooling, assume a concave form, from the pressure of the atmosphere, without which they cannot be air-tight, and the process has been unsuccessful. To determine this, the patentees expose the canisters, prepared as before, for at least a month in an apartment heated to about 100° F.; when, if the process has failed, putrefaction commences, and the ends of the cases, instead of remaining concave, bulge or become convex. This is called the 'test.' By this process, which was invented by M. Appert in France about the year 1808, fish, flesh, poultry, and vegetables may be preserved for years in any climate.

Goldner's process differs somewhat from the preceding, in the employment of a higher degree of heat, more hastily applied, and not prolonged or repeated after the tins are soldered up.

Gannal's process, having for its object the preservation of butchers' meat in the fresh state, depends on the peculiarly absorbent property of the flesh of recently killed animals, above referred to. This process consists in injecting a solution of sulphate of alumina, or, better, of chloride of aluminium, of the sp. gr. 1·070 to 1·085 (10°—12° Baumé), into the carotid artery, by means of a syphon, as soon as the blood ceases to flow from the slaughtered animal; both extremities of the jugular vein being previously tied. 9 to 12 quarts of the solution are sufficient for an ox, and a proportionate quantity for smaller animals. A less quantity is also required in winter than summer. When the animal has been well bled, and the injection skilfully performed, it is scarcely perceptible that the animal has undergone any preparation. The injected animal is cut up in the usual way; and when intended to be eaten within two or three weeks merely requires to be hung up in a dry, airy situation free from flies; but if it is to be kept for a longer period, it is directed to be washed with a mixed solution of common salt and chloride of aluminium at 10° Baumé, and then simply dried and packed in clean air-tight barrels, and kept in a cool dry place. If the air cannot be perfectly excluded, it should be packed in dry salt, not for the purpose of preserving it, but to prevent the vegetation of byssus, as, without this precaution, the meat becomes musty from exposure and the action of moisture. Meat preserved by this process may be kept for several years, and merely requires soaking for 24 hours in water, for the purpose of swelling its pores, to give it the appearance and taste of fresh meat, fit for either roasting or boiling. For hot climates a somewhat stronger solution, or a larger quantity of the usual one, may be injected. The use of the strong solutions ordered in some recent works, however, deprives the flesh of a portion of its apparent freshness, and makes it more nearly approach in flavour to that which has been slightly salted in the ordinary manner.

In addition to the above it may be added that both flesh and fish may be preserved by dipping them into, or brushing them over with, pyroligneous acid, and then drying them. This gives them a smoky flavour; but if pure acetic acid (Ph. L.) be used, no taste will be imparted. These fluids may be applied by means of a clean painter's brush, or even a stiff feather. A table-spoonful is sufficient to brush over a large surface. Fish and flesh so prepared will bear a voyage to the East Indies and back uninjured.

Fish may also be preserved in a dry state, and perfectly fresh, by means of sugar alone. Fresh fish may be thus kept for some days, so as to be as good when boiled as if just caught. If dried and kept free from mouldiness, there seems no limit to their preservation; and they are much more nutritious in this way than when salted. This process is particularly valuable in making what is called 'kippered salmon;' and the fish preserved in this manner are far superior in quality and flavour to those which are salted or smoked. A few table-spoonfuls of brown sugar are sufficient for a salmon of 5 or 6 pounds' weight; and if salt be desired, a teaspoonful or two may be added. Saltpetre may be used instead of salt, if it be wished to make the kipper hard.

The well-known property possessed by ether, alcohol, pyroxylic spirit, chloroform, and certain other hydrocarbons, of averting putrefaction, has been thus applied by M. Robin:—He encloses the meat or other substances to be preserved in a glass case, along with a sponge or a capsule containing the preservative liquid, which latter is continually evolved in a vaporous condition, and exercises the preservative agency. In this way the vapours of hydrocyanic acid are found to be very efficacious. Camphor is thus employed in the MUMMY CASES in the British Museum.

It has been asserted by Mr George Hamilton

that in an atmosphere of binoxide of nitrogen, in the dark, flesh preserves its natural colour and freshness for about five months; and eats well provided it be boiled in open vessels, to expel nitrous fumes. See CANDYING, EGG, FISH, FRUIT, MILK, PICKLES, POTTING, PRESERVES, SALTING, SMOKING, STUFFING, VEGETABLE SUBSTANCES, &c.

PUTTY. This name is given to the following preparations (when used alone, 'Glazier's putty' is generally indicated):

Putty, Glazier's. From whiting made into a stiff paste with drying oil. It is used to fix panes of glass in sashes, to fill holes and cracks in wood before painting it, &c.

Putty, Plasterer's. A fine cement used by plasterers, made of lime only. It differs from 'FINE STUFF' in the absence of hair.

Putty, Polisher's. *Syn.* PUTTY POWDER, CALCINE; CINERES STANNI, STANNI OXYDUM CRUDUM, L. A crude peroxide of tin, obtained by exposing metallic tin in a reverberatory furnace, and raking off the dross as it forms; this is afterwards calcined until it becomes whitish, and is then reduced to powder. Another method is to melt tin with rather more than an equal weight of lead, and then to rapidly raise the heat so as to render the mixed metal red-hot, when the tin will be immediately flung out in the state of 'putty' or 'peroxide.' The products of both these processes are very hard, and are used for polishing glass and japan work, and to colour opaque white enamel. See TIN.

Putty, To Soften. Take 1 lb. of American pearlash and 3 lbs. of quicklime. After slaking the lime in water add the pearlash, and let the mixture be made of a consistence about the same as that of paint. When required for use apply it to both sides of the glass, and let it remain in contact with the putty for 12 hours; after which the putty will have become so softened that the glass may be removed from the frame without any difficulty.

PUZZOLA'NA. PUOZZOLANA, POZZOLANA, or, more correctly, PUZZOLANA, is a volcanic ash found at Puzzuoli, near Naples, and over a large portion of Central Italy, especially in the Roman Campagna. The black variety is most esteemed, and next to this the bright red quality. When mixed with lime it forms an excellent hydraulic cement. A good FACTITIOUS PUZZOLANA may be made by heating a mixture of 3 bushels of clay and 1 bushel of fresh-slaked lime for some hours to redness (*M. Bruyère*). See CEMENT and MORTAR.

PYKNOMETER. *Syn.* PICNOMETER. The strength of a solution may be inferred from its specific gravity. The specific gravity is ascertained by comparing the weights of equal volumes of water and of the solution at the same temperature. For this purpose a light-stoppered bottle or *picnometer* is used, capable of containing about 2 fluid oz. This is thoroughly dried and counterpoised in a balance by placing in the opposite pan a piece of lead, which may be cut down to the proper weight. Suppose the strength of a solution of ammonia is required to be ascertained. The bottle is filled with the solution, the temperature observed with a thermometer and recorded, the stopper is inserted, the outside carefully dried, and the whole wiped. It is then well rinsed out, filled with distilled water, the temperature equalised with that of the ammonia by placing the bottle either in cold or warm water, and the weight ascertained as before. The specific gravity is obtained by dividing the weight of the ammonia solution by that of the weight of distilled water.

PYRI'TES.. A term applied to several native metallic sulphides. IRON PYRITES is the best known of these.

PY'RO-. The term is applied to several acids that are obtained by the action of heat on other substances; as PYROGALLIC ACID, PYROLIGNEOUS A., &c.

PYROACE'TIC SPIRIT. See SPIRIT (Pyroacetic).

PYROGAL'LIC ACID. $C_6H_3(OH)_3$. *Syn.* ACIDUM PYROGALLICUM, L. The old and pharmaceutical name of pyrogallin, pyrogallol, or trihydroxybenzene, as it is variously called.—*Prep.* 1. From either gallic or tannic acid, heated in a retort by means of an oil-bath, and steadily maintained at a temperature of about 420° F. as long as crystals are formed in the neck of the retort, or in the receiver, both of which should be kept well cooled. If a much higher heat is employed, the product consists chiefly of metagallic acid.

2. (*Stenhouse.*) By sublimation from the dry aqueous extract of nut-galls, in a Mohr's apparatus, in the same way that benzoic acid is obtained from benzoin resin, observing the precautions referred to in No. 1 (*above*). Nearly pure. The product is fully 10% of the weight of extract operated on.

3. Gallic acid dried at 100° C. is mixed with 3 times its weight of powdered pumice-stone, and distilled in a retort through which a slow stream of carbonic anhydride is passed, the heat being supplied by an oil-bath, and kept at 210°—220° C.

4. (As a developer in photography.) Heat 10 grms. of gallic acid with 80 c.c. of glycerin to 195° C. as long as carbonic acid anhydride is evolved, then make up to a litre with water.

Prop. Fine acicular crystals, which melt at 115° C., and boil to 210° C., and when perfectly pure, are quite white; freely soluble in water, alcohol and ether, 2½ parts, but the solution cannot be evaporated without turning black and suffering decomposition; it strikes a rich blackish-blue colour with the proto-salts of iron, and reduces those of the sesquioxide to the state of protoxide; when heated much above its boiling-point, it is converted into METAGALLIC ACID and water.

Uses, &c. Pure pyrogallic acid being a strong reducing agent, is now very extensively employed in photography as a developer. A solution of the crude acid mixed with a little spirit is used to dye the hair, to which it imparts a fine brown colour, but has the disadvantage of also staining the skin when applied to it. A mixture of potash and pyrogallic acid is employed to absorb oxygen in gas analysis. When heated with phthalic anhydride it yields gallein, which is used as a red dye.

PY'ROGEN ACIDS. Those generated by heat.

PYRO'LA. See WINTER-GREEN.

PYROLIG'NEOUS ACID. *Syn.* VINEGAR OF WOOD†, SPIRIT OF W.†, SMOKING LIQUOR†, ESSENCE OF SMOKE†; ACIDUM PYROLIGNOSUM, L. Impure acetic acid, obtained by the destructive distillation of wood in close vessels. It comes over along with tar, creasote, and other liquid and gaseous matters. In this state it contains much empyreumatic matter in solution; but by separation from the tar, saturation with slaked lime or chalk, defecation, and evaporation, an impure acetate of pyrolignate of lime is obtained, which, after being very gently heated, is again dissolved and defecated, and then treated with a solution of sulphate of soda, when a solution of acetate of soda and a precipitate of sulphate of lime are formed by double decomposition. The solution is next evaporated to dryness, the dry mass (pyrolignite of soda) dissolved in water, and the new solution filtered and recrystallised. The crystals of acetate of soda, obtained by the last process, yield nearly pure acetic acid by distillation along with sulphuric acid. See ACETIC ACID and VINEGAR.

PYROLIG'NEOUS SPIRIT. See SPIRIT (Pyroxylic).

PYROM'ETER. An instrument for measuring high degrees of heat. WEDGWOOD'S PYROMETER depends on the property which clay possesses of contracting when strongly heated. DANIEL'S PYROMETER consists, essentially, of a small rod or bar of platinum, which acts in a precisely opposite manner to the preceding, viz. by its expansion. None of the older forms give an exact measurement of temperature. Those now used are based upon the expansion of vapours and gases, the specific heat of solids, or upon the electrical properties of certain bodies. BECQUEREL'S ELECTRIC PYROMETER is an improved form of one devised by Pouillet. Two wires, each 2 m. in length, and 1 sq. mm. in cross-section, one being of platinum and the other of palladium, are firmly tied together for a distance of 1 cm. with fine platinum wire. The palladium wire is placed in a thin porcelain tube, the platinum wire being left outside, then the whole arrangement is enclosed in a larger porcelain tube. At one end of the outer tube is the junction of the wires, which is adjusted in the place the temperature of which is to be investigated. At the other end the platinum and palladium wires issue, and are soldered to the copper wires connected with a magnetometer. These wires at the junction are placed in a glass tube immersed in ice, so that being both at the same temperature they give rise to no current. The angular deflection of the magnetometer is observed, and the intensity of the current and the temperature of the junction of the palladium and platinum wires are deduced from pyrometric tables.

PYROPH'ORUS. *Syn.* LUFT-ZUNDER, Ger. Any substance that inflames spontaneously when exposed to the air.

Prep. 1. Neutral chromate of lead, 6 parts; sulphur, 1 part; triturate them with water, q. s. to form a paste, and make this into pellets; dry these perfectly by a gentle heat, then heat them in a closed tube until the sulphur is all driven off; lastly, transfer them to a stoppered phial.

2. (HOMBERG'S PYROPHORUS.) From alum

and brown sugar, equal parts; stir the mixture in an iron ladle over the fire until dry, then put it into an earthen or coated glass phial, and keep it at a red heat so long as the flame is emitted; it must then be carefully stopped up and cooled.

3. (*Dr Hare.*) Lampblack, 3 parts; burnt alum, 4 parts; carbonate of potassa, 8 parts; the last.

4. (*Gay Lussac.*) From sulphate of potassa, 9 parts; calcined lampblack, 5 parts; as No. 2.

5. Alum, 3 parts; wheat flour, 1 part; as No. 2.

6. (LEAD PYROPHORUS—*Göbel.*) Heat tartrate of lead to redness in a glass tube, and then hermetically seal it. See TARTRATE OF LEAD.

Obs. When the above are properly prepared, a little of the powder rapidly becomes very hot, and inflames on exposure to the air. The accession of the combustion is promoted by moisture, as a damp atmosphere or the breath. With the exception of the first and sixth, "they owe their combustibility to the presence of sulphide of potassium" (*Gay Lussac*).

PYROPHOSPHOR'IC ACID. See DIBASIC PHOSPHORIC ACID (Phosphorus).

PYRO'SIS. *Syn.* BLACK WATER, WATER BRASH, WATER QUALM. An affection of the stomach, attended by a sensation of heat and the eructation of a thin sour liquid, often in considerable quantity, especially in the morning.

The following pill will be found of service in this affection:—Powdered opium, ⅓ gr.; subnitrate of bismuth, 5 gr.; extract of gentian, sufficient to make into 2 pills. To be taken two or three times a day, before meals.

The solution of bismuth and citrate of ammonia (Liquor Bismuthi et Ammoniæ Citratis, B. P.), in doses of ½ dr. to 1 dr., taken as above, is another medicine which may be had recourse to, should the above fail to give relief.

PYROTARTAR'IC ACID. H₂C₅H₆O₄. Obtained by the destructive distillation of tartaric acid. See TARTARIC ACID.

PYROTECH'NY. The art of making fireworks. The three principal materials employed in this art are charcoal, nitre, and sulphur, along with filings of iron, steel, copper, or zinc, or with resin, camphor, lycopodium, or other substances, to impart colour, or to modify the effect or the duration of the combustion. Gunpowder is used "either in grain, half crushed, or finely ground, for different purposes. The longer the iron filings are, the brighter red and white spots they give; those being preferred which are made with a coarse file, and quite free from rust. Steel filings and cast-iron borings contain carbon, and afford a more brilliant fire, with wavy radiations. Copper filings give a greenish tint to flame; those of zinc, a fine blue colour; the sulphide of antimony gives a less greenish blue than zinc, but with much smoke; amber affords a yellow fire, as well as colophony (resin) and common salt; but the last must be very dry. Lampblack produces a very red colour with gunpowder, and a pink one with nitre in excess; it serves for making golden showers." When this substance is lightly mixed with gunpowder and put into cases, it throws out small stars resembling the rowel of a

spur; this composition has hence been called 'spur fire.' "The yellow sand, or glistening mica, communicates to fireworks golden radiations. Verdigris imparts a pale green; sulphate of copper and sal-ammoniac give a palm-tree green. Camphor yields a very white flame and aromatic fumes, which mask the bad smell of other substances. Benzoin and storax are also used, on account of their agreeable odour. Lycopodium burns with a rose colour and a magnificent flame; but it is principally employed in theatres to represent lightning, or to charge the torch of a Fury" (*Ure*). See FIRES (Coloured), FLAME COLOURS, GUNPOWDER, STARS, ROCKETS, &c.

The following substances are in requisition by the pyrotechnist:

ZINC. This metal is employed in the form of fine powder, which is obtained as follows:—The metal, scarcely melted, is poured into a hot mortar, where it is reduced to powder, being kept during the operation at a temperature of 401° F. It is then sifted to remove any particles which may have escaped contact with the pestle.

COPPER. This metal may be obtained in a state of minute division by precipitating it from a solution of sulphate of copper by means of iron, the precaution being taken of using a large quantity of iron. The precipitate, after being well washed, is dried between folds of blotting-paper, and kept in well-stoppered bottles.

IRON-SAND. A quantity of sulphur is melted in a crucible over a slow fire, and when it is quite fluid, iron filings are thrown in while the whole is being stirred. The crucible is removed from the fire, and the contents are rapidly stirred until cold. The material is then rolled on a board till it is broken up as fine as corned powder, after which the sulphur is sifted out.

SODA POWDER. This powder is prepared with the same precaution as ordinary gunpowder, the proportions which answer best being—

Nitrate of soda	630 parts.
Sulphur	125 „
Charcoal	125 „
	880 parts.

As the nitrate of soda is hygrometric, this powder must be preserved in close vessels from the moisture of the air.

LEAD POWDER. This mixture is also prepared like gunpowder, and the constituents are used in the following proportions:

Nitrate of lead	12 parts.
Nitrate of potash	2 „
Charcoal	3 „
	17 parts.

In the manufacture of this mixture on a large scale considerable care is necessary, since the mixture of nitrate of lead and charcoal is very liable to ignite by friction.

PREPARED BLOOD. 450 to 500 grms. of zinc is dissolved in 1340 grms. of hydrochloric acid 22° B., largely diluted with water, and filtered. This solution is again diluted with its own volume of water, and mixed with fresh blood. The whole is well stirred from time to time for 48 hours, and the clear liquor is siphoned off from the precipitate. The precipitate is well washed with water, dried, and reduced to powder, in which state it may be kept for any length of time.

TOUCH PAPER. This paper is prepared by immersing purple or blue paper in a solution of nitrate of potash in spirits of wine or vinegar, and carefully drying it.

When the touch paper is used with small articles, a piece is tied round the orifice with thread, leaving sufficient paper to form a small tube at the end. This tube is filled with gunpowder, and the paper twisted over it, when all is ready for firing.

Touch paper for capping every description of fireworks, such as squibs, crackers, Roman candles, &c., is prepared in the following manner:—Dissolve 2 oz. of the best saltpetre in 1 quart of warm water, and take care that the water is very clean.

After the mixture has stood for half an hour, pour off 1½ pints into a white basin, then cut your sheets of dark blue double-crown paper in half. The weight of the paper should be 12 or 14 lbs. per ream.

Place the paper on a slab sufficiently large to give you room to use a small piece of sponge, with which you use the liquor to wet your paper. Cover each half-sheet with the liquor as quickly as possible, on one side only, and immediately this is done place it on a line, the wet side outwards, and when nearly dry, if you have a great number of sheets, place them together as evenly as possible under a press for one hour, then lay them out to dry, after which they will be quite smooth and ready for use.

In pasting this paper on the work, take care that the paste does not touch that part which is to burn. To use this paper correctly, cut it in strips sufficiently long to go twice round the mouth of the case, or even more if requisite. When you paste on the strips, leave a little above the mouth of the case not pasted ; in small cases a little meal powder is put into the mouth, and then the paper is twisted to a point. In larger cases damp priming is used, and when dry, the capping process is proceeded with.

CRACKERS. The following mixtures are used for ordinary crackers:

Meal powder	. parts	5	15	6	8	16
Fine charcoal	„	1	4	—	2	17
Coarse charcoal	„	—	—	6	—	—
Sulphur . .	„	—	—	2	—	1
Saltpetre	„	—	—	16	1	7

Composition for crackers with Chinese fire :

Meal powder	. parts	. .	9	6	16
Saltpetre	„	. .	6	8	—
Sulphur . .	„	. .	1	2	3
Charcoal . .	„	. .	1½	1½	2
Fine iron . .	„	. .	5	—	7
Sand . .	„	. .	—	5	—

Composition for crackers with brilliant fire :

Meal powder	. parts	8	8	36	18	32
Sulphur . .	„	1	1½	1	1	3
Iron filings . .	„	2	2½	—	—	—
Litharge . .	„	—	—	—	2	—
Steel filings . .	„	—	—	3	3	12

The paper generally used for cartridge is that known as ' elephant ' or cartridge, the latter being the more frequently employed.

Cartridge paper is employed in the preparation of crackers, which vary from 12 to 15 inches, and 3½ inches diameter. One edge of the paper is folded down about ¾ inch in breadth, then the double edge is turned down about ¼ inch, and the single edge is bent back over the double fold so as to form a channel ¼ inch wide. This is filled with meal powder, which is then to be covered by the folds on each side, when the whole is to be pressed very smooth and close, by passing it over the edge of a flat ruler. The part containing the powder is to be gradually folded into the remainder of the paper, each fold being carefully pressed down. The cracker is then doubled backwards and forwards into as many folds of about 2½ inches as the paper will allow.

The whole is pressed together by means of a wooden vice, a piece of twine is passed twice round the middle across the folds, and the joinings are secured by causing the twine to take a turn round the middle at every turn. One of the ends of the folds may be doubled short under, which will produce an extra report, but the other must project a little beyond the rest, for the priming and capping with the touch paper. When these crackers are fired they give a report at every turn of the paper.

The crackers may also be made of two single cards, rolled over each other and covered with paper coated with paste. The crackers are partially filled with the composition by means of a tin funnel. Ordinary powder is then introduced, and the remaining space is filled with a little sawdust.

REVOLVING CRACKERS. These crackers are charged at each end with clay to a depth of two lines, and filled with a composition without gunpowder. The clay prevents the fire streaming out at the ends, and it escapes through two holes placed opposite each other. The two holes are united at the same time by connecting them by means of a quick-match, and a rotatory motion is thus communicated to the cylinder.

ENGLISH PIN WHEELS. Pin, or Catherine wheels are of very simple construction. A long wire about ₁⁄₁₆ of an inch in diameter is the former; on this wire are formed the pipes, which, being filled with composition, are afterwards wound round a small circle of wood so as to form a helix or spiral line. The cases are generally made of double-crown paper (yellow wove), and cut into strips to give the greatest length, and of width sufficient to roll about four times round the wire, and pasted at the edge so as to bite firmly at the end of the last turn. When a number of pipes are made and perfectly dry, they are filled with composition. These cases are not driven for filling, but are filled by means of a tin funnel with a tube ⅜ of an inch long, made to pass easily into the mouth of the case, which is gradually filled by lifting a wire up and down in this tube, the diameter of the charging wire being half that of the tube. The dry composition being placed in the funnel, the moment an action of the wire takes place the composition begins to fall into the case, which the charging wire compresses by continuous motion until you have filled the pipe to within ⅜ of an inch of the top. The pipe is then removed, and the mouth neatly twisted, which will be the point for lighting.

When a number of pipes are ready, place them on a damp floor, or in any damp situation, until they become very pliant, but by no means wet; then commence winding them round a circle of wood whose substance must be equal to the thickness of the diameter of the pipe; and when wound, secure the end with sealing-wax, to prevent its springing open; after winding the required quantity let them dry. Now cut some strips of crimson or purple paper $\frac{7}{16}$ of an inch wide, and in length twice the diameter of the wheel; then paste all over thoroughly. Take a strip and paste it across the wheel diametrically, rub it down, then turn the wheel over, and place the ends down to correspond with the opposite side; when dry, the wheel will be ready for firing. They may be fired on a large pin or held in the hand, but it is preferable to drive the pin into the end of a stick, which will prevent any accident, should a section of the wheel burst.

SQUIBS. These are either filled with grained powder, or with a mixture consisting of—Gunpowder, 8 parts; charcoal, 1 part; sulphur, 1 part. The cases, which are about 6 inches long, are made by rolling strips of stout cartridge paper three times round a roller, and pasting the last fold. They are then firmly tied down near the bottom, and the end is either dipped into hot pitch or covered with sealing-wax. The cases are filled by putting a thimble-full of the powder in, and ramming it tightly down with a roller, this operation being continued until the case is filled. It is then capped with touch paper.

SERPENTS (MARROON SQUIBS). A suitable case being ready, it is filled two thirds up with a powder consisting of—Saltpetre, 16 parts; sulphur, 8 parts; fine gunpowder, 4 parts; antimony, 1 part. This being rammed down into the case tolerably tight, the remainder of the space is filled with grained or corned powder.

SPARKS. These fireworks differ from stars in size, being very small and made without cases. The English method of preparing them is as follows. A mixture of—

Fine gunpowder . . .	1 part,
Powdered saltpetre . . .	3 parts,
Powdered camphor . . .	4 ,,

is placed in a mortar, and some weak gum-water in which a little gum tragacanth has been dissolved is poured over it, and the whole worked up into a thin paste. Some lint, prepared by boiling it in vinegar or saltpetre, and afterwards dried and unravelled, is placed in the composition so as to absorb the whole. This is then poured into balls about the size of a pea, dried, and sprinkled with fine gunpowder.

In Germany the following compositions are used:

	1	2	3	4	5	6	7	8	9	10	11	12	13	14
Chlorate of potash . parts	24	40	12	20	—	—	—	40	21	21	14	20	96	40
Chlorate of potash and copper . . ,,	—	—	—	—	—	—	—	23	23	—	—	—	—	—
Chlorate of baryta ,,	—	—	—	—	18	—	—	—	—	—	—	—	—	—
Nitrate of potash . ,,	—	—	—	—	—	12	26	—	—	—	—	—	—	—
Nitrate of lead . . ,,	24	—	—	—	—	—	—	—	—	—	—	—	—	—
Nitrate of baryta . ,,	—	—	—	40	—	—	—	—	—	—	—	—	—	—
Calomel . . . . ,,	—	—	—	13	7	—	—	28	12	12	4	8	18	—
Sulphide of copper ,,	—	—	—	—	—	—	—	28	—	12	6	4	—	—
Sulphate of strontia ,,	—	—	—	—	—	—	—	—	—	—	—	20	72	37
Oxalate of soda . ,,	—	16	10	—	—	—	—	—	—	—	—	—	—	—
Chalk . . . . ,,	—	—	—	—	—	—	—	—	—	5	—	—	—	—
Powdered zinc . . ,,	—	—	—	—	14	28	—	—	—	—	—	—	—	—
Powdered charcoal ,,	—	—	—	—	5	11	—	—	—	—	—	—	—	—
Sulphur . . . . ,,	12	—	1	13	—	—	—	—	—	—	6	8	—	—
Gum lac . . . . ,,	1	8	—	1	3	—	—	—	—	—	—	2	18	8
Soap . . . . . ,,	—	3	1	—	—	—	—	3	3	3	—	—	—	—
Starch . . . . ,,	—	—	—	—	—	—	—	—	10	—	—	—	—	—
Sugar . . . . ,,	—	—	—	—	—	—	—	—	—	4	4	—	—	—
Pine soot . . . ,,	—	—	—	—	—	—	—	—	—	—	—	—	1	—

The above mixtures are intended to give coloured sparks, according to the numbers.

No. 1 gives a bluish-white colour.
　,, 2 and 3 give yellow.
　,, 4 gives green.
　,, 5 gives green.
　,, 6, 7, 8, 9, and 10 give blue.
　,, 11 and 12 give violet.
　,, 13 gives red.
　,, 14 gives purple.

The materials are mixed with a small quantity of a solution of starch, so as to form a thick paste, which is forced through a perforated plate, the holes in which are twice as large as it is intended the sparks should be on drying. The small pieces fall on a pasteboard, to which the workman gives a rapid horizontal motion to round the grains. They are then dried, and those which are perfectly round are selected and separated by sieves of different meshes to collect those of the same size together.

The iron-sand is moistened with a little spirits of wine, and then mixed with the charcoal and saltpetre, which have been previously incorporated in another mortar.

CHINESE FIRE.

Red Chinese or Gerbe Fire.

Calibre of the case.	Saltpetre.	Sulphur.	Charcoal.	Iron-sand, 1st order.
12 to 16 lbs.	. . 1 lb.	. . . 3 oz.	. . . . 4 oz.	. . . . 7 oz.
16 to 22 „	. . 1 „	. . . 3 „	. . . 5 „	. . . 7 „ 8 drms.
22 to 36 „	. . 1 „	. . . 4 „	. . . 6 „	. . . 8 „

White Chinese Fire.

Calibre.	Saltpetre.	Bruised Powder.	Charcoal.	Iron-sand, 3rd order.
12 to 16 lbs.	. . 1 lb.	. . 12 oz.	. . 7 oz. 8 drms.	. . 11 oz.
16 to 22 „	. . 1 „	. . 11 „	. . 8 „	. . 11 „ 8 drms.
22 to 36 „	. . 1 „	. . 11 „	. . 8 „ 8 „	. . 12 „

SIMPLE STARS or FIREBALLS. These are generally used in combination with other arrangements, &c., and the composition of which they are made consists of—saltpetre, 16 parts; sulphur, 8 parts; fine gunpowder, 3 parts.

These materials are mixed with gum and as little spirits of wine as will suffice to make a very stiff paste. This paste is cut up into small squares, which are rolled up into balls on a board covered with gunpowder.

The gunpowder, which adheres, serves for the purpose of firing them. When perfectly dry they are ready for use.

	Ordinary.	Chinese.
Saltpetre . . . parts . .	16	4
Sulphur . . . „ . .	8	2
Fine charcoal . . „ . .	2	4
Pine soot . . . „ . .	2	—
Meal powder . „ . .	4	16

A portion of the cotton is softened in linseed oil, and the materials prepared in a mortar with water.

ROMAN CANDLES. These are made somewhat like gerbes and filled with the same materials, the only difference being that stars are placed between the different layers of substances. The materials must not be too tightly rammed down or the stars will be destroyed.

SIMPLE STARS or FIREBALLS. Take of saltpetre, 16 parts; sulphur, 8 parts; fine gunpowder, 3 parts; mix them with gum and only just enough spirits of wine to make a very stiff paste. Cut this up into small squares, and roll into balls covered with gunpowder. When properly dry they are ready for use.

MARROONS. These are small cubical boxes filled with an explosive composition which explodes suddenly, making a loud report. They are generally used in combination with other fireworks. The boxes are made of pasteboard, the corners being made tight by pasting paper over them, but leaving the top open until they are filled. They are filled with coarse gunpowder, when the top is closed with strong paper well cemented, and the whole box is wrapped round two or three times with lind cord dipped in strong glue. A hole is made in one of the corners, into which a quick-match is introduced, and the marroon is ready for action.

The reader who may be desirous of further information on the subject of Pyrotechny, cannot do better than consult the article on that subject in 'Knapp's Chemical Technology,' edited by Messrs. Richardson and Watts, vol. i, part 4, No. 1 (Baillière & Co.).

GERBES. These fireworks display themselves as luminous jets of fire somewhat resembling a waterspout. Previously to putting in the brilliant composition, put two scoops of first firing or preparatory fire, for which the following will suit, in cases not larger than ¼-lb. size:—16 oz. meal powder, 6 oz. saltpetre, 3 oz. sulphur, 3 oz. fine coal. It is important to see that the interior of the cases are quite smooth and free from wrinkles.

GOLD RAIN. The larger rockets are filled with this material, which consists of small squares made in the same way as the simple stars. It is composed as follows:

		Composition for immediate use.
Saltpetre . . . parts . .	4
Sulphur . . . „ . .	2
Fine small coals „ . .	1
Fine gunpowder „ . .	8
Coarse cast iron „ . .	4

To this work we are indebted for much of the material contained in the present papers. See COLOURED FIRES.

PYROXYLIC SPIRIT. See SPIRIT (Pyroxylic).

PYROXYLIN. Syn. FULMINATING COTTON, GUN-COTTON, CELLULOTRINITION. A highly inflammable and explosive compound, discovered by Schönbein. It is obtained by the action of nitric acid on cotton (cellulose, $C_6H_{10}O_5$), in the presence of sulphuric acid.

The action of nitric acid upon cellulose (e. g. cotton-wool, linen, paper, &c.) gives rise to several nitrates, mixtures of which are commonly called pyroxylin, but the hexanitrate prepared according to method 1 (see below) is the best for gun-cotton. Its composition is represented by the formula $C_{12}H_{14}(NO_3)_6O_{10}$.

Prep. 1. (Abel.) Purified, well-dried cotton-wool is placed in ten parts of a mixture of nitric acid (sp. gr. 1·5), 1 part; sulphuric acid (sp. gr. 1·85), 3 parts; where it is left for 24 hours. The wool is then thoroughly washed in such a washing machine as is used in the manufacture of paper, reduced to pulp, and then pressed into moulds ('Chem. News,' xxiv, p. 241).

2. Concentrated nitric acid (sp. gr. 1·500) and concentrated sulphuric acid (sp. gr. 1·845) are mixed together in about equal measures; when the mixture has become cold it is poured into a glass or wedgwood-ware mortar or basin, and clean dry carded cotton, in as loose a state as practicable, is immersed in it for 4 or 5 minutes, the action of the liquid being promoted by incessant stirring with a glass rod; the acid is next poured off, and the cotton, after being squeezed

as dry as possible, by means of the glass stirrer, or between two plates of glass, is thrown into a large quantity of clean soft water, and again squeezed to free it from superfluous moisture; it is then washed in a stream of pure water until it becomes perfectly free from acid, and is, lastly, carefully dried by the heat of hot water or steam, at a temperature not higher than about 180° F.

3. (*Schönbein.*) Nitric acid (sp. gr. 1·45 to 1·50), 1 part; sulphuric acid (sp. gr. 1·85), 3 parts (both by volume); proceed as above, but after the cotton has been squeezed from the acid, allow it to remain in a covered vessel for an hour before washing it, and after washing it dip it into a solution of carbonate of potash, 1 oz., in pure water, 1 gall., then squeeze, and partially dry it; next dip it into a weak solution of nitre, and dry it in a room heated by hot air or steam to about 150° F.

4. (*Von Lenk.*) The cotton, having been thoroughly cleansed and dried, is spun into loose yarn and steeped, as above, in a mixture of nitric and sulphuric acids (the strongest obtainable in commerce), squeezed as dry as possible, and immersed in a fresh mixture of strong acids, being allowed to remain in this second mixture 48 hours. It is then washed in a stream of water for several weeks, and finally treated with a solution of silicate of potash (soluble glass). This is the celebrated Austrian gun-cotton. The treatment with silicate of potash is adopted merely for the purpose of retarding the combustion.

5. ('Bulletin de St Pétersbourg.') *a.* Take of powdered nitre, 20 parts; sulphuric acid (1·830 to 1·835), 31 parts; dissolve in a glass vessel, and, whilst the solution is still warm (122° F.), add of dry carded cotton, 1 part, and agitate until this last is well saturated; then cover the vessel with a plate of glass, and let it stand for 24 hours at a temperature of about 86° F.; next well wash the cotton, as above, first with cold and afterwards with boiling water, and dry it carefully at a very low temperature.

5. From sulphuric acid (containing 3 equiv. of water), 13 parts; nitric acid (monohydrated), 12 parts; carded cotton, 1 part; the immersion being limited to one hour at a temperature of from 104° to 122° F. (See 'Pharm. Journ.,' vol. viii, No. 2.)

Prop., &c. Gun-cotton can scarcely be distinguished in appearance from that of raw cotton. It burns very rapidly when ignited, and explodes on percussion. It becomes powerfully electric on rubbing. Several modifications of pyroxylin are known, varying considerably in composition, though they are all more or less explosive except when wet. The hexanitrate is insoluble in a mixture of ether and alcohol, whilst others are readily dissolved, forming the glutinous solution which is used in surgery under the name of 'collodion,' and which is also extensively used in photography and in the manufacture of small balloons. The best gun-cotton is of no use whatever for making collodion. The pyroxylin prepared by the formula 5 *a* (*above*) is soluble in a mixture of 7 parts of ether and 1 part of alcohol; whilst the product of 5 *b*, if prepared by 2 hours' digestion instead of 1, is said to be even soluble in absolute alcohol.

Obs. Gun-cotton is a powerful, reliable, safe, portable, and convenient explosive, especially valuable for submarine operations. General von Lenk, Sir Frederick Abel, and others have overcome all the difficulties which have hitherto prevented gun-cotton being used in place of gunpowder. By spinning the gun-cotton into thread or yarn, and weaving this into webs, or by compressing the wet pulpy mass into moulds, cartridges can be made, which will produce the exact amount of force required. The time needed for the complete ignition of the cartridge can be diminished or increased at pleasure by varying the mechanical arrangement of the spun threads. Each kind of projectile requires a certain density of cartridge. In general it is found that the proportion of 11 lbs. of gun-cotton occupying 1 cubic foot of space produces a greater force than gunpowder of which from 50 to 60 lbs. occupies the same space, and a force of the nature required for ordinary artillery. See COLLODION, PHOTOGRAPHY, and XYLOIDIN; consult also Abel's researches in the 'Transactions of the Royal Society,' and the British Association Reports.

QUACK MED'ICINES. See PATENT MEDICINES, OINTMENT, PILLS, &c.

QUAIL. The *Coturnix vulgaris,* a gallinaceous bird, allied to the partridge, but of smaller size. Its flesh is highly esteemed by epicures. It is imported from Turkey, preserved in oil; and from Italy, potted with clarified butter.

QUARANTINE. The old laws of Quarantine, as the French derivation of the word indicates, compelled a vessel coming from the shores of a country liable to, or ravaged by, an infectious disease, such as plague, to those of a region free from contagion, to undergo 40 days' isolation before it was unladen, or its passengers were allowed to land at the healthy port.

In Europe these ancient enactments against the importation of infection are still more or less vexatiously enforced in Spain, Portugal, Greece, and Turkey; and in a modified form at Malta and some of the French and Italian ports. In the Mediterranean ports, ships coming from countries which lie in the southern or eastern shores of that sea are usually subjected to a quarantine of from 6 to 15 days, during which period the passengers are confined in a sort of barrack called a 'lazzretto,' the merchandise, letters, &c., of the vessel being in the meantime frequently fumigated, or otherwise disinfected.

The inconveniences to commerce and the necessary intercourse between nations attending the too rigorous carrying out of quarantine have, within the last 12 years, led to a series of sanitary international conferences between the European Governments, with the object of devising some methods which, without weakening the safeguards to the public health, should as much as possible reduce the inconveniences attending the enforcement of quarantine to a minimum. At the last of these conferences, which was held at Vienna in 1878, the members were almost unanimous in advising the abolition of quarantine on European rivers.

Until within the last 20 years the old quarantine laws were pretty strictly enforced in this country. Since this time, however, they have been considerably relaxed, or, we should rather say, superseded by the following ordinances, which, upon the authority of an order in council of July 31st, 1871, can be enforced in the case of suspected vessels.

This ordinance declares that it is lawful for a sanitary authority, having reason to believe that any ship arriving in its district comes from a place infected with cholera, to visit and examine the ship before it enters the port.

Art. 3 provides that the master of a cholera-infected ship, or one that has even been exposed to the infection of cholera, is to moor, anchor, or place her in such a position as from time to time the sanitary authorities shall direct.

Art. 4 provides that no person shall land from any such ship until after the examination.

Art. 5 provides for the proper examination of all persons on board by a legally-qualified practitioner, and permits those not suffering from cholera to land immediately.

Another order in council, dated August 3rd, 1874, empowers any custom-house officer, or other person having authority from the Commissioners or Board of Customs, at any time before the nuisance authority shall visit and examine the ship, to detain the ship.

"No person shall, after such detention, land from the ship, and the officer shall forthwith give notice of the detention, and of the cause thereof, to the proper nuisance (local) authority; and the detention shall cease as soon as the nuisance authority shall visit and examine the ship, or at the expiration of 12 hours after notice shall have been given to such nuisance authority."

Another order in council, dated August 5th, 1871, directs that the master of a vessel, in which cholera has existed, shall not be allowed to bring his vessel into port until he has destroyed the infected clothes and bedding.

Local Government Boards are also invested with considerable executive powers, by which they are enabled to enforce quarantine during the prevalence of any contagious disease in other countries. The main Act, however, relating to quarantine, is the 6th of Geo. IV., c. 78; and all vessels having on board any person or persons affected with a dangerous or infectious disorder, are to be deemed as coming within its provisions (see 'Public Health Act,' Schedule v, part 3). There is a land, as well as a sea quarantine. Thus, for instance, in some countries, particularly those of Eastern Europe, the former is still in force on the frontiers of contiguous States, to the great impediment of commerce and inconvenience of travellers.

The late outbreak of plague in Astrakan has led to its being established and very strictly carried out on the borders of Russia, Austria, Hungary, and Germany.

Hecker, writing on the probable origin of quarantine, remarks :—" The fortieth day, according to the most ancient notions, has always been regarded as the last of ardent diseases, and the limit of separation between these and those which are chronic. It was the custom to subject lying-in women for 40 days to a more exact super-intendence.

VOL. II.

"There was a good deal also said in medical works of 40 days' epochs in the formation of the foetus, not to mention that the alchemists always expected more durable revolutions in 40 days, which period they called the philosophical month. This period being generally held to prevail in natural processes, it appeared reasonable to assume and reasonable to establish it as that required for the development of latent principles of contagion, since public regulations cannot dispense with decisions of this kind, even though they should not be wholly justified by the nature of the case. Great stress has also been laid on theological and legal grounds, which were certainly of greater weight in the fifteenth century than in modern times ; such as the 40 days' duration of the flood ; the 40 days' sojourn of Moses on Mount Sinai ; our Saviour's fast for the same length of time in the wilderness ; lastly, what is called the Saxon term, which lasts for 40 days."

QUAR'TAN. Occurring every fourth day.

QUARTA'TION. The practice, among assayers, of alloying 1 part of gold with 3 parts of silver, before submitting it to the operation of 'parting ;' in order that its particles may be too far separated to protect the copper, lead palladium, silver, or other metals, with which it is contaminated, from the solvent action of the nitric or sulphuric acid, as the case may be. See ASSAYING.

QUARTZ. Pure native silica. It is an essential constituent of granite and many other rocks. Its crystalline, transparent varieties are known as rock crystal. See GLASS, POWDER, &c.

QUASS. Syn. POSCA VENALIS, L. Prep. Mix rye-flower and warm water together, and keep the mixture by the fireside until it has turned sour. Used as vinegar in Russia.

QUAS'SIA. Syn. QUASSIA ; QUASSIA LIGNUM, QUASSIA WOOD (B. P.). The "wood of Picraena (Picrasma) excelsa, Lindl.," or Jamaica quassia ; and also of the " Quassia amara, Linn." (Ph. E.), or Surinam quassia. The latter is the original quassia, but it is no longer imported. Quassia is characterised by its intense bitterness, due to a crystalline substance named quassin. It is reputed tonic and stomachic, assisting digestion, and giving tone and vigour to the system. Its name was given to it by Linnaeus, in honour of a negro slave who had long employed it as a remedy for the malignant endemic fevers of Surinam. When sliced, it forms the 'quassia chips' of the shops. It is generally taken in the form of infusion. This last, sweetened with sugar, forms a safe and effective poison for flies. Injected in the rectum it destroys thread worms. Sprayed on plants it destroys green fly and other insects.—Dose (in powder), 10 to 20 gr.

ROASTED QUASSIA, reduced to powder, is largely employed, instead of hops, to embitter porter ; and the unroasted powder is used for the same purpose in the adulteration of the bitter varieties of ale.

QUAS'SIN. Syn. QUASSITE, QUASSINA. A peculiar bitter principle, obtained by precipitating decoction of quassia with milk of lime, evaporating the filtrate, dissolving the residue in alcohol, treating with animal charcoal, again evaporating, dissolving in water, and crystallising. 3 lbs. of quassia chips yield 1 dr.

QUEBRACHO BARK. *Syn.* QUEBRACHO
CORTEX. WHITE QUEBRACHO BARK. The bark
of *Aspidosperma quebracho*, imported from
Chili. In pieces from ⅜ in. to 1 in. thick, reddish coloured, fissured and warty; taste aromatic,
bitter. A substance called *aspidospermine* is
sold as its active principle, but the researches of
Harnack, Hesse, and others show this to be a
mixture of *aspidosamin*, *quebrachin*, *quebrach-
amin*, *aspidospermatine*, *hypoquebrachin*, and *aspidospermine*. The bark has tonic and antipyretic powers, used also to relieve asthma.
Tincture 1 in 5 proof spirit.—*Dose*, ⅓ to 1 dr.

Quebracho Colorado. (*Quebrachia Lorentzii*,
Griesb). A tree abundant in the northern parts
of the Argentine Republic. The wood is valuable for building, as it resists water perfectly.
It is of a red colour, and is used for colouring
wines and also for tanning.

QUEEN'S BLUE. Thumb blue. See BLUE.

QUEEN'S MET'AL. A species of pewter used
for teapots, &c., made by fusing under charcoal a
mixture of tin, 9 parts, and antimony, bismuth,
and lead, of each, 1 part; or, tin, 100 parts;
antimony, 8 parts; copper, 4 parts; bismuth, 1
part. See BRITANNIA METAL and PEWTER.

QUEEN'S YEL'LOW. Subsulphate of mercury.

QUERCITRIN. The bark of the *Quercus tinctoria* yields a neutral substance, to which the
above name has been given. Quercitrin may be
prepared as follows by the process of Rochleder:—The bark is boiled with water, the decoction is left to cool, and the impure quercitrin
which separates is collected, then rubbed to a
pulp with alcohol of 35° B., heated over the
water bath, collected on linen, and pressed,
whereby the principal impurities are removed.
The residue is dissolved in a larger quantity of
boiling alcohol, the solution is filtered hot, and
water is added to it until it becomes turbid, so
that the greater part of the quercitrin separates
before the liquid is cold. It is then collected,
pressed, and purified by a repetition of the same
treatment.

Another process, by Zwenger and Dronke, is
this:—The bark, in small pieces, is exhausted
with boiling alcohol, the alcohol is distilled off,
and the residue, while still warm, is mixed with a
little acetic acid, and then with neutral acetate
of lead; the filtrate, freed from lead by sulphuric
acid, is evaporated, and the quercitrin which
crystallises is purified by repeated crystallisation
from alcohol.

" Hydrated quercitrin forms microscopic, rectangular, partly rhombic tablets, having their
obtuse lateral edges truncated; pale yellow when
pulverised. It is neutral, inodorous, tasteless in
the solid state, bitter in solution, permanent in
the air " (*Watts*).

QUER'CITRON. A yellow dye-stuff, composed
of the shavings and powder of the bark of *Quercus tinctoria*, or *Q·· nigra*, or *Q·· citrina*, a kind
of oak, a native of North America. It abounds
more particularly in Pennsylvania, Carolina, and
Georgia.

In America quercitron is used for tanning, and
in Europe for dyeing only. When employed for
the latter purpose it is used in the form of an
aqueous decoction, mordanted with alum or chlo-

ride of tin. Leeching states that a dye possessing greater colorific powder may be procured by
boiling the bark with dilute sulphuric or hydrochloric acid.

QUICK'SILVER. See MERCURY.

QUILLAI BARK. *Syn.* QUILLAY BARK, SOAP
BARK. The *Quillaia saponaria*, which yields this
bark, is an evergreen tree, growing in the mountainous parts of Chili, in South America.

It is believed to take its name from the native
word *quillay*, which signifies to wash. The inner
bark only is employed. When bruised and agitated in water it imparts a lather to the water,
in the same way that soap does. This quality has
been found to be due to the existence in the bark
of *sapotoxin* or *saponin*—the same principle which
confers a similar property on *Saponaria officinalis*.
The bark is free from any bitter principle, as well
as from tannic acid. It is very generally used
amongst the inhabitants residing on the western
side of South America, where it is employed for
removing grease from silk, and also in the form
of a wash for cleansing and preserving the hair.

When had recourse to for cleansing silks, quillai
bark is said not to change the colour of the fabric.
It is sometimes given as a febrifuge, and as a remedy for cold in the head. For this latter purpose the powder is snuffed up the nostrils, when
it occasions sneezing and profuse discharge from
the nose. A tincture, 1 in 10 of rectified spirit,
is used as a solvent for coal tar. It is a useful
emulsifying agent for fixed oils; a few drops of
the tincture shaken with cod-liver oil causes it to
readily mix with water.

QUILLS. *Prep.* 1. The quills or wing-feathers
of the goose (goose quills) are separately plunged,
for a few seconds, into hot ashes, cinders, or sand,
of a temperature about equal to that of boiling
water, after which they are scraped with a blunt
knife, strongly rubbed with a piece of flannel or
woollen cloth, and, gently ' stoved'; they are,
lastly, tied up in bundles by women or children.
A yellow tinge is often given to them by dipping
them for a short time into dilute hydrochloric or
nitric acid, or into an infusion of turmeric.

2. Suspend the quills in a copper over water
sufficiently high to nearly touch the nibs; then
close it steam tight, and apply three or four
hours' hard boiling; next, withdraw the quills,
and dry them, and in twenty-four hours cut the
nibs and draw out the pith; lastly, rub them with
a piece of cloth, and expose them to a moderate
heat in an oven or stove. Quills prepared in this
way are as hard as bone, without being brittle,
and nearly as transparent as glass. Crow quills
and swan quills may be cured in the same
manner.

QUINA. See QUININE.

QUINAMINE. *Syn.* QUINAMINA. $C_{19}H_{24}N_2O_2$.
This alkaloid was discovered by Hesse, in 1872,
in the bark of *Cinchona succirubra*, cultivated at
Darjiling, in British Sikhim.

Dr de Vrij gives the following process for the
preparation of quinamine:—The mixed alkaloids
obtained from the red bark are converted into
neutral sulphates, and the solution treated with
Rochelle salt, whereby the tartrates of quinine
and of cinchonidine are separated. After collecting these upon a filter, the filtered liquid is shaken

with caustic soda and ether. By this process the amorphous alkaloid and the quinamine are dissolved by the ether, with slight traces of cinchonine, whilst the bulk of this last alkaloid remains undissolved. After distilling the ethereal solution the residue is transformed into neutral acetate, and the solution of this mixed with a solution of sulphocyanate of potassium.

By this reaction the sulphocyanate of the amorphous alkaloid is precipitated in the shape of a yellow, soft, resinous substance, whilst the sulphocyanate of quinamine remains dissolved. After subsiding and filtering, the solution is clear and quite colourless, and by addition of caustic soda the quinamine is precipitated. It is then collected upon a filter, washed, and dried. It can now easily be obtained crystallised by dissolving it in boiling spirit, from which it crystallises in cooling. By this process the author obtained 0·38 per cent. of pure quinamine from samples of red cinchona quill bark, which he had received, through the Secretary of State for India, from the plantations in British Sikhim.

Quinamine fuses at 172° C., is sparingly soluble in boiling water, abundantly in boiling ether, benzol, or petroleum ether. Its solutions are not fluorescent, neither does it give the thallicoquin test. Moistened with nitric acid it assumes a yellow colour.

QUINCE. Syn. CYDONIA, L. The fruit of Cydonia vulgaris, or common quince tree. Its flavour in the raw state is austere, but it forms an excellent marmalade (quince marmalade), and its juice yields an agreeable and wholesome wine. The seed or pips (cydoniæ seminæ; cydonium—Ph. L.) abound in gummy matter, which forms a mucilage with water, and possesses the advantage of not being affected by the salts of iron or alcohol. See DECOCTION, FIXATURE, and JELLY.

QUINETUM. The alkaloid contained in the East Indian red bark (Cinchona succirubra) consist of a large percentage of cinchonidine, cinchonine, quinine, and amorphous alkaloid, besides a trace of quinidine, the preponderating alkaloid being cinchonidine.

Dr de Vrij, of the Hague, has devised a process by which these can be extracted in their entirety, and to the mixed alkaloids so obtained the name 'quinetum' has been given.

It is affirmed of quinetum that it possesses a remedial value as a tonic and antiperiodic that renders it, in many cases, superior to quinine, ague being one of these; also that it may be advantageously employed in affections in which quinine would be inadmissible. A medical correspondent informs us that he has used it with signal success in hay asthma. Another advantage it has over quinine is, it is much lower in price. Quinetum, according to Dr de Vrij's process, as well as a sulphate and hydrochlorate are prepared by Mr Whiffen, of Battersea.

QUINICINE. An alkaloid obtained in 1853 by Pasteur, by exposing quinine or quinidine, under favourable circumstances, to a temperature varying from 248° to 266° F., for several hours. It is very probable that this alkaloid is either identical, or in very close connection, with the amorphous alkaloid soluble in ether which occurs in all

barks, and particularly in the young barks of the plantations in India.

QUINIDINE. $C_{20}H_{24}O_2N_2.2Aq.$ Syn. QUINIDIA, CONCHININE, &c. An alkaloid contained in many species of cinchona, together with quinine and cinchonine, and therefore often found in the mother-liquors of quinine manufactures. It is identical with the β quinine of Van Heyningen, and was discovered, in 1833, by Henry and Delondre. As the cinchonidine discovered by Winckler, in 1848, has been unhappily denominated quinidine by this chemist, there is still a confusion about these alkaloids, and, therefore, the quinidine of commerce was very often a mixture of both, till Pasteur made, in 1853, a classical investigation of this matter. He maintained the name of quinidine for the alkaloid discovered by Henry and Delondre, because it is isomeric with quinine, and gives the same green colour when treated with chlorine followed by ammonia, and gave the name of cinchonidine to the alkaloid discovered by Winckler, because it is isomeric with cinchonine. He determined also the action of the solutions of these alkaloids on the plane of polarisation, and found that the quinidine turned this plane to the right, its molecular rotation in alcoholic solution being $[a] = 250·75°\ggg$, whilst he found that the cinchonidine turned this plane to the left, its molecular rotation in alcoholic solution being $[a] = 144·61°\lll$.

Prop., &c. Many of the salts of quinidine are very similar to those of quinine, but the normal salt with hydriodic acid is not only very different from that of quinine, but also from those of all the other cinchona-alkaloids. The normal hydriodate of quinidine is so very sparingly soluble in water that 1 part requires, at 60° F., not less than 1250 parts of water to be dissolved. Therefore the presence of sulphate of quinidine in the sulphate of quinine, which often occurs, either from that article being carelessly made or from wilful adulteration, can be easily detected by adding a few minims of solution of iodide of potassium to the saturated solution of sulphate of quinine in water of 60° F., whereby, if quinidine is present, its hydriodate will be separated either in the shape of a sandy precipitate or, if only traces are present, in the shape of striæ on the sides of the glass where this has been rubbed by a glass rod.

For an account of its medicinal properties the reader should consult the recent report from India upon the experiments made there by order of Government with all the four cinchona-alkaloids, which experiments are very favourable to the therapeutical action of quinidine compared with that of quinine.

QUININE. $C_{20}H_{24}N_2O_2 + 3H_2O.$ Syn. QUINA, QUINIA. Till recently it was found in the greatest quantity in good Calisaya bark, particularly in that from Bolivia, but since it has been found in great quantity in some other barks, especially in the bark of Cinchona officinalis, for instance, in the bark of that species grown in Ceylon. Red bark contains not only quinine and cinchonine, but also cinchonidine.

Prep. 1. By precipitating a solution of sulphate of quinine with a slight excess of ammonia, potassa, or soda, and washing and drying the precipitate. By solution in alcohol, sp. gr. ·815, and spontaneous evaporation, it may be procured in crystals. Crystals may also be obtained from "its solution in hot water with a little ammonia." (*Liebig.*)

2. (Direct.) By adding hydrate of lime, in slight excess, to a strong decoction of the ground bark made with water acidulated with sulphuric acid, washing the precipitate which ensues, and boiling it in alcohol; the solution, filtered while hot, deposits the alkaloid on cooling.

Prop., &c. Quinine, when prepared by precipitation, is an amorphous white powder, but when this precipitate is left in the liquor it assumes, after some time, the appearance of aggregated crystalline needles; when slowly crystallised from its solution, these needles are remarkably fine, and of a pearly or silky lustre. It is freely soluble in rectified spirit and in ether, and of all the cinchona-alkaloids it is the most soluble in ammonia. It is upon this fact that Kerner's method for testing the purity of sulphate of quinine is founded. Its normal salts, if dissolved in water, have a slightly alkaline reaction upon red litmus paper. It is only sparingly soluble in water, even when boiling; both the fixed and volatile oils dissolve it with the aid of heat, more especially when it has been rendered anhydrous, or is presented to them under the form of an ethereal solution. It fuses by a gentle heat at 57° C. without decomposition; forms crystallisable salts, which are only slightly soluble in water, unless it be acidulated, and, like the pure alkaloid, are extremely bitter. It is precipitated by the alkalies and their carbonates, by tannic acid, and by most astringent substances.

Pur. See QUININE, SULPHATE OF, and QUINOMETRY (*below*).

Tests. Quinine is recognised by—1. Its appearance under the microscope. 2. Its solubility in ether, and in pure ammonia water. 3. Its solubility in concentrated nitric acid, forming a colourless liquid, which does not become yellowish until it is heated. 4. The solubility of itself and salts, when pure, in concentrated sulphuric acid, forming colourless fluids, "which do not acquire any coloration upon being heated to the point of incipient evaporation of the sulphuric acid, but which afterwards become yellow, and finally brown." (*Fresenius.*) 5. Its solubility in concentrated sulphuric acid to which some nitric acid has been added, forming a colourless, or, at the most, only a faintly yellowish liquid. 6. It is wholly destroyed by heat.

A solution of quinine in acidulated water, and solutions of its salts, exhibit the following reactions:—1. Ammonia, potassa, and the alkaline carbonates, give white, pulverulent precipitates, becoming crystalline after some time (see *above*), and which are soluble in ammonia in excess, and which, when ether is added after the ammonia, and the whole is agitated, redissolve in the ether, whilst the clear liquid, on repose, presents two distinct layers. 2. Bicarbonate of soda (avoiding excess) gives a similar precipitate, both in acid and neutral solutions of quinine, either at once or after a short time. The precipitate is soluble in excess of the precipitant, and is again precipitated from the new solution upon protracted

ebullition. "Vigorous stirring of the liquid promotes the separation of this precipitate." (*Fresenius.*) 3. If recently prepared chlorine be added to it, and then ammonia, a beautiful emerald-green colour is developed. 4. A concentrated solution of ferrocyanide of potassium being added in excess, after the chlorine, instead of the ammonia, a dark red colour is instantly produced, which after some time passes into green, especially when freely exposed to the light. This reaction is not characteristic of quinine, for with quinidine one gets the same reaction. 5. If caustic potassa be used instead of ammonia (see *above*), the solution acquires a sulphur-yellow colour. "These reactions are restricted to this alkaloid" (*Dr Garrod*).

Flückiger (in 'Jahrb. f. Pharm.,' April, 1872, 136; 'Ph. Journ.,' 3rd series, ii, 901) says:— "The most characteristic test for ascertaining the presence of quinine is the formation of the splendid green compound called *thalleioquin*, which is produced if solutions of the alkaloid or its salts are mixed with chlorine water, and then a drop of ammonia added."

If one part of quinine is dissolved in 4000 parts of acidulated water, and then about $\frac{1}{10}$th of the volume of the liquid, of chlorine water, and a drop of ammonia added, a green zone will be readily formed if the liquids are cautiously placed in a flask without shaking.

If the solution of quinine contain no more than $\frac{1}{5000}$, the green of one may still be obtained, but in more diluted solutions the success becomes more and more uncertain.

From a practical point of view we may state that $\frac{1}{5000}$ of the alkaloid is the smallest quantity whose presence can thus be discovered with certainty; Kerner (1870) has succeeded with $\frac{1}{50000}$, but I was not able to corroborate this statement.

The author was also induced to try the action of bromine in place of chlorine. The *thalleioquin* is then, indeed, produced in solution which contain only $\frac{1}{50000}$ of quinine. Yet the behaviour of bromine displays some striking differences. Chlorine alone, as already stated, causes no immediate alteration of somewhat diluted solutions of quinine, whereas they became turbid on addition of bromine as long as there is about $\frac{1}{150}$ or more of quinine present. Now, the precipitate which is produced by bromine in solution of quinine does *not* turn green if a little ammonia is subsequently added, or, at least, the thalleioquin thus obtained is rather greyish. But in more dilute solutions of quinine bromine acts more readily than chlorine. An excess of bromine is to be carefully avoided.

This is easily performed if the vapour of bromine, not the liquid bromine itself, is allowed to fall down on the surface of the solutions of quinine; their superficial layer only must be saturated with bromine by gently moving the liquid. Then a drop of ammonia will produce the green or somewhat bluish zone, which is much more persistent than that due to chlorine.

Consequently, for demonstration of the test under notice, chlorine is to be used in comparatively concentrated solutions. In solutions containing so little quinine (less than $\frac{1}{5000}$) that it is no longer precipitated by the vapour of bromine, the thalleioquin test succeeds much better with bromine, and goes much further, as shown above.

The author also shows that morphine gives a dark, dingy brown colour with chlorine and ammonia, which is capable of more or less masking the reaction of quinine.

Another test for quinine is the formation of its iodosulphate, the so-called herapathite. For this purpose the quinine is dissolved in 10 parts of proof spirit, acidulated with $\frac{1}{10}$ part of sulphuric acid, and to this solution an alcoholic solution of iodine is carefully added, and the liquid in the meanwhile stirred with a glass rod. There appears either immediately or after some minutes a black precipitate of iodosulphate of quinine, which if redissolved in boiling proof spirit, forms in cooling the beautiful crystals of herapathite. 100 parts of this herapathite, if dried on a water-bath, represent 56·6 parts of pure quinine.

Dr de Vrij prefers the employment of the iodosulphate of chiniodine as a reagent for the detection and estimation of quinine. In a communication to the 'Pharmaceutical Journal' he writes (3rd series, vi, 461):—"In estimating quinine in a mixture of cinchona-alkaloids by means of an alcoholic solution of iodine the reagent requires to be added in slight excess, in order to ensure complete precipitation. An undue excess of the reagent, however, causes the formation of a compound richer in iodine and much more soluble in alcohol than herapathite, and thus renders the determination inaccurate." For this reason the author suggests the application of an alcoholic solution of iodosulphate of chiniodine (so-called sulphate of amorphous quinine) in place of free iodine. The reagent is made as follows:

Two parts of sulphate of chiniodine are dissolved in 8 parts of water containing 5% of sulphuric acid. To this clear solution, contained in a large capsule, a solution of 1 part of iodine and 2 parts of iodide of potassium, in 100 parts of water, is slowly added with continuous stirring, so that no part of the solution of chiniodine comes into contact with excess of iodine. By this addition an orange-coloured flocculent precipitate is formed of iodosulphate of chiniodine, which either spontaneously, or by a slight elevation of temperature, collapses into a dark brown, red-coloured, resinous substance, whilst the supernatant liquor becomes clear and slightly yellow coloured. This liquor is poured off, and the resinous substance is washed by heating it on a water-bath with distilled water. After washing, the resinous substance is heated on the water-bath till all the water has been evaporated. It is then soft and tenacious at the temperature of boiling water, but becomes hard and brittle after cooling. One part of this substance is now treated with 6 parts of alcohol of 92% or 94% until it is completely dissolved, and the solution allowed to cool. In cooling, a part of the dissolved substance is separated. The clear dark-coloured solution is evaporated on a water-bath, and the residue dissolved in 5 parts of cold alcohol. This second solution leaves a small part of insoluble

substance. The clear dark-coloured solution obtained by the separation of this insoluble matter, either by decantation or filtration, constitutes the reagent which the author has used for some time under the name of iodosulphate of chiniodine, both for the qualitative and quantitative determination of crystallisable quinine.

To determine a quantity of quinine contained in the mixed alkaloids obtained from a sample of cinchona bark, 1 part of the alkaloid is dissolved in 20 parts of alcohol, of 90% or 92%, containing 1·6% of sulphuric acid, to obtain an alcoholic solution of the acid sulphates of the alkaloids.

From this solution the quinine is separated by adding carefully, by means of a pipette, the above-mentioned solution of the iodosulphate of chiniodine, as long as a dark brown-red precipitate of iodosulphate of quinine-herapathite is formed. As soon as all the quinine has been precipitated, and a slight excess of the reagent has been added, the liquor acquires an intense yellow colour. The beaker containing the liquor with the precipitate is now covered by a watch-glass, and heated on a water-bath till the liquid begins to boil.

After cooling, the beaker is weighed, to ascertain the amount of liquid which is necessary, in order to be able to apply later the above-mentioned correction. For although quinine-herapathite is very little soluble in alcohol, it is not insoluble, and therefore a correction must be applied for the quantity which has been dissolved both by the alcohol used for the solution of the alkaloids and the alcohol contained in the reagents.

The liquor is now filtered to collect the iodosulphate of quinine on a small filter, where it is washed with a saturated solution of herapathite in alcohol. After the washing has been completed the weight of the funnel with the moist filter is taken, and the filter allowed to dry in the funnel. As soon as it is dry the weight is taken again, to ascertain the amount of solution of herapathite which remained in the filter, and which left the dissolved herapathite on the filter after the evaporation of the alcohol.

This amount is subtracted from the total amount of liquid, and for the remaining the correction is calculated with reference to the temperature of the laboratory during the time of the analysis. The dry iodosulphate of quinine is taken from the filter and dried on a water-bath in one of a couple of large watch-glasses closing tightly upon each other, so that the weight of the substance contained in the glass may be taken without the access of air.

When, after repeatedly ascertaining the weight, it remains constant, this weight is noted down, and to it is added the product of the calculated correction. The sum of this addition is the total amount of iodosulphate of quinine obtained from the mixed alkaloids subjected to the operation, and from this weight the amount of crystallisable quinine can be calculated by the use of Hauer's formula, $2C_{40}H_{24}N_2O_4$ (HO_1SO_3), 31 (old notation), which the author has found to be correct. According to this formula, 1 part of iodosulphate of quinine, dried at 100° C., represents 0·5509%

of anhydrous quinine, or 0·7345% of disulphate of quinine.

The accuracy of this determination may be seen from the following examples :

0·24 grm. of anhydrous crystallised quinine gave 0·541 grm. of herapathite dried at 100° C. − 0·298 grm. of quinine.

According to Hauer's formula, 0·5336 grm. of herapathite = 0·294 grm. of quinine, which ought to have been obtained.

1·048 grm. of bitartrate of quinine gave 1·224 grm. of herapathite = 0·674 grm. of quinine.

According to the formula of the bitartrate, $C_{20}H_{24}N_2O_{21}C_4H_6O_6$ + Aq. = 442 ; 1·048 of bitartrate represents 0·69 of quinine, so that 1·255 grm. of herapathite should have been obtained.

Notwithstanding the different circumstances in which the reagent was applied, the results are satisfactory.

. The two following experiments were made with pure quinine, dried at 100° C., at which temperature it still retains water under identical circumstances :

1·0664 grm. of hydrated quinine gave 1·7266 grm. of herapathite = 164·5%.

1·055 grm. of the same hydrated quinine gave 1·7343 grm. of herapathite = 164·3%.

The author further states that the iodosulphate of quinine and of quinidine prepared by means of his new reagent have an analogous composition, and are identical with the compound described by Herapath, whilst the iodosulphates of cinchonine and cinchonidine have a different composition from the former, and both require more iodine to be converted into the optical iodosulphates described by Herapath. Of all these iodosulphates that of quinine is by far the most insoluble in alcohol, and is precipitated first and free from the others by a judicious application of the iodosulphate of chinioidine.

Quinine is distinguished from both cinchonine and quinidine by its comparatively free solubility in ether ; the last of these being very sparingly soluble, and the other wholly insoluble, in that menstruum. The presence of cinchonine may also be positively determined by reference to the behaviour of that alkaloid. Quinidine is also distinguished from quinine by the different crystallisation, greater specific gravity, and freer solubility of its salts in cold water. An extremely elegant and highly sensitive method of testing for quinine and quinidine by means of the microscope, &c., is described at considerable length by Dr Herapath, in the 'Pharm. Journ.' for November, 1853.

Estim. See QUINOMETRY.

Uses, &c. Pure quinine is but rarely used in medicine, but several of its salts are employed as remedies on account of their great stimulant, tonic, and febrifuge powers. As a tonic in dyspeptic affections, and for restoring strength and vigour to morbidly weakened habits, and as an antiperiodic or agent to counteract febrile action, it appears to be superior to all other remedies, provided no abnormal irritability of the mucous membranes or of the circulatory organs exists. The dose of the salts of quinine, as a tonic, is ½ to 1 gr., twice or thrice daily ; as an antiperiodic, 2 to 5 gr., or even more, every second or third

hour, during the intervals of the paroxysms of ague, and of other intermittent or periodic affections ; also in acute rheumatism. The sulphate (disulphate) is the salt generally used ; this and other salts are most effective when taken in solution.

The nature of the influence exerted upon blood by quinine was, in 1872, made the subject of a fresh investigation by Schulte. ' N. Rep. Pharm.,' xx. 539 ('Pharm. Journ.,' 3rd series, ii, 629). Its extraordinary power of stopping fermentation and putrefaction, by destroying low organisms, such as bacteria and fungi, has been before pointed out. It is supposed to diminish the formation of pus in inflammation by arresting the motions and preventing the exit from the blood-vessels of the white blood-corpuscles, the accumulation of which, according to Cohnheim, constitutes pus.

By depriving the red blood-corpuscles of the power to produce ozone, it diminishes the change of tissue in the body, and thereby lessens the production of heat. Ranke and Kerner have shown the waste of tissue is reduced when large doses of quinine are administered, as indicated in the small proportion of uric acid and urea excreted.

With the object of ascertaining whether this effect is referable to the direct influence of quinine on oxidation in the blood, or to its indirect influence through the nervous system, Schulte employed a method, based upon the changes occurring in the alkalinity of the blood, observed by Zuntz, who had noticed that a considerable formation of acid takes place in freshly-drawn blood, and continues in a less degree till putrefaction commences.

The amount of acid formed was estimated from the diminished alkalinity of the blood, as comparatively shown by the quantity of dilute phosphoric acid required for exact saturation.

A sufficient quantity of chloride of sodium was added to the phosphoric acid to prevent the blood-corpuscles from being dissolved, and interfering with the reaction by their colouring matter. The point of saturation was fixed at the point of transient reddening of carefully prepared test paper by the carbonic acid. Schulte has thus been enabled to confirm the experiments of Zuntz and Scharrenbroich, showing that quinine and berberine lessen the production of acid, and that quinine can stop it both before and after coagulation ; that sodium nicropicrate has an action similar to, and nearly as powerful as, quinine ; while the action of cinchonine is much less energetic. Harley has shown that whilst quinine lessens oxidation in blood, some substances, such as snake poisons, increase it. Binz found that when putrid fluids were injected into the circulation of an animal, the temperature could be more or less prevented by the addition of quinine to the putrid liquid, or by the simultaneous injection of the quinine.

With respect to the influences of quinine on the change of tissues, Schulte gives the result of some careful experiments made by Zuntz, who found that after taking three 0·6 grm. doses of hydrochlorate of quinine for two days the amount of urine he excreted was increased by one third, and then decreased as much, the specific gravity falling from 1018 to 1012; the urea also showed a marked decrease.

The salts of quinine may be made by simply saturating the dilute acids with the base, so that part of the latter remains undissolved, and gently evaporating the solutions for crystals or to dryness.

Quinine, Ac'etate of. *Syn.* QUINÆ ACETAS, L. *Prep.* 1. (P. Cod.) Mix quinine, 2 parts, with water, 8 parts; heat the mixture, and add of acetate acid, q. s. to dissolve the alkaloid, and to render the solution slightly acid; lastly, decant or filter the solution whilst boiling hot, and set it aside to crystallise. The mother water, on evaporation, will yield a second crop of the acetate.

2. Effloresced sulphate of quinine, 17 parts, is dissolved in boiling water, and mixed with crystallised acetate of soda, 6 parts. The acetate of quinine crystallises.

Prop., &c. Satiny, acicular crystals, which are rather more suitable in water than those of the sulphate.—*Dose,* ⅓ to 5 gr.

Quinine, Arse"niate of. *Syn.* QUINÆ ARSENIAS, L. *Prep.* (*Bouricres.*) Arsenic acid, 1½ dr.; quinine, 5 dr.; distilled water, 6 fl. oz.; boil them together in a covered glass vessel until the alkaloid is dissolved, then set the solution aside to crystallise.

Uses, &c. Recommended by Dr Neligan, and others, as being more powerfully antiperiodic than the other preparations of quinine.—*Dose,* 1⁄12 to ¼ gr., made into pills; in agues, neuralgia, &c.; also in cancer.

Quinine, Ar'senite of. *Syn.* QUINÆ ARSENIS, L. *Prep.* Sulphate of quinine, 100 parts, is dissolved in alcohol, 600 parts, and boiled with arsenious acid, 14 parts. The liquid is then filtered. The poisonous salt is deposited in the crystalline form as the liquid cools.

Uses, &c. As the last.

Quinine, Chlo"ride of. Hydrochlorate of quinine (see *below*).

Quinine, Ci'trate of. *Syn.* QUINÆ CITRAS, L. *Prep.* 1. By mixing a hot solution of sulphate of quinine with a like solution of citrate of soda.

2. From quinine and citric acid, as the acetate. Needle-shaped prisms.—*Dose, &c.* As the sulphate or disulphate.

Quinine, Disulphate of. Sulphate of quinine (see *below*).

Quinine, Ferrocy'anide of. *Syn.* CYANIDE OF IRON AND QUININE; QUINÆ HYDROFERROCYANAS, QUINÆ FERRO-PRUSSIAS, L. *Prep.* (P. Cod.) Sulphate of quinine, 100 parts; ferrocyanide of potassium, 31 parts; distilled water, 5000 parts; boil for a few minutes, and, when cold, separate the impure salt which floats as an oily mass on the surface, wash it with a little cold water, and dissolve it in boiling alcohol; the solution will deposit crystals as it cools.

Obs. This compound is by some said to be the most efficacious of all the salts of quinia. Pelouze asserts that it is simply quinine mixed with some Prussian blue.—*Dose,* 1 to 6 gr.

Quinine, Ferrosul'phate of. See QUININE AND IRON, SULPHATE OF (*below*).

Quinine, Hydri'odate of. *Syn.* IODIDE OF QUININE; QUINÆ HYDRIODAS, Ǫ. IODIDUM, L.

Prep. 1. By adding, drop by drop, a concentrated solution of iodide of potassium to a like solution of acid sulphate of quinine, and drying the precipitate in the shade; or heat the liquid nearly to the boiling point, and allow it to crystallise.

2. (*Parrish.*) Effloresced sulphate of quinine, 5 parts, dissolved in alcohol, and decomposed by an alcoholic solution of 8 parts of iodide of potassium, precipitates sulphate of potassa, and yields, on cooling and evaporating, hydriodate of quinine in fine crystalline needles ("1 and 2 are not identical; 1 is an acid salt which readily crystallises, but 2 is a normal salt which I never saw crystallise but always like a fluid resin, quite amorphous."—(*De Vrij.*)

3. (IODURETTED — *Bouchardat.*) From an acid solution of quinia and a solution of iodide of iron, containing a slight excess of iron, as No. 1.

Obs. The above are reputed alterative, tonic, and antiperiodic.—*Dose,* 1 to 4 gr., in obstinate intermittents, and in the scrofulous affections of debilitated subjects.

Quinine, Hydrochlo"rate of. *Syn.* CHLORIDE OF QUININE, MURIATE OF QUININE†; QUINÆ HYDROCHLORAS, QUINÆ MURIAS, L. *Prep.* 1. By neutralising dilute hydrochloric acid with the base, as above.

2. (Ph. Bor.) Chloride of barium, 5 dr.; boiling water, 1 lb.; dissolve, add, gradually, of sulphate of quinine, 2 oz.; boil gently for a few minutes, filter the solution whilst hot, and set it aside that crystals may form.

3. (QUINÆ MURIAS—Ph. D.) Dissolve chloride of barium, 123 gr. in distilled water, 2 fl. oz.; add of sulphate of quinine, 1 oz., dissolved in boiling water, 1½ pint; mix, evaporate the solution to one half, filter, and again evaporate until spiculæ begin to appear; next allow the liquid to cool, collect the crystals, and dry them on bibulous paper. The mother liquor, by further concentration and cooling, will yield an additional product.

Obs. Hydrochlorate of quinine occurs in snow-white groups of feathery crystals, of a mother-of-pearl lustre, which are soluble in about 34 parts water. On account of its free solubility in water, without use of acid, it is preferred for eye lotions and antiseptic injections to bladders.

Quinine and Iodide of Iron. *Syn.* QUINÆ ET FERRI IODIDUM. (*Bouchardat.*) *Prep.* Pour a strong solution of acid sulphate of quinine into a fresh solution of iodide of iron; collect the precipitate, dry it quickly by pressing it between blotting paper, and keep it from the air.

Quinine, Ki'nate of. *Syn.* QUINÆ KINAS, L. *Prep.* By saturating a solution of kinic acid with quinine, and purifying by crystallisation out of alcohol. The kinate of quinine is obtained in crystalline warts, soluble in 4 parts of water and 8 parts of alcohol.

Quinine, Lac'tate of. *Syn.* QUINÆ LACTAS, L. *Prep.* As the ACETATE or CITRATE. By spontaneous evaporation fine crystals may be obtained. Said to agree better with dyspeptic patients than the other salts of quinine.

Quinine, Mu"riate of. Hydrochlorate of quinine (see *above*).

Quinine, Neutral Hydrobromate of. *Syn.* QUINÆ HYDROBROMAS. (*M. Boille.*) This salt is prepared by double decomposition of bromide of barium and neutral sulphate of quinia, and is thus easily obtained pure and free from chloride, the great solubility of bromide of barium in alcohol facilitating the removal of any chloride which is soluble.

The two salts are dissolved separately in alcohol and the solution filtered. The neutral sulphate of quinia solution is gradually added, in slight excess to the bromide of barium solution until a precipitate ceases to form,

The solutions, diluted with water, are distilled to recover the alcohol, afterwards filtered to separate the sulphate of quinia which has been precipitated by the water, and then concentrated sufficiently to induce rapid crystallisation. The addition of water is indispensable for the concentration and crystallisation; the hydrobromate, being soluble in alcohol of all proportions, redissolves as the alcoholic liquor is concentrated. M. Boille claims for his neutral hydrobromate of quinine its much readier solubility over the officinal sulphate, as well as its superior richness in quinine.

Quinine, Ni'trate of. *Syn.* QUINÆ NITRAS, L. *Prep.* As the HYDROCHLORATE, substituting dilute nitric acid, or nitrate of baryta (P. Cod.), for hydrochloric acid or chloride of barium.

Quinine, Phos'phate of. *Syn.* QUINÆ PHOSPHAS, L. As the ACETATE. Silky, needle-shaped crystals, with a pearly lustre. It has been highly recommended in intermittents, &c., associated with rickets and stomach affections.

Quinine, Salicylate of. *Syn.* QUINÆ SALICYLAS. This may be made by mixing an alcoholic solution of quinine with an alcoholic solution of salicylic acid to complete saturation, and afterwards allowing the alcohol slowly to evaporate.

Quinine, Sul'phates of. The salt often called 'disulphate of quinine' is now generally regarded as the normal sulphate, while the soluble salt, often called the 'neutral sulphate,' is considered to be an acid salt. This change in nomenclature results from doubling the atomic weight of the alkaloid quinine :

i. **Quinine, Acid Sulphate of.** $(C_{20}P_{24}N_2O_2, H_2SO_4. 7Aq.)$ *Syn.* SULPHATE OF QUININE†. NEUTRAL SULPHATE OF QUININE†, SOLUBLE S. OF Q.; QUINÆ SULPHAS SOLUBILIS, L. *Prep.* From sulphate of quinine, 1 oz., dissolved by the aid of heat, in water, ½ pint, previously acidulated with dilute sulphuric acid, 5 fl. dr. ; the solution affords crystals on cooling, and more on evaporation.

Obs. This salt possesses the advantage of being soluble in about 10 parts of water at 60° F.; but it is seldom used in the crystalline form; still, as the officinal sulphate ('disulphate') is generally prescribed along with a small quantity of dilute sulphuric acid to render it soluble, this acid sulphate is, in fact, the compound which is commonly given. It is the 'bisulphate,' 'supersulphate,' or 'acid sulphate of quinia' of Soubeiran and other Continental chemists.

ii. **Quinine, Sulphate of.** $([C_{20}H_{24}N_2O_2]_2H_2SO_4)_x$

$15H_2O$. *Syn.* NORMAL SULPHATE OF QUININE, DISULPHATE OF Q., QUININE ; QUINÆ DISULPHAS (Ph. L.), QUINÆ SULPHAS (Ph. E. D., & U. S., & P. Cod.), QUINÆ SULPHAS (B. P.), L ; SULPHATE DE QUININE, Fr. *Prep.* 1. (Ph. L. 1836.) Take of yellow cinchona bark, bruised, 7 lbs.; sulphuric acid, 4½ oz.; (diluted with) water, 6 galls.; boil them for 1 hour, and strain ; repeat this a second time for 1 hour, with a like quantity of acid and water, and again strain ; next boil the bark for 3 hours, in water, 8 galls., and strain; wash the residue with fresh quantities of boiling water; to the mixed decoctions and washings, add moist hydrated oxide of lead to saturation, decant the supernatant fluid, and wash the sediment with distilled water; boil down the liquor for 15 minutes, and strain, then precipitate the qnina with liquor of ammonia and wash the precipitate (with very cold water) until nothing alkaline is perceptible; saturate what remains with sulphuric acid, ½ oz , diluted with water, q. s.; digest with animal charcoal, 2 oz., and strain ; lastly, the charcoal being well washed, evaporate the mixed liquors, that crystals may form.

2. (Ph. E.) This process varies from the last one, in the bark (1 lb.) being first boiled in water (4 pints) along with carbonate of soda (4 oz.) ; the residuum, being pressed, is moistened with water, and again pressed, and this operation is repeated a second and a third time, the object being to remove, as much as possible, the acids, colouring matter, gum, and extractive, before proceeding to extract the alkaloid. Carbonate of soda is also used as the precipitant, instead of ammonia, and the precipitate is formed into a sulphate (disulphate) by being stirred with boiling water, 1 pint, to which sulphuric acid, 1 fl. scruple, or q. s., is subsequently added. The crystals, after digestion with prepared animal charcoal, 2 dr., are ordered to be dried at a heat not higher than 140° F.

3. (Ph. D.) Yellow bark, 1 lb., is macerated for 24 hours in water, 2 quarts, acidulated with oil of vitriol, 2 fl. dr. ; and then boiled for half an hour, after which the fluid is decanted; this is repeated a second and a third time with water, 2 quarts, and oil of vitriol, 1 fl. dr.; the decanted (or strained) liquors are evaporated to a quart, and filtered, and slaked lime, 1 oz., or q. s., added to the solution until it exhibits a decidedly alkaline reaction ; the precipitate is next collected on a calico filter, and, after having been washed with cold water, partially dried on porous bricks, and subjected to powerful pressure enveloped in blotting-paper, is boiled for 20 minutes in rectified spirit, 1 pint, and the liquid, after subsidence, decanted ; this is repeated a second and a third time with a fresh pint of spirit, and the residuum being well pressed, the mixed liquors are filtered and the spirit removed by distillation ; the brown viscid residuum is dissolved in boiling water, 16 fl. oz., boiled, and dilute sulphuric acid, ½ fl. oz., or q. s., added to render the solution neutral or only slightly acid ; animal charcoal, ½ oz., is next stirred in, the mixture boiled for about 5 minutes, filtered and set aside to crystallise ; the crystals are dried on blotting paper by mere exposure to a dry atmosphere.

4. (B. P.) Yellow cinchona bark, in coarse powder, 16 parts; hydrochloric acid, 8 parts; distilled water, a sufficiency; solution of soda, 80 parts; dilute sulphuric acid, a sufficiency. Dilute the hydrochloric acid with 10 pints of the water. Place the bark in a porcelain basin, and add to it as much of the diluted hydrochloric acid as will render it thoroughly moist. After maceration with occasional stirring, for 24 hours, place the bark in a displacement apparatus, and percolate with the diluted hydrochloric acid until the solution which drops through is nearly destitute of bitter taste. Into this liquid (hydrochlorate of quinine) pour the solution of soda, agitate well, let the precipitate (quinine) completely subside, decant the supernatant fluid, collect the precipitate on a filter, and wash it with cold distilled water until the washings cease to have colour. Transfer the precipitate to a porcelain dish containing a pint of distilled water, and, applying to this the heat of a water-bath, gradually add diluted sulphuric acid until very nearly the whole of the precipitate has been dissolved, and a neutral liquid has been obtained. (Or add about half the precipitated quinine to some water in an evaporating basin, warm the mixture and pour in diluted sulphuric acid until the precipitate has dissolved and the liquid is neutral or only faintly acid, then add the other half, stir well, and again heat liquid.) Filter the solution (sulphate of quinine), while hot, through paper, wash the filter with boiling distilled water, concentrate till a film forms on the surface of the solution, and set it aside to crystallise. The crystals should be dried on filtering paper without the application of heat.

5. Those who are well acquainted with the chemistry of the cinchona-alkaloids all agree with me in condemning the boiling of bark with dilute acids. I prefer the following method, which can also be used on a small scale for quinometry:

" Yellow bark, or any other bark in which quinine prevails, like, for instance, that of *Cinchona officinalis*, 1 lb., is mixed with milk of lime, made from 4 oz. of lime and 40 oz. of water. After drying this mixture it is exhausted with strong methylated spirit (the strongest possible) and the slightly coloured solution neutralized with sulphuric acid, so that the liquor has a slight acid reaction upon blue litmus paper. After filtering or subsiding, the clear liquid is distilled and the residue in the still dissolved in water, carefully neutralised, so that the solution has a slight alkaline reaction upon red litmus paper, treated with charcoal and crystallised, &c." (*De Vrij*).

(B. P. 1885.) The outlines only of a process is given as follows: " The sulphate of an alkaloid prepared from the powder of various kinds of cinchona and pemijia bark, by extraction with spirit after the addition of lime, or by the action of alkali on an acidulated aqueous infusion, with subsequent neutralisation of the alkaloid by sulphuric acid and purification of the resulting salt."

Prop. When pure, sulphate of quinine forms very light, delicate, flexible, white needles, which are efflorescent, inodorous, and intensely bitter; it is soluble in 740 parts of water at 60°, and in 30 parts at 212° F.; it takes about 60 parts of cold rectified spirit for its solution, but is freely soluble in boiling alcohol and in acidulated water; it melts at 240° F., and is charred and destroyed at a heat below that of redness. The crystals contain 76·1% of quinine, 8·7% of sulphuric acid, and 15·2% of water; of the last, they lose about three fourths by exposure to dry air, and nearly the whole when kept in a state of fusion for some time.

Pur. This may not be inferred from the form of its crystallisation, for the sulphates of quinidine and of cinchonidine may be obtained in the same form of crystallisation. As mentioned already, the reaction with chlorine and ammonia does not distinguish quinine from quinidine, as both give the same green colour. " It is entirely soluble in water (hot), and more readily so when an acid is present. Precipitated by ammonia, the residuary liquid, after evaporation, should not taste of sugar. By a gentle heat it loses 8% or 10% of water. It is wholly consumed by heat. If chlorine be first added, and then ammonia, it becomes green." Its solution in sulphuric acid gives with ammonia in excess a white precipitate of quinine soluble in ether and in large excess of ammonia. 25 gr. of the freshly made salts should lose 3·8 gr. of water by drying at 212° F. (100° C.). Ignited, with free access of air, it burns without leaving a residue.

Test for Cinchonidine and Cinchonine (B. P.). Heat 100 gr. of the sulphate of quinine in 5 or 6 oz. boiling water, with 3 or 4 drops of sulphuric acid. Set the solution aside until cold. Separate by filtration the purified sulphate of quinine which has crystallised out. To the filtrate, which should nearly fill a bottle or flask, add ether, shaking occasionally, until a distinct layer of ether remains undissolved. Add ammonia in very slight excess, and shake thoroughly, so that the quinine at first precipitated shall be redissolved. Set aside for some hours or during a night. Remove the supernatant, clear, ethereal fluid, which should occupy the neck of the vessel, by a pipette. Wash the residual aqueous fluid and any separated crystals of alkaloid, with a very little more ether, once or twice. Collect the separated alkaloid on a tared filter, wash it with a little ether, dry at 100° C., and weigh. 4 parts of such alkaloid correspond to 5 parts of crystallised sulphate of cinchonidine or of sulphate of cinchonine.

Test for Quinidine (B. P.). Recrystallise 50 gr. of the original sulphate of quinine as described in the previous paragraph. To the filtrate add solution of iodide of potassium and a little spirit of wine, to prevent the precipitation of amorphous hydriodates. Collect any separated hydriodate of quinidine, wash with a little water, dry, and weigh. The weight represents an equal weight of crystallised sulphate of quinidine.

Test for Cuprcine (B. P.). Shake the recrystallised sulphate of quinine obtained in testing the original sulphate of quinine for cinchonidine and cinchonine with 1 oz. of ether and ¼ oz. solution of ammonia, and to this ethereal solution separated, add the ethereal fluid and washings also obtained in testing the original sulphate for the two alkaloids just mentioned. Shake this ethereal liquor with ¾ fl. oz. of a 10% solution of caustic

soda, adding water if any solid matter separates.
Remove the ethereal solution, wash the aqueous
solution with more ether, and remove the ethereal
washings; add dilute sulphuric acid to the
aqueous fluid heated to boiling until the soda is
exactly neutralised. When cold, collect any sulphate of cupreine that has crystallised out on a
tared filter, dry, and weigh.
' Sulphate of quinine' should not contain much
more than 5% of sulphates of other cinchona
alkaloids.

Adult. Sulphate of quinine is said to be often
adulterated with starch, magnesia, gum, sugar,
cinchonine, quinidine, &c.; but, according to De
Vrij, those with starch, magnesia, gum, and sugar,
are very rare if ever they were really observed.
Very frequent are those with the sulphates of
the other cinchona-alkaloids, and these become
even still more frequent, as very different kinds
of bark are used for the manufacture of quinine.
Salicin is, if ever, but very seldom used for adulteration of quinine. The best practical test for
the purity of sulphate of quinine is the following:
—A saturated solution of the salt is made at 60°
F., and 1 part of this solution is mixed with 2 or
3 minims of a concentrated solution of iodide of
potassium, whilst another part is mixed with 2 or
3 minims of a concentrated solution of tartrate of
potash and soda. If the sulphate of quinine is pure
its solution will remain unaltered by both reagents,
even after rubbing the sides of the test tube with
a glass rod and standing many hours. But if it
contains one or more of the other cinchona-alkaloids there will appear either precipitates or striæ
on the glass where it has been rubbed by the glass
rod. Iodide of potassium indicates particularly
the presence of even traces of quinidine, but also
of cinchonidine and cinchonine, provided their
quantity be not too small. Tartarate of potash
and soda indicate, under these circumstances, only
the presence of cinchonidine. The first three remain undissolved when the salt is digested in
spirit; the fourth is dissolved out by cold water;
the fifth may be detected by its total insolubility
in ether; or by precipitating the quinine by solution of potassa, and dissolving the precipitate in
boiling alcohol; cinchonine crystallises out as the
solution cools, but the quinine remains in the
mother-liquor; and the last, by the greater solubility and sp. gr. of the salt, &c.

Uses, &c. The sulphate is more extensively
employed than any of the other salts of quinine,
and, indeed, to almost the exclusion of them. It
is the article intended to be used whenever 'sulphate' or 'disulphate' of quinine, or even 'quinine,' is ordered for medicinal purposes, unless
the name is qualified by some other term. It is
a most valuable stomachic, in doses of ¼ to 1
gr.; as a tonic, 1 to 3 gr.; and as a febrifuge, 2
to 20 gr.

Quinine, Sulpho-tar′trate of. *Syn.* QUINÆ
SULPHO-TARTRAS, L. *Prep.* From sulphate of
quinine, 4 parts; tartaric acid, 5 parts; distilled
water, 20 parts; mix, gently evaporate to dryness,
and powder the residuum.

Quinine, Tan′nate of. *Syn.* QUINÆ TANNAS,
L. *Prep.* Dissolve sulphate of quinine in
slightly acidulated water, and add a solution of
tannic acid as long as a precipitate forms; wash

this with a little cold water, and dry it. The
Ph. Græca orders infusion of galls to be used as
the precipitant. In intermittent neuralgia.

Quinine, Tar′trate of. *Syn.* QUINÆ TARTRAS,
L. *Prep.* (P. Cod.) From tartaric acid and
quinine, as the acetate.

Quinine, Vale″rianate of. *Syn.* QUINÆ VALE-
RIANAS (Ph. D.), L. *Prep.* 1. As the acetate
or citrate.

2. (Ph. D.) Valerianate of soda, 124 gr.; distilled water, 2 fl. oz.; dissolve; also dissolve
hydrochlorate of quinine, 7 dr., in distilled water,
14 fl. oz.; next heat each solution to 120° (not
higher), mix them, and set the vessel aside for 24
hours; lastly, press the mass of crystals thus obtained, and dry them, without the application of
artificial heat.

Prop., &c. Silky needles and prisms; its solution suffers decomposition when heated much
above 120° F. It is powerfully antispasmodic,
antiperiodic, and nervine.—*Dose,* ½ gr. every 2
hours, or 1 to 3 gr. twice or thrice daily; in epilepsy, hemicrania, hysteria, neuralgia, and other
nervous affections.

QUININE AND COD-LIVER OIL. *Syn.* COD-
LIVER OIL WITH QUININE, QUINIARETTED COD-
LIVER OIL; OLEUM MORRHUÆ CUM QUINÂ,
OLEUM JECORIS ASELLI CUM QUINÂ, L. This
medicine is a solution of pure anhydrous quinine
in pure cod-liver oil.

Prep. 1. Pure quinine (preferably recently
precipitated) is fused in a glass or porcelain capsule by the heat of an oil or sand bath, carefully
applied, by which it assumes a brown colour and
the appearance of a resin; it is then allowed to
cool out of contact with the air, after which it is
reduced to powder in a dry mortar, and added to
pure pale Newfoundland cod-liver oil, gently
heated in a closed glass vessel over a water-bath;
the solution of the alkaloid is promoted by constant agitation, and, when complete, the vessel,
still corked, is set aside in a dark situation to
cool; when the 'quiniaretted oil' is quite cold it
is put into bottles, in the usual manner, and preserved as much as possible from the light and air.

2. The anhydrous quinine is dissolved in a
little anhydrous ether before adding it to the oil,
which in this case need not be heated, as the
union is effected by simple agitation; should this
not take place, it may be gently warmed for a
few minutes.

3. The anhydrous quinine is dissolved in anhydrous alcohol, and after being added to the oil,
the whole is gently heated, in an open vessel, by
the heat of a water-bath, until the alcohol is expelled; agitation, &c., being had recourse to as in
No. 1.

Prop., &c. The above preparation resembles
ordinary cod-liver oil, except in having a pale
yellowish colour and a slightly bitter taste,
similar to that of cinchona bark. It is said to
possess all the properties of cod-liver oil combined with those peculiar to quinine, by which
the tonic, stomachic, and antiperiodic qualities of
the latter are associated, in one remedy, with the
genial supporting, and alterative action of the
other. The common strength is 2 gr. of quinine
per oz.

QUININE AND IRON. These two important

medicinal agents are combined together in various ways. The following compound salts are often prescribed :

Quinine and Iron, Cit'rate of. *Syn.* CITRATE OF IRON AND QUININE; FERRI ET QUINÆ CITRAS (B. P.), L. *Prep.* 1. (B. P.) Solution of persulphate of iron, 4½ parts; sulphate of quinia, 1 part; dilute sulphuric acid, 1½ parts; citric acid, 3 parts; solution of ammonia and distilled water, of each a sufficiency; mix 8 parts of the solution of ammonia with 40 parts of the water, and to this add the solution of persulphate of iron, previously diluted with 40 parts of the water, stirring them constantly and briskly. Let the mixture stand for 2 hours, stirring it occasionally, then put it on a calico filter, and when the liquid has drained away, wash the precipitate with distilled water until that which passes through the filter ceases to give a precipitate with chloride of barium. Mix the sulphate of quinia with 8 parts of the water, add the sulphuric acid, and when the salt is dissolved, precipitate the quinia with a slight excess of solution of ammonia. Collect the precipitate on a filter, and wash it with 30 parts of the water. Dissolve the citric acid in 5 parts of the water, and having applied the heat of a water-bath, add the oxide of iron, previously well drained; stir them together, and when the oxide has dissolved, add the precipitated quinia, continuing the agitation until this also has dissolved. Let the solution cool, then add, in small quantities at a time, 1½ parts solution of ammonia, dilute with 2 parts of the water, stirring the solution briskly, and allowing the quinia which separates with each addition of ammonia to dissolve before the next addition is made. Filter the solution, evaporate it to the consistence of a thin syrup, then dry it in layers on flat porcelain or glass plates, at the temperature of 100° F., remove the dry salt in flakes, and keep it in a stoppered bottle. Solubility, 2 in 1.—*Test.* Taste bitter as well as chalybeate. When burned with exposure to air, it leaves a residue (oxide of iron) which yields nothing to water. 50 gr., dissolved in an ounce of water, and treated with a slight excess of ammonia, gives a white precipitate (quinia) which, when collected on a filter and dried, weighs 8 gr. The precipitate is entirely soluble in pure ether, indicating absence of quinidia and cinchonia. When burned it leaves no residue. When dissolved by the aid of an acid it forms a solution which, after decolorisation by a little purified animal charcoal, turns the plane of polarisation strongly to the left (cinchona turns it to the right).—*Dose*, 5 to 10 gr. as a tonic, three times a day, in solution or in pill.

2. (Ph. U. S.) Triturate sulphate of quinine, 1 oz., with distilled water, 6 fl. oz., and having added sufficient diluted sulphuric acid to dissolve it, cautiously pour into the solution water of ammonia with constant stirring, until in slight excess. Wash the precipitated quinine on a filter, and having added solution of citrate of iron, 10 fl. oz., keep the whole at a temperature of 120° by means of a water-bath, and stir constantly until the alkaloid is dissolved. Lastly, evaporate the solution to the consistence of a syrup, and spread it on plates of glass, so that, on drying,

the salt may be obtained in scales.—*Dose*, 2 gr. to 5 gr.

Quinine and Iron, I'odide of. *Syn.* QUINÆ ET FERRI IODIDUM, L. *Prep.* From protiodide of iron, 2 parts; hydriodate of quinine, 1 part; rectified spirit, 12 parts; heat them together, and either evaporate to dryness or crystallise by refrigeration. A powder or crystalline scales.

Quinine and Iron, Sul'phate of. *Syn.* FERROSULPHATE OF QUININE; QUINÆ FERRO-SULPHAS, QUINÆ ET FERRI SULPHAS, L. *Prep.* From solutions of the sulphates of iron and quinine, in atomic proportions, mixed whilst hot, and the crystals which form as the liquid cools carefully dried and preserved from the air.

QUININE AND MERCURY. See MERCURIC AND QUININE CHLORIDE.

QUINOA. The seed of this plant (a species of *Chenopodium*) is largely consumed by the people who dwell in the elevated regions of Chili and Peru, in which countries it is found growing at a height of some 13,000 feet above the sea-level. Mr Johnston says there are two varieties of it, a sweet and a bitter one. It is a highly nutritious cereal, resembling ointment in properties. According to Voelker, quinoa has the following composition :

	Quinoa seeds dried at 212° F.	Quinoa flour.
Nitrogenous matter	22·86	19
Starch	56·80	60
Fatty matter	5·74	5
Vegetable fibre	9·53	—
Ash	5·05	—
Water	—	16

QUINOID'INE. *Syn.* AMORPHOUS QUININE, CHINOIDINE; QUINAAMORPHA, QUINA INFORMIS, QUINOIDA, QUINOIDINA, QUINOIDINUM, CHINOIDEUM, L. A few years after the discovery of the quinine by Pelletier and Caventou, Sertuerner, a German physician, obtained, by a peculiar method, from yellow bark, an amorphous alkaloid which was called by him Chinoidin (Sertuerner, ' Die neusten Entdeckungen in der Physik, Heilkunde, und Chemie,' 3ter Band, 2tes Heft, Seite 269 [1830]) (to which the name amorphous quinine is improperly given), and also fever-killer (Fiebertödter). He found that not only this alkaloid itself, but also all its compounds with acids, were equally amorphous. As recent investigations have proved that this amorphous alkaloid occurs in all cinchona barks, and is found particularly in many young Indian barks in great quantity, it is quite natural that in the manufacture of quinine the uncrystallisable sulphate of this alkaloid should accumulate in the mother liquors of the sulphate of quinine. From such liquors it is precipitated in an impure state by an alkali, and brought into commerce under the name of quinoidine. As this amorphous alkaloid has the property of preventing the crystallisation of the salts of the other alkaloids, particularly those of quinidine, it is clear that the quinoidine of commerce very often contains quinidine and also cinchonidine. Dr. de Vrij, for instance, found sometimes more than 20% of quinidine in some samples of quinoidine of commerce. Although commercial quinoidine contains many impurities which may have their origin partly in the adulteration of the cinchona-alkaloids, unadulterated quinoidine, no doubt, chiefly consists of the amorphous alkaloid discovered by Sertuerner.

The quinoidine of commerce ought never to be used in medicine unless purified. For this purification there are two methods: 1. That of Mr Bullock, which gives the purer but the more expensive product. Crude quinoidine is exhausted with ether, which, after defecation, is distilled off, leaving the purified quinoidine behind. This process has been patented in England by Mr Bullock. 2. That of Dr de ^{Vrij}, which consists in boiling 9 parts of crude quinoidine with a solution of 2 parts of oxalate of ammonium in water. By this process the alkaloids contained in the quinoidine are dissolved whilst the impurities, and amongst them the lime which is often contained in the crude quinoidine, remain undissolved. The solution is mixed with a large bulk of water, then filtered and the purified quinoidine precipitated by a slight excess of liquor of soda.

Prop., &c. In its crude form quinoidine somewhat resembles aloes; in its purest state it is a yellowish-brown resin-like mass, freely soluble in alcohol and ether, but nearly insoluble in water; with the acids it forms dark-coloured, uncrystallisable salts. It is powerfully febrifuge, but less so than either quinoidine or quinine, although it is identical in chemical composition with both of them.—*Dose*, 2 to 4 gr. for adults, ¼ to 1 gr. for children, given in wine, lemonade, or acidulated honey.

QUINOM'ETRY. *Syn.* CINCHONOMETRY. The art of estimating the quantity of quinine in cinchona bark, and in the commercial salts of this alkaloid. In addition to the following, other processes will be found under CINCHONA and QUININE SULPHATE.

Proc. 1. For BARK. *a.* (Ph. E.) A filtered decoction of 100 gr. of bark, in distilled water, 2 fl. oz., is precipitated with 1 fl. dr., or q. s. of a concentrated solution of carbonate of soda; the precipitate, after being heated in the fluid, so as to become a fused mass, and having again become cold, is dried and weighed. " It should be 2 gr. or more, and entirely dissolve in a solution of oxalic acid." To render the result strictly accurate, the product should be dissolved in 10 parts of proof spirit, containing $\frac{1}{10}$ of sulphuric acid, and to this solution carefully added an alcoholic solution of iodine as long as there appears a brown precipitate, which immediately turns black by stirring with a glass rod. This precipitate, collected upon a filter, washed with strong alcohol and dried on a water-bath, is Herapath's iodosulphate of quinine, of which 100 parts represent 56·5 parts of pure quinine.

b. (*Rebourdain.*) 100 gr. of the bark (coarsely powdered) are exhausted with acidulated water, and the filtered solution rendered alkaline with solution of potassa; it is next shaken with about one third of its volume of chloroform, and then allowed to repose for a short time; the chloroform holding the alkaloid in solution sinks to the bottom of the vessel in a distinct stratum, from which the supernatant liquid is separated by decantation; the chloroformic solution, either at once or after being washed with a little cold water, is allowed to evaporate; the residuum,

when weighed, gives the percentage richness of the sample.

Obs. A like result may be obtained with ether instead of chloroform, in which case the solution of quinine will form the upper stratum.

c. Instead of Rebourdain's process, Dr de Vrij prefers that of Charles ('Journal de Pharmacie et de Chimie,' 4e série, t. 12, p. 81, Août, 1870). so far as regards the separation of the total mixed alkaloids from the bark. To this mixture is applied the process mentioned above (*a*), viz. solution in acidulated proof spirit, &c.

2. For the SALTS. The above methods, as well as several others which have been devised for the purpose, may also be applied to the salts of quinine; but, unfortunately, they are inapplicable when great accuracy is required, owing to the non-recognition of the presence of quinidine as such, and which, consequently, goes to swell the apparent richness of the sample in quinine. The following ingenious method, invented by Dr Ure, not merely enables us to detect the presence of cinchonine and quinidine in commercial samples of the salts of quinine, but, with some trifling modifications, it also enables us to determine the quantity of each of these alkaloids present in any sample:—"10 gr. of the salts to be examined " (the sulphate is here more especially referred to) " is put into a strong test-tube furnished with a tight-fitting cork; to this are to be added 10 drops of dilute sulphuric acid (1 acid and 5 water), with 15 drops of water, and a gentle heat applied to accelerate solution. This having been effected, and the solution entirely cooled, 60 drops of officinal sulphuric ether, with 20 drops of liquor of ammonia, must be added, and the whole well shaken while the top is closed by the thumb. The tube is then to be closely stopped, and shaken gently from time to time, so that the bubbles of air may readily enter the layer of ether. If the salt be free from cinchonine and quinidine, or contain the latter in no greater proportion than 10%, it will be completely dissolved; while on the surface, where contact of the two layers of clear fluid takes place, the mechanical impurities only will be separated. After some time the layer of ether becomes hard and gelatinous, and no further observation is possible."

" From the above statement respecting the solubility of quinidine in ether, it appears that the 10 gr. of the salt examined may contain 1 gr. of quinidine, and still a complete solution with ether and ammonia may follow; but in this case the quinidine will shortly begin to crystallise in a layer of ether. The least trace of quinidine may be yet more definitely detected by employing, instead of the ordinary ether, some ether previously saturated with quinidine, by which means all of the quinidine contained in the quinine examined must remain undissolved. It is particularly requisite, in performing this last experiment, to observe (immediately), after the shaking, whether all has dissolved; for, owing to the great tendency of quinidine to crystallisation, it may become again separated in a crystalline form, and be a source of error."

" If more than 1-10th of quinidine or (any) cinchonine be present, there will be found an insoluble precipitate at the limits of the two layers of fluid. If this be quinidine it will be dissolved on the addition of proportionately more ether, while cinchonine will remain unaffected."

Note. To Dr Ure's test Dr de Vrij prefers, for several reasons, Dr Kerner's test, ' Zeitschrift für Analytische Chemie,' von Fresenius, 1st Jahrg., 1862; 'Ueber Die Prufung des Käuflichen Schwefelsauren Chinins auf fremde Alkaloides,' von Dr G. Kerner.

QUINOVIC ACID. $C_{24}H_{36}O_4$. This is insoluble in water, also in chloroform, and soluble with difficulty in alcohol. It can be obtained from the boiling alcoholic solution, by cooling in small crystals. In the leaves, bark, and wood of the cinchona tree this acid is contained, together with quinovin, and it is this mixture which has been recently applied in therapeutics, as a powerful tonic in cases of dysentery, &c. The mixture can easily be obtained from the leaves, bark, or wood of cinchona, and even from bark which has been exhausted by ebullition with water or diluted acids, by cold maceration with weak milk of lime by which it is dissolved, as it combines easily with bases. It is only the quinovate of lime which has till now been used in medicine.—*Doss,* 2 to 8 gr. every two hours.

QUINOVIN. $C_{30}H_{48}O_8$. *Syn.* CINCHOVIN, QUINOVIA. A very bitter amorphous substance contained in the genus Cinchona, and probably in many other allied genera. It is insoluble in water, very soluble in rectified spirit and in chloroform, with which last liquid it forms, in concentrated solutions, a jelly. If a current of hydrochloric gas is passed into its alcoholic solution the liquid becomes hot and the quinovin is split up into a peculiar kind of sugar.

QUIN'QUINA. Dr de Vrij states that the substance known under this name is a mixture of hydrochlorate of cinchonidine and of cinchonine. See CINCHONA.

QUIN'SY. See THROAT AFFECTIONS.

QUINTES'SENCE. *Syn.* QUINTA ESSENTIA, L. A term invented by the alchemists to represent a concentrated alcoholic solution of the active principles of organic bodies. It is still occasionally employed in perfumery and the culinary art. See ESSENCE, TINCTURE, &c.

QUITTOR. Generally shows itself at the top or coronet of the hoof of the horse, in the form of a fistulous opening (whence quittor is also called 'the pipes'), filled with a purulent discharge.

Quittor invariably points to the presence of an internal ulcer, abscess, or some other irritating cause, the discharge from which, accumulating under the hard hoof, slowly works its way to the surface. The origin of quittor is generally some injury to the hoof, such as a corn, a prick, or an inequality of tread.

The first thing to be done is to remove the animal's shoe, to cut away any dead or discoloured horn, so as to reach the seat of the suppuration, and to allow it to escape by a more direct outlet. Hot-water fomentations and poultices should afterwards be applied for a few days. Should the sores show an indisposition to heal, the parts should be washed with a tolerably strong solution of sulphate of zinc, or of bichloride of mercury—25 grms. of the latter to an ounce of water. The application of strong caustics is to be particularly deprecated.

QUOTID'IAN. Occurring or returning daily. See AGUE.

RAB'BIT. The *Lepus cuniculus,* Linn., of the Cuvian order RODENTIA. The domestic rabbit, when young, is a light and wholesome article of food, approaching in delicacy to the common barn-door fowl, but has less flavour than the wild animal. The fat is among the 'simples' of the Ph. L. 1618. Its hair and skin are made into cheap furs, gloves, hats, &c.

Composition of Rabbit's Flesh (BARTLETT, ' Lancet,' March 29th, 1878).

	Rabbit No. 1. Grains.	Rabbit No. 2. Grains.	Rabbit No. 3. Grains.	Average Grains.	Percentage. Grains.
Water	5,982	6,628	7,315	6,640	73·17
Fibrin and Syntonin	1,143	1,247	1,393	1,261	13·90
Gelatin	302	335	350	329	3·63
Fat	240	272	345	286	3·15
Albumen	276	305	340	307	3·38
Alcoholic extract, including salts . . .	106	119	135	120	1·32
Watery extract	102	108	125	112	1·23
Calcium phosphates	16	19	25	20	0·22
Edible portion	8,167	9,026	10,029	9,075	100·00
Additional gelatin from stewing bones .	215	232	251	233	2·06
Bones, &c., dissected out and stewed . .	1,501	1,674	1,854	}2,027	}17·88 waste.
Shank bones, fur and eyes, thrown away .	318	352	382		
	10,201	11,286	12,516	11,335	...

Rabbit Pie. Cut up two young rabbits, season with white pepper, salt, a little mace, and nutmeg, all in fine powder; add also a little cayenne. Pack the rabbit with slices of ham, forcemeat balls, and hard eggs, by turns in layers. If it is to be baked in a dish add a little water, but omit the water if it is to be raised in a crust. By the time it is taken out of the oven have ready a gravy of a knuckle of veal, or a bit of the scrag, with some shank bones of mutton, seasoned with herbs, onions, mace, and white pepper. If the pie is to be eaten hot, truffles, morels, or mushrooms may be added, but not if intended to be eaten cold. If it be made in a dish put as much gravy as will fill the dish, but in raised crusts the gravy must be carefully strained, and then put in cold as jelly.

Rabbit Pudding. Cut a rabbit into sixteen pieces, and slice a quarter of a pound of bacon; season with chopped sage, pepper, and salt; then add potatoes and onions according to the size of the family, and half a pint of water. Boil for two hours. The meat and vegetables must be well mixed. Rice may be substituted for potatoes if preferred.

Rabbit, Ragout of. "Wash and clean a good-sized Ostend rabbit; boil the liver and heart, chop them, and mix with veal stuffing; fill the rabbit, sew it up, and tie it into shape. Put a piece of fat beef and 1 lb. of bacon, cut in slices, into a saucepan, with 1 oz. of dripping; put in the rabbit to brown, turning it over to brown both sides; pour off the dripping, and put in 1 quart of water; let it simmer gently an hour and a half. A quarter of an hour before serving skim off all the fat, and thicken the gravy with a little corn flour; season with pepper and salt, and, if liked, stew a bunch of herbs and half an onion with it. Lay the rabbit on a dish with the bacon round it, and pour the gravy over " (*Tegetmeier*).

RACAHOUT. *Syn.* RACAHOUT DES ARABES. This is said to be farina prepared from the acorns of *Quercus ballota*, or Barbary oak, disguised with a little flavouring. The following is recommended as an imitation:—Roasted cacao or chocolate nuts, 4 oz.; tapioca and potato farina, of each, 6 oz.; white sugar, slightly flavoured with vanilla, ¼ lb. Very nutritious. Used as arrowroot.

RACEMIC ACID. *Syn.* PARATARTARIC ACID. This compound was discovered by Kestner in 1820, replacing tartaric acid in grape-juice of the Department of the Vosges. Racemic acid and tartaric acid are isomeric, and indeed, when exposed to heat, the same products; the racemates also correspond in the closest manner with the tartrates. Racemic acid crystallises in triclinic prisms containing 1 molecule of water, and fusing at 202° C.; it is rather less soluble than tartaric, and separates first from a solution containing the two acids. A solution of racemic acid precipitates a neutral salt of calcium, which is not the case with tartaric acid. A solution of racemic acid does not affect a ray of polarised light, while a solution of tartaric acid rotates the ray to the right.

Dessargnes and Jungfleisch found by experiment that, heated in a sealed tube to 175° C. with $\frac{1}{10}$ of its weight of water, ordinary tartaric acid

is readily transformed into inactive tartaric acid and racemic acid, and the latter chemist thought to find in this fact an explanation of the production of racemic acid.

" But observations continued through many years upon mother liquors from various tartaric acid factories showed that although more or less inactive tartaric acid was present in all of them; racemic acid was not, even when they had been subjected to prolonged treatment, and its occurrence in appreciable quantity was confined to a small number of specimens. In fact, some samples of mother liquor from factories where evaporation was carried on in a partial vacuum contained more racemic acid than others from factories where evaporation was carried on over a fire. Jungfleisch has since noticed that the liquors richest in inactive tartaric acid were also rich in alumina, and the suspicion that alumina favoured the conversion was confirmed by direct experiment; also that the neutral aluminium sulphate has but little action. Jungfleisch has come to the conclusion that when there is an accumulation of alumina in the mother liquor, the conditions are favourable for the production of a large proportion of inactive tartaric acid, and a small proportion of racemic acid, although when the latter is present in considerable quantity, it becomes the most manifest through its comparative insolubility. Examination of liquors from which racemic acid has been deposited has always shown them to contain much inactive tartaric acid. This theory does not exclude the probability that certain vines under particular conditions produce racemic acid " (*Pharmaceutical Journal*').

BACK'ING. See CIDER and WINES.

RAD'ICAL. According to the binary theory of the constitution of saline compounds, every salt is composed, like chloride of sodium (NaCl), of two sides or parts, which are termed its radicals. That part of a salt which consists of a metal, or of a body exercising the chemical functions of one, is called the metallic, basic, or basylous radical; while the other part, which, like chlorine, by combining with hydrogen would produce an acid, is designated the chlorous or acidulous radical. Every salt, therefore, consists of a basic and of an acid radical. Sometimes radicals are elementary in their nature, when they are called *simple*; and sometimes they are made up of a group of elements, when they are termed *compound*. Some radicals, both simple and compound, have been isolated, while many have but a hypothetical existence. In the following formulæ the vertical line separates the basic from the acid radicals, the former being on the left, the latter on the right:

H	F	Hydrofluoric acid (*Fluoride of hydrogen*).
Na	Cl	Chloride of sodium.
K	CN	Cyanide of potassium.
Ca	CO$_3$	Carbonate of calcium.
NH$_4$	Cl	Chloride of ammonium.
C$_2$H$_5$	NO$_2$	Nitrite of ethyl.

In organic chemistry the organic radical may be further defined as a group of elements which appear unchanged in the products of a reaction, and is, therefore, found on both sides of the equation.

RAD'ISH. The common garden radish (RA-

PHANUS, L.) is the root of *Raphanus sativus*, Linn., one of the CRUCIFERÆ. There are several varieties. They are all slightly diuretic and laxative, and possess considerable power in exciting the appetite. The seed is pressed for oil. The horseradish (ARMORACIA, L.) belongs to a distinct genus.

RAIN-GAUGE. *Syn.* OMBOMETER, PLUVIAMETER, UDOMETER. An instrument for determining the quantity of water, which falls as rain, at any given place. A simple and convenient raingauge for agricultural purposes is formed of a wide-mouthed funnel, or open receiver, connected with a glass tube furnished with a stopcock. The diameter of the tube may be exactly 1-100th that of the receiver, and if the tube be graduated into inches and tenths, the quantity of rain that falls may be easily read off to the 1-1000th of an inch. The instrument should be set in some perfectly open situation; and, for agricultural purposes, with its edge as nearly level with the ground as possible. Another form of gauge is furnished with a float, the height of which marks the amount of liquid. The diameter of the gauge should range between 4 and 8 inches. The quantity of water should be duly measured and registered at 9 a.m. daily.

Mr Symonds, F.R.M.S., has drawn the following code of instructions for the guidance of those registering the amount of rainfall at any locality:

1. *Site.* A rain-gauge should not be set on a slope or terrace, but on a level piece of ground, at a distance from shrubs, trees, walls, and buildings—at the very least as many feet from their base as they are in height.

Tall-growing flowers, vegetables, and bushes must be kept away from the gauges. If a thoroughly clear site cannot be obtained, shelter is most endurable from north-west, north, and east; less so from south, south-east, and west; and not at all from south-west or north-east.

2. *Old Gauges.* Old-established gauges should not be moved, nor their registration discontinued, until at least two years after a new one has been in operation, otherwise the continuity of the register will be irreparably destroyed. Both the old and the new ones must be registered at the same time, and the results recorded for comparison.

3. *Level.* The funnel of a rain-gauge must be set quite level, and so firmly fixed that it will remain so in spite of any gale of wind or ordinary circumstances. Its correctness in this respect should be ascertained from time to time.

4. *Height.* The funnel of gauges newly placed should be one foot above grass. Information respecting height above sea level may be obtained from G. J. Symonds, Esq., 64, Camden Square, N.W., London.

5. *Rust.* If the funnel of a japanned gauge become so oxidised as to retain the rain in its pores, or threatens to become rusty, it should have a coat of gas tar or japan-black, or a fresh funnel of zinc or copper should be provided.

6. *Float Gauges.* If the measuring rod is detached from the float it should never be left in the gauge; if it is attached to the float it should be pegged or tied down, and only allowed to rise

to its proper position at the time of reading. To allow for the weight of the float and rod these gauges are generally so constructed as to show 0 only when a small amount of water is left in them. Care must always be taken to set the rod to the zero or 0.

7. *Can and Bottle Gauges.* The measuring glass should always be held upright. The reading is to be taken midway between the two apparent surfaces of the water.

8. *Date of Entry.* The amount measured at 9 a.m. on any day is to be set against the previous one, because the amount measured at 9 a.m. of, say, the 17th, contains the fall during fifteen hours of the 16th, and only nine hours of the 17th. (The rule has been approved by the meteorological societies of England and Scotland, cannot be altered, and is particularly commended to the notice of observers.)

9. *Mode of Entry.* If less than one tenth (·10) has fallen, the cipher must always be prefixed; thus, if the measure is full up to the seventh line, it must be entered as ·07—that is, no inches, no tenths, and seven hundredths. For the sake of clearness it has been found necessary to lay down an invariable rule that there shall always be two figures to the right of the decimal point. If there be only one figure, as in the case of one tenth of an inch (usually written ·1), a cipher must be added, making it ·10. Neglect of this rule causes much inconvenience. All columns should be cast *twice*—once up and once down—so as to avoid the same error being made twice. When there is no rain a line should be drawn rather than a cipher inserted.

10. *Caution.* The amount should always be written down before the water is thrown away.

11. *Small Quantities.* The unit of measurement being ·01, observers whose gauges are sufficiently delicate to show less than that are, if the amount is under ·005, to throw it away; if it is ·005 to ·010 inclusive, they are not to enter it as ·01.

12. *Absence.* Every observer should train some one as an assistant; but where this is not possible, instructions should be given that the gauge should be emptied at 9 a.m. on the first of the month, and the water bottled, labelled, and tightly corked, to await the observer's return.

13. *Heavy Rains.* When very heavy rains occur it is desirable to measure immediately on their termination; and it will be found a safe plan, after measuring, to return the water to the gauge, so that the morning registration will not be interfered with. Of course, if there is the slightest doubt as to the gauge holding all the falls it must be emptied, the amount being previously written down.

14. *Snow.* In snow three methods may be adopted; it is well to try them all:—(1) Melt what is caught in the funnel by adding to the snow a previously ascertained quantity of warm water, and then, deducting this quantity from the total measurement, enter the residue as rain. (2) Select a place where the snow has not drifted, invert the funnel, and, turning it round, lift and melt what is enclosed. (3) Measure with a rule the average depth of snow, and take 1-12th as the equivalent of water. Some observers use in

snowy weather a cylinder of the same diameter as the rain-gauge, and of considerable depth. If the wind is at all rough all the snow is blown out of a flat-funnelled rain-gauge.

15. *Overflow*. It would seem needless to caution observers on this head, but as a recent foreign table contains *six instances in one day* in which gauges were allowed to run over, it is evidently necessary that British observers should be on the alert. It is not desirable to purchase any new gauge of which the capacity is less than four inches.

16. *Second Gauges*. It is often desirable that observers should have two gauges, and that one of them should be capable of holding eight inches of rain. One of the gauges should be registered daily, the other weekly or monthly, as preferred, but always on the first of each month. By this means a thorough check is kept on accidental errors in the entries, which is not the case if *both* are read daily.

17. *Dew and Fog*. Small amounts of water are at times deposited in rain-gauges by fog and dew. They should be added to the amount of rainfall, because (1) "they tend to water the earth and nourish the streams;" and not for that reason only, but (2) because in many cases the rain-gauges can only be visited monthly, and it would then obviously be impossible to separate the yield of snow, rain, &c.; therefore, for the sake of uniformity, all must be taken together.

18. *Doubtful Entries*. Whenever there is the least doubt respecting the accuracy of any observation, the entry should be marked with a ?, and the reason stated for its being placed there.

Obs. The height at which the rain-gauge is elevated from the ground is a matter of considerable moment. Thus one observer found the fall of rain at York for twelve months (1833–4) to be—at a height of 213 feet from the ground, 14·96 inches; at 44 feet, 19·85 inches; and on the ground, 25·71 inches.

Later experimentalists have confirmed this curious fact. Thus, Colonel Warde found the following to be the relative rainfall at different periods for the four years extending from 1864 to 1867:

	Inches.
On a level with the ground	1·07
At a height of 2 inches	1·05
„ 6 „	1·01
„ 1 foot	1·00
„ 2 feet	0·99
„ 3 „	0·98
„ 5 „	0·96
„ 10 „	0·95
„ 20 „	0·94

One of the causes that have been assigned for this singular phenomenon has been—the greater exposure in elevated situations of the rain to dispersive action of the wind; a surmise which derives some support from the circumstance that when a rain-gauge is placed on a building, the roof of which is flat, of large area, and with few, if any, chimneys to disturb the air currents, an amount of rain is collected equalling that obtained on the surface of the ground.

RAI'SINS. *Syn.* DRIED GRAPES; UVÆ (B.

P.), UVÆ SICCATÆ, UVA (Ph. L.), UVÆ PASSÆ (Ph. E. and D.), L. "The prepared fruit of *Vitis vinifera*," Linn. (Ph. L.). The grapes are allowed to ripen and dry on the vine. After being plucked and cleaned, they are dipped for a few seconds into a boiling lye of wood ashes and quicklime at 12° or 15° Baumé, to every 4 galls. of which a handful of culinary salt and a pint of salad oil have been added; they are then exposed for 12 or 14 days in the sun to dry; they are, lastly, carefully garbled, and packed for exportation. The sweet, fleshy kinds of grapes are those selected for the above treatment; and, in general, their stalks are cut about one half through, or a ring of bark is removed, to hasten their maturation.

Raisins are nutritious, cooling, antiseptic, and, in general, laxative; the latter to a greater extent than the fresh fruit. There are many varieties found in commerce. Their uses as a dessert and culinary fruit, and in the manufacture of wine, &c., are well known, and are referred to elsewhere. See GRAPES, WINES, &c.

RANCID'ITY. The strong, sour flavour and odour which oleaginous bodies acquire by age and exposure to the air. For its prevention, see FATS, OILS (Fixed), &c.

RAPE OIL. See OILS (Fixed).

RASH. Erasmus Wilson notices four different affections as included under this head:

1. ST ANTHONY'S FIRE, or ERYSIPELAS, the severest of them all, already referred to.

2. NETTLE-RASH, or URTICARIA, characterised by its tingling and pricking pain, and its little white elevations on a reddish ground, like the wheals caused by the sting of a nettle. This efflorescence seldom stays many hours, and, sometimes, not even many minutes, in the same place, and is multiplied or reproduced whenever any part of the skin is scratched or even touched. No part of the body is exempt from it, and when many of them occur together, and continue for an hour or two, the parts are often considerably swelled, and the features temporarily disfigured. In many cases these eruptions continue to infest the skin, sometimes in one place, and sometimes in another, for one or two hours together, two or three times a day, or, perhaps, for the greater part of the twenty-four hours. In some constitutions this lasts only a few days; in others several months.

There are several varieties of nettle-rash, or urticaria, noticed by medical writers, among which URTICARIA FEBRILIS, PERSISTENS, and EVANIDA are the principal.

The common cause of nettle-rash is some derangement of the digestive functions, arising either from the use of improper food or a disordered state of the nervous or other systems of the body. Lobsters, crabs, mussels, shrimps, dried fish, pork, cucumbers, mushrooms, and adulterated beer or porter, bear the character of frequently causing this affection. In childhood it commonly arises from teething. Occasionally, in persons of peculiar idiosyncrasy, the most simple article of food, as almonds, nuts, and even milk, rice, and eggs, will produce this affection.

The treatment may consist of the administration of gentle saline aperients, and in severer cases a gentle emetic, followed by the copious use of acidulated diluent drinks, as weak lemon-juice and water, effervescing potassa draughts, &c., and, when required, diaphoretics. The clothing should be light but warm, and the itching, when severe, may be allayed by the application of a lotion of water to which a little vinegar or camphorated spirit has been added; the latter must, however, be employed with caution. A hot knee-bath is useful in drawing the affection from the face and upper part of the body. A 'compress,' wrung out of cold water until it ceases to drip, and kept in contact with the stomach by means of a dry bandage, has been recommended to relieve excessive irritation of the stomach and bowels. It has been stated that decoction of Virginian snake-root is particularly useful in relieving chronic urticaria.

8. RED-RASH, RED-BLOTCH, or FIERY SPOT, is commonly the consequence of disordered general health, of dyspepsia, and particularly, in females, of tight lacing. Sometimes it is slight and evanescent; at others it approaches in severity to the milder forms of erysipelas, there being much

swelling and inflammation. Chaps, galls, excoriations, and chilblains are varieties of this disease produced by cold, excessive moisture, or friction. The treatment is similar to that of nettle-rash.

4. ROSE-RASH, FALSE MEASLES, or ROSEOLA, is an efflorescence, or rather a discoloration of a rose-red tint, in small irregular patches, without wheals or papulæ, which spread over the surface of the body, and are ushered in by slight febrile symptoms. There are several varieties. The causes are the same as those which produce the preceding affections, and the treatment may be similar. In all of them strict attention to diet, and a careful avoidance of cold applications or exposure to cold, so as to cause a retrocession, are matters of the first moment.

RASP'BERRY. Syn. HINDBERRY. The fruit of *Rubus idæus*, Linn., a small shrub of the Nat. Ord. ROSACEÆ. It is cooling, antiscorbutic, and mildly aperitive. It is frequently used to communicate a fine flavour to liqueurs, confectionery, wine, &c. See FRUITS and VEGETABLES.

Fresenius gives the following as the composition of raspberries:

Soluble Matter—		Wild Red.		CULTIVATED. Red.		White.
Sugar		3·597	. .	4·708	. .	3·703
Free acid (reduced to equivalent in malic acid) .		1·960	. .	1·356	. .	1·115
Albuminous substances		0·546	. .	0·544	. .	0·665
Pectous substances, &c.		1·107	. .	1·746	. .	1·397
Ash		0·270	. .	0·481	. .	0·318
Insoluble Matter—						
Seeds	}	8·460	. .	4·106	. .	4·520
Skins, &c.						
Pectose		0·180	. .	0·502	. .	0·040
[Ash from insoluble matter included in weights given]		[0·134]	. .	[0·296]	. .	[0·081]
Water		86·860	. .	86·557	. .	88·180
		100·000	. .	100·000	. .	100·000

RATAFI'A. Originally a liquor drunk at the ratification of an agreement or treaty. It is now the common generic name in France of liqueurs compounded of spirit, sugar, and the odoriferous and flavouring principles of vegetables, more particularly of those containing the juices of recent fruits, or the kernels of apricots, cherries, or peaches. In its unqualified sense this name is commonly understood as referring to cherry brandy or peach brandy.

Ratafias are prepared by distillation, maceration, or extemporaneous admixture, in the manner explained under the head LIQUEUR. The following list includes those which are commonly prepared by the French liquoristes:

Ratafia d'Angélique. From angelica seeds, 1 dr.; angelica stalks, 4 oz.; blanched bitter almonds, bruised, 1 oz.; proof spirit or brandy, 6 quarts; digest for 10 days, filter; add of water, 1 quart; white sugar, 3¼ lbs.; mix well, and in a fortnight decant the clear portion through a piece of clean flannel.

Ratafia d'Anis. See LIQUEUR, CORDIAL, ANISEED.

Ratafia de Baume de Tolu. From balsam of Tolu, 1 oz.; rectified spirit, 1 quart; dissolve, add water, 3 pints; filter, and further add of white sugar, 1½ lbs. Pectoral and traumatic.

Ratafia de Brou de Noix. From young walnuts with soft shells (pricked or pierced), 60 in number; brandy, 2 quarts; mace, cinnamon, and cloves, of each, 15 gr.; digest for 8 weeks; press, filter, add of white sugar, 1 lb.; and keeping it for some months before decanting it for use. Stomachic.

Ratafia de Cacao. Syn. R. DE CHOCOLAT. From Caracca cacao-nuts, 1 lb.; West Indian do., ½ lb. (both roasted and bruised); proof spirit, 1 gall.; digest for 14 days, filter, and add of white sugar, 2½ lbs.; tincture of vanilla, ½ dr. (or shred of vanilla may be infused with the nuts in the spirit instead); lastly, decant in a month, and bottle it.

Ratafia de Café. From coffee, ground and roasted, 1 lb.; brandy or proof spirit, 1 gall.; sugar, 2½ lbs.; (dissolved in) water, 1 quart; as last.

Ratafia de Cassis. From black currant juice, 1 quart; cinnamon, 1 dr.; cloves and peach kernels, of each, ½ dr.; brandy, 1 gall.; white sugar, 3 lbs.; digest for a fortnight, and strain through flannel.

Ratafia de Cerise. From Morella cherries, with their kernels bruised, 8 lbs.; brandy or proof spirit, 1 gall.; white sugar, 2 lbs.; as last.

Ratafia de Chocolat. Ratafia de cacao (see above).

Ratafia de Coings. From quince juice, 3 quarts; bitter almonds, 3 dr.; cinnamon and coriander seeds, of each, 2 dr.; mace, ½ dr.; cloves, 15 gr. (all bruised); rectified spirit (quite flavourless), ½ gall.; digest for a week, filter, and add of white sugar, 3½ lbs.

Ratafia de Crème. From crème de noyeau and sherry, of each, ½ pint; capillaire, ½ pint; fresh cream, 1 pint; beaten together.

Ratafia de Curaçoa. Curaçoa.

Ratafia de Framboises. Raspberry cordial.

Ratafia de Genièvre. From juniper berries (each pricked with a fork), ¼ lb.; caraway and coriander seed, of each, 40 gr.; finest malt spirit (22 u. p.), 1 gall.; white sugar, 2 lbs.; digest a week, and strain with expression.

Ratafia de Grenoble. From the small wild black cherry (with the kernels bruised), 2 lbs.; proof spirit, 1 gall.; white sugar, 3 lbs.; citron peels, a few grains; as before.

Ratafia de Grenoble de Teyssère. From cherries (bruised with the stones), 1 quart; rectified spirit, 2 quarts; mix, digest for 48 hours, then express the liquor, and heat it to boiling in a close vessel; when cold, add of sugar or capillaire, q. s., together with some noyeau, to flavour, and a little syrup of the bay laurel, and of galangal; in 3 months decant, and bottle it.

Ratafia de Noyeau. From peach or apricot kernels (bruised), 120 in number; proof spirit or brandy, 2 quarts; white sugar, 1 lb.; digest for a week, press, and filter.

Ratafia de Œillets. From clove-pinks (without the white buds), 4 lbs.; cinnamon and cloves, of each, 15 gr.; proof spirit, 1 gall.; macerate for 10 days, express the tincture, filter, and add of white sugar, 2½ lbs.

Ratafia d'Ecorce d'Orange. Crème d'Orange.

Ratafia de Fleurs d'Oranger. From fresh orange petals, 2 lbs.; proof spirit, 1 gall.; white sugar, 2½ lbs.; as last. Instead of orange flowers, neroli, 1 dr., may be used.

Ratafia à la Provençale. From striped pinks, 1 lb.; brandy or proof spirit, 1 quart; white sugar, ¼ lb.; juice of strawberries, ½ pint; saffron, 20 gr.; as before.

Ratafia des Quatre Fruits. From cherries, 30 lbs.; gooseberries, 15 lbs.; raspberries, 8 lbs.; black currants, 7 lbs.; express the juice, and to each pint add of white sugar, 6 oz.; cinnamon, 6 gr.; cloves and mace, of each, 3 gr.

Ratafia Rouge. From the juice of black cherries, 3 quarts; juice of strawberries and raspberries, of each, 1 quart; cinnamon, 1 dr.; mace and cloves, of each, 15 dr.; proof spirit or brandy, 2 galls.; white sugar, 7 lbs.; macerate, &c., as before.

Ratafia Sec. Take of the juice of gooseberries, 5 pints; juices of cherries, strawberries, and raspberries, of each, 1 pint; proof spirit, 6 quarts; sugar, 7 lbs.; as before.

Ratafia à la Violette. From orris powder, 3 oz.; litmus, 4 oz.; rectified spirit, 2 galls.; digest for 10 days, strain, and add of white sugar, 12 lbs.; dissolved in soft water, 1 gall.

RATS. The common or brown rat is the Mus decumanus, Linn., one of the most prolific and destructive species of the Rodentia. It was introduced to these islands from Asia, and has since spread over the whole country, and multiplied at the expense of the black rat (Mus rattus, Linn.), which is the old British species of this animal, until its inroads on our granaries, our stores, and dwelling-houses have increased to such an extent, that its extirpation has become a matter of serious, if not of national, importance.

For the destruction of these noxious animals two methods are adopted:

1. Trapping. To render the bait more attractive, it is commonly sprinkled with a little of one of the rat-scents noticed below. The trap is also occasionally so treated.

2. Poisoning. The following are reputed the most effective mixtures for this purpose:

ARSENICAL PASTE. From oatmeal or wheaten flour, 3 lbs.; powdered indigo, ½ oz.; finely powdered white arsenic, ¼ lb.; oil of aniseed, ½ dr.; mix, add of melted suet, 2½ lbs.; and beat the whole into a paste. A similar compound has the sanction of the French Government.

ARSENICAL POWDER. From oatmeal, 1 lb.; moist sugar, ½ lb.; white arsenic and rotten cheese, of each, 1 oz.; rat-scent, a few drops.

MILLERS' RAT POWDER. From fresh oatmeal, 1 lb.; nux vomica (in very fine powder), 1 oz.; rat-scent, 5 or 6 drops. This is highly spoken of by those who have used it.

MINERAL RAT POISON. From carbonate of baryta, ¼ lb.; sugar and oatmeal, of each, 6 oz.; oils of aniseed and caraway, of each, a few drops.

PHILANTROPE MUOPHOBON. A French preparation, which, according to Mr Beasley, consists of tartar emetic, 1 part, with farinaceous matter, 4 parts, and some other (unimportant) ingredients.

PHOSPHOR PASTE.

RAT-SCENTS. The following are said to be the most attractive:

a. Powdered cantharides steeped in French brandy. For traps. It is said that rats are so fond of this, that if a little be rubbed about the hands they may be handled with impunity.

b. From powdered assafœtida, 8 gr.; oil of rhodium, 2 dr.; oil of aniseed, 1 dr.; oil of lavender, ½ dr.; mix by agitation.

c. From oil of aniseed, ½ oz.; tincture of assafœtida, ¼ oz.

d. From oil of aniseed, ½ oz.; nitrous acid, 2 to 3 drops; musk (triturated with a little powdered sugar), 1 gr.

RAZORS. See PAPERS, PASTE, and SHAVING.

REA'GENTS. See TESTS.

REAL'GAR. This valuable red pigment is the bisulphide of arsenic. It is found native in some volcanic districts; but that of commerce is prepared by distilling, in an earthen retort, arsenical pyrites, or a mixture of sulphur and arsenic, of orpiment and sulphur, or of arsenious acid, sulphur, and charcoal, in the proper proportions. See DISULPHIDE OF ARSENIC.

RECOMMENDATIONS TO FARMERS. A series of valuable suggestions, intended for the guidance of farmers in the purchase of manures and cattle-feeding materials, have been issued by the Royal Agricultural Society of England. In substance they are as follows:—In the purchase of feeding-cakes, the guarantee of 'pure' should be insisted upon, since this means a legal warranty that the article is produced from good clean seed. The terms 'best' and 'genuine' are of no value, and should be objected to. Furthermore, the sample should be subjected to analysis. For this purpose a sample should be taken out of the middle of the cake, whilst the remainder of the cake from which the sample has been selected should be sealed up and set aside for reference in case of dispute.

The following advice is given to farmers about to purchase manures:—Raw bones or bone-dust should be purchased as 'pure,' whilst they should be guaranteed to contain not less than 45% of tribasic phosphate of lime, and 4% of ammonia. 'Boiled bones' should be purchased as 'pure' boiled bones, guaranteed to contain not less than 48% of tribasic phosphate of lime, and 1¼% of ammonia. Dissolved bones vary so greatly that the buyer should insist on a guarantee of quality under the heads of 'soluble phosphate of lime,' 'insoluble phosphate of lime,' and 'nitrogen,' or 'ammonia;' also for an allowance at current rates for each unit per cent. if the bones should prove on analysis to contain less than the guaranteed percentages, &c. It should be insisted that mineral superphosphates are delivered dry and in good condition, and be guaranteed to contain a certain percentage of soluble phosphates at a certain price per unit per cent. No value is to be attached to 'insoluble phosphates.' Compound artificial manures, which are rarely used, should be purchased on exactly the same terms. Nitrate of soda should be guaranteed to contain 94% to 95% of pure nitrate. Sulphate of ammonia should yield 35% of ammonia. Peruvian guano should be sold under that name, and guaranteed to be in a dry, friable condition, and to contain a certain percentage of ammonia.

In buying artificial manures the purchaser is recommended to obtain a guarantee that they shall be delivered in a sufficiently dry and powdery condition to allow of sowing by the drill.

Samples taken out of three or four bags should be well mixed together, and they should be analysed not later than three days after delivery. Two tins, holding about half a pound each, should be filled in the presence of a witness, sealed up, one sent to the analyst, and the other retained for future reference.

RECTIFICATION. The redistillation, &c., of a fluid, for the purpose of rendering it purer.

RED. A term denoting a bright colour, resembling blood. Red is a simple or primary colour, but of several different shades or hues, as scarlet, crimson, vermilion, orange-red, &c.

RED ANILINE. $C_{20}H_{19}N_3,H_2O$. *Syn.* ROSANILINE MAGENTA. This artificial base is prepared by the action of bichloride of tin, mercurial salts, arsenic acid, and many other oxidising agents, upon aniline. The aniline reds of commerce, now so largely used for dyeing, are salts, more or less pure, of rosaniline, with 1 equiv. of acid. These compounds are known under the names of 'magenta,' 'fuchsine,' 'roseine,' 'azaleine,' &c. In England the acetate of rosaniline is chiefly used. In France the hydrochlorate of rosaniline is most commonly employed. The free base crystallises in colourless plates, but its compounds with 1 equiv. of acid have, when dry, a beautiful green colour with golden lustre, and furnish with water and alcohol an intensely red-coloured solution. See PURPLE (Aniline) and RED DYE, also TAR COLOURS.

RED DYE. The substances principally employed for dyeing reds are cochineal, lac-dye, madder, and alizarin, which, under proper treatment, yield permanent colours of considerable brilliancy, the first and third more particularly so. Extremely beautiful but fugitive colours are also obtained from Brazil-wood, safflower, archil, and some other substances. For purple-red or crimsons (magenta, fuchsine, &c.), on silk or wool, the aniline reds (salts of rosaniline) are now extensively used (see TAR COLOURS). The mode of applying them is noticed under PURPLE DYE. SILK is usually dyed of a permanent red or scarlet with cochineal, safflower, or lac-dye; wool with cochineal and, still more frequently, with madder; and cotton with madder (chiefly), Brazil-wood, &c. The leading properties of these substances are given under their respective names, and the methods of employing them are generally referred to in the articles DYEING, MORDANTS, &c., and, therefore, need not be repeated here. The following may, however, be useful to the reader:

1. First, give the 'goods' a mordant of alum, or of alum and tartar; rinse, dry, and boil them in a bath of madder. If acetate of iron be used instead of alum the colour will be purple, and by combining the two as mordants any intermediate shade may be produced.

2. The yarn or cloth is put into a very weak boiling alkaline bath, then washed, dried, and 'galled' (or, when the calico is to be printed, for this bath may be substituted one of cow-dung, subsequent exposure to the air for a day or two, and immersion in very dilute sulphuric acid. In this way the stuff gets opened, and takes and retains the colour better). After the 'galling' the goods are dried and alumed twice; then dried, rinsed, and passed through a madder-bath, composed of ¼ lb. of good madder for every lb. weight of the goods; this bath is slowly raised to the boiling-point in the course of 50 or 60 minutes, more or less, according to the shade of colour required; after a few minutes the stuff is taken out and slightly washed; the operation is then repeated, in the same manner, with fresh madder; it is, lastly, washed and dried, or passed through a hot soap-bath, which carries off the fawn-coloured particles.

3. (ADRIANOPLE RED, TURKEY R.) This commences with cleansing or scouring the goods by alkaline baths, after which they are steeped in oily liquors brought to a creamy state by a little carbonate of soda; a bath of sheep's dung is next often used as an intermediate or secondary steep; the oleaginous bath, and the operation of removing the superfluous or loosely adhering oil with an alkaline bath, is repeated two or three times, due care being taken to dry the goods thoroughly after each distinct process; then follow the distinct operations of galling, aluming, maddering, and brightening, the last for removing the dun-coloured principle, by boiling at an elevated temperature with alkaline liquids and soap; the whole is generally concluded with treatment by stannic chloride. In this way the most brilliant reds on cotton are produced.

Obs. Wool takes from half its weight of madder to an equal weight to dye it red; cotton and linen take rather less. On account of the comparative insolubility of the colouring matter of madder this dye-stuff must be boiled along with the goods to be dyed, and not removed from the decoction, as is the practice in using many other articles. Other dye-stuffs are frequently added to the madder-bath to vary the shade of colour. Decoction of fustic, weld, logwood, quercitron, &c., are often thus employed, the mordants being modified accordingly. By adding bran to the madder-bath the colour is said to be rendered much lighter, and of a more agreeable tint. The red dyes of commerce are known under the names anisol red, Barwood red, claret red, Congo red, corallin red, fast red, French red, imperial red, Magdala red, neutral red, peony red, phenetol red, &c.

Red Dyes from Brazil'-wood (-zēle'-). *Syn.* BRAZILÌ; LIG'NUM BRAZILIEN'SE, L.; BOIS DE BRÉSIL, Fr. A dye-stuff furnished by several species of trees of the genus *Cæsalpinia*, and much used in dyeing various shades of red. The usual practice is to boil it for some hours in hard spring water, and to keep the resulting decoction for some time, or until it undergoes a species of fermentation, as it is thus found to yield more permanent and beautiful colours than when employed fresh. The following are examples of its application:

a. For COTTON. 1. The goods are first boiled in a bath of sumach, next worked through a weak mordant of solution of tin, and then run through the Brazil-bath lukewarm. This gives a bright red.

2. The goods are alumed, rinsed, next mordanted with solution of tin, rinsed again, and then turned through the Brazil dye-bath. This gives a rose colour.

b. For LINEN. This, for the most part, is similar to that adopted for cotton.

c. For SILK. The goods, after being alumed in the same way as wool, but at a lower temperature, are rinsed, and passed through the Brazil-wood bath lukewarm.

d. For WOOL. The goods are first steeped or boiled in a weak mordant of alum and tartar for one hour, and then allowed to lie in the cold liquor for two or three days, with frequent moving about; they are lastly boiled in the Brazil-wood bath for about half an hour.

Obs. The shades of colour given with Brazil-wood may be modified by varying the strength of the bath, the mordant, &c. The addition of a little alum turns it on the purple. A little alkali added to the bath, or passing the goods, after being dyed, through water holding a little alkali in solution, produces what is called false crimson. A deep crimson is obtained by adding a little logwood to the Brazil-wood bath. 1 lb. of Brazil-wood, ½ oz. of alum, and 2 oz. of tartar, are sufficient to dye from 20 to 28 lbs. of cotton, according to the depth of shade required.

RED GUM. *Syn.* EUCALYPTUS GUM; GUMMI RUBRI, EUCALYPTI GUMMI, L. A ruby-coloured exudation from the bark of *Eucalyptus rostrata*, imported from Australia. In properties and appearance it is very similar to gum kino; hence it is valued as an astringent and styptic. Red gum is distinguished from Botany Bay kino by its greater solubility in water, or by sticking to the teeth on chewing it.

Uses. Watery solution injected into the nose stays bleeding, or checks discharges from the vagina. Lozenges of the gum are valuable in congested and relaxed sore throats.—*Dose.* Of the gum, 2 to 10 gr.

RED-GUM. A slight eruptive disease of infancy, occasioned by teething, and, less frequently, by irritation from rough flannel worn next to the skin. See STROPHULUS.

RED LAV'ENDER. See TINCTURE OF LAVENDER (Compound).

RED LIQ'UOR. The crude solution of acetate or sulphoacetate of alumina employed in calico printing and cotton dyeing, as a mordant for producing alizarin reds. It is generally prepared by mixing crude sulphate of alumina with about an equal weight of crude acetate of lime, both being in the state of solution.

RED PIG'MENTS. The preparation of the principal red pigments is generally described under their respective names. The following list includes most of the reds in use:

Arme'nian Bole. *Syn.* BOLE ARMENIAN; BOLUS ARMENIÆ, L. Formerly imported from Armenia, Portugal, Tuscany, &c.; now generally made by grinding together a mixture of whiting, red oxide of iron, and red ochre, in nearly equal proportions.

Car'minated Lake.

Car'mine. A preparation from cochineal semi-permanent in water and fugitive in oil.

Crimson Lake. An extract of cochineal together with alumina or oxide of iron.

Lakes (Various).

Real'gar. Bisulphide of arsenic.

Red, Brown. A mixture of red oxide of iron and red ochre, in variable proportions.

Red, Chrome. *Syn.* DICHROMATE OF LEAD, RED CHROMATE OF L.; PLUMBI DICHROMAS, P. CHROMUS RUBRUM, L. *Prep.* 1. Boil pure carbonate of lead with chromate of potash, in excess, until it assumes a proper colour; then wash it well with pure water, and dry it in the shade.

2. Boil neutral chromate of lead with a little water of ammonia or lime water.

3. (*Liebig* and *Wöhler*.) Fuse saltpetre at a low red heat in a Hessian crucible, and throw in chromate of lead, by small portions at a time, as long as a strong ebullition follows upon each addition of the pigment, observing to stir the mixture frequently with a glass rod; after standing for a minute or two, pour off the fluid part, and, as soon as the solid residuum is cold, wash it with water, and dry it by a gentle heat.

Obs. Great care must be taken, in conducting the last process, not to employ too much heat, nor to allow the saline matter to stand long over the newly formed chrome red, as the colour is thus apt to change to a brown or orange. When well managed the product has a crystalline texture, and so beautiful a red colour that it vies with native cinnabar. The liquid poured from the crucible is reserved for manufacturing chrome yellow.

Red, In'dian. *Syn.* PURPLE OCHRE; OCHRA

PURPURA PERSICA, TERRA PERSICA, L. This is a native production, brought from Ormus. It is a hæmatite or peroxide of iron mixed with earthy matters. A factitious article is prepared by calcining a mixture of colcothar and red ochre.

Red, Light. From yellow ochre, by careful calcination. It works well with both oil and water, and produces an admirable flesh-colour by admixture with pure white. All the ochres, both red and yellow, are darkened by heat.

Red, Or'ange. Syn. SANDIX. Obtained from white-lead by calcination. Very bright.

Red, Vene'tian. Syn. BOLUS VENETA, L. A species of ochre, brought from Italy.

Red Bole. See ARMENIAN and VENETIAN BOLE (Ochres).

Red Chalk. A clay iron ore, much used for pencils and crayons, and, when ground, also for paints.

Red-lead. Syn. MINIUM. The finest red-lead is prepared by exposing ground and elutriated massicot, or dross of lead, in shallow iron trays (about 12 inches square, and about 4 or 5 inches deep), piled up on the hearth of a reverberatory furnace, to a heat of about 600° to 650° F., with occasional stirring, until it acquires the proper colour. The furnace employed for the preparation of massicot during the day usually possesses sufficient residuary heat during the night for this process, by which fuel is saved. Lead for the above purpose should be quite free from copper and iron. See LEAD OXIDES.

Red O'chre. A natural product abounding on the Mendip Hills.

Red Or'piment. Syn. RED ARSENIC. Tersulphide of arsenic.

Rose Pink. This is whiting coloured with a decoction of Brazil-wood to which a little pearlash has been added. A very pretty colour, but it does not stand. It is always kept in a damp state. The colour may be varied by substituting alum for pearlash, or by the addition of a little stannic chloride.

Vermil'ion. See under that word.

REDUC'TION. Syn. REVIVIFICATION. A term in its fullest sense applied to any operation by which a substance is restored to its neutral state; but now generally restricted, in chemistry, to the abstraction of oxygen, and hence frequently termed deoxidation. This change is effected by either heating the substance in contact with carbon or hydrogen, or by exposing it to the action of some other body, such as pyrogallic acid, &c., having a powerful affinity for oxygen.

REFI'NING. A term employed in commercial chemistry and metallurgy synonymously with purification. The separation of the precious metals from those of less value, as in the operation of parting, constitutes the business of the 'refiner.' See GOLD, SILVER, &c.

REFRAC'TION (of Light). The deviation of a ray of light from its original path on entering a medium of a different density. For the practical application of this property, see GEMS.

REFRI"GERANTS. Medicines or agents which tend to lessen the animal temperature without causing any marked diminution of sensibility or nervous energy. Among internal refrigerants, cold water, weak acidulous drinks, and saline aperients are those which are probably the best known and the most useful. Among external refrigerants are cold water, evaporating lotions, weak solutions of subacetate of lead, &c.

REFRIGERA'TION. The abatement of heat; the act or operation of cooling.

Among the purposes to which refrigeratory processes are applied in the arts, the principal are—the condensation of vapours, the cooling of liquids, the congelation of water, and the production of extreme degrees of cold in chemical operations. The first of these is referred to under the heads DISTILLATION, STILL, &c., and the second under WORT. It is, therefore, only necessary to notice here the third and fourth applications of cold, artificially produced, above referred to.

The refrigeratory processes at present employed depend upon the greater capacity for heat which the same body possesses as its density lessens, or its attenuation increases; as exhibited in the sudden liquefaction of solids, the rapid evaporation of liquids, and the almost instantaneous return of atmospheric air, or other gaseous body, from a highly condensed state to its normal condition. The loss of sensible heat in the first example is the basis of the various processes of producing cold by what are commonly called 'FREEZING' or 'FRIGORIFIC MIXTURES,' all of which act upon the principle of liquefying solid substances without supplying heat. The heat of liquidity being in these cases derived from that previously existing in the solid itself in a sensible state, the temperature must necessarily fall. The degree of cold produced depends upon the quantity of heat which is thus diffused through a larger mass, or which, as it were, disappears; and this is dependent on the quantity of solid matter liquefied, and the rapidity of the liquefaction. Saline compounds are the substances most frequently employed for this purpose; and those which have the greatest affinity for water, and thus liquefy the most rapidly, produce the greatest degree of cold.

Similar changes occur during the evaporation of liquids. When heat passes from the sensible to the insensible or latent state, as in the formation of vapour, cold is generated. This may be shown by pouring a few drops of ether or rectified spirit on the palm of the hand, when a strong sensation of cold is experienced. A still more familiar illustration of this fact is exhibited in the rapidity with which the animal body loses heat when enveloped in damp or wet clothing. The evaporation of water produces a degree of cold which is greater than that of other liquids, in exact proportion as the insensible or latent heat of its vapour exceeds theirs. In the attenuation or rarefaction of gases similar phenomena occur.

It has been found that evaporation proceeds much more rapidly from the surface of fluids in a vacuum than in the atmosphere. Water may be easily frozen by introducing a surface of sulphuric acid under the receiver of an air-pump, over which is placed a capsule filled with water, so that the vapour arising from the latter may be immediately absorbed by the former. After a few strokes of the piston the water is converted into a solid cake of ice. The acid operates by absorbing the aqueous vapours as soon as generated, and thus maintaining the vacuum. Professor Leslie found that, when air is thus rarefied 250 times, the surface of evaporation was cooled down 120° in winter, and when only 50 times, a depression of 80° or even 100° took place. Sulphuric acid, which has become diluted by the absorption of aqueous vapour, may be reconcentrated by heat. Any substance having a great tendency to absorb moisture may be substituted for the sulphuric acid. Fused chloride of calcium, quicklime, nitrate of magnesium, chloride of zinc, and oatmeal (dried nearly to brownness before a common fire) have been used for this purpose. Again, instead of employing an air-pump, a vacuum may be produced by the agency of steam, afterwards condensed by the affusion of cold water.

A pleasing illustration of the evaporative power of a vacuum is the 'CRYOPHORUS,' or 'FROST-

BEARER,' of Dr Wollaston. This instrument consists of two small glass globes united by a tube, one of which is partly filled with water. The whole apparatus is perfectly free from air, and is,

consequently, filled with attenuated aqueous vapours. No sooner is the pressure removed, as by plunging the empty ball into a freezing mixture (which condenses the vapour), than rapid evaporation commences, and the water in the other ball is frozen in two or three minutes.

In hot climates ice may be produced under favourable circumstances by evaporation. On the open plains, near Calcutta, this is effected by exposing a thin stratum of water to the atmosphere during the fine clear nights of December, January, and February. The pans are made of

TABLE *exhibiting a few of the most useful* FRIGORIFIC MIXTURES.

Ingredients.		Thermometer sinks.	Deg. F. of cold produced.
Snow *or* pounded ice . . . 2 parts Chloride of sodium . . . 1 ,,		. to – 5°	—
Snow *or* pounded ice . . . 5 ,, Chloride of sodium . . . 2 ,, Sal-ammoniac . . . . 1 ,,		. to – 12°	—
Snow *or* pounded ice . . . 12 ,, Chloride of sodium . . . 5 ,, Nitrate of ammonia . . . 5 ,,		. to – 25°	—
Snow . . . . . 8 ,, Hydrochloric acid (*concentrated*) 5 ,,		. From + 32° to – 27°	59°
Snow . . . . . 2 ,, Crystallised chloride of calcium . 3 ,,		. From + 32° to – 50°	82°
Sal-ammoniac . . . . 5 ,, Nitrate of potash . . . 5 ,, Water . . . . . 16 ,,		. From + 50° to + 10°	40°
Nitrate of ammonia . . . 1 ,, Water . . . . . 1 ,,		. From + 50° to + 4°	46°
Nitrate of ammonia . . . 1 ,, Carbonate of soda . . . 1 ,, Water . . . . . 1 ,,		. From + 50° to + 7°	57°
Phosphate of soda . . . 9 ,, Nitrate of ammonia . . . 6 ,, Diluted nitrous acid [1] . . . 4 ,,		. From + 50° to – 21°	71°
Sulphate of soda . . . 8 ,, Hydrochloric acid . . . 5 ,,		. From + 50° to 0°	50°
Snow . . . . . 3 ,, Diluted nitrous acid [1] . . . 2 ,,		. From 0° to – 46°	46°
Snow . . . . . 2 ,, Sulphuric acid [2] . . . 1 ,, Water . . . . . 1 ,,		. From – 20° to – 60°	40°
Snow . . . . . 1 ,, Crystallised chloride of calcium . 2 ,,		. From 0° to – 66°	66°
Snow . . . . . 1 ,, Crystallised chloride of calcium . 3 ,,		. From – 40° to – 73°	33°
Snow . . . . . 8 ,, Sulphuric acid . . . 5 ,, Water . . . . . 5 ,,		. From – 68° to – 91°	23°

[1] Fuming "nitrous acid," 2 parts; water, 1 part; by weight.
[2] Professor Pfaundler has shown that an acid containing 66·19 per cent. of H_2SO_4 is the most advantageous to employ for this purpose; one part of an acid of this strength with 1·097 parts of snow forming a refrigerating mixture which will reduce the temperature to – 37° C. (– 36° F.). For practical purposes it is suggested an excess of snow would be better, since the refrigerating value of the mixture is thereby largely increased, though the lowest temperature is not obtained. See ICE.

porous earthenware, and water is poured in to the depth of about 1½ in. A large number of these vessels are arranged in an excavation in the ground, 30 or 40 feet square and 2 feet deep, the bottom of which is covered, to the depth of 10 or 12 in., with sugar-canes or the stalks of Indian corn. At sunrise the pans are visited, the ice separated from the water, and packed as tight as possible in a deep cavity or pit, well screened from the heat.

Several machines have recently been invented by which water is frozen in large quantities by exposure to condensed air in the act of its subsequent expansion. They are worked by either hand or steam power. Others depend upon the liquefaction and evaporation of ammonia and similar substances.

For the production of an extremely low temperature, such as is required for the liquefaction of some gases, Faraday employed solid carbonic acid mixed with a little ether.

In the production of ice or an extreme degree of cold by saline mixtures, the salts should be in the crystallised state, and as rich as possible in water of crystallisation, but without being in the least damp. They should be coarsely pulverised at the time of using them, and should not be mixed until immediately before throwing them into the liquid ingredients. The mixture should be made in a thick vessel, well clothed, to prevent the accession of external heat; and the substance to be acted on should be contained in a very thin vessel, so as to expose it more fully to the action of the mixture. The preceding table, though founded on experiments made many years ago by Mr Walker, gives full and accurate information on the subject of freezing mixtures. See page 1446.

Obs. The materials in the first column are to be cooled, previously to mixing, to the temperature required in the second, by the use of other mixtures.

The following are some other very convenient freezing mixtures for laboratory purposes:

a. 8 oz. of sodium sulphate, and 4 fl. oz. of common hydrochloric acid.

b. 5 parts by weight of potassium sulphocyanide and 4 parts by weight of cold water.

c. Equal parts by weight of sal-ammoniac and nitre, dissolved in its own weight of water.

REGULATORS, GAS. There are many purposes for which artificial heat at a steady and uniform temperature is required—hot-air ovens for disinfection, incubators, cultivation chambers for bacteria, and other laboratory purposes. Where gas is available and the temperature required is not much above that of boiling water, regulators depending for their action on the expansion of mercury may be employed with great advantage,

Bunsen's Gas Regulator.

and, if properly constructed, perform the work required of them with absolute certainty. One of the earliest forms was Bunsen's (see *engr.*), which had, however, the very grave defect that not only was the supply of gas affected by the expansion of the mercury, but also by variations of atmospheric pressure acting on a volume of air in a closed space above the surface of the mercury. This might appear at first sight to be so trifling a matter as to be of no account, but in practice the irregularities produced are so serious as to render the instrument quite untrustworthy for scientific purposes.

The best and simplest form of gas regulator is undoubtedly Page's, the construction of which will be obvious from the annexed cut. The long bulb is filled with *clean* mercury up to within an inch of the T piece. On the top of the straight limb is fixed, by means of corks, a wide tube through which a fine tube passes into the straight limb of the

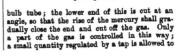

Page's Regulator.

bulb tube ; the lower end of this is cut at an angle, so that the rise of the mercury shall gradually close the end and cut off the gas. Only a part of the gas is controlled in this way ; a small quantity regulated by a tap is allowed to pass constantly to the burner, which thus never goes out, as might happen if the whole of the gas passed through the regulator.

The two tubes in the figure connected by the brass stopcock are best made of brass or other perfectly rigid material, and connected with the glass regulator by good rubber tubing.

To Set the Regulator. Bring the bath, incubator, &c., up to the required temperature, place the regulator in position freely immersed up to the T piece in the heated air or water, turn the stopcock until the burner shows a flame large enough to keep the bath within a little of the required temperature, then bring the internal tube down to the surface of the mercury until the end of the tube is just completely blocked and no gas passes, by applying a gentle screwing motion to the outer tube and corks; on being left to itself the temperature of the bath will fall to such a level as the gas, which finds its way to the burner by way of the stopcock, will allow. The mercury will then begin to contract, and so expose the end of the tube B, allowing a certain further supply of gas to find its way out through the T piece to the burner, thus maintaining a constant temperature. In large towns care must be taken to guard against the increased pressure of gas which is usually put on in the evening. For a regulator for use with paraffin lamps see INCUBATOR.

REG'ULUS. A term applied by the alchemists to various metallic matters obtained by fusion; as REGULUS OF ANTIMONY, ARSENIC, &c. It is now almost obsolete.

REL'ISHES. See SAUCES.

REMEDIES, FERRUGINOUS. Rob. Freygang:

STEEL BRANDY is an ordinary clear brownish brandy, containing a very little bitter matter, like the stomachic bitters of the apothecaries, and mixed with about 1% of sugar. 10,000 parts contain about 1½ parts oxide of iron.

STEEL STOMACHIC BITTERS. This is more aromatic, but otherwise similar to the steel brandy; 10,000 parts contain ½ part iron oxide.

STEEL LIQUEUR is a clear, agreeable-tasting liqueur, of the colour, and containing much of the juice of raspberries. 10,000 parts contain nearly 1200 of sugar and only 1 of iron oxide.

STEEL SYRUP (Syrup ferrugineux de Quinquina). A clear, slightly violet-coloured, thin, sweet fluid, containing spirit and sugar, of which cinchona bark may be an ingredient, though it is appreciable by neither taste nor tests. It contains 1½ parts iron in 10,000 parts.

STEEL BONBONS contain a trace of iron oxide. The iron present in the above preparations is in the form of citrate (*Hager*).

REMIT'TENT. A term applied to fevers, and other diseases, which exhibit a decided remission in violence during the twenty-four hours, but without entirely leaving the patient, in which they differ from intermittents or agues.

RENNET. *Syn.* RUNNET, PREPARED CALF'S MAW. The fourth or true digesting stomach of the calf, freed from the outer skin, fat, and useless membrane, washed, treated with either brine or dry salt for a few hours, and then hung up to dry. When well prepared, the dried 'vells' somewhat resemble parchment in appearance.

Uses, &c. Rennet is employed to curdle milk. A piece of the requisite size is cut off and soaked for some hours in whey or water, after which the whole is added to the milk for curdling, slightly warmed, and the mixture is slowly heated, if necessary, to about 122° F. In a short time after this temperature has been attained the milk separates into a solid white coagulum (curd), and into a yellowish, translucent, liquid whey. Two square inches from the bottom of a good 'vell' are sufficient for a cheese of 60 lbs. It is the gastric juice of the stomach that effects these changes. The stomachs of all sucking quadrupeds possess the same properties. See CHEESE.

Rennet, Essence of. *Prep.* 1. One calf's rennet; syrupy lactic acid, 1 dr.; glycerine, 1 oz.; sherry, 2 oz.; water, to 36 oz. Chop the rennet small, and macerate with the salt (about 3 oz.), used to preserve it, in the liquids for ten days; then filter, and colour with a little liquid cochineal.

2. Take 24 lbs. of dried rennet, cut small, freed from salt, and sifted—No. 4 sieve. To dry the rennets, take them and sprinkle well on both sides with salt, lay them singly on trays in a drying room heated to about 120° F., and they will be ready in a week. And also take salt, 12 lbs.; rectified spirit, 4 galls. 32 fl. oz.; sherry, 1 gall. 32 fl. oz.; aqua, 18 galls. Macerate seven days, shaking, drain on a fine sieve and filter; then filter again through fuller's-earth to brighten it. One teaspoonful to 1 pint of lukewarm milk will curdle in a few minutes.

Rennet, Liquid. *Syn.* ESSENCE OF RENNET. *Prep.* 1. From fresh rennet (cut small), 12 oz.; common salt, 3 oz.; knead them together, and leave the mixture at rest, in a cool place, for five or six weeks; then add of water, 18 oz.; good rum or proof spirit, 2 oz.; lastly, digest for 24 hours, filter, and colour the liquid with a little burnt sugar.

2. Fresh rennet, 12 oz.; salt, 2 oz.; proof spirit, 2 oz.; white wine, a quart; digest for 24 hours and strain. A quart of milk requires two or three teaspoonfuls. *Wielis* directs 10 parts of a calf's stomach; salt, 3 parts. The membrane of the stomach is to be cut with scissors and kneaded with the salt, and with the rennet found in the interior of that organ; the whole left in a cool place in an earthen pot till the cheesy odour is replaced by the proper odour of rennet, which will be in one or two months. Then add 16 parts of water and 1 of spirit. Filter and colour with burnt sugar.

3. The German Pharmacopœia gives the following formula for liquid rennet:—3 parts of the mucous membrane of fresh calf's rennet, macerated for three days in 26 parts of white wine, 1 part of table salt being added.

Obs. Two or 3 teaspoonfuls will curdle a quart of milk. Some persons use white wine instead of water, with simple digestion for a day or two.

RES'IN. *Syn.* RESINA, L. This name is applied to many vegetable principles composed of the elements carbon, hydrogen, and oxygen. The resins (RESINÆ) cannot be very accurately defined, but we may in a general way describe them as substances which are solid at ordinary temperatures, more or less transparent, inflammable, readily fusible, do not volatilise unchanged, become negatively electrified by rubbing; are insoluble in water, but soluble in alcohol; mostly inodorous, and readily incorporated with fatty bodies by fusion. Their sp. gr. varies from ·9 to 1·2. According to Liebig, they are oxidised essential oils. Common resin, rosin, or colophony, and the shellac of which sealing-wax is made, are familiar examples of these substances (see below).

Resin, Black. *Syn.* ROSIN‡, BLACK R.‡, COLOPHONY; RESINA NIGRA, COLOPHONIA, L. What remains of turpentine after the oil has been distilled. When this substance, whilst still fluid, is agitated with about 1-8th part of water, it forms the yellow resin of pharmacy. Used for violin bows, dark-coloured ointments, varnishes, &c.

Resin, Yel'low. *Syn.* YELLOW ROSIN‡, WHITE R.‡; RESINA FLAVA, RESINA (Ph. L.), L. Detergent. Used in ointments, plasters, &c. (see above).

RES'INOIDS. *Syn.* RESINOUS EXTRACTS, CONCENTRATED R.; EXTRACTA RESINÆ, L. Under this head the so-called 'Eclectics,' who form a numerous class among American physicians, place their most important 'concentrated remedies.' "Viewed as pharmaceutical preparations eligible for use in medicine, though not purified so as to rank as distinctive proximate principles, these are very appropriately named 'resinous extracts' or 'resins.' The term 'resinoid,' so commonly used, is less appropriate to the class, implying, as it does, a resemblance in resins, while all of these are either resins, oleo-resins, or more or less mixed proximate principles possessing no real resemblance to the class of resins" (*Parrish*). Most of them are prepared from plants indigenous to North America, by precipitating a strong alcoholic tincture with water. They are all brought to the condition of powder, those which are naturally soft and oily being mixed with a sufficient quantity of sugar of milk, or other dry material. Several of these eclectic remedies have been introduced into regular practice. See PODOPHYLLIN.

RESIN or ROSIN OIL. This is a product of the dry distillation of resin. The apparatus used consists of an iron pot, a head-piece, a condensing arrangement, and a receiver.

In distilling the resin a bright oil first comes over, and with it some acetic acid and water. As soon as a cessation in the flow of the distillate occurs the receiver is changed, and the heat is raised, when a red-coloured and heavy rosin oil comes over. The black residue remaining in the pot is used as pitch. The light oil, called 'pinoline,' is rectified, and the acetic acid water passing over with it is saturated with calcium hydrate, filtered and evaporated to dryness, the calcium acetate thus obtained being employed in the manufacture of acetic acid. The rosin oil, obtained after the light oil has passed over, has a dark violet-blue colour, and is called 'blue rosin oil.' The red oil is boiled for a day with water, the evaporated water being returned to the vessel; next day the water is drawn off, and the remaining rosin oil is saponified with caustic soda lye of 36° Baumé, and the resulting solid mass is distilled so long as oil passes over.

The product obtained is 'rectified rosin oil,' which is allowed to stand in iron vessels, protected by a thin layer of gypsum, whereby after a few weeks a perfectly clear oil is obtained free from water. Oil of the first quality is obtained by a repetition of the foregoing operation upon the once rectified oil. The residue of both operations is melted up with the pitch ('Dingler's Polytech. Journ.,' ccvi, 246).

Rosin oil is employed in the manufacture of axle grease, the oil being previously converted into a soap by heating with slaked lime.

Tests. A characteristic violet coloration with anhydrous stannic chloride. Allen recommends that the test be applied to the first fractions which come on when distilled if it is mixed with fatty oils. The presence of 10% of rosin oil in non-drying oils delays their solidification by the elaïdin test.

The admixture of rosin oil with mineral oil is detected by the polariscope (*vide* 'Dingler's Polytech. Journ.,' cccliii, p. 418). It is also shown by the increased solubility of the sample in glacial acetic acid. The differences in the iodine and tannin absorptions of resin oils and mineral oils are also distinguishing features.

RESOLV'ENTS. *Syn.* DISCUTIENTS; RESOLVENTIA, L. Substances or agents which discuss or resolve inflammatory and other tumours. See DIGESTIVES.

RESORCIN. *Syn.* META-DIHYDROXYBENZOL. $C_6H_4(OH)_2$. It was first obtained by Hlasiwetz and Barth by melting together gum-resins of ammoniacum, assafœtida, galbanum, &c., with potassium hydrate; from this circumstance arose its name (from *Orcinum resina*). It is now prepared directly from bensole itself by comparatively simple steps. When pure it forms colourless, or more often pale yellow, tabular or columnar crystals, with a faint urinous smell and an unpleasant tickling taste; m.-p. 118° C., b.-p. 276° C. It is readily and abundantly soluble in water, alcohol, and ether, but very sparingly in cold benzol or chloroform. In its therapeutical action resorcin closely resembles phenol, but is without its poisonous properties. Externally in substance or in concentrated solution it has been used as a painless caustic, particularly in diphtheria; in ointment form (1 : 6) for skin diseases; in solution (1% to 2%) for urethral injection; in dilute solution as an eye-douche. For the treatment of wounds it was employed in solution, and in wool or gauze form. Internally, it has been recommended as an antifermentative in acute and chronic disorders of the digestive tract. The dose is from 3 to 20 gr. several times a day in mixture, or powdered in wafers or capsules. Any brown spots formed on the skin by contact with resorcin can be removed by the application of citric acid.

Dr Unna recommends the treatment of erysipelas of the head by resorcin in the form of a lotion (1% or 2%). When the inflammation spreads beyond the scalp he recommends the edge of the inflammatory zone to be gently rubbed with equal parts of resorcin and zinc paste ('British Medical Journal').

RESPIRA'TION. The process by which air enters and leaves the chest for the oxygenation and purification of the blood in the lungs.

The chest is a closed cavity bounded behind by the spinal column, in front by the sternum, at the sides by the ribs extending between these two, below by the diaphragm, whilst above the cavity is closed by the approximation of the upper ribs to the structures which pass through, viz. the trachea, œsophagus, great blood-vessels, &c. The lungs are suspended in this cavity, lying against the wall of the thorax but not adhering to it, so that if the cavity of the thorax be enlarged the lung must follow it, and air will enter the lung through the windpipe. The chest is enlarged in three directions:

1. From before backwards by the raising of the ribs.

2. From side to side by the eversion of the ribs.

3. From above downwards by the descent of the diaphragm.

The number of air-vesicles in the lung is calculated at about 725,000,000, with a superficial area of about 90 square metres, or about 100 times the whole surface of the body.

The lungs are never completely emptied in respiration, and the air which moves in and out is classified as follows:

Residual air = the volume of air which remains in the chest after the most complete expiration (100 to 130 cub. in.).

Reserve air = volume of air expelled from the chest *after* a quiet expiration (about 100 cub. in.).

Tidal air = volume of air which passes in and out of the chest in quiet respiration (about 20 cub. in.).

Complemental air = air which can be forcibly taken into the chest over and above the amount taken in an ordinary quiet respiration.

Changes in composition of air by respiration:

	Oxygen.	Nitrogen.	Carbonic acid (by volume).
Atmospheric air	20·96	79·02	0·04
Respired air	16·03	79·02	4·00

Respired air is warmer and contains more moisture than atmospheric air.

As a result of this change of the air in the lung the 'venous blood' which entered the lungs from the right side of the heart has lost its dingy hue, and has acquired the rich florid colour which is characteristic of 'arterial blood.' In this state it is returned to the left side of the heart, and is propelled by that organ to every part of the body, from which it passes by the capillaries to the veins, and by these again to the heart and lungs, to undergo the same changes and circulation as before. The carbon and hydrogen of the blood, ultimately derived from the food, are, in this course, gradually converted into carbonic acid and water by a species of slow combustion; but how these changes are effected is not definitely ascertained.

Respiration, Artificial. See DROWNING.

REVALENTA ARABICA. A mixture of the red Arabian or Egyptian lentil with barley flour, and a little sugar or salt ('Lancet'). See LENTIL.

REVERB'ERATORY FURNACE. See FURNACE.

REVI"VER. *Prep.* 1. (BLACK REVIVER, PARIS'S ANTICARDIUM.) *a.* Blue galls (bruised),

4 oz.; logwood and sumach, of each, 1 oz.; vinegar, 1 quart; macerate in a closed vessel, at a gentle heat, for 24 hours, then strain off the clear, add iron filings and green copperas, of each, 1 oz., shake it occasionally for a week, and preserve it in a corked bottle.

b. Galls, 1 lb.; logwood, 2 lbs.; boil for 2 hours in water, 5 quarts, until reduced to a gallon, then strain, and add of green copperas, ¼ lb. Used to restore the colour of faded black cloth.

2. (BLUE REVIVER.) From soluble Prussian blue, 1 oz.; dissolved in distilled water, 1 quart. Used for either black or blue cloth.

RHAM'NIN. *Prep.* Express the juice from buckthorn berries scarcely ripe, which is to be rejected; boil the cake or residue with water, strain with pressure, and filter the liquid whilst hot; crude rhamnin will be deposited as the liquid cools, which, by solution in boiling alcohol and filtration, may be procured in crystals.

Obs. Buckthorn juice (succus rhamni), "the juice of the fruit of *Rhamnus catharticus*, Linn.," was officinal in the Ph. L.

RHAT'ANY. *Syn.* RHATANY ROOT; KRAMERIÆ RADIX (B. P.), KRAMERIA, RHATANE RADIX, L. "The root of *Krameria triandra*," and of *Krameria ixina.* It is stomachic, and powerfully astringent and styptic.—*Dose*, 20 to 60 gr., either in powder or made into a decoction or infusion. It is much employed in tooth powders, to fix the teeth when they become loosened by the recession of the gums, and also for improving the natural red colour of the lips and gums. A saturated tincture of fluid extract, made with brandy, forms the 'wine-colouring' used by the Portuguese to give roughness, colour, and tone to their port wine. Hard extract of rhatany is also much employed for the same purpose.

RHE'IN. *Syn.* CHRYSOPHANIC ACID. The yellow colouring principle of rhubarb.

RHEUMATIC and GOUT PILLS. (*W. Gross*, Cardiff.) Pills weighing 2 grms. rolled in lycopodium, the essential ingredients of which are quinine sulphate, gamboge, jalap, resin, and a little rhubarb (*Hager*).

RHEU'MATISM. *Syn.* RHEUMATISMUS, L. An affection of the joints, and of the external muscular, tendinous, and fibrous textures of the body, attended with swelling, stiffness, and great pain. Acute rheumatism or rheumatic fever,—arthritis, inflammation of the synovial membrane, or rheumatic gout,—sciatica, or rheumatism of the cellular envelope of the great sciatic nerve, affecting the hip,—and lumbago, or rheumatism of the loins, are varieties of this disease.

The treatment of rheumatism consists in the administration of purgatives and diaphoretics or sudorifics, accompanied by tonics, as bark, quinine, &c. Calomel with opium, and iodide of potassium, have also been frequently and successfully employed in this complaint. Of late years the administration of the bicarbonate, citrate, or nitrate of potass, in rather large doses, has been strongly recommended, and in numerous cases adopted with success. The salicylates of soda and potash have a most marked effect in acute rheumatism, reducing the temperature and relieving the pain in a very short time, and greatly diminishing the risk of subsequent heart troubles. Lemon juice, liberally taken, has also proved useful in suddenly cutting short severe attacks of certain forms of rheumatism. The compound powder of ipecacuanha, taken at night, will generally promote the ease and sleep of the patient, and, by its soporific action, tend considerably to hasten a cure. Where possible, a dry atmosphere and a regular temperature should be sought, since a damp atmosphere, and, indeed, exposure to damp under any form, are the principal causes of rheumatism. Stimulating embrocations, blisters, frictions, and, above all, the hot or vapour bath, are also frequently serviceable in rheumatism, especially in lumbago and casual attacks arising from cold. The daily use of oranges, or of lemon juice diluted with water, has been found, in the majority of cases, to lessen the susceptibility of those who employ them to attacks of rheumatism and rheumatic gout arising from a damp situation or exposure to the weather. See LEMON JUICE.

Rheumatic patients should abstain from ales, beers, stout, and champagne.

RHODIUM. A whitish metal discovered by Wollaston, in 1803, associated with palladium in the ore of platinum.

It is chiefly employed for tipping the nibs of metallic pens (' rhodium ' or ' everlasting pens '). A very small quantity added to steel is said to improve its closeness, hardness, and toughness, and to render it less easily corrodible by damp.

RHOPALOSIPHUM RIBIS, Koch. (From the Greek words signifying ' club ' and ' siphon,' or ' tube.') THE CURRANT APHIS. In the description of the cherry aphis, *Myzus cerasi*, it was shown that it also frequently was found upon currant bushes. There is yet another species of aphis which is common to these fruit bushes, namely, the *Rhopalosiphum ribis*, and as this is quite distinct in species, habit, and appearance, it is important to give its history, and point out the distinction between these two species of aphides.

The *Rhopalosiphum*, or currant aphis proper, makes galls or swellings form upon the upper surfaces of the leaves both of black and red currant bushes. These swellings look like blisters raised by the sun, and are mainly of a red colour. Upon examination of the under surface of the leaves companies of larvæ will be seen actively sucking away at the leaves, and although this blister, curl up, and eventually drop off. Although this aphis does not do so much harm as the *Myzus*, or not so much apparent harm, it often weakens the bushes considerably, so that the currants drop or ' run ' off, and the bunches ' shank ' like grapes in vineries from the exhaustion of the juices of the leaves, and consequently of the vital power of the bushes. It has been noticed that the bushes upon the poorer spots of the land of fruit plantations, or where the drainage is bad, or in what are known in Kent on the greensand soils as ' pinnocky places,' are more liable to receive injury from this aphis than those where the soil is good. This is probably because the bushes give up, or become exhausted sooner upon indifferent land.

As there is no honey-dew from these aphides, the larva being without the anal tubercles peculiar to many other species of aphides, their presence is often unsuspected and undetected, the galls and changed colouring of the leaves being attributed to conditions of weather or soil. Kaltenberg points out in his ' Pflanzenfeinde' that this aphis is well known in Germany as a foe to the currant bushes. Taschenberg also describes it as forming lumps (*Beulen*) upon their leaves, and making them curl up (' Praktische Insekten Kunde,' von Prof. E. L. Taschenberg). Prof. Lintner, in the ' First Annual Report of the Entomology of the State of New York,' alludes to the characteristic bulges and blister-like elevations upon currant leaves caused by it in American fruit plantations. Mr Saunders speaks of it as " an importation from Europe, where it has long been injurious to the currant " (' Insects Injurious to Fruit,' by W. Saunders, Philadelphia,' 1883).

Life History.—Somewhere about the 12th of May the wingless female, viviparous, or bringing forth living young, may be found upon the leaves, and, after the manner of aphides in general, soon begins the long and fertile series of parthenogenetic production. In an incredibly short space of time—in a day or two—the under surfaces of the leaves are covered with larvæ, whose continuous pumping with their club-shaped siphons disarranges the delicate economy of the leaf-tissue, and sucks out the very life-blood of the bush. Compared with the winged females and the winged males this progenitrix is large. It is of a yellowish or yellowish-green colour, and of a somewhat oval shape.

After a time, determined by circumstances not as yet accurately defined, the larvæ, or some of the larvæ, put on pupal form, and soon the winged female speeds from the colony upon long translucent wings to deposit living young on other currant bushes. It is prettily marked, having a yellow body with black and green bars and spots. The thorax is black. The legs are yellow with black extremities, while the antennæ are very long and black.

Very similar to this is the winged male, though rather smaller. The wingless, oviparous (egg-laying) female, with which the male pairs at the end of August, is rather darker in colour, and smaller than the wingless viviparous female, the direct product of the eggs. These are long, large, and peculiarly shaped, being fastened to the stems and twigs of currant bushes by a glutinous liquid, and carefully placed under the thin exfoliated layers of bark, or, more properly, skin, as it is so delicate.

Prevention. Black currant bushes infested with aphides must be cut very ' hard ' in the autumn, and all the cuttings should be removed far from the plantation. If eggs are found upon the stems that are left these should be washed over with a solution of soft soap and paraffin oil of thick consistency, or with a solution of soft soap and petroleum put on with a large paintbrush worked well up and down.

Red currant bushes may be treated similarly, care being taken to work the solution well into the joints between the ' snags,' or little twigs upon which the fruit comes.

Remedies. Washing or syringing with soft soap and quassia is the sole remedy that can be resorted to with any advantageous results, but this is obviously a delicate and a difficult operation, as the bunches of fruit hang immediately under the leaves, and in such thick clusters that the wash would drip into them and be retained, to the injury of the flavour (' Reports on Insects Injurious to Crops,' by Chas. Whitehead, Esq., F.Z.S.).

RHU'BARB. *Syn.* RHEI RADIX (B. P.), RHEUM, L. The root of *Rheum palmatum*, Linn.; *Rheum officinale*, Baillon; and probably other species. Collected and prepared in China and Thibet.

Three principal varieties of rhubarb are known in this country:

Russian or Turkey rhubarb is the produce of six-year-old plants of the mountain declivities of Chinese Tartary; and its principal excellence depends on its more careful preparation, and subsequent garbling, both before its selection for the Russian market, and after its arrival at Kiachta, and again at St Petersburg. At Kiachta all pieces of a porous, grey, or pale colour are rejected, the whole being pared and perforated, the better to determine the quality of the interior portion. At St Petersburg the pieces are again carefully examined and garbled, and are, finally, packed in close cases or chests, which are rendered air-tight by the application of pitch on the outside.

East India or Chinese rhubarb is the produce of the locality just referred to, as well as of other parts of China. It is obtained from younger plants, and its preparation and subsequent selection or garbling is conducted with less care.

English rhubarb is principally produced at Banbury, Oxfordshire, from the *Rheum rhaponticum*. It is cut and dressed up after the manner of Turkey rhubarb, for which it is sold by itinerant vendors, habited as Turks.

Adult. Dr Maisch ('American Journal of Pharmacy,' xliii, 259) says the presence of turmeric may be detected in powdered rhubarb by the following method:—A small quantity of the suspected rhubarb is agitated for a minute or two with strong alcohol, and then filtered, chrysophanic acid being sparingly soluble in this menstruum. The brown yellow colour of the filtrate is due to the resinous principles of rhubarb mainly; if adulterated with turmeric, the tincture will be of a brighter yellow shade; a strong solution of borax produces in both tinctures a deep red-brown colour.

If now pure hydrochloric acid be added in large excess, the tincture of pure rhubarb will instantly assume a light yellow colour, while the tincture of the adulterated powder will change merely to a lighter shade of brown-red.

The test is a very delicate one, and is based on the liberation of boracic acid, which imparts to curcumine a colour similar to that produced by alkalies, while the principles of rhubarb soluble in strong alcohol yield pale yellow solutions in acid liquids.

Qual. Russian or Turkey rhubarb occurs in irregular plano-convex or roundish lumps, perforated with a circular hole; if possesses a yellow colour outside; when recently broken, the inside presents a rich mottled appearance, and evolves a peculiar and somewhat aromatic odour. It is firm, compact, heavy, perfectly free from moisture, and easily grated. Its taste is bitter, slightly astringent, and subacid; and when chewed it feels gritty, and tinges the saliva of a beautiful yellow colour. It breaks with a rough, hackly fracture, is easily pulverised, and its powder is of a bright buff-yellow colour.

East India, Canton, or Chinese rhubarb is in flat pieces, seldom perforated, and its taste and odour are stronger than the other. It is also heavier, tinges the saliva of an orange-red hue, and when pulverised the powder is redder than that of Russian rhubarb.

English rhubarb possesses all the preceding qualities in a greatly less degree. It is light and spongy, does not feel gritty between the teeth, its taste is mucilaginous, and its powder has a peculiar pinkish hue not present in either of the other varieties of rhubarb. As a medicine it possesses little value, and is chiefly employed to adulterate East India and Turkey rhubarb.

Prop., &c. Rhubarb is astringent, stomachic, and purgative. In small doses its operation is principally or wholly confined to the digestive organs; in larger ones, it first acts as a mild aperient, and afterwards as an astringent; hence its value in diarrhœa. It has also been used externally to promote the healing of indolent sores. —*Dose.* As a stomachic, 1 to 5 gr.; as a purgative, 10 to 20 gr. It is most effective when chewed, or in the form of powder produced by grating it.

Rhubarb, Roast'ed. *Syn.* BURNT RHUBARB; RHEUM USTUM, L. *Prep.* 1. Rhubarb, in coarse powder, is carefully and regularly heated in a smooth shallow iron dish, with constant stirring, until its colour has changed to a moderately dark brown, when it is allowed to cool out of contact with the air; when cold, it is reduced to powder, and at once put into a well-closed bottle. 2. (*Hoblyn.*) Roast powdered rhubarb in an iron vessel, constantly stirring, until it becomes almost black; then smother it in a covered jar.—*Dose*, 5 to 10 gr.; as an astringent in diarrhœa, and a tonic in dyspepsia, &c. Professor Procter, the well-known American pharmaceutist, recommends the rhubarb to be only roasted to a 'light brown.'

RHYPOPH'AGON. *Prep.* From yellow soap, sliced, 1 oz.; soft soap (finest), 3 oz.; melt them by the heat of hot water, then allow them to cool a little, and stir in of oil of cloves, ½ dr.; essence of ambergris, 10 drops. It is kept a month before sale. Used for shaving.

RICE. *Syn.* ORYZA, L. The seed of *Oryza sativa*, a plant of the Nat. Ord. GRAMINACEÆ. Several varieties are known in commerce, distinguished by the name of the country or district which produces them. The finest is that imported from Carolina. It reaches this country in a decorticated condition. 'Paddy' is rice with the husk upon it. Dr Letheby estimates that it affords nourishment to not less than a hundred millions of people.

As an article of diet, rice is highly nutritious and wholesome when combined with fresh animal or other nitrogenised food; but, owing to the very small quantity of 'albuminoids' which it contains, and its comparative destitution in saline matter, it is totally unfit to form the principal portion of the diet of the working classes, or the poorly fed, at least in this climate. " It does not appear so well calculated for European constitutions as the potato, for we find the poor constantly reject it when potatoes can be had." This preference evidently depends on something more than mere whim or taste, for some years ago, when rice was substituted for potatoes in some of our union workhouses, the most serious consequences followed. In one of these, nine or ten deaths from scurvy and allied diseases occurred in a single fortnight. Large quantities of rice are annually imported into Britain, and used by distillers in the manufacture of spirits.

Letheby gives the following as the composition of rice:

Nitrogenous matter	.	. 6·3
Carbo-hydrates	.	. 79·5
Fatty matter .	.	. 0·7
Saline matter	.	. 0·5
Water .	.	. 13·0
		100·0

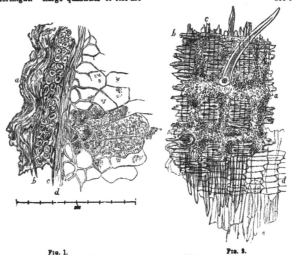

Fig. 1.   Fig. 2.

Microscopic Appearance of Rice.

Fig. 1.—Transverse section of the husk of rice.
Fig. 2.—Appearance of husk as seen in a transparent medium of glycerin and gum ;—a. Siliceous granules arranged in longitudinal and transverse ridges, perforated by openings—stomata, some having hairs over them. b, c. Transverse and longitudinal, brittle, rough-edged fibres. d. A fine membrane of transverse angular cells ; these overlie a very delicate membrane of large cells, e.

Payne gives the following as the composition of dried rice:

Nitrogenous matter	.	. 7·55
Starch .	.	. 88·65
Dextrin, &c.	.	. 1·00
Fatty matter	.	. 0·80
Cellulose	.	. 1·10
Mineral water	.	. 0·90
		100·00

Ash of rice:

Potash	.	. 18·48
Soda .	.	. 10·67
Lime .	.	. 1·27
Magnesia	.	. 11·69
Oxide of iron	.	. 0·45
Phosphoric acid	.	. 53·36
Chlorine	.	. 0·27
Silica .	.	. 3·35
		99·54

Rice, To Cook. If rice is boiled it should be

Microscopic appearance of ground rice-flour.

subjected to a low temperature. The best way of cooking rice, however, is by thoroughly steaming it. By this method, it is said, the loss of nitrogenous matter is prevented, and the grain consequently suffers no diminution of nutritive power, as in the case of boiling.

RICININ. The poisonous principle of castor-oil seeds. It is an albuminoid, one of the phytal-bumoses, and belongs to the class of unorganised ferments. Boiling destroys its activity. It can be prepared from the shelled seeds by percolating with a 10% salt solution. The percolate is saturated with sodium and magnesium sulphates, when there separates a white precipitate, which is purified by dialysis. The shiny mass is scraped from the septum and dried *in vacuo* over sulphuric acid. The product when powdered is white, and contains 10% to 20% of ash, which does not interfere with its physiological action. It is a very poisonous substance.

RICINO'LEIC ACID. A variety of oleic acid discovered in saponified castor oil.

RICK'ETS. *Syn.* RACHITIS, L. A disease, generally confined to childhood, characterised by a large head, prominent forehead, protruded breast-bone, flattened ribs, tumid belly, emaciated limbs, and great general debility. The bones, more particularly those of the spine and legs, become distorted, and exhibit a deficiency of earthy matter; the stools are frequent and loose, a slow fever succeeds, with cough, painful and difficult respiration, and, unless the child rallies, atrophy is confirmed, and death ensues. When recovery takes place there is always more or less deformity left.

The common causes of rickets are bad nursing, exposure to damp and cold, insufficient nutrition, and more especially improper food, *e. g.* meat, potatoes, and alcohol given to infants a few months old. Rickets, like caries of the bones, is a disease which is scarcely known amongst infants whose pap is made of pure wheaten bread, and whose mothers or nurses consume the same themselves.

The treatment of rickets depends more on proper domestic management than on direct medication. Careful nursing, warm dry clothing, thorough ventilation, moderate exercise, and, above all, a light nutritious mixed diet abounding in nitrogenous matter and the phosphates, will do much to effect a cure. To these may be added the administration of the milder chalybeate tonics, bark, or quinine, with occasional doses of some mild aperient, as phosphate of soda, or, when there is diarrhœa, of rhubarb or some other tonic purge. The administration of small doses of phosphate of lime or of dilute phosphoric acid, frequently repeated, is often useful. See BREAD, FARINA, NURSING, &c.

RING'WORM. *Syn.* SCALD-HEAD; PORRIGO, L. The common ringworm, the PORRIGO SCUTU-LATA of medical writers, is a disease that appears in circular patches of little pustules, which afterwards form scabs, leaving a red pimply surface, and destroying the bulbs of the hair in its progress. It spreads rapidly, and is very infectious, often running through a whole school. It chiefly affects the neck, forehead, and scalp of weakly children, and frequently arises without any apparent

cause, but in general may be traced to uncleanness, or contact with parties suffering from the disease.

The treatment of ringworm consists in shaving the part, and keeping it clean with soap and water, at the same time that an occasional mild saline aperient is administered, and a light, nutritious diet, of which the red meat and ripe fruits should form a portion, be rigorously adhered to. When the scabbing commences, dressings of tar ointment, or of the ointment of nitrate or red oxide of mercury, or a mixture of equal parts of the first and either the second or third, should be applied, in each case diluting the mixture with sufficient lard to adapt it to the state of irritability of the part. During this treatment the head should be covered with an ordinary nightcap, or some simple bandage, and not enveloped in a bladder or oil-skin case, as is commonly the practice, since the complete exclusion of atmospheric air tends to aggravate the disease.

RI'PENING. See BREWING, MALT LIQUORS, WINE, &c.

ROASTING. Alexis Soyer recommends, "as an invariable rule," that "all dark meats, such as beef and mutton, should be put down to a sharp fire for at least fifteen minutes, until the outside has acquired a coating of osmazome, or condensed gravy, and then removed back, and allowed to cook gently. Lamb, veal, and pork, if young and tender, should be done at a moderate fire. Veal should even be covered with paper.

"Very rich meat, if covered with paper, does not require basting. Fowls, &c., should be placed close to the fire, to set the skin, and in about ten minutes rubbed over with a small piece of butter, pressed in a spoon. Meats, whilst roasting, should be dredged with flour, just at the time when the gravy begins to appear; the flour absorbs it, and forms a coating which prevents any more coming out. Hares and small game should be treated in the same manner."

Under ordinary circumstances as to the fire, and the distance between it and the joint, beef, mutton, and veal take about ¼ hour per lb. in roasting. Lamb, poultry, and small game require only 12 to 14 minutes per lb.; whilst veal takes fully 15 minutes, and pork takes from ¼ hour to 20 minutes, as they must always be well done. The flesh of old animals requires more cooking than the flesh of young ones; and inferior, tough, and bony parts than the prime joints and pieces.

Roasting is not an economical method of cooking pieces of meat abounding in bone or tendinous matter, since the nutritious portion of these is either destroyed or rendered insoluble by the heat employed. Thus the raw bones from a joint are capable of affording a rich and excellent basin of soup, highly nutritious; whilst the bones from a corresponding joint which has been roasted are nearly worthless when so treated. The same applies with even greater force to the gristly and tendinous portions. A dry heat either destroys them or converts them into a horny substance, unfit for food; whilst by boiling they are transformed into a highly succulent and nutritious article of food, besides affording excellent soup or jelly. Hence the policy of 'boning' meat before roasting or baking it; or, at all events, of removing the bony portion which would be most exposed to the action of the fire. See BONE and JELLY.

ROB. *Syn.* ROOB. A term, derived from the Arabic, formerly applied to the inspissated juice of ripe fruit, mixed with honey or sugar to the consistence of a conserve of thin extract. Rob of elder-berries (ELDER ROB; ROOB SAMBUCI), juniper berries (JUNIPER ROB; ROOB JUNIPERI), mulberries (MULBERRY ROB; ROOB DIAMORUM), and walnuts (WALNUT ROB; ROOB DYACARYON), with a few others, are still found in some of the foreign Pharmacopœias.

ROCK. The popular name of a sweetmeat formed of sugar boiled to a candy, and then poured upon an oiled slab, and allowed to cool in the lump. It is variously flavoured.

ROCK CRYSTAL. Native crystallised silica. See QUARTZ.

ROCK OIL. See PETROLEUM.

ROCK SOAP. A native silicate of alumina; used for crayons, and for washing cloth.

ROC'KETS. (In *pyrotechny*.) *Prep.* The CASES. These are made of stout cartridge-paper, rolled on a mould and pasted, and then throttled a little below the mouth, like the neck of a phial. The diameter should be exactly equal to that of a leaden ball of the same weight, and the length should be equal to 3½ times the external diameter. Above the spindle there must be one interior diameter of composition driven solid. They are filled with the following mixtures, tightly driven in, and when intended for flight (SKY-ROCKETS) they are 'garnished,' and affixed to willow rods to direct their course.

The COMPOSITION. 1. (*Marsh.*) *a.* For 2-oz. rockets. From nitre, 54½ parts; sulphur, 18 parts; charcoal, 27½ parts; all in fine powder, and passed through lawn.

b. For 4-oz. do. From nitre, 64 parts; sulphur, 16 parts; charcoal, 20 parts; as the last.

c. For ½-lb. to 1-lb. do. From nitre, 62½ parts; sulphur, 15½ parts; charcoal, 21½ parts.

2. (*Ruggieri.*) *a.* For rockets of ½-inch diameter. From nitre, 16 parts; charcoal, 7 parts; sulphur, 4 parts.

b. For ½- to 1½-inch rockets, use 1 part more of nitre.

c. For 1½-inch rockets, use 2 parts more of nitre.

d. By using 1 part less of charcoal and adding respectively 3, 4, and 5 parts of fine steel filings, the above are converted into ' BRILLIANT FIRES.'

e. By the substitution of coarse cast-iron borings for filings, and a further omission of 2 parts of charcoal from each, the latter are converted into ' CHINESE FIRE.'

HAND-ROCKETS and GROUND-ROCKETS are usually loaded with nothing but very fine meal gunpowder and iron or zinc filings or borings.

After SKY-ROCKETS and WATER-ROCKETS are charged, a piece of clay is driven in, through which a hole is pierced, and the ' head' or ' garniture' filled with stars, and a little corn powder is then applied. See FIRES, STARS, and PYROTECHNY.

ROLL (Wine). *Prep.* Soak a French roll or sponge biscuit in raisin, marsala, or sherry wine, surround it by a custard or cream thickened with eggs, and add some spice and ornaments.

ROLLS. A variety of fancy bread, generally in the form of small semi-cylindrical cakes, prepared by the bakers, and intended to be eaten hot for breakfast. They differ from ordinary fine or French bread, as it is called, chiefly in containing more water. Some are wetted up with milk and water, and are hence called ' milk rolls.'

ROOT. *Syn.* RADIX, L. That part of a plant which imbibes its nourishment from the soil or medium in which it grows. In popular language, bulbs, corms, tubers, &c., are improperly included under this term.

ROPES AND KNOTS. The art of tying a knot under every imaginable condition, which shall be secure and perform the work required of it, is almost unknown except to sailors, or to those whose lives are spent in the constant use of ropes and cordage. There are very few accessible books in which any detailed description of knots is to be found, and, in the belief that such description will be of great practical utility, the following cuts, kindly lent by Messrs Cassell, are inserted. The reader who requires further information on the subject should consult the weekly periodical ' Work,' published by this firm, vol. iii, Nos. 109, 113, 117, &c. (See pages 1456—1459.)

RO'PINESS. See MALT LIQUORS and WINES.

ROSE. *Syn.* ROSA, L. The typical genus of the Nat. Ord. ROSACEÆ. It includes numerous species greatly prized as garden plants.

Rose, Cabbage. *Syn.* HUNDRED-LEAVED ROSE; ROSÆ CENTIFOLIÆ PETALA (B. P.), ROSA CENTIFOLIA (Ph. L. and E.), L. " The fresh petals " (Ph. L.) of this species are used in medicine. Odorous and slightly astringent and laxative. See WATER and SYRUP.

Rose, Dog. The *Rosa canina*, or wild briar. See HIPS.

Rose, French. *Syn.* RED ROSE; ROSÆ GALLICÆ PETALA (B. P.), ROSA GALLICA (Ph. L. E. and D.), L. " The fresh and dried unexpanded petals " (Ph. L.) of this species are officinal. The white claws of the petals are removed before drying them.

Uses, &c. The red rose is an elegant astringent and tonic, and, as such, is used as the basis of several pharmaceutical preparations. See CONFECTION, HONEYS, INFUSION, and SYRUP.

Rose of Jericho. *Syn.* ANASTATICA HIEROCHUNTICA, L. An annual plant from the deserts of Arabia and Egypt. After withering, its spreading branches roll themselves up in a ball, and the whole plant is detached and blown about by the wind, the branches expanding again with the first rainfall. By this means the seeds are easily dispersed.

ROSEMARY. *Syn.* ROSMARINUS (Ph. L. E. and D.). The flowering tops of *Rosmarinus officinalis*, Linn., or the common rosemary of our gardens, are officinal in the Ph. E. and D.; as is also the oil (oleum rosmarini) in the B. P. and Ph. L. The odour of both is refreshing, and they are reputed carminative, emmenagogue, and neurotic. The dried leaves are occasionally used by the hysterical and hypochondriacal as a substitute for China tea. The oil is an ingredient in Hungary water, and is much used in various cosmetic compounds, under the presumption of its encouraging the growth of hair and improving its quality.

OVERHAND KNOT.

FOURFOLD OVERHAND KNOT, made by passing the end of the rope several times through the bight, often termed a 'blood knot,' used for whip thongs, &c.

FLEMISH OR FIGURE-OF-EIGHT KNOT.

The Same drawn tight.

THE SAILOR'S KNOT, TRUE KNOT, or REEF KNOT for uniting two ropes *of the same size ;* will *not* answer when they are of different thicknesses.

REEF KNOT HAULED STRAIGHT.

REEF KNOT HALF MADE.

GRANNY KNOT.

WEAVER'S KNOT HALF MADE.

GRANNY KNOT CLOSED.

WEAVER'S KNOT CLOSED.

FISHERMAN'S KNOT.

EYE KNOTS.

CRABBER'S EYE.

RUNNING KNOT.

FISHERMAN'S EYE.

OPEN-HAND EYE.

FLEMISH EYE.

BOWLINE.

RUNNING BOWLINE ON BIGHT.

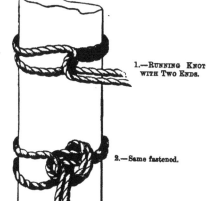

1.—RUNNING KNOT WITH TWO ENDS.

2.—Same fastened.

HANGMAN'S KNOT.

HITCHES AND BENDS.

Two Half Hitches.

Timber Hitch.

Builder's Knot.

Killick Hitch.

Magnus Hitch.

Fisherman's Bend.

Topsail Halliard Bend.

Carrick Bend.

Rolling Hitch.

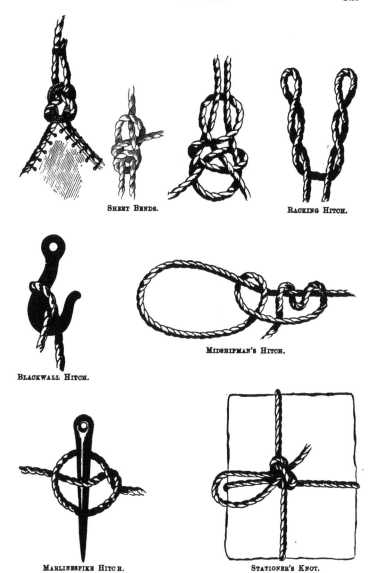

SHEET BENDS.

RACKING HITCH.

BLACKWALL HITCH.

MIDSHIPMAN'S HITCH.

MARLINESPIKE HITCH.

STATIONER'S KNOT.

ROSE PINK. See RED PIGMENTS.

RO'SIN. See RESIN.

ROSY-DROP. See ACNE.

ROT. *Syn.* GREAT-ROT, HYDROPHIC ROT, SHEEP-ROT, WET-ROT. A disease peculiar to sheep, produced by the presence in the liver of the *Distoma hepatica*, a parasite commonly known under the name of a 'fluke.' Rot prevails during very wet or rainy seasons. The leading symptoms are loss of flesh and vivacity; the lips and tongue look vivid, and the eyes sad and glassy; the pelt comes off on the slightest pull; the breath is fetid, and the urine highly coloured and scanty; and there is either black purging or obstinate costiveness. The treatment consists in a change to a dry, warm, elevated situation, and a dry diet, consisting of oats, barley meal, tail-wheat, &c., to which some turnips, carrots, or mangel-wurzel may be added, with a liberal supply of common salt, and a few grains of sulphur, daily. These last two substances form the active ingredients in Flesh's 'Patent Restorative.' See FLUKE, NEMATODA, WORMS.

ROT (in Timber). See DRY-ROT.

ROTA'TION (of Crops). The rotation or succession of crops is absolutely necessary for the successful and economical cultivation of the soils. Crops have been divided by agriculturists into exhausting crops, restoring crops, and cleaning crops. The most exhausting crops are usually considered to be those of corn, but all those that are allowed to ripen their seed and which are carried off the ground are also exhausting, but in different degrees. Even clover, tares, and grass cut green are considered as exhausting, but in a less degree than those that are allowed to ripen. Restoring crops are such as are allowed to decay upon the ground, or are consumed upon it by domestic animals. Cleaning crops are such as are grown in drills, and undergo the usual operations of weeding, hoeing, &c.; the majority of these may also be regarded as exhausting crops. An exhausting crop should always be followed by a restoring or a cleaning crop; or, where possible, by both combined. Crops should also succeed each other in such a way that the soil may not be exhausted of any one particular kind of nutriment. This is best effected by so rotating the crops that plants which are nearly allied should not succeed each other on the same soil, or, at all events, not more than once. See AGRICULTURE, SOILS, &c. :

ROTTEN-STONE. See TRIPOLI.

ROUGE. *Syn.* TOILET ROUGE; ROUGE VEGETAL, ROUGE D'ESPAGNE, Fr. *Prep.* Wash safflower (any quantity) until the water comes off colourless; dry and pulverise it, and digest the powder in a weak solution of crystallised carbonate of soda; then place same fine cotton-wool at the bottom of a porcelain or glass vessel, pour the filtered tinctorial solution on this, and throw down the colouring matter, by gradually adding lemon juice or white-wine vinegar, until it ceases to produce a precipitate; next wash the prepared cotton in pure cold water, and dissolve out the colour with a fresh solution of soda; to the new solution add a quantity of finely powdered talc or French chalk, proportionate to the intended quality of rouge; mix well, and precipitate with lemon juice, as before; lastly collect the powder, dry it with great care, with as little heat as possible, and triturate it with a very small quantity of oil of olives, to render it smooth and adhesive.

Obs. According to the best authorities, this is the only article which will brighten a lady's complexion without injuring the skin. The relative fineness and proportion of talc employed determines the quality of the rouge. It is applied by means of a camel-hair pencil, a small 'powder puff,' or a hare's foot. It is also employed under the form of 'pommade' and 'crepons.' The last of these consist of pieces of white woollen crape, upon which the colouring matter of the carthamus has been precipitated, instead of upon the talc, noticed above.

The following articles also pass under the name of rouge, and are used for the purposes named after each:

Rouge d'Athenes, Vert. *Syn.* PURE ROUGE. See CARTHAMINE.

Rouge, Brown-red. Jeweller's rouge.

Rouge, Chinese Card. This is said to be a 'carthamate of soda;' it is colourless when applied, but, being decomposed by the acid secretions of the skin, acquires a most beautiful rose-like tint (*O'Shaughnessy*).

Rouge, Indienne. The terra persica, or Indian red; imported from Ormus.

Rouge, Jeweller's. Sesquioxide of iron prepared by calcination. Used to polish gold, &c.

Rouge, Liquid. The red liquid left from the preparation of carmine; or a solution of carmine in weak carbonate of potash water, or of pure rouge in alcohol acidulated with acetic acid.

Rouge de Prusse. Light red or burnt yellow ochre. See RED PIGMENTS.

Rouge, Spanish Lady's. This is cotton-wool which has been repeatedly wetted with an ammoniacal solution of carmine, and dried. It is applied like 'rouge crepons.'

ROUGE'ENING. See WINES.

RUBBER, GUATEMALA and WEST INDIAN, from *Castilloa elastica*, Cerv. One of the largest forest trees of the north-east coast of Mexico, and found also in Honduras, Nicaragua, Guayaquil, &c. It is the Ulé of the natives. The plant has been introduced into India, Ceylon, and other countries.

RUBEFA'CIENTS. *Syn.* RUBEFACIENTIA, L. Substances or agents which, when applied for a certain time to the skin, occasion a redness and increase of heat, without blistering. They act as counter-irritants. Mustard, powdered ginger (both made into a paste with water), hartshorn and oil, and ether and spirit of wine (when their evaporation is prevented), are familiar examples of this class of remedies.

RUBE'OLA. See MEASLES.

RUBI'ACIN. An orange-coloured substance, obtained from madder.

RUBID'IUM. [Eng., L.] A metal belonging to the alkaline group discovered by Bunsen and Kirchhoff by means of spectrum analysis. It is found in many mineral waters associated with cæsium.

RU'BY. See GEMS and PASTES.

RUE. *Syn.* RUTÆ FOLIA, RUTA (Ph. L. & E.), L. "The leaf *Ruta graveolens*" (Ph. L.). A powerful antispasmodic, diuretic, and stimulant. It is also reputed nervine and emmenagogue. The fresh leaves are powerfully acrid, and even vesicant; but they become milder in drying.— *Dose.* Of the powder, 15 to 30 gr., twice or thrice daily; in hysteria, flatulent colic, &c. See IN-FUSION and OILS (Volatile).

RUM. *Syn.* SPIRITUS JAMAICENSIS, SPIRITUS SACCHARI, L. An ardent spirit obtained by distillation from the fermented skimmings of the sugar-boilers (syrup scum), the drainings of the sugar-pots and hogsheads (molasses), the washings of the boilers and other vessels, together with sufficient recent cane juice or wort, prepared by mashing the crushed cane, to impart the necessary flavour. The sweet liquor before fermentation commonly contains from 12% to 16% of saccharine, and every 10 galls. yields from 1 to 2 galls. of rum.

The average strength of rum, as imported into this country, is about 20 o. p. Like all other spirits, it is colourless when it issues from the still, but owing to the taste of the consumer the distiller is compelled to colour it before it leaves his premises.

Obs. Rum is imported from the West Indies. The best comes from Jamaica, and is hence distinguished by that name. Leeward Island rum is less esteemed. The duty on rum is 10s. 2d. per proof gallon if imported direct from any of the British colonies (colonial rum), but 10s. 5d. if from any other part of the world (foreign rum). The consumption of rum has long been declining in England, its place being chiefly supplied by gin. Rum owes its flavour to a volatile oil and butyric acid, a fact which the wary chemist has availed himself of in the manufacture of a butyric compound (essence of rum) for the especial purpose of enabling the spirit dealer to manufacture a factitious rum from malt or molasses spirit. In Jamaica it is usual to put sliced pine-apples into the puncheons containing the finer qualities of rum, which is then termed pine-apple rum. See ALCOHOL, SPIRIT, &c.

RUM, BAY (E. Rother's Formula for). According to an American authority, true bay rum is made from *Pimenta acris* (*Myrica acris*, Schwartz; *Myrtus acris*, Willd.), and not from *Laurus nobilis*, as commonly supposed; the method of its distillation not being known outside the West Indies, it has been customary to make it extemporaneously with the oil of bay distilled from the leaves of the former plant. This preparation is inferior in fragrance, however, to the genuine article. The following formula of R. Rother is said to give very good results:—Take of oil of bayberry, 1 fl. oz.; Jamaica rum, 1 pint; strong alcohol, 4 pints; water, 3 pints. Mix the rum, alcohol, and water, then add the oil; mix and filter.

RUMICIN. A resinous powdered extract obtained from the root of *Rumex crispus*, yellow dock. A tincture, 1 in 10 of proof spirit, is used as well as the resin, acting as tonic, astringent, and antiscorbutic.—*Dose.* Tincture, 2 to 10 minims; resin, 2 to 5 gr.

RUPERT'S DROPS. These are made by letting drops of melted glass fall into cold water. By this means they assume an oval form, with a tail or neck resembling a retort. They possess this singular property that, if a small portion of the tail is broken off, the whole bursts into powder with an explosion, and a considerable shock is communicated to the hand.

RUPIA. This is an affection of the skin attended by the formation on it of vesicles, that develop into ulcers which copiously discharge a foul, unhealthy, and reddish matter. After a time this matter hardens and forms a thick incrustation over the sores.

The best treatment is to put the patient upon a generous diet, including wine, and to administer iodide of potassium with sarsaparilla or quinine. The scabs should be poulticed.

RUPTURE. See SURGERY.

RUSKS. *Prep.* From 4 eggs; new milk and warm water, of each, ½ pint; melted butter and sugar, of each, ¼ lb.; yeast, 3 table-spoonfuls; beat well together with as much flour, added gradually, as will make a very light paste; let it rise before the fire for half an hour, then add a little more flour, form into small loaves or cakes 5 or 6 inches wide, and flatten them; bake these moderately, and, when cold, cut them into slices of the size of rusks, and put them into the oven to brown a little. A nice tea-cake when hot, or with caraways to eat cold. PLAIN RUSKS are made by simply cutting loaves of bread into slices, and baking them in a slow oven to the proper colour.

RUS'MA. An arsenical iron pyrites, found in Galatia, which, when reduced to powder, and mixed with half its weight of quicklime, is used by the Turkish ladies to make their 'PSILOTHRONS,' or compounds to remove superfluous hair. See DEPILATORY.

RUST. *Syn.* RUBIGO, L. The coating or film of oxide or carbonate which forms on the surface of several of the metals when exposed to a moist atmosphere; more particularly, that which forms on iron or steel (FERRI HYDRATE; HYDRATED SESQUIOXIDE OF IRON; FERRUGO, FERRI RUBIGO).

To prevent iron or steel goods rusting, it is merely necessary to preserve them from damp or moisture. In the shops, small articles in steel are, commonly, either varnished or enclosed in quicklime finely pulverised; large articles are generally protected with a coating of plumbago, or of boiled oil, or some cheap varnish, applied to them, previously gently heated. Surgical instruments are frequently slightly smeared with a little strong mercurial ointment with the same intention.

Spots of rust may be removed from the surface of polished iron or steel by rubbing them with a little tripoli or very fine emery made into a paste with sweet oil; or, chemically, by a mixture of polisher's putty powder with a little oxalic acid, applied with water. When the last is employed, the articles should be afterwards well rinsed in pure water, then wiped dry, and finished off with a warm and dry rubber, in order to remove every trace of acid.

RUTHE'NIUM. Ru = 103·5. A metal discovered by Claus, associated with iridium, in the residue from crude platinum, which is insoluble in aqua regia. It has recently been found in

Borneo in a mineral called laurite. It forms small angular masses, with a metallic lustre; is very brittle and infusible; resists the action of acids, but slowly oxidises when heated in the air. Sp. gr. 12·261 at 0°.

In Fremy's process for separating osmium from the residues of platinum ore, ruthenium occurs as the dioxide. By heating this dioxide in a current of hydrogen, the metal may be obtained in the form of a powder of dark grey colour.

With oxygen, ruthenium forms six compounds, of which the first three need only be noticed.

Ruthenium Monoxide. RuO. A dark grey powder insoluble in acids.

Ruthenium Dioxide. RuO₂. Small green crystals obtained in the extraction of osmium from the residue of a solution of platinum in aqua regia.

Ruthenium Trioxide. RuO₃. Syn. RUTHENIC ANHYDRIDE. Is known only in combination.

Ruthenic Sesquioxide. Ru₂O₃. Occurs in the anhydrous form when the metal is ignited in a current of air. It is the most stable of the basic oxides of the metal.

Ruthenium Sesquichloride. Ru₂Cl₆. By evaporating a solution of the corresponding hydroxide in hydrochloric acid. A deliquescent astringent mass.

Ruthenium Tetrachloride. RuCl₄. By evaporating a solution of the corresponding hydroxide in hydrochloric acid. A reddish-brown hygroscopic mass, soluble in water and alcohol, to which it imparts a bitter taste.

Tests, &c. Concentrated solution of potassium chloride and ammonium chloride precipitates the ruthenic salts dark red, and on boiling with water the characteristic black drug, finely divided oxychloride, is formed. Sulphuretted hydrogen first colours solutions of ruthenium compounds blue, and then the brown sulphide is thrown down; this is nearly insoluble in ammonium sulphide.

Ruthenium is estimated quantitatively as the metal like other members of the gold group.

RYE. Syn. SECALE, L. The seed of *Secale*

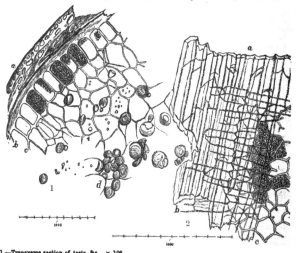

FIG. 1.—Transverse section of testa, &c. × 108.
FIG. 2.—Coats *in situ* from without. × 170. *a*, External; *b*, Middle; *c*, Internal coat; *d*, Starch grains. × 108.

cereale, a gramineous plant, the native country of which is undetermined. It is a more certain crop and requires less culture and manure than wheat, and is hence largely cultivated in Germany, Russia, and in the northern parts of Europe, where it is extensively employed for bread. When roasted it is occasionally used as a substitute for coffee. It furnishes an excellent malt for the distillation of spirit, and is much used in the making of hollands.

Rye bread is very likely to cause diarrhœa in those unaccustomed to partake of it. By continued use, however, this inconvenience disap-

pears. Rye bread is acid and dark in colour. It is about equal in nutritive power to wheat. It is less abundant than wheat in fibrin, but richer in casein and albumen.

The foregoing plate represents the microscopic appearance of rye.

Sommer recommends the microscopic examination of rye flour to be conducted as follows :— The flour is placed on a glass slide and moistened with water; a single drop of oil of vitriol is added, and a small disc is laid upon it. If, now, it be viewed with a magnifying power of 200, the starch grains of wheat and rye are seen to dissolve in a uniform manner, but the grains of barley starch, after losing their external coat, break up into a number of polyhedra before their solution is completed.

Rye, Spurred. See ERGOT.

SABADIL′LA. *Syn.* CEBADILLA, CEVADILLA, SABADILLA (B. P., Ph. E.), L. The dried fruit (*Asagræa officinalis*). A drastic and dangerous cathartic, occasionally used in tapeworm; and, externally, to destroy pediculi, but even for this purpose, when the scalp has been denuded or ulcerated, it has sometimes caused death. It is now used chiefly as a source of VERATRINE.

SA′BLE. The *Mustella Zibellina*, Linn., a small quadruped of the marten-cat family, found in Northern Asia. Its fur is remarkable for its fine quality and rich colour, and for the hairs turning with equal ease in every direction. The skins of the rabbit, cat, &c., dressed, painted, and lustred, are sold under the name of COMMON or MOCK SABLE.

SABOTIÈRE. [Fr.] An apparatus of peculiar construction, employed by the French confectioners for making ices. It consists of a pail to contain a freezing mixture, and an inner vessel for the creams to be iced. It may be used with a mixture of pounded ice and salt, or any other freezing mixture. The pail and cream vessels being loaded and closely covered, an alternate rotatory motion is given to the apparatus by means of the handle for ten or fifteen minutes, care being taken to occasionally scrape down the frozen portion of the cream from the sides by means of a wooden spoon. See ICES and REFRIGERATION.

SACCHAR′IC ACID. *Syn.* OXALHYDRIC ACID†. A compound resulting from the action of dilute nitric acid on sugar.

SACCHARIN. *Syn.* SACCHARINUM, BENZOYLSULPHONIC IMIDE, or BENZOIC SULPHIMIDE, GLUSIDUM, GLUSIDE. This compound is remarkable for its powerful sweetness. It is made from toluene, of which large quantities are produced in the manufacture of coal-gas. The toluene is treated with sulphuric acid, yielding ortho- and para-toluene sulphonic acids. These acids are converted into calcium salts, and, further, into sodium salts, by treating with sodic carbonate. The next step is to act on them with phosphorus trichloride and a current of chlorine, the product being a mixture of sulphonic chlorides.

These chlorides are separated by crystallisation, ortho-sulphonic chloride being retained. The ortho-compound is next treated with ammonium carbonate and steam, being thus converted into toluene-sulphonic imide, which, by oxidation with potassium permanganate, yields saccharin, having the formula

$$C_6H_4\binom{CO}{SO_2}NH.$$

Saccharin is a white minutely crystalline powder of intensely sweet taste. It is sparingly soluble in water; moderately soluble in alcohol, ether, chloroform, and glycerine. It unites with alkaline carbonates and hydrates to form salts, which are freely soluble in water, retaining their characteristic sweet taste. Fused with potash

saccharin gives salicylic acid, a solution of which gives a purple colour with ferric chloride.

It is generally estimated that the sweetening power of saccharin is 300 times that of sugar; if one grain of saccharin be dissolved in one gallon of water, sweetness is plainly perceptible.

Reports were published that the use of saccharin was injurious; but Dr Thos. Stevenson reports that (1) saccharin is quite innocuous when taken in quantities largely exceeding what would be taken in ordinary dietary; (2) saccharin does not interfere with or impede the digestive processes when taken in any practicable quantity.

Prep. 1. (Liquor saccharini; solution of saccharin.) Take of saccharin, 512 gr.; bicarbonate of sodium, 240 gr.; alcohol, 4 fl. oz.; water, sufficient to produce 16 fl. oz.

2. (Elixir saccharini, B. P. C. [the unofficial formulary of the British Pharmaceutical Conference]; elixir of saccharin.) Take of saccharin, 480 gr.; bicarbonate of sodium, 240 gr.; rectified spirit, 2½ fl. oz.; distilled water, a sufficiency. Rub the saccharin and bicarbonate of sodium in a mortar, with half a pint of distilled water gradually added. When dissolved, add the spirit, and filter with sufficient distilled water to produce one pint of elixir. Each fluid drachm represents three grains of saccharin.—*Dose*, 5 to 20 minims.

Uses. Mainly as a substitute for sugar. Especially useful in cases of diabetes for sweetening the patient's food. It also possesses an antiseptic action.

SACCHARINE FERMENTATION. This occurs during the germination and kiln-drying of grain in the operation of malting, and in the mashing of malt in brewing. The sweetening of bread during its exposure to heat in the oven is also included under this head by many writers.

The substance which most powerfully excites the sugar fermentation was first shown by Payen and Persoz to be a peculiar principle, to which they have given the name of 'DIASTASE.' This is always present in good malt, and possesses the singular property of converting STARCH successively into gum (dextrin) and sugar, at a temperature ranging between 149° and 168° F. During the action of this substance on starch it is itself decomposed; and when the sugar fermentation ceases it is found to have entirely disappeared. It is the presence of diastase in malt which alone converts the starch of the grain into sugar during the operation of mashing with hot water; and hence the absolute necessity of employing water at the proper temperature, as on this depends the strength and sweetness of the wort, and, consequently, its fitness for undergoing the vinous fermentation, and for making beer. Vegetable albumen and gluten also possess the property of exciting the saccharine fermentation, but in a considerably inferior degree to diastase.

The sugar formed during the germination of seeds containing starch results from the action of diastase, and disappears as soon as the woody fibre (lignin), which has a similar constitution, is developed, forming the skeleton of the young plant (*Liebig*). See BREWING, DIASTASE, DEXTRIN, &c.

SACCHAROM′ETER. An instrument similar

in principle to the common spirit hydrometer, but so weighted and graduated as to adapt it for the indication of the richness of malt worts in sugar or saccharine, expressed in pounds per barrel, or the excess of gravity over that of water, the last being taken at 1000. See BREWING, SYRUP, WORT, &c.

SACHET. *Syn.* SACCULUS, L. Sachets (SACCULI) are little bags containing dry substances, used for the external medication of parts, or for communicating agreeable perfumes to wearing apparel, drawers, furniture, &c. Those belonging to perfumery are commonly filled with mixtures of fragrant vegetable substances, reduced to coarse powder, and differ from those employed for *pot-pourri* chiefly in being used in the dry state. Sacculi are now seldom employed in this country in legitimate medicine. See POWDERS (Scented), &c.

Sachet, Ammoniacal. *Syn.* SACCULUS AMMONIACALIS. *Prep.* Equal parts of sal-ammoniac and quicklime are mixed, and sprinkled between cotton wadding, which is to be quilted in muslin.

Sachet, Anodyne. (Quincy.) *Syn.* SACCULUS ANODYNUS. *Prep.* Chamomiles, 1 oz.; bay berries, 1 oz.; lavender flowers, ½ oz.; henbane seed, 1 dr.; opium, 1 dr. To be dipped in hot spirits.

Sachet, Anti-phthisic. *Syn.* SACCULUS ANTIPHTHISICUS, L. *Prep.* Dissolve of aloes, 1 oz., in strong decoction of fresh rue, ¼ pint ; next fold a piece of soft muslin in eight folds large enough to cover the chest and part of the stomach ; steep this in the decoction, and dry it in the shade ; lastly, place in a small bag, one side of which is formed of scarlet silk or wool, and the other, intended to be worn next the skin, of the finest net or gauze. A celebrated domestic remedy for consumption and asthma. It is intended to be constantly worn on the chest.

Sachet, Resolv'ent. *Syn.* MELTING BAG; SACCULUS RESOLVENS, L. *Prep.* 1. (*Dr Breslau.*) Iodide of potassium, 1 part ; sal-ammoniac, 8 parts; dry, and reduce each separately to fine powder; mix them, and enclose ½ oz. to 1 oz. of the mixed powder in a small bag of linen or silk. Used as a resolvent to indolent tumours, especially goitres and scrofulous indurations. It should be worn on the part night and day for some time. The part next the skin should be well pricked with a needle, and the powder shaken up and readjusted every two or three days ; and it should be renewed about once a fortnight.

2. (*Trousseau* and *Reveil.*) Iodide of potassium, 1 part ; burnt sponge, 4 parts ; fine sawdust, 5 parts ; as before.

Sachet, Sponge. *Syn.* SACCULUS SPONGII, COLLIER DE MORAND. *Prep.* Muriate of ammonia, chloride of sodium, burnt sponge, of each, 1 oz.; mix; sprinkle the powder on a piece of cotton-wool, and quilt between muslin, in the form of a cravat. To be worn constantly in goitre or bronchocele, renewing it every month.

Sachet, Stomachic. (*Fuller.*) *Syn.* SACCULUS. *Prep.* Mint, 4 dr.; wormwood, thyme, red roses, each, 2 dr.; balastines, angelica root, caraway seed, nutmeg, mace, cloves, of each, 1

dr. Coarsely powder the ingredients, and put them into a bag, to be moistened with hot red wine when applied for flatulence.

SACK. [From SEC, Fr., dry.] A wine used by our ancestors, supposed by some to have been Rhenish or Canary ; but, with more probability, by others to have been dry mountain—vin d'Espagne, vin sec (Howell, 'Fr. and Eng. Dict.,' 1650). Falstaff (in Shakespeare's day sack was occasionally adulterated with lime, as we learn from Falstaff's speech to the Drawer : " You rogue, there's lime in this sack ") calls it 'sherris sack' (sherry sack), from Xeres, a sea town of Corduba, where that kind of sack (wine) is made (*Blount*). At a later period the term came to be used as a general name for all sweet wines.

SAFFLOWER. *Syn.* BASTARD SAFFRON, DYER'S S. ; CARTHAMUS, L. The florets of *Carthamus tinctorius*, a plant cultivated in Spain, Egypt, and the Levant. It contains two colouring principles, the one yellow, and the other red. The first is removed by water, and is rejected. The second is easily dissolved out by weak solutions of the carbonated alkalies, and is again precipitated on the addition of an acid. This property is taken advantage of in the manufacture of rouge, and in dyeing silk and cotton.

The most lively tints of cherry, flame, flesh, orange-red, poppy, and rose-colour are imparted to silk by the following process, modified to suit the particular shade required :—The safflower (previously deprived of its yellow colouring matter by water) is exhausted with water containing either carbonate of sodium or of potassium, in the proportion of about 5% of the weight of the prepared dye-stuff acted on ; the resulting liquid is next treated with pure lemon juice until it acquires a distinct and rich red colour ; the silk is then introduced and turned about as long as it is perceived to take up colour, a little more lemon juice being added as may appear necessary ; for deep shades this is repeated with one or more fresh baths, the silk being dried and rinsed between each immersion ; it is, lastly, brightened by turning it for a few minutes through a bath of warm water, to which a little lemon juice has been previously added. For flame-colour the silk should receive a slight shade with annotta before putting it into the safflower bath. For the deeper shades, when expense is an object, a little archil is commonly added to the first and second bath. See CARTHAMIN.

SAFFRON. *Syn.* CROCUS. The prepared stigmata or stigmas of the *Crocus sativus*, or saffron crocus. There are two principal varieties known in commerce :

1. (SAFFRON, HAY S. ; CROCUS IN FOENO, C. HISPANIOLUS, CROCI STIGMATA, CROCUS—B. P., Ph. L., E., and D.) This consisted of the stigmas, with part of the styles, carefully picked from the other parts of the flowers, and then dried on paper by a very gentle heat, generally in a portable oven constructed for the purpose. Saffron owes its value to a beautiful colouring matter called *polychroite*.

2. (CAKE SAFFRON ; CROCUS IN PLACENTÂ.) This, professedly, merely varies from the last, it being compressed into a cake after it has

become softened by the fire, and being then dried in that condition. The 'cake saffron' of commerce is now, however, mostly, if not entirely composed of safflower made into a paste with some sugar and gum water, rolled out on paper into oval cakes 10 to 12 inches long, 9 or 10 broad, and about 1-8th of an inch thick, and then dried. "I can detect neither saffron nor marigold in them" (Dr Pereira).

Pur. Saffron, of all the articles of commerce except French brandy, is, perhaps, the one most largely and constantly adulterated. Abroad it is frequently mixed with safflower, and in England with 'prepared marigolds,' or 'French (mock) saffron.' These frauds may be detected by the inferiority of the colour, and by soaking the leaves in water, when the stigmas of the *Crocus sativus* may be readily distinguished from the florets of safflower and the petals of marigolds. Winckler and Grüner proposed to detect these substances by means of a solution of nitrate of silver or of sesquichloride of iron. The infusion of true saffron is not altered by those reagents, but that of either of the above-mentioned adulterants is rendered opaque, and is at length precipitated. "It consists of tripartite filaments, of an orange-red colour, with the small filaments towards the apex dilated" (Ph. E.). Old and dry saffron is 'freshened up' by rubbing it between the hands slightly oiled, and then repicking it.

The late Mr D. Hanbury, F.R.S., found that the article known in commerce as alicante saffron was largely sophisticated with carbonate of lime, which he says had been made to adhere to the thread-like saffron without in the least altering its general appearance. To ascertain the amount of earthy matter thus fraudulently added, he subjected several specimens of saffron to incineration, each having in the first instance been dried in warm air until it ceased to lose its weight. The result indicated that while good Valentia saffron yields from 4% to 6% of ash, the alicante furnishes from 12% to 28%. The method of taking a sample of saffron for earthy adulteration which Mr Hanbury recommends is this:—Place in a watch-glass a small quantity (say 1 gr.) of the saffron, and drop upon it 8 or 10 drops of water; lightly touch the saffron with the tip of the finger, so as to cause the water to wet it. If the drug is free from earthy matter, a clear bright yellow solution will be immediately obtained; if adulterated, a white powder will instantly separate, causing the water to appear turbid; and if a drop of hydrochloric acid be now added, a brisk effervescence will take place.

Mr Hanbury says that saffron almost always contains a few of the pale yellow stamens, accidentally gathered; but the pollen from them which is detached when the drug is wetted, but which is minute in quantity, is easily distinguished from carbonate of lime by not dissolving when hydrochloric acid is added. Moreover the form of pollen-grains may be easily recognised under the microscope.

Mr Hanbury furthermore states that an effectual method of examination is to scatter a very small pinch of saffron on the surface of a glass of warm water. The stigma of the saffron crocus imme-

diately expands, and exhibits a form so characteristic that it cannot be confounded with the flowerets of safflower, marigold, or arnica, or with the stamens of crocus itself ('Pharm. Journ.').

Prop., &c. Saffron is anodyne, cordial, emmenagogue, and exhilarant; but is now seldom employed, except as an adjuvant, in medicine. Amongst cooks, confectioners, and liquorists it is largely used on account of its fine colour.

Saffron, Mead'ow. See COLCHICUM.

SAGAPE'NUM. This substance is described in the London Pharmacopœia as a gum-resin, the production of an uncertain species of *Ferula*. Its botanical source is unknown. The mass of the sagapenum sold to the retail trader is, however, a factitious article, formed by softening a mixture of assafœtida, 3 parts, and galbanum, 15 parts, over a water or steam bath, and then stirring in about 1-17th of their weight of oil of turpentine, with a little oil of juniper. This mixture is labelled 'Gum. Sagapeni Opt.,' an inferior sort being made by adding sundry portions of yellow resin and paste of gum tragacanth to the above.

PREPARED SAGAPENUM (SAGAPENUM PRÆPARA-TUM—Ph. L.) is ordered to be prepared in the same manner as 'prepared ammoniacum.'

Obs. Sagapenum is the feeblest of all the fetid gum-resins.—*Dose,* 5 to 15 gr., made into pills; as an antispasmodic and emmenagogue.

SA'GO. *Syn.* SAGO (Ph. L., E., and D.), L. "The fæcula (starch) from the stem of *Sagus lœvis, S. Rumphii,* and, perhaps, of other species of palms" (Ph. L.). It forms the principal portion of the pith of the sago palms, the Gommuti palm, the Talipot palm, and other allied trees. Its properties and uses, for the most part, resemble those of arrowroot. It is used for making puddings, jellies, &c.

Under the microscope the starch-grains of sago present an elongated form, rounded at the larger ends, and compressed at the smaller. They differ altogether in appearance from potato starch.

The hilum of the sago starch-grains is a point, or, more frequently, a crop, slit, or star, and is

Sago of commerce, magnified 147 times.

seated at the smaller end, whilst in the marsanta arrowroot the hilum is situated at the larger end. Rings are more or less clearly seen.

Sago, To Prepare. Wash an ounce of pearl sago in cold water; then boil it very gently in a pint of fresh water, stirring it frequently till dissolved.

It may be flavoured with wine, spices, and sugar. For children, and for consumptive and debilitated persons, it will be found advantageous to substitute milk for water. The common sago being in larger grains, more time is required to dissolve it, and it is usually steeped for some hours before boiling it.

Sa'go Milk. See *above*.

Sa'go Posset. (For invalids.) Macerate a table-spoonful of sago in a pint of water for 2 hours on the hob of a stove, then boil for 15 minutes, assiduously stirring. Add sugar, with an aromatic, such as ginger or nutmeg, and a table-spoonful or more of white wine. If white wine be not permitted flavour with lemon juice.

ST VITUS' DANCE. See CHOREA.

SAL. [L.] Salt. A word much used in compound names, handed down to us from the old chemists.

Sal Absin'thii. Carbonate of potassium.

Sal Acetosel'læ. Binoxalate and quadroxalate of potassium.

Sal Alem'broth. Ammoniated mercury (white precipitate).

Sal Ammo''niac. Chloride of ammonium.

Sal de Duobus. Sulphate of potassium.

Sal Diure'ticus. Acetate of potassium.

Sal Enix'um. Crude bisulphate of potassium.

Sal Gem'mæ. Rock or fossil salt (chloride of sodium).

Sal Mar'tis. Sulphate of iron.

Sal Mineralis. A mixture of salts representing the constituents of Carlsbad, Friedrichshall, Pullna, and other mineral waters.

Sal Mirab'ile. Sulphate of sodium.

Sal Perla'tum. Phosphate of sodium.

Sal Polychrest'um. Sulphate of potassium.

Sal Prunel'la. *Syn.* SORE-THROAT SALT, CRYSTAL MINERAL; POTASSÆ NITRAS FUSA, NITRUM TABULATUM, SAL PRUNELLÆ, L. From nitre fused in a Hessian crucible, and poured out on a smooth surface, or into moulds, to cool. Its usual form and size is that of an ordinary musket bullet, with the tail, in which state it is known in the drug trade as 'sal prunellæ globosum.' When in cakes it is often called 'sal. p. in placentis,' or 's. p. tabulatum.' A small portion allowed to dissolve slowly in the mouth, the saliva being slowly swallowed, often removes incipient inflammatory sore throat.

Sal Saturn'i. Sugar of lead (neutral acetate of lead).

Sal Seignette'. Rochelle salt (tartrate of potassium and sodium).

Sal Volat'ile. Sesquicarbonate of ammonia. The name is commonly used as an abbreviation of aromatic spirit of ammonia. See SPIRITS (Medicinal).

SAL'ADS are generally made of esculent vegetables, either singly or mixed, chosen according to taste or time of year, and 'dressed' with oil, vinegar, and salt, and sometimes also with mustard and other condiments. Sliced boiled egg is a common addition.

Sydney Smith's recipe for salad dressing:

To make this condiment your poet begs
The powdered yellow of two hard-boiled
    eggs;

Two boiled potatoes passed through kitchen
    sieve
Smoothness and softness to the salad give;
Let onion atoms lurk within the bowl,
And, half suspected, animate the whole;
Of mordant mustard add a single spoon
(Distrust the condiment that bites too soon);
But deem it not, thou man of taste, a fault
To add a double quantity of salt;
And, lastly, o'er the flavoured compound toss
A magic *soupçon* of anchovy sauce.
Oh, green and glorious! Oh, herbaceous treat!
'Twould tempt the dying anchorite to eat;
Back to the world he'd turn his fleeting soul,
And plunge his finger in the salad bowl;
Serenely full the epicure would say,
" Fate cannot harm me, I have dined to-day."
(The poet has inadvertently ignored the oil and vinegar.)

Another recipe for salad dressing :—Yolk of two eggs; table salt, ¼ oz.; salad oil, 4 oz.; mustard, ½ oz.; best vinegar, 6 oz.; isinglass, 1 dr.; soluble cayenne, 10 grms. (' Pharm. Journ.').

Cold meat, poultry, and game, sliced small, with some cucumber or celery, and a little onion or chopped parsley, or, instead of them, some pickles, make a very relishing salad. Fish are also employed in the same manner.

Mr C. J. Robinson, writing to 'Nature' (Aug. 18th, 1870) on our salad herbs, says :—" There is, perhaps, no country in the world so rich as England in native materials for salad making, and none in which ignorance and prejudice have more restricted their employment. At every season of the year the peasant may cull from the field and hedgerow wholesome herbs which would impart a pleasant variety to his monotonous meal, and save his store of potatoes from premature exhaustion. Besides, there can be no question that in hot seasons a judicious admixture of fresh green food is as salutary as it is agreeable. Much has been said lately about the advantage which the labouring man would derive from an accurate acquaintance with the various forms of fungus; he has been gravely told that the *Fistulina hepatica* is an admirable substitute for beef-steak, the *Agaricus gambosus* for the equally unknown veal cutlet.

" But deep-seated suspicion is not easily eradicated, and there will always be a certain amount of hazard in dealing with a class of products in which the distinctions between noxious and innocuous are not very clearly marked.

" There is not this difficulty with regard to salad herbs, and we conceive that the diffusion of a little knowledge as to their properties and value would be an unmixed benefit to our rural population.

" The first place must be assigned on the score of antiquity to the sorrel plant (*Rumex acetosa*), which in some districts still preserves the name of 'green sauce,' assigned to it in early times, when it formed almost the only dinner vegetable.

" Its acid is pleasant and wholesome, more delicate in flavour than that of the wood-sorrel (*Oxalis acetosella*), which, however, is used for table purposes in France and Germany. Chervil (*Anthriscus cerefolium*) is often found in a wild state, and is an admirable addition to the salad bowl; and it is unnecessary to enlarge upon the virtues of celery (*Apium graveolens*) when improved by cultivation."

John Ray, writing in 1668, says that "the Italians use several herbs for sallets, which are not yet, or have not been used lately, but in England, viz. *Selleri*, which is nothing else but the sweet smallage; the young shoots whereof, with a little of the head of the root, cut off, they eat raw with oil and pepper;" and to this we may add that the alexanders (*Smyrnium olusatrum*) is no bad substitute for its better known congener. The dandelion, which in France is blanched for the purpose, affords that *amaris aliquid* which the professed salad maker finds in the leaves of the endive, and the same essential ingredient may be supplied by the avens (*Geum urbanum*), the bladder campion (*Silene inflata*), and the tender shoots of the wild hop. Most people are familiar with the properties of the water-cress (*Nasturtium officinale*), garlic hedge-mustard (*Erysimum alliaria*), but it may not be generally known that the common shepherd's-purse (*Capsella bursa-pastoris*) and the lady's-smock (*Cardamine pratensis*) are pleasant additions, whose merits have long been recognised by our foreign neighbours. In fact, there is scarcely an herb that grows which has not some culinary virtue in a French peasant's eyes. Out of the blanched shoots of the wild chicory (*Cichorium intybus*) he forms the well-known *barbe de Capucin*, and dignifies with the title of *Salade de Chamoine* our own neglected corn-salad (*Fedia olitoria*). It would be very easy to extend the dimensions of our list of native salad herbs, for there are, perhaps, some palates to which the strong flavours of the chives (*Allium schœnoprasum*) and stonecrop (*Sedum reflexum*) may commend themselves; but enough has been said to show that Nature has not dealt niggardly with us, and that only knowledge is needful to make the riches she offers available.

If the British peasant can be taught to discover hidden virtues in these plants, with whose outward forms he has had lifelong familiarity, we do not despair of his acquiring the one secret of salad-making, viz. the judicious employment of oil, so as to correct the acrid juices of the plants, and yet preserve their several flavours unimpaired.

Salad, Lettuce. *Prep.* Take 2 large lettuces, remove the faded leaves and the coarser green ones; next cut the green tops off, pull each leaf off separately, rinse it in cold water, cut it lengthways, and then into 4 or 10 pieces; put these into a bowl, and sprinkle over them, with your fingers, 1 small teaspoonful of salt, ½ do. of pepper, 3 do. of salad oil, and 2 do. of English or 1 of French vinegar; then with the spoon and fork turn the salad lightly in the bowl until thoroughly mixed; the less it is handled the better. A teaspoonful each of chopped chervil and tarragon is an immense improvement.

Obs. The above seasoning is said to be enough for ¼ lb. of lettuce. According to Soyer, it is "such as the Italian count used to make some years since, by which he made a fortune in dressing salads for the tables of the aristocracy." The above may be varied by the addition of 2 eggs,

boiled hard and sliced, a little eschalot, or a few chives or young onions. Several other salad herbs, especially endive, water-cresses, and mustard-and-cress, may be 'dressed' in the same manner; always remembering that the excellence of a salad depends chiefly on the vegetables which compose them being recently gathered and carefully cleansed.

To improve the appearance of the above and other salads, when on the table or sideboard, before being used, the gay flower of the nasturtium or marigold, with a little sliced beetroot or radish, and sliced cucumber, may be tastefully intermixed with them.

Salad, Lobster. *Prep.* (*Soyer.*) "Have the bowl half filled with any kind of salad herb you like, as endive, lettuce, &c.; then break a lobster in two, open the tail, extract the meat in one piece, break the claws, cut the meat of both in small slices about a quarter of an inch thick, and arrange these tastefully on the salad; next take out all the soft part from the belly, mix it in a basin with 1 teaspoonful of salt, ½ do. of pepper, 4 do. of vinegar, and 4 do. of oil; stir these well together, and pour the mixture on the salad; lastly, cover it with 2 hard eggs, cut into slices, and a few slices of cucumber." "To vary this, a few capers and some fillets of anchovy may be added, stirred lightly, and then served either with or without some salad sauce. If for a dinner, ornament it with some flowers of the nasturtium and marigold."

SALEP. *Syn.* SALOP, SALOOP. The tuberous roots of *Orchis mascula*, and other allied species, washed, dried, and afterwards reduced to coarse powder. That imported from Persia and Asia Minor occurs in small oval grains, of a whitish-yellow colour, often semi-translucent, with a faint, peculiar smell, and a taste somewhat resembling gum tragacanth. It consists chiefly of bassorin and starch, is very nutritious, and is reputed aphrodisiac. It is employed in the same way as sago. A decoction of about 1 oz. of this substance in a pint of water was formerly sold at street stalls. A tea made of sassafras chips, flavoured with milk and coarse brown sugar or treacle, was also sold in the same way, and under the same name.

FRENCH SALEP is prepared from the potato. Dr Ure says that the *Orchis mascula* of our own country, properly treated, would afford an article of salep equal to the Turkey, and at a vastly lower price.

INDIAN SALEP is known in the Indian bazaars as *Salib misri*; it fetches a high price, and is derived from certain species of *Eulopia*.

SALICIN. $C_{13}H_{18}O_7$. A glucoside discovered by Le Roux and Buchner in the bark and leaves of several species of *Salix* and *Populus*. It occurs most abundantly in the white willow (*Salix alba*) and the aspen (*Salix helix*), but is also found in all the bitter poplars and willows. From willow bark which is fresh, and rich in salicin, it may be obtained by the cautious evaporation of the cold aqueous infusion.

Prep. 1. (*Merck.*) Exhaust willow bark repeatedly with water, concentrate the mixed liquors, and, while boiling, add litharge until the liquid is nearly decolourised; filter, remove the dis-

solved oxide of lead by treatment with sulphuretted hydrogen; filter and evaporate, that
crystals may form; the crystals must be purified
by re-solution and re-crystallisation.

2. To a strong filtered decoction of willow bark
add milk of lime, to throw down the colour;
filter, evaporate the liquor to a syrupy consistence,
add alcohol (sp. gr. 0·847), to separate the gummy
matter, filter, distil off the spirit, evaporate the
residuum, and set it aside in a cool place to crystallise; the crystals are purified by solution in
boiling water, agitation with a little animal charcoal, and re-crystallisation.

Prop., &c. Salicin forms bitter colourless
prisms, melting at 230° F., with decomposition;
burns with a bright flame; is soluble in about 30
parts of cold water; dissolves readily in alcohol,
but is insoluble in ether. When fermented by
emulsion or human saliva its aqueous solution
yields glucose and saligenin. It is tonic, like
sulphate of quinine, but less liable to irritate the
stomach. It is given in indigestion and intermittent diseases in from 5- to 10-gr. doses.

Salicin has lately been used with considerable
advantage in acute rheumatism.

Dr Maclagan ('Lancet,' March 4th and 11th,
1876) states that he found when administered in
doses of 10 gr. to ½ dr., every two to four hours,
the pain and fever ceased in the course of forty-
eight hours. The results are stated to have been
quite as favourable as those following the employment of salicylic acid. It was found to effect
with certainty a great reduction in the bodily
temperature.

Tests. 1. When strongly heated it is wholly
dissipated, and if kindled, burns with a bright
flame, leaving a bulky charcoal. 2. Its solution
is absolutely neutral to test-paper. 3. Concentrated sulphuric acid causes it to agglutinate into
resin-like lumps, with the accession of an intense
blood-red colour. 4. When its aqueous solution
is mixed with some hydrochloric acid or dilute
sulphuric acid, and the mixture is boiled for a
short time, the liquid suddenly becomes turbid,
and deposits SALIRETIN, under the form of a
granular crystalline precipitate. This is characteristic. 5. Gives a blue colour with ferric
chloride. See SALICYLIC ACID.

SAL'ICYL. C₇H₄O. A compound radical,
forming the basis of the so-called SALICYL COM-
POUNDS, or SALICYL SERIES. It is known only in
combination.

SALICYL'IC ACID. C₆H₄.OH.CO₂H. *Syn.*
ORTHO-HYDROXYBENZOIC ACID. Although this
acid was discovered fifty years ago by Piria (*vide*
'Annales de Chimie et de Physique,' vol. lxix), it
is only within comparatively recent times that its
value and most remarkable properties have been
recognised.

It occurs, according to Löwig, in the blossoms
of the meadow-sweet, *Spiræa ulmaria*, in certain
members of the genus *Viola*, and as the methyl-
ether in the oil of winter-green, which is extracted from *Gaultheria procumbens*, one of the
North American heaths. For a long time this
plant remained the only source of salicylic acid.

Prep. 1. From winter-green oil, by acting on
it with a strong and hot solution of potash, and
afterwards separating the acid by treatment with

excess of hydrochloric acid and subsequent crystallisation. This oil is salicylate of methyl.

2. Salicin, a glucoside extracted from the bark
of the willow, is melted with potassium hydrate,
and then converted into nitro-salicylic acid by the
action of dilute nitric acid; this is then melted
up with a fresh portion of potassium hydrate at a
gentle heat, and yields anthranilic acid. When
the temperature of the latter operation is raised
salicylic acid is produced (*Marchand* and *Ger-
hardt*, 'Ann. Chem. Pharm.,' lii, 343).

3. The greater part of the salicylic acid of
commerce is now obtained by a method invented by
Kolbe. This process, which consists in acting on
sodium carbolate with carbonic anhydride, is thus
described in the 'Archiv der Pharm.,' 3rd series,
v, 445 ('Pharm. Journal,' 3rd series, v, 421) :—In
a strong crude soda liquor of known strength is
dissolved a sufficiency of previously melted crystals of carbolic acid to saturate the soda.

The solution is then evaporated in an iron
capsule, stirring constantly, and brought to a dry
powder. The sodium carbolate so obtained is
gradually heated in a retort to a temperature of
from 180° to 200° C. in a continuous current of
dry carbonic anhydride.

The reaction is ended when at the above-mentioned temperature no more carbolic acid passes
over. It might have been expected that, the reaction going forward in this manner, a molecule
of carbonic anhydride would have been introduced into the molecule of sodium carbonate,
and thus a molecule of sodium salicylate be
formed.

This, however, is not the case, only half the
sodium carbolate being converted into salicylate.
The reaction proceeds according to the following
equation:

$$2NaC_6H_5O + CO_2 = Na_2C_7H_4O_3 + HC_6H_5O.$$

The disodic salicylate is dissolved in water and
decomposed by hydrochloric acid. Salicylic acid
then separates in crystalline films, and may be
purified by re-crystallisation out of its solution in
hot water.

Kolbe's process is the basis of most modern
methods of manufacturing salicylic acid; thus
Robbe's patent depends upon it; Lautermann has
also invented a synthetical process, for which
consult 'Ann. Chem. Pharm.,' cxv, 201.

Props. White acicular crystals; melts at from
155° to 156° C.; it dissolves in about 1800 parts of
cold water, but is more soluble in boiling water,
as well as in alcohol and in ether. When heated
it breaks up into phenol and carbonic anhydride.

Professor Kolbe surmised that from the constitution of salicylic acid, as revealed by his synthetical process, it would split up with heat
into carbonic anhydride and carbolic acid, and
hence that it might be employed as an antiseptic
and antiputrefactive agent. He quotes the following experiments as confirmatory of his views
on this point :—Mustard meal, which, in a few
minutes after being mixed with warm water, gave
off a strong smell of mustard oil, formed with
water a scentless mixture when a little salicylic
acid had been previously added. No fermentation was set up by yeast in a fermentation of
grape sugar to which salicylic acid had been
added; whilst in a sugar solution already in fer-

mentation the action stopped after the addition of some salicylic acid. The preservative influence of this acid upon fresh meat is also recorded.

The following, among other experiments, in their results illustrate the physiological action of salicylic acid:

Solution of amygdalin mixed with emulsion of sweet almonds developed no smell of bitter almonds if some salicylic acid was added. Beer, to which salicylic acid, in the proportion of 1 to 1000, was added, was thereby prevented from being spoiled by fungoid growth.

Fresh pure cow's milk, mixed with 0·04% of salicylic acid, and allowed to stand in an open vessel at a temperature of 18° C., curdled thirty-six hours later than a similar quantity of milk standing by the side of it, but containing no salicylic acid. The milk remained of a good flavour; the small quantity of salicylic acid present was not perceptible to the palate.

Some fresh urine was divided into two portions, and placed in separate vessels after some salicylic acid had been added to one portion. The urine containing the acid was on the third day still clear and free from ammoniacal odour, whilst the other portion was far advanced in putrefaction.

Professor Thiersch has investigated the antiseptic action of this acid specially in relation to surgery. He has found that as a powder, either alone or mixed with starch, it destroys for a long time the fœtid odour of cancerous surfaces or uncleansed wounds without setting up any inflammatory symptoms. A solution of 1 part of salicylic acid and 3 parts of sodium phosphate in 50 parts of water promotes the healing of granulated surfaces.

According to Dr Rudolph Wagner, salicylic acid may be applied to the following industrial purposes:

If a concentrated aqueous solution of salicylic acid be applied to fresh meat, and the meat be then placed in well-closed vessels, it will remain perfectly fresh for a long period. This solution is also very useful in the manufacture of sausages and such food. Butter containing a bitter salicylic acid will remain fresh for months, even in the hottest weather. The same acid prevents the moulding of preserved fruits. In the manufacture of vinegar this acid is of great utility.

The addition of a little salicylic acid renders all kinds of glue more tenacious and less liable to decompose. The acid also prevents decomposition in gut and parchment during their manufacture.

Skins to be used for making leather do not undergo decomposition if steeped in a dilute solution of salicylic acid.

Albumen may be preserved by the same means.

This acid is a very delicate test reagent for iron.

The methyl ether and amyl ether of salicylic acid are used as perfumes. The calcium salt on keeping and distilling with water yields a liquid which has a strong odour of roses.

In order to preserve syrups such as are used in the manufacture of aërated waters, &c., it is sufficient to add a quantity of salicylic acid

equivalent to 1·1000th the weight of sugar contained in the syrup. One part of salicylic acid in ten thousand parts of beer or wine is sufficient to prevent decomposition.

Tests. Dr Muter gives the following method for estimating the value of commercial samples of salicylic acid, and of detecting it in milk and similar organic solutions:—A standard solution of pure salicylic acid (1 gr. of acid dissolved in 1 litre of water, so that 1 c.c. represents 1 mgm. of acid) is prepared; the *indicator* solution consists of a solution of pure neutral ferric chloride, of such a strength that 1 c.c. added, drop by drop, to 50 c.c. of standard acid, just ceases to give any increase in intensity of colour; 1 grm. of the commercial sample is then dissolved in 1 litre of water, and 50 c.c. is put into a Nessler tube; to this 1 c.c. of ferric chloride solution is added, and the colour observed after standing for five minutes; some of the standard acid is also poured into another tube, and made up to 50 c.c. with water, and the 1 c.c. of ferric chloride added. When the colours are exactly alike in tint the amount of pure acid present in the sample is equal to the amount of pure acid added. All mineral acids should be absent; even acetic acid affects the reaction. To detect the presence of salicylic acid added to beer or milk, 4 oz. of these liquids are dialysed for twelve hours in a pint of distilled water; if after that time salicylic acid is still found to be present, the dialysis must be continued for forty-eight hours. The amount present is determined in the manner above stated.

Allen says, in his 'Commercial Organic Analysis,' vol. iii, part 1, published in 1889, "The most delicate reaction for salicylic acid is that of ferric chloride; which produces a beautiful violet colour." Nearly all the new methods are based upon the action of this reagent, the use of which was first noticed by Pagliani.

To detect the acid in beer, Hoorn recommends (*a*) that the suspected sample should be acidified with sulphuric acid, extracted with a mixture of light petroleum and ether, and the extract evaporated and examined with the ferric chloride test; (*b*) after acidifying, distil, and collect the *latter* portion of the distillate for examination as before ('Rec. Trav. Chim.,' vii, 41).

For the examination of wines, Wright recommends 50 c.c. of the sample to be agitated with 50 c.c. of chloroform; after standing, the layer of chloroform extract is separated off and passed through a dry filter. A perfectly fresh solution of ferric chloride having been prepared (the proportion being 1 in 1000), 10 c.c. is added to 30 c.c. of the chloroform extract, and the mixture shaken up, when, if salicylic acid be present, the usual violet coloration will be noticed. When it is desired to make this test a quantitative one, ether should be substituted for chloroform, and a Schwartz's extractor should be used ('Mitt. Chem. Physiol. Versuchs - stat.,' Klosternenberg bei Wien, v, 54).

A general plan devised by Elion is as follows:—The acid should be extracted with ether, about three times the volume of the solution under examination being used; the ether layer is then separated off and shaken with a little soda solution; this takes up the acid. The soda solution is next heated with a little hydrochloric acid, and then the ferric chloride test is applied ('Rec. Trav. Chim.,' vii, 211).

Obs. Pure salicylic acid is not known to be really poisonous in its effects; as much as 60 gr. have been taken by a patient in twenty-four hours. It is said by Kiersch and Godeffrey to be three times more powerful than carbolic acid in preventing putrefaction, and, since it is free from taste and smell, it is an excellent material for use in the preservation of food and drink.

There is no doubt that impure salicylic acid produces bad effects, and much of the artificial acid used to contain impurities; but this need not be the case now, and there can be no rational objection to the use of the pure acid, especially in the exceedingly small quantities which are found effective in retarding or preventing fermentation.

It will prevent the development of bacteria in fluids containing it, will kill or paralyse the action of torula, and will hinder most of those changes which are initiated by the action of vegetable ferment.

When salicylic acid is prepared from phenol without sufficient care, other homologous compounds derived from the higher phenols may contaminate it and render its use injurious.

Ewell and Prescott have devised two new processes for detecting foreign acids, &c. ('Transactions of the Pharmaceutical Society' [3], xix, 328).

a. Titrate 1 grm. against an alkali of centinormal strength, using phenol phthalein to indicate the 'end-action.' The foreign acids are calculated as hydrocytomeric acid.

b. Distil the sample with lime; this converts the salicylic acid and any of its homologues present into phenols, which are then estimated in the ordinary way. The percentage of other impurities is then calculated by difference.

SALI'VA. See MASTICATION.

SALIX NIGRA. *Syn.* BLACK WILLOW BARK, PUSSY WILLOW. A tree growing from 15 to 20 feet high, indigenous to the Southern States of America; has been used with success in cases of ovarian hyperæsthesia and uterine neuralgia. It acts as a powerful sexual sedative, similar to but without the depressing qualities of bromides.— *Dose.* Liquid extract, ⅓ to 1 dr.

SALM'ON. *Syn.* SALMO, L. The *Salmo salar*, Linn., a well-known, soft-finned abdominal fish. Its normal locality is at the mouth or estuary of the larger rivers of the northern seas, which, during the breeding season, it ascends, sooner or later, in the summer months, against all obstacles, for the purpose of depositing its spawn.

The salmon is an excellent and highly esteemed fish; but it is rich, oily, and difficult of digestion, and therefore ill adapted to the delicate or dyspeptic. When pickled, salted, or smoked, it is only fitted for persons of very strong stomachs, although in this state it is regarded as a great delicacy by epicures.

Salmon has the following composition:

Nitrogenous matter	. . .	16·1
Fat	. . . .	5·5
Saline matter	: . . .	1·4
Water	. . . . .	77·
		——
		100·0

Salmon is preferably cooked by boiling. One weighing 10 lbs. will require to be gently simmered for about an hour, reckoning from the time the water commences boiling. For fish of other weights, from 6 to 7 minutes per lb. may be allowed. See FISH, &c.

SALOL. A phenyl ether of salicylic acid $(HC_7H_4O_3)$, the radical phenyl (C_6H_5) replacing a hydrogen atom of the acid, thus giving C_6H_5. $C_7H_4O_3$, or more correctly $C_6H_4(OH)COO.C_6H_5$. Since the publication of a note on the substance in the 'Chemist and Druggist' (June 26th, p. 626) its action has been further investigated by Dr Sahli, of Berne, and also by Professor Löwenthal, who communicates a paper on the subject to the 'Semaine Médicale.' It is well known that salicylic acid and its salts produce many untoward effects when administered internally, which greatly militate against their undoubted value in the treatment of acute rheumatism, and many chemists and therapeutists have long aimed at producing a substance which will possess the better properties of salicylic acid, especially its power of warding off endocarditis, without its liability to produce secondary effects. It would appear from experiments which have been made, although they are yet imperfect, that salol is such a substance. As already stated, it is a white crystalline powder, of feebly aromatic odour (recalling oil of wintergreen) and almost tasteless. Merck states that it can also be obtained in rhombic crystals, melting at 42° C. to a clear, colourless liquid, which can be cooled much below that point without solidifying, unless it be touched with a glass rod. Its freedom from taste appears to be due to its insolubility in water; it dissolves, however, perfectly and rapidly in alcohol, benzol, and ether.

Salol has been administered in doses up to 8 grms. (ℨij) per day, without bad results, and noise in the ears has been seldom observed. The dose, however, must be modified to the case, just as that of salicylic acid is; for example, in phthisical cases doses of 0·5 grm. should be used at first, because it is undesirable in these cases to lower the temperature too quickly. It is supposed that salol is unaffected in its passage through the stomach, and that it is not decomposed until it reaches the duodenum and is acted upon by the pancreatic secretion. Certainly its administration is not followed by toxic symptoms, such as would be expected by rapid absorption of phenol by the stomach. When treated with pancreatic extract, the body is resolved into its component parts; moreover it acts as well when administered *per rectum.* The urine of patients is found to be almost black in colour, due to the presence of oxidation products of phenol, consequent on the absorption of phenol products into the blood and subsequent oxidation of these during circulation.

Salol may be applied externally as a dusting powder. Its insoluble nature makes it peculiarly applicable for this purpose, and it has been used with benefit for excoriated surfaces and fœtid wounds. It prevents the development of bacteria, but does not kill them. Bougies of salol are valuable for the treatment of gonorrhœa. As a mouth wash and as an injection it is used in solution with alcohol and water, but for the latter purpose it is more desirable to suspend the finely powdered salol in water with tragacanth or starch mucilage.

The reaction of the body with pancreatin may advantageously be employed as a test for its identity.

SALOOP'. Sassafras (chips) tea, flavoured with milk and sugar. A wholesome and useful drink in cutaneous and rheumatic affections. See SALEP.

SALT. *Syn.* SAL, L.; SEL, Fr. Salts may be regarded as acids in which one or more atoms of hydrogen, a constant constituent of all true acids, are replaced by a metal or other basic radical. This relationship between acids and salts will be better understood by reference to the subjoined list of acids and their corresponding potassium and ammonium salts:

ACIDS.		SALTS.
HCl (Hydrochloric acid)	—KCl (Chloride of potassium).
	—NH₄Cl (Chloride of ammonium).
HNO₃ (Nitric acid)	—KNO₃ (Nitrate of potassium).
	—NH₄NO₃ (Nitrate of ammonium).
H₂SO₄ (Sulphuric acid)	—K₂SO₄ (Sulphate of potassium).
	—(NH₄)₂SO₄ (Sulphate of ammonium).

Acids are, in fact, hydrogen salts. A basic salt is formed by replacing all the hydrogen in the acid, whilst an acid salt is formed by replacing part of the hydrogen in the acid by a metal or metallic radical. The so-called DOUBLE SALTS are, according to one view, combinations of two salts of the same acid, but of different basic radicals; thus common alum is a compound of sulphate of aluminum and sulphate of potassium. The salts are obtained by a variety of reactions, of which the following are the most important:

1. When certain metals are brought in contact with an acid, thus:

$$Zn + H_2SO_4 = ZnSO_4 + H_2.$$

2. When a basic acid or an hydroxide acts upon an acid or an acid-forming oxide, thus:

$$PbO + H_2SO_4 = PbSO_4 + H_2O.$$
$$Ba(OH)_2 + H_2SO_4 = BaSO_4 + 2H_2O.$$

The names of salts of acids having names ending in -ous terminate in -ite. Thus nitrous acid forms nitrites. The names of salts of acids having names ending in -ic terminate in -ate. Thus nitric acid forms nitrates.

The salts are a most important class of bodies, and their applications and uses in the arts of life and civilisation are almost infinite. See NOMENCLATURE, &c.

Salt of Bark. See EXTRACT OF BARK (Dried).
Salt, Bitter Pur'ging. Epsom salt.
Salt, Cathar'tic. Of GLAUBER, sulphate of sodium; ENGLISH or BITTER S., sulphate of magnesium (Epsom salt).
Salt, Common. *Syn.* CULINARY SALT. Chloride of sodium.
Salt, Diuret'ic. Acetate of potassium.
Salt, Ep'som. Sulphate of magnesium.
Salt, Feb'rifuge. Chloride of potassium.
Salt, Fu"sible. Phosphate of ammonium.

Salt, Glauber's. Sulphate of sodium.
Salt, Macquer's. Binarseniate of potassium.
Salt, Microcos'mic. Phosphate of sodium and ammonium.
Salt, Red. Common salt wetted with an infusion of beetroot or cochineal, or tincture of red sanders-wood, then dried, and rubbed through a sieve. Used to impart a colour to gravies, &c. Infusion of saffron also gives a beautiful colour for this purpose. It has been proposed to colour Epsom salt in this way to distinguish it from oxalic acid.
Salt, Rochelle. Tartrate of potassium and sodium.
Salt, Sea. Chloride of sodium.
Salt, Sed'ative. Boracic acid.
Salt, Smelling. See SALTS (below).
Salt, Sore throat. Sal prunella.
Salt, Taste'less. Phosphate of sodium.
Salt, Veg'etable. Tartrate of potassium.
Salt, Vol'atile. Common carbonate of ammonium.
Salt of Lem'ons. *Syn.* SAL LIMONUM, L. Citric acid. That sold in the shops for the removal of ink spots from linen is binoxalate or quadroxalate of potassium, either alone or mixed with one half its weight of cream of tartar.
Salt of Sor'rel. Binoxalate or quadroxalate of potassium.
Salt of Steel. Sulphate of iron.
Salt of Tar'tar. Carbonate of potassium.
Salt of Vit'riol. Sulphate of zinc.
Salt of Wormwood. Carbonate of potassium.
SALTING. *Syn.* PICKLING. This is an easy method of preserving butcher's meat, fish, and, indeed, most animal substances. It is performed in two ways.

1. (DRY SALTING.) This, as practised in Hampshire, Yorkshire, and in various large establishments elsewhere, consists in merely well rubbing ordinary culinary salt, mixed with a little saltpetre, into the meat, until every crevice is thoroughly penetrated, and afterwards sprinkling some over it, and placing it on a board or in a trough, in such a manner that the brine may drain off. On the small scale, in private families, a mixture of salt, 2 lbs., with saltpetre, 1½ or 2 oz., either with or without about an ounce of good moist sugar, is commonly used for the purpose, and imparts a fine flavour to the meat. In both cases the pieces are turned every day or every other day until sufficiently cured, a little fresh salt being added as required. Sometimes the fresh meat is packed at once in casks, with the best coarse-grained or bay salt. This method is that commonly adopted for sea stores.

2. (WET SALTING, or PICKLING IN BRINE.) When the meat is allowed to lie in the liquor that runs from it (see *above*), or is at once plunged into strong brine, it is said to be 'pickled,' or 'wet salted.' On the small scale this is most conveniently performed by rubbing the fresh meat with salt, &c., as above, and, after it has lain a few hours, putting it into a pickle formed by dissolving about 4 lbs. of good salt and 2 oz. of saltpetre in 1 gall. of water, either with or without the addition of ¼ to 1 lb. of moist sugar. This pickling liquor gets weaker by use, and should therefore be occasionally boiled down a little and skimmed, at

the same time adding some more of the dry ingredients. Three to ten days, depending on the size, is sufficiently long to keep meat in the brine. When it is taken out it should be hung up to dry, after which it may be packed in barrels with coarse-grained salt, or smoked, whichever may be desired. Saltpetre added to brine gives the meat a red colour, and brown sugar improves the flavour.

The sooner animal substances, more especially flesh, are salted after being killed, the better, as they then possess considerable absorbent power, which they gradually lose by age. See PUTRE-FACTION, SCURVY, SMOKING, &c.

SALTPE'TRE. Nitrate of potassium.

SALTS for producing Factitious Mineral Waters. AËRATED or CARBONATE WATERS. These require the aid of the powerful machine employed by soda-water manufacturers, to charge the waters strongly with carbonic acid gas. The gas is made from whiting and diluted sulphuric acid, and is forced by a pump into the watery solution. Sometimes the gas is produced by the mutual action of the ingredients introduced into the bottle of water, which must be instantly closed; but this method is found practically inconvenient, and is only adopted in the absence of proper apparatus. The quantity of gas introduced is directed, in the French and American pharmacopœias, in most cases, to be five times the volume of liquid. For chalybeate and sulphuretted waters the water should be previously deprived of the air it naturally contains by boiling, and allowing it to cool in a closed vessel.

There are various manufacturers of aërated water machines, and of syphon bottles for holding these waters when made. The names and addresses of these makers may be found in any trade directory.

SIMPLE AËRATED WATER. Carbonic acid gas water. Water charged with 5 or more vols. of carbonic acid gas, as above.

ALKALINE AËRATED WATERS. Aërated soda and potash waters should be made by dissolving a drachm of the carbonated alkali in each pint of water, and charging it strongly with carbonic acid gas. The soda water of the shops generally contains but little (or no) soda.

AËRATED MAGNESIA WATER. This is made of various strengths.

MURRAY'S and DINNEFORD'S FLUID MAGNESIA may be thus made:—To a boiling solution of 16 oz. of sulphate of magnesia in 6 pints of water add a solution of 19 oz. of crystallised carbonate of soda in the same quantity of water; boil the mixture till gas ceases to escape, stirring constantly; then set it aside to settle; pour off the liquid, and wash the precipitate on a cotton or linen cloth with warm water, till the latter passes tasteless. Mix the precipitate, without drying it, with a gallon of water, and force carbonic acid gas into it under strong pressure, till a complete solution is effected. The Eau magnésienne of the French Codex is about a third of this strength; and we have met with some prepared in this country not much stronger.

CARBONATED LIME WATER. Carrara water. Lime water (prepared from lime made by calcining Carrara marble) is supersaturated by strong pres-

sure with carbonic acid, so that the carbonate of lime at first thrown down is redissolved. It contains 8 gr. of carbonate of lime in 10 fl. oz. of water.

AËRATED LITHIA WATER. This may be conveniently made from the fresh precipitated carbonate, dissolved in carbonated water, as directed for fluid magnesia. Its antacid and antilithic properties promise to be useful.

SALINE CARBONATED WATERS.

The following afford approximate imitations of these waters. The earthy salts, with the salts of iron, should be dissolved together in the smallest quantity of water. The other ingredients to be dissolved in the larger portion of the water, and the solution impregnated with the gas. The first solution may be then added or be previously introduced into the bottle. The salts, unless otherwise stated, are to be crystallised.

BADEN WATER. Chloride of magnesium, 2 gr.; chloride of calcium, 40 gr.; perchloride of iron, ¼ gr. (or 3 minims of the tincture); chloride of sodium, 30 gr.; sulphate of soda, 10 gr.; carbonate of soda, 1 gr.; water, 1 pint; carbonic acid gas, 5 vols.

CARLSBAD WATER. Chloride of calcium, 8 gr.; tincture of chloride of iron, 1 drop; sulphate of soda, 50 gr.; carbonate of soda, 60 gr.; chloride of sodium, 8 gr.; carbonated water, 1 pint.

EGER. Carbonate of soda, 5 gr.; sulphate of soda, 4 scruples; chloride of sodium, 10 gr.; sulphate of magnesia, 3 gr.; chloride of calcium, 5 gr.; carbonated water, 1 pint. (Or it may be made without apparatus, thus:—Bicarbonate of soda, 30 gr.; chloride of sodium, 8 gr.; sulphate of magnesia, 3 gr.; water, a pint; dissolve and add a scruple of dry bisulphate of soda, and close the bottle immediately.)

EMS. Carbonate of soda, 2 scruples; sulphate of potash, 1 gr.; sulphate of magnesia, 5 gr.; chloride of sodium, 10 gr.; chloride of calcium, 3 gr.; carbonated water, a pint.

MARIENBAD. Carbonate of soda, 2 scruples; sulphate of soda, 90 gr.; sulphate of magnesia, 8 gr.; chloride of sodium, 15 gr.; chloride of calcium, 10 gr.; carbonated water, a pint. (Or, bicarbonate of soda, 50 gr.; sulphate of soda, 1 dr.; chloride of sodium, 15 gr.; sulphate of magnesia, 10 gr.; dissolve in a pint of water, add 25 gr. of dry bisulphate of soda, and cork immediately.)

MARIENBAD PURGING SALTS. Bicarbonate of soda, 5 oz.; dried sulphate of soda, 12 oz.; dry chloride of sodium, 1½ oz.; sulphate of magnesia (dried), 2 oz.; dried bisulphate of soda, 2½ oz. Mix the salts, previously dried, separately, and keep them carefully from the air.

PULLNA WATER. Sulphate of soda, 4 dr.; sulphate of magnesia, 4 dr.; chloride of calcium, 15 gr.; chloride of magnesium (dry), a scruple; chloride of sodium, a scruple; bicarbonate of soda, 10 gr.; water, slightly carbonated, 1 pint. One of the most active of the purgative saline waters.

PULLNA WATER WITHOUT THE MACHINE. Bicarbonate of soda, 50 gr.; sulphate of magnesia, 4 dr.; sulphate of soda, 3 dr.; chloride of sodium, a scruple; dissolve in a pint of water; add, lastly, 2 scruples of bisulphate of soda, and close the bottle immediately.

SALTS FOR MAKING PULLNA WATER. Dry bicarbonate of soda, 1 oz.; exsiccated sulphate of soda, 2 oz.; exsiccated sulphate of magnesia, 1 ½ oz.; dry chloride of sodium, 2 dr.; dry tartaric acid, ¾ oz. (or rather dry bisulphate of soda, 1 oz.).

SEIDLITZ POWDER. The common seidlitz powders do not resemble the water. A closer imitation would be made by using effloresced sulphate of magnesia instead of the potassio-tartrate of soda. A still more exact compound will be the following:—Effloresced sulphate of magnesia, 2 oz.; bicarbonate of soda, ½ oz.; dry bisulphate of soda, ½ oz.; mix, and keep in a close bottle.

SEIDLITZ WATER. This is usually imitated by strongly aërating a solution of 2 dr. of sulphate of magnesia in a pint of water. It is also made with 4, 6, and 8 dr. of the salt to a pint of water.

SEIDSCHUTZ WATER. Sulphate of magnesia, 3 dr.; chloride of calcium, nitrate of lime, bicarbonate of soda, of each, 8 gr.; sulphate of potash, 5 gr.; aërated water, 1 pint.

SELTZER WATER. Chloride of calcium and chloride of magnesium, of each, 4 gr.; dissolve these in a small quantity of water, and add it to a similar solution of 8 gr. of bicarbonate of soda, 20 gr. of chloride of sodium, and 2 gr. of phosphate of soda; mix, and add a solution of ½ gr. of sulphate of iron; put the mixed solution into a 20-oz. bottle, and fill up with aërated water. Much of the Seltzer water sold is said to be nothing more than simple carbonated water, containing a little chloride of sodium. An imitation of Seltzer water is also made by putting into a stone Seltzer-bottle, filled with water, 2 dr. bicarbonate of soda, and 2 dr. of citric acid in crystals, corking the bottle immediately. Sodaic powders are sometimes sold as Seltzer powders.

VICHY SALTS. Bicarbonate of soda, 1½ oz.; chloride of sodium, 15 gr.; effloresced sulphate of soda, 1 dr.; effloresced sulphate of magnesia, 1 scruple; dry tartarised potash and iron, 1 gr.; dry tartaric acid, 1 oz. (or dry bisulphate of soda); mix the powders, previously dried, and keep them in a close bottle.

VICHY WATER. Bicarbonate of soda, 1 dr.; chloride of sodium, 2 gr.; sulphate of soda, 8 gr.; sulphate of magnesia, 3 gr.; tincture of chloride of iron, 2 drops; aërated water, a pint. Dorvault directs 75 gr. of bicarbonate of soda, 4 gr. of chloride of sodium, ½ gr. sulphate of iron, 10 gr. sulphate of soda, 3 gr. sulphate of magnesia, to a pint of water. By adding 45 gr. (or less) of citric acid an effervescing water is obtained.

M. Soubeiran, relying on the analysis of Longchamps, imitates Vichy water by the following combination:—Bicarbonate of soda, 135 gr.; chloride of sodium, 2½ gr.; crystallised chloride of calcium, 12 gr.; sulphate of soda, 11½ gr.; sulphate of magnesia, 3¾ gr.; tartrate of iron and potash, ½ gr.; water, 2$\frac{1}{10}$ pints (1 litre); carbonic acid, 305 cub. inches (5 litres). Dissolve the salts of soda and iron in part of the water, and add the sulph. magnes. and then the chlor. calc. in the remaining water. Charge now with the carbonic acid gas under pressure.

SALINE WATERS, &C., NOT CARBONATED.

SEA WATER. Chloride of sodium, 4 oz.; sulphate of soda, 2 oz.; chloride of calcium, ½ oz.; chloride of magnesium, 1 oz.; iodide of potassium, 4 gr.; bromide of potassium, 2 gr.; water, a gallon. A common substitute for sea water as a bath is made by dissolving 4 or 5 oz. of common salt in a gallon of water.

The following mixture of dry salts may be kept for the immediate production of a good imitation of sea water:—Chloride of sodium (that obtained from evaporating sea water, and not recrystallised, in preference), 85 oz.; effloresced sulphate of soda, 15 oz.; dry chloride of calcium, 4 oz.; dry chloride of magnesium, 16 oz.; iodide of potassium, 2 dr.; bromide of potassium, 1 gr. Mix, and keep dry. Put 4 or 5 oz. to a gallon of water.

BALARUC WATER. Chloride of sodium, 1 oz.; chloride of calcium, 1 oz.; chloride of magnesium, ½ oz.; sulphate of soda, 3 dr.; bicarbonate of soda, 2 dr.; bromide of potassium, 1 gr.; water, a gallon. Chiefly used for baths.

SULPHURETTED WATERS.

SIMPLE SULPHURETTED WATERS. Pass sulphuretted hydrogen into cold water (previously deprived of air by boiling, and cooled in a closed vessel) till it ceases to be absorbed.

AIX-LA-CHAPELLE WATER. Bicarbonate of soda, 12 gr.; chloride of sodium, 25 gr.; chloride of calcium, 3 gr.; sulphate of soda, 8 gr.; simple sulphuretted water, 2½ oz.; water, slightly carbonated, 17½ oz.

BARÈGES WATER. (Cauterets, Bagnères de Luchon, Eaux Bonnes, St Sauveur, may be made the same.) Crystallised hydrosulphate of soda, crystallised carbonate of soda, and chloride of sodium, of each, 1½ gr.; water (freed from air), a pint. A stronger solution for adding to baths is thus made:—Crystallised hydrosulphate of soda, crystallised carbonate of soda, and chloride of sodium, of each, 2 oz.; water, 10 oz.; dissolve. To be added to a common bath at the time of using.

HARROGATE WATER. Chloride of sodium, 100 gr.; chloride of calcium, 10 gr.; chloride of magnesium, 6 gr.; bicarbonate of soda, 2 gr.; water, 18½ oz. Dissolve, and add simple sulphuretted water, 1½ oz.

NAPLES WATER. Crystallised carbonate of soda, 15 gr.; fluid magnesia, 1 oz.; simple sulphuretted water, 2 oz.; aërated water, 16 oz. Introduce the sulphuretted water into the bottle last.

CHALYBEATE WATERS.

SIMPLE CHALYBEATE WATER. Water freed from air by boiling, 1 pint; sulphate of iron, ½ gr.

AËRATED CHALYBEATE WATER. Sulphate of iron, 1 gr.; carbonate of soda, 4 gr.; water (deprived of air and charged with carbonic acid gas), a pint. Dr Pereira recommends 10 gr. each of sulphate of iron and bicarbonate of soda to be taken in a bottle of ordinary soda water. This is equivalent to 4 gr. of carbonate of iron.

BRIGHTON CHALYBEATE. Sulphate of iron, chloride of sodium, chloride of calcium, of each, 2 gr.; carbonate of soda, 3 gr.; carbonated water, 1 pint.

BUSSANG, FORGES, PROVINS, and other similar

waters may be imitated by dissolving from ½ to ¾ of a grain of sulphate of iron, 2 or 3 gr. of carbonate of soda, 1 gr. of sulphate of magnesia, and 1 gr. of chloride of sodium, in a pint of aërated water.

MONT D'OR WATER. Bicarbonate of soda, 70 gr.; sulphate of iron, ¾ gr.; chloride of sodium, 12 gr.; sulphate of soda, ½ gr.; chloride of calcium, 4 gr.; chloride of magnesium, 2 gr.; aërated water, a pint.

PASSY WATER. Sulphate of iron, 2 gr.; chloride of sodium, 3 gr.; carbonate of soda, 4 gr.; chloride of magnesium, 2 gr.; aërated water, a pint.

PYRMONT WATER. Sulphate of magnesia, 20 gr.; chloride of magnesium, 4 gr.; chloride of sodium, 2 gr.; bicarbonate of soda, 16 gr.; sulphate of iron, 2 gr.; Carrara water, a pint.

VARIOUS AËRATED MEDICINAL WATERS NOT RESEMBLING ANY NATURAL SPRING.

MIALHE'S AËRATED CHALYBEATE WATER. Water, a pint; citric acid, 1 dr.; citrate of iron, 15 gr.; dissolve, and add 75 gr. of bicarbonate of soda.

TROSSEAU'S MARTIAL AËRATED WATER. Potassio-tartrate of iron, 10 gr.; artificial Seltzer water, a pint.

BOUCHARDAT'S GASEOUS PURGATIVE. Phosphate of soda, 1½ oz.; carbonated water, a pint.

MIALHE'S IODURETTED GASEOUS WATER. Iodide of potassium, 15 gr.; bicarbonate of soda, 75 gr.; water, a pint; dissolve, and add sulphuric acid, diluted with its weight of water, 75 gr. Cork immediately.

DUPASQUIER'S GASEOUS WATER OF IODIDE OF IRON. Solution of iodide of iron (containing $\frac{1}{10}$ of dry iodide), 30 gr.; syrup of gum, 2½ oz.; aërated water, 17½ oz.

SALTS (Smelling). *Syn.* SAL VOLATILIS OLEOSUS, L. Sesquicarbonate of ammonia commonly passes under the name of 'SMELLING SALTS,' and, with the addition of a few drops of essential oil, is frequently employed to fill 'SMELLING BOTTLES;' but when a strong and durable pungency is desired, the carbonate should alone be used, as in one or other of the following formulæ:

1. Carbonate (not sesquicarbonate) of ammonia, 1 lb.; oil of lavender (Mitcham), 2 oz.; essence of bergamot, 1 oz.; oil of cloves, ½ oz.; rub them together, and sublime; keep the product in well-stopped bottles.

2. Carbonate of ammonia, 1 lb.; oil of lavender, 2 oz.; oils of bergamot and lemon, of each, 1 oz.; as the last.

3. Carbonate of ammonia, ½ lb.; essence of bergamot, 1 oz.; oil of verbena, ¼ oz.; otto of roses, 1 dr.; as before.

4. Carbonate of ammonia, ¾ lb.; essences of bergamot and lemon, of each, ½ oz.; essence de petit grain, ¼ oz.; oil of cloves, 1 dr.; as before.

5. (Extemporaneous.) *a.* From sal-ammoniac, 1 dr.; pure potassa, 3 dr.; grind them together, and add of essence of lemons, 15 drops; oil of cloves, 3 or 4 drops.

b. From carbonate or sesquicarbonate of ammonia (bruised), q. s.; volatile ammoniacal essence, a few drops.

According to Dr Paris, GODFREY'S SMELLING SALTS are made by resubliming volatile salt with subcarbonate of potassa and a little spirits of wine (and essential oil).

SALUFER. A fluosilicate of sodium is sold under this name, and used as an antiseptic wash and disinfectant. 1 gr. in 1 oz. of water is the usual strength for lotions.

SALVE. A name indiscriminately applied by the vulgar to any consistent, greasy preparation used in medicine.

Salve, Lip. *Syn.* CERATUM LABIALE, L. *Prep.* 1. (RED or PERUVIAN.) From spermaceti ointment, ½ lb.; alkanet root, ½ oz.; melt them together until sufficiently coloured, strain, and when the strained fat has cooled a little, add of balsam of Peru, 3 dr.; stir well, and in a few minutes pour off the clear portion from the dregs; lastly, stir in of oil of cloves, 20 or 30 drops. This never gets rancid.

2. (ROSE.) See CERATE.

3. (WHITE.) From the finest spermaceti ointment or cerate, 3 oz.; finely powdered white sugar, 1 oz.; neroli or essence de petit grain, 10 or 12 drops, or q. s.

Obs. Numerous formulæ are extant for lip salves, as for other like articles, but the preceding are those generally employed in trade. The perfumes may be varied at will, and the salve named after them. A very small quantity of finely powdered borax is occasionally added. FRENCH LIP SALVE is said to contain alum, in fine powder; and GERMAN LIP SALVE is said to be made of cacao butter. See CERATE, POMMADE, and OINTMENT.

SAND. *Syn.* ARENA, L. River and sea sand consist chiefly of finely divided siliceous matter, mixed occasionally with carbonate of lime. That of Lynn and Alum Bay is nearly pure silica, and is therefore selected for the manufacture of glass. Sand is used by moulders in metal, and as a manure for heavy land. It is a large and necessary portion of every fertile soil.

SAN'DAL-WOOD. *Syn.* RED SANDERS-WOOD, R. SAUNDERS-W.; LIGNUM SANTALI RUBRI, LIGNUM SANTALINUM RUBRUM, PTEROCARPUS, L. The wood of *Pterocarpus santalinus.* It is used in medicine as a colouring matter. It is also employed in dyeing, and to stain varnishes. WOOL may be dyed a carmine red by dipping it alternately into an infusion of this wood and an acidulous bath (*Trommsdorff*). Prepared with a mordant of alum and tartar, and then dyed in a bath of sandal-wood and sumach, it takes a reddish yellow (*Bancroft*). See SANTALIN.

WHITE AND YELLOW SANDAL-WOOD. There are more than a dozen species of the genus, which are chiefly restricted to Asia, Australia, and Oceania. The Indian species are *Santalum album* and *S. myrtifolium.* The Australian species are *S. cygnum*, *S. lanceolatum*, *S. oblongatum*, *S. obtusifolium*, *S. ovatum*, and *S. venosum.* The species found in the Pacific islands are *S. Austro-Caledonium*, Viell, which is superior to that of most other countries, owing to the strength and fineness of its odour; *S. ellipticum*, *S. Freycinetianum*, *S. paniculatum*, and *S. Yase.* But many of the species are not well determined, nor their localities clearly defined.

Santalum album, one of the Indian species, has long furnished the chief supply of wood, which is shipped from the Madras Presidency. In Mysore the sandal-wood trees form a Government monopoly, bringing in a revenue of about £40,000, the wood selling there at £35 to £40 a ton.

S. Freycinetianum is imported from Cochin China and the Pacific islands, but it is less esteemed, the colour of the wood being paler, and the odour less pronounced.

It is only the central portion of the tree which produces the scented yellow wood constituting the sandal-wood of commerce. The quality of the wood depends on the quantity of the oil contained in it, as indicated by the smell when freshly cut or burnt. The old trees produce the best, and of them that part of the wood near the root is the most prized. The distillation of oil from the roots in India is carried on chiefly at Mangalore. Five cwt. of wood yields about 80 lbs. of pure oil, thus giving a profit of nearly 37 per cent.

The essential oil is used as the basis of nearly all ottos manufactured in the country.

The wood is made into boxes, in which steel does not rust; curiously carved cases, fans, and other fancy articles; and it is also burnt in the temples.

The Mysore wood is divided into five classes. The first three go almost exclusively to China; the hollow fillets and the small broken pieces, which are not included in the five classes, going to Arabia, where they are either burnt whole, for the sake of the fragrant smell afforded, or ground up and used with other ingredients as incense. Of that sent to Surat the inferior descriptions of billets are burnt by the Parsees in their fire temples, and are also used at Hindoo funerals when the friends of the deceased are able to afford it. The wood, rubbed down with water and worked into a paste, is used by all Hindoos in their caste marks, and is also employed as an external application for headaches and some skin diseases. The powder of the roots and of the heart-wood is used by the Chinese against gonorrhœa, and is applied to wounds. They also consider it carminative, stomachic, and stimulant. The oil, which is yellow and of the consistency of castor oil, is much esteemed for its odour.

As opium ministers to the sensual gratification of the Chinese and others of the same class, so sandal-wood ministers to their superstition. Without it no religious ceremony can be conducted, and its absence is a mark of poverty, so that the Oriental of India and China will sacrifice anything rather than allow that, on the proper occasion, sandal-wood should not be burnt. The roots, which are richest in oil, and the chips go to the still, while Hindoos who can afford it show their wealth and respect for their departed relatives by adding sticks of sandal-wood to the funeral pile.

SANDARACH. *Syn.* SANDRAC, GUMS. A resin obtained from *Thuja articulata* and *Juniperus communis* (in warm climates). It is slightly fragrant, is freely soluble in rectified spirit, and has a sp. gr. of 1·05 to 1·09. It is used as incense, pounce, in varnishes, &c.

SANDERS-WOOD. See SANDAL-WOOD.

SANDIVER. *Syn.* GLASS GALL; FEL VITRI, SAL VITRI, L. The saline scum that swims on glass when first made. It is occasionally used in tooth powders.

SAND-PAPER. The 'American Builder' gives the following process for making sand-paper of superior quality, at almost nominal cost:

"The device for making sand-paper is simple and at hand to any one who has occasion to use the paper. A quantity of ordinary window glass is taken (that having a green colour is said to be the best) and pounded fine, after which it is passed through one or more sieves of different degrees of fineness, to secure the glass for coarse or fine paper. Then any tough paper is covered evenly with glue, having about one third more water than is generally employed for wood-work. The glass is sifted upon the paper, allowed a day or two in which to become fixed in the glue, when the refuse glass is shaken off, and the paper is fit for use.

SANGUIN'ARINE. *Syn.* SANGUINARINA, L. Obtained from the root of *Sanguinaria Canadensis*, Linn., or blood-root, by digesting it in anhydrous alcohol; exhausting it with weak sulphuric acid; precipitating by liquor of ammonia; dissolving out by ether, and precipitating sulphate of sanguinarine by the addition of sulphuric acid. The sulphate may be decomposed by ammonia, which precipitates the alkaloid as a white pearly substance, of an acrid taste, very soluble in alcohol, also soluble in ether and volatile oils. With acids it forms soluble salts, remarkable for their beautiful red, crimson, and scarlet colours. These salts are used in medicine as expectorants, in doses of fractions of a grain.

The 'sanguinarin' of the American 'Eclectics' is prepared by precipitating a saturated tincture of blood-root by water. It contains an uncertain proportion of the alkaloid, and is of a deep reddish-brown colour.

SANITARY AUTHORITIES AND SANITARY DISTRICTS. With the exception of the metropolis, the whole of England and Ireland is divided into urban and rural sanitary districts, which are respectively governed by urban and rural authorities.

The Public Health Act (sec. 6) defines an urban district and an urban authority in England as in table on the next page; provided that—

1. Any borough the whole of which is included in and forms part of a local government district or improvement act district, and any improvement act district which is included in and forms part of a local government district, and any local government district which is included in and forms part of an improvement act district, shall, for the purposes of this Act, be deemed to be absorbed in the larger district in which it is included, or of which it forms part; and the improvement commissioners, or local board, as the case may be, of such larger district, shall be the urban authority therein; and

2. Where an improvement act district is coincident in area with a local government district, the improvement commissioners, and not a local board, shall be the urban authority there; and

Urban Districts.	Urban Authority
Borough, constituted such either before or after the passing of this Act.	The mayor,aldermen, and burgesses, acting by the council.
Improvement act district, constituted such before the passing of the Public Health Act, 1872, and having no part of its area situated within a borough or local government district.	The improvement commissioners.
Local government district, constituted such either before or after the passing of this Act, having no part of its area situated within a borough, and not coincident in area with a borough or improvement act district.	The local board.

3. Where any part of an improvement act district is situated within a borough or local act district, or where any part of a local government district is situated within a borough, the remaining part of such improvement act district or of such local government district so partly situated within a borough, shall, for the purposes of this Act, continue subject to the like jurisdiction as it would have been subject to if this Act had not been passed, unless and until the Local Government Board by provisional order otherwise directs.

An English rural sanitary district and authority are thus defined by the Public Health Act (sec. 9):

" The area of any union which is not coincident in area with an urban district, nor wholly included in an urban district (in this section called a rural union), with the exception of those portions (if any) of the area which are included in any urban district, shall be a rural district, and the guardians of the union shall form the rural authority of such district; provided that—

" 1. An ex officio guardian resident in any parish or part of a parish belonging to such union, which parish or part of a parish forms or is situated in an urban district, shall not act or vote in any case in which guardians of such union act or vote as members of the rural authority, unless he is the owner or occupier of property situated in the rural district of a value sufficient to qualify him as an elective guardian for the union.

" 2. An elective guardian of any parish belonging to such union, and forming or being included within an urban district, shall not act or vote in any case in which guardians of such union act or vote as members of the rural authority.

" 3. Where part of a parish belonging to a rural union forms or is situated in an urban district, the Local Government Board may by order divide such parish into separate wards, and determine the number of guardians to be elected by such wards respectively, in such manner as to provide for the due representation of the part of the parish situated within the rural district; but

until such order has been made, the guardian or guardians of such parish may act and vote as members of the rural authority in the same manner as if no part of such parish formed part of, or was situated in, the urban district."

Where the number of elective guardians, who are not by this section disqualified from acting and voting as members of the rural authority, is less than five, the Local Government Board may from time to time by order nominate such number of persons as may be necessary to make up that number, from owners or occupiers of property situated in the rural district of a value sufficient to qualify them as elective guardians for the union; and the persons so nominated shall be entitled to act and vote as members of the rural authority, but not further or otherwise.

Subject to the provisions of this Act, all statutes, orders, and legal provisions applicable to any board of guardians shall apply to them in their capacity of rural authority under this Act for the purposes of this Act; and it is hereby declared that the rural authority are the same body as the guardians of the union or parish for or within which such authority act.

Sanitary districts in Ireland are—The City of Dublin, other corporate towns above 6000, and towns or townships having commissioners under local Acts.

And urban authorities are—In the City of Dublin, the Right Hon. the Lord Mayor, Aldermen, and Burgesses acting by the town council.

In towns corporate, the town council.

In towns exceeding 6000, having commissioners under the Lighting, Cleaning, and Watching Act of George IV; or having municipal commissioners under 8 and 4 Vict., c. 108; or town commissioners under the Towns Improvement (Ireland) Act (17 and 18 Vict., c. 103), the said commissioners, municipal or town councillors respectively.

In towns or townships having commissioners under local Acts, the town or township commissioners (37 and 38 Vict., c. 98, s. 3).

The Irish rural sanitary districts and authorities are exactly analogous to the English.

In Scotland, sanitary powers are exercised by town councils, police commissioners, and parochial boards, controlled and supervised by a board of supervision; but the names of urban and rural sanitary authorities have not yet been applied to them.

Under the English Public Health Act there may also be formed united districts; for example:

Where, on the application of any local authority of any district, it appears to the Local Government Board that it would be for the advantage of the districts, or any of them, or any parts thereof, or of any contributory places, in any rural district or districts, to be formed into a united district for all or any of the purposes following:

1. The procuring a common supply of water; or
2. The making a main sewer, or carrying into effect a system of sewerage for the use of all such districts or contributory places; or
3. For any other purposes of this Act, the Local Government Board may, by provisional order,

form such districts or contributory places into a united district.

All costs, charges, and expenses of and incidental to the formation of a united district are, in the event of the united district being formed, to be a first charge on the rates leviable in the united district in pursuance of section 279 of the Public Health Act.

Notice of the provisional order must be made public in the locality; and should the union be carried out, the incidental expenses thereto are a first charge on the sanitary rates of the united district. A united district is governed by a joint board consisting of such *ex officio*, and of such number of elective members as the provisional order determines.

The business arrangements of the joint board differ little from those of a sanitary authority.

The joint board is a body corporate having a name—determined by the provisional order,—a perpetual succession, and a common seal, and having power to acquire and hold lands without any licence in mortmain. The joint board has only business and power in matters for which it has been formed. With the exception of these special objects, the component districts continue as before to exercise independent powers.

Nevertheless the joint board may delegate to the sanitary authority of any component district the exercise of any of its powers, or the performance of any of its duties (Public Health Act, sec. 281).

Sanitary authorities and districts may be also combined for the execution and maintenance of works, for the prevention of epidemic diseases, as well as for the purpose of appointing a medical officer of health. Districts when once formed are not fixed and unvariable, the Local Government Board having the most extensive powers over the alterations of areas.

1. The Local Government Board, by provisional order, may dissolve any local government district, and may merge any such district in some other district, or may declare the whole or any portion of a local government or a rural district immediately adjoining a local government district to be included in such last-mentioned district, or may declare any portion of a local government district immediately adjoining a rural district to be included in such last-mentioned district; and thereupon the included area shall, for the purposes of the Public Health Act, be deemed to form part of the district in which it is included in such order; and the remaining part (if any) of such local government district or rural district affected by such order shall continue subject to the like jurisdiction as it would have been subject to if such order had not been made, unless and until the Local Government Board by provisional order otherwise directs.

2. In the case of a borough comprising within its area the whole of an Improvement Act district, or having an area co-extensive with such district, the Local Government Board, by provisional order, may dissolve such district, and transfer to the council of the borough all or any of the jurisdiction and powers of the Improvement Commissioners of such district, remaining vested in them at the time of the passing of the Public Health Act.

3. The Local Government Board may, by order, dissolve any special drainage district constituted either before or after the passing of the Public Health Act in which a loan for the execution of works has not been raised, and merge it into the parish or parishes in which it is situated; but in the cases where a loan has been raised the Local Government Board can only do this by provisional order (Public Health Act, sec. 271).

Disputes with regard to the boundaries of districts are to be settled by the Local Government Board after local inquiry (Public Health Act, sec. 278).

Where districts also are constituted for the purposes of main sewerage only in pursuance of the Public Health Act of 1848, or where a district has been formed subject to the jurisdiction of a joint sewerage board, in pursuance of the Sewage Utilisation Act of 1867, such districts or district may be dissolved by provisional order, and the Local Government Board may constitute it a united district, subject to the jurisdiction of a joint board (Public Health Act, sec. 323).

The Local Government Board may also declare by provisional order any rural district to be a local government district.

The Local Government Board has also the important power of investing a rural authority with urban powers as follows:

"The Local Government Board may, on the application of the authority of any rural district, or of persons rated to the relief of the poor, the assessment of whose hereditaments amounts at the least to one tenth of the net rateable value of such district, or of any contributory place therein, by order to be published in the 'London Gazette,' or in such other manner as the Local Government Board may direct, declare any provisions of this Act in force in urban districts to be in force in such rural district or contributory place, and may invest such authority with all or any of the powers, rights, duties, capacities, liabilities, and obligations of an urban authority under this Act, and such investment may be made either unconditionally or subject to any conditions to be specified by the board as to the time, portion of its district, or manner during, at, and in which such powers, rights, duties, liabilities, capacities, and obligations are to be exercised and attached, provided that an order of the Local Government Board made on the application of one tenth of the persons rated to the relief of the poor in any contributory place shall not invest the rural authority with any new powers beyond the limits of such contributory places" (Public Health Act, sec. 276).

Powers and Duties of Sanitary Authorities.
In England urban sanitary authorities have very extensive powers and duties under the Public Health Act of 1875, and in addition they have to carry out the Bakehouse Regulation Act, and the Artisans and Labourers' Dwellings Act. They also have power to adopt the Baths and Washhouses Acts, and the Labouring Classes' Lodginghouses Acts; but where adopted or in force, the powers, rights, duties, &c., of these Acts belong to the urban authority. The powers of any local Act for sanitary purposes (except a River Conservancy Act) are transferred to the urban authority.

The powers of an English rural authority are exercised principally under the Public Health Act, but they have also to carry out the Bakehouse Regulation Act.

The powers given by the Irish Public Health Act to Irish sanitary authorities are similar.

The Local Government Act is not in force there, and equal powers are given without distinction to urban and rural sanitary authorities.

The duties of sanitary authorities are to carry out the Acts which apply to them, and appoint certain officers, such as medical officers of health, inspectors of nuisances, clerk, treasurer, &c.

Speaking generally, it may be affirmed that all sanitary authorities are invested with ample powers for enforcing sanitary measures. Their duty consists in perfecting drainage, sewerage, and water supply. In towns they have the control of streets and houses, both private and public, and in all localities they possess ample powers to cause every species of nuisance to be abated, which is in the least inimical to health.

The Public Health Act contains a proviso for dealing with an authority which fails in its duty. Under these circumstances the Local Government Board is invested with compulsory powers, and may compel the due performance of whatever it may deem necessary.

SANITARY HERBAL BITTERS—Gesundheits-krauter-Bitter. An indispensable household remedy for every family, for colic, stomach-ache, cramp in the bladder, flatulence, loss of appetite, nausea, chronic liver diseases, constipation, and diarrhœa; also as a soothing agent for infants (*Gottschlich*). The fluid contains in 100 grms. the soluble portion of about ·8 grm. opium (*Hager*).

SANITARY LIQUEUR—Gesundheits Liqueur. Swedish elixir of life, with rhubarb in place of the aloes, made into a liqueur with sugar and spirit (*Hager*).

SANITARY, POPULAR, ERRORS. It is a popular sanitary error to think that the more a man eats the fatter and stronger he will become. To believe that the more hours children study the faster they learn. To conclude that, if exercise is good, the more violent the more good is done. To imagine that every hour taken from sleep is an hour gained. To act on the presumption that the smallest room in the house is large enough to sleep in. To imagine that whatever remedy causes one to feel immediately better is good for the system, without regard to the ulterior effects. To eat without an appetite; or to continue after it has been satisfied, merely to gratify the taste. To eat a hearty supper at the expense of a whole night of disturbed sleep and weary waking in the morning ('Sanitary Record ').

SANITARY RATAFIA—Gesundheits Ratafia. For removing all stomach, chest, and bowel complaints, indigestion, colic, diarrhœa, vomiting, flatulence, dysuria, and affections caused by chills. A clear brown schnapps, containing, in 250 grms. by weight, 75 grms. sugar, 105 grms. water, 100 grms. strong spirit, 40 grms. each of tincture of orange peel and tincture of orange berries, 2·5 grms. each tincture of cloves and tincture of wormwood, 1 drop oil of peppermint, 5 drops acetic ether, and some drops of caramel (*Dr Horn*).

SANITARY SOUL, Flowers of—Gesundheits-blumengeist. A mixture of spirit, 500 parts; tinct. aromatica, 5 parts; oils of bergamot, lavender, and rosemary, of each, 2 parts; oil of thyme, 3 parts; oil of spearmint, 1 part (*Hager*).

SANITAS. A powerfully oxidising liquid containing hydrogen peroxide, obtained by passing air and steam through oil of turpentine. Employed for disinfecting purposes.

SANITATION, DOMESTIC. Not one of the least creditable or important benefits conferred of late years, by the efforts of philanthropic and enlightened enterprise upon the poorer classes of this country, has been the erection—in cities and large towns more particularly—of healthy houses for them to dwell in. In the construction of these habitations the architects and designers have for the most part been guided by sound sanitary principles, the carrying out of which has been effected by means of legislative supervision, and, if needful, of legislative action.

The result of these measures has in most cases been to provide residences for our poorer brethren, wherein, amongst other advantages, they enjoy the two primary ones of pure air and water. That the richer, upper, and middle classes, whilst devising and achieving so much in the way of comfort and health for those beneath them, should themselves in so many cases live in houses notoriously unhealthy, and should fail to recognise the advantages of the compulsory enforcement of necessary hygienic arrangements, are anomalies so amazing as to be, at first sight, scarcely credible. Yet a few statistics may serve to discomfit those who are incredulous on this point. The average mortality in London is 24 persons in 1000. In the improved dwellings of the poor it is only 14 in the 1000.

This subject was ventilated in a very earnest and valuable paper read before the Social Science Congress at Brighton in 1875 by Mr H. H. Collins. In this paper Mr Collins refers only to the houses of the metropolis and its suburbs, and maintains that, as far as regards the enforcement of sanitary precautions in house - building, London and its suburbs are infinitely worse provided for than many second-rate provincial towns, most of which, he says, have the construction of their buildings and streets regulated by bye-laws issued under the powers of the Public Health Act, and sanctioned by the Home Secretary, whereas in London the various Acts of Parliament for this purpose have been inoperative. Mr Collins describes the insanitary condition of some of the high-rented houses he examined, and says the descriptions which follow equally apply to many others situated in the most aristocratic quarters of London.

Imagine one of our legislators who, perhaps, had been voting for the passing of the 'Nuisance Removals Act,' returning from his parliamentary duties to such a mansion as is portrayed by Mr Collins in the following extract:—" I have recently purchased on behalf of a client the lease of a mansion in Portland Place from a well-known nobleman, who had spent, as I was informed, a fortune in providing new drainage; indeed, I found the principal water-closet built out of the house altogether; the soil-pipe of it, however, was carried through the basement, where it was supposed to be connected with the drain. Upon removing the floor-boards to examine it, I found the ground surrounding the connection literally one mass of black sewage, the soil oozing through the joint even at the time of the examination, and the connection with the main drain laid in it at right angles. The 9-inch drain-pipes ran through the centre of the house, having a very slight gradient, and had evidently not been laid in many years, yet they were nearly full of consolidated sewage, and but little space was left for the passage of the fluid. With but a slightly increased pressure the joints would have given way, and the sewage would have flowed under the boards instead of into the sewer. The sinks, water-closets, and cisterns were all badly situated, and all more or less defective in sanitary arrangement. In the butler's pantry the sink was placed next to the turn-up bedstead of the butler, who must have inhaled draughts of impure atmosphere at every inspiration. The soil-pipes of the closets had indeed been ventilated with a zinc rectangular tube, but as this had been so placed as to let the sewer gas through an adjacent skylight into the house, and the odour being extremely disagreeable, it had been by his lordship's directions (as I am told) closed. Here was evidence that it had at all events been doing some service, and probably had only poisoned a few of the domestics. I found the bends of soil-pipes likewise riddled with holes, as described by Dr Leargus. There happened to be a house-maids' sink situated close to a bedroom, the waste from which had been carefully connected with the soil-pipe, so that probably had the closets been satisfactorily ventilated, this arrangement would have defeated the object in view. I should also mention that the best water-closet was situated on the bedroom floor under the stairs, and was lighted and ventilated through a small shaft formed of wood boarding and carried to the roof; it also opened by a window to the main or principal staircase. The gutter of the roof ran through the bedrooms and under the floors; at the time of examination it was full of black slimy filth. This is a fair specimen of the sanitary arrangements of a nobleman's town house, situated in one of the best streets of this great metropolis, in the year of grace 1875."

Let us take another example:—"A few years ago a client of mine, who resided in a large house in a wealthy suburb, informed me that his wife and two daughters had suffered in health ever since they had occupied their house; that he had consulted several medical men without beneficial result, and that he wished me to make a survey of the premises. He paid a rental of about £200 per annum. I found that the drainage was in every way defective, although he told me that he had spent a large sum of money in making it 'perfect;' the gradients were bad, the pipes choked, and the joints unsound. The servants' water-closet was adjacent to the scullery, which was in communication with the kitchen, the sink being directly opposite the kitchen range. The water-closet was supplied direct from the cistern, the waste from which entered the drain, although it was said to be trapped.: The waste of the sink was simply connected with the drains and trapped with an ordinary bell-trap, the cover or trap of which I found broken. Under the kitchen range hot-water tap I found a trapped opening, also leading into the drain. The domestics complained of frequent headaches and general depression, and I need not add that it excited no surprise, seeing that the kitchen fire was continuously drawing in from the sewers and house-drains a steady supply of sewer gas to the house and drinking-water cistern. In addition I found the basement walls damp, owing to the absence of a damp-proof course and the want of dry areas. The upper water-closets, house-closets, and cisterns were situated over each other, off the first-floor landing, and directly opposite the bedroom doors. The bath and lavatory were fixed in the dressing-room, communicating with the best bed-room, the wastes from which were carried into the soil-pipe of closets. This latter was unventilated, but was trapped with an S pipe at bottom. The water-closets were pan closets, and were trapped by D traps. The upper closet periodically untrapped the lower closet, and both traps leaving the impure air free access to the house and cistern, which latter was also in communication by means of its waste-pipe with the house-drains. The overflows from safes of the water-closets were practically untrapped. The peculiar nauseating odour of sewer gas was distinctly perceptible, and I had but little doubt but that atonic disease was rapidly making its inroads on the occupants. The landlord refused to recognise the truth of my report. My client, acting on my advice, relinquished his lease, took another house, the sanitation of which was carefully attended to, and his wife and children have had no recurrence of illness."

Mr Collins mentions a very alarming and unsuspected source of aërial poisoning in many town houses to be the existence of old disused cesspools in the centre of the buildings. These receptacles, which are frequently nearly filled with decaying fæcal substances, are very often found to be insecurely covered over with tiles, stones, or boarding. To ensure the construction of a healthy dwelling-house, Mr Collins regards attention to the following conditions as essential:—"All subsoil should be properly drained, proper thickness of the concrete should be applied to the foundations, damp-proof courses, should be inserted over footings, earth should be kept back from walls by dry areas properly. drained and ventilated, external walls should be, built of good hard well-burnt stock brickwork, of graduated thicknesses, and never less than 14, inches thick; internal divisions should be of brick in cement. The mortar and cement should be of good quality. All basement floors should have a concrete or cement bottom, with air flowing under the same, and the boarding thereof should be tongued so as to prevent draught and exhalation penetrating through the joints of the same. Ample areas back and front should be insisted on, the divisional or party fence walls of which should never be allowed to exceed 7 feet in height, to allow free circulation and to prevent the areas becoming wells or shafts for stagnant air. The main drains should be carried through the back yards, and, to prevent inconvenience to adjoining owners from any obstruction, they should be laid in subways, so that the sewer inspector could gain ready access thereto without entering any of the premises or causing any annoyance to the tenants. No basement should on any account be allowed to be constructed at such a level as will not permit of the pipes having good steep gradients to the sewer.

"All sinks should be placed next external walls, having windows over the same, and removed from the influence of the fire-grates. All wastes should discharge exteriorly over and not into trapped cesspits, all of which should be provided with splashing stones fixed round the same. The basement cisternage should be placed in convenient and accessible positions, protected from dirt and guarded from the effects of alternations of temperature. They should be of slate and galvanised iron, and never of lead or zinc. They should be fitted with overflows discharging over the sink, or over trapped cesses as just mentioned. They should be supplied with stout lead encased, block-tin pipe, the services therefrom for all drinking purposes should be of the same description, and should be attached to an ascending filter, so that water may be delivered free from lead or organic impurities. Lead poisoning is more frequent than is generally believed. Cupboards under stairs, under sinks, under dressers, or out-of-the-way places should be avoided, and when fitted up should always be well ventilated. All passages should be well lighted and ventilated. Borrowed lights are better than none at all. Every room should be furnished with a fire-place, and Comyn and Chingo ventilators over doors and windows should be freely disposed. It would conduce to the health of the house, without adding one shilling to its cost, to build next the kitchen flue a separate ventilating flue, and to conduct the products of combustion from gas and other impure or soiled air, &c., into the same, from ventilators placed in the centre of or close to the ceilings, as may be found most convenient. By carefully proportioning the inlet and outlet ventilation, the air will be kept moving without draught, and preserved in a pure and sweet condition for respiration. The windows and doors will then serve only their legitimate objects of admitting light, and of affording ingress and egress to the various apartments. The staircase should be made the main ventilator of the house, and it is essentially necessary to preserve the air surrounding the same uncontaminated, pure, and undefiled. It will be better to light and ventilate it from the top; and to prevent the Ethiopians or blacks of London finding their way into the house, an invisible gauze net may be placed under it, which can periodically be easily removed and cleansed; or it may be furnished with a moveable inner, ornamental, flat light.

"Under no circumstances must lavatories or sinks be brought in connection with the drains. Most people desire the bath-room to be in proximity to the bedrooms; whether so placed or not, all connection with main drainage must be studiously avoided. The hot and cold pipes, known as the flow and return pipes, should be of galvanised iron, with junctions carefully made with running joints in red-lead; on no account should these be in contact with any other pipes. The wastes from the bath safe (and lavatories, if any) should be carried through the front wall of the house, and should turn over and into rain-water head, covered with domical wire grating to prevent birds building their nests therein, and carried down to the basement area, where they must discharge over a trapped cesspit, as before described, surrounded with a splash-stone or curve to obviate the nuisance of the soapsuds flowing over the pavement. A brush passed up and down these waters now and then will effectually remove any soapy sediment which may cling to their surfaces. The waste from baths, &c., into heads should be furnished with a ground valve flap and collar to prevent draught, and the bath should be fitted with india-rubber seatings between the metal and wood framing. Mansards or sloping roofs should be avoided; they are injurious to the health of the domestics, whose sleeping chambers they are generally appropriated to; they are unhealthy, hot in summer, and prejudicially cold in winter, laying the basis for future disease for those least able to bear it. Gutters taken through roofs, known as 'trough,' should never be permitted; they congregate putrescent filth, which remains in them for years to taint and poison the atmosphere." Consult also, as supplementing this subject, the articles DRAINS, DUSTBINS, CESSPOOLS, TANKS, TRAPS, WATER-CLOSETS.

SANTAL VERT (*Croton* sp.). A dark green wood from Zanzibar. It is said to be exported from Zanzibar and Madagascar into India, where it is used for burning the bodies of Hindoos.

SANTALIN. The colouring principle of red sanders-wood.

SANTONIN. $C_{15}H_{18}O_3$. *Syn.* SANTONIC ACID; SANTONINUM, L. The crystalline and characteristic principle of the seed of several varieties of *Artemisia*.

Prep. (Ph. Baden, 1841.) Take of worm-seed, 4 parts; hydrate of lime, 1½ parts; mix, and exhaust them with alcohol of 90%; distil off 3-4ths of the spirit, and evaporate the remainder to one half, which, at the boiling temperature, is to be mixed with acetic acid in excess, and afterwards with water; on repose, impure santonin subsides; wash this with a little weak spirit, then dissolve it in rectified spirit, 10 parts, decolour by ebullition for a few minutes with animal charcoal, and filter; the filtrate deposits colourless crystals of santonin as it cools; these are to be dried, and kept in opaque bottles.

Mr W. G. Smith, M.B., states that two singular effects are known to result from the administration of santonin in moderate doses, viz. visual derangements and a peculiar alteration in the colour of the urine. He adds that three hours after taking 5 gr. of pure white santonin he became conscious, while reading, of a yellowish tint on the paper, and a yellow haze in the air. His own hands and the complexions of others appeared of a sallow, unhealthy colour; and the evening sky, which was really of a pale lavender colour, seemed to be light green. Vision was not perfectly distinct for some hours, and was accompanied by a certain vagueness of definition. Mr

Smith endorses the observations of previous observers, who had noticed that the urine of persons under the influence of santonin is tinged of a saffron-yellow or greenish colour. The coloured urine resembles that of a person slightly jaundiced, and, like this, permanently stains linen of a light yellow colour.

The best test for santonin in the urine is an alkali, upon the addition of which the urine immediately assumes a fine cherry-red colour, varying in depth according to the amount of santonin present. Potash was found to be the preferable alkali.

Prop., &c. Prismatic or tubular crystals; inodorous; tasteless, or only slightly bitter; fusible; volatilisable; soluble in 4500 parts of cold and about 250 parts of boiling water; soluble in cold alcohol and ether; freely soluble in hot alcohol. It is much esteemed as a tasteless worm medicine, and is especially adapted to remove lumbricales (large round-worms).—*Dose*, 2 to 6 gr., repeated night and morning, followed by a brisk purge.

Prep. (Ph. B.) Boil 1 lb. of santonica, bruised, with 1 gall. of distilled water and 5 oz. of slaked lime, in a copper or tinned iron vessel for an hour, strain through a stout cloth, and express strongly. Mix the residue with ⅓ gall. of distilled water and 2 oz. of lime, boil for half an hour, strain and express as before. Mix the strained liquors, let them settle, decant the fluid from the deposit, evaporate to the bulk of 2½ pints. To the liquor, while hot, add, with diligent stirring, hydrochloric acid, until the fluid has become slightly and permanently acid, and set it aside for five days that the precipitate may subside. Remove, by skimming, any oily matter which floats on the surface, and carefully decant the greater part of the fluid from the precipitate. Collect this on a paper filter, wash it first with cold distilled water till the washings pass colourless and nearly free from acid reaction, then with ¼ fl. oz. of solution of ammonia, previously diluted with 5 oz. of distilled water, and, lastly, with cold distilled water, till the washings pass colourless. Press the filter containing the precipitate between folds of filtering paper, and dry it with a gentle heat. Scrape the dry precipitate from the filter, and mix it with 60 gr. of purified animal charcoal. Pour on them 9 fl. oz. of rectified spirit, digest for half an hour, and boil for ten minutes. Filter while hot, wash the charcoal with 1 fl. oz. of boiling spirit, and set the filtrate aside for two days in a cool dark place to crystallise. Separate the mother-liquor from the crystals, and concentrate to obtain a further product. Collect the crystals, let them drain, redissolve them in 4 fl. oz. of boiling spirit, and let the solution crystallise as before. Lastly, dry the crystals on filtering paper in the dark, and preserve them in a bottle protected from the light.

SAP GREEN. See GREEN PIGMENTS.

SAPONIFICA'TION. See SOAP.

SAP'ONIN. *Syn.* SAPONINUM, L. A white, non-crystallisable substance, obtained by the action of hot diluted alcohol on the root of *Saponaria officinalis*, Linn., or soapwort.

Prop., &c. Saponin is soluble in hot water, and the solution froths strongly on agitation.

The smallest quantity of the powder causes violent sneezing.

SAPPAN WOOD (*Cæsalpinia Sappan*, Linn.). A red dye-wood, furnished by an East Indian tree growing to a height of 30 or 40 feet. It is imported from India, Siam, and Ceylon.

SARCOCOL'LA. A gum-resin supposed to be derived from one or more plants of the Nat. Ord. RHINEACEÆ, growing in Arabia and Persia. It somewhat resembles gum-arabic, except in being soluble in both water and alcohol, and in having a bitter-sweet taste. It was formerly used in surgery.

SAR'COSINE. $C_3H_7O_2N$. A feebly basic substance, obtained by boiling kreatine for some time with a solution of pure baryta. It forms colourless transparent plates, freely soluble in water, sparingly so in alcohol, and insoluble in ether; it may be fused and volatilised.

SARSAPARIL'LA. *Syn.* SARSÆ RADIX (B. P.), RADIX SARSÆ, RADIX SARSAPARILLÆ, SARSA (Ph. L. & E.), SARSAPARILLA (Ph. D. & U.S.), L. "Jamaica sarsa. The root of *Smilax officinalis*, Kunth" (Ph. L.), "and probably of other species" (Ph. E.).

The sarsaparillas of commerce are divided by Dr Pereira into two classes—'mealy sarsaparilla' and 'non-mealy sarsaparillas.' In the first are placed Brazilian or Lisbon, Caraccas or gouty Vera Cruz, and Honduras; the second includes Jamaica, Lima, and true Vera Cruz.

The mealy sarsaparillas are distinguished by "the mealy character of the inner cortical layers, which are white or pale-coloured. The meal or starch is sometimes so abundant that a shower of it, in the form of white dust, falls when we fracture the roots." The medulla or pith is also frequently very amylaceous.

The non-mealy sarsaparillas "are characterised by a deeply coloured (red or brown), usually non-mealy cortex. The cortex is red, and much thinner than in the mealy sorts." "If a drop of oil of vitriol be applied to a transverse section of the root of the non-mealy sarsaparillas, both cortex and wood acquire a dark red or purplish tint;" whilst in the preceding varieties the mealy coat, and sometimes the pith, is but little altered in colour. "The decoction of non-mealy sarsaparilla, when cold, is somewhat darkened, but does not yield a blue colour when a solution of iodine is added to it." The aqueous extract, when rubbed down with a little cold distilled water in a mortar, does not yield a turbid liquid, nor become blue on the addition of iodine. The reverse is the case with the decoction and extract of the mealy varieties.

The JAMAICA, RED JAMAICA, or RED-BEARDED SARSAPARILLA (SARSA JAMAICENSIS—Ph. D.) is the variety which should alone be used in medicine. This kind yields from 33% to 44% of its weight of extract (*Battley, Hennell, Pope*), and contains less starchy matter than the other varieties. It is distinguished by exhibiting the above peculiarities in a marked degree, by the dirty reddish colour of its bark, which "is not mealy," and by being "beset very plentifully with rootlets" (fibres—Ph. L.). Its powder has also a pale reddish-brown colour. The other varieties of sarsaparilla, viz. the Lisbon, Lima, Vera Cruz, and Honduras, are frequently substituted for the Jamaica by the druggist in the preparations of the decoctions and extracts of this drug; but the products are vastly inferior in quantity, colour, taste, and medicinal virtue to those prepared from the officinal sarsaparilla. Decoction of sarsaparilla, when made with the Honduras root, is very liable to ferment, even by a few hours' exposure, in hot weather. We have seen hogsheads of the strong decoction, after exposure for a single night, in as active a state of fermentation as a gyle of beer, with a frothy head, and evolving a most disagreeable odour, that was not wholly removed by several hours' boiling. When this occurs the decoction suffers in density, and the product in extract is, consequently, considerably lessened. Yet this is frequently allowed to occur in the wholesale laboratory, where the rule should be—always begin a 'bath of sarza' (as it is called), and, indeed, of other perishable articles, early in the morning, and finish it completely and entirely the same day.

Sarsaparilla has been recommended as a mild but efficacious alterative, diaphoretic, and tonic. It has long been a popular remedy in chronic rheumatism, rheumatic and gouty pains, scurvy, scrofula, syphilis, secondary syphilis, lepra, psoriasis, and several other skin diseases; and, especially, in cachexia, or a general bad habit of body, and to remove the symptoms arising from the injudicious use of mercurials, often falsely called 'secondary syphilis.' During its use the skin should be kept warm, and diluents should be freely taken. Its efficacy has been greatly exaggerated. It is, however, much more effective in warm than in northern climates.—*Dose.* In substance, ⅙ to 1 dr., three or four times daily; but preferably made into a decoction or infusion.

The articles so much puffed under the names of American or United States sarsaparilla and extract of sarsaparilla are "nothing more than the decoction of a common herb, a sort of 'aralia,' inhabiting the swamps and marshes of the United States. When cut up it has the appearance of chaff, but not the slightest resemblance in character, colour, or taste, to even the most inferior species of smilax (or sarza). The decoction is sweetened with a little sugar, flavoured with benzoin and sassafras, and, finally, preserved from decomposition by means of the bichloride of mercury." "I have heard of several cases of deadly sickness, and other dangerous symptoms, following its use." "We do not believe that a particle of real sarsaparilla ever entered into the composition of either of the articles referred to" ('Med. Circ.,' ii, 227). See DECOCTION and EXTRACT.

SARSAPARIL'LIN. *Syn.* PARILLIN, PARILLIC ACID, SALSAPARIN, SMILACIN. A white, crystallisable, odourless, and nearly tasteless substance, discovered by Pallotta and Folchi in sarsaparilla. *Prep.* The bark of Jamaica sarsaparilla is treated with hot rectified spirit, and the resulting tincture reduced to about one third by distilling off the spirit; the residual liquid is then filtered, whilst boiling, slightly concentrated by evaporation, and set aside to crystallise; the crystalline deposit is redissolved in either hot rectified spirit or boiling water, and decoloured by agitation with a little animal charcoal; the filtrate deposits crystals of nearly pure smilacin as it cools. It may also be extracted by boiling water. *Prop., &c.* A non-nitrogenised neutral body. Water holding a very small quantity of it in solution froths considerably on agitation. This is especially the case with infusion of Jamaica sarsaparilla, and this property has consequently been proposed as a test of the quality of sarsaparilla root. Its medicinal properties are similar to those of sarsaparilla. According to Pallotta, it is a powerful sedative, and diminishes the vital energies in proportion to the quantity taken.— *Dose,* 2 to 10 gr.; in the usual cases in which the root is given.

SAS'SAFRAS. *Syn.* SASSAFRAS RADIX (B. P.), SASSAFRAS RADIX, SASSAFRAS (Ph. L., E., & D.), L. "The root of *Sassafras officinale*, Nees; *Laurus sassafras*, Linn."—Ph. L. It has a fragrant odour and a sweetish aromatic taste. It has long been reputed a stimulating, alterative, diaphoretic, diuretic, and tonic; and an infusion of the chips (sassafras chips), under the name of sassafras tea, has been a popular 'diet drink' in various cutaneous affections, gout, chronic rheumatism, &c.

SATURA'TION. The state in which a body has taken its full dose, or chemical proportion, of any other substance with which it can combine, or which it can dissolve; as water with sugar or a salt, or an alkali with an acid, when the properties of both are neutralised.

SAUCE. A liquid or semi-liquid condiment or seasoning for food. The following receipts for sauces may be useful to the reader:

Sauce, Ancho'vy. 1. (Extemporaneous.) From 3 or 4 anchovies, chopped small; butter, 3 oz.; water, a wine-glassful; vinegar, 2 tablespoonfuls; flour, 1 do.; stir the mixture over the fire till it thickens, then rub it through a coarse hair sieve.

2. (Wholesale.) As essence of anchovies. Other fish sauces may be made in the same manner.

Sauce, Apple. From sharp apples, cored, sliced, stewed with a spoonful or two of water, and then beaten to a perfectly smooth pulp with a little good moist sugar. Tomato, and many other like sauces, may be made in the same manner.

Sauce, Aristocratique. From green-walnut juice and anchovies, equal parts; cloves, mace, and pimento, of each, bruised, 1 dr. to every lb. of juice; boil and strain, and then add to every pint 1 pint of vinegar, ⅛ pint of port wine, ¼ pint of soy, and a few shallots; let the whole stand for a few days, and decant the clear liquor.

Bech'amel. A species of fine white broth or consommée, thickened with cream and used as 'white sauce.'

Sauce, Caper. Put twelve tablespoonfuls of melted butter into a stewpan, place it on the fire, and, when on the point of boiling, add 1 oz. of fresh butter and one tablespoonful of capers; shake the stewpan round over the fire until the butter is melted, add a little pepper and salt, and serve where directed. Also as mint sauce.

Sauce, Chut'ney. 1. From sour apples (pared and cored), tomatoes, brown sugar, and sultana raisins, of each, 3 oz.; common salt, 4 oz.;

red chillies and powdered ginger, of each, 29 oz.; garlic and shallots, of each, 1 oz.; pound the whole well, add of strong vinegar, 3 quarts; lemon juice, 1 do.; and digest, with frequent agitation, for a month; then pour off nearly all the liquor, and bottle it. Used for fish or meat, either hot or cold, or to flavour stews, &c. The residue is the 'Chutney,' 'Chetney,' or 'Chitni,' which must be ground to a smooth paste with a stone and muller, and then put into pots or jars. It is used like mustard.

2. (BENGAL CHITNI.) As the last, but using tamarinds instead of apples, and only sufficient vinegar and lemon juice to form a paste.

Cor'atch. From good mushroom ketchup, ½ gall.; walnut ketchup, ¾ pint; India soy and chilli vinegar, of each, ½ pint; essence of anchovies, 5 or 6 oz.; macerate for a fortnight.

Sauce, Epicurienne. To the last add of walnut ketchup and port wine, of each, 1 quart; garlic and white pepper, of each (bruised), 4 oz.; chillies, (bruised), 1 oz.; mace and cloves, of each, ½ oz.

Sauce, Fish. From port wine, 1 gall.; mountain do., 1 quart; walnut ketchup, 2 quarts; anchovies (with the liquor), 2 lbs.; 8 lemons, 48 shallots, scraped horseradish, 1½ lbs.; flour of mustard, 8 oz.; mace, 1 oz.; cayenne, q. s.; boil the whole up gently, strain, and bottle.

Ketchup. See under that name.

Sauce, Kitchener's. *Syn.* KITCHENER'S RELISH. From salt, 3 oz.; black pepper, 2 oz.; allspice, horseradish, and shallots, of each, 1 oz.; burnt-sugar colouring, a wine-glassful; mushroom ketchup, 1 quart (all bruised or scraped); macerate for three weeks, strain, and bottle.

Lem'on Pickle. From lemon juice and vinegar, of each, 3 galls.; bruised ginger, 1 lb.; allspice, pepper, and grated lemon peel, of each, 8 oz.; salt, 3½ lbs.; cayenne, 2 oz.; mace and nutmegs, of each, 1 oz.; digest for 14 days.

Sauce, Lob'ster. From lobsters, as ANCHOVY SAUCE.

Sauce, Mint. From garden mint, chopped small, and then beaten up with vinegar, some moist sugar, and a little salt and pepper.

Sauce, On'ion. From onions boiled to a pulp, and then beaten up with melted butter and a little warm milk.

Sauce, Oys'ter. From about 12 oysters, and 6 or 7 oz. of melted butter, with a little cayenne pepper, and two or three spoonfuls of cream, stirred together over a slow fire, then brought to a boil, and served.

Sauce, Piquante. From soy and cayenne pepper, of each, 4 oz.; port wine, ½ pint; brown pickling vinegar, 1½ pints; mix, and let them stand for seven or eight days before bottling.

Sauce, Quin's. From walnut pickle and port wine, of each, 1 pint; mushroom ketchup, 1 quart; anchovies and shallots (chopped fine), of each, two dozen; soy, ½ pint; cayenne, ¼ oz.; simmer gently for ten minutes, and in a fortnight strain and bottle.

Sauce au Roi. From brown vinegar (good), 3 quarts; soy and walnut ketchup, of each, ½ pint; cloves and shallots, of each, 1 doz.; cayenne pepper, 1½ oz.; mix, and digest for fourteen days.

Sauce, Shrimp. From shrimps or prawns, as ANCHOVY SAUCE.

Soy. See that article.

Sauce, Superlative. From port wine and mushroom ketchup, of each, 1 quart; walnut pickle, 1 pint; soy, ½ pint; powdered anchovies, ½ lb.; fresh lemon peel, minced shallots, and scraped horseradish, of each, 2 oz.; allspice and black pepper (bruised), of each, 1 oz.; cayenne pepper and bruised celery seed, of each, ¼ oz. (or currie powder, ¾ oz.); digest for fourteen days, strain, and bottle. Very relishing.

Sauce, Toma'to. From bruised tomatoes, 1 gall.; good salt, ½ lb.; mix, in three days press out the juice, to each quart of which add of shallots, 2 oz; black pepper, 1 dr.; simmer very gently for twenty to thirty minutes, strain, and add to the strained liquor, mace, allspice, ginger, nutmegs, and cochineal, of each, ¼ oz.; coriander seed, 1 dr.; simmer gently for ten minutes, strain, cool, and in a week put it into bottles.

Sauce, Universal Relish. A sauce said to produce a first-class imitation of a celebrated one is thus prepared:—White-wine vinegar, Cxv; walnut ketchup, Cx; Madeira, Cv; mushroom ketchup, Cx; table salt, 25 lbs.; Canton soy, Civ; powdered allspice, 1 lb.; powdered coriander, 1 lb.; powdered mace, ½ lb.; powdered cinnamon, ½ lb.; assafœtida, ½ lb. (The latter dissolved in brandy, 1 gall.) Boil 20 lbs. of hog's liver with 10 galls. of water, renewing the water from time to time. Take out the liver, chop it, mix with a little more water, and work through sieve. Mix all together, let stand a month, and filter.

Sauce, Waterloo. From strong vinegar (nearly boiling), 1 quart; port wine, ¾ pint; mushroom ketchup, ½ pint; walnut ketchup, ½ pint; essence of anchovies, 4 oz.; 8 cloves of garlic; cochineal (powdered), ½ oz. (or red beet, sliced, 3 oz.); let them stand together for a fortnight or longer, occasionally shaking the bottle.

Sauce, White. *Syn.* BUTTER SAUCE, MELTED BUTTER. From good butter, 4 oz.; cream, 2½ oz.; salt (in very fine powder), ½ teaspoonful; put them into a pot or basin, set this in hot water, and beat the whole with a bone, wooden, or silver spoon, until it forms a perfectly smooth, cream-like mixture, avoiding too much heat, which would make it run oily. A table-spoonful of sherry, marsala, lemon juice, or vinegar is sometimes added, but the selection must depend on the dishes the sauce is intended for. Used either by itself, or as a basis for other sauces. Beaten up with any of the 'bottled sauces,' an excellent compound sauce of the added ingredient is immediately obtained.

Sauces, American. 1. White vinegar, 15 galls.; walnut ketchup, 10 galls.; Madeira wine, 5 galls.; mushroom ketchup, 10 galls.; table salt, 25 lbs. (troy); Canton soy, 4 galls.; powdered capsicum, 2 lbs. (troy); allspice (powdered), coriander powder, of each, 1 lb. (troy); cloves, mace, cinnamon, of each, ½ lb. (troy); assafœtida, ½ lb. (troy); dissolved in brandy, 1 galls.; 20 lbs. of hog's liver is boiled for twelve hours with 10 galls. of water, renewing the water from time to time. Take out the liver, chop it, mix it with water, and work it through a sieve; mix with the sauce.

2. White vinegar, 240 galls.; Canton soy, 36 galls.; sugar-house syrup, 30 galls.; walnut

ketchup, 50 galls.; mushroom ketchup, 50 galls.; table salt, 120 lbs. (troy); powdered capsicum, 15 lbs. (troy); allspice, coriander, of each, 7 lbs. (troy); cloves, mace, cinnamon, of each, 4 lbs. (troy); assafœtida, 2½ lbs. (troy); dissolved in St. Croix rum, 1 gall.

3. White vinegar, 1 gall.; Canton soy, molasses, of each, 1 pint; walnut ketchup, 1½ pints; table salt, 4 oz.; powdered capsicum, allspice, of each, 1 oz.; coriander, ½ oz.; cloves, mace, of each, ¼ oz.; cinnamon, 6 dr.; assafœtida, ¼ oz. in 4 oz. of rum.

SAUCERS (for Dyeing). *Prep.* 1. (BLUE.) From blue carmine, made into a paste with gum water, which is then spread over the inside of the saucers, and dried.

2. (PINK.) *a.* From pure rouge mixed with a little carbonate of soda, then made into a paste with thin gum water, and applied as the last.

b. Well-washed safflower, 8 oz.; carbonate of soda, 2 oz.; water, 2 galls.; infuse, strain, add of French chalk (scraped fine with Dutch rushes), 2 lbs.; mix well, and precipitate the colour by adding a solution of tartaric acid; collect the red powder, drain it, add a very small quantity of gum, and apply the paste to the saucers. Inferior. Both the above are used to tinge silk stockings, gloves, &c.

SAUERKRAUT. [Ger.] *Prep.* Clean white cabbages, cut them into small pieces, and stratify them in a cask along with culinary salt and a few juniper berries and caraway seeds, observing to pack them down as hard as possible without crushing them, and to cover them with a lid pressed down with a heavy weight. The cask should be placed in a cold situation as soon as a sour smell is perceived. Used by the Germans and other northern nations of Europe like our 'pickled cabbage,' but more extensively.

SAU'SAGES. From the fat and lean of pork (PORK SAUSAGES), or of beef (BEEF SAUSAGES), chopped small, flavoured with spice, and put into gut skins, or pressed into pots or balls (SAUSAGE MEAT). Crumb of bread is also added. Their quality is proportionate to that of the ingredients, and to the care and cleanliness employed in preparing them.

A pea sausage, composed of pea flour, fat pork, and a little salt, was largely consumed by the German soldiers during the Franco-German campaign. Dr Parkes found 100 parts of this sausage to consist of 16·2 parts of water, 7·19 of salts, 12·297 of albuminates, 33·65 of fat, and 30·663 of carbohydrates. It is ready cooked, but can be made into soup. Although much relished for a few days, the soldiers soon became tired of it. In some cases it gave rise to flatulence and diarrhœa. See MEAT.

Sausages, Painted. The following interesting particulars as to the manufacture of sausages are given in the report of the Dairy Commissioner of New Jersey for the year 1889. Twelve samples of Bologna sausages were examined, with the following result, says the ' British Medical Journal.' The analysis of the Bologna and the skin in which the meat was placed showed that some dye, probably one of the anilines, was used to colour the material, in order that some defect might be hidden or the article made to appear

better than it really was; also that some substance had been applied to the exterior of the sausage similar to varnish. Further analysis revealed the presence of triamidoazobenzine or Bismarck brown, one of the aniline colours; this was in the meat. The skin, or ' casing,' was coated with a varnish containing shellac. This discovery was the means of arriving at all the details of the process employed. The sausage in question was prepared in the following way: after the meat was chopped, and the sausage meat thus prepared put into the casings, the sausage was boiled in a bath containing a portion of the following colouring agent: Bismarck brown, 14 parts; garnet red, 2 parts; water, a pint and a half. This gave the sausage a brown colour. When this process was complete the sausages were coated with a varnish composed of shellac, resin, oil, and alcohol. In order that the small local manufacturers of sausage might engage in the practice of making dyed sausages, the compositions referred to above were offered for sale through the State, and the staining material was sold under the name of ' smokine ' or ' liquid smoke.' The sale of the article was checked by the official action of the inspectors throughout the State.

SAV'ELOYS. Pork sausages made in such a way that they keep good for a considerable time. —*Prep.* (*Mrs Rundell.*) Take of young pork, free from bone and skin, 3 lbs.; salt it with 1 oz. of saltpetre, and ½ lb. of common salt, for two days; then chop it fine, add 3 teaspoonfuls of pepper, 1 doz. sage leaves, chopped fine, and 1 lb. of grated bread; mix well, fill the skins, and steam them or bake them half an hour in a slack oven. They are said to be good either hot or cold.

SAV'INE. *Syn.* SAVIN; FOLIA SABINÆ, SABINA (Ph. L., E., and D.), L. " The recent and dried tops of *Juniperus sabina*, Linn.," or common savine (Ph. L.). It is a powerful stimulant, diaphoretic, emmenagogue, and anthelmintic; and, externally, rubefacient, escharotic, and vesicant. In large doses it is apt to occasion abortion, and acts as a poison. SAVINE POWDER mixed with verdigris is often applied to corns and warts. It is now chiefly used in the form of ointment.—*Dose,* 5 to 15 gr., twice or thrice daily (with care), in amenorrhœa and worms. See CERATE.

SAVONETTES. [Fr.] *Syn.* WASH-BALLS. These are made of any of the mild toilet soaps, scented at will, generally with the addition of some powdered starch or farina, and sometimes sand. The spherical or spheroidal form is given to them by pressure in moulds, or by first roughly forming them with the hands, and, when quite hard, turning them in a lathe. According to Mr Beasley, " they are formed into spherical balls by taking a mass of the prepared soap in the left hand, and a conical drinking-glass with rather thin edges in the right. By turning the glass and ball of soap in every direction the rounded form is soon given; when dry, the surface is scraped, to render it more smooth and even."

Prep. 1. Take of curd soap, 3 lbs.; finest yellow soap, 2 lbs. (both in shavings); soft water, ¼ pint; melt by a gentle heat, stir in of powdered starch (farina), 1½ lbs.; when the mass has considerably cooled, further add of essence of lemon or bergamot, 1 oz., and make it into balls.

2. (CAMPHOR.) Melt spermaceti, 2 oz.; add camphor (cut small), 1 oz.; dissolve, and add the liquid mass to white curd soap, 1½ lbs., previously melted by the aid of a little water and a gentle heat, and allowed to cool considerably as above. These should be covered with tin-foil.

3. (HONEY.) From the finest bright-coloured yellow soap, 7 lbs.; palm oil, ¼ lb.; melt, and add of oil of verbena, rose geranium, or ginger-grass, 1 oz.; as No. 1. Sometimes ½ oz. of oil of rosemary is also added.

4. (MOTTLED.) a. RED. Cut white curd or Windsor soap (not too dry) into small square pieces, and roll these in a mixture of powder bole or rouge, either with or without the addition of some starch; then squeeze them strongly into balls, observing to mix the colour as little as possible. b. BLUE. Roll the pieces in powdered blue, and proceed as before. c. GREEN. Roll the pieces in a mixture of powder blue and bright yellow ochre. By varying the colour of the powder, mottled savonettes of any colour may be produced.

5. (SAND.) From soap (at will), 2 lbs.; fine siliceous sand, 1 lb.; scent, q. s.; as No. 1. For the finer qualities, finely powdered pumice-stone is substituted for sand.

6. (VIOLET.) From palm-oil soap, 4 lbs.; farina, 2 lbs.; finely powdered orris root, 1 lb. Sometimes a little smalts, or indigo, is added.

SAWDUST, Preparation of Alcohol from. Zetterland ('Chemical News,' xxvi, 181) obtains alcohol from sawdust by the following process:—Into an ordinary steam boiler, heated by means of steam, are introduced 9 cwt. of very wet sawdust, 10·7 cwt. of hydrochloric acid (sp. gr. 1·18), and 30 cwt. of water; after eleven hours' boiling about 19·67% of grape-sugar is formed. The acid is next saturated with chalk, so as to leave in the liquid only a small quantity (½ degree by Ludersdorf's acid areometer); when the saccharine liquid has cooled down to 30° yeast is added, and the fermentation is finished in twenty-four hours. By distillation 26·5 litres of alcohol of 50% at 15°, quite free from any smell of turpentine, and of excellent taste, may be obtained. It appears that the preparation of alcohol from sawdust may be successfully carried on industrially. If all the cellulose present in sawdust could be converted into sugar, 50 kilos. of the former substance would yield, after fermentation, 12 litres of alcohol at 50%.

SCAB IN SHEEP. This disease, corresponding to mange in the dog, the horse, or in cattle, is caused by a species of acaris, a minute insect which burrows under the skin of the sheep, and sets up therein a considerable amount of irritation, which is followed by an irruption of pimples, accompanied with scurf, but frequently the wool comes off from the affected part.

The following are some of the numerous remedies employed for this disease:

1. Quicksilver, 1 lb.; Venice turpentine, ½ lb.; rub them together until the globules are no longer visible; then add ½ pint of oil of turpentine, and

4 lbs. of lard. The mode of applying this ointment is as follows:—Begin at the head of the sheep, and proceeding from between the ears along the back to the end of the tail, divide the wool in a furrow till the skin can be touched; and let a finger, slightly dipped in the ointment, be drawn along the bottom of the furrow. From this furrow similar ones must be drawn along the shoulders and thighs to the legs as far as the wool extends. And if much infected, two or more should also be drawn along each side, parallel with that on the back, and one down each side before the hind and fore legs. It kills the sheep-fag, and probably the tick and other vermin. It should not be used in very cold or wet weather (Sir Joseph Banks).

2. Tar oils. Tobacco juice. Stavesacre (Finlay Dun).

3. Strong mercurial ointment, 1 part; lard, 5 parts; mix (Youatt).

4. Quicksilver, 1 lb.; Venice turpentine, ½ lb.; spirit of turpentine, 2 oz.; lard, 4½ lbs.; to be made and used as No. 1. In summer 1 lb. of resin may be substituted for a like quantity of lard (Clater).

5. Strong mercurial ointment, 1 lb.; lard, 4 lbs.; oil of turpentine, 8 oz.; sulphur, 12 oz. (White).

6. Mild. Flowers of sulphur, 1 lb.; Venice turpentine, 4 oz.; rancid lard, 2 lbs.; strong mercurial ointment, 4 oz.; mix well (Clater).

7. Lard or other fat, with an equal quantity of oil of turpentine (Daubenton).

8. Without Mercury. Lard, 1 lb.; oil of turpentine, 4 oz.; flowers of sulphur, 6 oz. (White).

9. Strong mercurial ointment, 1 lb.; lard, 4 lbs.; Venice turpentine, 8 oz.; oil of turpentine, 2 oz. If mixed by heat, care must be taken not to use more heat than is necessary, and to add the oil of turpentine when the other ingredients begin to cool, and to stir till cold (M'Ewen).

10. Corrosive sublimate, 2 oz.; white hellebore, 3 oz.; fish oil, 6 quarts; resin, ½ lb.; tallow, ½ lb. The sublimate and then the hellebore to be rubbed with a portion of the oil till perfectly smooth, and then mixed with the other ingredients melted together (Stevenson).

11. The following once had considerable local celebrity, but it obviously requires to be used with caution:—Dissolve 2½ oz. of corrosive sublimate in the same quantity of muriatic acid, and beat up the solution with 6 lbs. of strong mercurial ointment; put in a large pan, and pour on it 19½ lbs. of lard, and 1½ lbs. of common turpentine, melted together and still hot, and stir the whole continually until it becomes solid.

12. Castor-oil seeds well bruised and steeped for 12 hours in sour milk, after straining rub the liquid briskly into the skin (an Indian remedy).

SCABIES. See ITCH.

SCAGLIOLA. A species of plaster or stucco, made of pure gypsum, with variegated colours, in imitation of marble. In general the liquid employed is a weak solution of Flanders glue; and the colours, any which are not decomposed or destroyed by admixture with sulphate of lime and exposure to the light. The composition is often applied upon hollow columns formed of wood, or even of laths nailed together, and the surface, when hard, is turned smooth in a lathe and polished.

SCALD-HEAD. See RINGWORM.

SCALDS. See BURNS AND SCALDS.

SCALES. A special article under the head of "BALANCE" has been devoted to the scales employed by the chemist and analyst.

But although these claimed, from their greater complexity of structure and the extreme delicacy of movement required of them, a separate notice, every pharmacist and apothecary will recognise the importance of bestowing an equal amount of attention upon his dispensing scales, and, to ensure accurate weighing by them, will take care to keep them scrupulously clean and properly poised.

For dispensing purposes scales fitted with glass pans (or at least with one glass pan, in which medicinal substances can be weighed) should always be employed. The beams should be of steel, and the attachments of one piece of brass only, in preference to chains or supports of silken thread. The beams are best cleaned by being rubbed with a little rotten-stone and oil used sparingly, and may be protected from rust by being wiped with a rag oiled with good Rangoon oil, such as is used for firearms.

SCALL. *Syn.* SCALD. The popular name of several skin diseases distinguished by scabs or scurfiness, whether dry or humid. See ERUPTIONS, RINGWORM, &c.

SCAMMONY. *Syn.* SCAMMONIUM (B. P., Ph. L., E., & D.), L. The "gum-resin emitted from the cut root of *Convolvulus scammonia*, Linn.," or Aleppo scammony plant.

There are three principal varieties or qualities of scammony known in the market, viz. VIRGIN (sp. gr. 1·21); SECONDS (sp. gr. 1·460 to 1·463); and THIRDS (sp. gr. 1·465 to 1·500). The best, and that only intended to be used in medicine, is imported from Aleppo.

Pur. Scammony is not only largely adulterated in the country of its production, but again after its arrival in England. SMYRNA SCAMMONY, a very inferior variety, is also commonly dressed up and sold as Aleppo scammony. In many cases substances are sold at the public sales in London and elsewhere as scammony, which contain only a mere trace of that article. This is all ground up to form the scammony powder of the shops (*vide* 'Evid. Com. Ho. Com.,' 1155). PURE SCAMMONY has a peculiar cheesy smell, and a greenish-grey colour. It is porous and brittle, and the freshly broken surface shines; hydrochloric acid being dropped on it, it emits no bubbles; nor does the powder digested in water, at a heat of 170° F., become blue by the simultaneous addition of iodide of potassium and dilute nitric acid. Out of 100 gr., 75 should be soluble in ether. The tincture of pure scammony is not turned green by nitric acid. If the powder effervesce with dilute acids it contains chalk.

Uses, &c. Pure scammony is a powerful drastic purgative and anthelmintic, inadmissible in inflammatory conditions of the alimentary canal, but well adapted for torpid and inactive conditions of the abdominal organs. Associated with calomel, rhubarb, or sulphate of potassa, it is useful in all cases in which an active cathartic or vermifuge may be required, especially for children.—*Dose* (for an adult), 5 to 15 gr. in powder, or made into a bolus or emulsion.

SCARLATINA. See SCARLET FEVER.

SCARLET DYE. *Proc.* (*Poirner.*) *a.* The 'Bouillon.' Take of cream of tartar, 1¾ oz.; water, q. s.; boil in a block-tin vessel, and when dissolved, add of solution of tin (made by dissolving 2 oz. of grain tin in a mixture of 1 lb. each of nitric acid and water, and 1½ oz. of sal-ammoniac), 1¾ oz.; boil for three minutes, then introduce the cloth, boil for two hours, drain it, and let it cool.—*b.* The 'Rougie.' Next take of cream of tartar, ¼ oz.; water, q. s.; boil, and add powdered cochineal, 1 oz.; again boil for five minutes, then gradually add of solution of tin, 1 oz., stirring well all the time; lastly, put in the goods and dye as quickly as possible. The quantities given are those for 1 to 1½ lbs. of woollen cloth. The result is a full scarlet. To make the colour turn on the 'ponceau' or poppy, a little turmeric is added to the bath.

Obs. A large number of scarlet dyes are now used, which are obtained by purely chemical processes. The following list, compiled from Dr Benedikt's 'The Chemistry of Coal Tar Colours,' gives some examples of these. See RED DYE.

Commercial Name.	Scientific Designation.
Scarlet 4 G B .	Benzeneazo-β-naphthol sulphonate of soda.
Scarlet 2 G . .	Benzeneazo-β-naphthol disulphonate of soda.
Scarlet G T . .	Tolueneazo-β-naphthol disulphonate of soda.
Wool scarlet R	Xyleneazo-α-naphthol disulphonate of soda.
Scarlet 3 R . .	Cumeneazo-β-naphthol disulphonate of soda.

SCARLET FEVER. *Syn.* SCARLATINA; Fr. SCARLATINE; Ger. SCHARLACH FIEBER.

Def. An infectious specific fever, characterised by deep redness of the throat; a finely diffused scarlet rash, most intense on the third day, beginning to fade on the fifth or sixth with some subsidence of fever, and followed by desquamation of the cuticle in both small and large flakes; and afterwards possibly by rheumatic or renal symptoms, with a tendency to severe effusions (*W. Squire*).

Symptoms. The onset of scarlet fever is abrupt. Slight pallor, languor, dizziness, restlessness at night, pains in the limbs, with sore throat and often vomiting. The pulse becomes very rapid, and the temperature rises considerably. The rash may make its appearance on the neck and chest soon after the sore throat is noticed, beginning in fine red points closely set over a considerable area; when pressed by the finger these red patches disappear, returning when the pressure is removed. The rash is at its height on the fourth day, fading on the fifth and sixth. The skin then begins to dry and desquamate from the sixth to the ninth day.

Treatment. No medicines will cut short the fever; simple salines, such as acetate of ammonia or chlorate of potash, are of great use if plenty of liquid be given. Tepid sponging of the whole body, part at a time, twice in twenty-four hours, affords the greatest relief, especially if a little aromatic vinegar be added to the water. Rubbing with carbolised oil is useful to prevent the particles of skin from flying off the body into the air.

The diet should at first be limited to milk and liquids, increased to eggs and beef-tea; and as soon as more food can be taken, fish or fowl with vegetables may be given. Meat should be given very cautiously.

General. The convalescence requires care. "Three weeks indoors for the disease to cease, and three weeks at home after that for restoration to health, is the safest rule for all." The disease is highly infectious, and every care must be taken to prevent its spread. All superfluous hangings, &c., should be taken out of the sick room at once. Everything which has been in contact with the patient should be disinfected. Stoving such materials as cannot be boiled is very effectual. The sick room after the patient is removed should be disinfected by burning sulphur in it at the rate of 1½ oz. for every 100 cubic feet; and the patient should not be allowed to mix with healthy persons for at least six weeks from the commencement of the disease, *however slight the attack.*

SCENE-PAINTING. A variety of distemper painting employed in theatres, &c., governed by perspective, and having for its object the production of striking effects when viewed at a distance. Water, size, turpentine, and the ordinary pigments are the materials used for the purpose.

SCENT-BAGS. See SACHETS.

SCENT-BALLS. *Syn.* PASTILLES DE TOILETTE ODORANTES, Fr. These are prepared from any of the materials noticed under POT POURRI, SCENTED POWDERS, and SACHETS, made into a paste with mucilage of gum tragacanth, and moulded into any desired forms, as that of balls, beads, medallions, &c. The larger ones are frequently polished.

SCENTED CASSOLETTES. See POT-POURRI, and *above.*

SCENTS (Pommade). *Prep.* 1. (COWSLIP.) From essence of bergamot, 8 oz.; essence of lemon, 4 oz.; oil of cloves, 2 oz.; essence de petit grain, 1 oz.

2. (JONQUILLE.) From essence of bergamot and lemon, of each, 8 oz.; oils of orange peel and cloves, of each, 2 oz.; oil of sassafras, 1 oz.; liquid storax, ½ oz.; digest, with warmth and agitation, for a few hours, and decant the clear portion in a week.

3. (MILLEFLEUR.) From essence of ambergris (finest), 4 oz.; essence of lemon, 3 oz.; oil of cloves and English oil of lavender, of each, 2 oz.; essence de petit grain, essence of bergamot, and balsam of Peru (genuine), of each, 1 oz.; as the last.

Obs. The above are employed to scent pomatums, hair oils, &c. 1 oz. of any one of them, dissolved in one pint of the strongest rectified spirit, produces a delicious perfume for the handkerchief.

SCENTS (Snuff). *Prep.* 1. Essence of bergamot, 2 oz.; otto of roses and neroli, of each, 1 dr.

2. Oil of lavender, 1 oz.; essence of lemon, 2 oz.; essence of bergamot, 4 oz.

3. To the last add of oil of cloves, 2 oz.

4. Essence of musk and ambergris, of each, 1 oz.; liquor of ammonia, ½ dr. See-SNUFF, &c.

SCHEELE'S GREEN. See GREEN PIGMENTS.

SCHIZONEURA LANIGERA, Hausmann. THE AMERICAN BLIGHT, or WOOLLY APHIS. Every apple-grower has frequently noticed knots or bunches of a downy or woolly substance on parts of the stems and branches of apple-trees, especially where any injury has been sustained, or where side shoots or branches have been pruned away in an unworkmanlike fashion, where the cuts have been delivered slanting upwards instead of slanting downwards, so that the wet has rested at the bottom of the cut and caused decay. By this means cracks are formed and gradually increase in width and depth. The edges of the outer layers of bark do not join together again, and a thin and tender tissue alone covers the exposed parts. Upon examination of the little groups of wool it will be seen that they are composed of larvæ having woolly or downy coverings, actively engaged in piercing the denudated surfaces with their suckers and extracting the sap, occasioning unhealthy and abnormal growths of tissue. Extravasation of sap occurs, giving rise to excrescences and warty lumps, which afford shelter and food for the numerous generations of larvæ, until the whole branch is injuriously affected and its vigour and fruitfulness are materially lessened by means of the hindrance to the due circulation of sap. Other branches soon become infested, and the small stems, and even the fruit-bearing twigs are attacked, and after a time are covered with swellings, which check both leaf and blossom development, and in time cause both stem and twigs to decay. In not a few cases apple-trees have died after a long and uninterrupted onslaught of these woolly aphides. In very many cases apple-trees might almost as well be dead, as their very life-sap is being systematically exhausted by the constant suckings of myriads of larvæ.

It is far too common to find apple-trees in all the apple-producing counties, whose branches are unsightly from warts and swellings, with the tops of their branches dead or decaying, and whose general appearance indicates extreme unhealthiness, due entirely to the unchecked influences of the woolly aphis. This state frequently is called "canker," and is attributed to over or injudicious pruning, or to unskilful pruning, or to something in the soil or subsoil that does not suit the trees. A little close observation demonstrates that it is quite different from canker, and is simply the effect of insects which have been permitted to increase and multiply for years.

It also very often happens that the bark, branches, and small twigs of apple-trees in many orchards are so enveloped by lichenous and mossy parasites, that the presence of the woolly aphides is unsuspected.

Young trees are seriously injured by these insects without any special wonder. Their bark is tender, and is easily pierced by the sharp beaks of the larvæ, and they cannot withstand their attacks so well or so long as old trees. I have seen trees of the sort known as Lord Suffield completely at a standstill, though they were only about sixteen years old, with their branches covered with swellings and swarming with larvæ. Yet it was stated persistently that the cause of the evil was canker, and that the roots had got down to something that did not suit them. If a young tree is permitted to be much injured by the woolly aphis it rarely recovers, and remains stunted, sickly, unfruitful. The Ribston Pippin is very liable to be attacked by this insect, as well as Cox's Orange Pippin, the Blenheim Orange, and other valuable kinds, both for dessert and culinary use, whose skins are comparatively tender, and not so cased in rugose layers of bark.

In the cider-making districts of Devonshire, Gloucestershire, Herefordshire, Somersetshire, and Worcestershire, incalculable mischief is wrought in the apple orchards by these insects, which increase and work unsuspected amid the lichens and mosses that clothe the branches, and the closely interlaced congeries of boughs and twigs. There is a record that the apple-trees in the Gloucestershire orchards were so attacked by the woolly aphis in 1810 that no cider was made in the county, and it was feared that this industry must be altogether abandoned.

Not only does the woolly aphis attack the trunks and branches of apple-trees, but it infests and injures their roots, also living upon them, and causing excrescences or swellings to form upon them. The injuries caused to old trees by the subterranean attack is not very apparent, though the constant action of series of generations of these insects must in time materially affect the health of the tree and diminish its constitutional vigour. It will, however, be easily understood that young trees are more liable to be seriously hurt than old well-established trees with large, wide-reaching roots. Very frequently it happens that young apple-trees languish and do so badly that it is said that the locality is not suited for apple-growing, whereas in fact it is the action of the woolly aphides, both upon their roots and branches, which is occasioning the evil.

This insect is also very destructive in French orchards, particularly in Normandy and in the south of France. In Germany it is well known, and in America and Canada the apple producers regard it with much apprehension. It was called American blight, as it was believed that it was first imported from America, but there is no proof of this. Dr Harris denies that it came originally from America, and states that there is good reason to believe that the miscalled American blight is not indigenous to America, but that it was introduced there first with fruit trees from European nursery-grounds.

Professor Asa Fitch also denies strongly that the woolly aphis was imported into America from Europe. Mr Buckton says that it was first noted in France in the Department of the Côte du Nord in 1812, and was seen in the garden of the École de Pharmacie in Paris in 1818. In 1822 it was common in the Departments of the Seine, the Somme, and the Aisne. It was first discovered in Germany in 1801, and in Belgium in 1812. However, it is possible that the woolly aphis may have been present in England and upon the Continent long before these dates, as entomological know-

ledge was very limited at the beginning of the present century, and practical observers were few and far between.

Life History. The woolly aphis belongs to the genus *Schizoneura* of the family *Aphididæ*, called *Schizoneura* from the peculiar neuration of their wings. It has the affix *lanigera* because the larvæ are covered with wool. It is not unfrequently confounded with the *Aphis mali*, which infests the leaves and blossoms of apple-trees, though it is quite distinct in its formation, in its habits, and its actions. The latter insect is furnished with tubercles at the end of its body for the secretion of honey-dew. The *Aphis lanigera* is entirely without these appendages. The perfect winged viviparous insect makes its appearance in the late summer, and has been seen as late as September. It is by no means common, as Mr Buckton remarks, or, being very minute, it possibly escapes all but the closest observation. It was held by some American entomologists that it never acquired wings; but this is a mistake, as winged specimens have been found in Kentish orchards, and abundantly in France. From this female some oviparous females are produced, which are without beaks or rostra, and therefore unable to feed. Mr Buckton and Professor Riley agree as to this curious formation or malformation, which is shared by the winged males. The life of both the sexes is necessarily very brief. Only one egg is laid by each female, and is placed under the bark or in crevices; this serves as a means of preserving the species when food fails, and to spread the plague of insects from tree to tree, though this is also continually brought about by the wind when trees are pretty close together, which bears the down-covered larvæ with its breezes.

Propagation, the regular continuity of existence, is, however, carried on by means of hibernating viviparous larvæ, as in the case of many species of the *Aphididæ.* These pass the winter upon the trunk of the trees, and on their boughs and branches, being protected in a degree by the bark, as well as by their woolly coverings. In the year 1886 larvæ were found alive upon the boughs of a Lord Suffield apple-tree in a Kentish orchard when snow was lying thick upon the ground, and the thermometer had registered 10 degrees of frost the previous night. These larvæ also hibernate under the surface of the ground upon the roots of the apple-trees, and in this case they are well preserved from the effects of severe weather, though, as has been shown, they are able to bear a considerable amount of cold. Mr Buckton confirms this, and remarks that he found the apterous larvæ alive and plentiful on the apple-branches when the thermometer stood at 21° F.

I have found the larvæ of the woolly aphis upon the roots of apple-trees in the spring and summer, but not in the winter. It may be that they go deeper down during this season. There is an opinion held by some persons that the subterranean generations differ from those which are upon the branches and bodies of the trees, in the same way as there are differences between the aërial and subterranean generations of the *Phylloxera vastatrix,* the terrible vine pest. But, from examination of the specimens of woolly aphides found upon the roots of apple-trees, there appears to be no distinction, and it is believed that these occupy the roots and branches indifferently, or according to the varying circumstances and conditions of food, weather, and seasons.

The winged generations of this insect are of a dark, somewhat shiny colour, with the under parts of their bodies approaching to chocolate, having large wings with a peculiar and distinctive venation. The wingless viviparous- females are more brown than the winged females, and have thin woolly coverings. Their offspring are still lighter in colour, between red and brown, being furnished with very long rostra. Mr Buckton states that when they are adult they exude from their pores long threads, which curve round a centre, and often form long spiral filaments round which they hide.

Prevention. Apple-trees should be kept from lichens and mosses, which serve as a shelter for the woolly aphides, as well as for many other kinds of injurious insects. Lichens and mosses can be killed by throwing quicklime up into the trees over the branches by means of scoops like flour scoops fixed to long poles. This should be done in damp weather, in a dripping November fog, and of course after the leaves have fallen. In a few days, if the operation has been thoroughly performed, the lichens and mosses turn rusty-coloured, and are washed away from the branches by the first heavy shower. These lichenous and mossy growths not only harbour insects of various kinds in various stages, but they also injure the trees by stopping up the pores of the rind and checking respiration. These parasites derive their sustenance from the air, and do not feed upon their host, as is very commonly supposed.

As the deep, extensive interstices in the thick layers of bark upon old apple-trees afford refuges for the woolly aphides as well as for other insects, it is desirable to scrape off the outer layers. This can be done with a scraper shaped like the instrument used for taking water and lather from horses, but of stouter materials. I have made a serviceable tool for this purpose by putting handles to semicircles of stout iron hooping, which has jagged edges, or edges sharpened somewhat upon a grindstone. In America and France, where scraping off the bark is much practised by apple-growers, a little tool is employed (consisting of a triangular plate of metal mounted on a short handle, after the manner of a triangular hoe), and found very effective. Unless the rough bark is removed it is almost useless to apply lime wash, soft-soap wash, or diluted paraffin, or other remedies, as they cannot be worked into the crevices. In California, Boards of Commissioners, appointed under an Act of the Legislature to protect and promote the horticultural interests of the States, have drawn up a set of rules. Rule 1 requires that every fruit-grower or owner of an orchard or orchards in which there are trees infested with insects should scrape all rough bark on such trees, and clean all crevices in the bark and 'crotches.'

In the case of young trees the stems and the branches should be rubbed with the hand cased in a glove composed of steel chains to remove rugosities and larvæ that may be *in situ*, or very young trees with tender bark may be brushed over with a stiff brush. A 'spoke' brush is very useful for this purpose, as it can be worked into the joints. The bark of trees may be kept in a healthy state by occasional use of the armed hand and brushes, just as friction is good for the skin of human beings and animals.

In grass orchards whose trees are persistently attacked by woolly aphides it would be well, in addition to the scrapings and cleanings of the trunk and branches, to water the roots round each tree with strong soapsuds. The roots of apple-trees do not go down very deep as a rule. Application of oil substances or of tar to the trunk, just above the ground, would prevent the passage of the larvæ from the roots upwards. It would be very efficacious to pen pigs round infested trees, as is frequently done for manurial purposes. These, by their rootings and disagreeable concomitants, make the roots very unpleasant quarters for the subterranean invaders. There is no doubt that young apple-trees become infested occasionally in nurseries, both with respect to their roots and branches. Professor Asa Fitch states that in America young trees often languish and die from the attacks of the woolly aphides on their roots soon after being planted from the nursery. Examination of roots, stems, and branches should be made before young trees are planted, and if signs of woolly aphis are present the roots should be washed with soft soap and quassia solutions.

Remedies. When young trees are attacked by woolly aphides they should be washed over with a mixture of soft soap and quassia, in the proportion of 15 to 20 lbs. of soft soap and 8 lbs. of quassia to 100 galls. of water. This may be put on with a whitewash brush or a large-sized paint brush, and a 'spoke' brush may be used for the forks of the stem and branches. Syringing with soft soap and quassia solutions by means of large garden engines like those used for washing hop plants is of some avail in the case of young trees, and especially those of an upright habit of growth. In the adoption of this process it is difficult to get at the insects on the upper parts of the branches. Still, if the pumps are worked well and the hose directed so that a good deal of liquid falls upon them from above, their quarters will be made unpleasant to them. Where trees are large, and of wide-spreading habit of growth, syringing will not be of much benefit, and the operation requires to be very carefully carried out. It would be well worth while for owners or occupiers of large acreage of orchard land to have special washing engines made to throw liquid well into and over large fruit trees.

Paraffin oil, mixed in the proportion of two to three wine-glasses to a pail of water, has been found efficacious worked into the bark and branches with brushes, and in some cases put on with garden engines. The oil must be held incorporated with the water, and the mixture should be kept stirred. Vegetable and animal oils, as linseed, whale, and neat's-foot oil, have been applied with good results, chiefly to the stems and lower branches. A mixture of linseed oil and powdered sulphur is successfully used in America upon the trunks and lower limbs of the trees after they have been well scraped.

Infested trunks should be scraped and washed over with lime wash, made with very quick lime. This will kill all the larvæ, and prevent migration from stems to roots. Hot lime thrown up by scoops in damp weather both removes lichens and mosses where these parasites are, and immediately destroys the larvæ of the woolly aphis. It need hardly be stated that where apple-trees have been long unpruned and neglected, and the small boughs are thickly intertwisted, it is most imperative that pruning should be done gradually but thoroughly, for many obvious reasons, and mainly that it will be more easy to stamp out the woolly aphis ('Reports on Insects Injurious to Crops,' by Chas. Whitehead, Esq., F.Z.S.).

SCHLIPPE'S SALT. Sulphantimoniate of sodium. See ANTIMONY SULPHANTIMONIATE.

SCHULTZE'S POWDER. A new sporting powder for guns, in which nitrated cellulose is the chief constituent.

SCHWARTZ'S DROPS. See DROPS, WORM.

SCIATICA. See RHEUMATISM.

SCIENCE. "Man," says Whewell, "is the interpreter of nature; science the right interpretation. The senses place before us the *characters* of the 'Book of Nature;' but these convey no knowledge to us till we have discovered the alphabet by which they are to be read."

Various classifications of the sciences have been proposed by different authors. Dr. Neil Arnott conceived that the object of all the sciences, viz. a knowledge of nature, might be best attained by the study of physics, chemistry, life, and mind, including under this latter the laws of society, or the modern science of sociology.

As to mathematics, he regarded it as a system of 'technical mensuration,' invented by the mind in order to enable it to study the other sciences. In his 'Cours de Philosophie Positive,' Comte traverses the entire circle of the theoretical, abstract, or fundamental sciences, and divides them into mathematics, astronomy, physics, chemistry, biology, and sociology. He admits no distinct science on psychology or the science of mind.

Mr Herbert Spencer proposes to classify the sciences into three groups; viz.—1. Abstract science, which treats of the forms of phenomena detached from their embodiments. 2. Abstract concrete science, or the phenomena of nature analysed into their separate elements. 3. Concrete science, or natural phenomena in their totalities.

"For the classification of the sciences it is convenient to prepare the way by distinguishing between theoretical sciences, which are the sciences properly so called, and practical science. A theoretical science embraces a distinct department of nature, and is so arranged as to give in the most compact form the entire body of ascertained (scientific) knowledge in that department, such as mathematics, chemistry, physiology, and zoology. A practical science is the application of scientifically obtained facts and laws in one or more departments to some practical end, which end rules the selection and arrangement of the whole; as, for example, navigation, engineering, mining, and medicine. Another distinction must be made before laying down the systematic order of the theoretical sciences.

"A certain number of these **sciences have for** their subject-matter each a separate department of natural forces or powers; thus **biology deals** with the department of organised beings, psychology with mind. Others deal **with the applica-** tion of powers elsewhere recognised to some region of concrete facts or phenomena. Thus geology does not discuss any natural powers not found in other sciences, but seeks to apply the laws of physics, chemistry, and biology to account for the appearance of the **earth's crust.**

"The sciences that embrace peculiar natural powers are called abstract, general, or fundamental sciences; those that apply to the powers treated of under these, to regions of concrete phenomena, are called concrete, derived, or applied sciences" ('Chambers's Encyclopædia').

The separation of the sciences into these two sections is that now generally accepted.

The first section, that of the abstract or theoretical sciences, is subdivided by almost common consent into mathematics, physics, chemistry, biology (vegetable and animal physiology), psychology (mind), and sociology (the laws of society).

The second section, that of the concrete sciences, includes meteorology, mineralogy, botany, zoology, geology, and geography.

The order in which the abstract sciences are arranged above has also been found to best illustrate the sequence in which they may be most advantageously studied.

SCIL'LITIN. *Syn.* SCILLITINA, SCILLITITE. A whitish, resinous, translucent, bitter, deliquescent substance, obtained by Vögel from squills. It is soluble in water, alcohol, and acetic acid, and is purgative, acrid, and poisonous.

SCOPOLA. Mr E. M. Holmes describes this plant as the *Scopola carniolica*, Jecquin. The plant grows in Eastern Germany and Hungary, and extends to Southern Russia. The plant resembles belladonna in many respects, but is not so tall or robust. It grows best in damp soil in the shade. Its root is a rhizome—not a rootstock, as belladonna is. Its leaves are larger, but thinner and greener. The flower is like that of belladonna, but the fruit more resembles henbane fruit. It flowers in March and April. It contains an alkaloid which Dunstan and Chaston identify with hyoscyamine; others have named it scopoleine, but the latter base is extracted from *S. Japonica*.

SCOPOLEINE. An alkaloid obtained from the root of *Scopola Japonica* (known as Japanese belladonna). It possesses mydriatic properties.

SCORBU'TUS. See SCURVY.

SCO'RIA. Dross; the refuse or useless part of any substance, more especially that left from bodies which have been subjected to the action of fire. It is frequently used in the plural (SCORIÆ).

SCOTT'S DROPS. See PATENT MEDICINES.

SCOUR'ING. The common method of cleaning cloth by beating and brushing it, unless it be very dirty, when it undergoes the operation of scouring. This is best done on the small scale, as with ARTICLES OF WEARING APPAREL, as follows: —A little curd soap is dissolved in water, and, after mixing it with a little clarified ox-gall, is applied to all the spots of grease, dirt, &c., and well rubbed into them with a stiff brush, until they appear to be removed; after which the article is well cleaned all over with a brush or sponge dipped into some warm water, to which the previous mixture and a little more ox-gall have been added. The cloth is next thoroughly rinsed in clean water, and hung up to dry. For dark-coloured cloths some fuller's-earth is often added to the mixture of soap and gall. When the article is nearly dry the nap is laid smooth, and it is carefully pressed (if with a hot iron, on the wrong side), after which a soft brush, moistened with a drop or two of olive oil, is frequently passed over it, to give it a finish and gloss.

Cloth is also cleaned in the dry way:—The spots being removed as above, and the wetted parts having become dry, clean damp sand is strewed over it, and beaten into it with a brush, after which the article is well gone over with a hard brush, when the sand comes out, and brings the dirt with it.

BUFF and DRAB CLOTH is generally cleaned by covering it with a paste made with pipe-clay and water, either with or without a little umber to temper the colour, which, when dry, is rubbed and brushed off.

When the article requires renovation as well as scouring, it is placed, whilst still damp, on a board, and the threadbare parts are rubbed with a half-worn hatter's card filled with flocks, or with a teasle, or a prickly thistle, until a nap is raised; it is next hung up to dry, after which it is 'finished off' as before. When the cloth is much faded it is usual to give it a 'dip,' as it is called, or to pass it through a dye-bath to freshen up the colour. BLACK and DARK BLUE CLOTH, if rusty or faded, is commonly treated to a coat of 're-viver,' instead of being 're-dipped,' and is then hung up until next day, before being pressed and finished off. See SPOTS and STAINS.

Scouring in Animals. See DIARRHŒA.

SCOURING DROPS. See DROPS.

SCROFULA. *Syn.* KING'S EVIL, STRUMA, STRUMOUS DISEASE. By modern pathologists scrofula is regarded as a constitutional tendency to form and deposit in various tissues and organs of the body a substance called *tubercle*. The *tendency* may, however, in some cases only exist without any actual tuberculous deposit taking place. Sir James Paget thus describes scrofula as generally understood to be a "state of constitution distinguished in some measure by peculiarities of appearance even during health, but much more by peculiar liability to certain diseases, including pulmonary phthisis. The chief of these 'scrofulous' diseases are various swellings of the lymphatic glands, arising from causes which would be inadequate to produce them in healthy persons.

"The swellings are due sometimes to mere enlargement, as from an increase of natural structure, sometimes to chronic inflammation, sometimes to an acute inflammation or abscess, sometimes to tuberculous disease of the glands.

"But besides these it is usual to reckon as 'scrofulous' affections certain chronic inflammations of the joints; slowly progressive carious inflammations of bones; chronic and frequent ulcers of the cornea; ophthalmia attended with extreme intolerance of light, but with little, if any, of the ordinary consequences of inflamma-

tion; frequent chronic abscesses; pustules or other cutaneous eruptions frequently appearing upon slight affection of the health or local irritation; habitual swelling and catarrh of the mucous membrane of the nose; habitual swelling of the upper lip."

Scrofula is a disease which almost always shows itself during childhood, and rarely after maturity has been attained.

Scrofulous children, or those of scrofulous diathesis, are frequently narrow-chested, or their chests present that projecting appearance known as 'pigeon-breasted;' their abdomens are also unnaturally large, and their limbs emaciated. Their circulation is languid, and they are very generally attacked with chilblains during inclement weather. They also suffer from obstinate indigestion. Bearing in mind the fact that scrofula is frequently induced, irrespective of hereditary taint, in the children of the poor by bad and damp air, insufficient food and clothing, exclusion from sunshine, and such like insanitary surroundings, the chief treatment that will suggest itself will consist in remedying these adverse conditions. Hence the patient should live on generous but digestible food, partaking of meat twice a day. Milk and eggs also form an excellent diet for the scrofulous. A scrofulous mother should refrain from suckling her offspring, and procure a wet-nurse for that purpose. Flannel should be worn both summer and winter. Various medicines have been employed in this disease, including cod-liver oil, sarsaparilla, bark, syrup of iodide of iron, the alkalies, and mineral acids. Of these, cod-liver oil and syrup of iodide of iron deservedly enjoy the highest reputation.

SCUDAMORE'S LOTION. See LOTION, GOUT.

SCURF. *Syn.* FURFURA. Scurf "is a natural and healthy formation, and though it may be kept from accumulating, it cannot be prevented. It is produced on every part of the body where hair is found, although, from the more active growth of hair on the scalp, the facilities for collecting, and the contrast of colour, it strikes the eye most disagreeably in that situation. This will show how futile any attempt must be which shall have for its object to prevent the formation of the scurf. It may be removed, and should be removed, every day, with the hair-brush; but prevention is impossible, inasmuch as it is opposed to a law of nature. Occasionally, as a morbid action, an unusual quantity of scurf is produced, in which case medical means may be adopted to bring the scalp into a more healthful state " (*Eras. Wilson*). In such cases the daily use of some mild stimulating or detergent wash, with due attention to the stomach and bowels, will generally abate this annoyance.

SCURF POWDER — Grindpulver. (*Mahon*, Paris.) Three powders which, according to Chevalier and Figuier, are nothing but wood ashes. Buchner found no alkalies, but announced the following composition:—Organic calcium carbonate (oyster shells, egg shells, crab shells), with a little gypsum, charcoal powder, and more or less brick-dust, powdered, mixed, and exposed to a moderate red heat in a covered crucible, till part of the chalk is converted into quicklime, and the gypsum reduced by the charcoal powder to calcium sulphide, which in its turn is gradually converted by the air into calcium sulphite. All three powders are made of the same ingredients, but in different proportions. No. 1 has more gypsum and charcoal powder; No. 2 less charcoal and more chalk; and No. 3 more brick-dust (*Wittstein*).

SCURF SALVE—Grimsalbe. In France it is generally a mixture of 2 parts slaked lime, 5 parts soda crystals, and 25 parts fat (*Hager*).

SCURVY. *Syn.* SCORBUTUS, L. This disease commences with indolence, sallow looks, debility, and loss of spirits; the gums become sore and spongy, the teeth loose, and the breath fœtid; the legs swell, eruptions appear on different parts of the body, and at length the patient sinks under general emaciation, diarrhœa, and hæmorrhages.

The treatment of ordinary cases of this disease mainly consists in employing a diet of fresh animal and green vegetable food, with mild ale, beer, and lemonade as beverages; scrupulously avoiding salted and dried meat. The fresh-squeezed juice of lemons is, perhaps, of all other substances the most powerful remedy in this disease in its early stages, and is useful in all of them. Effervescing draughts formed with the bicarbonate of potassa (not soda) are also excellent.

In former years, before the nature of this malady had been intelligently investigated, and when the proper preventive methods and remedial measures for combating it were unknown, scurvy was not only a very common but a very fatal disease in our own navy, as well as in the navies of other powers. Of 961 men who constituted the crews of Anson's fleet sent out during our war with Spain in 1742, 626 died of scurvy in nine months; whilst Sir Gilbert Blane records that in the year 1780, out of a fleet composed of between 7000 and 8000 men, more than 1000, or one in seven, perished from the same cause. Sir Richard Hawkins, one of the naval celebrities of Elizabeth and James's reign, affirmed that during twenty years he had known 20,000 sailors fall victims to scurvy alone; and a Portuguese writer, quoted by Sir Charles Blane, speaking of the number of victims from scurvy, during a naval exploring expedition of his own countrymen, says that "if the dead who from this cause had been thrown overboard between the coast of Guinea and the Cape of Good Hope, and between that Cape and Mozambique, could have had tombstones placed for them, each on the spot where he sank, the whole way would have appeared one continued cemetery" (*Dr Guy*).

The statistical report of the navy for 1871 offers a gratifying contrast to the above figures. From this document it appears that out of a total force of 4720 sailors, only four were affected with scurvy during that year. The much greater number of men attacked by the disease on board merchant ships appears to be due to the inferior or worthless character of the lime or lemon juice purchased by them.

Writing on the hygienic condition of the merchant marine in 1867, Mr Harry Leach says:

"We are prepared to maintain, from the following table (and other statistics from which these have been taken), that the want of good lime or lemon juice was distinctly the cause of scurvy in the vessels below mentioned.

Name of Ship.	No. of Hands (all told).	Cases of Scurvy.	Result of Examination of Lime-juice.
Hermione	17	5	Sulphuric acid.
Merrie England	29	10	Stinking.
Stirling Castle	32	6	Very weak.
Hoang-Ho	21	5	Acetic acid.
Blanche Moore	35	8	Musty and nauseous.
St Andrew's Castle	19	7	Citric acid.
Tamerlane	21	4	Nauseous.
Marlborough	23	8	Very weak.
Galloway	29	6	Short allowance.
Tamar	17	2	Very weak.
French Empire	27	7 or 8	Citric acid.
Eaglet	14	3	Thick and nasty.
Geelong	14	9	Taken irregularly.
Thorndean	35	2	Spoiled (short supply of provisions).

"Of direct causes this is undoubtedly first and foremost; but of indirect causes we have a few words to say. Dirt, bad provisions, and any form of disease to which sailors, in common with other men, are subject, will predispose to scurvy. This cannot and should not be denied, but it affords to parsimonious captains a very large peg whereon to hang sundry invectives as to the cry lately made about the continued prevalence of this disease in the mercantile marine. Such captains, with pardonable ignorance, consider scurvy a form of venereal disease, give the wretched subject thereof mercury, and bring him into port salivated as well as scorbutic."

Mr Leach further adds:

"In summing up statistics of scurvy for the past year (1867), we find that a total of 235 accredited cases were admitted into British hospitals, giving no account of those who convalesced in sailors' homes or elsewhere.

"To this we may add that seven sailors were left at St Helena, from a ship recently arrived in the Thames; that a vessel put into Falmouth on the 29th ult., with no less than sixteen severe cases of scurvy on board, and that between twenty and thirty cases have arrived in this port during the present month. It would be well (as a supplementary aid to the prevention of scurvy by inspection of lime-juice) that the dues levied for the St Helena Hospital should be abolished. It was stated to us some weeks ago by a very old inhabitant of that island that this fact alone caused many ships to pass without calling for needful supplies of antiscorbutic material.

"I would, however, remark that if the system proposed by the Seamen's Hospital Society were put in force, no such aid to the prevention of this disease would be required, inasmuch as every ship would then be supplied with good lime-juice."

The following figures, giving the number of patients suffering from scurvy admitted into the Seamen's Hospital, shows a decrease in the disease since the publication of the above:

In 1865, from British vessels, 101; foreign do. 1
" 1866 " " 96 " 5
" 1867 " " 90 " 4
" 1868 " " 64 " 10
" 1869 " " 31 " 9
" 1870 " " 30 " 21
" 1871 " " 24 " 16

SEALING-WAX. See WAX.

SEA-SICKNESS. The most effectual preventive of sea-sickness appears to be the prone position, and the application of ice-bags to the spine and back of the head. When there is much pain, after the stomach has been well cleared, a few drops of laudanum may be taken, or an opium plaster may be applied over the region of the stomach. Persons about to proceed to sea should put their stomach and bowels in proper order by the use of mild aperients, and even an emetic, if required, when it will generally be found that a glass of warm and weak brandy-and-water, to which 15 or 20 drops of laudanum, or, still better, 1 or 2 drops of creasote, have been added, will effectually prevent any disposition to sea-sickness, provided the bowels be attended to, and excess in eating and drinking be at the same time avoided. A spoonful of crushed ice in a wine-glassful of cold water, or weak brandy-and-water, will often afford relief when all other means fail. Smoking at sea is very apt to induce sickness. M. F. Curie, in the 'Comptes Rendus,' asserts that drawing in the breath as the vessel descends, and exhaling it as it ascends on the billows, by preventing the movements of the diaphragm acting abnormally on the phrenetic nerves, prevents sea-sickness. On this Mr Atkinson, at one of the meetings of the British Association, observed that if a person, seated on board ship, holding a tumbler filled with water in his hand, makes an effort to prevent the water running over, at the same time allowing not merely his arm, but also his whole body, to participate in the movements, he will find that this has the effect of prevent the giddiness and nausea that the rolling and tossing of the vessel have a tendency to produce in inexperienced voyagers. If the person is suffering from sickness at the commencement of his experiment, as soon as he grasps the glass of liquid in his hand, and suffers his arm to take its course and go through the movements alluded to, he feels as if he were performing them of his own free will, and the nausea abates immediately, and very soon ceases entirely, and does not return so long as he suffers his arm and body to assume the postures into which they seem to be drawn. Should he, however, resist the free course of his hand, he instantly feels a thrill of pain, of a peculiarly stunning kind, shoot through his head, and experiences a sense of dizziness and returning nausea.

Dr Doring, a Viennese physician, states that an ordinary dose of chloral hydrate is an unfailing remedy for sea-sickness. In various cases recorded by him it seems to have been of the greatest service, even during long sea voyages, ensuring a good night's rest, arresting violent sickness when it has set in, and preventing its return.

Every imaginable remedy has been suggested for this distressing malady, mostly narcotics in one form or another. Where it is possible plenty of physical exercise, e.g. rowing, is an excellent means of preventing it; and as there is little doubt that it is aggravated in persons by dread, it is well to be actively employed as much as possible during a sea voyage. Some persons never can overcome it. In most cases it will last three or four days, and then the nervous system gets accustomed to the motion, and the symptoms disappear.

SEDATIVE PILLS, Gunther's. These are composed of the following ingredients:—Assafœtida powder, 50 parts; extract of valerian, 50 parts; extract of belladonna, 3 parts; oxide of zinc, 1 part; castor oil, 2 parts. Make into a pill-mass, to be administered in doses of 3 to 10 grains, twice daily, in chorea, &c.

SEDATIVES. Syn. SEDATIVA, L. Medicines and agents which diminish the force of the circulation or the animal energy, and allay pain. Foxglove, henbane, tobacco, potassio-tartrate of antimony, and several of the neutral salts and acids, act as sedatives. Cold is, perhaps, the most powerful agent of this class.

SEED. Syn. SEMEN, L. The seeds of plants are conspicuous for their vast number and variety, and their extreme usefulness to man. The seeds of certain of the Graminaceæ furnish him with his daily bread; some of those of the Leguminosæ, in either the immature or ripe state, supply his table with wholesome esculents, or provide a nourishing diet for his domestic animals; whilst those of numerous other plants, dispersed through every class, order, and family, yield their treasures of oil, medicinals, or perfumes for his use.

SELANDRIA CERASI, Linn. (Tenthredo cerasi, Curtis). THE PEAR SAW-FLY. All fruit-growers and gardeners have remarked the action of a curious-looking insect resembling a slug, upon the leaves of apple, pear, cherry, plum, and damson trees. Though to all appearance it is but a mere lump of slime or dirt upon the leaves, it nevertheless soon eats away their soft green tissues, leaving only the nerves or ribs, so that the leaves become mere skeletons.

This insect, the larva of a saw-fly, is most repulsive to behold, having a disproportionately large head, and a dark, slimy, viscid fluid covers its body. It is not at present one of the often recurring pests of fruit trees like the Hyponomeuta padella and the gooseberry saw-fly, though serious attacks have been reported, and it constantly attacks single trees here and there, both standards in orchards and gardens, as well as espalier and wall trees in gardens.

A fruit-grower in Gloucestershire states that in

1881 the apple trees in an orchard facing to the south-east, lying low, had their leaves completely riddled by what he called 'snegs.' It was reported in this same year that "a lot of green slugs" were eating up the leaves of the pear trees in some orchards in Herefordshire. Also in 1881 reports of injury from 'slugs' came in from parts of Scotland and Yorkshire.

From inquiries made of fruit-growers in various parts of the country it is gathered that there was a deal of mischief caused by this insect in 1875, and that it seemed to increase year by year.

In the 'Introduction to British Entomology,' by Kirby and Spence, it is remarked that the pear saw-fly does not cause any material injury. But, seeing the rapid spread of many kinds of insects in these latter years, and of the general great increase of injury caused by such pests, it is as well to be forewarned by means of information as to the habits of those which may be dangerous, and forearmed with some means of preventing and of checking their onslaughts.

This pear saw-fly is well known in America, and has been known there for a long while. Harris says that it was so abundant in Massachusetts in 1797, that "small trees were covered with them and their foliage entirely destroyed; and even the air by passing through the trees became charged with a disagreeable and sickening odour given out by these slimy creatures" ('A Treatise on some of the Insects Injurious to Vegetation in New England,' by T. W. Harris, M.D.). Mr Cooke shows that this saw-fly is very troublesome, especially to pear trees, in the important and increasing fruit orchards of California. Dr Asa Fitch also speaks of it as injuring cherry trees in America; and Professor Lintner, the New York State Entomologist, mentions it as injurious to many kinds of fruit trees. Professor Saunders points out that in 1874 this saw-fly was "unusually abundant in the neighbourhood of London, Ontario, in many cases destroying the foliage so thoroughly that the trees looked as if they had been scorched by fire." It is also well known in Germany and in France; Réaumur called it la Tenthrède Limace.

By some it is supposed that this insect was imported into Europe from America in comparatively recent years, but Westwood wrote about it in the 'Gardener's Magazine' more than fifty years ago, and Réaumur described it one hundred and fifty years ago.

Life History. It could hardly be imagined upon looking at this creature, *monstrum horrendum, informe,* that its progenitrix was a little quiet-looking fly. However, it is so, and it belongs to the order HYMENOPTERA, and to the genus or sub-genus *Selandria* of the family *Tenthredinidæ.*

This fly is hardly the fourth of an inch in length. The wings are about eight lines, or three quarters of an inch, from tip to tip. Being dark, with its wings having dark markings, it has not a conspicuous or an attractive appearance. At the end of June the fly makes a slightly curved abrasion in the upper part of the leaf of an apple, pear, cherry, plum, or damson tree, with its admirable saw-like apparatus, like that of the *Nematus ribesii* and other saw-flies. A single

egg is placed upon each abrasion made thus by the saw for its reception. After about a fortnight a larva comes out, at first of a whitish colour—almost transparent, in fact; but after a short time it becomes of a green colour, and an olive-green slime issues from all parts of its body, covering it over, evidently as a protection against weather in its exposed position on the upper part of the leaf—a position, it may be said in passing, somewhat exceptional, as most larvæ live on the under sides of leaves. Though it has seven pairs of rudimentary feet upon its abdomen, and three pairs of distinct thoracic legs, it moves with uncommon slowness. With a large head and a body narrowing down towards the caudal extremity, it looks like a tadpole. When full grown it is nearly half an inch long.

Unlike many other larvæ, or caterpillars of flies and moths, it eats away the leaves from their centres and not from the outsides, and clears away the parenchyma between the nerves or ribs, leaving them bare as the framework of an un-covered parasol.

At the end of a month, after several moultings or castings of its skin, the larva loses its snail or slug-like form, and finally appears in an orange-yellow robe, and in shape more like a caterpillar. Giving up feeding, it crawls down the stem of the tree or falls to the ground, in which it ensconces itself and changes to a chrysalis, forming a little cell made of earth, glued together with a sticky material, in which it remains until late in the spring, and changing to a saw-fly goes forth to propagate its kind.

In America the saw-flies have two broods—the latter brood sometimes as late as September.

Prevention. As it is evident that the chrysalids cannot be far away from the fruit trees upon which the larvæ were reared, it would be well to dig the ground all round thoroughly, and to hoe it with prong-hoes well in the spring, taking care that the clods and lumps are well broken. Application of caustic substances and of offensive materials, as paraffin oil and carbolic acid, would not be very efficacious, it is thought, since the chrysalids are protected by their glue-cased cocoons. Upon grass orchards and upon cultivated land quicklime might be scattered round infested trees just before the general final change takes place, in order to kill them if they ventured into it.

Remedies. Fine particles powdered upon the slimy bodies of the larvæ of this saw-fly render existence burdensome to them. Thus very fine quicklime could be sent up by means of a machine like that used in Kent for putting sulphur upon hop plants for mildew—at least upon the smaller trees and the lower branches of the larger trees. Except in the case of cherries this would not affect the fruit to a great extent. And with regard to cherries, as a rule, these would be picked before the larvæ had done much mischief. In America powdered hellebore, mixed with water in the proportion of two pounds to 100 gallons, is syringed over the trees with good effect. Hellebore is, however, a far too deadly poison to be sprinkled upon fruit. If the fruit has been so injured as to be not worth picking, washing with soft soap and water, with the extract of two pounds of tobacco added to 100 gallons of water, put on with a washing engine, would effectually kill the larvæ, or make the leaves unpleasant for them. It is clearly most difficult to apply remedies to trees when the fruit upon them is plentiful and valuable ('Reports on Insects Injurious to Crops,' by Charles Whitehead, Esq., F.Z.S.).

SELE'NIUM. Se=79·0. A rare chemical element, discovered by Berzelius in 1817 in the refuse of a sulphuric acid manufactory near Fahlun, in Sweden, it having been derived from the pyrites employed in the manufacture of the acid. Hence the pyrites of Fahlun forms the chief source of this rare body, although it exists, but less abundantly, in combination with a few other metals, termed selenides. Selenium is chiefly interesting to the chemist from its remarkable analogy in chemical properties, natural history, and physical relations to sulphur. Like this latter element, it is capable of assuming three allotropic forms—the amorphous, the vitreous, and the crystalline, of which the first and last are the best understood.

The last variety of selenium, like the crystalline form of sulphur, dissolves in bisulphide of carbon, but much less readily. Selenium fuses at 100° C., boils at 680° (*Carnelly*), and becomes converted into a deep yellow vapour, which, when heated, is subject to the same anomalous expansion as sulphur vapour. It is not so combustible as sulphur, which it still further resembles by burning with a blue flame when ignited in the air. During combustion it gives off a peculiar and characteristic smell, resembling that of putrid horseradish. Heated with strong sulphuric acid, selenium forms a green solution. If this solution is poured into water, the selenium separates and is thrown down. Selenium is without taste or smell, is insoluble in water, and in its normal state is a non-conductor of heat and electricity. Selenium may be extracted from the Fahlun residue by the following process:—It should be first boiled with sulphuric acid, diluted with an equal volume of water, and nitric acid should then be added in small quantities until the oxidation of the selenium is accomplished, which may be known when red fumes cease to be evolved. The solution, which contains selenious and selenic acid, is then to be largely diluted with water, filtered, the filtrate mixed with about one fourth of its bulk of hydrochloric acid, and then concentrated a little by evaporation, when the hydrochloric acid reduces the selenic to selenious acid. A current of sulphurous acid being then passed through the solution, the selenium is precipitated in flakes, which form into a dense black mass when the liquid is gently heated.

Selenium Oxide. Like sulphur, selenium combines with oxygen and forms an anhydride corresponding to sulphurous anhydride. **SELENIOUS ANHYDRIDE** (SeO₂) may be obtained by burning selenium in a current of oxygen; it is, however, more easily prepared by boiling selenium with nitric acid or with aqua regia; the excess of acid being expelled by heat, the selenious anhydride is left as a white mass. When this is dissolved in water it yields a crystalline hydrate of selenious acid (H₂SeO₃).

Selenious Acid. H₂SeO₃. *Prep.* Formed when selenium is heated in nitric acid, or when 5 parts of the dioxide are dissolved in 1 part of hot water. *Prop., &c.* Long, colourless, prismatic crystals; strongly acid taste; decompose when heated. It forms not only acid and normal salts, but also salts containing selenites united with selenious acid.

The salts formed by selenious acid (selenites), with the exception of those of the alkali metals, are mostly insoluble in water. They are easily known by the peculiar odour of selenium which they give off when heated on charcoal in the reducing flame of the blowpipe; solutions of the selenites give a reddish-brown precipitate when treated with sulphurous acid.

Selenic Acid. H₂SeO₄. Discovered in 1827 by Mitscherlich.—*Prep.* By action of chlorine on selenium or selenious acid in the presence of water, or by treating a solution of sodium selenite with silver nitrate, and then acting upon the precipitate thus obtained with ammonia in the presence of water.

Prop. A colourless acid liquid which, by evaporation at 265° C. and afterwards under an air-pump, may be obtained of sp. gr. 2·627; this then contains 97·4% of selenic acid; dissolves gold and platinum. The selenates exhibit the closest analogy to the sulphates.

Seleniuretted Hydrogen. H₂Se. This may be obtained by acting on selenide of iron or potassium with diluted sulphuric or hydrochloric acid. Seleniuretted hydrogen is soluble in water, and precipitates many metals from their salts as selenides. The solution is feebly acid, and, like a solution of sulphuretted hydrogen, if exposed to the air, it absorbs oxygen and deposits selenium. The selenides of the alkali metals are soluble in water. The selenides of cerium, zinc, and manganese are flesh-coloured; most of the others are black. This gas is inflammable like sulphuretted hydrogen; it has, however, a still more offensive smell than this latter gas. Berzelius lost his sense of smell for several hours by the application to his nose of a bubble of seleniuretted hydrogen not larger than a pea.

Selenium Chlorides. There are two chlorides of selenium—a dichloride (Se₂Cl₂), a volatile liquid of a brown colour; and a tetrachloride (SeCl₄), which occurs as a white crystalline solid.

Selenium Sulphides. Selenium unites with sulphur, forming a bisulphide (SeS₂) and a tersulphide (SeS₃).

Obs., &c. A very curious physical property of selenium when exposed to the action of light was first noticed in 1873 by Mr May, who observed that a stick of crystallised selenium which had been used for some time in telegraphy, where high electrical resistance was required, offered a considerably less resistance to the current when exposed to the light than when kept in the dark. This discovery has since been amply corroborated by the observations and researches of many physicists, amongst them by Professor Werner Siemens, the result of whose experiments on this interesting subject we quote from a lecture delivered at the Royal Institution by his brother, Dr William Siemens, in February, 1876. After describing the method of arranging the selenium so that, when inserted in the galvanic current of a single Daniell's cell, the surface action produced by the light upon it attained a maximum effect, and thereby did away with the necessity of employing a large galvanic battery, and at the same time allowed an ordinary galvanometer to be used instead of a delicate one, as hitherto employed, Dr Siemens proceeded to illustrate the action of light upon the element by experiment. " I here hold," he said, " an element so prepared of amorphous selenium, which I place in a dark box, and insert in a galvanic circuit comprising a Daniell's cell and a delicate galvanometer, the face of which will be thrown upon the screen through a mirror by means of the electric light.

" In closing the circuit it will be seen that no deflection of the needle ensues. We will now admit light upon the selenium disc and close the circuit, when again no deflection will be observed, showing that the selenium in its present condition is a non-conductor both in the dark and under the influence of light. I will now submit a similar disc of selenium, which has been kept in boiling water for an hour and gradually cooled, to the same tests as before. In closing the circuit while the plate is in the dark a certain deflection of the galvanometer will be discernible, but I will now open the lid of the box so as to admit light upon the disc, when on again closing the circuit a slight deflection of the galvanometer needle will be observed. In closing the box against the light this deflection will subside, but will again be visible the moment the light is readmitted to the box. Here we have, then, the extraordinary effect of light upon selenium clearly illustrated.

" I will now insert into the same circuit another selenium plate which has been heated up to 210° C., and, after having been kept at that temperature for several hours, has been gradually cooled; it will be observed that this plate is affected to a greater extent than the former by the action of light; and other conditions, to which I shall presently allude, prove the selenium heated to a higher temperature to be in other respects dissimilar to the other two modifications of the same. These differences will be best revealed in describing my brother's experiment. He placed one of his amorphous preparations of selenium in an air-bath heated above the melting-point of selenium (to 260° C.), while the connecting wires were inserted in a galvanic circuit consisting of only one Daniell's element and a delicate reflecting galvanometer, and every five minutes the temperature and conductivity of the selenium were noted. Up to the temperature of 80° C. no current passed; from this point onward the conductivity of the material rapidly increased until it obtained its maximum at the temperature of 210° C., being nearly its melting-point, after which an equally rapid diminution of conductivity commenced, reaching a minimum at a temperature of about 240° C., when the conductivity was only such as could be detected by a most delicate galvanometer. In continuing to increase the temperature of the fluid selenium very gradually but steadily, its conductivity increased again.

" The interpretation of these experiments is as follows:—Amorphous selenium retains a very

large amount of specific heat, which renders it a non-conductor of electricity; when heated to 80° this amorphous solid mass begins to change its amorphous condition for the crystalline form, in which form it possesses a greatly reduced amount of specific heat, giving rise to the increase of temperature beyond that of surrounding objects when the change of condition is once set in. If care is taken to limit the rise of temperature of the selenium to 100° C., and if it is very gradually cooled after being maintained for an hour or two at that temperature, a mass is obtained which conducts electricity to some extent, and which shows increased conductivity under the influence of light. But in examining the conductivity of selenium so prepared at various temperatures below 80°, and without accession of light, it was found that its *conductivity increases with rise of temperature*, in which respect it resembles carbon, sulphides of metals, and electrolytes generally. This my brother terms his first modification of selenium.

" But in extending the heating influence up to 210°, and in maintaining that temperature by means of a bath of paraffin for some hours before gradually reducing the same, he obtained a second modification of selenium, in which its conductivity increases with fall of temperature, and in which modification it is, therefore, analogous to the metals. This second modification of selenium is a better conductor of electricity than the first, and its sensitiveness to light is so great that its conductivity in sunlight is fifteen times greater than it is in the dark, as will be seen from the following table, in which is given the effects of different intensities of light on selenium (Modification 2) obtained at Woolwich on the 14th of February, 1876:

Selenium in	Relative Conductivities.		Resistance in Ohms.
	Deflections.	Ratio.	
1. Dark	32	1·0	10,070,000
2. Diffused daylight	110	3·4	2,930,000
3. Lamplight .	180	5·6	1,790,000
4. Sunlight . . .	470	14·7	680,000

Unfortunately, however, the second modification is not so stable as the first; when lowered in temperature, parts of it change back into the first or metalloid modification by taking up specific heat, and in watching this effect a point is discovered at which ratio of increase of conductivity with fall of temperature changes sign, or where the electrolytic substance appears to predominate over the metallic selenium. If cooled down to 15° C., the whole of the metallic selenium is gradually converted back into the first variety. The physical conclusions here arrived at may be said to be an extension of Helmholtz's theory that the conductivity of metals varies inversely as the total heat contained in them. Helmholtz had only the sensible heat of temperature (counting from the absolute zero point) in view, but it has already been shown by Hittorf and Werner Siemens that it applies in the case of tin and some other metals, also to specific heat and to the latent heat of fusion. In selenium the specific heat is an extremely variable quantity, changing in the solid mass at certain temperatures, and, it is contended, under the influence of light. Aided by these experimental researches my brother arrives at the conclusion that the influence of light upon selenium may be explained by a '*change of its molecular condition near the surface, from the first or electrolytic into the second or metallic modification*, or in other words, by a *liberation of specific heat upon the illuminated surface of crystalline selenium*, which liberated heat is reabsorbed when the liberating cause has ceased to act.'" Professor Adams, who has likewise investigated this singular action of light upon selenium, ascribes it to a different cause. He says—

1. That the light falling on the selenium causes an electro-motive force in it in the same direction as the battery current passing through it, the effect being similar to the effect due to polarisation in an electrolyte, but in the opposite direction.

2. That the light falling on the selenium causes a change on its surface akin to the change which it produces on the surface of a phosphorescent body, and that in consequence of this change the electric current is enabled to pass more readily over the surface of the selenium.

SEM'OLA (Bullock's). This preparation consists of wheaten flour deprived of much of its starch by washings with water, and contains the largest amount (48%) of nitrogenous or albuminoid principles consistent with its adaptability to culinary purposes. It is specially intended as a food for infants, weakly children, and invalids.

SEMOLI'NA. *Syn.* SEMOULE, SEMOULINA. The large hard grains of wheat flour retained in the bolting machine after the fine flour has passed through its meshes. "The best semoule is obtained from the wheat of the southern parts of Europe. With the semoule the fine white Parisian bread called '*gruau*' is baked" (*Ure*).

SEN'EGA. *Syn.* SENEKA, SNAKE-ROOT, RATTLESNAKE R.; SENEGÆ RADIX (B. P.), SENEGÆ (Ph. L., E., and D.), RADIX SENEGÆ, L. "The root of the *Polygala senega*, Linn." (Ph. L.). A stimulating diaphoretic and expectorant; in large doses diuretic, cathartic, and emetic. In America it is used as an antidote to the bite of the rattlesnake. Drs Chapman and Hartshorne extol it as an emmenagogue. Dr Pereira says that it is an exceedingly valuable remedy in the latter stages of bronchial or pulmonary inflammation when this disease occurs in aged, debilitated, or torpid constitutions. Dr Ringer considers it of little value.—*Dose*, 10 to 30 gr., in powder or decoction (combined with aromatics, opium, or camphor), thrice daily.

According to Patrouillard senega is occasionally adulterated with the roots of *Asclepias vincetoxicum*. The branches of the latter root are cylindrical, very white, and almost devoid of taste; those of senega, on the contrary, are yellowish and twisted, and have a very acrid taste. The froth produced by shaking an infusion of senega keeps much longer than that produced by an infusion of the adulterant. In other respects

there is a great resemblance between the two roots.

SEN'EGIN. *Syn.* POLYGALIN, POLYGALIC ACID. A white odourless powder, discovered by Gehlin in the bark of seneka root (*Polygala senega*).

SEN'NA. *Syn.* SENNA, SENNÆ FOLIA, L. There are two principal varieties:

1. ALEXANDRIAN SENNA (SENNA ALEXANDRINA—B. P.). The dried leaflets of *Cassia acutifolia*, Delile. The leaves are "unequal at the base, ovate acute, or obovate mucronate." It is sometimes mixed with the leaves of *Solenostemma argel* (argel leaves), the presence of which is often the occasion of much griping. The leaf of argel is fully an inch long, warty, regular in its formation, and the lateral nerves are imperfectly seen on the under side; whilst that of the true Alexandrian senna never exceeds ⅝ inch in length, is oblique, and the nerves on the under side are very conspicuous.

2. INDIAN or TINNEVELLY SENNA is composed of the leaflets of *Cassia angustifolia*. These are pale green, thin, flexible, and from 1 to 2 inches long, and nearly ⅓ inch broad. This variety is equal in medicinal virtue to the best Alexandrian, and is to be preferred, on account of its being imported perfectly free from adulteration.

Senna is purgative in doses of 10 to 30 gr., either in powder or made into an infusion of tea with water, combined with ginger, caraways, or some other aromatic, to prevent griping. It acts chiefly on the small intestines, and generally effects its purpose within four hours after being taken.

SE'PIA. A pigment prepared from the 'ink' or black fluid secreted by *Sepia officinalis*, Linn., and several other varieties of cuttle-fish. The contents of the 'ink bags' are inspissated as soon as possible after collection, and then form the crude sepia of commerce. This is prepared for artists by boiling it for a short time in a weak lye of caustic alkali, precipitating the solution with an acid, and well washing and carefully drying the precipitate by a gentle heat. It possesses a fine brown colour, and is used like Indian ink.

SER'PENTARY. *Syn.* VIRGINIAN SNAKE-ROOT; SERPENTARIA RADIX (B.P.), SERPENTARIÆ RADIX, SERPENTARIA, L. The rhisome and rootlets of *Aristolochia serpentaria* and of *A. reticulata*. An excellent stimulating diaphoretic and tonic; in typhoid and putrid fevers, dyspepsia, &c. It is admirably suited to check vomiting and to tranquillise the stomach, particularly in bilious cases (*Dr Chapman*).—*Dose*, 10 to 20 gr. every third or fourth hour, its use being preceded by an aperient.

SE'RUM. *Syn.* SERALBUMEN. The clear pale fluid in which the blood-globules float, and which separates from blood during its coagulation. It is, essentially, a feebly alkaline solution of albumen. See ALBUMEN.

SESQUI-. See NOMENCLATURE.

SE'TON. *Syn.* SETACEUM. An artificial ulcer, made by passing a portion of silk or thread under the skin by means of a seton needle, a part of which is drawn through daily, and thus keeps up a constant irritation. Occasionally the thread is anointed with some irritating substance for the purpose of increasing the discharge.

SEVUM (Prepared). *Syn.* SEVUM PRÆPARA-TUM (B. P.), SEVUM MAGNETICUM, L. *Prep.* The internal fat of the abdomen of the sheep purified by melting and straining.

Used to make mercurial ointment. Triturated with 8, 12, or 16 times its weight of quicksilver, the globules are completely extinguished in from 11 to 15 minutes.

SEWAGE, Removal and Disposal of. The waste and putrescible refuse discharged from dwelling-houses by house-pipes and drains into sewers may be said, in general terms, to consist, besides human fæces and urine (in the drainage of some towns the fæces are not allowed to enter the sewers; this, however, is the exception), of the dirty water and soapsuds arising from washing our bodies, our houses, and linen, more or less foul, as well as the water which, having been used for cooking operations, necessarily contains variable quantities of mineral and vegetable matter.

The above statement will have prepared us not only for the complex nature of sewage water as shown in the following tables, but also for the variability in the amount of its constituents, this

Composition of Sewer Water (WAY).

	Grains per Gallon.			
	1.	2.	3.	4.
Organic matters (soluble) . .	19·40	41·03	12·30	} 9·20
„ (suspended) . .	39·10	17·00	24·37	
Lime	10·13	14·71	12·52	11·25
Magnesia	1·42	1·82	1·59	1·35
Soda	4·01	2·40	2·41	1·89
Potash	3·66	3·57	3·31	1·09
Chloride of sodium . .	26·40	22·61	34·30	5·58
Sulphuric acid . .	5·34	5·31	6·40	3·43
Phosphoric acid . .	2·63	5·76	2·48	0·64
Carbonic acid . .	9·01	8·92	11·76	} 4·77
Salicia { Oxide of iron } { Oxide of zinc } . .	6·20	13·55	6·46	
Ammonia	7·48	8·43	7·88	
	134·78	145·11	125·78	39·20

London Sewer Water (LETHEBY).

	Grains per Gallon.		
	Day Sewage.	Night Sewage.	Storm Sewage.
Soluble matters	55·74	65·09	70·26
Organic matters	15·08	7·42	14·75
Nitrogen	5·44	5·19	7·26
Mineral matters	40·66	57·67	55·71
Phosphoric acid . . .	0·85	0·69	1·03
Potash	1·21	1·15	1·61
Suspended matters . .	38·15	13·99	31·88
Organic	16·11	7·48	17·55
Nitrogen	0·78	0·29	0·67
Mineral	22·04	6·51	14·33
Phosphoric acid	0·89	0·64	0·98
Potash	8·08	0·04	0·16

latter condition depending upon locality, and, as experiment shows, the hour of the day at which the sewage was collected.

Letheby states that the sewer water in towns with water-closets has the following average composition per gallon:

Organic matter . . .	27·72
Nitrogen	6·21
Phosphoric acid . . .	1·57
Potash	2·03

Sewer water placed under the microscope reveals various dead decaying matters, besides swarms of bacteria, ciliated infusoria, amœbiform bodies, and fungi, consisting of spores and mycelium. The rotifera, diatoms, and desmids are few in number (*Parkes*). That a fluid having a composition such as sewage water has been shown to possess when mixed with solid excreta would, from the decomposition that so soon takes place in it, seriously endanger the health of those in whose habitations it was allowed to remain, is so self-evident to the sanitarian and pathologist that it is no wonder every civilised community should endeavour to get rid of this refuse from their habitations as speedily and effectively as possible. But the removal of the home sewage is a proceeding as illogical as it is imperfect if we afterwards neglect to dispose of it so as to render it innocuous or devoid of danger to the public health. The old method of getting rid of sewage (even when deprived of the fæcal matter) by turning it into rivers and streams has, more particularly since the report of the Rivers Pollution Commissioners in 1870, been gradually abandoned. That when sewer water passes into a river it undergoes a great amount of purification from oxidation, subsidence, and the agency of water plants is undeniable.

Letheby considered that if sewage mixed with twenty times its bulk of water flowed for nine miles it would be perfectly oxidised. It appears, however, from the experiments of Frankland, that, so far from sewage when mixed with twenty times its volume of water being oxidised during a flow of ten or twelve miles, scarcely two thirds of it would be so destroyed in a flow of 168 miles at the rate of one mile per hour, or after the lapse of a week. The results of Frankland's experiments led him to infer that there is no river in the United Kingdom of sufficient length to effect the destruction of sewage by oxidation; and he adds, "There is no process practicable on a large scale by which the noxious material (sewage matter) can be removed from water once so contaminated; and, therefore, I am of opinion that water which has been once contaminated by sewage or manure matter is thenceforth unsuitable for domestic use."

The discharge of sewage water, whether with or without solid excreta, into our springs and rivers, was a practice so dangerous and prejudicial to health that it is no cause for wonder the Legislature should, during the session of 1876, have passed a measure the object of which was, after the lapse of one year, to facilitate legal proceedings being instituted against persons who permitted sewage or other deleterious refuse to flow into rivers or streams. This measure, known as the 'Rivers Pollution Prevention Act,' is now in force, and permits offenders to be proceeded against; but it still leaves unsolved the important hygienic problem—How are we ultimately, and with safety to the community, to dispose of our sewage?

The numerous processes (the chief of which will be brought under notice) proposed for the attainment of this end have been divided by writers and authorities on sanitary science into—1. WET METHODS; 2. DRY METHODS.

1. WET METHODS. These comprise the removal of excreta—(1) By discharging it into running water. (2) By storage in tank with overflow. (3) By carrying it into the sea. (4) By precipitation. (5) By irrigation and filtration.

(1) *By discharging it into running water.* Our previous remarks have already shown in what respect this proposal is fallacious, and why it has, therefore, been discontinued.

(2) *By storage in tank with overflow.* In this process the sewage runs into a well-cemented tank fitted with an overflow pipe, which sometimes leads into a second tank arranged in the same manner; the solids subside, and are removed from time to time, whilst the liquid is allowed to run away. Instead of permitting the liquid to escape into a ditch or stream it has been proposed to carry it into drain-pipes, which are buried from half a foot to a foot in the subsoil, where it will be readily sucked up by the roots of grasses. This plan is only suited for small villages, or for a single house or mansion.

(3) *By carrying it into the sea.* The precautions to be observed in the working of this system are, wherever possible, to let the outlet or discharge-pipe, which conveys the sewage to the sea, be always under water, even at ebb tide, and to take special care that the wind does not blow up the sewers. A tide-flap, opening outwards, which is usually fixed by a hinge on the sewer at its outlet, will obviate this last contingency. At high water the tide will fill the outfall sewers to its own level, and to that extent will check the discharge of sewage, and thus cause a deposit in the sewers filled with mixed sea water and sewage. It is most important that this should be removed. "If the sewage cannot be got well out to sea, and if it issues in narrow channels, it may cause a nuisance, and may require to be purified before discharge" (*Parkes*).

(4) *By precipitation.* The simplest of the plans proposed for this method of removal is by subsidence only, and would afterwards permit the discharge of the supernatant sewage water into running water or over the land. The removal of the solid material is effected in a manner somewhat similar to that followed in plan No. 2; but as the thin water which runs off must, when poured into rivers or streams, be almost as dangerous as the sewage itself, the process of precipitation by settlement alone has little to commend it over the old rude and objectionable practice, a circumstance that in these days will doubtless lead to its entire prohibition.

In order to ensure greater purification the sewage in the subsiding tanks is now usually mixed with certain chemical reagents, which, it is believed, have the effect not only of speedily precipitating the solid materials, but also carrying down injurious matters suspended in the sewage water, thus rendering it sufficiently pure to be discharged without risk to health into any watercourse.

Of the numerous precipitants employed for this purpose, we may mention the following :

Lime and Salts of Lime. Quicklime, in the proportion of 8 gr. to a gall. of water, or 1 lb. to about 600 galls. of sewage; lime, with the addition of about a fortieth of its weight of chloride of lime; calcic phosphate dissolved in sulphuric acid; Whitehead's patent, which consists of a mixture of mono- and di-calcic phosphate; chloride of calcium.

Aluminous Compounds. Bird's process—A mixture of aluminous earths and sulphuric acid. Anderson's and Lenk's—Impure sulphate of alum; refuse of alum works, either alone or mixed with lime or charcoal. Scott's cement process—Clay mixed with lime; natural phosphate of aluminium dissolved by sulphuric acid and mixed with lime.

The quantities of the above substances when used as precipitants vary, in some of them 50, and in others 80 gr. to a gall. of sewer water being employed.

Magnesium Salts. Impure chloride of magnesium mixed with superphosphate of lime.

Carbon. As vegetable charcoal, peat, seaweed charcoal, carbonised tan, lignite, and Boghead coke.

Iron. In the form of sulphate. Ellerman's and Dale's—Perchloride; the sulphate is sometimes mixed with coal-dust.

Manganese. Condy's fluid.

Zinc. As sulphate and chloride.

Sillar's Process. The A. B. C. process, so called because composed of alum, blood, charcoal, and clay.

Hill's Process. Lime and tar are the precipitants. The effluent water is filtered through charcoal. The question now arises as to whether the sewer water after treatment with any of the above substances is in a fit condition to be poured into a stream or river. The River Pollution Commissioners in their first and second reports give a number of analyses, from which it appears that on an average the chemical treatment removes 89·8% of the matters suspended in the sewage waters, but only 36·6% of the organic nitrogen dissolved in them.

Of the A. B. C. process, Mr Crookes states that, when properly carried out, it removes all the phosphoric acid; and Professor Voelcker's analysis of the effluent water from sewage treated by the acid phosphate of alumina process gives more ammonia than the original sewer water, less organic nitrogen by one half, and less phosphoric acid. Such water is said by some authorities to be pure enough to be discharged into streams.

General Scott's Process. General Scott proposes to treat the sewer water with lime and clay, and instead of employing the precipitate obtained by this means as a manure, would, after burning it, use it as cement. He argues that the deposit contains so much combustible matter as to considerably reduce the quantity of coal usually expended in the manufacture of cement, and consequently the cement could be sold at a remunerative price.

This, like the 'carbonisation' process, possesses the merit of effectually destroying any noxious principles present in the deposit.

Commenting on the various precipitation processes, Dr Parkes writes:—"When the sewer water is cleared by any of these plans is it fit to be discharged into streams? In the opinion of some authorities, if the precipitate is a good one it may be so, and it appears certain that in many cases it is chemically a tolerably pure water, and it will no longer silt up the bed nor cause a nuisance. But it still contains, in all cases, some organic matter, as well as ammonia, potash, and phosphoric acid. It has, therefore, fertilising powers certainly, and possibly it has also injurious powers. No proof of this has been given, but also no disproof at present; and when we consider how small the agencies of the specific diseases probably are, and how likely it is that they remain suspended, we do not seem to be in a position to expect that the water, after subsidence of the deposit, will be safe to drink."

(5) *By Irrigation and Infiltration.* By this process is meant the passing of the sewer water over and through soil, with the object not only of effecting its purification to such an extent as to render it fit to be discharged into a river or stream, but also of employing it as a valuable manure. In the present article we shall treat only of the application of the process to the first of these purposes.

There is ample evidence to show that, if carried out with due attention to detail, no process for the treatment of effluent sewage water, so as to render it innocuous, is equal to that which subjects it to irrigation and filtration.

The Rivers Pollution Commissioners thus report on it:—"We are, therefore, justified in recommending irrigation as a safe as well as profitable and efficient method of cleansing town sewage."

The conditions necessary for the successful carrying out of this system are thus stated by Mr T. J. Dyke, in explaining "the process of the downward intermittent filtration of sewage at Troedyrhiw, near Merthyr Tydvil:"—"1. The soil of the land to be used must be porous. 2. A main effluent drain, which must not be less than six feet from the surface, must be provided. 3. The surface of the soil to be so inclined as to permit the sewage stream to flow over the whole land. 4. The filtering area should be divided into four equal parts, each part to be irrigated with the sewage for six hours, and then an interval of eighteen hours to elapse before a second irrigation takes place; each of the four parts would thus be used for six hours out of the twenty-four. An acre of land so prepared would purify 100,000 gallons of sewage per day." At Troedyrhiw the sewage has lime added to it, and the mixture is strained through cinders into tanks. From the tank it flows on to the conduit, from which it is conveyed to the filtering areas.

"These consist of about twenty acres of land, immediately adjoining the road on which the tanks are placed, and have been arranged into filtering areas or beds on a plan devised by Mr J. Bayley Denton. The land is a loamy soil, eighteen inches thick, overlying a bed of gravel. The whole of these twenty acres have been underdrained to a depth of from five to seven feet. The lateral drains are placed at regular distances from each other, and run towards the main or effluent drain. This is everywhere six feet deep. The surface of the land is formed into beds; these have been made to slope towards the main drain by a fall of 1 in 150.

"The surface is ploughed in ridges; on these vegetables are planted or seeds sown. The line of the ridged furrow is in the direction of the underdrain. Along the raised margin of each bed, in each area, delivering carriers are placed, one edge being slightly depressed.

"The strained sewage passes from the conduits into the delivery carriers, and as it overflows the depressed edges runs gently into and along the furrows down to the lowest and most distant part of the plot. The sewage continues to be so delivered for six hours, then an interval of rest of eighteen hours takes place, and again the land is thoroughly charged with the fertilising stream. The water percolates through the six feet of earth, and reaches the lateral drains, which convey it to the main effluent drain.

"The result of this plan of disposing of sewage by downward intermittent filtration may be seen in samples of the effluent water taken from the outlet of the main drain. Such water is bright, perfectly pellucid, free from smell, and tastes only of common salt. It may be safely drunk—in fact, is used by the workmen employed on the farm. During the process of irrigation no nuisance is caused, for the soil quickly absorbs all the fluids passed on to it; in fact, in two or three hours after the water has ceased to flow on the land, an observer would say that the ground had not been wetted for days. The workmen say that no unpleasant smell is noticed, nor has the health of the persons employed, in any one instance, been affected by any presumed poisonous exhalation.

" The only imperfection of the plan is that, at the end of the furrows nearest the lowest corner of a plot, a slight deposit of scum is formed. This scum is formed by the fine insoluble precipitate caused mainly by the addition of lime to the sewage stream."

The table given below, taken from the report of the Rivers Pollution Commissioners, gives the composition of the effluent water after it has passed through the soil.

If those results be compared with the condition of the supernatant sewage water, after treatment by any of the chemical precipitants already enumerated, the inferiority of these latter as methods of removal of the organic impurity of the sewage water will be evident.

The best of these precipitants give a removal of only 65·8% of organic nitrogen, whilst the A. B. C. process shows a diminution of 58·9% only. It appears from the first and second reports of the Rivers Pollution Commissioners, that on an average the precipitation processes remove 89·3% of the suspended matters, but only

36·6 per cent. of the organic nitrogen dissolved in the liquid.

The effects of a soil upon sewage water passing through it are the following :

1. The filtering property of the soil mechanically arrests and retains the suspended particles of the sewage.

2 and 3. The porosity and physical attraction of the soil lead to the oxidation of the organic matter contained in the sewage, as instanced in the discovery of nitrates and nitrites in the effluent water, which did not exist previous to filtration.

4. A chemical reaction takes place between the constituents of the sewage and those of the soil.

If the charge brought against the system of irrigation, viz. that it is detrimental to the health and comfort of those who reside near sewage farms, cannot be denied, it seems pretty certain that, in most cases, any ill effects arising from the method may be traced to its defective management. The selection of the soil which is to receive the sewage is a highly important consideration. The best for this purpose seems to be a loose marl, containing oxide of iron and alumina ; but sand, as well as chalk, is said to answer excellently.

If the soil be of a stiff clayey nature it must be broken up and mixed with sand, lime, or ashes. The upper parts must be comminuted and rendered porous, and it must be efficiently and deeply drained. At Troedyrhiw, as we have seen, the effluent drain is six feet deep.

Results of Irrigation, in parts per 100,000.	Percentage of dissolved Organic Pollution removed.		Percentage of suspended Organic Pollution removed.
	Organic Carbon.	Organic Nitrogen.	
On fallow land at Chorley (adhesive loam) .	62·3	70·2	100·
At Edinburgh (both sand and clay) . . .	45·3	81·1	84·9
Barking (gravelly soil)	65·8	86·2	100·
Aldershot (light sand)—			
Best result	91·8	87·8	99·7
Worst result	69·9	82·9	87·7
Average result	80·9	85·1	93·7
Carlisle (light loam)	77·9	59·8	100·
Penrith (light loam)	75·0	77·2	100·
Rugby (adhesive soil)	72·3	92·9	96·0
Banbury (principally clay)—			
Best result	87·8	91·3	96·0
Worst result	64·1	80·1	90·3
Average result	76·	85·7	93·2
Warwick (stiff clay)	71·7	89·6	100·
Worthing (loam)	42·7	85·3	100·
Bedford (light gravelly soil), average result .	71·6	81·3	100·
Norwood (clay), average result . . .	65·0	75·1	100·
Croydon (gravelly soil)—			
Best result	73·2	93·2	100·
Worst result	61·6	90·4	100·
Average result	67·4	91·8	100·

The sewer water should be poured over the land in as fresh a condition as possible, having been previously deprived of any solid or grosser parts by straining. At Carlisle, decomposition of the sewage during its flow is prevented by adding carbolic acid to it. Lastly, it is of the utmost consequence that the amount of land used as the filtering medium shall be large. Letheby has shown that where this precaution is neglected not only is the purification of the sewage incomplete, but the plan becomes a public nuisance. The amount of filtering earth should not be less than one cubic yard for eight gallons of sewage in twenty-four hours in properly prepared soils; in some soils more than a cubic yard is required.

The late Dr Parkes has given a summary of various reports that have from time to time been issued as to the effects of sewage farms upon the public health and comfort. He says:—" That sewage farms, if too near to houses, and if not carefully conducted, may give off disagreeable effluvia is certain; but it is also clear that in some farms this is very trifling, and that when the sewer water gets on the land it soon ceases. It is denied by some persons that more nuisance is excited than by any other mode of using manure. As regards health, it has been alleged that these farms may—1st, give off effluvia which may produce enteric fever or dysentery, or some allied affection; or, 2nd, end in the spread of entozoic diseases; or, 3rd, make ground swampy and marshy, and may also poison wells, and thus affect health."

The evidence of Edinburgh, Croydon, Aldershot, Rugby, Worthing, Romford, and the Sussex Lunatic Asylum is very strong against any influence in the production of typhoid by sewage farms effluvia. On the other hand, Dr Clouston's record of the outbreak of dysentery in the Cumberland Asylum is counter evidence of weight; and so is one of the cases noted by Dr Letheby of typhoid fever outbreak in Copley, when a meadow was irrigated with the brook water containing the sewage of Halifax.

The negative evidence is, however, so strong as to justify the view that the effluvia from a well-managed sewage farm do not produce typhoid fever or dysentery, or any affection of the kind. In a case at Eton, in which some cases of enteric fever were attributed to the effluvia, Dr Buchanan discovered that the sewer water had been drunk; this was more likely to have been the cause.

With regard to the second point, the spread of entozoic diseases by the carriage of the sewer water to the land has been thought probable by Cobbold, though as solid excreta from towns have been for some years largely employed as manure, it is doubtful whether the liquid plans would be more dangerous. The special entozoic diseases which, it is feared, might thus arise, are tapeworms, round-worms, trichina, Bilharzia, and Distoma hepaticum in sheep. Cobbold's latest observations show that the embryos of Bilharzia die so rapidly, that even were it introduced into England there would be little danger.

The trichina disease is only known at present to be produced in men by the worms in the flesh of pigs which is eaten, and it seems doubtful whether pigs receive them from the land. There remain, then, only tapeworms and round-worms for men and Distoma hepaticum for sheep to be dreaded. With regard to these the evidence at present is negative; and though much weight must be attached to any opinion of Cobbold's, this argument against sewage irrigation must be admitted to want evidence from experience.

The third criticism appears to be true. The land may become swampy and the adjacent wells poisoned, and disease (ague, and perhaps diarrhœa and dysentery) be thus produced. But this is owing to mismanagement, and when a sewage farm is properly arranged it is not damp, and the wells do not suffer ('Practical Hygiene').

The foregoing processes for the removal of excreta from dwellings necessitate the joint employment of sewers and large quantities of water. It may, however, sometimes happen that the adoption of either of these appliances may be not only difficult, but altogether impracticable; as, for instance, in localities where a sufficient fall cannot be obtained for the sewers, or where the supply of water is not adequate; or when the severity of the climate at certain times is such, that for months in the year the water is frozen. Under these conditions the excreta must either be allowed to accumulate about houses, or else be removed by methods other than those we have described at more or less short intervals. Of course their speedy removal is the best and safest; but in cases where they are permitted to accumulate, it is essential they should be mixed with deodorants, and confined in properly constructed receptacles (as far as possible from dwellings), from which category such pre-eminently unsanitary arrangements as cess-pools and dead wells must be excluded.

When excreta are got rid of from houses by other means than those of sewers and water, the processes employed are termed—

2. DRY METHODS. These comprise—

(1) Removal of the excreta without admixture.

(2) Removal of the excreta after treatment with deodorising and antiputrescent substances.

(1) *Removal without admixture.* In some cases boxes and tanks receive the ordure and urine, and these are changed more or less frequently.

In Glasgow the excreta from a part of the city containing eighty thousand people is thus collected and removed without admixture, except that from the garbage of the houses, daily.

In Edinburgh there are also many closets supplied with moveable metal pails, which are likewise removed daily. Many large dwelling-houses in this latter city are entirely without water-closet accommodation; hence the custom of placing pails full of excrement, urine, &c., outside the houses to be taken away by the scavenger. In Rochdale the excrement, &c., is collected in tubs, with tight-fitting lids, which are emptied twice or thrice a week. These tubs are manufactured out of disused paraffin casks. In Leeds, also, the excreta are collected in boxes without being subjected to admixture. In some towns in the north of England the excreta fall into receptacles constructed upon what is termed the 'Goux' principle. In this system the pails or receptacles are lined with some absorbent lining, which abstracts the urine. The refuse of cloth manufacturers is chiefly used for this purpose. Another contrivance is to have the receptacle fitted with a pipe or drain; the object in each case being to render the fæces drier and to delay their decomposition.

The pail or tub system (*fosses mobiles*), which is employed in Belgium, has, for its object the collection of the fæces in a state of purity, without admixture with water, in a clean and odourless condition.

The apparatus for carrying it out consists of—

1. *The seat.* This consists simply of a soil-pan of stoneware or *faience*, without woodwork, the soil-pan merely projecting from the top of the descent pipe. Its borders are furnished with a groove filled with water or sand, into which the raised rim of the lid fits.

2. *The connecting pipe.* This pipe is straight, without a syphon, and joins the descent pipe at the very acute angle of 22°, and is about 4 inches in diameter inside. It is, like the next, made of stoneware, glazed inside.

3. *Descent pipe.* This is from 6 to 8 inches in internal diameter; it is vertical, and is composed of a series of pipes, connected with each other by dry sand joints, without cements, fixed to the wall by iron bands.

It rests at the ground-floor level on a strong flagstone. Its prolongation through and below this stone consists of a sliding pipe of wrought copper capable of being lengthened or shortened, and solidly fixed to the stone by a cast-iron connector. A sort of circular shallow dish (*douille*), which can be hung under this last part of the descent pipe, serves at a given moment to shut its lower orifice.

4. *Tub (tonneau).* The excremental matters coming down the descent pipe fall into a tub of from 2 to 3 hectolitres (44 to 66 gallons), in a hole in the top of which the lower part of the pipe fits tightly. A cover fitted with a spring serves to shut and lute the tub when it is full. Placed on a stand furnished with wheels, the tub is easily managed.

When filled it is immediately replaced by another similar contrivance. If the tub is underground, the rails (on which the stand moves) should be placed on an incline, so that the removal and replacement may be easily effected. The underground chamber must be isolated, and the entrance to it placed outside the building. The thorough tarring of the interior of the tub not only preserves the staves, but also partly neutralises the effect of the mephitic gases which the excremental matters discharge.

Ventilation pipe. To prevent the smells and gases which are given off from the mouth of the tub from spreading themselves (in the house) by means of the opening in the privy seat, at the upper extremity of the descent pipe is fixed a ventilation pipe, which rises above the coping of the roof, and the action of which is increased by means of a vane, or any other contrivance producing the same effect (*Corfield*).

It is said that in the working of any of the above processes little or no nuisance ensues, if only ordinary care and intelligence are used. In many cases the excreta collected by the methods above specified are conveyed to manufactories and then converted into manure.

It does not appear that in England the health of the workmen employed in a manure manufactory or of those who live in the neighbourhood of it suffers in consequence.

(2) *Removal of the excreta after treatment with deodorising and antiputrescent substances.* This is the method usually adopted when the dry process is followed, the excreta mixed with the deodorising substance when removed from the house being at once applied to the land.

a. Coal and wood ashes. It is a common practice in the north of England to throw coal ashes on the excreta, which fall into closets made with hinged flaps or seats for the purpose of admitting the ashes, as at Manchester and Salford. Wood ashes are far more effective deodorisers than coal ashes, but they are seldom procurable. "In some towns there are receptacles called 'middens,' intended both for excreta and ashes; sometimes these are cemented, and there may be a pipe leading into a sewer so as to dry them. The midden system is a bad one; even with every care, the vast heaps of putrefying material which accumulate in some of our towns must have a very serious influence on the health, and the sooner the middens are abolished the better."

b. Deodorising powders. At some of the Indian stations deodorants, such as M'Dougall's or Calvert's carbolic acid powders, have been successfully employed, a comparatively small quantity being mixed with the excreta.

In Germany a mixture of lime, chloride of magnesium, and tar is largely used for the same purpose, and is known as 'Süverns' deodoriser.'

Another deodoriser (the Müller Schür), also used in the dry method, is composed of lime, 100 lbs.; powdered wood charcoal, 20 lbs.; peat powder or sawdust, 10 lbs.; and carbolic acid (containing 60% to 70% of real acid), 1 lb. After having been mixed, the mass is placed under cover for a night to avoid any chance of spontaneous ignition, and when dry it is packed in barrels.

c. Charcoal. The powerfully deodorising properties of charcoal obviously adapt it for the removal of excreta in the dry state, after the admixture with them. But the comparatively high price of animal charcoal, although nearly six times the value of dry earth as a deodorant, prohibits its being extensively used. Peat is, however, cheaper than animal charcoal. To obviate the objection of cost, Mr Stanford, in 1872, proposed to make charcoal for this purpose from seaweed. The charcoal obtained from this source is said to be cheap and of great service as an excretal deodoriser. The mixed charcoal and sewage is sufficiently odourless to be stored for some months in a convenient receptacle outside a dwelling-house.

After the seaweed charcoal has become thoroughly impregnated with fæces and urine, the mixture is recarbonised in a retort, and the charcoal can be again used; the distilled products (ammoniacal liquor, containing acetate of lime, tar, and gas) are sufficient to pay the cost, and it is said even to yield a profit. About the same time carbonisation of sewage in retorts, with or without previous admixture with charcoal, was proposed by Mr Hickey, of Darjeeling. There can be little doubt that, regarded from a purely sanitary point of view, carbonisation of sewage matter is an excellent plan. Mr Hickey proposed the utilisation of the ammoniacal products resulting from his process.

d. Dried earth. The Rev. Mr Moule was the

first to direct attention to the value of dried earth as a deodorant of excreta.

Mr Moule's 'earth closet' consists of a box with a receptacle below for the excreta. By pulling a plug dried earth, which is placed in a hopper above, enters the closet and falls upon the excreta, thus disinfecting and deodorising them. The consumption of earth averages from 1¼ lbs. to 1¼ lbs. a day. The slop water should not be thrown into the closet, but disposed of in some other way. In another plan, as in Taylor's improved closet, the urine is carried off without mixing at all with the fæces.

Clay, marl, and vegetable humus form the best kind of earths. When dried the clay may be easily reduced to powder. Chalk and sand are comparatively useless. The receptacle is emptied from time to time, the contents forming a valuable manure.

The earth closet is more particularly adapted for small villages and isolated mansions. One difficulty of its application by cottagers consists in the necessity of collecting, drying, and storing the earth; the limited space in the cottager's dwelling not permitting this. One great obstacle to the effective carrying out of this system amongst extensive communities is the difficulty of procuring the large supply of earth that its adoption necessitates. With proper supervision and care the 'earth system' answers admirably; if these are not bestowed on it, it as signally fails. It has been adopted with great success in many schools, barracks, and other large buildings.

"It is coming into great use in India, and is carried out with great attention to detail. In those European stations where water is not procurable Mr. Moule's invention has been a boon of great value, and medical officers say that nothing has been done in India of late years which has contributed so much to the health and comfort of the men. The plan of separating the urine from the fæces has been strongly advocated by Dr Cornish, of Madras, and would, no doubt, be attended with great advantages in India if there are means of disposal of the urine. The chief difficulty in the European barracks in India is felt during the rainy season, when the mixed excreta and earth cannot be kept sufficiently dry. In the case of natives of India, however, a serious difficulty arises in the use of the earth system, in consequence of the universal use of water for ablution after using the closet. Every native takes with him a small vessel holding ten to twenty ounces of water, so that a large amount of fluid has to be disposed of. The usual earth closet does not suffice for this. Mr Charles Turner, C.E., of Southampton, has contrived a closet suitable for the native family; it is unfortunately too costly, and possibly a simple iron box, with a pipe to carry off the urine and ablution water, would be better suited for the poorer classes" (*Parkes*).

e. Captain Lieurnur's pneumatic plan. This process, the invention of a Dutch engineer, is in use at Amsterdam, Leyden, Dordrecht, and a few other Continental towns. It is also known as the 'aspiration plan.' Its outlines are as follows:—
"The pipes and tubes leading from the various water-closets and privies peculiar to the system

are connected with street mains, which mains again communicate with underground horizontal cast-iron cylinders or tanks, these tanks being directly connected with a powerful air-pump worked by steam. Communication between the main and the tanks, as well as between the tanks and the pump, can be made or broken by means of stopcocks. Hence it follows that when access is allowed between one of the tanks and the air-pump, this latter will, when put into action, produce a vacuum in the tank, and if the stopcock of the main leading to the tank be then opened, the contents of all the privies and water-closets the pipes of which run into the main will be removed by being swept into the tank by pneumatic force. In this manner each tank is treated in succession. Similarly the sewage is carried to the large reservoirs of a manure manufactory. It is here mixed with a little sulphuric acid to prevent the formation of ammonia, and being evaporated down *in vacuo* becomes converted, when sufficiently dry, into poudrette. In Lieurnur's process all deodorants are dispensed with, and its mixture with water is prevented by means of porous drain-pipes laid above the sewers, by which contrivance the subsoil water is kept out of the sewers."

Sewage, Utilisation of. "Mr. Peregrine Birch read before the Institution of Surveyors a paper on 'The Use of Sewage by Farmers,' which embodied some facts that deserve to be noticed, as bearing on a question we have repeatedly discussed. It appears that there are at the present time 'upwards of one hundred owners and occupiers of land in Great Britain who use sewage for the sake alone of what they can get out of it by agricultural means.' Of these 'more than sixty are tenant farmers, who continue to use it although they have, annually at least, the option of ceasing to do so.' It seems five out of six of the tenant farmers purchase the sewage they employ, so that their adhesion to the method proves conclusively that it *pays*. Nearly four thousand acres of land are under regular cultivation with sewage. Mr Birch is of opinion that 'advocates of sewage precipitation processes should not regard sewage farmers as their rivals, for a chemical process might be very largely used with advantage when farmers are being persuaded or taught to use sewage. But this should be the distinct aim of all cultivation, for there is no chemical process that could not be worked to greater advantage during two months of the year than twelve, or applied to a small quantity of sewage at less cost than to a large.' Our primary interest is to see the utilisation of sewage generally adopted; the method employed must be determined by experience on the grounds of cheapness and expediency" ('Lancet').

A vast amount of very valuable information on this subject will be found in the 'First and Second Reports of the Royal Commission on Metropolitan Sewage Discharge, 1884' (C.—3842, 9d., and C.—4253, 8½d.).

SHADDOCK. A large species of orange, the fruit of *Citrus decumana*, Linn.

SHAGREEN'. This is prepared from the skins of the horse, wild ass, and camel, as follows:—
The skin is freed from epidermis and hair by soak-

ing in water, and, after dressing with the currier's fleshing-knife, is sprinkled over, whilst still wet and stretched, with the seeds of a species of *Chenopodium*, which are embedded in it by strong pressure, and in this state it is dried; the seeds are then shaken off, and the surface rubbed or shaved down, nearly to the bottom of the seed-pits or indentations; it is next soaked in water, by which the skin swells, and the recently depressed surface rises into a number of minute prominences; it is, lastly, dyed and smoothed off. Black is given to it with galls and copperas; blue, with a solution of indigo; green, with copper filings and sal-ammoniac; and red, with cochineal and alum. Shagreen was formerly very extensively used for covering the cases of watches, spectacles, surgical instruments, &c.

SHAKER EXTRACT. According to the manufacturers' statements, it represents an extraction of *Iris versicolor, Leptandra virginica, Stillingia officinalis, Juglans regia, Gaultheria procumbens, Taraxacum, Actæa racemosa, Gentiana rubra, Hydrastis canadensis, Euonymus atropurpureus, Capsicum annuum,* aloes, and sassafras, to which borax, hydrochloric acid, sugar, and podophyllin are added. Hager gives the following formula:—Ext. gentian, 5 dr.; ext. centaurii, marrubii, aurant. cort., tormentilli, of each, 2 dr.; aloes, 20 gr.; borax, 40 gr.; aq. cinnamomi, 2 oz.; aq. rosæ, 3 oz.; tinct. capsici, 1 oz.; ol. sassafras, 5 minims; ol. anisi stellat., 2 minims; acid. hydrochlor., 1 dr.

SHALLOT'. *Syn.* ESCHALOT. The *Allium ascalonicum,* Linn., a plant allied to the onion, the bulb of which is much used as a sauce or pot-herb.

SHAMPOOING. A practice common in the East, having for its object the increase or restoration of the tone and vigour of the body, or the mitigation of pain. It is applied either in the bath or immediately after quitting it, generally the latter, and consists in pressing and kneading the flesh, stretching and relaxing the knee-joints, and laboriously brushing and scrubbing the skin.

SHAMPOO LIQUID. *Prep.* 1. Sapo mollis (B. P.), 1 oz.; liquor potassæ, 2 oz.; rectified spirit, 2 oz.; perfume, q. s.; water, to 20 oz. Dissolve the soap in the water by aid of heat, add the potash, and when cold, the spirit and perfume.

2. Soft soap, ½ oz.; powdered borax, 1 dr.; ammonia, 1 dr.; eau de Cologne, ½ oz.; boiling distilled water, 20 oz. Dissolve the soft soap and borax in the water, and when cold add the ammonia and eau de Cologne.

SHAMPOO POWDER. *Prep.* Borax powder, 6 dr.; calcined soda, 1 oz.; quillayine, ½ oz.; perfume. It is very quickly dissolved by the use of 1 quart of warm water.

SHARPS. See FLOUR.

SHAVING. The following are Mr Mechi's instructions for this, to many persons, troublesome operation:—Never fail to well wash your beard with soap and cold water, and to rub it dry, immediately before you apply the lather, of which the more you use the easier you will shave. Never use warm water, which makes a tender face. Place the razor (closed, of course) in your pocket, or under your arm, to warm it. The

moment you leave your bed is the best time to shave. Always put your shaving-brush away with the lather on it.

The razor (being only a very fine saw) should be moved in a sloping or sawing direction, holding it nearly flat to your face, care being taken to draw the skin as tight as possible with the left hand, so as to present an even surface and throw out the beard. The practice of pressing on the edge of a razor in stropping generally rounds it; the pressure should be directed to the back, which must never be raised from the strop. If you shave from heel to point of the razor, strop it from point to heel; but if you begin with the point, then strop from heel to point. If you only once put away your razor without stropping or otherwise cleaning the edge, you must no longer expect to shave well, the soap and damp so soon rust the fine teeth or edge. A piece of plate leather should always be kept with the razors.

SHAVING FLUID. See ESSENCE OF SOAP.

SHAWLS, to Scour. Scrape 1 lb. of soap into thin shavings, and let it be boiled with as much water as will convert it into a thin jelly. When cold beat it with the hand, and mix with it three table-spoonfuls of oil of turpentine, and one of hartshorn. Let the shawl be well rubbed in this mixture, and afterwards rinsed in cold water, so as to get rid of the soap.

Next let the shawl be rinsed in salt and water, then wring out the water from it, and fold it between two sheets, being careful not to allow two folds of the shawl to lie together; finally mangle and iron with a cool iron.

SHEEP. *Syn.* OVIS, L. The *Ovis aries,* an animal domesticated almost everywhere. Its flesh supplies us with food, its skin with leather, its fleece with wool, and its intestines with catgut. Its fat (sevum) is officinal. See MUTTON, SUET, &c.

Sheep Washes. 1. Arsenious acid in powder, carbonate of potash, of each, 6 oz.; water, 14 galls. Boil together for half an hour.

2. Arsenious acid in powder, soft soap, and carbonate of potash, of each, 6 oz.; sulphur, 4 oz.; bruised hellebore root, 2 oz.; water, 14 galls. Boil the ingredients in a portion of the water for half an hour, or until the arsenic is dissolved, then add the remainder of the water, and strain through a coarse sieve. Mr Youatt says:—"More care than is usually taken should be exercised in order that the fluid may penetrate to every part of the skin, and which should be ensured by a previous washing in soap and water. The arsenic that necessarily remains about the wool when the water has dried away would probably destroy the acari as fast as they are produced. When a greater quantity of arsenic has been used, or the sheep has been kept too long in the water, fatal consequences have occasionally ensued."

3. A sheep-dipping composition employed on the Continent is—Arsenious acid, 1 lb.; sulphate of zinc, 10 lbs.; dissolved in 25 galls. of water.

4. The Australian sheep farmers use a weak solution of bichloride of mercury (1 oz. of the bichloride to 4 galls. of water).

5. Water, 40 parts, at the temperature of 50° to 57° C.; to this add 1 part of soluble glass

(the soluble silicates). This is recommended as a very efficient and perfectly safe sheep wash by Messrs Baerle and Co., of Worms. In washing the sheep with this preparation care should be taken to cover the eyes of the animal with a bandage, to perform the washing with the solution instantaneously, and to remove the surplus with tepid water.

"Yards into which newly clipped sheep are to be turned should be previously cleared of all green food, hay, and even fresh water; if perfectly empty they are still safer. When the dipping is finished they should be cleansed, washed, and swept, and any of the unused dipping solution at once poured down the drains. Dipped sheep should remain, if possible, in an open exposed place, as on a dry road, or in a large open yard. Over-crowding should be avoided, and every facility given for rapid drying, which is greatly expedited by selecting for the operation fine, clear, drying weather. On no account should sheep be returned to their grazings until they are dry, and all risk of dripping over" (*Finlay Dunn*).

SHELLAC. See LAC.

SHELL-FISH. The common name for the crustacean and molluscous animals that are used for food. 'Shell-fish' are extremely liable to disturb the functions of the stomach and bowels. The oyster (*Ostrea edulis*) and the cockle (*Cardium edule*) are, perhaps, the least objectionable. The crab (*Cancer pagurus*), the crayfish (*Astacus fluvialis*), the lobster (*Homarus vulgaris*) the mussel (*Mytilus edulis*), the prawn (*Palæmon serratus*), the periwinkle (*Littorina littorea*), and the shrimp (*Crangon vulgaris*), with the exception of the claws of the first three, are always suspicious, particularly in hot weather, and often absolutely poisonous. We have seen the most alarming, nay, fatal symptoms follow the use of mussels, even amongst those habitually accustomed to take them; whilst it is a well-known fact that the luscious bodies of the crab and lobster have too often formed the last supper of the epicure. See OYSTER, &c.

SHELLS (Prepared). *Syn.* TESTÆ PRÆPARATÆ (Ph. L. 1836), L. *Prep.* (Ph. L. 1836.) Wash oyster-shells (OSTREÆ—Ph. L.) with boiling water, having previously freed them from extraneous matters; then prepare them in the manner directed for chalk. The product is similar in constitution and properties to prepared chalk.

Shells, to Polish. 1. The surface of the shell should be first cleaned by rubbing it over with a rag dipped in hydrochloric acid till the outer dull skin is removed. It must be then washed in warm water, dried in hot sawdust, and polished with chamois leather. Those shells which are destitute of a natural polished surface may be either varnished or rubbed with a mixture of tripoli powder and turpentine, applied by means of a wash-leather, after which fine tripoli alone should be used, and, finally, a little olive oil, the surface being brought up with the chamois leather as before.

2. "The shells are first boiled in a strong solution of potash, then wound on wheels, sometimes through one stratum to show an underlying one,

then polished with hydrochloric acid and putty powder. In this operation the hands are in great danger. Shell-grinders are generally almost all cripples in their hands" (*Spon*).

SHER'BET. [Pers.] A cooling drink, used in the East, prepared with the juices of fruit, and water, variously sweetened and flavoured. The word has been, of late years, commonly employed in these countries in a similar manner. See LEMONADE, ORANGEADE, and POWDERS.

SHER'RY. *Syn.* SHERRY WINE, SHERRIS; VINUM XERICUM (Ph. L.), VINUM ALBUM (Ph. E.), VINUM ALBUM HISPANICUM (Ph. D.), L. This is the only wine ordered in the British Pharmacopœias. See WINES.

SHERRY-COBBLER. *Prep.* (*Redwood*.) Half fill a tumbler with clean pounded ice; add a table-spoonful of powdered white sugar, a few thin slices of lemon with the peel (or some strawberries or other similar fruit, bruised), and a wine-glassful or more of sherry wine; mix them together (lightly), and as the ice melts, suck the liquor through a straw (or a small tube of silver or glass).

Obs. A favourite American drink; very refreshing in hot weather.

SHIN"GLES. *Syn.* ZOSTER, HERPES ZOSTER, HERPES ZONA, L. A local variety of herpes or tetter, remarkable for forming a kind of belt round or partly round some part of the trunk of the body, chiefly the waist or abdomen. See TETTERS.

SHODDY. The epithet (we believe of American origin) is applied to the old, used-up wool and cloth, fraudulently mixed with fresh woollen fabrics. A plan for the examination of a fabric suspected of containing shoddy has been given by a German chemist, Herr Schlesinger, and is as follows:—Examine it with the microscope, and note if it contains cotton, silk, or linen, as well as wool. If so dissolve them by ammoniacal solution of copper. A qualitative examination is thus obtained. Then direct attention to the wool. In shoddy both coloured and colourless fibres are often seen, the fibres having been derived from different cloths which have been partially bleached; the colouring matter, if any, instead of consisting of one pigment, will be composed of two or three different kinds, such as indigo, purpurin, or madder. Again, the diameter of the wool is never so regular as in fresh wool, but is seen to vary suddenly or gradually in diameter, and suddenly widens again with a little swelling, and tapers off again, besides which the cross markings or scales are almost always absent. When shoddy wool is placed in liquor potasse it is much more speedily attacked than new wool.

SHOT METAL. *Prep.* From lead, 1000 parts; arsenic, 3 parts. When the lead is coarse, 6 to 8 parts of metallic arsenic are required to fit it for this purpose.

SHOW BOTTLES. The large ornamental carboys and jars filled with coloured liquids, and displayed in the shop windows of druggists, may be noticed under this head. They are striking objects when the solutions they contain are bright and of a deep pure tint, especially at night, when they are seen by transmitted light. The following formulæ for the solutions have been recommended by different persons:

AMBER. From dragon's blood (in coarse powder), 1 part; oil of vitriol, 4 parts; digest, and, when the solution is complete, dilute the mixture with distilled or soft water, q. s.

BLUE. a. From blue vitriol, 2 oz.; oil of vitriol, ½ oz.; water, 1 pint. b. A solution of indigo in sulphuric acid, diluted with water, q. s. c. A solution of soluble Prussian blue in either oxalic or hydrochloric acid, slightly diluted, and afterwards further diluted with water to the proper shade of colour.

CRIMSON. a. From alkanet root, 1 oz.; oil of turpentine, 1 pint. Used chiefly for the bull's-eyes of lamps. b. As PINK (b), below.

GREEN. a. From sulphate of copper, 2 oz.; bichromate of potash, 1 dr., or q. s.; water, 1 pint. b. A solution of sulphate of copper, 2 oz.; chloride of sodium, 4 oz.; water, 1 pint, or q. s. c. A solution of distilled verdigris in acetic acid, diluted with water, q. s. d. Dissolve blue vitriol in water, and add nitric acid until it turns green.

LILAC. a. Dissolve crude oxide of cobalt (zaffre) in nitric or hydrochloric acid, add sesquicarbonate of ammonia in excess, and afterwards sufficient ammonio-sulphate of copper to strike the colour. b. As the purple, but more diluted.

MAGENTA. Acetate of rosaniline, dissolved in water, q. s.

OLIVE. Dissolve sulphate of iron and oil of vitriol, equal weights, in water, and add of nitrate of copper, q. s. to strike the colour.

ORANGE. a. A solution of bichromate of potash in water, either with or without the addition of some hydrochloric or sulphuric acid. b. Dissolve gamboge or annotta in liquor of potassa; dilute with water, and add a little spirit.

PINK. a. To a solution of chloride or nitrate of cobalt in water, add sesquicarbonate of ammonia, q. s. to dissolve the precipitate at first formed. b. From madder (washed with cold water), 1 oz.; sesquicarbonate of ammonia, 4 oz.; water, 3 pints; digest, with agitation, for 24 hours, then dilute with more water, and filter.

PURPLE. a. A solution of sulphate of copper, 1 oz. in water, 1 quart, or q. s., with the addition of sesquicarbonate of ammonia, 1½ oz. b. To the last add a sufficient quantity of the first pink (above) to turn the colour. c. To an infusion of logwood, add carbonate of ammonia or of potassa, q. s. d. Sugar of lead, 3 oz.; powdered cochineal, 1 dr.; water, q. s. e. Add sulphate of indigo, nearly neutralised with chalk, to an infusion of cochineal till it turns purple.

RED. 1. a. Dissolve carmine in liquor of ammonia, and dilute with water. b. Digest powdered cochineal in a weak solution of ammonia or of sal-ammoniac, and afterwards dilute with water. c. Add oil of vitriol, 4 oz., to water, 1 gall., and digest dried red rose leaves, 8 oz., in the mixture for 24 hours. d. Dissolve madder lake in a solution of sesquicarbonate of ammonia, and dilute the solution with water.

2. a. Potassium sulphocyanide, 10 gr.; liq. ferri perchlor. fort. ♏x; water, 1 gall. (evanescent but beautiful). b. Cobalt nitrate, 1 oz.; carbonate of ammonia, q. s.; water, 1 gall. Dissolve the cobalt nitrate in 2 pints of water, and add a strong

solution of the ammonia salt until the precipitate formed is re-dissolved. Then dilute with the rest of the water. This is permanent.

VIOLET. To a solution of nitrate of cobalt in a solution of sesquicarbonate of ammonia, add solution of ammonio-sulphate of copper, q. s. to strike the colour.

YELLOW. a. A solution of sesquioxide or rust of iron, ½ lb., in hydrochloric acid, 1 quart, diluted with water. b. To a strong decoction of French berries add a little alum. c. A simple solution of chromate or bichromate of potash in distilled water. d. A solution of equal parts of nitre and either chromate or bichromate of potash in water.

Obs. Most of the above require filtering, which should be done through powdered glass, placed in a glass funnel, and never through paper. They usually need a second filtration, after being exposed to the light for some weeks; hence it is convenient always to make a little more of them than is required to fill the bottle, as several of them, when diluted after filtration, become again turbid. Distilled water or filtered rain water should be used.

SHRIMP. See SHELL-FISH.

SHRUB. A species of concentrated cold punch, prepared with lemon juice, spirit, sugar, and water. When the word is used in its unqualified form, RUM SHRUB is alluded to.

Shrub, Brandy. Prep. 1. Take of brandy, 1 gall.; orange and lemon juice, of each, 1 pint; peels of two oranges; do. of one lemon; digest for 24 hours, strain, and add of white sugar, 4 lbs., dissolved in water, 5 pints; in a fortnight decant the clear liquid for use.

2. As RUM SHRUB (below), but using brandy.

Shrub, Lemon. Syn. LEMONADE SHRUB. Concentrated lemonade, either with or without the addition of a little spirit. Used to make lemonade or lemon sherbet.

Shrub, Punch. Concentrated punch, made with equal parts of spirit and water. Used to make punch.

Shrub, Rum. Prep. 1. As BRANDY SHRUB, but substituting rum for brandy.

2. Take of rum, at proof, 84 galls. (or, if of any other strength, an equivalent quantity); essential oils of orange and lemon, of each, 2 oz., dissolved in rectified spirit, 1 quart; good lump sugar, 300 lbs., dissolved in water, 20 galls.; mix well by 'rummaging,' and gradually and cautiously add of Seville orange juice, or of a solution of tartaric acid in water, q. s. to produce a pleasant but scarcely perceptible acidity; next 'rummage' well for fifteen minutes, then add sufficient water to make the whole measure exactly 100 galls., and again 'rummage' well for at least half an hour; lastly, bung the cask down loosely, and allow it to repose for some days. In a fortnight, or less, it will usually be sufficiently 'brilliant' to be racked. The product is 100 galls. at 66 u. p.

Obs. Rum shrub is the kind in the greatest demand, and that having a slight preponderance of the orange flavour is the most esteemed. If wholly flavoured with lemon it is apt to acquire a kind of 'dead' or 'musty' flavour by long keeping. The substitution of a few gallons of brandy for a portion of the rum, or the addition, after racking, of about 1 oz. each of bruised bitter almonds, cloves, and cassia, the peels of about two dozen oranges, and a 'thread' of the essence of ambergris and vanilla, renders it delicious.

SIAL'OGOGUES. Medicines which increase the flow of saliva. Mercurials and pellitory of Spain belong to this class.

SICK'NESS. Nausea and vomiting frequently arise from the use of improper food, and other articles which offend the stomach; at other times it is symptomatic of some disease, as colic, cholera, dyspepsia, head affections, incipient fever, &c.; in which case the primary affection should be attended to. Nausea lowers the pulse, contracts the small vessels, occasions cold perspiration, severe rigors, and trembling; and diminishes, as long as it lasts, the actions, and even the general powers, of life. The act of retching, and vomiting more especially, on the contrary, rouses rather than depresses, puts to flight all the preceding symptoms, and often restores the system to itself.

The best remedies or palliatives in these affections are effervescing saline draughts, either with or without the addition of a few drops of tincture of henbane, or tincture of opium. A glass of genuine lemonade, iced, or a spoonful of crushed ice in a wineglassful of mint-water, is also very serviceable. Pepsin (*Bullock* and *Reynolds*) and oxalate of cerium are said to be most valuable remedies in the sickness of pregnancy. See DRAUGHTS, PREGNANCY, SEA-SICKNESS, &c.

SIFTING is to pulverulent substances what filtration is to liquids; but in this case the medium through which the substance passes is, usually, of a simpler and coarser description. Sieves are commonly employed for the purpose, which are fitted with silk or brass-wire gauze for fine purposes, and horsehair cloth or wire netting for coarser ones. Drum sieves are such as are furnished with covers and an enclosed space to receive the fine powder that passes through, by which dust and loss are prevented.

SIGHT, Effect of Gaslight on. The German Minister of Instruction issued some years ago a report on the influence of gaslight on the eye. The conclusion arrived at in this report—the result of frequent conference with well-known physicians—is that no evil results follow a moderate use of gas if the direct action of the yellow flame on the eye is prevented. For this purpose screens or shades are employed. Very great objections, however, exist to the use of zinc or lead shades, most evils affecting the eye being traceable to them. Their use, it is said, inevitably tends to blindness or inflammation, and other harmful effects. The milky-white glass shade is the best, as it distributes the light and has a grateful effect on the eyes. The burner should not be too close to the head, as congestions of the forehead and headaches result from the radiated heat. The glass plate below the gas, employed in some places, is especially useful for the purpose, as it causes an equal distribution of the light—necessary where a number are working at one burner,—prevents the radiation of heat, and tends to a steady illumination by shielding the flames from currents of air. In cases of highly inflamed eyes, dark blue globes can be very beneficially employed. With precautions of this kind no

evil effects from the burning of gas need be feared.

SIGNATURES (Fac-similes of). These may be readily obtained as follows:

1. Let the name be written on a piece of paper, and, while the ink is still wet, sprinkle over it some finely powdered gum-arabic, then make a rim round it, and pour on it some fusible alloy in a liquid state. Impressions may be taken from the plates formed in this way, by means of printing ink and the copper-plate press.

2. By the use of transfer ink and lithography.

SILBER LIGHT. This light is thus described in 'Dingler's Polytechnic Journal,' ccix, 79 ('Journ. Chem. Society,' vol. xi, new series, 1273). This mode of illumination is recommended where gas cannot be had.

The material used is oil, which is converted into gas before combustion takes place, whereby the combustion of the wick is greatly lessened (one wick may last a year), the accumulation of impurity is obviated, and the prevention of smell completely effected. The light is regular and uniform, and of a white colour. The light with a burner 1¼ inches wide, is equal to that of 28 sperm candles, each consuming 120 gr. per hour, and with one 1¾ inches wide a light is obtained equal to 50 such candles.

The burning apparatus consists of a row of concentrically enclosed double cylinders, perpendicularly arranged at definite intervals. The innermost cylinder contains the wick between its two walls, the hollow space in the interior serving to convey fresh air to the interior of the flame. The second cylinder conveys air to the outer side of the wick, and the third contains oil, and is in direct communication with both wick and reservoir. The mouths of all these chambers have a dome-shaped head, and from a suitable opening in this the gas streams forth in such a manner that it comes in contact with a current of air, and thus a complete combustion is attained.

According to the nature of the oil burnt the construction is somewhat varied in its minor details. Rape oil or light hydrocarbon oils are mentioned.

SIL'ICA. SiO₂. *Syn.* SILICIC ANHYDRIDE, SILICIC ACID, SILEX, SILICIOUS EARTH†, EARTH OF FLINTS†. This exists in quarts and rock crystal in a nearly pure state. Sand, flint, and almost all the scintillating stones chiefly consist of it.

Silica occurs under two conditions, the crystalline and the amorphous. The former variety has a sp. gr. of 2·642; the amorphous of 2·3. Some of our well-known native gems and precious stones consist almost wholly of one of the above forms of silica. In agate and chalcedony the two varieties are combined. Amethyst is silica coloured purple by ferric oxide. Onyx is formed of chalcedony arranged in layers of different colours. Carnelian is a red or brown variety of silica coloured with ferric oxide; whilst opal is amorphous silica combined with varying quantities of water. Silica is present in the stems of certain plants, such as wheat, and many grasses, to which the shining appearance of the stems is due, &c. The Italians polish marble with the ashes of burnt straw, the usefulness of which for such a purpose depends upon the silica contained in the straw; for similar reasons the Dutch rush is thus employed. Silica also occurs in solution in many natural waters. In the geysers, or boiling springs of Iceland, it exists in large quantity.

It may be obtained in a state of absolute purity by passing gaseous fluoride of silicon into water, collecting the resulting gelatinous precipitate on a calico filter, washing it with distilled water, drying it, and heating it to redness. Another method is to precipitate a solution of silicate of soda or potash (soluble glass) with dilute hydrochloric acid, and to treat the precipitate as before. Nearly pure silica may also be procured by heating colourless quartz to redness, and plunging it into cold water, by which treatment the quartz is rendered so friable as to be easily reducible to fine powder. Ordinary flints, subjected to this method, are found to yield silica in a condition approaching to purity. Amorphous silica is much more easily attacked by solvents than the crystalline variety. Most of the artificial forms of silica are amorphous, but crystals of quartz have recently been obtained by the action of water upon glass at a high temperature and pressure.

Tests, &c. The test for a silicate consists in fusing the suspected body with sodium or potassium carbonate, heating the residue with acid, and evaporating to dryness. If the residue be then treated with hot water the silica remains undissolved in the form of a white powder, which will yield a colourless bead when fused with sodium carbonate upon a piece of platinum-foil before the blowpipe flame. If silica be fused with borax it becomes slowly dissolved, forming a clear, colourless bead.

Chapman contests Plattner's opinion that, when silicates are fused with a phosphate, the 'silica skeleton' that results is especially due to the presence of alkalies or earthy bases.

Chapman says, "It is true enough that silicates in which these bases are present exhibit the reaction; but as other silicates—practically all, indeed—exhibit the reaction also, the inference implied in the above statement is quite erroneous. The opalescence of the glass arises entirely from precipitated silica.

"If some pure silica (or a silicate of any kind), in a powdered condition, be dissolved before the blowpipe flame in borax until the glass be saturated, and some phosphor salt be then added, and the blowing be continued for an instant, a precipitate of silicate will immediately take place, the bead becoming milky white (or, in the case of many silicates, opaque) on cooling. This test may be resorted to for the detection of silica in the case of silicates, which dissolve with difficulty in phosphor salt alone, or which do not give the pronounced 'skeleton' with that reagent" (Chapman on 'Blowpipe Reactions').

Prop., &c. A fine white, tasteless, infusible powder, insoluble in all acids, after being heated, except the hydrofluoric; requires the heat of the oxyhydrogen blowpipe for its fusion; approaches the precious stones in hardness; soluble in strong alkaline solutions; its salts are called SILICATES. The acid properties of this substance are very feeble. See GLASS; GLASS, SOLUBLE, &c.

Silica, Hydrates of. By pouring a dilute solution of sodium silicate into a considerable excess of hydrochloric acid the whole of the silica is retained in solution, together with the chloride of sodium formed by the action of the hydrochloric acid upon the soda. By subjecting this solution to dialysis (see DIALYSIS) the hydrochloric acid and chloride of sodium are removed, whilst the hydrate of silica is left behind, in solution, in the dialyser. Graham recommends a stratum of the liquid 4-10ths of an inch in depth, to be subjected for four or five days to dialysis, the water in the outer vessel to be changed every twenty-four hours.

If the solution so obtained be carefully evaporated down in a flask, any drying of the silicic acid at the edges of the liquid being prevented, a solution may be obtained containing 14% of silica. The solution has a very feebly acid reaction, and is without taste or colour. It cannot be preserved in the liquid state for more than a very few days, even in well-closed vessels, but becomes converted into a transparent gelatinous mass, which separates from the water. Hydrochloric acid, as well as small quantities of caustic potash or soda, retard the coagulation.

When the solution is evaporated *in vacuo* at 59° F. over sulphuric acid, a lustrous transparent glass is left behind, which consists of 22% of water, which closely accords with the formula SiO_3, H_2O.

By the action of moist air upon silicic ether a transparent glassy hydrate was obtained by Ebelmen, to which this chemist assigned the formula $2SiO_3, 3H_2O$. Two hydrates of silica were obtained by Fuchs, one having the formula $3SiO_3, H_2O$; the other, $4SiO_3, H_2O$. Evidence obtained recently appears to show that the solution left in the dialyser contains orthosilicic acid, H_4SiO_4.

Silicic Chloride. *Syn.* SILICIC TETRACHLORIDE. $SiCl_4$. This compound is rarely, if ever, obtained by the direct method, viz. by heating silicon in chlorine, but by the following indirect process:—A paste is made of finely divided silica, oil, and charcoal, and heated in a covered crucible. The fragments of the charred substance (consisting of silica and carbon) are then placed in a porcelain tube, which is raised to a red heat in a furnace, and during the ignition of the fragments a current of dry chlorine is passed over them; the silicic chloride which is thus formed being made to distil over into a bent tube surrounded by a freezing mixture of ice and salt, whereby it becomes condensed.

Silicic chloride is a strongly fuming gas, transparent and colourless, with an irritating and pungent smell. It is immediately decomposed by water into hydrochloric acid and hydrated silica, which deposits in the vessel.

Silicic Fluoride. *Syn.* SILICIC TETRAFLUORIDE. SiF_4. This gas is best prepared by heating in a capacious flask or retort equal parts of finely powdered fluor-spar and white sand, or glass, with ten or twelve times their weight of strong sulphuric acid. This gas must be collected over mercury, and in jars that are free from the least trace of moisture.

Silicic fluoride is a colourless gas, with a very

pungent odour, fuming strongly in the air, and neither burning nor supporting combustion. Faraday succeeded in liquefying it under great pressure, and Natterer states that at a temperature of —220° F. it may be solidified. By water it is partially decomposed and partially dissolved, yielding silicic acid and hydrofluosilicic acid.

Silicic Hydride. H₄Si. To procure this gas silicide of magnesium is decomposed with cold diluted hydrochloric acid.

The silicide of magnesium may be prepared as follows:—Mix intimately 40 parts of fused magnesium chloride, 35 parts of dried sodium silico-fluoride, and 10 parts of fused sodium chloride; these are mixed in a warm, dry tube, with 20 parts of sodium in small fragments, and thrown into a red-hot Hessian crucible, which is immediately covered, the operation being finished when the vapours of sodium cease to burn.

Silicic hydride becomes spontaneously ignited in the air, and in doing so gives off white fumes, which consist of amorphous silica (SiO₂). A cold body, such as a piece of porcelain or glass, introduced into the flame, becomes covered with a brown deposit of silicon. Passed into solutions of cupric sulphate, argentic nitrate, and palladium chloride, this gas throws down the metals, in most cases combined with silicon.

SILICO-FLUORIC ACID. See FLUOSILICIC ACID.

SIL'ICON. Si = 28·332 (Thorpe). Sym. SILICIUM. An elementary substance forming the base of silica. Next to oxygen it is the principal constituent of the earth's crust.

This element was first obtained by Sir Humphrey Davy, by acting upon silica with potassium. It is now procured much more easily by the decomposition of silico-fluoride of potassium, at an elevated temperature, with potassium or sodium. By heating a mixture of fluor-spar and ground flints with sulphuric acid a gaseous tetrafluoride of silicon is formed, which, being partially soluble in water, yields an acid solution of the tetrafluoride. Caustic potash is then added to the acid solution of the tetrafluoride until it becomes neutralised, and the sparingly soluble silico-fluoride of potassium thus formed is thoroughly dried and mixed in a glass or iron tube with 8- or 9-10ths of its weight of potassium or half its weight of sodium, and then heated. The following equation explains the reaction that takes place:

$$2KF,SiF_4 + 2K_2 = Si + 6KF.$$

The resulting mass, consisting of potassium fluoride, and silicon in partial combination with the excess of potassium, is treated with cold water, when a copious evolution of hydrogen gas ensues, owing to the decomposition of the water by the excess of potassium. The potassic fluoride is got rid of by washing with cold water, its entire removal being indicated by the water ceasing to have an alkaline reaction on test-paper, whilst amorphous silicon is left behind in the form of a brown powder.

Another method by which silicon may also be procured is by passing the vapour of silicic chloride over heated potassium or sodium, placed on a porcelain tray in a glass tube. In this operation it is advisable to protect the lining of the tube with thin plates of mica.

The silicon obtained by the above processes is known as amorphous silicon, and, as already stated, occurs as a brown powder. It is dull in colour, and being heavier than water, as well as insoluble in it, sinks in that fluid. It is a non-conductor of electricity, is unaffected by nitric or sulphuric acid, but dissolves readily in hydrofluoric acid, and in a warm solution of caustic potash. It burns with great brilliancy when heated in air or oxygen, and becomes converted into silica, which, owing to the great heat of combustion, fuses, and thus forms a superficial crust over the unburnt silicon.

A crystalline variety of silicon may be prepared by throwing a mixture of potassium silico-fluoride, 30 parts; granulated zinc, 40 parts; finely divided sodium, 8 parts, into a red-hot crucible kept at a temperature just below that of boiling zinc. The residue is then treated successively with hydrochloric acid, boiling nitric acid, and hydrofluoric acid, when dark glittering octahedral crystals remain behind.

By passing the vapour of silicic chloride over pure aluminium, placed on a porcelain tray, and raised to an intense heat, the aluminium becomes volatilised as aluminic chloride, whilst the silicon remains behind in crystals possessing a reddish lustre. These crystals occur in regular six-sided prisms, terminated by three-sided pyramids, derived from the octahedra, and are so hard that glass may be cut by them in the same way as by the diamond.

This variety will not take fire if heated strongly in the air or oxygen. Its density is such that it sinks in strong sulphuric acid, and hydrofluoric acid fails to dissolve it, although it is soluble in a mixture of hydrofluoric and nitric acids. It does not become oxidised, even if fused with potassium nitrate or chlorate, unless a white heat is obtained, when it burns brilliantly, giving rise, on so doing, to the formation of silica.

A graphitoid form of silicon, occurring in plates, has been described by Wöhler, who obtained this modification from an alloy of silicon and aluminium, which was treated in succession with boiling hydrochloric and hydrofluoric acids. The plates of silicon which are left have a metallic lustre, and a sp. gr. of 2·49. The graphitoid bears a great resemblance in properties to the crystalline silicon. It is a conductor of electricity. Like the crystalline variety, it dissolves in a mixture of hydrofluoric and nitric acids, although slowly, but, unlike the crystalline, it undergoes no change when heated to whiteness in a current of oxygen.

SILK. As an article of clothing, as far as "roundness of fibre, softness of texture, absence of attraction for moisture, and power of communicating warmth are concerned, silk is greatly superior to both linen and cotton; moreover, it gives the sensation of freshness to the touch which is so agreeable in linen. But, with all these advantages, silk (when worn next the body) has its defects; on the slightest friction it disturbs the electricity of the skin, and thus becomes a source of irritation. Sometimes, it is true, this irritation is advantageous, as causing a determination of blood to the surface; but when this action is not required it is disagreeable, and quite equal, in a sensitive constitution, to producing an eruption on the skin. I have seen eruptions occasioned in this manner, and, when they have not occurred, so much itching and irritation as to call for the abandonment of the garment" (*Eras. Wilson*).

Silk is characterised by its fibres appearing perfectly smooth and cylindrical, without depressions, even under a magnifying power of 160. Its fibres (even when dyed) acquire a permanent straw-yellow colour when steeped in nitric acid of the sp. gr. 1·20 to 1·30. The fibres of white or light-coloured silk are similarly stained by a solution of picric acid. A thread of silk, when inflamed, shrivels and burns with difficulty, evolves a peculiar odour, and leaves a bulky charcoal. By these properties silk is distinguished from cotton and linen.

Cotton, wool, and silk may be easily distinguished from each other by means of the microscope.

The cotton fibre will be seen to consist of only one cell; wool (as well as hair and alpaca) is made up of numerous cells in juxtaposition; whilst silk fibre is similar to the secreted matter of spiders and caterpillars.

The silk fibre (fig. 1) is smooth, cylindrical, devoid of structure, not hollow inside, and equally broad. The surface is glossy, and only seldom are any irregularities seen on it. If it is desired to detect in a woven fabric the genuineness of the silk, it is best to cut a sample to pieces, place it under water under the object-glass of a microscope magnifying 120 to 200 times, covering it with a thin piece of glass. The round, glazed, equally proportioned silk fibre (fig. 1) is easily distinguished from the unequalled and scaled wool fibre (w in fig. 2), and from the flat, bandlike, and spiral cotton fibre (b, fig. 3). Under the microscope also the mixture of inferior with superior fibres of silk can be easily detected.

Black silk, the weight of which has been

FIG. 1.   FIG. 2.   FIG. 3.

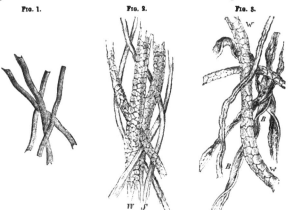

augmented by extensive sophistication, is not uncommon in English, French, and German markets. It is known as 'weighted' or 'shotted' silk, and very frequently contains no more—and frequently less—than one third of its weight of silk, the remaining two thirds consisting, according to Persoz, of a combination of iron salts with some astringent substance, salts of tin, and cyanides. It is easily distinguishable from genuine silk by its want of elasticity and tenacity, and its much greater combustibility. Persoz found a specimen of this adulterated silk to yield, upon incineration, more than 8 per cent. of ferric oxide.

The cleaning and renovation of articles of wearing apparel made of silk are matters requiring some care. No silk goods look well after being washed, however carefully it may be done; and this method should, therefore, never be resorted to but from absolute necessity. It is recommended to sponge faded silks with warm water and curd soap, then to rub them with a dry cloth

on a flat board, and afterwards to iron them on the wrong side with an ordinary smoothing iron. Sponging with spirit, benzol, or pure oil of turpentine, also greatly improves old silk, and is often preferable to any other method. The odour of the benzol passes off very quickly, that of the turpentine after exposure for a few days. When the ironing is done on the right side thin paper should be spread over the surface to prevent 'glazing.' See DYEING, GILDING, &c.

Silk Material, a New. The 'Textile Manufacturer' contains the following :—The utilisation of new substances as raw material for manufactures is a distinguishing feature of the scientific investigations of the nineteenth century. One of the most recent suggestions is the result of the researches of Herr Tycho Tulburg, an eminent German naturalist, on the products of the mussel. It will be remembered it was from one of the mussel species the famous purple dye was in past ages obtained, and this colour gained an imperishable renown from its being adopted by the Roman emperors, and the imperial purple became the symbol of sovereignty. In these latter days animal products have been displaced by aniline dyes, and there is no likelihood of their regaining their former celebrity. The researches of Tulburg have not, however, been in the direction of dyes, but in the adaptation of animal products other than the silkworm for silk yarns. The mussel (*Mytilus edulis*) fastens itself to the rocks by strong threads, called by naturalists *byssus*, and it is this substance which it is proposed to utilise for the manufacture of silk. The material is of a silky texture and very tough, and the experiments that have been made prove that it is well adapted to be made into yarn. Already the *Pinna*, one of the mussel tribe, has been manufactured into fabrics, although it is not of general use, nor at present of much commercial value, and the same obstacles to the use of the *byssus* of the common mussel are apparent. Notwithstanding the abundant supply of this popular shell-fish, it is difficult to see how a sufficient quantity of *byssus* can be collected to enable manufacturers to purchase the raw material at rates low enough for a marketable remuneration on the manufactured article. But the records of industrial progress testify to greater difficulties than these having been successfully overcome; and should the commercial value of the new material be satisfactorily demonstrated, there is no doubt some agency will be developed whereby the requisite supply may be obtained.

At present it is sufficient to notice the discovery that has been made, and to welcome another instance of the results of scientific labour being for the advantage of manufacturers.

SILKWORM, Diseases of. Silkworms are liable to a disease known as *pébrine*, which Pasteur has shown to be due to the presence, in the body, the egg, and in the blood of the insect, of peculiar parasitic corpuscles.

Pasteur states that the black specks which constitute these bodies are very easily distinguishable in the moth of the silkworm, but that in the earlier stages of its development, such as in the egg and worm condition, the detection of them becomes difficult, if not impossible. Pasteur further adds that sound moths produce sound eggs, and unsound moths the reverse; and that although the unsound eggs show no sign of disease, they never give rise to healthy worms.

Pasteur advises the silk cultivator, therefore, to ensure breeding from healthy moths at starting, and to abandon the old and useless precaution of hatching apparently healthy eggs.

The value of the cocoons grown in the whole world in 1870 was said to be as follows:—France, £4,334,000; Italy, £11,260,000; Spain and other European countries, £984,000; giving a total for Europe of £16,588,000. China, £17,000,000; India, £4,800,000; Japan, £3,200,000; Persia, £920,000; other Asiatic states, £2,192,000; giving a total for Asia of £28,112,000. Africa, £68,000. America, £20,000. Making a general total of £44,788,000.

The loss resulting from the silkworm disease in Italy may be seen from the following tables,

which are calculated for bales of 102 lbs. weight ('British Manufacturing Industries,' Stamford):

Average production prior to disease, 81,600.

1863	.	Bales, 50,600	.	Deficit, 38 per cent.
1864	.	„ 38,000	.	„ 53 „
1865	.	„ 38,700	.	„ 52 „
1866	.	„ 39,600	.	„ 51 „
1867	.	„ 44,000	.	„ 46 „
1868	.	„ 41,000	.	„ 49 „
1869	.	„ 47,300	.	„ 42 „
1870	.	„ 69,900	.	„ 14 „
1871	.	„ 76,300	.	„ 6 „
1872	.	„ 68,000	.	„ 16 „

Silkworm Gut. See GUT.

SIL'LABUB. *Prep.* Grate off the yellow peel of a lemon with lump sugar, and dissolve the sugar in ⅓ pint of wine; add the juice of ½ a lemon and a ¼ pint of cream; beat the whole together until of a proper thickness, and then put it into glasses.

Obs. ½ to 1 pint of new milk is often substituted for the cream, and strong cider or perry for the wine. Grated nutmeg is often added. When 'whipped' to a froth it is called 'WHIPPED SILLABUB.' See CREAM (Whipped).

SILPHA OPACA, Linn. (From Σιλφη, a grub or beetle.) THE BEET (CARRION) BEETLE. It is only comparatively recently that the mangel-wurzel plant has been cultivated to any important extent in this country. Its culture here has made a most rapid advance during the last thirty years, and therefore it is natural to find that insects injurious to this crop have increased proportionately. It seems also that certain insects have acquired the taste for this plant, so that they have forsaken their former food and have taken to it. *Steropus madidus* is one example of this. The insect which is the subject of this article, the *Silpha opaca*, or beet carrion beetle, is another instance of complete change of diet. It was supposed that this beetle lived upon carrion, as it and its larvæ were found in putrefying carcasses, as of moles, hedgehogs, birds, and other insects. Feeding on carrion is the attribute of the family of *Silphidæ*, to which this genus belongs, and yet it is clear beyond doubt that the larvæ of this, and at least another species, greedily devour the succulent young leaves of mangel-wurzel plants. Curtis reported this habit in 1844; Guérin Meneville also discovered the larvæ feeding upon beet plants in France in 1849.

Not much serious injury has been caused as yet to mangel-wurzel plants by this insect in this country. From time to time during the last ten years there have been occasional attacks, and in some cases it is believed that loss of plants caused by it has been placed to the credit of other insects.

Loss of plants was noticed in a field of mangels in the upper part of Kent. The leaves were eaten away, at least their juicy parts; their main ribs, or framework, being left. "Little black bugs like hop niggers," as the report stated, were found upon them, which turned out to be the larvæ of the *Silpha opaca*. These were very numerous, and were rapidly clearing away the plants.

Another intimation of damage to mangel-wurzel

plants by this insect was given with a note that the land had been manured with farmyard manure from an old mixen.

A friend in East Sussex sent some mangel plants in 1886 which were attacked by insects just after they had been singled. No specimens were forthcoming, but from the nature of the injuries seen on the leaves, and the description of the insects, there was not much doubt that they were *Silpha* larvæ.

It is certain that there are at least two kinds of *Silpha* which feed upon mangel leaves in this country, the *Silpha opaca* and the *Silpha atrata*, Linn. This is also Miss Ormerod's decided opinion. The *Silpha atrata* is known in Germany as injurious to beet. Taschenberg and Kaltenbach both state this. The latter says that in a district of Germany it reduced the leaves of beet plants to skeletons ('Die Pflansen-feinde,' von J. H. Kaltenbach, p. 509). Calwer speaks of *Silpha opaca*, Marsham, as being similar to *Silpha dispar*, Herbst, as known in Germany. He also describes *Silpha atrata* as synonymous with *S. rubrotus-data*, Stephens, and as stripping ("entblättern") beet plants of their leaves (C. J. Calwer, 'Käfer-buch,' pp. 92 and 93).

Silpha opaca is known in France as very destructive to beet plants. Brehm says that at different times these insects have made their appearance in the Département du Nord, where beet plants are largely cultivated for sugar-making purposes, and the larvæ were so thick upon these that they were quite black ('Les Merveilles de la Nature,' par A. H. Brehm).

The insect is not known in America, or, at least, there are no records of it.

Life History. The *Silpha opaca* is of the family *Silphidæ* of the order COLEOPTERA. It is a little black, egg-shaped beetle, five lines in length, having six black legs, with brownish claws. It is furnished with ample wings, and the body is covered with a silky grey pubescence, or down, which is soon rubbed off when touched. The beetle hibernates under stones and rubbish, and in the ground.

As soon as the weather becomes mild the female takes flight to suitable spots in which to place eggs, generally upon decaying leaves and other substances, and sometimes just under the surface of the ground. The larva hatches from the egg in about fourteen days, is black, about four and a half lines in length, and has twelve segments besides the head, and three pairs of feet. At the end of its body there are two points or spines. Its antennæ are long, and its jaws are furnished, Curtis remarks, with sharp minutely serrated teeth.

In the course of from fifteen days to three weeks the larva goes down deeply into the ground and constructs a cell, or shelter, in which it changes to a pupa, somewhat curved in form. At the end of ten days the perfect insect or beetle appears. It is said that there are two broods in a year, but this has not been proved.

Prevention. Manure intended for mangel-wurzel land should on no account be allowed to lie long in mixens after the heat has subsided, nor in lumps upon the ground before it is ploughed in. The beetles may be conveyed to the fields in this medium, or in early seasons the eggs might be carried out in the manure.

Farmyard manure should be well buried when put on for mangels, as it affords a convenient shelter and breeding-place for many insects.

Remedies. Soot applied when the dew is upon the plants is a valuable remedial measure against this insect. Quassia and soft soap washes would be most effective, it is fully believed. As the larvæ work quickly, it is very important that mangel-wurzel plants showing signs of failing should be promptly examined, and remedies tried at once ('Reports on Insects Injurious to Crops,' by Chas. Whitehead, Esq., F.Z.S.).

SILVER. Ag = 107·66. *Sys.* ARGENTUM, L. This metal, like gold, appears to have been as much valued in the remotest ages of antiquity of which we have any record as at the present time. It is found in nature, both in the metallic state and mineralised, in the state of alloy, and combined with sulphur, chlorine, bromine, &c. In Great Britain it is found in combination with lead. The largest supplies of silver come from the Mexican and Peruvian mines, but considerable quantities are furnished by Saxony, Hungary, and India. It is extracted from its ores principally by the process of amalgamation, founded on its easy solubility in mercury, and by subsequent cupellation.

Prep. 1. (*Liquation.*) The ore (generally argentiferous copper) is fused with three times its weight of lead, and the alloy cast into discs which are afterwards slowly heated upon a hearth which is so arranged that the lead, which melts more easily than the copper, flows off, carrying with it the silver. The lead and silver are afterwards separated by cupellation.

2. (*Cupellation.*) Founded upon the oxidation suffered by lead when heated in air, and the absence of any tendency upon the part of silver to combine directly with oxygen. The molten alloy is placed upon the hearth of a reverberatory furnace lined with moulded bone-ash, and then subjected to the action of a hot blast of air, which soon oxidises the lead, blowing it off as formed, leaving the silver pure.

3. (*Amalgamation.*) The silver is dissolved out from the crushed ore by mercury; the amalgam is placed in iron trays arranged one above the other, and covered with an iron till-shaped receiver standing over water. By heaping burning fuel round the upper part of the dome, its temperature is raised sufficiently to convert the mercury into vapour, which condenses again in the water, leaving the silver together with any lead and copper upon the trays. The lead and copper are then separated by cupellation, for the fused oxide of lead carries off with it the copper also in the form of oxide.

4. (*Net Process.*) Several have been proposed. Ziervogel's plan depends upon the fact that when copper argentiferous pyrites is soaked the copper and iron sulphides are converted into insoluble oxides, whilst the silver is converted into a soluble sulphate, which when the roasted ore is treated with hot water is dissolved; from this solution the silver is easily precipitated.

5. (*Chemically Pure.*) Ordinary silver is dissolved in pure nitric acid, the solution evaporated,

and the residue fused. The latter is then dissolved in weak ammonia solution, and the blue liquid which results diluted with water q. s. to bring the strength down to 2% of silver. Normal ammonium sulphite, q. s., is added to render the solution colourless on warming. After standing for 24 hours in a stoppered vessel, 1-3rd of the metal separates out in the crystalline form. The liquid, which is still blue, when cold, is passed off, and heated from 60° to 70° C., when the remainder of the metal is thrown down. The precipitate is washed, and then allowed to stand for several days in contact with strong ammonia; it is then again washed, dried, and fused in an unglazed porcelain crucible with 5% fused borax and 5% sodium nitrate, and finally cast in a mould lined with a mixture of burnt and unburnt kaolin. The bars are cleaned with sand and heated with potash, and finally washed in water (*Roscoe*).

Prop. Pure silver has a very white colour, a high degree of lustre, is exceedingly malleable and ductile, and is the best conductor of heat and electricity known. Its hardness is between that of copper and gold; its sp. gr. is 10·424 to 10·675; it melts at 1037° C.; is freely soluble in nitric acid, and dissolves in sulphuric acid by the aid of heat; it refuses to oxidise alone at any temperature, but, when strongly heated in open vessels, it absorbs many times its bulk of oxygen, which is again disengaged at the moment of solidification; its surface is rapidly tarnished by sulphuretted hydrogen and by the fumes of sulphur.

Tests. 1. The compounds of silver, mixed with carbonate of soda, and exposed on a charcoal support to the inner flame of the blowpipe, afford white, brilliant, and ductile metallic globules, without any incrustation of the charcoal. 2. The salts of silver are non-volatile and colourless, but most of them acquire a more or less black tint by exposure to full daylight.

The soluble salts of silver give—1. A white curdy precipitate with hydrochloric acid and the soluble metallic chlorides, which is soluble in ammonia, insoluble in nitric acid, and blackened by exposure to light. 2. White precipitates with solutions of the alkaline carbonates, oxalates, and ferrocyanides. 3. Yellow precipitates with the alkaline arsenites and phosphates. 4. With the arseniates, red precipitates. 5. With the caustic alkalies, brown precipitates. 6. With sulphuretted hydrogen and ammonium sulphide, a black precipitate, which is insoluble in dilute acids, alkalies, and cyanide of potassium, but readily soluble, with separation of sulphur, in boiling nitric acid. And 7. With phosphorus, and with metallic copper or zinc, pure reduced silver.

Assay. 1. The method of assaying silver by cupellation has been explained under ASSAY and CUPELLATION; and that method is alone applicable when the alloy contains a very small quantity of silver, as a few ounces only per ton. When the reverse is the case, as with the silver of commerce, the following is a much more accurate method.

2. *Humid Assay of Silver.* a. Dissolve 10 gr. of the silver for assay in 100 gr. of nitric acid, sp. gr. 1·28, by the aid of heat, the solution being

made in a tall stoppered glass tube, furnished with a foot; then place it in a very delicate balance, bring it into an exact state of equilibrium, and add the test solution (see *below*), gradually and cautiously, until the whole of the silver be thrown down; the number of grains now required to restore the equilibrium of the balance or scales gives the exact quantity of pure silver in 1000 parts of the sample.—*Obs.* To ensure accuracy, after each addition the stopper should be placed in the tube, and the latter violently agitated for a short time, when the liquor will rapidly clear and enable the operator to see when the operation is concluded. A small quantity of a solution of nitrate of silver must then be added to the liquor in the tube, after having first carefully taken the weight; this will serve as a check. If too much of the test liquor has been added, this will produce a fresh precipitate, and the assay cannot then be depended on. Instead of weighing the quantity of test liquor used, a tube graduated into 100 parts, and holding 1000 gr., may be employed, every division of which required to throw down the silver will represent 1-10th of a grain. See ALKALIMETRY and ACIDIMETRY.

b. The precipitate or chloride of silver may be collected in a paper filter, dried, washed, fused, and weighed. The previous weight of the paper, deducted from the gross weight of the filter and its contents, gives the quantity of chloride of silver present, which multiplied by 0·75278 gives the weight of the pure silver in the sample.

Test Solution. Dissolve 54·27 (54¼) gr. of sodium chloride in 9945·73 gr. (or 22 oz. and 320⅘ gr. avoirdupois) of distilled water; filter, and keep the filtrate in a stoppered bottle for use. Pure sodium chloride is obtained by boiling together for a few minutes, in a glass vessel, a solution of common salt with a little pure bicarbonate of soda; then adding to the filtered liquid sufficient hydrochloric acid to render it neutral to litmus and turmeric paper, and, lastly, evaporating and crystallising.

Obs. The presence of mercury, lead, or sulphide of silver interferes with the accuracy of the above assay. When mercury is present the precipitate blackens less readily by exposure to light; and when it contains 1/1000 or 1/100 of chloride of mercury it remains of a dead white; with 1/1000 it is not sensibly discoloured by the diffused light of a room; with 1/1000 only slightly darkened; with 1/1000 more so; but with pure chloride of silver the effect is very rapid and intense. When this metal is present—which is, however, seldom the case—the assay sample must be placed in a small crucible, and exposed to a full red heat before solution in the acid. Another method, proposed by Levol, and modified by Gay-Lussac, is to add to the nitric acid solution of the silver sufficient acetate of ammonia or crystallised acetate of soda to saturate all the nitric acid existing in the liquid, either in the free state or combined with the silver. When the alloy contains lead, shown by the precipitated chloride being partly soluble in water, it may either be laminated and subjected to the action of acetic acid before solution in the nitric acid, or the test solution of chloride of soda should be replaced by one of chloride of lead (189·355 gr. of the latter are equiv. to 58·782 gr. of the former). The presence of sulphide of silver is detected whilst dissolving the sample in nitric acid, by the black flakes which may be observed floating about in the liquor in an insoluble state. These flakes may be dissolved by fuming nitric acid, or by adding pure concentrated sulphuric acid to the solution, which should be then heated for about a ¼ hour in a steam-bath. When thus treated the precipitate produced by the test liquid represents the whole of the silver contained in the alloy.

Uses, &c. Metallic silver, unless in a state of very minute division, has no action on the human body. A plate of silver is ordered, in the Ph. L., as a test of the presence of nitric acid in the acetic and phosphoric acids; and metallic silver (preferably granulated) is employed by the other colleges in the preparation of the nitrate. Pure silver is used in volumetric analysis and in the preparation of laboratory utensils, for it is not attacked by fused caustic alkali. Its numerous applications in the arts are well known. The silver coinage of England contains 92·5% silver and 7·5% of copper.

Concluding Remarks. The researches of Tillet, D'Arcet, and Gay-Lussac have clearly shown that the percentage of silver in an alloy, as indicated by cupellation, is always below its real richness in that metal, owing to loss in the process; and that the cupelled button always retains a trace of lead and copper, the precise quantity of which is variable. The following table exhibits the additions to be made on this score when the quantity assayed (assay pound) is 20 gr. :

Weight after cupellation.	Actual richness in pure silver.	Percentage of richness in pure silver.
19·979	20	100
18·95	19	95
17·92	18	90
16·917	17	85
15·914	16	80
14·91	15	75
13·905	14	70
12·905	13	65
11·906	12	60
10·906	11	55
9·906	10	50
7·921	8	40
5·948	6	30
3·949	4	20
1·982	2	10

In assaying lead ores very poor in silver the best quantity to be taken for cupellation is 500 gr.; and from that quantity 0·0148 of silver, including compensation for loss, represents one ounce of silver to the ton. A cupel may absorb its own weight of lead. If the quantity of lead to be absorbed is more considerable, another cupel may be inverted, and the cupel in which the assay is to be made may be placed upon it. See ASSAY, and M. Gay-Lussac's elaborate memoir on the ' Humid Assay of Silver.'

For the recovery or reduction of silver from the chloride and its other compounds, several methods are employed.

a. The washed chloride is placed in a zinc or iron cup, along with a little water strongly acidulated with sulphuric acid; or in a glass or porcelain cup along with a zinc plate; the whole may then be left to itself for some hours—or, to hasten the reduction, gently heated; the precipitated silver is washed with pure water, and dried.

b. (*Hornung.*) Digest the chloride with ammonia and pure copper filings for twenty-four hours, then wash and dry the powder.

c. (*Levol.*) The washed chloride is mixed with an equal weight of sugar, and the mixture is digested in an excess of a moderately strong solution of caustic potash, with occasional agitation for twenty-four hours; the reduced silver is washed with distilled water.

d. (*Mohr.*) The dry chloride is mixed with 1·8rd of its weight of powdered black resin, and moderately heated in a crucible until the flame ceases to have a greenish-blue colour; the heat is then suddenly increased so as to melt the metal into a button or ingot.

e. (*Gay-Lussac.*) Take the chloride, dry it, and throw it, in successive portions, into twice its weight of carbonate of potash fused in a red-hot Hessian crucible; effervescence ensues, and the metal subsides to the bottom. If a 'soluble salt,' as the nitrate, acidulate the solution, and precipitate it by means of a polished plate of copper; the silver is then obtained in the form of powder. The products of the above processes, when the latter are carefully conducted, are almost pure silver.

Action of Light on Silver Salts. The observation of Boyle that silver chloride and other silver salts on being exposed to light darken (the chemical explanation of which was first given by Scheele) has led to the invention of the art of photography. See PHOTOGRAPHY.

Silver, Acetate of. AgC₂H₃O₂. *Syn.* ARGENTI ACETAS, L. *Prep.* By adding a solution of acetate of potash to a solution of nitrate of silver, washing the precipitate with cold water, redissolving it in a little hot water, and setting the solution aside to crystallise. Small colourless needles.

Silver, Ammonio-chloride of. 2AgCl + 3NH₃. *Syn.* ARGENTO-CHLORIDE OF AMMONIA; ARGENTI AMMONIO-CHLORIDUM, L. *Prep.* Add, gradually, chloride of silver (recently precipitated and well washed) to concentrated ammonia solution as long as it is dissolved on agitation, applying a gentle heat towards the end; then heat the solution to the boiling-point, concentrate a little, and allow it to cool very slowly; collect the crystals which form, dry them by pressure between folds of bibulous paper, and at once preserve them from the light and air.—*Dose*, 1/10 to ⅓ gr.

Silver, Ammoniuret of. See FULMINATING SILVER (Berthollet's, Nos. 1 and 2).

Silver, Ben'zoate of. AgC₇H₅O₂. Thin transparent plates, which are blackened by exposure to the light. See BENZOATE.

Silver, Bro'mide of. AgBr. Occurs in Chili and Mexico as the mineral bromargyrite.

Prep., &c. From silver nitrate and hydrobromic acid as the chloride; yellow octahedra, darken by light.

Silver, Car'bonate of. Ag_2CO_3. *Syn.* ARGENTI CARBONAS, L. A light yellow insoluble powder or needles obtained by precipitating a cold solution of nitrate of silver with another of carbonate of sodium. It is decomposed by heat, darkened by light.

Silver, Chlo"ride of. AgCl. *Syn.* ARGENTIC CHLORIDE. Occurs as the mineral cerargyrite (horn silver).—*Prep., &c.* Precipitate a solution of nitrate of silver by dilute hydrochloric acid or a solution of common salt; wash the precipitate, and dry it in the shade.—*Dose*, ⅓ to 8 gr., thrice daily; in epilepsy, chronic dysentery, cholera, diarrhœa, &c. Dr Perry regards it as preferable to the nitrate.

Silver, Cy'anide of. AgCN. *Syn.* ARGENTIC CYANIDE, HYDROCYANATE OF SILVER. *Prep., &c.* Add dilute hydrocyanic acid to a solution of nitrate of silver; wash the precipitate with distilled water, and dry it.

Prop., &c. Cyanide of silver is a white powder, soluble in ammonia, and decomposed by contact with vegetable substances; light turns it violet-coloured.—*Dose*, ¹⁄₁₆ to ½ gr.; in syphilis, &c. It has been proposed as a source of hydrocyanic acid (*Everitt*).

Silver, Hyposul'phite of. $Ag_2S_2O_3$. *Syn.* ARGENTI HYPOSULPHIS, L. A white unstable substance, insoluble in water, very soluble in the alkaline hyposulphites, forming compounds possessing an intensely sweet taste. See HYPOSULPHUROUS ACID and PHOTOGRAPHY.

Silver, I'odide of. AgI. *Syn.* ARGENTIC IODIDE; ARGENTI IODIDUM, L. Occurs as iodargyrite in Mexico, Spain, &c.

Prep. Precipitate a solution of nitrate of silver with another of iodide of potassium; wash the precipitate with distilled water, and dry it in the shade.

Prop., &c. Pale yellow powder or hexagonal plates; insoluble in water and in ammonia; soluble in a solution of hyposulphite of soda. It behaves abnormally with heat, contracting when heated from 10° to 70° C., and expanding on cooling. Used in some of the French hospitals in the stomach affections of scrofulous subjects; also in epilepsy.—*Dose*, ¹⁄₁₆ to 1 gr.

Silver, Ni'trate of. AgNO₃. *Syn.* ARGENTI NITRAS, L. This article is found in commerce under two forms.

1. CRYSTALLISED. *Prep.* By dissolving grain silver in nitric acid diluted with about twice its weight of water, evaporating the solution until it is strong enough to crystallise on cooling, and then allowing it to cool very slowly.

Prop., &c. Colourless; transparent, anhydrous rhombic plates; soluble in an equal weight of cold and in half their weight of boiling water; soluble in alcohol; fuse when heated, and at a higher temperature suffer decomposition; blackened by light, and by contact with organic substances. Its solution in distilled water is not sensibly darkened by light in the absence of organic matter. Used for solutions, and in photography, q. v.

2. FUSED (LUNAR CAUSTIC; ARGENTI NITRAS —B. P., Ph. L. & E.; A. N. FUSUM—Ph. D.).

Prep. (Ph. D.) Refined silver, 3 oz.; pure nitric acid, 4 fl. oz.; distilled water, 5 fl. oz.; mix in a glass flask, and apply a gentle heat until the metal is dissolved; transfer the solution to a porcelain dish, decanting it off a heavy black powder which appears at the bottom of the flask, and, having evaporated it to dryness, raise the heat (in a dark room) until the mass liquefies; then pour it into moulds furnished with cylindrical cavities of the size of a goose-quill, and which admit of being opened by a hinge; preserve the concreted salt in well-stoppered bottles, impervious to the light.

Obs. In preparing this salt care should be taken that the silver is free from copper. Pure nitrate of silver may, however, be prepared from silver containing copper by evaporating the nitric acid solution to dryness, and cautiously heating the mixed nitrates to fusion. A small portion of the melted mass is examined from time to time, until a little dissolved in water, and treated with ammonia in excess, ceases to strike a blue colour. When this point is arrived at, the fused nitrate is allowed to cool, when it is redissolved in water, filtered or decanted from the insoluble black oxide of copper, and evaporated in the usual way. The heat employed in preparing the fused nitrate should not exceed 420° F., and the fusion should be effected completely, but with moderate expedition, to prevent loss of nitric acid. The moulds should be gently heated before pouring the fused nitrate into them. Bensoin recommends moulds formed of white Bohemian talc or of English slate.

Pur., &c. Pure nitrate of silver, whether crystallised or fused, should be entirely soluble in water, yielding a colourless solution, from which metallic silver is precipitated by a piece of bright copper; both forms are originally white, but are darkened by exposure to light and contact with organic matter.

Uses, &c. Nitrate of silver is a powerful tonic, antispasmodic, astringent, and escharotic.—*Dose*, ¼ to 1 gr., gradually increased, twice or thrice a day, made into a pill with crum of bread; in cholera, epilepsy, &c., preceded by purgatives. It has been highly extolled by Mr Ross as a remedy in cholera. Its continued use permanently colours the skin. It is also extensively employed externally as a caustic. It is powerfully poisonous. A solution of common salt, emetics, and demulcents constitute the treatment in such cases. Nitrate of silver is much employed in the manufacture of hair dyes, of 'indelible ink' for linen, in chemical analysis, and in photography.

Silver, Oxide of. Ag_2O. *Syn.* SILVER HEMI-OXIDE, PROTOXIDE OF SILVER; ARGENTI OXYDUM, A. PROTOXYDUM, L. *Prep.* 1. (*Lane.*) Nitrate of silver, 2 parts; hydrate of potassium, 1 part; dissolve each separately in distilled water, mix the solutions, and, after frequent agitation during an hour, collect and wash the precipitate, and dry it by a gentle heat in the shade. A pale brown powder when moist, but black when dry.

2. Recently precipitated chloride of silver is boiled in a solution of hydrate of potassium of the sp. gr. 1·25, with frequent stirring and trituration, until, on testing a little of it, it is found to be entirely soluble in dilute nitric acid, when

it is washed and dried as before. A black and very dense powder. Chemically pure.

3. Nitrate of silver, ½ oz.; water, 4 fl. oz.; dissolve, and pour the solution into a bottle containing lime-water, 2 quarts, or q. s.; agitate the mixture well, collect and wash the sediment, and dry it at a heat not exceeding 212° F. Pure. *Prop.*, &c. Very soluble in solutions of ammonia and of the alkaline hyposulphites; slightly soluble in water; when moist it absorbs CO_2 from the air; reaction alkaline; decomposed by light; also when triturated with an easily oxidisable substance like amorphous phosphorus.—*Dose*, ⅓ to 2 gr.; in epilepsy, gastralgic irritations, &c. It is much used in France, and has been highly extolled in menorrhagia. By some, however, it is not considered superior to the nitrate.

Silver Tetratoxide. Ag_4O. *Syn.* ARGENTOUS OXIDE. *Prep.*, &c. From dry citrate of silver heated to 212° F., in a stream of hydrogen gas, until it turns dark brown, when it is dissolved in water; the solution is next treated with potash, and the precipitate is carefully washed and dried. A black powder, easily decomposed, and soluble in ammonia.

Silver Pencils, Nitrate of. According to A. Huber, very thin pencils of nitrate of silver, such as are sometimes required for intra-uterine applications, may be prepared in the following manner: —Silver nitrate is fused in a capsule, and then drawn up by slow and cautious suction into a glass tube, the calibre of which is a trifle larger than the required diameter of the pencil. Especial care is to be taken that no cavities filled with air-bubbles are produced in the contents of the tube. When entirely cold the tube is warmed by turning over a spirit lamp until the outer surface of the stick has become soft, when it may be easily pushed out by means of a knitting-needle.

Silver, Perox'ide of. Ag_2O_2. *Syn.* ARGENTI PEROXYDUM, L. A black crystalline substance which forms on the positive electrode when a galvanic current is passed from platinum electrodes through an aqueous solution of nitrate of silver.

Silver, Sul'phate of. Ag_2SO_4. *Syn.* ARGENTI SULPHAS, L. *Prep.* By boiling reduced silver in sulphuric acid, or by precipitating a solution of the nitrate by another of sulphate of sodium. It dissolves in 80 parts of hot water, and falls in small colourless needles as the solution cools.

Silver, Sul'phide of. Ag_2S. *Syn.* SULPHURET OF SILVER; ARGENTI SULPHURETUM, L. Occurs as argentite in Hungary, Norway, Mexico, &c.— *Prep.*, &c. Prepared by passing sulphuretted hydrogen through a solution of nitrate of silver, or by melting its constituents together. It possesses a brownish-black colour, and is a strong sulphur base. It forms the stains which tarnish silver on exposure to the air.

Silver, Ox'idised. The high appreciation in which ornamental articles in oxidised silver are now held renders a notice of the process followed interesting. There are two distinct shades in use; one produced by chlorine, which has a brownish tint, and the other by sulphur, which has a bluish-black tint. To produce the former it is only necessary to wash the article with a solution of sal-ammoniac. A much more beauti-

ful tint may, however, be obtained by employing a solution composed of equal parts of sulphate of copper and sal-ammoniac dissolved in vinegar. A fine black tint may be produced by a slightly warm solution of sulphide of potassium or of sodium ('Chem. Techn.').

SILVER DUST. *Syn.* SILVER POWDER; ARGENTI CROCUS, A. PULVIS, L. *Prep.* 1. Pure pulverulent silver, obtained by any of the methods given above. Used to coat pills by japanners, &c. 2. Heat oxide of silver to dull redness in a porcelain crucible, cool, triturate the powder in an agate mortar, and pass it through a fine sieve. Used at the hospital of Montpellier.

SILVER SHELLS. These are prepared and used like gold shells.

SIL'VERING. The art of covering the surfaces of bodies with a thin coating of silver. Leather, paper, wood, &c., are silvered by covering them with silver leaf, by a similar process to that employed for gilding them.

Silvering of Glass. Two distinct methods are adopted for this purpose—one of which consists in employing a layer of tin-foil and mercury, falsely called 'silvering;' the other in using a coating of real silver precipitated from a solution of that metal.

1. Plane surfaces, as those of mirrors, &c., are commonly silvered as follows:—A sheet of tin-foil corresponding to the size of the plate of glass is evenly spread on a perfectly smooth and solid marble table, and every wrinkle on its surface is carefully rubbed down with a brush; a portion of mercury is then poured on, and rubbed over the foil with a clean piece of very soft woollen stuff, or a hare's foot, after which two rules are applied to the edges, and mercury poured on to the depth of a crown piece, when any oxide on the surface is carefully removed, and the sheet of glass, made perfectly clean and dry, is slid along over the surface of the liquid metal, so that no air, dirt, or oxide can possibly either remain or get between them. When the glass has arrived at its proper position, gentle pressure is applied, and the table sloped a little to carry off the waste mercury, after which it is covered with flannel and loaded with heavy weights; in 24 hours it is removed to a wooden table and further slanted, and this position is progressively increased during a month, until it becomes perpendicular.

For silvering convex or concave surfaces a mould of plaster of Paris is employed, so that the amalgamated foil may be accurately fitted to the surface.

Globes and other hollow vessels are commonly silvered by the application of one of the silvering amalgams. See AMALGAM.

2. In the HUMID WAY. a. (*Drayton.*) A mixture is first made of nitrate of silver (in coarse powder), 1 oz.; ammonia, ½ oz.; and water, 2 oz.; which, after standing for 24 hours, is filtered (the deposit upon the filter, which is silver, being preserved), and an addition is made thereto of spirit (by preference, rectified spirit at 60% o. p.), or naphtha, 8 oz.; from 20 to 30 drops of oil of cassia are then added; and, after remaining for about 6 hours longer, the solution is ready for use. The glass to be silvered (first well cleaned and polished) is placed in a horizontal position,

and a wall of putty, or other suitable material, formed around it; the above solution is then poured over it to the depth of from ⅛ to ¼ inch; from 6 to 12 drops of a mixture of oil of cloves and spirit of wine (in the proportion of 1 part, by measure, of oil of cloves to 3 parts of spirit of wine) are next dropped into it, at different places; or the diluted oil of cloves may be mixed with the solution before it is poured upon the glass, a larger quantity, in both cases, increasing the rate of the deposit. When the glass is sufficiently silvered the solution is poured off; and as soon as the silver on the glass is perfectly dry it is varnished with a composition formed by melting together equal quantities of beeswax and tallow. The solution, after being poured off, is allowed to stand for three or four days in a close vessel, as it still contains silver, and may be again employed after filtration, and the addition of a sufficient quantity of fresh ingredients to supply the place of those which have been used. 18 gr. of nitrate of silver are sufficient for one square foot of glass. Hollow vessels may be silvered by pouring the solution into them. By the addition of a small quantity of oil of caraway, oil of cloves, or oil of thyme, the colour of the silver may be varied ('Patent Journ.').

b. (*Thomson* and *Mellish*.) Nitrate of silver, 2 oz.; water and rectified spirit, of each, 3 fl. oz.; dissolve, add of spirit of hartshorn or liquor of ammonia, 1 fl. oz.; mix, and after a short time filter the solution; to each ounce of this add of grape-sugar, ¼ oz., previously dissolved in a mixture of rectified spirit and water, of each, ½ pint; after 3 or 4 hours' repose it is fit for use. This solution is applied to the glass, heated to about 160° F., in a similar manner to the last. Patented.

c. The best plan of silvering plane or slightly curved surfaces is, however, the method employed for coating the specula of the silvered-glass Newtonian telescopes. This method is very easy, and has the advantages of giving a brilliant and durable surface on both sides, and the film is sufficiently firm to admit of being polished with rouge and fine wash-leather.

One half-ounce of pure nitrate of silver is dissolved in 4 oz. of distilled water, and divided into two equal portions. One is treated with dilute ammonia until the brownish precipitate is entirely redissolved; and to this clear solution ¼ oz. of pure hydrate of potassium, dissolved in 8 oz. of water, added; and the brown precipitate and grey sediment that remains after the brown precipitate disappears dissolved by the cautious addition of ammonia, stirring well all the time. The remaining nitrate of silver solution is now added, stirring well until it gives a greyish precipitate that does not disappear after well stirring. The bulk of the solution is next made up to 100 oz., and allowed to settle, when the clear solution is poured off for use.

The reducing solution is prepared by dissolving ½ oz. of pure milk-sugar in 10 oz. of hot water, and adding 10 minims of pure alcohol.

This quantity of silvering solution will coat over two square feet of glass surface with a brilliant film of pure silver. The glass must be perfectly clean, and is to be suspended face down-

wards on the surface of the solution, and allowed to stand one hour, the temperature of the solution being best about 80° F.

d. (*E. Siemens.*) As a reducing agent, acetic aldehyde is used in the form of aldehyde ammonia, prepared by passing dry ammoniacal gas into aldehyde. Four grms. of silver nitrate and 2½ grms. of aldehyde ammonia are separately dissolved in a litre of distilled water, and the solutions mixed and filtered. The article to be silvered, after washing out with potassium carbonate, and then with spirits of wine and distilled water, to remove every trace of grease, is filled with this solution (as far as it is desired to silver), and then hung up in the water-bath.

It is now gradually heated, and as soon as the temperature reaches 50° C. the separation of the silver mirror begins, and soon spreads over the whole inner glass surface. Its formation is soon finished, usually between 55° and 60°. When the beauty of the silver surface reaches a maximum it is time to withdraw the article from the water-bath, and pour off the contents, or the brilliancy of the mirror will be impaired. The article is finally rinsed in distilled water.

e. (*Martin.*) M. Martin makes use of four liquids, viz., first, a 10% solution of nitrate of silver; second, liquor ammonia, sp. gr. ·970; third, a 4% solution of caustic soda; and fourth, a 12½% solution of white sugar, to which he adds 2½% of nitric acid, and after twenty minutes' boiling he adds to it 25 parts of alcohol, and water to make up the bulk to 250. The silvering solution is made by mixing together 12 parts of solution No. 1, 8 parts of No. 2, 20 parts of No. 3, and 60 parts of distilled water, and finally, in twenty-four hours, 10 parts of No. 4. The object to be silvered is then immersed, when it will be covered with a film of reduced silver, which in ten minutes' time will be sufficiently thick for use. After having been washed with distilled water and dried the surface may be polished with chamois leather and rouge.

Silvering Glass. (*Böttger*, 'Chem. Centr.') The silvering liquid employed is made by dissolving 4 grms. of pulverised silver nitrate in strong ammonia, adding 1 grm. ammonium sulphate and 350 c.c. water. A solution of 1·2 grms. starch⁰ or grape-sugar with 3 grms. caustic potash in 350 c.c. of distilled water forms the reducing liquid. When used, equal volumes of the two liquids are mixed together, and applied to the surface of the substance to be coated.

Silvering of Metals. 1. (LEAF SILVERING.) This is performed with leaf silvering in the way described under GILDING for the gilding of polished metals.

2. (COLD SILVERING.) Mix chloride of silver, 1 part, with pearlash, 3 parts, common salt, 1½ parts, and whiting, 1 part, and well rub the mixture on the surface of the brass or copper (previously well cleaned) by means of a piece of soft leather, or a cork moistened with water and dipped into the powder. When properly silvered the metal should be well washed in hot water slightly alkalised, and then wiped dry.

3. (ELECTRO-SILVERING.) This is described under ELECTROTYPE.

Silver, a New Imitation of. A patent for an

alloy has been taken out by M. Lemarquand, which is said to bear a close resemblance to silver in appearance, and to be unaffected by atmospheric influences. It has the following composition:

Pure copper	. . .	750 parts.
Nickel	. . . .	140 „
Black oxide of cobalt	.	20 „
Tin, in sticks	. . .	18 „
Zinc	. . . .	72 „

SIMAROU'BA. *Syn.* SIMARUBA (Ph. E. and D.), L. The root-bark of *Simaruba amara* or *officinalis*, the mountain damson. Tonic, bitter, and astringent.—*Dose*, 20 to 30 gr.; in intermittents, obstinate diarrhœa, dysentery, and dyspepsia.

SIN'AMINE. $C_4H_6N_2$. A basic substance formed, along with sulphide of lead, when thiosinamine is treated with oxide of lead. It is very bitter tasted, has a powerful alkaline reaction, and, when slowly obtained from its concentrated aqueous solution, forms brilliant colourless crystals.

SIN'APISM. *Syn.* SINAPISMUS, L. A mustard poultice.

SIN'APOLINE. $C_7H_{12}N_2O$. A basic substance formed, along with carbonic acid, when the volatile oil of mustard, or sulphocyanide of allyl, is treated with oxide of lead. It is soluble in water and alcohol, has an alkaline reaction, and crystallises in colourless plates.

SINKS. "In no case," says Mr Eassie (' Healthy Houses,' by Wm. Eassie, C.E.—Simpkin, Marshall, & Co.), "should the waste-pipe of sink, laboratory, or bath lead direct into the drains; yet how frequently is this the case, and a special card sent out to disease and death!

"It must also be remembered it is every whit as dangerous if these waste conduits lead into the soil-pipe of a closet. Waste-pipes from the above-named places should be led down to within 12 or 18 inches from the ground, and should deliver on to the grating of a gully or yard trap."

This subject has been already treated in the article on ' Sanitation, Domestic,' wherein we have embodied the practical suggestions of Mr Collins, another sanitary house reformer, as well as in our article on ' Drainage,' in which will be found details for carrying out the system recommended by Messrs Eassie and Symonds, and thus preventing the admission into our dwelling-houses of the poisonous sewer gas.

The matter has so important a bearing upon health, that we shall make no apology for having thus reiterated and emphasized it by quoting Mr Eassie's words of warning. See TRAPS.

SIPHONOPHORA GRANARIA, Kirby. ·THE CORN APHIS. In some seasons this aphis is very injurious to wheat plants. It, with kindred species, is also found upon oats, barley, and rye, but the wheat plant is the chief object of its attack in this country. It is found upon the plants in the early spring, at this time usually in small numbers, wandering about restlessly and singly until the ear is formed, after which time, in favourable conditions, there is a rapid increase in its numbers. The ear, with the sweet juices destined for the support of the forming grains, is evidently its great attraction.

Upon an examination of ears infested with aphides, generations of all sizes, and in all stages —larvæ, pupæ, and perfect insects, commonly known as flies—will be seen actively engaged in sucking the juices from the stems within the ears and from the bases of the grain clusters.

Directly the plant begins to change for ripening and its tissues harden, the aphides cease because they are compelled to cease from active effect upon it, but their excrement and exuviæ mixed with ' honey-dew ' (' honey-dew ' is a glutinous sweet liquid secreted by many species of aphides, and ejected by them upon the plants they infest) hinder respiration, and in a degree affect the development and tend to spoil the colour of the grains. Aphis-affected ears of corn frequently have light and imperfectly shaped grains, and in bad and persistent attacks the sample is thin, shrivelled, and, especially in the case of white wheat, discoloured. In the season 1885 wheat plants were attacked by aphides in many parts of the country, as were many other agricultural and horticultural crops, with forest and ornamental trees, and in not a few localities much damage was sustained from loss of weight and imperfect shape of the corn, because in the abnormally cold weather in August the plants changed for ripening most slowly, so that the aphides had an unusually protracted time for work.

The corn aphis has been long known in this country. Curtis speaks of it as infesting wheat ears in 1797. Serious injury is also caused to wheat plants, as well as to oats, barley, and rye plants in America and Canada by aphides, which, according to the description of Fitch and Thomas, well-known American entomologists, and of Bethune, in Ontario, appear to belong to the same species as those in this country. A species of aphis is destructive to grain crops in Germany, described by Taschenberg (' Praktische Insekten Kunde ') as identical with the English corn aphis, minutely delineated, with an elaborate illustration, by Buckton in his ' Monograph of British Aphides.' As in America and Germany, so in England, aphides are found upon various corn plants and in many kinds of grasses, among which may be mentioned cock's-foot, *Dactylis glomerata*; soft-grass, *Holcus lanatus*; some of the Poas; rye-grass, *Lolium*; and couch-grass, *Triticum repens*.

Life History. The life history of this species of aphis, like that of many other species of aphis, is not yet completely clear. It has not yet been accurately ascertained as to whether the continuity of existence is maintained by eggs laid up during the winter, or by hibernating larvæ. It is believed that it is carried on by larvæ, because larvæ have been seen very early in the spring in the stems of the wheat plants, and upon the stems and blades of couch-grass, close to the ground. Drs C. Thomas and Asa Fitch and other American entomologists have also seen the larvæ of this aphis at the roots of wheat plants during the winter, together with the females producing them.

The winged female insect is light brown in colour, with the abdomen green, with legs of a dark yellow hue with black knees and feet. The eyes are red and the cornicles black. ·

In colour the larvæ or lice—wingless females bringing forth successive generations of live larvæ —differ from the perfect insects, being green or dark green with brown antennæ, having, however, legs of the same shade of yellow and black. Their beaks or rostra are short, as in the case of the perfect insects.

The winged, egg-laying female is not developed until late in the season, but winged females bringing forth live larvæ are generated at various times and intervals, as in the case of the hop aphis, *Aphis humuli*, when, as is commonly supposed, the food supply fails and the insects become too thick to thrive.

Prevention. After an attack of aphides the wheat stubbles should be scarified or cultivated and the rubbish burnt, or the land should at once be deeply ploughed. If the succeeding crop is to be tares, trifolium, potatoes, turnips, or mangels, thorough cleaning and destruction of couch and other grasses would be sufficient. A succeeding white straw crop should be avoided after a bad attack, as the aphides infest all crops of this character. Deep ploughing and thoroughly and deeply burying the stubble might prevent their recurrence. It would be safer to take another crop.

After an attack care should be taken to extirpate grasses and grassy growths from the fields and from the outsides of fields. It is usual in some counties for wheat to follow rye-grass and clover ley, or one or two years old, after oats, or after wheat. Should aphides have infested the previous oat or wheat crops, they might be carried on, and would probably be carried on, to the next corn crop by the rye-grass and other grasses in the leys. This would be detected by observation, and if aphides were found measures should be taken to circumvent them by altering the rotation, or by closely feeding the ley with sheep and treading it well before it was ploughed.

It may be suggested here that a strong magnifying lens for the pocket is a necessary part of a practical farmer's equipment in these days when insects are so numerous and rampant, and that it is as requisite to carefully examine the roots and lower stems of plants and the surface soil around them when walking round the farm, as to observe their upper parts within more convenient reach, or as to watch and note the signs of the weather.

Remedies. Obviously it would be impracticable, at all events to farmers generally, to apply remedies for aphides actually *in situ* upon corn plants when in ear. In cases where watchful and well-timed observations have shown that larvæ—lice—were present on the blades and stems close to the ground early in the spring, dressings of soot, guano, gas lime, or agricultural salt would check their progress. Where the plants were not too forward harrowings and rollings would interrupt them considerably. Feeding off with sheep would be remedial where the state of the plants and the condition and nature of the land allowed this to be done.

Lady-birds, *Coccinella*, called by the French peasants *Bêtes à Dieu*, or *Vaches à Dieu*, and in Italy *Bestioline del Signore*, are the natural and inveterate destroyers of these and all other species of aphides, as well as of other insects. These

should be regarded as sacred by all agriculturists and cultivators of every description. It is firmly believed that they eat fungi also, as they have been found upon hop cones and rose leaves affected with mildew which they appeared to be eating. The enormous benefits conferred upon agriculturists by the *Coccinellidæ* as devourers both of insects and fungi injurious to crops, and as scavengers of refuse, are described in a most graphic and interesting manner by Professor Forbes, State Entomologist, Illinois, U.S.A., in a paper entitled 'The Food Relations of the Carabidæ and Coccinellidæ.'

The grain aphis has fortunately dangerous enemies even more destructive than the lady-birds, in two parasites, ichneumon flies, known respectively as *Aphidius avenæ* and *Ephedrus plagiator*. These parasitic flies have long ovipositors with which they insert their eggs into the bodies of the aphides. In a short time the eggs become larvæ, and feed upon their bodies until nothing but empty skins remain. They lay many eggs and only one in each larva, so that they deal destruction wide-spread among these foes to the wheat crop ('Reports on Insects Injurious to Crops,' by Chas. Whitehead, Esq., F.Z.S.).

SITFASTS. These hard tumours, possessing but little sensibility, are situated in those superficial parts of the horse's body which have been exposed to the unequal pressure of the collar, the saddle, or the harness. The tumour should be removed by the veterinary surgeon if the previous application of either blisters, biniodide of mercury ointment, or a seton has been tried and failed to disperse it. The precursor of the sitfast is always a swelling filled with serum and lymph, caused, as before stated, by badly fitting harness. Hence the soundest treatment is to prevent its development into the hard form, by proper means, directly it shows itself, the best remedies being the application of salt and water or Goulard water, and correcting the defects of the harness.

SITONES CRINITUS, Oliv.; **SITONES LINEATUS,** Linn. **THE PEA AND BEAN WEEVILS.** These are two species of weevils very destructive to pulse crops. Farmers and gardeners have constantly noticed that the leaves of pea and bean plants are full of holes and notches, and so much so as to affect their growth most materially in some seasons. These weevils cause this, and are most dangerous when the plants are young; commencing their depredations in March, or as soon as the weather becomes spring-like, they work until the end of July.

It is said that they do not attack the common pea that is grown principally for pigs and sheep; but this is not correct, for complaints have been made from several parts showing that these have not by any means escaped. From observation it is clear that they eat all kinds of peas readily, in field and garden, as well as Mazagan, tick, and broad beans. In some seasons, and when the seed is sown late, they fairly prevent the plants from starting, eating off the leaves directly they appear from the cotyledons. A large grower of peas for seed reported that in 1883 he sustained considerable losses by the onslaughts of the pea weevils, especially upon the Early Sunrise sort.

Clover is much destroyed by these weevils, as

well as by a closely allied and almost identical species, known as *Sitones puncticollis*. The weevils eat the leaves, and the grubs or larvæ devour the roots of the clover. In 1883 there were many complaints made of clover dying in patches in various parts of England in October. It was thought at first this was due to clover sickness, or to a fungus. Upon close examination small maggots were found at the roots, which were living upon the juicy succulent parts. Again, in the early spring following, the mischief was continued. *Trifolium incarnatum* is also attacked frequently. Plant is lost quickly and mysteriously. It is said that the 'worm' is in it. In Kent, in 1882, this happened in a large piece of trifolium sown upon wheat stubble without ploughing. After the plants had nearly all disappeared the cause of the loss was traced to the larvæ of the *Sitona*.

Life History. The perfect insect—*Sitona lineata*—is about four lines or the third of an inch in length, rather narrow in shape. It is of an earthy colour, with light stripes or lines down its back. The head is dark coloured. The wings are large. *Sitones crinitus* is hardly so large as the *Sitones lineatus*, and is of a somewhat lighter colour, and without any stripes or lines, but has hairs or bristles on its body.

When disturbed these insects get on to the ground, either by falling or jumping, and remain perfectly still. Being similar in colour to the earth it is difficult to detect them. The eggs are white and numerous. The larvæ are found at the roots of clover plants from October until March, and of peas towards the end of May, and change to pupæ in the ground during June. The larvæ are nearly a quarter of an inch long, white, without legs, and having strong jaws.

Prevention. After an attack the land should be cleaned from all rubbish and deeply ploughed at once, as the larvæ remain in the soil during the winter. A dressing of lime would be most advantageous in serious cases. Care should be taken not to sow another leguminous crop after an attack.

Clover-fields literally swarm with these insects in some seasons. It would be highly dangerous to put peas, or beans, or tares in after clover in these circumstances, but this is an unusual course of cropping. As the weevils have been found in wheat stubbles after harvest, in land sown with wheat after clover, it is desirable not to put trifolium in after wheat without cultivation, as is often done. Trifolium crops have been materially injured by these *Sitona*.

Ashes, sawdust, or earth saturated with paraffin diluted in the proportion of 2 table-spoonfuls to 10 galls. of water put into the drills or rows when peas and beans are sown might be used upon a large scale as a preventive.

Remedies. A dressing of 2½ cwt. of guano per acre has been found to help peas and beans suffering from the attacks of the *Sitones*. If put on early when the dew is on the plants, or after a slight shower, this manure sticks to the leaves, and renders them distasteful to the weevils, and helps the plants along at the same time. In market gardens, and in gardens, it is very efficacious to send men and boys to walk with a foot on

either side of each row of plants, to press the earth tightly and firmly close to the plants in order to prevent the beetle from moving again easily. Many are killed by this process. This might be extended to larger cultivation, as a gang of men would get over a good deal of ground in a day. Horse hoeing cannot be done too often, and side hoeing will be found very useful ('Reports on Insects Injurious to Crops,' by Chas. Whitehead, Esq., F.Z.S.).

SIZE. Obtained, like glue, from the skins of animals, but is evaporated less, and kept in the soft state. See GOLD and GOLD SIZE.

Size, Oil. This may be made by grinding yellow ochre or burnt red ochre with boiled linseed oil, and thinning it with oil of turpentine.

SKATE. The *Raia batis*, Linn. Other varieties of *Raia* also pass under the name. It is a coarse fish, and is principally salted and dried for exportation.

SKIN (The). *Syn.* CUTIS, DERMIS, PELLIS, L. Every person must be familiar with the external appearance and general properties of the skin; but there are many of our readers who may not be aware of its peculiar compound character. The skin, then, although apparently a single membrane, is composed of three distinct layers, each of which performs its special duties :—1. The exterior of these is called the cuticle, epidermis, or scarf-skin. It is an albuminous tissue, possessing no sensibility, and is found thickest on those parts of the body most exposed to friction or injury. 2. The mucous net, or rete mucosum, which is a thin layer of rounded cells, which lies immediately under the cuticle, and is supposed to be the seat of the colour of the skin. 3. The dermis, cutis vera, or true skin, is a highly sensitive, vascular, gelatinous texture, the third and last in succession from the surface of the body. It is this which, when the scarf-skin and hair have been removed, is converted by the process of tanning or tawing into leather.

The skin, because of its tough, elastic, flexible nature and its underlying layer of fat, is admirably adapted for covering the various internal parts and organs, as well as for bodily movement and exertion. Besides this, it exercises, in common with the lungs, the liver, and the kidneys, the important function of a depurator, and may, with the organs above specified, be regarded as one of the main outlets for the waste products of the body; the effete and noxious matters of which, when in a healthy condition, it effects the removal, are those contained in the perspiration, and in addition carbonic acid.

The perspiration is variable in amount, owing to various causes, such as temperature, the amount of exercise taken, the more or less hygroscopic condition of the surrounding atmosphere, the quantity of fluid swallowed, the season of the year, &c. With the exception of that which occurs under the armpits and upon the soles of the feet, it has generally an acid reaction, due to the presence in it of uncombined organic acids. Under ordinary conditions of life it averages daily about 2 lbs. in quantity, being, as might be expected, more abundant than the urine in summer, and less in winter. The perspiration is of very complex composition, and contains lactates, butyrates, and acetates of sodium and ammonium, sodic chloride, phosphate of calcium, and sulphates—these latter, however, occurring in but small quantities.

Various observers have arrived at different conclusions respecting the amount of carbonic acid exhaled from the skin. Professor Scharling believed it to be from a fortieth to a sixtieth the amount given off by the lungs. Recent observations seem, however, to have shown that this

Sudoriparous Gland from the palm of the hand, magnified 40 diam.:—*s*, *s*, contorted tubes composing the gland, and uniting in two excretory ducts, *b*, *b*, which unite into one spiral canal that perforates the epidermis at *c*, and opens on its surface at *d*; the gland is embedded in fat-vesicles, which are seen at *e*, *e*.

estimate was too high. Dr Edward Smith, operating upon himself by placing every part of his body except the head in a caoutchouc bag, and subsequently collecting the evolved carbonic acid (the experiment being performed in the summer-time), found the quantity evolved to be 6 gr. per hour, or about a hundredth part of that passing off from the lungs.

Aubert's experiments led him to the conclusion that it was about half the amount given by Smith; whilst Reinhart estimated it at 34 or 35 gr. a day.

These excretory processes of the skin are effected by means of very minute vessels called the *sudoriparous* or *sweat glands*. These glands abound in almost every part of the human skin. They are of largest size under the axillæ or armpits, where perspiration is most profuse. They are also very abundant upon the palms of the hands. Professor Erasmus Wilson says that as many as 3528 of these sweat-glands exist in a square inch of surface on the palm of the hand; and as every tube, when straightened out, is about a quarter of an inch in length, it follows that, in a square inch of skin from the palm of the hand, there exists a length of tube equal to 882 inches, or 73½ feet. These glands, as we have seen, vary in number for different parts of the human body; but if we take Professor Wilson's average for the superficial area of a man of ordinary stature, viz. 2800 of them to the square inch, it follows "the total number of pores on such a man's skin would be about *seven millions*, and the length of perspiratory tubing would then be 1,750,000 inches, or 145,833 feet, or 48,611 yards, or nearly 28 miles" (Carpenter's 'Human Physiology').

In addition to the *sudoriparous*, the skin in those parts where hair is found also possesses *sebaceous* glands, which stud almost every part of its surface except the palms of the hands and the soles of the feet. The sebaceous glands secrete a semi-fluid, greasy kind of substance, the office of which is probably to lubricate the hair, these glands always opening into the air-follicles, generally in pairs. A parasite known as the *Acarus folliculorum* infests the sebaceous glands. In the cartilaginous part of the external passage of the ear are other glands, the *ceruminous*, which secrete the wax that forms a protective film for the membrane of the tympanum or drum, and guards it against dust, insects, &c. See EXERCISE, PERSPIRATION.

SKIN BALSAM, Glycerin—Glycerin Haut Balsam. A mixture of 1000 parts glycerin, 120 parts orange-flower water, 1 part each oils of neroli and bitter almonds (*Hager*).

SKIN COSMETICS. The simplest, cheapest, and most generally employed cutaneous cosmetics are soap and water, which at once cleanse and soften the skin. Soap containing a full proportion of alkali exercises a solvent power upon the cuticle, a minute portion of which it dissolves; but when it contains a small preponderance of oily matter, as the principal part of the milder toilet soaps now do, it mechanically softens the skin and promotes its smoothness. Almond, Naples, and Castile soaps are esteemed for these properties; and milk of roses, cold cream, and almond powder (paste) are also used for a similar purpose. To produce an opposite effect, and to harden the cuticle, spirits, astringents, acids, and astringent salts are commonly employed. The frequent use of hard water has a similar effect. The application of these articles is generally for the purpose of strengthening or preserving some particular part against the action of cold, moisture, &c.; as the lips, or mammæ, from chapping, or the hands from contracting chilblains; but in this respect oils, pommades, and other oleaginous bodies are generally regarded as preferable.

Another class of cutaneous cosmetics are employed to remove freckles and eruptions. Among the most innocent and valuable of these is Gowlland's lotion, which has long been a popular article, and deservedly so, for it not only tends to impart a delightful softness to the skin, but is a most valuable remedy for many obstinate eruptive diseases which frequently resist the usual methods of treatment. Bitter almonds have been recommended to remove freckles (*Celsus*), but moistening them with a lotion made by mixing 1 fl. oz. of rectified spirit, and a teaspoonful of hydrochloric acid with 7 or 8 fl. oz. of water, is said to do this more effectually. A safe and excellent cosmetic is an infusion of horseradish in cold milk (*Withering*).

Hermann prescribes the following lotion:— Blanched almonds, 2 oz.; rose water, 8 oz.; orange-flower water, 2 oz. Make an emulsion, strain, and add sal-ammoniac, 1 dr.; simple tincture of benzoin, 2 dr.

Skin paints and skin stains are employed to give an artificial bloom or delicacy to the skin. Rouge and carmine are the articles most generally used to communicate a red colour. The first is the only cosmetic that can be employed, without injury, to brighten a lady's complexion. The other, though possessing unrivalled beauty, is apt to impart a sallowness to the skin by frequent use. Starch powder is employed to impart a white tint, and generally proves perfectly harmless. The American ladies, who are very fond of painting their necks white, use finely powdered magnesia, another very innocent substance. Several metallic compounds, as the trisnitrate, chloride, and oxide of bismuth (pearl white, Fard's white, &c.), carbonate of lead (flake white), white precipitate, &c., are frequently used to revive faded complexions; but they are not only injurious to the skin, but act as poisons if taken up by the absorbents. Trisnitrate of bismuth (pearl white), probably the least injurious of these articles, has been known to cause spasmodic tremblings of the muscles of the face, ending in paralysis ('Voght. Pharm.'). The employment of liquid preparations containing sugar of lead, which are commonly sold under the name of milk of roses, cream of roses, &c., is equally injurious. Another disadvantage of these metallic preparations is that they readily turn black when exposed to the action of sulphuretted hydrogen gas, or the vapours of sulphur, such as frequently escape into the apartment from coal fires. There are many instances recorded of a whole company being suddenly alarmed by the pearly complexion of one of its belles being thus transformed into a sickly grey or black colour.

In conclusion, it may be remarked that the best purifiers of the skin are soap and water, followed by the use of a coarse but not a stiff cloth, in opposition to the costly and smooth diapers that are commonly employed; and the best beautifiers are health, exercise, and good temper.

Skin, Gold'beater's. See GOLDBEATER'S SKIN.

SKINS (of Animals). The preparation and preservation of skins is noticed under PELTRY; the preparation and uses of the skins of the larger animals under LEATHER, TANNING, TAWING, &c.

SLACK. Small coal, such as is used for kilns.

SLAG. The semi-vitrified compounds produced, on the large scale, during the reduction of metallic ores by fluxes. Those from iron and copper works are often used for building materials, mending roads, &c.

According to Egleston (Dingler's 'Polytech. Journ.,' 'Journ. Chem. Society') the following are some of the industial applications to which the slag from blast-furnaces is put.

When required for building stones the slag is run from the blast-furnace into a semicircular vessel on moving wheels, and having its bottom covered 3 cm. deep with sand and coke dust. By means of a bent iron instrument the slag is mixed with sand and coke dust till the escape of gases has nearly ceased and the mass is sufficiently tough. With the same tool it is next pressed into a mould furnished with a lid, which is forced down as soon as the escape of gas ceases. The red-hot stone is then placed in the cooling oven, covered with coke dust, and allowed to remain three or four days to cool completely. These stones are impervious to damp, and make good foundations. According to another method the slag, which should contain from 38% to 44% of silica, is run down a shoot into a large cavity, and then covered over with sand and ashes, and left to cool from five to ten days, when it is distributed in moulds, and there hardens. In certain parts of Belgium slag is poured upon iron plates and cooled by water, and thus a kind of glass is manufactured.

In other districts the slag is granulated as it flows from the blast-furnace by means of a stream of water. The granulated slag is preferred by the puddlers to the sand for the moulds of pig iron. The slag gravel may be advantageously substituted for sand in mortar making, a more rapid hardening being thus secured, a matter of great moment in building foundation walls.

Artificial stone is also manufactured from the granulated slag, and used for building purposes, furnishing warm dry houses of handsome appearance.

When stones for building with enamelled surfaces are required they are obtained in some parts of Europe as follows:—The unburnt bricks are covered with granulated slag, and after drying are burnt in a furnace where they do not come in contact with carbon. The stones are completely glazed, and according to the different kinds of slag used are tinted of different colours. This operation is also employed advantageously with tiles, pipes, and earthenware.

If, in the preparation of fireproof bricks, a certain proportion of mixture of clay and granulated slag be added to the mixture, very hard and durable fire-bricks are obtained. These have been tested in a brass furnace, and experiments are being tried as to their applicability to building puddling furnaces. This granulated slag may also be advantageously used for manure. Blast-furnace slag has also been drawn out in fine threads or filaments, furnishing the so-called 'furnace wool.' This substance, being a very bad conductor of heat, has suggested various household and other uses. A cheap and valuable cement, said to be equal to Portland cement, has been prepared from the finely granulated slag, which will also resist well the action of acids.

Mr Britten in 1876 patented a process for the manufacture of glass from blast-furnace slag. Large works for the purpose of carrying out this invention, under the title of 'Britten's Patent Glass Company,' have been erected at Finedon in Northamptonshire, and are, we believe, successfully worked in manufacturing glass bottles. The method consists in removing molten slag in a ladle from the blast-furnace, and pouring into a Siemens' furnace, when certain amounts of carbonate of sodium and silica are added, depending upon the quality of the slag used, and of the glass required.

SLATE. The excellence of this material for water cisterns deserves a passing notice here. Irish slate (*Lapis Hibernicus*) is an argillaceous mineral, said to contain iron and sulphur, found in different parts of Ireland. It is a common remedy, among the vulgar, for internal bruises, taken in a glass of gin.

SLATE WASTE (Utilisation of). Much has already been accomplished in the utilisation of basic slags from the steel and iron works, and experiments have shown that what has hitherto been called 'waste' in slate quarries can be manufactured into bricks and tiles. In Italy this new departure has already met with considerable success.

SLEEP. During the period of our waking hours the exercise of the animal functions entails a waste or destruction of tissue in the organs performing them, which, unless duly repaired, would soon lead to the enfeeblement and consequent failure of the powers of the organs themselves. For the animal economy, therefore, to be maintained in a state of efficiency the repair of the reduced tissues is a necessity; and this essential condition is effected by the agency of sleep, during which respiration, circulation, digestion, &c., continue to be carried on simultaneously with assimilative processes which end in the regeneration of the impaired tissue.

A proper amount of sleep is therefore as great or even a greater necessity than a proper supply of food; and any one failing to obtain it soon perishes of exhaustion. Thus it is that any great mental emotion—such as intense remorse, grief, anxiety, or the depressing effect of a reverse of fortune—so frequently expedites death. Like Macbeth, "it murders sleep," one of the great needs of man's existence.

Infants and children, it is well known, require much more sleep than adults. In these latter the organism, being already matured, demands only so much sleep as will enable it to make up for the daily waste of the body, which waste falls very far below the amount of nutrition required by the growing infant. In a still earlier state of development, viz. the fœtal one, life may be said to be passed entirely in slumber; whilst children prematurely born scarcely ever wake except for food. We may assume that, as a general rule, infants take treble the amount of sleep that

adults do; and that very young infants thrive the better the larger the amount of sleep they get, is borne out by the experience of medical practitioners, who affirm that they have known many children who were born small and weakly, but who slept the greatest part of their early existence, afterwards become strong and healthy; whilst those children, on the contrary, who, being born large and strong, were not good sleepers. As regards the sleep of adults, if the slumber has been of average length, or the subject of it awakes fully refreshed therefrom, a second sleep, instead of being conducive, is prejudicial to health, and should never be encouraged.

During sickness a patient, if in a very helpless and enfeebled state, may often be exposed whilst asleep to great peril, unless the nurse who attends him exercises intelligence and a proper amount of vigilance. In his work on 'Household Medicine' Dr Gardner has pointed out the dangers that beset the sleeping patient, and the means by which they may be avoided. "Having disposed," he says, " of the patient in bed in the best manner, be careful that no part of the pillow can project over the mouth or nose, and that the bedclothes do not cover the mouth.

" The attendant should be particularly attentive to these points when a narcotic has been taken, when the disease is paralysis, fever, head diseases, bronchitis, or any pulmonary complaint. The patient should be watched until he sleeps, and during his sleep, if a nurse is not constantly present, should be visited frequently, to observe whether the mouth and nostrils are free, and nothing obstructs the breathing.

" Very little suffices for an obstruction in such cases, which may extinguish life. Hundreds, perhaps we may say thousands of persons die prematurely from suffocation during sleep, in a low condition of the vital energies.

" How often does it happen that a patient left in a calm sleep is found dead upon being visited an hour or two after! Soft yielding pillows, in which the head and face get buried, are the instruments of suffocation to weakly persons very, very often."

The larger amount of sleep indulged in by the very old, over adults, is referable to the incapacity of the aged for exercise, and to their enfeebled powers of nutrition. Besides age, temperament, habit, and surrounding circumstances exercise considerable influence on the amount of sleep necessary for man. Persons of lymphatic temperament are generally great sleepers; whilst those of a nervous and active nature are mostly the reverse. The late Earl Russell was, we believe, in the years of his active political life a very small sleeper, his slumbers seldom extending over five hours. So likewise was the Duke of Wellington; General Elliott, the defender of Gibraltar, seldom slept more than four hours out of the twenty-four. As a contrast to these cases may be mentioned that of Dr Reid, the metaphysician, of whom it is stated that he could take as sufficient food and afterwards as much sleep as would suffice for an ordinary man for two days.

Several well-attested cases of excessive slumber

are on record in which the sleep lasted in some cases for weeks, and in others even for months.

In the 'Comptes Rendus' for 1864, Dr Blanchet records the case of one of his patients, a lady of 24 years of age, who had slept for 40 days when she was 18 years of age. Two years later she had a sleep lasting 50 days. Upon a subsequent occasion she fell asleep on Easter Sunday, 1862, and did not wake till March, 1863. She was fed during this period with milk and soup. She continued motionless and insensible, the pulse was low, the breathing scarcely perceptible, there were no evacuations, and she betrayed no signs of wasting away, whilst her complexion is described as florid and healthy.

This, however, as well as other cases of a similar kind, must not be regarded as an extreme instance of healthy slumber, but as a form of lethargy or coma, as indicative of disease, as the opposite condition of sleeplessness, that is frequently an accompaniment of certain forms of fevers, inflammatory affections, and brain disorders.

SLEEPLESSNESS AND COLD FEET. The relation between cold feet and sleeplessness is much closer than is commonly imagined. Persons with cold feet rarely sleep well, especially women. Yet the number of persons so troubled is considerable. We now know that if the blood-supply to the brain be kept up, sleep is impossible. An old theologian, when weary and sleepy with much writing, found that he could keep his brain active by immersing his feet in cold water; the cold drove the blood from the feet to the head.

Now, what this old gentleman accomplished by design is secured for many persons much against their will. Cold feet are the bane of many women. Tight boots keep up a bloodless condition of the feet in the day, and in many women there is no subsequent dilatation of the blood-vessels when the boots are taken off. These women come in from a walk and put their feet to the fire to warm—the most effective plan of cultivating chilblains. At night they put their feet to the fire, and have a hot bottle in bed. But it is all of no use; their feet still remain cold. How to get their feet warm is the great question of life with them—in cold weather. The effective plan is not very attractive at first sight to many minds. It consists in first driving the blood-vessels into firm contraction, after which secondary dilatation follows. See the snowballer's hands: the first contact of the snow makes the hands terribly cold; for the small arteries are driven thereby into firm contraction, and the nerve-endings of the finger-tips feel the low temperature very keenly. But as the snowballer perseveres, his hands commence to glow; the blood-vessels have become secondarily dilated, and the rush of warm arterial blood is felt agreeably by the peripheral nerve-endings. This is the plan to adopt with cold feet. They should be dipped in cold water for a brief period; often just to immerse them, and no more, is sufficient; and then they should be rubbed with a pair of hair flesh-gloves, or a rough Turkish towel, till they glow, immediately before getting into bed. After this a hot-water bottle will be successful enough in maintaining

the temperature of the feet, though without this preliminary it is impotent to do so. Disagreeable as the plan at first sight may appear, it is efficient; and those who have once fairly tried it continue it, and find that they have put an end to their bad nights and cold feet. Pills, potions, lozenges, 'night-caps,' all narcotics, fail to enable the sufferer to woo sleep successfully; get rid of the cold feet, and then sleep will come of itself ('British Medical Journal').

SMALLPOX. See POX.

Smallpox in Sheep. *Syn.* VARIOLA OVINA. This disease, although bearing the same name as that which attacks the human subject, is a perfectly distinct malady, and incapable of being communicated to man either by inoculation or contagion. In about ten days from the time of the animal's having imbibed the contagion feverish symptoms set in, accompanied with a mucous discharge of a purulent character from the nose. Red inflammatory pimples then begin to develop, first appearing where the skin is thin. After the pimples have been out about three days they assume a white appearance, and are filled with serum and pus. "Some of the vessels dry up, leaving brown scabs; others, especially in the severer cases, run together, and the scarf-skin is detached, leaving an ulcerated surface. It is in this ulcerated stage that the prostration reaches its height, and that most sheep die. The mortality from smallpox in sheep ranges from 25% to 90%." (*Finlay Dun*).

The disease being a very infectious one, the affected animals must be kept separate from the healthy ones. Thirty grains of chlorate of potash should be given three times a day, whilst the food should be nutritious and such as to tempt the animal's appetite. It may consist of bruised oil-cake, bran, and steeped oats. Professor Simonds recommends inoculation as a prophylactic measure.

Smallpox Marks (Prevention of). 1. For preventing disfigurement from smallpox marks, Dr Bernard suggests that the pustules as soon as they have acquired a certain size should be punctured with a fine needle, and then repeatedly washed with tepid water.

2. Dr Thorburn Patterson prescribes the following ointment:—Carbonic acid, 20 to 30 minims; glycerine, 1½ dr.; ointment of oxide of zinc, 6 dr.

3. Cream smeared on the pustules frequently during the day with a feather. See also OINTMENTS.

SMALTS. *Syn.* AZURE, POWDER BLUE, SILICEOUS B., SMALT; AZURUM, SMALTA, L. This consists, essentially, of glass coloured by fusing it with oxide of cobalt.

Prep. 1. Cobalt ore is roasted, to drive off the arsenic, then made into a paste with oil of vitriol, and heated to redness for an hour; the residuum is powdered, dissolved in water, and the ferric oxide precipitated with carbonate of potassium, gradually added, until a rose-coloured powder begins to fall; the clear portion is then decanted and precipitated with a solution of silicate of potassium (prepared by fusing together, for 5 hours, a mixture of 10 parts of potash, 15 parts of finely ground flints, and 1 part of charcoal); the precipitate after being dried is fused, and reduced to a very fine powder. A very rich colour.

2. Roasted cobalt ore and carbonate of potassium, of each, 1 part; siliceous sand, 3 parts; fuse them together, and cool and powder the residuum. Used as a blue pigment, also to colour glass, and for 'blueing' the starch used to get up linen. See BLUE PIGMENTS.

SMELL'ING SALTS. See SALTS, SMELLING.

SMELT. A beautiful little abdominal fish abounding in the Thames, and a few other rivers, between the months of November and February. It is esteemed a great delicacy by epicures, but sometimes proves offensive to the delicate and dyspeptic.

SMOKE PREVEN'TION. Although the full consideration of this subject belongs to public hygiene and civil engineering, its immediate application and advantages are interesting and important to everybody.

The history of smoke burning scarcely commences before the year 1840, at which date Mr Charles Wye Williams obtained a patent for this purpose. Since that time a 'thousand and one' schemes, either patented or non-patented, professedly for the same object, have been brought before the public. Most of these have been supported by the most reckless statements regarding their value, made by interested parties; and the most serious inconvenience and losses have often followed their adoption. Williams's method is to admit an abundant supply of cold air through a large number of small perforations in the door and front part of the furnace. Lark's method is based on the admission of heated air, under due regulation, both through the door and at the bridge or back of the furnace, by which means combustion is rendered more complete, and smoke thereby prevented.

Ivison's plan consists in the introduction of steam by minute jets over the fire, which is thus greatly increased in intensity without the production of smoke, and with a saving of fuel. In Jucke's arrangement the grate bars of a furnace are replaced by an endless chain web, which is carried round upon two rollers, in such a way that each part of the fuel is exposed to conditions most favourable for perfect combustion. Other inventions are based upon supplying fuel to the fires from beneath, so that the products of combustion must pass through the incandescent coals above.

For household fires, the smokeless grate, invented by Dr Arnott, will be found entirely successful, and most economical. Its general introduction would be a great advance in both domestic and public hygiene; and, being hence of national importance, should be enforced by law.

SMO'KING. This is done, on the large scale, by hanging up the articles (previously more or less salted) in smoking rooms, into which smoke is very slowly admitted from smothered dry-wood fires, kindled in the cellar, for the purpose of allowing it to cool and deposit its cruder part before it arrives at the meat. This process requires from six days to as many weeks to perform it properly, and is best done in winter. In farmhouses, where dry wood is burnt, hams, &c., are often smoked by hanging them up in some cool part of the kitchen chimney. When the meat is cut into slices, or scored deeply with a knife, to allow the smoke to penetrate it, it is called 'BUCANING.'

"The quality of the wood has an influence upon the smell and taste of the smoke-dried meat, smoke from beech wood and oak being preferable to that from fir and larch. Smoke from the twigs and berries of juniper, from rosemary, peppermint, &c., impart somewhat of the aromatic flavour of these plants" (*Ure*). The occasional addition of a few cloves or allspice to the fuel gives a very agreeable flavour to the meat.

Hung beef, a highly esteemed variety of smoked beef, is prepared from any part, free from bone and fat, by well salting and pressing it, and then drying and smoking it in the usual manner. It is best eaten shredded. See PUTREFACTION, SALTING, &c.

SNAKE-POISONING, Mortality from. The 'Lancet' (August 11th, 1870), quoting a letter from T. B. Beighton, Esq., of the Bengal Civil Service, magistrate of the Culna district of the Burdwan province of Bengal, says:—"The Culna district comprises, we presume, 80 or 100 square miles, and has a population of about 300,000. Mr Beighton says that deaths from snake-bite are singularly common in the subdivision. An average of one per day is reported through the police. The actual deaths are probably double the number reported. If this daily average is meant to apply the whole year round, we should thus get in a comparatively small district the frightful death of 700 persons from snake-bite. It is lamentable to think that, despite the supposed remedial discoveries in this direction, we still seem to be without an agent to neutralise the effects of the bites of poisonous snakes."

SNAKE-ROOT. See SENEGA. For 'Virginian snake-root' see SERPENTARY. Snakeweed (*Bistorta radix*) is the root of *Polygonum bistorta*, Linn.

SNIPE. The *Scolopax gallinago*, a well-known bird indigenous to this country. It is fine-flavoured, but rather indigestible.

SNOW, Foreign Bodies in. M. Bondier ('Journ. Ch. Soc.'), having lately made an examination of snow, records that of the solid matters floating in the air and retained in the snow, the most abundant was found to be sand; next some cells of *Protococcus viridis*, and spores and filaments of other cryptogams; then granules of starch and cells and fibres of various plants. Epithelial cells and hairs of animals were also present, as well as fibres of wool and silk. These last, being dyed, indicated the presence of man, as did also the fibres of hemp, cotton, and indigo. The amount of foreign matter was greatest in the snow collected at the lowest levels, especially in the vicinity of human habitations and of woods, which are both fertile sources of floating particles. Permanganate of potassium was used to estimate the amount of organic matter dissolved by the filtered snow water.

Immediately mixed with the soot were the ferruginous corpuscles observed by Tissandier. These are regarded by the author, in opposition to the opinion of Tissandier, as of terrestrial origin.

SNUFF. *Syn.* PULVIS TABACI, L.; TABAC EN POUDRE, Fr. A powder prepared from tobacco, for the purpose of being sniffed up the nose as a stimulant or intoxicant.

The finer kinds of snuff are made from the soft portions of the best description of manufactured leaf-tobacco, separated from the damaged portion; but the ordinary snuffs of the shops are mostly prepared from the coarser and damaged portions, the midribs, stems, or stalky parts that remain from the manufacture of 'shag tobacco,' the dust or powder sifted from the bales, and the fragments that are unfit for other purposes.

Prep. The proper materials being chosen, and if not in a sufficiently mature state rendered so by further fermentation, they are sufficiently dried by a gentle heat or exposure to the air to admit of being pulverised. This is performed on the large scale in a mill, and on the small scale with a kind of pestle and mortar. During the operation the tobacco is frequently sifted, that it may not be reduced to too fine a powder, and is several times slightly moistened with rose or orange-flower water, or eau d'ange, which are the only liquids fit for the superior kinds of snuff. In preparing the dry snuffs no moisture is used. The scent or other like matters are next added, and, after thorough admixture, the snuff is packed in jars or canisters.

Adult. During the grinding of tobacco it is frequently mixed with dark-coloured rotten wood, various English leaves, colouring, and other matter. Ammonia, hellebore, euphorbium, and powdered glass are common additions to snuffs to increase their pungency. We have seen powdered sal-ammoniac sent by the hundredweight at one time to a certain celebrated London tobacconist. The moist kinds of snuff are generally drugged with pearlash, for the triple purpose of keeping them damp and increasing their pungency and colour. The dry snuffs, especially 'Scotch' and 'Welsh,' are commonly adulterated with quicklime, the particles of which may be occasionally distinguished even by the naked eye. This addition causes their biting and desiccating effect on the pituitary membrane. "We were once severely injured by taking snuff which, after our suspicions were awakened, we found to contain a mixture of red-lead and umber" (*Cooley*).

The following circumstance, related by Dr Garrod ('Lancet') in a lecture at King's College Hospital, leads to the inference that the custom of packing snuff in lead is not free from danger. The doctor says:—A gentleman, a resident in India, began to suffer some time since from nervous exhaustion, anæmia, and debility of both extremities; he was a great snuff-taker, taking on an average as much as an ounce in the course of a day. He consulted several medical men in India, and they attributed his symptoms to inordinate snuff-taking. He, however, continued to take snuff and to get worse, and at last came to England to seek further advice. When Dr Garrod saw him he discovered a blue line on the gums. His suspicions were directed to the snuff, which he found to contain a considerable quantity of lead. To ascertain whether or not the presence of lead in this case was an accidental circumstance, six packets were ordered from the house in Calcutta with which the gentleman had been in the habit of dealing. The snuff was contained in sheet-lead packages, which were all found to contain lead to about the same extent as the first specimen. Dr Garrod exhibited a solution, which he tested in the following way:—Ten grains of snuff were burned in a platinum crucible, and the ash was treated with nitric acid, the crystallised result was dissolved in water with the addition of a small quantity of acetic acid, and then tested with iodide of potassium, which threw down an abundant precipitate of yellow iodide of lead. The leaden packages were labelled 'best brown rappee,' and bore the name of a well-known English firm, from which they had been exported to India. The snuff itself was rather moist. Where it adhered to the sides of the case it was dotted with white spots, probably consisting of carbonate of lead, formed by, Dr Garrod suggests, the fermentation of the damp snuff.

Since Dr Garrod's attention has been directed to this subject he has spoken to a medical man recently returned from Calcutta, who told him that he had quite lately met with three cases of lead-poisoning, which, on investigation, were found to be due to the use of snuff.

Var. Snuffs are divided into two kinds—DRY SNUFFS, as 'Scotch,' 'Irish,' 'Welsh,' and 'Spanish snuff,' 'Lundyfoot,' &c.; and MOIST SNUFFS, or RAPPEES, including 'black' and 'brown rappee,' 'carrotte,' 'Cuba,' 'Hardham's mixture,' 'prince's mixture,' 'princess,' 'queen's snuff,' &c. The last three also come under the denomination of SCENTED SNUFFS.

The immense variety of snuffs kept in the shops, independently of the above-named conditions, depend for their distinguishing characteristics on the length of the fermentation, the fineness of the powder, the height to which they are dried, and the addition of odorous substances. Tonquin beans, essence of tonquin bean, ambergris, musk, civet, leaves of *Orchis fusca*, root and oil of *Calamus aromaticus*, powder and essence of orris root, and the essences or oils of bergamot, cedra, cloves, lavender, petit grain, neroli, and roses (otto), as well as several others, either alone or compounded, are thus employed. TABAC PARFUMÉ AUX FLEURS is perfumed by putting orange flowers, jasmines, tuberoses, musk roses, or common roses, to the snuff in a close chest or jar, sifting them out after 24 hours, and repeating the treatment with fresh flowers as necessary. Another way is to lay paper, pricked all over with a large pin, between the flowers and the snuff.

MACOUBA SNUFF is imitated by moistening the tobacco with a mixture of treacle and water, and allowing it to ferment well.

SPANISH SNUFF is made from unsifted 'Havannah snuff,' reduced by adding ground Spanish nut-shells, sprinkling the mixture with treacle water, and allowing it to sweat for some days before packing.

YELLOW SNUFF is prepared from ordinary pale snuff moistened with a mixture of yellow ochre diffused in water, to which a few spoonfuls of thin mucilage have been added; when dry, the colour that does not adhere to the snuff is separated with a fine sieve.

RED SNUFF. As last, but using red ochre.

Snuff, Asarabac'ca. *Syn.* CEPHALIC SNUFF, COMPOUND POWDER OF ASARABACCA; PULVIS ASARI COMPOSITUS, L. *Prep.* 1. (Ph. D. 1826.) Asarabacca leaves, 1 oz.; lavender flowers, 1 dr. (both dried); mix and powder them.

2. (Ph. E. 1817.) Asarabacca leaves, 3 dr.; leaves of marjoram and flowers of lavender, of each, 1 dr.; as before. Both are used as errhines in headaches and ophthalmia. See SNUFF, CEPHALIC, ASARABACCA, &c.

Snuff, Cephal'ic. *Prep.* 1. From asarabacca leaves and Lundyfoot snuff, of each, 2 oz.; lavender flowers, ¼ oz.; essence of bergamotte and oil of cloves, of each, 2 or 3 drops; mixed and ground to a powder, the perfume being added last.

2. (*Boeli's.*) From tobacco or pure snuff and valerian root, of each, ½ oz.; reduced to powder, and scented with the oils of lavender and marjoram, of each, 5 or 6 drops. *Obs.* The first formula is an excellent one; and the product is very useful in nervous headaches, dimness of sight, &c. See SNUFF, ASARABACCA (*above*).

Snuff, Eye. *Prep.* From finely levigated tribasic sulphate of mercury ('Turpeth mineral'), ½ dr.; pure dry Scotch or Lundyfoot snuff, 1 oz.; triturate them well together. A pinch of this, occasionally, has been recommended in inflammation of the eyes, dimness of sight, headache, polypus, &c.; but it should be used with caution, and not too often.

SOAP. *Syn.* SAPO, L.; SAVON, Fr. SPANISH, CASTILE, or HARD SOAP, made with olive and soda (SAPO, SAPO EX OLIVE OLEO ET SODÂ CONFECTUS—Ph. L.; SAPO DURUS—B. P., Ph. E. & D.), and SOFT SOAP, made with olive oil and potash (SAPO MOLLIS—B. P., Ph. L. & E.; SAPO EX OLIVE OLEO ET POTASSÂ CONFECTUS—Ph. L.), are the only kinds directed to be employed in medicine. The former is intended whenever 'soap' is ordered, and is the one which is principally employed internally; the latter is used in ointments, &c., and in some of the officinal pills. *Definition.* Chemically speaking, a soap is

$$C_3H_5(C_{18}H_{35}O)_3.O_3 + 3NaHO = 3Na(C_{18}H_{35}O)O + C_3H_8O_3$$
Stearin.   Soda.   Sodium stearate.   Glycerin.

$$C_3H_5(C_{18}H_{33}O)_3.O_3 + 3NaHO = 3Na(C_{18}H_{33}O)O + C_3H_8O_3$$
Olein.   Soda.   Sodium oleate.   Glycerin.

Raw Materials used in Soap-making. Besides tallow, many other kinds of fatty matter are used, namely, bone grease (fresh bones bruised, boiled in water, and the fat skimmed off when cold), lard, kitchen waste, glue fat; seal, sperm, fish, and whale oils, spermaceti, castor, cotton-seed, dill, hemp-seed, linseed, and sunflower-seed oils; colza, illipe, olive, cocoa-nut, almond, and beech-nut oils; cacao butter, shea butter, palm oil or butter, palm-kernel oil, and ground-nut oil; rosin or colophony, recovered grease from the washings of woollen works.

Recovery of Alkali (*Tessié de Mothay*). In the print works of Alsace, where an immense quantity of egg-albumen is consumed, there collect, as a necessary result, enormous quantities of the yolks of egg. Amongst other purposes to which these are applied, that of soap-making is one. According to Kingzett, the olein is not the

produced whenever a *metallic base* is combined with a *fatty acid*, such as the acids of the general formula $C_nH_{m-2}O_p$, occurring in or obtainable from the natural fats or fixed oils; and hence, besides the ordinary commercial soaps, we have the *lead soap*, or lead plaster of pharmacy, and also *manganese, copper, mercury, zinc, tin, silver, aluminium,* and other *metallic soaps.* But in ordinary language by *soap* we understand a compound of an *alkali* and a fatty acid, the alkali potash affording, when so combined, *soft* soap, and the alkali soda forming *hard* soap.

Kingzett ('The Alkali Trade,' p. 173) gives this definition: "Soap, considered commercially, is a body which on treatment with water liberates alkali."

Prep. (HARD SOAP.) The fatty or oleaginous matter is boiled with a weak alkaline lye (soap lye), prepared by decomposing soda-ash with slaked lime and decanting the clear lye, and portions of stronger lye are added from time to time, the ebullition being still continued until these substances, reacting on each other, combine to form a tenacious compound, which begins to separate and rise to the surface of the water; to promote this separation and the granulation of the newly formed soap some common salt is generally added, and the fire being withdrawn, the contents of the boiler are allowed to repose for some hours, in order that the soap may collect into one stratum and solidify; when this happens it is put into wooden frames or moulds, and when it has become stiff enough to be handled it is cut into bars or pieces, and exposed to the air, in a warm situation, to further harden and to dry.

Chemistry of the Process. Tallow, a typical soap-material, contains two fatty substances, one of which, *stearin* ($C_{57}H_{110}O_6$), is solid; and the other, *olein* ($C_{57}H_{104}O_6$), liquid,—the quantity of the former being about three times more than that of the latter. The action of soda decomposes these fats into stearic and palmitic acids, which combine with the base to form soap, whilst a peculiar sweet substance (glycerin) passes into solution.

only ingredient of the yolk which reacts upon the soda or potash, and thus produces soap; but the yolk also contains another body (lecithine, $C_{44}H_{84}NPO_9$), which, absorbing water under the influence of the bases, splits up into oleic and margaric acids.

The soap water is decomposed by calcium, barium, or magnesium carbonate, and then carbonic acid is passed through the liquid. The bicarbonate formed precipitates organic matter and other impurities, and these settle down. The solution is then evaporated or treated with baryta water, which precipitates the last portion of foreign matters, and leaves a solution of caustic alkali. At a particular stage of the process an acid is used in order to hasten the separation of the resinous substances, and, in certain cases, of the sulphides of sodium and calcium, or barium and calcium and ferric oxide, and then carbonic acid passes into the liquid. The precipitated metallic substances carry down with them the humus-like substances present.

Recovery of the Glycerin from Spent Lyes. (*Foresmann's process:* 'Chemical News,' June 24, 1881.) A great many methods have been proposed. Cameron (*vide* 'Soap and Candle,' p. 176) gives twelve. Of these one of the most recent is as follows:

a. The lye is evaporated with heat until the salts which it contains begin to crystallise out.

b. The liquor is then cooled and filtered to get rid of gelatinous and albuminous matters.

c. Carbonic acid is sent through the liquor; this precipitates sodium bicarbonate, which is separated off.

d. Gaseous hydrochloric acid is next passed into the liquor until any remaining sodium carbonate is converted into chloride and precipitated as such.

e. The chloride of sodium is separated; the liquor, which now consists of water, glycerin, and hydrochloric acid, is evaporated to get rid of the acid, which is absorbed in water for use again.

f. The dilute glycerin is purified by filtration through animal charcoal, concentration, and, finally, by distillation.

Var. The principal varieties of soap found in commerce are—

ALMOND SOAP (SAPO AMYGDALINUS), made from almond oil and caustic soda, and chiefly used for the toilet.

The P. Codex gives the following formula for its preparation:—Solution of caustic soda (1·334), by weight, 10 oz.; oil of almonds, by weight, 21 oz.; add the lye to the oil in small portions, stirring frequently; leave the mixture for some days at a temperature of from 64° to 68° F., stirring occasionally, and when it has acquired the consistence of a soft paste put it into moulds until sufficiently solidified. It should be exposed to the air for one or two months before it is used.

ANIMAL SOAP (SAPO ANIMALIS, CURD SOAP—B. P.). A soap made with soda and a purified animal fat, consisting principally of stearin (P. Cod.). Put 5 parts of beef marrow with 10 parts of water into a porcelain or silver basin, heat, and when melted add by portions, with constant stirring, 2½ parts of liquor soda (1·38); when saponified add 1 part of salt; stir, remove the soap from the surface, drain it, melt it with a gentle heat, and pour it into moulds.

CASTILE SOAP (SPANISH s., MARSEILLES s.; SAPO CASTILLENSIS, SAPO HISPANICUS). An olive-oil soda soap, kept both in the white and marbled state. The former is said to be the purest, the latter the strongest. Olive oil contains, besides olein, a solid fat called margarin, which is really composed of palmitin and stearin; hence the soap made from it is a compound of oleate, palmitate, and stearate of soda.

CURD SOAP, made with tallow (chiefly) and soda (see *above*).

MEDICATED SOAPS, containing various active ingredients. The chief of these are noticed below.

MOTTLED SOAP, made with refuse kitchen stuff, &c.

An impure soda, containing sulphides, is preferred for the lye, and about 8 oz. of ferrous sulphate (green vitriol) is added for each cwt. of oil at the end of the preliminary boiling. This sulphate is precipitated partly as iron oxide and sulphide, and partly as an insoluble iron soap. The soap is worked with a rake before moulding, in such a way as to preserve and arrange the colouring matters yielded by the sulphate and the sulphides, &c., in series, so as to present a 'marbled' or 'mottled' appearance. By exposure to the air the iron gets oxidised to the state of sesquioxide, and a reddish tint called *manteau Isabelle* is diffused over the bluish mottled mass. Mottling in blue, grey, and red is also produced by Blake and Maxwell's process.

SOFT SOAP (of commerce), made with whale, seal, or cod oil, tallow, and caustic potash. The fish oils contain chiefly olein, which, when saponified with potash, gives potassium oleate, and this is the chief constituent of 'soft' soap.

N. Gräger gives the following method for the easy determination of the fat and alkali in soft (potash) soaps:—25 to 50 grms. of soap are dissolved in 150 c.c. of water by aid of heat, cooled, and mixed with an excess of salt, so that a soda soap separates out; the latter is washed on a paper filter with a saturated solution of salt. In the filtrate the free alkali is estimated by normal acid. The precipitate is decomposed by warming with excess of normal acid, and the quantity of acid neutralised by the combined alkali determined by a standard soda solution. The cake of fat which separates in the last operation is dried and weighed after adding to it, while melted, a known weight of stearin or paraffin to give it hardness.

TOILET SOAPS, prepared from any of the preceding varieties, and variously coloured and scented. Formulæ are given below.

YELLOW SOAP (RESIN SOAP), made with inferior tallow, 3 parts; resin, 1 part; and caustic soda, the resin being added shortly before the soap is 'finished.' Soluble glass is now largely employed in place of resin.

"When yellow soap is made with the cheaper kinds of fat it will hardly acquire a sufficient degree of firmness or hardness to satisfy the thrifty washerwoman. It melts away too rapidly in hot water, a defect which may be well remedied by the introduction into the soap of a little (1-20th) fused sulphate of soda; and this salt concreting gives the soap a desirable hardness, whilst it improves its colour, and renders it a more desirable article for the washing tub" (*Ure*).

See SOAPS (Medicated and Toilet).

Soaps are also divided into SOFT or POTASH SOAPS, and HARD or SODA SOAPS.

Assay. (*Filsinger's scheme*—'Chemiker Zeitung,' April 17th, 1884).

1. *Water.* Five grms. of hard soap scraped from the sides and centre of a fresh section are gently warmed over a water-bath, and finally dried at 100° C. until the weight is constant.

Ten grms. of soap are taken, spread in a thin layer over a large watch-glass, and treated in the same way.

2. *Unsaponifiable Fatty Matters.* The dry residue from (1) is finely powdered, and washed on a filter three or four times with lukewarm petro-

leum ether. The filtrates are collected in a weighed beaker, evaporated, dried, and weighed.

3. *Free Alkali.* The residue from (2) is digested for a short time with alcohol (95%), slightly warmed, filtered, the residue on the filter washed with warm alcohol, and the filtrate, to which a few drops of a phenol phthalein solution are added, triturated with decinormal sulphuric acid.

4. *Foreign Matters.* These are found by the usual method, together with the chlorides, sulphate, and carbonate on the filter in (3).

5. *Fatty Acid.* The neutralised alcoholic solution from (3) is mixed with water in a moderate-sized porcelain basin, the fatty acids precipitated by sulphuric acid, and after melting and settling 5 grms. of dry wax are added. When the whole is cool the fat acid wax is removed, washed with water and alcohol, dried without melting, and cooled. The weight − 5 grms. = the quantity of fatty acids.

6. *Glycerin.* The liquid from the cake of fatty acids is treated with a small excess of barium carbonate, heated, filtered, the filter washed with hot water, and the filtrate evaporated to dryness. The residue is repeatedly washed with alcoholic ether, the filtrate evaporated in a porcelain dish, dried at a temperature of 70° C., and weighed.

7. *Total Alkali.* Ten grms. of another portion of soap prepared as in (1) are dried in a platinum dish, and then heated till all the fatty acids have been destroyed. The porous carbonaceous residue is boiled with water, filtered into a ½-litre flask, and the filter washed with hot water till the washings cease to give an alkaline reaction. The bulk is then made up, the whole well mixed, and 25 c.c. (=1 grm. soap) of the solution are titrated with sulphuric acid. The result represents the amount of total alkali, and, after deducting the quantity of free alkali found by (3), the remainder is the proportion of alkali combined with fatty acids, and existing as carbonate and silicate.

8. *Chlorine.* The neutral titrated solution from (7) may be used for the determination of chlorine by decinormal silver solution.

9. *Silicic Acid.* Seventy-five c.c. of the solution from (7) are treated with excess of hydrochloric acid evaporated to dryness, treated with water, filtered, and the residue ignited and weighed as silica.

10. *Sulphuric Acid.* The filtrate from (9) is boiled, and, while boiling, barium chloride is added, the precipitated barium sulphate washed, dried, and weighed, and calculated as sodium or potassium sulphate, according to the nature of the soap under examination.

11. Potash and soda, if both are present, must be determined in the usual way by platinum chloride.

ANOTHER METHOD OF SOAP ASSAY (*Moffit*). The constituents to be determined in an analysis of soap are alkalies (combined and free), carbonates, fatty acids, resin, glycerin, salts, colouring matters, and water.

Three portions of the finely divided soap are weighed off, containing respectively 10 grms., 20 grms., and 40 grms. Ten grms. are digested with alcohol on the water-bath and filtered. The residue, containing carbonates and other salts, colouring matter, &c., is dried at 100°, weighed,

digested with water, and titrated with normal oxalic acid. Every c.c. of acid used indicates 0·053 Na_2CO_3.

Regard must be had to a slight precipitate of calcium oxalate. The weight of Na_2CO_3 found is subtracted from the total residue insoluble in alcohol; the difference is the weight of the salts and foreign matters. The filtrate is subjected to a stream of carbonic acid, filtered, and the precipitate dissolved in water and triturated with oxalic acid. Each c.c. of acid indicates 0·031 free soda, or 0·042 free potash. No precipitate shows the absence of free alkalies. The filtrate from the precipitate produced by the carbonic acid is, after the addition of 15 c.c. of water, evaporated to remove the alcohol. The aqueous solution, treated with normal oxalic acid to acid reaction, shows for every c.c. of acid 0·031 soda, or 0·042 potash in combination.

Sulphuric acid is then added, and the whole is heated on a water-bath with pure beeswax to separate the fatty acids and resin, which are then weighed, the weight of the beeswax being subtracted.

Forty grms. of the soap are next dissolved in water and mixed with sulphuric acid as long as any precipitate is formed. On standing, the fatty acids separate, and can be dried and weighed. These fatty acids are digested with a mixture of equal volumes of water and alcohol till the liquid on cooling ceases to appear milky. The solid layer is again dried and weighed, and the difference between the weight and that obtained above shows the weight of the resin.

The melting-point of the acids is next determined. Ten grms. are then dissolved in alcohol, and sulphuric acid mixed with alcohol is added till a precipitate is no longer formed. The liquid is filtered, mixed with barium carbonate, and again filtered. The sweet residue left after evaporation of the alcohol is glycerin. The weights of the carbonates, salts, and foreign matters, free and combined alkalies, fatty acids, resin, and glycerin are added together, and the sum subtracted from 10 grms. gives the weight of the water.

See also "Soap Analysis," 'Chem. News,' xxxv, 2. The article is too long to allow of insertion here.

Uses, &c. The common uses of soap need not be enumerated. As a medicine it acts as a mild purgative and lithontriptic, and it has been thought by some to be useful in certain affections of the stomach arising from deficiency of bile. Externally it is stimulant and detergent.—*Dose,* 3 to 20 or 30 gr., made into pills, and usually combined with aloes or rhubarb.

Concluding Remarks. Prior to the researches of Chevreul no correct ideas were entertained as to the constitution of soap. It was long known that the fixed oils and fats, in contact with caustic alkaline solutions at a high temperature, undergo the remarkable change which is called saponification; but here the knowledge of the matter stopped. Chevreul discovered that if the soap thus produced be afterwards decomposed by the addition of an acid, the fat which separates is found to be completely changed in character; to have acquired a strong acid reaction when applied in a melted state to test-paper, and to have become soluble with the greatest facility in warm alcohol; —in other words, that a new substance capable of forming salts, and exhibiting all the characteristic properties of an acid, has been generated out of the elements of the neutral fat under the influence of the base. Stearin, when thus treated, yields stearic acid, palmitin gives palmitic acid, olein gives oleic acid, and common animal fat, which is a mixture of several neutral bodies, affords, by saponification by an alkali and subsequent decomposition of the soap, a mixture of the corresponding fatty acids. These bodies are not, however, the only products of saponification; the change is always accompanied by the formation of a very peculiar sweet substance called glycerin, which remains in the mother-liquor from which the acidified fat has been separated. The process of saponification itself proceeds with perfect facility even in a closed vessel; no gas is disengaged; the neutral fat, of whatsoever kind, is simply resolved into an alkaline salt of the fatty acid, which is soap, and into glycerin, a neutral body resembling syrup, and miscible with water in every proportion. Liebig ('Familiar Letters on Chemistry,' letter xi, p. 129), referring to the extraordinary development for which the soap industry is remarkable, said, "The quantity of soap consumed by a nation would be no inaccurate measure whereby to estimate its wealth and civilisation. Of two countries with an equal amount of population, we may declare with positive certainty that the wealthiest and most highly civilised is that which consumes the greatest weight of soap."

Soap, Arsen'ical. *Syn.* SAPO ARSENICALIS, L. *Prep.* (Bécœurs.) From carbonate of potash, 12 oz.; white arsenic, white soap, and air-slaked lime, of each, 4 oz.; powdered camphor, ¼ oz.; made into a paste with water, q. s. Used to preserve the skins of birds and other small animals.

Soap, Black. *Syn.* SAPO NIGER, S. MOLLIS COMMUNIS, L. A crude soft soap, made of fish oil and potash; but the following mixture is usually sold for it :—Soft soap, 7 lbs.; train oil, 1 lb.; water, 1 gall.; boil to a proper consistence, adding ivory-black or powdered charcoal, q. s. to colour. Used by farriers.

SOAPS (Med'icated). A few only of these deserve notice here:

Soap, Antimo"nial. *Syn.* SAPO ANTIMONIALIS, SAPO STIBIATUS, L. *Prep.* (Hamb. Cod. 1845.) Golden sulphuret of antimony, 2 dr.; solution of caustic potassa, 6 dr. (or q. s.); dissolve and triturate the solution with medicated (Castile) soap (in powder), 1½ oz., until the mass assumes a pilular consistence. It should be of a greyish-white colour.

Soap, Cod-liver Oil. *Syn.* SAPO OLEI JECORIS (Deschamps). *Prep.* Cod-liver oil, 2 oz.; caustic soda, 2 dr.; water, 5 dr.; dissolve the soda in the water, and mix it with the oil. An iodurated soap is made by mixing with the above 1 dr. of iodide of potassium dissolved in 1 dr. of water.

Soap, Cro'ton. *Syn.* SAPO CROTONIS, L. *Prep.* From croton oil and liquor of potassa, equal parts; triturated together in a warm mortar until they combine. Cathartic.—*Dose*, 1 to 3 gr.

Soap of Gamboge. *Syn.* SAPO GAMBOGIE. (Soubeiran.) *Prep.* Mix 1 part of gamboge with 2 of soap, dissolve it with a little spirit, and evaporate to a pilular consistence.

Soap, Glycerine. The manufacture of transparent glycerine soap does not present any especial difficulty; there are nevertheless points which it is essential to observe. To produce first-class soap of this kind good materials are indispensable, and the proper proportions must be strictly adhered to. Tallow and stearin are the most useful hard fats, palm oil imparts to the soap its lathering qualities, and castor oil gives transparency. Great transparency is obtained by the use of spirit and of sugar-water, both of which bodies assist the normal saponification. The lye must be pure and clear as water; its proportion should not exceed 20% of lye per lb. of fatty matter, for an excess of alkali would make the soap too detergent, and a considerable excess of unsaponified fat would make it too weak and greasy. In one case the soap injures the skin in use, in the other the soap would soon lose its agreeable smell and finally become rancid. Even an excess of glycerin is to be avoided, as it renders the soap less transparent, and also too soft. No filling whatever except sugar solution is permissible. The following proportions are recommended :

White Alabaster Soap. Stearin, 13 lbs.; palm oil, 22 lbs.; glycerin, 13 lbs.; 38° lye, 18 lbs.; 96% alcohol, 26 lbs. The stearin and palm oil are to be heated to 65°, saponified with the lye, the alcohol added, and when the combination, which takes place at once, is complete, the glycerin is put in. When clear, the kettle is covered and the contents are allowed to stand at 45° R. The soap is run into the moulds and perfumed with bergamot oil, 120 grms.; geranium oil, 30 grms.; neroli oil, 25 grms.; citron oil, 20 grms.

As this is a white soap no colour is added.

Soap of Gua'iacum. *Syn.* SAPO GUAIACI, SAPO GUAIACINUS, L. *Prep.* (Ph. Bor.) Liquor of potassa, 1 oz.; water, 2 oz.; mix in a porcelain capsule, apply heat, and gradually add of resin of guaiacum (in powder), 6 dr., or as much as it will dissolve; next decant or filter, and evaporate to a pilular consistence.—*Dose*, 10 to 30 gr.; in chronic rheumatism, various skin diseases, &c.

Soap, I'odine. *Syn.* SAPO IODURATUS, L. *Prep.* From Castile soap (sliced), 1 lb.; iodide of potassium, 1 oz.; (dissolved in) water, 3 fl. oz.; melt them together in a glass or porcelain vessel over a water-bath. Excellent in various skin diseases; also as a common soap for scrofulous subjects.

Soap of Jal'ap. See JALAP, SOAP OF.

Soap, Larch. *Syn.* SAPO LARICIS (Dr Moore). *Prep.* Dissolve 12 oz. of white curd soap in 24 oz. of rose water on a steam-bath. Infuse 4 oz. of wheat bran in 10 oz. of cold water for twenty-four hours, and express. Add to the last 3 oz. of pure glycerine. Dissolve 6 dr. of extract of larch bark in 1 oz. of boiling water. Mix these solutions with the dissolved soap, evaporate over a steam-bath to a proper consistence, and pour into moulds to cool. For the local treatment of psoriasis.

Soap, Marine. (Patent.) This is made by substituting cocoa-nut oil for the fats and oils used in the manufacture of common soap. It has the advantage of forming a lather with salt water.

Soap, Mercu"rial. *Syn.* SAPO HYDRARGYRI (*M. Herbert*). 1. *Prep.* Dissolve 4 oz. of quicksilver in its weight of nitric acid without heat; melt in a porcelain basin by water-bath 18 oz. of veal suet, and add the solution, stirring the mixture till the union is complete. To 5 oz. of this ointment add 2 oz. of solution of caustic soda (1·33), porphyry slag till a soap is formed which is completely soluble in water. For external use, alone or dissolved in water, in some cutaneous diseases.

2. SAPO MERCURIALIS, L. (SAPO SUBLIMATIS CORROSIVI). *Prep.* From Castile soap (in powder), 4 oz.; corrosive sublimate, 1 dr.; (dissolved in) rectified spirit, 1 fl. oz.; beaten to a uniform mass in a porcelain or wedgwood-ware mortar.

3. (SAPO HYDRARGYRI, PRECIPITATI ALBI— *Sir H. Marsh.*) *Prep.* Beat 12 oz. of white Windsor soap in a marble mortar, add 1 dr. of rectified spirit, 2 dr. of white precipitate, and 10 drops of otto. Beat the whole to a uniform paste.

4. SAPO HYDRARGYRI, PRECIPITATI RUBRI (*Sir H. Marsh*). From white Windsor soap, 2 oz.; nitric oxide of mercury (levigated), 1 dr.; otto of roses, 6 or 8 drops; (dissolved in) rectified spirit, 1 to 2 fl. dr.; as the last. Both the above are employed as stimulant detergents and repellents in various skin diseases; also as SAVON ANTISYPHILITIQUE.

Soap Powder, Borax. Curd soap, 5 parts; soda ash, 3 parts; silicate of soda, 2 parts; borax, 1 part; mix.

Soap Powder, London. Yellow soap, 6 parts; soda crystals, 3 parts; pearlash, 1½ parts; sulphate of soda, 1½ parts; palm oil, 1 part; mix, spread out to dry, and powder.

Soap, Resin. Mr. H. Collier states the soap is made by boiling 180 gr. of common yellow resin and 300 gr. of caustic soda in a pint of water for two hours, at the end of which time it is reduced to a yellow pasty mass, which is to be heated to dryness and powdered. The product resembles powdered resin very closely, but dissolves readily in water with the aid of a little heat. The solution is saponaceous, but never gelatinises. The soap dissolves freely in rectified spirit also. If mercury is shaken with the aqueous solution (20 gr. to 1 oz.) the metal is broken up into minute globules, which do not run together again, as each globule is coated with the soap. Chloroform is transformed into a creamy liquid. In both cases the soap acts much better than tincture of quillaia; it gives more viscosity than the latter. Working on the B. P. C. formulary lines, an excellent *liquor carbonis* is obtained by dissolving 2 oz. of the soap in a pint of S. V. R. by heat, adding 4 oz. of purified coal-tar, heating to 120° F., setting aside for two days, and decanting or filtering. The soap gives splendid results with cod-liver, almond, olive, and castor oils. The plan is to dissolve 10 gr. of the soap in 1 oz. of water, and shake up with 1 oz. of the oil. Castor oil is less easily emulsified than the others,

but they do perfectly with 5 gr. of the soap. Essential oils are also very readily emulsified. There is an opening in this direction for making inhalation emulsions, the soap taking the place of the magnesia of the T. H. P. formulæ. Thus 10 gr. of the soap, 2 dr. of the volatile oil, and water to 3 oz., make an emulsion which mixes very well with water. Creosote requires 20 gr. of the soap. 5 gr. added to a drachm of spirit of camphor makes a preparation which mixes perfectly with water, the camphor not separating. So also 20 gr. with 1 oz. of tincture of tolu. Thymol, 18 gr., resin soap, 20 gr., spirit, 3 oz., is a good formula for a preparation which may be diluted, and for such oils as santal and copaiba we get good emulsions with a drachm of the oil, 10 gr. of the soap, and 2 oz. of water.

Soap, Sand. 100 lbs. of cocoa-nut oil are saponified with about 200 lbs. of lye at 20° B. The soap is then hardened by the addition of about 8 lbs. salt dissolved in water to a density of 15° B., with the addition of 6 to 8 lbs. soda ash. The soap is now covered up and the foam allowed to subside. After standing five to six hours the fob is skimmed and the soap is run off into the coolers. Whilst this process has been going on the sand has been dried and sifted, and the soap now being thoroughly crutched, the sand is sifted over it until 100 to 150 lbs. have been added. The crutching must be continued until the mass is perfectly cooled. The soap is very firm and hard, and must be cut as soon as cooled. To perfume the mixture add of essential oil of lavender, thyme, and coriander, 100 grms. each.

Soap, Scouring. Take 2 lbs. soda, 2 lbs. yellow bar soap, and 10 quarts water. Cut the soap in thin slices, and boil together two hours; strain, and it will be fit for use. The clothes should be soaked the night before washing, and to every pailful of water used to boil them add a pound of soap. They will need no rubbing, but merely rinsing ('Scientific American').

Soap, Sul'phuretted. *Syn.* SAPO SULPHURIS, SAPO SULPHURATUS, L. *Prep.* (*Sir H. Marsh.*) From white soap, 2 oz.; sublimed sulphur, ¼ oz.; beaten to a smooth paste in a marble mortar with 1 or 2 fl. dr. of rectified spirit strongly coloured with alkanet root, and holding in solution otto of roses, 10 or 12 drops. In itch and various other cutaneous diseases.

Soap, Tar. *Syn.* SAPO PICIS LIQUIDE, SAPO PICEUS, L. *Prep.* From tar, 1 part; liquor of potassa and soap (in shavings), of each, 2 parts; beat them together until they unite. Stimulant. Used in psoriasis, lepra, &c.

Soap, Tur'pentine. *Syn.* STARKEY'S SOAP; SAPO TEREBINTHINÆ, S. TEREBINTHINATUS, L.; SAVON TÉRÉBINTHINE, Fr. *Prep.* (P. Cod.) Subcarbonate of potash, oil of turpentine, and Venice turpentine, equal parts; triturate them together, in a warm mortar, with a little water, until they combine; put the product into paper moulds, and in a few days slice it, and preserve it in a well-stoppered bottle.

SOAPS (Toilet). Of toilet soaps there are two principal varieties:

1. (Hard.) The basis of these is, generally, a mixture of suet, 9 parts, and olive oil, 1 part saponified by caustic soda; the product is vari-

ously scented and coloured. They are also made of white tallow, olive, almond, and palm-oil soaps, either alone or combined in various proportions, and scented.

2. (Soft.) The basis of these is a soap made of hog's lard and potash, variously scented and coloured.

3. Guido Schnitzer, writing to 'Dingler's Journal' (cciii, 129—132; 'Journ. of Chem. Soc.,' new series, vol. x), says that the use of sodium silicate (ordinary water-glass) has proved of great value in the manufacture of palm oil and cocoa-nut oil soaps, as it increases their alkalinity, and gives to them greater hardness and durability. It is for these reasons the silicate is much used in the manufacture of toilet soaps.

He states that during the American war, when the price of resin soap reached a high figure, sodium silicate was much used as a substitute in soap-making. The soap is found to be the more active and durable in proportion to the amount of silica in the silicate.

Schnitzer made a series of experiments in order to discover a mixture which, on fusing, will yield a silicate as rich as possible in silica without being insoluble in boiling water, and he found the following proportions yielded on fusion the best silicate for the above purposes:

100 parts of soda ash (containing 91% of Na₂CO₃), and 180 of sand. In the solution of silicate obtained on treatment with boiling water, the proportion of the Na₂O to the SiO₂ would then be as 1 to 2·9.

After long boiling with water, there ordinarily remains a slimy residue, which on boiling up with fresh dilute soda lye for a long time, furnishes a concentrated solution of silicate. This residue, consisting of silica, with insoluble higher silicates, was boiled with soda solution at 6° Baumé, and the solution concentrated to 40° Baumé, when the proportion therein of Na₂O to SiO₂ was found to be as 1 to 1·4, and on cooling there crystallised out sodium silicate, of the formula Na₂SiO₃ + 8H₂O, in white foliated crystals.

On the small scale the perfume is generally added to the soap, melted in a bright copper pan by the heat of a water-bath; on the large scale it is mixed with the liquid soap at the soap-maker's before the latter is poured into the frames.

The following are examples of a few of the leading toilet soaps. See also SAVONETTES.

Soap, Bitter Al'mond. Sys. SAVON D'AMANDE, Fr. Prep. From white tallow soap, 56 lbs.; essential oil of almonds, ½ lb.; as before.

Savon au Bouquet. [Fr.] Prep. From tallow soap, 80 lbs.; olive-oil soap, 10 lbs.; essence of bergamot, 4 oz.; oils of cloves, sassafras, and thyme, of each, 1 oz.; pure neroli, ½ oz.; brown ochre (finely powdered), ¼ lb.; mixed as the last.

Soap, Cin'namon. Prep. From tallow soap, 14 lbs.; palm-oil soap, 7 lbs.; oil of cinnamon (cassia), 3 oz.; oil of sassafras and essence of bergamot, of each, ½ oz.; levigated yellow ochre, ¼ lb.

Soap, Float'ing. Prep. From good oil soap, 14 lbs.; water, 3 pints; melted together by the heat of a steam or water bath, and assiduously beaten until the mixture has at least doubled its volume, when it must be put into the frames, cooled, and cut into pieces. Any scent may be added.

Soap, Glycerin. Prep. 1. Any mild toilet soap being liquefied, glycerin is intimately mixed with it in the proportion of from a 20th to a 25th of the weight of the soap. Sometimes a red, and at others an orange tint is given to it. The scent usually consists of bergamot or rose geranium, mixed with a little oil of cassia, to which sometimes a little oil of bitter almonds is added.

2. (Spos.) 40 lbs. of tallow, 40 lbs. of lard, and 20 lbs. of cocoa-nut oil are saponified with 45 lbs. of soda lye, and 5 lbs. of potash lye, of 40° Baumé, when the soap is to be made in the cold way. To the paste then add pure glycerin, 6 lbs.; oil of Portugal, ¾ oz.; oil of bergamot, ½ oz.; bitter-almond oil, 5 oz.; oil of vitivert, 3 oz.

Soap, Hon'ey. Prep. 1. From palm-oil soap and olive-oil soap, of each, 1 part; curd soap, 3 parts; melted together and scented with the oil of verbena, rose geranium, or ginger-grass.

2. From the finest bright-coloured yellow soap, scented with the oils of ginger-grass and bergamot.

Soap, Liquid Glycerin—Glycerinseife, Flussige. Sesame or cotton-seed oil is saponified with sufficient caustic potash, and while moist is dissolved in six times its weight of spirit of wine. The solution is filtered, five sixths of the spirit is distilled from a water-bath, and the cool residue is reduced to the consistence of thin honey with a mixture of 2 parts glycerin and 1 part spirit. It is then perfumed.

Soap, Musk. Prep. 1. A good ox suet or tallow soap is generally used for the basis of this. The scent is composed of a mixture of essence of musk, with small quantities of the oils of bergamot, cinnamon, and cloves. The quantity of musk must be regulated by the amount of fragrance required. The soap is usually coloured with caramel.

2. Another kind is made with tallow and palm-oil soap, to which is added a mixture of the powders of cloves, roses, and gilliflowers, oil of bergamot, and essence of musk. The colouring matter is brown ochre.

Soap, Naples. From olive oil and potash.

Soap, Orange-flower. As SAVON à LA ROSE, with oil of neroli or essence de petit grain, supported with a little of the essence of ambergris and Portugal for perfume.

Soap, Palm-oil. Sys. VIOLET SOAP. Made of palm oil and caustic soda lye. It has a pleasant odour of violets and a lively colour.

Soap, Pearl. Sys. ALMOND CREAM; CRÈME D'AMANDES, Fr. Prep. From a soap made of lard and caustic potash lye; when quite cold it is beaten in small portions at a time in a marble mortar until it unites to form a homogeneous mass, or 'pearls,' as it is called; essence of bitter almonds, q. s. to perfume, being added during the pounding.

Soap, Rondeletia. This is merely cinnamon soap scented with the essence made with mixed essential oils, &c., known as rondeletia. It is coloured with brown or yellow ochre.

Savon à la Rose. [Fr.] Prep. From a mixture of olive-oil soap, 36 lbs.; best tallow soap, 24 lbs. (both new and in shavings) ; water, 1 quart; melted in a covered bright copper pan by the heat of a water-bath, then coloured with vermilion (finely levigated), 2½ oz. ; and, after the mixture has cooled a little, scented with otto of roses, 3 oz. ; essence of bergamot, 2½ oz. ; oil of cloves and cinnamon, of each, 1 oz.

Soap, Sha"ving. See PASTE (Shaving).

Soap, Transpa"rent. *Prep.* From perfectly dry almond, tallow, or soft soap, reduced to shavings, and dissolved in a closed vessel or still, in an equal weight of rectified spirit, the clear portion, after a few hours' repose, being poured into moulds or frames ; after a few weeks' exposure to a dry atmosphere the pieces are 'trimmed up' and stamped as desired. It may be scented and coloured at will by adding the ingredients to it while in the soft state. A rose colour is given by tincture of archil, and yellow by tincture of turmeric or annotta. It does not lather well.

Soap, Windsor. *Syn.* SAPO VINDSORR, S. VINDSORIENSIS, L. *Prep.* 1. (WHITE; S. V. ALBUS.) The best 'English' is made of a mixture of olive oil, 1 part, and ox tallow or suet, 9 parts, saponified by caustic soda. 'French Windsor soap' is made of hog's lard with the addition of a little palm oil. That of the shops is merely ordinary curd soap scented with oil of caraway, supported with a little oil of bergamot, lavender, or origanum. To the finer qualities a little of the essences of musk and ambergris is occasionally added. 1½ lbs. of the mixed scents is the common proportion per cwt.

2. (BROWN ; S. V. FUSCUS.) This merely differs from the last in being coloured with burnt sugar, or (less frequently) with umber. Originally it was the white variety, that had become mellow and brown with age.

SO'DA. See SODIUM.

SO'DIUM. Na = 22·99. *Syn.* NATRIUM. The metallic base of soda. It never occurs free, but its compounds are abundantly and universally diffused. It was first obtained by Sir H. Davy, in 1807, by means of a powerful galvanic battery; but it may be more conveniently and cheaply procured, in quantity, by the method described under POTASSIUM. The process, when well conducted, is, however, much easier and more certain than that for the last-named metal.

Prep. 1. (*Deville's Improvement on Brunner's Method.*) 30 parts by weight common soda ash, 13 parts small coal, and 3 parts chalk are placed in an iron cylinder which is lined with fire-clay, and heated in a reverberatory furnace to whiteness. The ends of the cylinder are closed, and one end is perforated with an iron pipe; through this pipe the gas and sodium vapours escape, the latter being condensed in an iron receiver as in the manufacture of potassium ('Ann. Chim. Phys.' [3], xliii, 5).

2. (*Castner's New Process.*) Fused sodium hydrate is distilled with a mixture of carbon and finely divided iron (prepared by reducing hæmatite with CO or H, mixing it with tar and coking). The carbon reduces the sodium, and the iron keeps it below the surface in direct contact with the fused hydrate. The residues are lixiviated with warm water, and the solution evaporated to recover the sodium carbonate formed,

and the iron is dried, mixed with tar, and used over again. The distillation is carried on in cast-iron crucibles heated in a gas furnace at 1000° C. Yield 90% of the sodium present.

Prop., &c. Sodium is a soft, lustrous, silver-white metal, scarcely solid at common temperatures, fuses at 95·6° C., boils at 861°—945° C., and volatilises at a red heat; it oxidises very rapidly in the air; when placed on the surface of cold water, it decomposes that liquid into free hydrogen and caustic soda with great violence, but generally without flame, in which it differs from potassium; on hot water or viscid solutions it burns with a bright yellow flame. Sp. gr. 0·9735 at 13·5° C. (*Baumhauer*) ; it is more malleable than any other metal, and may be easily reduced into very thin leaves (*Ure*) ; it conducts heat and electricity better than any metal except gold, silver, and copper, and is very electro-positive. Its other properties resemble those of potassium, but are of a feebler character. With oxygen it forms two oxides ; with chlorine, bromine, iodine, fluorine, &c., chloride, bromide, iodide, fluoride, &c., all of which may be obtained by similar processes to the respective compounds of potassium, which for the most part they resemble.

Uses. Until recently sodium has been regarded as a mere mechanical or philosophical curiosity ; it has now, however, become of great practical importance from being employed in the manufacture of silicon and borax, and of the metals aluminium, magnesium, &c., the price of which the production of cheap sodium has greatly lowered. An amalgam is employed with great advantage in extracting gold and silver from their ores.

Tests. Sodium salts are recognised by their solubility in water, and by their giving a precipitate with none of the ordinary reagents. They give a rich yellow colour to the colourless Bunsen or the pale blue blowpipe flame. They can, to a certain extent, be also distinguished from potassium salts by the carbonate being an easily crystallisable salt, effervescing in dry air ; the carbonate of potassium being crystallised with difficulty, and deliquescent. Platinum chloride does not give a precipitate with sodium chloride ; neither does picric acid, perchlorate of ammonium, nor tartaric acid.

Estim. Sodium generally occurs with potassium in qualitative analysis, for the separation of which see POTASSIUM. It may be determined directly by Bunsen's method. The alcoholic solution of the soluble double chloride of sodium and platinum is evaporated in a flask exposed to the light and filled with hydrogen gas. Sodium chloride, hydrochloric acid, and metallic platinum are formed. The latter is filtered off, and the filtrate evaporated to dryness after having previously transformed the sodium chloride into sodium sulphate. The weight of the salt is ascertained after gentle ignition.

Sodium, Acetate of. NaC₂H₃O₂,3H₂O. *Syn.* ACETATE OF SODA ; SODÆ ACETAS (B. P., Ph. D.), L. *Prep.* From carbonate of sodium by neutralisation with acetic acid ; but the resulting solution is evaporated, and set aside to crystallise.

Prop., &c. Its crystals are striated oblique

rhombic prisms; it effloresces slightly in the air, and is soluble in 4 parts of water at 60° F. Its solution in water forms one of the best examples of a supersaturated solution, in which state it is used for filling the foot-warmers for railway carriages, on account of the continuous evolution of heat during its crystallisation; for this purpose it is said to be four times as effective as an equal volume of water. Diuretic.—*Dose*, 20 to 40 gr.

Sodium Aluminate. $3Na_2O.Al_2O_3$. This salt has of late been in extensive demand by the calico printer and dyer. In France it is obtained from bauxite, a native hydrate of aluminate, by treatment with caustic soda or the carbonate. If caustic soda be employed, the powdered bauxite is boiled with a solution of the alkali, whereas if carbonate of soda be used, it is fused with the bauxite in a reverberatory furnace. By the first process the resulting aluminate of soda is dissolved in water, and, evaporated to dryness, forms the commercial article. If prepared by ignition, the semi-fused mass is lixiviated with water, and then evaporated to dryness.

Prop., &c. White powder, of a greenish-yellow hue. It is equally soluble in both hot and cold water, and readily decomposed by carbonic and acetic acids, bicarbonate and acetate of soda, chloride of ammonia, &c. Used for the preparation of lake colours, the induration of stone, in the manufacture of artificial stone, and for the saponification of fats in the manufacture of stearin candle manufacture, also in the preparation of an opaque, milky-looking glass.

Aluminate of soda may likewise be procured from cryolite, as described under ALUM.

Sodium Arsenates. 1. $Na_3HAsO_4.7H_2O$. *Syn.* COMMON ARSENATE OF SODA. *Prep.* From soda and arsenious oxide, the resulting arsenite being afterwards heated with sodium nitrate, from which it takes up oxygen.

Prop., Uses, &c. Feebly alkaline. Largely used in calico printing as a substitute for the dung-baths formerly employed.

2. HNa_2AsO_4. *Syn.* HYDRO-DISODIUM ARSENATE, ARSENATE OF SODA.

Prop., &c. From white arsenic dissolved in caustic soda solution, to which sodium nitrate is afterwards added; the liquid is then evaporated to dryness, the residue heated to redness, dissolved in water, and crystallised out. Crystallises with 12Aq., but the salt commonly sold in the shops contains 7Aq.

Sodium Arsenite. Na_4AsO_4. *Prep.* By dissolving 2 parts of white arsenic and 1 part of sodium carbonate in water. Used as a 'sheep-dipping' composition, in the manufacture of an arsenical soap, and for preventing incrustations in steam boilers.

Sodium Benzoate. $C_6H_5.CO_2Na$. *Syn.* SODÆ BENZOAS (B. Cod.). *Prep.* Heat gently benzoic acid and water, and add caustic soda to neutralise the acid. Filter, evaporate, and crystallise over sulphuric acid under a bell-glass.

Sodium Bisulphate. $NaHSO_4.H_2O$. *Syn.* ACID SULPHATE, SODIUM HYDROGEN S.; SODÆ BISULPHIS, L. *Prep., &c.* Dissolve crystallised carbonate of soda in twice its weight of water, and pass sulphuric acid in excess through the solution. Set it aside to crystallise. Prisms decomposed by heat into water and pyrosulphate. Its solution is used as a preservative.

Sodium Borate. See BORAX.

Sodium, Bromide of. NaBr. *Syn.* SODII BROMIDUM. Prepared as bromide of potassium. Monoclinic prisms containing 2Aq.

Sodium, Carbonate of. $Na_2CO_3.10Aq$. *Syn.* CARBONATE OF SODA, MONOCARBONATE OF SODA, SUBCARBONATE OF S.†, SALT OF BARILLA†, SODÆ CARBONAS (B. P., Ph. L., E., & D.), L. The carbonate of soda of commerce (WASHING SODA, BARILLA) was formerly prepared from the ashes of seaweed, and other marine vegetables, in a somewhat similar manner to that by which carbonate of potassium is obtained, and was chiefly imported from Spain, &c.; but it is now usually obtained from chloride of sodium by the action of heat, sulphuric acid, and carbonaceous matter, and by the ammonia-soda or 'Solvay' process.

Prep. 1. (*Leblanc Process, 'Salt-cake' Process.*) The salt is placed in an iron pan upon the hearth of a reverberatory furnace (see *engr.*), and mixed with an equal weight of sulphuric acid; this

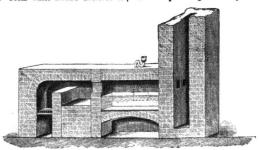

converts it into sodium bisulphate, hydrochloric acid gas being given off. The latter is condensed by contact with water. The flame of the furnace is allowed to play upon the mixture of salt and sulphuric acid until it has become dry. The residue is then called 'salt-cake.' This is now broken up, mixed with an equal weight of limestone, and rather more than half its weight of small coal. The mixture is again heated upon the hearth of a reverberatory furnace. Carbonic oxide gas is given off and a residue is formed, consisting of sodium carbonate, lime, and calcium sulphide; this residue is termed 'black ash.' The reaction may be represented by the following equations:

i. $H_2SO_4 + NaCl = HNaSO_4 + HCl.$
ii. $HNaSO_4 + NaCl = Na_2SO_4 + HCl.$
iii. $Na_2SO_4 + C_4 = Na_2S + 4CO.$
iv. $CaCO_3 + C = 2CO + CaO.$
v. $Na_2S + CaO + CO_2 = Na_2CO_3 + CaS.$

The ' black ash ' is now treated with water, which dissolves out the sodium carbonate, leaving the calcium sulphide and lime behind. Any sodium sulphide that may have formed is oxidised by blowing air through the liquid. The solution is finally evaporated to dryness, ordinary ' soda ash ' being thus obtained. Since this product contains as impurities common salt, sodium sulphate, and caustic soda, the last being formed by the action of the lime upon the sodium carbonate, it must be purified. The crude ' soda ash ' is mixed with small coal or sawdust, and again heated. Carbonic anhydride is given off, and this converts the caustic soda into sodium carbonate, and eliminates the other impurities. The mass is then lixiviated with water and the solution evaporated, when oblique rhombic prisms of common 'washing soda,' containing 10Aq, separate out. The hydrochloric acid obtained as a byeproduct in this process is used in the preparation of bleaching powder, &c.

Obs. During the wars of the French revolution the price of barilla (commercial carbonate of soda) rose very high, and since this substance was of prime importance in several of the leading French industries, Napoleon offered a premium for the discovery of a process by which it could be manufactured at home. This reward was obtained by an apothecary named Leblanc, who hit upon the principle of the process described above.

In making the ' salt-cake ' 16 cwt. of common salt are usually taken, and 123·5 galls. of chamber acid (crude sulphuric acid), of sp. gr. 1·42.

An average sample of ' salt-cake ' has the following composition:

Normal sodium sulphate .	. 95·275
Sodium-hydrogen sulphate	. 1·481
Sodium-chloride sulphate .	. 1·354
Calcium sulphate . . .	. 0·923
Ferric oxide and insoluble matter	. 0·321
Water	. 0·187
	99·541

Open roasters are commonly employed, but Deacon's close roasters are now coming into use. All hard labour is avoided in the modified process proposed by Jones and Walsh. Cammack and Walker's new process is said to moderate the violent action of the decomposition, and in Hargreaves' process the salt-cake is manufactured direct from salt, sulphur dioxide, and water.

In making the ' black ash ' the proportion of the materials taken are, in the Lancashire chemical works, salt-cake, 224 lbs.; limestone, 224 lbs.; coal-dust, 140 lbs.; in the Tyne works, salt-cake, 196 lbs.; limestone, 252 lbs.; coal-dust, 126 lbs.

Kynaston's analysis of English black ash is as follows:

Sodium carbonate	.	.	36·88
„ sulphate .		.	0·39
„ chloride .		.	2·53
„ silicate .		.	1·18
„ aluminate		.	0·69
Calcium sulphide.		.	28·68
„ carbonate		.	3·31
„ sulphite .		.	2·18
Lime .	.	.	9·27
Ferric oxide.	.	.	2·66
Coal .	.	.	7·00
Magnesia	.	.	0·25
Alumina	.	.	1·13
Water .	.	.	0·22
Ferrous sulphide .	.	.	0·37
Sand .	.	.	0·90
Ultramarine	.	.	0·96
			98·60

In Maclear's improved ' black ash ' process the proportion of limestone to salt-cake is 7 : 10, and in addition 6·5 parts of quicklime are used.

Improvements in the lixiviation of the black ash have been devised by Shanks.

2. (*Solvay's Process.*) Another process for the preparation of commercial carbonate of sodium, known as the ' ammonia process,' has of late years met with considerable adoption. The history of this process, together with the process itself, are thus described by Dr R. Wagner (' Journal of Applied Chemistry '):—" Six years ago [he was writing in 1873], when the international jury at the Paris Exhibition expressed their opinion upon the state of the soda industry at that time, all the judges, whether practical or theoretical men, believed that Leblanc's process (that previously described) would hold the field for a long time yet. This seemed still more probable since a process had just been introduced for recovering the sulphur from the soda residues. At that time all the soda in use was prepared by this process, excepting a comparatively small amount obtained from Chili saltpetre and cryolite, although there were already tangible indications that soda could be made on a larger scale by another method, which would be cheaper than Leblanc's process.

" The chemical section of the international jury at the Vienna Exhibition, under the presidency of Professor A. W. Hofmann, constituted a congress of chemical technology. This congress of scientific men was able to authenticate the very important fact that although Leblanc's process might in the future possess some importance for certain branches of the industry, yet in most places another soda process would be introduced in the immediate future, and entirely supersede that of Leblanc. Since the time of the Paris Exhibition this new process has grown from a small germ to a strong tree.

" The process in question, and which is called by Professor Hofmann ' the ammonia process,' is not new, from either a chemical or scientific point of view. It belongs to the same methods as those in which oxide of lead, bicarbonate of magnesia, quicklime, alumina, silicate of alumina, oxide of chromium, or fluosilicic acid are employed to decompose chloride of sodium, and convert it directly into soda or its carbonates. None of these attempts met with a success deserving of notice, although for a century past efforts have been made to render them practically operative. The new process is founded upon a reaction noticed over thirty years ago—that of bicarbonate of ammonia upon a strong solution of common salt. The greater part of the sodium is precipitated as bicarbonate, while chloride of ammonium remains in solution, from which the ammonia for a second operation is expelled by quicklime. The carbonic acid necessary to convert the ammonia into bicarbonate of ammonia, and thus make the process a continuous one, is obtained by heating the bicarbonate of soda to convert it into the simple carbonate.

"The sensation which the ammonia process has created in industrial circles will render a brief history of its development not uninteresting.

"So far as I know, Harrison, Dyer, Grey, and Hemming were the first to patent the ammonia process in Great Britain in 1838. Great expectations were excited by it, but it soon sank into oblivion.

"Thirty or forty years ago the manufacture of soda was by no means at the head of the great branches of industry; at that time, too, ammonia was not to be had cheaply or in immense quantities, and that branch of machine-building which has furnished the necessary apparatus for chemical industries did not exist. Besides this, Anton, of Prague, in 1840, claimed to have proved that in the ammonia process a very considerable portion of the common salt still remained undecomposed.

"After a sleep of sixteen years the ammonia process again entered the field. On the 26th of May, 1854, Turck took out a patent in France, and on the 21st June, the same year, Schlœsing, chemist of the Imperial Tobacco Factory at Paris, took out a patent for France and Great Britain. The mechanical portion and machinery for Schlœsing's process were designed by Engineer E. Rolland, director of the tobacco factory. In 1855 a company was organised to work this process. An experimental manufactory was started at Puteaux, near Paris, but, owing to its situation and arrangements, as well as the salt monopoly, it could not produce soda cheap enough to compete with the other process, and hence in 1858 the experiment was abandoned. Schlœsing and Rolland were of the opinion that sooner or later the new process must come into use in making soda.

"It must here be noticed that in 1858 Professor Heeren, of Hanover, subjected the ammonia process to a very careful test in his laboratory. From his experiments and calculations it was ascertained that this process was better adapted to the manufacture of the bicarbonate than of the simple protocarbonate of soda.

"To render this sketch more complete and historically true, it must be mentioned that T. Bell, of England, took out a patent, Oct. 13th, 1857, for a new soda process, which in principle and practice was almost literally the same as that of Dyer.

"It was known when the jury was working at

VOL. II.

Paris in 1867 that essential improvements had been introduced into the ammonia process by the efforts of Marguerite and De Sourdeval, of Paris, and James Young, of Glasgow. A more important fact, however, is that Solvay & Co., of Conillet, in Belgium, actually exhibited at the Paris Exhibition carbonate of soda prepared by this new process.

"Since that time the ammonia process has been developed and perfected to such an extent, especially by Solvay, Honigmann, and Prof. Gerstenhoefer, that as early as February, 1873, A. W. Hofmann, in his introduction to the third group of the catalogue of the Exhibition of the German Empire, was able to make this remark : 'At all events, the ammonia process is the only one which threatens to become an important competitor of the now almost exclusively employed process of Leblanc.'

"There are now large works in England, Hungary, Switzerland, Westphalia, Thuringia, and Baden, which employ the improved ammonia process.

"The advantages of the new process over that of Leblanc are very evident. The chief advantage consists in the direct conversion of salt into carbonate of soda, and next from the fact that from a saturated brine only the sodium is precipitated, with none of the other metals of the motherliquor. Besides this, the product is absolutely free from all sulphur compounds ; the soda is of a high grade ; the apparatus and utensils are very simple, there is a great saving of labour and fuel ; and no noxious gases and waste products are produced, which is of importance from a sanitary point of view. The only weak point of the ammonia process is the loss of chlorine, which is converted into worthless chloride of calcium.

"The effect which the general introduction of the new soda process will exert upon large chemical industries in general, and especially upon the consumption of sulphur, the manufacture of sulphuric acid, and chloride of lime, cannot be overlooked."

Obs. This important process, which would entirely supersede the Leblanc process were it not for the hydrochloric acid produced as a bye-product of the latter, depends upon the following reaction :

$$NaCl + NH_4.HCO_3 = NH_4Cl + NaHCO_3.$$

The most approved form of working it is as follows :

A saturated solution of common salt is mixed with about ½ its volume of ammonia liquor (sp. gr. 0·88) ; carbonic anhydride is then passed into the mixture, and this precipitates the bicarbonate of sodium. Some of the CO_2 escapes absorption, and this along with the NH_3 passes through a tall vertical cylinder with perforated shelves, through which trickles a solution of common salt, which absorbs the CO_2 and NH_3, forming more bicarbonate, which collects on the shelves. The sodium bicarbonate is heated to convert it into carbonate, and supply the CO_2.

$$2NaHCO_3 = Na_2CO_3 + H_2O + CO_2.$$

The ammonium chloride in solution is decomposed by heating with lime to recover the ammonia, which is used again.

$$2NH_4Cl + CaO = 2NH_3 + H_2O + CaCl_2.$$

97

Ordinary salt-cake (sodium sulphate) is now being employed also instead of common salt, thus:

$$Na_2SO_4 + 2NH_3 + 2CO_2 + 2H_2O = 2NaHCO_3 + (NH_4)_2SO_4.$$

The ammonium sulphate is fused with more salt-cake, and steam injected.

$$(NH_4)_2SO_4 + Na2SO_4 = 2NH_3 + 2NaHSO_4.$$

The ammonia is recovered and used over again, and the sodium bisulphate is converted into 'salt-cake' by fusion with common salt.

$$NaHSO_4 + NaCl = Na_2SO_4 + HCl.$$

3. Another method for the direct preparation of soda and potash from their chlorides is described in the "Bayerisches Industrie und Gewerbe Blatt.," 'New Remedies,' 1878, 4. The process is thus described by its author, Herr E. Bohlig:

Magnesium oxalate (freshly prepared when newly starting, but after the first operation obtained as a dry product in the next step) is allowed to drain, and then mixed in a large vat with the proper quantities of sodium chloride, or concentrated brine and hydrochloric acid, after which it is allowed to stand a few hours. Decomposition takes place almost instantaneously; all the magnesium goes into solution in the form of syrupy magnesium chloride, while all the sodium and oxalic acid are deposited as a crystalline acid salt (binoxalate of sodium).

Since the magnesium oxalate is always obtained of the same composition and in the same quantity, it is sufficient to determine its weight once for 3¹], and to take each time the previously common amounts of salt. The acid need not be weighed either; it must be added in just sufficient quantity to destroy the milky appearance which the mixture first assumes.

The reaction is as follows:

$$MgC_2O_4 + HCl + NaCl = NaHC_2O_4 + MgCl_2$$
Magnesium+Hydro- + Sodium=Sodium bin- + Magnesium
oxalate. chloric acid. chloride.   oxalate.   chloride.

The crystalline powder of sodium binoxalate is transferred to large draining filters, washed with water until the acid solution of magnesium chloride is removed, and worked up, as below described, while still moist.

The acid solution of magnesium chloride is made use of several times in succession as so much hydrochloric acid, together with a quantity of fresh acid sufficient for the reaction. Finally, when the magnesium chloride has inconveniently accumulated, it is worked up by itself into magnesia and hydrochloric acid.

In order to obtain the soda, the sodium binoxalate is brought together with an equivalent quantity of magnesium carbonate and water in a closed cylinder. As soon as the remaining air has been nearly expelled by the generated carbonic acid gas, the cylinder is closed, and a stirring mechanism set in motion.

A pressure-gauge attached to the cylinder indicates a gradual rise of the pressure to two atmospheres, but, on continual stirring, this diminishes, until, finally, the gauge stands again at 0°. The cylinder now contains a concentrated solution of sodium bicarbonate and a precipitate of magnesium oxalate, which latter, being coarsely granular, is easily separated from the liquid, and is used over again, after washing, for a new operation.

The solution of sodium bicarbonate is boiled for a short time with magnesia, obtained in distilling magnesium chloride, and both are thereby converted into simple carbonates. Both reactions are shown in the following scheme:

1. $NaHC_2O_2 + MgCO_3 = NaHCO_3 + MgC_2O_4$
    Sodium + Magnesium = Sodium + Magnesium
    binoxalate.   carbonate.   bicarbonate.   oxalate.

2. $2NaHCO_3 + MgO = Na_2CO_3 + MgCO_3 + H_2O$
    Sodium   + Magnesia = Sodium + Magnesium + Water.
    bicarbonate.   carbonate. carbonate.

As the solution of sodium carbonate, after concentration to 40° B., is incapable of dissolving or retaining in solution any sodium oxalate, it follows that the whole of the oxalic acid is recovered. The magnesia which is required for the purpose is obtained by distilling magnesium chloride, which thereby splits up into hydrochloric acid and magnesia. One half of the latter receives, as we have seen, its carbonic acid by boiling with sodium bicarbonate; the other half is placed, whilst still moist, upon trays, through which the gases of the furnace pass, and is thereby carbonated.

The process may also be so modified that the sodium binoxalate is first decomposed by caustic magnesia, and that magnesium carbonate is afterwards added. The whole mixture is then transferred to a stirring cask, provided with openings for the passage of cooled furnace gases, whereby the caustic soda present is very soon carbonated.

As soon as a large quantity of magnesium chloride solution has accumulated, it is tested as follows:—A small sample is mixed, while boiling, with magnesium oxalate, as long as the latter is dissolved, and then allowed to cool. There should be no crystalline deposit of sodium binoxalate formed, a proof that the solution does not contain any sodium chloride in excess, and is fit for distillation. It is first neutralised by adding some more magnesia, and evaporated over a fire in large kettles to a pasty consistence, short of driving off the hydrochloric acid. It is then transferred into the ordinary soda furnace, where it is distilled with a moderate fire. The eliminated hydrochloric acid is condensed in the usual manner.

The residuary mass should not be heated red-hot, so as not to impair its porosity or its ready affinity for carbonic acid. If, however, the first-mentioned test shows the magnesium chloride to contain sodium chloride the whole mass must be mixed with magnesium oxalate, and after removal of the precipitated sodium oxalate, saturated with magnesia and distilled. The same process, in all its details, may also be employed for the manufacture of potash and its carbonate.

Another method of manufacture of commercial soda is by heating the mineral cryolite (a double fluoride of sodium and aluminium) with chalk. The aluminate and caustic soda being both soluble in water, a stream of carbonic acid is passed through the solution containing them, whereby all the soda becomes converted into carbonate, whilst the alumina is thrown down as an insoluble precipitate. This process is largely used in Germany.

Various other processes for the manufacture of commercial soda have been devised, some of which are still followed, whilst others, being impracticable, have collapsed.

Prop., &c. Carbonate of sodium forms large, transparent, monoclinic prisms, which, as ordinarily met with, are of the formula $Na_2CO_3.10Aq$; but by particular management may be had with fifteen, nine, seven, or sometimes with only one molecule of water of crystallisation (*Fownes*); it is soluble in twice its weight of water at 60°, and less than an equal weight at 212° F. As a medicine it is deobstruent and antacid, and is given in doses of 10 to 30 gr. It is also, occasionally, used to make effervescing draughts. When taken in an overdose it is poisonous. The antidotes are the same as for carbonate of potassium. The crude carbonate is largely employed in the manufacture of soap, paper, textile fabrics, glass, &c.

When ANHYDROUS CARBONATE OF SODIUM is required (SODÆ CARBONAS EXSICCATA, B. P., Ph. L.; SODÆ CARBONAS SICCATUM, Ph. E. and D.), the crystallised carbonate is heated to redness, and, when cold, powdered; sp. gr. 2·5.

Fifty-three gr. of the dried carbonate are equal to 148 gr. of the crystallised salt. The medicinal properties of both are similar. It has, however, the disadvantage of being difficultly soluble in water.

The ordinary carbonate of sodium generally contains either sulphates or chlorides, or both; and these may be detected as under CARBONATE OF POTASSIUM. "When supersaturated with nitric acid, it precipitates only slightly, or not at all, chloride of barium or nitrate of silver; and 148 gr. require at least 960 grain-measures of solution of oxalic acid" (B. P.).

Sodium, Bicarbonate of. NaHCO₃. *Syn.* SESQUICARBONATE OF SODA; SODÆ BICARBONAS (B. P., Ph. L., E., and D.), L. *Prep.* This salt can be prepared in exactly the same manner as the corresponding salt of potassium. Other methods are as follows:—(1) Crystallised carbonate of sodium, 1 part; dried carbonate of sodium, 2 parts; triturate them well together, and surround them with an atmosphere of carbonic acid gas under pressure; let the action go on until no more gas is absorbed, which will generally occupy 10 to 14 hours, according to the pressure employed; then remove the salt, and dry it at a heat not above 120° F.

(2) The bicarbonate may be more simply prepared by passing carbonic acid gas through a solution of common salt mixed with ammonia, as in the manufacture of the normal carbonate by the Solvay or ammonia-soda process.

Prop., &c. A crystalline white powder; it is soluble in 10 parts of water at 60° F., but it cannot be dissolved in even warm water without partial decomposition; it has a more pleasant taste and is more feebly alkaline than the normal carbonate. When absolutely pure it does not darken turmeric paper, or only very slightly. The dose is from 10 to 40 gr., as an antacid and absorbent. It is much employed in the preparation of effervescing powders and draughts, for which purpose—

20 gr. of commercial bicarbonate of sodium are taken with

18 gr. of crystallised tartaric acid;
17 gr. of crystallised citric acid; or
¼ fl. oz. of lemon juice.

The quantity of bicarbonate any given sample contains may be approximately determined by well washing 100 gr. of the salt with an equal weight of cold water, and filtering the solution. The residuum left upon the filter, dried at a heat of 120° F., and weighed, gives the percentage of pure bicarbonate of sodium present (very nearly). The solution of this in water should give only a very trifling white precipitate with corrosive sublimate; whilst the filtered portion, which was used to wash the salt, will give a red one if it contains the simple carbonate of sodium. Hager's test for the normal carbonate, which nearly always occurs in commercial samples, is to shake in a stoppered bottle 1 grm. of the salt, 0·5 grm. of mercurous chloride, and 1·5 grms. water. If the normal salt be absent the liquid remains white for 24 hours, if present it turns grey.

Sodium Chlorate. NaClO₃. *Prep.* By boiling chlorate of potash (9 parts) with sodium silico-fluoride (7 parts), and crystallising out.

Prop., &c. Regular tetrahedrons, very soluble in water and alcohol. Largely manufactured for the use of calico printers in the production of aniline black.

Sodium, Chloride of. NaCl. *Syn.* SODII CHLORIDUM (B. P., Ph. L. and D.), SODÆ MURIAS (Ph. E.), L. This important and wholesome compound appears to have been known in the earliest ages of which we have any record. It is mentioned by Moses (Gen. xix, 26), and by Homer in the 'Iliad' (lib. ix, 214). In ancient Rome it was subjected to a duty (*vectigal salinarium*); and even at the present day a similar tax furnishes no inconsiderable portion of the revenue of certain nations. Sodium chloride occurs as rock salt in large deposits in various geological strata, in solution in sea water, brine springs, &c. The Trias formations yield the chief supplies of salt in Europe. The most important localities in which deposits of salt occur are in Cheshire (England), in Galicia, the Tyrol, and Stassfurt. The Cheshire salt beds are two in number, each about 180 feet thick, separated from each other by about 30 feet of clay, and extending over an area 16 miles long and 10 miles wide. Common salt forms no small portion of the mineral wealth of England, and has become an important article of commerce in every part of the known world. The principal portion of the salt consumed in this country is procured by the evaporation of the water of brine springs. It is also prepared by the evaporation of sea water (hence the term 'sea salt'), but this process has been almost abandoned in England, being more suited to hot dry climates or to very cold ones.

Var. BAY SALT; SAL MARINUS, SAL NIGER; imported from France, Portugal, and Spain, and obtained from sea water evaporated in shallow ponds by the sun; large-grained and dark-coloured. BRITISH BAY SALT, CHESHIRE LARGE-GRAINED S.; by evaporating native brine at a heat of 130° to 140° F.; hard cubical crystals. Both of the above are used to salt provisions for hot climates, as they dissolve very slowly in the brine as it grows weaker. CHESHIRE STOVED SALT, LUMP S., BASKET S.; obtained by evaporating the brine of salt springs; small flaky crystals. LONDON'S PATENT SOLID SALT; Cheshire rock salt, melted and ladled into moulds. ROCK SALT, FOSSIL S.; SAL GEMMÆ, SAL FOSSILIS; found in mineral beds in Cheshire, France, Galicia, &c.; has commonly a reddish colour.

Prop. Pure chloride of sodium crystallises in anhydrous cubes, which are often grouped into pyramids or steps; dissolves in about 2½ parts of water at 60° F.; its solubility is not increased by heat; it is slightly soluble in proof spirit, insoluble in alcohol; decrepitates when heated, fuses at a red heat, and volatilises at a much higher temperature.

Pur., &c. The common salt of commerce contains small portions of chloride of magnesium, chloride of calcium, and sulphate of calcium; and hence has commonly a slightly bitter taste, and deliquesces in the air. To separate these, dissolve the salt in four times its weight of pure water, and drop into the filtered solution first chloride of barium, then carbonate of sodium, as long as any precipitate falls; filter, and evaporate the clear fluid very slowly until crystals form, which are pure chloride of sodium ('Thomson's Chem.,' ii, 377). For medical purposes the Ph. E. orders the salt to be dissolved in boiling water, and the solution to be filtered and evaporated over the fire, skimming off the crystals as they form, which must then be quickly washed in cold water and dried. A solution of pure salt is not precipitated by a solution of carbonate of ammonium, followed by a solution of phosphate of sodium; a solution of 9 gr. in distilled water is not entirely precipitated by a solution of 26 gr. of nitrate of silver (Ph. E.).

Uses. Common salt is stimulant, antiseptic, and vermifuge, and is hence employed as a condiment, and for preserving animal and vegetable substances. It is also occasionally used in medicine, in clysters and lotions.

Sodium, Dried Sulphate of. Na₂SO₄. *Syn.* SODÆ SULPHAS EXSICCATA, EFFLORESCHD GLAUBER SALT. Expose the crystals to a warm dry air till they fall into powder. They lose half their weight. The dose is reduced in like proportion.

Sodium, Effervescing Citro-tartrate of. *Syn.* SODÆ CITRO-TARTRAS EFFERVESCENS (B. P.), L. *Prep.* Mix thoroughly powdered bicarbonate of soda, 17 oz.; tartaric acid, 8 oz.; and citric acid, 6 oz.; place in a dish or pan of suitable form, heated to between 200° and 220° F., and when the particles begin to aggregate, stir assiduously till they assume a granular form. By means of suitable sieves separate the granules of uniform and most convenient size. Preserve in well-closed bottles.

Sodium, Ethylate. C₄H₆NaO. Prepared as POTASSIUM ETHYLATE, substituting sodium for potassium. Properties similar to ethylate of potassium.

Sodium, Hydrate of. NaHO. *Syn.* HYDRATE OF SODA, SODIUM HYDRATE, CAUSTIC SODA; SODÆ HYDRAS, L. *Prep.* Exactly in the same manner from carbonate of sodium as potassium hydrate is prepared from carbonate of potassium. The solid caustic soda of commerce is generally obtained in the Leblanc process for manufacturing the carbonate of sodium; the solution obtained by treating the black ash with water is evaporated, so that the carbonate, sulphate, and chloride of sodium may crystallise out, leaving the more soluble hydrate in the concentrated liquid. The latter, which still retains a compound of sodium and iron sulphides which give it a red colour, is mixed with sodium nitrate, which oxidises these sulphides; the liquor is then evaporated down until a fused mass of sodium hydrate remains, and this is then poured into iron moulds. In another method the black ash liquor is allowed to filter through a column of coke against a current of air, when the sulphide of sodium is oxidised and the sulphide of iron deposited on the coke. After being mixed with a little chloride of lime to oxidise any remaining sulphides, the liquor is concentrated, carbonate and ferrocyanide of sodium separate out, and the remaining liquor, which contains the hydrate, is then concentrated till it solidifies on cooling.

Sodium hydrate is also made from the carbonate very cheaply by decomposing a dilute solution with quicklime thus:—Soda crystals, 3 parts; boiling water, 15 parts; and milk of lime (obtained by slaking lime, 1 part, with water, 3 parts) is added gradually to the boiling solution. When the liquid is found to be free of carbonic acid, it is concentrated till the soda fuses, and then the mass is moulded into sticks.

The 'Pharmaceutical Journal' (3rd series, i, 65) states that a pure hydrate of sodium is now manufactured from metallic sodium by the following method:—A deep silver vessel, of a hemispherical form, and capable of holding about 4 gallons of water, is employed. Into this vessel, which is cooled externally with a current of cold water, is placed a very little water, and upon the water is placed a cube of metallic sodium of about half an inch in diameter.

The vessel is made to revolve, so as continually to bring fresh portions of liquid into contact with the metal, and by this means explosion is avoided. When the first cube of metal has dissolved, and yielded a thick syrupy liquid, a little more water and a second tube of metal are added, and the reaction allowed to take place as before, the vessel being kept in motion all the time. In this manner several pounds of sodium may be worked up into pure soda.

The thick syrup so resulting is next evaporated down, heated to redness, fused, and poured into a mould.

The danger of explosions (which, however, are not likely to occur if proper care is taken) necessitates the employment of skilled labour in this manufacture, and constitutes a very serious drawback to the commercial success of the process.

Prop. White, semi-translucent, fibrous, deliquescent masses, very soluble in water, and bearing a very great resemblance to the corresponding potassium compound. Sp. gr. 2·13 (*Filhol*). It absorbs carbonic acid gas when moist. When heated to the melting-point of cast iron it is resolved into its elements, HO and Na.

Monoclinic crystals of the formula 2NaHO. 7H₂O, melting at 6° C., have recently been obtained.

Uses, &c. It is met with in commerce as a highly concentrated aqueous solution or in the solid condition. It is largely used in the manufacture of soap, the refining of paraffin and petroleum, in the preparation of silicate of soda and artificial stone, &c.; it is an invaluable reagent in the chemical laboratory, and is much used for absorbing carbonic acid gas; it acts as a painful cautery in surgery.

Schiff (vide 'Ann. Pharm.,' cvii, 300) has prepared the following table, showing the sp. gr. of soda lye at 15° C.:

Percentage of NaOH.	Specific Gravity.
1	1·012
5	1·059
10	1·115
15	1·170
20	1·225
25	1·279
30	1·332
35	1·384
40	1·437
45	1·488
50	1·540
55	1·591
60	1·643

Sodium, Hypochlo'rite of. NaOCl. *Syn.* CHLORINATED SODA, CHLORIDE OF SODA‡; SODA CHLORINATA, L. *Prep.* (*Christison.*) Dried carbonate of sodium, 19 parts, are triturated with water, 1 part, and the mixture placed in a proper vessel and exposed to the prolonged action of chlorine gas, generated from a mixture of chloride of sodium, 10 parts; binoxide of manganese, 8 parts; sulphuric acid, 14 parts; (diluted with) water, 10 parts.

Sodium Hypophosphate. See PHOSPHORUS.

Sodium Hypophosphite. NaH₂PO₂ + H₂O. *Prep.* From calcium hypophosphite in solution, by adding sodium carbonate and allowing the liquid to evaporate *in vacuo*.

Prop., &c. Pearly tabular crystals, deliquescent, soluble in alcohol. Employed in medicine for the same purpose as phosphorus.

Sodium Hyposulphite. NaHSO₂. *Prep.* Place a concentrated solution of acid sulphite of sodium in a well-corked bottle along with zinc clippings or turnings. Keep the mixture cool for half an hour, then decant the clear solution into three times its volume of strong alcohol. The bottle is then filled completely with this alcoholic liquid and tightly corked. As soon as all the zinc-sodium sulphite has fallen down the liquid is poured off into well-stoppered bottles, and allowed to stand and crystallise in a cool place; the crystals are dried between filter-paper. Purify by recrystallisation from alcohol.

Prop. Fine acicular crystals.

Use. As a reducing agent for indigo in dyeing and calico printing, and in the laboratory for estimating free oxygen, or that element in substances which easily evolve it.

Sodium, Iodide of. NaI or NaI.2H₂O. *Syn.* SODII IODIDUM. *Prep.* As IODIDE OF POTASSIUM. This, as well as the bromide, crystallises in clear or whitish cubes, deliquescent, and soluble in water. The hydrated form is deposited from water at the ordinary temperature. Used in medicine in the same manner as the corresponding potassium salts.

Sodium and Iron, Pyrophosphate of. *Syn.* SODAE ET FERRI PYROPHOSPHAS, NATRUM FYROPHOSPHORICUM FERRATUM (Ph. G.), L. *Prep.* Dissolve 20 oz. of pyrophosphate of soda in 40 oz. of cold distilled water, and add gradually to the solution, and with constant stirring, 8 oz. (by weight) of solution of perchloride of iron (Ph. G.), previously diluted with 22 oz. of distilled water, as long as the precipitate is redissolved. Filter, and to the clear, bright green liquid thus obtained pour in 100 oz. (by weight) of rectified spirit, wash the precipitate with more spirit, press it between blotting-paper, and dry by a gentle heat.

Sodium Lactate. NaC₃H₅O₃. *Syn.* SODII LACTAS, L. *Prep.* Let lactic acid be diluted with three parts of water; saturate whilst boiling with sodium carbonate; then evaporate.

Prop. Flattened prismatic crystals and stellar groups of needles, very deliquescent; when heated with metallic sodium it is converted into disodium lactate, Na₂C₃H₄O₃.

Sodium Manganate. Na₂MnO₄. *Prep.* By heating sodium hydrate and manganese dioxide together, freely exposed to the air.

Prop., Uses, &c. Green saline mass. Employed as a bleaching agent; in the preparation of oxygen at a cheap rate, and in solution in water as 'Condy's green disinfecting fluid.' If the water contains no potash or soda, to be pure the manganate is decomposed into the red permanganate.

Sodium, Ni'trate of. NaNO₃. *Syn.* CHILI SALTPETRE, CUBIC NITRE; SODAE NITRAS, L. This salt occurs native like ordinary nitre, and is chiefly imported into England from South America. It is refined by solution and crystallisation. It is largely employed as a manure, especially as a top-dressing for barley, in the preparation of nitric acid, and, recently, in the manufacture of fireworks, on account of the comparative slowness with which it burns. It is deliquescent,very soluble in water, and crystallises in oblique rhombohedrons; sp. gr. 2·26.

Sodium, Nitrite of. NaNO₂. *Syn.* SODAE NITRIS. Mix nitrate of soda, 1 lb., and charcoal, recently burned and in fine powder, 1½ oz., thoroughly in a mortar, and drop the mixture in successive portions into a clay crucible, heated to a dull redness. When the salt has become quite white, raise the heat so as to liquefy it, pour on to a clean flagstone, and when it has solidified, break into fragments, and keep in a stoppered bottle.

Sodium, Oxide of. Na₂O. *Syn.* ANHYDROUS SODA. *Prep., &c.* By burning dry metallic sodium in air, or (pure) by heating the hydroxide with the metal. A grey mass with conchoidal fracture, melting at dull red heat, volatilising at higher temperature; sp. gr. 2·805; very deliquescent, strongly basic, and soluble in water, forming pure sodium hydrate.

Sodium Peroxide. Na₂O₂. *Syn.* S. DIOXIDE. *Prep., &c.* By heating the metal in oxygen gas, or by igniting the nitrate. White powder, yellow when heated. When exposed to the air it deliquesces, then absorbs carbonic acid, and ultimately forms a solid mass of carbonate. Thrown into water, heat is evolved, and caustic soda and oxygen are formed.

Sodium, Phosphates of. There are three orthophosphates, namely—
i. Normal sodium orthophosphate, $Na_3PO_4 + 12H_2O$.
ii. Hydrogen-disodium orthophosphate, $Na_2HPO_4 + 12H_2O$.
iii. Dihydrogen-sodium orthophosphate, $NaH_2PO_4 + 4H_2O$.

These are tribasic, and give yellow precipitates with silver nitrate.

There are two pyrophosphates, namely—
i. Normal sodium pyrophosphate, $Na_4P_2O_7 + 10H_2O$.
ii. Dihydrogen-sodium pyrophosphate, $Na_2H_2P_2O_7$.

These are tetrabasic, and give white precipitates with silver nitrate, and do not precipitate albumen.

Five metaphosphates are known, namely—
i. Sodium monometaphosphate, $NaPO_3$.
ii. Sodium dimetaphosphate, $Na_2P_2O_6 + 2H_2O$.
iii. Sodium trimetaphosphate, $Na_3P_3O_9 + 6H_2O$.
iv. Sodium tetrametaphosphate, $Na_4O_4.P_4O_8 + 4H_2O$.
v. Sodium hexametaphosphate, $Na_6O_6.P_6O_{12}$.

These are monobasic, and give white precipitates with silver nitrate, and precipitate albumen.

Of these, two only will be noticed, namely, hydrogen-disodium orthophosphate (common phosphate of soda) and normal sodium pyrophosphate.

Sodium, Common Phos'phate of. Na_2HPO_4. 12Aq. Syn. HYDROGEN-DISODIUM ORTHOPHOSPHATE, RHOMBIC P. OF S.; SODÆ PHOSPHAS (B. P., Ph. L., E., & D.), L. Prep. 1. Take of powdered bone ashes, 10 lbs.; sulphuric acid, 44 fl. oz.; mix, add gradually of water, 6 pints, and digest for three days, replacing the water which evaporates; then add 6 pints of boiling water, strain through linen, and wash the residue on the filter with boiling water; mix the liquors, and, after defecation, decant and evaporate to 6 pints; let the impurities again settle, and neutralise the clear fluid, heated to boiling, with a solution of carbonate of sodium in slight excess; crystals will be deposited as the solution cools, and by successively evaporating, adding a little more carbonate of sodium to the mother-liquor till it is feebly alkaline, and cooling, more crystals may be obtained; these must be kept in close vessels.

2. (Funcke.) To ground calcined bones, diffused through water, add a little dilute sulphuric acid to saturate any carbonate of calcium present; when effervescence ceases, dissolve the whole in nitric acid, q. s.; to this solution add as much sulphate of sodium as the bone ash used, and distil the whole to recover the nitric acid; the residuum is treated with water, and the resulting solution filtered, evaporated, and crystallised.

Prop., &c. It forms large, transparent, monoclinic prisms, which effloresce in the air, dissolve in about 4 parts of cold water and in 2 parts at 212° F., and fuse when heated. As a medicine it is mildly aperient, in doses of ½ to 1 oz., or even more; and antacid in doses of 20 to 30 gr., frequently repeated. It has a purely saline taste, resembling that of culinary salt, and its aqueous solution turns red litmus blue.

Sodium Pyrophosphate (Normal). $Na_4P_2O_7$. $10H_2O$. Syn. SODÆ PYROPHOSPHAS CRYSTALLISATA (P. Cod.), L. Prep. This salt may be obtained by heating, gently at first, and afterwards to a red heat, common phosphate of sodium in a platinum crucible until all the water is driven off, and the salt has become fused. Dissolve the fused mass in water, filter, and concentrate until it has a density of 1·20, and crystallises in the cold. Monoclinic prisms.

Sodium, Salicylate of. $C_6H_4(OH).CO_2Na$. Syn. SODÆ SALICYLUS. Prep. Made by neutralising a solution of pure salicylic acid with caustic soda, and evaporating to dryness. It must be purified by crystallisation from alcohol. Antipyretic; given in acute rheumatism.—Dose, 10 to 20 grms.

Sodium Santonate. Syn. SODÆ SANTONAS. This salt is made by digesting an alcoholic solution of santonic acid with carbonate of soda, evaporating, redissolving in strong alcohol, and crystallising.

Sodium, Sesquicarbonate of. $Na_2CO_3.2NaHCO_3$. $3H_2O$. A salt found native in South America and on the banks of the soda lakes of Sotrena, in Africa, whence it is exported as 'Trona.' When strongly heated it is resolved into $Na_2CO_3.CO_2$ and H_2O.

Sodium Silicate. See GLASS, SOLUBLE.

Sodium, Stan'nate of. $Na_2SnO_3.4H_2O$. Prep. (Greenwood & Co.) Caustic soda, 22 lbs., is heated to low redness in an iron crucible, when nitrate of sodium, 8 lbs., and common salt, 4 lbs., are added; when the mixture is at a 'fluxing heat,' 10 lbs. of feathered block tin is stirred in with an iron rod, both the stirring and heat being continued until the mass becomes red-hot and 'pasty,' and ammoniacal fumes are given off. The product may be purified by solution and crystallisation. Used as a mordant by calico printers.

Sodium, Stan'nite of. Na_2SnO. Prep. (Greenwood & Co.) From caustic soda, 18½ lbs.; feathered block tin and common salt, of each, 4 lbs.; as the last. Used to prepare tin mordants (about 12 oz. to water, 1 gall.).

Sodium, Normal Sulphate of. Na_2SO_4. Syn. SODÆ SULPHIS. Occurs native as thenardite and in the mineral springs at Friedrichshall, &c. It is prepared on an enormous scale as 'salt-cake' in the manufacture of carbonate of soda (q.v.).

Prep. 1. Common salt is decomposed by the action of sulphuric acid, or by the combined action of sulphur dioxide, air, and aqueous vapour; purify by recrystallisation from water.

2. As a residue in the manufacture of nitric acid from Chili saltpetre, and in other chemical operations.

3. By exposing the hydrated sulphate (Glauber's salts) to the air, or by heating a saturated aqueous solution of this salt.

Prop., &c. Rhombic crystals, insoluble in alcohol, possessing a saline, bitter taste, and a neutral reaction. Used internally for sarcina ventriculi, and externally as an application in skin diseases of fungous origin. Hyposulphate of soda is employed in the same cases.

Sodium, Sul'phate of (Hydrated). Na₂SO₄,10Aq. *Syn.* GLAUBER'S SALT; SODÆ SULPHAS (B. P., Ph. L., E., & D.), SAL CATHARTICUS GLAUBERI†, L. Obtained by crystallisation from an aqueous solution of the normal sulphate at ordinary temperature.

Prop. Large, colourless, monoclinic prisms, which effloresce in the air and fall to an opaque white powder; soluble in about 3 parts of water at 60°, but at a higher temperature its solubility rapidly lessens; insoluble in alcohol; fuses when heated. It is seldom wilfully adulterated. When pure the solution is neutral to test-paper; nitrate of silver throws down scarcely anything from a dilute solution; nitrate of baryta more, which is not dissolved by nitric acid. It loses 55·5% of its weight by a strong heat. It readily forms supersaturated solutions. These crystallise suddenly when a small crystal of the salt is dropped in, and the solidification is accompanied by a rise of temperature.

Uses. It is purgative, but being extremely bitter tasted, is now less frequently used than formerly. Its nauseous flavour is said to be covered by lemon juice.—*Dose,* ¼ to 1 oz. The dried salt (SODÆ SULPHAS EXSICCATA) is twice as strong. LYMINGTON GLAUBER'S SALT is a mixture of the sulphates of soda and potash obtained from the mother-liquor of sea salt.

Sodium, Sul'phide of. Na₂S. *Prep.* (P. Cod.) Saturate a solution of caustic soda (sp. gr. 1·200) with sulphuretted hydrogen, closely cover up the vessel, and set it aside that crystals may form; drain, press them in bibulous paper, and at once preserve them in a well-closed bottle.

Prop., &c. Used in the commercial preparation of soluble glass and sodium thiosulphate, to make mineral waters, and in certain skin diseases. Dr Ringer says, " It possesses the property of preventing and arresting suppuration and stopping the formation of pus." Given for boils and carbuncles, it also produces excellent results.—*Dose.* For adults, ₁/₁₀ of a grain, mixed with sugar of milk, every hour or two on the tongue.

Sodium, Sulphocarbolate. *Syn.* SODÆ SULPHOCARBOLAS (*Pereira*). *Prep.* Mix 2 vols. of pure carbolic acid with 1 vol. of sulphuric acid in a flask, and heat the mixture to 280° or 290° F. for five minutes. Cool, dilute, and saturate with carbonate of soda, evaporate, and crystallise. The other sulphocarbolates may be prepared in the same manner.—*Dose,* 10 to 30 gr., in phthisis and zymotic diseases; externally, as a lotion in ozœna and fetid ulcers.

Sodium, Sulphosalicylate of. *Syn.* SODÆ SULPHOSALICYLAS, L. *Prep.* (*Williams.*) By treating very pure salicylic acid with about twice its weight of sulphuric acid, then adding carbonate of barium, and decomposing the sulphosalicylate of barium by sulphate of soda (see 'Pharm. Journ.,' Sept. 30th, 1876).

Sodium Sulphovinate. *Syn.* SODII SULPHOVINAS, L. *Prep.* Sulphovinic acid is first prepared by pouring gradually, with great care, and increasingly stirring with a glass rod, 1000 grms.

of 60° sulphuric acid into 1000 grms. of rectified 96° alcohol. The mixture is left for some hours in contact, then diluted with 4 litres of distilled water, and afterwards saturated with pure barium carbonate. When the saturation is complete the barium sulphate is filtered off. The solution of barium sulphovinate is then decomposed with pure carbonate of soda until it ceases to give a precipitate. The liquid, evaporated in a water-bath, is left to crystallise. If necessary the crystals are purified by recrystallisation. They should be kept in well-closed flasks.

Prop., &c. Sodium sulphovinate crystallises in hexagonal tables, which are slightly unctuous to the touch, and very soluble in water and in alcohol. If heated they give off, at 120°, the alcohol which they contain in combination; and they become gradually deprived of bitterness. Sodium sulphovinate ought not to contain sulphuric acid, nor have an acid taste. It should not be precipitated by barium chloride, and especially by soluble sulphates. The possession of either of these properties is a proof of faulty preparation, and that a portion of the sulphovinic acid has been decomposed.

Sulphovinate of soda is said to be a very effective, and by no means unpleasant, saline aperient, and to be unattended with subsequent constipation. The dose is from 5 to 6 dr.

Sodium, Tartrate of, and Potassium. NaKC₄H₄O₆.4Aq. *Syn.* TARTRATE OF POTASSA AND SODA, ROCHELLE SALT, SEIGNETTE'S S., TARTARISED SODA†; SODÆ TARTARATA (B. P.), SODÆ POTASSIOTARTRAS (Ph. L.), SODÆ ET POTASSÆ TARTRAS (Ph. E. & D.), SODA TARTARIZATA†, L. *Prep.* Take of carbonate of sodium, 12 oz.; boiling water, 2 quarts; dissolve, and add, gradually, of powdered bitartrate of potassium, 16 oz.; strain, evaporate to a pellicle, and set it aside to crystallise; dry the resulting crystals, and evaporate the mother-liquor that it may yield more of them.

Prop., &c. Large, transparent, hard, right rhombic prisms, often occurring in halves; slightly efflorescent; soluble in 5 parts of water at 60° F. Its " solution neither changes the colour of litmus nor of turmeric. On the addition of sulphuric acid, bitartrate of potassium is thrown down; on adding either nitrate of silver or chloride of barium nothing is thrown down, or only what is redissolved by the addition of water " (Ph. L.). By heat it yields a mixture of the pure carbonates of potassium and sodium.

Potassio-tartrate of sodium is a mild and cooling laxative.—*Dose,* ¼ to 1 oz., largely diluted with water. It forms the basis of the popular aperient called SEIDLITZ POWDERS.

Sodium, Thiosul'phite of. Na₂S₂O₃ + 5H₂O. *Syn.* SODÆ HYPOSULPHIS, L. *Prep.* 1. Dried carbonate of sodium, 1 lb.; flowers of sulphur, 10 oz.; mix, and slowly heat the powder in a porcelain dish until the sulphur melts; stir the fused mass freely to expose it to the atmosphere until the incandescence flags, then dissolve the mass in water, and immediately boil the filtered liquid with some flowers of sulphur; lastly, carefully concentrate the solution for crystals.

2. A stream of well-washed sulphurous anhydride gas is passed into a strong solution of carbonate of sodium, which is then digested with

sulphur at a gentle heat during several days; by evaporating the solution at a moderate temperature, the salt is obtained in large and regular crystals.

3. (*Capaun's process.*) Boil a dilute solution of caustic soda with sulphur to saturation, then pass sulphurous acid gas into the solution until a small portion, when filtered, is found to have a very pale yellow colour; when this is the case, it must be filtered and evaporated as before.

4. A cheap recently devised process is to decompose the soluble calcium thiosulphate obtained by the oxidation of 'alkali waste' by means of the sulphate or carbonate of sodium; the solution containing the thiosulphate is then drawn off from the precipitated sulphate or carbonate of calcium, and set aside to evaporate and crystallise.

Prop., &c. Crystallises in large transparent four-sided prisms, which in the dry state are unalterable in the air; it is freely soluble in water, and possesses a cooling taste. It may be perfectly freed from sulphide of sodium by agitating it with about half its weight of alcohol; the alcohol dissolves out the sulphide, which may then be easily separated. This salt is now very extensively used in the practice of photography as a solvent for unaltered silver chloride; also as an 'antichlor,' to extract the last traces from paper pulp.

Sodium Urate. $HNaC_5N_2.N_4O_3$. *Syn.* HYDRO-SODIUM URATE. Occurs in the gouty concretions termed 'chalk-stones,' and sometimes as a deposit from urine.

Sodium, Vale''rianate of. $NaC_5H_9O_3$. *Syn.* SODÆ VALERIANAS (Ph. D.), L. *Prep.* (Ph. D.) Dilute oil of vitriol, 6½ fl. oz., with water, ½ pint; then dissolve of powdered bichromate of potassium, 9 oz., in hot water, 3½ pints; when both solutions have cooled, put them into a matrass, and having added of fusel oil, 4 fl. oz., shake them together repeatedly until the temperature, which first rises to 150°, has fallen to 80° or 90° F.; a condenser having been connected, next apply heat so as to distil over about 4 pints of liquid; saturate this exactly with a solution of caustic soda, separate the liquid from the oil which floats upon the surface, and evaporate it until the residual salt is partially liquefied; the heat being now withdrawn, and the salt concreted, this last, whilst still warm, is to be divided into fragments, and preserved in well-stopped bottles.

Obs. This salt is intended to be used in the preparation of the VALERIANATES OF IRON, QUININE, and ZINC.

SOILS. These are classified by agriculturists, according to their chief ingredients, as loamy, clayey, sandy, chalky, and peaty soils. Of these the first is the best for most purposes, but the others may be improved by the addition of the mineral constituents of which they are deficient. Sand and lime or chalk and coal ashes are the proper additions to clayey soils, and clay gypsum, or loam, to sandy and gravelly ones. Clayey soils are expensive to bring into a fertile state; but when this is once effected, and they are well manured, they yield immense crops of wheat, oats, beans, clover, and most fruits and flowers of the rosaceous kinds.

The fertilisation of soils is suggested partly by chemical analysis, practical experience, and geological observations. In cases where a barren soil is examined with a view to its improvement, it is, when possible, compared with an extremely fertile soil in the same neighbourhood, and in a similar situation; the difference given by their analyses indicates the nature of the manure required, and the most judicious methods of cultivation; and thus a plan of improvement is suggested, founded upon scientific principles.

The analysis of soils may be briefly and generally described as follows:

1. The general character of the soil, as loamy, sandy, stony, rather stony, &c., being noted, 3 or 4 lbs. of it, fairly selected as an average specimen, may be taken during a period of ordinary dry weather. From this, after crushing or bruising the lumps with a piece of wood, all stones of a larger size than that of a filbert may be picked out, and their proportion to the whole quantity duly registered.

2. 1000 gr. of the remainder may be next dried by the heat of boiling water, until the mass ceases to lose weight, and afterwards exposed to a moist atmosphere for some time. The loss of weight in the first case, and the increase of weight in the second, indicate the absorbent powers of the soil.

3. The matter from No. 2, freed from siliceous stones by garbling, may be gradually heated to dull redness in a shallow open vessel, avoiding waste from decrepitation, &c. The loss of weight, divided by 10, gives the percentage quantity of vegetable or organic matter present (nearly).

4. Another 1000 grains (see No. 1) may be next washed with successive portions of cold water as long as anything is removed. The residuum, after being dried, indicates the proportion of sand and gravel (nearly).

5. Another portion of the soil (100, 200, or more gr., according to its character) is tested in the manner described under CARBONATE and ALKALIMETRY. The loss of weight in carbonic acid indicates the quantity of carbonate of lime present in the sample examined, 22 gr. of the former being equal to 50 gr. of the latter.

6. Another like portion of the soil may be gently boiled for four or five hours, along with dilute hydrochloric acid, in a flask furnished with a long glass tube passing through the cork, to prevent loss (see ÆTHER); after that time the whole must be thrown upon a filter, and what refuses to pass through (silica) washed with distilled water, dried, ignited, and weighed. The filtrate and washings from No. 6 are next successively treated for alumina (pure clay), lime, phosphate of lime, phosphoric acid, oxide of iron, alkalies (potasse or soda), ammonia (both ready formed and latent), &c. &c., in the manner noticed under GLASS, GUANO, and the names of the respective substances referred to. See MANURES, &c.

SOL'ANINE. *Syn.* SOLANI, SOLANINA, L. A peculiar basic substance, obtained from the leaves and stem of *Solanum dulcamara*, or bitter-sweet, and other species of the *Solanaceæ*.

SOLDERING. The union of metallic surfaces by means of a more fusible metal fluxed between them. The method of autogenous soldering, invented by M. De Richmont, is an exception to this definition. In all the cases surfaces must be perfectly clean, and in absolute contact, and the air must be excluded, to prevent oxidation. For this last purpose the brazier and silversmith use powdered borax made into a paste with water; the coppersmith, powdered sal-ammoniac; and the tinman, powdered resin. Tin-foil applied between the joints of fine brass-work, first wetted with a strong solution of sal-ammoniac, makes an excellent juncture, care being taken to avoid too much heat. See SOLUTION (Soldering), and below.

SOL'DERS. *Prep.* 1. (For copper, iron, and dark brass.) From copper and zinc, equal parts; melted together. For pale brass more zinc must be used.

2. (Fine solder.) From tin, 2 parts; lead, 1 part. Melts at 350° F. Used to tin and solder copper, tin plates, &c.

3. (For German silver.) From German silver, 5 parts; zinc, 4 parts; melted together, run into thin flakes, and then powdered. Also as No. 7.

4. (Glazier's.) From lead, 3 parts; tin, 1 part. Melts at 500° F.

5. (For gold.) Gold, 12 dwts.; copper, 4 dwts.; silver, 2 dwts.

6. (For lead and zinc.) From lead, 2 parts; tin, 1 part.

7. (For pewter, Britannia metal, &c.) From tin, 10 parts; lead, 5 parts; bismuth, 1 to 3 parts.

8. (For silver.) From fine brass, 6 parts; silver, 5 parts; zinc, 2 parts.

9. (For tin plate.) From tin, 2 parts; lead, 1 part. The addition of bismuth, 1 part, renders it fit for pewter.

SOLE. The *Solea vulgaris*, a well-known fish. It is, perhaps, more frequently eaten than any other flat fish, and, when skilfully cooked, exceeds them all in delicacy, nutritiousness, and flavour.

SOLU'TION. *Syn.* SOLUTIO, L. Under the head of solutions (SOLUTIONES), in pharmacy, are properly included only those liquids which consist of water, or an aqueous menstruum, in which has been dissolved an appropriate quantity of any soluble substance to impart to the liquor its peculiar properties. When spirit is the menstruum the liquid receives the name of alcoholic solution, spirit, or tincture. In the B. P. and the Ph. L. & D. aqueous solutions are named LIQUORS (LIQUORES), whilst in the Ph. E., and in the old pharmacopœias generally, they are termed WATERS (AQUÆ).

The following list embraces all the solutions of the British pharmacopœias, with a few others likely to be useful to the reader. Some other preparations to which the name has been given will be found under LIQUORS, TINCTURES, &c.

Solution of Ac'etate of Ammo"nium. *Syn.* LIQUOR AMMONII ACETATIS (B. P.); MINDERERUS' SPIRIT. *Prep.* Strong solution of acetate of ammonium, 4 oz.; distilled water to produce 20 oz. The solution should be stored in bottles free from lead. Sp. gr. 1·022.

Solution of Acetate of Ammonium (Strong). *Syn.* LIQUOR AMMONII ACETATIS FORTIOR, L. *Prep.* Carbonate of ammonium, 17½ oz.; acetic acid, 50 oz., or sufficient; add the former to the latter until the product is neutral; lastly, add distilled water to make 3 pints. Store in bottles free from lead. Sp. gr. 1·073.

Prop., &c. Free from colour and odour. It changes the colour neither of litmus nor turmeric. Sulphuretted hydrogen being dropped in, it is not discoloured, neither is anything thrown down on the addition of chloride of barium. What is precipitated by nitrate of silver is soluble in water, but especially so in nitric acid. Potassa being added, it emits ammonia; and sulphuric acid being added, it gives off acetic vapours. The fluid being evaporated, what remains is completely destroyed by heat.

Uses, &c. Solution of acetate of ammonia is a very common and excellent febrifuge and diaphoretic, and, in large doses, aperient saline liquor. Taken warm, in bed, it generally proves a powerful sudorific; and as it operates without heat it is much used in febrile and inflammatory disorders. Its action may likewise be determined to the kidneys by walking about in the cold air.—*Dose,* 2 to 6 dr. of the ordinary, ½ to 1 dr. of the strong, twice or thrice daily, either by itself or along with other medicines. Externally, as a discutient and refrigerant lotion; and diluted (1 oz. to 9 oz. of water) as a collyrium in chronic ophthalmia. For this last purpose it must be free from excess of ammonia.

(Concentrated.) Saturate acetic acid, sp. gr. 1·038, ½ gall., with carbonate of ammonia (in powder), 2½ lbs., or q. s., carefully avoiding excess.

Obs. This article is in great demand in the wholesale drug trade, under the name of 'concentrated liquor of acetate of ammonia' (LIQ. AMMON. ACET. CONC.). It is very convenient for dispensing.

Solution of Acetate of Lead. See SOLUTION OF DIACETATE OF LEAD.

Solution of Acetate of Morphia. *Syn.* LIQUOR MORPHIÆ ACETATIS (B. P., Ph. L. & D.), L. *Prep.* 1. (Ph. L.) Acetate of morphia, 4 gr.; acetic acid, 15 drops; distilled water, 1 pint; proof spirit, ½ pint; mix and dissolve. Sixty drops (minims) contain 1 gr. of acetate of morphia.—*Dose,* 5 to 15 or 20 drops.

2. (B. P.) Acetate of morphine, 9 gr.; diluted acetic acid, 18 minims; rectified spirit, ½ oz.; distilled water, 1½ oz.; dissolve in the mixed liquids.—*Dose,* 10 to 60 minims.

3. (Ph. D.) Acetate of morphia, 32 gr.; rectified spirit, 5 fl. oz.; distilled water, 15 fl. oz. 120 drops (minims) contain 1 gr. of the acetate.—*Dose,* 10 to 45 or 50 drops, or similar to that of tincture of opium.

4. (*Magendie.*) Each fl. dr. contains 1½ gr. of acetate (nearly).—*Dose,* 5 to 15 drops. Anodyne, hypnotic, and narcotic; in those cases in which opium is inadmissible. See MORPHIA.

Solution of Aconitia. *Syn.* SOLUTIO ACONITIÆ (*Dr Turnbull*). *Prep.* Aconitia, 1 gr.; rectified spirit, 1 dr. To be applied externally by means of a sponge in neuralgic and rheumatic affections.

Solution of Al'um (Compound). *Syn.* BATE'S ALUM WATER; LIQUOR ALUMINIS COMPOSITUS

(Ph. L.), AQUA ALUMINOSA COMPOSITA†, L. *Prep.* (Ph. L.) Alum and sulphate of zinc, of each, 1 oz.; boiling water, 3 pints; dissolve and filter (if necessary). Detergent and astringent. Used as a lotion for old ulcers, chilblains, excoriations, &c.; and, largely diluted with water, as an eye-wash and injection.

Solution of Ammo″nia. See LIQUOR OF AMMONIA.

Solution of Ammo″nio-ni′trate of Sil′ver. *Syn.* HUME'S TEST; SOLUTIO ARGENTI AMMONIATI (Ph. E.), L. *Prep.* (Ph. E.) Nitrate of silver (pure crystallised), 44 gr.; distilled water, 1 fl. oz.; dissolve and add ammonia water, gradually, until the precipitate, at first thrown down, is very nearly but not entirely redissolved. Used as a test for arsenic.

Solution of Ammo″nio-sul′phate of Cop′per. *Syn.* LIQUOR CUPRI AMMONIO-SULPHATIS (Ph. L.), CUPRI AMMONIATI SOLUTIO (Ph. E.), C. A. AQUA, L. *Prep.* (Ph. L.) Ammonio-sulphate of copper, 1 dr.; water, 1 pint; dissolve and filter. Stimulant and detergent. Applied to indolent ulcers, and, when largely diluted, to remove specks on the cornea; also used as a test for arsenic.

Solution for Anatom′ical Preparations, &c. *Syn.* ANTISEPTIC SOLUTION. *Prep.* 1. Nearly saturate water with sulphurous acid, and add a little creasote.

2. Dissolve chloride of tin, 4 parts, in water, 100 parts, to which 3% of hydrochloric acid has been added.

3. Dissolve corrosive sublimate, 1 part, and chloride of sodium, 3 parts, in water, 100 parts, to which 2% of hydrochloric acid has been added.

4. Mix liquor of ammonia (strong) with 3 times its weight (each) of water and rectified spirit.

5. Sal-ammoniac, 1 part; water, 10 or 11 parts. For muscular parts of animals.

6. Sulphate of zinc, 1 part; water, 15 to 25 parts. For muscles, integuments, and cerebral masses.

7. (*Dr Babington.*) Wood naphtha, 1 part; water, 7 parts; or wood naphtha undiluted, as an injection.

8. (*Sir W. Burnett.*) Concentrated solution of chloride of zinc, 1 lb.; water, 1 gall. The substances are immersed in the solution for 2 to 4 days, and then dried in the air.

9. (*Gannal.*) Alum and culinary salt, of each, ½ lb.; nitre, ¼ lb.; water, 1 gall.

10. (*Goadsby.*) a. From bay salt, 2 oz.; alum, 1 oz.; bichloride of mercury, 1 gr.; water, 1 pint. For ordinary purposes.

b. To the last add of bichloride of mercury, 1 gr.; water, 1 pint. For very tender tissues, and where there is a tendency to mouldiness.

c. From bay salt, ¼ lb.; bichloride of mercury, 1 gr.; water, 1 pint. For subjects containing carbonate of lime.

d. From bay salt, ¼ lb.; arsenious acid, 10 gr.; water, 1 pint; dissolve by heat. For old preparations.

e. To the last add of bichloride of mercury, 1 gr. As the last, when there is a tendency to the softening of parts; and, diluted, for mollusca. These solutions are approved of by Professor Owen.

11. (*M. Réboulet.*) Nitre, 1 part; alum, 2 parts; chloride of lime, 4 parts; water, 16 or 20 parts; to be afterwards diluted according to circumstances. For pathological specimens.

12. (*Dr Stapleton.*) Alum, 2½ oz.; nitre, 1 dr.; water, 1 quart. For pathological specimens.

13. (For FEATHERS—Beasley.) Strychnia, 16 gr.; rectified spirit, 1 pint.

14. Borax, 25 parts; salicylic acid, 5 parts; boric acid, 10 parts; glycerin, 30 parts; water, 40 parts. Place the whole in a porcelain dish, and heat until a clear solution is formed. This solution is suitable for diluting with 10 to 20 times its volume of water, as a preservative for meat. A teaspoonful of it is sufficient to add to a gallon of milk.

Obs. These fluids are used for preserving ANATOMICAL PREPARATIONS, OBJECTS OF NATURAL HISTORY, &c., by immersing them therein, in close vessels; or, for temporary purposes, applying them by means of a brush or piece of rag. The presence of corrosive sublimate is apt to render animal substances very hard. See PUTREFACTION.

Solution, Antiseptic. See *above*.

Solution of Arseniate of Ammonia. *Syn.* LIQUOR ARSENIATIS AMMONIÆ, L. *Prep.* (Hosp. of St Louis.) Arseniate of ammonia, 4 gr.; distilled water, 4 oz.; spirit of angelica, 2 dr.—*Dose*, 12 to 30 minims. There are other formulæ for the solution, differing in strength from the above.

Dr Neligan gives us Biett's:—Arseniate of ammonia, 1½ gr.; distilled water, 3 oz.; spirit of angelica, 6 dr.—*Dose*, 1 to 3 dr.

Bouchardat says 6 gr. to 8 oz. of distilled water. —*Dose*. From 12 drops to 1 dr.

Solution of Arseniate of Soda. *Syn.* LIQUOR ARSENIATIS SODÆ; PEARSON'S ARSENICAL SOLUTION. *Prep.* Arseniate of soda, 4 gr.; distilled water, 4 oz.—*Dose*, 12 minims to 30.

Solution of Arseniate of Soda. (B. P.) *Syn.* LIQUOR SODÆ ARSENIATIS. *Prep.* Dissolve arseniate of soda (rendered anhydrous by a heat not exceeding 300° F.), 4 gr., in distilled water, 1 oz.—*Dose*, 5 to 10 minims.

Solution, Arsenical. *Syn.* MINERAL SOLUTION; SOLUTIO ARSENICALIS, SOLUTIO MINERALIS, L. *Prep.* 1. (*Devergie.*) As SOLUTION OF ARSENITE OF POTASSA, Ph. L., but of only 1-50th the strength, flavoured with compound spirit of balm, and coloured to a deep rose with cochineal.

2. (*Pearson.*) Arseniate of soda, 4 gr.; water, 4 fl. oz.; dissolve.—*Dose*, 10 to 30 drops during the day (see *below*).

Solution of Arse″nious Acid. See DROPS, AGUE, and ARSENIOUS ACID.

Solution of Ar′senite of Potas′sa. *Syn.* FOWLER'S MINERAL SOLUTION; LIQUOR POTASSÆ ARSENITIS (Ph. L.), LIQUOR ARSENICALIS (B. P.), L. *Prep.* (B. P.) Arsenious acid, coarsely powdered, and carbonate of potassa, of each, 87 gr.; distilled water, 1 pint; boil until dissolved, and add, to the cold solution, compound tincture of lavender, 5 fl. dr.; water, q. s. to make the whole exactly measure a pint. Tonic, antiperiodic, and alterative.—*Dose*, 2 to 8 drops, gradually and cautiously increased; in agues and various scaly skin diseases. It is preferably taken soon after a meal. See ARSENIOUS ACID, &c.

Solution of Auro-chloride of Gold. *Syn.* SOLUTIO AURI AMMONIO-CHLORIDI (*Fwrasri*). *Prep.* Ammonio-chloride of gold, 8 gr. ; distilled water and rectified spirit, of each, 10 oz.—*Dose.* A teaspoonful morning and evening in sugared water for dysmenorrhœa and amenorrhœa.

Solution of Bimeconate of Morphine. *Syn.* LIQUOR MORPHINÆ BIMECONATIS (B. P.), L. *Prep.* Hydrochlorate of morphine, 9 gr. ; solution of ammonia, a sufficiency ; meconic acid, 6 gr.; rectified spirit, ½ oz. ; distilled water, a sufficiency. Dissolve the morphine in a little water by heat, then add ammonia until the morphine ceases to precipitate. Cool, filter, wash and drain the precipitate. Add it to enough water to produce 1½ oz. Finally add the spirit and meconic acid. Dissolve. —*Dose,* 5 to 40 minims.

Solution, Blistering. *Syn.* LIQUOR EPISPASTICUS (B. P.) ; LINIMENT OF CANTHARIDES. *Prep.* Powdered Spanish fly, 5 oz. ; acetic ether, a sufficiency. Percolate with the acetic ether until 20 oz. are obtained.

Solution, Brandish's. See SOLUTION OF POTASSA.

Solution of Bromine. *Syn.* LIQUOR BROMINII (*Pourche*). *Prep.* Bromine, 1 part; distilled water, 40 parts.—*Dose,* 5 or 6 drops, three times a day. A stronger solution (1 part to 10) is sometimes used externally.

Solution, Burnett's. A solution of chloride of zinc. See SOLUTION FOR ANATOMICAL PREPARATIONS (*above*), also DISINFECTING COMPOUNDS.

Solution of Camphor, Carbonated. *Syn.* SOLUTIO CAMPHORÆ CARBONICA (*Swediaur*). *Prep.* Water saturated with carbonic acid gas, 2 lbs. ; powdered camphor, 3 dr.

Solution of Camphor and Chloroform. *Syn.* SOLUTIO CAMPHORÆ ET CHLOROFORMI (*Messrs Smith*). *Prep.* Camphor, 3 dr.; chloroform, 1 fl. dr. Dissolve. For exhibiting camphor with yolk of egg in emulsions.

Solution of Carbolic Acid. (FOR THE TOILETTE.) *Prep.* Crystallised carbolic acid, 10 parts; essence of millefleur, 1 part ; tincture of *Quillaia saponaria,* 50 parts; water, 1000 parts. Mix. The saponine replaces soap with advantage. The above should be employed diluted with ten times its bulk of water, for disinfecting the skin, for washing the hands, after any risk of contagion, inoculation, &c.

Solution of Carbon (Detergent). *Syn.* LIQUOR CARBONIS DETERGENS. This name is applied to an alcoholic solution of coal-tar. Properly diluted it is used externally in skin diseases.

Solution of Carbonate of Magnesium. *Syn.* LIQUOR MAGNESII CARBONATIS (B. P.). *Prep.* Dissolve separately, each in half a pint of distilled water, sulphate of magnesium, 2 oz.; and carbonate of sodium, 2½ oz. Heat the solution of sulphate of magnesium to the boiling-point, add the solution of carbonate of soda, and boil together until the precipitated carbonate of magnesium, and wash until what passes ceases to give a precipitate with chloride of barium. Mix the precipitate with a pint of distilled water, and in a suitable apparatus, charge with pure washed carbonic acid gas. Retain excess of carbonic acid under pressure for twenty-four hours. Filter to remove undissolved carbonate, and again pass carbonic acid into the solution. Keep in a bottle securely closed (this contains about 10 gr. of carbonate of magnesium in each fluid ounce).

Solution of Chloride of Antimony. *Syn.* ANTIMONII CHLORIDI LIQUOR (B. P.). *Prep.* Dissolve black sulphide of antimony in boiling hydrochloric acid. Used as an escharotic, and in the preparation of oxide of antimony.

Solution of Chlo"ride of Ar'senic. *Syn.* LIQUOR ARSENICI HYDROCHLORICUS (B. P.), LIQUOR ARSENICI CHLORIDI (Ph. L.), L. *Prep.* 1. (Ph. L.) Arsenious acid (in coarse powder), ½ dr.; hydrochloric acid, 1½ fl. dr.;, distilled water, 1 fl. oz.; boil until the solution of the arsenious acid is complete, and, when cold, add enough distilled water to make the whole exactly measure a pint.— *Dose,* 4 to 5 drops.

2. (B. P.) Arsenious acid, 87 gr. ; hydrochloric acid, 2 dr.; distilled water, 20 oz.; boil the two acids with 4 oz. of the water until a solution is effected, then add sufficient distilled water to make up 20 oz.—*Dose,* 2 to 8 minims.

Solution of Chloride of Ba"rium. *Syn.* SOLUTION OF MURIATE OF BARYTA† ; LIQUOR BARII CHLORIDI (Ph. L. and D.), SOLUTIO BARYTÆ MURIATIS (Ph. E.), L. *Prep.* (Ph. L. and E.) Dissolve chloride of barium, 1 dr. (1 oz.—Ph. D.), in water, 1 fl. oz. (8 oz.—Ph. D.), and filter the solution. Sp. gr. (Ph. D.) 1·088.—*Dose,* 5 drops, gradually increased to 10 or 12, twice or thrice daily ; in scrofula, scirrhous affections, and worms ; *externally,* largely diluted, as a lotion in scrofulous ophthalmia.

Solution of Chloride of Cal'cium. *Syn.* SOLUTION OF MURIATE OF LIME† ; CALCII CHLORIDI LIQUOR (Ph. D.), CALCIS MURIATIS SOLUTIO (Ph. E.). *Prep.* 1. (Ph. L. 1836.) Fused chloride of calcium, 4 oz. (crystals, 8 oz.—Ph. E.) ; water, 12 fl. oz. ; dissolve and filter.

2. (Ph. D.) Fused chloride of calcium, 3 oz.; water, 12 oz. Sp. gr. 1·225.—*Dose,* 10 drops to 1 dr., or more ; in scrofulous and glandular diseases, &c.

Solution of Chloride of Zinc. *Syn.* LIQUOR ZINCI CHLORIDI (B. P.). *Prep.* Granulated zinc, 8 parts; hydrochloric acid, 22 parts; solution of chlorine, q. s.; carbonate of zinc, ½ part ; distilled water, 10 parts. Mix the acid and water in a porcelain dish, add the zinc, and apply a gentle heat to promote the action until gas is no longer evolved; boil for half an hour, supplying the water lost by evaporation, and allow the product to cool. Filter it into a bottle, and add solution of chlorine by degrees, with frequent agitation ; now add the carbonate of zinc until a brown sediment appears. Filter the liquid into a porcelain basin, and evaporate until it is reduced to the bulk of 20.

Solution of Chlo"rinated Lime. *Syn.* BLEACHING LIQUID, SOLUTION OF CHLORIDE OF LIME‡, S. OF HYPOCHLORITE OF LIME ; SOLUTIO CALCIS HYPOCHLORIS, S. CALCIS CHLORIDI, CALCIS CHLORINATÆ LIQUOR (Ph. D.), L. *Prep.* 1. (Ph. D.) Chlorinated lime ('chloride of lime'), ½ lb.; water, ¼ gall.; triturate them together, then transfer the mixture to a stoppered bottle, and shake it repeatedly for the space of 3 hours ; lastly, filter through calico, and preserve it in a well-stoppered bottle.

2. Chloride of lime (dry and good, and rubbed to fine powder), 9 lbs.; tepid water, 6 galls.; mix in a stoneware bottle capable of holding 8 galls., agitate frequently for a day or two, and after 2 or 3 days' repose decant the clear portion, and keep it in well-corked bottles in a cool situation. If filtered, it should be done as rapidly as possible, and only through coarsely powdered glass in a covered vessel.

3. LIQUOR CALCIS CHLORINATE (B. P.). *Prep.* Blend well together, by trituration in a large mortar, 1 lb. of chlorinated lime with 1 gall. of water, transfer the mixture to a stoppered bottle, and shake it frequently for the space of 3 hours; pour it on a calico filter, and let the solution which passes through be kept in a well-stoppered bottle. Sp. gr. 1·085.

Obs. The last is the usual strength sold in trade, under various attractive names, to give it importance. It is used as a disinfectant, bleacher, and fumigation; and, diluted with water, as a lotion, injection, or collyrium, in several diseases. See HYPOCHLORITE OF CALCIUM.

Solution of Chlorinated Lime, Spirituous. *Syn.* SOLUTIO CALCIS CHLORIDI SPIRITUOSA (*Chevallier*), L. *Prep.* Chloride of lime, 3 dr.; distilled water, 2 oz.; rectified spirit, 2 oz. Mix and filter.

Solution of Chlorinated Potas'sa. *Syn.* SOLUTION OF CHLORIDE OF POTASH‡, S. OF HYPOCHLORITE OF POTASSA, JAVELLE'S BLEACHING LIQUID; SOLUTIO POTASSÆ HYPOCHLORIS, LIQUOR POTASSÆ CHLORIDI, L. POTASSÆ CHLORINATE, L.; EAU DE JAVELLE, Fr. *Prep.* 1. Dissolve carbonate of potassa, 1 part, in water, 10 parts, and pass chlorine gas through the solution to saturation.

2. Chloride of lime (dry and good), 1 part; water, 15 parts; agitate them together for an hour; next dissolve of carbonate of potassa, 2 oz., in water, ½ pint; mix the two solutions, and after a time either decant or filter.—*Uses, &c.* As the last.

Solution of Chlorinated Soda. *Syn.* SOLUTION OF CHLORIDE OF SODA‡, S. OF HYPOCHLORITE OF SODA, LABARRAQUE'S DISINFECTING LIQUID; SOLUTIO SODÆ HYPOCHLORIS, HYPOCHLORIS SODICUS AQUA SOLUTUS (P. Cod.), LIQUOR SODÆ CHLORINATE (Ph. L. and D.), L. *Prep.* 1. (Ph. L.) Carbonate of soda (in crystals), 1 lb.; water, 1 quart; dissolve, and pass through the solution the chlorine evolved from a mixture of common salt, 4 oz.; binoxide of manganese, 3 oz.; sulphuric acid, 2½ fl. oz. (4 oz.—Ph. L. 1836); diluted with water, 3 fl. oz.; placed in a retort, heat being applied to promote the action, and the gas being purified by passing through 5 fl. oz. of water before it enters the alkaline solution.

2. (Ph. D.) Chlorinated lime, ½ lb., and water, 3 pints, are triturated together in a marble mortar, after which the mixture is transferred to a stoppered bottle, agitated frequently during three hours, and then filtered through calico; in the mean time carbonate of soda (cryst.), 7 oz., is dissolved in water, 1 pint; the two solutions are next mixed, and, after agitation for about ten minutes, the whole is filtered as before. The filtrate is to be preserved in a well-stoppered bottle.

3. (B. P.) Chlorinated lime, 16 oz.; carbonate of sodium, 24 oz.; water, 1 gall. Dissolve the soda in 2 pints of the water, triturate the lime salt with 6 pints of the water, and filter; well mix the solution, again filter. Keep in a stoppered bottle in a cool dark place.—*Dose*, 10 to 20 minims.

Obs. This solution is used as an antiseptic, disinfectant, and bleaching liquid, also in scarlet fever, sore throat, &c.; it is also made into a lotion, gargle, injection, and eye-water. Meat in a nearly putrid state, unfit for food, is immediately restored by washing or immersion in this liquid.

Solution of Chlo"rine. *Syn.* CHLORINE WATER; SOLUTIO CHLORINII, LIQUOR CHLORINI (Ph. L. & D.), CHLORINII AQUA (Ph. E.), L. *Prep.* 1. (Ph. L.) On binoxide of manganese (in powder), 2 dr., placed in a retort, pour hydrochloric acid, 1 fl. oz., and pass the chlorine in distilled water, ¾ pint, until it ceases to be evolved.

2. (Ph. E.) Muriate of soda (common salt), 60 gr.; red oxide of lead, 350 gr.; triturate them together, and put them into 8 fl. oz. of distilled water, contained in a stoppered bottle; then add of sulphuric acid, 2 fl. dr.; and having replaced the stopper, agitate the whole occasionally, until the oxide of lead turns white; lastly, after subsidence, pour off the clear liquid into another stoppered bottle.

3. (Ph. D.) Introduce into a gas bottle peroxide of manganese (in fine powder), ½ oz.; add of hydrochloric acid, 3 fl. oz., (diluted with) water, 2 fl. oz.; apply a gentle heat, and cause the evolved gas to pass through water, 2 fl. oz., and then into a three-pint bottle containing distilled water, 20 fl. oz., and whose mouth is loosely plugged with tow; when the air has been entirely displaced by the chlorine, cork the bottle loosely, and shake it until the chlorine is absorbed; it should now be transferred to a pint stoppered bottle, and preserved in a dark and cool place.

(B. P.) LIQUOR CHLORI. *Prep.* Put 1 oz. of black oxide of manganese, in fine powder, into a gas bottle, and having poured upon it 6 fl. oz. of hydrochloric acid, diluted with 2 oz. of distilled water, apply a gentle heat, and by suitable tubes cause the gas, as it is developed, to pass through 2 oz. of distilled water placed in an intermediate small phial, and thence to the bottom of a 3-pint bottle containing 30 oz. of distilled water, the mouth of which is loosely plugged with tow. As soon as the chlorine ceases to be developed let the bottle be disconnected from the apparatus in which the gas has been generated, corked loosely, and shaken until the chlorine is absorbed. Lastly, introduce the solution into a green bottle furnished with a well-fitting stopper, and keep it in a cool and dark place. Sp. gr. 1·008. One fluid ounce contains 2·66 grains of chlorine.

Prop., &c. Irritant and acrid, but, when largely diluted, stimulant and antiseptic.—*Dose*, ½ to 2 fl. dr., in ½ Pint of water, sweetened with a little sugar, in divided doses during the day; in scarlatina, malignant sore throat, &c. On the large scale liquid chlorine may be procured by passing the gas obtained by any of the methods named under CHLORINE into water, until it will absorb no more.

SOLUTION 1549

Solution of Chromic Acid. *Syn.* LIQUOR ACIDI
CHROMICI, L. (B. P.) Chromic acid, 1 part;
water, 3 parts. Used locally as a caustic.
Solution of Citrate of Ammonium. *Syn.* LI-
QUOR AMMONII CITRATIS, L. (B. P.) Strong
solution of citrate of ammonium, 5 oz., distilled
water to make 20 oz.—*Dose*, 2 to 6 dr.
Solution of Citrate of Ammonium (Strong).
Syn. LIQUOR AMMONII CITRATIS FORTIOR.
Citric acid, 12 oz.; strong solution, 11 oz., or a
sufficiency; distilled water, a sufficiency. Neutralise the acid with the ammonia, make up to a
pint with water.—*Dose*, ½ to 1½ dr.—*Uses.* Same
as acetate of ammonium.
Solution of Citrate of Bismuth and Ammonium.
(B. P.) *Syn.* LIQUOR BISMUTHI ET AMMONII
CITRATIS, LIQUOR BISMUTHI, L. Citrate of
bismuth, 800 gr.; solution of ammonia, distilled
water, of each, a sufficiency. Rub the citrate of
bismuth to a paste with a little water, gradually
stir in ammonia until the bismuth salt dissolves.
Dilute with distilled water to make 1 pint.—*Dose*,
½ to 1 dr.
Solution of Citrate of Magne'sium. *Syn.* SO-
LUTIO MAGNESII CITRATIS. See MAGNESIUM,
CITRATE OF.
(B. P.) Carbonate of magnesia, 100 gr.;
citric acid, 200 gr.; syrup of lemons, ½ fl. oz.;
bicarbonate of potassium in crystals, 40 gr.;
water, q. s. Dissolve the citric acid in 2 oz. of
the water, and having added the carbonate of
magnesium, stir until it is dissolved. Filter the
solution into a strong half-pint bottle, add the
syrup and water sufficient to nearly fill the bottle,
then introduce the bicarbonate of potassium, and
immediately close the bottle with a cork, which
should be secured with string or wire; afterwards
shake till the bicarbonate has dissolved.—*Dose*,
5 to 10 fl. oz.
Solution of Citrate of Morphia. *Syn.* LIQUOR
MORPHIAE CITRATIS, SOLUTIO M. C., L. *Prep.*
(*Magendie*.) Pure morphia, 13 gr.; citric acid,
8 or 10 gr.; water, 1 fl. oz.; tincture of cochineal,
2 fl. dr.—*Dose*, 3 to 12 drops.
Solution of Coal-tar. (B. P. C.) *Syn.* LIQUOR
PICIS CARBONIS. Quillaia bark, in No. 20 powder,
2 oz.; rectified spirit, a sufficient quantity.
Moisten the powder with a suitable quantity of
the menstruum and macerate for twenty-four
hours in a closed vessel. Then pack in a percolator,
and gradually pour rectified spirit upon it until
one pint of percolate is obtained. To this add
prepared coal-tar, 4 oz.; digest at a temperature
of 120° F. for two days, allow to become cold, and
decant or filter.
The product is an imitation of LIQUOR CAR-
BONIS DETERGENS.
Uses. 1 oz. to a pint of water, or a pound of
lard, yields a lotion or ointment useful as an antiseptic and stimulant in a variety of skin diseases.
A soap is made from it which is an excellent
cleanser.
Solution of Copai'ba. See SOLUTION, SPECIFIC.
Solution of Corro'sive Sub'limate. *Syn.* SOLU-
TION OF CHLORIDE OF MERCURY; LIQUOR
HYDRARGYRI BICHLORIDI† (Ph. L.), L. *Prep.* 1.
(Ph. L.) Corrosive sublimate and sal-ammoniac,
of each, 10 gr.; water, 1 pint; dissolve.—*Dose.*
As an alterative, 10 to 30 drops; as an antisy-

philitic, ½ to 2 fl. dr., in simple or sweetened water.
It must not be allowed to touch anything metallic.
It also forms a most useful lotion in various skin
diseases.
2. See LOTION, MERCURIAL.
Solution of Cyanide of Potassium. *Syn.* LI-
QUOR POTASSI CYANIDI (*Laming*). *Prep.* Cya-
nide of potassium, 22 gr.; proof spirit, 9 fl. dr.
This is the strength of his hydrocyanic acid,
which contains 1 gr. of real acid in 1 fl. dr.
Magendie's medicinal hydrocyanate of potash
consists of cyanide of potassium dissolved in 8
times its weight of distilled water.
Solution of Delphinia. *Syn.* SOLUTIO DEL-
PHINIAE (*Dr Turnbull*). Delphinia, 1 scruple;
rectified spirit, 2 oz. For outward use.
Solution of Diac'etate of Lead. See SOLUTION
OF SUBACETATE OF LEAD.
Solution of Dialysed Iron. (B. P.) *Syn.* LIQUOR
FERRI DIALYSATUS, L. Strong solution of perchloride of iron, 7 oz.; solution of ammonia and
water, of each, a sufficiency. Mix 6 oz. of the iron
with 3 pints of water, add ammonia, with constant
stirring, until the product has a distinct ammoniacal odour. Filter, wash and press the precipitate, add it to the remaining iron, dissolve with a
gentle heat; filter, and place the fluid in a covered
dialyser and float it on water, which is constantly
running, until it is almost tasteless. Make it to
measure 28 oz.—*Dose*, 10 to 30 minims.
Solution, Donovan's. See SOLUTION OF HY-
DRIODATE OF ARSENIC AND MERCURY (*below*).
Solution, Escharotic (Freyburg's). *Syn.* SOLU-
TIO ESCHAROTICA, L. *Prep.* From camphor,
30 gr.; corrosive sublimate, 60 to 100 gr.; rectified spirit, 1 fl. oz.; dissolve. In syphilitic vege-
tations, and especially condylomes. It is spread
over the diseased surface, either at once or after
the application of a ligature.
Solution of Ethylate of Sodium. (B. P.) *Syn.*
LIQUOR SODII ETHYLATIS. Sodium, 22 gr.;
ethylic alcohol, 1 oz. Dissolve the sodium in the
alcohol kept cool in a stream of cold water. It
should be freshly made, as it darkens by keeping.
Solution of Flints. *Syn.* LIQUOR OF FLINTS;
LIQUOREM SILICUM, LIQUOR POTASSAE SILICATIS,
L. *Prep.* 1. Soluble glass dissolved in water.
2. (*Bate*.) Powdered quartz, 1 part; dry
carbonate of potash, 2 parts (3 parts—*Turner*);
triturate them together, fuse the mixture in a
Hessian crucible, and allow the resulting glass to
deliquesce by exposure in a damp situation.—*Dose*,
5 or 6 to 30 drops; in gouty concretions, stone,
&c. "It resolves the stone, and opens obstructions." See SOLUBLE GLASS.

Solution, Gannal's. } See SOLUTION FOR ANA-
Solution, Goadsby's. } TOMICAL PREPARA-
                     } TIONS.
Solution, Goulard's. See SOLUTION OF SUB-
ACETATE OF LEAD.
Solution of Gutta Percha. (B. P.) *Syn.* LIQUOR
GUTTA PERCHA. Gutta percha, 1 part; chloro-
form, 8 parts; carbonate of lead, 1 part; mix
and dissolve; set aside for a few days, and decant
the clear fluid.
Solution, Hahnemann's Prophylac'tic. *Syn.*
LIQUOR BELLADONNAE, SOLUTIO PROPHYLACTICA,
L. *Prep.* From extract of belladonna (alcoholic),
3 gr.; distilled water, 6 fl. dr.; rectified spirit, 2

fl. dr.; dissolve. Used against scarlet fever.—
Dose, 2 or 3 drops for a child under 12 months,
and an additional drop for every year above that
age to maturity.

Solution of Hartshorn, Succinated. *Syn.* LI-
QUOR CORNU CERVI SUCCINATUS (P. Cod.).
Neutralise true spirits of hartshorn (or a solution
of 1 oz. of salt of hartshorn in 1 oz. of water)
with acid of amber.

Solution of Hydri'odate of Ar'senic and Mer'-
cury. *Syn.* DONOVAN's SOLUTION; SOLUTIO
ARSENICI ET HYDRARGYRI IODIDI, ARSENICI ET
HYDRARGYRI HYDRIODATIS LIQUOR (Ph. D.), L.
Prep. 1. (*Donovan.*) Triturate metallic arsenic,
6·08 gr., mercury, 15·38 gr., and iodine, 50 gr.,
with alcohol, 1 fl. dr., until dry; to this add,
gradually, of distilled water, 8 fl. oz., and again
well triturate; next put the whole into a flask,
add of hydriodic acid, ½ fl. dr., and boil for a few
minutes; lastly, when cold, add distilled water,
q. s. to make the whole measure exactly 8 fl. oz.
2. (Ph. D.) Pure arsenic (in fine powder),
6 gr.; pure mercury, 16 gr.; pure iodine, 50½
gr.; alcohol, ½ fl. dr.; triturate as before; add,
gradually, of water, 8 fl. oz., heat the mixture
until it begins to boil, and afterwards make up
the cold and filtered solution to exactly 8 fl. oz.
6 fl. dr.
3. (Wholesale.) From metallic arsenic, 61 gr.;
iodine, 500 gr.; mercury, 154 gr.; rectified spirit,
1½ fl. oz.; distilled water, 2 quarts; hydriodic
acid, 5 fl. dr.; as No. 1, the product being made
up with distilled water so as to measure exactly
4 pints, or 80 fl. oz., or to weigh 5 lbs. 1½ oz.
(av.), when cold.
4. (B. P.) Iodide of arsenicum, 45 gr.; red
iodide of mercury, 45 gr.; dissolve in distilled
water enough to make 10 oz.
Obs. Great care must be taken that the whole
of the arsenic be dissolved, which can only be
effected by the most careful trituration. Soubeiran
recommends the employment of 1 part, each, of
the respective iodides, with 98 parts of water,
as furnishing a simpler and equally effective product, proportions which are almost exactly those
employed by Mr Donovan.—*Dose*, 10 to 30 drops,
twice or thrice a day, preferably soon after a meal;
in lepra, psoriasis, lupus, and several other scaly
skin diseases. It is a most valuable medicine in
these affections.

Solution of Hydrochlorate of Cocaine. (B. P.)
Syn. LIQUOR COCAINE HYDROCHLORATIS, L. Hydrochlorate of cocaine, 33 gr.; salicylic acid, ½ gr.;
distilled water, to produce 6 dr.; dissolve.

Solution of Hydrochlo''rate of Mor'phine. *Syn.*
SOLUTION OF MURIATE OF MORPHIA; LIQUOR
MORPHLÆ HYDROCHLORATIS (Ph. L.), SOLUTIO
MORPHLÆ MURIATIS (Ph. E.), MORPHLÆ MU-
RIATIS LIQUOR (Ph. D.), L. *Prep.* 1. (Ph. L.)
Hydrochlorate of morphia, 4 dr.; proof spirit, ½
pint; distilled water, 1 pint; dissolve by the aid
of a gentle heat. 60 drops (minims) of this
solution contain 1 gr. of hydrochlorate of morphia.
—*Dose*, 5 to 15 or 20 drops.
2. (Ph. E. & D.) Muriate of morphia, 90 gr.;
rectified spirit, 5 fl. oz.; distilled water, 15 fl. oz.
107 drops (minims) contain 1 gr. of the hydrochlorate.—*Dose*, 10 to 30 or 40 drops, or nearly as
laudanum.

3. (Apothecaries' Hall.) Muriate of morphia,
16 gr.; rectified spirit, 1 fl. dr.; water, 1 fl. oz.;
30 drops (minims) contain 1 gr.—*Dose*, 3 to 10
drops. See SOLUTION OF ACETATE OF MOR-
phia, &c.
4. (B. P.) Hydrochlorate of morphine, 9 gr.;
dilute hydrochloric acid, 18 minims; rectified spirit,
½ oz.; distilled water, 1½ oz.; mix and dissolve.
—*Dose*, 10 to 60 minims.

Solution of Hydrochlorate of Strychnine. (B. P.)
Syn. LIQUOR STRYCHNIÆ HYDROCHLORATIS,
LIQUOR STRYCHNIÆ. Strychnine, 9 gr.; dilute
hydrochloric acid, 14 minims; rectified spirit,
½ oz.; distilled water, 1½ oz. Mix the strychnine,
acid, and water, dissolve by aid of heat, then add
the spirit.—*Dose*, 5 to 10 minims.

Solution of Hypochlo''rite of Lime. Solution of
chlorinated lime.

Solution of Hypophosphites (Compound).
(B. P. C.) *Syn.* LIQUOR FERRI HYPOPHOS-
PHITIS COMPOSITUS. Hypophosphite of calcium,
320 gr.; hyposulphate of sodium, 320 gr.; hyposulphate of magnesium, 160 gr.; strong solution
of hypophosphite of iron, 6 fl. oz.; hypophosphorous
acid, 30%, ½ fl. oz.; distilled water, a sufficient
quantity. Dissolve the hypophosphites of calcium,
sodium, and magnesium in 12 fl. oz. of distilled
water; add the solution of hypophosphite of iron
and the hypophosphorous acid. Filter, and make
up to 1 pint by the addition of distilled water.
Each fl. dr. contains about 2 gr. each of hypophosphite of sodium and calcium, 1 gr. of hypophosphite of magnesium, and 1½ gr. of hypophosphite of iron.—*Dose*, ½ to 2 fl. dr.

Solution of I'odide of Ar'senic. *Syn.* LIQUOR
ARSENICI PERIODIDI, L. *Prep.* (*Wackenroder.*)
Each dr. contains ½ gr. of teriodide of arsenic,
equivalent to ½ gr. of metallic arsenic and ₇₀ gr.
(nearly) of iodine.

Solution of Iodide of Iron. *Syn.* LIQUOR FERRI
IODIDI (Ph. U. S.). *Prep.* Mix 2 oz. (troy) of
iodine with 5 oz. of water, and add 1 oz. (troy) of
iron filings, stir frequently, and heat the mixture
gently till it assumes a greenish colour; then filter
into a glass bottle containing 12 oz. of powdered
sugar, and after it has passed, pour distilled water
on the filter until the filtered liquor, including
the sugar, measures 20 oz., last shake the bottle
till the sugar is dissolved.—*Dose*, 15 minims to
1 dr.

Solution of Iodide of Mer'cury and Potas'sium.
Syn. LIQUOR IODHYDRARGYRATIS POTASSII
IODIDI, L. *Prep.* (*Dr Channing.*) Iodide of
potassium, 8½ gr.; binoxide of mercury, 4½ gr.;
distilled water, 1 fl. oz.; dissolve.—*Dose*, 2 to 5
or 6 drops, three times a day, much diluted; in
dyspepsia, indurations, enlargement of the spleen,
dropsy, &c.

Solution of Iodide of Potas'sium (Compound).
Syn. IODURETTED WATER, COMPOUND SOLUTION
OF IODINE; LIQUOR POTASSII-IODIDI COMPOSI-
TUS (Ph. L. & D.), LIQUOR IODINEI COMPOSITUS
(Ph. E.), L. *Prep.* 1. (Ph. L. & D.) Iodide of
potassium, 10 gr.; iodine, 5 gr.; water, 1 pint;
dissolve.—*Dose*, 1 to 6 dr.; in the usual cases
where iodine is employed.
2. (Ph. E.) Iodide of potassium, 1 oz.; iodine,
2 dr.; water, 16 fl. oz. This is 30 times as strong
as the preceding.—*Dose*, 5 to 20 drops.

Solutions of Iodine. *Syn.* LIQUOR IODI (B. P.). *Prep.* Dissolve 22 gr. of iodine and 33 gr. of iodide of potassium in 1 oz. of distilled water. (*Lugol's.*) *Syn.* SOLUTIONES IODINII VEL IODURETE. *Prep.* Ioduretted waters, Nos. 1, 2, and 3; iodine, 1½ gr., 2 gr., and 2½ gr.; water, 1 pint. *Drops.*—Iodine, 1 scruple; iodide of potassium, 2 scruples; water, 9 dr. *Lotions, &c.*—Iodine, 1½ gr. to 3 gr.; iodide of potassium, 3 gr. to 6 gr.; water, 1 pint. *Rubefacient.*—Iodine, 1 part; iodide of potassium, 2 parts; water, 12 parts. *Caustic.*—Iodine, 1 part; iodide of potassium, 1 part; water, 2 parts.

Solution of Iodine with Hemlock. *Syn.* SOLUTIO IODINII CUM CONIO; DE SCUDAMORE'S SOLUTION. (For inhaling.) Iodine, 6 gr.; iodide of potassium, 6 gr.; rectified spirit, 2 dr.; water, 5 oz. 6 dr. From ½ dr. to 5 dr. of this solution with ½ dr. of tincture of hemlock to be added to warm water at 120° F. in a glass inhaler, and used twice a day. Two thirds of the ingredients are first put into the inhaler, and the rest added when half the time for inhaling has elapsed.

Solution of I'ron (Alkaline). *Syn.* LIQUOR FERRI ALKALINI, L. *Prep.* (Ph. L. 1824.) Iron filings, 2½ dr.; nitric acid, 2 fl. oz.; water, 6 fl. oz.; dissolve, decant, gradually add of solution of carbonate of potash, 6 fl. oz., and in 6 hours decant the clear portion. This was intended as an imitation of Stahl's Tinctura Martis Alkalina. It is tonic, emmenagogue, &c.—*Dose*, 20 to 60 drops.

Solution of Iron and Alum. *Syn.* SOLUTIO FERRI ALUMINOSA, L. (*Swediaur.*) *Prep.* Calcined sulphate of iron, 10 scruples; alum, 5 scruples; water, sufficient to dissolve them; sulphuric acid, 15 drops.—*Dose*, 10 to 15 drops. Once a celebrated nostrum in Germany, under the name of *Tinctura nervosa.*

Solution, Javelle's. See SOLUTION OF CHLORINATED POTASH.

Solution, Labarraque's. See SOLUTION OF CHLORINATED SODA.

Solution of Lime. *Syn.* LIME WATER; SOLUTIO CALCIS HYDRATIS, LIQUOR CALCIS (Ph. L. & D.), AQUA CALCIS (Ph. E.), L. *Prep.* (Ph. L.) Upon the lime, ¼ lb., first slaked (by sprinkling it) with a little of the water, pour the remainder of water, 12 pints, and shake them well together (for 5 minutes—Ph. D.); immediately cover the vessel, and set it aside for 3 hours; then keep the solution with the remaining lime (equally divided) in stoppered glass vessels, and, when it is to be used, decant the required portion from the clear solution (replacing it with more water, and agitating briskly, as before—Ph. E.).

LIQUOR CALCIS (B. P.). *Syn.* LIME WATER. *Prep.* Wash 2 oz. of slaked lime with water to free it from chlorides, put it into a stoppered bottle containing 1 gall. of distilled water, and shake well for 2 or 3 minutes. After 12 hours the excess of lime will have subsided, and the clear solution may be drawn off with a syphon as it is required for use, or transferred to a green glass bottle furnished with a well-ground stopper.

Obs. Cold water dissolves more lime than hot water. 1 pint of water at 32° F. dissolves 13½ gr., at 60° it dissolves 11½ gr., but at 212° only 5½ gr. (*Phillips*).

Uses, &c. Lime water is antacid, astringent, antilithic, tonic, and vermifuge.—*Dose.* A wineglassful, or more, 2 or 3 times a day, in milk or broth; in dyspepsia, diarrhœa, calculous affections, &c.; and, externally, as a detersive and discutient lotion.

Solution of Lime (Saccharated). (B. P.) *Syn.* LIQUOR CALCIS SACCHARATUS. *Prep.* Slaked lime, 1 part; refined sugar (in powder), 2 parts; distilled water, 20 parts; digest for some hours and strain.—*Dose*, 15 to 16 minims in milk.

Solution of Lithia, Effervescing. *Syn.* LIQUOR LITHIÆ EFFERVESCENS (B. P.). *Prep.* Mix 10 gr. of carbonate of lithia and 1 pint of water in a suitable apparatus, and charge with carbonic acid gas under a pressure of 7 atmospheres. Keep in bottles securely corked.

Solution, Mackenzie's. *Prep.* From nitrate of silver, 20 gr., dissolved in distilled water, 1 fl. oz. Used to wash the throat and fauces, and to sponge the trachea, in affections of those parts.

Solution of Magne'sia. *Syn.* AËRATED MAGNESIA WATER, CARBONATED M. W., FLUID MAGNESIA, CONDENSED SOLUTION OF M., CONCENTRATED S. OF M.; LIQUOR MAGNESIÆ CARBONATIS, AQUA M. C., L.; EAU MAGNÉSIENNE, Fr. *Prep.* (*Dinneford's.*) Water and Howard's heavy carbonate of magnesia, in the proportion of 17½ gr. of the latter to every fl. oz. of the former, are introduced into a cylindrical tinned copper vessel, and carbonic acid, generated by the action of sulphuric acid on whiting, is forced into it by steam power for 5½ hours, during the whole of which time the cylinder is kept in motion. Sir J. Murray's is similar. The Paris Codex orders recently precipitated carbonate of magnesia to be used while still moist. Antacid and laxative. See FLUID MAGNESIA.

Solution, Min'eral. See SOLUTION OF ARSENITE OF POTASSA.

Solution of Mor'phine. See SOLUTIONS OF ACETATE, HYDROCHLORATE, and SULPHATE.

Solution of Myrrh, Alkaline. *Syn.* SOLUTIO MYRRHÆ ALKALINA (*Swediaur.*) *Prep.* Carbonate of soda, 1 dr.; myrrh, 2 oz.; boiling water, 8 oz. Digest in a water-bath for 2 days, frequently stirring, and strain.

Solution of Nitrate of Mercury (Acid). *Syn.* LIQUOR HYDRARGYRI NITRATIS ACIDUS (B. P.). *Prep.* Mercury, 3 parts; nitric acid, 5 parts; distilled water, 1½ parts; mix the nitric acid with the water in a flask, and dissolve the mercury in the mixture without the application of heat. Boil gently for 15 minutes, cool, and preserve the solution in a stoppered bottle. Used alone, as a caustic; 1 to 2 minims to 1 oz. of water as a gargle; and 1 minim to 2 oz. of water as an injection in gonorrhœa.

Solution of Nitrate of Mercury and Ammonia. *Syn.* SOLUTIO HYDRARGYRI ET AMMONIÆ NITRATIS; WARD'S WHITE DROP. *Prep.* Nitrate of ammonia and mercury in crystals, 1 part; rose water, 3 parts; digest till dissolved.

Solution of Nitrate of Sil'ver. *Syn.* LIQUOR ARGENTI NITRATIS (Ph. L.), SOLUTIO A. N. (Ph. E.), L. *Prep.* (Ph. L.) Nitrate of silver (cryst.), 1 dr. (40 gr.—Ph. E.); distilled water, 1 fl. oz. (1600 gr.—Ph. E.); dissolve. Used as an escharotic, &c. It should be kept from the light. See LOTION, NITRATE OF SILVER, &c.

Solution of O'pium (Sed'ative). See LIQUOR.

Solution of Perchloride of Iron. Syn. LIQUOR FERRI PERCHLORIDI (B. P.). Prep. Stronger solution of perchloride of iron (see below), 1 part; distilled water, 3 parts.—Dose, 10 to 30 minims.

Solution of Perchloride of Iron (Stronger). Syn. LIQUOR FERRI PERCHLORIDI FORTIOR (B. P.). Prep. Iron wire, 4 oz.; hydrochloric acid, 20½ oz.; nitric acid, 1½ oz.; water, a sufficiency. Place the wire in a flask, add 12½ parts of hydrochloric acid and 7 parts of water, heat gently until effervescence ceases; filter from undissolved iron; add to the filtrate 7 oz. hydrochloric acid; mix, and pour the solution in a slow stream into 1½ oz. nitric acid, heating to assist the evolution of red fumes. Evaporate until a precipitate begins to form, then add 1 oz. of hydrochloric acid, and water to produce 17½ oz. Used as an application to diphtheritic patches, for injecting nævi, as a powerful styptic, and in the preparation of SOLUTION OF PERCHLORIDE OF IRON (see above).

Solution of Perchloride of Mercury. Syn. LIQUOR HYDRARGYRI PERCHLORIDI (B. P.). Prep. Corrosive sublimate, 10 gr.; chloride of ammonium, 10 gr.; distilled water, 20 oz.; dissolve.— Dose, 30 to 120 minims.

Solution of Perchloride of Mercury (Compound). Syn. LIQUOR HYDRARGYRI, PERCHLORIDI COMPOSITUS, L.; LIQUOR MERCURIELLE NORMALE (Mialhe), Fr. Prep. Distilled water, 16 oz.; chloride of sodium, 16 gr.; chloride of ammonium, 16 gr.; white of 1 egg; perchloride of mercury, 4 gr. Beat the white of egg with the water, filter, dissolve the salts in the liquid, and filter again.

Solution of Permanganate of Potassium. Syn. LIQUOR POTASSII PERMANGANATIS (B. P.), L. Prep. Permanganate of potassium, 88 gr.; distilled water, 1 pint; dissolve. Diluted with 40 parts of water, it is used as a gargle or as a cleansing wash for diseased surface.—Dose, 2 to 4 dr.

Solution of Perni'trate of Iron. (B. P.) Syn. SOLUTION OF PERSESQUINITRATE OF IRON; FERRI PERNITRAS LIQUOR (Ph. D.), SOLUTIO PERSESQUINITRAS FERRI (Kerr), L. Prep. (Ph. D.) Take of pure nitric acid, 4½ fl. oz.; water, 16 fl. oz.; mix, add fine iron wire, 1 oz.; dissolve, and to the clear solution add as much water as will make the whole measure 1½ pints. Sp. gr. 1·107.— Dose, 5 or 6 to 30 drops, or more; in passive hæmorrhages, mucous discharges, chronic diarrhœa with prostration, &c.

Solution of Persulphate of Iron. Syn. LIQUOR FERRI PERSULPHATIS (B. P.). Prep. Sulphate of iron, 8 parts; sulphuric acid, ½ part; nitric acid, ½ part; distilled water, 12 parts. Add the sulphuric acid to 10 parts of the water, and dissolve the sulphate of iron in the mixture with the aid of heat. Mix the nitric acid with the remaining 2 parts of the water, and add the dilute acid to the solution of sulphate of iron. Concentrate the whole by boiling until, by the sudden evolution of ruddy vapours, the liquid ceases to be black, and acquires a red colour. A drop of the solution is now to be tested with ferricyanide of potassium, and if a blue precipitate be formed, a few additional drops of nitric acid should be added, and the boiling renewed, in order that the whole may be converted into persulphate of iron. When the solution is cold, make up the quantity to 11 parts by the addition, if necessary, of distilled water. Used in making several preparations of iron; it is also a good styptic.

Solution of Phosphoric Ether. Syn. SOLUTIO PHOSPHORI ÆTHEREA, L. Prep. Sliced phosphorus, 5 gr.; rectified ether, 1 oz.; mix, set the bottle in a dark place for 3 or 4 days, shaking occasionally, and decant.

Solution for Plate. Syn. PLATE LIQUOR; SOLUTIO PRO ARGENTO, L. Prep. From alum, cream of tartar, and common salt, of each, 1 oz.; water, ½ gall.; dissolve. Used to increase the lustre and whiteness of silver plate, the articles being boiled in it.

Solution of Potas'sa. Syn. SOLUTION OF HYDRATE OF POTASSA, LIQUOR OF POTASSA, POTASH WATER, CAUSTIC P. W.; LIQUOR POTASSÆ (B. P., Ph. L.), AQUA POTASSÆ (Ph. E.); POTASSÆ CAUSTICÆ LIQUOR (Ph. D.), AQUA KALI PURI†, LIXIVIUM SAPONARIUM†, AQUA KALI CAUSTICUM†, LIXIVIUM CAUSTICUM†, L. Prep. 1. (Ph. L.) Lime (recently burnt), 8 oz.; boiling distilled water, 1 gall.; sprinkle a little of the water on the lime in an earthen vessel, and, when it is slaked and fallen to powder, add of carbonate of potassa, 15 oz., dissolved in the remainder of the water; bung down and shake frequently, until the mixture is cold, then allow the whole to settle, and decant the clear supernatant portion into perfectly clean and well-stoppered green glass bottles. Sp. gr. 1·063. It contains 6·7% of pure potassa.

2. (Ph. E.) Carbonate of potassa (dry), 4 oz.; quicklime, 2 oz.; water, 45 fl. oz.; boiling briskly for a few minutes after each addition of the milk of lime; to yield at least 35 fl.oz. by decantation, after 24 hours' repose in a deep, narrow glass vessel. Sp. gr. 1·072.

3. (Ph. D.) Pure carbonate of potassa, 1 lb.; distilled water, 1 gall.; dissolve, heat the solution to the boiling-point in a clean iron vessel, gradually add to it of fresh quicklime, 10 oz., previously slaked with water, 7 fl. oz.; and continue the ebullition for 10 minutes, with constant stirring; next allow it to cool out of contact with the air, and, when perfectly clear, decant it by means of a syphon, and bottle it as before. Sp. gr. 1·068.

4. (B. P.) Carbonate of potash, 2 parts; slaked lime, 1½ parts; distilled water, 20 parts; dissolve the carbonate of potash in the water, and having heated the solution to the boiling-point in a clean iron vessel, gradually mix the washed slaked lime, and continue the ebullition for 10 minutes with constant stirring; decant the clear liquid.—Dose, 15 to 60 minims 3 times a day in beer, milk, or Mistura Amygdalæ.

5. (Wöhler.) Nitrate of potassa, 1 part, is mixed in alternate layers with clippings of sheet copper, 2 or 3 parts, and then heated to moderate redness for about ½ an hour in a copper or iron crucible; when cold the potassa is washed out with distilled water, and the solution, after repose in a closed vessel, decanted as before. Not a trace of copper can be detected in the liquid. The clippings may be again used if mixed with a little fresh metallic copper.

6. (Wholesale.) From carbonate of potash (kali), 1 lb., and quicklime, ½ lb., to each gall. of water.

7. (BRANDISH'S ALKALINE SOLUTION; LIQUOR POTASSÆ BRANDISHII.) From American pearl-ashes, 6 lbs.; quicklime and wood ashes (from the ash), of each, 2 lbs.; boiling water, 6 galls. (old meas.); to each gall. of the clear product is added 12 or 15 drops of oil of juniper. This 'solution' is much asked for in trade. Ordinary liquor of potassa is generally sold for it.

Pur. "Nothing, or scarcely anything, is thrown down from this solution on the addition of lime water; and when it has been first saturated by nitric acid no precipitate falls on the addition of carbonate of soda, chloride of barium, or nitrate of silver. What is thrown down by bichloride of platinum is yellowish" (Ph. L.).

Uses, &c. Liquor of potassa is antacid, diuretic, resolvent, and lithontriptic.—*Dose,* 10 to 30 or 40 drops, in any bland diluent (not acidulous); in heartburn, gout, calculi, indurations, scrofula, lepra, psoriasis, &c.

Obs. Quicklime fails to abstract the carbonic acid from the alkaline carbonates in solutions much stronger than those above referred to. Weaker solutions may, however, be easily concentrated by evaporation in iron vessels. See POTASSIUM, HYDRATE of, and *below.*

Solution of Potas'sa (Effervescing). *Syn.* LIQUOR POTASSÆ EFFERVESCENS (B. P.); EFFERVESCING POTASH WATER, SUPERCARBONATE OF POTASSA W.; AQUA POTASSÆ EFFERVESCENS (Ph. E.), A. P. SUPERCARBONATIS, L. *Prep.* 1. (Ph. L. & E.) Bicarbonate of potash, 1 dr.; distilled water, 1 pint; dissolve, force in carbonic acid gas in excess, and keep it in a well-stoppered bottle. Resembles soda water, but sits better on the stomach. It is almost specific in the early stages of scurvy.

2. (B. P.) Dissolve 30 gr. of bicarbonate of potash in 1 pint of distilled water, filter, pass in washed carbonic acid (obtained by the action of sulphuric acid on chalk) up to a pressure of 4 atmospheres. Keep in bottles closely secured.

Obs. An excellent substitute for this preparation is to pour a bottle of soda water into a tumbler containing 20 gr. of powdered bicarbonate of potash, and to drink it immediately.

Solution of Potas'sio-tar'trate of An'timony. *Syn.* SOLUTIO ANTIMONII POTASSIO-TARTRATIS, ANTIMONII TARTARIZATI LIQUOR (Ph. D.), L. *Prep.* (Ph. D.) Tartarised antimony, 1 dr.; rectified spirit, 7 fl. oz.; distilled water, 1 pint; dissolve. Strength, doses, and uses similar to those of antimonial wine (which *see*), than which it keeps better.

Solution, Prophylac'tic. See HAHNEMANN'S SOLUTION.

Solution of Protonitrate of Mercury. *Syn.* LIQUOR HYDRARGYRI NITRICI (PROTONITRATIS) (G. Ph.). *Prep.* Protonitrate of mercury, 1 oz.; distilled water, 9 oz.; nitric acid (1·185), 46 gr.; filter.—*Dose,* 1 to 5 drops.

Solution of Sil'icate of Potas'sa. See SOLUTION OF FLINTS.

Solution of So'da. *Syn.* SOLUTION OF HYDRATE OF SODA, LIQUOR OF SODA, CAUSTIC SODA WATER; VOL. II.

LIQUOR SODÆ (B. P., Ph. L.), SODÆ CAUSTICÆ LIQUOR (Ph. D.), L. *Prep.* 1. (Ph. L.) Carbonate of soda (cryst.), 32 oz.; lime, 9 oz.; boiling distilled water, 1 gall.; proceed as for solution of potassa. "In 100 gr. is contained 1 gr. of (pure) soda" (Ph. L.). Sp. gr. 1·061.

2. (Ph. D.) Carbonate of soda (cryst.), 2 lbs.; fresh-burned lime, 10 oz.; water, 1 gall. 7 fl. oz.; as liquor of potassa. Sp. gr. 1·056.

3. (B. P.) Carbonate of soda, 7 parts; slaked lime, 3 parts; distilled water, 40 parts; dissolve the carbonate in the water, boil in a clean iron vessel, gradually mixing the washed lime, and stirring constantly for ten minutes; decant into a green glass bottle with air-tight stopper. Sp. gr. 1·047.—*Dose,* ½ to 1 dr.

Solution of Soda (Effervescing). *Syn.* SODA WATER; LIQUOR SODÆ EFFERVESCENS, AQUA S. E. (Ph. E.), A. S. SUPERCARBONATIS, SODÆ CARBONATIS AQUA ACIDULA, L. *Prep.* (Ph. E.) Bicarbonate of soda, 1 dr.; distilled water, 1 pint; dissolve, and force carbonic acid gas into the solution under pressure. Used as an antacid and grateful stimulant, often proving gently laxative. The soda water of the shops cannot be substituted for this preparation, as, in opposition to its name, it is usually made without soda. (B. P.) Half the strength.

Solution, Sol'dering. *Prep.* Dissolve zinc in hydrochloric acid nearly to saturation, add 1-5th part of powdered sal-ammoniac, and simmer for five minutes. Used to make solder flow easily and take well; applied with a feather. See SOLDERING.

Solution, Speci'fic (Frank's). *Syn.* SPECIFIC SOLUTION OF COPAIBA; LIQUOR COPAIBÆ ALKALINA, L. *Prep.* Take of balsam of copaiba, 3 parts; liquor of potassa (Ph. L.), 3 parts; water, 7 parts; boil the mixture for 2 or 3 minutes, put it into a separator, and allow it to stand for 5 or 6 days; then draw it off from the bottom, avoiding the upper stratum of oil, and to the clear liquid add of sweet spirit of nitre (perfectly free from acid), 1 part; should it turn foul or milky, a very little liquor of potassa will usually brighten it; if not, place it in a clean separator, and let it stand, closely covered, for a few days, and then draw it off from the bottom as before, when it will be perfectly transparent, without filtering. Some persons add the sweet spirit of nitre whilst the solution is still warm, mix it in as rapidly as possible, and immediately cork or fasten up the vessel. This is a good way when the article is wanted in a hurry, but is objectionable from the loss of spirit thereby occasioned, and the danger, without care, of bursting the separator.

Obs. A receipt for this article, upon the authority of Battley, has been going the round of the pharmaceutical works for many years. It is as follows:—Take 12 oz. of balsam of copaiba, and 6 oz. of calcined magnesia; rub together, add a pint of proof spirit, filter, and then add ¼ oz. of sweet spirits of nitre ('Gray's Supplement'). The product of this formula, utterly unlike 'Frank's specific solution,' is a colourless tincture, scarcely flavoured with copaiba, and holding very little of the active matter of the balsam in solution, owing to the compound formed with the

98

magnesia being insoluble in spirit. Such is the affinity of this earth for copaiba (copaibic acid), that it will even take it from caustic potassa. See COPAIBA and its preparations.

Solution of Subac'etate of Lead. *Syn.* LIQUOR OF SUBACETATE OF LEAD, L. OF DIACETATE OF L.†, GOULARD'S EXTRACT; LIQUOR PLUMBI, L. PLUMBI DIACETATIS (Ph. L.), PLUMBI DIACETATIS SOLUTIO (Ph. E.), PLUMBI SUBACETATIS LIQUOR (Ph. D.), L. *Prep.* 1. (Ph. L.) Acetate of lead, 27 oz.; litharge, in fine powder, 16 oz.; water, 3 quarts; boil for ½ an hour, constantly stirring, and then add enough distilled water to make the whole measure 3 quarts; lastly, filter, if required, and keep it in a closed vessel. The proportions ordered in the Ph. E. are similar. Sp. gr. 1·260.

2. (B. P.) Acetate of lead, 6 oz.; litharge, 3½ oz.; distilled water, 1 pint; boil, &c., as before; to produce 1 pint.

3. (Wholesale.) From finely powdered litharge, 32 lbs.; distilled vinegar, 32 galls.; boil in a perfectly bright copper pan for 2 hours, cool, add water to make up 32 galls., again simmer for 1 minute, cover up the vessel, and in an hour decant the clear portion. Common trade strength. (See *below*.)

Solution of Subacetate of Lead (Dilute). *Syn.* GOULARD, GOULARD'S LOTION, G.'S WATER; LIQUOR PLUMBI DIACETATIS DILUTUS (Ph. L.), PLUMBI SUBACETATIS LIQUOR COMPOSITUS (Ph. D.), L. *Prep.* 1. (Ph. L.) Liquor of diacetate of lead, 1½ fl. dr.; proof spirit, 2 fl. dr.; distilled water, 1 pint; mix.

2. (Ph. D.) Solution of subacetate of lead and proof spirit, of each, 2 fl. oz.; distilled water, ½ gall.; mix, filter, and preserve it in a well-stoppered bottle.

3. (B. P.) Solution of subacetate lead, 2 of fl. dr.; rectified spirit, 2 fl. dr.; distilled water, 19½ oz. Filter through paper.

Obs. Both the above preparations were formerly made with common vinegar, and hence were coloured, but those of the Pharm. are white. If wanted coloured, a little spirit colouring may be added. The stronger liquor is only used diluted, and the dilute solution is now seldom prepared by the wholesale druggist. The lead (diluted solution) is employed as a sedative, refrigerant, and astringent wash, in various affections. Both are poisonous. For the antidotes see LEAD.

Solution of Sulphate of Atropine. *Syn.* LIQUOR ATROPINÆ SULPHATIS (B. P.), L. *Prep.* Sulphate of atropia, 9 gr.; camphor water, 16½ dr.; dissolve.—*Dose*, 1 to 2 minims.

Solution of Sulphate of Copper. *Syn.* LIQUOR CUPRI SULPHATIS COMPOSITUS, AQUA STYPTICA, L. (Ph. L. 1746.) Sulphate of copper, 3 oz.; alum, 3 oz.; sulphuric acid, 2 oz.; (by weight,) water, 24 oz. For external use.

Solution of Sulphate of Indigo. *Syn.* LIQUOR INDIGO SULPHATIS, L. *Prep.* Digest 1 part of powdered indigo in 10 parts of sulphuric acid; when dissolved dilute it with water. Used as a test.

Solution of Sulphate of Morphine. (B. P.) *Syn.* LIQUOR MORPHINÆ SULPHATIS, L. Sulphate of morphine, 1 part; rectified spirit, 25

parts; water, to produce 100 parts.—*Dose*, 10 to 60 minims.

Solution of Sulphate of Zinc (Compound). See SOLUTION OF ALUM, COMPOUND.

Solution of Sul'phuret of Potassium. *Syn.* SOLUTION OF HYDROSULPHATE OF POTASSA; SOLUTIO POTASSII SULPHURETI, LIQUOR POTASSÆ HYDROSULPHATIS, AQUA POTASSÆ SULPHURETI (Ph. D.), L. *Prep.* Take of washed sublimed sulphur, 1 part; water of caustic potassa, 11 parts; mix, boil for 10 minutes, filter, and keep the solution in well-closed bottles. Sp. gr. 1·117. The product is a mixed solution of hydrosulphate and hyposulphate of potassa.—*Dose*, 10 to 60 drops, diluted in water; and, externally, made into a lotion; in itch, and several other eruptive diseases.

Solution, Swan's. *Syn.* SOLUTIO SODÆ HYPOPHOSPHITIS, L. *Prep.* Mr Squire says this contains 3 gr. of the salt in a drachm.

Solution of Tartrate of Magnesia. *Syn.* LIQUOR MAGNESIÆ TARTRATIS (*Airat*), L. *Prep.* Tartaric acid, 15½ oz. troy; distilled water, 20 pints; fresh calcined magnesia, diffused in 16 oz. of distilled water, 3 oz. troy and 1 dr.—*Dose.* As a purgative, 15 oz.

Solution of Trinitrine. (B. P.) *Syn.* LIQUOR TRINITRINÆ, L. NITRO-GLYCERINI, L. GLYCERINI, L. Nitro-glycerine, 1 part (by weight); rectified spirit, to produce 100 fluid parts.—*Dose*, ½ to 2 minims.

Solution of Veratria. *Syn.* SOLUTIO VERATRIÆ, L. *Prep.* Veratrine, 1 gr.; distilled water, 2½ oz. Dr Turnbull's solution, for external use, is—veratria, 1 scruple; rectified spirit, 2 oz.

SOL'VENT. *Syn.* MENSTRUUM, L. The liquid in which any substance is dissolved. The substance dissolved is, occasionally, called the 'solvend' (*Kirwan*).

Solvent, Glazier's. *Syn.* GLAZIER'S PICKLE. From soft soap dissolved in thrice its weight of strong soapers' lye; or from freshly slaked lime made into a thin paste or cream with twice its weight of pearlash dissolved in a little water. Very caustic. Used to soften old putty, and to remove old paint.

SOMNAL. This is described as an ethylised combination of urethane and chloral. It is prepared from chloral, alcohol, and urethane, and answers to the formula $C_5H_{12}Cl_3O_2N$, thus differing from the chloral-urethane, hitherto used, by the addition of 2 atoms of carbon and 4 atoms of hydrogen. Somnal has a melting-point of 40° C., and boils *in vacuo* at about 145° C. It is not influenced by the addition of nitrate of silver, nor by the action of acids. It is administered in doses of 30 gr., preferably with liquid extract of liquorice, or with syrup of raspberry, as follows:—Somnal, 2½ dr.; aq. destill., ad 3 oz.; ext. glycyrrhis. liq., 5 dr. One tablespoonful at night. Such a 30-gr. dose of somnal is said to create within half an hour of its administration a sound sleep of six to eight hours' duration, and without any injurious by or after effect. It is claimed for somnal that it does not affect the digestion, the breathing, or the temperature of the body, and that it possesses all the advantages of urethane and chloral hydrate without any of their ill effects.

SOMNAMBULISM. Children are most subjected to sleep-walking. When adults are affected with it the cause may generally be traced to mental exhaustion, over-excitement, or emotional feeling. The most preferable method of awakening a somnambulist, if this be desirable, is by dashing cold water on the face. It is well to occasionally administer an aperient, and also to rectify any errors of diet, if necessary, and to remove by the exercise of judicious and kindly advice, and change of scene, undue excitement or morbid feeling.

The other precautions, such as securing the feet, &c., during sleep, guarding the windows and the exits of the bedchamber, are so obvious as to need no further notice.

SOOT. *Syn.* FULIGO. Wood soot was formerly officinal, and reputed vermifuge and antiseptic. The soot from pit-coal contains, besides empyreumatic matter, sulphate of ammonia; hence it is valuable as a manure when not too freely applied. It is also employed by gardeners to kill insects.

SOPORIFICS. Hypnotics (which *see*).

SORBITE. A crystalline saccharine substance resembling mannite, obtained by Boussingault from the berries of the mountain ash. It was obtained from the liquid containing the undecomposed saccharine matter remaining after the juice of the berries had been subjected to fermentation.

SOU'JEE. *Syn.* SOOJEE. A species of semolina. Semoletta (*Semola rarita*) is a still smaller variety of pearled wheat, separated from the others by means of a sieve. 'Baster's soojee' is said to be a mixture of ordinary wheat-flour and sugar.

SOUP. A strong decoction of flesh, properly seasoned with salts, spices, &c., for the table. The different tastes of people require more or less of the flavour of spices, salt, garlic, butter, &c., which can, therefore, never be ordered by general rules. If the cook has not a good taste, and attention to that of his or her employers, not all the ingredients which nature and art can furnish will give an exquisite flavour to the dishes. The proper articles should be always at hand, and must be proportioned until the true zest be obtained. A variety of flavours may be given to different dishes served at the same time, or even to the same soup, by varying the condiments and spices. At a Parisian restaurant one caldron is made to produce almost every imaginable variety of soup.

Soup, Cabbage, Cheap. Wash a large cabbage and cut it into narrow strips, throwing them into ½ a gallon of boiling water containing 2 oz. of butter. Let it boil for an hour and a half; then add ½ a pint of milk, and flavour with pepper and salt. Serve when hot.

Soup, Carrot. INGREDIENTS REQUIRED. 4 quarts of liquor in which a leg of mutton or beef has been boiled, a few beef bones, 6 large carrots, 2 large onions, 1 turnip, seasoning of salt and pepper to taste, 3 lumps of sugar, and cayenne.

Mode. Put the liquor, bones, onions, turnips, pepper, and salt into a stewpan, and simmer for 3 hours. Scrape and cut the carrots thin, strain the soup on them, and stew them till soft enough to pulp through a hair sieve or coarse cloth; then boil the pulp with the soup, which should be of the consistency of pea soup. Add cayenne. Pulp only the red part of the carrot, and make this soup the day before it is wanted.

Time, 4½ hours. Seasonable from October to March. Sufficient for eight persons.

Soup, Celery. INGREDIENTS. 9 heads of celery, 1 teaspoonful of salt, nutmeg to taste, 1 lump of sugar, ½ pint of strong stock, 1 pint of cream, and 2 quarts of boiling water.

Mode. Cut the celery into small pieces, throw it into the water, seasoned with the nutmeg, salt, and sugar. Boil it till sufficiently tender; pass it through a sieve, add the stock and simmer it for ½ an hour. Now put in the cream, bring it to the boiling-point, and serve immediately.

Time, 1 hour.

Soup, Giblet. Scald and carefully clean 3 or 4 sets of goose or duck giblets; let them stew well, a pound or two of gravy beef, scrag of mutton, or the bone of a knuckle of veal, an ox-tail, or some shanks of mutton, with 3 onions, a large bunch of sweet herbs, a teaspoonful of white pepper, and a large spoonful of salt. Add 5 pints of water and simmer till the gizzards (which must be each in four pieces) are quite tender; skim nicely, and add a ½ pint of cream, 2 teaspoonfuls of mushroom powder, and 1 oz. of butter mixed with a dessert-spoonful of flour. Let it boil a few minutes, and serve with the giblets. Instead of cream, two glasses of sherry or Madeira, a large spoonful of ketchup, and some cayenne may be used for the seasoning. Add salt when the soup is in the tureen.

For the larger part of the above culinary preparations we are indebted to the excellent cooking manuals of Miss Acton and Mrs Beeton.

Soup, a Good Family. INGREDIENTS. Remains of a cold tongue, 2 lbs. of shin of beef, any cold pieces of meat or beef bones, 2 turnips, 2 carrots, 2 onions, 1 parsnip, 1 head of celery, 4 quarts of water, ½ teacupful of rice, salt and pepper to taste.

Mode. Put all the ingredients in a stewpan, and simmer gently for 4 hours, or until all the goodness is drawn from the meat. Strain off the soup and let it stand till cold. The kernels and soft part of the tongue must be saved. When the soup is wanted for use, skim off all the fat, put in the kernels and soft parts of the tongue, slice in a small quantity of fresh carrot, turnip, and onion; stew till the vegetables are tender, and serve with toasted bread.

Time, 5 hours. Seasonable at any time. Sufficient for eight persons.

Soup, Gravy. INGREDIENTS. 4 lbs. of shin of beef, a piece of the knuckle of veal weighing 4 lbs., a few pieces of trimmings of meat or poultry, 3 slices of nicely flavoured lean ham, ¼ lb. of butter, 2 onions, 4 carrots, 1 turnip, nearly a head of celery, 1 blade of mace, 6 cloves, a bunch of savoury herbs, seasoning of salt and pepper to taste, 3 lumps of sugar, 5 quarts of boiling soft water. It can be flavoured with ketchup, Leamington sauce, or Harvey's sauce and a little soy.

Mode. Slightly brown the meat and ham in the butter, but do not let them burn. When this is done, pour in to it the water, put in the salt, and as the scum rises take it off; when no more appears, add all the other ingredients, and let the soup simmer slowly by the fire for 6 hours without stirring it any more from the bottom; take it off, and pass it through a sieve. When perfectly cold and settled all the fat should be removed, leaving the sediment untouched, which serves nicely for thick gravies, hashes, &c. The flavourings should be added when the soup is heated for table.

Time, 7 hours. Seasonable all the year. Sufficient for twelve persons.

Soup, Green Pea. INGREDIENTS. 3 pints of green peas, ¼ lb. of butter, 2 or 3 thin slices of ham, 4 onions sliced, 4 shredded lettuces, the crumb of 2 French rolls, 2 handfuls of spinach, 1 lump of sugar, 2 quarts of medium stock.

Mode. Put the butter, ham, 1 quart of the peas, onions, and lettuces, to a pint of stock, and simmer for an hour; then add the remainder of the stock, with the crumb of the French rolls, and boil for another hour. Now boil the spinach, squeeze it very dry, and rub it, with the soup, through a sieve, to give the preparation a good colour. Have ready a pint of young peas boiled; add them to the soup, put in the sugar, give one boil, and serve.

Time, 2½ hours. Seasonable from June to the end of August. Sufficient for six persons.

** It will be well to add, if the peas are not quite young, a little more sugar; where economy is essential, water may be used instead of stock for this soup, boiling in it likewise the pea-shells, and using rather a larger quantity of vegetables.

Soup, Hare. Cut down a hare into joints, and put it into a soup-pot or large stewpan, with about 1 lb. of lean ham, in thick slices, 3 moderately sized mild onions, 3 blades of mace, a fagot of thyme, sweet marjoram, and parsley, with about 3 quarts of good beef stock. Let it stew very gently for fully 3 hours from the time of its first beginning to boil, and more if the hare be old. Strain the soup, and pound together very fine the slices of ham and all the flesh of the back, legs, and shoulders of the hare, and put this meat into a stewpan with the liquor in which it was boiled, the crumb of two French rolls, and ¾ a pint of port wine. Set it on the stove to simmer 20 minutes; then rub it through a sieve, place it again on the stove till very hot, but do not let it boil; season it with salt and cayenne, and send it to table directly.

INGREDIENTS. Hare, 1; ham, 12 to 16 oz.; onions, 3 to 6; mace, 3 blades; fagot of savoury herbs; beef stock, 3 quarts; 2 hours. Crumb of 2 rolls; port wine, ¾ pint; little salt and cayenne; 20 minutes.

Soup, Hare, a less Expensive. Pour on two pounds of neck or shin of beef, and a hare well washed and carved into joints, one gallon of cold water, and when it boils and has been thoroughly skimmed, add 1½ oz. of salt, 2 onions, 1 large head of celery, 3 moderate-sized carrots, a tablespoonful black peppercorns, and 6 cloves.

Let these stew gently for 3 hours, or longer, should the hare not be perfectly tender. Then take up the principal joints, cut the meat from them, mince, and pound it to fine paste, with the crumb of two penny rolls (or 2 oz. of crumb of household bread), which has been soaked in a

little of the boiling soup, and then pressed very dry in a cloth; strain, and mix smoothly with it the stock from the remainder of the hare; pass the soup through a strainer, season it with cayenne, and serve it when at the point of boiling; if not sufficiently thick, add to it a tablespoonful of arrowroot, moistened with a little broth, and let the soup simmer for an instant afterwards. Two or three glasses of port wine and two dozen of small forcemeat balls may be added to this soup with good effect.

INGREDIENTS. Beef, 2 lbs.; hare, 1; water, 1 gall.; salt, 1½ oz.; onions, 2; celery, 1 head; carrots, 3; bunch of savoury herbs; peppercorns, 1 teaspoonful; cloves, 6; 3 hours or more. Bread, 2 oz.; cayenne, arrowroot (if needed), 1 tablespoonful.

Soup, Haricot Bean. Take a quart of haricot beans and let them soak all night in cold water. Then pour on them 2½ pints of cold water, add 1 onion, and put on the fire, and when the liquid begins to boil, let them continue to boil for three hours. Then remove from the fire and strain through a wire sieve, after which return to the saucepan, and season with pepper and salt; next add 2 oz. of butter and a little milk. Then just boil up and serve. An economical and nutritious soup for the poor.

Soup, Julienne. INGREDIENTS. ½ pint of carrots, ½ pint of turnips, ½ pint of onions, 2 or 3 leeks, ½ head of celery, 1 lettuce, a little sorrel and chervil if liked, 2 oz. butter, 2 quarts of medium stock.

Mode. Cut the vegetables into strips about 1¼ in. long, and be particular they are all the same size, or some will be hard whilst the others will be done to a pulp. Cut the lettuce, sorrel, and chervil into larger pieces; fry the carrots in the butter, and pour the stock boiling to them. When this is done, add all the other vegetables thereto, and stew gently for nearly an hour. Skim off all the fat, pour the soup over thin slices of bread cut round, about the size of shilling, and serve.

Time, 1½ hours. Seasonable all the year. Sufficient for 7 or 8 persons.

** In summer, green peas, asparagus tops, French beans, &c., can be added. When the vegetables are very strong, instead of frying them in butter at first, they should be blanched, and afterwards simmered in the stock.

Soup, Macaroni. Throw 4 oz. of fine, fresh, mellow Naples macaroni into a pan of fast-boiling water, with about 1 oz. of fresh butter, and a small onion stuck with 3 or 4 cloves (the onion must be omitted for white soups). When it has swelled to its full size, and become tender, drain it well, cut it into half-inch lengths, and slip it into a couple of quarts of clear gravy soup; let it simmer for a few minutes, when it will be ready for table. Observe that the macaroni should be boiled quite tender; but it should by no means be allowed to burst, nor to become pulpy. Serve grated Parmesan cheese with it.

INGREDIENTS. Macaroni, 4 oz.; butter, 1 oz.; 1 small onion; 5 cloves; ¾ hour or more. In soup, 5 to 10 minutes.

Soup, Mock Turtle. INGREDIENTS. Half a calf's head, ¾ lb. butter, ¾ lb. of lean ham, 2 tablespoonfuls of minced parsley, a little minced lemon thyme, sweet marjoram, basil, 2 onions, a few chopped mushrooms (when obtainable), 2 shalots, 2 tablespoonfuls of flour, 2 glasses of madeira or sherry, forcemeat balls, cayenne, salt and mace to taste, the juice of one lemon and 1 Seville orange, 1 dessert-spoonful of pounded sugar, 3 quarts of best strong stock.

Mode. Scald the head with the skin on, remove the brain, tie the head up in a cloth, and let it boil for an hour. Then take the meat from the bones, cut it into small square pieces, and throw them into cold water. Now take the meat, put it into a stewpan, and cover it with stock; let it boil gently for an hour, or rather more if not quite tender, and set it on one side. Melt the butter in another stewpan, and add the ham, cut small, with the herbs, parsley, onions, shalots, mushrooms, and nearly a pint of stock; let these simmer slowly for 2 hours, and then dredge in as much flour as will dry up the butter. Fill up with the remainder of the stock, add the wine, let it stew gently for 10 minutes, rub it through a sieve, and put it to the calf's head; season with cayenne, and, if required, a little salt; add the juice of the orange and lemon; and when liked, ¼ teaspoonful of pounded mace, and the sugar. Put in the forcemeat balls, simmer 5 minutes, and serve very hot.

Time, 4½ hours. Seasonable in winter. Sufficient for 10 persons.

*** The bones of the head should be well stewed in the liquor it was first boiled in, and will make good white stock, flavoured with vegetables.

Soup, Ox-tail. A very inexpensive and nutritious soup may be made of ox-tails, but it will be insipid without the addition of a little ham, knuckle of bacon, or a pound or two of other meat.

Wash and soak 2 tails, pour on them a gallon of cold water, let them be brought gradually to boil, throw in 1½ oz. of salt, and clear off the scum carefully as soon as it forms upon the surface; when it ceases to rise add four moderate-sized carrots, from 2 to 4 onions according to the taste, a large fagot of savoury herbs, a head of celery, a couple of turnips, 6 or 8 cloves, and ½ a teaspoonful of peppercorns. Stew these gently from 3 to 3½ hours if the tails be very large; lift them out, strain the liquor, and skim off all the fat; divide the tails into joints, and put them into a couple of quarts or rather more of the stock; stir in, when these begin to boil, a thickening of arrowroot or rice flour, mixed with as much cayenne and salt as may be required to flavour the soup well, and serve it very hot.

INGREDIENTS. Ox-tails, 3; water, 1 gall.; salt, 1½ oz.; carrots, 4; onions, 2 to 4; turnips, 2; celery, 1 head; cloves, 8; peppercorns, ½ teaspoonful; fagot of savoury herbs; 3 to 3½ hours. For a richer soup, 5 to 6 hours.

Soup, Ordinary Pea. Well wash a quart of good split peas, and float off such as remain on the surface of the water; soak them for one night, and boil them with a bit of soda the size of a filbert, in just sufficient water to allow them to break to a mash. Put them into from 3 to 4 quarts of good beef broth, and stew them in it gently for an

hour; then work the whole through a sieve, heat afresh as much as may be required for table, season it with salt, or cayenne, or common pepper; clear it perfectly from scum, and send it to table with fried or toasted bread. Celery sliced and stewed in it will be found a great improvement.

INGREDIENTS. Peas, 1 quart; soaked one night, boiled in 2 quarts or rather more of water, 2 to 2½ hours. Beef broth, 3 to 4 quarts; 1 hour. Salt and cayenne or pepper, as needed; 3 minutes.

Soup, Portable. Syn. GLAZE. From shin of beef, or other like part, the soup being gently simmered until reduced to the consistence of a thin syrup, and then poured into small upright jelly-pots with covers, or upon flat dishes, to lie about ¼ inch deep. The latter, when set, is divided into pieces, which are dried. Used to make extemporaneous soup and glazes. A similar article, prepared on the large scale, now generally forms part of every ship's stores.

Soup, Potato. Mash to a smooth paste 3 lbs. of good mealy potatoes, which have been steamed or boiled very dry; mix with them, by degrees, 2 quarts of boiling broth, pass the soup through a strainer, set it again on the fire, add pepper and salt, and let it boil for five minutes. Take off entirely the black scum that will rise upon it, and serve it very hot with fried or toasted bread. Where the flavour is approved 2 oz. of onions, minced and fried a light brown, may be added to the soup, and stewed in it for ten minutes before it is sent to the table.

INGREDIENTS. Potatoes, 3 lbs.; broth, 2 quarts; 5 minutes. With onions, 2 oz., 10 minutes.

Soup, Spanish Onion. Peel two large Spanish onions and cut them into rings; fry them with a little dripping in a stewpan. When the onions have browned add 2½ pints of boiling water, and let them boil for two hours and a half; add pepper and salt to flavour, and a little vinegar. Thicken with oatmeal or bread crumbs (oatmeal is the more nourishing); let the mixture boil for another half-hour, and serve. A good, cheap, wholesome soup.

Soup, Turnip (Cheap). Wash and wipe the turnips, pare and weigh them; allow 1½ lbs. for every quart of soup, cut them in slices about ¼ inch thick. Melt 4 oz. of butter in a clean stewpan, and put in the turnips before it begins to boil; stew them gently ¾ hour, taking care that they shall not brown; then have the proper quantity of soup ready boiling, pour it on them, and let them simmer in it for ½ hour. Pulp the whole through a coarse sieve or soup strainer, put it again on the fire, keep it stirred until it has boiled three or four minutes, take off the scum, add salt or pepper if required, and serve it very hot.

INGREDIENTS. Turnips, 3 lbs.; butter, 4 oz.; ½ hour. Soup, 2 quarts; ½ hour. Last time, 3 minutes.

Soup, Vermicelli. Drop, very lightly and by degrees, 6 oz. of vermicelli, broken rather small, into 3 quarts of boiling bouillon, or clear gravy soup; let it simmer for half an hour over a gentle fire, and stir it often.

INGREDIENTS. Bouillon or gravy soap, 3 quarts; vermicelli, 6 oz.; 30 minutes. Or soup, 3 quarts; vermicelli, 4 oz.; blanched in boiling water, 5 minutes; stewed in soup, 10 to 15 minutes.

SOURING. See MALT LIQUORS and WINES.

SOUR-KROUT. See SAUER-KROUT.

SOY. Genuine soy is a species of thick black sauce imported from China.—*Prep.* Take of the seeds of *Soja hispida* (white haricots or kidney beans may be used for them), 1 gall.; boil them in water, q. s., until soft; add of bruised wheat, 1 gall., and keep the mixture in a warm place for 24 hours; then add of common salt, 1 gall.; water, 2 galls.; put the whole into a stone jar, and bung it up loosely for two or three months, shaking it very frequently during the whole time; lastly, press out the liquor and bottle it; the residuum may be treated afresh with water and salt for soy of an inferior quality.

Obs. The soy of the shops is, in nine cases out of ten, a spurious article made in this country by simply saturating molasses or treacle with common salt. A better and a really wholesome imitation is made as follows:—Malt syrup, 1 gall. (or 13½ lbs.); treacle, 5 lbs.; salt, 4½ lbs.; mushroom juice, 1 quart; mix, with a gentle heat, and stir until the union is complete; in a fortnight decant the clear portion.

SOZOIODOL. This substance is manufactured under a patent, and appears in the form of a white crystalline powder, which does not melt until heated to a temperature over 200° C. It is odourless, has a slightly acid taste, and dissolves to the extent of 7% in cold water, being more soluble in hot water.

The characteristic reactions of sozoiodol are that it readily gives off iodine vapour on heating, and if a few drops of sulphuric acid are added to a hot solution of the substance iodine is separated, and may be recognised in the usual way. Ferric chloride imparts a dark violet colour to the aqueous solution; silver nitrate throws down a white precipitate, soluble in sulphuric acid, proving that the precipitate is not silver chloride, but a silver salt of sozoiodol. The barium precipitate is soluble on heating, and from strong solutions separates in crystalline form. Sozoiodol is being extensively tried in the General Hospital of Vienna as a substitute for iodoform and salicylic acid. Being devoid of odour it has a great advantage over iodoform, and as it contains 42% of iodine it is hoped that it will be as active therapeutically. It is used in the form of an ointment made with lanoline, or as a dusting powder with Venetian talc as a diluent.

SPAN'ISH FLIES. See CANTHARIDES.

SPAR'ADRAP. *Syn.* SPARADRAPUM, L. Originally a cerecloth; now applied to spread plasters; as SPARADRAPUM COMMUNE, common strapping or adhesive plaster; s. VESICATORIUM, blistering plaster or tissue, &c.

The following are in occasional demand by the pharmacist:

Sparadrap, Opium. *Syn.* SPARADRAPUM OPII (*M. Schœufele*). *Prep.* On a piece of black sarcenet of a close and strong texture, properly stretched, spread, with a brush, 8 layers of extract of opium, softened with water, to the consistence of treacle, and mixed with a sixth part of powdered gum. Keep the plaster dry.

Sparadrap, Thapsian. *Syn.* SPARADRAPUM THAPSIÆ (P. Cod.). *Prep.* Yellow wax, 4½ oz.; resin, 1½ oz.; Burgundy pitch, 1½ oz.; boiled turpentine, 1½ oz.; Swiss turpentine, ½ oz.; glycerin, ½ oz.; honey, ½ oz.; resin of thapsia, ½ oz. Melt the first five substances together and strain through linen. Keep them liquefied and add the glycerin, the honey, and the resin. When well mixed, and of a proper consistence, spread on strips of linen cloth.

Sparadrap, Wax. *Syn.* SPARADRAPUM CUM CERA, TOILE DE MAI (P. Cod.). *Prep.* White wax, 8 oz.; (by wt.,) oil of almonds, 4 oz.; (by wt.,) Swiss turpentine, 1 oz. Melt together and dip into it strips of linen cloth, which are to be passed between wooden rollers, to remove the superfluous plaster. Spread on paper it forms waxed paper.

SPAR'TEINE. *Syn.* SPARTEINA, L. A volatile oily liquid, possessing basic properties, obtained from *Cytisus scoparius* or broom. It is highly poisonous, and resembles conine and nicotine in its general properties.

Sparteine Sulphate. Made by neutralising sparteine with sulphuric acid, crystallising.

Uses. Has a tonic action on the heart, and is a valuable diuretic.—*Dose,* ½ to 4 grs.

SPASMS. *Syn.* CRAMP; SPASMUS, L. An involuntary contraction of the muscles, generally of the extremities, accompanied with pain more or less severe. Spasms are distinguished into clonic spasms or convulsions, in which the contractions and relaxations are alternate, as in epilepsy; and into tonic spasms, in which there is continued rigidity, as in locked-jaw. That form which commonly attacks the muscles of the legs and feet, especially after great exertion or exposure to cold, is commonly called cramp. The best treatment for this is immediately to stand upright, and to well rub the part with the hand. The application of strong stimulants, as spirits of ammonia, or of anodynes, as opiate liniments, has been recommended. When spasm or cramp occurs in the stomach, a teaspoonful of sal volatile in water, or a teaspoonful of good brandy, may be swallowed immediately. When cramp comes or during cold bathing, the limb should be thrown out as suddenly and violently as possible, which will generally remove it, care being also taken not to become flurried or frightened, as presence of mind is very essential to personal safety on such an occasion. A common cause of spasm is indigestion, and the use of acescent liquors; these should, therefore, be avoided, and bitters and absorbents had recourse to. See ANTISPASMODICS, and the names of the principal spasmodic diseases.

SPEAR'MINT. See MINT.

SPE'CIES. (In *pharmacy*.) Mixtures of dried plants, or parts of plants, in a divided state, which, for convenience, are kept mixed for use. The dry ingredients of pills, conserves, electuaries, mixtures, &c., that do not keep well when made up, or which are in little demand, may be economically and conveniently preserved in this state. The word, thus applied, is obsolete out of the pharmaceutical laboratory.

Species, Anthelmin'tic. *Syn.* SPECIES AN-

THELMINTICÆ, L. The dried flowering tops of tansy and wormwood, and the flowers of chamomile, equal parts; mix, and keep them in a close vessel (P. Cod.).

Species, Aperitive. See SPECIES, DIURETIC (below).

Species, Aromat'ic. Syn. AROMATIC POWDER; SPECIES AROMATICÆ, L. Prep. (Ph. Bor.) Leaves of balm and curled-leaf mint (Mentha crispa), of each, 4 oz.; lavender flowers, 2 oz.; cloves, 1 oz.; dry them by a gentle heat, and then powder them.

Species, Astrin'gent. Syn. SPECIES ASTRINGENTES, L. The roots of bistort and tormentil, and bark of pomegranate, equal parts (P. Cod.).

Species, Bechics. (P. Cod.) 1. Leaves of Canadian maidenhair, ground-ivy, hart's-tongue, speedwell, hyssop tops, and poppy capsules (freed from seed), of each, equal parts. Cut and mix. 2. Dried flowers of mallow, catsfoot, coltsfoot, and petals of red poppy, of each, 1 oz.; mix. The Fructûs Bechici are—Dates (stoned), 1 oz.; jujubes, 1 oz.; figs, 1 oz.; raisins, 1 oz.

Species, Bitter. Syn. THREE BITTER HERBS; SPECIES AMARÆ, HERBÆ AMARÆ, L. The leaves of germander, and dried tops of lesser centaury and wormwood, equal parts (P. Cod.).

Species, Cap'illary. Syn. FIVE CAPILLARY HERBS; HERBÆ QUINQUE CAPILLARES, L. Hart's-tongue, black maidenhair, white do., golden do., and spleenwort, equal parts (Ph. L. 1720).

Species, Carminative. Syn. SPECIES CARMINATIVÆ (P. Cod.). Prep. Equal parts of aniseeds, caraway seeds, coriander seeds, and fennel seeds.

Species, Cor'dial. Syn. FOUR CORDIAL FLOWERS; SPECIES CORDIALES, L. The flowers of borage, bugloss, roses, and violets, equal parts (Ph. L. 1720).

Species for Decoction Woods. Syn. SPECIES AD DECOCTUM LIGNORUM (G. Ph.). Prep. Rasped guaiacum wood, 4 oz.; cut burdock root, 2 oz.; ononis root, 2 oz.; cut liquorice, 1 oz.; cut sassafras, 1 oz.; mix.

Species, Diuret'ic. Syn. APERIENT ROOTS, APERITIVE SPECIES; SPECIES DIURETICÆ, L. 1. (FIVE GREATER APERITIVE ROOTS—P. Cod., and Ph. E. 1744.) The dried roots of asparagus, butcher's-broom, parsley, smallage, and sweet fennel, equal parts. 2. (FIVE LESSER APERITIVE ROOTS.) Those of caper, dog-grass, eryngo, madder, and restharrow.

Species, Emol'lient. Syn. SPECIES EMOLLIENTES, L. 1. (THREE EMOLLIENT MEALS; FARINÆ EMOLLIENTES.) The meal of barley, linseed, and rye, equal parts (P. Cod.). 2. (FIVE EMOLLIENT HERBS; HERBÆ QUINQUE EMOLLIENTES.) a. The dried leaves of groundsel, common mallow, marsh-mallow, great mullein, and wall pellitory, equal parts (P. Cod.). b. The leaves of mallow, marsh-mallow, French mercury, pellitory of the wall, and violet (Ph. E. 1744).

Species of Ene'mas. Syn. HERBS FOR CLYSTERS; HERBÆ PRO ENEMATE, L. Mallow leaves, 2 parts; chamomile flowers, 1 part.

Species of the Five Herbs. Syn. SPECIES DIETÆ QUINQUE HERBÆ; FIVE CAPILLARY HERBS (Ph. L. 1720). Prep. Black and white maidenhair, spleenwort, hart's-tongue, and golden maidenhair.

Species for Fomenta'tions. Syn. SPECIES PRO FOTU, HERBÆ PRO FOTU, L. Leaves of southernwood, tops of sea-wormwood, and flowers of chamomile, of each, 2 parts; bay leaves, 1 part.

Species, Hot. 1. (FOUR GREATER HOT SEEDS.) The seeds of anise, caraway, cumin, and fennel. 2. (FOUR LESSER HOT SEEDS.) The seeds of bishopsweed, smallage, stone-parsley, and wild carrot.

Species, Lax'ative. Syn. ST GERMAIN LAXATIVE POWDER; SPECIES LAXANTES ST GERMAIN, L. Prep. (Ph. Bor.) Senna leaves (exhausted with spirit), 4 oz.; elder flowers, 2½ oz.; aniseed and fennel seed, of each, 1½ oz.; reduce them to coarse powder, and, when dispensing, add of powdered cream of tartar, 1 dr. to each 1½ oz. of the mixture.

Species, Narcotic. Syn. FOUR NARCOTIC HERBS; SPECIES NARCOTICÆ, L. Dried leaves of belladonna, black nightshade, henbane, and thorn-apple, equal parts.

Species, Pec'toral. Syn. SPECIES BECHICÆ, SPECIES AD INFUSUM PECTORALES, L. Mallow root, 4 oz.; coltsfoot leaves, 2 oz.; liquorice root, 1½ oz.; aniseed, great mullein flowers, and red poppy flowers, of each, 1 oz.; orris root, ½ oz. (Ph. Bor.).

Species, Purging. Syn. SPECIES PURGANTES, L.; THÉ DE SANTÉ, THÉ DE ST GERMAIN (P. Cod.), Fr. Senna, 12 dr.; elder flowers, 5 dr.; fennel seeds, 3 dr.; aniseed, 5 dr.; cream of tartar, 3 dr. Eighty grains in a cup of boiling water for a dose; said to be very serviceable and largely used in France for habitual constipation.

Species, Refri'gerant. 1. (FOUR COLD SEEDS.) The seeds of cucumber, gourd, melon, and watermelon. 2. (FOUR LESSER COLD SEEDS.) The seeds of endive, lettuce, purslane, and succory.

Species, Resol'vent. Syn. FARINÆ RESOLVENTES, L. The meal of the seeds of barley, bean, tare, and white lupine.

Species, Vulnerary. Syn. SPECIES VULNERARIÆ, L.; THÉ SUISSE, Fr. Prep. Leaves and tops of wormwood, betony, bugle, calamint, germander, hyssop, ground-ivy, milfoil, origanum, periwinkle, rosemary, self-heal, sage, hart's-tongue, water-germander, thyme, speedwell, flower of Arnica, flower of catsfoot, flower of coltsfoot, of each, equal parts. Cut and mixed.

SPECIFIC GRAVITY. In order to define this term we must consider what is meant by density. The mass or quantity of matter in a given body, as measured by its inertia, depends first on the density of its material, and secondly, on its size or volume; and the relation is expressed by the formula—

$$\text{Density} = \frac{\text{mass}}{\text{volume}};$$

or more simply, density is the mass of unit volume, or the mass of any volume divided by that volume. Similarly, we may define specific gravity as the *weight of unit volume*, or the weight of any volume divided by that volume, *i. e.*—

$$s = \frac{w}{v}.$$

In determining the specific gravity of any body, the weight of a certain volume of it is compared with that of the same volume of some standard substance. This standard is pure distilled water for liquids and solids, and atmospheric air for gaseous bodies and vapours. By modern chemists *hydrogen*, the lightest substance in nature, is taken as the standard for the specific gravity of gases and vapours. In England the sp. gr., unless when otherwise expressed, is always taken at 60° F. (15·5° C.); but in France it is taken at 32° F. (0° C.), or the temperature of melting ice. In the 'British Pharmacopœia,' whenever specific gravity is mentioned, the substance spoken of is supposed to be of the temperature of 60° F. In most cases, however, it is sufficient merely to note the temperature, and to apply a correction, depending on the known density of the standard substance, at the different degrees of the thermometric scale.

To determine the specific gravity of a solid, we weigh it first in the air, and then in water. In the latter case it loses, of its weight, a quantity precisely equal to the weight of its own bulk of water; and hence, by comparing this weight with its total weight, we find its specific gravity. The rule is—Divide the total weight by the loss of weight in water; the quotient is the specific gravity.

The specific gravity of a substance lighter than water may be determined by attaching it to some substance, as a piece of lead, the sp. gr., &c., of which are known. In this way, by deducting the loss in weight of the two substances, when weighed in water, from the loss sustained by the lead alone, when so weighed, we obtain a difference (*a*) which, added to the weight of the substance taken in air (*b*), gives the respective densities. From these the sp. gr. is found by the rule of three:

$$(a + b) : 1 :: b : sp. gr.$$

The specific gravities of substances soluble in water are taken in pure oil of turpentine, rectified spirit, olive oil, or some other liquid, the density of which is exactly known. Sometimes, for rough purposes, the article is covered with a coating of mastic varnish. This last method answers for mercurial pill.

The specific gravity of a substance in fragments, or in powder, may be found by putting a portion into a sp. gr. bottle, filling the latter with distilled water, and then weighing it. The weight of water which it is found to contain, deducted from 1000 (the weight of the bottle when filled with distilled water), gives a difference (*a*) which bears the same relation to the sp. gr. of water (1·000) as the weight of the powder (*b*) put into the bottle does to the required sp. gr. Or—

$$a : 1·000 :: b : sp. gr.$$

The specific gravity of alloys and mixtures, when no condensation has occurred, is equal to the sum of the weights divided by the sum of the volumes, compared to water reckoned as unity, and is not merely the arithmetical mean between the two numbers denoting the two sp. gr., as is frequently taught. See BEADS (*Love's*), HYDROMETER, MIXTURES (Arithmetic of), &c.

For the mode of determining the specific gravity of vapours the reader is referred to the works on chemistry of Miller and Fownes.

The specific gravity of a gas is determined by filling a large glass globe with the gas to be examined, in a perfectly dry and pure state, at a known temperature, and at a pressure equal to that of the atmosphere at the time of the experiment. The globe so filled is weighed; it is then exhausted with an air-pump, and again weighed; lastly, it is filled with dry air at a known temperature and pressure, and its weight once more determined. If the pressure and temperature have remained the same throughout the experiment the specific gravity of the gas is obtained by simply dividing the weight of the gas by that of the air; but if these conditions have varied corrective factors have to be introduced.

The specific gravity of a liquid is found by weighing it in a sp. gr. bottle, glass flask, or other vessel of known capacity, and dividing that weight by the weight of the same bulk of pure water at the same temperature; the quotient is, as before, the specific gravity. A bottle of the capacity of 1000 water-grains (sp. gr. bottle) gives the density of a liquid at once by simply filling it to the given mark, and then accurately weighing it.

We reprint from the 'Journal of the Chemical Society' (2, xi, 577) a method of determining the specific gravity of liquids, which is said by Dr H. Sprengel, the chemist who devised it, to be both expeditious and accurate:

"The form of my instrument, as shown in the accompanying fig. 1, is that of an elongated U-tube, the open ends of which terminate in two capillary tubes, which are bent at right angles in opposite directions. The size and weight of this instrument should be adapted to the size and capability of the balance in which it is to be weighed. As our usual balances indicate $\frac{1}{10}$ milligram when loaded with 50 grms., the U-tube,

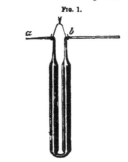

FIG. 1.

when charged with the liquid, should not be heavier than 1000 gr. (= 64·799 grms.).

"The instrument which served for my determinations, mentioned below, had a length of 17·7 cm. (7 inches), and was made of a glass tube, the outer diameter of which was 11 mm. ($\frac{7}{10}$ of an inch). It need hardly be mentioned that the U-shape is adopted for the sake of presenting a large surface, and so rendering the instrument sensitive to changes of temperature. The point, however, I wish to notice more particularly (for reasons explained below) is the different calibre of the two capillary tubes. The shorter one is a good deal narrower (at least towards the end) than the longer one, the inner diameter of which is about $\frac{1}{2}$ mm. The horizontal part of this wider tube is marked near the bend with a delicate line (b). This line and the extremity of the opposite capillary tube (a) are the marks which limit the volume of the liquid to be laid.

"The filling of the instrument is easily effected by suction, provided that the little bulb apparatus (as represented in fig. 2) has previously been attached to the *narrow* capillary tube by means of a perforated stopper, *i. e.* a bit of an india-rubber tube tightly fitting the conical tubules of the bulb. On dipping the wider and longer capillary tube into a liquid, suction applied to the open end of the india-rubber tube will produce a partial vacuum in the apparatus, causing the liquid to enter the U-tube. As the partial vacuum maintains itself for some time (on account of the bulb, which acts as an air-chamber), it is not necessary to continue the suction of the end if the india-rubber tube be closed by compression between the fingers. When bulb and U-tube have about equal capacity it is hardly necessary during the filling to repeat the exhaustion more than once.

"Without such a bulb the filling of the U-tube through these fine capillary tubes is found somewhat tiresome; the emptying the U-tube is effected by reversing the action, and so compressing the air. After the U-tube has been filled it is detached from the bulb, placed in water of the standard temperature almost up to the bends of the capillary tubes, left there until it has assumed this temperature, and, after a careful adjustment of the volume, is taken out, dried, and weighed.

"Particular care must be taken to ensure the correctness of the standard temperature, for a mistake of 0·1° causes the weight of 10 c.c. of water to be estimated either too high or too low by 0·14 milligram, giving rise to an error in the fifth decimal, or making 100,000 parts 100001·4 parts. These determinations have been made in Dupré's apparatus, which, when furnished with a sensitive thermometer, allows the fluctuations of temperature to be fixed within the limits of 0·01°. If many determinations had to be made, I should avail myself of Scheibler's (' *Zeitschrift für Analytische Chemie*,' vol. vii, p. 88, 1868) electro-magnetic regulator for maintaining a constant temperature.

"A peculiar feature of my instrument is the ease and precision with which the measurement of the liquid can be adjusted at the moment it has taken the standard temperature ; for it will be found that the liquid expands and contracts only in the direction of the least resistance. The narrow capillary tube remains always completely filled. Supposing the liquid reaches beyond the mark b, it may be re-

duced through capillary force by touching the point *a* with a little roll of filtering-paper. Supposing, however, that in so doing too much liquid is abstracted, capillary force will redress the fault if point *a* be touched with a drop of the liquid under examination ; for this gentle force acts instantly through the whole mass of the liquid, causing it to move forward again to or beyond the mark.

"As the instrument itself possesses the properties of a delicate thermometer, the time when it has reached the standard temperature of the bath may be learned from the stability of the thread of liquid inside the wider tube. The length of this thread remains constant after the lapse of about *five minutes*.

"In wiping the instrument (after its removal from the bath) care should be taken not to touch point *a*, as capillarity might extract some of the liquid; otherwise the handling of the liquid requires no especial precaution.

FIG. 2.

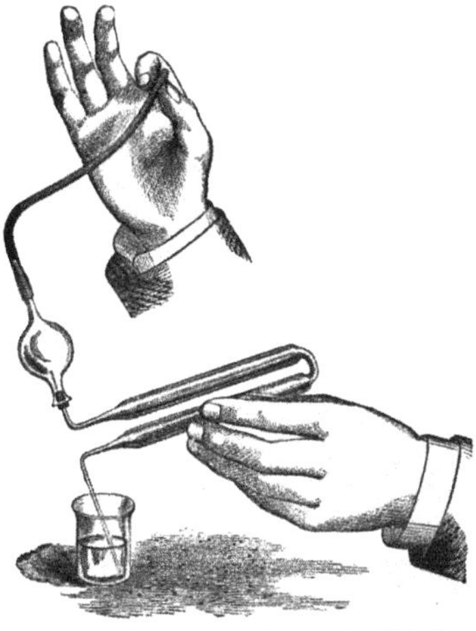

"The capillary tubes need not be closed for the purpose of arresting evaporation, at least that of water. I have learned from the mean of several determinations that the error arising from this source amounts in one hour to $\frac{1}{30}$ of a milligram.

"In cases where the temperature of the balance-room is high, and the expansion co-efficient of the liquid to be examined is considerable, it may be found necessary to put a small cap (bead-shaped and open at both ends) over the extremity of the *wider* capillary tube, for the purpose of retaining the liquid, which during the time of weighing might otherwise be lost, owing to its expansion. When a cap is used the wider capillary tube need not be longer than the narrow one."

The 'Comptes Rendus' (lxxxvi, 350—352, ' Journ. Chem. Soc.') describes a new specific gravity apparatus, the invention of M. Pisani,

The apparatus in question consists of a glass vessel about 5 c.c. capacity, closed with a perforated stopper like an ordinary specific gravity bottle. To the side of the vessel is joined a tube, coming off at an angle of about 45°, about 25 cm. long, and 4 mm. internal diameter, and graduated at 50ths of a c.c. The vessel is filled with water, the level of which is read off in the tube held vertically, the finger being held over the hole in the stopper; 2 or 3 grams of a mineral are then placed in the flask, the stopper is replaced, care being taken to lose no water, and the level is again read off in the graduated tube, held vertically as before. The difference in the two readings gives the volume of the mineral taken.

SPEC'TACLES. See EYE, VISION, &c.

SPEC'TROSCOPE. An instrument devised for examining the spectra of flames in spectrum analysis (see below).

SPECTRUM ANALYSIS. When a ray of sunlight is allowed to pass through a small round hole in the closed window-shutter of a dark room a round white spot of light will appear, exactly in the direction of the ray, upon a screen placed opposite the hole in the shutter. If, however, the ray of light be made to fall upon a prism of glass, it is at once deflected from its straight course upwards; that is to say, towards the base of the prism (the latter being placed with one angle pointing directly downwards), and away from the sharp edge of the refracting surface. On emergence it no longer forms a single ray, but is separated into many monocoloured rays, which as they diverge form upon the screen an elongated band of brilliant colours instead of the former round white image of the sun.

In the brilliant band the individual colours blend gradually one into the other, beginning at that end lying nearest the direction of the inci-

dent ray, with the least refrangible colour, **dark** red; this passes imperceptibly into **orange, and** orange again into bright yellow; a **pure green** succeeds, which is shaded off into a brilliant **blue,** and this gives place to a deep indigo; a **delicate** purple leads finally to a soft violet, by which **the** range of the visible ray is terminated.

The prism analyses the white light (R), separating it into the coloured rays of which it is composed, forming a coloured image (R'), which is called the *spectrum*.

By a similar experiment it can be shown that the light of the celestial bodies, the electric spark, and of all ordinary flames is of a compound nature. In order to observe this phenomenon with accuracy, and to study its variations according to the kind of light employed, an instrument has been devised called the spectroscope.

Every spectroscope is composed essentially of a slit through which the light passes from its source, and the width of which can be regulated; a collimating lens for making the rays which have passed through the slit parallel; and a prism which may be either of solid glass, or may be hollow and filled with some refracting liquid. Since the spectrum emerging from the prism is very small, a telescope is usually added through which to examine it; the prism is also usually enclosed in a tube, and other devices may be resorted to in order to exclude all light except that which is to be analysed.

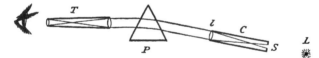

In the *engr.* L is the source of light, C a tube carrying the collimating lens l and the adjustable slit S; P is the prism, and T the telescope.

Since the coloured rays composing the spectrum form an angle with the incident rays as they enter the prism, it is necessary either that the tube of the telescope and the prism should be capable of adjustment, or that a compound prism should be adapted as in the case of direct vision instruments, convenient forms of which are described at the end of this article.

When a substance is gradually heated, a temperature will in time be reached when that substance becomes luminous. As the temperature continues to rise the substance will give off vapours or gases which glow with some definite coloured light. The light varies in character according to the substance examined; thus when potassium, sodium, and lithium are heated in a

sufficiently hot flame, luminous rays are evolved which are respectively lilac, yellow, and red.

Now when in the case of a metal, for instance, the quantity present is extremely minute, and the luminous rays proportionately scanty, this colour may escape notice; and this is especially the case when several metals are present at the same time. It is in such cases that the spectroscope can with advantage be resorted to.

Suppose that a flame contains several metals in a state of being vaporised. If now the light which proceeds from this flame be allowed to pass through a very narrow slit at A, collected by a lens, and transmitted through a prism of clear flint glass, or through a hollow prism B, filled with bisulphide of carbon, all the rays of any one colour will be refracted in a definite direction, so that when an observer looks through the telescope at C, he will perceive as many images of the slit as there are colours in the flame. The prism may be slowly moved round by a handle attached to a stage on which it rests, in order

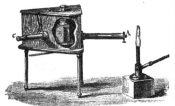

that these images (the different parts of the spectrum) may be successively brought into sight.

Any flame that emits white light—such, for instance, as an ordinary gas flame—will give what is called a *continuous spectrum;* that is to say, a series of overlapping images of the slit in all the colours of which white light is composed. If, however, a good Bunsen flame be employed, a single image of the slit will be seen in the form of a bright yellow line in the same place where the brightest yellow line is seen in the continuous spectrum. Were the air in the neighbourhood of the flame pure and completely free from dust, this line would not appear; it is due to the presence of traces of sodium derived from the dust in the air, and it becomes very intense if a little sodium chloride be held in the flame on a loop of platinum wire.

Hot sodium vapour emits yellow light only. By means of the spectroscope the eye can detect with the greatest ease less than 1-3,000,000 of a milligram of a sodium salt. But many other hot metallic vapours emit more than one kind of light. Hence they give two or more coloured images of the slit in different parts of the spectrum; this is called the *bright-line* spectrum. Thus the heated vapour of lithium emits a mixture of red and yellow rays, the former predominating; and hence the spectrum of a flame containing this vapour exhibits a very bright band of red light and a comparatively dull band of yellow light. Potassium vapour under the same conditions gives a darker red band and a feeble violet band of light.

If the solar spectrum be examined many *dark* lines placed parallel to the edge of the prism will be noticed; these are known as Fraunhofer's lines, though Dr Wollaston discovered them. Some, by reason of their strength and their relative positions, always be easily recognised, and serve as references. Sources of light which do not contain volatile constituents furnish continuous spectra exhibiting no such lines; but if such constituents are present, well-defined *bright* lines are observed, the breadth of which is limited by that of the slit, and these spectra are called *line-spectra,* or *bright line spectra.*

The spectra of the non-metals, which are of course obtained at much lower temperatures, are made up of bright bands, not lines, and the breadth of these is independent of the shape of the slit; these spectra are called *channelled-space spectra.* Chemical compounds, such as calcium chloride, which can be volatilised without decomposition, yield spectra which consist of a series of differently coloured broad bands.

Bunsen in his first memoir says, "Those who become acquainted with the various spectra by repeated observation do not need to have before them an exact measurement of the single lines in order to be able to detect the presence of the various constituents. The colour, relative position, peculiar form, variety of shade, and brightness of the bands are quite characteristic enough to ensure exact results even in the hands of persons unaccustomed to such work. These special distinctions may be compared with the difference of outward appearance presented by the various precipitates which we employ for detecting substances in the wet way. Just as it holds good as a character of a precipitate that it is gelatinous, pulverulent, flocculent, granular, or crystalline, so the lines of the spectrum exhibit their peculiar aspects, some appearing sharply defined at their edge, others blending off at either one side or both sides, either similarly or dissimilarly; or some, again appearing broader, others narrower; and just as in ordinary analyses we only make use of those precipitates which are produced with the smallest possible quantity of the substance supposed to be present, so in analysis with the spectrum we employ only those lines which are produced by the smallest possible quantity of the substance, and require a moderately high temperature. In these respects both analytical methods stand on an equal footing, but analyses with the spectra possess a great advantage over all other methods, inasmuch as the characteristic differences of colour of the lines serve as the distinguishing feature of the system. . . . In spectrum analysis the coloured bands are unaffected by alteration of physical conditions or by the presence of other bodies. The positions which the lines occupy in the spectrum give rise to chemical properties as unalterable as the combining weights themselves, and which can therefore be estimated with an almost astronomical precision."

Absorption Spectra. Every incandescent body is capable of absorbing at the same temperature the same kind of light which it emits; hence a body which yields under such conditions a continuous spectrum exhibits at the same temperature a continuous absorption spectrum, and a body whose emission spectrum is discontinuous yields under similar conditions a discontinuous or broken spectrum. This selective absorption is very general. The place of the bright lines is taken by black lines (Fraunhofer's lines) in absorption spectra, or in other words the spectra are reversed.

Selective absorption is exhibited by certain bodies at ordinary temperatures, and this serves as a means of detecting the presence of the substance in question. As an example we may cite the spectrum reaction of blood.

It is possible to detect by means of these bands the presence of carbon monoxide and various foreign substances in the blood, and to determine the identity of a blood-stain several years old.

The ordinary sodium spectrum gives two bright

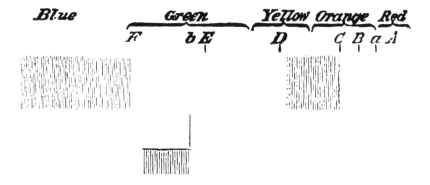

separately under a little black flux; next incorporate them thoroughly by stirring, and run the metal into the moulds, so that the face of the intended mirror may be downwards; lastly, allow the whole to cool very slowly.

2. Pure copper, 2 parts; pure tin, 1 part. Used to make the mirrors of reflecting telescopes. The addition of a little metallic arsenic, zinc, and silver renders it harder and more susceptible of a high polish.

SPELTER. See ZINC.

SPERMACETI. *Syn.* CETACEUM (B. P., Ph. L., E., & D.), L. The solid fat which is dissolved in sperm oil in the cephalic cavity of the sperm whale (*Physeter macrocephalus*), and which after death separates as a solid.

Prep. The oil is filtered off, the fat heated with potash, and then melted down.

Prop. White, inodorous, scaly, brittle mass, neutral to test-papers, and when pure, nearly tasteless. Sp. gr. 0·943 at 15° C. Melting-point, 38°—47° C. Spermaceti chiefly consists of cetylic palmitate, $C_{16}H_{31}O_2(C_{16}H_{33})$. It is demulcent and emollient, and is chiefly used in ointments and cerates.

SPHAGNUM. *Syn.* TURF MOSS, BOG MOSS. Used as a surgical dressing for absorbing discharge from wounds, also urinary discharges in diseases of the bladder and kidneys. When dry it absorbs eight times its weight of water.

SPHEROID'AL STATE. It is found that water, or any other volatile liquid, thrown on a metallic plate heated to dull redness, is not resolved into vapour, but, assuming a somewhat globular form, remains intact until the temperature becomes sufficiently lowered to allow of contact between the liquid and the heated surface. It is then immediately volatilised. M. Boutigny, who fully investigated this subject, has also shown that the same thing happens when a solid body containing water is substituted for the liquid in the above and similar experiments. Thus the finger or hand, under certain restrictions, may be thrust, with perfect impunity, into a stream of molten metal, and ice may be produced [by throwing water into a red-hot crucible. This last experiment, as performed by MM. Boutigny and Prevostaye, is essentially as follows:—A thick platinum crucible, of the capacity of 1 fl. oz., is heated to redness over a powerful spirit lamp, and some liquid anhydrous sulphurous acid (a very volatile substance) poured into it by means of a pipette; the acid assumes a spheroidal form, and does not evaporate; a few drops of water are now introduced into the sulphurous acid in the same way; the diluted and slightly cooled acid instantly flashes off in vapour, and, robbing the water of its caloric, leaves the latter in a frozen state; and, if the operator seizes the right moment, a solid lump of ice may be thrown out of the red-hot crucible.

By substituting for anhydrous sulphurous acid a mixture of solid carbonic anhydride and ether, and for water a few grains of quicksilver, this latter may be reduced to the solid condition, and may be turned out of the red-hot crucible in the form of a small frozen mass.

The spheroidal condition of "liquids is a complicated result of at least four distinct causes. Of these the most influential is the repulsive force which heat exerts between objects which are closely approximated towards each other. When the temperature reaches a certain point actual repulsion between the particles ensues.

"Besides this repulsive action occasioned by heat, the other causes which may be mentioned as tending to produce the assumption of the spheroidal condition by the liquid are these:

"1. The temperature of the plate is so high that it immediately converts any liquid that touches it into vapour, upon which the spheroid rests as on a cushion.

"2. This vapour is a bad conductor of heat, and prevents the rapid conduction of heat from the metal to the globule.

"3. The evaporation from the entire surface of the liquid carries off the heat as it arrives, and assists in keeping the temperature below the points of ebullition. The drop assumes the spheroidal form as a necessary consequence of the action of cohesion among the particles of the liquid, and the simultaneous action of gravity on the mass" (*Miller*).

Boutigny found that, when a liquid in a state of ebullition was brought into contact with a surface heated to such a degree as to cause the liquid to assume the spheroidal state, its temperature immediately fell 3° or 4° C. below the boiling-point.

All liquids are capable of assuming the spheroidal condition; but, as the temperature necessary for this purpose varies with the boiling-point of each liquid (the lower the boiling-point the lower the temperature necessary, and *vice versâ*), it follows that the conducting surface requires to be differently heated for each liquid. The exact temperature to which the plate should be heated to produce the spheroidal condition in any liquid depends partly upon the conducting power of the plate, and partly upon the latent heat of the vapour; the less this is, the more nearly the temperature of the plate approximates to the boiling-point of the liquid.

Boutigny believed that the temperature of each liquid, when in the spheroid condition, was as invariable as that of its boiling-point; but Boutan has demonstrated that this is a not quite accurate statement, since the temperature of the same liquids, when assuming the spheroidal form, is liable to slight divergence.

The following table, showing the lowest temperature of the plate and the temperature of the spheroid for certain liquids, is given by Boutigny:

Liquid employed.	Temperature of Plate.		Temperature of Spheroid.	
	° F.	° C.	° F.	° C.
Water	340	171	205·7	96·4
Alcohol . . .	273	134	167·9	75·5
Ether	142	61	93·6	34·2
Sulphurous anhydride . . .	...	...	13·1	10·5

Solids may also be made to assume the spheroidal condition, as when, for instance, some crystals of iodine are thrown upon a red-hot platinum disc, or into a platinum crucible similarly heated.

The nature of the plate or crucible employed appears to be immaterial, provided it be a good conductor. Platinum, silver, copper, and iron answer equally well; indeed, Tomlinson has shown that one liquid may even be made to assume the spheroidal state on the surface of another, as when water, alcohol, and ether are placed upon hot oil. If the experiment be conducted with water it must be carefully managed, since, if the water be allowed to sink in the oil, it soon becomes converted into steam, with the result that the hot oil is scattered to the danger of the operator.

Boutigny has advanced the opinion that the property of water to assume the spheroidal state under the conditions we have specified will account for certain cases of explosion in steam boilers. Thus we can imagine a boiler which has run dry of water to have become intensely overheated. Under these circumstances, when fresh water was admitted, it would at first assume the spheroidal state; and as more cold water flowed into it the boiler would become thereby reduced in temperature until it reached the point at which its conversion into steam would take place; the sudden generation, large volume, and elastic force of which would lead to the rupture of the boiler, accompanied with explosive violence.

SPICE. A general name for vegetable substances possessing aromatic and pungent properties, and employed for seasoning or flavouring food.

Spice, Cattle. The following formula gives an excellent condiment, a teacupful of which should be given in a bran mash once a week, or oftener:—Linseed cake, 14 lbs.; gentian, 1 lb.; liquorice, 1 lb.; fenugreek, ¾ lb.; nitre, ½ lb.; ginger (African), ¼ lb.; anise, 6 oz.; coriander, 6 oz.; sulphur, 6 oz.; cinchona, 4 oz. (all ground). Mix all the ingredients except the linseed meal, then incorporate it gradually.

Spice, Horse. *Syn.* COW SPICE; SPECIES EQUINUS, L. *Prep.* 1. Aniseed, allspice, cumin seed, ginger, liquorice, and turmeric, equal parts.
2. Turmeric and cumin seed, of each, 5 lbs.; ginger, 2½ lbs. Used by farriers.

Spice, Kit'chen. *Syn.* MIXED SPICE, KITCHEN PEPPER, &c. *Prep.* From black pepper, 2 lbs.; ginger, 1 lb.; cinnamon, allspice, and nutmegs, of each, 8 oz.; cloves, 1 oz.; dry salt, 6 lbs.; well ground together. Useful to flavour gravies, soup, &c.

Spice, Mixed. As the last, omitting half the salt.

Spice, Pease. See POWDER.

Spice, Ragoût. *Prep.* From dry salt, 1 lb.; flour of mustard, black pepper, and grated lemon peel, of each, ¼ lb.; cayenne pepper, 2 oz.; allspice and ginger, of each, 1 oz.; nutmeg, ½ oz.; all separately powdered.

Spice, Sausage (French). *Syn.* ÉPICE FINES, Fr. *Prep.* From black pepper, 5 lbs.; ginger, 2½ lbs.; cloves and nutmegs, of each, 1 lb.; aniseed and coriander seeds, of each, ½ lb.; powder and mix them.

Spice, Sa'voury. *Prep.* 1. (*Kidder's.*) From cloves, mace, nutmegs, pepper, and salt, equal parts. Used by cooks.
2. (*Dr Kitchener's.*) See SPICE, RAGOÛT (*above*).

Spice, Soup. *Syn.* KITCHENER'S SOUP-HERB POWDER, KITCHENER'S VEGETABLE RELISH, &c. *Prep.* From parsley, lemon thyme, sweet marjoram, and winter savoury, of each, dried, 2 oz.; sweet basil and yellow peel of lemon, of each, dried, 1 oz.; mix and powder.

Spice, Sweet (*Kidder's*). *Prep.* From cinnamon, cloves, mace, nutmegs, and sugar, equal parts. Used in pastry.

SPIGE'LIA. *Syn.* CAROLINA PINK-ROOT; SPIGELIA (U. S. P.), L. The root of *Spigelia marilandica*. It is purgative, narcotic, and vermifuge.—*Dose*, 10 to 40 gr., in powder or infusion night and morning, until the worms are expelled. Rhubarb or calomel is commonly added to it.

SPIRIT. *Syn.* SPIRITUS, L. Under this term are included all the inflammable and intoxicating liquors obtained by distillation, and used as beverages, as BRANDY, GIN, RUM, &c., each of which is noticed in its alphabetical order. Spirit may also be obtained by fermentation and distillation from all vegetable juices or solutions that contain sugar.

The spirit used in pharmacy and chemistry is distinguished by names which have reference to its richness in alcohol. See TABLES on next page.

Spirituous liquors, like all other fluids at common temperatures, expand when they are heated, and diminish in volume when they are cooled. It is found that 1000 galls. of proof spirit, measured at the temperature of 50° F., will, if re-measured at 59°, be found to have increased in bulk to full 1004¼ galls.; whilst 1000 galls. of the same spirit, measured at 77° F., will be only equal to 991½ galls. at 59°. These changes are still more marked at higher strengths and at extreme temperatures, and, from not being recognisable by the hydrometer, often lead to serious losses in trade, and to serious fluctuations in 'stock,' which, to those unaware of the action of temperature, are perfectly unaccountable. A gallon of proof spirit only weighs 9½ lbs. at 60° F. At a higher temperature it will weigh less—at a lower one more; but as this weight constitutes the standard gallon at the temperature the proof is calculated for, it is manifest that any variations from it must result in loss either to the buyer or seller. Hence the equity of buying and selling liquors by weight instead of by measure. The stock-keeper in every wholesale house should be aware of this fact, and on 'taking stock' should as regularly enter the temperature of his liquors in his stock-book as he does the 'dip' or 'wet inches.' See ALCOHOL, ALCOHOLOMETRY, SPECIFIC GRAVITY, SPIRITS (Medicinal), SPIRITS (Perfumed), &c.

Spirit of Acetic Ether. *Syn.* SPIRITUS ÆTHERIS ACETICI (Pruss. Ph.), L. *Prep.* Acetic ether, 1 oz.; rectified spirit, 3 oz.

Spirit, Alexiterius. *Syn.* SPIRITUS ALEXITERIUS, AQUA ALEXITERIA SPIRITUOSA (Ph. L. 1746). *Prep.* Mint, ½ lb.; angelica root, 4 oz.; tops of sea-wormwood, 4 oz.; proof spirit, 1 gall. (old wine measure); water, a sufficient quantity. Distil 1 gall.

I. TABLE of the Pharmacopœial Spirits.

	Sp. gr.			Sp. gr.
Alcohol, Ph. B. (absolute) . .	0·797	nearly pure Alcohol.	Rectified Spirit, Ph. D. . . .	0·840 or 54¾% o.-p.
" Ph. E.	0·796		Proof Spirit (Spiritus Tenuior),	0·920 " —
" Ph. D. 1826 . . .	0·810 or 70% o.p.		Ph. B.	
" Ph. L. 1836 . . .	0·815 " 68% "		Alcohol (absolute), P. Cod. .	0·797 " —
Stronger Spirit (Spiritus Fortior), Ph. D.	} 0·818 " 66% "		" (at 40°) " .	0·810 " 70% "
*Rectified Spirit (Spir. of Wine; Spiritus Rectificatus), B. P.	} 0·838 " 56% "		" (du commerce), P. Cod.	0·863 " 41% "
			" (faible) "	0·923 " 2½% u.-p.

* "This spirit can be reduced to the standard of the weaker (or proof) spirit by adding, to every 5 pints of it, 3 pints of distilled water at 62° F." (Ph. L.).

II. TABLE of the Principal Spirituous Liquors sold in England, with their usual Strengths, &c.

Denomination.	Revenue Mark.	Import Strength.	Legal Limits of Strength.	Usual Selling Strength.			Specific Gravity at 60° F.
				By Permit.	Contains Alcohol of 0·825.	Contains absolute Alcohol.	
*Gin (strongest) . . .	X (17 u. p.)	...	Not stronger than 25 o. p.	17 u. p.	...	40%	0·9395
*Do. (best ordinary) . .	X (22 u. p.)	...	do.	22 u. p.	...	37·4%	0·9445
†Do. (cordial) . . .	X (22 u. p.)	...	do.	22 u. p.	...	do.	‖0·
†Do.	X (24 u. p.)	...	do.	24 u. p.	...	36·5%	‖0·
‡Peppermint	X mint	...	do.	60 u. p.	...	21%	‖0·
‡Do.	do.	...	do.	64 u. p.	...	18%	‖0·
‡Cloves . . . . . ⎫							
‡Bitters							
‡Raspberry							
‡Noyau							
‡Cinnamon							
‡Tent . . . . . . ⎬ X (64 u. p.)		...	do.	64 u. p.	...	do.	‖1·065 to 1·080
‡Aniseed							
‡Caraway							
‡Lovage							
‡Usquebaugh . . .							
‡Orange cordial . .							
‡Citron . . . . . ⎭							
Rum	R.	About 10 o. p. to 43 o. p.	No limit	11 u. p.	...	43%	0·9329 to 0·8597
‡Rum shrub	R. Sh.	...	do.	64 u. p.	...	18%	‖0·
‡Do.	do.	...	do.	60 u. p.	...	21%	‖0·
French brandy . . .	F.	About 5 o. p. to 8 or 10 u. p.	do.	10 u. p.	...	44%	0·9318
§Spirit of wine . . .	S. W.	...	Not less than 43 o. p.	54 to 64 o. p.	...	...	0·8415 to 0·8221
Malt, grain, or molasses spirit (sent out by British distillers) . . .	P. S.	...	Not stronger than 25 o. p.	...	...	...	0·8669 to 0·9318
Hollands	Geneva	...	No limit	...	51·60%	40·5%	0·9358
Whiskey (Irish) . . .	P. S.	... ⎫	Not stronger	⎰ ...	54%	50%	
Do. (Scotch) . . .	P. S.	... ⎬	than 25 o. p.	⎱ ...	54·3%	50·2%	

* Frequently retailed at 25 to 35 u. p.   † Though 'permitted' at 22 to 24, are generally from 25 to 35 u. p., or even weaker.   ‡ These, though 'permitted' at 60 or 64 u. p., are generally 75 or 80 u. p.   § Usual strength 54 to 60 o. p.   ‖ The specific gravity is no guide when sugar is present, as in compounds.

Spirit, Amy'lic. See FUSEL OIL.

Spirit of Angelica. *Syn.* SPIRITUS ANGELICE. *Prep.* Sliced angelica root, 2 oz.; sliced valerian, ½ oz.; bruised juniper berries, ¼ oz. Put into a retort, and pour on 9 oz. of rectified spirit by weight, and 15½ oz. of water, and macerate for 24 hours; then draw out 12½ oz. (by weight), in which dissolve ¼ oz. of camphor.

Spirit of Ants. SPIRITUS FORMICARUM (Ph. G.), L. *Prep.* Ants freshly collected and bruised, 2 lbs.; spirit of wine (at ·830), 3 lbs.; water, 3 lbs. Macerate for 2 days. Distil 4 lbs.—*Dose,* 20 to 60 drops; also used outwardly.

Spirit, Blue. *Syn.* SPIRITUS OCERULEUS (Ham. Ph.). *Prep.* Wormwood, scordium, savin, lavender flowers, of each, 2½ oz.; proof spirit, 5 pints; distil 2½ pints, and add 6 dr. of verdigris, and water of ammonia, 9 oz. For outward use.

Spirit of Bryony (Compound). *Syn.* SPIRITUS BRYONIE COMPOSITUS (Ph. E. 1744). *Prep.* Bryony, ¼ lb.; valerian, 2 oz.; pennyroyal, 3 oz.; rue, 3 oz.; mugwort feverfew flowers, savin tops, of each, 4 dr.; orange peel, 1 oz.; lovage seeds, 1 oz.; brandy, 1 gall.; distil. Without the bryony this preparation is known as *Aqua hysterica.*—*Dose,* 1 oz.

Spirit of Cajeput. *Syn.* SPIRITUS CAJEPUTI (B. P.). *Prep.* Dissolve 1 fl. oz. of oil of cajeput in 49 fl. oz. of rectified spirit.

Spirit of Cardamom. *Syn.* SPIRITUS CARDAMOMI (Ph. L. 1746). *Prep.* Cardamom seed, 5 troy oz.; proof spirit, 1 gall.; water, a sufficient quantity. Distil 1 gall.

Spirit of Chloroform. *Syn.* SPIRITUS CHLOROFORMI (B. P.). *Prep.* Dissolve 1 fl. oz. of chloroform in 19 fl. oz. of rectified spirit. Sp. gr. ·871.

Spirit of Cloves. *Syn.* SPIRITUS CARYOPHYLLI (P. Cod.). *Prep.* Cloves, 10 oz.; spirit (·864), 80 oz., by weight; draw over all the spirituous part.

Spirit of Coriander. *Syn.* SPIRITUS CORIANDRI (P. Cod.). *Prep.* Coriander seed, 1 oz.; spirit (·0864), by weight, 8 oz. Distil to dryness.

Spirit, Dyer's. See TIN MORDANTS.

Spirit, Febrifuge, of Clutton. *Syn.* SPIRITUS FEBRIFUGUS CLUTTONI. See SPIRIT OF HYDROCHLORIC ETHER. *Prep.* The original form is—oil of sulphur by the bell, oil of vitriol, and sea salt, of each, 1 oz.; spirit of wine, 6 oz. Let them digest for a month, then distil to dryness.

Spirit, Fioravanti. *Syn.* SPIRITUS FIORAVANTI (P. Cod.). *Prep.* Swiss turpentine, 5 oz.; elemi, 1 oz.; resin of tacamahaca, 1 oz.; amber, 1 oz.; liquid styrax, 1 oz.; galbanum, 1 oz.; myrrh, 1 oz.; aloes, ½ oz.; bay berries, 1 oz.; galanga root, ½ oz.; ginger, ½ oz.; sedoary root, ½ oz.; cinnamon, ½ oz.; cloves, ½ oz.; nutmeg, ½ oz.; leaves of cretum marum, ½ oz. Macerate 6 days and distil over a water-bath till 35 oz. come over.

Spirit of Flower Sage. *Syn.* SPIRITUS SALVIE. *Prep.* Flower sage, 1 lb.; rectified spirit, 3 lbs.; water, 1 lb. Distil 3 lbs.

Spirit of French Wine. *Syn.* SPIRITUS VINI GALLICI; BRANDY.

Spirit of Lemon Peel. *Syn.* SPIRITUS CITRI

CORTICIS. With lemon peel as SPIRIT OF ORANGE PEEL.

Spirit of Marjoram. *Syn.* SPIRITUS MARJORANE. *Prep.* Sweet marjoram, 1 lb.; rectified spirit, 3 lbs.; water, 1 lb. Distil 3 lbs.

Spirit of Mastic (Compound). *Syn.* SPIRITUS MASTICHES COMPOSITUS. *Prep.* Mastic, 1 oz.; myrrh, 1 oz.; olibanum, 1 oz.; rectified spirit, 1 pint; distil.

Spirit, Meth'ylated. Spirit of wine to which one tenth of its volume of wood naphtha (strength not less than 60% o. p.) has been added, the object of such addition being that of rendering the mixture unpotable through its offensive odour and taste. The purification of this mixed spirit, or the separation of the two alcohols, though often attempted, has always proved a failure. It might be supposed that, owing to the low boiling-point of methylic alcohol, simple distillation would effect this; but experience has shown that both spirits distil over simultaneously. This is, no doubt, due to the difference of their vapour densities. Methylated spirit, being sold duty free, can be employed by the chemical manufacturer as a solvent in many processes for which, from its greater cost, duty-paid spirit would be commercially inapplicable. But in the preparation of medicines containing spirit, as the vehicle or menstruum by which more active substances are administered, the employment of methylated spirit is highly improper. The Council of the Pharmaceutical Society obtained from the Pharmacopœia Committee of the Medical Council the decided opinion that "the substitution of 'methylated' for 'rectified' spirit in any of the processes of the Pharmacopœia should be strictly prohibited."

The use of methylated spirit in the preparation of tinctures, sweet spirit of nitre, common ether, or any medicine to be used internally, is now prohibited by law. Certain new regulations as to the substances to be added to spirits to be sold as 'methylated' came into force in September, 1891. In addition to the usual naphtha, a certain quantity of crude benzol is required to be added. This interferes with the dilution test for the presence of resinous matter.

Spirit of Orange Peel. *Syn.* SPIRITUS AURANTII (P. Cod.). *Prep.* The yellow part of fresh orange peel, 1 lb.; spirit of wine (·864), 6 lbs.; macerate for 2 days, and distil by water-bath to dryness.

Spirit of Origanum. *Syn.* SPIRITUS ORIGANI. *Prep.* Wild marjoram, 1 lb.; rectified spirit, 3 lbs.; water, 1 lb. Distil 3 lbs.

Spirit of Para Cress. *Syn.* SPIRITUS SPILANTHI (Beral). *Prep.* Bruised Para cress (*Spilanthes oleracea*), in flower, 1 part; spirit (·863), 2 parts. Macerate 2 or 3 days, and distil 2 parts.

Spirit, Proof. See ALCOHOL.

Spirit, Pyroace'tic. *Syn.* ACETONE; SPIRITUS PYROACETICUS, L.; ACÉTONE, ESPRIT PYROACÉTIQUE, Fr. An inflammable volatile liquid obtained with carbonic acid and other products when the metallic acetates in an anhydrous state are subjected to destructive distillation. The acetate of lead is the most eligible salt for this purpose.

Prep. 1. Dried acetate of lead is carefully distilled in a large earthen or coated glass retort, by a heat gradually raised to redness, the volatile products being passed through a condenser well supplied with cold water. The distillation is continued until nothing but finely divided lead (lead pyrophorus) remains in the retort. The receiver contains crude acetone, which is to be saturated with carbonate of potassa, and afterwards rectified in a water-bath from chloride of calcium.

2. By passing the vapour of strong acetic acid through an iron tube heated to dull redness, and condensing the acetone thus formed.

Obs. In both of the above processes carbonic acid and other permanent gases are produced, consequently the receiver must not fit too closely to the tube of the condenser.

Prop. Colourless, limpid, of peculiar odour, and very inflammable, giving a brilliant flame without smoke; boiling-point, 132° F.; sp. gr. ·792. It dissolves resins and essential oils. See MESITILOL, MESITYL, METACETONE, &c.

Spirit, Pyroxyl'ic. *Syn.* PYROLIGNEOUS SPIRIT, WOOD S., MEDICINAL NAPHTHA, WOOD N., HYDRATED OXIDE OF METHYL; SPIRITUS PYROXYLICUS (Ph. D.), L. A light volatile liquid, discovered by P. Taylor, in 1812, among the limpid products of the distillation of dry wood. It has been shown by Dumas and Peligot to be "really a second alcohol, forming an ether, and a series of compounds (METHYL SERIES) exactly corresponding with those of vinous spirit, and in some points even more complete than the latter."

Prep. Crude pyroligneous acid (which contains about 1% of the spirit) is subjected to distillation, and the first or more volatile portion which passes over is neutralised with hydrate of lime. After repose the clear liquid is separated from the oil which floats on the surface, and from the sediment at the bottom of the vessel; this, when redistilled, forms the wood spirit of commerce. It may be strengthened in the same manner as ordinary alcohol, by rectification, and ultimately rendered pure by careful distillation from quicklime by the heat of a water-bath. Berzelius recommends the crude spirit to be agitated with a fatty oil, to remove empyreumatic matter, and then to rectify it, first from recently burnt charcoal, and next with chloride of calcium.

Prop., &c. Pure pyroxylic spirit is a transparent, colourless liquid, having a penetrating ethereal smell and a hot, disagreeable taste; it is very inflammable, burning with a pale blue flame. It is neutral to test-paper; mixes with water, alcohol, and ether, in all proportions; and boils at 152° F.; sp. gr. ·798 at 68° F. (*Regnault* and *Liebig*). Dr Ure states the sp. gr. to be ·824 at 60°; the Dublin College makes it ·846. That of the latter must, therefore, have contained a little water. It does not dissolve india rubber and gutta percha, like mineral or true naphtha.

Pyroxylic spirit is distinguished from acetone or pyroacetic spirit by the character of its flame, and by freely dissolving chloride of calcium, which is quite insoluble in the latter. In a mixture of these two liquids two distinct strata are formed when this substance, either in powder or concentrated solution, is added.

Pyroxylic spirit is distinguished from vinous spirit by Nessler's test (which *see*), by its forming a solid crystalline salt (methylic oxalate) when distilled with an oxalate and sulphuric acid, and by its lower boiling-point. The presence of alcohol in a mixture of the two is readily detected by distilling the suspected sample with sulphuric acid. The formation of common ether indicates ethylic alcohol, and, from the amount formed, the proportion of alcohol may be determined.

Uses, &c. Chiefly to dissolve resins and volatile oils, especially shellac, and as a substitute for alcohol in spirit lamps. As a medicine it is anodyne and sedative, and has been beneficially employed by Drs Christison, Hastings, and Neligan, to allay the harassing cough, troublesome vomiting, and excessive expectoration in phthisis and some other affections.—*Dose,* 5 to 30 drops, thrice a day, in water.

Spirit, Rai'sin. *Prep.* From raisins fermented along with water, and the wash distilled by a quick fire. One gall. added to 150 galls. of plain spirit, along with some colouring and a little catechu, either with or without a little acetic ether, makes a very decent 'British brandy.'

Spirit of Raspberries. *Syn.* SPIRITUS RUBRI IDÆI. *Prep.* Raspberries, 3 lbs.; rectified spirit, 2 lbs.; distil 2 lbs.

Spirit, Rec'tified. See ALCOHOL, and Table I, under SPIRIT.

Spirit of Salt†. Hydrochloric acid.

Spirit of Sassafras. *Syn.* SPIRITUS SASSAFRAS. *Prep.* Sassafras, 1 troy lb.; rectified spirit (·863), 8 lbs. Macerate 4 days, and distil nearly to dryness.

Spirit of Scurvy-grass. *Syn.* SPIRITUS COCHLEARIÆ (Ph. G.). *Prep.* Fresh leaves of flowering scurvy-grass, 28 lbs.; rectified spirit, 8 lbs.; water, 8 lbs. Distil 4 lbs.

Spirit of Scurvy-grass (Compound). *Syn.* SPIRITUS COCHLEARIÆ COMPOSITUS (P. Cod.). *Prep.* Fresh scurvy-grass, 5 lbs.; spirit (·0864), 6 lbs.; horseradish, 10½ oz. Distil 5 lbs.

Spirit of Soot. *Syn.* SPIRITUS FULIGINIS. An empyreumatic spirit was formerly distilled from wood soot, in the same manner as hartshorn. An alcoholic spirit is also made from 1 part of wood soot, 5 of proof spirit, 15 of water. Distil 4 parts.

Spirit of Soup Herbs. As essence of soup herbs, but substituting 1 quart of brandy or proof spirit for the rectified spirit.

Spirit of Sweet Flag Root. *Syn.* SPIRITUS CALAMI (P. Cod.). *Prep.* Calamus, 1 lb. troy; spirit of wine (·863), 8 lbs.; macerate 4 days, and distil nearly to dryness.

Spirit of Thyme. *Syn.* SPIRITUS THYMI. From lime, as spirit of sage.

Spirit of Turpentine, Ethereal. *Syn.* SPIRITUS TEREBINTHINÆ ÆTHEREUS (*Van Mons*). *Prep.* Spirit of nitric ether, with as much rectified oil of turpentine as it will dissolve. Rectified oil of turpentine is also termed 'Ethereal spirit of turpentine.'

Spirit of Vanilla. See ESSENCE OF VANILLA.

Spirit of Wine. See ALCOHOL, and Table I, under SPIRIT.

Spirit of Wormwood (Compound). *Syn.* SPIRITUS ABSINTHII COMPOSITUS VEL AQUA (Ph. L. 1720). *Prep.* Dried wormwood, ½ lb.; cardamom seed, ½ oz.; coriander seed, 1½ oz.; brandy, 1 gall. Distil.

SPIRITS (Medic'inal). *Syn.* SPIRITUS MEDICINALES, *L.* The spirits of pharmacy are either prepared by macerating the bruised seeds, flowers, herbs, &c., in the spirit for 2 or 3 days before distillation, and then drawing it off by a gentle heat; or extemporaneously, by adding a proper proportion of essential oil to pure spirit of the prescribed strength (in the British Pharmacopœia, for most distilled spirits is substituted a solution of 1 part of volatile oil in 49 parts of rectified spirit—ED.). This latter plan is very generally adopted in the Ph. D. In the first method, when a naked fire is employed, a little water is put into the still along with the spirit, to prevent empyreuma. These spirits are principally employed as aromatics and stimulants, or as adjuvants in draughts and mixtures.

The following are the principal medicinal spirits:

Spirit of Ammo"nia. *Syn.* SPIRITUS AMMONIÆ (Ph. E.). *Prep.* 1. (Ph. E.) Take of quicklime, 12 oz.; shake it with water, 6½ fl. oz.; add of finely powdered chloride of ammonium, 3 oz.; and distil in a glass retort furnished with a tube reaching nearly to the bottom of a bottle containing rectified spirit, 2 pints, and kept well cooled. A sand heat is to be employed, and the distillation continued as long as anything passes over. The product has a sp. gr. about ·845, and should not effervesce with acids. The alkali is here in the caustic state, and in this respect it resembles the spirit of ammonia, Ph. U. S., and Dzond's caustic spirit of ammonia, Ph. Bor.

2. (Ph. L. 1836.) Chloride of ammonium, 10 oz.; carbonate of potassa, 16 oz.; rectified spirit and water, of each, 3 pints; mix, and let 3 pints distil.

3. (Ph. D. 1826.) Dissolve 3½ oz. of carbonate of ammonia in rectified spirit, 3 wine pints. *Obs.* The ammonia in the last two preparations exists in the carbonated state. They are chiefly employed to make other preparations.

4. (ANISATED SPIRIT OF AMMONIA; LIQUOR AMMONIÆ ANISATUS, SPIRITUS A. A.—Ph. Bor.) Rectified spirit, 12 oz.; oil of aniseed, 3 dr.; dissolve, and add of caustic solution of ammonia (·960), 3 oz.

5. (AROMATIC SPIRIT OF AMMONIA, SPIRIT OF SAL VOLATILE; SPIRITUS AMMONIÆ AROMATICUS —B. P., Ph. L., E., and D.) *a.* (Ph. L.) Take of hydrochlorate of ammonia, 6 oz.; carbonate of potassa, 10 oz.; cinnamon and cloves, of each, bruised, 2½ dr.; fresh lemon peel, 5 oz.; rectified spirit and water, of each, 2 quarts; mix, and distil 3 quarts. Sp. gr. ·918.

b. (Ph. E.) Spirit of ammonia, 8 fl. oz.; oil of rosemary, 1½ fl. dr.; oil of lemon peel, 1 fl. dr.; mix.

c. (Ph. D.) Rectified spirit, 8 pints; oil of lemon, ½ fl. oz.; oil of nutmeg, 2 fl. dr.; oil of cinnamon, ½ fl. dr.; dissolve, and add of stronger solution of ammonia, 6 fl. oz. Sp. gr. ·852.

d. (B. P.) Carbonate of ammonium, 4 oz.; strong solution of ammonia, 8 oz.; volatile oil of nutmeg, 4½ dr.; oil of lemon, 6½ dr.; rectified spirit, 6 pints; water, 3 pints; mix the oils with the spirit and water, distil 7 pints, then distil an

additional 9 oz.; in the 9 oz. dissolve the ammonia and carbonate of ammonium, and gradually mix it with the 7 pints of spirit. The product should measure 1 gall.—*Dose*, 20 to 60 minims in camphor water.

Obs. The ammonia exists in the state of neutral carbonate in the product of the *a* formula, but in the caustic state in those of the others.— *Dose*, ½ to 1 fl. dr., in water or any bland liquid; as a diffusible stimulant and antacid in debility, low spirits, dyspepsia, heartburn, flatulent colic, hysteria, &c. The spirit of sal volatile of the shops is generally a spurious compound of little more than half the above strength.

6. (FETID SPIRIT OF AMMONIA; SPIRITUS AMMONIÆ FŒTIDUS—B. P., Ph. L., E., and D.) *a.* (Ph. L.) Hydrochlorate of ammonia, 10 oz.; carbonate of potassa, 16 oz.; assafœtida, 5 oz.; rectified spirit and water, of each, 3 pints; mix well, then slowly distil 3 pints. Sp. gr. ·861.

b. (Ph. E.) Spirit of ammonia, 10½ fl. oz.; assafœtida (broken small), ½ oz.; digest for 12 hours, then distil 10½ fl. oz. by the heat of a vapour (water) bath.

c. (Ph. D.) Assafœtida, 1½ oz.; rectified spirit, 1½ pints; digest for 24 hours, then distil off the whole of the spirit, and mix the product with stronger solution of ammonia, 3 fl. oz. Sp. gr. ·849.

d. (B. P.) Strong solution of ammonia, 3 parts; assafœtida, in small pieces, 1½ parts; rectified spirit, sufficient; macerate the assafœtida in 15 of the spirit for 24 hours, distil, add the distillate to the ammonia, and make up with spirit to 20 parts.—*Dose*, ½ to 1 dr.

Obs. The dose, &c., are the same as those of the last, but it is preferred for hysterical and spasmodic affections.

Spirit, Amyl'ic. *Syn.* ALCOHOL AMYLICUM (Ph. D.), *L.* See FUSEL OIL.

Spirit of An'iseed. *Syn.* SPIRITUS ANISI (Ph. D.), *L.* *Prep.* 1. (Ph. L.) Oil of aniseed, 3 fl. dr.; proof spirit, 1 gall.; dissolve. Carminative. —*Dose*, ½ fl. dr. to 4 fl. dr.

2. (ESSENTIA ANISI—Ph. D.) Oil of aniseed, 1 fl. oz.; rectified spirit, 9 fl. oz.; mix with agitation. Chiefly used to make aniseed water.

3. (COMPOUND SPIRIT OF ANISEED; SPIRITUS ANISI COMPOSITUS—Ph. D. 1826.) Aniseed and angelica seed, of each, ½ lb.; proof spirit, 1 gall.; water, q. s.; distil 1 gall. When coloured with saffron, or sap green, it closely resembles the Irish usquebaugh (*Montgomery*).—*Dose*, 1 to 4 fl. dr.

Spirit, Arquebusade'. See VULNERARY SPIRIT (*below*).

Spirit of Balm (Compound). *Syn.* BALM WATER, CARMELITE W.; AQUA MELISSÆ COMPOSITA, SPIRITUS M. COMPOSITUS, *L.*; EAU DES CARMES, EAU DE MELISSE DES CARMES, Fr. *Prep.* (P. Cod.) Fresh flowering tops of balm, 24 oz.; fresh lemon peel, 4 oz.; cinnamon, cloves, and nutmegs, of each, 2 oz.; coriander seed and dried angelica root, of each, 1 oz.; rectified spirit, 8 lbs.; macerate for eight days and distil in a water-bath to dryness. The spirit is much esteemed in France as a stomachic, a cosmetic, and a stimulant.

Spirit, Bath'ing. Soap liniment.

Spirit of Cam'phor. *Syn.* CAMPHORATED SPIRIT; SPIRITUS CAMPHORÆ (B. P., Ph. L.), TINCTURA CAMPHORÆ, SPIRITUS CAMPHORATUS, L. *Prep.* 1. (Ph. L.) Camphor, 5 oz.; rectified spirit, 1 quart; dissolve.

2. (B. P.) Camphor, 1 part; rectified spirit, 9 parts; dissolve.—*Dose,* 10 to 30 minims, in milk or on sugar. Used as an application to chilblains, and in chronic rheumatism, cholera, &c. See ESSENCE and TINCTURE.

Spirit of Car'away. *Syn.* SPIRITUS CARUI (Ph. L. and E.), L. *Prep.* 1. (Ph. L.) Oil of caraway, 2 fl. dr.; proof spirit, 1 gall.; dissolve.

2. (Ph. E.) Caraway seeds (bruised), ½ lb.; proof spirit, 7 pints; macerate for two days in a covered vessel, then add of water, 1½ pints, and distil 7 pints. Aromatic and carminative.—*Dose,* 1 to 4 fl. dr. A similar spirit, 'sweetened with sugar,' is drunk in Germany as a dram (KÜMMELLIQUEUR, KÜMMELBRANDTWEIN).

3. (ESSENTIA CARUI—Ph. D.) Oil of caraway, 1 fl. oz.; rectified spirit, 9 fl. oz. Used to make caraway water.

Spirit of Cas'sia. *Syn.* SPIRITUS CASSIÆ (Ph. E.), L. *Prep.* From coarsely powdered cassia, 1 lb.; proof spirit, 7 pints; water, 1½ pints, or q. s.; draw off 7 pints.—*Dose,* &c., as the last. It is almost universally substituted for spirit of cinnamon.

Spirit of Cin'namon. *Syn.* SPIRITUS CINNAMOMI (Ph. L. and E.), L. *Prep.* 1. (Ph. L.) Oil of cinnamon, 2 fl. dr.; proof spirit, 1 gall.; dissolve.

2. (Ph. E.) From cinnamon, as spirit of cassia.—*Dose,* 1 to 4 fl. dr.

3. (B. P.) Oil of cinnamon, 1 part; rectified spirit, 49 parts.

Spirit of E'ther. *Syn.* SPIRIT OF SULPHURIC ETHER, SWEET SPIRIT OF VITRIOL†; SPIRITUS ÆTHERIS (B. P.), SPIRITUS ÆTHERIS SULPHURICI (Ph. E.), L. *Prep.* 1. *a.* (Ph. E.) Sulphuric ether, 1 part; rectified spirit, 2 parts. Sp. gr. ·809.—*Obs.* This preparation should be neutral to test-paper, mix (clear) with water, and, when shaken with twice its volume of concentrated solution of chloride of calcium, 28% of ether should separate.—*Dose,* ½ to 2 or 3 fl. dr.; as a stimulant and anodyne.

b. (B. P.) Ether, 1 part; rectified spirit, 2 parts; mix.—*Dose,* 30 to 60 minims.

2. COMPOUND SPIRIT OF ETHER, HOFFMANN'S ANODYNE LIQUOR; SPIRITUS ÆTHERIS COMPOSITUS (Ph. L.), S. ÆTHEREUS OLEOSUS (Ph. D.), L. *a.* (Ph. L.) Ether, 8 fl. oz.; rectified spirit, 16 fl. oz.; ethereal oil, 3 fl. dr.; mix.

b. (Ph. D.) Mix, in a glass matrass, oil of vitriol, 1½ pints, with rectified spirit, 1 pint; connect this with a Liebig's condenser, apply heat, and distil until a black froth begins to rise; then separate the upper stratum of the distilled liquid, and, having exposed it to the air for twenty-four hours, let the oil be transferred to a moist paper filter, and washed with a little cold water; lastly, dissolve it in a mixture of rectified spirit, ½ pint; sulphuric ether, 5 fl. oz.

Obs. This compound is anodyne and antispasmodic, and was once held in very great repute.—*Dose,* ½ to 2 fl. dr.

3. AROMATIC SPIRIT OF ETHER, A. S. OF SULPHURIC E., SWEET ELIXIR OF VITRIOL† ; SPIRITUS ÆTHERIS AROMATICUS, L. *Prep.* (Ph. L. 1824.) Bruised cinnamon, 3 dr.; cardamoms, 1½ dr.; long pepper and ginger, of each, 1 dr.; rectified spirit, 10 fl. oz.; sulphuric ether, 5 fl. oz.; mix, and digest fourteen days. The last two preparations are also frequently called 'sweet elixir of vitriol.'

Spirit of Harts'horn. *Syn.* LIQUOR OR SPIRITUS VOLATILIS CORNU CERVI, L. Originally distilled from hartshorn. Dilute liquor ammonia is now generally sold for spirit of hartshorn.

Spirit of Horserad'ish (Compound). *Syn.* SPIRITUS ARMORACIÆ COMPOSITUS (B. P., Ph. L.), L. *Prep.* 1. (Ph. L.) Sliced horseradish and dried orange peel, of each, 20 oz.; bruised nutmegs, ½ oz.; proof spirit, 1 gall.; water, 3 pints; distil 1 gall. Stimulant and diuretic.—*Dose,* 1 to 4 fl. dr.; in dropsies, when there is much debility. It is usually combined with infusion of juniper berries or foxglove.

2. (B. P.) Fresh root, sliced, 20 parts; dried orange peel, 20 parts; nutmeg (bruised), ½ ; proof spirit, 160 parts; water, 60 parts; mix, and distil over 160 parts.—*Dose,* 1 to 3 dr.

Spirit of Hydrochlo"ric Ether. *Syn.* SPIRIT OF MURIATIC ETHER, CLUTTON'S FEBRIFUGE SPIRIT; ÆTHER HYDROCHLORICUS ALCOHOLICUS, SPIRITUS ÆTHERIS MURIATICI, L. *Prep.* 1. From hydrochloric ether and rectified spirit, equal parts, mixed together.

2. (Ph. E. 1744.) Hydrochloric acid, 1 part; rectified spirit, 3 parts; digest some days, and distil in a sand-bath.—*Dose,* ½ to 3 fl. dr.; in dyspepsia, liver complaints, hectic fever, &c.

Spirit of Ju"niper. *Syn.* SPIRITUS JUNIPERI (B. P.), L. *Prep.* English oil of juniper, 1 part; rectified spirit, 49 parts; dissolve.—*Dose,* 30 to 60 minims.

Spirit of Juniper (Compound). *Syn.* SPIRITUS JUNIPERI COMPOSITUS (Ph. L., E., & D.), L. *Prep.* 1. (Ph. L.) Oil of juniper, 1½ fl. dr.; oils of caraway and fennel, of each, 12 drops; proof spirit, 1 gall.; dissolve.

2. (Ph. L. 1836.) Juniper berries, bruised, 15 oz.; caraway and fennel seed, of each, bruised, 2 oz.; proof spirit, 1 gall.; water, 1 quart, or q. s.; distil 1 gall.

Obs. This spirit is stimulant and diuretic.—*Dose,* 2 to 4 fl. dr. Mixed with twice or thrice its weight of proof spirit, and sweetened with a little sugar, it makes no bad substitute for Hollands gin.

Spirit of Lav'ender. *Syn.* SPIRITUS LAVANDULÆ (B. P., Ph. E.), L. *Prep.* 1. From fresh lavender, 2½ lbs.; rectified spirit, 1 gall.; water, 1 quart, or q. s.; distil 1 gall. (7 pints—Ph. E.).

2. (Wholesale.) From Mitcham oil of lavender, 3 oz.; rectified spirit, 1 gall.; dissolve. Cordial and fragrant.

3. (B. P.) English oil of lavender, 1 part; rectified spirit, 49 parts; dissolve.—*Dose,* 30 to 60 minims. See SPIRITS (Perfumed), TINCTURE, &c.

Spirit of Ni'tric Ether. *Syn.* SPIRIT OF NITROUS ETHER, SWEET SPIRIT OF NITRE, NITROUS ETHEREAL SPIRIT, NITRE DROPS; SPIRITUS ÆTHERIS NITRICI (B. P., Ph. L. & E.), SPIRITUS

ÆTHERIS NITROSUS (Ph. D.), L. *Prep.* 1. (Ph. L.) Take of rectified spirit, 1 quart; nitric acid, 3½ fl. oz.; add the acid by degrees to the spirit; then mix them, and let 28 fl. oz. distil over. An earthenware still and condensing worm should be employed. Sp. gr. ·834.

2. (Ph. E.) Pure hyponitrous ether (Ph. E.), 1 part; rectified spirit, 4 parts (both by volume); mix. Sp. gr. ·847.

3. (Ph. D.) Nitrous or hyponitrous ether (which has been washed with half of its volume of liquor of ammonia), 4 fl. oz.; rectified spirit "in 42 fl. oz.; mix, and preserve the compound in small, strong, and accurately stoppered bottles."

4. (B. P.) Nitric acid (sp. gr. 1·42), 3 parts; sulphuric acid, 2 parts; copper, in fine powder (No. 25), 2 parts; rectified spirit, a sufficiency; to 20 parts of the spirit add gradually the sulphuric acid, stirring them together; then add to this, also gradually, 2½ parts of the nitric acid. Put the mixture into a retort or other suitable apparatus, into which the copper has been introduced, and to which a thermometer is fitted. Attach now an efficient condenser, and, applying a gentle heat, let the spirit distil at a temperature commencing at 170° and rising to 175°, but not exceeding 180°, until 12 parts have passed over and been collected in a bottle kept cool, if necessary, with ice-cold water; then withdraw the heat, and, having allowed the contents of the retort to cool, introduce the remaining half of nitric acid, and resume the distillation as before, until the product has been further increased to 14 parts. Mix this with 40 parts of the rectified spirit, or as much as will make the product correspond to the tests of specific gravity and percentage of ether indicated in the B. P. Preserve it in well-closed vessels.

Char. and Tests. Transparent and nearly colourless, with a very light tinge of yellow, mobile, inflammable, of a peculiar penetrating apple-like odour, and sweetish, cooling, sharp taste. It effervesces feebly, or not at all, when shaken with a little bicarbonate of soda. When agitated with solution of sulphate of iron and a few drops of sulphuric acid it becomes deep olive-brown or black. Sp. gr. ·840 to ·845.—*Dose*, 1 to 2 fl. dr.

Pur., &c. Pure spirit of nitric ether boils at about 160° F., scarcely reddens litmus paper, and "gives off no bubbles of carbonic acid gas on the addition of carbonate of soda" (Ph. L.). "When agitated with twice its volume of concentrated solution of chloride of calcium 12% of ether slowly separates" (Ph. E.).—*Dose*, ½ to 3 fl. dr., as a febrifuge, a diaphoretic, diuretic, antispasmodic, &c.; in various affections.

Obs. The mass of the sweet spirits of nitre of the shops is of very inferior quality, and is scarcely, if ever, made directly from spirit that has paid the duty. One, and a very large portion is obtained from Scotland; another from the manufacturers of fulminating mercury; and a third—and, in fact, the principal part—from certain persons in the neighbourhood of the metropolis, who employ contraband spirit for its preparation, as this article is not under the excise. Recently methylated spirit has been employed for the purpose.

Sweet spirits of nitre, sp. gr. ·850, is now commonly and publicly sold in quantity at a price which is only about 2-3rds that of the spirit in it if the latter had paid duty. The spirit obtained from the manufacturers of fulminating mercury frequently contains no inconsiderable quantity of hydrocyanic acid.

The mere admixture of nitric or hyponitrous ether with alcohol does not afford an officinal SPIR. ÆTHER. NITR., as this always contains aldehyde, which, according to Prof. Liebig, is an essential constituent of the officinal compound.

Spirit of Nitrous Ether. *Syn.* SPIRITUS ÆTHERIS NITROSI (B. P.). See SPIRIT OF NITRIC ETHER.

Spirit of Nut'meg. *Syn.* SPIRITUS MYRISTICÆ (B. P., Ph. L. and E.), S. NUCIS MOSCHATÆ, L. *Prep.* 1. (Ph. L. and E.) Bruised nutmegs, 2½ oz.; proof spirit, 1 gall.; water, 1 pint, or q. s.; distil a gallon. Cordial and carminative. —*Dose*, 1 to 4 fl. dr.; chiefly used to flavour mixtures and draughts.

2. ESSENTIA MYRISTICÆ MOSCHATÆ (Ph. D.). Oil of nutmegs, 1 fl. oz.; rectified spirit, 9 fl. oz. Used in dispensing.

3 (B. P.) Volatile oil of nutmeg, 1 part; rectified spirit, 49 parts; dissolve.—*Dose*, 30 to 60 minims.

Spirit of Pennyroy'al. *Syn.* SPIRITUS PULEGII (Ph. L.), S. MENTHÆ PULEGII, L. *Prep.* 1. (Ph. L.) Oil of pennyroyal, 3 fl. dr.; proof spirit, 1 gall.; dissolve. Stimulant, antispasmodic, and carminative.—*Dose*, ½ to 2 fl. dr.

2. ESSENTIA MENTHÆ PULEGII (Ph. D.). Oil of pennyroyal, 1 fl. oz.; rectified spirit, 9 fl. oz. Used chiefly in dispensing.

Spirit of Pep'permint. *Syn.* SPIRITUS MENTHÆ PIPERITÆ (B. P., Ph. L.), S. MENTHÆ (Ph. E.) L. *Prep.* 1. (Ph. L.) Oil of peppermint, 3 fl. dr.; proof spirit, 1 gall.; dissolve.

2. (Ph. E.) Green peppermint, 1½ lbs.; proof spirit, 7 pints; macerate two days; add of water, q. s., and distil 7 pints.—*Dose*, ½ to 2 dr.

3. ESSENTIA MENTHÆ PIPERITÆ (Ph. D.). Oil of peppermint, 1 fl. oz.; rectified spirit, 9 fl. oz. See ESSENCE OF PEPPERMINT.

4. (B. P.) English oil of peppermint, 1 part; rectified spirit, 49 parts; dissolve.—*Dose*, 30 to 60 minims, or for children under five years, 1 to 3 minims.

Spirit of Pimen'to. *Syn.* SPIRIT OF ALLSPICE; SPIRITUS PIMENTÆ (Ph. L. and E.), L. *Prep.* 1. (Ph. L.) Oil of pimento, 2 fl. dr.; proof spirit, 1 gall.; dissolve.

2. (Ph. E.) From pimento (bruised), ½ lb.; and proof spirit, 7 pints; as SPIRIT OF CARAWAY. Carminative and stomachic.—*Dose*, 1 to 4 fl. dr.; in flatulent colic, dyspepsia, &c.

3. ESSENTIA PIMENTÆ (Ph. D.). Oil of pimento, 1 fl. oz.; rectified spirit, 9 fl. oz. Used to make pimento water, and in dispensing.

Spirit of Pine-tops. *Syn.* SPIRITUS TURIONUM PINI, L. See BALSAM, RIGA.

Spirit of Rose'mary. *Syn.* SPIRITUS ROSMARINI (B. P., Ph. L. and E.), L. *Prep.* 1. (Ph. L.) As SPIRIT OF PIMENTO.

2. (Ph. E.) Rosemary tops, 2½ lbs.; rectified spirit, 1 gall.; as SPIRIT OF LAVENDER. Fragrant and stimulant.

3. ESSENTIA ROSMARINI (Ph. D.). As ESSENCE OF PIMENTO.

4. (B. P.) Oil of rosemary, 1 part; rectified spirit, 49 parts; dissolve.—*Dose*, 10 to 30 minims.

Spirit of Spear'mint. *Syn.* SPIRITUS MENTHÆ VIRIDIS (Ph. L.), S. MENTHÆ SATIVÆ, L. *Prep.* 1. (Ph. L.) As SPIRIT OF PEPPERMINT (Ph. L.).

2. ESSENTIA MENTHÆ VIRIDIS (Ph. D.). As ESSENCE OF PEPPERMINT (Ph. D.). The uses and doses are also the same.

Spirit of Sulphu'ric E'ther. See SPIRIT OF ETHER (*above*).

Spirit of Vitriol (Sweet). See AROMATIC SPIRIT OF ETHER (*above*).

Spirit, Vul'nerary. *Syn.* VULNERARY WATER, ARQUEBUSADE; SPIRITUS VULNERARIUS, L.; EAU D'ARQUEBUSADE, Fr. *Prep.* 1. Dried tops of sage, wormwood, fennel, hyssop, marjoram, savory, thyme, rosemary, calamint, balm, peppermint, and scordium, fresh leaves of angelica and basil, and lavender flowers, of each, 4 oz.; proof spirit, 2 galls.; digest for fourteen days, and distil over 1½ galls.

2. Rosemary leaves, 1½ lbs.; leaves of thyme and summits of milfoil, of each, ½ lb.; juniper berries, 8 oz.; proof spirit, 2 galls.; distil over 5 quarts.

Obs. This preparation is stimulant and vulnerary, and is in great repute on the Continent as a cosmetic and cordial.

SPIRITS (Perfumed). *Syn.* SPIRITUS ODORIFERI, ODORES SPIRITUOSI, L. The odoriferous spirits of the perfumer are, for the most part, prepared from various aromatic and odorous substances, by a similar process to that described under ESSENCES and SPIRITS (Medicinal); but in this case a perfectly pure, flavourless, and scentless spirit must be employed. The distillation should also be preferably conducted by steam, or the heat of a water-bath, and the distilled spirit should be kept for some time in a cellar, or other cold situation, previously to being used. When simple solution of an essential oil in the spirit is adopted, care should be taken that the oil is pale and new; or, at least, has not been much exposed to the air; as in that case it would contain resin, which would make the perfumed spirit, or essence, liable to stain delicate articles of clothing to which it may be applied. Most of the 'eaux' and 'esprits' of the perfumers are prepared by one or other of the above methods. It is found, however, that the perfumed spirits of some of the more delicate flowers cannot be well obtained by either infusion or distillation, or by the simple solution of their essential oils in spirit; or, at least, they are not usually so prepared by the foreign perfumers. The spirits of orange flowers, jasmine, tuberose, jonquil, roses, and of some other flowers, and of cassia, vanilla, &c., are commonly prepared by digesting pure rectified spirit for three or four days on half its weight of the respective pommades or oils, obtained by infusion or contact. The operation is performed in a closed vessel placed in a water-bath, and frequent agitation is employed for three or four days, when the perfumed spirit is decanted into a second digester,

containing a like quantity of oil to the first. The whole process is repeated a second and a third time, after which the spirit is allowed to settle and is then decanted. It now forms the most fragrant and perfect odoriferous spirit (extrait) of the Continental perfumer. The product is called 'esprit' or 'extrait of the first infusion.' The three portions of oil are then treated again with fresh spirit in the same manner, and thus spirits or essences of inferior quality are obtained, which are distinguished by the perfumers as Nos. 2, 3, 4, &c., or 'esprits' or 'extraits of the first, second, third,' &c., operation or infusion. In some, though only a very few cases, the spirits are afterwards distilled.

The strength of the spirit for the concentrated essences should not be less than 56 o. p. (sp. gr. ·8376); that for eaux, esprits, and extraits, not less than 35 o. p. (sp. gr. ·8723). The strength of the second quality of the last three must be fully proof (sp. gr. ·920). See ALCOHOL, DISTILLATION, ESSENCE, OILS, POMMADE, &c., and below.

Eau d'Ambre Royale. [Fr.] From essences of ambergris and musk, of each, 1 fl. oz.; spirit of ambrette and orange-flower water, of each, 1 pint; rectified spirit, 1 quart; mix.

Eau d'Ange. [Fr.] From flowering tops of myrtle (bruised), 1½ lbs.; rectified spirit, 7 pints; water, 3 pints; digest a week, add of common salt, 2 lbs., and distil 1 gall.

Eau d'Arquebusade. [Fr.] See VULNERARY SPIRIT (*above*).

Eau de Bouquet. [Fr.] From spirits of rosemary and essence of violets, of each, 1 fl. oz.; essences of bergamot and jasmine, of each, 1 fl. dr.; oils of verbena and lavender, of each, ½ fl. dr.; orange-flower water, 1 fl. oz.; eau de rose, ½ pint; rectified spirit, 1 quart; mix.

Eau de Bouquet de Flore. [Fr.] From spirits of rosemary and roses and essence of violets, of each, ½ fl. oz.; oil of cedrat and essence of ambergris, of each, 1 fl. dr.; orange-flower water, 5 fl. oz.; rectified spirit, 1 pint.

Eau des Carmes. [Fr.] See SPIRIT OF BALM (COMPOUND).

Eau de Cologne. [Fr.] *Syn.* COLOGNE WATER; AQUA COLONIENSIS, A. C. SPIRITUOSA, SPIRITUS COLONIENSIS, L. For the production of good eau de Cologne it is absolutely essential that the spirit be of the purest description, both tasteless and scentless, and that the oils be not only genuine, but recently distilled, as old oils are less odorous, and contain a considerable quantity of resin and camphor, which prove injurious. When flowers and the flowering tops of plants are ordered, it is also necessary that they be either fresh gathered or well preserved, without drying them. To produce an article of the finest quality, distillation should be had recourse to. A very excellent eau de Cologne may, however, be produced by simple solution of the oils or essences in the spirit, provided they be new, pale-coloured, and pure. The mass of the eau de Cologne prepared in England, some of which possesses the most delicate fragrance, and is nearly equal to the best imported, is made without distillation. In the shops two kinds of this article are generally kept—French and German. That prepared by Farina of Cologne is esteemed the best, and is preferred in the fashionable world.

Prep. 1. From essences of bergamot and lemon, of each, 1 fl. dr.; oil of orange, ½ dr.; oil of neroli, 20 drops; oil of rosemary, 10 drops; essence of ambergris and musk, of each, 1 drop; rectified spirit, ½ pint; mix.

2. Essence of bergamot, 3 fl. oz.; essence of lemon, 3 fl. dr.; essence of cedrat, 2 fl. dr.; oils of neroli and rosemary, of each, 1½ fl. dr.; oil of balm, ½ fl. dr.; rectified spirit, 1½ galls.; mix.

3. (*Cadet Gassincourt.*) Take of pure neroli, essences (oils) of cedrat, orange, lemon, bergamot, and rosemary, of each, 24 drops; lesser cardamom seeds, ¼ oz.; spirit at 32° Baumé (sp. gr. ·869), 1 quart; digest a few days and then distil 1½ pints.

4. (*Farina.*) Take of rectified spirit, 5 galls.; calamus aromaticus, sage, and thyme, of each, ½ dr.; balm-mint and spearmint, of each, 1 oz.; angelica root, 10 gr.; camphor, 15 gr.; petals of roses and violets, of each, 3 dr.; lavender flowers, 1½ dr.; orange flowers, 1 dr.; wormwood, nutmeg, cloves, cassia lignea, and mace, of each, 20 gr.; oranges and lemons, sliced, of each, 2 in number; bruise or slice the solids, macerate with agitation for 48 hours, then distil off 2-3rds, and add to the product—essences of lemon, cedrat, balm-mint, and lavender, of each, 1 fl. dr.; pure neroli and essence of the seeds of anthos, of each, 20 drops; essences of jasmine and bergamot, of each, 1 fl. oz.; mix well and filter, if necessary.

5. (P. Cod.) Oils of bergamot, lemon, and cedrat, of each, 3 oz.; oils of rosemary, lavender, and neroli, of each, 1½ oz.; oil of cinnamon, ¾ oz.; spirit of rosemary, 1 quart; compound spirit of balm (eau de melisse des Carmes), 3 pints; rectified spirit, 3 galls.; digest for 8 days, then distil 3 galls.

6. (*Dr A. T. Thomson.*) Oils of bergamot, orange, and rosemary, of each, 1 fl. dr.; cardamom seeds, 1 dr.; rectified spirit and orange-flower water, of each, 1 pint; mix, digest for a day, and then distil a pint.

7. (*Trommsdorff.*) Oils of neroli, citron, bergamot, orange, and rosemary, of each, 12 drops; Malabar cardamoms, bruised, 1 dr.; rectified spirit of wine, 1 quart; mix, and, after standing 2 or 3 days, distil a quart.

Obs. Eau de Cologne is principally used as a perfume, but a very large quantity is consumed by fashionable ladies as a cordial and stimulant. For this purpose it is dulcified with sugar. A piece of linen dipped in Cologne water, and laid across the forehead, is a fashionable remedy for headache.

Eau d'Élégance. [Fr.] From spirit of jessamine, 1 pint; rectified spirit and spirits of hyacinth and storax, of each, ½ pint; tinctures of star-anise and tolu, of each, 2 fl. oz.; tincture of vanilla, 1 fl. oz.; essence of ambergris, ½ dr.; mix, and in a week decant the clear portion.

Eau de Framboises. [Fr.] *Prep.* From strawberries (bruised), 16 lbs.; rectified spirit, 1 gall.; digest, and distil to dryness in a salt water or steam bath.

Eau d'Héliotrope. [Fr.] *Prep.* From essence of ambergris, ½ fl. dr.; vanilla, ½ oz.; orange-flower water, ½ pint; rectified spirit, 1 quart; digest a week, and filter.

Eau d'Hongrie. [Fr.] *Syn.* HUNGARY WATER; AQUA HUNGARICA, SPIRITUS ROSMARINI COMPOSITUS, L.; EAU DE LA REINE D'HONGRIE, Fr. A fragrant stimulant and cosmetic. Sweetened with sugar it is also used as a liqueur.

Prep. 1. Rosemary tops (in blossom), 4 lbs.; fresh sage, ½ lb.; bruised ginger, 2 oz.; rectified spirit, 1½ galls.; water, ¼ gall.; macerate for 10 days, add of common salt, 3 lbs., and then distil 11 pints.

2. From oil of rosemary (genuine), 1½ fl. dr.; oil of lavender, ½ dr.; orange-flower water, ½ pint; rectified spirit, 1½ pints; mix. SPIRIT OF ROSEMARY (see *above*) is now commonly sold for it.

Eau d'Ispahan. [Fr.] *Prep.* From oil of the bitter orange, 2 fl. oz.; oil of rosemary, 2 dr.; oils of cloves and neroli, of each, 1 fl. dr.; oil of spearmint, ½ fl. dr.; eau de rose, 1 pint; rectified spirit, 7 pints; mix. It is better for distillation. Used as eau de Cologne.

Eau de Jasmin. [Fr.] See ESPRIT DE JASMIN ODORANTE (*below*).

Eau de Lavande. [Fr.] *Syn.* LAVENDER WATER, DOUBLE DISTILLED L. W.; AQUA LAVANDULÆ, A. L. ODORIFERA, SPIRITUS L., L. *Prep.* 1. From the flowering tops of lavender (freshly and carefully picked), 7 lbs.; rectified spirit, 2 galls.; macerate for a week, add of water, ½ gall.; (holding in solution) common salt, 3 lbs.; and distil 2 galls.

2. From Mitcham oil of lavender, 8 oz.; essence of musk, 4 oz.; essence of ambergris and oil of bergamot, of each, 1½ oz.; rectified spirit, 2 galls.; mix well. Very fine.

3. (*Brande.*) Oil of lavender, 20 oz.; oil of bergamot, 5 oz.; essence of ambergris (finest), ½ oz.; rectified spirit, 5 galls.; mix.

Obs. The products of the last two formulæ are better for distillation; but in that case the essences of ambergris and musk should be added to the distilled spirit. The oils should be of the best quality and newly distilled, and the spirit should be perfectly scentless.

It may be useful to observe here that the common lavender water, double distilled lavender water, or spirit of lavender of the shops, is made with spirit at proof, or even weaker; hence its inferior quality to that of the more celebrated perfumers. 1 oz. of true English oil of lavender is all that will properly combine with 1 gall. of proof spirit without rendering it muddy or cloudy.

Eau de lavande is a most agreeable and fashionable perfume. The article produced by the second formula has received the commendation of Her Majesty and many of the nobility.

Eau de Lavande de Millefleurs. *Prep.* To each quart of the ordinary eau de lavande (No. 2 or 3) add of oil of cloves, 1½ fl. dr.; essence of ambergris, ½ fl. dr.

Eau de Lavande (Ammoniacal). *Prep.* 1. To lavender water, 1 pint, add of liquor of ammonia, ½ fl. oz.

2. (P. Cod.) English oil of lavender, 1 oz.; spirit of ammonia, 2 lbs.; dissolve. Used as a stimulating scent in fainting. See PERFUMES (Ammoniated).

Eau de Luce. [Fr.] See TINCTURE OF AMMONIA, COMPOUND.

Eau de Maréchale. [Fr.] *Syn.* EXTRAIT DE MARÉCHALE, Fr. *Prep.* 1. From ambergris and grain musk, of each, 20 gr.; oils of bergamot, lavender, and cloves, of each, 1 oz.; oils of sassafras and origanum, of each, ½ fl. dr.; rectified spirit, 2 quarts; macerate with agitation for a week.

2. Rectified spirit, 1 pint; essence of violets, 1 oz.; essences of bergamot and œillets, of each, ½ oz.; orange-flower water, ½ pint; mix.

Eau de Mélisse. [Fr.] See SPIRIT OF BALM, COMPOUND.

Eau de Miel. [Fr.] *Syn.* HONEY WATER, SWEET-SCENTED H. W.; AQUA MELLIS, A. M. ODORIFERA, L. *Prep.* 1. Take of spirit of roses (No. 3—see *above*), 2 quarts; spirit of jasmine and rectified spirit, of each, 1 quart; essence of Portugal, 1 fl. oz.; essences of vanilla and musk, of each (No. 3), 4 fl. oz.; flowers of benzoin, 1½ dr.; mix, agitate, and add of eau de fleurs d'oranges, 1 quart. Delightfully fragrant.

2. Honey (finest), ½ lb.; essence of bergamot, ½ oz.; essence of lemon, ¼ oz.; oil of cloves, 12 drops; musk, 12 gr.; ambergris, 6 gr.; orange-flower and rose water, of each, 1 quart; rectified spirit, 1 gall.; macerate for 14 days, with frequent agitation, and filter.

Obs. The last is often coloured with 20 or 30 gr. of saffron, and made into a ratafia with sugar. HONEY WATER FOR THE HAIR is a different article from the above. It is obtained by the dry distillation of honey, mixed with an equal weight of clean sand, a gentle heat only being employed. The product is yellowish and acidulous, from the presence of acetic acid. This last is used to promote the growth of the hair.

Eau de Millefleurs. [Fr.] *Syn.* EXTRAIT DE MILLEFLEURS, Fr. *Prep.* 1. From grain musk, 12 gr.; ambergris, 20 gr.; essence of lemon, 1½ oz.; oils of cloves and lavender (English), of each, 1 oz.; neroli and oil of verbena, of each, ½ dr.; rectified spirit, 2 quarts; macerate in a closed vessel and a warm situation for a fortnight.

2. Balsam of Peru (genuine) and essence of cloves, of each, 1 oz.; essences of bergamot and musk, of each, 2 oz.; essences of neroli and thyme, of each, ¼ oz.; eau de fleurs d'oranges, 1 quart; rectified spirit, 9 pints; mix well. Very fine.

3. Essence of bergamot, ½ oz.; eau de lavande and essence of jasmine, of each, 1 oz.; orange-flower water, 8 fl. oz.; rectified spirit, 1 pint; mix.

Eau de Mousselline. [Fr.] From eau de fleurs d'oranges and spirit of clove gilliflower, of each, 1 quart; spirit of roses (No. 3—see *above*), spirit of jasmine (No. 4), and spirit of orange flowers (No. 4), of each, 2 quarts; essences of vanilla and musk, of each (No. 3), 2 fl. oz.; sanders-wood, ½ oz. Very fine.

Eau de Naphe. [Fr.] See WATERS (Perfumed).

Eau sans Pareille. [Fr.] 1. From essence of bergamot, 5 dr.; essence of lemon, 8 dr.; essence of citron, 4 dr.; Hungary water, 1 pint; rectified spirit, 6 quarts; mix and distil.

2. Grain musk, 20 gr.; ambergris, 25 gr.; oils of lavender and cloves, of each, 1 oz.; essence of bergamot, ½ oz.; oils of sassafras and origanum, of each, 20 drops; rectified spirit, 1 gall.; macerate for 14 days.

Eau, Romain. [Fr.] From essence of ambergris, 1 fl. oz.; tincture of benzoin, 4 fl. oz.; spirit of tuberose, ½ pint; spirit of acacia flowers and tincture of vanilla, of each, 1 pint; spirit of jasmine, 3 pints; mix.

Eau de Rosières. [Fr.] From spirit of roses, 1 pint; spirits of cucumber, angelica root, and celery seeds, of each, ½ pint; spirits of jasmine and orange flowers, of each, ¼ pint; tincture of benzoin, 2 fl. oz.; mix.

Eau de Violette. [Fr.] See ESPRIT DE VIOLETTES (*below*).

Esprit d'Ambrette. [Fr.] See ESSENCE.

Esprit de Bergamotte. [Fr.] From essence (oil) of bergamot (best), 5 oz.; essence of ambergris (pale), 2 fl. oz.; essence of musk, ½ fl. oz.; oil of verbena, 2 fl. dr.; rectified spirit, 1 gall.; mix.

Esprit de Bouquet. [Fr.] From Mitcham oil of lavender, 1 oz.; oils of cloves and bergamot, of each, 3 fl. dr.; essence of musk, 1 fl. dr.; otto of roses, 10 drops; rectified spirit, 1 quart.

Esprit de Fleurs. [Fr.] See SPIRIT OF THE FLOWERS OF ITALY (*below*).

Esprit de Jasmin. [Fr.] *Syn.* EAU DE JASMIN, Fr.

Esprit de Jasmin Odorante. [Fr.] From spirit of jasmine and rectified spirit, of each, 1 pint; essence of ambergris, 1 fl. dr.

Esprit de Jonquille. [Fr.]

Esprit de la Reine. [Fr.] From oil of bergamot, 1 fl. oz.; essence of ambergris, 2 fl. dr.; otto of roses, 1 fl. dr.; rectified spirit, 1 quart.

Esprit de Rondeletia. [Fr.] *Syn.* EXTRAIT DE RONDELETIA, Fr. From Mitcham oil of lavender, 3 oz.; oil of cloves, 1½ oz.; oil of bergamot, 1 oz.; essences of musk and ambergris, of each, 2 fl. dr.; rectified spirit, 3 pints.

Esprit de Rose. [Fr.] 1. From spirit of roses, 1 pint; essence of ambergris and oil of rose geranium, of each, ½ fl. dr.

2. From otto of roses, 2 dr.; neroli, ½ dr.; rectified spirit, 1 gall.; dissolve, add of chloride of calcium (well dried and in powder), 1½ lbs.; agitate well, and distil 7 pints. Very fine.

Esprit de Suave. [Fr.] From the essences of cloves and bergamot, of each, 1½ fl. dr.; neroli, ½ fl. dr.; essence of musk, 1 fl. oz.; spirit of tuberose and rectified spirit, of each, 1 pint; spirits of jasmine and cassia, of each, 1 quart; dissolve, then add of eau de rose, 1 pint, and mix well.

Esprit de Taiz. [Fr.] *Syn.* SPIRIT OF LEMON THYME; SPIRITUS THYMI, L. From tops of lemon thyme, 2 lbs.; proof spirit, 1 gall.; distil 7 pints.

Esprit de Violettes. [Fr.] *Syn.* SPIRIT OF VIOLETS, ESSENCE OF V., E. OF ORRIS. From Florentine orris root, reduced to coarse powder, ½ lb.; rectified spirit, 1 pint; by simple maceration for a fortnight. A stronger and finer article (ESSENCE OF VIOLETS) is prepared from orris root, 5 lbs., to rectified spirit, 1 gall.; by percolation.

Extrait de Bouquet. [Fr.] Extract of nose-gay.

Extrait de Maréchale. [Fr.] See EAU DE MARÉCHALE (above).

Extrait de Millefleurs. [Fr.] See EAU DE MILLEFLEURS (above).

Extrait de Rondeletia. [Fr.] See ESPRIT (above).

Odeur, Délectable. [Fr.] From oils of lavender, bergamot, rose geranium, and cloves, of each, 1 fl. dr.; eaux de rose and fleurs d'orange, of each, ¼ pint; rectified spirit, 1½ pints.

Odeur Suave. [Fr.] See ESPRIT (above).

Spirit of Cytherea. From the spirits of violets, tuberose, clove gillyflower, jasmine (No. 2—see above), roses (No. 2), and Portugal, of each, 1 pint; orange-flower water, 1 quart; mix.

Spirit of the Flowers of Italy. Syn. ESPRIT DE FLEURS, Fr. From the spirits of roses (No. 1—see above), jasmine (No. 2), oranges (No. 3), and cassia (No. 2), of each, 4 pints; orange-flower water, 3 pints; mix.

Victoria Perfume. See ESPRIT DE LA REINE (above).

SPIRONE. English proprietary article for the cure of consumption. Contains chloroform, glycerine, iodide of potassium, and an odorous substance which has not yet been defined. Price is said to be £3 for 2½ dr.

SPITTING OF BLOOD. See HÆMOPTYSIS.

SPLINT. This is the common name given to an enlargement of the bone in horses; which generally occurs below the knee, between the large and small splint-bones, usually on the inside of the limb. It mostly results from fast driving or riding, or from the animal having been much worked while young, or made to unduly traverse hard or paved roads. The splint is a frequent cause of lameness if it develops just under the knee, since it interferes with and circumscribes the free movement of the joint. It is very essential to have recourse to prompt measures directly this affection shows itself.

The treatment usually prescribed is the constant application to the part of cold water if the splint be accompanied by much tenderness or inflammation. This may be accomplished by bandages soaked in cold water, taking care to renew the cold water as soon as it becomes warm. Mr Finlay Dun advises the horse, where practicable, to stand for an hour several times a day up to the knees in a stream or pool of water. In addition he prescribes rest for ten days or a fortnight, and when the heat and tenderness have been subdued the application of a blister, or of biniodide of mercury ointment, or the hot iron.

SPONGE. Syn. SPONGIA, S. OFFICINALIS, L. Sponge is a cellular fibrous structure, produced by marine animals of the humblest type, belonging to the sub-kingdom Protozoa. The finest quality is imported from Smyrna, and is known as TURKEY SPONGE; another, called WEST INDIAN or BAHAMA SPONGE, is much less esteemed, being coarse, dark-coloured, and very rotten.

Sponge, as collected, and also as generally imported, contains many impurities, more especially sand, most of which may be removed by beating it, and by washing it in water. Amusing disputes often arise between the smaller importers and the wholesale purchasers on this subject—the privilege of beating it before weighing it, the number of minutes so employed, and even the size of the stick, being often made important matters in the 'haggling.'

1. BLEACHED SPONGE (WHITE SPONGE; SPONGIA DEALBATA) is prepared by soaking ordinary sponge in very dilute hydrochloric acid, to remove calcareous matter, then in cold water, changing it frequently, and squeezing the sponge out each time, and next, in water holding a little sulphuric or sulphurous acid, or, still better, a very little chlorine, in solution; the sponge is, lastly, repeatedly washed and soaked in clean water scented with rose or orange-flower water, and dried.

2. The sponges are first soaked in hydrochloric acid to remove the lime; they are then washed in water, and afterwards placed for ten minutes in a 2% solution of permanganate of potassium. When taken out they have a brown appearance; this is owing to the deposition of manganous oxide, and may be removed by steeping the sponge for about two minutes in a 2% solution of oxalic acid, to which a little sulphuric acid has been added. As soon as the sponges appear white they are well washed out in water to remove the acid. Strongly diluted sulphuric acid may be used instead of oxalic acid.

3. Sponges can be bleached by first soaking them in hydrochloric acid, diluted with 1½ parts of water, until no more carbonic acid is given off; then wash in pure water, and afterwards leave in a bath composed of 2 lbs. of hyposulphite of soda, 12 lbs. of water, and 2 lbs. of hydrochloric acid. If the sponge be afterwards dipped in glycerin and well pressed, to remove excess of liquid, it remains elastic, and can be used for mattresses, cushions, and general upholstery. Sponge mattresses prepared in this way are now finding great favour. It is, of course, not necessary to bleach the sponge where it is intended to be used for such purposes ('Pharmacist').

BURNT SPONGE (SPONGIA USTA—Ph. D.) is prepared by heating the cuttings and unsaleable pieces in a closed iron crucible until they become black and friable, avoiding too much heat, and allowing the whole to cool before exposing it to the air. It was formerly in great repute in bronchocele and scrofulous complaints.—Dose, 1 to 3 dr., in water, or made into an electuary or lozenge. When good, burnt sponge evolves violet fumes of iodine on being heated in a flask along with sulphuric acid.

COMPRESSED or WAXED SPONGE (SPONGIA CERATA, S. COMPRESSA) is sponge which has been dipped into melted wax and then compressed between two iron plates until cold. When cut into pieces it forms 'SPONGE TENTS,' which are used by surgeons to dilate wounds.

The following notes on sponge are abstracted from a paper by H. B. Marks:

The sponge is an animal belonging to the Porifera class, and is formed of organic matter around a horny skeleton. When first taken out of the sea sponges are covered with a bluish-black skin, and discharge a very offensive, thick, milky fluid. To purify them pressure is first resorted to. They are then scraped with a knife to rid them of the skin, and afterwards thoroughly washed in the sea, when only the skeleton remains. If the cleansing process be carefully done the sponge in its skeleton state becomes elastic, and proves pleasant in use; but should the process be only partially attended to, no amount of after-cleansing will avail, and the sponge will always remain more or less sticky and disagreeable.

Many experiments have been made on the growing sponges by artificial means; but although from a scientific point of view success has been achieved, no result of commercial value has yet been obtained. Pieces cut from the living sponge and replanted, it has been found, will continue to grow, and two pieces cut from the same species of sponge will unite if placed together; but parts from different species fail to unite, however closely they may be fixed together. The four important sponge-fishing grounds of the world are the Mediterranean Sea, Florida, the West Indies and Bahamas, and Cuba. At the end of the first year's growth sponges attain the size of a small lemon, at the end of the second that of a large orange, while at the end of the third year they are twice or three times the last-mentioned size. Beyond this no positive information has been obtained, but it is thought that they are very slow of growth, and that the very large species are probably very ancient.

The introduction of diving apparatus, and the largely increased demand for the article, has caused much deterioration in the Mediterranean sponges, as, by the improved apparatus enabling the diver to stay a considerable time under the water, every year large and small sponges are gathered indiscriminately, and before the latter have time to grow. To remedy this, efforts have been made to bring about laws to prevent the fishing being carried on in the same spot, except after a lapse of three years. There are four methods of sponge-fishing in vogue at the present time. First, by means of the native naked diver; second, by the diving apparatus; third, by net-fishing; and fourth, by the harpoon. After describing at length the different modes, and tracing the extent of the various fishing-grounds, the lecturer went on to speak of sponge-buying. It was, he said, no easy task, and many years of experience had proved that the oldest buyers were very often seriously at fault in their speculations. A buyer, to be successful, must be well acquainted with the large number of classes and their variations in value.

The bleaching of sponges is a process which requires great attention, the success of the operation depending to a large extent on the care bestowed. The general method is by steeping the sponges successively in preparations of acid and permanganate of potash.

From the statistics which had been gathered the Mediterranean sponge fishery showed the largest yield, with an annual production amounting to £250,000, and employing from 4500 to 5000 men. Florida stood second with from £50,000 to £70,000, bringing employment to 1200 men. The Bahamas showed an output worth £60,000, employing 4500 men; and in Cuba the employment of 700 men produced a yield of £60,000. In round numbers the total amount reached £500,000 worth for the whole world.

Spongia Decolorata. Sys. DECOLOURISED SPONGE, BLEACHED SPONGE. Prep. Sponge, permanganate of potassium, hyposulphite of sodium, hydrochloric acid, and water, of each, a sufficient quantity. Free the sponge from sand and any other obvious impurities or damaged portions by beating, washing, and trimming; then soak it for about fifteen minutes in a sufficient quantity of solution of permanganate of potassium, containing 120 gr. to the pint, wringing the sponge out occasionally and replacing it in the liquid. Then remove it and wash it with water until the latter runs off colourless. Wring out the water, and then place the sponge into a solution of hyposulphite of sodium containing 1 troy oz. to the pint. Next add, for every pint of the last-named solution used, 1 fl. oz. of hydrochloric acid diluted with 4 fl. oz. of water. Macerate the sponge in the liquid for about fifteen minutes, expressing it frequently and replacing it in the liquid. Then remove it, wash it thoroughly with water, and dry it. In the case of large and dark-coloured sponges this treatment may be repeated until the colour has been removed as far as possible.

Sponge, To Clean a. There is nothing more pleasant for washing the skin than a fresh good sponge, or the reverse when not kept thoroughly clean. Without the greatest care a sponge is apt to get slimy long before it is worn out. It may be made almost as good as—in fact, often better than new, by the following process:—Take about 2 or 3 oz. of carbonate of soda, or of potash; dissolve in 2½ pints of water; soak the sponge in it for twenty-four hours, then wash and rinse it in pure water. Then put it for some hours in a mixture, 1 glassful of muriatic acid to 3 pints of water; finally, rinse in cold water, and dry thoroughly. A sponge should always be dried, if possible, in the sun every time it has been used.

SPONGES employed in Washing Wounds, Purification of. M. Leriche advises the sponge to be first saturated with a solution of 4 parts of permanganate of potassium in 100 parts of water; then passed through a solution of sulphurous acid, and finally washed thoroughly with water. The sponges are said to become perfectly disinfected and deodorised, whilst the tissue is not affected by the treatment.

SPOROKTON. See SULPHUROUS ANHYDRIDE.

SPOTS and STAINS. 1. OIL and GREASE SPOTS on boards, marble, &c., when recent, may be removed by covering them with a paste made of fuller's-earth and hot water, and the next day, when the mixture has become perfectly dry, scouring it off with hot soap and water. For old spots, a mixture of fuller's-earth and soft soap, or a paste made of fresh-slaked lime and pearlash, will be better; observing not to touch the last with the fingers.

2. RECENT SPOTS of OIL, GREASE, or WAX, on woollen cloth or silk, may be removed with a little clean oil of turpentine or benzol; or with a little fuller's-earth or scraped French chalk, made into a paste with water, and allowed to dry on them.

They may also be generally removed by means of a rather hot flat-iron and blotting-paper or spongy brown paper, more especially if the cloth, or one of the pieces of paper, be first slightly damped. OLD OIL and GREASE SPOTS require to be treated with ox-gall or yolk of egg, made into a paste with fuller's-earth or soap. PAINT SPOTS, when recent, generally yield to the last treatment. Old ones, however, are more obstinate, and require some fuller's-earth and soft soap made into a paste with either ox-gall or spirit of turpentine.

The 'American Chemist' gives the following method for extracting grease spots from books or paper :—Gently warm the greased or spotted part of the book or paper, and then press upon it pieces of blotting-paper one after another, so as to absorb as much of the grease as possible. Have ready some fine, clear, essential oil of turpentine heated almost to a boiling state (this operation ought to be very carefully accomplished, as the turpentine is a highly inflammable body); warm the greased leaf a little, and then with a soft, clean brush, wet with the heated turpentine both sides of the spotted part. By repeating this application the grease will be extracted. Lastly, with another brush dipped in rectified spirits of wine go over the place, and the grease will no longer appear, neither will the paper be discoloured.

FRUIT and WINE STAINS, on linen, yield easily to hot soap and water. If not, they must be treated as those below.

INK SPOTS and RECENT IRONMOULDS on washable fabrics may be removed by dropping on the part a little melted tallow from a common candle before washing the articles ; or by the application of a little lemon juice, or of a little powdered cream of tartar made into a paste with hot water. Old ink spots and ironmoulds will be found to yield almost immediately to a very little powdered oxalic acid, which must be well rubbed upon the spot previously moistened with boiling water, and kept hot over a basin filled with the same.

Boettger recommends the use of pyrophosphate of soda for the removal of ink stains from coloured woven tissues, to be applied in the form of a concentrated solution. The recent ink stains are readily removed, but older stains require washing and rubbing with the solution for a long time.

STAINS arising from ALKALIES and ALKALINE LIQUORS, when the colours are not destroyed, give way before the application of a little lemon juice; whilst those arising from the weaker acids and acidulous liquids yield to the fumes of ammonia, or the application of a little spirit of hartshorn or sal volatile.

STAINS of MARKING INK may be removed by soaking the part in a solution of chloride of lime, and afterwards rinsing it in a little solution of ammonia or of hyposulphate of soda; or they may be rubbed with the tincture of iodine, and then rinsed as before.

NITRIC ACID STAINS, TO REMOVE. The yellow stain left by nitric acid can be removed either from the skin or from brown or black woollen garments by moistening the spots for a while with permanganate of potash, and rinsing with water. A brownish stain of manganese remains, which may be removed from the skin by washing with

aqueous solution of sulphurous acid. If the spots are old they cannot be entirely removed. See BALLS, CLOTHES, HANDS, SCOURING, STAINS, &c.

SPRAIN. *Syn.* SUBLUXATIO, L. An injury of a joint, in which it has been strained or twisted in an unnatural manner, without actual dislocation. Pain, swelling, and inflammation are the common consequences, which must be combated by repose with refrigerant lotions, or warm fomentations, according to circumstances. Where there is simple stiffness and weakness, exercise is often serviceable.

Treatment for the Horse and other Animals. Foment. Apply lead lotion and refrigerants.

IF FOR CURB use counter-irritants, or red iodide of mercury ointment, or the firing iron; and if for a horse a high-heeled shoe.

SPRAT. The *Clupea sprattus*, Linn., a small fish of the herring family, abounding on our coasts. Gutted, coloured, and pickled, it is sold for anchovies, or as British anchovies, and much used to make the sauce of that name. Sprats contain about 6% of fat.

SPRENGEL'S PUMP. See AIR-PUMP.

SPRINKLES. See BOOKBINDING.

SPRUCE. See BEER, ESSENCE, and POWDERS.

SPUNK. See AMADOU.

SQUILL. *Syn.* SCILLA (B. P., Ph. L., E., & D.), L. The bulb of '*Urginea scilla*,' sliced and dried. In small doses, squill acts as a stimulating expectorant and diuretic ; in larger ones, as an emetic and purgative. With the first intention it is generally given in substance (powder), in doses of 1 to 3 or 4 gr. ; with the latter, either made into vinegar or oxymel (which *see*). It is an excellent remedy in coughs, &c., after the inflammatory symptoms have subsided.

STAGGERS. There are two varieties of the disease known under this name by which horses are affected, viz. stomach staggers, and grass or sleepy staggers. The first, which occasionally kills the horse in twelve or fifteen hours after the attack, is generally induced by an overladen stomach and improper food. The animal has perhaps partaken largely and rapidly, and after too long a fast, of some diet to which it is unaccustomed, such as vetches, clover, or grass. These undergo decomposition within the stomach and intestines, and give rise to such an evolution of gas as either to set up inflammation of the stomach and intestines, or to lead to their rupture, in which latter case the result is, of course, fatal. The symptoms are a quick and feeble pulse, attempts at vomiting, a staggering gait, whilst very frequently the animal sits on its haunches like a dog. Sleepy staggers, which is a more chronic manifestation of the disease, is most common during the summer and autumn months, and generally occurs amongst horses fed on tough and indigestible food, such as vetches or rye-grass, from which circumstances the complaint has been called 'grass staggers.' Both kinds of the disease require the same treatment.

Mr Finlay Dun prescribes a brisk purge, consisting of 6 dr. of aloes in solution, with a dr. of calomel and 2 oz. of oil of turpentine; also the injection every hour of clysters, consisting of salt, soap, or tobacco smoke, the abdomen being at the same time diligently rubbed and fomented with water nearly boiling. To ward off stupor he recommends the frequent administration of 2 or 3 dr. of carbonate of ammonia, with an ounce or two of spirit of nitrous ether, or of strong whiskey toddy, combined with plenty of ginger. To guard against a return of the attack light and easily digestible food should be administered every four or five hours, and occasional mild purgatives should be given.

Horses are also subject to another form of staggers, called 'mad staggers.' This disease originates, however, in causes wholly dissimilar from those just stated, being the result of phrenitis or inflammation of the brain. The animal is frequently very furious and excited, and seems wholly unable to control itself, throwing itself madly about, and attempting to run down anybody that comes in its way; it is also frequently unable to keep on its legs, and when it falls, plunges and struggles violently.

The treatment recommended is prompt and copious bloodletting, combined with active purges and enemas, with refrigerant lotions to the head.

STAINED GLASS. The art of painting or staining glass resembles enamel painting, the effect being produced by fluxing certain metallic substances, as oxides or chlorides, on its surface, by means of heat applied in a suitable furnace. The operations it embraces are difficult, and require great promptitude and experience to prove successful. The colours or compounds employed are, for the most part, similar to those noticed under ENAMEL and PASTE.

STAINS. Discolorations from foreign matters. Liquid dyes are also frequently termed 'stains.' See SPOTS, &c., and below.

Stains, Blood. Spots of dried blood on wood, linen, &c., however old, are easily recognised by the microscope; but simple stains or marks of blood of a slight character, especially those occurring on iron or steel, are recognised with greater difficulty. To obviate this, H. Zollikofer adopts the following plan:—The spot is removed, by scraping, from the surface of the metal, and the resulting powder is digested in tepid water, when a liquid is obtained which exhibits the following reactions:

1. The liquid is neutralised with acid, and heated to ebullition, when opalisation occurs, or a dirty red coagulum forms.

2. The coagulum is dissolved in hot liquor of potassa; the solution, if blood (hæmatin) be present, is diachromatic, or appears green by transmitted light and red by reflected light.

3. By the addition of concentrated chlorine water, in excess, to either solution, white flocks of albumen and chlorhæmatin separate, which are free from iron, as tested by sulphocyanide of potassium.

Obs. The last two reactions are said to be characteristic. Very old spots must be boiled in water containing a little liquor of potassa. See Dr Taylor's 'Medical Jurisprudence,' and BLOOD.

Fresh blood-stains should be treated with a weak solution of common salt, ½; this will generally remove them effectually.

Stains, Bookbinder's. See LEATHER, MARBLING, &c.

Stains, Confectioner's. These are similar to those noticed under LIQUEUR. Mineral colours, especially mineral blues, greens, and yellows, must on no account be used, as they are nearly all dangerous poisons; nor is there any inducement to use them, since the vegetable substances referred to afford, by proper management, every shade that can be possibly required. These stains are also used for cakes and pastry.

Stains, Liqueur. See LIQUEUR.

Stains, Map. See MAPS, VELVET COLOURS, &c.

Stain Remover (for textile fabrics). 1. Soap bark extract, 1 oz.; borax, 1 oz.; fresh ox-gall, 4 oz.; tallow soap, 15 oz. Mix the borax, extract, and gall together by triturating in a mortar, then incorporate the soap so as to produce a plastic mass, which may be moulded or put up in boxes. 2. Oleic acid, 1 part; borax, 2 parts; fresh ox-gall, 5 parts; tallow soap, 20 parts. Mix the borax and ox-gall, then incorporate the soap, and lastly mix in the oleic acid.

STAMMERING. *Syn.* BLÆSITAS, L. Occasionally this depends on some organic affection, or slight malformation of the parts of the mouth or throat immediately connected with the utterance of vocal sounds; but, much more frequently, it is a habit resulting from carelessness, or acquired from example or imitation. When the latter is the case, it may be generally removed by perseveringly adopting the plan of never speaking without having the chest moderately filled with air, and then only slowly and deliberately. Hasty and rapid speaking must not be attempted until the habit of stammering is completely subdued. Nervous excitement and confusion must be avoided as much as possible, and the general health attended to, as circumstances may direct. This variety of stammering is commonly distinguished by the person being able to sing without hesitation. Stammering depending on elongation of the uvula, and other like causes, may be generally removed by a simple surgical operation.

STANNIC ACID. Peroxide of tin.

STARCH. $C_6H_{10}O_5$. *Syn.* AMYLACEOUS FECULA; AMYLUM, L. One of the most important and widely diffused of the proximate principles of vegetables, being found in greater or less quantity in every plant. The mealy and farinaceous seeds, fruits, roots, and the stem-pith of certain trees consist chiefly of starch in a nearly pure state. Wheat contains about 75% and potatoes about 15% of this substance. From these sources the fecula is obtained by rasping or grinding to pulp the vegetable structure, and washing the mass upon a sieve, by which the torn cellular tissue is retained, whilst the starch passes through with the liquid, and eventually settles down from the latter as a soft, white, insoluble powder, which, after being thoroughly washed with cold water, is dried in the air, or with a very gentle heat.

WHEAT STARCH (AMYLUM, B.P., Ph. L., E., & D.) is commonly prepared by steeping the flour in water for a week or a fortnight, during which time the saccharine portion ferments, and the starch granules become freed, for the most part, from the glutinous matter which envelops them, by the disintegrating and solvent action of the lactic acid generated by the fermentation. The sour liquor is then drawn off, and the feculous residue washed on a sieve; what passes through is allowed to settle, when the liquid is again drawn off, and the starch thoroughly washed from the slimy matter; it is then drained in perforated boxes, cut up into square lumps, placed on porous bricks to absorb the moisture, and, lastly, air- or stove-dried.

In the preparation of starch from potatoes (potato starch) and other like vegetable substances, the roots or tubers, after being washed and peeled, either by hand labour or by machinery, are rasped by a revolving grater, and the pulp washed on hair sieves until freed from feculous matter. Successive portions of the pulp are thus treated until the vessel over which the sieves are placed, or into which the washings run, is sufficiently full. The starch held in suspension in the water having subsided to the bottom, the water is drawn off, and the starch stirred up with fresh water, and again allowed to subside. This operation is repeated several times with fresh water until the starch is rendered sufficiently pure for commercial purposes, when it is washed and dried as before. The waste fibres and the washing waters are used as manure.

The starch manufactory at Hohenziatz treated 1216 tons of potatoes for starch between the 4th October, 1874, and the 6th February, 1875. The waste water, after passing through precipitating vats, &c., for the purpose of collecting all the particles of starch, was conducted into a reservoir and mixed with spring water. This water was conducted over a meadow of 18·5 acres, and then passed to a meadow of 4·95 acres, and from this to the third and last, which contained 6·19 acres. The 29·64 acres received the water from 1064 tons of potatoes, or for each acre 4·38 cwt. of potash, 1·26 cwt. of phosphoric acid, and 1·27 of nitrogen.

The following table shows analyses (1) of potato water; (2) of the same diluted; (3) of water from the first meadow; (4) water from the second meadow; one litre contained—

	1. mg.	2. mg.	3. mg.	4. mg.
Whole solid matter	1857·8	323·8	322·8	262·0
Organic matter .	1134·2	101·8	38·0	78·8
Inorganic matter .	723·8	222·0	348·8	183·2
Potash	212·5	55·0	41·2	8·2
Phosphoric acid .	56·6	5·5	trace	trace
Nitrogen	140·7	12·0	4·0	9·1
Ammonia . . .	37·4	0	0	0
Nitric acid . . .	3·8	trace	trace	trace

The disappearance of ammonia and phosphoric acid in 2 is accounted for by the precipitation of phosphate of magnesium and ammonium on the addition of the spring water.

The harvest in hay before the use of potato water was 19·13 cwt. per acre, and afterwards 31·88 cwt. The composition of the hay is better than before, as will be seen by the following comparative table:

	1.	2.
Moisture	15·00	15·00
Woody matter	22·66	22·88
Mineral matter	7·64	8·69
Soluble in ether	2·00	2·30
Albumen	10·89	15·85
Extractable matter not containing nitrogen	41·81	35·34
	100·00	*100·00

* 'Dingl. Polyt. Journ.,' ccxxv, 394—396 (' Journ. Chem. Soc.').

In the manufacture of starch from rice and Indian corn (rice starch, maize starch) a very dilute solution of caustic soda, containing about 200 gr. of alkali to each gallon of liquid, is employed to facilitate the disintegration and separation of the gluten and other nitrogenised matters. A weak solution of ammonia, or sesquicarbonate of ammonia, is also similarly employed with advantage. The gluten may be recovered by saturating the alkali with dilute sulphuric acid. Such starch does not require boiling, and is less apt than wheat starch to attract moisture from the atmosphere. Most of the so-called 'wheaten starch' of commerce used by laundresses is now prepared from rice.

To whiten the starches made from damaged roots and grains, and the coarser portions of those from sound ones, a little solution of chloride of lime is occasionally added to the water, followed by another water containing a very little dilute sulphuric acid; every trace of the last being afterwards removed by the copious use of pure soft or spring water.

The bluish-white starch used by laundresses is coloured with a mixture of smalts and alum in water, and is regarded as unfit for medicinal purposes.

Prop., &c. Starch is insoluble in cold water, and in alcohol and most other liquids, but it readily forms a gelatinous compound (amidin) with water at about 175° F.; alcohol and most of the astringent salts precipitate it from its solutions; infusion of galls throws down a copious yellowish precipitate, containing tannic acid, which is redissolved by heating the liquid; heat and dilute acids convert it into dextrin and grape-sugar; strong alkaline lyes dissolve it, and ultimately decompose it. Sp. gr. 1·53.

To the naked eye it presents the appearance of a soft, white, and often glistening powder; under the microscope it is seen to be altogether destitute of crystalline structure, but to possess, on the contrary, a kind of organisation, being made of multitudes of little rounded transparent bodies, upon each of which a series of depressed parallel rings, surrounding a central spot or hilum, may be traced. The starch granules from different plants vary both in magnitude and form. Those of potato starch and canna starch (tous-les-mois) are the largest, and those of rice and millet starch the smallest, the dimensions ranging from ⅟₁₀₀ to ⅟₁₀₀₀₀ of an inch. The granules of arrowroot and tous-les-mois are ovoid, those of potato starch both oblong and circular, those of tapioca muller-shaped, and those of wheat starch circular.

Identif. One of the commonest frauds practised upon the profession and the public is the admixture of the cheaper kinds of starch, chiefly potato farina, with arrowroot, and the vending of manufactured for genuine tapioca, sago, and other articles of diet used for invalids and children. These sophistications are most easily detected with a good microscope. Drawings of the principal starches will be found under the substances from which they are obtained, as 'arrowroot,' &c.

The following is an outline of the process followed at Messrs Orlando Jones and Co.'s factory:

Rice is bought in the husk, husked in the rice-mills on the premises, and separated into large and small grains; the large are sold for use as rice, the small, as a matter of economy, converted into starch. It is first ground, then treated with a very dilute solution of caustic soda to dissolve out the gluten. This solution is run off and wasted. Many experiments have been made, but without success, to discover a method of recovering the gluten. After various washings with dilute soda the mixture is run into settling-tubs, and kept gently agitated by revolving stirrers to allow the fibre to settle, while the starch is still in suspension. The starch and water is then syphoned to other tubs, where it is allowed to deposit. The thick pasty mass is dried in centrifugal machines until of sufficient consistency to be made into cubes of 6 or 8 inch side. These are dried to a certain degree in stoves heated by steam, and are then transferred to the hands of girls, who with large knives trim off the slightly discoloured surface. One or two strokes of the knife are enough to slice off the whole of a side. The cubes are then wrapped in glazed paper and stacked in store-rooms maintained at a gentle heat, where they remain for at least three weeks, gently drying. When the parcels are opened after this period, it is found that the starch, without any further treatment, has broken up into the well-known columnar pieces. It is noticeable that the ends of the columns are at the surface, the fissures extending at right angles thereto. The process is not one of crystallisation, properly so called, but is apparently the same that produced the curious structure of the basalt at the Giant's Causeway, and may be seen in action on mud banks drying when the river is low.

Starch Glass. *Prep.* Borax, 10 oz.; starch, 20 oz.; stearic acid, ½ oz.; absolute alcohol, 6 dr. Dissolve the stearic acid in the alcohol and mix with the starch; expose to the air until dry, then add the borax and sift.

Starch, Glazing. *Prep.* Melt 5 parts of stearic acid, add 5 parts of absolute alcohol, and triturate the mixture with 95 parts of wheat starch. Starch prepared from this takes easily a fine polish. The effect is the same as adding a piece of stearin to the starch before the boiling water is poured upon it.

Starch Polish. *Prep.* Lard, 7 oz.; white wax, ½ oz.; glycerin, ¼ oz.; strong solution of ammonia, 1 oz.; citronella oil, 5 drops. Melt the wax and lard together and stir constantly until of a creamy consistence, then add the perfume, and incorporate the glycerin and ammonia previously mixed.

Starch, Iodide of. *Syn.* AMYLI IODIDUM, AMYLI IODATUM, L. *Prep.* (Ph. Castr. Ruthena.) Iodine, 24 gr.; rectified spirit, a few drops; rub them to a powder; then add of starch, 1 oz., and again triturate until the mass assumes a uniform colour. Recommended by Dr A. Buchanan, of Glasgow, as producing the alterative effects of iodine, without the usual irritant action of that medicine.—*Dose.* A teaspoonful, or more, in water-gruel, or any bland liquid, twice or thrice a day.

Starch, Soluble Iodide of. (*Petit.*) *Prep.* Iodine, 12 grms.; starch, 100 grms.; ether, q. s. Dissolve the iodine in the ether, pour the resulting solution over the starch, and triturate until the ether has sufficiently evaporated. Put the product in a porcelain capsule and expose it to the heat of a boiling water bath for half an hour, with occasional stirring. This treatment is sufficient to render it entirely soluble in hot water.

Dr Bellini strongly recommends iodide of starch as a valuable antidote in cases of poisoning by caustic alkalies, alkaline or earthy sulphides, and vegetable alkaloids. The advantages attending its employment, he says, are—that it may be administered in large doses; that it does not possess the irritating properties of free iodine; and that it readily forms harmless compounds with the substances named. To avoid the subsequent decomposition of the latter, he advises its administration to be followed by an emetic. As an antidote to alkaline and earthy sulphides, the author thinks it preferable to all others. In cases of poisoning by ammonia, caustic potash, or soda, it is applicable when acid drinks are not on hand.

STARCHING (Clear). Muslins, &c., are 'clear-starched' or 'got up' by laundresses in the following manner:—Rinse the articles in three waters, dry them, and dip them into thick-made starch, which has been previously strained through a piece of muslin; squeeze them, shake them gently, and again hang them up to dry; when they are dry, dip them twice or thrice into clear water, squeeze them, spread them on a linen cloth, roll them up in it, and let them lie an hour before ironing them. Some persons put a morsel of sugar into the starch, to prevent its sticking whilst ironing, and others stir the starch with a candle to effect the same end; both these practices are as injurious as unnecessary. The best plan to prevent sticking is simply to use the best starch, and to make it well, and to have the irons quite clean and highly polished. Mr W. B. Tegetmeier recommends the addition of a small piece of paraffin (a piece of paraffin candle-end) to the starch, to increase the glossiness of the ironed fabric.

STARS (in *pyrotechny*). *Prep.* 1. (Brilliant—*Marsh.*) Nitrate, 52½ parts; sulphur and black antimony, of each, 13 parts; reduce them to powder, make them into a stiff paste with isinglass, 1½ parts, dissolved in a mixture of vinegar, 6½ parts; and spirits of wine, 13 parts; lastly, form this into small pieces, and whilst moist roll them in meal gunpowder.

2. (WHITE—*Ruggieri.*) Nitre, 16 parts; sulphur, 7 parts; gunpowder, 4 parts; as the last.

8. (GOLDEN RAIN.) *a.* (*Ruggieri.*) Nitre and gunpowder, of each, 16 parts; sulphur, 10 parts; charcoal, 4 parts; lamp-black, 2 parts; mix, and pack it into small paper tubes.

b. (*Ruggieri.*) Nitre, 16 parts; sulphur and gunpowder, of each, 8 parts; charcoal and lamp-black, of each, 2 parts; as the last.

c. (*Marsh.*) Mealed gunpowder, 66¼ parts; sulphur, 11 parts; charcoal, 22½ parts; as before. Used for the 'garniture' of rockets, &c. See PYROTECHNY.

STAVES'ACRE. *Syn.* STAVESACRE SEEDS; STAPHISAGRIÆ SEMINA, STAPHISAGRIA (Ph. L. and D.), L. "The seed of *Delphinium staphisagria,* Linn." (Ph. L.). This article is powerfully emetic and cathartic, but is now scarcely ever used internally. Mixed with hair powder, it is used to kill lice. An infusion or ointment made with it is said to be infallible in itch, but its use requires some caution.

STAYS. *Syn.* CORSET. Stays, "before womanhood, are instruments of barbarity and torture, and then they are needed only to give beauty to the chest. It is the duty of every mother, and every guardian of children, to inquire the purpose for which stays were introduced into female attire. Was it for warmth? If so, they certainly fulfil the intention very badly, and are much inferior to an elastic woollen habit, or one of silk quilted with wool. Was it to force the ribs, while yet soft and pliable, into the place of the liver and stomach, and the two latter into the space allotted for other parts, to engender disease and deformity to the sufferer and her children for generations? Truly, if this were the object, the device is most successful, and the intention most ingeniously fulfilled" (*Eras. Wilson*).

"Only observe," exclaimed Dr John Hunter—"only observe, if the statue of the Medicean Venus were to be dressed in stays, and her beautiful feet compressed into a pair of execrably tight shoes, it would extort a smile from an Heraclitus, and a horse-laugh from a Cynic."

"The Turkish ladies express horror at seeing Englishwomen so tightly laced" (*Lady M. W. Montague*). See DISTORTIONS.

STEAM. The application of steam of the laboratory, as a source of heat, is commonly effected by means of double pans, to the space between which steam, at a moderate pressure, is introduced, the arrangements being such as to permit of the condensed steam, or distilled water, being removed, by means of a cock, nearly as soon as formed, or as may be desirable. Another plan is to place coils of metal pipe along the bottom of cisterns, vats, &c., formed either of wood or metal, and to keep them supplied with high-pressure steam.

"It is quite susceptible of positive proof, that by no arrangement yet discovered can more than two thirds of the heat generated by a given quantity of coal, during combustion, be fairly absorbed and utilised in any of our manufactories; and, moreover, there are undeniable facts which demonstrate that seldom, in the burning of coal, are more than three fourths of the total heat, which might be eliminated, actually obtained; thus justifying the supposition that one half of all the coal now consumed is virtually wasted and lost to society." To lessen, as much as possible, this loss various improvements have been made, "which, for the most part, have consisted in lengthening the flues, and exposing a larger surface of the boiler to the action of the heated air passing from the furnace to the chimney." "Remembering that air is an extremely bad conductor of heat, and that water about to be converted into steam is also a bad conductor, it is evident that time must form an important element in the perfect transmission of heat from one of these to the other; and hence, with a great velocity of current existing in the flues, very little heat would pass from air, however high its temperature, to water contained in a boiler, and so circumstanced with respect to its all but gaseous condition." The results of the experiments on fuel made at the Museum of Practical Geology by Sir H. De la Beche and Dr Lyon Playfair go clearly to show that "to open the damper of a steam-boiler furnace is pretty generally to diminish the effective power of the fuel." "Great waste of coal now arises from this simple circumstance; and much of the heat of the fire, which ought to go to the boiler, is lost by its [too] hasty transmission up the chimney. If, however, there be thus far room for improvement in the direction just indicated, still wider is the vacant space caused by imperfect combustion, or, in technical phrase, 'bad stoking,' merely because the stoker, to economise his labour and to avoid trouble, throws on to the bars of his furnace a thick layer of fuel, by which loss is caused in two or three directions. These are, principally, imperfect combustion, and the volatilisation of fuel, as smoke, &c., from an insufficient supply of air, and from a mass of mere red-hot coke or cinder, two or three inches thick, lying between the boiler and the hottest part of the furnace; which last, according to Dr Kennedy, is about one inch above the fire-bars. Besides which, "in passing over this red-hot coke the carbonic acid would be converted into carbonic oxide, and thus not only remove a quantity of carbon equal to its own, without yielding any additional heat, but actually with the production of cold, or, in other words, the absorption of heat" ('Dict. Arts, Manuf., and Mines'). This points to the evident policy of using a smoke-consuming furnace, as noticed elsewhere.

Another matter worthy of remark is the constant waste of heat, and, consequently, of fuel, in laboratories and manufactories in which steam is employed, owing to the exposed condition of the pipes, boilers, and pans. All of these should be well 'clothed' or covered by some non-conducting medium, to prevent loss of heat by radiation, and by contact with the atmosphere. Not only does economy dictate such a course, but the health and comfort of the workpeople demand that the atmosphere in which they labour should be as little heated and poisoned as possible.

A cubic inch of water, during its conversion into steam, under the ordinary pressure of the atmosphere, expands into 1696 cubic inches, or nearly a cubic foot.

TABLE of corresponding Pressure and Temperatures of Steam. By ARAGO and DULONG.

Pressure in Atmospheres.[1]	Temperature, F.	Pressure in Atmospheres.[1]	Temperature, F.
	Degrees.		Degrees.
1	212·	13	380·66
1½	234·	14	386·94
2	250·5	15	392·86
2½	263·8	16	398·48
3	275·2	17	403·83
3½	285·	18	408·92
4	293·7	19	413·78
4½	300·3	20	418·46
5	307·5	21	422·96
5½	314·24	22	427·23
6	320·36	23	431·42
6½	326·26	24	435·56
7	331·7	25	439·34
7½	336·86	30	457·16
8	341·78	35	472·73
9	350·78	40	486·59
10	358·88	45	499·14
11	366·85	50	510·6
12	374·		

[1] Estimating 14·6 lbs. = 1 atmosphere.

One part, by weight, of steam, at 212° F., when condensed into cold water, is found to be capable of raising 5·6 parts of the latter from the freezing to the boiling point. See FUEL, PIT-COAL, SMOKE, &c.

STEAR'IC ACID. $C_{17}H_{35}.CO_2H$. Syn. STEARIN (Commercial). This is obtained from stearin (see below) by saponification.

Prep. 1. Repeatedly dissolve and crystallise commercial stearic acid in hot alcohol, until its melting-point becomes constant at not less than 158° F. Pure.

2. (Chevreul.) Saponify mutton suet with caustic potash, and dissolve the soap in 6 times its weight of hot water; to the solution add 40 or 50 parts of cold water, and set the mixture aside in a temperature of about 52° F.; after a time separate the pearly matter (stearate and palmitate of potash) which falls, drain and wash it on a filter, and dissolve it in 24 parts of hot alcohol of sp. gr. 0·820; collect the stearate of potash which falls as the liquid cools, recrystallise it in alcohol, and decompose it, in boiling water, with hydrochloric acid; lastly, wash the disengaged stearic acid in hot water, and dry it.

3. (Commercial.) Ordinary tallow is boiled in large wooden vessels by means of high-pressure steam, with about 16% of hydrate of lime (equiv. to 11% of pure lime), for 3 or 4 hours, or until the combination is complete, and an earthy soap is formed, when the whole is allowed to cool; the product (stearate of lime) is then transferred to another wooden vessel, and decomposed by adding to it 4 parts of oil of vitriol (diluted with water) for every 3 parts of slaked lime previously employed, the action being promoted by steam heat and brisk agitation; after repose, the liberated fat is decanted from the sediment (sulphate of lime) and water, and is then well washed with water, and by blowing steam into it; it is next allowed to cool, when it is reduced to shavings by machinery, and in this divided state is placed in canvas bags and submitted to the action of a powerful hydraulic press, by which a large portion of the oleic acid which it contains is expelled; the pressed cakes are then a second time exposed to the action of steam and water, again cooled, and coarsely powdered, and again submitted to the joint action of steam and pressure; they are, lastly, melted, and cast into blocks for sale.

Obs. The commercial product is a more or less impure mixture of stearic acid and other fatty bodies, particularly the so-called 'margaric acid,' now generally regarded as a mixture of palmitic and stearic acids. The hard, fatty acids of vegetable origin, now so extensively used as candle materials, are obtained from the natural oils and butters by the process known as 'sulphuric acid saponification,' which consists in treating the fatty bodies with 5% or 6% of concentrated sulphuric acid at a high temperature (about 350° F., produced by superheated steam), and distilling the resulting mass by the aid of steam heated to about 560° F. Frequently the operations of hot and cold pressing are resorted to in order to free the product from the softer fats.

By a patent process employed at Price's candle works the natural vegetable fats are decomposed into their constituents (fatty acids and glycerin) by the action of superheated steam alone, without previous 'saponification' with alkali or sulphuric acid.

Another method for the preparation of commercial stearic acid is that of Messrs Moinier and Boutigny. This process is thus described in the 'Chemical Technology' of Messrs Ronalds and Richardson:—2 tons of tallow and 900 galls. of water are introduced into a large rectangular vat of about 270 feet capacity. The tallow is melted by means of steam admitted through a pipe coiled round the bottom, and the whole kept at the boiling heat for an hour, during which a current of sulphurous acid is forced in. At the end of this period 6 cwt. of lime, mixed with 350 galls. of water, are added. The mixture soon becomes frothy and viscid. The whole is now agitated in order to prevent the sudden swelling up of the soapy materials. The pasty appearance of the lime soap succeeds, and it then agglomerates into small nodular masses.

The admission of sulphurous acid is now stopped; but the injection of the steam is continued until the small masses become hard and homogeneous. The whole period occupies 8 hours, but the admission of the sulphurous acid is discontinued at the end of about 3 hours. The water containing the glycerin is run off through a tube into cisterns prepared to receive it.

Retorts are used for preparing sulphurous acid, into which are put sulphuric acid and pieces of wood; upon the application of heat the sulphurous acid passes off, and is conveyed by leaden pipes into the vessels containing the tallow. The lime soap formed is then moistened with 12 cwt. of sulphuric acid at 150° F., diluted with 50 galls. of water. The whole is thoroughly agitated and the steam cautiously admitted, so as not to dilute the acid too much until the decomposition is general at all points. This occupies about 3 hours, and in 2 or 3 hours more the sulphate of lime has collected at the bottom, while the fatty acids are floating on the surface of the solution of the bisulphate of lime. Several processes of washing with steam and water are necessary to ensure the removal of the sulphate of lime, &c., and after settling for 4 hours the fatty acids are forced through a fixed siphon into a vat, where they are again washed with water ; they are then siphoned at last into a trough lined with lead, on the bottom of which are placed leaden gutters, pierced below by long pegs of wood. The fatty acids are then placed in bags and subjected to pressure in the stearin cold press.

In 1871 Prof. Bock, of Copenhagen, after a careful microscopic and chemical investigation, discovered that the neutral fats were composed of a congeries of little globules enclosed in albuminous envelopes. To the presence of these latter substances in the fat he attributed the difficulty of eliminating the fatty acids from it by means either of sulphuric acid, except in excess, or of alkali, except under great pressure; conceiving that both these agents, as employed under the usual methods, were expended in rupturing and destroying the albuminous coverings.

The inconveniences arising from the above processes are, in the case of the excess of the sulphuric acid, a considerable destruction of the fatty acid, as well as the necessity of its distillation, and the consequent danger of conflagration; whilst in the case of the alkali, this must either be used in quantities much greater than theory requires, or else be heated under great pressure, at the risk of giving rise to an explosion.

In Prof. Bock's process these dangers, together with the waste of material, are avoided. By submitting the fat for a limited time and at a given temperature to the action of a small quantity of sulphuric acid, the albuminous envelopes are broken and partly destroyed. The neutral fat thus liberated is then placed in open tanks in water, by which, after the expiration of several hours, it becomes decomposed. When this is completely effected the glycerin, dissolved in the water used for the decomposition, is removed; the fatty acids which remain behind, and which amount to 94% of the original fat, being at this stage of the operation dark brown or blackish in colour.

In this condition they are placed in open tanks, and dilute solutions of certain reagents are poured upon them, whereby the albuminous débris as well as the colouring matters with which they are associated become oxidised, whilst the specific gravity of these latter is in consequence so increased as to cause them to subside to the bottom of the tank, leaving the fatty acids, now greatly whitened, in the upper part of the liquid.

The acids, after being washed 2 or 3 times with dilute acid and water, are then cooled, and hot-pressed in the usual manner, and the stearic acid thus obtained is said to have a higher melting-point and to be larger in yield than that obtained by any other method, an oleic acid of excellent quality being at the same time produced.

In a French patent carbon disulphide is employed to increase the fluidity of the oleic acid, so that the warm pressure of the crude stearic acid

is avoided. The addition of the carbon disulphide may be made either before or after the cold pressing of the stearic acid. The crude fatty acid is melted in a special apparatus, and 20% of the disulphide is mixed with it whilst in the fluid state. It is then left to cool and subjected to cold pressure. The stearic acid thus obtained should be free from oleic acid.

Prop., &c. Pure stearic acid crystallises in milk-white needles of the same specific gravity as water, and freezing at 69° C., which are soluble in ether and in cold alcohol, and form salts with the bases, called stearates. The commercial acid is made into candles. For the method of estimating stearic acid see PALMITIC ACID. See CANDLES, FAT, OILS (Fixed), and TALLOW.

STEARIN. $C_3H_5(C_{18}H_{35}O_2)_3$. The solid portion of fats which is insoluble in cold alcohol. There are three stearins or glyceryl stearates.

Prep. Pure strained mutton suet is melted in a glass flask along with seven or eight times its weight of ether, and the solution allowed to cool ; the soft, pasty, semi-crystalline mass is then transferred to a cloth and is strongly pressed as rapidly as possible, in order to avoid unnecessary evaporation ; the solid portion is then redissolved in ether, and the solution allowed to crystallise as before.

Prop., &c. White ; semi-crystalline ; insoluble in water and cold alcohol ; soluble in 225 parts of cold ether, and freely so in boiling ether. It melts at 130° F. The 'stearin' of commerce is stearic acid.

STEAROPTENE. The name given by Herberger to the solid crystalline compound which separates in the cold from certain volatile oils. Bizio calls it stereusin.

STEEL. This important material may be defined as iron chemically combined with sufficient carbon to give it extreme toughness and hardness without brittleness. According to one of our greatest authorities on metallurgy, steel is a combination of iron with from 0·1% to 1·8% of carbon, these numbers referring respectively to the softest and the hardest varieties.

The influence of traces of foreign matters is very important in the economy of steel, and a great deal of research is at the present time being done in this country by Professor Roberts-Austen, in France by the brothers Le Chatelier Demond, and by other scientists, and it is expected that very soon some important deduction will be made from their experiments.

By Pourcel and other authorities silicon in small quantities is supposed to be a useful ingredient in steel, and to increase its capacity for being hardened ; an opinion dissented from by others, who hold that its presence has a tendency to diminish the malleability and ductility of the metal.

Faraday and Stodart believed that the addition of small quantities of chromium and iridium to steel served to improve its quality, and the same has been asserted of tungsten and titanium ; but on these points there is still a divergence of opinion, and no satisfactory decision has yet been arrived at concerning them.

Manganese has also been credited by Hadfield and others with the property of improving steel, but as it has been found that only a very minute quantity of the manganese is taken up by the steel, an indirect influence may possibly be exercised by it, viz. its power of carrying away any prejudicial excess of sulphur and phosphorus with it; and in this manner it may contribute to the increased purity of the metal. Steel that contains manganese is always harder, stronger, and more ductile than steel which does not. The addition of manganese to cast steel constitutes Heath's patent, the chief advantage of which is that blistered steel made from British bar iron can be substituted for the much more expensive Swedish and Russian iron in certain branches of iron manufacture.

Among the various substances which are frequently present in malleable iron and in cast iron, those which are more prejudicial to the quality of steel are sulphur, phosphorus, and copper. 0·05% of sulphur in steel renders the metal brittle and 'red-short;' 0·1% of phosphorus renders steel 'cold-short,' i.e. unworkable at ordinary temperatures; 0·5% of copper renders steel decidedly red-short, and for this reason iron smelted from ores containing copper pyrites is not suitable for making steel.

Within the last few years great attention has been paid to the investigation of the chemistry of steel. The researches of Despretz and Fremy tend to the conclusion that nitrogen exercises a very important influence over the phenomena of 'steeling,' and that carbon plays a less necessary part; while those of Carron and Deville still refer the formation of steel to the chemical combination of iron with carbon. There is no test of the value of steel beyond its elasticity and temper, and the fineness, equality, and smoothness of its grain.

Cast iron, wrought iron, and steel are all combinations of iron and carbon, differing in the amount they contain of the latter element. As cast iron contains a larger and wrought iron a smaller proportion of carbon than steel, it follows that to convert the cast iron into steel its excess of carbon must be removed; whilst conversely, to make the wrought iron into steel, the requisite amount of carbon must be added to it.

Thus it is that the various processes for the manufacture of steel (with the exception of those which propose to obtain it direct from the ores) are directed to one or other of these ends, viz. the decarburation of cast or pig iron, and the carburation of wrought or malleable iron.

1. In the first, or decarburation method, the oxygen of the air plays an important part. Best wrought iron is heated with coal or charcoal, in some works on the refining hearth, in others upon the bed of the puddling furnace. The oxygen of

The following TABLE, *from ' Payen's Industrial Chemistry,' gives the Composition of several kinds of Steel.*

Kind of Steel	Locality	Fe	Mn	Cu	Carbon		Si	S	P	Authority
					Combined	Graphitic				
Natural steel	Siegen	...	...	0·379	1·698	...	0·038	...	...	Karsten.
,, ,,	Solingen	...	...	...	1·570	...	0·020	...	...	Lampadius
Puddled steel	Hartz	...	0·012	...	1·380	...	0·006	(Al 0·12)	trace	Brauns.
Cement steel	English	...	...	...	1·807	...	0·100	...	...	Berthier.
,, ,,	German	...	...	...	0·416	0·080	...	...	...	Bromeis.
Cast steel	Sheffield	...	...	...	0·950	0·220	...	...	...	,,
,, ,, ,,	,,	...	...	...	1·758	...	...	...	...	Karsten.
,, ,,	French	...	...	...	0·650	0·040	...	...	...	...
Sword steel	Damascus	...	0·070	...	1·069	...	...	(Ni 0·07 Wo 0·01)	...	...
,, ,,	,,	...	trace	...	0·775	...	...	(Ni 0·21 Co trace Wo trace)	...	...
Wootz	Indian	...	...	...	1·500	0·600	...	...	...	...
,, ,,	,,	98·092	...	...	1·333	0·312	0·045	(As 0·037)	...	Henry.
Cast steel	German	...	trace	0·300	1·180	0·330	...	(Ni 0·12)	0·020	...
,, ,,	English	...	0·024	0·066	1·275	0·213	...	(As 0·007)	...	...
Bessemer steel	Dowlais	...	0·576	0·025	0·490	...	0·009	0·003	0·036	...
,, ,,	Sweden	...	trace	...	0·085	...	0·008	trace	0·025	Brusewitz.
,, ,,	...	...	0·179	...	0·300	...	0·044	,,	0·033	,,
,, ,,	...	...	0·256	...	0·700	...	0·032	,,	...	,,
,, ,,	...	...	0·464	...	0·950	...	0·047	,,	0·032	,,
,, ,,	...	...	0·355	...	1·050	...	0·067	,,	...	,,
Wired	Barrow-in-Furness	...	0·214	...	0·200	...	0·179	0·030	0·026	,,
Rail heads	German	...	0·386	...	0·138	...	0·306	0·040	0·034	,,
Rails	...	...	0·264	...	0·150	...	0·091	0·025	0·032	,,
,, ,,	...	...	0·638	...	0·046	...	0·634	0·045	0·093	,,
Boiler plates	...	...	0·136	...	0·250	...	0·016	0·010	...	,,
,, ,,	...	...	0·273	...	0·300	...	0·056	0·040	0·041	,,

the air burns off the excess of carbon from the iron, and steel is left. Payen says that when the iron contains slag, the ferrous silicate present in this takes part in the reaction.

The steel obtained by this method is called *natural steel*. It is afterwards subjected to forging, and is employed in the manufacture of springs for machinery, railway carriages, wheel tyres, ploughs, and other farming implements.

Krupp's cast steel, manufactured at Essen, near Cologne, is a natural steel, being made on the bed of a puddling furnace. It is obtained from hæmatite and spathic ores, coke being used for the smelting. The proportion of carbon in Krupp's steel is about 1·2%. When required for ordnance it is fused with a little bar iron in pots, each of which holds 30 lbs. It sometimes happens that in the manufacture of a huge gun or cannon the contents of as many as 1200 of these pots are required. When this is the case the pots are emptied of their molten contents simultaneously into a channel leading to the cast, 400 well-drilled men being required to carry out the operation.

It is very essential that castings of such magnitude should be allowed to cool very gradually. They are therefore enveloped in hot cinders for two or three months, after which they are ready for the forging.

2. *The carburation method.* This is generally effected by the process known as 'cementation,' which is carried out as follows :—Two chests, made of fire-brick or stone, one narrow end of each of which is shown in the accompanying plate, are so fixed in a dome-shaped furnace that the flames from the hearth beneath can effectually play around them.

The process renders it necessary that the temperature of the furnace should be steadily maintained for some days ; and this is achieved by surrounding the furnace with a conical wall of brick-work, as shown in the *engr*. The chests are usually about 10 or 12 feet in length, 3 feet in height, and 3 feet in depth. A layer of charcoal of a fineness to pass through a sieve of a ¼-inch mesh, or of soot, is placed on the bottom of each chest, and upon this the bars of wrought iron which are intended for conversion into steel. The bars inside must be of iron of the best quality and generally about 3 inches broad and ¾ of an inch thick. When arranged regularly a little distance apart, the interstices between them are filled up with charcoal, with which they are then covered to a depth of about an inch. Similar layers of bars, similarly arranged, succeed this first one, until the chests are filled. They are then covered in to a depth of 6 inches with a luting of damp clay or sand. Each chest when thus filled contains from 5 to 6 tons of iron. One of the bars projects through an opening at the end of the chest, to facilitate an inspection of it from time to time, so that the progress of the operation may be judged. The materials of which the chests are composed render it important that

the temperature of the furnace should be carefully and gradually increased, as a too sudden accession of heat would lead to damage. The temperature necessary to effect the carburation of the iron has been found to be that required for the melting of copper, viz. 1996° F. When this temperature is reached it is maintained for eight or ten days, or even longer, the period depending upon the thickness of the iron, and the degree of hardness it is desired to possess. Six or eight days are sufficient to yield steel of a moderate degree of hardness. At the end of the requisite time the fire is gradually put out, and the chests as gradually cooled, a process which occupies about another ten days.

The effect of the treatment to which the iron bars have been subjected has been, in the first place, to entirely alter their interior structure; for if they are broken asunder at any part, instead of showing the fibrous arrangement observable in bar iron, they present a closely granular one. In the second place, chemical analyses demonstrate that the iron has combined with about 1% of carbon, and that this combination has not only taken place on the surface of the bar, but has extended throughout its whole substance. It is because of this perfect impregnation of the iron by the solid carbon that the process by which it has thus been converted into steel is called 'cementation.' The converting furnace in the cementation is usually of the form shown in the *engr*. (see next page).

N, N are two fire-brick boxes, 3 × 4 × 12 feet, open at their upper surface. C is the fireplace, which runs the whole length of the furnace. A is a vault of fire-brick covered in by the arch B. D, D are flues. E, E, short chimneys. F, a dome of brick, which encloses the whole. G is the man-hole, through which the boxes N, N are charged.

Two suggestions have been offered in explanation of the blistered surface presented by the steel. One of these, the theory of Mr T. H. Henry, is that part of the carbon, in penetrating into the body of the bar iron, has combined with the small quantity of sulphur present in the iron, and that the bisulphide of carbon thus formed, becoming vaporised by the elevated temperature in escaping through the soft surface of the metal, has caused its blistered condition. The second conjecture is that the blebs have arisen from the extrication of carbonic oxide, which had been formed in the bar by the union of the carbon with the small quantity of oxide of iron or slag accidentally remaining in it.

Graham, has shown that soft iron has the power of absorbing or occluding at a low red heat 4·15

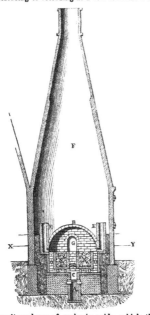

times its volume of carbonic oxide, which the metal, when it becomes cold, retains, but which it parts with when subjected to a temperature such as that which prevailed in the cementation box. This fact seems to offer a reasonable confirmation of the reaction it has been surmised takes place during the cementation process, and which is supposed to be as follows:

The small quantity of atmospheric oxygen remaining in the chest unites with the carbon to form carbonic oxide. This carbonic oxide gives up half its carbon to the iron (which thereby becomes converted into steel), and in doing so changes to carbonic acid, which becomes reduced to carbonic oxide by the absorption of more carbon from the charcoal, which carbon the carbonic oxide again transfers to the iron.

The above reaction may not improbably occur throughout the substance of the bar. By some chemists cyanogen compounds are supposed to be present in the cementation powder, and the cyanogen contained in these is supposed to be the carrier of the carbon to the iron.

"The blistered steel obtained by this process is, as would be expected, far from uniform, either in composition or texture; some portions of the bar contain more carbon than others, and the interior contains numerous cavities. In order to improve its quality it is subjected to a process of fagoting similar to that employed in the case of bar iron; the bars of blistered steel, being cut into short lengths, are made up into bundles, which are raised to a welding heat, and placed under a tilt hammer weighing about 2 cwt., which strikes 200 or 300 blows in a minute; in this way the several bars are consolidated into one compound bar, which is then extended under the hammer till of the required dimensions. The bars, before being hammered, are sprinkled with sand, which combines with the oxide of iron upon the surface and forms a vitreous layer, which protects the bar from oxidation.

"The steel which has been thus hammered is much denser and more uniform in composition; its tenacity, malleability, and ductility are greatly increased, and it is fitted for the manufacture of shears, files, and other tools. It is commonly known as shear steel. Double shear steel is obtained by breaking the tilted bars in two, and welding these into a compound bar.

"The best variety of steel, which is perfectly homogeneous in composition, is that known as cast steel, to obtain which about 30 lbs. of blistered steel are broken into fragments, and fused in a fire-clay or plumbago crucible, heated in a wind-furnace, the surface of the metal being protected from oxidation by a little glass melted upon it. The fused steel is cast into ingots, several crucibles being emptied simultaneously into the same mould. Cast steel is far superior in density and hardness to shear steel, but, since it is exceedingly brittle at a red heat, great care is necessary in forging it. It has been found that an addition to 100 parts of the cast steel, of one part of a mixture of charcoal and oxide of manganese, produces a very fine grained steel, which admits of being cast on to a bar of wrought iron in the ingot mould, so that the tenacity of the latter may compensate for the brittleness of the steel; when the compound bar is forged, the wrought iron forming the back of the implement, and the steel its cutting edge" (Bloxam's 'Chemistry, Inorganic and Organic').

Another distinct method from the cementation one, by which the carburation of iron is effected, is that in which scrap or malleable iron is mixed with pig or cast iron, this latter being fused with the scrap iron in quantity sufficient to afford such an amount of carbon as is necessary to convert the mixture into steel. Steel made by this operation is entirely homogeneous; the tilting process which precedes the casting of the steel obtained by cementation is therefore unnecessary. The pig-iron is placed on the bed (made of refractory sand) of one of Siemens' regenerative furnaces, heated by gaseous fuel. The temperature in this furnace is so intense that the pig-iron becomes perfectly liquid, and, when in this condition, the scrap iron, which has been previously heated to redness in an adjoining refractory furnace, is added, and becomes dissolved by it.

In the manufacture of Bessemer steel both the carburation and decarburation processes are practised. From 1 to 5 tons of pig or cast iron in a molten state are run from a continuous blast, cupola, or reverberatory furnace, with an apparatus known as a converter, which is previously heated to redness by means of coke. The converter is figured under two aspects in the annexed engraving.

This vessel, which is generally made of boiler-plates of sheet iron, has an inside lining, consisting of a siliceous fireproof material, and is perforated at the bottom with a number of concentric little openings, which are the orifices of as many little tubes or tuyères, that lead into an outside main tube, as shown in the engr. By

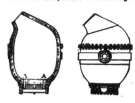

means of these tubes condensed air is forced into the mass of melted metal, which is soon thrown into violent commotion, and sends out a shower of ignited sparks. The oxide of iron formed at the same time, being set into active movement by the incoming blast of air, is brought into intimate contact with every particle of the carbon and silicon contained in the cast iron, and converts the former into carbonic oxide, which burns with its characteristic flame at the mouth of the converter, and the silica into silicic acid, which enters into the slag, and floats, in the form of foam, on the top of the heavier molten iron.

The removal of the carbon (which is recognised by the discontinuance of the carbonic oxide flame) being thus accomplished, the iron has next to be submitted to the carburetting operation. This is performed by running into the liquid iron in the converter such a quantity of molten pig or cast iron as contains the required proportion of carbon.

The pig-iron used for this purpose generally contains, in addition to a large amount of carbon, a very perceptible quantity of manganese. The converter is then by means of trunnions tilted, so that its contents can be run into a ladle and transferred to the necessary moulds. The time of conversion occupies from ten to twenty minutes.

By Bessemer process the sulphur present in the pig-iron is almost entirely eliminated; the greater part of the silicon is also separated, together with the carbon, and almost in the same proportion; but the phosphorus is not removed, and, owing to the oxidation of some iron, the amount is actually greater in the finished steel than in the pig-iron (Paven's 'Industrial Chemistry,' edited by B. H. Paul, Ph.D.). Bessemer steel is in large demand, and is excellently suited for rails for railroads, cannon, boiler-plates, armour-plates, and similar heavy material, for the manufacture of which it has largely supplanted wrought iron; but it is not at

all adapted for the manufacture of knives, razors, lancets, or similar instruments, in which a sharp or keen edge is desirable.

Latterly attempts have been made to obtain steel direct from the ores. The efforts made have been greatly stimulated by the invention of the regenerating furnace of Siemens. In these furnaces, in which an intense temperature is obtained by means of the combustion of inflammable gases (chiefly consisting of carbonic oxide, hydrogen, and carburetted hydrogen), the ore, after (in one process) being melted in hoppers by means of the burning gases, runs down, and is gradually dissolved in some melted pig-iron placed on the hearth of the furnace. When this latter has been sufficiently diluted with the decarbonised iron the operation is complete.

Properties of Steel. The effects of temperature upon steel are remarkable, and a knowledge of them has proved of great practical utility in the manufacture of the various steel-ware articles that are so indispensable to our every-day wants and needs. If forged and soft steel is heated, and then suddenly cooled, it becomes hard, the hardness varying with the temperature and the rapidity with which this has been reduced. The higher the temperature and the more rapidly it is cooled, the greater will be its hardness. Steel which has been heated until white-hot, and then suddenly plunged into a bath of cold mercury, acquires a hardness nearly equalling that of the diamond. That, however, which the steel gains in hardness it loses in pliancy and elasticity, becoming exceedingly brittle.

Soft steel, which has been made hard by heating it to redness, and by subsequent sudden immersion in cold water, may be reconverted into soft steel by again heating it to redness and allowing it to cool slowly. By stopping short, however, of heating it to redness, its hardness may be proportionally modified.

Hence steel articles, varying as much in the qualities of hardness and elasticity as a lancet and watch-spring, are made either by 'heating down' hard steel to requisite temperature and allowing it to cool, or by 'heating up' soft steel to the necessary point and also letting it gradually cool. When steel is so treated it is said to be *tempered* or *annealed.* If polished steel be heated over a flame to a temperature of 430° F. its surface becomes of a very pale yellow colour; the colour passes through different shades of yellow and blue with each successive increase of temperature, until when raised to 600° F. it becomes blackish blue.

These effects are due to the formation on the surface of the steel of films of oxide of different degrees of thickness, and to the action of the light on these. They are precisely analogous to those which are caused when a ray of reflected light falls upon any other body, the surface of which is composed of thin layers, which are continually changing in thickness, such as a soap-bubble, or a thin coating of tar or oil swimming on water, and which are exemplified in Newton's rings.

As each shade of colour is an index of the temperature of the steel, and as this determines its adaptability for various purposes, all that the workman has to do, when he requires it for any special object, is to heat it by the proper methods (such as a bath of oil, or tallow, or melted metal) until it acquires the desired colour, and then to allow it to gradually cool.

The following table, exhibiting the different melting-points of steel when employed in the manufacture of different kinds of works, together with the corresponding colours, the composition of the metallic baths, &c., is from Dr Wagner's 'Handbook of Chemical Technology,' edited by W. Crookes, Esq., F.R.S.

	Composition of Metallic Bath. Lead.	Tin.	Melting-point.	Colour.
Lancets	7	4	220° C.	Hardly pale yellow.
Razors	8	4	228° „	Pale yellow to straw yellow.
Penknives	8½	4	232° „	Straw yellow.
Pairs of scissors	14	4	254° „	Brown.
Clasp-knives, joiners' and carpenters' tools	19	4	265° „	Purplish colour.
Swords, cutlasses, watch-springs	48	4	288° „	Bright blue.
Stilettos, boring tools, and fine saws	50	2	292° „	Deep blue.
Ordinary saws	in boiling linseed oil		316° „	Blackish blue.

Steel is of a greyish-white colour, and has a sp. gr. varying from 7·6224 to 7·8131 (*Karsten*). During hardening its physical and even its chemical properties are modified, and it experiences a slight increase of volume. The property that steel possesses of becoming hard after being heated to redness, and suddenly chilled, does not belong to pure iron, such as may be obtained by electrolysis. Unlike pure iron, too, steel presents a granular instead of a fibrous structure when broken, the best samples closely resembling silver in this respect. The chemical difference between hard and soft steel appears to consist in the much more intimate combination of the carbon with the iron in the hard variety than in the soft. In this latter kind the carbon seems to be only mechanically mixed, for if it be immersed in hydrochloric acid the iron is dissolved, and leaves the carbon behind. Steel is the most tenacious of all metals; its tenacity varies with its temper. Some kinds require a load of seventy tons per square inch to break it. Its melting-point, about 1800° C., is between that of pig-iron and of malleable iron; it is less easily magnetised, but its magnetism is more permanent than is that of pure iron; it is less oxidisable on exposure to moist air than is malleable iron. In elasticity steel is superior to malleable iron.

What is termed *case-hardening* (which *see*) is a process by which small articles of iron, such as keys, gun-locks, &c., are superficially converted into steel. It is performed by heating the articles in contact with powdered charcoal. Another method is to make the iron red-hot, and then to sprinkle powdered potassium ferrocyanide all over it.

STEREOCHROMY. This is a branch of the pictorial art confined to the embellishment of walls and monuments. In the operations by which it is accomplished it will be seen that the soluble silicates (water-glass) play an important part.

The foundation for the future picture or coloured design must be of some durable stone or imperishable cement. Over this is first placed a layer of lime mortar, to which is applied, when it is dry and has become sufficiently hard, a solution of water-glass, by which all the interstices of the mortar are filled up. Another coating of mortar made of sharp sand and a lye of chalk is next laid on, and this, after it has been carefully smoothed, properly levelled on the surface, and become quite dry, is washed over and thoroughly impregnated with water-glass solution. When this last layer has become dry it is ready to receive the painting, which must be executed in water colours. After laying on these colours may be permanently fixed by covering them with water-glass. The following are the colours used:—Zinc white, chrome green, chrome oxide, cobalt green, chrome red, zinc yellow, oxide of iron, sulphide of cadmium, ultramarine, ochre, &c. Vermilion is inadmissible, since, in fixing, it turns from red to brown. Cobalt ultramarine, on the contrary, increases greatly in brilliancy upon the application of the fixing solution. Stereochromatic paintings are found to be very durable, and impervious to damp, smoke, or variations of temperature.

STE'REOTYPE METAL. See TYPE METAL.

STER'LING. The truth of the old proverb, that "all is not gold which glitters," is often painfully experienced by the purchaser of modern jewellery.

Sterling Value of Gold of different degrees of 'Fineness.'

Carats fine.		Value per oz. Troy. £	s.	d.
24 *carats*		4	4	11½
23 „		4	1	5
22 „	(*British standard*)	3	17	10½
21 „		3	14	4
20 „		3	10	9½
19 „		3	7	3
18 „	(*lowest Hall-mark*)	3	3	8½
17 „		3	0	2
16 „		2	16	7½
15 „		2	13	1
14 „		2	9	6½
13 „		2	6	0
12 „		2	2	5½
11 „		1	18	11
10 „		1	15	4½
9 „		1	11	10
8 „		1	8	3½
7 „		1	4	9
6 „		1	1	2½
5 „		0	17	8
4 „		0	14	2
3 „		0	10	7½
2 „		0	7	1
1 *carat*		0	3	6

The foregoing table will, therefore, prove highly useful to the reader in determining the value of articles in gold, provided he ascertain the 'fineness' of the metal, either by examination or written warranty.

STEREO METAL. A remarkable alloy recently invented by Baron de Rosthorn, of Vienna, and used in place of ordinary gun-metal. It consists of copper and spelter, with small proportions of iron and tin, and to these latter its peculiar hardness, tensile strength, and elasticity are attributed.

STEROPUS MADIDUS, Fabricius. THE NIGHT-FEEDING GROUND BEETLE. It was fondly imagined that this beetle was altogether useful to agriculturists by destroying other insects. Curtis had this opinion, and believed that it was the natural enemy of wireworms. Like many others of the *Geodephaga*, it feeds upon the roots of plants as well probably as upon insects. Westwood says of the *Carabidæ* that "some of the species generally found in corn-fields are clearly ascertained to feed upon growing grain." There can be no doubt that the *Steropus madidus* feeds eagerly upon mangel-wurzel plants, as it has been caught frequently *flagrante delicto*. It attacks these plants just under the ground, at the point where the root begins, and bites away the soft substance. Sometimes the plant is bitten through and through, or all round the collar of the root, so that it is completely killed. In other cases it is partially cut through, and cannot develop a large or healthy root.

The beetle begins its operations directly after the plants have been singled, and their roots have begun to swell, and for three or four weeks it is able to do inconceivable mischief. Like many of the beetles of this family it works at night, and is therefore difficult to detect actually at its work of destruction.

In 1885 a report was received from Shropshire of a strange attack upon mangel plants. These were a capital plant when hoed out, but signs of failing were noticed soon after this, and they fell away one by one. Upon pulling the leaves the plants gave way just at the junction between the leaves and roots, and there was evident proof that at this point the plants had been bitten round by some insect. The farmer stated that no insect could be found. He was advised to hunt in the very early morning, and soon forwarded specimens of *Steropus madidus* taken in the very act of gnawing the plants.

Two or three different attacks were reported in 1885. In 1886 injury to mangel-wurzel plants of a somewhat serious nature was traced, after some patient watching, to this insect in a large field in Kent. It was stated that the soil of this field was light, and that there were a good many stones in the soil, which was on the Lower Greensand formation. Reports of damage by this insect came also from a farm in another part of Kent, where flint stones abound, and from one near Salisbury in Wilts, upon which flints are plentiful.

This insect is common in England, and is known also in Germany, Switzerland, France, and Belgium, according to C. G. Calwer ('Käferbuch').

Life History. This beetle belongs to the genus *Steropus*, a subdivision of the family *Feroniidæ*, of the section *Geodephaga* of the COLEOPTERA. It is black in colour, and has no wings. It is eight lines long. The female lays her eggs under the ground, generally under stones. The larvæ which come from the eggs in about eight days are as long as the beetle when full grown, and are dark, with six legs and a pair of spines or bristly points at the end of the body. They change to pupæ, and pass the winter in this form. The larvæ do not injure mangel-wurzel plants.

Prevention. As these beetles have no wings their range of mischief is limited. After an attack the land should be deeply ploughed, and mangels should not be planted in fields adjacent.

Remedies. When mangel plants fail, and the cause is ascertained to be the *Steropus madidus*, ashes, sawdust, or sand saturated with paraffin should be scattered on both sides of the drills, or roots of plants, and lightly dropped in close to the plants. Soot would be serviceable if fresh, pungent, and pure. Frequent horse-hoeings and side-hoeings would disturb the insect ('Reports on Insects Injurious to Crops,' by Chas. Whitehead, Esq., F.Z.S.).

STETHOSCOPE. An instrument employed in auscultation. It consists of a tube (usually made of wood, sometimes of gutta percha) widening considerably at one end, and but slightly at the other. The wide end is applied to the chest or other part of the patient, the physician putting his ear at the other end; and from the sounds emitted by the heart, lungs, &c., the state of these parts is ascertained.

STEWING. A method of cooking food intermediate to frying and boiling, performed by simmering it in a saucepan or stewpan, with merely sufficient water to prevent burning, and to effect the object in view; the whole being served up to form the 'dish.' It is undoubtedly the most simple and economical, and, when skilfully conducted, one of those best calculated to develop the flavour and nutritious qualities of animal food. The following is one of the most popular stews:

Stew, Irish. *Prep.* (*Soyer.*) Take about 3 lbs. of scrag or neck of mutton; divide it into ten or twelve pieces, and lay them in the pan; add 8 large potatoes and 4 onions cut into slices; season with 1½ teaspoonfuls of pepper, and 3 do. of salt; cover all with water, put it into a slow oven, or on a stove, for two hours, then stir it all up well, and serve it up in deep dishes. If a little more water is added at the commencement, you can take out, when half done, a nice cup of broth.

STIGMATA MAIDIS ('*corn silk*,' the stigmata of maize, *Zea mays*). Demulcent and diuretic, in catarrhal affections of the kidneys and bladder. Gives the best results in cases of uric acid phosphatic gravel, chronic cystitis, and mucous or purulent catarrh. Fluid extract.—*Dose*, 1 dr.

STILL. A vessel or apparatus employed for the distillation of liquids on the large scale. The forms of still, and the materials of which they are made, vary according to the purposes for which they are intended, some being exceedingly simple, whilst others are equally elaborate and

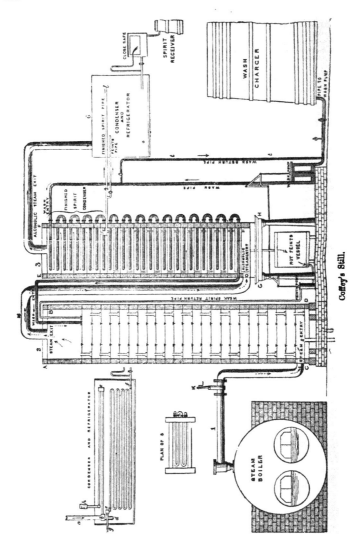

Coffey's Still.

complicated. The *engr.* represents the most common and useful apparatus of this kind, and the one almost exclusively employed in the laboratory. It is used as follows:—After the fluid and other matters (if any) are put into the still, the head is put on and connected with the worm-tub or refrigerator, and the joints are all securely luted.

a. Body of still, which may be either placed in a steam jacket or in a brick furnace.
b. Still head or capital.
c. Worm-tub.
d. Pewter worm or refrigerator.
e. Cold-water pipe.
f. Waste-pipe.
g. Receiver.

For ordinary liquids, a stiff paste made with linseed meal and water, to which a little chalk may be added, answers well for this purpose. For corrosive liquids, nothing is better than elastic

to a very extended heated surface; whilst it affects the evaporation of the alcohol from the wash by passing a current of steam through it.

The wash is pumped from the 'wash charger' into the worm-tub, which passes from top to bottom of the rectifier. In circulating through this tube it experiences a slight elevation of temperature. Arrived at the last convolution of the tube in the rectifier, the wash passes by the tube M in at the top of the 'analyser.' It falls, and collects on the top shelf till this overflows, whence it falls on the second shelf, and so on to the bottom. All the time this operation is going on steam is passed up from the steam boiler through fine holes in the shelves, and through valves opening upwards. As the wash gradually descends in the analyser it becomes rapidly weaker in alcohol, partly from condensation of steam which is passed into it, and partly from loss of alcohol, either evaporated or expelled by the steam, till when it arrives at the bottom it has parted with the last traces of spirits.

At the same time the vapour, as it rises through each shelf of the analyser, becomes constantly richer in alcohol, and contains less and less water because of its condensation; it then passes from the top of the analyser in at the bottom of the lower compartment of the rectifier. Here it ascends in a similar way, bubbling through the descending wash, until it arrives at F, above which it merely circulates round the first windings of

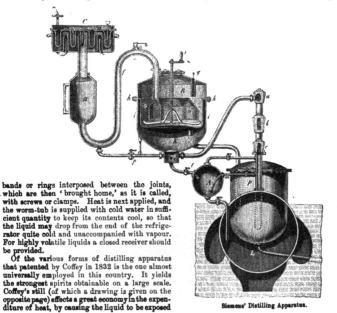

Siemens' Distilling Apparatus.

bands or rings interposed between the joints, which are then 'brought home,' as it is called, with screws or clamps. Heat is next applied, and the worm-tub is supplied with cold water in sufficient quantity to keep its contents cool, so that the liquid may drop from the end of the refrigerator quite cold and unaccompanied with vapour. For highly volatile liquids a closed receiver should be provided.

Of the various forms of distilling apparatus that patented by Coffey in 1832 is the one almost universally employed in this country. It yields the strongest spirits obtainable on a large scale. Coffey's still (of which a drawing is given on the opposite page) effects a great economy in the expenditure of heat, by causing the liquid to be exposed the wash-pipe, the low temperature of which condenses the spirits; which, collecting on the shelf at F, flows off by the tube into the finished spirit condenser.

To still further effect a saving of heat, the water for supplying the boiler is made to pass through a long coil of pipe, immersed in boiling spent wash, by which means its temperature is raised before it enters the boiler.

Another variety of distillatory apparatus is that of Siemens (see page 1593), much employed in the distillation of brandy.

It consists of two mash stills set in a boiler, and capable of being alternately used by means of the

Deroane's Distilling Apparatus.

three cocks (a, b, and c). L is the boiler; P one of the mash retorts; K is the low wine receiver; R the fore warmer, a reservoir in which the condensed water intended as feed water of the boiler is collected; C is the dephlegmator; B a reservoir for the vapours condensed in c.

From the dephlegmator the vapour passes to a condenser, not shown in the engr., page 1592.

The mash warmer consists of a cylindrical portion (i i), the lower part of which has an indentation (c). In the cylinder is placed a smaller portion (o o) of the real mash, containing a vessel, fitted with the heading tube (f n). The upper part of the fore warmer is fitted to the lower part by means of the flange (h h); r is a stirring apparatus, which is frequently set in operation during the process of distillation. The vapours from the second still are carried into the depression (c) under the fore warmer, which, in order that the vapours may come into contact with the phlegma, is covered with a sieve.

The vapours surround the under part of the mash reservoir, and enter into the tube (f), through which they pass to the lower cylinder of the dephlegmator. The condensed water of the dephlegmator is conducted into the reservoir (A). The upper and under parts of the fore warmer are made of cast iron, but the interior bottom and heating surfaces are made of copper. This kind of fore warmer has the advantage of uniformly distributing the heat, while it can be easily cleansed.

The dephlegmator (c) is so contrived that the rectified vapour can be conveyed to the condenser by two separate pipes placed in an opposite direction to each other, which are joined again in close proximity to the condenser.

The remainder of the details will be seen by studying the engr.

Another distilling apparatus is that known as Deroane's, which is an improvement upon one invented by Cellier-Blumenthal. This apparatus is only designed for the distillation of wine.

The engr. annexed gives a representation of it.

It consists of two stills (A and A'); the first rectifier (B); the second rectifier (C); the wine warmer and dephlegmator (D); the condenser (F); the regulator (E); a contrivance for regulating the flow of the wine from the cistern (G).

The still A', which, as well as the still A, is filled with wine, acts as a steam boiler. The low wine vapours evolved when they have arrived in the rectifiers come in contact with an uninterrupted stream of wine, whereby dephlegmation is effected; the vapour, thus enriched in alcohol, becomes stronger in the vessel (D), and thus arrives at the cooling apparatus (F). In order that a real rectification should take place in the rectifiers the stream of wine should be heated to a certain temperature, which is imparted to it by the heating of the condensed water. The steam from the still A' is carried by means of the pipe (z) to the bottom of the still A.

Both stills are heated by the fire of the same furnace. By means of the tube B' the liquid contained in the still A can be run into the still A'. The first rectifier (B) contains a number of semicircular discs of unequal size, placed one above the other, and which are so fastened to a vertical centre rod that they can be easily removed and cleansed. The larger discs, perforated in the manner of sieves, are placed with their concave surfaces upwards.

In consequence of this arrangement the vapours ascending from the stills meet with large surfaces moistened with wine, which, moreover, trickles downward in the manner of a cascade from the discs, and comes, therefore, into very intimate contact with the vapours. The second rectifier (c) is fitted with six compartments; in the centre of each of the partition walls (iron or copper plates) a hole is cut, and over this hole, by means of a vertical bar, is fastened an inverted cup, which nearly reaches to the bottom of the compartment wherein it is placed. As a portion of the vapours are condensed in these compartments, the vapours are necessarily forced through a layer of low wine, and have to overcome a pressure of a column of liquid 2 cm. high. The fore warmer and dephlegmator (D) is a horizontal cylinder made of copper fitted with a worm, the convolutions of which are placed vertically. The tube M communicates with this worm, the other end of which passes to o. A phlegma collects in the convolutions of this tube, which is richer in alcohol in the foremost windings, and weaker in those more remote; this fluid, collecting in the lower part of the spirals, may be drawn off by means of small tubes, thence to be transferred, either all or in part, by the aid of another tube and stopcocks to the tube o, or into the rectifier.

By means of the tube L the previously warmed wine of the dephlegmator can be run into the rectifier. The condenser (F) is a cylindrical vessel closed on all sides, and containing a worm communicating with the tube o.

The other end of the condensing tube carries the distillate away. On the top of this portion of the apparatus the tube K is placed, by means of which wine is run into the dephlegmator. The cold wine flows into the cooling vessel by the tube I.

Another variety of distillatory apparatus, invented by Langier, is that represented in the accompanying engr.

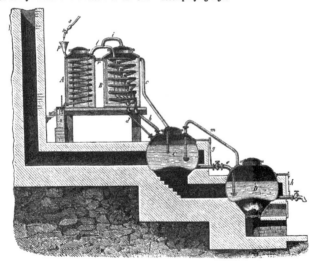

The fluid intended for distillation flows from the tube s into the funnel p, thence into the vessel A, entering its lower part, and serving to condense the alcoholic vapour. From this vessel the warmed fluid passes by means of the tube r into the lower part of the second vessel (B), where dephlegmation takes place by means of a condensing tube. From B the fluid flows through the tube *c* into the second still (C), which is heated by the hot gases evolved from the fire kept burning under the first still (C); in the still C the fluid undergoes a rectification, and the vinasses flow by the tube *e* into the still D; *m* is the pipe for conveying the hot vapour from D

FIG. 1.

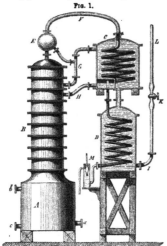

into C; the tube *b* carries the alcoholic vapours into the dephlegmator. The tube *d* conveys the phlegma into the still C; *g* and *h* are glass gauging tubes for indicating the height of the fluid in the interior of the stills; the tube *l* conveys the uncondensed vapours from the dephlegmator into the condensing apparatus, while *i* carries the vapours formed in the vessel B into the condensing apparatus.

The alcohol condensed in the cooling apparatus flows into the vessel *o*, provided with a hydrometer, which shows the strength of the liquid. The cooling apparatus of the vessel B consists of seven compartments or sections formed by wide spirals, to each of which, at its lower level, is attached a narrow tube, all of which tubes are connected to the tube *d*, which latter conveys the condensed fluids back into the still.

Fig. 1 is a very simple form of apparatus. A is a cylinder made of cast iron or copper, in which the fluid to be distilled is heated by a spiral tube made of copper. The inlet of this tube is shown at *b*, and the outlet at *a*; *c* serves to carry off the vinasses; B is the dephlegmator, through which the fluid to be distilled continually flows in a downward direction, while the vapour of the low wine evolved in A ascends uninterruptedly.

The dephlegmator is so constructed as to have as large a surface and as many points of contact as possible. The vapour ascends to the reservoir (E), and passes into the rectifier (C) by the tube F. The condensed portion returns through the tube H to the dephlegmator, whilst the uncondensed vapour passes on to the condenser of the vessel (D), where it becomes condensed, and is carried off through M. The liquid intended for distillation is kept in a tank (not shown in the *engr.*) placed above the apparatus, and is conveyed to the latter by the tube I, fitted with the stopcock K, so that the liquid arrives first in D, is next conveyed to C, thence through G into the dephlegmator, and finally into the cylinder.

FIG. 2.

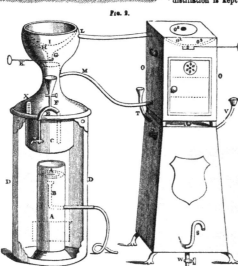

Divers adaptations for heating by steam have been arranged, in a very convenient form, by Mr Coffey. His so-called Es-CULAPIAN STILL affords the pharmaceutical chemist the means of conducting the processes of ebullition, distillation, evaporation, desiccation, &c., on a small scale. The annexed *engr.* (fig. 2) represents his apparatus.

B, a burner supplied with gas by a flexible tube; C, the boiler or still; I, an evaporating pan fixed over the boiler, and forming the top of the still head; X, a valve for shutting off the steam from I when it passes through the tube M, otherwise it would pass through L, and communicate heat to the drying closet (O O), and from thence to the condenser (T T); O is a second evaporating pan over the drying closet.

For further information on the subject of stills consult 'Ure's Dictionary,' 'Illustrated Chemistry,' and Wagner's 'Chemical Technology.'

STIM'ULANTS. *Syn.* STIMULANTIA, L. Medicines or agents which possess the power of exciting vital action. They are divided into general stimulants, or those which effect the whole system, as mercury or bark; and local or topical stimulants, or those which affect a particular organ or part only, as mustard applied as a poultice. Diffusible stimulants are general stimulants the effects of which are rapid but fugacious, as ether or alcohol. "Much discrimination and caution are required in the administration of articles of this class, because, if given when inflammation is present, they are liable to create more mischief than benefit; but they are called for when, on the decline of that condition of an organ or organs, a state of relaxation or torpidity exists. In this state of things a gentle stimulation materially assists the functions, and is productive of much benefit."

STINGS. See BITES.

STIR-ABOUT. Thick gruel formed of oatmeal and water boiled together. When eaten with cold milk, it forms the porridge of the Scotch; and when mixed with the liquor in which meat or vegetables have been boiled, it is called beef brose, kale brose, &c.

STOCK, among cooks, is condensed soup or jelly, used to make extemporaneous soup, broth, &c.

STOM'ACH AFFEC'TIONS. Those of a character to admit of being usefully noticed in a popular work are referred to under the heads APPETITE, DYSPEPSIA, SICKNESS, &c.

Dr Budd recommends small doses of ipecacuanha as a remedy for those cases of indigestion in which digestion is slow, and the food lies heavily on the stomach, and there is an inability for mental or bodily exertion for some time after meals. He says it should be given in the morning, fasting, and in quantity barely sufficient to occasion a slight feeling of vermiculating motion in the stomach, but without causing any sensation of pain or nausea. The dose to produce this effect varies from ¼ to 2 gr. He thinks there is no other medicine which appears so effectual in removing the affections in question. Small doses of rhubarb, ginger, and cayenne pepper have a similar kind of action, and may be given singly or together for the same purpose. "I generally prescribe from ¼ to 1 gr. of ipecacuanha, in a pill, with 3 or 4 gr. of rhubarb. With many, a favourite remedy for the discomfort resulting from slow digestion is a grain of cayenne pepper, with 3 or 4 gr. of rhubarb. The best time for giving these medicines is shortly" (say half an hour) "before any meal after which a sense of oppression is usually felt."

STOPP'ERS, when obstinately immoveable in bottles, are the most safely treated by patiently hitting them upwards alternately on opposite sides with a piece of wood. When this fails the part may be dipped into hot water.

"Another method of removing a bottle-stopper is to insert its head into a chink, and then endeavouring to turn the bottle with both hands. If the neck of the stopper breaks, the hand is out of the way of danger. An upright board, such a one as supports the ends of a set of shelves, should be selected in a convenient situation in the laboratory, and a vertical slit cut through it about a foot in length, an inch in width above, but gradually decreasing in size, so as to be about one third of an inch at the bottom. The top of the hole may be about the height of the breast. This aperture will in one part or another receive and retain the head of almost any stopper, and prevent its turning with the bottle. Then by wrapping a cloth about the bottle and grasping it with both hands, the attempt to turn it round so as to move the stopper may be made with any degree of force which it may be thought safe to exert. The force employed should never be carried so far as to cause fracture anywhere, but the attempts, if unavailing with the application of a moderate degree, should be desisted. Another and very successful method of removing a stopper is to turn the bottle round when held horizontally over the small flame of a spirit lamp or candle applied to the neck. The heat should be applied only to the part round the plug of the stopper, and in a few moments, when that has become warm, the stopper should be tapped with the piece of wood as before stated. As soon as the stopper moves by tapping it is to be taken out, and must not be replaced till the glass is cold.

"The application of heat in this manner must be short, and the operation altogether, to be successful, must be a quick one. If the contents of the bottle are fluid, it should be so inclined that they must not become heated; if they are volatile this method should be tried very carefully, lest the vapour formed within should burst the bottle.

"It is often advantageous to put a little olive oil round the edge of the stopper at its insertion, allowing it to soak in for a day or two. If this be done before the heat be applied, it frequently penetrates by increased facility; by oil, heat, and tapping very obstinate stoppers may be removed.

"When a stopper has been fixed by crystallisation from solution, water will sometimes set it free, and it is more efficacious in such cases than oil, because it dissolves the cement. When the cementing matter is a metallic oxide or sub-salt, a little muriatic acid may be useful if there be no objection to its application arising from the nature of the substance involved" (*Faraday*).

A writer in 'New Remedies' suggests that, in attempting to extricate the fixed stopper by means of knocking with a piece of wood, the motion given to it when putting it in should be reversed, that is, the stopper should be knocked from *right* to *left*.

STORM-GLASS. A philosophical toy, consisting of a thin glass tube about 12 inches long and ¾ inch in diameter, about three fourths filled with the following liquid, and covered with a brass cap having an almost capillary hole through it, or else tied over with bladder.

The solution. Take of camphor, 2 dr.; nitre, 1½ dr.; sal-ammoniac, 1 dr.; proof spirit, 2½ fl. oz.; dissolve, and place it in the tube above referred to. Used to foretell changes of the weather.

STOVES. In England the open grate or fire-place, because of its cheerful appearance and the sense of comfort it suggests when filled with glowing coal, is the favourite and general receptacle for the fuel with which we warm our apartments. The cosy appearance, however, of our old-fashioned English grate, constitutes its chief, if not its only merit; for it not only fails in uniformly warming and effectively ventilating our apartments, but it more or less sets into circulation a number of draughts of cold air, and besides occasionally filling our rooms with smoke, and spoiling our furniture by the deposition of soot and dust, wastes our fuel, by allowing it to escape unconsumed in the shape of smoke, and thus pollutes the atmosphere of our cities and towns.

In France, Germany, Belgium, Russia, and other European countries, as well as in Canada and other parts of America, the stove or closed fireplace is used. The domestic stove of these countries is made either of sheet or cast iron, or fire-clay. The iron stoves, being mostly composed of thin plates, soon absorb and radiate the heat; and although this property enables them to rapidly warm an apartment, it has the disadvantage, if the stove becomes red-hot, of allowing the escape through the heated metal into the surrounding air of the carbonic acid generated in the stove; and furthermore, in its immediate vicinity converts a portion of it into carbonic oxide. Such stoves must necessarily be unsafe unless used in well-ventilated apartments. (Dr Bond has suggested coating them with soluble glass as a remedy for this.) Another effect of the over-heating of the stove is to desiccate or parch the air, and to render it irritating when breathed. The fire-clay stoves are free from these drawbacks, and continue to radiate from their surfaces a large amount of heat, even when the fuel with which they have been supplied is consumed. But although we exclude the close stove from our sitting-rooms and dormitories, it is in frequent requisition in halls, picture galleries, churches, theatres, lecture-rooms, and the like.

'Stove literature,' if such a term may be applied to the various treatises descriptive of the multitude of patterns in use which have emanated alike from inventors and their critics, is so voluminous that it is impossible for us to attempt to give even a list of the numberless stoves in use, to say nothing of a commentary on their relative value. Of close stoves suitable for heating spaces other than dwelling or sleeping rooms, mention may be made of Arnott's stove, and one known as 'the Belfast.' These stoves are serviceable when it is desirable to keep up a fire for some time, as in heating a lobby. They have the advantage of requiring little, if any, attention after the fuel has been placed in them and ignited.

Of late years gas stoves, both for heating and cooking purposes, have come largely into use. One of those for the former purpose is called the "Pyropneumatic." The inner part of this apparatus is formed of lumps of fire-clay traversed by vertical air-passages which communicate with the external air by a special channel. The air becomes heated as it passes through the lumps of fire-clay, and, rising to the top of the stove, escapes therefrom by an outlet into the room. Another so-called 'ventilating' warming gas-stove is Mr George's 'Culirogen.' It consists of a stove made of thin-rolled iron, inside of which is a coil of wrought-iron tubing open at the top of the stove. The lower end of this tubing is in connection with an iron pipe which is carried through the wall of the apartment, and fed with air from without. Gas is the fuel generally used to heat the inside of the stove. The continuous current of air as it rushes into the iron pipe from without thus becomes warmed as it ascends into the coil, which it leaves to become diffused into the surrounding apartment, whilst the products of combustion of the gas used as fuel are, by means of a pipe attached to the stove, carried into the chimney, as with coal fires.

A gas cooking apparatus possesses many advantages over an ordinary coal fire. In the first place, it is more cleanly; in the second, it affords a much more uniform and equable temperature; in the third, it forms no smoke; and in the fourth, it is more economical, as well as expeditious.

Mr Eassie gives the following practical suggestions to intending purchasers of gas-stoves:

"It is not necessary here to enter into a description of any of the numberless common patterns extant, but it might be well to record the opinion of the best engineers, that the simplest gas-stove is the best. They should not be surrounded by a non-conducting material, as that affords no advantage, but the contrary. An Argand or fish-tail burner should also be used instead of rings pierced for so many separate jets; and where practicable the Bunsen burner should be employed, as the mixture of common air with the gas not only prevents the formation of soot, but also intensifies the heat."

STRABIS'MUS. *Syn.* SQUINTING. This need not be described. When one eye only is affected, an excellent plan is to blindfold the sound eye during several hours each day, until the affection be removed. When both eyes are affected, a projecting piece of pasteboard, in the line of the nose, may be worn as much as possible with the same object. In bad cases of squinting inwards, as it is called, the division of the internal rectus muscle of the eyeball by a skilful surgeon is said to often relieve the deformity.

STRANGULATION. See HANGING.

STRAP'PING. Spread adhesive plaster. Used to dress wounds, &c.

STRASS. See ENAMEL.

STRAW'BERRY. *Syn.* FRAGARIA, L. The fruit of *Fragaria vesca*, Linn., or strawberry plant. Strawberries are, perhaps, the mildest of all the cultivated fruits; they are cooling, and slightly laxative and diuretic; rubbed on the teeth, they dissolve the tartar, and whiten them. They were formerly in repute in gout, stone, and consumption. The root of the plant is aperient.

Strawberry Essence, Factitious. Nitric ether, 1 part; acetate of ethyl, 5 parts; formiate of ethyl, 5 parts; butyrate of ethyl, 5 parts; salicylate of methyl, 1 part; acetate of amyl, 3 parts; butyrate of amyl, 3 parts; glycerin, 2 parts; alcohol, 100 parts ('Pharm. Journ.').

STRAW PLAIT, and the articles made of it, are bleached by exposing them to the fumes of burning sulphur in a close chest or box; or by immersing them in a weak solution of chloride of lime, and afterwards well washing them in water. Water acidulated with oil of vitriol or oxalic acid is also used for the same purpose. Straw plait may be dyed with any of the simple liquid dyes.

STRINGHALT. The same as CHOREA, which see.

STRON'TIUM. Sr. The metallic base of the earth strontia. It was discovered by Sir H. Davy in 1808. It closely resembles barium, but is less lustrous. With chlorine it combines to form a chloride of strontium, a somewhat deliquescent salt, soluble in 2 parts of cold and in less of boiling water, and freely soluble in alcohol. With oxygen it forms an oxide.

Test. Strontium salts are precipitated by sulphuric acid and alkaline carbonates and sulphate. They are distinguished from barium by not giving such a decided precipitate with sulphates, and by not being precipitated by bichromate of potassium. From calcium, by sulphates of calcium solution giving a precipitate, and by concentrated solutions giving a precipitate with chromate of potassium. It is distinguished from magnesium by the insolubility of its sulphate.

Strontium, Oxide of. SrO. *Syn.* PROTOXIDE OF STRONTIUM, STRONTIA. *Prep.* Quite pure crystalline nitrate of strontium.

Prop. Greyish-white powder, uniting with water to form a white, somewhat soluble substance, the hydrate of strontium, Sr(HO)₂.

With acids it forms various salts, of which the carbonate is a white insoluble powder, and the nitrate a white crystalline salt, soluble in 5 parts of cold water, and in alcohol; communicating a brilliant red colour to flame.

STROPHANTHIN. C₃₀H₃₄O₁₀. A glucoside obtained by T. R. Fraser from strophanthus seeds.

Prep. To extract of strophanthus in water, add tannic acid, collect the precipitate, and digest with fresh lead oxide. Dissolve the residue in alcohol and filter; to the solution add ether in large excess, when the strophanthin will be slowly precipitated. Dissolve the precipitate in weak alcohol, pass carbonic acid gas through the liquid to remove lead, filter, evaporate in a vacuum.

Thus obtained it is colourless and imperfectly crystalline, freely soluble in water and alcohol, insoluble in ether and chloroform. Intensely bitter. Acids quickly change it to glucose.

Uses. Similar to strophanthus.—*Dose,* ₅₀₀ to ₇₀ gr.

STROPHANTHUS. The ripe seeds of *Strophanthus hispidus* DC., var. *Kombé,* Oliver, freed from their awns (B. P.). Nat. Ord. APOCYNACEÆ.

The plant yielding these seeds is a creeper found growing in various parts of Africa, where it tops the highest trees in the forest. The natives prepare from it the Kombé arrow poison, which in the Gabon district is called Inée, Onaye, or Onage. The natives pound the seed to an oily

mass, with which they coat the stem of their arrows. The effect of the poison on wild animals is to cause stupor, foaming at the mouth, and death. The flesh of the beast is eaten after cutting away the part near the wound. The seeds are contained in two pod-like follicles, bearing a plumose or feathery awn of great beauty. Their colour is greenish fawn covered with adpressed hairs; a ridge runs along one side. Length about ⅔ in., breadth ⅛ in. Kernel white and oily, taste very bitter. The active principle of the seeds is *strophanthin,* a bitter glucoside and powerful poison. Strophanthus is a valuable cardiac tonic, and strengthener of the heart-muscle; small doses increase the systole and slow the contractions; it exerts a more powerful action upon the heart, and a less powerful action on the blood-vessels than digitalis. It does not produce the digestive troubles and cumulative action observed with digitalis, and does not lose its effects by the system becoming habituated to it.—*Dose,* tincture 2—10 minims.

STROPH'ULUS. A papular eruption peculiar to infants. There are several varieties:—In strophulus intertinctus, red-gum, or red gown, the pimples rise sensibly above the level of the cuticle, possess a vivid red colour, and are usually distinct from each other; they commonly attack the cheeks, forearm, and back of the hand, and, occasionally, other parts of the body. In strophulus albidus, or white-gum, there are a number of minute whitish specks, which are sometimes surrounded by a slight redness. The two preceding varieties commonly occur during the first two or three months of lactation. In strophulus confertus, rank red-gum, or tooth-rash, which usually appears about the fourth or fifth month, the pimples usually occur on the cheeks and sides of the nose, sometimes on the forehead and arms, and still less frequently on the loins. They are smaller, set closer together, and less vivid, but more permanent than in the common red-gum. In strophulus volaticus small circular patches or clusters of pimples, each containing from six to twelve, appear successively on different parts of the body, accompanied with redness; and as one patch declines, another patch springs up near it, by which the efflorescence often spreads gradually over the whole face and body. In strophulus candidus the pimples are larger than in the preceding, and are pale, smooth, and shining; it principally attacks the upper parts of the arms, the shoulders, and the loins. The last two varieties commonly appear between the third and ninth month.

The treatment of the above affections consists chiefly in removing acidity and indigestion and duly regulating the bowels by an occasional dose of magnesia or rhubarb, or both combined. Diarrhœa may be met by the warm bath and the daily use of arrowroot (genuine), to which a teaspoonful or two of pure port wine has been added; and itching and irritation may be alleviated by the use of a lotion consisting of water, to which a little milk, lemon juice, borax, or glycerin has been added.

STRYCH'NINE. C₂₁H₂₂N₂O₂. *Syn.* STRYCH-NINA, STRYCHNIA (B. P., Ph. L., E., and D.), L. *Prep.* 1. Dissolve hydrochlorate or sulphate of

strychnine in distilled water, and throw down
the alkaloid with ammonia, carefully avoiding
excess; redissolve the precipitate in hot rectified
spirit, and collect the crystals which form as the
liquid cools.

2. (Ph. D.) Nux vomica (in powder), 1 lb., is
digested for 24 hours in ½ gall. of water acidulated with 2 fl. dr. of sulphuric acid, after which
it is boiled for half an hour, and the decoction
decanted; the residuum is boiled a second and a
third time with a fresh ½ gall. of water acidulated with 1 fl. dr. of the acid, and the undissolved matter is finally submitted to strong
expression; the decoctions are next filtered and
concentrated to the consistence of a syrup, which
is boiled with rectified spirit, 3 pints, for about
20 minutes, hydrate of calcium, 1 oz., or q. s.,
being added in successive portions during the
ebullition, until the solution becomes distinctly
alkaline; the liquid is then filtered, the spirit distilled off, and the residuum dissolved in diluted
sulphuric acid, q. s.; ammonia, in slight excess, is
added to the filtered solution, and the precipitate
which falls is collected upon a paper filter, and
dried; it is next redissolved in a minimum of
boiling rectified spirit, and digested with ½ oz.
of animal charcoal for 20 minutes; the filtered liquid, as it cools, deposits strychnine in
crystals.

3. (Ph. B.) Nux vomica, 1 lb.; acetate of
lead, 180 gr., solution of ammonia, q. s.; rectified
spirit, q. s.; distilled water, q. s. Subject the
nux vomica for two hours to steam in any convenient vessel; chop or slice it; dry it in a water-
bath or hot-air chamber, and immediately grind
it in a coffee mill. Digest the powder at a
gentle heat for 12 hours with 2 pints of the
spirit and 1 pint of the water; strain through
linen, express strongly, and repeat the process
twice. Distil off the spirit from the mixed fluid,
evaporate the watery residue to about 16 oz., and
filter when cold. Add now the acetate of lead,
previously dissolved in distilled water, so long as
it occasions any precipitate; filter; wash the precipitate with 10 oz. of cold water, adding the
washings to the filtrate; evaporate the clear fluid
to 8 oz., and when it has cooled add the ammonia
in slight excess, stirring thoroughly. Let the
mixture stand at the ordinary temperature for
12 hours; collect the precipitate on a filter, wash
it once with a few ounces of cold distilled water,
dry it in a water-bath or hot-air chamber, and
boil it with successive portions of rectified spirit,
till the fluid scarcely tastes bitter. Distil off
most of the spirit, evaporate the residue to the
bulk of about ½ oz., and set it aside to cool.
Cautiously pour off the yellowish mother-liquor
(which contains the brucia of the seeds) from the
white crust of strychnia which adheres to the
vessel. Throw the crust on a paper filter, wash
it with a mixture of two parts of rectified spirit
and one of water, till the washings cease to
become red on the addition of nitric acid;
finally, dissolve it by boiling it with 1 oz. of
rectified spirit, and set it aside to crystallise.
More crystals may be obtained by evaporating
the mother-liquor. [Strychnine is more readily
obtained, and in greater purity, from St Ignatius's bean.] The usual dose of strychnia and

its salts to commence with is from 1-30th to
1-12th of a grain, to be very slowly increased,
carefully watching its effects. Magendie says
the salts are more active than their base.

Prop. A white inodorous powder; or small,
but exceedingly brilliant, transparent, colourless,
octahedral crystals; soluble in about 7000 parts
of water at 60°, and in 2500 parts at 212° F.;
freely soluble in hot rectified spirit; insoluble in
absolute alcohol, ether, and solutions of the
caustic alkalies; imparts a distinctly bitter taste
to 600,000 times its weight of water (1 part in
1,000,000 parts of water is still perceptible—
Fownes); exhibits an alkaline reaction; and
forms salts with the acids, which are easily prepared, are crystallisable, and well defined.

Tests. 1. Potassium hydrate and the carbonate produce, in solutions of the salts of
strychnia, white precipitates, which are insoluble
in excess of the precipitant, and which, when
viewed through a lens magnifying 100 times,
appear as aggregates of small crystalline needles.
In weak solutions the precipitate only separates
after some time, in the form of crystalline
needles, which are, however, in this case, perfectly visible to the naked eye. 2. Ammonia
gives a similar precipitate, which is soluble in
excess of the precipitant. 3. Bicarbonate of sodium produces, in neutral solutions, a like white
precipitate, which is insoluble in excess, but
which redissolves on the addition of a single drop
of acid; in acid solutions no precipitate occurs
for some time in the cold, but immediately on
boiling the liquid. 4. Nitric acid dissolves pure
strychnia and its salts to colourless fluids, which
become yellow when heated. Commercial strychnine, from containing a little brucine, is reddened by this test. 5. A minute quantity of
strychnine being mixed with a small drop of concentrated sulphuric acid, placed on a white capsule or slip of glass, forms a colourless solution,
but yields, on the addition of a very small crystal
of bichromate of potassium, or a very minute
portion of chromic acid, a rich violet colour,
which gradually changes to red and yellow, and
disappears after some time. The 1-1000th of a
grain yields very distinct indications. 6. Pure
oxide or peroxide of lead produces a similar
reaction to the last, provided the sulphuric acid
contain about 1% of nitric acid.

Pois. The characteristic symptom is the
special influence exerted upon the nervous
system, which is manifested by a general contraction of all the muscles of the body, with
rigidity of the spinal column. A profound calm
soon succeeds, which is followed by a new tetanic
seizure, longer than the first, during which the
respiration is suspended. These symptoms then
cease, the breathing becomes easy, and there is
stupor, followed by another general contraction.
In fatal cases these attacks are renewed, at intervals, with increasing violence, until death ensues. One phenomenon which is only found in
poisonings by substances containing strychnine
is, that touching any part of the body, or even
threatening to do so, instantly produces the
tetanic spasm.

Treat. The stomach should be immediately
cleared by means of an emetic, tickling the fauces,

&c. To counteract the asphyxia from tetanus, &c., artificial respiration should be practised with diligence and care. The patient may be kept fully under chloroform or ether; chloral hydrate and bromide of potassium may be given. "If the poison has been applied externally, we ought immediately to cauterise the part, and apply a ligature tightly above the wound. If the poison has been swallowed for some time, we should give a purgative clyster, and administer draughts containing sulphuric ether or oil of turpentine, which in most cases produce a salutary effect. Lastly, injections of chlorine and decoction of tannin are of value."

According to Ch. Gunther, the greatest reliance may be placed on full doses of opium, assisted by venesection, in cases of poisoning by strychnia or nux vomica. His plan is to administer this drug in the form of solution or mixture, in combination with a saline aperient.

Uses, &c. It is a most frightful poison, producing tetanus and death in very small doses. Even $\frac{1}{16}$ gr. will sometimes occasion tetanic twitchings in persons of delicate temperament. $\frac{1}{2}$ gr. blown into the throat of a small dog produced death in six minutes. In very minute doses it acts as a useful tonic in various nervous diseases, chronic diarrhœa, leucorrhœa, &c.; in slightly larger ones it has been advantageously employed in certain forms of paralysis, in tic-douloureux, impotence, &c.—*Dose,* $\frac{1}{16}$ to $\frac{1}{12}$ gr. (dissolved in water by means of a drop of acetic or hydrochloric acid), gradually and cautiously increased until it slightly affects the muscular system. Externally, $\frac{1}{8}$ to $\frac{1}{4}$ gr. at a time.

The Edinburgh College ordered the nux vomica to be exposed for two hours to steam, to soften it, then to chop or slice it, next to dry it by the heat of a vapour-bath or hot air, and, lastly, to grind it in a coffee mill. In the process of the Ph. L. 1836 magnesia was employed to effect the precipitation. In the last Ph. L. strychnine appears in the Materia Medica. Most of that of commerce is now obtained from St Ignatius's bean, which, according to Geiseler, yields 1½% of it; whereas 8 lbs. of nux vomica produce little more than 1 dr. Commercial strychnine may be freed from brucine by digesting the powder in dilute alcohol.

The salts of strychnine, which are occasionally asked for in trade, are the acetate (strychniæ acetas), hydrochlorate or muriate (s. murias—Ph. D.), hydriodate (s. hydriodas), nitrate (s. nitras), phosphate (s. phosphas), and sulphate (s. sulphas). All of these may be easily formed by simply neutralising the acid, previously diluted with 2 or 3 parts of water, with the alkaloid, assisting the solution with heat; crystals are deposited as the liquid cools, and more may be obtained by evaporating the mother-liquor.

STRYCHNOS. See NUX VOMICA, and BEAN, ST IGNATIUS'S.

STUC'CO. The name of several calcareous cements or mortars. Fine stucco is the third or last coat of three-coat plaster, and consists of a mixture of fine lime and quartzose sand, which, in application, is "twice hand floated and well trowelled." See CEMENTS.

STUFFING. Seasoning, placed in meat, poul-

try, game, &c., before dressing them, to give them an increased relish. The same materials formed into balls are added to soups, gravies, &c., under the name of FORCEMEAT.

Prep. 1. (For fowls, &c.) Shred a little ham or gammon, some cold veal or fowl, some beef suet, a small quantity of onion, some parsley, a very little lemon peel, salt, nutmeg, or pounded mace, and either white pepper or cayenne, and bread crumbs, pound them in a mortar, and bind it with 1 or 2 eggs.

2. (For hare, or anything in imitation of it—*Mrs Rundell.*) The scalded liver, an anchovy, some fat bacon, a little suet, some parsley, thyme, knotted marjoram, a little shallot, and either onion or chives, all chopped fine, with some crumbs of bread, pepper, and nutmeg, beaten in a mortar with an egg.

3. (For goose.) From sage, onion, suet, and crumb of bread. Geese are now, however, more commonly stuffed with veal stuffing.

4. (For veal—*Soyer.*) Chop $\frac{1}{2}$ lb. of suet, put it into a basin with $\frac{1}{4}$ lb. of bread crumbs, a tea-spoonful of salt, a $\frac{1}{4}$ do. of pepper, a little thyme or lemon peel chopped, and 8 whole eggs; mix well.

Obs. 1 lb. of bread crumbs and one more egg may be used; they will make it cut firmer. This, as well as No. 1, is now commonly employed for poultry and meat. Ude, a great authority in these matters, observes that "it would not be amiss to add a piece of butter, and to pound the whole in a mortar." "Grated ham or tongue may be added to this stuffing" (*Rundell*). This is also used for turkeys, and for 'forcemeat patties.'

STUFFING (Birds, &c.). The skins are commonly dusted over with a mixture of camphor, alum, and sulphur, in about equal quantities; or they are smeared with Bécœur's arsenical soap, noticed under SOAP. According to Crace Calvert, carbolic acid, which is worth only about 2s. per gall., is superior to all other substances for preserving the skins of birds and animals as well as corpses. See TAXIDERMY, PRACTICAL.

STURDY. This disease, known also by the name of GAD, which attacks cattle and sheep, but more particularly the latter, is caused by the presence in the brain of the animal of a hydatid —a creature enclosed in a sac of serous fluid. This hydatid develops from the ova of the tape-worm in the animal's body, whence it has gained the grass which constitutes the cattle or sheep's food, upon which it has been voided by dogs and other animals.

It is most common in sheep of from six to eight months old, and, as might be expected, with those which feed in damp meadows. The animals attacked by it turn round and round in one position, lose their gregarious habits, seem dazed, and refuse their food; which latter circumstance frequently causes death by inducing starvation.

As regards the treatment of this disease, Mr Finlay Dun writes:—"A stout stocking wire thrust up the nostrils has long been used with occasional success to get rid of the hydatid; but the use of the trocar and canula now sold by most surgical instrument makers is much safer and better. The sheep is placed with its feet tied upon a table or bench, and the head carefully examined, when a soft place may often be detected, indicating that the hydatid lies underneath. A portion of the skin is dissected back and the trocar and canula introduced, when the hydatid will often come away as the trocar is withdrawn." Mr Dun says that, "should the trocar fail to extract it, it must be drawn to the surface by a small syringe made for the purpose. Furthermore, the wound, after the removal of the hydatid, must be treated with a cold water dressing."

All cattle similarly affected should be treated as above.

STUR'GEON. Several species of *Acipenser* pass under this name. The common sturgeon is the *Acipenser sturio*, Linn. The roe is made into 'caviare,' the swimming-bladder into 'isinglass.'

STY. *Syn.* STYE, STIAN; HORDEOLUM, L. A small inflamed tumour or boil at the edge of the eyelid, somewhat resembling a barleycorn. It is usually recommended to promote its maturation by warm applications, since "the stye, like other furunculous inflammations, forms an exception to the general rule that the best mode in which inflammatory swellings can end is resolution."

STYP'TICS. *Syn.* STYPTICA, L. Substances which arrest local bleeding. Creasote, tannic acid, alcohol, alum, and most of the astringent salts belong to this class.

Styptic, Brocchieri's. A nostrum consisting of the water distilled from pine tops.

Styptic, Eaton's. A solution of sulphate disguised by the addition of some unimportant substances. "Helvetius's styptic was for a long time employed under this title" (*Paris*).

Styptic, Helvetius's. *Syn.* STYPTICUM HELVETII, L. Iron filings (fine) and cream of tartar, mixed to a proper consistence with French brandy. See POWDER, HELVETIUS'S.

Styptic Euspinis. Tannic acid, 5 parts; brandy, 10 parts; rose-water, 120 parts; dissolve.

STY'RAX. *Syn.* STORAX, STORAX BALSAM; STYRAX (Ph. L. & E.), L. A balsam prepared from the inner bark of *Liquidambar orientalis*. Two or three varieties are known in commerce—liquid storax (*styrax liquida*); lump of red storax (*s. in massis*), which is generally very impure; storax in tears (*s. in lachrymis*); and storax in reeds (*s. calamita*). The last are now seldom met with in trade.

PREPARED STORAX (*styrax colata*, *s. præparata*, B. P., Ph. L.) is obtained by dissolving storax, 1 lb., in rectified spirit, 4 pints, by a gentle heat, straining the solution through linen, distilling off greater part of the spirit, and evaporating what is left to a proper consistence by the heat of a water-bath. It is less fragrant than the raw drug.

Storax is stimulant, expectorant, and nervine. It was formerly much used in menstrual obstructions, phthisis, coughs, asthmas, and other breath diseases. It is now chiefly used as a perfume.—*Dose*, 6 to 20 or 30 gr. (10 to 20 gr. twice a day—B. P.).

A factitious strained storax is made as follows:—1. Balsam of Peru, 1 lb.; balsam of tolu, 4 lbs.; mix.

2. Gum benzoin, 8 lbs.; liquid storax, 6 lbs.; balsam of tolu and Socotrine aloes, of each, 3 lbs.; balsam of Peru, 2 lbs.; N.S.W. yellow gum, 7 lbs.; rectified spirit, 7 galls.; digest, with frequent agitation, for a fortnight; strain and distil off the spirit (about 5½ galls.) until the residuum has a proper consistence.—*Prod.*, 28 lbs.

3. Liquid storax, 1 oz.; Socotrine aloes, ½ lb.; balsam of tolu, 2 lbs.; rectified spirit, q. s.

SUB-. See NOMENCLATURE and SALTS.

SU'BERIC ACID. $C_6H_{10}(CO_2H)_2$. Obtained by boiling rasped cork, palm oil, or castor oil, for some time in nitric acid.

SUBLIMA'TION. The process by which volatile substances are reduced to the state of vapour by heat, and again condensed in the solid form. It differs from ordinary distillation in being confined to dry solid substances, and in the heat employed being, in general, much greater.

SUB'STANTIVE COLOURS, in the art of dyeing, are such as impart their tints to cloth and yarns without the intervention of a mordant; in contradistinction to adjective colours, which require to be fixed by certain substances which have a joint affinity for the colouring matter and the material to be dyed.

SUCCIN'IC ACID. $C_4H_6O_4$. *Syn.* ACIDUM SUCCINICUM. There are two modifications, ordinary succinic acid and isosuccinic acid. The former need only be considered. It occurs ready formed in amber and certain lignites, and occasionally in the animal organism.—*Prep.* 1. From amber, in coarse powder, mixed with an equal weight of sand, and distilled by a gradually increased heat; or from the impure acid obtained during the distillation of oil of amber; the product in both cases being purified by wrapping it in bibulous paper, and submitting it to strong pressure, to remove the oil, and then resubliming it.

2. From malic acid, by fermentation. The juice of mountain ash berries is neutralised with chalk, and the calcium malate thus obtained is mixed with water and yeast or decaying cheese in an earthen jar, and kept warm for a few days. The calcium succinate thus obtained is decomposed with dilute sulphuric acid, and the succinic acid purified by recrystallisation from water and by sublimation.

Prop., &c. Colourless, inodorous, monoclinic prisms; soluble in 23 parts of water at 20° C., and in 4 parts of boiling water; fusible at 180° C., and volatile, without decomposition. Its salts are called 'succinates,' most of which are soluble. Succinate of ammonium is used as a test for iron. Succinic acid is distinguished from benzoic acid by its greater solubility, and by giving a brownish or pale red bulky precipitate with ferric chloride in neutral solutions; whereas that with benzoic acid is paler and yellower.

Uses, &c. Succinic acid is antispasmodic, stimulant, and diuretic, but is now seldom used.—*Dose*, 5 to 15 gr.

SUC'CORY. Chicory, or wild endive. See CHICORY.

SUDORIF'ICS. See DIAPHORETICS.

SU'ET. *Syn.* SEVUM, L. This is prepared from the fat of the loins of the sheep or bullock by melting it by a gentle heat, and straining the liquid fat. In this state it forms the ADEPS OVILLUS (Ph. D.), SEVUM (Ph. L. & E.), SEVUM OVILLUM, or SEVUM PRÆPARATUM, employed in medicine and perfumery, as the basis of ointments, cerates, plasters, pommades, &c. Suet, Mel'ilot. *Syn.* SEVUM MELILOTI, L. *Prep.* From suet, 8 lbs.; melilot leaves, 2 lbs.; boil until the leaves are crisp, strain, and allow it to cool very slowly, so that it may 'grain well.' Used by farriers, and to make melilot plaster.

SUFFOCA'TION. The treatment varies with the cause. See ASPHYXIA, CHARCOAL, DROWNING, HANGING, SULPHURETTED HYDROGEN, &c.

SUG'AR. $C_{12}H_{22}O_{11}$. *Syn.* CANE SUGAR, SACCHAROSE, SACCHAROBIOSE; SACCHARUM, L.; SUCRE, Fr. This well-known and most useful substance is found in the juice of many of the canes or grasses, in the sap of several forest trees, in the nectar of flowers, in many seeds, and in the roots of various plants. In tropical climates it is extracted from the sugar-cane (*Saccharum officinarum*), in China from the sweet sorgho (*Sorghum saccharatum*), in North America from the sugar maple (*Acer saccharinum*), and in France, Germany, Russia, and Belgium from white beetroot (*Beta vulgaris*, var. *alba*).

Until of late years the ordinary sugar consumed in this country was that chiefly sent from the West Indian islands, South America, the Mauritius, &c., and was the produce of the sugar-cane; recently, however, large and increasing quantities of beetroot sugar have found their way into the English markets from the Continental factories, at such a price and quality as to seriously threaten the future of the sugar-cane industry.

The *Saccharum officinarum*, or sugar-cane, of which there are several varieties, ranges in height from 6 to 15 feet, and in diameter from 1 to 2 inches. In order to obtain the saccharine juice contained in it, the cane, stripped of its leaves, is cut just before the commencement of inflorescence, the period in which it is richest. As the sap or juice is found to abound most in sugar when taken from the lower part of the stem, the cane is cut off nearly close to the ground.

The stump which remains develops into a fresh plant, and one plant thus treated will last several years, not, however, without a gradual diminution in the size and quality of the successive crops.

In South America and the West Indies a variety known as the Otaheite cane is extensively cultivated, since it is very productive, and yields a large amount of juice.

The annual average produce in raw sugar per acre of land is in—

Demerara	. .	4480 lbs.
Louisiana	. . .	1200 „
Mauritius	. 3500 to 5500 „
Jamaica	. .	1344 „
India	. .	896 „
Rio Janeiro	. .	2100 „
Java	. . .	3360 „

Sugar-cane growing in the below-mentioned places has, according to the analysis of the three chemists whose names are appended, the following composition:

	(a) Péligot.	(b) Dupuy.	(c) Icery.
Sugar	18·0	17·8	20·0
Water	. 72·1	72·0	69·0
Cellulose	. 9·9	9·8	10·0
Salts .	. —	0·4	0·7—1·2

The cane, therefore, may be said to yield 90% of juice, which latter contains from 18 to 20 parts of pure sugar. However, the actual quantity obtained is rarely if ever more than 1 lb. of sugar to a gallon of juice, or 10%, and much more frequently only 8%.

A large part of this loss is due to the prolonged exposure of the cane juice to heat during its repeated boilings, whereby a large proportion of its crystallisable sugar is converted into the uncrystallisable variety, which passes away in the form of molasses or treacle. Another important cause of loss is the retention of a large amount of juice by the cane.

The following figures will convey an idea of the enormous quantities of sugar consumed yearly throughout the globe. They are taken from the 1888 edition of Messrs Lock and Newlands Brothers' 'Handbook for Planters and Refiners.'

	Year.	Aggregate consumption.	lbs.per head.
United Kingdom . .	1875	18,374,543	62·80
Holland . . .	1874	8,000,000	25·03
Belgium . . .	1874	1,000,000	28·19
Hamburgh (imports) .	1873	2,223,733	—
Germany . . .	1874	6,120,000	16·60
Denmark . . .	1873	533,881	33·30
Sweden . . .	1873	630,741	16·90
Norway . . .	1873	193,086	12·70
France . . .	1874	5,000,000	15·50
Austria and Hungary .	1874	3,400,000	15·10
Switzerland . .	1873	381,295	15·90
Portugal . . .	1874	300,000	8·40
Spain . . .	1873	81,817	0·54
Russia and Poland .	1874	4,000,000	5·40
Turkey . . .	1874	500,000	3·80
Greece . . .	1871	86,800	6·60
Italy . . .	1873	865,350	3·60
United States . .	1873	13,040,500	37·80
British America .	1875	1,721,386	51·40
Brazil . . .	1874	642,857	8·00
Peru . . .	1874	570,000	5·61
River Plate States .	1874	1,000,000	43·90
Other Southern and Central American States	1874	500,000	—
West Indian Islands (British and foreign) .	1874	1,000,000	—
North and South Africa .	1874	1,000,000	—
Australia . . .	1874	1,718,142	85·90
India, China, and the Eastern and Pacific Islands . . .	—	25,000,000	—

The late Dr Edward Smith found that 98% of indoor operatives partook of 7½ oz. of sugar per adult weekly; that 96% of Scotch labourers use it, and 80% of Irish. He further states that in Wales sugar is commonly used to an average extent of 6 oz. per adult weekly; but that there is a marked difference in the rate of consumption in the northern and southern portions of the Principality. In North Wales, for example, the average amount per head is 11¼ oz., whereas in South Wales it is only 3 oz.

The manufacture of sugar is exclusively conducted on the large scale. The recently cut canes are crushed between rollers, and the expressed juice is suffered to flow into a suitable vessel, where it is slowly heated to nearly the boiling-point, to coagulate albuminous matter. The crushed canes generally supply the fuel needed for this purpose. The ashes left after the combustion of the canes are carefully collected and used as a manure for future crops of sugarcane, as they are rich in potash, &c.

The cut below represents a press for the extraction of the juice from the canes. By means of the screws (i i), the rollers are adjusted to the proper distance apart; the upper roller is half the size of the two lower ones, and all are moved by cogged wheels fitting on to the axes of the rollers. The sugar-canes are transferred from the slate gutter (d d) to the rollers (a c), which press them a little; and from thence they are carried over the arched plate (n) to the rollers (o b). The pressed sugar-canes fall over the gutter (f), the expressed juice collecting in g g, and running off through h. A small quantity of milk of lime is then added to the juice to remove mechanical impurities, and the skimmed and clarified juice, after being sufficiently concentrated by rapid evaporation in open pans, is transferred to coolers, and thence into upright casks perforated at the bottom, and so placed that the syrup, or uncrystallisable portion, may drain off into a tank or cistern from the newly formed sugar. During the period of crystallisation it is frequently agitated, in order to hasten the change, and to prevent the formation of large crystals. The solid portion of the product forms moist, raw, or muscovado sugar; the uncrystallisable syrup, molasses or treacle. The term 'molasses' is usually restricted to the drainings from raw sugar, and 'treacle' to the thicker syrup which has drained from refined sugar in the moulds.

Raw sugar is refined by redissolving it in water, adding to the solution albumen, in the form of serum of blood or white of egg, and, sometimes, a little lime-water, and heating the whole to the boiling-point; the impurities are then removed by careful skimming, and the syrup is decolourised by filtration through recently burnt animal charcoal. The clear decolourised syrup is next evaporated to the crystallising point in vacuo, and at once transferred into conical earthenware moulds, where it solidifies, after some time, to a crystallised mass; this,

when drained, washed with a little clean syrup and dried in a stove, constitutes ordinary loaf lump, or refined sugar. Sometimes, in washing the crystallised mass for the purpose of removing the coloured syrup which is mingled with it, the process known as 'claying' is followed.

In this case, instead of white syrup being used, a layer of a thin paste of clay is poured into the mould on to the base of the inverted sugar cone, through which the water escaping from the clay paste permeates, and carries with it the coloured syrup. Neither the mud nor the clay mixes with the sugar, but remaining on the top they soon become hard, when they are removed. As the syrup running from the moulds still contains a large quantity of crystalline cane sugar, this is recovered as follows:

The syrup, after being sufficiently concentrated by boiling in a vacuum pan, is removed and allowed to cool, when it assumes the appearance of a crystalline magma known as 'crushed sugar.'

Crushed sugar is a mixture of a large quantity

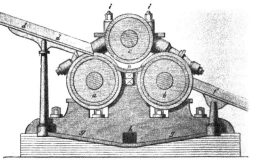

of sugar crystals with uncrystallisable syrup. To get rid of this latter from the crystals, the mass is placed in quantities of 3 or 4 cwt. at a time in a 'centrifugal machine.' This, of which an engraving is given below, consists, as will be seen, of a drum fixed on a vertical axis. The walls of the drum are made of perforated metal, or are formed of meshed wire-work, and the drum itself enclosed in an outer metal cylinder, which is fixed, and, of course, unperforated. When the drum is made to revolve on its axis at the rate of 1000 or 1200 revolutions in a minute, the syrup flying off by centrifugal action, and escaping through the perforation at the sides of the drum, is received into the outer cylinder, whence it escapes by a trough into a proper receptacle, leaving behind the crystals in the interior of the drum.

a is an open drum of fine-meshed wire-work, caused to revolve in the cast-iron vessel (b b), by means of the bevel wheels (c d), gearing with a motive power. The motion of the drum can be stopped by means of the brake (e), and regulated by the weights placed at o.

When the crystallisation of sugar is allowed to take place quietly and slowly, the product is sugar-candy. The evaporation at a low temperature in vacuum pans has the effect of diminishing the yield of treacle. One half of the sugar

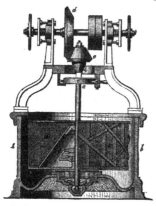

contained in the cane usually remains in the molasses. To obtain this many methods have been prepared, but one of the most important depends upon the formation of basic strontium saccharosate, which being insoluble in syrup can be separated; it is then decomposed by the action of carbonic acid, and caused to yield crystallisable sugar.

Prop. Pure cane sugar crystallises from aqueous solution in hard, transparent, monoclinic prisms, which when broken emit a curious bluish light. Sugar requires for its solution only 1-3rd of its weight of cold and still less of boiling water; it is practically insoluble in cold rectified spirit; aqueous alcohol dissolves it, and Schiebler has constructed tables showing this solubility in alcohol of different strengths—a very important matter in the manufacture of liqueurs (*vide* ' Berichte deutsch. Chem. Ges.,' v, 348). It melts at 160°—161° C., and cools to a glassy amorphous mass (barley-sugar); at higher temperatures it suffers rapid decomposition, and fuses to a brown, uncrystallisable mass (caramel, $C_{12}H_{18}O_9$, which is used for tinting liquids brown). At still higher temperatures sugar becomes carbonised, giving off aldehyde, acetic acid, carbon monoxide, carbon dioxide, marsh gas, &c.; long boiling with water lessens its tendency to crystallise. When sugar is boiled with dilute acids, *e.g.* hydrochloric acid, it is transformed into invert sugar, which is a mixture of equal quantities of dextrose and levulose; it takes up water thus:

$$C_{12}H_{22}O_{11} + H_2O = C_6H_{12}O_6 + C_6H_{12}O_6$$
Cane sugar + Water = Dextrose + Levulose.

Certain ferments also have the power of effecting the transformation. When yeast is added to a solution of cane sugar moist sugar is formed on standing, and this is then broken down into carbonic acid and alcohol by the action of the yeast-cells. Cane sugar acts as an antiputrescent, and is therefore used in the preservation of fruits, &c. Cane sugar stands to dextrose and levulose in the same relation as the ethers do to their corresponding alcohols. Its aqueous solution dissolves alkalies, earths, and many metallic oxides with facility. The presence of cane sugar in solutions containing certain metallic salts prevents the precipitation of their oxides by alkalies. The oxides of copper and iron are amongst those thus kept in solution. Sugar also possesses the power of effecting the partial or complete reduction of many metallic oxides if boiled with their salts. The first result is exemplified in the case of the chromates; for if a chromate be added to a solution of sugar, and to the mixture a few drops of free acid, the chromic acid suffers reduction to chromic oxide, which, dissolving in the excess of acid, imparts a green colour to the liquid. Mercuric salts become reduced to mercurous, whilst the salts of gold throw down a precipitate of the metal in fine powder. The action of strong oil of vitriol on cane sugar is very energetic. The sugar is instantly reduced to a black charred mass, whilst carbonic and formic acids are given off. The same effects are produced by exposing it to dry chlorine at a temperature of 100° C. By nitric acid of sp. gr. 1·25 cane sugar is converted into saccharic acid; if a stronger acid be employed, oxalic acid is produced. Triturated with 8 parts of lead oxide the mixture takes fire. When a mixture of concentrated nitric and sulphuric acids is poured on to cane sugar, an explosive compound, resembling gun-cotton, is produced. This body is known as 'nitro-sugar.' Cane sugar combines with several oxides and hydroxides, forming metallic compounds; these are called saccharosates, and amongst them those of sodium, ammonium, calcium, strontium, lead, and iron are the best known. Sp. gr. 1·593 at 3·9° C.

Pur. Moist or muscovado sugar and crushed lump sugar are occasionally adulterated with chalk, plaster, sand, potato flour, and other fecula; but frequently with starch sugar (fecula), sugar, or potato sugar (*see* further on). These frauds may be detected as follows:

Tests. 1. Pure cane sugar dissolves freely and entirely in both water and proof spirit, forming transparent colourless solutions, which are unaffected by either sulphuretted hydrogen or dilute sulphuric acid.

2. Its solution bends the luminous rays in circumpolarisation to the right, whereas grape sugars bend it to the left. (Of late years, owing to the little difference in price between the two, this form of adulteration has been abandoned.—Ed.)

3. (*Chevallier.*) Boiled for a short time in water containing 2% or 3% of caustic potash, the liquid remains colourless; but it turns a more or less intense brown, according to the quantity, if starch sugar is present. Even 2% or 3% of starch sugar may be thus detected.

4. (*E. Krauts.*) A filtered solution of 33 gr. of cane or beet sugar in 1 fl. oz. of water, mixed with 3 gr. of pure hydrate of potassium, and then agitated with 1½ gr. of sulphate of copper in an air-tight bottle, remains clear, even after the lapse of several days; but if starch or potato sugar be

present a red precipitate is formed after some time; and if it is present in considerable quantity the copper will be wholly converted into oxide within twenty-four hours, the solution turn-ing first blue or green, and then entirely losing its colour.

5. (*Fehling*.) A solution of cane sugar is mixed with a solution of sulphate of copper, and hydrate of potassium added in excess; a blue liquid is obtained, which, on being heated, is at first but little altered; a small quantity of red powder falls after a time, but the liquid long retains its blue tint. When grape sugar or fecula sugar is thus treated, the first application of heat throws down a copious yellowish precipitate, which rapidly changes to scarlet, and eventually to dark red, leaving a nearly colourless solution. This is an excellent test for distinguishing the two varie-ties of sugar, or discovering an admixture of grape sugar with cane sugar. The $\frac{1}{1000}$ part of grape sugar may be thus detected. The proportion of oxide of copper produced affords a good criterion, not only of the purity of the sugar, but also of the extent of the adulteration. Tables should be consulted when this test is used quantitatively.

6. (*Trommer*.) Copper sulphate, caustic soda, and alkaline acetate solution, added in this order to a solution of cane sugar containing grape sugar, will precipitate copious oxide on warming.

7. Riffard ('Journ. de Pharm. et de Chimie,' 1874, 49—'Pharm. Year-book,' 1874), taking advantage of the fact that sugar, like tartaric, malic, citric acid, and albumen, prevents the pre-cipitation of iron by ammonia, employs iron as a means for estimating sugar. A solution contain-ing sugar and iron in a certain proportion, when saturated with ammonia, will form a compound of a fine red colour, which remains clear if no alkaline earthy metals are present. Riffard has applied to sugar the method proposed by Juette for the estimation of tartaric acid. He observed that a neutral or acid solution of crystallised perchloride of iron, when heated for a consider-able time to 100° C., requires 2·710 grms. of sugar, if 100 milligrammes of iron are to remain in solution in the presence of ammonia. If, on the other hand, the solution is prepared simply by dissolving crystallised perchloride of iron in pure water, without the addition of an acid, 100 milli-grammes of iron only require 2·587 grms. of sugar to remain dissolved. In this case the liquid is perfectly clear, and remains so; but if a smaller quantity of sugar be added, it is turbid, and deposits peroxide of iron. To estimate the sugar by this process, 25·870 grms. of the substance to be tested are dissolved, the solution mixed with a few drops of oxalate of ammonia to precipitate the lime, filtered and made up with water to 250 c.c.; 25 c.c. of this mixture require the addition of as many milligrammes of iron as there are per cents. of pure sugar in the example under examination, and by two tests the following results will be arrived at :—With *s* milligrammes of iron the solution is clear *;* with *s* + 1 milli-grammes of iron the solution is precipitated *; s* re-presenting the number of per cents. of sugar contained in the sample.

8. Perrot's method for the determination of sugars by means of normal solutions is as follows :

—He prepares a standard solution of copper by dissolving 39·275 grms. of sulphate of copper, very pure, and dried between several folds of fil-tering paper, and makes it up with distilled water to 1000 c.c. Each c.c. of this solution contains 0·01 grm. of copper. On the other hand, he dissolves about 25 grms. of pure cyanide of potassium in one litre of distilled water. Of this solution 10 c.c. are taken and put in a flask, to which about 20 c.c. of ammonia are added, and the liquid is kept at a tempera-ture of 60° or 70° C. He pours in the copper solution drop by drop by means of a burette graduated into tenths of a c.c. until there appears the blue tint characteristic of salts of copper in an ammoniacal solution. The number of degrees of the burette are then read off, and indicate the quantity of copper which has been required to produce the reaction. The solution of the sugar in question (previously inverted if it is required to determine crystalline sugar) is then placed in contact with an excess of Fehling's liquor, and reduced in the water-bath. The whole is filtered in order to collect the precipitate of suboxide, which is first well washed with hot water, and dissolved in nitric acid, diluted with an equal volume of water, and a few fragments of chlorate of potassa are added. This solution is effected on the filter, which is then carefully washed in acidulated water. The filtrate to which the washings are added is then mixed with water enough to make up 100 or 150 c.c., and is then poured by means of the burette into 10 c.c. of cyanide, mixed with 20 c.c. of ammonia as above, stopping when the blue colour appears, and read-ing off the quantity of copper employed. From the former experiment it is known how much copper 10 c.c. of the cyanide solution require. Hence it is easy to calculate the total amount of copper which has been present as suboxide. The amount of sugar is then found from the data that 9298 parts of copper equal 5000 of crystalline sugar, or 5263 of glucose ("Comptes Rendus"—'Chem. News,' January 5th, 1877).

9. The specific gravities and crystalline forms offer other means of distinguishing the varieties of sugar. For other methods consult Tucker's 'Manual of Sugar Analysis,' p. 287.

Concluding Remarks. Refined sugar (SAC-CHARUM—Ph. L., S. PURUM—Ph. E., S. PURIFI-CATUM—Ph. D.), raw sugar (S. COMMUNE—Ph. E.), and molasses or treacle (SACCHARI FÆX—Ph. L. and E.) were officinal.

The relative sweetening power of cane sugar is estimated at 100; that of pure grape sugar at 60; that of common fecula or starch sugar at 30 to 40.

Several improved processes have been intro-duced during the last few years for treating the juice. Most of these refer to the machinery. Mr Fryer's ' concretor ' very quickly evaporates the clarified juice, and turns it at once into a solid mass which can easily be packed. Large vacuum pans are now a great feature of the sugar re-finery. Heckmann's pan will boil down at one time 20 tons of juice, while that of Messrs Adam, of Greenock, will deal with 27 tons.

The strontia process is now very widely used. It consists in precipitating the sugar from the boiling by adding strontium hydrate; the precipitate solution, $C_{12}H_{22}O_1(SrO)$, is well washed with hot water, and afterwards suspended in boiling water and allowed to cool, when most of the strontia is deposited as hydrate, and the remainder is precipitated by blowing in carbonic acid gas.

TABLE *showing the Specific Weight of Sugar Solutions with the corresponding percentage of Cane Sugar at 17·5° C.*—GERLACH.

Percentage Cane Sugar.	Specific Weight of Sol.	Percentage Cane Sugar.	Specific Weight of Sol.
75	1·383,342	37	1·164,056
74	1·376,822	36	1·159,026
73	1·370,345	35	1·154,032
72	1·363,910	34	1·149,073
71	1·357,518	33	1·144,150
70	1·351,168	32	1·139,261
69	1·344,860	31	1·134,406
68	1·338,594	30	1·129,586
67	1·332,370	29	1·124,800
66	1·326,188	28	1·120,048
65	1·320,046	27	1·115,330
64	1·313,946	26	1·110,646
63	1·307,887	25	1·105,995
62	1·301,868	24	1·101,377
61	1·295,890	23	1·096,792
60	1·289,952	22	1·092,240
59	1·284,054	21	1·087,721
58	1·278,197	20	1·083,234
57	1·272,379	19	1·078,779
56	1·266,600	18	1·074,356
55	1·260,861	17	1·069,965
54	1·255,161	16	1·065,606
53	1·249,500	15	1·061,278
52	1·243,877	14	1·056,982
51	1·238,293	13	1·052,716
50	1·232,748	12	1·048,482
49	1·227,241	11	1·044,278
48	1·221,771	10	1·040,104
47	1·216,339	9	1·035,961
46	1·210,945	8	1·031,848
45	1·205,589	7	1·027,764
44	1·200,269	6	1·023,710
43	1·194,986	5	1·019,686
42	1·189,740	4	1·015,691
41	1·184,531	3	1·011,725
40	1·179,358	2	1·007,788
39	1·174,221	1	1·003,880
38	1·169,121	0	1·000,000

The presence of certain saline bodies in a solution of cane sugar exercises a very prejudicial effect upon it, since these, by combining with the sugar, give rise to compounds which contribute to the more or less reduction of the sugar to the uncrystallisable condition, and to a consequent increase of the molasses.

One of the chief constituents of the sugarcane that possesses this objectionable property is potash in combination with acids, both organic and inorganic. A patent for the removal of these potash salts has been taken out by Messrs Newlands. The patentees proceed upon the facts that the solubility of alum in water is very trifling, and that it contains only 1-10th part of its weight of potash. They add to a concentrated syrup a strong solution of sulphate of alumina (having by a previous examination of the syrup determined the quantity required). Sulphate of

potash is thereby formed, and this uniting with the sulphate of alumina, the resulting alum after a time deposits in a crystalline form at the bottom of the vessel containing the sugar solution. This being run off into another receptacle, the free acid of which it now contains a large quantity is neutralised with lime or chalk, boiled, filtered, and passed through charcoal.

The addition of the lime also throws down the alumina liberated by the reaction, which carries with it and removes certain injurious nitrogenous principles previously present in the saccharine liquid.

Some few years back Messrs Dubrunfaut and Péligot, being cognizant of the fact of the insolubility, in boiling water, of the compounds of sugar with lime, based upon it a method of separating crystallisable sugar from treacle. Péligot has obtained from common treacle one fourth of its weight of crystallised sugar by dissolving the precipitated sugar lime in water, and separating the lime by passing into the mixture a stream of carbonic acid; but the strontia process is now largely used for this also.

Sugar may be obtained from nearly all sweet vegetable substances by a process essentially similar to that described above.

Sugar, Al'um. *Syn.* SACCHARUM ALUMINATUM, ALUMEN SACCHARINUM, L. *Prep.* From alum and white sugar, in fine powder, equal parts, formed into minute sugar-loaf shaped lumps with mucilage of gum-arabic made with rose-water. Used to make astringent lotions and eye-waters.

Sugar, Bar'ley-. *Syn.* SACCHARUM HORDEATUM, PENIDIUM, SACCHARUM PENIDIUM, L. When sugar is melted in a little water (barley-water was formerly used) it cools to a glassy mass (barley-sugar) enclosing a little water.—*Prep.* Take of saffron, 12 gr.; hot water, q. s.; sugar, 1 lb.; boil to a full 'candy height,' or that state called 'crack,' or 'crackled sugar,' when 2 or 3 drops of clear lemon juice or white vinegar must be added, and the pan removed from the fire and set for a single minute in cold water, to prevent its burning; the sugar must be then at once poured out on an oiled marble slab, and either cut into pieces or rolled into cylinders and twisted in the usual manner. One drop of oil of citron, orange, or lemon will flavour a considerable quantity. White barley-sugar is made with a strained decoction of barley instead of water, or starch is added to whiten it.

Sugar, Beetroot. *Syn.* SACCHARUM BETÆ, L. Sugar obtained from the white beet. Chemically speaking it is cane sugar or saccharose, and when well refined is quite equal to and indistinguishable from the refined produce of the sugar-cane.

Countries where Factories exist.	Production in 1877–8.	Factories existing.
Germany	375,000,000 kilos.	380
France	325,000,000 „	513
Austria, Hungary	245,000,000 „	248
Russia, Poland	250,000,000 „	288
Belgium	50,000,000 „	153
Holland, Sweden, Denmark	25,000,000 „	42
	1,270,000,000 „	1574

FIG. 1.

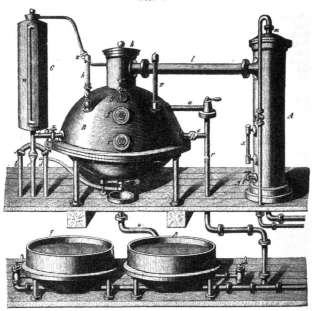

FIG. 2.

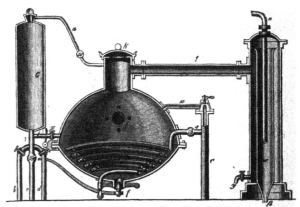

In the foregoing table the names of the countries in which this plant is chiefly cultivated are given, together with the amount of sugar produced in each in 1877–8.

The white beet is used in preference to the red varieties, not only because of the colour of its juice, but also in consequence of its being richer in sugar. The roots vary in their yield of sugar according to quality and the season of the year. They are generally in best condition in October. According to Wagner the constituents of the sugar-beet are as follows :

Water . . . . 82·7
Sugar . . . . 11·3
Cellulose . . . . . . 0·8
Albumen, casein, and other bodies . 1·5
Fatty matter . . . . . 0·1
Organic substances, citric acid, pectin, and pectic acid. Asparagin, aspartic acid, and betain, a substance having, accord. ing to Schiebler, the formula $C_{12}H_{28}N_2O_8$
Organic salts, oxalate and pectate of cal- } 3·7
cium, oxalate and pectate of potassium and sodium . . . . .
Inorganic salts, nitrate and sulphate of potash, phosphate of lime and magnesia

Twelve and a half cwt. of beet yield on an average 1 cwt. of raw sugar, or 8 per cent.

With this analysis compare the following analysis of the sugar-cane by Dr Phipson :

Water . . . . . 71·04 ⎫
Sugar . . . . 18·02 ⎪
Cellulose . . . . 9·56 ⎬ Derived from
Albuminous matter . . 0·55 ⎪ the air.
Fatty and colouring matter 0·35 ⎭
Salts soluble in water . 0·12 ⎫ Derived from
„ insoluble „ . 0·16 ⎬ the soil.
Silica . . . . 0·20 ⎭
                  ———
                  100·00

The first operation in the manufacture of beet-root sugar after washing and cleansing the roots (an operation which sometimes reduces their weight 10% or 20%) is the extraction from them of the juice. This may be effected either by—
1. Pressure.
2. Centrifugal power.
3. Dialysis.

1. *Pressure.* The roots being put into a proper crushing machine are reduced to a uniform pulp, which in some manufactories is subjected to pressure, wrapped in linen cloths, under stone or iron rollers, and in others is placed in bags and placed under an hydraulic press, the resulting juice being collected in proper receptacles.

2. *Centrifugal Power.* This method is that gene. rally employed for separating the juice from the pulp, which thus yields between 50% and 60% of juice. A weak saccharine solution, also used in sugar manufacture, is afterwards obtained by mixing the residue of the pulp with water, and subjecting it to the same process.

3. *Dialysis.* The application of the principle of diffusion for the extraction of the sugar from the beetroot originated with M. Robert. The fresh roots, cut into thin slices, are immersed in a little more than their own weight of water heated to about 120° F. The crystalloid sugar

thus diffuses out through the coil membrane which encloses it into the surrounding water, leaving the pectous and colloid matters, such as albumen, gum, &c., behind. The operation, which is so managed as to bring the same water into contact with successive quantities of root, yields a saccharine solution of nearly the same strength as the natural juice. The solution so obtained is, after concentration, &c., converted into sugar. The same process is said to have been tried with cane sugar, and with equally satisfactory results.

The succeeding stages of the manufacture of beet sugar, such as refining, liming, decolourising, &c., are the same as those already described under CANE SUGAR.

Beet sugar is in every respect identical with cane. It was discovered in 1747 by Marggraf, of Berlin, but it did not come into use until about the beginning of the present century, its manufacture at this period in France being necessitated by an edict of the first Napoleon's, which prohibited the importation of cane sugar into that country.

The engrs. on opposite page represent a common type of vacuum pan used in the sugar refineries.

Fig. 1 gives a perspective, and fig. 2 a sectional view of this evaporating pan.

The boiling pan (B) consists of two air-tight hemispheres, surmounted by a funnel, connected by the tube *l* with the condenser (A). The apparatus is supplied by steam by *r s*, the steam circulating in the boiling pan by means of the pipes (*q*), fig. 2. By opening the lever valves (*f*) the juice can be run by means of the pipe *o* into the pan (*p*). When the pan, after continued boiling, requires to be refitted, the pipes *l* and *w* are connected to an air-pump. The manometer (*h*) shows the state of the air pressure, which can be regulated by opening the pipes connected to the vacuum chamber. By means of the gauge cylinder (G) the quantity of syrup in the boiling pan can be ascertained, the gauge cylinder being connected to the boiling pan by the pipes *a* and *i*, and the height read off from the gauge-tube (*n*). The syrup can be removed, for the purpose of ascertaining its consistency, from the gauge cylinder by means of either of the three pipes (*b*, *c*, *d*). By *u* steam can be admitted to the boiling pan and condenser. *e* is generally of stout glass, and enables the state of the juice to be seen. *g* is the grease cock ; *f* the manhole. The condenser consists of the jacket (B), arranged to prevent the mixing of the juice with the water used for condensation. *x* is the gauge. The pipe (*m*) conveying water to the condenser terminates in a rose. *x* is a thermometer showing the interior temperature of the boiling pan.

The air-pump being set in operation the tube *e* is opened, and the gauge cylinder filled by the juice rising from *q*. By closing *m* and opening *s* the juice is admitted to the boiling pan. When this is half full the steam pipe (*s*) is opened, the steam quickly heating the contents of the pan to the boiling-point. The condenser is then placed in working ; by opening the pipe *l* the steam of the juice passes into the condenser, where it is speedily condensed, passing with the water through β.

Sugar, Diabetic. Grape sugar found in the

urine of persons suffering with *diabetes mellitus*. In *diabetes insipidus* a substance having the general properties of a sugar, but destitute of a sweet taste, appears to be produced (*Thénard*).

Sugar, Gel'atin. See GLYCOCINE.

Sugar, Grape. $C_6H_{12}O_6.H_2O$. *Syn.* GLUCOSE, DEXTROSE, FRUIT SUGAR; SACCHARUM UVÆ, S. FRUCTUS, L. This substance is found in the juice of grapes and other fruit, associated with levulose as invert sugar, in the urine of diabetic patients, and in the liquid formed by acting on starch and woody fibre with dilute sulphuric acid.

Prep. 1. From the juice of ripe grapes or an infusion of the dried fruit (raisins), by saturating the acid with chalk, decanting the clear liquid, evaporating to a syrup, clarifying with white of egg or bullock's blood, and then carefully evaporating to dryness; it may be purified for chemical purposes by solution and crystallisation in either water or boiling alcohol. Like other sugar, it may be decolourised by animal charcoal.

2. From honey, by washing with cold alcohol, which dissolves the fluid syrup and leaves the solid crystallisable portion.

On the large scale it is prepared from starchy matter. See STARCH SUGAR.

Prop. It is less sweet and less soluble than cane sugar, requiring 1½ parts of cold water for its solution; the sweetness of cane sugar : dextrose :: 5 : 3; instead of bold crystals, it forms granular warty masses, without distinct crystalline faces; it reduces Fehling's solution; it does not easily combine with either oxide of calcium or oxide of lead; with heat, caustic alkaline solutions turn it brown or black; with chloride of sodium it forms a soluble salt, which yields large, regular, and beautiful crystals. It rotates the plane of polarisation to the right + 112° when fresh, + 56° after standing. For uses see STARCH SUGAR.

The various fruits contain grape sugar in the following proportions:

	Per Cent.
Peach . . .	1·57
Apricot . .	1·80
Plum . . .	2·12
Raspberry .	4·00
Blackberry .	4·44
Strawberry .	5·73
Bilberry .	5·78
Currant . .	6·10
Plum . . .	6·26
Gooseberry .	7·16
Cranberry .	7·45 (*Fresenius*)
Pear . . .	8·02 to 10·8 (*E. Wolff*)
Apple . . .	8·37 (*Fresenius*)
„ . . .	7·28 to 8·04 (*E. Wolff*)
Sour cherry .	8·77
Mulberry . .	9·19
Sweet cherry .	10·79
Grape. . .	14·93

Obs. Cane sugar is converted into grape sugar during the process of fermentation, and by the action of acids. See SUGAR, and SUGAR, STARCH (*below*).

Sugar, Invert. A mixture of about equal parts of dextrose and levulose. It exists to some extent in most fruits, in the sap of many plants and trees, in honey, &c.

Prep. Cane sugar is converted into invert sugar by hydrolysis, that is, by boiling with very dilute sulphuric or hydrochloric acid.

Sugar, Maple. *Syn.* SACCHARUM ACERINUM, L. From the juice of the sugar maple, *Acer saccharinum*. It is identical with cane sugar, but is always impure.

In the United States and the British colonies of North America considerable quantities of this sugar are made. The juice is obtained through the bark of the tree to a depth of about a quarter or half an inch. Each tree has generally two perforations made in it, and they are always made on that side of the tree which faces towards the south, and at a distance of about 20 inches from the ground. The juice flows into suitable vessels, into which it is conducted by reeds placed under the perforations. The period chosen for tapping the trees is that during which it is known the sap is ascending, from March to May. Sometimes the tree undergoes a second tapping in the autumn, but this is not generally practised, inasmuch as it is injurious to the tree. A daily yield of 6 galls. of juice from each incision is looked upon as a 'good run,' and if these 6 galls. be the produce of an old tree or 'old bush' they will yield 1 lb. of sugar. In a young tree or 'young bush' the yield of sugar from the same quantity of sap is only half. By proper care the same tree may be tapped 20 or 30 years following. Unlike the sugar-cane, the juice in the maple is the richest in sugar the higher it is found from the ground. The thick saccharine liquid is concentrated every 24 hours. The raw crystallised sugar undergoes no refining, and being made into blocks is then sent to market.

Sugar, Milk. $C_{12}H_{22}O_{11}.H_2O$. *Syn.* SUGAR OF MILK, LACTOSE, LACTOBIOSE; SACCHARUM LACTIS (Ph. D.), L. *Prep.* Gently evaporate pure whey until it crystallises on cooling, and purify the crystals by digestion with animal charcoal and repeated crystallisations.

Prop., &c. White, translucent, very hard cylindrical masses or four-sided prisms; soluble in about 6 parts of cold and in 2 parts of boiling water; nearly insoluble in alcohol and ether; ammoniacal plumbic acetate precipitates it from its solutions. When an alkaline solution of grape sugar is boiled with the salts of copper, silver, or mercury, it reduces them; it produces right-handed rotation of a ray of polarised light (sp. rotatory power = + 52·53° at 20°); by boiling with dilute acid it is converted into *galactose* ($C_6H_{12}O_6$) treated with nitric acid it yields mucic acid, with small quantities of saccharic, oxalic, and tartaric acid. Milk sugar is not susceptible of the vinous fermentation, except under the action of dilute acids, which convert it into grape sugar; in solution it is converted by fermentation into lactic or butyric acid by the action of caseine and albuminous matter. Milk contains about 5% of it (*Boussingault*).

Obs. Sugar of milk is chiefly imported from Switzerland. In this country it is chiefly used as a vehicle for more active medicines, especially among the homœopathists.

Sugar, Starch. *Syn.* POTATO SUGAR, FÆCULA
s. This is grape sugar obtained by the action
of diastase on starch, in the manner noticed under
·Gum (British), or by the action of dilute sulphuric
acid on starch, or of the strong acid on lignin, or
on substances containing it.

Prep. 1. From corn. The corn is first steeped
in soda lye; it is then ground wet and passed
through revolving sieves to separate the husks
and gluten. The starch is carried through long
troughs, in which are placed transverse pieces of
wood, against which the solid particles of starch
lodge, and are thus separated from the washing
waters. These wash-waters run into a large cistern, where it undergoes fermentation into weak
vinegar. The starch in the wet state is then put
into a mash-tub, and treated for from 3 to 8
hours with 1% of sulphuric acid. The acid
liquor is neutralised with chalk and evaporated
in vacuum pans, and after being separated
from the sulphate of lime it is run into barrels and allowed to crystallise. The sugar is
sometimes manufactured in blocks 6 inches
square and dried on plaster plates in a current
of dry air, as hot air would discolour it.
Large quantities of grape sugar are now produced in Germany, France, and in the United
States, particularly in New Orleans, Buffalo, and
Brooklyn.

Uses. Starch sugar (glucose, potato sugar,
&c.) is much used by brewers and distillers for
making alcohol, as well as by confectioners; dyers
and calico printers use it to reduce indigo. When
specially prepared for the use of brewers the
blocks are crushed into small pieces about the size
of malt grains. Our excise authorities prohibit
the entrance of glucose into a brewer's premises
in the liquid state. In the brewing of pale sparkling ales grape is esteemed more than either cane
sugar or malt, and is said to yield a more sound
and wholesome liquor, and one free from the
acidity, impurity, and treacly sweetness frequently
found in beers brewed from raw or inferior sugars.
Glucose may also be obtained from cellulose, but
the process is too expensive to admit of being
practically worked.

2. Potato starch, 100 parts; water, 100 parts;
sulphuric acid, 6 parts; mix, boil for 35 or 40
hours, adding water to make up for evaporation;
then saturate the acid with lime or chalk, decant
or filter, and evaporate the clear liquor. Under
pressure the conversion is more rapid.—*Prod.*,
105%.

3. " The starch of potatoes can be converted
into glucose by digestion for a few hours with
parings of the potato. This operation is largely
practised by German farmers in the preparation
of food for fattening hogs. An excellent starch
sugar can be prepared from Indian corn, which
will yield alcohol one eighth cheaper, and quite as
pure as that from cane sugar" ('Journ. of
Applied Chemistry ').

4. Shreds of linen or paper, 12 parts; strong
sulphuric acid, 17 parts (*Braconnot*; 5 parts of
acid and 1 part of water—*Vogel*); mix in the cold;
in 24 hours dilute with water, and boil it for 10
hours; then neutralise with chalk, filter, evaporate to a syrup, and set the vessel aside to
crystallise.—*Prod.*, 114%. Sawdust, glue, &c.,

also yield grape sugar by like treatment. See
LIGNIN.

Sugar, Uncrystallisable Fruit (levulose—left-
handed sugar). $C_6H_{12}O_6$. *Prep.* Invert cane
sugar by heating it with dilute sulphuric acid.
Allow it to stand in the light, and strain off the
clear syrup which forms.

Prop. Much sweeter than dextrose. Under
proper conditions this sugar can be got to crystallise from alcohol. It does not ferment so
readily as dextrose; it also reduces cupric solution
less readily. It rotates the plane of polarisation to
the left.

Sugar from other Sources. Considerable quantities of East Indian cane sugar or jaggery are
yielded by certain Indian palms, the principal
of which are the *Arenga saccharifera* and the
Phœnix sylvestris, or wild date. Another source
whence large quantities of cane sugar are procured is the *Sorghum saccharatum*, or sugar-grass.
This plant is extensively grown in Ohio, and
yields annually more than 15,000,000 galls. of
juice, which is made into sugar.

The preparation of syrup from the melon
(*Cucumis melo*) is fast assuming some importance
in America. The juice of the fruit is stated to be
free from those non-saccharine substances which
make the extraction of beet and cane sugars such
an expensive matter.

The following sugars, besides those which have
been dealt with, are known:

Galactose. $C_6H_{12}O_6$. From milk sugar by
boiling it with dilute sulphuric acid; also from
plum gum.

a-Acrose. $C_6H_{12}O_6$. By the action of baryta
water on acrolein dibromide.

Mannose. $C_6H_{12}O_6$. By the action of platinum-
black on mannitol.

Formose. $C_6H_{12}O_6$. By the action of lime on
paraformaldehyde.

Raffinose. $C_{18}H_{32}O_{16} + 5H_2O$. From beet sugar
molasses.

Melesitose. $C_{72}H_{22}O_{11}$. From larch manna.

Mycose or *trehalose.* $C_{12}H_{22}O_{11}$. From Turkish
manna.

Melitose. $C_{12}H_{24}O_{12}$. From the eucalyptus.

Maltose. $C_{12}H_{24}O_{12}$. From malt.

Eucalyn. $C_6H_{12}O_6$. By fermentation of melitose, a substance found in eucalyptus manna.

Sorbinose. $C_6H_{12}O_6$. From the berries of the
mountain ash tree.

*Effects of the Varieties of Sugar on Polarised
Light.* Both sucrose or cane sugar and dextrose
produce rotation upon a ray of polarised light.
The plane of rotation is rotated to the right by
sucrose rather more powerfully than by dextrose.
It is remarkable that the uncrystallisable sugar of
fruits produces an opposite rotation, *viz.* to the
left. Since the degree of rotation is proportionate
in columns of equal length to the quantity of
sugar present, it has been proposed to employ this
property in order to determine the quantity of
sugar present in syrups (*Miller*). The following,
according to Berthelot, are the rotatory powers of
the different varieties of sugar if equal weights
of each are dissolved in an equal bulk of water;
the quantities of each sugar are calculated for the
formulæ annexed:

Variety.	Formula.	Rotation.	Tempera- ture.	
			° F.	° C.
Sucrose (cane sugar)	(C₁₂H₂₂O₁₁)	Right 73·8°		
Melezitose .	(C₁₈H₂₂O₁₁)	„ 94·1°		
Mycose .	(C₁₂H₂₂O₁₁)	„ 193°		
Melitose . .	(C₁₂H₃₄O₁₂)	„ 102°		
Dextrose (grape sugar)	(C₆H₁₂O₆)	„ 57·4°		
Malt sugar .	(C₆H₁₂O₆)	„ 172°		
Levulose (fruit sugar)	(C₆H₁₂O₆)	Left 106°	56	13·3
Eucalyn . .	(C₆H₁₂O₆)	Right 50°		
Sorbin . .	(C₆H₁₂O₆)	Left 46·9°		
Lactose (milk sugar)	(C₆H₁₂O₆)	Right 56·4°		
Glucose of ditto (galactose)	(C₆H₁₂O₆).	„ 83·3°		
Inverted cane sugar	(C₆H₁₂O₆)	Left 28°	57	13·9

SUGAR-BOILING. The art or business of the confectioner or sugar-baker; the candying of sugar. The stages are as follow :—Well-clarified and perfectly transparent syrup is boiled until a 'skimmer' dipped into it, and a portion 'touched' between the forefinger and thumb, on opening them, is drawn into a small thread, which crystallises and breaks. This is called a 'weak candy height.' If boiled again, it will draw into a larger string, and if bladders may be blown through the 'drippings' from the ladle, with the mouth, it has acquired the second degree, and is now called 'bloom sugar.' After still further boiling, it arrives at the state called 'feathered sugar.' To determine this re-dip the skimmer, and shake it over the pan, then give it a sudden flirt behind, and the sugar will fly off like feathers. The next degree is that of 'crackled sugar,' in which state the sugar that hangs to a stick dipped into it, and put directly into a pan of cold water, is not dissolved off, but turns hard and snaps. The last stage of refining this article reduces it to what is called 'carmel sugar,' proved by dipping a stick first into the sugar, and then into cold water, when, on the moment it touches the latter, it will, if matured, snap like glass. It has now arrived at a 'full candy height.' Care must be taken throughout that the fire is not too fierce, as, by flaming up against the sides of the pan, it will burn and discolour the sugar; hence the boiling is best conducted in steam jacketed pans.

Any flavour or colour may be given to the candy by adding the colouring matter to the syrup before boiling it, or the flavouring essences when the process is nearly complete. See STAINS, &c.

SUGAR-CANDY. *Syn.* SACCHARUM CANDIDUM, S. CRYSTALLINUM, S. CRYSTALLIZATUM, L. Sugar crystallised by leaving the saturated syrup in a warm place (90° to 100° F.), the shooting being promoted by placing sticks, or threads, at small distances from each other in the liquor; it is also deposited from compound syrups, and does not seem to retain much of the foreign substances with which they are loaded. Brown sugar-candy is prepared in this way from raw sugar; white do., from refined sugar; and red do., from a syrup of refined sugar which has been coloured red by means of cochineal.

Sugar-candy is chiefly used as a sweetmeat; and, being longer in dissolving than sugar, in coughs, to keep the throat moist; reduced to powder, it is also blown into the eye, as a mild escharotic in films or dimness of that organ.

SUGAR OF LEAD. Acetate of lead.

SUGAR-PLUMS. *Syn.* BONBONS, DRAGÉES, Fr. These are made by various methods, among which are those noticed under DROPS (Confectionery), LOZENGES, and PASTILS, to which may be added the following :—Take a quantity of sugar syrup, in the proportion to their size, in that state called a 'blow' (which may be known by dipping the skimmer into the sugar, shaking it, and blowing through the holes, when parts of light may be seen), and add a drop or two of[any esteemed flavouring essence. If the 'bonbons' are preferred white, when the sugar has cooled a little, stir it round the pan till it grains and shines on the surface. When all is ready, pour it through a funnel into little clean, bright, leaden moulds, which must be of various shapes, and be previously slightly moistened with oil of sweet almonds; it will then take a proper form and harden. As soon as the plums are cold, take them from the moulds; dry them for two or three days in the air, and put them upon paper. If the bonbons are required to be coloured, add the colour just as the sugar is ready to be taken off the fire.

CRYSTALLISED BONBONS are prepared by dusting them with powdered double-refined lump sugar before drying them.

LIQUEUR BONBONS, now so beautifully got up by the Parisian confectioners, are obtained by pressing pieces of polished bone or metal into finely powdered sugar, filling the hollow spaces so formed with saturated solutions of sugar in the respective liqueurs, and then spreading over the whole an ample layer of powdered sugar. In the course of three or four days the bonbons may be removed, and tinted by the artist at will. Instead of white powdered sugar ordered above, coloured sugar may be used. These bonbons are found to be hollow spheres, containing a small quantity of the spirit or liqueur employed, and will bear keeping for many months. See SWEETMEATS, &c..

SUGARS (Medicated). *Syn.* SACCHARIDES; SACCHARA MEDICATA, L.; SACCHAROLÉS, SACCHARURES, Fr. Some of these are prepared by moistening white sugar with the medicinal substance, then gently drying it, and rubbing it to powder; in other cases they are obtained in the manner noticed under PULVERULENT EXTRACTS, or OLEOSACCHARUM. The most valuable preparation of this class in British *pharmacy* is the saccharated carbonate of iron (FERRI CARBONAS CUM SACCHARO—Ph. L.).

SUINT. Sheep while browsing abstract a considerable amount of potash, which, after having passed into the blood, &c., is sweated through the skin and deposited on the wool as *suint*. The substance constitutes about one third part of the weight of crude merino wool (*Chevreul*). The fatty acids of suint are compounds of oleic, stearic, and palmitic acids (*Reich* and *Ulbricht*). Suint is used as a manure in agriculture, and as a source of potassium salts and illuminating gas.

Suint, Gas from. By this is understood a gas prepared from the fatty materials present in the soap-suds used in washing raw wool and spun yarns. The water containing the suint and soap-suds is run into cisterns, and is there mixed with milk of lime, and left to stand for twelve hours. A thin precipitate is formed, which, after the supernatant clear liquor has been run off, is put upon coarse canvas for the purpose of draining off any impurities, sand, hair, &c., while the mass which runs through the filter is put into a tank, in which it forms, after six or eight days, a pasty mass, which, having been dug out and moulded into bricks, is dried in open air. At Rheims the first wash-water of the wool is used for making both gas and potash, because the water contains no soap and only suintate of potash, potassium sudorate. Havrez, at Verviers, has recently proposed to employ suint—which, by-the-bye, is very rich in nitrogen—for the purpose of making ferro-cyanide of potassium.

The dried brick-shaped lumps are submitted to distillation, yielding a gas which does not require purification, and which possesses an illuminating power three times that of good coal-gas. The wash-water of a wool-spinning mill with 20,000 spindles yields daily, when treated as described, about 500 kilos. of dried suinter, as the substance is technically called. One kilo. of this substance yields 210 litres of gas. Annually about 150,000 kilos. of suinter are obtained, and this quantity will yield 31,500,000 litres = 1,112,485 cubic feet of gas. Every burner consuming 35 litres of gas per hour, and taking the time of burning at 1200 hours, the quantity of gas will suffice for 750 burners, and as a spinning-mill of 20,000 spindles only requires 500 burners, there is an excess of gas supply available for 250 other burners.

SUL'PHATE. *Syn.* SULPHAS, L. A salt of sulphuric acid. There are three kinds of sulphates. In *normal* sulphates both atoms of hydrogen in sulphuric acid are replaced by a metal, as N_2SO_4; in acid sulphates one atom only is so replaced, as $KHSO_4$; and in double sulphates the hydrogen atoms are replaced by two different metals, as $KAl(SO_4)_2$.

SUL'PHIDE. A salt consisting of sulphur and a metal or other basic radical. See SULPHURETTED HYDROGEN.

SULPHINDYL'IC ACID. $C_{16}H_9(SO_3.OH)_2N_2O_2$. *Syn.* INDIGOTIN DISULPHONIC ACID. An intensely blue pasty mass, formed by dissolving 1 part of indigo in about 15 parts of concentrated sulphuric acid. Used in dyeing Saxony cloth. See SULPHATE OF INDIGO.

SUL'PHITE. A salt of sulphurous acid. Sulphurous acid forms two classes of salts: the normal sulphites, in which both atoms of hydrogen are replaced by a metal, as Na_2SO_3; and the acid sulphites, in which one atom only is replaced, as $KHSO_3$.

SULPHOCARBOLIC ACID. $C_6H_4(OH)SO_3H$. (SULPHOCARBOLATES.) Carbolic acid, when acted

upon by bases, yields a class of salts termed carbolates. These compounds are very unstable; they readily absorb water from the air, which sets free carbolic acid; they usually have the powerful odour of the latter. When, however, equivalent weights of carbolic and sulphuric acids are mixed, union takes place, a definite double acid results, and the salts formed by this double acid with the various bases are entirely different from the simple salts of carbolic acid. They are very stable, very soluble, possess neither odour nor taste of carbolic acid, and are singularly beautiful in crystalline form.

Prop. Sulphocarbolic acid crystallises in long colourless needles; unlike carbolic acid, it is soluble in water, alcohol, and ether, in any proportions.

Sulphocarbolate of Calcium ($Ca[C_6H_5SO_4]_2 +$ Aq) is obtained in very long, fine, densely interlacing crystals, which form in bulk, by their interlacement, a porous mass. Unlike the usual lime salts, this is exceedingly soluble. This fact overcomes the great difficulty of treatment when in disease there is a deficiency of lime in the body, especially in rickets, in which disease the want of lime in the bones gives rise to distortions.

Sulphocarbolate of Copper ($Cu[C_6H_5SO_4]_2$) forms fine prismatic crystals of a blue colour. It is used as the zinc sulphocarbolate, chiefly as a lotion and dressing, in the proportion of 3 to 10 gr. to the ounce of distilled water.

Sulphocarbolate of Iron ($Fe[C_6H_5SO_4]_2$) forms colourless or pale green rhombic plates. It is readily administered, and seems in some instances to be preferred to other salts of iron. It seems to have been of especial use in the skin diseases of children, wherein there is much formation of matter.

Sulphocarbolate of Sodium ($Na[C_6H_5]SO_4.Aq$) forms brilliant, clear, rhombic prisms. The salt is very soluble in water. This salt can be administered as a medicine in doses of 20 to 60 gr.; it is slowly decomposed, carbolic acid being evolved. It thus becomes a very simple means of obtaining the beneficial effects of the administration of this antiseptic without the difficulties and dangers which attend it in its uncombined irritant and caustic form. It has proved of great service in the treatment of infectious diseases. Administered in the severest cases of diphtheria, malignant scarlet fever, typhoid, erysipelas, &c., the remedy has proved of extreme value.

Sulphocarbolate of Zinc ($Zn[C_6H_5SO_4]_2$) is chiefly employed in solution as a lotion. By high surgical authorities it is considered to answer all the purposes of the antiseptic dressing of carbolic acid. It is inodorous, and has very slight irritating action.

The Sulphocarbolates of Potassium ($KC_6H_5SO_4$) and Ammonium ($NH_4C_6H_5SO_4$) are also brilliant crystals; they are freely soluble, administered with the greatest ease, and have been used with success as remedial agents.

SULPHOCYAN'OGEN. CNS. A well-defined salt radical, containing sulphur united to the elements of cyanogen. Its compounds are the sulphocyanides, most of which may be formed by directly saturating hydrosulphocyanic acid with the oxide or hydrate of the base; or, from the sulphocyanide of potassium and a soluble salt of the base, by double decomposition.

SULPHOFORM. *Syn.* SULPHOFORMUM, L. An oily liquid obtained by distilling 1 part of iodoform with 3 of sulphide of mercury.

SULPHONAL. $C_7H_{16}S_2O_4$. It is chemically diethylsulphondimethylmethane, and occurs in colourless prismatic crystals slightly soluble in water, more so in alcohol or a mixture of alcohol and ether. Professor Olt, of Prague, gives his experience of sulphonal as being an excellent hypnotic. In most cases tranquil and prolonged sleep supervened after the administration of 1 to 2 grms.

Dr Perregaux, of Montreux, finds this drug particularly suitable in cases of nervous insomnia, and is also very useful in cases of severe neuralgic pain. He gives doses varying from 1·5 to 2 grms., which is sufficient to produce quiet and refreshing sleep of five to six hours' duration. He, however, observed some accessory cerebral symptoms following the use of this remedy, such as initial mental excitement and atactic disturbances of certain fine movements of the hands. Hence the question arises whether prolonged use of this new hypnotic can be altogether harmless in regard to the cerebral cortical function ('British Medical Journal').

Mr Conolly Norman, Medical Superintendent to the Richmond Asylum, Dublin, reports ('Dublin Journal Medical Science') several cases of insanity and mental disorder in which sulphonal proved most valuable. As also Dr Lojacono, assistant in Professor Bianchi's Clinique for mental diseases at Palermo, has recently reported the results of experiments made with sulphonal on patients suffering from mental diseases. The drug was administered in syrup of doses of 1 to 4 grms. In no case did the drug fail to give tranquil and refreshing sleep, besides having a marked effect as a general sedative. In cases of acute maniacs of hysterical delirium and epileptiform convulsions the paroxysms were either prevented or greatly mitigated.

Dr Kisch, of Berlin, gives his experience of this drug as producing most rapid effects as an hypnotic on individuals suffering from nervous complaints in doses of 7 to 15 gr. at bedtime. This produced sleep during the whole or greater part of the night. In cases of neuralgia of rheumatic origin 15 to 30 gr. produced good results, whilst one of gastric neuralgia was unaffected.

SULPHOPHE'NIC ACID. A synonym of sulphocarbolic acid. See SULPHOCARBOLATES.

SULPHOVIN'IC ACID. $C_2H_5.HSO_4$. *Syn.* SULPHETHYLIC ACID; ACIDUM SULPHOVINICUM, L. This substance is formed by the action of heat on a mixture of alcohol (1 part by weight) and sulphuric acid (2 parts by weight); it is the intermediate product which is developed in the preparation of ether. The salts are called sulphovinates or sulphethylates.

SUL'PHUR. S = 31·98. [Eng., L.] *Syn.* BRIMSTONE; SOUFRE, Fr. Occurs in the free state in many volcanic districts, e.g. Iceland, California, Italy, New Zealand. In Sicily, from which the chief supplies come, it exists in beds associated with blue clay. As sulphuretted hydrogen it is found in many mineral waters; in combination with metals it forms numerous ores known as sulphides or sulphurets, e.g. galena, cinnabar, iron and copper pyrites combined with oxygen and a metal. It is plentifully distributed as sulphates, e.g. gypsum, Glauber's salt, &c.; in this latter form it occurs also in the tissues of animals and plants.

Prep. 1. As 'rough' sulphur from ores which contain more than 12% by mixing them in furnaces with a little fuel, igniting and smothering with earth; the sulphur liquefies and is caught in wooden moulds, or high-pressure steam is applied to the ores contained in an iron vessel.

2. The ores are heated in a boiler with a 66% solution of calcium chloride at 120° C.

3. By extraction with carbon disulphide.

4. By driving off the sulphur in the form of vapour by the application of heat, and condensing the vapour in a cooled receiver.

Var. The principal of these are—

AMORPHOUS SULPHUR, BROWN s.; SULPHUR AMORPHUM, S. FUSCUM, S. INFORME, S. RUBRUM, L. Prepared from sublimed sulphur, by melting it, increasing the heat to from 320° to 350° F., and continuing it at that temperature for about half an hour, or until it becomes brown and viscid, and then pouring it into water. In this state it is ductile, like wax, may be easily moulded in any form, is much heavier than usual, and when it has cooled does not again become fluid until heated to above 600° F. The same effect is produced more rapidly by at once raising the temperature of the melted mass to from 430° to 480° F. The soft mass soon becomes brittle again.

PRECIPITATED SULPHUR, HYDRATE OF SULPHUR, MILK OF S.; SULPHURIS HYDRAS, LAC SULPHURIS, SULPHUR PRECIPITATUM (Ph. L.), L. *Prep.* 1. From sublimed sulphur, 1 part; dry and recently slaked lime, 2 parts; water, 25 parts, or q. s.; boil for 2 or 3 hours, dilute with 25 parts more of water, filter, and precipitate with dilute hydrochloric acid; drain, and well wash the precipitate, and dry it by a gentle heat. Resembles sublimed sulphur in its general properties, but is much paler, and in a finer state of division.

2. (B. P.) Sublimed sulphur, 5 oz.; slaked lime, 3 oz.; hydrochloric acid, 3 fl. oz., or q. s.; distilled water, q. s. Heat the sulphur and lime, previously well mixed, in 1 pint of water, stirring diligently with a wooden spatula, boil for 15 minutes, and filter. Boil the residue again in ½ pint of water and filter. Let the united filtrates cool, dilute with 3 pints of water, and in an open place, or under a chimney, add in successive quantities the hydrochloric acid previously diluted with 1 pint of water until effervescence ceases, and the mixture acquires an acid reaction. Allow the precipitate to settle, decant off the supernatant liquid, pour on fresh distilled water, and continue the purification by affusion of distilled water and subsidence, until the fluid ceases to have an acid reaction, and does not precipitate with oxalate of ammonia. Collect the precipitated sulphur on a calico filter, wash it once with distilled water, and dry it at a temperature not exceeding 120° F.

Prop. A greyish-yellow powder free from

grittiness, and with no smell of sulphuretted hydrogen.

Obs. Many pharmacists regard LAC SULPHURIS and SULPHUR PRECIPITATUM as distinct substances, and assume that by milk of sulphur is intended a preparation made by an old pharmacopœial process, in which sulphuric acid being employed, the sulphur so precipitated contains from 50% to 75% of sulphate of lime. Pareira, Royle, Attfield, and some other authorities, hold that LAC SULPHURIS and SULPHUR PRECIPITATUM are synonymous; whilst others, including Professor Redwood (one of the compilers of the B. P.), entertain a contrary opinion.

ROLL SULPHUR, CANE S., STICK S.; SULPHUR IN BACCULIS, S. IN ROTULIS, S. ROTUNDUM, L. This is crude sulphur, purified by melting and skimming it, and then pouring it into moulds. That obtained during the roasting of copper pyrites, and which forms the common roll sulphur of England, frequently contains arsenic.

SUBLIMED SULPHUR, FLOWERS OF SULPHUR; FLORES SULPHURIS, SULPHUR (Ph. L.), SULPHUR SUBLIMATUM (B. P., Ph. E. & D.), L. Prepared by subliming sulphur in iron vessels. For medical purposes it is ordered to be well washed with water, and dried by a gentle heat. " A slightly gritty powder, of a fine greenish-yellow colour, without taste and without odour till heated " (B. P.).

Octahedral Sulphur. In this state it is often found in nature in large transparent crystals, which, however, are best obtained from a solution of flowers of sulphur in carbon disulphide, which at the ordinary temperature dissolves about 1-3rd of its weight, and on slow evaporation deposits octahedra.

SULPHUR VIVUM, BLACK SULPHUR, CRUDE S., HORSE BRIMSTONE; SULPHUR NIGRUM, S. CABALLINUM, S. GRISEUM, L. This is crude native sulphur. It is a grey or mouse-coloured powder. The residuum in the subliming pots from the preparation of flowers of sulphur is now commonly substituted for it. It generally contains much arsenic, and is consequently very poisonous.

Pur. The sublimed sulphur of the shops is now, in general, of respectable quality, but the precipitated sulphur frequently contains about 2-3rds of its weight of sulphate of lime (plaster of Paris), owing to the substitution of sulphuric acid for hydrochloric acid in its manufacture (see PRECIPITATED SULPHUR, *above*). This is readily detected by strongly heating a little of the suspected sample in an iron spoon or shovel, when the sulphur is burnt or volatilised, and leaves behind the sulphate of lime as a white salt; this, when mixed with water and gently dried, gives the amount of the adulteration. A still simpler plan is to dissolve out the sulphur in the sample with a little hot oil of turpentine; the undissolved portion is foreign matter.

Prop. Sulphur melts to a clear thin fluid at 115° C., and in open vessels rapidly takes fire at 260° C., burning with a bluish flame. It is easily electrified. It is insoluble in both water and alcohol; it is soluble in oil of turpentine and the fatty oils, and freely so in bisulphide of carbon. About thirty different crystallographic modifications of sulphur are known to exist. With oxygen it unites to form sulphurous anhydride, and with the metals to form sulphides. Sp. gr. 1·982 to 2·015, according to its state.

Detect. and Estim. Sulphur is most easily detected in a substance by mixing it with pure sodium carbonate, and fusing before the blowpipe on a piece of sound charcoal; the fused mass is placed upon a bright silver coin and moistened with water; the smallest quantity of sulphur is recognised by the formation of a brown or black stain upon the bright surface. The proportion of sulphur is best determined by oxidising a known weight of the substance by strong nitric acid, or by fusing it in a silver vessel with ten or twelve times its weight of pure hydrate of potash and about half as much nitre. The sulphur is thus converted into sulphuric acid; the quantity of which can be determined by dissolving the fused mass in water, acidulating the solution with nitric acid, adding barium chloride, and weighing the resulting sulphate. See ORGANIC SUBSTANCES.

Uses, &c. Sulphur is extensively used in the manufacture of gunpowder, in bleaching, sulphuric acid, &c. When swallowed, it acts as a mild laxative and stimulating diaphoretic; and has hence been long taken in various chronic skin diseases, in pulmonary, rheumatic, and gouty affections, and as a mild purgative in piles, prolapsus ani, &c. Externally, it is extensively used in skin diseases, especially the itch, for which it appears to be a specific.—*Dose,* 20 to 63 gr., in sugar, honey, treacle, or milk.

Obs. Sulphur is now extensively recovered from alkali waste, which contains large quantities of calcium sulphide. Various processes have been proposed, but one of them, the 'Chance process,' is the most important, as it promises to prolong the struggle between the 'Leblanc' and 'ammonia soda' manufacturers of alkali till such time as the chlorine shall be recovered in the latter.

Chance's sulphur-recovery process consists in pumping carbonic acid gas through the 'vat waste;' this causes sulphuretted hydrogen to be given off, which is subsequently used either for making sulphuric acid or sulphur. In the latter case the sulphuretted hydrogen is mixed with a carefully regulated supply of air and passed through a kiln containing some porous material, when the hydrogen only is burnt, and the sulphur is deposited in condensing chambers.

The theory of the operation is illustrated by the following formulæ:

$$CaS + H_2O + CO_2 = CaCO_3 + H_2S.$$
$$CaS + H_2S = Ca(HS)_2.$$
$$Ca(HS)_2 + CO_2 + H_2O = CaCO_3 + 2H_2S.$$
$$H_2S + O = H_2O + S.$$

Sulphur, Chlo"rides of. Three of these compounds exist. 1. MONOCHLORIDE. S_2Cl_2. *Prep.* By passing dry chlorine gas over the surface of sulphur melted in a bulbed tube or small retort connected with a well-cooled receiver. The product is a deep orange-yellow and very mobile liquid, which possesses a disagreeable odour, and boils at 138° C.; sp. gr. 1·7055. It is soluble in bisulphide of carbon, and in benzol, without decomposing. It dissolves sulphur in large quantities at the ordinary temperature. This property has been largely used for the purpose of vulcanising caoutchouc.

2. SULPHUR DICHLORIDE. SCl₂. *Prep.* By passing chlorine gas into the monosulphide at a low temperature, and then removing the excess of Cl with dry CO₂. A brown liquid.

3. SULPHUR TETRACHLORIDE. SCl₄. *Prep.* By saturating the dichloride with chlorine gas at 22° C. A light, mobile, yellowish-brown liquid.

Sulphur, I'odide of. S₂I₂. *Syn.* BINIODIDE OF SULPHUR; SULPHURIS IODIDUM (Ph. L.), SULPHUR IODATUM (Ph. D.), L. *Prep.* Into a glass flask put 1 part of sublimed sulphur, and over it place 4 parts of iodine; insert the cork loosely, and place the flask in a water-bath; as soon as its contents melt, stir them with a glass rod, replace the cork, remove the bath from the fire, and let the whole cool together. When cold, break the iodide into pieces, and place it in a wide-mouthed stoppered bottle. Beautiful semi-crystalline, dark grey mass, resembling native sulphide of antimony.

Uses, &c. It is stimulant and alterative. An ointment made of it has been recommended by Biett and others in tuberculous affections of the skin, in lepra, psoriasis, lupus, prurigo, &c. Iodide of sulphur stains the skin like iodine, and is readily decomposed by contact with organic substances.

SULPHURATION. The process by which silk, cotton, and woollen goods, straw plait, &c., are subjected to the fumes of burning sulphur, or sulphurous acid, for the purpose of bleaching or decolourising them. On the large scale this is effected in closed apartments, called 'sulphuring rooms,' to which sufficient air only is admitted to keep up the slow combustion of the sulphur. On the small scale, as for straw hats, bonnets, &c., a large wooden chest is frequently employed in the same way.

SUL'PHURET. *Syn.* SULPHIDE; SULPHURETUM, SULPHIDUM, L. See SULPHIDE.

SUL'PHURETTED HY'DROGEN. H₂S. *Syn.* HYDROGEN SULPHIDE, DIHYDRIC SULPHIDE, HYDRIC SULPHIDE, HYDROSULPHURIC ACID. Sulphuretted hydrogen occurs in nature amongst the gaseous products given off by volcanoes, as well as in many mineral waters, amongst which may be instanced those of Harrogate, in England, of Moffat, in Scotland, and of Barèges, Eaux Bonnes, St Sauveur, &c., in the Pyrenees. It is also evolved from decaying animal matter containing albumen, such as white of egg, as well as from putrefying animal and vegetable substances when in contact with a soluble sulphate, and is always one of the gases present in the air of drains and sewers.

Prep. 1. Sulphuretted hydrogen may be procured by the direct union of hydrogen and sulphur, as by passing hydrogen into boiling sulphur. But this method of procuring it is rarely adopted.

2. The much readier process of acting upon a metallic sulphide by an acid constitutes the means by which the chemist almost invariably obtains this gas.

About an ounce of ferrous sulphide, previously reduced to small pieces, is placed in a bottle, and then there is poured on to it a fluid ounce of sulphuric acid diluted with 8 times its bulk of water, when the following reaction ensues:—
FeS + H₂SO₄ = H₂S + FeSO₄.

The following sketch shows a convenient form of apparatus.

The cooled diluted acid is poured through the thistle funnel upon the ferrous sulphide, and the evolved gas passing through the small intermediate wash-bottle into the bottle at the reader's right hand, is absorbed by the water therein contained, the operation being continued until the water has become saturated with the gas. The glass tubes are connected with vulcanised india rubber, as shown in the above *engr.* Diluted hydrochloric acid is frequently substituted for sulphuric. In this process, the gas obtained, owing to the contamination of the iron sulphide, is more or less impure.

3. When sulphuretted hydrogen is required in a state of purity, 1 oz. of antimonious sulphide must be employed instead of the iron sulphide, and instead of sulphuric 3 or 4 parts of hydrochloric acid. As heat must be applied to the mixture, it will be necessary to substitute a flask for the larger bottle, and to support it on a retort stand. In other respects the apparatus needs no alteration.

4. Sulphuretted hydrogen is also obtainable when paraffin is heated at a moderately elevated temperature with sulphur, the reaction being attended with an abundant evolution of the gas, and a simultaneous separation of carbon.

Obs. The solution of sulphuretted hydrogen, which is so indispensable to the chemist, and consequently in such constant requisition in the laboratory, unfortunately very quickly decomposes into water and sulphur, which deposits at the bottom of the vessel containing it. To diminish as much as possible the tendency to deterioration, the solution should be made either with boiled water, or with the clear spoilt solution.

When a constant supply of sulphuretted hydrogen gas is required it is best to use a Kipp's apparatus, such as is shown in the accompanying *engr.* The two globes *b* and *d* are connected by a narrow neck, whilst the tubulus of the third globe *e* passes air-tight through the neck of *b* nearly to the bottom of *d*.

The middle globe *b* contains lumps of sulphide of iron, and dilute sulphuric acid is poured through the bent thistle funnel until the lowest globe *d* is filled, and a portion of the acid has come in contact with the sulphide. The gas passes off at *e* through the wash-bottle to the right of the figure. When it is desired to stop the current of gas the stopcock is closed, and the gas accumulating in *b* forces the acid in *d* up the tubulus into the upper globe.

Qualities, &c. Sulphuretted hydrogen is a colourless inflammable gas, somewhat heavier than air, its sp. gr. being 1·174. It possesses a sweet taste, and an odour like that of rotten eggs. When ignited, it burns with a bluish flame, to water and sulphurous anhydride if the

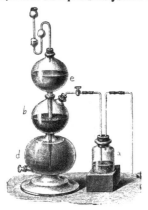

combustion take place in a sufficient quantity of air, but if the supply of air be too limited, sulphur is deposited. Under a pressure of 17 atmospheres it is condensed to a colourless and very mobile fluid, which boils at —61·8° C., and freezes at —85° C. to a transparent solid. Both the gas and its aqueous solution exercise a feebly acid reaction on litmus.

Sulphuretted hydrogen is highly poisonous; when inhaled in any quantity it causes fainting; and in smaller quantities, even when considerably diluted by air, if breathed for any length of time, it acts as a dangerous depressant and insidious poison. Upon the lower animals it acts with fatal rapidity, even if diluted with 800 or 1000 parts of atmospheric air. Transmitted through tubes heated to redness, sulphuretted hydrogen becomes partially decomposed into its elements, hydrogen and sulphur. Water at 32° F. takes up 4·37 times its bulk of this gas, and at 59°, 3·23 times its bulk; hence the importance of collecting it over warm water if required in the gaseous form.

In the presence of moisture, sulphurous anhydride and sulphuretted hydrogen, if equivalent quantities of each react upon each other, become decomposed into sulphur, water, and pentathionic acid; hence the value of sulphurous acid as a disinfectant. The deposited sulphur is found always to occur in the electro-positive condition. Chlorine, bromine, and iodine also decompose sulphuretted hydrogen with deposition of sulphur, and formation of hydrochloric, hydrobromic, and hydriodic acids. Sulphuretted hydrogen turns silver black, hence egg-spoons are gilded to prevent their being tarnished.

Hydrosulphates or Sulphides. Sulphuretted hydrogen or hydrosulphuric acid, as it is sometimes called, when brought into contact with bases in solution, gives rise to compounds, which by some chemists are regarded as hydrosulphates or

combinations of the base with hydrosulphuric acid; and by others as sulphides or combinations of the metal with sulphur, the latter reaction being attended with the elimination of water, as when a base is acted upon by hydrochloric acid. By those who hold the former view the reaction would be as follows:

$$K_2O + H_2S = K_2O.H_2S.$$

In the latter case it would be thus represented:

$$K_2O + H_2S = K_2S + H_2O.$$

The latter is the more general opinion, and it receives support from the fact that when sulphuretted hydrogen is passed into the solution of a metallic salt, an insoluble precipitate of a sulphide of the metal is thrown down. Thus, when the gas is passed into a solution of cupric sulphate, the precipitate consists of hydrated cupric sulphide; the liberated sulphuric acid renders the liquid, which was before neutral, acid. The larger number of sulphides so formed, combining with water at the instants of their precipitation, occur as hydrates.

There is also a class of sulphides known as hydrosulphides, sulphydrates, or double sulphides, in which an equivalent of the metal is replaced by an equivalent of hydrogen. Examples of these are the potassic hydrosulphide (KHS), sodic hydrosulphide (NaHS), and ammonic hydrosulphide (H_4NHS). No such combinations occur with hydrogen and the metals of the earth proper, and of the iron group.

Tests. Many of the hydrosulphates or sulphides may be detected by dropping on them some hydrochloric acid, when the characteristic smell of sulphuretted hydrogen will be immediately evolved from them. Very small quantities of a sulphide may be detected as follows:—Place the suspected sulphide in a small test-tube, on the upper part of which is inserted a piece of blotting-paper moistened with a solution of plumbic acetate; then carefully pour some hydrochloric acid on to the substance, when, if it be a sulphide, the paper will become browned or blackened.

Many small quantities of the soluble sulphides are revealed in neutral or alkaline solutions by the rich purple colour which they form on the addition of a solution of sodic nitro-prusside. Most of them, when heated before the blowpipe, give off the smell of sulphurous acid.

The quantitative determination of free sulphuretted hydrogen, or of a soluble sulphide in any solution is conducted as follows:—The liquid to be tested is mixed with a small quantity of a cold solution of starch, made slightly acid with acetic acid. A solution of iodine of known strength, dissolved in potassium iodide, is then added, until the liquid just begins to turn blue from the action of the excess of iodine on the starch. In this process the sulphuretted hydrogen converts the iodine into hydriodic acid, whilst sulphur is liberated.

Of course the quantity of sulphuretted hydrogen is calculated from the quantity of iodine employed. The reaction is—

$$2H_2S + 2I_2 = 4HI + S_2.$$

The value of sulphuretted hydrogen as a reagent has already been alluded to. It throws down most of the metals from solutions of their salts in the form of insoluble sulphides, the sulphides so produced in many cases being distinguished from the others by a characteristic colour. The sulphuretted hydrogen thus presents the metal in a form in which it can, in many instances, be easily and with certainty recognised. Thus sulphide of lead is black, of arsenic yellow, of antimony orange, of manganese salmon colour, and of zinc white.

By means of sulphuretted hydrogen, also, the chemist is enabled to separate the metals into groups. For instance, from solutions containing certain metallic salts, sulphuretted hydrogen throws down the metals as sulphides, provided the solution has been previously made slightly acid. Copper, arsenic, tin, and cadmium are some of the metals thrown down under these conditions.

The salts of iron, nickel, cobalt, and certain others, although they do not yield precipitates under like circumstances, are found to do so if their solutions are made alkaline instead of acid. Again, there are other salts, those of the alkalies and alkaline earths, which, when sulphuretted hydrogen is passed through their solutions, give no precipitates either in acid or alkaline solution. The chemist, therefore, in the course of an analysis, frequently avails himself of a knowledge of these facts to separate certain metals from each other.

Hydrogen, Persulphide of. *Syn.* HYDRIO PERSULPHIDE, HYDROGEN DISULPHIDE. *Prep.* One part slaked lime, 2 parts flowers of sulphur, and 16 parts of water are boiled together, and the clear cold solution decanted into some dilute hydrochloric acid; the persulphide subsides at the bottom as an oily fluid. Hydric persulphide has a great resemblance to hydric peroxide in qualities. It bleaches organic matters, and is decomposed with violence when brought into contact with the oxides of manganese and silver. It easily decomposes into sulphur and sulphuretted hydrogen; its vapour attacks the eyes.

SULPHU'RIC ACID. H₂SO₄. *Syn.* OIL OF VITRIOL, BRITISH O. OF V., VITRIOLIC ACID†; ACIDUM SULPHURICUM (B. P., Ph. L. & E.), ACIDUM SULPHURICUM VENALE (Ph. D.), ACIDUM VITRIOLICUM†, L. This acid, in a concentrated form, was discovered by Basil Valentine towards the end of the 15th century. At first it was obtained by the distillation of green vitriol, but is now made from sulphurous anhydride, obtained by the combustion either of sulphur or of certain sulphides. In consequence of the growing demand for sulphur in the manufacture of gunpowder, ultramarine, and for the destruction of the vine parasites in the vineyards of France, Italy, and Spain, sulphuric acid is now seldom made by burning sulphur, but, with few exceptions, by roasting iron pyrites or bisulphide of iron.

Since 1746, when Dr Roebuck proposed the use of leaden chambers, the production of sulphuric acid has been steadily increasing, until at the present time upwards of 100,000 tons are annually consumed in Great Britain, and a very large quantity is exported. The price is now about 1¼d. per lb., whereas the original oil of sulphur prepared by Valentine cost 2s. 6d. per lb.

Of the other sulphides employed in vitriol making may be mentioned galena, or native sulphide of lead, which, when roasted, gives up half its sulphur. The chief consumption of this mineral is in the Harz. Copper pyrites is also used in the Harz, as well as in Swansea and Glasgow. Blende, or native sulphide of zinc, is also occasionally had recourse to.

In addition to the above sulphides the vitriol maker in England, France, and Germany has lately largely availed himself of a compound known as 'Laming's mixture,' which is an impure oxide of iron that has been used in purifying coal-gas from sulphur. Laming's mixture is consequently rich in this element.

Associated with the pyrites in small quantities are various substances, some of which, becoming volatilised when the ore is burnt, enter the chambers with the mixed gases, and thus find their entrance into the acid, whilst others remain behind in the iron residue of impure ferric oxide left on the hearth of the furnace after roasting. The former of these foreign bodies, which are found in most commercial acids, are described below under the section 'Purification.' Amongst the solid non-volatile matters, the extraction of which from the burnt iron has been found in many works to yield a profit, are zinc, copper, silver, and thallium.

At Wolcrum, in Germany, the zinc which exists in the residue in the form of sulphate is extracted by lixiviation, and then treated with common salt, the reaction giving rise to the production of sulphate of soda and chloride of zinc. The soda obtained is sufficient to pay for the working of the operation, whilst a good profit is made by the sale of the large quantities of chloride of zinc which are thus yielded ("Development of the Chemical Arts during the Last Ten Years," by Dr A. W. Hofman—'Chemical News,' vol. xxiv, 1879).

The copper, which in some residues is met with to the amount of 4%, also pays for extraction, and is sold to the smelter. It is first converted into chloride, and then precipitated by iron. The silver is recovered by Claudet's process, which consists in precipitating it from a saline solution, in which it is in the state of a soluble chloride, by iodide of potassium.

In the Widnes Copper Works the silver so extracted yields an annual profit of £3000 (*ibid.*).

Thallium is found in the fine dust caused by the combustion of the pyrites, which dust deposits in the flues between the furnace and the chambers. The metal is extracted from the dust by treating this latter with dilute sulphuric acid. The resulting sulphate is converted into chloride, and again reconverted alternately into sulphate and chloride several times, the sulphate last obtained being reduced by metallic zinc (*ibid.*).

Selenium is also a frequent constituent in the fine dust. Some ores, after being subjected to roasting, yield iron capable of being worked. This is more particularly the case with the Spanish and Portuguese pyrites.

Prep. The following is an outline of the process by which sulphuric acid is obtained, and of the chemical changes which occur during its manufacture:

The sulphur or sulphide being placed on the hearth of the furnace, shown at A in the accompanying *engr.* (see p. 1620), when heated from below, soon takes fire, and combining with the oxygen of the atmospheric air, the admission of which into the furnace is regulated by an experienced workman, by the door shown in the plate, forms sulphurous anhydride. An iron pot, standing on the hearth of the furnace, contains a mixture of nitrate of soda and oil of vitriol, and this becoming heated by the burning sulphur, decomposition of the salt ensues, and fumes of nitric acid are given off. The sulphurous anhydride and nitric acid gases thus formed together with air are carried into large leaden chambers, standing on, and supported by, massive frameworks of stout timber. Steam is admitted continuously by several jets (see *engr.*) into these chambers, which are covered at the bottom with water to a depth of about 2 inches.

As soon as the mixed gases enter the chamber and come into contact with the steam, the sulphurous anhydride acts on the nitric acid, forming sulphuric acid, which falls into and is absorbed by the water on the floor of the chamber; and nitric oxide, which is liberated in the chamber.

The following equation will illustrate the reaction:

$$2HNO_3 + 3SO_2 + 2H_2O = 3H_2SO_4 + 2NO.$$

170 parts by weight of nitrate of soda are required to oxidise to sulphuric acid 96 parts of sulphur, whereas rarely more, and frequently less than 5 parts of soda are required by the vitriol maker. This saving of material is effected by the function performed in the chamber by the nitric oxide resulting from the decomposition of the nitric acid.

The nitric oxide reacting upon the air in the chamber abstracts oxygen from it and becomes converted into nitric peroxide, thus:

$$2NO + O_2 = 2NO_2.$$

Nitric peroxide is a very unstable compound, and directly it comes into contact with the fresh sulphurous anhydride entering the chamber, it oxidises it in the presence of water to sulphuric acid, thus:

$$NO_2 + SO_2 + H_2O = H_2SO_4 + NO.$$

This deportment of the nitric oxide being continuous, it will be seen it acts the part of a carrier of oxygen from the atmospheric air contained in the chamber to the sulphurous acid, and by so doing (theoretically) renders any further supply of nitrate of soda than that required to start the process unnecessary.

As soon as the sulphuric acid on the floor of the leaden chambers has acquired the sp. gr. of 1·6, and contains 70% H_2SO_4, it is drawn off, and concentrated by boiling in shallow leaden pans to the density of about 1·72, and contains about 80% H_2SO_4, after which it is further concentrated in green glass or platinum retorts until the sp. gr. reaches 1·84, and contains about 98% H_2SO_4. When of sp. gr. from 1·35 to 1·50 it is called chamber acid, and is used in the manufacture of salt-cake, sulphate of ammonia, some kinds of manure, and nitric acid. Sulphuric acid of sp. gr. 1·72 is mostly employed in the preparation of superphosphate of lime. After concentration to 1·84, the clear acid is put into large globular bottles of green glass (carboys), surrounded with straw and basket-work, and is sent into the market under the name of 'oil of vitriol.'

The leaden chambers in which the chemical changes take place, that result in the formation of the acid, vary greatly in dimensions in different works, being sometimes as much as 12 or 15 feet high, 15 or 20 wide, and from 150 to 300 feet long. They are mostly partially divided by incomplete leaden partitions, known as curtains, so arranged on the roof and the floor as to cause the currents of mixed gases to come into collision, and bring about their admixture. Where there are a number of small separate chambers they are connected by means of leaden tubes. A chamber having a capacity of 25,000 cubic feet will yield 10 tons of acid weekly.

The sheets of lead used in the construction of the chambers are united by the melting together of their edges. If cement were used, it would be speedily attacked and destroyed by the acid and gaseous products.

In 1746 the first vitriol factory in Great Britain was set up at Birmingham by Dr Roebuck, with whom originated the idea of the leaden chambers. The continuous process for the manufacture of sulphuric acid above described was devised in 1774 by a calico printer of Rouen, and improved by Chaptal.

Various attempts have been at different times unsuccessfully made to supersede the old process. Of these we may mention—

1. The proposal to oxidise sulphurous acid by means of chlorine in the presence of steam.

2. Persoz's method to oxidise sulphurous acid by means of nitric acid, and to regenerate the nitric oxide resulting from the reduction of the acid by the oxygen of the air in the presence of steam.

3. *a*, by the decomposition of gypsum by superheated steam at a red heat; or *b*, by decomposing the gypsum by chloride of lead.

The failure of the above and other efforts has led to the chemist turning his attention to the elaboration and perfection of the old process, in the working of which considerable improvements have been introduced within the last ten or fifteen years; improvements resulting not only in a diminished cost of production, but in the manufacture of a purer, and therefore better acid.

The proper construction of the furnaces, ovens, and grates on which the firing of the sulphur or pyrites takes place, together with the flues, is an important condition in the manufacture of the acid; and to this end a great deal of scientific knowledge and experience have lately been applied with excellent effect. Of the many improvements in this direction for burning poor ores of pyrites is a contrivance much used in Germany, where the furnace

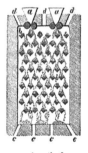

on which it is carried out is known as Gersten-höfer's oven. It is shown in the preceding drawing.

The furnace is fitted inside with a number of little fire-clay projections, arranged as shown in the plate, in banks or terraces, the function of which is to prolong the exposure of the pyrites to heat. The furnace having been previously raised to a red heat, by means of a coal or wood fire (which is then extinguished), the pyrites are admitted into it through the hoppers (a). At the base of the hoppers are grooved iron rollers, which crush the lumps of ore as they enter the chambers, and by thus reducing their size, expose a larger amount of surface to the action of heat. The greater part of the sulphur of the pyrites is thus burnt off, as the lumps pass from terrace to terrace, the heat at the same time generated by their combustion being sufficient to keep up that of the furnace. A moderate blast of air is ad-mitted at c, whilst the sulphurous acid formed ascends through d into the leaden chambers, the spent pyrites falling out through the apertures at e.

Another improved furnace is Perret's, which is largely used in France. In this small lumps of pyrites are placed on horizontal plates, and ex-posed to the hot gases generated in kilns below. The gases, on their way to the chambers, sweep over the pyrites and rob them of their sulphur.

The most important and noticeable improve-ment, however, of late years in sulphuric acid manufacture is the addition to the plant of the Gay-Lussac and Glover towers. Previous to this invention, the sulphuric acid of commerce, amongst other impurities, always contained ap-preciable quantities of certain oxides of nitrogen, the results of which were not only the contamina-tion of the acid, but a waste of substances which, properly utilised, are essential for the conversion of the sulphurous and sulphuric acid, and the loss of which leads to an increased consumption of nitrate of soda. Under the old method, these valuable oxides of nitrogen, which, with a large amount of nitrogen and a small quantity of oxygen, constituted the spent air of the last leaden chamber, were carried off into the air, and consequently lost. Now, instead of being allowed to diffuse into the atmosphere, they are made to pass through a tower or chamber (shown at c) filled with coke, through which a thin stream of sulphuric acid is made to trickle. In passing through the coke, therefore, the expiring spent gases come into contact with the sulphuric acid, to which they give up their oxides of nitrogen. From the tower (c) the acid flows into a cistern (D), whence it is pumped up to the top of another tower (E), called a Glover's tower, which is either filled with broken flints or arranged with inclined shelves, as shown in the plate. In this tower the acid meets with a current of hot sulphurous acid and air coming up from the furnace, which de-

prive it of the oxides of nitrogen, and the gaseous mixture enters the chambers, whilst the denitrified acid flows off into a suitable reservoir.

Since the introduction of the above, the consumption of nitrate of soda is sometimes lessened by more than one half.

Another very recent improvement, the invention of a German chemist named Sprengel, is the substitution of water spray, blown in by steam, for steam jets, in the leaden chambers. By this method a saving of coal to the extent of one third is said to be effected.

In theory, 1 molecule of sulphur requires only 3 molecules of oxygen to convert it into sulphuric acid, viz. 2 to form sulphurous anhydride, and 1 to convert the latter into sulphuric anhydride, which combines with 1 molecule of water to form the acid. Thus 1 kilogram of sulphur requires 1500 grms. or 1055 litres of oxygen, which is equivalent to 5275 litres of air containing 4220 litres of nitrogen; when pyrites is used a far larger quantity of air is required, for the obvious reason that the pyrites becomes converted into ferric peroxide. One kilogram of pyrites requires for its combustion nearly 6600 litres of air.

In well-regulated works the spent and escaping gases should not contain more than 2% of oxygen. If from 100 kilograms of sulphur 306 kilograms of strong acid of sp. gr. 1·84 be obtained, the result is regarded as very satisfactory; more frequently the product from 100 kilograms of sulphur does not exceed 280 or 290 kilograms.

Purif. Commercial sulphuric acid frequently contains nitrous acid and other oxides of nitrogen, arsenic, lead, and saline matter. The nitrous acid may be removed by adding a little sulphate of ammonia, and heating the acid to ebullition for a few minutes. Both nitric and nitrous acid are thus entirely decomposed into water and nitrogen gas. The arsenic may be got rid of by adding a little sulphide of barium to the acid, agitating the mixture well, and, after repose, decanting and distilling it. Lead, which exists as sulphate, may be separated as a white precipitate by simply diluting the acid with water. Saline matter may be removed by simple rectification. A good way of purifying oil of vitriol is to heat it nearly to the boiling-point, and pass a current of hydrochloric acid through it; the arsenic is thus carried over as the volatile chloride of arsenic, while the nitrous and nitric acids are expelled almost completely. To obtain a perfectly pure acid it should be distilled after the removal of the nitrous acid and arsenic by the methods indicated above. The distillation is most conveniently conducted, on the small scale, in a glass retort containing a few platinum chips, and heated by a sand-bath or gas flame, rejecting the first ½ fl. oz. that comes over.

According to Dr Ure the capacity of the retort should be from four to eight times as great as the volume of the acid, and connected with a large tubular receiver by a loosely fitting glass tube, 4 feet long and 1 to 2 inches in diameter.

The receiver should not be surrounded with cold water. We find that fragments of glass, or of rock crystals, may be advantageously substituted for platinum-foil, to lessen the explosive

violence of the ebullition; and it is better to heat the retort at the sides rather than at the bottom. Sulphuric acid which has become brown by exposure may be decolourised by heating it gently, the carbon of the organic substances being thus converted into carbonic acid.

The following table exhibits the sp. gr. of different degrees of concentration of sulphuric acid. In consequence of the discovery of errors in Kolb's table (the one usually resorted to), G. Lunge and M. Isler have made a fresh determination with great care (*vide* ' Zeit. ang. Chem.,' 1890, 129—135; and ' Chem. Soc. Journ.,' 1891, vol. lx, p. 150).

Sp. gr. at 15°/4° in vacuo.	Percentage of H_2SO_4.	Sp. gr. at 15°/4° in vacuo.	Percentage of H_2SO_4.
1·000	0·09	1·600	68·51
1·020	3·03	1·620	70·32
1·040	5·96	1·640	71·99
1·060	8·77	1·660	73·64
1·080	11·60	1·680	75·42
1·100	14·35	1·700	77·17
1·120	17·01	1·720	78·92
1·140	19·61	1·740	80·68
1·160	22·19	1·760	82·44
1·180	24·76	1·780	84·50
1·200	27·32	1·800	86·90
1·220	29·84	1·820	90·05
1·240	32·28	1·824	90·80
1·260	34·57	1·826	91·25
1·280	36·87	1·828	91·70
1·300	39·19	1·830	92·10
1·320	41·50	1·832	92·52
1·340	43·74	1·834	93·05
1·360	45·88	1·836	93·80
1·380	48·00	1·838	94·60
1·400	50·11	1·840	95·60
1·420	52·15	1·8405	96·95
1·440	54·07	1·8410	97·00
1·460	55·97	1·8415	97·70
1·480	57·83	1·8410	98·20
1·500	59·70	1·8405	98·70
1·520	61·59	1·8400	99·20
1·540	63·43	1·8395	99·45
1·560	65·08	1·8390	99·70
1·580	66·71	1·8385	99·95

Prop. Commercial sulphuric acid (oil of vitriol) is a colourless, oily-looking, odourless, and highly corrosive liquid, the general properties of which are well known. Its sp. gr. at 60° should never be greater than 1·848, or less than 1·840. The acid purified by distillation contains about 2% of water. It is immediately coloured by contact with organic matter. It attracts water so rapidly from the atmosphere when freely exposed to it, as to absorb 1-3rd of its weight in 24 hours, and 6 times its weight in a few months. When 3 volumes are suddenly mixed with 2 of water, the temperature of the mixture rises more than 180° F. Its freezing-point appears to be about 60° below that of water (Miller and Odling give that of the rectified acid as —30° F.; Apjohn and Abel and Bloxam, —29° F.). It boils at 338° C. (640° F.); solidifies at 34° C.

It rapidly corrodes the skin and other organic textures, usually blacking them at the same time. It dissolves melted sulphur, converting it into sulphur dioxide. All the ordinary metals except gold and platinum are acted upon by the acid; the metal being oxidised by one portion of the acid, which is thus converted into SO_3, and this reacts with another part of the acid to form a sulphate.

Pur. Free from colour and odour. Sp. gr. 1·854 at 0° C., and 1·834 at 24° C. (*Roscoe*). "100 gr. are saturated by 285 gr. of crystallised carbonate of soda" (Ph. L.). "What remains after the acid is distilled to dryness does not exceed 1-400th part of its weight. Diluted sulphuric acid is not discoloured by sulphuretted hydrogen" (Ph. L. 1836). "Diluted with its own volume of water, only a scanty muddiness arises, and no orange fumes escape. Sp. gr. 1·840" (Ph. E.). "The rectified acid (ACIDUM SULPHURICUM PURUM—Ph. E. and D.) is colourless; dilution causes no muddiness; solution of sulphate of iron shows no reddening at the line of contact when poured over it. Sp. gr. 1·845" (Ph. E.; 1·846—Ph. D.; 1·843—B. P.; 1·842— *Urs*).

Uses, &c. The uses of sulphuric acid are so numerous that it would be impossible to mention all of them, sulphuric acid being to chemical what iron is to mechanical industry. Sulphuric acid is employed in preparing a great many other acids —among them, nitric, hydrochloric, sulphurous, carbonic, tartaric, citric, phosphoric, stearic, oleic, and palmitic. It is used in making superphosphates, soda, sulphate of ammonia, alum, sulphates of copper and iron, in paraffin and petroleum refining, silver refining, manufacture of madder preparations, manufacture of glucose from starch, to dissolve indigo, in the manufacture of blacking, vegetable parchment, and on account of its hygroscopic properties for the purpose of drying gases, &c. In the diluted state it is used in medicine. When swallowed, it acts as a violent corrosive poison. The antidotes are chalk, whiting, magnesia, carbonate of soda, or carbonate of potash, mixed with water or any bland diluent, and taken freely, an emetic being also administered.

Tests. The addition of barium chloride solution to a liquid containing sulphuric acid or a sulphate throws down a heavy white precipitate of barium sulphate, insoluble in dilute hydrochloric acid. Free sulphuric acid may be detected in a liquid c. of vinegar by evaporating it on a water-bath with a small quantity of sugar, when if present a black residue will be obtained.

Estim. The strength of sulphuric acid is most correctly ascertained by its power of saturating bases. In commerce it is usually determined from its sp. gr. The quantity of sulphuric acid present in a compound may be determined by weighing it under the form of sulphate. See ACIDIMETRY.

Concluding Remarks. According to most of our standard works on chemistry, British oil of vitriol, when purified and brought to its maximum strength by distillation, is a definite chemical compound, having the formula H_2SO_4, and designated normal sulphuric acid by Odling. Marignac, however, asserts that the distilled acid always contains an excess of water, and that the true monohydrate can only be obtained by sub-

mitting fuming sulphuric acid ("Nordhausen s. a.') to congelation. According to this chemist, the true monohydrate readily freezes in cold weather, and remains solid up to 51° F. Several other definite hydrates of sulphuric acid are now generally recognised by chemists, of which we may notice bihydrated sulphuric acid (" glacial s. a.'), having a sp. gr. of 1·78, freezing at about 40° F. (47°—*Miller*), and boiling at about 435° F. (*Aojohn*; 401° to 410°—*Odling*); and terhydrated sulphuric acid, having a sp. gr. of 1·632, and the boiling-point 348° F. See also SULPHURIC ACID, NORDHAUSEN (*below*).

Sulphuric Acid, Al'coholised. *Syn.* ACIDUM SULPHURICUM ALCOHOLISATUM, L.; EAU DE RABEL, Fr. *Prep.* (P. Cod.) To rectified spirit, 3 parts, add, very gradually, sulphuric acid, 1 part. It is generally coloured by letting it stand over a little cochineal. Refrigerant, and, externally, escharotic.—*Dose*, ½ fl. dr. to water, 1 pint; as a cooling drink in fevers, &c.

Sulphuric Acid, Anhy'drous. SO_3. *Syn.* SULPHURIC ANHYDRIDE, DRY SULPHURIC ACID; ACIDUM SULPHURICUM SINE AQUÂ, L. *Prep.* 1. By heating Nordhausen acid to about 100° F. in a glass retort connected with a well-cooled receiver.

2. By passing a mixture of sulphur dioxide and oxygen over heated platinum-sponge or platinised asbestos.

3. (*Barreswill.*) 2 parts of the strongest oil of vitriol are gradually added to 3 parts of anhydrous phosphoric acid, contained in a retort surrounded by a freezing mixture; when the compound has assumed a brown colour the retort is removed from the bath, and connected with a receiver which is set there in its place; a gentle heat is now applied to it, when white vapours pass over into the receiver, and condense there under the form of beautiful silky crystals. The product equals in weight that of the phosphorus originally employed. "If a few drops of water be added, a dangerous explosion ensues."

Prop. Transparent prisms, deliquescing rapidly, and fuming in the air; put into water, it hisses like a red-hot iron; it melts at 14·8° C., and boils at 46° C.; it does not redden dry litmus paper; sp. gr. 1·97 at 20° C.

A second modification is said to be obtained when the melted binoxide is kept below 25° C.; it crystallises in silky needles which melt at 50° C.

Sulphuric Acid, Aroma'tic. *Syn.* ELIXIR OF VITRIOL, ACID N. OF V.; ACIDUM SULPHURICUM AROMATICUM (B. P., Ph. E. & D.), L. *Prep.* 1. (Ph. E. & D.) Oil of vitriol, 3½ fl. oz.; rectified spirit, 1½ pints; mix, add of powdered cinnamon, 1½ oz.; powdered ginger, 1 oz.; digest for 6 days (7 days—Ph. D.), and filter. Sp. gr. 0·974 (Ph. D.).

2. (Wholesale.) From compound tincture of cinnamon, 1 gall.; oil of vitriol, 1 lb.; mix, and in a week filter.—*Dose*, 10 to 30 drops, in the same case as the dilute acid.

3. (B. P.) Sulphuric acid, 3 parts; rectified spirit, 40 parts; cinnamon, in powder, 3 parts; ginger, in powder, 1½ parts; mix the acid gradually with the spirit, add the powders, macerate for seven days, and filter.—*Dose*, 5 to 30 minims.

Sulphuric Acid, Dilute'. *Syn.* SPIRIT OF VITRIOL; ACIDUM SULPHURICUM DILUTUM (B. P., Ph. L., E., & D.), L. *Prep.* 1. (Ph. L.) Take of sulphuric acid, 15 fl. dr., and dilute it gradually with distilled water, q. s. to make the whole exactly measure a pint. Sp. gr. 1·103. 1 fl. oz. of this acid is exactly saturated by 216 gr. of crystallised carbonate of soda.

2. (Ph. E.) Sulphuric acid, 1 fl. oz.; water, 13 fl. oz. Sp. gr. 1·090.

3. (Ph. D.) Pure sulphuric acid, 1 fl. oz.; distilled water, 13 oz. Sp. gr. 1·084.

4. (B. P.) Sulphuric acid, 3 parts; distilled water, q. s. to measure 35½ parts; mix by adding the acid gradually to the water.—*Dose*, 4 to 20 minims.

Prop., &c. Antiseptic, tonic, and refrigerant. —*Dose*, 10 to 30 drops, largely diluted with water, several times daily; in low typhoid fevers, passive hæmorrhages, profuse perspiration, in various skin diseases to relieve the itching, in dyspepsia, &c. It is also used externally.

Sulphuric Acid, Nordhausen. *Syn.* FUMING SULPHURIC ACID; ACIDUM SULPHURICUM FUMANS, L. *Prep.* By distilling calcined ferrous sulphate ('green vitriol') in earthen retorts. The retorts, which are shown at A in the accompanying *engr.*, after the 'green vitriol' has been put into them, are placed in a galley-furnace, as shown below, the necks passing through the wall of the furnace, and being properly secured to the necks of the receivers (B B). Into each of the flasks 2¼ lbs. of green vitriol are put; on the first application of heat only sulphurous acid and weak hydrated sulphuric acid come over, and are usually allowed to escape, the receivers not being securely luted until white vapours of anhydrous sulphuric acid are seen. Into each of the receiving flasks 30 grms. of water are poured, and the distillation continued for 24 to 36 hours. The retort flasks are then again filled with raw material, and the operation repeated four times before the oil of vitriol is deemed strong enough. The residue in the retorts is red (per)oxide of iron, still retaining some sulphuric acid. The product is a brown oily liquid, which fumes in the air, is intensely corrosive, and has a sp. gr. about 1·9. When heated to about 100° F. the anhydrous acid is given off, and ordinary oil of vitriol is left. According to Marignac crystals of normal sulphuric acid (H₂SO₄) are formed in this acid when it is submitted to a low temperature. Nordhausen acid is so called from the place of its manufacture in Saxony. It may be regarded as a mixture or compound of H_2SO_4 and SO_3. In England it is now made by dissolving sulphuric anhydride in about twice its weight of oil of vitriol. It is chiefly used for dissolving indigo; also in making alizarine, and is a convenient source of sulphuric anhydride.

SULPHURIC ANHYDRIDE. See SULPHURIC ACID, ANHYDROUS.

SULPHURIC E'THER. See ETHER.

SUL'PHUROUS ACID. H_2SO_3. This substance is only known in aqueous solution. It is formed by dissolving the corresponding anhydride in

water. It forms two series of salts, termed sulphites, q. v.

Sulphurous Acid, Anhydrous. SO_2. *Syn.* SULPHUROUS ANHYDRIDE; ACIDUM SULPHUROSUM (B. P.), L. This compound is freely evolved in the gaseous form when sulphur is burnt in air or oxygen, and when the metals are digested in hot sulphuric acid; and, mixed with carbonic acid, when charcoal, chips of wood, cork, and sawdust are treated in the same way.

Prep. 1. By heating together sulphur and strong sulphuric acid.

2. By the action of sulphuric acid, 4 fl. oz., on chippings of copper, 300 gr., at a gentle heat. Pure.

3. (*Berthier.*) By heating, in a glass retort, a mixture of black oxide of manganese, 100 parts, and sulphur, 12 or 14 parts. Pure. The gas evolved should be collected over mercury, or by downward displacement.

4. (*Redwood.*) Pounded charcoal, ½ oz.; oil of vitriol, 4 fl. oz.; mix in a retort, apply the heat of a spirit lamp, and conduct the evolved gases by means of a bent tube into a bottle containing water. The sulphurous acid is absorbed, whilst the carbonic acid gas passes off.

Prop., &c. Heavy, colourless gas, smelling of burning brimstone. Water absorbs 43·5 times its volume of this gas. Pure liquid sulphurous

acid can only be obtained by passing the pure dry gas through a glass tube surrounded by a powerful freezing mixture. Its sp. gr. is 1·45 at 20° C.; boiling-point, 14° F.; it causes intense cold by its evaporation. Sulphurous acid forms salts called sulphites. Its disposition to absorb oxygen and pass into sulphuric acid renders it a powerful reducing agent.

Uses. To bleach silks, woollens, straw, sponge, isinglass, baskets, &c., and to remove vegetable stains and ironmoulds from linen. For these purposes it is prepared from sawdust or any other refuse carbonaceous matter, and the articles to be bleached must be moistened with water.

Distilled water, saturated with sulphurous anhydride, is used as a deoxidiser, disinfectant, and antiseptic. Diluted with from 1 to 2 parts of water it is employed as a lotion for wounds, cuts, ulcers, bedsores, scalds, and burns; with from 1 to 5 parts of water it is used as a gargle, also as a lotion in parasitic skin diseases; from ½ to 1 dr., in a wine-glassful of water, 3 times a day, relieves constant sickness.

Several preparations containing sulphurous acid have recently been introduced to the public as agents in sanitation under the name of *Sporokton* (germ-killer). To understand the nature and merits of these preparations it is desirable to explain the true and individual meanings of 'deodoriser,' 'antiseptic,' and 'disinfectant' —words which are too often improperly employed as if they had the same signification, and as if, in fact, they were convertible terms.

A deodoriser is a substance which will absorb or destroy bad smells; an antiseptic is an agent which will prevent or retard putrefaction; and a disinfectant is an agent which will render harmless the virus of smallpox, scarlet fever, measles, diphtheria, influenza, pleuro-pneumonia, cattle plague, glanders, distemper in dogs, and other infectious or contagious diseases.

Now, medical authorities and sanitarians are of opinion that the most potent disinfectant with which we are acquainted is sulphurous acid, a gas which has been used for ages as a fumigator. Sulphurous acid has not, however, been so generally employed for disinfecting purposes as one might from these circumstances have expected, on account of the difficulties and inconveniences which formerly attended its generation.

To remove these drawbacks, and to render sulphurous acid, both as a gas and in solution, easily and cheaply available for the above-named and many other applications, sporokton has been invented. Several varieties are made; they are as follows:

Liquid No. 1. This preparation consists of a colourless solution of a non-volatile antiseptic, usually a salt of zinc, impregnated with 80 times its bulk of sulphurous acid gas; in other words, 1 pint of the liquid contains 10 galls. of gas. Liquid sporokton is, in fact, a combination of one of the most powerful antiseptics with the disinfectant; the former ingredient will effectually prevent the putrefaction of any solid or liquid animal or vegetable matter with which it may come in contact, while the sulphurous acid will rapidly pass off in the gaseous state into the surrounding air, and act as an energetic destroyer of noxious atmospheric impurities.

Liquid sporokton absorbs ammonia and sulphuretted hydrogen, destroys bad smells, and prevents the spread of infectious diseases; it is, consequently, a valuable agent for the deodorisation and disinfection of wards of hospitals, sick rooms, dairies, larders, ship, stables, cowhouses, kennels, piggeries, slaughter-houses, urinals, water-closets, privies, cesspools, sewers, drains, and other similar buildings and places.

After it has parted with the whole of its sul-

phurous acid gas, liquid sporokton remains as an odourless, non-volatile antiseptic and absorber of ammonia and sulphuretted hydrogen.

Liquid sporokton evolves its sulphurous acid by simple exposure to air, without the aid of heat, so that no risk of fire attends its use, as is the case when rooms, buildings, holds of ships, &c., are fumigated with this gas by the old plan; it will not stain or in any other way injure undyed woollen, linen, or cotton goods. It is consequently well adapted for the disinfection of underclothing, sheets, blankets, bed-furniture, &c.

Liquid sporokton may be employed for the instantaneous preparation of a bath or lotion of sulphurous acid, to be used, under medical direction, in the treatment of itch, ringworm, chronic eczema, lepra, psoriasis, impetigo, pityriasis, &c., in man, as well as mange, scab, and other skin affections in the lower animals.

Liquid sporokton is clean, it requires no skill in using it, and its action is perfectly controllable.

Liquid No. 2. This preparation is specially made for the disinfection and purification of old beer barrels, wine casks, and the like. It is similar in composition to, and may be used for the same purpose as, No. 1; except, however, that as No. 2, unlike No. 1, is liable, from its containing iron instead of zinc, to stain linen, wood, &c., it should not be employed for disinfecting clothing or sprinkling over floors, decks of ships, and the like.

Solid. This is a powder, usually a mixture of calcium sulphite and ferric chloride, which, by simple exposure to air, will slowly and steadily, or, when sprinkled with water, rapidly give out 25% of its weight of sulphurous acid, and leave no unpleasant smell behind it.

Sulphurous acid gas, unlike non-volatile disinfectants, quickly mingles with the air, and seeks out, as it were, the noxious atmospheric impurities it is capable of destroying.

Solid sporokton, in addition to evolving sulphurous acid, contains an excess of ferric chloride which, together with this gas, renders it a most useful and efficient antiseptic.

Sulphurous Anhydride. See SULPHUROUS ACID (Anhydrous).

SUMACH. This dye-stuff is chiefly used as a substitute for galls. With a mordant of acetate of iron it gives grey or black; with tin or acetate of alumina, yellow; and with sulphate of zinc, a yellowish brown; alone it gives a greenish-fawn colour.

SUMBUL. *Syn.* MUSK-ROOT, JATAMANSI, SUMBUL-ROOT; SUMBUL RADIX (B. P.). A substance introduced to British medicine by Dr A. B. Granville, in 1850. It occurs in circular pieces, varying from 1 to 3 or 4 inches in diameter; has a musk-like odour, and a sweet balsamic taste. It acts as a powerful stimulant, especially of the nervous system. In India and Persia it has long been used as a medicine, a perfume, and as incense.—*Dose*, 15 gr. to 1 dr., either masticated, or made into an infusion, electuary, or tincture; in cholera, hysteria, neuralgia, epilepsy, low fevers, and various other spasmodic and nervous disorders.

SUMMER DRINKS. See LEMONADE, SHERBET, &c.

SU'PER-. See NOMENCLATURE.

SUP'PER. The evening meal; the last meal of the day. Supper is generally an unnecessary meal, and when either heavy, or taken at a period not long before that of retiring to rest, proves nearly always injurious, preventing sound and refreshing sleep, and occasioning unpleasant dreams, nightmare, biliousness, and all the worst symptoms of imperfect digestion. The last meal of the day should be taken at least three hours before bedtime. Even when it consists of some 'trifle,' as a sandwich or biscuit, an interval of at least an hour should elapse before retiring to rest. In this way restlessness and unpleasant dreams will become rare.

SUPPOSITORY. *Syn.* SUPPOSITORIUM, L. A conical shaped preparation used to insert in the rectum for the purpose of affecting the lower intestine, or, by absorption, the system generally. Suppositories are rounded, usually elongated masses, having the active medicine combined with some substance which will retain the proper shape, as stearin, gelatin, or cacao butter. The last substance is, perhaps, the best vehicle for remedies prescribed in this form. It is, however, rather too soft to be used without admixture. According to Dorvault, the addition of one eighth part by weight of wax imparts the proper hardness.

All difficulty of removing suppositories from the mould may be obviated by having the moulds ice-cold and immersed in soapy water.

The mode of proportioning the doses of active ingredients has been noticed in the article ENEMA.

Suppositories of Assafœtida. *Syn.* SUPPOSITORIA ASSAFŒTIDÆ (Ph. U. S.). *Prep.* Tincture of assafœtida, 1 oz.; oil of theobroma, 390 gr. Let the tincture evaporate by exposure to the air until of the consistence of a thick syrup, and proceed as for suppositories of carbolic acid.

Suppository, Astringent. *Syn.* SUPPOSITORIUM ASTRINGENS (*Reuss*). *Prep.* Powdered oak-bark, 2 dr.; tormentil, 2 dr.; honey, q. s. For 8 suppositories.

Suppositories of Carbolic Acid with Soap. *Syn.* SUPPOSITORIA ACIDI CARBOLICI CUM SAPONE (B. P.). *Prep.* Carbolic acid, 12 gr.; curd soap, in powder, 180 gr.; starch, q. s.; mix the carbolic acid with the soap, and add starch, q. s. to make of a suitable consistency; divide into twelve equal parts, and make each suppository into a conical or other convenient form.

Suppository, Emollient. *Syn.* SUPPOSITORIUM EMOLLIENS. *Prep.* Butter of cacao and spermaceti in equal parts, melted together.

Suppositories of Glycerin. *Syn.* SUPPOSITORIA GLYCERINI. Gelatin, ½ oz.; glycerin, 2½ oz.; water, q. s. Soak the gelatin in water 1 or 2 minutes, pour off the water, leave the gelatin to rest till quite soft, and dissolve in the glycerin, until the product weighs 1560 gr. The product may be moulded into 30, 60, or 120 gr. suppositories.

Suppository of Iodide of Potassium. *Syn.* SUPPOSITORIUM POTASSII IODIDI (*Mr Stafford*). *Prep.* Iodide of potassium, 1 gr. to 4 gr.; extract of henbane, 6 gr.; extract of hemlock, 6 gr. In enlarged prostate.

Suppositories of Iodoform. *Syn.* SUPPOSITORIA IODOFORMI. Iodoform, 36 parts; oil of theobroma, 144 parts. Melt, mix, and divide into 15 gr. suppositories.

Suppository, Irritant. *Syn.* SUPPOSITORIUM IRRITANS (*Richard*). *Prep.* Butter of cacao, 2 dr.; aloes, 4 gr.; tartarised antimony, 1 gr. To restore the hæmorrhoidal flux.

Suppository of Lead (Compound). *Syn.* SUPPOSITORIUM PLUMBI COMPOSITUM (B. P.). *Prep.* Acetate of lead, in powder, 36 parts; opium, in powder, 12 parts; oil of theobroma, 132 parts; melt the oil of theobroma with a gentle heat, then add the other ingredients, previously rubbed together in a mortar, and, having mixed them thoroughly, pour the mixture while it is fluid into suitable moulds of the capacity of 15 gr. The above makes 12 suppositories.

Suppository of Mercury. *Syn.* SUPPOSITORIUM HYDRARGYRI (B. P.), L. *Prep.* Ointment of mercury, 60 gr.; oil of theobroma, 120 gr.; melt together, stir till well mixed, and immediately pour into moulds of the capacity of 15 gr. The above makes 12 suppositories.

Suppository of Morphine. *Syn.* SUPPOSITORIUM MORPHINÆ (B. P.), L. *Prep.* Hydrochlorate of morphine, 6 gr.; oil of theobroma, 174 gr.; melt the oil of theobroma with a gentle heat, then add the hydrochlorate of morphine, and mix all the ingredients thoroughly; pour the mixture, while it is fluid, into suitable moulds of the capacity of 15 gr., or the fluid mixture may be allowed to cool, and then be divided into 12 equal parts, each of which should be made into a conical form.

Suppositories of Morphine with Soap. *Syn.* SUPPOSITORIA MORPHINÆ CUM SAPONE (B. P.). *Prep.* Hydrochlorate of morphine, 6 gr.; glycerin of starch, 50 gr.; curd soap, in powder, 100 gr.; starch, q. s. Mix the hydrochlorate with the glycerin of starch and soap, and add starch, q. s. to form a paste of suitable consistence. Divide into 12 equal parts, each of which is to be made into a conical or other convenient form of suppository.

Suppository for Piles. *Syn.* SUPPOSITORIUM HÆMORRHOIDALE, S. SEDATIVUM, L. *Prep.* 1. Powdered opium, 2 gr.; finely powdered galls, 10 gr.; spermaceti cerate, 1 dr.

2. (*Ellis.*) Powdered opium, 2 gr.; soap, 10 gr.; mix.

3. (*Richard.*) Extracts of opium and stramonium, of each, 1 gr.; cacao butter, 2 dr. Used when the piles are very painful.

Suppository, Pur'gative. *Syn.* SUPPOSITORIUM CATHARTICUM, L. *Prep.* 1. Soap, 1 dr.; elaterium, 1 to 2 gr.; mix. As a strong purge.

2. (*Niemann.*) Soap, 2 dr.; common salt, 1 dr.; honey, q. s.; mix. As a mild cathartic.

Suppository of Quinine. *Syn.* SUPPOSITORIUM QUINÆ (*Boudin*), L. *Prep.* Sulphate of quinine, 15 gr.; butter of cacao, 1½ dr.; mix.

Suppository, Resol'vent. *Syn.* SUPPOSITORIUM RESOLVENS, L. *Prep.* (*Stafford.*) Iodide of potassium, 3 to 4 gr.; extracts of henbane and hemlock, of each, 6 gr. In enlargement or induration of the prostate gland.

Suppository of Rhatany. *Syn.* SUPPOSITORIUM RHATANIÆ (P. Cod.), L. *Prep.* Butter of cacao, 1 dr.; extract of rhatany, 15 gr., for 1 suppository.

Suppository of Santonin. *Syn.* SUPPOSITORIUM SANTONINI. Santonin, 2 or 5 gr.; oil of theobroma to make 15 gr. Very useful to destroy thread-worms.

Suppository, Sed'ative. *See above.*

Suppository of Tannic Acid. *Syn.* SUPPOSITORIUM ACIDI TANNICI (B. P.). *Prep.* Tannic acid, 36 gr.; benzoated lard, 44 gr.; white wax, 10 gr.; oil of theobroma, 90 gr.; melt the wax and oil with a gentle heat, then add the tannic acid and benzoated lard, previously rubbed together, and mix thoroughly. Pour the mixture while it is fluid into suitable moulds of the capacity of 15 gr. The above makes twelve suppositories.

Suppositories of Tannic Acid with Soap. *Syn.* SUPPOSITORIA ACIDI TANNICI CUM SAPONE (B. P.). *Prep.* Tannic acid, 36 gr.; glycerin of starch, 50 gr.; curd soap, in powder, 100 gr.; starch, q. s. Mix the tannic acid with the glycerin of starch and soap, and add starch, q. s. to form a paste of suitable consistence; divide into 12 equal parts, each of which is to be made into a conical or other convenient form of suppository.

Suppositories, Vaginal. *Syn.* SUPPOSITORIA VAGINALE (*Guadrist*). *Prep.* Liquid chloride of zinc, 5 minims; sulphate of morphia, ½ gr.; mix with 2 dr. of the following paste:—Thick mucilage of tragacanth, 6 parts; white sugar, 3 parts; starch, 9 parts. Mr. Druitt prescribes in leucorrhœa—Tannin, 10 gr., with mucilage of tragacanth, q. s.

Suppository, Ver'mifuge. *Syn.* SUPPOSITORIUM ANTHELMINTICUM, S. VERMIFUGUM, L. *Prep.* (*Swediaur.*) Aloes, 44 dr.; common salt, 3 dr.; flour, 2 dr.; honey, q. s. to make a stiff mass; divide into proper shaped pieces, weighing about 15 gr. each. One to be used after each motion.

SURVEYING. The act of measuring land for various purposes. It is not proposed in this article to deal with surveying in all its details, but merely to indicate the means whereby maps and plans may be constructed with sufficient accuracy for ordinary purposes.

Where very accurate measurements are required the chain or measuring tape and the theodolite must be employed; to the traveller or explorer such methods are useless, as involving great consumption of time and labour, the carriage of heavy instruments, and, above all, skilled assistance. The methods in use for military purposes, reconnaissance, &c., can be made by practice to yield very excellent results. The instruments required are few and simple, easily carried and easily mastered, and the principles on which military surveys are conducted are of such general utility that it is a matter of regret that they are not more widely known.

Scales. All maps and plans are drawn to scale; that is to say, there is a fixed and definite relation between the distances as measured on the ground and as measured on paper. The scale of a map may be expressed in two ways. (1) By means of a fraction ("the representative fraction") which expresses the relation between the plan and the ground it represents, e.g. on the

scale of 1 : 1000, every inch on the map would represent 1000 inches on the ground, or one metre would represent a kilometre. (2) The other way is by a comparison of units, e.g. the scale of one inch to the mile means that every inch on the map represents one mile of ground. The former method of expressing the scale is by far the most convenient, as no question need ever arise as to the distances on a foreign map if we are ignorant of the measures of the country. It is easy to convert the one system into the other in any given case; e.g. there are 63,360 inches in the statute mile; the scale of one inch to the mile may, therefore, be expressed by the fraction $\frac{1}{63360}$, i. e. one inch on the map represents 63,360 inches on the ground. If we have a foreign map on the scale of $\frac{1}{10000}$ we need not concern ourselves with the measures of the country, for we know that one inch or other unit of length on the map represents 10,000 inches or other units on the ground.

In deciding on the scale on which a map is to be constructed we must consider the amount of detail required, and the size of the paper which it will ultimately cover. Twenty miles of country on the scale of six inches to the mile would give a map 10 feet long, which would be quite unmanageable, though for some purposes it might be necessary in order to get in all the details required to be shown; 10 yards on the scale would be represented by $\frac{1}{10}$ inch, whereas on the scale of 1 inch to the mile 10 yards would be expressed by $\frac{1}{176}$ inch, a length too small to be appreciated or measured by ordinary instruments. For military purposes the 6-inch scale is used for maps of positions, &c., where detail is of great consequence; and a 2 or 1 inch scale for large areas, long stretches of roads or rivers, &c.

To make a Scale. Suppose a scale of 15 inches to the mile to show yards is required, a line 5 inches long would represent ¼ of a mile (586 yards); a round number is more convenient, say 600 yards; then by proportion—

$$\underset{\substack{\text{Yards in a} \\ \text{mile.}}}{1760} \; : \; \underset{\substack{\text{Yards in} \\ \text{scale.}}}{600} \; :: \; \underset{\substack{\text{Inches repre-} \\ \text{senting one mile.}}}{15} \; : \; \underset{\substack{\text{Length of} \\ \text{scale.}}}{5·11}$$

Draw a line 5·11 inches long on paper by means of a scale of inches, and divide this into six equal parts of 100 yards each.

This may be done most easily by geometry as follows:

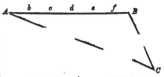

Let *A B* be the line as measured 5·11 inches long. Draw another line *A C* at any angle with *A B*, and by means of a scale divide it or any part of it into six equal parts. If *A C*, for example, be exactly 6 inches long, and the inches be marked off upon it, this is all that is required; in this case join *B C*, and through each of the divisions on *A C* indicating inches, draw a line parallel with *B C*, and cutting the line *A B* at *b, c, d, e,* and *f,* the line *A B* will be divided into six equal parts. As *A B* represented 600 yards, each of these divisions will represent 100 yards. By a similar process the division *A b* may be subdivided into 10 equal parts, each of which will represent 10 yards. This scale having been constructed, we are in a position to translate any distance on our map into yards.

Measurement of Distance. The first requirement in surveying is the accurate measurement of distances; this is generally done with a measuring tape or 'chain.' 'Gunter's chain' is 22 yards long, so that 10 chains = 1 mile, 80 chains = 1 mile; the chain is divided into 100 links, each 7·92 inches long. Two persons are required in order to use a chain, the 'leader' and the 'follower.' The leader has 10 iron arrows when the measuring begins; he walks on from the starting-point till the chain is taut, the follower holding the other end, and sticks an arrow into the ground at his end of the chain; the chain is dragged on until the 'follower' reaches this arrow, when he holds the end of the chain close to it, waits till the 'leader' pulls it tight, and then removes the arrow; this operation is proceeded with till all 10 arrows have been picked up by the 'follower,' who returns them to the 'leader,' and the operation begins again.

For military sketching and other like purposes distances are measured by *pacing, i. e.* by walking over the ground and counting the number of paces taken; if the length of the pace is known, the distance can be calculated. It is not difficult to pace yards with a little practice, and this should be learned by measuring 100 or 200 yards on level ground, and pacing it frequently until the art is acquired. On rough ground the length of the pace is less than on smooth, and it is well for the surveyor to learn how far the length of his pace is affected by irregularities of surface. If the regulation pace of 30 inches be used the number of yards is found by deducting ⅙ from the number of paces, *e. g.* 36 paces = 30 yards. Pacing cannot be relied upon on slopes of more than 15°, and it must be remembered that the horizontal distance between any two points on uneven ground is less than the distance paced between them up and down the slopes; two sides of any triangle are greater than the third, and the distance between two points with a hill between is represented by the length of a tunnel through this hill, and *not* by the distance traversed in going up the hill and down the other side. This is a common source of error when the ground is uneven.

Pacing may be done on horseback, the length of pace of a well-broken average horse at the walk being 83 inches. Young horses pace 39 or 40 inches.

For rapid work the measurement of distance by the eye is of great importance. This can be acquired by practice.

The simplest method of keeping count of the paces is to close one of the fingers of the left hand for every hundred paces walked. The count should begin again at every halt or change of direction.

Triangulation. If two sides of a triangle and the angle between them are known, the third side

and the remaining angles can always be found by calculation. Suppose that two points, *A* and *B*, are accessible, and the length of the line *A B* is accurately known, also the distance from *A* to *C*,

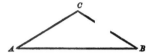

the distance from *B* to *C* can be calculated by trigonometrical formulæ.

Or if the line *A B* is known, and the point *C* can be seen from both *A* and *B*, we may by means of instruments observe the angle *B A C*. This tells us that the point *C* lies somewhere on the line *A C*. If now we observe from *B* the angle *A B C*, we know that *C* must lie somewhere on the line *B C*; it also lies on the line *A C*; *C* must, therefore, be situate at the point where these two lines intersect. The length of *A B* being known, and the two angles *C A B* and *A B C* being also known, the length of the two sides *A C* and *B C* can be calculated, and the position of *C* determined. Other triangles may be built on the line *A B*, with other points for their vertices, and thus, by accurate observation of the length of one line, *A B*, and of a number of angles, the relative position of a number of points with regard to *A* and *B* may be determined. This process is called *triangulation*, and the line *A B* is called the *base*. It is clear that any error in the measurement of the base will affect the whole work. In actual survey the length of this line is always determined with the greatest exactitude, and its position is so selected as to afford the greatest possible advantages with regard to neighbouring points of importance. Surveying is thus seen to depend upon the accurate measurement of distance and angular direction. The instruments by which this is effected for the kind of survey under consideration are few and simple.

FIG. 1.—Plane Table.

The plane table consists of a board on which a sheet of drawing-paper is stretched, attached to a tripod stand by a screw in such a way that it may be revolved to any position. The table being placed in position and fixed, a ruler with sights attached, *e.g.* a slit at the eye end and a hair or wire at the other, is laid upon the paper and pivoted against a needle driven into the paper at a point corresponding to the observer's position.

The eye, the sight, and some distant object being all placed in a line, it is obvious that the *direction* of this object with regard to the observer can be marked on the paper. The method of use is briefly as follows:—A base line, *A B*, is selected and measured, and set off on the paper to the required scale; the table is placed at *A* and levelled; by means of the sighted ruler the direction of points *D, E, F, G,* &c., is found and marked by fine lines on the paper. The table is now taken to *B*, the needle driven into the corresponding point on the board, and the ruler directed on the various points as before, beginning with the line *A B*, care being taken to turn the board so that the point *A* on the paper is towards the actual point on the ground. The points in which the new set of direction lines intersect those previously made indicate the position of the distant object at which the ruler was aimed.

The Compass. The compass card is divided

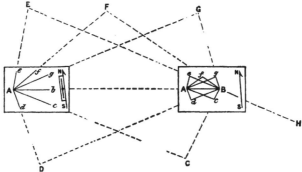

FIG. 2.—Mode of using Plane Table.

into 360 degrees; and it will be obvious that if, instead of using the sighted ruler as above, the compass bearings of all the other points were accurately taken at *A* and *B*, the same result would be obtained as with the ruler, if the angles observed were set out on the paper. In practice, however, this plan is not adopted, though a compass is usually let into one corner of the plane table, in order that the bearings of the sketch may be laid down.

The relation of the direction given by the compass needle to the true meridian is very frequently misunderstood. The compass shows *magnetic north*, which may differ considerably from the true or *sidereal* north.

The angular difference between the two meridians is known as the magnetic 'variation' or 'declination.'

If the bearing of the true north as found by compass is 355° 15', the variation would be 4° 45' east (see fig. 3). This variation is of serious consequence, and varies considerably in different places in the world and in the same place from year to year: for example, in the year 1576 the declination in London was 11° *east*; in 1660 it was reduced to zero; in the year 1814 it reached its maximum of 24° 20' *west*, since which time the needle has been slowly returning northwards. The pole-star affords the simplest means of determining the true meridian, and thence the magnetic variation.

The heavens apparently revolve around an invisible point (the pole) only 1° 27' distant from the pole-star, and which can easily be found by drawing an imaginary line through the 'pointers' of the constellation of the Great Bear (*not* the line shown in the figure). Twice in every twenty-four hours the pole-star in revolving round the pole comes into the same vertical plane with, and either above or below it. This occurs when the star ζ of the Great Bear (second from the end of the line) is in the same vertical plane with the pole-star (fig. 4). If the bearing

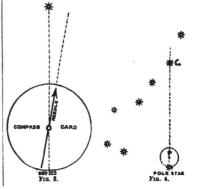

FIG. 3.   FIG. 4.

of the star be observed with the prismatic compass when this occurs, the difference between
the reading of the compass and the true north
(360°) will give the variation. If the observation be taken when the two
stars (pole-star and ζ Ursæ Ma-
joris) are in the same horizontal plane, the quantity 1° 27'
must be added if the Great Bear
be on the east, and subtracted
if on the west of the pole-star.

*To find the true north with-
out instruments,* support a pole
as shown in the figure by means
of a forked stick with its end
pointing approximately to the
north.

About half an hour *before*
noon mark the extremity of the
shadow (*a*), then with centre
at a point found by plumb-line
vertically under the end of the
pole, and with radius equal to
the length of the shadow observed at *a*, describe an arc.
When the shadow after getting
shorter and shorter again
lengthens till it touches the
arc, mark the spot (*b*); draw
the arc *a b*; the point *c*, where
a line from the centre of the
circle bisecting the arc cuts the circle, is *true*
north of the centre.

The Prismatic Compass (fig. 6, I). One of
the most useful instruments for surveying consists of a magnetic needle balanced on a pivot,
and carrying a card divided into 360°, showing

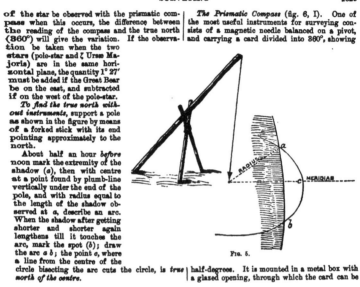
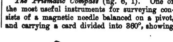

Fig. 5.

half-degrees. It is mounted in a metal box with
a glazed opening, through which the card can be

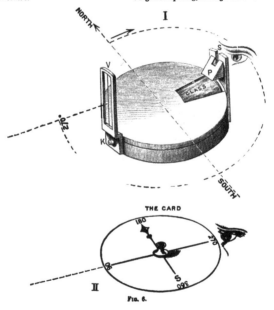

THE CARD

II

Fig. 6.

observed by means of a prism, and also a sight vane V. When the card is steady, and the eye, the object, and the wire of the sight vane are all in a line, the figure seen on the card is read off and noted. The graduation of the card commences at south (fig. 6, II), so that the degree showing *the bearing of the object* shall appear under the eye.

The use of the prismatic compass is very simple. The observer standing at any point on his map may determine with considerable accuracy the angular bearing of any object visible from this point; or, marching along a winding road, he may determine its direction, and by laying down his paces to scale on his sketch at the angle formed a correct representation of the general direction of the road will be obtained.

The Protractor (fig. 7) is in its best form a thin rectangular slab of ivory about 6 × 1¾ inches, with bevelled edges.

The degrees of the semicircle from 0° to 180°, *i. e.* from north to south by east, are marked on the outer edge of the protractor; the centre is marked on the lower edge by a vertical line or arrow. The sketch is prepared with parallel lines about one third of an inch apart, to represent magnetic north and south, and with relation to these the bearings are protracted. The protractor is laid on the sketch with its centre at the point where the bearing is to be drawn, and its edges set north and south or parallel to these lines. If the bearing is *under* 180°, the graduated edge of the protractor is laid *to the right*; if *over* 180°, *to the left*, the north margin of the sketch being uppermost.

Traversing with the Compass. The operation of observing the direction and measuring the length of a number of lines is called 'traversing.' The observations may be either laid down at once on the sketch (plotted), or recorded in a field book and plotted afterwards. This book is a large pocket-book, with a column about three quarters of an inch wide ruled down the centre of each page lengthwise; in it are booked the *forward* angles or direction of the traverse lines and the forward distances thus:

Column

186

142

Chain

Fᴵɢ. 8.

The chain column represents a line having breadth, so that a line representing, say, a fence or ditch crossing the road traversed, should meet the chain column on one side, and leave it at a point exactly opposite on the other.

Winding roads should be traversed with as few observations as possible, in order to diminish error (fig. 9).

It is also very desirable whenever possible not only to take the forward bearing, *e. g.* from *A* to *b*, but also on arriving at *b* to turn round and

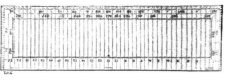

Fᴵɢ. 7.

take the bearing of *A*. The two observations should agree perfectly, and the sum of the angles

Fᴵɢ. 9.

should be 180°. In practice it is usual to judge the distance of objects right and left of the traverse line, and mark them at the proper point in the side column of the field book. These measurements *perpendicular to the traverse line* are called *offsets*, and in regular surveying would be carefully measured by a tape or chain. In military surveying offsets as a rule are not taken to points more distant than 100 yards from the traverse line.

For very accurate surveying, circular protractors fitted with folding arms and verniers are used; at the end of each arm is a point with which, when the instrument is set, a minute hole is punctured in the paper (see fig. 10).

The Pocket Sextant. The following description of the instrument and mode of using it is taken from Colonel W. H. Richards' 'Text-book of Military Topography.'

The instrument is shown in fig. 11; it is used for measuring the angle which the distance between two visible objects subtends. When required for use the cover is taken off and screwed on underneath the sextant as represented. The parts of the instrument are—L, the index mirror; H, the horizon glass, the upper half only of which is silvered; V, the index arm which moves with the index mirror, and shows its position by a vernier on the graduated arc A; C, the milled head, by means of which the index mirror is moved; K, the adjusting key; E, the eye-hole through which the observation is taken. Some instruments are provided with a dark glass, which can be slid into the place of E, for observations of the sun; others have a small telescope which fits in the same place. D, a magnifying glass for examining the vernier; it should always be placed directly over the latter, so that it may be viewed perpendicularly to the plane of the instrument; L, M, squares to which the key is applied when altering the adjustments.

To measure the angle between two objects, hold the sextant in the left hand. Look through E and the opening O at the left-hand

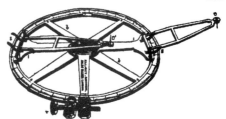

FIG. 10.—Elliot's Circular Protractor.

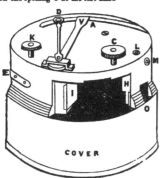

FIG. 11.

object. Turn E until the right-hand object reflected from the index mirror appears on the silvered part of the horizon glass coinciding with the left-hand object, as shown in fig. 12. The angle may then be read.

There is a certain amount of knack in bringing the two images into contact in the sextant; the beginner should practise at first with two objects which are near each other.

Adjustment of the Pocket Sextant. The 'index error,' as it is called, of the sextant is the result of incorrect adjustment of the horizon glass, so that the point of the graduation from which the arc angles are measured is not zero.

If this error be considerable it may be removed by adjustment, thus:

Set the index exactly at 0° on the arc. Select a well-defined point, such as a chimney-top, at

some distance. First observe whether the direct and reflected images are on the same level; if not, apply the key K to the square L (fig. 11), and turn it until they are so. The effect of this correction is to set the horizon glass truly perpendicular to the plane of the instrument.

Next, if the reflected image overlaps that seen directly, or does not exactly cover it, apply the key to the square M (fig. 11), and turn it slowly until they coincide. The effect of this correction is to set the horizon glass parallel to the index mirror. It is generally safer to note and apply the index error in correction of observations than to alter the adjustments.

Theory of the Sextant. It will be noticed that the actual angle between the graduations O and 120 on the arc of the sextant is only 60°. The reason for doubling the graduation may be thus demonstrated:

Let P and X be the distant objects between which the angle P E X is measured; M, the horizon glass of the sextant; E, the moveable index mirror which was in the position $e\ e$, parallel to M, when the index was at O; E X was then also perpendicular to $e\ e$. Let E R be perpendicular to E. By a well-known principle of optics, the angle of incidence, P E R, is equal to the angle of reflection, X E R.

From the right angles X E O, R E V, take the common angle X E V, leaving X E R = V E O. But P E X = 2X E R = V E O. Hence the ray of light being reflected successively from two mirrors, the total angular deviation of the ray is double the angle of inclination of the mirrors. Therefore V E O, though only 15°, is read 30° on the arc (fig. 13). For the sake of simplicity the observer's eye is supposed to be at E in the figure.

The sextant may be used in any position from which the required objects are visible, and is not affected by local attraction like the compass. Its disadvantages are—It is not of general use like the compass, but can only be employed for observ-

FIG. 12.—Objects coinciding in the Sextant by Reflection.

ing triangulation, determining heights and distances, and observing for latitude; it cannot be conveniently used for traversing, or the more ordinary operations of surveying; any difference of level between the points observed is liable to cause error.

The Representation of Hills. Two distinct methods of representing hills are in use at the present time, 'hâchure shading' and 'contouring.'

In *hâchure shading* the degree of slope is represented by short strokes called hâchures, the thickness and number of which are regulated by a 'scale of shade.' For military purposes, and all others in which accurate indication of the slope is essential, the system of *contouring* is now adopted.

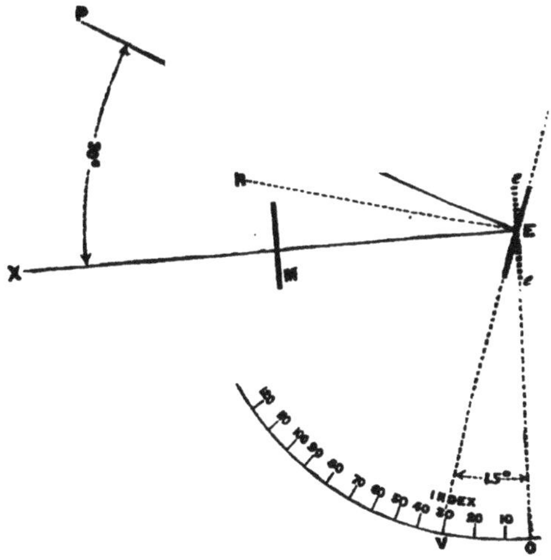

A contour is the line of intersection of a hill by a horizontal plane. Supposing a cone representing a hill, 20 inches in height, to be built up of 20 circular pieces of wood, each 1 inch thick, piled one on the top of the other, the upper or lower edge of each piece would represent a contour, and if these outlines were projected on paper they would appear as a series of concentric circles; and knowing the scale, it would be possible to reconstruct the cone from these circles. By this plan all the errors and necessary defects of shading are avoided, and it is possible from a properly contoured map to reproduce a scale model of the country by cutting out a model of each contour in wood or other material of definite thickness, and fastening them in their proper position one on the top of the other.

A little consideration will show that all contour lines should be continuous, and should enclose a space so that where they appear to run together there must be a vertical cliff in which, instead of sloping, the contours are vertically over one another, and, seen from above, as they are supposed to be in a map, they cover one another and appear as one line. The first contour, i. e. the lowest, is generally taken at sea level or some other established datum line. The intervals between the contours may vary somewhat with the nature of the country. It will be obvious that in a flat country, unless the vertical interval be small, few or no contours may appear over very large areas. Much will depend on the scale of the map also; 10 or 20 foot contours may be necessary for some purposes, while 50 or 100 feet of interval may suffice in others. There are two ways in which a slope may be described:

(1) The plan adopted on railways, in which the *gradient* is expressed as a fraction, *e. g.* $\frac{1}{292}$, signifying that there is a rise or fall of one foot in every 292 feet.

(2) The angle of the slope may be expressed in *degrees* of elevation or depression above or below

Fig. 13.

the horizontal plane. This method is the one most generally in use.

On a slope of 1°, a difference of level of one foot will occur in 57·3 feet, which may be expressed as 1 in 57·3, 57·29 being the natural cotangent of the angle 1°.

On a slope of 2° the difference of level will occur in half the distance, and so on. For rough work the round number 60 is taken instead of 57·3, and we get the simple relations of angle to difference of level as follows:

A slope of 1° is equivalent to 1 in 60
" 2° " " 1 " 30
" 4° " " 1 " 15
" 5° " " 1 " 12

As distances are usually measured by yards, it is well to remember that the horizontal equivalent for *one foot vertical on a slope of one degree is 19·1 yards nearly.*

The scale of slopes adopted in this country as a standard is constructed on the condition of *horizontal equivalents for 20 feet vertical intervals on the scale of 6 inches to the mile.* Thus for 1° the horizontal equivalent for 20 feet is 19·1 × 20 = 382 yards; and it is usual to set out on every map a scale of slopes showing the horizontal equivalent for 20 feet at all angles from 1° up to 20°. The eye soon becomes accustomed to the value of the scale, and learns to recognise the degree of slope in any part of the map at once.

In sketching hill country an instrument known as a *clinometer* is used. Briefly this consists of a circle divided with degrees and half-degrees, so arranged with a prism and sight vane, as in the prismatic compass, that the degrees may be read when the graduated card is in the vertical position. The card is so weighted that when the eye and sight vane are perfectly horizontal the read-ing is 0°, and by holding it in such a position that it is parallel with the ground the degree of slope can be read off with considerable accuracy.

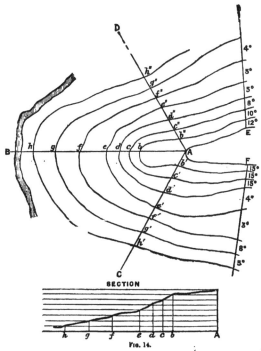

Fig. 14.

The above figures will serve to illustrate the procedure in sketching a simple spur of a hill.

The scale is supposed to be 12 inches to a mile, and the normal scale of horizontal equivalents is used; but, inasmuch as this is constructed for the scale of 6 inches to the mile, the vertical interval between the contours will be in inverse ratio to the scale, i. e. the larger the scale the smaller the vertical interval. As the scale is doubled in this case the vertical interval will be halved, and in this case will be 10 feet instead of 20 feet.

Standing at A, the slope is observed with the clinometer to be 2° in the direction of B. At this inclination the next contour, 10 feet below A, will occur 95 yards down the slope. The observer paces this distance to b, and marks it on his sketch to scale; at b the slope is 10°, and the horizontal equivalent for 10 feet is 19 yards, which is paced and set out on the sketch as before. The slope is observed to be the same at d and e; the proper distances are paced and contours marked. At e the slope decreases to 8°; the horizontal equivalent is 63 yards, which

brings the observer to f, the slope remaining constant as far as the stream, the position of which is supposed to be marked already on the sketch. The remaining contours may be put in without pacing. The hill is surveyed in a similar manner from A in other directions, A C, A D, the contours marked as before, and drawn on the sketch so as to pass round the hill. Small irregularities may be put in by eye. In this way an accurate map of a hill may be made—not, of course, so accurate as one compiled from the data obtained by careful levelling, but accurate enough for all ordinary purposes.

The clinometer requires some practice in order to obtain good results, and the conformation of the ground may be such as to require considerable judgment in determining *where* to begin. The reader who desires further information is referred to 'A Text-book of Military Topography,' by Colonel W. H. Richards, printed under the super-intendence of Her Majesty's Stationery Office, from which the above descriptions are very largely drawn, and which contains a mass of information on the whole subject of the very greatest practical value.

DETERMINATION OF HEIGHTS AND DISTANCES.

The following problems from Colonel Richards' work above referred to will be found useful under many circumstances:

I. *To trace a Right Angle or a Perpendicular to a given Line.*

A method suitable to any situation, and not requiring the aid of a second person, is the following:—A perpendicular to the line A B is required at A. Stick a peg in the ground at any point C 10 or 12 feet from A, slip a noose of a cord over the peg, measure the distance A C on the cord, find the point *d* in A B such that C *d* = A C, and the point *e* in the line *d* C produced, *d* A *e*, is a right angle by Euclid, iii, 31.

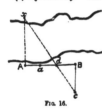

Fig. 15.

II. *To find the Breadth of a River or the Distance of an Inaccessible Point.*

(1) *By Geometry.* The distance of *x* from A being required, place a mark at A, and another at *a* perpendicular to A *x*; in the line A *a* produced place a picket B, and another at a convenient point *c* perpendicular to B A. Standing at *c*, direct an assistant in placing picket *d* in the line *c x*, he at the same time placing it in the alignment A *a*. The triangles *x* A *d*, *d* B *c*, are similar. Measure *c* B, B *d*, *d* A; then—

$$d\,B\,:\,B\,c\,::\,d\,A\,:\,A\,x.$$

The length of the lines constructed should be more or less proportional to the distance of the object, otherwise errors occur.

(2) *Second Method.* Place a mark at B in pro-

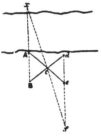

Fig. 16.

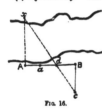

Fig. 17.

longation of *x* A, another at a convenient point *c*; make *c d* equal to *c* B, and *c e* equal to A *c*; place *d* and *e* in the prolongation of B *e* and A *c* respectively. Find *f* where *x c* and *d c* would intersect when produced. The triangles *x* A *c*, *f e c*, are similar and equal; and *ef* equal to A *x* may be measured.

With the Pocket Sextant. Set the index at 90°,

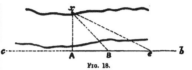

Fig. 18.

and find some distant object *b*, perpendicular to A *x*. Set the index at 45°, and move along this line towards *b* until A and *x* coincide by reflection. Then A B = A *x*.

III. *To find the Length of a Line Accessible only at the Extremities, as x y (fig. 19).*

Place a mark at a convenient point A; another at *a* in the prolongation of and equal to A *y*; and a third at *b* in the prolongation of and equal to A *x*.

The triangle *a* A *b* is equal and similar to *x* A *y*, and the distance *a b* may be measured.

Fig. 19.

IV. *To find the Length of an Object Accessible at the Base on Level Ground.*

By Geometry. At any convenient spot plant a pole vertically, and a picket in line with the object at A; observe and mark the points *b c*, in which the lines A B, A C, intersect the pole. Then, the triangles A *b c*, A B C, being similar,

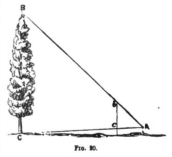

Fig. 20.

measure A *c*, *c b*, A C—and A *c* : *b c* :: A C : B C.

With the Pocket Sextant. Make a mark on

the wall *m* at the height of the eye. Set the index at one of the angles given in the table of tangents below. Find a place where *m* and B coincide by reflection in the sextant. Measure the distance from this place to the foot of the wall under *m*; multiply or divide this distance by the figure given for the angle. To this add the length of the eye above the ground. The angle 45° will give the distance equal to the length.

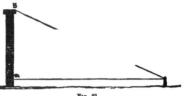

Fig. 21.

Table of Natural Tangents.

Multiplier.	Angle.	Divisor.	Angle.
1	45°	1	45°
2	63°26′	2	26°34′
3	71°34′	3	18°26′
4	75°58′	4	14°2′
5	78°41′	5	11°19′
6	80°32′	6	9°28′
8	82°52′	8	7°8′
10	84°17′	10	5°43′
12	85°14′	12	4°46′
15	86°11′	15	3°49′
18	86°49′	18	3°11′
20	87°8′	20	2°52′
25	87°42′	25	2°18′
30	88°6′	30	1°54′

V. *To find the Height of an Object, Inaccessible at the Base, on Level Ground.*

By Clinometer. With the clinometer the tangent tables may be employed as above explained, the observer advancing towards or retiring from the object until its top appears to coincide with the required degree.

By Geometry. Similarly to the method shown under Problem IV; but the object being inaccessible, the distance A C must be found by Problem II. The triangle A *b c* will usually be so much smaller than the similar triangle A B C, that the slightest error in the verticality of the pole or in the measurements of A *c*, *c b*, will produce a large error in C B.

VI. *To find the Distance between Two Inaccessible Objects, and the Difference of Level between them.*

Let C and D be the two objects; measure a base A B of convenient length midway between them. Observe with the theodolite at A the angles B A C, B A D, also the vertical angles of C and D.

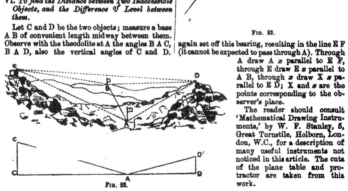

Fig. 22.

At B observe the angles A B C, A B D. To find the sides A C, A D, we have in the triangles A B C, A B D, two known angles and the side A B. To find C D (the distance between the inaccessible points) there are now two known sides, A C, A D, and the included angle C A D.

To find the Heights. In the right-angled triangles A C C′ and A D D′ the sides A C, A D, and the vertical angles are given; to find the perpendiculars C C′ and D D′, the difference between them is the difference of height required.

VII. *To find with the Prismatic Compass the Place on the Ground with reference to an Inaccessible Point.*

Take the bearing of the inaccessible point A, set it off the reverse way from A, giving the line A B, in some part of which is the observer's place. Take the bearing of any point E, and set it off at any point D in A B; measure D E. At E observe A;

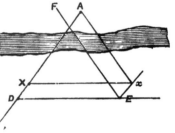

Fig. 23.

again set off this bearing, resulting in the line E F (it cannot be expected to pass through A). Through A draw A *x* parallel to E F, through E draw E *x* parallel to A B, through *x* draw X *x* parallel to E D; X and *x* are the points corresponding to the observer's place.

The reader should consult 'Mathematical Drawing Instruments,' by W. F. Stanley, 5, Great Turnstile, Holborn, London, W.C., for a description of many useful instruments not noticed in this article. The cuts of the plane table and protractor are taken from this work.

SUSPENDED ANIMATION. See ASPHYXIA.

SWALLOW. Three or four species of *Hirundo* (Linn.) pass under this name. It was once held in great repute in medicine. Even the excrement was included among the simples of the Ph. L., 1618. The swallow is an insectivorous bird, but, like the sparrow and rook, is much persecuted for its good services. It has been calculated that, directly and indirectly, a single swallow is the humble means of lessening the race of one kind of insect alone to the extent of 560,970,489,000,000,000 of its race in one year.

SWEEPING. Before commencing to sweep, the floor should be strewn with a good amount of damp tea-leaves, saved for the purpose; these collect the dust and thereby save the furniture, which as far as practicable should be covered up during the process. Tea-leaves may also be advantageously used upon druggets and short-piled carpets. Light sweeping and soft brooms are desirable if these latter are to be operated upon. Many a carpet is prematurely worn out by overviolent sweeping.

In sweeping thick-piled carpets, such as Axminster and Turkey carpets, the servant should always be instructed to brush the way of the pile; by following this advice the carpets may be kept clean for years; but if the broom is used in a contrary direction, all the dust will be forced into the carpet, and soon spoil it.

SWEET BALLS. *Prep.* Take of Florentine orris root, 3 oz.; cassia, 1 oz.; cloves, rhodium wood, and lavender flowers, of each, ½ oz.; ambergris and musk, of each, 6 gr.; oil of verbena, 10 or 12 drops; beat them to a paste, form this into balls with mucilage of gum tragacanth made with rose water; pierce them, whilst soft, with a needle, and, when they are quite dry and hard, polish them. Worn in the pocket as a perfume. Some persons varnish them, but that keeps in the smell.

SWEET BAY. *Syn.* LAUREL; LAURUS NOBILIS (Linn.), L. The fruit (LAURI BACCÆ; LAURUS—Ph. L.), as well as the leaves (LAURI FOLIA), is reputed aromatic, stimulant, and narcotic. They were formerly very popular in coughs, colic, hysteria, suppressions, &c.; and, externally, in sprains, bruises, &c.

SWEETBREAD. The thymus gland of the calf. When boiled, it is light and digestible; but when highly dressed and seasoned it is improper both for dyspeptics and invalids (*Pereira*).

SWEET FLAG. *Syn.* ACORUS CALAMUS, L. A plant of the Natural order ORONTIACEÆ. The rhizome ('root') is an aromatic stimulant, and is regarded by some as a valuable medicine in agues, and as a useful adjunct to other stimulants and bitter tonics. It is sometimes employed by the rectifiers of gin. The volatile oil obtained from it by distillation is employed for scenting snuff, and in the preparation of aromatic vinegar.

SWEETMEATS. Under this head are properly included confections, candies, and preserves in sugar; but, as generally employed, the word embraces all the sweet compounds of the confectioner.

Sweetmeats, as well as cakes, blancmange, and jellies, are not unfrequently coloured with deleterious substances, the consequences of which are always pernicious, and in many instances have proved fatal. Gamboge, a drastic cathartic; chrome yellow, red-lead, orpiment, emerald green, and various other pigments containing lead, arsenic, copper, or other poisons, have been thus employed. The whole of these may be readily detected by the tests and characteristics appended to their respective names.

The colours and stains which may be safely employed to increase the beauty of these articles are noticed under STAINS and LIQUEUR.

SWEETS. Home-made wines; British wines.

SWINE-POX. See POX.

SYDENHAM'S LENITIVE. *Prep.* Take of rhubarb (recently grated or powdered), 3 dr.; tamarinds, 3 oz.; senna, ½ oz.; coriander seeds (bruised), 3 dr.; boiling water, 1 pint; macerate for 3 hours in a covered vessel, and strain. An excellent stomachic and laxative.—*Dose,* ½ to 1 wineglassful.

SYLVIC ACID. *Syn.* SILVIC ACID. The portion of common resin or colophony which is the least soluble in cold and somewhat dilute alcohol.

SYMBOLS, in *chemistry,* are representations of one atom of each of the elementary bodies by the capital initial letter, with or without the addition of a small letter, of their Latin names: as C, for *carbon;* Fe (*ferrum*), iron; O, *oxygen,* &c.

Symbols, Alchemical.—The following list of alchemical and botanical symbols and abbreviations is a reprint of that contained in the 'Lexicon of Terms used in Medicine and the Allied Sciences,' now being published by the New Sydenham Society, under the editorship of Henry Power, M.B., and Leonard W. Sedgwick, M.D.

Acetum	⚗
Acetum destillatum	⚗ ♒ 1
Acidum	+
Aër	⊿
Aerugo	⊕
Æther	℞
Alembic	☒
Alumen	○
Amalgama	⚱
Ammonium	⊕
Aqua	▽
Aqua fortis	▽ⁿ
Aqua pluvialis	♆
Aqua regia	♆
Arena	∴
Argentum	☽
Arsenicum	o—o
Aurantium	⊙ rant.
Auripigmentum	o=o
Aurum	☉

Baln. arenæ	B ∴	Plumbum		♄
Baln. mariæ	BM	Precipitare		
Baln. vaporis	BV	Preparare		PP
Baryta	♈	Pulvis		
Bismuth	♉	Regulus		
Borax		Regulus		
Calcaria	♅	Resina		
Calcaria usta	♅ va	Retorta		
Camphora	≋	Saccharum		ff
Cancer	69	Sal		⊖
Caput mortuum	☠	Sal ammoniac		⊖
Carbo		Sal kali		⊕
Carbonicum	♣	Sal medius		⊜
Carduus benedictus	C·B·	Sapo		☐
Card. Marianus	C·M·	Spiritus		⌒
Cera		Spiritus rectificatissimus		℞ss.
Cinis	♇	Spiritus rectificatus		℞
Cinis clavellatum	♇	Spiritus vini		V̇ V̇
Cinnabar	55	Stannum		♃
Cornu cervi	C.C	Stibium		♂◇
Cristalli	XIIG	Stratum super stratum		S, S, S.
Crucibulum	♒	Sublimare		
Cuprum	♀	Succinum		⊕
Distillare	♌ re	Sulphur		♄
Ferrum	♂	Tartarus		♉
Fictile	Fict.	Terra		∇
Fixum	♈	Terra foliata		∇
Flores	FI·	Tinctura		℞
Gummi	55	Urine		☐
Hora	×⃝	Ustare		∽
Hydrargyrum	☿	Vitriolum		⊕
Hydr. chloridum	☿ ♈ l.	Vitrum		♏
Hydr. corrosivum	☿ ♌ lcor.	Volatile		Λ
Ignis	△	Zincum		♌

Symbols and Abbreviations, Botanical.

Kali	♅ v.
Lapis	∇
Lithargyrum	
Magnesia	♅
Magnet	♌
Menstruum	⊠
Natrum	⊕ m.
Nitrum	◑
Oleum	°°
Oxidatum	Xdal:
Oxidulatum	Xdul·
Per deliquium	P d·

⊙ *Monocarp.* A plant which produces seed only once during its life. The symbol representing the sun.

A, ① *Annual.* A monocarp which dies in the same year that it germinated, e. g. *Mustard.*

B, ② *Biennial.* A monocarp which produces leaves only the first year, and perfects its seed the next, e. g. *Mullein.*

P *Perennial.* A plant which produces seed for an indefinite number of years, e. g. *Apple.*

♃ *Rhizocarp.* A perennial, the stems of which die down to the ground every year, e. g. *Rhubarb, Mint.* The symbol representing Jupiter, which has a period of revolution round the sun of twelve years.

♄ *Caulocarp.* A perennial, the stems of which are persistent throughout the whole of its life, e. g. *Apple.* The symbol representing Saturn, the period of revolution of which round the sun is thirty years.

H *Herb.* A plant, the stems of which remain soft or succulent, e. g. *Mint* or *Rhubarb.*

S, ♃ *Shrub.* A plant in which the stems are woody, and which usually divide near the ground into numerous branches and twigs, e. g. *Lilac.*

♃ *Under-shrub.* A small shrub; one that does not grow more than three feet in height, e. g. *Gooseberry.*

T, ♃ *Tree.* A plant which grows to twenty feet or more in height, having a woody stem forming a distinct trunk, e. g. *Oak.*

⟲ A climbing plant which follows the sun, e. g. *Hop.*

⟳ A climbing plant which moves against the sun, e. g. *Scarlet runner.*

♂ Flowers having stamens only (unisexual, staminiferous, or male), e. g. male flowers of *Box.* The symbol representing Mars, the period of revolution of which is two years.

♀ Flowers having pistils only (unisexual, pistillate, or female), e. g. female flowers of *Box.* The symbol representing Venus.

⚥ Flowers having both stamens and pistils (bisexual or hermaphrodite), e. g. *Buttercup.*

♂ Abortive staminiferous flowers (neuter).

♀ Abortive pistillate flowers (neuter), e. g. the florets of the ray in *Daisy.*

♂ – ♀ Monœcious plants, producing male and female flowers upon the same individual, e. g. *Box.*

♂ : ♀ Diœcious plants, producing male and female flowers, but upon separate individuals, e. g. *Willow.*

⚥ ♂ ♀ Polygamous plants, which produce hermaphrodite and unisexual flowers upon the same or different individuals, e. g. *Atriplex.*

∞ Indefinite in number; applied to stamens and other parts of flowers.

O = Cotyledons accumbent, radicle lateral.

O || „ incumbent, „ dorsal.

O ≫ „ conduplicate, „ „

O || || „ twice folded, „ „

O || || || „ thrice folded, „ „

3/ Trimerous, applied to flowers when the whorls of the flower are multiples of three, as in most endogens.

5/ Pentamerous, applied to flowers when the whorls of the flower are multiples of five, as in exogens generally.

Bab., Babington.
Berk., Berkeley.
Br., Brown.
Cal., calyx.
Caul., caulis, stem.
Cl., classis, class.
Cor., corolla.
Cuv., Cuvier.
DC., or De Cand., De Candolle.
Endl., Endlicher.
Fam., family.
Fr., fructus, fruit.
Gen., genus.
Hook., Hooker.
Juss., Jussieu.
L. or Linn., Linnæus.
Lindl., Lindley.
Nat. Ord., Natural order.
O. or Ord., ordo, order.
Per., perianthus, perianth.
Rad., radix, root.
Rich., Richard.
Sp. or Spec., species.
Subk., subkingdom.
Subord., suborder.
Var., varietas, variety.
V. s. c., vidi siccam cultam, a dry cultivated plant seen.
V. s. s., vidi siccam spontaneam, a dried specimen seen.
V. v. c., vidi vivam cultam, a living cultivated plant seen.
V. v. s., vidi vivam spontaneam, a living wild plant seen.
Willd., Willdenow.
With., Withering.

SYMPATHETIC INK. See INK.

SYN'APTASE. *Syn.* EMULSIN. The name given by Robiquet to the EMULSIN, a nitrogenised or albuminoid principle existing in both the bitter and sweet almond. It possesses the remarkable property of converting amygdalin, in the presence of water, into hydrocyanic acid and the essential oil of bitter almonds; 100 gr. of amygdalin yield, under the influence of synaptase and water, 47 gr. of raw oil, and 5·9 gr. of anhydrous hydrocyanic acid (*Liebig*).

SYN'COPE. See FAINTING.

SYRINGIN. A glucoside obtained from the bark of *Syringa vulgaris* (lilac) and *Ligustrum vulgare*, occurring in white crystalline needles, readily soluble in hot water and in alcohol, but insoluble in ether. It does not appear to be poisonous, and has been recommended as a febrifuge in malaria; but little is known at present as to its properties, dose, &c.

SYR'UP. *Syn.* SIRUP, SIROP; SYRUPUS, L. A saturated, or nearly saturated, solution of sugar in water, either simple, flavoured, or medicated.

In the preparation of syrups care should be taken to employ the best refined sugar, free from lime and ultramarine, and either distilled water or filtered rain-water; by which they will be rendered much less liable to spontaneous decomposition, and will be perfectly transparent, without the trouble of clarification. When inferior sugar is employed, clarification is always necessary. This is best done by dissolving the sugar in the water, or other aqueous menstruum, in the cold, and then beating up a little of the cold syrup with some white of egg, and an ounce or two of cold water, until the mixture froths well;

this must be added to the syrup in the boiler, and the whole 'whisked up' to a good froth; heat should now be applied, and the scum which forms removed from time to time with a clean 'skimmer.' As soon as the syrup begins to slightly simmer it must be removed from the fire, and allowed to stand until it has cooled a little, when it should be again skimmed, if necessary, and then passed through clean flannel. When vegetable infusions or solutions enter into the composition of syrups, they should be rendered perfectly transparent by filtration or clarification before being added to the sugar.

M. Magnes-Lahens ('Germ. Pharm. Chem.,' 4th Series, xv, 140; 'Year-book Phar.,' 1872) describes below a process for the clarification of syrups, the originator of which was M. Demarest, a pharmacien. The process is as follows:—White unsized paper is beaten up into a pulp with a portion of the syrup, and then mixed with the bulk. The proportion of paper should be one gram to every litre of syrup; and the latter should be maintained at a temperature of 35° to 40° C.

A filter of moleskin capable of holding about one third of the volume of the syrup, and having the form of an inverted sugar-loaf, is supported over a suitable receptacle; the syrup with the pulp is poured rapidly into it, so as to fill it as quickly as possible; and the filter is kept full so long as any of the syrup remains. When the greater part has run through, and but little remains in the filter, and consequently the 'felting' of the paper pulp is complete, the syrup which has already run through is again poured into the filter. The liquid which now passes is perfectly bright, and may be collected. In pouring the syrup into the filter, the stream should be directed into the middle, and not upon the sides, so as to avoid disarranging the felt, which would interfere with the success of the operation.

The author very strongly recommends this method for the clarification of all kinds of syrups; its advantages being that it results in a perfectly limpid liquid, and that it involves neither trouble nor loss of time or material. He states that in 4 or 5 hours, with a filter of 8 litres in capacity, 24 litres of syrup may be clarified.

The small quantity of syrup retained in the filter and pulp may be recovered by pouring on a sufficient quantity of warm water, pressing strongly, evaporating the liquid to a syrupy consistence, beating up with a little paper pulp, and passing it again through a small filter.

The proper quantity of sugar for syrups will, in general, be found to be 2 lbs. (avoir.) to every imperial pint of water or thin aqueous fluid. These proportions, allowing for the water that is lost by evaporation during the process, are those best calculated to produce a syrup of the proper consistence, and possessing good 'keeping qualities.' They closely correspond to those recommended by Guibourt for the production of a perfect syrup, which, he says, consists of 30 parts of sugar to 16 parts of water.

In the preparation of syrups it is of great importance to employ as little heat as possible, as a solution of sugar, even when kept at the temperature of boiling water, undergoes slow decomposition. The plan which we adopt is to pour the water (cold) over the sugar, and to allow the two to lie together for a few hours in a covered vessel, occasionally stirring, and then to apply a gentle heat (preferably that of steam or water bath) to finish the solution. Some persons (falsely) deem a syrup ill prepared unless it has been allowed to boil well; but if this method be adopted, the ebullition should be only of the gentlest kind ('simmering'), and should be checked after the lapse of 1 or 2 minutes.

Mr Orynski recommends the preparation of all syrups without the application of heat, as follows:

Introduce 30 or 32 oz. of sugar (according to the temperature) into a percolator, in which has been previously introduced a piece of lint or sponge, well adjusted, and gradually pour on 16 oz. of liquid, so as to make the percolate (syrup) pass drop by drop. If the first liquid is turbid, pour it back into the percolator till the syrup passes clear.

The advantages claimed for this process are—First, the syrups are clear, and there is no necessity for purifying them.

Secondly, they possess their medicinal properties unaltered; since many drugs may be injured by heat, more especially aromatics, and those containing readily volatile substances; and—

Thirdly, the syrups will neither crystallise nor ferment; and may be prepared in large quantity, provided the vessels or bottles are clean before filling them with syrup.

When it is necessary to thicken a syrup by boiling, a few fragments of glass should be introduced, in order to lower the boiling-point.

To make highly transparent syrups, the sugar should be in a single lump, and, by preference, taken from the bottom or broad end of the loaf, as, when taken from the smaller end, or if it be powdered or bruised, the syrup will be more or less cloudy.

A fl. oz. of SATURATED SYRUP weighs 577½ gr.; a gall. weighs 13½ lbs. (avoir.); its sp. gr. is 1·380, or 35° of Baumé's aerometer; its boiling-point is 221° F., and its density at the temperature of 212° is 1·260 to 1·261, or 30° Baumé. The syrups prepared with the juices of fruits, or which contain much extractive matter, as those of sarsaparilla, poppies, &c., mark about 2° or 3° more on Baumé's scale than the other syrups.

In most pharmaceutical works directions are given to completely saturate the water with sugar. Our own experience, which is extensive, leads us to disapprove of such a practice, since we find that, under all ordinary circumstances, a syrup with a very slight excess of water keeps better than one fully saturated. In the latter case a portion of sugar generally crystallises out on standing, and thus, by abstracting sugar from the remainder of the syrup, so weakens it that it rapidly ferments and spoils. This change proceeds with a rapidity proportionate to the temperature. Saturated syrup kept in a vessel that is frequently uncorked or exposed to the air soon loses sufficient water, by evaporation from its surface, to cause the formation of minute crystals of sugar, which, falling to the bottom of the vessel, continue to increase in size at the expense of the sugar in the solution. We have seen a single 6-gall. stone bottle, in which syrup has been kept for some time, the inside of which, when broken, has been found to be entirely cased with sugar-candy, amounting in weight to 16 or 18 lbs. On the other hand, syrups containing too much water also rapidly ferment, and become acescent; but of the two this is the lesser evil, and may be more easily prevented. The proportions of sugar and water given above will form an excellent syrup, provided care be taken that an undue quantity be not lost by evaporation.

The decimal part of the number denoting the sp. gr. of a syrup, multiplied by 26, gives the number of pounds of sugar it contains per gall. very nearly (Ure).

In boiling syrups, if they appear likely to boil over, a little oil, or rubbing the edges of the pan with soap, will prevent it.

Syrups may be decoloured by agitation with, or filtration through, recently burnt animal charcoal. Medicated syrups should not, however, be treated in this way.

The preservation of syrups, as well as of all other saccharine solutions, is best promoted by keeping them in a moderately cool, but not a very cold, place. "Let syrups be kept in vessels well closed, and in a situation where the temperature never rises above 55° F." (Ph. L.). They are better kept in smaller rather than in large bottles, as the longer a bottle lasts the more frequently it will be opened, and, consequently, the more it will be exposed to the air. By bottling syrups whilst boiling hot, and immediately corking down and tying the bottles over with bladder perfectly air-tight, they may be preserved, even at a summer heat, for years, without fermenting or losing their transparency.

The 'candying,' or crystallisation, of syrup, unless it be over-saturated with sugar, may be prevented by the addition of a little acetic or citric acid (2 or 3 dr. per gall.).

The fermentation of syrups may be effectually prevented by the addition of a little sulphite of potassium or of calcium. M. Chereau recommends the addition of some (about 3% to 4%) sugar of milk with the same intention. Fermenting syrups may be immediately restored by exposing the vessel containing them to the temperature of boiling water.

Syrup of Ac'etate of Mor'phia. *Syn.* SYRUPUS MORPHIÆ ACETATIS, L. *Prep.* (Ph. D.) Solution of acetate of morphia, 1 fl. oz.; simple syrup, 15 fl. oz.; mix. Each fl. oz. contains ⅛ gr. of acetate.—*Dose,* ½ to 2 teaspoonfuls.

Syrup of Al'mond. *Syn.* BARLEY SYRUP, ORGEAT; SYRUPUS AMYGDALÆ, L.; SIROP D'ORGEAT, Fr. *Prep.* 1. Sweet almonds, 1 lb.; bitter almonds, 1 oz.; blanch, beat them to a smooth paste, and make an emulsion with barley-water, 1 quart; strain, to each pint add of sugar, 2 lbs., and a table-spoonful or two of orange-flower water; put the mixture into small bottles, and preserve it in a cool place. Some persons add a little brandy.

2. (Ph. Bor.) Sweet almonds, 8 oz.; bitter almonds, 2 oz.; blanch them after cold maceration, then beat them in a marble mortar, with a

wooden pestle, to a paste, adding gradually of water, 16 fl. oz.; orange-flower water, 3 fl. oz.; after straining through flannel dissolve 3 lbs. of sugar in each pint of the emulsion. An agreeable pectoral and demulcent.

Syrup of Aniseed. *Syn.* SYRUPUS ANISI, L. *Prep.* Infuse ½ oz. of bruised aniseed in 4 oz. of hot water, strain, and add 2 dr. of sugar.

Syrup, Antiscorbutic. *Syn.* SYRUPUS ANTISCORBUTICUS, L. (P. Cod.). *Prep.* Scurvy-grass, watercresses, horseradish, all fresh, of each, 10 oz.; buckbean, 1 oz.; bitter orange peel, 2 oz.; cinnamon, ½ oz.; white wine, 40 oz. (by weight); macerate 2 days, and distil off 10 oz. (by weight); then add to the distillate, sugar, 26 oz.; strain the residue left in the retort, decant and make into a syrup with another 25 oz. of sugar; clarify with white of egg, and when cold add to it the former syrup.—*Dose,* 4 dr.

Syrup of Balsam of Peru. *Syn.* SYRUPUS BALSAMI PERUVIANI (Ph. G.), L. *Prep.* Balsam of Peru, 1 oz.; boiling water, 11 oz.; digest with frequent agitation till cold, and form 10 oz. of the filtered liquid into a syrup with 18 oz. of sugar.

Syrup of Bark. *Syn.* SYRUPUS CINCHONÆ (P. Cod.), L. *Prep.* Calisaya bark, 1 oz.; percolate with 10 oz. of proof spirits ('996), and then with water, so as to yield 10 oz. of liquid; distil off spirit, filter, and add 10 oz. of sugar; reduce by a gentle heat, so as to obtain 15½ oz. (by weight) of product.

Syrup of Bark, Vinous. *Syn.* SYRUPUS CINCHONÆ VINOSUS (P. Cod.), L. *Prep.* Soft extract of bark, 1 oz.; white wine, 2 pints 3 oz.; dissolve, filter, add 3½ lbs. of white sugar, and dissolve by a water-bath.

Syrup of Belladonna. *Syn.* SYRUPUS BELLADONNÆ (P. Cod.), L. *Prep.* Tincture of belladonna (P. Cod.), ½ oz. (by weight); syrup, 10 oz. (by weight).

Syrup, Boyle's. See SYRUP, SYMPHYTIC.

Syrup of Bromide of Iron. *Syn.* SYRUPUS FERRI BROMIDI (B. P. C.), L. *Prep.* Iron wire, free from oxide, ⅛ oz.; bromine, 553 gr.; refined sugar, 14 oz.; distilled water, q. s. Dissolve the sugar in 6 oz. of distilled water by the heat of a water-bath; put the iron wire with 4 oz. of distilled water into a glass flask, having a capacity of at least 1 pint, and surround it with cold water; then add the bromine in successive quantities; shake occasionally until the froth becomes white and the reaction is complete; filter the solution into the warm syrup, and add, if necessary, distilled water sufficient to produce 1 pint. Each fl. dr. contains about 4½ gr. of bromide of iron. —*Dose,* ½ to 1 fl. dr.

Syrup of Buck'thorn. *Syn.* SYRUPUS RHAMNI (B. P., Ph. L. & E.), S. RHAMNI CATHARTICI, L. *Prep.* 1. (Ph. L.) Juice of buckthorn, defecated by 3 days' repose, 2 quarts; ginger and allspice, of each (bruised), 6 dr.; macerate the spice in 1 pint of the juice, at a gentle heat, for 4 hours, and filter; boil the remainder of the juice to 1½ pints, mix the liquors, dissolve therein of white sugar, 6 lbs., and add to the (nearly cold) syrup 6 fl. oz. of rectified spirit. In the Ph. E. the spirit is omitted.

2. (Wholesale.) *a.* Take of buckthorn juice,

3 galls.; bruised pimento and ginger (sifted from the dust), of each, ½ lb.; simmer for 15 minutes, strain, and add of sugar, 44 lbs.

b. Take of buckthorn juice, 3 galls.; boil to 2 galls.; add of bruised pimento and ginger gruffs (free from dust), of each, ¾ lb.; boil to 1 gall., strain, add molasses, 72 lbs., and finish the boiling.

Obs. Syrup of buckthorn is a brisk but unpleasant cathartic. It is now chiefly used in veterinary practice.—*Dose,* ½ fl. oz. to 1 fl. oz. Should the colour be dull, the addition of a few grains of citric or tartaric acid will brighten it.

Syrup of Butyl-chloral. *Syn.* SYRUPUS BUTYL-CHLORAL (B. P. C.), L. *Prep.* Hydrate of butyl-chloral, 320 gr.; syrup, sufficient to produce 1 pint; dissolve the hydrate of butyl-chloral in the syrup, previously made hot.—*Dose,* 1 to 4 fl. dr.

Syrup of Cabbage-tree Bark. *Syn.* SYRUPUS GEOFFROYAE (*Dr Wright*), L. *Prep.* Decoction of cabbage-tree bark made into a syrup with twice its weight of sugar. Vermifuge.—*Dose,* 1 to 4 table-spoonfuls.

Syrup of Cahinca. *Syn.* SYRUPUS CAHINCAE (*Soubeiran*), L. *Prep.* Alcoholic extract of cahinca, 64 gr.; syrup, 16 oz.; dissolve the extract in a little water, and add the solution to the boiling syrup.—*Dose,* 1 oz. daily.

Syrup of Capillaire'. *Syn.* SYRUP OF MAIDEN-HAIR; SYRUPUS ADIANTI, SYRUPUS CAPIL-LORUM VENERIS, L.; CAPILLAIRE, SIROP DE CAPILLAIRE, Fr. *Prep.* (P. Cod.) Canadian maidenhair (*Adiantum pedatum*, Linn.), 4 oz.; boiling water, 2½ pints; infuse, strain, add of white sugar, 5 lbs., and pour the boiling clarified syrup over 2 oz. more of maidenhair; re-infuse for 2 hours, and again strain.

Obs. Demulcent. Clarified syrup flavoured with orange-flower water or curaçoa is now commonly sold for CAPILLAIRE. It is usually 'put up' in small bottles of a peculiar shape, known in the trade as 'capillaires.' It is now chiefly used to sweeten and flavour grog. See CAPIL-LAIRE.

Syrup of Car'rageen. *Syn.* SYRUP OF ICE-LAND MOSS. *Prep.* Boil horehound, 1 oz.; liver-wort, 6 dr., in water, 4 pints, for 15 minutes; express and strain; then add carrageen (previously softened with cold water), 6 dr.; again boil for 15 minutes, strain through flannel, and add sugar, 1 lb., to each pint. An agreeable demulcent in coughs.

Syrup of Cascara Sagrada. *Syn.* SYRUPUS CASCARA SAGRADA (B. P. C.), L. Take of liquid extract of cascara sagrada, 4 fl. oz.; liquid extract of liquorice, 3 fl. oz.; carminative tincture, 2 fl. dr.; syrup, sufficient to produce 1 pint. Mix.—*Dose,* 1 to 4 fl. dr.

Syrup of Castor, Compound. *Syn.* SYRUPUS CASTORII COMPOSITUS (*Lebron*), L. *Prep.* Valerian water, 5 oz.; cherry laurel water, 2½ oz.; castor (dissolved in a sufficient quantity of spirit), 3 dr.; white sugar, 15 oz. In spasmodic asthma.

Syrup of Catechu. *Syn.* SYRUPUS CATECHU (P. Cod.), L. *Prep.* Extract catechu, 2½ oz.; syrup, 6 lbs.; dissolve the extract in double its weight of water, and add to the syrup.

Syrup of Chamomile. *Syn.* SYRUPUS ANTHE-

MIDIS (P. Cod.), L. *Prep.* Chamomile flowers, dried, 1 lb.; boiling water, 10 lbs.; macerate, strain with expression, and form the infusion into a syrup with twice its weight of sugar.

Syrup of Chloral Hydrate. *Syn.* SYRUPUS CHLORALIS HYDRATIS (B. Ph.), L. *Prep.* Hydrate of chloral, 80 gr.; distilled water, 1½ fl. dr.; syrup, to measure 1 fl. oz.—*Dose,* ½ fl. dr. to 2 fl. dr.

Syrup of Chloride of Lime. *Syn.* SYRUPUS CHLORIDI CALCIS (*Dr Reid*), L. *Prep.* Liquid chloride of lime, 1 dr.; mucilage, 2 dr.; syrup of orange peel, 10 dr.

Syrup of Cinchonine. *Syn.* SYRUPUS CINCHO-NINE (P. Cod.), L. *Prep.* Sulphate of cinchonine, 20 gr.; syrup, 16 oz. (by weight).

Syrup of Citrate of Caffeine. *Syn.* SYRUPUS CAFFEINAE CITRATIS (*Hannon*), L. *Prep.* Citrate of caffeine, 1 scruple; syrup, 1 oz.

Syrup of Citrate of Iron and Ammonia. *Syn.* SYRUPUS FERRI ET AMMONIAE CITRATIS (*Boral*), L. *Prep.* Ammonio-citrate of iron, ½ oz.; syrup, 9½ oz. (by weight); cinnamon water, ½ oz.

Syrup of Citrate of Iron and Quinine. *Syn.* SYRUPUS FERRI ET QUINIAE CITRATIS, L. A syrup is prepared by Mr Bullock under this name, but its composition has not been made known. Another form is citrate of iron and quinine, 1 oz.; syrup of orange peel, 1 pint (*Beasley*).

Syrup of Citric Acid. *Syn.* SYRUPUS ACIDI CITRICI (Ph. D.), L. *Prep.* (Ph. D.) Take of citric acid (in powder) and distilled water, of each, 2½ oz.; dissolve, add the solution, together with tincture of lemon peel, 5 fl. dr., to simple syrup, 3 pints, and mix with agitation. An agreeable refrigerant. Used for sweetening barley-water, &c., and for flavouring water to be used as a beverage in fevers and other inflammatory diseases.

Syrup of Cloves. *Syn.* SYRUPUS CARYOPHYLLI (Ph. E.), L. *Prep.* Clove July flowers, 1 troy oz.; boiling water, 4 oz.; macerate for 12 hours, strain, and add sugar, 7 oz.; make a syrup. Used for its colour and flavour.

Syrup of Cochineal. *Syn.* SYRUPUS COCCI-NELLE, S. COCCI (Ph. L.), L. *Prep.* (Ph. L.) Take of cochineal (bruised), 80 gr.; boiling distilled water, 1 pint; boil for 15 minutes in a closed vessel, strain, and add of sugar, 3 lbs., or twice that of the strained liquor; lastly, when the syrup has cooled, add of rectified spirit, 2½ fl. oz., or ½ fl. dr. to each fl. oz. of syrup. Used as a colouring syrup, and often sold for SYRUP OF CLOVE PINKS.

Syrup of Cochineal, Alkaline. *Syn.* SYRUPUS COCCI ALKALINUS, L. *Prep.* Cochineal, in powder, 2 scruples; carbonate of potash, in powder, 4 scruples; triturate, and add distilled water, 16 oz.; strain, and add 4 oz. of sugar-candy. A popular domestic remedy for whooping-cough.—*Dose,* From a teaspoonful to a tablespoonful, according to the age of the child, 3 or 4 times a day.

Syrup of Codeine. *Syn.* SYRUPUS CODEINAE (B. P. C.), L. *Prep.* Take of codeine, in powder, 20 gr.; proof spirit, 1½ fl. oz.; distilled water, 1½ fl. oz.; dissolve, and add syrup, sufficient to produce 1 pint.—*Dose,* ½ to 2 fl. dr.

Syrup of Cod-liver Oil. *Syn.* SYRUPUS OLEI

MORRHUÆ (*Duolos*), L. *Prep.* Mix 5 parts of powdered gum with 4 parts of simple syrup; add 8 parts of cod-liver oil, triturate till perfectly mixed, gradually adding 12 parts of water; lastly, dissolve in the emulsion 24 oz. of sugar by means of a gentle heat. In the same manner may be prepared syrups from the oil of skate, castor oil, &c.

Syrup of Coffee. *Syn.* SYRUPUS CAFFRÆ, L. *Prep.* Concentrated infusion of fresh-roasted coffee, 4 oz.; refined sugar, 8 oz.; dissolved in a closed vessel by a gentle heat.

Syrup of Colchicum. *Syn.* SYRUPUS COLCHICI (Ph. E. 1817), L. *Prep.* Fresh colchicum, 1 oz.; vinegar, 16 oz.; macerate for 2 days, and strain with gentle expression; add to the clear liquor 26 oz. of sugar, and boil.

Syrup of Colts'foot. *Syn.* SYRUPUS TUSSILAGINIS, L. *Prep.* (P. Cod.) Flowers of colts-foot, 1 lb. (or dried flowers, 2 oz.); boiling water, 2 lbs.; macerate for 12 hours; strain, press, filter, and add of white sugar, 4 lbs. A popular remedy in coughs, colds, &c.—*Dose*, 1 to 2 table-spoonfuls, *ad libitum*.

Syrup of Copaiba. *Syn.* SYRUPUS COPAIBÆ (*Puche*), L. *Prep.* Triturate 2 oz. of copaiba with ½ oz. of powdered gum and 1½ oz. of water; add 32 drops of essence of peppermint, and 12 oz. of simple syrup.

Syrup of Corsican Moss. *Syn.* SYRUPUS HELMINTHOCORTI (P. Cod.), L. Macerate 1 lb. of cleansed Corsican moss in 1 lb. of boiling water; in six hours strain. Macerate the residue in sufficient boiling water, so as to obtain, including the product of the first maceration, 2½ lbs., in which dissolve 5 lbs. of sugar.

Syrup of Cream. *Prep.* Finely powdered lump sugar mixed with an equal weight of fresh cream. It will keep for a long time if put into bottles, and closely corked and sealed over. It is commonly placed in 2-oz. wide-mouthed phials, and taken on long voyages, a fresh phial being opened at every meal.

Syrup of Dittany. *Syn.* SYRUPUS DICTAMNI, L. From dittany of Crete as SYRUP OF HYSSOP.

Syrup of Dulcamara. *Syn.* SYRUP OF BITTERSWEET; SYRUPUS DULCAMARÆ, L. As SYRUP OF CORSICAN MOSS.

Syrup, Easton's. See SYRUP OF THE PHOSPHATES OF IRON, QUININE, and STRYCHNINE.

Syrup of Emetina. *Syn.* SYRUPUS EMETINÆ (P. Cod.), L. *Prep.* Coloured emetine, 12 gr.; syrup, 17½ oz. (by weight); mix.

Syrup, Empyreumat'ic. Treacle.

Syrup of Ergotine. *Syn.* SYRUPUS ERGOTINÆ (*Bonjean*), L. *Prep.* Ergotine (watery extract of ergot), 2 dr.; orange-flower water, 1 oz.; dissolve, and add the solution to 16 oz. (by weight) of boiling syrup.—*Dose*, 2 to 4 spoonfuls in the day.

Syrup of E'ther. *Syn.* SYRUPUS ÆTHERIS, S. Æ. SULPHURICI, L.; SIROP D'ÉTHER, Fr. *Prep.* (P. Cod.) Sulphuric ether, 1 part; white (simple) syrup, 16 parts; place them in a glass vessel having a tap at the bottom, shake them frequently for 5 or 6 days, and then draw off the clear syrup into small bottles.—*Dose*, ½ to 3 fl. dr.

Syrup of Eucalyptus. *Syn.* SYRUPUS EUCALYPTI GLOBULI, L. *Prep.* 1. 100 gr. of the chopped leaves are infused for 6 hours in 1 litre of boiling water, the liquid expressed, and after allowing it to deposit, it is made into a syrup by the addition of 190 grms. of sugar for 100 grms. of the clear liquid.

2. (*Dorvault.*) Distilled water of eucalyptus, 50 parts; sugar, 95 parts; dissolve.

Syrup, Euston's. See SYRUP OF PHOSPHATE OF IRON WITH QUININE AND STRYCHNIA.

Syrup of Fennel. *Syn.* SYRUPUS FŒNICULI (Ph. G.), L. *Prep.* Infuse bruised fennel seed, 2 oz., in 12 oz. of boiling water for 3 hours; strain off 10 oz., and dissolve it in 18 oz. sugar.

Syrup of Foxglove. *Syn.* SYRUPUS DIGITALIS (P. Cod.), L. *Prep.* Tincture of foxglove (P. Cod.), ½ oz. (by weight); syrup, 20 oz. (by weight).

Syrup of Fumitory. *Syn.* SYRUPUS FUMARIÆ (P. Cod.), L. *Prep.* Clarified juice of fumitory, 1 lb.; white sugar, 2 lbs.; boil to a syrup.

Syrup of Garlic. *Syn.* SYRUPUS ALLII (Ph. U. S.), L. *Prep.* Garlic, 6 oz. (troy); distilled vinegar, 16 oz. (o. m.); macerate for 4 days; express, and form a syrup with the clear liquor and sugar, 2 troy lbs.

Syrup of Garlic, Compound. *Syn.* SYRUPUS ALLII COMPOSITUS, L.; DR. WILL'S SYRUP. *Prep.* Garlic (cut small), ½ oz.; bruised aniseed, ½ oz.; elecampane root, 3 dr.; liquorice root, 2 dr.; brandy, 24 oz.; digest for 2 or 3 days, strain, and form a syrup with 1½ lbs. of sugar.

Syrup of Gin'ger. *Syn.* SYRUPUS ZINGIBERIS (B. P., Ph. L., E., & D.), L. *Prep.* 1. (Ph. L.) Bruised ginger, 2½ oz.; boiling water, 1 pint; macerate for 4 hours, strain, and add of white sugar, 2½ lbs., or q. s.; and rectified spirit, as directed for syrup of cochineal. The Ph. E. omits the spirit.

2. (Ph. D.) Tincture of ginger, 1 fl. oz.; simple syrup, 7 fl. oz.; mix. Stimulant and carminative. Chiefly used as an adjuvant, in mixtures.

3. (B. P.) Strong tincture of ginger, 1 part; syrup, 25 parts; mix.—*Dose*, 1 to 4 dr.

Syrup of Guaiacum Wood. *Syn.* SYRUPUS GUAIACI LIGNI (P. Cod.), L. *Prep.* Boil rasped guaiacum wood, 3 oz., twice, and for an hour each time, in 30 oz. of water; strain through a thick cloth; mix the two liquids, and concentrate until they are reduced to 6 oz. (by weight); let cool, filter through paper, and add 10 oz. of sugar.

Syrup of Guarana. *Syn.* SYRUPUS PAULLINIÆ, S. GUARANÆ, L. *Prep.* Extract of guarana, 2½ dr.; syrup, 32 oz.

Syrup of Gum. *Syn.* SYRUPUS ACACIÆ, L.; SIROP DE GOMME, Fr. *Prep.* (P. Cod.) Dissolve pale and picked gum-arabic in an equal weight of water, by a gentle heat, add the solution to 4 times its weight of simple syrup, simmer for 2 or 3 minutes, remove the scum, and cool. A pleasant demulcent. The addition of 1 or 2 fl. oz. of orange-flower water to each pint greatly improves it.

Syrup of Gum Ammoniacum. *Syn.* SYRUPUS GUMMI AMMONIACI (Wurt. Ph.), L. *Prep.* Dissolve 2 oz. of gum ammoniacum in 8 oz. of white wine, by the heat of a water-bath, and add sugar, 16 oz.

Syrup of Gum Tragacanth. *Syn.* SYRUPUS

GUMMI TRAGACANTHÆ (*Mouchon*), L. *Prep.* Gum tragacanth, 1 oz.; water, 32 oz.; macerate for 48 hours, press through a linen cloth, and mix the mucilage with 8 lbs. of syrup, heated to 176° F., and strain through coarse cloth.

Syrup of Hedge-mustard. *Syn.* SYRUPUS ERYSIMI, L. *Prep.* From the expressed juice of hedge-mustard (clarified), 1 lb.; sugar, 2 lbs.; make into a syrup.

Syrup of Hedge-mustard, Compound. *Syn.* SYRUPUS ERYSIMI COMPOSITUS, L.; SIROP DE VÉLAR, SIROP DE CHANTRE (P. Cod.), Fr. *Prep.* Pearl barley, raisins, liquorice root, of each, ¾ oz.; cut and dried leaves of borage and chicory, of each, 1 oz.; fresh hedge-mustard, 15 oz.; dried elecampane root, 1 oz.; maidenhair, ¼ oz.; dried lavender and rosemary tops, of each, ¼ oz.; green aniseeds, ¼ oz.; sugar, 20 oz.; honey, 5 oz. Boil the pearl barley in the water until it bursts, add the raisins and the sliced liquorice, the borage and the chicory, and after just boiling strain and press. Then pour the strained liquid on to the other substances properly bruised and cut, and let the mixture digest for 24 hours over a water-bath; then distil, drawing over 2¼ oz. liquid (by weight); on the other hand, press and strain the liquor that remains in the retort, clarify with white of egg, add the sugar and honey, and make into a syrup that, when boiling, shall have a sp. gr. of 1·29; when nearly cold add the 2¼ oz. of distilled liquid, and strain.—*Dose*, ½ to 2 oz.

Syrup of Henbane. *Syn.* SYRUPUS HYOSCYAMI (P. Cod.). From the tincture (P. Cod.) as syrup of belladonna.

Syrup, Hive. Compound syrup of squills.

Syrup of Horehound. *Syn.* SYRUPUS MARRUBII, L.; SIROP DE PRASSIO, Fr. *Prep.* 1. (P. Cod.) Dried horehound, 1 oz.; horehound water, 2 lbs.; digest in a water-bath for 2 hours, strain, and add of white sugar, 4 lbs.

2. White horehound (fresh), 1 lb.; boiling water, 1 gall.; infuse for 2 hours, press out the liquor, filter, and add of sugar, q. s.

Obs. A popular remedy in coughs and diseases of the lungs.—*Dose.* A table-spoonful, *ad libitum.* "It is sold for any syrup of herbs that is demanded, and which is not in the shop" (*Gray*).

Syrup of the Hydrobromates of Iron and Quinine. *Syn.* SYRUPUS FERRI ET QUININÆ HYDROBROMATUM (B. P. C.), SYRUPUS FERRI BROMIDI CUM QUININÂ, L. *Prep.* Take of acid hydrobromate of quinine, 160 gr.; diluted hydrobromic acid, 1 fl. oz.; distilled water, 1 fl. oz. Mix the diluted hydrobromic acid with the distilled water, and in the mixture dissolve the acid hydrobromate of quinine. Then add syrup of bromide of iron, sufficient to produce 1 pint. Each fluid drachm contains 1 gr. of acid hydrobromate of quinine, and about 4 gr. of bromide of iron.—*Dose*, ½ to 1 fl. dr.

Syrup of the Hydrobromates of Iron, Quinine, and Strychnine. *Syn.* SYRUPUS FERRI QUININÆ ET STRYCHNINÆ HYDROBROMATUM (B. P. C.), SYRUPUS FERRI BROMIDI CUM QUININÂ ET STRYCHNINÂ, L. *Prep.* Take of strychnine, in powder, 2¼ gr.; acid hydrobromate of quinine, 160 gr.; diluted hydrobromic acid, 1 fl. oz.; distilled water, 1 fl. oz. Mix the diluted hydrobromic acid with the distilled water, and in the mixture dissolve the strychnine and acid hydrobromate of quinine, by the aid of a gentle heat. Then add syrup of bromide of iron, sufficient to produce 1 pint. Each fluid drachm contains $\frac{1}{57}$ gr. of strychnine, 1 gr. of acid hydrobromate of quinine, and about 4 gr. of bromide of iron.—*Dose*, ½ to 1 fl. dr.

Syrup of Hydrochlorate of Apomorphine. *Syn.* SYRUPUS APOMORPHINÆ HYDROCHLORATIS, L. *Prep.* Take of hydrochlorate of apomorphine, 5 gr.; dilute hydrochloric acid, 2 fl. dr.; rectified spirit, 7 fl. dr.; distilled water, 7 fl. dr.; syrup, 18 fl. oz. Mix the rectified spirit and distilled water, dissolve the hydrochlorate of apomorphine in the mixture by agitation; add the hydrochloric acid, and mix with the syrup.—*Dose*, ½ to 1 fl. dr.

Syrup of Hydrochlo″rate of Mor′phine. *Syn.* SYRUP OF MURIATE OF MORPHINE; SYRUPUS MORPHIÆ MURIATIS (Ph. D.), L. *Prep.* (Ph. D.) Solution of muriate of morphia, 1 fl. oz.; simple syrup, 17 fl. oz.; mix. Each fl. oz. contains ¼ gr. of the muriate.—*Dose*, ½ to 2 teaspoonfuls.

Syrup of Hydrocyanic Acid. *Syn.* SYRUPUS ACIDI HYDROCYANICI, L. The Paris Codex orders a syrup in which 200 parts (by weight) contain 1 part (by weight) of officinal medicinal hydrocyanic acid, containing 10% of anhydrous acid.

Syrup of the Hypophosphites. *Syn.* SYRUPUS HYPOPHOSPHITICUS, L. *Prep.* Hypophosphite of lime, potash, and soda, 1 part each; dissolved with heat in syrup, 100 parts.—*Dose*, 1 dr.

Syrup of Hypophosphites, Compound. *Syn.* SYRUPUS HYPOPHOSPHITUM COMPOSITUS (B. P. C.), L. Take of quinine (alkaloid), 20 gr.; strychnine, 1 gr.; hypophosphorous acid, 30%, 2 fl. dr.; strong solution of hypophosphite of iron, 3 fl. oz.; dissolve, and add hypophosphite of calcium, 80 gr.; hypophosphite of manganese, 40 gr.; hypophosphite of potassium, 40 gr.; dissolve, filter, and add syrup, sufficient to produce 1 pint. Mix. Each fl. dr. contains $\frac{1}{128}$ gr. of strychnine and ⅛ gr. of quinine.—*Dose*, ½ to 2 fl. dr.

Syrup of Hypophosphite of Iron. *Syn.* SYRUPUS FERRI HYPOPHOSPHITIS, L. *Prep.* Strong solution of hypophosphite of iron, 4 oz.; syrup, 16 oz.—*Dose*, ½ to 2 dr.

Syrup of Hyssop. *Syn.* SYRUPUS HYSSOPI (P. Cod.), L. *Prep.* As syrup of coltsfoot.

Syrup of Iceland Moss. *Syn.* SYRUPUS LICHENIS, L. Iceland moss deprived of its bitterness, 1 oz.; syrup, 32 oz. Make a concentrated decoction of the moss, strain, and add the syrup, and boil to a proper consistence.

Syrup of Indian Sarsaparil′la. *Syn.* SYRUPUS HEMIDESMI (B. P., Ph. D.), L. *Prep.* 1. (Ph. D.) Indian sarsaparilla (*Hemidesmus Indicus*, Brown), bruised, 4 oz.; boiling water, 1 pint; infuse for 4 hours, and to the strained and defecated infusion add twice its weight of sugar. Tonic, diuretic, &c.—*Dose*, 2 to 4 fl. dr.; in nephritic complaints, and in some others, instead of sarsaparilla.

2. (B. P.) Hemidesmus, bruised, 1 part; refined sugar, 7 parts; boiling distilled water, 5 parts; infuse 4 hours, strain, add the sugar, and dissolve. The product should weigh 10½ oz., and measure 8. Sp. gr. 1·335.—*Dose*, 1 to 4 dr.

Syrup of I′odide of Iron. *Syn.* SYRUPUS FERRI IODIDI (B. P., Ph. L., E., & D.), L. *Prep.* 1.

(Ph. L.) Mix iodine, 1 oz., and iron wire, 3 dr., with distilled water, 8 fl. oz.; and heat the solution until it assumes a greenish colour; then strain it, evaporate it to about 4 fl. oz., and add to it of white sugar, 10 oz.; lastly, when the syrup has cooled, add as much water as may be necessary, that it may measure exactly 15 fl. oz., and keep it in a well-stoppered black glass bottle. The formulæ of the Ph. E. & D. are nearly similar, a fl. dr. of each containing about 5 gr. of the pure dry iodide. This syrup is tonic and re-solvent, and hæmatinic.—*Dose*, 15 or 20 drops to 1 fl. dr.; in anæmia, debility, scrofula, &c.

2. (B. P.) Iron wire, 1 part; iodine, 2 parts; refined sugar, 28 parts; distilled water, 13 parts. Make a syrup with the sugar and 10 parts of the water, and keep it hot. Put into a strong soda-water bottle, covered with a cloth, the iron wire, the iodine, and 3 parts of the water; shake them together until the froth of the mixture becomes white; add now 2 oz. of the syrup, and boil for ten minutes; filter whilst still hot into the syrup. The product should be made up by water to weigh 43 parts, or to measure 31½ parts. Sp. gr. 1·385. —*Dose*, 20 to 60 minims.

Syrup of Iodide of Iron, Compound. *Syn.* SYRUPUS FERRI IODIDI COMPOSITUS (*Ricord*), L. *Prep.* This may be made by adding 1 oz. of the syrup to 9 oz. of compound syrup of sarsaparilla, both by weight.

Syrup of Iodide of Iron and Quinine. *Syn.* SYRUPUS FERRI ET QUINIÆ IODIDI (*Bouchardat*), L. *Prep.* Digest 1 dr. of iodine with ½ dr. of iron filings and 4 dr. of water, with a gentle heat and frequent agitation, till the solution is colourless. Filter it rapidly into a vessel containing 28 oz. of simple syrup. Dissolve also 12 gr. of sulphate of quinine in 2 dr. of water acidulated with sulphuric acid, and add to the former.

Syrup of Iodide of Manganese. *Syn.* SYRUPUS MANGANESII IODIDI (*M. Hannon*), L. Pure hydrated carbonate of manganese, 1 dr.; concentrated hydriodic acid, q. s. to dissolve it. Mix this solution with 16½ oz. of sudorific syrup.— *Dose.* From 2 to 6 tablespoonfuls daily.

Syrup of Iodide of Potassium. *Syn.* SYRUPUS POTASSII IODIDI (P. Cod.), L. *Prep.* Iodide of potassium, ½ oz.; water, ¼ oz.; syrup, 9½ oz. (by weight).

Syrup of Iodine. *Syn.* SYRUPUS IODINII (*Foy*), L. *Prep.* Compound tincture of iodine, 4 dr.; mint water, 4 oz.; syrup, 16 oz.

Syrup of Iodo-hydrargyrate of Potassium. *Syn.* SYRUPUS IODO-HYDRARGYRATIS POTASSII (*Puche*), L. *Prep.* Iodo-hydrargyrate of potassium, 16 gr.; tincture of saffron, 2½ dr.; syrup, 16 oz.

Syrup of Ipecacuanha. *Syn.* SYRUPUS IPE-CACUANHÆ, L. *Prep.* (Ph. E.) Ipecacuanha (in coarse powder), 4 oz.; rectified spirit, 15 fl. oz.; digest for 24 hours at a gentle heat, and strain; add of proof spirit, 14 fl. oz., and again digest and strain, and repeat the process with water, 14 fl. oz.; distil off the spirit from the mixed liquors, evaporate to 12 fl. oz., and filter; next add to the residuum rectified spirit, 5 fl. oz., and simple syrup, 7 pints, and mix well.—*Dose.* As an emetic for infants, ½ teaspoonful; as an expectorant, 1 to 3 teaspoonfuls.

Syrup of Ipecacuanha (Acetic). *Syn.* SYRUPUS IPECACUANHÆ ACETICUS (B. P. C.), L. *Prep.* Take of vinegar of ipecacuanha, 1 pint; refined sugar, 2½ lbs. Dissolve by the aid of a gentle heat. Sp. gr. about 1·33.—*Dose*, ½ to 2 fl. dr.

Syrup of Iron and Iodide of Potassium. *Syn.* SYRUPUS FERRI ET POTASSII IODIDI, L. *Prep.* Dissolve 1 oz. of iodide of potassium in 6 oz. of hot water; add 12½ oz. (fl.) of syrup of iodide of iron, and sufficient simple syrup to make up 1½ pints.

Syrup of Jalap. *Syn.* SYRUPUS JALAPINII (P. Cod.), L. Jalap, 10 dr.; coriander, ½ dr.; fennel seed, ½ dr.; water, 12 oz.; heat to 212° F. for twenty minutes, let it stand 24 hours; strain, and make a syrup with 24 oz. of sugar.

Syrup of Kermes. *Syn.* SYRUPUS KERMETIS, SYRUPUS ANTIMONIATUS, L. *Prep.* Kermes mineral, 20 gr.; syrup of squills, 1½ oz.; syrup of marsh-mallow, 1½ oz. Mix.

Syrup of Lactate of Iron. *Syn.* SYRUPUS FERRI LACTATIS (*Cap*), L. *Prep.* Lactate of iron, 1 dr.; boiling distilled water, 6 oz.; pure sugar, 12 oz.—*Dose*, 2 to 4 dr.

Syrup of Lacto-phosphate of Lime. *Syn.* SYRUPUS CALCIS LACTO-PHOSPHATIS (*P. Vincent*, 'Pharm. Journ.'), L. *Prep.* Burnt bones, 155 gr.; hydrochloric acid, 310 gr.; liquid ammonia, 200 gr.; concentrated lactic acid, distilled water, of each, q. s.; sugar, 18½ oz.; leave together for some time the bone ash and the acid until effervescence ceases, then add distilled water, 500 gr.; precipitate with the ammonia, filter, and well wash the precipitate with distilled water, until the washings cease to give a precipitate with nitrate of silver. Leave to drain for 12 hours, after gently heat in a porcelain capsule, and add sufficient lactic acid to dissolve the precipitate; add sufficient distilled water to make the product weigh 9½ oz.; filter, and add the sugar; make dissolve with a gentle heat.

Syrup of Lactucarium. *Syn.* SYRUPUS LAC-TUCARII (U. S.), L. *Prep.* Lactucarium, 1 troy oz.; syrup, 14 oz. (o. m.); proof spirit, q. s.; rub the lactucarium with the proof spirit to a syrupy consistence, transfer to a percolator and percolate with proof spirit until 8 oz. (o. m.) of tincture have been obtained. Evaporate this portion in a water-bath at 160° F. to 2 oz. (o. m.). Mix it with the syrup made hot, and strain immediately.

Syrup of Le'mon. *Syn.* SYRUPUS LIMONIS (B. P.), SYRUPUS LIMONUM (Ph. L. & E.), SYRUPUS CITRI MEDICÆ, L. *Prep.* 1. (Ph. L.) Lemon juice (strained or defecated), 1 pint; white sugar, 2½ lbs.; dissolve by a gentle heat, and set it aside; in 24 hours remove the scum, decant the clear portion, and add of rectified spirit, 2½ fl. oz. The Edinburgh College omits the spirit. A pleasant refrigerant syrup in fevers, &c.—*Dose*, 1 to 4 fl. dr., in any diluent. With water it forms an excellent extemporaneous lemonade.

2. (B. P.) Fresh lemon peel, 2 parts; lemon juice, strained, 20 parts; refined sugar, 36 parts. Heat the lemon juice to the boiling-point, and having put it into a covered vessel with the lemon peel, let them stand until they are cold, then filter and dissolve the sugar in the filtered liquid

with a gentle heat. The product should weigh 56 parts and measure 41 parts.—*Dose*, 1 to 2 dr.

3. An excellent syrup for making lemonade is prepared in the following manner, according to a German paper:—Fresh lemon peel is steeped for 24 hours in an equal quantity, by weight, of alcohol, after which the latter is drawn off by distillation. This spirit of lemon is used as required by adding 30 parts to a syrup made from 500 parts of sugar and 250 parts of water; this syrup is mixed with a solution of 15 parts of citric acid in 30 parts of orange-flower water. Such a lemon syrup is said to be far superior, both in flavour and durability, to that made either from the freshly expressed juice or from citric acid and oil of lemons.

Syrup of Lettuce. *Syn.* SYRUPUS LACTUCÆ (P. Cod.), L. *Prep.* Dissolve 2 oz. of extract of lettuce in 8 times its weight of cold water, filter, and add 6 lbs. 2 oz. of syrup, which, when boiling, has a specific gravity of 1·26.

Syrup of Lime. *Syn.* SYRUPUS CALCIS (*Trousseau*), L. *Prep.* Slake 2½ dr. of quicklime with 3 oz. of water, and add it to 32 fl. oz. of simple syrup; boil 10 minutes and filter. This is usually diluted with 4 parts of syrup. Given in diarrhœa.

Syrup of Liquorice. *Syn.* SYRUPUS GLYCYRRHIZÆ, L. *Prep.* Liquorice root, 4 oz.; boiling water, 16 oz.; digest, strain, and make a syrup with sugar.

Syrup of Lobelia. *Syn.* SYRUPUS LOBELIÆ (*Mr Proctor*), L. *Prep.* Vinegar of lobelia, 6 oz.; sugar, 12 oz. Dissolve in a gentle heat.

Syrup of Malate of Manganese. *Syn.* SYRUPUS MANGANESII MALATIS (*M. Hannon*), L. *Prep.* Malate of manganese, 1 oz.; simple syrup, 16 oz.; spirit of lemon peel, 2 dr.—*Dose*, ½ dr. to 1 dr.

Syrup of Manna. *Syn.* SYRUPUS MANNÆ (Ph. G.), L. *Prep.* Dissolve 8 oz. of manna in 12 oz. of water, strain, and filter; then add 16 oz. of sugar, and make it into a syrup.

Syrup of Marsh-mal'low. *Syn.* SYRUPUS ALTHÆÆ (Ph. L. & E.), L. *Prep.* 1. (Ph. L.) Marsh-mallow root, fresh and sliced, 1½ oz.; distilled water (cold), 1 pint; macerate for 12 hours, press out the liquor, strain it through linen, and add to the strained liquor twice its weight of white sugar (about 3 lbs.); dissolve by a gentle heat, and, when cold, add of rectified spirit, 2½ fl. oz., or q. s. See SYRUP OF COCHINEAL.

2. (Ph. L. 1836.) Take of fresh marsh-mallow root, bruised, 8 oz.; water, 4 pints; boil down to one half, and express the liquor when it is cold; set it aside for 24 hours, that the fæces may subside, then decant off the clear liquid, and, having added to it of sugar, 2½ lbs., boil the whole to a proper consistence. The formula of the Ph. E. is similar.

Obs. This is a popular demulcent and pectoral.—*Dose*, 1 to 4 fl. dr.; in coughs, &c., either alone or added to mixtures.

Syrup of Mercury. *Syn.* SYRUPUS HYDRARGYRI, L. *Prep.* "There are several forms for mercurial syrups, but they all appear liable to serious objection. Plenk:—Quicksilver, 1 dr.; powdered gum-arabic, 3 dr.; syrup, 2 oz.; triturate, and gradually add 1 oz. of water. Larry:—Sudorific syrup, 1 pint; bichloride of mercury, 5 gr.; muriate of ammonia, 5 gr.; extract of opium, 5 gr.; Hofman's anodyne liquor, ½ dr.—*Dose*, ½ oz. to 1 oz. Creron's syrup consists of mercurial ether (4 gr. of sublimate to 2 dr. of ether), 2 dr.; syrup, 8 oz." (*Beasley*).

Syrup of Milk. *Syn.* SYRUPUS LACTIS, L. Reduce skimmed milk by gentle evaporation to one half, and add twice its weight of sugar.

Syrup of Monosulphide of Sodium. *Syn.* SYRUPUS SODII MONOSULPHIDI (P. Cod.), L. *Prep.* Crystallised monosulphide of sodium, 44 gr.; distilled water, 1 oz.; syrup, 94 oz. (by weight).

Syrup of Mugwort. *Syn.* SYRUPUS ARTEMISIÆ, L. As SYRUP OF WORMWOOD.

Syrup of Mugwort, Compound. *Syn.* SYRUPUS ARTEMISIÆ COMPOSITUS (P. Cod.), L. *Prep.* Take of fresh tops of mugwort, pennyroyal, catmint, and savine, of each, 2 oz.; fresh roots of elecampane, lovage, and fennel, of each, 88 gr.; fresh tops of wild marjoram, hyssop, feverfew, rue, and basil, of each, 1 oz.; aniseed, ½ oz.; cinnamon, ½ oz., all properly divided; rectified spirit, 2½ oz. (by weight); water, 30 oz.; syrup of honey, 12½ oz. (by weight). Put the plants in a vessel over a water-bath, pour on the water mixed with the spirit, let it stand 24 hours, and then distil over 3½ oz. (by weight). On the other hand, press the residue of the distillation, clarify with white of egg, and add sugar, 25 oz.; then make into a syrup, which, when boiling, has the sp. gr. 1·26. Take the weight and evaporate until it has lost weight equal to that of the distilled liquid, then add the syrup of honey, and, lastly, when nearly cold, the distilled liquid, and strain. —*Dose*, 2 to 12 dr.

Syrup of Mul'berries. *Syn.* SYRUPUS MORI (B. P., Ph. L.), L. *Prep.* 1. (Ph. L.) Juice of mulberries, strained, 1 pint; sugar, 2½ lbs.; dissolve by a gentle heat, and set the solution by for 24 hours; then remove the scum, decant the clear liquid, and add of rectified spirit, 2½ fl. oz. Used as a colouring and flavouring when alkalies and earths are not present. Syrup of red poppies (*rhœados*), slightly acidulated with tartaric or dilute sulphuric acid, is very generally sold for it.

2. (B. P.) Mulberry juice, 20 parts; refined sugar, 36 parts; rectified spirit, 2½ parts; heat the juice to the boiling-point, and, when it has cooled, filter it; dissolve the sugar in the filtered liquid by a gentle heat, and add the spirit. The product should weigh 54 parts. Sp. gr. 1·33.—*Dose*, 1 to 2 dr.

Syrup of Mu''riate of Morphia. See SYRUP OF HYDROCHLORATE OF MORPHINE.

Syrup of Opium. *Syn.* SYRUPUS OPII (P. Cod.), L. *Prep.* Extract of opium, 87½ gr., dissolve in 6 dr. of cold water, and mix with sufficient syrup to make up 6½ lbs. (1 in 500).

Syrup of Orange Flowers. *Syn.* SYRUPUS AURANTII FLORIS (B. P.), L. *Prep.* Orange-flower water, 8 parts; refined sugar, 48 parts; distilled water, 16 parts, or a sufficiency; heat the sugar and water together, strain, and when nearly cold add the orange-flower water. When finished should weigh 72 parts and measure 54 parts. Sp. gr. 1·33.—*Dose*, 1 to 2 dr.

Syrup of Orange Juice. *Syn.* SYRUPUS E SUCCO AURANTIORUM (Ph. E., 1744), L. *Prep.* Orange juice, 1 lb.; sugar, 2 lbs. Dissolve by heat.

Syrup of Orange Peel. *Syn.* SYRUPUS AU-RANTII (B. P., Ph. L., E., & D.), S. CITRI AU-RANTII, S. E CORTICIBUS AURANTIORUM, L. *Prep.* 1. (Ph. L.) Dried orange peel, 2½ oz.; boiling distilled water, 1 pint; macerate for 12 hours in a covered vessel, press out the liquor, simmer it for 10 minutes, and then complete the process as directed for SYRUP OF COCHINEAL. In the Ph. E. & D., and Ph. L. 1836, no spirit is ordered.

2. (B. P.) Tincture of orange peel, 1 part; syrup, 7 parts; mix.—*Dose*, 1 to 2 dr.

3. (Wholesale.) *a.* From fresh orange peel, 18 oz. (or dried, ¾ lb.); sugar, 18 lbs.; water, q. s.

b. From tincture of orange peel, 1 fl. oz.; simple syrup, 19 fl. oz.; mix. An agreeable flavouring and stomachic.—*Dose*, 1 to 4 fl. dr.

Syrup, Pectoral. *Syn.* SYRUPUS PECTORALIS (Ph. L., 1746), L. *Prep.* Black maidenhair, 5 oz.; liquorice root, 4 oz.; boiling water, 4 pints; macerate for some hours, strain, add to the infusion twice its weight of sugar, and make a syrup.

Syrup of Pepsine. *Syn.* SYRUPUS PEPSINÆ (*Corvisart*), L. *Prep.* 6 parts of pepsine in 20 parts of cold water, and added to 70 parts of acidulated syrup of cherries.

Syrup of Persulphuret of Iron. *Syn.* SYRUPUS FERRI PERSULPHURETI (*Bouchardat*), L. *Prep.* Reduce 10 oz. of syrup by evaporation to 9 oz., and add 2 oz. hydrated persulphuret of iron in a gelatinous state; mix, and keep in a close bottle. —*Dose.* A teaspoonful two or three times a day in scrofulous and cutaneous affections. As an antidote for poisoning by the salts of lead, mercury, and copper. Give a teaspoonful frequently.

Syrup of Phosphate of Iron. *Syn.* SYRUPUS FERRI PHOSPHATIS (B. P.), L. *Prep.* Granulated sulphate of iron, 224 gr.; phosphate of soda, 200 gr.; bicarbonate of sodium, 56 gr.; concentrated phosphoric acid, 1½ oz.; refined sugar, 8 oz.; distilled water, 8 oz. Dissolve the sulphate of iron in 4 oz. of the water, and the phosphate and bicarbonate of soda in the remainder; mix the two solutions, and, after careful stirring, transfer the precipitate to a calico filter, and wash it with distilled water till the filtrate ceases to be affected with chloride of barium. Then press the precipitate strongly between folds of bibulous paper, and add to it the phosphoric acid. As soon as the precipitate has dissolved, filter the solution, add the sugar, and dissolve without heat. The product should measure exactly 12 fl. oz. 1 fl. dr. contains 1 gr. of phosphate of iron.

In the preparation of this syrup as much expedition as possible should be used in washing and pressing the precipitate of phosphate of iron formed. It is best washed by decantation. The water employed should be just previously boiled to expel oxygen; the protosulphate of iron should be entirely free from persulphate, and clear crystals of phosphate of soda should be chosen. Mr W. H. Jones ('Pharm. Journ.,' 3rd series, vol. v, p. 541) gives a process for the preparation of this syrup, which consists in dissolving metallic iron

in phosphoric acid and water, and then adding the solution to syrup.

Syrup of Phosphate of Iron, Compound. *Syn.* PARRISH'S CHEMICAL FOOD, SYRUP OF THE COMPOUND PHOSPHATES; SYRUPUS FERRI PHOSPHATIS COMPOSITUS, S. PHOSPHATICUS (*Mr E. Parrish*, U.S.), L. *Prep.* 1. Dissolve sulphate of iron, 10 dr., in boiling water, 2 oz.; and phosphate of soda, 12 dr., in boiling water, 4 oz. Mix, and wash the precipitated phosphate of iron. Dissolve phosphate of lime, 12 dr., in 4 oz. of boiling water, with enough hydrochloric acid to make a clear solution; precipitate with liquid ammonia, and wash precipitate. Add to the fresh precipitates glacial phosphoric acid, 20 dr., dissolved in 4 dr. of water; when clear, add carbonate of soda, 2 scruples, and carbonate of potassa, 1 dr., and then sufficient hydrochloric acid to dissolve the precipitate. Now add water to make the solution measure 22 oz. (old measure), and add powdered cochineal, 2 dr.; mixed sugar, 32 troy oz.; apply heat, and, when the syrup is formed, strain it and add orange-flower water, 1 oz.—*Dose.* A teaspoonful. In addition to phosphate of iron and phosphate of lime this syrup contains smaller quantities of the alkaline phosphates. Dr Howie points out that Parrish is incorrect in stating that this syrup contains 1 gr. of phosphate of iron and 2½ gr. phosphate of lime in the fl. dr., if this statement be compared with his formula, which by calculation will be found to give ·715 gr. of phosphate of iron and 2 gr. of phosphate of lime for the fl. dr., even if none of the former were wasted in the process. Mr Howie deprecates the use of hydrochloric acid sometimes had recourse to in preparing the syrup, and he adds that the purest sugar only should be used, and that made from beetroot should be carefully avoided. See a valuable paper by Mr Howie on this subject, ' Pharm. Journ.,' 3rd series, vol. vi, p. 804.

2. (B. P. C.) Iron wire (free from rust), 37½ gr.; concentrated phosphoric acid (sp. gr. 1·5), 1 fl. oz.; distilled water, 5 fl. dr. Put these into a glass flask, so that the liquid completely covers the iron wire, plug the neck with cotton wool, and heat gently till dissolved. Add this solution to the following when the latter has cooled :— Precipitated carbonate of calcium, 120 gr.; concentrated phosphoric acid, 4 fl. dr.; distilled water, 2 fl. oz.; mix and add bicarbonate of potassium, 9 gr.; phosphate of sodium, 9 gr.; filter and set aside. Then take of cochineal, 30 gr.; water, 7½ oz. Boil 15 minutes and filter to 7 oz.; add 14 oz. sugar, heat till dissolved, and strain. When cold add the former filtrate set aside, and a sufficient quantity of water to make the whole measure 1 pint.—*Dose*, ½ to 2 dr.

Syrup of Phosphate of Iron and Manganese. *Syn.* SYRUPUS FERRI PHOSPHATIS ET MANGANESII, L. *Prep.* Dissolve 6 dr. of glacial phosphoric acid in a small quantity of water, add 72 gr. of phosphate of iron, and 48 gr. of phosphate of manganese; apply heat to dissolve, then add sugar, 10 oz., and water up to measure of 12 oz.—*Dose*, 1 to 4 dr.

Syrup of the Phosphates of Iron, Quinine, and Strychnine. *Syn.* SYRUPUS FERRI, QUININÆ, ET STRYCHNINÆ PHOSPHATUM (B. P. C.), L. (Easton's syrup.) *Prep.* Take of strychnine, in powder, 5 gr.; concentrated phosphoric acid, sp. gr. 1·5, 75 minims; distilled water, 225 minims; dissolve, and add phosphate of quinine, 120 gr. Dissolve by the aid of a gentle heat, and add syrup of phosphate of iron, sufficient to produce 1 pint; mix thoroughly. Each fluid drachm contains 1 gr. of phosphate of iron, ¾ gr. of phosphate of quinine, and $\frac{1}{32}$ gr. of strychnine.—*Dose*, ½ to 1 fl. dr.

Syrup of Phosphate of Manganese. *Syn.* SYRUPUS MANGANESII PHOSPHATIS (*M. Hannon*), L. *Prep.* Phosphate of manganese, ½ dr.; spirit of tolu, 3 oz. 3 dr.; syrup of bark, 5 oz.; spirit of lemon peel, 1½ dr.; powder of tragacanth, 10 gr. Mix quickly and preserve in a well-stoppered bottle.

Syrup of Phosphate of Quinine. *Syn.* SYRUPUS QUININÆ PHOSPHATIS, L. *Prep.* Phosphate of quinine, 96 gr.; water, 18½ fl. dr.; syrupy phosphoric acid (sp. gr. 1·500), 2¼ fl. dr.; syrup, 10 fl. dr. Mix the acid with the water, add the quinine, and filter into the syrup.

Syrup of Phosphate of Zinc. *Syn.* SYRUPUS ZINCI PHOSPHATIS, L. *Prep.* Phosphate of zinc, 192 gr.; water, 11 fl. dr.; syrupy phosphoric acid (sp. gr. 1·500), 5 fl. dr.; syrup, 10 fl. oz. Rub the phosphate with the water, add the acid, and filter into the syrup.

Syrup of Pomegranate-root Bark. *Syn.* SYRUPUS CORTICIS RADICIS GRANATI (*Guibourt*), L. *Prep.* Obtain from 1 lb. of powdered bark of pomegranate root 4 lbs. of infusion by percolation. Boil this with 28½ oz. of syrup till reduced to 2 lbs.

Syrup of Pop'pies. *Syn.* SYRUP OF WHITE POPPIES; SYRUPUS PAPAVERIS (B. P., Ph. L. and E.), S. P. SOMNIFERI, L. *Prep.* 1. (B. P.) Poppy capsules, coarsely powdered, free from seeds, 36 parts; rectified spirit, 16 parts; refined sugar, 64 parts; boiling distilled water, a sufficiency; macerate the poppy capsules in 80 parts of the water. Infuse for 24 hours, then pack in a percolator, and, adding more of the water, allow the liquor slowly to pass until 320 parts have been collected or the poppies are exhausted; evaporate the liquor by a water-bath until it is reduced to 60 parts; when quite cold add the spirit; let the mixture stand for 12 hours and filter. Distil off the spirit, evaporate the remaining liquor to 40 parts, and then add the sugar. The product should weigh 104 parts, and measure 78¾ parts. Sp. gr. 1·32.—*Dose*, 1 dr.; 10 to 20 minims for children, increasing cautiously.

2. (Ph. L.) Poppy heads, dried, bruised, and without the seed, 3 lbs.; boiling water, 5 galls.; boil down to 2 galls., press out the liquor, evaporate the expressed liquid to 2 quarts, strain it whilst hot, and set it aside for 12 hours; next decant the clear portion from the fæces, boil this down to 1 quart, and dissolve in it sugar, 5 lbs.; lastly, when cold, add of rectified spirit, 5 fl. oz. "Each fl. oz. is equivalent to 1 gr. of dry extract." In the Ph. E. and Ph. D. 1826 no spirit is ordered.

3. (Wholesale.) Extract of poppies, 1½ lbs.; boiling water, 2½ galls.; dissolve, clarify, or filter, so that it may be perfectly transparent when cold; then add of white sugar, 44 lbs., and dissolve.

Obs. Syrup of poppies is anodyne and soporific.—*Dose.* For an infant, ¼ to ½ teaspoonful; for an adult, 2 to 4 fl. dr. According to M. Chereau, its tendency to fermentation is prevented by the addition of 32 parts of sugar of milk to every 1000 parts of the syrup.

Syrup of Pyrophosphate of Iron. *Syn.* SYRUPUS FERRI PYROPHOSPHATIS (*Parrish*), L. *Prep.* Pyrophosphate of iron in scales, 16 gr.; syrup, 1 fl. oz.

Syrup of Pyrophosphate of Iron and Ammonia. *Syn.* SYRUPUS FERRI PYROPHOSPHATIS ET AMMONIÆ (P. Cod.), L. *Prep.* Pyrophosphate of iron with citrate of ammonia, 1 dr.; water, 2 dr.; syrup, 12 oz.

Syrup of Quinine with Coffee. *Syn.* SYRUPUS QUINIÆ CUM COFFEÂ, L. *Prep.* Prepare 1½ pints of clear infusion from 4 oz. of roasted coffee; dissolve it in 5 lbs. of refined sugar, and add to the syrup 1½ dr. of sulphate of quinine dissolved in a little water, with the addition of a few drops of sulphuric acid.

Syrup of Raspberry. *Syn.* SYRUPUS ACETI RUBI IDÆA (P. Cod.), L. *Prep.* Raspberry vinegar, 10 oz. (by weight); sugar, 17½ oz.; boil them together.

Syrup of Red Pop'pies. *Syn.* SYRUPUS RHŒADOS (Ph. L. and E.), S. PAPAVERIS RHŒADOS, L. *Prep.* 1. (Ph. L.) Petals of the red poppy, 1 lb.; boiling water, 1 pint; mix in a water-bath, remove the vessel, macerate for 12 hours, press out the liquor, and, after defecation or filtering, complete the process as directed for SYRUP OF COCHINEAL.

2. (Wholesale.) From dried red poppy petals, 3 lbs.; boiling water, q. s.; white sugar, 44 lbs.; as the last.

Obs. Syrup of red poppies is chiefly employed for its fine red colour. A little acid brightens it. The colour is injured by contact with iron, copper, and all the common metals.

Syrup of Red Roses. *Syn.* SYRUPUS ROSÆ (B. P.), SYRUPUS ROSÆ GALLICÆ (Ph. E. and D.), L. *Prep.* 1. (Ph. E.) Dried petals of the red rose, 2 oz.; boiling water, 1 pint; pure sugar, 20 oz.; as the last.

2. (Ph. D.) Dried petals of the Gallic rose, 2 oz.; boiling water, 1 pint; boil in a glass or porcelain vessel until the colour is extracted, strain with expression, and, after defecation, add to the clear decanted liquor twice its weight of white sugar. Astringent and stomachic; chiefly used as an adjunct in mixtures, &c.

3. (B. P.) Dried rose petals, 1 part; refined sugar, 15 parts; boiling distilled water, 10 parts; infuse the petals in the water 2 hours, squeeze through calico, heat the liquor to the boiling-point, and filter; add the sugar and dissolve with heat. The product should weigh 23 parts, and measure 17½ parts. Sp. gr. 1·335.—*Dose*, 1 to 2 dr.

Syrup of Rhatany. *Syn.* SYRUPUS KRAMERIÆ (P. Cod.), L. As syrup of catechu.

Syrup of Rhu'barb. *Syn.* SYRUPUS RHEI (B. P.), L. *Prep.* 1. (B. P.) Rhubarb, in coarse powder, 2 parts; coriander fruit, in powder, 2 parts; refined sugar, 24 parts; rectified spirit, 8 parts; distilled water, 24 parts. Mix the rhubarb and coriander, pack them in a percolator, pass the spirit and water, previously mixed, slowly through them, evaporate the liquid that has passed until it is reduced to 13 parts, and in this, after it has been filtered, dissolve the sugar with a gentle heat.—*Dose*, 1 to 4 dr.

2. (P. Cod.) Bruised rhubarb, 3 oz.; cold water, 16 fl. oz.; macerate for 12 hours, filter, and add of white sugar, 32 oz.

3. (Ph. U. S.) Take of rectified spirit, 8 fl. oz.; water, 24 fl. oz.; rhubarb (coarsely powdered), 2 oz.; (mixed with) sand, an equal bulk, or q. s.; make a tincture by percolation, evaporate this, over a water-bath, to 13 fl. oz., and dissolve it in 2 lbs. of white sugar. An excellent formula.

4. (Wholesale.) Rhubarb (bruised), 1½ lbs.; cold water, q. s.; sugar, 20 lbs.; as No. 1. Stomachic and purgative.—*Dose*. For an infant, ¼ to 1 teaspoonful; for an adult, ½ to ¾ fl. oz., or more.

Syrup of Rhubarb (Spiced). *Syn.* SYRUPUS RHEI AROMATICUS, L. *Prep.* (Ph. U. S.) Rhubarb, 2½ oz.; cloves and cinnamon, of each, ½ oz.; nutmeg, ¼ oz. (all bruised); proof spirit, 32 fl. oz.; macerate for 14 days (or percolate), strain, gently evaporate to 16 fl. oz., filter whilst hot, and mix the liquid with simple syrup (gently warmed), 4½ pints. A cordial laxative.—*Dose*, ½ to 1 teaspoonful; in infantile constipation, diarrhœa, &c.

Syrup of Rhubarb and Senna. *Syn.* SYRUPUS RHEI ET SENNÆ (Ph. E., 1745), L. *Prep.* Rhubarb, 1 oz.; senna, 2 oz.; fennel seed, 2 dr.; cinnamon, 2 dr.; boiling water, 2½ pints; macerate for 12 hours, strain, and boil with 3 lbs. of sugar to a syrup.

Syrup of Ro"ses. *Syn.* SYRUPUS ROSÆ (Ph. L.), SYRUPUS ROSÆ CENTIFOLIÆ (Ph. E.), L. *Prep.* 1. (Ph. L.) Dried petals of damask roses (*Rosa centifolia*), 7 oz.; boiling water, 3 pints; macerate for 12 hours, filter, evaporate in a water-bath to 1 quart, and add of white sugar, 6 lbs.; and, when cold, rectified spirit, 5½ fl. oz.

2. (Wholesale.) From rose leaves, 1 lb.; sugar, 19 lbs.; water, q. s.; as the last. Gently laxative.—*Dose*, ½ to 1 fl. oz. It is usual to add a few drops of dilute sulphuric acid, to brighten the colour. Alkalies turn it green.

Syrup of Rue. *Syn.* SYRUPUS RUTÆ, L. *Prep.* Take of oil of rue, 12 to 15 drops; rectified spirit, ½ fl. oz.; dissolve, and add it to simple syrup, 1 pint.—*Dose*, ½ to 1 teaspoonful; in the flatulent colic of children. An infusion of ½ oz. of the herb is sometimes substituted for the solution of the essential oil.

Syrup of Saffron. *Syn.* SYRUPUS CROCI (Ph. L., E., & D.), L. *Prep.* 1. (Ph. L.) Hay saffron, 5 dr. (10 dr.—Ph. E.; ¼ oz.—Ph. D.); boiling water, 1 pint; macerate in a covered vessel for 12 hours, then strain the liquor, and add of white sugar, 3 lbs., or q. s., and rectified spirit, 2½ oz., or q. s., in the manner directed under SYRUP OF COCHINEAL. The Ph. E. & D. omit the spirit.

2. (Wholesale.) Hay saffron, 6 oz.; boiling water, 6 quarts; white sugar, 24 lbs.; as the last. Used for its colour and flavour; the first is very beautiful.

Syrup of Salicin. *Syn.* SYRUPUS SALICINI, L.

Prep. Salicin, 1 dr.; boiling water, 1 oz.; sugar, 2 oz.

Syrup of Santonate of Soda. *Syn.* SYRUPUS SODÆ SANTONATIS, L. This formula is recommended because of the adaptability of its administration to children, the syrup being of very pleasant taste. It is made as follows:—Powdered santonate of soda, 5 grms.; simple syrup, 900 grms.; syrup of orange flower, 100 grms. Suspend the santonate in 250 grms. of the syrup, and heat it over a spirit lamp until dissolved; add the remainder of the syrup, then the syrup of orange flower, and mix carefully. A tablespoonful or 20 grms. of this syrup will contain 10 centigrams of santonate, or the equivalent of 5 centigrams of santonin. For adults the dose might be double, or a syrup made containing 20 centigrams to the tablespoonful.

Syrup of Sarsaparil'la. *Syn.* SYRUPUS SARZÆ (Ph. L. & E.), SYRUPUS SARSAPARILLÆ, L. *Prep.* 1. (Ph. L.) Take of sarsaparilla (sliced), 2½ lbs.; boil it in water, 2 galls., down to one half; pour off the liquor, and strain it whilst hot; again boil the sarsaparilla in another gall. of water down to one half, and strain; evaporate the mixed liquors to 1 quart, and in these dissolve of white sugar, 8 oz.; lastly, when the syrup has cooled, add to it of rectified spirit, 2 fl. oz.

2. (Ph. E. & Ph. L. 1836.) Sarsaparilla (sliced), 15 oz.; boiling water, 1 gall.; macerate for 24 hours, boil to 2 quarts, strain, add of sugar, 15 oz., and boil to a syrup.

3. (Wholesale.) Take of extract of sarsaparilla, 3 lbs.; boiling water, 3 quarts; dissolve, strain, and add of white sugar, 12 lbs. Alterative and tonic.—*Dose*, 2 to 4 dr. See SARSAPARILLA.

Syrup of Sarsaparil'la, Compound. *Syn.* SYRUPUS SARZÆ COMPOSITUS, L.; SIROP DE CUISINIER, Fr. *Prep.* (Ph. U. S.) Sarsaparilla (bruised), 2 lbs.; guaiacum wood, rasped, 3 oz.; damask roses, senna, and liquorice root, bruised, of each, 2 oz.; diluted alcohol (proof spirit), 10 wine pints (1 gall. imperial); macerate for 14 days, express, filter through paper, and evaporate in a water-bath to 4 wine pints (3½ pints imperial); then add of white sugar, 8 lbs.; and, when cold, further add of oils of sassafras and aniseed, of each, 5 drops, and oil of partridge-berry (*Gaultheria procumbens*), 3 drops, previously triturated with a little of the syrup.

Obs. This is an excellent preparation, but the rose leaves might be well omitted.—*Dose*, ½ fl. oz. three or four times a day, as an alterative, tonic, and restorative. The syrup of the P. Cod. is made with water instead of spirit, and is inferior as a remedy to the preceding.

Syrup of Sarsaparilla, Ioduretted. *Syn.* SYRUPUS SARZÆ IODURETI (*Ricord*), L. *Prep.* Syrup of sarsaparilla, 31 parts; iodide of potassium, 1 part.

Syrup of Senega. *Syn.* SYRUPUS SENEGÆ (U. S.), L. *Prep.* Senega in moderately fine powder, 4 troy oz.; sugar, 15 troy oz.; proof spirit, 2 pints (o. m.). Introduce the senega into a percolator, and pour on the proof spirit; when finished, evaporate the percolate by a water-bath at 160° F. to 8 oz. (o. m.); filter, add the sugar, dissolve by a gentle heat, and strain whilst hot.

Syrup of Sen'na. *Syn.* SYRUPUS SENNÆ (B. P.,

Ph. L. & E.), L. *Prep.* 1. (Ph. L.) Take of senna, 3½ oz.; fennel seed (bruised), 10 dr.; boiling water, 1 pint; macerate for 6 hours with a gentle heat; then strongly press out the liquid through linen, and dissolve in it of manna, 3 oz.; next add this solution to treacle, 3 lbs., previously evaporated over a water-bath until a little of it, on being cooled, almost concretes, and stir them well together.

2. (Ph. E.) Senna, 4 oz.; boiling water, 24 fl. oz.; infuse, strain, add of treacle, 48 oz., and evaporate to a proper consistence. Aperient.—*Dose*, 1 to 4 dr.

3. (B. P.) Senna (broken small), 8 oz.; oil of coriander, 1½ minims; refined sugar, 12 oz.; distilled water, 50 oz., or a sufficiency; rectified spirit, 1½ oz.; digest the senna in three quarters of the water 24 hours at a temperature of 120°, press, and strain; digest the marc in the remainder of the water 6 hours, press, and strain; evaporate the mixed liquors to 5 oz.; when cold add the rectified spirit containing the oil of coriander; filter, and wash the filter with water to make up to 8 oz.; add the sugar and dissolve with gentle heat. The product should weigh 21 oz., and measure 16 oz. Sp. gr. 1·310.—*Dose*, 1 to 2 dr.

Syrup of Senna with Manna. *Syn.* SYRUPUS SENNÆ CUM MANNÂ (Ph. G.), L. *Prep.* Infuse 10 oz. of senna leaves and 1 oz. of bruised fennel seeds for some hours in 205 pints of hot water; strain, and dissolve in the strained liquor 15 oz. of manna. Pour off 5½ oz. (by weight) of liquid from the sediment, and dissolve it in 3 lbs. 2 oz. of sugar.

Syrup, Simple. *Syn.* SYRUPUS (B. P., Ph. L.), S. SIMPLEX (Ph. E. & D.), L. *Prep.* 1. (Ph. L.) White sugar, 3 lbs.; distilled water, 1 pint; dissolve by a gentle heat.

2. (Ph. E. & Ph. L. 1836.) Pure sugar, 10 lbs.; boiling water, 3 pints.

3. (Ph. D.) Refined sugar (in powder—crushed), 5 lbs.; distilled water, 1 quart.

4. (B. P.) Refined sugar, 5 lbs.; distilled water, 2 pints; dissolve the sugar in the water with the aid of heat, and when cool add water to make the product weigh 7½ lbs. Sp. gr. 1·33.

5. (Wholesale.) Finest double refined sugar, 44 lbs.; distilled water, 2¾ galls.; make a syrup.

Obs. This preparation should be as white and transparent as water. Used as capillaire, &c., and to give cohesiveness and consistence to pulverulent substances in the preparation of electuaries, pills, &c.

Syrup of Snails. *Syn.* SYRUPUS LIMACIBUS (P. Cod.), L. *Prep.* Vine snails, deprived of their shells and of the black portions, and cut up, 2 oz.; wash in cold water, and then boil with 10 oz. of water to 7 oz.; then add 10 oz. of sugar.

Syrup of Soapwort. *Syn.* SYRUPUS SAPONARIÆ, L. From the root, the same as SYRUP OF COLTSFOOT.

Syrup of Squills. *Syn.* SYRUPUS SCILLÆ (B. P., Ph. E. & D.), L. *Prep.* 1. (Ph. E.) Vinegar of squills, 3 pints; white sugar (in powder), 7 lbs.; dissolve by a gentle heat.

2. (Ph. D.) Vinegar of squills, 8 fl. oz.; refined sugar (in powder), 1 lb.; dissolve.

3. (B. P.) Vinegar of squills, 20 parts; refined sugar, 40 parts; dissolve with the aid of heat.—*Dose*, ½ to 1 dr.

4. (Wholesale.) Take of vinegar of squills (perfectly transparent), 14 lbs.; double refined sugar, 28 lbs.; dissolve in a stoneware vessel, in the cold, or at most by a very gentle heat.

Obs. This syrup, like the last, should be as clear as water, and nearly colourless.—*Dose*, 1 to 2 fl. dr., as an expectorant; in chronic coughs and asthma. In large doses it proves emetic.

Syrup of Squills, Compound. *Syn.* HIVE SYRUP; SYRUPUS SCILLÆ COMPOSITUS, L. *Prep.* (Ph. U. S.) Squills and senega, of each (bruised), 5 oz.; water, ½ gall.; boil to a quart; add of sugar, 4½ lbs.; evaporate to 3 pints, or a proper consistence, and dissolve in it, whilst hot, of potassio-tartrate of antimony (in powder), 1 dr.

Obs. This syrup is a popular expectorant in the United States, where it is known as hive syrup.—*Dose.* As an expectorant, 20 to 30 drops for adults; for children, 5 to 10 drops; in croup, 10 drops to ½ fl. dr., repeated until it vomits.

Syrup of Stramonium. *Syn.* SYRUPUS STRAMONII, L. From the tincture, as SYRUP OF BELLADONNA.

Syrup of Strychnia. *Syn.* SYRUPUS STRYCHNIÆ (P. Cod.), L. The Paris Codex orders a syrup containing ¼ gr. of sulphate of strychnia in 1000 gr. of syrup.

Syrup of Subchloride of Iron. *Syn.* SYRUPUS FERRI SUBCHLORIDI, L.; SYRUP OF FERROUS CHLORIDE. Iron wire, 300 gr.; hydrochloric acid, 2 oz.; citric acid, 10 gr.; water, 10 dr.; syrup, a sufficiency. Dissolve the iron in the hydrochloric acid with a gentle heat, add the citric acid, and filter into 10 oz. of the syrup; wash the filter with the remaining water; make up to 1 pint with syrup.

Syrup, Sudorific. *Syn.* SYRUPUS SUDORIFICUS (*Foy*), L. *Prep.* Sarsaparilla, 6 oz.; guaiacum raspings, 6 oz.; water, 3 pints. Macerate for 24 hours, evaporate to 1½ pints, strain, and make into a syrup with 2½ lbs. of sugar.

Syrup of Sulphate of Iron. *Syn.* SYRUPUS FERRI SULPHATIS (*Willis*), L. *Prep.* Sulphate of iron, 1 dr.; water, 2 dr.; syrup, 16 oz.

Syrup of Sulphate of Quinine. *Syn.* SYRUPUS QUININÆ SULPHATIS (P. Cod.), L. *Prep.* Dissolve 30 gr. of sulphate of quinine in 4 dr. of water with ½ dr. of dilute sulphuric acid, and mix the solution with 13 oz. of white syrup.

Syrup of Sulphuret of Potassium. *Syn.* SYRUPUS POTASSII SULPHURETI (P. Cod.), L. *Prep.* Liver of sulphur, 8 gr.; water, 16 gr.; syrup, 1 oz.

Syrup of Superphosphate of Iron. *Syn.* SYRUPUS FERRI SUPERPHOSPHATIS (*Mr Greenish*), L. *Prep.* Superphosphate of iron, 2 scruples; simple syrup, 1 fl. oz.

Syrup, Symphytic. *Syn.* SYRUPUS SYMPHYTI, L.; BOYLE'S SYRUP (Ph. E. 1745). *Prep.* Fresh comfrey root, ½ lb.; plantain leaves, ½ lb.; bruise, express the juice, boil to half, and make a syrup with an equal weight of sugar.

Syrup of Tannin. *Syn.* SYRUPUS TANNINI (*Foy*), L. *Prep.* Tannin, 2 oz.; water, 16 oz.; sugar, 32 oz.

Syrup of Tar. *Syn.* SYRUPUS PICIS (P. Cod.),

L. *Prep.* Tar water, 5½ oz.; sugar, 10 oz. Dissolve by water-bath, and filter through paper.

Syrup of Tar. *Syn.* SYRUPUS PICIS LIQUIDÆ. *Prep.* Washed wood-tar, 6 parts; boiling water, 50 parts; stir for 15 minutes; set aside 36 hours, stirring from time to time; decant, filter, and add 60 parts sugar.

Syrup of Tartaric Acid. *Syn.* SYRUPUS ACIDI TARTARICI (P. Cod.), L. *Prep.* Tartaric acid, 1 oz.; water, 2 oz.; sugar, 6 lbs. 1 oz. Dissolve in the cold.

Syrup of Tartrate of Manganese. *Syn.* SYRUPUS MANGANESII TARTRATIS, L. Made with tartrate of manganese, as syrup of malate of manganese.

Syrup of Tolu'. *Syn.* BALSAMIC SYRUP; SYRUPUS TOLUTANUS (B. P., Ph. L., E., & D.), L. *Prep.* 1. (Ph. L.) Balsam of Tolu, 10 dr. (1 oz.—Ph. D.); boiling distilled water, 1 pint; boil in a covered vessel for ½ an hour, frequently stirring, then cool, strain, and dissolve in the liquor sugar, 2½ lbs.

2. (Ph. E.) Simple syrup (warm), 2 lbs.; tincture of Tolu, 1 oz.; mix by degrees, and agitate them briskly together in a closed vessel.

3. (B. P.) Balsam of Tolu, 1¼ parts; sugar, 32 parts; water, 20 parts; boil the balsam half an hour, adding water when required; when cold make up to 16 parts, filter, add the sugar, and dissolve. The product weighs 48 parts, and measures 36 parts. Sp. gr. 1·33.—*Dose*, 1 to 2 dr.

4. (Wholesale.) To warm water, 28 lbs., add tincture of Tolu gradually, until it will bear no more without becoming opaque; then cork down the bottle, and occasionally agitate until cold; when quite cold filter it through paper, and add of the finest double-refined sugar, 44 lbs.; lastly, promote the solution, in a closed vessel, by a gentle heat in a water-bath.

Obs. This syrup should be clear and colourless as water; but, as met with in the shops, it is usually milky. It is strange that the London College should have omitted from their formula the usual addition of rectified spirit, although this syrup, perhaps more than any other, would be benefited by it.

Syrup of Tolu is pectoral and balsamic.

Syrup of Valerian. *Syn.* SYRUPUS VALERIANÆ (P. Cod.), L. *Prep.* Infuse 1 lb. of bruised valerian in 4 lbs. of boiling water for six hours; strain and press; then pour upon the marc 2 lbs. more of boiling water, or q. s., so as to obtain 4½ lbs. of infusion, including the product of the first infusion; filter, and add 1 lb. of valerian water, and then dissolve in it, by the aid of a water-bath, 10 lbs. of sugar.

Syrup of Vanilla. *Syn.* SYRUPUS VANILLÆ. *Prep.* Vanilla, 2 oz.; white sugar, 18 oz.; water, 9 oz. Beat the vanilla with a few drops of spirit, then with part of the sugar, and water q. s. to form a soft paste; add the rest of the sugar and water, and digest for 18 or 20 hours in a glass vessel placed in a water-bath. Strain and clarify with white of egg if required.

Syrup, Velno's Vegetable. According to Dr Paris and Sir B. Brodie, this celebrated nostrum is prepared as follows:—Young and fresh burdock root, sliced, 2 oz.; dandelion root, 1 oz.; fresh spearmint, senna, coriander seed, and bruised liquorice root, of each, 1½ dr.; water, 1½ pints; boil down gently to a pint, strain, add of lump sugar, 1 lb.; boil to a syrup, and, lastly, add a small quantity of corrosive sublimate, previously dissolved in a little spirit. Used as an alterative and purifier of the blood.

Syrup of Vin'egar. *Syn.* SYRUPUS ACETI, L. *Prep.* (Ph. E.) Take of vinegar (French in preference), 11 fl. oz.; white sugar, 14 oz.; and make a syrup.—*Dose*, 1 dr. to 1 fl. oz.; as an expectorant in coughs and colds, or, diffused through any mild diluent, as a drink in fevers. A more agreeable preparation is that of the P. Cod., made by dissolving 30 parts of sugar in 16 parts of raspberry vinegar.

Syrup of Vi'olets. *Syn.* SYRUPUS VIOLARUM, SYRUPUS VIOLÆ (Ph. L. & E.), L. *Prep.* 1. (Ph. L.) Macerate violet flowers, 9 oz., in boiling water, 1 pint, for twelve hours; then press, strain, and set aside the liquid, that the fæces may subside; afterwards complete the process with sugar, 3 lbs., and rectified spirit, 2½ fl. oz. (or as much of each as may be necessary), in the way which has been ordered concerning syrup of cochineal.

2. Violet flowers, 100 parts; alcohol, 50 parts; digest eight hours, press, and make up to 100 vol. with water; filter. To 1 part of the filtered product add 9 parts of strong simple syrup (*C. Bernbeck*).

3. (Ph. E.) Fresh violets, 1 lb.; boiling water, 2½ pints; infuse for 24 hours in a covered vessel of glass or earthenware, strain off the liquor (with gentle pressure), filter, and dissolve in the liquid white sugar, 7½ lbs.

4. (Wholesale.) From double-refined white sugar, 66 lbs.; 'anthokyan' (the expressed juice of violets, defecated, gently heated in earthenware to 192° F., then skimmed, cooled, and filtered; a little spirit is next added, and the next day the compound is again filtered), 11 lbs.; water, 22 lbs., or q. s.; dissolve in earthenware.

Uses. Syrup of violets is gently laxative.—*Dose.* For an infant, a teaspoonful.

Obs. Genuine syrup of violets has a lively violet-blue colour, is reddened by acids, turned green by alkalies, and both smells and tastes of the flowers. It is frequently used as a test. A spurious sort is met with in the shops, which is coloured with litmus, and slightly scented with orris root. The purest sugar, perfectly free from either acid or alkaline contamination, should alone be used in the manufacture of this syrup. The Ph. E. orders the infusion to be strained without pressure; and the P. Cod., and some other Ph., direct the flowers to be first washed in cold water.

Syrup of Wild Cherry Bark. *Syn.* SYRUPUS PRUNI VIRGINIANÆ (U. S.), L. *Prep.* 1. Moisten 5 troy oz. of coarsely powdered bark of wild cherry and water; let it stand 24 hours, then put it into a percolator, adding water till 16 oz. (o. m.) of liquid are obtained. To this add 2½ troy lbs. of sugar in a bottle, and agitate until it is dissolved.

2. (B. P. C.) Take of wild cherry bark, in No. 20 powder, 3 oz.; refined sugar, in coarse powder, 15 oz.; glycerine, 1½ fl. oz.; distilled water, a sufficient quantity. Moisten the powder with dis-

tilled water, and macerate for 24 hours in a closed vessel, then pack it in a percolator, and gradually pour distilled water upon it until 9 fl. oz. of percolate are obtained. Dissolve the sugar in the liquid by agitation, without heat, add the glycerine, strain, and, if necessary, pour sufficient distilled water over the strainer to produce one pint of syrup.—*Dose*, ¼ to 2 fl. dr.

Syrup, Wilks's. See SYRUP OF GARLIC, COMPOUND.

Syrup of Worm'wood. *Syn.* SYRUPUS ABSINTHII, L.; SIROP D'ABSINTHE, Fr. *Prep.* (P. Cod.) Tops of wormwood (dried), 1 part; boiling water, 8 parts; infuse for 12 hours, strain with expression, and dissolve in the liquor twice its weight of sugar. Bitter, tonic, and stomachic. —*Dose*, 1 to 3 fl. dr.

Syrups for Aërated Waters. 1. *Lemon Syrup.*
a. Dissolve 1 oz. of citric acid in 4 oz. of water, and add to 9 pints of simple syrup; also add 4 fl. oz. of mucilage of acacia, and half a fluid ounce of tincture of lemon.

b. Grate off the yellow rind of lemons, and beat it up with a sufficient quantity of granulated sugar. Express the lemon juice; add to each pint of juice 1 pint of water and 3½ lbs. of granulated sugar, included that rubbed up by the rind. Warm until the sugar is dissolved, and strain.

c. Dissolve 6 dr. of tartaric acid and 1 oz. of gum-arabic in pieces in 1 gall. of simple syrup, then flavour with 1½ fl. dr. of best oil of lemon. Or flavour with the saturated tincture of the peel in Cologne spirits.

2. *Orange Syrup.* *a.* To be prepared from the fruit in the same manner as *b*, Lemon Syrup.

b. Dissolve 6 dr. of citric acid in 1 gall. of simple syrup, and add 2 fl. dr. of fresh oil of orange in 2 oz. of alcohol, or, instead of the alcohol solution of the oil, use the saturated tincture, obtained by macerating the fresh peel for ten days in sufficient Cologne spirits to cover. The lemon and orange syrups made from the fruit, after being strained, may be diluted with an equal bulk of simple syrup. One dozen of the fruit is sufficient to make 1 gall. of finished syrup.

3. *Vanilla Syrup.* See SYRUP.
4. *Syrup of Coffee.* See SYRUP.
5. *Strawberry and Raspberry Syrups.* Mash the fresh fruit, express the juice, and to each quart add 3½ lbs. of granulated sugar. The juice, heated to 180° F. and strained or filtered previous to dissolving the sugar, will keep for an indefinite time. See also STRAWBERRY ESSENCE, FACTITIOUS.

6. *Pine-apple Syrup.* Expressed juice of pine-apple, 1 pint; sugar, 2 lbs. Boil gently, and when cold, filter.

7. *Nectar Syrup.* Mix 3 parts of vanilla syrup with 1 part each of pine-apple and lemon syrup.

8. *Sherbet Syrup.* Mix equal parts of orange, pine-apple, and vanilla syrup.

9. *Grape Syrup.* Mix ¼ pint of brandy, ½ oz. of tincture of lemon, and sufficient tincture of red sanders, with 1 gall. of syrup.

10. *Cream Syrup.* Condensed milk, 1 pint; water, 1 pint; sugar, 1½ lbs. Heat to boiling, and strain.

11. *Orgeat Syrup.* Cream syrup and vanilla

syrup, of each, 1 pint; oil of bitter almonds, a minims.

12. *Ginger Syrup.* Syrup, 7½ fl. oz.; essence of ginger (1 part of ginger to 4 parts of spirit), ¼ oz.

13. *Syrup of Chocolate.* Chocolate, 8 oz.; syrup, sufficient; water, ¼ pint; white of one egg. Grate the chocolate, and rub it in a mortar with the egg. When thoroughly mixed, add the water gradually, and triturate till a uniform mixture is obtained. Finally, add syrup to 4 pints, and strain.

Syrups, Mineral Water. The following American forms are taken from Remington's 'Practice of Pharmacy.'

Vanilla. Fluid extract of vanilla, 2 oz.; syrup, to make 32 oz.

Ginger. Tincture of ginger, 4 oz.; syrup, to make 128 oz.

Lemon. Solution of citric acid (1 in 10), 3 oz.; spirit of lemon, 1½ oz.; syrup, 8 pints; tincture of curcuma, enough to colour.

Orange. Oil of orange, 10 minims; citric acid, 120 gr.; syrup, 64 oz.

Strawberry. Strawberry juice, 32 oz.; sugar, 128 oz.; water, 32 oz.

Raspberry. Same as strawberry, using raspberry juice.

Pine-apple. Same as strawberry, using pine-apple juice.

Nectar. Vanilla syrup, 40 parts; pine-apple syrup, 8 parts; strawberry syrup, 16 parts.

Chocolate. Best chocolate, 8 oz.; sugar, 64 oz.; water, 32 oz.

Coffee. Coffee, 8 oz.; boiling water, 8 pints; sugar, 112 oz. Make an infusion, filter, and add the sugar.

TABASHEER. A deposit chiefly composed of silica, found in the joints of the bamboo. When dry it is opaque, but possesses the property of becoming transparent when placed in water. Its deposition in the nodes and joints of the bamboo appears to be due to a diseased condition of these parts. Tabasheer is much and unduly prized by the natives of India as a tonic and constitutional restorative, and is chewed mixed with betel. It has the least refractive power on light of any body known.

TABELLÆ. *Syn.* TABLETS. Small discs variously medicated, and weighing from 2 to 10 gr. Any simple substance may be used as the basis, such as sugar and gum, or black currant paste; but cocoa or chocolate is generally preferred. All the drugs must be intimately incorporated, so as to form a stiff paste; this is rolled into an even sheet, and cut by means of a punch into tablets of the required weight. The following method of preparing tablets is described by H. Wyatt :—The cacao and other ingredients, including the medicine to be administered, are rubbed together in a mortar, massed, in the same way as a pill-mass, with the liquid excipient, and cut into pills on a pill machine. Each pill is then taken, dusted with a powder of equal parts powdered sugar and arrowroot to prevent sticking, and placed in a tube of brass or wood standing vertically on a tile, an accurately fitting piston of wood giving a round form to the lozenge on being forced down the tube on the top of the pill.

The tablets, Mr Wyatt says, may be turned out equally well without the mould, by simply placing the mass on the cutter of the pill machine after piping, and pressing down the upper cutter upon it, oblong or square tablets resulting, according to the amount of mass used.

For most medicines this process answers admirably, yet there are some which could be administered in lozenge form were it not for their nauseous taste, which requires an amount of cacao and sugar to disguise scarcely compatible with the dimensions of an ordinary lozenge. In such cases Mr Wyatt recommends glycyrrhizin, the glucoside from liquorice root, and saccharin as substitutes for the sugar. He finds, for example, that 5 gr. of antipyrin are rendered nearly unnoticeable by ½ gr. of glycyrrhizin, and that the intense bitterness of strophanthus is covered by the addition of ¼ gr. of saccharin to every 5 minims of tincture in the tablet.

Tablets sweetened with saccharin may be used freely in diabetes and other diseases in which the administration of ordinary sugar lozenges would be attended with injurious results.

Another variation is in the use of a warm mortar so as to melt the cacao, in which state the powders can be more easily incorporated. Other cases in which medicines may be administered in the form of tabellæ may be suggested by the following formulæ:

Tabellæ Acidi Arseniosi. Trituration of arsenious acid (1 in 100), 48 gr.; cacao, 70 gr.; saccharin, 1 gr.; tragacanth powder, 24 gr.; rectified spirit, 30 minims; essence of vanilla (1 in 10), 24 minims; distilled water, 30 minims. Place the cacao in a warm mortar, when liquefied add the powders, previously well rubbed together, and after the mass has set powder it with the help of the spirit and essence of vanilla, massing with the distilled water and cutting into 48 tablets. Each contains $\frac{1}{100}$ gr. arsenious acid, equal to 1 minim of liquor arsenicalis.

Tabellæ Aconiti. Tincture of aconite, 60 minims; cacao, 170 gr.; tragacanth powder, 20 gr.; saccharin, ½ gr.; distilled water, 40 minims. Add the tincture to the cacao, melted, stir until the spirit is evaporated. Add the powders, mass with water, and cut into 24 tablets. Each contains 2½ minims tincture of aconite.

Tabellæ Antipyrin. Antipyrin, 120 gr.; cacao, 120 gr.; tragacanth powder, 24 gr.; glycyrrhizin, 8 gr.; rectified spirit, a sufficiency; distilled water, 40 minims. Make as tabellæ acidi arseniosi, massing with the water in which the glycyrrhizin has been dissolved, and cut into 24 tablets. Each contains 5 gr. antipyrin.

Tabellæ Belladonnæ. Tincture of belladonna, 120 minims; cacao, 170 gr.; tragacanth powder, 20 gr.; saccharin, ¼ gr.; distilled water, a sufficiency. Make as tabellæ aconiti, and divide into 24 tablets. Each contains 5 minims of tinct. belladonnæ.

Tabellæ Caffeinæ. Caffeine, 24 gr.; cacao, 160 gr.; tragacanth powder, 24 gr.; saccharin, 2 gr.; essence of vanilla (1 in 10), 12 minims; distilled water, 50 minims. Make as tabellæ acidi arseniosi,

and divide into 24 tablets. Each contains 1 gr. caffeine.

Tabellæ Cerii et Bismuthi. Cerium oxalate, 48 gr.; ammonio-citrate of bismuth, 48 gr.; cacao, 120 gr.; tragacanth powder, 24 gr.; saccharin, 1 gr.; cherry laurel water, 96 minims; distilled water, a sufficiency. Make as tabellæ acidi arseniosi, and cut into 24 tablets. Each contains 2 gr. cerium oxalate, 2 gr. bismuth ammonio-citrate.

Tabellæ Gelsemii. Fluid extract of gelsemium, 60 minims; cacao, 170 gr.; tragacanth powder, 24 gr.; saccharin, 2 gr.; distilled water, a sufficiency. Make as tabellæ aconiti, and cut into 24 tablets. Each contains 2½ minims extract of gelsemium.

Tabellæ Hamamelidis. Fluid extract of hamamelis, 120 minims; cacao, 170 gr.; tragacanth powder, 24 gr.; saccharin, 2 gr. Make as tabellæ aconiti, and cut into 24 tablets. Each contains 5 minims ext. hamamelidis.

Tabellæ Nitro-glycerini. Solution of nitroglycerin (1 gr. in 100 minims), 24 minims; cacao, 100 gr.; tragacanth powder, 18 gr.; saccharin, ¼ gr.; distilled water, a sufficiency. Make as tabellæ aconiti, and cut into 24 tablets. Each contains $\frac{1}{100}$ gr. nitro-glycerin.

Tabellæ Strophanthi. Tincture of strophanthus (1 in 20), 120 minims; cacao, 90 gr.; saccharin, 4 to 8 gr.; tragacanth powder, 24 gr.; essence of vanilla (1 in 10), 24 minims; distilled water, a sufficiency. Make as tabellæ aconiti, and cut into 48 tablets. Each contains 2½ minims of tinct. strophanthi.

TABES DORSALIS. A disease of the posterior column of the spinal cord, resulting in incoordination of the movements of the legs, sometimes spreading to the upper limbs, so that the patient in walking throws out the legs with a jerk, and brings them down violently upon the heels. Such patients are popularly called 'Stampers.'

TACAMAHACA. The resinous substance known by this name is believed to be obtained from the *Fagara octandra* (of Linnæus), a large tree growing in the island of Curaçoa and in Venezuela. The juice, which exudes from the tree spontaneously, becomes hard upon exposure. The commercial article varies greatly in size, sometimes occurring in irregular-shaped pieces of one or two inches in diameter, whilst at others it is met with no larger than a mustard seed. The pieces are usually of a reddish-brown or light yellow colour. They have a resinous agreeable odour, with a balsamic, bitter, slightly acrid taste. Tacamahaca dissolves partially in alcohol, and entirely so in ether and fixed oils. It is composed of resin and a little volatile oil. There are several varieties of this substance. At one time tacamahaca enjoyed a high reputation as an internal remedy for urinary and scorbutic affections. It is now only occasionally employed in medicine as an ingredient in ointments and plaster. Sometimes it enters into the composition of incense. In properties it is very similar to the turpentines.

TAFFETAS. Plasters on silk are occasionally so called. For TAFFETAS ANGLICUM, see COURT PLASTER; for TAFFETAS VESICANS, se VESICANTS.

TALC. *Syn.* FOLIATED TALC; URRUC. A transparent, foliated, siliceous magnesian mineral, flexible, but not elastic, found in Scotland, the Tyrol, and elsewhere. It is used as a cosmetic, to impart a silky whiteness to the skin; also in the composition of *rouge végétal*, and to give a flesh-like polish to alabaster figures, and in the extraction of grease from cloth, &c. A second and harder species of this mineral (FRENCH CHALK, SOAPSTONE, STEATITE; CRETA GALLICA) is employed as a crayon by carpenters, glaziers, and tailors, and forms the boot-powder of the bootmakers. Writing executed with it on glass, even after being apparently removed by friction, becomes again visible when breathed upon.

TALLOW. The name given to the fat separated from the membranes which enclose it in the suet or solid fat of ruminating animals, *e.g.* cow, sheep, &c. From this it is "rendered by heating it in an open copper; the heat bursts the membranes and liberates the fat as oil. The fat is separated from the waste matters by straining and pressing; what is left is called the 'greaves.'" Tallow is commonly purified by the following methods:—1. Melting it along with water, passing the mixed fluids through a sieve, and letting the whole cool slowly, when a cake of cleansed fat is obtained. 2. Keep the tallow melted for some time, along with about 2% of oil of vitriol, largely diluted with water, employing constant agitation, and allowing the whole to cool slowly; then to remelt the cake with a large quantity of hot water, and to wash it well. 3. Blow steam for some time through the melted fat. By either this or the preceding process a white, hard tallow may be obtained. Some persons add a little nitre to the melted fat, and afterwards a little dilute nitric or sulphuric acid, or a solution of bisulphate of potash. Others boil the fat along with water and a little dilute nitric or chromic acid, or a mixture of bichromate of potash and sulphuric acid, and afterwards wash it thoroughly with water. These methods answer well for the tallow or mixed fats of which ordinary candles are made.

Tallow converted into stearic acid by saponification is readily hardened and bleached if moderately pure. A mixture composed of 1 part of oxalic acid and 2000 parts of water is sufficient to bleach 1000 parts of stearic acid. The mode of operating is as follows:—Throw the stearic acid, cut into small pieces, into a vessel of cold water, and turn on steam; as soon as it has melted and assumed a turbid appearance, add the solution of oxalic acid, and boil the mixture. After boiling for ½ hour, long threads appear in the liquid; the liquid itself, which previously was of a greyish colour, becomes black, and the threads unite together. The boiling must now be discontinued, and the contents of the vessel, having been allowed to settle for three or four hours, must be drawn off into the coolers.

As commercial stearic acid frequently contains undecomposed tallow, as well as various foreign matters, this process is occasionally unsuccessful. To obviate the inconveniences connected with the use of an impure material, the candle may be run at two operations, as follows:—"The stearic acid, treated as above, is exposed for a month to

the sun, by which means the foreign matters are oxidised, and the bleached stearic acid acquires a dirty yellow colour; the oxidised blocks are then melted in water containing a little sulphuric acid, at about 150° F.; an addition of about 10% of good white wax (or spermaceti) is next made, and the whole boiled for half an hour; the white of an egg, previously beaten up in a quart of water, is then added to each 1 cwt. of stearic acid, the temperature of the mass having been reduced to 100°, or at most 120° F., after which the mixture is again well stirred and boiled, when the liquid soon becomes clear, which is seen by the dark colour it assumes.

"This mixture of stearic acid and wax or spermaceti is very suitable for forming the exterior coating of the candle; it is transparent, and of perfect whiteness, and, as it is devoid of oxalic acid, it does not injure the moulds; whilst at the same time, as it is less fusible than pure stearic acid, candles made with it do not run. The first coating may be run hot without crystallising; the interior of the candle, being protected from without against too sudden a cooling, may also be run somewhat hot; by this means the candle acquires a whiteness and a transparency which cannot be realised by other processes" ('Le Moniteur Industriel').

The sulphuric acid saponification of inferior tallow and other solid or semi-solid fatty bodies is now carried out on a very large scale for producing the cheaper varieties of 'stearine candles.' For this purpose the tallow or fat is mixed with 5% or 6% of concentrated sulphuric acid, and exposed to a steam heat of 350° to 360° F. After cooling, the black mass thus obtained crystallises to a tolerably solid fat, which is well washed once or twice with water, or high-pressure steam, and is then submitted to distillation by the aid of steam heated to about 560° F. The product of the distillation is beautifully white, and may be at once used for making candles. It is better, however, to first submit it to the processes of cold and hot pressing, whereby a much more solid fat is obtained.

Russian tallow has long been considered the best, but the imports from America and New Zealand bid fair to displace it.

Chemically, tallow consists of stearin, palmitin, and olein, the stearin predominating, but varying according to the species of the animal, its age, and food.

Good tallow should be white or nearly yellowish in colour; it should be practically free from water and mineral matters; its melting-point should not vary outside 115°—121° F.

An oil corresponding to the oil of lard, and called tallow oil, is obtained from tallow by pressure; it is very useful in the manufacture of the finer kinds of soap.

According to Pohl, palm oil, or, as it is often called, palm tallow, is most easily purified by simple exposure to a high temperature, provided it has been first well defecated. When quickly heated to about 465° F., and kept at that temperature for from 5 to 15 minutes, it is completely decolourised. The product has a slight empyreumatic odour, but this disappears by age, exposure, or saponification, and the natural violet odour of the oil returns. Cast-iron pans should be employed in the process, and should be only 2-3rds filled, and well covered during the operation.

By the distillation of sulphurated palm oil in closed vessels, at a heat ranging from 570° to 600° F., from 68% to 75% of a mixture of palmitic and palm-oleic acid passes over, of which 25% to 30% is colourless, hard, and crystalline, and the rest darker and softer (*Pohl*). The residuum in the still is a fine hard pitch. See CANDLES, FAT, GLYCERIN, OILS (Fixed), STEARIC ACID, &c.

TAMAR INDIEN. Dr H. Hager states this preparation is prepared as follows:—Tamarind pulp, 450 parts; sugar, 40 parts; sugar of milk, 80 parts; glycerine, 50 parts; senna powder, 50 parts; anise powder, 10 parts; oil-sugar of lemon, 3 parts; tartaric acid, 3 parts. Mix well to form a plastic mass, divide into oblong tablets 1½ inches by ⁷⁄₁₀ inch, and sprinkle with the following powders:—Cream of tartar, 5 parts; sugar, 35 parts; sugar of milk, 35 parts; tragacanth, 3 parts; tartaric acid, 2 parts; red saunders powder, 25 parts. Dry the tablets and wrap in tin-foil.

TAMA'RA. A mixed spice used in Italian cookery, consisting of cinnamon, cloves, and coriander, of each 2 parts; aniseed and fennel seed, of each 1 part.

TAM'ARIND. *Syn.* TAMARINDUS (B. P., Ph. L., E., & D.), L. The pulp or preserved fruit or pod of the *Tamarindus indica*, or tamarind tree.

Tamarind pulp is refrigerant and gently laxative. Mixed with water, it forms a grateful acidulous drink in fevers.—*Dose*, ¼ oz. and upwards.

Composition of the Tamarind (Vanquelin).

Citric acid .	.	9·40 per cent.
Tartaric acid	. .	1·55 „
Malic acid	. .	0·45 „
Bitartrate of potash	.	3·25 „
Sugar	. . .	12·50 „

Besides gum, vegetable jelly, parenchyma, and water.

TANKS. The difference between water-tanks and cisterns is not very obvious. Perhaps the definition the most nearly representing the general idea respecting them would be, that whilst both were receptacles for water, in tanks water would be stored for a longer period than in cisterns, which, supplying the constantly recurring needs of a house or a building of any kind, would be more frequently filled and emptied; although in many instances there might be no such distinction between them, and they might be regarded as synonymous. In whatever sense the terms may be understood, the remarks that follow as to their construction and management have a common application.

The materials for tanks and cisterns for the reception of water consist of stone, cement, brick, slate, iron, zinc, and lead. Of these materials, the best, although the dearest, is slate. The slate cistern, however, is occasionally liable to leakage, a defect mostly arising from the employment of mortar instead of cement for joining the slabs.

Wrought-iron cisterns and tanks, as well as the pipes in connection with them, are in very general use. The tendency of both to corrosion by the action of the water is considerably reduced by coating the insides with Portland cement or a vitreous glaze.

Mr Burn advocates the employment of a compound of tar, which, he says, most effectually protects them. Zinc, although cheap, and little acted upon by water, is seldom employed for cisterns. Dr Osborne says he has seen several cases of zinc poisoning caused by drinking water that had passed through zinc pipes, or had stood in zinc pails. Equal if not greater risk is incurred when drinking-water is kept in lead cisterns, or is made to run through lead pipes. In setting up cisterns or tanks made of stone or cement, common mortar must not be used, as lime is taken up and the water is rendered hard in consequence.

In seasons of drought it is by no means an unusual occurrence for many rural districts to lack a sufficiency of water, the limited supply of which entails considerable suffering, sometimes terminating fatally upon farm stock, with frequent loss to the owners. Few persons, perhaps, can form a correct idea of the immense quantity of water that in the shape of rain falls even in the least humid portions of our islands. If this rain, which is now allowed to run waste, were properly collected and stored, it would form a valuable resource in times and at places where there was a dearth or scarcity of this necessary element. Mr Bayley Denton, writing on this subject, says:—"Take an ordinary middle-class house in a village with stabling and outbuildings, the space of ground covered by the roofs will frequently reach 10 poles, while the space covered by a farm labourer's cottage and outbuildings will be 2½ poles.

"Assuming that the roof is slate, and the water dripping from it is properly caught by eave-troughing, and conducted by down-pipes and impervious drain-pipes into a water-tight tank sufficiently capacious to prevent overflow under any circumstances, and that by this method 20 inches of water from rain and dew are collected in the course of the year, the private houses will have the command of 28,280 galls., and the cottage 7070 galls. in a year. . . . A tank 16 feet long and 10 feet wide will hold 1000 galls. in every foot of depth, and where the water is not wanted for drinking, it need not be covered, except with a common boarded floating roof of half-inch boards fastened together. This floating roof keeps the water clean, and prevents evaporation."

Leakage of pipes of any kind into a cistern or tank should be particularly guarded against. Another important precaution claiming adoption is to see that the overflow-pipe is not directly connected with the sewer, for if it be the sewer gases will rise through it, and, being prevented escaping from the cistern because of its covering, will become absorbed by the water. To obviate this the overflow-pipe is curved, so as to form a syphon trap; but this device conduces to a sense of false security, since it mostly fails owing to the evaporation of the water in it, or to the gases forcing their way through it. The overflow-pipe, therefore, should never have direct communication with the sewer, but should always end above ground, and discharge over a trapped grating into it. For similar reasons the same tanks or cisterns should never supply the water used for culinary or drinking purposes, and also the water-closets.

To the water in the tanks attached to these latter some disinfectant substance should from time to time be added, more particularly during hot weather.

Unless a cistern be efficiently protected, particularly if it be placed in an exposed situation, various disgusting and filthy substances, such as the ordure of birds, cats, rats, dead insects, &c., will be liable to fall into it and foul its contents. This must not only be guarded against by the proper means, but even where the contamination may not be suspected, or likely to occur, the cistern should be frequently examined and periodically cleansed; part of the proper carrying out of which should consist in always running off the water remaining in it, and renewing it with fresh.

The London and General Water Purifying Company have adopted an excellent idea in connection with tanks and cisterns; they fit them with filters, so that the water drawn from the pipes shall have been submitted to filtration previous to delivery.

TAN'NATE. A salt of tannic acid.

TAN'NER'S BARK. The best is oak bark; but the bark of the chestnut, willow, larch, and other trees which abound in tannin, are also used for preparing leather.

TAN'NIC ACID. $C_{18}H_8O_7.CO_2H$. *Syn.* TAN, TANNIN, GALLO-TANNIC ACID†; TANNINUM, ACIDUM TANNICUM (B. P., Ph. L., D., & U. S.), L. A peculiar vegetable principle, remarkable for its astringency and its power of converting the skins of animals into leather.

Prep. From galls, in moderately fine powder, 240 grms.; by percolation, in a closed vessel, with sulphuric ether, 1800 c.c., that has been previously agitated with water, 150 c.c. After some time the percolated liquid will be found divided into two distinct portions, the lower and heavier one being a watery solution of tannic acid, and the upper one an ethereal solution of gallic acid and colouring matter. Fresh ether must be passed through the powder as long as the lower stratum of liquid continues to increase. The two fluids are now placed in a separate funnel and carefully separated, and after the heavier one has been well washed with ether, it is gently evaporated to dryness, *in vacuo*, or over sulphuric acid. The ether may be recovered unaltered from the ethereal solution by cautious distillation in a retort connected with a Liebig's condenser supplied with ice-cold water.—*Prod.* About 40%.

Prop., &c. Pure tannic acid is perfectly white, but as ordinarily met with it has a slight yellowish colour, owing to the action of the air; it is uncrystallisable, and usually occurs as shining scales; it possesses a powerful astringent taste, without bitterness; is freely soluble in water, less so in alcohol, and only very slightly in anhydrous ether; it reddens vegetable blues; when boiled with acids it assimilates water and splits into gallic acid and grape-sugar; when heated in the dry state it suffers decomposition, metagallic and pyrogallic acids being formed; it unites with the

bases, forming salts called **tannates**, which are characterised by striking a bluish black with ferric chloride (ink), and forming a white precipitate with gelatin.

E. Schmidt (' Chem. News,' from 'Bull. de la Soc. Chem. de Paris') gives the following comparative method of determining tanning materials, stating, preliminarily, " that the question to be solved is, knowing that a certain weight of pure tannin is required to obtain a certain result, how much of another tanning body, *e.g.* the extract of a wood, is required to produce the same result?"

He proposes a modification of Pibram's method with sugar of lead, the modification being as follows:

A. *Preparation of the Test Liquor.* Fifty grms. neutral acetate of lead are dissolved in 400 grms. of alcohol of 92%, and distilled water is added to make up 1 litre.

On the other hand, 1 grm. of tannin is dissolved in 40 grms. of alcohol of the same strength, and the solution is made up with water to the bulk of 100 c.c. This being done, 10 c.c. of the tannin solution are mixed with 20 c.c. of water, and heated to 60°. The lead liquor is then run into the hot solution from a burette, graduated to tenths of a c.c., so long as a precipitate is formed. At this temperature, and with these alcoholised liquids, the precipitate forms and settles rapidly. Iodide of potassium may be used as an indicator to show excess of lead, proceeding in the same manner as is done with ferrocyanide in titrating phosphates with nitrate of uranium. If we suppose that to precipitate 10 c.c. of the tannin solution 28° of the lead liquor have been required, then 2·8 c.c. of the latter = 0·10 grm. of tannin.

B. *Preparation of the Sample to be tested.* Suppose that chestnut bark is to be examined. It is coarsely powdered, and 10 grms. are mixed with an equal volume of washed sand, and exhausted with water at 50° or 60° C. The filtered liquid is evaporated to dryness in a water-bath in a tared porcelain capsule. After evaporation the capsule is weighed, which shows the yield of the bark in aqueous extract. This is taken up in 40 grms. of alcohol at 92°, and water is added to make up 100° c.c. The liquid is filtered if needful. In this manner the resinous, albuminoid, pectic, and gummy matters are got rid of.

C. *Titration.* The liquid thus prepared is divided into two parts. The first, one third of the entire volume, serves for direct determination of the acetate of lead. Suppose that a grm. of the dry extract of bark has required for 10 c.c. of the tannin liquor, in three successive experiments, 16°, 17°, and 16° of the burette, which corresponds to 57% of tannin. This figure 57 represents not only tannin, but every other substance capable of precipitating acetate of lead.

The tannin is then absorbed with bone-black, previously washed with hydrochloric acid, and dried at 100° C. in the following manner:—We act with bone-black upon the tannin liquor, and on a solution of pure tannin, prepared at a standard somewhat lower than that indicated for the extract by the first direct titration. In the present case this solution of tannin should be prepared at 55%.

From one and the same glass tube, about 1 cm. in diameter, we cut off two lengths of 20 cm. each, and we draw out each at one of its ends. The two tubes are fixed perpendicularly with the points downwards, and plugged with a little carded cotton. Into each is put 10 grms. of the bone-black, pouring into one of them the second part of the tannin liquor under examination, and into the other the same volume of the pure solution of pure tannin at 55%.

Twenty c.c. of the tannin liquor (which will be found to have retained its original brown colour in spite of the bone-black) are now heated to 60° C., and the standard lead liquor is added from the burette as before. Two successive trials show 16°=8° for 10 c.c. in place of the 16° found for 10 c.c. on direct titration. On the other hand, 20 c.c. of the solution of pure tannin require 14°, or 7° for 10 c.c. Thus we see that in the chestnut extract there is a certain quantity of matter which acts upon the standard lead solution like tannin, corresponding to 1° of the lead liquor, i. e. to 357 thousandths of a centigram of tannin ; 28°, therefore, correspond to 10 centigrams. The figure 57, obtained by direct titration, is, therefore, too high by 3·57%, and the extract contains 57 — 3·57 = 53·43% of tannin.

The best method of determining tannin is a modification of Löwenthal's method by von Schroeder (' Zeits. Anal. Chem.,' xxv, 121).

A solution of ferrous acetate is prepared from iron alum in the following proportions :—Iron ammoniacum alum, 48·2 grms. ; crystallised sodium acetate, 25 grms. ; acetic acid, containing 50% hydrate, 40 c.c., dissolved in one litre. 10 c.c. of this solution are added to 50 c.c. of the tannin solution to be determined, which is prepared, according to Schroeder, of 5 grms. per litre. The mixture is examined at the end of fifteen minutes, to see if the iron is in excess, which should always be the case, made up to 100 c.c., and filtered ; 20 c.c. of the filtrate, equal to 10 c.c. of the original solution, are then titrated with standard potassium permanganate after addition of 20 c.c. in digestive solution. The difference between the amount of iron added and that found on titration gives the amount precipitated by tannin (' Chem. News,' 1887).

Uses, &c. The value of substances containing tannin in the preparation of leather is well known. In its pure form it is used as an astringent in medicine ; internally, in diarrhœa, hæmorrhages, as a tonic in dyspepsia, &c. ; externally, made into a gargle, injection, or ointment.—*Dose,* 1 to 10 gr., in the form of pills or solution. See GALLIC ACID, &c.

TAN'NIN. See TANNIC ACID.

TAN'NING. When the skin of an animal, carefully deprived of hair, fat, and other impurities, is immersed in a dilute solution of tannic acid, the gelatin gradually combines with that substance as it penetrates inwards, forming a perfectly insoluble compound, which resists putrefaction completely ; this is tanned leather. In practice lime water is used for cleansing and preparing the skin, water acidulated with oil of vitriol for ' raising ' or opening the pores, and an infusion of oak bark, sumach, galls, wattle bark, or other astringent matter, as the source of tannic acid. The process itself is necessarily a slow one, as dilute solutions only can be safely used. Skins intended for the curriers, to be dressed for ' uppers,' commonly require about 3 weeks ; and ' thick hides' from 12 to 18 months.

Of late years various ingenious contrivances have been adopted, with more or less success, to hasten the process of tanning skins and hides. Among these may be mentioned the employment of stronger tan solutions ; the application of a gentle heat ; puncturing the skins to afford more ready access for the liquid to their interior parts ; and maceration in the tan liquor under pressure, either at once or after the vessel containing them has been exhausted of air by means of an air-pump. On the merit of these several methods it has been remarked that " the saturated infusions of astringent barks contain much less extractive matter, in proportion to their tannin, than the weak infusions; and when the skins are quickly tanned in the former, common experience shows that it produces leather which is less durable than leather slowly formed" (*Sir H. Davy*). " One hundred lbs. of skin, quickly tanned in a strong infusion of bark, produce 137 lbs. of leather ; while 100 lbs. slowly tanned in a weak infusion produce only 117½ lbs." " Leather thus highly (and hastily) charged with tannin is, moreover, so spongy as to allow moisture to pass readily through its pores, to the great discomfort and danger of persons wearing shoes made of it " (*Ure*).

According to Mr G. Lee, much of the original gelatin of the skin is wasted in the preliminary processes to which they are subjected, more especially the ' liming' and ' bating.' He says that 100 lbs. of, perfectly dry hide, cleaned from extraneous matter, should, on chemical principles, afford at least 180 lbs. of leather.

KID for gloves undergoes a ' tanning' operation, the chief chemical features of which are the removal of excess of lime, and opening the pores by steeping in a sour bran-bath, impregnation with aluminium chloride, and kneading with alum flour and the yolks of eggs.

MOROCCO LEATHER is prepared from goat or sheep skins, which, after the action of lime water and a bath of sour bran or flour, are slighly tanned in a bath of sumach. They are subsequently dyed, grained, polished, &c. Red morocco, which is dyed before tanning, is steeped first in alum or chloride of tin, and afterwards in an infusion of cochineal. Black morocco is dyed with acetate of iron, which combines with the tannic acid. The aniline dyes are now much used.

RUSSIA LEATHER is generally tanned with a decoction of willow bark, after which it is dyed and curried with the empyreumatic oil of the birch tree. It is the last substance which imparts to this leather its peculiar odour and power of resisting mould and damp. See LEATHER, TANNIC ACID, TAWING, &c.

TANTALIC ACID. *Syn.* TANTALIC ANHYDRIDE, COLUMBIC ACID. Rose believed this substance to be a dioxide, to which he gave the formula TaO_2; but the subsequent researches of Marignac, and the crystalline form of potassic tantalic fluoride, $2KF.TaF_5$, seem to show that it is to be regarded rather as Ta_2O_5.

TANTALUM. Ta=182. *Syn.* COLUMBIUM. A rare metal discovered by Hatchett, in 1801, in a mineral from Massachusetts, and by Ekeberg in 1808 in tantalite, a mineral found in Sweden; it exists in most of its ores in combination with oxygen. The yttrotantalite of Sweden is an ore which consists of tantalic acid combined with ferrous and manganous oxides.

TAPEWORM. See WORMS.

TAPEWORM CURE (*Bloch*, Vienna). Coarsely powdered pomegranate-root bark, 125 grms., boiled for half an hour in 800 grms. water. To this add solution of ammonia, 5 grms.; boil again for a quarter of an hour. Add kousso flowers, 25 grms. Again boil for a few minutes, and when cold add citric acid, 1 grm.; and alcohol, 30 grms. Press, filter, and set aside. The product should weigh about 500 grms. Klinger says this remedy is merely a concentrated essence of pomegranate-root bark, and contains neither ammonia nor citric acid.

Tapeworm Cure (*Jacoby*, Berlin). A box containing 20 grms. of kousso powder and directions for use (*Hager*).

Tapeworm Cure (*Mix*). (*a*) A mixture containing 3 decigrms. of sulphate of quinine, with a few drops of hydrochloric acid to dissolve it in 200 grms. of water. To be taken in the course of three days. (*b*) A box with 12 grms. kousso powder. A teaspoonful to be taken each morning in black coffee (*Schädler*).

Tapeworm Cure (*Richard Mohrmann*, Frankenberg, Saxony). This Mohrmann travels about in the fashion of the old charlatans to sell his medicines. These consist of two varieties, the first being 10 grms. of extract of male fern, the second a mixture of 8 grms. each of raspberry juice and castor oil. These remedies have been used for tapeworm for almost 100 years. The doctor's directions for use are to mix 30 grms. of the extract with the castor oil and raspberry compound, and 30 drops of the mixture to be taken every quarter of an hour until purging occurs.

Tapeworm Cure (*Mork*, Berlin). A decoction of about 110 grms. of pomegranate-root bark, yielding 400 grms. of liquid, and mixed with 1 grm. of extract of male fern. The directions order that on one day 1 or 2 table-spoonfuls of castor oil should be taken, a herring salad in the evening, and the following morning, after coffee, a third of the contents of the bottle, another third half an hour later, and the remainder in yet another half-hour (*Hager*).

Tapeworm Cure for Children and Adults (*E. Karig*, Berlin). Burnt oxide of copper, 1 grm.; cassia powder, 1½ grms.; sugar of milk, 10 grms. Divide in 24 powders (*Schädler*).

Tapeworm Pills, Laffon's, are compounded of the ethereal extract of the root of *Aspidium lonchitis, Asp. helveticum*, and *Asp. filix-mas*, together with the alcoholic extract of the flowers of *Achillea mutellina* and *moschata*, and the powder of the flowers of *Arnica doronicum* (*Wittstein*).

Tapeworm Pills, Peschier's. Ethereal extract and powder of the rhizome of male fern, of each, grm. 1·6; make 20 pills. Take 10 at night and 10 in the morning.

TAPIO'CA. *Syn.* TAPIOCA (Ph. E. & D.), L. The fecula of the root of *Janipha manihot* (*Jatropha manihot*, Linn.), which has been well washed in water, and dried on hot plates, by which it assumes the appearance of warty-looking granules.

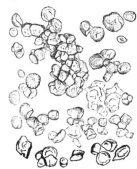

Microscopic appearance of tapioca.

Pure tapioca is insipid, inodorous, only slightly soluble in cold water, but entirely soluble in boiling water, forming a translucent and highly nutritious jelly. Its granules are muller-shaped, about $\frac{1}{1000}$ of an inch in diameter, and display very marked hilums. It is used in a similar manner to sago and arrowroot. See CASSAVA.

TAPS, WOODEN (to prevent their cracking). The taps are placed in mother-paraffin, heated to from 110° to 120°; by this means the water is eliminated from the wood, and the latter becomes thoroughly impregnated with paraffin. The taps are heated in this bath until all the aqueous vapour has been expelled, and are left in it, after the removal of the vessel from the fire, up to the very moment the paraffin begins to solidify. Wooden taps thus prepared are very durable, do not become soaked with liquids, keep very tight, and are not liable to become mouldy. The excess of paraffin is wiped off with care, and the taps are then rubbed clean with a piece of flannel (*Kopp,* 'Chemical News').

TAR. *Syn.* PIX LIQUIDA (B. P., Ph. L., E., & D.), L. A brown-black viscid liquid produced together with gaseous and watery products in the dry distillation of organic bodies and bituminous minerals. It is a mixture of various substances, acid, alkaline, and neutral, and its composition varies according (*a*) to the nature of the original substances placed under distillation, and (*b*) to the temperature at which the distillation is carried on. Tar obtained from vegetable substances has an acid reaction, but coal tar and the tar of animal substances is alkaline.

The principal groups of compounds contained in tars are liquid and solid hydrocarbons, alcohols, ethers, acids, and bases, together with resins and empyreumatic products of indeterminate composition. On subjecting tar to repeated distillation the more volatile and liquid hydrocarbons, together with the alcohols and ethers, pass over first, while the less volatile oils consist chiefly of acid and basic compounds, and the last portions which distil over contain the solid hydrocarbons. The residue left after about half the tar has distilled over—called pitch, and likewise asphalte when obtained from coal tar—also contains solid hydrocarbons, together with resinous compounds. The volatile constituents of tar can be separated by fractional distillation into portions of constant boiling-point. The oily portions are also treated with dilute acids to remove their basic constituents, and with alkalies to remove their acid constituents (*Watts*).

Tar, Barbadoes. *Syn.* PIX LIQUIDA BARBADENSIS, PETROLEUM BARBADENSE, PETROLEUM (Ph. L. & E.), L. "Black liquid bitumen, exuding spontaneously from the earth" (Ph. L.). Its medicinal properties are stimulant, diuretic, sudorific, and vermifuge.—*Dose*, 10 to 30 drops; in asthma, chronic coughs, tapeworm, &c. Externally, in chilblains, chronic and rheumatic pains, &c. See PETROLEUM.

Tar, Coal. The black liquid obtained in the destructive distillation of coal, peat, lignite, or bituminous shale. The more volatile portions are called *light oil* or coal naphtha, and consist mainly of benzene and its homologues, together with a number of bases of the formula $C_nH_{2n-5}N$. When obtained from cannel and Boghead coal it is chiefly composed of alcoholic hydrides, homologous with marsh-gas, together with olefines and homologues of benzene. The less volatile portion, or *dead oil*, of tar contains phenol, creosote, aniline, picoline, chinoline, and other bases; also naphthalene, anthracene, and other solid hydrocarbons.

The preparation of this tar from coal, shale, peat, &c., has attained great importance of late years for obtaining illuminating and lubricating oils; it is also a source of aniline colours, picric acid, lampblack, &c.

Tar, Wood (Stockholm tar). Chiefly prepared from the wood of *Pinus sylvestris* and by dry distillation.

The chief liquid constituents are methyl acetate, acetone, hydrocarbons (toluene, xylol, cumene), methol, eupione, creosote, and a number of indefinite oxidised substances. The solid portions consist for the most part of resinous matters which resemble colophony, paraffin, naphthalene, anthracene, chrysene, retene, &c.

It possesses powerful antiseptic properties, due to the creasote which it contains; hence it is much used in the preservation of wood and in shipbuilding.

TARAXACUM. See DANDELION.

TAR COLOURS. *Syn.* COAL-TAR COLOURS, ANILINE COLOURS, &c. Coal tar, the source of the aniline colours, consists of the oily fluid obtained in the destructive distillation of coal, during the manufacture of ordinary illuminating gas, and collected in a tank from the hydraulic main and condensers.

The composition of coal tar is highly complex, the most important constituents being, however, a series of homologous hydrocarbons obtained by distilling coal tar, and known as 'coal naphtha.' Naphtha, by rectification between 180° and 250° F. (82° and 121° C.), yields a light yellow oily liquid, of sp. gr. 0·88, the benzol of commerce.

By the action of a mixture of nitric and sul-

phuric acids on benzol, nitro-benzol, a heavy oily liquid with an odour of oil of bitter almonds, is obtained. In commerce this substance is made in large cast-iron pots, fitted with tight covers, and provided with stirrers worked by steam power. By means of pipes the reagents are admitted and the nitrous fumes are carried off, while the nitro-benzol and the spent reagents are drawn off from the bottom. The entire charge of benzol is first placed into the vessels, and the mixed acids are, as the reaction is very energetic, cautiously run in, the whole being well stirred throughout. When finished, the contents are drawn off, and the nitro-benzol collected, washed with water, and, if necessary, neutralised with a solution of soda. See BENZOL.

Nitro-benzol is converted into *aniline* in a similar apparatus, but it should be provided with means of admitting a current of superheated steam, and condensing the aniline as it distils over. Into the vessel iron borings are placed, and acetic acid and nitro-benzol cautiously run in as the reduction is violent, stirring well all the time. A current of superheated steam is passed through, and the aniline collected as it distils over as a pale, sherry-coloured, oily liquid, boiling at 184° C., and of sp. gr. 1·036. See ANILINE.

MAUVE, INDISINE, VIOLINE, PHENAMINE, the first-discovered coal-tar or aniline colour, was obtained by Perkin during some experiments directed towards the artificial formation of quinine, and was also first practically manufactured by him in 1856. Commercially, mauve is made as follows:

Aniline and sulphuric acid in proper proportions are dissolved in water in a vat by aid of heat, and when cold a solution of bichromate of potassium added, and the whole allowed to stand a day or two, when a black precipitate is obtained, which, after collecting on shallow filters, is washed and well dried. This black resinous substance is digested with dilute methylated spirit in a suitable apparatus, to dissolve out the mauve, and the major portion of the spirit distilled off. The mauve is precipitated from the aqueous solution left behind by hydrate of sodium, and after washing is either drained to a paste or dried.

The amount of mauve thus obtained is but small in comparison with the raw material, coal tar, as 100 lbs. of coal yield 10 lbs. 12 oz. of coal tar; 8½ oz. of mineral naphtha, 2½ oz. of benzol, 4½ oz. of nitro-benzol, 2¼ oz. of aniline, and ¼ oz. of mauve. Mauve is usually sent into the market in paste or solution, the expense of the crystals being heavy, and offering no corresponding advantages.

Other salts than the bichromate of potassium have been employed to convert aniline into mauve, as chloride of copper, permanganate of potassium, &c.; but experience has shown none to possess the same advantages as the bichromate of potassium.

MAUVEINE, the organic base of mauve or aniline purple, is a black crystalline powder, of the formula $C_{27}H_{24}N_4$, yielding a dull violet solution. The moment, however, mauveine is brought in contact with an acid, it turns a magnificent purple colour. The salts of mauveine form beautiful crystals possessing a splendid green metallic lustre, soluble very readily in alcohol, and less so in water. The commercial salt, or mauve, is the acetate, or sometimes the hydrochlorate.

MAGENTA. *Syn.* **ANILINE RED, ROSEINE, FUCHSINE, AZALEINE, SOLFERINO, TYRALINE.** Various processes have been proposed and patented for the preparation of this commercially important coal-tar colour. Amongst these processes are—

1. Gerber-Keller's. By this the aniline is treated with mercuric nitrate.

2. Lauth and Depouilly used nitric acid.

3. Medlock, Nicholson, and Messrs Girard and De Laire, in 1860, separately patented the use of arsenic acid. This process, being the one now almost exclusively employed, is thus described in Crace Calvert's work, 'Dyeing and Calico Printing,' edited by Messrs Stenhouse and Grove :— "The manufacture of magenta, as it is now conducted in the large colour works, is a comparatively simple process, the apparatus employed consisting of a large cast-iron pot set in a furnace, provided with means of carefully regulating the heat. It is furnished with a stirrer, which can be worked by hand or by mechanical means, the gearing for the stirrer being fixed to the lid, so that by means of a crane the lid may be removed, together with the stirrer and gearing. There is also a bent tube passing through the lid for the exit of the vapours, which can be easily connected or disconnected with a worm at pleasure. Lastly, there are large openings at the bottom of the pot, closed by suitable stoppers, so that the charge can be removed with facility as soon as the reaction is complete. Into this apparatus, which is capable of holding about 500 galls., a charge of 2740 lbs. of a concentrated solution of arsenic acid, containing 72% of the anhydrous acid, is introduced, together with 1600 lbs. of commercial aniline. The aniline selected for this purpose should contain about 35% of the toluidine.

"After the materials have been thoroughly mixed by the stirrer the fire is lighted, and the temperature gradually raised to about 360° F. In a short time water begins to distil, then aniline makes its appearance along with the water, and, lastly, aniline alone comes over, which is nearly pure, containing, as it does, but a very small percentage of toluidine. The operation usually lasts about eight or ten hours, during which time about 170 gallons of liquid pass over, and are condensed in the worm attached to the apparatus; of this about 150 lbs. are aniline. The temperature should not exceed 380° F. at any period during the operation. When this is complete, steam is blown in through a tube, in order to sweep out the last traces of the free aniline, and boiling water is gradually introduced in quantity sufficient to convert the contents into a homogeneous liquid. When this occurs the liquid is run out of the openings at the bottom into cisterns provided with agitators; here more boiling water is added in the proportion of 300 galls. to every 600 lbs. of crude magenta, and also 6 lbs. of hydrochloric acid. The mass is then boiled for four or five hours by means of steam pipes, the agitators being kept in constant motion. The solution of hydrochloride, arsenite, and arseniate

of rosaniline thus obtained is filtered through woollen cloth, and 720 lbs. of common salt added to the liquid (which is kept boiling) for each 600 lbs. of crude magenta. By this means the whole of the rosaniline is converted into hydrochloride, which, being nearly insoluble in the strong solution of arseniate and arsenite of sodium produced in the double decomposition, separates and rises to the surface; a further quantity is deposited from the saline solution on allowing it to cool and stand for some time. In order to purify the crude rosaniline hydrochloride it is washed with a small quantity of water, redissolved in boiling water slightly acidulated with hydrochloric acid, filtered, and allowed to crystallise."

If in the treatment of aniline with arsenic acid the latter be considerably beyond the proportion of aniline employed, VIOLET and BLUE dyes may be formed. The production of such has been patented by Girard and De Laire.

4. Laurent and Casthélaz have obtained aniline red direct from benzol, without the preliminary isolation of aniline. Nitro-benzol is treated with twice its weight of iron finely divided, and half its weight of concentrated hydrochloric acid. The colouring matter obtained by this process is said to be inferior in beauty to that procured from aniline.

5. Messrs Renard Brothers include in their patent the ebullition of aniline with stannous, stannic, mercurous, and mercuric sulphates, with ferric and uranic nitrates and nitrate of silver, and with stannic and mercuric bromides.

6. Messrs Dale and Curo's patent (dated 1860) consists in the treatment of aniline or hydrochlorate of aniline with nitrate of lead.

7. Mr Smith claims the ebullition of aniline with perchloride of antimony, or the action of antimonic acid, peroxide of bismuth, stannic, ferric, mercuric, and cupric oxides, upon hydrochlorate or sulphate of aniline, at the temperature of 180°.

Coupier's process for the manufacture of magenta without the use of arsenic acid is as follows :—He heats together pure aniline, nitrotoluene, hydrochloric acid, and a small quantity of finely divided metallic iron, to a temperature of about 400° F. for several hours. The pasty mixture soon solidifies to a friable mass resembling crude aniline red—ordinary commercial aniline.

The above processes are for the preparation of crude aniline red only. The crude colours contain some undecomposed aniline, mostly in the form of salts. They are also contaminated with tarry matters, some insoluble in water and dilute acids; others soluble in bisulphide of carbon, naphtha, or in caustic or carbonated alkalies. If, therefore, the crude red be boiled with an excess of alkali the undecomposed aniline is expelled, the acid which exists in the product being fixed. On treating the residue with acidulated boiling water the red is dissolved, while certain tarry matters remain insoluble. If now the boiling solution be filtered, and then saturated with an alkali, the colouring matter is precipitated in a tolerable state of purity. By redissolving the precipitated red in an acid, not employed in excess, a solution is obtained which frequently crystallises, or from which a pure red may be thrown down by a new addition of chloride of sodium or other alkaline salt.

Dr Hofmann and Mr Nicholson have demonstrated that pure aniline, from whatever source obtained, is incapable of furnishing a red dye, but that it does so when mixed with its homologue toluidine—toluidine by itself being equally incapable of yielding it. From this it will be evident that an aniline rich in toluidine is an essential condition for obtaining aniline red.

Magenta consists of brilliant large crystals, having a beautiful golden-green metallic lustre, and soluble in water to an intense purplish-red solution. It is a salt of a colourless base, rosaniline, which is prepared from magenta by boiling with hydrate of potassium, and allowing the solution to cool, when it crystallises out in colourless crystals, having the formula $C_{20}H_{19}N_3.H_2O$. All the salts of rosaniline have a splendid purple-red colour, and that usually met with as magenta is the hydrochlorate, although the nitrate, oxalate, and acetate are also to be obtained.

Sugar, previously dyed with magenta, is sometimes used as an adulterant of crystallised magenta. If present, the larger crystals of dyed sugar may be readily detected by their colour being paler at the edges, when the suspected sample is spread out on a sheet of white paper in the sunshine. One of the best methods of testing magenta is to make a comparative dyeing experiment with the sample under examination, and with one of known purity, using white woollen yarn.

From magenta or hydrochlorate of rosaniline a large number of colouring matters are produced, the most important of which will be briefly described below.

ANILINE BLACK. "Dissolve 20 parts of potassium chlorate, 40 parts of sulphate of copper, 16 parts of chloride of ammonium, and 40 parts of aniline hydrochloride, in 500 parts of water, warming the liquid to about 60°, and then removing it from the water-bath. In about 3 minutes the solution froths up and gives off vapours which strongly attack the breathing organs. If the mass does not become quite black after the lapse of a few hours it is again heated to 60°, and then exposed in an open place for a day or two, and afterwards carefully washed out till no salts are found in the filtrate. For use in printing, the black paste is mixed with a somewhat large quantity of albumen, and the goods after printing are strongly steamed. The paste can be pressed into moulds, and used as a substitute for Indian ink" (A. Müller). "Mix equal weights of aniline (containing toluidine), hydrochloric acid, and potassium chlorate, with a minute quantity of cupric chloride and a sufficient quantity of water, and leave the mixture to evaporate spontaneously, when a black powder will be obtained" (Rheineck).

ANILINE BLUE, or BLEU DE LYONS. This dye is prepared by heating a mixture of magenta, acetate of sodium, and aniline in iron pots, provided with stirrers, &c., in an oil-bath, to 370° F. (190° C.), and the excess of aniline distilled over. When a good blue has been obtained the heat is removed, and the thick treacly fluid purified. This is effected for the commoner varieties by

treating the crude product with hydrochloric acid, to dissolve all the excess of aniline, and the various red and purple impurities; but for the better qualities by mixing the crude product with methylated spirit and pouring the whole into water acidulated with hydrochloric acid, and then thoroughly washing the colouring matter that is precipitated, with water and drying.

This blue, like magenta, is a salt of a colourless base, which has been named Triphenyl-rosaniline, $C_{20}H_{16}(C_6H_5)_3N_3$. Aniline blue, or Lyons blue, is sent into the market either as a coarse powder or a coppery lustre, or in alcoholic solution; as it is insoluble in water, which necessitates it being added to the dye-bath in solution in spirit, a great drawback.

Mr Nicholson, by treating Lyons blue in the same manner as indigo is converted into sulpindigotic acid, has succeeded in rendering it soluble; dissolving in alkalies to form colourless salts, and decomposed by acids into its original blue colour. By a modification of this method 'NICHOLSON'S BLUE' is prepared, a fine soluble blue dye.

Another colouring matter, called Paris blue, or bleu de Paris, was obtained by heating stannic chloride with aniline for 30 hours at a temperature of 356° F. (180° C.). It is a fine pure blue, soluble in water, and crystallising in large blue needles with a coppery lustre.

Another method pursued in the manufacture of this colour on a large scale is carried out by allowing a mixture of a salt of rosaniline, with an excess of aniline, to digest at a temperature of 150° to 160° C. for a considerable time. A mixture of 2 kilogrammes of dry hydrochlorate of rosaniline, and 4 kilogrammes of aniline be employed, the operation is completed in 4 hours. The crude blue is purified by treating it successively with boiling water, acidulated with hydrochloric acid, and with pure water, until it is of the purest blue colour. 'Nicholson's blue' is obtained by digesting triphenyl-rosaniline monosulphonic acid (made by dissolving triphenyl-rosaniline hydrochloride in strong sulphuric acid, and heating the solution for five or six hours; on the addition of water, the acid is obtained as a dark blue precipitate, and dried at 100° C.), with a quantity of soda lye not quite sufficient for saturation, filtering the solution and evaporating. It is dried at 100° C. Wool dipped into a hot aqueous solution of Nicholson's blue, especially if borax or water-glass be added, extracts it in a colourless state, and holds it so fast that it cannot be washed out with water, but on dipping the wool thus prepared into an acid the salt is decomposed, and the colouring matter is set free.

ANILINE BLUE FOR PRINTING. Blumer-Zweefel gives the following process:—"Mix 100 parts of starch with 1000 parts of water, and add to it while warm 40 parts of potassium chlorate, 3 to 4 parts of ferrous sulphate, and 10 parts of sal-ammoniac. The well-mixed paste, when quite cold, is mixed with 70 parts of aniline hydrochloride, or an equivalent quantity of tartrate, and immediately used. The printed goods are oxidised, then passed through warm or faintly alkaline water, whereby the blue colour is developed."

VIOLET IMPERIAL. If the action of the aniline and magenta in the process of manufacturing aniline blue be stopped before it is finished, and the resulting product treated with dilute acid, a colouring matter called violet imperial is obtained. It is now, however, replaced by the Hofmann violets.

Mr Nicholson obtains another violet from aniline red, by heating it in a suitable apparatus to a temperature between 200° and 215° C. The resulting mass is exhausted with acetic acid, and the deep violet solution diluted with enough alcohol to give the dye a convenient strength. Aniline violet, although it resists the action of light to a very considerable extent, has been shown by Chevreul to be inferior in this particular to either madder, cochineal, or indigo.

HOFMANN VIOLETS. Primula, red violet 5 R. extra. On a large scale these violets are produced in deep cast-iron pots, surrounded by a steam jacket, and provided with a lid, having a perforation for distilling over the excess of reagents.

These vessels are charged with a solution of magenta in methylated or wood spirit, and iodide of ethyl or methyl, in proportions according to the shade required, and the whole heated by steam for five or six hours, when the excess of alcohol and iodide of ethyl is distilled over. The resulting product is dissolved in water, filtered, precipitated with common salt, and well washed. Like most of the other colours, Hofmann violets are salts of colourless bases. That of a red shade has a formula $C_{20}H_{19}(C_2H_5)N_3$; of a true violet shade, $C_{20}H_{17}(C_2H_5)_3N_3$; and of a blue shade of violet, $C_{20}H_{16}(C_2H_5)_4N_3$; but these and other methyl derivatives. They are all moderately fast on wool and silk, although less so on cotton, and as they can be produced in nearly every shade of violet, are in great use, having replaced most of the other violets.

The following processes have also been proposed for the production of aniline violet :

1. Oxidation of an aniline salt by means of a solution of permanganate of potassium (*Williams*).

2. Oxidation of an aniline salt by means of a solution of ferricyanide of potassium (*Smith*).

3. Oxidation of a cold and dilute solution of hydrochlorate of aniline by means of a dilute solution of chloride of lime (*Bolley, Beale,* and *Kirkmann*).

4. Oxidation of a salt of aniline by means of peroxide of lead under the influence of an acid (*Price*).

5. Oxidation of a salt of aniline in an aqueous solution of peroxide of manganese (*Kay*).

6. Oxidation of a salt of aniline by free chlorine or free hypochlorous acid (*Smith*).

DAHLIA. This is prepared from mauve and iodide of ethyl, in a manner analogous to that of the Hofmann violets, and is a purple-red violet. It is a good colour, but the expense precludes its general use.

BRITANNIA VIOLET. This is obtained in the same manner as the Hofmann violets, by acting on an alcoholic solution of magenta, with a thick, viscid, oily fluid of the formula $C_{10}H_{15}Br_3$, obtained by cautiously treating oil of turpentine with bromine. It is a beautiful violet, capable of being manufactured of every shade, from purple to blue, and most extensively used.

ALDEHYDE GREEN. Prepared by dissolving one part of rosaniline in three parts of sulphuric acid, diluted with one part of water, adding by degrees one and a half pint of aldehyde, and heating the whole on a water-bath until a drop put in water turns a fine blue. It is then poured into a large quantity of hot water containing three parts of hyposulphite of sodium, boiled and filtered. The filtrate contains the green, which can either be kept in solution or be precipitated by means of tannic acid or acetate of sodium. Like the other colours, this green is a salt of a colourless base containing sulphur, the formula of which is not known, and is principally used for dyeing silk, being very brilliant in both day and artificial light.

IODIDE GREEN. Produced during the manufacture of the Hofmann colours, and is now used for dyeing cotton and silk, its colour being bluer and more useful than that of aldehyde green. Iodide green, not being precipitated by carbonate of sodium, is usually sold in alcoholic solution.

PERKIN GREEN. This is also a magenta derivative, and a salt of a powerful colourless base. It resembles the iodide green, but is precipitated by alkaline carbonates and picric acid. This colour is used chiefly for calico printing, and is quite as fast as the Hofmann colours.

ANILINE GREEN. When treated with chlorate of potassium, to which a quantity of hydrochloric acid has been added, aniline assumes a rich indigo-blue colour. The same result occurs if the aniline be treated with a solution of chlorous acid. Similar blues have been obtained by Crace Calvert, Lowe, and Clift. Most of these blues possess the property, when subjected to the action of acids, of acquiring a green tint, called EMERALDINE. Calvert obtained this colour directly upon cloth by printing with a mixture of an aniline salt and chlorate of potassium, and allowing it to dry. In about 12 hours the green colour is developed. This colour may be converted into blue by being passed through a hot dilute alkaline solution, or through a bath of boiling soap.

ANILINE YELLOW. Amongst the secondary products obtained during the preparation of aniline red, there occurs a well-defined base of a splendid yellow colour, to which the name *chrysaniline* has been given. It is prepared by submitting the residue from which the rosaniline has been extracted to a current of steam for some time, when a quantity of the chrysaniline passes into solution. By adding nitric acid to the solution the chrysaniline is thrown down in the form of a difficultly soluble nitrate. The intimate relation between chrysaniline, rosaniline, and leucaniline has been shown by Hofmann.

Chrysaniline, $C_{20}H_{17}N_3$.
Rosaniline, $C_{20}H_{19}N_3$.
Leucaniline, $C_{20}H_{21}N_3$.

SAFRANINE. SAFRANINE T., PINK. This dyestuff is of a bright red-rose colour. Mené says it may be prepared commercially by treatment of heavy aniline oils successively with nitrous and arsenic acids; or two parts of the aniline may be heated with one of arsenic acid, and one of an alkaline nitrate for a short time, to 200° or 212° F. The product is extracted with boiling water, neutralised with an alkali, filtered, and the colour thrown down by common salt. According to a more recent and improved process, safranine is now made by oxidising a mixture of monamines and diamines. The aniline oil is converted into amido-azobenzene and amido-azo-orthotoluene, as in Mené's process, and this mixture is then heated with zinc and hydrochloric acid. The product of the reaction is then diluted with water, one molecular weight of toluidine hydrochloride is added, and the whole oxidised with potassium bichromate.

It comes into the market as a brown-red powder.

Besides the above products obtained from aniline, a series of colours have been obtained from phenol, or carbolic acid, another substance obtained from coal tar.

PICRIC ACID. TRINITRO-PHENOL. $C_6H_2(NO_2)_3$, OH. This is obtained by treating in a suitable apparatus, with proper precautions, carbolic acid with nitric acid. It is a pale yellow crystalline acid, melting at 122·5° C., forming dark orange explosive salts, and dyeing silk a fine yellow.

ISOPURPURATE OF POTASSIUM. GRENATE BROWN. GRENATE SOLUBLE. By treating picric acid with cyanide of potassium a very explosive salt is obtained, used to dye wool a dark maroon colour.

AURINE, or ROSOLIC ACID. $C_{19}H_{14}O_3$. This is obtained by heating a mixture of sulphuric, oxalic, and carbolic acids, and purifying the products. It is a beautiful reddish, resinous substance, with a pale green lustre, and yielding an orange-coloured solution, changed by alkalies to a splendid crimson. Owing to the difficulty in using it, however, it is not very extensively employed.

PEONINE, or CORALLINE. This dye is formed when rosolic acid and ammonia are heated to between 248° and 284° F. (120° to 140° C.). It is a fine crimson dye, forming shades similar to safranine on silk, but, owing to the bad effects of acids, not much used.

AZULINE. Prepared by heating coralline and aniline together. A coppery-coloured resinous substance, soluble in alcohol, and with difficulty in water, and dyeing silk a blue colour. The aniline blues, however, have superseded it to a great extent.

There are other substances obtained from coal tar that have been employed to form dyes, but of which we shall only refer to one—naphthalin. By treating this in exactly the same manner as benzol is converted into aniline, a solid crystalline white base, termed naphthylamine, is produced. From this substance a large series of dyes is obtained; the following may be given as an example.

DINITRONAPHTHOL, or MANCHESTER YELLOW. Alpha-naphthol treated at 212° F. (100° C.) with a mixture of sulphuric and nitric acids yields a nitro-compound, which is precipitated by water; or sulphuric acid alone is used, and the alpha-naphthol converted into the monosulphuric acid and then nitrated; then the sulpho-group is removed, and substituted by NO_2. Dinitronaphthol forms yellow needles of the formula $C_{10}H_5(NO_2)_2$, OH, melting at 138° C., insoluble in water. It is a strong acid, forming yellow or orange salts. The salt employed in commerce is the beautiful yellow crystalline calcium salt which dyes silk and wool a magnificent golden-yellow colour.

PRIMULINE. The trade name of the sodium salt of the monosulphuric acid of a complex base discovered by A. G. Green in 1887 (*vide* ' J. Soc. Chem. Ind.,' 1888, 179).

Prep. By heating paratoluidine with sulphur in the proportion of from 4 to 5 atoms of the latter to 2 molecules of the former. This process yields 2 bases, from one of which primuline is obtained.

Prop. A bright yellow powder, extremely soluble in water; dyes unmordanted cotton a primrose-yellow; capable of being diazotised within the fibre, and of combining with various amines and phenols.

Uses. It is largely employed in cotton dyeing on account of the great range of fast shades that may be obtained with it. Commercial preparations of it are put up under the names ' polychromine,' ' thiochromogen,' ' sulphine,' ' aureoline,' ' chameleon yellow,' ' carnoline,' &c. Quite recently (*vide* ' J. Soc. Arts,' Jan. 23rd, 1891) Messrs Green, Bevan, and Cross have applied primuline in a new method invented by them, which is likely to work a radical change in the colouring of the photographs of the future.

The diazo-compound of dehydrothiotoluidine and its condensed derivatives which form the dyes of the primuline group can be used for photographic purposes, as the sensitiveness of the compounds is increased by combination with the complex colloids which constitute animal or vegetable textile fabrics. The sensitive surface is prepared (at present) by colouring a cotton or silk fabric with primuline (1% to 2%), and then diazotising. Such a surface will give a complete positive picture after 40—180 seconds exposure; that is to say, in the bright lights the diazo-compound is completely, in the half lights only partially, decomposed, so that a reproduction of the original is obtained in the form of diazoprimuline. The picture can be developed with any of the various amines or phenols which form a dye with the diazo-compound.

TARPAULIN. *Syn.* TARPAWLING. Canvas covered with tar or any composition which will render it waterproof.

TARRAS. *Syn.* TERRAS. A volcanic product resembling puzzuolano, that imparts to mortar the property of hardening under water. Several other argillo-ferruginous minerals possess the same power, and are used under this term.

TARTAR. *Syn.* ARGOL, ORGOL; TARTARUM, TARTARUS, L. Impure bitartrate of potash. Crude tartar is the concrete deposit formed upon the sides of the casks and vats during the fermentation of grape juice. That obtained from white wine is white argol; that from red wine, red argol. After purification it forms cream of tartar.

Tartar, Ammo"niated. $C_4H_4K(NH_4)O_6$. *Syn.* AMMONIO-TARTRATE OF POTASSA, SOLUBLE TARTAR (AMMONIATED); TARTARUS AMMONIATUS, TARTARUM SOLUBILE AMMONIATUM, L. *Prep.* Neutralise a solution of cream of tartar with ammonia in slight excess, then evaporate and crystallise. Very soluble in water. A favourite laxative on the Continent.

Tartar, Bo"raxated. *Syn.* SOLUBLE CREAM OF TARTAR, BORO-TARTRATE OF POTASSA AND SODA ;

TARTARUM BORAXATUM, CREMOR TARTARI SOLU-BILIS, POTASSÆ ET SODÆ TARTRAS BORAXATA, L. *Prep.* From borax, 2 lbs.; cream of tartar, 5 lbs. (both in powder); dissolved in water, evaporated, and crystallised. See POTASSIUM BORO-TARTRATE.

Tartar, Chalyb'eated. Potassio-tartrate of iron.

Tartar, Cream of. $C_4H_5KO_6$. Bitartrate of potash. *Prep.* From crude tartar (argol, q. v.), dissolved in hot water and treated with a little pipeclay and animal charcoal, to remove the colouring matter derived from the wine; the filtered solution is set aside to crystallise.

Prop. Irregular groups of small transparent or translucent prisms; soluble in boiling water, less soluble in cold water; 1 part of the latter takes up about $\frac{1}{180}$; heat decomposes it into potassium carbonate, carbon, and inflammable gases, and evolves an odour of burnt sugar.

Tartar Emet'ic. $2C_4H_4K(SbO)O_6 + H_2O$. Potassio-tartrate of antimony, potassio-antimonious tartrate. *Prep.* By boiling antimony trioxide, 5 oz., in a solution of cream of tartar, 6 oz., in water, 2 pints.

Prop. Transparent rhombic octahedral crystals; soluble in 15 parts of cold and 3 parts of boiling water without decomposition. The solution and the salt have an extremely acid, metallic, and disagreeable taste. The crystals lose their water of crystallisation at the temperature of boiling water, but regain it on re-solution and recrystallisation.

Tartar, Oil of. Deliquesced carbonate of potash.

Tartar, Reduced. *Syn.* CREMOR TARTARI REDUCTUS, L. An article is sold under the name of 'British cream of tartar,' which contains ½ its weight or more of bisulphate of potash.

Tartar, Salt of. Carbonate of potash.

Tartar, Sol'uble. Neutral tartrate of potash.

Tartar, Spirit of. $C_4H_6O_4$. Pyrotartaric acid. Four modifications of this acid exist, viz. methylsuccinic acid, glutaric acid, ethylmalonic acid, diethylmalonic acid. Neither is of much importance except chemically.

TARTAR'IC ACID. $H_2C_4H_4O_6$. *Syn.* ACID OF TARTAR, ESSENTIAL SALT OF T.†; ACIDUM TARTARICUM (B. P., Ph. L., E., & D.), SAL ESSEN-TIALE TARTARI†, L. *Prep.* 1. (Ph. L. 1836.) Take of cream of tartar, 4 lbs.; boiling water, 2 galls.; dissolve by boiling; add, gradually, of prepared chalk, 12 oz. 7 dr. (made into a milk with water), and, when the effervescence ceases, add another like portion of prepared chalk dissolved in hydrochloric acid, 26½ fl. oz., or q. s., diluted with water, 4 pints; collect the precipitate ('tartrate of lime'), and, after well washing it with water, boil it for 15 minutes in dilute sulphuric acid, 7 pints and 17 fl. oz.; next filter, evaporate the filtrate (to the density of 1·38), and set it aside to crystallise; redissolve the crystals in water, concentrate the solution by evaporation, and recrystallise a second and a third time. The Edinburgh formula is nearly similar. In the Ph. L. & D. tartaric acid is placed in the Materia Medica.

2. (*Gatty.*) The solution of argol or tartar is first neutralised with carbonate of potash, and to every 300 galls. of the clear liquid, at 5° Twaddell, 34 galls. of milk of lime (1 lb. of lime per

gall.) are added; carbonic acid gas is then forced in, with agitation; decomposition ensues, with the formation of 'bicarbonate of potash' and 'tartrate of lime;' the last is converted into tartaric acid in the usual manner, and the former is evaporated in iron pans, and roasted in a reverberatory furnace for its potash.

Prop. Tartaric acid forms inodorous, scarcely transparent, monoclinic prisms, more or less modified, which are permanent in the air; it possesses a purely sour taste, dissolves in about 3 parts of water at 17·5° C., and in about its own weight of boiling water; it is slightly soluble in alcohol; the aqueous solution exhibits right-handed polarisation, and suffers gradual decomposition by age.

Heated to 135° C. it fuses and becomes an amorphous deliquescent mass of metatartaric acid, which is isomeric with it; at 145° it becomes tartralic acid; at 180° it yields tartrelic acid and tartaric anhydride, which is isomeric with it. All these are reconverted into tartaric acid by solution in water. On further heating it undergoes destructive distillation, yielding acetic, pyroracemic, pyrotartaric, pyrotritartaric, and formic acids; also dipepotetracetone, which has a peculiar odour like burnt sugar, acetone aldehyde, carbonic oxide, and carbon dioxide.

Tests. 1. Tartaric acid is known to be such by its solution giving white precipitates with solutions of caustic lime, baryta, and strontia, which dissolve in excess of the acid. 2. A solution of potash causes a white granular precipitate of cream of tartar, soluble by agitation in excess of the precipitant. 3. Nitrate of silver and acetate of lead give white precipitates, which, when heated, fume, and yield the pure metal. 4. If to a solution of tartaric acid, or a tartrate, solution of a ferric or aluminium salt be added, and subsequently ammonia or potash, no precipitate is formed. 5. At about 570° F. all the tartrates are blackened, and yield a peculiar and characteristic odour.

Uses, &c. Tartaric acid is chiefly employed in calico printing, and in medicine as a substitute for citric acid and lemon juice in the preparation of cooling drinks and saline draughts. For the latter purpose bicarbonate of soda is the alkaline salt commonly employed.—*Dose*, 10 to 30 gr.

Concluding Remarks. On the large scale the decomposition of the tartar is usually effected in a copper boiler, and that of the tartrate of lime in a leaden cistern. This part of the process is often performed by mere digestion for a few days without the application of heat. Leaden or stoneware vessels are used as crystallisers. Good cream of tartar requires 26% of chalk and 28·5% of dry chloride of calcium for its perfect decomposition. Dry tartrate of lime requires 75% of oil of vitriol to liberate the whole of its tartaric acid. A very slight excess of sulphuric acid may be safely and advantageously employed. Some manufacturers bleach the coloured solution of the first crystals by treating it with animal charcoal; but for this purpose the latter substance should be first purified by digesting it in hydrochloric acid, and afterwards by lixiviating it with water, and exposing it to a dull red heat in a covered vessel. The general management of this manufacture resembles that of citric acid. To obtain a large product care must be taken that the whole of the tartrate of lime be thoroughly decomposed, a matter not always effected by clumsy manipulators, who do not adapt their quantities or practice to the circumstances before them.

TARTRATE. A salt of tartaric acid.

TARTS. These may be regarded as miniature pies, consisting of fruit, either fresh or preserved, baked or spread on puff-paste.

To make an apple tart take about 2 lbs. of apples, peel them, cut each into 4 pieces, and remove the cores; then let each of the quarters be subdivided into 2 or 3 pieces, according to the size of the apple. Having done this, put half the pieces into a pie-dish, press them evenly down, and sprinkle over them 2 oz. of brown sugar; then add the remaining apples, and afterwards another 2 oz. of sugar, so that the apples shall form a kind of dome, the centre of which is about 2 inches above the sides; now add a wine-glassful of water, and cover the top over with short paste. Let bake in a moderately heated oven from half to three quarters of an hour.

The quantity of sugar must depend upon the quality, and the degree of sweetness, or the reverse, of the apples used. If they are of the sweet kind or very ripe, use less sugar, but a double quantity of water; in the latter case a little of the juice of lemon will improve the flavour. Chopped lemon peel, or cinnamon, or cloves, may also be added to the tart with advantage.

On making green rhubarb or greengage tarts it will be necessary to use a little more sugar, and to proceed as for apple tart, taking care, however, to omit the lemon juice and peel, cinnamon, or cloves. Tarts of ripe currants, raspberries, cherries, damsons, and mulberries, may be made in the same manner as rhubarb tart. Pink rhubarb does not require peeling.

TAU′RIN. $C_9H_7NSO_5$. Obtained when purified bile is boiled for some hours with an excess of hydrochloric acid. By filtration, evaporation, and dissolving the dry residuum in about 6 parts of boiling alcohol, nearly pure taurin crystallises out as the solution cools. It forms with crystalline needles, which are soluble in water, and sparingly soluble in alcohol. It is remarkable for containing fully 25% of sulphur.

TAUROCHOLALIC ACID. See CHOLEIC ACID.

TAW′ING. In the preparation of the TAWED LEATHER used for gloves, housings, &c., the skins are first soaked, scraped, and hung in a warm room until they begin to exhale an ammoniacal odour, and the wool readily comes off; they are then dehaired, and soaked in water with some quicklime for several weeks, the water being changed two or three times during that period; they are then again beamed, smoothed, and trimmed, after which they are rinsed, and resoaked in a vat of bran and water, where they are kept in a state of gentle fermentation for some weeks (in this state they are called 'pelts'); the skins are next well worked about in a warm solution of alum and salt, again fermented in bran and water for a short time, and are then stretched on hooks and dried in a stoveroom; they are, lastly, again soaked in water and trodden or worked in a pail or tub containing some yelks of eggs beaten to a froth with water, after which they are stretched and dried in a loft,

and are smoothed with a warm smoothing-iron. Sometimes the process is shortened by soaking the skins in the following mixture after the first steep with bran:—Common salt, 3½ lbs.; alum, 8 lbs.; boiling water, q. s.; dissolve, add of wheaten flour, 21 lbs.; yelks of 9 dozen eggs; make a paste. For use, a portion is to be largely diluted with water.

CHAMOIS or **SHAMMY LEATHER** is generally prepared from either sheep- or doe-skins, which, after dressing, liming, &c., are well oiled on the grain side, then rolled into balls, and thrown into the trough of the fulling-mill, where they are beaten for 2, 3, or 4 hours. They are next aired, and again oiled and fulled, and this is repeated a third time, or oftener, as circumstances may direct. The oiled skins are then exposed to a fermenting process, or heating in a close chamber, and are afterwards freed from redundant oil by being scoured in a weak alkaline lye. They are, lastly, rinsed in clean water, wrung at the peg, dried, and 'finished' at the stretcher-iron.

TAWED LEATHER differs from **TANNED LEATHER** in yielding size or glue under the influence of heat and moisture, in nearly the same way as the raw skins.

TAXIDERMY, Practical. The following review on the excellent work on this subject is from the 'Bazaar:'—'Practical Taxidermy: a manual of instruction to the amateur in collecting, preserving, and setting up Natural History Specimens of all kinds. By Montagu Browne.'

"The author of this little book begins at the beginning, and, before detailing the process of skinning, preserving, and mounting any given Vertebrate, he starts with a chapter on 'Trapping and Decoying Birds and Animals.' In this chapter descriptions are given of various forms of springs, snare, 'figure-4 trap,' clap-net, glade-net, bow-net, and box trap; and some wrinkles are imparted which may be useful to wildfowl shooters, to say nothing of gamekeepers, whose livelihood depends on their success in destroying what they are pleased to regard as 'vermin.' Chapter 3 is devoted to 'Necessary Tools,' of which not only descriptions, but figures are given, and those who have no knowledge of taxidermy will probably be surprised to learn how few tools are really necessary for the purpose.

"On the subject of preservative soaps and powders Mr Browne has a good deal to say, and gives no less than seventeen different receipts. Many of these, however, are only noticed to be condemned, for the author has been a great experimentalist, and has tested the efficiency or otherwise of all the preparations he names, with a view of ascertaining the best, and at the same time that which is most harmless to the operator. For the preservation of birds he pins his faith to the fourth formula (p. 46), which is a preservative soap for the inside of the skin, composed of 1½ lbs. whiting or chalk, 1 lb. of soft soap, and 2 oz. of chloride of lime, finely pounded. These ingredients are boiled together in a pint of water, and the mixture, when properly applied, is said to be so efficacious as to completely supersede arsenical paste or soap. Presuming that it is used only for such specimens as are to be immediately cased up in air-tight cases, nothing further is needed; but as regards such specimens as are left exposed or uncased, 'a wash of benzoline, liberally applied from time to time—say twice a year—to the outside,' is recommended.

" We have long since proved the efficacy of this fluid, not only in repelling the attacks of moths, and the larvæ of destructive beetles such as *Dermestes lardarius*, but in killing them in skins that have been already attacked by them. We can therefore endorse Mr Browne's remarks, and may supplement them by a 'wrinkle' which he has probably discovered by this time, although he does not refer to it. It is this. If the amateur in search of benzoline applies to the nearest chemist for it, he will be served with the 'rectified' fluid, and will be asked three shillings or three shillings and sixpence for an ordinary medicine bottle-full. If he proceeds, however, to an oil and colour shop, and asks for it as supplied for burning in the sponge lamps, he may get a pint for about a tithe of the cost.

" The best way to apply it is to pour some out in a saucer, saturate a pinch of cotton wool with it, and dab it on all over the fur or feathers. The great advantage about it is that, while killing or repelling insects, it does not in the least injure the specimen to which it is applied. We have seen a mounted specimen of a bird almost saturated with it, so that the feathers looked quite draggled; but as soon as the moisture had evaporated, the feathers all resumed their former shape and glossy appearance. Care should be taken not to use it by candle-light, as the vapour is inflammable.

" Amongst the seventeen receipts for ' preservatives' of different kinds, we do not see a very simple powder which we have used with success for very small bird-skins, which were almost too delicate to stand the application of a brush and paste, or soap. It is composed of burnt alum and sugar of lead, and mixed in the proportion of two thirds of the former to one third of the latter. The alum dries, the sugar of lead preserves, and if the specimen while being skinned be dusted with this mixture, it will absorb all moisture as it arises, for which purpose plaster of Paris is usually employed. A very little goes a long way, and it is desirable not to apply too much, lest the astringent nature of the alum should cause the skin to become brittle and crack.

" Possibly Mr Browne has not referred to this mixture on account of the poisonous nature of the sugar of lead. He justly remarks that too much care cannot be exercised in the employment of poisonous preservatives, and we fully agree with him.

" On the subject of his instructions for skinning birds we have not much criticism to offer, except perhaps as regards his mode of filling out the skin after the preservative has been applied, and before the skin is sewn up. We have found by experience that, after the artificial neck of tow or cotton wool has been inserted, and the wing-bones tied inside, the skin is much better filled by degrees with little bits of cotton wool inserted piecemeal, instead of with an artificial body ' as nearly as possible shaped to the original body of the bird.' The advantage of the former plan is that it is much more quickly executed,

and a nice soft skin is the result, instead of a comparatively hard one. By introducing the wool piecemeal, too, scarcely any portion of the skin is left without support on the inside, as is often the case when a ' made body ' is inserted. This is material; for if there be any want of inside support at a given point, pressure upon that point from the outside will cause the skin to crack. These remarks, however, must be taken to apply only to such specimens as are intended to be preserved as skins, and are not to be set up.

" Few amateurs, probably, give much time to mounting their specimens, for they can get them so well done by professional taxidermists, at prices varying to suit all purses. The art of skinning wild animals and birds, however, and curing or dressing animal hides, should be acquired by every sportsman and naturalist who intends to travel and collect and bring home trophies. To such a one we may specially recommend the chapters which are devoted to this portion of the subject. In this, as in other respects, Mr Browne's book is a ' practical ' manual of taxidermy."

TAXINE. A poisonous alkaloid present in the leaves and seeds of the yew (*Taxus baccata*).

TEA. *Syn.* **THEA,** L. The dried leaves of the Chinese tea plants (*Thea Bohea* and *Thea viridis*).

It was formerly supposed that BLACK TEAS could only be obtained from *T. Bohea*, and GREEN TEAS from *T. viridis*, but Fortune and others have proved that both sorts may be made from either species, and that the differences in colour and flavour depend chiefly on the age of the leaves and the treatment they undergo in the drying process. Another species, named *Thea Assamica*, furnishes ASSAM TEA.

Mulder gives the following as the composition of tea :

	Black Tea.	Green Tea.
Essential oil	0·60	0·79
Chlorophyll	1·84	2·22
Wax	0·00	0·28
Resin	3·64	2·22
Gum	7·28	8·56
Tannin	12·88	17·80
Theine	0·46	0·43
Extractive matter . .	21·36	22·80
Colouring substances .	19·19	23·60
Albumen	2·80	3·00
Fibre	28·33	17·80
Ash (mineral substances)	5·24	5·56

Dr Walter Blyth, commenting upon the above, says the amount of theine is certainly understated.

Pur. The chief adulteration of tea which is extensively practised at the present day is mixing it with a certain portion of exhausted tea-leaves, which have been redried and curled. The collection and preparation of these occupy several hundred persons, chiefly women and children, in and about London. The leaves which have been found in the possession of the manufacturers of imitation tea are those of the aloe tree, ash tree, elder bush, and whitethorn. According to Mr Warrington, a most extensive system of adulterating tea is practised in China. Many samples

105

directly imported from that country, examined by him, did not contain a single grain of tea, being made up entirely of other leaves. The ordinary green teas he found, for the most part, spurious, being manufactured out of the cheaper black teas. These are 'faced up' or 'painted' with various colouring substances, powdered porcelain, clay, &c., which are readily perceived under the microscope, and even admit of being separated and chemically examined.

It is a general practice among the grocers in this country to impart what they call a 'bloom' to their green teas by 'rouncing' them up with a little calcined magnesia, or finely powdered talc or French chalk. The quantity that adheres to the tea is very trifling, but it greatly improves its appearance. Black teas are 'faced,' in a similar manner, with finely powdered plumbago or blacklead.

Pure China tea is not turned black by being put into water impregnated with sulphuretted hydrogen gas, nor does it tinge a solution of ammonia blue. The infusion is amber-coloured, and is not reddened by the addition of an acid. The ashes left from the combustion of genuine tea are white, and do not exceed 5% to 5½%. If they exceed this they may be chemically examined with the usual tests for alumina, chromate of lead, copper, cyanide of potassium, gypsum, lime, magnesia, &c. Many of these substances may be detected by simply agitating the tea with a little cold water, when they will be detached from its surface, and render the water turbid, or, by their gravity, sink to the bottom.

Mr A. H. Allen ('Chemical News,' xxix, 123, 167, 189, 221; and xxx, 2) arranges the adulteration of tea under four heads, giving at the same time directions for their detection:

1. *Mineral Additions for increasing Weight or Bulk.* (a) Magnetic matter. Detected by drawing a magnet under a weighted portion of the tea spread upon paper, whereby the magnetic matter is separated from the tea and may be weighed.

(b) Siliceous matter. The ash must be estimated by igniting a weighed portion of the tea. The ash of genuine tea varies from 5·24% to 6·0%. The ash is then boiled with water, the insoluble part again treated with hydrochloric acid, and the silica collected and weighed. Genuine tea does not contain, on an average, more than 0·80% of ash insoluble in acid; adulterated teas sometimes contain as much as 10%.

2. *Organic Adulterations for increasing Weight or Bulk.* (a) Exhausted tea-leaves. Best detected by estimating the tannin, gum, soluble ash, insoluble matter, &c.

a. Tannin. 5 gr. of lead acetate are dissolved in 1 litre of water, and the solution filtered after standing; 5 mgrms. of pure potassium ferricyanide are dissolved in 5 c.c. of water, and an equal bulk of strong ammonia solution is added. The lead solution is standardised by diluting 10 c.c. to 100 c.c. with boiling water, and adding to it from a burette a solution of 0·1 pure tannin in 100 c.c. of water, until a few drops, when allowed to fall through a filter on to a drop of the ferricyanide solution, spotted on a slab, produce a pink colour. A solution of the tea is made by

repeatedly boiling about 2 grms. of the finely powdered sample with 80 c.c. of water until it is completely exhausted. The solution is filtered and made up to 250 c.c., and used as already described.

The amount of tannin in genuine black tea averages about 10%. A small quantity of tannin, about 2%, remains in the exhausted leaves. The percentage of exhausted leaves, E, in a sample may be estimated, when the percentage of tannin, T, is known, by the equation—

$$E = \frac{(10 - T) \, 100}{8}.$$

β. Insoluble matter is best estimated by boiling the pounded sample repeatedly with water, and drying the residue at 120° C., until the weight is constant. The insoluble matter in black tea varies from 46·7% to 53·6%, while in previously infused leaves it varies between 72% and 75%.

γ. Gum. The aqueous decoction is evaporated nearly to dryness, the residue treated with methylated spirit, filtered, washed with spirit, rinsed off the filter with hot water, the liquid evaporated at a steam heat, weighed, ignited, and weighed again. The loss represents gum.

δ. Soluble ash. The aqueous solution of the ash is evaporated, *gently* ignited, and weighed. Genuine tea contains not less than 3% of soluble ash, while in exhausted leaves this item falls as low as 0·52%. If S represent the percentage of soluble ash, the percentage of exhausted leaves, E, may be approximately found, in the absence of foreign leaves, by the equation—

$$E = (6-2S) \, 20.$$

(b) Foreign leaves. The presence of leaves other than those of the tea plant may be detected with some accuracy by estimating the insoluble matter, tannin, gum, and ash; but the microscope must decide this question.

3. *Adulterants for imparting a Fictitious Strength.* (a) Extraneous tannin matters, such as catechu, &c., are detected by an unusually high percentage of tannin, as indicated by the lead process. Tea adulterated with catechu gives an infusion which quickly becomes muddy on cooling. 1 grm. of the sample and 1 grm. of pure tea are each infused in 100 c.c. of water, and the solutions poured off from the leaves are precipitated while boiling, with a slight excess of neutral lead acetate, filtered, and tested as follows :—About 20 c.c. of the pure tea infusion, when gently heated with a few drops of silver nitrate, give a slight cloudiness only; while tea containing catechu gives a copious brownish precipitate, and the liquid acquires a distinct yellow tinge. One drop of ferric chloride gives a light green colour if catechu is present, and a greyish-green precipitate on standing; the solution from pure tea gives a reddish colour with ferric chloride, due to acetate, and no precipitate on standing. These tests are applicable only when catechu is present in tolerably large quantities.

(b) Lie tea, when thrown into hot water, falls to powder, because the gum or starch used to keep it in a compact form is dissolved. The liquid may be acidified with sulphuric acid, decolourised with permanganate, and tested for starch. The ash of lie tea is often as high as 30% or 40%.

(c) Caper tea is made into little glossy masses by the aid of gum or starch; it is usually much adulterated. The insoluble matter is usually much less than in genuine tea; the gum amounts to 15% or 20%. The soluble ash often falls below 2%.

(d) Soluble iron salts are added to give an appearance of strength by the formation of tannate of iron. They are detected by shaking the powdered leaves with cold dilute acetic acid, filtering, and testing for iron in the filtrate.

(e) Alkaline carbonates are sometimes added to tea. The soluble ash gives the yellow sodium flame if sodium salts have been added; the alkalinity may also be determined in the soluble ash. The average amount of potash (K_2O) in tea is about 1·62%.

4. *Facing and Colouring Materials.* These may be detected under the microscope, or the leaves may be washed with warm water, the colouring matter collected and examined. Indigo is best detected by the microscope; Prussian blue by boiling with caustic alkali, filtering and testing for ferrocyanide by ferric chloride. The re-

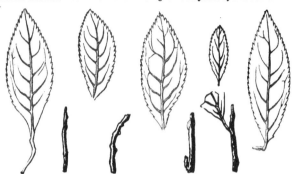

Leaves and stalks of best tea brought from China (1861) by private hand. Natural size.
Generally in commercial tea the leaves are much larger and thicker, and often are, cut transversely into two or three parts. Some stalks and remains of flowers are found in all tea, even the best.

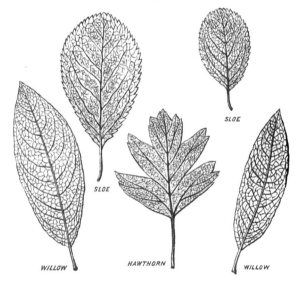

SLOE

SLOE

WILLOW

HAWTHORN

WILLOW

sidue, insoluble in alkali, is fused with alkaline carbonate, evaporated to dryness with hydrochloric acid; the residue tested for silica, and the filtrate tested for lime and magnesia.

Moisture varies from 6% to 8% . Among domestic substitutes for tea are the leaves of speedwell, wild germander, black currant, syringa or mock orange, purple-spiked willow-

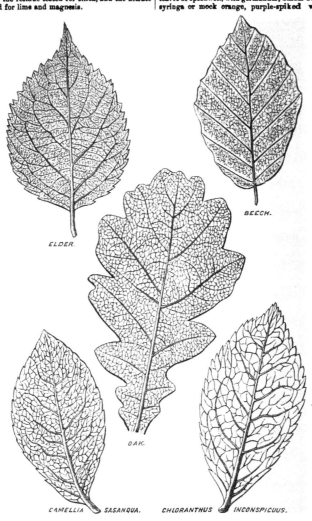

BEECH.

ELDER.

OAK.

CAMELLIA SASANQUA.   CHLORANTHUS INCONSPICUUS.

Leaves used in the adulteration of tea—the sloe, willow, oak, beech, elder, and hawthorn, have been nature-printed and then lithographed. The drawings of the *Chloranthus inconspicuus* and the *Camellia sasanqua*, which are said to be used by the Chinese, are copied from Hassall. The leaves of the elm, poplar, and plane are said to be sometimes used in England. Falsification with any kind of leaf is, however, now decidedly uncommon in this country.

herb, winter-green, sweet-briar, cherry tree, sloe, &c., all of which are used for tea, either singly or mixed. The addition of a single bud of the black currant to the infusion of ordinary black tea imparts to it a flavour closely resembling that of green tea.

The brownish-coloured powder vended under the name of 'la veno beno' is a mixture of 3 parts of tea-dust with 5 parts of powdered catechu or terra japonica. A few grains of this substance thrown into the teapot are described in the advertisements as being capable of more than doubling the strength of the beverage.

Tea, Lie. Of this compound Dr Hassall says: "It is so called because it is a spurious article, and not tea at all. It consists of dust of tea-leaves, sometimes of foreign leaves and sand, made up by means of starch or gum into little masses, which are afterwards painted and coloured, so as to resemble either black or green gunpowder. The skill exhibited in the fabrication of this spurious article is very great, and we have met with at least a dozen varieties of it, differing from each other in the size and colouring of the little masses."

The once notorious ' PARAGUAY PLANT,' sold in packets, was simply new meadow hay that had been wetted with a strong infusion of catechu, then dried, chopped small, and strongly compressed. See THEINE and CAFFEINE.

Tea. "The tea is not a meal; when it is properly used it should not be a meal; but it has a special purpose to fulfil, which I will now explain. Tea—and under the generic term tea I include coffee,—tea is usually taken three hours after dinner. This is the moment which corresponds with the completion of digestion, when, the food having been conveyed away from the stomach, nothing remains behind but the excess of the acid juices employed in digestion; these acid juices create an uneasy sensation at the stomach, and a call is made for something to relieve the uneasiness; tea fulfils that object." "On the same principle, after the business of the dining-room, the antacid and refreshing beverage, either in the shape of tea or coffee, is prepared in the drawing-room. In taking either, the nearer they approach to the simple infusion the better; little milk or cream, and less sugar, should be the principle. But, seeing the purpose of tea, how unreasonable to make it the excuse for a meal, to conjoin with it toast, muffins, bread and butter, and *id genus omæ !* " "Three meals a day may be taken as the standard of habit and custom, tea and coffee having a specific place and purpose as a beverage, but none as a meal" (*Eras. Wilson*). See MEALS, &c.

Although tea is undoubtedly prejudicial to children and to adults of nervous and irritable temperament, there can be no question that, if its use be not abused, it possesses valuable physiological properties. On the nervous system it acts as a pleasant stimulant and restorative, its moderate use not being followed by depression. Dr Parkes says these effects are in some measure due to the warmth of the infusion. According to the same authority its use is followed by very little quickening of the pulse, whilst there is an increase in the amount of perspiration, and a slightly

diminished action on the bowels. Cases, however, are not uncommon in which this latter effect is reversed.

Dr Edward Smith says that tea increases the excretion of pulmonary carbonic acid. The contention that the elimination of urea is lessened does seem to have been not satisfactorily established. If so, the diminution is very trifling. Sir Ranald Martin says tea is most useful against excessive fatigue, especially in hot climates. The traveller in the Australian bush speaks highly of its renovating effects at the end of a long day passed in the saddle.

A cup of strong green tea without milk or sugar is a popular and frequently by no means inefficient remedy for a severe nervous headache. According to Liebig, tea and coffee resemble soup in their effect on the system. Lehmann's experiments seem to show that they lessen the waste of tissue in the human body. Tea taken too continuously, or in excess, produces indigestion, flatulence, and constipation, besides rendering its votaries anæmic and depressed in spirits.

It is a fallacy to suppose that soft water makes the best tea. It certainly yields a darker infusion than that made from moderately hard water, but this is owing to the soft water taking up a large quantity of bitter, physiologically inert, extractive matter from the tea, the delicate flavour of which becomes thereby greatly impaired. This is why connoisseurs object to an infusion of too dark a colour. Moderately hard *boiling* water, on the contrary, fails to dissolve this objectionable ingredient, and hence produces a beverage in which the characteristic taste of the pleasant aromatic principle of the tea is not masked by the bitter substance. London water, which, when boiled, has a hardness of about 5 degrees (equal to 5 gr. of lime salts to the gall.), makes excellent tea—better, in fact, than a water of half the hardness, the latter yielding a slightly bitter infusion. In the use of moderately hard water, it is essential that it should be allowed to remain on the tea sufficiently long. The Chinese never employ either very soft or immoderately hard water, but *a water of medium hardness*.

"Experimentally it is found that infusions of tea and coffee are strong enough when the former contains 0·6% of extractive matter, and the latter 3%, so that a moderate-sized cup (5 oz.) should contain about 18 gr. of the extract of tea, or 66 gr. of coffee. These proportions will be obtained when 263 gr. of tea (about 2½ teaspoonfuls) or 2 oz. of freshly roasted coffee are infused in a pint of boiling water; and the amounts of the several constituents dissolved are about as follows:

" Constituent.	Tea. gr.	Coffee. gr.
Nitrogenous matters	17·2	44·0
Fatty matter	—	3·0
Gum, sugar, and extractive	31·7	103·2
Mineral matters	9·1	22·8
Total extracted	58·8	173·0

So that tea yields, to a pint of fresh water, about 2% of its weight, and coffee about 20%. Lehmann found that only 15½% of tea was dissolved by water, whereas Sir Humphry Davy estimated it at 33½%. No doubt the quality of the water, as well as that of the tea, affects the results, for cold distilled water will extract from 40% to 44% of black tea, and nearly 50% of green; but for all this, about 22% is a good average with boiling water" (*Letheby*, 'Lectures on Food,' Longmans).

Dr Edward Smith has shown in the following table that, when the usual custom of measuring tea into the teapot by the spoonful is followed, very varying weights of tea are employed. Thus he found that the weight of a spoonful of tea was for—

Black Teas.

Oolong	. .	39 grains.
Congou (inferior)	.	52 „
Flowery Pekoe	. .	62 „
Souchong	. .	70 „
Congou (fine)	.	87 „

Green Teas.

Hyson	. . . .	66 grains.
Twankay	. .	70 „
Fine Imperial	. .	90 „
Scented Caper	.	103 „
Fine Gunpowder	.	123 „

The attempt to make good tea will prove a failure unless the water employed is *boiling*. Previously to making the infusion, the teapot should always be warmed by means of boiling water. The kettle should be filled from the *tap*, and not the boiler. It should also be borne in mind that neither good tea nor coffee can be obtained if they are made with water that has been in the kettle for many hours. The tea is ready to be drunk after the boiling water has stood on it for five minutes.

Tea, Algerian. *Syn.* THÉ ARABE. The flowers of *Paronychia argentea*, Lam., and *P. nivea*, DC., Nat. Ord. ILLECEBRACEÆ. Used as a medicinal tea in Algeria, and sold in Paris.

Tea, Beef. *Syn.* INFUSUM CARNIS BUBULÆ, JUSCULUM CUM CARNE BOVIS, L. This is merely a very concentrated soup formed of lean beef. According to the common plan, lean beef, 1 lb., is gently simmered in water, 1 quart, for about half an hour, when spices, salt, &c., are added, and in a few minutes the whole is strained for use. The following are other formulæ:

1. (*Dr A. T. Thomson.*) Take good rump steak, ¼ lb.; cut it into thin slices, spread these over a hollow dish, sprinkle a little salt on them, add a pint of boiling water, and place the dish (covered) near the fire for half an hour; then remove the whole to a saucepan, and boil it gently for 15 minutes; lastly, strain through a hair sieve.

2. (*Professor Liebig.*) Beef, free from fat, 1 lb., is to be minced very small, mixed with an equal weight of cold water, and, after digestion and agitation in the cold for about half an hour, heated slowly to boiling; when it has boiled for a minute or two, strain it through a cloth. It may be coloured with roasted onion or burnt sugar, and spiced and salted to taste.

Obs. Similar preparations are ordered in some foreign Pharmacopœias from calves' lights, crayfish, frogs, mutton, pullets, snails, tortoise, veal, &c. In the Ph. L., 1746, a form was given for viper broth (JUSCULUM VIPERINUM). See ESSENCE OF BEEF, EXTRACT OF MEAT, BEEF TEA, &c.

Tea, Brouss. *Vaccinium arctostaphylos*, Linn. Used at Broussa, and sold at about 8*d.* per lb.

TEETH (The). *Syn.* DENTES, L. An object very subservient to health, and which merits due attention, is the preservation of the teeth; the care of which, considering their importance in preparing the food for digestion, is, in general, far from being sufficiently appreciated. Those who abhor a fetid breath, rotten teeth, and the toothache, would do well to invariably clean their teeth before retiring to rest. With smokers this practice is almost obligatory. Washing the mouth frequently with cold water is not only serviceable in keeping the teeth clean, but in strengthening the gums, the firm adhesion of which to the teeth is of the greatest importance in preserving them sound and secure. Some persons think it serviceable to add a few drops of spirit or essence of camphor to the water thus employed, a plan we certainly approve of. See BREATH, DENTIFRICES, PASTES, POWDERS, TOOTH CEMENTS, WASHES, &c.

Teeth, Stoppings for. See DENTISTRY.

TEETHING. *Syn.* DENTITION. Children are sometimes born with one or more teeth; but, in general, the teeth at birth consist of mere pulpy rudiments buried in the gum. Their development is gradual. About the third or fourth month they begin to assume shape and hardness. At this period children become fretful, the saliva flows copiously, the gums grow turgid, and there is a fondness for biting hard cold objects. In nearly all cases there is more or less fever, frequently a cough or diarrhœa, and a rash commonly appears, which is called by nurses the 'red-gum.' These symptoms generally abate after a fortnight or three weeks, and the child remains undisturbed until the seventh or eight month. About this period the gums again become red, tender, and swollen, and often extremely sensitive and painful. The upper part of the gum gradually becomes attenuated and pale, and, just before the tooth appears, even covered with a blister. These changes are usually attended by an increased flow of saliva, or 'drivelling,' and a lax state of the bowels, both of which are regarded as favourable symptoms. Sometimes, however, the diarrhœa is excessive, when it may be cautiously restrained by a dose or two of rhubarb and magnesia, with a little dill or peppermint water; or, better, by the daily use of a little arrowroot, to which a few drops of pure port wine may be added. Sometimes the local irritation is considerable, or there are spasms or convulsions, in which case the practice is to lance the gums. Sluggishness of the bowels may be removed by a little castor oil. Excessive irritability, without other marked symptoms, is best combated by a drop or two of tincture of hops in sweetened water. Throughout the whole period of dentition the use of warm dry clothing, freedom from tight bandages, with thorough ventilation, good nursing, exercise, fresh air without undue exposure, abundance of crawling on the carpet, and frequent warm baths, will be found most advantageous. Indeed, the last, without other treatment, are often sufficient to subdue the most distressing convulsions and the most obstinate diarrhœa, and in no case can they do harm. See NURSING, STROPHULUS, &c.

TELEPHONE. Within the memory of the present generation Sir Charles Wheatstone made some experiments on the transmission of sound, which were subsequently repeated and enlarged upon by Professor Henry in America. Connecting together by means of a bar of wood the sounding boards of two pianos placed in houses on opposite sides of the street, Henry found that when the piano on one side of the street was played upon the musical sounds it gave out were reproduced by that on the other side. The next research in this direction was that of Page, in 1837, who, setting up vibrations in bars of iron by rapidly magnetising and demagnetising them, elicited from them musical notes corresponding with the velocity of the vibration. Similar effects, but more marked in character, were produced by De la Rive, in 1843, by means of a succession of electric currents transmitted through a copper wire stretched through a cylinder made of insulated copper wire.

In 1861, Reiss, of Friedrichsdorf, perfected an instrument which, by means of the vibrations of a diaphragm alternately completing and breaking the continuity of a galvanic circuit, reproduced musical sounds in an iron bar at a distance.

Varley, in 1870, obtained similar results to Reiss by the rapid charging and discharging of a condenser.

In the first of these experiments—viz. Henry's—the sound was mechanically conducted along the bar of wood from the strings of one piano to those of the other, which being thrown into similar vibratory movements gave rise to similar sounds.

In the other experiments, on the contrary, the sounds were not due to the chemical conduction at all, but to currents of electricity. It has been explained that Reiss's instrument was capable of reproducing musical sounds at a distance from their origin. Reiss's may, therefore, be regarded as the original telephone. But, although able to reproduce a musical note or sound originating at a distance, this instrument failed altogether in the case of a word or a sentence, for the simple reason that the current of electricity which passes through the wires is an intermittent one. Musical sounds differ in tone, in intensity, and quality. The tone depends upon the number of vibrations produced in the air per second; when these are less than sixteen no sound is produced. The intensity is due to the extent or amplitude of the vibrations; and the quality, or *timbre*, to the form of the undulations made by the vibrating particles of the atmosphere. Now, of all these qualities or varieties of sound, the first only, or the tone, can be reproduced by a current of intermittent electricity, so that Reiss's is a *tone* telephone, and as such is only capable of redelivering a number of musical notes. To Professor A. Graham Bell alone belongs the merit of having invented an *articulating* or *speaking* telephone, or an apparatus by means of which not only tone, but intensity and *timbre* of sound—in short, speech in its entirety—can be electrically conveyed from one point to another, no matter how distant. The practical result of this is that a conversation can be carried on, the distance by which the speakers may be separated being of no import. To the particular species of

electricity by which this is accomplished Professor Bell has given the name 'undulatory,' in contradistinction to 'intermittent' or 'pulsatory.' The annexed plate, which is half the actual size of Bell's articulating telephone, represents that instrument in section.

m m is a permanent bar magnet, to the upper end of which is attached a soft iron core, which becomes magnetised by the permanent magnet. Surrounding the iron core is a coil of very fine insulated copper wire (*b*), the two ends of which are carried to the terminals (*t t*), by means of which one is connected with the line wire, and the other with the earth. *d* is a disc of thin iron plate, either tinned or japanned, about the size of a crown piece; and *c* is the cavity or mouth-piece. Upon applying the lips to this and speaking into

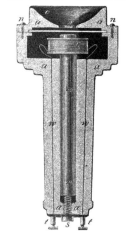

it, the iron disc *d* vibrates towards the soft iron core, the result being that a current of induced electricity is set up in the coil *b*, which being in connection by means of the telegraph wire with a precisely similar arrangement at the other end of the line, reproduces there the spoken words by means of a corresponding disc. The magnet with attachments is enclosed in a wooden case (*a a, a a, a a*); *s s* are screws which secure the iron disc *b*; *s* is a screw for adjusting the distance between the pole of the magnet and the disc *b*.

The extreme simplicity of Professor Bell's telephone was the outcome of several antecedent experiments, worked out by forms of apparatus gradually diminishing in complexity.

The German physicist Helmholtz had previously shown that, by the agency of a current of intermittent electricity passed through a tuning-fork, he could produce simultaneous vibrations in a number of other forks connected with the first by a wire, and that by varying the loudness of these vibrations by means of resonators, so as to combine the musical notes in different proportions, the resulting sound was an imitation of certain vowel sounds, or a copy of the *timbre* of sound.

Professor Bell's first telephone was an extension of Helmholtz's device for producing vowel or composite sounds. A number of steel wires of different pitch were made into a harp, and connected by a powerful permanent magnet, the same arrangement being repeated at the other end of the circuit. In the magnetic field of the permanent magnet was an electro-magnet. When a permanent magnet is vibrated in the neighbourhood of an electro-magnet, this latter will have a current of electricity generated in it, the intensity of which will vary with the velocity of the vibrations in the permanent magnet, whilst it will be either positive or negative according to the direction of these same vibrations, so that a vowel sound, if produced by causing a number of the rods of the harp to vibrate at the same time, can be transmitted by a current of electricity, and will be reproduced by the harp at the other end of the connecting wire. If a piano were sung into whilst on the pedal was down, not only would the pitch of the voice be echoed back, but an approach to the quality of the vowel would also be obtained; and theory teaches that if the piano had a very much larger number of strings to the octave, we should get not only an approximation to, but an exact vocal reproduction of the vowel. If, therefore, in the harp there were a large number of steel rods to the octave, and you were to speak in the neighbourhood of such a harp, the rods would be thrown into vibration with different degrees of amplitude, producing currents of electricity, and would throw into vibration the rods at the other end with the same relative amplitude, and the *timbre* of the voice would be reproduced.

The effect when you vibrate more than one of these rods simultaneously is to change the shape of the electrical undulation, and a similar effect is produced when a battery is included in the circuit. In this case the battery current is thrown into waves by the action of the permanent magnets. Hence you will see that the resultant effect on the current of a number of musical tones is to produce a vibration which corresponds in every degree to the moving velocity of the air. Suppose, for instance, you vibrate two rods in the harp, you have two musical notes produced; but of course if you pay attention to a particle of air, it is impossible that any particle of air can vibrate in two directions at the same time; it follows the resultant form of vibration. One curve would show the vibration of a particle of air for one musical tone, the next one for another, and the third the resulting motion of a particle of air when both musical tones are sounded simultaneously. You have by the harp apparatus the resultant effect produced by a current of electricity, but the same resultant effect could be produced in the air. There is an instrument called the phonautograph. It consists of a cone which, when spoken into, condenses the air from the voice. At the small end of the cone there is a stretched membrane which vibrates when a sound is produced, and in the course of its vibration it controls the movement of a long style of wood, about one foot in length. If a piece of

glass with a smoked surface is rapidly drawn before the style during its movement, a series of curves will be drawn upon the glass. I myself uttered the vowels *e, ay, eh, ah, aah*. These vowels were sung at the same pitch and the same force, but each is characterised on the glass by a shape of vibration of its own. In fact, when you come to examine the motion of a particle of air, there can be no doubt that every sound is characterised by a particular motion. It struck me that if, instead of using that complicated harp, and vibrating a number of rods tuned to different pitches, and thus creating on the line of wire a resultant effect, we were at once to vibrate a piece of iron, to give to that piece of iron not the vibration of a musical tone, but to give it the resultant vibration of a vowel sound, we could have an undulatory current produced directly, not indirectly, which would correspond to the motion of the air in the production of a sound.

The difficulty, however, was how to vibrate a piece of iron in the way required.

The following apparatus gave me the clue to the solution of the problem in the attempt to improve the phonautograph. I attempted to construct one modelled as nearly as possible on the mechanism of the human ear, but on going to a friend in Boston, Dr Clarence J. Blake, an aurist, he suggested the novel idea of using the human ear itself as a phonautograph, and this apparatus we constructed together. It is a human ear. The interior mechanism is exposed, and to a part of it is attached a long style of hay. Upon moistening the membrane and the little bones with a mixture of glycerin and water, the mobility of the parts was restored, and on speaking into the external artificial ear a vibration was observed, and after many experiments we were enabled to obtain tracings of the vibration on a sheet of smoked glass drawn rapidly along. This apparatus gave me the clue to the present form of the telephone. What I wanted was an apparatus that should be able to move a piece of iron in the way that a particle of air is moved by the voice. (From Professor Bell's lecture at the Society of Arts, November 28th, 1877, published in the Journal of the Society, vol. xxvi, p. 17.)

We need not follow Professor Bell through the various stages by which he arrived at his most successful solution of this problem further than to state that the simplicity of construction exhibited in the present form of instrument did not characterise the earlier articulating telephones. Amongst the causes contributing to this simplicity may be mentioned the abandonment of an animal membrane attached to the iron plate, the diminution of the coil of insulated wire, and the substitution for the galvanic battery, which formerly formed part of the circuit, of the permanent magnet.

Professor Bell records the curious fact that hardly any difference is observable in the results by varying the size, thickness, and force of the permanent magnet, and that beyond a remarkable effect in the quality of the voice, distinct articulations might be obtained from iron plates of from 1 inch to 2 feet in diameter, and from $\frac{1}{32}$ to $\frac{1}{4}$ inch in thickness. With plates of uniform thickness, but of varying diameter, he obtained the following results. With a plate of small diameter the articulation was perfectly distinct, but the sound emitted was as if a person were speaking through the nose. By gradually enlarging the diameter of the plate this nasal effect as gradually disappeared, until, when a certain diameter was attained, a very good quality of voice manifested itself.

By continuing to enlarge the diameter a coarse, hollow, drum-like effect was produced, until when the diameter became very large the sound re-sembled that one hears when the head is inside a barrel, and was accompanied with a reverberating sound. By reversing the above conditions—that is, by keeping the diameter constant and varying the thickness—it was found that with a very thin plate the drum-like sound was produced; by gradually increasing the thickness this effect passed off; then followed distinct articulation, until at a certain increase of thickness the peculiar nasal quality again developed itself.

In practice it has been found desirable, in establishing speaking communication between two distant places, to employ two telephones instead of a single one, one being applied to the mouth and the other to the ear during a conversation.

With one telephone it was no unusual occurrence for confusion to arise in consequence of the two speakers talking or listening at the same time.

So faithful is the transmission by the telephone of every variety of sound, that Mr Preece states, when in telephonic communication with Prof. Bell through a quarter of a mile, he has heard him "laugh, sneeze, cough, and, in fact, make any sound the human voice can produce." It must be borne in mind, however, that the transmitted speech can only be distinctly heard in the immediate vicinity of the receiving apparatus; the keenest hearing fails to detect it at the distance of little more than a foot away. Hence, when a message is expected, the recipient has to place his ear to the mouth-piece of the instrument, and use it as an ear-trumpet.

A circumstance tending to impair the satisfactory working of Bell's telephone is that the line wire to which the ends of the coil are attached becomes inductively affected by the currents of electricity passing through the parallel and contiguous telegraph wires, the effect, on a line where there is an active transmission of telegraphic messages, being that the telephone "emits sounds that are very like the pattering of hail against a window, and which are so loud as to overpower the effects of the human voice" (Preece).

This inconvenience can, however, it is stated, be remedied.

If all the arrangements of the instrument were perfect there should be no limits to the distance through which speech could be conveyed by the telephone. Prof. Bell says that in laboratory experiments "no difficulty has been found in using an apparatus of this construction through a circuit of 6000 miles;" and that he had found it act efficiently between New York and Boston, a distance of 258 miles, subject to the condition that the neighbouring telegraph wires were not in action.

Mr Preece has carried on conversations between Dublin and Holyhead, a distance of 100 miles. The telephone is now (1891) in regular use between London and Paris.

Two useful applications of the telephone are recorded by Prof. Bell, the one its employment in connection with the diving-bell, the other as a means of communication between those above and below ground in mines. It has been largely adopted in extensive factories and in commercial houses, both in America and in this country, supplementing, because of its much greater simplicity and easy application, the electric telegraphs previously in use in such establishments.

We extract the following from the 'Journal of the Society of Arts' (vol. xxvi, p. 887):

"THE TELEPHONE AND THE TORPEDO.

"A novel application of the Bell telephone is one which has been made in connection with torpedoes by Captain C. A. M'Evoy, of 18, Adam Street, Adelphi. The torpedoes to which the telephone has been applied are those of the buoyant contact class—that is, floating torpedoes—which are used for the protection of rivers and harbours. These torpedoes are held in position beneath the surface of the water by mooring lines and anchors, and it is necessary to ascertain from time to time that these deadly agents are in active working order. They are, of course, connected to the shore by electric wires, by which they may be exploded. They are also arranged so that they may be exploded electrically by contact with passing vessels. For this latter purpose they are fitted with what is known as a circuit closer, which is placed in the middle of the charge within the torpedo. The testing is ordinarily performed by sending a current of electricity through the torpedo and fuse; but, in order that the fuse may not be fired, and the torpedo consequently exploded during the process of testing, an extremely weak current has to be used in connection with a sensitive galvanometer. The consequence is that the indications received are so very delicate that they are not always to be relied on. Now, what Captain M'Evoy does is to supplement the electrical test by the test of sounds, and to this end he encloses an ordinary Bell telephone in each torpedo. The telephone is so placed that the vibrating diaphragm is in a horizontal plane, and upon it are laid a few shot or particles of metal, and these are boxed in. Every motion of the torpedo causes the shot to shift their position upon the face of the diaphragm, and to cause a slight noise, which is distinctly heard in the receiving telephone on shore. Thus each torpedo, two or three miles away in the restless waters of a channel, is continually telling the operator on shore of its own condition in language sometimes excited, according to the state of calmness or agitation of the water at the time. Should the torpedoes be sunk they would lie motionless on the bottom, and the silence of the telephone would indicate the fact of their inoperativeness. The telephones are connected to the ordinary electric wires of the torpedoes, but this does not prevent them from being tested in the usual way from the battery on shore."

TELLURIUM. Te—125. A rare greyish-white elementary substance, found only in small quantities, associated with gold, silver, lead, and bismuth, in the gold mines of Transylvania, California, Brazil, &c. It has often been described as a metal, but is now commonly classed with the non-metals and in the sulphur group.

Prep. 1. Tellurium may be obtained from the bismuth ore (the telluride of bismuth) by strongly heating the ore with an equal weight of carbonate of potash, having first rubbed the mixture into a paste with oil. A sodium telluride is formed which dissolves in water, forming a solution of a purplish-red colour, from which the tellurium deposits on exposure of the liquid to the atmosphere. The tellurium after washing and drying is distilled in a current of hydrogen (*Berzelius*).

2. (*Schrötter.*) Graphic or black tellurium ore is heated with dilute hydrochloric acid as long as any carbonic acid is evolved, then with strong acid until sulphuretted hydrogen ceases to come off. The liquid is decanted from the residue, which is washed with hydrochloric acid and hot water, then boiled with aqua regia until the insoluble matter is white. From the aqua regia solution any gold that may be present is precipitated by means of ferrous sulphate, and afterwards the tellurium is thrown down in the clear filtrate by passing in sulphur dioxide gas.

Prop. Tellurium bears a great resemblance to bismuth in appearance, having a pinkish metallic lustre, a bluish-white colour, and in being crystalline (rhombohedral) and brittle. Its sp. gr. = 6·24, and it fuses at 452° C.; at a high temperature it becomes converted into a yellow vapour. It burns in air, when strongly heated, with a blue flame having a green rim, and giving off white fumes of tellurium dioxide that have a peculiar odour. When taken internally, even in very minute quantities, tellurium imparts to the breath an offensively powerful odour of garlic. Tellurium dissolves in cold fuming sulphuric acid, to which it imparts a rich purple-red colour; it is insoluble in water and carbon bisulphide. If the acid solution be diluted with water the tellurium precipitates unchanged.

Tellurous acid, H_2TeO_3, is obtained by pouring a solution of tellurium in nitric acid of 1·25 into water, when the tellurous acid is precipitated as a bulky white hydrate. This hydrate is slightly soluble in water, and reddens litmus. It forms two classes of salts, called tellurites.

Telluric Acid. H_2TeO_4. When tellurium or tellurous acid is gently heated with nitre and potassium carbonate a potassic tellurate is formed, this being decomposed by barium chloride, whilst the resulting barium tellurate is in its turn decomposed, and the telluric acid separated by sulphuric acid.

Prop., &c. Hexagonal prismatic crystals, which, when heated usually to redness, become converted into telluric anhydride. It forms salts with metals, termed tellurates.

TELLURIUM CHLORIDES. *The dichloride*, $TeCl_2$, is formed with the tetrachloride when chlorine gas is passed over melted tellurium; it is separated from the tetrachloride by distillation as a greenish-yellow powder.

The tetrachloride, $TeCl_4$, is formed by the continued action of chlorine on the dichloride. A white crystalline substance.

Telluretted Hydrogen, or *Dihydric Telluride*. H_2Te. *Prep.* 1. By heating tellurium with zinc, and decomposing the zinc telluride with hydrochloric acid.

2. By heating tellurium in hydrogen gas.

Prop., &c. This compound presents a striking analogy to seleniuretted and sulphuretted hydrogen. Like both of these it is gaseous, but resembles the latter in smell more than the former. It burns with a blue flame, reddens litmus, and when dissolved in water forms a colourless solution, which becomes brownish red by exposure to the air, owing to the oxidation of hydrogen and the deposition of tellurium. The salts of most of the metals are decomposed when a current of telluretted hydrogen is passed through their solutions, from which the metals are then thrown down as tellurides. These tellurides present a close resemblance to the corresponding sulphides. The tellurides of the alkali metals, like the sulphides, are soluble in water.

Tests. The most distinctive character of tellurium compounds is the reddish-purple solution of potassium telluride they furnish when fused with potassium carbonate and charcoal and treated with water.

TEMPERATURE. In English pharmacy it is customary to measure the degree of heat by Fahrenheit's thermometer. When a boiling heat is directed, 212° is meant. A gentle heat is that which is denoted by any degree between 90° and 100° F.

Whenever specific gravity is mentioned, the substance spoken of is supposed to be of the temperature of 62° F. (Ph. L.).

In the B. P., Ph. E. and D., and in chemical works in this country generally, the specific gravities of bodies are taken at, or referred to, the temperature of 60° F. See THERMOMETERS.

The following data may be of use to the pharmacist:

Degree of F.

2786	.	.	Cast iron melts (Daniell).
2016	.	.	Gold melts (Daniell).
1996	.	.	Copper melts (Daniell).
1873	.	.	Silver melts (Daniell).
1750	.	.	Brass (containing 25% of zinc) melts (Daniell).
1000	.	.	Iron, bright cherry red (Poillet).
980	.	.	Red heat, visible in daylight (Daniell).
941	.	.	Zinc begins to burn (Daniell).
773	.	.	Zinc melts (Daniell).
644	.	.	Mercury boils (Daniell; 662—Graham).
640	.	.	Sulphuric acid boils (Magrignac; 620—Graham).
630	.	.	Whale oil boils (Graham).

Degree of F.

617	. . .	Pure lead melts (Rudberg).
600	. . .	Linseed oil boils.
518	. . .	Bismuth melts (Gmelin).
442	. . .	Tin melts (Crichton).
380	. . .	Arsenious acid volatilises.
356	. . .	Metallic arsenic sublimes.
315	. . .	Oil of turpentine boils (Kaure).
302	. . .	Etherification ends.
257	. . .	Saturated sol. of sal-ammoniac boils (Taylor).
256	. . .	„ „ acetate of soda boils.
239	. . .	Sulphur melts (Miller; 226—Fownes).
238	. . .	Saturated sol. of nitre boils.
221	. . .	„ „ salt boils (Paris Codex).
220	. . .	„ „ alum, carb. soda, and sulph. zinc boil.
218	. . .	„ „ chlorate and prussiate potash boil.
216	. . .	„ „ sulph. iron, sulph. copper, nitrate of lead, boil.
214	. . .	„ „ acetate of lead, sulph. and bitartrate potash, boil.
213 (or 213·5)	. . .	„ „ water begins to boil in glass.
212	. . .	Water boils in metal, barometer at 30°.
211	. . .	Alloy of 5 parts bismuth, 3 parts tin, 2 parts lead, melts.
207	. . .	Sodium melts (Regnault).
201	. . .	Alloy of 8 parts bismuth, 5 parts lead, 3 parts tin, melts (Kaue).
185	. . .	Nitric acid 1·52 begins to boil.
180 (about)	. . .	Starch forms a gelatinous compound with water.
176	. . .	Rectified spirit boils, benzol distils.
173	. . .	Alcohol (sp. gr. ·796 to ·800) boils.
151	. . .	Beeswax melts (Kane; 142—Lepage).
150	. . .	Pyroxylic spirit boils (Scanlan).
145	. . .	White of egg begins to coagulate.
141·8	. . .	Chloroform and ammonia of ·945 boil.
132	. . .	Acetone (pyroacetic spirit) boils (Kane).
122	. . .	Mutton suet and styracin melt.
116	. . .	Bisulphuret of carbon boils (Graham).
115	. . .	Pure tallow melts (Lepage; 92—Thomson).
112	. . .	Spermaceti and stearin of lard melt.
111	. . .	Phosphorus melts (Miller).
98	. . .	Temperature of the blood.
95	. . .	Ether (·720) boils.
95	. . .	Carbolic acid crystals become an oily liquid.
88	. . .	Acetous fermentation ceases, water boils in vacuo.
77	. . .	Vinous fermentation ends, acetous fermentation begins.
64·4	. . .	Oil of anise liquefies.
59	. . .	Gay Lussac's Alcoolomètre graduated at.
55	. . .	Syrups to be kept at (Ph. L.).
44·5	. . .	Potassium melts (Bunsen).
32	. . .	Water freezes.
30 (about)	. . .	Olive oil becomes partially solid.
5	. . .	Cold produced by snow, 2 parts, and salt, 1 part.
—37·9	. . .	Mercury freezes.

TENT. A piece of lint, or compressed sponge, used to dilate openings, wounds, &c.

TEPHRITIS ONOPORDINIS, Fabricius. (Tephritis, from τέφρος, ash-grey, because so many of these insects are of this colour; and Onopordon, a tribe of plants of which the cotton thistle is the English representative.) THE CELERY FLY. This small fly belongs to the order DIPTERA and its family Muscidæ, to the same family as the common house-fly. This family of Muscidæ comprises a very large number of species. Professor Westwood says that in this country 700 or 800 species have already been recorded ('An Introduction to the Modern Classification of Insects,' by J. W. Westwood, F.L.S.). Among these are the meat-fly, or 'blow-fly,' several kinds of Anthomyia, particularly injurious to vegetables, the fly which deposits maggots in cheese, the Oscinis and Chlorops, which ravage wheat and barley fields, and many other destructive and offensive insects.

Though this celery fly is small, it is intensely mischievous, frequently ruining crops of celery, and also doing much harm to parsnips in fields and gardens.

It may be urged that celery and parsnips hardly come within the definition of root crops, and that insects injurious to these should not have been included in those monographs which are intended to relate only to farm crops. But celery is grown on an extensive scale by market gardeners and market-garden farmers in the home counties and near cities and large towns, and being a profitable crop its cultivation will probably be extended. It is not by any means unusual for market gardeners and market-garden farmers to have as much as 50 acres of celery (see 'Report upon the Market Garden and Market-garden Farm Competition in connection with the Royal Agricultural Society's Show at Kilburn, in 1879,' by Charles Whitehead). Parsnips also are grown extensively by farmers, as well as by market gardeners, for cattle feeding.

In some seasons this celery fly is not very troublesome, while in others—as in 1883, for example—it causes most serious injuries in many parts of the United Kingdom. This variation in the degree of attack is seen in the case of all insects which affect the crops of the farm and garden, and is due to the influences of weather, as well as to the supply of parasitic insects and to other circumstances. Nature has provided for a balance or natural equipoise in the animal as in the vegetable kingdom ; but the exigencies of civilisation and of artificial systems of cropping and of cultivation have tended to interfere with natural laws, whose action the finite mind of man, with all its wisdom and all its intelligence, has not been able to supplement. This is proved by the fact that the number of the various insects injurious to crops has greatly increased during the last fifty years, in which the science of agriculture has made marvellous progress in every branch, and that new and most destructive insects have appeared in swarms upon the scene.

It is greatly to be desired that accurate observations should be made and careful records kept of the raids of insects upon crops, in order to show when and in what circumstances they may be expected.

In 1879 there was some injury occasioned by the *Tephritis onopordinis* both in England and Scotland, but not nearly so much as was caused in 1882 and 1883. It will be remembered that 1879 was a very wet year, while 1882 was fairly dry, and 1883 was very dry, and in the latter year more damage was caused to celery and parsnips than had ever been known before. This corroborates the opinion of all economic entomologists that this fly likes dry warm weather, and cannot propagate freely and rapidly in wet seasons.

Very serious harm was done by it in 1883 in the large market gardens and market-garden farms in Kent, Middlesex, Essex, Surrey, Hertford, Bedford, Lancashire, and elsewhere. One market-garden farmer reported that he had ten acres of parsnips almost entirely spoilt by its action. Another, near Bedford, only obtained about the third of a crop from nine acres planted with this vegetable. A large market-garden farmer in Essex planted forty acres with celery plants, and he calculated that he lost at least half of the crop from the persistence of the fly. He remarked that the leaves of the plants looked just as if they had been scorched by fire. Many reports of a similar nature came from celery and parsnip producers in various parts of the kingdom. Looking back at the outbreaks of this insect during the last ten years, it will be seen that they have been far worse when the summer season has been dry.

It is only since 1865, or about 20 years, that this insect has been known to occasion serious harm, at least upon anything like a large scale. Curtis, who wrote about the celery fly in 1859, does not speak of any amount of mischief attributed to it. Kirby and Spence do not allude to it. Rusticus does not mention it in his letters. Like many other insects in Great Britain, and indeed in all parts of the world, its power of affecting crops has become intensified during the last few years.

As there are no references to this insect by either of the German economic entomologists, Kaltenbach and Taschenberg, and as Köllar does not mention it, we may conclude that it is not known in Germany. Neither do French writers speak of it, and it appears to be unknown in America and Canada.

Life History. The celery fly is close upon the eighth of an inch in length of body, and has a wing expanse of not quite half an inch. It is of a tawny brown colour, or honey-yellow, as Meigen terms it ('Europaischen zweiflügeligen Insekten,' von J. W. Meigen), with the under part of the body light coloured.

Its halteres, or poisers, are yellowish, and its wings have lines of rusty-looking or brownish spots running obliquely on their upper parts, while the lower parts are hyaline. There are six legs, of a dark yellow hue. When at rest upon the plant its wings are carried upright, after the manner of some moths and butterflies, and of the *Hemerobidæ* or lacewing flies. The female fly is furnished with a long ovipositor, with which it places its eggs in the outer cuticle of the upper side of the leaves of celery and parsnip plants.

Many eggs are laid by each female. These are hatched in about six days, and from them proceed pale greenish larvæ, or maggots, which quickly grow and attain a length of about a quarter of an inch. The maggot has a dark line under the skin traversing the length of the body. It has no legs, and its mouth is small, with pointed jaws well suited for mining in the parenchyma of the leaves, and feeding upon its soft juicy substance. Very soon the leaf contracts, and its colour becomes bronzed. After a time it shrivels up, and is utterly useless to the plant as a medium of respiration and of receiving food from the air.

In the case of a celery plant thus affected, the underground leaves or stalks cannot grow and fill out properly with their surfaces which are above the ground in an unhealthy condition. The blanching process is also interfered with, so that the celery is green and small. In some cases of very bad attack the plant is killed outright. With respect to parsnips assailed by the fly, the injury to the leaves is the same as in the case of celery plants, and the roots are small, forked, and badly shaped.

Larvæ or maggots have been found upon celery and parsnip plants as late as November.

After about fourteen days the larva changes into a pupa, which either remains in the leaf or falls on to the ground beneath. From this, which is of a yellow colour, oval, and much wrinkled, gradually becoming brown or dark yellow, the fly emerges in the early summer in a few days, and commences a new cycle of existence. There are at least two generations in a season.

In fine dry summers there are more. The pupæ of the last generation, which is determined by conditions of weather and of food supply, pass the winter in their pupal form in the ground, and in the *débris* of the leaves.

Prevention. As many of the pupæ are in the ground, it is most essential when the celery crop has been taken from the trenches that the earth should be well levelled and dug deeply, and care should be taken to put the upper surface well underneath.

A good dressing of lime, or of lime ashes, might be applied with much advantage. Every particle of leafage must be buried deeply, or the leaves should be collected carefully and burnt. This should be done directly the celery is dug up. It is very important to destroy the leaves, so that the pupæ which are left within them may be prevented from changing into flies in the spring. If the leaves are merely put in a lump with other decaying vegetable matter, or upon composts, or mixens not in active fermentation, it is very probable that the pupæ within them may be carried out with manure for celery or parsnip crops, or to land near where these are grown. Celery and parsnip growers will not eradicate this pest unless they are most particular in destroying the leaves of plants that have been in the least degree affected. Weeds should be cleared away in the neighbourhood of celery and parsnip plants, as the fly has been bred from maggots from thistle leaves. Meigen says it is occasionally found in the thistle, and Curtis states that he has seen the fly come from blisters in the leaves of a weed known commonly as alexanders (*Smyrnium olusatrum*), formerly eaten like celery.

Remedies. Pinching the blistered leaves between the finger and thumb certainly kills the maggots. It also injures the leaves considerably, and if it is necessary to pinch a leaf in two or three different places it will do almost as much harm to it as the maggots. Besides, if the attack is extensive and general, it would be impossible to adopt this drastic remedy. Nor could it be carried out where the cultivation of celery and parsnips was on a large scale.

Soot scattered over the plants in the early morning while the dew is yet upon the leaves, or after a shower of rain when the leaves are wet, is very likely to prevent the flies from laying eggs upon them, and to keep the eggs from being hatched. Finely powdered guano may also be applied in this manner. Fine quicklime has also been known to be of benefit.

Even when the leaves are blistered, it has been found that these applications are of use.

It is desirable to force rapid leaf growth where there is a bad attack. Nitrate of soda sown broadcast at the rate of 1 cwt. to 2 cwt. per acre, with 2 cwt. of agricultural salt, will do this and keep the plants moving. In dry seasons, which are favourable for the fly, celery and parsnips, loving moisture, are most benefited by dressings of nitrate of soda and common salt, which give off moisture.

Natural Enemies. There are at least two parasitic flies which keep the celery fly in check. One of these is a fly smaller than the celery fly,

called by Curtis *Alysia apii.* It places its eggs in the larvæ of the celery fly, and from these larvæ quickly come, which feed upon the other larvæ and destroy them ('Reports on Insects Injurious to Crops,' by Chas. Whitehead, Esq., F.Z.S.).

TERBIUM. A rare metal found by Professor Mosander, associated with erbium and yttrium in ordinary yttria. See ERBIUM and YTTRIUM.

TEREBENE. *Syn.* TEREBENA, L. A compound body consisting of various isomers of turpentine. It is made by the action of strong sulphuric acid on turpentine, and distillation of the product. Amongst other substances it contains borneol, camphene, cymene, and terpilene. Its odour is refreshing, and very like that of pitch pine sawdust. It is volatile, inflammable, and insoluble in water; sp. gr. about ·864; soluble in alcohol, ether, and chloroform.

Uses. A common cheap form of terebene is used as a disinfectant. The pure colourless variety is given for the relief of winter cough, and forms a useful sedative and antiseptic inhalation for phthisis.—*Dose,* 5 or 6 drops on sugar every four hours, or placed in an inhaler with a pint of water at 140° F. and the vapour inhaled.

TERPIN HYDRATE. An interesting compound, formed when oil of turpentine is allowed to stand for a length of time over a small quantity of water. It is obtained in larger quantity in the presence of alcohol or of acids.

Terpene hydrate ($C_{10}H_{18}.2H_2O + H_2O$) forms large rhombic crystals, without colour or smell. It dissolves in 32 parts of water at 100° C., or 250 at 15° C., in 10 of alcohol, 100 of ether, 200 of chloroform, also in carbon disulphide and bensole; it is very little taken up by turpentine. In the capillary tube it melts between 116° and 117° C., and giving off water forms terpin ($C_{10}H_{18}.2H_2O$), which boils without decomposition at 258° C.

Recommended as a good expectorant in chronic and subacute bronchitis; also given in nephritis. In small doses it increases the secretion of the bronchial mucous membrane, and, in larger doses, of the kidneys. It was primarily anticipated that it would prove particularly active against diphtheria, but experience has not fulfilled this expectation. In bronchial catarrh of emphysematic and phthisical patients, where the secretion is small in quantity and tenacious, it has proved efficacious. Quite recently it has been recommended by Dr Manassé in the hooping-cough of children; he used it in doses of from 7 to 15 gr. three times a day, and observed a decrease in the number and intensity of the attacks.

TERRA. [L.] Earth. TERRA JAPONICA, catechu; TERRA PONDEROSA, sulphate of baryta; &c.

TERRA-COTTA. Literally, baked clay; a term applied to statues, architectural ornaments, &c., made of pure white clay, fine sand, and powdered potsherds, slowly dried, and baked to a strong hardness.

TEST. *Syn.* REAGENT. Any substance employed to determine the name or character of any other substance, or to detect its presence in compounds.

TEST SOLUTIONS. The test solutions here given are those of the British Pharmacopœia, which are used for determining the strength of various Pharmacopœial preparations by volumetric analysis. In the Pharmacopœia it is stated the processes for volumetric estimations may be performed either with British or with metrical weights and measures, and the solutions are so arranged that they will be of the same strength, and the same indications will be obtained in using them, whichever system is employed, without the necessity of altering any of the figures by which the quantities of the substances tested, or of the test solutions required in the process, are expressed.

According to the British system, the quantities of the substances to be tested are expressed in grains by weight, whilst the quantities of the test solutions employed in testing are expressed in grain-measures, the grain-measure being the volume of a grain of distilled water.

According to the metrical system, the quantities of the substances to be tested are expressed in grammes by weight, whilst the quantities of the test solutions employed in testing are expressed in cubic centimètres, the cubic centimètre being the volume of a gramme of distilled water.

As the cubic centimètre bears the same relation to the gramme that the grain-measure bears to the grain, the one system may be substituted for the other, with no difference in the results excepting that, by the metrical system, all the quantities will be expressed in relation to a weight (the gramme) which is more than fifteen times as great as the British grain.

In practice it will be found convenient, in substituting metrical for British weights and measures, to reduce the values of all numbers to one tenth by moving the decimal points, and this has been done in the tables appended to the descriptions of the volumetric solutions. The quantities indicated in the Pharmacopœia, which in grains and grain-measures can be conveniently used, would be found inconveniently large if the same numbers of grammes and cubic centimètres were employed.

The following apparatus is required in the preparation and use of these solutions.

For British weights and measures:

1. A flask which, when filled to a mark on the neck, contains exactly 10,000 grains of distilled water at 60° F. (15·5° C.). The capacity of the flask is therefore 10,000 grain-measures.

2. A graduated cylindrical jar which, when filled to O, holds 10,000 grains of distilled water, and is divided into 100 equal parts.

3. A burette. A graduated glass tube which, when filled to O, holds 1,000 grains of distilled water, and is divided into 100 equal parts. Each part, therefore, corresponds to 10 grain-measures.

For metrical weights and measures:

1. A glass flask which, when filled to a mark on the neck, contains 1 litre, or 1,000 cubic centimètres.

2. A graduated cylindrical jar which, when filled to O, contains 1 litre (1,000 cubic centimètres), and is divided into 100 equal parts.

3. A burette. A graduated tube which, when filled to O, holds 100 cubic centimètres, and is divided into 100 equal parts.

(One cubic centimètre is the volume of one gramme of distilled water at 4° C. [39·2° F.] [it is customary to make the measurements with metrical apparatus at 60° F. (15·5° C.)]. 1000 cubic centimètres equal 1 litre.)

Volumetric solutions, before being used, should be shaken in order that they may be throughout of uniform strength. They should also be preserved in stoppered bottles. All measurements should be made at 60° F. (15·5° C.).

VOLUMETRIC SOLUTION OF BICHROMATE OF POTASH (bichromate of potash, $K_2Cr_2O_7 = 295$). Take of bichromate of potash, 147·5 gr.; distilled water, a sufficiency.

Put the bichromate of potash into the 10,000-grain flask, and, having half filled the flask with water, allow the salt to dissolve; then dilute the solution with more water, until it has the exact bulk of 10,000 grain-measures: 1000 grain-measures of this solution contain 14·75 gr. of the bichromate (1-20th of $K_2Cr_2O_7$ in grains), and, when added to a solution of a ferrous salt, acidulated with hydrochloric acid, are capable of converting 16·8 gr. (1-20th of 6Fe in grains) from the ferrous to the ferric state.

Grammes and cubic centimètres may be employed instead of grains and grain-measures, but for convenience 1-10th of the numbers should be taken. Thus 14·75 grammes of bichromate of potash should be made to 1000 cubic centimètres of solution. 100 cubic centimètres of this solution contain 1·475 grammes of the bichromate (1-200th of $K_2Cr_2O_7$ in grammes), and, when added to a solution of a ferrous salt acidulated with hydrochloric acid, are capable of converting 1·68 grammes of iron (1-200th of 6Fe in grammes) from the ferrous to the ferric state.

This solution is used for determining the proportion of protoxide of iron in the following preparations. It is known that the whole of the ferrous salt has been converted into a ferric salt when a minute drop of the liquid, placed in contact with a drop of very dilute solution of potassium ferricyanide on a white plate, ceases to strike with it a blue colour.

		British Weights and Measures.				Metrical Weights and Measures.		
		Grains weight of substance.	=	Grain-measures of vol. sol.	or	Grams weight of substance.	=	C.C. of vol. sol.
Ferri arsenias	. .	100·0	=	225	or	10·0	=	22·5
„ carb. sacch.	. .	80·0	=	287·5	„	8·0	=	28·75
„ phosphas	. .	30·0	=	279	„	8·0	=	27·9
„ sulphas	.	42·1	=	500	„	4·21	=	50·0
„ „ exsiccata	.	10·0	=	191	„	1·0	=	19·1
„ „ granulata	.	41·7	=	500	„	4·17	=	50·0

VOLUMETRIC SOLUTION OF HYPOSULPHITE OF SODA (hyposulphite of sodium crystallised, $Na_2S_2O_3$, $5H_2O = 248$). Take of hyposulphite of soda, in crystals, 280 gr.; distilled water, a sufficiency.

Dissolve the hyposulphite of soda in 10,000 grain-measures of water. Fill a burette with this solution and drop it cautiously into 1000 grain-measures of the volumetric solution of iodine until the brown colour is just discharged. Note the number of grain-measures (*n*) required to produce this effect; then put 8000 grain-measures of the same solution into a graduated jar, and augment this quantity by the addition of distilled water until it amounts to $\dfrac{8000 \times 1000}{n}$ grain-measures. If, for example, *n* ≈ 950, the 8000 grain-measures of solution should be diluted to the bulk of $\dfrac{8000 \times 1000}{950} = 8421$ grain-measures. 1000

grain-measures of this solution contain 24·8 gr. of the hyposulphite ($\frac{1}{10}$ of $Na_2S_2O_3.5H_2O$ in grains), and therefore correspond to 12·7 gr. of iodine ($\frac{1}{10}$ of an atomic weight in grains).

Grammes and cubic centimètres may be employed instead of grains and grain-measures, but for convenience $\frac{1}{10}$ of the numbers should be taken. 100 c.c. of this solution contain 2·48 grms. of the hyposulphite ($\frac{1}{100}$ of $Na_2S_2O_3.5H_2O$ in grammes), and therefore correspond to 1·27 gr. of iodine ($\frac{1}{100}$ of atomic weight in grammes).

This solution is used for testing the following substances. In each case, except that of iodine, a solution of iodide of potassium and hydrochloric acid is added to the substance, and the amount of iodine so liberated is indicated by this solution.

		British Weights and Measures.		or	Metrical Weights and Measures.	
		Grains weight of substance.	= Grain-measures of vol. sol.		Grams weight of substance.	= C. C. of vol. sol.
Calx chlorinata	. .	5·0	= 467	or	0·50	= 46·7
Iodum	. . .	12·7	= 1000	„	1·27	= 100·0
Liq. calc. chlorinatæ	.	80·0	= 450	„	8·00	= 45·0
„ chlori	. .	439·0	= 750	„	43·90	= 75·0
„ sodæ chlorinatæ	.	70·0	= 500	„	7·00	= 50·0

VOLUMETRIC SOLUTION OF IODINE (iodine, I ‒ 127). Take of iodine, 127 gr.; iodide of potassium, 180 gr.; distilled water, a sufficiency.

Put the iodide of potassium and the iodine into the 10,000-grain flask, fill the flask to about two thirds its bulk with distilled water, gently agitate until solution is complete, and then dilute the solution with more water, until it has the exact volume of 10,000 grain-measures. 1000 grain-measures of this solution contain $\frac{1}{10}$ of an equivalent in grains (12·7 gr.) of iodine, and therefore correspond to 1·7 gr. of sulphuretted hydrogen,

3·2 gr. of sulphurous anhydride, and 4·95 gr. of arsenious anhydride.

Grammes and cubic centimètres may be employed instead of grains and grain-measures, but for convenience $\frac{1}{10}$ of the numbers should be taken. 100 c.c. contain 1·27 grms. of iodine, and correspond to 0·17 grm. of sulphuretted hydrogen, 0·32 grm. of sulphurous anhydride, and 0·495 grm. of arsenious anhydride. This solution is for testing the following substances. It is dropped from the burette into the liquid to be tested, until free iodine begins to appear in the solution.

		British Weights and Measures.		or	Metrical Weights and Measures.	
		Grains weight of substance.	= Grain-measures of vol. sol.		Grams weight of substance.	= C. C. of vol. sol.
Acid. arseniosum	. .	4·0	= 808	or	0·40	= 80·8
„ sulphurosum	.	64·0	= 1000	„	6·40	= 100·0
Liquor arsenicalis	. .	442·0	= 875	„	44·20	= 87·5
„ arsenici hydrochloricus }	442·0	= 875	„	44·20	= 87·5
Sodii hyposulphas	. .	24·8	= 1000	„	2·48	= 100·0

VOLUMETRIC SOLUTION OF NITRATE OF SILVER (nitrate of silver, $AgNO_3 = 170$). Take of nitrate of silver, 170 gr.; distilled water, a sufficiency.

Put the nitrate of silver into the 10,000-grain flask, and having filled half the flask with water, allow the salt to dissolve; then dilute the solution with more water until it has the exact bulk of 10,000 grain-measures.

The solution should be kept in an opaque

stoppered bottle. 1000 grain-measures of this solution contain $\frac{1}{10}$ of a molecular weight in grains of nitrate of silver (or 1·70 gr.). Grammes and cubic centimètres may be employed instead of grains and grain-measures, but for convenience $\frac{1}{10}$ of the numbers should be taken. 100 c.c. contain $\frac{1}{100}$ of a molecular weight in grammes of nitrate of silver (or 1·7 grms.).

It is used in testing the following substances:

		British Weights and Measures.		or	Metrical Weights and Measures.	
		Grains weight of substance.	= Grain-measures of vol. sol.		Grams weight of substance.	= C. C. of vol. sol.
Acid. hydrocyan. dil.	.	270	= 1000	or	27·0	= 100·0
Ammonii bromidum.	.	5	= 508·5 to 514·5	„	0·5	= 50·85 to 51·45
Potassii „	.	10	= 838 „ 850	„	1·0	= 83·8 „ 85·0
„ cyanidum	.	10	= 730	„	1·0	= 73·0
„ iodidum	.	10	= 602	„	1·0	= 60·2
Sodii bromidum	.	10	= 960	„	1·0	= 96·0
„ iodidum	.	10	= 660	„	1·0	= 66·0

VOLUMETRIC SOLUTION OF OXALIC ACID (crystallised oxalic acid, $H_2C_2O_4.2H_2O = 126$). Take of purified oxalic acid in crystals, quite dry, but not effloresced, 630 gr.; distilled water, a sufficiency.

Put the oxalic acid into the 10,000-grain flask, fill the flask to about two thirds of its bulk with water, allow the acid to dissolve, and then dilute the solution with more water until it has the exact volume of 10,000 grain-measures. Fill a burette with the solution, and add it gradually to a solution of 10·6 gr. of pure carbonate of sodium (which may be obtained by heating the pure bicarbonate of sodium to redness in a platinum crucible for 15 minutes), containing a few drops of solution of litmus, until the red colour produced ceases to change to blue on boiling. Note the number of grain-measures used (n), then put 9000 grain-measures of the solution of oxalic acid into a graduated jar, and increase this quantity by the addition of distilled water until it amounts to

$\dfrac{9000 \times 200}{n}$ grain-measures. 1000 grain-measures of this solution contain half a molecular weight in grains (63 gr.) of oxalic acid, and are therefore capable of neutralising 1 molecular weight in grains of an alkali, such as potash, &c.; or half the molecular weight in grains of such salts as anhydrous carbonate of sodium, Na_2CO_3, crystallised carbonate of sodium, $Na_2CO_3.10H_2O$, &c. Grammes and cubic centimètres may be employed instead of grains and grain-measures, but for convenience $\frac{1}{10}$ of the numbers should be taken. 100 c.c. contain $\frac{1}{10}$ of a molecular weight in grammes (6·3 grms.) of oxalic acid, and will neutralise $\frac{1}{10}$ of a molecular weight in grammes of an alkali. The following substances are tested with this solution:

	British Weights and Measures.			Metrical Weights and Measures.	
	Grains weight of substance.	= Grain measures of vol. sol.	or	Grams weight of substance.	= C. C. of vol. sol.
Ammonii carb.	52·3	= 1000	or	5·23	= 100·0
Borax .	191·0	= 1000	„	19·10	= 100·0
Liq. ammon.	85·0	= 500	„	8·50	= 50·0
„ „ fort.	52·3	= 1000	„	5·23	= 100·0
„ calcis .	4375·0	= 180	„	437·5	= 18·0
„ „ sacchar.	460·2	= 254	„	46·02	= 25·4
„ plumbi subacet.	284·5	= 500	„	28·45	= 50·0
„ potassæ	462·9	= 483	„	46·29	= 48·2
„ „ efferves.	4375·0	= 150	„	437·50	= 15·0
„ sodæ .	458·0	= 470	„	45·80	= 47·0
„ „ efferves.	4375·0	= 178	„	437·50	= 17·8
Plumbi acetas .	38·0	= 200	„	3·80	= 20·0
Potassa caustica .	56·0	= 900	„	5·60	= 90·0
Potassii bicarb. .	50·0	= 500	„	5·00	= 50·0
„ carb.	83·0	= 980	„	8·30	= 98·0
„ citras .	102·0	= 1000	„	10·20	= 100·0
„ tartras .	122·0	= 990	„	12·2	= 99·0
„ „ acida .	204·0	= 1000	„	20·40	= 100·0
Soda caustica	40·0	= 900	„	4·00	= 90·0
„ tartarata .	141·0	= 990	„	14·10	= 99·0
Sodii bicarb. .	84·0	= 960	„	8·40	= 100·0
„ carb. .	143·0	= 960	„	14·30	= 96·0

VOLUMETRIC SOLUTION OF SODA (hydrate of soda, NaHO = 40).

Take of solution of soda, a sufficiency; distilled water, a sufficiency.

Fill a burette with the solution of soda, and cautiously drop this into 1000 grain-measures of the volumetric solution of oxalic acid, until the acid is exactly neutralised as indicated by litmus.

Note the number of grain-measures (n) of the solution used, and having then introduced 9000 grain-measures of the solution of soda into a graduated jar, augment this quantity by the addition of water until it becomes—

$\dfrac{9000 \times 1000}{n}$ grain-measures.

If, for example, $n = 930$, the 9000 grain-measures should be augmented to—

$\dfrac{9000 \times 1000}{930} = 9677$ grain-measures.

One thousand grain-measures of this solution contain one molecular weight in grains (40 gr.) of hydrate of soda, and will therefore neutralise one molecular weight in grains of any monobasic acid.

Grammes and cubic centimètres may be employed instead of grains and grain-measures; but, for convenience, $\frac{1}{10}$ of the number should be

taken. 1000 cubic centimètres contain $\frac{1}{10}$ of a molecular weight in grammes (4 grms.) of hydrate of soda, and will neutralise $\frac{1}{10}$ of a molecular weight in grammes of an acid.

This solution is used for testing the following substances (see next page).

INDICATORS OF THE TERMINATION OF REACTIONS IN VOLUMETRIC OPERATIONS. The most important of these are as follows:

MUCILAGE OF STARCH, which gives an intensely blue colour with iodine. It may be used with the following substances:

Acidum arseniosum.
 „ sulphurosum.
Calx chlorinata.
Iodum.
Liquor arsenicalis.

Liquor arsenici hydrochloricus.
Liquor calcis chlorinatæ.
 „ sodæ „
 „ chlori „
Sodii hyposulphis.

SOLUTION OF FERRICYANIDE OF POTASSIUM, which gives an intensely blue precipitate with ferrous salts, but none with ferric. It is used with the following substances:

Ferri arsenias.
 „ carbonas saccharata.
 „ phosphas.

Ferri sulphas.
 „ „ exsiccata.
 „ „ granulata.

	British Weights and Measures.			or	Metrical Weights and Measures.		
	Grains weight of substance.	=	Grain measures of vol. sol.		Grams weight of substance.	=	C. C. of vol. sol.
Acetum	445·4	=	402	or	44·54	=	40·2
Acid. acet.	182·0	=	1000	„	18·20	=	100·0
„ „ dil.	440·0	=	813	„	44·00	=	81·3
„ „ glac.	60·0	=	990	„	6·00	=	99·0
„ citric	70·0	=	1000	„	7·00	=	100·0
„ hydrochloric	114·8	=	1000	„	11·48	=	100·0
„ „ dil.	345·0	=	1000	„	34·50	=	100·0
„ nitric	90·0	=	1000	„	9·00	=	100·0
„ „ dil.	361·3	=	1000	„	36·13	=	100·0
„ nitro-hydrochlor. dil.	352·4	=	883	„	35·20	=	88·3
„ sulph.	50·0	=	1000	„	5·00	=	100·0
„ „ arom.	195 0	=	500	„	19·50	=	50·0
„ „ dil.	359·0	=	1000	„	35·90	=	100·0
„ tart.	25·0	=	330	„	2·50	=	33·0

Solution of litmus, which gives a red colour with acids, and a blue colour with alkalies. It may be used with the following substances:

Acidum hydrochloricum.	Liquor potassæ.
„ „ dilutum.	„ „ effervescens.
Acidum nitricum.	„ sodæ.
„ „ dilutum.	„ „ effervescens.
„ nitro-hydrochlor. dil.	Potassa caustica.
	Potassii bicarbonas.
„ sulphuricum.	„ carbonas.
„ „ arom.	„ citras.
„ „ dil.	„ tartras.
Ammonii carbonas.	„ „ acida.
Borax.	Soda caustica.
Liquor ammoniæ.	„ tartarata.
„ „ fortior.	Sodii bicarbonas.
„ calcis.	„ carbonas.
„ „ saccharatus.	

Solution of yellow chromate of potassium, which gives a red colour with nitrate of silver, but not until any soluble bromide or iodide present is entirely decomposed. It may be used with the following substances:

Ammonii bromidum.	Potassii iodidum.
Potassii „	Sodii bromidum.

Tincture of phenol phthalein, which gives an intense red colour with potash or soda. It may be used with the following substances:

Acetum.	Acidum aceticum glaciale.
Acidum aceticum.	
„ „ dilutum.	Acidum citricum.
	„ tartaricum.

Test solutions for qualitative work, &c., which require special preparation (from the B. P.):—

Acetate of Copper. Take of subacetate of copper (powdered), ½ oz.; acetic acid, 1 fl. oz.; aqua dest., a sufficiency. Dissolve the acid with ½ a fl. oz. of the water; digest the subacetate of copper in the mixture at a temperature not exceeding 212° F. (100° C.) with repeated stirring, and continue the heat until a dry residue is obtained. Digest this in 4 oz. of boiling distilled water, and by the addition of more of the water make up the solution to 5 fl. oz. Filter it.

Acetate of Potassium or Sodium. Take of acetate of potassium or sodium, ½ oz.; aqua dest., 5 fl. oz. Dissolve and filter.

Solution of Albumen. Take of the white of one egg; aq. dest., 4 fl. oz. Mix by trituration in a mortar, and filter through clean tow, first moistened with distilled water.

This solution must be neatly prepared.

Ammonio-nitrate of Silver. Take of nitrate of silver, in crystals, ¼ oz.; solution of ammonia, ½ fl. oz., or a sufficiency; aq. dest., a sufficiency. Dissolve the nitrate of silver in 8 fl. oz. of the water, and to the solution cautiously add the ammonia until the precipitate first proved is nearly dissolved. Clear the solution by filtration, and then add distilled water, so that the bulk may be 10 fl. oz.

Ammonio-sulphate of Copper. Take of sulphate of copper, in crystals, ½ oz.; solution of ammonia, a sufficiency; aq. dest., a sufficiency. Proceed as directed for ammonio-nitrate of silver.

Ammonio-sulphate of Magnesium. Take of sulphate of magnesium, 1 oz.; chloride of ammonium, ½ oz.; solution of ammonium, ½ fl. oz.; aq. dest., a sufficiency. Dissolve the sulphate of magnesium and chloride of ammonium in 8 fl. oz. of the water, and to the solution add the ammonia and as much distilled water as will make up the bulk to 10 fl. oz. Filter it.

Boric Acid. Take of boric acid, 50 gr.; rectified spirit, 1 fl. oz. Dissolve and filter.

Bromine Water. Take of bromine, 10 minims; aq. dest., 5 fl. oz. Place the bromine in a bottle furnished with a well-fitting stopper, pour in the water, and shake several times. Keep it excluded from light.

Carbonate of Ammonium.—Take of ammonium carbonate, in small pieces, ½ oz.; solution of ammonia, ¾ fl. oz.; aq. dest., 10 fl. oz. Dissolve and filter.

Chloride of Ammonium. Take of ammonium chloride, 1 oz.; aq. dest., 10 fl. oz. Dissolve and filter.

Chloride of Barium. Take of barium chloride, in crystals, 1 oz.; aq. dest., 10 fl. oz. Dissolve and filter.

Ferricyanide of Potassium, or *Ferrocyanide of Potassium.* Take of ferri- or ferro-cyanide of potassium, ½ oz.; aq. dest., 5 fl. oz. Dissolve and filter.

Iodide of Potassium. Take of potassium iodide, 1 oz.; aq. dest., 10 fl. oz. Dissolve and filter.

Solution of Isinglass. Take of isinglass, in shreds, 50 gr.; warm distilled water, 5 fl. oz. Mix and digest an hour in a water-bath with repeated shaking, and filter through clean tow moistened with distilled water.

Solution of Litmus. Take of litmus, in powder, 1 oz.; rectified spirit, 10 fl. oz.; aq. dest., 10 fl. oz. Boil the litmus with 4 fl. oz. of the spirit for 1 hour, and pour away the clean fluid; repeat the operation with 3 oz. of the spirit, and a third time with the remainder of the spirit. Digest the residual litmus in distilled water, and filter.

Oxalate of Ammonium. Take of ammonium oxalate, ½ oz.; warm aq. dest., 1 pint. Dissolve and filter.

Phosphate of Sodium. Take of sodium phosphate, in crystals, 1 oz.; aq. dest., 10 fl. oz. Dissolve and filter.

Platinum Tetrachloride. Take of platinum-foil (thin), ½ oz.; nitric acid, a sufficiency; hydrochloric acid, a sufficiency; aq. dest., 7 fl. oz. Mix 1 fl. oz. of the nitric acid with 4 fl. oz. of the hydrochloric acid and 2 fl. oz. of the water; pour the mixture into a small flask containing the platinum, and digest with a little heat, adding more of the acids mixed in the same proportion should this be necessary, until the metal is dissolved. Transfer the solution to a porcelain dish, add to it 1 fl. dr. of hydrochloric acid, and evaporate on a water-bath until acid vapours cease to be given off. Let the residue be dissolved in the remaining 5 oz. of aq. dest. Filter, and [preserve it in a stoppered bottle.

Potassio-mercuric Iodide (Nessler's reagent). Take of potassium iodide, 135 gr.; perchloride of mercury, a sufficiency; caustic soda, 2 oz.; aq. dest., 1 pint. Dissolve the iodide of potassium and 100 gr. of the perchloride of mercury in 15 fl. oz. of boiling aq. dest. To this fluid add more aqueous solution of the perchloride of mercury until the precipitate produced no longer continues to disappear on well stirring, and a slight permanent precipitate remains. Then add the caustic soda. When the latter has dissolved, add a little more of the aqueous solution of perchloride of mercury; shake, allow to settle, and dilute the whole with distilled water to the volume of 1 pint. The solution should be kept in a stoppered bottle.

Stannous Chloride. Take of granulated tin, 1 oz.; hydrochloric acid, 3 fl. oz.; aq. dest., a sufficiency. Dilute the acid in a flask with 1 fl. oz. of the water, and, having added the tin, apply heat gently until gas ceases to be evolved. Add as much of the water as will make up the bulk to 5 fl. oz., and transfer the solution, together with the undissolved tin, to a bottle fitted with an accurately grooved stopper.

Sulphate of Indigo. Take of indigo, dry and in fine powder, 5 gr.; sulphuric acid, 10 fl. oz. Mix the indigo with 1 fl. dr. of the sulphuric acid in a small test-tube, and heat on a water-bath for an hour. Pour the blue liquid into the remainder of the acid, agitate the mixture, and, when the undissolved indigo has subsided, decant the clear liquid into a stoppered bottle.

Sulphate of Iron. Take of granulated sulphate of iron, 10 gr.; boiling aq. dest., 1 fl. oz. Dissolve and filter. The solution should be recently prepared.

Sulphate of Lime. Take of sulphate of calcium, ½ oz.; aq. dest., 1 pint. Put the sulphate of calcium in a porcelain mortar for a few minutes with 2 oz. of the water, introduce the mixture thus obtained into a pint bottle containing the rest of the water, shake well several times, and allow the undissolved sulphate to subside. Filter.

Sulphuretted Hydrogen. Take of sulphide of iron, ¼ oz.; water, 4 fl. oz.; sulphuric acid, a sufficiency. Place the sulphide of iron and the water in a gas-bottle, closed with a cork perforated by two holes, through one of which passes air-tight a funnel tube of sufficient length to dip into the water, and through the other a tube for giving exit to the gas. Through the former pour from time to time a little of the acid, so as to develop the sulphuretted hydrogen as it may be required. Allow the gas to bubble into a small bottle of distilled water very slowly until the water is saturated with the gas. Close the bottle with a well-ground stopper.

Sulphydrate of Ammonium. Take of solution of ammonium, 5 fl. oz. Put 3 fl. oz. of the ammonia into a bottle, and conduct into this a stream of sulphuretted hydrogen as long as the gas continues to be absorbed; then add the remainder of the ammonia, and transfer the solution to a green glass bottle furnished with a well-ground stopper.

Tartaric Acid. Take of tartaric acid, in crystals, 1 oz.; aq. dest., 8 fl. oz.; rectified spirit, 2 fl. oz. Dissolve the tartaric acid in the water, add the rectified spirit, and preserve the solution in a stoppered bottle.

Tincture of Phenol Phthalein. Take of phenol phthalein, 1 gr.; proof spirit, 500 gr. Dissolve. The solution should be colourless.

Turmeric Tincture. Take of turmeric, bruised, 1 oz.; rectified spirit, 6 fl. oz. Macerate for 7 days in a closed vessel, and filter.

Yellow Chromate of Potassium. Take of red chromate of potassium, 295 gr.; bicarbonate of potassium, 200 gr.; aq. dest., 10 fl. oz. Dissolve the red chromate in the water, and exactly neutralise the solution with the bicarbonate, evolution of all carbonic acid being ensured by ebullition. Filter.

TE'TANUS. Spasm with rigidity. When it affects the under jaw it is called TRISMUS, or locked-jaw; when the body is drawn backward by the contraction of the muscles it is called OPISTHOTONOS; when the body is bent forward, EMPROSTHOTONOS; and when the body is drawn to one side, PLEUROSTHOTONOS.

TETRANYCHUS TELARIUS. THE RED SPIDER (spinning mite). (*As affecting Hop Plants.*) This is a species of the order *Acarina*, or mites, in which are included many familiar and unpleasant creatures, as ticks, cheese mites, itch mites, among others. A familiar but disagreeable acquaintance of country people—the harvest bug, *Tetranychus autumnalis*—is another species of the genus of spinning mites.

Gardeners know the *red spider*, as it is commonly called, which is found on the under leaves of many plants both in the open air and in greenhouses and frames, and make lamentations over the great mischief it causes. It is not a spider, being essentially different in form, though it spins a kind of web upon the under surfaces of the leaves of the plants it infests, for its protection, and it has a peculiar arrangement of stiff hairs with round terminations, for the purpose of spreading and fixing this web. Nor is it always red. Its colour is also sometimes green, sometimes brown or brick-red, varying, as some naturalists think, according to its food. Upon hop leaves it has been found of many shades of colour, ranging from green, with tiny black specks on the sides of the mite, all through the variations between brown and bright red, upon the same leaf; rather indicating that the differences in colour are hardly attributable to diet, but to degrees of age. It is difficult to detect the presence of red spiders even when they are bright red, and almost impossible to see them when green or brown without a glass, so that casual observers or persons in any degree short-sighted do not discover them upon hop plants until considerable injury has been done.

Indeed, for some time the work of these mites upon hop plants was mistaken for the effect of drought and heat, particularly as the injury was first noticed upon badly drained spots, where drought would naturally show its results. This supposed disorder was called *Fireblast*, because the leaves turn bronze-coloured at first, then they become yellowish red, as if they had been burnt. In Germany the hop planters term it *Kupferbrand*. As red spider is only troublesome to hop plants in hot, dry seasons, it is easy to understand that its action may be mistaken for that of heat and drought.

Many would think it impossible that this tiny mite could work much harm upon the masses of vegetation in hop plantations. In 1868, when the summer was excessively hot and almost without rain, the crop was utterly ruined by red spiders, upon thousands of acres in England and upon the Continent. In Tasmania it is frequently very troublesome, so that the planters irrigate the hop land to destroy it in the ground before it can get to the plants. There was a sharp attack in England also in certain localities in 1872. During the late cycle of wet summers there was no sign of red spider. In 1884 there were clear indications of an attack in parts of Kent, Worcestershire, and Herefordshire. The weather changed, however, becoming damp and cool, and the mites could not work.

In 1868, at the beginning of July, the lower leaves of the hop plants became discoloured, 'fireblasted,' as the labourers said. This discoloration rapidly spread upwards, extending even to the lateral shoots. After a time the leaves fell off, the plants being quite exhausted of sap, and it was impossible for them in most cases to form any hop cones. Where these were formed they quickly shrivelled up and dropped off. Upon close examination of the leaves they were found to be desiccated. Their juices had been sucked out by myriads of mites, whose fine webs covered their under surfaces with countless filaments. Many plantations, which in June were green and flourishing, looked at the end of July as if a scorching fire had passed over them.

Not only do the mites exhaust the juices of the plants by means of the barbed suckers with which their mouths are fitted, but they hinder their respiration with their webs and excrements.

Life History. The red spiders pass the winter in the perfect state, either in the ground near where they have fallen with the leaves they have injured, or in other convenient places of shelter. They have frequently been found under stones, Kaltenbach states. In the case of hop plantations they also retire into the cracks of the poles, and they have been found upon the hop bines after they have been stacked for litter, as well as upon the ends of bines left in the hills or stocks during the winter. The females lay eggs which are rather large, spherical, and colourless, and are glued to the silky webs under the leaves. These are hatched in seven or eight days. The larva has six legs, but after the pupa or nymph stage there are eight legs, the full complement.

Means of Prevention. As English hop planters cannot irrigate the hop land, as is done in Tasmania, the only means of prevention are to apply hot lime or other caustic and pungent substances, as soot or lime, round the hop stocks in the late autumn after an attack, taking care that this should be put over the stocks and pieces of bine left on them. After an attack it would be of course desirable that the poles should be treated with a solution of paraffin or petroleum to kill the mites in their cracks. Practically, however, as hop planters would agree, this is almost impossible.

In the case of poles that are fixtures in the ground to carry wires or strings, according to the new methods of training hops, so much adopted in Germany and extending in this country, it would be well after an attack of red spiders to wash these poles with a strong solution of soft soap and water, with quassia added, or with paraffin or petroleum solutions brushed well into the crevices.

Poles should be well shaved before they are set up, as their bark harbours these mites and many insects injurious to hop plants.

Remedies. Kaltenbach, the German entomologist, says that washing with water containing solutions of sulphur and tobacco may be advantageously employed. This was tried in 1868 in England without much benefit. The only effectual remedy would appear to be washing the plants by means of hand or horse engines, with a composition of water, soft soap, and quassia, in the following proportions:—100 galls. of water, 4 to 6 lbs. of soft soap, 4 to 6 lbs. of quassia (extract after well boiling).

Water alone would be effectual, only it runs off the web-covered leaves. The soap fixes it on them, and the bitter of the quassia makes them unpleasant to the tastes of the red spiders.

TETRANYCHUS TELARIUS. THE RED SPIDER. (*As affecting Plum and Damson Trees.*) The red spider is most destructive to plum and damson trees, and this was most plainly shown in the season of 1886. Though by many it is regarded quite as a new scourge to these fruit trees, it has been observed previously upon them as well as upon other kinds, and has in many cases escaped detection on account of its very minute size. The changed colours of the leaves of plum and damson trees from deep green to rusty brown, which have sometimes been noted, have been often set down to the influences of weather, to the east wind and scorching sun; while the red spiders, mere specks, the causes of the evil, have been overlooked or regarded as accessories after the fact.

In the spring of the year 1886 the plum and damson trees were full of blossoms. These set well, and there was a prospect of a large crop; but a change came over the trees in some places. The green colour of the trees became duller and duller in hue, and many fell off, while the fruit became stunted, and some of it dropped off. There was no doubt at all as to the cause of this. Upon examination of the under surfaces of the leaves tiny, moving, reddish specks might be seen even with the naked eyes congregated by the sides of the midribs and the lateral ribs of the leaves. With the aid of a pocket glass experienced persons could see plainly that they were spinning mites—red spiders—as their webs were apparent. Under the microscope they were found, without any shadow of doubt, to be of the species *Tetranychus telarius*, as the globular terminations of their feet, or claws, illustrated by Mr Andrew Murray ('Economic Entomology,' by Andrew Murray), and other distinctions peculiar to this species, were clearly seen. Deep in the hollows by the sides of the midribs of the leaves were rows of eggs, some distance apart from each other, very large in proportion to the mites. These were spherical, and some were of a white colour, whilst others were of a semi-transparent golden hue.

A somewhat unusual and noteworthy circumstance in connection with this attack of red spiders was that the mites were full grown by the 25th of May, and had been evidently laying eggs for some time, and were perfectly vigorous and very numerous, although the weather had been very cold at times. Cold and moisture have been held generally to be most unfavourable to red spiders, and their mischief to hop plants has been done in abnormally hot, dry summers—as in 1868, for example.

There are two new features in connection with red spiders.

First. Their early appearance, and their rapid and injurious spread upon damson and plum trees.

Second. Their alarming increase and hurtful action upon plants in normal spring weather.

These new features should cause much anxiety to hop planters and fruit-growers, and to all cultivators who know how very rapidly these spinning mites breed, and what mischief they occasion by sucking the juices—the life-blood—from the leaves of trees and plants, as well as by obstructing their respiration.

Life History. In Californian fruit plantations the red spider is very troublesome, causing just the same injury to fruit trees and most other deciduous trees as that described above. In other parts of the United States and in Canada it is also a recognised pest to fruit trees. It is also well known as injurious in France and Germany, as well as in Tasmania.

Prevention. The stems and larger branches of damson and plum trees that have been infested with red spiders should be washed over with a strong solution of soft soap and quassia. A mixture of soft soap and paraffin is a valuable wash, in the proportion of 15 lbs. of soft soap and 2 quarts of paraffin to 100 galls. of water, put on with a whitewash brush or a large paint brush and worked well into the stems and branches. Or a wash composed of 15 lbs. of soft soap and 12 lbs. of the finest flowers of sulphur to 100 galls. of water may be applied with much advantage, as red spiders do not like sulphur (the very finest flowers of sulphur must be used, and the liquid should be well stirred when the sulphur is added —C. W.). The petroleum soap described before would also be a valuable application.

The leaves and cuttings from infested trees should be burnt, and the stakes used as supports for young trees should be treated with washes in the same manner as the stems.

Remedies. Syringing the trees with soft soap and quassia effectually drives away or kills the mites themselves. The eggs, however, are not affected. In the present season, 1886, it was necessary to syringe again at the end of a fortnight, in order to kill the broods that came from the eggs that were upon the leaves at the first syringing. This washing was of much benefit, and undoubtedly saved the crop upon a good many of the trees, and helped the next crop, as the trees would hardly bear after a bad attack in the preceding year.

Flowers of sulphur may be put into the soft soap wash, either with or without the quassia, at the rate of 8 lbs. of soft soap and 10 lbs. of sulphur to 100 galls. of water ('Reports on Insects Injurious to Crops,' by Chas. Whitehead, Esq., F.Z.S.).

TETTERS. The popular name of several cutaneous diseases, the treatment of which can only be properly undertaken by the experienced medical man.

THALLIN SULPHATE. The base thallin, first prepared in 1885 by Skraup, has the systematic name 'tetrahydroparachinanisol.' The salt, $(C_{10}H_{13}NO)_2H_2SO_4$, forms a yellowish-white crystalline powder, with a cumarin-like smell, and an acid, saline, somewhat bitter taste. It dissolves in 7 parts of cold or ½ a part of boiling water, also in 100 parts of alcohol; in chloroform it is very slightly, and in ether practically not at all soluble It melts at 100° C., and when strongly heated decompose; and forms a deep black swollen mass, which burns away without residue. It possesses antipyretic, antiseptic, and antifermentative properties, and has been used internally in various febrile conditions, mostly in solution with water or wine and syrup of orange peel. If used in small doses no unpleasant secondary symptoms appear, but larger quantities should be prescribed with caution. Externally its antiseptic properties have been particularly requisitioned for the treatment of gonorrhœa, in which, applied by means of bougies, it has attained conspicuous success.

THALLIUM. Tl=203·6. A heavy metal, belonging to the mercury, silver, and lead group, discovered by Crookes in the early part of 1861, and displayed by him as 'a new metallic element' at the opening of the International Exhibition, on the 1st of May, 1862. Thallium is a widely diffused metal, being found in many minerals, particularly in iron- and copper-pyrites and native sulphur. It has recently been obtained in comparatively large quantities from the dust of the flues leading to sulphuric acid chambers. The spectrum of thallium consists of a single most characteristic line of a beautiful green colour. The spectrum produced when the metal is burnt in the electric arc is, however, more complicated, and consists of several green, blue, and other lines.

Thallium melts at 550° F., and at a less heat may be readily welded, a property that has hitherto been regarded as peculiar to iron and platinum. Its specific gravity varies from 11·8 to 11·9, according to the mode of preparation. When freshly cut it has a dull white colour, destitute of the brilliancy of polished silver. Exposed to the air it tarnishes rapidly, a straw-coloured oxide making its appearance on the surface. The oxides Tl_2O, Tl_2O_3, are alkaline and caustic to the taste, and much more soluble than the oxides of silver and lead. The metal is remarkable for its strongly marked diamagnetic characters, resembling bismuth in this respect. The alloys of thallium are very remarkable. Copper, alloyed with only ½% of thallium, becomes quite brittle; but the alloy with tin is malleable. Crookes has prepared a great number of the salts of this interesting metal; all these are poisonous, and may be readily detected by the beautiful green colour which they impart to a non-luminous flame. These need not further be described here, as they have not yet been applied to any use in the arts. See SULPHURIC ACID.

THALLOGENS. Thallogens or thallogenous plants are structurally the simplest of the acotyledonous or flowerless plants, consisting simply of a collection of cellular tissue, called a *thallus*. They are entirely destitute of woody fibre. The *Algæ, Characeæ, Fungi*, and Lichens are thallogenous plants.

THEBA'INE. $C_{19}H_{21}NO_3$. *Syn.* THEBAIA, PARAMORPHIA. A crystalline substance obtained by Thibourméry from an infusion of opium that has had its morphia extracted by acting on it by an excess of lime.

THE'INE. $C_8H_{10}N_4O_2$. *Syn.* THEINA, L. An alkaloid extracted from tea. It is identical with caffeine, and may be obtained from tea in the same manner as that substance is from coffee. The best 'gunpowder tea' contains fully 6% of theine, about one half of which is lost in the present careless mode of making infusion of tea for the table.

Mr Lewis Thompson, M.R.C.S., in a contribution to the ' Medical Times and Gazette ' for 1871, directs attention to the value of theine as a therapeutic agent, as well as gives an easy method for its preparation. He writes as follows :—" I wish to direct the attention of the medical profession to a valuable agent which has hitherto escaped notice, although its powers are most unquestionable, and its cost price very trivial. The article to which I allude is theine, a substance existing in tea and coffee, and, as I believe, in many other vegetable products.

" As a medicine, theine is powerfully tonic and stimulant, and appears to possess the tonic virtues of the disulphate of quina united to the stimulating power of wine, but with this difference, that the stimulus from theine is not followed by depression, as in the case of wine and alcohol.

" Theine seems to act chiefly on the great sympathetic or ganglionic system of nerves, and but slightly on the brain. I have used it in doses of from 1 to 5 grms. with very marked advantage in the low stage of typhoid fevers, confluent small-pox, and that form of mortification of the toes which is so singularly fatal to old people. But, in addition to this, different medical friends of mine have found it useful in hemicrania, neuralgia, and what has been called relapsing fever ; and in the case of an overdose of opium it appeared to relieve the narcotic symptoms speedily.

" With regard to the cost of this medicine, I have discovered that in the ordinary process of roasting coffee the whole of the theine is driven off before the torrefaction of the coffee is completed ; and thus theine may be cheaply collected by making the axis of the coffee roaster tubular. If, instead of a solid axis, we employ at one end of the roasters a tube passing away to the distance of about three feet, the theine is condensed in this tube by the refrigerating power of the atmosphere, and may afterwards be easily dissolved out by a little water and purified in the manner about to be indicated.

" As the result' of much experience, I have obtained on an average 75 gr. of theine from the roasting of 1 lb. of raw coffee ; and when we reflect than in Great Britain alone there are more than 13,000 tons of coffee roasted annually, we see that about 140 tons of theine are wasted and lost every year by sheer ignorance. It may, perhaps, be thought that the saving of the theine will damage the flavour of the coffee, but from experience I know that it has no such effect ; and, in point of fact, it is an advantage to the flavour of the coffee to make both the axes of the roaster tubular, and to cause a gentle current of air to pass through the apparatus during the roasting of the coffee, so as to expel the empyreumatic products that are formed. I will now relate the fact upon which the purification of theine depends ; and when this is once clearly understood, the manufacture of theine from either tea or coffee becomes an extremely simple matter. Theine is absolutely insoluble in a concentrated solution of the carbonate of potash, and thus we may precipitate it from its admixture with sugar, mucilage, and vegetable extract. If, then, by means of the subacetate of lead we have removed from a vegetable infusion the tannin, malic acid, &c., we have only to evaporate the filtered solution to a small bulk, and add to it its own weight of dry carbonate of potash, and the whole of the theine becomes at once insoluble ; so that having collected this insoluble product, and boiled it in rectified spirits of wine, we have a solution of pure theine, which, after distilling off the spirit, furnishes crystals fit for immediate use. In conclusion, I will merely mention a distinctive test for theine, sufficiently delicate to detect the one thousandth of a grain of that substance. Dissolve the theine in a small quantity of water, and pass through this a stream of euchlorine, then allow the fluid to evaporate at a steam heat ; a blood-coloured substance will remain, which, on the application of a few drops of cold water, forms a beautiful scarlet solution like red ink. It is, I apprehend, almost unnecessary for me to say that euchlorine gas is formed by the action of hydrochloric acid upon the chlorate of potash.

"I ought, perhaps, to add that theine collected as a waste product of coffee, and purified by myself, has cost me less than threepence per ounce troy."

THENARD'S BLUE. See ULTRAMARINE (Cobaltic).

THEOBROMÆ OLEUM. *Syn.* CACAO BUTTER. A concrete oil obtained by expression and heat from the ground seeds of *Theobroma cacao.* Occurs in cakes of a yellowish colour, of a pleasant cacao odour. Does not become rancid from exposure to air. Contained in all the suppositories. *Not Official.* The following form good bases for suppositories:—Theobroma oil, when melted, begins to solidify at 72° F.; stearine of cocoa-nut oil at 75° F.; 4 parts of stearine and 2 parts of mutton fat at 77° F.; 4 parts of stearine and 1 part of spermaceti at 80° F. Stearine alone is, perhaps, a better substance than cacao butter for making suppositories. It begins to solidify at 78° F., but there is stearine that solidifies at 120° F.; this will not answer for suppositories.

THEOBRO'MINE. A peculiar principle, closely resembling caffeine or theine, found by Woskresensky in the seed of the *Theobroma cacao,* or the nuts from which chocolate is prepared. Its form is that of a light, white, crystalline powder, which is rather less soluble than caffeine. It is obtained like caffeine. See COCOA.

THERI'ACA. A name given in ancient pharmacy to various compound medicines, chiefly electuaries or confections, employed as antidotes to poisons or infection. The THERIACA ANDROMACHI, Ph. L. 1746, contained above 60 ingredients. Mithridate and Venice treacle are examples of this class. See TREACLE.

THERMOM'ETERS. FAHRENHEIT'S scale is the one generally employed in England, while that of CELSIUS, or the CENTIGRADE scale, is principally used on the Continent. REAUMUR'S is another scale occasionally employed. DE LISLE'S thermometer was formerly used in Russia and some other parts of the north of Europe. As references to these scales are frequently met with in books, it is useful to know their relative value and the method of reducing the one to the other. The boiling-point of water is indicated by 212° on Fahrenheit's scale, 100° on the Centigrade scale, 80° on that of Reaumur, and 0° on that of De Lisle; the freezing-point of water marks 32° Fahrenheit, and 0°, or zero, on the Centigrade and Reaumur, and −150° on the scale of De Lisle. The 0° or zero of Fahrenheit is 32° below the freezing-point of water.

1. To reduce Centigrade degrees to those of Fahrenheit multiply them by 9, divide the product by 5, and to the quotient add 32; that is—

$$\frac{\text{Cent.}° \times 9}{5} + 32 = \text{Fahr.}°$$

2. To reduce Fahrenheit's degrees to Centigrade:

$$\frac{\text{Fahr.}° - 32 \times 5}{9} = \text{Cent.}°$$

3. To reduce Reaumur's to Fahrenheit's:

$$\frac{\text{Reau.}° \times 9}{4} + 32 = \text{Fahr.}°$$

4. To convert Fahrenheit's to Reaumur's :-

$$\frac{\text{Fahr.}° - 32 \times 4}{9} = \text{Reaumur}°$$

Thermometers intended to register extreme degrees of heat are called PYROMETERS (which see).

THIBANT'S BALSAM. FOR WOUNDS. Digest flowers of St John's-wort, one handful, in ½ pint of rectified spirit; then express the liquor, and dissolve in it myrrh, aloes, and dragon's-blood, of each, 1 dr., with Canada balsam, ½ oz.

THIOCAMF. This is a colourless, syrupy liquid, for which the formula $(C_{10}H_{16}O)_2SO_2$ has been suggested. It is prepared by bringing sulphurous acid gas in contact with camphor contained in a vessel surrounded by ice. Analysis shows that the camphor absorbs more than three times its volume of the gas, probably forming a definite though weak chemical compound. When exposed to the air it gives off free sulphur dioxide, and becomes covered with a pellicle of camphor. By virtue of this property, and because it is not apparently liable to spontaneous change, thiocamf was introduced first in England by Dr E. Reynolds, and a few months later recommended in America by Parke, Davis, and Co. as an efficient disinfectant. It seems that the best method of using is to pour it on paper and suspend these in the apartment to be disinfected, but it can also be used in water (1 : 40) like other liquid disinfectants.

THO'RIUM. Th. *Syn.* THORINUM. A very rare element, belonging to the group of earthy metals. Metallic base of thoria. It is obtained by the action of potassium on the chloride of thorium, and washing the resulting mass in water.

THORN-APPLE. See DATURA.

THRIPS CEREALIUM, Haliday. THE CORN THRIPS. Although very small indeed, this little creature does an infinity of harm to wheat, oat, and barley plants in some seasons and in some localities. It is that tiresome insect which gets on the face and hands, and occasions much annoyance, and even irritation of the skin, by running over these in the months of July and August.

It belongs to the order *Thysanura,* and to its family *Physopoda,* or bladder-footed, so called, from the shape of its feet.

Being so tiny, its action upon corn plants is frequently unnoticed, and the results are attributed to other than insect agencies; or they are frequently called blight, or supposed to be due to an abnormal state of the plants.

Upon close examination of affected plants it will be found that the thrips have taken up positions under the covering, or case, or corolla of the seed of corn, within the slits of the seeds, and are sucking the milky juices from them with their short stout beaks. They seem only to enter the ears of corn just previous to the blossoming period. It has been supposed that they are attracted by the pollen, but it is certain that their chief attraction is the sweet fluid of developing seeds. Their action upon the grains of corn renders them light and shrivelled. This insect is known in America, Germany, and France as very destructive to corn crops, and it has been unusually troublesome to wheat plants in England during the summer of 1888.

Life History. The perfect insect is only about a line—the twelfth of an inch—in length. It is of a blackish or darkish brown colour, with long wings having long thick cilia or fringes. The antennæ are also fringed. The males are wingless. It is believed by Taschenberg that they pass the winter in decayed roots and in stubble in the perfect state, emerging thence in early spring, and laying eggs on grasses and on corn plants, and producing many generations in the course of the summer. The larvæ are of a bright orange-yellow colour, and may be distinguished from the larvæ of the *Cecidomyia tritici* by being of a rather brighter colour, and not quite so large, and by the end of the abdomen being dark coloured, as well as by both the larvæ and pupæ being furnished with three pairs of claw feet.

Prevention. After an attack of thrips upon corn the stubble should be burnt or removed to cattle yards, and undergo fermentation in mixens. Grass must be brushed close on the outsides of the corn-fields and the rubbish burnt.

Remedies. There appears to be no remedial measures against these numerous and almost microscopic insects. Earwigs feed upon them, but earwigs are not frequently found in white straw crops ('Reports on Insects Injurious to Crops,' by Chas. Whitehead, Esq., F.Z.S.).

THRUSH. *Syn.* APHTHA, L. A disease of infancy, which, in its common form, is marked by small white ulcers upon the tongue, palate, and gums. In some cases it extends through the whole course of the alimentary canal, and, assuming a malignant form, proves fatal. The treatment consists of a gentle emetic of ipecacuanha wine, followed by an occasional dose of rhubarb and magnesia, to keep the bowels clear, and to arrest diarrhœa. The ulcerations may be touched with a little honey or borax; and if they assume a dark colour, or there be much debility, astringents and tonics should be had recourse to. In all cases the diet should be light, but supporting, as imperfect nutrition is a common cause of the disease.

In Animals. Topical applications of alum or borax, glycerin, Condy's fluid; laxatives. The food should be cooling and digestible.

THYMOL. *Syn.* THYMIC ACID. $C_{10}H_{14}O$. This substance is the oxygenated constituent of the essential oils of thyme (*Thymus vulgaris*), horse-mint (*Monarda punctata*), and *Carum ajowan*, a common umbelliferous plant grown in India. Thymol is isomeric with cymilic alcohol, and homologous with phenyl.

Thymol may be procured from either of the above sources by treatment with caustic potash or soda, as described below, or by submitting the essential oils to a low temperature for some days. When prepared by the first process thymol occurs as an oily fluid, and when by the second as a crystalline solid.

The following are the details of the preparation of the liquid variety of thymol as given by the Paris Pharmaceutical Society in their formulæ for new remedies published in 1877:

"Treat essential oil of thyme with an equal volume of an aqueous solution of potash or soda, and shake several times to facilitate combination. The thymol dissolves, forming a soluble compound, whilst the thymene, a carbide of hydrogen that accompanies it in the essence, does not combine with the alkali and separates. Filter the solution obtained and treat with an acid—hydrochloric acid, for example—which sets free the thymol. The product should be purified by washing, dried, and distilled. Thymol was obtained in fine tabular crystals by Flückiger and Hanbury, who exposed oil of ajowan to a temperature of 0° C.; the oil so treated yielded 85% of its weight of crystallised thymol. Mr Gerrard says it is stated that oil of thyme yields as much as 50%.

"As found in commerce, thymol consists of irregular broken crystals, nearly transparent and colourless; the taste is burning and aromatic; sp. gr. 1·028, but lighter than water when fused; its melting-point is about 44° C. When once completely fused, and allowed to cool to the ordinary temperature, it will maintain itself in the fluid condition for several days, but the contact of a crystal will at once cause it to crystallise. It is freely soluble in alcohol, ether, chloroform, benzol, carbon bisulphide, fats, and oils, and but sparingly in water and glycerin. The alkaline hydrates of potash and soda are powerful solvents of thymol; ammonia dissolves it but sparingly.

"The potash and soda solutions are spoken of by some authors as chemical combinations; but the following test will demonstrate them otherwise. When shaken with ether the thymol can be entirely removed, and obtained as a neutral volatile residue."

With sulphuric acid thymol forms a crystallisable colligated acid, the thymol sulphuric having the formula $HC_{10}H_{13}SO_4$. Undiluted thymol is an energetic caustic. According to Buchola, thymol possesses ten times the septic power of carbolic acid, over which it also has the advantage of being non-poisonous, and of giving off an agreeable odour. Although considerably dearer than carbolic acid, the much smaller quantity required to produce an equivalent effect nearly equalises it in point of cost. It is said to have been successfully employed in the antiseptic treatment of wounds, in destroying the fœtor arising from ulcerated surfaces and carious bones; in the form of spray during surgical applications, as well as for certain throat affections; and as an ointment and lotion in psoriasis and other skin diseases. When thymol is to be used for lotions, injections, inhalations, or spray solutions, the Paris Pharmaceutical Society recommends 1 part of thymol to be dissolved in 4 parts of alcohol at 90°, and this to be added to 995 parts of distilled water.

Dr Crocker, of University College Hospital, strongly recommends thymol lotion to be prepared with glycerin, which, he says, obviates the drying effect upon the skin produced by aqueous or spirituous solutions of the thymol alone. According to Mr Gerrard ('Thymol and its Pharmacy,' by A. W. Gerrard, F.C.S., 'Ph. Journ.,' vol. viii, third series, p. 645), this lotion is prepared by dissoving 1 part of thymol in 120 parts of glycerin, and reducing by water to 600 parts. Dr Symes says he finds milk to be an excellent solvent for thymol, of which it will take up readily to nearly 10% of its weight. In cases, therefore, in which solutions are required of greater strength than aqueous ones he recommends the employment of the fluid.

An ointment varying in strength from 1 to 5 parts of thymol to 100 of lard is said by Mr Gerrard to be employed in our hospitals. In the preparation of this ointment it is of importance to first dissolve the thymol in a few drops of spirits, and then to mix it with the lard. The neglect of this precaution causes the undissolved particles of thymol present in the ointment to act as a caustic irritant on the skin, and to eat little holes in it. Mr Gerrard found vaseline an unsuitable and objectionable vehicle for the application of thymol, since, after a few days, an ointment prepared with it had its surface covered with minute crystals of thymol.

The 'Medical Times' contains the following formula for the preparation of thymol gauze for dressing wounds:—" Bleached gauze, 1000 parts; spermaceti, 500 parts; resin, 50 parts; thymol, 16 parts." This is said to yield an extremely soft and pliant preparation, excellently adapted for wounds, fitting accurately to them, and absorbing, at the same time, the blood and secretions from them like a sponge would do. Dr Ranke has pointed out that, in consequence of the great reduction in the amount of secretion from wounds caused by the use of thymol, the consequent consumption of bandages becomes so much less as to more than compensate for the great difference in price between thymol and carbolic acid.

Another advantage possessed by thymol over carbolic acid is that the redness, vesication, and eczema, frequently induced when dressings of the latter agent are used, does not follow the application of thymol dressings.

Mr Squire prepares an antiseptic adhesive plaster containing 1 part of thymol to 1000 parts of plaster.

Mr Gerrard, in operating upon nine different samples of commercial oil of thyme (so-called oil of origanum) by means both of caustic soda and refrigeration, states that, except in one doubtful case, he was unable to obtain the slightest trace of thymol. From this circumstance Mr Gerrard infers that thymol is not present in the English oils of thyme of commerce, from which it must have been removed in the countries where it is produced, the residual cymene and thymene being sent to us as an oil of thyme.

Large quantities of thymol are prepared in Germany, principally from the seeds of the *Ptychotis ajowan*. One firm of chemical manufacturers residing in Leipzig is reported to have sent out during the months of September and November last year more than a ton of it. Thymol wadding is also in extensive demand.

Thymol Gauze. Thymol, 16 parts; resin, 50 parts; spermaceti, 500 parts. Mix by fusion and heat. Impregnate cotton gauze with the fluid.

Thymol Solution (Volckmann's). Thymol, 1 part; alcohol, 20 parts; glycerin, 20 parts; water, to make 1000 parts. Used as an antiseptic lotion or spray.

Thymol Spirits. Thymol, 1 part; rectified spirit, 9 parts. Used for medicating wools and lint.

Thymol Vapour. Thymol, 6 gr.; rectified spirit, 1 dr.; carbonate of magnesium, 3 gr.; water, to 1 oz. A teaspoonful to a pint of water at 140° F. Used in pharyngitis and laryngitis.

TIC-DOULOUREUX'. [Fr.] According to a writer in one of the medical periodicals, a solution of atropia, 2 gr., in water, 1 fl. dr., to which nitric acid, 1 drop (minim), has been previously added, applied as a paint, by means of a camel-hair pencil, to the part of the face over the spot affected, immediately and completely subdues the pain within three to five minutes in all accidental cases, and affords considerable relief in others. The application is to be continued until some relief is experienced. The solution, being very poisonous, must not be taken internally, nor applied to the skin when broken. See ATROPIA and NEURALGIA.

TILIACIN is a new glucoside extracted from the leaves of the linden tree by Mr Latschinow, and found also in *Cirsium arvense* and *Phlox paniculata*. It is similar to hesperidin, and among other products yields anisic acid on decomposition.

TIN. Sn = 118·8. *Sym.* STANNUM (Ph. E. & D.), L. This metal has been known from the most remote antiquity, being mentioned in the books of Moses (Numb. xxxi, 22), and by Homer ('Iliad,' x, 25) and other early writers. The ancients obtained it principally, if not solely, from Cornwall. The Phœnicians traded with England for this metal at least 1000 years before the birth of Christ.

Tin occurs in nature in the state of oxide and, more rarely, as sulphide (TIN PYRITES). In Cornwall it is found under the form of peroxide (MINE TIN, TINSTONE), associated with copper ore, in the slate and granite rocks, and as an alluvial deposit (STREAM TIN) in the beds of rivers.

Prep., &c. The ore is first reduced to powder in stamping-mills, washed to remove earthy matter, and then roasted to expel arsenic and sulphur; it is next deoxidised or reduced by smelting it with about 1-6th of its weight of powdered culm and a little slaked lime; it is, lastly, refined by 'liquation,' followed by a second smelting of the purer portion, which, after being treated in a state of fusion for some time with billets of green wood, or 'tossed,' as the workmen call it, is allowed to settle, and is then cast into large blocks, which, after being assayed, receive the stamp of the duchy. Two varieties of commercial tin are known, called respectively grain tin and bar tin. The first is the best, and is prepared from the stream ore.

Prop. Tin approaches silver in whiteness and lustre; in hardness it is intermediate between gold and lead; it is very malleable when pure, but the presence of a very small quantity of any other metal, particularly lead, deprives it of this property; it exhibits a fibrous fracture, and can be easily rolled or hammered into foil; at 100° C. it can be drawn out into a brittle wire; when rubbed it evolves a peculiar odour, and when bent backwards and forwards it emits a peculiar crackling noise; it melts at 232·7° C., volatilises at a white heat, and when heated above its melting-point, with free access of air, is speedily converted into a yellowish-white powder, which is the peroxide, or the 'putty powder' of polishers. Sp. gr. 7·293 at 13°.

Pur. It is almost entirely dissolved by hydrochloric acid, yielding a colourless solution; the precipitate thrown down by hydrate of potassium is white, and soluble in excess of the precipitant. If it contain arsenic, brownish-black flocks will be separated during the solution, and arseniuretted hydrogen evolved, which may be inflamed and tested in the usual manner. The presence of other metals in tin may be detected by treating the hydrochloric solution with nitric acid (sp. gr. 1·16), first in the cold, and afterwards with heat, until all the tin is thrown down in the state of insoluble stannic oxide. The decanted acid solution from pure tin leaves no residuum on evaporation. If, after all the acid has been dissipated by heat, dilution with water occasion a heavy white precipitate, the sample contained bismuth; if, after dilution, a solution of sulphate of ammonium or of sodium produce a similar white precipitate (sulphate of lead), it contained lead; if ammonia, added in excess, occasion reddish-brown flocks, or if ferricyanide of potassium give a blue precipitate, it contained iron; and, if the clear supernatant liquid leave a residuum on evaporation, copper.

Tests. The stannous salts are characterised as follows:—1. Potash gives a bulky white precipitate, readily soluble in excess of the precipitate; on concentrating the solution the precipitate is changed from stannous hydrate into stannic hydrate, which remains in solution, and metallic tin, which separates in brown flakes. 2. Ammonia, and the carbonates of potassium, sodium, and ammonium, give white precipitates, insoluble in excess. 3. Sulphuretted hydrogen gives, in neutral and acid solutions, a dark brown precipitate, which is soluble in potash, in the alkaline sulphides (especially when they contain an excess of sulphur), and in strong hot hydrochloric acid; and insoluble in nitric acid, even when boiling. 4. Sulphide of ammonium produces a similar brown precipitate, soluble in excess of the precipitant, provided the latter contains an excess of sulphur. 5. Terchloride of gold gives, in the cold, on the addition of a little nitric acid, a precipitate of the purple of Cassius. 6. Mercuric chloride gives a black precipitate, but in excess it produces a white one. The stannic salts are precipitated yellow by sulphuretted hydrogen, and the sulphide is readily soluble in ammonium sulphide. Alkalies precipitate a white hydroxide, which dissolves in an excess of the precipitant.

Separation. From other metals precipitated by sulphuretted hydrogen by treating the well-washed precipitate with yellow ammonium sulphide, filtering and acidifying the filtrate with cold dilute hydrochloric acid; again well wash with water, digest with solid ammonium carbonate, heat the residue with hydrochloric acid, and place in the solution a slip of platinum-foil upon which a piece of zinc rests; the tin is deposited upon the zinc, and is then dissolved off with hydrochloric acid.

Determination. Oxidise (if in the state of metal or alloy) with pure strong nitric acid, well wash, and then ignite the residue. From solution it is precipitated with ammonia as hydroxide; if present, however, in the *stannous* condition it must be first oxidised with chlorine or hydrochloric acid and potassium chlorate. The precipitate from ammonia is then dissolved in a very small quantity of pure hydrochloric acid, and heated with a strong solution of Glauber's salt; this precipitates the hydroxide, which may then be washed, ignited, and weighed. Each grain of stannic oxide well washed and dried is equivalent to 0·78365 gr. of pure tin.

Uses. Tin is used for a large number of purposes, for the preparation of vessels for technical and household use, for the manufacture of tin-foil, for tinning copper and iron, and especially in the preparation of alloys of tin. These alloys are amongst the most useful and important that are known. Tin and lead form 'pewter' and 'solder.' Tin, copper, and arsenic give 'gun-metal' and 'speculum metal.' Bronze consists of copper, tin, zinc, and sometimes lead; bell-metal of copper and tin; phosphor-bronze of copper, tin phosphide, and lead. With mercury tin forms an amalgam which is largely used for 'silvering' mirrors.

Stannous Chloride. SnCl₂. *Syn.* PROTO-CHLORIDE OF TIN. *Prep.* (ANHYDROUS.) Distil a mixture of tin and mercuric chloride. Grey, resin-like, solid, fusible, and volatile.

(HYDRATED; TIN-SALT, TIN-CRYSTALS, SnCl₂, 2Aq.) Boil an excess of tin in hydrochloric acid. A powerful deoxidising agent. It is somewhat extensively used as a mordant by dyers and calico printers, and for imparting a fine golden colour to sugar.

Stannous Hydrate. Sn(HO)₂. *Syn.* HYDRATED OXIDE OF TIN. *Prep.* Precipitate stannous chloride with carbonate of potassium, well wash, and dry under 196°. Greyish-white powder, soluble in acids and alkaline hydrates, except ammonia.

Stannous Iodide. SnI₂. *Syn.* PROTIODIDE OF TIN. Heat tin and iodine together. Yellowish needles, slightly soluble in water.

Stannous Nitrate. Sn(NO₃)₂. By the action of dilute nitric acid on the metal ammonium nitrate is also produced.

Stannous Oxide. SnO. *Syn.* PROTOXIDE OF TIN. *Prep.* Ignite the oxalate in an atmosphere of carbonic anhydride. Olive-coloured powder, inflammable in air, turning black on exposure to sunlight, and insoluble in acids.

Stannous Sulphate. SnSO₄. By dissolving the metal or the hydrated oxide in dilute sulphuric acid. Granular crystals, more soluble in hot than in cold water.

Stannous Sulphide. SnS. *Syn.* PROTOSULPHIDE OF TIN. A brittle bluish-grey substance, obtained by heating tin and sulphur.

Stannic Ammonium Chloride. (NH₄)₂SnCl₆. From a mixture of concentrated solution of ammonium chloride and stannic chloride. Small regular octahedra; formerly much used by calico printers under the name of pink salt as a mordant for madder-red colours. Its use has been lately greatly superseded by the crystalline pentahydrated stannic chloride.

Stannic Chloride. SnCl₄. *Syn.* BICHLORIDE OF TIN, TETRACHLORIDE OF TIN, PERCHLORIDE OF TIN, PERMURIATE OF T.†; STANNI BICHLO-

RIDUM, STANNI PERMURIAS, L. *Prep.* 1. (*Liebig.*) By dissolving grain tin in a mixture of hydrochloric acid, 2 parts; nitric acid and water, of each, 1 part (all by volume); observing to add the tin by degrees, and to allow one portion to dissolve before adding another, as without this precaution the action is apt to become violent, and stannic oxide of tin to be deposited.

2. (ANHYDROUS; LIBAVIUS'S FUMING LIQUOR.) By heating stannous chloride in chlorine gas; or by distilling a mixture of powdered tin, 1 part, with corrosive sublimate, 3 parts (5 parts—*Fownes*). A very volatile, colourless, mobile liquid, which fumes in the air, and boils at 113·9° C., sp. gr. = 2·234 at 15° C.; when mixed with 1-3rd of its weight of water, it solidifies to a crystalline mass (' butter of tin').

Obs. Solution of stannic chloride is much used by dyers, under the names of 'SPIRITS OF TIN,' 'DYKES' SPIRITS,' 'TIN MORDANT,' &c., the proportions of the ingredients and the state of dilution being various, according to circumstances or the caprice of the manufacturer. Drebbel discovered that by aid of it a *permanent* red dye can be obtained from cochineal. Dyers now usually use the crystalline pentahydrate (oxymuriate of tin), $SnCl_4 + 5H_2O$. A process which has been highly recommended, and which seems preferable to all others, is to prepare a simple solution of the stannous chloride, and to convert it into a solution of the stannic chloride, either by the addition of nitric acid and a gentle heat, or by passing chlorine through it. See TIN MORDANTS.

Stannic Hydrate. H_2SnO_3. *Syn.* HYDRATED PEROXIDE OF TIN, STANNIC ACID. *Prep.* By adding hydrate of potassium or an alkaline carbonate to a solution of stannic chloride. A glassy mass, soluble in acids and pure alkalies. Its compound with the latter are sometimes call STANNATES. Of these the most important is sodium stannate, Na_2SnO_3. This is manufactured very extensively for calico printers under the name of 'preparing salts.'

Stannic Iodide. SnI_4. By dissolving stannic hydrate in hydriodic acid. Yellow, silky crystals.

Stannic Oxide. SnO_2. *Syn.* BINOXIDE OF TIN, PEROXIDE OF TIN. Occurs in nature as tinstone or cassiterite as quadratic crystals. *Prep.* By the action of nitric acid on metallic tin, the resulting white powder being well washed with water; or by heating metallic tin above its melting-point in the air. Yellow amorphous powder; anhydrous; insoluble. It can be obtained crystalline by heating in a current of hydrochloric acid.

Obs. Frémy has given the name of METASTANNIC ACID to the oxide prepared by the action of nitric acid on metallic tin; the hydrate he calls STANNIC ACID. See POLISHERS' PUTTY.

Stannic Sulphide. SnS_2. *Syn.* BISULPHIDE OF TIN, BRONZE POWDER, MOSAIC GOLD; AURUM MUSIVUM, AURUM MOSAICUM, STANNI BISULPHURETUM, L. *Prep.* 1. To pure tin, 12 oz., melted by a gentle heat, add of mercury, 6 oz.; to the powdered mass when cold, add of chloride of ammonium, 6 oz.; flowers of sulphur, 7 oz.; and after thorough admixture place the compound in a glass flask, and gradually heat it, on a sand-bath,

to low redness, and continue the heat for several hours, or until white fumes cease to be disengaged; the 'aurum musivum' remains at the bottom of the vessel, under the form of soft and very brilliant gold-coloured flakes.

2. (*Berzelius.*) Stannic oxide and sulphur, of each, 2 parts; chloride of ammonium, 1 part; mix, and expose it to a low red heat, in a glass or earthenware retort, until sulphurous fumes cease to be evolved.

Used as a metallic gold colour, or substitute for powdered gold, in bronzes, varnish work, sealing-wax, &c.

TIN FI'LINGS. See TIN POWDER (*below*).

TINFOIL, Lead in. Tinfoil very rarely consists of pure tin; generally it contains more or less lead. According to the recent analysis of August Vogel, who has examined a great number of samples from very different sources, it contains from 1 to 19% of lead. There are, however, specimens of tinfoil which contain so little lead that it hardly gives a reaction with the appropriate tests.

Since tinfoil is so much used for covering articles of diet, of confectionery, or of perfumery, it was a matter of some interest to determine whether or not there was any danger of transference of lead from the wrapper to the contents. A number of experiments upon soap, chocolate, and different kinds of dry sugar, which had been enveloped in tinfoil very highly charged with lead, showed that there was no contamination with lead. Cheese, on the other hand, on account of its being moist, and being closely in contact with the foil, did take up lead.

Of course the lactic acid of the cheese would also favour the taking up of the metal. A point worthy of being recorded in connection with this matter is the rapid diminution of the lead toward the centre of the cheese. Often plenty of lead was found in the rind, and none a little way in the cheese (' Repertorium für Pharmacie,' Von Buchner).

TIN GLASS†. See BISMUTH.

TIN MOR'DANTS. *Syn.* DYERS' SPIRIT, SOLUTION OF TIN, SPIRIT OF T., NITRO-MURIATE OF T.† These, as noticed above, vary greatly in their composition and character.

Prep. 1. Take of aquafortis, 8 parts; sal-ammoniac or common salt, 1 part; dissolve, and add very gradually of grain tin, 1 part; and, when dissolved, preserve it in stoppered bottles from the air. This is the common 'SPIRIT OF TIN' of the dyers.

2. (*Berthollet.*) Nitric acid, at 30° Baumé, 8 parts; sal-ammoniac, 1 part; dissolve, then add by degrees, of tin, 1 part; and when dissolved, dilute the solution with 1-4th of its weight of water.

3. (*Dambourney.*) Hydrochloric acid, at 17° Baumé, 4 parts; nitric acid, at 30° Baumé, 1 part; mix, and add by degrees, of Molucca tin, 1 part.

4. (*Hellot.*) Nitric acid and water, of each, 1 lb.; sal-ammoniac, 1 oz.; nitre, ½ oz.; dissolve, then add, by degrees, of granulated tin, 2 oz.

5. (*Poerner.*) Nitric acid and water, of each, 1 lb.; sal-ammoniac, 1½ oz.; dissolve, then add, by very slow degrees, of pure tin beaten into ribands, 2 oz.

6. (*Schoeffer.*) Nitric acid and water, of each, 2 lbs.; sal-ammoniac, 2 oz.; pure tin, 4½ oz.; as last. All the above are used chiefly for dyeing scarlet, more particularly with cochineal.

7. (LAC SPIRIT.) From grain tin, 1 lb.; slowly dissolved in hydrochloric acid (sp. gr. 1·19), 20 lbs. Recommended as a solvent for lac dye. For use, ¾ to 1 lb. of the liquid is digested on each lb. of the dye for 5 or 6 hours, before adding it to the dye-bath.

8. Hydrochloric acid, 6½ lbs.; aquafortis, ½ lb.; grain tin, gradually added, 1 lb. Recommended for lac dye.

TIN'NING. *Proc.* 1. Plates or vessels of brass or copper, boiled with a solution of stannate of potass, mixed with turnings of tin, become, in the course of a few minutes, covered with a firmly attached layer of pure tin.

2. A similar effect is produced by boiling the articles with tin filings and caustic alkali or cream of tartar.

Obs. By either of the above methods chemical vessels made of copper or brass may be easily and perfectly tinned.

3. The following method for tinning copper, brass, and iron in the cold, and without apparatus, is by F. Stolba ('The Pharmacist,' iv, 86). The requisites for accomplishing this object are—1st. The object to be coated with tin must be entirely free from oxide. It must be carefully cleaned, and care be taken that no grease spots are left; it makes no difference whether the object be cleaned mechanically or chemically. 2nd. Zinc powder; the best is that prepared artificially by melting zinc, and pouring it into an iron mortar. It can be easily pulverised immediately after solidification; it should be about as fine as writing sand. 3rd. A solution of protochloride of tin containing 5% or 10%, to which as much pulverised cream of tartar must be added as will go on to the point of a knife. The object to be tinned is moistened with the tinned solution, after which it is rubbed hard with the zinc powder. The tinning appears at once. The tin salt is decomposed by the zinc, metallic tin being deposited. When the object tinned is polished brass or copper, it appears as beautiful as if silvered, and retains its lustre for a long time. 4th. (*C. Pawl*, 'Dingl. Polyt. J.,' ccviii, 47—49, 'Journ. Chem. Soc.') The zinc or iron articles are immersed in a mixture of 1 part sulphuric or nitric acid with 10 parts of water; a solution of copper sulphate or acetate is then slowly added. After the deposition of a thin layer of copper, the articles are removed, washed, moistened with a solution of 1 part 'tin crystals' in 2 parts water and 2 parts hydrochloric acid, and then shaken up with a mixture of fine chalk and copper. Ammonium sulphate is prepared by dissolving 1 part of copper sulphate in 16 parts of water, and adding ammonia until a clear dark blue liquid is obtained.

The articles may now be tinned by immersion in a solution of 1 part of tin crystals with 3 parts white argol in water. Brass, copper, or nickel goods, also iron and zinc articles which have been copper-plated, may be silvered by treatment (after thorough cleansing) with a solution of 14 grms. silver in 28 grms. of nitric

acid, to which is added a solution of 120 grms. of potassium cyanide in 1 litre water, and also 28 grms. of finely powdered chalk.

TIN-PLATE. Iron-plate covered with a coating of tin, by dipping it into a bath of that metal. The best kind, known as 'block tin,' is that which is covered with the thickest layer of tin, and afterwards hammered upon a polished anvil in order to consolidate the coating and make it adhere more firmly.

TIN-PLATE, To Crystallise. Crystallised tin-plate is made as follows:—Place the tin-plate, slightly heated, over a tub of water, and rub its surface with a sponge dipped in a liquid composed of 4 parts of nitric acid and 2 parts of water, containing 1 part of common salt or sal-ammoniac in solution. When the crystalline spangles seem to be thoroughly brought out, the plate must be immersed in water, washed carefully, dried, and coated with a lacquer varnish, otherwise it loses its lustre in the air. If the whole surface is not plunged at once in cold water, but is partially cooled by sprinkling water on it, the crystallisation will be finely variegated with large and small figures.

TIN POW'DER. *Syn.* TIN FILINGS, TIN DUST; STANNI PULVIS (Ph. E. and D.), L. *Prep.* 1. (Ph. E.) Melt grain tin in an iron vessel, pour it into an earthenware mortar heated a little above its melting-point, and triturate briskly as the metal cools; lastly, sift the product, and repeat the process with what remains in the sieve.

2. (Ph. D.) Melt grain tin in a black-lead crucible, and, whilst it is cooling, stir it with a rod of iron until it is reduced to powder; let the finer particles be separated by means of a sieve, and then, after having been several times in succession shaken with distilled water, the decanted liquor appears quite clear, let the product be dried for use.

Obs. Powdered tin is also prepared by filing and rasping.—*Dose*, 2 to 4 dr., as a vermifuge. POLISHERS' PUTTY, coloured with ivory-black, is frequently substituted for this powder, and hence arise the ill effects that sometimes follow its use.

TINS, To Clean. All kinds of tins, moulds, measures, &c., may be cleaned by being well rubbed with a paste made of whiting and water. They should then be rubbed with a leather, and any dust remaining on them should be removed by means of a soft brush. Finally they must be polished with another leather. Always let the inside of any vessel be cleaned first, since in cleaning the inside the outside always becomes soiled. For very dirty or greasy tins, grated Bath brick and water must be used.

TINC'TURE. *Syn.* TINCTURA, L.; TEINTURE, Fr. Tinctures (TINCTURE; ALCOOLÉS, ALCOOLATURES) are solutions of the active principles of bodies, obtained by digesting them in alcohol more or less dilute. ETHEREAL TINCTURES (TINCTURE ÆTHEREE; ÉTHÉROLÉS, ÉTHÉROLATURES) are similar solutions prepared with ether.

Prep. "Tinctures are usually prepared by reducing the solid ingredients to small fragments, coarse powder, or fine powder, macerating them for 7 days, or longer, in proof spirit or rectified spirit, straining the solution through linen or calico (or paper), and finally expressing the residuum strongly, to obtain what fluid is still retained in the mass. They are also advantageously prepared by the method of displacement or percolation." "All tinctures should be prepared in closed glass (or stoneware) vessels, and be shaken frequently during the process of maceration." Cooper's patent jars are very convenient for the preparation of tinctures, as they are made with wide mouths large enough to admit the hand, and yet may be closed in an instant, with as much ease and certainty as an ordinary stoppered bottle.

Tinctures are better clarified by repose than by filtration, as in the latter case a considerable portion is retained by the filtering medium, and lost by evaporation. The waste in this way is never less than 10% of spirit. In all ordinary cases it is sufficient to allow the tincture to settle for a few days, and then to pour off the clear supernatant portion through a funnel loosely choked with a piece of sponge or tow, or absorbent wool; after which the remaining foul portion of the liquid may be filtered through bibulous paper in a covered funnel. The filtration should be conducted as rapidly as possible, for the double purpose of lessening the amount lost by evaporation and the action of the air on the fluid. Tinctures which have been long exposed to the air frequently lose their transparency within a few days after being filtered, owing to the oxidisation and precipitation of some portion of the matter previously held in solution, a change which occurs even in stoppered bottles. Resinous and oily tinctures, as those of myrrh, tolu, and lavender (comp.), may be generally restored to their former brightness by the addition of a quantity of rectified spirit equal to that which they have lost by evaporation; but many tinctures resist this mode of treatment, and require refiltering.

Ethereal tinctures are best prepared by percolation, and should be both made and kept in stoppered bottles.

Mr Umney says it must always be remembered that the quantity of spirit required to make the measure of tinctures to a given bulk will only be strictly uniform in so far as the operators proceed under precisely the same circumstances.

No causes will be found to influence results more than the manufacture of tinctures upon a small as compared with a large scale, and the use of the screw as compared with the hydraulic press, in the final removal of the spirit from the marc; even the temperature of summer and winter may cause a variation in the results.

Qual. The tinctures of the shops are usually very uncertain and inferior preparations, owing to their manufacture being carelessly conducted, and refuse drugs and an insufficient quantity of spirit being employed in their production. It is a general practice among the druggists to substitute a mixture of equal parts of rectified spirit and water, or a spirit of about 26 u. p., for proof spirit; and a mixture of 2 galls. of water with 5 galls. of rectified spirit for rectified spirit.

Assay. 1. The RICHNESS in ALCOHOL may be readily determined by Brande's method of alcoholometry, but more accurately by the method of M. Gay-Lussac (see ALCOHOLOMETRY). That of tinctures containing simple extractive, saccharine, or like organic matter in solution may be approximately found from the boiling-point, or from the temperature of the vapour of the boiling liquid.

2. The QUANTITY of SOLID MATTER per cent. may be ascertained by evaporating to dryness 100 gr.-measures in a weighed capsule, by the heat of boiling water.

3. The QUANTITY of the INGREDIENTS used in the preparation of tinctures may be inferred from the weight last found, reference being had to the known percentage of extract which the substances employed yield to spirit of the strength under examination. When the ingredients contain alkaloids, or consist of saline or mineral matter, an assay may be made for them.

Uses, &c. Tinctures, from the quantity of alcohol which they contain, are necessarily administered in small doses, unless in cases where stimulants are indicated. The most important and useful of them are those that contain very active ingredients, such as the tincture of opium, foxglove, hemlock, henbane, &c. In many instances the solvent, even in doses of a few fluid drachms, acts more powerfully on the living system than the principles it holds in solution; and, when continued for some time, produces the same deleterious effects as the habitual use of ardent spirits. When the action of a substance is the reverse of stimulant it cannot with propriety be exhibited in this form, unless the dose be so small that the operation of the spirit cannot be taken into account, as with the narcotic tinctures. Hence this class of remedies are in less frequent use than formerly.

The following list embraces all the formulæ of the tinctures of the London, Edinburgh, Dublin, and British Pharmacopœias, with a few others likely to be useful to the reader. These will furnish examples for the preparation of others in less general use, care being had to proportionate the ingredients with due reference to the proper or usual dose of tinctures of that class.

Tincture of Ac'etate of I'ron. *Syn.* TINCTURA FERRI ACETATIS, L. *Prep.* Strong solution of acetate of iron, 5 oz.; acetic acid, 1 oz.; rectified spirit, 5 oz.; water, 9 oz.; mix.—*Dose,* 5 to 30 minims.

Tincture of Acetate of Zinc. *Syn.* TINCTURA ZINCI ACETATIS, L. *Prep.* (Ph. D. 1826.) Acetate of potash and sulphate of zinc, of each, 1 oz.; rub them together, then add of rectified spirit, 16 fl. oz.; macerate for a week, and filter. Astringent. Diluted with water, it is used as a collyrium and injection.

Tincture of Ac'onite. *Syn.* TINCTURA ACONITI (Ph. L.), TINCT. ACONITI RADICIS (B. P., Ph. D.), L. *Prep.* 1. (B. P.) Powdered root, 1 part; rectified spirit to percolate, 8 parts; macerate for 48 hours with three fourths of the spirit, agitating occasionally; pack in a percolator and let it drain, then pour on the remaining spirit; when it ceases to drop, press the marc and add spirit to make up 8 parts.—*Dose,* 5 to 15 minims,twice or thrice a day.

2. (Ph. L.) Take of aconite root, coarsely powdered, 15 oz. (20 oz.—Ph. D.); rectified spirit, 1 quart; macerate for 7 days, press, and filter.

Obs. These tinctures differ materially in strength.—*Dose.* Of the Ph. L., 5 to 10 drops; of the Ph. D., 3 to 6 drops, two or three times daily (carefully watching its effects); in rheumatism, gout, syphilis, &c., where a narcotic sedative is indicated. Diluted with water, it forms an excellent embrocation in rheumatism, neuralgia, &c. It should be applied by means of a small sponge tied to the end of a stick or glass rod. The Ph. D. formula is nearly the same as that for Dr Turnbull's concentrated tincture of aconite root, and that given by Dr Pereira. The TINCTURA ACONITI FOLIORUM of the Ph. U. S. is made with 1 oz. of the dried leaves to 8 fl. oz. of rectified spirit.

Tincture of Aconite, Ethereal. *Syn.* TINCTURA ACONITI ÆTHEREA (P. Cod.), L. *Prep.* Powdered aconite, 4 oz.; sulphuric ether, 16 oz. (by weight). It is best prepared by percolation.

Tincture of Ailanthus Bark. *Syn.* TINCTURA AILANTHI CORTICIS, L. *Prep.* Take of ailanthus bark, bruised, 1½ oz.; proof spirit, 1 pint; macerate for 7 days in a closed vessel with occasional agitation; then strain, press, filter, and add sufficient spirit to make 1 pint.—*Dose.* From ½ to 2 fl. dr.

Tincture of Al'oes. *Syn.* TINCTURA ALOËS (B. P., Ph. L. & E.), L. *Prep.* 1. (B. P.) Socotrine aloes, 1 part; extract of liquorice, 3 parts; proof spirit, 40 parts; macerate 7 days, press, and wash the marc with spirit to make 40 parts.—*Dose.* 1 to 2 dr.

2. (Ph. L.) Socotrine or hepatic aloes, coarsely powdered, 1 oz.; extract of liquorice, 3 oz.; water, 1½ pints; rectified spirit, ½ pint; macerate for 7 days, and filter. The formula of the Ph. E. is nearly similar. Purgative and stomachic.—*Dose.* ½ to fl. oz.

Tincture of Aloes, Alkaline. *Syn.* TINCTURA ALOËS ALKALINA (*Swediaur*), L. *Prep.* Aloes, ½ oz.; extract of liquorice, 1½ dr.; cinnamon water, 8 oz.; proof spirit, 8 oz.; carbonate of soda, 1 oz. Digest and strain.—*Dose.* 1 dr. to 4 dr.

Tincture of Aloes (Compound). *Syn.* TINCTURE OF ALOES AND MYRRH; TINCTURA ALOËS COMPOSITA (Ph. L.), TINCTURA ALOËS ET MYRRHÆ (Ph. E.), ELIXIR ALOËS†, L. *Prep.* 1. (Ph. L. & E.) Socotrine or hepatic aloes, coarsely powdered, 4 oz.; hay saffron, 2 oz.; tincture of myrrh, 1 quart; macerate for 7 days, with occasional agitation, and strain. The Dublin College (1826) omits the saffron.

2. (Wholesale.) From aloes, 1 lb.; myrrh, ¾ lb.; hay saffron, 2 oz.; rectified spirit, 5 pints; water, 3 pints; as the last. Purgative, stomachic, and emmenagogue.—*Dose.* ½ to 2 fl. dr.

Tincture of Amber. *Syn.* TINCTURA SUCCINI (P. Cod.), L. *Prep.* Amber, in fine powder, 1 oz.; rectified spirit, 6 oz. Digest for 6 days and filter.—*Dose.* 20 to 30 drops.

Tincture of Amber, Alkaline. *Syn.* TINCTURA SUCCINI ALKALINA (Ph. E. 1744), L. *Prep.* Rub 2 oz. of amber with a sufficient quantity of carbonate of potash to form a soft paste; dry this, and digest it in 16 oz. of rectified spirit for 8 days.

Tincture of Ambergris. *Syn.* TINCTURA AMBERGRISEÆ (P. Cod.), L. *Prep.* Ambergris,

1 part; rectified spirit, 10 parts. Macerate for 10 days.

Tincture of Amm"onia (Compound). *Syn.* TINCTURA AMMONIÆ COMPOSITA (Ph. L.), L. *Prep.* 1. (Ph. L.) Mastic, 2 dr.; rectified spirit, 9 fl. dr.; digest until dissolved, decant, add of oil of lavender, 14 drops; stronger solution of ammonia, 1 pint; and mix well.

2. (Ph. L. 1836; AQUA LUCIÆ; EAU DE LUCE.) As the last, but adding 4 drops of oil of amber along with the oil of lavender.

Obs. This preparation is reputed antacid, antispasmodic, and stimulant.—*Dose,* 10 to 20 drops, in water; in hysteria, low spirits, &c. In the East Indies eau de luce is regarded almost as a specific for the bite of the cobra di capello and other venomous reptiles.

Tincture of Ammo"nio-chlo"ride of I'ron. *Syn.* AMMONIATED TINCTURE OF IRON, MYNSIGHT'S A. T. OF I.; TINCTURA FERRI AMMONIO-CHLORIDI (Ph. L.), TINCTURA FERRI AMMONIATI, L. *Prep.* (Ph. L.) Ammonio-chloride of iron, 4 oz.; proof spirit and distilled water, of each, 1 pint; dissolve.—*Dose,* 20 to 60 drops, or more; as a stimulant, chalybeate tonic. "A fl. oz. of this, on potassa being added, yields 5·8 gr. of sesquioxide of iron" (Ph. L.).

Tincture of Ammoniacum. *Syn.* TINCTURA GUMMI AMMONIACI (P. Cod.), L. *Prep.* Gum ammoniacum, 4 oz.; rectified spirit, 20 oz. (by weight). Digest 10 days and strain.

Tincture of Angelica. *Syn.* TINCTURA ANGELICÆ (Aust. Ph.), L. *Prep.* Dried angelica root, 1 oz.; proof spirit, 6 oz. Digest and filter. —*Dose,* 1 dr.

Tincture of Angostu'ra. Tincture of cusparia.

Tincture, Antiscorbutic. *Syn.* TINCTURA ANTISCORBUTICA, TINCTURA ARMORACIÆ COMPOSITA (P. Cod.), L. *Prep.* Fresh horseradish root, 8 oz.; black mustard seed, 4 oz.; muriate of ammonia, 2 oz.; proof spirit, 16 oz. (by weight) compound spirit of scurvy-grass, 16 oz. (by weight). Macerate 10 days.

Tincture of Ants. *Syn.* TINCTURA FORMICARUM (Ph. G.), L. *Prep.* Ants recently collected, cleaned, and bruised, 2 oz.; rectified spirit, 3 oz. (by weight). Digest 8 days.

Tincture of Ar'nica. *Syn.* TINCTURA ARNICÆ; T. A. FLORUM, L. *Prep.* (Ph. Bor. and Hamb. Cod.) Flowers of *Arnica montana*, 1½ oz.; spirit, sp. gr. ·900 (15½ o. p.), 1 lb.; digest for 8 days, and strain with expression.—*Dose,* 10 to 30 drops; in diarrhœa, dysentery, gout, rheumatism, paralysis, &c.

Tincture of Arnica Root. *Syn.* TINCTURA ARNICÆ (B. P.), TINCTURA ARNICÆ RADICIS, L. *Prep.* 1. (B. P.) Bruised root, 1 part; rectified spirit to percolate, 20 parts; macerate 48 hours with 15 parts of the spirit, agitating occasionally; pack in a percolator, and, when it ceases to drop, pour on the remaining spirit, let it drain, wash the marc, press, filter, and make up to 20 parts. —*Dose,* 1 to 2 dr.

2. From arnica root, 2 oz.; proof spirit, 1 pint; as the last.

Tincture, Aromatic. *Syn.* TINCTURA AROMATICA (G. Ph.), L. *Prep.* Cinnamon, 4 oz.; cardamoms, 1 oz.; cloves, 1 oz.; galangal root, 1 oz.; ginger, 1 oz.; all in coarse powder; proof spirit, 3 lbs. 2 oz. (by weight). Macerate 8 days, and strain.

Tincture, Aromat'ic. Compound tincture of cinnamon.

Tincture of Artichoke. *Syn.* TINCTURA CYNARÆ, L. *Prep.* Fresh artichoke leaves, bruised, 2 lbs.; rectified spirit, 1 lb. Digest for 7 days, express, and filter.

Tincture of Assafœtida. *Syn.* TINCTURA ASSAFŒTIDÆ (Ph. L., E., & D.), L. *Prep.* 1. (B. P.) Assafœtida (small fragments), 1 part; rectified spirit, 8 parts; macerate 7 days, strain, filter, and add spirit to make 8 parts.—*Dose*, ½ dr. to 1 dr.

2. (Ph. L.) Assafœtida (small), 5 oz.; rectified spirit, 1 quart; macerate for 7 days (14 days —Ph. D.), and filter. "It cannot be made by percolation with delay" (Ph. E.).

3. (Wholesale.) Assafœtida, 2½ lbs.; boiling water, 2 quarts; dissolve, add of rectified spirit, 1½ gall.; agitate well for 3 or 4 days, then let it settle, and decant the clear portion.—*Dose*, ½ to 2 fl. dr.; in hysteria, flatulent colic, &c.

Tincture of Assafœtida (Ammo"niated). See FETID SPIRIT OF AMMONIA.

Tincture of Assafœtida, Ethereal. *Syn.* TINCTURA ASSAFŒTIDÆ ETHEREA (P. Cod.), L. *Prep.* Assafœtida, 1 part; alcoholised ether, 5 parts (by weight). Macerate for 10 days. The ether is made by mixing equal weights of ether and rectified spirit.

Tincture, Asthmat'ic. Compound tincture of camphor.

Tincture, Astringent. *Syn.* TINCTURA ASTRINGENS (*Dr Copland*), L. *Prep.* Catechu, ½ oz.; myrrh, ½ oz.; Peruvian bark, 2 dr.; balsam of Peru, 1½ dr.; spirit of horseradish, 1½ oz.; rectified spirit, 1½ oz. Digest. For sponginess of the gums.

Tincture, Balsamic. *Syn.* TINCTURA BALSAMICA (Ph. E. 1744), L. *Prep.* Copaiba, 1 oz.; balsam of Peru, 3 dr.; balsam of Tolu, 2 dr.; benzoin, ½ dr.; saffron, 1 scruple; rectified spirit, 16 oz.; digest 4 days in a sand-bath, and strain.

Tincture, Balsam of Copaiba. *Syn.* TINCTURA BALSAMI COPAIBÆ (*Guibourt*), L. *Prep.* One part of copaiba to 8 parts of alcohol. Digest and filter.

Tincture of Balsam of Gilead. *Syn.* TINCTURA BALSAMI GILEADENSIS (*Guibourt*), L. *Prep.* One part of balsam to 8 parts of rectified spirit.

Tincture of Bal'sam of Peru. *Syn.* TINCTURA BALSAMI PERUVIANI, L. *Prep.* (Ph. L. 1788.) Balsam of Peru, 4 oz.; rectified spirit, 16 fl. oz.; dissolve. Pectoral, stimulant, and fragrant.—*Dose*, 10 to 30 drops.

Tincture of Balsam of Tolu. Tincture of Tolu.

Tincture of Bark. Tincture of cinchona.

Tincture of Belladon'na. *Syn.* TINCTURA BELLADONNÆ (B. P., Ph. L. and D.), L. *Prep.* 1. (B. P.) The dried leaves in coarse powder, 1 part; proof spirit, 20 parts; macerate 48 hours in 15 parts of the spirit, agitating occasionally; pack in a percolator, and when it ceases to drop, add the remaining spirit, let it drain, wash and press the marc; filter and make up 20 parts.—*Dose*, from 5 to 20 minims.

2. (Ph. L.) Dried leaves of belladonna, 4 oz. (5 oz. in coarse powder—Ph. D.); proof spirit, 1 quart; macerate for 7 days (14—Ph. D.), press, and filter.

3. (Wholesale.) From the dried leaves, 1 lb.; proof spirit, 1 gall.; macerate 14 days.—*Dose*, 5 to 10 drops, gradually increased; also externally, diluted with water.

Tincture of Benzoin. *Syn.* TINCTURA BENZOINI (Ph. G.), L. *Prep.* Benzoin, 2 oz.; rectified spirit, 10 oz. (by weight). Digest for 8 days, frequently shaking; then filter.

Tincture of Benzoin (Simple). *Syn.* TINCTURA BENZOINI SIMPLEX (B. P. C.), L. *Prep.* Take of benzoin, in powder, 2 oz.; rectified spirit, 1 pint. Macerate for 24 hours with frequent agitation, then filter, and add sufficient rectified spirit, if required, to produce 1 pint.

Tincture of Benzoin (Compound). *Syn.* FRIAR'S BALSAM, TRAUMATIC B., BALSAM FOR CUTS, COMMANDER'S BALSAM, VERVAIN'S B., WOUND B., JESUITS' DROPS, WADE'S D.; TINCTURA BENZOINI COMPOSITA (B. P., Ph. L. & E.), TINCT. BENZOÏS COMP., BALSAMUM TRAUMATICUM, L. *Prep.* 1. (B. P.) Benzoin, 8 parts; prepared storax, 6 parts; balsam of Tolu, 2 parts; Socotrine aloes, 1½ parts; rectified spirit, 80 parts; macerate 7 days, filter, and wash the marc with spirit to make up 80 parts.—*Dose*, ½ to 1 dr., triturated with mucilage or yolk of egg.

2. (Ph. L.) Gum benzoin, coarsely powdered, 3½ oz.; prepared storax, 2½ oz.; balsam of Tolu, 10 dr.; Socotrine or hepatic aloes, in coarse powder, 5 dr.; rectified spirit, 1 quart; macerate, with frequent agitation, for 7 days, and strain.

3. (Ph. E.) Benzoin, 4 oz.; balsam of Peru, 2½ oz.; East Indian (hepatic) aloes, ½ oz.; rectified spirit, 1 quart.

Obs. Either of the above formulæ produces a most beautiful tincture, truly balsamic. The following is, however, very generally employed by the wholesale druggists, and the product, though possessing a very rich colour, is thin and watery.

4. (Wholesale.) From gum benzoin, 4 lbs.; aloes (lively coloured), 1½ lbs.; liquid storax, 1 lb.; balsam of Tolu, ½ lb.; powdered turmeric (best), 9 oz.; rectified spirit, 5½ galls.; digest with frequent agitation for 10 days, then add of hot water, 1½ galls., again digest for 4 days, and, after 24 hours' repose, decant the clear portion.

Dose, 10 drops to 2 fl. dr.; as a stimulating expectorant, in chronic coughs and various breath affections. It is also employed to stop the bleeding from cuts, &c., and promote their healing.

Tincture, Bitter. *Syn.* TINCTURA AMARA (Ph. G.), L. *Prep.* Urnipe oranges, 2 oz.; centaury, 2 oz.; gentian root, 2 oz.; zedoary root, 1 oz.; proof spirit, 35 oz. (by weight). Digest 8 days and strain.

Tincture, Bitter Stomach'ic. Tincture of gentian.

Tincture of Black Snake-root. (B. P.) *Syn.* TINCTURA CIMICIFUGÆ, T. ACTÆÆ RACEMOSÆ, L. *Prep.* Bruised root of black snake-root, 2½ oz.; proof spirit, 20 oz.—*Dose*, 1 to 2 dr.

Tincture of Blessed Thistle. *Syn.* TINCTURA CARDUI BENEDICTI (Ph. Bruns.), L. *Prep.* Blessed thistle, 6 oz.; rectified spirit, 2 pints.

Tincture of Blood-root. *Syn.* TINCTURA SANGUINARIÆ (Ph. U. S.), L. *Prep.* Blood-root in moderately fine powder, 4 oz.; proof spirit, 32 oz.; made by percolation.—*Dose.* As a stimulant and alterative, 30 to 60 drops; as an emetic, 3 to 4 dr.

Tincture, Brandish's. Alkaline tincture of rhubarb.

Tincture of Bryony. *Syn.* TINCTURA BRYONIÆ (B. P. C.), L. *Prep.* Fresh bryony root, rectified spirit, and distilled water, of each, a sufficient quantity. Ascertain the percentage of moisture in the root by drying 100 gr. of it over a water-bath. Bruise the remainder, after having calculated the moisture it contains, and reckon this as part of the water to form, with rectified spirit, a mixture equal in strength to proof spirit. Produce a tincture by macerating for 7 days of such a strength as that 10 fl. oz. shall represent 1 oz. of the dried root.—*Dose,* 1 to 10 minims.

Tincture of Buchu. *Syn.* TINCTURA DIOSMÆ, T. BUCKU (Ph. E.), T. BUCHU (B. P., Ph. D.), L. *Prep.* 1. (B. P.) Buchu, bruised, 1 part; proof spirit, 8 parts; macerate for 48 hours with ⅔ of the spirit, pack in a percolator and let it drain, then pour on the rest of the spirit; when it ceases to drop, press and wash the marc, filter and make up to 8 parts.—*Dose,* 1 to 2 dr.

2. (Ph. E.) Buchu leaves, 5 oz.; proof spirit, 1 quart; macerate 7 days (14 days—Ph. D.); or proceed by the method of percolation.—*Dose,* 1 to 4 fl. dr.; as a tonic, sudorific, and diuretic. It is inferior to the fresh infusion.

Tincture of Calum'ba. *Syn.* TINCTURA CALUMBÆ (B. P., Ph. L. & E.), T. COLOMBÆ (Ph. D.), L. *Prep.* 1. (B. P.) Bruised calumba, 1 part; proof spirit, 8 parts; macerate 48 hours with 6 parts of the spirit, agitating occasionally; pack in a percolator, and let it drain; then pour on the remaining spirit; when it ceases to drop, press, and wash the marc with spirit to make up 8 parts.—*Dose,* ½ to 2 dr.

2. (Ph. L.) Calumba root, finely sliced, 3 oz.; proof spirit, 1 quart; macerate a week (14 days —Ph. D.), press, and filter. "Or, more conveniently, by percolation, allowing the calumba, in moderately fine powder, to first soak in a little spirit for 6 hours" (Ph. E.).

Tincture of Cam'phor. *Syn.* SPIRIT OF WINE AND CAMPHOR, CAMPHORATED SPIRIT; TINCTURA CAMPHORÆ (Ph. E. & D.), SPIRITUS CAMPHORÆ (Ph. L.), S. CAMPHORATUS, L. *Prep.* 1. (Ph. E.) Camphor, 2¼ oz.; rectified spirit, 1 quart; dissolve. This is only one half as strong as the Ph. L. preparation.

2. (Ph. D.) Camphor, 1 oz.; rectified spirit, 8 fl. oz. Stimulant and anodyne.—*Dose,* 10 to 60 drops. Also as a liniment for sprains, bruises, chronic rheumatism, &c. For the Ph. L. formula see SPIRIT.

Tincture of Camphor (Compound). *Syn.* CAMPHORATED TINCTURE OF OPIUM, ASTHMATIC ELIXIR, PAREGORIC E., ASTHMATIC TINCTURE; TINCTURA CAMPHORÆ COMPOSITA (B. P., Ph. L.), T. OPII CAMPHORATA (Ph. E. & D.), ELIXIR PAREGORICUM, L. *Prep.* 1. (B. P.) Opium, in coarse powder, 40 gr.; benzoic acid, 40 gr.; camphor, 30 gr.; oil of anise, ½ dr.; proof spirit, 20 oz.; macerate 7 days, strain, wash the marc with spirit, and filter 20 oz.—*Dose,* 15 to 60 minims.

2. (Ph. L.) Camphor, 50 gr.; powdered opium and benzoic acid, of each, 72 gr.; oil of aniseed, 1 fl. dr.; proof spirit, 1 quart; macerate for 7 days, and filter. The formulæ of the Ph. E. & D. are nearly similar. The oil of aniseed, probably one of the most useful and characteristic of the ingredients, was omitted in the Ph. L. 1824, but was restored in that of 1836.

3. (Wholesale.) From powdered opium, 3 oz.; benzoic acid, camphor, and oil of aniseed, of each, 2 oz.; rectified spirit and water, of each, 3 galls.; as before.

Obs. This tincture is a popular and excellent pectoral and anodyne where there are no inflammatory symptoms.—*Dose,* ½ to 2 fl. dr.; in troublesome coughs, &c. ½ fl. oz. contains about 1 gr. of opium.

Tincture of Canthar'ides. *Syn.* TINCTURA CANTHARIDIS (B. P., Ph. L., E., & D.), TINCTURA LYTTÆ, L. *Prep.* 1. (B. P.) Cantharides, in coarse powder, 1 part; proof spirit, 80 parts; macerate, agitating occasionally, for seven days, in a closed vessel, strain, press, filter, and add sufficient proof spirit to make up 80 parts.—*Dose,* 5 to 20 minims.

2. (Ph. L.) Powdered cantharides, 4 dr. (½ oz.—Ph. D.), and strain with expression.

3. (Wholesale.) From powdered cantharides, 2¼ oz.; rectified spirit and water, of each, ½ gall.; as the usual.—*Dose,* 10 drops, gradually raised to 1 fl. dr., in any bland liquid; in fluor albus, gleets, incontinence of urine, lepra, &c. It should be used with caution. The Ed. College recommends it to be prepared by displacement.

Tincture of Cantharides (Ethereal). *Syn.* TINCTURA CANTHARIDIS ETHEREA (P. Cod.), L. *Prep.* Powdered cantharides, 1 oz.; acetic ether, 10 oz. (by weight). Macerate for 10 days in a stoppered bottle, express, and filter.

Tincture of Capsicum. *Syn.* TINCTURE OF CAYENNE PEPPER; TINCTURA CAPSICI (B. P., Ph. L., E., & D.), L. *Prep.* 1. (B. P.) Capsicum, bruised, ¾ part; rectified spirit, 20 parts; macerate 48 hours with three fourths of the spirit, agitating occasionally, pack in a percolator and let it drain, then pour on the remaining spirit; as soon as it ceases to drop wash the marc with spirit to make up 27 parts.—*Dose,* 10 to 20 minims.

2. (Ph. L.) Capsicum, bruised, 10 dr.; proof spirit, 1 quart; digest 14 days (or percolate—Ph. E).—*Dose,* 10 to 60 drops; in atonic dyspepsia, scarlet fever, ulcerated sore throat, &c. It is also made into a gargle.

3. (Ph. D.) Cayenne pods, bruised, 1½ oz.; proof spirit, 1 pint; macerate for 14 days. This is of fully twice the strength of the preceding.

4. (B. P. C.) Capsicum fruit, in No. 40 powder, 10 oz.; rectified spirit, a sufficient quantity. Moisten the powder with a suitable quantity of the menstruum, and macerate for 24 hours in a closed vessel; then pack in a percolator, and gradually pour rectified spirit upon it until a pint and a half of tincture is obtained.—*Dose,* 1 to 3 minims. Principally used externally.

Tincture of Capsicum (Concentrated). See ESSENCES.

Tincture of Capsicum with Veratria. *Syn.* TINCTURA CAPSICI CUM VERATRIÂ (Dr Turnbull), L. *Prep.* Dissolve 4 gr. of veratria in 1 oz. of concentrated tincture of capsicum.

Tincture of Card'amoms. *Syn.* TINCTURA CARDAMOMI (Ph. E.), TINCT. AMOMI REPENTIS, L. *Prep.* (Ph. L. 1836.) Cardamom seeds, 3½ oz. (4½ oz.—Ph. E.); proof spirit, 1 quart; digest for 14 days (or percolate—Ph. E.). *Obs.* The shells should be sifted from the seeds before maceration, and the latter are preferably ground in a pepper mill instead of being bruised in a mortar. Aromatic and carminative.—*Dose*, 1 to 2 fl. dr., as an adjunct to purgative mixtures.

Tincture of Cardamoms (Compound). *Syn.* STOMACHIC TINCTURE; TINCTURA CARDAMOMI COMPOSITA (B. P., Ph. L., E., & D.), TINCTURA STOMACHICA, L. *Prep.* 1. (B. P.) Cardamom seeds, freed from their pericarps, bruised, 1 part; caraway, bruised, 1 part; raisins, freed from their seeds, 8 parts; bruised cinnamon, 2 parts; cochineal, in powder, ½ part; proof spirit, 80 parts; macerate 48 hours with ¾ of the spirit, agitating occasionally, pack in a percolator and let it drain, pour upon it the remainder of the spirit, and, when it ceases to drop, press, and wash the marc with spirit to make up 80 parts.—*Dose*, ½ to 2 dr.

2. (Ph. L.) Cardamoms (without the shells), caraways, and cochineal, of each, bruised, 2½ dr.; cinnamon, bruised, 5 dr.; raisins, stoned, 5 oz.; proof spirit, 1 quart; macerate 7 days, then strain with expression.

3. (Ph. E., and Ph. L. 1836.) As the last, but using only 1 dr. of cochineal, and macerating 14 days; or "it may be prepared by the method of displacement" (Ph. E.).

4. (Wholesale.) From cardamoms and caraway seeds, of each, 4 oz.; cochineal (s. g.), 6 oz.; cassia, 8 oz.; sultana raisins, 5 lbs.; proof spirit, 4 galls. (or rectified spirit and water, of each, 2 galls.); macerate, &c., as before.

Tincture, Carminative. *Syn.* TINCTURA CARMINATIVA (B. P. C.), L. *Prep.* Cardamom seeds, bruised, 600 gr.; stronger tincture of ginger, 1½ fl. oz.; oil of cinnamon, 100 minims; oil of caraway, 100 minims; oil of clove, 100 minims; rectified spirit, sufficient to produce 1 pint. Macerate the cardamoms in 15 fl. oz. of the spirit for a week; decant, express, and dissolve the oils in the mixed tinctures, making up to 1 pint with rectified spirit.—*Dose*, 2 to 10 minims.

Tincture of Casca. *Syn.* TINCTURA ERYTHROPHLŒI (B. P. C.), L. *Prep.* Casca bark, in No. 20 powder, 2 oz.; rectified spirit, a sufficient quantity; moisten the powder with a suitable quantity of the menstruum, and macerate for 24 hours, then pack in a percolator, and gradually pour rectified spirit upon it until 1 pint of tincture is obtained.—*Dose*, 5 to 10 minims.

Tincture of Cascaril'la. *Syn.* TINCTURA CASCARILLÆ (B. P., Ph. L., E., & D.), L. *Prep.* 1. (B. P.) Cascarilla, bruised, 1 part; proof spirit, 8 parts; macerate 48 hours with 6 parts of the spirit, agitating occasionally; pack in a percolator, let it drain, and pour on the remainder of the spirit, and, when it ceases to drop, wash the marc, press, filter, and make up 8 parts.—*Dose*, ½ to 2 dr.

2. (Ph. L.) Cascarilla, bruised, 5 oz.; proof spirit, 1 quart; macerate for 7 days (14 days—Ph. D.; or percolate—Ph. E.). An excellent tonic and stomachic; chiefly employed as an adjunct to mixtures, &c.—*Dose*, 1 to 2 fl. dr.

Tincture of Cas'sia. *Syn.* TINCTURA CASSIÆ (Ph. E.), L. *Prep.* (Ph. E.) Cassia, 3½ oz.; proof spirit, 1 quart; macerate for 7 days, or percolate. Stomachic and carminative.—*Dose*, 1 to 2 fl. dr.

Tincture of Castor. *Syn.* TINCTURA CASTOREI (Ph. L. & E.), TINCT. CASTOREI ROSSICI, L. *Prep.* 1. Castor, in coarse powder, 1 part; spirit, 20 parts; macerate 7 days, strain, and wash the marc with spirit sufficient to make up to 20 parts.—*Dose*, ½ to 1 dr.

2. (Ph. L.) Castor, bruised, 2½ oz.; rectified spirit, 1 quart; macerate for 7 days (or percolate—Ph. E.).

Obs. The Dublin College ordered Russian castor in their Ph. of 1826; but the scarcity and high price of that variety, we fear, too often precludes its use. The tincture of the shops is usually made with only 8 oz. of castor to the gall. of proof spirit. Nervine and antispasmodic.—*Dose*, 20 drops to 2 fl. dr.; in hysteria, epilepsy, &c.

Tincture of Castor (Ammo"niated). *Syn.* ELIXIR FŒTIDUM, TINCTURA CASTOREI COMPOSITA, T. c. AMMONIATA (Ph. E.), L. *Prep.* (Ph. E.) Castor, bruised, 2½ oz.; assafœtida, in small fragments, 10 dr.; spirit of ammonia, 1 quart; digest 7 days in a well-closed vessel. Stimulant and antispasmodic.—*Dose* and *uses*, as the last. With the addition of ½ oz. of opium, it forms the Elixir Uterinum, or Elixir Castorei Thebaicum of foreign Pharmacopœias.

Tincture of Castor (Ethereal). *Syn.* TINCTURA CASTOREI ÆTHEREA (P. Cod.), L. *Prep.* Castor, in powder, 1 oz.; alcoholised ether (see ETHEREAL TINCT. OF ASSAFŒTIDA), 10 oz. (by weight).

Tincture of Castor Oil Seeds. *Syn.* TINCTURA RICINI, L. Castor oil seeds bruised are digested with five times their weight of rectified spirit. This tincture is stated to be four times the strength of the oil.

Tincture of Cat'echu. *Syn.* COMPOUND TINCTURE OF CATECHU; TINCTURA CATECHU COMPOSITA (Ph. L.), T. CATECHU (B. P., Ph. E. and D.), L. *Prep.* 1. (B. P.) Pale catechu, in coarse powder, 2½ parts; cinnamon, bruised, 1 part; proof spirit, 20 parts; macerate for seven days with agitation, strain, press, and filter, and add spirit to make up 20 parts.—*Dose*, ½ to 2 dr.

2. (Ph. L.) Catechu, in powder, 3½ oz. (4 oz.—Ph. D.); cinnamon, bruised, 2½ oz. (2 oz.—Ph. D.); proof spirit, 1 quart; macerate for seven days (or percolate—Ph. E.).

3. (Wholesale.) From catechu, 2 lbs.; oil of cassia, 3 fl. dr.; rectified spirit and water, of each, 1 gall.; macerate for 10 days.—*Dose*, 1 to 2 fl. dr., as a warm astringent; in diarrhœa, &c., either alone or combined with chalk.

Tincture of Cevadilla. *Syn.* TINCTURA SABADILLÆ (Dr *Turnbull*), L. *Prep.* Digest the seeds of cevadilla (freed from their capsules), and bruise for ten days in as much rectified spirit as will cover them; express and filter. For external use only in rheumatism.

Tincture of Chamomile. *Syn.* TINCTURA ANTHEMIDIS (Aust. Ph.), L. Dried chamomile flowers, 2 oz.; proof spirit, 1 pint.

Tincture of Chiret'ta. *Syn.* TINCTURA CHIRAYTÆ (B. P.), TINCTURA CHIRAYTA, T. CHI-

RETTÆ (Ph. D.), L. *Prep.* 1. (B. P.) Chiretta, cut small and bruised, 1 part; proof spirit, 8 parts; macerate 48 hours with 6 parts of the spirit, agitating occasionally, pack in a percolator, and let it drain, then pour on the remaining spirit; when it ceases to drop, press, and wash the marc with spirit to make up 8 parts.—*Dose,* 15 to 60 minims (B. Ph. dose ½ to 2 dr.).

2. (Ph. D.) Chiretta or chirayta (bruised), 5 oz.; proof spirit, 1 quart; macerate for 14 days. Tonic and stomachic.—*Dose,* ½ to 2 fl. dr.

Tincture of Chloroform (Compound). *Syn.* TINCTURA CHLOROFORMI COMPOSITA (B. P.), L. *Prep.* Mix 2 fl. oz. of chloroform with 8 fl. oz. of rectified spirit and 10 fl. oz. of compound tincture of cardamoms.—*Dose,* 20 to 40 minims.

Tincture of Chloroform and Morphine. *Syn.* TINCTURA CHLOROFORMI ET MORPHINÆ, L. Chloroform, 1 oz.; ether, 2 dr.; rectified spirit, 1 oz.; hydrochlorate of morphine, 8 gr.; diluted hydrocyanic acid, ½ oz.; oil of peppermint, 4 minims; liquid extract of liquorice, 1 oz.; treacle, 1 oz.; syrup, enough to make 8 oz.—*Dose,* 5 to 10 minims.

Tincture, Cholera. Bacc. capsici, 1 oz.; ol. menth. pip., 1½ oz.; camphora, 1½ oz.; opii g., ½ oz.; rhei r., ½ oz.; croci stig., 45 gr.; zingiberis r., 60 gr.; succ. solazzi, 90 gr.; 8. V. R., 40 oz.; aquæ, 10 oz. Boil opium, saffron, and solazzi in the water. Mix the liquor with the 8. V. R. in which the camphor and peppermint have been dissolved, and make with the mixture a tincture of the other ingredients.—*Dose,* 5 to 30 drops to be taken in water every ten or fifteen minutes, until the pain and purging cease.

Tincture of Cinchona. *Syn.* TINCTURE OF BARK; TINCTURA CINCHONÆ (B. P., Ph. L., E., and D.), T. CORTICIS PERUVIANI, T. C. P. SIM-PLEX, L. *Prep.* 1. (B. P.) Red cinchona bark, in coarse powder, 4 parts; proof spirit, 20 parts; macerate 48 hours with 15 parts of the spirit, agitating occasionally, pack in a percolator and let it drain, then pour on the remaining spirit, and when it ceases to drop, press, and wash the marc with spirit to make 30 parts.—*Dose,* 1 to 2 dr.

2. (Ph. L.) Yellow cinchona bark (bruised), 8 oz.; proof spirit, 1 quart; macerate for 14 days (or percolate—Ph. E.).

Obs. The Dublin College orders pale bark, and the Edinburgh either species, according to prescription.—*Dose,* 1 to 3 fl. dr.; as a tonic, stomachic, and febrifuge.

Tincture of Cinchona (Ammoniated). *Syn.* TINCTURA CINCHONÆ AMMONIATÆ (Ph. L. 1824), L. *Prep.* Peruvian bark, 4 oz.; aromatic spirit of ammonia, 32 fl. oz. Macerate for 10 days.— *Dose,* ½ dr. to 1 dr.

Tincture of Cinchona (Compound). *Syn.* COMPOUND TINCTURE OF BARK, HUXHAM'S T. OF B., FEVER TINCTURE; TINCTURA CINCHONÆ COMPOSITA (B. P., Ph. L., E., and D.), T. CORTICIS PERUVIANI COMPOSITA, L. *Prep.* 1. (B. P.) Red cinchona bark, in coarse powder, 4 parts; bitter orange peel, cut small and bruised, 2 parts; serpentary, bruised, 1 part; saffron, ½ part; cochineal, ½ part; proof spirit, 40 parts; macerate 48 hours with 30 parts of spirit, agitating occasionally, pack in a percolator and let it

drain, then pour on the remainder of the spirit; when it ceases to drop, press, and wash the marc with spirit to make up 40 parts.—*Dose,* ½ to 2 dr.

2. (Ph. L.) Pale bark, bruised, 4 oz.; dried bitter orange peel, 3 oz..(2 oz.—Ph. D.); serpentary root, bruised, 6 dr.; hay saffron, 2 dr.; cochineal, in powder, 1 dr.; macerate for 7 days (14 days—Ph. D.; or percolate—Ph. E.), press, and filter.

3. (Wholesale.) From pale bark, 3½ lbs.; dried orange peel, 2 lbs.; serpentary root, 4 oz.; hay saffron, 1 oz.; cochineal, ½ oz.; proof spirit, 4 galls. (or rectified spirit and water, of each, 2 galls.); macerate for 14 days.

Obs. In the Ph. E. yellow bark is ordered.— *Dose* and *use,* as the last.

Tincture of Cinchona (Pale). *Syn.* TINCTURE OF PALE BARK; TINCTURA CINCHONÆ PALLIDÆ (Ph. L.), L. *Prep.* From pale bark, as the last.

Tincture of Cin'namon. *Syn.* TINCTURA CIN-NAMOMI (B. P., Ph. L. and E.), L. *Prep.* 1. (B. P.) Cinnamon, in coarse powder, 1 part; rectified spirit, 8 parts; macerate 48 hours with 6 parts of the spirit, agitating occasionally, pack in a percolator and let it drain, then pour on the remaining spirit; when it ceases to drop, press, and wash the marc with spirit to make up 8 parts.

2. (Ph. L.) Cinnamon, bruised, 3½ oz.; proof spirit, 1 quart; macerate for 7 days (or percolate —Ph. E.). In the shops cassia is usually substituted for cinnamon, and spirit 26 u. p. for proof spirit.—*Dose,* 1 to 4 fl. dr.; as a cordial, aromatic, and stomachic.

Tincture of Cinnamon (Compound). *Syn.* AROMATIC TINCTURE; TINCTURA CINNAMOMI COMPOSITA (Ph. L., E., and D.), T. AROMATICA, L. *Prep.* 1. (Ph. L.) Cinnamon, bruised, 1 oz.; cardamoms (bruised, without the shells), ½ oz.; long pepper and ginger, of each, 2½ dr.; proof spirit, 1 quart; digest for 7 days (or percolate—Ph. E.). The Ph. E. omits the ginger, and uses ½ oz. more cardamoms.

2. (Ph. D.) Cinnamon, 2 oz.; cardamoms, 1 oz.; ginger, ½ oz.; proof spirit, 1 quart; macerate for 14 days. The following form is current in the wholesale houses.

3. Cassia, 1 lb.; cardamoms, 6 oz.; long pepper and ginger, of each, ¼ lb.; oil of cassia, 1½ fl. dr.; proof spirit, 4 galls. (or rectified spirit and water, of each, 2 galls.). Cordial, aromatic, stomachic. —*Dose,* 1 to 2 fl. dr.; in atonic gout, debility, flatulence, &c.

Tincture of Cloves. *Syn.* TINCTURA CARYO-PHYLLI (*Guibourt*), L. *Prep.* Cloves, 2 oz.; rectified spirit, 16 oz. Macerate 10 days.

Tincture of Coch'ineal. *Syn.* TINCTURA COCCI CACTI (Ph. D.), L. *Prep.* 1. (B. P.) Cochineal, in powder, 1 part; proof spirit, 8 parts; macerate 7 days; strain, and wash the marc with spirit to make up 8 parts. *Dose,* 30 to 90 minims twice a day. Used chiefly for colouring medicines.

2. (Ph. D.) Cochineal, in fine powder, 2 oz.; proof spirit, 1 pint. Antispasmodic and sedative, but chiefly employed for its colour.—*Dose,* ½ to 2 fl. dr.

Tincture of Cochineal (Ammoniated). *Syn.* TINCTURA COCCI AMMONIATA (*Dr Eberle*), L. *Prep.* Cochineal, ½ oz.; water of ammonia, ½ oz.; rectified spirit, 8 fl. oz.—*Dose*, 5 drops, in hooping-cough.

Tincture of Col'chicum. *Syn.* GOUT TINCTURE, TINCTURE OF MEADOW SAFFRON; TINCTURA COLCHICI SEMINUM (B. P.); TINCTURA COLCHICI (Ph. L. and E.), T. SEMINUM COLCHICI (Ph. D.), L. *Prep.* 1. Colchicum seed, bruised, 1 part; proof spirit, 8 parts; macerate 48 hours with 6 parts of the spirit, agitating occasionally, pack in a percolator and let it drain, then pour on the remainder of the spirit; when it ceases to drop, wash the marc with spirit to make up 8 parts.—*Dose*, 15 to 30 minims.

2. (Ph. L.) Seeds of meadow saffron (*Colchicum autumnale*), bruised (finely ground in a coffee-mill—Ph. E.), 5 oz.; proof spirit, 1 quart; macerate for 7 days (14 days—Ph. D.; or percolate—Ph. E.); then press and filter.—*Dose*, 15 to 20 drops to 1 fl. dr.; in gout, &c.

Tincture of Colchicum Bulbs. *Syn.* TINCTURA COLCHICI E RADICE (P. Cod.), L. *Prep.* Macerate 1 part of the bulbs in 5 parts (by weight) of proof spirit for 10 days.

Tincture of Colchicum (Compound). *Syn.* TINCTURA COLCHICI COMPOSITA (Ph. L.), SPIRITUS COLCHICI AMMONIATUS, L. *Prep.* (Ph. L.) Colchicum seeds, bruised, 5 oz.; aromatic spirit of ammonia, 1 quart; digest for 7 days, then press and filter.—*Dose*, 20 drops to 1 fl. dr.; in gout, &c.

Tincture of Colchicum Flowers. *Syn.* TINCTURA FLORUM COLCHICI; EAU MÉDICINALE D'HUSSON (*Dr Wilson*), L. *Prep.* Take of the fresh juice of colchicum flowers, 2 parts; French brandy (or proof spirit), 1 part; mix, and in a few days decant or filter, and preserve it in small bottles in a cool place.

Tincture of Colocynth. *Syn.* TINCTURA COLOCYNTHIDIS (Ph. G.), L. *Prep.* Colocynth, 1 part; rectified spirit, 10 parts.—*Dose*, 6 to 20 drops.

Tincture of Contrayerva. *Syn.* TINCTURA CONTRAYERVÆ (P. Cod.), L. *Prep.* Contrayerva root, 4 oz.; rectified spirit, 1 pint.

Tincture of Copai'ba (Alkaline). *Syn.* TINCTURA COPAIBÆ ALKALINA, L. *Prep.* (*Lewis Thompson*.) Dissolve carbonate of potassa, 2oz., in water, 1 pint, and add to this balsam of copaiba, in a thin stream, constantly stirring, until the mixture, at first white and milky, becomes clear, like jelly or amber, which will generally take place when about a pint of balsam has been added; set the mixture aside for 2 or 3 hours, then pour in of rectified spirit, 1 quart, and mix the whole together. Sweet spirit of nitre may be substituted for spirit of wine, provided it does not contain free acid.—*Dose*, 1 to 2 teaspoonfuls.

Tincture of Coto. *Syn.* TINCTURA COTO (B. P. C.), L. *Prep.* Take of coto bark, bruised, 2 oz.; rectified spirit, 1 pint. Macerate for 7 days, with occasional agitation; then press, filter, and add sufficient rectified spirit to produce 1 pint.—*Dose*, 10 to 30 minims.

Tincture of Croton. *Syn.* TINCTURA CROTONIS, L. *Prep.* Croton seed, 1 part; rectified spirit, 5 parts (*Beasley*).

Tincture of Cu'bebs. *Syn.* ESSENCE OF CUBEBS; TINCTURA CUBEBÆ (B. P., Ph. L. and D.), TINCTURA PIPERIS CUBEBÆ, L. *Prep.* 1. (B. P.) Cubebs, in powder, 1 part; rectified spirit, 8 parts; macerate 48 hours with 6 parts of the spirit, agitating occasionally, pack in a percolator, and let it drain; pour on the remaining spirit, and when it ceases to drop, wash the marc with spirit to make up 8 parts.—*Dose*, 1 to 2 dr.

2. (Ph. L.) Cubebs (bruised or ground in a pepper-mill), 1 lb.; proof spirit, 1 quart; macerate for 7 days, press out the liquor, and filter.—*Dose*, ½ to 1 fl. dr., three or four times a day; in gonorrhœa, &c.

3. (Ph. D., and Ph. L. 1836.) Cubebs, 5 oz.; rectified spirit, 1 quart (proof spirit—Ph. D. 1826); macerate for 14 days.—*Dose*, 1 to 2 fl. dr.

Tincture of Cuspa'ria. *Syn.* TINCTURA CUSPARIÆ (Ph. E.), T. ANGOSTURÆ, L. *Prep.* (Ph. E.) Angostura bark or cusparia, 4½ oz.; proof spirit, 1 quart; digest or percolate. Tonic, stimulant, and stomachic.—*Dose*, 1 to 2 fl. dr.

Tincture of Deadly Nightshade. Tincture of belladonna.

Tincture, De Coetlogon's. Haffenden's tincture.

Tincture of Digita'lis. Tincture of foxglove.

Tincture of Elaterium. *Syn.* TINCTURA ELATERII, L. *Prep.* Extract of elaterium, 8 gr.; rectified spirit, 8 fl. oz.—*Dose*, ½ to 2 dr.

Tincture of Elecampane'. *Syn.* TINCTURA INULÆ, T. HELENII, L. *Prep.* (P. Cod.) Powdered elecampane, 4 oz.; proof spirit, 1 pint; macerate for 15 days. Tonic, deobstruent, and expectorant.—*Dose*, ½ to 2 fl. dr.; in dyspepsia, palsy, dropsies, uterine obstructions, &c.

Tincture of Er'got. *Syn.* TINCTURA SECALIS CORNUTI, T. ERGOTÆ (B. P., Ph. D.), L. *Prep.* 1. (B. P.) Ergot, bruised, 1 part; proof spirit, 4 parts; macerate 48 hours with 3 parts of spirit, agitating occasionally; pack in a percolator, let it drain, then pour on the remaining spirit; when it ceases to drop, wash the marc with the spirit to make up 4 parts.—*Dose*, 15 to 60 minims.

2. (Apothecaries' Hall.) Ergot (ground in a coffee-mill), 2½ oz.; proof spirit, 1 pint; digest for 7 days.—*Dose*. A teaspoonful; to excite the action of the uterus in labour.

3. (Ph. D.) Ergot, 8 oz.; proof spirit, 1 quart; macerate for 14 days, and strain with expression.—*Dose*, 20 drops to 1 fl. dr.; as the last.

Tincture of Ergot (Ammoniated). *Syn.* TINCTURA ERGOTÆ AMMONIATA, L. *Prep.* Ergot (in No. 20 powder), 10 oz.; aromatic spirit of ammonia, a sufficient quantity. Moisten the powder with a suitable quantity of the menstruum, and macerate for 12 hours; then pack in a percolator, and gradually pour aromatic spirit of ammonia upon it until 1 pint of tincture is obtained.—*Dose*, 10 to 60 minims.

Tincture of Ergot (Ethereal). *Syn.* TINCTURA ERGOTÆ ÆTHEREA (Ph. L.), L. *Prep.* (Ph. L.) Ergot, bruised, 15 oz.; ether, 1 quart; macerate for 7 days, press, and filter.—*Dose*, 10 drops to 1 fl. dr.

Tincture of Eucalyptus. *Syn.* TINCTURA EU-

CALYPTI (B. P. C.), L. *Prep.* Take of eucalyptus leaves (in No. 20 powder), 4 oz.; rectified spirit, a sufficient quantity. Moisten the powder with a suitable quantity of the menstruum, and macerate for 24 hours; then pack in a percolator, and gradually pour rectified spirit upon it until 1 pint of tincture is obtained.—*Dose*, 15 minims to 2 fl. dr.

Tincture of Euphorbia. *Syn.* TINCTURA EUPHORBIÆ (B. P. C.), L. *Prep.* Take of euphorbia (in No. 20 powder), 4 oz.; proof spirit, a sufficient quantity. Moisten the powder with a suitable quantity of the menstruum, and macerate for 24 hours; then pack in a percolator, and gradually pour proof spirit upon it until 1 pint of tincture is obtained.—*Dose*, 10 to 30 minims.

Tincture, Febrifuge. *Syn.* TINCTURA FEBRIFUGA (*Dr Clutton*), L. *Prep.* Febrifuge spirit, ½ pint; angelica root, 1½ dr.; serpentary, 1½ dr.; cardamom seeds, 1½ dr. Digest and filter.

Tincture of Fleabane. *Syn.* TINCTURA ERIGERONIS, L. *Prep.* Dried Canada fleabane (*Erigeron canadense*), 4 oz.; proof spirit, 16 oz. Macerate, express, and filter.

Tincture of Foxglove. *Syn.* TINCTURA DIGITALIS (Ph. L., E., & D.), L. *Prep.* 1. (Ph. L.) Dried foxglove leaves, 4 oz. (5 oz.—Ph. D.); proof spirit, 1 quart; macerate for 7 days (14 days—Ph. D.; or percolate—Ph. E.); then press and strain.

2. (B. P.) Digitalis leaves, in coarse powder, 2½ oz.; proof spirit, 1 pint. Proceed as for tincture of aconite (B. P.).

Obs. This tincture is a powerful sedative, diuretic, and narcotic. The commencing dose should be 10 drops, gradually and cautiously increased to 30, or even 40; in asthmas, dropsies, fevers, phthisis, &c. "If 40 fl. oz. of spirit be allowed to pass (percolate) through the sp. gr. will be ·944; and the solid contents of 1 fl. oz. will amount to 24 gr." (Ph. E.).

Tincture of Galangal. *Syn.* TINCTURA GALANGÆ (Ph. Amst.), L. *Prep.* Galangal root, 1 oz.; proof spirit, 6 oz.—*Dose*, 30 to 60 drops.

Tincture of Gal'banum. *Syn.* TINCTURA GALBANI, L. *Prep.* (Ph. D. 1826.) Galbanum, 2 oz.; proof spirit, 32 fl. oz.; digest 7 days. Stimulant and antispasmodic.—*Dose*, 1 to 3 fl. dr. "If less nauseous than tincture of assafœtida, it is also less powerful" (*Dr A. T. Thomson*).

Tincture of Galls. *Syn.* TINCTURA GALLÆ (B. P., Ph. L. & D.), T. GALLARUM (Ph. E.), L. *Prep.* 1. (B. P.) Galls, bruised, 1 part; proof spirit, 8 parts; macerate for 48 hours with 6 parts of the spirit, agitating occasionally, pack in a percolator, let it drain, and then pour on the remaining spirit; when it ceases to drop, wash the marc with spirit to make up 8 parts.—*Dose*, ½ to 2 dr.

2. (Ph. L.) Galls, in powder, 5 oz.; proof spirit, 1 quart; macerate for 7 days (14 days—Ph. D.; or percolate—Ph. E.); then express the liquid, and filter it. Astringent and styptic.—*Dose*, ½ to 2 fl. dr. It is chiefly used as a test for iron.

Tincture of Garden Marigold. *Syn.* TINCTURA CALENDULÆ, L. *Prep.* A saturated tincture of the leaves and flowers of the garden marigold is prepared with whisky, and is reputed to be of service as an application for lacerated wounds.

Tincture of Garden Nightshade (Ethereal). *Syn.* TINCTURA SOLANI ÆTHEREA (P. Cod.), L. *Prep.* Powdered leaves of garden nightshade, 4 oz.; sulphuric ether, 16 oz. (by weight). Make by percolation.

Tincture of Gentian (Ammo"niated). *Syn.* TINCTURA GENTIANÆ AMMONIATÆ, L.; ÉLIXIR ANTISCROFULEUX, Fr. *Prep.* (P. Cod.) Gentian, 1 oz.; sesquicarbonate of ammonia, ½ oz.; proof spirit, 32 fl. oz. As the last, but preferred in acidity and low spirits.

Tincture of Gen'tian (Compound). *Syn.* BITTER STOMACHIC TINCTURE; TINCTURA GENTIANÆ COMPOSITA (B. P., Ph. L., E., & D.), T. AMARA, L. *Prep.* 1. (B. P.) Gentian, bruised, 1½ parts; bitter orange peel, bruised, ¾ part; cardamom seeds, bruised, ½ part; proof spirit, 20 parts; macerate for 48 hours with 15 parts of the spirit, agitating occasionally, pack in a percolator, let it drain, and then pour on the remaining spirit; when it ceases to drop, wash the marc with spirit to make up 20 parts.—*Dose*, 1 to 2 dr.

2. (Ph. L.) Gentian root, sliced and bruised, 2½ oz.; dried orange peel, 10 dr.; cardamoms, bruised, 5 dr.; proof spirit, one quart; macerate for 7 days (or percholate—Ph. E.). The Edinburgh College substitutes canella for cardamoms, and adds of cochineal, ½ dr.

3. (Ph. D.) Gentian root, 3 oz.; dried bitter orange peel, 1½ oz.; cardamoms, ½ oz.; proof spirit, 1 quart; macerate for 14 days.

4. (Wholesale.) Gentian, 2½ lbs.; dried orange peel, 1¼ lbs.; bruised cardamoms, 2½ lbs.; proof spirit, 4 galls. (or rectified spirit and water, of each 2 galls.); digest as last.

Obs. This is an excellent and popular stomachic bitter and tonic.—*Dose*, 1 to 2 fl. dr.; in dyspepsia, loss of appetite, &c.

Tincture of Geranium. *Syn.* TINCTURA GERANII, L. *Prep.* Dried roots of *Geranium maculatum*, 5 oz.; proof spirit, 2 pints. Astringent. Used chiefly in gargles.

Tinc'ture of Ginger. *Syn.* TINCTURA ZINGIBERIS (B. P., Ph. L., E., & D.), L. *Prep.* 1. (B. P.) Ginger, bruised, 1 part; rectified spirit, 8 parts; macerate the ginger 48 hours in 6 parts of the spirit, agitating occasionally; pack in a percolator, let it drain, pour on the remaining spirit, and when it ceases to drop, press, filter, and add spirit to make 8 parts.—*Dose*, 10 to 30 minims.

2. (Ph. L.) Ginger, bruised, 2½ oz.; rectified spirit, 1 quart; macerate for 7 days (or percolate—Ph. E.).

3. (Wholesale.) Coarsely powdered unbleached Jamaica ginger, 1½ lbs.; rectified spirit (or spirit distilled from the essence), 1½ galls.; water, ½ gall.; digest as above. Stimulant and carminative.—*Dose*, 1 to 2 fl. dr.

Obs. The formula of the Ph. D. 1826 resembles the above; that of the last Ph. D. orders 8 oz. of ginger to 1 quart of rectified spirit. The product is, consequently, of fully 3 times the strength of that of the others, and is similar to the common ESSENCE OF GINGER of the shops.

Tincture of Ginger (Stronger). *Syn.* TINCTURA ZINGIBERIS FORTIOR (B. P.), L. Pack tightly in a percolator, ginger in fine powder, 10 oz., and pour over it carefully ½ pint of rectified spirit. After two hours more add more spirit, and let it percolate slowly until 1 pint of tincture has been collected.—*Dose*, 5 to 20 minims.

Tincture of Gold-Thread. *Syn.* TINCTURA COPTIS (*Dr Wood*), L. *Prep.* Gold-thread, 1 oz.; proof spirit, 16 oz.—*Dose*, 1 dr. Tonic.

Tincture, Gout. *Syn.* TINCTURA ANTARTHRITICA, L. *Prep.* 1. (*Dr Graves's*.) Take of dried orange peel and powder of aloes and canella, of each, 2 oz.; rhubarb, 1 oz.; French brandy (or proof spirit), 1 quart; digest a week, and strain with expression.—*Dose*, 1 to 2 teaspoonfuls night and morning.

2. (*Dr Wilson's*.) Tincture of colchicum flowers.

3. Tincture of colchicum.

Tincture of Green Hellebore Root. *Syn.* TINCTURA VERATRI VIRIDIS (B. P.), L. *Prep.* Green hellebore root in coarse powder, 4 oz.; rectified spirit, 1 pint. Prepared as tincture of aconite (B. P.).

Tincture of Gua'iacum. *Syn.* TINCTURA GUAIACI (Ph. E. & D.), L. *Prep.* (Ph. E., & Ph. L. 1836.) Guaiacum resin (powdered), 7 oz. (8 oz.—Ph. D.); rectified spirit, 1 quart; digest for 14 days, and filter. An excellent sudorific; in chronic gout and rheumatism.—*Dose*, 1 to 3 fl. dr., taken in milk.

Tincture of Guaiacum (Alkaline). *Syn.* TINCTURA GUAIACI ALKALINA (*Dr Dewees*), L. *Prep.* Guaiacum, 5 oz.; carbonate of potash or of soda, 3 dr.; pimento, 2 oz.; proof spirit, 2 pints. —*Dose*. A teaspoonful 3 times a day in dysmenorrhœa.

Tincture of Guaiacum (Compound). *Syn.* AMMONIATED TINCTURE OF GUAIACUM, VOLATILE T. OF G., RHEUMATIC DROPS; TINCTURA GUAIACI COMPOSITA (Ph. L.), T. G. AMMONIATA (B. P., Ph. E.), L. *Prep.* 1. (B. P.) Guaiac resin, in fine powder, 4 parts; aromatic spirit of ammonia, 20 parts; macerate 7 days, filter, and wash the filter with the spirit to make up 20 parts. —*Dose*, ½ to 1 dr., with 1 dr. of mucilage or yolk of egg, to form an emulsion.

2. (Ph. L.) Guaiacum in coarse powder, 7 oz.; aromatic spirit of ammonia (spirit of ammonia—Ph. E.), 1 quart; digest for 7 days, and decant or filter. A powerful, stimulating sudorific and emmenagogue; in chronic rheumatism, gout, amenorrhœa, &c.—*Dose*, 1 to 2 fl. dr., in milk or some viscid liquid.

Tincture of Guaiacum Wood. *Syn.* TINCTURA GUAIACI LIGNI (P. Cod.), L. *Prep.* 1 part of the rasped wood to 5 parts by weight of proof spirit. Digest 10 days and strain.

Tincture of Guarana. *Syn.* TINCTURA PAULLINIÆ (*Dorvault*), L. *Prep.* Alcoholic extract of guarana, 1 oz.; proof spirit, 16 oz. Dissolve.

Tincture, Haffenden's Balsam'ic. *Syn.* DE COETLOGON'S BALSAMIC TINCTURE. This is a nostrum of many virtues, prepared from tincture of serpentary (of double strength), 1½ fl. oz.; compound tincture of benzoin, 1 fl. oz.; tinctures of Tolu and opium, of each, ½ fl. oz.; with rectified spirit, q. s. to render the mixture 'bright,' should it turn milky ('Anat. of Quackery').

Tincture of Hamamelis. *Syn.* TINCTURE OF WITCH-HAZEL; TINCTURA HAMAMELIDIS (B. P.), L. *Prep.* Hamamelis bark (in No. 20 powder), 2 parts; proof spirit, 20 parts. Macerate, filter, and make up to a pint with proof spirit. .

Tincture, Hatfield's. *Prep.* From gum guaiacum and soap, of each, 2 dr.; rectified spirit, 1 pint; digest for a week. Used as TINCTURE OF GUAIACUM; also externally.

Tincture of Hedge-hyssop. *Syn.* TINCTURA GRATIOLÆ (*Reece*), L. *Prep.* Dried hedgehyssop, 4 oz.; proof spirit, 32 oz.

Tincture of Hel'lebore. *Syn.* TINCTURE OF BLACK HELLEBORE; TINCTURA HELLEBORI (Ph. L.), TINCTURA HELLEBORI NIGRI, L. *Prep.* (Ph. L.) Black hellebore root, bruised, 5 oz.; proof spirit, 1 quart; macerate 7 days, then strain with expression.

Obs. This tincture is a powerful emmenagogue, and was a favourite remedy with Dr Mead in uterine obstructions and certain cutaneous affections.—*Dose*, 20 drops to one fl. dr. See TINCTURE OF VERATRUM.

Tincture of Hem'lock. *Syn.* TINCTURA CICUTÆ, T. CONII (Ph. L. & E.), T. CONII MACULATI, L. *Prep.* 1. (Ph. L.) Dried hemlock leaves, 5 oz.; proof spirit, 1 quart; digest a week, press, and filter. In the Ph. L. 1836, cardamom seeds, 1 oz., was added.

2. (Ph. E.) Fresh hemlock leaves, 12 oz.; express the juice, bruise the residuum, and treat it, by percolation, first with tincture of cardamoms, 10 fl. oz., and next with rectified spirit, 1½ pints; mix the liquids, and filter. Deobstruent and narcotic.—*Dose* of the Ph. L., 30 to 60 drops; that of the Ph. E. tincture is less, it being a much stronger and certain preparation. See HEMLOCK.

3. (B. P.) Hemlock fruit, bruised, 2½ oz.; proof spirit, 1 pint. Proceed as for tincture of aconite (B. P.).

Tincture of Hemp. Tincture of Indian hemp.

Tincture of Hen'bane. *Syn.* TINCTURA HYOSCYAMI (B. P., Ph. L., E., & D.), L. *Prep.* 1. (B. P.) Hyoscyamus leaves, dried and bruised, 1 part; proof spirit, 8 parts; macerate 48 hours with 6 parts of the spirit, pack in a percolator, and when it has drained pour on the remaining spirit, and when it ceases to drop, press, and wash the marc with spirit to make up 8 parts.—*Dose*, 15 to 60 minims.

2. (Ph. L.) Dried leaves of henbane, 5 oz.; proof spirit, 1 quart; macerate for 7 days (14 days —Ph. D.); or percolate—Ph. E.), then press and filter. Anodyne, sedative, soporific, and narcotic. —*Dose*, 20 drops to 2 fl. dr.

Obs. This, as well as the TINCTURES OF FOXGLOVE, HEMLOCK, HOPS, JALAP, LOBELIA INFLATA, RHATANY, SAVINE, SQUILLS, SENNA, VALERIAN, WORMWOOD, &c., is usually prepared by the druggists with 1 lb. of the dried leaves (or dried drugs) to each gall. of a mixture of equal parts of rectified spirit and water.

Tincture of Hops. *Syn.* TINCTURA LUPULI (B. P., Ph. L. & E.), TINCTURA HUMULI, L. *Prep.* 1. (B. P.) Hops, 1 part; proof spirit, 8 parts; macerate 48 hours in 6 parts of the spirit, agitating occasionally, pack in a percolator, let it drain, add the remaining spirit, and when fluid ceases to drop, wash the marc, filter, and make up 8 parts.—*Dose*, ½ to 2 dr.

2. (Ph. E.) Hops, 6 oz.; proof spirit, 1 quart; digest 7 days, then press and filter. Anodyne, sedative, and soporific.—*Dose*, ½ to 2 fl. dr. For

the formula of the Ph. E. and D., see TINCTURE OF LUPULIN.

Tincture of Hops (Compound). *Syn.* TINCTURA LUPULI COMPOSITA, L.; LIQUEUR DES TEIGNEUX (P. Cod.), Fr. *Prep.* Hops, 1 oz.; smaller centaury, 1 oz.; orange peel, 2 dr.; carbonate of potash, 12 gr.; proof spirit, 18 oz. (by weight).

Tincture of Horse-chestnut. *Syn.* TINCTURA HIPPOCASTANEI, L. *Prep.* Horse-chestnut bark, 4 oz.; proof spirit, 2 pints. Macerate for 10 days, and filter.

Tincture, Hudson's. Tooth tincture.

Tincture, Huxham's. Compound tincture of cinchona.

Tincture of Hydrastis. *Syn.* TINCTURE OF GOLDEN SEAL; TINCTURA HYDRASTIS, L. Hydrastis rhizome (in No. 60 powder), 2 parts; proof spirit, 20 parts. Pack in a percolator, and pass the spirit through till 20 parts are collected.—*Dose*, ½ to 1 dr.

Tincture of Indian Hemp. *Syn.* TINCTURA CANNABIS, T. C. INDICÆ (B. P., Ph. D.), L. *Prep.* 1. (B. P.) Extract of Indian hemp, 1 part; rectified spirit, 20 parts; dissolve.—*Dose*, 5 to 20 minims with 1 dr. of mucilage, adding 1 oz. of water.

2. (Ph. D.) Purified extract of Indian hemp, ½ oz.; rectified spirit, ½ pint; dissolve. 21 drops (minims) contain 1 gr. of the extract.

Obs. The formulæ of O'Shaughnessy and the Bengal Ph. are similar.—*Dose*, 10 drops every half-hour in cholera; 1 fl. dr. every half-hour in tetanus till the paroxysms cease, or catalepsy is induced.

Tincture of Indian Tobac'co. Tincture of lobelia.

Tincture of I'odine. *Syn.* TINCTURA IODINII (Ph. E.), TINCTURA IODINII, L. *Prep.* (Ph. E.) Iodine, 2½ oz.; rectified spirit, 1 quart; dissolve, and preserve it in well-closed bottles.—*Dose*, 5 to 30 drops, twice or thrice daily, where the use of iodine is indicated. Externally as a paint, &c.

Obs. The formulæ of Magendie, the Ph. U. S., and the Paris Codex are similar.

Tincture of Iodine (Colourless). *Syn.* TINCTURA IODI DECOLORATA (Ph. G.), L. *Prep.* 1. Iodine, 10 oz. Digest with gentle heat, occasionally shaking; and when the solution is completed, add liquor ammoniæ (·960), 16 oz. (by weight), shake together, and add rectified spirit, 75 oz. (by weight).

2. (*R. Rother.*) Iodine, 1 troy oz.; sodium sulphite, q. s.; ammonium carbonate, q. s.; alcohol, q. s.; water, 2 fl. oz. Upon 1 troy oz. of sodium sulphite and 200 gr. of ammonium carbonate, each previously powdered, pour the water, and then gradually add the iodine until its colour is no longer discharged. If now the effervescence has ceased, add ammonium carbonate in proportion to the remaining iodine, but if ammonium carbonate still predominates, then add sodium sulphite in proportion to the surplus of iodine, and continue the incorporation of the iodine until all has been added, and a faintly yellow solution results, whilst some sulphite and carbonate remain in excess. Now gradually add alcohol with constant stirring until the mixture measures 12 fl. oz. Pour this upon a muslin strainer, press the liquid out, and measure it. Mix the solid residue with enough alcohol to make the measure of a pint when united

with the first expression, then press the liquor out mix it with that first obtained, and filter the tincture through paper.

3. (B. P. C.) Iodine, 250 gr.; rectified spirit, 5½ fl. oz. Dissolve by the aid of a gentle heat. When cold transfer to a stoppered bottle, and add of stronger solution of ammonia, 10 fl. dr. Keep the mixture in a warm place until decolourised, after which dilute it with rectified spirit sufficient to produce 1 pint.

Tincture of Iodine (Compound). *Syn.* ANTI-SCROFULOUS DROPS; TINCTURA IODI (B. P.), TINCTURA IODINII COMPOSITA (Ph. L. & D.), L. *Prep.* 1. Iodine, ½ part; iodide of potassium, ¼ part; rectified spirit, 20 parts; dissolve.—*Dose*, 5 to 20 minims. Also an excellent application to the throat in diphtheria.

2. (Ph. L. & D.) Iodine, 1 oz.; iodide of potassium, 2 oz.; rectified spirit, 1 quart; dissolve.—*Dose*, &c., as the last.

Tincture, Iodine'(Ethereal). *Syn.* TINCTURA IODINII ÆTHEREA. *Prep.* Iodine, 2 scruples; sulphuric ether, 1½ fl. oz.

Tincture of Iodoform (Ethereal). *Syn.* TINCTURA IODOFORMI ÆTHEREA (*Odin* and *Lemaire*), L. *Prep.* Crystallised iodoform, 15 gr.; ether at 60° Baumé, 1 dr. (by weight).

Tincture of Ipecacuan'ha. *Syn.* TINCTURA IPECACUANHÆ, L. *Prep.* (Ph. Bor.) Ipecacuanha (coarsely powdered), 1 oz.; spirit sp. gr. ·897 to ·600 (16 to 17 o. p.), 8 oz.; macerate for 8 days. The tincture of the P. Cod. has twice its strength.—*Dose*, 10 or 12 drops to 2 fl. dr., according to the intention.

Tincture of Jaborandi. *Syn.* TINCTURA JABO-RANDI ('Ph. Journ.'), L. *Prep.* Powdered jaborandi leaves, 10 oz.; rectified spirit, q. s. Percolate until a pint of tincture is obtained.—*Dose*, 10 minims to 1 or 2 dr.

Tincture of Jal'ap. *Syn.* TINCTURA JALAPÆ (B. P., Ph. L., E., & D.), L. *Prep.* 1. (B. P.) Jalap, in coarse powder, 1 part; proof spirit, 8 parts; macerate for 48 hours in 6 parts of the spirit, agitating occasionally; pack in a percolator, and when the fluid ceases to pass, pour on the remaining spirit, press, filter, and add spirit to make 8 parts.—*Dose*, ½ to 2 dr.

2. (Ph. L.) Jalap, coarsely powdered, 5 oz. (10 oz.—Ph. L. 1836; 7 oz.—Ph. E.); proof spirit, 1 quart (1½ pints—Ph. D.); macerate for 7 days (or percolate—Ph. E.), then press and filter. Cathartic.—*Dose*, 1 to 4 fl. dr.

Tincture of Jalap (Compound). *Syn.* TINCTURA JALAPÆ COMPOSITA (Ph. E. 1744), L. *Prep.* Jalap root, 6 dr.; black hellebore root, 3 dr.; juniper berries, ½ oz.; guaiacum shavings, ½ oz.; French brandy, 24 oz.; digest for 3 days, and strain. The *Eau de Vie Allemande* of the Paris Codex is—Jalap, 8 oz.; turpeth root, 1 oz.; scammony, 2 oz.; proof spirit, 96 oz. (by weight). —*Dose*, 4 dr.

Tincture of Ki'no. *Syn.* TINCTURA KINO (B. P., Ph. L. & E.), L. *Prep.* 1. (B. P.) Kino in powder, 2 parts; rectified spirit, 12 parts; glycerine, 3 parts; water, 5 parts; macerate 7 days, filter, and make up to 20 parts.—*Dose*, ½ to 2 dr.

2. (Ph. L.) Powdered kino, 3½ oz.; rectified spirit, 1 quart; macerate for 7 days (or percolate —Ph. E.), and filter. Astringent.—*Dose*, 1 to 2

fl. dr., combined with chalk mixture; in diarrhœa, &c.

Tincture of Lactuca"rium. *Syn.* TINCTURA LACTUCARII, L. *Prep.* (Ph. E.) Powdered lactucarium, 4 oz.; proof spirit, 1 quart; digest or percolate. Anodyne, soporific, antispasmodic, and sedative.—*Dose,* 20 to 60 drops; in cases for which opium is unsuited. 10 drops (minims) contain 1 gr. of lactucarium.

Tincture of Larch. *Syn.* TINOTURA LARICIS (B. P.), L. *Prep.* Larch bark, in coarse powder, 2½ oz.; rectified spirit, 1 pint. Macerate the bark for 48 hours in 15 oz. of the spirit in a closed vessel, agitating occasionally; then transfer to a percolator, and when the fluid ceases to pass, continue the percolation with the remaining 5 oz. of spirit. Afterwards subject the contents of the percolator to pressure, filter the product, mix the liquid, and add rectified spirit, q. s. to make 1 pint. —*Dose,* 20 to 30 minims.

Tincture of Lav'ender (Compound). *Syn.* RED LAVENDER, RED LAVENDER DROPS, RED HARTS-HORN; TINCTURA LAVANDULÆ COMPOSITA (B. P., Ph. L. & D.), SPIRITUS LAVANDULÆ COMPOSITUS (Ph. E.). *Prep.* 1. (B. P.) English oil of lavender, 90 minims; English oil of rosemary, 10 minims; cinnamon, bruised, 150 gr.; nutmeg, bruised, 150 gr.; red sandal-wood, 300 gr.; rectified spirit, 40 oz.; macerate the cinnamon, nutmeg, and red sandal-wood in the spirit for 7 days, then press out and strain; dissolve the oils in the strained tincture, and add sufficient rectified spirit to make 40 oz.—*Dose,* ½ to 2 dr.

2. (Ph. L.) Cinnamon and nutmegs, of each, bruised, 2½ dr.; red sanders-wood, sliced, 5 dr.; rectified spirit, 1 quart; macerate for 7 days, then strain with expression, and dissolve in the strained liquid, oil of lavender, 1½ fl. dr.; oil of rosemary, 10 drops.

3. (Ph. L. 1836.) Spirit of lavender, 1½ pints; spirit of rosemary, ½ pint; red sanders-wood (rasped), 5 dr.; cinnamon and nutmegs (bruised), of each, 2½ dr.; macerate for 14 days.

4. (Ph. E.) Spirit of lavender, 1 quart; spirit of rosemary, 12 fl. oz.; cinnamon, 1 oz.; nutmeg, ½ oz.; red sanders, 3 dr.; cloves, 2 dr.; as No. 1.

5. (Ph. D.) Oil of lavender, 3 fl. dr.; oil of rosemary, 1 fl. dr.; cinnamon, 1 oz.; nutmegs, ½ oz.; cloves and cochineal, of each, ¼ oz.; rectified spirit, 1 quart; macerate for 14 days.

6. (Wholesale.) From oil of cassia, ⅝ fl. oz.; oil of nutmeg, 1 fl. oz.; oils of lavender and rosemary, of each, 4½ fl. oz.; red sanders (rasped), 3 lbs.; proof spirit, 6 galls. (or rectified spirit and water, of each, 3 galls.); digest 14 days. Should it be cloudy, add a little more proof spirit.

Obs. Compound tincture of lavender is a popular stimulant, cordial, and stomachic.—*Dose,* 1 to 3 teaspoonfuls (½ to 2 fl. dr.); in lowness of spirits, faintness, flatulence, hysteria, &c.

Tincture of Lem'ons. *Syn.* TINCTURA LIMO-NUM (Ph. L.), T. LIMONIS (B. P., Ph. D. & L.), L. *Prep.* 1. (B. P.) Fresh lemon peel, sliced thin, 1 part; proof spirit, 8 parts; macerate for 7 days in a closed vessel with occasional agitation, strain, press, filter, and make up with spirit to 8 parts.—*Dose,* ½ to 2 dr.

2. Fresh lemon peel, 3½ oz. (cut thin, 5 oz.—Ph. D.); proof spirit, 1 quart; macerate for 7 days (14 days—Ph. D.), then express the liquid and filter it. An aromatic bitter and stomachic. —*Dose,* ½ to 2 fl. dr.

Tincture of Lily of the Valley. *Syn.* TINC-TURA CONVALLARIÆ (B. P. C.), L. Take of lily of the valley flowers and stalks, dried in No. 20 powder, 2½ oz.; proof spirit, a sufficient quantity. Moisten the powder with a suitable quantity of the menstruum, and macerate for 24 hours; then pack in a percolator, and gradually pour proof spirit upon it until one pint of tincture is obtained.—*Dose,* 5 to 20 minims.

Tincture of Lobe"lia. *Syn.* TINCTURE OF INDIAN TOBACCO; TINCTURA LOBELIÆ IN FLATE, T. LOBELIÆ (B. P., Ph. L., E., & D.), L. *Prep.* 1. (B. P.) Lobelia, dried and bruised, 1 part; proof spirit, 8 parts; macerate 48 hours with 6 parts of the spirit, agitating occasionally; pack in a percolator, and let it drain; pour on the remaining spirit, and when it ceases to drop, press and wash the marc with spirit to make up 8 parts.— *Dose,* 10 to 30 minims, but 1 dr. may be given for dyspnœa; 4 dr. as an emetic.

2. (Ph. L.) Dried and powdered lobelia inflata, 5 oz.; proof spirit, 1 quart; macerate for 7 days (14 days—Ph. D.; or percolate—Ph. E.), press, and filter.—*Dose.* As an expectorant, 10 to 60 drops; as an emetic and antispasmodic, 1 to 2 fl. dr., repeated every third hour until it causes vomiting. It is principally employed in spasmodic asthma, and some other pulmonary affections.

Tincture of Lobelia (Ethereal). *Syn.* TINC-TURA LOBELIÆ ÆTHEREA (B. P., Ph. L. & E.), L. *Prep.* 1. Lobelia, dried and bruised, 1 part; spirit of ether, 8 parts; macerate 7 days, press, and strain 8 parts.—*Dose,* 10 to 30 minims, as an antispasmodic.

2. (Ph. L.) Indian tobacco, powdered, 5 oz.; ether, 1 fl. oz.; rectified spirit, 25 fl. oz.; macerate 7 days, press, and filter.

3. (Ph. E.) Dry lobelia, 5 oz.; spirit of sulphuric ether, 1 quart; by digestion for 7 days, or by percolation.—*Dose,* 6 or 8 drops to 1 fl. dr.

4. (*Whitlaw's.*) From lobelia, 1 lb.; rectified spirit and water of nitrous ether, of each, 4 pints; sulphuric ether, 4 oz.—*Dose,* &c., as the last.

Tincture of Lu'pulin. *Syn.* TINCTURE OF HOPS; TINCTURA LUPULI (Ph. E.), TINCTURA LUPULINÆ (Ph. D.), L. *Prep.* (Ph. D.) Lupulin (the yellowish-brown powder attached to the scales of hops, separated by friction and sifting), 5 oz.; rectified spirit, 1 quart; macerate for 14 days (or proceed by displacement—Ph. E.), press, and filter.—*Dose,* ½ to 2 fl. dr. See TINCTURE OF HOPS.

Tincture of Malate of Iron. *Syn.* TINCTURA FERRI MALATIS, T. FERRI POMATA (Ph. G.), L. *Prep.* Extract of malate of iron (see EXTRACT OF APPLES), 2 oz.; spirituous cinnamon water, 18 oz. Dissolve and filter.—*Dose,* 15 to 30 minims.

Tincture of Mastic. *Syn.* TINCTURA MAS-TICHES, L. *Prep.* Mastic, 2 oz.; rectified spirit, 9 fl. oz. Used in making eau de luce. If required for stopping hollow teeth, double the quantity of mastic must be used.

Tincture of Mat'ico. *Syn.* TINCTURA MATICO (Ph. D.), L. *Prep.* (Ph. D.) Matico leaves, in coarse powder, 8 oz.; proof spirit, 1 quart; macerate for 14 days, and strain with expression.— *Dose*, 1 to 2 fl. dr., as an internal astringent or hæmostatic. It is a very feeble remedy, as matico leaves are destitute of either tannin or gallic acid, and derive their power of stopping local bleeding from the peculiar mechanical construction of their surface.

Tincture of Mea'dow Saf'fron. Tincture of colchicum.

Tincture of Mone'sia. *Syn.* TINCTURA MONESIÆ, L. *Prep.* From monesia, 2½ oz.; proof spirit, 1 pint; macerate a week. Astringent.—*Dose*, ½ to 2 fl. dr.

Tincture of Musk. *Syn.* TINCTURA MOSCHI, L. *Prep.* (Ph. D. 1826.) Musk, 2 dr.; rectified spirit, 16 fl. oz.; digest 7 days. Antispasmodic; but principally used as a perfume, being too weak for medical use.

Tincture of Musk (Artificial). *Syn.* TINCTURA MOSCHI ARTIFICIALIS (*Van Mons*), L. *Prep.* Artificial musk, 1 dr.; rectified spirit, 2 oz.

Tincture of Musk Seed. *Syn.* TINCTURA ABELMOSCHI SEMINUM (*Dr Reece*), L. *Prep.* Musk seed, 2 oz.; proof spirit, 16 oz. Digest 7 days, and strain.—*Dose*, 1 fl. dr.

Tincture of Myrrh. *Syn.* GOLDEN TOOTH DROPS; TINCTURA MYRRHÆ (B. P., Ph. L., E., & D.), L. *Prep.* 1. (B. P.) Myrrh, in coarse powder, 1 part; rectified spirit, 8 parts; macerate 48 hours with 6 parts of the spirit, agitating occasionally; pack in a percolator, and when it ceases to drop, pour on the remaining spirit, wash the marc, press, and make up to 8 parts.—*Dose*, ½ to 1 dr. More frequently used mixed with water to form a gargle.

2. (Ph. L.) Myrrh, in powder, 3 oz. (3½ oz.—Ph. E.; 4 oz.—Ph. D.); rectified spirit, 1 quart; macerate for 7 days (14 days—Ph. D.; or by displacement—Ph. E.), and filter.

3. (Wholesale.) Myrrh, in coarse powder, 2½ lbs.; rectified spirit, 2 galls.; water, 1 gall.; as the last.

Obs. Tincture of myrrh is tonic and stimulant.—*Dose*, ½ to 1 fl. dr., as an adjuvant in mixtures, gargles, &c. Chiefly used diluted with water, as a dentifrice or wash for ulcerated and spongy gums.

Tincture of Myrrh (Alkaline). *Syn.* TINCTURA MYRRHÆ ALKALIZÆ (Ph. E. 1744). *Prep.* Powdered myrrh, 1½ oz.; solution of carbonate of potash, a sufficient quantity; mix into a soft paste, dry it, and add rectified spirit, 1 pint. Digest for 6 days, and strain.

Tincture of Myrrh (Compound). *Syn.* TINCTURA MYRRHÆ COMPOSITA, L. *Prep.* From myrrh and Socotrine aloes, of each, 2 lbs.; rectified spirit, 3 galls.; water, 2 galls.; digest for 14 days. This is frequently substituted for 'COMPOUND TINCTURE OF ALOES' in the wholesale trade.

Tincture of Myrrh and Borax. *Prep.* 1. Borax pulv., 2 oz.; glycerine, 4 oz. (by measure); tinct. myrrhæ, 32 oz.; spirit. rect., 1 gall.; aq. destillat., 1 gall.; eau de Cologne, 20 oz.; tinct. krameriæ, 1 oz.; syrup. simpl., 8 oz.

2. Tincture of myrrh, 1½ oz.; glycerine of borax, 3 dr.; ess. bouquet, 1 dr.; saccharin, 2 gr. Mix.

3. Glycerine boracis, 1½ oz.; tr. krameriæ, 3 dr.; tr. myrrh., 8 oz.; eau de Cologne, 8 oz. Mix.

Tincture of Nux Vom'ica. *Syn.* TINCTURA NUCIS VOMICÆ, L. *Prep.* 1. Extract of nux vomica, 133 gr.; water, 4 oz.; rectified spirit, enough to make 20 oz.—*Dose*, 10 to 20 minims.

2. (Ph. D. 1826.) Nux vomica (ground in a coffee-mill), 2 oz.; rectified spirit, 8 fl. oz.; macerate 7 (14) days.—*Dose*, 5 to 20 drops; in paralysis, &c. It is poisonous.

Tincture of Nux Vomica (Ethereal). *Syn.* TINCTURA NUCIS VOMICÆ ÆTHEREA (Ph. G.), L. *Prep.* Coarsely powdered nux vomica, 1 oz.; spirits of ether, 10 oz. (by weight). Macerate for 8 days.

Tincture, Odontal'gic. *Syn.* TOOTHACHE TINCTURE; TINCTURA ODONTALGICA, L. *Prep.* 1. Tincture of opium, 1 fl. dr.; ether, 2 fl. dr.; oil of cloves, 15 drops.

2. Rectified spirit, 3 fl. dr.; chloroform, 2 dr.; creasote, 1 dr.; mix.

3. (*Collier.*) Pellitory of Spain, 4 dr.; camphor, 3 dr.; opium, 1 dr.; oil of cloves, 2 fl. dr.; rectified spirit, 16 fl. oz.; digest for a week.

4. (*Niemann.*) Digest 60 or 80 common ladybirds (*Coccinella septempunctata*, Linn.) in rectified spirit, 1 fl. oz., for 8 days, and strain.

Obs. The above are commonly applied, on a small piece of lint, in toothache. For other formulæ see DROPS, TINCTURES OF MYRRH and PELLITORY, &c.

Tincture of O'pium. *Syn.* LAUDANUM, LIQUID L., ANODYNE TINCTURE, THEBAIC T.; TINCTURA OPII (B. P., Ph. L., E., & D.), THEBAICA, LAUDANUM LIQUIDUM, L. *Prep.* 1. (B. P.) Opium, in coarse powder, 1½ parts; proof spirit, 20 parts; macerate 7 days, strain, express, filter, and add spirit to make 20 parts.—*Dose*, 10 to 30 minims.

2. (Ph. L.) Powdered opium, 3 oz. (3 oz.—Ph. D.); proof spirit, 1 quart; macerate for 7 days (14 days—Ph. D.), and strain with expression.

3. (Ph. E.) Opium, sliced, 3 oz.; boiling water, 13½ fl. oz.; digest with heat, for 2 hours, break down the opium with the hand, strain, and express the infusion; then macerate the residuum for about 20 hours in rectified spirit, 1 pint 7 fl. oz.; next strain, press, mix the watery and spirituous infusions, and filter.

Obs. This preparation has a deep brownish-red colour, and the characteristic odour and taste of opium. 14 minims or measured drops of the London, and about 15 minims of the Edinburgh and Dublin tinctures, are equivalent to 1 gr. of dry opium, or 1·13 gr. of ordinary opium. 14 minims of this tincture are equal to about 25 drops of it poured from a bottle. Its sp. gr. is ·952 (*Phillips*).—*Dose*, 10 to 60 drops; as an anodyne, sedative, or hypnotic. The following form is substituted for that of the Pharmacopœia by many of the wholesale drug-houses:—Take of Turkey opium, 2½ lbs.; boiling water, 9 quarts; digest till dissolved or disintegrated, cool; add of rectified spirit, 2 galls.; and after repose for 24 hours, decant the clear portion.—*Prod.*, 4 galls.

Tincture of Opium (Ammo"niated). *Syn.* AMMONIATED TINCTURE OF OPIUM, SCOTCH PARE-

GORIC; TINCTURA OPII AMMONIATA (B. P., Ph-
E.), L. *Prep.* 1. (B. P.) Opium, in powder,
100 gr.; saffron, cut small, 180 gr.; benzoic acid,
180 gr.; oil of anise, 60 minims; strong solution
of ammonia, 4 oz.; rectified spirit, 16 oz.; macerate 7 days in a closed vessel, with occasional
agitation, strain, and add sufficient rectified spirit
to make up 20 oz.—*Dose,* ½ to 1 dr.
2. (Ph. E.) Benzoic acid and hay saffron, of
each, 6 dr.; opium, sliced, 4 dr.; oil of aniseed, 1
dr.; spirit of ammonia (Ph. E.), 1 quart; digest
for a week, and filter. Stimulant, antispasmodic,
and anodyne.—*Dose,* 20 to 80 drops; in hysteria,
hooping-cough, &c.
Obs. This preparation is called 'PAREGORIC,'
or 'PAREGORIC ELIXIR,' in Scotland, but should
be carefully distinguished from the compound
tincture of camphor, which passes under the same
names in England; as the former contains about
four times as much opium as the latter. 80
minims, or 145 poured drops, contain about 1 gr.
of opium.

Tincture of Opium (Cam'phorated). Compound
tincture of camphor.

Tincture of Opium (Ecard's or Bamberg's).
Syn. ECARD'S or BAMBERG'S THEBAIC TINCTURE;
TINCTURA OPII ECARDI, L. *Prep.* Opium, 2
oz.; cloves, 1 dr.; cinnamon water, 8 oz.; rectified
spirit, 4 oz. Digest in a warm room for 6 days,
and strain.

Tincture of Opium (Fetid). *Syn.* TINCTURA
OPII FŒTIDA (Ph. Fulda), L. *Prep.* Castor oil,
4 oz.; assafœtida, 2 oz.; salt of hartshorn, 1 oz.;
dry opium, 4 dr.; rectified spirit, 32 oz.—*Dose,*
15 minims to 1 dr.

Tincture of Opium (Odourless). *Syn.* TINCTURA
OPII DEODORATA (Ph. U. S.), L. *Prep.* Opium,
dried, and in moderately fine powder, 2½ troy oz.;
ether, rectified spirit, of each, 8 oz. (o. m.); water,
a sufficient quantity. Macerate the opium with ½
pint of the water for 24 hours, and express. Repeat
this operation twice with the same quantity of
water, mix the expressed liquids and evaporate to
4 oz. (o. m.); let cool, and shake repeatedly in a
bottle with the ether. When it has separated by
standing, pour off the ethereal solution, and evaporate the remaining liquid till all the ether has
disappeared. Mix the residue with 20 oz. (o. m.)
of water, and filter. When the liquid has ceased
to pass, add enough water to make the filtrate
measure 1½ pints (o. m.). Lastly, mix in the spirit.

Tincture of Orange with Iron. *Syn.* TINCTURA
FERRI AURANTIACA (Ph. Wurt.), L. *Prep.* Iron
fillings, 4 oz.; Seville oranges deprived of their
seeds, 4. Beat them together, leave them for
2 days; then add Madeira wine, 10 oz.; spirit of
orange peel, 2 oz. Digest, express, and filter.

Tincture of Or'ange Peel. *Syn.* TINCTURA
AURANTII (B. P., Ph. L., E., & D.), T. CORTICIS
AURANTII, L. *Prep.* 1. (B. P.) Dried bitter orange
peel, cut small and bruised, 1 part; proof spirit,
10 parts; macerate for 7 days in a closed vessel
with occasional agitation, then strain, press, and
filter, add sufficient proof spirit to make 10 parts.
—*Dose,* 1 to 2 dr.
2. (Ph. D.) Dried orange peel, 3½ oz. (4 oz.—
Ph. D.); proof spirit, 1 quart; digest for 7 days
(14 days—Ph. D; or by percolation—Ph. E.),
press, and filter. A grateful bitter stomachic.—

Dose, 1 to 3 fl. dr.; chiefly as an adjuvant in
mixtures, &c.

Tincture of Orange Peel (Fresh). *Syn.* TINCTURA
AURANTII RECENTIS (B. P.), L. *Prep.* Bitter
orange, rectified spirit, of each, a sufficient quantity.
Carefully cut from the orange the coloured part
of the rind in thin slices, and macerate 6 oz. of
this in 1 pint of spirit for a week, with frequent
agitation; then pour off the liquid, press the
dregs, mix the liquid products, and filter; finally,
add sufficient spirit to make 1 pint.—*Dose,* 1 to
2 dr.

Tincture of Orris Root. *Syn.* TINCTURA IRIDIS,
L. *Prep.* Freshly powdered orris root, 1 part;
proof spirit, 5 parts. Sold as *Esprit de Violette.*

Tincture of Ox-gall. *Syn.* TINCTURA FELLIS
BOVINI, L. *Prep.* Inspissated ox-gall, 2 oz.; proof
spirit, 1 pint. Digest until dissolved.

Tincture of Paracress. *Syn.* TINCTURA SPI-
LANTHI COMPOSITA. *Prep.* Paracress, dried and
bruised, 2 oz.; pyrethrum root, in coarse powder,
2 oz.; spirit, 10 oz. (by weight). Digest 8 days.
Sialogogue.

Tincture of Pareira Brava. *Syn.* TINCTURA
PAREIRÆ BRAVÆ (*Brodie*), L. *Prep.* Pareira
brava root, 2 oz.; French brandy, 1 pint. Digest
for 7 days.

Tincture of Pel'litory. *Syn.* TOOTHACHE TINC-
TURE; TINCTURA PYRETHRI (B. P.), L. *Prep.* 1.
Pellitory root, in coarse powder, 4 parts; rectified
spirit, 20 parts; macerate for 48 hours with 15
parts of the spirit, agitating occasionally; then
pack it in a percolator, let it drain, and pour on
the remaining spirit; when it ceases to drop,
press, filter, and make up to 20 parts.
2. Pellitory of Spain (bruised), 1 oz.; rectified
spirit, ½ pint; digest a week. In the P. Cod. a
tincture is ordered prepared with spirit about 41
o. p., and another prepared with spirit of sulphuric
ether.
3. (COMPOUND—Brande.) Pellitory root, 4 dr.;
camphor, 3 dr.; oil of cloves, 2 dr.; opium, 1 dr.;
rectified spirit, 6 fl. oz.; digest for 8 days. Both
the above are used for the toothache. See TINC-
TURE, ODONTALGIC.

Tincture of Pepper (Stomachic). *Syn.* TINCTURA
PIPERIS STOMACHICA, ESSENTIA STOMACHICA
POLYCHRESTA (*Spielman*), L. *Prep.* Capsicum, 1
oz.; black pepper, 2 dr.; long pepper, 2 dr.; white
pepper, 2 dr.; solution of acetate of potash,
6 oz.; spirit of ammonia, 1 oz. Digest and
filter.

Tincture of Perchloride of Iron. *Syn.* TINCTURA
FERRI PERCHLORIDI (B. P.), L. *Prep.* Mix 5 fl. oz.
of strong solution of perchloride of iron with 15
fl. oz. of rectified spirit.—*Dose,* 10 to 30 minims.

Tincture of Phos'phorus (Ethereal). *Syn.*
ÆTHER PHOSPHORATUS, TINCTURA PHOSPHORI
ÆTHEREA, L. *Prep.* 1. (Ph. Hann. 1831.)
Phosphorus (powdered by agitation with rectified
spirit), 16 gr.; ether, 2 oz.; macerate with agitation for 4 days, then decant the clear portion, and
preserve it in a stoppered bottle, in a cool dark
situation.
2. (P. Cod.) Phosphorus, cut small, 1 part;
ether, 50 parts; digest with occasional agitation
for 1 month, and decant the clear.—*Dose,* 5 to 15
drops, in any bland vehicle, thrice daily; in impotency, low sinking conditions of the system, &c.

Tincture of Phosphorus (Compound). *Syn.* TINCTURA PHOSPHORI COMPOSITA (B. P. C.), L. *Prep.* Take of phosphorus, 12 gr.; chloroform, 2½ fl. oz. Place in a stoppered bottle, and apply the heat of a water-bath until dissolved. Then add the solution to ethylic alcohol, 12½ fl. oz.

Tincture of Poplar Buds. *Syn.* TINCTURA POPULI (*Van Mons*), L. *Prep.* Poplar buds, 4 oz.; rectified spirit, 24 oz. Macerate and filter.

Tincture of Quas'sia. *Syn.* TINCTURA QUASSIÆ (B. P., Ph. E.), L. *Prep.* 1. (B. P.) Quassia in chips, ¾ part; proof spirit, 20 parts; digest 7 days, filter, and make up to 20 parts.—*Dose*, 1 to 2 dr.

2. (Ph. E.) Quassia, in chips, 10 dr.; proof spirit, 1 quart; digest 7 days. Bitter, tonic.—*Dose*, ½ to 2 fl. dr.; in dyspepsia, &c.

Tincture of Quassia (Compound). *Syn.* TINCTURA QUASSIÆ COMPOSITA (Ph. E.), L. *Prep.* (Ph. E.) Cardamoms and cochineal, of each, bruised, ½ oz.; powdered cinnamon and quassia chips, of each, 6 dr.; raisins, 7 oz.; proof spirit, 1 quart; digest for 7 days (or by percolation), then press and filter. Aromatic and tonic.—*Dose* and *use*, as the last.

Tincture of Quinine (Ammoniated). *Syn.* TINCTURA QUININÆ AMMONIATA (B. P.), L. *Prep.* Sulphate of quinine, 160 gr.; solution of ammonia, 2½ fl. oz.; proof spirit, 17½ oz.; dissolve the quinine in the spirit with a gentle heat, and add the solution of ammonia.—*Dose*, ½ to 2 dr.

Tincture of Quinine (Compound). *Syn.* FEVER DROPS; TINCTURA QUINÆ (B. P.), TINCTURA QUINÆ COMPOSITA (Ph. L.), L. *Prep.* 1. (B. P.) Hydrochlorate of quinia, 1 part; tincture of orange peel, 60 parts; dissolve with a gentle heat, digest for 3 days with occasional agitation, and strain.—*Dose*, 1 to 1½ dr.

2. (Ph. L.) Sulphate of quinine, 5 dr. 1 scrup. (or 320 gr.); tincture of orange peel, 1 quart; digest, with agitation, for 7 days, or until solution is complete.

Obs. Unless the tincture employed as the solvent be of the full strength some of the disulphate remains undissolved. It is an excellent medicine when faithfully prepared.—*Dose*, ½ to 2 fl. dr.; in debility, dyspepsia, &c.

Tincture of Red Gum. *Syn.* TINCTURA GUMMI RUBRI (*Mr Squire*), L. *Prep.* Red gum, 1 part; rectified spirit, 4 parts; digest and strain.—*Dose*, 20 to 40 minims.

Tincture of Red Lav'ender. Compound tincture of lavender.

Tincture of Rhat'any. *Syn.* TINCTURA KRAMERIÆ (B. P., Ph. D.), L. *Prep.* 1. (B. P.) Rhatany, bruised, 1 part; proof spirit, 8 parts; macerate 48 hours in 6 parts of the spirit, agitating occasionally; pack in a percolator; when it ceases to drop pour on the remaining spirit, and wash the marc with spirit to make up 8 parts.—*Dose*, 1 to 2 dr.

2. (Ph. D.) Rhatany root, in coarse powder, 8 oz.; proof spirit, 1 quart; macerate for 14 days, then press and filter. Astringent.—*Dose*, 1 to 2 fl. dr. The formula of the Ph. U. S. is similar.

Tincture of Rhododendron. *Syn.* TINCTURA RHODODENDRI (*Niemann*), L. *Prep.* Leaves of *Rhododendron chrysanthum*, 2 oz.; French brandy, ½ lb.; sherry, ½ lb. Digest for 15 days.

Tincture of Rhu'barb. *Syn.* TINCTURA RHEI (B. P., Ph. E.), L. *Prep.* 1. (B. P.) Rhubarb, bruised, 2 parts; cardamom seeds, bruised, ¼ part; coriander, bruised, ¼ part; saffron, ¼ part; proof spirit, 20 parts; macerate for 48 hours with 15 parts of the spirit, agitating occasionally; pack in a percolator, and when it ceases to drop pour on the remaining spirit; press and wash the marc, and add spirit to make up 20 parts.—*Dose*. As a stomachic, 1 to 2 dr.; as a purgative, ½ to 1 oz.

2. (Ph. E.) Powdered rhubarb, 3½ oz.; cardamom seeds, bruised, ½ oz.; proof spirit, 1 quart; digest, or proceed by the method of displacement.

3. (Ph. L. 1824.) Rhubarb, 2 oz.; cardamoms, 4 dr.; saffron, 2 dr.; proof spirit, 32 fl. oz. Both the above are cordial, stomachic, and laxative.—*Dose*, 1 fl. dr. to 1 fl. oz.

Tincture of Rhubarb (Aqueous). *Syn.* TINCTURA RHEI AQUOSA (Ph. G.), L. *Prep.* Rhubarb, 10 oz.; borax, 1 oz.; carbonate of potash, 1 oz.; boiling water, 85 oz.; infuse for ¼ hour, then add rectified spirit, 10 oz. (by weight); let it stand 2½ hours, and add cinnamon water, 15 oz.

Tincture of Rhubarb (Brandish's Al'kaline). *Syn.* TINCTURA RHEI ALKALINA BRANDISHII, L. *Prep.* From rhubarb, in coarse powder, 1½ oz.; Brandish's alkaline solution, 1 quart; macerate for a week. In the original formula only ½ oz. of rhubarb is ordered.—*Dose*, 20 drops to 2 fl. dr., in any bland liquid not acidulous; in acidities, dyspepsia, &c.

Tincture of Rhubarb (Compound). *Syn.* TINCTURA RHEI COMPOSITA (Ph. L. & D.), L. *Prep.* 1. (Ph. L.) Rhubarb, sliced, 2½ oz.; liquorice root, bruised, 6 dr.; ginger (bruised) and hay saffron, of each, 3 dr.; proof spirit, 1 quart; macerate for 7 days, then press and filter.

2. (Ph. D.) Rhubarb, 3 oz.; cardamoms, 1 oz.; liquorice root, ½ oz.; saffron, ¼ oz.; proof spirit, 1 quart; macerate for 14 days.

3. (Ph. L. 1824.) Rhubarb, 2 oz.; liquorice root, 4 dr.; ginger and saffron, of each, 2 dr.; proof spirit, 16 fl. oz.; water, 12 fl. oz.

Obs. This tincture is a popular remedy in diarrhœa and colic, and is an especial favourite with drunkards.—*Dose*. As a stomachic, 1 to 3 fl. dr.; as a purgative, ½ to 1½ fl. oz. The tincture of rhubarb of the shops is mostly inferior, being generally deficient both in rhubarb and spirit. The following forms we have seen extensively employed in the wholesale trade:—East Indian rhubarb, 20 lbs.; boiling water, q. s. to cover it; infuse for 24 hours; then slice the rhubarb, and put it into a cask with moist sugar, 14 lbs.; ginger, bruised, 3½ lbs.; hay saffron, 1 lb.; carbonate of potash, ½ lb.; bruised nutmegs, ½ lb.; rectified spirit, 19 galls.; water, 21 galls.; macerate with frequent agitation for 14 days, decant the clear portion, and press and filter the bottoms. Those houses that adhere to the Ph. L. for 1824 substitute cardamom seeds, 5 lbs., for the ginger. For the corresponding Ph. E. formula see the last article.

Tincture of Rhubarb and Aloes. *Syn.* SACRED ELIXIR†; TINCTURA RHEI ET ALOËS (Ph. E.), L. *Prep.* (Ph. D.) Rhubarb, in powder, 1½ oz. Socotrine or East Indian aloes, 6 dr.; cardamom seeds, bruised, 5 dr. ; proof spirit, 1 quart ; macerate 7 days, or percolate. A warm stomachic purgative.—*Dose*, ½ fl. oz. to 1 fl. oz.

Tincture of Rhubarb and Gen'tian. *Syn.* TINCTURA RHEI AMARA, TINCTURA RHEI ET GENTIANÆ (Ph. E.), L. *Prep.* (Ph. E.) Rhubarb, 2 oz. ; gentian, ½ oz. ; proof spirit, 1 quart ; proceed as for the last. Stomachic, tonic, and purgative.—*Dose*, 1 fl. dr. to 1 fl. oz.

Tincture of Rhubarb and Sen'na. *Syn.* WARNER'S GOUT CORDIAL ; TINCTURA RHEI ET SENNÆ (Ph. U. S.), L. *Prep.* (Ph. U. S.) Rhubarb, 1 oz. ; senna and red sanders-wood, of each, 2 dr. ; coriander and fennel seed, of each, 1 dr. ; saffron and extract of liquorice, of each, ½ dr. ; stoned raisins, 6 oz. ; proof spirit, 2½ pints ; macerate for 14 days. A popular stomachic and laxative.—*Dose*, ½ to 1½ fl. oz.

Tincture (Riemer's Nervous). *Prep.* From oil of juniper, 1 part ; volatile liquor of hartshorn, 4 parts ; rectified spirit, 16 parts.—*Dose*, 1 teaspoonful in water.

Tincture of Rose. *Syn.* TINCTURA ROSÆ, L. *Prep.* Dried red rose, 4 oz. ; proof spirit, 1 pint. Digest for 10 days.

Tincture of Rosemary. *Syn.* TINCTURA ROSMARINI (Ph. Bruns.), L. *Prep.* Flowering tops of rosemary, 1½ oz. ; spirit of rosemary, 6 oz. Digest, express, and filter.

Tincture, Ruspini's. See TINCTURE, TOOTH.

Tincture of Saffron. *Syn.* TINCTURA CROCI (B. P., Ph. E. & D.), T. C. SATIVÆ, L. *Prep.* 1. (B. P.) Saffron, 1 part ; proof spirit, 20 parts ; macerate 48 hours with 15 parts of the spirit, agitating occasionally ; pack in a percolator, let it drain, and then pour on the remaining spirit ; when it ceases to drop, wash the marc with spirit to make up 20 parts.—*Dose*, ½ to 2 dr.

2. (Ph. E.) Hay saffron, 2 oz. (2 oz.—Ph. D.) ; proof spirit, 1 quart (1 pint—Ph.D.) ; proceed either by maceration (for 14 days—Ph. D.) or by displacement. Stimulant and emmenagogue.—*Dose*, 1 to 2 fl. dr. Chiefly used for its colour and flavour.

Tincture of Saponin. *Syn.* TINCTURA SAPONINI, *Prep.* Bark of *Quillaia saponaria*, 1 part ; alcohol (90 per cent.), 4 parts. Heat to ebullition and filter.

Tincture of Sarsaparilla (Compound). *Syn.* TINCTURA SARZÆ COMPOSITA, L. ; LIQUEUR DÉPURATIVE (*François*), Fr. *Prep.* Sarsaparilla, guaiacum, china root, sassafras, of each, 1 oz. ; proof spirit, 16 oz.

Tincture of Savine. *Syn.* TINCTURA SABINÆ (B. P.), L. *Prep.* Savine tops, dried and coarsely powdered, 2½ oz. ; proof spirit, 1 pint. Proceed as for tincture of aconite.

Tincture of Scammony. *Syn.* TINCTURA SCAMMONII (P. Cod.), L. *Prep.* Scammony, 4 oz. ; rectified spirit, 20 oz. (by weight).

Tincture of Senega. *Syn.* TINCTURA SENEGÆ (B. P.), L. *Prep.* Senega, bruised, 2½ oz. ; proof spirit, 1 pint. Prepared as tincture of aconite.

Tincture of Senna (Compound). *Syn.* TINCTURE OF SENNA, ELIXIR OF HEALTH† ; TINCTURA SENNÆ (B. P.), TINCTURA SENNÆ COMPOSITA (Ph. L., E., & D.), L. *Prep.* 1. (B. P.) Senna, broken small, 5 parts ; raisins, freed from seeds, 4 parts ; cara-

way, bruised, 1 part ; coriander, bruised, 1 part ; proof spirit, 40 parts ; macerate the ingredients 48 hours in three fourths of the spirit, agitating occasionally ; pack in a percolator, and when it ceases to drop, pour on the remaining spirit ; press, filter, and make up 40 parts.—*Dose*, 2 to 8 dr.

2. (Ph. L.) Senna, 3½ oz. ; caraway seeds, bruised, 3½ dr. ; cardamom seeds, bruised, 1 dr. ; stoned raisins, 5 oz. ; proof spirit, 1 quart ; macerate for 7 days, press, and filter.

3. (Ph. E.) Senna and stoned raisins, of each, 4 oz. ; sugar, 2½ oz. ; corianders, 1 oz. ; jalap, 6 dr. ; caraways and cardamoms, of each, 5 dr. ; proof spirit, 1 quart ; digest, or proceed by percolation.

4. (Ph. D.) Senna, 4 oz. ; caraway and cardamom seeds, of each, bruised, ½ oz. ; proof spirit, 1 quart ; macerate for 14 days.

5. (Wholesale.) From senna, 6 lbs. ; treacle, 2 lbs. ; caraways, ½ lb. ; cardamoms, ½ lb. ; rectified spirit, and water, of each, 4 galls. ; as before. Carminative, stomachic, and purgative.—*Dose*, ½ to 1 fl. oz.

Obs. " If Alexandrian senna be used for this preparation, it must be freed from cynanchum (argol) leaves by picking " (Ph. E.).

Tincture of Ser'pentary. *Syn.* TINCTURE OF VIRGINIAN SNAKE-ROOT ; TINCTURA SERPENTARIÆ (B. P., Ph. L. & E.), L. *Prep.* 1. (B. P.) Serpentary, bruised, 1 part ; proof spirit, 8 parts ; macerate 48 hours with 6 parts of the spirit, agitating occasionally ; pack in a percolator, and let it drain ; pour on the remaining spirit, and when it ceases to drop, press and wash the marc to make up 8 parts.—*Dose*, ½ to 2 dr.

2. (Ph. L.) Serpentary, bruised, 3½ oz. (cochineal, 1 dr.—Ph. E.) ; proof spirit, 1 quart ; macerate for 7 days (or by percolation—Ph. E.), strain, and filter. Stimulant, tonic, and diaphoretic.—*Dose*, 1 to 3 fl. dr.

Tincture of Sesquichloride of Iron. *Syn.* TINCTURE OF MURIATE OF IRON, TINCTURE OF STEEL, STEEL DROPS ; TINCTURA FERRI PERCHLORIDI (B. P.), TINCTURA FERRI SESQUICHLORIDI (Ph. L. & D.), T. FERRI MURIATIS (Ph. E.), FERRI MURIATIS LIQUOR, L. *Prep.* 1. (B. P.) Strong solution of perchloride of iron, 5 parts ; rectified spirit, 5 parts ; water, 10 parts ; mix.—*Dose*, 10 to 30 minims in water.

2. (Ph. L.) Sesquioxide of iron, 6 oz. ; hydrochloric acid, 1 pint ; mix and digest in a sandbath, frequently shaking (with a gentle heat, for a day—Ph. E.), until solution is complete ; then add, when cold, of rectified spirit, 3 pints, and (in a short time) filter. Sp. gr. ·992. " 1 fl. oz. potash being added, deposits nearly 80 gr. of sesquioxide of iron " (Ph. L.).

3. (Ph. D.) Iron wire, 8 oz. ; pure hydrochloric acid, 1 quart ; distilled water, 1 pint ; mix, and dissolve by a gentle heat ; next add, in successive portions, of pure nitric acid, 18 fl. dr. ; evaporate by a gentle heat to a pint, and when cold, mix this in a bottle with rectified spirit, 1½ pints ; lastly, after 12 hours, filter. Sp. gr. 1·237.

Obs. This tincture is an active ferruginous tonic.—*Dose*, 10 to 30 drops, gradually increased, take in water, ale, or wine. In the old Tinctura Martis (Ph. L.) iron filings, and in the T. Ferri Muriatis (Ph. E.1817) black oxide of iron were used instead of the sesquioxide or carbonate. ' Bestuchef's nervine tincture' of the P. Cod. is prepared by dissolving 1 dr. of dry sesquichloride of iron in 7 dr. of spirit of sulphuric ether. See GOLDEN DROPS.

Tincture of Sesquini'trate of Iron‡. *Syn.* TINCTURA FERRI SESQUINITRATIS, L. *Prep.* (*Onion.*) Iron filings, ½ oz.; nitric acid (1·5), 2½ oz.; dissolve; add of hydrochloric acid (1·16), 6 dr.; simmer for 2 or 3 minutes, cool, add of rectified spirit, 8 oz., and filter. Proposed as a substitute for the last preparation, but the name misrepresents its chemical constitution.

Tincture of Soap. *Syn.* TINCTURA SAPONIS (P. Cod.), L. *Prep.* White soap, 3 oz.; carbonate of potash, 1 dr.; proof spirit, 5 oz. (by weight). Dissolve.

Tincture of Soap with Turpentine. *Syn.* TINCTURA SAPONIS TEREBINTHINATA, BAUME DE VIE EXTERNE, L. *Prep.* White soap, 3 oz.; oil of turpentine, 3 oz.; spirit of wild thyme, 2 lbs.; water of ammonia, 2 oz.

Tincture of Squills. *Syn.* TINCTURA BECHICA, TINCTURA SCILLE (B. P., Ph. L., E., and D.), L. *Prep.* 1. (B. P.) Dried squill, bruised, 1 part; proof spirit, 8 parts; macerate for 48 hours with 6 parts of the spirit, agitating occasionally; pack in a percolator, let it drain, and pour on the remaining spirit; when it ceases to drop, press, filter, and make up to 8 parts.—*Dose,* 15 to 30 minims.

2. (Ph. L.) Squills, recently dried and sliced (or in coarse powder), 5 oz.; proof spirit, 1 quart; macerate for 7 days (14 days—Ph. D.; or by percolation—Ph. E.), press, and filter. A stimulating expectorant and diuretic.—*Dose,* 10 to 30 drops; in chronic coughs, and other bronchial affections.

Tincture of Squills (Alkaline). *Syn.* TINCTURA SCILLE ALKALINA (Ph. G.), L. *Prep.* Squill, 8 parts; caustic potash, 1 part; rectified spirit, 50 parts; macerate 8 days.

Tincture of Stavesacre (Concentrated). *Syn.* TINCTURA STAPHISAGRIE CONCENTRATA (*Dr Turnbull*), L. *Prep.* Digest stavesacre seeds in twice their weight of rectified spirit. For external use in neuralgic and rheumatic affections, as a substitute for solutions of delphinia.

Tincture, Stomach'ic. Compound tincture of cardamoms. Compound tincture of gentian is also occasionally so called.

Tincture of Stramo"nium. *Syn.* TINCTURE OF THORN-APPLE; TINCTURA STRAMONII (B. P., Ph. D. and U. S.), L. *Prep.* 1. (B. P.) Stramonium seeds, bruised, 1 part; proof spirit, 8 parts; macerate 48 hours with 6 parts of the spirit, agitating occasionally; pack in a percolator, let it drain, and pour on the remaining spirit. When it ceases to drop, press, filter, and add spirit to make 8 parts.—*Dose,* 10 to 20 minims.

2. (Ph. D.) Stramonium or thorn-apple seeds (bruised), 5 oz.; proof spirit, 1 quart; macerate for 14 days (or by displacement—Ph. U. S.), then press and filter. Anodyne.—*Dose,* 10 to 20 drops; in neuralgia, rheumatism, &c. Said to be superior to laudanum.

Tincture of Strophanthus. *Syn.* TINCTURA STROPHANTHI, L. *Prep.* Strophanthus seeds in No. 80 powder, dried at 110° F., 1 oz.; pure

ether and rectified spirit, of each, a sufficiency. Percolate the seeds with ether until the fluid passes colourless. Remove the seeds, and dry them at 120° F., repack in the percolator, add slowly the rectified spirit until 10 oz. of tincture is obtained. Dilute with more spirit to make 1 pint.—*Dose,* 2 to 10 minims.

Tincture, Styptic. *Syn.* TINCTURA STYPTICA (Ph. L. 1746), L. *Prep.* Calcined sulphate of iron, 1 dr.; French brandy, coloured by the cask, 2 lbs.

Tincture, Sudorific. *Syn.* TINCTURA SUDORIFICA (Ph. E. 1744), L. *Prep.* Serpentary root, 5 dr.; cochineal, 4 dr.; castor oil, 1 dr.; saffron, 2 scruples; opium, 1 scruple; spirit of Mindererus, 16 oz. Digest for 3 days, and strain.

Tincture of Sum'bul. *Syn.* TINCTURA SUMBULI (B. P.), L. *Prep.* 1. (B. P.) Sumbul, bruised fine, 1 part; proof spirit, 8 parts; digest 7 days, and filter.—*Dose,* 15 to 30 minims.

2. From sumbul, bruised, 5 oz.; proof spirit, 1 quart; macerate for a week, and strain with expression. Stimulant and tonic.—*Dose,* 10 to 30 or 40 drops.

Tincture of Tartrated Iron. *Syn.* TINCTURA MARTIS TARTARIZATA, L. *Prep.* Pure iron filings, 100 parts; cream of tartar, 250 parts; rectified spirit, 50 parts (by weight). Put the filings and tartar into an iron kettle with sufficient water to form a paste, leave them for 24 hours, add 3000 parts of soft water and boil for 2 hours, stirring constantly and supplying the waste of water. Decant and filter the liquor, and evaporate it till it has the density of 1·286, and add the spirit.—*Dose,* 3 to 6 drops.

Tincture of Thorn-apple. Tincture of stramonium.

Tincture of Thuja. *Syn.* TINCTURA THUJÆ,. L. *Prep.* Fresh leaves of thuja, 1 part; spirit (90%), 10 parts. Macerate for 10 days, and filter.—*Dose,* 10 drops in water. The leaves of thuja are collected in June and July. Those of the *T. orientalis* and *T. occidentalis* have the reputation in Belgium of curing smallpox.

Tincture of Tobac'co. *Syn.* TINCTURA NICOTIANÆ, TINCTURA TABACI, L. *Prep.* From pure manufactured tobacco, 1½ oz.; proof spirit, 1 pint; macerate for 7 days. Compound spirit of juniper is often used instead of proof spirit. Sedative and narcotic.—*Dose,* 10 to 30 drops. A tincture is also made from the fresh leaves. See VEGETABLE JUICES and TINCTURES (Ethereal).

Tincture of Tolu'. *Syn.* TINCTURA TOLUTANUS (B. P.), TINCTURA TOLUTANA (Ph. L. and D.), T. BALSAMI TOLUTANI, T. TOLUIFERÆ BALSAMI, L. *Prep.* 1. (B. P.) Balsam of Tolu, 1 part; rectified spirit, 8 parts; dissolve, filter, and make up to 8 parts.—*Dose,* 15 to 30 minims, mixed with mucilage or syrup.

2. (Ph. L.) Balsam of Tolu, 2 oz. (3½ oz.— Ph. E.); rectified spirit, 1 quart (1 pint—Ph. D.); dissolve (by the aid of a gentle heat—Ph. E. and D.) and filter. *Obs.* This tincture is reputed pectoral and expectorant; but it is chiefly used as an adjunct in mixtures, on account of its flavour.—*Dose,* 10 to 60 drops.

Tincture, Tooth. *Prep.* 1. (*Greenhough's.*)

From bitter almonds, 2 oz.; Brazil-wood, cinnamon, and orris root, of each, ¼ oz.; alum, cochineal, and salt of sorrel, of each, 1 dr.; spirit of scurvy-grass, 2 fl. oz.; proof spirit, 1½ pints; macerate a week.

2. (*Hudson's*.) From the tinctures of myrrh and cinchona, and cinnamon water, equal parts, with a little arquebusade and gum-arabic.

3. (*Ruspini's*.) From orris root, 8 oz.; cloves, 1 oz.; ambergris, 20 gr.; rectified spirit, 1 quart; digested for 14 days. The above are used as cosmetics for the teeth and gums. The last has long been a popular dentifrice.

Tincture, Toothache. Odontalgic tincture. See also DROPS, ESSENCE, &c.

Tincture of Turmeric. *Syn.* TINCTURA CURCUMÆ (*Dr Wood*), L. *Prep.* Turmeric, 1 oz.; proof spirit, 6 oz.

Tincture of Turpentine. *Syn.* TINCTURA TEREBINTHINÆ (P. Cod.), L. *Prep.* Venice turpentine, 4 oz.; rectified spirit, 1 pint.

Tincture of Tuyaya. *Syn.* TINCTURA TUYAYÆ, L. *Prep.* Tuyaya root, in powder, 12 oz.; proof spirit, 36 oz.; macerate for 14 days. If for internal use it must be diluted with four times its volume of spirit.—*Dose*, 1 to 10 minims. In syphilis.

Tincture of Vale'rian. *Syn.* TINCTURA VALERIANÆ (B. P., Ph. L., E., & D.), L. *Prep.* 1. (B. P.) Valerian, bruised, 1 part; proof spirit, 8 parts; macerate the valerian 48 hours with 6 parts of the spirit, agitating occasionally; pack in a percolator, let it drain, pour on the remainder of the spirit; when it ceases to drop press and filter, washing the marc with spirit to make up 8 parts.—*Dose*, 1 to 2 dr.

2. (Ph. L.) Valerian root, bruised, 5 oz.; proof spirit, 1 quart; macerate 7 days (14 days—Ph. D.; or by percolation—Ph. E.), press, and filter. Tonic and antispasmodic.—*Dose*, 1 to 3 fl. dr.; in hysteria, epilepsy, &c.

Tincture of Valerian (Compound). *Syn.* AMMONIATED TINCTURE OF VALERIAN, VOLATILE T. OF V.; TINCTURA VALERIANÆ COMPOSITA (Ph. L.), T. V. AMMONIATA (B. P., Ph. E.), L. *Prep.* 1. (B. P.) Valerian, bruised, 1 part; aromatic spirit of ammonia, 8 parts; macerate the valerian 7 days, press, filter, and add spirit to make up 8 parts.—*Dose*, ½ to 1 dr.

2. (Ph. L.) Valerian root, bruised, 5 oz.; aromatic spirit of ammonia (simple—Ph. E.), 1 quart; macerate for 7 days (or by percolation—Ph. E.), then press and filter. Stimulant, tonic, and antispasmodic.—*Dose* and *use*, same as those of the simple tincture, than which it is thought to be more powerful. The tincture of the shops is generally made with only 1 lb. of the root to the gallon.

Tincture of Vanilla. *Syn.* TINCTURA VANILLÆ (Ph. G.), L. *Prep.* Vanilla pods, 1 part; rectified spirit, 5 parts; macerate 8 days.

Tincture of Veratria. *Syn.* TINCTURA VERATRIÆ (*Magendie*), L. *Prep.* Veratria, 4 gr.; rectified spirit, 1 dr.

Tincture of Vera'trum. *Syn.* TINCTURE OF WHITE HELLEBORE; TINCTURA VERATRI VIRIDIS, TINCTURA HELLEBORI ALBI, T. VERATRI (Ph. E.), L. *Prep.* 1. (B. P.) Green hellebore root, in coarse powder, 4 parts; rectified spirit, 20 parts;

macerate the powder with 15 parts of the spirit 48 hours, agitating occasionally; pack it in a percolator, let it drain, pour on the remainder of the spirit; when it ceases to drop, press, filter, wash the marc with spirit to make up 20 parts.—*Dose*, 5 to 20 minims.

2. (Ph. E.) White hellebore, 4 oz.; proof spirit, 1 pint; digest or percolate.—*Dose*, 10 drops, 2 or 3 times a day, gradually increased; in gout, rheumatism, &c., as a substitute for colchicum; also externally.

Tincture of Virginian Snake-root. Tincture of serpentary.

Tincture of Vittie-vayr. *Syn.* TINCTURA VETIVERIÆ, L. *Prep.* From vittie-vayr (roots of *Andropogon muricatis*), 2½ oz.; proof spirit, 1 pint; macerate a week. Stimulant, tonic, and sudorific.—*Dose*, 15 to 30 drops.

Tincture, Vulnerary. *Syn.* TINCTURA VULNERARIA (P. Cod.), L. *Prep.* Fresh leaves of wormwood, angelica, basil, calamint, fennel, hyssop, marjoram, balm peppermint, origanum, rosemary, rue, savory, sage, wild thyme, St John's-wort tops, lavender tops, of each, 1 oz. (all cut up), rectified spirit, 20 oz. (by weight); digest 10 days.

Tincture of Walnuts. *Syn.* TINCTURA JUGLANDIS (Ph. Dan.), L. *Prep.* Green shells of walnuts, 6 oz.; proof spirit, 24 oz. Digest for 6 days.

Tincture of Walnut Leaves. *Syn.* TINCTURA JUGLANDIS FOLIORUM (*Mr Ince*), L. *Prep.* Dried walnut leaves, 16 oz.; macerated for 7 days in a gallon of proof spirit.—*Dose*, 1 to 2 teaspoonfuls. To prevent sickness, or to cover the taste of cod-liver oil.

Tincture, Warburg's Fever. *Syn.* TINCTURA FEBRIFUGA WARBURGII, L. *Prep.* 1. The composition was for a long time kept secret; but in 1875 Dr W. published the following formula for it through Prof. Maclean:—Aloes (Soc.), lb. j; Rad. Rhei, Sem. Angelic., Conf. Damocrat., āā ʒiv; Rad. Helenii, Croci Sat., Sem. Fœnic., Cretæ ppt., āā ʒij; Rad. Gent., Rad. Zedoar., Pip. Cubeb., Myrrhæ, Camphoræ, Boleti Laric., āā ʒj. Digest with 500 oz. of proof spirit in a water-bath for 12 hours; express, and add Quinin. Disulph., ʒx. Then replace in water-bath until the quinine is dissolved. Filter when cool.—*Dose*, ½ oz. (undiluted), after an aperient.

2. Tincture of orange, 5 parts; compound tincture of aloes, 20 parts; alcohol, 15 parts; spirit of camphor, 2 parts; sulphate of quinine, 1 part.

Tincture of Wild Cherry Bark. *Syn.* TINCTURA PRUNI VIRGINIANÆ (B. P. C.), L. Wild cherry bark, in No. 20 powder, 4 oz.; distilled water, 7½ fl. oz.; macerate for 24 hours in a closed vessel, and add rectified spirit, 12½ fl. oz.; macerate for 7 days, then press, filter, and add proof spirit sufficient to produce 1 pint.—*Dose*, 20 to 60 minims.

Tincture of Winter Cherry. *Syn.* TINCTURA PHYSALIS ALKAKENGÆ, L. *Prep.* Take of the whole plant, dried, 4 oz.; rectified spirit, 1 pint. Digest for 10 days, strain, and filter. Diuretic and febrifuge.—*Dose*, ½ dr. to 1 dr.

Tincture of Wormwood. *Syn.* TINCTURA ABSINTHII (Ph. G.), L. *Prep.* Dried wormwood, 6 oz.; proof spirit, 30 oz. (by weight); macerate for 8 days, and strain.—*Dose*, 1 dr

Loss of Spirit in making Tinctures by the British Pharmacopœia.[1]—Mr UMNEY gives the following table as embodying the result of his experience:

	Quantity made.	Alcohol ·838 to make up measure.	Alcohol ·920 to make up measure.	Loss per cent. by volume.	Gain per cent. by volume.
	Galls.	Pints.	Pints.		
Tinct. aconit.	4	2·5	...	7·9	
„ arnicæ	10	3·0	...	3·8	
„ aurantii	10	...	5·0	6·3	
„ belladonnæ	2	...	·7	4·1	
„ benz. comp.	5	...	...	...	10·0
„ buchu	2	...	1·0	6·3	
„ calumbæ	5	...	3·0	7·1	
„ camphor. comp.	10	...	·5	·7	
„ cantharid.	4	...	·6	1·9	
„ capsici	5	1·0	...	2·5	
„ cardam. co.	20	...	...	...	2·5
„ cascarillæ	5	...	2·0	5·0	
„ castor	2	·5	...	3·2	
„ catechu	5	...	No loss		
„ chirettæ	2	...	2·0	12·5	
„ cinchonæ	10	...	12·0	15·0	
„ „ co.	10	...	9·0	11·3	
„ cinnam.	5	...	4·0	10·0	
„ cocci	1	...	·25	3·1	
„ colchici sem.	2	...	1·5	9·4	
„ conii	2	...	·75	4·6	
„ croci	1	...	·6	7·5	
„ cubebæ	2	·5	...	3·2	
„ digitalis	5	...	3·0	7·5	
„ ergotæ	2	...	2·0	12·5	
„ ferri acet.	1	1·2	...	15·0	
„ gallæ	1	...	·4	5·0	
„ gent. co.	10	...	7·0	8·7	
„ hyoscyami	10	...	6·0	7·5	
„ jalapæ	5	...	2·0	5·0	
„ kino	1	No loss	...	...	
„ krameriæ	5	...	2·0	5·0	
„ lavand. comp.	20	2·0	...	1·3	
„ limonis	2	...	1·5	9·4	
„ lobeliæ	3	...	2·5	10·5	
„ „ æther.	3	3·5[2]	...	14·5	
„ lupuli	5	...	4·0	10·0	
„ myrrhæ	5	2·5	...	6·2	
„ nuc. vom.	4	·0	...	6·2	
„ opii	10	...	·75	1·0	
„ „ ammon.	2·5	·5	...	2·5	
„ pyrethri	1	1·0	...	12·5	
„ quassiæ	1	...	·7	8·7	
„ rhei	10	...	4·0	5·0	
„ sabinæ	2	...	·8	5·0	
„ scillæ	5	...	3·0	7·5	
„ senegæ	1	...	·75	9·3	
„ sennæ	10	...	2·0	2·5	
„ serpentar.	1	...	·9	10·9	
„ stramon.	1	...	·5	6·3	
„ sumbul.	3	...	1·0	4·2	
„ valer. ammon.	5	3·5[3]	...	4·4	
„ valerian.	5	...	2·3	5·7	
„ veratri virid.	2	2·5	...	15·6	
„ zingiber.	5	2·5	...	6·3	
„ „ fort.	10	3·6	...	37·5	

[1] 'Pharm. Journ.,' 3rd series, i, 321—379.   [2] Sp. æther. sulph.   [3] Sp. ammon. aromat.

Tincture of Wormwood (Compound). *Syn.*
TINCTURA ABSINTHII COMPOSITA (P. Cod.), L.
Prep. Dried wormwood tops, 1 oz.; gentian,
bitter orange peel, of each, 1 oz.; germander,
1 oz.; rhubarb, ¾ oz.; aloes, 2 dr·; cascarilla, 2
dr.; proof spirit, 2 pints.

Tincture of Yellow Jasmine. *Syn.* TINCTURA
·GELSEMII SEMPERVIRENS. *Prep.* Yellow jasmine,
2½ oz.; proof spirit, 1 pint.—*Dose,* 10 to 30 minims.

Tincture of Zedoary. *Syn.* TINCTURA ZEDO-
ARIÆ (Ph. Amst.), L. *Prep.* Zedoary root, 1
part; rectified spirit, 8 parts. Digest and filter.

Tincture of Zedoary (Compound). *Syn.* TINC-
·TURA ZEDOARIÆ COMPOSITA, WEDEL'S ESSENTIA
CARMINATIVA, L. *Prep.* Zedoary, 4 oz.; calamus,
galangal, of each, 2 oz.; chamomile, aniseed, caraway seed, of each, 1 oz.; bay berries and cloves, of
each, 6 dr.; orange peel and mace, of each, 4 dr.;
peppermint water and rectified spirit, of each, 24
dr. In 6 days strain, and add hydrochloric ether,
4 oz.

TINCTURES (Concentrated). *Syn.* TINCTURÆ
CONCENTRATÆ HAENLI, L. *Prep.* (Ph. Baden.)
Digest 8 parts of the vegetable powder in 16 parts
of spirit of the sp. gr. ·857 (45 o. p.) for 4 days
at 72° F., with occasional agitation; then press
and filter; to the marc or residuum add as much
spirit as it has absorbed, and again press and
filter; the weight of the mixed liquors should be
16 parts. In this way are prepared concentrated
tinctures of aconite leaves; arnica and chamomile
flowers; belladonna, digitalis, hemlock, henbane,
peppermint, and savine leaves; ipecacuanha and
valerian roots, &c.

TINCTURES (Cu'linary). See ESSENCES,
SPIRITS, &c.

TINCTURES (Ethe"real). *Syn.* TINCTURÆ
ÆTHEREÆ, L. *Prep.* (P. Cod.) From the vegetable
substance, 1 oz.; sulphuric ether, 4 oz. (or 6 fl.
oz.); by maceration for 4 days in a well-closed
vessel; or preferably by percolation in a cylindrical glass vessel furnished with a stopper, and
terminating at the lower end in a funnel, obstructed
with a little cotton. The powder being introduced
over the cotton, pour on enough ether to moisten it,
put in the stopper, fix the tube into the neck of
a bottle, and leave it for 48 hours; then add,
gradually, the remaining portion of the ether, and,
lastly, enough water to displace the ether absorbed.
In this manner are prepared the ethereal tinctures
of aconite leaves, arnica flowers, belladonna,
hemlock, foxglove, tobacco, pellitory, solanum,
valerian, stramonium, &c., of the Paris Codex.
The ethereal tinctures of amber, ambergris,
assafœtida, cantharides (acetic ether), castor,
musk, tolu, &c., are prepared by maceration only.

TINCTURES (Odorif'erous). These are prepared from odoriferous substances by the usual
processes of digestion or percolation. See SPIRITS.

TINCTURES from Recent Vegetables. See
VEGETABLE JUICES.

TIN'DER (German). See AMADOU.

TINEA GRANELLA, Linn. THE CORN WOLF
MOTH. This pretty little moth belongs to the
family *Tineidæ* of the group *Tineina,* according
to Mr Stainton, the great authority upon this
division of LEPIDOPTERA. It is called the wolf,
and is so called because of its ravages to corn in
granaries and storehouses, and is known in every

part of the world. The manner of the injury
done by the larva of this moth is much the same
as that caused by the *Trogosita mauritanica,*
only that it appears to consume much more of
the grain. Its attack is often mistaken for that
of the Trogosita. This moth belongs to the same
genus as the clothes moth, the carriage moth,
and the fur moth.

Life History. The moth appears first towards
the middle of May, and is seen flying towards
dark in granaries and warehouses, not only of
corn but also of other commodities. Is is about
three sevenths of an inch across the wings, and
its body is less than half an inch in length. It
is of a dull white colour with dark spots on the
whitish wings. The female lays thirty or more
eggs, yellowish, and so small that they cannot be
seen without a glass. She places one or two
upon single grains of corn—wheat, barley, oats,
and rye. In the course of a fortnight tiny caterpillars with dark brown heads come forth and
attack the grain with their stout jaws. They are
of a light buff colour with reddish heads, having
thirteen segments, and are close upon one third of
an inch in length. They fasten several grains
together with a kind of web. Sometimes heaps
of corn that have been undisturbed for some time
are covered with these grey webs, which Curtis
believes are for their protection. In Kirby and
Spence's 'Introduction to Entomology' it is
stated. "On visiting corn granaries at Bristol we
found the barley lying on the floors covered
with a gauze-like tissue formed of the five silken
threads spun by the larvæ in traversing its surface." In due time the larvæ retire to chinks
and holes in rafters, beams, and ceilings, and
make cocoons covered with fine webs, and rest
until the warmth of the spring sun tempts them
forth. The larvæ of this moth are frequently
found in the fissures of the bark of oak trees and
of fruit trees, from whence the perfect insects fly
to the storehouses of grain.

Prevention. To prevent the attacks of this
destructive moth all rooms and buildings used for
storing grain must be kept well and constantly
swept, and the whole places—sides, ceilings, and
floors—cleansed. As it is said that the larvæ
bore into wood to make resting-places for their
transformation, it is important that all wood-
work should be scrubbed hard and well, so as to
let the soap and water into every cranny. No
lumps of dust or grain should be allowed to
remain in corners, or on the ledges or window-
sills. Ceilings should be carefully and frequently
whitewashed. Strong decoctions of quassia may
be mixed advantageously with the soap and water
used for cleansing.

Remedies. When corn in store is found to be
' moth-eaten,' and webs are seen upon the heap, it
must at once be moved, and frequently. If possible it should be run down through the winnowing machine. Should the injury be great and
evidently increasing, kiln drying should be resorted to. Corn in sacks should be frequently
examined, as the moth and larvæ work in sacks
as well as in the heaps (' Reports on Insects Injurious to Crops,' by Chas. Whitehead, Esq.,
F.Z.S.).

TIPULA OLERACEA, Linn. THE CRANE FLY

(DADDY LONG-LEGS). Every one knows this long-legged, awkward fly, called Daddy Long-legs. It is also known as the crane fly, as Curtis says, on account of its beaked head. It is a general destroyer of crops, and an omnivorous feeder upon farm and garden productions, attacking all kinds of corn, grass, turnips, mangels, clover, peas, cabbages, strawberries, and others. There is also another and a smaller species, known as *Tipula maculosa*, or the spotted crane fly, having spots on its body, which is injurious to various crops. Its habits and history are similar to those of the *Tipula oleracea*, and the methods of prevention and remedies against it are the same. There has been a large increase of these insects during the past few years, and the injury caused to corn crops in England and Scotland has been very great. The wet summer season previous to 1880 favoured their propagation, as they delight in moisture and revel in damp, marshy, boggy places, in which they prefer to deposit their eggs.

It is the larvae or grubs that injure plants of corn and grass by attacking them with their strong jaws, and eating into them just beneath the surface of the ground, so as either to kill them or to make them weak and sickly. In the early spring, if wheat plants which show signs of failing are examined, large ash-grey grubs, or maggots, will often be found close to the affected plants. Oats and barley are equally liable to harm from these grubs—not, perhaps, quite to such an extent as autumn-sown wheat, and especially wheat sown after clover leys. One instance may be given here of serious loss in a large wheat-field in Kent after clover ley, well ploughed and duly pressed, with a deal of sward turned in. The plant, which was forward and very vigorous, looked like yielding five quarters per acre. In February it began to fail, but the actual cause was not ascertained until the middle of March— too late for any effectual remedies. Only about four sacks per acre were obtained from this field. A field of oats sown on the 1st of March, after clover, was attacked by these grubs. Although it was an even strong plant it was soon nearly half devoured, and, instead of nine quarters per acre being obtained, as might have been expected from the state of the land and the circumstances of its cultivation, and the produce of other land hard by, only about four quarters per acre were grown. It is computed that the loss in this case amounted to £80. Upon a farm in Essex, in 1882, the bean crop was materially reduced by an attack of these crane fly grubs; and on a market-garden farm near Rainham, in Essex, early peas were almost entirely ruined by them. Grave complaints of great injury and of heavy losses to corn have been rife from many parts of England, Scotland, and Ireland during the last six years.

Pastures have also suffered alarmingly in some places. The grubs seem to select the best and most succulent grasses, and those of upright growth, such as cock's-foot (*Dactylis glomerata*). In pastures and meadows the amount of damage done by these insects cannot be estimated, as so much of it is unseen and unknown. It is stated that in pasture land known to be attacked by them as many as 200 grubs have been taken from a square

foot of turf. In 1884 Lord's Cricket Ground was seriously injured by the grubs of the Daddy Long-legs.

Life History. The life history of the crane flies, both of the *Tipula oleracea* and its close congener, *Tipula maculosa*, is simple. The eggs are small, oval, conical grains, shining and black as Curtis writes, forming a mass which occupies nearly the whole of the abdomen. As many as 300 have been found in one female. These are deposited in the autumn upon grass and herbage, and more frequently in the ground. Wet undrained meadows, and marshy and damp places are preferred by these insects, and the conditions of such spots are probably favourable to the preservation and the ultimate hatching of the eggs. This hatching takes place in the early spring, directly the weather becomes mild. Taschenberg reports that they have been found as early as January and February in mild winters.

Miss Ormerod relates that they may be found as early as February, for in 1880 they were destroying hundreds of acres of autumn-sown wheat on heavy land after clover at that date. After hatching, the maggots or larvae grow fast until they become an inch in length. Labourers call them 'leather jackets,' because of their tough skins. Their colour is of the earth, with a slight dash of grey or ash colour in it. Although they have no legs they are able to move rapidly from place to place, and burrow in the ground. It is in this grub form that they do mischief to crops, and they remain in this stage of their existence until the beginning of July, at which period they change into pupae under the surface of the soil. After a while the pupae work their way up to the light by means of the hooks or recurved spines, and in a short time the crane flies appear, and soon unfold their long wings, and fly away to commence a new series. Most persons are acquainted with these insects in their perfect forms, but it may be stated that the females are about an inch long, with wings two inches across. Their colour is light brown; they have six legs, and a long tapering body with nine divisions or segments. The male insects are not so large as the females. Pairing takes place in the beginning of August. In fields and meadows infested with crane flies thousands of empty pupa cases can be seen sticking half out of the ground. It should be mentioned that in some seasons the flies may be seen as late as October.

Prevention. Spots where there is long intertwisted herbage and weed-growth, wet ditches, the wet sides of hedgerows, damp headlands, undrained meadows, and marshes, are congenial habitations of the crane flies. These being their head-quarters and chief breeding-places, an obvious means of prevention is to keep ditches well brushed and cleaned out, to abolish hedgerows where possible, or to keep them well and closely trimmed up, also to drain wet land. Wetness and decay, as is affirmed by Curtis, Taschenberg, and Kaltenbach ('Die Pflanzenfeinde aus der Klasse der Insekten,' von J. H. Kaltenbach), are natural to them, and long immersion in water does not destroy their eggs.

In very many instances the attack of crane flies upon field crops is where these follow clover or artificial grasses, whose herbage has served as shelter for the eggs. The necessity of keeping clover and other leys down close before they are ploughed cannot be too strongly urged, not only as a means of prevention against crane flies, but also against many other injurious insects. However well ploughed, however well pressed down the land may be, some eggs will be left in circumstances in which they can be hatched out if there are long stalks and much herbage. It is also important to plough leys early, in order that the eggs may be buried deeply, so that they may be prevented from changing into larvæ. It goes almost without saying that weed-growth in fields serves equally as a harbour for the eggs. The clean and careful farmer, as a rule, is not so liable to attacks from insects as he who is slovenly. As the eggs and the grubs of these flies are without doubt carried out on to the land in farmyard manure, it is very desirable to keep old mixens clear from weeds, and to turn mixens that have lain some time three weeks or so before they are carted out, that the eggs and grubs may be destroyed by the renewed heating. Old mixens are a very fertile source of insect attacks of many other kinds. Weeds should not be allowed upon them. They should never be carried out when their heat has been long exhausted; but old mixens are altogether a mistake from all points of view.

Remedies. In the case of wheat plants suffering from a bad attack of crane-fly grubs the following treatment was adopted with much success in 1883 :— Early in February the grubs were found at work in numbers in a strong piece of wheat after clover. It was horse-hoed well and side-hoed as soon as the land was dry enough for these operations. A few days afterwards 2 cwt. of nitrate of soda was put on, and a heavy roller was applied. This checked the grubs, and gave time to the plants to grow away and produce a fair crop, though the grubs were very numerous as seen at first.

In 1882 a field of wheat was losing plant fast at the beginning of March; 1½ cwt. of nitrate of soda, mixed with 4 cwt. of soot, was broadcasted over the plants, and, as the soil was a little unkindly, it was ring-rolled both ways. Growth was stimulated; the soil was pressed well round the plants. Eventually a crop of 4½ quarters was grown. Oats attacked badly have been much helped by hand-hoeings and horse-hoeings, and dressings of soot, salt, nitrate of soda, and guano. Frequent horse-hoeings at the right time have materially benefited peas and beans which appeared to be giving way fast. Hand-picking by women and boys following hoers where labour is plentiful has been of great service. In market-garden farms and market gardens crops have been saved by careful hand-picking, also by means of stimulating manures hoed in. Chemical manures act as deterrents, but obviously it is difficult in large fields and in large cultivation to get them directly in the paths of the grubs. Applications should, if it is practicable, be sprinkled up the drills or rows of plants instead of being broadcasted in the usual way, in order that the application may be close to the grubs; not that these act directly upon their leathern jackets, but they are offensive to them.

Moles, rooks, starlings, peewits, plovers, and gulls are natural remedial agents, devouring these grubs wholesale, and should be encouraged and protected (' Reports on Insects Injurious to Crops,' by Chas. Whitehead, Esq., F.Z.S.).

TISANE. [Fr.] *Syn.* PTISAN; PTISANA, L. This form of medicine is much used in France. Tisanes may be readily prepared by slightly medicating barley, rice, or tamarind water, lemonade, &c. See DECOCTION, INFUSION, JULEP, PTISAN, &c., and *below.*

Tisane, Antimonial. (*Brera.*) Lemonade, 2 pints ; tartar emetic, 2 gr.; sugar, q. s.

Tisane, Antiscorbutic. An infusion of buck-bean and the fresh roots of horseradish.

Tisane, Antivenereal. Various compound decoctions of sarsaparilla are known by this name.

Tisane of Arnica. (P. Cod.) As elder-flower tisane.

Tisane of Arnica Flowers. (P. Cod.) *Prep.* Arnica flowers, ⅓ oz.; boiling water, 6 pints. Infuse half to an hour and filter through paper. Prepare in the same manner saffron tisane.

Tisane of Asparagus. (P. Cod.) Asparagus root, ¾ oz.; boiling water, 2 pints. Infuse 2 hours and strain. Tisanes are prepared in the same way from the roots of elecampane, comfrey, strawberries, rhatany, soapwort, spruce fir buds, Peruvian bark, dulcamara, and burdock root.

Tisane of Bread. *Syn.* DECOCTUM ALBUM (P. Cod.), L. *Prep.* Prepared hartshorn, 1 oz.; bread crumb, 2 oz.; gum-arabic, 1 oz.; water, sufficient quantity to yield 5 pints ; boil for half an hour, strain through a coarse sieve, and add white sugar, 6 oz.; orange-flower water, 1 oz.

Tisane of Cassia. (P. Cod.) *Prep.* Extract of cassia, 1 oz.; dissolve in 5 pints of water at 140° F.

Tisane, Common. A decoction of pearl barley and couch-grass, flavoured with liquorice root.

Tisane of Couch-grass. (P. Cod.) *Prep.* Root of couch-grass, sliced, 6 dr. Boil for half an hour with water sufficient to yield 2 pints.

Tisane of Elder Flowers. (P. Cod.) *Prep.* Elder flowers, 1 dr.; boiling water, 1½ pints. Macerate for half an hour, and strain.

Tisane of Gentian. (P. Cod.) *Prep.* Gentian, sliced, ⅓ oz.; cold water, 5 pints. Infuse 4 hours and strain. In the same manner prepare tisanes of quassia, simaruba, and rhubarb.

Tisane of Groats. From groats, as tisane of pearl barley.

Tisane of Guaiacum Wood. (P. Cod.) *Prep.* Guaiacum shavings, 1½ oz. Boil for 1 hour with sufficient water to yield 2 pints, and strain.

Tisane of Gum. *Syn.* EAU DE GOMME, Fr. *Prep.* Bruised gum-arabic, 1 oz.; water, 2½ pints. Dissolve without heat, and strain.

Tisane of Iceland Moss. (P. Cod.) *Prep.* Just boil 2½ dr. of Iceland moss in a little water and throw away the first decoction, then wash the remaining moss with cold water ; then add a fresh quantity of water and boil for half an hour, so as to obtain 2 pints.

Tisane of Irish Moss. (P. Cod.) *Prep.* Wash 90 gr. of carrageen in cold water ; then, after boiling 10 minutes, add water in sufficient quantity so as to yield 2 pints.

Tisane of Liquorice. (P. Cod.) *Prep.* Sliced

liquorice, 1 oz.; boiling water, 5 pints. Infuse 2 hours, and strain. Prepare in the same manner (but infusing for ½ hour) tisanes from the dried leaves of borage, wormwood, holy thistle, chicory, fumitory, ground-ivy, pellitory, wild pansy, soapwort, scabious, from the cones of the hop, aniseed, red rose leaves, lesser centaury tops, and linseed.

Tisane of Mezereon. (P. Cod.) *Prep.* Mezereon bark, 2 dr.; water, 2½ pints; boil to 1¾ pints, and strain.

Tisane of Orange Leaves. (P. Cod.) *Prep.* Leaves of the orange tree, ½ oz.; boiling water, 5 pints; infuse ½ hour, and strain. Prepare in the same manner tisanes from the leaves of wormwood, maidenhair, hyssop, balm, mint, and sage; and from the flowers of white mullein, chamomile, red poppy, mallow, marsh-mallow, lime, coltsfoot, and violet.

Tisane of Pearl Barley. (P. Cod.) *Prep.* Wash ¾ oz. of pearl barley in cold water; strain off water, and boil in a sufficient quantity of water, so as to yield 2 pints. Tisanes of groats and rice are made in the same manner.

Tisane, Pectoral. An infusion of the roots of liquorice and marsh-mallow, Canadian maidenhair, and the flowers of the red poppy and coltsfoot, in a decoction of rice.

Tisane of Rice. (P. Cod.) Prepared in the same manner as tisane of pearl barley.

Tisane of Rice with Lemon. (*Augustin.*) *Prep.* Washed rice, 1 oz.; water, 4 lbs.; boil, strain, add barley-sugar, ½ dr.; lemon juice, 1 oz.

Tisane of Roses with Milk. (P. Cod.) *Prep.* Conserve of roses, 1 oz.; new milk, 1 pint. Rub together and strain.

Tisane, Royal. *Prep.* From senna, fresh chervil, and sulphate of soda, of each, 4 dr.; aniseed and cinnamon, of each, 1 dr.; 1 lemon, sliced; cold water, 1¼ pints; macerate for 24 hours, stirring occasionally, then press and filter. Aperient.—*Dose.* A wineglassful or more, repeated every half-hour until it operates.

Tisane of Salep. (P. Cod.) *Prep.* Boil 1 dr. of salep in 16 oz. of water, and strain.

Tisane of Senega. (P. Cod.) *Prep.* Senega, 1 oz.; boiling water, 5 pints. Infuse for 2 hours, and strain. Prepare in the same manner tisanes of the roots of marsh-mallow and valerian.

Tisane of Sulphuric Acid. (P. Cod.) *Syn.* LIMONADE SULFURIQUE, Fr. *Prep.* Sulphuric acid (1·84), 72 minims; water, 4½ pints; syrup, 10 oz. (by weight); mix *s. a.* Prepare in the same manner nitric and phosphoric acid lemonade; the first with acid of sp. gr. 1·42; the second with acid of sp. gr. 1·45.

Tisane of Tamarinds. (P. Cod.) *Prep.* Pulp of tamarind, 1 oz.; boiling water, 2 pints. Infuse ½ hour.

Tisane, Tartaric. (P. Cod.) *Prep.* Syrup of tartaric acid, 2 oz.; water, 18 oz. Prepare in the same manner, with their respective syrups, lemonades of citric acid, gooseberries, cherries, and raspberries.

TIS'SUE (Blis'tering). See VESICANTS.

TITA'NIUM. A rare metal, discovered by Klaproth in 1794, and examined by Wollaston in 1822. It is occasionally found at the bottom of

the smelting-furnaces of iron-works, in combination with nitrogen and cyanogen, under the form of minute crystals, having a coppery lustre.

TOAD-IN-THE-HOLE. One to six ounces of flour, break the contents of one egg, and stir in by degrees one pint of milk, taking care to keep the mixture free from lumpiness. Place meat or ox kidney cut in slices in a greased pie dish or tin; then pour the batter over the meat after adding a pinch of salt, and let it bake for an hour to an hour and a quarter. The batter should be allowed to stand before being cooked.

TOAST (Essence of). This is liquid burnt sugar or spirit colouring. Used to make extemporaneous toast-and-water (3 or 4 drops to the glass), and to flavour soups, gravies, &c.

TOAST AND WATER. Toast a crust of bread, taking care not to char it, and put it into a pint of cold water in a covered vessel. After standing half an hour it will be ready for use.

TOBAC'CO. *Syn.* TABACUM (Ph. L., E., and D.), L.; TABAC, Fr. The prepared leaf of *Nicotiana tabacum*, Linn., or other species of the same genus. The name was given to this herb by the Spaniards, because it was first seen by them at Tabasco, or Tabaco, a province of Yucatan, in Mexico.

The tobacco of commerce is chiefly obtained from Virginia and other parts of the United States, and recently from Japan and California, but the finest varieties are imported from Havannah and from the East. The plants are gathered when mature, during hot dry weather, and are hung up in pairs, in sheds, to dry. When sufficiently dry, the leaves are separated from the stems, bound up in bundles, and these are formed into bales, or packed in hogsheads, for exportation.

Prep. To impart to the dried leaves the characteristic odour and flavour of tobacco, and to render them agreeable to smokers and snuffers, it is necessary that they should undergo a certain preparation, or kind of fermentation. If a fresh green leaf of tobacco be crushed between the fingers, it emits merely the herbaceous smell common to most plants; but if it be triturated in a mortar along with a very small quantity of quicklime or caustic alkali, it will immediately exhale the peculiar odour of manufactured tobacco. This arises from the active and volatile ingredients being liberated from their previous combination by the ammonia developed by fermentation, or the action of a stronger base. Tobacco contains a considerable quantity of chloride of ammonium, and this substance, as is well known, when placed in contact with lime or potassa, immediately evolves free ammonia. If we reverse the case, and saturate the excess of alkali in prepared tobacco by the addition of any mild acid, its characteristic odour entirely disappears. In the preparation of tobacco previously to its manufacture for sale, these changes are effected by a species of fermentation. Pure water, without any addition, is quite sufficient to promote and maintain the perfect fermentation of tobacco. The leaves soon become hot and evolve ammonia; during this time the heaps require to be occasionally opened up and turned over, lest they become too hot, take fire, or run into the putrefactive fermentation. The extent to which the process is allowed to proceed varies, for different kinds of snuff or tobacco, from one to three months.

Qual., &c. Tobacco is a powerful narcotic, sedative, and emetic; and is also cathartic and diuretic; but the last in a weaker degree than either squills or foxglove. Its action is violently depressing and relaxing, producing fainting, and even death, in comparatively small quantities. Toxicologists rank it among the more active narcotico-acrid poisons; and physicians, when they wish to produce sudden physical prostration, in accidents or spasmodic diseases, order an enema of the infusion or smoke of tobacco. Its deleterious properties depend on the presence of narcotine, one of the most frightful vegetable poisons known, of which ordinary Virginia tobacco contains from 6% to 7%.

"The chief sources of tobacco in Europe are Germany, Holland, Salonica, Hungary, and Russia; in Asia, the principal are China, Japan, the East Indies, Latakia, and other parts of Asiatic Turkey, Persia, Java, Syria, and Manilla; in Africa, Algiers; in South America, Varinas, Brazil, Uruguay, New Grenada, Paraguay, Cumana, and other fields are most productive; while the great tobacco districts of North America are the United States, Mexico, Cuba, Hayti, and Porto Rico. The extent to which these and others are severally laid under contribution by the manufacturers of this country is shown by the following partial analysis of the imports of 1873 and 1874" ('British Manufacturing Industries,' Sandford):

From	1873.	1874.
	lbs.	lbs.
Germany . . .	687,720	856,646
Holland	5,429,511	7,856,798
France	1,436,985	1,712,839
Greece	330,712	84,161
Turkey	1,430,572	696,132
British India . .	3,068,109	2,359,987
Philippine Islands .	171,803	780,098
China	2,136,637	1,398,467
Japan	4,846,892	2,948,086
Spanish West India Islands . . .	295,654	242,304
New Grenada (United States of Columbia) .	2,199,885	1,617,573
Argentine Republic .	340,787	663,940
United States of America	57,593,826	53,567,555
Other countries . .	1,404,640	1,890,679
	81,382,733	76,175,215

Most of the so-called Havannah cigars which arrive in England are shipped from German ports. It appears that a higher price is obtainable for dark than for light-coloured cigars, the demand for the former being about three times as large as for the latter. Unfortunately, however, owing in a great measure to the partial failure of the tobacco crops of late years, light-coloured tobacco is much more common than dark. In order,

therefore, to render the cigars made of light-coloured tobacco saleable at a higher price, and also to improve the appearance of old and faded cigars, if we are to believe a pamphlet recently published at Bremen, where there are several of these manufactories, various infusions have of late been prepared and largely sold, under the name of 'Havanna brown,' 'sap brown,' and 'condensed sauce.' All these preparations are now openly advertised, and directions given for using them. None of these infusions contain anything particularly injurious, most of them consisting of brown vegetable dyes; nevertheless they enable the manufacturer to give to cigars made of old and faded leaves the appearance of good Havannah cigars. A German paper states that if a piece of white blotting-paper, saturated with diluted sal-ammoniac, is passed a few times lightly over the cigar, the colouring matter, if any has been used, will come off on it, whereas the natural brown of the tobacco leaf will remain.

Tobacco Adulteration. The popular belief that bad cigars are made of cabbage leaves is not justified by the last official report on tobacco adulteration. This document contains a tabulated account of the seizures of spurious tobacco made in the United Kingdom since 1864; and in the whole paper there is no mention whatever of the much-suspected vegetable. Its place in the black list is supplied by a variety of ingredients large enough to rejoice the heart of any member of the Anti-Tobacco League. The dishonest dealer in things smokable is shown by the report to make use of three different sorts of materials besides that which he professes to employ. The first sort is required for the actual substance of the cigar; the second for improving its outward appearance; and the third for imparting to it what is supposed to be a better taste. In the former category the favourite substances seem to be the leaves of the lime tree, the husks of wheat and oats, cotton yarn, and Tonquin bean. But there are numerous cases where the ingredients have been much more curiously selected, and have included cocoa-nut fibre, small seeds, cotton, wood, and bread. At one establishment 50 lbs. of 'tobacco dust' were found and analysed, when it was shown to contain string, wood, nails, grindings of tobacco-pipe, dirt, and all sorts of refuse. Another large class of materials is apparently used for securing the adhesion and consistency of the cigar when made. Amongst these starch is the most prominent; but it includes gum and amidine blue, gum-arabic, glue, glycerin, and essential oils. The colour of the fabrication is the next thing to be attended to, and for this purpose resort is had to yellow ochre, lead sandal-wood, logwood, lampblack, and Venetian red. As for the flavour of the cigar, it is varied to suit the most diverse tastes; but the usual object seems to be to impart to it a pleasing sweetness of tone. Accordingly saccharine matter, and especially treacle, is very largely pressed into the service. For those who like a rather more decided taste, liquorice, salt, logwood, glycerin, and aniseed are used. It is in Dublin where the last ingredient is most fashionable, while Edinburgh is fondest of treacle and sugar, and East London is addicted to liquorice ('Pall Mall Gazette').

Tobacco, British. *Syn.* HERB TOBACCO; TABACUM ANGLICUM, Species STERNUTORIÆ, L. *Prep.* Take of thyme, marjoram, and hyssop, of each, 2 oz.; betony and eyebright, 3 oz.; rosemary and lavender, of each, 4 oz.; coltsfoot, 1 lb.; mix, press them together, and cut the mass in imitation of manufactured foreign tobacco. Some asthmatic subjects add 5 or 6 oz. of stramonium or thorn-apple leaves; and others add ½ lb. of genuine tobacco.

Tobacco, Indian. See LOBELIA.

TOD'DY. Obtained from various species of palms, by cutting off the end of the flowering bud, and collecting the sap. Used fresh as a cooling beverage; and, after fermentation, as an intoxicating one. Sweetened grog is so called in Cornwall, and in some other parts of England.

TOFFY. *Syn.* EVERTON TOFFY. A sweetmeat prepared by heating brown sugar in a saucepan or skillet, with about one half its weight of fresh butter, for 15 to 20 minutes, or until a 'little' of it dropped into cold water forms a lump that breaks crisply;' it is then poured into a little buttered tin mould.

TOLUOLE. *Syn.* TOLUOL; BENZOENE. C₇H. One of the hydrocarbons homologous with benzol, with which it occurs, associated with xylol and isocumole, in the light oil obtained from the distillation of coal tar. It is also one of the products of distillation of balsam of Tolu, and would seem to be identical with the *retinaphtha* obtained by Pelletier and Walter from the distillation of rosin. If oxidised by means of chromic acid it yields benzoic acid. Its boiling-point is 230° F., and its sp. gr. 0·87.

TOM'BAC. An alloy consisting of copper, 16 lbs.; tin, 1 lb.; zinc, 1 lb. Red tombac is composed of copper, 10 lbs.; zinc, 1 lb.

TONGA. A remedy used by the natives of the Tonga Islands, a group of the Fijis. It is a mixture of parts of three plants. Mr E. M. Holmes has shown two of them to be the scraped stem of *Raphidophora vitensis*, and bark of *Premna Taitensis*; the third substance is some unknown leaf. These are enclosed in a wrapper composed of the inner bark of the cocoa-nut tree. The drug is used for neuralgia, and seems harmless. A liquid extract is made, and sold as a proprietary article.—*Dose*, 1 dr.

TON'ICS. Medicines that increase the tone of the muscular fibre, and impart vigour to the system.

TON'QUIN REMEDY. *Syn.* PULVIS TRUNCHINENSIS, P. ALEXIPHARMICUS SINENSIS, L. *Prep.* From valerian, 20 gr.; musk, 6 gr.; camphor, 6 gr.; mix. Antispasmodic and alexiterian; in doses of 6 to 10 gr., in hoopingcough, &c.; 1 dr. in hydrophobia, exanthemata, and mania.

TOOTH'ACHE. *Syn.* ODONTALGIA, L. This annoying affection frequently arises from sympathy with a disordered stomach. In such cases a saline purgative should be administered, and an emetic if required. When cold is the cause, an excellent remedy is a hot embrocation of poppy heads, followed by the use of flannel and diaphoretics. When it arises from a hollow or decayed tooth, the best application is a piece of lint moistened with creasote, or a strong spirituous

solution of creasote, and closely rammed into the cavity of the tooth. Laudanum, the essential oils of cloves, caraway, and cajeput, and essence or tincture of pellitory of Spain, are also used in the same way. To prevent the recurrence of the latter kind of toothache, the cavity should be filled with an amalgam of gold, or with mineral marmoratum, or some other good cement. In many cases, chewing a piece of good ginger, or, still better, a small piece of pellitory, will afford relief in a few minutes. The celebrated John Wesley recommended a ' few whiffs' at a pipe containing a little caraway seed mixed with the tobacco. A slight 'shock' from a voltaic battery will often instantly remove the toothache after all other means have failed. See DROPS, ESSENCE, TINCTURE, &c.

TOOTH CEMENTS. See DENTISTRY. *Obs.* It is absolutely necessary for success that the teeth be well cleaned out, and wiped dry, before applying any of the above stoppings or cements.

TO'PAZ. See GEMS.

TOR'MENTIL. *Syn.* TORMENTILLÆ RADIX; TORMENTILLA (Ph. L. & E.), L. The root or rhizome of *Potentilla tormentilla*. It is astringent and febrifuge, without being stimulant.—*Dose*, 20 to 60 gr.; in agues, diarrhœa, &c.; also formerly in syphilis.

TORTOISESHELL, to Polish. Dip a soft linen rag into rouge powder, and rub the tortoiseshell with it, and finish off with the hand. Tortoiseshell combs will not lose their polish if they are rubbed with the hand after removal from the hair.

TOUCH-NEEDLES. See ASSAYING.

TOUCHWOOD. See AMADOU.

TOUS-LES-MOIS. The fecula of the roots of *Canna edulis* (Ph. D.); intended as a substitute for arrowroot. To the naked eye it closely resembles the finest quality of potato starch, but under the microscope its granules are found to be oblong, oval, with a concentric structure, and larger than those of the potato tuber.

Microscopic appearance of tous-les-mois.

TOXICOL'OGY. See POISON.

TRAG'ACANTH. See GUM.

TRANSPA"RENCIES. Water-colour pictures

on paper, linen, or calico, if executed in non-opaque or glazing colours, may be converted into transparencies by simply brushing over their backs with Canada balsam, thinned down, when necessary, with a little oil of turpentine. For coarse work, boiled oil may be employed.

TRAPS, HOUSE. With few exceptions, the endless varieties of traps advertised for house drains are all modifications of the older forms of the syphon, the midfeather, and the ball trap. The syphon trap consists of a bent tube with a deep curve, in which the water lies and acts as an hydraulic valve.

The following conditions are essential for its proper action:—The curve must be of such a depth as to ensure a height of not less than ¾ inch of water always standing above the highest level of the water in the curve. This water attached to the trap should not be too small nor have too sudden a fall as it leaves the trap, otherwise when 'running full' of water, all the water will be sucked out of it by the pipe beyond, owing to the too narrow bore and too perpendicular inclination of this latter.

The midfeather trap consists of a round or square box or receptacle, into the upper part of which, on one side, an inlet pipe discharges, whilst at a corresponding height on the opposite side there is an outlet pipe. The upper part of the box is divided by a partition, which dips at least ¾ inch below the surface of the water, always standing in the receptacle, at the level of the outlet pipe. The principle, therefore, of the midfeather is similar to the syphon trap. The receptacle is so arranged that any heavy substances collected at the bottom can from time to time be removed. A useful variety of the midfeather is 'Dean's patent drain-trap,' manufactured by Edwards, of Ruabon.

The ball trap is not in very general use. By this arrangement the drain is trapped by means of a hollow ball, which rises with the water in the drain until it is carried against and closes an orifice.

The common ball trap is inefficient and unsatisfactory. The facility with which it can be removed or placed out of gear often leads, in the hands of careless servants, to the untrapping of the drain altogether. A good description of common sink trap is Antel and Lock's, shown in the accompanying *engr.*, which explains itself.

Amongst the circumstances that impair the efficiency of house traps may be included the neglect to allow the passage of water through them sufficiently often, and with force enough to flush and cleanse the trap, and renew the water in it. The results are that the water becomes saturated with sewage exhalations, which escape into and contaminate the air in the house; and that the trap becomes either dry or choked up.

Another contingency to which house traps are exposed when the drains are made to form a continuous and disconnected system with the sewers is that of the water being sucked out of the trap, owing to the combined effect of the pressure of sewer air and the aspirating power of the house, into which the sewer gas would then pass unchecked.

In our articles DRAINAGE and SINKS we have pointed out the peril attaching to the intimate connection between the house drains and sewers, and given in the former practical directions for its avoidance.

To rest in the belief that the danger can be removed (although the risk may be slightly diminished) by the use of traps alone, is to entertain a very false sense of security.

TRAUMATIC BALSAM. Compound tincture of benzoin is known by this name. See TINCTURE.

TRAUMATICINE. This article, as manufactured by the Gutta Percha Company, is simply a solution of white and dry unmanufactured gutta percha in bisulphuret of carbon. A small portion dropped on a wound, or raw surface, almost instantly forms a pliable, waterproof, and air-tight defensive covering to the part. The only objection to the preparation is the fetid odour of the menstruum, which, however, is lost in a few seconds, or may be obviated by employing chloroform as the solvent.

When fresh it is a useful external application, and being painted over a given surface with a camel's-hair pencil, a thin covering of a brownish colour is deposited. It has been thought possible to increase the value of the preparation by dissolving in the chloroform solution a certain quantity of chrysophanic acid, in order to make the pellicle more adherent. So medicated, it has been employed with success in the treatment of psoriasis; but traumaticine can be used as a vehicle for a great number of remedies soluble in chloroform, or susceptible of being held in suspension by its means.

M. Auspitz introduces, in 8 parts of chloroform, 1 part of chrysophanic acid and 1 part of gutta percha, and this formula is stated to have given favourable results. M. Besnier prefers applying first chrysophanic acid in the chloroform solution, and covering the slight deposit formed with a varnish of gutta percha. He gives two formulæ:

1. Chrysophanic acid, 10 to 15 grms.; chloroform, 85 to 90 grms.

2. Gutta percha, purified, 10 grms.; chloroform, 90 grms.

TRAVELLER'S JOY. *Syn.* CLEMATIS VITALBA, L. The inner bark is used in Switzerland for straining milk and for other domestic purposes. The slender shoots are used in France to bind faggots, and their tips are sometimes pickled.

TREA'CLE. *Syn.* MOLASSES; THERIACA (B. P.), SACCHARI FÆX (Ph. L. & E.), L. The viscid, brown, uncrystallisable syrup which drains from moist sugar during its formation (molasses), and from the sugar-refining moulds (sugar-house molasses). The last, according to Dr Ure, has generally the sp. gr. 1·4, and contains about 75% of solid matter.

Treacle is more laxative than sugar, and always

contains more or less free acid. It is used as the vehiculum in some of the pill-masses of the Ph. L. See SUGAR.

Treacle, German. *Syn.* THERIACA GERMANICE, L. An evaporated infusion or decoction of juniper berries. It is sweet-tasted, aromatic, and diuretic.

Treacle, Venice. *Syn.* LONDON TREACLE; THERIACA, T. ANDROCHI, L. The theriaca of the Ph. L. 1746 consists of 61 ingredients, and contains 1 gr. of opium in 75 gr.; that of the Paris Codex consists of 72 ingredients, and contains 1 gr. of opium in 72 gr.; that of the Ph. E. 1744 consists of 10 ingredients, and contains 1 gr. of opium in every 100 gr. It is prepared as follows: —Take of serpentary root, 6 oz.; valerian and contrayerva roots, of each, 4 oz.; aromatic powder, 3 oz.; guaiacum resin, castor, and nutmeg, of each, 2 oz.; saffron and opium (dissolved in a little wine), of each, 1 oz.; clarified honey, 75 oz.; reduce all the dry ingredients to fine powder, then mix them. The confections or electuaries of catechu and opium are the representatives of the above polypharmic compounds in the modern British Pharmacopœias.

TRI-, TRIS-. See NOMENCLATURE.

TRIBASIC·PHOSPHATE OF LIME. *Syn.* TRI-CALCIC PHOSPHATE. $Ca_3(PO_4)_2$. Tricalcic phosphate occurs nearly pure in the mineral known as OSTEOLITE. See CALCIUM PHOSPHATE for its artificial preparation.

TRIMETHYLAMINE. *Syn.* TRIMETHYLIA.

C_3H_9N, or $\left\{\begin{array}{c}CH_3\\CH_3\\CH_3\end{array}\right\}$ N. An ammonia found in

large quantities in the roe of the herring. It also occurs in putrefying flour and urine, and is the ingredient which gives to the *Chenopodium vulvaria* its peculiar and disagreeable odour. It may also be obtained by distilling ergot of rye with caustic potash. Trimethylia is a volatile fluid, with a very pungent and ammoniacal fishy smell. It boils at about 41° F. It is metameric with propylamine.

TRINITRIN. *Syn.* GLONOINE, NITRO-GLY-CERIN (B. P.). A solution is official containing 1 part of trinitrin by weight, and rectified spirit to produce 100 fl. parts.—*Dose*, ½ to 2 minims. For properties see NITRO-GLYCERIN.

TRIPE. This is the paunch, or first portion of the ruminant stomach of the ox. It is nutritious and easy of digestion, except when very fat. Letheby gives the following as its composition:

Nitrogenous matter	.	13·2
Fat	. . .	16·4
Saline matter	. .	2·4
Water .	. . .	68·0
		100·0

Tripe fried in Batter. "Tripe is cut into pieces about 3 inches square, and dipped into a batter made of 6 oz. of flour, 1 table-spoonful of oil, or 1 oz. of butter, and ½ pint of tepid water. Mix the oil with the flour, add the water by degrees, whip the whites of 2 eggs to a stiff froth, stir into the batter, dip the tripe in, throw it into a saucepan of boiling fat, let it fry 3 or 4 minutes, take it out, and drain" (Tegetmeier's 'Scholar's Handbook of Cookery, &c.,' Macmillan and Co.).

TRIP'OLI. *Syn.* ROTTEN-STONE; ALANA, TERRA CARIOSA, L. A mineral employed as a polishing powder, originally imported from Tripoli, in Barbary. It consists almost entirely of silica, and is composed of the skeletons of minute infusoria, the precise character of which is readily distinguishable under the microscope.

TRIS'MUS. See TETANUS.

TRITICUM. The rhizome of *Triticum repens* or creeping couch-grass, dog-grass, or quitch gathered in the spring and deprived of its rootlets. Used in the form of decoction or liquid extract, as a diuretic and emollient in bladder and kidney affections.—*Dose*. Decoction, 4 to 8 oz.; liquid extract, 1 to 4 dr.

TRITURA'TION. *Syn.* TRITURA, TRITURATIO, L. The act of rubbing a solid body to powder. See PULVERISATION.

TRO'CHES. See LOZENGES.

TROGOSITA MAURITANICA, Linn. THE CORN BEETLE. This is another grain-boring insect, and belongs to the family *Tenebrionidæ*. Curtis says it was introduced from Africa, and that it is abundant in America, and in many European countries. In France it is called the Cadelle, and Olivier alludes to it as doing great harm to housed grain in the south of France. It is of the same family as the meal-worm and the worm which eats ship's biscuits. It is found in granaries and warehouses, and its larvæ sometimes greatly damage corn and other produce by biting the cuticle or skin, as it would seem in mere wanton mischief. At first sight it appears as if the corn lying in heaps had been nibbled by mice, but on close inspection the bran flakes are smaller, and bitten off differently. Sometimes when corn has been lying long, the quantity of bran which comes from the heap is surprising.

Life History. The beetle inhabits stores and warehouses among other places. It is about the third of an inch long, of a dark brown or chestnut colour, with six legs, and fairly sized wings. It is not known where the eggs are placed. The larvæ live in the corn heaps, in which they go from grain to grain, biting off the skin and consuming the flour. They are ½ of an inch in length, of a white colour, having dark brown heads. Their bodies are covered with short hairs or bristles, and have twelve divisions or segments, with six thoracic feet. The jaws are strong, pointed, and horny, adapted especially for biting hard substances. At the beginning of autumn they bury themselves in dust, and in cracks of floors, and lie there until the early spring, when they assume the pupa form, and from thence soon come forth in beetle shape.

Prevention. As this insect is most troublesome in foreign countries, particularly in hot climates, constant suspicion should be directed towards granaries and warehouses where foreign corn is stored. After the presence of these beetles has been detected the floors and boardings all round should be scrubbed with water, and strong solutions of soft soap well worked into the joints and cracks. All dust should be swept away and burnt, all ceilings whitewashed, and non-boarded sides must be washed with hot lime wash.

Remedies. When corn, English or foreign, is found to be infested with the *Trogosita* larvæ it should be frequently moved, and winnowed occasionally. If this does not prove effectual, kiln drying must be adopted to kill them ('Reports on Insects Injurious to Crops,' by Chas. Whitehead, Esq., F.Z.S.).

TRO'NA. A native carbonate of soda, found on the banks of the soda lakes of Sokena, in Africa.

TROPH'AZOME. A concentrated infusion of minced lean meat mixed with the fluid obtained from the residuum after being heated for 20 minutes in a water-bath, and flavoured with salt and spices, the whole being, lastly, simmered for a few minutes. Excellent for convalescents.

TROUT. The *Salmo fario* of Linnæus, a highly esteemed fish, found in most of the rivers and lakes of this country. Other members of the genus *Salmo* are also so called, as *S. eriox*, the bull or grey trout; *S. ferox*, the great grey or lake trout; *S. trutta*, the salmon trout, &c. All of these varieties are in the finest condition from the end of May to late in September.

The trout contains about 6 per cent. of fat. It is desirable to cook this fish as soon as convenient after taking it.

TRUSSING. This is a well-known operation performed on poultry or game previous to their being roasted or boiled. It simply consists in *drawing* or removing the intestines and other objectionable parts. In doing this, care must be taken to avoid rupturing the gall-vessel, which, if broken, would impart a very bitter flavour to the poultry, &c., extremely difficult of removal.

The cook should never take for granted that poultry or game, when it comes from the dealer, has been thoroughly cleansed inside, but, in order to be safe in this matter, should always make a point of cleansing it herself.

TRYPSIN. A ferment found in the pancreas. It has the power of changing proteids into peptones, both in alkaline and neutral media.

TULIPIN. An active principle, of which little is known at present, has been called *tulipin* from the fact of its having been extracted from the garden tulip when fully developed. It occurs probably in every part of the plant, even in the brilliantly coloured petals (according to Dr Phipson). Tulipin was discovered some little time back by Gerrard, and next alluded to by Dr Ringer in the 'Edinburgh Medical Journal.' It acts as a powerful sialogogue, producing a considerable flow of saliva without exerting any action on the pupil of the eye. Frogs poisoned by tulipin died with the heart in systole, and with the same symptoms as those exhibited by animals poisoned with veratrine. Chemically and therapeutically very little is yet known with regard to this new alkaloid. It is supposed to be a 'muscular poison' by some writers, acting also on the medulla and on the nerves of sensation. It is not improbable that it will be found to be more or less closely related to scillitin.

TULIP TREE (*Liriodendron tulipifera*, Linn.). A large tree of North America. Wood fine and even grained; used in America for cabinet work, door panels, &c. The bark is used as a stimulant tonic.

TUMOURS. Tumours, of which there are a great variety, are abnormal growths, occurring in different parts of the body. Sir Jas. Paget describes them as belonging to the class of overgrowths or hypertrophies, their most constant distinctive characters being—1. Their deviations, both in respect to size and shape, from the normal type of the body in which they are found. 2. Their apparently inherent power and method of growth. 3. Their development and growth being independent of those of the rest of the body, and continuing with no evident purpose when the rest of the body is only being maintained in its normal type.

Tumours are divided by pathologists into *malignant* and *innocent* or *benign*.

In the former division is included cancer. The most common varieties belonging to the second division are *cutaneous cysts, fatty tumours,* and *fibro-cellular tumours.* Cutaneous cysts, which may occur under any part of the skin, are most frequently met with in the scalp. They mostly arise from " the morbid growth of natural ducts or follicles, or by the enormous growth of elementary structures, which increase from the form of cells or nuclei, and become closed sacs with organised walls capable of producing other growths" ('Chambers' Cyclopædia').

The most commonly occurring tumour is the fatty one. It usually develops itself on the bodies of persons of from forty to fifty years of age. It seldom occasions inconvenience, and appears to be in no way prejudicial to health; occasionally, however, these tumours are very unsightly and unpleasant to look upon. The fat of which they are composed appears to differ in no respect from ordinary human fat. The uterus is the principal seat of the fibro-cellular tumour. It occurs also in the scrotum, the bones, and the subcutaneous tissue. These tumours sometimes attain great size and grow very rapidly. They are sometimes met with exceeding 40 lbs. in weight. Certain polypi belong to this class of tumour.

A pseudo-tumour is occasionally met with in surgical practice, which may often be mistaken for a real one by the unwary or inexperienced practitioner. This, which is known as a *phantom tumour,* appears to be caused by muscular contraction. Sir Jas. Paget, writing on this subject, says, " The abdominal muscles of hysterical women are most often affected, sometimes with intentional fraud. The imitation of a tumour may be so close as to require great tact for its detection; but chloroform, by relaxing the muscles, dissipates the swelling. Occasionally the apparent tumours move."

TUNGSTATE OF SODIUM. $Na_2WO_4 + 2H_2O$. This salt is used for rendering linen, cotton, and other fabrics uninflammable; also as a substitute for stannate of sodium as a mordant in dyeing. It may be prepared by adding 9 parts of finely powdered tungsten to 8 parts of fused carbonate of sodium, and continuing the heat for some time; on boiling the cooled and pulverised mass with water, evaporating the filtrate to dryness, and treating the residue with lukewarm water, the salt dissolves out. Muslin steeped in a 20% solution of this salt is perfectly uninflammable when dry, and the saline film left upon its surface is so smooth that the muslin may be ironed without difficulty.

TUNGSTEN. W = 183·5. *Syn.* TUNG-

STENUM, WOLFRAMIUM, L. A heavy, grey, brittle metal, discovered by Delhuyart.

The word *tungsten*, in Swedish, signifies 'heavy stone' (*tung-sten*), the name being applied to the element because the source from which it is obtained is a heavy mineral called *Wolfram*. Wolfram may be regarded as a variable double tungstate of iron and manganese, and the tungsten occurs in this. A native tungstate of copper has been discovered in Chili. Tungsten is also found in the mineral *scheelite*, a tungstate of lime.—*Prep.* By digesting finely powdered scheelite in hydrochloric acid. Chloride of calcium is formed, together with insoluble tungstic acid. Upon heating the acid to bright redness in a stream of hydrogen gas, the metal is left behind. When thus procured it is of a dark grey colour, but under the burnisher it may be made to assume a metallic lustre. Metallic tungsten may also be obtained by the reduction of tungstic acid, by means of charcoal at a white heat. When procured by this method it is unaffected by hydrochloric or diluted sulphuric acid, although it becomes reconverted into tungstic acid by the action of nitric acid.

When tungsten occurs in the pulverulent form, it burns easily into tungstic anhydride when heated in the air; and is oxidised and dissolved when boiled with the caustic alkalies or their carbonates in solution.

Uses. An alloy, possessed of extreme hardness, may be procured when tungsten is combined with steel, in the proportion of 1 part of tungsten to 10 parts of steel. This alloy is a good material for tools. The addition of from 1% to 2% of tungsten to nickel silver forms the alloy 'platinoid.'

Tests. A solution of a tungstate turns blue on treatment with zinc and hydrochloric acid. Fused with a small quantity of tin in the reduction flame with microcosmic salt, a tungsten compound if pure yields a blue bead, but if it contains iron the bead will be blood-red. Tungsten is determined quantitatively as the trioxide.

Dioxide or **Binoxide of Tungsten.** WO_2. This is obtained by treating tungstic acid with hydrogen at a low red heat. It occurs as a brown powder, which absorbs oxygen greedily from the air, and is dissolved by boiling with solution of caustic potash, hydrogen being evolved and potassium tungstate being formed.

Trioxide of Tungsten. WO_3. *Syn.* **TUNGSTIC ANHYDRIDE.** A bright canary-yellow coloured powder which may be obtained by decomposing wolfram with aqua regia, and evaporating to dryness. The resulting tungstic acid is dissolved in ammonia, and the ammonic tungstate purified by crystallisation. When this ammonic tungstate is heated in the air, it loses ammonia and water, pure tungstic trioxide being left behind. The trioxide combined with potassium or with sodium forms the so-called tungsten bronzes.

Tungstic Acid. H_2WO_4. This compound may be procured by adding an excess of hydrochloric acid to a boiling solution of the trioxide in any of the alkalies. It occurs as a yellow powder.

Tungstic Chloride. WCl_6. This may be

obtained by heating tungsten in chlorine, when it sublimes in bronze-coloured needles, which are decomposed by water. When gently heated in hydrogen, this chloride becomes converted into tetrachloride (WCl_4). Two other chlorides are known, viz. WCl_2 and WCl_5.

Bisulphide of Tungsten. WS_2. By the action of sulphur, sulphuretted hydrogen, or carbon disulphide on ignited metallic tungsten, a black crystalline substance, having the above composition, may be obtained, resembling plumbago in appearance. A trisulphide, WS_3, is also known.

Of the salts of tungsten, tungstate of baryta has been used as a substitute for white-lead in painting; but the most important of these is the tungstate of soda, described above. See also TUNGSTIC GLUE.

TUNGSTIC GLUE. Tungstic glue has been suggested as a substitute for hard india rubber, as it can be used for all the purposes to which this latter is applied. It is thus prepared :—Mix together a thick solution of glue with tungstate of soda and hydrochloric acid. A compound of tungstic acid and glue is precipitated, which, at a temperature of 86° to 104° F., is sufficiently elastic to be drawn out into very thin sheets.

TUNNY FISH, à la Parisienne. As a rule tunny fish is very indigestible, and may be described as "neither fish, flesh, fowl, nor good red herring;" nevertheless some of our readers may come across this fish, and will be glad to hear of a way in which to make it palatable and digestible. Take three or four pounds of fresh tunny fish; lard it with bacon as you would veal; cook it gently in its own gravy for three hours, with salt, pepper, sweet herbs, little onions, and a small quantity of water. When well cooked tunny fish makes a nice dish cold for breakfast.

TURBOT. The *Rhombus maximus*, Cuvier, said to be the best, and, excepting the halibut, the largest of our flat-fishes. Dutch turbots are the most esteemed.

Composition of the turbot :

Nitrogenous matter	18·1 per cent.
Fat	2·9 ,,
Saline matter	1·0 ,,
Water	78·0 ,,
	100·0

TURKEY. See POULTRY.

TURMERIC. *Syn.* **CURCUMA** (Ph. L. & D.), L. The rhizome of *Curcuma longa*. The best is imported from Ceylon. It is stimulant and carminative, but is chiefly used in dyeing yellow, and as an ingredient in curry powder; also as a test for alkalies. It gives a fugitive golden yellow with woad, and an orange tinge to scarlet. It dyes wool and silk, mordanted with common salt or salammoniac, a fugitive yellow.—*Dose*, 10 to 30 gr. See CURCUMINE.

Under the microscope turmeric presents a very characteristic structure, viz. "a cellular tissue containing large, loose, yellow cells, with here and there small but very distinct starch granules, similar in shape and size to those of *Curcuma arrowroot*, and some woody fibre and dotted ducts. The yellow granular cells can readily be identified whenever they occur" (*Dr Winter Blyth*).

Composition of an average sample of Curcuma
longa:

Water	. . .	14·249
Curcumine	. . .	11·000
Turmeric	. . .	12·075
Volatile oil	. . .	1·000
Gum	. . .	8·113
Starch	. . .	3·627
Extractive	. . .	3·388
Woody fibre	. .	46·548
Ash, included in above weights	. . .	[5·463]
		100·000

TURNBULL'S BLUE. *Syn.* FERRICYANIDE
OF IRON; FERRI FERRIDCYANIDUM, L. *Prep.*
Precipitate a solution of protosulphate of iron
with another of red prussiate of potash (ferricyanide of potassium).

Obs. This is a variety of Prussian blue remarkable for its beautiful colour, and may be
distinguished from the ordinary Prussian blue
of commerce by its action on the yellow prussiate of potash. When boiled in a solution of
the latter it is decomposed, a portion is dissolved,
and a grey residue remains.

TURNER'S YELLOW. See YELLOW PIG-
MENTS.

TURNIP. The *Brassica napus.* This vegetable possesses but little nutritive value, as will
be inferred from the annexed description of its
composition given by Letheby:

Nitrogenous matter	.	1·2
Starch	. . . .	5·1
Sugar	. . .	2·1
Salt	. . .	0·6
Water	. . ..	91·0
		100·0

Turnips should always be thoroughly cooked,
otherwise they are very liable to produce indigestion.

TURN'SOLE. See LITMUS.

TUR'PENTINE. *Syn.* TURPENTIN; TERE-
BINTHINA (Ph. L., E., & D.), L. "An oleo-resin
flowing from the trunk, the bark being removed,
of *Pinus palustris* (pitch or swamp pine) and
Pinus Tæda (loblolly, or old field pine)" (Ph. L.),
"From *Pinus sylvestris* (the Scotch fir)" (Ph. D.),
"From various species of *Pinus* and *Abies*" (Ph.
E.). It is viscid, of the consistence of honey,
and transparent; by distillation it is resolved into
oil of turpentine, which passes over into the
receiver; and into resin, which remains in the
still.

Turpentine, Bordeaux. *Syn.* FRENCH TUR-
PENTINE. From the *Pinus maritima*, or cluster
pine. Solidifies with magnesia (*Lindley*).

Turpentine, Chian. *Syn.* CHIO TURPENTINE,
CYPRUS T., SCIO T.; TEREBINTHINA CHIA (Ph. L.
& E.), L. "An oleo-resin flowing from the incised trunk of *Pistachia terebinthus*," Linn. (Ph.
L.). It is pale, aromatic, fragrant, and has a
warm taste, devoid of acrimony or bitterness. It
is much adulterated. A factitious article (terebinthina Chia factitia), made as follows, is also
very generally sold for it:—Black resin, 7 lbs.;
melt, remove the heat, and stir in of balsam of

Canada, 7 lbs.; oils of fennel and juniper, of each,
1 fl. dr.

Turpentine, Venice. *Syn.* TEREBINTHINA
VENETA (Ph. E.), L. Liquid resinous exudation
from the *Abies larix*, or larch tree. It is sweeter
and less resinous-tasted than common turpentine,
but is now scarcely ever met with in trade. That
of the shops is wholly a factitious article, made
as follows:—Black resin, 48 lbs.; melt, remove
the heat, and add of oil of turpentine, 2 galls.

TUR'PETH MINERAL. Basic sulphate of
mercury.

TURRET OF CHESTNUTS. A 'turret of chestnuts' (*tourelle de marrons*) is the name of a
most toothsome dish. Take rather over 2 lbs. of
chestnuts, peel, and cook them in water, with a
pinch of salt therein; then put them, whilst hot,
into a colander; beat into a paste, with a little
milk, sugar, and vanilla; put the mixture into
a mould in the form of a turret, about an inch
thick; when quite firm, open the mould and turn
out the contents carefully; glaze with syrup;
fill the middle with whipped cream, flavoured
with chocolate or vanilla.

TUR'TLE. *Syn.* GREEN TURTLE. The *Tes-
tudo midas*, Linn., a chelonian reptile, highly
esteemed for its flesh, eggs, and fat.

TUSSILA'GO. See COLTSFOOT.

TU'TENAG. A name sometimes applied to
German silver, at others to pale brass and bell
metal. "In India zinc sometimes goes under this
name" (*Brande*).

TUTTY. *Syn.* TUTIA, TUTHIA, IMPURE OXIDE
OF ZINC. The sublimate that collects in the
chimneys of the furnaces in which the ores of
zinc are smelting. Drying; astringent. Used
in eye-waters and ointments.

TYPE METAL. An alloy formed of antimony,
1 part; lead, 3 parts; melted together. Small
types are usually made of a harder composition
than large ones. A good stereotype metal is said to
be made of lead, 9 parts; antimony, 2 parts; bismuth, 1 part. This alloy expands as it cools, and
consequently brings out a fine impression.

TYPHOID FEVER. *Syn.* GASTRIC FEVER,
ENTERIC or INTESTINAL FEVER, LOW FEVER,
COMMON CONTINUED FEVER, INFANTILE REMIT-
TENT, ENDEMIC FEVER, PYTHOGENIC FEVER.
Although the term 'typhoid' expresses the fact
that this particular form of fever resembles typhus, the researches of later pathologists, including Perry, Lombard, Stewart, and Jenner, have
satisfactorily demonstrated that the two diseases
are altogether distinct.

"Typhus and typhoid fevers differ," says Sir
Thomas Watson, "notably and constantly in
their symptoms and course, in their duration, in
their comparative fatality, in the superficial
markings which respectively belong to them, and
which warrant our classing them among the exanthemata, in the internal organic changes with
which they are severally attended, and (what is
the most important and the most conclusive) in
their exciting causes."

About the beginning of the present century,
the French practitioners, after several *post-
mortem* examinations, were the first to point out
that the specifically distinguishing feature of this
disease was an internal exanthema. This salient

characteristic, coupled with the highly infectious nature of typhoid fever, has caused it to be defined by pathologists as "a contagious eruptive fever occurring on the mucous membrane of the intestines, and therefore removed from view."

The morbid appearance presented by the intestinal mucous membrane varies with the time that elapses between the period of seizure and death. If the patient dies within the week, the follicles on the membrane present a thickened appearance, and are raised above it, whilst they are seen to be filled with a yellowish, cheesy-looking substance. The result of these details is to give the bowels the appearance of being covered with pustules.

When death has occurred at a later period, ulceration more or less extensive has been observed to have set in.

The influence of age in predisposing to typhoid fever is forcibly illustrated in the following table extracted from Dr Walter Blyth's 'Dictionary of Practical Hygiene:'

Years of Age.	Per cent.
Under 5	0·98
From 5 to 9	9·44
„ 10 „ 14 . . .	18·16
„ 15 „ 19 . . .	26·86
„ 20 „ 24 . . .	19·69
„ 25 „ 29 . . .	10·15
„ 30 „ 34 . . .	5·36
„ 35 „ 39 . . .	3·40
„ 40 „ 44 . . .	2·09
„ 45 „ 49 . . .	1·08
„ 50 „ 54 . . .	0·60
„ 55 „ 59 . . .	0·33
„ 60 „ 64 . . .	0·33
„ 65 „ 69 . . .	0·08
„ 70 „ 79 . . .	1·33

Dr Murchison asserts that those under thirty are more than four times as liable to be attacked by typhoid fever as persons over that age. The practical bearing of the above figures is obvious. Typhoid patients should only be nursed by the middle-aged.

The season of the year also exercises an influence over the development and spread of this disease. In most countries it prevails with the greatest violence and is most general in autumn, and much more frequently follows a very hot and dry summer than a damp one. The carriers of the typhoid poison are the alvine and possibly the cutaneous and other excretions. The disease may therefore be conveyed by contact with the hands or skin of an infected patient, by his urine, by his body linen, the bedclothes, or by dissemination from these into the surrounding air. But the most fertile and unquestionable cause of propagation is the contamination of drinking-water by matter derived from the fæcal discharges of typhoid patients, which having soaked into the soil from the privy into which they had found their way, filtered from thence into a neighbouring well, or by means of drains proceeding from a privy or cesspool into a stream. We can easily understand that the disease, when traced to potable water, should always assume so virulent and frequently fatal a cha-

racter. The fever poison is thus directly conveyed into the stomach, and hence easily reaches the intestines, whence the disease originates. This will also account for the very small quantity of infected water which it has been found communicates the disease.

The outbreak of typhoid fever in Marylebone in 1874, which attacked some 500 persons, was traced to the milk vended by a certain company; this milk having been placed in cans that had merely been washed out and cooled with water obtained from a well into which it was discovered the excreta from a typhoid patient had percolated from an adjoining privy. After the statement of these facts, the need for thorough and efficient disinfection of all the excretions, immediately they leave the body of the patient, as well as of his body and bed linen, mattresses, &c., and also of the sick apartment, will be obvious.

The best method of effecting this will be to follow the instructions given by Dr William Budd for the prevention of the propagation of this disease, which are as follows:

" The means by which typhoid fever may be prevented from spreading are very simple, very sure, and their cost next to nothing.

" They are founded on the discovery that the poison by which this fever spreads is almost entirely contained in the discharges from the bowels.

" These discharges infect (1) the air of the sick room; (2) the bed and body linen of the patient; (3) the privy and the cesspool, or the drains proceeding from them.

" In these various ways, including the contamination of drinking water, already described, the infection proceeding from the bowel discharges often spreads the fever far and wide. The one great thing to aim at, therefore, is to disinfect these discharges on their very escape from the body, and before they are carried from the sick-room. This may be perfectly done by the use of disinfectants. One of the best is made of green copperas.

" This substance, which is used by all shoemakers, is very cheap, and may be had everywhere. A pound and a half of green copperas to a gallon of water is the proper strength. A teacupful of this liquid put into the night-pan every time before it is used by the patient renders the bowel discharge perfectly harmless. One part of Calvert's liquid carbolic acid in fifty parts of water is equally efficacious.

" To disinfect the bed and body linen, and bedding generally, chloride of lime, or Macdougal's or Calvert's powder, is more convenient.

" These powders should be sprinkled by means of a common dredger on soiled spots on the linen, and about the room, to purify the air.

" All articles of bed and body linen should be plunged, immediately on their removal from the bed, into a bucket of water containing a tablespoonful of chloride of lime, or Macdougal's or Calvert's powder, and should be boiled before being washed; a yard of thin white gutta percha, placed beneath the blanket, under the breech of the patient, by effectually preventing the discharges from soaking into the bed, is a great additional safeguard. The privy or closet, and all drains communicating with it, should be flushed twice daily with the green copperas liquid, or with carbolic acid diluted with water. [See SPOROXTON.]

"In towns and villages where the fever is already prevalent, the last rule should be put in force for all the houses, whether there be fever in them or not, and for all public drains.

"In the event of death, the body should be placed as soon as possible in a coffin sprinkled with disinfectants. Early burial is on all accounts desirable.

"As the hands of those attending on the sick often become unavoidably soiled by the discharges from the bowels, they should be frequently washed.

"The sick-room should be kept well ventilated day and night.

"The greatest possible care should be taken with regard to the drinking-water. When there is the slightest risk of its having become tainted with fever poison, water should be got from a pure source, or should at least be boiled before being drunk.

"Immediately after the illness is over, whether ending in death or recovery, the dresses worn by the nurses should be washed or destroyed, and the bed and room occupied by the sick should be thoroughly disinfected. These are golden rules.

"Where they are neglected the fever may become a deadly scourge; where they are strictly carried out it seldom spreads beyond the person first attacked."

No part of the globe appears to be exempt from the visitations of typhoid fever, since it occurs not only in all the older countries of Europe, Asia, and Africa, but in those also included in the North and South American continents, as well as in Australia, Tasmania, and New Zealand. It would appear also to have prevailed in the earliest ages, since it is evidently alluded to in the works of Hippocrates, Galen, and others. Later writers, including Sydenham and Hoffman, also constantly refer to it under a different name.

Pathologists differ as to the time that this disease lies dormant in the system before developing itself. Some practitioners contend that the usual period is from ten to fourteen days, whilst others think it is much less than this, and, in some instances, that it may not exceed one or two days. The late Dr Murchison entertained the latter opinion. The symptoms, when they show themselves, are as follows:—An irritable condition of the stomach, accompanied by sickness or vomiting; pain, with more or less tenderness, about the abdomen; sometimes the patient suffers from great constipation, at others from diarrhœa; he also experiences great prostration of strength, has a feeble pulse and a brown furred tongue; he is extremely restless, and at night frequently delirious; the lower limbs are frequently cold; he passes but little urine, and that of an offensive smell; the stools are dark, offensive, and very frequently bloody, this latter being a very characteristic accompaniment of typhoid fever. Bleeding from the nose sometimes occurs. The perspiration has a sour and fetid odour. After seven or eight days, small rose-coloured spots or petechiæ make their appearance on the skin.

Treatment. The abdomen should be leeched, and mustard poultices applied. If not too prostrated, the patient should be given a hot bath; but if he is not sufficiently strong to venture upon this, ablution of the whole of the body with hot water and soap should be had recourse to, the operation being performed by means either of a sponge or a flannel.

An effervescing draught, consisting of 20 gr. of carbonate of ammonia dissolved in water, to which a tablespoonful of lemon-juice should be added, ought to be administered, and drunk while effervescing, every three or four hours. The diet should consist of beef tea, nutritious broth, milk, and eggs.

The necessity of thorough ventilation of the patient's apartment, together with the methods of disinfection of the bodily discharges, the linen, &c., have been already emphasised in the directions given by Dr Budd for the prevention of the propagation of this fever.

It is, perhaps, needless to state that the outline of treatment given above is intended only for adoption by the emigrant, or of any one so unfortunately situated with regard to locality as to be unable to secure the services of a medical practitioner. Where these are obtainable, the patient or his friends should use all speed in procuring them.

Horses. Horses are occasionally attacked with typhoid fever, the symptoms of which bear a general resemblance to those which characterise the disease in the human subject. The appearances presented after death are also very similar, particularly in the lesions observable in the mucous membrane of the intestines. As in man, the disease is greatly aggravated by insanitary surroundings and depressing external agencies, and by the animals partaking of water containing decaying organic matters.

Upon the commencement of the attack give a few doses of calomel or laudanum, or of tincture of aconite, and if the bowels are costive two or three drachms of aloes, afterwards keeping up the laxative effect by mild clysters and mashes. Afterwards administer, three or four times a day, a drachm each of chlorate of potash and chloride of ammonium, adding to these an ounce of oil of turpentine or ether, or sweet spirit of nitre, if the animal exhibit dulness or weakness. If there is tenderness or pain about the abdomen, apply hot fomentations constantly; and should there be much flatulence give occasional drenches of ammonia, carbonate of ammonia, or whisky and water. The food of the animal must be nutritive and generous. He should be kept in a loose box, his legs should be bandaged in flannel, and warm rugs should envelop his body. Rest and quiet are essential. During convalescence let him have small doses of gentian, chloride of iron, with ale (*Finlay Dun*).

TYPHUS FEVER. This fever is known under various names, such as SPOTTED TYPHUS, JAIL FEVER, SHIP FEVER, CAMP FEVER, MILITARY FEVER, IRISH AGUE, FAMINE FEVER, BRAIN FEVER, PESTILENTIAL FEVER, MALIGNANT FEVER, OCHLOTIC FEVER, TYPHO-RUBEOLOID.

"1. Typhus prevails for the most part in great and wide-spread epidemics.

"2. The epidemics appear during seasons of

general scarcity and want, or amidst hardships and privations arising from local causes, such as warfare, commercial failures, and strikes among the labouring population. The statement that they always last for three years and then subside is erroneous.

"3. During the intervals of epidemics sporadic cases of typhus occur, particularly in Ireland, and in the large manufacturing towns of Scotland and England.

"4. Although some of the great epidemics of this country have commenced in Ireland, and spread thence to Britain, appearing first in those towns on the west coast of Britain where there was the freest intercourse with Ireland, it is wrong to imagine that all epidemics have commenced in Ireland, or that typhus is a disease essentially Irish. The disease appears wherever circumstances favourable to its development are present.

"5. In many epidemics typhus has been associated with relapsing fever, and the relative proportion of the two fevers has varied greatly.

"6. From the earliest times typhus has been regarded as a disease of debility, forbidding depletion, and demanding support and stimulation.

"7. The chief exception to the last statement originated in the erroneous doctrines taught in the early part of this century, according to which the disease was looked upon as symptomatic of inflammation or congestion of internal organs.

"8. The success believed at one time to follow the practice of venesection was only apparent. It was due to the practice, for the most part, having been resorted to in cases of relapsing fever and acute inflammations, and to the result having been compared with those of the treatment by stimulation of the much more mortal typhus.

"9. Although typhus fever varies in its severity and duration at different times, and under different circumstances, there is no evidence of any change in its type or essential characters. The typhus of modern times is the same as that described by Fracastorius and Cordames. The period during which epidemic fever was said to present an inflammatory type was that in which relapsing fever was most prevalent; and the times in which the type has been described as adynamic have been those in which relapsing fever has been scarce or absent" (*Murchison*).

In the article "TYPHOID FEVER" it has been stated that its propagation was mainly due, and had been very clearly traced, to the drinking of water contaminated by the alvine discharges of typhoid patients; in the dissemination of typhus, on the contrary, the air in the neighbourhood of the infected person appears to be the great medium for the conveyance of the disease, the poison, it is conceived, being disseminated into the surrounding atmosphere from the surface of the body or the lungs of the patient, or from the clothes, body linen, &c., worn and discarded by him. Hence it is we find, as we should expect, in the past no less than in the present, that the spread and degree of virulence of the malady have always been associated with overcrowding and bad ventilation. Although amongst the causes that predispose and induce susceptibility to its attacks, as shown above, are poverty, and consequent deficiency of food and clothing, and

squalor, it has been demonstrated that, with all these unfavourable conditions, patients may often recover from typhus provided they are supplied with a sufficiency of fresh air.

The fact that of late years typhus has rarely visited the inmates of our prisons, barracks, or shops, and that their comparative immunity from it has been coincident with improved ventilation and the avoidance of overcrowding, can lead to no other deduction than that previous to this reform these sanitary conditions were altogether neglected. We may narrate some of those outbreaks of typhus that have taken place previous to the application of hygienic principles to the treatment of the disease. During an assize held at Cambridge in 1522 the disease, which had broken out amongst the prisoners, spread to the justices, the bailiffs, and other officers, as well as to many people frequenting the court-house, with the result that many of those so seized died.

Another outbreak of a very malignant character occurred at Exeter in 1586. Some Portuguese were captured at sea, and (the words of the old historian who records the fact clearly indicate the cause of the virulent nature of the malady) were "cast into the deep pit and *stinking dungeon*." When brought into court they imparted the contagion to those around them. The judge and eleven out of the twelve jurymen who were thus attacked died, whilst the disease spread through and devastated the whole country.

A fourth case is recorded by Howard "at the Lent assizes in Taunton in 1730. Some prisoners who were brought there from Ivilchester jail infected the court, and Lord Chief Baron Pengelly, Sir James Shepherd, serjeant, John Pigot, Esq., sheriff, and some hundreds besides, died of the jail distemper."

Another eruption, which broke out during an assize held at the Old Bailey in 1750, resulted in the contraction of the disease by the judge, the Lord Mayor, and the alderman, and caused the death of forty persons who were present in the close and narrow court-house during the judicial proceedings. One circumstance recorded in connection with this last attack needs no comment. It is to the effect that "a hundred prisoners were put into two rooms measuring fourteen feet by eleven feet, and seven feet high." The instances above quoted explain why it was this disease acquired the name of *jail fever*.

During the present century six different epidemics of typhus have broken out amongst the convicts on board the Toulon Galleys. They occurred in 1820, 1829, 1833, 1845, 1855, and 1856. Although the above statement of facts indisputably points to the intimate connection existing between the prevalence and violent character of typhus and overcrowding, and consequent contamination and vitiation of the air breathed by the patient, it is still a moot point with pathologists whether the disease can be generated *de novo* by these conditions, or whether they merely assist to disseminate and intensify it. Dr Parkes, writing on this subject, says :—" With reference to the particular kind of fever at Metz, it may be noticed that an important argument against the production of exanthematic typhus from simple overcrowding has been drawn from the experience both of Metz and Paris. In both places during the sieges there was overcrowding, wretchedness, and famine, particularly at Metz; yet, as pointed out by Professor Chauffard to the Académie de Médecine, there was scarcely any or no typhus, as there had been in the wars of the first Napoleon. There was typhus in the German besieging force, but so strict was the blockade that it was not imported into Metz, and was not generated there" (Blyth).

The mortality which has been caused in large armies by the ravages of typhus has been enormous. During the thirty years' warfare that desolated Germany from 1619 to 1648, innumerable soldiers fell victims to it, the Bavarian army alone losing 20,000 men from this cause. Typhus also committed appalling havoc among the legions of the first Napoleon. The Bavarian contingent of the French army in the campaign in 1812 lost nearly 26,000 men from this cause; whilst in Mayence 25,000 of Napoleon's soldiers in garrison perished from the same cause in six months. More lately, viz. during the Crimean campaign (in 1856), typhus slew more than 17,000 French soldiers.

It frequently infested the German armies during the Franco-German war of 1869-70, and committed great havoc both amongst the hosts of Russia and Turkey in the late war between those countries.

When we turn to the civil population, we find that typhus has been no less ruthless, and has slain its myriads of these also. Confining our attention to our own country, we find it to have especially devastated Ireland, which has suffered from no less than eleven violent outbreaks of typhus within the last 130 years.

In one of these visitations, viz. that of 1840, 80,000 people are estimated to have died from the disease. The largest recorded epidemic of typhus within our islands during the present century was that of 1846. It extended over the whole of the British islands, and the number of persons attacked by it were nearly 1,400,000, out of which 1,000,000 occurred in Ireland. "The Irish flocked to England in thousands, bringing the pestilence with them. It therefore was extremely prevalent in Liverpool, no less than 10,000 persons dying of typhus in that city" (Blyth). The latest outbreaks in England have been in 1862 and 1869; they were principally confined to London.

Dr Murchison says that 14,000 persons were admitted to the London Fever Hospital during the two epidemics, and that amongst them a small number only were Irish.

All European countries, as well as North America and some parts of Asia, suffer from the ravages of this alarming disease. Africa, however, as well as Australia and New Zealand, is said to be exempt from it.

Symptoms. The symptoms of typhus are thus described by Dr Murchison:—"More or less sudden invasion, marked by rigors or chilliness; frequent compressible pulse; tongue furred, and ultimately dry and brown; bowels in most cases constipated; skin warm and dry; a rubeloid rash appearing between the fourth and seventh days, the spots never appearing in successive crops, at first slightly elevated, and disappearing on pressure, but after the second day persistent, and often becoming converted into true petechiae; great and early prostration; heavy flushed countenance; injected conjunctivae; wakefulness and obtuseness of the mental faculties, followed, at the end of the first week, by delirium, which is sometimes acute and noisy, but oftener low and wandering; tendency to stupor and coma, tremors, subsultus, and involuntary evacuations, with contracted pupils. Duration of the fever from ten to twenty-one days, usually fourteen. In the dead body no specific lesion, but hyperaemia of all the internal organs, softening and disintegration of the heart and voluntary muscles, hypostatic congestion of the lungs, atrophy of the brain, and oedema of the pia mater, are common."

Treatment. The following remarks, bearing on this branch of the subject, are suggested for adoption by the non-medical reader, in the event of his being precluded by circumstances from calling in the aid of the medical practitioner. The most important points to be observed are the isolation of the patient, and the thorough ventilation of his apartment by the continuous admission into it of fresh air without stint or hindrance. Dr Parkes recommends the patient to be put in the top room of the house or hospital, since there is strong evidence to show that the contagious virus is volatile and ascends through the atmosphere.

The forms of disinfection best suited for adoption in this disease, together with the method of employing them, will be found fully described in the article 'DISINFECTANTS.'

The body of the patient should also be frequently sponged with Condy's fluid, properly diluted, or covered with olive oil, to which has been added a small quantity of carbolic acid.

As an internal remedy, dilute hydrochloric or nitro-hydrochloric acid has been highly commended. Chlorate of potash, in large doses, was formerly much employed.

These remedies may be supplemented by the use of saline medicines, sudorifics, and moderate purgatives. The diet should consist largely of milk and water, beef tea, broth, and such like digestible and nutritious food.

UDDER, Inflammation of. *Syn.* GARGET; MAMMITIS, L. Amongst domestic animals, cows are the most frequently subject to this affection. It is most common amongst those cows that have lately calved or have been thoroughly milked. Heifers, and even young cows that have never had a calf, however, are not exempt from it, and occasionally suffer from its attacks. The inflammation varies in intensity, in some cases only showing itself in a slight, tenseness, heat and tenderness of the skin of the udder, whilst in others it is much more serious, and extends to the interior parts and vessels; in the latter case giving rise to hard lumps amongst the softer texture of the udder.

Inflammation of the udder appears to result from various causes—indigestion, over-driving, the too long retention of the milk in the udder, and cold; it is also very frequently associated with murrain, rheumatism, and swelling of the joints.

Treatment. If the disease be constitutional, as indicated by the suddenness of the attack, the best course would be to administer at once a good dose of Epsom or Glauber salts combined with a little ginger, and to give copious doses of nitre. A modified form of this treatment should be kept up for some little time by means of gentle aperients, and smaller doses of nitre. In the milder form of inflammation, viz. in that confined to the exterior of the udder, it will be best to have recourse to spirit lotions or refrigerant applications, such as ice water, or a mixture of chloride of ammonium and nitre applied immediately after mixing with water. The udder should be kept constantly cool by means of these. When the inflammation is deeper seated, as evidenced by the presence of lumps in the udder, the continuous application of warm water is advisable. Whether the cold or hot treatment be indicated, it should be diligently kept up for a day at least. In the adoption of either the cold or hot local remedies above specified, they should be combined with some means of support (by the agency of a proper bandage) to the udder. It is also important to have the milk removed every 3 or 4 hours; if the milking operation cause pain, a syphon should be used. The hard lumps will be found to disperse best under gentle friction applied by the hand twice a day, for an hour each time, the hand being previously anointed with lard. When the surface pits or becomes soft, and very hot, it may be assumed that suppuration has set in, in which case the confined pus must be liberated by means of the lancet.

ULCERATED SORE THROAT. This form of sore throat, in which ulcers develop themselves upon the tonsils, is a very frequent accompaniment of scarlatina, syphilis, and other diseases, in which cases to prescribe the method of treatment would be beyond our province. For ordinary ulcerated sore throat arising from cold, chronic inflammation of the part, or a low state of health, the best course will be to gargle the throat 4 or 5 times daily with either of the following gargles : —1. Alum, 30 gr.; infusion of rose, 6 oz. 2. Sulphate of zinc, 30 gr.; distilled water, 6 oz. 3. Hydrochloric acid, 1 dr.; water, 6 oz. Should these fail, touch the ulcers every morning with a solution of nitrate of silver, containing 10 gr. of the nitrate to an ounce of distilled water.

ULCERS. These are open sores, mostly accompanied by a discharge of pus, or serous matter. They differ from ordinary wounds by the edges showing no disposition to unite. When they extend or deepen, it is by a process of absorption ; while they heal by granulation, whereby they become filled up with little granular growths of flesh. Ulcers may appear on all parts of the body, but they most frequently attack the legs and arms.

In enfeebled states of the body, wounds, boils, and abscesses may degenerate into ulcers; they are also a consequence of enlarged or varicose veins, or the result of some specific poison in the system.

Ulcers may be classed into simple, irritable, indolent, and specific.

Treatment. When an ordinary wound or sore shows a disinclination to heal, but, on the contrary, extends or deepens, it should be poulticed with bread and water or linseed meal. Should these remedies be ineffectual, an old-fashioned but useful one, viz. a carrot poultice, may be applied. When the ulceration is irritable or painful, the poultices may be supplemented by the frequent use of a lotion consisting of 4 parts of water to 1 of tincture of opium, or of a warm decoction of poppy heads applied by means of a linen rag. Filling the cavity with prepared chalk has been recommended.

It sometimes happens that during poulticing, *proud flesh* may form in an ulcer. This may require the attention of a surgeon. When the ulcer has a bad or fetid odour, it should be washed with a lotion composed of 1 part of solution of chlorinated soda to 16 parts of water ; or it may be sprinkled over with charcoal powder, or with a mixture of starch and salicylic acid. The best application to bad smelling ulcers caused by varicose veins is a lotion consisting of nitric acid considerably diluted with water. Very irritable ulcers are often greatly relieved by the gentle application to them of lunar caustic, and indolent ones by dressing with yellow basilicon ointment, or by the judicious use of black wash. The general health should be attended to by the administration of tonics consisting of the mineral acids, gentle aperients, and a digestible and nourishing diet. Small ulcers on the mucous membrane of the mouth or on the gums may be made to disappear instantly upon touching them with a piece of lunar caustic. Where any difficulty is experienced in the healing of an ulcer, or if it be at all of a serious nature, the medical practitioner should be consulted.

ULEXINE. An alkaloid obtained by M. Gerrard from *Ulex europæus*, common gorse. Gerrard and Symons, who chemically examined the base, gives it the following formula : $\frac{1}{2}C_{26}H_{36}N_4O_2$. The physiological action of ulexine has been recently studied by Mr J. Rose Bradford. The drug acts first as a stimulant, and then as a depressor of the respiratory system, and, in large doses, paralyses the motor nerves of mammals. It has a powerful effect on the kidney, causing constriction, followed by a very large expansion of short duration. The physiological effects point to a possible diuretic action, and this is what really happens. It has been employed in University College Hospital with success in cases of dropsy due to heart disease.—*Dose*, $\frac{1}{10}$ to $\frac{1}{5}$ gr.

ULMIN, ULMIC ACID. By boiling sugar in dilute sulphuric acid for a long time, a brownish-black substance is produced. Boullay and Malaguti state that this is a mixture of two distinct bodies—ulmin (sacchulmin—*Liebig*) and ulmic acid (sacchulmic acid—*Liebig*). The first is insoluble in solutions of the alkalies ; the latter dissolves in them freely. A number of black uncrystallisable substances, produced by the action of powerful chemical agents upon vegetable matter, have been confounded under these names.

ULTRAMARINE'. *Syn.* LAPIS-LAZULI BLUE, ULTRAMARINE B.; CÆRULEUM ULTRAMONTANUM, L. This beautiful pigment is obtained from the blue mineral azure-stone, lazulite, or lapis lazuli, the finest specimens of which are brought from. China, Persia, &c.

Prep. Pure lapis lazuli (reduced to fragments about the size of a pea, and the colourless pieces rejected), 1 lb., is heated to redness, quenched in water, and ground to an impalpable powder; to this is added of yellow resin, 6 oz.; turpentine, beeswax, and linseed oil, of each, 2 oz.; previously melted together; the whole is next made into a mass, which is kneaded in successive portions of warm water as long as it colours it blue; from these it is deposited on repose, and is then collected, well washed with clean water, dried, and sorted according to its qualities. The first water, which is usually dirty, is thrown away; the second gives a blue of the first quality; and the third and following ones yield samples of less value. The process is founded on the property which the colouring matter of azure-stone has of adhering less firmly to the resinous cement than the foreign matter with which it is associated. When azure-stone has its colour altered by a moderate heat, it is reckoned bad or factitious.

Obs. Ultramarine is the most costly, but at the same time the most splendid and permanent of our blue pigments, and works well in oil.

Ultramarine, Artificial. *Syn.* Azure blue, Meissner ultramarine, Paris b., Vienna b.; Cæruleum ultramontanum factitium, L. From the researches of Clement, Desormes, and Robiquet, it has been inferred that the colour of ultramarine depends on the presence of sulphide of sodium in a peculiar state of combination with the silicates of soda and alumina; but, according to Elsner and Tirnmon, a minute quantity of sulphide of iron is also an essential ingredient. It is by heating mixtures of this kind that the artificial ultramarine of commerce is prepared. The finer specimens, thus obtained, are quite equal in durability and beauty of colour to those prepared from lazulite, while they are very much less expensive.

Prep. 1. Kaolin, 37 parts; sulphate of soda, 15 parts; carbonate of soda, 22 parts; sulphur, 18 parts; charcoal, 8 parts; intimately mixed and heated from twenty-four to thirty hours in large crucibles; the product is then heated again in cast-iron boxes, at a moderate temperature, till the required tint is obtained; it is, finally, pulverised, washed, and dried.

2. (*Gmelin.*) Sulphur, 2 parts; dry carbonate of soda, 1 part; mix well; gradually heat in a covered crucible to redness, or till the mixture fuses, then sprinkle in, by degrees, another mixture of silicate of soda and 'aluminate of soda' (containing 72 parts of silica and 70 parts of alumina), and continue the heat for an hour. The product contains a little free sulphur, which may be separated by water.

3. (*Robiquet.*) By exposing to a low red heat, in a covered crucible, as long as fumes are given off, a mixture of pure kaolin, 2 parts; anhydrous carbonate of soda and sulphur, of each, 3 parts. Some manufacturers who adopt this process use 1-3rd less carbonate of soda.

4. (*Tirnmon.*) Take of crystallised carbonate of soda, 1075 gr.; apply a gentle heat, and, when fused in its water of crystallisation, shake in of finely pulverised orpiment, 5 gr.; and, when partly decomposed, add as much gelatinous hydrate of alumina as contains 7 gr. of anhydrous alumina; finely sifted clay, 100 gr., and flowers of sulphur, 221 gr., are next to be added, and the whole placed in a covered crucible, and at first gently heated, to drive off the water; but as soon as this is effected, raised to redness, the heat being so regulated that the ingredients only 'sinter' together, without actually fusing; the mass is then to be cooled, finely pulverised, suspended in river water, and brought upon a filter; the product has now a very beautiful delicate green or bluish colour, but on being heated in a covered dish, and stirred about from time to time, until the temperature reaches that of dull redness, at which it must be kept for one or two hours, it changes to a rich blue. If the heat of the first calcination has been properly regulated, the whole of the mass taken from the crucible will have uniform colour; but if too little heat has been used, and the ingredients have not been properly mixed, there will be colourless parts, which should be rejected; if too much heat has been used, or the mass allowed to fuse, brown parts will appear, especially if the crucible is of a bad kind, or easily destroyed; these must also be rejected.

5. Heat to bright redness, in a covered crucible, three or four hours, an intimate mixture of pure kaolin, 100 parts, dried carbonate of soda, 100 parts, sulphur, 60 parts, and charcoal, 12 parts. The mass has now a green colour (green ultramarine). This is powdered finely, washed, dried, mixed with 1-5th its weight of sulphur, and gently wasted to a thin layer till the sulphur has burnt off; this operation being repeated with fresh additions of sulphur till the residue has a fine blue colour.

Ultramarine Ashes. *Syn.* Saunders blue. Obtained from the resinous mass from making ultramarine, by melting it with fresh oil, and kneading it in water containing a little potash or soda; or by burning away the wax and oil of the mass, and well grinding and washing the residue with water. Very permanent, but much less brilliant than ultramarine.

Ultramarine, Cobalt'ic. *Syn.* Chinese blue, Cobalt b., Louisa b., Höffner's b., Thénard's b. A very rich blue pigment, prepared by slowly drying and heating to dull redness a mixture of freshly precipitated alumina (freed from water as much as possible), 8 to 10 parts; arseniate or phosphate of cobalt, 1 part. By daylight it is of a pure blue, but by artificial light the colour turns on the violet. For other formulæ see Blue Pigments.

UMBER. A species of clay coloured with oxides of iron and manganese, used as a pigment; the commonest kind consists of limonite or brown hæmatite, and the hydrated oxide of manganese, mixed with clay. It occurs in beds associated with brown jasper in Cyprus and elsewhere. It may be used either raw or burnt as a brown pigment.

UPAS. The Javanese name for several deadly poisons. 'Bohun upas' is a gum-resin obtained from the bark of the *Antiaris toxicaria.* (See Antiarine.) The 'upas tieuté' is obtained from the *Strychnos tieuté*, and owes its fatal power to strychnine. They are both used to poison arrows and other deadly weapons.

URA''NIUM. U = 239. A rare metal, discovered by Klaproth in 1789. It occurs in the pitch-blende of Saxony and the uranite of Cornwall. Uranium is employed in the arts only in the state of compound.

Uranic Oxide (U_2O_3) may be obtained in the anhydrous state by heating the hydrated sesquioxide to a temperature of 572° F. It is capable of acting both as an acid and a base. UO_4 is also known. The uranic salts are yellow.

Uranous Oxide. UO. This may be procured by igniting uranium oxalate in a closed vessel, or in a stream of hydrogen gas. Acids are without action upon this oxide. When, however, it is obtained as a hydrate (which it may be by treatment of its chloride with ammonia), this latter is easily acted upon by acids, and gives rise to salts having a green colour, which rapidly absorb oxygen. Peligot proposed to call this oxide *Uranyl*, from the tendency it showed to follow the deportment of a metal when it combined with elementary bodies.

Chlorides of Uranium. Uranium forms two chlorides, U_2Cl_3 and UCl_2.

Uses. Its ores and oxides (U_2O_3 and $2UO.U_2O_3$) are used to colour glass yellow and porcelain black. This glass possesses a beautiful canary-green fluorescence.

U'RATES. Salts of uric acid.

U'REA. COH_4N_2. *Syn.* CARBAMIDE. A crystalline, colourless, transparent substance, discovered by Fourcroy and Vauquelin in urine, and by Wöhler as the first organic compound artificially produced.

Urea generally occurs in slender, striated, colourless prisms, as shown on next page. It is slightly deliquescent. It has a neutral reaction and a bitterish taste. It is extremely soluble in water and in hot alcohol, but very slightly so in ether. At about 248° F. it melts. At a little higher temperature it becomes decomposed into ammonia carbonate, cyanate of ammonium, and cyanuric acid, this last being left in the retort.

The ammoniacal odour acquired by urine after a few days is due to the conversion of the urea into carbonate of ammonia, as shown by the following equation:

Urea.	Water.	Carb. Ammonia.

$$CH_4ON_2 + 2H_2O = (NH_4)_2CO_3.$$

This change is effected by a minute organism, *Micrococcus ureæ*, present in the urine. A solution of pure urea may be kept at ordinary temperature, or even boiled, without undergoing alteration.

Urea occurs as an essential component of the urine of man and animals, being more particularly abundant in the urinary excretion of the flesh-eating mammalia; nor is it altogether absent from the urine of birds and amphibia. According to Bischoff and Voit, urea is the result of tissue metamorphosis. The greater number of inquirers, however, hold an opposite opinion, and believe that it is derived from the albuminous constituents of the food. [Accurate experiment on the human subject shows that the total daily nitrogen of the excreta corresponds, within limits of error, with that of the ingesta.—ED.]

Prep. (*Thénard.*) Fresh urine, gently evaporated to the consistence of a syrup, is treated with its own volume of nitric acid of sp. gr. 1·19; the mixture is shaken and immersed in an ice-bath, to solidify the crystals of nitrate of urea (p. 1728); these are washed with ice-cold water, drained, and pressed between sheets of blotting-paper; they are next dissolved in water, and the solution is decomposed and precipitated with carbonate of potassium (or carbonate of barium); the whole is then gently evaporated nearly to dryness, and the residuum is exhausted with pure alcohol, which dissolves the urea, which crystallises out as the solution cools.

Urea, Factitious. *Prep.* Mix 28 parts of well-dried ferrocyanide of potassium with 14 of black oxide of manganese (both in fine powder), and heat them to dull redness on an iron plate. Lixiviate with cold water, add 22½ parts of dry sulphate of ammonia, concentrate by evaporation with a heat not exceeding 212° F., decant the concentrated liquid, treat it with rectified spirit, and crystallise. This is intended as a cleanly substitute for the preceding.

Urea, Ni'trate of. *Syn.* UREÆ NITRAS, L. *Prep.* From urine, as described above; or it may be prepared by saturating artificial urea with nitric acid. Diuretic.—*Dose*, 2 to 5 gr., twice or thrice daily; in dropsy.

UREOMETER. See p. 1738.

URETHAN. *Syn.* URETHANE. Ethyl-urethan, which is commonly understood by the commercial name urethan, is $NH_2COOC_2H_5$, and forms colourless columnar or tabular crystals, melting between 47° and 50° C. It is practically tasteless, giving rise, when placed in substance on the tongue, to a feeble saline sensation. In water and in most media it is very readily soluble; the aqueous solution is neutral. It boils between 170° and 180° C. without decomposition; the vapours given off burn with a bluish flame. By Kobert, Schmiedeberg, and others it has been recommended as a useful hypnotic, distinguished from morphine, chloral hydrate, paraldehyde, &c., by the absence of unpleasant secondary symptoms, which makes it particularly valuable in the treatment of children, of delirium tremens, and acute mania.—*Dose*, 16 to 60 gr.

U'RIC ACID. $C_5H_4N_4O_3$. *Syn.* LITHIC ACID; ACIDUM LITHICUM, A. URICUM, L. A substance discovered by Scheele, and peculiar to the urine of certain animals, and the excrement of serpents and several birds. The fæces of the boa-constrictor consist of little else than urate of ammonium. It constitutes one of the commonest varieties of urinary calculi, and of the red gravel or sand which is voided in certain morbid states of the urine. Guano derives its principal value as a manure from the presence of urate of ammonium. The gouty concretions of the joints, popularly known as chalk-stones, consist chiefly of urate of sodium.

Prep. Dissolve the chalk-like excrement of serpents, reduced to fine powder, in a solution of caustic potassa, by boiling; then add hydrochloric acid in excess, again boil for 15 minutes, and well wash the precipitate with water.

Prop., &c. Brilliant, very minute, white and silky scales, which are tasteless, inodorous, slightly soluble in boiling water, and dissolve in strong sulphuric acid, but are again precipitated by water. It forms salts with the bases called urates, all of which are very sparingly soluble. The characteristic reaction of uric acid is, that when moistened with nitric acid and heated, it dissolves, and by evaporation yields a red compound, which, upon the addition of a drop or two of solution of ammonia, assumes a magnificent crimson colour, being converted into murexide.

Uric acid is a constituent of healthy human urine, in which it exists combined with bases in the form of urates, which, being in small quantity, are soluble in the urine. 1000 gr. of the urine contain from ¼ gr. to 1 gr. of the acid. Drs Beale and Thudichum respectively estimate the amount of uric acid excreted in 24 hours by a healthy adult man at from 5 to 8 gr. To determine the amount of uric acid in urine proceed as follows: —To a certain weight of the urine, hydrochloric or nitric acid is added, and the urine set aside for some hours, at the end of which time the insoluble crystals of uric acid which are formed are washed, dried, and weighed.

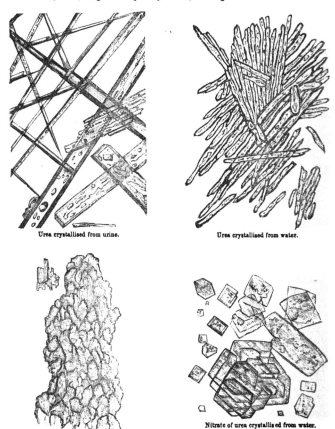

Urea crystallised from urine.

Urea crystallised from water.

Nitrate of urea formed in urine.

Nitrate of urea crystallised from water.

The majority of the cuts illustrating 'urea,' 'urinary diseases,' and 'urine,' are taken from Dr Beale's work on 'Kidney Diseases, Urinary Deposits, and Calculous Deposits,' by that gentleman's kind permission.

With the exception of the urates, uric acid is one of the deposits most frequently met with in abnormal urine, wherein it occurs as a small red-

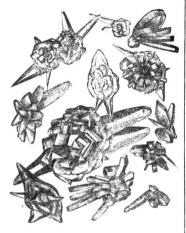

dish powder adhering to the bottom or sides of the containing vessel. As a urinary deposit, uric acid assumes a great variety of forms, that of most frequent occurrence being the rhombic, modified in many of the crystals by the rounding of two of the angles, as shown in the *engr.*

URINARY DISEASES. This class of disorders, which in general terms may be said to embrace affections of the kidneys, bladder, ureters, &c., comprises diseases of these parts varying greatly in character and pathological importance.

The most serious forms of kidney disease are CONGESTION of the kidney, a very frequent accompaniment of heart or lung disease; PYELITIS, or inflammation of the pelvis of the kidney; SUPPURATIVE NEPHRITIS, or inflammation of the substance of the kidney, which ends in suppuration; ACUTE NEPHRITIS, or ACUTE BRIGHT'S DISEASE, acute inflammation of the kidney, frequently arising from scarlatina or cold; CHRONIC NEPHRITIS, or CHRONIC BRIGHT'S DISEASE, a formidable and incurable variety of kidney affection, giving rise to dropsy, and, owing to the disintegration of the organ, to poisoning of the blood by urea. In advanced cases of this disease the urine contains a large quantity of albumen and casts of the urinary tubes.

CALCULUS OF THE KIDNEY. The most dangerous diseases of the bladder are those caused by the deposition in it of earthy and other concretions, known as URINARY CALCULI, which are described in the present work under CALCULUS. The presence of these calculi is indicated by acute pain in the bladder and urinary passages, extending to the adjacent parts, the pain being excruciating immediately after passing the urine. Sometimes

during the act of excretion the stream of water is suddenly stopped.

Inflammation is another dangerous disease of the bladder, calling for the prompt summoning of the medical practitioner wherever possible. The following particulars as to its symptoms and treatment are offered for adoption to emigrants and others, so placed as to be beyond the means of medical succour.

Inflammation of the bladder commences with pain in the region of that organ, the pain becoming continuous and increasing in violence, and being accompanied with a sense of burning heat and of tenderness on pressure. The urine is frequently voided. The inflammation is sometimes so acute as to give rise to suppuration, and the consequent discharge of pus with the urine. Sometimes the disease assumes a chronic character.

In the acute form of the disease recourse should be had to leeches, hot fomentations, and warm baths; a dose of calomel, to be followed by a brisk dose of castor oil, should likewise be administered. Alcoholic drinks of any kind must be carefully avoided, the patient being allowed to drink only cool demulcent beverages. With these should be combined effervescing draughts, frequently repeated, and small doses of Dover's powder. Great relief will also be derived from the use of enemas of gruel containing laudanum. Where inflammation of the bladder arises from gout or rheumatism it must be treated as for these diseases. A suppository, consisting of 2 gr. of opium combined with 20 gr. of soap, is frequently of great benefit.

Should the disease become chronic the best method of treatment will be the repeated use of mild aperients, the combined employment of uva ursi in infusion or powder, with either tincture of perchloride of iron or the mineral acids. Spirituous liquors of any kind must be avoided. Demulcent drinks form the best beverage, and a farinaceous or milk diet the most desirable food.

NEURALGIA OF THE BLADDER. The pain which attends this disease is unaccompanied either by inflammation or irritation, and is recurrent in character. It may generally be arrested by tincture of perchloride of iron, or of iron and quinine, administered three times a day.

IRRITATION OF THE BLADDER. The patient affected with this disorder gratifies the frequent desire he has to pass his urine, the operation being accompanied with pain and forcing, the most severe pain being experienced after the excretion has taken place. The tincture of perchloride of iron will also be found the best remedy for this disorder. It should be given in conjunction with the infusions of uva ursi, Pareira brava, or buchu. Mucilaginous drinks should also be had recourse to.

CATARRH OF THE BLADDER. The symptoms of this disease are irritation, and the presence of much mucus in the urine. The same treatment may be adopted as recommended for irritation of the bladder. If there be an absence of pain, spirits of nitre and copaiba balsam in moderate doses frequently afford relief.

STRANGURY. Constant micturition, only a few drops of urine passing at the time, occasioning burning and cutting pains around the parts. Strangury is generally due to some irritating cause, which should, if possible, be discovered and removed. Cantharides taken either internally, or applied externally, as in the form of a blister, will sometimes give rise to it. The patient should drink copiously of mucilaginous beverages, such as linseed tea, slippery elm bark, barley water, with gum-arabic dissolved in it. An injection consisting of thirty or forty drops of laudanum in a spoonful of gruel will be found to afford immediate relief. If the above means fail, a pill containing a grain of camphor in five grains of extract of henbane should be given, and a warm bath taken. See GRAVEL.

U'RINE. The density of the urine varies from 1·005 to 1·030 (from 1·015 to 1·025—*Beale*); the average, in health, being 1·020, when it contains about 380 gr. of solid matter in the pint. According to Berzelius, the proportion is about 6¾%, the rest being pure water. It exhibits a decidedly acid reaction, and is never alkaline, except during disease, or the use of large quantities of alkaline salts of the vegetable acids. The average quantity secreted during 24 hours may be taken at 2 pints to 3 pints; as might be supposed, a larger quantity is passed during the summer than in the winter months.

Miller gives the following as the composition of healthy urine:

				In 100 parts of solid matter.
Specific gravity			1·020	
Water			956·80	
Solid matters, 43·2.	Organic matters, 29·79	Urea	14·23	33·00
		Uric acid	0·37	0·86
		Alcoholic extract	12·53	29·03
		Watery extract	2·50	5·80
		Vesical mucus	0·16	0·37
	Fixed salts, 13·35	Sodic chloride	7·22	16·73
		Phosphoric anhydride	2·12	4·91
		Sulphuric anhydride	1·70	3·94
		Lime	0·21	0·49
		Magnesia	0·21	0·28
		Potash	1·93	4·47
		Soda	0·09	0·12
		Loss	0·03	
			1000·00	100·00

The presence of bile in urine, or other like fluids, may be detected as follows:—Put a small quantity of the suspected liquid into a test-tube, and add to it, drop by drop, strong sulphuric acid, until it becomes warm, taking care not to raise the temperature above 122° F.; then add from 2 to 5 drops of syrup (made with 5 parts of sugar to 4 parts of water), and shake the mixture. If the liquid contain bile, a violet coloration is observed. Acetic acid may be substituted for sugar.

Another test for bile consists in pouring a little of the suspected urine into a test-tube, and adding to it a few drops of tincture of iodine, when, if bile be present, the fluid becomes distinctly green. Rosenbach says that urine containing bile, when passed through white filtering-paper, imparts a yellow or brown colour to the paper. On allowing one drop of strong nitric acid to run down the side of the moist filter it leaves a yellow streak, soon changing to orange, with a violet border, on the outside of which blue and emerald-green zones may be observed. These colours remain visible for some time.

Dark-coloured urine, owing to substances other than bile, does not produce this play of colours.

The reagents most generally employed for detecting the presence of sugar in urine are Trommer's (see SUGAR) and Fehling's solutions. For the effective application of Fehling's test, Dr Roberts ('Urinary and Renal Diseases,' by Dr W. Roberts) advises the following method of procedure:—Pour some of the Fehling's solution into a narrow test-tube to the depth of ¾ of an inch; heat until it begins to boil; then add 2 or 3 drops of the suspected urine. If the sugar be abundant, a thick yellow opacity or deposit of yellow suboxide is produced (and this changes to a brick-red at once if the blue colour of the test remains dominant). If no such reaction ensue, go on adding the urine until a bulk nearly equal to the test employed has been poured in; heat again to ebullition, and, no change occurring, set aside without further boiling. If no milkiness is produced as the mixture cools, the urine may confidently be pronounced free from sugar, or, at any rate, it contains less than $\frac{1}{10}$%.

If the quality of sugar is very small, viz. from ½% to $\frac{1}{10}$%, the precipitation of the yellow or cuprous oxide does not take place immediately, but occurs after some time as the liquid cools, and the manner of the change is peculiar. First, the mixture loses its transparency, and passes from a clear bluish-green to a light greenish opacity, just as if some drops of milk had fallen into the tube. This green milky appearance is quite characteristic of sugar.

Before using the Fehling's solution it should be always examined previously to the addition of the urine, by being first boiled alone, when if it remains clear it may be pronounced in fit condition. On the contrary, should the preliminary boiling give a deposit, the solution must be discarded, and some freshly made employed instead.

Böttger has proposed the following quantitative test for the presence of sugar in urine:

He first adds some potash to the sample of urine, and then a small quantity of subnitrate of bismuth, and boils the mixture. If sugar is present, the suboxide is reduced, and metallic bismuth being liberated is precipitated as a black powder.

Another method of applying the bismuth test is as follows :—1 part of crystallised carbonate of soda is dissolved in 3 parts of water, and added to an equal quantity of the urine. A small quantity of basic nitrate of bismuth is then added to the mixture, which is then heated to the boiling-point. A black precipitate is formed if the urine contains sugar.

Horsley's test consists in boiling with the suspected urine a mixture of equal parts of neutral chromate of potash and solution of potash, when, if sugar be present, a green colour will be produced, owing to the formation of the sesquioxide of chromium.

In M. Luton's, which is a modification of Horsley's test, a solution of bichromate of potash is decomposed by excess of sulphuric acid, and upon the urine being boiled with the mixture, a splendid green colour is imparted to it. Urea, albumin, and the urates do not interfere with this reaction.

Vidau has observed that a mixture of equal parts of hydrochloric acid and oil of brune (oil of sesame), either in the cold, or when slightly heated, assumes a distinct rose colour in the presence of cane or grape sugar, provided 0·001 gramme of sugar is present for every c.c. of mixture.

One of the best methods for the accurate and quick estimation of the amount of sugar in urine is, perhaps, the volumetric, devised by Fehling, who employed a standard copper solution, known as 'Fehling's solution,' of the following composition :

Sulphate of copper . . . 90½ grains.
Neutral tartrate of potash . 364 ,,
Solution of caustic soda, sp. gr.
  1·12 . . . . . . 4 fl. oz.
Add water to make up exactly 6 ,,

Of this solution 200 gr. are exactly decomposed by 1 gr. of sugar.

The following is the mode of performing the analysis given by Dr Roberts ('Urinary and Renal Diseases,' by Dr W. Roberts) :—Measure off 200 gr. of the above standard solution in a 200-grain tube, pour this into a flask, and add about twice its volume of water ; then place over a spirit-lamp to boil. While the copper solution is being heated the urine to be analysed should be diluted with water to a known degree. In the case of ordinary diabetic urines the best dilution is 1 in 10. This is obtained by carefully filling a 6-oz. measure with water to the depth of 4½ oz., and then adding urine so as to make up exactly 5 oz. The mixture will then contain exactly ₁₀ of urine (when the quantity of sugar in the urine is very small, a dilution of 1 in 5, or even the undiluted urine may be employed). The next step is to fill a burette (which must be graduated to grains) with the diluted urine to 0. Then proceed to add it in successive small portions to the boiling copper solution until the blue colour has entirely disappeared. After each fresh addition from the burette, the mixture should be raised to the boiling-point, and then allowed to stand a few seconds, so that the precipitated copper may sub-

side, and the observer may see, by holding the flask between the eye and the light, whether the mixture still retains any blue colour.

As soon as the blue colour has disappeared the analysis is complete, and the quantity of diluted urine may be read off. The percentage of sugar in the urine can now be readily calculated. Suppose 125 gr. had been added from the burette, this represents ₁₀, or 12·5 gr. of undiluted urine, and contains exactly 1 gr. of sugar ; by dividing 12·5 into 100 the percentage of sugar is obtained.

$$\frac{100}{12 \cdot 5} = 8 \; ; \; \text{the urine contains 8\% of sugar.}$$

Another process for the quantitative determination of sugar in urine, called by its author, Dr Roberts, 'the differential density method,' is based upon the loss of density experienced by diabetic urine after all the sugar has been removed by fermentation. Dr Roberts says repeated examples derived from diabetic urine so treated, together with corresponding experiments made with solution of sugar of known strength in normal urine, and in pure water, as well as theoretical calculation, have warranted the conclusion, *that the number of degrees of density so lost indicates as many grains of sugar per fluid ounce.*

The method, which is extremely simple, is thus performed :—Into a 12-oz. bottle measure 4 fl. oz. of the diabetic urine, and drop into it a piece of fresh German yeast, about as large as a cobnut or walnut ; insert a cork in the bottle, and let the cork have a nick cut in the side, to allow of the escape of the carbonic acid. Then fill an ordinary 4-oz. bottle with the same sample of urine, omitting to add any yeast, and cork it in the ordinary manner. Place both bottles in a warm situation, where the temperature is about 80° or 90° F., for twenty or twenty-four hours ; at the end of which time, the fermentation being over, the scum will either have cleared off or subsided. The fermented urine is then poured into a proper urine-glass, and its specific gravity ascertained.

The specific gravity of the unfermented companion portion is also taken, and by comparing the two results the loss of density is thus arrived at. Before the respective densities are taken it is best to remove the two samples to a cool place, where they should remain for two or three hours, in order that they may acquire the temperature of the surrounding air.

The two following examples may serve as illustrations of the method.

	I.	II.
Density before fermentation	. 1053	. 1088
Density after fermentation	. 1004	. 1013
Degrees of density lost	. 49	. 25
Grains of sugar per fluid ounce	49	. 25

If it be desired to bring out the result as so much per cent., this is accomplished by multiplying the number indicating the 'density lost' by the coefficient 0·23. Thus, in the first of the above examples, 49 × 0·23 = 11·27 ; and in the second 25 × 0·23 = 5·69, which are amounts of sugar respectively per 100 parts (*Roberts*).

In taking the densities Dr Roberts advises the

operator to employ a urinometer having a long scale, since the degrees are much further apart than in the scales of the short-stemmed instruments, and are therefore more distinct and can be more easily read off.

The following are examples of diabetic urine:

No. 1 (*Simon*).

Specific gravity	1018·00
Water	957·00
Solid constituents	43·00
Urea	Traces.
Uric acid	Traces.
Sugar	39·80
Extractive matter and soluble salts .	2·10
Earthy phosphates	0·52
Albumen	Traces.

No. 2 (*Dr Percy*).

Specific gravity	1042·00
Water	894·50
Solid constituents	105·50
Urea	12·16
Uric acid	0·16
Sugar	40·12
Extractive matters and soluble salts	53·06

No. 3 (*Bouchardat*).

Water	887·58
Solid constituents	162·42
Urea	8·27
Uric acid	Not isolated.
Sugar	134·32
Extractive matters and soluble salts	20·34
Earthy phosphates . . .	0·38

"Diabetic urine usually possesses a peculiar smell, which has been compared with that of violets, apples, new hay, whey, horses' urine, musk, and sour milk. Such comparisons serve only to show how difficult it is to give by description a correct idea of a particular odour. The colour of diabetic urine is generally pale. Sometimes, but not usually until after two or three days, the surface becomes coloured with a whitish film, owing to the development of the *sugar fungus* and the *Penicillium glaucum*, and gradually the urine becomes opalescent in consequence of these fungi multiplying in great numbers in every part of the fluid. See URINARY DEPOSITS (FUNGI).

"Diabetic urine has a sweet taste, and often numbers of flies are attracted to it, which fact sometimes leads the patient to suspect that the urine is not healthy" ('Kidney Diseases, Urinary Deposits,' &c., Dr Lionel Beale).

White merino, that has been wet with a solution of bichloride of tin, is also said to form a ready test for sugar in urine.

Albumen in urine may be detected by the nitric acid, or by the heat test. The nitric acid test is performed as follows:—Fill a test-tube to about an inch with the urine, then incline the tube and pour in strong nitric acid down the side of the tube, so that the acid sinks to the bottom and displaces the urine, which by reason of its smaller specific gravity rests above it. Let the acid be added till it forms a stratum about a quarter of an inch thick at the bottom.

If the urine contain albumen three layers will be perceptible—one, perfectly colourless, of nitric

acid at the bottom; immediately above this an opalescent zone of the coagulated albumen; and, on the top, the unaltered urine.

In his work, 'Kidney Diseases and Urinary Deposits,' Dr Lionel Beale directs attention to the very important fact that "two or three drops of nitric acid to about a drachm of albuminous urine in a test-tube will produce a precipitate of albumen which will be *dissolved on agitation*; while, on the other hand, about half as much strong nitric acid as there is of urine will redissolve the precipitate of albumen, unless the quantity present be excessive. Albumen precipitated by nitric acid is *soluble in weak nitric acid*, and in a considerable excess of urine, and it is also *soluble in strong nitric acid. It is therefore necessary in employing the nitric acid test to add from ten to fifteen drops of the strong acid to about a drachm of the urine suspected to contain albumen.*"

Dr Roberts gives the following directions for applying the heat test:—If the urine have its usual acid reaction it becomes turbid on boiling when it contains albumen, and this turbidity persists after the addition of an acid. There are two points to be remembered on using heat alone as a test for albumen. First, that albumen is not coagulated by heat when the urine is alkaline; in such cases, therefore, it is necessary before boiling to restore the acidity by a few drops of acetic acid (carefully avoiding excess). Secondly, when the urine is neutral, or very feebly acid, it may become turbid on heating, from precipitation of the earthy phosphates, but turbidity from the cause is easily distinguished from albumen by a drop of nitric or acetic acid, which instantly causes the phosphates to disappear. It may sometimes happen that the patient whose urine is to be submitted to examination for albumen may be taking large doses of nitric or hydrochloric acid. Under these circumstances Dr Bence Jones recommends the addition of ammonia to the urine, nearly to the point of neutralisation.

Mr Louis Siebold proposes a modification of Dr Roberts's method of applying the heat test in acid states of the urine, which is as follows:—Add solution of ammonia to the urine until just perceptibly alkaline, filter, and add diluted acetic acid very cautiously until the urine acquires a faint acid reaction, avoiding the use of a single drop more than is necessary. Now place equal quantities of this mixture into two test-tubes of equal size, heat one of them to ebullition, and compare it with the cold sample contained in the other test-tube. The least turbidity is thus distinctly observed, and gives absolute proof of the presence of albumen, the error of confounding phosphates with albumen being out of the question, as they are precipitated by the ammonia and removed by filtration.

M. Galipe, 'Pharm. Zeitung für Russland,' xiv, 48 ('Pharm. Journ.'), says the following is a delicate as well as trustworthy test for albuminous urine. A few drops of the urine are carefully added to a solution of picric acid contained in a small conical test-glass. If albumen be present a well-marked turbidity will be produced at the point of contact between the two liquids. On applying heat the albumen agglutinates, and rises to the surface. Phosphates and urates are said not to interfere with this test.

In order to determine the quantity of albumen in urine proceed as follows:—Add a little acetic acid to the urine, and then heat it in a water bath until it boils. Or the albuminous urine may be dropped into boiling water acidulated with acetic acid. In either case collect the precipitate on a weighed filter, wash it well, dry it, and weigh it. The albumen must afterwards be incinerated, and the resulting residue, which consists of earthy salts, must be deducted from the dried precipitate.

Stolnikow, 'Chem. Centralb.' ('Pharm. Journ.'), adopts the following method for the quantitative estimation of albumen in urine:—The urine is diluted with water until a sample poured upon some nitric acid contained in a test-tube produces still a faint white ring at the point of contact after the lapse of forty seconds. The number of volumes of water added to the volume of urine (which may be taken as one) is divided by 250, and the quotient will be the percentage of albumen in the urine. This relation has been established and confirmed by gravimetric determinations.

It is sometimes desirable to remove the albumen from the urine before proceeding to search for other substances. There are several methods of accomplishing this. If the urine be boiled the albumen will become coagulated, but in many cases it may happen, owing to the urine being slightly alkaline or neutral, that a small quantity may remain in solution. Hence it will be advisable to add a little acetic acid to the urine before applying heat to it, to remove the precipitated matters by filtration, and to exactly neutralise the acid in the filtrate. If a few crystals of sulphate of soda be heated with albuminous urine, the albumen and allied matters may be entirely removed without injury to other organic matters dissolved, and without interfering with the employment of other reagents. When it is desirable to free the urine from albumen previous to testing for sugar, this latter method will be found the best and most convenient.

The following analyses represent the amount of albumen present in the urine of two patients suffering from Bright's disease:

No. 1 (*Simon*).

Specific gravity	. .	1014·
Water	. . .	966·10
Solid constituents	. .	33·90
Urea .	. . .	4·77
Uric acid	. . .	0·40
Fixed salts .	. .	8·04
Extractive matters	. .	2·40
Albumen	. . .	18·00

No. 2 (*Dr Percy*).

Specific gravity	. . .	1020·
Water .	. . .	946·82
Solid constituents .	. .	53·18
Urea .	. . .	7·68
Uric acid and indeterminate animal matter . }		17·52
Fixed soluble salts .	. .	5·20
Earthy phosphates .	. .	0·14
Albumen	. . .	22·64

Dr Parkes records the case of a patient suffering from albuminuria, who excreted 545 grains of albumen in twenty-four hours. See URATES.

Urine frequently contains an abnormally large quantity of urea. Such urine is of high specific gravity—1·080 or more. When present in large excess the urea becomes deposited in 'sparkling crystalline lamellæ' of the nitrate, if it be mixed with an equal quantity of strong nitric acid in the cold.

The crystals vary slightly in character, according to the amount of nitric acid employed and the degree of concentration of the urine. Urine which thus yields, without previous concentration, the nitrate, is said to contain an excess of urea. See page 1727.

The quantity of urea present in urine is best determined by a process invented by Liebig. When a solution of pernitrate of mercury is added to one of pure urea, the urea and mercuric salt unite and form an insoluble compound, of under-determined constitution. If, however, the chlorides of the alkalies and alkaline earths are present, this combination does not take place, owing to the decomposition of the mercuric nitrate, and the formation of bichloride of mercury, and a nitrate of the alkali or alkaline earth, both of which are soluble. When, however, the decomposition of the chloride has been completed, the urea may be entirely precipitated, provided a sufficient quantity of mercuric nitrate be added to the solution. In estimating the amount of urea in urine, therefore, it is only necessary to add to the urine a solution of the mercuric salt of known strength, since from the quantity of this latter, which has been employed in throwing down the urea, this can easily be calculated.

In performing this analysis, three special solutions are requisite:

1. A solution consisting of one part (by measure) of a cold saturated solution of barium nitrate (also by measure) in saturated baryta water. This serves for the removal of the phosphates and sulphates, the presence of which in the urine would interfere with the analysis.

2. A standard solution of mercuric nitrate, which is made as follows:—$\frac{77}{2}$ grains of red oxide of mercury placed in a beaker are dissolved in a sufficient quantity of nitric acid (sp. gr. 1·20) by a gentle heat, and evaporated over a water-bath until all *excess of free acid* is driven off. This may be known by the liquid becoming dense and syrupy in appearance. It is then poured into a properly graduated vessel and diluted to 10,000 grain-measures. Of this solution, 10 grain-measures = 0·1 grain of urea.

3. A solution of carbonate of soda in distilled water, 20 grains to the ounce. This solution is employed to indicate when the titration is complete, and to show the operator that all the urea has been precipitated by the mercuric salt.

The operation is thus performed:

(*a*) 400 grain-measures of the clear urine are mixed with 200 grain-measures of the baryta solution, No. 1. The mixture is poured into a filter, and of the clear filtrate which passes through 150 grain-measures are carefully measured off, and poured into a small beaker. This quantity, of course, contains two thirds, or 100 grain measures of wine.

(b) A graduated burette (each division of which equals a grain-measure of water) is next filled with the solution (No. 2) of mercuric nitrate, which is then dropped into the beaker containing the filtered urine, until the mixture becomes turbid. The quantity of solution that has been required to just reach the point of turbidity is then noted down; it shows that all the chloride of sodium has been decomposed, and that the urea is now beginning to precipitate.

(c) The solution (No. 2) is now added more liberally, and thoroughly mixed with the contents of the beaker by means of a glass rod; a copious white precipitate is being formed. The operation is completed when, of course, no more precipitate is thrown down.

(d) This point is ascertained by means of the solution of carbonate of soda (No. 3), to a few isolated drops of which dotted about a white plate or slab, or placed on a watch-glass, give, when mixed by means of the stirring rod with a drop of the turbid mixture from the beaker, a yellow tinge, owing to the formation of hydrated oxide of mercury.

(e) The quantity of solution of mercuric nitrate that it has taken to produce the above reaction is then noted down, and from this the portion used before the occurrence of the turbidity is deducted, the remainder, of course, being the amount required to precipitate the urea. By bearing in mind the statement already made that 10 grain-measures of the mercurial solution indicate 0·1 grain of urea, the quantity excreted in twenty-four hours may be arrived at by a very easy and obvious calculation.

Dr Davy's Method of estimating Urea. This consists in the decomposition of a known quantity of urine by sodium hypochlorite, the amount of urea being calculated from the resulting nitrogen. A glass tube, 12 or 14 inches in height, and graduated to tenths and hundredths of a cubic inch, is filled to more than a third of its length with mercury; a measured quantity of urine, varying from a quarter of a drachm to a drachm, is next poured into the tube, which is then filled up with a solution of sodium hypochlorite (the liquor sodæ chlorinatæ of the Dublin Pharmacopœia). This latter must be poured in quickly, and the open end of the tube immediately closed with the thumb. The tube is then shaken to ensure admixture between the urine and hypochlorite, and stood with the open end downwards in a cup filled with a saturated solution of common salt; the mercury escapes into the tube, its place being filled by the solution of salt, which being heavier than the mixture of urine and hypochlorite, retains them in the upper part of the tube. The urine becomes soon decomposed, the carbonic acid, which is one of the products of its decomposition, being absorbed by the excess of chloride of sodium present, whilst the liberated nitrogen bubbles up to the top of the tube. When no more evolution of gas takes place, the volume of nitrogen is read off, and from its amount the quantity of urea present in the amount of urine experimented upon is calculated: one fifth of a

grain of urine=0·3096 parts of a cubic inch of nitrogen at 60° F. and 30″ barometric pressure.

Determination of the Water. The amount of water in any sample of urine may be determined by weighing 1000 gr. of the recently excreted urine into a counterpoised platinum or porcelain dish, and ascertaining the loss it has undergone after evaporation to dryness. The operation should be performed as speedily as possible. The best plan is to concentrate the urine in a water-bath, the evaporation should be continued *in vacuo* over strong sulphuric acid, until the weight of the residue remains constant. By way of control, another sample of the same urine, consisting of 500 gr., may be operated upon at the same time, and under the same conditions.

URINARY DEPOSITS, &c. These differ from the albumen, sugar, bile, &c., previously described, in being insoluble in abnormal urine. Sometimes they are diffused throughout the whole body of the urine, when they give it an opaque appearance. Sometimes they may be met with floating on its surface; at others they are only partially diffused through the fluid, frequently in the form of a transparent or opaque cloud, when they occupy a considerable space; whilst very often they occur in a crystalline or granular form, deposited sometimes at the bottom and sometimes at the sides of the vessel holding the urine.

Of the numberless insoluble substances met with in urine, both in health and disease, our limits will only permit us to notice those which are most important and of frequent occurrence.

For the detection of the generality or these the microscope is indispensable. An instrument magnifying 200 to 220 diameters (½ of an inch objective) will generally be found sufficiently powerful, and in some instances an inch objective, magnifying 40 diameters (as in the larger forms of crystalline deposit) will answer all the purposes.

Some of the varieties of these deposits admit of a double examination, viz. a microscopical and chemical one. When this is the case, the particulars applying to each kind of investigation will be given.

Mucus. Mucus is always present in small quantity in healthy urine, in which it shows itself within a few hours after the urine has been excreted in the shape of a transparent cloud towards the bottom of the vessel containing the urine.

Pus. The presence of pus in urine is indicated by an opaque, more or less bulky, cream-like deposit at the bottom of the vessel holding the urine, to which some separated pus globules, finding their way to the supernatant liquid, give an appearance of slight turbidity. By shaking the vessel the whole of the liquid becomes turbid, owing to the equal dissemination through it of the pus globules. The pus again deposits on standing. A small quantity of albumen is always met with in the clear part of urine which contains pus; the albumen being derived from the *liquor puris*, the liquid by which the pus-corpuscles are surrounded.

Whenever it can be obtained in sufficient quantity, pus should always be examined chemically, as follows:—The supernatant urine being decanted, the suspected sediment is shaken up with liquor potassæ, when if it become converted into a gelatinous, viscid substance, incapable of being dropped from the tube, and when poured from it running as a slimy and almost continuous mass, it may be pronounced pus. This same gelatinous viscid mass is met with in alkaline urines containing pus, adhering to the sides of the vessel in which the urine is placed, where it has been formed by the action of the carbonate of ammonia (caused by the decomposition of the urea) upon the pus. The reaction upon the pus is the same as that which takes place when liquor potassæ is employed. The stringy viscid substance due to the last cause is frequently, but erroneously, termed *mucus*.

In urines containing pus, the clear portion should always be examined for albumen, since where this is found, except in small amount, some form of kidney disease may be suspected.

But it sometimes happens that the pus is present in such small quantity in the urine as to preclude its chemical examination. Under these circumstances, recourse must be had to the microscope. Dr Lionel Beale says, "Pus-globules, which have been long removed from the body, always have a granulated appearance in the microscope, and, when fresh, do not always exhibit a well-defined nucleus ; the outline is usually distinct and circular, but it is finely crenated. Upon the addition of acetic acid the globule increases somewhat in size, becomes spherical, with a smooth, faint outline, and from one to four nearly circular bodies are developed in the centre of each. If the pus-corpuscles have lain some days in the urine they will have undergone complete disintegration."

Epithelium. A great many varieties of epithelium, derived from different parts of the kidneys, ureters, bladder, urethra, vagina, &c., are more or less present in urine. A few of these are given in the accompanying *engr.* In the various diseases peculiar to the urinary and genito-urinary organs the quantity of epithelium present in the urine is frequently considerable, and as in some cases it presents itself in an imperfect or disintegrated form, its identification, except to the experienced microscopist and physiologist, becomes a matter of great difficulty.

Casts. Casts or moulds which have been formed in the tubes of the kidneys, or in the uterus and vagina, are constantly finding their way into the urine of persons affected with acute or chronic renal diseases and uterine affections. They are very varied both in character and appearance, and difficult of recognition, except by the skilled microscopist and pathologist.

Blood-corpuscles. These, when present in quiescent urine, occur as a sediment at the bottom of the vessel. Some few globules, however, are diffused throughout the supernatant urine, and impart to it a smoky appearance, if the fluid have a marked acid reaction ; whereas if the reaction be alkaline the corpuscles assume a bright red colour.

In the accompanying plate the three upper groups represent blood-corpuscles taken from the human body; the three lower, those found in urine. Of these latter some will be seen to have

lost their circular outline, and to have become jagged or crenated. In some cases, on the contrary, they swell and become much enlarged. These changes in appearance take place when the blood has remained for some time in the urine, and appear to be due to the forces of endosmose and exosmose.

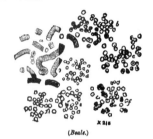

(*Beale.*)

Fungi. The chief vegetable organisms found in urine are the *sugar fungus* and the *Penicillium glaucum.* The sugar fungus is precisely the same as the yeast plant (the *Torula cerevisiæ*). The *Penicillium* is very frequently present in albuminous urine, with an acid reaction, as well as in diabetic.

Uric Acid. See *above.*

Urates. According to Bence Jones the soluble urates met with in healthy urine consist of uric acid, potassium, ammonium, and sodium.

In abnormal urine the urates of ammonium and sodium sometimes occur, the latter, which are the more general, presenting under the microscope the appearance shown below.

Urate of sodium is, however, much more common in the urine of children than of adults, when it presents itself in the form of spherical crystals.

In both cases the urates are associated with uric acid (resulting from their partial decomposition), represented by the small spiked crystals

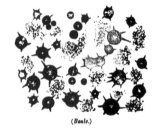

(*Beale.*)

protruding from the spheres in the form of needle-shaped crystals. Urate of sodium occurs as the concretions known as 'chalk-stones' in gout. But by far the most abundant kind of urates met with in abnormal urine is that known as *amorphous urates*, which constitute the most common variety of urinary deposits.

Urate of soda in a globular form, commonly found in the urine of children.

Heintz states that they are a mixture of urate of sodium with small quantities of the urates of ammonium, lime, and magnesium. They are very frequently seen in the urine of persons in excellent health, in which, owing perhaps to too abundant or nitrogenous diet and an insufficiency of muscular exercise, being in excess, they are thrown down when the urine cools.

An excess of the amorphous urates in urine, like the presence of pus and phosphates, is indicated by the bulky precipitate more or less diffused throughout the vessel containing the urine. A very easy test will decide as to which of the three classes of substances (if only one of them be present) the precipitate belongs. The supernatant fluid being decanted from the deposit, about an equal bulk of liquor potassæ is added to the latter, when one of three results will ensue:

1. If it be *pus*, and become viscid, it will exhibit the qualities already mentioned under the description of that substance.

2. If *phosphates*, no alteration will ensue.

3. If *amorphous urate*, it will at once dissolve.

When amorphous urates are uniformly distributed throughout the urine they give it a milky appearance, which may sometimes lead to its being mistaken for *chylous* urine, or urine throughout which fatty particles of chyle are diffused. This latter doubt, however, may be easily set at rest by gently heating it. If the turbidity is owing to the urate it will disappear; if to chyle it will remain.

If the amorphous urate be decomposed by a little hydrochloric acid, it will yield uric acid, easily recognised by its characteristic form under the microscope, or when treated with nitric acid and ammonia will answer to the annexed test.

It sometimes happens that in testing an acid urine suspected to contain albumen, the urine may contain so large an amount of uric acid in solution that, upon adding a drop of nitric acid to it, a bulky precipitate of uric acid exactly resembling albumen is thrown down, and it may be erroneously regarded as this substance if examined under the microscope immediately upon its formation. Upon being allowed, however, to stand some time, and then placed under the microscope, the well-known crystals of the acid will reveal themselves.

In such urine no precipitate takes place when the liquid is heated—another essential feature in which it differs from albumen.

Phosphates. The urinary earthy phosphates occur under two varieties, viz. the phosphate of ammonia and magnesia, known as the triple phosphate, and the phosphate of lime.

In the following *engravings* the principal crystalline forms of the triple phosphate are shown.

Of these the triangular prismatic, with the truncated extremities, is the most common. In some cases the prisms are so much reduced in length as to resemble the octahedral crystals of oxalate of lime, for which they are sometimes mistaken by the inexperienced. When any doubt exists on this point it must be set at rest by having recourse to the chemical tests given further on. The triple phosphate is rarely met with alone, urate of ammonia, and sometimes uric acid and oxalate of lime, being present, although generally occurring in neutral or alkaline urine. The triple acid is sometimes found in that which is acid.

When ammonia is added to fresh urine the triple phosphate is precipitated, and if it be then examined by the microscope it will be found to consist of beautiful stellate crystals, and to form a most attractive object. The presence of phosphoric acid can be demonstrated by the ordinary reagents.

FIG. 1.

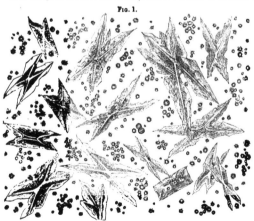

Crystals of triple phosphate, with spherules of urate of soda (*Beale*).

Phosphate of lime dissolves in strong acids without effervescence. The presence of lime, as well as of phosphoric acid, can easily be verified by the usual tests.

Oxalate of Lime. The principal crystalline forms of oxalate of lime, when it occurs as a urinary deposit, are the octahedral and the dumb-bell. Of these the most common is the octahedral. These octahedra (which have one axis much shorter than the other two) vary considerably in size; but there is reason to believe that the diversity in appearance which they exhibit is due to crystals of precisely the same shape occupying different positions as to the direction of their axes, when examined by the microscope. There are a great many diversities of the dumb-bell form of oxalate of lime, which seem to be derived from circular and oval crystals. The subjoined *engrs.* illustrate the varieties of crystalline oxalates the most generally met with.

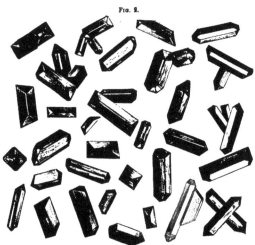

Fig. 2.

Crystals of triple phosphate, with triangular prisms with truncated extremities (*Beale*).

Oxalate of lime (*Beale*).

When the crystals of oxalate are extremely minute they are very liable to be overlooked, since they then appear as almost opaque cubes, and may not unnaturally be taken for urate of soda, to which they bear no slight resemblance, but from which they differ by being insoluble in potash or acetic acid, and not dissolving on the application of heat. We have already alluded to their resemblance to the dumb-bells of the earthy phosphates. Another distinctive feature is that the oxalates rarely sink to the bottom of the vessel, but are diffused through the mucous cloud, which forms in urine after a short time.

Cystine. Cystine is an occasional ingredient in urine, when it occurs as a whitish precipitate crystallised in hexagonal plates. At other times, but not so frequently, it is met with dissolved in the urine. It may be separated from the urine holding it in solution by the addition of an excess of acetic acid. Under the microscope cystine bears somewhat of a resemblance to uric acid, from which, however, it differs when under treatment with ammonia. When ammonia is added to cystine the cystine dissolves, but by the spontaneous evaporation of the ammonia remains behind in its original form; whilst, if the ammonia be allowed to escape under the same circumstances from the urate of ammonia which has been formed, this remains behind as an amorphous mass. Ammonia, therefore, dissolves the cystine without entering into chemical union with it. Potash also readily dissolves cystine, as do also oxalic acid and the strong mineral acids. It is, however, insoluble in boiling water, in weak hydrochloric acid, and, as we have seen, in acetic acid.

Obs. In the examination of urine it is important that the investigation should be conducted upon a portion taken from *the whole of the urine excreted during twenty-four hours*, and not on an isolated quantity voided at any particular time.

The compiler of the present article has to acknowledge his indebtedness to Dr Lionel Beale's very valuable and exhaustive work, 'Kidney Diseases, Urinary Deposits, &c.,' as well as to Dr W. Roberts's excellent book, 'Urinary and Renal Diseases,' to both of which volumes the reader, desirous of further and more explicit information on the subject, is referred.

Mr A. W. Gerrard communicates to the 'Lancet' a note on an improved form of burette called a glycosometer, for estimating the amount of sugar in urine, which is based upon a reversal of the usual method, the urine in this case being placed in the burette and the Fehling's solution in the boiling-dish. As 10 c.c. of Fehling's solution is equal to 0·5 grm. of glucose it follows that the following are equivalent figures:

Amount of sugar in urine. Per cent.	Volume of urine (1 in 20) required to reduce 10 cub. cent. of Fehling's solution. Cub. cent.
10·0	10·0
9·5	10·52
9·0	11·1
8·5	11·76
8·0	12·5
7·5	13·33
7·0	14·28
6·5	15·38
6·0	16·66
5·5	18·18
5·0	20·0
4·5	22·22
4·0	25·0
3·5	28·57
3·0	33·33
2·5	40·0
2·0	50·0
1·5	66·66
1·0	100·0

The burettes are graduated in percentages, and there is a pair of them clasped by a pair of swinging arms, supported by a shoulder fixed to an upright brass stand; the swinging arrangement allows the burettes to be moved at will, so as to be brought over the dish containing the Fehling's solution. To graduate the instrument it is filled with water and marked with a 0 line, then the volumes of water represented by the right-hand column of figures in the table are withdrawn in the order of their sequence, the levels after each withdrawal being marked with their proper percentage, as shown to the left. As the range of percentage from 1 to 10 would necessitate a very large burette, two are employed, one thin and narrow for high percentages, the other of larger capacity for low percentages.

URINOM'ETER. An hydrometer adapted to determining the density of urine. That of Dr Prout is the simplest and best. Urinometers should always be tested by placing them in distilled water at 60° F. from 1·015 to 1·025 (*Beale*).

UREOMETER, Gerrard's. This is a simple form of apparatus for estimating urea, the operation being completed in less than five minutes. The determinations are scientifically accurate, no calculations being needed, as the result of the analysis is read at once in per cents of urea (*vide* 'Diseases of the Kidney, and Morbid Conditions of the Urine, dependent on Functional Derangements,' by C. H. Ralfe, M.A., M.D.Cantab., F.R.C.P.Lond.).

Method of Using (see *engr.*). Pour into the tube 5 c.c. of the urine to be examined, and in the bottle (*a*) 25 c.c. or 6 fl. dr. of sodium hypobromite solution. Place the tube carefully inside the bottle, as shown in the illustration, avoiding spilling

any of the contents. Fill the glass tubes (*b, c*) with water, so that the level reaches the zero line, taking care that when this is done the tube *c* contains only a little water by being placed high—it having to receive what is displaced from *c* by the nitrogen evolved. Now connect the india-rubber tubing to the bottle, and noting, lastly, that the water is exactly at zero, upset the contents of the tube into the hypobromite solution. Nitrogen is evolved, and depresses the water in *b*. When this ceases, lower *c* until the level of the water in both tubes is equal. To be exact, dip *a* into cold water to cool the gas before taking a reading, and note the result, which shows percentage of urea.

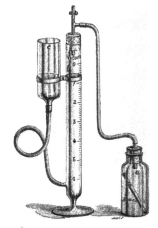

The solution of hypobromite of soda is made by dissolving 100 grms. of caustic soda in 250 c.c. of water, then adding 22 c.c. of bromine.

To avoid the danger of the bromine vapour, the bromine is sold in hermetically sealed glass tubes, containing 2·2 c.c.; one of these placed in the large bottle with 25 c.c. of the soda solution gives, when broken with a sharp shake, the exact quantity of hypobromite for one estimation of urea, and all bad odour is avoided.

UREN POWDER. Crocus martis, or jeweller's rouge.

URTICAR'IA. See RASH.

US'QUEBAUGH. *Syn.* ESCUBAC. Literally, mad water, the Irish name of which 'whisky' is a corruption. At the present time it is applied to a strong cordial spirit, much drunk in Ireland, and made in the greatest perfection at Drogheda.

Prep. 1. Brandy or proof spirit, 3 galls.; dates (without their kernels) and raisins, of each, bruised, ¼ lb.; juniper berries, bruised, 1 oz.; mace and cloves, of each, ½ oz.; coriander and aniseed, of each, ½ oz.; cinnamon, ¼ oz.; macerate, with frequent agitation, for 14 days, then filter, and add of capillaire or simple syrup, 1 gall.

2. Pimento and caraways, of each, 8 oz.; mace, cloves, and nutmegs, of each, 2 oz.; aniseed, corianders, and angelica root, of each, 8 oz.; raisins, stoned and bruised, 14 lbs.; proof spirit, 9 galls.; digest as before, then press, filter, or clarify, and add of simple syrup, q. s. Should it turn milky, add a little strong spirit, or clarify it with alum, or filter through magnesia.

Obs. Usquebaugh is either coloured yellow with saffron (about ¼ oz. per gall.), or green with sap-green (about ½ oz. per gall.); either being added to the other ingredients before maceration in the spirit.

UVA URSI. The *Arctostaphylos uva-ursi* (the Bearberry) is an indigenous plant, the leaves of which are employed in medicine. Bearberry leaves contain a large percentage of tannic acid, a small quantity of gallic acid, some resin, and a little volatile oil and extractive, together with a crystallisable principle named *arbutin*, which is said to be a very powerful diuretic. Another crystallisable resinous body named *arbutin* has also been discovered in them. Bearberry leaves, either in the form of powder, infusion, or extract, are chiefly used in chronic diseases of the bladder in which there is an abnormal secretion of mucus, such as *catarrhus vesicæ*, but neither acute nor active inflammation.

VACCINATION. See COW-POX (POX).

VACCINE MATTER. *Syn.* LYMPHIA VACCINLÆ, L. This is collected either upon the points of lancet-like pieces of ivory, or by opening the pustule, and applying a small glass ball and tube (like those called by the boys in London candle-pops, or fire-pops) to the orifice, expelling part of the air in the ball by bringing a lighted taper near it; then, withdrawing the taper, the matter is sucked into the ball, in which it may be sealed up hermetically or cemented, and thus kept for a length of time. It is, however, now generally preserved between two small pieces of glass, or in straight capillary glass tubes. It is said that cotton thread is a convenient and efficient vehicle. The matter may be liquefied with a little clean water before application. A degree of heat scarcely higher than that of the blood lessens its efficacy.

VACUUM. Empty space; a portion of space void of matter. For experimental and manufacturing purposes, a sufficient vacuum is produced either by means of the air-pump, or by filling an enclosed space by steam, which is then condensed by the application of cold. Evaporation proceeds much more rapidly, and liquids boil at much lower temperatures in an exhausted receiver than when exposed to the air. Thus, under ordinary circumstances, in the air, ether boils at 96°, alcohol at 177°, and water at 212° F.; but *in vacuo* water boils at about 88°, alcohol at 56°, and ether at —20° F. In the best vacuum obtainable by a powerful air-pump, water placed over oil of vitriol to absorb the aqueous vapour as it forms, will often enter into violent ebullition whilst ice is in the act of formation on its surface. The reduction of the boiling-point with reduced pressure is practically taken advantage of by the pharmaceutist in the preparation of extracts, by the sugar refiner in the evaporation of his syrups, by the

distiller in the production of certain liqueurs, and by the chemist in a variety of processes of interest or utility. See EXTRACTS, EVAPORATION, REFRIGERATION, &c.

VALE'RIAN. *Syn.* VALERIANÆ RADIX (B., P.), VALERIANÆ RADIX, VALERIANA (Ph. L., E., & D.), L. " The root of the wild plant *Valeriana officinalis*, Linn., or wild valerian " (Ph. L.). An excitant, antispasmodic, tonic, and emmenagogue, not only acting on the secretions, but exercising a specific influence over the cerebro-spinal system, and in large quantities producing agitation, mental exaltation, and even intoxication.—*Dose*, 10 to 30 or 40 gr., thrice daily; in hysteria, epilepsy, headache (affecting only one side), morbid nervous sensibility, &c. Even the odour of it exerts a species of fascination over cats.

VALERIAN'IC ACID. $C_5H_{10}O_2$. *Syn.* VALERIC ACID; ACIDUM VALERIANICUM, A. VALERICUM, L. Chemically speaking, there are four valeric acids, viz. normal valeric acid (propyl acetic), isovaleric acid (isopropyl acetic), secondary valeric acid (methyl-ethyl acetic), tertiary valeric acid (trimethyl acetic). The valerianic acid of the shops is the isovaleric acid. It is commonly prepared as follows:—1. A mixture of potato oil or corn-spirit oil (hydrated oxide of amyl) with about 10 times its weight of quick-lime and hydrate of potassium in equal proportions, placed in a glass flask, is kept heated to about 400° F., for 10 or 12 hours, by means of a bath of oil or fusible metal; the nearly white solid residuum is mixed with water, an excess of sulphuric acid added to the mixture, and the whole subjected to distillation; the distillate is supersaturated with potash, evaporated nearly to dryness, to dissipate any undecomposed potato oil, and then mixed with weak sulphuric acid in excess; a light oily liquid separates, which by cautious rectification yields at first water containing a little acid, and afterwards pure monohydrated valerianic acid, which is identical with that prepared from valerian root.

2. (Ph. D.) See VALERIANATE OF SODIUM. This is a most economical process.

Prop., &c. A limpid oily liquid, smelling strongly of valerian root; it has an acid taste and reaction, and leaves a sensation of sweetness and a white spot on the tongue; is inflammable; boils at 175° C.; is freely soluble in alcohol and ether; dissolves in 30 parts of water, and forms salts called valerianates, most of which have a sweetish taste, are soluble, and uncrystallisable; sp. gr. 0·947 at 0° C.; placed in contact with water, it absorbs a portion of it, and is converted into the terhydrated acid, with increase of sp. gr. and reduction of the boiling-point.

VALE'RIC ACID. See VALERIANIC ACID.

VALO'NIA. The cup of a large species of acorn, imported from the Levant. Used in tanning leather.

VANA'DIUM. V=51·2. A rare metal discovered by Sefstom, in 1830, in some Swedish iron extracted from an iron mine near Jönköping. It has since been found in a vanadinite lead ore met with in Scotland, Zimpanan in Mexico, and Chili, and in the iron slag of Staffordshire. Of late years a more abundant source of vanadium has been discovered by Professor Roscoe in the cupriferous stratum of the new red sandstone at Alderley Edge in Cheshire. There are four, and possibly five oxides of this element.

Vanadic Acid. V_2O_5. *Syn.* VANADIC ANHYDRIDE, TEROXIDE OF V.; ACIDUM VANADICUM, L. *Prep.* (*Johnston.*) From the native vanadate of lead, by dissolving it in nitric acid, passing sulphuretted hydrogen through the solution, to throw down lead and arsenic, filtering, and evaporating the resulting blue liquid to dryness; the residuum is then dissolved in a solution of ammonia, and a piece of sal-ammoniac, considerably larger than can be dissolved, introduced; as the latter dissolves, a pulverulent precipitate of vanadate of ammonium is formed, which must be washed, first in a solution of sal-ammoniac, and then in alcohol of 0·860; by exposing this salt, in an open platinum crucible, to a heat a little below redness, and keeping it constantly stirred, until it acquires a dark red colour, pure vanadic acid is obtained.

Prep., &c. Vanadic acid is orange-coloured, scarcely soluble in water, and forms, with the alkaline bases, soluble salts called vanadates; and with the other bases sparingly soluble salts. All of these have an orange or yellow colour. "Vanadate of ammonia mixed with solution of galls forms a black fluid, which is the best writing ink hitherto known. The quantity of salt required for this purpose is very small; the writing is perfectly black, and not obliterated by alkalies, acids, chlorine, or other reagents" (*Ure*).

Vanadic Oxychloride. *Syn.* VANADIC OXY-TRICHLORIDE. $VOCl_3$. Roscoe states there are several oxychlorides of vanadium, which, however, have not been studied. The most interesting of them is the oxytrichloride, which corresponds to the phosphorous oxychloride. This oxytrichloride is a yellow fuming liquid, which is instantly decomposed by water into vanadic and hydrochloric acids. The oxytrichloride may be obtained by heating vanadic anhydride and charcoal (mixed together) in a current of hydrogen, after which it is heated in a current of dry chlorine. An easier method is by passing dry chlorine over the sesquioxide of vanadium.

Vanadic Pentoxide. *Syn.* VANADIC ANHYDRIDE. V_2O_5. At a red heat this oxide fuses, and on cooling, crystallises in rhombic prisms. It is but little soluble in water; the aqueous solution, which is of a yellow tint, is strongly acid, and produces a marked reddening effect on litmus. Vanadic anhydride forms both normal and acrid salts. The ammonic vanadiate (Roscoe's meta-vanadiate) is the chief source of the acid. This salt may be obtained by adding pieces of sal-ammoniac to a crude solution of potassic vanadiate; the resulting ammonic vanadiate, being insoluble in a saturated solution of sal-ammoniac, is deposited in small crystalline grains. The vanadic anhydride may be obtained from the ammonic vanadiate by heating an aqueous solution of the salt in the open air, when the ammonia is driven off, and the vanadic anhydride is left behind. The acid ammonic vanadiate, mixed with tincture of galls, makes a very durable writing ink, unacted upon either by alkalies or chlorine. Acids turn such blue without, however, destroying it.

Vanadic Trioxide, V_2O_3, is the *Vanadyl* of

Roscoe, who obtained it in the form of a grey metallic-looking powder, by the transmission of a current of dry hydrogen charged with the vapours of oxychloride of vanadium through a tube containing ignited charcoal. It dissolves in dilute acids, with evolution of hydrogen. Solutions of its salts are lavender coloured. Berzelius regarded this oxide as a metal.

Tests. The vanadiates mostly occur of a red or yellow colour. When treated with sulphuretted hydrogen they yield a solution of a fine blue colour, a reaction that distinguishes them from the chromates, which, under similar treatment, would give a green liquid. When mixed with borax and exposed to the reducing flame of the blowpipe, compounds containing vanadium give a green gloss, which turns to yellow in the oxidising flame. Professor Roscoe, to whose researches we are indebted for all the chemical knowledge we possess respecting vanadium, says:—" All the main facts now established in connection with the chemical department of this element prove it to bear a strong analogy to the elements phosphorus and arsenic; in fact, it occupies a previously vacant place in a well-defined group of triad, or, as some chemists prefer to consider them, pentad elements. There is a property of vanadium in virtue of which it may ultimately obtain considerable importance in the arts, though in the present infancy of the history of the metal it is difficult to foretell this with any certainty. This property is the power of forming a permanent black for dyeing purposes. The black produced by the action of vanadium has the advantage over copper and aniline blacks, viz. that it is permanent, whereas the latter are liable to turn green. This application of an element that was first introduced into notice as a chemical curiosity furnishes one more example of the importance of original scientific investigation. However far a newly discovered substance may seem to be removed from purposes of practical utility, we never know at what moment it may be turned to account for the benefit of the human race."

VANIL'LA. *Syn.* VANILLE, Fr. The dried pods of various species of *Vanilla*, a genus of the Nat. Ord. ORCHIDACEÆ. It is chiefly used in the manufacture of chocolate and perfumery. As a medicine it is much employed on the Continent as an aromatic stimulant and narcotic.—*Dose*, 6 to 12 gr.; in asthenic fevers, hysteria, hypochondriasis, impotency, &c.

Vanilla is reduced to powder (PULVIS VANILLÆ; POUDRE DE VANILLE) by slicing it, and triturating the fragments with twice or thrice their weight of well-dried lump sugar. For SUCRE DE VA-NILLE, 11 parts of sugar are employed.

Vanilla Extract. Musk greatly improves the flavour of vanilla, and a good flavouring essence may be made as follows:—Take of vanilla, 1 oz.; musk, 2 gr.; carbonate of potash, 1 scruple; put into a pint flask and pour on 1 oz. of tepid water; when cold add 10 oz. of rectified spirit and 5 oz. of water. Allow to stand for at least 14 days before filtering.

The following table, given by Messrs Tiemann and Harmann in the 'Journal of the Berlin Chemical Society,' represents the quantities of vanillin (the aromatic principle of vanilla) contained in that substance, as obtained from different sources:

	Vanillin per cent.
Mexican vanilla (1873 harvest)	1·69
„ „ (1874 harvest)	1·86
„ „ (medium quality)	1·32
Bourbon, best quality (1874_75)	1·91
„ (1874_75)	1·97
„ „	2·90
„ small medium (1874_75)	1·55
Java, best quality (1873)	2·75
„ „ (1874)	1·56

VANILLIN. A crystallised substance obtained from pine juice by Messrs Tiemann and Harmann. It has been shown to be identical with the aromatic principle of vanilla.

In a paper read before the Royal Society the authors have described the process by which vanillin was artifically prepared by them. They state that the sap of the cambium of coniferous trees contains a beautiful crystalline glucoside, coniferine, which was discovered by Kartig, and examined some years ago by Kubel, who arrived at the formula $C_{24}H_{22}O_{12} + 3Aq$. A minute study of this compound leads us to represent the molecule of coniferine by the expression $C_{16}H_{22}O_9 + 2Aq.$, the percentages of which nearly coincide with the theoretical values of Kubel's formula.

Submitted to fermentation with emulsine, coniferine splits into sugar and a splendid compound, crystallising in prisms which fuse at 73°. This body is easily soluble in ether, less so in alcohol, almost insoluble in water; its composition is represented by the formula $C_{10}H_{12}O_3$. The change is represented by the equation—

$$C_{16}H_{22}O_9 + H_2O = C_6H_{12}O_6 + C_{10}H_{12}O_3.$$

Under the influence of oxidising agents, the product of fermentation undergoes a remarkable metamorphosis. On boiling it with a mixture of potassium bichromate and sulphuric acid, there passes with the vapour of water in the first place ethylic aldehyd, and subsequently an acid compound soluble in water, from which it may be removed by ether. On evaporating the ethereal solution, crystals in stellar groups are left behind, which fuse at 81°. These crystals have the taste and odour of vanilla.

An accurate comparative examination has proved them to be identical with the crystalline substance which constitutes the aroma of vanilla, and which is often seen covering the surface of vanilla pods.

On analysis, the crystals we obtained were found to contain $C_8H_8O_3$. This is exactly the composition which recent researches of Carles have established for the aromatic principle of vanilla. The transformation of the crystalline product of fermentation into vanillin is represented by the following equation:—

$$C_{10}H_{12}O_3 + O = C_2H_4O + C_8H_8O_3.$$

To remove all doubt regarding the identity of artificial vanillin with the natural compound, we have transformed the former into a series of salts, which have the general formula $C_8H_7MO_3$, and into two substitution products, $C_8H_7BrO_3$, and $C_8H_7TO_3$, both of which had previously been prepared by Carles from the natural compound.

VAN SWIETEN'S SOLUTION. Contains $\frac{1}{1000}$

part of its weight of corrosive sublimate, or ½ gr, per fl. oz.

VAPOUR. Vapours are really gases, and amenable to substantially the same physical laws; as ordinarily understood, however, the difference between a gas and a vapour is the following :— A gas is a form of matter which exists, at ordinary temperatures and pressures, in a state of vapour; whilst a vapour has been formed by the application of heat to a body usually existing in the solid or liquid form; gases, therefore, differ from vapours only in being derived from bodies which, in the solid or liquid form, boil at very much lower temperatures. See INHALATIONS.

VARICOSE VEINS. See VARIX.

VARIX. The permanent unequal dilation of a vein or veins, which are then said to be 'varicose.' It is known by the presence of a soft tumour, which does not pulsate, and often assumes a serpentine figure. Varicose veins of the groin and scrotum generally form a collection of knots. The treatment consists of cold applications and pressure from bandages. Some cases are relieved by ligature. When occurring in the legs much standing or walking should be avoided, and the use of the elastic stockings made for the purpose will be proper. Professional advice should be sought at once in all cases of varicose veins; much intolerable suffering may be saved by so doing.

VARNISH. *Syn.* VERNIS, Fr. Any liquid matter which, when applied to the surface of a solid body, becomes dry, and forms a hard, glossy coating, impervious to air and moisture.

Varnishes are commonly divided into two classes—FAT or OIL VARNISHES and SPIRIT VARNISHES. The fixed or volatile oils, or mixtures of them, are used as vehicles or solvents in the former, and concentrated alcohol in the latter (methylated spirit is now generally used for making spirit varnishes in place of duty-paid alcohol). The sp. gr. of alcohol for the purpose of making varnishes should not be more than ·8156 (=·67 o. p.), and it should be preferably chosen of even greater strength. A little camphor is often dissolved in it to increase its solvent power. The oil of turpentine, which is the essential oil chiefly employed for varnishes, should be pure and colourless. Pale drying linseed oil is the fixed oil generally used, but poppy oil and nut oil are also occasionally employed. Among the substances which are dissolved in the above menstrua are amber, anime, copal, elemi, lac, mastic, and sandarach, to impart body and lustre; benzoin, on account of its agreeable odour; annotta, gamboge, saffron, Socotrine aloes, and turmeric, to give a yellow colour; dragon's blood and red sandal-wood, to give a red tinge; asphaltum, to give a black colour and body; and caoutchouc, to impart toughness and elasticity.

In the preparation of spirit varnishes care should be taken to prevent the evaporation of the alcohol as much as possible, and also to preserve the portion that evaporates. On the large scale a common still may be advantageously employed, the head being furnished with a stuffing-box, to permit of the passage of a vertical rod, connected with a stirrer at one end and a working handle at the other. The gum and spirit being introduced, the head of the still closely fitted on and luted, and the connection made with a proper refrigerator, heat (preferably that of steam or a water-bath) should be applied, and the spirit brought to a gentle boil, after which it should be partially withdrawn, and agitation continued until the gum is dissolved. The spirit which has distilled over should be then added to the varnish, and after thorough admixture the whole should be run off, as rapidly as possible, through a silk-gauze sieve into stone jars, which should be immediately corked down, and set aside to clarify. On the small scale spirit varnishes are best made by maceration in closed bottles or tin cans, either in the cold or by the heat of a water-bath. In order to prevent the agglutination of the resin it is often advantageously mixed with clean siliceous sand or pounded glass, by which the surface is much increased, and the solvent power of the menstruum greatly promoted.

To ensure the excellence of oil varnishes one of the most important points is the use of good drying oil. Linseed oil for this purpose should be very pale, perfectly limpid or transparent, scarcely odorous, and mellow and sweet to the taste. 100 galls. of such an oil is put into an iron or copper boiler, capable of holding fully 150 galls., gradually heated to a gentle simmer, and kept near that point for about 2 hours, to expel moisture; the scum is then carefully removed, and 14 lbs. of finely pulverised scale litharge, 12 lbs. of red-lead, and 8 lbs. of powdered umber (all carefully dried and free from moisture) are gradually sprinkled in; the whole is then kept well stirred, to prevent the driers sinking to the bottom, and the boiling is continued at a gentle heat for about 3 hours longer; the fire is next withdrawn, and after 30 to 40 hours' repose, the scum is carefully removed, and the clear supernatant oil decanted from the 'bottoms.' The product forms the best boiled or drying oil of the varnish maker. Another method is to heat a hogshead of the oil gradually for 2 hours, then to gently simmer it for about 3 hours longer, and, after removing the scum, to add, gradually, 1 lb. of the best calcined magnesia, observing to mix it up well with the oil, and afterwards to continue the boiling pretty briskly for at least an hour, with constant agitation. The fire is then allowed to die away, and, after 24 hours, the oil is decanted as before. The product is called 'clarified oil,' and requires to be used with driers. It should be allowed to lay in the cistern for 2 or 3 months to clarify.

In the preparation of oil varnishes, the gum is melted as rapidly as possible, without discolouring or burning it; and when completely fused, the oil, also heated to nearly the boiling-point, is poured in, after which the mixture is boiled until it appears perfectly homogeneous and clear, like oil, when the heat is raised, and driers (if any are to be used) gradually and cautiously sprinkled in, and the boiling continued, with constant stirring, for 3 or 4 hours, or until a little, when cooled on a palette knife, feels strong and stringy between the fingers. The mixture is next allowed to cool considerably, but while still quite fluid, the turpentine, previously made moderately hot, is cautiously added, and the whole thoroughly incorporated. The varnish is then run through a filter or sieve into stone jars, cans, or other vessels, and set aside to clarify itself by subsidence. When no driers are used, the mixture of oil and gum is boiled until it runs perfectly clear, when it is removed from the fire, and, after it has cooled a little, the turpentine is added as before.

It is generally conceived that the more perfectly the gum is fused, or run, as it is called, the larger and stronger will be the product; and the longer the boiling of the 'gum' and oil is continued, within moderation, the freer the resulting varnish will work and cover. An excess of heat renders the varnish stringy, and injures its flowing qualities. For pale varnishes as little heat as possible should be employed throughout the whole process. Good body varnishes should contain 1½ lbs.; carriage, wainscot, and mahogany varnish, fully 1 lb.; and gold size and black japan, fully ½ lb. of gum per gall., besides the asphaltum in the latter. Spirit varnishes should contain about 2½ lbs. of gum per gall. The use of too much driers is found to injure the brilliancy and transparency of the varnish. Copperas does not combine with varnish, but only hardens it; sugar of lead, however, dissolves in it to a greater or less extent. Boiling oil of turpentine combines very readily with melted copal, and it is an improvement on the common process, to use it either before or in conjunction with the oil, in the preparation of copal varnish that it is desired should be very white. Gums of difficult solubility are rendered more soluble by being exposed, in the state of powder, for some time to the air.

Varnishes, like wines, improve by age, and should always be kept as long as possible before use.

From the inflammable nature of the materials of which varnishes are composed, their manufacture should be only carried on in some detached building of little value, and built of uninflammable materials. When a pot of varnish, gum, or turpentine catches fire, it is most readily extinguished by closely covering it with a piece of stout woollen carpeting, which should be always kept at hand, ready for the purpose.

An excellent paper, by Mr J. W. Niel, on the manufacture of varnishes, will be found in the 'Trans. of the Soc. of Arts,' vol. xlix. See also the articles ALCOHOL, AMBER, COPAL, OILS, &c., in this work.

Varnish, Am'ber. *Prep.* 1. Take of amber (clear and pale), 6 lbs.; fuse it, add of hot clarified linseed oil, 2 galls.; boil until it 'strings well,' then let it cool a little, and add of oil of turpentine, 4 galls., or q. s. Nearly as pale as copal varnish; it soon becomes very hard, and is the most durable of the oil varnishes; but it requires some time before it is fit for polishing, unless the articles are 'stoved.' When required to dry and harden quicker, drying oil may be substituted for the linseed oil, or 'driers' may be added during the boiling.

2. Amber, 4 oz.; pale boiled oil, 1 quart; proceed as last. Very hard.

3. Pale transparent amber, 5 oz.; clarified

linseed oil or pale boiled oil, and oil of turpentine, of each, 1 pint; as before.

4. Amber, broken small, 2 oz.; Venice turpentine, 2 oz.; pale linseed oil, 1½ oz. Dissolve, and thin with about 2 oz. of oil of turpentine.

Obs. Amber varnish is suited for all purposes where a very hard and durable oil varnish is required. The paler kind is superior to copal varnish, and is often mixed with the latter to increase its hardness and durability. The only objection to it is the difficulty of preparing it of a very pale colour. It may, however, be easily bleached with some fresh-slaked lime.

Varnish, Balloon. See VARNISH, FLEXIBLE (*below*).

Varnish, Bessemer's. This consists of a pale oil copal varnish, diluted with about six times its volume of oil of turpentine, the mixture being subsequently agitated with about 1-30th part of dry slaked lime, and decanted after a few days' repose. Five parts of the product mixed with 4 parts of bronze powder forms 'Bessemer's gold paint.'

Varnish, Black. *Prep.* 1. (BLACK AMBER VARNISH.) From amber, 1 lb.; fuse, add of hot drying oil, ½ pint; powdered black resin, 3 oz.; asphaltum (Naples), 4 oz.; when properly incorporated and considerably cooled add of oil of turpentine, 1 pint. This is the beautiful black varnish of the coachmakers.

2. (IRONWORK BLACK.) From asphaltum, 48 lbs.; fuse, add of boiled oil, 10 galls.; red-lead and litharge, of each, 7 lbs.; dried and powdered white copperas, 3 lbs.; boil for 2 hours, then add of dark gum amber (fused), 8 lbs.; hot linseed oil, 2 galls.; boil for 2 hours longer, or until a little of the mass, when cooled, may be rolled into pills, then withdraw the heat, and afterwards thin it down with oil of turpentine, 30 galls. Used for the ironwork of carriages, and other nice purposes.

3. (BLACK JAPAN, BITUMINOUS VARNISH.) *a.* From Naples asphaltum, 50 lbs.; dark gum anime, 8 lbs.; fuse, add of linseed oil, 12 galls.; boil as before, then add of dark gum amber, 10 lbs., previously fused and boiled with linseed oil, 2 galls.; next add of driers, q. s., and further proceed as ordered in No. 2. Excellent for either wood or metals.

b. From burnt umber, 8 oz.; true asphaltum, 4 oz.; boiled linseed oil, 1 gall.; grind the umber with a little of the oil; add it to the asphaltum, previously dissolved in a small quantity of the oil by heat; mix, add the remainder of the oil, boil, cool, and thin with a sufficient quantity of oil of turpentine. Flexible.

4. (BRUNSWICK BLACK.) *a.* To asphalt, 2 lbs., fused in an iron pot, add of hot boiled oil, 1 pint; mix well, remove the pot from the fire, and, when cooled a little, add of oil of turpentine, 2 quarts. Used to blacken and polish grates and ironwork. Some makers add driers.

b. From black pitch and gas-tar asphaltum, of each, 25 lbs.; boil gently for 5 hours, then add of linseed oil, 8 galls.; litharge and red-lead, of each, 10 lbs.; boil as before, and thin with oil of turpentine, 20 galls. Inferior to the last, but cheaper.

Varnish, Body. *Prep.* 1. From the finest African copal, 8 lbs.; drying oil, 2 galls.; oil of turpentine, 3½ galls.; proceed as for AMBER VARNISH. Very hard and durable.

2. Pale gum copal, 8 lbs.; clarified oil, 2 galls.; dried sugar of lead, ½ lb.; oil of turpentine, 3½ galls.; proceed as before, and mix the product, whilst still hot, with the following varnish:—Pale gum anime, 8 lbs.; linseed oil, 2 galls; dried white copperas, ¼ lb.; oil of turpentine, 3½ galls.; the mixed varnishes are to be immediately strained into the cans or cistern. Dries in about six hours in winter, and in about four hours in summer. Used for the bodies of coaches and other vehicles.

Varnish, Bookbinder's. *Prep.* Take of pale gum sandarach, 3 oz.; rectified spirit, 1 pint; dissolve by cold digestion and frequent agitation. Used by binders to varnish morocco leather book-covers. A similar varnish is also prepared from very pale shell-lac and wood naphtha.

Varnish for Boots and Shoes. See BOOTS and SHOES.

Varnish, Cabinet-maker's. French polish is occasionally so called.

Varnish, Carriage. *Prep.* 1. (SPIRIT.) Take of gum sandarach, 1½ lb.; very pale shellac, ¾ lb.; very pale transparent resin, ½ lb.; rectified spirit of ·8221 (64 o. p.), 3 quarts; dissolve, and add of pure Canadian balsam, 1½ lbs. Used for the internal parts of carriages, &c. Dries in 10 minutes or less.

2. (OIL.) *a.* (Best pale.) Take of pale African copal, 8 lbs.; fuse, add of clarified linseed oil, 2½ galls.; boil until very stringy, then add of dried copperas and litharge, of each, ¼ lb.; again boil, thin with oil of turpentine, 5½ galls.; mix whilst both are hot with the following varnish, and immediately strain the mixture into a covered vessel:—Gum anime, 8 lbs.; clarified linseed oil, 2½ galls.; dried sugar of lead and litharge, of each, ¼ lb.; boil as before, thin with oil of turpentine, 5½ galls. Dries in four hours in summer and six in winter. Used for the wheels, springs, and carriage parts of coaches and other vehicles, and by house painters, decorators, &c., who want a strong, quick-drying, and durable varnish.

b. (Second quality.) From gum anime ('sorts'), 8 lbs.; clarified oil, 3 galls.; litharge, 5 oz.; dried and powdered sugar of lead and white copperas, of each, 4 oz.; boil as last, and thin with oil of turpentine, 5½ galls. Used as the last.

Varnish, Chinese. *Prep.* From mastic and sandarach, of each, 2 oz.; rectified spirit (64 o. p.), 1 pint; dissolve. Dries in 6 minutes. Very tough and brilliant.

Varnish, Colourless, for Prints, &c. The 'Art Amateur' gives the following directions for making a colourless varnish suitable for prints, oil paintings, and hard white wood:—Dissolve 2½ oz. shellac in a pint of rectified spirits of wine; to this about 5 oz. of well-burnt animal charcoal that has been recently heated must be added, and the whole boiled for a few minutes. If, on filtering a small portion of the mixture through blotting-paper, it is not found to be perfectly colourless, more charcoal must be added, until the desired result is obtained. When this has been achieved, the mixture must be strained through a piece of silk and filtered through blotting-paper.

Varnish, Copal. *Prep.* 1. (OIL.) *a.* From pale hard copal, 2 lbs.; fuse, add of hot drying oil, 1 pint; boil as before directed, and thin with oil of turpentine, 3 pints, or q. s. Dries hard in 12 to 24 hours.

b. From clear and pale African copal, 8 lbs.; pale drying oil, 2 galls.; rectified oil of turpentine, 3 galls.; proceed as before, and immediately strain it into the store can or cistern. Very fine, hard, and durable.

2. (SPIRIT.) *a.* From coarsely powdered copal and glass, of each, 4 oz.; alcohol of 90% (64 o. p.), 1 pint; camphor, ¼ oz.; heat the mixture, with frequent stirring, in a water-bath, so that the bubbles may be counted as they rise until solution is complete, and, when cold, decant the clear portion.

b. From copal (which has been melted, dropped into water, and then dried and powdered), 4 oz.; gum sandarach, 6 oz.; mastic, 2 oz.; pure Chio turpentine, 3 oz.; powdered glass, 5 oz.; spirit of 90%, 1 quart; dissolve by a gentle heat. Dries rapidly.

3. (TURPENTINE.) To oil of turpentine, 1 pint, heated in a water-bath, add, in small portions at a time, of powdered copal (prepared as above), 3 to 4 oz.; dissolve, &c., as before. Dries slowly, but is very pale and durable.

4. (JAPANNER'S COPAL VARNISH.) From pale African copal, 7 lbs.; pale drying oil, ½ gall.; oil of turpentine, 3 galls.; proceed as in No. 1. Dries in 20 to 60 minutes, and may be polished as soon as hard, particularly if stoved. See JAPANNING.

Obs. All copal varnishes, when properly made, are very hard and durable, though less so than those of amber; but they have the advantage over the latter of being paler. They are applied on coaches, pictures, polished metal, wood, and other objects requiring a good durable varnish. Anime is frequently substituted for copal in the copal varnishes of the shops. See VARNISHES, BODY, CARRIAGE (above), and COPAL, &c.

Varnish, Crystal. *Prep.* 1. From genuine pale Canada balsam and rectified oil of turpentine, equal parts. Used for maps, prints, drawings, and other articles of paper, and also to prepare tracing paper, and to transfer engravings.

2. Mastic, 3 oz.; rectified spirit, 1 pint; dissolve. Used to fix pencil drawings.

Varnish, Drying. Spirit copal varnish.

Varnish, Dutch. Lac and toy varnishes are often so called.

Varnish, Etch'ing. See ETCHING.

Varnish, Fat. See OIL VARNISH.

Varnish, Flexible. *Syn.* BALLOON VARNISH, CAOUTCHOUC V., INDIA-RUBBER V. *Prep.* 1. From india rubber (cut small), 1½ oz.; chloroform, ether (washed), or bisulphuret of carbon, 1 pint; digest in the cold until solution is complete. Dries as soon as it is laid on. Pure gutta percha may be substituted for india rubber.

2. India rubber, in shavings, 1 oz.; rectified mineral naphtha or benzol, 1 pint; digest at a gentle heat in a closed vessel, and strain. Dries very badly, and never gets perfectly hard.

3. India rubber, 1 oz.; drying oil, 1 quart; dissolve by heat. Very tough; dries in about 48 hours.

4. Linseed oil, 1 gall.; dried white coppera and sugar of lead, of each, 3 oz.; litharge, 8 oz.; boil, with constant agitation, until it strings well, then cool slowly, and decant the clear portion. If too thick, thin it down with quick-drying linseed oil. The above are used for balloons, gas bags, &c. See BALLOON, CAOUTCHOUC, &c.

Varnish, Furniture. A solution of pure white wax, 1 part, in rectified oils of turpentine, 4 parts, frequently passes under this name. See VARNISHES, BODY, CARRIAGE, COPAL, &c.

Varnish, Gilder's. *Prep.* (*Watin.*) Pale gum-lac in grains, gamboge, dragon's blood, and annotta, of each, 12½ oz.; saffron, 3½ oz.; dissolve each resin separately in 5 pints of alcohol, of 90%, and make two separate tinctures of the dragon's blood and annotta, with a like quantity of spirit; then mix the solutions in the proper proportions to produce the required shade. Used for gilded articles, &c.

Varnish, Glass. A solution of soluble glass. Used to render wood, &c., fireproof.

Varnish, Gun-barrel. *Prep.* From shellac, 1½ oz.; dragon's blood, 3 dr.; rectified spirit, 1 quart. Applied after the barrels are 'browned.'

Varnish, Hair. *Prep.* From hog's bristles (chopped small), 1 part; drying oil, 10 parts; dissolve by heat. Said to be used to give cotton or linen cloth the appearance of horse-hair.

Varnish, India-rubber. See VARNISH, FLEXIBLE (*above*).

Varnish, Italian. *Prep.* Boil Scio turpentine until brittle, powder it, and dissolve this in oil of turpentine. Used for prints, &c.

Varnish, Japan. Pale amber or copal varnish. Used for japanning tin, papier mâché, &c.

Varnish, Label or Paper. *Prep.* African copal, 60 grms.; powdered glass, 60 grms.; camphor, 15 grms.; ether, 250 grms.; absolute alcohol, 60 grms. Reduce the copal to fine powder, and mix the glass with it; place both in a 500-gramme bottle with the camphor and the ether, close well and set aside for a month, shaking occasionally. At the end of this time add the alcohol, and after shaking well, set aside for 14 days; then pour off the clear portion of the varnish. Before using this varnish it is advisable to size the paper surface with a solution of isinglass in spirit, 1 part; and water, 3 parts.

Varnish, Lac. *Prep.* 1. Pale seed-lac (or shellac), 8 oz.; rectified spirit, 1 quart; dissolve.

2. Substitute lac bleached with chlorine for seed-lac. Both are very tough, hard, and durable, but quite inflexible. Wood naphtha may be substituted for spirit. Used for pictures, metal, wood, or leather, and particularly for toys.

Varnish, Lac (Aqueous). *Prep.* From pale shellac, 6 oz.; borax, 1 oz.; water, 1 pint; digest at nearly the boiling-point until dissolved; then strain. Equal to the more costly spirit varnish for many purposes; it is an excellent vehicle for water-colours, inks, &c.; when dry, it is waterproof.

Varnish, Lac (Coloured). *Syn.* LACQUER, BRASSWORK VARNISH. *Prep.* 1. Take of turmeric (ground), 1 lb.; rectified spirit, 2 galls.;

macerate for a week, strain with expression, and add to the tincture, gamboge, 1½ oz.; pale shellac, ¾ lb.; gum sandarach, 3½ lbs.; when dissolved, strain, and further add of good turpentine varnish, 1 quart. Gold coloured.

2. Seed-lac, 3 oz.; turmeric, 1 oz.; dragon's blood, ¼ oz.; rectified spirit, 1 pint; digest for a week, frequently shaking, then decant the clear portion. Deep gold coloured.

3. Spanish annotta, 3 lbs.; dragon's blood, 1 lb.; gum sandarach, 3¼ lbs.; rectified spirit, 2 galls.; turpentine varnish, 1 quart; as before. Red coloured.

4. Gamboge, 1 oz.; Cape aloes, 3 oz.; pale shellac, 1 lb.; rectified spirit, 2 galls.; as before. Pale brass coloured.

5. Seed-lac, dragon's blood, annotta, and gamboge, of each, ¼ lb.; gum sandarach, 2 oz.; saffron, 1 oz.; rectified spirit, 1 gall. Resembles the last.

Obs. Lacquers are used upon polished metals and wood, to impart to them the appearance of gold. Articles in brass, tin plate, and pewter, or which are covered with tinfoil, are more especially so treated. As lacquers are required of different depths and shades of colour, it is best to keep a concentrated solution of each of the colouring ingredients ready, so that it may be added at any time to produce any desired tint.

Varnish, Mahogany. *Prep.* From gum anime ('sorts'), 8 lbs.; clarified oil, 3 galls.; litharge and powdered dried sugar of lead, of each, ¼ lb.; proceed as for body varnish, and thin with oil of turpentine, 5 galls. or q. s.

Varnish, Mastic. *Syn.* PICTURE VARNISH, TURPENTINE V., TINGRY'S ESSENCE V. *Prep.* 1. Take of pale and picked gum mastic, 5 lbs.; glass (pounded as small as barley, and well washed and dried), 3 lbs.; finest newly rectified oil of turpentine (lukewarm), 2 galls.; put them into a clean 4-gall. tin, bottle, or can, bung them securely, and keep rolling it backwards and forwards pretty smartly on a counter, or any other solid place, for at least 4 hours, when, if the gum is all dissolved, the varnish may be decanted, strained through muslin into another bottle, and allowed to settle; if the solution is still incomplete, the agitation must be continued for some time longer, or a gentle warmth applied as well. Very fine.

2. (Second quality.) From mastic, 4 lbs.; oil of turpentine, 2 galls.; dissolve with heat.

Obs. Mastic varnish is much used for pictures, &c.; when good, it is tough, hard, brilliant, and colourless. It greatly improves by age, and, when possible, should never be used before it has been made at least a twelvemonth. Should it get 'chilled,' 1 lb. of well-washed siliceous sand should be made moderately hot and added to each gallon, which must then be well agitated for 5 minutes, and afterwards allowed to settle.

Varnish, Oak. *Syn.* WAINSCOT VARNISH, COMMON TURPENTINE V. *Prep.* 1. Clear pale resin, 3½ lbs.; oil of turpentine, 1 gall.; dissolve.

2. To the last add of Canada balsam, 1 pint. Both are cheap and excellent common varnishes for wood or metal.

Varnish, Oil. The finer qualities are noticed under AMBER, BODY, CARRIAGE, and COPAL VARNISH; the following produces the ordinary oil varnish of the shops:—Take of good clear resin, 3 lbs.; drying oil, ½ gall.; melt, and thin with oil of turpentine, 2 quarts. A good and durable varnish for common work.

Varnish, Painter's. See CARRIAGE, COPAL, MAHOGANY, OAK, OIL, and other varnishes; the selection depending greatly on the colour and quality of the work.

Varnish, Patent Leather. This is carefully prepared drying oil. The skins being stretched on a board, and every trace of grease being removed from them by means of a mixture of fuller's-earth and water, they are ready to receive the varnish, which is then spread upon them very thinly by means of a species of scraper. The first coat varnish consists of pale Prussian blue (that containing some alumina), 5 oz.; drying oil, 1 gall.; boiled to the consistence of single size, and when cold, ground with a little vegetable black; it is stoved and afterwards polished with fine-grained pumice. The second coating resembles the first, excepting in having a little pure Prussian blue mixed with it. The third coat varnish consists of a similar mixture, but the oil is boiled until it strings well, and a little more pure Prussian blue and vegetable black are added. The last coat varnish, or finish, is the same as the third, but must contain ½ lb. of pure dark-coloured Prussian blue, and ¼ lb. of pure vegetable black, per gall., to which a little oil copal or amber varnish is often added; each coat being duly stoved and pumiced before the next is applied. The heat of the stove or oven is commonly 120° F. for 'enamelled skins,' as those of the calf and seal, intended for 'uppers;' and 175° to 180° for stout 'japan leather;' the exposure in the stove is commonly for 6 to 10 hours. The skins are next oiled and grained. The 'graining' of the 'enamelled skins' is done by holding the skin in one hand, and with a curved board lined with cork (graining stick), lightly pressed upon the fleshy side, working it up and down until the proper effect is produced.

Varnish, Picture. Several varnishes, especially mastic varnish, are called by this name. Pale copal or mastic varnish is generally used for oil paintings; and crystal, white hard spirit, or mastic varnish, for water-coloured drawings on paper.

Varnish, Printer's. Diluted with twice its volume of oil of turpentine, it forms a good common varnish.

Varnish, Sealing-wax. Black, red, or any coloured sealing-wax, broken small, with enough rectified spirit (or methylated spirit) to cover it, digest till dissolved. A most useful varnish for wood-work of electrical or chemical apparatus, for tops of corks, &c.

Varnish, Spirit. *Prep.* 1. (BROWN HARD.) *a.* From gum sandarach, 3 lbs.; pale seed-lac or shell-lac, 2 lbs.; rectified spirit (65 o. p.), 2 galls.; dissolve, and add of turpentine varnish, 1 quart; agitate well, strain (quickly) through gauze, and in a month decant the clear portion from the sediment. Very fine.

b. From seed-lac and yellow resin, of each, 1½ lbs.; rectified spirit, 5 quarts; oil of turpentine, 1½ pints; dissolve. Inferior to the last.

2. (WHITE HARD.) *a.* From gum sandarach, (picked), 5 lbs.; camphor, 2 oz.; washed and dried coarsely pounded glass, 3 lbs.; rectified spirit (65 o. p.), 7 quarts; proceed as in making

mastic varnish; when strained, add of pure Canada balsam, 1 quart. Very pale, durable, and brilliant.

b. From gum sandarach and gum mastic, of each, picked, 4 oz.; coarsely powdered glass, 8 oz.; rectified spirit, 1 quart; dissolve and add of pure Strasburg turpentine, 3 oz. Very fine.

3. (SOFT BRILLIANT.) From sandarach, 6 oz.; elemi (genuine), 4 oz.; anime, 1 oz.; camphor, ½ oz.; rectified spirit, 1 quart; as before.

4. (SCENTED.) To the preceding add some gum benzoin, balsam of Peru, balsam of Tolu, oil of lavender, or the essence of musk or ambergris. The first two can only be employed for dark varnishes.

Obs. The above varnishes are chiefly applied to articles of the toilette, as work-boxes, card-cases, &c.; but are also suitable to other articles, whether of paper, wood, linen, or metal, that require a brilliant and quick-drying varnish. They dry almost as soon as applied, and are usually hard enough to polish in 24 hours. They are, however, much less durable, and more liable to crack, than oil varnishes.

Varnish, Stopping-out. *Syn.* PETIT VERNIS, Fr. From lamp-black, made into a paste with turpentine. Used by engravers. See ETCHING.

Varnish, Tingry's. See MASTIC VARNISH.

Varnish, Toy. Similar to common spirit varnish, but using carefully rectified wood naphtha as the solvent. See VARNISHES, LAC and SPIRIT.

Varnish, Transfer. *Syn.* MORDANT VARNISH. *Prep.* From mastic (in tears) and sandarach, of each, 4 oz.; rectified spirit, 1½ pints; dissolve, and add of pure Canada balsam, ⅓ pint. Used for transferring and fixing engravings or lithographs on wood, and for gilding, silvering, &c. See VARNISH, CRYSTAL.

Varnish, Turpentine. See VARNISHES, MASTIC and OAK.

Varnish, Wainscot. See VARNISH, OAK.

Varnish, Waterproof, for Boots. *Prep.* 1. Ozokerite (hard paraffin), 1 part; castor oil, 2 parts; lamp-black, 1 part. Mix.

2. Salad oil, 1 pint; mutton suet, 4 oz.; white wax and spermaceti, of each, 1 oz. Melt together, and apply to the boots warmed.

3. Spermaceti, 3 oz.; melt and add india rubber in thin shavings, ⅓ oz.; when dissolved add tallow, 8 oz.; lard, 2 oz.; amber varnish, 4 oz. Mix well, and while still warm apply with a brush.

Varnish, Wax. *Syn.* MILK OF WAX; EMULSIO CERÆ SPIRITUOSA, L. *Prep.* 1. Take of white wax (pure), 1 lb.; melt it with as gentle a heat as possible, add of warm rectified spirit, sp. gr. ·830 (60 o. p.), 1 pint; mix perfectly, and pour the liquid out upon a cold porphyry slab; next grind it with a muller to a perfectly smooth paste, adding more spirit as required; put the paste into a marble mortar, make an emulsion with water, 3½ pints, gradually added, and strain it through muslin. Used as a varnish for paintings; when dry, a hot iron is passed over it, or heat is otherwise evenly applied, so as to fuse it, and render it transparent, after which, when quite cold, it is polished with a clean linen cloth. The most protective of all varnishes.

2. Wax (pure), 5 oz.; oil of turpentine 1 quart; dissolve. Used for furniture. See VARNISH, SEALING-WAX.

Varnish, White. See VARNISH, SPIRIT, 2, *a* and *b.* ·

VARNISHING. To give the highest degree of lustre to varnish after it is laid on, as well as to remove the marks of the brush, it undergoes the operation of polishing. This is performed by first rubbing it with very finely powdered pumice-stone and water, and afterwards with an oiled rag and tripoli, until the required polish is produced. The surface is, last of all, cleaned with soft linen cloths, cleared of all greasiness with powdered starch, and then rubbed bright with the palm of the hand.

In varnishing great care must be taken that the surface is free from grease or smoke; as, unless this be the case, the best oil or turpentine varnish in the world will not dry and harden. Old articles are usually washed with soap and water, by the painters, before being varnished, to prevent any misadventure of the kind alluded to.

VASELINE. See COSMOLINE.

VEAL. "The grain should be close, firm, and white, and the fat of a pinkish white, not a dead white, and the kidneys well covered with a thick white fat" (*Soyer*).

Veal, like pork, requires to be well dressed, to develop its nutritive qualities. It should also be eaten fresh, as a peculiar principle is generated in it when improperly kept, which acts as a malignant poison. See ROASTING, &c.

VEG′ETABLE AL′KALI†. Potassa.

VEGETABLE JUICES. See *below.*

VEGETABLES. Vegetables are organic beings, which are distinguished from animals by a number of characteristics, but, like them, are composed of certain proximate principles, or compounds, which possess a high degree of scientific interest, and in many cases are invaluable to man. Among the most important of these are—albumen, gluten, gum, lignin, starch, sugar, tannin, wax, the fixed and volatile oils, the resins and gum-resins, the alkaloids, and innumerable forms of extractive matter. Several of these substances are noticed under their respective names.

The method of propagating plants from their seeds, depending on their simple exposure, at the proper season, to warmth and moisture, under the protection of the soil, is well known; that by propagation from 'slips' and 'cuttings,' which will doubtless prove interesting to the amateur gardener, is noticed below.

The choice of slips and cuttings should be made from the side shoots of trees and plants, and, when possible, from such as recline towards the ground, observing, when they are removed by the knife, to leave a little wood of a former year or season's growth attached to them, as such are found to take root more readily than when they are wholly composed of new wood. The time to take slips or cuttings is as soon as the sap gets into full motion. Before setting them the latter should be cut across, just below an eye or joint, with as smooth a section as possible, observing not to injure the bud. The superfluous leaves may be removed, but a sufficient number should be left on for the purposes of vegetation. The common practice of removing all or nearly all the leaves of cuttings is injudicious.

In some cases leaves alone will strike root. When cuttings are set in pots, they should be so placed as to reach to the bottom and touch the sides throughout their whole length, when they will seldom fail to become rooted plants. In the case of tubular-stalked plants it is said to be advantageous to insert both ends into the soil, each of which will take root, and may then be divided, when two parts will be produced instead of one. An equable temperature, a moist atmosphere, a shady situation, and a moderate supply of water, are the principal requisites to induce speedy rooting. Excess of any of these is prejudicial. When the size of the cuttings admits, it is better to place them under a hand- or bell-glass, which will preserve a constant degree of heat, and prevent evaporation from the surface of the leaves, which is the most common cause of their dying, especially in hot dry weather.

Qual. The vegetable kingdom furnishes by far the larger portion of the food of man, and indirectly, perhaps, the whole of it. The great value of culinary vegetables and fruit in a mixed diet need not be insisted on, since it is a fact which is almost universally known and appreciated.

In the choice of culinary vegetables observe that if they are stiff and break freely and crisply, they are fresh, and fit for food; if, 'on the contrary, they have a flabby appearance, or are soft or discoloured, they are stale, and should be rejected.

The dose of the generality of vegetable substances that exercise no very marked action on the human frame is about ½ to 1 dr. of the powder, night and morning; or 1 oz., or q. s. to impart a moderately strong colour or taste, may be infused or boiled in 1 pint of water, and a wineglassful or thereabouts taken 2 or 3 times a day.

Collection and Pres. The following general directions are given in the London Pharmacopœia for the collection and preservation of vegetable substances (vegetabilis—Ph. L.):

" Vegetables are to be collected in dry weather, and when neither wet with rain nor dew; they are to be collected annually, and are not to be kept beyond a year.

" Barks are to be collected at that season in which they can be most easily separated from the wood." Spring is the season here alluded to; as at this time, after the sap begins to ascend, the bark is, in general, very easily separated.

" Flowers are to be collected recently blown." The red rose, however, must be gathered before the buds are expanded.

" Fruits and seeds are to be collected when ripe.

" Herbs and leaves are to be gathered after the flowers have expanded, and before the seeds are mature.

" Roots and rhizomes (underground stems), for the most part, are to be dug up after the old leaves and stalks have fallen, and before the new ones appear." (" Roots which are required to be preserved fresh should be buried in dry sand "—Ph. L. 1836.)

" Seeds are to be collected when they are ripe, and before they drop from the plant." (" They ought to be preserved in their seed-vessels "—Ph. L. 1836.)

" The different parts of vegetables are to be kept dried for use, except where we shall otherwise direct. Expose those you wish to dry, within a short time after they have been gathered, in shallow wicker baskets, to a gentle heat, in a dark place, and where there is a current of air. Then, the moisture being driven off, gradually increase the heat to 150° F., in order that they may be dried. Finally, preserve the more delicate parts, viz. flowers and leaves, in black glass vessels, well closed, and keep the rest in proper vessels, preventing the access of light and moisture."

Fruits, culinary vegetables, and vegetable juice, of every class, may be preserved for any length of time by several of the methods described under PUTREFACTION. On the small scale the following method is often adopted :—The substances to be preserved are put into strong glass or stoneware bottles, with necks of a proper size, which are then corked with the greatest care, tied or wired, and luted with a mixture of lime and soft cheese, or with a paste formed of linseed meal and water, spread on rags; or tin cases are employed, and are soldered up instead of being corked. The bottles are then placed in an oven, the temperature of which is cautiously raised to fully 212° F.; or they are enclosed, separately, in canvas bags, and put into a copper of water, to which some salt has been added, which is then gradually heated until it boils, and thus kept for 15 or 20 minutes; the whole is next left to cool, when the bottles are taken out and carefully examined before being laid by, lest they should have cracked or the lute have given way.

Herbs and flowers are now generally preserved for distillation by means of common salt. The objection which is raised against the use of fresh aromatic plants is thus obviated, whilst the odours of the distilled products are rendered superior to those obtained from either the recent or dried plant, fruit, or flower, without the great loss, inconvenience, or trouble attending the common methods. Besides, many aromatic and odorous substances almost entirely lose their properties by drying; while most of them yield more oil, and that of a finer quality, in the fresh than in the dried state. The odours of roses, elder flowers, and a variety of others are vastly improved by this treatment, and these flowers may thus be preserved with ease and safety from season to season, or even longer, if required. The process simply consists in intimately mixing the flowers or other vegetables, soon after being gathered, with about 1-4th their weight, or less, of good dry salt, and ramming down the mixture as tightly as possible in strong casks. The casks are then placed in a cold cellar, and covered with boards, on which heavy weights are put, to keep the mass tight and close. See FRUITS, PUTREFACTION, &c.

Vegetables, Juices of. Prep. 1. (EXPRESSED VEGETABLE JUICES, SIMPLE, V. J.; SUCCI EXPRESSI, L.) These are obtained by bruising the fresh leaves, or other vegetable matter, in a marble mortar, or in a mill, and expressing the liquid portion by means of a powerful screw press. After defecation for 12 or 14 hours in a cold situation, the juice is either decanted or filtered from the feculous sediment, and is next heated for some

minutes to about 185° F., to coagulate albuminous matter. The clear portion is subsequently
separated as before, and the product preserved
for use in well-closed and well-filled bottles, in a
cool situation. Some plants, as borage, cabbage,
&c., require the addition of 1-8th of water before
being pressed. The expression of the juice of
lemons, oranges, quinces, &c., is facilitated by
previously mixing the pulp with clean chopped
straw. Buckthorn berries, mulberries, &c., after
being crushed between the hands, are commonly
left for 3 or 4 days to undergo a slight fermentation before pressing them.

The expression of the juices of the narcotic
plants, and of some other vegetables, has lately
assumed considerable interest, from these juices
being now extensively used in pharmacy for the
preparation of extracts and the preserved juices
noticed below. It appears that the juice of young
plants just coming into flower yields only 2-3rds
the amount of extract which may be obtained from
the same quantity of juice expressed from the
matured plant, or when the flowers are fully
blown, and the strength of the product is also inferior ; the case appears to be best met by selecting
the plants when more than half the flowers are
fully blown. The leaves alone should be preferably
employed, and should be exclusively of the second
year's growth when the plants are biennials
(Squire). The homœopathists commonly employ
the whole flowering herb.

The INSPISSATED VEGETABLE JUICES (SUCCI
SPISSATI) are now included among the extracts.

The principal simple vegetable juices of commerce are—

BUCKTHORN JUICE (SUCCUS RHAMNI—Ph. L.),
from the fruit of Rhamnus cartharticus, or
buckthorn berries.

CITRON JUICE (SUCCUS CITRI), chiefly imported
from Italy in large casks.

LEMON JUICE (SUCCUS LIMONUM—Ph. L.), from
lemons that spoil before they can be sold; also
imported.

MULBERRY JUICE (SUCCUS MORI—Ph. L.),
from the fruit of the mulberry.

ORANGE JUICE (SUCCUS AURANTII), obtained
from the same source as that of lemons.

CONCENTRATED ORANGE JUICE (SUCCUS SPIS-
SATUS AURANTII vel AURANTIORUM) and CON-
CENTRATED LEMON JUICE (SUCCUS SPISSATUS
LIMONUM) are prepared by evaporating the fresh
juices of oranges and lemons, either alone or
mixed with sugar, and are employed as substitutes for the fruit where the latter cannot be
obtained.

2. (ALCOHOLISED VEGETABLE JUICES, PRE-
SERVED V. J., TINCTURES OF RECENT PLANTS;
SUCCI ALCOHOLATI, L.; ALCOOLATURES, Fr.)
a. The juice, obtained by powerful pressure,
in the manner noticed above, is allowed to remain for 24 hours in a cold place, when the
clear portion is decanted from the feculous matter
which has subsided, and is then agitated with one
half its volume of rectified spirit (56 o. p.); after
another 24 hours the clear portion is again
decanted and, if necessary, filtered through
bibulous paper or linen. In this way are now
generally prepared the preserved juices of aconite,

belladonna, colchicum (corms), hemlock, henbane,
foxglove, elaterium, lactuca virosa, taraxacum,
&c., sold in this country.

b. (P. Cod.) To the fresh leaves, bruised in a
marble mortar, is added an equal weight of rectified spirit, and after maceration for 15 days the
whole is pressed, and the resulting tincture filtered. In this manner are prepared tinctures of
the fresh leaves of aconite (tinctura aconiti cum
foliis recentibus), belladonna, foxglove, hemlock,
henbane, strong-scented lettuce (Lactuca virosa),
stramonium, trailing poison oak (Rhus toxicodendron), mugwort (Artemisia vulgaris), colchicum (corms), squirting cucumber, white poppy,
taraxacum, &c., of the Paris Codex.

Obs. These tinctures are much more powerful,
and more certain in their operation, than those
prepared from the dried plants. The commencing
dose is from 2 to 5 drops, the effects of which
should be carefully watched. The products of
the first of the above formulæ keep as well as
the ordinary tinctures, and there is less waste of
spirit than with the second. That of the P. Cod.
is, however, preferred by M. Soubeiran, as affording more uniform products—an opinion which is
questionable. Béral orders equal weights of juice
and spirit; Mr Squire recommends ¼ part,
Messrs Bentley & Davenport ½ part (both by
volume), and Mr Gieseke only ¼ part (by weight)
of spirit to 1 part of the expressed juice. The
homœopathists generally go with M. Béral.
"Our own experience, which has been very considerable, and extends over upwards of sixteen
years, leads us to prefer the proportions given in
formula a, which are similar to those of Mr
Squire. If less spirit be employed, the product is
apt to suffer rapid deterioration when kept in a
warm shop or surgery" (Cooley).

3. (ETHERISED VEGETABLE JUICES; SUCCI
ÆTHERIZATI, L.; ÉTHÉROLATURES, SUCS ÉTHÉ-
RÉS, Fr.) For these we are indebted to M.
Bouchardat. They are prepared as follows :—
Ether is gradually added to the depurated freshly
expressed juice until, after active agitation, a thin
layer of it rises to the surface, on the mixture
being allowed to repose for a minute or two; the
whole is then set aside for 24 hours, when the
supernatant ether is expertly removed by means
of a pipette or syringe, and the juice is filtered;
lastly, the decanted ether is returned to the filtrate, and the etherised juice is at once put into
well-stoppered bottles. For use, one of the bottles
is reversed, and the dose taken from the lower
part, so that the ether remains behind. We find
in practice that decantation, carefully conducted,
may be substituted for filtration; and this not only
rendering the process less costly, but ensuring a
more uniform product.

The etherised juices are said to retain their
active properties for an indefinite period. The
method has been successfully applied to the juices
of aconite, anemone, black hellebore, and hemlock, and is probably applicable to many others;
but, we think, not to the juices of all the narcotic
plants, as has been asserted.

Vegetable Fibres. The following method for
the identification of vegetable fibres is intended
to supplement the information previously given
on this subject. Its originator, M. Vetillard, ap-

plies it for distinguishing the fibres of linen, hemp, cotton, jute, China grass, and New Zealand flax. The following extract, descriptive of the process, is from the 'Journal of Applied Chemistry:'

"If a woven or spun fibre is to be examined it must first be disintegrated into the single fibres, and any colour or finish must be removed as completely as possible. Vertical and longitudinal microscopic sections are next made. These are rendered transparent by glycerin or chloride of calcium, and treated with tincture of iodine, made by simply dissolving iodine in a solution of iodide of potassium. The excess of this tincture is removed, a drop of dilute sulphuric acid added, and the sections examined by the aid of the microscope."

Linen Fibre. Bundles of similar fibres, with a fine canal in the centre, long, uniformly thick, and pointed at the ends. Longitudinal section: the fibres are coloured blue, the canal yellow. Cross section: regular polygons, loosely connected, coloured blue; centres yellow.

Hemp. Fibres aggregated; each fibre covered with a thin skin; coloured yellow. They are thick and less uniform than the linen fibres. The ends are thick and of the shape of spatulas, and become blue or greenish blue with iodine. Cross section: irregular polygons, firmly connected; rim yellow, the mass blue, the centre colourless.

Cotton. Longitudinal section: single fibres spirally wound on their own axes, with a central canal and broad ends; coloured blue by iodine. The cross sections are rounded in the shape of kidneys and coloured blue, with yellow spots interspersed.

China Grass. Longitudinal section: fibres separated lengthwise, of varying thickness. The interior canal is often filled with a yellow granular substance, which is coloured brown by iodine. The fibre is turned blue by iodine. Cross section: irregular, with re-entrant angles, and little cohesion. The fibres are stouter than all other fibres, and are turned blue by iodine.

Jute. Fibres strongly coherent, the ends undulating and difficult to separate; central canal wide, empty, and gently rounded at the ends; coloured yellow. Cross section: polygons strongly coherent and regular, much like those of hemp, but the central opening is larger; coloured yellow, darker at the rim.

New Zealand Flax. Bundles of cells of the leaves easily separated with a needle into stiff little fibres, provided with a canal of uniform width. The sides are rolled inwards, coloured yellow. The cross section resembles that of jute, but the corners of the polygons are rounded off. They are coloured yellow by iodine tincture.

VEGETATION. Vegetation (which is here employed in the sense of plants in general) is very unequally distributed over the earth's surface. Thus towards the poles plants are found not only in diminished numbers compared to their occurrence in warmer and more temperate regions, but also of much smaller size or stunted growth. No plants at all, are met in the regions of eternal frost and snow, whilst in equatorial climes they attain to the most gigantic proportions, and are possessed of the most exquisite colours and perfumes, and yield the finest fruits. The habitat of a plant will, of course, be that on which it finds the conditions favourable to its existence and growth, in the shape of soil, climate, moisture, geographical position, &c.

VEGETATION (Metallic). This name has been fancifully applied to the following:

LEAD TREE; ARBOR SATURNI. Take of sugar of lead, 1 oz.; distilled water, 1½ pints; acetic acid, a few drops; dissolve, place the liquid in a clear white glass bottle, and suspend a piece of zinc in it, by means of a fine thread.

SILVER TREE; ARBOR DIANÆ. From nitrate of silver, 20 gr.; water, 1 fl. oz.; dissolve in a phial, and add about ½ dr. of pure mercury.

TIN TREE; ARBOR JOVIS. From chloride of tin, 3 dr.; nitric acid, 10 to 15 drops; distilled or rain water, 1 pint; dissolve in a white glass bottle, and hang in it, by a thread, a small rod of zinc.

Obs. In the above experiments the metals are precipitated in a very beautiful arborescent form. It is curious to observe the laminæ shoot out, as it were, from nothing, assuming forms resembling real vegetation. This phenomenon results from voltaic action being set up between the liquid and the metal.

VEGETO-AL'KALI. See ALKALOID.

VELLUM, PREPARING. The skins used are those of calves, kids, and stillborn lambs. These are unhaired either by steeping with lime, by sweating (i. e. by hanging in a smokehouse heated by a smouldering fire till fermentation sets in), or by soaking with dilute acids. As you seek a cleanly method you may, perhaps, prefer the last. The hair, &c., is scraped off with a two-handed unhairing knife. After this the skin is stretched in a 'herse' (merely a square frame of four sticks joined at corners); strings from the edges of the skin to this frame allow of its being made quite tight, and it is well scraped with a half-moon knife to clear away all fleshy particles, dirt, &c. Next it is ground. The grain side is merely ground over with a flat pumice-stone, but the flesh side is rubbed over with powdered chalk before grinding. The half-moon knife is now passed over the skin to drain it; this makes it look whiter. Fine chalk is then rubbed over both sides, and it is put to dry. It has next to be pared down to a proper thickness—probably about one half—with a sharp circular knife, and then pumiced smooth where required. Lastly, it is glazed with albumen—white of egg.

The skins of sheep are commonly used for parchment; those of he-goats and wolves for drum-heads; and those of the ass for battledores. The species of vellum used for church services by binders is said to be prepared from pig-skins. See POUNCE.

VEL'VET COLOURS. *Syn.* MAP STAINS, PAPER STAINS; LACCA FLUIDA, L. *Prep.* 1. (BLUE.) *a.* Dissolve litmus in water, and add ¼ of spirit of wine. *b.* Dilute Saxon blue or sulphate of indigo with water. If required for delicate work, neutralise the acid with chalk. *c.* To an aqueous infusion of litmus add a few drops of vinegar, until it turns of a full blue.

2. (GREEN.) *a.* Dissolve crystallised verdigris in water. *b.* Dissolve sap green in water, and add a little alum. *c.* Add a little salt of tartar to a blue or purple solution of litmus, until it turns green. *d.* Dissolve equal parts of crystallised verdigris and cream of tartar in water.

3. (PURPLE.) *a.* Steep litmus in water, and strain the solution. *b.* Add a little alum to a strained decoction of logwood. *c.* Add a solution of carmine (red) to a little blue solution of litmus or Saxon blue.

4. (RED.) *a.* Macerate ground Brazil wood in vinegar, boil a few minutes, strain, and add a little alum and gum. *b.* Add vinegar to an infusion of litmus until it turns red. *c.* Boil or infuse powdered cochineal in water containing a little ammonia or sal volatile. *d.* Dissolve carmine in liquor of ammonia, or in weak carbonate of potash water; the former is superb.

5. (YELLOW.) *a.* Dissolve gamboge in water, and add a little alum. *b.* Dissolve gamboge in equal parts of proof spirit and water. Golden coloured. *c.* Steep French berries in boiling water, strain, and add a little alum. *d.* Steep turmeric, round zedoary, gamboge, or annotta, in a weak ley of subcarbonate of soda or potash.

Obs. The preceding, thickened with a little gum, are used as inks for writing, as colours to tint maps, foils, paper, artificial flowers, &c., and to paint on velvet. Some of them are very beautiful. Those containing litmus are, however, fugitive. It must be observed that those made with strong spirit do not mix well with gum water, unless somewhat diluted with water. Any other transparent colours or stains may be employed for painting on velvet, as well as the above.

VELVET LEAF. *Syn.* PAREIRA BRAVA, PAREIRA (Ph. L., E., and D.), L. "The root of *Cissampelos Pareira*" (Ph. L.), white pareira or velvet leaf. It is tonic, aperient, and diuretic.— *Dose,* 20 to 60 gr.; in chronic and purulent inflammation, and extreme irritability of the bladder; in leucorrhœa, dropsy, ulceration of the kidney, &c.

VENESECTION. The practice of venesection, bloodletting, or phlebotomy, as it is variously denominated, has within the last thirty or forty years been nearly banished from medical practice. It seems very evident that prior to the above period medical practitioners were in the habit of resorting to venesection to an unwise extent, and in cases which the progress of modern pathology has shown it to be wholly inapplicable.

There are, we believe, some practitioners who, whilst admitting the evils arising from its misapplication and abuse, still advocate its occasional and judicious employment.

Because of the dangers that beset the operation when performed by a tyro, we forbear to give any particulars as to the method of carrying it out. The veins of the arm are those invariably opened in venesection, although the operation may be performed on many other superficial veins.

VEN'ISON. The flesh of several species of deer. That from good land, killed at the proper season, and eaten in a moderately fresh state, is most easily digestible, and, perhaps, the most wholesome, of all the red meats; but when it is 'high,' or in a state of incipient putrefaction, it is far from wholesome, and often poisonous.

VENO BENO (La). See TEA.

VENOM. Drs Brunton and Fayrer, who have devoted many years to the study of the nature and physiological action of snake poisons, state that there appears to be some resemblance in the action of the venom or virus of the cobra, *Naja tripudians*, and of curara, the arrow-poison of the Indians; both poisons causing death by paralysing the respiratory organs.

Dr Armstrong, who has analysed the cobra poison, has not been enabled to isolate from it any crystalline principle. From its reactions he concludes that its chief ingredient is an albuminoid substance.

Dr Armstrong obtained a white precipitate from the poison by treating it with absolute alcohol, and also prepared an alcoholic extract from it.

He gives the following as the composition of the three substances. The albumen is appended for comparison:

	Crude poison.	Alcoholic precipitate.	Alcoholic extract.	Albumen.
Carbon	43·55	45·76	43·04	53·5
Nitrogen	43·30	14·30	12·45	15·7
Hydrogen	...	6·60	7·00	7·1
Sulphur	...	2·50	...	...
Ash	...	traces.		

"But although there is little difference between the composition of the alcoholic precipitate and extract, there is an immense difference between their physiological action, the extract being a virulent poison, the precipitate almost inert. This is notably different from what has been stated by Dr Weir Mitchell respecting the poison of the rattlesnake, viz. that the alcoholic precipitate is active, whilst the extract is inert" (Royal Society's Proceedings, ' Pharm. Journ.').

VENTILA'TION. The proper ventilation of our habitations, as well as of other buildings in which we pass any considerable portion of our time, is quite as necessary to health as food and clothing. Lavoisier, writing in the middle of the last century, remarks: "It is certain that mankind degenerate when employed in sedentary manufactures, or living in crowded houses, or in the narrow lanes of large cities; whereas they improve in their nature and constitution in most of the country labours which are carried on in the open air." Yet many persons, by the care which they take to shut out fresh air, and to prevent the escape of that which their own bodies, by pulmonary and surfacial respiration, have contaminated, would seem to hug to themselves the discomfort of breathing over and over again an impure and unrefreshing atmosphere, and to be anxious to finish their career by lingering suicide. The almost universal indifference to the subject, considering its importance, is unaccountable.

The first step towards effecting and maintaining a liberal supply of fresh air, is either by means of ventilators or by regularly opening the windows for stated periods daily. During the colder portions of the year, when fires are kept burning, and there is an up-current in the chimney, nothing is so simple and effective as the well-known chimney-valve of Dr Arnott; and, indeed, without this, open fires are powerful instruments of ventilation. In cold weather, where expense is not an object, the apartments may be supplied with air that has been previously warmed by passing through a heated chamber, on the principle recommended by Dr Reid; but care must be taken that, in warming the air, we do not overheat it, nor contaminate it.

A sufficient supply of light, another powerful sanitary agent, is now regarded as nearly as essential as thorough ventilation, and the two are commonly treated of together. According to Palladio, the opening of windows should not exceed a fourth, nor be less than a fifth, of the length of the side of a room, and should be in height two and one sixth times the width. Mr Gwilt, another high authority on this subject, has given as a definite rule, that we should allow 1 square foot of glass to every 100 cubic feet of space in any apartment or inclosure. A great deal must, however, depend on the shape of the apartment; but, in all cases, care should be taken that the windows are placed at the longest side of the room, and not at the narrowest, or the end of it. A southern aspect affords the most light and heat; a northern one the most diffused and least variable light, and is hence usually chosen by artists for their studios.

VERA′TRINE. *Syn.* VERATRIA, VERATRINA, SABADILLINE; VERATRIA (B. P., Ph. L. & E.), L. An alkaloid, or mixture of alkaloids, discovered by Pelletier and Caventou, in the seeds of *Schœnacaulon officinale* (sabadilla), and in the rhizomes of *Veratrum album* (white hellebore).

Prep. 1. (Ph. E.) Digest sabadilla seeds in boiling water for 24 hours, then squeeze them, dry them thoroughly by a gentle heat, beat them in a mortar, and separate the seeds from the capsules by agitation in a deep and narrow vessel; next grind the seeds in a coffee-mill, and extract them by percolation with rectified spirit; concentrate the resulting tincture by distillation so long as no deposit forms, and pour the residuum, whilst still hot, into 12 times its volume of cold water; then filter through calico, and wash the residuum on the filter as long as the washings yield a precipitate with ammonia; unite the filtered liquid with the washings, add ammonia in excess, collect the precipitate on a filter, wash it slightly with cold water, and dry it first by imbibition with filtering paper, and then in the vapour-bath. "The product is not pure, but sufficiently so for medical use. From this coloured substance it may be obtained white, but at considerable loss, by solution in very weak hydrochloric acid, decolorisation with animal charcoal, and reprecipitation with ammonia."

2. (Ph. L. 1836.) This is the same in principle as the last; a tincture is formed by boiling the seeds in rectified spirit, which is then evaporated to a syrup, dissolved in very dilute sulphuric acid, the veratrine precipitated with magnesia, redissolved in very dilute acid, treated with animal charcoal, the filtrate again evaporated to a syrup, and precipitated with ammonia; it is, lastly, washed and dried.

3. By means of ether, as noticed under ALKALOID and ACONITINE. This is by far the best method.

4. (B. P.) Cevadilla, 2 lbs.; distilled water, q. s.; rectified spirit, q. s.; solution of ammonia, q. s.; hydrochloric acid, q. s.; purified animal charcoal, 60 gr. Macerate the cevadilla with half its weight of boiling distilled water in a covered vessel for 24 hours. Remove the cevadilla, squeeze it, and dry it thoroughly with a gentle heat. Beat it now in a mortar and separate the seeds from the capsules by brisk agitation in a deep narrow vessel, or by winnowing it gently on a table with a sheet of paper.

Grind the seeds in a coffee-mill, and form them into a thick paste with rectified spirit.

Pack this firmly in a percolator, and pass rectified spirit through it till the spirit ceases to be coloured. Concentrate the spirituous solution by distillation so long as no deposit forms, and pour the residue, while hot, into twelve times its volume of cold distilled water. Filter through calico, and wash the residue on the filter with distilled water, till the fluid ceases to precipitate with ammonia. To the united filtered liquid add the ammonia in slight excess, let the precipitate completely subside, pour off the supernatant fluid, collect the precipitate on a filter, and wash it with distilled water till the fluid passes colourless. Diffuse the moist precipitate through 12 oz. of distilled water, and add gradually, with diligent stirring, sufficient hydrochloric acid to make the fluid feebly but persistently acid.

Then add the animal charcoal, digest at a gentle heat for 20 minutes, filter, and allow the liquid to cool. Add ammonia in slight excess, and when the precipitate has completely subsided, pour off the supernatant liquid, collect the precipitate on a filter, and wash it with cold distilled water till the washings cease to be affected by nitrate of silver accidental with nitric acid. Lastly, dry the precipitate, first by imbibition with filtering paper, and then by the application of a gentle heat.

Prop. Pure veratrine is perfectly white; but, as usually met with, it is a grey powder; it is highly acrid; scarcely soluble in water, soluble in ether, and freely soluble in hot alcohol; heated to about 125° F., it fuses like wax, and solidifies, upon cooling, to a transparent yellow mass. With the dilute acids it forms salts, which are either amorphous or difficultly crystallisable. The smallest possible portion of its powder causes violent sneezing.

Tests. 1. Potassa, ammonia, and their carbonates, give flocculent white precipitates which at first are not crystalline under the microscope, but which, after some minutes, assume the appearance of small scattered clusters of short prismatic crystals; they are insoluble in excess of potassa and its carbonate, and only very slightly so in excess of ammonia. 2. With sulphuric acid it strikes an intense red colour, changing afterwards to crimson, and finally to violet. 3. A dilute acetic solution of veratrine is turned to a superb red by strong sulphuric acid.

Veratrine is distinguished from brucine and the other alkaloids by its fusibility, by the crystalline form of its precipitate with potassa, and by its reaction with oil and vitriol.

Uses, &c. "As an external application, it has been efficaciously employed by Magendie in France, and by Dr Turnbull in this country; but the extravagant eulogies of the latter have not tended to confirm the reputation of this remedy" (*Dr A. T. Thompson*). From 6 to 12 gr. dissolved in 1 fl. oz. of rectified spirit, as a liniment; or 30 gr., mixed with 1 dr. of olive oil and 1 oz. of lard, as an ointment, have been occasionally found very serviceable in neuralgia, and other like painful affections, and in gouty and rheumatic paralysis. As an internal remedy it possesses no advantage, as it merely acts as a violent and depressing cathartic. —*Dose*, $\frac{1}{32}$ to $\frac{1}{16}$ gr. In larger doses it acts as a powerful irritant poison. For antidotes, &c., see ALKALOID.

VERA'TRUM. See WHITE HELLEBORE.

VER'DIGRIS. *Syn.* ÆRUGO, L.; VERT-DE-GRIS, Fr. This is a mixture of several basic acetates of copper which have a green or blue colour. It is obtained in the wine districts of the south of Europe, by the action of refuse grapes, from which the juice has been expressed, on thin sheets of copper. When pure it should dissolve almost entirely, and without effervescence, in dilute sulphuric acid. It is very poisonous; for antidotes, see COPPER.

An inferior quality of verdigris is now prepared from pommage, or apple marc, in the cider districts of England.

Verdigris, Distilled. *Syn.* CRYSTALLISED VERDIGRIS. This name is applied to the normal acetate of copper, which is prepared in the wine districts by dissolving ordinary verdigris, 1 part, in good distilled vinegar, 2 parts; the operation being performed in a copper vessel by the aid of a gentle heat and agitation; the solution is afterwards slowly evaporated until a pellicle begins to form on the surface, when it is transferred into glazed earthen pans ('oulas'), in each of which are placed two or three cleft sticks, and it is then left in a warm apartment for 14 or 15 days to crystallise.

A spurious article is often prepared by adding a solution of sulphate of copper, 12½ lbs., to a solution of sugar of lead, 19 lbs., or q. s., and filtering, evaporating, and crystallising the mixture.

There is an acetate of copper and lime which resembles distilled verdigris in colour. It was manufactured pretty extensively in Scotland some years ago, and fetched a high price, till Dr Ure published an analysis of it in the 'Edin. Phil. Trans.' It is much inferior for all uses in the arts.

Pure distilled verdigris is entirely soluble in water, and is not precipitated on the addition of sulphuric acid or of ammonia in excess.

Verdigris, English. *Prep.* Blue vitriol, 24 lbs.; white vitriol, 16 lbs.; sugar of lead, 12 lbs.; alum, 2 lbs. (all coarsely powdered); mix, and heat them in a pot over the fire until they unite into a mass. Sold by fraudulent dealers for foreign verdigris.

VER'DITER. *Syn.* BLUE VERDITER, REFINER'S VERDITER; CENDRES BLEUES, Fr. A blue pigment, obtained by adding chalk, whiting, or milk of lime, to a solution of copper in nitric acid; or by triturating recently precipitated and

still moist carbonate of oxide of copper with hydrate of lime.

Prep. A quantity of whiting or milk of lime is put into a tub, and upon this the solution of copper is poured; the mixture is stirred every day for some hours together, until the liquor loses its colour; it is then poured off, and more solution of copper added; this is repeated until the whiting or lime has acquired the proper colour; the whole is then washed with water, drained, spread on chalk stones, and dried in the sun.

Obs. The cupreous solution employed in the above process is made by neutralising the nitric solution obtained from the refiners of gold and silver by beating it along with metallic copper. For the finer qualities of verditer the lime should be of the purest kind, and the cupreous precipitate should be carefully triturated with it, after it is nearly dry, by which a fine velvety appearance is produced. The 'cendres bleues en pâtes' of the French differ from the above mainly in a solution of chloride of copper being employed, and in the resulting green precipitate being turned blue by the action of carbonate of potassa. Verditer is made into crayons whilst moist, or dried into a powder, or it is used as a water-colour in the moist state.

Verditer, Green. *Syn.* BREMEN GREEN. The process for refiner's verditer frequently miscarries, and a green colour is produced instead of a blue one. It may also be obtained directly by omitting the 'blueing up' with carbonate of potassa, mentioned above.

VER'JUICE. *Syn.* AGRESTA, OMPHACIUM, L. The expressed juice of unripe grapes. The term is also often extended to the expressed juice of the wild or crab apple. It was formerly used as an astringent and refrigerant in medicine; but it is now principally employed as an ingredient in sauces, ragoûts, &c.

VERMICEL'LI. This, like macaroni, is prepared from a stiff paste made of a peculiar fine kind of granular wheat flour, called sémoule, which is mixed up with hot water, and, after being well kneaded, is formed into small ribands, cylinders, or tubes, by being placed in a vertical cylinder press, the bottom of which is filled with proper-shaped holes, through which it is driven by an iron plate or 'follower' being forced down by a powerful screw. The pieces that protrude are broken off, twisted into any desired shape upon paper, and dried. Those in the form of fillets or ribands are called 'lasagnes.' Vermicelli contains a large amount of gluten, and is extremely nutritious, although slightly less digestible than the ordinary wheaten foods. See MACARONI.

VER'MIFUGES. *Syn.* ANTHELMINTICS; ANTHELMINTICA, HELMINTHAGOGA, VERMIFUGA, L. Medicines employed to destroy or expel intestinal worms. Some of these, as coarsely powdered tin and iron filings and cowhage, act as mechanical agents, by irritating the worms; others have a specific action upon worms, as male fern, kousso, santonin, &c.; others, again, owe their power to their action as purgatives, as calomel, gamboge, jalap, &c. See WORMS.

VERMIL'ION. *Syn.* FACTITIOUS CINNABAR, RED SULPHIDE OF MERCURY, RED SULPHURET

OF MERCURY. This article may be prepared both in the moist and dry way; that of commerce is almost entirely obtained by the latter; it consists, for the most part, of mercuric sulphide, HgS.

Prep. · 1. By sublimation. Take of pure mercury, 202 parts; pure sulphur, 33 parts; fuse them together by a gentle heat, observing not to allow the mass to take fire; when fused, cover over the vessel, and, when the whole has become cold, powder the mass, and sublime it in a closed vessel, so placed in a furnace that the flame may freely circulate and play upon it to about half its height, the heat being at first gradually applied, and afterwards augmented until the lower part of the subliming vessel becomes red-hot; the cold sublimate is broken into pieces, ground along with water to a fine powder, elutriated, passed through a sieve, and dried.—*Prod.* Fully 112% of the weight of the mercury employed.

2. In the humid way (*Brunner*). Take of pure quicksilver, 300 parts; pure sublimed sulphur, 114 parts; triturate them together for several hours, until a perfectly black product is formed; add gradually of caustic potash, 75 parts, (dissolved in) water, 400 parts; continue the trituration for some time longer, then gently heat the mixture in an iron vessel, at first constantly stirring, but afterwards only from time to time, observing to keep the heat at about 120° F. (49° C.), and to add fresh water to compensate for the portion evaporated. When the colour begins to redden, great caution is requisite to preserve the mixture at the lower temperature, and to keep the sulphide of mercury perfectly pulverulent; as soon as the colour becomes nearly 'fine,' the process must be conducted with increased caution, and at a lower heat for some hours, or until a rich colour is produced, when the newly formed vermilion must be elutriated with water, to separate any particles of metallic mercury, and carefully dried.—*Prod.*, 332 parts of vermilion, equal in brilliancy to the finest Chinese.

Obs. It has been said that the rich tone of Chinese vermilion may be imitated by adding to the materials 1% of sulphide of antimony, and by digesting the ground sublimate, first in a solution of sulphide of potassium, and next in diluted hydrochloric acid, after which it must be well edulcorated with water, and dried. Our own belief is that the finer qualities of vermilion owe their superiority of shade more to the care bestowed on their sublimation, and the extent to which their division is limited, than to anything else. The sublimed vermilion is generally inferior to that obtained by the wet process.

Vermilion is a beautiful and permanent red pigment, and works and covers well both in oil and water.

VERMIN. This term has rather a large application, since, although it is generally understood to be applied to rats, mice, and certain parasitic insects infesting the dwellings and sometimes the bodies of men, it is extended by the farmer, the gardener, and the breeder of game, to those creatures from the depredations of which these three classes suffer pecuniary loss. Hence it embraces not only foxes and polecats, but weasels, stoats, hedgehogs, owls, hawks, kites, carrion crows, magpies, wood-pigeons, hares, rabbits, rooks, moles, and small birds.

Whilst the attempted partial destruction of any of the classes of animals or birds above specified may be regarded as of doubtful value, there can be no question about the practice when it is carried to the verge of extermination.

In this latter case the balance of Nature is interfered with, and the system of checks which she has established for the prevention of the undue preponderance of one tribe of the animal kingdom over the other being interfered with, the result will be the undue propagation of particular species inimical to the operations of the husbandman, &c.

As illustrating this we may mention the destruction to various crops in France caused some years ago by the ravages of certain grubs and insects, the unusual increase in the numbers of which was clearly traced to the foolish practice, amongst French farmers, of shooting all the small birds. See BUG, LOUSE, RATS.

VERMOUTH, or VERMUTH. This preparation is one of the many unofficial wines which require for their sale a wine licence. It appears to have recently become an article of commerce in this country, but it has been used for a number of years as a kind of liqueur on the Continent, especially in Italy and France, and is taken whenever a pick-me-up would be called for by us. It is a stomachic, and composed of the following ingredients in the proportions named :

1. Chamædrys, 12 parts; inula root, 12 parts; calamus rhizome, 12 parts; cinchona bark, 12 parts; cinnamon, 12 parts; elder flowers, 16 parts; cloves, 20 parts; coriander, 20 parts; star anise, 20 parts; tansy, 16 parts; orange peel, 24 parts; blessed thistle, 16 parts; common centaury, 16 parts; wormwood, 16 parts; quassia, 8 parts; nutmeg, 4 parts; galangal, 4 parts; white wine, containing 11% or more alcohol, 8000 parts. The whole is macerated for 8 days and completed in the usual way. The soluble matter of 1 part of the solid ingredients is present in 33·3 parts of finished product, and the medicinal dose would be about a table-spoonful, diluted with an equal quantity of water.

2. Wormwood, 4 oz.; gentian, 2 oz.; angelica root, 2 oz.; blessed thistle, 4 oz.; calamus aromaticus, 4 oz.; elecampane root, 4 oz.; century leaves, 4 oz.; germander leaves, 4 oz.; nutmegs, No. 15; oranges, sliced, No. 6; alcohol of 85°, 9 pints; sweet white wine, 20 galls. Macerate 15 days and filter.

VERTI'GO. Dizziness and swimming of the head. In its more serious forms there is more or less mental confusion, the objects around the patient appear in motion, the ears are oppressed with strange sounds, and visible illusions are experienced, whether the eyes be closed or open, and in darkness as well as in the light. The causes are fulness of the vessels of the head, nervous derangement, general debility, hæmorrhage, the use of narcotics, an overloaded stomach, and, in some cases, an empty one. It is also frequently symptomatic of fevers and inflammations, and of a condition threatening apoplexy. The treatment must be varied, according to the cause and the peculiar habit or condition of the patient.

VES'ICANTS. *Syn.* EPISPASTICS; EPISPASTICA, VESICANTIA, L. Substances which vesicate or raise blisters. Among these are the cantharis or blistering fly, mezereon, croton oil, boiling water, &c., the first only of which is now in common use in England.

"It is a principle sufficiently established with regard to the living system, that, where a morbid action exists, it may often be removed by inducing an action of a different kind, in the same or a neighbouring part. On this principle is explained the utility of blisters in local inflammation and spasmodic action, and it regulates their application in pneumonia, gastritis, hepatitis, phrenitis, angina, rheumatism, colic, and spasmodic affections of the stomach—diseases in which they are employed with the most marked advantage. A similar principle exists with respect to pain; exciting one pain often relieves another. Hence blisters often give relief in toothache, and some other painful affections. Lastly, blisters, by their operation, communicate a stimulus to the whole system, and raise the vigour of the circulation. Hence, in part, their utility in fevers of the typhoid kind, though in such cases they are used with still more advantage to obviate or remove local inflammation" ('Med. Lex.').

Blisters are commonly prepared with cantharides plaster, or with some other preparation of cantharides; and, in the former case, are usually lightly covered with the powdered fly. In order to prevent the action of the cantharides upon the mucous membrane of the bladder, blistering plasters are often sprinkled with a little powdered camphor, or, better still, are moistened with camphorated ether, which leaves a thin layer of camphor. In all these cases the layer should not be too thick, for in that case the plaster would not take effect.

When it is not wished to maintain a discharge from the blistered part, it is sufficient to make a puncture in the vesicle, to let out the fluid; but when the case requires the blister to be 'kept open,' as it is called, the whole of the detached cuticle is carefully removed with a pair of scissors, and the part is dressed with either the ointment of cantharides or of savine, at first more or less diluted with lard, or simple ointment, with an occasional dressing of resin cerate. According to Mr Crowther, the blistered surface is best kept clean by daily fomentation with warm water.

Of late years, to obviate the unpleasant effects occasionally arising from the common blister, various compounds having cantharides for their base have been brought before the public. Of these, the vesicating collodion noticed under COLLODION is the most convenient and effective. The following also deserve notice:

1. Take of cantharides, in fine powder, 2 parts; spermaceti, 2 parts; olive oil, 4 parts; white wax, 8 parts; water, 10 parts; simmer, with constant agitation, for 2 hours, strain through flannel, separate the plaster from the water, gently remelt it with common turpentine, 1 part, and spread the mass whilst still fluid. This nearly resembles the form recommended by MM. Henry and Guibourt.

2. (P. Cod.) Distil off the ether from a concentrated ethereal tincture of cantharides, melt the oily residue with twice its weight of white wax, and spread the mixture on thin oiled silk, or on cloth prepared with wax plaster.

3. (*Oettinger.*) Cantharidal ether (prepared from cantharides, 1 part; ether, 2 parts) and sulphuric ether, of each, 10 dr.; turpentine and black resin, of each, 2½ dr.; mix, dissolve, and apply it to the surface of stretched silk or taffeta which has been previously prepared with two coatings of a solution of isinglass.

4. (CHARTA EPISPASTICA, B. P.) Digest 4 oz. of white wax, 1½ oz. spermaceti, 2 oz. fluid of olive oil, 3 oz. of resin, 1 oz. of cantharides in powder, and distilled water, 6 oz., in a waterbath for two hours, stirring constantly, strain, and separate the plaster from the watery liquid. Mix ½ fl. oz. of Canada balsam with the plaster, melted in a shallow vessel, and pass strips of paper over the hot liquid, so that one surface of the paper shall receive a thin coating of plaster. It may be convenient to employ paper ruled in square inches.

5. (VESICATING SINAPIS, B. P.) Black mustard seeds, in powder, 1 oz.; solution of gutta percha, 2 oz., or q. s. Mix so as to make a semi-fluid, and having poured this into a shallow flat-bottomed vessel, such as a dinner-plate, pass strips of cartridge paper over its surface, so that one side of the paper shall receive a thin coating of the mixture. Then lay the paper on a table, with the coated side upwards, and let it remain exposed to the air until the coating has hardened. Before being applied let the mustard paper be immersed for a few seconds in tepid water.

6. (VESICATING SPARADRAP, P. Cod.) Gum elemi, 1 oz.; olive oil, ½ oz.; basilicon ointment, 2½ oz.; resin, 1 oz.; yellow wax, 3½ oz.; cantharides, in fine powder, 4½ oz. Melt the first five substances together, and stir in the cantharides; when sufficiently cold, and well mixed, spread on waxed strips of linen.

Obs. The above compounds are spread on leather, linen, paper, silk, oiled silk, taffeta, &c., and then form the numerous compounds vended under the names of blistering tissue, rannus vesicatorius, papier épispastique, sparadrapum vesicatorium, taffetas vesicans, tela vesicatoria, &c.

Acetic extract of cantharides, croton oil, or extract of mezereon is sometimes substituted for the ethereal extract ordered in the above formulæ.

The 'papier épispastique' of Vée is prepared of three strengths, which are respectively distinguished by the colours white, green, and red. The composition is made by boiling powdered cantharides for an hour with water, lard, and green ointment, or with lard coloured with alkanet root, adding white wax to the strained fats, and spreading the mixture whilst fluid. No. 1 is made with 10 oz. of cantharides to 4 lbs. of lard; No. 2 of 1 lb. of cantharides to 8 lbs. of green ointment; and No. 3 of 1½ lbs. of flies to 8 lbs. of reddened lard. To each are added 2 lbs. of white wax (*Dorvault*).

The magistral blister of Valleix is a revival of the vesicating epithem. See BLISTER, CANTHARIDES, COLLODION, &c., and *below*.

VESICATION. The formation of a blister is a vital process, and its success may be taken as a proof of the presence of life. Hence a French physician, Dr Mandl, has suggested such a stimulation of the skin as would ordinarily cause a blister as a test of life, in those cases of long-continued trance which we occasionally hear of, where all the functions of life seem to be extinct. Dr Mandl's plan is to apply a stick of lunar caustic. The application of a little strong vinegar of cantharides, or other cantharidal blister, of the size of a sixpenny piece, or of 2 or 3 spoonfuls of boiling water, by means of a bent tube of like diameter, is, however, more certain and satisfactory.

VESICATORIN. *Syn.* CANTHARIDIN, CANTHARIDINA, CANTHARIDES CAMPHOR. The blistering principle of Spanish flies, discovered by M. Robiquet.

Prep. 1. (P. Cod.) Exhaust powdered cantharides with concentrated alcohol by percolation; distil off the spirit from the filtered tincture, and leave the residuum to deposit crystals; these may be purified by dissolving them in boiling alcohol, digestion with animal charcoal, filtration whilst hot, and crystallising by refrigeration.

2. (*Thierry.*) Macerate cantharides (in coarse powder) for several days in ether, in a closed displacement apparatus; then, after the whole of the soluble matter has been extracted by the addition of fresh portions of ether, pour on sufficient water to displace the retained ether; next distil off the ether, dissolve the remaining extract in boiling alcohol, filter while hot, and abandon the filtrate to spontaneous evaporation. —*Prod.,* 5%.

3. Digest the aqueous extract of cantharides in hot alcohol, filter, evaporate to dryness, digest the residuum in sulphuric ether, evaporate, and slightly wash the resulting crystals with cold alcohol.

Prop., &c. Micaceous plates resembling spermaceti; fusible; vaporisable; insoluble in water; soluble in ether, oils, acetic acid, and hot alcohol; powerfully vesicant and poisonous. Its vapour, even at ordinary temperatures, frequently produces temporary blindness. The 1-100th part of a gr., placed on a piece of paper, and applied to the edge of the lower lip, caused small blisters in 15 minutes, which, when rubbed with a little simple cerate, extended over a large surface, and covered both lips with blisters (*Robiquet*).

VETCH. The common name of various leguminous plants of the genera *Vicia* and *Ervum*, now much cultivated as green fodder for milch cows and working stock. The seeds (tares) were formerly reputed detersive and astringent. Those of "the Canadian variety make good bread" (*Lindley*).

VETERINARY MEDICINES. The common form of medicine for horses is that popularly known as horse-balls. They are usually prepared by mixing the dry ingredients, in the state of powder, with a sufficient quantity of treacle, or syrup bottoms, to give the mass a proper consistence for rolling into balls; adding, when necessary, linseed meal, or any other simple powder, to increase the bulk. The usual practice among the veterinary druggists is to keep a compound known in the trade as 'ball-mass' or 'common mass,' ready prepared to give form and bulk to more active ingredients. This is usually made of about equal parts of linseed meal and treacle, together with a little palm or lard, thoroughly incorporated by kneading with the hands; and it is kept in a cool situation, tied over to prevent it drying and hardening. For use, the ball-masses are either rolled or moulded into small cylinders of about 1½ to 1¾ oz. in weight; and in size from 2 to 2¾ inches long, and from about ½ to ¾ of an inch in diameter; and they are wrapped in soft paper, which is administered with them. Those for dogs are commonly formed into large boluses or nut-like pieces. The common practice in some houses of adding a little salt of tartar or acetate of potassa to ball-masses kept in stock, for the purpose of preserving them in a soft state, is not to be commended, since these articles decompose many of the saline and mineral compounds which are subsequently added to them.

Medicines for neat cattle are always administered in a liquid form, popularly called drenches. A similar plan is adopted with small cattle, as sheep and goats. For these, however, the quantity should seldom exceed ½ pint. In all cases, drenches should be very slowly administered.

The following are a few useful horse-balls:

ALTERATIVE BALLS. 1. Levigated sulphide of antimony, sulphur, and linseed meal, of each, 3 oz.; nitre, 4 oz.; palm oil, q. s. to form a mass; for 12 balls. One to be taken every day, or every other day.

2. (*Bell.*) Sulphide of antimony, nitre, sulphur, and Æthiops mineral, of each, 3 oz.; soft soap, 10 oz.; oil of juniper, ½ oz.; for 12 balls. As the last.

3. (*White.*) Sulphide of antimony, caraways, and treacle, of each, ¼ oz.; for one ball. As the last.

CORDIAL BALLS. 1. (*Blaine.*) Coriander seed, caraway, and anise, of each, 8 oz.; ginger, 4 oz.; oil of aniseed, ½ oz.; honey or palm oil, q. s. to form a mass. Cordial, warming, and stomachic.—*Dose,* 1½ oz.

2. (*Hill.*) Anise, caraway, and cumin seed, of each, 4 lbs.; ginger, 2 lbs.; treacle, q. s.; divide into 1½-oz. balls.—*Prod.,* 21 lbs.

COUGH BALLS. 1. (*Blaine.*) Ipecacuanha, 1 dr.; camphor, 2 dr.; honey, q. s. to form a ball. One night and morning.

2. (*Bracy Clark.*) Emetic tartar and benzoin, of each, 2 dr.; squills, 4 dr.; spermaceti and balsam of copaiba, of each, 1 oz.; elecampane and sulphur, of each, 2 oz.; syrup of poppies, q. s. to mix; for 8 balls. As the last.

DIURETIC BALLS. 1. (*Bracy Clark.*) Nitre and common turpentine, of each, 1 lb.; Castile soap, ¼ lb.; barley meal, 2½ lbs., or q. s. For common-sized balls.

2. (*Morton.*) Digitalis, 1 oz.; aloes, 3 oz.; liquorice, 13 oz.; honey or Barbadoes tar, q. s. to mix; for 1-oz. balls. One, twice a day, with care.

PHYSIC BALLS, PURGING B., CATHARTIC B. 1. Aloes and hard soap, of each, 5 oz.; salt of tartar and cayenne pepper, of each, 1 oz. melt together. For 8 balls.

2. (Vet. Coll.) *a.* (Common physic ball.) Aloes, 8 oz.; treacle, 3 oz.; olive oil, 1 oz.; melted together.—*Dose,* 1 to 1½ oz.

b. (Stronger ball.) To each dose of the last add of croton oil, 4 to 8 drops.

Obs. The *dose* of the above is 1 ball, fasting in the morning, preceded by a bran mash, on one or two successive nights, and followed by gentle exercise until the ball begins to operate.

WORM BALLS. **1.** Barbadoes aloes, 5 dr.; calomel and ginger, of each, 2 dr.; oil of cloves, 12 drops; treacle, q. s. for a ball.

2. (*J. Bell & Co.*) Barbadoes aloes, 5 to 8 dr.; powdered tin, Æthiops mineral, and ginger, of each, 2 dr.; oils of aniseed and savine, of each, 20 drops; treacle, q. s. for a ball.

3. (*Clater.*) Sulphur and emetic tartar, of each, 1 dr.; linseed meal, 4 dr.; palm oil, q. s. to form a ball. One every morning, having prepared the animal with a physic ball containing 1 dr. of calomel. See BALLS; also Tuson's 'Veterinary Pharmacopœia.'

VIBRIO TRITICI. (*Tylenchus tritici,* Bastian.) THE EAR COCKLES (OR PURPLES) WORM. Although this is not an insect it causes frequent and serious injury to corn plants and grasses, and should, it is considered, be treated of in this report. Curtis describes it in 'Farm Insects,' and Taschenberg also in his 'Praktische Insekten Kunde,' so that there are eminent precedents for this course.

The chief and most dangerous consequences of the attack of this nematoid, or thread-worm, a species of the order NEMATOIDEA, is the replacement which it causes of the grains of corn by a black or dark brown substance known as 'ear cockle' or 'purples,' of a round shape. It is commonly supposed that it is the grain that is actually converted into this dark mass of cellular tissue; but it is in fact a foreign body, an excrescence, or gall, and contains what appears to be a mass of small fibres. This is, as the microscope reveals, a cluster of tiny thread-worms.

Corn plants are also attacked by this, or, as Mr Carruthers, the Consulting Botanist of the Royal Agricultural Society, thinks, by another, species of nematoid, in their stems, so that they are seriously weakened, and in some instances prevented from flowering (Dr Bastian found specimens of *Plectus tritici* between the lower part of the sheaths of wheat-stalks at Broadmoor).

In the case of the ordinary attack in the ears of corn not only is the crop reduced but the sample is more or less spoilt, and, besides, there is the great danger of the extensive reproduction of nematodes from the galls being sown with the seed-corn.

Life History. The galls containing these worms are sown with the seed-corn, or are carried by some means into the fields. Upon being moistened, either by rain or the natural dampness of the earth, they become active and penetrate into the stems of the corn plants. They were formerly supposed to permeate the tissues of the plants, as the worms of the trichinæ permeate the flesh of animals; but it is held by Davaine, Bastian, and others, that the worms ascend with the growth of the plants to the flowers. Mr Carruthers believes that the worms

attack the flowers before their development is perfect, and suggests that, as in other cases, the gall is a simple excrescence caused by the abnormal flow of sap, probably to heal the injury occasioned to the plant.

After the gall is formed, and is yet soft or green, the worms within it pair, and eggs are laid in strings of five and six together, and worms are hatched from these. When the gall in course of time becomes harder, or ripe, no more eggs are laid, and when it is quite hard and dry, the worms are apparently lifeless. Upon moisture being applied they begin to show signs of vitality, and wriggle about. They retain vitality for a long time when the galls are kept dry, Taschenberg says for many years. He also shows that they may be exposed to a heat of 122° F. without being affected. It is well known that the hardest frosts do not destroy them. They will also live for months in water. Those who wish to study the history of these and other nematoids are referred to Baner's descriptions, illustrated by most elaborate drawings, in the Department of Botany, British Museum, Natural History; and Dr Bastian's paper on the Anguillulidæ, in part 2 of the 'Transactions of the Linnean Society' for 1865, gives admirable accounts and figures of them.

Prevention. Each gall may supply tens of thousands of worms ready to prey upon the young corn plants. It is therefore most important to keep the seed wheat free from galls. To accomplish this careful screening or winnowing is essential. Corn intended for seed should be 'run down' as long as any of the black galls or cockles remain in the samples. Even when corn is not intended for seed these should be eliminated, as they may be the sources of disease to human beings or animals.

All the tail corn, and 'chogs' especially, should be burnt, and not by any means given to the chickens, nor allowed to get into the yard where manure is made, as there would be the risk of the galls being carried out to the corn-fields. In thrashing corn affected the chaff had better be destroyed, and the returns from the larger screens put through the fine screens of the hand winnower with the greatest care.

Another species of *Vibrio* is peculiar to oat plants, which it injures by attacking their stems and causing the main stem to die. New shoots are thrown out, and these are in turn destroyed. Much injury was occasioned to oat plants in 1884 and 1885 by these *Vibrios* in various parts of this country. Mr Carruthers reported upon this in December, 1885, to the Council of the Royal Agricultural Society, and remarked that the species was different from that which attacks wheat plants, and was in his opinion peculiar to oat plants. Dr Bastian recounts in the paper alluded to above that he found several species of nematoids lying between the inner sheaths of various kinds of grasses, among which was *Festuca elatior* ('Reports on Insects Injurious to Crops,' by Chas. Whitehead, Esq., F.Z.S.).

VIBURNUM. *Syn.* BLACK HAW. The bark of *Viburnum prunifolium.* A sedative and tonic to the uterine system; has a good reputation for preventing abortion. Is also astringent and antispasmodic, relieving most forms of cramp. —*Dose*, of liquid extract (U. S. P.), ⅙ to 1 dr.

VICTORIA WATER-LILY (*Victoria regia*, Lindl.). A native of Guiana and Brazil. The leaves sometimes measure 12 feet across, and the expanded flowers about a foot in diameter. The seeds are eaten by the Indians.

VIN'EGAR. *Syn.* ACETUM, L.; VINAIGRE, Fr. Dilute acetic acid, more or less contaminated with gum, sugar, and vegetable matter.

Prep. 1. MALT VINEGAR, ACETUM, BRITISH VINEGAR (B. P.); ACETUM BRITANNICUM (Ph. L. & E.), L. This is the ordinary coloured vinegar consumed in this country, and is correctly described in the Ph. L. as "impure (dilute) acetic acid, prepared by fermentation from an infusion of malt (malt wort)."

In the manufacture of MALT VINEGAR, a mixture of malt and barley is mashed with hot water, and the resulting wort fermented, as in the common process of brewing. The liquor is then run into barrels, placed endways, tied over with coarse canvas, and arranged side by side in darkened chambers, moderately heated by a stove, and freely supplied with air. Here it remains till the acetous fermentation is nearly complete, which usually occupies several weeks, or even months. The newly formed vinegar is next run off into two large tuns, furnished with false bottoms, on which some 'rape' (the pressed cake from making domestic wines, or the green twigs or cuttings from vines) is placed. One of these vessels is wholly, and the other only about 3-4ths, filled. The fermentation recommences, and the acetification proceeds more rapidly in the latter than in the former tun, and the liquor it contains consequently matures the sooner. When fit for sale, a portion of the vinegar is withdrawn from the smaller quantity, and its place supplied with a like quantity from the full tun, and this in its turn is refilled from the barrels before noticed. This process is carried on with a number of tuns at once, which are all worked in pairs.

Prop., &c. The general properties of malt vinegar are well known. Its pleasant and refreshing odour is chiefly derived from acetic acid and acetic ether. Its strength is distinguished by the makers as Nos. 18, 20, 22, and 24; the last of which, also called ' proof vinegar,' is the strongest, and usually contains about 4·6% of real or about 5% of glacial acetic acid. Its density varies according to the quantity of foreign matter which it contains. Sp. gr. 1·017 to 1·019 (B. P.). This vinegar usually contains a small quantity of sulphuric acid. The presence of 1-1000th part of this acid is allowed by law.

Pur. "Brownish; of a peculiar odour. Its sp. gr. is 1·019. 1 fl. oz. of the acid is saturated by 1 dr. of the crystals of carbonate of soda. If, after 10 minims of solution of chloride of barium have been added to the same quantity, more of the chloride be poured into the filtered acid, nothing further is thrown down. The colour is not changed by the addition of hydrosulphuric acid " (Ph. L.).

2. WINE VINEGAR, FRENCH V.; ACETUM GALLICUM (Ph. E. & D.), A. VINI, L.; VINAIGRE D'ORLÉANS, Fr. This is prepared, in wine countries, from grape-juice and inferior new wines,

worked up with wine-lees, by a nearly similar process to that adopted for malt vinegar. That prepared from white wine (WHITE-WINE VINEGAR) is the most esteemed. It is purer and pleasanter than malt vinegar. Sp. gr. 1·014 to 1·022 (Ph. E.; 1·016—*Phillips*). It usually contains from 5% to 6% of acetic acid. "100 parts of good Orleans vinegar should require 10 parts of dry carbonate of potassa for saturation " (*Soubeiran*).

3. GERMAN, OR QUICK METHOD OF MAKING VINEGAR; PROCESS OF HAM. This method is based upon the fact that acetification is the mere oxidation of alcohol in contact with organic matter. Hence, by employing dilute alcohol, or liquors containing it, and by vastly enlarging the surface of the liquid exposed to the air at a proper temperature, we may reduce the period occupied in acetification from weeks to as many hours. In practice this is effected by causing the dilute spirit, previously mixed with 1-1000th part of sugar or malt extract, or the fermented and clarified malt wort, to slowly trickle down through a mass of beech shavings steeped in vinegar, and contained in a vessel called a vinegar generator (Essigbilder), or graduation vessel. This is an oaken tub, narrower at the bottom than at the top, furnished with a loose lid or cover, below which is a perforated shelf (colander or false bottom), having a number of small holes, which are loosely filled with pack-thread about 6 inches long, and prevented from falling through by a knot at the upper end. The shelf is also perforated with four open glass tubes as air-vents, each having its ends projecting above and below the shelf. This arrangement is repeated a second and a third time, or even oftener, according to the size of the vessel. The tube or graduator at its lower part is pierced with a horizontal row of eight equidistant round holes, to admit atmospheric air. One inch above the bottom is a syphon-formed discharge pipe, whose upper curvature stands one inch below the level of the air-holes in the side of the tub. The floors or partitions of the tub or generator being covered with birch twigs or beech chips to the depth of a few inches, the alcoholic liquor (first heated to between 75° and 83° F.) is introduced at the upper part of the apparatus. This immediately commences trickling slowly down through the holes by means of the packthreads, diffuses itself over the chips or twigs forming the respective strata, slowly collects at the bottom of the tub, and then runs off by the syphon-pipe. The air enters by the circumferential holes, circulates freely through the tub, and escapes by the glass tubes. As the acetification proceeds, the temperature of the liquid rises to 100° or 105° F., and remains stationary at that point while the action goes on favourably. The alcoholic solution or wort requires to be passed three or four times through the cask before acetification is complete, which is, in general, effected in from 24 to 36 hours.

Obs. For the production of a superior vinegar by this process, it is necessary that the spirit employed be sufficiently pure not to contaminate the product with its flavour or odour, and that the malt wort should be fermented and treated with all the care usually employed in the production of beer. The best English manufacturers who have adopted this process are in the habit of filtering or clarifying their fermented wash, and also of storing it away for several months before they subject it to acetification in the graduator. The most favourable temperature for the process is about 90° F., and this should be preserved, as much as possible, by artificial means.—*Prod.* A malt wort of the sp. gr. 1·072, or, in "technical language, weighing about 26 lbs. and barrel," afforded a vinegar containing 5·4% of pure acetic acid, and a residuary extract of 10 lbs. for 36 galls. The former of these would indicate 35 lbs. of sugar, or 13·7 lbs. per barrel of gravity; whilst the latter shows 3·8 lbs. per barrel; the two united being only 17·5 lbs. instead of 26 lbs., the original weight. The loss, therefore, has been 8·5 lbs., or from a sp. gr. of 1·072 to less than 1·050" (*Ure*). Thus about one third of all the extractive matter of the malt is lost or dissipated during the processes of fermentation and acetification. According to Knapp, a mixture of about 80 galls. of water, 9 galls. of spirit of from 44% to 45% Tralles (18 or 20 u. p.), and 3 galls. of vinegar containing 3·5% of real acid, forming together 92 galls., yields, on an average, an almost equal quantity of vinegar, or about from 90 to 91 galls. of the above-stated strength.

4. WOOD VINEGAR. See PYROLIGNEOUS ACID.

5. OTHER VARIETIES OF VINEGAR, of minor importance, chiefly domestic, and commonly 'worked' as malt vinegar—ALE VINEGAR, ALEGAR; ACETUM CEREVISIÆ. From strong pale ale which has soured.—ARGOL VINEGAR; ACETUM EX TARTARO. From white argol or cream of tartar, 1 lb.; dissolved in boiling water, 2 galls.; with the addition, when cold, of proof spirit of whiskey, 3 pints.—CRYSTAL VINEGAR. Pickling vinegar, discoloured with fresh-burnt animal charcoal.—CIDER VINEGAR. From cider, 'worked' as malt vinegar.—GERMAN HOUSEHOLD VINEGAR. From soft water, 7½ galls.; honey or brown sugar, 2 lbs.; cream of tartar, 2 oz.; corn spirit, or whiskey, 1 gall.—GOOSEBERRY VINEGAR. From bruised gooseberries and brown sugar, of each, 1½ lbs.; water, 1 gall. Other fruits may be substituted for gooseberries.—PICKLING VINEGAR. The strongest pale malt vinegar.—RAISIN VINEGAR. From the marc left from making raisin wine, 1 cwt. to every 12 or 15 galls. of water, along with a little yeast.—SUGAR VINEGAR. From brown sugar, 4 lbs. to each gallon of water.—WHISKEY VINEGAR. From whiskey, 1 pint; sugar, 2 oz.; yeast, a dessertspoonful.

Pur., Tests, and Assay. These are, for the most part, rather fully noticed under ACETIC ACID, ACIDIMETRY, and above. The following additional tests, &c., may, however, be useful:—1. Paper written on or smeared with pure vinegar is not charred when strongly warmed before the fire; if it is, the sample examined contains fully 2% of oil of vitriol. 2. A small porcelain capsule or china cup, dipped into a solution of sugar in 30 times its weight of water, and then heated to a temperature equal to that of boiling water, is not materially discoloured when a drop of pure vinegar is poured on it; but a spot of an intensely brown or black colour is formed if the sample contains only 1-300th part of sulphuric acid; if it contains only 1-1000th part the spot is olive-green;

and if a less quantity, then only of a pale green colour. 3. The heavy white precipitate given with chloride of barium (see *above*) shows the presence of sulphuric acid; each grain, after being dried and gently ignited, represents ·344 gr. of dry sulphuric acid. If the precipitate from 1000 gr. of the vinegar exceeds 2½ gr. it contains an illegal quantity of this acid. 4. If a solution of nitrate of silver gives a cloudy white precipitate hydrochloric acid is present. 5. If, after the addition of 2 or 3 gr. of carbonate of potash, and evaporation of the sample to dryness, the residuum deflagrates when ignited, the sample under examination contains nitric acid. 6. If the vinegar be blackened by sulphuretted hydrogen or hydrosulphuret of ammonia, it contains either lead or copper. If it gives a yellow precipitate with iodide of potassium or chromate of potash, the metal is lead. If ferrocyanide of potassium gives a bronze-brown coloured precipitate, or a little olive oil, when agitated with some of the vinegar, be turned green, the metal is copper. 7. If a small sample, gently evaporated to dryness, leaves more than 1% of residuum, and this has a sweet taste, it is undecomposed sugar. The presence of acrid substances, as capsicum, chillies, grains of paradise, mustard seed, pellitory of Spain, pepper, &c., may be detected by neutralising the acidity of the vinegar with carbonate of soda, when the acrid taste of the adulterant will be readily perceived.

Vinegar, Antihyster'ic. *Syn.* ACETUM ANTIHYSTERICUM, L. *Prep.* Castor, 2 dr.; galbanum, 4 dr.; rue, 1 oz.; vinegar, 3 lbs.; macerate and strain.

Vinegar, Aromat'ic. *Syn.* ACETUM AROMATICUM, L. *Prep.* 1. Glacial acetic acid, 1 lb.; oil of cloves, 1½ dr.; oil of rosemary, 1 dr.; oils of bergamot, cinnamon, pimento, and lavender, of each, ½ dr.; neroli, 20 drops; camphor, 2½ oz.; rectified spirit, 2 fl. oz.; mix. Very fine.

2. (*Henry's.*) From glacial acetic acid, strongly scented with the oils of cloves, lavender, rosemary, and *Calamus aromaticus*, to which the usual quantity of camphor is added. This is the formula adopted at Apothecaries' Hall.

3. (Extemporaneous.) From acetate of potash (dry), 1 dr.; oil of vitriol, 20 drops; oils of lemon and cloves, of each, 3 drops.

Vinegar, Aromatic (*for sick rooms*). *Prep.* Camphor, 8 parts; oil of cassia, 1 part; oil of pimento, 1 part; oil of bergamot, 1 part; oil of cloves, 1 part; oil of lavender, 2 parts; acetic acid, 24 parts; rectified spirit, 24 parts; glacial acetic acid to 128 parts.

Obs. Aromatic vinegar is used as a pungent and refreshing perfume, in faintness, &c. For this purpose it is generally dropped on a small piece of sponge placed in a stoppered bottle or a vinaigrette. It is highly corrosive, and should therefore be kept from contact with the skin and clothes.

Vinegar, Camp. *Prep.* Take of sliced garlic, 8 oz.; Cayenne pepper, soy, and walnut ketchup, of each, 4 oz.; 36 chopped anchovies; vinegar, 1 gall.; powdered cochineal, ½ oz.; macerate for a month, strain, and bottle.

Vinegar, Cam'phorated.

Vinegar of Canthar'ides. *Syn.* BLISTERING

VINEGAR; ACETUM CANTHARIDIS (B. P., Ph. L., E., and D.), L. *Prep.* 1. (Ph. L.) Cantharides, in powder, 2 oz.; acetic acid, 1 pint; macerate, with agitation, for 8 days, then press and strain.

2. (Ph. E.) Cantharides, 3 oz.; euphorbium, ½ oz.; acetic acid, 5 fl. oz.; pyroligneous acid, 15 fl. oz.; macerate a week.

3. (Ph. D.) Spanish flies, 4 oz.; strong acetic acid, 4 fl. oz.; commercial acetic acid (sp. gr. 1·044), 16 fl. oz.; macerate, as before, for 14 days.

4. (B. P.) Cantharides, in powder, 2 parts; glacial acetic acid, 2 parts; acetic acid (28%), 18 parts, or a sufficiency; add the glacial acetic acid to 13 parts of acetic acid, and in this mixture digest the cantharides for two hours at a temperature of 200° F.; when cold, place them in a percolator, and when the liquid ceases to drop, pour over the residuum the remaining 5 parts of acetic acid, and when the percolation is finished, press and make the whole liquid up to 20 parts.

Uses, &c. As a counter-irritant, and to raise blisters. For the last purpose it is applied on a piece of lint, evaporation being prevented with a piece of oiled skin or thin sheet gutta percha. The last is the best, and, indeed, the only effective form, the others being too weak. "If the acetic acid be strong, a blister will be as rapidly raised without the cantharides as with them" (*Dr A. T. Thomson*).

Vinegar of Cap'sicum. *Syn.* ACETUM CAPSICI. *Prep.* Capsicum, 1 oz.; vinegar, 24 oz. Used as gargle.

Vinegar of Col'chicum. *Syn.* ACETUM COLCHICI (Ph. L., E., and D.), L. *Prep.* 1. (Ph. L.) Dried corms of colchicum or meadow saffron, 3½ dr.; dilute acetic acid, 1 pint; macerate for 3 days, then press out the liquor, and, after defecation, add to the strained liquid proof spirit, 1½ fl. oz.

2. (Ph. E.) Fresh colchicum bulbs (dried), 1 oz.; distilled vinegar, 16 fl. oz.; proof spirit, 1 fl. oz.

3. (Ph. D.) Dried colchicum bulbs, 1 oz.; acetic acid (1·044), 4 fl. oz.; distilled water, 12 fl. oz.; as before, but prolonging the maceration for 7 days.

Obs. Vinegar of colchicum is chiefly used in gout.—*Dose*, 20 drops to 1 fl. dr. The Dublin preparation is about three times as strong as the others, and the dose must therefore be proportionately less.

Vinegar, Cur'rie. *Prep.* From currie powder, ½ lb.; vinegar, 1 gall.; infuse for a week. Used as a flavouring. Other like vinegars may be made in the same way.

Vinegar, Distilled. *Syn.* ACETUM DESTILLATUM (Ph. L. and E., and Ph. D. 1826), L. *Prep.* 1. (Ph. L.) Vinegar, 1 gall.; distil in a sand-bath 7 pints. Sp. gr. 1·0065.

2. (Ph. E.) Vinegar (preferably French), 8 parts; distil over with a gentle heat 7 parts; and dilute the product, if necessary, with distilled water, until the sp. gr. is 1·005.

Pur., &c. "1 fl. oz. is saturated by 57 gr. of crystallised carbonate of soda" (Ph. L.). 100 gr. are saturated by 13 gr. of crystallised carbonate of soda. It contains about 4·6% of real acetic acid. If a pewter worm is used, a portion of lead is dissolved, and the product becomes cloudy and poisonous. Distilled vinegar is more agreeable than pure dilute acetic acid of the same strength.

Vinegar of Fox'glove. *Syn.* ACETUM DIGITALIS (Ph. G.), L. *Prep.* Dried foxglove, 1 oz.; vinegar, 9 oz. (by weight); rectified spirit, 1 oz. (by weight). Macerate for 8 days, press, and fil. ter.—*Dose*, 30 minims.

Vinegar of Gar'lic. *Syn.* ACETUM ALLII. *Prep.* Fresh garlic, 1 oz.; distilled vinegar, 12 oz.

Vinegar of Lav'ender. *Syn.* ACETUM LAVANDULÆ (P. Cod.), L. *Prep.* Digest 1 troy oz. of dried lavender flowers with 12 oz. of vinegar for 10 days. The vinegars of other flowers are made in the same manner.

Vinegar of Lobelia. *Syn.* ACETUM LOBELIÆ, L. *Prep.* Lobelia in moderately coarse powder, 4 troy oz. Diluted acetic acid, 2 pints (o. m.). Macerate for 7 days.

Vinegar, Marseilles. *Syn.* VINEGAR OF THE FOUR THIEVES, PROPHYLACTIC VINEGAR; ACETUM PROPHYLACTICUM, A. ANTISEPTICUM, A. THERIACALE, A. QUATUOR FURUM, L.; VINAIGRE DES QUATRE VOLEURS, Fr. *Prep.* Take of the summits of rosemary and flowers of sage (dried), of each, 4 oz.; dried lavender flowers, 2 oz.; cloves, 1 dr.; distilled vinegar, 1 gall.; digest for 7 days, press, and strain. Used as a corrector of bad smells, and formerly as a prophylactic against the plague and other contagious diseases. It is said to have been a favourite preventive with Cardinal Wolsey, who always carried some with him. The original formula also contained of garlic, ½ oz.; fresh rue, 1½ oz.; and camphor, dissolved in spirit, 1 oz.

Vinegar of Mus'tard. *Syn.* ACETUM SINAPIS (Beral), L. *Prep.* Mustard, 1 oz.; vinegar, 12 oz.; distil 8 oz. For outward use, as a counter-irritant.

Vinegar of O''pium. *Syn.* ACETUM OPII (Ph. E. and D.), L. *Prep.* 1. (Ph. E.) Opium, sliced, 4 oz.; distilled vinegar, 16 fl. oz.; macerate for 7 days, press, and filter.—*Dose*, 5 to 20 drops.

2. (Ph. D.) Opium, in coarse powder, 1½ oz.; dilute acetic acid, 1 pint; macerate for 7 days.—*Dose*, 10 or 12 to 60 drops.

Obs. These were intended to supersede the old 'black drop,' which they closely resemble in their action.

Vinegar, Rasp'berry. *Syn.* ACETUM RUBI IDÆI, L.; VINAIGRE FRAMBOISE, Fr. *Prep.* 1. Bruised ripe raspberries and white wine vinegar, of each, 3 pints; macerate for 3 days, press, strain, and to each pint add of white sugar, 1 lb.; boil, skim, cool, and at once bottle. Some persons add 2 fl. oz. of brandy to each pint.

2. (P. Cod.) French raspberries, picked from their calices, 3 lbs. (1 lb.—Ph. Bor.); good vinegar, 2 lbs.; macerate in glass for a fortnight, then strain, without pressure.

Obs. In a similar manner may be made cherry vinegar, strawberry v., and the vinegars of all other like fruits.

Vinegar of Rue. *Syn.* ACETUM RUTÆ (Ed.

Ph. 1744). *Prep.* Rue, 1 lb. troy; vinegar, 1 gall.

Vinegar of Squills. *Syn.* ACETUM SCILLÆ, ACETUM SCILLITICUM, L. *Prep.* 1. (B. P.) Take of squills, recently dried and bruised, 2½ oz.; dilute acetic acid, 1 pint; macerate in a covered vessel for 7 days, then press out the liquor, and filter.

2. (Wholesale.) From squills, 7 lbs.; distilled vinegar, 6 galls.; macerate in the cold for 10 days, press, and filter. Expectorant and diuretic. —*Dose*, ½ to 1½ fl. dr.; in chronic pulmonary affections, dropsies, &c.

Vinegar (Culinary). *Prep.* 1. BLACK PEPPER VINEGAR, CAPER V., CAPSICUM V., CELERY-SEED V., CHILLIE V., CRESS-SEED V., GARLIC V., GINGER V., HORSERADISH V., ONION V., RED-ROSE V., SEVILLE ORANGE-PEEL V., SHALLOT V., TRUFFLE V., WHITE PEPPER V., with several others of a like kind, are made by steeping about an oz. of the respective articles in a pint of good vinegar for 14 days, and straining.

2. BASIL VINEGAR, BURNET V., CELERY V., CHERVIL V., ELDER-FLOWER V., GREEN-MINT V., TARRAGON V., with several others from like substances, are prepared from 2 to 3 oz. of the leaves to each pint of vinegar; the whole being frequently shaken for 14 days, then strained and bottled. They are used in cookery. The culinary vinegars may also be prepared in the same manner as the 'culinary spirits' and 'tinctures,' by simply substituting strong pickling vinegar for the spirit.

Vinegars (Perfumed). *Syn.* ACETA ODORIFERA, L. *Prep.* From the dried flowers, 1 to 2 oz., or the fresh flowers, 2 to 4 oz.; strongest distilled vinegar, 1 pint; digest for a week, strain with pressure, and repeat the process with fresh flowers if necessary. They may also be made by adding 15 to 20 drops, or q. s. of the respective essential oils to the vinegar. In a similar way are prepared the vinegars of clove gillyflowers, elder flowers, lavender flowers (vinaigre distillé de lavande), musk roses, orange flowers (fresh), Provence roses, red roses (vinaigre de rose; acetum rosatum), rosemary flowers (vinaigre de rosmarin; acetum anthosatum), tarragon flowers, &c. &c. Another excellent plan is to add 1 fl. oz. of glacial acetic acid to each pint of the respective perfumed spirits. This answers admirably for acetic eau de Cologne and like perfumes.

Vinegars (Spiced). The following are given by the 'Mineral Water Trades Review:'

For French Beans. Distilled or very pale malt vinegar, 1 gall.; white peppercorns, 4 oz.; bleached ginger (sliced), 2 oz.; chillies, 1 oz. Into ½ gall. of the vinegar place the whole of the spices and allow to macerate for twelve hours; then simmer (do not boil) gently for one hour in an enamelled pan, covering the top. To be used hot.

For Gherkins. Good malt vinegar, 1 gall.; black peppercorns, 6 oz.; sliced ginger, 4 oz.; chillies, 1 oz.; garlic, in slices, 1 oz. Boil the spices and garlic gently in half the vinegar for half an hour, strain through a sieve, and add the rest of the vinegar to the spices and again strain. To the remnant spices add 2 oz. of salt and 1 pint of water, and boil for half an hour. After removing from the fire add 1 pint of vinegar, and again

strain into the spiced vinegar, which when perfectly cold may be poured over the gherkins.

For Walnuts (to be used hot). Good malt vinegar, 2 galls.; black peppercorns, ½ lb.; ginger, unbleached, 6 oz.; mustard seed, 1 lb.; cloves, 2 oz.; mace, 2 oz.; garlic, in slices, 2 oz. In 1 gall. of vinegar boil the whole of the spices, and having strained, pour the hot liquor over the walnuts, then boil the remaining gallon of vinegar and pour over spices, &c. This pickle takes some time to mature, but if properly prepared should be ready for use in three months.

VI'NOUS FERMENTA'TION. *Syn.* ALCOHOLIC FERMENTATION. The peculiar change by which sugar, in solution, is converted into carbonic acid, which is eliminated, and into alcohol, which remains in solution in the fermented liquor.

The presence of a 'ferment' is essential to excite the vinous fermentation, as a solution of absolutely pure water remains unaltered, even though exposed to the conditions most favourable to its accession. In the juices of the sweet fruits, and in those vegetable solutions that spontaneously run into a state of fermentation, the ferment is supplied by nature, and is intimately associated with the saccharine matter. In the juice of those grapes which produce the more perfect wines, the relative proportions of the exciters of fermentation and the sugar are so accurately apportioned, that the whole of the former are decomposed, and nearly the whole of the latter is converted into alcohol; so that the liquid (wine) is left in a state but little liable to future change. An infusion of malt, however, in which the nitrogenised matters (gluten, vegetable albumen, &c.) are absent, or at least present in too small quantities to vigorously excite the vinous fermentation, undergoes a mixed species of decomposition, with the formation of products widely different from those that result from the true vinous fermentation; or, in other words, the liquid becomes spoiled. But if a ferment (yeast) be added to this infusion of malt under the above circumstances, and in the proper proportion to the sugar present, the true vinous fermentation speedily commences, and the liquid becomes converted into beer. This is what actually takes place in the process of brewing, and the scientific brewer endeavours to employ a proper quantity of ferment to decompose the whole of the saccharine matter of his wort; but, at the same time, as equally endeavours to avoid the use of an excess.

The chief product of the vinous fermentation is alcohol, but there are other substances simultaneously produced, and which remain associated with the fermented liquor. Among the principal of these are œnanthic acid, œnanthic ether, fusel oil (oil of potato spirit, oil of grain), &c.; none of which exist previously to fermentation, and are generally supposed to result from the action of the nitrogenised matters of the solution on the sugar. Under certain circumstances these extraneous products are formed in much larger quantities than under others; and as these substances injure the value of the alcohol with which they are associated, a knowledge of the peculiar circumstances favourable and unfavourable to their production is a desideratum to the brewer and distiller.

According to MM. Colin and Thénard, Frémy, Rousseau, and others, the essential condition of a ferment, to be able to excite the pure vinous fermentation, is to be sufficiently acidulous to act on coloured test-paper; and this acidity should arise from the presence of certain vegetable acids and salts, capable of conversion into carbonic acid and carbonates by their spontaneous decomposition. Those acids and salts which are found to pre-exist in fermentable fruits and liquors, as their tartaric, citric, malic, and lactic acids, and their salts should be chosen for this purpose; preference being given to the bitartrate of potassa, on account of its presence in the grape. The addition of any of these substances to a saccharine solution renders its fermentation both more active and complete. The favourable influence of cream of tartar on fermentation was first pointed out by Thénard and Colin, and the addition of a little of this article has been adopted in practice, with manifest advantage, by the manufacturers of British wine.

There is good reason for supposing that each variety of sugar which is susceptible of the alcoholic fermentation is first converted into grape sugar by contact with the ferment, and that this variety of sugar is alone capable of yielding carbonic acid and alcohol.

The circumstances most favourable to this fermentation are a certain quantity of active ferment, and its due distribution through the liquor. The temperature of from 68° to 77° F. is usually regarded as the most propitious for the commencement and progress of fermentation; but it has been ably shown by Liebig that, at this temperature, the newly formed alcohol slowly undergoes the 'acetous fermentation,' forming vinegar, by which the vinous character of the liquor is lessened. This conversion of alcohol into vinegar proceeds most rapidly at a temperature of 95° F., and gradually becomes more languid, until, at about 46° to 50° F. (8 to 10 C.), it ceases altogether, while the tendency of the nitrogenous substances to absorb oxygen at this low temperature is scarcely diminished in a perceptible degree. " It is therefore evident that if wort (or any other saccharine solution) is fermented in wide, open, shallow vessels, as is done in Bavaria, which afford free and unlimited access to the atmospheric oxygen, and this in a situation where the temperature does not exceed 46° to 50° F., a separation of the nitrogenous constituents, *i. e.* the exciters of acidification, takes place simultaneously on the surface, and in the whole body of the liquid " (*Liebig*). By this method wine or beer is obtained which is invariably far superior in quality to that fermented in the usual manner. See FERMENTATION.

The symptoms of a perfect fermentation of malt wort, according to the usual English system with top yeast (*oberhefe*), have been thus described by a well-known practical writer on brewing:—1. A cream-like substance forms round the edges of the gyle tun, which gradually extends itself, and ultimately covers the whole surface of the liquor. 2. A fine curly or cauliflower head in a similar way extends itself over the surface, and indicates to the experienced brewer the probable quality of the fermentation. 3. The 'stomach,' or 'vinous

odour,' is next evolved, and continues to increase with the attenuation of the wort. The peculiar nature of this odour is also an indication of the state of the fermentation. 4. The cauliflower head changes, or rises to a fine 'rocky' or 'yeasty' head, and ultimately falls down. 5. In this stage the head assumes a peculiar yeasty appearance, called by brewers 'close-yeasty,' and the gas is evolved in sufficient quantity to blow up little bells or bubbles, which immediately burst, and are followed by others, at intervals depending on the activity and forwardness of the fermentation. These bells should be bright and clear; as, if they appear opaque and dirty, there is something the matter with the wort (*Black*).

It is often of the utmost importance to brewers, wine merchants, sugar refiners, druggists, &c., to be able to lessen the activity of the vinous fermentation, or to stop it altogether, or to prevent its accession to syrups and other saccharine and vegetable solutions. Whatever will still the motion of the molecules of the nitrogenous matter forming the ferment will render them inoperative as exciters of fermentation. Among the simplest means of effecting this object, and such as admit of easy practical application, may be mentioned exposure to either cold or heat. At a temperature below 50° F. the acetous fermentation is suspended, and the alcoholic fermentation proceeds with diminished activity as the temperature falls, until at about 38° F. it ceases altogether. In like manner the rapid increase of the temperature of a fermenting liquid arrests its fermentation, and is preferable to the action of cold, as it is of easier application, and perfectly precipitates the ferment in an inert state. For this purpose a heat of about 180° F. is sufficient; but even that of boiling water may be employed with advantage. In practice fluids are commonly raised to their boiling-point for this purpose, or they are submitted to the heat of a warm bath (207½° F.). In this way the fermentation of syrups and vegetable solutions and juices is commonly arrested in the pharmaceutical laboratory.

Among substances that may be added to liquids to arrest fermentation the most active are—the volatile oil of mustard, coarsely powdered mustard seed, or pure flour of mustard, sulphurous acid or the fumes of burning sulphur, sulphuric acid, sulphite of lime, tincture of catechu, strong spirit, strong acetic acid, chlorate of potassa, sugar of milk, bruised horse-radish, garlic, and cloves, and their essential oils, and all the other volatile oils that contain sulphur, and most of the salts that readily part with their oxygen. These substances arrest fermentation by rendering the yeast inoperative, and they possess this power nearly in the order in which they stand above. In practice, mustard, the fumes of burning sulphur, sulphite of lime, and chlorate of potassa, are those most adapted for beer, cider, wines, syrups, &c.; but some of the others are occasionally used, though less active. For arresting or preventing the fermentation of the vegetable juices and solutions, and the medicated syrups employed in pharmacy, mustard seed, either alone or combined with a little bruised cloves, may be safely used, as the addition of acids or salts would lead to the decomposition of their active principles.

For this reason such liquids should be kept in a sufficiently low temperature to prevent fermentation; and should they pass into that state it should be preferably arrested by the application of heat or cold, as above explained. Sugar of milk is also very effective for certain syrups, if not all of them.

To prevent, or rather to lessen, the production of fusel oil, it has been proposed to add a certain quantity of tartaric acid or bitartrate of potassa to the wort, or to arrest the fermentative process somewhat before the liquid has reached its utmost degree of attenuation. The best means of depriving the spirit of this and other substances of a similar nature is to largely dilute it with water, and to redistil it at a gentle heat. Agitation with olive oil, decantation, dilution with a large quantity of water, and redistillation, have also been recommended. An excellent method is filtration through newly burnt and coarsely powdered charcoal. This plan succeeds perfectly with moderately diluted spirit. On the Continent, the addition of about 10% of common vinegar, and a very little sulphuric acid, followed by agitation, repose for a few days, and redistillation, is a favourite method. A solution of chloride of lime is also employed for the same purpose, and in the same way. In both these cases a species of ether is formed, which possesses a very agreeable odour. In the first, acetate of oxide of amyl (essence of jargonelle) is produced; and in the other, chloride of amyl, which also possesses a pleasant ethereal smell and taste. The affinity of the hydrated oxide of amyl (fusel oil) for acetic acid is so great that they readily unite without the intervention of a mineral acid (*Doebereiner*). Thus the oil of vitriol mentioned above, though always used in practice, might be omitted without any disadvantage.

According to Messrs Bowerbank, the distillers quoted by Dr Pereira, 500 galls. of corn-spirit yield about 1 gall. of corn-spirit oil. See ACETIFICATION, ALCOHOL, BREWING, DISTILLATION, FERMENTATION, FUSEL OIL, SPIRIT, VINEGAR, VISCOUS FERMENTATION, YEAST, &c.

VI'OLET. *Syn.* PURPLE VIOLET, SWEET V.; VIOLA (Ph. L. & E.), L. "The recent petals of *Viola odorata*, Linn." (Ph. L.). It is chiefly used on account of its colour. See SYRUP.

VIOLET DYE. Violet, like purple, is produced by a mixture of red and blue colouring matter, applied either together or in succession. The 'aniline colours' are now almost exclusively used for obtaining violet on silk and wool (see ANILINE, PURPLE, and TAR COLOURS). With the old dye-stuffs, violet may thus be obtained:—A good violet may be given to silk or wool by passing it first through a solution of verdigris, then through a decoction of logwood, and lastly through alum water. A fast violet may be given by first dyeing the goods a crimson with cochineal, without alum or tartar, and, after rinsing, passing them through the indigo vat. Linens and cottons are first galled with about 18% of gall-nuts, next passed through a mixed mordant of alum, iron liquor, and sulphate of copper, working them well, then through a madder-bath made with an equal weight of root, and, lastly, brightened with soap or soda. Another good method is to pass cloth, previously dyed Turkey red, through the blue vat. Wool, silk, cotton, or linen, mordanted with alum and dyed in a logwood-bath, or a mixed bath of archil and Brazil, takes a pretty but false violet.

VIS'COUS FERMENTATION. *Syn.* MUCILAGINOUS FERMENTATION, MUCOUS F. The peculiar change by which sugar, in solution, is converted into gummy matters, and other products, instead of into alcohol.

When the expressed juice of the beet is exposed to a temperature of 90° to 100° F. for a considerable time, the sugar it contains suffers this peculiar kind of fermentation. Gases are evolved which are rich in hydrogen, instead of being exclusively carbonic acid; and when the sugar has, for the most part, disappeared, mere traces of alcohol are found in the liquid, but, in place of that substance, a quantity of lactic acid, mannite, and a mucilaginous substance resembling gum-arabic, and said to be identical with gum in composition. By boiling yeast or the gluten of wheat in water, dissolving sugar in the filtered solution, and exposing it to a tolerably high temperature, the viscous fermentation is set up, and a large quantity of the gummy principle generated, along with a ferment of a globular texture, like that of yeast, but which is capable of producing only the viscous fermentation in saccharine solutions.

The peculiar cloudy, stringy, oily appearance of wine and beer, called by the French 'graisse,' and the English 'ropiness,' depends on the accession of the viscous fermentation. The mineral acids and astringent substances, especially the sulphuric and sulphurous acids, and tannin, precipitate the viscous ferment, and are, hence, the best cures for this malady of fermented liquors. It is the large amount of tannic acid in the red wines and well-hopped beer which is the cause of their never being attacked with 'graisse,' or 'ropiness.' See VINOUS FERMENTATION, WINES, &c.

VI"SION. The following means of preserving and restoring the sight may be appropriately inserted here:

For NEAR-SIGHTEDNESS. Close the eyes and press the fingers very gently, from the nose outward, across the eyes. This flattens the pupil, and thus lengthens or extends the angle of vision. This should be done several times a day, or at least always after washing the face, until short-sightedness is overcome.

For LOSS OF SIGHT BY AGE, such as require magnifying glasses, pass the fingers or towel from the outer corners of the eyes inwardly, above and below the eyeballs, pressing very gently against them. This rounds them up, and preserves or restores the sight.

It is said that many persons, by this last means, have preserved their sight so as to read fine print at 80 years of age; others, whose sight has been impaired by age, by carefully manipulating the eyes with their fingers, from their external angles inwardly, have restored their sight, and been able to dispense with glasses, and have since preserved it by a continuance of the practice. To be successful, or safe, these practices must be applied with great gentleness and caution.

The 'Lancet' remarks that "there is good reason to believe that chicory (the coffee of the Londoners), from its narcotic character, exerts an injurious effect on the nervous system. So convinced of this is Professor Beer, of Vienna, a most celebrated oculist, that he has enumerated chicoried coffee among the causes of amaurotic blindness."

To strengthen the eyes, to relieve them when swollen or congested, and to remove chronic ophthalmia, purulent discharges, &c., nothing is equal to frequently bathing them with water, at first tepid, but afterwards lowered in temperature to absolute coldness.

VITRIOL. A common name for sulphuric acid and for several of its salts (see *below*).

Vitriol, Blue. *Syn.* ROMAN VITRIOL. Commercial sulphate of copper.

Vitriol, Green. Commercial sulphate of iron.

Vitriol, White. Commercial sulphate of zinc.

VITTIE-VAYR. *Syn.* VETIVER. The Tamool name of the odorous and fibrous roots of the *Andropogon muricatus* sold by the perfumers.

VOLTAIC ELECTRICITY. *Syn.* GALVANIC E., GALVANISM, VOLTAISM. That branch of electrical science which has reference to the phenomena attendant on the development of electricity by chemical action. Electricity thus developed may be made to show itself in the 'static' condition, so as to produce the effects of frictional electricity, but it is much more easily obtained in the 'dynamic' condition—in other words, as a 'voltaic current'—when it is especially remarkable for its chemical and magnetic effects. If a plate of zinc and a plate of platinum be immersed in dilute sulphuric acid, and connected outside the liquid by a wire, a current of electricity will immediately be set up, and will continue as long as the conducting circuit is complete and the action of the acid on the zinc goes on. The current of 'positive' electricity passes from the zinc, through the liquid, to the platinum, and thence through the wire to the zinc. The arrangement of two dissimilar metals immersed in a liquid which acts upon one of them is called a voltaic couple. By uniting a number of couples together in regular order, a voltaic pile or battery is formed.

The older forms of the voltaic battery, viz. VOLTA'S PILE, CRUIKSHANK'S TROUGH, and WOLLASTON'S BATTERY, are now but little used. They all consist of a series of couples of zinc and copper, excited by an acid liquid, generally a mixture of water with $\frac{1}{20}$ of its bulk of sulphuric acid, and $\frac{1}{20}$ of nitric acid.

One of the most useful forms of the voltaic battery is that proposed by the late Professor Daniell, and commonly known by his name. Its peculiar advantages arise from its action continuing without interruption for a long time; hence the name of 'constant battery' that has been applied to it. The following figure will explain the construction of each couple.

One of these couples is sufficient for electrotyping; six of them form a circle of considerable power, and about twenty produce one sufficiently strong for most experiments of demonstration and research.

In arranging these, as well as other batteries, when intensity, or travelling power, is desired, the metallic communication is made between the opposite metals (the zinc of one couple being

A. A copper cylinder, filled with a saturated solution of sulphate of copper.
B. A smaller porous cylinder (earthenware or membrane), containing a mixture of 1 measure of strong sulphuric acid, and about 8 measures of water.
C. A rod of amalgamated zinc, supported in the smaller cylinder by the cross-piece *i.*
D. A shelf full of small holes, for supporting crystals of sulphate of copper, to keep up the strength of the solution.
s and *f.* Screws and caps to connect the wires *g* and *h* with the battery.
g. The negative wire, connected with the zinc.
h. The positive wire, connected with the copper.

united with the copper of another); but when simple quantity without intensity is required, the zinc of one battery is united with the zinc of the other, and the copper of the one with the copper of the other—an effect which is equally attainable with a single battery of enlarged dimensions.

Another useful apparatus is GROVE'S BATTERY, in which the positive metal consists of amalgamated zinc immersed in sulphuric acid, diluted with ten times its bulk of water; and the negative metal of platinum immersed in strong nitric acid. The two liquids are kept separate by the use of porous vessels, as in 'Daniell's battery.' This is an extremely powerful arrangement, but not so constant as Daniell's, owing to the reduction of the nitric acid to lower oxides of nitrogen. After this battery has been in action for about an hour, copious red nitrous fumes are given off, which cause great annoyance.

In place of platinum, compact charcoal or coke, prepared by a rather troublesome process, may be used, and the arrangement then constitutes a BUNSEN'S BATTERY. Other substitutes for the costly platinum have been proposed, as lead coated with gold or platinum, and iron rendered 'passive' by immersion in strong nitric acid. Callan has obtained very good results with amalgamated zinc and cast iron immersed in diluted sulphuric acid, without the use of nitric acid (MAYNOOTH BATTERY).

In SMEE'S BATTERY, which is much used in the arts, pairs of amalgamated zinc and platinised silver (or platinised platinum) are immersed in dilute sulphuric acid (1 part acid to 7 parts water). The plates of zinc are usually bent double, and the platinised plates interposed between the two surfaces formed by the bend. See PLATINISING (p. 1337).

In every voltaic combination the passage of the electricity (*i.e.* the positive modification of the force) in the liquid is from the active element to the inactive element; in the case of a simple zinc-and-copper couple, for instance, it is from the zinc to the copper. If this simple fact be borne in mind, it will decide in every case the question which confuses so many, namely, which is the positive, and which the negative end of a battery? The positive is the end where the electricity leaves the battery; the negative where it re-enters it. For further information connected with the subject of voltaic electricity, see articles on ELECTRICITY, ELECTROLYSIS, ELECTROTYPE, ETCHING, &c.

VOLUME'TRIC ANALYSIS. Quantitative chemical analysis by measure. This method of analysis "consists in submitting the substance to be estimated to certain characteristic reactions, employing for such reactions liquids of known strength, and from the quantity of the liquid employed determining the weight of the substance to be estimated by means of the known laws of equivalence." As an example of this method we give the following from the Introduction in Sutton's excellent 'Handbook of Volumetric Analysis:'—" Suppose that it is desirable to know the quantity of pure silver contained in a shilling. The coin is first dissolved in nitric acid, by which means a bluish solution, containing silver, copper, and probably other metals, is obtained. It is a known fact that chlorine combines with silver in the presence of other metals to form chloride of silver, which is insoluble in nitric acid. The proportions in which the combination takes place are 35·46 of chlorine to every 108 of silver; consequently, if a standard solution of pure chloride of sodium is prepared by dissolving 58·46 grains of the salt (i. e. 1 equiv. sodium = 23, 1 eq. chlorine = 35·46 = 1 eq. chloride of sodium 58·46) in so much distilled water as will make up exactly 1000 grains by measure, every single grain of this solution will combine with 0·108 grain of pure silver to form chloride of silver, which precipitates to the bottom of the vessel in which the mixture is made. In the process of adding the salt solution to the silver, drop by drop, a point is at last reached when the precipitate ceases to form. Here the process must stop. On looking carefully at the graduated vessel from which the standard solution has been used, the operator sees at once the number of grains which have been necessary to produce the complete decomposition. For example, suppose the quantity used was 520 grains; all that is necessary to be done is to multiply 0·108 grain by 520, which shows the amount of pure silver present to be 56·16 grains." For grains of course grammes may be substituted, and it is convenient to reckon the number of cubic centimetres of standard solution used, it being known how much each cubic centimetre is equivalent to. The volumetric method is much less troublesome than the ordinary method of analysis (by separating the constituents of a mixture and weighing them), and is admirably adapted for the examination of substances used in arts and manufactures. Most of the processes described under ACIDIMETRY and ALKALIMETRY are examples of this method. See those articles, also EQUIVALENTS, TEST SOLUTIONS, &c.

WADE'S DROPS. Compound tincture of benzoin.

WA'FER PAPER. See WAFERS, in Cookery (below).

WA'FERS. Thin adhesive discs, used for securing letters or sticking papers together.

Prep. 1. (WAFERS, FLOUR W.) The finest wheaten flour is mixed with water, either pure or coloured, to a smooth pap or batter, which, after being passed through a sieve, to remove clots or lumps, is poured into the 'wafer-irons' (previously warmed and greased with butter or olive oil), and in this state exposed to the heat of a clear charcoal fire; the whole is then allowed to cool, when the irons are opened, and the thin cake, which has become hard and brittle, is cut into wafers by means of sharp annular steel punches made exclusively for the purpose.

2. (GELATINE WAFERS, TRANSPARENT W.) Good gelatine or glue is dissolved, by the heat of a water-bath, in just sufficient water to form a consistent mass on cooling; it is then poured, whilst hot, upon the surface of a warm plate or mirror glass, slightly oiled, and surrounded with a border of card paper (laid flat); a similar plate, also warmed and oiled, is next laid upon the gelatine, and the two plates pressed into as close contact as is permitted by the card paper; when quite cold the thin sheet of gelatine is removed, and cut into wafers with punches, as before. 1 to 2 oz. of sugar is commonly added to each lb. of gelatine.

3. (MEDALLION WAFERS.) A sheet of metal or glass, having designs sunk in it corresponding to the raised part of seals, being provided, the hollows are filled up with a mixture formed of any appropriate coloured powder, made into a paste with gum water or size, leaving the flat part clear; melted coloured glue is then poured on the plate, and the process is otherwise conducted as before. For use, the paper is wetted where the wafer is to be applied.

Obs. Care must be taken that no poisonous colours be employed. For gelatine wafers, transparent colours only can be used. Those noticed under LIQUEURS and STAINS (Confectioner's) are appropriate. To these may be added plumbago, sesquioxide of iron (crocus martis), smalts, levigated vegetable charcoal, and vermilion.

Wafers (in Cookery). Prep. Make fine flour, dried and sifted, into a smooth thin batter with good milk, or a little cream-and-water; add about as much white wine as will make it thick enough for pancakes, sweeten it with a little loaf sugar, and flavour it with powdered cinnamon. When thus prepared, have the wafer-irons made ready by being heated over a charcoal fire; rub them with a piece of linen cloth dipped in butter; then pour a spoonful of the batter upon them, and close them almost immediately; turn them upon the fire, and pare the edges with a knife if any of the batter oozes out. A short time will bake them when the irons are perfectly heated. The wafers must be curled round whilst warm when they are for ornaments. 'Wafer paper' is prepared in a similar way to the above; but when intended to be kept for some time, the milk must be omitted. Used by cooks, &c.; and recently, as an envelope for nauseous medicines.

Wafers, DE SILVA's. These nostrums were introduced to the public some time ago, as though they were prepared from the formulæ of a celebrated physician whose name was affixed to them (for an exposition of the Da Silva quackery, with Dr Locock's letter on the subject, see the 'Anat. of Quackery,' or the 'Med. Circ.,' ii, 106—126). There are three varieties, which are said to be prepared as follows:

1. APERIENT or ANTIBILIOUS WAFERS. From sugar and extract of liquorice (Spanish juice), equal parts; senna and jalap, of each, in fine powder, about ½ dr. to every oz. of sugar employed; made into a mass with a concentrated infusion of senna, and divided into 12-gr. lozenges or squares with the corners rounded off.

2. FEMALE WAFERS. From sugar, horehound candy (or honey), and aperient wafer mass, equal parts; beaten to a proper consistence with weak gum water, to which a little orange-flower water has been added, and divided into 8-gr. tabellæ, as before.

3. PULMONIC WAFERS. From lump sugar and starch, of each in powder, 2 parts; powdered gum, 1 part; made into a lozenge-mass with vinegar of squills, oxymel of squills, and ipecacuanha wine, equal parts, gently evaporated to 1·6th their weight, with the addition of lactucarium in the proportion of 20 to 30 gr. to every oz. of the dry powders, the mass being divided into half-inch squares, weighing about 7½ gr. each (when dry), as before.

WAI-FA. Flower-buds of Saphore japonica, used by the Chinese for dyeing yellow, or rather for rendering blue cottons and silks green.

WALNUT. The Juglans regia, a tree of the Nat. Ord. JUGLANDACEÆ. The sap yields sugar; the fruit is the walnut; the kernels of the latter are eaten and, pressed for their oil; the peel or husks are used for 'rooting' or dyeing brown; the unripe fruit is pickled, and its juice is used as a hair-dye; the leaves are reputed diaphoretic and antisyphilitic; and the wood is esteemed for cabinet work.

WARBURG'S FEVER DROPS. See TINCTURE, WARBURG'S FEVER.

WARD'S RED DROP. A strong solution of emetic tartar in wine.

WARTS. Syn. VERRUCÆ, L. These chiefly attack the hands, and may be removed by the daily use of a little nitrate of silver, nitric acid, or aromatic vinegar, as directed under CORNS. The first of the above applications produces a black stain, and the second a yellow one; both of which, however, wear off after the lapse of some days. Acetic acid scarcely discolours the skin. Erasmus Wilson, the eminent surgeon and talented author of several works on the skin, mentions the case of a gentleman who removed an entire crop of warts from his knuckles and fingers by subjecting them to a succession of sparks from one of the poles of an electrical machine. "He was in the habit, as is usual, of trying the amount of electric fluid collected in his machine by placing his knuckle near the brass knob, and receiving a spark. Observing that an odd sensation was produced whenever the spark struck a wart, he was tempted to give them a round of discharges. When his attention was next directed to his hands he found, to his surprise and satisfaction, that all

the warts had disappeared." Dr Pees, of Wiesbaden, recommends the internal use of carbonate of magnesia in cases of warts.

The papular eruption which covers the hands of some persons, and which is occasionally called 'soft warts,' is best removed by the daily use of Goulard's lotion.

WASH. The fermented wort of the distiller.

WASH-BALLS. See SAVONETTES.

WASH-BOTTLE. The principle of this very common and indispensable laboratory utensil, by which precipitates are washed, will be readily understood by reference to No. 1 of the engrs.

The bottle being two thirds filled with distilled water by blowing into the shorter tube, b, a small jet of water is forced through the nozzle of the longer tube, c. We give the following directions for the construction of a WASH-BOTTLE, from Mr Clowes' excellent little manual, entitled 'An Elementary Treatise on Practical Chemistry' (J. and A. Churchill, New Burlington Street):—

"A thin, flat-bottomed flask is chosen, of 16 or 18 ounces capacity; the neck must not be less than an inch in diameter. Procure a sound cork, which is slightly too large to enter the neck, soften the cork by placing it upon the floor and rolling it backwards and forwards under the foot with gentle pressure; when thus softened, the cork must fit tightly into the flask. (A vulcanized india-rubber stopper is much more durable for this and most other chemical processes. It is perforated by a sharp, well-wetted cork bore, or by a wetted round file.) Two pieces of glass tubing rather longer than would be required for the tubes a and b are then bent into the form shown in Fig. 1. The ends of the tubes are, if necessary, cut off to the right length, and their sharp edges rounded by holding them in the Bunsen flame, or the tip of the blowpipe flame.

"Two parallel holes are then bored in the cork by means of a round file, or by a proper size corkborer; the holes must be rather smaller than the glass tubes, and must not run into one another, or to the outside of the cork. They are slightly enlarged, if necessary, by the round file. Into these holes the tubes a and b are then pushed with a twisting motion; if the holes have been made of the proper size the tubes must enter somewhat stiffly, but without requiring much pressure. Upon the upper end of a is fitted a small piece of india-rubber tubing, about an inch and a half in length, and into the other end of this is a finished short jet (c) made by drawing out a piece of glass tubing in the flame; its nozzle may be constructed, if necessary, by holding it perfectly dry in the flame for some time. The neck of the bottle should then be bound round with twine, like the handle of a cricket bat, or tightly covered with a piece of flannel. This prevents the fingers from being burnt when the bottle contains boiling water."

We append below some varieties of washing bottles. The round-bottomed are in more general requisition than the flat-bottomed description; although this latter presents the advantage of standing more firmly, and, if boiling water be required, of furnishing it more quickly than the bottle with the round base.

Fig. 1.

Fig. 2.

Fig. 3.

In some laboratories earthenware bottles are in use. These are not so easily broken as those made of glass, but, unlike these latter, water cannot be boiled in them, neither can we see whether they be full or empty.

WASHERWOMAN'S SCALL. See PSORIASIS.

WASHES. The familiar name of lotions, more especially of those employed as cosmetics. See FRECKLES, LOTION, MILK OF ROSES, SKIN COSMETICS, &c., and the following.

Washes, Hair. *Prep.* 1. From rosemary tops, 2 oz.; boiling water, 1 pint; infused together in a teapot or jug, either with or without the addition of rectified spirit, 1 fl. oz. (or rum, 2 fl. oz.) to the cold strained liquor.

2. Box leaves, a small handful; boiling water, 1 pint; digest for an hour, simmer 10 minutes, and strain. Both are used to improve the growth of and to strengthen the hair.

3. To clean the 'partings,' remove scurf, &c. *a.* (ANTIPITYRIENNE.) From sesquicarbonate of ammonia, 1 oz.; spirit of rosemary, ½ pint; rose or elder water, 1½ pints.

b. (DETERGENT ESSENCE.) From honey, 2 oz.; borax, 1 oz.; cochineal (bruised), ¼ oz.; camphor, 1 dr.; (dissolved in) rectified spirit, 2 fl. oz.; soft water, ½ pint; oil of rosemary, 20 drops.

c. (VEGETABLE EXTRACT.) Take of salt of tartar, 1 oz.; rosemary water, 1 pint; burnt sugar, q. s. to tinge it brown; dissolve, filter, and add of essence of musk, 10 drops.

4. To darken the hair. *a.* From pyrogallic acid, ¼ oz.; distilled water, orange-flower water, and rectified spirit, of each, 1½ fl. oz.

b. (LA FOREST'S COSMETIC LOTION or LIQUID HAIR DYE.) Boil, for a few minutes, chloride of sodium, 1 dr., and sulphate of iron, 2 dr., in red wine, 1 lb.; then add of verdigris, 1 dr.; in 2 or 3 minutes remove it from the fire, and further add of powdered galls, 2 dr.; the next day filter. For use, moisten the hair with the liquid; in a few minutes dry it with a cloth, and afterwards wash the skin with water.

5. To prevent the hair falling off. *a.* (AMERICAN SHAMPOO LIQUID.) Take of carbonate of ammonia, ½ oz.; carbonate of potash, 1 oz.; water, 1 pint; dissolve, and add the solution to a mixture of tincture of cantharides, 5 fl. oz.; rectified spirit, 1 pint; good rum, 3 quarts. Used

to strengthen the hair and to remove dandruff, by moistening it with the mixture, rubbing so as to form a lather, and then washing with cold water.

b. (BALM OF COLUMBIA.) As the last, omitting the potash, quadrupling the carbonate of ammonia, and adding some perfume.

c. (*Eras. Wilson.*) Eau de Cologne (strongest), 8 fl. oz.; tincture of cantharides, 1 fl. oz.; oils of rosemary and lavender, of each, ½ fl. dr.

d. (DE LOCOCK'S LOTION.) From expressed oil of mace (nutmeg), 1 oz., liquefied, at a gentle heat, with olive oil, ½ oz.; and, when cold, formed into an emulsion by agitation, with rose water, ½ pint; spirit of rosemary, 2½ fl. oz.; stronger liquor of ammonia, 1½ fl. dr. For other formulæ, see BALDNESS, HAIR DYES, LOTION, &c.

Washes, Medicinal. See LOTION, &c.

Washes, Mouth. *Syn.* TOOTH WASHES; COLLUTORIA, L. *Prep.* 1, Take of camphor (cut small), ¼ oz.; rectified spirit, 2 fl. oz.; dissolve. A few drops to be added to a wine-glassful of water, to sweeten the breath and preserve the teeth.

2. Chloride of lime, ½ oz.; water, 2 fl. oz.; agitate well together in a phial for ½ an hour, filter, and add of rectified spirit, 2 fl. oz.; rose or orange-flower water, 1 fl. oz. Used, highly diluted with water, as the last, by smokers and persons having a foul breath.

3. Mastic (in powder), 2 dr.; balsam of Peru, ½ dr.; gum, 2 dr. or q. s.; orange-flower water, 6 fl. oz.; tincture of myrrh, 2 fl. dr.; for an emulsion. In loose teeth, &c.

4. Tannin, ½ dr.; tincture of tolu, 2 fl. dr.; tincture of myrrh, 6 fl. dr.; spirit of horseradish, 2 fl. oz.; mix. In spongy gums, scurvy, &c.; diluted with tepid water.

5. (*Swediaur.*) Borax, ¼ oz.; water and tincture of myrrh, of each, 1 fl. oz.; honey of roses, 2 oz. In tender or ulcerated gums.

6. Balsam of Peru, 2 dr.; camphor, ½ dr.; essence of musk and liquor of ammonia, of each, ½ fl. dr.; tincture of myrrh, 3 fl. dr.; spirit of horseradish, 1½ fl. oz. To sweeten and perfume the breath; a teaspoonful in ½ a wine-glassful of tepid water to rinse the mouth with.

Washes for the Nose. *Syn.* NASAL DOUCHES, COLLUNARIA. The following formulæ medicinally employed for the purpose of washing or rinsing out the nostrils are from the 'Pharmacopœia of the Throat Hospital.'

In applying them it is directed that "not more than twenty ounces of fluid should ever be used for a nasal douche, and ten ounces are generally sufficient. If an apparatus on the syphon principle be applied, it should be placed only just above the level of the patient's head, in order to avoid too great force of current. The temperature of the fluid should be about 90° F."

NASAL DOUCHE OF ALUM. Syn. COLLUNARIUM ALUMINIS, L. Prep. Alum, 4 gr.; water, 1 oz.; dissolve.—Use. As a mild astringent.

NASAL DOUCHE OF PERMANGANATE OF POTASH. Syn. COLLUNARIUM POTASSÆ PERMANGANATIS, L. Prep. Solution of permanganate of potash (B. P.), 6 minims; water to 1 oz.; mix.—Use. Detergent.

NASAL DOUCHE OF QUININE. Syn. COLLUNARIUM QUINLÆ, L. Prep. Sulphate of quinine, ½ gr.; water, 1 oz. Dissolve by the aid of a gentle heat.

The solution is occasionally useful in hay-fever. It is generally sufficient to place a little in the palm of the hand and draw it up through the nose.

NASAL DOUCHE OF SULPHOCARBOLATE OF ZINC. Syn. COLLUNARIUM ZINCI SULPHOCARBOLATIS, L. Prep. Sulphocarbolate of zinc, 2 gr.; water, 1 oz.; dissolve. — Use. Antiseptic.

NASAL DOUCHE OF TANNIC ACID. Syn. COLLUNARIUM ACIDI TANNICI, L. Prep. Tannic acid, 3 gr.; water, 1 oz.; dissolve.—Use. Astringent.

Washes, Tooth. See above.

WASHING (as applied in Chemistry). In the chemical laboratory the washing of precipitates is an operation of constant occurrence, and the accurate result of the quantitative analysis in which the process of precipitation is had recourse to, essentially depends upon the manner in which the washing has been carried out. In washing a precipitate the object is, of course, to entirely free it from all extraneous matter, so as to ensure, after proper drying, its being weighed in an absolutely pure and uncontaminated state. To arrive at a correct knowledge as to when a precipitate has been properly washed, the operator must never trust to guess-work, but to ocular demonstration, by testing a minute portion, such as a drop or so of the washings, from time to time.

This may be done either by adding—1. A very minute quantity of the proper precipitant to the washings; or—2. By evaporating a drop of the latter on a platinum knife, or a blade of platinum foil; when, if in the former case no turbidity is caused and in the latter no fixed residue remain, the precipitate may be pronounced perfectly washed. The operator, however, instead of not sufficiently washing his precipitate, is frequently liable to fall into another dilemma, which consists not so much in overwashing it as in washing it with an unsuitable liquid, or one in which the precipitate is, to a greater or lesser extent, soluble.

It may not unfrequently happen that the best available precipitant may be one in which the precipitate is soluble to some small extent. Under these circumstances, before throwing down the precipitate, the liquid should, as far as practicable, be removed by evaporation.

Many precipitates which are not altogether insoluble in water may, by the addition of some other liquid to the water, be rendered much less so. Thus the double chloride of platinum and ammonium, which is incompletely thrown down in water, is perfectly precipitated if alcohol be added to the water, as are also chloride of lead and sulphate of lime; whilst the basic phosphate of magnesium and ammonium may be rendered insoluble in water by the addition of ammonia to the water.

The precipitate having subsided to the bottom of the fluid in which it was suspended, the supernatant liquid may be removed from it either by filtration or decantation. In some cases both processes are had recourse to. To wash a precipitate which has been separated by filtration, and which in a moist condition more or less fills the paper filter inserted in a proper funnel, the wash-bottle described below is employed. In using this apparatus the jet of water that is made to issue from the bottle should be directed upon the sides of the filter, and never in the centre, since this would cause a splashing and a consequent loss of the precipitate. The same contingency would be liable to follow if the waters were propelled too violently from the bottle. On no account must the wash-water be allowed to reach to the top of the filter. Another precaution to be guarded against is the formation in the precipitate of fissures or channels; if these are not prevented, the water will not permeate all the parts of the precipitate, and it will be only very insufficiently washed. When such channels form, it will be best to stir up the precipitates with a glass rod or a platinum spatula, taking care, however, to avoid tearing or making a hole in the filter.

Precipitates that are washed by decantation ought to consist of such substances as readily subside from the liquid in which they are suspended and are practically insoluble in water, since a very much larger quantity of this menstruum has to be employed than when filtration is resorted to. The process is generally carried out in deep vessels. The supernatant liquid being removed, the vessel is filled up with water, and the precipitate well stirred up with a glass rod; after it has again fallen down fresh water is added, and the process is continued until the washings cease to show the presence of any soluble matter. The several washings being collected, are let stand some 12 or 24 hours; after which time, should no precipitate show itself, they are thrown away. Should any deposit form in the washing, it is carefully removed either by filtration or decantation, and its amount being determined, the result is added to that obtained from the bulk of the precipitate. Where the nature of the precipitate is in no way influenced by hot water, this latter should always be used in washing precipitates, as it greatly facilitates and expedites the operation. Many precipitates require to stand a long time before they entirely subside from the fluid in which they are suspended. Most gela-

·tinous, pulverulent, and crystalline precipitates are of this nature. The separation of the precipitate should not be attempted until after the liquid containing the precipitate has stood several hours.

WASHING FLUIDS. Solutions of carbonate of soda, rendered caustic with quicklime.

WASHING LIQUIDS. Various fluids are sold under this name; they are employed by dyers, cleaners, and laundresses. The addition of turpentine to any of them will exercise a bleaching effect on linen.

Prep. 1. Carbonate of soda, dissolved in water, and made caustic by shaking with slaked lime.

2. Alcohol, 1 pint; spirits of turpentine, 1 pint; strongest solution of ammonia, 2 oz.; mix. Put 3 or 4 table-spoonfuls to 1 pint of soft soap, or 1 lb. of hard soap. The clothes should be soaked overnight if possible before using this mixture, but if soaked an hour or two it will aid much.

3. Washing fluid for fine linen, laces, &c.:— Borax, 4 oz.; water, 5 galls. For crinoline, or any other stiff fabric, increase the quantity of borax to 6 oz.

4. Nottingham washing liquor:—Water, 1 gall.; white soap, 3 oz.; pearlash, 3 dr.

5. Hull washing liquor:—Yellow soap, 1½ oz.; water, 1 gall.; strongest solution of ammonia, 4 oz.

6. Yorkshire wash:—Strongest solution of ammonia, 1 oz.; common water, 1 pint.

7. Silicate of soda or potash or water-glass is in itself a good detergent. It is added to cheap soaps to allow of the retention of large quantities of water in the finished product. The retail chemist should not attempt to manufacture the article. It is purchased in casks, and is a thick, viscid, translucent mass, flowing very slowly. When dissolved in hot water it forms a solution which unites with certain kinds of soap very readily (curd soap, yellow soap, and soaps containing resin. Probably a useful washing liquor could be made from this substance.

WASHING POWDERS. See POWDERS.

WATCHFULNESS. *Syn.* SLEEPLESSNESS; AGRYPNIA, L. The common causes of watchfulness are thoughtfulness or grief, disordered stomach or bowels, heavy and late suppers, and a deficiency of outdoor exercise. The best treatment, in ordinary cases, simply consists in an attention to these points. The method of producing sleep recommended by a late celebrated hypnotist consists in merely adopting an easy recumbent position, inclining the head towards the chest, shutting the eyes, and taking several deep inspirations with the mouth closed. Another method, recommended by an eminent surgeon, and which appears infallible if persevered in with proper confidence, and which is suitable either to the sitting or recumbent posture, consists in tying a decanter cork with a bright metallic top, a pencil case, or any other bright object on the forehead, in such a position that the eyes must be distorted or strained to be capable of seeing it. By resolutely gazing in this way for a short time, without winking, with the mind fully absorbed in the effort, the muscles of the eyes gradually relax, and the experimenter falls asleep. Gazing in a similar manner on any imaginary bright spot in the dark, as at night, exerts a like effect. A tumblerful of cold spring water, either with or without a few grains of bicarbonate of potash in it, taken just before lying down, will frequently succeed with the dyspeptic and nervous, when all other means fail.

The following valuable advice to those who suffer from unnatural wakefulness is abridged from the late Dr Tanner's valuable work on the 'Practice of Medicine' ('The Practice of Medicine,' by Thomas Hawkes Tanner, M.D.; Renshaw, London).

At his starting-point, Dr Tanner enjoins the practice of taking a proper amount of exercise daily. A digestible diet, such as is not liable to cause acidity or flatulence, must also be adopted, and tea and coffee must be abstained from in the after part of the day. Early dinners and light suppers are also recommended. The reading of any thrilling work of fiction previous to retiring to rest is also prohibited. The patient is advised to seek his bed at an early and regular hour, and it is desirable to have his sleeping chamber well ventilated, and if the weather be chilly the bedroom fire should be lighted. Feather beds should be abandoned for mattresses; there should not be too many blankets on the bed, the pillows should be firm and high, and no curtains or hangings should be allowed. Should the above means fail to produce the required sleep, before going to bed the patient is advised to try a tumbler of port-wine negus, or of mulled claret, or of white-wine whey, the last thing. The aged are recommended (should the above methods be unsuccessful) to imbibe a glass of spirit and water, which is said to be all the more effective if drunk when in bed. In some cases, attended by a hot or dry skin, a glass of cold water has been found useful. Another remedy is the use of a bath, for about three or five minutes, just before getting into bed, at temperature varying from 90° to 96° F.

Rapid sponging of the body with tepid water is also recommended, as also the use of a warm footbath, at a temperature of 100° F., or of a hot-water bottle in the bed, or putting the feet in cold water for a minute, and then vigorously rubbing them.

For those whose sleeplessness is caused by their prosecuting literary work till a late hour, a short brisk walk, just before retiring to bed, is recommended.

If the wakefulness can be traced to any bodily ailment, this, of course, must be removed by the proper means. Constipation, which is not at all an unfrequent cause of insomnia, must be combated by the methods described under that article. If there be headache it will be best removed by applying a rag dipped in cold water to the scalp, or a bladder containing ice.

Should the adoption of any of the above suggestions fail, all kinds of mental labour and excitement during the day must be greatly diminished, and physical exercise must replace them. Sedatives should be had recourse to with great caution, and under medical supervision only. Because of the hazard attending their use, and of the ready tendency their adoption has to degenerate into a pernicious ineradicable habit, we have forborne to specify the medicinal agents Dr Tanner prescribes for sleeplessness, strongly recommending the patient, before he has recourse to them, to exhaust the category of suggestions given by Dr Tanner; and should these unhappily be found to fail, and he is drawn to soporifics, we again reiterate, let him take them only under medical supervision.

Another method, adopted by professional hypnotists, consists in gently moving, in opposite directions, a finger of each hand over the forehead just above the eyebrows. A soothing and drowsy effect is said to be thereby produced, which ends in tranquil slumber.

Dr Ainslie Hollis contributes some excellent hints on the treatment of wakefulness to the practitioner. He classifies the treatment under two heads—first, the induction of natural sleep; and secondly, the production of narcosis of artificial rest. The application of mustard plasters to the abdomen generally brings about the first result, producing, according to Schuler, first dilatation, and subsequently contraction of the vessels of the pia mater. Dr Preyer, of Jena, on the supposition that sleep may be induced by the introduction of the fatigue products of the body, advocates the administration of a solution of lactate of soda. When sleeplessness is the result of brain exhaustion Dr Hollis advocates a tumbler of hot claret negus. The alkalies and alkaline earths, says the 'Boston Journal of Chemistry,' are useful when acid dyspepsia is associated with the insomnia. In hot weather sprinkling the floor of the sleeping apartment with water lessens the irritant properties of the air, adding much to the comfort of the sleepers; possibly the quantity of ozone is at the same time increased. When sleep is broken by severe pain, opium or morphia is of value, bringing not only relief, but producing anæmia of the cerebral vessels; when neuralgia is the cause an injection of morphia under the skin, near the branch of the affected nerve, will have more effect than by administering it by the mouth. Again, when wakefulness is due to defective cardiac power, digitalis may be useful. Chloral hydrate is supposed to owe its hypnotic effect to its power of diminishing the amount of blood in the brain, and therefore it may be used when sleeplessness arises from the pains of muscular spasm. The bromides, although undoubtedly sedatives, possess very doubtful hypnotic properties. See SUPPER, &c.

WATER. H₂O. Syn. OXIDE OF HYDROGEN, PROTOXIDE OF H.; AQUA, L.; EAU, Fr.; WASSER, Ger.; ὕδωρ, Gr. The ancients regarded water as a simple substance, and as convertible into various mineral and organic products. Earth, air, fire, and water were at one time conceived to be the elementary principles or essences of matter from which all form and substance derived their existence. The true constitution of water was not discovered until about the year 1781, when Cavendish and James Watt, independently and nearly simultaneously, showed it to be a compound of hydrogen and oxygen. Five years, however, before this time (1776), the celebrated Macquer, assisted by Sigaud de la Fond, obtained pure water by the combustion of hydrogen in the air. It has since been satisfactorily demonstrated that

hydrogen and oxygen exist in water in the proportion of 1 to 8 by weight, or 2 to 1 by volume; the sp. gr. of hydrogen being to that of oxygen as 1 to 16. One cubic inch of perfectly pure water at 62° F., and under an atmospheric pressure represented by 30 inches of the barometer, weighs 252·458 gr., by which it will be seen that it is 770 times heavier than atmospheric air. Its sp. gr. is 1·0, it being made the standard by which the densities of all solid and liquid bodies are estimated. The sp. gr. of frozen water (ice) is 0·9175, water being 1·0 (Dufour); that of aqueous vapour (steam), 0·6252, air being in this case taken as the unit. Water changes its volume with the temperature; its greatest density is about 39½° F. (4° C.), and its sp. gr. decreases from that point, either way. Water is nearly incompressible. By subjecting water to a pressure of 705 atmospheres, Cailletet found the compressibility to be at the rate of 0·0004451 for each atmosphere. Water evaporates at all temperatures; but at 212° F. (100° C.), under ordinary circumstances, this takes place so rapidly that it boils, and is converted into vapour (steam), whose bulk is nearly 1700 times greater than that of water. Var. Of these the following are the principal:

DISTILLED WATER; AQUA DESTILLATA (B. P., Ph. L., E., & D.), L. Obtained by the distillation of common water through a block-tin worm, rejecting the first and last portions that come over. The still employed for this operation should be used for no other purpose; and when great nicety is required, the distillation should be performed in glass or earthenware vessels. It remains clear on the addition of lime water, chloride of barium, nitrate of silver, oxalate of ammonium, or hydrosulphuric acid. It is the only kind of water that should be employed in chemical and pharmaceutical operations. When distilled water is not at hand, clean filtered rain water is the only kind that can be successfully substituted.

NATURAL WATERS. In respect of wholesomeness, palatability, and general fitness for drinking and cooking, natural waters may be classified in order of excellence as follows ('Rivers Pollution Commissioners' Sixth Report'):

Wholesome	1. Spring water . . . 2. Deep well water . . 3. Upland surface water 4. Stored rain water .	Very palatable. Moderately palatable.
Suspicious	5. Surface water from cultivated lands .	
Dangerous	6. River water to which sewage gains access 7. Shallow well water .	Unpalatable.

The average composition of the four classes of unpolluted waters is given by the same authorities as follows. Their estimations are in parts per 100,000, but may be converted into grains per gallon by multiplying by 7 and dividing by 10.

RAIN WATER in the country contains, among natural waters, the smallest amount of solid matter in solution. From the columns headed "Organic carbon" and "Organic nitrogen" it will be seen that even rain collected with special precautions, away from any large town, is by no means free from organic matter. Rain water

	Total solid impurity.	Organic carbon.	Organic nitrogen.	Ammonia.	Nitrogen as nitrates and nitrites.	Total combined nitrogen.	Previous sewage or animal contamination.	Chlorine.	Hardness.		
									Temporary.	Permanent.	Total.
Rain water . . .	2·95	0·070	0·015	0·029	0·003	0·042	42	0·22	0·4	0·5	0·3
Upland surface water	9·67	0·322	0·032	0·002	0·009	0·042	10	1·13	1·5	4·3	5·4
Deep well water . .	43·78	0·061	0·018	0·012	0·495	0·522	474	5·11	15·8	9·2	25·0
Spring water . . .	28·20	0·056	0·013	0·001	0·388	0·396	3559	2·49	11·0	7·5	18·5

collected from roofs, and stored in underground tanks, is often very impure.

SURFACE WATERS form the main supply of rivers. If collected from high uncultivated districts they are usually unpolluted with animal matter. The organic matter is usually peaty, is sometimes very small, but is liable to considerable variations with the season, and is occasionally present in excessive quantities, discolouring the water, and rendering it unpalatable. From their softness these waters are admirably adapted for manufacturing purposes. The amount of solid matter in solution ranges from 2 gr. to 7 gr. per gallon.

SURFACE WATER from *cultivated land* contains on an average less organic matter than upland surface water, but the pollution, being derived from manure and other objectionable matter, is more harmful.

RIVER WATER consists of the above, aided by springs, and most frequently the drainage of towns on its banks. The amount of solid matter varies from 10 gr. to 30 gr. per gallon. In Thames water there are on the average about 20 gr.

WELLS, if *shallow*, are usually a most undesirable supply. Unless far from any house they are contaminated by drainage, and sometimes, from proximity to cesspools, contain more animal matter than ordinary town sewage. They are, as a class, hard waters, the polluted ones excessively so.

Wells of 100 feet deep and upwards are, as a class, very superior waters, the filtration and oxidation of so great a depth of soil having removed the greater part of the organic matter. The hardness varies with the strata, but, as a class, the deep wells are softer than the shallow.

SPRING WATER greatly resembles deep well water, possessing all its good qualities in a higher degree. Spring and deep well water are very uniform in quality, and little affected by climatic changes.

SEA WATER. The characteristic of this variety is its saltness. Its density is about 1·0274, and the average quantity of saline matter which it contains is about 3½%, of which about ¼ are chloride of sodium, and the remainder chiefly chloride of magnesium and sulphate of magnesium.

The average proportion of organic carbon and nitrogen in 23 samples of sea water was 0·278 carbon, 0·165 nitrogen, as compared with Thames water averages of 0·203 parts carbon, 0·033 nitrogen, in 100,000 parts of water.

Analysis of sea water (British Channel), by Dr Schweitzer, of Brighton:

1000 gr. contained—		Grains.
Water	.	963·745
Chloride of sodium	.	28·059
„ of potassium	.	0·766
„ of magnesium	.	3·666
Bromide of magnesium	.	0·029
Sulphate of magnesium	.	2·296
„ of calcium	.	0·406
Carbonate of calcium	.	0·033
		1000·

Pur. Pure water is perfectly transparent, odourless, and colourless, and evaporates without residue, or even leaving a stain behind. The purest natural water is that obtained by melting snow or frozen rain that has fallen at some distance from any town. Absolutely pure water can only be obtained by the union of its gaseous constituents; but water sufficiently pure for all purposes may be procured by the careful distillation of common water.

Among the methods adopted for improving the quality of water are—

(*a*) *For reducing the amount of organic and suspended matter.* 1. Filtration through or agitation with coarsely powdered, freshly burnt charcoal, either animal or vegetable, but preferably the former. *When in good condition* a filter of animal charcoal will not only remove suspended matter in water, but will considerably reduce the amount of organic matter, and also the calcareous and gaseous impurities held in solution; but it, however, loses its power of removing lime in a week or two, and of abstracting the organic matter in about three to four months, and then becomes foul, and requires to be recharged. Spongy metallic iron is more energetic in its action than charcoal, and remains serviceable for a twelvemonth. 2. Free exposure to the action of the air, by which the organic matters become oxidised and insoluble, and speedily subside. This may be easily effected by agitating the water in contact with fresh air, or by forcing air through it by means of bellows. 3. The addition of a little sulphuric acid has a like effect; 15 or 20 drops are usually sufficient for a gallon. This addition may be advantageously made to water intended for filtration through charcoal, by which plan at least 2-3rds of the latter may be saved (*Lowitz*). 4. An ounce of powdered alum (dissolved), well agitated with a hogshead or more of foul water, will purify it in the course of a few hours, when the clear portion may be decanted. When the water is very putrid about ¼ dr. (or even 1 dr.) per gall. may be employed; any alum that may be left in solution may be precipitated by the cautious addition of an equivalent proportion of carbonate of sodium. 5. A solution of ferric sulphate acts in the same way as alum; a few drops are sufficient for a gallon. 6. Agitation with about ½% to 1% of finely powdered black oxide of manganese has a similar effect to the last. 7. The addition of a little aqueous chlorine, or chlorine gas, to foul water, cleanses it immediately. This method has the advantage of the water being capable of being perfectly freed from any excess of the precipitant by heat.

(*b*) *For reducing the amount of inorganic matter.* 1. Distillation separates all non-volatile matter, including organised bodies. It is used to obtain a potable water from sea water. The waste heat of the cook's galley is amply sufficient for this purpose. There are several patent contrivances for the distillation of water on shipboard. 2. Hard water may be softened by adding carbonate of soda to the water so long as it turns milky. The precipitation of the hardening ingredients, lime and magnesia, is most rapid when the water is heated. The water cannot be used for drinking purposes, from the unpleasant flavour of the carbonate of soda. When used on a hard water intended for washing, it effects a saving of soap equal to about fifteen times its own cost. Sea water can be made fit for washing by this means. It removes both the 'temporary' hardness, due to carbonates of calcium and magnesium, and the 'permanent,' due to the sulphates, chlorides, and nitrates of these metals. 3. Hard water may be both aërated and softened by the addition of a few grains of bicarbonate of potassium per gallon, followed by half as much lime juice or tartaric acid as is sufficient to saturate the alkali in the carbonate thus added. 4. The 'temporary' hardness may be nearly removed by ebullition, or, as recommended by Professor Clarke, by mixing the hard water with lime water, when the calcium combines with the excess of carbonic acid, which previously rendered the carbonate of calcium soluble, and is precipitated as carbonate (chalk), together with the carbonate originally present. This method removes, at the same time, much of the organic matter, and carries down suspended matter. The water is often made more palatable than before. The directions are—For every degree of hardness on Clarke's scale each 1000 gallons of water to be softened requires 1 ounce of quicklime. Slake the lime and work up to a thin cream with water and pour into the cistern, which already contains at least 50 gallons of water to be softened. Then add the remainder of the 1000 gallons in such a way as to stir up and mix uniformly with the contents of the cistern. In about three hours the milky water is clear enough for washing. After twelve hours' rest the water is fit to drink. If the exact hardness of the water is not known, water may be added to the milk of lime till, on adding a drop of nitrate of silver to a cupful of the cistern water, the brown tint indicative of an excess of lime is replaced by a very faint yellow. 5. To save boilers from scaling, water intended for steam purposes is sometimes treated with lime to remove carbonates, and then the sulphate of calcium (which forms a very tenacious scale) is decomposed with baric chloride (Haen's pro-

cess). The precipitated mineral matter may also be prevented from forming a scale or fur by adding organic substances, such as potatoes, sound or otherwise, swedes, mangolds, or other vegetable. Oak bark, spent tan, sawdust, and their decoctions are efficacious on account of the tannic acid they contain, but they attack the boiler-plates at the same time. Zinc suspended in the water is said to answer well. It has been recommended to polish the inside of the boiler-plates with black-lead or coat it with linseed oil and dissolved india rubber. Numerous chemical preparations, most of which do more harm than good, are also sold. A recent plan is to suspend a small bundle of fibrous material in the boiler; upon this the solid matter tends to accumulate.

Tests (Physical). 1. To observe colour, stand in tall colourless glass cylinder on white ground. If very turbid allow to settle, and examine sediment by microscope for evidence of sewage contamination (linen fibres, hairs, epithelium) and for moving organisms. Slight turbidity is best noted by filling a clean quart flask, and holding it towards the light with some dark object as a window-pane between. Taste and odour are most marked when the water is made lukewarm. 2. For poisonous metals add one drop of strong colourless ammonium sulphide to about 1000 grains of water in glass cylinder, and observe if the liquid darkens. If the coloration or precipitate disappears on adding acid, it is iron; if it remains, lead or copper is present, either of which condemns the water. 3. For chlorine add a couple of drops of nitric acid to a little of the water, and a crystal or drop of solution of nitrate of silver. If the water turns very milky it is a bad sign; make, if possible, a comparative experiment with water of known composition. 4. The residuum, if any, of evaporation is impurity; if it be organic matter, smoke and a peculiar odour will be evolved as the residue becomes dry and charred. 5. Neither litmus, syrup of violets, nor turmeric is discoloured or affected when moistened with pure water; if the first two are reddened, it indicates an acid; if the litmus is turned blue or the turmeric is turned brown, an alkali is present. 6. If a precipitate is formed, or a fur or crust deposited on the vessel during ebullition, it indicates the presence of carbonates of calcium, magnesium, or iron. 7. Calcium salts produce a white precipitate with oxalate of ammonium. 8. The liquid filtered off from 7, on being treated with a solution of phosphate of sodium and ammonium (microcosmic salt), and allowed to stand, gives a white precipitate if magnesium is present. 9. Tincture or infusion of galls turns water containing iron black. When this takes place both before and after the water has been boiled, the metal is present under the form of sulphate; but if it only occurs before boiling, then ferrous carbonate may be suspected, and it will be precipitated as a reddish powder by exposure to air and heat. 10. Ferrocyanide of potassium gives a dark blue precipitate in water containing a ferric salt; and a white one, turning blue by exposure to the air, in water containing a ferrous salt. 11. If sulphuric acid be run into water and allowed to cool, and a crystal of sulphate of iron dropped into the water, a dark brown cloud round the crystal indicates nitrates; the bleaching of indigo added to the hot mixture of equal parts water and pure oil of vitriol also indicates the presence of these salts. 12. Sulphuric acid or sulphate is indicated by a soluble salt of barium throwing down a white precipitate insoluble in nitric acid.

Water, Quantitative Analysis of. The quantitative analysis of potable water is confined to the following : total residue, hardness, temporary and permanent, chlorine, ammonia, nitrates and nitrites, and organic matter.

Of these, all but the first two are intended to throw light on the organic contamination of the water. Chlorine, ammonia, and nitrates and nitrites are in themselves innocuous substances, but are estimated because they supplement the somewhat imperfect information obtained from the organic matter itself. A sewage-polluted supply being an agent in propagating zymotic diseases, a knowledge of the source of the organic matter in a water is of the highest importance.

Before passing to the mode of estimating the above items it may be desirable to explain the object of each analysis, and the interpretation which may be placed on the results.

Total solid residue includes all the substance, organic or mineral, dissolved in the water. Everything beyond the two gases which enter into the combination of the water being useless, the 'residue' of a water is sometimes called the 'total solid impurity.' The less residue left by a water on evaporation the better, but a water need not be objected to for drinking purposes till the residue reaches 40 grains per gallon. For raising steam a water should not contain more than 20 grains, and should be, if possible, much less.

The hardness, or soap-wasting power of a water, is chiefly determined on economic grounds. Unless the hardness is very excessive, the hardness or softness of the water does not appear to materially affect the health of the consumer. Hardness is caused by salts of lime and magnesia. If the property of hardness be caused by the presence of bicarbonates of the above substances, the water is said to be ' temporarily ' hard, for by boiling or adding lime as above described the hardness may be reduced without affecting the potability of the supply ; but when the hardness is due to calcium or magnesium sulphates it is called ' permanent' hardness, for it is not then practicable to remove the hardening ingredients without adding some more objectionable substance. The average hardness of the four classes of pure water is shown in the analysis given above. Thames water has a total hardness of 15°. Loch Katrine water, as supplied to Glasgow, 0·70 on Clarke's scale.

Chlorine. Except in places near the sea, or in salt-bearing strata, an unpolluted water does not contain more than the merest trace of chlorine. Sewage, however, contains a large quantity of chlorine as sodic chloride (common salt), derived from the salt used in cooking, &c. Hence a mixture of sewage with water becomes known by the quantity of chlorine present. It is not safe to drink a water containing such an excessive

quantity of chlorine as 4 grains per gallon. The chlorine in Ullswater and Thames water is 0·7 and 1·1 grains per gallon respectively. Sewage has about 8 grains on the average.

Ammonia. This determination acquires significance because it is one of the early substances produced by the decomposition of animal matter. It therefore indicates, when present in large quantities, *recent* contamination by sewage. Rain always contains a small amount of ammonia, and deep wells occasionally show ammonia derived from the reduction of nitrates by the oxygen-seeking organic matter. The above inferences must, therefore, be applied with caution.

Nitrates and *Nitrites* result from the oxidation of animal matter. Vegetable substances, under like conditions, yield none or but mere traces of these compounds. The presence of nitrates is a most unfavourable sign in a shallow well or river water, because the conditions to which these waters are subjected are so variable that there is a constant liability of the purifying processes diminishing, and allowing the sewage, now only represented by innoxious nitrates, to appear in its dangerous, unoxidised condition.

Dr Frankland takes the sum of the nitrogen existing in the water as ammonia and as nitrites and nitrates, as a sort of measure of the minimum amount of animal or sewage.matter destroyed. The amount due to sewage or animal matter is considered to be all over 0·032 part per 100,000 (or 0·022 gr. per gallon), which is the average of ' inorganic nitrogen ' natural to unpolluted rain water. Dr Frankland also expresses this ' previous sewage or animal contamination ' in terms of London sewage containing 10 parts of nitrogen in 100,000 parts of liquid, by multiplying the above-named corrected sum by 10,000. Thus a water containing 1 part per 100,000 (0·7 gr. per gall.) of ' inorganic nitrogen ' would have a ' previous sewage or animal contamination ' of 9680 parts per 100,000, for. it would have re-

$$\text{quired } 100,000 \frac{(1-0\cdot032)}{10} = 9680 \text{ parts of}$$

London sewage to produce an amount of nitrogen equal to that found by analysis. A water which contains over 20,000 parts of previous sewage contamination (1·5 grains of inorganic nitrogen) is said to be dangerous. All other waters containing more inorganic nitrogen than in rain are said to be ' doubtful,' except springs and deep well waters containing less than 10,000 parts of previous sewage contamination per 100,000, and such shallow wells and running water which from their source may be taken to be free from sewage.

Organic Matter. There is no method by which the actual weight of organic matter can be determined, still less is it possible to say how much is likely to be actually injurious organic matter, but there are several means of measuring the proportionate amount of organic contamination. Dr Frankland determines the amount of carbon and nitrogen in the organic matter. The smaller the amount of these elements, the better the water; and the less the amount of nitrogen, especially in proportion to organic carbon, the less chance of *animal* matter. A good drinking water will not have more than 0·2 part in 100,000 (0·14 gr. per gall.) of carbon, or 0·08 part of organic nitrogen in 100,000 parts (0·02 gr. per gall.) of the water. The amount of putrescent matter may be estimated by the amount of oxygen consumed in destroying it. Dr Tidy ('Chem. Soc. Journ.,' January, 1879) considers, speaking generally, waters requiring 0·05 part per 100,000 (0·035 gr. per gall.) to be of great organic purity; 0·15 part (0·1 gr. per gall.), waters of medium purity; waters of doubtful purity, from 0·15 to 0·21 part per 100,000 (0·15 gr. per gall.). Impure waters, all above 0·15 gr. per gall.

The proportion of albuminous substances present is measured by Mr Wanklyn by the amount of ammonia set free by alkaline permanganate. A water containing over 0·15 part per million albuminoid ammonia condemns a water absolutely ('Wanklyn's Water Analysis,' 4th edit., p. 54); 0·10 part per million with little free ammonia, or 0·05 part albuminoid ammonia with much free ammonia, is 'suspicious.' A water with less than 0·05 part albuminoid ammonia belongs to the class of very pure waters.

Of course the above data are not hard and fast lines, but serve as aids to a judgment which may be modified by other circumstances connected with the analysis, and the source of the water.

Methods of Analysis : Total Solid Residue. 1000 grains are evaporated to dryness in a platinum dish over a water-bath, and residue dried in an oven at 212° F. for an hour, or until the weight is constant. The increase in weight of the platinum vessel multiplied by 70 gives the number of grains of total solid residue per gallon.

Hardness is determined by a solution of soap, of which 320 grain-measures will soften a water of 16° of hardness. Each degree of hardness represents an amount of soap-destroying matter equivalent to 1 grain of chalk per gallon. 1000 measured grains of the water are measured into a narrow-mouthed six or eight ounce stoppered bottle, then well shaken, and the air sucked out by means of a piece of glass tube. The standard soap solution is now run in, 10 grains at a time, shaking well between each addition until there is formed over the whole surface a lather, which, when the bottle is placed upon its side, shall last just five minutes. The number of grain-measures used will indicate the hardness of the water by reference to Table A. Should, however, the permanent lather not be formed before 320 measures of soap solution have been added, a second trial must be made, in which only 500 grain-measures of the water are taken, to which a like amount of recently boiled distilled water is added. The degree of hardness now obtained must be multiplied by 2. With very hard waters it is necessary to dilute still further, say 250 grains to 750 of distilled, and multiplying the result by 4. If the number of soap measures does not correspond with any degree on the table, observe which number it falls between. The degree corresponding to the lower of these soap volumes will be the whole number in the answer; the fraction will be the difference between the observed number of measures and the next lower on the table, divided by the difference (given in column 3) between the figure above and below it. Thus

if 14 measures were used the hardness would be 6·2°, 13·6 measures being equivalent to 6 degrees, and the fraction being $\dfrac{14-13\cdot6}{13\cdot6-11\cdot6} = \dfrac{4}{20} = \cdot2$.

The hardness of the water in the natural state is the 'total hardness.' By boiling for an hour and making up loss by evaporation with boiled distilled water and again determining the hardness, the 'permanent hardness' is found. That which has been removed by the boiling is the temporary hardness.

TABLE A.

Soap Test Measures corresponding to one thousand measures of water of each degree of hardness.

Degree of hardness.	Soap test measures.	Difference.
0	14	18
1	32	22
2	54	22
3	76	20
4	96	20
5	116	20
6	136	20
7	156	19
8	175	19
9	194	19
10	213	18
11	231	18
12	249	18
13	267	18
14	285	18
15	303	17
16	320	—

The standard water of 16° of hardness is thus made :—Pure carbonate of calcium (Iceland spar) is weighed out into a porcelain or platinum dish in the proportion of 16 gr. for a gallon of solution. It is dissolved in weak hydrochloric acid, and the whole cautiously evaporated to dryness over a water-bath, then re-dissolved in water, and again evaporated to drive off any excess of acid. The dish is covered with a glass during the operation to prevent loss by spirting. The resulting neutral chloride of calcium is dissolved in a gallon of pure distilled water if 16 gr. were weighed out, or a proportionate quantity in other cases. The soap solution can be made by dissolving good curd soap in weak methylated spirit in the proportion of 1 oz. of soap to the gallon. A potash soap made as follows is, however, less liable to change : —150 gr. of lead plaster (Emplastrum Plumbi— R. P.) and 40 gr. of dry potassium carbonate are rubbed together in a mortar and repeatedly extracted with small portions of methylated spirit, triturating the mass meanwhile, till about a pint of spirit has been used; filter and add an equal bulk of recently boiled distilled water. Whichever method is followed, the clear solution has now to be standardised by the 'water of 16° of hardness.' 1000 gr. of the water of 16° of hardness are placed into a bottle, and this soap solution is run in from a burette until a permanent lather is formed. The soap solution must be fortified by strong soap solution or diluted with alcohol till 320 measures produce a lather permanent for five minutes in 1000 grain-measures of water of 16° of hardness.

Chlorine. To 1000 gr. of the water add a drop or two of neutral chromate of sodium, so as to tinge the water yellow; run in standard nitrate of silver till the liquid acquires a very faint red tinge, showing that all the chlorine has been precipitated and that red silver chromate is beginning to be formed. The number of grains of standard solution divided by 100 will give the grains of chlorine in 1 gall. of the water.

The standard solution is prepared by dissolving pure nitrate of silver in the proportion of 47·90 gr. to 1 gall. of distilled water.

Ammonia estimations are always carried out as described in the account of Messrs Wanklyn and Chapman's process.

Nitrates and Nitrites. These substances can be most expeditiously estimated by the indigo process as follows:—200 grain-measures of the water are placed in a flask, and a little of a standard solution of indigo added thereto; twice the volume of pure sulphuric acid is then suddenly poured in from a measuring cylinder, and the whole shaken. The temperature rises immediately to about 270° F., and the blue colour will probably be immediately discharged; more indigo, therefore, must be rapidly run in till a brown-green tint shows itself. This gives the trial estimation, but the maximum amount of indigo is only used up when all the indigo is added previous to the addition of acid; hence a second experiment is now started, and an amount equal to that previously used run in at once, and on it is poured exactly twice as much sulphuric acid as there is water and indigo in solution. The second result will be somewhat higher than the first. If the solutions below mentioned be used, the amount of indigo required by the 200 gr. of water divided by the number of grains of indigo required to bleach 200 c.c. of standard nitre represents the grains per gallon of nitrogen as nitrates and nitrites. The standardising of the indigo with the nitrate solution is performed exactly as for an actual water. The requisites are—1. A solution of pure potassium nitrate of known strength, say 14·442 gr. of nitre (equivalent to 2 gr. of nitrogen or 9 gr. of nitric acid) in a gallon of distilled water. 2. A solution of indigo made by dissolving soluble indigo carmine in distilled water in such a proportion that 200 gr. is about equal to 200 gr. of nitre solution. 3. Strong pure oil of vitriol; it must be free from nitrous compounds, not become turbid when diluted, and its sp. gr. not be less than 1·84. It is important to maintain the same proportion of acid, and not to allow the temperature to fall below 250° F. throughout the experiment.

Messrs Wanklyn and Chapman's aluminium method is also a very convenient process. 2000 gr. of the water are placed in a retort, and half as much of a solution of 10% soda added. The soda solution is made from sodium soda, and the absence of nitrates is secured by boiling the liquid with a piece of aluminium. Half the contents of the retort are distilled over, and the residue cooled. A piece of aluminium-foil of about 6 square inches area is tied to a piece of clean glass rod and sunk in the liquid. The neck of the retort is guarded by a tube containing fragments of glass moistened with hydrochloric acid; it is sloped, so that any liquid spurted into the neck will flow back into

the retort. After resting several hours the neck of the retort is washed down with pure distilled water, the contents of the tube are transferred to the retort, and the contents distilled over, down to about an ounce in 2 or 3 oz. water placed as a receiver. The contents of the receiver are made up to 200 gr., and the ammonia is estimated in one half by Nessler's test as below described.

An exceedingly accurate eudiometric method has also been devised by Dr Frankland based on Crum's observation that a highly concentrated solution of nitrates, when vigorously agitated with mercury and an excess of concentrated pure sulphuric acid, yields all its nitrogen from the nitrates and nitrites as nitric oxide, a compound occupying twice the volume of the nitrogen as nitrates. The weight of gas is easily calculated from the volume measured (' Journal Chem. Soc.,' March, 1868).

Organic Contamination: means of estimating. Messrs Wanklyn and Chapman's method is most generally employed. It depends on the conversion of the nitrogen of the organic matter into ammonia, and the employment of Nessler's test to estimate this ammonia.

Nessler's Test. 500 gr. of iodide of potassium are dissolved in a small quantity of hot distilled water, and to this is gradually added a cold saturated solution of mercuric chloride till the precipitate produced ceases to be dissolved upon stirring. To render this alkaline add 2000 gr. of potassic hydrate, and dilute the volume to 10,000 gr. measures. A little more saturated mercuric chloride is added, and the whole allowed to settle, and the clear liquid decanted off. The test should have a slightly yellowish tint. If colourless it is not sensitive, and more mercuric chloride must be added.

Standard Ammonia Solution. Dissolve 27·164 gr. of pure sulphate of ammonium in 1 gall. of distilled water. For use dilute 100 gr. to 1000 gr. It will then contain 1 gr. of ammonia in 100,000 of water.

In order to estimate ammonia several six-ounce tall glass cylinders, free from colour, are graduated at 1000 gr. One of these is filled up to the graduation mark with the ammonia to be estimated, and about 30 gr. of Nessler's reagent added from a pipette. The coloration produced is noted, a second cylinder is filled nearly to the mark with distilled water, and what is thought sufficient ammonia to produce a similar colour to the first run in, and the whole made up to 1000 gr., and 30 gr. of Nessler added; if after standing five minutes the colour in the second is the same as in the water examined, the quantity of ammonia they contain will be equal; but if this is not the case a second trial must be made, using more or less standard ammonia as the intensity of colour is greater or less than the first. After a little experience more than two trials are rarely necessary.

Examination. (*a*) *Free Ammonia.* 7000 gr. (a deci-gallon) of the water to be analysed is placed in a tubulated retort, and to it is added half an ounce of a supersaturated solution of carbonate of soda, made by dissolving ignited carbonate of soda in water free from ammonia. The contents are distilled over in two portions of 1000 gr. each, and the second Nesslerised : if it contains no ammonia, the distillation may be stopped ; if it does, the distillation must be continued and tested in portions of 500 gr. till the ammonia no longer can be detected. If there is much ammonia in the cylinder of the second 1000 gr. the first will probably contain too much to be conveniently estimated, and therefore an aliquot part diluted to 1000 gr. with distilled water free from ammonia should be used. The sum of the ammonia in these different portions, multiplied by 10, gives grains per gallon.

(*b*) *Albuminoid Ammonia.* To the retort, after all the free ammonia has been driven off, 1 oz. of a solution of hydrate and permanganate of potassium of a strength of 2000 gr. of hydrate of potassium and 80 gr. of permanganate to 10,000 gr. of water is added, and the distillation continued until no more ammonia comes over, collecting the distillate in portions of 1000 c.c. as before. The sum is the albuminoid ammonia derived from the nitrogenous organic matter.

It is, of course, essential that the utmost care be taken to remove by rinsing or distilling all traces of ammonia from the apparatus employed. Water which has been distilled till free from ammonia should alone be used in estimations and preparations of solutions, and the alkaline permanganate should be boiled for a short time when made to expel ammonia.

'*Oxygen*' *Process.* This is a useful process when comparing waters of similar origin. It is probably a more reliable measure of the *putrescent* matter present than the *total* organic contamination. It is essential that the oxidising agent, potassium permanganate, be added in excess and allowed to stand three hours. The following method is very delicate (*vide* Dr Tidy on " Potable Waters," ' Chem. Soc. Journ.,' January, 1879).

Cleanse with sulphuric acid and with tap-water two flasks, and place in one 500 septems (₁⁄₁₀ gall.) of the water, in the other an equal quantity of distilled water. Add to each 20 septems (140 gr.) of sulphuric acid (1 part pure acid to 3 parts of distilled water) and 20 septems of potassium permanganate, and allow to rest for 3 hours. Then add to each flask a couple of drops of an aqueous solution of potassium iodide (1 in 10), when iodine is liberated equivalent to the amount of permanganate unacted on by the waters. Observe the amount of a sodium hyposulphite solution (5·4 gr. in 7000 gr.) which must be added to each to remove this free iodine (judging of the exact point by adding towards the end of the experiment a few drops of starch). The strength of the potassium permanganate solution is 2 gr. of the salt in 7000 gr. to ₁⁄₁₀ gall. ; therefore the 20 septems will contain 0·04 gr. permanganate, equivalent to 0·01 of available oxygen. The experiment (A), with the amount of hyposulphite used up for the blank distilled water, shows the amount of hyposulphite equivalent to 20 septems, or 0·01 gr. of oxygen. Therefore the amount of oxygen unconsumed in the

water (B) to be examined was $\dfrac{B}{A} \times 0·01$, and the

amount (C) actually used up was $\dfrac{A-B}{A} \times 0·01$ for

500 septems (₁⁄₁₀ gall.). Then the oxygen consumed per gallon would be $\dfrac{A-B \times 0·2}{A}$. It is necessary to perform this standardising of hyposulphite with every series of experiments, on account of its tendency to change. Dr Tidy recommends that in addition to the three hours' experiment one of a single hour duration be executed. The higher the proportion of oxygen consumed in one hour to the oxygen consumed in three hours, the worse the water.

Nitrites, sulphuretted hydrogen, and ferrous salts interfere with this test, and there appears to be a different ratio between the oxygen consumed and the amount of organic matter, according to the amount of oxidation that has already taken place. The organic matter of deep wells is proportionately least acted upon.

Combustion Methods. The 'Frankland and Armstrong process' consists in burning with oxide of copper *in vacuo* the residue left on evaporating the water, and collecting and measuring in a suitable gas apparatus the carbonic acid, nitrogen, and nitric oxide proceeding from the organic matter. From these estimations are calculated the organic carbon and nitrogen.

This method, though forming the most accurate means of measuring organic contamination, is not in general use in consequence of the difficulties attending Dr Frankland's method of analysis. Professor Dittmar and Drs Dupré and Hake have lately introduced processes by which the same results may be obtained without necessitating the use of expensive gas apparatus.

Dittmar's Carbon Process. Concentrate a suitable quantity (say 10,000 gr.) in a pear-shaped flask, and, after adding some saturated solution of sulphurous acid to expel carbonates and nitrates, evaporate to dryness in a glass dish on a water-bath. Transfer the residue from the dish to a porcelain or platinum boat, and introduce it into the tail end of a combustion tube, filled three fourths of its length with oxide of copper, and having a roll of silver gauze in the front part of the tube. Previous to the boat being put in, this tube is heated to redness, and a stream of air freed from carbonic acid passed through it till the gas which comes out no longer renders clear baryta water turbid. The combustion tube has attached in front a small V-shaped tube charged with chromic acid dissolved in 60% sulphuric acid. To it is permanently fixed a small tube filled with calcium chloride, and in front of all is a small weighed U-tube, the first three fourths of which is filled with soda lime, and the other fourth with calcium chloride. On turning the gas on gradually from the front to the tail the residue is at last reached, and burnt in the stream of pure air. The carbonic acid given off, after being freed from sulphurous anhydride by passing through the chromic acid solution, and of moisture by the calcium chloride, passes into the soda lime tube and is absorbed. The increase in weight multiplied by ₃⁄₁₁ gives the amount of carbon and the amount of water taken.

Dittmar's Nitrogen Process. An amount of water, about half that taken for the carbon, is evaporated in a similar way. The residue is transferred to a large copper or silver boat, and mixed with about fifty grains of soda made from pure sodium, or with a mixture of soda and baryta, and burnt in a stream of hydrogen in a short combustion tube, which is closed in front by a nitrogen absorption bulb charged with exceedingly weak acidulated water. The amount of ammonia given off is estimated by the Nessler test as described under 'Ammonia.' Subtracting the amount of inorganic ammonia, the residue multiplied by $\frac{14}{17}$ yields the quantity of organic nitrogen in that volume of water.

A few blank experiments must be made to observe and allow correction for the amount of experimental error.

Carbon Method of Drs Dupré and Hake ('Chem. Soc. Journ.,' March, 1879). This method appears to be very accurate, but it necessitates a number of minute precautions, which cannot here be particularised. A residue is obtained by evaporating the water either in the ordinary hemispherical glass dish, or in an exceedingly thin silver one, which after being ignited is supported in a platinum hemisphere of convenient size. At the close of the evaporation this dish is crumpled up without being handled and introduced into a combustion tube, similar to that described under Dittmar's process. The carbonic acid is absorbed in clear barium hydrate solution, and the precipitated barium carbonate is, with suitable precautions to prevent access of impure air, collected on a filter and washed. It is dried and weighed. The result divided by 19·4 gives the weight of organic carbon. As another method of estimating the carbon the authors propose to compare the turbidity produced by the carbonic acid evolved from the combustion of the residue in solutions of basic acetate of lead with that produced by known quantity of carbonic acid.

Pres. The preservation of rain water in a state of purity necessitates the greatest care in constructing the tanks, especially if the latter are under ground. Of eight samples of stored rain water examined by the River Commissioners only one was fit for domestic use; the others were all polluted by animal matter. Storage room sufficient to hold 120 days' supply will be found sufficient for the driest district. The small cisterns for service water should not be placed in positions where it can receive the emanation of water-closets or sleeping apartments. They should be frequently cleaned out. The best are made of enamelled slate or properly painted iron. Wherever possible a water service should be on the constant supply system.

For wells the chief precaution necessary is to keep out surface and drainage water by maintaining the walls waterproof for a considerable depth. On shipboard water is preserved in iron tanks, or in casks well charred on the inside. Water cannot be safely kept in copper or leaden vessels, and it receives a calcareous impregnation by contact with lime, mortar, slate, or stone containing lime. The addition of $\frac{1}{2}$% to 1% of finely powdered binoxide of manganese materially promotes preservation, especially at sea, where the motion of the vessel and the subsequent agitation of the water increases the points of contact. Water never putrefies in iron vessels, or when some fragments of metallic iron are immersed in it.

Distilled water should be preserved in glass bottles or carboys. See LOTION, SPIRITS, WATER, DISTILLED EYE WATER, PERFUMED WATER, and the articles below.

Water, Soda. Each bottle of this liquid should contain at least 15 grains of carbonate of sodium, but that of the shops is usually nothing else but water highly charged with carbonic anhydride. Not a particle of soda enters into its composition, on which account it cannot be substituted for the preparation of the Pharmacopœias.

To produce a superior article of soda water, the possession of a powerful aërating and bottling machine is absolutely necessary. The water employed must also be of the purest quality, the carbonic anhydride well washed with water, and the corks so prepared that they will not impart their peculiar flavour to the beverage. See POWDERS, SOLUTION, WINES, and LEAD IN AËRATED WATER.

Water, Tar. See INFUSION OF TAR.

Water-glass. Strong solutions of both sodium and potassium silicate are sold under this name. The solutions are very viscid, transparent, and soluble in water. The soda salt contains about 10% caustic soda and 20% silica. Surgeons use both preparations to impregnate bandages, for mounting fractured limbs, in place of starch; in very dilute solution it is antiseptic, and employed as injection for gonorrhœa, cystitis, leucorrhœa, and uterine ulceration.

WATERS. *Syn.* AROMATIC WATERS, ODORIFEROUS W., PERFUMED W.; AQUÆ (Ph. L.), AQUÆ · DESTILLATÆ (Ph. E. & D.), L. Pure water charged, by solution or distillation, with volatile, odorous, and aromatic principles.

Prep. 1. (Ph. L.) *a.* 2 galls. of water are put into the still along with the vegetable matter (bruised if necessary), but only 1 gall. is drawn over. In the Ph. L. 1836, 7 fl. oz. of proof spirit were added before distillation.

b. Take of the essential oil of the plant, 2 fl. dr.; powdered silex, 2 dr.; triturate them diligently together, and then with distilled water, 1 gall., gradually added; lastly (after briskly agitating the whole for some time), strain the solution.

2. (Ph. E.) As 1, *a*, but adding of rectified spirit, 3 fl. oz., before distillation.

3. (Ph. D.) From the respective essences, 1 fl. oz.; distilled water, 2 quarts; agitated well together, and then filtered through paper.

The following are the AQUÆ DESTILLATÆ of the British Colleges, with some others, the quantities referring to a product of 1 gall., to be prepared as above when not otherwise directed.

ANGELICA WATER; AQUA ANGELICÆ (P. Cod.), L. Bruised seed, 1 lb.; water, q. s.: distil 4 lbs.

ANISEED WATER; AQUA ANISI (P. Cod.), L. From seeds, as AQUA ANGELICÆ.

BALM WATER; AQUA MELISSÆ (P. Cod.), L. Fresh tops, 12 lbs.

BERGAMOT WATER; AQUA BERGAMII (Ph. L. 1746), L. Bergamot peel, 5 oz.

BITTER ALMOND WATER; AQUA AMYGDALÆ AMARÆ, A. AMYGDALARUM AMARARUM (P. Cod.), L. Bitter almond cake (from which the oil has been expressed), 5 lbs.; macerate for 24 hours, and filter the distilled product through paper previously wetted with pure distilled water. Poisonous.—*Dose*, 10 to 60 drops, as a substitute for hydrocyanic acid.

BLACK MUSTARD-SEED WATER; AQUA SINAPIS NIGRÆ (*Guibourt*), L. Mix 1 part of ground black mustard seed with 8 parts of water; macerate for 12 hours, and distil 4 parts, by means of steam conducted by a tube from a boiler to the bottom of the still. Filter through moistened paper to separate the oil. Used externally as a rubefacient.

BORAGE WATER; AQUA BORAGINIS (P. Cod.), L. Fresh leaves, 12 lbs.

WATER OF BOTOT; EAU DE BOTOT, Fr. The following is Winkler's formula:—Tincture of cedar wood, 500 grms.; tincture of myrrh, 125 grms.; tincture of rhatany, 125 grms.; oil of peppermint, 5 drops. It is employed as a tooth and mouth wash.

CAMPHOR WATER; AQUA CAMPHORÆ (B. P.), MISTURA C., L. Enclose 1 oz. of camphor, broken into pieces, in a muslin bag, and attach this to one end of a glass rod, to keep it at the bottom of a bottle containing 1 gall. of distilled water. Macerate for two days, then pour off the solution as required.

CARAWAY WATER; AQUA CARUI (B. P., Ph. L. & D.), L. Caraway, bruised, 1 part; water, 20 parts; distil 10 parts.

CASCARILLA WATER; AQUA CORTICIS CASCARILLÆ (P. Cod.), L. Cascarilla, bruised, 3 lbs.

CASSIA WATER; AQUA CASSIÆ (Ph. E.), L. Cassia, bruised, 1½ lbs.

CASTOR WATER; AQUA CASTOREI, L. Castor, 4 oz.

CHAMOMILE WATER; AQUA ANTHEMIDIS (Ph. G.), L. Dried chamomile flowers, 2 lbs.; water, q. s.; distil 20 lbs.

CHERRY-LAUREL WATER; AQUA LAURO-CERASI (B. P., Ph. E. & D.), L. *Prep.* (B. P.) Fresh leaves of cherry laurel, 16 parts; water, 50 parts; chop the leaves, crush them in a mortar, and macerate them in the water for 24 hours; distil 20 parts of the liquid, shake the product, filter through paper; adjust the strength of the finished product, either by addition of hydrocyanic acid, or by diluting with water so that 810 gr. of it shall correspond, on testing with silver nitrate, to 0·1% of real hydrocyanic acid.—*Dose*, 10 to 60 drops, as a substitute for hydrocyanic acid. It is commonly imitated in trade by dissolving 75 drops (minims) of the oil of bitter almonds in 2½ fl. oz. of rectified spirit, agitating the mixture with warm distilled water, 1 gall., and filtering.

CHLOROFORM WATER; AQUA CHLOROFORMI (B. P.), L. Chloroform, 1 part; water, 200 parts. Shake till dissolved.

CINNAMON WATER; AQUA CINNAMOMI (B. P., Ph. L., & D.), L. *Prep.* 1. Cinnamon, bruised, 18 oz., or oil, 2 fl. dr.

2. (B. P.) Cinnamon, bruised, 1 part; water, 16 parts; distil 8 parts.

CLOVE WATER; AQUA CARYOPHYLLI (P. Cod.), L. Cloves, bruised, 3 lbs.

CORIANDER WATER; AQUA CORIANDRI, L. As ANGELICA WATER.

DILL WATER; AQUA ANETHI (B. P., Ph. L. & E.), L. *Prep.* 1. Bruised seed, 1½ lbs., or essential oil, 2 fl. dr.

2. (B. P.) Bruised fruit, 1 part; water, 20 parts; distil 10 parts.

DISTILLED WATER; AQUA DESTILLATA (B. P.), L. Take of water, 10 galls.; distil from a copper state, connected with a block-tin worm; reject the first ½ gall., and preserve the next 8 galls. It should remain clear on the addition of either lime water, chloride of barium, nitrate of silver, oxalate of ammonia, or hydrosulphuric acid (sulphuretted hydrogen).

ELDER-FLOWER WATER; AQUA SAMBUCI (B. P., Ph. L. & E.), L. *Prep.* 1. Fresh elder flowers, 10 lbs.

2. (B. P.) Fresh elder flowers, separated from the stalks, 1 part; water, 2 parts; distil 1 part.

EUCALYPTUS WATER; AQUA EUCALYPTI, L. Dry leaves, 1 part; add sufficient water to yield 4 parts of product.

FENNEL WATER; AQUA FŒNICULI (B. P., Ph. L., E., & D.), L. As DILL WATER.

HONEY WATER; AQUA MELLIS, L. Ol. caryophylli, 2½ dr.; ol. bergamot, 10 dr.; ol. lavandulæ ang., 2½ dr.; moschi, 4 gr.; ol. santal. flav., 15 minims; spt. vini rectificat., 32 oz.; aq. rosæ, 8 oz.; aq. flor. aurantii, 8 oz.; mellis opt. aug., 2 oz. Macerate the whole for at least 3 weeks, agitating occasionally. Allow to settle, decant, and filter if necessary.

HYSSOP WATER; AQUA HYSSOPI (P. Cod.), L. Fresh tops, 12 lbs.

HYSTERIC WATER; AQUA HYSTERICA, L. Compound of spirit of bryony, omitting the bryony.

JUNIPER WATER; AQUA BACCÆ JUNIPERI (P. Cod.), L. Berries, bruised, 3 lbs.

LAVENDER WATER; AQUA LAVANDULÆ (P. Cod.), L. Flowering tops, 3 lbs.

LEMON-PEEL WATER; AQUA LIMONIS (E. 1817), L. Fresh lemon peel, 2 lbs.; water, q. s.; distil 10 lbs.

LETTUCE WATER; AQUA LACTUCÆ (P. Cod.), L. Fresh lettuces, bruised, 12 lbs.

LILY WATER; AQUA LILIORUM CONVALLARIUM (Ph. Bruns.), L. Flowers of lily of the valley, 1 lb.; water, 4 lbs.; distil 2 lbs.

LIME-TREE FLOWER WATER; AQUA TILIÆ, L. From lime flowers, as MELILOT WATER.

MELILOT WATER; AQUA MELILOTI (P. Cod.), L. Dried flowers, 3 lbs.

MINT WATER, SPEARMINT W.; AQUA MENTHÆ VIRIDIS (B. P., Ph. L., E., & D.), L. *Prep.* 1. Dried herb, 2 lbs.; or fresh herb, 4 lbs.; or essential oil, 2 fl. dr.

2. (B. P.) English oil of spearmint, 1½ dr.; water, 1½ galls.; distil 1 gall.

MYRTLE-FLOWER WATER; AQUA MYRTI, L. Myrtle flowers, 3 lbs.; water, q. s.; distil 1 gall.

OPIUM WATER; AQUA OPII (Ph. G.), L. Opium, sliced and dried, 1 oz. Put into a glass retort with 10 oz. of water, and distil 5 oz.

ORANGE-FLOWER WATER; AQUA AURANTII FLORIS (B. P., Ph. L.), A. FLORUM AURANTII, L. "Water distilled from the flowers of *Citrus bigaradia*, Risso, and *Citrus aurantium*, DC." (Ph. L.). Orange flowers, 10 lbs.; proof spirit, 7 fl. oz. (Ph. L. 1836).

ORANGE-PEEL WATER; AQUA CORTICIS AU-RANTII (Ph. L. 1746), L. Rind of oranges, 5 oz. ORIGANUM WATER; AQUA ORIGANI (P. Cod.), L. Dried flowers, 3 lbs.

PARSLEY-SEED WATER; AQUA PETROSELINI (P. Cod.), L. From parsley seed, as ANGELICA WATER.

PEACH WATER; AQUA PERSICÆ (P. Cod.), L. Fresh leaves, chopped small, 12 lbs.; as CHERRY-LAUREL WATER.

PEACH-LEAF WATER; AQUA PERSICÆ (P. Cod.), L. Fresh peach leaves, cut small, 2 lbs.; water, 4 lbs.; distil gently 3 lbs.

PENNYROYAL WATER; AQUA PULEGII (Ph. L. & E.), A. MENTHÆ PULEGII (Ph. D.), L. As MINT WATER (above).

PEPPERMINT WATER; AQUA MENTHÆ PIPERITÆ (B. P., Ph. L., E., & D.), L. As MINT WATER (above).

PIMENTO WATER; AQUA PIMENTÆ (B. P., Ph. L., E., & D.), L. Prep. 1. Pimento, bruised, 1 lb.; or oil, 2 fl. dr.

2. (B. P.) Pimento, bruised, 1 part; water, 23 parts (nearly); distil one half.

PLANTAIN-LEAF WATER; AQUA PLANTAGINIS (P. Cod.), L. From fresh plantain leaves, as LETTUCE WATER.

RASPBERRY WATER. Fresh raspberries, 6 lbs.

RED ANT WATER; AQUA FORMICARUM, L. Distilled from red ants with water, q.s.

RHODIUM WATER; AQUA RHODII, L. Rhodium wood, 1 part; water, 8 parts; macerate, and distil 4 parts.

ROSE WATER; AQUA ROSÆ (B. P., Ph. L., E., & D.), L. Damask or hundred-leaved rose, 10 lbs. (Ph. L. & E.). Otto, 40 drops (Ph. D.). Fresh cabbage-rose petals, 1 part; water, 2 parts; distil 1 part (B. P.).

ROSEMARY WATER; AQUA ROSMARINI, A. ANTHOS, L. Rosemary, in flower, 1 lb.; infuse 24 hours; distil 1 gall.

RUE WATER; AQUA RUTÆ, L. Fresh rue, 1 lb.; macerate 24 hours; distil 1 gall.

SAGE WATER; AQUA SALVIÆ (P. Cod.), L. As LAVENDER WATER (above).

SASSAFRAS WATER; AQUA LIGNI SASSAFRAS (P. Cod.), L. Sassafras chips, 3 lbs.

SASSAFRAS WATER; AQUA SASSAFRAS (P. Cod.), L. From sassafras, as MELILOT WATER.

SCURVY-GRASS WATER; AQUA COCHLEARIÆ (P. Cod.), L. Fresh scurvy-grass, 8 lbs.

SPEARMINT WATER. See MINT WATER.

SPIRITUOUS WATERS. Many of the distilled spirits were formerly termed waters.

SPRUCE FIR WATER; AQUA ABIETIS (P. Cod.), L. Bruised buds of spruce fir, 2 lbs.

STINKING GOOSEFOOT WATER; AQUA CHENO-PODII VALVARIÆ, L. Stinking goosefoot, 1 lb.; water, 6 lbs.; distil 3 lbs.—Dose, 1 to 2 oz.; in hysteria.

STRAWBERRY WATER; AQUA FRAGARIÆ, L. Strawberries, 3 lbs.; water, q.s.; distil 3 lbs.

TANSY WATER; AQUA TANACETI (P. Cod.), L. Flowering tops, 6 lbs.

THYME WATER; AQUA THYMI (P. Cod.), L. As the last.

VALERIAN WATER; AQUA VALERIANÆ, A. RADICIS V. (P. Cod.), L. Root, bruised, 3 lbs.

VANILLA WATER; AQUA VANILLÆ, L. Va-

nilla, coarsely powdered, 1 lb.; salt, 5 lbs.; water, 2½ galls.; macerate for 24 hours in a covered vessel, then distil 1 gall.

VIOLET WATER; AQUA VIOLÆ, L. Violets, 1 part; water, 4 parts; after 6 hours distil 2 parts.

WORMWOOD WATER; AQUA ABSINTHINI (P. Cod.), L. Wormwood tops, 4 lbs.

Uses, &c. Distilled waters are mostly employed as vehicles or perfumes. A few, as bitter almond, cherry-laurel, and peach water, are poisonous in doses larger than a few drops. The dose of the aromatic or carminative waters, as those of dill, caraway, peppermint, pennyroyal, &c., is a wine-glassful ad libitum.

Concluding Remarks. In the preparation of distilled waters for medical purposes the utmost care should be taken to prevent contamination from contact with either copper, lead, or zinc, since these metals are gradually oxidised and dissolved by them. In preparing them from the essential oils, silica, in impalpable powder, is the best substance that can be employed to promote the division and diffusion of the oil, as directed in the Ph. L. Magnesia and sugar, formerly used for the purpose, are objectionable; as the first not only decomposes a portion of the oil, but the water is apt to dissolve a little of it, and is hence rendered unfit to be used as a solvent for metallic salts, more especially for corrosive sublimate and nitrate of silver; whilst the other causes the water to ferment and acetify.

In the distillation of waters intended for perfumery the utmost care is requisite to produce a highly fragrant article. The still should be furnished with a high and narrow neck, and the heat of steam, or a salt-water bath, should alone be employed. The first 2 or 3 fl. oz. of the runnings should be rejected, except when spirit is used, and the remainder collected until the proper quantity be obtained, when the whole product should be mixed together, as distilled waters progressively decrease in strength the longer the process is continued. When a very superior article is desired, the waters may be redistilled by a gentle heat, the first two thirds only being preserved. The herbaceous odour of recently distilled waters is removed by keeping them for some months, loosely covered, in a cold cellar.

When distilled waters have been carefully prepared, so that none of the liquor in the still has 'spirted' over into the condensing worm, they keep well, and are not liable to change; but when the reverse is the case, they frequently become ropy and viscid. The best remedy for this is to redistil them. Waters which have acquired a burnt smell in the 'stilling' lose it by freezing. Distilled waters may be prevented from turning sour by adding a little calcined magnesia to them, and those which have begun to spoil may be recovered by adding 1 gr. each of borax and alum to the pint. The doctoring is not, however, to be recommended, and should never be adopted for those used in medicine. A drop of solution of terchloride of gold added to these waters shows whether they contain any uncombined essential oil, by forming, in that case, a fine metallic film on the surface. After distilled waters have acquired their full odour, they should be carefully preserved in well-stopped bottles. Some houses keep a separate still for each of the more delicate perfumed waters, as it is extremely difficult to remove any odour that adheres to the body of the still and worm. The addition of the small quantity of spirit ordered in the Ph. E. and Ph. L. 1836, in the preparation of their waters, in no way tends to promote their preservation.

In general, the druggist draws off 2 galls., or more, of water from the quantities of the herbs, barks, seeds, or flowers, ordered in the Pharmacopœias; hence the inferior quality of the waters of the shops. They do, however, very well for vehicles. The perfumers, on the contrary, use an excess of flowers, or at least reserve only the first and stronger portion of the water that distils over, the remainder being collected and used for a second distillation of fresh flowers.

The most beautiful distilled waters are those prepared in the south of France, and which are imported into England under the French names. Thus eau de rose, eau de fleurs d'oranges, &c., are immensely superior to the best English rose or orange-flower water, &c. The water that distils over in the preparation of the essential oils is usually of the strongest and finest class. See ESSENCE, OILS (Volatile), SPIRITS (Perfumed), VEGETABLES, &c.

WATERS (Eye). *Syn.* COLLYRIA, L. *Prep.*
1. From distilled vinegar, 1 fl. oz.; distilled water, ½ pint. Half a fl. oz. of rectified spirit, or 1 fl. oz. of brandy, is often added. In simple chronic ophthalmia, blear eyes, &c.; also to remove particles of lime from the eyes.

2. Sugar of lead, 10 gr.; pure vinegar, ½ teaspoonful; distilled water, ½ pint. In ophthalmia, as soon as active inflammation ceases; also as the last.

3. Wine of opium, 2 fl. dr.; sulphate of zinc, 20 gr.; distilled water, ½ pint. Astringent and anodyne; in painful ophthalmia and extreme irritability.

4. Opium, 15 gr.; boiling water, 8 fl. oz.; when cold, add of solution of acetate of ammonia, 2½ fl. oz., and filter. As the last.

5. Sulphate of zinc, 20 gr.; distilled water, ½ pint; dissolve. An excellent astringent water in chronic ophthalmia, weak and irritable eyes, &c.

6. Sulphate of copper, 10 gr.; camphor mixture (julep), ½ pint; dissolve. In the purulent ophthalmia of infants.

7. Camphor julep, 5 fl. oz.; solution of acetate of ammonia and rose-water, of each, 2½ fl. oz.; mix. For weak or swollen eyes, particularly after ophthalmia.

8. Chloride of barium, 30 gr.; distilled water, ½ pint. In the ophthalmia of scrofulous and syphilitic habits.

9. (*Bate's.*) From blue vitriol, 15 gr.; camphor, 4 gr.; hot water, ½ pint; agitate in a corked bottle, and, when cold, make it up to 4 pints, and filter. In purulent ophthalmia and blear eyes.

10. (*Goulard's.*) From solution of diacetate of lead, 16 drops; distilled water, ½ pint; mix. As No. 2.

11. (*Krimer.*) Hydrochloric acid, 20 drops; mucilage, 1 dr.; water, 2 fl. oz. To remove particles of iron or lime from the eye.

12. (*Marshall's* 'EYE DROPS.') Nitrate of silver, 2 gr.; dilute nitric acid, 2 drops; pure soft or distilled water, 1 fl. oz.; dissolve; add powdered gum, 15 gr.; agitate until dissolved, and the next day decant the clear portion.

13. (P. Cod.) Extract of opium, 4 gr.; rose-water, 1 fl. oz.; dissolve. In painful ophthalmia.

WATERS (Mineral). *Syn.* SALINE WATERS; AQUÆ MINERALES, L. Our space will not permit a description of these individually. The tables given on pages 1780-1, exhibiting their composition, will, however, enable the reader, with a little attention, to produce artificial waters more closely resembling the natural ones than can be done by adopting any of the numerous formulæ published for the purpose. The 'aërated waters' are charged with 5 or 6 times their volume of carbonic acid gas, by means of the apparatus employed by the soda-water manufacturers. On the small scale the gas is often produced by the reaction of the ingredients on each other, in which case, on the introduction of the latter, the bottle must be instantly closed and inverted. Distilled water, or filtered rain water, should alone be employed in their composition; and for the chalybeated and sulphuretted waters it should be first boiled, and allowed to cool out of contact with the air.

In addition to the tables it may be remarked that traces of iodine have been found in the water of Cheltenham (old well), traces of 'bromine in the water of Epsom, and traces of both bromine and iodine in that of Leamington (royal pump). Manganese has been found in the waters of Tunbridge, Carlsbad, Spa, Pyrmont, Marienbad, Saidschütz, &c. Traces of phosphoric and fluoric acid have also been found in some mineral waters. It is the opinion of many high authorities that the medicinal virtues of these waters depend more on the minute quantities of the above substances, and the high state of dilution in which they are held, than on their more abundant saline ingredients.

SUPPLEMENTARY TABLE OF MINERAL WATERS.

BADEN-BADEN. In 16 oz.

Chloride of sodium	. . . .	16·520 gr.
Bicarbonate of lime	. . .	1·273 „
Bicarbonate of magnesia	. .	0·042 „
Bicarbonate of protoxide of iron	.	0·037 „
Bicarbonate of protoxide of manganese	. . . . .	traces
Bicarbonate of ammonia	. .	0·060 „
Sulphate of lime	. . .	1·556 „
Sulphate of potash	. . .	0·017 „
Phosphate of lime	. . .	0·021 „
Arseniate of iron	. . .	traces
Chloride of magnesium	. .	0·097 „
Chloride of potassium	. .	1·258 „
Bromide of sodium	. . .	traces
Silica	. . . . . .	0·914 „
Alumina	. . . . .	0·008 „
Nitrates	. . . . .	traces
		22·098 gr.
Free carbonic acid	. . .	0·299 „
		(*Bunsen.*)

I. TABLE exhibiting the Composition of several of the more Celebrated MINERAL WATERS. 16 fl. oz. contain the following Ingredients:

WATERS.	Nitrogen in cubic inches.	Carbonic anhydride in cubic inches.	Sulphuretted hydrogen in cubic inches.	Carbonate of sodium in grains.	Carbonate of magnesium in grains.	Carbonate of calcium in grains.	Sulphate of sodium in grains.	Sulphate of magnesium in grains.	Sulphate of calcium in grains.	Chloride of sodium in grains.	Chloride of magnesium in grains.	Chloride of calcium in grains.	Ferric oxide.	Silica.	Temperature.	Total of saline contents.	AUTHORITY.
CARBONATED.																	
Seltzer		17·		4·	5·	8·				17·					Cold	29·	Bergman.
Pyrmont		26·		1·5	10·	4·5			8·5	1·5			0·6		ditto	30·6	ditto.
Spa		18·		5·	4·5	1·5	8·5	6·5		0·2			0·6		ditto	8·3	ditto.
Carlsbad		80·		10·	1·2	1·2				4·5			a trace	0·3	165°	19·6	Klaproth.
Pouges		22·			0·5	11·5							2·5	0·5	Cold	28·4	Hassenfratz.
Saint Parize									18·	2·2					ditto	25·	ditto.
CHALYBEATE.																	
Tunbridge	0·59	1·	trace of oxygen			0·08			0·17	0·80	0·08	0·05	0·28		ditto	0·66	Scudamore.
Cheltenham chalybeate		2·5					22·7	6·	2·5	41·8			0·8		ditto	78·8	Parkes & Brande.
Brighton		2·2	0·5	0·5					4·	8·	0·75		1·4	0·14	ditto	9·29	Marcet.
SALINE.																	
Seidlitz								180·	5·		4·5				ditto	192·8	Bergman.
Cheltenham pure saline	0·2					0·8	16·	11·	4·5	50·	1·				ditto	80·5	Parkes & Brande.
Bristol		8·5				1·5	1·5		1·5	0·5			0·8	0·2	74°	6·	Carrick.
Buxton						1·3			0·8	0·2					82°	1·88	Pearson.
Bath		1·2			2·5	0·8	1·5		9·	8·8					116°	14·8	Phillips.
Scarborough			uncertain			a trace	30·		9·	0·5			a trace ditto		Cold	29·	Saunders.
Barèges				2·2		ditto	2·3		a trace	1·5					120°	8·	ditto.
Plombières		8·5	8·64?			0·3		87·	5·5	2·6			a trace	0·8	?	6·6	Vauquelin.
Kilburn	0·4	a trace	a trace		0·5	1·	13·	7·	1·4	68·	6·5	0·2	0·8		Cold	64·2	Schweisser.
Leamington New Bath	0·8	ditto					19·		18·	41·	1·5				ditto	88·3	Lamb.
Leamington Old Bath		ditto					7·5								ditto	78·5	ditto.
SULPHUROUS.																	
Harrogate	0·8	1·	2·8		0·7	2·5		1·8		77·	11·	1·5			ditto	94·	Garnet.
Moffat	0·5	0·8	1·2							4·5					ditto	4·5	ditto.
Aix-la-Chapelle			5·5			4·2	23·5	5·	1·2	35·					143°	21·2	Bergman.
Cheltenham Sulp. Spring			1·5										0·8		Cold	65·	Parkes & Brande.

II. TABLE exhibiting the Composition of the principal MINERAL WATERS of GERMANY and of the SARATOGA CONGRESS SPRING of AMERICA, rearranged expressly for this Work.

Grains of Anhydrous Ingredients in one pound Troy.	Adelheids-Quelle.	Anadowitz Ferdinands-brunnen.	Carlsbad.	Eger Franzens-brunnen.	Ems.	Pachingen.	Kissengen Ragoczi.	Kreuznach Elisen-brunnen.	Marienbad Kreuthr.	Püllna.	Pyrmont.	Saratoga Congress Spring.	Obersalz-brunnen Schlesischer.	Seidschütz.	Seltera.	Spa Pouhon.
Carbonate of soda	5·3443	4·5976	7·2719	3·9914	8·0836	19·3338			5·8499			0·8961	7·6311		4·6168	0·5631
" lithia	0·0902	0·0607	0·0160	0·0263	0·0405				0·0868							
" baryta	0·0094				0·0058							0·0073			0·0014	
" strontia	0·0387	0·0040	0·0065	0·0058	0·0590		0·0699	0·2068	2·9459	5·7776	4·7781	5·6631	0·0170	5·1046	0·0144	0·7287
" lime	0·4708	3·0065	1·7775	1·3501	0·8555	1·8647	4·6180	1·1313	0·0390	4·8046		4·1155	1·5464	9·9256	1·4004	0·6431
" magnesia (proto) manganese	0·9980	3·2887	1·0375	5·6040	2·1911	1·9963	1·8186						1·5496		1·5000	
" (proto) iron	0·0012	0·0098	0·0098	0·0384	0·0098		0·0131	0·0073	2·0298		0·0864	0·0096	0·0096	0·0096		0·0699
Subphosphate of lime	0·0131	0·9995	0·0908	0·1763	0·0130	0·0061	0·1897	0·1495	0·1819	0·0098	0·9313	0·0173	0·0345	0·0094	0·0007	0·8213
" alumina			0·0013	0·0173							0·0110		0·0117		0·0050	0·0103
Sulphate of potassa	0·0064	0·0040	0·0019	0·0093	0·0014	0·1897	1·5540				0·0314	0·1879	0·0168	0·0088	0·9978	0·0064
" soda		16·9098	14·9019	18·9765	0·4060		5·5485		98·5868	3·6000	1·0099		0·5160	3·6706		0·0635
" lithia										1·9600	0·0067	1·9579	5·9106	17·0590		0·0581
" lime										69·3146	5·9385					
" strontia					0·0538		0·0864	0·7287			0·0164	0·1004	0·0164	1·1897	0·2494	
" magnesia	0·1846							64·0917			9·9684	0·0856		0·0847	19·9690	0·2871
Nitr. of magnesia	98·4408		5·9590	6·9599	5·7966	8·5397	89·9783	0·0569	10·1797			1·9666	0·3689	69·9455		
Chlor. of ammonium								9·7388				19·9663		5·9603		
" potassium																
" sodium	0·9060					3·8699		0·2866	14·7496				1·2528			0·0899
" lithium	0·1400							0·9404								
" calcium								0·2304								
" magnesium			0·0184			0·5381		0·0094				0·1613	0·0061	1·2528		
" barium												0·0046				
" strontium																
Bromide of sodium	0·0165	0·0060	0·0164	0·3648	0·0014	0·0587	0·1609	0·0098	0·0098	0·1330	0·9460	0·0089	0·0061		0·0013	0·7389
Iodide of sodium	0·1983	0·1400	0·4559		0·3104			0·8835	0·9908		0·8727	0·1113	0·2498	0·0900	0·1566	
Fluoride of calcium												0·0069				
Alumina																
Silica		0·6038														
Total saline contents	34·4739	34·4719	31·4408	31·6870	16·0836	18·9300	34·7136	63·0190	51·6417	188·9606	16·4281	33·7443	14·7309	98·0133	21·2936	3·9691
Carbonic acid gas in 100 cubic inches	10	144	58	154	61	136	96	13	106	7	160	114	98	90	136	136
Temperature, F.	58°	49°	Sprn. 165° Neub.138° Mühl 138° Ther. 138°	54°	Kaa. 117° Krän. 84°	50°	6 °	47°	68°	68°	· 56°	50°	58°	58°	58°	58°
Authorities	Struve.	Steinm.	Berzelius.	Berzelius.	Struve.	Bischof.	Struve.	Struve.	Berzelius.	Struve.	Struve.	Schweiz.	Struve.	Struve.	Struve.	Struve.

APOLLINARIS BRUNEN.

Carbonate of soda	9·65	grains.
" of magnesia	3·39	"
" of lime	0·45	"
Chloride of sodium	3·57	"
Sulphate of soda	2·30	"
Oxide of iron } Alumina }	0·15	"
Silica	0·06	"
	19·59	
Carbonic acid	47·04	

The above are the contents of 16 oz. Temp. 70° F. (*Bish*).

FRIEDRICHSHALL. In 16 oz.

Sulphate of soda	46·51 gr.
Sulphate of magnesia	39·55 "
Chloride of sodium	61·10 "
Chloride of magnesium	30·25 "
Bromide of magnesium	0·37 "
Sulphate of potash	1·52 "
Sulphate of lime	10·34 "
Carbonate of lime	0·11 "
Carbonate of magnesia	1·16 "
Silica	0·33 "
	190·25 gr.
Carbonic acid	5·32 c.i.

(*Liebig*.)

TOEPLITZ. In 16 oz. Temp. 14° F.

Sulphate of potash	0·098 gr.
Sulphate of soda	0·290 "
Carbonate of soda	2·635 "
Phosphate of soda	0·014 "
Fluoride of silicon	0·351 "
Chloride of sodium	0·433 "
Carbonate of lime	0·330 "
Carbonate of strontia	0·027 "
Carbonate of magnesia	0·088 "
Carbonate of protoxide of iron	0·019 "
Carbonate of protoxide of manganese	0·021 "
Sulphate of alumina	0·020 "
Silica	0·443 "
Crenic acid	0·034 "
	4·804 gr.

(*Wolf*.)

VICHY (Grand Grille).

Temp. 106° F. In a litre.

Carbonic acid	0·908
Bicarbonate of soda	4·883
Bicarbonate of potash	0·352
Bicarbonate of magnesia	0·303
Bicarbonate of strontia	0·003
Bicarbonate of lime	0·434
Bicarbonate of protoxide of iron	0·004
Bicarbonate of protoxide of manganese	a trace
Sulphate of soda	0·291
Phosphate of soda	0·130
Arseniate of soda	0·002
Borate of soda	a trace
Chloride of sodium	0·534
Silica	0·070
Organic matter, bituminous	a trace
	Grms. 7·914

WOODHALL (Lancashire).

Iodine and bromine, with chlorides of calcium, magnesium, potassium; more than ½ gr. of bromide of sodium and ¼ gr. of iodide of sodium.

190 gr. in 20 oz. Strongly impregnated with carbonic acid.

WATERS (Perfumed). *Syn.* AQUÆ ODORIFERÆ, L. The simple distilled waters of the perfumer have been already noticed (see p. 1777). They may be prepared from any substance which imparts its fragrance to water by distillation. The compound waters (eaux) employed as perfumes consist of very pure rectified spirit, holding in solution essential oils or other odorous matter, and resemble the esprits, essences, and spirits before noticed. They differ from extraits in being mostly colourless, or nearly so, and in being generally prepared by distillation, or by the addition of the pure essential oils or essences to carefully rectified and perfectly scentless spirit; whereas the extraits are mostly and preferably prepared by macerating the flowers, &c., in the spirit, or by digesting the spirit with the oils, in the manner noticed under SPIRITS (Perfumed). Extraits are preferred to eaux and esprits as the basis of good perfumery, where the colour is not objectionable.

The following are a few additional formulæ and remarks:

ANGEL WATER, PORTUGAL W. From orange-flower and rose water, of each, 1 pint; myrtle water, ½ pint; essence of ambergris, ⅛ fl. oz.; essence of musk, ¼ fl. oz.; shake them well together for some hours, then filter the mixture through paper.

EAU D'ANGE, Fr.; AQUA MYRTI, L. From myrtle flowers, 3½ lbs.; water, 2 galls.; distil a gallon. A pleasant perfume.

EAU D'ANGE BOUILLÉE, Fr. From rose-water and orange-flower water, of each, 3 pints; benzoin, ½ lb.; storax, ¼ lb.; cinnamon, 1 oz.; cloves, ⅛ oz.; 3 fresh-emptied musk bags; digest in a securely covered vessel, at nearly the boiling heat, for 2 hours, then allow it to cool; strain off the clear, press the remainder, and filter for use. Very fragrant.

EAU D'ANGE DISTILLÉE, Fr. From benzoin, 4 oz.; storax, 2 oz.; cloves, ½ oz.; calamus and cinnamon, of each, ¼ oz.; coriander seeds, 1 dr. (all bruised); water, 3 quarts; distil 2 quarts. Eau d'ange distillée et musquée is made by adding a little essence of musk to the distilled product. Both are highly fragrant.

EAU DE LAVANDE, LAVENDER WATER. See SPIRITS (Perfumed).

EAU DE NAPHRE, EAU DE NAPHE, Fr.; AQUA NAPHÆ, L. This article is distilled in Languedoc from the leaves of the bigarade, or bitter orange tree, but the preparation sold in England under this name is often prepared as follows:—Orange flowers, 7 lbs.; fresh yellow peel of the bigarade or Seville orange, ½ lb.; water, 2 galls.; macerate 24 hours, and distil 1 gall. In many cases ordinary orange-flower water is sold for eau de naphe.

ROSE-WATER. From otto of roses, 3 dr.; rectified spirit (warm), 1 pint; dissolve, add of hot water, 10 galls.; mix in a 12-gallon carboy, cork, and well agitate the whole until quite cold.

This makes the ordinary rose-water of the shops, and is really excellent, but it is better for distillation. See WATERS (Distilled).

UNPARALLELED WATER; EAU INCOMPARABLE, Fr. From oil of lemon, 4 dr.; oil of bergamot, 2½ dr.; oil of cedrat, 2 dr.; rectified spirit, 3½ pints; Hungary water, ½ pint; mix, and distil all but 9 oz. (*Guibourt*).

WATER-BRASH. See PYROSIS.

WATER-CLOSET. There are a number of conditions necessary to be observed in the construction and arrangement of the water-closet if we wish to prevent its becoming a nuisance and a source of danger to the health of the inmates of a dwelling-house. 1. As regards situation there can be no doubt that, upon strict sanitary principles, the closet, instead of forming part of the house, should, whilst within easy access to it, be entirely detached. Owing to various causes, however, this isolation is frequently impossible.

Under such circumstances, the closet, whilst forming part of the dwelling, should be built out from it, so as to have as little connection as possible with the rooms, corridors, &c. To still further accomplish this end the approach to the closet should be through a small vestibule or passage connecting the closet with the corridor, and opening into the latter by means of a door. Where there are more than one closet, they should be built upon the plan just proposed, and one over the other. The basement of a house is a particularly objectionable locality for a water-closet, since the warm house acts as an aspirator, and thus draws any fetid and poisonous gases there may be in the closet into the house, and causes them to be diffused throughout it. The water-closet should, therefore, always be placed in the higher parts of a building. 2. As regards construction, &c., it would be impossible for us to attempt to canvass the merits or the reverse of the numerous designs, patents, &c., that relate to this part of our subject. We shall indicate, therefore, only the more important desiderata, which are—That the pan should be nearly cone-shaped, and not round, like a half-circle. It is mostly made of earthenware, sometimes of metal, and occasionally of enamelled iron. The preferable substance is earthenware: the pan should always be ventilated, and there should likewise be a sufficient flow and force of water to sweep everything out of it and thoroughly cleanse it.

The cistern supplying the closet should be kept solely for this purpose, and not, as is sometimes the case, be taken from the house cistern, as this latter practice may lead to the contamination of the drinking water, owing to the gases rising from the closet.

The bottom of the pan is attached to the soil-pipe which discharges into the drain. The soil-pipe is mostly trapped by means of a syphon valve; and it is important that the points of junction between the pipe and the syphon valve, and the pipe and the main drain, should be thoroughly secure and air-tight. Furthermore it is imperative, if we wish to prevent an influx into the pan of the gases and foul air which rise through the syphon as the water runs off, that the soil-pipe should be ventilated. This may be effected by attaching a small pipe having connection with the outer air to the discharge-pipe just below the syphon, and carrying it up to the top of the house. Another advantage arising from ventilating the soil-pipe, besides the prevention of the escape of sewer gas into the house, is that there is no danger of its corrosion (if it be of lead) by the action of the pent-up sulphuretted vapours. The seat, which is mostly of wood, should be so arranged as to be easily moveable, and thus allow of easy inspection of the different parts should they get out of order.

The seat as well as the closet should always be ventilated. A good and simple method for the ventilation of the latter is to carry a tube from the top of the closet into the outer air. "If the closet is in a bad situation, it should be heated by a gas jet" (*Parkes*).

The lid attached to the seat should have a hole cut in it, so as to allow of the handle being pulled up when the pan is covered, which, strange to say, in perhaps ninety-nine cases out of every hundred it never is, after being used. Of course, in the absence of the ventilation of the pan and soil-pipe, the result of keeping the seat covered over would only be to fill the pan with malodorous and more or less dangerous gases, which would escape into the closet when the lid was again raised.

3. *Precautions.* The use of unduly large pieces of paper, such as cause stoppage and obstruction in the discharge-pipe, should be particularly avoided. Any defect or impediment in the working of the closet should be remedied at once. As a general rule, servants are very careless in all matters connected with the water-closet; so much so that the masters of many houses are themselves compelled to exercise supervision over it.

During very hot weather, or the prevalence of an infectious disease in a dwelling-house or in the neighbourhood of the house, some disinfectant should be added to the water that supplies the closet. A substance that will very satisfactorily answer this purpose is the commercial sulphate of iron known as green vitriol. A pound of it should be put into the tank when filled with water.

The same disregard of sanitary obligations so frequently shown in the construction, site, &c., of water-closets is more obvious in the case of privies. The Public Health Act not only renders unlawful the erection or rebuilding of any dwelling-house without "a sufficient water-closet, earth-closet, or privy, and an ash-pit, furnished with proper doors and coverings;" but also requires that, "if a house within the district of a local authority appears to such authority by the report of their surveyor or inspector of nuisances to be without a sufficient water-closet, earth-closet, or privy and ash-pit, furnished with proper doors and coverings, the local authority shall, by written notice, require the owner or occupier of the house within a reasonable time therein specified to provide a sufficient water-closet, earth-closet, or privy and an ash-pit furnished as aforesaid, or either of them, as the case may require."

Although in many large towns and cities a more or less effectual supervision may be exercised by the sanitary inspector in the above direction, every one's experience of the usual outdoor privy of a small English country town or village will suggest to him the extreme toleration prevailing amongst the sanitary authorities in many provincial and rural districts in this particular. Ventilation is as essential for the privy as the water-closet; so also is the thorough trapping of the exit-pipe from the pan, as well as the cleansing and flushing of this latter by water directly after it has been used. Yet how rarely do we find not only all, but even one of these conditions fulfilled in the arrangement of the ordinary privy; but instead an untrapped, immoveable pan (and in some cases even this is wanting), covered with filth, and no contrivance of any kind for a constant water supply!

No wonder, therefore, that the atmosphere of an ordinary privy should be so foul and noisome as it invariably is.

The following specification for a useful description of privy is published by Messrs Knight and Co., 90, Fleet Street, London:

Specification. The privy and dustbin to be built of 4½-inch brickwork, in well-ground mortar of approved quality. Two rows of 4½-inch and 3-inch bond timber to be built in at back of privy for securing ventilating shafts. The ventilating shafts to be 7 by 4½ inches, inside measurement, of best red deal boards, 1 inch thick, closely put together with strong white-lead paint, and well nailed and carefully seamed to the 4½-inch and 3-inch bond timber. These shafts to have coats of boiled tar, both inside and out.

The lid of refuse-bin to be of best 1-inch red deal boards, with two strong ledges or battens across them, to be hung with three strong band-hinges to the sides of the ventilating shafts, and the making-up piece between the same. A circular orifice to be made in centre of lid, between the battens, 10 inches wide. The lid to have two coats of boiled tar, both inside and out. A 4½-inch and 3-inch frame of red deal to be securely fixed on top of the dustbin as a seat for the lid. A lid over the privy seat to be hinged on at the back, with a child's seat over centre of large one. The larger seat to be provided with an earthenware circular rim beneath. The earth compartment to be without lid, and provided with a pint scoop for each occupant to throw in a pint of the stored dry earth or dry ashes through the seat into the galvanised iron pail, the contents of which must be scattered over the garden or put in the dustbin before the pail becomes full. A loose foot-block may be furnished where there are young children (the earth-closet is described under Sewage, Removal of).

The dustbin may be placed at side of the privy if required. The floor of dustbin to be at the ground level, slightly inclined outwards, and paved with brick. See Sewage, Removal and Disposal of; Drains, Tanks, Cesspools.

WATER-COLOUR CAKES. These are prepared from any of the ordinary pigments that work well in water, made into a stiff and perfectly smooth paste with gum water or isinglass size, or a mixture of the two, and then compressed in polished steel moulds and dried. See Painting, and the respective pigments.

WATERCRESS. The *Nasturtium officinale*, a well-known plant of the Nat. Ord. Cruciferæ.

It is alterative and antiscorbutic, and was formerly used in medicine, but now chiefly as a salad, or a refreshing relish at breakfast.

WATER-GAS. By forcing steam through fire-clay, or iron retorts filled with red-hot charcoal or coke, the steam is decomposed into a mixture of hydrogen, carbonic oxide, and carbonic anhydride (possibly a small quantity of marsh-gas is also present).

To this mixture, after it has been purified, the name of 'water-gas,' owing to the source from whence it has been derived, has been given.

According to some chemists the purified gas (obtained by passing the crude gaseous product sometimes over lime, sometimes over crystallised carbonate of soda) consists solely of hydrogen gas. Langlois's analysis, however, has led to the conclusion that it is a compound of hydrogen and carbonic oxide gases. Water-gas, obtained as above, possesses no illuminating power. This is imparted to it by impregnating the gas with the vapour of certain hydrocarbons, a plan suggested by Jobbard, of Brussels, in 1832. Another but less usual method, originating with Gengembre and Gillard, is to place on the burners which consume the gas small platinum cylinders. When these become white-hot a strong and brilliant light is produced. See Platinum Gas.

WATER-POX. See (Pox) Chicken-pox.

WATERPROOFING. Cloth is 'waterproofed' as follows:

1. Moisten the cloth on the wrong side, first with a weak solution of isinglass, and, when dry, with an infusion of nut-galls.

2. As the last, but substitute a solution of soap for isinglass, and another of alum for galls.

3. (Hancock's Patent.) By spreading the liquid juice of the caoutchouc tree upon the inner surface of the goods, and allowing them to dry in the air. Absolutely chimerical.

4. (Potter's Patent.) The cloth is first imbued on the wrong side with a solution of isinglass, alum, and soap, by means of a brush; when dry, it is brushed on the same side against the grain, and then gone over with a brush dipped in water. Impervious to water, but not to air.

5. (Sievier's Patent.) By applying first a solution of india rubber in oil of turpentine, and afterwards another india-rubber varnish, rendered very dry by the use of driers. On this, wool or other material of which the fabric is made, cut into proper lengths, is spread, and the whole passed through a press, whereby the surface acquires a nap or pile.

6. A simple method of rendering cloth water-proof, without being airproof, is to spread it on any smooth surface, and to rub the wrong side with a lump of beeswax (perfectly pure and free from grease) until it presents a light, but even, white or greyish appearance; a hot iron is then to be passed over it, and, the cloth being brushed whilst warm, the process is complete. When the operation has been skilfully performed a candle may be blown out through the cloth, if coarse, and yet a piece of the same, placed across an inverted hat, may have several glassfuls of water poured into the hollow formed by it, without any of the liquid passing through. Pressure or fric-

tion will alone make it do so. "We have shown this to numerous cloth manufacturers, waterproofers, tailors, and others, several of whom have adopted the method very extensively and with perfect success" (*Cooley*).

7. About the year 1862 a patent was taken out by Dr Stenhouse for employing paraffin as a means of rendering leather waterproof, as well as the various textile and felted fabrics; and in August, 1864, an additional patent was granted to him for an extension of and improvement on the previous one, which consisted chiefly in combining the paraffin with various proportions of drying oils, it having been found that paraffin alone, especially when applied to fabrics, became to a considerable extent detached from the fibre of the cloth after a short time, owing to its great tendency to crystallise. The presence, however, of even a small quantity of drying oil causes the paraffin to adhere much more firmly to the texture of the cloth, from the oil gradually becoming converted into a tenacious resin by absorption of oxygen.

In the application of paraffin for waterproofing purposes it is first melted along with the requisite quantity of drying oil and cast into blocks. This composition can then be applied to fabrics by rubbing them over with a block of it, either cold or gently warmed, or the mixture may be melted and laid on with a brush, the complete impregnation being effected by subsequently passing it between hot rollers. When this paraffin mixture has been applied to cloth such as that employed for blinds or tents, it renders it very repellent to water, although still pervious to air.

Cloth paraffined in this manner forms an excellent basis for such articles as capes, tarpaulins, &c., which require to be rendered quite impervious by subsequently coating them with drying oil, the paraffin in a great measure preventing the well-known injurious effect of drying oil on the fibre of the cloth. The paraffin mixture can also be advantageously applied to the various kinds of leather. One of the most convenient ways of effecting this is to coat the skins or manufactured articles, such as boots, shoes, harness, pump-buckets, &c., with the melted composition, and then to gently heat the articles until it is entirely absorbed. When leather is impregnated with the mixture, it is not only rendered perfectly waterproof, but also stronger and more durable. The beneficial effects of this process are peculiarly observable in the case of boots and shoes, which it renders very firm without destroying their elasticity. It therefore not only makes them exceedingly durable, but possesses an advantage over ordinary dubbing in not interfering with the polish of these articles, which, on the whole, it rather improves.

The superiority of paraffin over most other materials for some kinds of waterproofing consists in its comparative cheapness, in being easily applied, and in not materially altering the colour of fabrics, which in the case of light shades and white cloth is of very considerable importance.

8. A waterproof packing cloth which does not break may be made by covering the fabric with the following varnish:—2 lbs. of soft (potash)

soap is dissolved in water and mixed with an aqueous solution of sulphate of iron. The washed and dried soap is dissolved in 3 lbs. of linseed oil, in which ¼ lb. of caoutchouc has been previously dissolved.

WATERPROOF LIQUID. *Prep.* 1. India rubber, in fragments, 1 oz.; boiled oil, 1 pint; dissolve by heat, carefully applied, then stir in of hot boiled oil, 1 pint, and remove the vessel from the fire.

2. Boiled oil, 1 pint; beeswax and yellow resin, of each, 2 oz.; melt them together.

3. Salad oil, 1 pint; mutton suet, ¼ lb.; white wax and spermaceti, of each, 1 oz.; as the last. For 'ladies' work.'

4. Bisulphide of carbon, 2 oz.; gutta percha, ¼ oz.; asphaltum, 2 oz.; brown amber, ½ oz.; linseed oil, 1 oz.; mix. Dissolve the gutta percha in the bisulphide of carbon, the asphalte and amber in the oil, and mix well.

Obs. The above are used for boots, shoes, harness, leather straps, leather trunks, &c., applied warm before the fire.

WAX. *Syn.* BEESWAX, YELLOW W.; CERA (Ph. L.), CERA FLAVA (B. P., Ph. E. & D.), L. The substance which forms the cells of bees; obtained by melting the comb in water, after the honey has been removed, straining the liquid mass, remelting the defecated portion, and casting it into cakes.

Pure beeswax has a pleasant ceraceous odour, a pale yellowish-brown colour, and the sp. gr. ·960 to ·965. It is brittle at 32°, softens and becomes plastic at 88° or 90°, and melts at 154° to 155° F. "It becomes kneadable at about 85°, and its behaviour while worked between finger and thumb is characteristic. A piece the size of a pea being worked in the hand till tough with the warmth, then placed upon the thumb, and forcibly stroked down with the forefinger, curls up, following the finger, and is marked by it with longitudinal streaks" (*B. S. Proctor*). It is very frequently adulterated with farina, resin, and mutton suet or stearin. Dr Normandy ('Chem. Central,' 1872, No. 29) met with a sample containing 23% of effloresced sulphate of soda. The first may be detected by oil of turpentine, which dissolves only the wax; the second, by its solubility in cold alcohol, and by its terebinthinate taste; the third and fourth, even when forming less than 2% of the wax, may be detected by it affording sebacic acid on distillation. When greasy matter is present in any considerable quantity, it may also be detected by the suspected sample having an unctuous feel and a disagreeable taste. A spurious beeswax met with in the American markets is described in 'New Remedies' for 1877, and is said to have been a very clever imitation externally of the genuine substance, which it closely resembled in appearance, colour, fracture, bitterness, pliability, and odour. Upon analysis it was found to be composed of 60 parts of paraffin and 40 parts of yellow resin covered with a thin coating of true-beeswax. The specific quantity of the counterfeit article was identical with that of many samples of genuine beeswax. Saline matter may be detected by the loss of weight when a weighed quantity of the wax is boiled in water. Heavy substances, as chalk, plaster of Paris, white-lead, oxide of zinc, &c., may also be thus separated, since they subside, owing to their superior gravity, to the bottom of the vessel. The rough mealy fracture of pure wax is rendered finer grained, smoother, and duller by the addition of lard or spermaceti, and becomes sparkling and more granular by the addition of resin (*Proctor*).

Wax, Bleached. See WAX, WHITE (*below*).

Wax, Carnauba ('Ph. Journal,' vol. vi, 3rd series, p. 746). The leaves of the carnauba tree (*Copernica cerifera*), a South American palm, have lately become a very important source for the supply of large quantities of vegetable wax. Carnauba wax is extensively used in the manufacture of candles. Mr Consul Morgan, in a paper laid before Parliament in 1876, on the trade and commerce of Brazil, states "that the exportation of this wax is calculated at 871,400 kilos.; exceeding in value reis 1,500,000, or £162,500."

Wax, Etching. See ETCHING GROUND and VARNISH.

Wax, Facti"tious. *Syn.* CERA FLAVA FAC-TITIA, L. A spurious compound, sold by the farriers' druggists for veterinary purposes.

Prep. 1. From yellow resin, 16 lbs.; hard mutton suet or stearin, 8 lbs.; palm oil, 2½ lbs.; melted together.

2. As last, but substituting turmeric, 1 lb., for the palm oil.

3. Best annotta, 6 oz., or q. s.; water, 1 gall.; boil; add of hard mutton suet or stearin, 36 lbs.; yellow resin, 70 lbs.; again boil, with constant agitation, until perfectly mixed and of a proper colour, and, as soon as it begins to thicken, pour it out into basins to cool. When cold, rub each cake over with a little potato starch.

Wax, Gilder's. See GILDING.

Wax, Mod'elling. *Prep.* Take of beeswax, lead plaster, olive oil, and yellow resin, equal parts; whiting, q. s. to form a paste; mix well, and roll it into sticks. Colours may be added at will.

Wax, Refined. Crude wax, especially that imported, is generally loaded with dirt, bees, and other foreign matter. To free it from these substances it undergoes the operation of 'refining.' This is done by melting the wax along with about 4% or 5% of water in a bright copper or stoneware boiler, preferably heated by steam, and after the whole is perfectly liquid, and has boiled for some minutes, withdrawing the heat, and sprinkling over its surface a little oil of vitriol, in the proportion of about 5 or 6 fl. oz. to every cwt. of wax. This operation should be conducted with great care and circumspection; as, when done carelessly, the melted wax froths up, and boils over the sides of the pan. The acid should also be well scattered over the whole surface. The melted wax is next covered over, and left for some hours to settle, or until it becomes sufficiently cool to be drawn off for 'moulding.' It is then very gently skimmed with a hot ladle, baled or decanted into hot tin 'jacks,' and by means of these poured into basins, where it is left to cool. Great care must be taken not to disturb the sediment. When no more clear wax

can be drawn off, the remainder in the melting-pan is allowed to cool, and the cake, or 'foot,' as it is called, is taken out, and the impurities (mostly bees) scraped from its under surface. The scraped cake is usually reserved for a second operation; but if required, it may be at once remelted, and strained through canvas into a mould.

Much of the foreign wax has a pale, dirty colour, which renders it, no matter however pure, objectionable to the retail purchaser. Such wax undergoes the operation of 'colouring' as well as 'refining.' A small quantity of the best roll annotta, cut into slices (¼ lb., more or less, to wax, 1 cwt., depending on the paleness of the latter), is put into a clean boiler with about a gallon of water, and boiled for some time, or until it is perfectly dissolved, when a few ladlefuls of the melted wax are added, and the boiling continued until the wax has taken up all the colour, or until the water is mostly evaporated. The portion of wax thus treated has now a deep orange colour, and is added, in quantity as required, to the remainder of the melted wax in the larger boiler, until the proper shade of colour is produced when cold; the whole being well mixed, and a sample of it cooled now and then, to ascertain when enough has been added. The copper is next brought to a boil, and treated with oil of vitriol, &c., as before. Some persons add palm oil (bright) to the wax, until it gets sufficient colour, but this plan is objectionable from the quantity required for the purpose being often so large as to injure the quality of the product; besides which the colour produced is inferior, and less transparent and permanent than that given by annotta.

Another method of refining crude wax, and which produces a very bright article, is to melt it in a large earthen or stoneware vessel, heated by steam or a salt-water bath, then to cautiously add to it about 1% of concentrated nitric acid, and to continue the boiling until nitrous fumes cease to be evolved, after which the whole is allowed to settle, and is treated as before.

Obs. The great art in the above process is to produce a wax which shall at once be 'bright,' or semi-translucent in thin places, and good coloured. The former is best ensured by allowing the melted mass to settle well, and by carefully skimming and decanting the clear portion without disturbing the sediment. It should not be poured into the moulds too warm, as in that case it is apt to 'separate,' and the resulting cakes to be 'streaky,' or of different shades of colour. Again, it should be allowed to cool very slowly. When cooled rapidly, especially if a current of air fall upon its surface, it is apt to crack, and to form cakes full of fissures. Some persons, who are very nice about their wax, have the cakes polished with a stiff brush when quite cold and hard. It is absolutely necessary that the 'jacks' or cans, ladles, and skimmers, used in the above process, be kept pretty hot, as without this precaution the wax cools, and accumulates upon them in such quantity as to render them inconvenient, and often quite useless, without being constantly scraped out.

Wax, Seal'ing-. *Prep.* 1. (RED.) *a.* Take of

shellac (very pale), 4 oz.; cautiously melt it in a bright copper pan over a clear charcoal fire, and when fused, add of Venice turpentine, 1¼ oz.; mix, and further add of vermilion, 3 oz.; remove the pan from the fire, cool a little, weigh it into pieces, and roll them into circular sticks on a warm marble slab by means of a polished wooden block; or it may be poured into moulds whilst in a state of fusion. Some persons polish the sticks with a rag until quite cold.—*b.* From shellac, 3 lbs.; Venice turpentine, 1¼ lbs.; finest cinnabar, 2 lbs.; mix as before. Both the above are 'fine.'—*c.* As the last, but using half less of vermilion. Inferior.—*d.* Resin, 4 lbs.; shellac, 2 lbs.; Venice turpentine and red-lead, of each, 1¼ lbs.; as before. Common.

2. (BLACK.) *a.* From shellac, 60 parts; finest ivory-black, reduced to an impalpable powder, 30 parts; Venice turpentine, 20 parts; Fine.—*b.* Resin, 6 lbs.; shellac and Venice turpentine, of each, 2 lbs.; lampblack, q. s. Inferior.

3. (GOLD-COLOURED.) By stirring gold-coloured mica spangles or talc, or aurum musivum, into the melted resins just before they begin to cool. Fine.

4. (MARBLED.) By mixing two or three different coloured kinds just as they begin to grow solid.

5. (SOFT.) *a.* (Red.) Take of beeswax, 8 parts; olive oil, 5 parts; melt, and add of Venice turpentine, 15 parts; red-lead, to colour.—*b.* (Green.) As the last, but substituting powdered verdigris for red-lead. Both are used for sealing official documents kept in tin boxes; also as a cement.

6. (BOTTLE WAX.) *a.* (Black.) From black resin, 6¼ lbs.; beeswax, ½ lb.; finely powdered ivory-black, 1¼ lbs.; melted together.—*b.* (Red.) As the last, but substitute Venetian red or red-lead for ivory-black.

Obs. All the above forms for 'fine' wax produce 'superfine' by employing the best qualities of the ingredients; and 'extra superfine,' or 'scented,' by adding 1% of balsam of Peru or liquid storax to the ingredients when considerably cooled. The 'variegated' and 'fancy coloured kinds' are commonly scented with a little essence of musk or ambergris, or any of the more fragrant essential oils. The addition of a little camphor, or spirit of wine, makes sealing-wax burn easier. Sealing-wax containing resin, or too much turpentine, runs into thin drops at the flame of a candle.

Wax, Testing of. Mr Barnard S. Proctor gives the following as a convenient mode of working when it is desired to operate on several samples at once:—A little fragment of pure wax is placed between two slips of glass, such as are used for mounting microscopic objects, and heated till the wax is fused; a small india-rubber band is then placed round the slips to hold them together. Fragments of commercial cake (white wax), spermaceti, and cerasin are mounted in the same way, and then the four samples are arranged vertically round the inside of a small beaker containing warm water; a bung placed in the centre serves to steady the glass slips by pressing them against the sides of the beaker. A thermometer

is immersed in the water, and heat gradually applied by means of a Bunsen flame of very small dimensions kept an inch or two below the bottom of the beaker. All the samples are thus kept conveniently under observation while the thermometer gradually rises, and the melting-points noted and read off as the different films, one after the other, become transparent; and then the congealing points are noted in the same manner, as the films become dull after the beaker is removed from the source of heat. The following figures give the best results which the author obtained in a couple of experiments of this description:—Spermaceti melts at 112° F., congeals at 109°; white wax (commercial cake) melts at 130°, and congeals at 128°; pure beeswax melts at 150°, congeals at 142°; and cerasin melts at 160°, and congeals at 146°. These results are fairly concordant when the experiment is repeated, but are not quite so accurate as when the wax is rubbed on to the bulb of the thermometer itself, and the latter placed in a test-tube which plunges into the warm water. In this method of proceeding pure beeswax is found to melt at 146° F., and to congeal at 143°. The Food and Drugs Act requires pure beeswax to have a specific gravity = 0·95 to 0·97, and to melt at 146° F., congealing a few degrees lower.

Wax, White. *Syn.* BLEACHED WAX; CERA ALBA (B. P., Ph. L., E., & D.), L. *Prep.* From pure beeswax, by exposing it in thin flakes to the action of the sun, wind, and rain, frequently changing the surface thus exposed, by remelting it, and reducing it again into thin flakes. Used in making candles, and in white ointments, pommades, &c., for the sake of its colour. Block white wax (CERA ALBA IN MASSIS) is the above when cast into blocks; the best foreign is always in this form. Virgin wax (CAKE WHITE WAX; CERA ALBA IN OFFIS) should be the last made into round flat cakes; but this is seldom the case, the mixture sold under the name generally containing from 1-3rd to half its weight of spermaceti. The 'white wax' supplied by certain wholesale druggists to their customers is often totally unfit for the purposes to which it is applied. Spermaceti is constantly added to the white wax of commerce to improve its colour. Mr B. S. Proctor states that wholesale houses of the highest reputation supply an article as white cake wax which is in many cases half spermaceti, and in some as much as two thirds spermaceti to one of wax (see articles on "Adulteration of Wax," and "Substitutes for Wax," in 'Chemist and Druggist,' vol. iv, 1863).

WEATHER, Effects of, on Health. The 'Medical Press and Circular' says:—"We are in the midst of a severe winter (1878), and as hygiene is the order of the day, we cannot be too particular in impressing upon the public certain facts which are too often disregarded. Few are aware of the killing powers of intense cold and great heat, even in this comparatively temperate climate. Those who have been in the habit, as we have, of watching the returns of the Registrar-General, well know how quickly the death-rate rises during even a short continuance of cold weather. Now that the increase in the mortality affects chiefly the young and the old, as well as those who are either suffering from, or are predisposed to, affections of the chest and throat, indicates the class of people who should be especially careful to protect themselves against the inclemency of the weather. With regard to children, the system of 'hardening' them, by allowing them to go thinly clad, and exposing them to all sorts of weather, is a delusion from which the minds of some parents are even now not altogether free. It is thought that if their chest is kept warm, there is no need of caring about their arms and legs. But that is a great mistake. In proportion as the upper and lower extremities are well clothed will the circulation be kept up and determined to the surface of those parts; and in proportion to the quickness and equable distribution of the circulation will be the protection against those internal congestions which are but the first stage of the most fatal diseases of infancy and childhood. The same observation holds good with respect to grown-up people who are predisposed to pulmonary complaints. There is no exaggeration in saying that the mortality from these and other affections would be considerably diminished were people to avoid that 'catching cold' of which they so often and so lightly speak; and it is a matter of surprise to us that this fact, of which most of us are aware, does not lead to more precautions being taken by those who are anxious about either their own health or that of others. To take care that the body is thoroughly warm and well clothed just before going out in very wet or very cold weather—to keep up the circulation and warmth of the body rather by exercise of some kind than by sitting over great fires or in over-heated rooms —to be sure that the temperature of the sleeping apartments is not ever so many degrees below that of the sitting-room,—these are three golden maxims, attention to which would prevent thousands from catching that 'chill' or 'cold' to the results of which so many valuable lives have been prematurely sacrificed."

WEATHER PLANT. The Kew 'Bulletin of Miscellaneous Information,' 1890, contains a report made by Dr Oliver, of University College, on a series of experiments carried out in the Jodrell Laboratory of the Royal Gardens last autumn on Herr Joseph Nowack's renowned weather plant, of whose marvellous properties as a forecaster of weather, earthquakes, and fire-damp we heard so much towards the close of the summer of 1888. The proprietor attended at Kew Gardens in person, and superintended the experiments, he himself preparing the forecasts, and Dr Oliver and Mr Weiss making a close inspection of the movements of the plants and noting the actual weather experienced from day to day, so as to check the correctness or otherwise of the predictions. The plant is the well-known tropical legume, *Abrus precatorius*, a shrubby climber, originally a native of the East Indies, but now scattered to the Mauritius, West Indies, and other tropical countries. Dr Oliver enters minutely into a description of the ingenious devices by which Herr Nowack professed to be able to ascertain what would occur at some future date, the weather forty-eight hours hence, and earthquakes and fire-damp days or weeks hence anywhere up to a

distance of many hundreds of miles; also an explanation of the mode of fixing on the day to which the forecasts referred (for it was found in practice not to be limited strictly to forty-eight hours); if the results were not favourable the final determination of the day for which the forecast was made out was only made after the event. In this way less than one half the forecasts were two days ahead, the others being one, three, four, &c., days. The dates for some of them were altered twice, so that every opportunity was afforded to select the most suitable day to agree with the forecasts. But, notwithstanding these unexampled facilities, Herr Nowack made a very poor show of what it was possible to attain in the matter of forecasting anything. There were numerous changes in the weather during October, but although there were over 140 predictions, Dr Oliver states that only one change was anticipated by Herr Nowack. The predictions of earthquakes, *schlagwetter* (fire-damp in coal mines), and the positions of areas of high and of low barometer within the limits of the Meteorological Office Daily Weather Charts (the last-mentioned idea having occurred to Herr Nowack since his arrival in England) were submitted to Mr R. H. Scott, and were found to be as unsuccessful as the weather forecasts. Of nine earthquake predictions one was correct and eight failures; of nine *schlagwetter* two were correct, two nearly so, and five failures. Between fifty and sixty barometric charts were drawn up, and on placing them side by side with those prepared from the facts at the Meteorological Office "no accordance was found between the successive pairs of maps." The result of the inquiry, therefore, has been to show that the plant is not to be relied on as a substitute for the ordinary systems of weather prediction.

WEIGHT. The quantity of a body determined by means of a balance, and expressed in terms having reference to some known standard; the measure of the force of gravity, from which the relative quantity of a body is inferred. The relation between the weight and volume of a body, compared to a given standard taken as unity constitutes its specific gravity.

For the purpose of weighing, a balance or lever is required, which, when accurately suspended in a state of equilibrium, will be affected, in precisely an equal manner, by like weights applied to its extremities. Hence the construction of such an instrument is not more difficult than its application is important in chemical and philosophical research. Oertling, the most celebrated maker of the chemical balance, constructs this important instrument in seven different varieties, more or less elaborate. The largest of these, with a 16-inch beam, will carry 2 lbs. in each pan, and yet turn with $\frac{1}{100}$ of a gr. A balance with arms of unequal length or weight will weigh as accurately as another of the same workmanship with equal arms, provided the substance weighed be removed and standard weights placed in the same scale until the equilibrium be again restored, when the weights so employed, being exactly in the same condition as the substance previously occupying the scale, will, of course, indicate its proper weight. A knowledge of this fact is useful, as it enables any one to weigh correctly with unequal scales, or with any suspended lever.

Small weights may be made of thin leaf brass, or, preferably, of platinum foil. Quantities below the $\frac{1}{100}$ of a gr. may be either estimated by the position of the index, or shown by actually counting rings of wire, the value of which has been previously determined. The readiest way to subdivide small weights consists in weighing a certain quantity of very fine wire, and afterwards cutting it into such parts, by measure, as are desired; or the wire may be wrapped close round two pins, and then cut asunder with a knife. By this means it will be divided into a great number of equal lengths, or small rings. An elaborate essay on the BALANCE, in Watt's 'Dict. of Chemistry,' gives minute directions for weighing, with rules for the elimination of errors. See BALANCE.

The following tables represent the values of the weights legally employed in this country for the sale of gold, silver, and articles made thereof, as well as platinum, diamonds, and other precious metals and stones; also for drugs when sold by retail (see WEIGHTS AND MEASURES ACT, 1878, and MEASURES):

1. Troy Weight.

Grains. gr.	Pennyweights. dwt.	Ounces. oz.	Pound. lb. or lb.
24	1		
480	20	1	
5760	240	12	1

₊ The standard of the above measure is 1 cubic inch of distilled water, which, at 62° F. and 30 inches of the barometer, weighs 252·458 troy grains.

The carat used in weighing diamonds is 3¼ grains (nearly). Troy weight is employed in weighing gold, jewelry, &c., and, under a somewhat modified form, in prescribing and dispensing medicines (see below).

2. Apothecaries' Weight.
(Modified Troy Weight.)

Grains (Troy). gr.	Scruples 3	Drachms. 3	Ounces. 3	Pounds. lb	Equivalent in French grammes.	Equiv. in minims or measured drops.	Equivalent in cubic inches.	Equivalent in Avoirdupois weight.
1·	·05	·01666	·002083	·0001736	·06475	1·09	·008961055	1· gr.
20·	1·	·3333	·0416	·008472	1·295	21·94	·07922109	20· ,,
60·	3·	1·	·1250	·0104166	3·885	65·82	·23766329	60· ,,
480·	24·	8·	1·	·0833333	31·08	526·62	1·90130635	1 oz. 42·5 ,,
5760·	288·	96·	12·	1·	372·96	6319·54	22·81567609	13 ,, 72·5 ,,

₊ Apothecaries' weight is employed in prescribing and dispensing medicines according to the Ph. L., E., and U. S. But in the last Ph. D. and the new Brit. Ph. it has been superseded by avoirdupois weight.

Troy.
1 lb is equivalent to . . . . . .
1 oz. ,, ,, ...

Avoirdupois.
0·822857 lb.
1·097143 oz.

WEIGHTS, FOREIGN.

Binary Weights. (Système usuel.) French.

French grain.	Scruple.	Gros.	Ounce.	Livre.	Kilo-gramme.	Equivalent in grammes métrique.	Round number of the Codex in grammes.	Equivalent in Avoirdupois weight.
								lb. oz. gr.
1·	...	...	...	...	...	·0542	·05	... ... 0·837
24·	1·	...	...	...	...	1·30	1·30	... ... 20·1
72·	3·	1·	...	...	...	3·906	4·	... ... 60·284
576·	24·	8·	1·	...	...	81·25	32·	... 1 45·
9216·	384·	128·	16·	1·	...	500·	500·	1 1¼ 61·
18432·	768·	256·	32·	2·	1·	1000·	1000·	2 3¼ 13·

₊ The old French grain is equal to ·820 of an imperial troy grain; hence 1 troy grain is equal to 1·21 old French grains. The gros, once, and other multiples of the grain, are, of course, proportionate. The new French grain (of 1812) is equal to ·0542 gramme, or ·8365228 gr. troy. It is said, in some works, to be equal to ·878 gr. troy, or, in round numbers, ·9, but this is much too high.

CONTINENTAL MEDICINAL WEIGHTS in Troy Grains.

(From Dr Christison's ' Dispensatory.')

Country.	Pound.	Ounce.	Drachm.	Scruple consisting of		Grain.
				24 med. grs.	20 med. grs.	
French	5670·5	470·50	59·10	19·7	...	0·820
Spanish	5326·3	443·49	55·14	18·47	...	0·769
Tuscan	5240·3	436·67	54·58	18·19	...	0·758
Roman	5235·0	436·25	54·58	18·17	...	0·757
Austrian . . .	6495·1	541·25	67·65	...	22·5	1·127
German . . .	5524·8	460·40	57·55	...	19·18	0·960
Russian . . .	5524·8	460·40	57·55	...	19·18	0·960
Prussian . . .	5415·1	451·26	56·40	...	18·80	0·940
Dutch	5695·8	474·64	59·33	...	19·78	0·988
Belgian . . .	5695·8	474·64	59·33	...	19·78	0·988
Swedish . . .	5500·2	458·34	57·29	...	19·09	0·954
Piedmontese . .	4744·7	395·39	49·45	...	16·48	0·824
Venetian . .	4661·4	388·45	48·55	...	16·18	0·809

WEIGHTS AND MEASURES ACT, 1878. On the 1st of January, 1879, there came into force an Act to consolidate throughout the United Kingdom the law relating to weights and measures. Legislation on this subject had been long rendered necessary from the extreme inconvenience and friction to commerce of all kinds arising from the adherence to local standards of weight or measurements; and from the divergent values in different parts of the kingdom, and in places more or less contiguous to each other, of weights and measures often bearing the same name. Thus, previous to the passing of the above Act, there were twelve different markets in this country in which when corn was sold by the bushel, the weight of the bushel varied in each; and six different localities in which the same thing occurred when vended by the quarter and the load. In some places a score of grain would imply 20 lbs., but often less, whilst in others it was not an infrequent transaction for wheat to be sold by one measure, delivered by another, and eventually paid for by weight. And the same perplexing and arbitrary conditions attached to the sale of numberless other commodities.

We give below the most important sections of the Weights and Measures Act of 1878:

LAW OF WEIGHTS AND MEASURES.

Uniformity of Weights and Measures.

The same weights and measures shall be used throughout the United Kingdom.

Standards of Measure and Weight.

The bronze bar and the platinum weight, more particularly described in the first part of the First Schedule of this Act, and at the passing of this Act, deposited in the Standards Departments of the Board of Trade, in the custody of the Warden of the Standards, shall continue to be the imperial standard of measure and weight, and the said bronze bar shall continue to be the imperial standard for determining the imperial standard yard for the United Kingdom, and the said platinum weight shall continue to be the imperial

standard for determining the imperial standard pound for the United Kingdom.

Imperial Measures of Length.

The straight line or distance between the centres of the two gold plugs or pins (as mentioned in the First Schedule to this Act further on), in the bronze bar by this Act declared to be the imperial standard for determining the imperial standard yard, measured when the bar is at the temperature of sixty-two degrees of Fahrenheit's thermometer, and when it is supported on bronze rollers placed under it in such a manner as best to avoid flexure of the bar, and to facilitate its free expansion and contraction from variations of temperature, shall be the legal standard measure of length, and shall be called the imperial standard yard, and shall be the only unit or standard measure of extension from which all measures of extension, whether linear, superficial, or solid, shall be ascertained.

One third part of the imperial standard yard shall be a foot, and the twelfth part of such foot shall be an inch, and the rod, pole, or perch in length shall contain five such yards and a half, and the chain shall contain twenty-two such yards, and the furlong two hundred and twenty such yards, and the mile one thousand seven hundred and sixty such yards.

The rood of land shall contain one thousand two hundred and ten square yards according to the imperial standard yard, and the acre of land shall contain four thousand eight hundred and forty such square yards, being one hundred and sixty square rods, poles, or perches.

Imperial Measures of Weight and Capacity.

The weight in vacuo of the platinum weight (mentioned in the First Schedule to this Act), and by this Act declared to be the imperial standard for determining the imperial standard pound, shall be the legal standard measure of weight, and of measure having reference to weight, and shall be called the imperial standard pound, and shall be the only unit or standard measure of weight from which all other weights and all measures having reference to weight shall be ascertained.

One sixteenth part of the imperial standard pound shall be an ounce, and one sixteenth part of such ounce shall be a dram, and one seven thousandth part of the imperial standard pound shall be a grain.

A stone shall consist of fourteen imperial standard pounds, and a hundredweight shall consist of eight such stones, and a ton shall consist of twenty such hundredweights.

Four hundred and eighty grains shall be an ounce troy.

All the foregoing weights except the ounce troy shall be deemed to be avoirdupois weights.

The unit or standard measure of capacity from which all other measures of capacity, as well as for liquids as for dry goods, shall be derived, shall be the gallon containing ten imperial standard pounds weight of distilled water weighed in air against brass weights, with the water and air at the temperature of sixty-two degrees of Fahrenheit's thermometer, and with the barometer at thirty inches.

The quart shall be one fourth part of the gallon, and the pint shall be one eighth part of the gallon.

Two gallons shall be a peck, and eight gallons shall be a bushel, and eight such bushels shall be a quarter, and thirty-six such bushels shall be a chaldron.

A bushel for the sale of any of the following articles, namely, lime, fish, potatoes, fruit, or any other goods and things which before (the passing of the Weights and Measures Act, 1835, that is to say) the ninth day of September, one thousand eight hundred and thirty-five, were commonly sold by heaped measure, shall be a hollow cylinder having a plane base, the internal diameter of which shall be double the internal depth; and every measure used for the sale of any of the above-mentioned articles which is a multiple of a bushel, or is a half bushel or a peck, shall be made of the same shape and proportion as the above-mentioned bushel.

In using an imperial measure of capacity, the same shall not be heaped, but either shall be stricken with a round stick or roller, straight, and of the same diameter from end to end, or if the article sold cannot from its size or shape be conveniently stricken, shall be filled in all parts as nearly to the level of the brim as the size and shape of the article will admit.

Metric Equivalents of Imperial Weights and Measures.

The table in the Third Schedule to this Act shall be deemed to set forth the equivalents of imperial weights and measures and of the weights and measures therein expressed in terms of the metric system, and such table may be lawfully used for computing and expressing, in weights and measures, weights and measures of the metric system.

Use of Imperial Weights and Measures.

Every contract, bargain, sale, or dealing, made or had in the United Kingdom for any work, goods, wares, or merchandise, or other thing which has been or is to be done, sold, delivered, carried, or agreed for by weight or measure, shall be deemed to be made and had according to one of the imperial weights or measures ascertained by this Act, or to some multiple or part thereof, and if not so made or had shall be void; and all tolls and duties charged or collected according to weight or measure shall be charged and collected according to one of the imperial weights or measures ascertained by this Act, or to some multiple or part thereof.

Such contract, bargain, sale, dealing, and collection of tolls and duties as is in this section mentioned is in this Act referred to under the term 'trade.'

No local or customary measures, nor the use of the heaped measure, shall be lawful.

Any person who sells by any denomination of weight or measure other than one of the imperial weights or measures, or some multiple or part thereof, shall be liable to a fine not exceeding forty shillings for every such sale.

All articles sold by weight shall be sold by avoirdupois weight; except that—

(1) Gold and silver, and articles made thereof, including gold and silver thread, lace, or fringe, also platinum, diamonds, and other precious metals or stones, may be sold by the ounce troy or by any decimal parts of such ounce; and all contracts, bargains, sales, and dealings in relation thereto shall be deemed to be made and had by such weight, and when so made or had shall be valid; and

(2) Drugs, when sold by retail, may be sold by apothecaries' weight.

Every person who acts in contravention of this section shall be liable to a fine not exceeding five pounds.

A contract or dealing shall not be invalid or open to objection on the ground that the weights or measures expressed or referred to therein are weights or measures of the metric system, or on the ground that decimal subdivisions of imperial weights and measures, whether metric or otherwise, are used in such contract or dealing.

Nothing in this Act shall prevent the sale, or subject a person to a fine under this Act for the sale, of an article in any vessel, where such vessel is not represented as containing any amount of imperial measure, nor subject a person to a fine under this Act for the possession of a vessel where it is shown that such vessel is not used nor intended for use as a measure.

Any person who prints, and any clerk of a market or other person who makes, any return, price list, price current, or any journal or other paper containing price list or price current, in which the denomination of weights and measures quoted or referred to denotes or implies a greater or less weight or measure than is denoted or implied by the same denomination of the imperial weights and measures under this Act, shall be liable to a fine not exceeding ten shillings for every copy of every such return, price list, price current, journal, or other paper which he publishes.

Every person who uses or has in his possession for use for trade a weight or measure which is not of the denomination of some Board of Trade standard, shall be liable to a fine not exceeding five pounds, or in the case of a second offence ten pounds, and the weight or measure shall be liable to be forfeited.

Unjust Weights and Measures.

Every person who uses or has in his possession for use for trade any weight, measure, scale, balance, steelyard, or weighing machine which is false or unjust, shall be liable to a fine not exceeding five pounds, or in the case of a second offence ten pounds, and any contract, bargain, sale, or dealing made by the same shall be void, and the weight, measure, scale, balance, or steelyard shall be liable to be forfeited.

Where any fraud is wilfully committed in the using of any weight, measure, scale, balance, steelyard, or weighing machine, the person committing such fraud, and every person party to the fraud, shall be liable to a fine not exceeding five pounds, or in the case of a second offence ten pounds, and the weight, measure, scale, balance, or steelyard shall be liable to be forfeited.

A person shall not wilfully or knowingly make or sell, or cause to be made or sold, any false or unjust weight, measure, scale, balance, or weighing machine.

Every person who acts in contravention of this section shall be liable to a fine not exceeding ten pounds, or in the case of a second offence fifty pounds.

MISCELLANEOUS.

Every inquisition which, in pursuance of any Act hereby repealed, has been taken for ascertaining the amount of contracts to be performed or rents to be paid in grain or malt, or in any other commodity or thing, or with reference to the measure or weight of any grain, malt, or other commodity or thing, and the amount of any toll rate or duty payable according to any weight or measure in use before the passing of the said Act, and has been enrolled of record in Her Majesty's Court of Exchequer, shall continue in force, and may be given in evidence in any legal proceeding, and the amount ascertained by such inquisition shall, when converted into imperial weights and measures, continue to be the rule of payment in regard to all such contracts, rents, tolls, rates, or duties.

Standards and Definitions.

Nothing in this Act shall affect the validity of the models of gas holders verified and deposited in the Standards Department of the Board of Trade, in pursuance of the Act of the session of the twenty-second and twenty-third years of the reign of Her present Majesty, chapter sixty-six, intituled 'An Act for regulating measures used in sales of gas,' and of the Acts amending the same, and the provisions of this Act with respect to Board of Trade standards shall apply to such models; and the provisions of this Act with respect to defining the amount of error to be tolerated in local standards when verified or reverified, shall apply to defining the amount of error to be tolerated in such copies of the said models of gas holders as are provided by any justices, council, commissioners, or other local authority in pursuance of the said Acts.

Nothing in this Act shall extend to prohibit, defeat, injure, or lessen the rights granted by charter to the master, wardens, and commonalty of the mystery of the Founders of the City of London.

Nothing in this Act shall prohibit, defeat, injure, or lessen the rights of the mayor and commonalty and citizens of the City of London, or of the Lord Mayor of the City of London for the time being, with respect to the stamping or sealing of weights and measures, or with respect to the gauging of wine or oil, or other gaugeable liquors.

APPLICATION OF ACT TO SCOTLAND.

This Act shall apply to Scotland with the following modifications:

In the application of this Act to Scotland the expression 'rents and tolls' includes all stipends, feu duties, customs, casualties, and other demands whatsoever, payable in grain, malt, or meal, or any other commodity or thing.

The fair's prices of all grain in every county shall be struck by the imperial quarter, and all other returns of the prices of grain shall be set forth by the same, without reference to any other measure whatsoever.

APPLICATION OF ACT TO IRELAND.

This Act shall apply to Ireland with the following modifications:

In Ireland every contract, bargain, sale, or dealing—

For any quantity of corn, grain, pulses, potatoes, hay, straw, flax, roots, carcasses of beef or mutton, butter, wool, or dead pigs, sold, delivered, or agreed for:

Or for any quantity of any other commodity sold, delivered, or agreed for by weight (not being a commodity which may by law be sold by the troy ounce or by apothecaries' weight), shall be made or had by one of the following denominations of imperial weight, namely, the ounce avoirdupois; the imperial pound of sixteen ounces; the stone of fourteen pounds; the quarter hundred of twenty-eight pounds; the half hundred of fifty-six pounds; the hundredweight of one hundred and twelve pounds; or the ton of twenty hundredweight; and not by any local or customary denomination of weight whatsoever, otherwise such contract, bargain, sale, or dealing shall be void:

Provided always, that nothing in the present section shall be deemed to prevent the use in any contract, bargain, sale, or dealing of the denomination of the quarter, half, or other aliquot part of the ounce, pound, or other denomination aforesaid, or shall be deemed to extend to any contract, bargain, sale, or dealing relating to standing or growing crops.

In Ireland every article sold by weight shall, if weighed, be weighed in full net standing beam; and for the purposes of every contract, bargain, sale, or dealing, the weight so ascertained shall be deemed the true weight of the article, and no deduction or allowance for tret or beamage, or on any other account, or under any other name whatsoever, the weight of any sack, vessel, or other covering in which such article may be contained alone excepted shall be claimed or made by any purchaser on any pretext whatever under a penalty not exceeding five pounds.

FIRST SCHEDULE.

PART I.—IMPERIAL STANDARDS.

The following standards were constructed under the direction of the Commissioners of Her Majesty's Treasury, after the destruction of the former imperial standards in the fire at the Houses of Parliament.

The imperial standard for determining the length of the imperial standard yard is a solid square bar, thirty-eight inches long, and one inch square in transverse section, the bar being of bronze or gun-metal; near to each end a cylindrical hole is sunk (the distance between the centres of the two holes being thirty-six inches) to the depth of half an inch; at the bottom of this hole is inserted in a smaller hole a gold plug or pin, about one tenth of an inch in diameter, and upon the surface of this pin there are cut three fine lines at intervals of about the one hundredth part of an inch transverse to the axis of the bar, and two lines at nearly the same interval parallel to the axis of the bar; the measure of length of the imperial standard yard is given by the interval between the middle transversal line at one end and the middle transversal

SECOND SCHEDULE.

DENOMINATIONS OF STANDARDS OF APOTHECARIES' WEIGHT AND MEASURE.

1. *Apothecaries' Weight.*

Denomination.		Weight in grains in terms of the Imperial Standard Pound, which contains 7000 such grains.
Ounces.	10 ounces . .	4800 grains.
	8 „ . .	3840 „
	6 „ . .	2880 „
	4 „ . .	1920 „
	2 „ . .	960 „
	1 ounce . .	480 „
Drachms.	4 drachms or half an ounce .	} 240 „
	2 drachms	120 „
	1•drachm . .	60 „
Scruples.	2 scruples. .	40 „
	1½ „ or half a drachm .	} 30 „
	1 scruple . .	20 „
	½ „ . .	10 „
	6 grains . .	6 „
	5 „ . .	5 „
	4 „ . .	4 „
	3 „ . .	3 „
	2 „ . .	2 „
	1 grain	1 grain.
	½ „ . .	0·5 „

2. *Apothecaries' Measure.*

Denomination.		Containing the following weight of distilled water. Temperature = 62° F. Barometer = 30 inches. Imperial Pound = 7000 gr.
A fluid ounce and the multiples thereof from 1 to 40 fluid ounces . Half a fluid ounce . .	}	One fluid ounce contains 437·5 grains weight, or $\frac{1}{160}$ imperial gallon.
A fluid drachm and the multiples thereof from 1 to 16 fluid drachms . Half a fluid drachm . .	}	One fluid drachm equals ⅛ fluid ounce.
A minim and the multiples thereof from 1 to 60 minims	}	One minim equals $\frac{1}{80}$ fluid drachm.

line at the other end, the part of each line which is employed being the point midway between the longitudinal lines; and the said points are in this Act referred to as the centres of the said gold plugs or pins; and such bar is marked 'Copper 16 oz., tin 2½ oz., zinc 1 oz. Mr Baily's metal. No. 1 standard yard at 62·30° Fahrenheit. Cast in 1845. Troughton and Simms, London.'

The imperial standard for determining the weight of the imperial standard pound is of platinum, the form being that of a cylinder nearly 1·35 inches in height and 1·15 inches in diameter, with a groove or channel round it, whose middle is about 0·34 inch below the top of the cylinder, for insertion of the points of the ivory fork by which it is to be lifted; the edges are carefully rounded off, and such standard pound is marked 'P.S. 1844, 1 lb.'

The following new and additional denominations of standards of apothecaries' weights and measures were created under the Weights and Measures Act, by an order in Council, dated the 14th August, 1879 (published in the 'London Gazette,' August 15th, 1879).

THIRD SCHEDULE.

PART I.—METRIC EQUIVALENTS.

Table of the values of the principal denominations of measures and weights on the metric system, expressed by means of denominations of imperial measures and weights, and of the values of the principal denominations of measures and weight of the imperial system, expressed by means of metric weights and measures.

Measures of Surface.

Metric Denominations and Values.		Equivalents in Imperial Denominations.		
	Square Metres.	Acres.	Square Yards.	Decimals.
Hectare, *i. e.* 100 ares .	10,000	{ 2 or	2280·3326	
			11960·3326	
Decare, *i. e.* 10 ares	1000	...	1196·0333	
Are	100	...	119·6033	
Centiare, *i. e.* $\frac{1}{100}$ are .	1	...	1·1960	

Measures of Length.

Metric Denominations and Values.		Equivalents in Imperial Denominations.				
	Metres.	Miles.	Yards.	Feet.	Inches.	Decimals.
Myriametre	10,000	{ 6 or	376 10,936	0 0	11·9 11·9	
Kilometre	1000	...	1093	1	10·79	
Hectometre	100	...	109	1	1·079	
Decametre	10	...	10	2	9·7079	
Metre	1	...	1	0	3·3708	
Decimetre	1/10	...	...	...	3·9371	
Centimetre	1/100	...	...	...	0·3937	
Millimetre	1/1000	...	...	...	0·0394	

Measures of Capacity.

Metric Denominations and Values.		Equivalents in Imperial Denominations.						
	Cubic Metres.	Quarters.	Bushels.	Pecks.	Gallons.	Quarts.	Pints.	Decimals.
Kilolitre, i. e. 1000 litres . .	1	3	2	2	0	0	0·77	
Hectolitre, i. e. 100 litres .	1/10	...	3	3	0	0	0·077	
Decalitre, i. e. 10 litres . .	1/100	...	...	1	0	0	1·6077	
Litre	1/1000	...	...	...	...	...	1·76077	
Decilitre, i. e. 1/10 litre . .	1/10000	...	...	...	...	...	0·176077	
Centilitre, i. e. 1/100 litre .	1/100000	...	...	...	...	...	0·0176077	

Weights.

Metric Denominations and Values.		Equivalents in Imperial Denominations.					
	Grams.	Cwt.	Stones.	Pounds.	Ounces.	Drams.	Decimals.
Millier	1,000,000	19	5	6	9	15·04	
Quintal	100,000	1	7	10	7	6·304	
Myriagram	10,000	...	1	8	0	11·8304	
Kilogram	1,000	{ ... (or 15432·3487 grs.)		2	8	4·3830	
Hectogram	100	...	...	...	3	8·4383	
Decagram	10	...	...	...	...	5·6438	
Gram	1	...	...	...	...	0·56438	
Decigram	1/10	...	...	...	...	0·056438	
Centigram	1/100	...	...	...	...	0·0056438	
Milligram	1/1000	...	...	...	...	0·00056438	

Measures of Length.

Imperial Measures.	Equivalents in Metric Measures.			
	Millimetre.	Decimetre.	Metre.	Kilometre.
Inch	= 25·39954			
Foot, or 12 inches . . .	...	= 3·04794	= 0·30479	
Yard, or 3 feet, or 36 inches. . .	...	...	= 0·91428	
Fathom, or 2 yards, or 6 feet . .	...	...	= 1·82877	
Pole, or 5½ yards	...	...	= 5·02911	
Chain, or 4 poles, or 22 yards .	...	...	= 20·11644	
Furlong, 40 poles, or 220 yards .	...	...	= 201·16437	= 0·20116
Mile, 8 furlongs, or 1760 yards .	...	...	=1609·31493	= 1·60931

Measures of Surface.

Imperial Measures.	Equivalents in Metric Measures.			
	Square Decimetres.	Square Metres.	Ares.	Hectares.
Square inch	= 0·06451			
Square foot, or 144 square inches . .	= 9·28997	= 0·092900		
Square yard, or 9 square feet, or 1296 square inches	= 83·60971	= 0·836097		
Pole or perch, or 30¼ square yards .	...	=25·291939		
Rood, or 40 perches, or 1210 square yards	...	...	=10·116776	
Acre, or 4 roods, or 4840 square yards .	...	...	...	= 0·40467
Square mile, or 640 acres . .	...	...	...	=258·98945

Measures of Capacity.

Imperial Measures.	Equivalents in Metric Measures.			
	Decilitres.	Litres.	Decalitres.	Hectolitres.
Gill	= 1·41983	= 0·14198		
Pint, or 4 gills	= 5·67932	= 0·56798		
Quart, or 2 pints	...	= 1·13587		
Gallon, or 4 quarts	...	= 4·54346		
Peck, or 2 gallons	...	= 9·08692	= 0·90869	
Bushel, or 8 gallons, or 4 pecks . .	...	...	= 3·63477	
Quarter, or 8 bushels . . .	...	...	...	= 2·90781

Cubic Measure.

Imperial Measures.	Equivalents in Metric Measures.		
	Cubic Centimetres.	Cubic Decimetres.	Cubic Metres.
Cubic inch	16·38618		
Cubic foot, or 1728 cubic inches	...	28·31531	
Cubic yard, or 27 cubic feet	...	...	0·76451

Weights.

Imperial Weights.	Equivalents in Metric Weights.			
	Grams.	Decagrams.	Kilograms.	Millier or Metric Ton.
Grain	= 0·06479895			
Dram	= 1·77185			
Ounce avoirdupois, or 16 drams, or 437·5 grains	= 28·34954	= 2·83495		
Pound, or 16 ounces, or 256 drams, or 7000 grains	=453·59265	=45·35927	= 0·45359	
Hundredweight, or 112 lbs.	...	...	= 50·80238	
Ton, or 20 cwt.		...	=1016·04754	=1·01605
Ounce troy, or 480 grains	= 31·103496	= 3·11035		

USEFUL REFERENCE TABLE OF METRIC WEIGHTS AND MEASURES AND EQUIVALENTS.

Length.

Unit of Measurement.	Approximate Equivalent.	Accurate Equivalent.
1 inch	2½ centimetres	2·539
1 centimetre (1/100 metre)	0·4 inch	0·393
1 yard	1 metre	0·914
1 metre (39·37 inches)	1 yard	1·093
1 foot	30 centimetres	30·479
1 kilometre (1000 metres)	⅝ mile	0·621
1 mile	1⅗ kilometres	1·609
Weight.		
1 gramme	15½ grains	15·432
1 centigramme	between ⅙ and ⅐ grain	0·154
1 milligramme	1/65 grain	0·015
1 grain	0·064 gramme	0·064
1 kilogramme (1000 grammes)	2⅕ pounds avoirdupois	2·204
1 pound avoirdupois	½ kilogramme	0·453
1 ounce avoirdupois (437½ grains)	28⅓ grammes	28·348
1 ounce troy or apothecary (480 grains)	31 grammes	31·103
Bulk.		
1 cubic centimetre	0·06 cubic inch	0·061
1 cubic centimetre (1 c.c.)	17 minims	16·932
1 cubic inch	16⅓ cubic centimetres	16·386
1 litre (1000 cubic centimetres)	35 fluid ounces	35·275
1 gallon	4½ litres	4·535
1 fluid ounce	28⅓ cubic centimetres	28·348
Surface.		
1 hectare (10,000 square metres)	2½ acres	2·471
1 acre	⅖ hectare	0·404

WELD. *Syn.* WOAD. The *Reseda luteola*, Linn., an herbaceous annual employed by the dyers. A decoction of the stems and leaves gives a rich yellow to goods mordanted with alum, tartar, or muriate of tin. See YELLOW PIGMENTS.

WELSH RAREBIT (vulg. rabbit). *Prep.* Cut slices of bread, toast and butter them; then cover them with slices of rich cheese, spread a little mustard over the cheese, put the bread in a cheese-toaster before the fire, and in a short time serve it up very hot.

WEN. The popular name of pulpy, encysted, and fleshy tumours of the face and neck.

WET (to keep out from Gun Locks). In giving hints to sportsmen going to Norway, Mr Lock, in his book on 'Sport in Norway,' gives some capital advice on this subject, which would be equally serviceable in wet weather in England. Sportsmen will do well, he says, to remove the locks from their rifle and gun, oil them with a little Rangoon oil, lay them on the hob of the fireplace until they are quite hot, and then wipe them as dry as possible with a little cotton waste, so that there will be no superfluous oil left to clog the works. While the locks are getting hot get a little beeswax and melt it in a cup, and with the tip of a penknife carefully pay, as though you were using putty to place in a pane of glass, though more sparingly, the wooden ledges where the lock-plates rest when in their places, in such a manner that none of the wax gets into the places hollowed out to receive the works of the lock. When the warm locks are put back in their places, and screwed up tight, the wax will adhere to the edge of the lock-plates and the wood wherein they bed, and effectually render them impervious to wet. The sportsman can afterwards, when stalking, push his rifle through wet grass, and use his fowling-piece when the water, after a shower, drops from the trees upon him as he forces his way between the wet branches, without fear of the wet making its way into the locks.

WHEAT. *Syn.* TRITICUM, L. The ripe seed or fruit of several varieties of *Triticum vulgare*, Linn., of which the principal are *Triticum æstivum*, or spring wheat; *Triticum hybernum*, or winter wheat; and *Triticum turgidum*, or turgid wheat, the last two of which include several red and white sub-varieties. Of all the cereal grains wheat appears to be that best adapted for bread-corn, not merely on account of its highly nutritious character, but also on account of the power it possesses, from its richness in gluten, of forming a light and agreeable loaf by the process of fermentation.

According to Sir H. Davy, good English wheat contains of gluten, 19%; starch, 77%; soluble matter, 4% to 5%.

The average weight of good wheat per bushel is from 58 to 60 lbs.; and its average yield of flour is fully 12½ lbs. for every 14 lbs. The weight of the straw is said to be about double that of the grain. The produce per acre varies from 12 to 60, or even 64 bushels an acre. See FLOUR, STARCH, &c.

Buckwheat. *Syn.* FAGOPYRUM, L. The seed of *Fagopyrum esculentum*, a plant of the Nat. Ord. POLYGONACEÆ. It makes excellent cakes, crumpets, and gruel. In North America, buckwheat cakes, or rather fritters, are in general use at breakfast, eaten with molasses. In England, buckwheat is cultivated as food for pheasants.

Wheat, Indian. See MAIZE.

Wheat, Steeps for. Quicklime, sulphate of zinc or white vitriol, sulphate of copper or blue vitriol, and arsenious acid or white arsenic, are the substances chiefly employed for this purpose. About 5 lbs. of the first (slaked and made into a milk with water), 1½ lbs. of the second, 1 lb. of the third, and 3 or 4 oz. of the last, are regarded as sufficient for each sack of seed. The method of applying them is either to dissolve or mix them with just sufficient water to cover the seed, which is then to be soaked in the mixture for a few hours, or a less quantity of water is employed, and the more concentrated solution is, at intervals well sprinkled, by means of a 'watering pot,'

over the seed wheat spread upon the barn floor, the action being promoted by occasional stirring. *Obs.* The first two substances above named have been separately proved to be amply sufficient to destroy the 'smut' in seed wheat, and are perfectly harmless in their effects, which renders them greatly preferable to arsenic, or even to sulphate of copper. Nearly all the numerous advertised 'anti-smuts,' or nostrums to prevent the smut in wheat, contain one or other of the last three of the above substances.

WHEY. *Syn.* SERUM LACTIS, L.; PETIT LAIT, Fr. The liquid portion of milk after the curd has been separated. It consists chiefly of water, holding in solution 3% or 4% of sugar of milk. A pound of milk mixed with a tablespoonful of proof spirit allowed to become sour, and the whey filtered from the sediment, yields, in the course of a few weeks, a good vinegar (whey vinegar), free from lactic acid (*Scheele*). Skimmed milk may be used.

Whey, Al'um. *Syn.* SERUM LACTIS ALUMINATUM,** L. *Prep.* Take of powdered alum, 1 dr.; hot milk, 1 pint; simmer a few seconds, let it repose for a short time, and strain the whey from the coagulum. Used in diarrhœa, &c.; a wine-glassful after every motion. Acid whey (SERUM LACTIS ACIDUM) may be prepared in a similar manner by substituting ½ dr. of tartaric or citric acid for the alum. Orange whey and lemon whey are prepared from the juice of the respective fruits, with a little of the yellow peel to impart flavour.

WHEY POWDER. *Prep.* 1. From whey gently evaporated to dryness, and powdered along with about one third of its weight of lump sugar.

2. Sugar, 7 oz.; sugar of milk, 2 oz.; gum-arabic, 1 oz. (all in fine powder); mix well. 1 oz. dissolved in 1½ pints of water forms extemporaneous whey.

WHIS'KY. Dilute alcohol obtained from the fermented wort of malt or grain. That from the former is the most esteemed. The inferior qualities of this spirit are prepared from barley, oats, or rye, a small portion only of which is malted, the other potatoes mashed with a portion of barley malt, the resulting wash being carelessly fermented and distilled, and purposely suffered to burn, to impart the peculiar empyreumatic or smoky flavour so much relished by the lower orders of whisky drinkers. The malt whisky, sold as such, of the principal Scotch and Irish distillers is fully equal in quality to London gin, from which it merely differs in flavour. The peculiar flavour of whisky may be imitated by adding a few drops each of pure creosote and purified fusel oil to 2 or 3 gallons of good London gin; and the imitation will be still more perfect if the liquor be kept for some months before drinking it.

We are indebted to 'Land and Water' for the following interesting particulars relating to Irish whisky:

" Genuine unadulterated Irish whisky has, of late years, become a great desideratum as a wholesome and agreeable beverage, and in the article produced by the large and successful company whose premises and business I am about to describe, the consuming public have every guarantee of its excellence and purity, as far as can be insured by the use of the very best materials, great skill and care in the manufacturing processes, and the valuable and extensive buildings in which the spirit is stored until it attains the maturity and mellowness which age alone can confer.

"No blending process of new whiskies can effect this, no distiller who has not very extensive bonded warehouses is to be trusted. Acre after acre of cellars, vault after vault, corridor after corridor, each and all dim, damp, and dark, and guarded by the exciseman's talismanic padlock—all these are necessary for the soundness of the distillery. For to secure age and quality, the effect of several years' storage in these vaults is required. If you wish to see such store-rooms to perfection, go to Cork, which may be considered the capital of the Irish whisky trade. Even Dublin, with its Jamiesons, its Powers, and its Roes, must bow down before it.

"But what is most singular of all, one company represents that important branch of manufacture, and have therefore a good right to their title of 'Cork Distilleries Company.' A little over a quarter of a century ago there were five distilleries in Cork—Wise's, Hewett's, Daly's, Murphy's, and Waters'. In 1867, however, an amalgamation took place, and the present company was started, and the work of the five distilleries was concentrated into three—the North Mall, still known as Wise's (that proprietor wisely allowing himself to be bought out, after having made one of the largest private fortunes in Ireland); the Midleton, situated at a pretty village of that name, about ten miles from Cork, and the Watercourse, in the north-western suburb of the city. The three distilleries are capable of producing 1,000,000 gallons each per annum, which represents an annual duty of one million and a half pounds sterling. Their paid-up capital is a quarter of a million, and a very large rest fund. Their works and property are insured for over three quarters of a million sterling, and they find employment for about 1000 men.

"*The Brewing Process.* I shall have occasion to describe each of these three distilleries during the course of this paper, but it would perhaps be as well to run hurriedly through the several processes of whisky distillation. It may be divided roughly into brewing and distilling. Malt and barley are, of course, the ingredients used. Barley as it comes from the market is distinguished by the appellation 'green.' This is either steeped and converted into malt, or kiln-dried and ground. It is then removed to the mash-tuns, where water is added, and the whole mixed by revolving machinery. After some hours' steeping, the water has soaked all the desired properties from the grain, and is known as wort. This is led away or pumped by a complicated series of pipes to the top of the manufactory, where it undergoes a cooling process. When of the desired temperature it is conducted to the fermenting vats—vast wooden vessels of imposing appearance ranged in rows. The brewing processes end with this fermentation.

"*The Distilling Processes.* When this is done, which generally takes five days, the fermented liquor is conducted to the 'wash' charger, and from thence pumped to the intermediate charger, where it is heated before undergoing the first process of distillation, which now takes place. The still is a vast copper vessel, shaped exactly like an inverted funnel, with the pipe leading to the roof. The 'wash' or liquor from the charger is conducted into this vessel. Beneath it are two furnaces, which soon raise the temperature of the vessel to boiling-point. When evaporation commences the steam (which is the spirit, and is technically known as 'low wines') is conducted up the copper pipe into a refrigerator, known as the 'worm.' This worm is, in reality, a continuation of the pipe of the still twisted into regular coils in and about a vessel filled with the coldest water obtainable. By this means the steam is converted into liquor. This liquor passes into the close safe, a glass vessel somewhat like an aquarium tank. The distiller stands by and watches the running liquor, and his practised eye and educated palate immediately detect any fault in the distillation. He is not allowed to open his tank, however, except by notice in the presence of the excise officer, one or more of whom are always present in every distillery. Through this tank it runs into 'low wines' receiver, a large tank placed below, and from these it again passes to 'feints chargers' en route to the 'low wines still,' where the second distillation takes place. I forgot to say that the refuse liquor left after the first distillation is much valued by farmers for its milk-producing qualities, and is bought up by them for cow-food. The refuse liquor from the second distillation, however, is only water, and the refuse liquor from the third and final distillation is water also.

"The second distillation is like the first—the same process of 'worm' cooling, conducting, and charging is carried on. The third still is known as the spirit still. The spirit is now considered perfect, and is led off to the large vats in the spirit stores, where it is reduced to desired strength, racked off into casks, and removed to bonded warehouses for maturity. Such are the processes carried on here—such are the processes carried on by all honest distillers during the last century; but modern science has discovered that many very common—tasteless, I grant, but easily flavoured—vegetables will yield ardent spirits, and there are not wanting those who will take advantage of the discovery.

"*Within a Distillery.* But the distillery itself, who can describe it—its story upon story of granaries—its kilns floored with perforated tiles—its steeping vats and its low-roofed malting sheds—its roaring mills—its terrible and mysterious tanks—its inextricable machinery—its innumerable rafters and false roofs—its ladders perched up in inaccessible places—its bewildering passages—and far away, above all, its immense chimneys, towering up to the sky? But this is not all—the bonded warehouses have to be gone through. The excise officer has to be called, and the sealed lock has to be broken, and you enter into the vast cool place. Black as night is everything around you; the lamps which the attendants hold are utterly incapable of dissipating the darkness, and only cast a strong orange glare upon the faces of the men who hold them. To show one the dimensions of the place a man is sent to the opposite end. Away he goes, only traceable by the lamp he bears, and before he waves it to show that the opposite end of the vault is reached, it has become a scarcely discernible glimmer. As we become more used to the darkness we see straight passages leading in every direction, and lined on every side by barrels piled almost to the ceiling.

"*The North Mall.* The first of the Cork distilleries I visited was the one at North Mall, formerly, and, in fact, still known as Wise's. It is in a western suburb of the town, out among the meadows. The Lee winds its silvery course between tall alders close by it, and a branch stream is made to do much of the work of the immense manufactory. On approaching it, it has a picturesque effect. It lies underneath a tall bank, over which the road to Sunday's Well leads. Looking down from this road the whole of its vast dimensions can be taken in at a glance. The extensive yard, where one would imagine enough coal was stored to supply the whole city, is being raised from the adjoining fields. I was looking over an old history of Cork, published by a certain Dr Smith, over a century ago, and I find that formerly on this spot a Franciscan monastery stood. Such discipline was preserved here that it was called the Mirror of Ireland, and their sacerdotal character was so great that they had the power of curing sore eyes. The only remains of this ancient edifice now visible is a carved stone built into the wall of the great bonded warehouses in the Sunday's Well Road. It was here that the noted water oozed out of the redstone rock. Whether it is ever now used in making the agreeable beverage manufactured from the old whisky stored below I did not ascertain. These old Franciscan fathers had, doubtless, a good cellar of their own; but what would they have said of the vast, well-filled vaults which now are found upon perhaps the identical spot? But, large as these are, they are not large enough for the requirements of the distillery, and other extensive premises have been secured in Leitrim Street, which are now used as bonded warehouses.

"The whisky produced at this distillery is, if possible, still better now than it was in Wise's time; the same distiller who worked the concern for him for twenty years is still there, and none but the very finest description of malt and barley (a large proportion of the former) is used. Its production, as well as those of the other two distilleries of the company, gained a first-class medal last year at Philadelphia, and the jurors described it as 'very fine, full flavour, and good spirit.' As a natural consequence, there is a demand for this whisky all over the world, and there are very few large towns in either hemisphere where it is not represented by an agent.

"*The 'Watercourse.'* By-the-bye, these lie on our way to the celebrated Watercourse Distillery, the second of those used by this great firm. Entering through the broad portals, long ranges of old-fashioned buildings spread out on every side. Here is the mill, gaunt and square and solid; those jealously guarded doors to the right are the bonded warehouses; the buildings across the yard are devoted to the coopers' and smiths' work, which in all three distilleries is done on the premises.

"That tall black and white building far away on the opposite side is the grain store; this, nearer to you, with the irregular roofs, the complicated piping and open-walled structures running away overhead, is the distillery proper. This distillery is about of equal size to that at North Mall. It turns out as good and extensive work, and, like it, is not satisfied with the extensive storing facilities at its command, but must needs go abroad to an old unused distillery, farther in the suburbs, where it hides most of its rich and treasured productions. A picturesque old place is this; the ruins of the old works are still standing, and their architecture is such that it only requires a mantle of ivy to transform it into a remnant of feudal savagery.

"*The Midleton Distillery.* The company have handsome and extensive offices on Morrison's Island, in the centre of the city, and close to the water's edge. Here the directors sit day after day, and the scores of clerks attend to the interests of 4000 customers. But I cannot linger here, for I have another distillery to visit. Another, the brightest of all, far out in the beautiful country, at the town of Midleton, situate at the north-east extremity of Cork's magical harbour. Approaching the distillery from the town, it has somewhat the appearance of a fortress. A massive stone gateway bars the entrance, and heavy walls encompass it. But when once admittance is gained the sternness of the approach vanishes. Great buildings loom aloft, but they have all a bright look; trees are on every side, and handsome garden plots, and clinging ivy, relieve the monotony of the high square structures. Here, I believe, is the largest still in the world—certainly the largest in Ireland. No work was in progress at the time of my visit, save the work of repairs and the storage of coal. Here, as at North Mall, water gives considerable aid in driving the machinery, a canal having been raised after considerable engineering difficulties and much expense. The vast works of Midleton Distillery cover over eight acres. It was a hot July day when I paid my visit, not at all the day to attempt remarkable pedestrian feats. Will it be forgiven me, therefore, if I forsook Irish whisky for Irish hospitality? Under the very shadow of the tall manufactory, yet altogether hidden from it, there is a luring lawn, a cool shrubbery, and an elegant villa radiant with flowers. Is it not more pleasant to lounge through conservatories than to climb staircases, to drink iced claret cup than to sip raw spirit, or to examine the points of a horse than to note the intricacies of machinery? Beyond the garden and the tennis court and the conservatory is a grotto, so cunningly placed that none but the initiated can find it; the air there is deliciously cool, a luxuriant growth of honeysuckle and dog-rose and fern surrounds you, and at your feet is a spring of as pure water as ever mortal tasted. With pleasant society, and chat and gossip to while away the time, will it be deemed strange that I stayed there until it was impossible to see more of the distillery, and that it would only be possible to catch my last train by a hard and almost breakneck gallop?" See GIN, SPIRITS, and USQUEBAUGH.

WHITE AR'SENIC. See ARSENIOUS ACID.

WHITEBAIT. . The *Clupea catulus* (*Clupea alba*, Yarrell), a very small and delicate fish, common in the brackish waters of the Thames from April to September. When fried in oil it is esteemed a great luxury by epicures.

WHITE COPPER. See GERMAN SILVER and PACKFONG.

WHITE HEL'LEBORE. *Syn.* VERATRUM; VERATRI ALBI RADIX, L. "The rhizome of *Veratrum album*, Linn., or white hellebore." A powerful acrid cathartic, emetic, and sternutatory. It is now seldom exhibited internally, and its external use over a large or ulcerated surface is not unaccompanied with danger.—*Dose*, ¼ to 2 gr. of the powder made into a pill; in gout, mania, &c.; or 1 to 3 gr., carefully triturated with 12 or 15 gr. of liquorice powder, as an errhine, in amaurosis, &c.

WHITE-LEAD. *Syn.* FINE WHITE, FLAKE W., CARBONATE OF LEAD, CERUSE, MAGISTERY OF LEAD; CERUSSA, PLUMBI CARBONAS (B. P., Ph. E. & D.), L. *Prep.* 1. By suspending rolls of thin sheet lead over malt liquor or pyroligneous acid in close vessels, the evaporation from the acid being kept up by the vessels being placed in a heap of dung or a steam-bath. This is the Dutch method.

2. A new process consists in passing carbonic acid gas into an intimate mixture of litharge with about 1% of lead acetate and water.

3. Another new process is to grind together for some hours a mixture of common salt, litharge, and water, and then to pass carbonic acid gas into the creamy liquid until it is neutral.

4. Quite lately a process for manufacturing 'white-lead' has been patented, the product of which is not poisonous, but analysis shows it to have the composition of the sulphate, not the carbonate.

Obs. Commercial carbonate of lead, however prepared, is not the pure carbonate of lead, but always contains a certain proportion of hydrate. The usual composition of white-lead is represented by the formula $Pb(OH)_2.2Pb.CO_3$. It is generally largely adulterated with native sulphate of baryta ('heavy spar'), and sometimes with chalk. The former may be detected by its insolubility in dilute nitric acid, and the latter by the nitric solution yielding a white precipitate with dilute sulphuric acid, or a solution of oxalic acid or oxalate of ammonia, after having been treated with sulphuretted hydrogen, or a hydrosulphuret, to throw down the lead. "Pure carbonate of lead does not lose weight at a temperature of 212° F.; 68 gr. are entirely dissolved in 150 minims of acetic acid diluted with 1 fl. oz. of distilled water; and the solution is not entirely precipitated by a solution of 60 gr. of phosphate of soda" (Ph. E.). The solution in nitric acid should not yield a precipitate when treated with a solution of sulphate of soda. Used as a superior white paint, and, in medicine, as an external astringent, refrigerant, and desiccant. It is very poisonous.

The particles of carbonate of lead prepared by precipitation, or by any of the quick processes, are in a somewhat crystalline and semi-translucent condition, and hence do not cover so well as that just noticed. The following are some of the varieties of ' white-lead ' found in commerce:

1. (DUTCH WHITE-LEAD.) *a.* (Finest.) From flake white, 1 cwt.; cawk, 3 cwt.—*b.* (Ordinary.) Flake white, 1 cwt.; cawk, 7 cwt. These form the best white-lead of the shops.

2. (ENGLISH WHITE-LEAD.) Flake white lowered with chalk. Covers badly, and the colour is inferior to the preceding.

3. (FRENCH WHITE-LEAD; BLANC DE PLOMB, Fr.) From litharge dissolved in vinegar, and then thrown down by a current of carbonic acid gas from coke. Does not cover so well as flake white.

4. (GRACE'S WHITE-LEAD.) Made from sheet lead, with the refuse water of the starch-makers, soured brewer's grain, &c.

5. (HAMBURG WHITE, HAMBURG WHITE-LEAD.) From flake white, 1 cwt.; cawk, 2 cwt. Also sold for best Dutch white-lead.

6. (VENETIAN WHITE, VENETIAN WHITE-LEAD; CERUSA VENETA, L.) From flake white, or pure white-lead and cawk, equal parts. (See *below.*)

White Precip'itate of Lead. *Syn.* MINIATURE PAINTER'S WHITE, SULPHATE OF LEAD. From an acetic or nitric solution of litharge, precipitated by adding dilute sulphuric acid, and the white powder washed and dried. The clear liquid decanted from the precipitate is poured on fresh litharge, when a second solution takes place; and this may be repeated for any number of times. Used in miniature painting, being a beautiful and durable white.

White, Wilkinson's. From litharge ground with sea water until it ceases to whiten, and then washed and dried.

White, Zinc (Hubbuck's). A hydrated oxide of zinc. It possesses the advantage of being innocuous in use, and not being blackened by sulphuretted hydrogen, like white-lead.

Whi"ting. Another name for ground chalk, but prepared more carelessly, in horse-mills.

WHITE PIG'MENTS. *Syn.* PIGMENTA ALBA, L. The following list embraces the more important white pigments of commerce.

White, Alum. *Syn.* BAUME'S WHITE. Take of powdered Roman alum, 2 lbs.; honey, 1 lb.; mix, dry, powder, calcine in a shallow dish to whiteness, cool, wash, and dry. A beautiful and permanent white, both in oil and water.

White, Chinese. *Syn.* ZINC WHITE. See *above.* Often adulterated with chalk, kaolin, or starch.

White, Cremnitz. Green white-lead.

White, Derbyshire. From cawk or heavy spar, by grinding and elutriation.

White, Flake. The finer kinds of white-lead are so called.

White, Min'eral. Precipitated carbonate of lead.

White, Newcastle. White-lead made with molasses vinegar.

White, Nottingham. White-lead made with alegar (sour ale). Permanent white is now commonly sold for it.

White, Pearl. *Syn.* FARD'S SPANISH WHITE. Trisnitrate of bismuth.

White, Per'manent. Artificial sulphate of baryta, prepared by precipitating chloride of barium with dilute sulphuric acid, or a solution of Glauber's salts. A good fast white, unchanged by sulphurous fumes. Used to mark jars and bottles for containing acids or alkalies, as it is affected by very few substances; also to adulterate white-lead.

White, Spanish. *Syn.* BLANC D'ESPAGNE, BLANC DE TROYES, Fr. The softest and purest white chalk, elutriated, made into balls, and well dried. Used as a cheap white paint.

WHITE SWELL'ING. *Syn.* HYDRARTHRUS, L. A variety of indolent, malignant, scrofulous tumours, attacking the knee, ankle, wrist, and elbow, especially the first.

WHITES (Sharp). *Prep.* 1. From wheaten flour and powdered alum, equal parts, ground together.

2. (STUFF; BAKER'S STUFF.) From alum, ground to the coarseness of common salt, 1 lb.; common salt, 3 lbs.; mix together. Both the above are used by bakers for the purpose of clandestinely introducing alum into their bread.

WHITE'WASH. Whiting is made into a milk with water, and a small quantity of melted size or dissolved glue added. It is applied to walls or ceilings with a broad, flat brush, worked in a uniform direction. Should the surface have been previously whitewashed, it is requisite first to remove the dirt by washing it with a brush and abundance of clean water.

"LIME-WASHING is, from the cleansing action of the quicklime, much the more effectual mode of purification, but is less frequently had recourse to, from the general ignorance respecting the proper mode of preparing the lime-wash. If glue is employed, it is destroyed by the corrosive action of the lime, and, in consequence, the latter easily rubs off the walls when dry. This is the case also if the lime be employed, as is often absurdly recommended, simply slaked in water, and used without any binding material. Lime-wash is prepared by placing some freshly burned quicklime in a pail, and pouring on sufficient water to cover it; 'boiled oil' (linseed) should then be immediately added, in the proportion of a pint to a gallon of the wash. For coarser work, any common refuse fat may be used instead of the boiled oil. The whole should then be thinned with water to the required consistency, and applied with a brush. Care should be taken not to leave the brush in the lime-wash for any length of time, as it destroys the bristles" (*W. B. Tegetmeier*). For conservatory roofs, &c., by adding a small quantity of potassium bichromate solution to the whitewash and size just before use the size is rendered insoluble on exposure to light, and the work is more permanent, and will resist the weather for a long time (ED.).

WHI″TING. See WHITE PIGMENTS.

WHITING. The *Gadus merlangus*, Linn., a member of the cod family of fishes. It is a very light and nutritious fish, and well adapted to dyspeptics and invalids; but it has too little flavour to be a favourite with gourmands.

WHIT'LOW. *Syn.* WHITLOE; PARONYCHIA, L. A painful inflammation, tending to suppuration and abscess at the ends of the fingers, and mostly under or about the nails. Emollient poultices are useful in this affection; extreme tension and pain may be relieved by an incision, so as to allow the exit of the pus or matter from under the nail. The treatment must also be directed to establish the general health, as without this local remedies often fail.

WHITWORTH BOTTLE. A remedy much used in the neighbourhood of Whitworth, Lancashire.—*Prep.* Camphor, 6 parts; oil of origanum, 6 parts; anchusa root, 1 part; methylated spirit, 80 parts.

WHOOP'ING-COUGH. *Syn.* CHIN-COUGH, HOOPING-C., KIN-C.; PERTUSSIS, L. A convulsive, strangling cough, characterised by peculiar sonorous or whooping inspirations, from which its popular name is taken. It comes on in fits, which are usually terminated by vomiting. It is infectious, chiefly attacks children, and, like the smallpox, only occurs once during life. The treatment of whooping-cough consists chiefly in obviating irritation, and in exciting nausea and occasional vomiting. From the first, aperients and sedatives (hemlock or henbane), in small doses, may be given; for the second intention an extremely weak sweetened solution of tartarised antimony, or a mixture containing squills or ipecacuanha, may be administered in small doses every hour or two, according to the effect produced. In full habits blisters and leeches may be resorted to; and in all cases opiate and stimulating embrocations may be applied to the chest and spine with advantage. Whenever the head is affected the use of narcotics is contra-indicated. The hot bath is often serviceable. Other medicinal agents employed in pertussis are alum, bromide of ammonium, sulphate of zinc, belladonna, tincture of myrrh, carbolic acid, and lobelia. Abundance of homemade lemonade, not too sweet, is an excellent thing for children with whooping-cough. See ANTIMONIALS, DRAUGHTS, MIXTURE, OXYMEL, SYRUP, WINES, &c.

WHOR'TLEBERRY (Bear's). *Syn.* UVÆ URSI FOLIA (B. P.), UVA URSI (Ph. L., E., & D.), L. The leaf of *Arctostaphylos uva-ursi*, trailing arbutus, or bearberry. Astringent.—*Dose*, 10 to 30 gr. of the powder, thrice daily. See DECOCTION and EXTRACT.

WIK'ANA. *Syn.* WACAKA DES INDS, Fr. *Prep.* (*Guibourt.*) Roasted chocolate nuts (ground), 2 oz.; powdered cinnamon, 2 dr.; powdered vanilla, ½ dr.; ambergris, 3 gr.; musk, 1½ gr.; sugar, 6 oz.; well mixed together. A teaspoonful is boiled with ½ pint of milk or arrowroot, as a stimulating diet for convalescents.

WILD CHERRY. The *Prunus Virginiana*, a beautiful tree growing wild in the western States of America. The inner bark (wild cherry bark) is officinal in the Ph. U. S., and is a valuable sedative tonic. It is specially adapted for the alleviation of the distressing cough which is so harassing to patients with pulmonary disease. See INFUSION.

WILD'FIRE RASH. *Strophulus volaticus.*

WILLOW. *Syn.* SALIX, L. The barks of *Salix alba* or white willow, *Salix fragilis* or crack willow, and *Salix caprea* or great round-leaved willow (WILLOW BARKS; SALICIS COR-TICIS—Ph. E.), were officinal in the Ph. D. 1826; and, with that of *Salix Russelliana* and other species, are rich in salicin, and hence possess considerable febrifuge power.—*Dose*, ½ to 1 dr., either in powder or made into a decoction; as a substitute for Peruvian bark, in agues, hectics, debility, dyspepsia, &c.

WIN'DOWS. A prismatic or crystalline appearance may be imparted to windows by several expedients.

1. Mix a hot solution of sulphate of magnesia (Epsom salt) with a clear solution of gum-arabic, and lay it on hot. For a margin, or for figures, wipe off the part you wish to remain clear with a wet towel as soon as the surface has become cold and hard. The effect is very pretty, and may be varied by substituting oxalic acid, red or yellow prussiate of potash, or any other salt (not efflorescent), for the sulphate of magnesia. Sulphate of copper gives a very beautiful crystallisation of a blue colour.

2. Evenly cover the surface of the glass with a layer of thin gum water, and sprinkle any of the saline crystals before noticed over it whilst wet. The gum water may be tinged of any colour to vary the effect.

A blinded appearance more or less resembling ground glass may be given as follows :

1. By evenly dabbing the surface with a piece of soft glazier's putty.

2. A coating of stained rice jelly, laid on with a painter's brush (sash tool), and afterwards dabbed with a duster brush applied endways.

3. Tissue paper, either white or coloured, applied by means of clear gum water or some pale varnish. The pattern may be lined with a pencil, and, when the whole is somewhat dry, but not hard, the lines may be cut through, and the pattern stripped off with the flat point of a knife.

4. The surface of the glass being coated with mucilage or any pale varnish, as before, coarsely powdered glass or quartz, reduced to a uniform state of grain by a sieve, may be sprinkled over it ; when dry, the loose portion should be removed with a soft brush.

WINE. *Syn.* VINUM, L. ; VIN, Fr. The fermented juice of the grape. The general characters and quality of wine are principally influenced by climate, soil, and aspect, the nature and maturity of the grape, and the method of conducting the fermentation. The sp. gr. of the 'must' varies from 1·063 to 1·285, from which the proportion of saccharine matter and the ultimate alcoholic richness of the wine resulting from its fermentation may be inferred. That of Rhenish grapes seldom exceeds 1·095 to 1·100. Want of space compels us to confine our remarks chiefly to the properties, uses, and management of grape-juice after it has passed through the stage of fermentation, or, in reality, become wine.

Officinal Wine. The only wine ordered by the British Colleges is sherry (WHITE WINE ; VINUM XERICUM—B. P., Ph. L.; VINUM ALBUM—Ph. E. ; VINUM HISPANICUM—Ph. D.); but several other wines are employed in medicine, as tonics, stimulants, antispasmodics, and restoratives, according to the circumstances of the case or the

I. TABLE *of the Quantity of Alcohol in Wine.* By Dr CHRISTISON.

Name, &c.		Alcohol of ·7937 per cent. by weight.	Proof spirit per cent. by volume.
Port . .	Weakest .	14·97	31·31
	Mean of 7 samples .	16·20	34·91
	Strongest	17·10	37·27
	White .	14·97	31·31
Sherry .	Weakest .	13·98	30·84
	Mean of 13 wines, excluding those very long kept in cask	15·37	33·59
	Strongest	16·17	35·12
	Mean of 9 wines long kept in cask in the East Indies .	14·72	31·30
	Madre da Xeres .	16·90	37·06
Madeira (long kept in cask in the East Indies) { Strongest	16·90	37·06
	Weakest	14·09	30·86
Teneriffe (long in cask at Calcutta)	13·84	30·21
Cercial .	15·45	33·65
Lisbon (dry) .	16·14	34·71
Shiraz .	12·95	28·30
Amontillado .	12·63	27·60
Claret (a first growth of 1811) .	7·72	16·95
Château-Latour (ditto 1825) .	7·78	17·06
Rosan (second growth of 1825) .	7·61	16·74
Ordinary Claret (Vin Ordinaire) .	8·99	18·96
Rivesaltes .	9·31	22·35
Malmsey .	12·86	28·17
Rüdesheimer. 1st quality .	8·40	18·44
„ Inferior .	6·90	15·19
Hambacher. Superior quality .	7·35	16·15

II. *Quantity of Alcohol* (sp. gr. ·825 at 60° F.) *in* 100 *parts of Wine by volume.*

Alcohol of ·825 contains 92·6% of real or anhydrous alcohol; or, in the language of the Excise, is about 62½% o.p., and in round numbers may be said to be of about twice the strength of brandy or rum as usually sold.

Names of Wines.	Alcoholic content.	Authority.	Names of Wines.	Alcoholic content.	Authority.
Alba Flora . . .	17·26	Brande.	Lunel	15·52	Brande.
Barsac	13·86	do.	Madeira (average) .	22·27	do.
Bucellas . . .	18·49	do.	Ditto (do.) .	21·20	Prout.
Burgundy (average) .	14·57	do.	Malaga	17·26	Brande.
Ditto	12·16	Prout.	Ditto	18·94	do.
Calcavella (average) .	18·69	Brande.	Malmsey Madeira .	16·40	do.
Cape Madeira (do.) .	20·51	do.	Marsala (average) .	25·09	do.
Cape Muschat . .	18·25	do.	Ditto (do.) .	18·40	Prout.
Champagne (average)	12·61	do.	Nice	14·63	Brande.
Ditto	12·20	Fontenelle.	Orange (average) .	11·26	do.
Claret (average) . .	15·10	Brande.	Port (do.) .	20·64	Prout.
Colares	19·75	do.	Ditto (do.) .	22·96	Brande.
Constantia (White) .	19·75	do.	Raisin (do.) .	25·41	do.
Ditto (Red) . .	18·92	do.	Ditto (do.) .	15·90	Prout.
Ditto (average)	14·50	Prout.	Red Madeira (do.) .	20·35	Brande.
Côte Rôtie . .	12·32	Brande.	Roussillon (do.) . .	18·13	do.
Currant . . .	20·55	do.	Sauterne . . .	14·22	do.
Elder	8·79	do.	Shiras . . .	15·52	do.
Frontignac (Rivesalte)	12·79	do.	Sherry (average)	19·17	do.
Gooseberry . .	11·84	do.	Ditto (do.) . .	23·80	Prout.
Grape (English) .	18·11	do.	Syracuse . . .	20·00	do.
Hermitage (Red) .	12·32	do.	Ditto	15·28	Brande.
Ditto (White) . .	17·43	do.	Teneriffe . . .	19·79	do.
Hock (average) . .	12·08	do.	Tent	13·30	do.
Lachryma Christi .	19·70	do.	Tokay	9·88	do.
Lisbon	18·94	do.	Vidonia . . .	19·25	do.
Lissa (average) . .	25·41	do.	Vin de Grave . .	13·94	do.
Ditto (do.) . .	15·90	Prout.	Zante	17·05	do.

taste of the patient. In pharmacy, the less expensive Cape or marsala, or even raisin wine, is usually substituted for sherry in the preparation of the medicated wines of the Pharmacopœias.

Varieties, Characteristics, &c. The preceding Tables will convey much useful information on this subject in a condensed form.

Composition. The constituents of wine are—alcohol, which is one of its principal ingredients, and on which its power of producing intoxication depends; sugar, which has escaped the process of fermentation, and which is most abundant in the sweet wines, as tokay, tent, frontignac, &c.; extractive, derived chiefly from the husk of the grape, and is extracted from it by the newly formed alcohol; tartar, or bitartrate of potassa, which constitutes the most important portion of the saline matter of wine; odoriferous matter, imparting the characteristic vinous odour, depending chiefly upon the presence of œnanthic acid and ether; and bouquet, arising from essential oil or amyl compounds, probably existing under the form of ethers. Besides these, small quantities of tannin, gum, acetic and malic acid, acetic ether, lime, &c., are found in wine. The specific gravity of wine depends on the richness and ripeness of the grapes used in its manufacture, the nature of the fermentation, and its age. It varies from about ·970 to 1·041.

Purity. The most frequent species of fraud in the wine trade is the mixing of wines of inferior quality with those of a superior grade. In many cases the inferior kinds of foreign wines are flavoured and substituted for the more expensive ones. This is commonly practised with Cape wine, which, after having a slight 'nuttiness' communicated to it by bitter almonds or peach kernels, a lusciousness or fulness by honey, and additional strength by a little plain spirit or pale brandy, is made to undergo the operation of 'fretting in,' and is then sold for 'sherry.' Formerly it was a common practice of ignorant wine dealers to add a little litharge or acetate of lead to their inferior wines to correct their acidity; but it is believed that this highly poisonous substance is now never employed in this country, 'salt of tartar' being made to perform the same duty. The lead which is frequently detected in bottled wine, and which often causes serious indisposition, may be generally traced to shot being carelessly left in the bottles, and not to wilful fraud. Sherry is commonly coloured in Spain by the addition of 'must' boiled down to 1-5th of its original volume; and in England, by burnt brown sugar, or spirit colouring. Amontillado (a very nutty wine) is frequently added to sherries deficient in flavour. Various other ingredients, as the essential oil of almonds, bitter almonds in substance, cherry-laurel leaves, cherry-laurel water, &c., are also employed for a like purpose. In Portugal the juice of elderberries is very commonly added to port wine to increase its colour, and extract of rhatany for the double purpose of improving its colour and imparting an astringent taste. In England beetroot, Brazil-wood, the juices of elderberries and bilberries, the pressed cake of elder wine, extract of logwood, &c., are frequently added to port to deepen its colour; and oak sawdust, kino, alum, and extract of rhatany, to increase its astringency. But the most common adulterant of port wine, both in Portugal and this country, is 'jerupiga,' or 'gerupiga,' a compound of elder juice, brown sugar, grape juice, and crude Portuguese brandy. That imported here contains about 45% of proof spirit, and is allowed by the Custom-house authorities to be mixed with port wine in bond. A factitious bouquet is also commonly given to wine by the addition of sweet-briar, orris root, clary, orange flowers, elder flowers, esprit de petit grain, &c.

Tests. These, for the most part, are applicable to all fermented liquors:

1. Richness in alcohol. This may be found by any of the methods noticed under ALCOHOLO-METRY, PORTER, and TINCTURE.

2. SACCHARINE and EXTRACTIVE MATTER. The sp. gr. corresponding to the alcoholic strength, last found, is deducted from the real sp. gr. of the sample; the difference divided by ·0025, or multiplied by 400, gives the weight of solid matter (chiefly sugar) in oz. per gallon (nearly).

3. NARCOTICS. These may be detected in the manner already noticed.

4. LEAD. The presence of lead or litharge in wine may be readily detected by sulphuretted hydrogen, or a solution of any alkaline sulphydrate, which will, in that case, produce a black precipitate. See WINE TESTS.

5. POTASSA or SODA improperly present. A portion of the wine is evaporated nearly to dryness, and then agitated with rectified spirit; the filtered tincture, holding in solution acetate of potassa, is then divided into two portions, one of which is tested for acetic acid, and the other for the alkali.

6. ALUM. A portion of the wine is evaporated to dryness and ignited; the residuum is then treated with a small quantity of hydrochloric acid, the mixture evaporated to dryness, again treated with dilute hydrochloric acid, and tested with liquor of potassa. If a white bulky precipitate forms, which is soluble in an excess of caustic potassa, and which is reprecipitated by a solution of sal-ammoniac, the sample examined contained alum.

7. OIL OF VITRIOL. *a.* A drop or two of the suspected wine may be poured upon a piece of paper, which must then be dried before the fire. Pure wine at most only stains the paper, but one containing sulphuric acid causes it to become charred and rotten. The effect is more marked on paper which has been previously smeared with starch paste.

b. According to M. Lassaigne, pure red wine leaves, by spontaneous evaporation, a violet or purple stain on paper; whilst that to which sul-

phuric acid has been added, even in quantity only equal to $\frac{1}{1000}$ to $\frac{1}{1000}$ part, leaves a pink stain in drying.

8. SPURIOUS COLOURING MATTER. *a.* Genuine red wine yields greenish-grey precipitates with sugar of lead, and greenish ones with potassa; but those coloured with elderberries, bilberries, litmus, logwood, and mulberries, give deep blue or violet precipitates, and those coloured with Brazil-wood, red sanders-wood, or red beet, give red ones.

b. Pure red wine is perfectly decoloured by agitation with fresh hydrate of lime.

c. Dissolve a piece of caustic potash in a small quantity of the liquid to be experimented upon. If no deposit is formed, and the wine assumes a greenish shade, there is no artificial coloration. A violet-coloured deposit indicates the presence of elderberries or mulberries, a red one indicates the presence of beetroot for Brazil-wood, red violet that of logwood. If the deposit is blue violet, privet berries have been employed; and if of a pale violet the coloration is due to litmus.

d. For the detection of the principal colouring matters employed in the sophistication of wines, M. Chancel proceeds as follows:—He takes 10 c.c. of wine, and adds 3 c.c. of a dilute solution of subacetate of lead, allowing the mixture to subside for a few minutes to make sure that the precipitation is complete. If this is not the case a slight excess of the reagent is added.

After stirring and heating for a few moments it is thrown on a very small filter, the filtrate collected in a test-tube, and the precipitate washed three or four times in hot water. If the filtrate is coloured magenta is present, and may be sought for by the aid of the spectroscope. But if the wine contains a mere trace of this colour, it is retained in the precipitate, and is sought for in the manner directed below. To discover the colouring matter which may be contained in the lead precipitate, it is treated upon the filter with a few c.c. of a solution of carbonate of potassa (2 parts of the dry salt to 100 parts of water), taking care to repass *the same solution* several times through the precipitate. Any magenta present is thus extracted, along with carminamic (ammoniacal cochineal) and sulphindigotic acid. The colouring matters of logwood and of alkanet remain undissolved.

With a genuine wine the alkaline liquid takes a very faint yellow or greenish-yellow tint. For the detection of magenta the filtrate is mixed with a few drops of acetic acid, and it is then shaken up with amylic alcohol. The magenta dissolves in this alcohol with a fine rose tint, and its presence is proved by spectroscopic examination. Carminamic and sulphindigotic acids remain in the aqueous solution, and are decanted off. A couple of drops of sulphuric acid are added, and the mixture is again shaken up with amylic alcohol, which now dissolves the ammoniacal cochineal. It may be detected by the spectroscope. The sulphindigotic acid remains undissolved in the amylic alcohol, and may be found in the blue aqueous residual liquor by means of the spectroscope. Logwood is most conveniently sought for in a fresh portion of the wine by digestion with a little precipitated carbonate of lime, adding a few drops of lime-water, and filtering. In a natural wine the filtrate has a faint greenish-yellow colour, but if logwood is present it takes a fine red shade, and the absorption bands of logwood may be detected with the spectroscope. On treating the lead precipitate above mentioned with an alkaline sulphide, washing with boiling water, and then treating with alcohol, the colouring matter of alkanet, if present, is dissolved, and may be detected by spectroscopic examination ("Comptes Rendus," February 19th, 1877; 'Chem. News,' xxxv, 106).

e. (*Dr Dupré.*) The colouring matter of pure red wine does not pass through the dialyser. The dialysate from pure wine is therefore colourless, or shows but a slight purplish coloration, such as water would assume on the addition of a small quantity of the wine. A yellow or brownish-yellow dialysate indicates an adulteration with logwood, Brazil-wood, or cochineal, the colouring matters of which may be identified by the chemical and optical tests employed for this purpose. The ammoniacal solution of the colouring matter of cochineal yields three well-marked absorption bands.

f. For the detection in wine of fuchsine only, the following methods are given by M. E. Jacquemin:— (1) A small quantity of gun-cotton is heated for a few minutes in 10—20 c.c. of the wine, and then washed with the water. The nature of the coloration (if any) imparted to the cotton is now identified by means of solution of ammonia, which decolourises rosaniline, but turns archil violet. (2) 100 c.c. of the wine are boiled to expel the alcohol, and then boiled for some time with white Berlin wool, previously moistened with water. The colour imparted to the wool by fuchsine is retained after washing, and may be distinguished from archil by ammonia. (3) 100—200 c.c. of the wine are boiled to expel the alcohol, then allowed to cool, mixed with ammonia in excess, and shaken with ether. By immersing white wool in the ethereal solution, and evaporating the latter, the wool acquires the characteristic colour of fuchsine.

9. ARTIFICIAL FLAVOURING. This can only be detected by a discriminating and sensitive palate.

10. ARTIFICIAL BOUQUET. The substances added for this purpose may often be readily detected by a comparison of the sample with another of known purity.

Uses. The uses of wine as a beverage are too well known to require description. As a medicine, port wine is most esteemed as an astringent and tonic; and sherry and Madeira as stimulants and restoratives, in diseases where the acidity of the former would be objectionable; champagne is reputed diuretic and excitant, but its effects are not of long duration; and the Rhenish wines are regarded as refrigerant, diuretic, and slightly aperient. Claret, Rhenish, and Moselle wines are said to be the most wholesome. In *pharmacy*, wine is used as a menstruum.

MANAGEMENT OF WINE.

Age. The sparkling wines are in their prime in from 18 to 30 months after the vintage, depending on the cellaring and climate. Weak wines, of inferior growths, should be drunk within 12 or 15 months, and be preserved in a very cool cellar. Sound, well-fermented, full-bodied still wines are improved by age, within reasonable limits, provided they be well preserved from the air, and stored in a cool place, having a pretty uniform temperature. See *Maturation* (*below*).

Bottling. The secret of bottling wine with success consists in the simple exercise of care and cleanliness. The bottles should be all sound, clean, and dry, and perfectly free from the least mustiness or other odour. The corks should be of the best quality, and immediately before being placed in the bottles should be compressed by means of a 'cork-squeezer.' For superior or very delicate wines, the corks are usually prepared by placing them in a copper or tub, covering them with weights to keep them down, and then pouring over them boiling water holding a little pearlash in solution. In this state they are allowed to remain for 24 hours, when they are well stirred about in the liquor, drained, and re-immersed for a second 24 hours in hot water, after which they are well washed and soaked in several successive portions of clean and warm rain water, drained, dried out of contact with dust, put into paper bags, and hung up in a dry place for use. The wine should be clear and brilliant, and if it be not so, it must undergo the process of 'fining' before being bottled. In fact, it is a common practice with some persons to perform this operation whether the wine require it or not; as, if it has been mixed and doctored, it "amalgamates and ameliorates the various flavours." The bottles, corks, and wine being ready, a fine clear day should be preferably chosen for the bottling, and the utmost cleanliness and care should be exercised during the process. Great caution should also be observed to avoid shaking the cask so as to disturb the 'bottoms.' The remaining portion that cannot be drawn off clear should be passed through the 'wine-bag,' and, when bottled, should be set apart as inferior to the rest. The coopers, to prevent breakage and loss, place each bottle, before corking it, in a small bucket, having a bottom made of soft cork, and which is strapped on the knee of the bottler. They thus seldom break a bottle, though they 'fig in' the corks very hard. When the process is complete the bottles of wine are stored in a cool cellar, and on no account upright, or in damp straw, but on their sides, in sweet, dry sawdust or sand.

Bouquet. See *Flavouring* and *Perfuming.*

Brandying. Brandy is frequently added to weak or vapid wines, to increase their strength or to promote their preservation. In Portugal, one third of brandy is commonly added to port before shipping it for England, as without this addition it generally passes into the acetous fermentation during the voyage. A little good brandy is also usually added to sherry before it leaves Spain. By the regulation of the Customs of England, 10% of brandy may be added to wines in bond, and the increased quantity is only charged the usual duty on wine. The addition of brandy to wine injures its proper flavour, and hence it is chiefly made to port, sherry, and other wines whose flavour is so strong as not to be easily injured. Even when brandy is added to wines of the latter description, they require to be kept for some time to recover their natural flavour. To promote this object, the wine-doctors employ the process called ' fretting in,' by which they effect the same change in three or four weeks as would otherwise require some months at the very least.

Cellaring. A wine-cellar should be dry at bottom, and either covered with good hard gravel, or be paved with flags. Its gratings or windows should open towards the north, and it should be sunk sufficiently below the surface to ensure an equable temperature. It should also be sufficiently removed from any public thoroughfare, so as not to suffer vibration from the passing of carriages. Should it not be in a position to maintain a regular temperature, arrangements should be made to apply artificial heat in winter, and proper ventilation in summer.

Colouring. Wines are as commonly doctored in their colour as their flavour. A fawn-yellow and golden-sherry yellow are given by means of tincture or infusion of saffron, turmeric, or safflower, followed by a little spirit colouring, to prevent the colour being too lively. All shades of amber and fawn, are given by burnt sugar. Cochineal (either alone or with a little alum) gives a pink colour; beetroot and red sanders give a red colour; the extracts of rhatany and logwood, and the juice of elderberries, bilberries, &c., give a port-wine colour.

Crusting. To make port wine form a crust on the inside of the bottles, a spoonful of powdered catechu, or half a spoonful of finely powdered cream of tartar, is added to each bottle before corking it, after which the whole is well agitated. It is also a common practice to put the crust on the bottle before putting the wine into it, by employing a hot saturated solution of red tartar, thickened with gum and some powdered tartar.

Deacetification. This is effected by the cautious addition of either salt of tartar or carbonate of soda. Wine so treated soon gets insipid by exposure and age; and, without care, the colour of red wines is thus frequently spoiled.

Deacidification. See *Detartarization* (below).

Decanting. Wine only refers to small quantities of wine ready for consumption. In decanting wine care must be taken not to shake or disturb the crust when moving it about or drawing the cork, particularly of port wine. Never decant wine without a wine-strainer, with some clean fine cambric in it, to prevent the crust and bits of cork going into the decanter. In decanting port wine do not drain it too close, as there are generally two thirds of a wine-glassful of thick dregs in each bottle, which ought to be rejected. In white wine there is not much settling; but it should nevertheless be poured off very slowly, the bottle being raised gradually.

Decolouring. The colour of wine is precipitated by age and by exposure to the light. It is also artificially removed by the action of skimmed milk, lime water, milk of lime, and fresh-burnt charcoal. Wine merchants avail themselves of this property for the purpose of whitening wines that have acquired a brown colour from the cask,

or which are esteemed pale, and also for turning ' pricked ' red or dark-coloured wines into white wines, in which a small degree of acidity is not so much perceived. In this way brown sherry is commonly converted into pale or gold-coloured sherry. For the latter purpose 2 to 3 pints of skimmed milk are usually sufficient; but to decolour red wine 2 to 3 quarts or more will be required, according to the nature and intensity of the colour, or the shades of paleness desired. Charcoal is seldom used, as it removes the flavour as well as colour; but a little milk of lime may sometimes be advantageously substituted for milk when the wine has much acidity, more particularly for red wines, which may even be rendered quite colourless by it.

Detartarization. Rhenish wines, even of the most propitious growths, and in the best condition, besides their tartar, contain a certain quantity of free tartaric acid, on the presence of which many of their leading properties depend. The excess of tartar is gradually deposited during the first years of the vatting, the sides of the vessels becoming more and more encrusted with it; but, owing to the continual addition of new wine and other causes, the liquid often gains such an excess of free tartaric acid as to acquire the faculty of redissolving the deposited tartar, which thus again disappears after a certain period. The taste and flavour of the wine are thus exalted, but the excess of acid makes the wine less agreeable in use, and probably less wholesome. Amateurs and manufacturers should therefore welcome a means of taking away the free tartaric acid without altering, in any respect, the quality of the wine. This is pure neutral tartrate of potash. When this salt, in concentrated solution, is added to such a fluid as the above, the free acid combines with the neutral salt, and separates from the liquid under the form of the sparingly soluble bitartrate of potash. " If to 100 parts of a wine which contains 1 part of free tartaric acid we add 1½ parts of neutral tartrate of potash, there will separate, 'on repose at 76°—75° F., 2 parts of crystallised tartar, and the;wine will then contain only ¼ part of tartar dissolved, in which there is only ·2 part of the original free acid, ·8 part of the original free acid having been withdrawn from the wine" (Liebig's 'Annalen '). This method is particularly applicable to recent must, and to wines which do not contain much free acetic acid; but when this last is the case so much acetate of potash is formed as occasionally to vitiate the taste of the liquor.

Fining. Wine is clarified in a similar manner to beer. White wines are usually fined by isinglass in the proportion of about 1½ oz. (dissolved in 1½ pints of water, and thinned with some of the wine) to the hogshead. Red wines are generally fined with the whites of eggs, in the proportion of 15 to 20 to the pipe. Sometimes hartshorn shavings, or pale sweet glue, is substituted for isinglass.

Flatness. This is removed by the addition of a little new brisk wine of the same kind, or by rousing in 2 or 3 lbs. of honey, or by adding 5 or 6 lbs. of bruised sultana raisins and 3 or 4 quarts of good brandy, per hogshead. By this treatment the wine will usually be recovered in about a fortnight, except in very cold weather. Should it be wanted sooner, a table-spoonful or two of yeast may be added, and the cask removed to a warmer situation.

Flavouring. Various ingredients are added to inferior wines to give them the flavour of others more expensive, and to British wines to make them resemble those imported. Substances are also added in a similar manner to communicate the aroma of the high-flavoured grape wines. Among the first are bitter almonds, almond cake, or the essential oil of almonds, or, preferably, its alcoholic solution, which are used to impart a 'sherry' or 'nutty' taste to weak-flavoured wines, as poor sherry, white Cape, and malt, raisin, parsnip, and other similar British wines; rhatany, kino, oak sawdust and bark, alum, &c., to convey astringency; and tincture of the seeds of raisins, to impart a 'port wine' flavour. Among the substances employed to communicate the bouquet of the finer wines may be mentioned orris root, eau de fleurs d'oranges, neroli, essence de petit grain, ambergris, vanilla, violet petals, essence of cedrat, sweet-briar, clary, and elder flowers, quinces, cherry-laurel water, &c. By the skilful, though fraudulent, use of the above flavouring substances and perfumes the experienced wine-brewer manages to produce, in the dark cellars of London, from white Cape, currant, gooseberry, raisin, rhubarb, parsnip, and malt wine, very excellent imitations of foregn wine, and which pass current among the majority of English wine-drinkers as the choicest productions of the grape, 'genuine as imported.' A grain or two of ambergris, well rubbed down with sugar, and added to a hogshead of claret, gives it a flavour and bouquet much esteemed by some connoisseurs.

Fretting in. See *Sweating in* (*below*).

Improving. This is the cant term of the wine trade, under which all the adulteration and 'doctoring' of wine is carried on. A poor sherry is improved by the addition of a little almond flavour, honey, and spirit; a port deficient in body and astringency, by the addition of some red tartar (dissolved in boiling water), some rhatany, kino, or catechu, and a little honey or foots, and brandy. See *Mixing* (*below*).

Insensible Fermentation. See *Maturation* (*below*).

Insipidity. See *Flatness* (*above*).

Maturation. The natural maturation or 'ripening' of wine and beer by age depends upon the slow conversion of the sugar which escaped decomposition in the 'gyle tun,' or fermenting vessel, into alcohol. This conversion proceeds most perfectly in vessels which entirely exclude the air, as in the case of wine in bottles; as when air is present, and the temperature sufficiently high, it is accompanied by slow acetification. This is the case of wine in casks, the porosity of the wood allowing the very gradual permeation of the air. Hence the superiority of bottled wine over draught wine, or that which has matured in wood. Good wine, or well-fermented beer, is vastly improved by age when properly preserved; but inferior liquor, or even superior liquor, when preserved in improper vessels or situations, becomes acidulous, from the conversion of its alcohol into vinegar. Tartness or acidity is consequently very generally, though wrongly, regarded by the ignorant as a sign of age in liquor. The peculiar change by which fermented liquors become mature or ripe by age is termed the 'insensible fermentation.' It is the alcoholic fermentation impeded by the presence of the already formed spirit in the liquor, and by the lowness of the temperature. See *Ripening* (*below*).

Mixing. Few wines are sold without admixture. It is found that the intoxicating properties of wine are increased by mixing them with other wines of a different age and growth. In many cases the flavour is at the same time improved. Thus a thin port is improved by the addition of a similar wine having a full body, or by a little Malaga, Teneriffe, or rich sherry; and an inferior old sherry may be improved by admixture with a little full-bodied wine of the last vintage. In this consists the great art of 'cellar management,' and to such an extent is this carried, both abroad and in England, that it may be confidently asserted that few wines ever reach the consumer in an unmixed or natural state.

Mustiness. This may generally be removed by violently agitating the wine for some time with a little of the sweetest olive oil or almond oil. The cause of the bad taste is the presence of an essential oil, which the fixed oil seizes on, and rises with to the surface, when it may be skimmed off; or the liquor under it may be drawn off. A little coarsely powdered fresh-burnt charcoal, or even some slices of bread toasted black, will frequently have a like effect. A little bruised mustard seed is also occasionally used for the same purpose.

Perfuming. This is chiefly performed on British wines for family use. For its application to foreign wines, see *Flavouring* (*above*). Wines may be perfumed by the simple addition of any odorous substances previously well mixed with a little of the wine, or dissolved in a few fluid ounces of rectified spirit.

Racking. This should be performed in cool weather, and preferably early in the spring. A clean syphon, well managed, answers better for this purpose than a cock or faucet. The bottoms, or foul portion, may be strained through a wine-bag, and added to some other inferior wine.

Ripening. To promote the maturation or ripening of wine, various plans are adopted by the growers and dealers. One of the safest ways of hastening this, especially for strong wines, is not to rack them until they have stood fifteen or eighteen months upon the lees; or, whether 'crude' or 'racked,' keeping them at a temperature ranging between 50° and 60° F., in a cellar free from draughts and not too dry. Another method is to remove the corks or bungs, and to substitute bladder tied or fastened air-tight over the openings. Bottled wine, treated in this way, ripens very quickly in a temperate situation. Some dealers add a little dilute sulphuric acid to the coarser wines for the same purpose; but a small quantity of concentrated acetic acid or tartaric acid would be preferable, since these acids are found in all wines. Four or five drops of the former, added to a bottle of some kinds of new wine, immediately give it the appearance of being two or three years old.

Ropiness, Viscidity; Graisse. This arises from the wine containing too little tannin or astringent matter to precipitate the gluten, albumen, or other azotised substance occasioning the malady. Such wine cannot be clarified in the ordinary way, because it is incapable of causing the coagulation or precipitation of the finings. The remedy is to supply the principle in which it is deficient. M. François, of Nantes, prescribes the bruised berries of the mountain ash (1 lb. to the barrel) for this purpose. A little catechu, kino, or, better still, rhatany, or the bruised footstalks of the grape, may also be conveniently and advantageously used in the same way. For pale white wines, which are the ones chiefly attacked by the malady, nothing equals a little pure tannin or tannic acid dissolved in proof spirit. See VIS-COUS FERMENTATION, MALT LIQUORS, &c.

Roughening. See *Flavouring (above).*

Second Fermentation; La Pousse. Inordinate fermentation, either primary or secondary, in wine or any other fermented liquor, may be readily checked by sulphuration, or by the addition of mustard seed or sulphite of lime ; 1 oz. of brimstone, ¼ to 1 lb. of bruised mustard seed, and about 4 to 8 oz. of sulphite of lime, are fully sufficient for a hogshead. This substance seldom fails of arresting the fermentation. In addition to the above remedies, a little sulphuric acid is sometimes employed, and the use of black oxide of manganese, or chlorate of potash, has been proposed on theoretical grounds.

Souring. This is either occasioned by the wine having been imperfectly fermented, or from its having been kept in too warm a cellar, where it has been exposed to draughts of air or to continual vibrations, occasioned by the passage of loaded vehicles through the adjoining thoroughfare. The remedy commonly recommended in books for this purpose is to saturate the acid with chalk, milk of lime, or calcined oyster-shells ; but such additions, made in sufficient quantity to effect this object, destroy the character of the wine, and render it sickly and vapid. The best and only safe remedy is a little neutral tartrate of potash, cautiously added ; or it may be mixed with a considerable portion of full-bodied new wine of its class, adding at the same time a little brandy, and in two or three weeks fining it down, when it should be either at once put into bottles, or consumed as soon as possible. See *Deacetification* and *Detartarization (above).*

Sparkling, Creaming, and Briskness. These properties are conveyed to wine by racking it into closed vessels before the fermentation is complete, and while there still remains a considerable portion of undecomposed sugar. Wine of this description, which has lost its briskness, may be restored by adding to each bottle a few grains of white lump sugar or sugar-candy. This is the way in which champagne is treated in France. The bottles are afterwards inverted, by which means any sediment that forms falls into the necks, when the corks are partially withdrawn, and the sediment is immediately expelled by the elastic force of the compressed carbonic

acid. If the wine remains muddy, a little solution of sugar and finings are added, and the bottles are again placed in a vertical position, and, after two or three months, the sediment is discharged as before.

Sweating in. The technical terms 'sweating in' and 'fretting in' are applied to the partial production of a second fermentation, for the purpose of mellowing down the flavour of foreign ingredients (chiefly brandy) added to wine. For this purpose 4 or 5 lbs. of sugar or honey, with a little crude tartar (dissolved), are commonly added per hogshead ; and when the wine is wanted in haste, a spoonful or two of yeast, or a few bruised vine leaves, are also mixed in, the cask being placed in a moderately warm situation until the new fermentation is established, when it is removed to the wine cellar, and, after a few days, 'fined down.'

Taste of Cask. The remedies for this malady are the same as those for mustiness.

** For further information connected with the nature and management of wines, and other fermented liquors, see BREWING, FERMENTATION, MALT LIQUORS, PORTER, SUGAR, SYRUP, VINOUS FERMENTATION, VISCOUS F., WORT, YEAST, &c., and *below.*

Wine, British. The various processes in British wine-making depend upon the same principles, and resemble those employed for foreign wine.

The FRUIT should be preferably gathered in fine weather, and not until mature, as evinced by its flavour ; for if it be employed whilst unripe, the resulting wine will be harsh, disagreeable, and unwholesome, and a larger quantity of sugar and spirit will be required to render it palatable. The common practice of employing unripe gooseberries for the manufacture of British champagne arises from a total ignorance of the scientific principles of wine-making. On the other hand, if ordinary British fruit be employed in too ripe a state, the wine is apt to be inferior, and deficient in the flavour of the fruit.

The fruit, being gathered, at once undergoes the operation of picking or garbling, for the purpose of removing the stalks and unripe or damaged portions. It is next placed in a tub, and is well bruised, to facilitate the solvent action of the water. Raisins are commonly permitted to soak about twenty-four hours previously to bruising them, but they may be advantageously bruised or minced in the dry state. The bruised fruit is then put into a vat or vessel with a guard placed over the tap-hole, to keep back the husks and seeds of the fruit when the must, juice, or extract is drawn off. The water is now added, and the whole is allowed to macerate for thirty to forty hours, more or less, during which time the magma is frequently roused up with a suitable wooden stirrer. The liquid portion is next drawn off, and the residuary pulp is placed in hair bags, and undergoes the operation of pressing, to expel the fluid which it contains. The sugar, tartar (in very fine powder or in solution), &c., are now added to the mixed liquors, and the whole is well stirred or 'rummaged' up for some time. The temperature being suitable, the vinous fermentation soon commences, when the liquor is frequently skimmed (if necessary), and well 'roused' up, and, after 3 or 4 days of this treatment, it is run into casks, which should be quite filled, and left purging at the bung-hole. In about a week the flavouring ingredients, in the state of coarse powder, are commonly added, and well stirred in; and in about another week, depending upon the state of the fermentation, and the attenuation of the must, the brandy or spirit is added, and the cask is filled up and bunged down close. In four or five weeks more the cask is again filled up, and, after some weeks (the longer the better), it is 'pegged' or 'spiled,' to ascertain if it be fine or transparent; if so, it undergoes the operation of racking; but if, on the contrary, it still continues muddy, it must be either again bunged up, and allowed to repose for a few weeks longer, or it must pass through the process of fining. Its future treatment is similar to that already noticed under FOREIGN WINE (see *above*).

The must of many of the strong-flavoured fruits, as black currants, mulberries, &c., is improved by being boiled before being made into wine. The flavour and bouquet of the more delicate fruits are either greatly diminished or utterly dissipated by boiling.

General Formula for the *Preparation* of BRITISH WINES:

1. From ripe saccharine fruits. Take of the ripe fruit, 4 to 6 lbs.; clear soft water, 1 gall.; sugar, 3 to 5 lbs.; cream of tartar (dissolved in boiling water), 1½ oz.; brandy, 2% to 3% flavouring, as required. If the full proportions of fruit and sugar are used, the product will be good without the brandy, but better with it; 1½ lbs. of raisins may be substituted for each pound of sugar.

In the above way are made the following wines:—Gooseberry wine ('British champagne'); currant wine (red, white, or black); mixed fruit wine (currants and gooseberries, or black, red, and white currants, ripe black-heart cherries, and raspberries, equal parts), a good family wine; cherry wine; Colepress's wine (from apples and mulberries, equal parts); elder wine; strawberry wine; raspberry wine; mulberry wine (when flavoured, makes 'British port'); whortleberry wine (bilberry wine), makes a good factitious 'port;' blackberry wine; damson wine (makes good factitious 'port'); morella wine; apricot wine; apple wine; grape wine, &c.

2. From dry saccharine fruit (as raisins). Take of the dried fruit, 4½ to 7½ lbs.; clear soft water, 1 gall.; cream of tartar (dissolved), 1 oz.; brandy, 1½% to 4%. Should the dried fruit employed be at all deficient in saccharine matter, 2 to 3 lbs. of it may be omitted, and half that quantity of sugar, or two thirds of raisins added. In the above way are made date wine, fig wine, raisin wine, &c.

3. From ACIDULOUS, ASTRINGENT, or SCARCELY RIPE FRUITS, or those which are deficient in saccharine matter. Take of the picked fruit, 2½ to 3½ lbs.; sugar, 3½ to 5½ lbs.; cream of tartar (dissolved), ½ oz.; water, 1 gall.; brandy 2% to 6%.

In the above way are made gooseberry wine ('British champagne'); bullace wine (which makes an excellent 'factitious port'); damson wine, &c.

4. From FOOTSTALKS, LEAVES, CUTTINGS, &c. By infusing them in water, in the proportion of 3 to 6 lbs. to the gall., or q. s. to give a proper flavour, or to form a good saccharine liquor; and adding 2½ to 4 lbs. of sugar to each gall. of the strained liquor. 1½ lbs. of raisins may be substituted for each lb. of sugar.

In the above way are made grape wine (from the pressed cake of grapes); English grape wine; rhubarb wine ('Bath champagne,' 'patent c.'), from garden rhubarb; celery wine, &c.

5. From SACCHARINE ROOTS and STEMS OF PLANTS. Take of the bruised, rasped, or sliced vegetable, 4 to 6 lbs.; boiling water, 1 gall.; infuse until cold, press out the liquor, and to each gall. add of sugar, 3 to 4 lbs.; cream of tartar, 1 oz.; brandy, 2% to 5%. For some roots and stems the water must not be very hot, as they are thus rendered troublesome to press.

In the above way are made beetroot wine ('British Roussillon'); parsnip wine ('British malmsey'); turnip wine, &c.

6. From FLOWERS, SPICES, AROMATICS, &c. These are prepared by simply infusing a sufficient quantity of the bruised ingredient for a few days in any simple wine (as that from sugar, honey, raisins, &c.) after the active fermentation is complete, or, at all events, a few weeks before racking them.

In the above way are made clary wine ('muscadel'), from flowers, 1 quart to the gall.; cowslip wine (flowers), 2 quarts to the gall.; elder-flower wine ('Frontignac'), flowers of white-berried elder, ¾ pint, and lemon juice, 3 fl. oz. to the gall.; ginger wine (1½ oz. of ginger to the gall.); orange wine (1 dozen sliced oranges per gall.); lemon wine (juice of 12 and rinds of 6 lemons to the gall.); spruce wine (½ oz. of essence of spruce per gall.); juniper wine (berries, ¾ pint per gall.); peach wine (4 or 5 sliced, and the stones broken, to the gall.); apricot wine (as peach wine, or with more fruit); quince wine (12 to the gall.); rose clove gillyflower, carnation, lavender, violet, primrose, and other flower wines (distilled water, 1½ pints, or flowers, 1 pint to the gall.); balm wine (balm tops, 4 oz. per gall.), &c.

7. From SACCHARINE JUICES, or INFUSIONS, or from fermented liquors. Take of the juice or liquor, 1 gall.; honey or sugar, 2 to 3 lbs. (or raisins, 3 to 5 lbs.); cream of tartar, 1½ oz.; brandy, 2% to 4%.

In this way are made English grape wine; mixed fruit wine; pine-apple wine; cider wine; elder wine; birch wine (from the sap, at the end of February or beginning of March); sycamore wine (from the sap); malt wine ('English Madeira'), from strong wort; and the wines of any of the saccharine juices of ripe fruit.

8. From SIMPLE SACCHARINE MATTER. Take of sugar, 3 to 4 lbs.; cream of tartar, ½ oz.; water, 1 gall.; honey, 1 lb.; brandy, 2% to 4%. A handful of grape leaves or cuttings, bruised, or a pint of good malt wort, or mild ale may be substituted for the honey. Chiefly used as the basis for other wines, as it has little flavour of its own, but makes a good 'British champagne.'

Obs. In all the preceding formulæ lump sugar is intended when the wines are required very pale, and good Muscovado sugar when this is not the case. Some of the preceding wines are vastly improved by substituting good cider, perry, or pale ale or malt wort, for the whole or a portion of the water. Good porter may also be advantageously used in this way for some of the deep-coloured red wines. When expense is no object, and very strong wines are wanted, the expressed juices of the ripe fruits, with the addition of 3 or 4 lbs. of sugar per gall., may be substituted for the fruit in substance, and the water.

Examples of BRITISH IMITATIONS OF FOREIGN WINES:

AMERICAN HONEY WINE. From good honey, 21 lbs.; cider, 12 galls.; ferment, then add of rum, 5 pints; brandy, 2 quarts; red or white tartar (dissolved), 6 oz.; bitter almonds and cloves, of each (bruised), ¼ oz.; powdered capsicum, 3 dr. This is also called 'mead wine.' With the addition of 3 oz. of unbleached Jamaica ginger (finely grated), it forms the best American ginger wine.

BRITISH BURGUNDY. By adding a little lemon juice, and a 'streak' of orris or orange-flower water, to 'British port,' the ingenious wine-brewer converts it into 'British Burgundy.' It is also made by mixing together equal parts of 'British port' and claret.

BRITISH CAPE. 1. (White.) Raisin wine well attenuated by fermentation, either alone or worked up with a little cider and pale malt wort.

2. (Red.) British white cape, sound rough cider, and mulberry wine, equal parts; well mixed and fined down.

BRITISH CHAMPAGNE. 1. From stoned raisins, 7 lbs.; loaf sugar, 21 lbs.; water, 9 galls.; crystallised tartaric acid, 1 oz.; cream of tartar, ½ oz.; Narbonne honey, 1 lb.; sweet yeast, ½ pint; ferment, skimming frequently, and, when the fermentation is nearly over, add of coarsely powdered orris root, 1 dr.; eau de fleurs d'oranges, ¼ pint; and lemon juice, 1 pint; in 3 months fine it down with isinglass, ¼ oz.; in 1 month more, if not sparkling, again fine it down, and in another fortnight bottle it, observing to put a piece of double-refined white sugar, the size of a pea, into each bottle; lastly, wire down the corks, and cover them with tin-foil, after the manner of champagne.

2. As the preceding, but substituting 32 lbs. of double-refined sugar for the sugar and raisins therein ordered, with the addition of 3 galls. of rich pale-coloured brandy.

3. From amber hairy champagne gooseberries, English grape juice, or the stalks of garden rhubarb, and lump sugar; with a little sweet-briar, orris, or orange-flower water, to impart a slight bouquet. The last forms what is know as 'patent' or 'Bath champagne.'

4. (Pink.) To either of the preceding add red currant juice, q. s. to colour; or 1 oz. of coarsely powdered cochineal to each 10 or 12 galls. at the time of racking.

Obs. It is notorious that two bottles of wine out of every three sold for 'genuine champagne' in England is of British manufacture. "We have ourselves seen sparkling gooseberry, rhubarb, and white sugar wines, sold for imported champagne, at 7*s.* 6*d.* per bottle, and the fraud has passed undetected, even by habitual wine drinkers" (*Cooley*).

BRITISH CLARET. 1. Rich old cider or perry and port wine, equal parts.

2. To each gall. of the last add of cream of tartar (genuine), 3 dr., with the juice of 1 lemon. Sometimes ¼ pint of French brandy is also added.

Obs. If these mixtures are well fined down, and not bottled for at least a month or 5 weeks, they closely resemble good 'Bordeaux.' A mixture of 4 parts of raisin wine, with 1 part each of raspberry and barberry or damson wine, also forms, when so treated, an excellent factitious 'claret.'

BRITISH CYPRUS. From the juice of white elderberries, 1 quart, and Lisbon sugar, 4 lbs., to water, 1 gall.; together with ½ dr. each of bruised ginger and cloves. When racked, add minced raisins and brandy, of each, 2 oz.

BRITISH HOCK, BRITISH RED HOCK. From cream of tartar, 1½ oz.; tartaric acid, ½ oz. (both in extremely fine powder); juices of the purple plum, ripe apples, and red beet, of each (warmed), 5 pints; lemon juice, 1 pint; with white sugar, 2½ lbs. per gall.

BRITISH MADEIRA. From very strong pale malt wort, 36 galls.; sugar-candy, 28 lbs.; and cream of tartar, 3 oz.; fermented with yeast, 2 lbs.; adding, when the fermentation is nearly finished, raisin wine, 2½ galls.; brandy and sherry wine, of each, 2 galls.; rum and brandy, of each, 3 pints; after 6 or 9 months, fine it down, and in another month bottle it. See BRITISH SHERRY (*below*).

BRITISH MALMSEY. From sliced or grated parsnips, 4 lbs.; boiling water, 1 gall.; when cold, press out the liquor, and to each gallon add of cream of tartar, ½ oz., and good Muscovado sugar, 3 lbs.; ferment, rack, and add of brandy, 3% to 5%. Good Malaga raisins may be substituted for the sugar.

BRITISH RED MOSELLE. The last, coloured with clarified elderberry juice.

BRITISH SPARKLING MOSELLE. From rich cider apples (carefully peeled and garbled), pressed with 1-4th of their weight of white magnum-bonum plums (previously stoned), and the juice fermented with 2½ lbs. of double-refined sugar per gall., as champagne.

BRITISH MUSCADEL. As 'British sparkling Moselle,' with some infusion of clary, or of the musk plant, to flavour it.

BRITISH PORT, LONDON P., SOUTHAMPTON P. 1. From red cape, 2 galls.; damson or elder wine, 1 gall.; brandy, ½ pint; powdered kino, ½ oz.

2. Strong old cider, 6 galls.; elderberry juice, 4 galls.; sloe juice, 3 galls.; sugar, 28 lbs.; powdered extract of rhatany, 1 lb.; at the time of racking add brandy, ½ gall.; good port wine, 2 galls.

3. Good port, 12 galls.; rectified spirit, 6 galls.; French brandy, 3 galls.; strong rough cider, 42 galls.; mix in a well-sulphured cask ('Publican's Guide').

4. Port wine, 8 galls.; brandy, 6 galls.; sloe juice, 4 galls.; strong rough cider, 45 galls.; as the last ('Licensed Victualler's Companion').

5. Cider, 24 galls.; juice of elderberries, 6 galls.;

sloe juice, 4 galls.; rectified spirit, 3 galls.; brandy, 1½ galls.; powdered rhatany, 7 lbs.; isinglass, 4 oz., dissolved in a gallon of the cider; bung it down; in three months it will be fit to bottle, but should not be drunk until the next year; if a rougher flavour is required the quantity of rhatany may be increased, or alum, 5 or 6 oz. (dissolved), may be added.

BRITISH SHERRY. 1. From cape or raisin wine, slightly flavoured with a very little bitter almond cake, or, what is more convenient, a little of the essential oil dissolved in alcohol (essence of bitter almonds). A mere 'streak' or 'thread' of sweet-briar, eau de fleurs d'oranges, orris, is occasionally added by way of bouquet; but care must be taken not to overdo it.

2. To each gallon of strong raisin must be added, when racking, 1 Seville orange and 3 or 4 bitter almonds, both sliced. By omitting the almonds, and adding 1 green citron to each 2 or 3 galls., this forms ' British madeira.'

3. Very strong pale malt wort, 36 galls.; finest Muscovado sugar, 1 cwt.; yeast, 1 pint; ferment; on the third day add of raisins, stoned, 14 lbs., and in another week add, of rectified spirit, 1 gall.; rum, ½ gall.; and bitter almonds, grated, 1½ oz.; bung down for 4 months, then draw it off into another cask, add of brandy, 1 gall., and in 3 months bottle it.

4. Teneriffe, slightly flavoured with cherry-laurel or bitter almonds, forms an excellent ' British sherry,' either alone or diluted with an equal quantity of cape or raisin wine, or good perry.

BRITISH TOKAY. To good cider, 18 galls., add of elderberry juice, ½ gall.; honey, 28 lbs.; sugar, 14 lbs.; red argol (powdered), ¾ lb.; crystallised tartaric acid, 3 oz.; mix, boil, ferment, and, when the active fermentation is complete, add of brandy, 1 gall., and suspend in the liquor, from the bung-hole, a mixture of cassia and ginger, of each, ½ oz.; cloves and capsicum, of each, ¼ oz.; the whole bruised and loosely enclosed in a coarse muslin bag. It will be ripe in 12 months.

Obs. Some of the preceding formulæ, by skilful management, produce very good imitations of some of the imported wines; but (prejudice aside) many of the British fruit wines possess an equally agreeable flavour, and are frequently more wholesome. All British wines require to be kept at least a year to 'mellow.' Much of the superiority of foreign wines arises from their age.

WINES (Culinary). *Syn.* WINES FOR KITCHEN USE. These are prepared in a similar manner to the MEDICATED WINES noticed below.

Wine, Basil. *Prep.* From green basil leaves, 4 or 5 oz.; sherry, cape, or raisin wine, 1 pint; digest for 10 days, press, and strain. Used to give a turtle flavour to soups and gravies. In a similar way may be made the wines of celery leaves, celery seed, sage, shallots, and the various green and dried herbs used in cookery.

Wine, Cayenne. *Prep.* From capsicum or cayenne, 1 oz.; cape, 1 pint; steep for a fortnight, and strain.

Obs. In a similar way may be made currie (powder), ragout (spice), and several other similar wines used in the kitchen.

WINES (Medicated). *Syn.* IMPREGNATED WINES; VINA MEDICATA, L. The medicated wines of pharmacy are prepared by cold maceration, in well-closed vessels, in precisely the same way as the tinctures. In the Ph. L. of 1824 a diluted spirit was substituted for wine, without altering the name of the preparation; but the use of wine (sherry) was restored in that of 1836. The druggists commonly use cape or raisin wine as a menstruum, from its being cheaper than sherry, and perhaps scarcely less power as a solvent. The ' vinum ' of the Ph. U. S. was formerly Teneriffe. Dr B. Lane's process for preparing medicated wines by fermentation is noticed at the end of the alphabetical list given below.

" Medicated wines should be kept in stoppered glass vessels, and be frequently shaken during maceration" (Ph. L.).

The following are the principal medicated wines at present in use:

Wine of Acetate of Iron. *Syn.* VINUM FERRI ACETATIS (*Soubeiran*), L. *Prep.* Acetate of iron, 32 gr.; white wine, 16 oz.

Wine, Alkaline Diuretic. *Syn.* VINUM ALKALINUM DIURETICUM (*Sydenham*), L. *Prep.* Ashes of broom, 12 oz.; Rhenish wine, 4 pints.—*Dose*, 3 oz. twice a day.

Wine of Al'oes. *Syn.* VINUM ALOËS (B. P.), TINCTURA SACRA†, TINCT. HIERÆ-PICRÆ†, L. *Prep.* 1. (B. P.) Socotrine aloes, 1½ oz.; ginger, in coarse powder, 80 gr.; cardamom seeds, bruised, 80 gr.; sherry, 40 oz.; digest seven days, strain, and make it up to 40.—*Dose*, 1 to 2 dr.

2. (Ph. L.) Powdered Socotrine or hepatic aloes, 2 oz.; powdered canella, ¼ oz.; sherry, 1 quart; macerate for 14 days, and filter. In the Ph. E. cardamoms and ginger, of each, 1½ dr., are substituted for canella.—*Dose.* As a purgative, ½ to 2 fl. oz.; as a stomachic, 1 to 2 fl. dr.

Wine of Aloes (Al'kaline). *Syn.* VINUM ALOËS ALKALINUM, L. *Prep.* (Dr *A. T. Thomson*.) Carbonate of soda, 3 oz.; myrrh and extract of aloes, of each, 6 dr.; sesquicarbonate of ammonia, 4½ dr.; sherry, 24 fl. oz. (say 1½ pints); macerate as before. In dyspepsia, chlorosis, &c.—*Dose.* As the last.

Wine, Antimo"nial. *Syn.* TARTAR EMETIC WINE, WINE OF POTASSIO-TARTRATE OF ANTIMONY; VINUM ANTIMONII POTASSIO-TARTRATIS (Ph. L.), V. ANTIMONIALE (B. P., Ph. E.), L. *Prep.* 1. (B. P.) Tartarated antimony, 2 gr.; sherry, 1 oz.—*Dose*, 10 to 60 minims. (In consequence of the insolubility of the tartarated antimony in the sherry, Squire recommends it to be dissolved in about ten times its weight of hot water, and that the wine be added to the solution.)

2. (Ph. L. & E.) Potassio-tartrate of antimony, 40 gr.; sherry, 1 pint; dissolve. Each fl. oz. contains 2 gr. of emetic tartar.—*Dose.* As a diaphoretic and expectorant, 10 to 30 drops, frequently; as a nauseant, 1 to 2 fl. dr.; as an emetic, 2 to 4 fl. dr. The corresponding compound of the Ph. D. is ANTIMONII TARTARIZATI LIQUOR. See SOLUTION OF POTASSIO-TARTRATE OF ANTIMONY.

Wine, Antiscorbutic. *Syn.* VINUM ANTISCORBUTICUM (P. Cod.), L. *Prep.* Fresh horseradish root, 3 oz.; scurvy-grass, 1½ oz.; water-cress leaves, 1½ oz.; buckbean, 1½ oz.; mustard seed, 1½ oz.; chloride of ammonium, 5½ dr.; wine, 5 pints; compound spirit of scurvy-grass, 1½ oz.

Wine, Aromatic. *Syn.* VINUM AROMATICUM (P. Cod.), L. *Prep.* Aromatic species, 1 oz.; vulnerary tincture, 1 oz.; red wine, 10 oz. For outward use. M. Ricord sometimes adds from 1 to 6 per cent. of tannin.

Wine of Bark. *Syn.* VINUM CINCHONE (P. Cod.), L. Yellow bark, 3 oz.; proof spirit, 6 oz. (by weight). Macerate 24 hours, and add red wine, 5 pints. Macerate for 10 days, shaking it occasionally; strain with expression, and filter.

Wine of Bark (Compound). *Syn.* VINUM CINCHONE COMPOSITUM (P. Cod.), L. *Prep.* Yellow bark, 1 oz.; bitter orange peel, 44 gr.; chamomiles, 44 gr.; alcohol ('864), 1 oz. (by weight); white wine, 9 oz. (by weight). Macerate for 10 days.

Wine of Bark, Muriated. *Syn.* VINUM CIN-CHONE MURIATUM, L. *Prep.* Ammonio-citrate of iron, ½ oz.; wine of pale Peruvian bark, 5 pints (made with double the quantity of bark contained in the yellow); dissolve the ammonio-citrate in twice its weight of distilled water, and add to the wine.

Wine of Beef and Iron ('Pharmaceutical Era'). *Prep.* Extract of beef, 2 tr. oz.; phosphate of iron (soluble scale), 4½ tr. oz.; tincture of orange, 2 fl. oz.; essence of lemon, ½ fl. oz.; syrup (simple), 26 fl. oz.; alcohol, 21 fl. oz.; hot water, q. s.; wine (native), to make 128 fl. oz. Dissolve the extract of beef and the phosphate of iron, each separately, in about 8 oz. or more of hot water. Mix the solutions, and when cold add the wine, tincture of orange, and essence of lemon, and filter. To the filtrate add the syrup and alcohol, previously mixed.—*Dose*, ½ to 2 oz.

Wine of Beef, Iron, and Cinchona. *Syn.* VINUM CARNIS, FERRI, ET CINCHONE, L. *Prep.* Extract of beef, 256 gr.; citrate of iron and ammonium, 64 gr.; sulphate of quinine, 16 gr.; sulphate of cinchonine, 8 gr.; citric acid, 6 gr.; water, ½ fl. oz.; simple elixir, 4 fl. oz.; angelica wine, enough to make 16 fl. oz. Dissolve the phosphate of iron in ½ fl. oz. of boiling water, add to the hot solution the citric acid and the sulphates of quinine and cinchonine, and, when they are dissolved, pour the hot solution upon the extract of beef contained in a mortar or other suitable vessel. Triturate the liquid and the extract until they form a smooth mixture; then gradually add, while stirring, the simple elixir, and transfer the mixture to a graduated vessel, using enough angelica wine to rinse the mortar and to make the product measure 16 fl. oz. Allow the mixture to stand during a few hours, then filter.—*Dose*, ½ to 2 oz.

Wine of Buchu. *Syn.* VINUM BUCHU (*Brandes*), L. *Prep.* Buchu leaves, 2½ oz.; white wine, 1 pint.

Wine, Camphorated. *Syn.* VINUM CAMPHO-RATUM (Ph. G.), L. *Prep.* Camphor and gum acacia, in powder, of each, ½ oz. Mix accurately and gradually. Add 24 oz. (by weight) of white wine.

Wine of Cascarilla. *Syn.* VINUM CASCA-

RILLE (*Bernardeau*), L. *Prep.* Cascarilla, 1 oz.; Malaga wine, 1 pint.—*Dose*, 1 oz., twice a day in consumption.

Wine of Catechu. *Syn.* VINUM CATECHU (*Soubeiran*), L. *Prep.* Tincture of catechu, 1 part; red wine, 12½ parts. Mix, and after a few days filter.

Wine of Centaury (Compound). *Syn.* VINUM CENTAURII COMPOSITUM, HOFFMAN'S ELIXIR VIS-CERALE, L. *Prep.* Centaury, orange peel, extract of blessed thistle, gentian, myrrh, cascarilla, of each, 1 dr.; sherry, 2 pints.

Wine, Chalyb'eate. See WINE OF IRON.

Wine of Cinnamon. *Syn.* VINUM CINNAMOMI (*Beral*), L. *Prep.* Cinnamon, 1 oz.; alicant wine, 16 oz.; macerate and filter. Sugar is sometimes added.

Wine of Citrate of Iron. *Syn.* VINUM FERRI CITRATIS (B. P.), L. *Prep.* Dissolve 160 gr. of citrate of iron and ammonia in 1 pint of orange wine; let the solution remain for 3 days in a closed vessel, shaking occasionally; then filter.

Wine, Coca. *Syn.* VINUM COCA, L. *Prep.* Coca leaves, 3 oz.; brandy, 1½ oz.; sherry, 24 oz.; tokay wine, 6 oz.; macerate for about a week, press, and to the liquor add 8 gr. of citric acid; set aside for a few days more, and filter.

Wine of Col'chicum. *Syn.* WINE OF COL-CHICUM ROOT; VINUM COLCHICI (B. P.), VINUM RADICIS COLCHICI (Ph. L. & E.), L. *Prep.* 1. (B. P.) Colchicum corms, dried and sliced, 4 parts; sherry, 20 parts; macerate 7 days, and strain.—*Dose*, 20 to 30 minims.

2. (Ph. L.) Dried corms of meadow saffron (sliced), 8 oz.; sherry wine, 1 quart; macerate 7 days, and strain (press strongly the residuum and filter the mixed liquor—Ph. E.). A powerful sedative and purgative.—*Dose*, ½ to 1 fl. dr.; in gout, acute rheumatism, and other painful and inflammatory and nervous affections.

Obs. The celebrated EAU MÉDICINALE of M. Husson (AQUA MEDICINALIS HUSSONII) resembles, in composition and action, the above preparation in every point except its strength, which, we believe, is much above that of the wine of the British Colleges.

Wine of Colchicum Seed. *Syn.* VINUM SEMINIS COLCHICI, V. SEMINUM C., L. *Prep.* (Ph. U. S. and *Dr Williams*.) Seeds of meadow saffron (preferably ground in a coffee-mill), 2 oz.; sherry, 16 fl. oz.; macerate for 14 days.—*Dose*, 1 to 1½ fl. dr.; in gout, &c.

Wine of Colocynth. *Syn.* VINUM COLOCYNTHI-DIS (*Van Mons*), L. *Prep.* Colocynth, 2 oz. white wine, 24 oz.; macerate for 8 days, an filter.

Wine of Elecampane. *Syn.* VINUM INULÆ, As wine of wormwood.

Wine, Emet'ic. See ANTIMONIAL WINE, W OF IPECACUANHA, &c.

Wine of Ergot. *Syn.* VINUM ERGOTE (U. S.), L. *Prep.* Fluid extract of ergot (Ph. S.), 2 oz. (o. m.); white wine, 14 oz. (o. m.); r and filter.

Wine of Foxglove. *Syn.* VINUM DIGITA (Ph. Port.), L. *Prep.* Dried foxglove, 1 o good white wine, 32 oz.; macerate for 4 days, strain.

Wine of Gen'tian. *Syn.* BITTER WINE, To

W.; **Vinum Amara,** V. **Gentiane** (Ph. E.),
L. *Prep.* (Ph. E.) Gentian, in coarse powder,
½ oz.; yellow bark (do.), 1 oz.; dried orange peel,
2 dr.; canella, in coarse powder, 1 dr.; proof
spirit, 4½ fl. oz.; digest for 24 hours, then add of
sherry, 1 pint and 16 fl. oz., and further digest for
7 days. Tonic and stomachic.—*Dose,* ½ to ½ fl. oz.

Wine of Hedge-hyssop. *Syn.* **Vinum Gratiole**
(*Niemann*), L. *Prep.* Hedge-hyssop, 2 dr.; white
wine, 16 dr.; digest at a gentle heat for 4 hours,
and strain.—*Dose,* 1 oz.; frequently in hypochondriasis.

Wine of Hellebore. See **Wine of White
Hellebore.**

Wine of Holly. *Syn.* **Vinum Ilicis** (*Rousseau*),
L. *Prep.* Powdered holly leaves, 2 dr.; white
wine, 6 oz.; infuse for 12 hours.

Wine of Iodide of Iron. *Syn.* **Vinum Ferri
Iodidi** (*Pierquin*), L. *Prep.* Iodide of iron, 4 dr.;
Bordeaux wine, 1 pint.

Wine of Ipecacuan'ha. *Syn.* **Emetic Wine;
Vinum Ipecacuanhæ,** L. *Prep.* 1. (B. P.) Ipecacuanha, 1 part; acetic acid, 1 part; distilled
water, a sufficiency; sherry, 20 parts; macerate
the ipecacuanha in the acid 24 hours, transfer
to a percolator, and pass water through to produce 1 pint. Evaporate to dryness, dissolve residue in the sherry, and filter.—*Dose,* 10 to 40
minims as an expectorant, 3 to 6 dr. as an
emetic.

2. Ipecacuanha root, bruised, 2½ oz.; sherry,
1 quart; macerate for 7 days (14 days, and strain
with expression—Ph. D.). This is a mild and
excellent preparation.—*Dose.* As a diaphoretic
and expectorant, 10 to 40 drops, in coughs, diarrhœa, dysentery, dyspepsia, &c.; as an emetic,
2 fl. dr. to 1 fl. oz., in divided doses; as an emetic
for infants and young children, ½ teaspoonful
every 10 or 15 minutes until it operates.

Wine of I'ron. *Syn.* **Chalybeate Wine, Steel
Wine; Vinum Ferri** (B. P., Ph. L.), L. *Prep.*
1. (B. P.) Fine iron wire (No. 35), 1 oz.; sherry,
20 oz.; digest 30 days with frequent agitation.
The bottle to be corked, but the wire not wholly
immersed.—*Dose,* 1 to 4 dr.

2. (Ph. L.) Iron wire, 1 oz.; sherry, 1 quart;
digest, with frequent agitation, for 30 days, and
strain. Each fl. oz. contains less than 1½ gr. of
metallic iron.

3. Ammonio-tartrate of iron (*Aikin's*), 1½ dr.;
sherry, 1 pint; dissolve. Frequently substituted
for the last, especially when the preparation is
required in a hurry.—*Dose,* 1 to 5 fl. dr.; as a
mild chalybeate.

Obs. The formula for **Wine of Iron** was modified in the Ph. L. 1824, omitted in that of 1836,
and restored, in its original character, in that
of 1851.

Wine of Liquorice. *Syn.* **Vinum Glycyrrhizæ,**
L.; **Fuller's Sweet Tincture.** *Prep.* Liquorice
(Italian juice), 1 oz.; cochineal, 2 scruples;
canary wine, 2 pints. Sometimes 1 dr. of saffron
is added.

Wine of Malate of Iron. Iron wire steeped in
cider.

Wine of Mea'dow Saf'fron. See **Wine of Colchicum.**

Wine of O''pium. *Syn.* **Sydenham's Liquid
Laudanum†; Vinum Opii** (B. P., Ph. L., E., &

D.), **Tinctura Thebaica†, Laudanum Liquidum
Sydenham†,** L. *Prep.* 1. (B. P.) Extract of
opium, 1 oz.; cinnamon bark, 75 gr.; cloves,
75 gr.; sherry wine, 20 oz.; macerate for 7 days,
and filter.—*Dose,* 10 to 40 minims.

2. (Ph. L.) Extract of opium (Ph. L.), 2½ oz.;
cinnamon and cloves, of each, bruised, 2½ dr.;
sherry, 1 quart; macerate for 7 days (14 days—
Ph. D.), and filter. In the Ph. E., opium, 3 oz.;
and in the Ph. D., opium, in coarse powder, 3 oz.,
are ordered, instead of extract of opium. The
Dublin College also omits the aromatics.

3. (Wholesale.) From extract of opium, 11 oz.;
oil of cassia, 25 drops; oil of cloves, 20 drops;
wine, 1 gall. (or rectified spirit, 1½ pints; water,
6½ pints; colouring, q. s.); digest, with agitation, until dissolved. Milder than the tincture.—
Dose, 10 to 40 drops; as an anodyne and hypnotic.

Wine of Opium (Fermented). *Syn.* **Rousseau's
Laudanum, Black Drop; Vinum Opii Fermen-
tatione Paratum, Gutta Nigra,** L. *Prep.*
(P. Cod.) Opium, 4 oz.; boiling water, 5 lbs.;
dissolve; add of honey, 1 lb.; yeast, 2 dr.; keep
it at 86° F. for a month, or until the fermentation
is complete; then press, filter, distil off 16 oz.,
and evaporate the residuum to 10 oz.; distil the
16 oz. of spirit obtained above until 12 oz. have
passed over, and from this, by a third distillation,
obtain 4½ oz., which add to the evaporated solution (10 oz.), and filter. About four times as
strong as tincture of opium. See **Drop, Black.**

Wine of Orange. *Syn.* **Vinum Aurantii** (B.
P.), L. *Prep.* Made in Britain by the fermentation of a saccharine solution, to which the
fresh peel of the bitter orange has been added.
Contains 12% of alcohol, and is but slightly acid
to test-paper.

Wine of Pepsin. *Syn.* **Vinum Pepsini** (Ph.
G.), L. *Prep.* 1. Remove by hard scraping, by
means of a bone knife, the pepsin from the
mucous membrane of a previously washed, freshly
killed pig's or ox's fourth stomach, and mix 10
dr. of it with 5 dr. (by weight) of glycerin diluted
with 5 dr. of water; put into a large flask and
shake up vigorously with 13½ oz. (by weight) of
white wine, and ½ dr. (by weight) of hydrochloric
acid. Macerate for 3 days at 68° F., frequently
shaking, and filter.

2. Two stomachs are sufficient for a pint of the
wine; open these and wash slightly, then scrape
off the mucous surface and macerate for 2 days
in a mixture of hydrochloric acid, 1 dr.; water, 5
oz.; and glycerin, 2 oz.; then add 12 oz. of sherry
and 1 oz. rectified spirit; macerate for 5 or 6
days. Most of the wine can be poured off perfectly bright, and the rest may be filtered in the
ordinary manner.

Wine of Potas'sio-tartrate of Antimony. See
Wine, Antimonial (above).

Wine of Quinine. *Syn.* **Vinum Quinæ** (B.
P.), L. *Prep.* 1. (B. P.) Sulphate of quinine,
20 gr.; citric acid, 30 gr.; orange wine, 20 oz.;
dissolve the citric acid and then the sulphate of
quinina in the wine; digest 3 days and filter.—
Dose, ½ to 1 oz.

2. (*Magendie.*) Sulphate of quinine, 14 gr.;
sherry, 1 quart; agitate frequently for some time.
"The sulphate of quinine requires to be dissolved

in a little dilute sulphuric acid before it is added to the wine" (*Dr Hayes*).—*Dose*, 1 wine-glassful, as a tonic and stomachic.

Wine of Quinine, Aromatic. *Syn.* VINUM QUINIÆ AROMATICUM, L.; DR COLLIER'S AROMATIC QUININE WINE. *Prep.* Disulphate of quinine, 18 gr.; citric acid, 15 gr.; sound orange wine, 1 bottle (24 fl. oz.).

Wine of Rhu'barb. *Syn.* VINUM RHEI (B. P., Ph. E. & D.), TINCTURA RHEI VINOSA, L. *Prep.* 1. (B. P.) Rhubarb in coarse powder, 1½ parts; canella bark, ½ part; sherry, 20 parts; macerate 7 days, filter, and make up to 20 parts. —*Dose*, 1 to 2 dr.

2. (Ph. E.) Rhubarb, in coarse powder, 5 oz.; canella, in coarse powder, 2 dr.; proof spirit, 5 fl. oz.; sherry, 1½ pints; macerate for 7 days, press, and filter.

3. (Ph. D.) Rhubarb, 3 oz.; canella, 2 dr.; sherry, 1 quart; macerate 14 days. Weaker than the last.—*Dose*. As a stomachic, 1 to 3 fl. dr.; as a purgative, ½ to 1 fl. oz., or more. It does not keep well.

Wine of Sarsaparilla. *Syn.* VINUM SARSAPARILLÆ (*Beral*), L. *Prep.* Alcoholic extract of sarsaparilla, 1 oz.; white wine, 16 oz.

Wine of Senna. *Syn.* VINUM SENNÆ (Ph. Swed.). *Prep.* Senna, 4 oz.; coriander seed, 2 dr.; fennel seed, 2 dr.; sherry, 2½ lbs. Digest for 8 days, add stoned raisins, 3½ oz. Macerate for 24 hours, and strain with expression.

Wine of Squills. *Syn.* VINUM SCILLÆ (P. Cod.), L. *Prep.* Dried squills, 3 oz.; Malaga wine, 2½ pints. Macerate for 10 days.

Wine of Squills, Bitter. *Syn.* VINUM SCILLITICUM AMARUM (P. Cod.), L. *Prep.* Pale Peruvian bark, 6 oz.; winter's bark, 6 oz.; lemon peel, 6 oz.; swallowwort, 1½ oz.; angelica root, 1½ oz.; squill, 1½ oz.; wormwood, 3 oz.; balm, 3 oz.; juniper berries, 1½ oz.; mace, 1½ oz.; white wine, 2½ galls.; proof spirit, 1 pint. Macerate for 10 days.

Wine of Squills, Compound. *Syn.* VINUM SCILLÆ COMPOSITUM (*Richter*), L. *Prep.* Dried squills, 1 oz.; orange peel, 3 dr.; juniper berries, 2 dr.; white wine, 2½ pints. Digest for 3 days, filter, and add 2 oz. of oxymel of squills.

Wine of Stramonium. *Syn.* VINUM STRAMONII (Ph. Bat.), L. *Prep.* Stramonium seeds, 2 oz.; Malaga wine, 8 oz.; rectified spirit, 1 oz. Digest and filter.

Wine of Tobac'co. *Syn.* VINUM TABACI (Ph. E.), L. *Prep.* (Ph. E.) Tobacco, 3½ oz.; sherry, 1 quart; digest 7 days, strain, with strong pressure, and filter. A powerful sedative and diuretic. —*Dose*, 10 to 30 drops; in dropsy, lead colic, ileus, &c.

Wine of White Hel'lebore. *Syn.* VINUM VERATRI (Ph. L.), TINCTURA VERATRI ALBI†, L. *Prep.* (Ph. L.) White hellebore, sliced, 8 oz.; sherry wine, 1 quart; digest for 7 days, press, and filter. —*Dose*, 10 drops, gradually increased to 25 or 30; as a substitute for colchicum, in gout and rheumatism, &c. It is less manageable than wine of colchicum, and is now seldom employed.

Wine of White Hellebore (Opiated). *Syn.* MOORE'S EAU MÉDICINALE; VINUM VERATRI OPIATUM, L. *Prep.* From wine of white hellebore, 3 fl. dr.; tincture of opium, 1 fl. dr.—*Dose*. As the last.

Wine of Wormwood. *Syn.* VINUM ABSINTHII (P. Cod.), L. *Prep.* Dried wormwood leaves, 3 oz.; white wine, 5 pints; proof spirit, 6 oz. Macerate the leaves in the spirit, in 24 hours add the wine, macerate for 10 days, and strain.

Wines, Medicated (Dr B. Lane's). *Syn.* VINOUS ESSENCES; ESSENTIÆ VINOSÆ, LIQUORES VINOSI, L. *Prep.* From an infusion or solution of the drug, of about 3 or 4 times the usual strength, fermented with a little yeast, and about 3 or 4 lbs. of sugar per gall.; the fermented liquor being afterwards set in a cool cellar until fit for bottling. Compounds of CALUMBA, CASCARILLA, GENTIAN, OPIUM, RHUBARB, SENNA, and VALERIAN, have been thus prepared. That of OPIUM is made of only twice the strength of the common tincture.

WINE-STONE. CRUDE TARTAR OF ARGOL.

WINE TESTS. *Prep.* 1. (*Hahnemann's*.) From quicklime, 1 oz.; flowers of sulphur, 1½ oz.; mix, and heat them in a covered crucible for 5 or 6 minutes; put 2 dr. of the product and an equal weight of tartaric acid (separately powdered) into a stoppered bottle, with a pint of water, and shake them well; let the liquid settle, pour off the clear portion, and add of tartaric acid, 1½ dr.

2. (*Dr Paris's*.) From sulphide of calcium and cream of tartar, of each (in powder), ½ oz.; hot water, 1 pint; agitate, &c., as before; decant the cold clear liquid into 1-oz. phials, and add 20 drops of hydrochloric acid to each of them.

Obs. The above tests will throw down the least quantity of lead from wines, as a very sensible black precipitate. As iron might be accidentally contained in the wine, the hydrochloric acid is added to the last test, to prevent the precipitation of that metal.

WIN'TER-GREEN (American). *Syn.* PIPSISSEWA; CHIMAPHILA (Ph. L. & E.), PYROLA (Ph. D.), L. The herb of *Chimaphila umbellata*. It is astringent, diuretic, tonic, and stomachic; and has been successfully administered in loss of appetite, dyspepsia, dropsy, chronic affections of the urinary organs, scrofula, &c. It must not be confounded with ordinary winter-green (box berry, chequer b., partridge b., mountain tea), which is the *Gaultheria procumbens*, a plant belonging to the *Ericaceæ*, whilst the former plant belongs to a genus of the *Pyrolaceæ*. See DECOCTION, EXTRACT, and OILS (Essential).

WIREWORM. Dr Spencer Cobbold, F.R.S., made the following communication on this subject to Dr Tuson:—"Dear Professor Tuson,—You asked me about the remedies for wireworm. Although a great deal has been said on the subject, yet it is not easy to advise. I believe the *best plan* is to 'catch-'em-alive' by means of sliced potatoes, turnips, or carrots laid in rows, women and children being employed every morning to pick up the slices, and brush off the larvæ into a jar (the slices being replaced). Mr Hogg (the Ettrick Shepherd) found lettuce leaves very serviceable when laid as a bait in a similar way. Pheasants are very destructive to them. As agriculturists do not like the trouble and expense of this baiting method (by far the best if persevered in) some have recommended deep ploughing, &c. The following extract taken from the 'Journal of the Agricultural Society of Victoria,' bears on the question at issue. Trusting it may be found useful, believe me, yours faithfully, T. SPENCER COBBOLD."

"*Remedy for Wireworm.*—Having seen in your issue of the 24th ult. that 'B.' would be glad if any one could give any information as to a remedy for the ravages of the wireworm, which plays such havoc in our corn fields during the early part of the growth of our cereal crops, I beg to offer a few observations on the subject. I have for years paid particular notice as to any remedy or preventive, and it is with regard to the latter that I shall chiefly confine my remarks, as there is positively no known remedy when once the insects have attacked the crop. Some persons recommend the application of lime or salt, but it is a well-known fact that if either of these is applied in such quantity as to destroy the worm, it will likewise destroy vegetation, and consequently the crop will be entirely lost; and not alone this crop, but the soil will be poisoned to such an extent as to injure succeeding crops. What I have found most successful is deep ploughing, not what is ordinarily called deep ploughing, 7 or 8 inches, but to the depth of at least 10 or 12, where the soil will admit of it. The wireworm lives not more than 4 inches below the surface, and by burying it 10 or 12 inches it is found that it cannot again make its way to the surface, and consequently can do no injury to surface-rooted plants, such as the grain crops. The operation of ploughing should be performed as follows: a strong skim coulter is attached to the beam of an ordinary strong plough, which is drawn by three horses. The skim coulter pares off the surface, which is buried underneath the sod turned over by the mould-board. Or it is sometimes performed in a different way. A small plough, drawn by one horse, precedes the ordinary plough, skimming off the surface exactly the same as the skim coulter. So much for a preventive. As to remedy, what I have found most effectual is heavy rolling, using, if possible, such a roller as that called the Crosskill, which crushes the insects, killing some, and preventing others doing much damage until the crop is sufficiently far advanced as not to be affected by the insect.—JOHN THOMAS, 32, Capel Street, Dublin."

WITCH MEAL. *Syn.* VEGETABLE SULPHUR, LYCOPODIUM. The spores of *Lycopodium clavatum*, or club-moss.

WOAD. *Syn.* DYER'S WOAD; PASTEL, Fr. The *Isatis tinctoria.* To prepare them for the dyer, the leaves are partially dried and ground to a paste, which is made into balls; these are placed in heaps, and occasionally sprinkled with water, to promote the fermentation; when this is finished, the woad is allowed to fall down into lumps, which are afterwards reground and made into cakes for sale. On mixing the prepared woad with boiling water, and, after standing for some hours in a closed vessel, adding about 1-20th its weight of newly slaked lime, digesting in a gentle warmth, and stirring the whole together every 3 or 4 hours, a new fermentation begins; a blue froth rises to the surface, and the liquor, though it appears itself of a reddish colour, dyes woollens of a green, which, like the green from indigo, changes in the air to a blue. This is said to be one of the nicest processes in the art of dyeing, and does not well succeed on the small scale. Woad is now mostly used in combination with indigo. 50 lbs. of woad are reckoned equal to 1 lb. of indigo.

WOLFRAM. See TUNGSTEN.

WOLFSBANE. See ACONITE.

WOOD is polished by carefully rubbing down the grain with fine glass paper, or pumice-stone, and then rubbing it, first with finely-powdered pumice-stone and water, and afterwards with tripoli and linseed oil, until a proper surface is obtained. For common purposes, glass paper, followed by a metal burnisher, is employed.

Wood is stained by the application of any of the ordinary liquid dyes employed for wool or cotton. They sink deeper into the wood when they are applied hot. When the surface is properly strained and dried, it is commonly cleaned with a rag dipped in oil of turpentine or boiled oil, after which it is either varnished or polished with beeswax. Musical instruments, articles of the toilette, &c., are usually treated in this way.

Wood is preserved by any agents which destroy the tendency to putrefaction of the matter within its pores, or which enables it to resist the attacks of insects, or renders it unsuited to the growth of minute fungi. See DRY-ROT.

WOOD-APPLE (*Feronia elephantum*, Corr). A large Indian tree, the wood is used for house building, agricultural implements, &c.

WOOD NAPHTHA. See SPIRIT (Pyroxylic).

WOOD OIL. See BALSAM, GURGUN.

WOODS. List of some of those most used, and the purposes to which they are applied.

Acacia or **Locust Wood.** From India, the West Indies, and tropical regions of Africa. A dark coloured wood resembling mahogany.

African Oak. Formerly used in shipbuilding, but its use has died out. It is a hard wood of a dark red colour.

Almond Wood. From the North of Africa and parts of Europe and Asia bordering on the Mediterranean. Is a very hard, dense wood like lignum vitæ. Used for teeth and bearings of cogwheels.

Aloes Wood, Calambak, or **Green Sandal-wood.** From tropical countries.

Amboyna Wood. Sometimes called *Kaibooca wood*, from the Moluccas. It has the appearance of being the burr of a large tree, used in inlaying and veneering.

Bird's-eye Maple. From Prince Edward's Island. Used chiefly for picture frames.

Black Birch. North America. Used for cabinet-making and household furniture.

Black Walnut. Used for fret-sawing, and all kinds of cabinet work.

Botany Bay Wood or **Beefwood.** From Botany Bay. Is a dense, hard, heavy wood of a black colour. Chiefly used for ornamental turning.

Box. From Turkey. Used by turners for fancy articles, lathe chucks, &c.

Butternut. From United States. Used for cabinet work.

Camphor Wood. Used for boxes for preserving furs, &c., from the attacks of moths.

Canary Wood. From South America; a straight grained wood of yellow colour, used by turners and cabinet makers.

Cane Wood. From Southern Africa; is very hard, and of a close, fine texture; colour, a rich reddish brown. Used for ornamental turning.

Cocabola Wood. A hard and resinous wood of a red colour, chiefly used for inlaying.

Cocus Wood. From the West Indies; is a hard wood, and of a black colour when cut, and exposed to the action of the air; used in turnery.

Coral Wood. From tropical countries; is a hard and close-grained wood of a yellow colour when first cut, but changes to a rich coral red. Used in turning and fancy cabinet work.

Coromandel Wood. From Southern India and Ceylon; colour, a rich hazel-brown streaked with black. Used in cabinet work.

Ebony (Black). Africa, East Indies, and the Mauritius; used by turners and cabinet makers.

Fustic. A wood used for dyeing, and for turning and inlaying. Colour, a greenish yellow.

Greenheart. From the West Indies and Brazil; is a coarse and heavy wood used in shipbuilding; is of a brownish-green colour when first cut, but darkens on exposure to the air.

Hickory. From the United States; is a tough and elastic wood used for shafts of carriages, spokes of wheels, &c.

Hornbeam. An American wood; very strong and tough, colour white, hard, and close-grained. Used for teeth of cog-wheels, &c., for skittle-pins. The English hornbeam is a different species.

Ironbark. From Australia. A species of the *Eucalyptus*, used formerly in shipbuilding.

Ironwood. From South America, East and West Indies; is very hard and straight grained. Frequently used for making ramrods, colour reddish brown.

Jarrak. From Australia. Used in Australia for railway work, &c.

Kingwood. From Brazil; is a hard and durable wood. Used for turning and small cabinet work. Sometimes called violet wood through being beautifully streaked in tints of violet.

Lancewood. From the West Indies. A very tough wood. Used for shafts of carriages in general. A good wood for bending when steamed.

Lignum Vitæ. From Central America; is a kind of box-tree; is heavy, hard, and close grained. The medicinal resin called guaiacum is obtained from it. Used for sheaves of pulleys, &c.

Logwood. From South America; is the heavy, red heartwood of a tree used for dyeing. Sometimes called campeachy wood.

Mahogany. From the West Indies and Central America. Used by joiners, cabinet-makers, and turners.

Nettle Tree. From Southern Europe. A close grained wood used by musical instrument makers.

Olive Wood. Of a close, fine grain. Used for fretwork, carved work, and ornamental cabinet work.

Partridge-wood. From Brazil; very hard and heavy, close grain. Used for walking-sticks and handles of umbrellas and parasols.

Pomegranate Wood. Is of a brownish-green colour, and used chiefly by musical instrument makers.

Purple Heart. The wood of *Copaifera Martii*, Hayne, var. *pubiflora*. A large timber tree of British Guiana, where the wood, which is of a beautiful purple colour when freshly cut, is used for structural purposes on account of its great strength and durability.

Red Satinwood. From the East Indies; is a hard wood used for marquetry; colour, a beautiful reddish purple, with veins and markings of a darker tint.

Rosetta Wood. From the East Indies; is of a bright, reddish orange colour, marked with veins and streaks of a darker tint; used for ornamental cabinet work, but is rather scarce.

Rosewood. Rio Janeiro, East Indies, and Canary Islands. Used for ornamental furniture, and by turners.

Sandal-wood. From the East Indies; is a highly scented wood, like cedar. Used for ornamental boxes for furs, gloves, &c.

Satinwood. From the East Indies; is of a soft and lustrous appearance, with a yellowish tint. Used for marquetry and inlaying.

Service Wood. Has a hard and close grain. Used for making and handling joiners' tools.

Teak. From the East Indies; is a very hard and durable wood. Used chiefly in shipbuilding.

Tulip Wood. From Brazil; resembling rose-wood, of a reddish colour striped with darker shades. Used for marquetry.

West Indian Ebony. Furnished by *Brya ebenus ?*, A. DC., a small tree of Jamaica. It takes a beautiful polish, and is used for making walking-sticks, inlaying, &c. Cocus wood, used for making flutes and flageolets, is supposed to be produced by this plant.

Whitewood. From North America. Used for ordinary cabinet work and fret-sawing.

WOODY FIBRE. See LIGNIN.

WOODY NIGHT'SHADE. BITTER-SWEET, *Solanum dulcamara.*

WOOL. *Syn.* LANA, L. Wool is a fine, soft, elastic variety of hair, cellulated in its structure. Its filaments are cylindrical, like those of silk; but the surface is covered with thin scales or epidermic cells. In the finer qualities, these filaments vary in thickness from $\frac{1}{1100}$ to $\frac{1}{1000}$ of an inch; and under a good microscope distinctly exhibit, at intervals of about $\frac{1}{750}$ of an inch, a series of serrated rings, imbricated towards each other, "like the joints of equisetum, or, rather, like the scaly zones of a serpent's skin." These appearances render it almost impossible to mistake wool for silk, linen, or cotton. From experiments made by different competent authorities, it is found that wool is one of the worst conductors of heat known. This property renders woollen fabrics particularly adapted for clothing in cold, damp, and changeable climates, since it enables them to maintain the surface of the body at a proper and equable temperature. Wool, sometimes, however, proves too irritative for highly sensitive skins, and, moreover, disturbs the electricity of the cutaneous surface, on friction, even more than silk. On these accounts there are persons who find it unpleasant to wear woollen garments, of any description, next the skin; in which cases all the advantages that can be derived from their use may be obtained by wearing them outside one of linen or cotton. According to Erasmus Wilson, this method "is preferable in warm weather, since the linen absorbs the perspiration, while the woollen garment preserves the warmth of the body, and prevents the inconvenience resulting from its evaporation." See FLANNEL.

Identif. 1. By the microscope. (See *above*.)

2. Its fibres, when inflamed, shrivel up, and burn with difficulty, and evolve the peculiar and characteristic odour of hair when similarly treated, leaving a bulky charcoal; whereas cotton and linen burn rapidly, leave no charcoal, and evolve little or no odour. Silk acts in nearly the same manner as wool.

3. Nitric acid, picric acid, and gaseous chlorine, stain the fibres of wool and silk yellow. Dr J. J. Pohl recommends an aqueous or alcoholic solution of picric acid as well adapted for a test. After immersion of a small portion of the fabric or yarn for from 5 to 10 minutes in the solution, it is to be taken out and washed in warm water. The linen and cotton in it will then appear white, but the wool, silk, or other animal fibre, will retain its yellow colour. In stuffs, cloths, flannel, &c., the mixed threads may be readily observed by means of a pocket lens, and their relative numbers may be ascertained by means of a 'thread counter.' This test acts best with white, but is also applicable to many other colours. For dark colours nitric acid is preferable.

4. A square inch of the fabric or a small portion of the yarn is boiled for a short time in a solution of caustic soda or potass, and then withdrawn and washed. If it be of pure wool, it will be dissolved, and wholly disappear; if any threads or fibres are left undissolved, they consist of either cotton or linen; of these, such as have acquired a dark yellow tinge, are linen, whilst those which have retained their whiteness, or which are only slightly discoloured, are cotton. The relative proportion of the adulteration may be ascertained as before. See COTTON, LINEN, and SILK.

Wool and woollen goods exhibit a greater affinity for colouring matter than either cotton or linen, and in many cases this exceeds even that of silk. The most difficult dye to impart to wool is a rich, deep, and permanent black. See DYEING, MORDANTS, &c.

Woollen goods are cleaned and scoured in the manner noticed under BLEACHING and SCOURING.

Wool, Spanish. Rouge-crepons.

Wool, Styptic. Dr Erle, of Isny, prepares this by boiling the finest carded wool for half an hour or an hour in a solution containing 4% of soda, then thoroughly washes it out in cold spring water, wrings it, and dries it. The wool is thus effectually purified, and is now capable of imbibing fluids uniformly. It is then to be dipped two or three times in fluid chloride of iron diluted with one third of water, expressed and dried in a draught of air, but not in the sun or by the aid of high heat; finally, it is carded out. Thus prepared, it is of a beautiful yellowish-brown colour, and feels like ordinary dry cotton wool.

As it is highly hygroscopic, it must be kept dry, and when required to be transported, must be packed in caoutchouc or bladder. Charpie may be prepared in a similar manner, but on account of its coarser texture, is not so effective as cotton wool, presenting a less surface for producing coagulation. When the wool is placed on a bleeding wound, it induces moderate contraction of the tissue, coagulation of the blood that has escaped, and subsequently coagulation of the blood that is contained within the injured vessels, and thus arrests the hæmorrhage. The coagulating power of the chloride of iron is clearly exalted by the extension of its surface that is in this way affected. The application of the prepared wool is not particularly painful, whilst by sucking up the superfluous discharge, and preventing its decomposition, it seems to operate favourably on the *process of the wound* ('Lancet').

Wool Work, Woollen Shawls, &c., to Clean. Boil a large piece of soap in rain-water. Put it into an earthenware pan and add a teacupful of ox-gall, which any butcher will supply. Put in the work to be cleaned, and rub it briskly, as you would a pocket handkerchief, lifting it up and down. Wash in two waters, if very dirty; then rinse quickly in cold water, lay a cloth over it, and fold it tightly. Iron it immediately on the wrong side with hot, heavy irons.

WOORA'RA. *Syn.* OURARI, WOURALI, L. A deadly poison employed in Guiana, obtained from the *Strychnos toxifera.* See UPAS.

WOOTZ. The Indian name of steel; applied in this country to the steel imported from Bengal.

WORM BARK. *Syn.* CABBAGE-TREE BARK, or GEOFFRÆYA INERMIS, L. The bark of *Andira inermis* (*Geoffraya i.* of some botanists). A powerful astringent, purgative, anthelmintic, and narcotic.—*Dose*, 10 to 30 gr. In larger doses, or if cold water be drunk during its action, it is apt to occasion sickness, vomiting, and delirium. The remedy for this is copious draughts of warm water.

WORM-SEED. *Syn.* SEMEN CONTRA, SEMEN CINÆ, L. The broken peduncles, mixed with the calyces and flower-buds, of several species of *Artemisia* imported from the Levant.—*Dose*, 10 to 30 gr., in powder; as a vermifuge. See SANTONIN.

WORMS (Intes'tinal). *Syn.* VERMES, INTESTINALIA (*Cuvier*), ENTOZOA (*Rudolphi*), L. The principal parasites which are generated and nourished in the human intestinal canal are—the *Acaris lumbricoides*, Gmelin, or long round-worm, found in the small intestines, and which is generally of the thickness of a goose-quill, and varies in length from 10 to 15 inches; *Ascaris vermicularis*, Gmelin, maw, or thread-worms, which is thread-like in appearance, varies from 1¼ to 5 lines in length, and confines itself chiefly to the rectum; *Tricocephalus hominis*, Gmelin, or long thread-worm, varying from 1¼ to 2 inches in length, and found chiefly in the cæcum; *Tænia solium*, Gmelin, or common tapeworm, having a flattened riband-like appearance, varying in length from 3 or 4 to 15 or 20 feet, and occupying the small intestines; *Bothriocephalus latus*, or broad tapeworm, a variety seldom found in this country, but common in Switzerland and the north of

Europe; and *Tænia mediocanellata*, another large species, described by Küchenmeister.

Causes. A debilitated state of the digestive organs, improper food, sedentary habits, impure air, bad water, and, apparently, an occasional hereditary tendency to worms.

Symp. Griping pains, especially about the navel; acid eructations; slimy stools; occasional nausea and vomiting, without any manifest cause; heat and itching about the anus; tenesmus; emaciation; disturbed dreams; grinding of the teeth during sleep; pallor of countenance; discolouration round the eyes; feverishness; headache; vertigo, &c. In many cases these symptoms are often highly aggravated, and mistaken for primary diseases. The only absolutely positive evidence of the existence of worms is, however, their being seen in the fæces.

Treat. In common cases, an occasional moderately strong dose of calomel overnight, followed by a smart purgative the next morning, is an excellent remedy, where the use of mercurials is not contra-indicated. Cowhage, made into an electuary with honey or treacle, is also an excellent vermifuge. Oil of turpentine is useful against nearly every variety of worms, and, when taken in sufficient doses to reach the rectum, is almost specific in ascarides (thread-worms). When this is inconvenient, an occasional enema of oil of turpentine is even more effective. Enemas of aloes are also very useful in such cases. Scammony, under the form of basilic powder, has long been employed to expel worms in children. Aloes, castor-oil, worm-seed, tin-fillings, and sulphur, are likewise popular remedies. Cabbage-tree bark is a powerful anthelmintic; but its use requires caution. Most of the quack vermifuges contain either aloes or gamboge, or calomel and jalap. The substances which have been most highly extolled for the destruction or expulsion of tapeworm (tænia) are kousso, oil of turpentine, male fern, pomegranate, and tin-filings. The first two are those on which the most dependence may be placed. Madame Nouffer's celebrated 'Swiss remedy' for tapeworm, for which Louis XVI gave 18,000 francs, consisting of 2 or 3 dr. of powdered male fern, taken in ¼ pint of water, in the morning, fasting, followed in two hours by a bolus made of calomel and scammony, of each, 10 gr.; gamboge, 6 or 7 gr. 'Swain's vermifuge' is prepared from worm-seed, 2 oz.; valerian, rhubarb, pink-root, and white agaric, of each, 1½ oz.; boiled in water, q. s. to yield 3 quarts of decoction, to which 30 drops of oil of tansy, and 45 drops of oil of cloves, dissolved in a quart of rectified spirit, are added. All purgatives may be regarded as vermifuges. Besides our efforts to destroy and expel the worms, the tone of the primæ viæ should be raised by the use of stomachics and tonics, by which the tendency to their equivocal generation will be either removed or lessened. See ASCARIS LUMBRICOIDES, DECOCTION, ENEMA, PILLS, PATENT MEDICINES, VERMIFUGES, &c.; and also the several vermifuges under their respective names.

Obs. Parasitic worms as existing in animals are so remarkably prevalent and so widely diffused that probably no creature can be said to be secure against their attack. Among domestic animals, sheep often suffer to a most serious extent from these parasites, and more especially from the nematoid known as *Strongylus bronchialis.* In some years lambs are lost by hundreds from the complications of disease which attend upon the presence of these worms within the windpipe and the bronchial tubes. Their existence is marked by great wasting of the body, hurried breathing, and distressing cough. After a time diarrhœa sets in, which quickly carries off the animal. It has been found that lambs fed on clover, and other allied plants, which had been pastured the year previously with sheep, suffer the most, and are far more likely to be affected than those which are differently managed. Remedial measures too often prove ineffectual, especially when structural disease of the lungs has followed as a consequence. The exhibition of oil of turpentine in doses of about half an ounce, mixed with an equal quantity of linseed oil, is sometimes found to be beneficial; but it must be conjoined with a corn diet, the free use of salt, and also sulphate of iron mixed with the manger food, tincture of assafœtida, and the essential oil of savin, in small doses, are remedies greatly extolled by some persons. The inhalation of diluted chlorine gas or of sulphurous acid gas is often exceedingly beneficial; but remedies of this kind ought always to be confided to the superintending care of the veterinary surgeon.

WORM TEA. A preparation sold in the shops of the United States, and much used, consisting, according to Brande, of spigelia or pink-root, savin, senna, and manna.

WORMWOOD. *Syn.* ABSINTHIUM (Ph. L. & E.), L. The flowering herb of *Artemisia absinthium*, a well-known plant, indigenous to this country, and largely cultivated for medicinal purposes. It is a bitter tonic and stomachic, and also anthelmintic.—*Dose*, 20 to 40 gr. of the dried herb, either in powder or made into a tea or infusion; in dyspepsia, dropsy, scurvy, sympathetic epilepsy, &c. See ABSINTHINE and ABSINTHIC ACID.

WORT. The technical name for the fermentable infusion of malted grain.

The strength of worts is ascertained by means of an instrument termed a saccharometer. "Brewers, distillers, and the excise, sometimes denote by the term 'gravity' the excess of weight of 1000 parts of a liquid by volume above the weight of a like volume of distilled water, so that if the specific gravity be 1045, 1070, 1090, &c., the gravity is said to be 45, 70, or 90; at others, they thereby denote the weight of saccharine matter in a barrel (36 galls.) of wort; and again, they denote the excess in weight of a barrel of wort over a barrel of water equal to 36 galls., or 360 lbs. This and the first statement are identical, only 1000 is the standard in the first case, and 360 in the second" (*Ure*). The last is that commonly adopted by the brewers.

According to Dr Ure, the solid dry extract of malt, or 'saccharine,' has the specific gravity 1·264, and the specific volume ·7911; "that is, 10 lbs. of it will occupy the volume of 7·911 lbs. of water. The mean sp. gr., by computation of a solution of that extract in its own weight of water, is 1·116; but by experiment, the sp. gr. of that solution is 1·216, showing considerable condensation of volume in the act of combination with water." The quantity of solid saccharine or sugar in a wort may be determined in the manner mentioned under SYRUP.

"According to the compilers of the tables accompanying Field's alcoholometer, 1·8 lb. of saccharine is decomposed for the production of 1% of proof spirit; but according to our experiments, the proportion of saccharine named is rather below the true equivalent" (Cooley).

The rapid cooling of worts is an important object with the brewer and distiller. On the large scale, the old system in which shallow coolers are employed, with all its numerous inconveniences and accidents, is now for the most part abandoned, being supplanted by the method introduced by Mr Yandall nearly forty years ago. This consists in the use of a 'refrigerator,' which is an apparatus so constructed that any hot liquid may be cooled by about its own volume of cold water in a very short space of time. The principle is that of passing the two fluids through very shallow and very long passages, in opposite directions, being essentially that of a 'Liebig's condenser' on a gigantic scale. The apparatus may consist of zigzag passages, flattened tubes or convoluted curves, of any convenient shape, so that they possess little capacity in one direction, but great breadth and length. A refrigerator, having the passages for the fluids 1-8th of an inch thick, is said to require a run of about 80 feet. The success of this method is such as to leave nothing more to desire. See BREWING, FERMENTATION, MALT, &c.

WOUND. Syn. VULNUS, L. A solution of continuity in any of the soft parts of the body, arising from external violence.

Wounds are distinguished by surgeons into CONTUSED WOUNDS, INCISED W., LACERATED W., POISONED W., PUNCTURED W., &c.; terms which explain themselves. Sword-cuts are incised wounds; gun-shot wounds lacerated and contused ones. Slight wounds, and, indeed, all those not demanding material surgical assistance, after dirt and foreign substances have been removed, may be treated in the manner noticed under CUTS and ABRASIONS.

WOUND BAL'SAM. See TINCTURE OF BENZOIN.

WRIT'ING, executed in the ordinary tanno-gallic ink, and which has been rendered illegible by age, may be restored by carefully moistening it, by means of a feather, with an infusion of galls, or a solution of ferrocyanide of potassium slightly acidulated with hydrochloric acid, observing to apply the liquid so as to prevent the ink spreading.

WRITING FLU'IDS. A term commonly applied, of late years, to easy-flowing inks, adapted for metallic pens, in contradistinction to the old tanno-gallic compounds at one time exclusively employed for writing.

Prep. 1. Dissolve pure basic or soluble Prussian blue in pure distilled water, and dilute the resulting solution with pure water until the desired shade of colour is obtained. Very permanent and beautiful. It is not affected by the addition of alcohol, but is immediately precipi-

tated by saline matter. The precipitate, however, still possesses the property of dissolving in pure water.

2. From the soluble ferrocyanide of potassium and iron, dissolved in pure water. Resembles the last, but it is precipitated from its solution by alcohol.

3. Powdered Prussian blue, 1 oz.; concentrated hydrochloric acid, 1¼ fl. oz.; mix in a matrass or glass bottle, and, after 22 or 30 hours, dilute the mass with a sufficient quantity of water.

4. Dissolve sulphindylate of potassa or ammonia in hot water, and, when cold, decant the clear portion. It is an intense blue, and dries nearly black; is perfectly incorrosive, and very permanent and easy flowing.

5. (Horning.) Perchloride of iron, 4 parts; water, 750 parts; dissolve, add of cyanide of potassium, 4 parts, dissolved in a little water; collect the precipitate, wash it with several effusions of pure water, allow it to drain until it weighs about 200 parts, then add of oxalic acid, 1 part; and promote solution by agitating the bottle or vessel containing it.

6. (Mohr.) Pure Prussian blue, 6 parts; oxalic acid, 1 part; triturate with a little water to a perfectly smooth paste, then dilute the mass with a proper quantity of soft water. The product resembles Stephen's ' patent-blue ink.'

7. (Rev. J. B. Reade—patented). a. A solution of his patent soluble Prussian blue in distilled water. Blue.

b. Prepared by adding to good gall ink a strong solution of his soluble Prussian blue. This addition "makes the ink, which was previously proof against alkalies, equally proof against acids, and forms a writing fluid which cannot be erased from paper by any common method of fraudulent obliteration without the destruction of the paper." This ink writes greenish-blue, but afterwards turns intensely black. Stephen's ' patent ink,' which does the same, is a similar compound.

8. (Prof. Runge—CHROMIC INK.) Logwood, in fine chips, ¼ lb.; boiling water, 3 pints; digest for 12 hours, then simmer the liquid down gently to 1 quart, carefully observing to avoid dust, grease, and smoke; when cold, decant the decoction, and add to it of yellow chromate of potash, 20 gr.; dissolve by agitation, after which it will be fit for use. Cheap and good. It resists the action of all ordinary destructive agents better than the tanno-gallic inks; it may be washed after use with a wet sponge, or steeped for twenty-four hours in water, or even tested with dilute acids, and yet preserve its original blackness. It is perfectly liquid, it scarcely thickens by age, and neither deposits a sediment nor corrodes steel pens.

9. (Ure.) From vanadate of ammonia decomposed with infusion of galls. It is of a perfectly black hue, flows freely from the pen, is rendered blue by acids, is unaffected by dilute alkaline solutions, and resists the action of chlorine.

Obs. The preceding formulæ, under proper management, produce excellent products, all of which are extremely mobile, and most of them of a more or less beautiful colour. The blue ones, when concentrated, dry of a blue-black, whilst two or three of the others, though at first pale, rapidly pass into a deep black when exposed to the air. Care must be taken in all cases that the ingredients be pure. The Prussian blue, except when directly prepared for the purpose, should be washed in dilute hydrochloric acid before attempting its solution by means of oxalic acid. Unless these precautions are attended to, success is unlikely. A little gum may be added, if required, to prevent the fluid spreading on the paper; but in most cases the addition is no improvement. Most of the blue fluids may be used as 'indelible ink' to mark linen, and will be found very permanent, provided the part be first moistened with alum water, and dried.

XYLOID'IN. When starch is immersed in concentrated nitric acid (sp. gr. 1·45 to 1·50), it is converted, without disengagement of gas, into a colourless, tremulous jelly, which, when treated with an excess of water, yields a white, curdy, insoluble substance, which after being edulcorated with pure water, until every trace of acid is removed, is xyloïdin. Paper, sugar, gum, mannite, and several other substances, treated in the same manner, become in great part changed to xyloidin or analogous compounds.

Obs. Pure xyloidin differs but slightly from pyroxylin, or pure gun-cotton.

XY'LOL. A hydrocarbon homologous with benzol, found in wood-tar and coal-gas naphtha. Useful as a solvent for Canada balsam in the mounting of objects for the microscope.

YAWS. *Syn.* FRAMBŒSIA, L. A peculiar disease of the skin, common in the Antilles and some parts of Africa. It is characterised by mulberry-like excrescences, which discharge a watery humour. The treatment chiefly consists in alleviating urgent symptoms (if any), and the adoption of a temperate diet and regimen, until the eruptions, having run their course, begin to dry, when tonics and alteratives, as cinchona bark, quinine, and sarsaparilla, with occasional small doses of mercurials, generally prove advantageous. The master (or principal) yaw, which frequently remains troublesome after the others have disappeared, may be dressed with the ointment of red oxide of mercury, or of nitrate of mercury, diluted with an equal weight of lard. The yaws is not a dangerous, although a very disgusting disease. It is contagious by contact, and, like the smallpox, only occurs once during life.

YEAST. *Syn.* BARM, FERMENT, ZUMINE; FERMENTUM, L. Yeast, which consists almost entirely of minute vegetable cells, termed *Torula cerevisiæ,* is either the froth or the deposit of fermenting worts, according to the character of the fermentation.

The top yeast, or superficial ferment, which covers the surface of fermenting worts, is called 'oberhefe' by the Germans; and the bottom yeast, or the ferment of deposit, is termed 'unterhefe.' The first is the common yeast of the English brewer; the other, that used in Bavaria for the fermentation of worts from below (untergährung). Both varieties yield their own kind under proper conditions. Wort fermented

with top yeast, at from 46° to 50° F., yield both varieties, and each of these furnishes its own kind, nearly pure, by a second fermentation. See BREWING, FERMENTATION, &c.

Pres. 1. Ordinary beer yeast may be kept fresh and fit for use for several months, by placing it in a close canvas bag, and gently and gradually squeezing out the moisture until the press until the remaining matter acquires the consistence of clay or soft cheese, in which state it must be preserved in close vessels, or wrapped in waxed cloth. This is the method generally adopted for the best Flanders and German yeast.

2. Whisk the yeast until it forms a uniform liquid mass, and then lay it with a clean and soft painter's brush evenly and thinly on flat dishes, or any convenient surface, on which it can be exposed to the sun or air; this operation must be repeated as soon as the first coat is sufficiently solid, and so on, until the layers acquire a proper thickness, when it must be detached and preserved as before. If rendered quite dry, its power of exciting fermentation will be destroyed.

3. By employing strips of clean new flannel (well washed), as above, and, when sufficiently dry, rolling these up, and covering them with waxed cloth or paper, or placing them in tin canisters or boxes. For use, a few inches of one of the strips is cut off, and soaked in lukewarm water, when the barm leaves the flannel, and mixes with the water, which may then be stirred up with the flour.

Yeast, Artifi"cial. "Although the conversion of a small into a large quantity of yeast is a very easy thing, yet to produce that substance from the beginning is very difficult" (*Berzelius*). Both cases are met in the formulæ below.

Prep. 1. (*Without a ferment—Fownes.*) Wheat flour is to be mixed with water into a thick paste, which is to be slightly covered, in a moderately warm place; about the third day it begins to emit a little gas, and to exhale a disagreeable sour odour; about the sixth or seventh day the smell changes, much gas is evolved, accompanied by a distinct and agreeable vinous odour, and it is then in a state to excite either to vinous or panary fermentation, and may be either at once employed for that purpose or formed into small and very thin cakes, dried in the air, and preserved for future use. Wort fermented with it in the ordinary way yields a large quantity of yeast, of excellent quality, which is found at the bottom of the vessel. "This is a revival of a method which, although Mr Fownes seems to regard it as new, is to be found in the 'Chemistry' of Boerhaave" ('Lancet'). It is, indeed, a mere modification of the mode of preparing leaven, as practised from the most remote ages of antiquity; but is not the less valuable on that account.

2. (*With a ferment.*) *a.* Take of bean flour, ¼ lb.; water, 6 quarts; boil for ¼ an hour, pour the decoction into any suitable vessel, add of wheat flour, 3½ lbs.; stir the whole well together, and, when the temperature reaches 55° F., add of beer yeast, 2 quarts; mix well, and in 24 hours after the commencement of the fermentation add of barley or bean flour, 7 lbs.; make a

uniform dough by thorough kneading, form it into small cakes, as above, and then preserve these in a dry situation. For use, one of these discs is to be broken into pieces, laid in tepid water, and set in a warm place during 12 hours, when the soft mass will serve the purpose of beer yeast.

b. (PATENT YEAST.) Take of hops, 6 oz.; water, 3 galls.; simmer for 3 hours, strain, and in 10 minutes stir in of ground malt, ¼ peck; next reboil the hops in water, as before, and let the strained liquor run into the first mash, which must then be well stirred up, covered over, and left for 4 hours; after that time drain off the wort, and, when the temperature has fallen to 90° F., set it to work with yeast (preferably patent), 1 pint; after standing for 20 to 24 hours in a warm place, take off the scum, strain it through a coarse hair-sieve, and it will be fit for use. 1 pint is said to be enough for 1 bushel of bread.

Obs. The preparation of artificial yeast, and substitutes for yeast, has long engaged the attention of both the scientific chemist and the practical tradesman. The subject is, undoubtedly, a great importance to emigrants and voyagers. The above processes, by good management, yield products which are all that can be desired.

YEL'LOW DYES. The following substances impart a yellow to goods, either at once or after they have been mordanted with alumina or tin:—annotta, barberry root, dyer's broom, French berries, fustic, fustet, quercitron bark, and turmeric. Goods mordanted with acetate of lead, and afterwards passed through a bath of chromate of potash, acquire a brilliant chrome-yellow colour; solution of sulphate or acetate of iron, followed by immersion in potash or lime water, gives a buff or orange; orpiment dissolved in ammonia water, imparts a golden yellow (see the above-named substances, in their alphabetical places). An aniline yellow (chrysaniline) has recently been obtained by Mr Nicholson, which is said to be a most valuable dye-stuff, comparable, indeed, with the aniline reds and purples.

YELLOW FE'VER. The bilious remittent fever of hot climates. It is very common in the West Indies and the Southern States of America. New Orleans has been several times nearly depopulated by it.

YELLOW PIG'MENTS. Of these the principal are:

Brown Pink. *Prep.* Take of French berries and pearlash, of each, 1 lb.; fustic chips, ½ lb.; water, 2 galls.; boil in a tin or pewter vessel, and strain the decoction through flannel whilst hot; then dissolve alum, 1¼ lbs., in hot water, 2½ galls.; add the solution to the strained decoction as long as a precipitate falls, which must afterwards be washed, drained, and dried. Some manufacturers omit the fustic. A good glazing colour, when ground in linseed, and used with drying oil.

Yellow, Chrome. *Syn.* CHROMATE OF LEAD, YELLOW C. OF L.; PLUMBI CHROMAS, PLUMBI CHROMAS FLAVUM, L. The preparation of the pure salt is noticed under CHROMIUM and LEAD; that of the commercial pigment is as follows:

1. Add a filtered solution of nitrate or ace-

tate of lead to a like solution of neutral chromate of potash as long as a precipitate falls; then collect this, wash it well with clean soft water, and dry it out of the reach of sulphuretted vapours.

2. To the lye of chromate of potash, prepared by roasting the chrome ore with nitre, and lixiviation with water, add a solution of acetate of lead, and otherwise proceed as before.

3. Dissolve acetate of lead in warm water, and add of sulphuric acid, q. s. to convert it into sulphate of lead; decant the clear liquid (vinegar), wash the residuum with soft water, and digest it, with agitation, in a hot solution of neutral (yellow) chromate of potash, containing 1 part of that salt to every 3 parts of sulphate of lead operated on; afterwards decant the liquid, which is a solution of sulphate of potash, and carefully drain, wash, and dry the newly formed pigment. The product contains much sulphate of lead, but covers as well, and has as good a colour, as pure chromate of lead, whilst it is much cheaper. The shade may be varied by increasing or lessening the quantity of the chromate (Armengaud's ' Génie Industriel,' April, 1858).

Obs. Four shades of this beautiful pigment are met with in the shops, viz. pale yellow or straw colour, yellow, deep yellow, and orange. The former are made by adding a little alum or sulphuric acid to the solution of the chromate before mixing it with the solution of lead; the last, by the addition of a little subacetate of lead (tribasic acetate), or by washing the precipitate with a weak alkaline lye. The darker colour appears to arise from a little ' dichromate ' being thrown down intimately mixed with the neutral chromate, and the paler shades from a slight excess of acid, or from the presence of water-sulphate of lead, and, occasionally, alumina. The colour is also influenced by the temperature of the solutions at the time of admixture. Anthon has found that, when hot solutions of equal equivalents of acetate of lead (190 parts) and chromate of potash (100 parts, both neutral and in crystals) are mixed, the yellow precipitate, when dried, is anhydrous; but when the mixture is made at ordinary temperatures, the precipitate has a paler yellow, and, when dried contains 1 eq., or nearly 5½% of water ('Buch.' Rept.'). It thus appears that the shades of colour of chrome yellow may be varied, without any foreign addition. In practice, the third formula will be found very satisfactory. See ORANGE CHROME and CHROME RED.

Dutch Pink. *Prep.* Take of French berries, 1 lb.; turmeric, ¼ lb.; alum, ¼ lb.; water, 1¼ galls.; boil half an hour, strain, evaporate to 2 quarts, adding of whiting, 3 lbs., and dry by a gentle heat. Starch, or white lead, is sometimes employed instead of whiting to give it a body. Golden yellow. Used as a pigment; but will not glaze like brown pink.

English Pink. *Syn.* LIGHT PINK. As the last, but using 5 lbs. of whiting.

Indian Yellow. See PURREE.

King's Yellow. Factitious tersulphuret of arsenic.

Naples Yellow. *Syn.* MINERAL YELLOW. *Prep.* 1. Take of metallic antimony, in powder,

3 lbs.; red-lead, 2 lbs.; oxide of zinc, 1 lb.; mix, calcine, well triturate the calx, and fuse it in a covered crucible; the fused mass must be reduced to an impalpable powder by grinding and elutriation.

2. Flake white, 1¼ lbs.; diaphoretic antimony, ¼ lb.; calcined alum, 1 oz.; sal-ammoniac, 2 oz.; calcine in a covered crucible with a moderate heat for 8 hours, so that at the end it may be barely red hot. More antimony and sal-ammoniac turn it on the gold colour.

3. (*Guimel.*) Washed diaphoretic antimony, 1 part; pure red-lead, 2 parts; grind them to a paste with water, and expose this mixture to a moderate red heat for 4 or 5 hours, as before. Used in oil, porcelain, and enamel painting. Chrome has now nearly superseded it for ordinary purposes.

Patent Yellow.   *Syn.*   CASSEL YELLOW, MONTPELLIER Y., TURNER'S Y.   *Prep.*   1. Take of dry chloride of lead, 28 parts; pure carbonate of lead, 27 parts; grind them together, fuse, and powder.

2. Common salt, 1 part, and litharge, 4 parts, are ground together with water, and digested at a gentle heat for some time, water being added to supply the loss by evaporation; the carbonate of soda formed is then washed out with more water, and the white residuum heated until it acquires a fine yellow colour. Works well in oil. Chiefly used in coach-painting. See OXYCHLORIDE OF LEAD.

Weld Yellow. Prepared from a decoction of weld brightened with a little alum, in the same manner as Dutch pink. Used chiefly for paper hangings.

YTTRIUM. Y = 89·0. The oxide of this metal (yttria), a rare, white earth, was discovered by Gadolin, in 1794, in a mineral from Ytterby, in Sweden, since called gadolinite. Yttrium was obtained by Wöhler in 1828, as a brittle, dark-grey metal, from the chloride by the action of sodium. Its salts have in general a sweet taste, and the sulphate and several others an amethystine colour. Its solutions are precipitated by pure alkalies, but alkaline carbonates, especially carbonate of ammonium, dissolve it in the cold. They are distinguished from glucinium salts by the colour of the sulphate by being insoluble in pure alkalies, and by yielding a white precipitate with ferrocyanide of potassium. Yttria may be obtained from gadolinite by a similar process to that by which glucina is extracted from the beryl.

According to Professor Mosander, ordinary yttria is a mixture of the oxides of not less than three metals—yttrium, erbium, and terbium. These metals differ from each other in many important particulars. The first is a powerful base, and the others are said to be weak ones. They are separated with extreme difficulty, and are only interesting in a scientific point of view.

ZAFFRE.   *Syn.*   SAFFRA, SAFFLOR, ZAFFER. Crude oxide of cobbalt, obtained by roasting cobalt ore, reduced to an impalpable powder, and then ground with 2 or 3 parts of very pure quartzose or siliceous sand. Used as a blue colour by enamellers and painters on porcelain and glass. Chiefly imported from Saxony. See SMALTS.

ZE'RO.   See THERMOMETER.

ZESTS.   See POWDERS, SAUCE, SPICE, &c.

ZEUZERA ÆSCULI, Latreille. THE WOOD LEOPARD MOTH. This is a very handsome, beautifully marked, large moth. Fortunately for the fruit producers it is not very common, as its larvæ can do much harm to apple and pear trees, together with forest trees of various kinds, by boring into their trunks and boughs, and living upon their substance in the same manner as the larvæ of the goat moth, described in the preceding monograph. Pear trees are especial favourites of the *Zeuzera*, and large boughs are frequently detached from them by their own weight, or by the wind, showing upon close examination plain traces of having been occupied by the caterpillars of this insect. Apple trees also suffer severely in this way, and the injury done to them is often attributed to the goat moth or to *Scolyti*.

The difference, however, can be easily detected, as the burrows of the goat moth are much larger and longer than those made by the *Zeuzera*. Though it derives its affix *Æsculi* from *Æsculus hippocastanum*, the horse-chestnut, it rarely attacks this tree. So in Germany it is called *Ross-kastanien*, of the horse-chestnut; but Köllar remarks that it is found in this tree less than in many others.

Similarly in France it is termed *la Zeuzère du Marronnier*, although it by no means confines its attention to chestnut trees, but it is very troublesome to apple, pear, and walnut trees.

In an orchard in Herefordshire in 1879, it was noticed that several pear and apple trees were drooping, and that their branches fell off from time to time. Upon investigation it was seen that the trunks, and in many cases the branches also, were quite honeycombed with passages running all manner of ways. Caterpillars of the *Zeuzera* were found *flagrante delicto*. From one tree as many as seventy-six were taken, and there were signs that many more had been lodging there.

Life History. Like the goat moth, the wood leopard moth belongs to the *Lepidopterous* family or group of *Hepialidæ*. It is nocturnal in its habits, and occasionally comes heavily against the lamps in sitting-rooms through the open windows in summer.

The female moth measures from 30 to 32 lines across the wings when fully expanded. Its fore wings are of [a light colour, almost white, with many black or bluish-black spots dotted all over them, here and there, irregularly. Upon the light-coloured hind, or posterior wings, the dots or marks are not so dark coloured. Upon the upper part of its body, between the wings, there are six or seven blue-black spots on a white ground. Lower down, on the abdomen, there are dark-coloured rings.

As is often the case among moths and butterflies, the male *Zeuzera* is smaller and less handsome than the female, which is made showy, brilliant, and conspicuous, a very ' cynosure of neighbouring eyes.' To compensate it, however, for diminished size and glory the male has fine recurved antennæ, feathered or pectinated, toothed like a comb, about half way up their length. The female has mere commonplace feelers, without fringes or curves of beauty.

For the deposition of its dark yellow or orange-coloured eggs deep into the bark or rind of trees, the female is furnished with a long horny ovipositor, with the assistance of which she places from 200 to 300 eggs in various parts of the stem and branches. Egg-laying takes place from the end of June to the last week in July, and the caterpillars are produced in a few days after oviposition, beginning to push their heads into the wood immediately. They live for two years in this condition, depending upon the trees entirely for their support, and constantly labouring at their task of excavation.

In colour, the caterpillar is yellow, with black dots upon its smooth body. Its head is black, and the segment next to the head, and the last or posterior segment are shiny black. Like the goat-moth caterpillar, it is not a looper, having six pectoral, eight abdominal, and two anal feet. Upon the abdominal segments there are rows of recurved spines, which evidently serve for the retrogression of the caterpillar in its tunnel.

Prevention. There are scarcely any means of preventing the approach of this insect to place its eggs upon fruit trees. Birds are very fond of it in its moth state, and of its eggs. The titmouse and hedge sparrow are constantly upon the watch for these latter dainties.

Remedies. Plugs of cement, or of some substance that will dry hard, should be put into the holes from which sap is flowing, and little bits of wood mixed with excrement appear on the outsides. Pieces of wire may be introduced with effect so long as the tunnels are not tortuous (' Reports on Insects Injurious to Crops,' by Chas. Whitehead, Esq., F.Z.S.).

ZINC. Zn = 65·1. *Syn.* ZINK, SPELTER; ZINCUM (Ph. L., E., & D.), L. This metal was first noticed by Paracelsus, in the sixteenth century, who called it ' zinetum ;' but its ores must have been known at a much earlier period, as the ancients were acquainted with the manufacture of brass. It has been found native near Melbourne, Australia; but it occurs chiefly as smithsonite or zinc spar in Belgium, Spain, North America, Great Britain, and as zinc blende and calamine in England, Saxony, Bohemia, North America, &c.

Prep. The zinc of commerce is obtained from the native sulphide (zinc blende), or carbonate (calamine), by roasting those ores, and distilling the calx with carbonaceous matter in a covered earthen crucible, having its bottom connected with an iron tube, which terminates over a vessel of water situated beneath the furnace. The first portion that passes over contains cadmium and arsenic, and is indicated by what is technically called ' brown blaze ;' but when the metallic vapour begins to burn with a bluish-white flame, or the ' blue blaze' commences, the volatilised metal is collected.

The following method, by which several pounds of chemically pure zinc may be obtained in about ¼ of an hour, will be found very useful:—Melt the zinc of commerce in a common crucible, and granulate it by throwing it into a tolerably deep vessel of water, taking care that the metal be very hot at the time; dry the metallic grains, and dispose them by layers in a Hessian crucible with ¼ of their weight of nitrate of potassium, using the precaution to place a slight excess at the top and at the bottom; cover the crucible and secure the lid, then apply heat; after the vivid deflagration which occurs is over, remove the crucible from the fire, separate the dross with a tube, and, lastly, run the zinc into an ingot mould. This zinc, tested in Marsh's apparatus during entire days, has never given any stain, and in solution the most sensitive reagents, such as hydro-sulphocyanic acid, have never indicated the least atom of iron (' Journ. de Pharm.').

Prop. Zinc is a bluish-white metal, having the sp. gr. 6·9; tough (under some circumstances, brittle) when cold, ductile and malleable at from 100° to 150° C.; brittle and easily pulverised at 205°; fuses at 433°; at a white heat it boils, and sublimes unchanged in close vessels; heated to whiteness in contact with the air, it burns with a brilliant green light, and is converted into oxide. It is very soluble in dilute sulphuric and hydrochloric acid, with the evolution of hydrogen gas. It is little acted on by the air, even when moist. The salts of zinc are colourless.

Tests. 1. The solutions of zinc give a gelatinous white precipitate with the caustic alkalies and carbonate of ammonium, which is completely redissolved by an excess of the precipitant. 2. The carbonates of potassium and sodium give a white precipitate of carbonate of zinc. All the above precipitates acquire a lemon-yellow colour when dried and heated, but again become white on cooling. 3. Sulphide of ammonium gives, in neutral solutions, a white precipitate, insoluble in excess of the precipitant, or in solutions of hydrate of potassium or ammonium, but freely soluble in the dilute mineral acids. 4. Sulphuretted hydrogen, in neutral and alkaline solutions, also gives a like white precipitate. 5. Ferrocyanide of potassium gives a gelatinous white precipitate.

Assay. a. 100 gr. are digested in dilute hydrochloric acid in excess, and the insoluble portion dried and weighed.

b. The acid solution (see a) is next treated with a current of sulphuretted hydrogen until it smells very strongly of that gas; the whole is then left for some time in a warm situation. The precipitate which subsides consists of the sulphides of arsenic, cadmium, copper, lead, &c., if any of these metals were present in the sample.

c. The filtrate from b, after being boiled, is treated with a little nitric acid, after which it is again boiled, and, when cold, is precipitated with carbonate of barium added in excess ; the precipitate (ferric hydrate) is then collected, dried, ignited, and weighed. The weight, in grains, multiplied by 0·7, gives the percentage of iron in the sample examined.

d. The filtrate from c is next precipitated with dilute sulphuric acid, and solution of carbonate of sodium is added in excess to the filtered liquid ; the whole is then boiled, after which the new precipitate is washed, dried, gently ignited for some time, and then cooled and weighed. The weight in grains, multiplied by 0·80,247, gives the percentage of pure zinc in the sample.

Uses. Zinc is used in the process of desilverising lead, in galvanic batteries, for covering sheet-iron (galvanising) destined for roofing and other purposes which require lightness and durability; in the chemical laboratory in reduction operations, especially in the form of *zinc dust,* which is a mixture of the metal with a certain amount of oxide; in the form of zinc dust also for the reduction of indigo blue. Zinc is an important ingredient in several alloys, *e. g.* brass.

Zinc, Amalgamated, which is employed for voltaic batteries, is prepared as follows:—The plates, having been scoured with emery, are immersed for a few seconds in dilute sulphuric acid, then rinsed in clean soft water, and drained. They are then dipped into a strong solution of either mercuric nitrate or chloride, or into equal parts of a mixture of saturated solutions of mercuric chloride and acetate of lead; the plates are, lastly, dipped into water, and then rubbed with a soft cloth. Another and simpler method is to rub mercury over the plates while wet with dilute sulphuric acid.

Zinc, Granulated. *Syn.* ZINCUM GRANULATUM (B. P.), L. Fuse commercial zinc in a crucible, pour it in a very thin stream into a bucket of cold water, and afterwards dry the zinc.

Zinc, Acetate of. $Zn(C_2H_3O_2)_2 2H_2O.$ *Syn.* ZINCI ACETAS, L. *Prep.* 1. Acetate of lead, 1 lb., is dissolved in distilled water, 2½ pints, and the solution being placed in a cylindrical jar, sheet zinc, 4 oz., rolled into a coil, is immersed therein; after 24 hours the liquid is decanted, evaporated to 15 oz., and solution of hypochlorite of calcium added drop by drop, until a reddish precipitate ceases to form; the liquid is then filtered, acidulated by the addition of a few drops of acetic acid, reduced by evaporation to 10 fl. oz., and set aside to crystallise; the crystals are dried on bibulous paper set on a porous brick, and then preserved in a well-stopped bottle.

2. Add 2 oz. of carbonate of zinc in successive portions to 5 fl. oz. of acetic acid, previously mixed with 6 fl. oz. of distilled water, in a flask; heat gently, add by degrees 2 fl. oz. of acid, or q. s., till the carbonate is dissolved; boil for a few minutes, filter while hot, and set it aside for two days to crystallise. Decant the mother liquor, evaporate to one half, and again set it aside for two days to crystallise. Place the crystals in a funnel to drain, then spread them on filtering paper on a porous tile; and dry them by exposure to the air at ordinary temperatures.

Prop., &c. Efflorescent, white, hexagonal plates, having a powerful styptic taste; very soluble in water; less soluble in alcohol; decomposed by heat. It is tonic, antispasmodic, and emetic.— *Dose,* 1 to 2 gr.; as an emetic, 10 to 20 gr.; externally, 2 or 3 gr. to water, 1 fl. oz., as an astringent lotion or injection.

Zinc, Bro'mide of. $ZnBr_2.$ *Syn.* ZINCI BROMIDUM, L. Prepared by passing bromine vapour over the red hot metal; white needles.

Zinc, Carbonate of. $ZnCO_3.$ *Syn.* ZINCI CARBONAS PURUM, ZINCI CARBONAS (B. P., Ph. D.), L. Found in nature as calamine. *Prep.* 1.

Solution of chloride of zinc (Ph. D.), 1 pint is added, in successive portions, to a solution of crystallised carbonate of sodium of commerce, 2 lbs., dissolved in boiling distilled water, 6 pints, and the whole is boiled until gas ceases to be evolved; the precipitate is then washed, and dried, at first on blotting-paper, and, finally, by a steam or water heat.

2. (B. P.) Dissolve 10½ oz. carbonate of soda with 1 pint of boiling water in a capacious porcelain vessel, and pour into it 10 oz. of sulphate of zinc, also dissolved in 1 pint of water, stirring diligently. Boil for 15 minutes after effervescence has ceased, and let the precipitate subside. Decant the supernatant liquor, pour on the precipitate 3 pints of boiling distilled water, agitating briskly; let the precipitate again subside, and repeat this process till the washings are no longer precipitated by chloride of barium. Collect the precipitate on calico, let it drain, and dry at a gentle heat.

Obs. When a solution of zinc vitriol is precipitated by an excess of acid potassium carbonate a white precipitate of the hydrated normal carbonate is obtained; but if normal sodium carbonate be employed for precipitation, hydrated basic zinc carbonates of variable composition are thrown down; these become more basic in proportion as the temperature is raised and the water increased.

Zinc, Chlo'ride of. $ZnCl_2.$ *Syn.* BUTTER OF ZINC, MURIATE OF z.†; ZINCI CHLORIDUM (B. P., Ph. L.), Z. MURIAS†, L. *Prep.* 1. By heating metallic zinc in chlorine.

2. (Ph. L.) Hydrochloric acid, 1 pint; water, 1 quart; and zinc (in small pieces), 7 oz.; when the effervescence is nearly finished, apply heat until bubbles cease to be evolved; decant the clear liquid and evaporate to dryness; fuse the resulting mass in a lightly covered crucible, by a red heat, pour it out on a flat smooth stone, and, when cold, break it into small pieces, or cast it into rods in iron moulds, and preserve it in a well-stoppered bottle.

3. (B. P.) Put 16 oz. of granulated zinc into a porcelain basin, add by degrees 44 fl. oz. of hydrochloric acid previously mixed with 1 pint of distilled water, and aid the action by gently warming it on a sand-bath until gas is no longer evolved. Boil for half an hour, supplying the water lost by evaporation, and allow it to stand on the cool part of the sand-bath for 24 hours, stirring frequently. Filter the product into a gallon bottle, and pour in a solution of chlorine, q. s. by degrees, with frequent agitation, until the fluid acquires a permanent odour of chlorine. Add ½ oz. or a sufficient quantity of carbonate of zinc, in small quantities at a time, and with renewed agitation, until a brown sediment appears. Filter through paper into a porcelain basin,?and evaporate until a portion of the liquid, withdrawn on the end of a glass rod and cooled, forms an opaque white solid. Pour it out now into proper moulds, and when the salt has solidified, but before it has cooled, place it in closely-stoppered bottles.

4. (IN SOLUTION.) a. (LIQUOR ZINCI CHLORIDI —Ph. D.) Hydrochloric acid and water, of each, 2½ pints; sheet zinc, 1 lb.; dissolve, filter through

calico, add of hyperchlorite of calcium, 1 fl. oz., and evaporate, by boiling, to a pint; when cold, pour it into a bottle, add of prepared chalk, 1 oz., and water, q. s. to make the whole measure 1 quart; agitate occasionally for 24 hours, decant or filter, and preserve the liquid in a stoppered bottle. Sp. gr. 1·593. See SOLUTION.

b. (*E. Parish.*) Granulated zinc, 4 lbs.; hydrochloric acid, 4 lbs., or q. s.; water, 9 quarts; dissolve, avoiding excess of acid. The solution contains 1 in 12 of chloride of zinc. Recommended as of the proper strength for a disinfectant.

Prop., &c. When pure, a colourless, amorphous mass or crystals; generally a whitish-grey, semi-transparent mass, having the consistence of wax; fusible at 100° C., volatile at a strong heat, condensing in acicular crystals; freely soluble in alcohol, ether, and water; highly deliquescent; coagulates albumen and gelatin, and corrodes animal substances. The solution possesses the same properties in a minor degree.

Pur. From the aqueous solution, hydrosulphuric acid or ferrocyanide of potassium being dropped in, a precipitate is thrown down. What is thrown down by ammonia or potash from the same solution is white, and is redissolved by either precipitant in excess. The precipitate thrown down by the carbonate of either ammonium or potassium is also white, but is not redissolved when these are added in excess.

Uses, &c. Dry chloride of zinc is chiefly used as a caustic, for which it is highly recommended by Voght, Canquoin, and others. It is more powerful than chloride of antimony, and its action extends deeper than does nitrate of silver, whilst it exercises an influence over the vital actions of neighbouring parts. The sore is generally healthy after the separation of the eschar, and no constitutional disorder ensues. It has been given in scrofula, epilepsy, chorea, &c.; and, combined with hydrocyanic acid, in facial neuralgia.—*Dose,* ¼ gr. to 2 gr.; externally, as a lotion, 2 to 3 gr. to water, 1 oz. In large doses it is poisonous.

The solution is also used as a caustic, but chiefly as a disinfectant and deodoriser, *e. g.* as Sir Wm. Burnett's Fluid and Professor Tuson's ' Sporokton,' of which it is one of the very best, possessing, as it does, the power of rapidly decomposing sulphide of ammonium and of rendering inert the virus of infectious diseases. It is also employed on the large scale in 'weighting' cotton goods. When a solution (sp. gr. 1·7) is boiled with an excess of oxide, a liquid is obtained which possesses the property of dissolving silk; this is used for separating silk fibres from those of wool, cotton, or linen, all of which dissolve in *normal* zinc chloride.

Zinc, Cy'anide of. ZnCy₂. *Syn.* CYANURET OF ZINC; ZINCI CYANIDUM, ZINCI CYANURETUM, L. *Prep.* (P. Cod.) Add a solution of cyanide of potassium to another of pure sulphate of zinc; wash, and dry the precipitate. It is insoluble in water and alcohol, but dissolves in solutions of the cyanides of ammonium and potassium.—*Dose,* ¼ to 1 gr., twice a day; in epilepsy, hysteria, and other nervous affections, heartburn, worms, &c.; and as a substitute for hydrocyanic acid.

Zinc, Ferrocy'anide of. Zn₂FeCy₆. *Syn.* ZINCI FERROCYANIDUM, L. *Prep.* By adding a hot solution of ferrocyanide of potassium to a hot and strong solution of pure sulphate of zinc, and washing and drying the precipitate (*White*).—*Dose,* 1 to 4 gr.; in the same cases as the last.

Zinc, I'odide of. ZnI₂. *Syn.* HYDRIODATE OF ZINC†; ZINCI IODIDUM, ZINCI HYDRIODAS, L. *Prep.* 1. (*Duflos.*) Iodine, 2 parts; granulated zinc, 1 part; water, 4 parts; proceed as for ferrous iodide, but employ a glass or porcelain vessel.

2. (*Magendie.*) Iodine, 17 parts; zinc (in powder), 20 parts; mix, and sublime in a matrass.

Prop., &c. Deliquescent; colourless; octohedrous. Chiefly used externally; 15 gr. to water, 6 fl. oz., as a collyrium in scrofulous inflammation of the eye (*Poulet*); 1 dr. to lard, 1 oz., as a powerful resolvent in scrofulous and other glandular swellings, rubbed on the part twice a day (*Ure*).

Zinc, Lac'tate of. Zn(C₃H₅O₃)₂. *Syn.* ZINCI LACTAS, L. Prepared from zinc in the same way as ferrous lactate is from iron.

Zinc, Ox'ide of. ZnO. *Syn.* PROTOXIDE OF ZINC; ZINCI OXYDUM (B. P., Ph. L., E., & D.), L. *Prep.* 1. Sulphate of zinc (pure), 1 lb.; carbonate of ammonium, 6½ oz.; dissolve each separately in 6 quarts of water, filter, mix the solutions, well wash the precipitate with water, and calcine it for two hours in a strong fire. The Ph. E. is nearly similar.

2. Place carbonate of zinc in a covered clay crucible, and expose it to a very low red heat, until a portion taken from the centre of the mass ceases to effervesce on being dropped into dilute sulphuric acid.

Prop., &c. A white, tasteless powder; insoluble in water; freely soluble in acids, the solution yielding colourless and easily crystallisable salts; strongly basic.

Uses, &c. It is tonic and antispasmodic, and has been advantageously used in chorea, epilepsy, and other nervous and spasmodic affections.— *Dose,* 2 to 10 gr. It is also used as a dusting powder, and to make an ointment. It has been proposed as a substitute for white-lead in painting, than which it covers better, but dries slower, and hence requires the addition of dried white vitriol. This oxide is the only compound which zinc forms with oxygen.

Zinc Phosphide. Zn₃P₂. Fragments of pure distilled zinc are introduced into a tubulated stoneware retort, so as to occupy about one fourth of its capacity; the retort is placed in an ordinary furnace, and a current of dry carbonic acid is passed into it through the neck. Over the tubulure is placed a crucible cover, so as to close the orifice incompletely, and allow the carbonic acid, after traversing the retort, to escape. When the zinc enters into ebullition small fragments of previously dried phosphorus are successively thrown in through the tubulure, the cover being removed and returned after each addition, to prevent loss of phosphorus. From time to time it is necessary to break the crust of phosphorus formed, in order to expose a new layer of metal to its action.

The operation is terminated by increasing the heat strongly—a precaution that is indispensable, in order to separate as completely as possible the zinc phosphide from the metallic button of nearly pure zinc which collects at the bottom of the retort. The product should be reduced to very fine powder, and the fragments which resist the action of the metal, however slightly, should be reserved for another operation. In the pure state it resembles iron reduced by hydrogen, and only thus should it be used by pharmaceutists. It is completely soluble in hydrochloric acid (from ' Formulæ for New Medicaments,' adopted by the Paris Pharmaceutical Society).

Zinc, Sulphate of. $ZnSO_4$. Syn. WHITE COP-PERAS*, WHITE VITRIOL*; ZINCI SULPHAS (B. P., Ph. L., E., & D.), L. Long known to the alchemists as white vitriol. Prep. 1. (Ph. L.) Granulated zinc, 5 oz.; diluted sulphuric acid, 1 quart; dissolve and filter. Evaporate to a pellicle, and set it aside to crystallise.

2. (Ph. D.) Zinc (laminated or granulated), 4 oz.; sulphuric acid, 3 fl. oz.; water, 1 pint; mix in a porcelain capsule, and, when gas ceases to be evolved, boil for 10 minutes, filter through calico, and evaporate the filtrate to dryness; dissolve the dry salt in water, 1 pint; frequently agitate the solution, when cold, during 6 hours, with prepared chalk, ½ oz.; next filter, acidulate the filtrate with nitric acid and dilute sulphuric acid, of each, 1 fl. dr.; evaporate until a pellicle forms on the surface, and set it aside to crystallise; dry the crystals on bibulous paper without heat.

3. (B. P.) Pour 12 fl. oz. of sulphuric acid, previously mixed with 4 pints of distilled water, on 16 oz. of granulated zinc contained in a porcelain basin, and when effervescence has nearly ceased, aid the action by a gentle heat. Filter the fluid into a gallon bottle, and add gradually, with constant agitation, chlorine water, until the fluid acquires a permanent odour of chlorine. Add now, with continued agitation, ½ oz. or q. s. of carbonate of zinc, until a brown precipitate appears; let it settle, filter the solution, evaporate till a pellicle forms on the surface, and set aside to crystallise. Dry the crystals by exposure to air on filtering paper placed on porous tiles.

4. The common sulphate of zinc of commerce frequently contains copper, cadmium, lead, iron, and manganese, and nearly always one or more of them. By digesting its concentrated solution for some time with metallic zinc it may be freed from copper, lead, and cadmium, for these metals are all reduced and precipitated in a metallic state; or the acidulated solution may be treated with sulphuretted hydrogen as long as any precipitate forms. In order to separate the iron, chlorine gas may be passed into the solution, by which the iron is converted into the ferrous chloride; if this solution be exposed to the air for a length of time, it absorbs oxygen, and oxide of iron is deposited as a yellow powder, from which the solution must be filtered. When the sulphate contains manganese, which is not very often the case, the solution must be boiled up a few times with purified charcoal, filtered, and evaporated (' Journ. für prakt. Chem.'). The product of each of the above formulæ is nearly chemically pure.

5. (Commercial.) The crude sulphate of zinc (white copperas, or white vitriol, of the shops) is prepared by roasting native sulphide of zinc (zinc blende) in a reverberatory furnace, exposing the calcined mass to the air and humidity for some time, then lixiviating it, and evaporating the resulting solution until it forms a white semi-crystalline mass on cooling.

Prop. Pure sulphate of zinc forms inodorous, colourless, transparent, quadrangular prisms, closely resembling in appearance those of Epsom salt, which effloresce slightly in the air, and contains 7 equiv. of water; sp. gr. 1·95; it has a slightly acidulous and very styptic metallic taste; the crystals dissolve in 2½ parts of cold and in less than their own weight of boiling water; they are insoluble in alcohol. The crude sulphate of zinc of commerce (white vitriol) occurs in irregular granular masses, which somewhat resemble loaf sugar.

When a solution of this salt, in 6 parts of water, is boiled with a little nitric acid, and a solution of ammonia is then added until the oxide of zinc at first precipitated is all redissolved, no yellow precipitate remains, or a trace only, and the solution is colourless.

Uses. In medicine, as a tonic, antispasmodic, &c.; in doses of 1 to 2 gr., twice daily; as an emetic, 10 to 30 gr. In large doses it is poisonous. It has been employed with benefit in dyspepsia, fluor albus, chorea, epilepsy, hooping-cough, and other convulsive and nervous affections, generally combined with bitters, foxglove, hemlock, henbane, or opium. As an emetic, it acts almost immediately, and is therefore well suited to empty the stomach at the commencement of a fit of ague, and in cases of poisoning, &c. It is used externally to form astringent and repellent collyria, injections, and lotions. It is used also very extensively in dyeing.

Zinc Sulphide. ZnS. Occurs as blende, an ore of zinc. When pure it is yellow and transparent; it usually, however, is contaminated with iron and other metals, which cause it to assume a red, brown, or black tint.

Zinc Sulphocarbolate. $Zn(C_6H_5SO_4)_2.H_2O$. The acid, prepared as in sulphocarbolate of soda (which see), is saturated by aid of a gentle heat with oxide of zinc, filtered, and crystals allowed to form. The crystals should be dried by exposure to the air.

Zinc, Valerianate of. $Zn(C_5H_9O_2)_2$. Syn. ZINCI VALERIANAS (B. P., Ph. D.), L. Prep. (Ph. D.) Valerianate of sodium, 2½ oz., and sulphate of zinc, 2 oz. 7 dr., are each separately dissolved in distilled water, 1 pint; the solutions are then heated to 200° F., mixed, and the resulting crystals skimmed off; the liquid is next evaporated at a temperature not higher than 200°, until it measures 4 fl. oz., the crystals, as they form, being removed from the surface; the salt thus obtained is steeped for an hour in distilled water, just sufficient to cover it, after which the whole is transferred to a paper filter, on which it is at first drained, and then dried at a heat not exceeding 100°.

Prop., Vale'rianate of. Brilliant white, pearly tabular crystals; very light; astringent; smells feebly of valerianic acid; only slightly soluble in cold water, more so in hot water, and freely soluble in alcohol and ether; exposure to heat rapidly decomposes it; exposure to the air also decomposes it, but more slowly. It is regarded as powerfully antispasmodic and tonic.—*Dose*, 1 to 3 gr., thrice daily, made into pills; in neuralgia, tic-douloureux, nervous headaches (more particularly hemicrania), hysteria, palpitation of the heart, vertigo, chorea, epilepsy, &c.

Obs. Butyrate of zinc, scented with valerianic acid, which is often sold for the above compound, may be detected by distilling it with sulphuric acid; the distillate, tested with a strong solution of acetate of copper, gives a bluish-white precipitate if it contains butyric acid. The valerianate is distinguished from the other salts of zinc by its extreme lightness.

ZINC ETHYL. $Zn(C_2H_5)_2$. A curious liquid body, discovered by Dr Frankland, and formed, along with iodide of zinc, when iodide of ethyl is heated with pure zinc in a sealed glass tube. The mixed white product, by distillation in a current of hydrogen, yields pure zinc ethyl. It is a highly volatile liquid, having a rather disagreeable odour, and so rapidly decomposed by contact with the air that it takes fire. Water resolves it into hydride of ethyl and other products.

ZINC METHYL. $Zn(CH_3)_2$. Obtained by the action of zinc upon iodide of methyl, as zinc ethyl. It takes fire on coming in contact with the air.

ZINCKING. *Syn.* ZINKING. Vessels of copper and brass may be covered with a firmly adherent layer of pure zinc by boiling them in a solution of chloride of zinc, pure zinc turnings being at the same time present in considerable excess. The same object may be effected by means of zinc and a solution of chloride of ammonium or hydrate of potassium.

The variety of zincked iron commonly known by the name of 'galvanised iron' is prepared by immersing the sheets of metal, previously scoured and cleaned with dilute hydrochloric acid, in a bath of melted zinc covered with powdered sal-ammoniac, and moving them about until they are sufficiently coated.

ZINCOG'RAPHY. A process of printing closely resembling lithography, in which plates of zinc are substituted for slabs of stone.

ZIR'CON. See GEMS.

ZIRCONIUM. $Zr = 90$. The oxide of this metal, a white pulverulent earth, was discovered in the mineral zircon of Ceylon, by Klaproth, in 1789. It has since been found in hyacinth, a mineral found in Ceylon.

Prep. The ore is calcined and thrown into cold water, and then powdered in an agate mortar; the power is mixed with 9 parts of pure hydrate of potassium, and the mixture very gradually shaken into a red-hot crucible, care being taken that each portion is fused before another is added; after fusion, with an increased heat, for an hour and a half, the whole is allowed to cool; the calcined mass is next powdered, and boiled in water; the insoluble portion is then dissolved in hydrochloric acid, and the solution heated, so that the silicic acid may fall down, after which zirconia is precipitated with hydrate of potassium; or the zirconia may be precipitated with carbonate of sodium, and the carbonic acid expelled by heat.

From this, metallic zirconium is obtained by heating it with magnesium, and then treating the residue with dilute hydrochloric acid, when the insoluble zirconium remains behind (*Phipson*).

Prop., &c. An iron-grey powder; it acquires a feeble metallic lustre under the burnisher, and takes fire when heated in the air.

Oxide of zirconium, or zirconia, ZrO_2, is a white tasteless powder, is insoluble in water, and forms salts with the acids. It is distinguished from all the other earths, except thorina, by being precipitated when any of the neutral salts of zirconium are boiled with a saturated solution of sulphate of potassium. The salts of zirconium are distinguished from those of aluminium and glucinum by being precipitated by all the pure alkalies, and by being insoluble when they are added in excess. The precipitated hydrate and carbonate are readily soluble in acids. Zirconia cylinders have recently been successfully substituted for lime in producing a strong light by the oxyhydrogen flame.

With 596 Illustrations, royal 8vo., 32s.

CHEMICAL TECHNOLOGY

BY

RUDOLF VON WAGNER

TRANSLATED AND EDITED BY

WILLIAM CROOKES, F.R.S.,

FROM THE THIRTEENTH ENLARGED GERMAN EDITION AS REMODELLED BY

DR. FERDINAND FISCHER

CONTENTS.

SECTION I.—TECHNOLOGY OF FUEL.

Fuel and its Treatment—Thermometry—Determination of the Value of Fuels—Manufacture of Wood Charcoal—Peat—Lignite (Brown Coal, Bovey Coal)—Coal—Coke—Degasifying, Gasifying, Combustion—Heating Arrangements—Lighting-gas—Mineral Oil—Paraffine and Solar Oil Industry—Production of Light—Photometry—Lighting with Candles—Lighting with Lamps—Gas Lighting—Electric Light.

SECTION II.—METALLURGY.

Iron—Crude Iron—Examination of Iron and Steel—Iron Founding—Wrought or Bar Iron—Steel—Manganese—Cobalt—Nickel—Copper—Lead—Silver—Gold—Platinum—Tin—Bismuth—Antimony—Arsenic—Mercury—Zinc—Cadmium—Potassium and Sodium—Aluminium—Magnesium.

SECTION III.—CHEMICAL MANUFACTURING INDUSTRY.

Water and Ice—Artificial Mineral Waters—Sulphur—Sulphuric Acid—Properties of Sulphuric Acid—Potassium Salts—Common Salt and Salt Works—Soda—Natural Soda—Soda from Plants—Soda obtained by Chemical Means—Chlorine, Chloride of Lime, and Chlorates—Bromine—Iodine—Nitric Acid and Nitrates—Nitric Acid—Explosives—Ammonia—Phosphorus—Matches: Production of Fire—Phosphates: Manures—Boric Acid and Borax—Salts of Aluminium—Ultramarine—Compounds of Tin and Antimony—Compounds of Antimony—Compounds of Arsenic—Compounds of Gold, Silver, and Mercury—Compounds of Copper—Compounds of Zinc and Cadmium—Compounds of Lead—Compounds of Manganese and Chromium—Iron Compounds, including Ferrocyanogen—Conspectus of Inorganic Pigments—Thermo-chemistry.

SECTION IV.—THE ORGANIC CHEMICAL MANUFACTURES.

Alcohols and Ethers—Organic Acids—Treatment of Coal Tar—Organic Colouring Matters—Tar Colours—Benzol Colours—Examination of Colouring Matters—Artificial Colours soluble in Water—Solid or Pasty Colours insoluble in Water.

SECTION V.—GLASS, EARTHENWARE, CEMENT, AND MORTAR.

Glass Manufacture—Earthenware or Ceramic Manufacture—Mortars, &c.

SECTION VI.—ARTICLES OF FOOD AND CONSUMPTION.

Starch and Dextrine—Sugar—Fermentation Arts—Wine Making—Beer Brewing—The Manufacture of Spirits—Flour and Bread—Milk, Butter, and Cheese—Meat—Nutrition.

SECTION VII.—CHEMICAL TECHNOLOGY OF FIBRES.

Wool—Silk—Vegetable Fibres—Bleaching—Dyeing and Tissue-printing—Paper Manufacture.

SECTION VIII.—MISCELLANEOUS ORGANO-CHEMICAL ARTS AND MANUFACTURES.

Tanning—Glue, Size, Gelatine—Sizes—Bones—Fats—Soap—Stearine and Glycerine—Essential Oils and Resins—Preservation of Wood.

LONDON: J. & A. CHURCHILL, 11, NEW BURLINGTON STREET.

Lightning Source UK Ltd.
Milton Keynes UK
UKHW050913041218
333390UK00036B/736/P